2026 Release

Prescott's Microbiology

Joanne M. Willey
HOFSTRA UNIVERSITY

Kathleen M. Sandman

Sarah N. Salm
BOROUGH OF MANHATTAN COMMUNITY COLLEGE

Nathan W. Rigel
KEAN UNIVERSITY

McGraw Hill

PRESCOTT'S MICROBIOLOGY

This book is printed on acid-free paper.

26 CPI 2

ISBN 978-1-265-82717-5
MHID 1-265-82717-6

Cover Image: *(Scientist) Kristopher Grunert/Corbis/Glow Images; (Fungus) Jannicke Wiik-Nielsen/iStock/Getty Images; (Yeast) Steve Gschmeissner/Science Photo Library/Getty Images; (Influenza) Cynthia Goldsmith/Centers for Disease Control and Prevention; (Staphylococcus) Mediscan/Alamy Stock Photo; (Bacillus) Medical-on-Line/Alamy Stock Photo*

mheducation.com/highered

Brief Contents

About the Authors xiv

Part One Introduction to Microbiology

1 The Evolution of Microorganisms and Microbiology 1
2 Microscopy 23
3 Bacterial Cell Structure 44
4 Archaeal Cell Structure 74
5 Eukaryotic Cell Structure 85
6 Viruses and Other Acellular Infectious Agents 103

Part Two Microbial Nutrition, Growth, and Control

7 Bacterial and Archaeal Growth 121
8 Control of Microorganisms in the Environment 155
9 Antimicrobial Chemotherapy 170

Part Three Microbial Metabolism

10 Introduction to Metabolism 190
11 Catabolism: Energy Release and Conservation 208
12 Anabolism: The Use of Energy in Biosynthesis 244

Part Four Microbial Molecular Biology and Genetics

13 Bacterial Genome Replication and Expression 265
14 Regulation of Cellular Processes 299
15 Eukaryotic and Archaeal Genome Replication and Expression 326
16 Mechanisms of Genetic Variation 343
17 Microbial DNA Technologies 362
18 Microbial Genomics 380

Part Five The Diversity of the Microbial World

19 Archaea 401
20 Nonproteobacterial Gram-Negative Bacteria 414
21 Proteobacteria 436
22 Gram-Positive Bacteria 454
23 Protists 470
24 Fungi 490
25 Viruses 504

Part Six Ecology and Symbiosis

26 Exploring Microbes in Ecosystems 528
27 Microbial Interactions 543
28 Biogeochemical Cycling and Global Climate Change 555
29 Microorganisms in Marine and Freshwater Ecosystems 569
30 Microorganisms in Terrestrial Ecosystems 587

Part Seven Pathogenicity and Host Response

31 Innate Host Resistance 606
32 Adaptive Immunity 633
33 The Microbe-Human Ecosystem 666
34 Infection and Pathogenicity 684

Part Eight Microbial Diseases, Detection, and Their Control

35 Epidemiology and Public Health Microbiology 700
36 Clinical Microbiology and Immunology 718
37 Human Diseases Caused by Viruses and Prions 737
38 Human Diseases Caused by Bacteria 769
39 Human Diseases Caused by Fungi and Protists 811

Part Nine Applied Microbiology

40 Microbiology of Food 836
41 Biotechnology and Industrial Microbiology 851
42 Applied Environmental Microbiology 862

Appendix 1 A Review of the Chemistry of Biological Molecules A-1
Appendix 2 Common Metabolic Pathways A-9
Appendix 3 Microorganism Pronunciation Guide A-17

Glossary G-1
Index I-1

McGraw Hill connect®

Your complete course platform

Connect enables you to build deeper connections with your students through cohesive digital content and tools, creating engaging learning experiences. We are committed to providing you with the right resources and tools to support all your students along their personal learning journeys.

90% student satisfaction

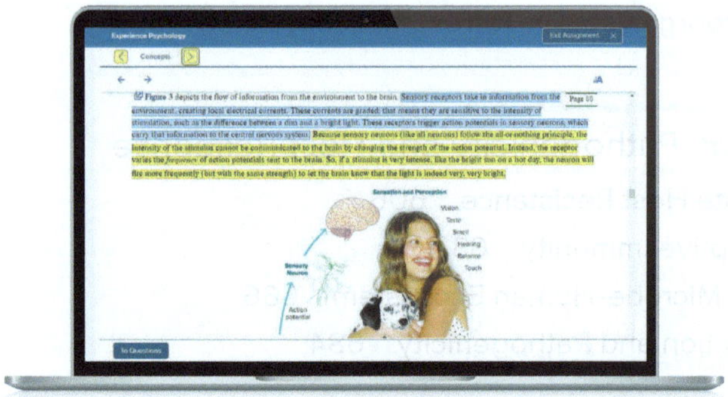

Laptop: Getty Images; Woman/dog: George Doyle/Getty Images

Every learner is unique

In Connect, instructors can assign an adaptive reading experience with SmartBook®. Rooted in advanced learning science principles, SmartBook delivers a personalized experience, focusing students on their learning gaps and ensuring the time they spend studying is time well spent.
mheducation.com/highered/connect/smartbook

Study anytime, anywhere

Encourage your students to download the free ReadAnywhere® app so they can access their online eBook, SmartBook, or Adaptive Learning Assignments when it's convenient, even when they're offline. Because the app automatically syncs with students' Connect accounts, all their work is available every time it is opened.
mheducation.com/readanywhere

"I really liked this app—it made it easy to study when you don't have your textbook in front of you."

Jordan Cunningham, a student at *Eastern Washington University*

Effective tools for efficient studying

Connect is designed to help students be more productive with simple, flexible, intuitive tools that maximize study time and meet students' individual learning needs and prepare them with real-world application. Get learning that works for everyone with Connect.

SUPPORT AT every step

Education for all

McGraw Hill works directly with Accessibility Services departments and faculty to meet the learning needs of all students. Please contact your Accessibility Services Office, and ask them to email **accessibility@mheducation.com**, or visit **mheducation.com/about/accessibility** for more information.

Affordable solutions, added value

Make technology work for you with LMS integration for single sign-on access, mobile access to the digital textbook, and reports to quickly show you how each of your students is doing. And with Affordable Access programs, you can provide all these tools at the lowest available market price to your students. Ask your McGraw Hill representative for more information.

Solutions for your challenges

A product isn't a solution. Real solutions are affordable, reliable, and come with training and ongoing support when you need it and how you want it. Visit **supportateverystep.com** for videos and resources both you and your students can use throughout the term.

Updated and relevant content

Our Evergreen™ Delivery Model provides you with the most relevant and updated content, tools, and accessibility. Engage students and freshen up assignments with up-to-date coverage of select topics and new questions, all without having to switch editions or build a new course.

Digital Tools for Your Success

Virtual Labs and Lab Simulations

While the biological sciences are hands-on disciplines, instructors are often asked to deliver some of their lab components online: as full online replacements, supplements to prepare for in-person labs, or make-up labs.

These simulations help each student learn the practical and conceptual skills needed, then check for understanding and provide feedback. With adaptive pre-lab and post-lab assessment available, instructors can customize each assignment.

From the instructor's perspective, these simulations may be used in the lecture environment to help students visualize complex scientific processes, such as DNA technology or Gram staining, while at the same time providing a valuable connection between the lecture and lab environments.

Relevancy Modules for Microbiology

With the help of our Relevancy Modules within McGraw Hill Connect, students can see how microbiology actually relates to their everyday lives. For select microbiology titles, students and instructors can access the Relevancy Modules eBook at no additional cost. Auto-graded assessment questions that correlate to the modules are also available within Connect. Each module consists of videos, an overview of basic scientific concepts, and then a closer look at the application of these concepts to the relevant topic. Some topics include microbes and cancer, fermentation, vaccines, biotechnology, global health, SARS-CoV-2, antibiotic resistance, and several others.

Proctorio

Remote Proctoring & Browser-Locking Capabilities

Remote proctoring and browser-locking capabilities, hosted by Proctorio within Connect, provide control of the assessment environment by enabling security options and verifying the identity of the student.

Seamlessly integrated within Connect, these services allow instructors to control the assessment experience by verifying identification, restricting browser activity, and monitoring student actions.

Instant and detailed reporting gives instructors an at-a-glance view of potential academic integrity concerns, thereby avoiding personal bias and supporting evidence-based claims.

ReadAnywhere® App

Read or study when it's convenient for you with McGraw Hill's free ReadAnywhere app. Available for iOS or Android smartphones or tablets, ReadAnywhere gives users access to McGraw Hill tools including the eBook and SmartBook® or Adaptive Learning Assignments in McGraw Hill Connect®. Students can take notes, highlight, and complete assignments offline—all their work will sync when connected to Wi-Fi. Students log in with their Connect username and password to start learning—anytime, anywhere!

Online Learning Consortium–Aligned Courses

Implementing High-Quality Instruction and Assessment Through Preconfigured Courseware

In consultation with the Online Learning Consortium (OLC) and our certified Faculty Consultants, McGraw Hill has created preconfigured courseware using OLC's quality scorecard to align with best practices in online course delivery. This turnkey courseware contains a combination of formative and summative assessments, homework, and application activities, and can easily be customized to meet an individual's needs and course outcomes. For more information, visit mheducation.com/highered/olc.

Evergreen™

Content and technology are ever-changing, and we know how important it is for instructors to keep courses up to date. That's why *Prescott's Microbiology* uses an Evergreen delivery model—providing the latest content, tools, and accessibility updates hassle-free.

With Evergreen, you can engage students and freshen up assignments with up-to-date coverage of select topics and new questions—all without switching editions or building a new course.

Writing Assignment

Available within McGraw Hill Connect®, the Writing Assignment tool delivers a learning experience to help students improve their written communication skills and conceptual understanding. Assign, monitor, grade, and provide feedback on writing more efficiently and effectively.

Digital Tools for Your Success

Test Builder in McGraw Hill Connect®

Available within Connect, Test Builder is a cloud-based tool that enables instructors to format tests that can be printed, administered within a Learning Management System, or exported as a Word document. Test Builder offers a modern, streamlined interface for easy content configuration that matches course needs, without requiring a download.

Test Builder allows you to:

- access all test bank content from a particular title.
- easily pinpoint the most relevant content through robust filtering options.
- manipulate the order of questions or scramble questions and/or answers.
- pin questions to a specific location within an exam.
- determine your preferred treatment of algorithmic questions.
- choose the layout and spacing.
- add instructions and configure default settings.

Test Builder provides a secure interface for better protection of content and allows for just-in-time updates to flow directly into assessments.

Prep for Microbiology

Prep for Microbiology is designed to get students ready for a forthcoming course by quickly and effectively addressing prerequisite knowledge gaps that may cause problems down the road. This question bank highlights a series of questions, including Fundamentals of Science, Fundamentals of Math and Statistics, Fundamental Skills for the Scientific Laboratory, and Student Success, to give students a refresher on the skills needed to enter and be successful in their course. Prep maintains a continuously adapting learning path individualized for each student and tailors content to focus on what the student needs to master in order to have a successful start in the new class.

Polling

Every learner has unique needs. Uncover where and when you're needed with the new Polling tool in McGraw Hill Connect®! Polling allows you to discover where students are in real time. Engage students and help them create connections with your course content while gaining valuable insight during lectures. Leverage polling data to deliver personalized instruction when and where it is needed most.

Create®

Your Book, Your Way

McGraw Hill's Content Collections Powered by Create is a self-service website that enables instructors to create custom course materials—print and eBooks—by drawing upon McGraw Hill's comprehensive, cross-disciplinary content. Choose what you want from our high-quality textbooks, digital products, articles, cases, and more. Combine it with your own content quickly and easily, and tap into other rights-secured, third-party content such as cases, articles, readings, cartoons, and labs. Content can be arranged in a way that makes the most sense for your course, and you can select your own cover and include the course name and school information as well. Choose the best format for your course: color print, black-and-white print, or eBook. The eBook can be included in your McGraw Hill Connect® course and is available on the free ReadAnywhere® app for smartphone or tablet access as well. When you are finished customizing, you will receive a free digital copy to review in just minutes! Visit McGraw Hill Create®—mcgrawhillcreate.com—today and begin building!

McGraw Hill GO

McGraw Hill GO is an easy-to-use, quick-to-set-up eBook+ that lives within the Learning Management System (LMS). GO equips instructors and students with essential course materials: trusted, assignable content and autograded chapter questions that sync directly into the LMS. GO makes it easier to keep up with student progress and direct attention where it matters.

A Modern Approach to Microbiology

Evolution as a Framework

Introduced immediately in chapter 1 and used as an overarching theme throughout, evolution helps unite microbiological concepts and provides a framework upon which students can build their knowledge.

An Introduction to the Entire Microbial World

Covered in chapters 3–6, separate chapters on the structure and function of bacteria and archaea are followed by the discussion of eukaryotic cells and viruses.

Broad Coverage of Microbial Ecology

The importance and multidisciplinary nature of microbial ecology are demonstrated by content that ranges from global climate change to the human microbiome.

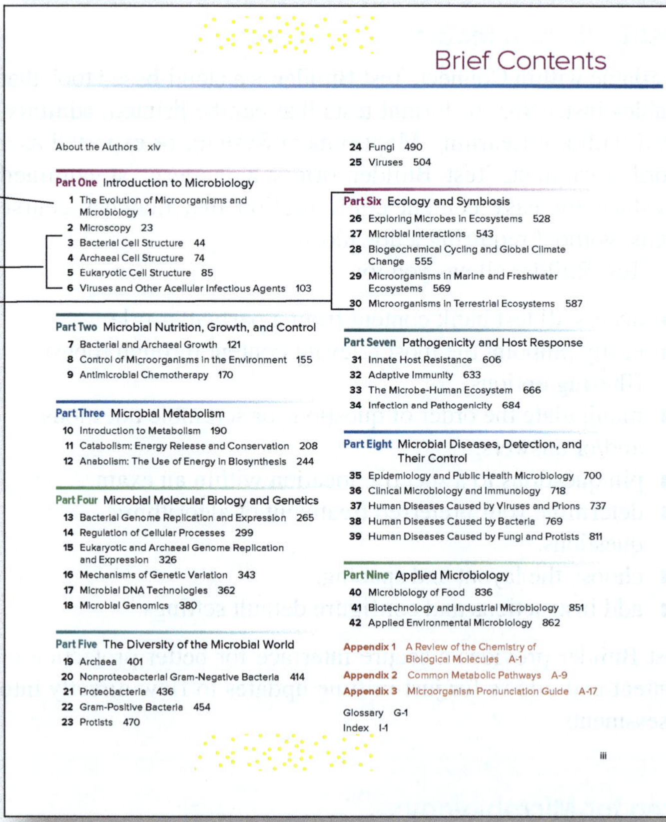

Brief Contents

About the Authors xiv

Part One Introduction to Microbiology
1 The Evolution of Microorganisms and Microbiology 1
2 Microscopy 23
3 Bacterial Cell Structure 44
4 Archaeal Cell Structure 74
5 Eukaryotic Cell Structure 85
6 Viruses and Other Acellular Infectious Agents 103

Part Two Microbial Nutrition, Growth, and Control
7 Bacterial and Archaeal Growth 121
8 Control of Microorganisms in the Environment 155
9 Antimicrobial Chemotherapy 170

Part Three Microbial Metabolism
10 Introduction to Metabolism 190
11 Catabolism: Energy Release and Conservation 208
12 Anabolism: The Use of Energy in Biosynthesis 244

Part Four Microbial Molecular Biology and Genetics
13 Bacterial Genome Replication and Expression 265
14 Regulation of Cellular Processes 299
15 Eukaryotic and Archaeal Genome Replication and Expression 326
16 Mechanisms of Genetic Variation 343
17 Microbial DNA Technologies 362
18 Microbial Genomics 380

Part Five The Diversity of the Microbial World
19 Archaea 401
20 Nonproteobacterial Gram-Negative Bacteria 414
21 Proteobacteria 436
22 Gram-Positive Bacteria 454
23 Protists 470
24 Fungi 490
25 Viruses 504

Part Six Ecology and Symbiosis
26 Exploring Microbes in Ecosystems 528
27 Microbial Interactions 543
28 Biogeochemical Cycling and Global Climate Change 555
29 Microorganisms in Marine and Freshwater Ecosystems 569
30 Microorganisms in Terrestrial Ecosystems 587

Part Seven Pathogenicity and Host Response
31 Innate Host Resistance 606
32 Adaptive Immunity 633
33 The Microbe-Human Ecosystem 666
34 Infection and Pathogenicity 684

Part Eight Microbial Diseases, Detection, and Their Control
35 Epidemiology and Public Health Microbiology 700
36 Clinical Microbiology and Immunology 718
37 Human Diseases Caused by Viruses and Prions 737
38 Human Diseases Caused by Bacteria 769
39 Human Diseases Caused by Fungi and Protists 811

Part Nine Applied Microbiology
40 Microbiology of Food 836
41 Biotechnology and Industrial Microbiology 851
42 Applied Environmental Microbiology 862

Appendix 1 A Review of the Chemistry of Biological Molecules A-1
Appendix 2 Common Metabolic Pathways A-9
Appendix 3 Microorganism Pronunciation Guide A-17

Glossary G-1
Index I-1

iii

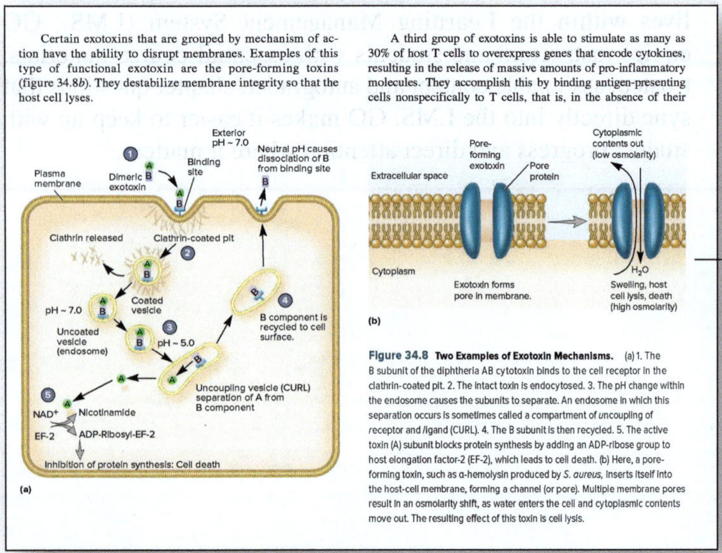

Certain exotoxins that are grouped by mechanism of action have the ability to disrupt membranes. Examples of this type of functional exotoxin are the pore-forming toxins (figure 34.8*b*). They destabilize membrane integrity so that the host cell lyses.

A third group of exotoxins is able to stimulate as many as 30% of host T cells to overexpress genes that encode cytokines, resulting in the release of massive amounts of pro-inflammatory molecules. They accomplish this by binding antigen-presenting cells nonspecifically to T cells, that is, in the absence of their

Figure 34.8 Two Examples of Exotoxin Mechanisms. (a) 1. The B subunit of the diphtheria AB cytotoxin binds to the cell receptor in the clathrin-coated pit. 2. The intact toxin is endocytosed. 3. The pH change within the endosome causes the subunits to separate. An endosome in which this separation occurs is sometimes called a *c*ompartment of *u*ncoupling of *r*eceptor and *l*igand (CURL). 4. The B subunit is then recycled. 5. The active toxin (A) subunit blocks protein synthesis by adding an ADP-ribose group to host elongation factor-2 (EF-2), which leads to cell death. (b) Here, a pore-forming toxin, such as α-hemolysin produced by *S. aureus*, inserts itself into the host-cell membrane, forming a channel (or pore). Multiple membrane pores result in an osmolarity shift, as water enters the cell and cytoplasmic contents move out. The resulting effect of this toxin is cell lysis.

Molecular Microbiology and Immunology

Updates on genetic mechanisms, molecular microbiology, and the human microbiome are included. An up-to-date discussion of immunity, with enhanced detail between innate and adaptive linkages, helps students grasp the complexity and specificity of immune responses. The microbiome and its interaction with immunity and its impact on human homeostasis are introduced in chapter 33, The Microbe-Human Ecosystem.

A Modern Approach to Microbiology

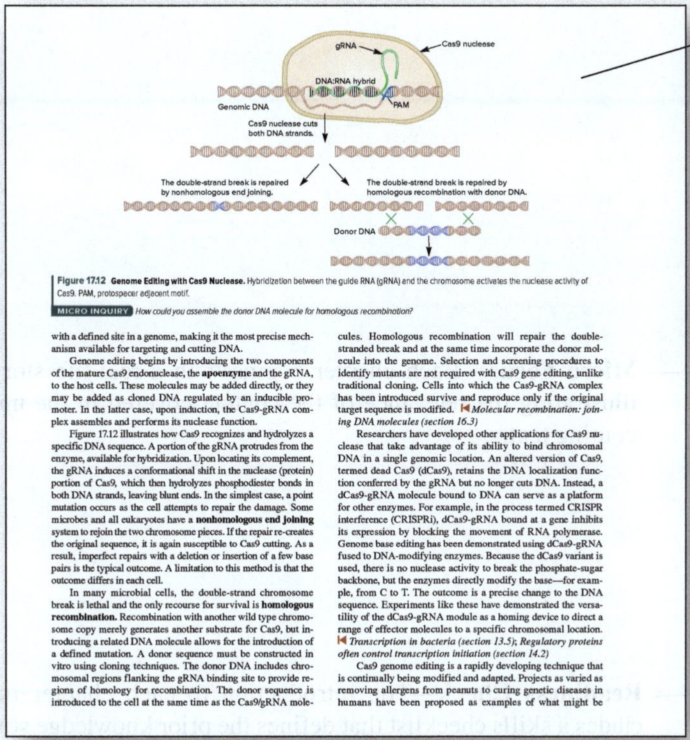

Figure 17.12 Genome Editing with Cas9 Nuclease. Hybridization between the guide RNA (gRNA) and the chromosome activates the nuclease activity of Cas9. PAM, protospacer adjacent motif.

MICRO INQUIRY *How could you assemble the donor DNA molecule for homologous recombination?*

with a defined site in a genome, making it the most precise mechanism available for targeting and cutting DNA.

Genome editing begins by introducing the two components of the mature Cas9 endonuclease, the **apoenzyme** and the gRNA, to the host cells. These molecules may be added directly, or they may be added as cloned DNA regulated by an inducible promoter. In the latter case, upon induction, the Cas9-gRNA complex assembles and performs its nuclease function.

Figure 17.12 illustrates how Cas9 recognizes and hydrolyzes a specific DNA sequence. A portion of the gRNA protrudes from the enzyme, available for hybridization. Upon locating its complement, the gRNA induces a conformational shift in the nuclease (protein) portion of Cas9, which then hydrolyzes phosphodiester bonds in both DNA strands, leaving blunt ends. In the simplest case, a point mutation occurs as the cell attempts to repair the damage. Some microbes and all eukaryotes have a **nonhomologous end joining** system to rejoin the two chromosome pieces. If the repair re-creates the original sequence, it is again susceptible to Cas9 cutting. As a result, imperfect repairs with a deletion or insertion of a few base pairs is the typical outcome. A limitation to this method is that the outcome differs in each cell.

In many microbial cells, the double-strand chromosome break is lethal and the only recourse for survival is **homologous recombination.** Recombination with another wild type chromosome copy merely generates another substrate for Cas9, but introducing a related DNA molecule allows for the introduction of a defined mutation. A donor sequence must be constructed in vitro using cloning techniques. The donor DNA includes chromosomal regions flanking the gRNA binding site to provide regions of homology for recombination. The donor sequence is introduced to the cell at the same time as the Cas9/gRNA molecules. Homologous recombination will repair the double-stranded break and at the same time incorporate the donor molecule into the genome. Selection and screening procedures to identify mutants are not required with Cas9 gene editing, unlike traditional cloning. Cells into which the Cas9-gRNA complex has been introduced survive and reproduce only if the original target sequence is modified. *Molecular recombination: joining DNA molecules (section 16.3)*

Researchers have developed other applications for Cas9 nuclease that take advantage of its ability to bind chromosomal DNA in a single genomic location. An altered version of Cas9, termed dead Cas9 (dCas9), retains the DNA localization function conferred by the gRNA but no longer cuts DNA. Instead, a dCas9-gRNA molecule bound to DNA can serve as a platform for other enzymes. For example, in the process termed CRISPR interference (CRISPRi), dCas9-gRNA bound at a gene inhibits its expression by blocking the movement of RNA polymerase. Genome base editing has been demonstrated using dCas9-gRNA fused to DNA-modifying enzymes. Because the dCas9 variant is used, there is no nuclease activity to break the phosphate-sugar backbone, but the enzymes directly modify the base—for example, from C to T. The outcome is a precise change to the DNA sequence. Experiments like these have demonstrated the versatility of the dCas9-gRNA module as a homing device to direct a range of effector molecules to a specific chromosomal location. *Transcription in bacteria (section 13.5); Regulatory proteins often control transcription initiation (section 14.2)*

Cas9 genome editing is a rapidly developing technique that is continually being modified and adapted. Projects as varied as removing allergens from peanuts to curing genetic diseases in humans have been proposed as examples of what might be

Twenty-first-Century Microbiology

Prescott's Microbiology leads the way with text devoted to CRISPR genome engineering, global climate change, and microbial plastic degradation. For more details, see chapters 17, 28, and 42.

Metagenomics and the Human Microbiome

Expanded coverage of metagenomics and its importance in understanding the role of microbes in all environments and in exploring symbionts of invertebrates is threaded throughout the text. Chapter 33, The Microbe-Human Ecosystem, explores the human microbiome and its role in health and disease.

Laboratory Safety

Reflecting recommendations from the Centers for Disease Control and Prevention, along with the American Society for Microbiology, chapter 36 provides specific guidance for laboratory best practices to help instructors provide safe conditions during the teaching of laboratory exercises.

Special Interest Essays

Organized into four themes—Microbial Diversity & Ecology, Techniques & Applications, Historical Highlights, and Disease—these focused and interesting essays provide additional insight into relevant topics.

TECHNIQUES & APPLICATIONS

17.1 Animal-Free Foods

The meaty flavor of that grilled plant-based burger and the cheesy vegan topping that you melted onto it—all derive from fungal fermentation! Food science meets consumer demand for animal-free foods and the result is ingredients produced by precision fermentation.

Pichia pastoris and *Trichoderma reesei* are workhorses of synthetic biology: They grow to a high cell density in fermentation vats on inexpensive media and are popular platforms for assembling the DNA components for a vast collection of biological molecules. These fungi have few secreted proteins and low toxicity and allergenicity. As eukaryotic cells, they are capable of many posttranslational modifications required for recombinant protein activity.

Leghemoglobin (LegH) from soybeans is the not-so-secret ingredient to the burger's meaty flavor. It's a monomeric protein with an iron-containing heme group (*see figure 10.10*). When cooked, LegH denatures and releases its heme, which then reacts with other components to generate the flavor and aroma that we associate with meat. *Rhizobia are symbionts of leguminous plants (section 30.3)*

It's far more cost-effective to produce LegH in *P. pastoris* than to isolate it from soybeans. This requires cloning the gene encoding the protein portion and the nine enzymes that assemble the heme group into the yeast. All these genes need to be overexpressed to optimize LegH production. Because glycine starts the heme biosynthetic pathway, its levels must be balanced between cell growth and heme synthesis. This complex regulatory system was worked out by **metabolomics,** which identified *P. pastoris* genes to be up- or down-regulated. Genome editing technologies make these changes feasible.

As for the vegan cheese, the milk protein casein is singularly responsible for cheese melting and stretching when heated. *T. reesei* has been engineered to synthesize and secrete recombinant casein. When taken through the cheese-making process, this casein results in a nondairy cheese indistinguishable from the original. Because it is nondairy, the product is a good option for consumers with lactose intolerance. While the details of recombinant casein production in *T. reesei* remain proprietary, multiple phosphorylation and glycosylation reactions occur after casein synthesis, and it is reasonable to expect co-expression of enzymes that modify the recombinant casein substrate.

DENNIS KUNKEL MICROSCOPY/SCIENCE PHOTO LIBRARY/Alamy Stock Photo

RHJPhtotos/Shutterstock

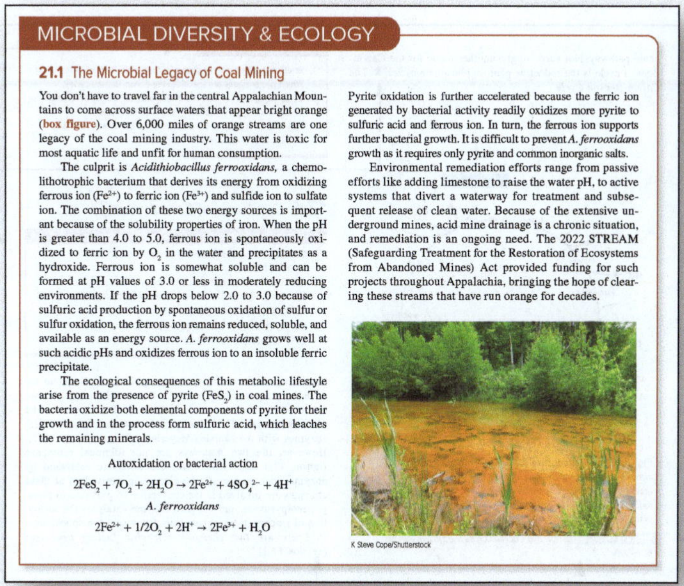

MICROBIAL DIVERSITY & ECOLOGY

21.1 The Microbial Legacy of Coal Mining

You don't have to travel far in the central Appalachian Mountains to come across surface waters that appear bright orange (**box figure**). Over 6,000 miles of orange streams are one legacy of the coal mining industry. This water is toxic for most aquatic life and unfit for human consumption.

The culprit is *Acidithiobacillus ferrooxidans,* a chemolithotrophic bacterium that derives its energy from oxidizing ferrous ion (Fe^{2+}) to ferric ion (Fe^{3+}) and sulfide ion to sulfate ion. The combination of these two energy sources is important because of the solubility properties of iron. When the pH is greater than 4.0 to 5.0, ferrous ion is spontaneously oxidized to ferric ion by O_2 in the water and precipitates as a hydroxide. Ferrous ion is somewhat soluble and can be formed at pH values of 3.0 or less in moderately reducing environments. If the pH drops below 2.0 to 3.0 because of sulfuric acid production by spontaneous oxidation of sulfur or sulfur oxidation, the ferrous ion remains reduced, soluble, and available as an energy source. *A. ferrooxidans* grows well at such acidic pHs and oxidizes ferrous ion to an insoluble ferric precipitate.

The ecological consequences of this metabolic lifestyle arise from the presence of pyrite (FeS_2) in coal mines. The bacteria oxidize both elemental components of pyrite for their growth and in the process form sulfuric acid, which leaches the remaining minerals.

Autoxidation or bacterial action

$$2FeS_2 + 7O_2 + 2H_2O \rightarrow 2Fe^{2+} + 4SO_4^{2-} + 4H^+$$

A. ferrooxidans

$$2Fe^{2+} + 1/2O_2 + 2H^+ \rightarrow 2Fe^{3+} + H_2O$$

Pyrite oxidation is further accelerated because the ferric ion generated by bacterial activity readily oxidizes more pyrite to sulfuric acid and ferrous ion. In turn, the ferrous ion supports further bacterial growth. It is difficult to prevent *A. ferrooxidans* growth as it requires only pyrite and common inorganic salts.

Environmental remediation efforts range from passive efforts like adding limestone to raise the water pH, to active systems that divert a waterway for treatment and subsequent release of clean water. Because of the extensive underground mines, acid mine drainage is a chronic situation, and remediation is an ongoing need. The 2022 STREAM (Safeguarding Treatment for the Restoration of Ecosystems from Abandoned Mines) Act provided funding for such projects throughout Appalachia, bringing the hope of clearing these streams that have run orange for decades.

K Steve Cope/Shutterstock

Student-Friendly Organization

CHAPTER

31

Innate Host Resistance

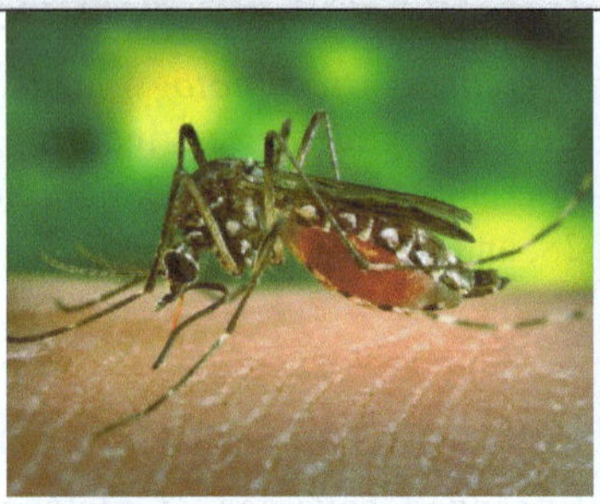

James Gathany/CDC

Vaccines for a Changing Planet

Think about a mosquito proboscis slicing into your skin. Next the mosquito injects its saliva, which contains a complex mixture of over 100 different molecules that promote feeding by suppressing blood clotting and dilating blood vessels. Although you may think your only defense is to swat the mosquito, all animal hosts (including humans) possess innate immunity: a set of immune chemicals, cells, and tissues at the ready to shield against the invasion of all foreign molecules, including those from blood-sucking insects. However, because such arthropods coevolved with their hosts, their saliva also includes proteins that modulate the animal's innate immune response, setting up a host-mosquito interaction that leaves that familiar, itchy welt.

What if that mosquito's saliva also carries protozoa that cause malaria or viruses that cause diseases like yellow fever, Zika virus disease, or dengue (like the mosquito *Aedes aegypti* shown above)? As you may be aware, or even have experienced, the last decade has seen an increase in mosquito-borne diseases. As the climate warms, the breeding season and geographical distribution of mosquitos that transmit these diseases are growing, thereby putting ever larger numbers of human populations at risk. New approaches to controlling disease transmission are needed.

One innovative strategy targets mosquito saliva, aiming to block transmission at its source. The specific goal of saliva vaccines is to train the host's immune system to block the activity of salivary proteins, which in turn would prevent the mosquito-borne pathogen from gaining a foothold in the human host. Such vaccines are in development and offer broad-spectrum defense by targeting proteins found in multiple mosquito species. Thus a single vaccine could protect against many mosquito-borne diseases.

Early trials of one such anti–mosquito saliva vaccine illustrate the importance of innate immunity. An experimental vaccine targeting four different salivary proteins was prepared in two ways: Some vaccines stimulated both innate immunity as well the more specific, adaptive arm of the immune system, while other vaccine triggered just an adaptive response. When tested on human volunteers, only those who received the vaccine that stimulated both arms of the immune system showed a clear immune response against salivary proteins, demonstrating that, in addition to innate immunity serving as our first line of defense, it is critical in calling up the responses of adaptive immunity. Although saliva vaccines are still in the early stage of development, they represent the kind of groundbreaking approaches needed in a world of rapidly evolving infectious diseases. To better understand how infectious agents cause disease, we first explore the components of innate immunity in this chapter and learn about adaptive immunity in chapter 32.

Readiness Check:

Based on what you have learned previously, you should be able to:

✓ Explain basic eukaryotic cell biology (sections 5.1 through 5.6)

✓ Diagram the flow of information in eukaryotes from DNA to protein and regulation of this flow (chapter 15)

606

Micro Focus—Each chapter begins with a real-life story illustrating the relevance of the content covered in the upcoming text.

Readiness Check—The introduction to each chapter includes a skills checklist that defines the prior knowledge students need to understand the material that follows.

phate pathway. Not surprisingly, another name for the Calvin-Benson cycle is the reductive pentose phosphate cycle. *The Calvin-Benson Cycle*

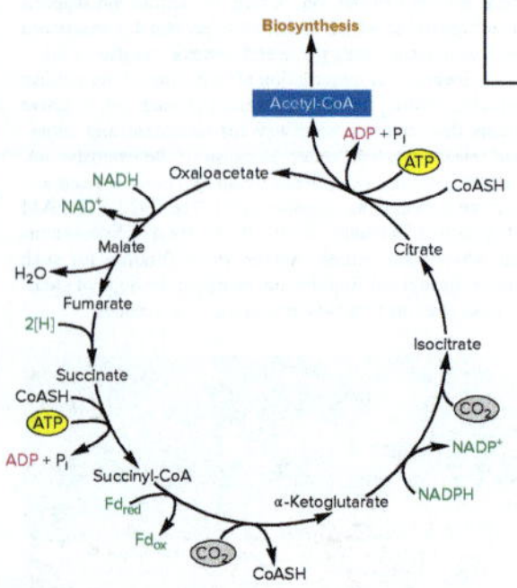

Figure 12.4 The Reductive TCA Cycle. Two CO_2 molecules are incorporated into organic carbon in this pathway. The cycle runs in the opposite direction (i.e., counterclockwise) as the TCA cycle. ATP and reducing power (NADH, NADPH, and [H]) drive the reversal. The product of the pathway is acetyl-CoA.

MICRO INQUIRY *How might acetyl-CoA be used in biosynthesis (see figure 12.2)?*

Comprehension Check

1. Briefly describe the three phases of the Calvin-Benson cycle.
2. Which two enzymes are specific to the Calvin-Benson cycle?
3. Compare figures 12.4 and 11.8. Which enzyme is likely not shared between the TCA cycle and the reductive TCA cycle?
4. Why is it important that the product(s) of a CO_2-fixation pathway be a precursor metabolite?

12.4 Synthesis of Carbohydrates

After reading this section, you should be able to:

a. Compare and contrast gluconeogenesis and the Embden-Meyerhof pathway
b. Describe the role of ATP and UTP in the synthesis of monosaccharides (other than glucose) and polysaccharides
c. Outline the major steps in peptidoglycan synthesis
d. Identify the steps in which peptidoglycan biosynthesis can be targeted by antibiotics

Many microbes need to synthesize glucose 6-phosphate for anabolic purposes (figure 12.2). The synthesis of glucose 6-phosphate from noncarbohydrate precursors is called **gluconeogenesis.** The gluconeogenic pathway shares several enzymes with the Embden-Meyerhof pathway (**figure 12.5**). However, the two pathways are not identical (compare figures 12.5 and 11.5). Three reactions are catalyzed by enzymes that are specific for gluconeogenesis. Two of these enzymes are involved in the conversion of pyruvate to phosphoenolpyruvate, and the third enzyme catalyzes the formation of fructose 6-phosphate from fructose 1,6-bisphosphate. *There are two chemoheterotrophic fueling processes (section 11.2)*

Comprehension Check—Questions within the narrative of each chapter help students master section concepts before moving on to other topics.

Animation Icon—This symbol indicates that material presented in the text is accompanied by an animation within Instructor Resources in Connect. Create a file attachment assignment in Connect to have your students view the animation, or post it to your Learning Management System for students.

Learning Outcomes—Every section in each chapter begins with a list of content-based activities students should be able to perform after reading.

Cross-Referenced Notes—In-text references refer students to other parts of the book to review.

Micro Inquiry—Selected figures in every chapter contain probing questions, adding another assessment opportunity for the student.

Student-Friendly Organization

Concept Maps—Organizational frameworks of foundational content are provided by concept maps, helping learners grasp the big picture.

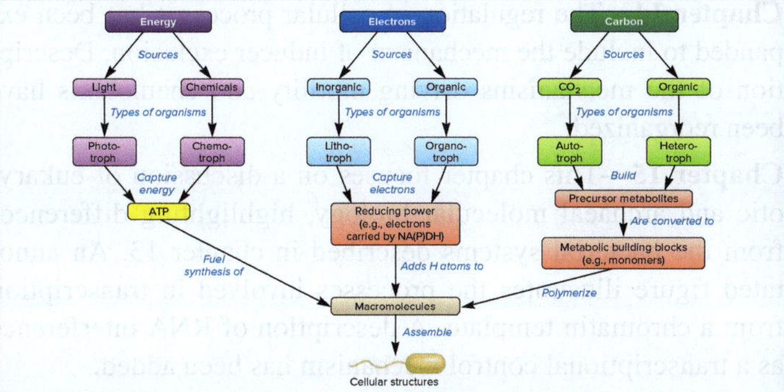

Figure 11.2 A Concept Map of Cellular Synthesis Requirements for Energy, Electron, and Carbon Sources. ATP and reducing power (e.g., NADPH) are used in anabolic reactions that convert precursor metabolites into monomers and other building blocks. Monomers are used to generate macromolecules such as proteins, which are assembled into various cell structures.

Annotated Figures—All key metabolic pathways and molecular processes are annotated, so each step is clearly illustrated and explained.

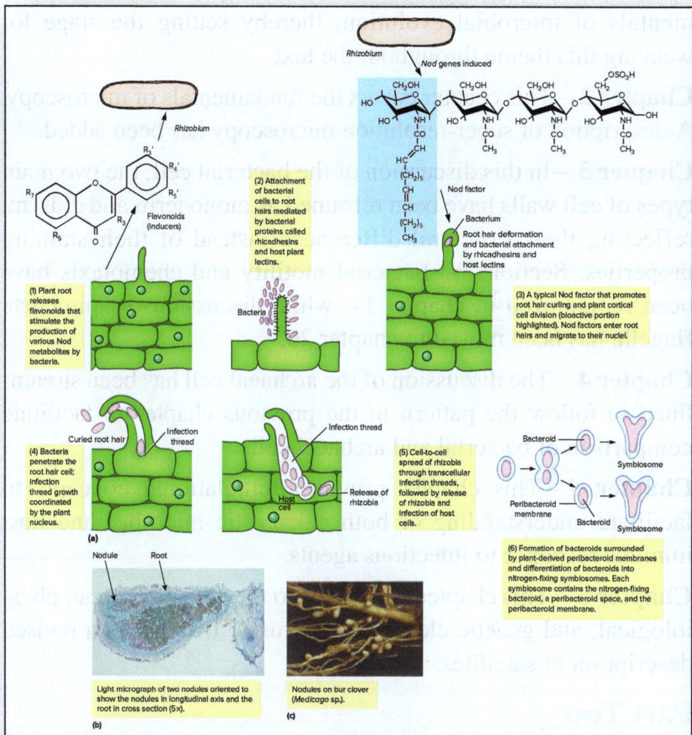

Figure 30.9 Root Nodule Formation by a *Rhizobium* Species. (a) Initiation of bacterial penetration into the root hair cell, infection thread development, and differentiation of bacterial cells into bacteroids. (b) Light micrograph of two nodules that develop by cell division (×5). This section is oriented to show the nodules in longitudinal axis and the root in cross section. (c) Nodules on bur clover (*Medicago* sp.). (b) Biology Pics/Science Source; (c) Hugh Spencer/Science Source

Key Concepts—At the end of each chapter, organized by numbered headings, this feature distills the content to its essential components with cross-references to figures and tables.

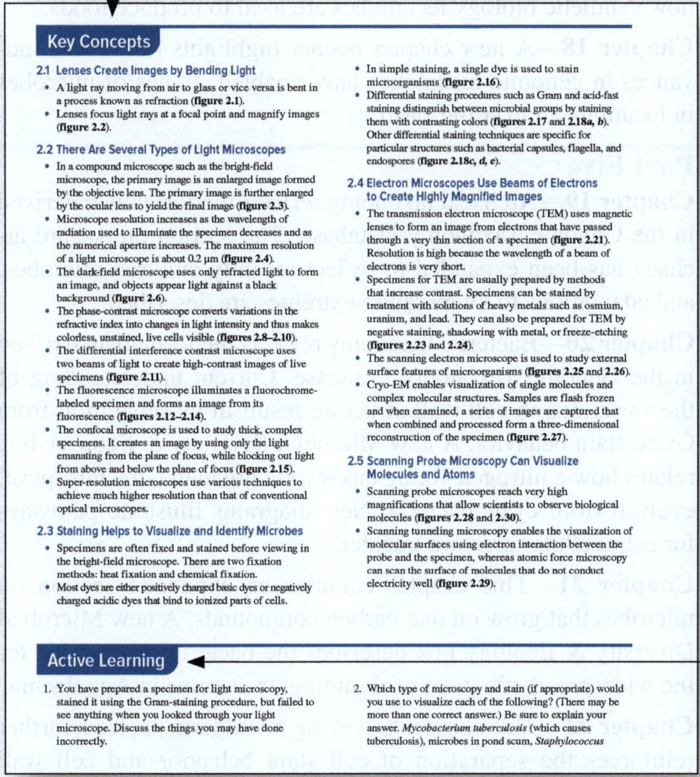

Key Concepts

2.1 Lenses Create Images by Bending Light

- A light ray moving from air to glass or vice versa is bent in a process known as refraction (**figure 2.1**).
- Lenses focus light rays at a focal point and magnify images (**figure 2.2**).

2.2 There Are Several Types of Light Microscopes

- In a compound microscope such as the bright-field microscope, the primary image is an enlarged image formed by the objective lens. The primary image is further enlarged by the ocular lens to yield the final image (**figure 2.3**).
- Microscope resolution increases as the wavelength of radiation used to illuminate the specimen decreases and as the numerical aperture increases. The maximum resolution of a light microscope is about 0.2 μm (**figure 2.4**).
- The dark-field microscope uses only refracted light to form an image, and objects appear light against a black background (**figure 2.6**).
- The phase-contrast microscope converts variations in the refractive index into changes in light intensity and thus makes colorless, unstained, live cells visible (**figures 2.8–2.10**).
- The differential interference contrast microscope uses two beams of light to create high-contrast images of live specimens (**figure 2.11**).
- The fluorescence microscope illuminates a fluorochrome-labeled specimen and forms an image from its fluorescence (**figures 2.12–2.14**).
- The confocal microscope is used to study thick, complex specimens. It creates an image by using only the light emanating from the plane of focus, while blocking out light from above and below the plane of focus (**figure 2.15**).
- Super-resolution microscopes use various techniques to achieve much higher resolution than that of conventional optical microscopes.

2.3 Staining Helps to Visualize and Identify Microbes

- Specimens are often fixed and stained before viewing in the bright-field microscope. There are two fixation methods: heat fixation and chemical fixation.
- Most dyes are either positively charged basic dyes or negatively charged acidic dyes that bind to ionized parts of cells.
- In simple staining, a single dye is used to stain microorganisms (**figure 2.16**).
- Differential staining procedures such as Gram and acid-fast staining distinguish between microbial groups by staining them with contrasting colors (**figures 2.17** and **2.18*a*, *b***). Other differential staining techniques are specific for particular structures such as bacterial capsules, flagella, and endospores (**figure 2.18*c*, *d*, *e***).

2.4 Electron Microscopes Use Beams of Electrons to Create Highly Magnified Images

- The transmission electron microscope (TEM) uses magnetic lenses to form an image from electrons that have passed through a very thin section of a specimen (**figure 2.21**). Resolution is high because the wavelength of a beam of electrons is very short.
- Specimens for TEM are usually prepared by methods that increase contrast. Specimens can be stained by treatment with solutions of heavy metals such as osmium, uranium, and lead. They can also be prepared for TEM by negative staining, shadowing with metal, or freeze-etching (**figures 2.23** and **2.24**).
- The scanning electron microscope is used to study external surface features of microorganisms (**figures 2.25** and **2.26**).
- Cryo-EM enables visualization of single molecules and complex molecular structures. Samples are flash frozen and when examined, a series of images are captured that when combined and processed form a three-dimensional reconstruction of the specimen (**figure 2.27**).

2.5 Scanning Probe Microscopy Can Visualize Molecules and Atoms

- Scanning probe microscopes reach very high magnifications that allow scientists to observe biological molecules (**figures 2.28** and **2.30**).
- Scanning tunneling microscopy enables the visualization of molecular surfaces using electron interaction between the probe and the specimen, whereas atomic force microscopy can scan the surface of molecules that do not conduct electricity well (**figure 2.29**).

Active Learning

1. You have prepared a specimen for light microscopy, stained it using the Gram-staining procedure, but failed to see anything when you looked through your light microscope. Discuss the things you may have done incorrectly.
2. Which type of microscopy and stain (if appropriate) would you use to visualize each of the following? (There may be more than one correct answer.) Be sure to explain your answer. *Mycobacterium tuberculosis* (which causes tuberculosis), microbes in pond scum, *Staphylococcus*

Active Learning—These questions are designed to stimulate analytical problem-solving skills.

List of Content Changes

- Each chapter has been thoroughly reviewed and updated.
- Artwork updated to reflect all learners and abilities.

Part One

Chapter 1—We open the text with an emphasis on the fundamentals of microbial evolution, thereby setting the stage for weaving this theme throughout the text.

Chapter 2—This chapter covers the fundamentals of microscopy. A description of super-resolution microscopy has been added.

Chapter 3—In this discussion of the bacterial cell, the two main types of cell walls have been reframed as monoderm and diderm, reflecting their structural differences instead of their staining properties. Sections on bacterial motility and chemotaxis have been reorganized in chapter 14, while discussion of spirochete flagella has been moved to chapter 20.

Chapter 4—The discussion of the archaeal cell has been streamlined to follow the pattern in the previous chapter to facilitate comparison of bacterial and archaeal cells.

Chapter 5—This chapter provides foundational concepts to facilitate understanding of both eukaryotic microbes and host immune response to infectious agents.

Chapter 6—This chapter introduces to the morphological, physiological, and genetic elements of viruses. It includes a revised description of satellites.

Part Two

Chapter 7—A new chapter opener highlights the challenges in cultivating a diverse array of microbes in the laboratory.

Chapter 8—Microbial control is organized into physical, chemical, and biological methods. A new Techniques & Applications box highlights how changes to materials' surfaces can inhibit microbial colonization.

Chapter 9—The discussion of β-lactamase inhibitors and anti-HIV drugs has been updated. In addition to reviewing the structure and mechanism of action of antimicrobial classes, the growing threat of antimicrobial resistance is emphasized. The challenges in bringing new antimicrobial drugs to market are also discussed.

Part Three

Chapter 10—This chapter provides the foundation for understanding energy conservation and biosynthesis.

Chapter 11—Catabolic pathways and energy conservation emphasize bacterial and archaeal processes. Concept maps provide overviews of microbial catabolic strategies such as aerobic versus anaerobic respiration, chemolithotrophy, and phototrophy.

Chapter 12—Biosynthetic pathways are illustrated in detail in this chapter, including archaeal variations. Multiple figures have been redrawn for added clarity.

Part Four

Chapter 13—A new section describes how diderms assemble outer membrane proteins (OMPs) and is accompanied by a new figure of the Bam complex.

Chapter 14—The regulation of cellular processes has been expanded to include the mechanism of inducer exclusion. Description of the mechanisms driving motility and chemotaxis have been reorganized.

Chapter 15—This chapter focuses on a discussion of eukaryotic and archaeal molecular biology, highlighting differences from the bacterial systems described in chapter 13. An annotated figure illustrates the processes involved in transcription from a chromatin template. A description of RNA interference as a transcriptional control mechanism has been added.

Chapter 16—Mutation and repair discussions are updated with new figures. Mechanisms of gene transfer have been expanded and reorganized. A new section on the mobilome describes some of the hybrid genetic elements that populate microbial genomes.

Chapter 17—This chapter introduces students to the common laboratory techniques for manipulating DNA, including gene cloning, PCR, heterologous gene expression, and CRISPR/Cas9 gene editing. A new Techniques & Applications box illustrates how synthetic biology techniques are used to produce foods.

Chapter 18—A new chapter opener highlights how recent advances in genomic techniques have enabled studies of microbes in locations all over the earth.

Part Five

Chapter 19—Archaeal taxonomy reflects the formalism derived in the Genome Taxonomy Database. The section on Asgard archaea has been expanded to reflect recently cultured microbes, and adaptations to temperature extremes are described.

Chapter 20—Bacterial taxonomy reflects the formalism derived in the Genome Taxonomy Database. Current understanding of the variations in cell wall structure result in its separation from Gram stain behavior. A new Microbial Diversity & Ecology box relates how a nitrogen-fixing eukaryotic organelle, the nitroplast, evolved from cyanobacteria. New diagrams illustrate pathways for extracellular electron transfer.

Chapter 21—This chapter contains an expanded section on microbes that grow on one-carbon compounds. A new Microbial Diversity & Ecology box describes the bacteria responsible for the widespread effects of coal mining on streams in Appalachia.

Chapter 22—This chapter surveying monoderm bacteria further reinforces the separation of cell stain behavior and cell wall structures.

List of Content Changes

Chapter 23—Protist classification based on recent phylogenomic analysis introduces clades of medically and environmentally important protists.

Chapter 24—This chapter presents six major fungal groups based on recent phylogenomic evidence.

Chapter 25—The life cycles of representative RNA and DNA viruses are discussed in this chapter. The HIV life cycle has been revised and is accompanied by an updated figure.

Part Six

Chapter 26—This chapter has been extensivey revised and focuses on the study of microbial communities in the broader context of ecology. The influence of metagenomic sequencing is apparent in new sections on the techniques for assessing microbial communities. New techniques featured include BONCAT, substrate analog probing, and marker gene analysis.

Chapter 27—This chapter on microbial interactions features new and improved diagrams, notably of the organisms in a coral ecosystem.

Chapter 28—An introduction to nutrient cycling and biogeochemical cycling precedes the review of major elemental cycles. The chapter builds upon these concepts to explain the role of microbes in an updated discussion of climate change.

Chapter 29—Discussions of microbial adaptation to the marine environment and the importance of the oceans in global climate change have been updated. Coverage of freshwater microbiology has also been revised, emphasizing anthropogenic impacts.

Chapter 30—This chapter complements chapter 27 with discussions of mycorrhizal fungi and nitrogen-fixing bacteria. The role of metagenomics in advancing our understanding of soil microbiology is stressed.

Part Seven

Chapter 31—This chapter provides a concise introduction of innate immunity, including advances in our understanding of the role of the inflammasome and innate lymphoid cells.

Chapter 32—Revised and updated, this discussion of adaptive immunity and immunopathologies provides a current overview to introduce students to the dynamics of human immunity. Many figures have been revised for clarity.

Chapter 33—The rapidly expanding field of the human microbiome is introduced. This chapter follows those on immunology for a complete discussion of the role of human microbiota in immune function, as well as their role in maintaining system homeostasis.

Chapter 34—This chapter provides a broad overview of infectious disease from pathogen transmission to pathogenicity. A new figure illustrating the cycle of a bacterial pathogen infection has been added.

Part Eight

Chapter 35—This chapter covers fundamental epidemiological concepts. A figure summarizing the leading causes of healthcare-associated infections in the United States replaces two previous figures.

Chapter 36—This chapter provides students with an overview of key microbiological and immunological techniques enabling the identification of clinical samples. A description of isothermal amplification has been added, with an accompanying figure.

Chapter 37—This chapter covers important human diseases caused by viruses and prions. Information has been revised and updated.

Chapter 38—Students are introduced to bacterial diseases, including pathogenesis, prevalence, and clinical presentation. Where applicable, the importance of vaccine prevention is stressed.

Chapter 39—This chapter provides an overview of fungal and protozoan diseases of local and global significance. The global burden of key diseases such as Chagas and malaria is emphasized.

Part Nine

Chapter 40—This chapter has been updated to reflect that methods for food testing now rely on the use of molecular methods and whole-genome sequencing.

Chapter 41—The growing reach of biotechnology is illustrated in several examples, including an expanded discussion of industrial enzymes derived from microbes and vaccine design strategies.

Chapter 42—The discussion of water safety has been expanded to include modern sewage treatment strategies. A new Microbial Diversity & Ecology box tracks the changes in microbial community structure following a chemical spill.

About the Authors

Photographer: Joanne Willey

Joanne M. Willey has been a professor at Hofstra University on Long Island, New York, since 1993, where she is the Leo A. Guthart Professor of Biomedical Science and Chair of the Department of Science Education at the Donald and Barbara Zucker School of Medicine at Hofstra/Northwell. Dr. Willey received her B.A. in Biology from the University of Pennsylvania, where her interest in microbiology began with work on cyanobacterial growth in eutrophic streams. She earned her Ph.D. in biological oceanography (specializing in marine microbiology) from the Massachusetts Institute of Technology–Woods Hole Oceanographic Institution Joint Program in 1987. She then went to Harvard University, where she spent her postdoctoral fellowship studying the filamentous soil bacterium *Streptomyces coelicolor.* Dr. Willey has coauthored a number of publications that focus on its complex developmental cycle. She is an active member of the American Society for Microbiology (ASM), and served on the editorial board of the journal *Applied and Environmental Microbiology* for nine years and as Chair of the Division of General Microbiology. Dr. Willey taught microbiology to biology majors for 20 years and now teaches microbiology and infectious disease to medical students. She has taught courses in cell biology, marine microbiology, and laboratory techniques in molecular genetics. Dr. Willey is an avid runner and enjoys skiing, hiking, rock climbing, and reading. She can be reached at joanne.m.willey@hofstra.edu.

Photographer: Kim Tyndall

Kathleen M. Sandman received her B.A. in Biology from La Salle University and her Ph.D. in Cellular and Developmental Biology from Harvard University. She was inspired to a career in science by her older brother's experience as an organic chemist and by the developing technology in recombinant DNA in the 1970s. Her graduate work used a transposable element as a mutagen in *Bacillus subtilis* to study gene expression during endospore formation. She continued in *Bacillus* genetics with a postdoctoral year studying *Bacillus thuringiensis* at the University of Cambridge in the United Kingdom. Another postdoctoral opportunity at The Ohio State University provided an introduction to the emerging field of archaeal molecular biology, where Dr. Sandman discovered archaeal histones and continued research in the structural biology of archaeal chromatin for about 20 years. She served the National Science Foundation as a research grant reviewer and panelist for the Life in Extreme Environments program, and has organized conference sessions on archaeal molecular biology and proteins from extremophiles. Dr. Sandman has taught microbiology to hundreds of students, both at the introductory level and in an advanced molecular microbiology laboratory. She has worked as a consultant in a variety of industries, including industrial microbiology, environmental geomicrobiology, and technical publishing. She lives with her husband in Columbus, Ohio, and enjoys biking, fabric arts, reading, and genealogy, and can be reached at kathleenmsandman@gmail.com.

About the Authors

Photographer: Sandy Coetzee

Sarah N. Salm Sarah Salm is a professor at the Borough of Manhattan Community College (BMCC), of the City University of New York. She earned her B.S., M.S., and Ph.D. degrees at the University of the Witwatersrand in Johannesburg, South Africa. Her doctoral work focused on the epidemiology and characterization of ryegrass mosaic virus (RMV), a mite-transmitted member of the Potyviridae that infects pasture species, resulting in substantial crop losses. Dr. Salm then moved to the United States, where she had the opportunity to do postdoctoral work at the NYU Grossman School of Medicine. Her research there focused on molecular aspects of prostatic stem cells. She subsequently joined the faculty of BMCC, where she has been teaching Microbiology for 20 years. Dr. Salm lives in Manhattan but frequently leaves the city to go hiking. She enjoys traveling, triathlons, reading, and bird watching. She can be reached at ssalm@bmcc.cuny.edu.

Photographer: Lisa Wong

Nathan W. Rigel is the Director of Research Compliance in the Office of Research and Sponsored Programs at Kean University in Union, New Jersey. Dr. Rigel completed his undergraduate studies in Microbiology at the Pennsylvania State University. He then moved south to begin his research career studying bacterial protein export. In 2008, Dr. Rigel earned his Ph.D. in Microbiology and Immunology at the University of North Carolina at Chapel Hill while studying the accessory SecA2 protein export system found in mycobacteria. In 2009, Dr. Rigel returned north to study assembly of outer membrane proteins in *Escherichia coli* as a postdoctoral research associate in the Department of Molecular Biology at Princeton University. He opened his own lab at Hofstra University in 2013, where he focused his studies on *Acinetobacter baylyi*. While at Hofstra, Dr. Rigel taught courses in microbiology, cell and molecular biology, bioinformatics, and bacterial genetics. In 2024, Dr. Rigel made a career pivot to become a research administrator, supporting the rapidly growing research enterprise at Kean University. He can be reached at nathanrigel@gmail.com.

Acknowledgements

In the preparation of each release, we are guided by the collective wisdom of reviewers who are expert microbiologists and excellent teachers. They represent experience in community colleges, liberal arts colleges, comprehensive institutions, and research universities. We have followed their recommendations, while remaining true to our overriding goal of writing readable, student-centered content. Each feature incorporated into this release has been carefully considered in terms of how it may be used to support student learning in both the traditional and the flipped learning environment.

The authors wish to extend their gratitude to our team at McGraw Hill, including Lauren Vondra, Darlene Schueller, Tami Hodge, Vicki Krug, Lori Hancock, Allison Campbell, Sandy Ludovissy, Betsy Blumenthal, and Rajesh Negi. We would like to acknowledge our friend and colleague Dorothy Wood, who has led the effort on the accompanying online content. Thanks to Rebecca E. Steiner for her help with this Release as well. Finally, we thank our spouses and children, who provided support and tolerated our absences (mental, if not physical) while we completed this demanding project.

Contents

Part One Introduction to Microbiology

1 The Evolution of Microorganisms and Microbiology **1**

Micro Focus: The Microbial Universe 1

1.1 Members of the Microbial World 2

1.2 Microbes Have Evolved and Diversified for Billions of Years 4

Microbial Diversity & Ecology 1.1 Hydrothermal Vents: Did Life Begin Under the Sea? 7

1.3 Microbiology Advanced as New Tools for Studying Microbes Were Developed 14

1.4 Microbiology Encompasses Many Subdisciplines 20

2 Microscopy **23**

Micro Focus: Anthrax Bioterrorism Attack 23

2.1 Lenses Create Images by Bending Light 24

2.2 There Are Several Types of Light Microscopes 24

2.3 Staining Helps to Visualize and Identify Microbes 33

2.4 Electron Microscopes Use Beams of Electrons to Create Highly Magnified Images 35

2.5 Scanning Probe Microscopy Can Visualize Molecules and Atoms 40

3 Bacterial Cell Structure **44**

Micro Focus: Bacteria Use Rapid Transport 44

3.1 Bacteria Are Diverse but Share Some Common Structural Features 45

3.2 Bacterial Plasma Membranes Control What Enters and Leaves the Cell 47

3.3 Cell Walls Have Many Functions 52

3.4 Extracellular Vesicles Emerge from Bacterial Membranes 58

3.5 The Cell Envelope Often Includes Layers Outside the Cell Wall 59

3.6 The Bacterial Cytoplasm Is Complex 60

Microbial Diversity & Ecology 3.1 Organelles Without Membranes? 63

3.7 External Structures Are Used for Attachment and Motility 66

3.8 Bacterial Endospores Are a Survival Strategy 68

4 Archaeal Cell Structure **74**

Micro Focus: Methane—The Other Greenhouse Gas 74

4.1 Archaea Are Diverse but Share Some Common Features 75

4.2 Archaeal Cell Envelopes Are Structurally Diverse 77

4.3 Archaeal Cytoplasm Is Similar to Bacterial Cytoplasm 80

4.4 Many Archaea Have External Structures Used for Attachment and Motility 81

5 Eukaryotic Cell Structure **85**

Micro Focus: Red Means Dead 85

5.1 Eukaryotic Cells Are Diverse but Share Some Common Features 86

5.2 Eukaryotic Cell Envelopes 87

5.3 The Eukaryotic Cytoplasm Contains a Cytoskeleton and Organelles 88

5.4 Several Organelles Function in the Secretory and Endocytic Pathways 90

5.5 The Nucleus and Ribosomes Are Involved in Genetic Control of the Cell 94

5.6 Mitochondria, Related Organelles, and Chloroplasts Are Involved in Energy Conservation 95

Microbial Diversity & Ecology 5.1 There Was an Old Woman Who Swallowed a Fly 98

5.7 Many Eukaryotic Microbes Have External Structures Used for Motility 99

6 Viruses and Other Acellular Infectious Agents **103**

Micro Focus: Viruses to the Rescue 103

6.1 Viruses Are Acellular 104

Contents

6.2 Virion Structure Is Defined by Capsid Symmetry and Presence or Absence of an Envelope 104
6.3 Viral Life Cycles Have Five Steps 108
6.4 There Are Several Types of Viral Infections 113
6.5 Virus Cultivation and Enumeration 115
6.6 Viroids and Satellites: Nucleic Acid–Based Subviral Agents 117
6.7 Prions Are Composed Only of Protein 118

Part Two Microbial Nutrition, Growth, and Control

7 Bacterial and Archaeal Growth 121

Micro Focus:
Nurturing Microbes in the Lab 121
7.1 Many Bacteria and Archaea Reproduce by Binary Fission 122
7.2 Bacterial Cell Cycles Are Divided into Three Phases 123
7.3 Archaeal Cell Cycles Are Unique 127
7.4 Growth Curves Consist of Five Phases 128
7.5 Environmental Factors Affect Microbial Growth 131

Microbial Diversity & Ecology 7.1
Microbial Sculptors 134
7.6 Microbial Growth in Natural Environments 138

Historical Highlights 7.1
VBNC: An Important Part of the Bacterial Lifespan 140
7.7 Laboratory Culture of Microbes Requires Conditions That Mimic Their Normal Habitats 143
7.8 Microbial Population Size Can Be Measured Directly or Indirectly 148
7.9 Chemostats and Turbidostats Are Used for Continuous Culture of Microorganisms 151

8 Control of Microorganisms in the Environment 155

Micro Focus:
Scalpel, Please 155
8.1 Microbial Growth and Replication: Targets for Control 156
8.2 Microbes Can Be Controlled by Physical Means 158
8.3 Microorganisms Are Controlled with Chemical Agents 162

Techniques & Applications 8.1
Nanosurfaces to Repel Microbes 166
8.4 Antimicrobial Agents Must Be Evaluated for Effectiveness 166
8.5 Microorganisms Can Be Controlled by Biological Methods 167

9 Antimicrobial Chemotherapy 170

Micro Focus:
A Gift from Traditional Chinese Medicine 170
9.1 Antimicrobial Chemotherapy Evolved from Antisepsis Efforts 171
9.2 Antimicrobial Drugs Have Selective Toxicity 172
9.3 Antimicrobial Activity Can Be Measured by Specific Tests 173
9.4 Antibacterial Drugs 175
9.5 Antiviral Drugs 181
9.6 Antifungal Drugs 182
9.7 Antiprotozoan Drugs 183

Disease 9.1
Chloroquine and COVID-19: A Cautionary Tale 184
9.8 Antimicrobial Drug Resistance Is a Public Health Threat 185

Part Three Microbial Metabolism

10 Introduction to Metabolism 190

Micro Focus:
Flushed Away 190
10.1 Metabolism: Important Principles and Concepts 191
10.2 ATP: The Major Energy Currency of Cells 192
10.3 Redox Reactions: Reactions of Central Importance in Metabolism 194
10.4 Electron Transport Chains: Sets of Sequential Redox Reactions 196

Contents

10.5 Biochemical Pathways: Sets of Linked Chemical Reactions 198
10.6 Enzymes and Ribozymes Speed Up Cellular Chemical Reactions 199
10.7 Metabolism Must Be Regulated to Maintain Homeostasis 203

11 Catabolism: Energy Release and Conservation 208

Micro Focus:
Solar, Wind, and ... Microbes? 208
11.1 Metabolic Diversity and Nutritional Types 209
11.2 There Are Two Chemoorganotrophic Fueling Processes 211
11.3 Aerobic Respiration Starts with Glucose Oxidation 212
11.4 Electron Transport and Oxidative Phosphorylation Generate the Most ATP 219
11.5 Anaerobic Respiration Uses the Same Steps as Aerobic Respiration 224
11.6 Fermentation Does Not Involve an Electron Transport Chain 225
11.7 Catabolism of Organic Molecules Other Than Glucose 228
11.8 Chemolithotrophy: "Eating Rocks" 230
11.9 Flavin-Based Electron Bifurcation 233
11.10 Phototrophy 234

12 Anabolism: The Use of Energy in Biosynthesis 244

Micro Focus:
Building Penicillin 244
12.1 Principles Governing Biosynthesis 245
12.2 Precursor Metabolites: Starting Molecules for Biosynthesis 246
12.3 CO_2 Fixation: Reduction and Assimilation of CO_2 Carbon 247
12.4 Synthesis of Carbohydrates 248
12.5 Synthesis of Amino Acids Consumes Many Precursor Metabolites 251
12.6 Synthesis of Purines, Pyrimidines, and Nucleotides 257
12.7 Lipid Synthesis 258

Part Four Microbial Molecular Biology and Genetics

13 Bacterial Genome Replication and Expression 265

Micro Focus:
Making Code 265
13.1 Experiments Using Bacteria and Viruses Demonstrated That DNA Is the Genetic Material 266
13.2 Nucleic Acid and Protein Structure 268
13.3 DNA Replication in Bacteria 271
13.4 Bacterial Genes Consist of Coding Regions and Other Sequences Important for Gene Function 277
13.5 Transcription in Bacteria 279
13.6 The Genetic Code Consists of Three-Letter "Words" 282
13.7 Translation in Bacteria 285
13.8 Coordination of Gene Expression Processes 290
13.9 Protein Maturation and Translocation 292

14 Regulation of Cellular Processes 299

Micro Focus:
Promoting Expression 299
14.1 Bacteria Use Many Regulatory Strategies 300
14.2 Regulation of Transcription Initiation Saves Considerable Energy and Materials 300
14.3 Attenuation and Riboswitches Stop Transcription Prematurely 305
14.4 RNA Secondary Structures Control Translation 308
14.5 Mechanisms Used for Global Regulation 309
14.6 Bacteria Combine Several Regulatory Mechanisms to Control Complex Cellular Processes 316

15 Eukaryotic and Archaeal Genome Replication and Expression 326

Micro Focus:
Pharming 326
15.1 Genetic Processes in the Three Domains 327
15.2 DNA Replication 327
15.3 Transcription 331

Contents

15.4 Translation and Protein Maturation and Localization 335
15.5 Regulation of Cellular Processes 338

16 Mechanisms of Genetic Variation 343

Micro Focus:
Antimicrobial Resistance 343
16.1 Mutations: Heritable Changes in a Genome 344
16.2 DNA Repair Maintains Genome Stability 347
16.3 Microbes Use Mechanisms Other Than Mutation to Create Genetic Variability 350
16.4 Environmental DNA Brings New Genes into Cells 352
16.5 Mobile Genetic Elements Move Genes Within and Between Cells 353
16.6 Evolution in Action: The Development of Antibiotic Resistance in Bacteria 359

17 Microbial DNA Technologies 362

Micro Focus:
Spinning Stronger Silk 362
17.1 Cloning Technologies Developed from DNA Structure and Enzymology 363
17.2 Polymerase Chain Reaction Amplifies Targeted DNA 367
17.3 Genomic and Metagenomic Libraries: Cloning Genomes in Pieces 370
17.4 Expressing Foreign Genes in Host Cells 371
17.5 Cas9 Nuclease Is a Programmable Tool for Genome Editing 373
17.6 Biotechnology Develops Custom Microbes for Industrial Use 375
Techniques & Applications 17.1
Animal-Free Foods 377

18 Microbial Genomics 380

Micro Focus:
Life on the Microbial Frontier 380
18.1 DNA Sequencing Methods 381
18.2 Genome Sequencing 385
18.3 Metagenomics Provides Access to Uncultured Microbes 387
18.4 Bioinformatics: What Does the Sequence Mean? 388
18.5 Functional Genomics Links Genes to Phenotype 390
18.6 Systems Biology: Making and Testing Complex Predictions 395
18.7 Comparative Genomics 396

Part Five The Diversity of the Microbial World

19 Archaea 401

Micro Focus:
Taking the Pledge 401
19.1 Overview of Archaea 402
19.2 Many Archaea Are Known Primarily from Metagenomics 405
19.3 Phylum Thermoproteota Features Temperature Adaptations 407
19.4 Phyla Methanobacteriota and Halobacteriota: Methanogens, Halophiles, and Others 408

20 Nonproteobacterial Gram-Negative Bacteria 414

Micro Focus:
From Food Waste to Fuel 414
20.1 Bacterial Taxonomy and Cell Envelopes 415
20.2 Thermotogota and Deinococcota Live in Extreme Conditions 415
20.3 Photosynthetic Bacteria Are Diverse 416
Microbial Diversity & Ecology 20.1
UCYN-A: From PCR Fragment to Genome to Organelle 423
20.4 PVC Superphylum (Planctomycetota and Verrucomicrobiota): Atypical Cell Division 423
20.5 Phylum Spirochaetota: Bacteria with a Corkscrew Morphology 426
20.6 Phylum Bacteroidota Includes Important Gut Microbiota 427
20.7 Phylum Fusobacteriota: Commensal Anaerobes 429
20.8 Phylum Desulfobacterota: Anaerobic Sulfate/Sulfur or Metal Reducers 429
20.9 Phyla Bdellovibrionota and Myxococcota: Bacterial Predators 431
20.10 Phylum Campylobacterota: Human and Animal Commensals 433

Contents

21 Proteobacteria **436**

Micro Focus:
Welcome back! 436

21.1 Class Alphaproteobacteria Includes Many Oligotrophs 437

21.2 Gammaproteobacteria Is the Largest Bacterial Class 443

Microbial Diversity & Ecology 21.1
The Microbial Legacy of Coal Mining 446

22 Gram-Positive Bacteria **454**

Micro Focus:
Antibiotic Production: Is It Actually Bacterial Chitchat? 454

22.1 Phylum Actinomycetota 455

22.2 Phylum Bacillota 460

22.3 Gram-Positive Taxonomic Groups Include Some Bacteria with Outer Membranes 467

23 Protists **470**

Micro Focus:
Human Evolution Shaped by a Protist 470

23.1 Protist Diversity Reflects Broad Phylogeny 471

23.2 Discoba–Metamonada Clade 473

23.3 Amoebozoa Clade Includes Protists with Pseudopodia 475

23.4 TSAR Clade: Protists of Global Importance 477

23.5 Haptista Clade 486

23.6 Archaeplastida Clade Includes Green and Red Algae 487

24 Fungi **490**

Micro Focus:
The Complex Story of Caterpillar Fungus 490

24.1 Fungal Biology Reflects Vast Diversity 491

24.2 Zoosporic Fungi Form Motile Spores 494

24.3 Zygomycetous Fungi Have Coenocytic Hyphae 495

24.4 Dikarya Is the Most Diverse Fungal Group 497

Disease 24.1
White-Nose Syndrome Is Decimating North American Bat Populations 500

25 Viruses **504**

Micro Focus:
Disrupting the Viral Life Cycle 504

25.1 Virus Phylogeny Relies on Genomics 505

25.2 Double-Stranded DNA Viruses Infect All Cell Types 506

25.3 Single-Stranded DNA Viruses Use a Double-Stranded Intermediate in Their Life Cycles 514

25.4 Double-Stranded RNA Viruses: RNA-Dependent RNA Polymerase Replicates the Genome and Synthesizes mRNA 515

25.5 Positive-Strand RNA Viruses: Genomes That Are Translated upon Entry 517

25.6 Negative-Strand RNA Viruses: RNA-Dependent RNA Polymerase Is Part of the Virion 521

25.7 Retroviruses: Positive-Strand Viruses That Use Reverse Transcriptase in Their Life Cycles 523

25.8 Reverse Transcribing DNA Viruses 524

Part Six Ecology and Symbiosis

26 Exploring Microbes in Ecosystems **528**

Micro Focus:
Can Soil Have Memory? 528

26.1 The Vocabulary of Ecology Applies to Microbial Environments 529

26.2 Microbial Identification Is Largely Based on Molecular Characterization 529

26.3 Microbial Ecology Also Relies on Cultures 531

Microbial Diversity & Ecology 26.1
The Case of the Endosymbiotic Hypothesis 533

26.4 Assessing Microbial Communities 534

26.5 Assessing Microbial Community Activity 538

27 Microbial Interactions **543**

Micro Focus:
Microbes in Community 543

27.1 Many Types of Microbial Interactions Exist 544

27.2 Mutualism: Obligatory Positive Interaction 545

27.3 Cooperation: Nonobligatory Positive Interaction 549

Contents

27.4 Antagonistic Interactions Prompt Microbial Responses 550

Microbial Diversity & Ecology 27.1
Wolbachia: The World's Most Infectious Microbe? 552

28 Biogeochemical Cycling and Global Climate Change 555

Micro Focus:
Global Climate Change; Infectious Disease Change 555
28.1 Biogeochemical Cycling Sustains Life on Earth 556
28.2 Microbes Mediate Nutrient Cycling 558
28.3 Global Climate Change: Biogeochemical Cycling Out of Balance 565

29 Microorganisms in Marine and Freshwater Ecosystems 569

Micro Focus:
Ocean Death Coming Soon to a Coast Near You 569
29.1 Water Is the Largest Microbial Habitat 570
29.2 Microorganisms in Marine Ecosystems 571
29.3 Microorganisms in Freshwater Ecosystems 580

Microbial Diversity & Ecology 29.1
Attention All Dog Owners! 585

30 Microorganisms in Terrestrial Ecosystems 587

Micro Focus:
Bread for a Hungry World 587
30.1 Soils Are an Important Microbial Habitat 588
30.2 The Soil Microbiome 590
30.3 The Plant MIcrobiome 591
30.4 The Subsurface Biosphere Is Vast 601

Disease 30.1
Citrus Greening and the Power of "Why?" 603

Part Seven Pathogenicity and Host Response

31 Innate Host Resistance 606

Micro Focus:
Vaccines for a Changing Planet 606
31.1 Immunity Arises from Innate Resistance and Adaptive Defenses 607
31.2 Innate Resistance Starts with Barriers 607
31.3 Innate Resistance Relies on Chemical Mediators 609
31.4 Each Type of Innate Immune Cell Has a Specific Function 616
31.5 Organs and Tissues of the Immune System Are Sites of Host Defense 620
31.6 Phagocytosis Destroys Invaders 624
31.7 Inflammation Unites All Components of Immunity 629

32 Adaptive Immunity 633

Micro Focus:
Killing Cancer, Immunologically 633
32.1 Adaptive Immunity Relies on Recognition and Memory 634
32.2 Antigens Elicit Immunity 634
32.3 Adaptive Immunity Can Be Earned or Borrowed 635
32.4 Recognition of Foreignness Is Critical for a Strong Defense 636
32.5 T Cells Are Critical for Immune Function 639
32.6 B Cells Make Antibodies 644
32.7 Antibodies Bind Specific 3-D Antigens 646
32.8 Antibodies Doom Antigens 650

Historical Highlights 32.1
Convalescent Plasma: An Old Treatment for a New Disease 651
32.9 Generating Populations of Diverse Lymphocytes 652
32.10 The Immune System Can Malfunction 655

Techniques & Applications 32.2
Monoclonal Antibody Therapy 656

33 The Microbe-Human Ecosystem 666

Micro Focus:
Embrace Your Gut Flora 666
33.1 Humans Are Holobionts 667
33.2 The Microbiome Develops from Birth to Adulthood 667
33.3 A Functional Core Microbiome Is Required for Host Homeostasis 672
33.4 Many Diseases Have a Connection with Dysbiosis 677
33.5 Microbiome Manipulation Can Be Therapeutic 681

Contents

34 Infection and Pathogenicity **684**

Micro Focus:
The Unlikely Tale of Miasmas, Bras, and Masks 684

34.1 The Process of Infection 685

34.2 Transmission and Entry into the Host 686

Historical Highlights 34.1
The First Indications of Person-to-Person Spread of an Infectious Disease 687

34.3 Surviving the Host Defenses 692

34.4 Damage to the Host 693

Part Eight Microbial Diseases, Detection, and Their Control

35 Epidemiology and Public Health Microbiology **700**

Micro Focus:
Protecting the Herd 700

35.1 Epidemiology Is an Evidence-Based Science 701

Historical Highlights 35.1
John Snow, the First Epidemiologist 702

35.2 Epidemiology Is Rooted in Well-Tested Methods 702

35.3 Infectious Disease Is Revealed Through Patterns Within a Population 705

Historical Highlights 35.2
"Typhoid Mary" 706

35.4 Infectious Diseases and Pathogens Are Emerging and Reemerging 708

35.5 Healthcare Facilities Harbor Infectious Agents 710

35.6 Coordinated Efforts Are Required to Prevent and Control Epidemics 711

Historical Highlights 35.3
The First Immunizations 712

35.7 Bioterrorism Readiness Is an Integral Component of Public Health Microbiology 715

Historical Highlights 35.4
1346—Early Biological Warfare Attack 715

36 Clinical Microbiology and Immunology **718**

Micro Focus:
Ebola and Global Health Security 718

36.1 The Clinical Microbiology Laboratory Detects Infectious Agents and Protects Its Workers 719

36.2 Identification of Microorganisms from Specimens 721

36.3 Immune Responses Can Be Exploited to Detect Infections 729

37 Human Diseases Caused by Viruses and Prions **737**

Micro Focus:
Remembering HIV/AIDS 737

37.1 Viruses Can Be Transmitted by Airborne Routes 738

37.2 Arthropods Can Transmit Viral Diseases 747

37.3 Direct Contact Diseases Can Be Caused by Viruses 749

37.4 Food and Water Are Vehicles for Viral Diseases 760

Historical Highlights 37.1
A Brief History of Polio 762

37.5 Zoonotic Diseases Arise from Human-Animal Interactions 763

37.6 Prion Proteins Transmit Disease 765

38 Human Diseases Caused by Bacteria **769**

Micro Focus:
The Plague Family Tree 769

38.1 Bacteria Can Be Transmitted by Airborne Routes 770

38.2 Arthropods Can Transmit Bacterial Diseases 779

38.3 Direct Contact Diseases Can Be Caused by Bacteria 782

Disease 38.1
Syphilis and the Tuskegee Study 789

Disease 38.2
Biofilms 790

38.4 Food and Water Are Vehicles for Bacterial Diseases 795

Techniques & Applications 38.3
Clostridial Toxins as Therapeutic Agents: Benefits of Nature's Most Toxic Proteins 798

38.5 Zoonotic Diseases Arise from Human-Animal Interactions 802

38.6 Opportunistic Diseases Can Be Caused by Bacteria 804

Contents

39 Human Diseases Caused by Fungi and Protists **811**

Micro Focus:
One Health for Tropical Diseases That Should Not Be Neglected 811

39.1 Relatively Few Fungi and Protists Are Human Pathogens 812

39.2 Fungi Can Be Transmitted by Airborne Routes 813

39.3 Arthropods Can Transmit Protozoal Disease 815

Disease 39.1
A Brief History of Malaria 816

39.4 Direct Contact Diseases Can Be Caused by Fungi and Protists 822

39.5 Food and Water Are Vehicles of Protozoal Diseases 825

39.6 Opportunistic Diseases Can Be Caused by Fungi and Protists 829

Part Nine Applied Microbiology

40 Microbiology of Food **836**

Micro Focus:
The Art, Science, and Genetics of Brewing Beer 836

40.1 Microbial Growth Can Cause Food Spoilage 837

40.2 Environmental Factors Control Food Spoilage 838

40.3 Food-Borne Illnesses 840

40.4 Detection of Food-Borne Pathogens Requires Government-Industry Cooperation 842

40.5 Microbiology of Fermented Foods: Beer, Cheese, and Much More 844

Techniques & Applications 40.1
Chocolate: The Sweet Side of Fermentation 845

41 Biotechnology and Industrial Microbiology **851**

Micro Focus:
Where Are the New Antibiotics? 851

41.1 Microbes Are the Source of Many Products of Industrial Importance 852

41.2 Microbial Cells Are Customized for Specific Tasks 854

41.3 Growing Microbes in Industrial Settings Presents Challenges 856

41.4 Agricultural Biotechnology Repurposes Plant and Insect Pathogens 858

41.5 Biofuel Production Is a Dynamic Field 859

42 Applied Environmental Microbiology **862**

Micro Focus:
Viruses Down the Drain 862

42.1 Purification and Sanitary Analysis Ensure Safe Drinking Water 863

42.2 Wastewater Treatment Maintains Human and Environmental Health 865

42.3 Microbial Fuel Cells: Batteries Powered by Microbes 869

42.4 Biodegradation and Bioremediation Harness Microbes to Clean the Environment 871

Microbial Diversity & Ecology 42.1
Microbial First Responders 873

Appendix 1 A Review of the Chemistry of Biological Molecules A-1

Appendix 2 Common Metabolic Pathways A-9

Appendix 3 Microorganism Pronunciation Guide A-17

Glossary G-1

Index I-1

CHAPTER

1

The Evolution of Microorganisms and Microbiology

Denis Belitsky/Shutterstock

The Microbial Universe

If you have ever gazed at the night sky on a cloudless evening in a region far from light pollution, you have probably been amazed and perhaps a little humbled by the vast number of stars. It's hard to estimate just how many stars are out there; best estimates start with our own Milky Way Galaxy, which has roughly 100 billion (1×10^{11}) stars. Astronomers figure there are something like 10 trillion (1×10^{13}) galaxies in the universe. Assuming all galaxies are roughly the size of the Milky Way, you wind up with a number around 1×10^{24} stars in the universe—not the most accurate number, but nonetheless daunting.

If you turn your gaze back to Earth, you will find even more awesome abundance. For example, if you stacked all of the 1×10^{31} viruses on Earth one on top of the other, they would stretch about 100 million light-years. That's about 43 times farther away than the Andromeda Galaxy. And next time you take a dip in the sea, consider that there are at least a hundred thousand (1×10^{5}) more bacteria in the ocean (about 1.3×10^{29}) than stars in the universe. Or perhaps you prefer staying on land, where a teaspoon of soil has about a billion (1×10^{9}) microorganisms.

What are all these microorganisms doing? The short answer is, making life for the rest of us possible. How? Starting about 2.4 billion years ago, bacteria called cyanobacteria started releasing oxygen in abundance into the atmosphere. This has been dubbed the "great oxidation event," and it set the stage for oxygen-consuming organisms (like us) to evolve. Microorganisms have another starring role in the evolution of life because bacteria can fix nitrogen—that is, take gaseous nitrogen and convert it to organic nitrogen used by plants, animals, and other microbes. Finally, can you imagine what life on Earth would look like if dead organic material were not degraded? Probably best not to. Much better to think of foods like beer, wine, chocolate, cheese, and yogurt—all microbial products.

But, of course, not all microorganisms make life possible, or even easier. Prior to the COVID-19 pandemic, about 13 million (1.3×10^{7}) people died from infectious disease annually; many of these deaths were preventable by either vaccination or antibiotic treatments. As of early 2025, over 7 million (7×10^{6}) people were known to have died from COVID-19 worldwide. This is considered an underestimate, with the real number thought to be between 20 and 50 million ($2–5 \times 10^{7}$). Although we know the most about disease-causing microorganisms, there is a lot left to learn about all microbes. In fact, like the number of stars, the number of microbial species (including bacteria, viruses, fungi, and protists) is debated. What is certain is that microbes are important for all other life on Earth.

Our goal in this chapter is to introduce you to the amazing world of microorganisms and outline their evolution and history of discovery. Much of microbiology is similar to what you have learned in other biology classes that focus on large organisms. But microbes have unique properties that often require specialized approaches to understand them. But before you delve into this chapter, check to see if you have the background needed to get the most from it.

Readiness Check:

Based on what you have learned previously, you should be able to:

✓ List two major features of eukaryotic cells that distinguish them from other cell types

✓ Describe the basic structure of the macromolecules nucleic acids, proteins, carbohydrates, and lipids (*see appendix I*)

✓ Explain the terms *genome, genotype,* and *mutation*

1.1 Members of the Microbial World

After reading this section, you should be able to:

a. Define the term *microbiology*
b. Explain Carl Woese's contributions in establishing the three-domain system for classifying cellular life
c. Determine the type of microbe (bacterium, fungus, etc.) when given a description of a microorganism
d. Provide an example of the importance to humans of each of the major types of microbes

Microorganisms—those organisms too small to be seen clearly by the unaided eye (**figure 1.1**)—are fabulously diverse and unimaginably abundant. It is difficult to accurately count the number of microbes on Earth, but estimates are about 10^{30} microbial cells in habitats as diverse as termite guts and sediments deep beneath the seafloor (**figure 1.2**). There are roughly an equal number of viruses and other acellular entities across these ecosystems.

Although microbes are generally 1 millimeter or less in diameter, some, such as molds, are visible without microscopes. Some macroscopic microorganisms are multicellular. They are distinguished from other multicellular life forms such as plants and animals by their lack of highly differentiated tissues. In addition, a variety of acellular biological entities, including viruses and subviral agents, are also called *microorganisms* and *microbes*. This is not without controversy because these agents cannot reproduce independently.

The diversity of microorganisms has always presented a challenge to microbial taxonomists. Early descriptions of cellular microbes as either plants or animals were too simple. For instance, some microbes are motile like animals but also have cell walls and are photosynthetic like plants. An important breakthrough in microbial taxonomy arose from studies of their cellular architecture, when it was discovered that cells exhibited one of two possible "floor plans." Cells that came to be called **prokaryotic cells** (Greek *pro,* before; *karyon,* nut or kernel) have an open floor plan. That is, their contents are not divided into compartments by membranes. Only **eukaryotic cells** (Greek *eu,* true) have a nucleus and other membrane-bound organelles (e.g., mitochondria, chloroplasts) that separate some cellular materials and processes from others.

These observations eventually led to the development of a classification scheme that divided organisms into five kingdoms: Monera, Protista, Fungi, Animalia, and Plantae. Microorganisms (except for viruses and other acellular infectious agents) were placed in the first three kingdoms. In this scheme, all organisms with prokaryotic cell structure were placed in Monera. However, the five-kingdom system is no longer accepted by microbiologists. This is because prokaryotes are too diverse to be grouped together in a single kingdom.

Classifying microbes has benefited from progress in three technologies. First, electron microscopy reveals the detailed structure of microbial cells. Second, methods that measure the biochemical and physiological characteristics of many different microorganisms demonstrate their similarities and differences. Third, advances in genomics enables the analysis of nucleic acid and protein sequences from a wide

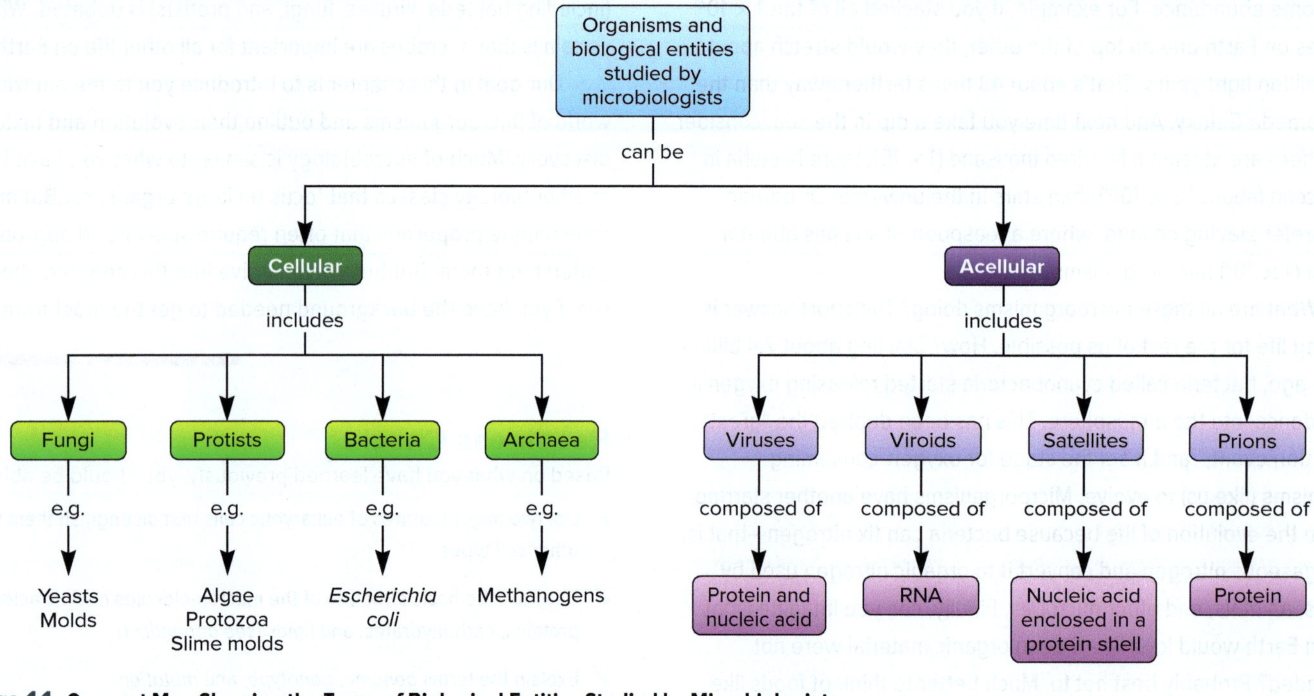

Figure 1.1 Concept Map Showing the Types of Biological Entities Studied by Microbiologists.

MICRO INQUIRY *How would you alter this concept map so that cellular organisms are differentiated by their key features?*

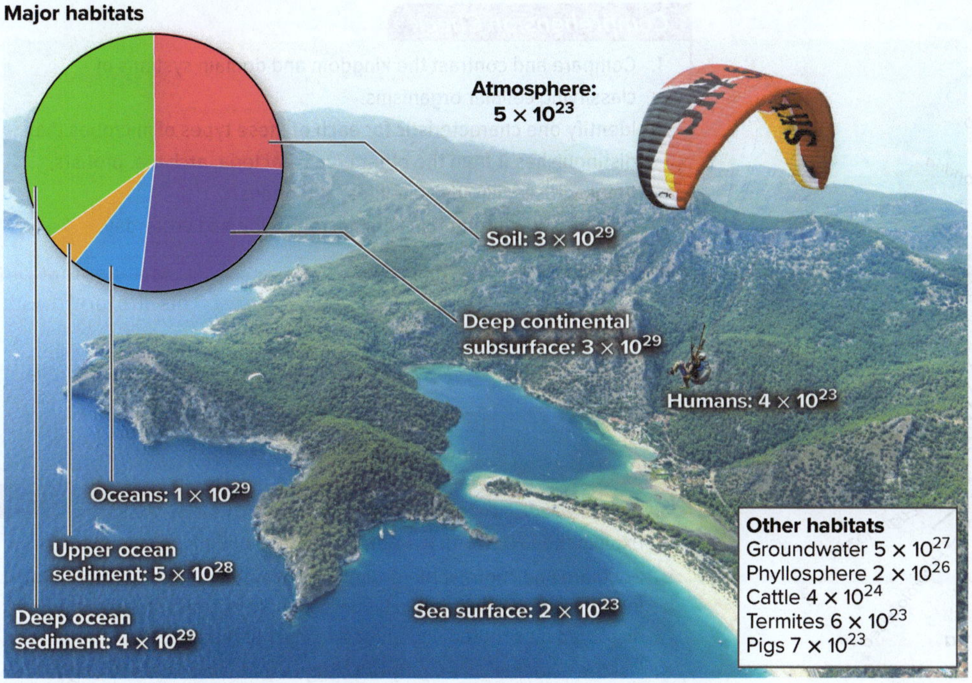

Figure 1.2 Bacterial and Archaeal Habitats and Abundance. Numbers indicate the number of microbial cells in each habitat. The majority of bacteria and archaea live in oceans and sediments, either within the Earth's crust or deep below the crust (subsurface). The discovery of viable microbes so deep within our planet is a recent and exciting development. Other habitats include the phyllosphere (aboveground portions of plants), livestock, and humans.

Fotout/Shutterstock

variety of organisms. The comparison of ribosomal RNA (rRNA) nucleic acid sequences, begun by Carl Woese (1928–2012) in the 1970s, transformed our understanding of the term *prokaryote*. It was discovered that there are two very different groups of organisms with prokaryotic cell morphology: Bacteria and Archaea. Because of these discoveries and additional analyses, Norman Pace proposed in 2006 that the term *prokaryote* be abandoned. Many microbiologists, including the authors of this textbook, agree.

Among eukaryotic microbes, studies show that Protista is not a cohesive taxonomic unit (i.e., taxon) and that it should be divided into multiple groups. These studies and others led many taxonomists to reject the five-kingdom system in favor of one that divides cellular organisms into three domains: Bacteria, Archaea, and Eukarya (all eukaryotic organisms) (**figure 1.3**). ▶| *Nucleic acids (appendix I); Proteins (appendix I)*

Members of domain **Bacteria** are usually single-celled organisms.[1] Most have cell walls that contain the structural molecule peptidoglycan. Despite popular belief, most bacteria do not cause disease. In fact, bacteria are major inhabitants of our bodies, forming the human **microbiome.** Indeed, about as many microbial cells are found in and on the human body as there are human cells, where they help maintain our health and well-being. For example, microbes contribute to the development of the body's immune system and those that inhabit the large intestine produce vitamins and help digest food. ▶| *Overview of bacterial cell wall structure (section 3.3); A Functional Core Microbiome is Required for Host Homeostasis (section 33.3)*

Unfortunately some bacteria do cause disease, and some of these diseases can have a huge impact. In 1347 the plague (Black Death), caused by bacteria living in fleas, struck Europe with brutal force, killing one-third of the population within 4 years. Over the next 80 years, the disease struck repeatedly, eventually wiping out roughly half of the European population. The resulting labor shortage gave workers more power, ultimately eliminating serfdom, and preparing the way for the Renaissance.

Members of domain **Archaea** are distinguished from bacteria by many features, most notably their distinctive rRNA sequences, cell walls, and membrane lipids. Some have unique metabolic characteristics, such as the ability to generate methane (natural) gas. Some archaea are found in extreme environments, including those with high temperatures and high concentrations of salt. Archaea do not appear to directly cause disease in humans.

Domain **Eukarya** includes plants, animals, and microorganisms classified as protists or fungi. **Protists** are generally unicellular but larger than most bacteria and archaea. They have traditionally been divided into **protozoa,** which have an animal-like metabolism, and **algae,** which are photosynthetic. However, these terms lack taxonomic value because algae and protozoa do not form distinct groups, each with a single evolutionary history. Nonetheless, for convenience, we use these terms here. ▶| *Protist diversity reflects broad phylogeny (section 23.1)*

Fungi are a diverse group of microorganisms that range from unicellular forms (yeasts) to multicellular molds and mushrooms. Because of their metabolic capabilities, many fungi play beneficial roles, including making bread dough rise, producing antibiotics, and decomposing dead organisms. Some fungi associate with plant roots to form mycorrhizae. Mycorrhizal fungi transfer nutrients to the roots, improving growth of the plants, especially in poor soils. Other fungi cause plant diseases (e.g., rusts, powdery mildews, and smuts) and diseases in humans and other animals. ▶| *Fungal biology reflects vast diversity (section 24.1)*

1 In this text, the term *bacteria* (s., *bacterium*) is used to refer to those microbes belonging to domain Bacteria, and the term *archaea* (s., *archaeon*) is used to refer to those that belong to domain Archaea. In some publications, the term *bacteria* is used to refer to all cells having prokaryotic cell structure. That is not the case in this text.

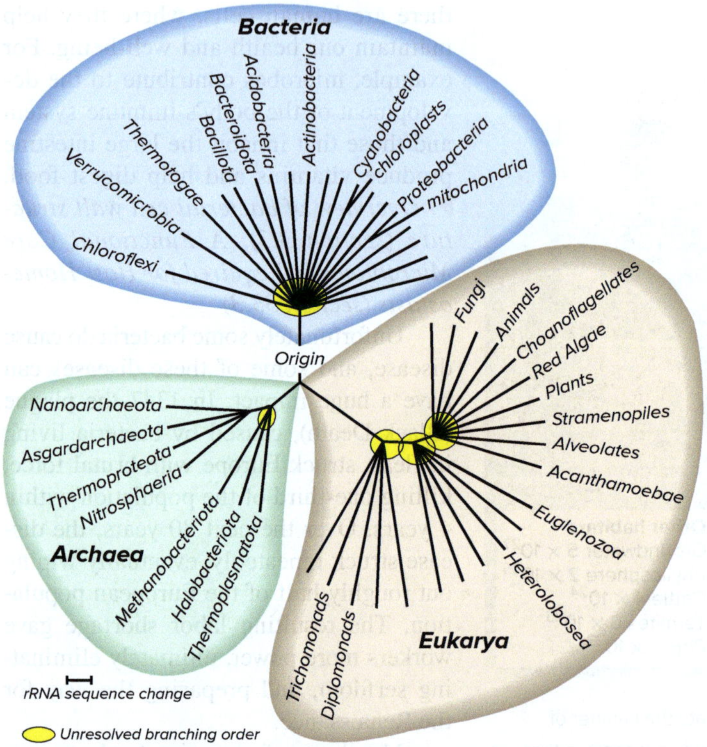

Figure 1.3 Universal Phylogenetic Tree. Only representative lineages have been identified. All Eukarya except fungi, animals, and plants are considered protists.

MICRO INQUIRY *How many of the taxa listed in the figure include microbes?*

The microbial world also includes numerous acellular infectious agents. **Viruses** are acellular entities that must invade a host cell to multiply. The simplest virus particles (also called virions) are composed only of proteins and a nucleic acid, and can be extremely small (the smallest is 10,000 times smaller than a typical bacterium). However, their small size belies their power. They cause many animal and plant diseases and, as we saw most recently with COVID-19, can trigger epidemics and pandemics that shape human history. Other viral diseases include rabies, influenza, AIDS, the common cold, and some virus-related cancers. Viruses are also important in aquatic environments, where they play a critical role in shaping microbial communities. **Viroids** are infectious agents composed only of ribonucleic acid (RNA). They cause numerous plant diseases. **Satellites** are composed of a nucleic acid enclosed in a protein shell. They must coinfect a host cell with a virus, called a helper virus, to complete their life cycle. Satellites and their helper viruses cause both plant and animal diseases. Finally, **prions,** infectious agents composed only of protein, are responsible for causing neurological diseases such as scrapie and mad cow disease. ▶| *Viruses and other acellular infectious agents (chapter 25)*

Comprehension Check

1. Compare and contrast the kingdom and domain systems of classifying cellular organisms.
2. Identify one characteristic for each of these types of microbes that distinguishes it from the other types: bacteria, archaea, protists, fungi, viruses, viroids, satellites, and prions.
3. Describe one interaction with microbes you had yesterday.

1.2 Microbes Have Evolved and Diversified for Billions of Years

After reading this section, you should be able to:

a. Explain the RNA world hypothesis and the evidence that supports it
b. Compare and contrast the evolution of mitochondria and chloroplasts
c. Interpret the evolutionary relationships between microorganisms displayed on a rooted and unrooted phylogenetic tree
d. List the taxonomic hierarchy of an organism from domain through species

A review of figure 1.3 reminds us that microbes are the dominant organisms on Earth. How has microbial life been able to radiate to such an astonishing level of diversity? To answer this question, we must consider microbial evolution. The field of microbial evolution, like any other scientific endeavor, is based on the formulation of hypotheses, the gathering and analysis of data, and the reformation of hypotheses based on newly acquired evidence. That is to say, the study of microbial evolution is based on the scientific method. To be sure, it is difficult to amass evidence when considering events that occurred millions, and often billions, of years ago, but the application of molecular methods has revealed a living record of life's ancient history. This section describes the outcome of this scientific research.

Theories of the Origin of Life Depend Primarily on Indirect Evidence

Dating meteorites through the use of radioisotopes places our planet at an estimated 4.5 to 4.6 billion years old. However, conditions on Earth for the first 100 million years or so were far too harsh to sustain any type of life. Eventually bombardment by meteorites decreased, water appeared on the planet in liquid form, and gases were released by geological activity to form Earth's atmosphere. These conditions were amenable to the origin of the first life forms. But how did this occur, and what did these life forms look like?

To find evidence of life and develop hypotheses about its origin and subsequent evolution, scientists must be able to define life. Although even very young children can examine an object and correctly determine whether it is living or not, defining life succinctly is actually not that easy. Most definitions of life consist of a set of attributes. The attributes of particular importance to paleobiologists are an orderly structure, the ability to obtain and use energy (i.e., metabolism), and the ability to reproduce. Microbiologists use **extant organisms,** those organisms present today, to explore the origin of life. Some extant organisms have structures and molecules that represent relics of ancient life forms. These can provide clues about the type of evidence to seek when testing hypotheses.

The best direct evidence for the nature of primitive life would be a fossil record. There have been reports of microbial fossil discoveries since 1977. These have always met with skepticism because some objects that look like cells can be formed by geological forces that occurred as the rock was formed. The result is that the fossil record for microbes is sparse and always open to reinterpretation.

Despite these problems, most scientists agree that life was present on Earth about 3.5 to 3.9 billion years ago (**figure 1.4**). To reach this conclusion, biologists rely on indirect evidence such as molecular fossils. These are chemicals found in rock or sediment that are chemically related to biological molecules. For instance, the presence of molecules called hopanes in a rock indicates that bacteria were present when the rock was formed. This conclusion is reached because hopanes are formed from hopanoids, which are found in the plasma membranes of extant bacteria. As you can see, no single piece of evidence can stand alone. Instead many pieces of evidence are put together in an attempt to get a coherent picture to emerge, as with a jigsaw puzzle.

Early Life Was Probably RNA-based

Before there was life, most evidence suggests that Earth was a very different place: hot and **anoxic** (without oxygen), with an atmosphere rich in water vapor, carbon dioxide, and nitrogen. In the oceans, hydrogen, methane, and carboxylic acids were formed by geological and chemical processes. Areas near hydrothermal vents may have provided the conditions that allowed chemicals to react with one another, randomly testing the usefulness of the reaction and the stability of its products. Some reactions generated molecules that functioned as catalysts, some aggregated with other molecules to form the predecessors of modern cell structures, and others were able to replicate and act as units of hereditary information (**Microbial Diversity & Ecology 1.1**).

How did early cells, sometimes called *probionts*, arise? Extant cells require three types of molecules: catalysts, structural molecules, and hereditary molecules. Proteins include catalytic proteins, or **enzymes,** and structural proteins, which have myriad functions, such as transport, attachment, and motility. DNA stores hereditary information that is replicated and passed on to the next generation. RNA converts the information stored in DNA into protein. Any hypothesis about the origin of life must account for the evolution of these molecules, but their relationships to each other in modern cells complicates attempts to imagine how they evolved. Proteins can do cellular work, but their synthesis involves other proteins and RNA, and uses information stored in DNA. DNA cannot do cellular work, and proteins are needed for its replication. RNA synthesis requires both DNA as the template and proteins as catalysts.

Based on these considerations, it is hypothesized that at some time in the evolution of probionts, there must have been a single molecule with both catalytic and hereditary activities. This idea was supported in 1981 when Thomas Cech discovered an RNA molecule in a protist (*Tetrahymena* sp.) that, like a protein enzyme, also had catalytic activity. Since then, other catalytic RNA molecules have been discovered, including RNA found in ribosomes that is responsible for forming peptide bonds—the bonds that hold together amino acids, the building blocks of proteins. Catalytic RNA molecules are now called **ribozymes.**

The discovery of ribozymes suggested that RNA at some time was capable of storing, copying, and expressing genetic information, as well as catalyzing other chemical reactions. In 1986 Nobel laureate Walter Gilbert coined the term **RNA world** to describe this precellular stage in the evolution of life. However, for this precellular RNA-based stage to proceed to the evolution of cellular life forms, a lipid membrane must have formed around the RNA (**figure 1.5**). This important evolutionary step is easier to imagine than other events in the origin of cellular life forms because lipids, major structural components of the membranes of modern organisms, spontaneously form liposomes—vesicles bounded by a lipid bilayer. The notion of an RNA world has caused some scientists to look for evidence on Mars, where conditions are thought to have been frozen in the prebiotic era. ▶| *Lipids (appendix I)*

Back here on Earth, Jack Szostak, also a Noble laureate, is a leader in experimentally simulating how protobionts containing only RNA may have formed. When his group created liposomes using simpler fatty acids than those found in membranes today, the liposomes were leaky. These leaky liposomes allowed single RNA nucleotides to move into the liposome, but prevented large RNA chains from moving out. Furthermore, researchers could prod the liposomes into growing and dividing. Dr. Szostak's group has also been able to create conditions in which an RNA molecule could serve as a template for synthesis of a complementary RNA strand. In 2024, they showed that such ribozymes could evolve to RNAs with improved function. These experiments may have recapitulated early steps in the evolution of cells. As seen in figure 1.5, several other processes need to occur to reach the level of complexity found in extant cells.

Apart from its ability to perform catalytic activities, the function of RNA suggests its ancient origin. Consider that much of the cellular pool of RNA in modern cells exists in the ribosome, a structure that consists largely of ribosomal RNA (rRNA) and uses messenger RNA (mRNA) and transfer RNA

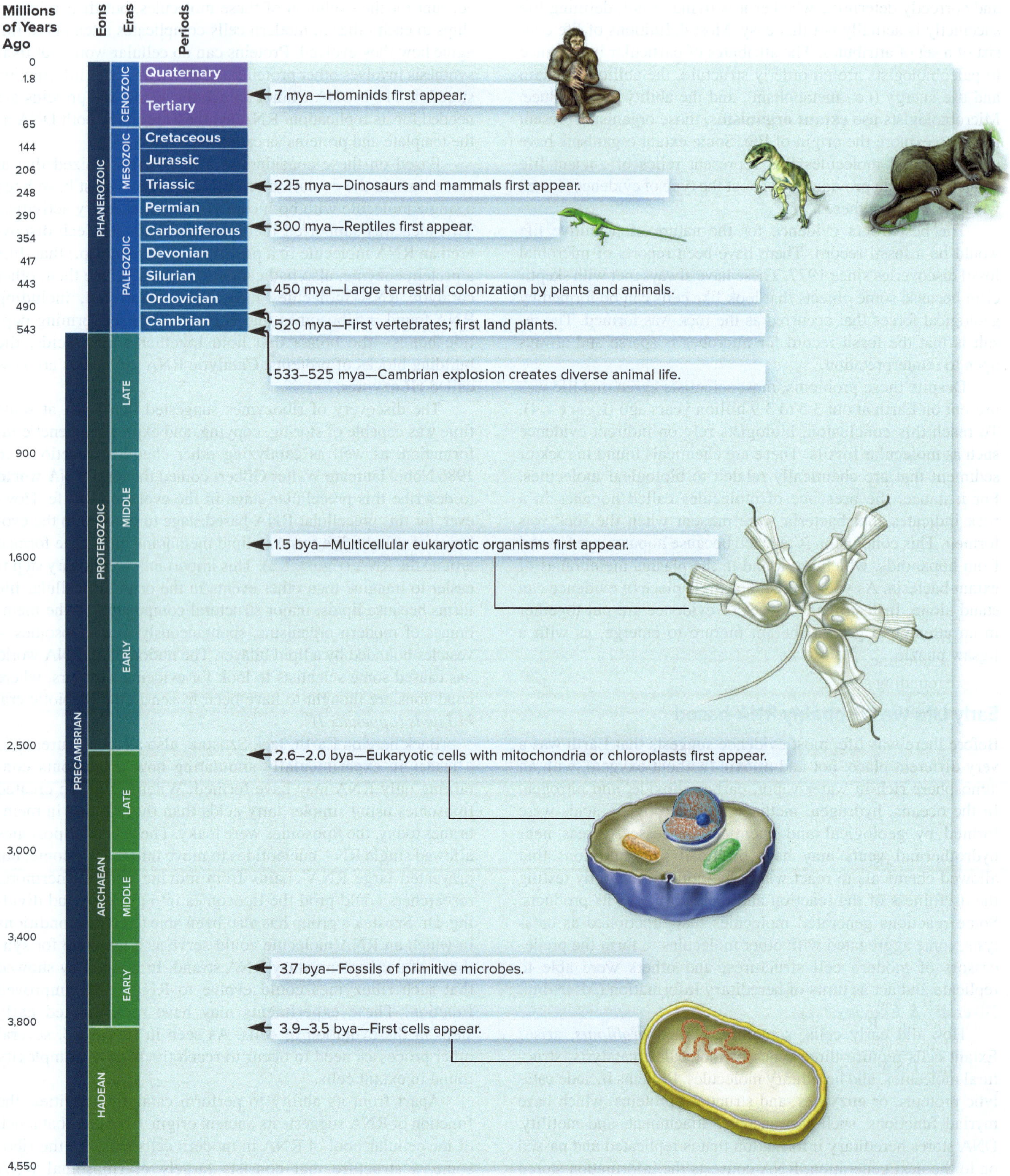

Figure 1.4 An Overview of the History of Life on Earth. mya = million years ago; bya = billion years ago.

MICROBIAL DIVERSITY & ECOLOGY

1.1 Hydrothermal Vents: Did Life Begin Under the Sea?

Whether or not early life was RNA-based, one thing is clear: The origin of life needed energy to synthesize biomolecules. So, perhaps the most fundamental evolutionary question is "Where did biomolecules and the energy needed to build them come from?" Three hypotheses have been suggested. First, the *panspermia theory* speculates that meteorites bombarded our planet, bringing with them other-worldly biomolecules. Second, the more familiar *primordial soup theory* suggests that organic molecules spontaneously assembled with an input of energy, such as lightning strikes. The last theory, which has gained evidence in recent years, hypothesizes that both the energy and the molecules originated in hydrothermal vents. Let's explore the *hydrothermal vent theory.*

Hydrothermal vents are geothermally active deep-sea chasms thousands of meters below the surface of the ocean. Their discovery in 1977 sparked tremendous excitement as images of entirely new ecosystems with mysterious organisms captured the attention of scientists and the public (*see section 27.2*). These vents pump 400°C sulfide-rich water into cold ambient water, causing the sulfide to instantly precipitate, so these chimneylike structures are dubbed "black smokers." In 2000, scientists made yet another deep-sea discovery with a different kind of vent system. These are cooler (45–90°C) and alkaline (pH 9–11). When these waters mix with the surrounding seawater (pH about 8.0), calcium carbonate precipitates, forming white chimneys, as seen in the Lost City vents (**box figure**).

This pH gradient is critical to the hypothesis that a vent system, such as Lost City, could be the origin of biomolecules. As you may have learned when studying mitochondria or batteries, the separation of positive and negative charges captures potential energy (remember that energy can't be created). In Lost City vents, the thin walls of the chimneys serve to separate these fluids with as much as a 3-unit pH difference. The question now being asked is "Was this potential energy tapped to convert CO_2 in seawater to simple carbon-based molecules, such as amino acids, short hydrocarbons, and others?"

D.Kelley, University of Washington.

If the answer is yes, it is likely relevant that a mixture of molecules called single-chain amphiphiles (SCAs), which are simpler versions of more familiar phospholipids, can form vesicles in hot, alkaline pH seawater that mimics that of Lost City. Putting this together, we can hypothesize a series of events that occurred 3.7–4.0 billion years ago. First, the presence of the pH gradient across geological barriers in the Lost City drove the formation of random organic molecules, some of which were SCAs. These SCAs accumulated and formed vesicles that entrapped fluids preserving the pH gradient. These vesicles had the energy to test the formation of different molecules. Was one of them RNA?

(tRNA) to construct proteins. Also rRNA itself catalyzes peptide bond formation during protein synthesis. Thus RNA seems to be well poised for its importance in the development of proteins. Because RNA and DNA are structurally similar, RNA could have given rise to double-stranded DNA. It is suggested that once DNA evolved, it became the storage facility for genetic information because it provides a more chemically stable structure. Two other pieces of evidence support the RNA world hypothesis: the fact that the energy currency of cells, ATP, is a ribonucleotide and the discovery that RNA can regulate gene expression. ▶| *ATP: the major energy currency of cells (section 10.2); Riboswitches: effector-mRNA interactions regulate transcription (section 14.3); Translational riboswitches (section 14.4)*

However, the RNA world hypothesis is not without problems, and more recent experiments suggest the first nucleic acids may have been a mix of DNA and RNA molecules. Another area of research also fraught with considerable debate is the evolution of metabolism. Early Earth was a hot environment that lacked oxygen. Thus cells that arose there must have been able to use the available energy sources under these harsh conditions. Today there are heat-loving archaea capable of using inorganic molecules such as FeS as a source of energy. Some suggest that this interesting metabolic capability is a remnant of

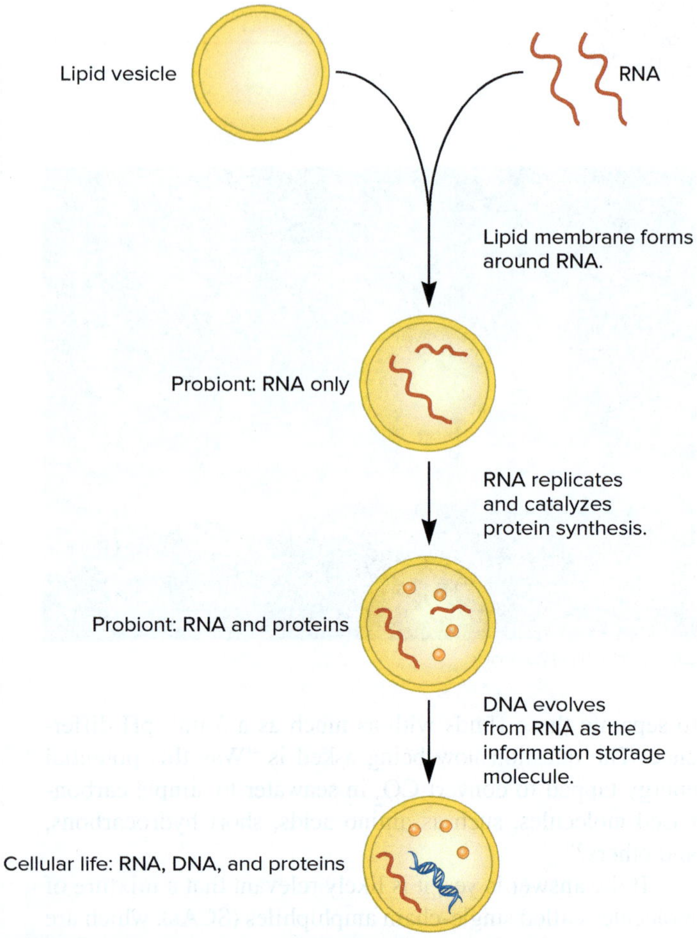

Figure 1.5 The RNA World Hypothesis for the Origin of Life.

MICRO INQUIRY *Why are the probionts pictured above not considered cellular life?*

the first form of energy metabolism. Another metabolic strategy, oxygen-releasing photosynthesis (oxygenic photosynthesis), appears to have evolved perhaps as early as 2.7 billion years ago. This is supported by the discovery of ancient stromatolites (**figure 1.6**). Stromatolites are layered rocks formed by the incorporation of mineral sediments into layers of cyanobacteria growing in thick mats on surfaces. The oxygen released by these early cyanobacteria is thought to have altered Earth's atmosphere to its current oxygen-rich state, allowing the evolution of additional energy-capturing strategies such as aerobic respiration, the oxygen-consuming metabolic process used by many microbes and animals.

Evolution of the Three Domains of Life

Look closely at figure 1.3 and find a line labeled "Origin." This is where data indicate the ***last universal common ancestor* (LUCA)** to all three domains should be placed. LUCA is the most recent organism from which all three types of life—bacterial, archaeal, and eukaryotic—arose. On this tree of life, LUCA is on the bacterial branch, which means that Archaea and Eukarya evolved independently, separate from Bacteria.

Figure 1.6 Stromatolites. Modern stromatolites from Western Australia. Each stromatolite is a rocklike structure, typically 1 m in diameter, containing layers of cyanobacteria.

Horst Mahr/imagebroker/age fotostock

The evolutionary relationship of Archaea and Eukarya is still a matter of considerable debate. According to the **universal phylogenetic tree** (figure 1.3), Archaea and Eukarya shared common ancestry but diverged and became separate domains. Recent evidence supports the notion that Eukarya evolved from Archaea *(see Microbial Diversity & Ecology 26.1)*. The close evolutionary relationship of these two forms of life is still evident in the manner in which they process genetic information. For instance, certain protein subunits of archaeal and eukaryotic RNA polymerases, the enzymes that catalyze RNA synthesis, resemble each other to the exclusion of those of bacteria. However, archaea have other features that are most similar to their counterparts in bacteria (e.g., mechanisms for conserving energy). This has further complicated and fueled the debate. The evolution of the nucleus and endoplasmic reticulum is also controversial. However, hypotheses regarding the evolution of other membrane-bound organelles are more widely accepted and are considered next.

Mitochondria, Mitochondria-Like Organelles, and Chloroplasts Evolved from Endosymbionts

The **endosymbiotic hypothesis** is generally accepted as the origin of several eukaryotic organelles, including mitochondria, chloroplasts, and hydrogenosomes. **Endosymbiosis** is an interaction between two organisms in which one organism lives inside the other. One hypothesis actively being explored is that archaea belonging to the Asgard group may have engulfed a bacterial partner. If the intracellular bacterium used aerobic respiration, it became a mitochondrion. If the endosymbiont was a cyanobacterium and therefore performed photosynthesis, it became a chloroplast (**figure 1.7**).

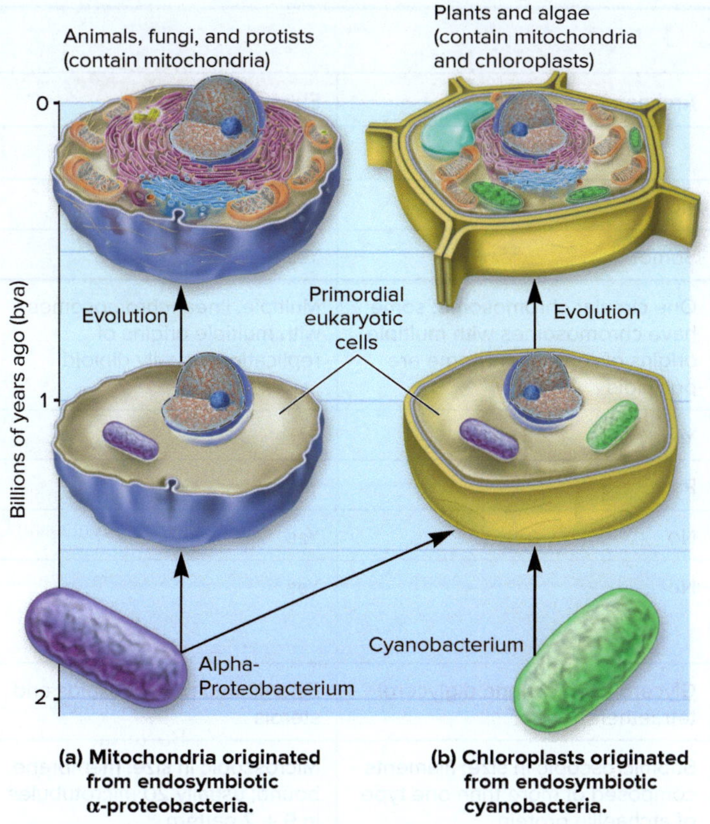

Figure 1.7 The Endosymbiotic Theory. (a) According to this hypothesis, mitochondria derived from a bacterium in the phylum Pseudomonadota in the class Alphaproteobacteria. (b) A similar phenomenon occurred for chloroplasts, which derived from cyanobacteria.

Although the mechanism by which the endosymbiotic relationship was established is unclear, there is considerable evidence to support this hypothesis. Mitochondria and chloroplasts contain DNA and ribosomes that are similar to bacterial DNA and ribosomes. Peptidoglycan, the unique bacterial cell wall molecule, has even been found between the two membranes that enclose the chloroplasts of some algae. Indeed, inspection of figure 1.3 shows that both organelles belong to the bacterial lineage. More specifically, mitochondria are most closely related to bacteria called Alphaproteobacteria. The chloroplasts of plants and green algae are thought to have descended from an ancestor of the cyanobacterial genus *Prochloron,* which contains species that live within marine invertebrates. ▶| *Phylum Cyanobacteriota: oxygenic photosynthetic bacteria (section 20.3); The origin of mitochondria (section 21.1)*

The endosymbiotic hypothesis for mitochondria has been refined by the **hydrogen hypothesis.** This asserts that the endosymbiont was an anaerobic bacterium that produced H_2 and CO_2 as end products of its metabolism. Over time, the host became dependent on the H_2 produced by the endosymbiont. Ultimately the endosymbiont evolved into one of several organelles (*see figure 5.14*). Some endosymbionts evolved into organelles such as a hydrogenosome—an organelle found in some extant protists that produces ATP by a process called fermentation *(see figure 5.16).*

Evolution of Cellular Microbes

Although the history of early cellular life forms may never be known, it's clear that once microbial cells arose, they were subjected to the same evolutionary processes as modern organisms. The ancestral bacteria, archaea, and eukaryotes possessed genetic information that could be duplicated, lost, or mutated in other ways. While some mutations led to the death of the microbe, others allowed new functions and characteristics to evolve. Mutations that allowed the organism to increase its rate of reproduction or survival were selected and passed on to subsequent generations. In addition to selective forces, geographic isolation of populations allowed some groups to evolve separately from others. Thus selection and isolation led to the eventual development of new collections of genes (i.e., genotypes) and new species.

In addition to mutation, other mechanisms exist for reconfiguring genomes and therefore creating genetic diversity. Most eukaryotic species increase their genetic diversity by reproducing sexually, giving rise to offspring with a mixture of parental genes and a unique genotype. Bacteria and archaea do not reproduce sexually. They increase their genetic diversity by mutation and horizontal gene transfer (HGT). During HGT, genetic information from a donor organism is transferred to a recipient, creating a new genotype in the recipient. In this way genetic information is passed between individuals of the same generation and even between species found in different domains of life. Genome sequencing has revealed that HGT has played an important role in the evolution of all microbial species. Importantly, HGT still occurs in bacteria and archaea, leading to the rapid evolution of microorganisms with antibiotic resistance, new virulence properties, and novel metabolic capabilities. The outcome of HGT is that most microbes have mosaic genomes composed of bits and pieces of the genomes of other organisms. ▶| *Horizontal gene transfer: creating genetic variation the asexual way (section 16.3)*

Phylogenetic or **phyletic classification systems** compare organisms on the basis of evolutionary relationships. The term **phylogeny** (Greek *phylon,* tribe or race; *genesis,* generation or origin) refers to the evolutionary development of organisms. As discussed, microbial phylogeny relies on comparisons of multiple features found in extant organisms. These include cell wall structure, biomolecules such as fatty acids, and certain housekeeping proteins (proteins used to maintain cellular life, therefore found in many different organisms), and nucleotide sequences, particularly of small subunit rRNA molecules (SSU rRNA) (**table 1.1**). ▶| *Bacterial ribosomes (section 3.6); Archaeal ribosomes (section 4.3); Eukaryotic ribosomes (section 5.5)*

Phylogenetic Trees

Figure 1.3 is an example of a **phylogenetic tree.** The goal of phylogenetic tree construction is to display the evolutionary

Table 1.1 Comparison of Bacteria, Archaea, and Eukarya

Property	Bacteria	Archaea	Eukarya
Organization of Genetic Material			
Membrane-bound nucleus	No	No	Yes
DNA complexed with histones	No	Some	Yes
Chromosomes	Usually one circular chromosome; chromosomes have single origin of replication; some are polyploid	One circular chromosome; some have chromosomes with multiple origins of replication; some are polyploid	Multiple, linear chromosomes with multiple origins of replication; usually diploid
Plasmids	Yes	Yes	Rare
Introns in genes	Rare	Rare	Yes
Nucleolus	No	No	Yes
Mitochondria, Chloroplasts, Endoplasmic Reticulum, Golgi, and Lysosomes Observed	No	No	Yes
Plasma Membrane Lipids	Ester-linked phospholipids; some have sterols	Glycerol diethers and diglycerol tetraethers	Ester-linked phospholipids and sterols
Flagella	Submicroscopic in size; filament composed of single type of flagellin protein	Submicroscopic in size; filaments composed of more than one type of archaellin protein	Microscopic in size; membrane bound; usually 20 microtubules in 9 + 2 pattern
Peptidoglycan in Cell Walls	Yes	No	No
Ribosome Size and Structure	70S; 3 rRNAs; 49–59 ribosomal proteins	70S; 3 rRNAs; 58–68 ribosomal proteins	80S; 4 rRNAs; 78–80 ribosomal proteins
Cytoskeleton	Rudimentary	Rudimentary	Yes

relationships between different organisms. Trees are built by comparing the sequences of amino acids or nucleotides from diverse organisms. Amino acid sequences are often compared because nucleotide changes may not alter the resulting protein and therefore have little or no impact on evolution. Historically, the rRNAs from small ribosomal subunits (16S from bacteria and archaea and 18S from eukaryotes) were the molecules of choice for inferring microbial phylogenies and making taxonomic assignments for several reasons (**figure 1.8**). Like housekeeping proteins (proteins with essential functions), SSU rRNAs play the same role in all microorganisms and are absolutely necessary for survival. Therefore neither housekeeping proteins nor genes encoding SSU rRNAs tolerate large changes in sequence. The utility of SSU rRNAs is extended by the presence of certain sequences within SSU rRNA genes that vary among organisms as well as other regions that are quite similar. The variable regions enable comparison between closely related microbes, whereas the stable sequences allow the comparison of more distantly related microorganisms. The stability of SSU rRNA sequences means that they evolve very slowly, so analysis of housekeeping genes is more frequently used to infer the evolution of more closely related organisms.

Comparative analysis of SSU rRNA sequences from thousands of organisms has demonstrated the presence of **oligonucleotide signature sequences.** These are short, conserved nucleotide sequences specific for phylogenetically defined groups of organisms. Thus the signature sequences found in bacterial rRNAs are rarely or never found in archaeal rRNAs, and vice versa. Likewise, the 18S rRNA of eukaryotes bears signature sequences that are specific to the domain Eukarya.

Approaches to building a phylogenetic tree can be divided into two broad categories: a distance-based approach and a character-based approach. *Distance-based* approaches are the most intuitive. Here the differences between the aligned sequences are counted for each pair and summarized in a single statistic (**figure 1.9**). This value serves as a measure of the evolutionary distance between the organisms; the more differences counted, the greater the evolutionary distance. The evolutionary distances from many comparisons are anyalyzed using computer software to construct the tree. The tip of each branch in the tree (called a *node*) represents one of the organisms used in the comparison. The distance from one node to another is the evolutionary distance between the two organisms.

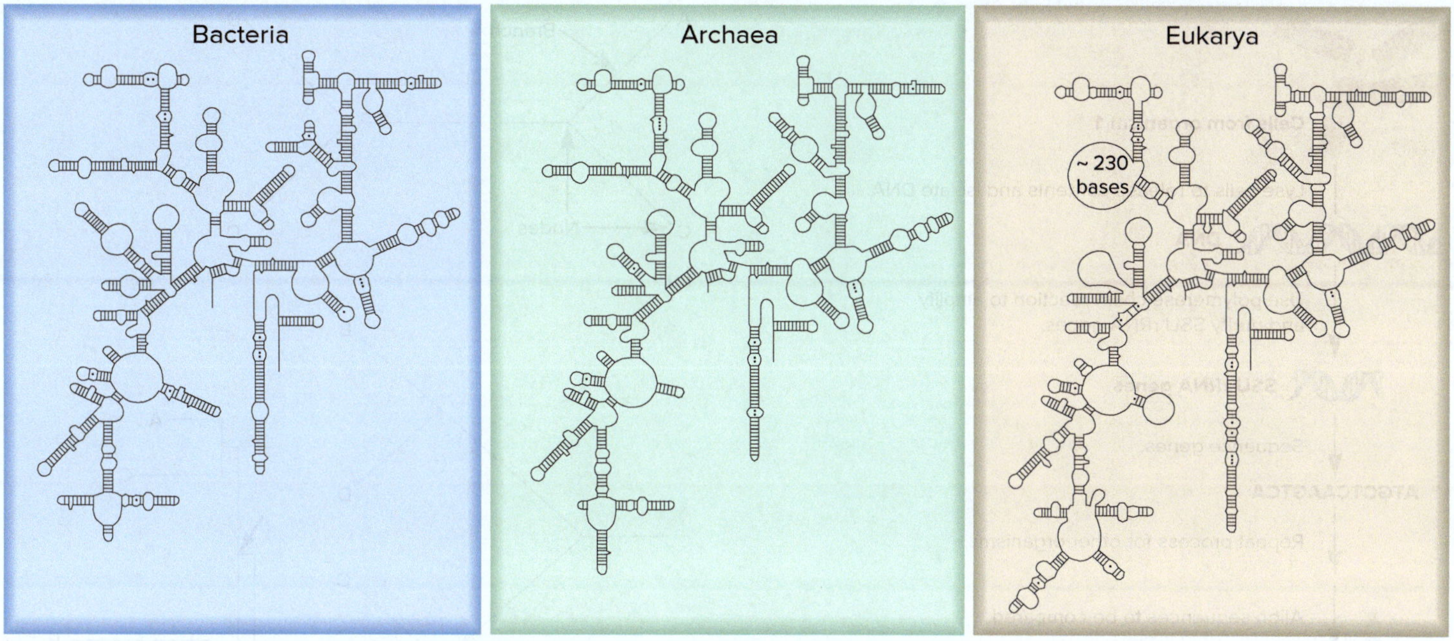

Figure 1.8 Signature rRNA Help Identify Microbes. Representative examples of SSU rRNA secondary structures from the three domains: Bacteria *(Escherichia coli)*, Archaea *(Methanococcus vannielii)*, and Eukarya *(Saccharomyces cerevisiae)*.

An algorithm analyzes this information to generate a tree. One algorithm called cluster analysis serially links pairs that are ever more distantly related (i.e., start with those with the least number of sequence differences and move to those with the most). Although this is the easiest to understand, it can generate trees based on artifact. Neighbor joining is another distance-based method that uses a different matrix to modify the distance between each pair of nodes based on the average divergence from all other nodes.

Character-based approaches for phylogenetic tree building are more complicated but generate more robust trees. These methods start with assumptions about the pathway of evolution, infer the ancestor at each node, and choose the best tree according to a specific model of evolutionary change. Sometimes called tree-searching, these methods include maximum parsimony, which assumes that the fewest number of changes occurred between ancestor and extant organisms. Another approach is called maximum likelihood. This requires a large data set because the probability (i.e., the likelihood) for each possible tree is calculated based on certain evolutionary and molecular information. In this way, the tree with the greatest probability based on these criteria is selected. Bayesian inference is another character-based approach. Rather than looking at a single tree, Bayesian inference analyzes multiple potential trees and calculates the probability that each branch would appear based on this comparison.

Once a tree is constructed, it is important to get a sense of whether the placement of its branches and nodes is legitimate. There are a variety of methods to assess the strength of a tree, but the most common is bootstrapping. Bootstrapping involves reanalyzing a randomly selected subset of the data presented on the tree. A bootstrap value is the percent of times in which that particular branch was found. Typically, bootstrap values of 70% or greater are thought to support a tree. Of note is that Bayesian inference values are also reported as percentages, but they are not directly comparable to bootstrap values. Only values greater than 95% are acceptable when Bayesian inference is used.

Two things should be kept in mind when examining phylogenetic trees. The first is that they are molecular trees, not organismal trees. In other words, they represent, as accurately as possible, the evolutionary history of a molecule (e.g., rRNA). Second, the distance between nodes is a measure of relatedness, not of time. If the distance along the lines is very long, then the two organisms are more evolutionarily diverged (i.e., less related). However, we do not know when they diverged from each other. This concept is analogous to a printed map that accurately shows the distance between two cities but because of many factors (traffic, road conditions, etc.) cannot show the time needed to travel that distance.

Importantly, a tree may be unrooted or rooted. An unrooted tree (**figure 1.10*a***) represents phylogenetic relationships but does not indicate which organisms are more primitive relative to the others. Figure 1.10*a* shows that A is more closely related to C than it is to either B or D, but it does not indicate which of the four species might be the oldest. In contrast, the rooted tree (figure 1.10*b*) includes a node (taxonomic unit) that serves as the common ancestor and shows the development of the four species from this root. It is much more difficult to develop a rooted tree. For example, there are 15 possible rooted trees that connect four species but only three possible unrooted trees.

An unrooted tree can be rooted by adding data from an outgroup—a species known to be very distantly related to all the species in the tree (figure 1.10*c*). The root is determined by the point of the tree where the outgroup joins. This provides a point of reference to identify the oldest node on the tree, which is the node closest to the outgroup. So, for example, in figure 1.10*c*,

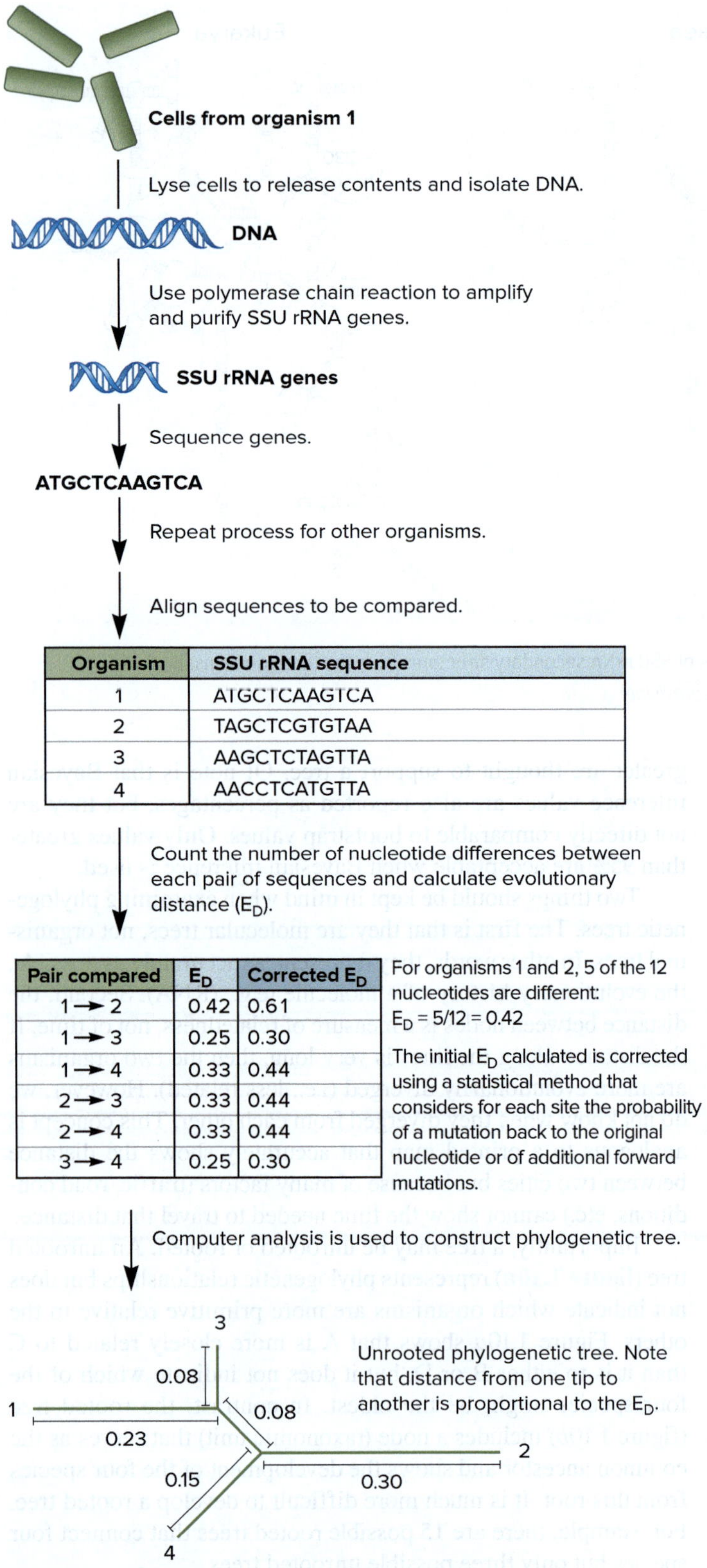

Organism	SSU rRNA sequence
1	ATGCTCAAGTCA
2	TAGCTCGTGTAA
3	AAGCTCTAGTTA
4	AACCTCATGTTA

Pair compared	E_D	Corrected E_D
1 → 2	0.42	0.61
1 → 3	0.25	0.30
1 → 4	0.33	0.44
2 → 3	0.33	0.44
2 → 4	0.33	0.44
3 → 4	0.25	0.30

Figure 1.9 The Construction of a Phylogenetic Tree Using a Distance Method. The polymerase chain reaction is described in chapter 17.

MICRO INQUIRY *Why does the branch length indicate amount of evolutionary change but not the time it took for that change to occur?*

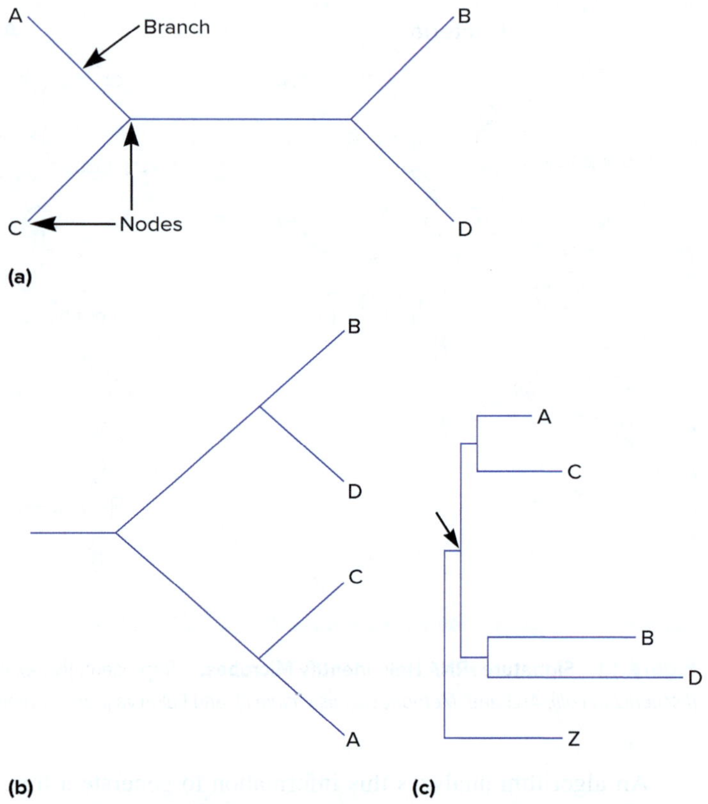

Figure 1.10 Phylogenetic Tree Topologies. (a) Unrooted tree joining four taxonomic units (A, B, C, D). (b) Rooted tree. (c) An unrooted tree can be rooted by adding an outgroup, represented by Z in this example. The arrow indicates a speciation event that resulted in the development of new species from an ancestral organism.

organism Z is the outgroup and the oldest node on the tree is marked with an arrow.

Microbial Taxonomy

The science of classifying living things is called **taxonomy.** Taxonomy consists of three separate but interrelated parts: classification, nomenclature (naming), and identification. A taxonomic scheme is used to arrange organisms into groups called taxa (s., **taxon**) based on mutual similarity. Microbes are placed in taxonomic levels arranged in a nonoverlapping hierarchy so that each level includes not only the traits that define the rank above it but also a new set of more restrictive traits (**figure 1.11**). Thus within each domain—Bacteria, Archaea, or Eukarya—each organism is assigned (in descending order) to a phylum, class, order, family, genus, and species epithet or name. Some microbes are also given a subspecies designation. Microbial groups at each level have a specific suffix that indicates rank or level.

Microbiologists name microbes using the binomial system of the eighteenth-century biologist and physician Carl Linnaeus. The Latin, italicized name consists of two parts. The first part, which is capitalized, is the generic name (i.e., the name of the

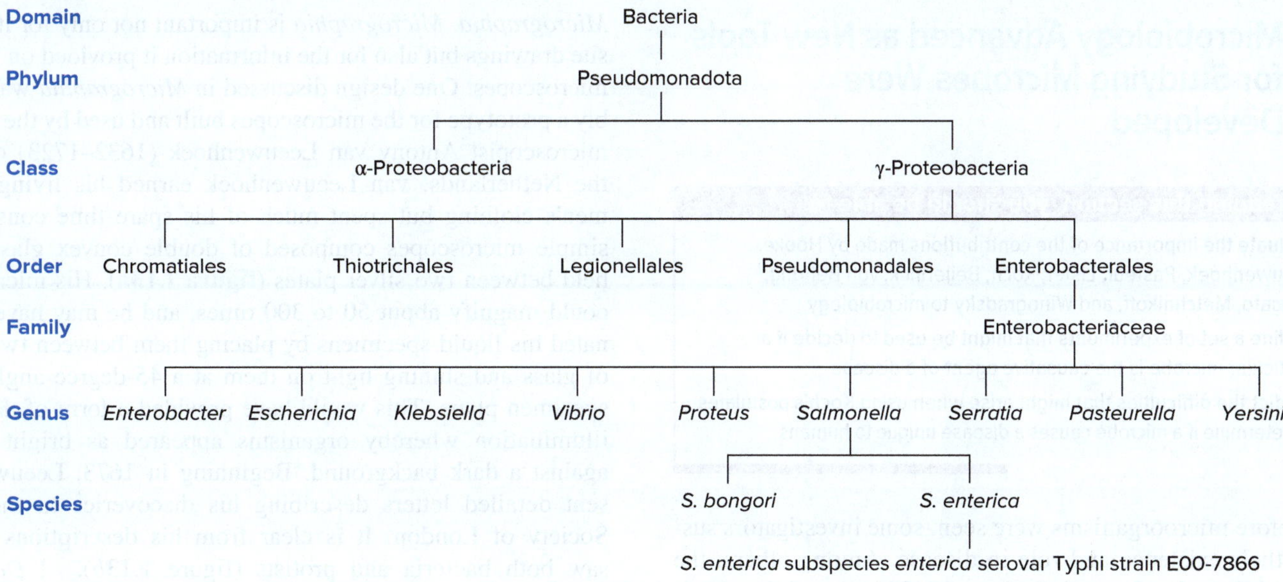

Figure 1.11 Hierarchical Arrangement in Taxonomy. In this example, members of the genus *Salmonella* are placed within higher taxonomic ranks. The species *S. enterica* is further classified into subspecies, serovar (based an surface proteins), and strain (based on minor genetic differences). Not all classification possibilities are given for each rank to simplify the diagram. Note that *-ales* denotes order and *-ceae* indicates family.

genus to which the microbe belongs), and the second is the uncapitalized species epithet. For example, the bacterium that causes plague is called *Yersinia pestis*. Often the name of an organism will be shortened by abbreviating the genus name with a single uppercase letter (e.g., *Y. pestis*).

This straightforward organizational approach is complicated by the fact that bacteria and archaea do not reproduce sexually. You may recall from a general biology class that plant and animal **species** are defined as a group of interbreeding or potentially interbreeding natural populations reproductively isolated from other groups. This definition also is appropriate for the many eukaryotic microbes that reproduce sexually. However, bacterial and archaeal species cannot be defined by this criterion because they do not reproduce sexually. Therefore comparisons of genome sequences are often used to distinguish one species from another. An appropriate definition is currently debated. A common definition is that bacterial and archaeal species are a collection of strains that share many stable properties and differ significantly from other groups of strains. A **strain** consists of the descendants of a single, pure microbial culture. Strains within a species may be described in a number of different ways. Biovars are variant strains characterized by biochemical or physiological differences, morphovars differ morphologically, serovars have distinctive properties that can be detected by antibodies, and pathovars are pathogenic strains distinguished by the plants in which they cause disease.

Although microbiologists continue to use Linnaeus's classification system, the ongoing explosion in metagenomic analysis has had an impact on this historically accepted hierarchy. Within the last few decades, thousands of 16S rRNA genes and protein-coding genes have been sequenced that do not belong to any previously defined taxa. These data have led to the recent development of the taxonomic classification **superphylum,** below domain and above phylum (e.g., in figure 1.11, superphylum would be placed between Bacteria and Pseudomonadota). A superphylum includes organisms of several phyla that share a number of distinctive characteristics, such as unusual morphological or metabolic features.

While we focus here on how microbes can be classified, it is important—and exciting—to point out that the vast majority of microorganisms are considered **microbial dark matter**. Like dark matter in physics, which describes the unseen mass in the universe inferred indirectly by gravitational forces, microbial dark matter is life known to exist only from nucleotide sequencing of environmental samples. Such studies have revealed a staggering diversity of unique microbes with potentially novel functions in ecosystems, as well as applications for human health and biotechnology.

Comprehension Check

1. Describe two reasons RNA is thought to be the first self-replicating biomolecule.
2. Explain the endosymbiotic hypothesis of the origin of mitochondria, hydrogenosomes, and chloroplasts. List two pieces of evidence that support this hypothesis.
3. What is the goal of phylogenetic tree construction?
4. In general terms, what is the difference between a distance-based and character-based tree? Does a rooted or unrooted phylogenetic tree provide more information? Explain your answer.
5. What is the difference between mutation and horizontal gene transfer?
6. What is the correct way to write this microbe's name: *bacillus subtilis*, Bacillus subtilis, *Bacillus Subtilis*, or *Bacillus subtilis*? Identify the genus name and the species epithet.

1.3 Microbiology Advanced as New Tools for Studying Microbes Were Developed

After reading this section, you should be able to:

a. Evaluate the importance of the contributions made by Hooke, Leeuwenhoek, Pasteur, Lister, Koch, Beijerinck, von Behring, Kitasato, Metchnikoff, and Winogradsky to microbiology
b. Outline a set of experiments that might be used to decide if a particular microbe is the causative agent of a disease
c. Predict the difficulties that might arise when using Koch's postulates to determine if a microbe causes a disease unique to humans

Even before microorganisms were seen, some investigators suspected their existence and role in disease. Among others, the Roman philosopher Lucretius (about 98–55 BCE) and the physician Girolamo Fracastoro (1478–1553) suggested that disease was caused by invisible living creatures. However, until microbes could actually be seen and studied in some other way, their existence remained a matter of conjecture. Therefore **microbiology** is defined not only by the organisms it studies but also by the tools used to study them. The development of microscopes was the critical first step in the evolution of the discipline. However, microscopy alone is unable to answer the many questions scientists ask about microbes. A distinct feature of microbiology is that microorganisms are usually removed from their normal habitats and grown in isolation, apart from all other microbes. This is called a **pure** or **axenic culture.** Although the development of techniques for isolating microbes in pure culture was another critical step in microbiology's history, it is now recognized as having limitations. Microbes in pure culture are in some ways like animals in a zoo; just as a zoologist cannot fully understand animals by studying them in zoos, microbiologists cannot fully understand microbes by studying them in pure culture. Today molecular genetic techniques and genomic analyses are providing new insights into the lives of microbes and the communities they inhabit.

Here we describe how the tools used by microbiologists have influenced the development of the field. As microbiology evolved as a science, it contributed greatly to the well-being of humans. The historical context of some of the important discoveries in microbiology is shown in **figure 1.12.**

Microscopy Led to the Discovery of Microorganisms

The earliest microscopic observations of organisms appear to have been made between 1625 and 1630 on bees and weevils by the Italian Francesco Stelluti (1577–1652), using a microscope probably supplied by Galileo (1564–1642). Robert Hooke (1635–1703) is credited with publishing the first drawings of microorganisms in the scientific literature. In 1665 he published a highly detailed drawing of the fungus *Mucor* in his book *Micrographia. Micrographia* is important not only for its exquisite drawings but also for the information it provided on building microscopes. One design discussed in *Micrographia* was probably a prototype for the microscopes built and used by the amateur microscopist Antony van Leeuwenhoek (1632–1723) of Delft, the Netherlands. van Leeuwenhoek earned his living selling men's clothing but spent much of his spare time constructing simple microscopes composed of double convex glass lenses held between two silver plates (**figure 1.13a**). His microscopes could magnify about 50 to 300 times, and he may have illuminated his liquid specimens by placing them between two pieces of glass and shining light on them at a 45-degree angle to the specimen plane. This would have provided a form of dark-field illumination whereby organisms appeared as bright objects against a dark background. Beginning in 1673, Leeuwenhoek sent detailed letters describing his discoveries to the Royal Society of London. It is clear from his descriptions that he saw both bacteria and protists (figure 1.13*b*). ▶| *Dark-field microscope: bright object, dark background (section 2.2)*

Culture-based Methods for Studying Microorganisms Were a Major Development

As important as Leeuwenhoek's observations were, the development of microbiology essentially languished for almost the next 200 years until techniques for isolating and culturing microbes in the laboratory were formulated. During this time, scientists grappled with the conflict over the theory of spontaneous generation. This conflict and the subsequent studies on the role played by microorganisms in causing disease ultimately led to what is now called the golden age of microbiology.

Spontaneous Generation

From earliest times, people had believed in **spontaneous generation**—that living organisms could develop from nonliving matter. This view was challenged by the Italian physician Francesco Redi (1626–1697), who carried out a series of experiments on decaying meat, which was thought to produce maggots (fly larvae) spontaneously. Using covered and uncovered containers of meat, Redi clearly demonstrated that maggots on decaying meat resulted from the presence of fly eggs. Other experiments helped discredit the theory for larger organisms. However, Leeuwenhoek's communications on microorganisms renewed the controversy. Some proposed that microbes arose by spontaneous generation but larger organisms did not. They pointed out that boiled extracts of hay or meat gave rise to microorganisms after sitting for a while. In 1748, the English priest John Needham (1713–1781) suggested that organic matter in these extracts contained a "vital force" that could confer the properties of life on nonliving matter.

A few years after Needham's experiments, the Italian priest and naturalist Lazzaro Spallanzani (1729–1799) sealed glass flasks that contained water and seeds and then placed the flasks in boiling water for about 45 minutes. He found that no growth took place as long as the flasks remained sealed. He proposed

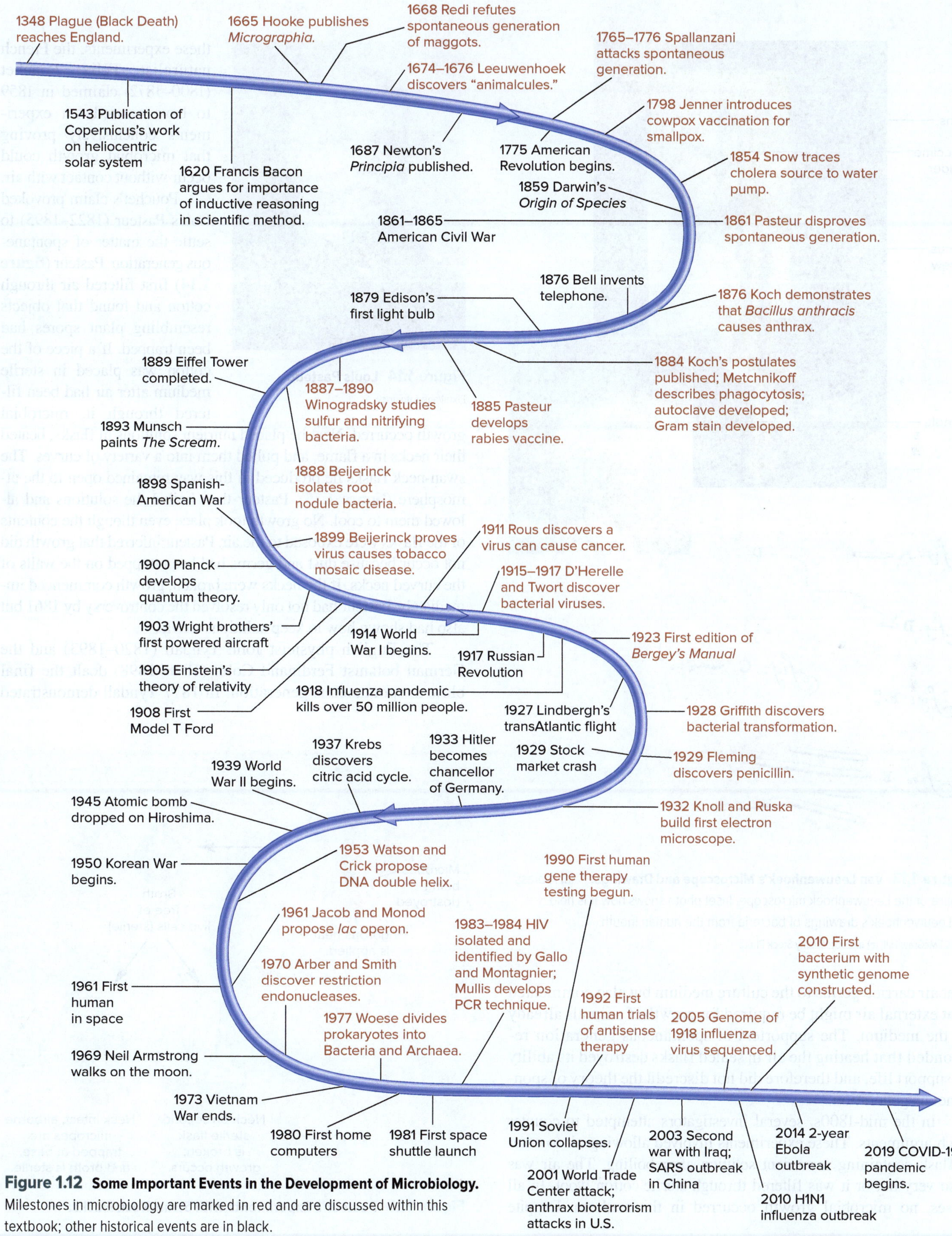

Figure 1.12 Some Important Events in the Development of Microbiology. Milestones in microbiology are marked in red and are discussed within this textbook; other historical events are in black.

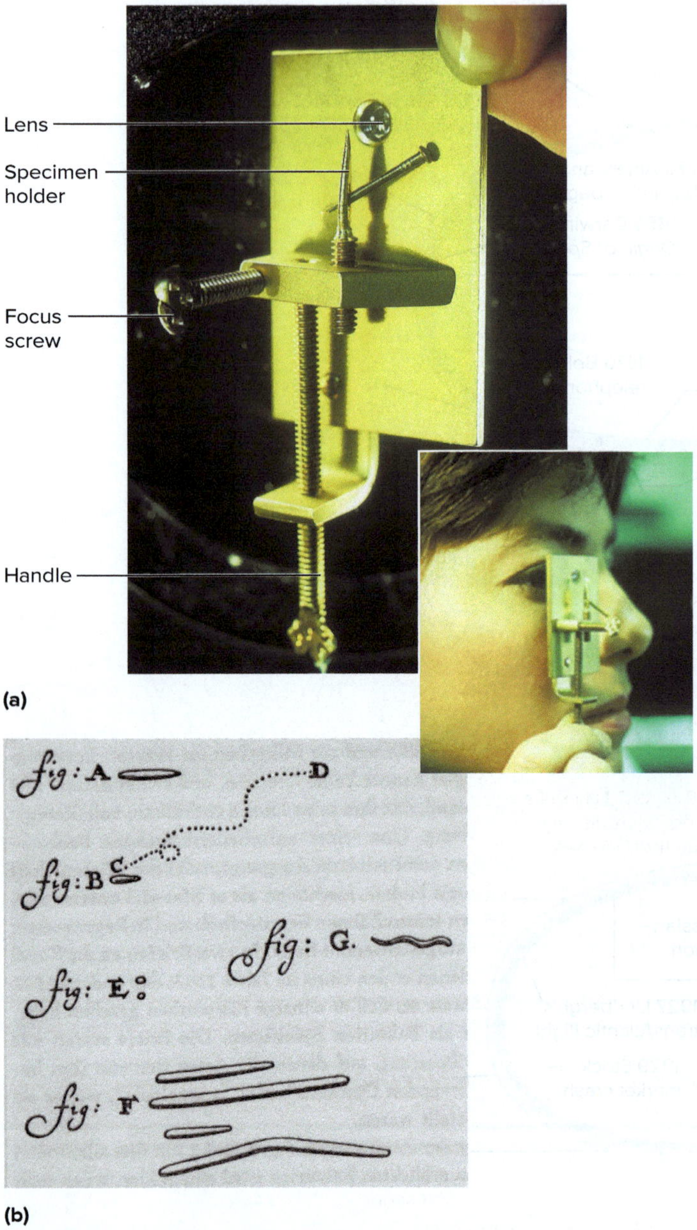

Figure 1.13 van Leeuwenhoek's Microscope and Drawings. (a) A brass replica of the Leeuwenhoek microscope. Inset photo shows how it is held. (b) Leeuwenhoek's drawings of bacteria from the human mouth.
(a, 1-2) McGraw Hill; (b) Interfoto/Alamy Stock Photo

Figure 1.14 Louis Pasteur.
Pixtal/age fotostock

these experiments, the French naturalist Felix Pouchet (1800–1872) claimed in 1859 to have carried out experiments conclusively proving that microbial growth could occur without contact with air.

Pouchet's claim provoked Louis Pasteur (1822–1895) to settle the matter of spontaneous generation. Pasteur (**figure 1.14**) first filtered air through cotton and found that objects resembling plant spores had been trapped. If a piece of the cotton was placed in sterile medium after air had been filtered through it, microbial growth occurred. Next he placed nutrient solutions in flasks, heated their necks in a flame, and pulled them into a variety of curves. The swan-neck flasks he produced in this way remained open to the atmosphere (**figure 1.15**). Pasteur then boiled the solutions and allowed them to cool. No growth took place even though the contents of the flasks were exposed to the air. Pasteur inferred that growth did not occur because dust and germs had been trapped on the walls of the curved necks. If the necks were broken, growth commenced immediately. Pasteur had not only resolved the controversy by 1861 but also had shown how to keep solutions sterile.

The English physicist John Tyndall (1820–1893) and the German botanist Ferdinand Cohn (1828–1898) dealt the final blow to spontaneous generation. In 1877 Tyndall demonstrated that air carried germs to the culture medium but also commented that external air might be required for growth of animals already in the medium. The supporters of spontaneous generation responded that heating the air in sealed flasks destroyed its ability to support life, and therefore did not discredit the theory of spontaneous generation.

In the mid-1800s, several investigators attempted to counter such arguments. These experiments involved allowing air to enter a flask containing a nutrient solution after boiling. The air was also very hot or it was filtered through sterile cotton wool. In all cases, no microbial growth occurred in the medium. Despite

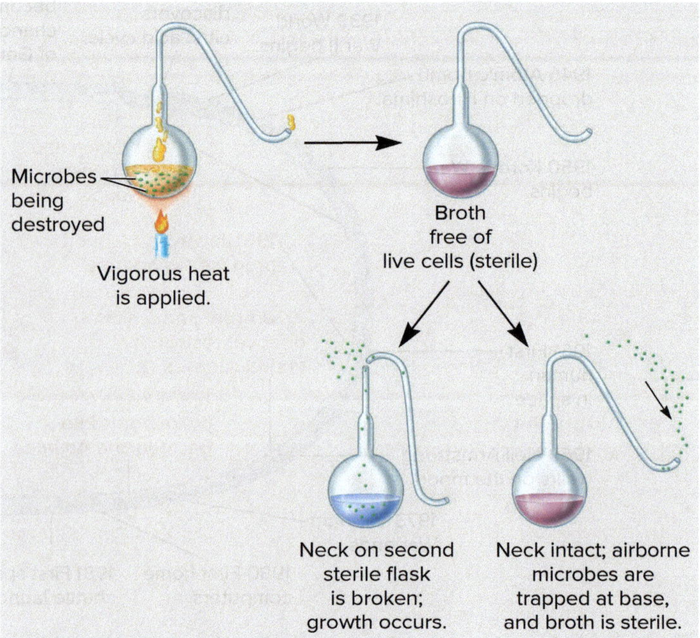

Figure 1.15 Pasteur's Experiments with Swan-Neck Flasks.

that dust did indeed carry germs and that if dust was absent, broth remained sterile even if directly exposed to air. During the course of his studies, Tyndall provided evidence for the existence of exceptionally heat-resistant forms of bacteria. Working independently, Cohn discovered that the heat-resistant bacteria recognized by Tyndall were species capable of producing bacterial endospores. Cohn later played an instrumental role in establishing a classification system for bacteria based on their morphology and physiology. ▶| *Bacterial endospores are a survival strategy (section 3.8)*

These early microbiologists disproved spontaneous generation, thereby contributing to the birth of microbiology. They developed liquid media and the methods for sterilizing it so that microbes could be cultured. These techniques were next applied to understanding the role of microorganisms in disease.

Microorganisms and Disease

For hundreds of years, most people believed that disease was caused by supernatural forces, poisonous vapors, and imbalances among the four humors thought to be present in the body. The role of the four humors (blood, phlegm, yellow bile [choler], and black bile [melancholy]) in disease had been widely accepted since the time of the Greek physician Galen (129–199). Support for the idea that microorganisms cause disease—that is, the germ theory of disease—began to accumulate in the early nineteenth century from diverse fields. Agostino Bassi (1773–1856) demonstrated in 1835 that a silkworm disease was due to a fungal infection. In 1845 M. J. Berkeley (1803–1889) proved that the great potato blight of Ireland was caused by a protozoan (then thought to be a fungus), and in 1853 Heinrich de Bary (1831–1888) showed that fungi caused crop diseases.

Pasteur also contributed to this area of research in what may seem an unlikely way. Pasteur was trained as a chemist and spent many years studying the fermentations that yield ethanol used in the production of wine and other alcoholic beverages. When he began his work, leading chemists were convinced that fermentation was due to a chemical instability in sugars that resulted in their breakdown into alcohol. Pasteur did not agree; he believed that fermentations were carried out by living organisms.

In 1856 M. Bigo, an industrialist in Lille, France, where Pasteur worked, requested Pasteur's assistance. His business produced ethanol from the fermentation of beet sugars, and the alcohol yields had recently declined and the product had become sour. Pasteur discovered that the fermentation was failing because the yeast normally responsible for alcohol formation had been replaced by bacteria that produced acid rather than ethanol. In solving this practical problem, Pasteur demonstrated that fermentations were due to the activities of yeasts and bacteria.

Pasteur was also called upon by the wine industry in France for help. For several years, poor-quality wines had been produced. Pasteur referred to the wines as diseased and demonstrated that particular wine diseases were linked to particular microbes contaminating the wine. He eventually suggested a method for heating the wines to destroy the undesirable microbes. The process is now called *pasteurization*.

Indirect evidence for the germ theory of disease came from the work of the English surgeon Joseph Lister (1827–1912) on the prevention of wound infections. Lister, impressed with Pasteur's studies on fermentation, developed a system of antiseptic surgery designed to prevent microorganisms from entering wounds. Instruments were heat sterilized, and phenol was used on surgical dressings and at times sprayed over the surgical area. The approach was remarkably successful in reducing the rate of surgical infection. It also provided strong indirect evidence for the role of microorganisms in disease.

Koch's Postulates

The first direct demonstration that bacteria cause disease came from the study of anthrax by the German physician Robert Koch (pronounced "Koke"; 1843–1910). Koch (**figure 1.16**) used the criteria proposed by his former teacher Jacob Henle (1809–1885) and others to establish the relationship between *Bacillus anthracis* and anthrax. In these studies, published in 1876, Koch used mice as his model or test organism. Following the steps outlined in **figure 1.17**, he next used guinea pigs to show that *Mycobacterium tuberculosis* causes tuberculosis (TB), which at that time was a leading cause of death in Europe. In 1905 Koch was awarded the Nobel Prize in Physiology or Medicine and his criteria for proving the causal relationship between a microorganism and a specific disease are known as **Koch's postulates.** Koch's postulates

Figure 1.16 Robert Koch. Koch examining a specimen in his laboratory.
Bettmann/Getty Images

Postulate	Experimentation
1 The microorganism must be present in every case of the disease but absent from healthy organisms.	Koch developed a staining technique to examine human tissue. *Mycobacterium tuberculosis* could be identified in diseased tissue.
2 The suspected microorganisms must be isolated and grown in a pure culture.	Koch grew *M. tuberculosis* in pure culture on coagulated blood serum.
3 The same disease must result when the isolated microorganism is inoculated into a healthy host.	Koch injected cells from the pure culture of *M. tuberculosis* into guinea pigs. The guinea pigs subsequently died of tuberculosis.
4 The same microorganisms must be isolated again from the diseased host.	Koch isolated *M. tuberculosis* in pure culture on coagulated blood serum from the dead guinea pigs.

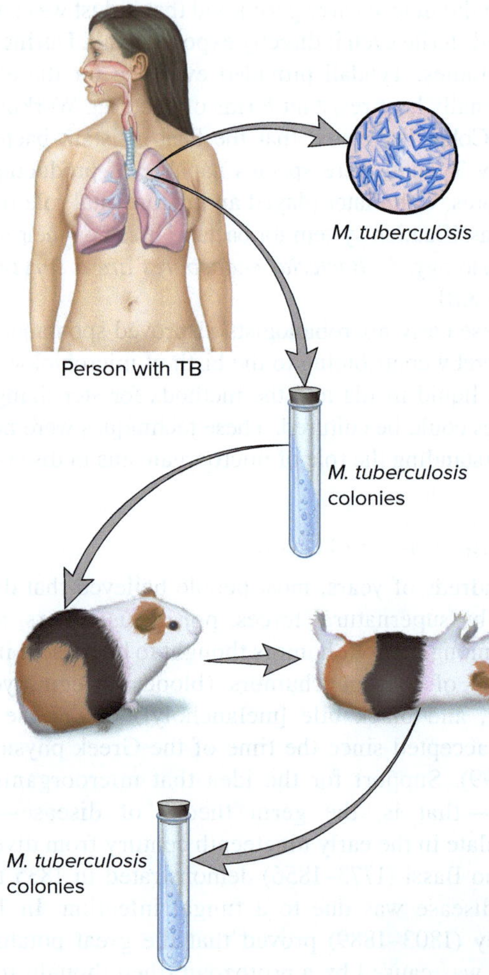

Figure 1.17 Koch's Postulates Applied to Tuberculosis.

MICRO INQUIRY *Why is the fourth postulate necessary?*

have since been used to discover the causative microorganisms for many infectious diseases.

While Koch's postulates are still widely used, their application is at times not feasible. For instance, microbes that must live within host cells such as viruses and *Mycobacterium leprae,* the bacterium that causes leprosy, cannot be isolated in pure culture. Some human diseases lack an appropriate animal model so the postulates cannot be fully met. To avoid some of these difficulties, microbiologists sometimes use molecular and genetic evidence. For instance, molecular methods might be used to detect the nucleic acid of a microorganism in body tissues, rather than isolating it, or the genes thought to be associated with the virulence of a disease-causing microbe (pathogen) might be mutated. In this case, the mutant organism should have decreased ability to cause disease. Introduction of the normal gene back into the mutant should restore the pathogen's virulence.

Our focus thus far has been on the discovery of bacteria, fungi, and protists. But viral pathogens were also being studied during this time. The discovery of viruses and their role in disease was made possible when Charles Chamberland (1851–1908), one of Pasteur's associates, constructed a porcelain filter to remove bacteria. Dimitri Ivanowski (1864–1920) and Martinus Beijerinck (pronounced "by-a-rink"; 1851–1931) used the filter to study tobacco mosaic disease. They found that plant extracts and sap from diseased plants were infectious, even after being filtered with Chamberland's filter. Because the infectious agent passed through a filter that trapped bacterial cells, they reasoned that the agent must be something smaller than a bacterium. Beijerinck proposed that the agent was a "filterable virus" (Latin *virus,* slimy liquid, poison). Eventually viruses were shown to be tiny, acellular infectious agents.

Immunology

The ability to culture microbes also played an important role in early immunological studies. During studies on the bacterium that causes chicken cholera, Pasteur and Pierre Roux (1853–1933) discovered that bacteria incubated in cultures for long periods of time lost their ability to cause disease. These bacteria were said to be attenuated or weakened. When chickens were injected with attenuated bacteria, they were surprisingly able to resist the disease when exposed to virulent strains of the same bacteria. Pasteur called the attenuated bacteria a vaccine (Latin *vacca,* cow) in honor of Edward Jenner (1749–1823) because, many years earlier, Jenner had used material from cowpox lesions to protect people against smallpox (*see Historical Highlights 35.3*). Shortly after this, Pasteur and Chamberland developed an attenuated anthrax vaccine. ▶| *Vaccines immunize susceptible populations (section 35.6)*

Pasteur also prepared a vaccine using an attenuated strain of rabies virus. During the course of these studies, a nine-year-old boy named Joseph Meister was bitten by a rabid dog and was brought to Pasteur. Because Joseph's death was certain in the absence of treatment, Pasteur agreed to try vaccination. Joseph was injected 13 times over the next 10 days with increasingly virulent preparations of the attenuated virus (**figure 1.18**). His survival marked a huge advance in the use of vaccines. To thank Pasteur, people from around the world contributed to the construction of the Pasteur Institute in Paris, France. One of the initial tasks of the institute was vaccine production.

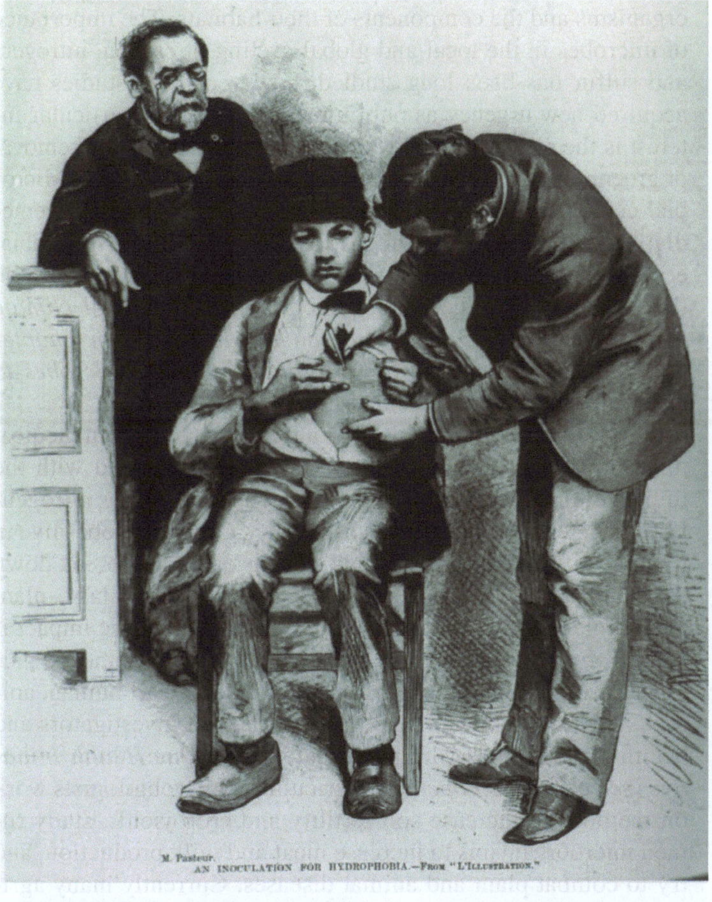

Figure 1.18 The First Rabies Vaccine. Louis Pasteur (standing) looks on as a physician administers the rabies vaccine developed by Pasteur to 9-year-old Joseph Meister.

Courtesy of the National Library of Medicine

These early advances in immunology were made without understanding how the immune system works. Immunologists now know that white blood cells and the chemicals they produce play a central role in immunity. Among the chemicals are soluble proteins called antibodies, found in blood, lymph, and other body fluids. The role of antibodies in preventing disease was recognized by Emil von Behring (1854–1917) and Shibasaburo Kitasato (1852–1931). After the discovery that diphtheria was caused by a toxin produced by bacteria, they injected inactivated diphtheria toxin into rabbits. The inactivated toxin induced rabbits to produce antibodies against the toxin, known as antitoxin, which protected against the disease (*see Historical Highlights 32.1*). The first immune system cells were discovered when Élie Metchnikoff (1845–1916) found that some white blood cells could engulf disease-causing bacteria. He called these cells phagocytes and the process phagocytosis (Greek *phagein,* eating).

Microbial Ecology

Early microbial ecologists studied microbial involvement in the carbon, nitrogen, and sulfur cycles. The Russian microbiologist Sergei Winogradsky (1856–1953) made many contributions to soil microbiology, including the discovery that soil bacteria could oxidize iron, sulfur, and ammonia to obtain energy and that many of these bacteria could incorporate CO_2 into organic matter much as photosynthetic organisms do. Winogradsky also isolated anaerobic nitrogen-fixing soil bacteria and studied the decomposition of cellulose. Martinus Beijerinck made fundamental contributions to microbial ecology as well as virology. He isolated several kinds of nitrogen-fixing bacteria and sulfate-reducing bacteria. Beijerinck and Winogradsky also developed enrichment culture techniques and selective media, which have been of great importance in microbiology. ▶| *Enrichment cultures (section 7.7); Biogeochemical cycling sustains life on Earth (section 28.1)*

Comprehension Check

1. Why did Pasteur's experiments to disprove spontaneous generation succeed while others failed? In general, do you think it is easier to prove or disprove an idea?
2. Draw a timeline illustrating the contributions of Lister, Pasteur, and Koch to the germ theory of disease and the treatment or prevention of diseases.
3. What role did the ability to grow bacteria in pure culture play in the development of Koch's postulates?
4. What did Jenner, Pasteur, von Behring, Kitasato, and Metchnikoff contribute to the development of immunology?
5. How did Winogradsky and Beijerinck contribute to the study of microbial ecology?

1.4 Microbiology Encompasses Many Subdisciplines

After reading this section, you should be able to:

a. Construct a concept map, table, or drawing that illustrates the diverse nature of microbiology and how it has improved human conditions
b. Discuss the opinion held by many microbiologists that microbiology is experiencing its second golden age

Microbiology today is as diverse as the organisms it studies. It has both basic and applied aspects. The basic aspects are concerned with the biology of microorganisms themselves. The applied aspects are concerned with practical problems such as disease, water and wastewater treatment, food spoilage and food production, and industrial uses of microbes. Despite this apparent dichotomy, the basic and applied aspects of microbiology are intertwined. Basic research is often conducted in applied fields, and applications often arise out of basic research.

An important development in microbiology is the increasing use of molecular and genomic methods to study microbes and their interactions with other organisms. These methods have led to a time of rapid advancement that rivals the golden age of microbiology. Indeed, many feel that microbiology is in its second golden age. ▶| *Microbial DNA technologies (chapter 17); Microbial genomics (chapter 18); Exploring microbes in ecosystems (chapter 26)*

Major Fields in Microbiology

Microbiology is commonly divided into subdisciplines based on the type of microbe studied. Thus microbiology encompasses bacteriology and virology as well as other microbe-specific fields. Microbiology can also be divided based on the activities of microbes—for instance, environmental microbiology and agricultural microbiology. Finally, microbiologists may study only one aspect of the biology of microbes, leading to subdisciplines such as microbial genetics and microbial physiology.

One of the most active and important fields in microbiology is medical microbiology, which deals with diseases of humans. Medical microbiologists investigate agents causing infectious diseases and measures for their control and elimination. They are involved in tracking down new, unidentified pathogens such as the virus that causes COVID-19. These microbiologists also study how microorganisms cause disease. Clinical laboratory scientists, the microbiologists who work in hospitals and other clinical laboratories, use culture and molecular techniques to provide information needed by physicians to diagnose and treat infectious disease.

Major epidemics have regularly affected human history. For example, the human immunodeficiency virus/acquired immune deficiency syndrome (HIV/AIDS) has infected about 88 million people since the beginning of the AIDS pandemic in 1981. The COVID-19 pandemic demonstrated to the world that public health microbiology is concerned with the control and spread of such communicable diseases. Public health microbiologists and epidemiologists monitor the amount of disease in populations to detect outbreaks and epidemics as they begin, and implement appropriate control measures. They also conduct surveillance for new diseases as well as bioterrorism events. Public health microbiologists working for local governments monitor community food establishments and water supplies to ensure they are safe and free from pathogens. ▶| *Epidemiology and public health microbiology (chapter 35)*

To understand, treat, and control infectious disease, it is important to understand how the immune system protects the body from pathogens; this question is the concern of immunology. Immunology is one of the fastest growing areas in science. Many advances emerged through the study of HIV, which specifically targets cells of the immune system. Immunology also deals with the nature and treatment of allergies and autoimmune diseases such as rheumatoid arthritis. ▶| *Innate host resistance (chapter 31); Adaptive immunity (chapter 32)*

Microbial ecology is another important field in microbiology. Microbial ecologists employ a variety of culture and molecular approaches to describe the vast diversity of microbes in terms of their morphology, physiology, and relationships with organisms and the components of their habitats. The importance of microbes in the local and global cycling of carbon, nitrogen, and sulfur has been long studied; however, these studies have acquired new urgency as our climate changes. Of particular interest is the role of microbes in both the production and removal of greenhouse gases such as carbon dioxide and methane. Microbial ecologists also are employing microorganisms in bioremediation to reduce pollution. An exciting frontier in microbial ecology is the study of the microbes normally associated with the human body—the human microbiome. ▶| *Global climate change (section 28.3); The microbe-human ecosystem (chapter 33); Biodegradation and bioremediation harness microbes to clean the environment (section 42.4)*

Agricultural microbiology is a field related to both medical microbiology and microbial ecology. It is concerned with the impact of microorganisms on food production, such as nitrogen-fixing bacteria, which affect soil fertility. Other microbes live in the digestive tracts of ruminants such as cattle and break down the plant materials these animals ingest. There are also plant and animal pathogens that have significant economic impact if not controlled. Furthermore, some pathogens of domestic animals also cause human disease. The intersection of human, animal, and environmental health is the focus of investigators and health professionals who participate in the *One Health Initiative (see chapter 39 opener)*. Agricultural microbiologists work on methods to increase soil fertility and crop yields, study rumen microorganisms to increase meat and milk production, and try to combat plant and animal diseases. Currently many agricultural microbiologists are studying the use of bacterial and viral insect pathogens as substitutes for chemical pesticides. ▶| *Microorganisms in terrestrial ecosystems (chapter 30)*

In addition to agricultural microbiology, food microbiology has contributed to the ready supply of high-quality foods. Food microbiologists study microbes used to make food and beverages (e.g., yogurt, cheese, beer) as well as microbes that cause food spoilage or are pathogens that are spread through food. For example, periodic outbreaks of *Listeria monocytogenes* and certain *Escherichia coli* strains can have lethal consequences. To protect the public, the specific strain must be traced back to the contaminated food source. Food microbiologists also work to prevent microbial spoilage of food and conduct research on the use of microorganisms as nutrient sources for livestock and humans. ▶| *Microbiology of food (chapter 40)*

Industrial microbiology involves the use of microbes to make products helpful to humans. An important advance occurred in 1929 when Alexander Fleming rediscovered that the fungus *Penicillium* sp. produced what he called penicillin, the first antibiotic that could successfully control bacterial infections. Although it took World War II for scientists to learn how to mass-produce penicillin, microbiologists soon found other microbes capable of producing additional antibiotics. Today industrial microbiologists also use microorganisms to make products such as vaccines, steroids, alcohols and other solvents, vitamins, amino acids, enzymes, and biofuels. These alternative fuels are renewable and may help reduce fossil fuel dependence. ▶| *Antimicrobial chemotherapy (chapter 9); Biofuel production is a dynamic field (section 41.5); Microbial fuel cells: batteries powered by microbes (section 42.3)*

The advances in medical microbiology, agricultural microbiology, food and dairy microbiology, and industrial microbiology are outgrowths of basic research in areas such as microbial physiology, microbial genetics, molecular biology, and bioinformatics. Microbial physiologists study many aspects of the biology of microorganisms, including their diverse metabolic capabilities. They also study the synthesis of antibiotics and toxins, the ways in which microorganisms survive harsh environmental conditions, and the effects of chemical and physical agents on microbial growth and survival. Microbial geneticists, molecular biologists, and bioinformaticists study the nature of genetic information and how it regulates the development and function of cells and organisms. The bacteria *E. coli* and *Bacillus subtilis*, the yeast *Saccharomyces cerevisiae* (baker's yeast), and bacterial viruses such as T4 and lambda continue to be important model organisms used to understand biological phenomena.

The future of microbiology is bright. Genomics has revolutionized biology, as scientists are now beginning to understand organisms holistically, rather than in a reductionist, piecemeal manner. How the genomes of microbes evolve, the nature of host-pathogen interactions, the minimum set of genes required for an organism to survive, and many more topics are aggressively being examined by molecular and genomic analyses. This is an exciting time to be a microbiologist. Enjoy the journey.

Comprehension Check

1. Briefly describe the major subdisciplines in microbiology. Which do you consider to be applied fields? Which are basic?
2. Log all the microbial products you use in a week. Be sure to consider all foods and medications (including vitamins). Consider how your life might be different without these items.
3. List all the activities or businesses you can think of in your community that directly depend on microbiology.

Key Concepts

1.1 Members of the Microbial World

- Microbiology studies microscopic cellular organisms that are often unicellular or, if multicellular, do not have highly differentiated tissues. Microbiology also focuses on biological entities that are acellular (**figure 1.1**).
- Microbiologists divide cellular organisms into three domains: Bacteria, Archaea, and Eukarya (**figure 1.3**).
- Microbes in the domains Bacteria and Archaea lack organelles. Eukaryotic microbes (protists and fungi) are placed in Eukarya. Viruses, viroids, satellites, and prions are acellular entities not placed in any domain and are classified by a separate system.

1.2 Microbes Have Evolved and Diversified for Billions of Years

- Earth is approximately 4.5 billion years old. Within the first 1 billion years of its existence, life arose (**figure 1.4**).
- The RNA world hypothesis posits that the earliest self-replicating entity had RNA enclosed in a lipid bilayer. RNA stores genetic information and conducts cellular processes (**figure 1.5**).
- The last universal common ancestor (LUCA) is placed on the bacterial branch of the universal phylogenetic tree (**figure 1.3**). Thus Bacteria are thought to have diverged first, and Archaea and Eukarya arose later.
- Mitochondria, chloroplasts, and hydrogenosomes are thought to have evolved from bacterial endosymbionts of ancestral cells in the eukaryotic lineage (**figure 1.7**).
- Comparisons of housekeeping proteins and small subunit (SSU) rRNA genes have been useful in establishing phylogenetic relationships (**figure 1.8**).
- Phylogenetic relationships often are shown as branched diagrams called phylogenetic trees. Trees can be constructed using distance-based or character-based approaches (**figure 1.9**).
- Trees may be either rooted or unrooted. Unrooted trees can be rooted by including an outgroup when the tree is constructed (**figure 1.10**).

- Taxonomic ranks are arranged in a nonoverlapping hierarchy (**figure 1.11**). Species are named using the binomial system of Linnaeus.
- It is difficult to define a bacterial or archaeal species because these microbes do not reproduce sexually.

1.3 Microbiology Advanced as New Tools for Studying Microbes Were Developed

- Microbiology is defined not only by the organisms it studies but also by the tools it uses. Microscopy and culture-based techniques play important roles in the evolution of the discipline.
- Antony van Leeuwenhoek used simple microscopes and was the first person to extensively describe microorganisms (**figure 1.13**).
- Culture-based techniques for studying microbes began to develop as scientists debated the theory of spontaneous generation. Experiments by Francesco Redi and others disproved the theory of spontaneous generation of larger organisms. The spontaneous generation of microorganisms was disproved by Louis Pasteur and others (**figure 1.15**).
- Koch's postulates are used to establish a direct relationship between a suspected pathogen and a disease. Robert Koch and his coworkers developed techniques to grow bacteria on solid media and isolate pure cultures of pathogens (**figure 1.17**).
- Viruses were discovered following the invention of a bacterial filter by Charles Chamberland. Dimitri Ivanowski and Martinus Beijerinck were important contributors to the field of virology.
- Microbiologists founded the field of immunology; they created vaccines and discovered antibodies and phagocytic cells.
- Microbial ecology grew out of the work of Sergei Winogradsky and Beijerinck. They studied the role of microorganisms in carbon, nitrogen, and sulfur cycles and developed enrichment culture techniques and selective media.

1.4 Microbiology Encompasses Many Subdisciplines

- There are many fields in microbiology. These include medical, public health, industrial, and food and dairy microbiology. Microbial ecology, physiology, and genetics are important subdisciplines of microbiology.

Active Learning

1. Why aren't viruses, viroids, satellites, and prions included in the three-domain system?
2. Some individuals can be infected by a pathogen yet not develop disease. In fact, some become chronic carriers of the pathogen. How does this observation affect Koch's postulates? How might the postulates be modified to account for the existence of chronic carriers?
3. Support this statement: "Vaccinations against various childhood diseases have contributed to the entry of females, particularly those with children at home, into the full-time workplace."
4. For many years, the oldest stromatolites with compelling evidence of ancient life on Earth were found in Western Australia. The microbial communities that built these sedimentary structures were dated at roughly 3.5 billion years old. However, in 2016 stromatolites in Greenland were reported to be 3.7 billion years old. Interestingly, the complexity and morphology of the Greenland stromatolites are similar to those found in Australia. Why is it of interest to document the oldest evidence of life on Earth? What might this suggest about the evolution of life during the intervening 200 million years? What is the significance, if any, of the geographic distribution of these stromatolites?
5. The RNA world hypothesis suggests that ribonucleotides came together (condensed) to form RNAs, some of which had catalytic activity, as do current day ribozymes. It has been suggested that modified or unusual (called noncanonical) ribonucleotides may have been essential in creating these early ribozymes. In fact, RNA molecules with noncanonical ribonucleotides have been shown to establish the synthesis of short proteins (peptides) directly on RNA. What is the significance of noncanonical ribonucleotides and how does this observation impact our understanding of the RNA world hypothesis?

CHAPTER

2

Microscopy

Kenneth Lambert/AP Images

Anthrax Bioterrorism Attack

While on a trip to North Carolina in late September 2001, a photojournalist from Florida began feeling ill. His muscles ached, he felt nauseous, and he had a fever. After returning home, his symptoms worsened, and on October 2 he awoke with vomiting and mental confusion. When his wife took him to a local emergency room, doctors suspected meningitis and performed a spinal tap to collect cerebrospinal fluid (CSF). The fluid was sent to the hospital laboratory, where it was stained using the Gram-staining technique. Much to the surprise of the clinical lab scientists, the CSF contained long, Gram-positive rods about 1 micrometer (µm) wide by 3–5 µm long, arranged in chains. This morphology is unlike that of typical meningitis-causing bacteria. Based on the Gram-stain results, doctors made an initial diagnosis of inhalation anthrax, caused by the bacterium *Bacillus anthracis*. This was very surprising because inhalation anthrax is extremely rare in the United States. Doctors immediately began treatment with antibiotics, but the photojournalist's condition worsened, and he died on October 5.

Thus began the first anthrax bioterrorism attack in the United States. Over the next few weeks, 16 other people developed inhalation anthrax, all after opening mail that contained *B. anthracis* spores. Four people died. Much was learned about the disease from this attack. Much was also learned about the nation's readiness to deal with bioterrorism. This event is a reminder of the importance of microscopy in microbiology and in diagnosing disease.

In this chapter, we introduce some of the most commonly used types of microscopy. We begin with light microscopes and then describe other common types of microscopes, as well as how specimens are prepared for examination by microscopes.

Readiness Check:

Based on what you have learned previously, you should be able to:

✓ Explain why microscopy is a major tool used by microbiologists (sections 1.1 and 1.3)

✓ Define magnification

✓ Express the size of organisms using the metric system (**table 2.1**)

Table 2.1 Common Units of Measurement

Unit	Abbreviation	Value
1 centimeter	cm	10^{-2} meter
1 millimeter	mm	10^{-3} meter
1 micrometer	µm	10^{-6} meter
1 nanometer	nm	10^{-9} meter
1 Angstrom	Å	10^{-10} meter

2.1 Lenses Create Images by Bending Light

After reading this section, you should be able to:

a. Relate the refractive indices of glass and air to the path light takes when it passes through a prism or convex lens
b. Correlate lens strength and focal length

Light microscopes were the first microscopes invented and they continue to be the most commonly used. The goal of all microscopes is to accurately enlarge, or magnify, the image of an object. To understand light microscopy, we must consider the way lenses bend and focus light to form images. When a ray of light passes from one medium to another, refraction occurs; that is, the ray is bent at the interface. The **refractive index** is a measure of how much a substance slows the velocity of light; the direction and magnitude of bending are determined by the refractive indices of the two media forming the interface. For example, when light passes from air into glass, which has a greater refractive index, it is slowed and bent toward the normal, a line perpendicular to the surface (**figure 2.1**). As light leaves glass and returns to air, a medium with a lower refractive index, it accelerates and is bent away from the normal. Thus a prism bends light because glass has a different refractive index from air and the light strikes its surface at an angle.

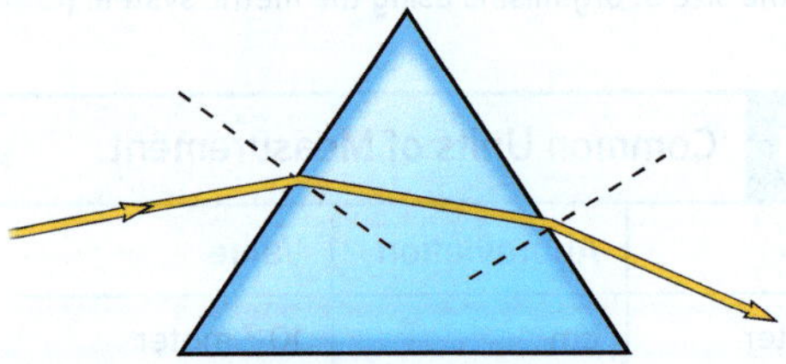

Figure 2.1 The Bending of Light by a Prism. Lines perpendicular to the surface of the prism are called normal; they are indicated by dashed lines. As light enters the glass, it is bent toward the first normal. When light leaves the glass and returns to air, it is bent away from the second normal.

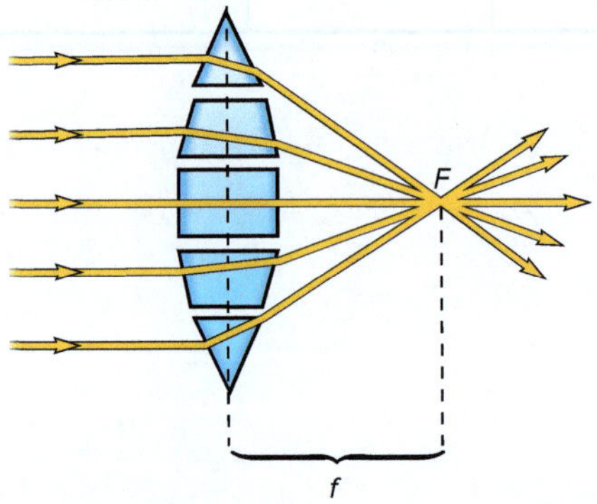

Figure 2.2 Lens Function. A lens functions like a collection of prisms. Light rays from a distant source are focused at the focal point *F*. The focal point lies at a distance *f*, the focal length, from the lens center.

MICRO INQUIRY *How would the focal length change if the lens shown here were thicker?*

Lenses act like a collection of prisms operating as a unit. When the light source is distant so parallel rays of light strike the lens, a convex lens focuses the rays at a specific point, the focal point (*F* in **figure 2.2**). The distance between the center of the lens and the focal point is called the *focal length* (*f* in figure 2.2).

Our eyes cannot focus on objects nearer than about 25 cm (i.e., about 10 inches). This limitation may be overcome by using a convex lens as a simple magnifier (or microscope) and holding it close to an object. A magnifying glass provides a clear image at much closer range, and the object appears larger. Lens strength is related to focal length; a lens with a short focal length magnifies an object more than a lens having a longer focal length.

Comprehension Check

1. How do refraction, refractive index, focal point, and focal length differ?
2. Draw the path of a light ray through a thin and a thick lens.
3. If the lenses in the above question were in corrective eyeglasses, which would be used for reading and which would improve distance vision?

2.2 There Are Several Types of Light Microscopes

After reading this section, you should be able to:

a. Evaluate the parts of a light microscope in terms of their contributions to image production and use of the microscope
b. Predict the relative degree of resolution based on light wavelength and numerical aperture of the lens used to examine a specimen
c. Create a table that compares and contrasts the various types of light microscopes in terms of their uses, how images are created, and the quality of images produced

Microbiologists use a variety of light microscopes in their work, each designed for specific applications. Modern microscopes are compound microscopes; that is, they have two sets of lenses. The **objective lens** is the lens closest to the specimen. It forms a magnified image that is further enlarged by one or more additional lenses.

Bright-Field Microscope: Dark Object, Bright Background

The **bright-field microscope** is routinely used to examine both stained and unstained specimens. It forms a dark image against

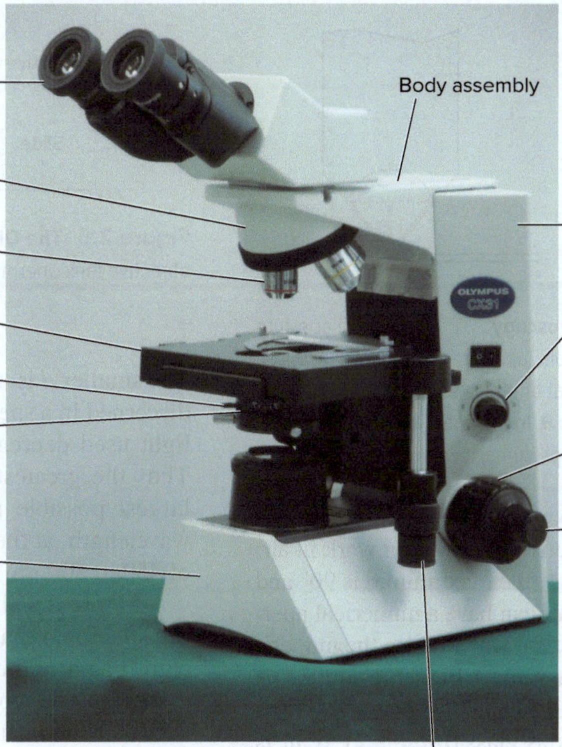

Figure 2.3 A Bright-Field Microscope.
James Redfearn/McGraw Hill

a lighter background; thus it has a bright field. The microscope consists of a metal stand composed of a base and an arm to which the remaining parts are attached (**figure 2.3**). A light source is located in the base. Two focusing knobs, the fine and coarse adjustment knobs, are located on the arm and move either the stage or the nosepiece vertically to focus the image.

The stage is positioned about halfway up the arm. Microscope slides are clipped to the stage, which can be moved during viewing by rotating control knobs. The **condenser lens** (or, simply, condenser) is within or beneath the stage and focuses a cone of light on the slide. Its position may be fixed in simpler microscopes but can be adjusted vertically in more complex models.

The curved upper part of the arm holds the body assembly, to which a nosepiece and one or more **ocular lenses** (also called **eyepieces**) are attached. Binocular microscopes have eyepieces for both eyes. The body assembly contains a series of mirrors and prisms so the barrel holding the eyepiece may be tilted for ease in viewing. The nosepiece holds three to five objective lenses of differing magnifying power and can be rotated to change magnification. Ideally a microscope should be **parfocal;** that is, the image should remain in focus when objective lenses are changed.

The image seen when viewing a specimen with a compound microscope is created by the objective and ocular lenses working together. Light from the illuminated specimen is focused by the objective lens, creating an enlarged image within the microscope. The ocular lens further magnifies this primary image. The total magnification is calculated by multiplying the objective and eyepiece magnifications. For example, if a 45× objective lens is used with a 10× eyepiece, the overall magnification of the specimen is 450×.
Parts and Care of the Microscope; Use of the Microscope

Better Microscope Resolution Means a Clearer Image

The most important part of the microscope is the objective lens, which must produce a clear image, not just a magnified one. **Resolution** is the ability of a lens to distinguish between two objects that are close together, rather than detecting them as a single larger object. At best, the resolution of a bright-field microscope is 0.2 μm, which is about the size of a very small bacterium. Why is this the case? Resolution is in part dependent on the **numerical aperture** ($n \sin \theta$) of a lens. Numerical aperture is defined by two components: n is the refractive index of the medium in which the lens works (e.g., air = 1) and θ is ½ the angle of the cone of light entering an objective lens (**figure 2.4**). A cone with a narrow angle does not adequately separate the rays of light emanating from closely packed objects, so the images are not resolved. A cone of light with a very wide angle is able to separate the rays, and the closely packed objects appear widely separated and resolved. Recall that the refractive indices of all materials through which light waves pass determine the direction of the light rays

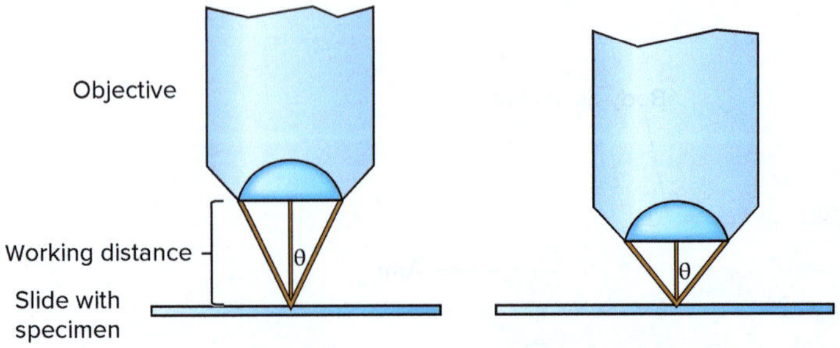

Figure 2.4 Numerical Aperture in Microscopy. The numerical aperture of a lens is related to a value called the angular aperture (symbolized by θ), which is ½ the angle of the cone of light that enters a lens from a specimen. The equation for numerical aperture is $n \sin \theta$. In the right-hand illustration, the lens has larger angular and numerical apertures; its resolution is greater and its working distance smaller.

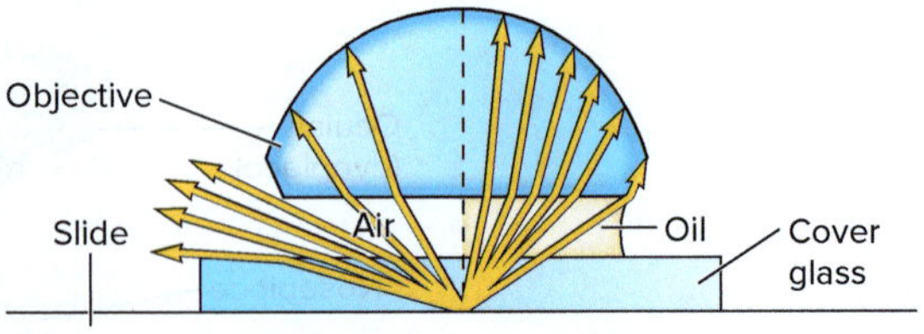

Figure 2.5 The Oil Immersion Objective Lens. An oil immersion objective lens operating in air and with immersion oil.

emanating from the specimen. Objective lenses often work in air; because sin θ cannot be greater than 1 (the maximum θ is 90° and sin 90° is 1.00), no lens working in air can have a numerical aperture greater than 1.00. The only practical way to raise the numerical aperture above 1.00, and therefore achieve higher resolution, is to increase the refractive index with *immersion oil,* a colorless liquid with the same refractive index as glass (**table 2.2**). If air is replaced with immersion oil, many light rays that would have otherwise escaped due to reflection and refraction at the surfaces of the lens and slide now enter the objective lens (**figure 2.5**). This results in an increase in numerical aperture and resolution.

Resolution is described mathematically by an equation developed in the 1870s by Ernst Abbé (1840–1905), a German physicist responsible for much of the optical theory underlying microscope design. The Abbé equation states that the minimal distance (d) between two objects that reveals them as separate entities depends on the wavelength of light (λ) used to illuminate the specimen and on the numerical aperture of the lens ($n \sin \theta$), which is the ability of the lens to gather light.

$$d = \frac{0.5\,\lambda}{n \sin \theta}$$

The smaller d is, the better the resolution, and finer detail can be discerned in a specimen; d becomes smaller as the wavelength of light used decreases and as the numerical aperture increases. Thus the greatest resolution is obtained using a lens with the largest possible numerical aperture and light of the shortest wavelength, at the blue end of the visible spectrum (in the range of 450 to 500 nm).

Figure 2.4 shows that one objective lens is much closer to the specimen than the other. Specifically, the objective on the right has a higher numerical aperture and a shorter working distance. The working distance of an objective is the distance between the surface of the lens and the surface of the cover glass (if one is used) or the specimen when it is in sharp focus. As illustrated here, objectives with large numerical apertures and great resolving power have short working distances (table 2.2).

Numerical aperture is also an important feature of a microscope's condenser lens. The condenser is a large, light-gathering lens used to project a wide cone of light through the slide and into the objective lens. Although most condensers have a numerical aperture between 1.2 and 1.4, the numerical aperture will not exceed about 0.9 unless the top of the condenser lens is oiled to the bottom of the slide. During routine microscope operation, the condenser usually is not oiled, and this limits the overall resolution, even with an oil immersion objective lens.

The most accurate calculation of a microscope's resolving power considers both the numerical aperture of the objective lens

Table 2.2 Properties of Objective Lenses

	OBJECTIVE			
Property	**Scanning**	**Low Power**	**High Power**	**Oil Immersion**
Magnification	4×	10×	40–45×	90–100×
Numerical aperture	0.10	0.25	0.55–0.65	1.25–1.4
Approximate focal length (f)	40 mm	16 mm	4 mm	1.8–2.0 mm
Working distance	17–20 mm	4–8 mm	0.5–0.7 mm	0.1 mm
Approximate resolving power with light of 450 nm (blue light)	2.3 µm	0.9 µm	0.35 µm	0.18 µm

and that of the condenser, as is evident from the following equation, where NA is the numerical aperture.

$$d_{\text{microscope}} = \frac{\lambda}{(\text{NA}_{\text{objective}} + \text{NA}_{\text{condenser}})}$$

However, in most cases the limit of resolution of a light microscope is calculated using the Abbé equation, which considers the objective lens only. We now see why the maximum theoretical resolving power of a microscope when viewing a specimen using an oil immersion objective (numerical aperture of 1.25) and blue-green light is approximately 0.2 μm.

$$d = \frac{(0.5)(530 \text{ nm})}{1.25} = 212 \text{ nm or } 0.2\ \mu\text{m}$$

Given the limit of resolution of a light microscope, the largest useful magnification—the level of magnification needed to increase the size of the smallest resolvable object to be visible with the light microscope—can be determined. Our eye can just detect a speck 0.2 mm in diameter. When the acuity of the eye and the resolution of the microscope are considered together, it is calculated that the useful limit of magnification is about 1,000 times the numerical aperture of the objective lens. Most standard microscopes have 10× eyepieces and have an upper limit of about 1,000× with oil immersion. A 15× eyepiece may be used with good objective lenses to achieve a useful magnification of 1,500×. Any further magnification does not enable a person to see more detail. Indeed, a light microscope can be built to yield a final magnification of 10,000×, but it would simply be magnifying a blur. Only a type of microscope called the electron microscope provides sufficient resolution to make higher magnifications useful, as we discuss in section 2.4.

Visualizing Live, Unstained Microbes

Bright-field microscopes are probably the most common microscope found in teaching, research, and clinical laboratories. However, many microbes are unpigmented so are not clearly visible because there is little contrast between the cells, subcellular structures, and water. As we discuss in section 2.3, one solution to this problem is to stain cells before observation. Unfortunately staining procedures usually kill cells. But what if an investigator wishes to view living cells? Three types of light microscopes create detailed, clear images of living specimens: dark-field microscopes, phase-contrast microscopes, and differential interference contrast microscopes.

Dark-Field Microscope: Bright Object, Dark Background

The **dark-field microscope** produces detailed images of living, unstained cells and organisms by changing the way in which they are illuminated. A hollow cone of light is focused on the specimen in such a way that unreflected and unrefracted rays do not enter the objective. Only light that has been reflected or refracted by the specimen forms an image (**figure 2.6**). The field surrounding a specimen appears black, while the object itself is brightly illuminated (**figure 2.7**). The dark-field microscope can be used to visualize certain bacteria such as the thin and distinctively shaped *Treponema pallidum*, the causative agent of syphilis (figure 2.7*a*). It can also reveal considerable internal structure in larger eukaryotic microorganisms (figure 2.7*b*).

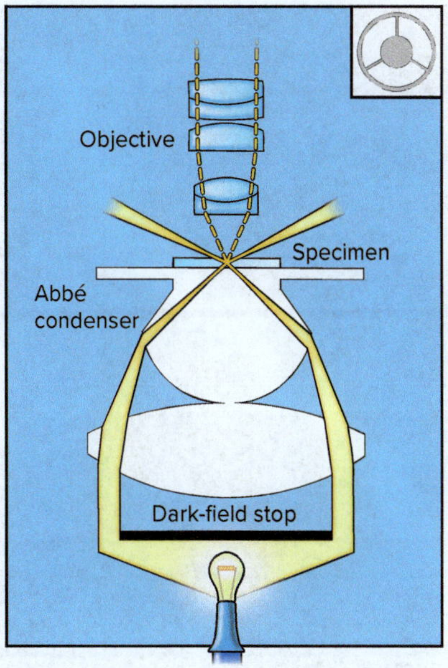

Figure 2.6 Dark-Field Microscopy. In dark-field microscopy, a dark-field stop (inset) is placed underneath the condenser lens system. The condenser then produces a hollow cone of light so that the only light entering the objective is reflected or refracted by the specimen.

Phase-Contrast Microscope: Dark Object, Bright Background

To understand phase-contrast microscopy, consider a bacterium in a drop of water on a microscope slide (i.e., a wet mount) as illustrated in **figure 2.8**. The refractive indices of bacterial cell structures are greater than that of water. Therefore, light waves passing through a cell structure will be diffracted and slowed (deviated) more than light waves passing through the water inside and outside the cell (undeviated). Because the deviated light waves are slowed relative to the undeviated light waves, they are said to be out of phase. That is, the crests and troughs of the deviated and undeviated waves do not align. Typically the deviated light waves are slowed by about ¼ wavelength compared to the undeviated light (figure 2.8).

Phase-contrast microscopes take advantage of this phenomenon to create differences in light intensity that provide contrast to allow the viewer to see a clearer, more detailed image of the specimen (**figure 2.9**). Two components of these microscopes allow the light manipulation needed for phase contrast to occur: a condenser annulus and a phase plate

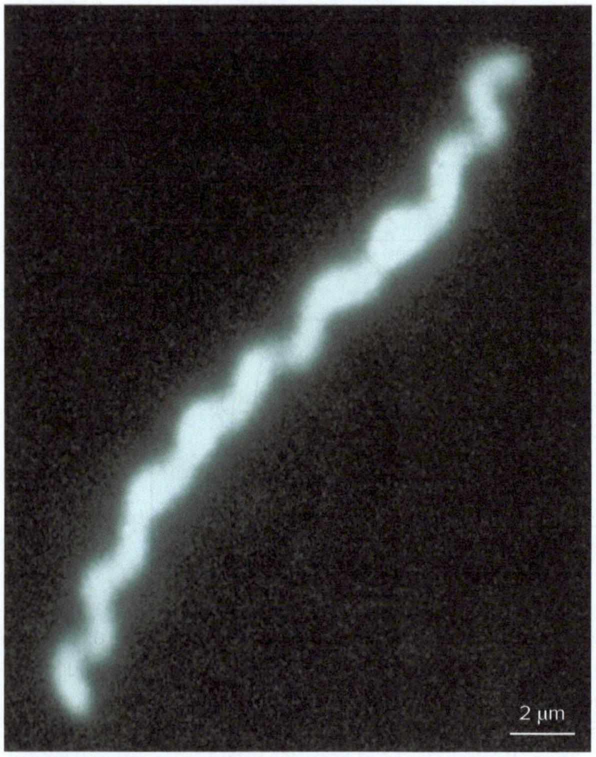

(a) *T. pallidum*

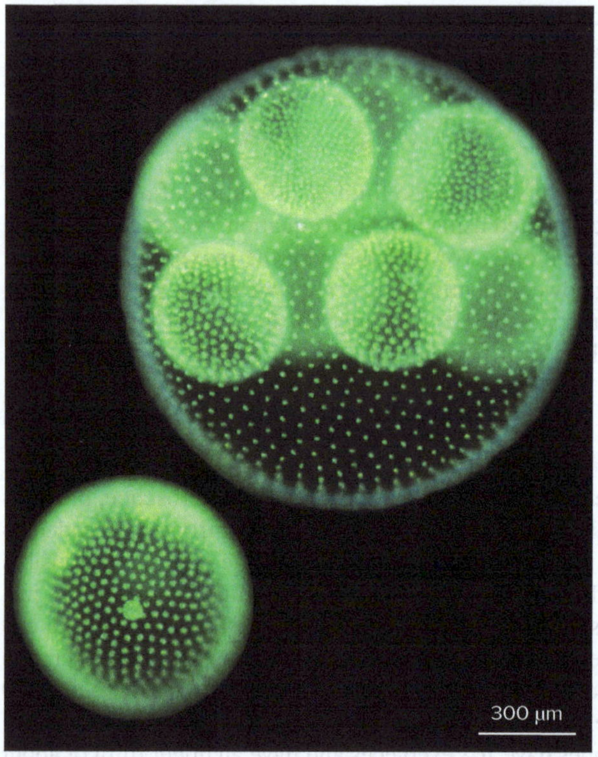

(b) *Volvox*

Figure 2.7 Examples of Dark-Field Microscopy. (a) *Treponema pallidum,* the spirochete that causes syphilis. (b) The protist *Volvox*. Note daughter colonies within the mature *Volvox* colony.

(a) Schwartz/CDC; (b) Stephen Durr

(**figure 2.10**). The condenser annulus is an opaque disk with a thin transparent ring. Light is directed through the condenser annulus to the condenser, which focuses it onto the specimen, as shown in figure 2.10. Because light waves can pass through the transparent ring but not through the rest of the condenser annulus, the specimen is illuminated with a hollow cone of light. As the light passes through the specimen, some of it is deviated because of the uneven surface of the the specimen. Light directly from the light source, however, remains undeviated. Both deviated and undeviated light then pass through the objective toward the phase plate. The phase plate is usually glass, with a thin ring etched in it through which the undeviated light is focused (figure 2.8). In the most common type of phase-contrast microscopy (positive phase contrast), the ring is coated with a metallic film that advances the phase of the undeviated light by ¼ wavelength. The deviated light is focused on the rest of the phase plate, through which it passes unchanged. After leaving the phase plate, the deviated and undeviated light are now out of phase by ½ wavelength (figure 2.8).

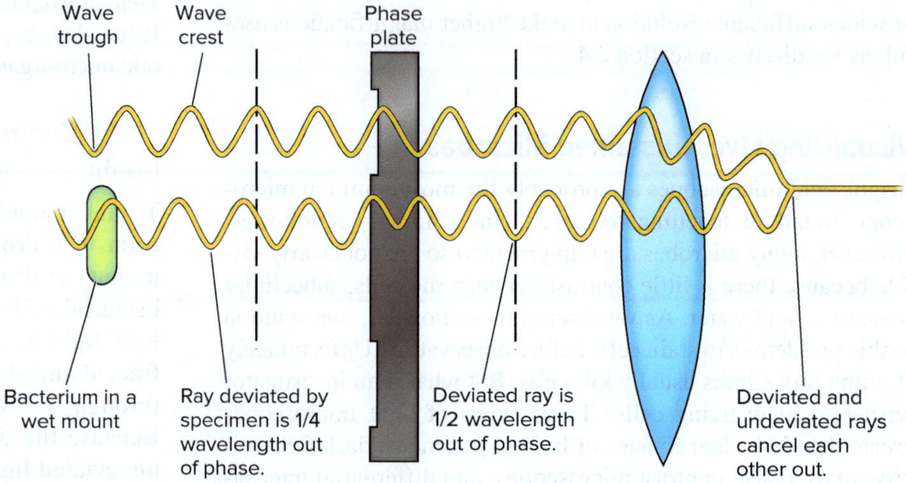

Figure 2.8 The Production of Contrast in Phase-Contrast Microscopy. The behavior of deviated and undeviated (i.e., undiffracted) light rays in the positive-phase-contrast microscope. Because the light rays tend to cancel each other out, the image of the specimen will be dark against a brighter background.

When the two rays of light recombine to form an image, they cancel each other out, a phenomenon called *destructive interference*. In our example, the resulting image consists of a darker bacterium against a lighter background.

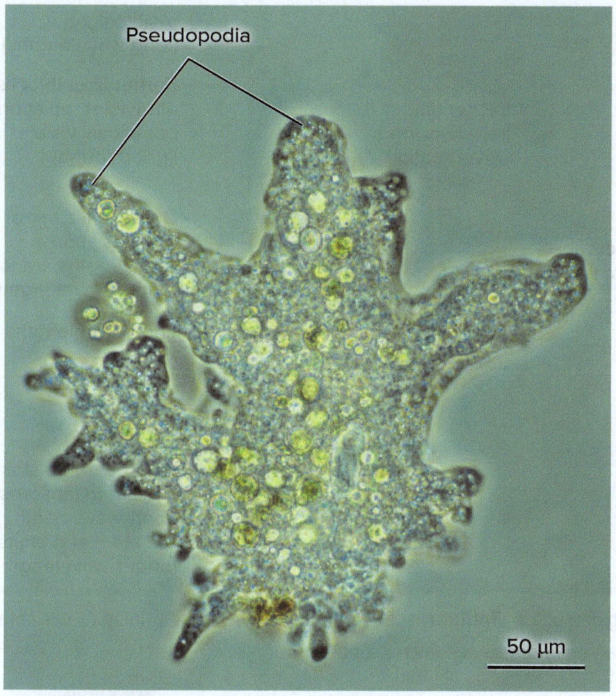

(a) An amoeba

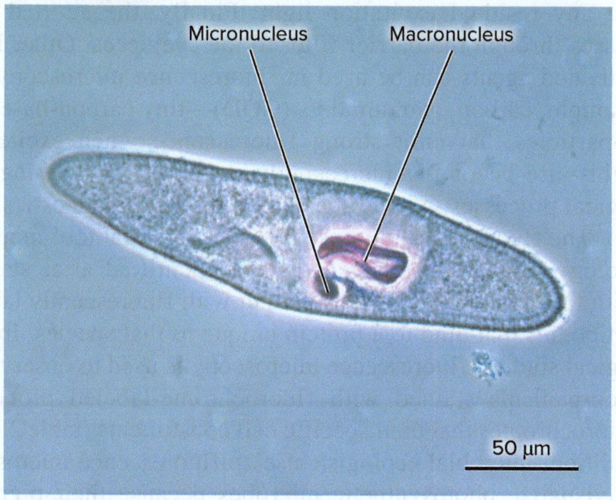

(b) *Paramecium* sp.

Figure 2.9 Example of Phase-Contrast Microscopy. (a) An amoeba, a eukaryotic microbe that moves by means of pseudopodia, which extend out from the main part of the cell body. (b) *Paramecium* sp. stained to show a large central macronucleus with a small spherical micronucleus at its side.

(a) Stephen Durr; (b) James Redfearn/McGraw Hill

Phase-contrast microscopy is especially useful for studying microbial motility, determining the shape of living cells, and detecting bacterial structures such as endospores and inclusions. These are clearly visible because they have refractive indices markedly different from that of water. Phase-contrast microscopes also are widely used to study eukaryotic cells. In 1953, Frits Zernike was awarded the Nobel Prize in Physics for inventing this type of microscope. ▶| *Inclusions (section 3.6); Bacterial endospores are a survival strategy (section 3.8)*

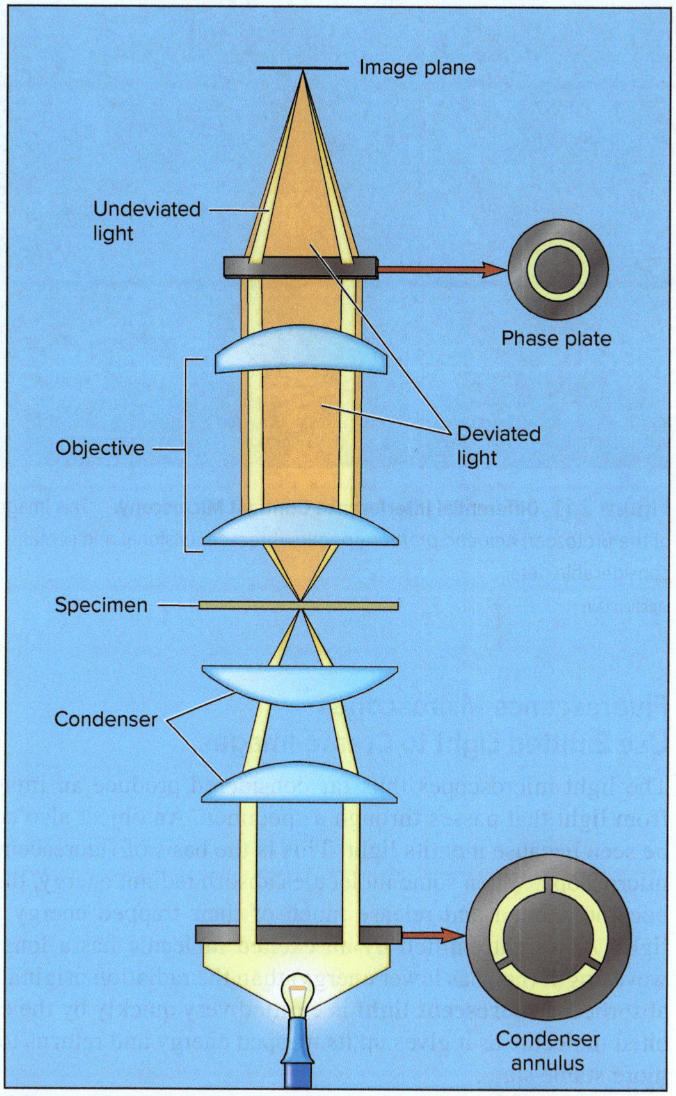

Figure 2.10 Phase-Contrast Microscopy. The optics of a positive-phase-contrast microscope.

MICRO INQUIRY *What is the purpose of the condenser annulus in a phase-contrast microscope?*

Differential Interference Contrast Microscope

The **differential interference contrast (DIC) microscope** is similar to the phase-contrast microscope in that it creates an image by detecting differences in refractive indices and thickness. Two beams of plane-polarized light at right angles to each other are generated by prisms. In one design, the object beam passes through the specimen, while the reference beam passes through a clear area of the slide. After passing through the specimen, the two beams combine and interfere with each other to form an image. A live, unstained specimen appears brightly colored and seems to pop out from the background, giving the viewer the sense that a three-dimensional image is being viewed (**figure 2.11**). Structures such as cell walls, endospores, granules, vacuoles, and nuclei are clearly visible.

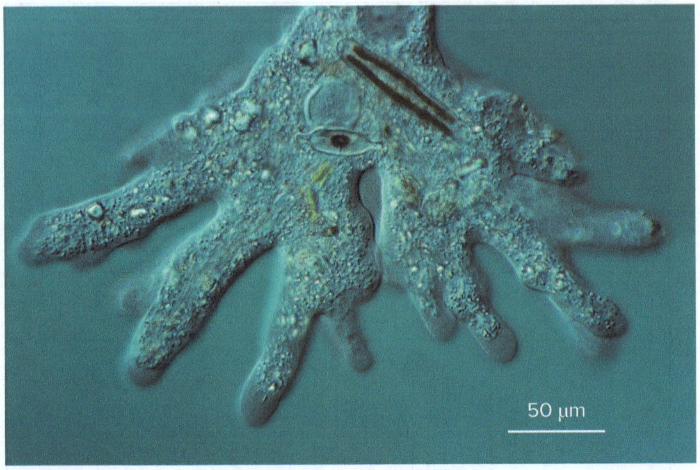

Figure 2.11 Differential Interference Contrast Microscopy. This image of the protozoan *Amoeba proteus* appears three-dimensional and contains considerable detail.

Stephen Durr

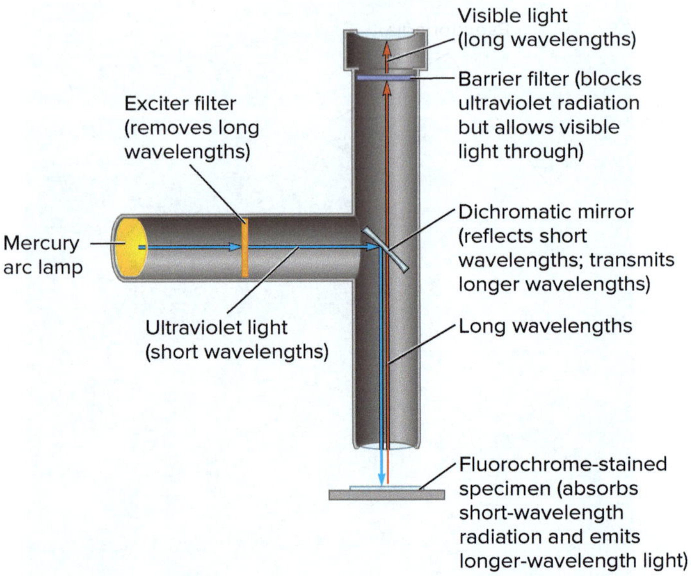

Figure 2.12 Epifluorescence Microscopy. The principles of operation of an epifluorescence microscope.

Fluorescence Microscopes Use Emitted Light to Create Images

The light microscopes thus far considered produce an image from light that passes through a specimen. An object also can be seen because it emits light. This is the basis of fluorescence microscopy. When some molecules absorb radiant energy, they become excited and release much of their trapped energy as light. Any light emitted by an excited molecule has a longer wavelength (i.e., has lower energy) than the radiation originally absorbed. **Fluorescent light** is emitted very quickly by the excited molecule as it gives up its trapped energy and returns to a more stable state.

The **fluorescence microscope** excites a specimen with a specific wavelength of light that triggers the object to emit fluorescent light. The most commonly used fluorescence microscopy is epifluorescence microscopy, also called incident light or reflected light fluorescence microscopy. Epifluorescence (Greek *epi*, upon) microscopes illuminate specimens from above. The objective lens also acts as a condenser (**figure 2.12**). A light-emitting diode (LED) or other light source produces an intense beam of light that passes through an exciter filter. The exciter filter transmits only the desired wavelength of light. This is directed down the microscope by the dichromatic mirror, which reflects light of shorter wavelengths (i.e., the excitation light) but allows light of longer wavelengths to pass through. The excitation light continues down, passing through the objective lens to the specimen, which is stained with molecules called **fluorochromes** (**table 2.3**). The fluorochrome absorbs light energy from the excitation light and emits visible fluorescent light that travels up through the objective lens into the microscope. Because the emitted fluorescent light has a longer wavelength, it passes through the dichromatic mirror to a barrier filter, which blocks out any residual excitation light. Finally, the emitted light passes through the barrier filter to the eyepieces. Other light-activated agents can be used in fluorescence microscopy. For example, carbon quantum dots (CQD)—tiny carbon-based nanoparticles that emit strong fluorescence when excited by light—are becoming increasingly common for tracking biological processes in real time, and in medical diagnostics.

The fluorescence microscope is an essential tool in microbiology. Bacterial pathogens can be identified after staining with fluorochromes or tagging them with fluorescently labeled antibodies that bind to a protein unique to that species. In ecological studies, fluorescence microscopy is used to observe microorganisms stained with fluorochrome-labeled probes or fluorochromes that bind specific cell constituents (table 2.3). In addition, microbial ecologists use epifluorescence microscopy to visualize photosynthetic microbes because their pigments

Table 2.3 Commonly Used Fluorochromes

Fluorochrome	Uses
Acridine orange	Stains nucleic acids
Diamidino-2-phenyl indole (DAPI)	Stains DNA
Fluorescein isothiocyanate (FITC)	Often attached to DNA probes or to antibodies that bind specific cellular components
Tetramethyl rhodamine isothiocyanate (TRITC or rhodamine)	Often attached to antibodies that bind specific cellular components

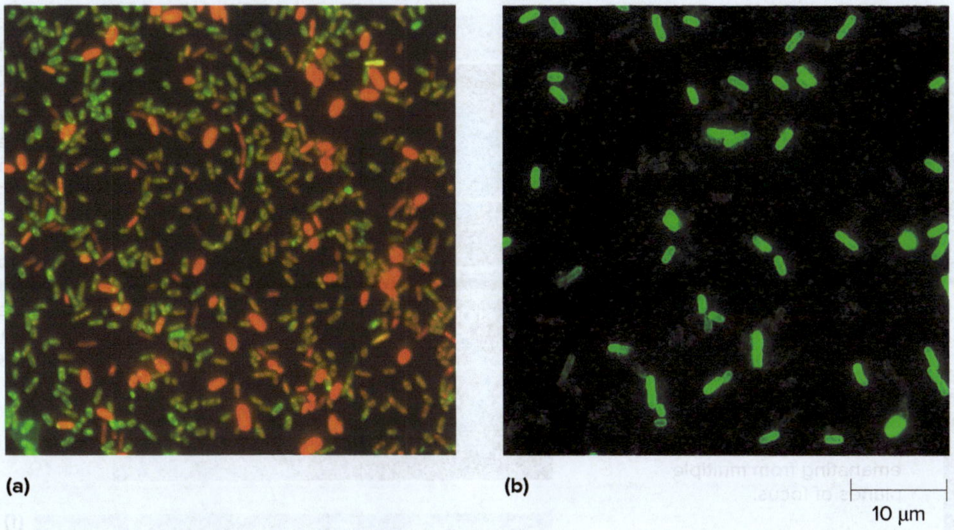

Figure 2.13 Fluorescent Dyes and Tags. (a) A sample of bacteria stained with dyes that cause live cells to fluoresce green and dead ones red. (b) Fluorescent antibodies tag specific molecules. In this case, the antibody binds to a molecule that is unique to *Yersinia pestis,* the bacterium that causes plague.
(a) Dr. Rita B. Moyes; (b) Larry Stauffer/Oregon State Public Health Laboratory/CDC

MICRO INQUIRY *How might the fluorescently labeled antibody used in figure 2.13b be used to diagnose an infectious disease?*

naturally fluoresce when excited by light of specific wavelengths. It is even possible to distinguish live bacteria from dead bacteria by the color they fluoresce after treatment with a specific mixture of stains (**figure 2.13a**). ▶| *Assessing microbial communities (section 26.4); Identification of microorganisms from specimens (section 36.2)*

Another important use of fluorescence microscopy is the localization of specific proteins within cells. One approach is to use genetic engineering techniques that fuse the gene for the protein of interest to a gene isolated from jellyfish belonging to the genus *Aequorea.* This jellyfish gene encodes a protein that naturally fluoresces green when exposed to light of a particular wavelength and is called green fluorescent protein (GFP). Thus when the protein is made by the cell, it is fluorescent. GFP has been used extensively in studies on bacterial cell division and related phenomena (**figure 2.14**). The 2008

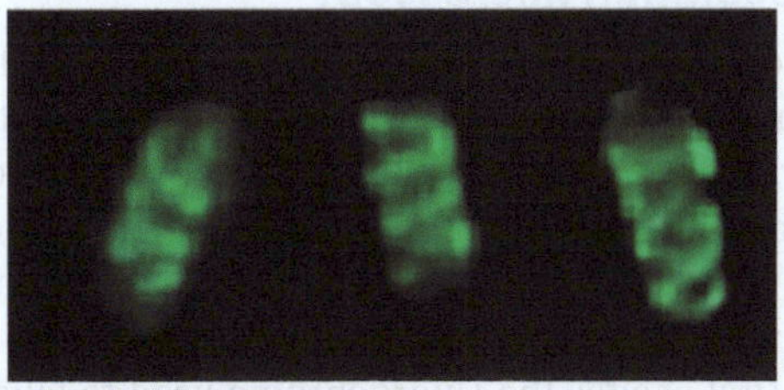

Figure 2.14 Green Fluorescent Protein. Visualization of Mbl, a cytoskeletal protein of *Bacillus subtilis.* The Mbl protein has been fused with green fluorescent protein and therefore fluoresces green.
Centre for Bacterial Cell Biology/Newcastle University/Jeff Errington

Nobel Prize in Chemistry was awarded to Osamu Shimomura of Japan and Americans Martin Chalfie and Roger Tsien for their development of this important tool. ▶| *Fluorescence labeling (section 17.4)*

Confocal Microscopy Creates Images That Appear 3-Dimensional

Like the large and small beads illustrated in **figure 2.15a**, biological specimens are three-dimensional. When three-dimensional objects are viewed with traditional light microscopes, light from all areas of the object, not just the plane of focus, enters the microscope and is used to create an image. The resulting image is blurry (figure 2.15*b*). This problem has been solved by the confocal scanning laser microscope (CSLM), or simply, **confocal microscope.** The confocal microscope uses a laser beam to illuminate a fluorescently stained specimen. A major component of this microscope is an aperture (an opening) placed above the objective lens. The aperture eliminates stray light from parts of the specimen that lie above and below the plane of focus. Thus the only light used to create the image is from the plane of focus, and a much sharper image is formed.

To generate a confocal image, a computer interfaced with the confocal microscope receives digitized information from each plane in the specimen. This information can be used to create a clear and detailed composite image (figure 2.15*c*) or a three-dimensional reconstruction of the specimen

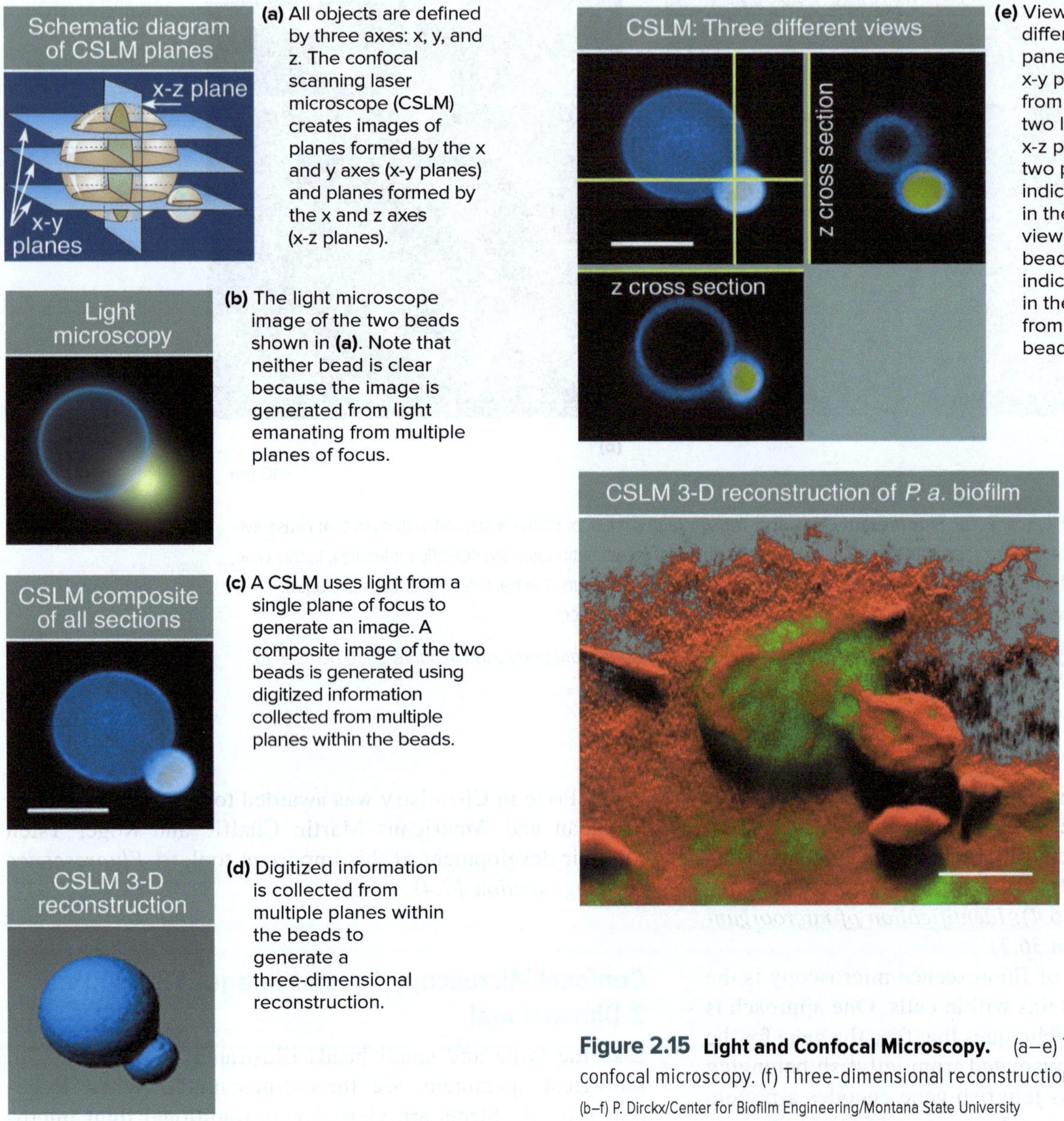

(a) All objects are defined by three axes: x, y, and z. The confocal scanning laser microscope (CSLM) creates images of planes formed by the x and y axes (x-y planes) and planes formed by the x and z axes (x-z planes).

(b) The light microscope image of the two beads shown in **(a)**. Note that neither bead is clear because the image is generated from light emanating from multiple planes of focus.

(c) A CSLM uses light from a single plane of focus to generate an image. A composite image of the two beads is generated using digitized information collected from multiple planes within the beads.

(d) Digitized information is collected from multiple planes within the beads to generate a three-dimensional reconstruction.

(e) Views are also generated using different planes. The top left panel is the image of a single x-y plane (i.e., looking down from the top of the beads). The two lines represent the two x-z planes imaged in the other two panels. The vertical line indicates the x-z plane shown in the top right panel (i.e., a view from the right side of the beads) and the horizontal line indicates the x-z plane shown in the bottom panel (i.e., a view from the front face of the beads).

(f) A three-dimensional reconstruction of a *Pseudomonas aeruginosa* biofilm. The biofilm was exposed to an antibacterial agent and then stained with dyes that distinguish living (green) from dead (red) cells. The cells on the surface of the biofilm have been killed, but those in the lower layers are still alive. This image clearly demonstrates the difficulty of killing all the cells in a biofilm.

Figure 2.15 Light and Confocal Microscopy. (a–e) Two beads examined by light and confocal microscopy. (f) Three-dimensional reconstruction of a biofilm.

(b–f) P. Dirckx/Center for Biofilm Engineering/Montana State University

(figure 2.15*d*). Images of x-z plane cross sections of the specimen can also be generated, giving the observer views of the specimen from three perspectives (figure 2.15*e*). Confocal microscopy has numerous applications. One is the study of biofilms, which can form on many different types of surfaces, including indwelling medical devices such as hip joint replacements. As shown in figure 2.15*f* it is difficult to kill all cells in a biofilm. This makes biofilms a particular concern because their formation on medical devices can result in infections that are difficult to treat. ▶| *Biofilms are common in nature (section 7.6)*

Super-Resolution Microscopy

Super-resolution microscopy (SRM) allows much higher resolution with an optical microscope than is typically attained with these instruments. SRM depends on the use of fluorochromes with different chemical and physical properties. There are several SRM techniques including patterned resolution and single molecule imaging. In the first approach, two beams of light are used. The first excites the fluorochrome used to label the specimen of interest, creating an excitation spot. The second beam creates a cone of light that quenches all fluorescence at the edge of the excitation spot, reducing its size and resulting in higher resolution within it. An example of this approach is *st*imulation *e*mission *d*epletion microscopy (STED). In single molecule imaging, special fluorochromes are used that can be induced by light to switch on and off (photoswitchable fluorochromes), allowing flashes of fluorescence to be detected at different times. Thousands of images are captured before being combined to build one image; examples of this method include *p*hoto*a*ctivation *l*ocalization *m*icroscopy (PALM) and *st*ochastic *o*ptimization *r*econstruction *m*icroscopy (STORM). In 2014, Eric Betzig, Stefan W. Hell, and William E. Moerner won the Nobel Prize in Chemistry for their contributions to the field of SRM.

Comprehension Check

1. Compare and contrast the function of the condenser, objective, and eyepiece lenses.
2. What happens to resolution if each of the following increases: wavelength of light, refractive index, and numerical aperture? How are resolution and magnification related?
3. What might a microscopist see through an oil immersion objective if no immersion oil is used?
4. Why don't most light microscopes use 30× ocular lenses for greater magnification?
5. Construct a table that compares how dark-field, phase-contrast, differential interference contrast, epifluorescence, and confocal microscopes work and the kind of image or images provided by each. Give a specific use for each type.
6. Identify which type of microscope(s) would be used to examine the following specimens: photosynthetic bacteria, swimming bacteria, GFP-labeled microorganisms, unstained microbes, and pond scum. Explain each of your answers.

2.3 Staining Helps to Visualize and Identify Microbes

After reading this section, you should be able to:

a. Recommend a fixation process to use when the microbe is a bacterium or archaeon and when the microbe is a protist
b. Plan a series of appropriate staining procedures to describe an unknown bacterium as fully as possible
c. Compare what happens to Gram-positive and Gram-negative bacterial cells during each step of the Gram-staining procedure

As noted in section 2.2, specimens examined by bright-field microscopy are often fixed and stained before being examined. Such preparation serves to increase the visibility of the microorganisms, accentuate specific morphological features, and preserve them for future study. Importantly, some staining procedures help microbiologists identify the organism being examined.

Fixation

Stained cells seen using a microscope should resemble living cells as closely as possible. **Fixation** is the process by which the internal and external structures of specimens are preserved and fixed in position. It inactivates enzymes that might disrupt cell morphology and toughens cell structures so that they do not change during staining and observation. A microorganism usually is killed and attached firmly to the microscope slide during fixation.

There are two fundamentally different types of fixation: heat fixation and chemical fixation. **Heat fixation** is routinely used to observe bacteria and archaea. Typically, a film of cells (a smear) on a microscope slide is dried and gently heated. Heat fixation attaches cells to the slide, preserves overall morphology, and inactivates enzymes. However, it also destroys proteins in subcellular structures, which may distort their appearance. **Chemical fixation** is used to protect fine cellular substructure as well as morphology. It is used when examining microorganisms by electron microscopy. Chemical fixatives penetrate cells and react with cellular components, usually proteins and lipids, to render them inactive, insoluble, and immobile. Common fixative mixtures contain ethanol, acetic acid, mercuric chloride, formaldehyde, and glutaraldehyde.

Dyes and Simple Staining

The many types of dyes used to stain microorganisms have two features in common. First, they have **chromophore groups**—chemical moieties with conjugated double bonds that give the dye its color. Second, they bind cells by ionic, covalent, or hydrophobic bonds.

Dyes that bind cells by ionic interactions are probably the most commonly used dyes. They can be divided into two general classes based on the nature of their charged (ionized) group: acidic dyes have anionic (negatively charged) chromophores, whereas basic dyes have cationic (positively charged) chromophores (**table 2.4**). The staining effectiveness of ionizable dyes may be altered by pH because the nature and number of the charged moieties on cell components change with pH. Thus acidic dyes stain best under acidic conditions (low pH) when proteins and many other molecules carry a positive charge; conversely, basic dyes are most effective at higher pH values.

Dyes that bind through covalent bonds or that have certain solubility characteristics are also useful. For instance, DNA can be stained by the Feulgen procedure in which the staining compound (Schiff's reagent) is covalently attached to the deoxyribose sugars of DNA. Sudan III (Sudan Black) selectively stains lipids because it is lipid soluble but does not dissolve in aqueous portions of the cell.

Microorganisms can be stained by **simple staining,** in which a single dye is used (**figure 2.16**). The value of simple staining is its ease of use. The heat-fixed smear is covered with stain for a short time, excess stain is washed off with water, and the slide is blotted dry. Basic dyes such as crystal violet, methylene blue, and

Table 2.4	Ionizable Dyes	
Type of Dye	**Examples**	**Chromophore Characteristics**
Basic dyes	Methylene blue, basic fuchsin, crystal violet, safranin, malachite green	Have positively charged groups; bind to negatively charged molecules such as nucleic acids, many proteins, and the surfaces of bacterial and archaeal cells
Acidic dyes	Eosin, rose bengal, and acid fuchsin—possess groups such as carboxyls (—COOH) and phenolic hydroxyls (—OH)	Have negatively charged groups; bind to positively charged cell structures

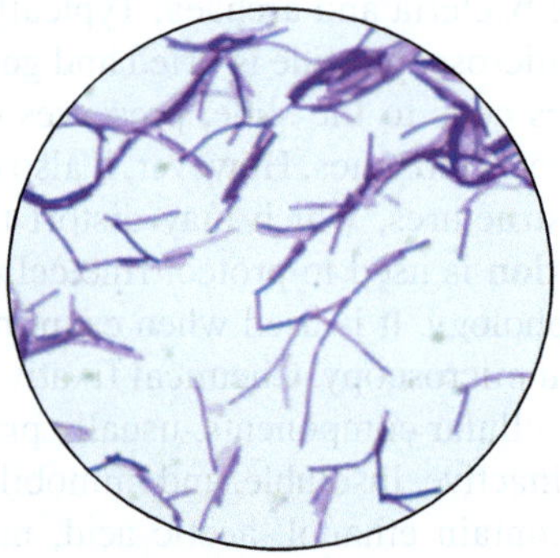

Figure 2.16 Simple Staining Illustrates Cell Size and Morphology.
Dr. Rita B. Moyes

carbol fuchsin are frequently used in simple staining to determine the size, shape, and arrangement of bacterial and archaeal cells.

While most dyes directly stain the cell or object of interest, some dyes (e.g., India ink and nigrosin) are used in **negative staining.** In negative staining, the background is stained, not the cell; instead, the unstained cells appear as bright, clear objects against a dark background.

Differential Staining

The **Gram stain,** developed in 1884 by the Danish physician Christian Gram, is the most widely used staining method in bacteriology. The Gram stain is an example of **differential staining**—procedures that distinguish organisms based on their staining properties. Use of the Gram stain classifies most bacteria into one of two groups—Gram negative or Gram positive—based on the composition of their cell walls.

The Gram-staining procedure is illustrated in **figure 2.17.** The result of Gram staining is that Gram-negative bacteria are stained pink to red and Gram-positive bacteria are stained dark purple (**figure 2.18*a***). *Cell walls have many functions (section 3.3)* *Gram Staining*

Acid-fast staining is another important differential staining procedure. It is used to identify mycobacteria such as *Mycobacterium tuberculosis* and *M. leprae* (figure 2.18*b*), the pathogens that cause tuberculosis and leprosy, respectively. These bacteria have cell walls containing lipids constructed from mycolic acids, a group of branched-chain hydroxy fatty acids, which prevent dyes from readily binding to the cells (*see figure 22.7*). A commonly used staining procedure, the Ziehl-Neelsen method, uses heat to drive the primary stain, carbol fuchsin, into mycobacterial cells. Phenol in the stain also aids in driving it into the cells. The slide is rinsed with acid alcohol and cells that resist decolorization are acid-fast. Non-acid-fast cells lose the primary color during this step and are counterstained with a different dye, frequently methylene blue. *Acid Fast Staining*

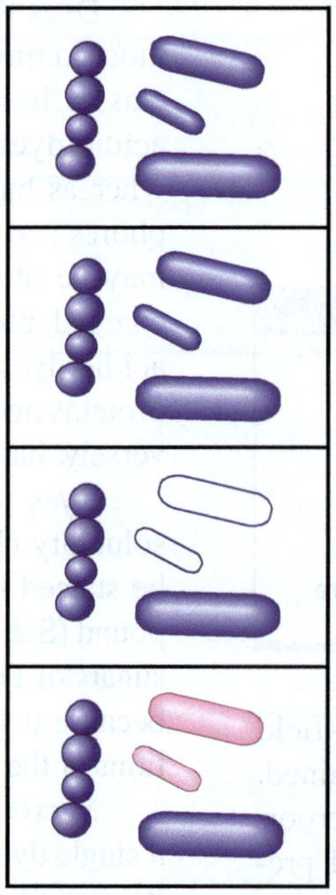

Steps in Staining	State of Bacteria	Explanation
Step 1: Crystal violet (primary stain) for 1 minute. Water rinse.	Cells stain purple.	The dye is taken up equally well by all cells present.
Step 2: Iodine for 1 minute. Water rinse.	Cells remain purple.	Iodine is a mordant, a compound that enhances binding of dye to cell walls.
Step 3: Alcohol (decolorizer) for 10–30 seconds. Water rinse.	Gram-positive cells remain purple. Gram-negative cells become colorless.	Gram-negative cell walls are thin and cannot retain the dye.
Step 4: Safranin (counterstain) for 30–60 seconds. Water rinse. Blot dry.	Gram-positive cells remain purple. Gram-negative cells appear pink or red.	Only unstained Gram-negative cells can take up the dye; crystal violet in the Gram-positive cells repels the counterstain.

Figure 2.17 Gram Stain. Steps in the Gram-staining procedure.

MICRO INQUIRY *Why is the decolorization step considered the most critical in the Gram-staining procedure?*

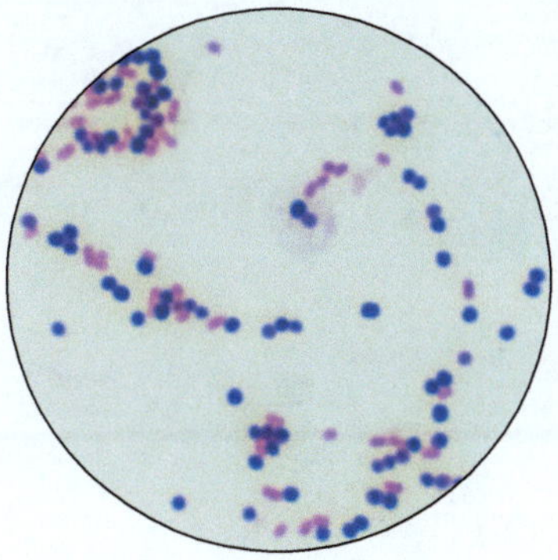

(a) Gram stain
Purple cells are Gram positive.
Pink cells are Gram negative.

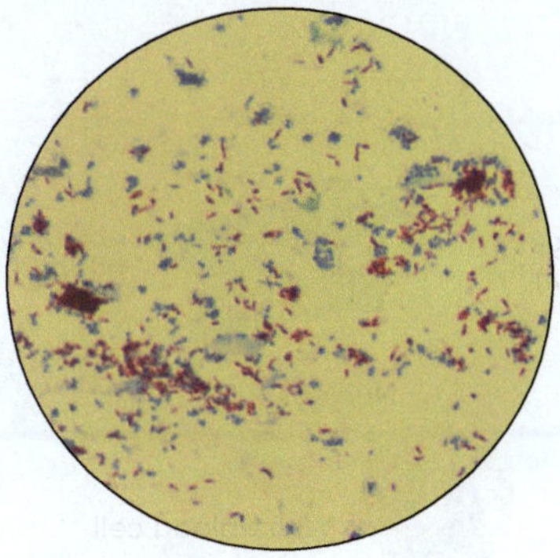

(b) Acid-fast stain
Red cells are acid-fast.
Blue cells are nonacid-fast.

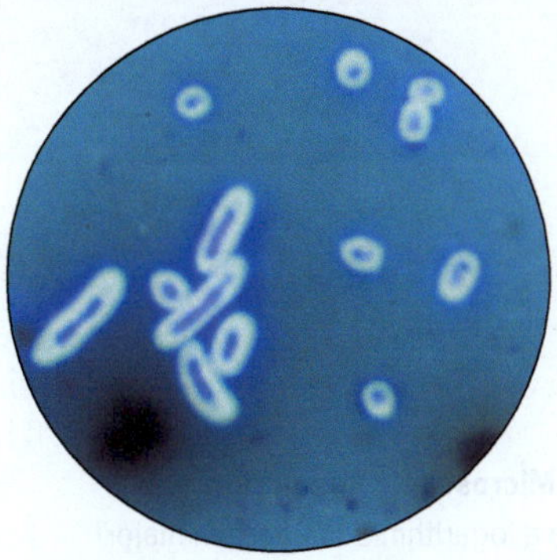

(c) Capsule stain of *Klebsiella pneumoniae*

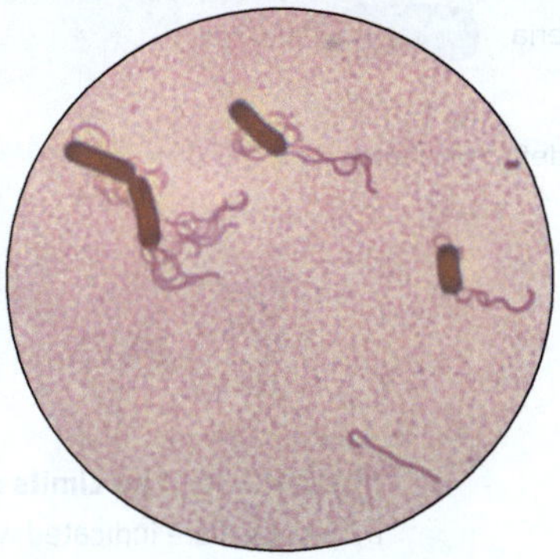

(d) Flagellar stain of *Bacillus brevis*

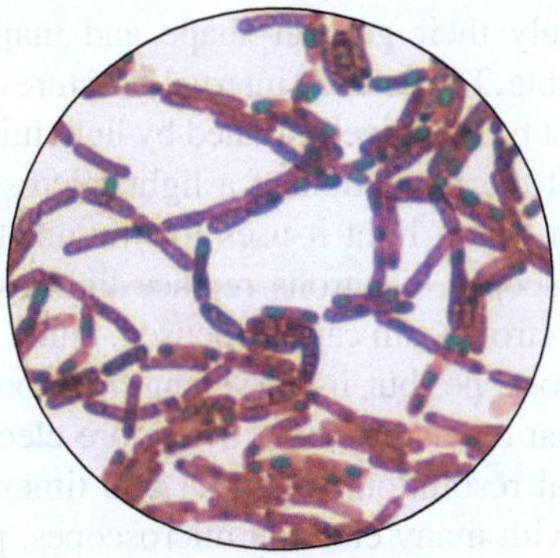

(e) Endospore stain
Endospores are green.
Vegetative cells are pink.

Figure 2.18 Differential Stains.

(a, c) Lisa Burgess/McGraw Hill; (b) James Redfearn/McGraw Hill; (d) Dr. William A. Clark/CDC (e) Larry Stauffer/Oregon State Public/CDC

One of the simplest staining procedures is **capsule staining** (figure 2.18*c*), a technique that reveals the presence of capsules—a layer of well-organized material, usually polysaccharides, that surrounds many bacteria and some fungi. Cells are mixed with India ink or nigrosin dye and spread out in a thin film on a slide. After air-drying, cells appear as bodies surrounded by a colorless halo in the midst of a blue-black background because ink and dye particles cannot penetrate the capsule. Thus capsule staining is an example of negative staining. There is little distortion of cell shape, and the cell can be counterstained for even greater visibility. *Capsules and slime layers (section 3.5)* *Capsule Staining*

Flagella staining provides taxonomically valuable information about the presence and distribution pattern of flagella on bacterial and archaeal cells (figure 2.18*d; see also figure 3.34*). Their flagella are fine, threadlike organelles of locomotion that are so slender (about 10 to 30 nm in diameter) they can only be seen directly using an electron microscope (although bundles of flagella can be visualized by dark-field microscopy). To observe bacterial flagella with a light microscope, their thickness is increased by coating them with mordants such as tannic acid and potassium alum and then staining with a dye such as basic fuchsin. *Bacterial flagella (section 3.7)*

Endospore staining (figure 2.18*e*) allows the detection of bacterial endospores. It is a differential stain in which the primary stain, malachite green, is driven into the endospores by heat. After a washing step to remove the primary stain from everything except the endospores, colorless vegetative cells are counterstained using safranin. *Bacterial endospores are a survival strategy (section 3.8)* *Endospore Staining*

Comprehension Check

1. Describe the two general types of fixation. Which would you use when Gram staining a bacterium? Which would you use before observing the organelles of a protist?
2. Why are basic dyes more effective under alkaline conditions?
3. What procedural error(s) might be responsible if you Gram stained a mixture of bacteria known to be Gram-positive and Gram-negative and see only red cells?
4. Are capsular and flagellar staining differential staining procedures? Explain your answer.

2.4 Electron Microscopes Use Beams of Electrons to Create Highly Magnified Images

After reading this section, you should be able to:

a. Create a concept map, illustration, or table that compares transmission electron microscopes (TEMs) to light microscopes
b. Decide when it would be best to examine a microbe by TEM, scanning electron microscopy (SEM), or cryo-electron microscopy

As we have learned, light microscopes have a resolution limit of about 0.2 μm, which greatly compromises their usefulness for detailed studies of many microorganisms (**figure 2.19**). Viruses, for example, are too small to be seen with light microscopes (with the exception of some recently discovered giant viruses). Bacteria and archaea can be observed, but because they are usually only 1 to 2 μm

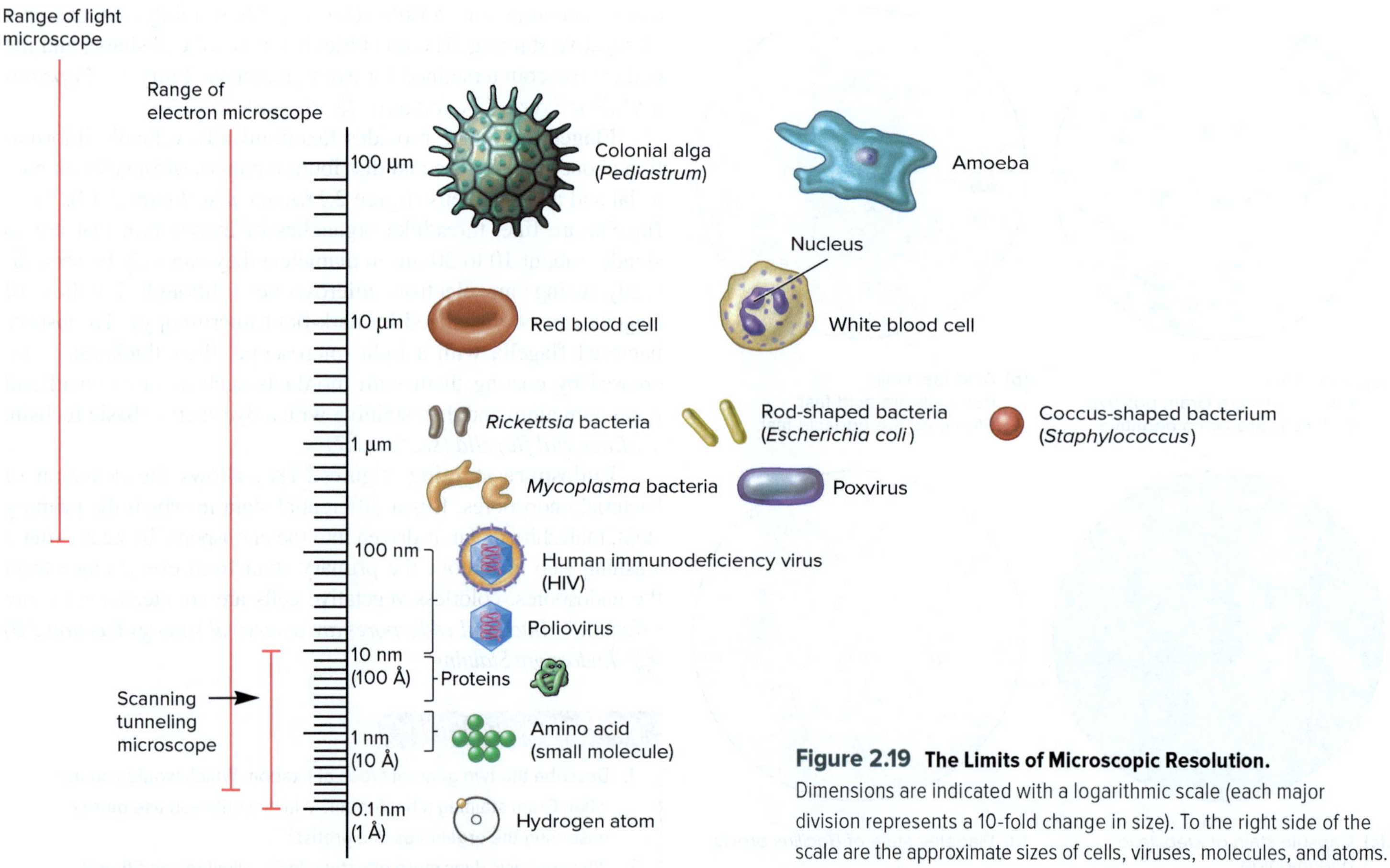

Figure 2.19 The Limits of Microscopic Resolution. Dimensions are indicated with a logarithmic scale (each major division represents a 10-fold change in size). To the right side of the scale are the approximate sizes of cells, viruses, molecules, and atoms.

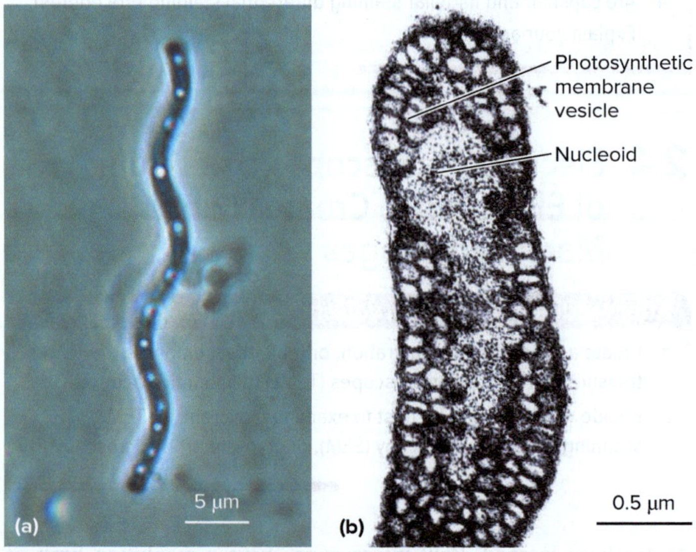

Figure 2.20 Light and Electron Microscopy. A comparison of light and electron microscopic resolution. (a) The proteobacterium *Spirillum volutans* in a phase-contrast light microscope. (b) A thin section of another spiral-shaped proteobacterium, *Rhodospirillum rubrum,* in a transmission electron microscope.

(a) James Redfearn/McGraw Hill; (b) Biology Media/Science Source

in diameter, only their general shape and major morphological features are visible. The detailed internal structure of microorganisms therefore cannot be effectively studied by light microscopy.

Recall that the resolution of a light microscope increases as the wavelength of the light it uses for illumination decreases. In electron microscopes, electrons replace light as the illumination source. The electron beam can be focused, much as light is focused in a light microscope, but its wavelength is about 100,000 times shorter than that of visible light. Therefore electron microscopes have a practical resolution roughly 1,000 times better than light microscopes; with many electron microscopes, points closer than 0.2 nm can be distinguished, and the useful magnification is well over 100,000×, meaning that microbial morphology can be studied in great detail (figure 2.19). The value of the electron microscope is evident on comparison of the micrographs in **figure 2.20.**

Transmission Electron Microscope

A **transmission electron microscope** (**TEM**) uses a heated tungsten filament in a component called the *electron gun* to generate a beam of electrons. The electron beam is focused on the specimen by a condenser (**figure 2.21**). Because electrons cannot pass through a glass lens, doughnut-shaped electromagnets called *magnetic lenses* focus the beam. The column containing the lenses and specimen must be under vacuum because electrons are deflected by collisions with air molecules. The specimen scatters some electrons, but those that pass through it are used to form an

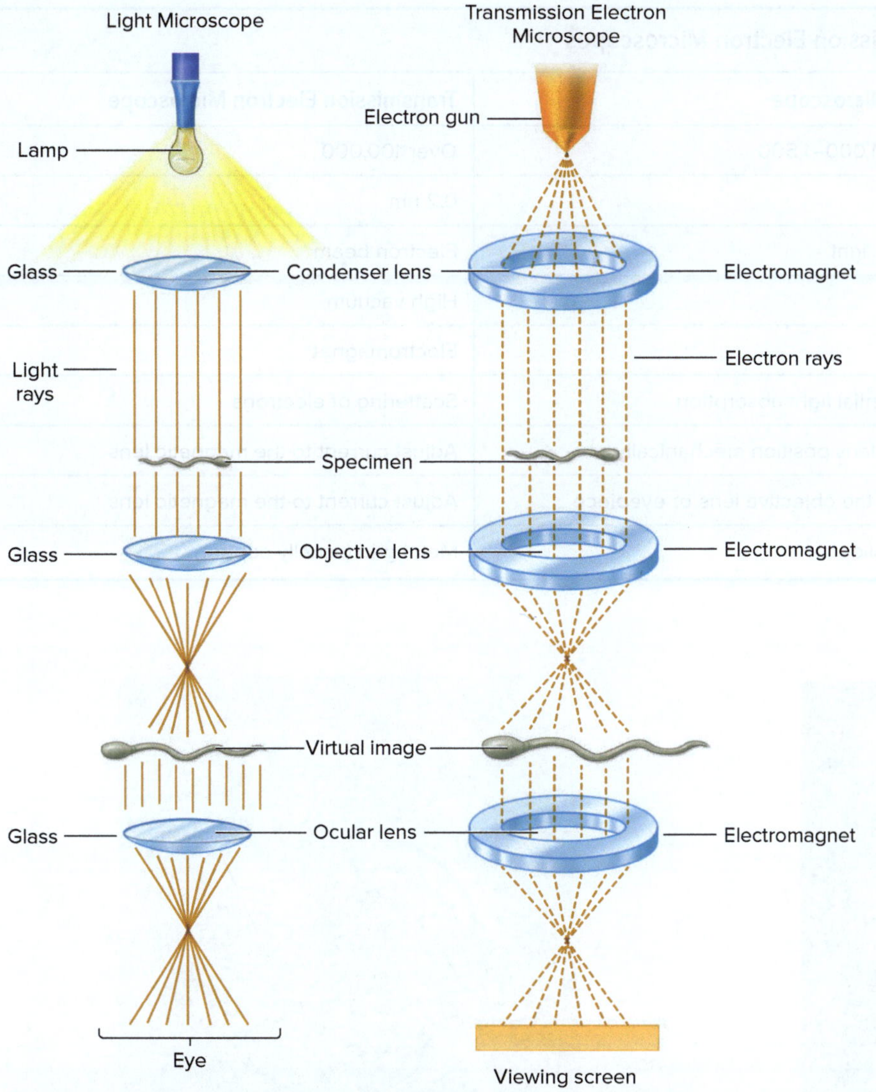

Figure 2.21 A Comparison of Light and Transmission Electron Microscopes. For comparison purposes, the light microscope has been turned upside down.

Figure 2.22 A Transmission Electron Microscope. The electron gun is at the top of the central column, and the magnetic lenses are within the column. The image on the fluorescent screen is also viewed on a computer monitor.
Anamaria Mejia/Shutterstock

enlarged image of the specimen on a fluorescent screen that interfaces with a computer monitor (**figure 2.22**). Denser regions in the specimen scatter more electrons and therefore appear darker because fewer electrons strike that area of the screen; these regions are said to be "electron dense." In contrast, electron transparent regions are brighter.

Table 2.5 compares some of the important features of light and transmission electron microscopes. The TEM has distinctive features that place harsh restrictions on the nature of samples that can be viewed and the means by which those samples must be prepared. Specimens must be sliced into very thin sections (20–100 nm) before being placed in the vacuum to allow the electron beam to penetrate the specimen for image formation. To slice specimens so thinly, they are first fixed with chemicals such as glutaraldehyde and osmium tetroxide to stabilize cell structure. The specimen is then dehydrated with organic solvents (e.g., acetone or ethanol). Next the specimen is soaked in unpolymerized, liquid epoxy plastic until it is completely permeated, after which the plastic is hardened using UV light or heat to form a solid block. Thin sections are then skillfully cut from the block with a glass or diamond knife using a device called an ultramicrotome.

As with bright-field light microscopy, cells are usually stained so they can be seen clearly with a TEM. The probability of electron scattering is determined by the density (atomic number) of atoms in the specimen. Biological molecules are composed primarily of atoms with low atomic numbers (H, C, N, and O), and electron scattering is fairly constant throughout an unstained cell or virus. Therefore specimens are further prepared by soaking thin sections with solutions of heavy metal salts such as lead citrate or uranyl acetate. The lead and uranium ions bind to structures in the specimen and make them more electron opaque, thus increasing contrast in the material. Heavy osmium atoms from the osmium tetroxide fixative also stain specimens and increase their contrast. The stained thin sections are then mounted on tiny copper grids and viewed.

Two other important techniques for preparing specimens for TEM include negative staining and shadowing. In negative staining, the specimen is spread out in a thin film with either phosphotungstic acid or uranyl acetate. Just as in negative staining for light microscopy, the specimen appears bright against a dark background, in this case because the heavy metals do not penetrate biological material. Negative staining enables visualization of viruses and cellular microbes, but unlike thin sections, internal structures cannot be discerned (**figure 2.23a**). In shadowing, a specimen is coated with a thin film of platinum or other heavy metal by evaporation at an angle of about 45 degrees from horizontal so that the metal strikes the microorganism on only one side. In one commonly used imaging method, the area coated with metal appears dark in photographs, whereas the uncoated side and

Table 2.5 Characteristics of Light and Transmission Electron Microscopes

Feature	Light Microscope	Transmission Electron Microscope
Highest practical magnification	About 1,000–1,500	Over 100,000
Best resolution[1]	0.2 µm	0.2 nm
Radiation source	Visible light	Electron beam
Medium of travel	Air	High vacuum
Type of lens	Glass	Electromagnet
Source of contrast	Differential light absorption	Scattering of electrons
Focusing mechanism	Adjust lens position mechanically	Adjust current to the magnetic lens
Method of changing magnification	Switch the objective lens or eyepiece	Adjust current to the magnetic lens
Specimen mount	Glass slide	Metal grid (usually copper)

[1] The resolution limit of a human eye is about 0.2 mm.

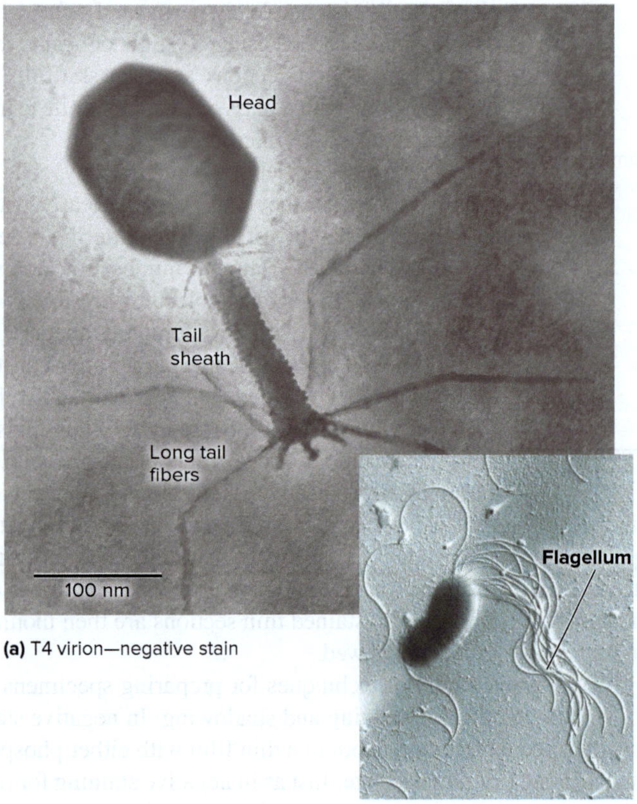

(a) T4 virion—negative stain

(b) *P. fluorescens*—after shadowing

Figure 2.23 Stained Microorganisms Visualized by TEM. (a) T4 is a virus that infects *Escherichia coli*. (b) *Pseudomonas fluorescens* with its polar flagella.

(a) Ami Images/Science Source; (b) Dr. Tony Brain/SPL/Science Source

MICRO INQUIRY *Why are all electron micrographs black and white (although they are sometimes artificially colorized after printing)?*

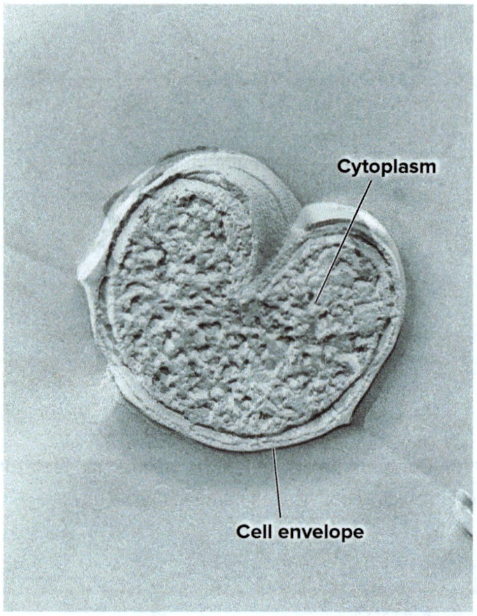

Figure 2.24 Example of Freeze-Etching. A freeze-etched preparation of the bacterium *Nitrospira* sp.

Dr. Kari Lounatmaa/SPL/Science Source

the shadow region created by the object are light (figure 2.23*b*). This technique is particularly useful in studying virus particle morphology, bacterial and archaeal flagella, and DNA.

The process of chemical fixation and dehydration can introduce artifacts that can alter cellular morphology. This can be minimized or avoided by using a freeze-etching procedure. When cells are rapidly frozen in liquid nitrogen, they become very brittle and can be broken along lines of greatest weakness, usually down the middle of internal membranes (**figure 2.24**). The exposed surfaces are then shadowed and coated with layers of platinum and carbon to form a replica of the surface. After the specimen has been removed chemically, this replica is studied in the TEM, providing a detailed view of intracellular structure.

Scanning Electron Microscope

Transmission electron microscopes form an image from radiation that has passed through a specimen. The **scanning electron microscope (SEM)** produces an image from electrons released from an object's surface. The surfaces of microorganisms are visualized in great detail; most SEMs have a resolution of about 10 nm.

Specimen preparation for SEM is relatively easy, and in some cases, air-dried material can be examined directly. However, microorganisms usually must first be fixed, dehydrated, and dried to preserve surface structure and prevent the collapse of cells when they are exposed to the SEM's vacuum. Before viewing, dried samples are mounted and coated with a thin layer of metal to prevent the buildup of an electrical charge on the surface and to give a better image.

To create an image, the SEM scans a narrow, tapered electron beam back and forth over the specimen (**figure 2.25**). When the beam strikes a particular area, surface atoms discharge a tiny shower of electrons called secondary electrons,

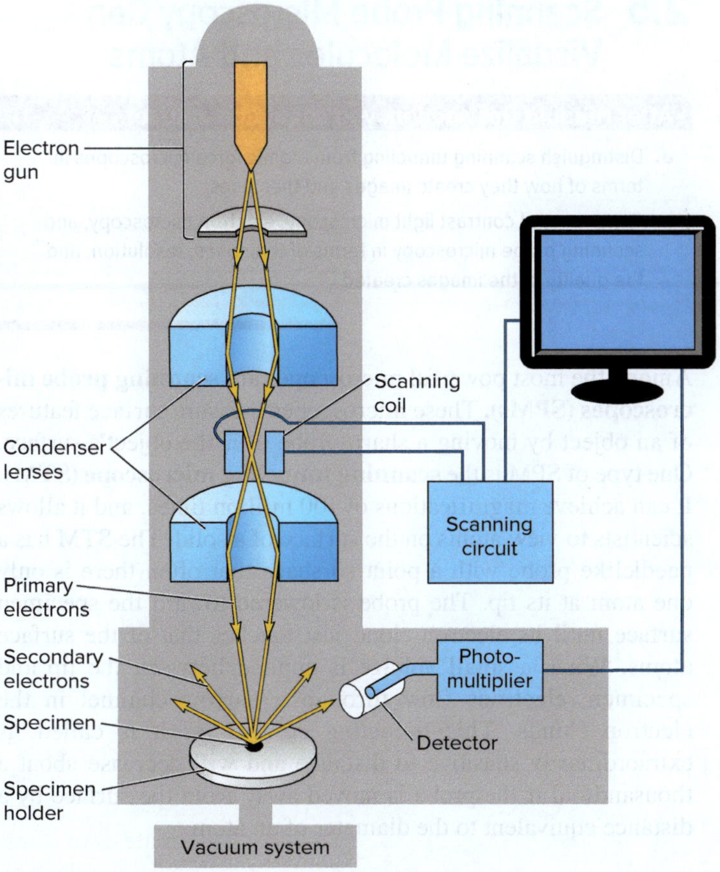

Figure 2.25 The Scanning Electron Microscope.

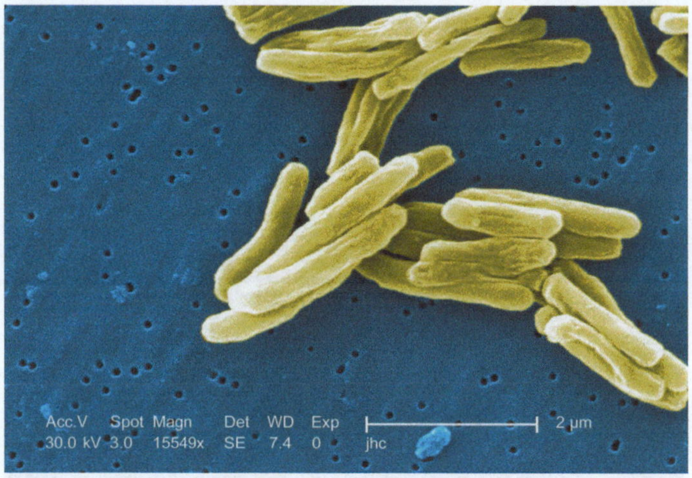

Figure 2.26 Scanning Electron Micrograph of *Mycobacterium tuberculosis*. Colorized image.

Janice Carr/CDC

and these are trapped by a detector. Secondary electrons strike a material in the detector that emits light when struck by electrons (the material is called a scintillator). The flashes of light are converted to an electrical current and amplified by a photomultiplier. The signal is digitized and sent to a computer, where it can be viewed.

The number of secondary electrons reaching the detector depends on the nature of the specimen's surface. When the electron beam strikes a raised area, a large number of secondary electrons enter the detector; in contrast, fewer electrons escape a depression in the surface and reach the detector. Thus raised areas appear lighter on the screen and depressions are darker. A realistic three-dimensional image of the microorganism's surface results (**figure 2.26**). A single instrument can house both transmission and scanning electron microscopes (S/TEM).

Cryo-Electron Microscopy

In 2017, the Nobel Prize in Chemistry was awarded to three scientists for their contributions to the development of **cryo-electron microscopy,** or **cryo-EM.** The Nobel committee recognized the advances this technique has made in visualizing biomolecules, from proteins to complex molecular structures. While protein crystallization and X-ray diffraction (called X-ray crystallography) had long been the workhorse of resolving protein structure, not all proteins can be crystallized, and the crystallization process can introduce changes to protein structure. Furthermore, it is difficult to study the interaction between molecules by X-ray crystallization. To understand how cryo-EM works, let's explore the contribution of each Nobel laureate.

Richard Henderson invented an approach to see proteins using TEM. As we have discussed, TEM typically is used to look at either whole cells and viruses, or thin slices of them to examine their interior. Dr. Henderson packed many molecules of the

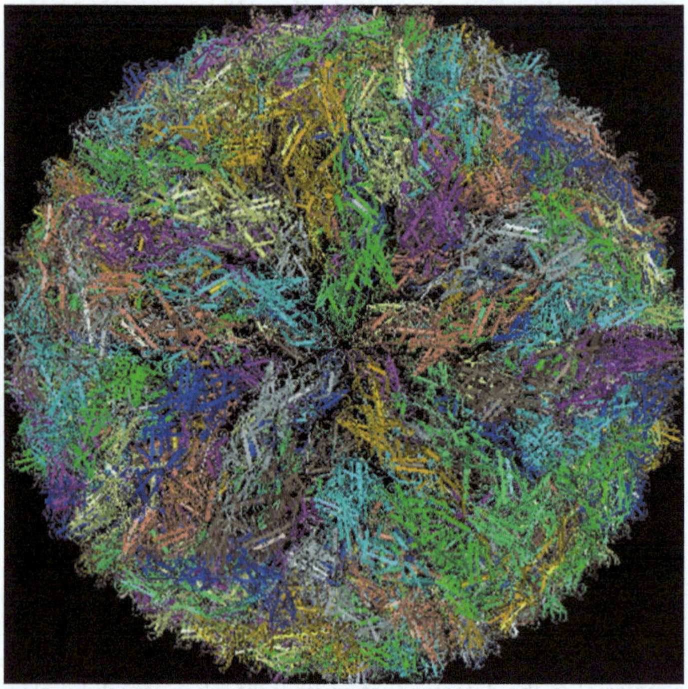

Figure 2.27 Cryo-Electron Micrograph. The Zika virus structure was resolved to 3.8 angstroms by cryo-EM. It was determined that the shell of the virus contains 180 glycoproteins. Colors are added to the image during computer analysis.
National Center for Biotechnology Information

archaeal protein archaerhodopsin into a single sample and took images from multiple angles. By combining the images, he was able to generate a 3-D image. Ultimately, he was able to refine this technique to obtain images that in some cases had the same resolution as X-ray crystallography. Joachaim Frank then built upon Henderson's work by developing computational methods that assembled many images of a single sample into 3-D images.
▶| *Halobacteriota (section 19.4)*

The third laureate, James Dubochet, knew that as long as molecules had to be dehydrated before visualization, their true structure could not be revealed. His solution was to rapidly freeze samples so water could not form crystalline ice, which would diffract electron beams. Instead, the water would form a glasslike film called *vitreous ice*. This is accomplished by plunging samples into ethane cooled to −196°C by liquid nitrogen. These vitrified molecules retain their 3-D shape and can be visualized by TEM while still frozen. The application of Henderson's and Frank's image capture and analysis algorithms creates stunning, high-resolution photomicrographs (**figure 2.27**).

Since the development of cryo-EM, advances in both software and hardware technologies have helped create images that are extremely sharp and, unlike X-ray diffraction, include complex multi-molecular structures such as ribosomes and even biological membranes, which are very unstable. Additional advances, including a method to ensure all electrons in the EM beam travel at similar speeds before hitting a sample, have resulted in atomic resolution, enabling the visualization of individual atoms in stable proteins. An image bank with thousands of cryo-EM images has been collected. It is anticipated that in the coming decade, cryo-EM will overtake X-ray crystallography as the method of choice for high-resolution structural analysis.

Comprehension Check

1. Why does an electron microscope have much greater resolution than a light microscope?
2. Where is the electron gun located relative to the sample? Why must a TEM use high vacuum and very thin sections?
3. How are thin sections for TEM typically stained for visualization? Under what circumstances would sections be prepared by negative staining? Shadowing? Freeze-etching?
4. How does a scanning electron microscope differ from a TEM? If you were shown a micrograph, how would you know if it were a TEM, SEM, or cryo-EM?
5. What advantages does cryo-EM have over X-ray crystallography?

2.5 Scanning Probe Microscopy Can Visualize Molecules and Atoms

After reading this section, you should be able to:

a. Distinguish scanning tunneling from atomic force microscopes in terms of how they create images and their uses
b. Compare and contrast light microscopy, electron microscopy, and scanning probe microscopy in terms of their uses, resolution, and the quality of the images created

Among the most powerful microscopes are **scanning probe microscopes (SPMs).** These microscopes measure surface features of an object by moving a sharp probe over the object's surface. One type of SPM is the **scanning tunneling microscope (STM).** It can achieve magnifications of 100 million times, and it allows scientists to view atoms on the surface of a solid. The STM has a needlelike probe with a point so sharp that often there is only one atom at its tip. The probe is lowered toward the specimen surface until its electron cloud just touches that of the surface atoms. When a small voltage is applied between the tip and specimen, electrons flow through a narrow channel in the electron clouds. This *tunneling current,* as it is called, is extraordinarily sensitive to distance and will decrease about a thousandfold if the probe is moved away from the surface by a distance equivalent to the diameter of an atom.

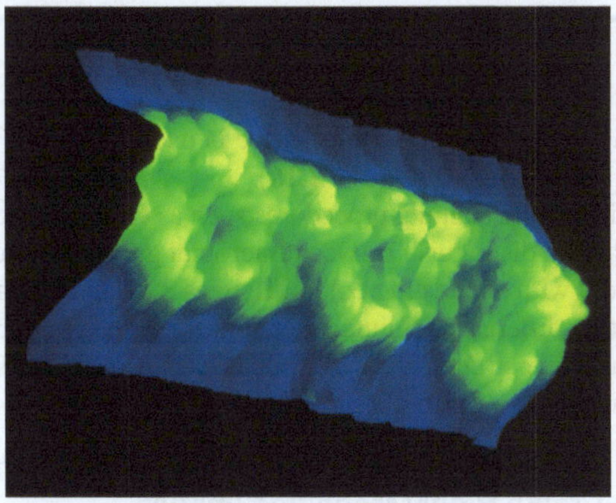

Figure 2.28 Scanning Tunneling Microscopy of DNA. The DNA double helix with approximately three turns shown (false color; ×2,000,000).
Driscoll, Youngquist & Baldeschwieler, California Institute of Technology/Science Source

MICRO INQUIRY *Compare the magnification of this image with that shown in figure 2.26 (SEM). How many times greater is the magnification shown here?*

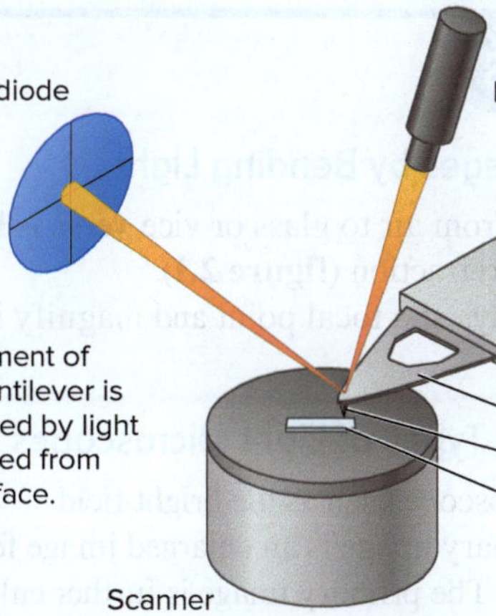

Figure 2.29 The Basic Elements of an Atomic Force Microscope. The tip used to probe the specimen is attached to a cantilever. As the probe passes over the "hills and valleys" of the specimen's surface, the cantilever is deflected vertically. A laser beam directed at the cantilever is used to monitor these vertical movements. Light reflected from the cantilever is detected by the photodiode and used to generate an image of the specimen.

The arrangement of atoms on the specimen surface is determined by scanning the probe tip back and forth over the surface while keeping the probe at a constant height above the specimen. As the tip follows the surface contours by moving up and down, its motion is recorded and analyzed by a computer to create an accurate three-dimensional image of the surface atoms. The surface map can be displayed on a computer screen or plotted on paper. The resolution is so great that individual atoms are observed easily. Even more exciting is that the microscope can examine objects when they are immersed in water. Therefore it can be used to study biological molecules such as DNA (**figure 2.28**). The microscope's inventors, Gerd Binnig and Heinrich Rohrer, shared the 1986 Nobel Prize in Physics for their work, together with Ernst Ruska, the designer of the first TEM.

Another type of SPM is the **atomic force microscope (AFM),** which moves a sharp probe on a cantilever over the specimen surface while keeping the distance between the probe tip and the surface constant. It does this by exerting a very small amount of force on the tip, just enough to maintain a constant distance but not enough to damage the surface. The vertical motion of the probe tip as it moves over the specimen surface is detected by a photosensor that measures the deflection of a laser beam that strikes the cantilever (**figure 2.29**). Unlike the STM, the AFM can be used to study surfaces that do not conduct electricity well. It has been used to study the interactions of proteins, to follow the behavior of living bacteria and other cells, and to visualize membrane proteins such as aquaporins (**figure 2.30**).

Comprehension Check

1. Briefly describe each of the most popular versions of the scanning probe microscope. Discuss a structure that might be visualized with each microscope.
2. Scanning probe microscopy can be applied to samples that have not been dehydrated, but electron microscopy cannot. Why is this an important advantage?

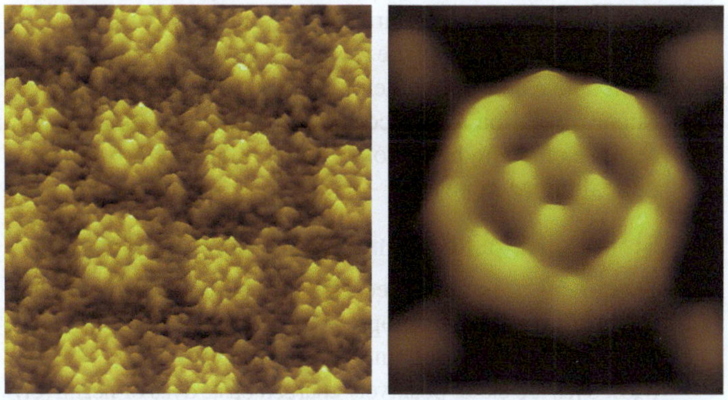

(a) (b)

Figure 2.30 The Membrane Protein Aquaporin Visualized by Atomic Force Microscopy. Aquaporin is a membrane-spanning protein that allows water to move across the membrane. (a) Surface view of aquaporin proteins. (b) A single aquaporin protein observed in more detail and at higher magnification.
(a–b) Simon Scheuring

Key Concepts

2.1 Lenses Create Images by Bending Light

- A light ray moving from air to glass or vice versa is bent in a process known as refraction (**figure 2.1**).
- Lenses focus light rays at a focal point and magnify images (**figure 2.2**).

2.2 There Are Several Types of Light Microscopes

- In a compound microscope such as the bright-field microscope, the primary image is an enlarged image formed by the objective lens. The primary image is further enlarged by the ocular lens to yield the final image (**figure 2.3**).
- Microscope resolution increases as the wavelength of radiation used to illuminate the specimen decreases and as the numerical aperture increases. The maximum resolution of a light microscope is about 0.2 μm (**figure 2.4**).
- The dark-field microscope uses only refracted light to form an image, and objects appear light against a black background (**figure 2.6**).
- The phase-contrast microscope converts variations in the refractive index into changes in light intensity and thus makes colorless, unstained, live cells visible (**figures 2.8–2.10**).
- The differential interference contrast microscope uses two beams of light to create high-contrast images of live specimens (**figure 2.11**).
- The fluorescence microscope illuminates a fluorochrome-labeled specimen and forms an image from its fluorescence (**figures 2.12–2.14**).
- The confocal microscope is used to study thick, complex specimens. It creates an image by using only the light emanating from the plane of focus, while blocking out light from above and below the plane of focus (**figure 2.15**).
- Super-resolution microscopes use various techniques to achieve much higher resolution than that of conventional optical microscopes.

2.3 Staining Helps to Visualize and Identify Microbes

- Specimens are often fixed and stained before viewing in the bright-field microscope. There are two fixation methods: heat fixation and chemical fixation.
- Most dyes are either positively charged basic dyes or negatively charged acidic dyes that bind to ionized parts of cells.
- In simple staining, a single dye is used to stain microorganisms (**figure 2.16**).
- Differential staining procedures such as Gram and acid-fast staining distinguish between microbial groups by staining them with contrasting colors (**figures 2.17** and **2.18*a, b***). Other differential staining techniques are specific for particular structures such as bacterial capsules, flagella, and endospores (**figure 2.18*c, d, e***).

2.4 Electron Microscopes Use Beams of Electrons to Create Highly Magnified Images

- The transmission electron microscope (TEM) uses magnetic lenses to form an image from electrons that have passed through a very thin section of a specimen (**figure 2.21**). Resolution is high because the wavelength of a beam of electrons is very short.
- Specimens for TEM are usually prepared by methods that increase contrast. Specimens can be stained by treatment with solutions of heavy metals such as osmium, uranium, and lead. They can also be prepared for TEM by negative staining, shadowing with metal, or freeze-etching (**figures 2.23** and **2.24**).
- The scanning electron microscope is used to study external surface features of microorganisms (**figures 2.25** and **2.26**).
- Cryo-EM enables visualization of single molecules and complex molecular structures. Samples are flash frozen and when examined, a series of images are captured that when combined and processed form a three-dimensional reconstruction of the specimen (**figure 2.27**).

2.5 Scanning Probe Microscopy Can Visualize Molecules and Atoms

- Scanning probe microscopes reach very high magnifications that allow scientists to observe biological molecules (**figures 2.28** and **2.30**).
- Scanning tunneling microscopy enables the visualization of molecular surfaces using electron interaction between the probe and the specimen, whereas atomic force microscopy can scan the surface of molecules that do not conduct electricity well (**figure 2.29**).

Active Learning

1. You have prepared a specimen for light microscopy, stained it using the Gram-staining procedure, but failed to see anything when you looked through your light microscope. Discuss the things you may have done incorrectly.
2. Which type of microscopy and stain (if appropriate) would you use to visualize each of the following? (There may be more than one correct answer.) Be sure to explain your answer. *Mycobacterium tuberculosis* (which causes tuberculosis), microbes in pond scum, *Staphylococcus*

aureus from a skin wound, measles virus, bacterial flagella, *Borreliella burgdorferi* (the spirochete that causes Lyme disease), and isolated bacterial ribosomes.

3. Stimulated emission depletion (STED) microscopes, described in section 2.2, are super-resolution microscopes that use two highly focused beams of light rather than one, as in typical confocal microscopy. The first beam uses a wavelength that excites the fluorochrome with which a specimen is labeled. This creates a fluorescent spot where the fluorochrome is located. The second beam is a hollow cone. It strikes the outermost edges of the fluorescent spot illuminated by the first beam, quenching that light so that only the center of the fluorescent spot is visible. The result is that the resolution of STED microscopes is greater and objects smaller than 200 nm can be resolved. If you could use STED microscopy, what questions in biological science would you want to ask and why would you ask these questions? Refer to figures 2.15 and 2.19 to help formulate your answer.
4. When using the SEM to examine microbes, specimens are chemically fixed, coated with a thin layer of metal, and then viewed in a vacuum. The environmental scanning electron microscope (ESEM) makes it possible to get images of specimens that are wet and/or uncoated and are thus in their natural state. The ESEM differs from conventional SEM in many ways, including the fact that the specimen chamber pressure can be controlled, and specimens can be viewed in a variety of gaseous environments and at different temperatures. Consider the benefits of the ESEM over the SEM and describe microbiology studies that might be made using this microscope.

CHAPTER

3

Bacterial Cell Structure

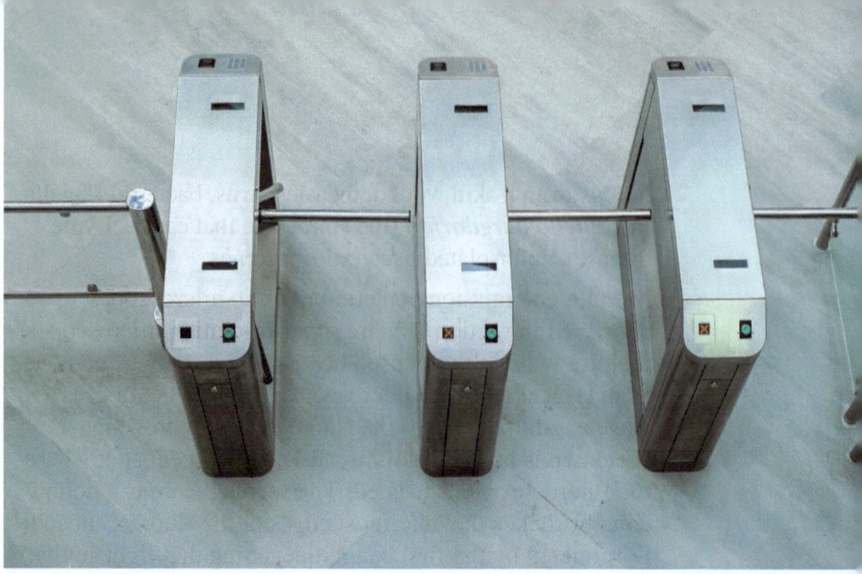

Roberto Sorin/Shutterstock

Bacteria Use Rapid Transport

In the popular imagination, the term *bacteria* is often preceded by *flesh-eating*. Tabloid headlines aside, only a small minority of skin infections progress to such a severe outcome. Although there are several microbes that can cause these infections, the most common culprit is the group A streptococci (GAS). If that name sounds familiar, you may be thinking of strep throat, another outcome of GAS interaction with humans. Scientists have puzzled over how a tiny microbe with a small genome can multiply so rapidly and cause such damage in different human body tissues. The answer may well lie in its diversity of transporters: molecular gates that control which molecules enter or leave the cell.

Entry to the bacterial cell is controlled by transport proteins embedded in the cell's membrane. Transporters are specific for one type of molecule: ions, amino acids, vitamins, or sugars. The cell membrane is stuffed with transporters so nutrients are efficiently imported. Other transporters export molecules for communication, toxicity, and waste products. In a given niche of the human body—saliva, for example—GAS employs the transporters best suited to the nutrients available in that environment.

Which transporters do GAS use when flesh-eating? The data are quite clear that GAS requires amino acids in significant quantities when growing in skeletal muscle (flesh). This makes sense because muscle is made largely from protein. Most of the transporters that GAS produces for use in this tissue are responsible for amino acid transport into the cell. How does this information help to treat flesh-eating GAS infections? Two avenues of research are opened here: First, molecules that plug up amino acid transporters would slow GAS multiplication, and second, because the transporters are on the cell surface, they make good candidates for a vaccine.

In this chapter, we explore the basic structures of a bacterial cell. Establishing boundaries like the cell membrane or the cell wall and conditions for crossing those boundaries is a significant part of an organism's lifestyle. Although bacterial cells look like simple structures in the light microscope, a closer look reveals astounding complexity.

Readiness Check:

Based on what you have learned previously, you should be able to:

✓ Identify the following structures or regions of a plant or animal cell and describe their functions: cell wall, plasma membrane, cytoplasm, mitochondria, chloroplasts, and ribosomes

✓ Define and give examples of essential nutrients; describe how they are used by cells

3.1 Bacteria Are Diverse but Share Some Common Structural Features

After reading this section, you should be able to:

a. Distinguish a typical bacterial cell from a typical plant or animal cell in terms of cell shapes and arrangements, size, and subcellular structures
b. Discuss the factors that determine the size and shape of a bacterial cell

Much of this chapter is devoted to a discussion of individual cell components. Therefore a preliminary overview of the features common to many bacterial cells is in order. We begin by considering overall cell morphology and then move to subcellular structures.

Shape, Arrangement, and Size

It might be expected that because bacterial cells are small and relatively simple, they would be uniform in shape and size. This is not the case, as the microbial world offers considerable variety in terms of morphology. However, the two most common shapes are cocci and bacilli (**figure 3.1**). **Cocci** (s., **coccus**) are roughly spherical cells. They can exist singly or can be associated in characteristic arrangements that are useful in their identification. **Diplococci** (s., **diplococcus**) arise when cocci divide and remain together to form pairs. Long chains of cocci result when cells adhere after repeated divisions in one plane; this pattern is seen in the genera *Streptococcus, Enterococcus,* and *Lactococcus.* Members of the genus *Staphylococcus* divide in random planes to generate irregular, grapelike clusters (figure 3.1*a*). Divisions in two or three planes can produce symmetrical groupings of cocci. Bacteria in the genus *Micrococcus* often divide in two planes to form square groups of four cells called tetrads. In the genus *Sarcina,* cocci divide in three planes, producing cubical packets of eight cells.

Legionella pneumophila is an example of a bacterium with a rod shape (figure 3.1*b*). Rods, sometimes called **bacilli** (s., **bacillus**), differ considerably in their length-to-width ratio, the coccobacilli being so short and wide that they resemble cocci. The shape of the rod's end often varies between species and may be flat, rounded, football-shaped, or bifurcated. Although many rods occur singly, some remain together after division to form pairs or chains.

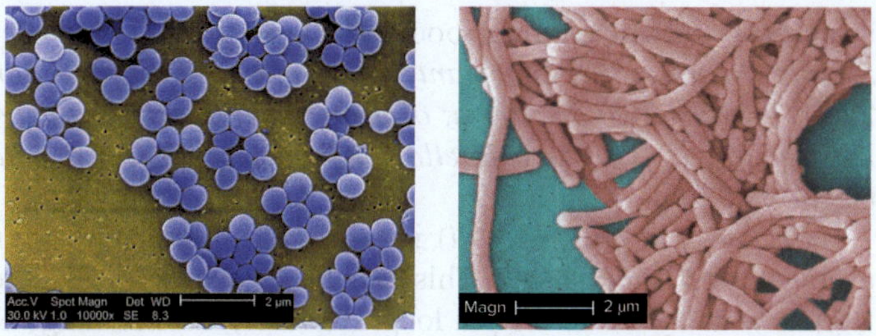

(a) *S. aureus*—cocci in clusters **(b)** *L. pneumophila*—bacilli in chains

Figure 3.1 Cocci and Bacilli Are the Most Common Bacterial Shapes. These images are color-enhanced scanning electron micrographs. (a) *Staphylococcus aureus.* (b) *Legionella pneumophila*, the cause of Legionnaires' disease.
Janice Haney Carr/CDC

There are several less common cell shapes and arrangements. **Vibrios** are comma-shaped; **spirilla** are rigid, spiral-shaped cells; and **spirochetes** are flexible, spiral-shaped bacteria (**figure 3.2*a,b***). Other bacteria are **pleomorphic,** being variable in shape and lacking a single, characteristic form. ▶| *Phylum Spirochaetota (section 20.5)*

Some bacteria can be thought of as multicellular. Many actinobacteria form long filaments called **hyphae** (figure 3.2*c*). The hyphae form a network called a **mycelium,** and in this sense, they are similar to eukaryotic filamentous fungi. Many cyanobacteria, a group of photosynthetic bacteria, are also filamentous. Being filamentous allows some degree of differentiation among cells in the filament. For instance, some filamentous cyanobacteria form specialized cells within the filament, called heterocysts, that carry out nitrogen fixation (*see figure 20.8a*). Myxobacteria are also morphologically complex. These bacteria sometimes aggregate to form

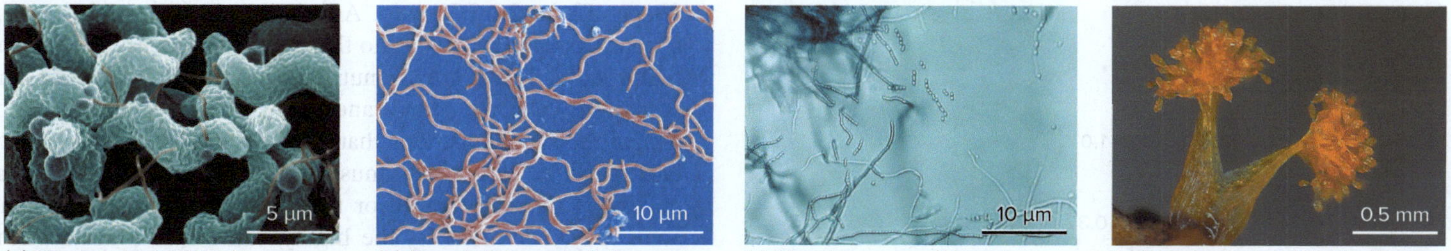

(a) *Campylobacter jejuni*—spiral-shaped **(b)** *Borreliella burgdorferi*—a spirochete **(c)** *Streptomyces*—a filamentous bacterium **(d)** Myxobacteria fruiting bodies

Figure 3.2 Other Cell Shapes and Aggregations. (a) *Campylobacter jejuni,* an organism that can cause food poisoning; SEM. (b) *Borreliella burgdorferi,* the spirochete that causes Lyme disease; SEM. (c) *Streptomyces* sp., a filamentous bacterium; light micrograph. (d) Fruiting body typical of myxobacteria, composed of thousands of cells; light micrograph. Note that the scale bar in (d) is in mm, not µm.
(a) De Wood, Chris Pooley/USDA; (b) Janice Haney Carr/CDC; (c) Dr. David Berd/CDC; (d) Simia Attentive/Shutterstock

structures called fruiting bodies (figure 3.2*d*). ▶| *Order Streptomycetales: an important source of antibiotics (section 22.1); Phylum Cyanobacteria: oxygenic photosynthetic bacteria (section 20.3); Phyla Bdellovibrionota and Myxococcota (section 20.9)*

Escherichia coli (*E. coli*) is an excellent example of an average-sized bacterial cell. This rod-shaped bacterium is 1.1 to 1.5 μm wide by 2.0 to 6.0 μm long. However, the size range of bacterial cells extends far beyond this average (**figure 3.3**). Near the small end of the size continuum are members of the genus *Mycoplasma* (0.3 μm in diameter). Parasitic bacteria in the size range of 0.2–0.4 μm can only survive when growing attached to their hosts, which provide nutrients to the symbionts (**figure 3.4*a***). At the other end of the continuum are bacteria such as some spirochetes, which can reach 500 μm in length. For instance, *Epulopiscium fishelsoni* grows as large as 600 by 80 μm, a little smaller than a printed hyphen and clearly larger than the eukaryote *Paramecium* (figure 3.4*b*). An even larger bacterium, *Thiomargarita namibiensis,* lives in ocean sediment (*see figure 21.18*). Thus a few bacteria are much larger than the average eukaryotic cell (typical plant and animal cells are around 10 to 50 μm in diameter).

The variety of sizes and shapes exhibited by bacteria raises a fundamental question: What causes a bacterial species to have a particular size and shape? For many years it was thought that microbes had to be small to increase the surface area-to-volume ratio (S/V ratio). As this ratio increases, the uptake of nutrients and the diffusion of these and other molecules within the cell become more efficient, which in turn facilitates rapid growth. Shape affects the S/V ratio. A rod with the same volume as a coccus has a higher S/V ratio than does the coccus. This means that a rod can have greater nutrient flux (i.e., nutrient transfer) across its plasma membrane. However, the discovery of *E. fishelsoni* demonstrates that bacteria can be very large. For bacteria to be large, they must have other characteristics that maximize their S/V ratio, or their size must be beneficial in some way. For instance, the highly convoluted plasma membrane of *E. fishelsoni* has a much greater surface area than a typical plasma membrane. This increased surface area thereby increases the S/V ratio. In addition, large cells are less likely to be eaten by predatory protists. Cells that are filamentous, have stalks, or are oddly shaped are also less susceptible to predation. ▶| *Predators come in all sizes (section 27.4)*

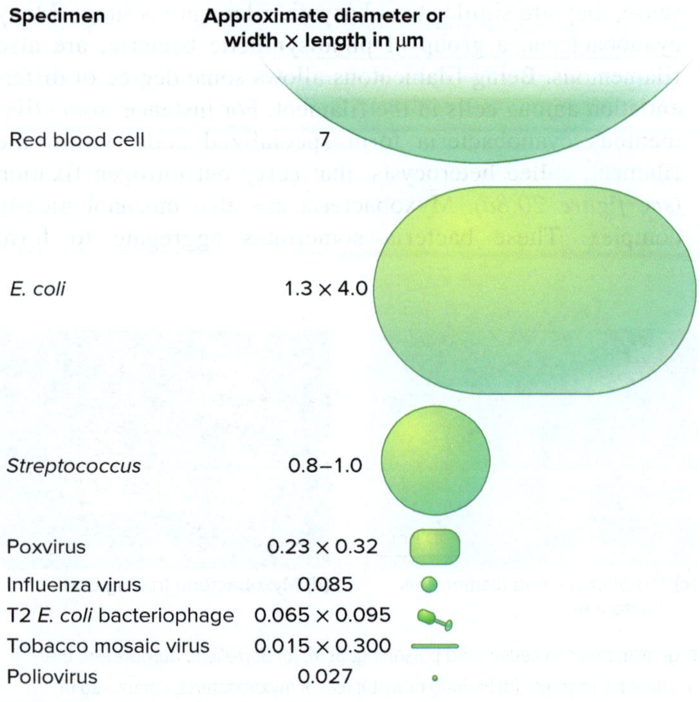

Figure 3.3 Sizes of Bacteria Relative to a Red Blood Cell and Viruses. The larger viruses are comparable in size to the smaller bacteria.

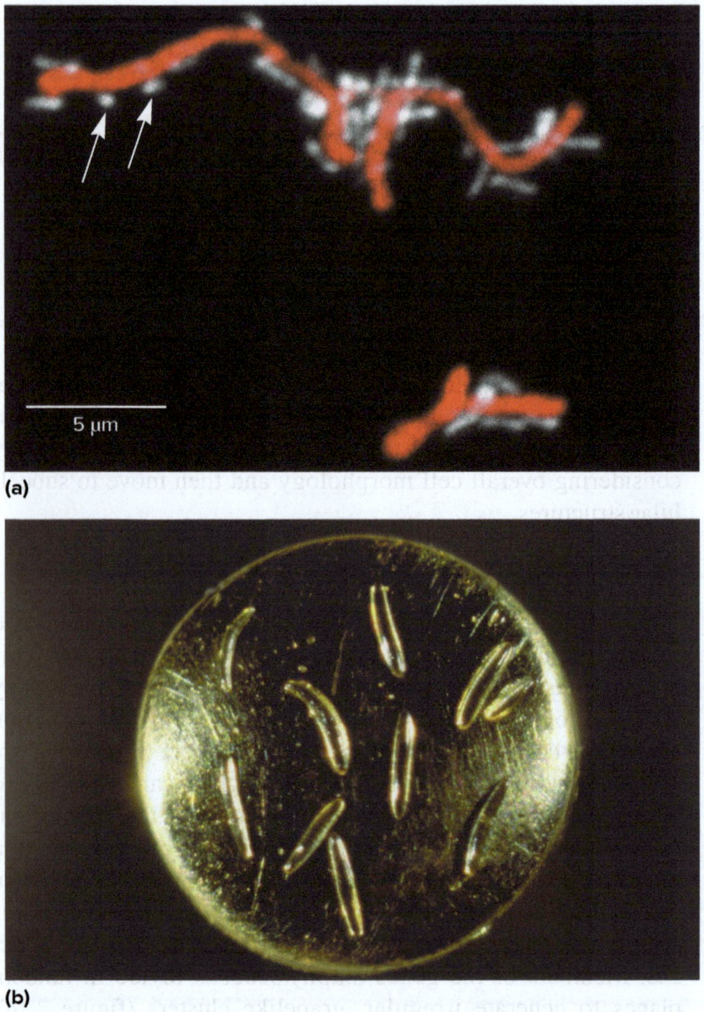

Figure 3.4 Size Extremes in Bacteria. (a) In this fluorescence micrograph, the ultrasmall *Candidatus* Saccharibacteria (white) grows as an obligate parasite on *Actinomyces odontolyticus* (red) in a sample obtained from a human mouth. (b) This micrograph shows *Epulopiscium fishelsoni* on the head of a pin.

(a) Batbileg Bor; (b) Charles O'Rear/Corbis Documentary/Getty Images

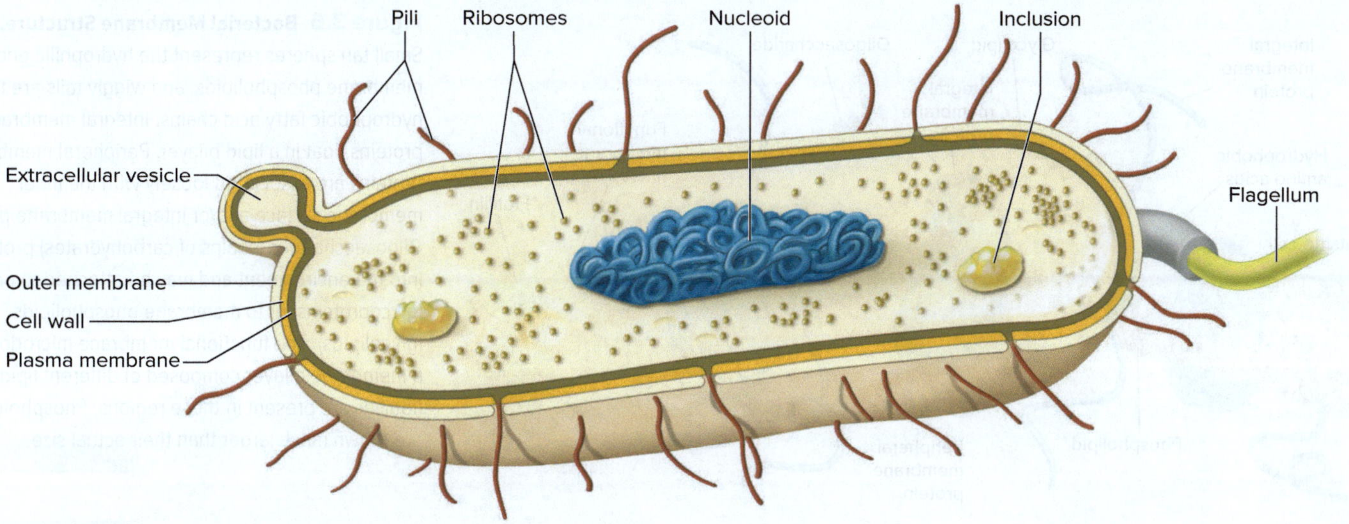

Figure 3.5 Structure of a Typical Bacterial Cell. Crossection of a typical rod-shaped bacterial cell.

Cell Organization

Structures often observed in bacterial cells are summarized and illustrated in **figure 3.5**. No single bacterium possesses all of these structures at all times. Some are found only in certain cells in particular conditions or in specific phases of the life cycle.

There are several common features of bacterial cell structure. Bacterial cells are surrounded by several layers, which are collectively called the cell envelope. The most common cell envelope layers are the plasma membrane, cell wall, and capsule or slime layer. The innermost layer of the cell envelope is the plasma membrane, which surrounds the cytoplasm. Most bacteria have a chemically complex cell wall, which surrounds the plasma membrane. Many bacteria encase the cell wall with a capsule or slime layer. Because most bacteria do not contain internal, membrane-bound organelles, their interior appears morphologically simple. The genetic material is localized in a discrete region called the **nucleoid** and is not separated from the surrounding cytoplasm by a membrane. **Ribosomes** and larger masses called **inclusions** are scattered about the cytoplasm. Filamentous structures termed **pili** (s. **pilus**) may protrude from the surface, facilitating gene transfer or attachment to surfaces. Finally, many bacteria use **flagella** (s. **flagellum**) for locomotion. In the remaining sections of this chapter, we describe these major structures observed in bacterial cells in more detail.

Comprehension Check

1. What characteristic shapes can bacteria assume? Describe the ways in which bacterial cells cluster together.
2. What advantages might the ability to form multicellular arrangements (e.g., clusters or chains) have that are not afforded single cells?
3. What is the relevance of the surface area-to-volume ratio?

3.2 Bacterial Plasma Membranes Control What Enters and Leaves the Cell

After reading this section, you should be able to:

a. Describe bacterial membrane structure and identify the types of lipids typically found there
b. Distinguish macronutrients from trace elements (micronutrients) and provide examples of each
c. Provide examples of growth factors needed by some microorganisms
d. Compare and contrast passive diffusion, facilitated diffusion, active transport, and group translocation, and provide examples of each
e. Discuss the challenge of iron uptake and describe how bacteria overcome this difficulty

The **cell envelope** is defined as the plasma membrane and all the surrounding layers external to it. The cell envelopes of many bacteria consist of the plasma membrane, cell wall, and at least one additional layer (e.g., capsule or slime layer). The **plasma membrane** encompasses the cytoplasm and defines the cell. If it is removed or damaged, the cell's contents spill into the environment and the cell dies. Importantly, the plasma membrane serves as a platform for many of the biochemical reactions required to sustain cellular growth. Thus we begin our consideration of bacterial cell structure by describing the plasma membrane.

To survive, cells must interact in a selective fashion with their environment, acquire nutrients, and eliminate waste. A primary role of all plasma membranes is to serve as selectively permeable barriers: They allow particular molecules to pass into or out of the cell, while altogether preventing the

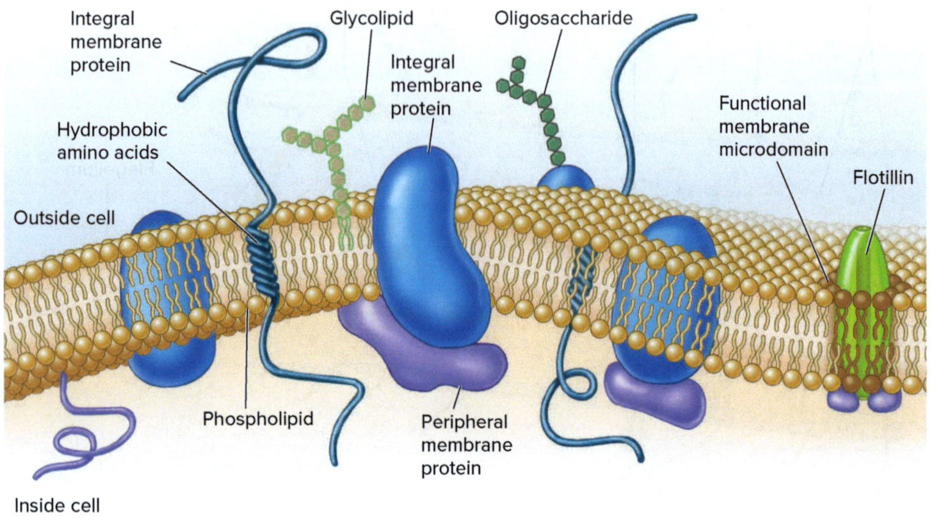

Figure 3.6 Bacterial Membrane Structure. Small tan spheres represent the hydrophilic ends of membrane phospholipids, and wiggly tails are the hydrophobic fatty acid chains. Integral membrane proteins float in a lipid bilayer. Peripheral membrane proteins are associated loosely with the inner membrane surface and/or integral membrane proteins. Oligosaccharides (chains of carbohydrates) protrude into the environment and may be attached to proteins (glycoproteins) or to membrane phospholipids (glycolipids). The functional membrane microdomain is a membrane bilayer composed of different lipids; flotillins are present in these regions. Phospholipids are drawn much larger than their actual size.

movement of other molecules. Cells also have to maintain their interior in a constant, highly organized state in the face of external changes. Plasma membranes are an absolute requirement for all living organisms because they are involved in carrying out these cellular tasks.

Plasma Membrane Structure Is Dynamic

Plasma membranes are very thin structures, about 7 to 8 nm thick, that look like two dark lines on either side of a light interior when imaged by transmission electron microscopy (TEM). This characteristic appearance is evidence that the membrane is composed of two sheets of lipid molecules arranged end-to-end (**figure 3.6**). All membranes share this same basic structure. Bacterial membranes have roughly equal amounts of lipids and proteins. Cleavage of membranes by freeze-etching, a technique that reveals fine detail, exposes the proteins lying within the membrane lipid bilayer. ◀ *Electron microscopes use beams of electrons to create highly magnified images (section 2.4); Scanning probe microscopy can visualize molecules and atoms (section 2.5)*

The chemical nature of membrane lipids is critical to their ability to form bilayers. Most membrane-associated lipids (e.g., the phospholipids shown in figure 3.6) are **amphipathic:** They are structurally asymmetric, with polar and nonpolar ends (**figure 3.7**). The polar ends interact with water and are **hydrophilic;** the nonpolar **hydrophobic** ends are insoluble in water and tend to associate with one another. In aqueous environments, amphipathic lipids spontaneously interact to form a bilayer. The outer surfaces of the bilayer are hydrophilic, whereas hydrophobic ends are buried in the interior away from the surrounding water (figure 3.6). Individual phospholipid molecules associate through noncovalent interactions and are in continuous motion in a membrane. ▶ *Lipids (appendix I)*

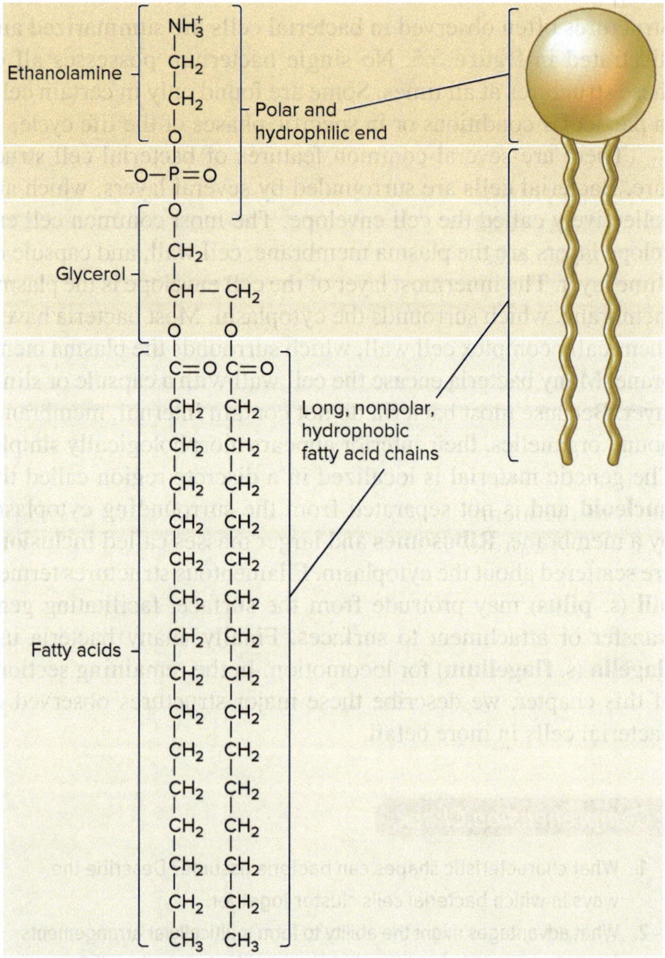

Figure 3.7 The Structure of a Phospholipid. Phosphatidylethanolamine, a phospholipid often found in bacterial membranes.

Two types of membrane proteins have been identified based on their ability to be separated from the membrane. **Peripheral membrane proteins** are loosely connected to the membrane and can be easily removed (figure 3.6). They are soluble in aqueous solutions and make up about 20 to 30% of total membrane protein. The remaining proteins are **integral membrane proteins** whose hydrophobic regions are buried in membrane lipids, whereas the hydrophilic portions project from the membrane surface (figure 3.6). These are not easily extracted from membranes and are insoluble in aqueous solutions when freed of lipids. Integral membrane proteins, like membrane lipids, are amphipathic.

Integral membrane proteins carry out some of the most important functions of the membrane. Many are transport proteins used to move materials either into or out of the cell. Others are involved in energy-conserving processes, such as the proteins found in electron transport chains. Those integral membrane proteins with regions exposed to the outside of the cell enable the cell to interact with its environment. ▶| *Proteins (appendix I)*

Neither the lipid bilayers nor the membrane proteins are homogeneous. Several related phospholipids comprise the bulk of the plasma membrane, and these are accompanied by smaller amounts of lipids like steroids, hopanoids, and carotenoids (*see figure 11.29*). **Hopanoids** are similar in structure to cholesterol found in eukaryotic membranes (**figure 3.8**), and their rigid planar structure makes them more hydrophobic than phospholipids. Because of their hydrophobic nature, these lipids easily insert into membranes, but because they are not amphipathic, they distort the regular bilayer structure. This affects fluidity and shape in membrane regions and, in turn, determines where certain integral membrane proteins can reside. These form **functional membrane microdomains** that serve as platforms for large protein complex assembly. For example, integral membrane proteins called **flotillins** organize secretion systems for transporting molecules out of the cell and complexes that transmit signals from the environment to molecules in the cytoplasm (figure 3.6). Within the plasma membrane, regions of lipid variation differentiate parts of the membrane and allow for the subcellular localization of some proteins.

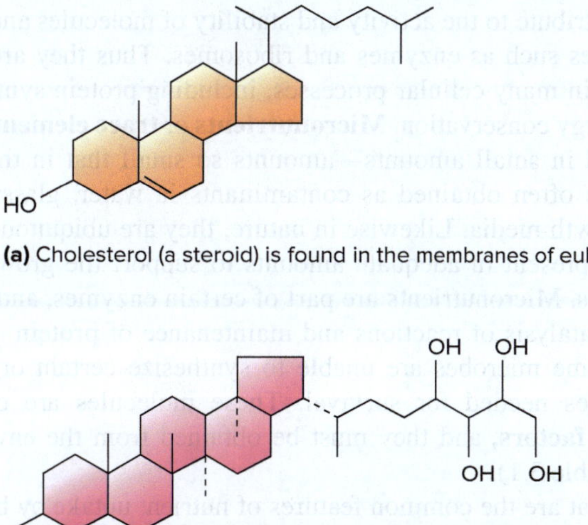

(a) Cholesterol (a steroid) is found in the membranes of eukaryotes.

(b) Bacteriohopanetetrol (a hopanoid) is found in many bacterial membranes.

Figure 3.8 Membrane Steroids and Hopanoids.

Bacteria Use Many Mechanisms to Bring Nutrients into the Cell

All plasma membranes function as barriers. Yet they must also allow movement of nutrients into the cell. If a microbe does not obtain nutrients from its environment, it will quickly exhaust its supply of amino acids, nucleotides, and other molecules needed to survive. In addition, if a microbe is to thrive and reproduce, it must have a source of energy. The energy source is used to generate the cell's major energy currency: the high-energy molecule adenosine triphosphate (ATP). Clearly, obtaining energy and nutrient sources is one of the most important jobs an organism has, and it is primarily a function of the plasma membrane. Here we discuss nutrient uptake, but first let's define some terms used to describe the nutrients needed by cells.

The six **macronutrients** are required in relatively large amounts (**table 3.1**). They are found in organic molecules such as proteins, lipids, nucleic acids, and carbohydrates. Other macronutrients exist as ions and generally are associated with

Table 3.1 Nutrients Required by Bacterial Cells

Macronutrients	Ionic Macronutrients	Micronutrients	Growth Factors
Carbon	Potassium	Manganese	Amino acids
Oxygen	Calcium	Zinc	Purines
Hydrogen	Magnesium	Cobalt	Pyrimidines
Nitrogen	Iron	Molybdenum	Vitamins
Sulfur	Chlorine	Nickel	
Phosphorus		Copper	

and contribute to the activity and stability of molecules and cell structures such as enzymes and ribosomes. Thus they are important in many cellular processes, including protein synthesis and energy conservation. **Micronutrients** or **trace elements** are required in small amounts—amounts so small that in the lab they are often obtained as contaminants in water, glassware, and growth media. Likewise in nature, they are ubiquitous and usually present in adequate amounts to support the growth of microbes. Micronutrients are part of certain enzymes, and they aid in catalysis of reactions and maintenance of protein structure. Some microbes are unable to synthesize certain organic molecules needed for survival. These molecules are called **growth factors,** and they must be obtained from the environment (table 3.1).

What are the common features of nutrient uptake by bacteria? In view of the enormous variety of nutrients and the complexity of the task, it is not surprising that bacteria use several different transport mechanisms: passive diffusion, facilitated diffusion, primary and secondary active transport, and group translocation.

Passive Diffusion

Passive diffusion, often called diffusion or simple diffusion, is the process by which molecules move from a region of higher concentration to one of lower concentration; that is, the molecules move down the concentration gradient. The rate of passive diffusion depends on the size of the concentration gradient between a cell's exterior and its interior (**figure 3.9**). A large concentration gradient is required for adequate nutrient uptake by passive diffusion (i.e., the external nutrient concentration must be high while the internal concentration is low). The rate of diffusion decreases as more nutrient accumulates in the cell. This occurs if the nutrient is not used immediately upon entry.

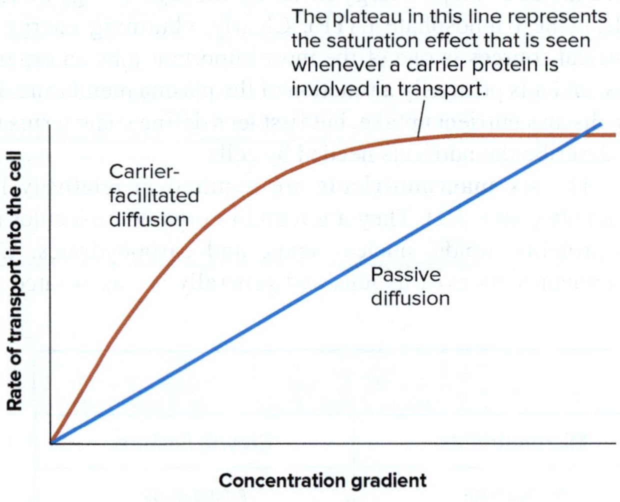

Figure 3.9 Passive and Facilitated Diffusion. The rate of diffusion depends on the size of the solute's concentration gradient (the ratio of the extracellular concentration to the intracellular concentration). This example of facilitated diffusion involves a carrier protein that can be saturated. Sometimes facilitated diffusion is mediated by a channel. Channels often do not exhibit a saturation effect.

MICRO INQUIRY *How would the curve appear for a solute that can cross the membrane both by passive diffusion and through a carrier?*

Most substances cannot freely diffuse into a cell. However, water and some nonpolar molecules, including O_2 and CO_2, easily cross the plasma membrane by passive diffusion. Larger molecules, ions, and polar substances must enter the cell by other mechanisms, all of which involve specialized proteins that are referred to as transport proteins. *How Diffusion Works*

Facilitated Diffusion

During **facilitated diffusion,** substances move across the plasma membrane with the assistance of transport proteins that are either **channels** or **carriers.** Channels, as their name indicates, are proteins that form pores in membranes through which substances can pass; they are often involved in facilitated diffusion. Channels show some specificity for the substances that pass through them, but this is considerably less than that shown by carriers, which are far more substrate specific. The rate of facilitated diffusion is greater than that of passive diffusion (figure 3.9). An example of channel-mediated facilitated diffusion involves aquaporins (*see figure 2.30*), which transport water. When the transporter is a carrier, the diffusion rate reaches a plateau above a specific gradient value because the carrier protein is saturated; that is, it is transporting as many solute molecules as possible. The resulting curve resembles an enzyme-substrate curve (*see figure 10.16*) and is different from the linear response seen with passive diffusion.

Although facilitated diffusion relies on transport proteins, it is truly diffusion because a concentration gradient spanning the membrane drives the movement of molecules, and no energy is used. If the concentration gradient disappears, net inward movement ceases. The gradient can be maintained by converting the transported nutrient to another compound as soon as it enters the cytoplasm, as occurs when a nutrient is metabolized.

Facilitated diffusion has been documented in some bacteria, but it does not seem to be the major uptake mechanism for these microbes. Recall that many bacteria live in environments where nutrient concentrations are low, and facilitated diffusion cannot concentrate nutrients inside cells. Therefore energy-dependent transport mechanisms capable of concentrating nutrients are significantly more important uptake mechanisms for bacterial cells.

Primary and Secondary Active Transport

Active transport is the transport of solute molecules to higher concentrations (i.e., against a concentration gradient) with the input of energy. Three types of active transport are observed: primary active transport, secondary active transport, and group translocation. They differ in terms of the energy used to drive transport and whether or not the transported molecule is modified as it enters.

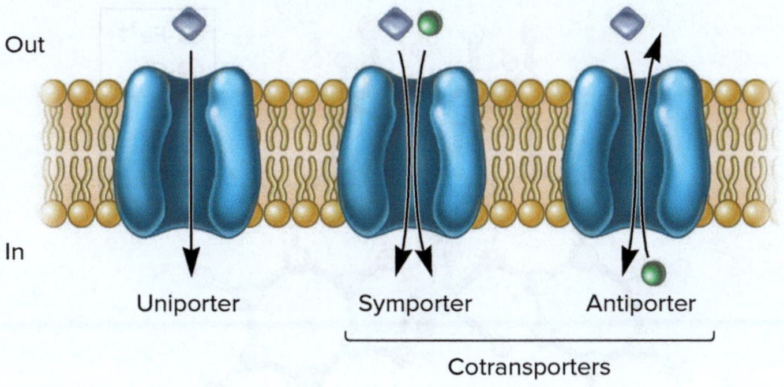

Figure 3.10 Carrier Proteins Can Be Uniporters or Cotransporters.

Active transport resembles facilitated diffusion in that it involves carrier proteins. Recall that carrier proteins bind particular solutes with great specificity. Active transport is also characterized by the carrier saturation effect at high solute concentrations (figure 3.9). Nevertheless, active transport differs from facilitated diffusion because it uses energy to concentrate substances within the cell.

Primary active transport is mediated by carriers called primary active transporters. They use energy provided by ATP hydrolysis to move substances against a concentration gradient without modifying them. Primary active transporters are **uniporters;** that is, they move a single molecule across the membrane (**figure 3.10**). ***ATP-Binding Cassette transporters* (ABC transporters)** are important primary active transporters, responsible for moving substrates such as vitamins, ions, and sugars. Our focus here is on those ABC transporters used for import rather than for export. Export transporters are described in chapter 13. *Protein maturation and secretion (section 13.9)*

Most ABC transporters consist of two hydrophobic membrane-spanning regions (domains) with two ATP-binding domains facing the cytoplasm (**figure 3.11**). The membrane-spanning domains form a pore in the membrane, and the ATP-binding domains bind and hydrolyze ATP to drive uptake.

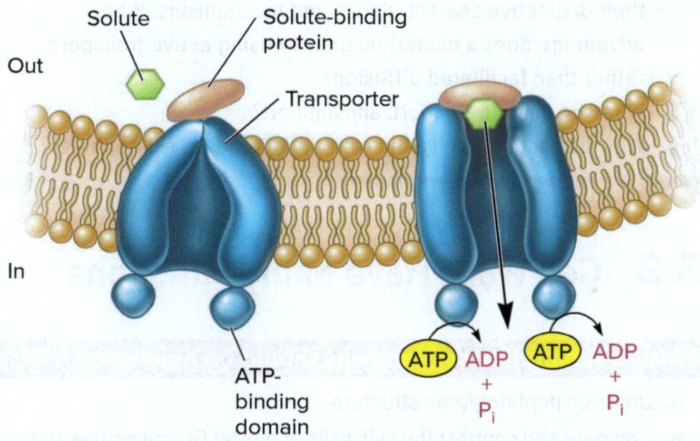

Figure 3.11 ABC Transporter Function. Solute-binding proteins may be associated with the plasma membrane, associated with the transporter, or even fused to the transporter.

Most ABC transporters employ solute-binding proteins to deliver the molecule to be transported to the transporter.

Secondary active transport couples the potential energy of ion gradients to transport substances without modifying them. Secondary active transporters are cotransporters because they move two substances simultaneously: the ion whose gradient powers transport and the substance being moved across the membrane (figure 3.10). When the ion and other substance both move in the same direction, it is called **symport.** When they move in opposite directions, it is called **antiport.** *Cotransport (Symport and Antiport)*

The ion gradients used by secondary active transporters arise primarily in three ways. The first results from bacterial metabolic activity. During energy-conserving processes, electron transport generates a proton gradient in which protons are at a higher concentration outside the cell than inside. The proton gradient is used to do cellular work, including secondary active transport. Some bacteria use the second method, in which an enzyme called a V-type ATPase hydrolyzes ATP and uses the energy released to create either a proton gradient or a sodium gradient across the plasma membrane. (Although F-type ATP synthases are more familiar to microbiologists, they *use* a proton gradient rather than *generate* a proton gradient.) Finally, a proton gradient can be used to create another ion gradient such as a sodium gradient. This is accomplished by an antiporter that brings protons in as sodium ions are moved out of the cell. The sodium gradient can then be used to drive uptake of nutrients by a symport mechanism. *Electron transport and oxidative phosphorylation (step 3) generate the most ATP (section 11.4)*

The lactose permease of *E. coli* is a well-studied symport secondary active transporter. It is a single protein that transports a lactose molecule inward as a proton simultaneously enters the cell. As the proton moves down a proton gradient, the energy released drives lactose transport. X-ray diffraction studies show that the carrier protein exists in outward- and inward-facing conformations. When lactose and a proton bind to separate sites on the outward-facing conformation, the protein changes to its inward-facing conformation, and the sugar and proton are released into the cytoplasm. *Lactose operon: negative transcriptional control of inducible genes (section 14.2)*

Bacteria often have more than one transport system for a nutrient. *E. coli* has at least five transport systems for the sugar galactose, three systems each for the amino acids glutamate and leucine, and two potassium transport complexes. When several transport systems exist for the same substance, the systems differ in such properties as their energy source, their affinity for the solute transported, and the nature of their regulation. This diversity gives the bacterium a competitive advantage in a changing environment.

Group Translocation

The distinguishing characteristic of **group translocation** is that a molecule is chemically modified as it is brought into the cell. The best-known group translocation system is the **phosphoenolpyruvate: sugar phosphotransferase system (PTS),** which is observed in many bacteria. The PTS imports a variety

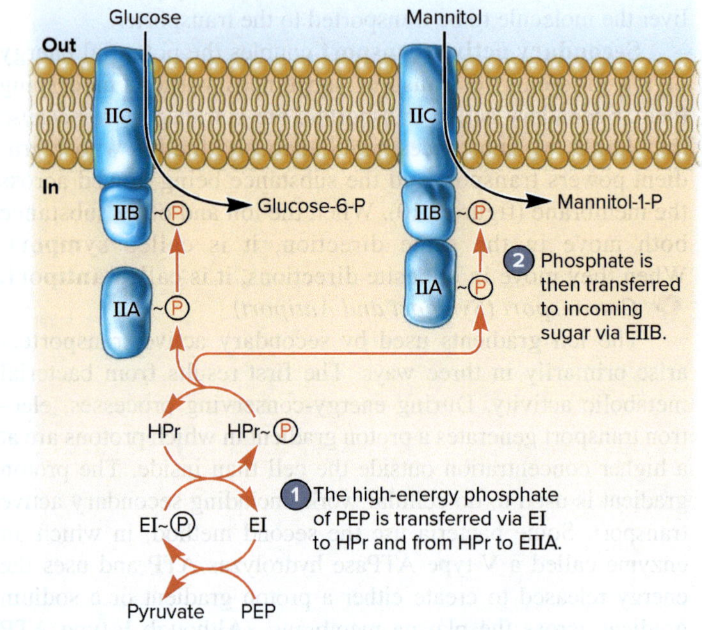

Figure 3.12 Group Translocation: Bacterial PTS Transport. Two examples of the phosphoenolpyruvate:sugar phosphotransferase system (PTS) are illustrated. The following components are involved in the system: phosphoenolpyruvate (PEP), enzyme I (EI), the low molecular weight heat-stable protein (HPr), and enzyme II (EII). EIIA is attached to EIIB in the mannitol transport system and is separate from EIIB in the glucose system.

of sugars while phosphorylating them, using phosphoenolpyruvate (PEP) as the phosphate donor. PEP is a common metabolic intermediate found in many bacteria.

The transfer of phosphate from PEP to the incoming sugar molecule involves several proteins and is an example of a **phosphorelay system.** In *E. coli* and *Salmonella,* the PTS consists of two enzymes and a low molecular weight heat-stable protein (HPr). A phosphate is transferred from PEP to enzyme II with the aid of enzyme I and HPr (**figure 3.12**). Enzyme II is an integral membrane protein that phosphorylates the sugar molecule as it crosses the membrane. Many different PTSs exist, and they vary in terms of the sugars they transport. The specificity lies with the type of Enzyme II used in the PTS. ▶| *Enzymes and ribozymes speed up cellular chemical reactions (section 10.6)*

Active Transport by Group Translocation

Iron Uptake

Almost all microorganisms require iron for building molecules important in energy-conserving processes (e.g., cytochromes), as well as for the function of many enzymes. Iron is unique among the ions transported into the cell (table 3.1) because of the extreme insolubility of ferric iron (Fe^{3+}) and its derivatives. Simply put, bacteria can only transport dissolved molecules; they cannot transport rocks! Rather than using typical membrane transporters, many bacteria secrete **siderophores** (Greek for iron bearers).

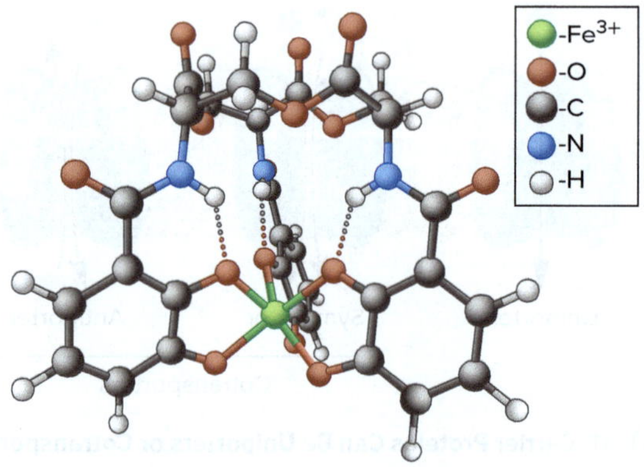

Figure 3.13 Enterobactin: A Siderophore Produced by *E. coli*. Ball-and-stick model of enterobactin complexed with Fe^{3+}.

Siderophores are small organic molecules that bind ferric iron and supply it to the cell (**figure 3.13**). ▶| *Electron transport chains: sets of sequential redox reactions (section 10.4)*

Microorganisms secrete siderophores when iron is scarce in the environment. Once the iron-siderophore complex reaches the cell surface, it binds to a siderophore-receptor protein. Then either the iron is released to enter the cell directly or the whole iron-siderophore complex is transported inside by an ABC transporter. Iron is so crucial to microorganisms that more than one route of iron uptake may be used to ensure an adequate supply.

Comprehension Check

1. List the functions of bacterial plasma membranes. Why must their plasma membranes carry out more functions than the plasma membranes of eukaryotic cells?
2. On what basis are elements divided into macronutrients and trace elements?
3. What is a growth factor?
4. Compare and contrast facilitated diffusion, primary and secondary active transport, and group translocation in terms of their distinctive characteristics and mechanisms. What advantage does a bacterium gain by using active transport rather than facilitated diffusion?
5. What are uniport, symport, and antiport?
6. Why did bacteria evolve the capacity to produce siderophores?

3.3 Cell Walls Have Many Functions

After reading this section, you should be able to:

a. Describe peptidoglycan structure
b. Compare and contrast the cell walls of typical Gram-positive and Gram-negative bacteria
c. Explain how the cell wall protects against osmotic stress

The **cell wall** is the layer that lies just outside the plasma membrane. It is one of the most important structures for several reasons: It helps maintain cell shape and protect the cell from osmotic lysis; it can protect the cell from toxic substances; and it can contribute to the ability of a pathogen to cause disease. Cell walls are so important that most bacteria have them. Those that lack them have other features that fulfill cell wall function. Bacterial cell wall synthesis is an important target for many antibiotics. *Antibacterial drugs (section 9.4)*

Overview of Bacterial Cell Wall Structure

After Christian Gram developed the Gram stain in 1884, it soon became evident that most bacteria could be placed into one of two groups based on their response to the Gram-staining procedure. Gram-positive bacteria stained purple, whereas Gram-negative bacteria were pink or red (*see figure 2.17*). The true structural difference between these two groups did not become evident until TEM experiments revealed the relationship between staining behavior and the cell wall. *Differential staining (section 2.3)*

The cell walls of typical Gram-positive bacteria consist of a single, 20- to 80-nm-thick layer of **peptidoglycan (murein)** lying outside the plasma membrane (**figure 3.14**). In contrast, the cell walls of typical Gram-negative bacteria consist of a 2- to 7-nm-thick peptidoglycan layer. An additional feature of Gram-negatives is that the cell wall is covered by a 12- to 14-nm-thick **outer membrane.**

One important feature in typical Gram-negative bacteria is a space between the plasma membrane and the outer membrane. It also is sometimes observed between the plasma membrane and cell wall in typical Gram-positive bacteria. This space is called the **periplasmic space.** The aqueous substance that occupies the periplasmic space is the **periplasm.**

For decades, the Gram-negative/Gram-positive model of cell wall structure guided taxonomists and became a cornerstone in identifying bacteria in clinical settings. More recently, however, it has become clear that these long-held models do not hold true for all bacteria. Iain Sutcliffe has proposed that microbiologists should more precisely describe bacterial cell envelopes by the number of membranes surrounding the cytoplasm. **Monoderms** (Greek *mono,* one; *derma,* skin) have a single membrane as seen in typical Gram-positive bacteria, while **diderms** (Greek *di,* two; *derma,* skin) have both a plasma membrane surrounding the cytoplasm and an outer membrane surrounding the periplasm, as seen in typical Gram-negative bacteria.

Although we continue to use the Gram-positive and Gram-negative terminology, we are mindful that it oversimplifies the diversity of bacterial cell envelope architectures. We highlight organisms that are best described by the monoderm/diderm classification where they appear in this book. *Bacterial taxonomy and cell envelopes (section 20.1).*

Peptidoglycan Structure

The feature common to nearly all bacterial cell walls is the presence of peptidoglycan, which forms an enormous meshlike structure

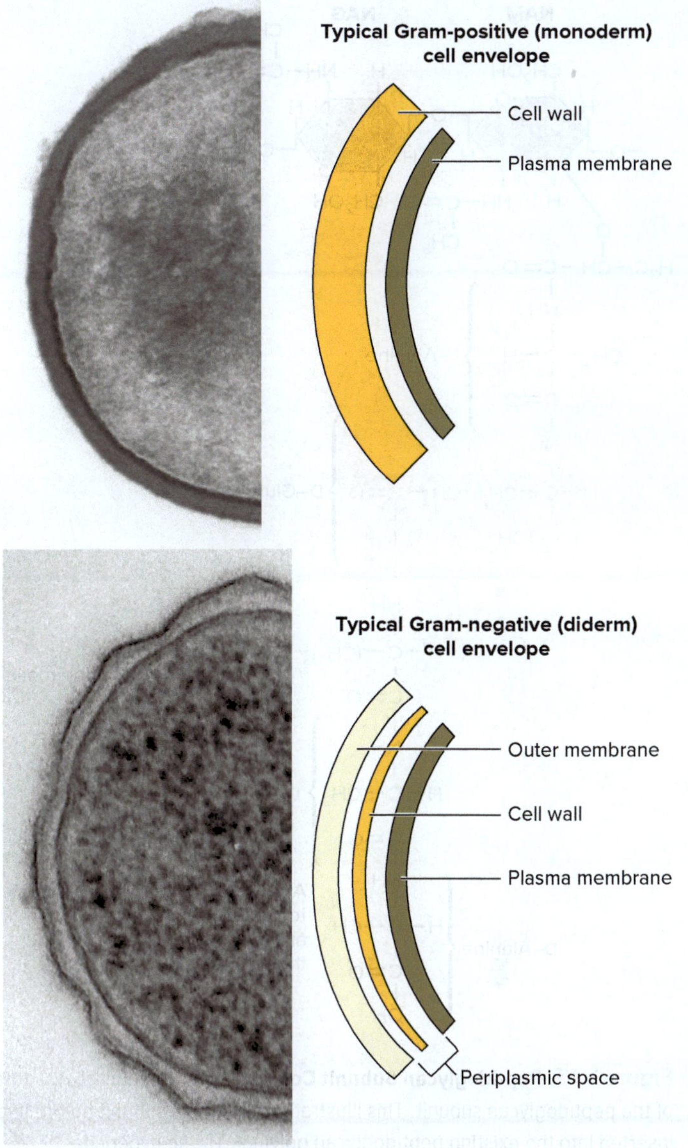

Figure 3.14 Cell Envelopes of Typical Gram-Positive (Monoderm) and Gram-Negative (Diderm) Bacteria. Cell envelopes consist of the plasma membrane and any layers (e.g., cell wall) exterior to it. *Staphylococcus aureus* (top) has a typical Gram-positive cell envelope that consists primarily of a plasma membrane and a peptidoglycan cell wall. *Myxococcus xanthus* (bottom) has a typical Gram-negative cell envelope consisting of a plasma membrane, a thin layer of peptidoglycan, an outer membrane, and the periplasmic space.
Egbert Hoiczyk

often referred to as the peptidoglycan sacculus. Peptidoglycan, a single molecule covering the entire plasma membrane, is composed of many identical subunits. Each subunit within the sacculus contains two sugar derivatives, *N-a*cetyl*g*lucosamine (NAG) and *N-a*cetyl*m*uramic acid (NAM), and several different amino acids. The amino acids form a short peptide, sometimes called the stem peptide, consisting of four alternating D- and L-amino acids; the peptide is connected to the carboxyl group of NAM (**figure 3.15**). Three of the amino acids are not found in proteins: D-glutamic acid,

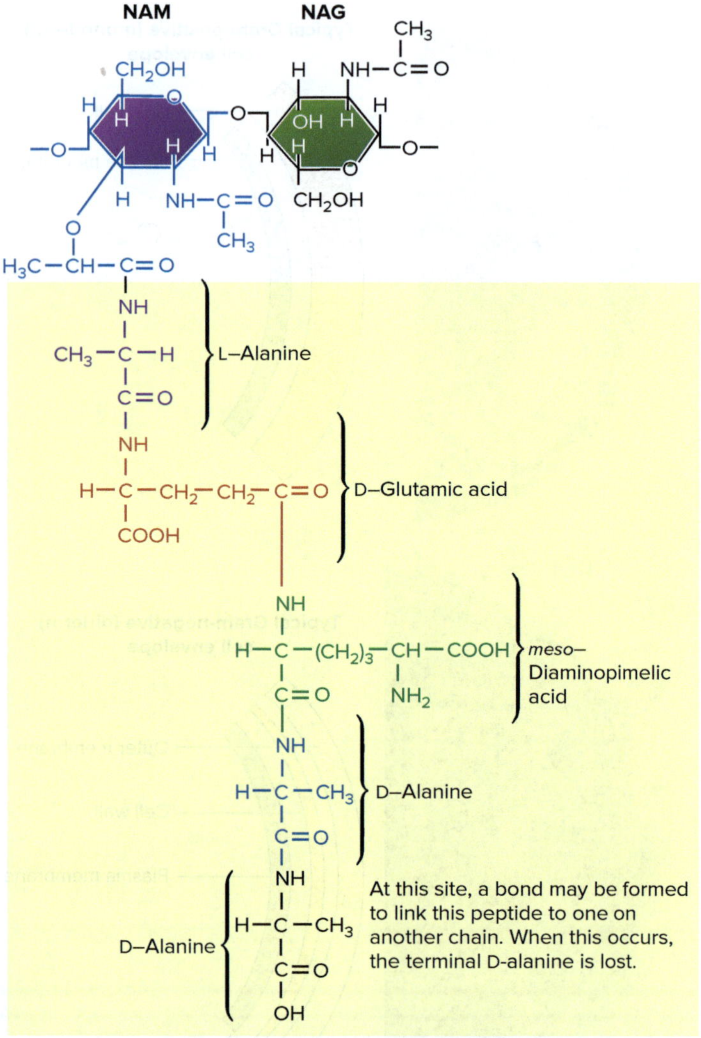

Figure 3.15 Peptidoglycan Subunit Composition. Chemical structure of the peptidoglycan subunit. This illustration shows the subunit before it is inserted into the existing peptidoglycan polymer. The stem peptide is composed of alternating D- and L-amino acids; it terminates with two D-alanines. The amino acids are shown in different colors for clarity. NAG, *N*-acetylglucosamine; NAM, *N*-acetylmuramic acid.

D-alanine, and *meso*-diaminopimelic acid. The presence of D-amino acids in the stem peptide protects against degradation by most peptidases, which recognize only the L-isomers of amino acid residues. ▶| *Carbohydrates (appendix I); Proteins (appendix I); Proteins are polymers of amino acids (section 13.2)*

The peptidoglycan sacculus is formed by linking the sugars of the peptidoglycan subunits together to form a strand; the strands are then cross-linked to each other by covalent bonds formed between the stem peptides extending from each strand. As seen in **figure 3.16**, the backbone of each strand is composed of alternating NAG and NAM residues. The strand is helical, and the stem peptides extend out from the backbone in different directions. There are two types of cross-links: direct and indirect.

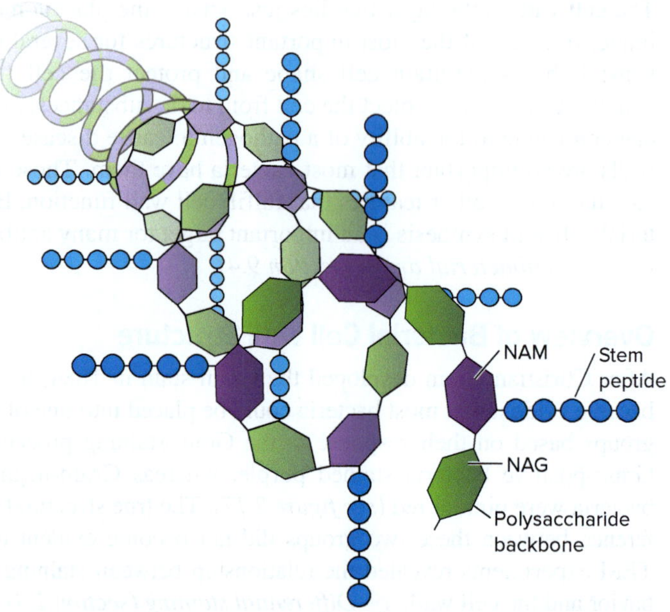

Figure 3.16 A Helical Peptidoglycan Strand. Because of the strand's helical nature, the stem peptides project out in different directions from the NAM-NAG backbone at approximately 90-degree angles.

A direct cross-link is characterized by connecting the carboxyl group of an amino acid in one stem peptide to the amino group of an amino acid in another stem peptide. For instance, many bacteria cross-link the strands by connecting the carboxyl group of the D-alanine at position 4 of the stem peptide directly to the amino group of diaminopimelic acid (position 3) of the other peptidoglycan strand's stem peptide. The position 5 D-alanine is removed as the cross-link is formed. Bacteria that have indirect linkage use a **peptide interbridge**, a short chain of amino acids linking the stem peptide of one peptidoglycan strand to that of another (**figure 3.17**). Cross-linking results in a dense, interconnected network of peptidoglycan strands ▶| *Synthesis of peptidoglycan occurs in the cytoplasm, at the plasma membrane, and in the periplasmic space (section 12.4)*

The peptidoglycan sacculus is strong but elastic. It is able to stretch and contract in response to osmotic pressure. This is due to the rigidity of the backbone coupled with the flexibility of the cross-links. Peptidoglycan sacculi are also porous, allowing globular proteins having a molecular weight as large as 50,000 to pass through, depending on whether the sacculus is relaxed or stretched; thus only extremely large proteins are unable to pass through peptidoglycan.

Variants of peptidoglycan are found, particularly among typical Gram-positive bacteria. For example, some substitute the diamino acid lysine for *meso*-diaminopimelic acid (**figure 3.18**) and cross-link chains via peptide interbridges. These peptide interbridges can vary considerably (figure 3.17). Peptidoglycan can also vary in terms of the length of the NAM-NAG strands and the amount of cross-linking. Bacteria that stain Gram positive tend to have much more cross-linking, whereas those that stain Gram negative have considerably less.

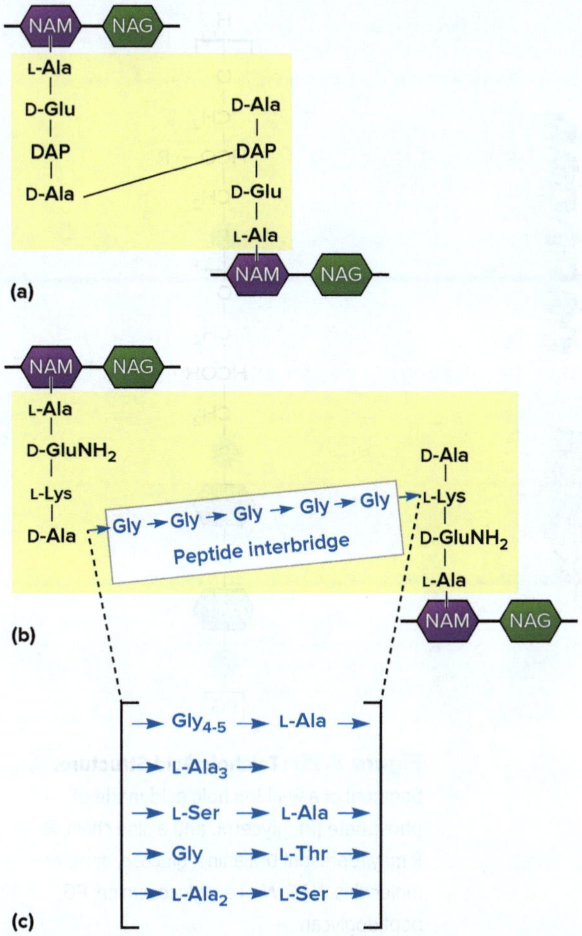

Figure 3.17 Peptidoglycan Cross-Links Can Be Direct or Indirect via a Peptide Interbridge. (a) *E. coli* peptidoglycan with direct cross-linking, typical of many Gram-negative bacteria. (b) *Staphylococcus aureus* peptidoglycan with an interbridge. *S. aureus* stains Gram positive. (c) Most variation in peptidoglycan structure occurs in the composition of the stem peptide and the method by which peptidoglycan strands are cross-linked. The bracket contains five typical Gram-positive interbridges. Gly, glycine; D-GluNH$_2$, D-glutamic acid with an amino group attached to the α carbon (the carbon next to the carboxyl group).

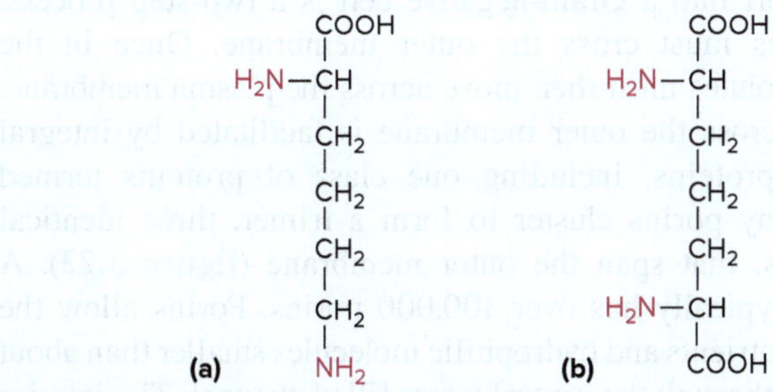

Figure 3.18 Diamino Acids Present in Peptidoglycan. (a) L-lysine, (b) *meso*-diaminopimelic acid. Amino groups are shown in red font.

Bacterial Cell Walls Consist Primarily of Peptidoglycan

Most cultured bacteria that stain Gram positive belong to only two phyla, Actinomycetota and Bacillota. Most of these bacteria have thick cell walls composed of peptidoglycan and large amounts of other polymers such as teichoic acids (**figure 3.19**). **Teichoic acids** are polymers of glycerol or ribitol joined by phosphate groups. Some teichoic acids are covalently linked to peptidoglycan and are referred to as wall teichoic acids (**figure 3.20**). Others are covalently connected to the plasma membrane; they are called lipoteichoic acids. Wall teichoic acids extend beyond the surface of the peptidoglycan. The negatively charged phosphates give the cell wall its negative charge.

Teichoic acids have several important functions. They help create and maintain the structure of the cell envelope by anchoring the wall to the plasma membrane. They are important during cell division, and they protect the cell from harmful substances in the environment (e.g., antibiotics and host defense molecules). In addition, they function in ion uptake and are involved in binding pathogenic species to host tissues, thus initiating the infectious disease process.

The vast majority of bacteria belonging to the remaining phyla stain Gram negative, making the diderm architecture far more widespread than the monoderm. Even a brief inspection of figure 3.14 shows that typical Gram-negative cell walls are more complex than typical Gram-positive walls. One of the most striking differences is the paucity of peptidoglycan. The peptidoglycan layer is very thin (2 to 7 nm, depending on the bacterium) and sits within the periplasmic space.

The periplasmic space is much larger than that of a typical Gram-positive cell, ranging from about 30 to 70 nm wide (**figure 3.21**). Some studies indicate that it may constitute about 20 to 40% of the total cell volume. The periplasmic space is home to a variety of proteins. Some periplasmic proteins participate in nutrient acquisition—for example, hydrolytic enzymes and transport proteins. Other periplasmic proteins are involved in peptidoglycan synthesis and modification of toxic compounds that could harm the cell.

Gram-Negative Cell Envelopes Include a Second Membrane

One of the defining features of a Gram-negative cell is the presence of an outer membrane. The **outer membrane** lies outside the thin peptidoglycan layer. In some Gram-negative cells, the outer membrane is connected to the cell wall by Braun's lipoprotein, the most abundant protein in the outer membrane, present at over a million copies per cell (figure 3.21). This small lipoprotein is covalently joined to both the outer membrane and the peptide chain of peptidoglycan.

In contrast to the symmetry of the plasma membrane, the most distinctive structural feature of the outer membrane is the difference between the inner leaflet (facing the periplasm) and the outer leaflet (facing the environment). The inner leaflet is composed of phospholipids as found in the plasma

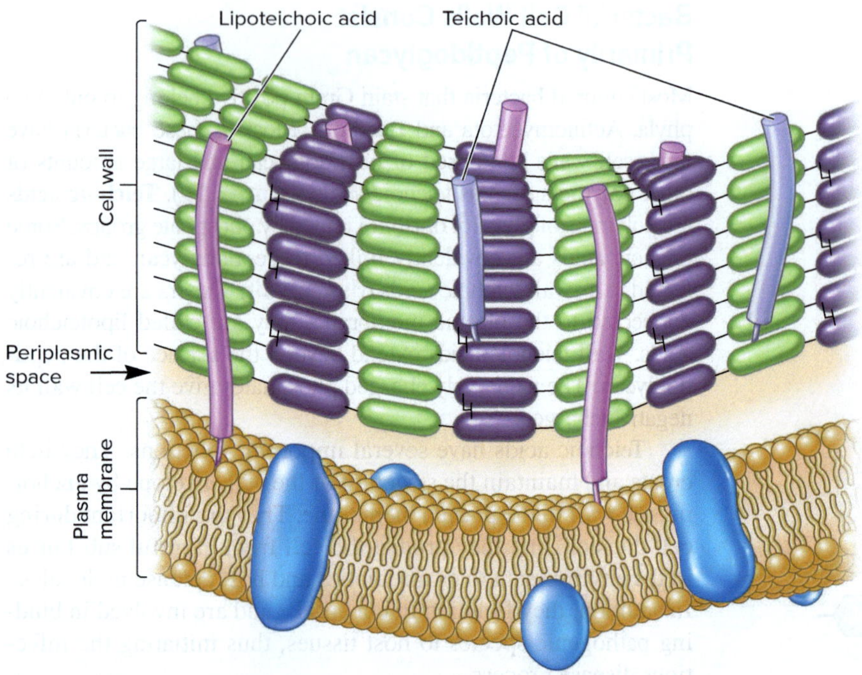

Figure 3.19 Typical Gram-Positive (Monoderm) Cell Envelope.

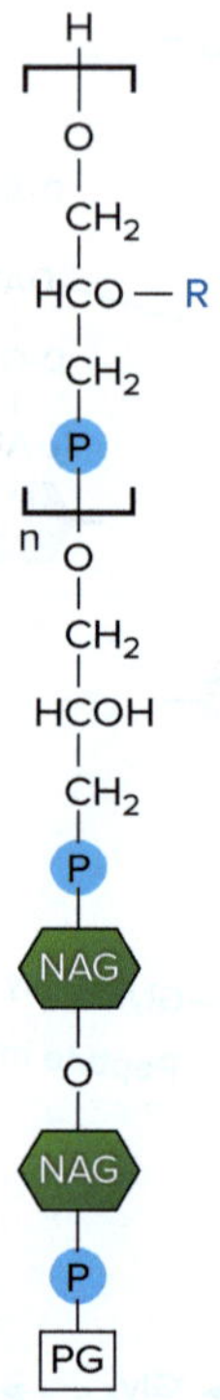

Figure 3.20 Teichoic Acid Structure. Segment of a wall teichoic acid made of phosphate (P), glycerol, and a side chain, R. R may represent D-alanine, glucose, or other molecules. NAG, *N*-acetylglucosamine; PG, peptidoglycan.

membrane, whereas the external layer of the Gram-negative cell is comprised of **lipopolysaccharides (LPS)**. These large, complex molecules contain both lipid and carbohydrate, and consist of three parts: (1) lipid A, (2) the core polysaccharide, and (3) the O antigen (**figure 3.22**). **Lipid A** contains two glucosamine sugar derivatives, each with fatty acids and phosphate attached. The fatty acids of lipid A comprise the hydrophobic component of the outer leaflet, whereas the remainder of the LPS molecule projects from the surface. The **core polysaccharide** is joined to lipid A and is constructed of 10 sugars, many of them unusual in structure. The **O antigen** is a polysaccharide chain extending outward from the core. It has several unusual sugars and varies in composition between bacterial strains.

LPS has many important functions: (1) It contributes to the negative charge on the bacterial surface because the core polysaccharide usually contains charged sugars and phosphate (figure 3.22). (2) It helps stabilize outer membrane structure because lipid A is a major constituent of the exterior leaflet of the outer membrane. (3) It helps create a permeability barrier. (4) LPS helps protect pathogenic bacteria from host defenses. The O antigen of LPS is so named because it elicits an immune response by an infected host. For example, microbiologists refer to specific strains of Gram-negative bacteria using the O antigen, such as *E. coli* O157; here the O antigen is the antigenic type, number 157. Unfortunately, many bacteria can rapidly change the antigenic nature of their O antigens, thus thwarting host defenses. (5) Importantly, the lipid A portion of LPS can act as a toxin and is called *endotoxin*. If LPS or lipid A enters the bloodstream, a form of septic shock develops for which there is no direct treatment. ▶| *Endotoxins (section 35.4); Antibodies bind specific 3-D antigens (section 32.7)*

Like the plasma membrane, the outer membrane is an effective barrier for the cell, and its asymmetry contributes significantly to this effect. Within the LPS, the negatively charged phosphate groups on the core polysaccharide interact with calcium ions to stabilize and tightly pack the LPS molecules. This results in an impermeable barrier that excludes small molecules, including many antibiotics and toxins. In addition, although integral proteins move freely in the plasma membrane, movement in the outer membrane is significantly more constrained.

Transport into a Gram-negative cell is a two-step process. First, solutes must cross the outer membrane. Once in the periplasm, solutes must then move across the plasma membrane. Transport across the outer membrane is facilitated by integral membrane proteins, including one class of proteins termed **porins.** Many porins cluster to form a trimer, three identical polypeptides, that span the outer membrane (**figure 3.23**). A single cell typically has over 100,000 porins. Porins allow the passage of nutrients and hydrophilic molecules smaller than about 600 daltons through the central water-filled channel. The interior of the channel narrows to a segment where the amino acid side chains set up an electric field. This electric field orients water

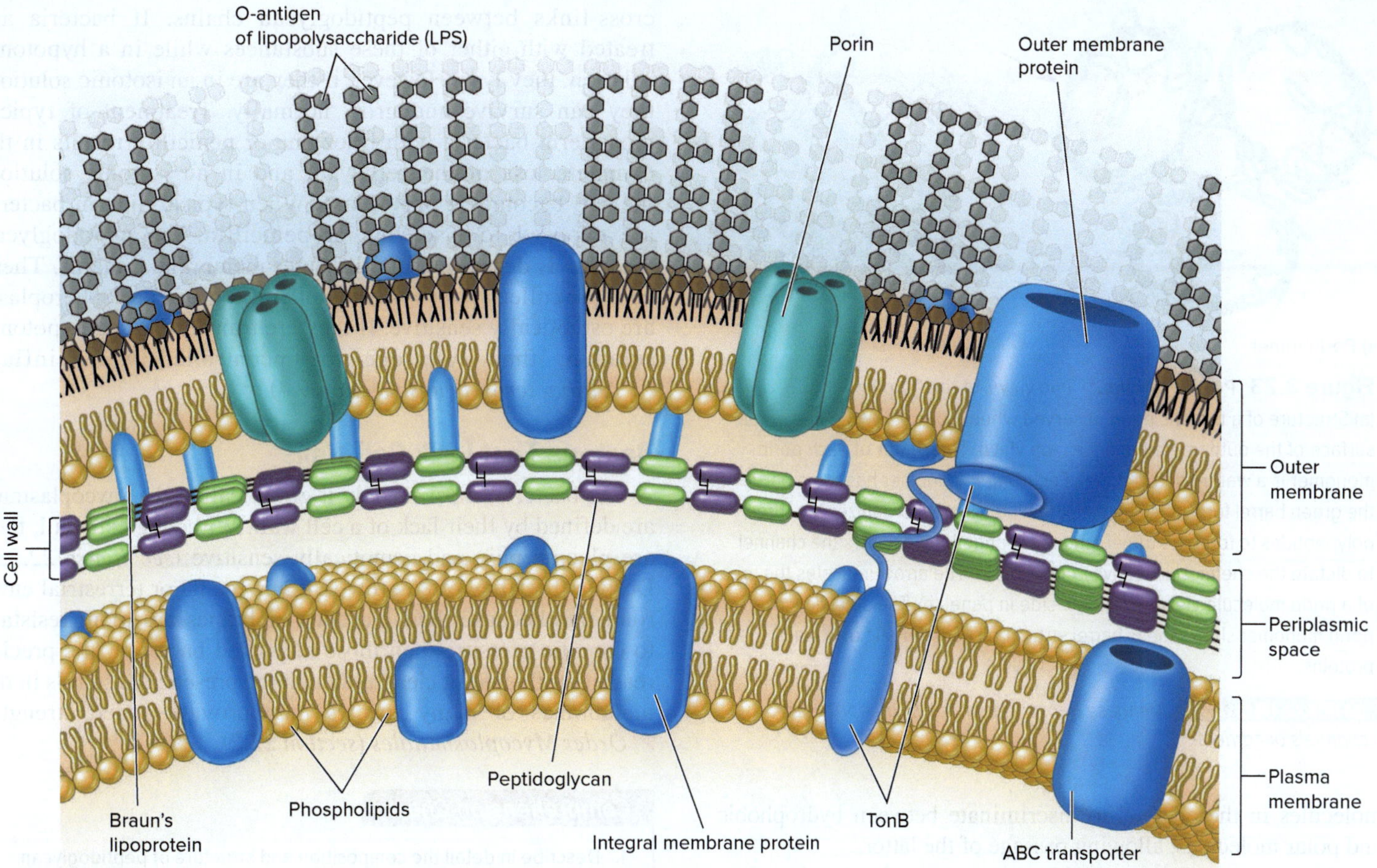

Figure 3.21 Typical Gram-Negative (Diderm) Cell Envelope. These bacteria are bounded by two membranes, the plasma membrane and the outer membrane.

MICRO INQUIRY *How does the outer membrane differ from the plasma membrane?*

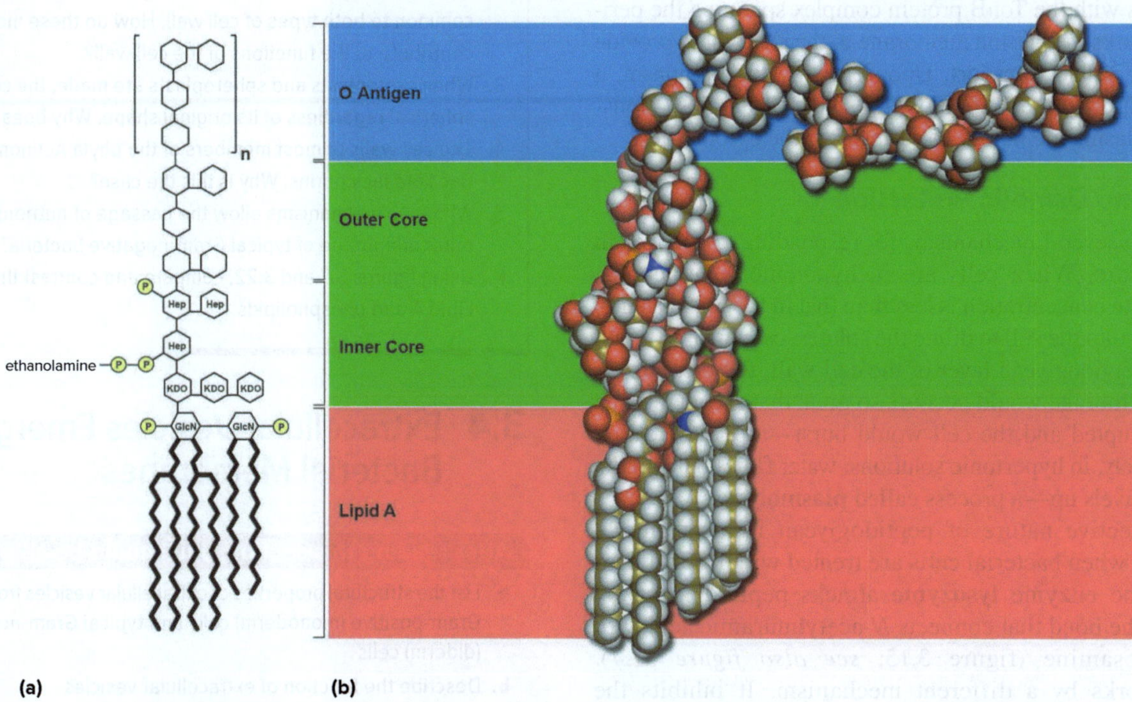

Figure 3.22 Lipopolysaccharide (LPS) Structure. (a) A simplified diagram of LPS. Lipid A is buried in the outer leaflet of the outer membrane. (b) Molecular model of a lipopolysaccharide. The lipid A and core polysaccharide are straight; the O antigen is bent at an angle in this model. GlcN, glucosamine; Hep, heptulose; KDO, 2-keto-3-deoxyoctonate; P, phosphate.

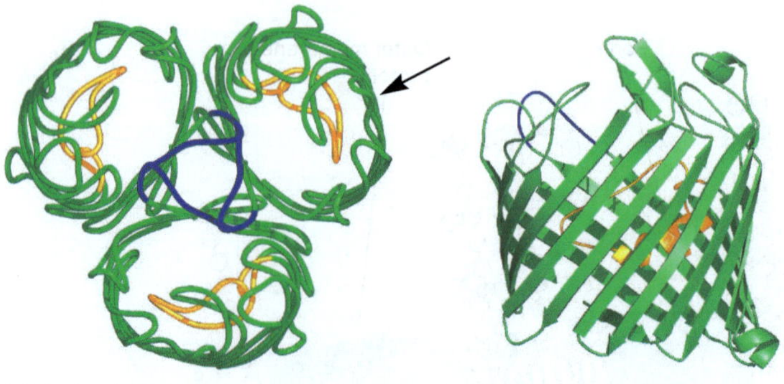

Figure 3.23 Porin Proteins. Two views of the OmpF porin of *E. coli.* (a) Structure of a trimeric porin observed when looking down at the outer surface of the outer membrane (i.e., top view). The center of each porin monomer is a water-filled channel. Each OmpF monomer has three parts: the green barrel forms the channel, the blue segment organizes the three polypeptides to form the trimer, and the orange loop narrows the channel to dictate the chemical selectivity of the porin. The arrow indicates the area of a porin molecule viewed from the side in panel (b). The side view of a porin monomer shows the β-barrel structure characteristic of porin proteins.

MICRO INQUIRY *Are these transporter proteins categorized as channels or carriers?*

molecules in the channel to discriminate between hydrophobic and polar molecules, allowing passage of the latter.

In contrast to the nonspecific transport mechanism based on porins, larger molecules such as vitamin B_{12} and iron-siderophore complexes cross the outer membrane through the action of specific receptors (figure 3.21). A solute-bound receptor in the outer membrane interacts with the TonB protein complex spanning the periplasm. TonB links the plasma membrane to the receptor to provide the energy for solute transport. Upon entering the periplasm, a different transport protein shuttles the solute to an ABC-transporter in the plasma membrane for transfer to the cytoplasm.

Cell Walls and Osmotic Protection

Microbes have several mechanisms for responding to changes in osmotic pressure. When cells are in hypotonic solutions—in which the solute concentration is less than that in the cytoplasm—water diffuses into the cell to dilute the solutes, causing it to swell. Without the peptidoglycan layer of the cell wall, the pressure on the plasma membrane would become so great that the membrane would be disrupted and the cell would burst—a process called **lysis.** Conversely, in hypertonic solutions, water flows out and the cytoplasm shrivels up—a process called **plasmolysis.**

The protective nature of peptidoglycan is most clearly demonstrated when bacterial cells are treated with lysozyme or penicillin. The enzyme **lysozyme** attacks peptidoglycan by hydrolyzing the bond that connects *N*-acetylmuramic acid with *N*-acetylglucosamine (figure 3.15; *see also figure 31.4*). Penicillin works by a different mechanism. It inhibits the enzyme transpeptidase, which is responsible for making the cross-links between peptidoglycan chains. If bacteria are treated with either of these substances while in a hypotonic solution, they lyse. However, if they are in an isotonic solution, they can survive and grow normally. Treatment of typical monoderm bacteria with lysozyme or penicillin results in the complete loss of the cell wall, and in an isotonic solution, the cell becomes a **protoplast.** When typical diderm bacteria are exposed to lysozyme or penicillin, the peptidoglycan sacculus is destroyed, but the outer membrane remains. These cells are called **spheroplasts.** Both protoplasts and spheroplasts are osmotically sensitive. If they are transferred to a hypotonic solution, they lyse due to uncontrolled water influx. ▶| *Antibacterial drugs (section 9.4)*

Bacteria That Lack Cell Walls

A few bacteria, most notably a group called the mycoplasmas, are defined by their lack of a cell wall. Without a cell wall, they are pleomorphic and osmotically sensitive (*see figure 22.23*). Despite this, they often grow in dilute media or terrestrial environments because their plasma membranes are more resistant to osmotic pressure than those of walled bacteria. The precise reason for this is not clear, although the presence of sterols in the membranes of many species may provide added strength. ▶| *Order Mycoplasmatales (section 22.2)*

Comprehension Check

1. Describe in detail the composition and structure of peptidoglycan. Why does peptidoglycan contain the unusual D-isomers of alanine and glutamic acid rather than the L-isomers observed in proteins?
2. Draw a Venn diagram to classify the major molecules that make up typical Gram-positive and Gram-negative cell walls and those common to both types of cell wall. How do these molecules contribute to the functions of the cell wall?
3. When protoplasts and spheroplasts are made, the cell becomes spherical regardless of its original shape. Why does this occur?
4. The cell walls of most members of the phyla Actinomycetota and Bacillota lack porins. Why is this the case?
5. What two mechanisms allow the passage of nutrients across the outer membrane of typical Gram-negative bacteria?
6. Using figures 3.7 and 3.22, compare and contrast the structures of Lipid A and phospholipids.

3.4 Extracellular Vesicles Emerge from Bacterial Membranes

After reading this section, you should be able to:

a. List the structural properties of extracellular vesicles from typical Gram-positive (monoderm) cells and typical Gram-negative (diderm) cells
b. Describe the function of extracellular vesicles

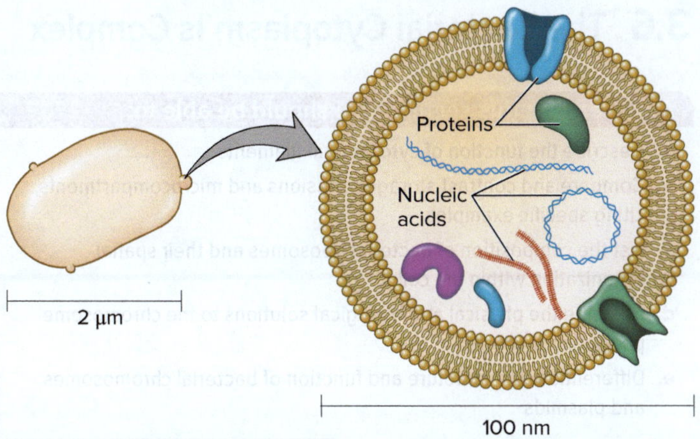

Figure 3.24 Extracellular Vesicle from a Monoderm Cell. Extracellular vesicles are membrane-bound nanoparticles consisting of a sample of the cytoplasm and its contents.

Extracellular vesicles (EVs), also called membrane vesicles, are small membrane-bound particles, about 20–400 nm in size, that have been observed by electron microscopy on a wide variety of bacterial cells (**figure 3.24**). They develop when a membrane buds out, pinches off, and is released from a cell.

EVs formed by a typical monoderm cell are composed of the plasma membrane surrounding a small amount of cytoplasm. Osmotic pressure in monoderm cells appears to force EVs through pores in the peptidoglycan. EVs that form from a typical diderm cell are composed of the lipopolysaccharide-containing outer membrane surrounding a sample of periplasm, and these structures are also referred to as **outer membrane vesicles (OMVs).** The outer membrane needs to dissociate from Braun's lipoprotein before it can grow outward. Other particles originating from diderm bacteria contain both plasma and outer membranes enclosing cytoplasm and periplasm, respectively.

Although they are approximately the size of some small symbiotic cells (figure 3.4*a*), EVs are not cells because they do not reproduce. They may carry some ATP, but they do not have the capacity to conserve energy. Despite these limitations, EVs actively influence a broad range of cell–cell interactions, including the ability to transfer toxin molecules during infection and the ability to transfer genetic material between cells. Because EVs are protected by a lipid bilayer, they persist in the environment longer than naked macromolecules. EVs deliver their contents to eukaryotic cells after membrane fusion or endocytosis (*see figure 5.10*). Similarly, OMVs fuse with the outer membrane of other diderm cells and release the contents to the periplasm.

Vesicle production increases in response to cell stress, and EVs vary in size and cargo with cellular physiology. Although the general phenomenon of EVs is well-documented, the specific molecular details of how they form is currently under debate.

3.5 The Cell Envelope Often Includes Layers Outside the Cell Wall

After reading this section, you should be able to:

a. List the structures found in all the layers of bacterial cell envelopes
b. Identify the functions and major component molecules in cell envelope structures

Figures 3.19 and 3.21 show the bacterial envelope as consisting solely of the plasma membrane and cell wall. However, many bacteria have another layer in their cell envelopes that lies outside the cell wall. This layer is given different names depending on its makeup and how it is organized.

Capsules and Slime Layers

Capsules are well-organized layers that are most often composed of polysaccharides. In typical diderm bacteria, these polysaccharides are covalently bonded to outer membrane lipids, and therefore are not easily washed away. Some capsules are constructed of other materials. For example, *Bacillus anthracis* (the Gram-positive anthrax bacterium) has a proteinaceous capsule composed of poly-D-glutamic acid (**figure 3.25**). Capsules are clearly visible in the light microscope when negative stains or specific capsule stains are employed (*see figure 20.7c*).

Capsules are not required for growth and reproduction in laboratory cultures. However, they confer several advantages when bacteria grow in their normal habitats. They help pathogenic bacteria resist phagocytosis by host phagocytes. *Streptococcus pneumoniae,* which causes ear infections, pneumonia, and other diseases, provides a dramatic example (*see figure 13.1*). In the absence of a capsule, it is phagocytosed easily and does not cause disease. In contrast, the capsulated variant commonly causes disease. Capsules also protect against desiccation because they contain a great deal of water. In addition, they exclude viruses and most hydrophobic toxic materials such as detergents.

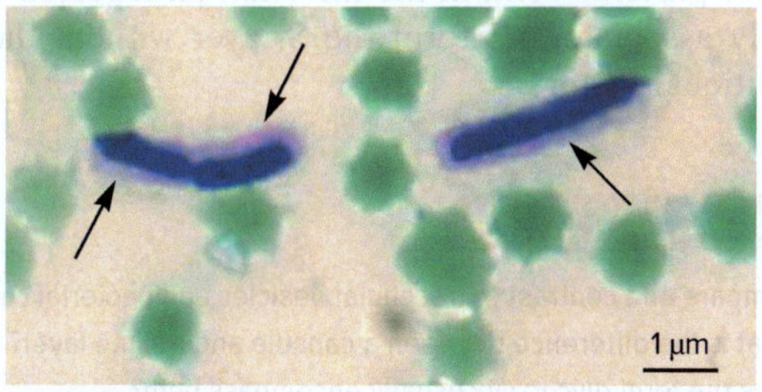

B. anthracis

Figure 3.25 Bacterial Capsule. Capsules on *Bacillus anthracis* (arrows) appear as a pink layer surrounding the purple cells. Bright field light microscopy.
Larry Stauffer/Oregon State Public Health Laboratory/CDC

Figure 3.26 The S-Layer. An electron micrograph of the hexagonal S-layer of the bacterium *Viridilinea mediisalina.*

Dr. Vasil Gaisin, Professor, Martin Pilhofer, Dr. Romain Kooger

A **slime layer** is also a polysaccharide layer, but it differs from a capsule in being more diffuse and unorganized, and more easily removed from the cell. It is not as readily observed by light microscopy. Gliding bacteria often produce slime, which in some cases has been shown to facilitate motility.

The term **glycocalyx** also describes a polysaccharide layer extending from the cell surface. The term encompasses both capsules and slime layers because both are composed of polysaccharides. The glycocalyx aids in attachment to solid surfaces, including tissue surfaces in plant and animal hosts.

S-Layers

Many bacteria have an ordered surface covering called an **S-layer.** The S-layer has a regular geometric pattern something like floor tiles and is composed of protein or glycoprotein (**figure 3.26**). In most bacteria with an S-layer, many copies of a single protein form this two-dimensional crystalline surface. In typical diderm bacteria, the S-layer adheres noncovalently to the outer membrane; it is associated with the peptidoglycan surface of typical monoderm cell walls.

S-layers are of considerable interest not only for their biological roles but also in nanotechnology. Their biological roles include protecting the cell against ion and pH fluctuations, osmotic stress, or enzymes. The S-layer also helps maintain the shape and envelope rigidity of some cells, and it can promote adhesion to surfaces. Finally, the S-layer seems to protect some bacterial pathogens against host defenses, thus contributing to their virulence. The potential use of S-layers in nanotechnology is due to their ability to self-assemble; that is, S-layer proteins contain the information required to spontaneously associate and form the S-layer without the aid of any additional factors.

Comprehension Check

1. Compare and contrast extracellular vesicles and bacterial cells.
2. What is the difference between a capsule and a slime layer? Why does the term *glycocalyx* usually encompass both?
3. S-layers and some capsules are composed of proteins. How does an S-layer differ from a proteinaceous capsule?

3.6 The Bacterial Cytoplasm Is Complex

After reading this section, you should be able to:

a. Describe the function of cytoskeletal filaments
b. Compare and contrast storage inclusions and microcompartments, citing specific examples
c. List the composition of bacterial ribosomes and their spatial organization within the cell
d. Describe the physical and biological solutions to the chromosome packaging problem
e. Differentiate the structure and function of bacterial chromosomes and plasmids

The plasma membrane and everything within is called the **protoplast.** The **cytoplasm** is the material bounded by the plasma membrane; thus the cytoplasm is a major part of the protoplast. The cytoplasm is divided into three components: the liquid **cytosol,** structures that float in the cytosol, and the solutes in the cytosol. The high concentration of these macromolecules and their precursors and metabolites creates a phenomenon known as *macromolecular crowding*. Abundant solutes result in less available space around each molecule, a familiar experience to anyone who travels around a city at rush hour. The bacterial cytoplasm is estimated to be about 10 times more viscous than water, and physical and chemical processes like diffusion are affected.

Despite their small size and apparent simplicity, there are clear patterns of organization within the bacterial cell. Imaging by both electron microscopy and fluorescence microscopy have contributed to the emerging field of bacterial cell biology. In this section, we consider the variety of structures in the cytoplasm.

Bacterial Cytoskeleton

The **cytoskeleton** is composed of several types of protein filaments. Individual cytoskeleton polypeptides polymerize to form functional filaments that often extend to the full inner dimensions of the cell. They play important roles in cell division, maintaining cell shape, and positioning and segregating inclusions and plasmid DNA (**table 3.2**). Some bacterial and archaeal cytoskeleton proteins share common ancestors with actin and tubulin, components of the eukaryotic cytoskeleton. It is therefore likely that the evolution of the cytoskeleton was an early event in the history of life on Earth.

The morphology of rod-shaped bacterial cells is conferred by their rigid cell walls. The cell wall is sculpted by the actin homologue MreB and related proteins, whose filaments associate with the plasma membrane and position the enzymes that synthesize peptidoglycan as the cell grows (**figure 3.27**). In the curved bacterium *Caulobacter crescentus,* the CreS protein localizes to the inner surface of the curve (*see figure 7.8*). CreS is

Table 3.2 Some Bacterial Cytoskeleton Proteins		
Function	**Examples**	**Eukaryotic Homolog**
Cell Morphology. This protein filament positions cell wall synthesis machinery, conferring a rod shape to cells.	MreB	actin
Cell Morphology. Filaments of this protein localize to the membrane on the inner side of a curved cell.	CreS	intermediate filaments
Cell Division. These protein filaments mark the position of the new cell wall after division and constrict to separate dividing cells.	FtsZ	tubulin
	FtsA	actin
DNA Segregation. Filaments of these proteins push plasmid copies toward opposite cell poles during cell division.	ParM, AlfA	actin
DNA Segregation. Filaments of these proteins push bacteriophage genomes toward opposite cell poles during cell division.	TubZ, PhuZ	tubulin
Magnetotaxis. This protein organizes and orients magnetosomes within the cells of magnetotactic bacteria.	MamK	actin

a homologue of eukaryotic intermediate filaments. *Cellular growth and determination of cell shape (section 7.2)*

The tubulin homologue FtsZ is one of the best studied cytoskeletal proteins in bacteria. Its filaments form a ring at the center of a dividing cell that constricts as the daughter cells separate (*see figure 7.6*). The Z-ring is a structural platform for the enzymes that synthesize the new cell wall. *Chromosome replication and partitioning (section 7.2)*

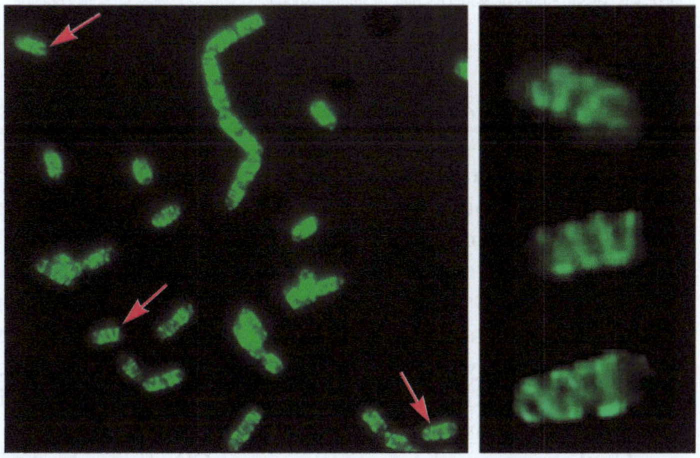

(a) Mbl (b) Mbl

Figure 3.27 The Bacterial Cytoskeleton. (a) The MreB-like cytoskeletal protein (Mbl) of *Bacillus subtilis;* Mbl-GFP in live cells was examined by fluorescence microscopy. With this method, Mbl seems to form helices (arrows). More sensitive methods show that Mbl forms patches that move perpendicular to the long axis of the cell. (b) Three of the cells from (a) are shown at a higher magnification.

Centre for Bacterial Cell Biology/Newcastle University/Jeff Errington

Intracytoplasmic Membranes

Although bacteria do not contain complex membranous organelles like mitochondria or chloroplasts, internal membranous structures are observed in some cells (**figure 3.28**). These can be extensive and

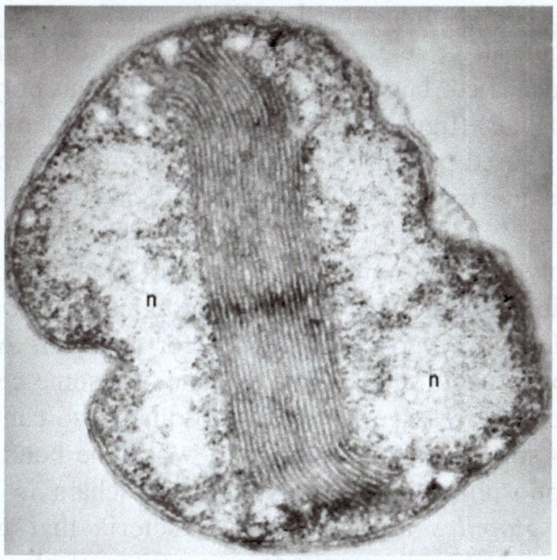

Figure 3.28 Internal Bacterial Membranes. The nitrifying bacterium *Nitrosococcus oceani* has parallel membranes traversing the whole cell. Note the nucleoid (n).

Dr. Robert G.E. Murray

complex in bacteria with high respiratory activity, such as nitrifying bacteria and photosynthetic bacteria. The internal membranes of the photosynthetic cyanobacteria are called thylakoids and are analogous to the thylakoids of chloroplasts. ▶| *Mitochondria, related organelles, and chloroplasts are involved in energy conservation (section 5.6); Photosynthetic bacteria are diverse (section 20.3); Nitrifying bacteria oxidize ammonium or nitrite to gain energy and electrons (section 21.1)*

The internal membranous structures observed in bacteria may be aggregates of spherical vesicles, flattened vesicles, or tubular membranes (*see figure 21.1*). They are often connected to the plasma membrane and arise from it by invagination. These internal membranes differ from the plasma membrane by being enriched for molecules involved in energy conservation. For instance, the thylakoids of cyanobacteria contain the chlorophyll and photosynthetic reaction centers responsible for converting light energy into ATP. Thus the function of internal membranes is to provide a larger membrane surface for greater metabolic activity. Regardless of their function, all membrane structures are driven by the same hydrophilic and hydrophobic interactions as described in section 3.2.

Inclusions

Inclusions are common in all cells. They are formed by the aggregation of substances that may be either organic or inorganic. The primary function of inclusions is to segregate certain cellular components so they do not diffuse freely in the cytoplasm. They function as storage sites, locations to sequester enzymatic reactions, or guides for cell movement. Inclusions can take the form of granules, crystals, or globules; some are amorphous. The most common storage inclusions are glycogen inclusions, polyhydroxyalkanoate (PHA) granules, sulfur globules, and polyphosphate granules.

Carbon is often stored as PHA granules, also termed carbonosomes (**figure 3.29*a***). Several types of PHA granules have been identified, but the most common contain polyhydroxybutyrate (PHB) (figure 3.29*b*). PHA granules are surrounded by a single-layered protein shell. Much of the interest in PHA granules is due to their industrial use in making biodegradable plastics.

Polyphosphate granules and sulfur globules are inorganic inclusions observed in many microorganisms. *Polyphosphate granules* store phosphate needed for synthesis of important cell constituents such as nucleic acids. In some cells, they act as an energy reserve, and polyphosphate also can serve as an energy source in some reactions, when the bond linking the final phosphate in the polyphosphate chain is hydrolyzed. *Sulfur globules* are formed by bacteria that use reduced sulfur-containing compounds as a source of electrons during their energy-conserving metabolic processes (figure 3.29*c*). For example, some photosynthetic bacteria use hydrogen sulfide (rather than water) as an electron donor and accumulate the resulting sulfur either externally or internally. ▶| *Light reactions in anoxygenic photosynthesis (section 11.10); Order Chromatiales: purple sulfur bacteria perform anoxygenic photosynthesis (section 21.2)*

Some bacterial inclusions called *microcompartments* are relatively large polyhedra. Because bacteria lack membrane-bound organelles, microcompartments are often bounded by a protein shell (**Microbial Diversity & Ecology 3.1**). Enclosed within the protein shell are one or more enzymes. These enzymes are sequestered because of the nature of the chemical reaction or pathway that they catalyze. In some cases, the reaction produces a toxic intermediate, and in others the reactant is a gas that could easily pass through a lipid membrane.

Carboxysomes are present in many CO_2-fixing bacteria and are about 100 nm in diameter. Two enzymes critical for CO_2 fixation, carbonic anhydrase (CA) and ribulose 1,5-bisphosphate carboxylase/oxygenase (RuBisCO), are located in carboxysomes. CA releases CO_2 into the protein shell so that it accumulates to a high concentration, where RuBisCO catalyzes the conversion of CO_2 to sugar. ▶| *CO_2 fixation: reduction and assimilation of CO_2 carbon (section 12.3); Phylum cyanobacteria: oxygenic photosynthetic bacteria (section 20.3)*

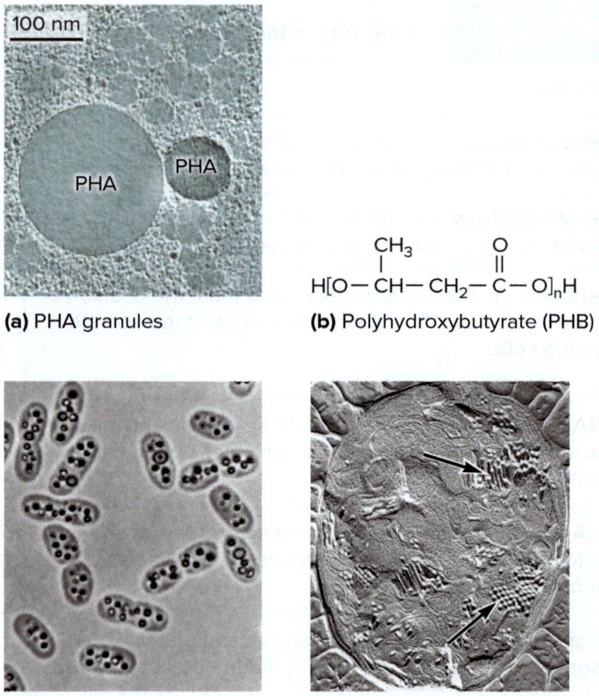

(a) PHA granules (b) Polyhydroxybutyrate (PHB)

(c) Sulfur globules (d) Gas vacuoles

Figure 3.29 Bacterial Cell Inclusions. (a) Spherical PHA granules in the cytoplasm of Chloroflexota. (b) Chemical structure of polyhydroxybutyrate. (c) *Allochromatium vinosum,* a purple sulfur bacterium, with intracellular sulfur globules. (d) A freeze-fracture preparation of *Dolichospermum flosaquae* showing gas vesicles and gas vacuoles. Both longitudinal and cross-section views of gas vesicles are indicated by arrows.

(a) Dr. Vasil Gaisin, Professor, Martin Pilhofer, Dr. Romain Kooger; (c) James T. Staley; (d) Daniel Branton

MICROBIAL DIVERSITY & ECOLOGY

3.1 Organelles Without Membranes?

One of the defining characteristics of eukaryotic cells is the presence of membrane-bound organelles. As selectively permeable barriers, membranes limit the movement of many types of molecules. In bacterial cells that lack membrane-bound organelles, some cytoplasmic inclusions use a novel boundary mechanism. ▶| *Cell organization (section 5.1)*

Cyanobacteria and other CO_2-fixing organisms segregate the critical enzyme RuBisCO in carboxysomes. Electron micrographs show carboxysomes as discrete bodies within the cytoplasm, lacking a membrane (**box figure**). Biochemically, they are entirely protein. So how do carboxysomes assemble around active enzymes?

RuBisCO is joined in the carboxysome by scaffold proteins that assemble by a mechanism termed **liquid-liquid phase separation (LLPS).** Both RuBisCO subunits and scaffold subunits have regions that are structurally disordered; that is, their amino acid chains are not fixed in any of the typical protein configurations. These disordered protein regions interact and aggregate. It is difficult to imagine how this is possible, so think of two Lego building blocks with random snaps that magically align when they are brought close together. In a similar fashion (but without magic), RuBisCO subunits and scaffold subunits assemble the carboxysome polyhedron. ▶| *Proteins (Appendix I)*

LLPS is increasingly recognized as a means for organizing protein complexes within bacterial cells. The complexes are known as **biomolecular condensates,** and while they are a compartment distinct from the cytoplasm, they are separated not by a membrane, but by the biophysical properties of the molecules at high concentration. Notably, the components of a condensate are dynamic, moving between the condensed phase of the inclusion and the cytoplasmic diffuse phase.

In addition to carboxysomes, other biomolecular condensates may organize the cell poles in asymmetrically dividing cells like *Caulobacter (see figure 21.6)*, or regulate the septation machinery in cytokinesis *(see figure 7.6)*.

Carboxysomes in the bacterium *Halothiobacillus neapolitanus*.
Michael Schmid

Two inclusions, gas vacuoles and magnetosomes, are involved in bacterial movement. The **gas vacuole** provides buoyancy to some aquatic bacteria, many of which are photosynthetic. Gas vacuoles are aggregates of enormous numbers of small, hollow, cylindrical structures called **gas vesicles** (figure 3.29*d*). Gas vesicle walls are composed of many copies of a single small protein. These protein subunits assemble to form a rigid cylinder that is impermeable to water but freely permeable to atmospheric gases. Cells with gas vacuoles regulate their buoyancy to float at the depth necessary for proper light intensity, oxygen concentration, and nutrient levels. They descend by simply collapsing vesicles and float upward when new ones are constructed.

Aquatic magnetotactic bacteria use **magnetosomes** to orient themselves in Earth's magnetic field and swim to nutrient-rich sediments or locate the optimum depth in freshwater and marine habitats. Magnetosomes are intracellular chains of magnetite (Fe_3O_4) or greigite (Fe_3S_4) particles (**figure 3.30**). They are around 35 to 125 nm in diameter and enclosed within invaginations of the plasma membrane. The invaginations contain distinctive proteins and thus constitute functional

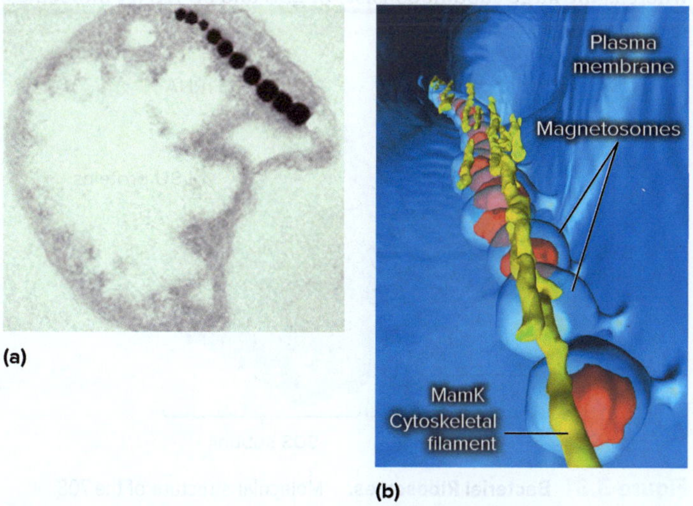

Figure 3.30 Magnetosomes. (a) *Magnetococcus marinus* cells are roughly spherical with chains of magnetite crystals. (b) Three-dimensional reconstruction of the magnetosomes of *Phaeospirillum magneticum*.
(a) Dennis Bazylinski; (b) Grant J. Jensen

membrane microdomains. For the cell to move properly within a magnetic field, magnetosomes must be arranged in a chain. The cytoskeletal protein MamK establishes a framework upon which the chain can form (table 3.2).

Bacterial Ribosomes

Ribosomes are the site of protein synthesis, and large numbers (10,000 to 20,000) are found in all cells. The cytoplasm of rapidly growing bacterial cells is often packed with ribosomes, and additional ribosomes may be loosely attached to the plasma membrane. The cytoplasmic ribosomes synthesize proteins destined to remain within the cell, whereas plasma membrane-associated ribosomes make proteins that will reside in the cell envelope or be transported to the outside.

Translation, the process of protein synthesis, is amazingly complex and is discussed in detail in chapter 13. This complexity is evidenced in part by the structure of ribosomes, which are made of dozens of proteins and several ribonucleic acid (RNA) molecules. The ribosomes of all three domains of life share similarities. However, there are important differences. Here we focus on bacterial ribosomes. We compare bacterial and archaeal ribosomes in chapter 4 and compare all three types of ribosomes in chapter 5.

Bacterial ribosomes are called 70S ribosomes and are constructed of a 50S and a 30S subunit (**figure 3.31**). The S in these values stands for **Svedberg unit.** This is the unit of the sedimentation coefficient, a measure of sedimentation velocity in a centrifuge; the faster a particle travels when centrifuged, the greater its Svedberg value. The sedimentation coefficient is a function of a particle's molecular weight, volume, and shape. Heavier and more compact particles normally have larger Svedberg numbers.

Bacterial ribosomes are composed primarily of ribosomal RNA (rRNA) molecules. The small subunit contains 16S rRNA, whereas the large subunit consists of 23S and 5S rRNA molecules.

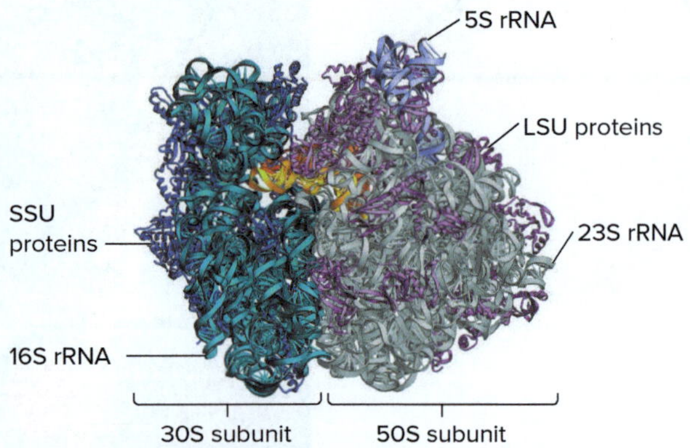

Figure 3.31 Bacterial Ribosomes. Molecular structure of the 70S ribosome of *Thermus thermophilus.* The 50S subunit (LSU) includes 23S rRNA (gray) and 5S rRNA (lavender), while 16S rRNA (turquoise) is found in the 30S subunit (SSU). A molecule of tRNA (gold) is shown.

Approximately 55 proteins make up the rest of the mass of the ribosome: 21 in the small subunit and 34 in the large subunit.

Nucleoid

The **nucleoid** is an ellipsoidal region that contains the cell's chromosome and numerous proteins and occupies about 20% of the cell volume (**figure 3.32*a***). Although the nucleoid is not separated from the cytoplasm by a membrane, it is a distinct region of the cell, as observed by microscopy. Ribosomes are excluded from the nucleoid.

The chromosomes of most bacteria are a circle of double-stranded **deoxyribonucleic acid (DNA)**, but a few bacteria have a linear chromosome. Many bacteria carry a single copy of their chromosome (monoploid), but others are polyploid, often with more than 10 copies per cell.

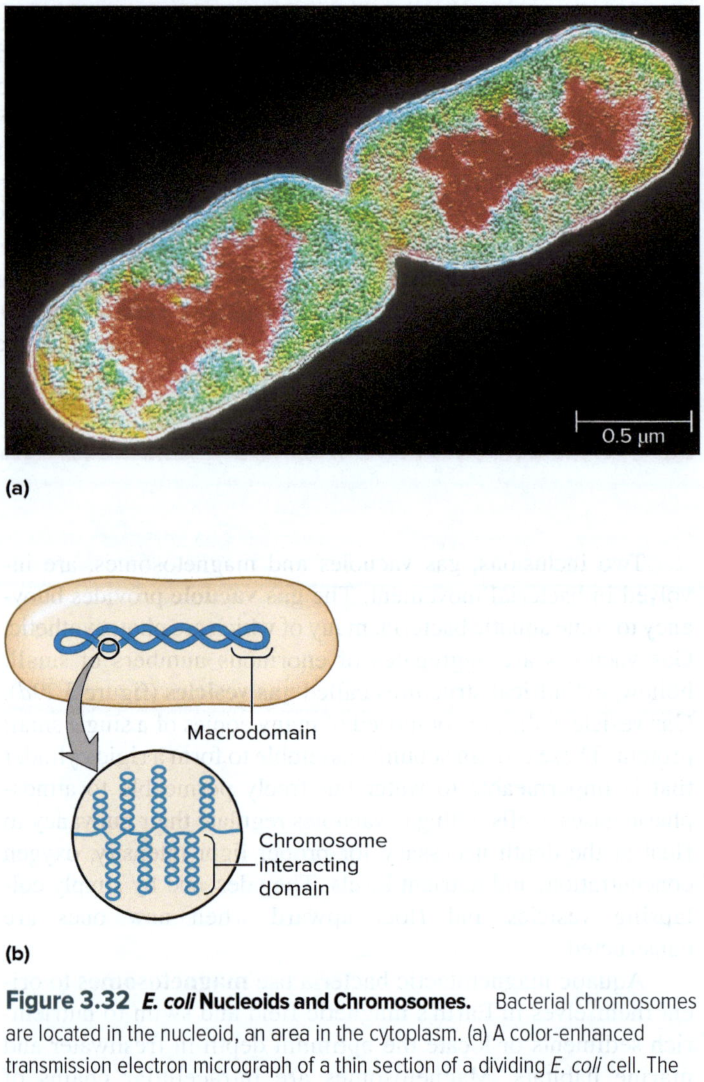

Figure 3.32 *E. coli* Nucleoids and Chromosomes. Bacterial chromosomes are located in the nucleoid, an area in the cytoplasm. (a) A color-enhanced transmission electron micrograph of a thin section of a dividing *E. coli* cell. The red areas are the nucleoids present in the two daughter cells. (b) Diagram illustrating the levels of chromosome organization within the nucleoid.

(a) CNRI/Science Source

Bacterial chromosomes are typically 0.5 to 10 million base pairs (0.5–10 Mbp), and this represents between 500 and 10,000 genes. Bacterial chromosomes are longer than the cell itself. Therefore, an important question is how bacteria fit their chromosomes into the relatively small space occupied by the nucleoid. For instance, *E. coli*'s circular chromosome of about 4.6 Mbp measures approximately 1,400 μm, or about 230–700 times longer than the cell. Thus the chromosome must be organized and compacted in a manner that decreases its overall size yet retains easy accessibility to the genetic information. ▶| *DNA is a polymer of deoxyribonucleotides (section 13.2)*

Physical factors like macromolecular crowding and supercoiling contribute to nucleoid organization. The chromosome itself has an orientation and several levels of structure within the nucleoid. At the grossest level, chromosomes have a few macrodomains of about 1 Mbp, and these are partitioned to a few dozen **chromosome interaction domains** (figure 3.32*b*). On a still smaller scale, a few thousand bp comprise a topological domain constrained by supercoiling.

Architectural proteins also contribute to nucleoid structure. Nucleoid-associated proteins (NAPs) are small, abundant DNA binding proteins that cause the chromosome to bend and fold (**figure 3.33**). NAPs do not form covalent bonds with DNA, but rather bind by electrostatic interaction. NAPs have a track of positively charged amino acids on their surface that aligns with the negatively charged DNA backbone. This type of interaction allows the proteins to move on and off the DNA as the DNA engages in its other cellular roles. All cells contain multiple different NAPs that may act at different scales of chromosome structure. On short segments of DNA, certain NAPs bend or stiffen the chromosome (figure 3.33*a,b,c*); on a global level, other NAPs bridge nonadjacent regions of the chromosome (figure 3.33*d*). NAPs are particularly important during cell division, when they further compact the chromosome. This extra level of packing is important for proper segregation of daughter chromosomes during cell division.

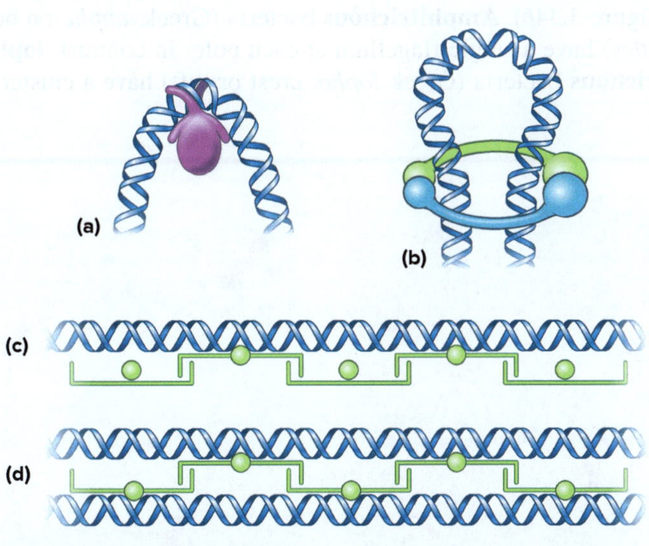

Figure 3.33 Selected nucleoid-associated proteins from *E. coli* and their interaction with DNA. (a) HU has two arms extending from its globular core to contact DNA in multiple places. HU binding introduces a sharp bend in the DNA. (b) SMC encircles DNA and extrudes a loop. (c,d) H-NS stiffens DNA when bound to one duplex, or can link two duplexes.

Plasmids

In addition to the genetic material present in the nucleoid, many bacteria contain additional small DNA molecules called plasmids. The genome of an organism includes all DNA, including plasmids. **Plasmids** are double-stranded DNA molecules that can exist independently of the chromosome. Indeed, most bacterial genomes sequenced thus far include plasmids. In some cases, numerous different plasmids within a single species have been identified. For instance, *Borreliella burgdorferi* carries 12 linear and 9 circular plasmids. Plasmids play many important roles in the lives of the organisms that have them. They also have proved invaluable to microbiologists in constructing and transferring new genetic combinations and in cloning genes, as described in chapter 17.

Both circular and linear plasmids have been documented, but most known plasmids are circular. Plasmids have relatively few genes, generally less than 30. Many plasmids carry genes that confer a selective advantage to the bacterium in certain environments. Plasmids can differ in terms of the number of copies found within the cell.

Plasmids use the cell's DNA-synthesizing machinery to replicate, but their replication is not linked to any particular stage of the cell cycle. Thus both plasmid and chromosomal replication are regulated independently. However, some plasmids are able to integrate into the chromosome. Such plasmids are called **episomes** and when integrated are replicated as part of the chromosome. Plasmids are inherited stably during cell division, but some require a partitioning system involving cytoskeletal elements to ensure they are equally apportioned into daughter cells and not lost (table 3.2).

Comprehension Check

1. Briefly describe the nature and function of the cytoplasm, and the regions and structures within it. How is the cytosol different from the cytoplasm?
2. Compare and contrast the most common kinds of bacterial inclusions with eukaryotic organelles.
3. How do plasmids differ from chromosomes? What is an episome?

3.7 External Structures Are Used for Attachment and Motility

After reading this section, you should be able to:

a. Distinguish pili (fimbriae) and flagella
b. Illustrate the various patterns of flagella distribution
c. Describe the assembly of pili and flagella

Bacteria must constantly respond to their changing environments. Sometimes it is advantageous to attach to a surface. Other times it is better to move toward or away from something in the environment. Many bacteria have appendages that are involved in either attachment to surfaces or motility. In addition, these appendages can function in protection and horizontal gene transfer. Several are discussed in this section.

Bacterial Pili and Fimbriae

Many bacteria have fine, hairlike appendages that are thinner and typically shorter than flagella. These are called **fimbriae** (s., **fimbria**) or **pili** (s., **pilus**). Certain structures are historically called pili (e.g., sex pili), while others are called fimbriae. A cell may be covered with up to 1,000 pili, but they are only visible in an electron microscope due to their small size (**figure 3.34*a***). They are slender tubes composed of helically arranged protein subunits and are about 3 to 10 nm in diameter and up to several micrometers long. Pili grow by adding protein subunits to their base (*see figure 20.13*). Several different types have been identified. Those termed fimbriae function to attach cells to solid surfaces such as rocks in streams and host tissues. **Type IV pili** are involved in motility and two gene transfer mechanisms: transformation and conjugation. ▶| *Conjugation requires cell cell contact (section 16.5); Transformation is the uptake of free DNA (section 16.6)*

Many bacteria have up to 10 **sex pili** per cell. These structures differ from other pili in the following ways. Sex pili often are larger than other pili (around 9 to 10 nm in diameter). They are genetically encoded on plasmids and are required for conjugation. Some bacterial viruses attach specifically to sex pili at the start of their infection cycle.

Bacterial Flagella

Many motile bacteria move by means of **flagella** (s., **flagellum**), threadlike locomotor appendages extending outward from the plasma membrane and cell wall. Although the main function of flagella is motility, they can be involved in attachment to surfaces, and in some bacteria, they are virulence factors; that is, they contribute to the ability of the bacterium to cause disease.

Bacterial flagella are slender, rigid structures about 20 nm across and up to 15 μm long. Flagella are so thin they cannot be observed directly with a light microscope but must be stained with techniques designed to increase their thickness. The detailed structure of a flagellum can only be seen in the electron microscope. |◀ *Differential staining (section 2.3); Electron microscopes use beams of electrons to create highly magnified images (section 2.4)*

Bacterial species often differ in their patterns of flagella distribution, and these patterns are useful in identifying bacteria. **Monotrichous** bacteria (Greek *trikhos,* hair) have one flagellum; if it is located at an end, it is said to be a **polar flagellum** (figure 3.34*b*). **Amphitrichous** bacteria (Greek *amphi,* on both sides) have a single flagellum at each pole. In contrast, **lophotrichous** bacteria (Greek *lopho,* crest or tuft) have a cluster of

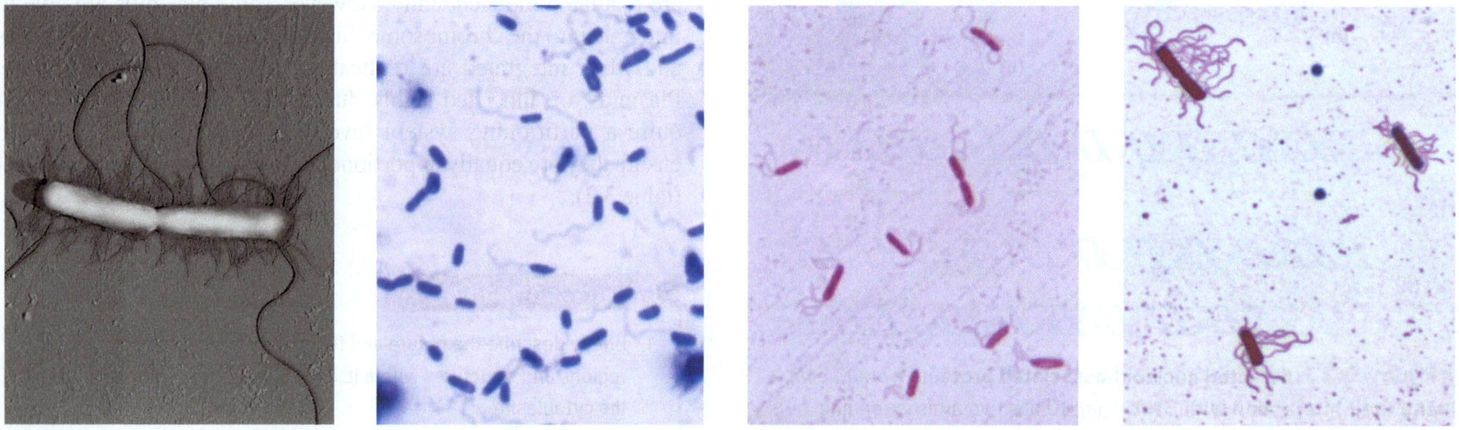

(a) (b) (c) (d)

Figure 3.34 Flagellar Distribution. Examples of various patterns of flagellation as seen in the light microscope. (a) The long flagella and numerous shorter pili are evident in this SEM of *Proteus vulgaris.* (b) *Pseudomonas aeruginosa*—monotrichous polar flagellation. (c) *Comamonas terrigena*—lophotrichous flagellation. (d) *Bacillus cereus*—peritrichous flagellation.

(a)Thomas Deerinck,NCMIR/Science Source; (b) Lisa Burgess/McGraw Hill; (c) R. E. Weaver, MD, PhD/CDC; (d) Dr. William A. Clark/CDC flagella at one or both ends (figure 3.34c). Flagella are spread evenly over the whole surface of **peritrichous** (Greek *peri,* around) bacteria (figure 3.34d).

Transmission electron microscope studies have shown that the bacterial flagellum is composed of three parts (**figure 3.35**). (1) The *filament* is the longest and most obvious portion. It extends from the cell surface to the tip. (2) The *basal body* is embedded in the cell envelope; and (3) a short, curved segment, the *hook,* links the filament to its basal body and acts as a flexible coupling.

The filament is a hollow, rigid cylinder constructed of subunits of the protein **flagellin,** which ranges in molecular mass from 30,000 to 60,000 daltons, depending on the species. The filament ends with a capping protein. Some bacteria have sheaths surrounding their flagella. For example, *Vibrio cholerae* flagella have sheaths that are an extension of the outer membrane.

The hook and basal body are quite different from the filament (figure 3.35). Slightly wider than the filament, the hook is made of different protein subunits. The basal body is the most complex part of a flagellum. The basal bodies of most typical diderm bacteria have four rings: L, P, MS, and C, which are connected to a central rod (figure 3.35*a*). The L, P, and MS rings are embedded in the *l*ipopolysaccharide, *p*eptidoglycan, and plasma *m*embrane, respectively, and the C ring is on the *c*ytoplasmic side of the MS ring. Typical monoderm bacteria have only two rings: an inner ring connected to the plasma membrane and an outer one probably attached to the peptidoglycan (figure 3.35*b*).

Bacterial flagella assembly is complex and involves more than 20 different proteins. Besides flagellin, there are 10 or more hook and basal body proteins; other proteins control flagellar construction or function. Many components of the flagellum lie outside the cell envelope and must be transported out of the cell for assembly. Individual flagellin subunits are transported through the hollow filament. When the subunits reach the tip, they spontaneously aggregate under the direction of a protein called the filament cap; thus the filament grows at its tip rather than at the base (**figure 3.36**). Filament synthesis, like S-layer formation, is an example of self-assembly that occurs without the aid of enzymes or other factors.

Comprehension Check

1. What are the functions of fimbriae and sex pili?
2. What terms are used to describe the different flagella distribution patterns observed among bacteria?
3. What is self-assembly? Why does it make sense that the filament of a flagellum is assembled in this way?

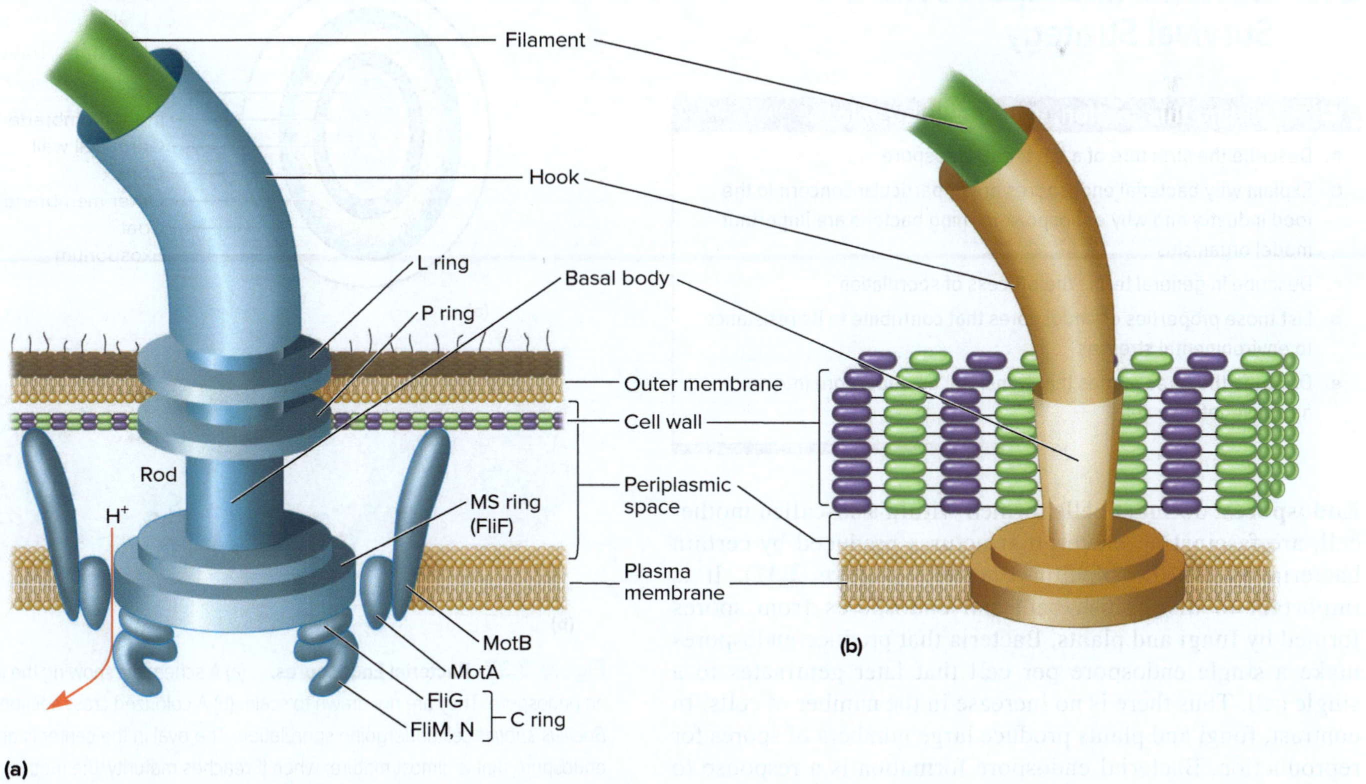

Figure 3.35 The Ultrastructure of Bacterial Flagella. (a) This diagram of a typical diderm flagellum shows some of the more important protein components of the basal body, and the arrow indicates the flow of protons that drives rotation. (b) The basal body, filament, and hook components of a typical monoderm flagellum.

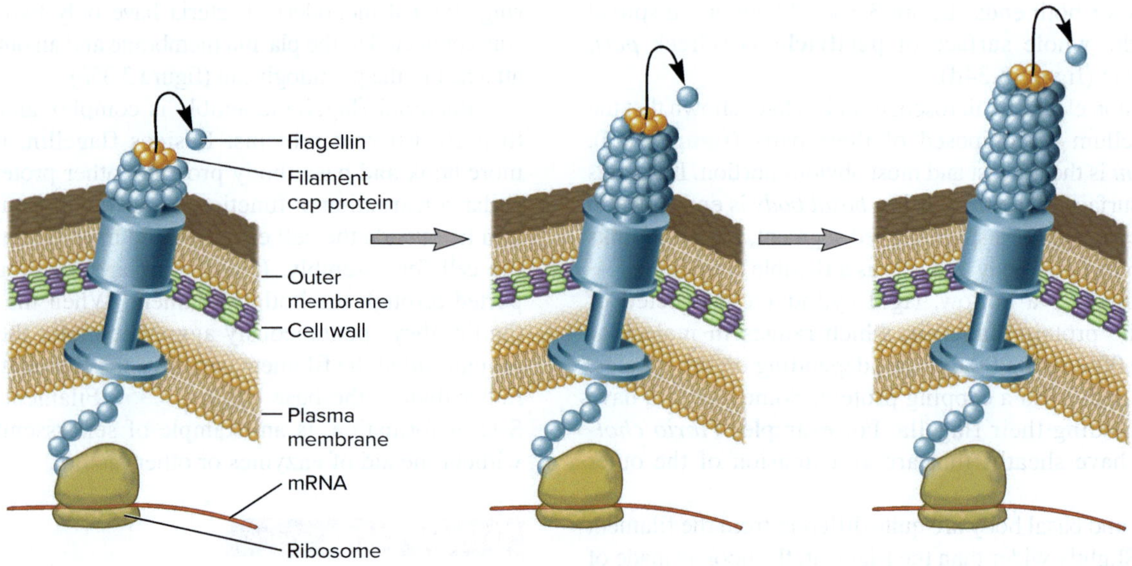

Figure 3.36 Flagellar Filaments Grow at the Tip. Flagellin subunits travel through the hollow flagellum and attach to the growing tip. Their attachment is directed by the filament cap protein.

MICRO INQUIRY *How does flagellum growth compare to pilus growth?*

3.8 Bacterial Endospores Are a Survival Strategy

After reading this section, you should be able to:

a. Describe the structure of a bacterial endospore
b. Explain why bacterial endospores are of particular concern to the food industry and why endospore-forming bacteria are important model organisms
c. Describe in general terms the process of sporulation
d. List those properties of endospores that contribute to its resistance to environmental stresses
e. Describe the three stages that transform an endospore into an active vegetative cell

Endospores, dormant cells formed within a so-called mother cell, are fascinating bacterial structures produced by certain bacteria within the phylum Bacillota (**figure 3.37**). It is important to distinguish bacterial endospores from spores formed by fungi and plants. Bacteria that produce endospores make a single endospore per cell that later geminates to a single cell. Thus there is no increase in the number of cells. In contrast, fungi and plants produce large numbers of spores for reproduction. Bacterial endospore formation is a response to nutrient depletion, and the endospores represent a dormant stage of the organism.

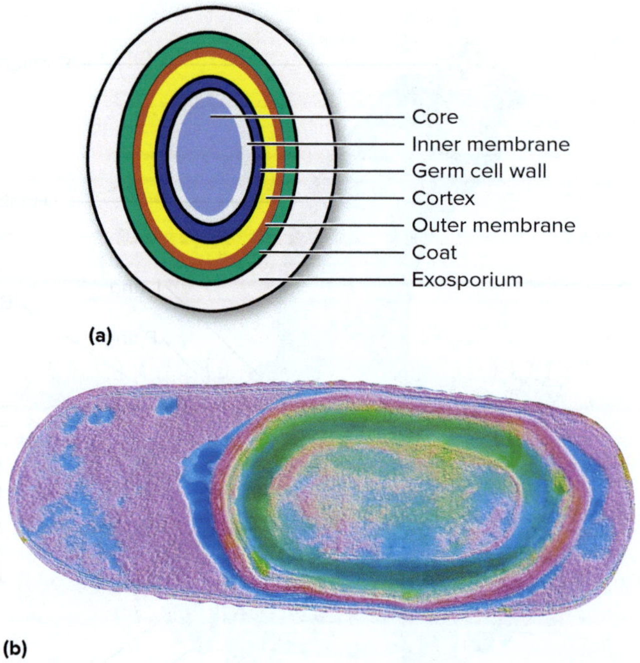

Figure 3.37 Bacterial Endospores. (a) A schematic showing the layers of an endospore. They are not drawn to scale. (b) A colorized cross section of a *Bacillus subtilis* cell undergoing sporulation. The oval in the center is an endospore that is almost mature; when it reaches maturity, the mother cell will lyse to release it.
CNRI/Science Source

There are several reasons why endospores have long held the interest of microbiologists. The primary reason is that endospores are extraordinarily resistant to environmental stresses that damage living cells. Endospores can withstand greater exposure to heat, ultraviolet radiation, gamma radiation, chemical disinfectants, and desiccation. In fact, some endospores have remained viable for around 250 million years. Another important reason is that several species of endospore-forming bacteria are dangerous pathogens. For example, *Clostridium botulinum* causes botulism, a food-borne disease that results from ingestion of botulinum toxin, the deadliest toxin known. The extreme heat resistance of *C. botulinum* endospores is a major concern of the food industry. Other medically important endospore-forming bacteria include *Bacillus anthracis* (causes anthrax), *C. tetani* (causes tetanus), *C. perfringens* (causes gas gangrene and food poisoning), and *Clostridioides difficile* (causes colitis). Endospores also are of considerable interest to scientists because of their complex structures and the equally complex process that generates them. We discuss endospore structure here, but save the consideration of regulating endospore formation for chapter 14. ▶| *Sporulation in* B. subtilis *(section 14.6)*

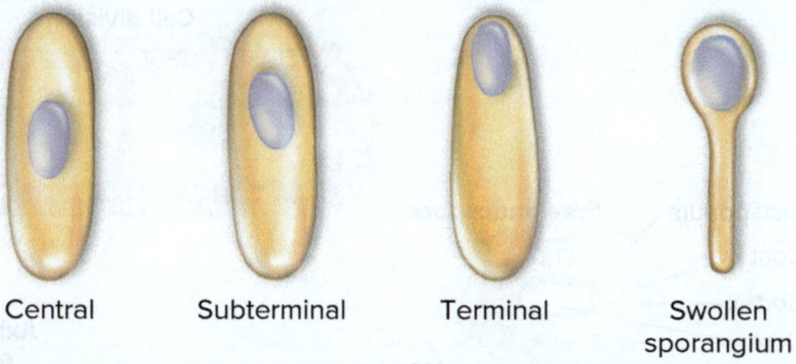

Figure 3.38 **Examples of Endospore Location and Size.**

Sporulation: Making Endospores

Endospore-forming bacteria are common in soil, where they must be able to withstand fluctuating nutrient levels. **Sporulation** commences when growth slows due to nutrient limitation. It is thus a survival mechanism that allows the bacterium to produce a dormant cell that can persist until nutrients are available and vegetative growth can resume. These bacteria cycle between two states: vegetative growth and survival as an endospore. Vegetative growth is the normal, continuous cycle of growth and cell division. By contrast, sporulation is a complex process that occurs in a highly organized fashion over several hours. The mature endospore occupies a characteristic location in the mother cell (referred to as the sporangium), depending on the species of bacteria. Endospores may be centrally located, close to one end (subterminal), or terminal (**figure 3.38**). Sometimes an endospore is so large that it swells the sporangium. One of the best-studied endospore formers is *Bacillus subtilis,* which is an important model organism.

Electron microscopy has been an important tool for dissecting endospore structure. Electron micrographs show that the endospore consists of a core surrounded by several distinct layers (figure 3.37*b*). The core has normal cell structures such as ribosomes and a nucleoid but a low water content.

To describe the endospore structure in detail, let's follow the process of its formation. Bacteria that form endospores have typical monoderm cell walls: a plasma membrane covered by a thick peptidoglycan (PG) layer. The shift from vegetative growth to endospore formation involves two side-by-side cells transitioning to one engulfing the other. The folding and arrangement of the cell envelopes in these two cells give the endospore its multiple layers.

Sporulation is divided into seven stages in *Bacillus subtilis* (**figure 3.39**). In stage I, the chromosome replicates, followed by cell division. The division septum is set down away from the center of the cell, resulting in two unequal cells (stage II). The smaller is termed the *forespore* and will become the endospore. The larger cell supports endospore construction and is termed the *mother cell* or *sporangium.* The endospore core begins as the forespore cytoplasm and its plasma membrane is termed the *inner membrane.* The endospore core, therefore, contains the chromosome and cytoplasmic contents, including ribosomes, proteins, and small molecules. The core also accumulates a complex of *ca*lcium and *dip*icolinic *a*cid (Ca-DPA) which is responsible for dehydrating the cytoplasm.

Following septum formation, the mother cell membrane continues to grow to engulf the forespore in a second membrane (stage III), termed the *outer membrane.* (Although this term is also applied to the diderm cell envelope, there is no similarity between the two membranes.) Between the forespore and mother cell membranes, PG is synthesized (stage IV). The inner layer, the germ cell wall, is chemically similar to vegetative cell wall PG. The outer layer, the cortex, has PG with fewer cross-links. These PG layers are critical in counteracting osmosis as the core dehydrates.

Atop the outer membrane is the coat, a complex structure composed of more than 80 different proteins highly cross-linked to each other. In stage V, the coat proteins are synthesized in the mother cell and deposited on the forespore surface. Stage VI is characterized by continued dehydration that results in changes in the forespore resistance properties, such that physical and chemical insults are no longer lethal. Finally, in stage VII, the mother cell sporangium lyses to release the mature endospore. In some species, endospores are enclosed by a thin glycoprotein covering, the *exosporium.* The exposporium is hydrophobic, which allows the endospores to adhere to surfaces.

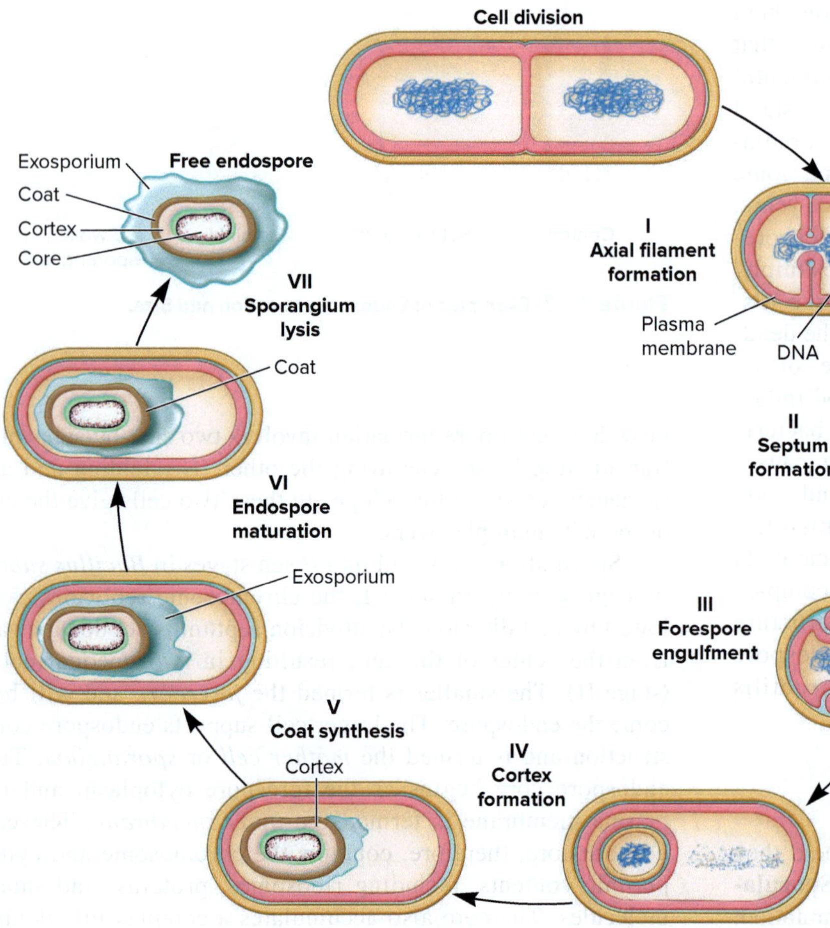

Figure 3.39 Endospore Formation. The stages are indicated by Roman numerals.

Endospore Resistance

The ability of the endospore to tolerate heat, radiation, and damaging chemicals requires that its enzymes and DNA be protected and the layers of the endospore contribute to this resistance. The dense protein coat protects the endospore from chemicals and lytic enzymes such as lysozyme. The inner membrane is extremely impermeable to chemicals, including those that cause DNA damage. The core has low water content, high amounts of Ca-DPA, and a slightly lower pH, all of which contribute to the endospore's resistance. The water content of the core is low enough to prevent rotation of enzymes and other proteins. However, protein denaturation can still occur. It has been suggested that immobilization prevents proteins from interacting and becoming entangled. Thus, even though they may denature, they are able to refold to their active structure when the endospore germinates. The endospore DNA is protected by two mechanisms. Ca-DPA complexes insert between the nitrogenous bases, which stabilizes the DNA. In addition, *s*mall, *a*cid-*s*oluble DNA-binding *p*roteins (SASPs) saturate endospore DNA. SASPs alter the three-dimensional structure of DNA in the endospore core, converting it from the common B form to A form. The A form DNA is less susceptible to damage by ultraviolet (UV) light, which partially accounts for the ability of endospores to tolerate UV exposure. *DNA is a polymer of deoxyribonucleotides (section 13.2).* *Bacterial Endospore Formation*

Endospore to Vegetative Cell

The transformation of dormant endospores into active vegetative cells is almost as complex as sporulation. It occurs in three stages: activation, germination, and outgrowth. Activation prepares endospores for germination and results from treatments such as brief heating to a sublethal temperature, which may cause conformational changes in germination proteins. This is followed by **germination,** the breaking of the endospore's dormant state. It begins when proteins called *germinant receptors*, located in the inner membrane and the cortex, detect small molecules such as sugars and amino acids. These activated receptors trigger the release of the Ca-DPA complexes and water uptake. Cortex peptidoglycan is broken down, allowing the core to expand and hydrate further. Eventually water levels inside the germinating cell reach those characteristic of vegetative cells, causing enzymes in the core to reactivate. This allows the endospore to synthesize molecules required for outgrowth and return to a vegetative state. The SASPs coating the DNA are broken down to free the chromosome and to supply amino acids for new protein synthesis. Because germination of some species' endospores results in disease or toxin production, inhibiting germination by blocking germinant receptors is an active area of research.

Comprehension Check

1. Describe the structure of the bacterial endospore using a labeled diagram.
2. Briefly describe endospore formation and germination. What is the importance of the endospore?
3. What features of the endospore contribute to its resistance to harsh conditions?

Key Concepts

3.1 Bacteria Are Diverse but Share Some Common Structural Features

- Rods (bacilli) and spheres (cocci) are common bacterial shapes. In addition, bacteria may be comma-shaped (vibrio), spiral (spirillum and spirochetes), or filamentous; they may form fruiting bodies; or they may have no characteristic shape (pleomorphic) (**figures 3.1 and 3.2**).
- Some cells remain together after division to form pairs, chains, and clusters of various sizes and shapes.
- A typical coccus cell measures about 1 micron in diameter, whereas a typical rod measures a few microns in length.
- Frequently observed bacterial cell structures include a cell wall, plasma membrane, cytoplasm, extracellular vesicles, a nucleoid, pili, inclusions, capsule, ribosomes, and flagella (**figure 3.5**).

3.2 Bacterial Plasma Membranes Control What Enters and Leaves the Cell

- The cell envelope consists of the plasma membrane and all external coverings, including the cell wall and other layers (e.g., capsules).
- The bacterial plasma membrane fulfills many roles, including acting as a semipermeable barrier, carrying out respiration and photosynthesis, and detecting and responding to chemicals in the environment.
- Cell membranes are lipid bilayers in which integral membrane proteins are buried. Peripheral membrane proteins are loosely associated with the membrane (**figure 3.6**).
- Bacterial membranes are bilayers composed of phospholipids constructed of fatty acids connected to glycerol by ester linkages (**figure 3.7**). Bacterial membranes include microdomains characterized by an altered lipid composition, in which large protein complexes are assembled by flotillins.
- Microorganisms require nutrients, materials that are used in energy conservation and biosynthesis. Macronutrients are needed in relatively large quantities (**table 3.1**). Micronutrients (trace elements) are used in small amounts.
- In passive diffusion, a substance moves through the membrane on its own. Movement is down a concentration gradient and does not require an input of energy. Only a few substances enter bacteria by passive diffusion.
- Transport proteins are divided into two major types: channels and carriers. Channels create a pore through which a substance moves. Carriers are so called because they carry the molecule across the membrane.
- In facilitated diffusion, a transport protein (either a channel or a carrier) helps move substances in the direction of decreasing concentration; no energy is required. Facilitated diffusion is not an important uptake mechanism for bacterial cells.
- Gradients of ions drive solute uptake in secondary active transport systems. Secondary active transporters are either symporters (two substances are transported in the same direction) or antiporters (two substances are transported in opposite directions) (**figure 3.10**).
- Active transport systems use energy and carrier proteins to concentrate substances by transporting them against a gradient. ATP is the energy source for ABC transporters, a type of primary active transporter (**figure 3.11**). ABC transporters are uniporters (one substance is transported).
- Group translocation is a type of active transport in which bacterial cells transport organic molecules while modifying them (**figure 3.12**).
- The secretion of siderophores, which bind ferric iron, enables the accumulation of iron within bacterial cells (**figure 3.13**).

3.3 Cell Walls Have Many Functions

- The vast majority of bacteria have a cell wall outside the plasma membrane to help maintain their shape and protect them from osmotic stress.
- Bacterial walls are chemically complex and usually contain peptidoglycan (**figures 3.14–3.18**). Typical Gram-positive (monoderm) walls have thick, homogeneous layers of peptidoglycan and teichoic acids

(**figures 3.19** and **3.20**). Typical Gram-negative (diderm) bacteria have a thin peptidoglycan layer surrounded by a complex outer membrane containing lipopolysaccharides (LPSs) and other components (**figures 3.21** and **3.22**). The lipid A portion of LPS is also called endotoxin. Its release into the human body can lead to septic shock.

3.4 Extracellular Vesicles Emerge from Bacterial Membranes

- Extracellular vesicles (EVs) form when the plasma membrane pinches off and is released from a cell. Their contents reflect the composition of the cytoplasm or periplasm in the cell before release (**figure 3.24**).

3.5 The Cell Envelope Often Includes Layers Outside the Cell Wall

- Capsules, slime layers, and glycocalyces are layers of material lying outside the cell wall. They can protect cells from certain environmental conditions, allow cells to attach to surfaces, and protect pathogenic bacteria from host defenses (**figure 3.25**).
- S-layers are the external-most layer in some bacteria. They are composed of proteins or glycoprotein and have a characteristic geometric shape (**figure 3.26**).

3.6 The Bacterial Cytoplasm Is Complex

- The bacterial cytoplasm is a highly concentrated and crowded solution of biochemicals.
- The bacterial cytoplasm contains proteins that are similar in structure and function to the cytoskeletal proteins observed in eukaryotes (**figure 3.27** and **table 3.2**).
- Some bacteria have internal membrane systems that support photosynthetic and respiratory machinery (**figure 3.28**).
- Inclusions are observed in all cells (**figure 3.30**). Most are used for storage (e.g., PHA inclusions and polyphosphate granules), but some are used for other purposes (e.g., magnetosomes and gas vacuoles). Microcompartments such as carboxysomes contain enzymes that catalyze important reactions (e.g., CO_2 fixation).
- Bacterial ribosomes are 70S in size. They are composed of numerous proteins and several rRNA molecules (**figure 3.31**).
- The genetic material of bacterial cells is located in an area within the cytoplasm called the nucleoid. The nucleoid is not enclosed by a membrane (**figure 3.32**). The chromosome usually consists of a double-stranded, covalently closed, circular DNA molecule. Nucleoid-associated proteins participate in structuring the chromosome.
- Plasmids are extrachromosomal DNA molecules found in many bacteria. Some are episomes—plasmids that exist freely in the cytoplasm or that can be integrated into the chromosome. Although plasmids are not required for survival in most conditions, they can encode traits that confer selective advantage in some environments.

3.7 External Structures Are Used for Attachment and Motility

- Many bacteria have hairlike appendages called fimbriae or pili. Fimbriae function primarily in attachment to surfaces, but type IV pili are involved in twitching motility. Sex pili participate in the transfer of DNA from one bacterium to another (**figure 3.34**).
- Many bacteria are motile, often by means of threadlike, locomotory organelles called flagella. Bacterial species differ in the number and distribution of their flagella (**figure 3.34**). Each bacterial flagellum is composed of a filament, hook, and basal body (**figure 3.35**).

3.8 Bacterial Endospores Are a Survival Strategy

- Some bacteria from the phylum Bacillota survive adverse environmental conditions by forming endospores, dormant structures resistant to heat, desiccation, and many chemicals (**figure 3.37**).
- Both endospore formation and germination are complex processes made in response to certain environmental signals (**figure 3.39**).

Active Learning

1. Propose a model for the assembly of a flagellum in a typical Gram-positive cell envelope.
2. Review figure 2.17 describing the Gram stain, and relate cell wall structure to staining behavior.
3. Do you think there is now enough evidence for an entrenched system like Gram-positive/Gram-negative to be replaced with monoderm/diderm? Explain your reasoning.
4. The peptidoglycan of bacteria has been compared with the chain mail worn beneath a medieval knight's suit of armor. It provides both protection and flexibility. Describe other structures in biology that have an analogous function. How are they replaced or modified to accommodate the growth of the inhabitant?
5. Why might a microbe have more than one uptake system for certain substrates?

6. Design a demonstration to illustrate the cell wall's role in protecting against lysis.
7. Compare the plasma membrane and the outer membrane in terms of their thickness. How does this affect their respective transporters/porins?
8. What would you expect to observe if you were able to transplant CreS into a rod-shaped bacterium such as *Bacillus subtilis?*
9. Develop a hypothesis to explain why gas vacuoles are bounded by proteins rather than a lipid bilayer membrane.
10. Support or refute the statement that the periplasm is an organelle because it is a membrane-bound compartment of the bacterial cell.
11. When viewed in the microscope, ultrasmall bacteria (figure 3.4*a*) appear similar to extracellular vesicles (figure 3.24). How could you determine which of the two you might be observing in a specimen?
12. Photosynthetic microbes experience a daily cycle, in which their energy source, sunlight, is available for only part of the day. How might the presence of storage vesicles (e.g., polyphosphate) in the cells also vary each day?

CHAPTER

4

Archaeal Cell Structure

Heath Johnson/Shutterstock

Methane—The Other Greenhouse Gas

One of the driving factors in climate change is the accumulation of greenhouse gases in Earth's atmosphere. These gases (carbon dioxide (CO_2) and methane (CH_4) among others) insulate the planet, thereby raising the temperature globally. While CO_2 is more abundant, CH_4 traps more heat and is therefore a greater concern to control. What's the source of the CH_4? Livestock accounts for about 20% of the CH_4 generated. Ruminant animals, namely cattle, sheep, and goats, burp out about 200 million metric tons of methane annually. Add in the 18 million metric tons generated from their manure, and the impact of these 3.5 billion animals is a serious concern.

But don't blame the animals; they are not directly responsible for generating methane. Rather, the guts of ruminant animals host a unique group of archaea—the methanogens—that make methane from hydrogen, carbon dioxide, and acetate. These substrates arise in the animals as their food is digested with the help of a complex community of microbes living in their rumen. There is considerable interest in figuring out how to decrease methane production. Possibilities include genetically modified livestock that host fewer methanogens, and feed additives to control methane production.

Methanogens are only one group in the domain Archaea. Other archaea have attracted attention because of their ability to live in extreme environments. All are of interest because of their chimeric natures: They have some features that are similar to bacteria and some that are similar to eukaryotes. Archaeal cells are morphologically similar to bacterial cells, but the molecules used to construct archaeal cells are often unique or are similar to those found in eukaryotes. As we describe in chapters 15 and 19, their physiologies exhibit this dichotomy. In general, the processes and molecules used to conserve energy are like those of bacteria, whereas the processes and molecules used to replicate and express their genomes are like those of eukaryotes. In this chapter, we describe some of the molecular and structural aspects of archaeal cells that set them apart.

Readiness Check:

Based on what you have learned previously, you should be able to:

- ✓ Summarize the structures observed in bacterial cells and their molecular makeup (chapter 3)
- ✓ Describe the mechanisms used by bacteria to obtain nutrients (section 3.2)
- ✓ Describe the structural features of the membranes and cell walls of bacteria (sections 3.2 and 3.3)

4.1 Archaea Are Diverse but Share Some Common Features

After reading this section, you should be able to:

a. Describe a typical archaeal cell
b. Discuss key differences between bacterial and archaeal cells

Recall from Chapter 1 that, for many years, **archaea** and bacteria had been jointly referred to as prokaryotes. Today Archaea and Bacteria are recognized as distinct taxa. Their separation based on 16S rRNA sequences correlated with the observation that they each have unique and distinguishing characteristics. Some of these are summarized in **table 4.1**.

Many archaea inhabit niches that constitute extreme environments from a human perspective, and in many cases, these environments were not previously believed to support life. Once microbiologists broadened their exploration of Earth's microbial habitats, archaea have been found to be ubiquitous. Surveys of nonextreme environments have established that archaea reside in most habitats, but often in small numbers. Our understanding of archaeal cell structure derives from experiments performed on a few model organisms that can be grown in the laboratory, and these represent only a small fraction of the known archaeal diversity. Methanogens (microbes that generate methane gas) are killed by exposure to oxygen, and must be handled in sealed containers. Extreme halophiles (salt-lovers) require high salt conditions. Another commonly studied archaeon is *Sulfolobus,* which requires both high temperature and low pH for growth. Because of the technical challenges in culturing archaea, many are known only in the sense that their 16S rRNA sequence or genome sequence has been determined. We begin our discussion of archaeal cells by considering overall cell morphology and organization and then specific cell structures.

Shape, Arrangement, and Size

Archaeal cells, like bacterial cells, exhibit a variety of shapes. **Cocci** and **bacilli** (rods) are common (**figure 4.1*a, b***). Both usually exist singly, but some cocci form clusters and some rods form chains. Curved and spiral shapes have been observed, but no spirochete-like forms have yet been discovered. However, some archaea exhibit unique shapes, such as the branched form of *Thermoproteus tenax* (figure 4.1*c*) and the flat, postage-stamp-shaped *Haloquadratum walsbyi,* an archaeon that lives in salt ponds and measures about 2 µm by 2 to 4 µm and is only 0.25 µm thick (*see figure 19.12*). Recently cultured Asgard archaea include organisms featuring a central body with multiple long protrusions (figure 4.1*d*). These atypical shapes

Table 4.1 Comparison of Bacterial and Archaeal Cells

Cellular Component	Bacteria	Archaea
Plasma membrane lipids	Ester-linked phospholipids	Glycerol diethers, diglycerol tetraethers
Cell wall constituents	Peptidoglycan present in nearly all; some lack cell walls.	Most commonly a protein S-layer; some include polysaccharides or protein or both; some lack cell walls. Peptidoglycan absent.
Extracellular vesicles	Yes	Yes
Capsule or slime layers	Common	Rare
Cytoskeleton	Multiple homologues of tubulin and actin	Multiple homologues of tubulin and actin
Ribosomal RNA	16S, 23S, 5S in size, but specific to bacteria	16S, 23S, 5S in size, but specific to archaea
Ribosomal proteins	Half universal in all three domains, half unique to bacteria	Half universal in all three domains, half common with eukaryotes
Chromosome structure	Circular, double-stranded (ds) DNA	Circular, ds DNA
Plasmids	Circular and linear ds DNA	Circular, ds DNA
Fimbriae or pili	Common	Common
Inclusions	Yes	Yes
Motility structure	Flagellum	Archaellum

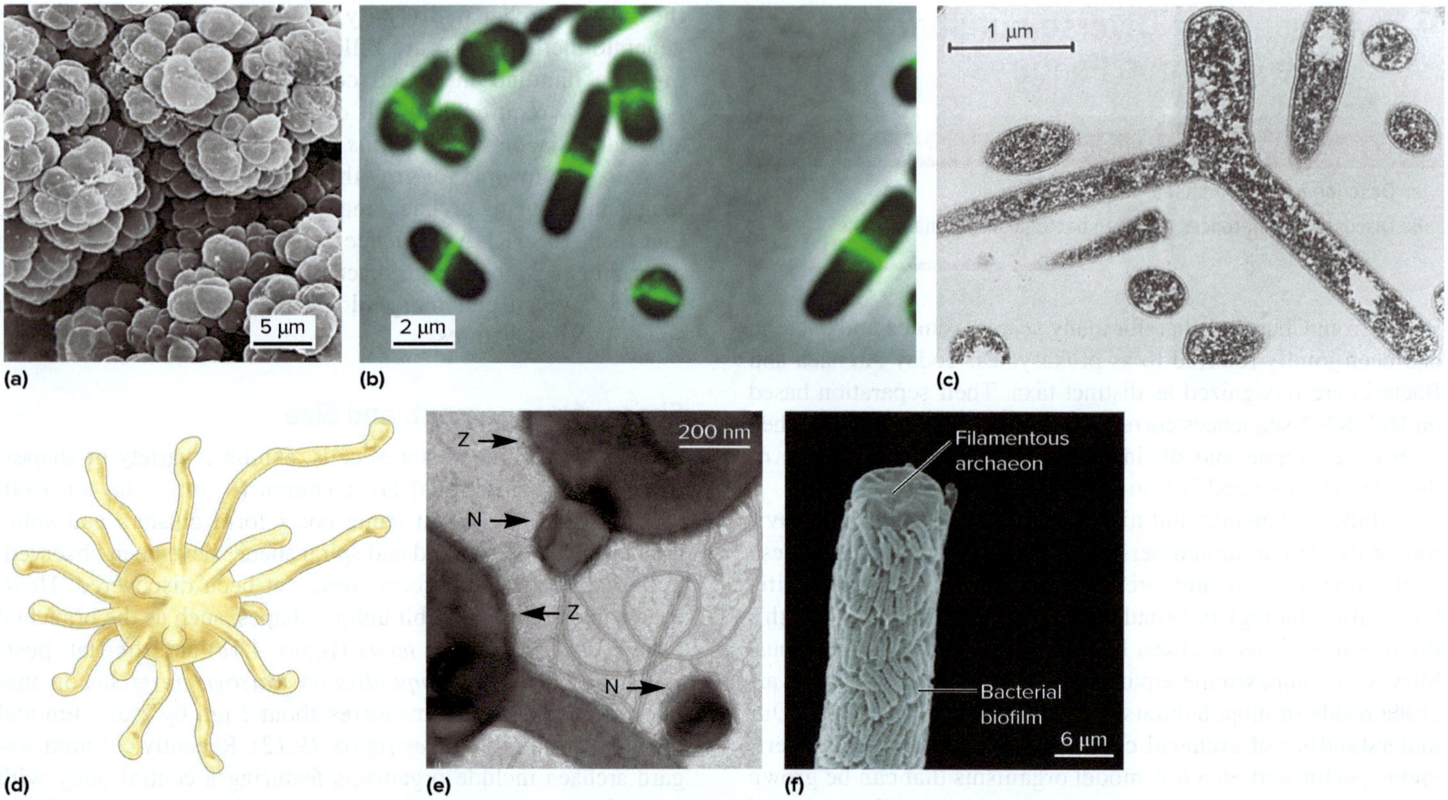

Figure 4.1 Archaeal Cell Morphology. Note the difference in size among the scale bars in each panel. (a) *Methanococcus mazei*, a coccus that forms clusters; SEM. (b) *Haloferax volcanii*; fluorescence micrograph of dividing cells with the contractile ring labeled green. Cells are rods or disk-shaped. (c) *Thermoproteus tenax*, a branched cell; TEM. (d) *Lokiarchaeum ossiferum*, artist interpretation of a cell with many projections. The cell body diameter is 1-2 μm. (e) *Nanoclepta minutus* (N) lives in a stable association with its host archaeon *Zestosphaera tikiterensis* (Z); TEM. (f) *Giganthauma karukerense* forms filaments covered with a bacterial biofilm; SEM.

(a) James T. Staley; (b) Dr. Iain Duggin; (c) James T. Staley (e) St. John, E. et al. A new symbiotic nanoarchaeote (*Nanoclepta minutus*) and its host (*Zestosphaera tikiterensis* gen. nov.; sp. nov.) from a New Zealand hot spring. *Syst. Appl. Microbiol.* 42, 94–106 (2019).; (f) Prof. Olivier Gros.

have the advantage of greatly increasing the surface area-to-volume (S/V) ratio, which in turn increases the efficiency of nutrient uptake, the diffusion of molecules, and growth rate. Pleiomorphic archaea are exemplified by the halophile *Haloferax volcanii* (figure 4.1*b*). These cells exhibit two forms, depending on the growth stage, switching from a flattened disk shape to a rod. It has been suggested that the rod shape is required for the assembly of the motility organelle, the archaellum. ◀ *Bacteria are diverse but share some common structural features (section 3.1)*

Archaeal cells also vary in size as much as in shape. Typical rods are 1 to 2 μm wide by 1 to 5 μm long; cocci are 1 to 3 μm in diameter. However, extremely small and extremely large archaea have been identified. At the small end of the size continuum are symbiotic nanoarchaea (0.2 μm) that live in conjunction with typically sized archaeal host cells (figure 4.1*e*). These miniature cells are characterized by both their small cell size and small genomes. They are unable to grow independently and require import of many metabolites from the host cell. At the other extreme is a giant archaeon that forms long, macroscopic filaments up to 30 mm in length. The filament is composed of cells measuring 8 to 10 μm wide by 20 to 24 μm long. An interesting characteristic of this archaeon is that its filaments are coated with a biofilm formed by bacteria of a single species (figure 4.1*f*). It is thought that the archaeon is the host and the bacteria are symbionts.

Comprehension Check

1. Which cell shapes are observed in members of both Bacteria and Archaea? Which are unique to bacteria? Which to archaea?
2. Archaea was first defined as a distinct taxon by comparisons of ribosomal RNA sequences. Identify two other molecules that could be used to determine if a microbe having a prokaryotic morphology is a bacterium or an archaeon.

4.2 Archaeal Cell Envelopes Are Structurally Diverse

After reading this section, you should be able to:

a. Draw an archaeal cell envelope and identify the component layers
b. Compare and contrast archaeal and bacterial cell envelopes in terms of their structure, molecular makeup, and functions
c. Compare and contrast nutrient uptake mechanisms observed in bacteria and archaea

We define the **cell envelope** as the plasma membrane and any layers external to it. For bacteria, these layers include cell walls, S-layers, capsules, and slime layers. One of the most distinctive features of archaeal cells is the nature of their cell envelopes. Their uniqueness begins with their plasma membranes. Recall from chapter 3 that bacterial membranes are lipid bilayers within which proteins float (*see figure 3.6*). Imagine the surprise of microbiologists when they found that some archaea have monolayer membranes that function like bilayers. This structural difference is due to the presence of unique lipids in archaea. Archaeal cell envelopes also differ in terms of organization. For many archaea, an S-layer is the major, and sometimes only, component of the cell wall. Slime layers and capsules have been observed in some archaea where they appear to mediate cell-cell interactions, but little is known about their composition and regulation.

Archaeal Plasma Membranes Are Composed of Unique Lipids but Function Like Bacterial Membranes

Archaeal membranes are composed primarily of lipids that differ from bacterial and eukaryotic lipids in two ways. First, they contain hydrocarbons derived from isoprene units—5-carbon, branched molecules (**figure 4.2**). This affects the way the lipids pack together, which in turn affects the fluidity and permeability of the membrane. This is especially important for extremophilic archaea for which membrane fluidity and permeability could be compromised by harsh growth requirements. Second, the hydrocarbons are attached to glycerol by ether links rather than ester links (figure 4.2). Ether links are more resistant to chemical attack and heat than ester links.

Two major types of archaeal lipids have been identified: glycerol diethers and diglycerol tetraethers. **Glycerol diether lipids** are formed when two hydrocarbons are attached to glycerol (**figure 4.3**, lipid 4). Usually, the hydrocarbon chains in glycerol diethers are 20 carbons in length. **Diglycerol tetraether lipids** are formed when two glycerol residues are linked by two long hydrocarbons that are 40 carbons in length (figure 4.3, lipids 5 and 6). Tetraethers are more rigid lipids than diethers.

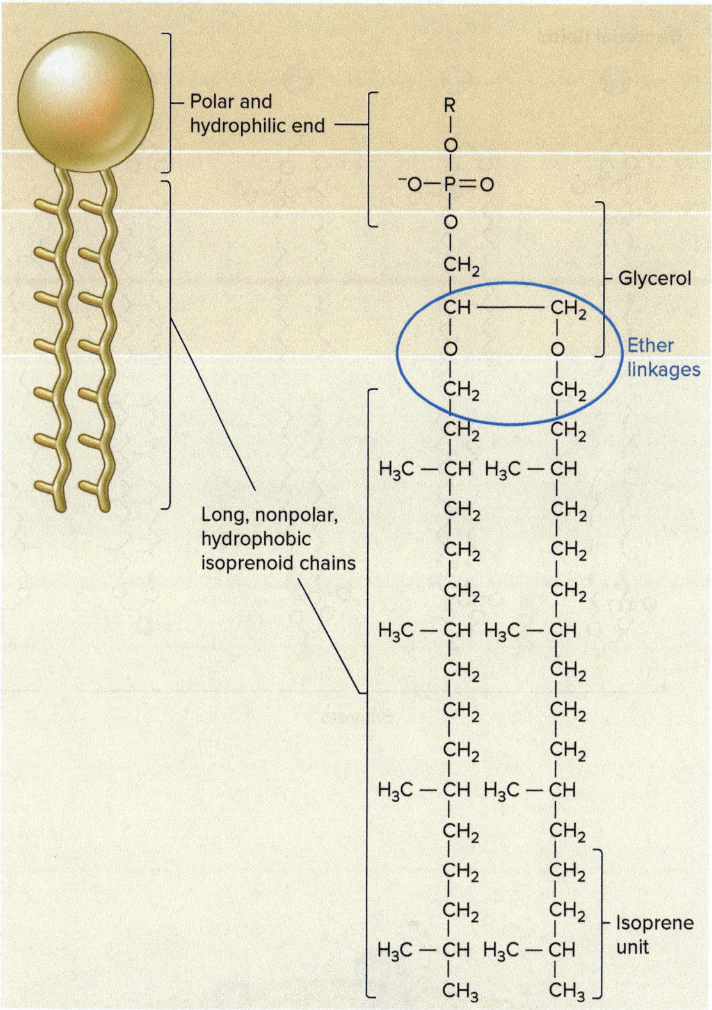

Figure 4.2 The Structure of an Archaeal Phospholipid. Phosphatidylglycerol archaeol, a lipid found in archaeal membranes.

MICRO INQUIRY *Compare this figure with figure 3.7. What are the chemical differences between this lipid and the phospholipids common in bacteria?*

Cells can adjust the overall length of the tetraethers by cyclizing the chains to form pentacyclic rings (figure 4.3, lipid 5). Phosphorus-, sugar-, and amino-containing groups can be attached to the glycerol moieties in the diethers and tetraethers, just like the phospholipids observed in bacterial and eukaryotic membranes.

Despite the structural differences in membrane lipids, the basic design of archaeal membranes is similar to that of bacterial and eukaryotic membranes: There are two hydrophilic surfaces and a hydrophobic core. When C_{20} diethers are used, a typical bilayer membrane is formed (**figure 4.4*a***). When the membrane is constructed of C_{40} tetraethers, a monolayer membrane with much more rigidity is formed (figure 4.4*b*).

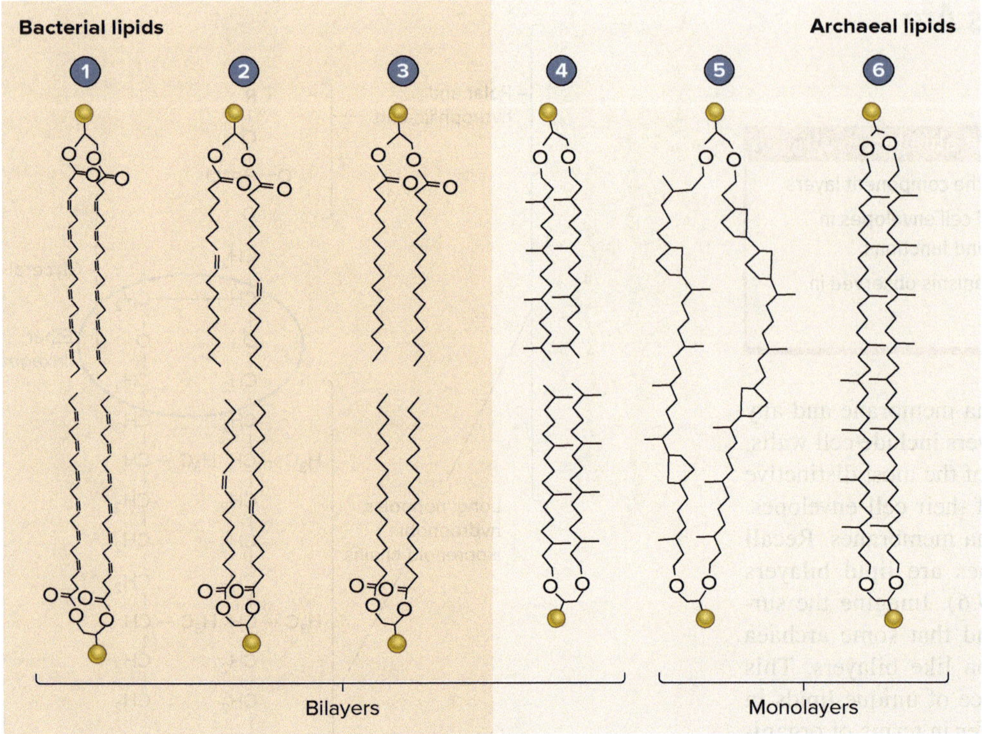

Figure 4.3 Bacterial and Archaeal Lipids. Carbon skeletons are illustrated for six types of lipids. Gold balls represent the polar hydrophilic region in each lipid. Molecules 1, 2, and 3 are bacterial lipids, and molecules 4, 5, and 6 are archaeal lipids. Molecule 4 is a glycerol diether, while 5 and 6 are diglycerol tetraethers. Note that some archaeal lipids (5, 6) form monolayers (figure 4.4*b*), whereas all bacterial lipids form bilayers.

MICRO INQUIRY *Identify the two archaeal-specific chemical signatures in these lipids: the ether linkages between glycerol and the side chains and the isoprene subunits of the hydrophobic side chains.*

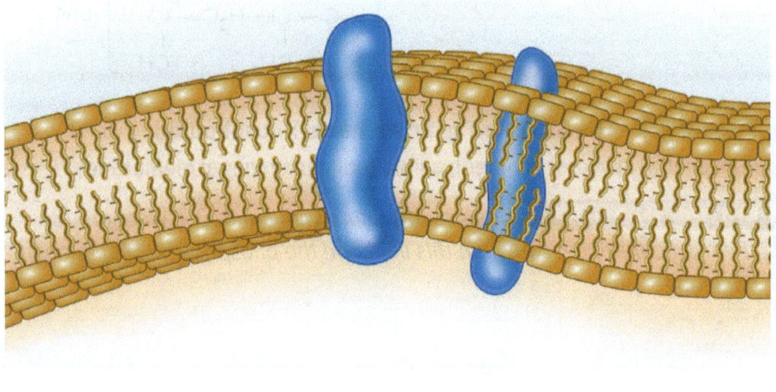

(a) Bilayer of C_{20} diethers

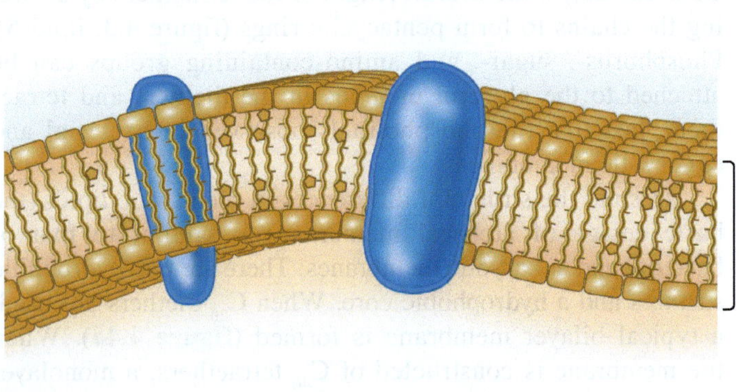

(b) Monolayer of C_{40} tetraethers

Figure 4.4 Examples of Archaeal Plasma Membranes.

The addition of pentacyclic rings further increases this rigidity. As might be expected from their need for stability, the membranes of extreme thermophiles are up to 90% tetraether monolayers. Most archaea have membranes containing some regions with monolayers and some with bilayers with the exception of the extreme halophiles that have entirely bilayer membranes. ▶| *Lipids (section 41.1)*

Like the lipids in bacterial membranes, archaeal lipids move laterally within the plane of the membrane, rendering the membrane a dynamic platform as it fulfills its role as a barrier. Membrane fluidity is dramatically affected by the relative proportions of the more rigid monolayers and the more flexible bilayers. The presence of both types of lipids within an archaeal plasma membrane partitions it into **functional membrane microdomains.** Assembly of membrane-bound protein complexes depends on the association of specific lipid types and protein subunits. As a consequence, the cell partitions certain functions to distinct regions of the cell envelope, as exemplified by the polar position of the archaellum.

As we discuss in chapter 3, organisms have several options for obtaining the nutrients they need from their environment. In methanogenesis, nonpolar gases like CO_2, H_2 (growth substrates), and CH_4 (product) cross membranes by passive diffusion. Methanogens are typically found in association with bacteria that release H_2 and CO_2, positioning themselves to exploit the concentration gradient. Other archaea use active transport for nutrient uptake. Both primary (e.g., ABC transport) and secondary (e.g., symport and antiport) active transport systems have been identified in archaeal cells, often by searching the genome for the genes encoding components of these systems (i.e., genomic analysis). These systems tend to be similar to those seen in bacteria. The group translocation system phosphoenolpyruvate:sugar phosphotransferase system (PTS) also functions in some halophiles. ◀ *Bacteria use many mechanisms to bring nutrients into the cell (section 3.2)* ▶ *Microbial genomics (chapter 18)* ▶ *Methanogens (section 19.4)*

There Are Many Different Types of Archaeal Cell Walls

Archaeal cell walls perform the same essential functions as those in bacteria: determining shape, stabilizing the cell against osmotic pressure, conferring mechanical strength, and forming a permeability barrier. The molecules used to construct archaeal cell walls are entirely different from those used in bacteria. Peptidoglycan is universally absent from archaea.

The most common type of archaeal cell wall is an **S-layer** composed of many copies of one or two proteins (**figure 4.5**, structure 1). The S-layer may be as thick as 70 nm. The S-layer in *Sulfolobus* spp. resembles a protein canopy tethered to the plasma membrane surface (**figure 4.6**). The surface geometric patterns observed in electron microscopy (*see figure 3.26*) arise from the manner in which the individual polypeptides pack into the two-dimensional surface. The lattice style in the S-layer varies with the species and determines the size of the pores. In an analogy to bacterial cell wall structure, the space between the S-layer and the plasma membrane is termed the quasiperiplasm. Archaeal S-layer proteins are extensively decorated with carbohydrates that protect and stabilize the S-layer. These sugars may also partially penetrate the pores, allowing for regulation of molecules crossing the S-layer.

Some archaea have additional layers of material in their cell envelope. For instance, *Methanospirillum* spp. enclose dozens of cells in a single protein sheath external to the S-layer. *Methanosarcina* spp. have a polysaccharide layer

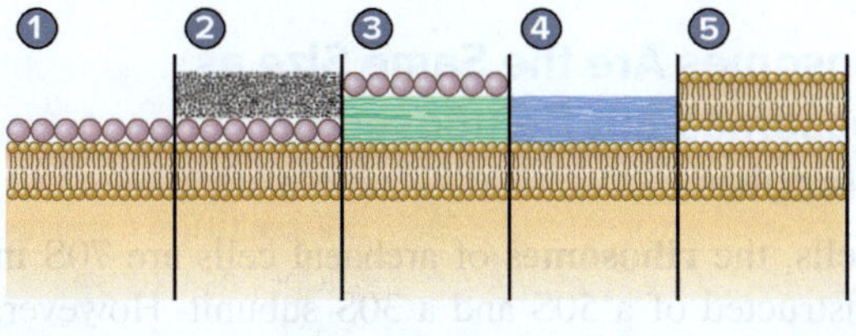

Figure 4.5 Archaeal Cell Envelopes. (1) The most common envelope is a simple S-layer. (2) Some S-layers are covered with an additional protein sheath or a carbohydrate layer. (3) Other archaea have a carbohydrate layer beneath the S-layer or (4) in place of an S-layer. (5) A few have a double membrane.

MICRO INQUIRY *Can the monoderm/diderm terminology be used for archaeal cell walls?*

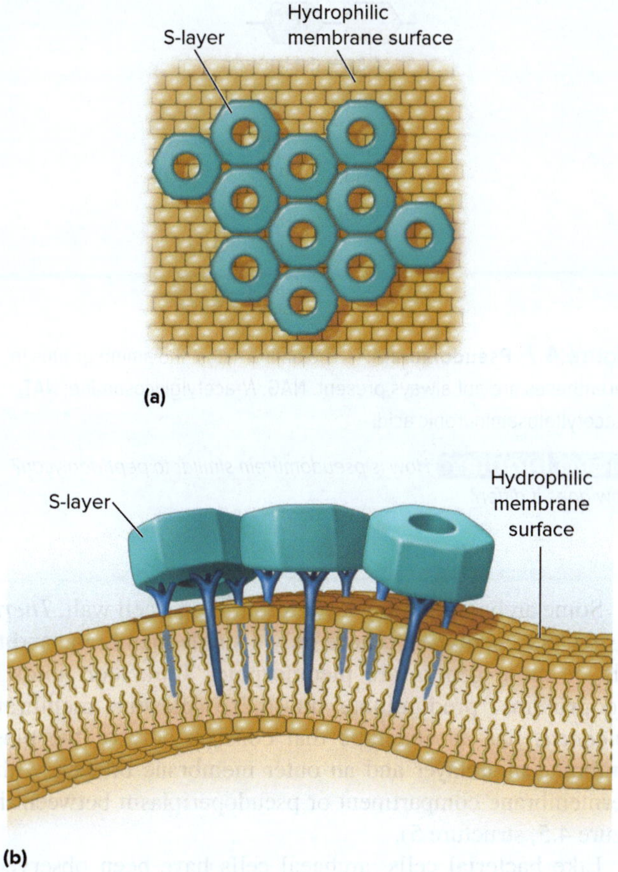

Figure 4.6 Archaeal S-Layer Structure. The two panels represent structures visible at 90-degree angles to one another, on the surface of the cell (a) and looking at the membrane (b). (a) Example of a geometric pattern formed by the two-dimensional network of S-layer proteins. Hexagonal structures form pores through which solutes pass. (b) Diagram of how the S-layer polypeptides attach to the plasma membrane via hydrophobic stalks.

covering the S-layer (figure 4.5, structure 2). This material, called methanochondroitin, is similar in composition to the chondroitin sulfate of animal connective tissue.

In some methanogens, the S-layer is separated from the plasma membrane by a peptidoglycan-like molecule called pseudomurein (figure 4.5, structure 3). **Pseudomurein** differs from peptidoglycan in that it has L-amino acids instead of D-amino acids in its cross-links, *N*-acetyltalosaminuronic acid (NAT) instead of *N*-acetylmuramic acid (NAM), and $\beta(1\rightarrow3)$ glycosidic bonds instead of $\beta(1\rightarrow4)$ glycosidic bonds joining the two sugars (**figure 4.7**). These differences mean that lysozyme, penicillin, and other chemicals that affect peptidoglycan structure and synthesis in bacterial cell walls have no effect on archaeal cell walls. ◀ *Cell walls and osmotic protection (section 3.3)*

Another type of archaeal cell wall found in some extreme halophiles consists of a single, thick homogeneous layer resembling that in monoderm bacteria (figure 4.5, structure 4). These archaea lack an S-layer and their wall chemistry varies from species to species but usually consists of complex polysaccharides.

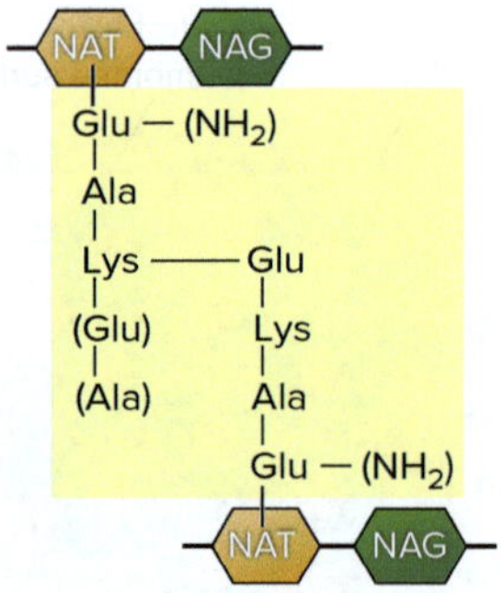

Figure 4.7 Pseudomurein. The amino acids and amino groups in parentheses are not always present. NAG, *N*-acetylglucosamine; NAT, *N*-acetyltalosaminuronic acid.

MICRO INQUIRY *How is pseudomurein similar to peptidoglycan? How does it differ?*

Some archaea lack any layer resembling a cell wall. *Thermoplasma* and *Ferroplasma* have only a single membrane and thus cells of these genera are pleomorphic. Tetraether lipids and glycoproteins predominate in these plasma membranes. *Ignicoccus* has an envelope that consists only of the plasma membrane monolayer and an outer membrane bilayer, with an intermembrane compartment or pseudoperiplasm between them (figure 4.5, structure 5).

Like bacterial cells, archaeal cells have been observed to produce extracellular vesicles (EVs) (**figure 4.8**). These vesicles are composed of the plasma membrane and enclose cytoplasmic contents including proteins and nucleic acids. How EVs form in archaea with tetraether membranes and how EVs are released through the varied archaeal cell walls remain open questions. It has been proposed that EVs may be an important mechanism for intercellular gene transfer in thermophiles because the lipid coating confers stability against high temperatures that would denature naked DNA.

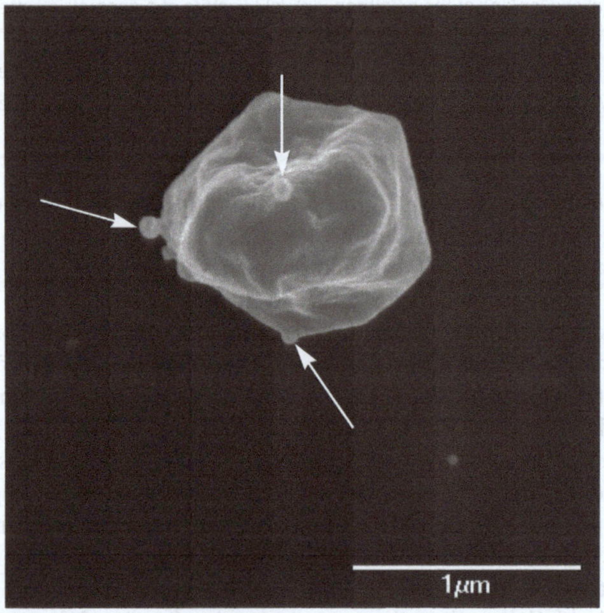

Figure 4.8 Archaeal Extracellular Vesicles (EVs). Arrows indicate EVs budding from a *Sulfolobus* cell.

Gill S, Catchpole R, Forterre P. Extracellular membrane vesicles in the three domains of life and beyond. *FEMS Microbiol Rev.* 2019;43(3):273-303. doi:10.1093/femsre/fuy042

Comprehension Check

1. Identify three features that distinguish archaeal plasma membranes from those of bacteria.
2. Both bacteria and archaea can have S-layers. How does their use in the cells differ?
3. Compare and contrast archaeal and bacterial extracellular vesicles.

4.3 Archaeal Cytoplasm Is Similar to Bacterial Cytoplasm

After reading this section, you should be able to:

a. Compare and contrast the cytoplasm of bacterial and archaeal cells
b. Discuss the organization of archaeal DNA and similarities to both bacteria and eukaryotes

Overall, the physical properties of archaeal cytoplasm are similar to those of bacteria. Inclusions are found within the cytoplasm —polyhydroxyalkanoates, polyphosphate granules, glycogen granules, and gas vesicles; ribosomes; a nucleoid; and plasmids. Protein filaments that might form a **cytoskeleton** have also been identified (**table 4.2**). These proteins include homologues of those found in bacteria, as well as some unique to archaea. In several cases, the similarities have been detected from DNA sequences, and the actual structures have not been visualized in cells. In this section, we focus on cytoplasmic structures.

Archaeal Ribosomes Are the Same Size as Bacterial Ribosomes but Composed of Different Molecules

Like bacterial cells, the **ribosomes** of archaeal cells are 70S in size and are constructed of a 50S and a 30S subunit. However, their shape is somewhat different and their component molecules are not the same. Both have **ribosomal RNA (rRNA)** molecules of similar size: 16S in the small subunit, and 23S and 5S in the large subunit (table 4.1). However, as we discuss in chapter 1, the differences in nucleotide sequences of these molecules were the initial basis for establishing the taxon Archaea.

Table 4.2 Some Archaeal Cytoskeleton Proteins

Function	Example	Homologs
Cell Division. These protein filaments form a contractile ring to separate daughter cells during cell division.	FtsZ	tubulin (eukaryotic) FtsZ (bacterial)
Cell Morphology. Filaments of this protein control archaeal cell shape (phylum Methanobacteriota).	CetZ	tubulin (eukaryotic)
Cell Morphology. Filaments of this protein span the cell length (phylum Thermoproteota).	crenactin	actin (eukaryotic) MreB (bacterial)
Cell Morphology. Filaments of this protein are present in membrane protrusions (phylum Asgardarchaeota) (*see figure 4.1d*).	lokiactin	actin (eukaryotic)

The protein composition of bacterial and archaeal ribosomes also differs. Given the conserved function of ribosomes, it should be no surprise that more than half of the ribosomal proteins (r-proteins) are present in all three domains. Archaeal ribosomes have more r-proteins than bacterial ribosomes, and these share homology with eukaryotic r-proteins. The difference in bacterial and archaeal ribosomes correlates with the fact that archaeal ribosomes are unaffected by antibiotics that bind and inhibit bacterial ribosomes. ▶| *Protein synthesis inhibitors (section 9.4)*

Nucleoid

The **nucleoid** provides another example of the difference between archaea and bacteria. This region in the cytoplasm contains the **chromosome** and nucleoid-associated proteins (NAPs). The chromosomes of all known archaea are circular, double-stranded **deoxyribonucleic acid (DNA).** However, some archaea are polyploid, carrying multiple chromosomes, whereas others are monoploid, with a single copy. It is believed that this difference in chromosome number reflects the mechanism for chromosome partitioning to daughter cells during cell division. Whereas some archaea have a highly regulated chromosome segregation mechanism, the process appears to be random in others. Maintaining multiple copies of the genetic information is good insurance for an intact copy to end up in each daughter cell. ▶| *Archaeal cell cycles are unique (section 7.3)*

As in bacteria, archaeal chromosomes appear to be organized into chromosome interaction domains within the nucleoid. In the model archaeon *Sulfolobus* the chromosome occupies two distinct compartments that partition regions of high and low gene expression. A NAP related to SMC (*see figure 3.33*), termed coalescin, helps to establish these domains.

The problem of compacting a genome to fit into a nucleoid is common to archaea and bacteria, and the solution is similar. Both supercoiling and the presence of NAPs on the genomic DNA contribute to chromosome organization. As in bacteria, there are multiple NAPs, but they bear little resemblance to bacterial NAPs. The common features of these proteins are small size and positive charge.

Many archaea have NAPs termed **histones** associated with their chromosomes. These histones organize the chromosome in **nucleosomes** that are related to the nucleosomes observed in eukaryotes (*see figure 5.11*). In *Haloferax volcanii,* the nucleosomes differ from eukaryotic nucleosomes in that they consist of four histones rather than eight and they organize a shorter length of DNA. In *Thermococcus kodakarensis,* however, histones assemble into complexes containing from two histones to more than a dozen. It is hypothesized that, in thermophiles, histones help prevent chromosome denaturation. |◀ *Nucleoid (section 3.6);* ▶| *Nucleus (section 5.5)*

4.4 Many Archaea Have External Structures Used for Attachment and Motility

After reading this section, you should be able to:

a. Compare and contrast bacterial and archaeal pili
b. Compare and contrast bacterial flagella and archaella in terms of their structure and function

Like bacteria, many archaea have appendages that extend beyond the cell envelope, namely pili and flagella. These external structures and their functions are discussed in this section.

Pili

Archaeal **pili** exhibit a range of lengths and diameters, some solid and others hollow. Some pili are single-protein polymers, and others are composed of bundles of individual filaments. In all cases, pili function in the adherence of cells to other cells or to nonliving surfaces.

Sulfolobus acidocaldarius can display at least four distinct types of pili. The first, the archaellum, confers motility and is described in detail below. The *a*rchaeal *a*dhesive *p*ilus (Aap) confers gliding motility as it attaches and retracts from a surface. The *u*ltraviolet (UV)–inducible *p*ilu*s* (Ups) is only produced following exposure of the cells to UV light. As UV light damages DNA, the Ups mediates cell aggregation and promotes DNA transfer between cells. The fourth pilus, termed *thread*, has several polymers aligned into filaments that can interact with other threads on the same cell or neighboring cells. ◀ *Bacterial pili and fimbriae (section 3.7)*

Two archaeal pili are particularly noteworthy: cannulae and hami. **Cannulae** are hollow, tubelike structures observed on the surface of thermophiles belonging to the genus *Pyrodictium* (**figure 4.9*a***). The function of cannulae is unknown, but daughter cells arising from cell division remain connected to each other by cannulae. After many rounds of cell division, a network of cells is formed. **Hami** are of interest because of their shape. They resemble tiny grappling hooks, which suggests they might function in attachment (figure 4.9*b*). Indeed, the cells that produce hami, *Altiarchaeum hamiconexum* are members of biofilm communities in which cells connect via hami.

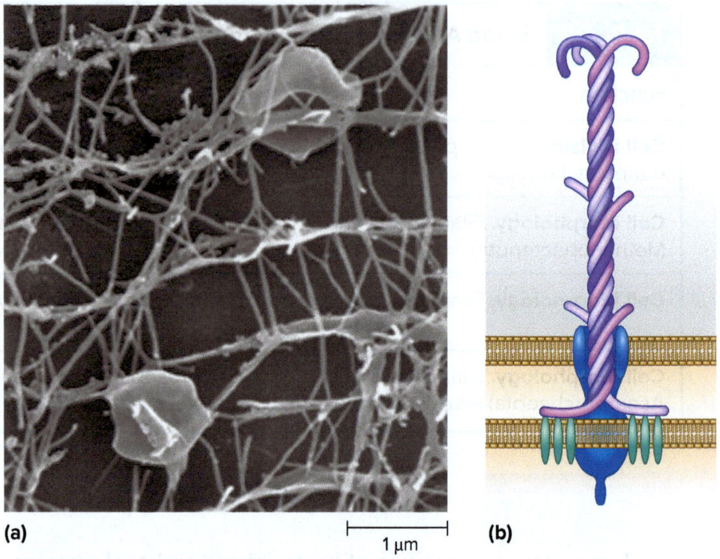

Figure 4.9 Archaeal Specialized Pili. (a) Cannulae are tubular structures about 25 nm in diameter. They connect cells, ultimately forming a dense network. (b) Hami are hooklike structures that are thought to promote surface adherence. They are often several times longer than the cell.
(a) Karl O. Stetter

Archaella and Motility

Archaeal flagella, more precisely termed **archaella** have been studied in detail in a few model organisms. They are superficially similar to their bacterial counterparts, but important differences exist. Archaella are best described as a type IV pilus, which denotes their mechanism of assembly within the cell envelope. Archaella are thinner than bacterial flagella (10 to 14 nm rather than 18 to 22 nm) and most are composed of more than one type of protein subunit (**figure 4.10**). Archaella may be up to 10 μm in length. The filament is not hollow, so its assembly differs from bacterial flagella. Archaellin (pilin) monomers are exported through the motor complex (figure 4.10) and polymerized at the base of the filament. Additional proteins anchor the archaellum in the S-layer. ◀ *Bacterial flagella (section 3.7)*

Archaella operate in a manner similar to bacterial flagella: Filament rotation propels the cell. Again, there are some important differences. First, rotation is powered by ATP hydrolysis rather than by proton motive force. Second, when the direction of rotation switches, it causes the cell to move in either the forward or reverse direction, as seen for some bacteria having polar flagella.

Archaea exhibit chemotaxis, thermotaxis, and phototaxis, and the proteins responsible for these behaviors are homologous to those in bacteria. Chemoreceptors are clustered in the plasma membrane near the cell poles, where the sensory input can be relayed to the motor driving archaellum rotation.

Archaea have been observed to engage in two types of swimming behaviors, termed a relocate-and-seek strategy. Fast speeds in a straight line (up to ~500 μm/sec) result in a rapid change of location. A slower zigzag motion (~50 to 100 μm/sec) constitutes the seek, a scan for favorable conditions. The fastest known swimmers, *Methanocaldococcus*

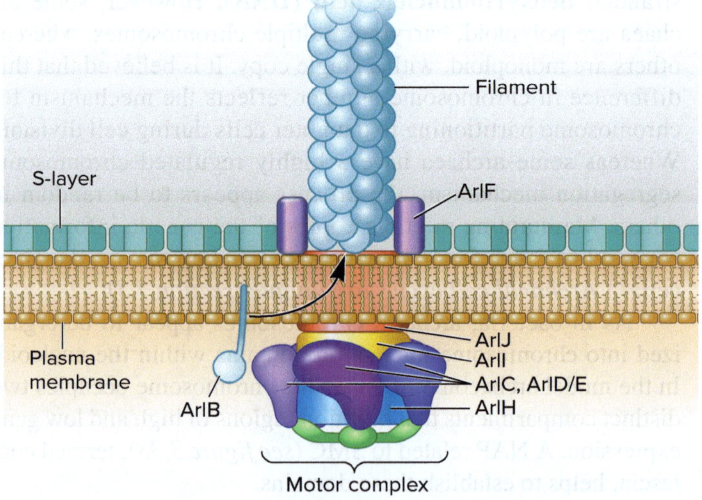

Figure 4.10 Structure of the Archaellum, Based on *Pyrococcus*. The archaellum is anchored to the plasma membrane and bound to the S-layer by ArlF. Assembly of the filament occurs by the addition of ArlB monomers through ArlJ. ArlCDE and ArlHI complexes comprise the motor. ArlI is the ATPase that generates torque to rotate the filament.

spp., can travel at about 500 body lengths per second (bps), compared to 20 bps for *E. coli* and 11 bps for humans. This extreme swimming behavior is an adaptation to their deep sea hydrothermal vent environment. Temperature gradients force these microbes to inhabit a narrow zone between the 400°C vent fluid and the surrounding 2°C seawater. When fast swimming identifies a suitable niche (50° to 90°C), the microbes use the archaella for attachment.

In addition to their role in motility, archaella are involved in adhesion to substrates, biofilm formation, and cell-cell interactions.

Comprehension Check

1. What observations about cannulae and hami suggest that they allow archaeal cells to adhere to surfaces, including other cells?
2. List three aspects of archaella and their motility that are similar to bacterial flagella and their motility.
3. Defend the proposal that the archaellum is a type of pilus.
4. Archaea have some features similar to bacteria and some similar to eukaryotes. Based on what you have learned, provide two examples that illustrate the similarity of archaea to bacteria; list two examples of their similarity to eukaryotes.

Key Concepts

4.1 Archaea Are Diverse but Share Some Common Features

- Members of Bacteria and Archaea share a common cell architecture. However, they have characteristics that define them as distinct taxa (**table 4.1**).
- Many archaea have been identified only through 16S rRNA, partial genome sequences, and metagenomes.
- Rods and cocci are the most common archaeal shapes. Archaea also can be curved rods, branched, square, and pleomorphic. Most archaea are similar in size to bacteria. Archaea that are extremely small or are extremely large have also been identified (**figure 4.1**).

4.2 Archaeal Cell Envelopes Are Structurally Diverse

- The cell envelope consists of the plasma membrane and all external coverings, including cell walls and other layers. Archaeal cell envelopes usually consist of only the plasma membrane and the cell wall.
- Archaeal membranes are composed of glycerol diether and diglycerol tetraether lipids (**figure 4.3**). Membranes composed of glycerol diether are lipid bilayers. Membranes composed of diglycerol tetraethers are lipid monolayers. The overall structure of a monolayer membrane is similar to that of the bilayer membrane in that the membrane has a hydrophobic core and its surfaces are hydrophilic (**figure 4.4**).
- Gaseous substrates enter archaeal cells by passive diffusion. Archaea use active transport systems to move nutrients up a concentration gradient.
- Archaeal cell walls do not contain peptidoglycan and exhibit diversity in their makeup. The most common type of cell wall is one consisting of an S-layer only (**figures 4.5** and **4.6**).
- Archaeal cells release extracellular vesicles comprised of the plasma membrane and S-layer (**figure 4.8**).

4.3 Archaeal Cytoplasm Is Similar to Bacterial Cytoplasm

- Numerous inclusions are observed in archaeal cells, including gas vesicles and various types of granules.
- Cytoskeletal proteins have been identified in archaeal cells. These are homologues of eukaryotic cytoskeletal proteins (**table 4.2**).
- Bacterial and archaeal ribosomes are 70S in size but differ slightly in their morphology. The rRNA molecules differ in sequence because these are the molecules used to determine phylogeny. About half of archaeal ribosomal proteins are found in all three domains, whereas the rest are homologues of the eukaryotic ribosomal proteins.
- The genetic material of archaeal cells is located in the nucleoid, which is not enclosed by a membrane. All known archaeal chromosomes consist of a double-stranded, covalently closed, circular DNA molecule.
- Like bacteria, archaea use nucleoid-associated proteins (NAPs) to organize their chromosomes, notably including histones that are homologues of eukaryotic histones.

4.4 Many Archaea Have External Structures Used for Attachment and Motility

- Many archaea have pili, which assume many roles, including attachment to living and nonliving surfaces and motility.
- Many archaea are motile by means of pili called archaella (**figure 4.10**). They are rigid protein helices that rotate, and the direction of rotation determines if the cell moves forward or backward. Rotation is powered by ATP hydrolysis.
- Motile archaea exhibit chemotaxis. Some are also phototactic.
- Archaea engage in relocate-and-seek swimming behaviors that may aid in positioning cells optimally in a temperature gradient.

Active Learning

1. Archaea with cell walls consisting of a thick, homogeneous layer of complex polysaccharides often retain crystal violet dye when stained using the Gram-staining procedure. Suggest a reason for this staining reaction.
2. Isoprene serves as a building block not only for the hydrocarbons observed in archaeal membranes but also for sterols, carotenoids, retinal, and quinones. Use any resources necessary to identify the function of these other isoprene-based molecules and to determine their distribution in nature. What does the use of isoprene to make this diverse array of molecules suggest about the nature of the last universal common ancestor (LUCA)?
3. Archaeal cells have shapes in common with bacteria, for example, cocci and bacilli. Most cells need an expansive surface area for importing nutrients and exporting wastes. Consider some of the archaeal cell shapes described in this chapter (figure 4.1) and explain how they optimize cell surface area.
4. Identify some challenges to working with archaea in the laboratory, given their habitats in locations like salt ponds, acid mine drainage sites, and deep sea hydrothermal vents.
5. A recent study of *Sulfolobus* cell surface structures used cryo-electron microscopy to examine how three types of pili assemble external to the S-layer. The S-layer pores have a diameter of 48Å, and the pili (called thread, Aap, and the archaellum) have diameters of 40Å, 80Å, and 110Å, respectively. Refer to figure 4.6 and draw how *Sulfolobus* might distribute its S-layer proteins to accommodate all three types of pili.

CHAPTER

5

Eukaryotic Cell Structure

U.S. Department of Agriculture (USDA)

Red Means Dead

Anyone driving down a highway in the Rocky Mountains of Canada and the United States will observe an astonishing sight: large stands of rust-colored, dead pine trees in once verdant mountain forests. In British Columbia alone, over half the commercial lodgepole pine trees have been killed, while in northern Colorado and southern Wyoming, these trees have been completely eliminated.

This destruction is typically attributed to the mountain pine beetle, but that's not the whole story. Each beetle hosts three fungi, *Grosmannia clavigera, Leptographium longiclavatum,* and *Ophiostoma montium*. How they kill is an interesting example of a mutualistic relationship—one in which both host and microbes benefit. The fungi are helped by the beetle, which acts as a mode of transportation from one tree to another. The beetle is helped by the fungi, which protect it from the tree's defense mechanisms and nourishes its young.

The killing process begins when the beetle chews through the bark of a pine tree to reach underlying tissues. The tree fights back by producing a resin toxic to the beetle. However, the fungi degrade the toxic resin, using it as a carbon and energy source, thereby protecting the beetle. This enables the beetle to continue its attack, ultimately laying its eggs in the tree and carrying fungi inside the tree. The fungi produce masses of filaments that help feed the beetle's progeny. These filaments are lethal to the tree because they literally clog its circulatory system. Nutrients and water can no longer flow through the tree, so it dries up and starves to death.

Much of this text is devoted to discussions of bacteria and archaea. But as this story of deforestation shows us, eukaryotic microbes also deserve our attention. They are prominent members of ecosystems, important model organisms, and exceptionally useful in industrial microbiology. Some are also major human pathogens.

Eukaryotic microbes can be divided into two major groups: protists and fungi. **Protists** are a group of very diverse unicellular eukaryotes. Some use organic molecules as energy sources just as animals do and are called **protozoa. Algae** are photosynthetic protists. Because protists did not descend from a single common ancestor, the term *protist* is a common name for these microbes rather than a valid taxon. Fungi, on the other hand, all share a common evolutionary history, so they are placed in one taxon (Fungi).

The division of eukaryotes into appropriate taxa is currently being reevaluated, as is their evolution. The taxonomy of eukaryotic microbes is discussed more completely in chapters 23 and 24, and their role as pathogens is covered in chapter 39. In this chapter, we introduce eukaryotic cell structures and their functions.

Readiness Check:

Based on what you have learned previously, you should be able to:

- ✓ List the three domains of life (section 1.1)
- ✓ Describe the application of small subunit (SSU) rRNA analysis to the establishment of the three-domain classification system proposed by Carl Woese (section 1.2)
- ✓ Identify the following components of a plant or animal cell and describe their functions: cell wall, plasma membrane, cytoplasm, mitochondria, chloroplasts, and ribosomes
- ✓ Identify the major cell structures observed in bacterial and archaeal cells, and describe the function and molecular makeup of each (chapters 3 and 4)

5.1 Eukaryotic Cells Are Diverse but Share Some Common Features

After reading this section, you should be able to:

a. Compare and contrast eukaryotic, bacterial, and archaeal cells in terms of their use of membranes, size, morphological diversity, and organelles
b. Define the term *organelle* and explain the role of intracellular membranes in eukaryotic cells

We begin our discussion of the structure of eukaryotic cells with an overview. Here we make broad generalizations about eukaryotic cells but provide examples of deviations from common features.

Shape and Size

One hallmark of eukaryotic cells is their morphological diversity. This is especially true for eukaryotic microbes, which are found in numerous habitats and have evolved morphological adaptations accordingly. Some of the diverse shapes of eukaryotic microbes are shown in **figure 5.1.**

In general, eukaryotic microbes are larger than bacterial and archaeal cells (*see figure 2.19*). However, just as some bacteria and archaea are very large, even larger than some of the largest protists (*see figure 3.4*), so too are some protists smaller than many bacteria and archaea. These small eukaryotic microbes include *Ostreococcus tauri,* which, at about 0.8 μm in diameter, is the smallest free-living alga known. In fact, these and other picoeukaryotes (Spanish *pico,* little bit) in marine environments can reach densities as high as 10^3 to 10^4 per milliliter.

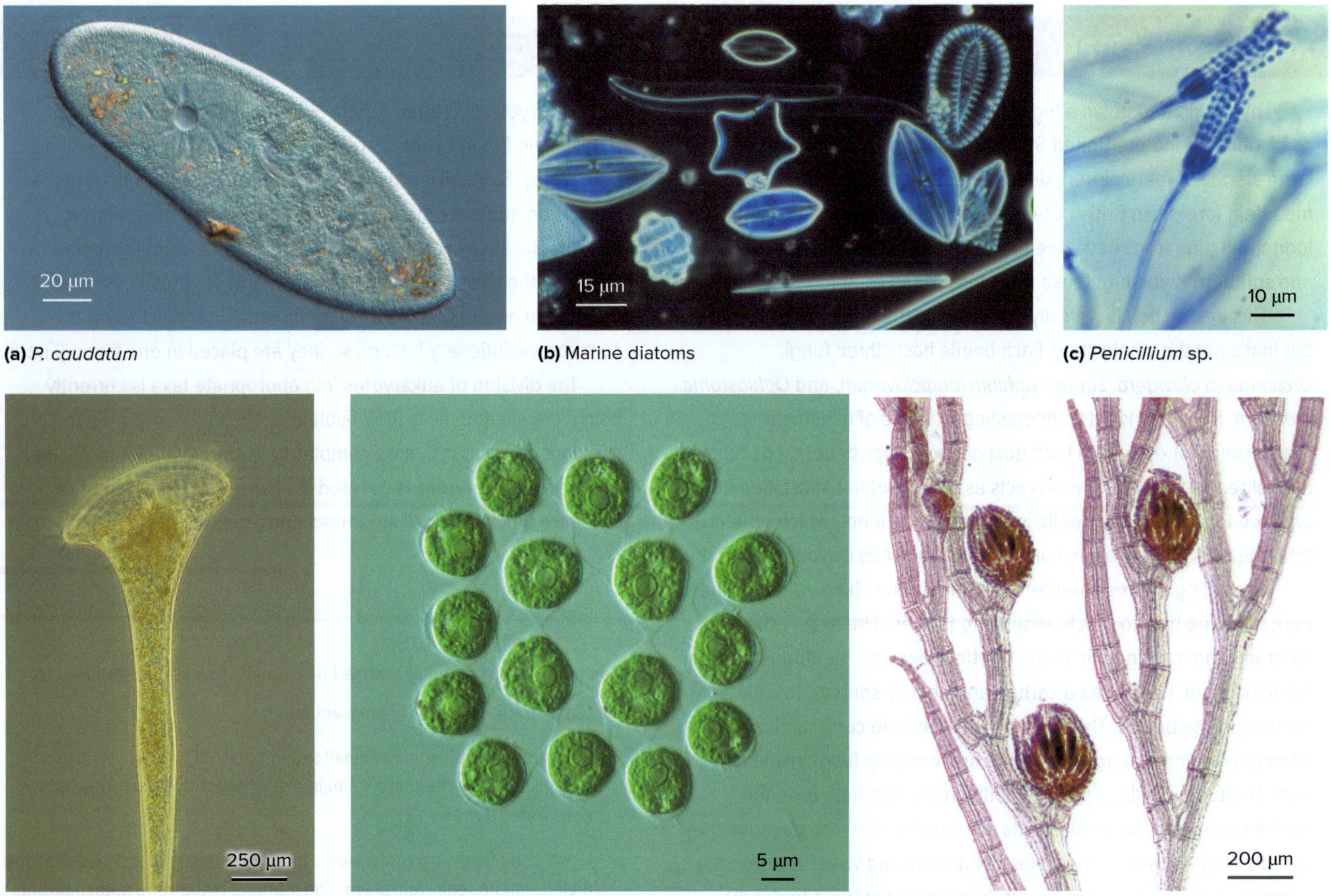

(a) *P. caudatum* **(b)** Marine diatoms **(c)** *Penicillium* sp. **(d)** *Stentor* sp. **(e)** *Gonium* sp. **(f)** *Polysiphonia* sp.

Figure 5.1 Some Eukaryotic Microorganisms. (a) *Paramecium caudatum,* a protist; differential interference-contrast (DIC) microscopy. (b) Mixed diatom frustules; dark field microscopy. (c) *Penicillium* sp., the mold's hyphae and conidia; phase contrast microscopy. (d) *Stentor* sp., a ciliated protist in its extended, actively feeding form; phase-contrast microscopy. (e) *Gonium* sp., a freshwater, colonial alga; DIC microscopy. (f) *Polysiphonia* sp., a filamentous, red alga; phase contrast microscopy.

(a) micro_photo/iStock/Getty Images; (b) Oxford Scientific/Photodisc/Getty Images; (c) Dr. Lucille K. Georg/CDC; (d–e) Stephen Durr; (f) Steven P. Lynch

Cell Organization

Eukaryotic cells are distinct from bacterial and archaeal cells because of their use of membranes (**figures 5.2** and **5.3**; *compare these figures with figure 3.5*). They have membrane-bound nuclei, and membranes play a prominent part in the structure of many other organelles. **Organelles** are intracellular structures that perform specific functions in cells analogous to the functions of organs in the body of a multicellular organism. The partitioning of the eukaryotic cell interior into organelles makes possible the placement of different cellular functions in separate compartments, so they can more easily take place simultaneously under independent control and proper coordination. Large membrane surfaces make possible greater respiratory and photosynthetic activity because these processes are located exclusively in membranes. The intracytoplasmic membrane complex also serves as a transport system to move materials between different cell locations.

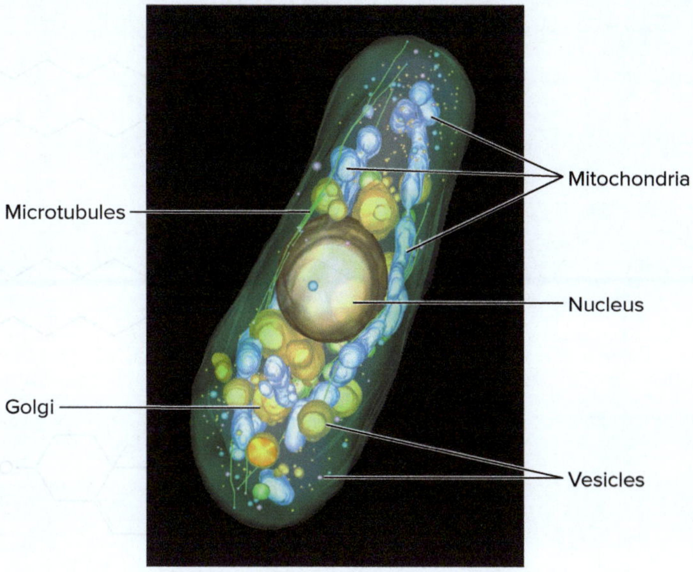

Figure 5.2 Eukaryotic Cell Ultrastructure. A cryo-electron micrograph of the yeast *Schizosaccharomyces pombe*. This image is the first high-resolution, three-dimensional reconstruction of a complete eukaryotic cell.
Johanna Hoog/European Molecular Biology Laboratory/Science Source

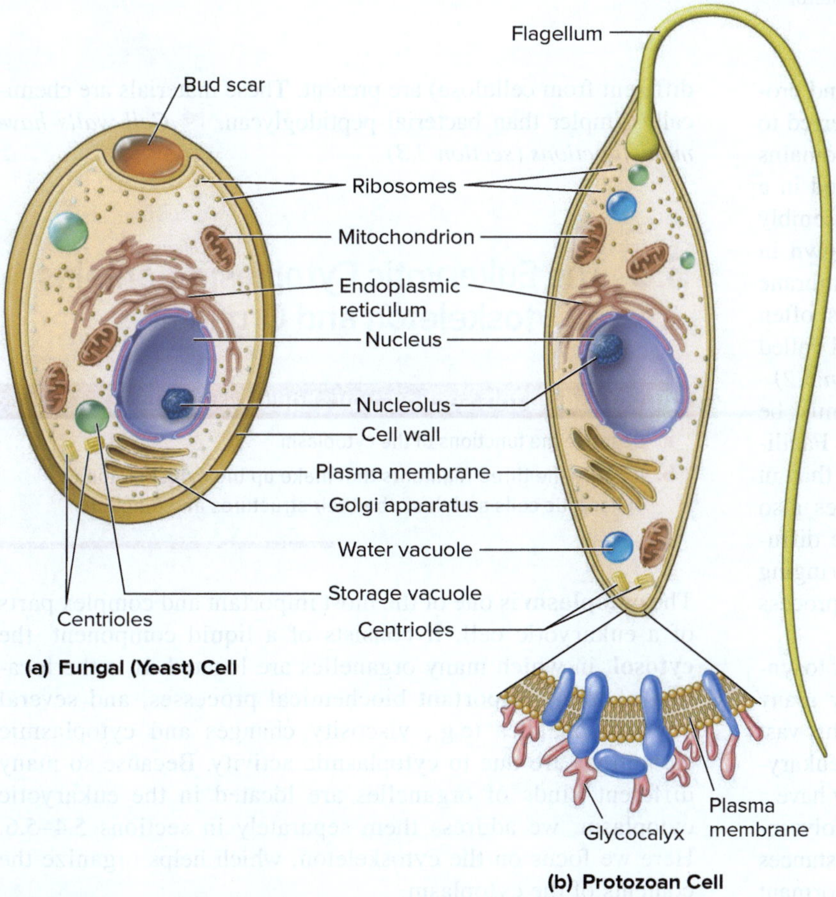

Figure 5.3 The Structure of Two Representative Eukaryotic Microbes. Illustrations of (a) a yeast cell (fungus) and (b) the flagellated protozoan *Peranema*.

MICRO INQUIRY *In addition to separating each organelle from the cytoplasm, what other functions do internal membranes serve?*

5.2 Eukaryotic Cell Envelopes

After reading this section, you should be able to:

a. Describe lipids commonly found in eukaryotes
b. Identify the types of eukaryotic microbes that have cell walls and describe their composition

The cell envelope consists of the plasma membrane and all coverings outside it. Many eukaryotic microbes have only a plasma membrane without a cell wall. However, when cell walls are present, they are quite different from those of bacteria and archaea.

The plasma membrane of eukaryotes is a lipid bilayer composed of a high proportion of sphingolipids and sterols (e.g., cholesterol and ergosterol) in addition to the phospholipids observed in bacterial membranes (**figure 5.4**). In phospholipids and sphingolipids, the fatty acids are attached to glycerol with ester links. Sphingolipids and sterols can pack together very tightly, so they affect the fluidity and permeability of the plasma membrane. Eukaryotic cells therefore alter the quantity of sterols and sphingolipids in response to stresses such as changes in temperature.

The distribution of lipids in the plasma membrane is asymmetric. This means that the lipids in the outer leaflet differ from those of the inner leaflet. Although most lipids in each leaflet mix

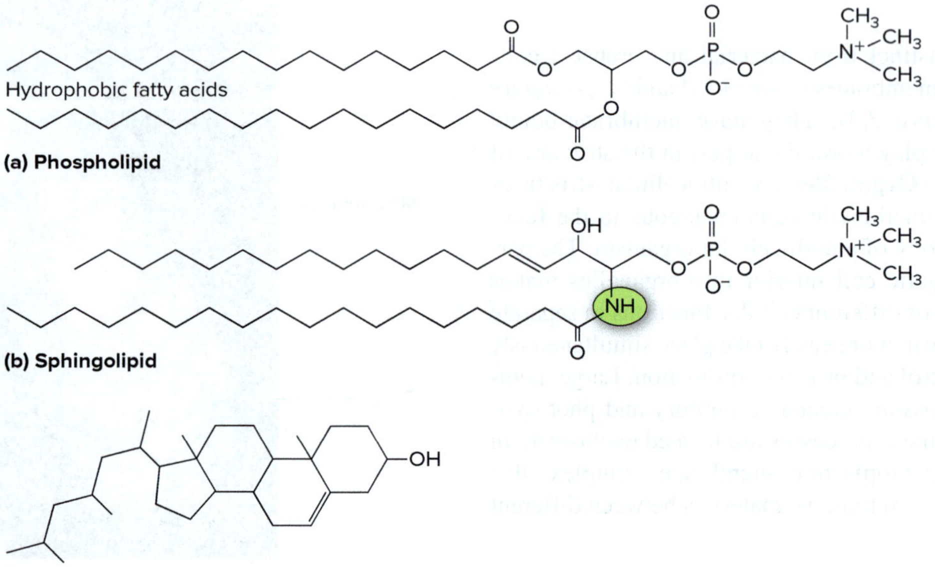

Figure 5.4 Examples of Eukaryotic Membrane Lipids. (a) Phosphatidylcholine, a phospholipid. (b) Sphingomyelin, a sphingolipid. The amine group is highlighted as this differentiates a sphingolipid from a phospholipid. (c) Cholesterol, a sterol.

freely with each other, microdomains that differ in lipid and protein composition are found. These microdomains are referred to as *lipid rafts* and are equivalent to membrane microdomains found in bacteria and archaea. They have been implicated in a variety of cellular processes (e.g., signal transduction, assembly and release of virus particles, and endocytosis). As shown in figure 5.3*b*, many eukaryotic cells have integral membrane proteins and lipids decorated with carbohydrates. This often forms a carbohydrate-rich layer on the surface of the cell called the **glycocalyx.** ◀ *Membrane structure is dynamic (section 3.2)*

Like bacterial and archaeal cells, eukaryotic cells must be able to move materials across their plasma membranes. Facilitated diffusion is much more common in eukaryotic cells than in bacterial and archaeal cells (*see figure 3.9*). Eukaryotes also use primary and secondary active transport and passive diffusion. In addition, eukaryotes have another option for bringing materials into cells: endocytosis. We focus on this unique process in section 5.4.

Eukaryotic microbes that have cell walls include photosynthetic protists (commonly called algae), cysts made by many types of protists, and most fungi. Not surprisingly given this vast diversity, the chemical composition of the cell walls of eukaryotic microbes varies considerably. Algal cell walls usually have a layered appearance and contain large quantities of the polysaccharides cellulose and pectin. In addition, inorganic substances such as silica or calcium carbonate may be present. Dormant cysts formed by many protists consist of multiple layers of chitin. Fungal cell walls are rigid. Their exact composition varies with the organism; usually chitin and glucan (a glucose polymer different from cellulose) are present. These materials are chemically simpler than bacterial peptidoglycan. ◀ *Cell walls have many functions (section 3.3)*

5.3 The Eukaryotic Cytoplasm Contains a Cytoskeleton and Organelles

After reading this section, you should be able to:

a. Describe the functions of the cytoplasm
b. Identify the three filaments that make up the cytoskeleton of eukaryotic cells and describe their structures and functions

The **cytoplasm** is one of the most important and complex parts of a eukaryotic cell. It consists of a liquid component, the **cytosol,** in which many organelles are located. It is the location of many important biochemical processes, and several physical changes (e.g., viscosity changes and cytoplasmic streaming) are due to cytoplasmic activity. Because so many different kinds of organelles are located in the eukaryotic cytoplasm, we address them separately in sections 5.4–5.6. Here we focus on the cytoskeleton, which helps organize the contents of the cytoplasm.

The **cytoskeleton** is a vast network composed of three types of interconnected filaments: actin filaments (also called microfilaments), intermediate filaments, and microtubules

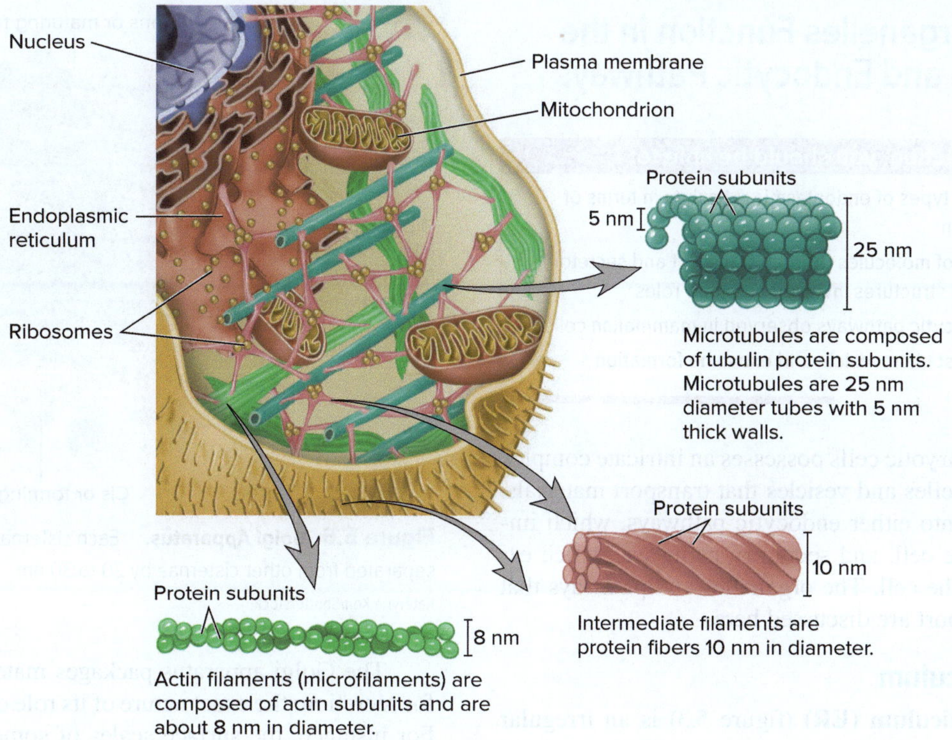

Figure 5.5 The Eukaryotic Cytoplasm and Cytoskeleton. The cytoplasm of eukaryotic cells contains many organelles. The cytoskeleton helps form a scaffold within which the organelles lie. The cytoskeleton is composed of three elements: actin filaments, microtubules, and intermediate filaments.

MICRO INQUIRY *Which cytoskeletal filament is made of the proteins α- and β-tubulin?*

(**figure 5.5**). The cytoskeletal network is like a highway that guides the movement of cellular structures. Three motor proteins (myosin, kinesin, and dynein) associated with actin filaments and microtubules track along the cytoskeletal filaments, helping to move cell structures from one location to another.

Actin filaments are minute protein filaments organized to form a variety of structures depending on the cell type (figure 5.5). They are composed of an actin protein that is similar to the actin contractile protein of muscle tissue. In addition to moving cellular structures, actin filaments help cells change shape; thus they are involved in amoeboid movement, endocytosis, and cytokinesis.

Intermediate filaments (IFs) are flexible yet very strong. They are not associated with motor proteins and therefore play a variety of structural roles. They are assembled from a group of proteins that can be divided into several classes (e.g., keratin and vimentin). IFs having different functions are assembled from one or more of these classes of proteins. Fungi and plant cells lack IFs. In animal cells and certain protists, IFs form the *nuclear lamina,* a structure that provides support for the nuclear envelope. Additional IFs in animal cells help position organelles and help link cells together to form tissues.

Microtubules are shaped like thin cylinders constructed of two globular protein subunits—α-tubulin and β-tubulin (figure 5.5). The two proteins are the same molecular weight and differ only slightly in terms of their amino acid sequence and tertiary structure. Each tubulin monomer is approximately 4 to 5 nm in diameter. These subunits form a helical cylinder with an average of 13 subunits in one turn.

Microtubules have several important functions. They form the spindle apparatus that separates chromosomes during mitosis and meiosis. They form tracks along which numerous types of organelles and vesicles move about the cell, and are found in cilia and flagella, which confer motility. Microtubules also are found in cell structures requiring support such as long, slender, rigid pseudopodia (called axopodia) observed in some protists (*see figure 23.7*).

Comprehension Check

1. What are the major lipids in eukaryotic membranes? Which other domain of life possesses similar lipids?
2. How do the cell envelopes of eukaryotic organisms differ from those of bacterial and archaeal cells?
3. List the name and function of the three cytoskeletal elements observed in eukaryotic cells. Do all eukaryotic microbes have each of the three elements?

5.4 Several Organelles Function in the Secretory and Endocytic Pathways

After reading this section, you should be able to:

a. Differentiate the two types of endoplasmic reticulum in terms of structure and function
b. Outline the pathway of molecules through transport and secretory pathways, noting the structures involved and their roles
c. Describe three endocytic pathways observed in mammalian cells
d. Compare and contrast microvesicle and exosome formation

The cytoplasm of eukaryotic cells possesses an intricate complex of membranous organelles and vesicles that transport materials. We divide transport into either endocytic pathways, which import materials into the cell, and secretory pathways, which export molecules from the cell. The organelles and pathways that accomplish this transport are discussed here.

Endoplasmic Reticulum

The **endoplasmic reticulum (ER)** (figure 5.3) is an irregular network of branching and fusing membranous tubules, around 40 to 70 nm in diameter, and many flattened sacs called *cisternae* (s., cisterna). The nature of the ER varies with the functional and physiological status of the cell. In cells synthesizing a great deal of protein to be secreted, much of the ER is studded on its outer surface with ribosomes and is called **rough endoplasmic reticulum (RER).** Other cells, such as those producing large quantities of lipids, have ER without ribosomes. This is **smooth endoplasmic reticulum (SER).**

The endoplasmic reticulum is an important hub of molecular synthesis and transport. Lipids and proteins are synthesized by ER-associated enzymes and ribosomes. Polypeptide chains synthesized on RER-bound ribosomes may be inserted either into the ER membrane or into its lumen for transport elsewhere. The ER is also a major site of cell membrane synthesis.

The Golgi Apparatus Prepares Materials for Secretion

The **Golgi apparatus** is composed of flattened, saclike cisternae (**figure 5.6**). A complex network of tubules and vesicles is located at the edges of the cisternae. In many eukaryotes, the cisternae are stacked, forming a structure called a *dictyosome*. There are usually around four to eight cisternae in a stack. In many cells, multiple stacks are linked to form a Golgi ribbon. Most eukaryotic microbes have just a single stack of cisternae, and in some, the cisternae are not in stacks (e.g., the yeast *Saccharomyces cerevisiae*).

A stack of cisternae has two faces that are quite different. The sacs on the cis or forming face are closest to the ER and often associated with it. The sacs on the trans or maturing face are farthest from the ER. The two faces of the Golgi differ in thickness, enzyme content, and degree of vesicle formation.

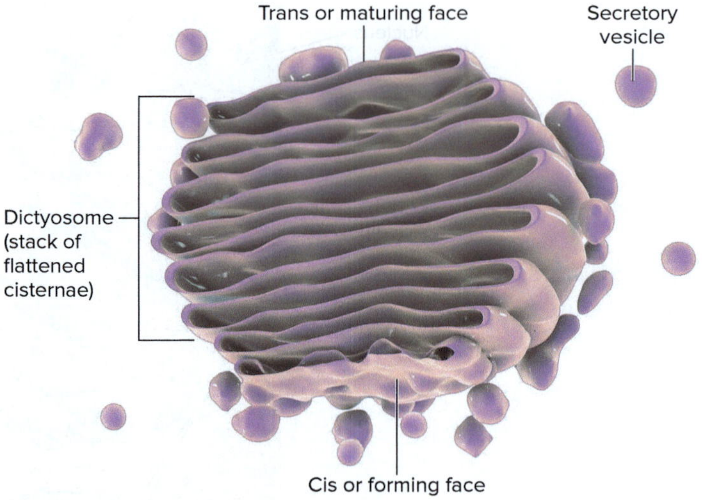

Figure 5.6 Golgi Apparatus. Each cisterna is 15 to 20 nm thick and separated from other cisternae by 20 to 30 nm.
Kateryna Kon/Shutterstock

The Golgi apparatus packages materials and prepares them for secretion; the exact nature of its role depends on the organism. For instance, the surface scales of some flagellated protists are constructed within the Golgi apparatus and then transported to the surface in vesicles. The Golgi often participates in the development of cell membranes and the packaging of cell products. The growth of some fungal hyphae occurs when Golgi vesicles contribute their contents to the wall at the hyphal tip. ▶| *Protists (chapter 23); Fungi (chapter 24)*

Lysosomes Are Degradative Organelles that Function in Endocytic Pathways

Lysosomes are found in animal cells because they are involved in digesting nutrients. They are roughly spherical, are enclosed in a single membrane, and average about 500 nm in diameter but range from 50 nm to several micrometers in size **figure 5.7**. Lysosomes contain the enzymes needed to digest all types of macromolecules. These enzymes, called *hydrolases,* catalyze the hydrolysis of molecules and function best under slightly acidic conditions

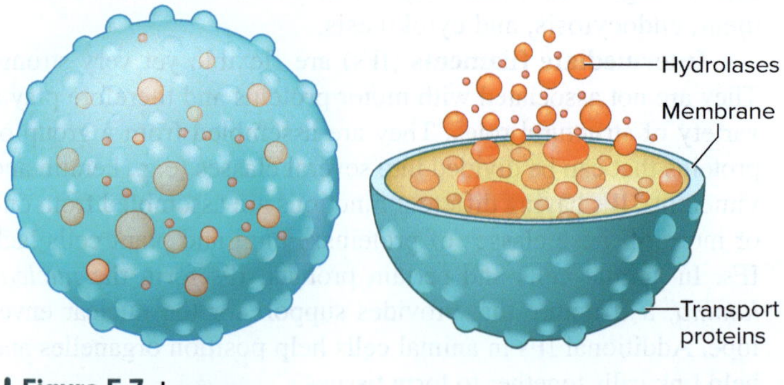

Figure 5.7 Lysosome.
Designua/Shutterstock

MICRO INQUIRY *How and why do lysosomes maintain an acidic environment?*

(usually around pH 3.5 to 5.0). Lysosomes maintain an acidic environment by pumping protons into their interior. *Lysosomes*

Organelles with the same function as lysosomes are found in fungal and protist cells, where they instead may be called vacuoles, phagocytic vacuoles, or food vacuoles. In addition to functioning in intracellular digestion, they may have other functions, including storage of calcium ions, phosphate, and amino acids.

Transport and Secretory Pathways Move Molecules to Other Locations, Including Outside the Cell

Various pathways are used to move materials within a cell. The early steps in these pathways are the same as those used to move materials into the plasma membrane or outside the cell. Thus they are often described together and referred to as the **secretory pathway.** The process of moving materials such as proteins and lipids involves numerous molecules that help target the protein or lipid to its proper location. The movement of proteins is of particular importance and is our focus.

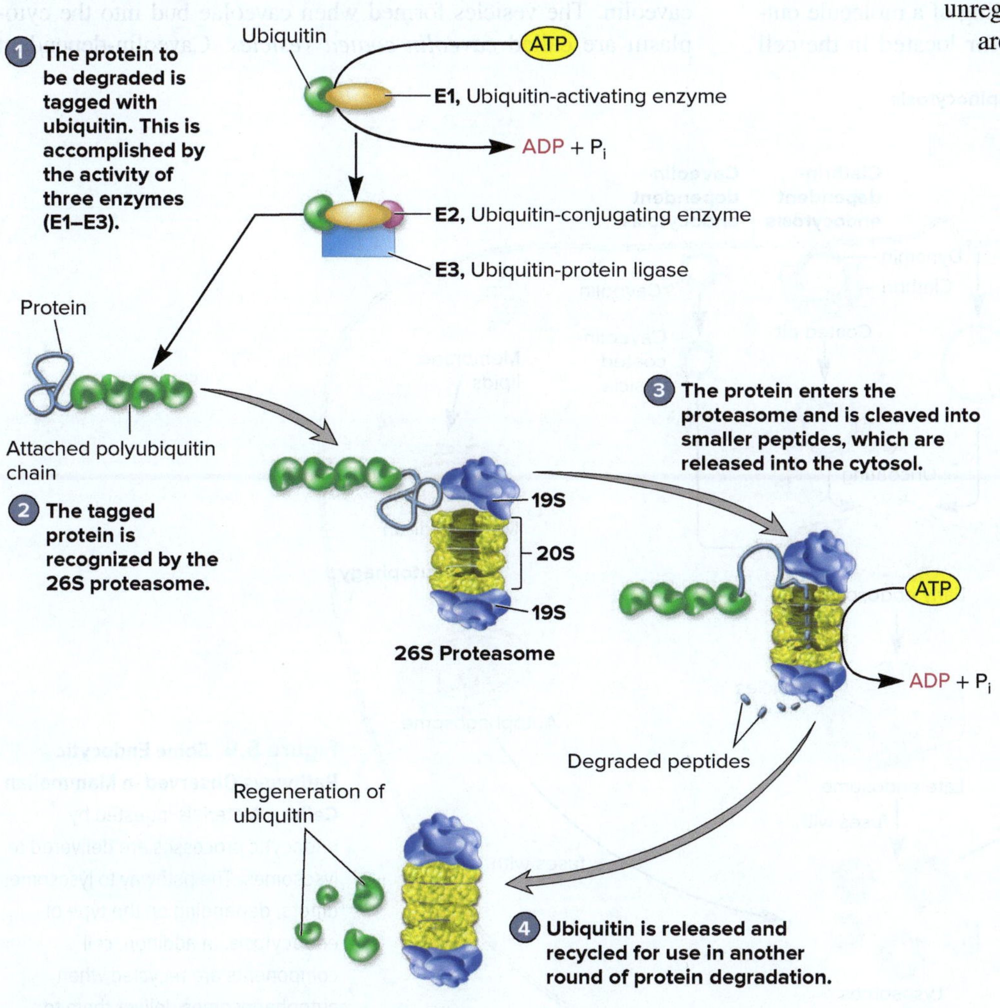

Figure 5.8 Proteasome Degradation of Proteins.

MICRO INQUIRY *Why must proteins be unfolded when they enter the proteasome? What do you think happens to the peptides after exiting the proteasome?*

Initially, proteins destined for a cellular target or secretion follow a similar path. Proteins synthesized by ribosomes attached to the rough endoplasmic reticulum (RER) have sequences of amino acids that target them to the lumen of the RER. These proteins move through the RER lumen until they are released in small vesicles that bud from the ER. On their journey through the ER, sugars are often added to the proteins—a process known as *glycosylation.* *Protein localization and secretion in eukaryotes (section 15.4)*

Vesicles released from the ER travel to the cis face of the Golgi apparatus (the side closest to the ER; figure 5.6), to which they may fuse. The proteins then proceed from the cis to the trans side of the Golgi (the side farthest from the ER) and are further modified. Some of these modifications target the proteins for their final location.

Transport vesicles released from the trans face of the Golgi deliver their contents to various locations within the cell. If the destination is the cell surface or the external environment, secretory vesicles are released and one of two pathways is used. The *constitutive secretory pathway* delivers proteins in an unregulated manner. In this pathway, proteins are released to the outside of the cell as the secretory vesicle fuses with the plasma membrane. The other pathway, called the *regulated secretory pathway,* requires a signal before its secretory vesicles fuse with the plasma membrane and release their contents.

It is important that only properly made proteins reach their final destination. Therefore, the cell has a built-in quality assurance mechanism to make sure proteins that fail to fold or are misfolded are not transported to their final destination. Instead they are released into the cytosol, where they are destroyed. However, because there are many normal proteins in the cytosol, proteins with mistakes must be labeled so only they are eliminated. This is accomplished by attaching several small polypeptides, called ubiquitin, to the protein (**figure 5.8**). The addition of ubiquitin marks a protein for degradation by a huge, cylindrical complex called a 26S **proteasome.** The ubiquitin-tagged protein is fed into the 26S proteasome much like a log into a wood chipper. It is broken down to smaller peptides using energy supplied by adenosine triphosphate (ATP) hydrolysis. As the protein is broken down, ubiquitin polypeptides are released and reused. Proteasomes and proteasome-like structures have also been identified in some bacteria and archaea. However, they are simpler in structure and the proteins to be degraded are targeted to the degradation machinery by mechanisms other than ubiquitin.

Endocytic Pathways Bring Materials into the Cell

Endocytosis is used by all eukaryotic cells to bring materials into the cell from the outside. During endocytosis, a cell takes up solutes (pinocytosis) or particles (phagocytosis) by enclosing them in vesicles pinched off from the plasma membrane. Endocytosis occurs regularly in all eukaryotic cells as a mechanism for recycling molecules in the membrane. The cellular steps that accomplish endocytosis are called **endocytic pathways.** Some eukaryotic cells have specialized endocytic pathways that allow them to concentrate materials outside the cell before bringing them in. Others use endocytic pathways as a feeding mechanism. Endocytosis is not found in bacteria and archaea; on the contrary, eukaryotic cells use endocytosis to engulf many viruses and other intracellular pathogens.

When does a cell "know" when to undergo endocytosis? Endocytosis is often triggered by the binding of a molecule outside the cell, called a **ligand,** to a **receptor** located in the cell membrane. This is referred to as **receptor-mediated endocytosis.** The nature of ligands varies (e.g., hormones or molecules on the surface of a virus). Sometimes receptors are concentrated in certain areas of the membrane, and receptor-mediated endocytosis serves to concentrate the ligand.

Mammalian cells have several types of endocytic pathways (**figure 5.9**). All generate an endocytic vesicle of some type, and the substances in the vesicle are eventually delivered to a lysosome for degradation. **Phagocytosis** involves the use of protrusions from the cell surface to surround and engulf particulates. The endocytic vesicles formed by phagocytosis are called **phagosomes. Clathrin-dependent endocytosis** begins with specialized membrane regions coated with the protein clathrin on the cytoplasmic side; these are called *coated pits*. The endocytic vesicles formed when coated pits invaginate are then called *coated vesicles*. Clathrin-dependent endocytosis is used to internalize hormones, growth factors, iron, and cholesterol. **Caveolin-dependent endocytosis** involves caveolae ("little caves"), tiny, cup-shaped invaginations of the plasma membrane (about 50 to 80 nm in diameter) enriched in cholesterol and the membrane protein caveolin. The vesicles formed when caveolae bud into the cytoplasm are called *caveolin-coated vesicles*. Caveolin-dependent

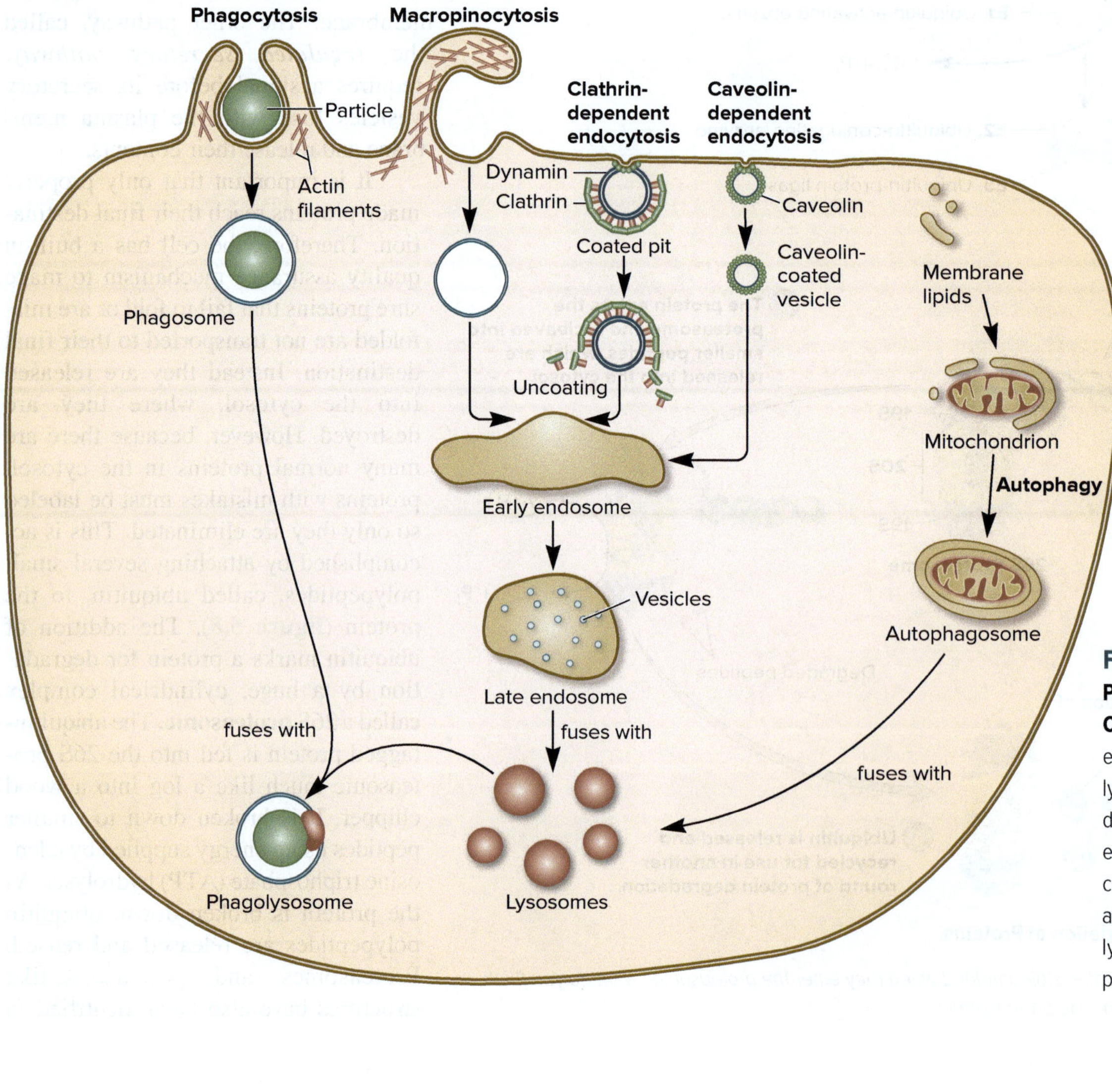

Figure 5.9 Some Endocytic Pathways Observed in Mammalian Cells. Materials ingested by endocytic processes are delivered to lysosomes. The pathway to lysosomes differs, depending on the type of endocytosis. In addition, cell components are recycled when autophagosomes deliver them to lysosomes for digestion. This process is called autophagy.

endocytosis has been implicated in signal transduction, transport of small molecules such as folic acid, as well as transport of macromolecules. Caveolae also are used by many viruses, bacteria, and protozoa to enter host cells.

The next step in degradation of endocytosed material is delivery to lysosomes. How this proceeds depends on the endocytic pathway. Clathrin-coated vesicles and caveolin-coated vesicles first deliver their contents to small organelles containing hydrolytic enzymes. These organelles are called early **endosomes** (figure 5.9). Early endosomes mature into late endosomes and then fuse with lysosomes, which deliver more degradative compounds. Phagosomes skip the early endosome stage and simply fuse directly with lysosomes (figure 5.9) to form a phagolysosome.

Endocytosis in eukaryotic microbes has not been studied as extensively as it has in mammals. It is known that some eukaryotic microbes carry out phagocytosis and probably macropinocytosis (figure 5.9). Clathrin-dependent endocytosis has been observed in fungi and is best characterized in the yeast *Saccharomyces cerevisiae*.

Lysosomes are involved in another important process called **autophagy** (Greek *auto,* self; *phagia,* to eat). Several types of autophagy have been identified. Some selectively sequester and destroy bacteria, viruses, aggregated proteins, and mitochondria. However, these selective processes are less well studied than **macroautophagy,** which nonselectively digests and recycles cytoplasmic components (including organelles such as mitochondria). This is a normal process that helps maintain cellular homeostasis.

During macroautophagy, materials to be digested are surrounded by a double membrane to form a structure called an **autophagosome** (figure 5.9). Autophagosome formation begins with creation of an isolation membrane. Numerous autophagy-related proteins and lipids are involved in the formation of the isolation membrane and its subsequent maturation into the autophagosome. The mature autophagosome ultimately fuses with a lysosome.

No matter the route taken, digestion occurs upon lysosome fusion. The lytic enzymes delivered by the lysosome generate small products of digestion, which enter the cytosol, where they are used as nutrients or for other purposes. The resulting vacuole containing undigested material is sometimes called a **residual body.** In some cases, the residual body releases its contents to the cell exterior.

Extracellular Vesicles

Like bacteria and archaea, all types of eukaryotic cells release **extracellular vesicles (EVs)** containing a wide variety of molecules, including proteins, lipids, and nucleic acids. EVs in mammalian cells transport molecules that contribute to cell signaling and the development of an immune response. Fungi and protist pathogens produce EVs that carry molecules that can harm the host and suppress immune responses.

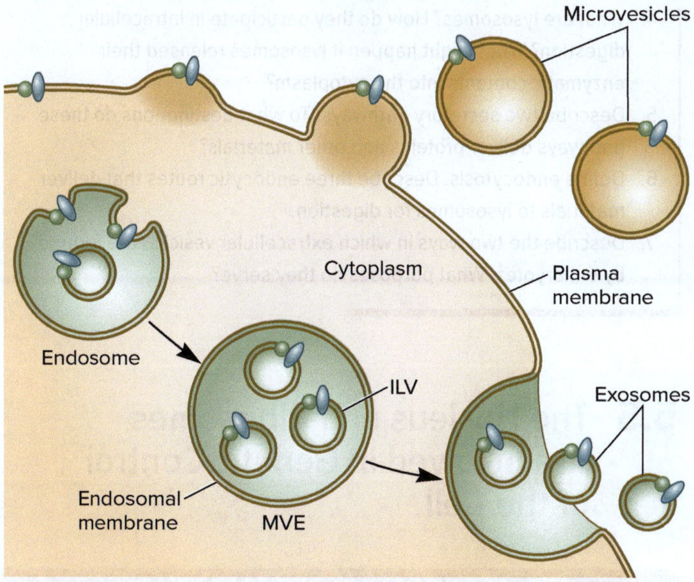

Figure 5.10 Extacellular Vesicle (EV) Formation. Microvesicles are EVs formed by budding from the plasma membrane. Exosomes arise when a region of endosome membrane buds inwardly, forming intra-luminal vesicles (ILVs). An endosome filled with ILVs is called a multivesicular endosome (MVE). Exosomes are released when an MVE and plasma membranes fuse. Note that microvesicles are often much larger than exosomes, but both carry a diverse range of molecules.

All EVs are formed by pinching off membrane that then surrounds their molecular cargo. Because eukaryotes have intracellular membranes, it follows that their EVs can arise from plasma or organelle membranes (**figure 5.10**). To distinguish these origins, the term *microvesicles* (50–1,000 nm in diameter) is used to refer to EVs released following outward budding of the plasma membrane. EVs generated within endosomes are called *exosomes* (30–100 nm). Exosome production is a multistep process. First, the endosomal membrane inwardly buds, forming intraluminal vesicles (ILVs). As the endosome fills up with ILVs, it is termed a multivesicular endosome (MVE). MVEs migrate to the cytoplasmic face of the plasma membrane, where the MVE and plasma membranes fuse and mature exosomes are expelled. Fungi and algae present an interesting dilemma for EVs: traversing the cell wall to finally reach the outside environment. How this is accomplished is not well understood.

Comprehension Check

1. How do RER and SER differ in terms of structure and function? List the processes in which the ER is involved.
2. What is a proteasome? Why is it important to the proper functioning of the ER?
3. Describe the structure of a Golgi apparatus. How do the cis and trans faces of the Golgi apparatus differ? List the major functions of the Golgi.

4. What are lysosomes? How do they participate in intracellular digestion? What might happen if lysosomes released their enzymatic contents into the cytoplasm?
5. Describe two secretory pathways. To what destinations do these pathways deliver proteins and other materials?
6. Define endocytosis. Describe three endocytic routes that deliver materials to lysosomes for digestion.
7. Describe the two ways in which extracellular vesicles are formed by eukaryotes. What purposes do they serve?

5.5 The Nucleus and Ribosomes Are Involved in Genetic Control of the Cell

After reading this section, you should be able to:

a. Describe the structure and function of the nucleus, chromosomes, nucleolus, and ribosomes
b. Differentiate chromosomes from chromatin
c. Compare and contrast the chromosomes and ribosomes of bacterial, archaeal, and eukaryotic cells

Perhaps the most profound way in which eukaryotic cells differ from bacterial and archaeal cells is the eukaryotic nucleus and the way DNA is stored and used. In this section, the organelles involved with these important cellular functions are discussed.

Nucleus

The nucleus is by far the most visually prominent organelle in eukaryotic cells. It was discovered early in the study of cell structure and was shown by Robert Brown (1773–1858) in 1831 to be a constant feature of eukaryotic cells. The **nucleus** is home to the cell's DNA; RNA is made in the nucleus but is then exported.

A cell's nucleus is a membrane-bound spherical body about 5 to 7 μm in diameter (figures 5.2 and 5.3). The exact number of chromosomes within a given nucleus depends on the organism, cell type, and stage in the life cycle. Each eukaryotic **chromosome** is composed of chromatin. **Chromatin** is a complex of DNA and proteins, including histones. Histones are small proteins rich in the basic amino acids lysine and arginine. As basic proteins, histones are positively charged; recall that DNA is acidic and negatively charged. This allows histones to associate closely with DNA. Histones provide structural support and condense chromosomes so they fit in the nucleus. Some histones serve a spool-like function, with DNA wrapped around the proteins. Eight histone molecules plus their associated DNA form a "beads-on-a-string" appearance (**figure 5.11**). Each bead is called a **nucleosome.** *Nucleoid (sections 3.6 and 4.3)*

The nucleus is surrounded by the **nuclear envelope,** a complex structure consisting of two lipid bilayer membranes, the inner and outer membranes, separated by a perinuclear space. The nuclear envelope is continuous with the ER at several points, and its outer membrane is covered with ribosomes. Many nuclear pores penetrate the envelope, and each pore is formed by about 30 proteins; each pore plus the associated proteins is called a **nuclear pore complex** (**figure 5.12**). Pores are about

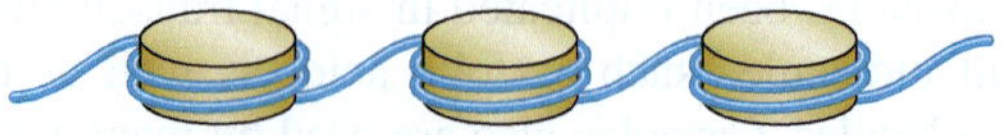

Figure 5.11 Nucleosomes. DNA is wrapped around an ellipsoid composed of eight histone proteins (two each of H2A, H2B, H3, and H4). In most eukaryotes, the histone H1 associates with the nucleosome.

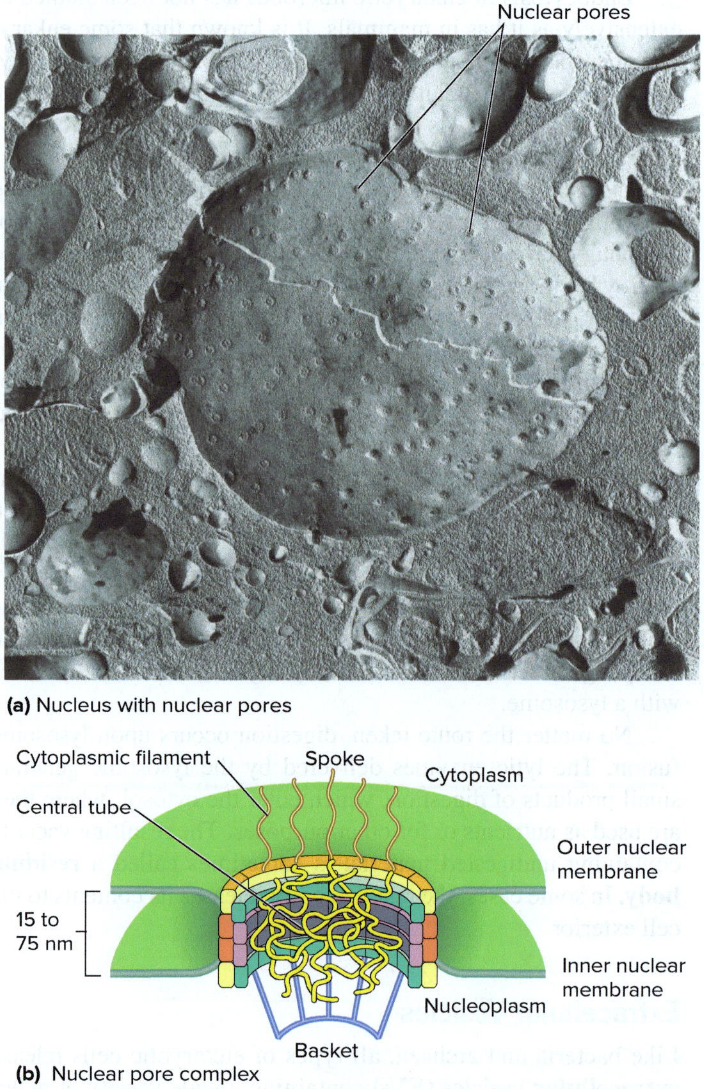

Figure 5.12 The Nucleus and Nuclear Pore Complex. (a) A transmission electron micrograph (TEM) of a freeze-etch preparation of a cell, showing the large, convex nuclear surface with pores scattered over it. (b) Schematic of the nuclear pore complex. Note that the nuclear envelope consists of two lipid bilayer membranes: outer nuclear membrane and inner nuclear membrane.

(a) Biophoto Associates/Science Source

70 nm in diameter and collectively occupy about 10 to 25% of the nuclear surface. The nuclear pore complexes serve as transport routes between the nucleus and surrounding cytoplasm. Small molecules move through the nuclear pore complex unaided. Large molecules are transported with the help of nuclear pore complex proteins.

Often the most noticeable structure within the nucleus is the **nucleolus**—the site of ribosomal RNA (rRNA) synthesis (figure 5.3). Recall that rRNA is needed to build ribosomes, the site of protein synthesis. Therefore, depending on how many ribosomes the cell needs, a single nucleus may contain one to many nucleoli. Although the nucleolus is not membrane-enclosed, it is considered a complex organelle with separate granular and fibrillar regions. ◀ *Microbial diversity & ecology 3.1*

The DNA that encodes the rRNA is located on a highly condensed chromosomal region called the *nucleolar organizer.* This RNA is synthesized in a single long piece that is cut to form the final rRNA molecules. The processed rRNAs combine with ribosomal proteins (which have been synthesized in the cytoplasm and transported to the nucleus) to form partially completed ribosomal subunits. Immature ribosomal subunits then leave the nucleus, by way of the nuclear pore complexes, and mature in the cytoplasm.

Eukaryotic Ribosomes Are Different from Bacterial and Archaeal Ribosomes

Eukaryotic ribosomes are larger than bacterial and archaeal 70S ribosomes with the exception of those found in mitochondria and chloroplasts. Each ribosome is a dimer of a 60S and a 40S subunit. The 60S subunit is composed of three rRNA molecules (5S, 28S, and 5.8S rRNAs) and about 50 proteins. The 40S subunit is composed of an 18S rRNA and about 30 proteins. Each ribosome is about 22 nm in diameter and has a sedimentation coefficient of 80S and a molecular weight of 4 million (**figure 5.13**). Eukaryotic ribosomes are either associated with the endoplasmic reticulum or free in the cytoplasm. When bound to the endoplasmic reticulum to form rough ER, they are attached through their 60S subunits.

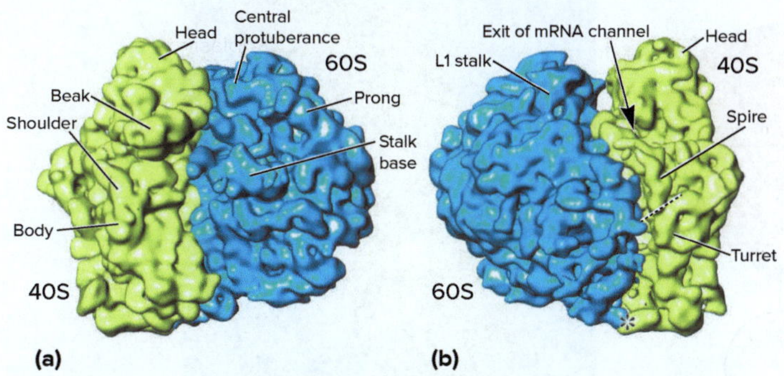

(a) (b)

Figure 5.13 Eukaryotic Ribosome. This image showing two views of the 80S ribosome of *Trypanosoma cruzi* was generated using cryo-electron microscopy. The image is a density map with a resolution to 12 Å. Landmarks of the ribosome are shown; compare these to bacterial 70S ribosomes shown in figure 3.31.

MICRO INQUIRY *Which subunit attaches to the rough ER?*

Both free and ER-bound ribosomes synthesize proteins. Recall that proteins made on the ribosomes of the RER are often secreted or are inserted into the ER membrane as integral membrane proteins. Free ribosomes are the sites of synthesis for nonsecreted and nonmembrane proteins. Some proteins synthesized by free ribosomes are transported to organelles such as the nucleus, mitochondrion, and chloroplast. As we discuss in chapters 13 and 15, proteins called molecular chaperones aid the proper folding of proteins after synthesis. They also assist the transport of proteins into eukaryotic organelles such as mitochondria.

Comprehension Check

1. What proteins contribute to the folding of eukaryotic chromosomes?
2. Why is the nucleolus important for protein synthesis?
3. Describe the structure of the eukaryotic 80S ribosome and contrast it with bacterial and archaeal ribosomes.
4. How do free ribosomes and those bound to the ER differ in function?

5.6 Mitochondria, Related Organelles, and Chloroplasts Are Involved in Energy Conservation

After reading this section, you should be able to:

a. Draw a mitochondrion and identify its component parts
b. Compare and contrast mitochondria and hydrogenosomes in terms of their structure and the chemical processes they carry out
c. Draw a chloroplast and identify its component parts

Three important energy-conserving organelles are the focus of this section: mitochondria, hydrogenosomes, and chloroplasts. They are of scientific interest not only for their role in cellular energy metabolism but also because of their evolutionary history. All are thought to be derived from bacterial cells that invaded or were ingested by early ancestors of eukaryotic cells (*see figure 1.7*). ◀ *Evolution of the three domains of life (section 1.2)* ▶ *The origin of mitochondria (section 21.1)*

Aerobic Mitochondria Belong to the Family of Mitochondrial-Related Organelles

Eukaryotic microbes contain numerous organelles related to mitochondria. It is thought they arose from a single endosymbiotic event, as we describe in chapter 1, but then different evolutionary paths were taken to yield at least five different mitochondrial-related organelles, or MROs (**figure 5.14**). In this section we describe two MROs: the aerobic mitochondrion and the hydrogenosome.

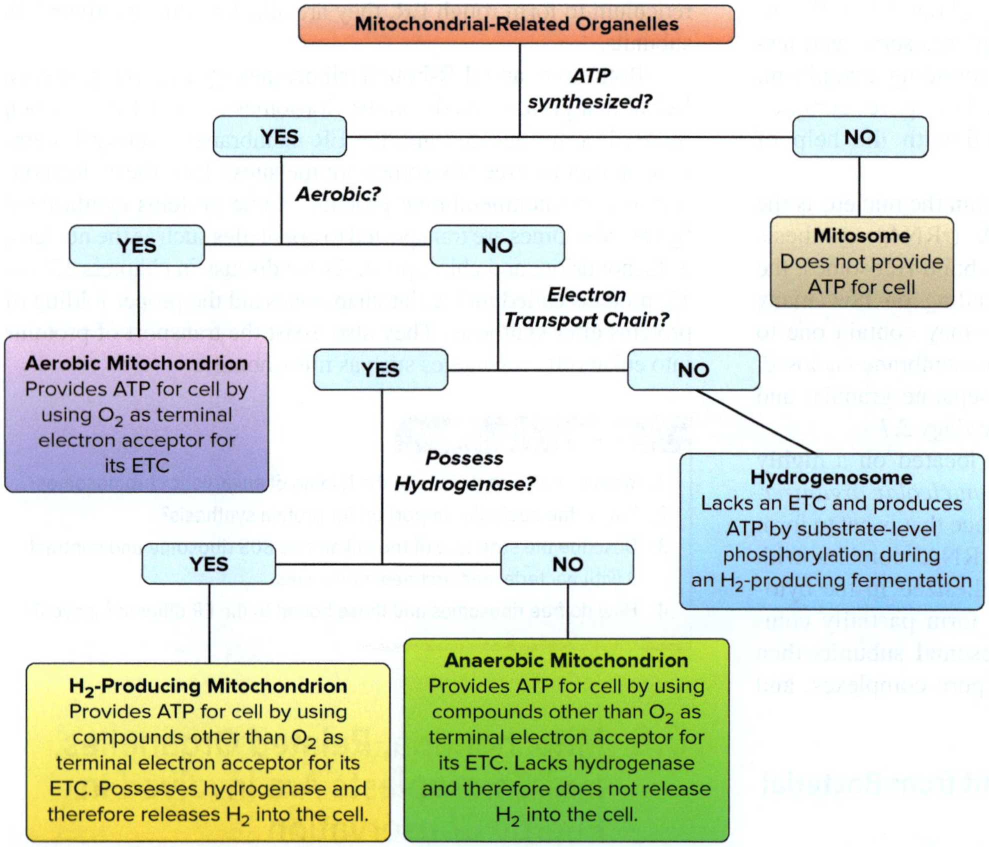

Figure 5.14 Mitochondrial-Related Organelles. It is thought that the ancestral endosymbiont evolved in different cell lineages to give rise to these five different organelles. Descriptions of the metabolic processes noted are described in chapter 11. ETC, electron transport chain.

The aerobic mitochondrion is the organelle in this family most familiar to biology students. For our discussion, we will simply refer to it as the mitochondrion. Found in most eukaryotic cells, these **mitochondria** (s., **mitochondrion**) frequently are called the powerhouses of the cell (**figure 5.15**). Metabolic processes such as the tricarboxylic acid cycle and generation of ATP, the major energy currency of all life forms, take place here, but only when oxygen is available. Many eukaryotes have cylindrical mitochondria that measure approximately 0.3 to 1.0 μm by 5 to 10 μm. Some cells possess 1,000 or more mitochondria; some yeasts, unicellular algae, and trypanosome protists have a single, giant, tubular mitochondrion twisted into a continuous network permeating the cytoplasm. ▶| *Electron transport and oxidative phosphorylation generate the most ATP (section 11.4)*

Like diderm bacteria, mitochondria are bounded by two membranes: an outer membrane separated from an inner membrane by a 6- to 8-nm intermembrane space (figure 5.15). The outer mitochondrial membrane contains porins also like diderm bacteria (*see figure 3.23*). The inner membrane has infoldings called **cristae** (s., **crista**), which greatly increase its surface area. The shape of cristae differs in mitochondria from various organisms. Platelike (laminar) cristae, cristae shaped like disks, tubular cristae, and cristae in the shape of vesicles have all been observed. The inner membrane encloses the mitochondrial matrix, a dense material containing ribosomes, DNA, and often large calcium

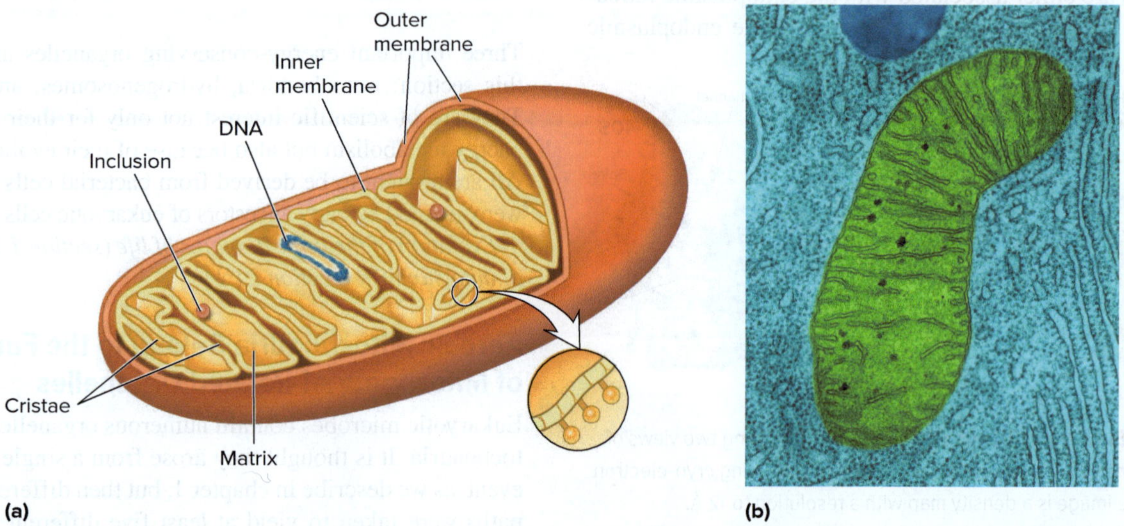

Figure 5.15 Mitochondrial Structure. (a) A diagram of mitochondrial structure. The insert shows the ATP-synthesizing enzyme ATP synthase lining the inner surface of the cristae. (b) Colorized transmission electron micrograph of a mitochondrion showing the cristae.

(b) Science History Images/Alamy Stock Photo

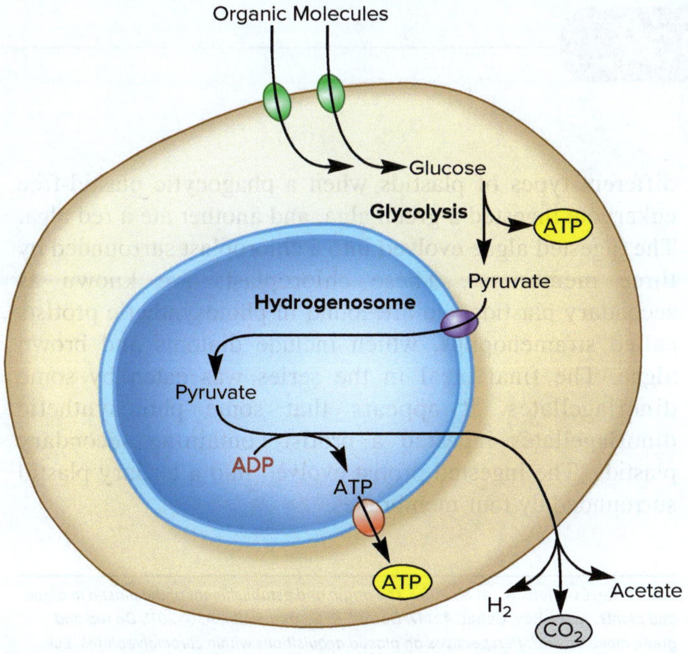

Figure 5.16 **The Metabolic Activities of Hydrogenosomes.**

phosphate granules. In many organisms, mitochondrial DNA is a closed circle, like most bacterial DNA. However, in some protists, mitochondrial DNA is linear.

Each mitochondrial compartment has a characteristic chemical and enzymatic composition. For example, the inner mitochondrial membranes possess enzymes and electron carriers involved in electron transport and oxidative phosphorylation (the formation of ATP during respiration). Enzymes of the tricarboxylic acid cycle (*see figure 11.8*) and those involved with the catabolism (breaking down) of fatty acids are located in the matrix. ▶| *Lipid catabolism (section 11.7)*

Mitochondria are unlike other organelles because they possess their own DNA and ribosomes. This reflects their bacterial ancestry; however, mitochondria synthesize only some of their own proteins. Most mitochondrial proteins are manufactured under the direction of the nucleus and must be transported into the mitochondrion. In another similarity to bacteria, mitochondria reproduce by binary fission. ▶| *Bacterial cell cycles are divided into three phases (section 7.2); Protein localization and secretion in eukaryotes (section 15.4)*

Hydrogenosomes are small organelles involved in energy capture in some anaerobic protists (**figure 5.16**). Like mitochondria, hydrogenosomes are bound by a double membrane. They differ from mitochondria in terms of the method used to generate ATP. ATP is made during fermentation so hydrogenosomes do not require extensive internal membranes and lack cristae. In some hydrogenosome-bearing protists, the metabolic products are consumed by symbiotic bacteria and archaea living within the protist. ▶| *Fermentation does not involve an electron transport chain (section 11.6); Mutualism: obligatory positive interactions (section 27.2)*

Chloroplasts

Plastids are cytoplasmic organelles found in algae (photosynthetic protists) and plants. They often possess pigments such as chlorophylls and carotenoids (which capture light energy), and are the sites of synthesis and storage of food reserves. The most important type of plastid is the chloroplast. **Chloroplasts** contain chlorophyll and use light energy to convert CO_2 and water to carbohydrates and O_2; that is, they are the site of photosynthesis. Two major types of chloroplasts have been identified: those that evolved from a primary endosymbiotic event and those that evolved from a secondary (or tertiary) event (**Microbial Diversity & Ecology 5.1**). Plant and some algal chloroplasts are primary plastids and are the focus of this discussion.

Chloroplasts are quite variable in size and shape, but they share many structural features. Most are ellipsoid with dimensions of 2 to 4 µm by 5 to 10 µm, but some algae possess a single chloroplast that fills much of the cell. Like mitochondria, chloroplasts are enclosed within two membranes (**figure 5.17**). The inner membrane surrounds the stroma, which contains DNA, ribosomes, lipid droplets, and starch granules. Also located within the stroma is a complex internal membrane system that features flattened, membrane-bound sacs called **thylakoids.** Clusters of two or more thylakoids are dispersed within the stroma of most algal chloroplasts (figure 5.17). In some photosynthetic protists, several disklike thylakoids are stacked on each other like coins to form grana (s., granum).

Photosynthesis is defined as the combination of light-driven ATP and NADPH production, called the light reaction, and the use of that ATP and NADPH to convert CO_2 and water to carbohydrate, termed the dark reaction. The light reaction takes place in the thylakoid membranes, where chlorophyll and electron transport components are also found. The dark reaction of photosynthesis takes place in the stroma. ▶| *Phototrophy (section 11.10)*

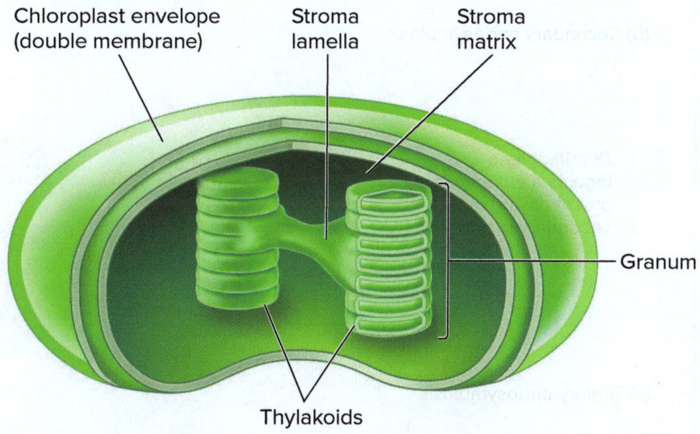

Figure 5.17 **Diagram of Chloroplast Structure.**

MICRO INQUIRY *In what regions of the chloroplast do light reactions occur? Dark reactions?*

MICROBIAL DIVERSITY & ECOLOGY

5.1 There Was an Old Woman Who Swallowed a Fly

The children's song "There Was an Old Woman Who Swallowed a Fly" describes the unusual eating habits of an old woman. Her story began when she ate a fly but continued as she ate a series of other animals, all in an attempt to get rid of the fly. The evolutionary history of chloroplasts also appears to involve a series of meals. Molecular studies of the chloroplasts of a wide variety of photosynthetic protists have revealed that some chloroplasts arose when a phagocytic ancestor of a eukaryotic cell engulfed the ancestor of a cyanobacterium (**box figure**). This evolutionary event eventually gave rise to plants and the photosynthetic protists commonly called green algae and red algae (*see chapter 23*). The chloroplasts that evolved from this endosymbiosis are the typical chloroplasts that biology students are familiar with. These primary plastids have two membranes. It is thought that these primary plastids gave rise to at least two different types of plastids when a phagocytic plastid-free eukaryote ingested a green alga, and another ate a red alga. The ingested algae evolved into a chloroplast surrounded by three membranes. These chloroplasts are known as secondary plastids and are found in photosynthetic protists called stramenopiles, which include diatoms and brown algae. The final meal in the series was eaten by some dinoflagellates. It appears that some photosynthetic dinoflagellates ingested a protist containing secondary plastids. The ingested protist evolved into a tertiary plastid surrounded by four membranes.

Source: Reyes-Prieto, A., et al. 2007. The origin and establishment of the plastid in algae and plants. Annu. Rev. Genet. *41:147; Dorrell, R. G., and Smith, A. G. 2011. Do red and green make brown?: Perspectives on plastid acquisitions within chromalveolates.* Euk. Cell *10:856.*

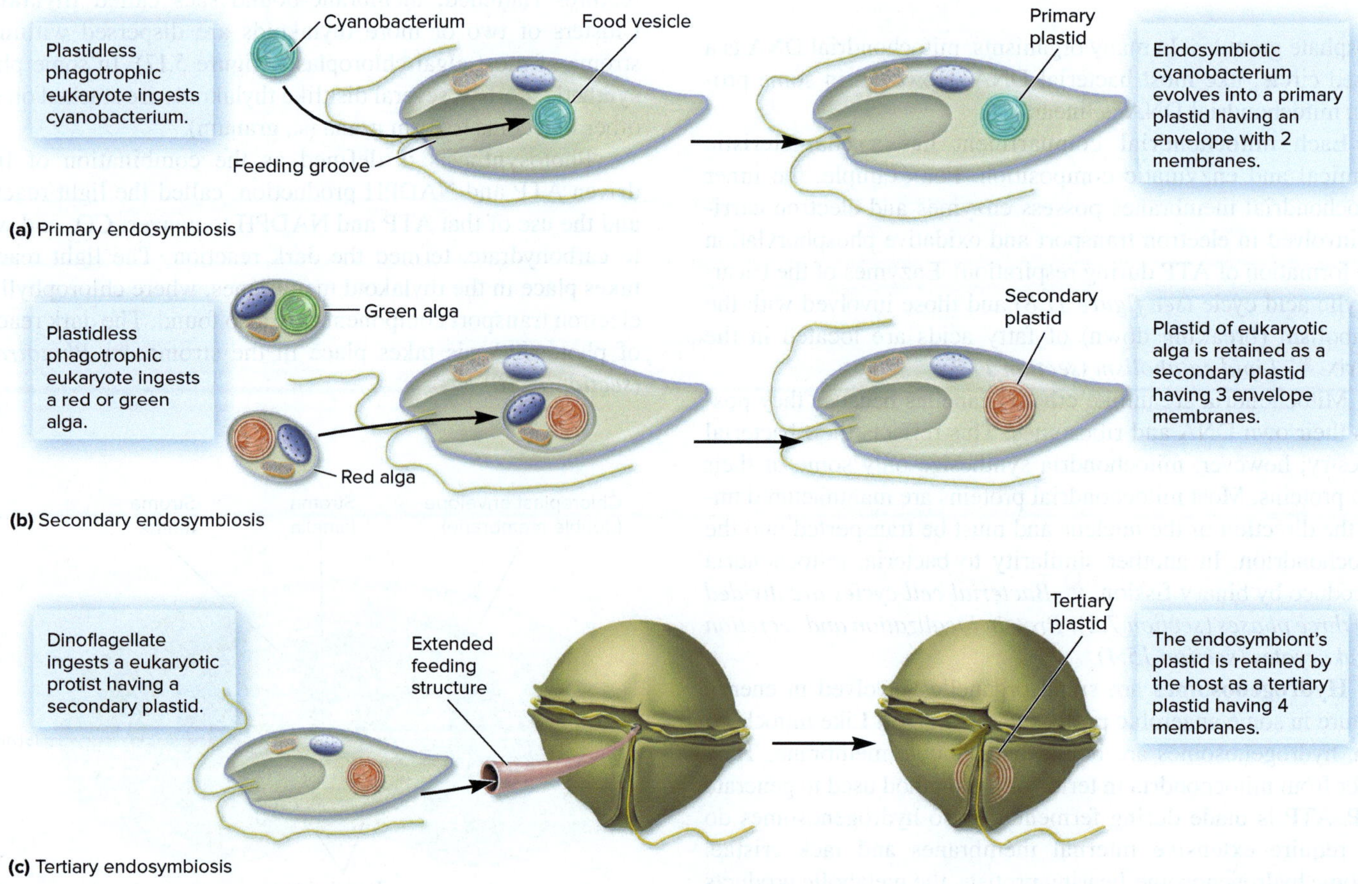

Comprehension Check

1. Differentiate between aerobic mitochondria, hydrogenosomes, and chloroplasts. Where are the different components of mitochondria and chloroplast energy-trapping systems located?
2. Why do hydrogenosomes lack cristae?
3. What is the role of mitochondrial DNA?

5.7 Many Eukaryotic Microbes Have External Structures Used for Motility

After reading this section, you should be able to:

a. Explain why the terms flagella, archaella, and cilia have been proposed for bacteria, archaea, and eukaryotes, respectively
b. Describe the structure of cilia, in general terms
c. Discuss the types of motility observed in eukaryotic microbes

Many microorganisms use extracellular appendages to move through their environments. Traditionally, these appendages were called flagella, regardless of the microbe. However, genomic, biochemical, and biomechanical analysis has demonstrated that "flagella" are evolutionarily and structurally distinct among bacteria, archaea, and eukaryotes. To reflect these differences, it has been proposed that the term *flagella* be reserved to describe appendages on bacteria, *archaella* for archaea, and *cilia* for eukaryotes (s., flagellum, archaellum, cilium, respectively). This nomenclature has not been universally accepted; the term **flagella** in eukaryotes continues to be used to describe long (100–200 mm) whiplike filaments, while the term **cilia** often refers to shorter (5–20 mm) hairlike structures.

The use of a single term for eukaryotes has been justified as cilia and eukaryotic flagella are structurally identical. Both are membrane-bound cylinders about 0.2 μm in diameter with hundreds of small, ATP-dependent motors called *dynein* that run the length of the appendage. Dynein is housed within a central filament called an *axoneme*. An axoneme usually consists of nine microtubule doublets arranged in a circle around two central microtubules. This is called the 9 + 2 pattern of microtubules. A basal body lies in the cytoplasm at the base of each cilium. Cilia bend because adjacent microtubule doublets slide along one another while maintaining their individual lengths. ATP powers the movement of cilia.

Despite their identical structure, eukaryotic flagella and cilia have distinct patterns of movement, making it useful to differentiate between them (**figure 5.18**). The beating of flagella causes them to undulate and generate planar or helical waves from the base to the tip, or vice versa. If the wave moves from base to tip, the cell is pushed along; a beat traveling from the tip toward the base pulls the cell through the water. Some flagella have lateral hairs called flimmer filaments (thicker, stiffer hairs are called mastigonemes). These filaments change flagellar action so that a wave moving down the filament toward the tip pulls the cell along instead of pushing it. Such flagella often are called *tinsel flagella,* whereas naked flagella are referred to as *whiplash flagella* (**figure 5.19**). Cilia, on the other hand, normally have a beat with two distinctive phases. In the effective stroke, the cilium moves through the surrounding fluid like an

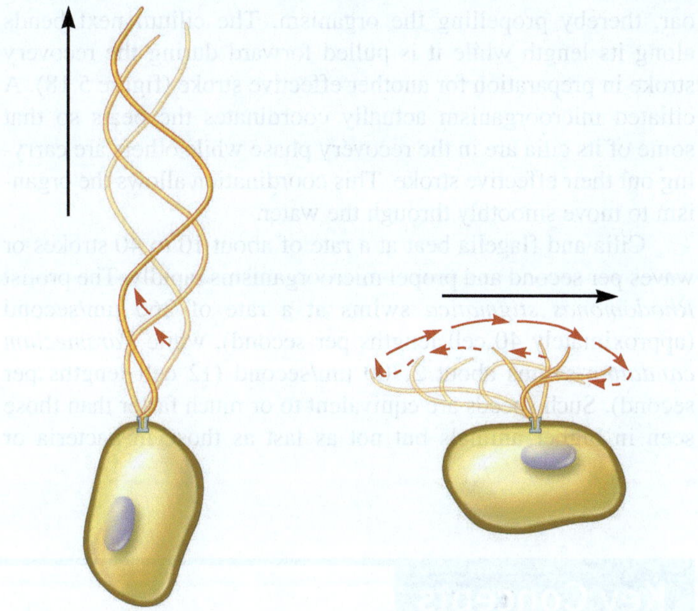

Figure 5.18 Patterns of Flagellar and Ciliary Movement. Flagellar and ciliary movement take the form of waves. To propel an organism, flagella (left illustration) move either from the base of the flagellum to its tip or in the opposite direction. The beat of a cilium (right illustration) may be divided into two phases. In the effective stroke, the cilium remains fairly stiff as it swings through liquid. This is followed by a recovery stroke in which the cilium bends and returns to its initial position. The black arrows indicate the direction of liquid movement in these examples.

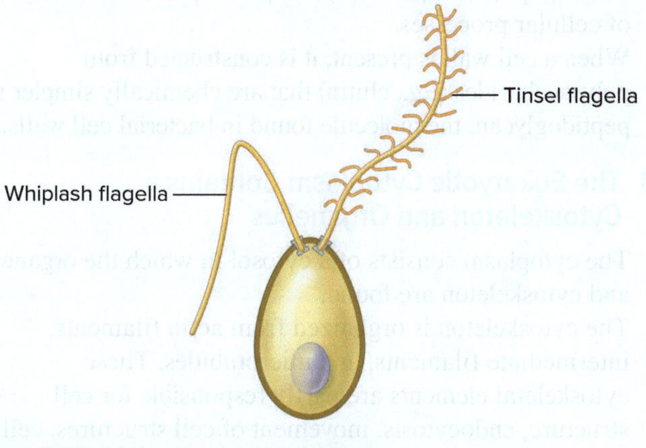

Figure 5.19 Whiplash and Tinsel Flagella.

oar, thereby propelling the organism. The cilium next bends along its length while it is pulled forward during the recovery stroke in preparation for another effective stroke (figure 5.18). A ciliated microorganism actually coordinates the beats so that some of its cilia are in the recovery phase while others are carrying out their effective stroke. This coordination allows the organism to move smoothly through the water.

Cilia and flagella beat at a rate of about 10 to 40 strokes or waves per second and propel microorganisms rapidly. The protist *Rhodomonas stigmatica* swims at a rate of 260 µm/second (approximately 40 cell lengths per second), while *Paramecium caudatum* swims about 2,700 µm/second (12 cell lengths per second). Such speeds are equivalent to or much faster than those seen in higher animals but not as fast as those in bacteria or archaea. ◀ *Archaella and motility (section 4.4);* ▶ *Motility and chemotaxis (section 14.6)*

Comprehension Check

1. What is the rationale changing the nomenclature of "flagella" as it pertains to bacteria, archaea, and eukaryotes?
2. Contrast the ways in which flagella and cilia propel eukaryotic microorganisms through water.
3. Calculate your running speed relative to that of *R. stigmatica:* 40 cell lengths per second. Is an equivalent running speed (body length per second) humanly possible?

Key Concepts

5.1 Eukaryotic Cells Are Diverse but Share Some Common Features

- Eukaryotic cells are generally larger than bacterial and archaeal cells, and they exhibit considerably more morphological diversity.
- The eukaryotic cell has a membrane-bound nucleus and many membranous organelles (**figures 5.2** and **5.3**).
- The membranous organelles form compartments in the cytoplasm of the cell. This allows the cell to carry out a variety of biochemical reactions simultaneously. It also provides more surface area for membrane-associated activities such as respiration.

5.2 Eukaryotic Cell Envelopes

- Eukaryotic membranes are similar in structure and function to bacterial membranes. The two differ in terms of their lipid composition (**figure 5.4**).
- Eukaryotic membranes contain lipid rafts enriched for certain lipids and proteins, and participate in a variety of cellular processes.
- When a cell wall is present, it is constructed from polysaccharides (e.g., chitin) that are chemically simpler than peptidoglycan, the molecule found in bacterial cell walls.

5.3 The Eukaryotic Cytoplasm Contains a Cytoskeleton and Organelles

- The cytoplasm consists of a cytosol in which the organelles and cytoskeleton are found.
- The cytoskeleton is organized from actin filaments, intermediate filaments, and microtubules. These cytoskeletal elements are partly responsible for cell structure, endocytosis, movement of cell structures, cell division, and motility (**figure 5.5**).
- Actin filaments and microtubules are found in eukaryotic microbes. Actin filaments are composed of actin proteins; microtubules are composed of α-tubulin and β-tubulin.
- Intermediate filaments are assembled from a heterogeneous family of proteins. They are found in animal cells and some protozoa but have not been identified in fungi or plant cells.
- Uniquely eukaryotic organelles are numerous and include the endoplasmic reticulum, Golgi, mitochondria, and chloroplasts.

5.4 Several Organelles Function in the Secretory and Endocytic Pathways

- Specific organelles and vesicles are involved in the synthesis and secretion of materials (secretory pathways). Some are involved in the uptake of materials from the extracellular milieu (endocytic pathways).
- The endoplasmic reticulum (ER) is an irregular network of tubules and flattened sacs (cisternae). Rough ER has ribosomes attached to it. Smooth ER lacks ribosomes.
- The ER donates some materials to the Golgi apparatus, an organelle usually composed of one or more stacks of cisternae (**figure 5.6**). The Golgi prepares and packages cell products for secretion.
- The Golgi apparatus buds off vesicles that deliver hydrolytic enzymes and other proteins to lysosomes and lysosome-like organelles. Lysosomes contain hydrolytic enzymes that aid in intracellular digestion of materials delivered to them by endocytosis and macroautophagy (**figure 5.7**).
- Misfolded proteins are marked for destruction by the addition of a polypeptide called ubiquitin and then degraded by proteasomes (**figure 5.8**).
- Eukaryotic cells use several kinds of endocytosis (**figure 5.9**). These include phagocytosis, clathrin-dependent endocytosis,

and caveolin-dependent endocytosis. Some macromolecules are bound to receptors prior to endocytosis in a process called receptor-mediated endocytosis.

- Eukaryotic cells release two types of extracellular vesicles. Microvesicles arise from the plasma membrane and exosomes are generated within endosomes that then fuse with the plasma membrane (**figure 5.10**).

5.5 The Nucleus and Ribosomes Are Involved in Genetic Control of the Cell

- The nucleus is a large organelle containing the cell's chromosomes. It is surrounded by a double-membrane envelope perforated by pores for the nuclear import and export of cellular materials (**figure 5.12**).
- The nucleolus lies within the nucleus and participates in the synthesis of ribosomal RNA and ribosomal subunits.
- Eukaryotic ribosomes are found either free in the cytoplasm or bound to the ER. They are 80S in size and differ from bacterial and archaeal ribosomes in their size and molecular make-up (**figure 5.13**).

5.6 Mitochondria, Related Organelles, and Chloroplasts Are Involved in Energy Conservation

- The family of mitochondrial-related organelles consists of at least five members, including aerobic mitochondria and hydrogenosomes (**figure 5.14**).
- Aerobic mitochondria are organelles bounded by two membranes, with the inner membrane folded into cristae (**figure 5.15**). They are the site of energy conservation by cell respiration, which involves electron transport and oxidative phosphorylation.
- Hydrogenosomes have a double membrane and carry out fermentation that generates hydrogen gas (**figure 5.16**). They are found in certain protozoa.
- Chloroplasts are pigment-containing organelles that serve as the site of photosynthesis (**figure 5.17**). The trapping of light energy takes place in the thylakoid membranes of the chloroplast, whereas CO_2 fixation occurs in the stroma.

5.7 Many Eukaryotic Microbes Have External Structures Used for Motility

- Cilia and flagella are membrane-enclosed organelles that provide motility. Both contain nine microtubule doublets surrounding two central microtubules.
- Cilia and flagella are structurally identical, although cilia are shorter and more numerous. Both exhibit whiplike motion. This occurs when the microtubule doublets slide along each other, causing the cilium or flagellum to bend.

Active Learning

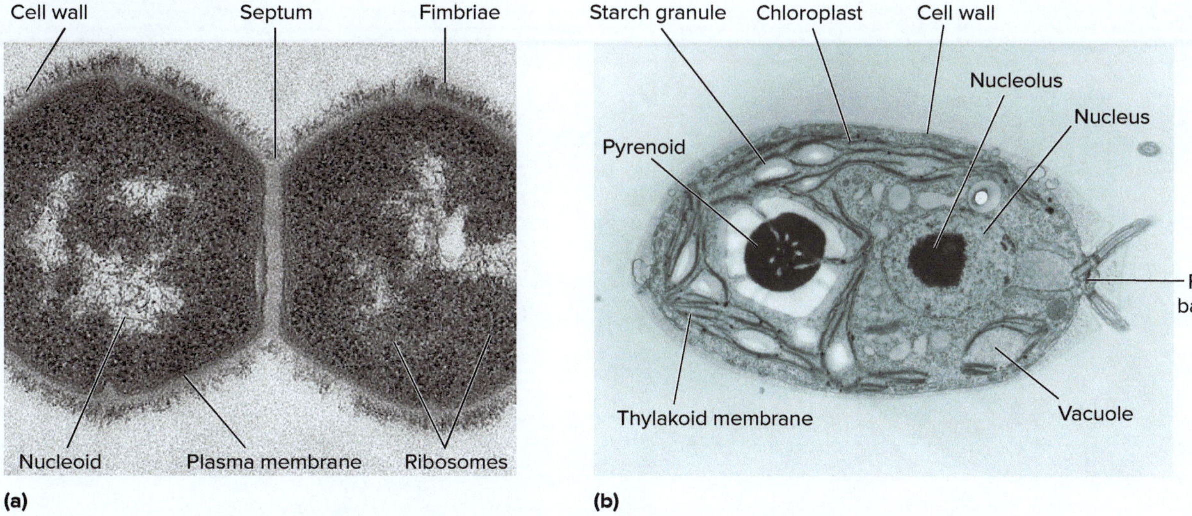

Comparison of Bacterial and Eukaryotic Cell Structure. (a) TEM of the bacterium *Streptococcus pyogenes* undergoing cell division. The septum separates the two daughter cells. Each cell is about 1 µm in diameter. (b) TEM of the eukaryotic alga *Chlamydomonas reinhardtii,* which measures about 10 µm in diameter.

(a) ©Vincent A. Fischetti, Ph.D/Rockefeller University, www.rockefeller.edu/vaf; (b) Biophoto Associates/Science Source

1. Examine the figure above, which shows transmission electron micrographs of a bacterial and algal cell. Draw a Venn diagram that includes all the shared and unique morphological and functional features you can think of for the two microoganisms.
2. Bacterial and archaeal cell sizes are limited by the rate at which nutrients diffuse from one place in the cell to another. Are eukaryotic microorganisms similarly limited? Why or why not?
3. Compare the mechanisms by which most eukaryotic cells acquire nutrients with those used by bacteria and archaea. What are the implications in terms of survival in different habitats and competition between eukaryotic microbes such as protists and bacteria and archaea?
4. Many pathogenic protozoa release extracellular vesicles (EVs) during infection, although EV contribution to disease processes has only recently been explored. One such microbe, *Acanthamoeba castellanii,* can cause eye and brain infections. *A. castellanii* EVs contain about 130 different proteins within them. The majority of these proteins are hydrolases (degradative enzymes). Propose a hypothesis that predicts the role of *A. castellanii* EVs in causing disease and outline an experiment or series of experiments designed to test your hypothesis. (Don't forget to include controls!)

CHAPTER

6

Viruses and Other Acellular Infectious Agents

Klaus Tiedge/Blend Images LLC

Viruses to the Rescue

Imagine that to treat a bacterial infection, instead of going to the pharmacist for an antibiotic, you might go to the local virologist for a sample of a bacteriophage—a virus that infects bacteria. Using bacteriophages to treat bacterial infections, an approach called phage therapy, holds promise at a time when drug-resistant infections require new types of treatments.

Antibiotics and bacteriophages both destroy bacteria, but they differ in specificity. Bacteriophages are like snipers that eliminate a single species of bacteria, unlike many antibiotics that obliterate multiple species. Both therapeutics were developed in the early twentieth century, but by 1955, Western scientists lost interest in phage therapy, instead amassing a medicine cabinet full of antibiotics. Meanwhile, Soviet scientists collected bacteriophages specific for certain pathogens, but their expertise remained unknown outside Eastern Europe due to Cold War politics.

With the worldwide rise of antimicrobial resistance, alternative methods like bacteriophage therapy are gaining new attention. The challenge is finding compatible bacteriophage-host pairings that will target a specific bacterium. In 2018, a young patient with cystic fibrosis in Great Britain developed an antibiotic-resistant *Mycobacterium abscessus* infection following a lung transplant. The infection had spread to many organs, and after conventional treatments proved ineffective, her physicians sought out bacteriophages that might destroy the bacteria. The best candidates were found in a freezer in Pittsburgh; the phages required some genetic modifications but effectively destroyed the *M. abscessus* cells. After months of treatment, the patient became free of infection, establishing a milestone in the clinical use of bacteriophages. Phage therapy is routinely used in various European countries today. In the United States, however, it has not yet been approved for general use by the Food and Drug Administration (FDA); rather, it is restricted to clinical trials and for compassionate use in those for whom standard treatment has failed .

Viruses as useful agents may come as a surprise to many. Typically we think of them as major causes of disease. However, viruses are significant for other reasons. They are vital members of aquatic ecosystems, where they interact with cellular microbes and contribute to nutrient recycling. A number of animal viruses, including poliovirus and herpes simplex virus, are being modified to target and destroy cancer cells. The human virome—the collection of viruses found in and on the human body—is relatively unexplored, but our expanding knowledge indicates that the bacteriophages in our guts are important in regulating the bacterial members of our microbiome. Finally, viruses are important model organisms. In this chapter, we introduce viruses and other acellular infectious agents. *Microorganisms can be controlled by biological methods (section 8.5); Marine viruses: mortality at sea (section 29.2)*

Readiness Check:

Based on what you have learned previously, you should be able to:

✓ Define the term *acellular*

✓ Compare and contrast in general terms viruses, viroids, satellites, and prions (section 1.1)

6.1 Viruses Are Acellular

After reading this section, you should be able to:

a. Define the terms *virology, bacteriophage,* and *phage*
b. List organisms that are hosts to viruses

The discipline of **virology** studies **viruses,** infectious agents unique in their simple, acellular organization and pattern of multiplication. Despite this simplicity, viruses are major causes of disease. In 2020, the entire world learned about the impact viruses can have on human health as SARS-CoV-2, the virus that causes COVID-19, reached pandemic proportions. However, the simplicity of viruses also has made them useful as models for understanding DNA replication, RNA synthesis, and protein synthesis. In this way, virology has contributed significantly to the discipline of molecular biology.

Viruses can exist either extracellularly or intracellularly. When extracellular, they are inactive because they cannot reproduce outside of living cells. When intracellular, viruses commandeer host cells and use them to synthesize viral components from which mature progeny viruses are assembled and eventually released.

Viruses can infect all cell types. Numerous viruses infect bacteria. They are called **bacteriophages,** or **phages.** Fewer archaeal viruses have been identified. Most known viruses infect eukaryotic organisms, including plants, animals, protists, and fungi.

In this chapter, we introduce the structure and multiplication strategies of viruses. Their taxonomic diversity is presented in chapter 25, their ecological relevance is discussed in chapter 29, and their role in human disease is the topic of chapter 37.

6.2 Virion Structure Is Defined by Capsid Symmetry and Presence or Absence of an Envelope

After reading this section, you should be able to:

a. State the size range of virions
b. Identify the parts of a virion and describe their function
c. Distinguish enveloped viruses from nonenveloped viruses
d. Describe the types of capsid symmetry
e. List the forms that viral genomes may take

A mature virus particle is called a **virion.** Virion morphology has been intensely studied because of the importance of viruses and because it can help answer fundamental biological questions. Although our knowledge is incomplete due to the vast diversity of viruses, we can discuss the general nature of virion structure.

General Structural Properties

Virions can be extraordinarily tiny (about 20 nm in diameter) to about the same size as a rod-shaped bacterial cell (1.5 × 0.5 μm)

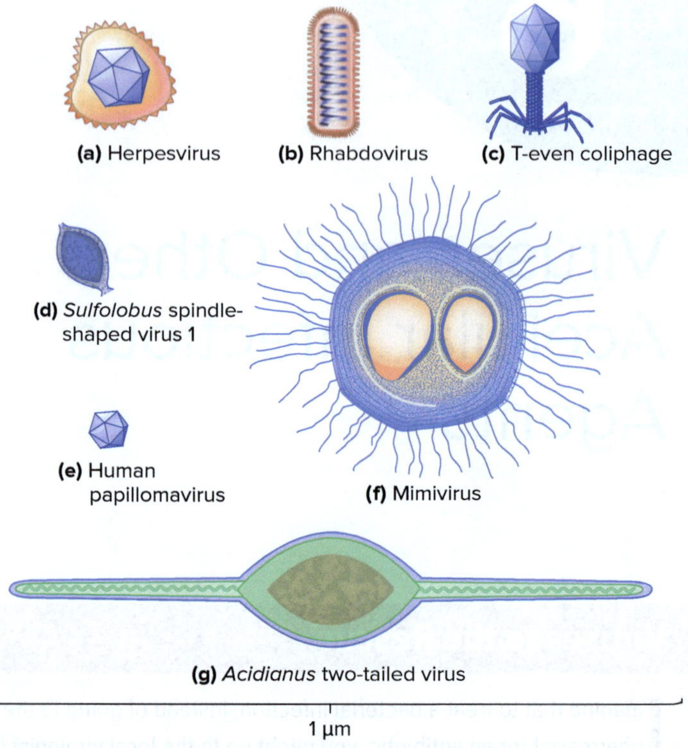

Figure 6.1 The Size and Virion Morphology of Selected Viruses. The virions are drawn to scale. For context, a *Staphylococcus aureus* cell is about 1 μm in diameter.

MICRO INQUIRY *Which capsids are icosahedral? Which are helical? Which have complex symmetry?*

(**figure 6.1**). The smallest are a little larger than ribosomes, whereas mimiviruses, among the largest viruses known, can be seen using the light microscope. However, most virus particles must be viewed with electron microscopes. ◀ *Electron microscopes use beams of electrons to create highly magnified images (section 2.4)*

The simplest virions consist only of a **nucleocapsid,** which is composed of a nucleic acid, either DNA or RNA, and a protein coat called a **capsid** (**figure 6.2**). The capsid surrounds the viral

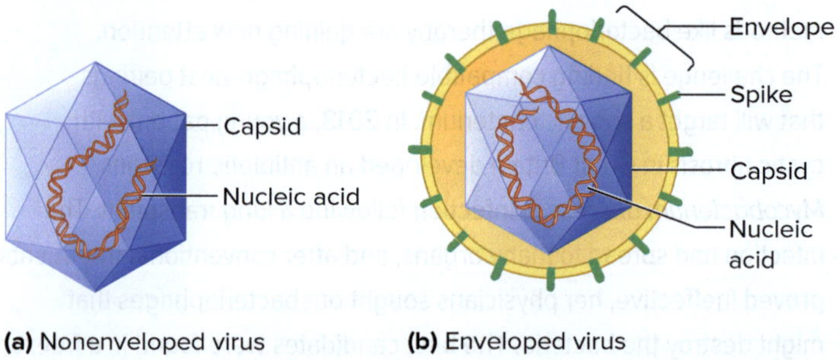

Figure 6.2 Generalized Structure of Virions. (a) The simplest virion is that of a nonenveloped virus, consisting of a capsid assembled around its nucleic acid (nucleocapsid). (b) Virions of enveloped viruses are composed of a nucleocapsid surrounded by a membrane called an envelope. The envelope usually has viral proteins called spikes inserted into it.

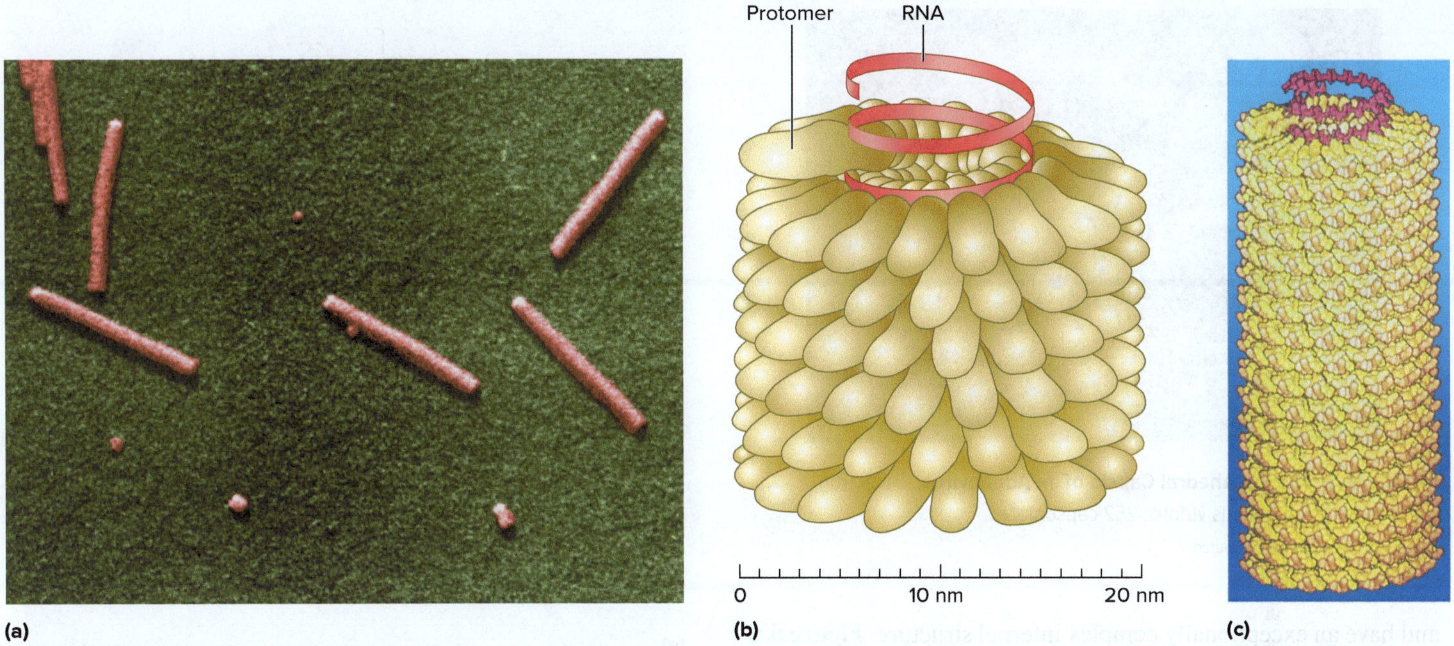

Figure 6.3 Tobacco Mosaic Virus (TMV) Virions. (a) An electron micrograph of negatively stained helical capsids. Virions are ~ 20 nm in diameter and 300 nm long. (b) Illustration of a TMV nucleocapsid composed of a helical array of protomers with RNA spiraling on the inside. (c) Model of a TMV nucleocapsid.

(a) Photo Researchers/Science History Images/Alamy Stock Photo

nucleic acid, thereby protecting the genome. Some virions are covered by a lipid membrane, and these are termed **enveloped viruses,** whereas those lacking a membrane are called **nonenveloped** or **naked viruses.**

Capsids are composed of many subunits of a protein termed a **protomer,** and thousands of protomers self-assemble to form the capsid (**figure 6.3**). Nonenveloped viruses construct a capsid from many protomers and a few additional proteins. In contrast, enveloped viruses require both nucleocapsid proteins and additional proteins, called *matrix proteins,* to anchor the membrane. Some viruses use noncapsid proteins as scaffolding upon which the capsids are assembled. Probably the most important advantage of this design strategy is that the viral genome is used with maximum efficiency. For example, the *t*obacco *m*osaic *v*irus (TMV) capsid is constructed using a single type of protomer. Recall that the building blocks of proteins are amino acids and that each amino acid is encoded by three nucleotides. The TMV protomer is 158 amino acids in length. Therefore only about 474 nucleotides are required to code for the coat protein. The entire TMV genome consists of only 6,400 nucleotides. Thus only a small fraction of the genome is used to code for the capsid.

Notice what is missing from viruses: ribosomes for protein synthesis and a mechanism for generating ATP. Cytoplasm is absent, and while a few enzymes may be found, there are not enough to sustain cellular processes.

Helical Capsids

Helical capsids are shaped like hollow tubes with protein walls. Tobacco mosaic virus is a well-studied example of helical capsid structure (figure 6.3). The self-assembly of TMV protomers into a helical arrangement produces a rigid tube. The capsid encloses an RNA genome, which is wound in a spiral and lies within a groove formed by the protein subunits. ▶| *Tobacco mosaic virus (section 25.5)*

The size of a helical capsid is influenced by both its protomers and the viral genome. The diameter of the capsid is a function of the size, shape, and interactions of the protomers. The length of the capsid appears to be determined by the size of the viral genome.

Icosahedral Capsids

An icosahedron is a regular polyhedron with 20 equilateral triangular faces and 12 vertices, or corners (figure 6.1*a,e*). **Icosahedral capsids** are the most efficient way to enclose a space. They are constructed from ring- or knob-shaped assemblages (groups) of five or six protomers; the assemblages are called **capsomeres** (**figure 6.4**). The vertices of icosahedral capsids are composed of capsomeres with five protomers (pentons), while the triangular faces contain capsomeres with six protomers (hexons). Although most icosahedral capsids contain both pentons and hexons, some have only pentons.

Capsids of Complex Symmetry

Most viruses have either icosahedral or helical capsids, but some viruses do not fit into either category. Poxviruses and large bacteriophages are two examples.

Poxvirus virions are among the largest of the animal viruses (about 400 by 240 by 200 nm in size; *see figure 3.3*) and can be seen with a light microscope. They are ovoid- to brick-shaped

Figure 6.4 The Icosahedral Capsid of an Adenovirus. Electron micrograph of adenovirus virions, 252 capsomeres.
Biophoto Associates/Science Source

and have an exceptionally complex internal structure. **Figure 6.5** shows the morphology of a poxvirus called molluscum contagiosum virus that causes small benign skin lesions, primarily in humans. The mature virion has an outer envelope surrounding a central biconcave core that contains a double-stranded DNA genome. Between the core and an enclosing core membrane are two lateral bodies that contain proteins, some of which modulate the host's antiviral responses.

Some large bacteriophages have virions that are even more elaborate than those of poxviruses. The virions of T2, T4, and T6 phages (T-even phages) that infect *Escherichia coli* are said to have **binal symmetry** because they have a head that resembles an icosahedron and a helical tail. The icosahedral head is elongated by one or two rows of hexons in the middle and contains

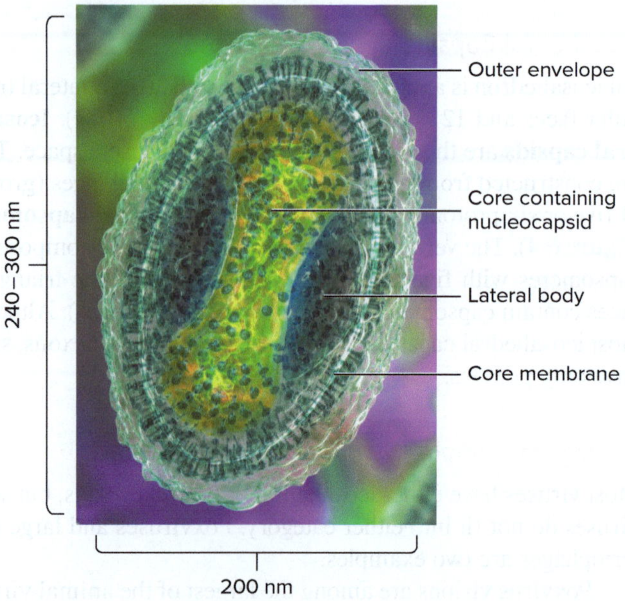

Figure 6.5 Morphology of Molluscum Contagiosum Virus Virions (illustration).
Kateryna Kon/Shutterstock

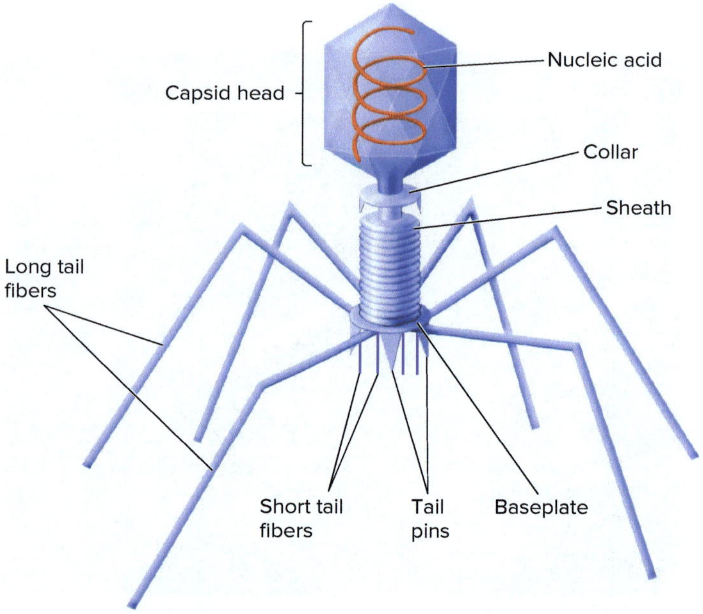

(a)

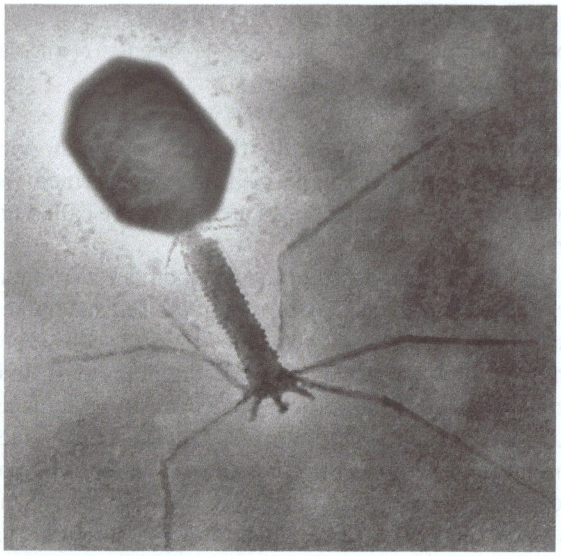

(b)

Figure 6.6 T4 Phage of *E. coli*. (a) The structure of T4 bacteriophage virion. (b) Electron micrograph of the virion. The head and the tail sheath are each about 100 nm in length.
(b) Ami Images/Science Source

MICRO INQUIRY *Why is T4 said to have binal symmetry?*

the DNA genome (**figure 6.6**). The tail is composed of a collar joining it to the head, a central hollow tube, a sheath surrounding the tube, and a complex baseplate. In T-even phages, the baseplate is hexagonal and has a pin and a jointed long tail fiber and a short tail fiber at each corner. ▶| *Bacteriophage T4: a virulent bacteriophage (section 25.2)*

Viral Envelopes and Enzymes

The nucleocapsids of many animal viruses, some plant viruses, and at least two related bacteriophages are surrounded by an outer

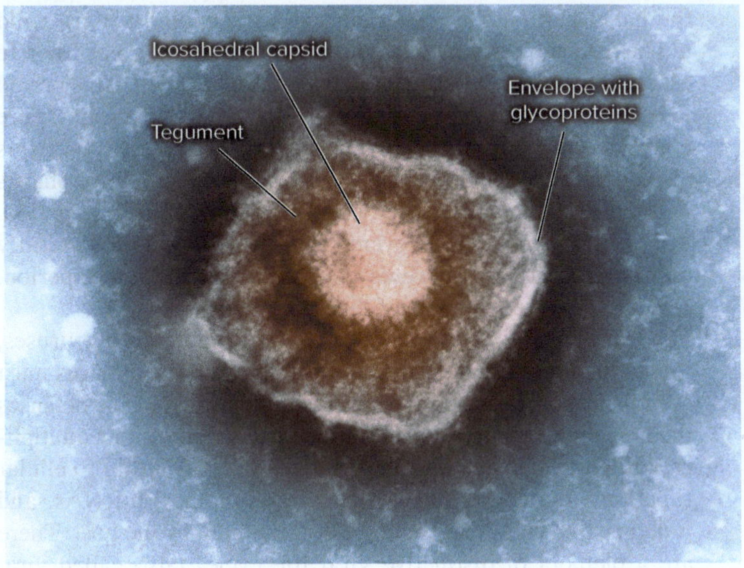

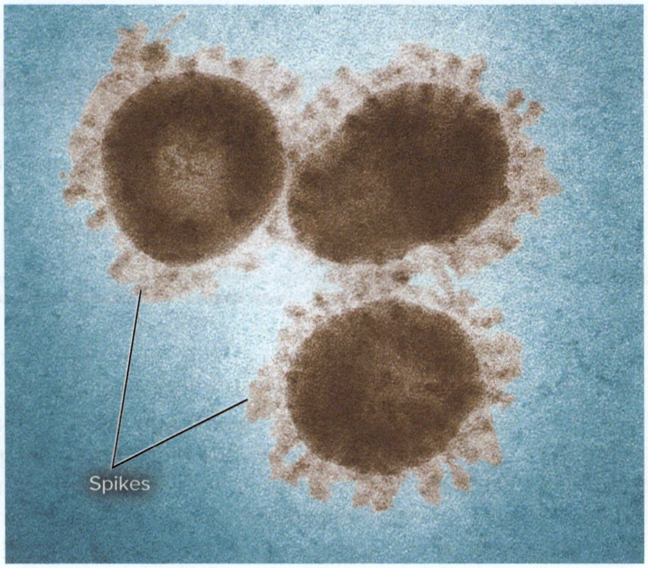

Figure 6.7 Examples of Enveloped Viruses. (a) Varicella-zoster virus particle (~200 nm in diameter), the causative agent of chickenpox and shingles. (b) Virions of infectious bronchitis virus, a coronavirus, illustrating the surface spikes that protrude from the envelope and inspired its name. Each virion is ~100 nm in diameter. Images are artificially colorized transmission electron micrographs.

(a) Heather Davies/Science Source; (b) Dr. Fred Murphy/Sylvia Whitfield/CDC

membranous layer called an **envelope** (**figure 6.7**). Animal virus envelopes usually arise from the plasma or organelle membranes of the host cell. Envelope lipids and their attached carbohydrates are therefore acquired from the host. In contrast, envelope proteins are coded for by viral genes and often project from the envelope surface as **spikes,** also called **peplomers** (figure 6.7*b*). In many cases, spikes are involved in virion attachment to the host cell surface. Some surface proteins also have enzymatic activity needed for entry into or exit from the host cell. Because spikes differ among viruses, they also can be used to identify some viruses.

In addition to enzymes associated with the envelope or capsid, some viruses have enzymes within their capsids. Such enzymes are usually involved in nucleic acid replication. For example, many viruses with an RNA genome carry RNA-dependent RNA polymerase, an enzyme that synthesizes RNA using an RNA template (*see figure 25.23*). Thus although viruses lack true metabolism and cannot reproduce independently of living cells, their virions may carry one or more enzymes essential to the completion of their life cycle.

Viral Genomes Are Structurally Diverse

One clear distinction between cellular organisms and viruses is the nature of their genomes. Cellular genomes are always double-stranded (ds) DNA. Viruses, on the other hand, employ all four possible nucleic acid types: dsDNA, single-stranded (ss) DNA, ssRNA, and dsRNA. Most plant viruses have ssRNA genomes, and most bacterial and archaeal viruses have dsDNA. Animal viruses use any of the four types. The size of viral genomes also varies greatly. Very small genomes are around 4,000 nucleotides—just large enough to code for three or four proteins. Some viruses save space by using overlapping genes. At the other extreme are the genomes of giant viruses such as pandoraviruses, which infect protists. They are over 2.5 million nucleotides long, exceeding some bacteria and archaea in coding capacity.

Most DNA viruses use dsDNA as their genetic material. However, some have ssDNA genomes (e.g., ϕX174 and parvovirus). In both cases, the genomes may be either linear or circular. The ends of linear viral genomes may be covalently closed (the ends of the DNA are joined together by covalent bonds), attached to a protein, or otherwise masked. Relatively few RNA viruses have dsRNA genomes. More common are viruses with ssRNA genomes. Coronaviruses, tobacco mosaic virus, rabies virus, and human immunodeficiency viruses (HIV) are all examples of ssRNA viruses. ▶| *Single-stranded DNA viruses use a double-stranded intermediate in their life cycles (section 25.3)*

Some RNA viruses have **segmented genomes**—genomes that consist of multiple pieces (segments) of RNA. In many cases, each segment codes for one protein and there may be 10 to 12 segments, usually enclosed in the same capsid. Influenza viruses are an important example of segmented viruses.

Comprehension Check

1. How are viruses similar to cellular organisms? In what fundamental way do they differ?
2. What is the difference between a nucleocapsid and a capsid?
3. What is an envelope? What are spikes?
4. All four nucleic acid forms can serve as viral genomes. Describe each. What is a segmented RNA genome?

6.3 Viral Life Cycles Have Five Steps

After reading this section, you should be able to:

a. Describe the five steps common to the life cycle of all viruses
b. Discuss the role of receptors, capsid proteins, and envelope proteins in the life cycles of viruses
c. Describe the two most common methods for virion release from a host cell

In the early years of virology, one of the fundamental questions was how progeny viruses are made. The bacteriophage T4 one-step growth experiment devised in 1939 by Max Delbrück (1906–1981) and Emory Ellis (1906–2003) answered this question. They knew that T4 lysed its host, *E. coli,* and released progeny phages. In their experiment, *E. coli* cells were mixed with T4. After a short interval, the mixture was greatly diluted so that any virions released upon host cell lysis would not encounter and infect other cells. The diluted culture was then incubated, and over time samples were removed to determine the number of infectious phage particles in the culture. A plot of the number of phage particles versus time shows several distinct periods in the resulting growth curve (**figure 6.8**, red line). The latent period occurs immediately following addition of the phages. During this period, no virions are present in the medium. The rise period follows and is characterized by the rapid release of infective phages. Finally, a plateau is reached and no more virions are produced.

The one-step growth experiment led to a fundamental question: What is occurring during the latent period? In subsequent experiments, infected cells were artificially lysed during

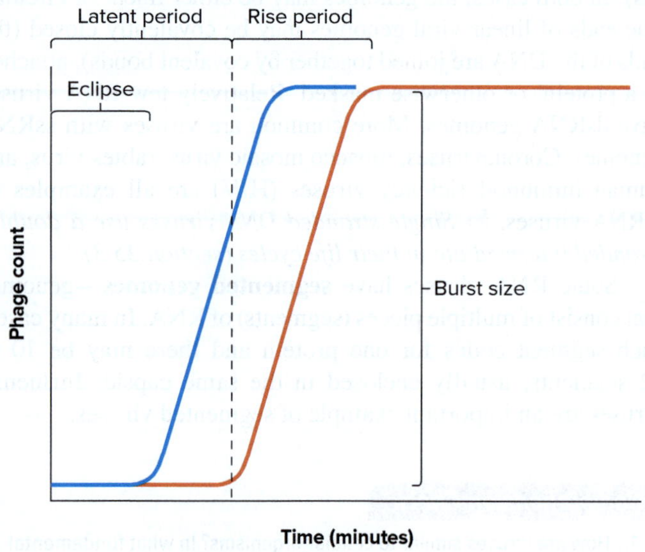

Figure 6.8 The One-Step Growth Curve. The red line represents the number of extracellular viruses, which is the sum of those that have not infected host cells and those released by host cell lysis. The blue line represents both intracellular and extracellular viruses collected after host cells were experimentally lysed.

the latent period. This revealed that intracellular virions could not be detected early in the latent period (figure 6.8, blue line). In essence, the phages disappeared once inside the cell. This period is called the *eclipse period* because the infecting virions were concealed or eclipsed within the host cell. These experiments also demonstrated that the number of complete, infective phages within the host cell increases after the end of the eclipse period. Still other experiments eventually showed that a carefully orchestrated series of events occurs during the latent period. These events are the focus of this section.

If a virus is to multiply and give rise to new progeny viruses, it must find and use a host cell. To accomplish this, a virus must access an appropriate host, enter that host, and avoid any defenses the host might use to rid itself of the virus or prevent its multiplication. Once inside a host cell, a virus takes control of cellular functions, thereby ensuring that viral genomes, mRNAs, and proteins are synthesized using ATP generated by the host. There are a plethora of distinctive viral life cycles (also called viral replicative cycles), and these are usually related to the structure of the viral genome.

Despite the diversity of viral life cycles, a general pattern of viral replicative cycles can be discerned; it is divided into five steps (**figure 6.9**). We now review each of these steps taking a general approach.

Attachment (Adsorption)

Because viruses need a host cell in which to multiply, the first step is attachment, often called adsorption (Latin *sorbere,* to suck in), to a host. Attachment is accomplished by specific interactions between molecules on the surface of the virion (ligands) and molecules on the surface of the host cell called **receptors.** In many cases, the viral surface ligand is a protein or glycoprotein, but some animal viruses use lipids derived from the cell membrane as ligands. Viruses have evolved to use host receptors that are important for normal host cell function. Some bacteriophages attach to cell wall lipopolysaccharides and proteins, while others use teichoic acids, flagella, or pili. A recurring theme among archaeal viruses is the presence of virion appendages for host cell attachment. These include beardlike fiber clusters and structures resembling claws. Binding of an animal virus to its receptor often causes conformational changes in virion proteins that facilitate interaction with secondary receptors and entry into the host cell.

Receptor specificity is at least partly responsible for the preferences viruses have for a particular host. Bacteriophages not only infect a particular bacterial species but often infect only certain strains within a given species. In complex animals like humans, the presence of receptor molecules on a cell surface varies with the organ and tissue. Viruses have a **tropism;** that is, particular cell types are infected based on the distribution of surface receptors in tissues. ◀ *Eukaryotic cell envelopes (section 5.2)*

Plant viruses are a notable exception to attachment based on receptor binding because no receptors have been identified for

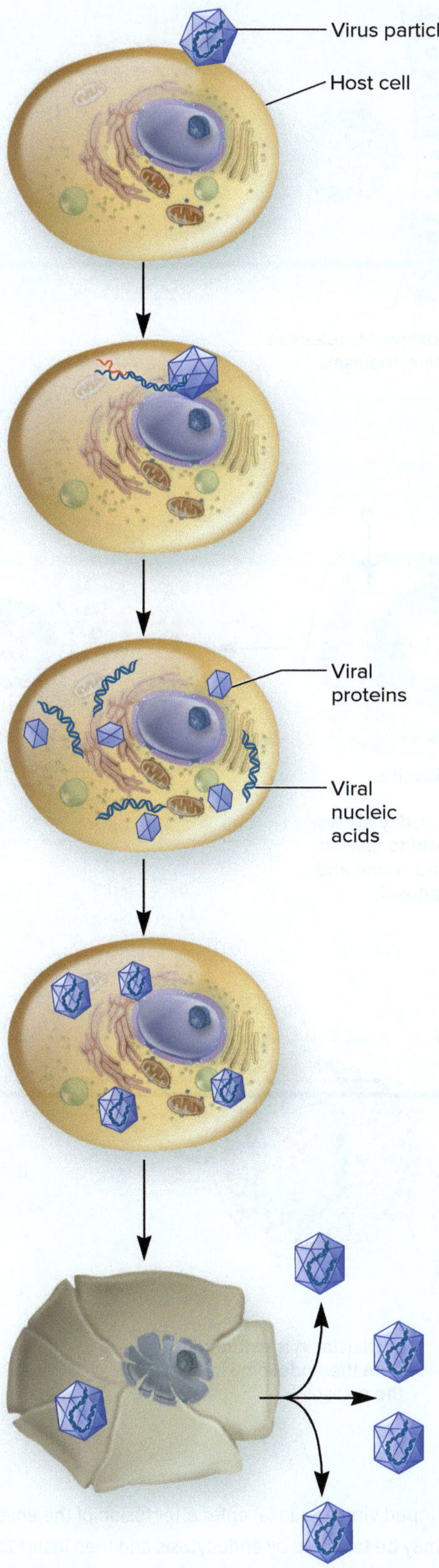

Figure 6.9 Generalized Illustration of Virus Multiplication. The life cycle of a virus that infects an animal cell is shown. Figure 6.15 illustrates the generalized life cycle of many bacterial and archaeal viruses. Virions are not drawn to scale; they are much smaller than illustrated here.

plant viruses. Rather, damage of host cells is required for the virus particles to enter the host. Infection is often achieved by plant-eating insects that carry virions from one plant to another. The virions are deposited in plant tissues as the insect devours the plant. Interestingly, some viruses may alter their plant hosts to promote activity of the insects and thereby foster transmission to new plants.

Entry into the Host

After attachment to the host cell, the virus's genome or the entire nucleocapsid enters the cytoplasm. For many bacteriophages only their genome enters the host's cytoplasm, leaving the capsid outside and attached to the cell. In contrast, many eukaryotic viral nucleocapsids enter the cytoplasm with the genome still enclosed. Once inside the cytoplasm, some shed their capsid proteins in a process called *uncoating,* whereas others remain encapsidated. Because penetration and uncoating are often coupled, we consider them together.

The mechanisms of penetration and uncoating vary with the type of virus. One of three different modes of entry is usually used by animal viruses: fusion of the viral envelope with the host cell's plasma membrane, entry by endocytosis, or release of viral nucleic acid into the cytoplasm of the host cell (**figure 6.10**). *Entry of Animal Viruses into Host Cells*

Penetration by fusion involves interaction between viral envelope glycoproteins or phospholipids with proteins in the plasma membrane of the host cell (figure 6.10*a*). This interaction sets into motion events that allow the nucleocapsid to enter. In the second mode of entry, virions of nonenveloped viruses and some enveloped viruses enter cells by one of the endocytic pathways, including clathrin-dependent endocytosis and macropinocytosis (figure 6.10*b; see figure 5.9*). The resulting endocytic vesicle contains the virion and fuses with an endosome. Depending on the virus, escape of the nucleocapsid or its genome from the endocytic vesicle may occur either before or after fusion with an endosome. Endosomal enzymes aid in virion uncoating, and low pH often triggers the process. For some enveloped viruses, the viral envelope fuses with the endosomal membrane, and the nucleocapsid is released into the cytosol. Once in the cytosol, the viral nucleic acid may be released from the capsid upon completion of uncoating or may function while still attached to capsid components. The third mode of entry is used by some nonenveloped animal viruses that cannot use the membrane fusion mechanism for release from the endosome (figure 6.10*c*). In this case, the low pH of the endosome causes a conformational change in the capsid. The altered capsid contacts the endosome membrane and either releases the viral nucleic acid into the cytosol or ruptures the membrane to release the intact nucleocapsid. *There are several endocytic pathways (section 5.4)*

Synthesis

This stage of the viral life cycle differs dramatically among viruses because the genome of a virus dictates the events that occur. For dsDNA viruses, the synthesis stage is similar to the typical

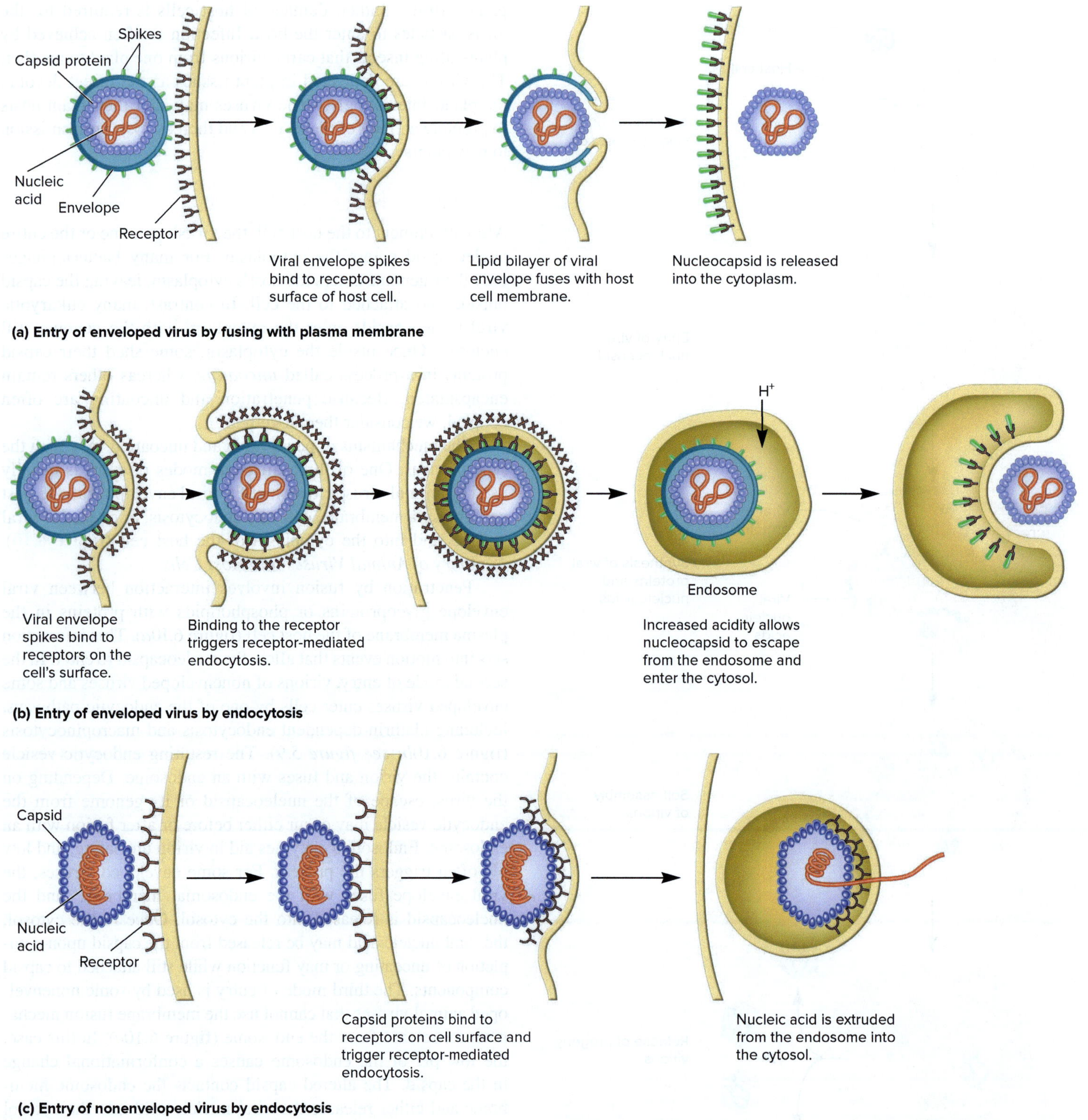

Figure 6.10 Animal Virus Entry. Examples of animal virus attachment and entry into host cells. Enveloped viruses can (a) enter after fusion of the envelope with the plasma membrane or (b) escape from an endosome after endocytosis. (c) Nonenveloped viruses may be taken up by endocytosis and then insert their genome into the cytosol through the vesicle membrane. It also is possible that they insert the genome directly through the plasma membrane.

MICRO INQUIRY *Which animal virus entry mechanism involves the production of a vesicle within the host cell? Which involves the interaction between viral proteins and host cell membrane proteins?*

flow of information in cells. That is, the genetic information is stored in DNA and replicated by enzymes called DNA polymerases, recoded as mRNA (transcription), and decoded during protein synthesis (translation). Because of this similarity, many dsDNA viruses depend solely on their host cells' biosynthetic machinery to replicate their genomes and synthesize their proteins.

The same cannot be true for RNA viruses. Cellular organisms (except for plants) lack the enzymes needed to synthesize RNA from an RNA template. Therefore the nucleocapsids of RNA viruses must carry the enzymes needed to complete the synthesis stage, or the enzymes must be synthesized during the infection process. This is discussed in more detail in chapter 25.

Some animal and plant viruses carry out the synthesis stage and subsequent assembly step within the host's cytoplasm. To protect these processes from host defenses, some viruses reorganize host cell membranes (e.g., membranes of the endoplasmic reticulum, Golgi apparatus, or lysosomes) to form membranous structures that enclose the machinery needed for genome replication, transcription, and protein synthesis (**figure 6.11**). The structures are called *viral replication complexes,* and they appear as vesicles, tubular structures, and other forms in electron micrographs of infected cells. Other viruses carry out synthesis and assembly in defined areas within the cytoplasm that are not enclosed by membranes. These areas of concentrated viral genomes, mRNAs, and proteins are called *viroplasms,* and they are also visible in electron micrographs of infected cells. Both viral replication complexes and viroplasms are sometimes referred to as *virus factories.*

One important feature of the synthesis stage is the tight regulation of gene expression and protein synthesis. Genes are often referred to as early, middle, or late genes based on when they are expressed. The proteins they encode are likewise referred to as early, middle, or late proteins. Many early proteins are involved in taking over the host cell. Middle proteins often participate in replication of the viral genome or activation of expression of late genes. These and other proteins not incorporated into the virion are often called *nonstructural (NS) proteins.* Late proteins usually include capsid proteins and other proteins involved in self-assembly and release. Proteins present in the virion are termed *structural proteins.*

Assembly

Several kinds of late proteins are involved in the assembly of mature virions. Some are nucleocapsid proteins, some are not incorporated into the nucleocapsid but participate in its assembly, and still other late proteins are involved in virion release. In addition, proteins and other factors synthesized by the host may be involved in assembling mature virions.

The assembly process can be quite complex with multiple subassembly lines functioning independently and converging in later steps to complete nucleocapsid construction. As shown in **figure 6.12**, the baseplate, tail fibers, and head components

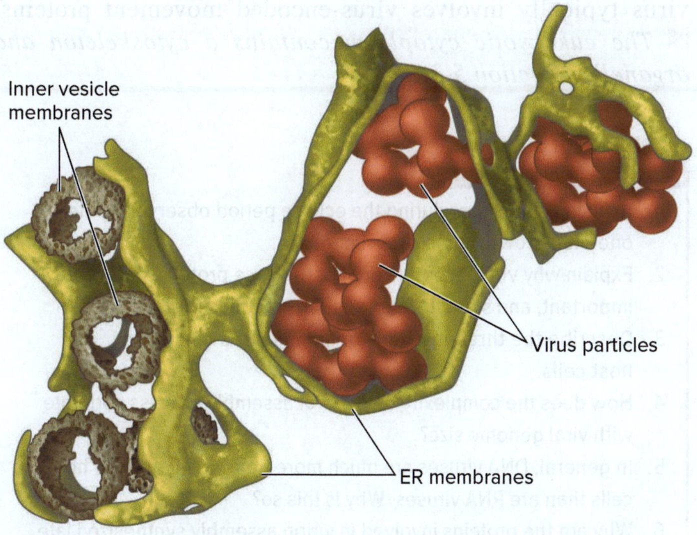

Figure 6.11 A Viral Replication Complex. In this example, severe acute respiratory syndrome coronavirus (SARS-CoV) has formed a viral replication complex derived from endoplasmic reticulum (ER) membranes (gold). The membranes of vesicles (brown) are connected to the ER. Virus particles are red.

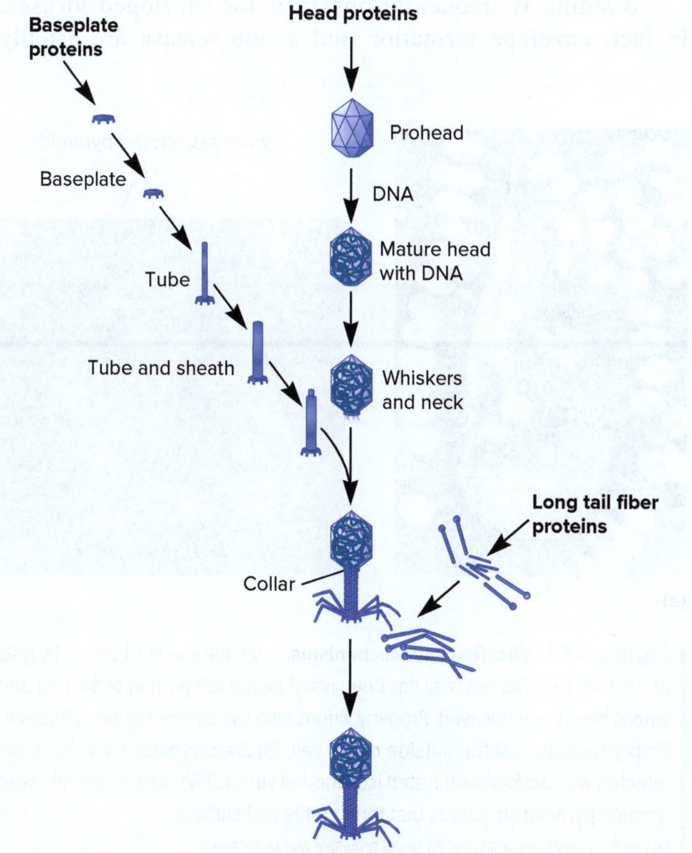

Figure 6.12 The Assembly of T4 Virions. Note the subassembly lines for the baseplate, tail tube and sheath, long tail fibers, and head.

of bacteriophage T4 are assembled separately. Once the baseplate is finished, the tail tube is built on it and the sheath is assembled around the tube. The prohead (procapsid) is constructed with the aid of scaffolding proteins that are degraded or removed after assembly is complete. DNA is incorporated into the prohead by a complex of proteins sometimes called the *packasome.* The packasome consists of a protein called the *portal protein,* which is located at the base of the prohead, and an enzyme called *terminase,* which moves DNA into the prohead. The movement of DNA uses energy in the form of ATP, which is supplied by the metabolic activity of the host bacterium. After the head is complete, it spontaneously combines with the tail assembly.

Virion Release

The two most common viral release mechanisms are host cell lysis and release by budding. Lysis is especially common for bacterial viruses and some nonenveloped animal viruses. This process involves the activity of viral proteins that puncture host plasma membranes and, in the case of bacteriophages, degrade the host peptidoglycan (**figure 6.13*a***). Viral enzymes that hydrolyze the bacterial cell wall result in host cell lysis through osmotic stress.

Budding is frequently observed for enveloped viruses; in fact, envelope formation and virion release are usually concurrent processes. All envelopes of animal viruses are derived from host cell membranes by a multistep process. First, virus-encoded proteins are incorporated into the membrane. Then the nucleocapsid is simultaneously released and the envelope formed by membrane budding (**figure 6.14**). In several virus families, a matrix protein attaches to the plasma membrane and aids in budding. Most envelopes arise from the plasma membrane. The endoplasmic reticulum, Golgi apparatus, and other internal membranes may also be used to form envelopes. When virions are released by budding, the host cell may continue releasing virions for some time before cell death. *Mechanism for Releasing Enveloped Virions*

A novel release mechanism has been elucidated for some archaeal viruses. The virions are assembled in the cytoplasm, and a seven-sided pyramid structure forms near the cell envelope. These virus-associated pyramids puncture the envelope and open up much like the petals of a flower (figure 6.13*b*). This allows escape of the progeny virions, leaving the empty cell ghosts behind.

Interestingly, some viruses are not released from their host cell into the surrounding environment. Rather, their virions move from one host cell directly to another host cell. Most fungal viruses lack an extracellular phase in their replicative cycles. Instead they are transmitted by cell division, by spore formation, or during mating. Vaccinia viruses elicit the formation of long actin tails that propel nucleocapsids through the plasma membrane, directly into an adjacent cell. In this way, the virus avoids detection by the host immune system. The genomes or nucleocapsids of many plant viruses also move directly from cell to cell through small connections called plasmodesmata that link adjacent cells. This spread of the virus typically involves virus-encoded movement proteins. *The eukaryotic cytoplasm contains a cytoskeleton and organelles (section 5.3)*

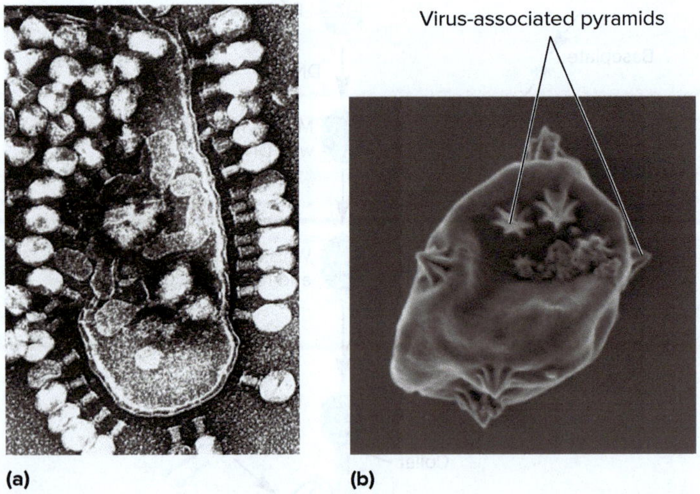

Figure 6.13 Viral Release Mechanisms. (a) Release of T4 virions by lysis of the host cell. The host cell has been lysed (upper left portion of the cell) and virions have been released. Progeny virions also can be seen in the cytoplasm. Empty capsids coat the outside of the cell. (b) *Saccharolobus solfataricus* infected with *Sulfolobus* turreted icosahedral virus (STIV). Virions are released through pyramid structures that form on the cell surface.

(a) Lee D. Simon/Science Source; (b) Susan Brumfield and Mark Young

MICRO INQUIRY *Why do the empty capsids remain attached to the cell after the viral genome enters the host cell?*

Comprehension Check

1. Explain what occurs during the eclipse period observed in the one-step growth curve experiment.
2. Explain why viral receptors are host surface proteins that serve important, and sometimes essential, functions for the host cell.
3. Describe the three methods of animal virus entry into host cells.
4. How does the complexity of the viral assembly process correlate with viral genome size?
5. In general, DNA viruses are much more dependent on their host cells than are RNA viruses. Why is this so?
6. Why are the proteins involved in virion assembly synthesized late in the viral life cycle?
7. Compare two methods of virus release from a host cell. Discuss the unique exit method of some archaeal viruses.

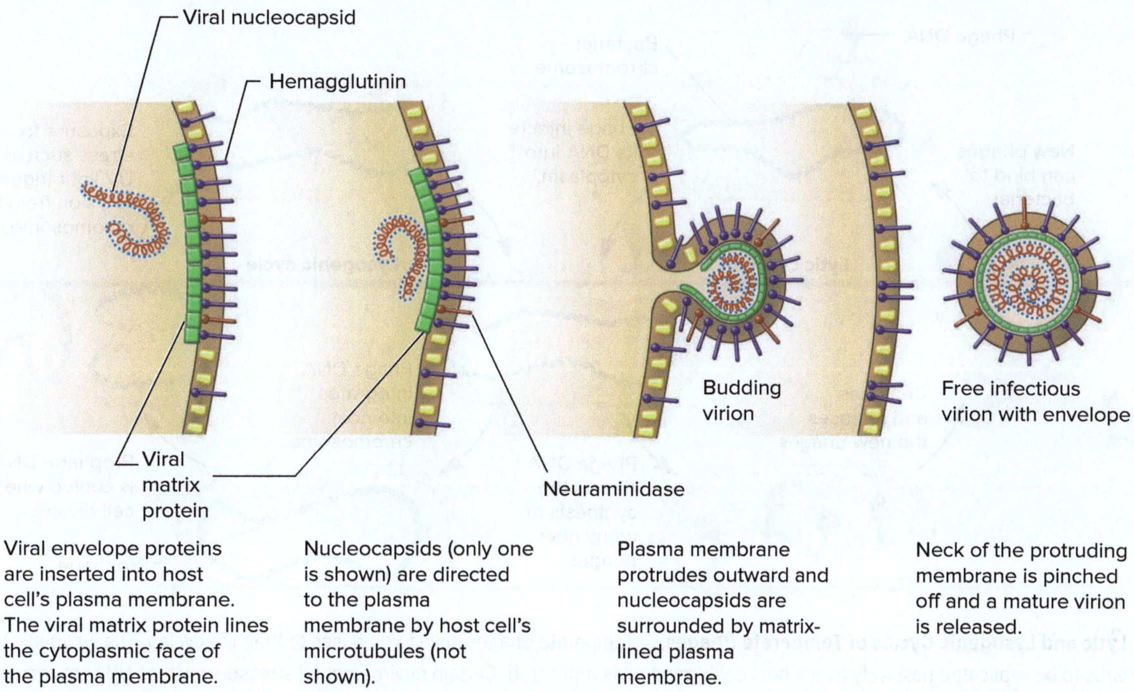

Figure 6.14 Release of Influenza Virus Virions by Budding. Hemagglutinin and neuraminidase are influenza spike proteins that must be inserted into the host cell membrane before budding.

6.4 There Are Several Types of Viral Infections

After reading this section, you should be able to:

a. Compare and contrast the major steps of the life cycles of virulent phages and temperate phages
b. Define lysogenic conversion
c. Differentiate among the types of viral infections of animal cells

The dependence of viruses on their host cells has many consequences—as anyone who has experienced a viral illness well understands. Next we discuss the interaction between virus and host cell.

Lytic and Lysogenic Infections Are Common for Bacterial and Archaeal Cells

Most bacteriophages are either virulent or temperate. **Virulent phages** begin to multiply immediately upon entering a bacterial host, and are released from the host by lysis. T4 is an example of a virulent phage. **Temperate phages** have two options: upon entry into the host cell, either they can multiply like virulent phages and lyse the host cell, or they can remain within the host without destroying it (**figure 6.15**). Bacteriophage lambda is an example of a temperate phage. *Steps in the Replication of T4 Phage in* E. coli; *Lambda Phage Replication Cycle*

The relationship between a temperate phage and its host is called **lysogeny.** The form of the virus that remains in the host is called a **prophage.** This either integrates into the host's chromosome, or more rarely, exists in the cytoplasm as a circular molecule called an *episome.* The infected bacteria are called **lysogens** or **lysogenic bacteria.** Lysogenic bacteria have two distinct characteristics. First, they cannot be reinfected by the same virus; that is, they have immunity to superinfection. Second, the prophage is replicated and inherited within the genome of progeny cells. This can continue for many generations until conditions arise that cause the prophage to initiate synthesis of phage proteins and to assemble new virions, a process called **induction.** Induction is commonly caused by cellular stress such as changes in growth conditions or ultraviolet irradiation. As a result of induction, the **lysogenic cycle** ends and the **lytic cycle** commences; the host cell lyses and progeny phage particles are released.

Another important outcome of lysogeny is **lysogenic conversion.** This occurs when the prophage changes the phenotype of its bacterial host. For example, *E. coli* infected with Stx phages produce shiga toxins; these strains (such as *E. coli* O157:H7) may cause bloody diarrhea and sometimes a life-threatening condition called hemolytic uremic syndrome (HUS) that affects the

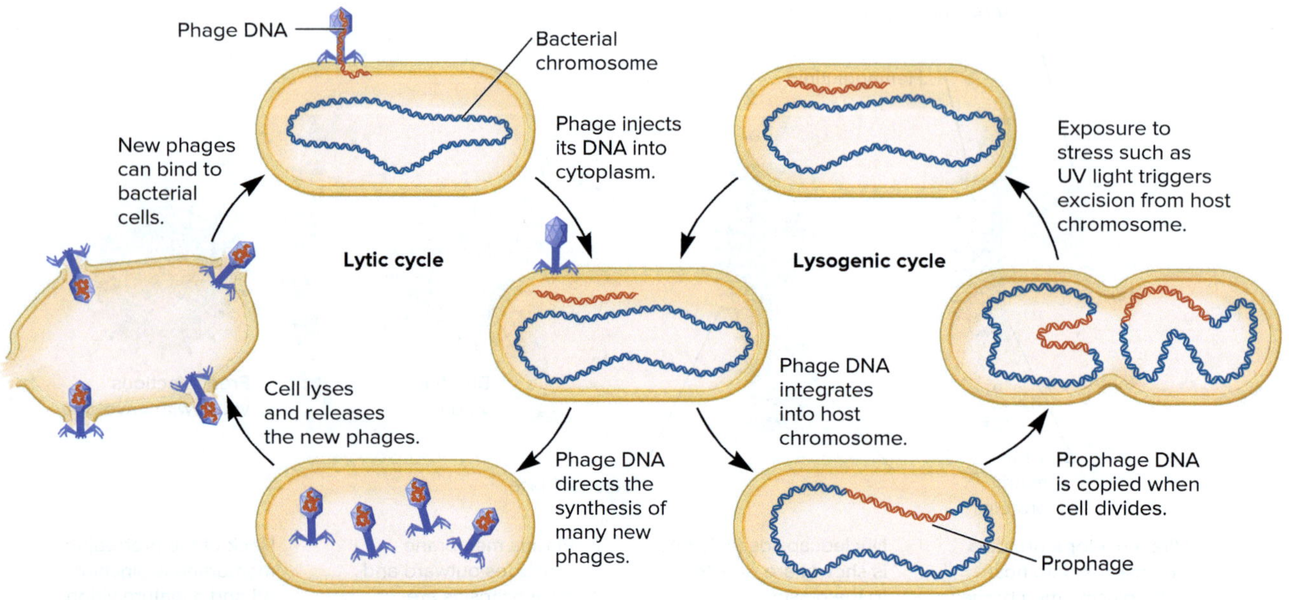

Figure 6.15 Lytic and Lysogenic Cycles of Temperate Phages. Temperate phages have two phases to their life cycles. The lysogenic cycle allows the genome of the virus to be replicated passively as the host cell's genome is replicated. Certain environmental stressors such as UV light can cause a switch from the lysogenic cycle to the lytic cycle. In the lytic cycle, new virus particles are made and released when the host cell lyses. Virulent phages are limited to just the lytic cycle.

MICRO INQUIRY *Why is a lysogen considered a new or different strain of a given bacterial species?*

kidneys and blood-clotting functions. In this case, lysogeny converts a nonpathogenic strain of *E. coli* into one that causes severe disease.

Clearly the infection of a bacterium by a temperate phage has significant impact on the host, but why would viruses evolve this alternate cycle? Two advantages of lysogeny have been recognized. The first is that lysogeny allows the viral nucleic acid to be maintained within its host. Bacteria often become dormant due to nutrient deprivation, and while in this state, they do not synthesize nucleic acids or proteins. In such situations, a prophage would survive but most virulent bacteriophages would not be replicated, as they require active cellular biosynthetic machinery. Furthermore, their genome would be degraded as the host cell entered dormancy. The second advantage arises when there are many more phages in an environment than there are host cells, a situation virologists refer to as a high multiplicity of infection (MOI). In these conditions, lysogeny enables the survival of infected host cells within a population that has few uninfected cells. When MOI is high, a virulent phage would rapidly destroy the available host cells in its environment. However, a prophage will be replicated as the host cell reproduces.

Archaeal viruses can also be virulent or temperate. In addition, many archaeal viruses establish long-lasting infections. Unfortunately, little is known about the mechanisms they use to regulate their replicative cycles.

Infections of Eukaryotic Cells

Viruses can harm their eukaryotic host cells in many ways. An infection that results in cell death is a *cytocidal* infection. As with bacterial and archaeal viruses, this can occur by lysis of the host (**figure 6.16*a***). However, infection does not always result in lysis of host cells. Some viruses (e.g., herpesviruses) can establish persistent infections lasting many years (figure 6.16*b,c*). Viruses that infect eukaryotic cells can cause microscopic or macroscopic degenerative changes or abnormalities in host cells and in tissues that are distinct from lysis. These are called **cytopathic effects (CPEs).** Viruses use a variety of mechanisms to cause cytopathic and cytocidal effects. One mechanism of particular note is that some viruses cause the host cell to be transformed into a malignant cell (figure 6.16*d*). ▶| *Viral hepatitides (section 37.3)*

Comprehension Check

1. Define the terms *lysogeny, temperate phage, lysogen, prophage, immunity,* and *induction.*
2. What advantages might a phage gain by being capable of lysogeny?
3. Describe lysogenic conversion and its significance.
4. How does a latent infection differ from a chronic infection?
5. What is a cytocidal infection? What is a cytopathic effect?

Figure 6.16 Types of Viral Infections and Their Effects on Animal Cells. (a) Lytic infections of host cells can lead to disease states in animal hosts called acute infections. There are two types of persistent infections: (b) latent infections and (c) chronic infections. (d) Some infections of animal cells cause the cell to be transformed into a malignant cell, which can cause cancer in the animal host.

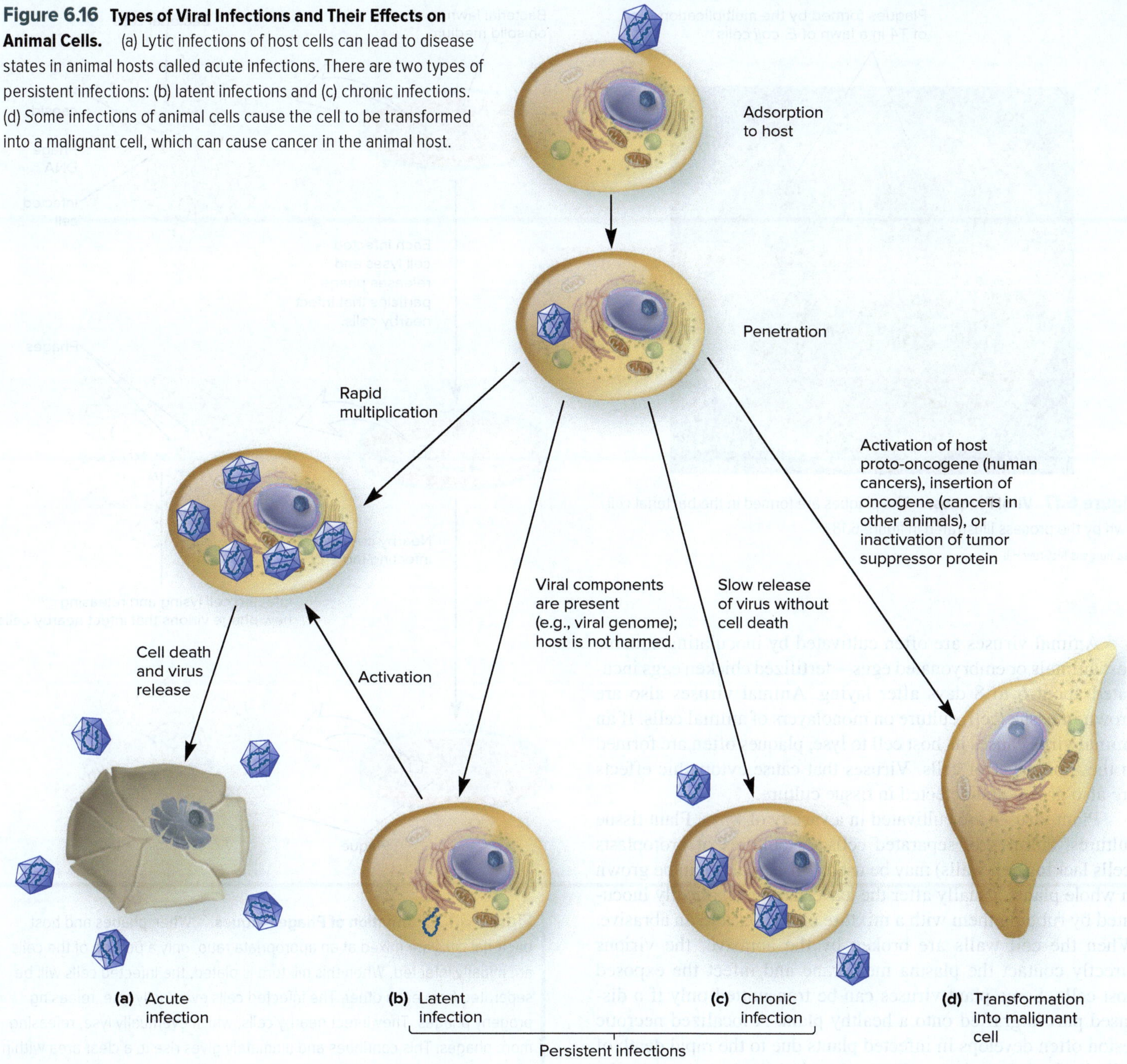

6.5 Virus Cultivation and Enumeration

After reading this section, you should be able to:

a. List the approaches used to cultivate viruses, noting which types of viruses are cultivated by each method
b. Define *plaque-forming unit* and explain its role in the quantification of viruses
c. Outline the events that lead to formation of a plaque in a lawn of bacterial cells

Because they are unable to reproduce outside of living cells, viruses cannot be cultured in the same way as cellular microorganisms. It is relatively simple to cultivate lytic bacterial and archaeal viruses as long as the host cell is easily grown in culture. Cells and viruses are simply mixed together in a broth culture. Over time, more and more cells are infected and lysed, releasing virions into the broth. Culturing temperate viruses requires an additional step of inducing the lytic phase of their life cycles. When cultured with cells on a solid medium, lysis of the host cells is observed as **plaques** on the bacterial lawn (**figure 6.17**).

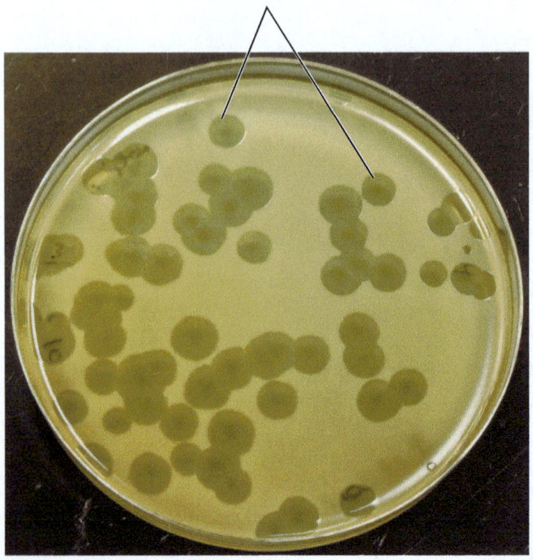

Figure 6.17 Viral Plaques. The plaques are formed in the bacterial cell lawn by the process illustrated in figure 6.18.
Lisa Burgess/McGraw Hill

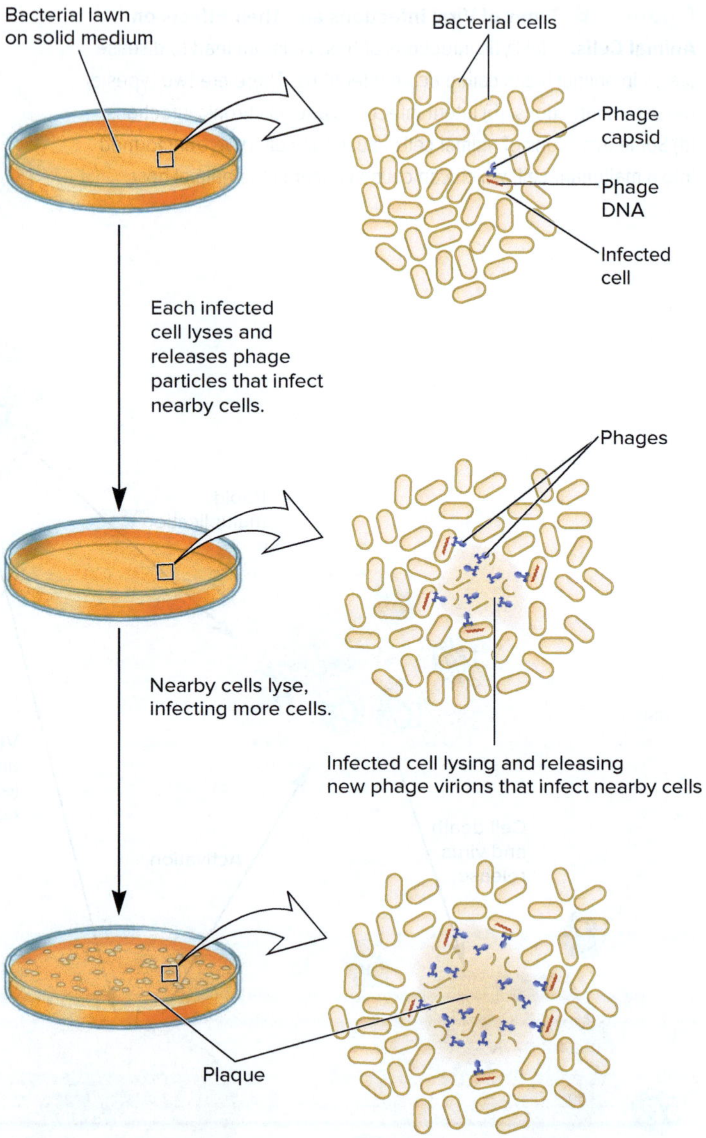

Figure 6.18 Formation of Phage Plaques. When phages and host bacterial cells are mixed at an appropriate ratio, only a portion of the cells are initially infected. When this mixture is plated, the infected cells will be separated from each other. The infected cells eventually lyse, releasing progeny phages. They infect nearby cells, which eventually lyse, releasing more phages. This continues and ultimately gives rise to a clear area within a lawn of bacteria. The clear area is a plaque.

MICRO INQUIRY *What would happen to a plate with individual plaques if it were returned to an incubator so that the viral infection could continue?*

Animal viruses are often cultivated by inoculating suitable host animals or embryonated eggs—fertilized chicken eggs incubated about 6 to 8 days after laying. Animal viruses also are grown in tissue (cell) culture on monolayers of animal cells. If an animal virus causes its host cell to lyse, plaques often are formed in the monolayer of cells. Viruses that cause cytopathic effects are also grown and detected in tissue culture.

Plant viruses are cultivated in a variety of ways. Plant tissue cultures, cultures of separated cells, or cultures of protoplasts (cells lacking cell walls) may be used. Viruses also can be grown in whole plants, usually after the leaves are mechanically inoculated by rubbing them with a mixture of viruses and an abrasive. When the cell walls are broken by the abrasive, the virions directly contact the plasma membrane and infect the exposed host cells. Some plant viruses can be transmitted only if a diseased part is grafted onto a healthy plant. A localized necrotic lesion often develops in infected plants due to the rapid death of cells in the infected area. Even when lesions do not occur, the infected plant may show symptoms such as changes in pigmentation or leaf shape.

A common approach to enumerating viruses is the **plaque assay.** Here several dilutions of a sample containing virus particles are plated with appropriate host cells. Theoretically, when the multiplicity of infection is very low, each plaque in a layer of host cells arises from the multiplication of a single virion (**figure 6.18**). Therefore a count of the plaques produced can be used to estimate the number of virions in the original sample. The resulting value is expressed as **plaque-forming units (PFU)** rather than as virions, for several reasons. First, not all virions may be infective. Furthermore, even though there are far fewer viruses than host cells, it is still possible for more than one virion to infect the same cell (figure 6.13). Finally, it is possible for two infected cells to be plated to the same area, which would then give rise to a single plaque. Despite these caveats, the number of PFU is proportional to the number of viruses: A preparation with twice as many virions will have roughly twice the plaque-forming units.

Comprehension Check

1. Discuss the ways that viruses can be cultivated. Define the terms *plaque* and *necrotic lesion*.
2. Given that in many cases viruses must be cultivated to make vaccines against viral diseases, discuss the advantages and disadvantages that might occur with each approach (embryonated egg versus cell culture) to growing animal viruses.

6.6 Viroids and Satellites: Nucleic Acid–Based Subviral Agents

After reading this section, you should be able to:

a. Describe the structure of a viroid and discuss the practical importance of viroids
b. Distinguish satellite viruses from satellite nucleic acids

Although viruses are exceedingly simple, even simpler infectious agents exist. **Viroids** are subviral infectious agents that cause over 20 different plant diseases, including potato spindle-tuber disease and chrysanthemum stunt disease. They consist of covalently closed, circular ssRNAs, about 250 to 430 nucleotides long and they do not encode any proteins. The viroid genome does however carry motifs that interact with host proteins to facilitate host specificity, replication, organelle entry and exit, cell-to-cell movement, and systemic spread (**figure 6.19**).

Viroids are divided into two families. The RNA of those in family Pospiviroidae exists in a rodlike shape due to intrastrand base pairing, which forms double-stranded regions with single-stranded loops (figure 6.19*a*). The RNA of viroids in the family Avsunviroidae is rod-shaped with a highly branched structure at one end (figure 6.19*b*). Each branch is formed by intramolecular base pairing of the RNA that creates a stem-loop structure. The two types of viroids replicate in different locations within an infected plant cell. Pospiviroids replicate in the nucleus, whereas avsunviroids replicate in chloroplasts. In both groups, replication involves DNA-dependent RNA polymerase (DdRP), an enzyme normally used by the host cell to copy messenger RNA from the DNA genome. Interestingly, viroids redirect DdRP to copy their genomic RNA. It has been suggested that they may represent survivors of the RNA world that existed before DNA. ◀ *Early life was probably RNA-based (section 1.2)*

A plant may be infected with a viroid without showing symptoms; that is, it may have a latent infection. However, the same viroid in another host species may cause severe disease. The pathogenicity of viroids is not well understood, but it is known that particular regions of the RNA are required; studies have shown that removing these regions (e.g., P and T_L) blocks the development of disease (figure 6.19*a*). Some data suggest that viroids cause disease by triggering a response called **RNA interference (RNAi),** which normally functions to protect the plant against infection by RNA viruses. During RNAi, the cell detects the presence of dsRNA and cuts it into fragments called small interfering RNAs (siRNAs). These are then used to target and destroy viral mRNA molecules or prevent their translation. Viroids may usurp this response by hybridizing to specific host mRNA molecules to which they have a complementary nucleotide sequence. Formation of the viroid-host hybrid dsRNA molecule is thought to elicit RNAi, leading to the destruction of host mRNA and therefore silencing of host genes. Failure to express required host genes leads to disease in the host plant. ▶ *Eukaryotes use RNA interference mechanisms for control (section 15.5)*

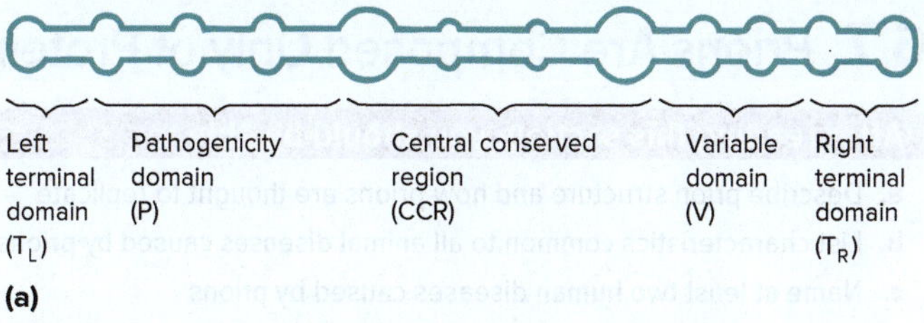

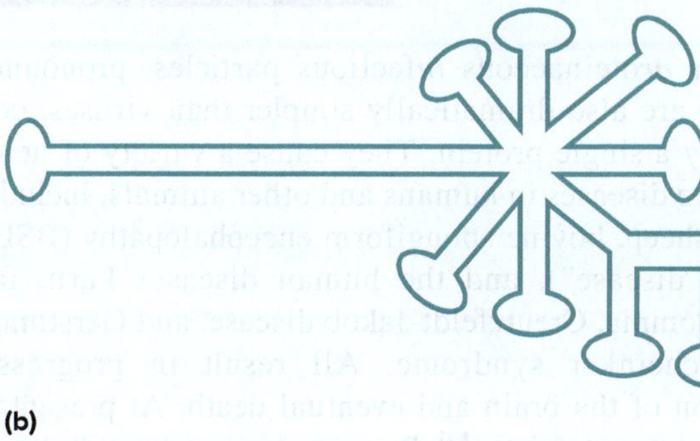

Figure 6.19 Viroid Structure. (a) This schematic diagram shows the general organization of a viroid belonging to the family Pospiviroidae. The closed single-stranded RNA circle has extensive intrastrand base pairing and interspersed unpaired loops. Also shown are the five domains identified in the molecule. Most changes in viroid pathogenicity arise from variations in the P and T_L domains. (b) Schematic diagram of a viroid belonging to Avsunviroidae. These viroids lack the central conserved region observed in pospiviroids.

Satellites are another type of subviral agent. They depend on helper viruses (HV) for replication and can be classified as either satellite viruses or satellite nucleic acids, based on their ability to encode capsid proteins. Satellite viruses can synthesize their own capsid proteins, which are distinct from those of the HV. By comparison, satellite nucleic acids—which can be either RNA or DNA—are encapsidated by HV-encoded proteins. There is little or no homology between the nucleic acid of the satellite and that of its HV (i.e., a satellite is not a defective version of its HV). Most satellites are associated with plant infections and use plant viruses as their HVs. However, an important example of a human satellite virus is hepatitis delta virus (HDV), for which the HV is hepatitis B virus (HBV). Coinfection with both HBV and HDV may result in liver disease that usually resolves within a few months but is occasionally fatal.

6.7 Prions Are Composed Only of Protein

After reading this section, you should be able to:

a. Describe prion structure and how prions are thought to replicate
b. List characteristics common to all animal diseases caused by prions
c. Name at least two human diseases caused by prions

Prions (for *pro*teinaceous *in*fectious particles, pronounced PREE-ons) are also dramatically simpler than viruses, comprising only a single protein. They cause a variety of neurodegenerative diseases in humans and other animals, including scrapie in sheep, bovine spongiform encephalopathy (BSE or "mad cow disease"), and the human diseases kuru, fatal familial insomnia, Creutzfeldt-Jakob disease, and Gerstmann-Strässler-Scheinker syndrome. All result in progressive degeneration of the brain and eventual death. At present, no effective treatment exists. *Prion proteins transmit disease (section 37.6)*

The disease state is caused by an abnormal form of a cellular protein called PrP^{C} (for *c*ellular *prion* *p*rotein). In healthy brain cells, this protein is localized to the plasma membrane of neurons, and is involved with several aspects of brain development and function. The abnormal form, PrP^{SC} (*sc*rapie-associated *pr*ion *p*rotein), differs from the normal form only in its three-dimensional conformation. Contact between the two protein forms causes PrP^{C} to irreversibly convert to PrP^{SC} (**figure 6.20**). Two important characteristics of PrP^{SC} are its stability and its aggregation with other PrP^{SC} molecules to form fibrils that accumulate in infected tissue. At present, it is uncertain if disease results from the loss of normal PrP^{C}, from the accumulation of PrP^{SC} aggregates, or from both. *Prion Diseases*

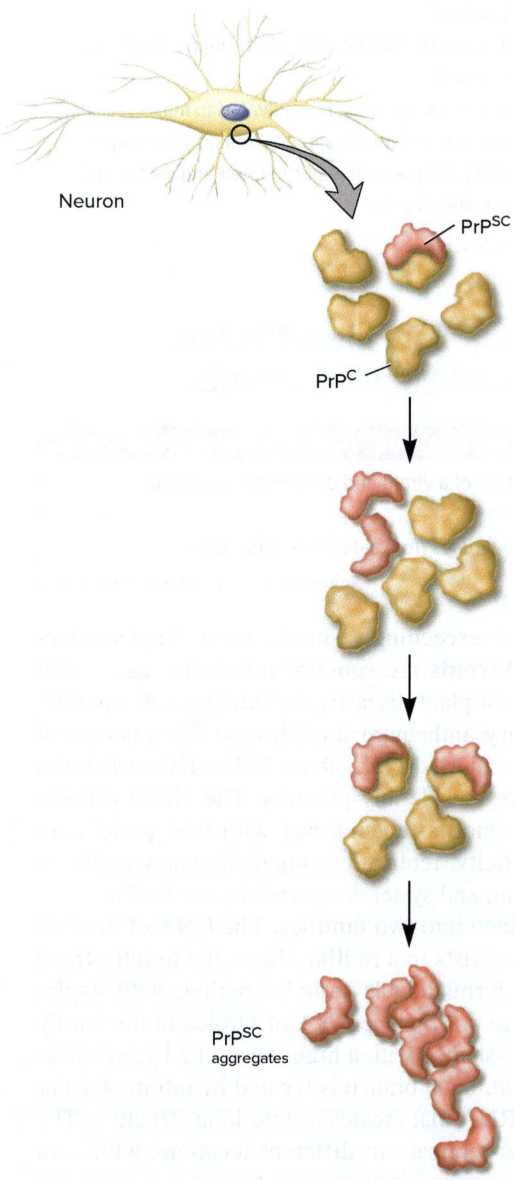

Figure 6.20 One Proposed Model for Prion Replication. The normal (PrP^{C}) and abnormal (PrP^{SC}) prion proteins differ in their tertiary structures. The abnormal prion protein acts as a template to catalyze the conversion of the normal prion protein to the abnormal form.

Lachina Publishing Services/McGraw Hill

Comprehension Check

1. What are viroids and why are they of great interest?
2. How does a viroid differ from a virus? From a satellite?
3. What is a prion? In what way does a prion differ fundamentally from viruses and viroids?
4. Prions are difficult to detect in host tissues. Why do you think this is so? Why do you think it has been challenging to develop effective treatments for prion-caused diseases?

Key Concepts

6.1 Viruses Are Acellular

- Virology is the study of viruses and other acellular infectious agents.
- Viruses cannot reproduce independently of living cells.

6.2 Virion Structure Is Defined by Capsid Symmetry and Presence or Absence of an Envelope

- All virions have a nucleocapsid composed of a nucleic acid, either DNA or RNA, held within a protein capsid made of one or more types of protein subunits called protomers (**figure 6.2**).
- Helical capsids resemble hollow protein tubes and may be either rigid or flexible (**figure 6.3**).
- Icosahedral capsids are usually constructed from two types of capsomeres: pentamers (pentons) at the vertices and hexamers (hexons) on the edges and faces of the icosahedron (**figure 6.4**).
- Complex virions (e.g., poxviruses and large phages) have complicated morphology not characterized by icosahedral and helical capsid symmetry (**figure 6.5**). Virions of large phages often have binal symmetry—that is, their heads are icosahedral and their tails are helical (**figure 6.6**).
- Some viruses have a membranous envelope surrounding their nucleocapsid. The envelope lipids usually come from the host cell; in contrast, many envelope proteins are viral and may project from the envelope surface as spikes (**figure 6.7**).
- Viral nucleic acids can be either single stranded (ss) or double stranded (ds), DNA or RNA. Most DNA viruses have dsDNA genomes that may be linear or closed circles. RNA viruses usually have ssRNA. Some RNA genomes are segmented.
- Although viruses lack true metabolism, some contain a few enzymes necessary for their multiplication.

6.3 Viral Life Cycles Have Five Steps

- The one-step growth experiment devised by Delbrück and Ellis was an important first step in understanding how viruses multiply (**figure 6.8**). This type of experiment showed that there is an initial latent period, during which no virions are released, followed by a rise period where virions are released. Early in the latent period is an eclipse period, in which no virus particles are present in infected cells.
- Viral life cycles can be divided into five steps: attachment to host; entry into host; synthesis of viral nucleic acids and proteins; self-assembly of nucleocapsids; and release from host (**figures 6.9–6.14**).
- The synthesis stage of a virus's life cycle depends on the nature of its genome. DNA viruses may use enzymes that are similar to host enzymes. In some cases, they rely solely on their hosts for synthesis of their nucleic acids and proteins.
- RNA viruses must either encode the enzymes they need to make mRNA and replicate their genomes upon entry into the host or carry these enzymes in their capsids.

6.4 There Are Several Types of Viral Infections

- Virulent bacteriophages and archaeal viruses lyse their host. In addition to host cell lysis, temperate viruses can enter the lysogenic cycle, in which they remain dormant in the host cell. This is often accomplished by integrating the viral genome into that of the host (**figure 6.15**).
- Viruses that infect eukaryotic cells can cause either host cell lysis or a more insidious cellular death: such viruses cause cytocidal infection (**figure 6.16**). Viral infection may result in cytopathic effects, or abnormalities in the host cell.

6.5 Virus Cultivation and Enumeration

- Viruses are cultivated using tissue cultures, embryonated eggs, bacterial cultures, and other living hosts.
- Phages produce plaques in bacterial lawns. Sites of animal viral infection may be characterized by cytopathic effects such as plaques (**figures 6.17** and **6.18**).

6.6 Viroids and Satellites: Nucleic Acid–Based Subviral Agents

- Viroids are infectious agents of plants that consist only of circular ssRNA molecules (**figure 6.19**). They do not encode any proteins.
- Satellite viruses are subviral infectious agents that encode their own capsid proteins but require a helper virus (HV) for replication. Satellite nucleic acids (RNA or DNA) are encapsidated by HV proteins.

6.7 Prions Are Composed Only of Protein

- Prions are small proteins associated with at least six degenerative nervous system disorders: scrapie, bovine spongiform encephalopathy, kuru, fatal familial insomnia, Gerstmann-Strässler-Scheinker syndrome, and Creutzfeldt-Jakob disease.
- Prion proteins exist in two forms: the infectious, abnormally folded form and a normal cellular form. The interaction between the abnormal form and the cellular form converts the cellular form into the abnormal form (**figure 6.20**).

Active Learning

1. Many classification schemes are used to identify bacteria. These start with Gram staining, progress to morphology and arrangement characteristics, and include a battery of metabolic tests. Build an analogous scheme that could be used to identify viruses. You might start by considering the host, or you might start with viruses found in a particular environment, such as a marine filtrate.
2. The origin and evolution of viruses are controversial. Discuss whether you think viruses evolved before the first cell or whether they have coevolved and are perhaps still coevolving with their hosts.
3. Consider the separate stages of an animal virus life cycle. Assemble a short list of structures and processes that are unique to the virus and would make good drug targets for an antiviral agent. Explain your rationale for each choice.
4. Consider the origin of viral envelopes and suggest why enveloped viruses that infect bacteria or plants are rare.
5. How might temperate and virulent bacteriophages differ upon encountering a population of susceptible bacteria? Consider both the first and subsequent rounds of infection.

CHAPTER

7

Bacterial and Archaeal Growth

frans lemmens/Alamy Stock Photo

Nurturing Microbes in the Lab

Learning how to grow microbes in the laboratory was a fundamentally important breakthrough in the history of microbiology. While many students (and professors!) might take them for granted, the stacks of Petri dishes that fill microbiology labs take a great deal of effort to prepare. Part of that effort requires knowing the correct media needed to support growth of your desired microbe. This is particularly challenging when trying to grow microbes that are fastidious or grow in so-called extreme environments.

One notably fastidious bacterium is *Legionella pneumophila*, which causes a pneumonia-like illness called Legionnaire's disease (see section 38.1). Culturing this microbe requires a nutritionally rich medium to support its growth. This medium must also be supplemented with charcoal, something that was not commonly used when *L. pneumophila* was first discovered in the 1970s. Even with the right media, this organism grows very slowly compared to commonly studied bacteria like *Escherichia coli* or *Staphylococcus aureus*. We now know that *L. pneumophila* is an intracellular pathogen, meaning that it grows inside eukaryotic host cells. This likely explains why it is hard to devise a chemical formulation to support rapid growth of *L. pneumophila* in the laboratory.

Another notoriously slow-growing genus is *Mycobacterium*. *M. tuberculosis* is the causative agent of tuberculosis, a disease that infects one-third of the world's population. While methods exist to culture *M. tuberculosis* in the laboratory, it takes three weeks of incubation for colonies to appear on a plate! Cultivating *M. leprae*, which causes leprosy in humans, is even more difficult as no known media formulation exists. In such cases, animal hosts are often employed. As it turns out, *M. leprae* is found naturally in the nine-banded armadillo native to the southeastern United States. Most microbiology labs are not equipped with armadillo farms, making studies of *M. leprae* extraordinarily challenging.

Readiness Check:

Based on what you have learned previously, you should be able to:

- ✓ Describe the eukaryotic cell cycle
- ✓ Define and list examples of essential nutrients, and describe how they are used by cells
- ✓ Distinguish macronutrients, micronutrients, and trace elements; list examples of each; and describe how they are used (section 3.2)
- ✓ Provide examples of growth factors needed by some microorganisms (section 3.2)

7.1 Many Bacteria and Archaea Reproduce by Binary Fission

After reading this section, you should be able to:

a. Describe binary fission as observed in bacteria and archaea
b. Compare binary fission to other reproductive strategies

Most bacterial and archaeal cells reproduce by **binary fission** (**figure 7.1**). During binary fission, the cell elongates as new cell envelope material is synthesized. The nucleoid must be replicated and partitioned into each half of the elongated cell during binary fission. By contrast, subcellular structures such as ribosomes and inclusions are dispersed throughout the cytoplasm and are easily partitioned to daughter cells when a septum (cross wall) is formed at midcell. Thus the division of a parent cell into two progeny cells results in each having its own nucleoid and a complement of other cellular constituents.

Several other reproductive strategies have been identified in bacteria (**figure 7.2**). Some reproduce by forming a bud. Others undergo multiple fission events with the progeny cells being held within the parent cell until they mature (*see figure 20.18*). Some cyanobacteria form **baeocytes,** reproductive cells produced via multiple fissions. Other bacteria, such as members of the genus *Streptomyces,* form multinucleoid filaments that eventually divide to form uninucleoid spores. These spores are readily dispersed, much like the dispersal spores formed by filamentous fungi. ▶| *Order Streptomycetales: an important source of antibiotics (section 22.1)*

Despite the diversity of reproductive strategies, certain features are shared. In all cases, the genome must be replicated and segregated to form distinct nucleoids. Each nucleoid and its surrounding cytoplasm must become enclosed within its own plasma membrane. Failure to correctly replicate and partition chromosomes between progeny cells can be lethal.

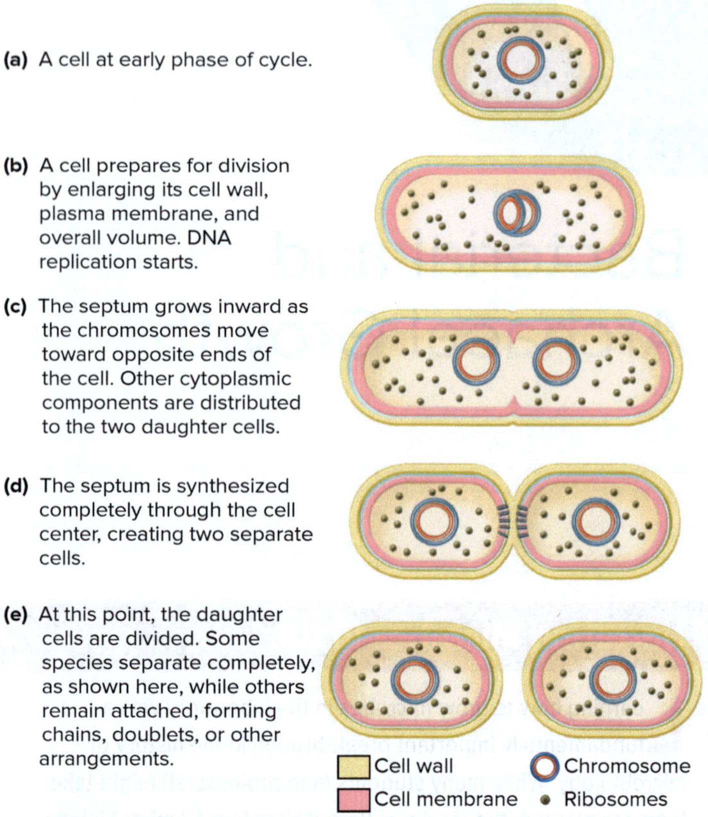

Figure 7.1 Binary Fission.

MICRO INQUIRY *In addition to chromosomes, what other cytoplasmic contents must be partitioned between daughter cells?*

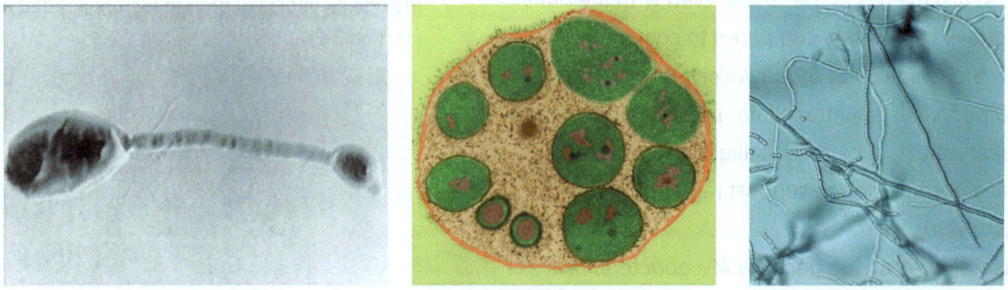

(a) Budding *Hyphomicrobium* **(b)** Multiple fission in *Dermocarpa* **(c)** *Streptomyces* spore formation

Figure 7.2 Some Bacteria Reproduce by Methods Other than Binary Fission. (a) Transmission electron micrograph of a budding *Hyphomicrobium* cell. (b) Baeocytes are daughter cells produced by the cyanobacterium *Dermocarpa.* (c) Spore formation by *Streptomyces* spp. Branched filaments, aerial hyphae, and chains of spores are visible in the light micrograph.

(a) McGraw Hill; (b) Dennis Kunkel Microscopy/Science Source; (c) Smith Collection/Gado/Archive Photos/Getty Images

7.2 Bacterial Cell Cycles Are Divided into Three Phases

After reading this section, you should be able to:

a. Summarize the three phases in a typical bacterial cell cycle
b. Summarize molecular mechanisms for chromosome partitioning and cytokinesis
c. State the functions of cytoskeletal filaments during cytokinesis and in determining cell shape

The **cell cycle** encompasses the formation of a new cell through the next round of cell division. This fundamentally important biological process has immense practical importance as well. For instance, synthesis of peptidoglycan during the cell cycle is the target of numerous antibiotics used to treat bacterial infections. ▶| *Inhibitors of cell wall synthesis (section 9.4)*

The cell cycles of several bacteria—*Escherichia coli, Bacillus subtilis,* and *Caulobacter vibrioides*—have been examined extensively, and our understanding of the bacterial cell cycle is based largely on these studies. The bacterial cell cycle consists of three phases: (1) a period of growth after the cell is born, which is similar to the G1 phase of the eukaryotic cell cycle; (2) chromosome replication and partitioning period, which functionally corresponds to the S and mitosis events of the M phase of the eukaryotic cycle; and (3) cytokinesis, during which a septum and daughter cells are formed (**figure 7.3**). Recall that in the eukaryotic cell cycle, the S phase is separated from the M phase by the G2 period. In G2, chromosome replication is completed and some time passes before chromosome segregation occurs. This is not the case for bacteria. As you will see in the discussion that follows, chromosome replication and partitioning occur concurrently. Furthermore, the initial events of cytokinesis actually occur before chromosome replication and partitioning are complete. Finally, some bacteria are able to initiate new rounds of replication before the first round of

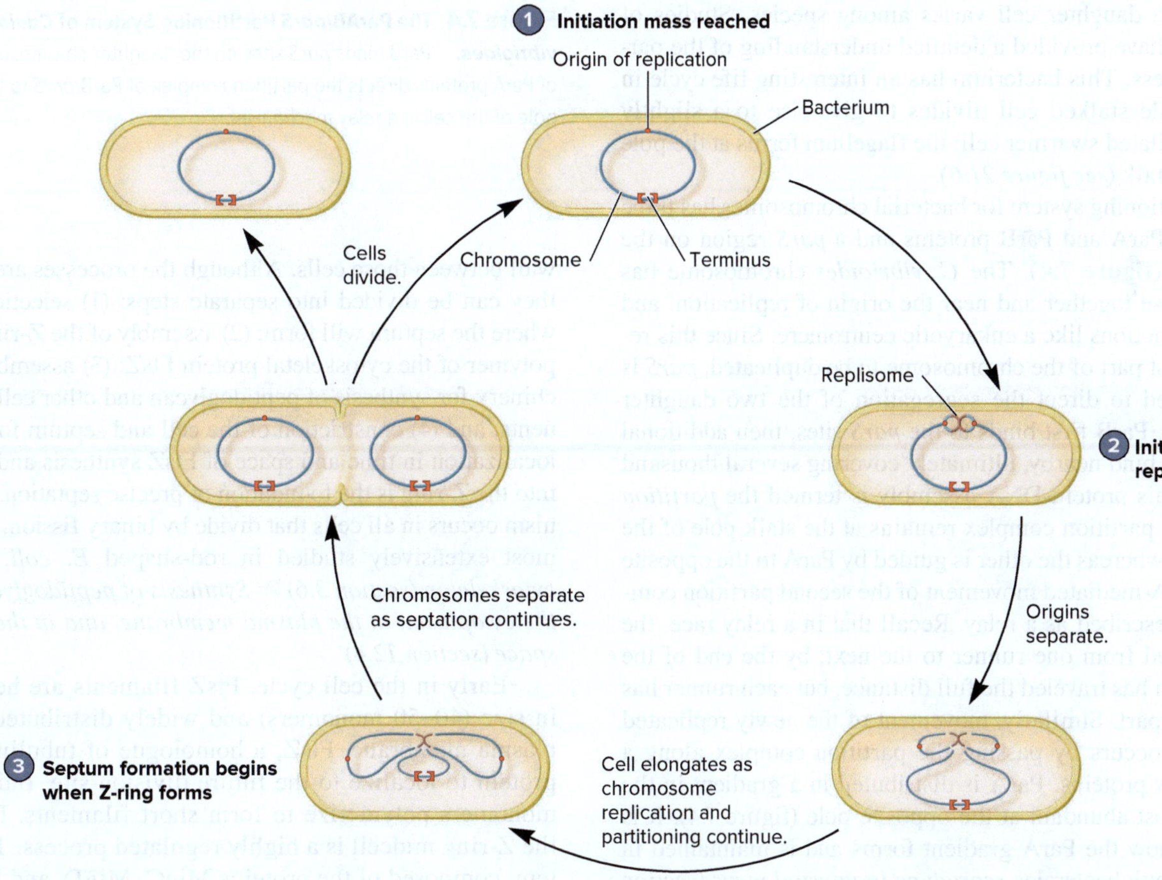

Figure 7.3 Cell Cycle of *E. coli*. An increase in cell mass triggers replication initiation. The origin of replication migrates to the center of the cell and proteins that make up the replisome assemble. In this illustration, one round of DNA replication is completed before the cell divides. In rapidly growing cultures, second and third rounds of DNA replication initiate before the original cell division is complete. Thus the daughter cells inherit partially replicated DNA.

MICRO INQUIRY *Why is it important that the origin of replication migrate to the center of the cell prior to replication?*

replication and cytokinesis is finished. Although chromosome replication and partitioning overlap with cytokinesis, we consider them separately.

Chromosome Replication and Partitioning

Most bacteria have one circular chromosome. Each chromosome has a single site at which replication starts called the **origin of replication,** or simply the origin (figure 7.3). Replication is completed at the **terminus,** located opposite the origin. In a newly formed *E. coli* cell, the chromosome is compacted and oriented so that the origin and terminus are in opposite halves of the cell. Early in the cell cycle, the origin and terminus move to midcell, and proteins needed for chromosome replication assemble at the origin. This DNA synthesizing machinery is called the **replisome,** and DNA replication proceeds bidirectionally (in both directions) from the origin. As the chromosome is replicated, the two origins move toward opposite ends of the cell, which facilitates separation of the newly replicated chromosomes from one another. ▶| *DNA replication in bacteria (section 13.3)*

The mechanism by which bacterial chromosomes are partitioned to each daughter cell varies among species. Studies of *C. vibrioides* have provided a detailed understanding of the partitioning process. This bacterium has an interesting life cycle in which a sessile stalked cell divides to give rise to a slightly smaller, flagellated swarmer cell; the flagellum forms at the pole opposite the stalk (*see figure 21.6*).

The partitioning system for bacterial chromosomes has three components: ParA and ParB proteins and a *parS* region on the chromosome (**figure 7.4**). The *C. vibrioides* chromosome has *parS* sites close together and near the origin of replication, and this region functions like a eukaryotic centromere. Since this region is the first part of the chromosome to be duplicated, *parS* is well-positioned to direct the segregation of the two daughter chromosomes. ParB first binds to the *parS* sites, then additional ParB proteins bind nearby, ultimately covering several thousand base pairs. This protein-DNA assembly is termed the *partition complex.* One partition complex remains at the stalk pole of the dividing cell, whereas the other is guided by ParA to the opposite pole. The ParA-mediated movement of the second partition complex is best described as a relay. Recall that in a relay race, the baton is passed from one runner to the next; by the end of the race, the baton has traveled the full distance, but each runner has traveled only part. Similarly, movement of the newly replicated chromosome occurs by passing the partition complex along a series of ParA proteins. ParA is distributed in a gradient in the cell and is most abundant at the opposite pole (figure 7.4). It is still unclear how the ParA gradient forms and is maintained in the cell. ▶| Caulobacterales *reproduce in unusual ways (section 21.1)* *Binary Fission; Bidirectional DNA Replication*

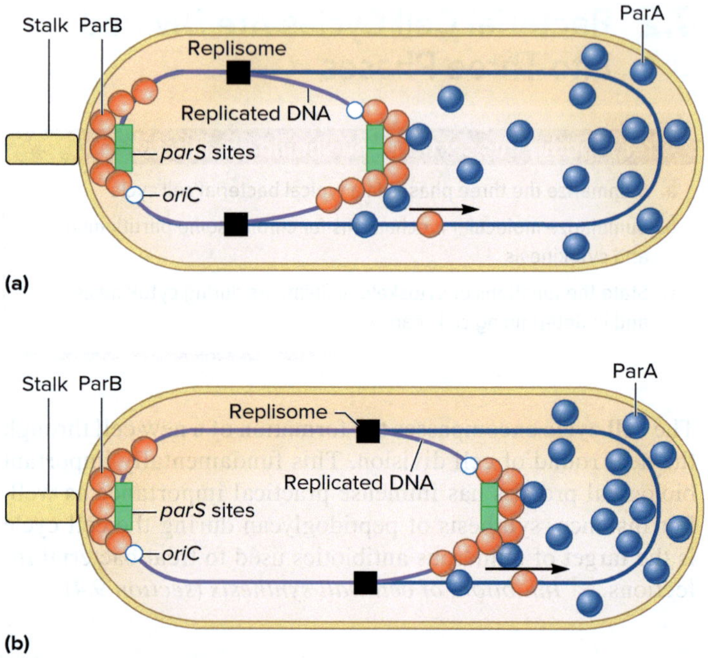

Figure 7.4 The ParAB/*parS* Partitioning System of *Caulobacter vibrioides.* ParB binds *parS* sites on the daughter chromosomes. A gradient of ParA proteins directs the partition complex of ParB/*parS* to the opposite pole of the cell in a relay mechanism.

Cytokinesis

Cytokinesis describes the formation of two daughter cells following cell division, while **septation** is the process of forming a cross wall between those cells. Although the processes are interrelated, they can be divided into separate steps: (1) selection of the site where the septum will form; (2) assembly of the Z-ring, which is a polymer of the cytoskeletal protein FtsZ; (3) assembly of the machinery for synthesis of peptidoglycan and other cell wall constituents; and (4) constriction of the cell and septum formation. The localization in time and space of FtsZ synthesis and its assembly into the Z-ring is the foundation of precise septation. This mechanism occurs in all cells that divide by binary fission, and has been most extensively studied in rod-shaped *E. coli.* |◀ *Bacterial cytoskeleton (section 3.6)* ▶| *Synthesis of peptidoglycan occurs in the cytoplasm, at the plasma membrane, and in the periplasmic space (section 12.4)*

Early in the cell cycle, FtsZ filaments are heterogeneous in size (40–50 monomers) and widely distributed around the plasma membrane. FtsZ, a homologue of tubulin, is the first protein to localize to the future division site. Individual FtsZ monomers polymerize to form short filaments. Placement of the Z-ring midcell is a highly regulated process. The Min system, composed of the proteins MinC, MinD, and MinE, determines where the Z-ring will form (**figure 7.5**). These proteins do this not by placing FtsZ at the correct site, but by preventing its polymerization in the wrong places. The Min proteins oscillate from end to end along the length of the cell. Over time, the highest concentration of MinC is found nearest the poles. MinC is a strong inhibitor of FtsZ polymerization and its

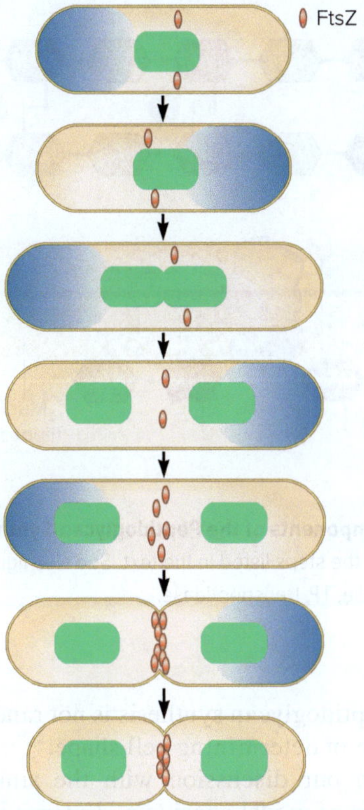

Figure 7.5 Mechanisms to Establish the Site of Septum Formation. The Min system (blue) oscillates between the poles to prevent FtsZ polymerization at the poles. Nucleoid (green) occlusion prevents Z-ring formation until the two nucleoids have separated. Both mechanisms restrict Z-ring formation to the center of the cell.

MICRO INQUIRY *What would be the outcome if FtsZ formed a Z-ring that was not in the center of the cell? Consider both the morphology of the daughter cells and chromosome partitioning.*

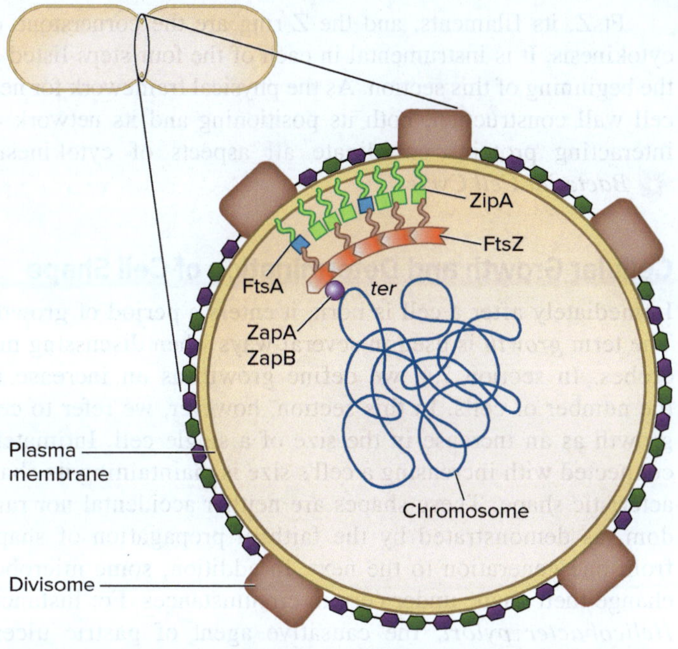

Figure 7.6 Cross-sectional View of a Dividing Bacterial Cell. FtsZ filaments attach to the plasma membrane through anchoring proteins FtsA and ZipA, which also coordinate the enzymes that synthesize and remodel peptidoglycan in the developing septum. ZapA and ZapB link the Z-ring to the chromosome replication terminus (*ter*) to ensure that septum constriction follows the conclusion of replication.

Source: Du and Lutkenhaus. At the Heart of Bacterial Cytokinesis: The Z Ring. *Trends in Microbiology.* 2019. Figure 4. https://doi.org/10.1016/j.tim.2019.04.011

location at the cell poles thus allows accumulation of FtsZ filaments at midcell.

The timing of Z-ring formation is also important. If done too early, the Z-ring could constrict around the replicating chromosome, preventing proper partitioning. Therefore, coordination of FtsZ filaments to form the mature Z-ring occurs only after the replicated chromosomes have almost completed partitioning to the poles. This process is called **nucleoid occlusion**, whereby chromosome movement is coordinated with cell separation. The protein SlmA coats the chromosome except at the replication termination region (*ter*). Similar to MinC, SlmA inhibits FtsZ polymerization. This ensures that the SlmA-decorated chromosome has moved away from midcell prior to Z-ring formation (figure 7.5).

The Z-ring is a dense aggregate of FtsZ filaments that associate discontinuously and randomly at midcell. It is a dynamic structure 80–100 nm wide positioned just inside the plasma membrane (**figure 7.6**). The Z-ring is indirectly attached to the membrane through interactions with several adaptor proteins. Within the Z-ring, individual filaments move around the interior circumference of the membrane by a process called treadmilling. Treadmilling results in the removal of individual FtsZ monomers from one end of a filament and repolymerization of new monomers at the other end. The net result is a moving filament, with energy provided by GTP hydrolysis. (GTP is related to ATP and provides energy in the same manner.) ▶| *ATP: The major energy currency of cells (section 10.2)*

The Z-ring has a connection to the *ter* region of the chromosome (figure 7.6). This interaction both stabilizes the dynamic Z-ring and prevents constriction until the two new chromosomes have moved away from midcell.

In describing Z-ring assembly and its coordination with chromosome partitioning, we have yet to address its primary function. The Z-ring is the scaffold upon which the enzymes that synthesize the layers of the cell wall assemble (figure 7.6). The proteins that anchor FtsZ to the membrane also coordinate **divisome** formation. The divisome is a complex of more than 30 proteins that catalyze peptidoglycan remodeling to split the sacculus. Peptidoglycan strands are linked by short peptides (*see figure 3.17*), and these strands must be separated and rejoined to one daughter sacculus or the other. Ultimately, the Z-ring circumference is reduced, the two cytoplasms (and their contents) separate, and two daughter cells each begin a new cell cycle.

FtsZ, its filaments, and the Z-ring are the cornerstone of cytokinesis. It is instrumental in each of the four steps listed at the beginning of this section. As the physical framework for new cell wall construction, both its positioning and its network of interacting proteins coordinate all aspects of cytokinesis. *Bacterial Cell Cycle*

Cellular Growth and Determination of Cell Shape

Immediately after a cell is born, it enters a period of growth. The term *growth* is used in several ways when discussing microbes. In section 7.4, we define growth as an increase in the number of cells. In this section, however, we refer to cell growth as an increase in the size of a single cell. Intimately connected with increasing a cell's size is maintaining its characteristic shape. These shapes are neither accidental nor random, as demonstrated by the faithful propagation of shape from one generation to the next. In addition, some microbes change their shape under certain circumstances. For instance, *Helicobacter pylori,* the causative agent of gastric ulcers and stomach cancer, changes from a helical shape to straight rods, cocci, or filaments depending on the environmental conditions.

To consider the shape of a cell, we must first review cell wall function. The cell wall constrains the turgor pressure exerted by the cytoplasm, preventing the cell from swelling and bursting. Turgor pressure describes the force pushing against the cell wall because of the osmolarity of the cytoplasmic contents. Peptidoglycan is a rigid but elastic structure that protects the cell from lysis. In fact, the peptidoglycan sacculus is a single molecule that constrains the plasma membrane and its contents. The strength of the existing sacculus must be maintained as new peptidoglycan subunits are added. Thus understanding peptidoglycan synthesis is critical to understanding cell growth and the determination of cell shape. *Peptidoglycan structure (section 3.3)*

Peptidoglycan synthesis involves many proteins that sculpt the sacculus during growth and cell division. **Figure 7.7** illustrates some of the components of the peptidoglycan-synthesizing machinery and outlines a general scheme of peptidoglycan synthesis. (1) Synthesis of the NAG-NAM-pentapeptide building block is completed in the cytoplasm while it is attached to a lipid carrier, **bactoprenol,** located in the plasma membrane. (2) The carrier-bound building block is then flipped across the membrane by a protein called MurJ. (3) Upon its release into the periplasmic space, the NAG-NAM pentapeptide is inserted into a peptidoglycan strand by enzymes termed peptidoglycan glycotransferases (GTases). Strands are crosslinked by enzymes termed transpeptidases (TPases). TPases are also called penicillin-binding proteins, reflecting their discovery as molecules that physically associated with penicillin. Maintenance of cell shape requires both of these proteins, along with scaffold proteins to direct them to the appropriate location in the plasma membrane. The cellular

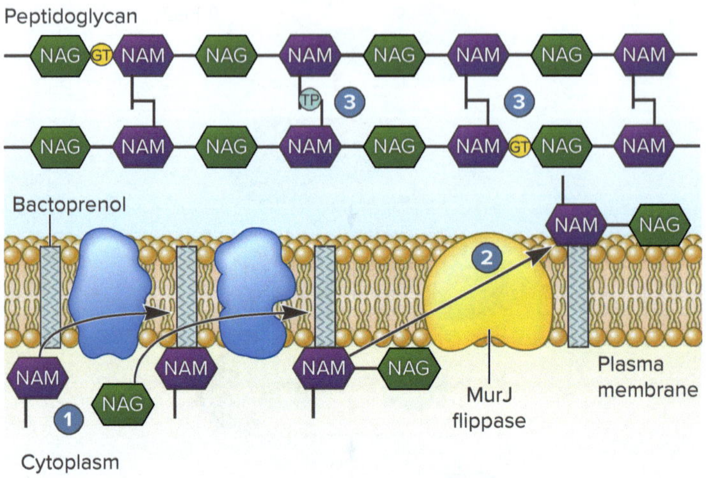

Figure 7.7 Components of the Peptidoglycan-Synthesizing Machinery. Numbers refer to the steps listed in the text. See also figures 9.6 and 12.8. GT, glycosyltransferase; TP, transpeptidase.

location of peptidoglycan synthesis is not random and plays an important role in determining cell shape.

We begin our discussion with the simplest shape, the coccus. Studies of model cocci (e.g., *Enterococcus faecalis* and *Staphylococcus aureus*) show that new peptidoglycan forms only at the central septum (**figure 7.8*a***). When daughter cells separate, each has one new and one old hemisphere. As in most bacteria, proper placement of the septum depends on FtsZ localization. Thus FtsZ placement determines the site of cell wall growth by recruiting enzymes needed for peptidoglycan synthesis to the divisome.

In contrast to cocci, maintaining the rod shape requires a molecular machine called the **elongasome.** It shares many enzymes with the divisome, but its scaffold is the actin homologue MreB. In a manner that echoes FtsZ and the Z-ring, MreB proteins polymerize, creating filaments along the cytoplasmic face of the plasma membrane (figure 7.8*b*). MreB directs the assembly of elongasome proteins including GTases and TPases on the periplasmic face of the plasma membrane. Cell wall growth occurs in numerous bands around the circumference of the cell. The bands are positioned along the length of the cell but not at the poles (*see figure 3.27*). The importance of MreB in determining cell shape is demonstrated by two observations. First, rod-shaped cells in which MreB has been depleted assume a spherical shape. In addition, while almost all rod-shaped bacteria and archaea synthesize at least one MreB homologue, cocci lack proteins in the MreB family.

The last cell shape we consider is that of curved (vibroid) cells, as studied in *C. vibrioides.* In addition to MreB and FtsZ, these cells (and other vibroid-shaped cells) produce a cytoskeletal protein called crescentin. Filaments of this protein localize to one side of the cell, where it slows the insertion of new

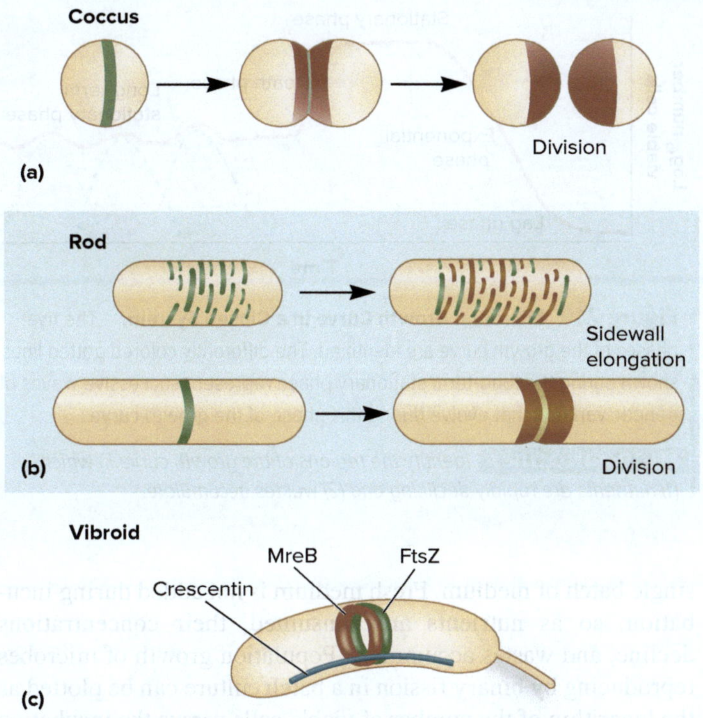

Figure 7.8 Cell Wall Biosynthesis and Determination of Cell Shape in Spherical, Rod-Shaped, and Curved Cells.

MICRO INQUIRY *Which step in the development of rod-shaped cells is essential in determining cell morphology?*

peptidoglycan units into the sacculus (figure 7.8*c*). The resulting asymmetric cell wall growth gives rise to the curvature that characterizes the vibroid shape (*see figure 21.5*).

It is clear from these examples that common microbial cell shapes require cytoskeletal filaments that bear a startling similarity with those of eukaryotes. However, some cell shapes appear to be determined by other mechanisms. For example, the spirochete *Borreliella burgdorferi,* which causes Lyme disease, relies on its periplasmic flagella to confer its spiral shape. As another example, *Spiroplasma,* a citrus pathogen that has no cell wall or flagella, relies on contractile cytoplasmic fibrils to give its spiral shape. ▶| *Phylum Spirochaetota: bacteria with a corkscrew morphology (section 20.5)*

Comprehension Check

1. Describe the three phases of a bacterial cell cycle. The overlapping of cytokinesis and chromosome partitioning could potentially create problems for a cell during the cell cycle. What mechanisms does the cell use to prevent problems?
2. How does the bacterial cell cycle compare with the eukaryotic cell cycle? List two ways they are similar and two ways they differ.
3. Do you think the Min system and nucleoid occlusion function in coccoid-shaped cells? Explain your answer.

7.3 Archaeal Cell Cycles Are Unique

After reading this section, you should be able to:

a. Compare and contrast the *Sulfolobus* spp. cell cycle and the typical eukaryotic cell cycle
b. Compare and contrast the *Sulfolobus* spp. cell cycle and a bacterial cell cycle

As is true with many other aspects of archaeal biology, understanding of archaeal cell cycles lags behind that of bacteria. However, studies of model archaea such as *Sulfolobus* spp. have yielded intriguing results. Their cell cycle is reminiscent of a mitotic cell cycle. After a growth phase (G1), their DNA is replicated (S) using machinery similar to that of eukaryotes. Of note is the fact that unlike bacteria, the circular chromosome of *Sulfolobus* spp. has three origins of replication. Furthermore, once replicated, the daughter chromosomes remain unsegregated for some time (G2); the *Sulfolobus* G2 phase consumes more than 50% of the cell cycle. G2 is followed by segregation of the chromosome and then cytokinesis occurs. ▶| *Archaea use homologues of eukaryotic replisome proteins (section 15.2).*

Two proteins involved in chromosome segregation have been identified: SegA and SegB. SegA is similar to bacterial ParA, while SegB is structurally unique but appears to be the functional equivalent of bacterial ParB. SegA forms filaments and is thought to exert the force, either pulling or pushing, that segregates the chromosomes. The role of SegB is unclear, although it binds DNA and interacts with SegA, enhancing filament formation.

Although chromosome segregation appears to be bacteria-like, cytokinesis in *Sulfolobus* spp. is most similar to eukaryotic processes. These archaea undergo cytokinesis using cell *division* proteins termed CdvA, CdvB, and CdvC. CdvA, found only in *Sulfolobus,* binds the membrane and forms a noncontractile ring at midcell (**figure 7.9**). CdvA recruits CdvB and CdvC to the site

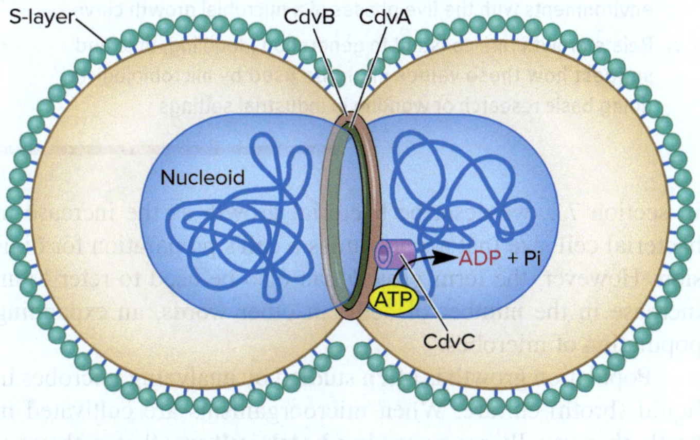

Figure 7.9 Cytokinesis in *Sulfolobus* spp.

Source: Haertel and Schwille. ESCRT-III mediated cell division in *Sulfolobus acidocaldarius*, a reconstitution perspective. 2014. Figure 1B. https://doi.org/10.3389/fmicb.2014.00257

of cell division. The CdvB ring constricts to separate the daughter cells. Widespread among archaeal phyla, CdvB and CdvC are homologues of eukaryotic proteins ESCRT-III and Vsp4, respectively. In eukaryotes, these proteins encircle membranes to bring about membrane scission.

Other archaea have cell cycles comparable to the bacterial cell cycle, featuring growth, DNA replication, and cytokinesis. FtsZ is present in these cells, often in two related forms. Recall that all archaea lack peptidoglycan, so unlike the bacterial model where the Z-ring is physically associated with the peptidoglycan synthesizing enzymes, the archaeal Z-ring associates with new S-layer proteins and possibly other cell wall components.

As we note in chapter 4, some archaea are polyploid, often with a dozen or more chromosome copies. It is possible that a genome segregation system does not exist, and the large number of chromosome copies may compensate for this deficiency. *Nucleoid (section 4.3)*

Comprehension Check

1. What elements of the *Sulfolobus* spp. cell cycle are similar to the bacterial cell cycle? What elements are similar to the eukaryotic cell cycle?
2. Many archaeal genomes encode an FtsZ homologue. Describe how FtsZ might function in an archaeal cell cycle.
3. Review figure 4.6 and the variety of cell wall types in archaea. How might these coordinate with the cytokinesis proteins to sculpt the cell wall during cell division?

7.4 Growth Curves Consist of Five Phases

After reading this section, you should be able to:

a. Describe the five phases of a microbial growth curve observed when microbes are grown in a batch culture
b. Correlate changes in nutrient concentrations in natural environments with the five phases of a microbial growth curve
c. Relate growth rate constant to generation (doubling) time and suggest how these values might be used by microbiologists doing basic research or working in industrial settings

In section 7.2, we describe bacterial growth as the increase in bacterial cell size that accompanies a cell's preparation for division. However, the term *growth* can also be used to refer to an increase in the number of cells; in other words, an expanding population of microbes.

Population growth is often studied by analyzing microbes in liquid (broth) culture. When microorganisms are cultivated in broth, they usually are grown in a **batch culture;** that is, they are incubated in a closed vessel such as a test tube or a flask with a

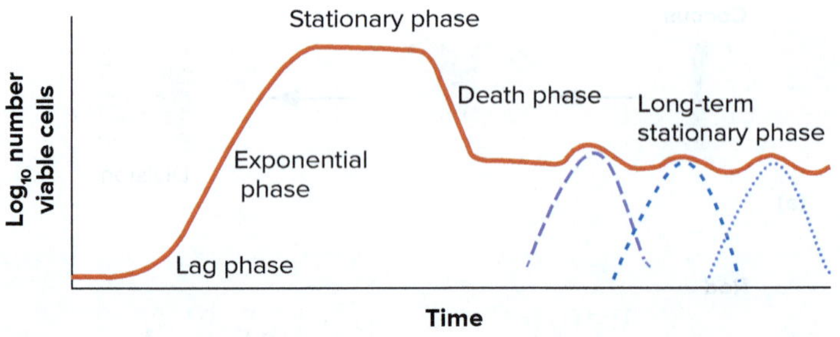

Figure 7.10 Microbial Growth Curve in a Closed System. The five phases of the growth curve are identified. The differently colored dotted lines shown during the long-term stationary phase represent successive waves of genetic variants that evolve during this phase of the growth curve.

MICRO INQUIRY *Identify the regions of the growth curve in which (1) nutrients are rapidly declining and (2) wastes accumulate.*

single batch of medium. Fresh medium is not added during incubation, so as nutrients are consumed, their concentrations decline, and wastes accumulate. Population growth of microbes reproducing by binary fission in a batch culture can be plotted as the logarithm of the number of viable cells versus the incubation time. The resulting curve has five distinct phases (**figure 7.10**), which we examine in this section. Although this is "life in the lab," microbes do encounter conditions in their natural environments that mimic what occurs in a batch culture. Furthermore, humans routinely create artificial environments for microbes (e.g., the fermentation vessel in a pharmaceutical plant) that are batch cultures (*see figure 41.5*). Therefore understanding the growth curve is of paramount importance.

When microorganisms are introduced into fresh culture medium, usually no immediate increase in cell number occurs. This period is called the **lag phase.** It is not a time of inactivity; rather cells are synthesizing new components. This can be necessary for a variety of reasons. The cells may be old and depleted of ATP, essential cofactors, and ribosomes; these must be synthesized before growth can begin. The medium may be different from the one the microorganism was growing in previously. In this case, new enzymes are needed to use different nutrients. Eventually, however, the cells replicate their DNA, increase in mass, and divide. As a result, the number of cells in the population increases.

During the **exponential phase,** microorganisms grow and divide at the maximal rate possible given their genetic potential and culture conditions. Their growth rate is constant during the exponential phase as cells are completing the cell cycle and doubling in number at regular intervals (figure 7.10). The population is most uniform in terms of chemical and physiological properties during this phase; therefore exponential phase cultures are usually used in biochemical and physiological studies. The growth rate during the exponential phase depends on several factors, including nutrient availability. When microbial

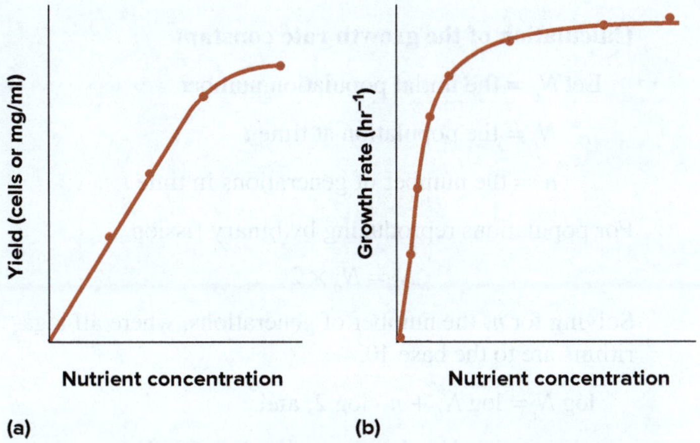

Figure 7.11 Nutrient Concentration and Growth. (a) The effect of changes in limiting nutrient concentration on total microbial yield. At sufficiently high concentrations, total growth will plateau. (b) The effect of nutrient availability on growth rate.

growth is limited by the low concentration of a required nutrient, the final net growth or yield of cells increases with the initial amount of the limiting nutrient present (**figure 7.11*a***). The growth rate also increases with nutrient concentration (figure 7.11*b*), but it saturates, much like what is seen with many enzymes (*see figure 10.16*). The shape of the curve is thought to reflect the rate of nutrient uptake by microbial transport proteins. At sufficiently high nutrient levels, the transport systems are saturated, and the growth rate does not rise further with increasing nutrient concentration. *Bacteria use many mechanisms to bring nutrients into the cell (section 3.2)*

In a closed system such as a batch culture, population growth eventually ceases and the growth curve becomes horizontal (figure 7.10). This **stationary phase** is attained by some bacteria at a population level of around 10^9 cells per milliliter. Final population size depends on nutrient availability and other factors, as well as the type of microorganism. In stationary phase, the total number of viable microorganisms remains constant. This may result from a balance between cell division and cell death, or the population may simply cease to divide but remain metabolically active. Although we are talking about the entire population of microbes in a culture, remember that the individual microbes within a stationary phase culture are heterogeneous. Although some cells are dying, others are still reproducing. Experimental means of enumerating cells in a population do not distinguish the metabolic activity of individuals within that population.

Microbes enter the stationary phase for many reasons. One important reason is nutrient limitation. When an essential nutrient is depleted, population growth slows and eventually stops. Aerobic organisms often are limited by O_2 availability. Oxygen is not very soluble and may be depleted so quickly that only the surface of a culture will have an O_2 concentration adequate for growth. Population growth also may cease due to the accumulation of toxic waste products. This seems to limit the growth of many cultures growing in the absence of O_2. For example, streptococci can produce so much acid from sugar fermentation that growth is inhibited (*see figure 11.18*). Finally, growth may cease when a critical population level is reached. In reality, entrance into the stationary phase may result from several factors operating in concert.

There are a variety of mechanisms by which nutrient availability helps regulate growth. For example, DnaA, the protein that binds to the chromosome's origin to initiate replication, becomes less active in stationary phase. Ongoing replication is completed, but no further initiation occurs. This is one of many ways the cell conserves energy by eliminating processes that are not essential to survival.

Cells growing in batch culture cannot remain in stationary phase indefinitely. Eventually they enter a phase known as the **death phase** (figure 7.10). During this phase, the equilibrium between dying and reproducing cells shifts and the number of viable cells declines exponentially, with cells dying at a constant rate. Nutrient deprivation and the buildup of toxic wastes cause irreparable harm to the cells.

Long-term growth experiments reveal that after a period of exponential death, some microbial cultures have a long period where the population size remains more or less constant. This **long-term stationary phase** (also called extended stationary phase) can last months to years (figure 7.10). During this time, the bacterial population continually evolves so that actively reproducing cells are those best able to use the nutrients released by their dying brethren and most tolerant to accumulated toxins. This dynamic process is marked by successive waves of genetically distinct variants. Thus natural selection can be witnessed within a single culture vessel.

Mathematics of Growth

Knowledge of microbial growth rates during the exponential phase is indispensable to microbiologists. Growth rate studies contribute to basic physiological and ecological research, and are applied in industry. As growth is typically measured in batch cultures, only the exponential phase features a population acting in synchrony. For this reason, the quantitative aspects of growth are only valid for cultures in exponential phase and only apply to microorganisms that divide by binary fission.

During the exponential phase, each microorganism divides at a constant interval. Thus the population doubles in number during a specific length of time called the **generation (doubling) time (*g*)**. This can be illustrated with a simple example. Suppose that a broth culture is inoculated with one cell that divides every 20 minutes (**table 7.1**). The culture will have 2 cells after 20 minutes, 4 cells after 40 minutes, and so forth. Because the number of cells doubles every generation, the increase in population is always 2^n, where n is the number of generations. The resulting population increase is exponential (**figure 7.12**).

Table 7.1 An Example of Exponential Growth[1]

Time[2] (min)	Generation Number (n)	2^n	Number of cells[3] $(N_0 \times 2^n)$	$\log_{10} N_t$
0	0	$2^0 = 1$	1	0.000
20	1	$2^1 = 2$	2	0.301
40	2	$2^2 = 4$	4	0.602
60	3	$2^3 = 8$	8	0.903
80	4	$2^4 = 16$	16	1.204

[1]Definitions of n, N_0, N_t are as in figure 7.13.
[2]The hypothetical culture begins with one cell having a 20-minute generation time.
[3]Number of cells in the culture.

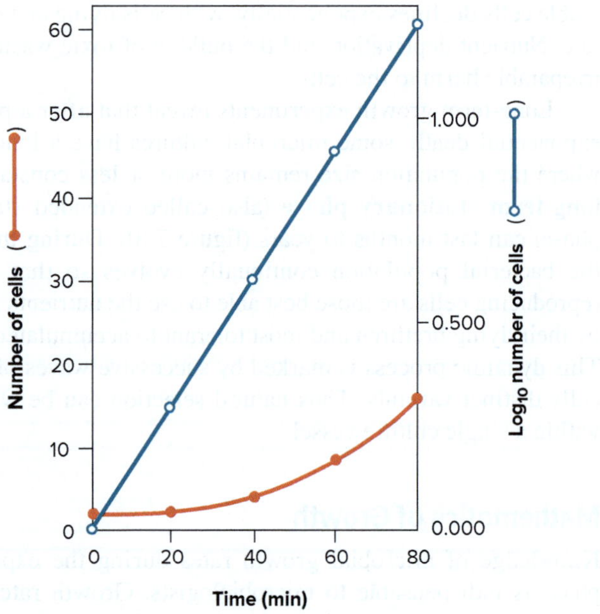

Figure 7.12 Exponential Microbial Growth. Four generations of growth are plotted on a linear scale (—) and on a logarithmic scale (○—○). The linearity of the log plot indicates that the population is in exponential phase.

The mathematics of growth during the exponential phase is illustrated in **figure 7.13**, which shows the calculation of two important values. The **growth rate constant (k)** is the number of generations per unit time and is often expressed as generations per hour (hr^{-1}). Larger values for k indicate more rapid growth. k can be used to calculate g. As can be seen in figure 7.13, g is simply the reciprocal of k. g can also be determined directly from a semilogarithmic plot of growth curve data (**figure 7.14**). Generation times vary markedly between different microbial species and environmental conditions (**table 7.2**).

Calculation of the growth rate constant

Let N_0 = the initial population number

N_t = the population at time t

n = the number of generations in time t

For populations reproducing by binary fission

$$N_t = N_0 \times 2^n$$

Solving for n, the number of generations, where all logarithms are to the base 10,

$$\log N_t = \log N_0 + n \cdot \log 2, \text{ and}$$

$$n = \frac{\log N_t - \log N_0}{\log 2} = \frac{\log N_t - \log N_0}{0.301}$$

The growth rate constant (k) is the number of generations per unit time $\left(\frac{n}{t}\right)$. Thus

$$k = \frac{n}{t} = \frac{\log N_t - \log N_0}{0.301t}$$

Calculation of generation (doubling) time

If a population doubles, then

$$N_t = 2N_0$$

Substitute $2N_0$ into the growth rate constant equation and solve for

$$k = \frac{\log (2N_0) - \log N_0}{0.301g} = \frac{\log 2 + \log N_0 - \log N_0}{0.301g}$$

$$k = \frac{1}{g}$$

The generation time is the reciprocal of the growth rate constant.

$$g = \frac{1}{k}$$

Figure 7.13 Calculation of the Growth Rate Constant and Generation Time. The calculations are only valid for the exponential phase of growth, when the growth rate is constant.

Comprehension Check

1. Define *microbial growth*.
2. Describe the phases of the growth curve and discuss the causes of each.
3. Why must k and g be calculated from exponential phase measurements?
4. Why would vigorously growing cells have a shorter lag phase than those that have been stored in a refrigerator when inoculated into fresh culture medium?
5. Calculate the growth rate constant and generation time of a culture that increases in the exponential phase from 5×10^2 to 1×10^8 in 12 hours.
6. An archaeal culture has 10^3 cells at the start of exponential phase and a 90-minute generation time. How many cells will there be after 8 hours of exponential growth?

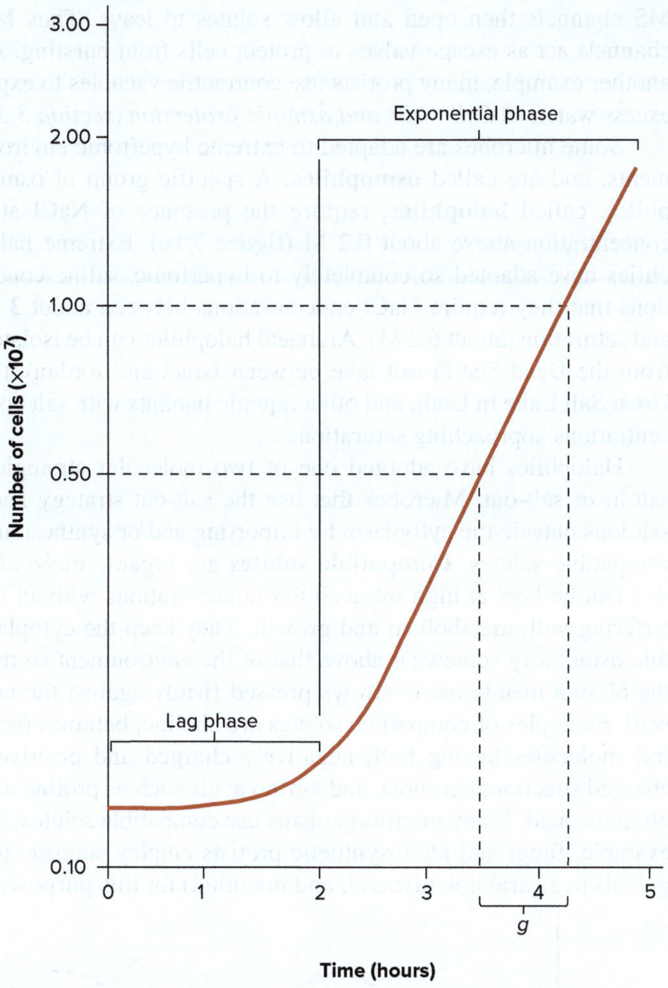

Figure 7.14 Generation Time Determination. The generation time (*g*) can be determined from a microbial growth curve. The population data are plotted with the number of cells on a logarithmic scale. The time to double the population number is then read directly from the plot.

7.5 Environmental Factors Affect Microbial Growth

After reading this section, you should be able to:

- **a.** Use terms that describe a microbe's growth range or requirement for each of the factors that influence microbial growth
- **b.** Summarize the adaptations of extremophiles to their habitats
- **c.** Summarize the strategies used by nonextremophiles to acclimate to changes in their environment
- **d.** Describe enzymes that protect microbes from toxic O_2 products

Almost every year at Yellowstone National Park, a visitor is seriously burned or killed after falling into one of the park's hot springs. Yet a rich microbial community lives in these same springs, as well as in the hot pots and other thermal features of the park. Clearly, the adaptations of some microorganisms to what humans perceive as inhospitable, **extreme environments** are truly remarkable. Indeed, microbes are thought to be present nearly everywhere on Earth. Bacteria such as *Bacillus infernus* are able to live over 2.4 km below Earth's surface, without oxygen and at temperatures above 60°C. Other microbes live under crushing pressure deep in the ocean or in lakes such as the Great Salt Lake in Utah (USA) that have high concentrations of sodium, chloride, and many other ions. Microorganisms that thrive in such harsh conditions are called **extremophiles.** Whether an extremophile or not, all microbes must respond to changes in their environment. However, if the conditions exceed their ability to respond, they will not grow and eventually may die. Thus for each environmental parameter, all microbes have a characteristic range at which growth occurs, defined by maximum and minimum values beyond which the microbe cannot survive. Within the range is an optimal value at which growth is best. **Figure 7.15** illustrates this as a graph of the growth rate constant (k) over a range of an environmental factor. Recall from the previous section that a higher value of k indicates faster growth. This general type of plot illustrates that microbes grow fastest at their characteristic optimum condition and that growth slows and ultimately stops as conditions deviate in both directions.

To study the ecological distribution of microbes, it is important to understand the strategies they use to survive. An

Table 7.2 Examples of Generation Times[1]

Microorganism	Incubation Temperature (°C)	Generation Time (Hours)
Bacteria		
Escherichia coli	40	0.35
Staphylococcus aureus	37	0.47
Mycobacterium tuberculosis	37	~12
Treponema pallidum	37	33
Archaea		
Pyrococcus abyssi	90	0.67
Sulfurisphaera tokodaii	75	6
Nitrososphaera viennensis	37	45
Eukaryotes[2]		
Euglena gracilis	25	22
Ceratium tripos	20	48–72
Saccharomyces cerevisiae	30	2
Monilinia fructicola	25	30

[1]Generation times differ depending on the growth medium and environmental conditions used.
[2]Generation times are for asexual growth.

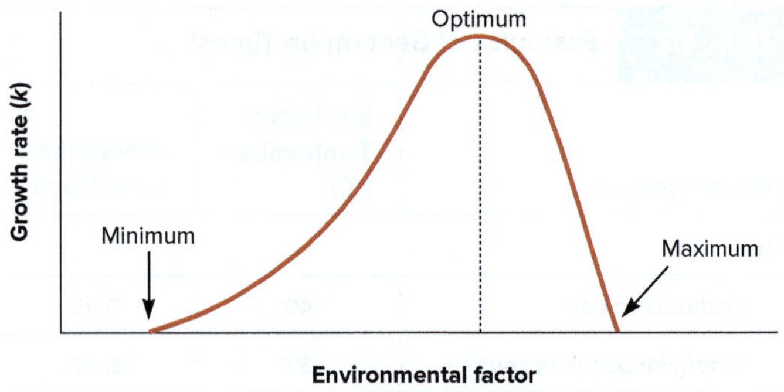

Figure 7.15 Growth over a Range of an Environmental Factor. The effect of an environmental factor on growth rate. The y-axis is the growth rate constant, *k*.

understanding of the environmental influences on microbes and their activity also aids in the control of microbial growth. In this section, we briefly review the effects of the most important environmental factors on microbial growth. Major emphasis is given to solutes and water activity, pH, temperature, oxygen level, pressure, and radiation. The adaptations that microbes have evolved to live in their environments also are discussed.

Solutes Affect Osmosis and Water Activity

Water is critical to the survival of all organisms, but water can also be destructive. Solutes in an aqueous solution alter the behavior of water. One way this occurs is by osmosis, which is observed when two solutions are separated by a semipermeable membrane that allows movement of water but not solutes. If the solute concentration of one solution is higher than the other, water moves to equalize the concentrations. In other words, water moves from solutions with lower solute concentrations to those with higher solute concentrations. (Solute concentration is also referred to as osmotic concentration or osmolarity.) Because a selectively permeable plasma membrane separates a cell's cytoplasm from its environment, microbes can be affected by changes in the solute concentration of their surroundings. If a microorganism is placed in a hypotonic solution (one with a solute concentration that is lower outside the cell than inside), water will enter the cell and cause it to burst unless something prevents the influx of water or inhibits plasma membrane expansion. Conversely, if the microbe is placed in a hypertonic solution (one with a higher osmolarity outside the cell than inside), water will flow out of the cell. In microbes that have cell walls, the membrane shrinks away from the cell wall. Dehydration of the cell in hypertonic environments may damage the plasma membrane and cause the cell to become metabolically inactive.

Because of the potential damaging effects of uncontrolled osmosis, microbes must be able to respond to changes in the solute concentrations of their environment. Recall that the cytoplasm is a crowded environment. As a consequence, most microbes live in hypotonic environments. They are protected in part by their cell wall, which prevents overexpansion of the plasma membrane. Microbes use several mechanisms to lower the solute concentration of their cytoplasm. For example, some bacteria have mechanosensitive (MS) channels in their plasma membrane. In a hypotonic environment, the membrane stretches due to an increase in hydrostatic pressure and cellular swelling. MS channels then open and allow solutes to leave. Thus MS channels act as escape valves to protect cells from bursting. As another example, many protists use contractile vacuoles to expel excess water. ◀ *Cell walls and osmotic protection (section 3.3)*

Some microbes are adapted to extreme hypertonic environments, and are called **osmophiles.** A specific group of osmophiles, called **halophiles,** require the presence of NaCl at a concentration above about 0.2 M (**figure 7.16**). Extreme halophiles have adapted so completely to hypertonic, saline conditions that they require NaCl concentrations between about 3 M and saturation (about 6.2 M). Archaeal halophiles can be isolated from the Dead Sea (a salt lake between Israel and Jordan), the Great Salt Lake in Utah, and other aquatic habitats with salt concentrations approaching saturation.

Halophiles have adopted one of two molecular strategies: salt-in or salt-out. Microbes that use the salt-out strategy keep salt ions outside the cytoplasm by importing and/or synthesizing compatible solutes. **Compatible solutes** are organic molecules that can be kept at high intracellular concentrations without interfering with metabolism and growth. They keep the cytoplasmic osmolarity somewhat above that of the environment so that the plasma membrane is always pressed firmly against the cell wall. Examples of compatible solutes are choline, betaines (neutral molecules having both negatively charged and positively charged functional groups), and amino acids such as proline and glutamic acid. Many microorganisms use compatible solutes; for example, fungi and photosynthetic protists employ sucrose and polyols (e.g., arabitol, glycerol, and mannitol) for this purpose.

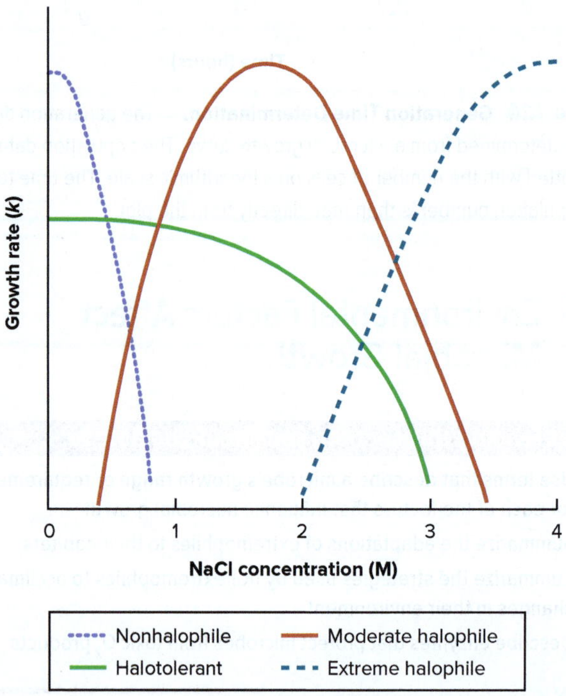

Figure 7.16 The Effects of Sodium Chloride on Microbial Growth. Four different patterns of microbial dependence on NaCl concentration are depicted. The curves are only illustrative and are not meant to provide precise shapes or salt concentrations required for growth.

MICRO INQUIRY *What is the difference between halophilic and halotolerant?*

Archaeal halophiles use the salt-in strategy. They accumulate cytoplasmic potassium and chloride ions to remain hypertonic to their environment; the internal potassium concentration may reach 4 to 7 M. Their enzymes, ribosomes, and transport proteins are unusual because they require high potassium levels for stability and activity. In addition, their plasma membranes and cell walls are stabilized by high concentrations of sodium ion. If the sodium concentration decreases too much, the wall and plasma membrane disintegrate. Although extreme halophiles have successfully adapted to environmental conditions that would destroy most organisms, they have become so specialized that they lack ecological flexibility. ▶| *Halobacteriota (section 19.4)*

Solutes can also decrease the availability of water to microbes. When solutes such as salts and sugars are present, the water is "tied up" by its interaction with the solutes. Microbiologists express the degree of water availability quantitatively by determining **water activity (a_w).** The water activity of a solution is 1/100 the relative humidity of the solution (when expressed as a percent). It is also equivalent to the ratio of the solution's vapor pressure (P_{soln}) to that of pure water (P_{water}). Distilled water has an a_w of 1, milk has an a_w of 0.97, a saturated salt solution has an a_w of 0.75, and the a_w of dried fruits is only about 0.5. Most microorganisms only grow well at water activities around 0.98 (the approximate a_w for seawater) or higher. This is why drying food or adding large quantities of salt or sugar effectively prevents food spoilage. ▶| *Environmental factors control food spoilage (section 40.2)*

To survive in a habitat with a low a_w value, microorganisms must maintain a high internal solute concentration to retain water. These **osmotolerant** microorganisms grow over wide ranges of water activity but optimally at higher levels. Osmotolerant organisms can be found in all domains of life. For example, the bacterium *Staphylococcus aureus* is osmotolerant, can be cultured in media containing up to about 3 M sodium chloride, and is well adapted for growth on human skin. The yeast *Zygosaccharomyces rouxii* grows in sugar solutions with a_w values as low as about 0.65. The protist *Dunaliella viridis* tolerates sodium chloride concentrations from 1.7 M to a saturated solution. In contrast, **xerotolerant** microbes can withstand either high solute concentrations or the effects of desiccation. These microbes thrive in desert regions, household dust, and preserved foods.

Comprehension Check

1. How do microorganisms adapt to hypotonic and hypertonic environments?
2. Review transport mechanisms in section 3.2. Which type is used by mechanosensitive channels?
3. What are halophiles and why do they require sodium and potassium ions?
4. Define *water activity.* Why is it difficult for microorganisms to grow at low a_w values?

pH

pH is a measure of the relative acidity of a solution and is defined as the negative logarithm of the hydrogen ion concentration (expressed in terms of molarity).

$$pH = -\log [H^+] = \log(1/[H^+])$$

The pH scale extends from pH 0.0 (1.0 M H^+) to pH 14.0 (1.0×10^{-14} M H^+), and each pH unit represents a 10-fold change in hydrogen ion concentration. **Figure 7.17** shows that microbial habitats vary widely in pH.

Each species has a pH growth range and pH growth optimum. **Acidophiles** have their growth optimum between pH 0 and 5.5; **neutrophiles,** between pH 5.5 and 8.0; and **alkaliphiles** (alkalophiles), between pH 8.0 and 11.5. In general, different microbial groups have characteristic pH preferences. Most bacteria and protists are neutrophiles. Most fungi prefer more acidic surroundings, about pH 4 to 6; photosynthetic protists also seem to favor slight acidity. Some archaea are acidophiles. For

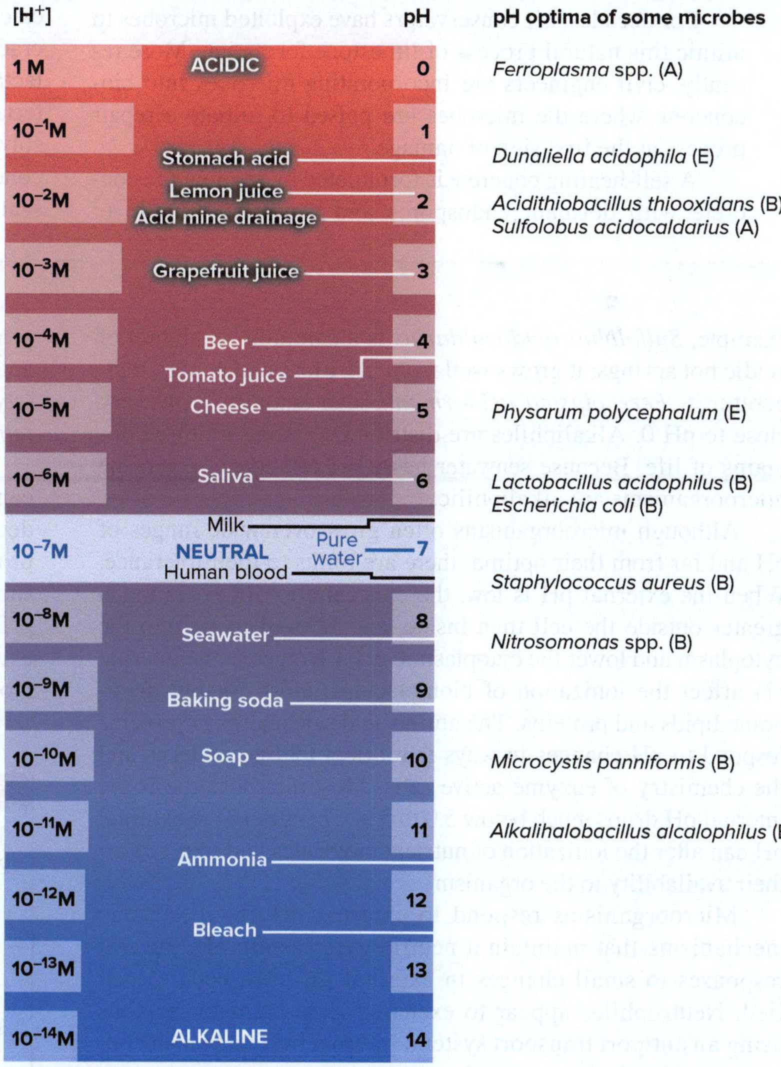

Figure 7.17 The pH Scale. Examples of substances with different pH values. Several microorganisms are placed at their growth optima.
A, archaeon; E, eukaryote; B, bacterium.

MICROBIAL DIVERSITY & ECOLOGY

7.1 Microbial Sculptors

Wanted: stone masons. Must be able to tolerate alkaline and high salt working conditions. Sleeping on the job encouraged.

Limestone (calcium carbonate, $CaCO_3$) formation requires only four things:

- Solution with a high calcium concentration
- Solution with a high concentration of inorganic carbon
- High pH (basic) conditions
- Nucleation sites

If provided with the first three items, some bacteria can fulfill the job of nucleating $CaCO_3$ precipitation, thereby depositing new limestone. Certain species of *Bacillus* have cell walls that effectively precipitate $CaCO_3$. An important career path is open to microbes with this skill set: conservation of decaying monuments (**box figure**) and concrete repair.

DrObjektiff/Shutterstock

For decades, art conservators have exploited microbes to mimic this natural process of limestone formation. More recently, civil engineers are incorporating microbes into bioconcrete where the microbes are poised to initiate a repair process at the first sign of damage.

A self-healing concrete is formulated by seeding the concrete with dormant endospores and a nutrient supply of calcium lactate. Upon a breach of the concrete structure, cracks form and water enters. The water dissolves the calcium lactate and initiates endospore germination. The revived bacteria produce CO_2 as a by-product of metabolism, which dissolves as carbonate in the high pH environment of the concrete. The system is primed to precipitate $CaCO_3$, thereby sealing cracks before structural damage occurs.

example, *Sulfolobus acidocaldarius* is a common inhabitant of acidic hot springs; it grows well from pH 1 to 3 and at high temperatures. *Ferroplasma acidarmanus* can actually grow very close to pH 0. Alkaliphiles are distributed among all three domains of life. Because seawater has a pH of about 8.1, marine microorganisms are alkaliphilic.

Although microorganisms often grow over wide ranges of pH and far from their optima, there are limits to their tolerance. When the external pH is low, the concentration of H^+ is much greater outside the cell than inside, and H^+ will move into the cytoplasm and lower the cytoplasmic pH. Changes to the internal pH affect the ionization of biomolecules, most notably membrane lipids and proteins. The amino acid side chains in proteins respond to pH changes in ways that affect folding patterns and the chemistry of enzyme active sites. Most microbes die if the internal pH drops much below 5.0 to 5.5. Changes in the external pH can alter the ionization of nutrient molecules and thus reduce their availability to the organism.

Microorganisms respond to external pH changes using mechanisms that maintain a neutral cytoplasmic pH. Several responses to small changes in external pH have been identified. Neutrophiles appear to exchange potassium for protons using an antiport transport system. Internal buffering also contributes to pH homeostasis. Chaperone proteins prevent protein denaturation and aid in refolding denatured proteins in acidic conditions. ◀ *Primary and secondary active transport (section 3.2)* ▶ *Molecular chaperones: proteins that help proteins fold (section 13.9)*

What about microbes that live at pH extremes? Acidophiles can transport cations (e.g., potassium ions) into the cell, thus decreasing the movement of H^+ into the cell; they can also use proton transporters that pump H^+ out of the cell. Alkaliphiles such as *Alkalihalobacillus alcalophilus* maintain their internal pH close to neutrality by exchanging internal sodium ions for external protons. Tolerance of high pH conditions can be exploited for biotechnology purposes, as illustrated in **Microbial Diversity & Ecology 7.1**.

Comprehension Check

1. Define *pH, acidophile, neutrophile,* and *alkaliphile*.
2. Using figure 7.17, classify each of the following organisms as an alkaliphile, a neutrophile, or an acidophile: *Staphylococcus aureus, Microcystis panniformis, Sulfolobus acidocaldarius,* and *Escherichia coli* (see figure 7.17). Which might be human pathogens? Explain your choices.
3. Describe the mechanisms microbes use to maintain an internal neutral pH. Explain how extreme pH values might harm microbes.

Temperature

Microorganisms are particularly susceptible to external temperatures because they cannot regulate their internal temperature. The range of temperatures over which microbes grow reflects their critical dependence on liquid water. The boundaries of microbial life currently extends from −15°C (*Planococcus halocryophilus,* a halophilic bacterium active in permafrost) to 113°C (*Pyrolobus fumarii,* an archaeon that lives in deep-sea hydrothermal vents). Chemical and physical mechanisms to maintain water in its liquid state are required for microbial life at these temperatures: High salt concentration depresses freezing, and high pressure inhibits boiling.

An important factor influencing growth is the temperature sensitivity of enzyme-catalyzed reactions. Each enzyme has a temperature at which it functions optimally. At temperatures below the optimum, it ceases to be catalytic. As the temperature rises from this low point, the rate of catalysis increases to that observed for the optimal temperature. The velocity of the reaction roughly doubles for every 10°C rise in temperature. When all enzymes in a microbe are considered together, as the rate of each reaction increases, metabolism as a whole becomes more active and the microorganism grows faster. However, beyond a certain point, further temperature increase actually slows growth, and sufficiently high temperatures are lethal. High temperatures denature enzymes, as well as transport and other proteins. Temperature also has a significant effect on membranes. At very low temperatures, membranes solidify. At high temperatures, the lipid bilayer simply melts and disintegrates. Thus when organisms are above or below their optimum temperature, both cell function and structure are affected. ◀ *Plasma membrane structure is dynamic (section 3.2)*

Because of these opposing temperature influences, microbial growth has a characteristic temperature dependence with distinct **cardinal temperatures**—minimum, optimum, and maximum growth temperatures (**figure 7.18*a***). Although the shape of temperature dependence curves varies, the temperature optimum is always closer to the maximum than to the minimum. The cardinal temperatures are not rigidly fixed. Instead, they depend to some extent on other environmental factors such as pH and available nutrients. For example, *Crithidia fasciculata,* a flagellated protist living in the gut of mosquitoes, grows in a simple medium at 22° to 27°C. However, if its medium is supplemented with extra metals, amino acids, vitamins, and lipids, it can grow at 33° to 34°C.

The cardinal temperatures vary greatly among microorganisms (figure 7.18*b*). The growth temperature range for a particular microorganism usually spans about 30°. Some species (e.g., *Neisseria gonorrhoeae*) have a small range; others grow over a wide range of temperatures. Some bacteria and archaea can grow at much higher temperatures than eukaryotes. It has been suggested that eukaryotes are not able to manufacture stable and functional organellar membranes at temperatures above 60°C.

Several terms describe microbes based on their temperature range for growth (figure 7.18*a*). Microbes that grow in cold

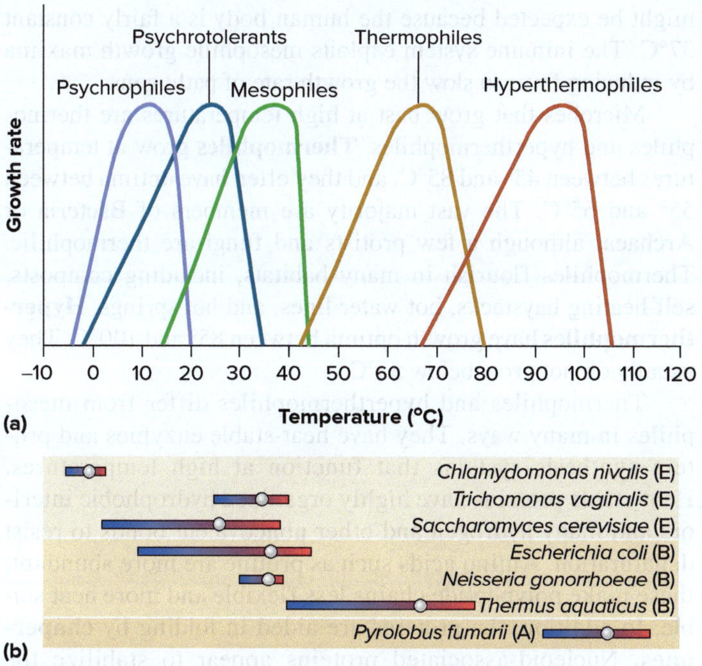

Figure 7.18 Temperature Ranges for Microbial Growth.
(a) Microorganisms are classified based on their temperature ranges for growth. (b) Temperature ranges for representative microbes. Minimum and maximum temperatures are the blue and red ends of the bar; optimum temperature is the white dot. E, eukaryote; B, bacterium; A, archaeon.

MICRO INQUIRY Which organisms in panel (b) may be human pathogens?

environments are either psychrotolerants or psychrophiles. **Psychrotolerants** (sometimes called **psychrotrophs**) grow at 0°C or higher and typically have maxima at about 35°C. **Psychrophiles** (sometimes called **cryophiles**) grow well at 0°C and have an optimum growth temperature of 15°C; the maximum is around 20°C. They are readily isolated from Arctic and Antarctic habitats. Oceans constitute an enormous habitat for psychrophiles because most ocean water is 5°C or colder. The psychrophilic protist *Chlamydomonas nivalis* can actually turn a snowfield or glacier pink with its bright red spores (a phenomenon called watermelon snow). Psychrophiles are widespread among bacterial taxa and have adapted to their environment in several ways. The membrane lipids of psychrophilic microorganisms have high levels of unsaturated fatty acids and remain semifluid when cold. Indeed, many psychrophiles begin to leak cellular constituents at temperatures higher than 20°C because of membrane disruption. Two strategies used by psychrophiles to decrease the freezing point of the cytosol are to accumulate compatible solutes and to synthesize antifreeze proteins and ice-structuring proteins. Psychrophilic and psychrotolerant bacteria and fungi are major causes of refrigerated food spoilage.

Mesophiles are microorganisms that grow in moderate temperatures. They have growth optima around 20° to 45°C and often have a temperature minimum of 15° to 20°C and a maximum of about 45°C. Almost all human pathogens are mesophiles, as might be expected because the human body is a fairly constant 37°C. The immune system exploits mesophilic growth maxima by inducing fever to slow the growth rate of pathogens.

Microbes that grow best at high temperatures are thermophiles and hyperthermophiles. **Thermophiles** grow at temperatures between 45° and 85°C, and they often have optima between 55° and 65°C. The vast majority are members of Bacteria or Archaea, although a few protists and fungi are thermophilic. Thermophiles flourish in many habitats, including composts, self-heating haystacks, hot water lines, and hot springs. **Hyperthermophiles** have growth optima between 85° and 100°C. They usually do not grow below 55°C.

Thermophiles and hyperthermophiles differ from mesophiles in many ways. They have heat-stable enzymes and protein synthesis systems that function at high temperatures. Heat-stable proteins have highly organized hydrophobic interiors and many hydrogen and other noncovalent bonds to resist denaturation. Amino acids such as proline are more abundant; these make polypeptide chains less flexible and more heat stable. In addition, the proteins are aided in folding by **chaperones.** Nucleoid-associated proteins appear to stabilize the DNA of thermophiles. Hyperthermophiles have an enzyme called reverse DNA gyrase that changes the topology of their DNA and enhances its stability. The membrane lipids of thermophiles and hyperthermophiles are also quite temperature stable. They tend to be more saturated, more branched, and of higher molecular weight. Recall that archaeal membrane lipids have ether linkages making them resistant to hydrolysis at high temperatures. Furthermore, the tetraethers observed in some archaeal membranes form a rigid, stable monolayer. ◀ *Archaeal cell envelopes are structurally diverse (section 4.2)* ▶ *Proteins (appendix I)* ▶ *Translation and protein maturation and localization (section 15.4)*

Comprehension Check

1. What are cardinal temperatures?
2. Why does the growth rate rise with increasing temperature and then fall again at higher temperatures?
3. Identify environments where you could find microbes considered psychrophilic, psychrotolerant, mesophilic, thermophilic, and hyperthermophilic.
4. Compare and contrast the metabolic and structural adaptations for extreme temperatures found in psychrophiles and thermophiles.

Oxygen Concentration

For the first billion years or so of microbial evolution, microbes lived without oxygen (O_2). The success of cyanobacteria at conserving light energy and producing O_2 resulted in increasing levels of O_2 in the Earth's atmosphere. However, using O_2 causes oxidative stress through the formation of toxic derivatives. Therefore the O_2-containing atmosphere represented an extreme environment. Enzymes that counteract O_2 damage evolved as microbes adapted to conditions with increasing levels of O_2. ◀ *Microbes have evolved and diversified for billion of years (section 1.2);* ▶ *Phototrophy (section 11.10)*

The influence of O_2 on the growth of an organism correlates with the processes it uses to conserve energy and its capacity to detoxify reactive oxygen species. Almost all energy-conserving metabolic processes involve the movement of electrons through a series of membrane-associated electron carriers called an electron transport chain (ETC). For organisms that use chemical energy sources, an externally supplied terminal electron acceptor is critical to the functioning of the ETC. The ability of an organism to use O_2 as its terminal electron acceptor determines its response to O_2. ▶ *Electron transport chains (section 11.4)*

Traditionally, five types of relationships to O_2 have been described. Almost all multicellular organisms and many microbes are completely dependent on atmospheric O_2 for growth; that is, they are **obligate aerobes.** O_2 is the terminal electron acceptor for the ETC in the metabolic process called aerobic respiration. **Microaerophiles** are damaged by the atmospheric level of O_2 (20%) and require O_2 levels in the range of 2 to 10% for growth. **Facultative anaerobes** can grow with or without O_2, but grow faster in its presence where they perform aerobic respiration. **Aerotolerant anaerobes** grow equally well whether O_2 is present or not; they can tolerate O_2, but they do not use it. The ability to grow in both oxic (oxygen-rich) and anoxic (oxygen-free) environments provides considerable flexibility and yields a competitive advantage for facultative and aerotolerant anaerobes. In contrast, for strict or **obligate anaerobes,** O_2 is toxic, and they are usually killed by prolonged exposure. Strict anaerobes do not generate energy through aerobic respiration and employ other metabolic strategies that don't require O_2. Although obligate anaerobes are killed by O_2, they may be recovered from habitats that appear to be oxic. In such cases, they associate with facultative anaerobes that use up the available O_2 and thus make the growth of obligate anaerobes possible. ▶ *Aerobic respiration starts with glucose oxidation (section 11.3); Anaerobic respiration uses the same steps as aerobic respiration (section 11.5); Fermentation does not involve an electron transport chain (section 11.6)*

The nature of bacterial O_2 responses can be readily determined by growing the microbe in a medium such as thioglycollate broth, which contains a reducing agent to lower O_2 levels (**figure 7.19**). All five responses are found among bacteria, archaea, and protists. Most fungi are aerobic, but a number of species—particularly among yeasts—are facultative anaerobes. Photosynthetic protists are usually obligate aerobes.

The different relationships with O_2 are due to several factors. Enzymes can be inactivated when sensitive groups such as sulfhydryls are oxidized. However, even molecules that have evolved to function aerobically can be damaged by O_2. This is because O_2 has two unpaired electrons in the outer shell, whereas both superoxide and hydroxyl radicals have only one, making them extremely unstable (*see figure AI.2*). Toxic O_2 derivatives called **reactive oxygen species (ROS)** are formed when cellular

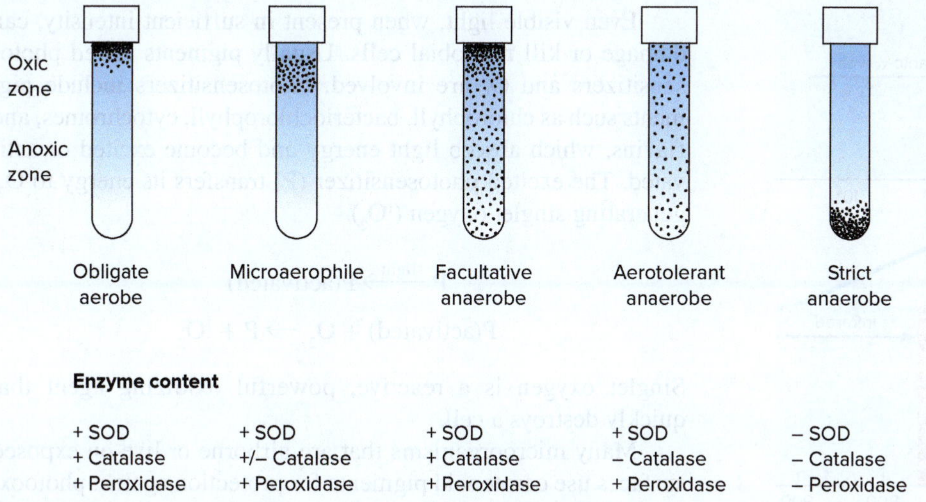

Figure 7.19 Oxygen and Bacterial Growth. Each dot represents an individual bacterial colony within the medium or on its surface. The surface, which is directly exposed to atmospheric oxygen, is oxic. The oxygen content of the medium decreases with depth until the medium becomes anoxic toward the bottom of the tube. The presence and absence of the enzymes superoxide dismutase (SOD), catalase, and peroxidase for each type are shown.

MICRO INQUIRY *Why do facultative anaerobes grow best at the surface of the tube while aerotolerant anaerobes demonstrate a uniform growth pattern throughout the tube?*

proteins such as flavoproteins transfer electrons to O_2. ROS can damage proteins, lipids, and nucleic acids, and include the superoxide radical, hydrogen peroxide, and the most dangerous hydroxyl radical.

$$O_2 + e^- \rightarrow O_2\bar{\bullet} \text{ (superoxide radical)}$$

$$O_2\bar{\bullet} + e^- + 2H^+ \rightarrow H_2O_2 \text{ (hydrogen peroxide)}$$

$$H_2O_2 + e^- + H^+ \rightarrow H_2O + OH\bullet \text{ (hydroxyl radical)}$$

A microorganism must be able to protect itself against ROS or it will be killed. Indeed, neutrophils and macrophages, two important immune system cells, produce ROS to destroy invading pathogens. ▶| *Redox reactions: reactions of central importance in metabolism (section 10.3); Phagocytosis destroys invaders (section 31.6)*

Many microorganisms possess enzymes that neutralize toxic O_2 products (figure 7.19). **Superoxide dismutase (SOD)** catalyzes the destruction of superoxide radicals. **Catalase** and **peroxidase** use different mechanisms, but both detoxify hydrogen peroxide.

$$2O_2\bar{\bullet} + 2H^+ \xrightarrow[\text{dismutase}]{\text{superoxide}} O_2 + H_2O_2$$

$$2H_2O_2 \xrightarrow{\text{catalase}} 2H_2O + O_2$$

$$H_2O_2 + \text{NADH} + H^+ \xrightarrow{\text{peroxidase}} 2H_2O + \text{NAD}^+$$

Strict anaerobes lack these enzymes or have them in low concentrations and therefore cannot tolerate O_2.

Pressure

Organisms that spend their lives on land or the surface of water are always subjected to a pressure of 1 atmosphere (atm; 1 atm is ~0.1 megapascal). Other organisms, including many bacteria and archaea, live in the deep sea (ocean depths of 1,000 m or more), where the hydrostatic pressure can reach 600 to 1,100 atm and the temperature is about 2° to 3°C. Many of these microbes are **barotolerant**—that is, increased pressure adversely affects them but not as much as it does nontolerant microbes. Some are truly **piezophilic (barophilic).** A piezophile is an organism that requires high pressure for growth. For instance, a piezophile recovered from the Mariana trench near the Philippines (depth about 10,500 m) grows only at pressures between 400 and 500 atm when incubated at 2°C.

An important adaptation observed in piezophiles is that their membrane lipids change in response to increasing pressure. For instance, membranes composed of lipids that are shorter and more unsaturated are better able to resist higher pressure. Piezophiles are thought to play important roles in nutrient cycling in the deep sea and include both bacteria and archaea. ▶| *Microbial communities in benthic marine sediments are enormous and mysterious (section 29.2)*

Radiation

Our world is bombarded with electromagnetic radiation of various types (**figure 7.20**). Radiation behaves as if it were composed of waves like those traveling on the surface of water. The distance between two wave crests or troughs is the wavelength. As the wavelength of electromagnetic radiation decreases, the energy of the radiation increases; gamma rays and X rays are much more energetic than visible light or infrared waves. Electromagnetic radiation also acts like a stream of energy packets called photons, each photon having a quantum of energy whose value depends on the wavelength of the radiation.

Sunlight is the major source of radiation on Earth. It includes visible light, ultraviolet (UV) radiation, infrared rays, and radio waves. Visible light is a conspicuous and important aspect of our environment because most life on Earth depends on the ability of photosynthetic organisms to trap the energy it holds. The visible spectrum consists of the ROYGBV wavelengths that most of us memorized for early science classes. Despite the

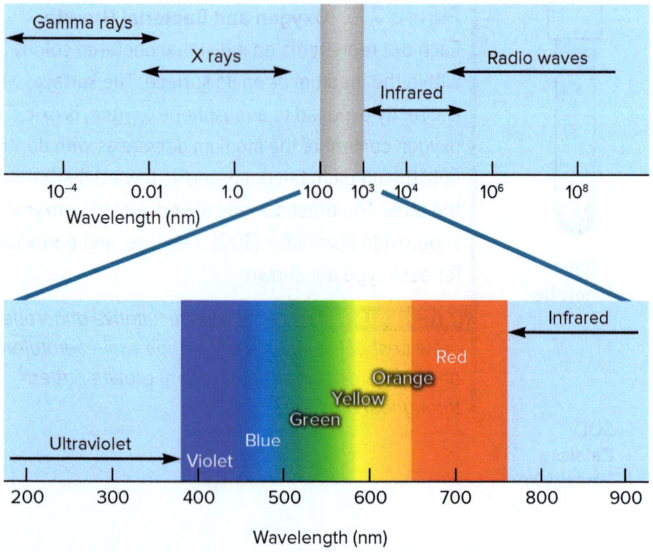

Figure 7.20 The Electromagnetic Spectrum. A portion of the spectrum showing the visible light range and the nearby ultraviolet and infrared wavelengths is expanded at the bottom of the figure.

importance of visible light, it is not the predominant component of sunlight. Infrared rays hold that honor, making up almost 60% of the sun's radiation. Infrared radiation is the major source of Earth's heat. Only about 3% of the light reaching Earth's surface is UV radiation. ▶| *Phototrophy (section 11.10)*

Many forms of electromagnetic radiation are harmful to microorganisms. One of the most damaging is **ionizing radiation,** radiation of very short wavelength and high energy (figure 7.20), which causes atoms to lose electrons (ionize). Two major forms of ionizing radiation are X rays, which are artificially produced, and gamma rays, which are emitted during natural radioisotope decay. Ionizing radiation causes a variety of changes in cells. It breaks hydrogen bonds, destroys ring structures, and polymerizes some molecules, but the most severe effect is protein oxidation. Low levels of ionizing radiation produce mutations that indirectly result in death, whereas higher levels are lethal. Indeed, ionizing radiation can be used to sterilize inanimate objects like laboratory supplies, medical devices, and foods. However, bacterial endospores and microbes such as *Deinococcus radiodurans* are extremely resistant to large doses of ionizing radiation. ▶| *Radiation (section 8.2)*

Ultraviolet (UV) radiation can kill microorganisms due to its short wavelength (approximately from 10 to 400 nm) and high energy. The most lethal UV radiation has a wavelength of 260 nm, the wavelength most effectively absorbed by DNA. The damage caused by UV light can be repaired by DNA repair mechanisms, as we discuss in chapter 16. However, excessive exposure to UV light outstrips the organism's ability to repair the damage and death results. ▶| *Mutations: heritable changes in a genome (section 16.1)*

Even visible light, when present in sufficient intensity, can damage or kill microbial cells. Usually pigments called photosensitizers and O_2 are involved. Photosensitizers include pigments such as chlorophyll, bacteriochlorophyll, cytochromes, and flavins, which absorb light energy and become excited or activated. The excited photosensitizer (P) transfers its energy to O_2, generating singlet oxygen ($^1O_2$).

$$\text{P} \xrightarrow{\text{light}} \text{P(activated)}$$

$$\text{P(activated)} + O_2 \longrightarrow \text{P} + {}^1O_2$$

Singlet oxygen is a reactive, powerful oxidizing agent that quickly destroys a cell.

Many microorganisms that are airborne or live on exposed surfaces use carotenoid pigments for protection against photooxidation. Carotenoids effectively quench singlet oxygen; indeed, they absorb energy from singlet oxygen and convert it back into the unexcited ground state. Both phototrophic and nonphototrophic microorganisms employ pigments in this way.

Comprehension Check

1. Describe the five types of O_2 relationships seen in microorganisms.
2. What are the toxic effects of O_2? How do aerobes and other oxygen-tolerant microbes protect themselves from these effects?
3. Where would you expect to find barotolerant and piezophilic microbes? Explain your answer.
4. List the types of electromagnetic radiation in the order of decreasing energy or increasing wavelength.
5. How do ionizing radiation, ultraviolet radiation, and visible light harm microorganisms? How do microorganisms protect themselves against damage from UV and visible light?

7.6 Microbial Growth in Natural Environments

After reading this section, you should be able to:

a. Discuss the mechanisms used by microbes to survive starvation
b. Predict how the presence of viable but nonculturable cells in food or water systems might impact public health
c. Distinguish sessile and planktonic microbial life styles
d. Describe the formation of biofilms and summarize their importance in natural environments, industrial settings, and health
e. Define *quorum sensing* and provide examples of cellular processes regulated by quorum sensing

Natural microbial environments—those not created by humans—are complex and constantly changing. They expose microbes to many overlapping gradients of nutrients and environmental factors. These include the environmental parameters we describe in section 7.5, as well as inhibitory substances that limit microbial growth. Furthermore, microbes do not exist alone. Microbial habitats contain both micro- and macroorganisms. Microbes often associate with each other and they frequently form relationships with plants and animals. In this section, we explore three aspects of life in nature: stress-induced growth arrest, biofilms, and the mechanisms used by microbes to communicate among and between species.

Most Microbes Live in Growth-Arrested States

Relatively few microbes are lucky enough to live in nutrient-rich **eutrophic** environments. Rather, most inhabit **oligotrophic** environments—ones in which nutrient levels are low—and some live in "feast or famine" environments. For microbes in oligotrophic or nutrient-labile environments, the ability to survive starvation conditions is paramount. Fortunately, microbes have evolved numerous responses to starvation and environmental stress.

A few bacteria respond to nutrient limitation with obvious morphological changes such as endospore formation. Recall that endospores are not only resistant to a variety of stressors (e.g., high temperatures and radiation) but are also metabolically dormant. Other bacteria and archaea become dormant without such a drastic reorganization of the cell. Many eukaryotic microbes and some bacteria form cysts (*see figures 21.3 and 23.2*). *Bacterial endospores are a survival strategy (section 3.8)*

Most commonly, nutrient and environmental conditions leave microbes in a state termed **growth arrest.** They are not actively dividing, nor are they dead (**figure 7.21**). They are stressed by lack of nutrients or by deviation from optimal conditions. Inspection of figure 7.10 illustrates this phenomenon in the growth curve. Entry into stationary phase is triggered by nutrient limitation or waste accumulation, and cells stop growing as they adjust to new conditions. As time goes on, cells may die, but the long-term stationary phase illustrates that many cells remain viable even though the population is not increasing. Natural populations, too, are comprised of a heterogeneous group of microbes, most of which are in some form of growth arrest.

Cells must maintain their integrity during growth arrest to be ready to respond when conditions improve. Energy-intensive processes such as cell division are halted to conserve ATP and redirect it to cellular maintenance. Resources such as storage inclusions are tapped first to provide nutrients. Prolonged starvation induces the use of cellular components as nutrients. For example, many growth-arrested cells decrease in size. This results from the loss of membrane lipids and ribosomes as their components are recycled within the cell. A further benefit of cell size decrease is an increase in the surface area-to-volume ratio, which in turn promotes more efficient nutrient transport.

In *E. coli*, two molecules that regulate the transition from rapid growth to growth arrest are RpoS and ppGpp. RpoS is a protein subunit of RNA polymerase, the enzyme responsible for binding DNA and initiating RNA synthesis (transcription). RpoS directs RNA polymerase to transcribe genes encoding proteins that help the bacterium survive starvation, as well as proteins involved in a general stress response. ppGpp is related to the guanine nucleotides in DNA and RNA. It is a signaling molecule that functions in a regulatory network called the stringent response, which we describe in more detail in chapter 14. ppGpp is often referred to as an **alarmone** because it is synthesized in response to many stressors and regulates the activity of numerous genes and proteins so that the cell can survive. *Transcription in bacteria (section 13.5); Stringent response in* E. coli *(section 14.5)*

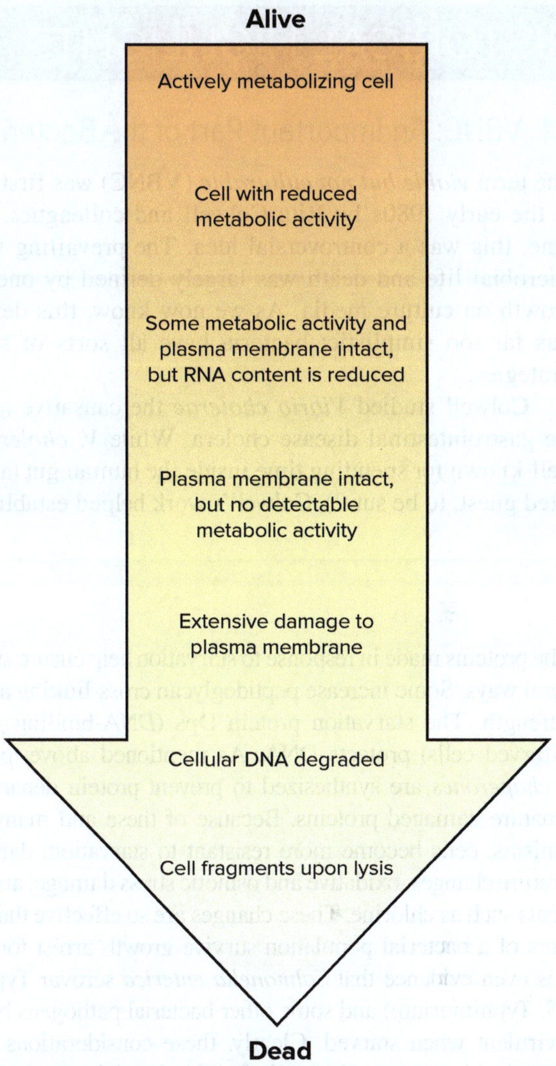

Figure 7.21 Death Trajectory.
Microbiologists have traditionally defined microbes as dead when they could not be cultured. However, cells may be inactive or damaged and therefore unable to reproduce temporarily. With time and appropriate conditions, the cells may recover and begin to reproduce. Various criteria have been used to establish these states; however, microbiologists are still unable to define the point at which a microbe is truly dead and cannot be resuscitated.

HISTORICAL HIGHLIGHTS

7.1 VBNC: An Important Part of the Bacterial Lifespan

The term *viable but not culturable* (VBNC) was first coined in the early 1980s by Rita Colwell and colleagues. At the time, this was a controversial idea. The prevailing view of microbial life and death was largely defined by one test—growth on culture media. As we now know, this definition was far too simplistic; bacteria have all sorts of survival strategies.

Colwell studied *Vibrio cholerae* the causative agent of the gastrointestinal disease cholera. While *V. cholerae* was well-known for spending time inside the human gut (an uninvited guest, to be sure!), Colwell's work helped establish how this microbe survives in aquatic environments. By combining agar plating methods with microscopy techniques, Colwell's team showed that *V. cholerae* cells could enter a state where they were alive and capable of metabolism yet could not be grown using traditional culture techniques. Elucidating this part of *V. cholerae*'s lifestyle helped reshape how public health officials responded to cholera outbreaks. These and similar studies sparked development of sensitive molecular techniques to detect viable microbes in wide-ranging environments without the cumbersome task of formulating an appropriate growth medium in the laboratory.

The proteins made in response to starvation help ensure survival in several ways. Some increase peptidoglycan cross-linking and cell wall strength. The starvation protein Dps (*D*NA-binding *p*rotein from *s*tarved cells) protects DNA. As mentioned above, proteins called *chaperones* are synthesized to prevent protein denaturation and renature damaged proteins. Because of these and many other mechanisms, cells become more resistant to starvation, damaging temperature changes, oxidative and osmotic stress damage, and toxic chemicals such as chlorine. These changes are so effective that some members of a bacterial population survive growth arrest for years. There is even evidence that *Salmonella enterica* serovar Typhimurium (*S.* Typhimurium) and some other bacterial pathogens become more virulent when starved. Clearly, these considerations are of great practical importance in medical and industrial microbiology.

After an extended time in growth arrest, cells may enter a state termed **viable but not culturable (VBNC).** This is defined as the transient inability to grow under conditions that previously permitted routine growth. Once the appropriate conditions are available (e.g., a change in temperature or passage through an animal), VBNC microbes resume growth. VBNC microorganisms could pose a public health threat as many assays that test food and drinking water are culture-based.

Some cells, called **persisters,** represent another type of growth arrest. Persisters are identified by their ability to survive exposure to an antibiotic (often multiple antibiotics) even though they do not harbor antibiotic resistance genes. These nongrowing cells are characterized by low ATP levels. Because antibiotics target growing bacteria, nongrowing cells are unaffected. Persisters are of considerable interest for several reasons, including their potential to cause recurrent infections after antibiotic treatment. ▶| *Antimicrobial drug resistance is a public health threat (section 9.8)*

Why does growth arrest occur? Evidence shows that in some cases, small subsets of cells in a population of growing cells spontaneously become growth arrested, even when nutrients are readily available. These cells can be thought of as the population's insurance against future possible starvation. There is also increasing evidence that growth-arrested cells arise in response to various triggers, with starvation probably being the most important.

Biofilms Are Common in Nature

Many introductory microbiology students have scraped their teeth and examined the microbes in this sampling, thus replicating the way van Leeuwenhoek (*see figure 1.13*) reportedly entertained dinner guests. What was sampled by all was dental plaque, an important example of a biofilm (**figure 7.22**). Ecologists observed as early as the 1940s that more microbes in aquatic environments were found on surfaces (**sessile**) than were free-floating (**planktonic**). These complex, slime-encased communities of microbes are called **biofilms.** They are ubiquitous in nature, and it is estimated that 40–80% of the bacterial and archaeal cells on the

Figure 7.22 Dental Plaque Biofilm. Biofilms form on almost any surface exposed to microorganisms as shown in this colorized scanning electron micrograph.
Steve Gschmeissner/SPL/Getty Images

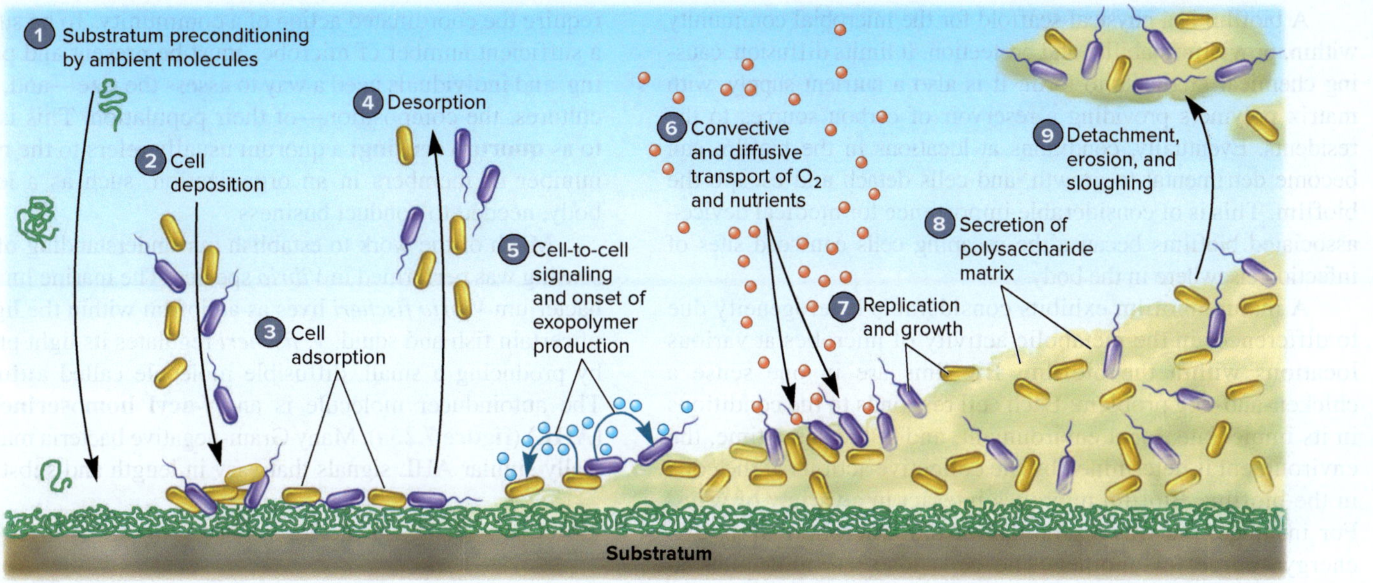

Figure 7.23 Biofilm Formation.

MICRO INQUIRY *What biomolecules make up the extracellular polymeric matrix, and what functions does the matrix serve?*

planet are in biofilms. Biofilms predominate in all environments except the open ocean surface. Of major concern is biofilm formation on medical devices such as hip and knee implants and indwelling catheters. These biofilms often cause serious illness and failure of the device. Biofilms can also form within wounds and delay healing. Biofilm formation is apparently an ancient ability among microbes, as evidence for biofilms is found in the fossil record from about 3.4 billion years ago.

Biofilms can form on virtually any living or nonliving surface. Cells recognize an interface for colonization, often through chemotaxis toward attractants on the surface; for example, soil microbes moving toward plant roots (*see figure 30.9*). Surface contact is typically through flagella or pili (**figure 7.23**). For example, when a rotating flagellum strikes a surface, the force feeds back to the stator proteins in the flagellar motor. This initiates a process to transform mechanical signals to chemical signals that prepare the cell to enter a sessile phase. Changes include the synthesis of adhesin proteins and the biofilm matrix molecules that enable the cell to stick to the substratum. The matrix is made up of various polymers, depending on the microbes. The polymers are collectively called **extracellular polymeric substances (EPS),** and they include polysaccharides, proteins, glycoproteins, glycolipids, and DNA. The composition of biofilm EPS varies with the species present, nutrient availability, and the physical properties of the substratum and the surrounding environment. EPS allows the microbes to stick more stably to the surface. As the biofilm thickens and matures, some cells remain at the substratum, whereas others move toward the surface to expand the biofilm in three dimensions. Most cells in biofilms are embedded in the EPS matrix. The matrix is continuously formed and remodeled by the resident microbes, making it highly heterogeneous (**figure 7.24**).

Bacterial flagella (section 3.7)

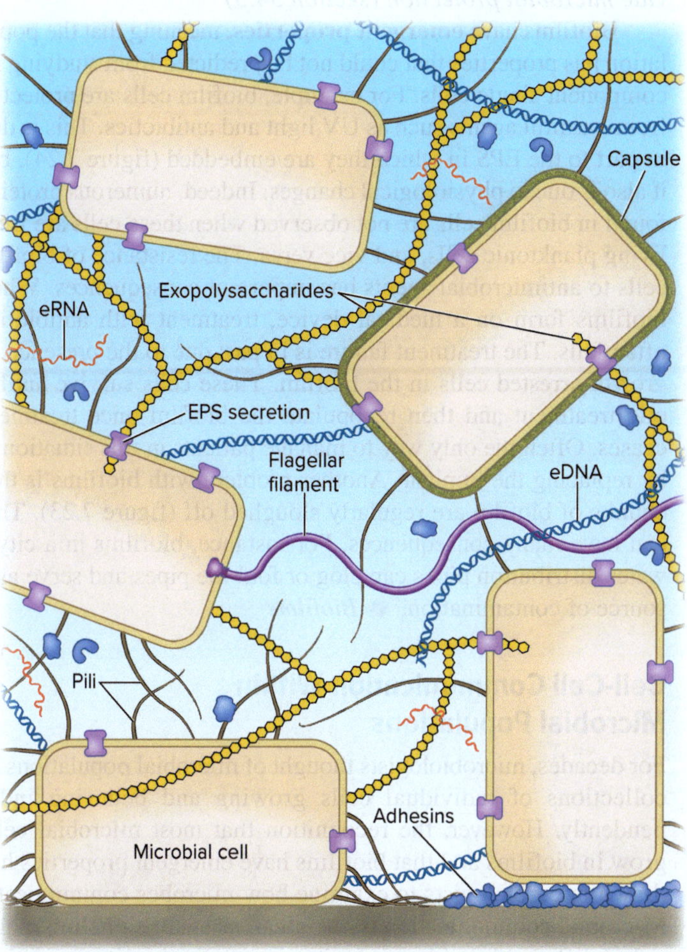

Figure 7.24 Biofilm Heterogeneity. eDNA, extracellular DNA; eRNA, extracellular RNA.

A biofilm is a physical scaffold for the microbial community within, providing stability and protection. It limits diffusion, causing chemical gradients to form. It is also a nutrient supply, with matrix polymers providing a reservoir of carbon sources to the residents. Eventually, conditions at locations in the biofilm can become detrimental to growth, and cells detach and escape the biofilm. This is of considerable importance for medical device–associated biofilms because the escaping cells can seed sites of infection elsewhere in the body.

A mature biofilm exhibits considerable heterogeneity due to differences in the metabolic activity of microbes at various locations within the biofilm. Biofilms are in one sense a chicken-and-egg problem: Each cell responds to the conditions in its immediate, local environment, and at the same time, the environment is determined by the collective actions of the cells in the biofilm. Biofilm microbes interact in a variety of ways. For instance, the waste products of one microbe may be the energy source for another. The cells also use molecules to communicate with each other, as we describe next. Finally, DNA present in the EPS can be taken up by members of the biofilm community. Thus genes can be transferred from one cell (or species) to another. ▶| *Horizontal gene transfer: creating variability the asexual way (section 16.3); Biofilms provide microbial protection (section 34.3)*

Biofilms have **emergent properties,** meaning that the population has properties that could not be predicted from studying its component single cells. For example, biofilm cells are protected from harmful agents such as UV light and antibiotics. This is due in part to the EPS in which they are embedded (figure 7.24), but it also is due to physiological changes. Indeed, numerous proteins found in biofilm cells are not observed when these cells are free-living planktonic cells, and vice versa. The resistance of biofilm cells to antimicrobial agents has important consequences. When biofilms form on a medical device, treatment with antibiotics often fails. The treatment failure is in part due to the presence of growth-arrested cells in the biofilm. These cells survive antibiotic treatment and then repopulate the biofilm once treatment ceases. Often the only way to manage patients in this situation is by replacing the implant. Another problem with biofilms is that chunks of biofilm are regularly sloughed off (figure 7.23). This can have many consequences. For instance, biofilms in a city's water distribution pipes can clog or foul the pipes and serve as a source of contamination. *Biofilms*

Cell-Cell Communication Within Microbial Populations

For decades, microbiologists thought of microbial populations as collections of individual cells growing and behaving independently. However, the recognition that most microbial cells grow in biofilms and that biofilms have emergent properties has drawn microbiologists to examine how microbes communicate. Microbial communication is chemical, as small signaling molecules diffuse through an environment. Emergent properties, such as the ability to initiate pathogenesis or bioluminescence, require the coordinated action of a community. To be successful, a sufficient number of microbes must be present and participating, and individuals need a way to assess the size—and, in mixed cultures, the composition—of their population. This is referred to as **quorum sensing;** a quorum usually refers to the minimum number of members in an organization, such as a legislative body, needed to conduct business.

Much of the work to establish our understanding of quorum sensing was performed in *Vibrio* species. The marine luminescent bacterium *Vibrio fischeri* lives as a biofilm within the light organ of certain fish and squid. *V. fischeri* regulates its light production by producing a small, diffusible molecule called **autoinducer.** The autoinducer molecule is an **N-acyl homoserine lactone (AHL)** (**figure 7.25a**). Many Gram-negative bacteria make chemically similar AHL signals that vary in length and substitution at

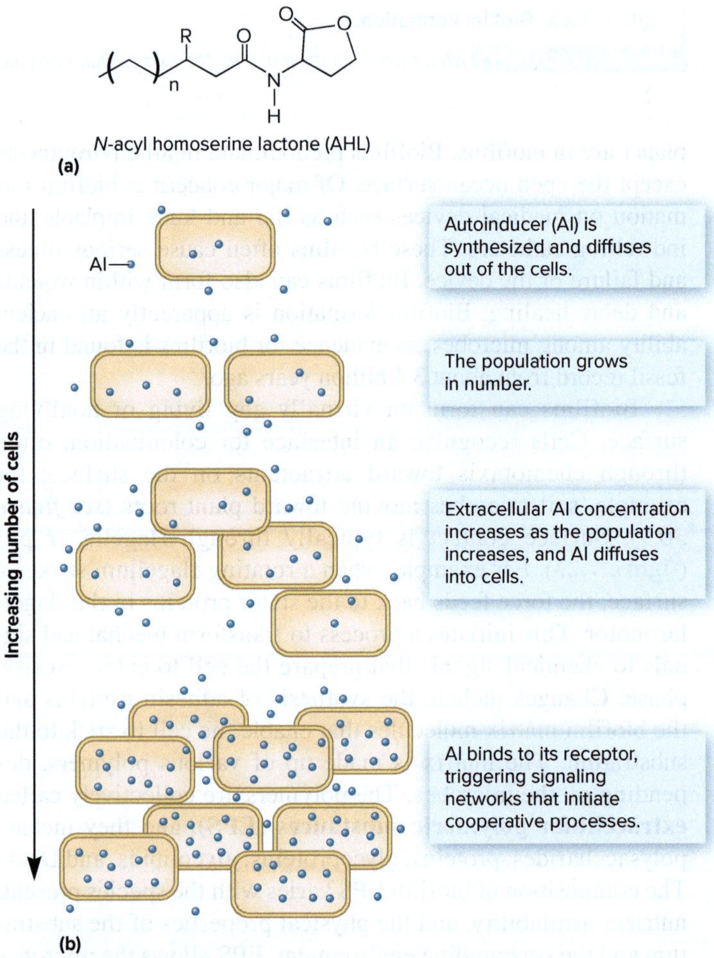

Figure 7.25 Quorum Sensing. (a) Chemical structure of N-acyl homoserine lactone. R and n indicate the regions of the molecule that vary in different microbes. (b) Mechanism of quorum sensing. Autoinducer (AI) freely diffuses across the plasma membrane into the environment. When the external AI concentration reaches a high level (a "quorum"), it diffuses back into cells to prompt individual cells to simultaneously initiate specific activities.

Figure 7.26 Squid–*Vibrio* Symbiosis. This squid is colonized by *Vibrio fischeri.* When autoinducer accumulates to a threshold concentration, light production is triggered, as evidenced by the glowing eyes.
Cdascher/E+/Getty Images

the third position of the acyl side chain. In many of these species, AHL is freely diffusible across the plasma membrane. Thus at a low cell density, the diffusion gradient favors movement of AHL from the cytoplasm to the outside of the cell (figure 7.25*b*). However, when the cell population increases, the diffusion gradient is reversed to favor movement of AHL into the cell. Because the influx of AHL is cell-density dependent, it enables individual cells to assess population density. When AHL reaches a threshold level inside the cell, it induces the expression of genes that regulate a number of functions, depending on the microbe. These functions are most effective only if a large number of microbes are present. For instance, the light produced by one *V. fischeri* cell would be invisible and energetically costly, so bioluminescence is inhibited. However, cell densities within the light organ of marine fish and squid reach 10^{10} cells per milliliter. At such densities, light is clearly visible (**figure 7.26**). Thus quorum sensing works to promote bioluminescence at high densities. This provides fish with a flashlight effect. In squid, bioluminescence provides countershading, which protects the squid from predation as it hunts at night. In return, the microbes have a safe and nutrient-enriched habitat. ▶| *Quorum sensing by* Vibrio *spp. (section 14.6)*

Many of the processes regulated by quorum sensing involve host-microbe interactions, including pathogenicity. For instance, the Gram-negative, opportunistic bacterial pathogens *Burkholderia cepacia* and *Pseudomonas aeruginosa* use AHLs to regulate biofilm formation and the expression of virulence factors. These bacteria cause debilitating infections in people who are immunocompromised and are important pathogens in patients with cystic fibrosis.

Gram-positive bacteria communicate using short (5–17 amino acid) peptides called **autoinducing peptides (AIPs).** Examples include *Enterococcus faecalis,* whose AIP signal is used to determine the best time to conjugate (transfer genes). AIP communication by *Staphylococcus aureus* and *Bacillus subtilis* is also used to trigger the uptake of DNA from the environment.

We now know that cell-cell communication is important for regulating cellular processes in microbes. For instance, while only Gram-negative bacteria are known to make AHLs, many bacteria and archaea make autoinducer-2 (AI-2), a universal quorum-sensing signal. The soil microbe *Streptomyces griseus* produces a γ-butyrolactone known as A-factor. This small molecule regulates both morphological differentiation and the production of the antibiotic streptomycin. Eukaryotic microbes also rely on cell-cell communication to coordinate key activities within a population. For example, the pathogenic fungus *Candida albicans* secretes farnesoic acid to govern morphology and virulence. These examples of cell-cell communication demonstrate multicellular behavior in that many individual cells communicate and coordinate their activities to act as a unit.

Comprehension Check

1. As a cell adapts to starvation conditions during growth arrest, which cell division molecules are no longer needed?
2. What is a biofilm? List two ways life in a biofilm is advantageous for microbes.
3. What medical challenges do biofilms present?
4. What is quorum sensing? Describe how it occurs, and briefly discuss its importance to microorganisms.

7.7 Laboratory Culture of Microbes Requires Conditions That Mimic Their Normal Habitats

After reading this section, you should be able to:

a. Describe the importance of culturing microbes to the study of microorganisms
b. Distinguish defined (synthetic) media from complex media and the uses of liquid versus solid growth media
c. Compare and contrast supportive (general purpose), enriched, selective, and differential media, listing examples of each and describing how each is used
d. Discuss the use of enrichment cultures in isolating microbes
e. Differentiate the streak-plate, spread-plate, and pour-plate methods for isolating pure cultures

For decades, studying microbes in their natural habitats has proven to be a significant hurdle for microbiologists. For that reason, microbiology research has depended largely on the ability to grow and maintain microorganisms in the laboratory. Unfortunately, it is estimated that less than 5% of all microbes are currently culturable. Recently nonculture-based alternatives have become available as genome sequencing techniques become more advanced. Yet even with these techniques, an important goal is to eventually isolate microbes from their normal habitat. Thus culturing microbes continues to be an important tool used by microbiologists. ▶| *Metagenomics provides access to uncultured microbes (section 18.3), Microbial ecology also relies on cultures (section 26.3),*

Culture Media

Culturing microbes is possible only if suitable media are available. A **culture medium** is a solid or liquid mixture of nutrients and other compounds used to grow, transport, and store microorganisms. To be effective, the medium must contain all the growth factors the microorganism requires. Specialized media are essential in the isolation and identification of microorganisms, the testing of antibiotic sensitivities, water and food analysis, industrial microbiology, and other activities. Although all microorganisms need sources of energy and macro- and micronutrients, the precise composition of a satisfactory medium depends on the species being cultivated due to the great variety of nutritional requirements among microbes. Knowledge of a microorganism's normal habitat is useful in selecting an appropriate medium because its nutrient requirements reflect its natural surroundings. Frequently a medium is used to select and grow specific microorganisms or to help identify a particular species. Media can also be designed to facilitate the growth of only one type of microbe present in a sample from nature. |◀ *Bacteria use many mechanisms to bring nutrients into the cell (section 3.2)*

Culture media are classified on the basis of their chemical composition, their physical nature, and their function (**table 7.3**). The types of media defined by these parameters are described here.

Chemical and Physical Types of Culture Media

A **defined** or **synthetic medium** is one in which each ingredient can be described with a chemical formula. It can be in liquid form (broth) or solidified by an agent such as agar. Many organisms that use reduced organic molecules as carbon and energy sources can be grown in defined media (for example, a medium with glucose as a carbon source and an ammonium salt as a nitrogen source). While some defined media have only a few ingredients, others are constructed from dozens of components. Defined media are used widely in research because it is often desirable to know exactly what the microorganism is metabolizing.

Complex media contain at least one ingredient of nonspecific chemical composition such as peptones, meat extract, or yeast extract. Peptones are prepared by partial proteolytic digestion of protein sources. They serve as sources of carbon, energy, and nitrogen. Beef extract and yeast extract are aqueous extracts that contain amino acids, peptides, nucleotides, organic acids, vitamins, and minerals. Yeast extract is an excellent source of B vitamins as well as nitrogen and carbon compounds.

Complex media are useful because a single complex medium may be able to meet all the nutritional requirements of many different microorganisms. In addition, they often are needed because the nutritional requirements of a particular microorganism are unknown, and thus a defined medium cannot be formulated. Complex media are also used to culture *fastidious* microbes, those with complicated nutritional or cultural requirements.

Both liquid and solidified media are routinely used in laboratories. However, solidified media are particularly important because they can be used to isolate different microbes to establish pure cultures. As we discuss in chapter 1, this is a critical step when using Koch's postulates to demonstrate the relationship between a microbe and a disease (*see figure 1.17*). **Agar** is the most commonly used solidifying agent. It is a sulfated polymer composed mainly of D-galactose, 3,6-anhydro-L-galactose, and D-glucuronic acid, usually extracted from red algae. Agar is well suited as a solidifying agent for several reasons. First, it melts at about 90°C but, once melted, does not harden until it cools to about 45°C. Thus after being melted in boiling water, it can be cooled to a temperature that is tolerated by human hands as well as microbes. Furthermore, microbes growing on agar medium can be incubated at a wide range of temperatures. Finally, agar is an excellent hardening agent because most microorganisms cannot degrade it.

Table 7.3 Types of Media

Basis for Classification	Types
Chemical composition	Defined (synthetic), complex
Physical nature	Liquid, semisolid, solid
Function	Supportive (general purpose), enriched, selective, differential

Functional Types of Media

Media such as tryptic soy broth and tryptic soy agar are called general purpose or **supportive media** because they sustain the growth of many types of microorganisms. Blood and other nutrients may be added to supportive media to encourage the growth of fastidious microbes. These fortified media (e.g., blood agar) are called **enriched media** (**figure 7.27*a,b***).

Selective media allow the growth of particular microorganisms, while inhibiting the growth of others. For instance, many Gram-negative bacteria grow on media containing bile salts or dyes such as basic fuchsin and crystal violet; however, the growth of Gram-positive bacteria is inhibited.

Differential media distinguish among different groups of microbes and even permit tentative identification of microorganisms based on their biological characteristics. Some types of media have multiple functions. Blood agar is both a differential medium and an enriched one. It differentiates hemolytic and nonhemolytic bacteria. Some hemolytic bacteria (e.g., many streptococci and staphylococci) produce clear zones around their colonies because of red blood cell destruction (figure 7.27*b*). Blood agar is an enriched growth medium in that blood (usually from sheep) provides protein, carbohydrate, lipid, iron, and a number of growth factors and

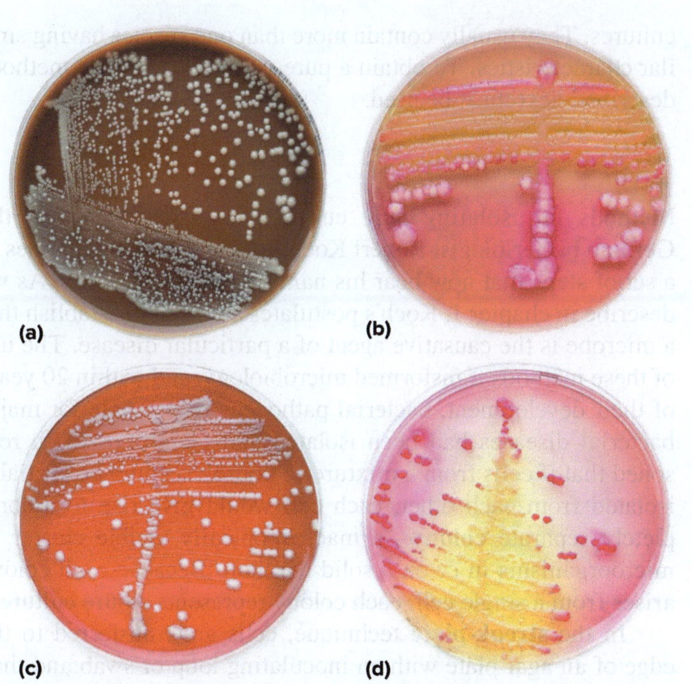

Figure 7.27 Enriched, Selective, and Differential Media. (a) Chocolate agar is used to grow fastidious organisms. It is called chocolate agar because of its color, the result of heated and lysed red blood cells added to the medium. (b) MacConkey agar includes dye to differentiate bacteria. Colonies of lactose fermenters are distinguished by pink pigmentation. (c) If bacteria grown on blood agar produce a hemolysin, the subsequent lysis of red blood cells produces a clearing in the agar. (d) Mannitol salt agar has a high salt concentration that selects for growth of staphylococci. Shown here is *Staphylococcus aureus*, which can ferment mannitol to acid, as detected by the pH indicator turning yellow.
Jarun Ontakrai/Shutterstock

MICRO INQUIRY *Classify these media as enriched, selective, or differential. Explain.*

vitamins necessary for the cultivation of fastidious organisms. Eosin methylene blue agar and MacConkey agar are differential and selective. Both media contain lactose and pH indicator dyes. Bacteria that consume lactose by fermenting it release acidic waste products that trigger color change in the dyes (figure 7.27*c*). Both media are widely used for the detection of *E. coli* and related bacteria in water supplies and elsewhere. *Blood Agar; EMB; MacConkey Agar (MAC)*

Cultivation of Aerobes and Anaerobes

Because aerobes need O_2 and anaerobes are killed by it, radically different approaches must be used when they are cultivated. When large volumes of aerobic microorganisms are cultured in liquid media, they must be shaken or sterile air must be pumped through the culture vessel to aerate the culture medium. Without aeration, the low solubility of O_2 in liquid prevents aerobic microbes from obtaining an adequate supply.

Precisely the opposite problem arises with anaerobes, in that all O_2 must be excluded. This can be accomplished in several ways:

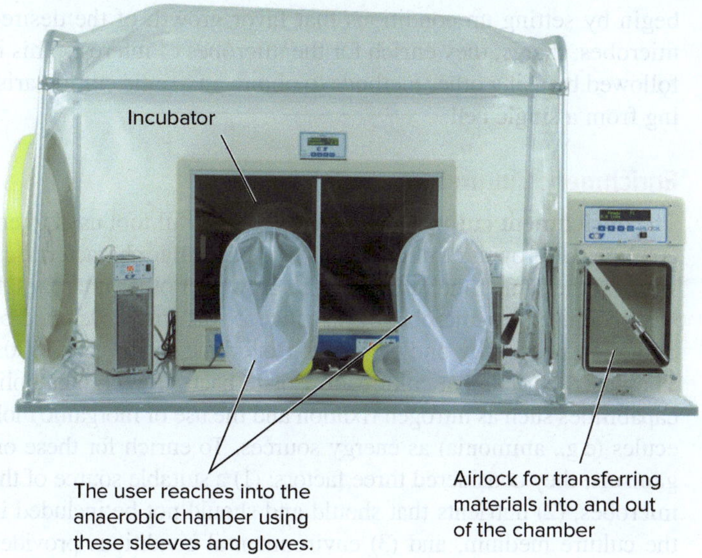

Figure 7.28 An Anaerobic Workstation and Incubator. This apparatus contains an oxygen-free work area and an incubator. The airlock compartment (right) allows materials to be transferred without exposing the interior to oxygen.
Coy Laboratory Products, Inc.

(1) Anaerobic media containing reducing agents such as thioglycollate or cysteine may be used. The medium is boiled during preparation to dissolve its components and drive off oxygen. The reducing agents eliminate any residual dissolved O_2 in the medium so that anaerobes can grow beneath its surface (figure 7.19). (2) Oxygen also may be eliminated from an enclosed work area, often called an anaerobic chamber or anaerobic workstation. Most of the air is removed with a vacuum pump followed by purges with nitrogen gas (**figure 7.28**). A gas mix containing hydrogen (at less than 3%) and nitrogen is then introduced into the workstation. In the presence of a palladium catalyst, the hydrogen and any remaining O_2 react to form water, maintaining an anoxic environment. Often CO_2 is added to the chamber because many anaerobes require a small amount of CO_2 for optimal growth. (3) Hard transparent containers that can be sealed tightly are often used to incubate Petri dishes. After adding the Petri dishes, a catalyst envelope is opened and placed in the container, which is promptly sealed. The catalyst reacts with oxygen to remove most or all of the gas, resulting in an anaerobic or microaerophilic atmosphere in the container. (4) A candle jar operates similarly but less expensively. A lit candle is included in the jar before sealing. The fire consumes all available O_2 and is extinguished when the atmosphere has no remaining O_2. *Oxygen Requirements*

Enrichment and Isolation of Pure Cultures

In natural habitats, microorganisms grow in complex communities containing many different species. This presents a problem for microbiologists because a single type of microorganism cannot be studied adequately in a mixed culture. To characterize an individual species, a population of cells arising from a single cell, called a **pure** or **axenic culture,** is needed. When isolating microbes from soil or aquatic habitats, microbiologists often begin by setting up conditions that favor growth of the desired microbes; that is, they enrich for the microbes of interest. This is followed by using other methods to obtain an axenic culture arising from a single cell.

Enrichment Cultures

The **enrichment culture** technique is a powerful tool used to encourage the growth of microbes having particular characteristics, while at the same time inhibiting the growth of other microbes. It has been used by microbiologists since it was first developed by Sergei Winogradsky and Martinus Beijerinck in the late 1800s. They used enrichment cultures to isolate bacteria with metabolic capabilities such as nitrogen fixation and the use of inorganic molecules (e.g., ammonia) as energy sources. To enrich for these organisms, they considered three factors: (1) a suitable source of the microbes, (2) nutrients that should and should not be included in the culture medium, and (3) environmental conditions provided during the incubation period. *Microbial ecology (section 1.3)*

Consider the following: You want to isolate an oil-degrading microbe and study it in pure culture. A logical source of such an organism (i.e., an inoculum) would be soil contaminated with oil. The assumption is that this source would have exposed all microbes in the soil to the oil but the population of those able to use oil as a sole source of energy and carbon will expand, or become enriched. To select for the oil-degrading microbes, a medium would be formulated that contained inorganic nitrogen, sulfur, and phosphorus sources but no source of carbon and energy other than the types of molecules typically found in oil. Finally, a temperature, pH, and oxygen level similar to that of the soil environment would be used during incubation.

Despite its utility, the enrichment culture technique is not without problems. It is sometimes difficult to mimic the proper selective conditions to successfully isolate the desired organisms. Furthermore, even if microbes are successfully cultivated, they may not adequately represent the population as a whole at a sampling site. Generally, enrichment cultures yield the organisms that grow fastest, and these may not be the predominant organisms in the habitat. Finally, enrichment cultures are not pure cultures. They usually contain more than one species having similar characteristics. To obtain a pure culture, one of the methods described next must be used.

Streak Plate

Methods for isolating pure cultures were developed by the German bacteriologist Robert Koch. He used these techniques in a set of steps that now bear his name: Koch's postulates. As we describe in chapter 1, Koch's postulates are used to establish that a microbe is the causative agent of a particular disease. The use of these methods transformed microbiology, and within 20 years of their development, bacterial pathogens responsible for major bacterial diseases had been isolated and identified. Koch reasoned that if cells from a mixture of microbes could be spatially isolated from each other, each cell would give rise to a completely separate **colony**—a macroscopically visible cluster of microorganisms in or on a solid medium. Because each colony arises from a single cell, each colony represents a pure culture.

In the **streak-plate** technique, cells are transferred to the edge of an agar plate with an inoculating loop or swab and then streaked across the surface in one of several patterns (**figure 7.29**). After the first sector is streaked, the inoculating loop is sterilized and an inoculum for the second sector is obtained from the first sector. A similar process is followed for streaking the third sector, except that the inoculum is from the second sector. Thus the number of organisms is diluted as each sector is streaked. Eventually very few cells will be on the loop, and single cells will drop from it as it moves across the agar surface. These develop into separate colonies. *Streak Plate*

Spread Plate and Pour Plate

Spread-plate and pour-plate techniques are similar in that they both dilute a sample of cells before separating them spatially. They differ in that the spread plate supports the cells on the surface of the agar, whereas the pour plate embeds them within the agar. Both methods can be used to determine the number of microorganisms in a sample.

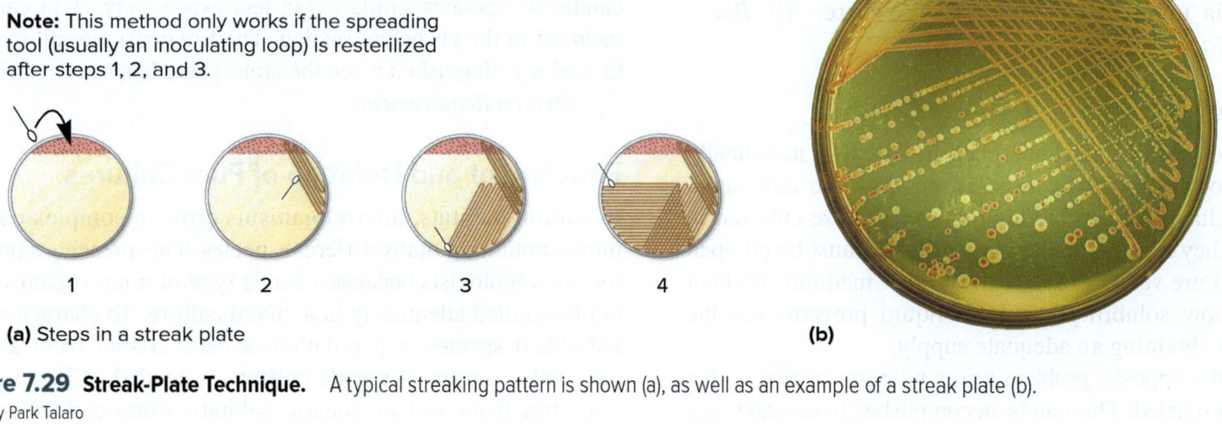

Figure 7.29 Streak-Plate Technique. A typical streaking pattern is shown (a), as well as an example of a streak plate (b).
(b) Kathy Park Talaro

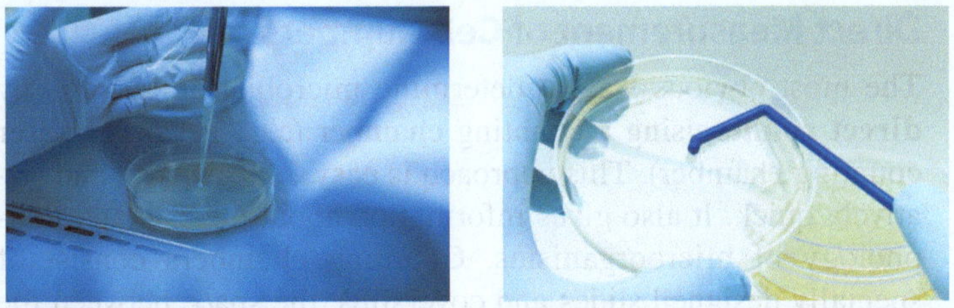

A small amount of the sample is pipetted to the center of a solidified medium.

A sterile spreader is used to spread the sample evenly over the surface of the medium.

(a)

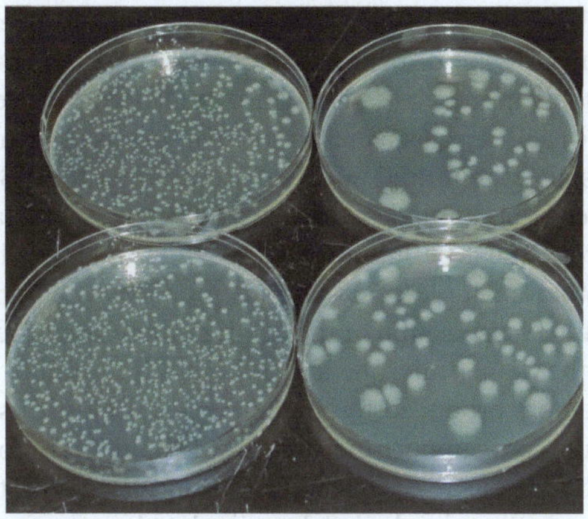

(b)

Figure 7.30 Spread-Plate Technique. (a) The preparation of a spread plate. (b) Typical result of spread-plate technique.

(a,1) Choksawatdikorn/Shutterstock; (a,2) unoL/Shutterstock; (b) Lisa Burgess/McGraw Hill

For the **spread plate,** a small volume of a diluted mixture is transferred to the center of an agar plate and spread evenly over the surface with a sterile bent rod (**figure 7.30**). The dispersed cells develop into isolated colonies during incubation. The dilutions for a spread plate are usually made in the same way as for a pour plate—by preparing serial dilutions (**figure 7.31**). *Standard Plate Count*

The **pour plate** is extensively used with bacteria, archaea, and fungi. It is particularly useful when sampling a heterogeneous population of microbes, some of which might produce colonies that overgrow an agar surface if isolated by the streak-plate or spread-plate methods. In the pour-plate method, the original sample is serially diluted to reduce the microbial population sufficiently to obtain separate colonies when plating (figure 7.31). Then small volumes of several diluted samples are mixed with liquid agar that has been cooled to about 45°C, and the mixtures are poured immediately into sterile Petri dishes. Most microbes survive a brief exposure to the warm agar. Each cell becomes fixed in place to form an individual colony after the agar hardens and the plate is incubated.

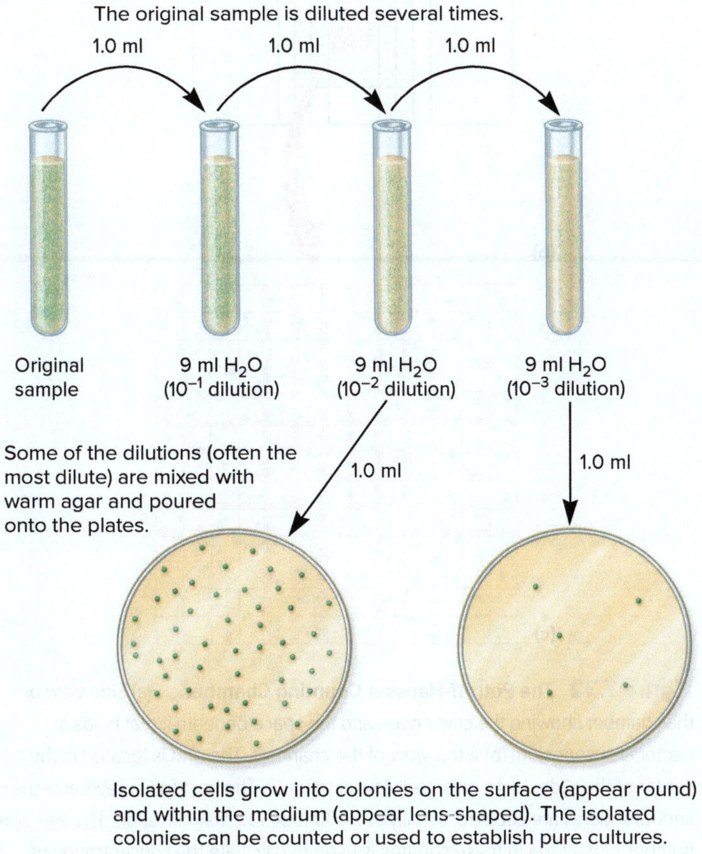

Figure 7.31 Serial Dilution and the Pour-Plate Technique.

New Approaches to Culturing Microbes

The importance of culturing even those microbes that are the most recalcitrant to growth in the lab has led microbiologists to devise new techniques. As we have noted, traditional methods for culturing microbes are based on constructing artificial habitats in a flask or Petri dish that mimic the microbes' natural environments. However, it is often difficult to know all the environmental conditions required for growth. One way to solve this problem is to formulate many types of media and incubate those media in many different conditions, an approach termed **culturomics.** This approach uses miniaturized cultures and robotics because of the vast number of manipulations, but it has succeeded in greatly expanding the number of cultured microbes from the human gut environment. Not surprisingly, many of the media that permit cultivation are enriched by the addition of mammalian blood and blood serum.

A different approach is to bring part of the natural environment into the lab. For instance, an aquarium containing seawater, beach sand, marine flora, and macroscopic organisms collected from a natural habitat can be used to culture marine microbes. This works very well for maintaining a mixed population in the lab, but it still leaves the problem of isolating particular microbes. One solution is to create a diffusion chamber, an enclosure within the natural habitat that allows diffusion of nutrients and other factors from the surroundings but retains the microbes. In this example, a dilute suspension of the seawater theoretically containing a single cell would be used to inoculate the enclosure. Thus an axenic culture would form in the enclosure.

Another approach is useful for microbes that require the presence of a different species to survive. In this case the microbe of interest is co-cultured with its required partner. Sometimes the partner is an animal, animal tissue, a plant, or plant tissue. The microbe must be maintained growing in or on that organism. Another example of co-culturing occurs in a Petri dish on which a dilute suspension of the desired microbe has been spread and then the helper microbe spotted onto the medium at one location. Factors produced by the helper microbe diffuse through the medium and support the growth of the microbe of interest. This method is particularly useful for long-term maintenance of cultures once the microbe has been isolated in axenic culture.

Microbiologists are exploring even more methods for culturing the uncultured. In the future it is expected that with these new techniques microbiologists will gain a better understanding of the diversity of microbes and their roles in many environments. *Microbial ecology also relies on cultures (section 26.3)*

Comprehension Check

1. Describe the following kinds of media and their uses: defined media, complex media, supportive (general purpose) media, enriched media, selective media, and differential media. Give an example of each.
2. What are peptones, yeast extract, beef extract, thioglycollate, and agar? In what kinds of media are they used?
3. List four ways in which anaerobes may be cultured.
4. What are pure cultures and why are they important? How are spread plates, streak plates, and pour plates inoculated?

7.8 Microbial Population Size Can Be Measured Directly or Indirectly

After reading this section, you should be able to:

a. Evaluate direct cell counts, viable counting methods, and cell mass measurements for determining population size
b. Explain why plate count results are expressed in terms of colony forming units (CFU)
c. Design appropriate approaches for measuring the population size of different types of samples

There are many ways to measure microbial growth to determine growth rate constants and generation times. Either cell number or mass can be followed because growth leads to increases in both. Here the most commonly employed techniques for determining population size are examined and the advantages and disadvantages of each noted. No single technique is always best; the most appropriate approach depends on the experimental situation.

Direct Measurement of Cell Numbers

The most obvious way to determine microbial numbers is by **direct counts** using a counting chamber (e.g., Petroff-Hausser counting chamber). This approach is easy, inexpensive, and relatively quick. It also gives information about the size and morphology of microorganisms. Counting chambers consist of specially designed slides and coverslips; the space between the slide and coverslip creates a chamber of known depth. On the bottom of the chamber is an etched grid that facilitates counting the cells (**figure 7.32**). The number of microorganisms in a sample can be calculated from the chamber's volume and any dilutions made of the sample before counting. One disadvantage of

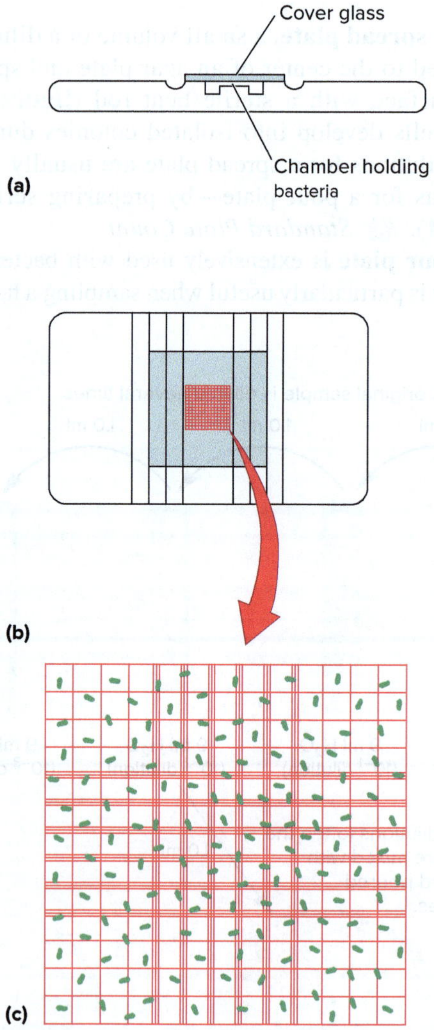

Figure 7.32 The Petroff-Hausser Counting Chamber. (a) Side view of the chamber showing the cover glass and the space beneath it that holds a bacterial suspension. (b) A top view of the chamber. The grid is located in the center of the slide. (c) An enlarged view of the grid. The bacteria in several of the central squares are counted, usually at ×400 to ×500 magnification. The average number of bacteria in these squares is used to calculate the concentration of cells in the original sample.

using counting chambers is that to determine population size accurately, the microbial population must be relatively large and evenly dispersed because only a small volume of the population is sampled.

The number of bacteria in aquatic samples is frequently determined from direct counts after the bacteria have been trapped on membrane filters. In the membrane filter technique, the sample is first filtered through a black polycarbonate membrane filter. Then the bacteria are stained with nucleic acid fluorescent dyes such as acridine orange or DAPI and observed microscopically (*see figure 26.1*). Alternatively, fluorescently labeled dyes specific for members of a given taxon may be used (*see figure 3.4a*). The stained cells are easily observed against the black background of the membrane filter and can be counted when viewed with an epifluorescence microscope. Direct cell counts of environmental samples almost invariably result in higher cell densities than do methods that rely on culturing. This is because only a small fraction of cells growing in nature can be cultivated in the laboratory. ◀ *Fluorescence microscopes use emitted light to create images (section 2.2)*

Flow cytometry is increasingly being used to directly count microbes and to gain detailed information about them. A flow cytometer creates a stream of cells so narrow that one cell at a time passes through a beam of laser light. As each cell passes through the beam, the light is scattered. Scattered light is detected by the flow cytometer. Because cells are separated in space, each light-scattering event is detected independently. Thus the number of light-scattering events represents the number of cells in the sample (*see figure 26.4*). Cells of differing size, internal complexity, and other characteristics within a population can also be counted. This usually involves the use of fluorescent dyes or fluorescently labeled antibodies. These more sophisticated uses of flow cytometry can provide valuable information about characteristics of the population of cells, which can allow them to be sorted.

Microorganisms also can be directly counted with electronic counters such as the Coulter counter. In the Coulter counter, a microbial suspension is forced through a small hole. Electrical current flows through the hole, and electrodes placed on both sides of the hole measure electrical resistance. Every time a microbial cell passes through the hole, electrical resistance increases (i.e., the conductivity drops), and the cell is counted.

Traditional methods for directly counting microbes in a sample usually yield cell densities that are much higher than the plating methods described next, in part because direct counting procedures do not distinguish dead cells from culturable cells. Newer methods for direct counts help alleviate this problem. Fluorescent dyes can differentiate between live and dead cells, making it possible to count the number of live and dead microorganisms in a sample (*see figures 2.13a and 26.3*). Unfortunately, the method is not completely accurate. Just as scientists have difficulty defining life, so too they have difficulty deciding when a cell is truly dead (figure 7.21).

Viable Counting Methods

Several plating methods can be used to determine the number of viable microbes in a sample. These are referred to as either **viable counting methods** or **standard plate counts** because they count only those cells able to reproduce when cultured. Two commonly used procedures are the spread-plate and the pour-plate techniques, which are described in section 7.7. When using these methods to determine population size, the samples should yield between 25 and 250 colonies per plate for accurate counting. Once the number of colonies is known, the original number of viable microorganisms in the sample can be calculated from that number and the sample dilution. For example, if 1.0 milliliter of a solution diluted by a factor of 1×10^6 yields 150 colonies (i.e., 1.5×10^2 colonies), then the original sample contained around 1.5×10^8 cells per milliliter. However, because it is not possible to be certain that each colony arose from a single cell, the results are often expressed in terms of **colony forming units (CFU),** rather than the number of microorganisms.

Another commonly used plating method first traps bacteria in aquatic samples on a membrane filter. The filter is then placed on an agar medium or on a pad soaked with liquid media (**figure 7.33**) and incubated until each cell forms a separate colony. A colony count gives the number of microorganisms in the filtered sample, and selective media can be used to enumerate specific microorganisms. This technique is especially useful in analyzing water purity. ▶ *Purification and sanitary analysis ensures safe drinking water (section 42.1)*

Plating techniques are simple, sensitive, and widely used for viable counts of bacteria and other microorganisms in samples of food, water, and soil. However, several problems can lead to inaccurate counts. The population will be underestimated if clumps of cells are not broken up and microorganisms well dispersed. The hot agar used in the pour-plate technique may injure or kill sensitive cells; thus spread plates sometimes give higher counts than pour plates. A major problem is that plate counts will be artificially low if the medium employed cannot support growth of all the viable microorganisms present. This is a common problem encountered by microbial ecologists trying to understand the makeup of microbial communities. In such studies, the number of cells observed by microscopy is often much higher than the population size determined by plate counts. This discrepancy is referred to as *the great plate count anomaly* and is discussed in detail in chapter 26. ▶ *Microbial ecology also relies on cultures (section 26.3)*

Measurement of Cell Mass

Techniques for measuring changes in cell mass also can be used to determine population size. One approach is the determination of microbial dry weight. Cells growing in liquid medium are collected by centrifugation, washed, dried, and weighed. This is an especially useful technique for measuring the growth of filamentous fungi. However, it is time consuming and not very sensitive. Because bacteria weigh so little, large volumes of culture may be needed to collect a sufficient quantity.

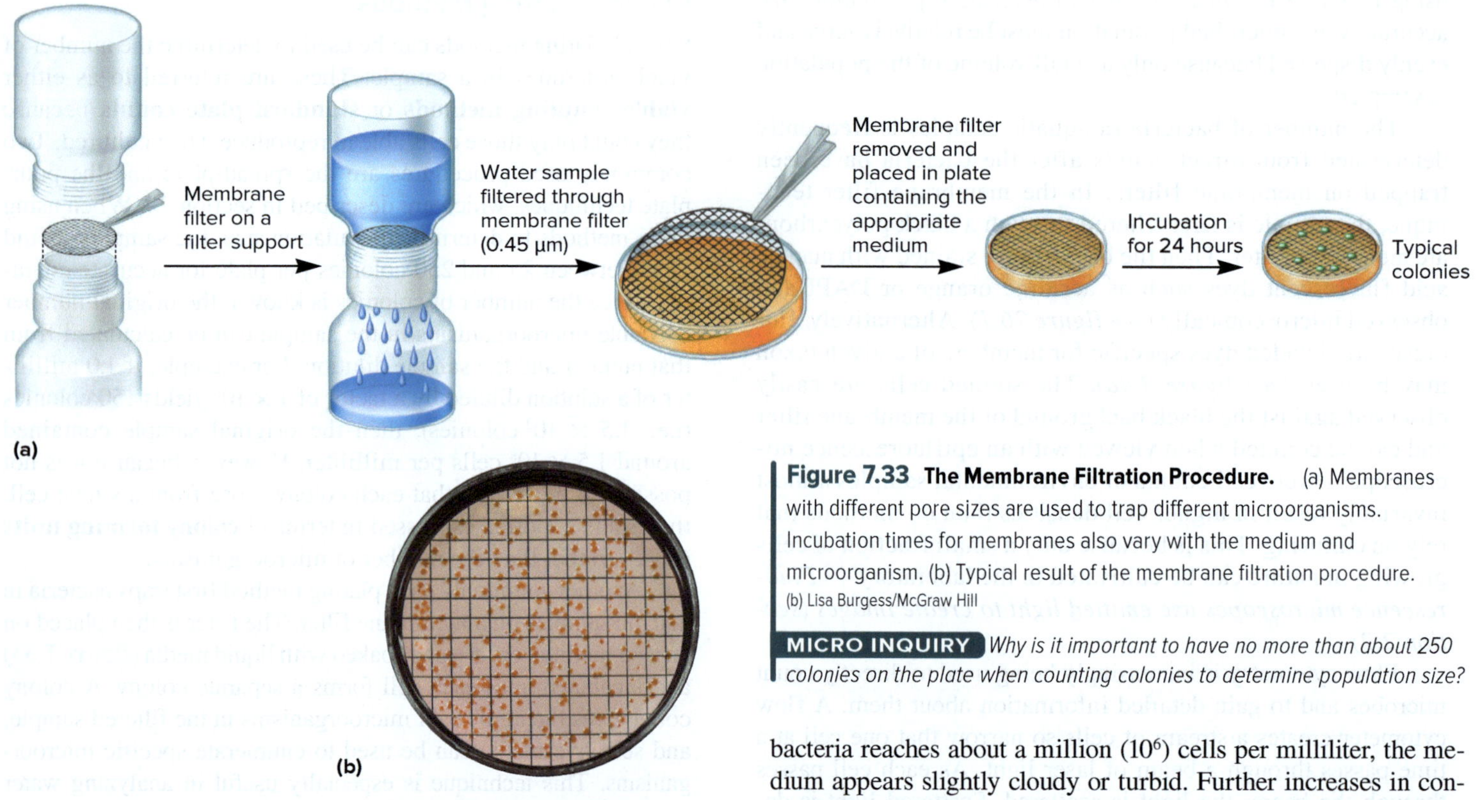

Figure 7.33 The Membrane Filtration Procedure. (a) Membranes with different pore sizes are used to trap different microorganisms. Incubation times for membranes also vary with the medium and microorganism. (b) Typical result of the membrane filtration procedure.
(b) Lisa Burgess/McGraw Hill

MICRO INQUIRY *Why is it important to have no more than about 250 colonies on the plate when counting colonies to determine population size?*

A more rapid and sensitive method for measuring cell mass is spectrophotometry (**figure 7.34**). Spectrophotometry depends on the fact that microbial cells scatter light. Because microbial cells in a pure culture are of roughly constant size, the amount of scattering is directly proportional to the biomass of cells present and indirectly related to cell number. When the concentration of bacteria reaches about a million (10^6) cells per milliliter, the medium appears slightly cloudy or turbid. Further increases in concentration result in greater turbidity, and less light is transmitted through the medium. The extent of light scattering (i.e., decrease in transmitted light) can be measured by a spectrophotometer and is called the absorbance (optical density) of the medium. Absorbance is almost linearly related to cell concentration at absorbance levels less than about 0.5. If the sample exceeds this value, it must first be diluted. Thus population size can be easily measured as long as the population is high enough to give detectable turbidity.

Cell mass can also be estimated by measuring the concentration of some cellular substance, as long as its concentration is constant in each cell. For example, a sample can be analyzed for total protein. An increase in the microbial population will be reflected in higher total protein levels. Similarly, chlorophyll determinations can be used to measure phototrophic protist and cyanobacterial populations, and the quantity of ATP can be used to estimate the amount of living microbial mass.

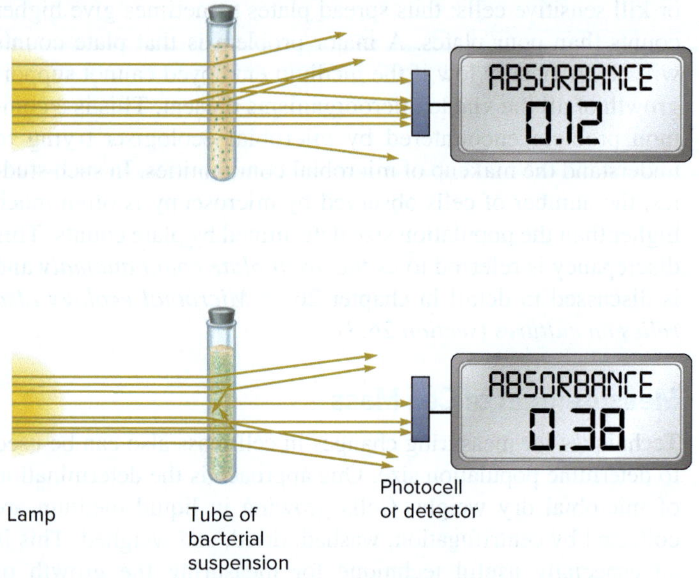

Figure 7.34 Turbidity and Microbial Mass Measurement. Determination of microbial mass by measurement of light absorption. As the population and turbidity increase, more light is scattered and the absorbance reading given by the spectrophotometer increases.

Comprehension Check

1. Briefly describe each technique by which microbial population numbers may be determined and give its advantages and disadvantages.
2. When using direct cell counts to follow the growth of a culture, it may be difficult to tell when the culture enters the death phase. Why?
3. Why are plate count results expressed as colony forming units?
4. For each of the following, which enumeration technique would you use? Explain your choice. (a) A pure culture of *Staphylococcus aureus;* (b) a water sample that needs to be checked for *E. coli* contamination; (c) a sample of yogurt.

7.9 Chemostats and Turbidostats Are Used for Continuous Culture of Microorganisms

After reading this section, you should be able to:

a. Distinguish batch culture and continuous culture
b. Differentiate chemostats and turbidostats
c. Discuss the relationship between the dilution rate of a chemostat and population size and growth rate

In section 7.4, our focus is on closed systems called batch cultures in which nutrients are not renewed nor wastes removed. Exponential growth lasts for a few generations and soon stationary phase is reached. In contrast to this approach, it is possible to grow microorganisms in a system with constant environmental conditions maintained by supplying nutrients and removing wastes. Such a system is called a **continuous culture system.** These systems maintain a microbial population in exponential growth at a known rate and at a constant biomass concentration for extended periods. Continuous culture systems make possible the study of microbial growth at very low nutrient levels, concentrations close to those present in natural environments. These systems are essential for research in many areas, particularly microbial ecology. Two major types of continuous culture systems in common use are chemostats and turbidostats.

Chemostats

A **chemostat** is constructed so that the rate at which sterile medium is fed into a culture vessel is the same as the rate at which the medium containing microorganisms is removed (**figure 7.35**). The culture medium for a chemostat has a limited quantity of an essential nutrient (e.g., a vitamin). Because one nutrient is limiting, the growth rate is determined by the rate at which fresh medium is fed into the growth chamber; the final cell density depends on the concentration of the limiting nutrient. The rate of nutrient exchange is expressed as the dilution rate (D), the rate at which medium flows through the culture vessel relative to the vessel volume, where f is the flow rate (milliliter/hr) and V is the vessel volume (milliliter).

$$D = f/V$$

For example, if f is 30 milliliters/hr and V is 100 milliliters, the dilution rate is 0.30 hr^{-1}.

Both population size and generation time are related to the dilution rate (**figure 7.36**). When dilution rates are very low, only a limited supply of nutrient is available and the microbes can conserve only a limited amount of energy. Much of that energy must be used for cell maintenance, not for growth and reproduction. Slightly higher dilution rates make more nutrients available to the microbes. When the dilution rate is high enough to provide enough nutrients for both maintenance and reproduction, the cell density will begin to rise. In other words, the growth rate can increase

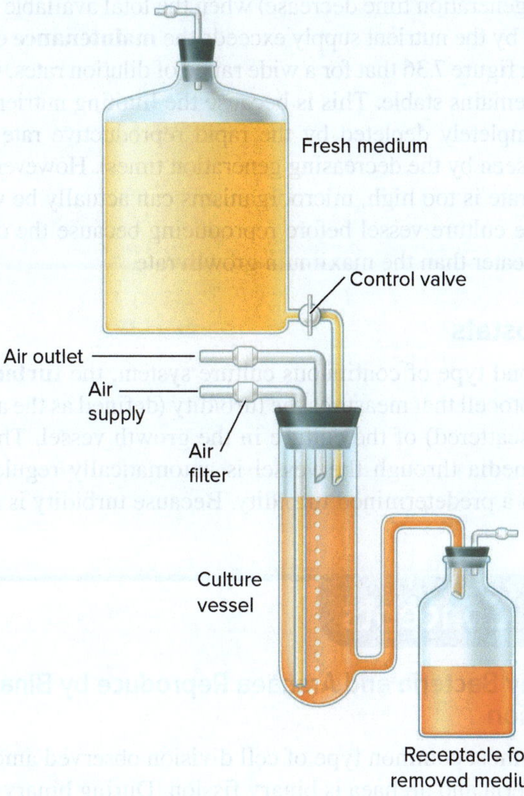

Figure 7.35 A Chemostat. The fresh medium contains a limiting amount of an essential nutrient. Although not shown, usually the incoming medium is mechanically mixed with that already in the culture vessel.

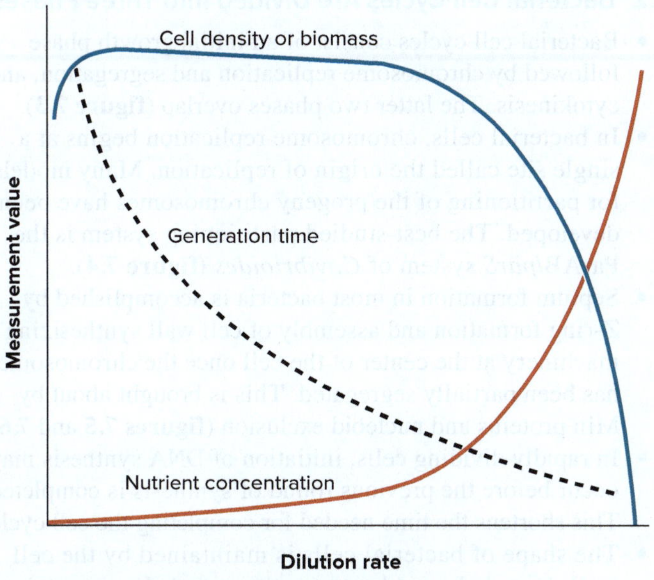

Figure 7.36 Chemostat Dilution Rate and Microbial Growth. The effects of changing the dilution rate in a chemostat.

MICRO INQUIRY *Why does the cell density stay relatively constant over such a large range of dilution rates?*

(and the generation time decrease) when the total available energy provided by the nutrient supply exceeds the **maintenance energy.** Notice in figure 7.36 that for a wide range of dilution rates, the cell density remains stable. This is because the limiting nutrient is almost completely depleted by the rapid reproductive rate of the cells (as seen by the decreasing generation times). However, if the dilution rate is too high, microorganisms can actually be washed out of the culture vessel before reproducing because the dilution rate is greater than the maximum growth rate.

Turbidostats

The second type of continuous culture system, the **turbidostat,** has a photocell that measures the turbidity (defined as the amount of light scattered) of the culture in the growth vessel. The flow rate of media through the vessel is automatically regulated to maintain a predetermined turbidity. Because turbidity is related to cell density, the turbidostat maintains a desired cell density. Turbidostats differ from chemostats in several ways. The dilution rate in a turbidostat varies, rather than remaining constant, and a turbidostat's culture medium contains all nutrients in excess; that is, none of the nutrients is limiting. A turbidostat operates best at high dilution rates; a chemostat is most stable and effective at lower dilution rates.

Comprehension Check

1. How does a continuous culture system differ from a closed culture system (i.e., a batch culture)?
2. Describe how chemostats and turbidostats operate. How do they differ?
3. What is the dilution rate? What is maintenance energy? How are they related?

Key Concepts

7.1 Many Bacteria and Archaea Reproduce by Binary Fission

- The most common type of cell division observed among bacteria and archaea is binary fission. During binary fission, the cell elongates, the chromosome is replicated, and then it segregates to opposite poles of the cell prior to completing the septum, which divides the cell into two progeny cells (**figure 7.1**).
- Other bacterial reproductive strategies include budding, baeocytes, and spore formation (**figure 7.2**).

7.2 Bacterial Cell Cycles Are Divided into Three Phases

- Bacterial cell cycles consist of an initial growth phase followed by chromosome replication and segregation, and cytokinesis. The latter two phases overlap (**figure 7.3**).
- In bacterial cells, chromosome replication begins at a single site called the origin of replication. Many models for partitioning of the progeny chromosomes have been developed. The best-studied partitioning system is the ParAB/*parS* system of *C. vibrioides* (**figure 7.4**).
- Septum formation in most bacteria is accomplished by Z-ring formation and assembly of cell wall synthesizing machinery at the center of the cell once the chromosome has been partially segregated. This is brought about by Min proteins and nucleoid exclusion (**figures 7.5** and **7.6**).
- In rapidly dividing cells, initiation of DNA synthesis may occur before the previous round of synthesis is completed. This shortens the time needed for completing the cell cycle.
- The shape of bacterial cells is maintained by the cell wall. In rod-shaped bacteria, the cytoskeletal proteins FtsZ and MreB are involved in determining where peptidoglycan will be synthesized. When rod-shaped cells are lengthening prior to cell division, MreB localizes along the length of the cell, and peptidoglycan synthesis occurs there as a result. During cell division, FtsZ localization confines peptidoglycan synthesis to the midcell (**figures 7.7** and **7.8**). In cocci, FtsZ and peptidoglycan synthesis localize to the midcell.

7.3 Archaeal Cell Cycles Are Unique

- *Sulfolobus* spp. have a cell cycle that consists of G1-, S-, and G2-like phases followed by partitioning and cytokinesis. The partitioning process is similar to the ParAB/*parS* system of *C. vibrioides*. However, these archaea use homologues of eukaryotic ESCRT proteins for cytokinesis (**figure 7.9**).
- The cell cycles of other archaea are more similar to the bacterial model. Many have homologues of FtsZ, which are thought to function in cytokinesis.

7.4 Growth Curves Consist of Five Phases

- Growth is an increase in cellular constituents and results in an increase in cell size, cell number, or both.
- When microorganisms are grown in a batch culture, the resulting growth curve usually has five phases: lag, exponential, stationary, death, and long-term stationary phase (**figure 7.10**).
- In the exponential phase, the number of cells undergoing binary fission doubles at a constant interval called the generation or doubling time (**figure 7.13**). The growth rate constant (k) is the reciprocal of the generation time (g) (**figure 7.14**).

7.5 Environmental Factors Affect Microbial Growth

- Solute concentrations can adversely affect microbes by causing lysis or shrinkage of the cytoplasm. Most bacteria, photosynthetic protists, and fungi have rigid cell walls that help protect them from osmotic stress. By synthesizing cytoplasmic compatible solutes, their cytosol remains hypertonic relative to their habitats.

- Solutes also affect microbes because they interact with water, making it less available for microbial activities. The amount of water available to microorganisms is expressed as the water activity (a_w). Most microorganisms do not grow well at water activities below 0.98; osmotolerant organisms survive and even flourish at low a_w values. Halophiles require high sodium chloride concentrations for growth (**figure 7.16**).
- Each microbial species has an optimum pH for growth and can be classified as an acidophile, neutrophile, or alkalophile (**figure 7.17**).
- Microorganisms have distinct temperature ranges for growth with minima, maxima, and optima—the cardinal temperatures. These ranges are determined by the effects of temperature on the rates of catalysis, protein denaturation, and membrane disruption. There are five major classes of microorganisms with respect to temperature preferences: psychrophiles, psychrotolerants, mesophiles, thermophiles, and hyperthermophiles (**figure 7.18**).
- Microorganisms can be placed into five different categories based on their O_2 requirements: obligate aerobes, microaerophiles, facultative anaerobes, aerotolerant anaerobes, and obligate anaerobes (**figure 7.19**).
- Oxygen can become toxic because of the production of hydrogen peroxide, superoxide radical, and hydroxyl radical. These are destroyed by the enzymes superoxide dismutase, catalase, and peroxidase.
- Most deep-sea microorganisms are barotolerant, but some are piezophilic and require high pressure for optimal growth.
- High-energy or short-wavelength radiation harms organisms in several ways. Ionizing radiation (X rays and gamma rays) ionizes molecules and destroys DNA and other cell components. Ultraviolet (UV) radiation induces DNA damage. Visible light can provide energy for the formation of reactive singlet oxygen, which will destroy cells.

7.6 Microbial Growth in Natural Environments

- Microbial growth in natural environments is profoundly affected by nutrient limitation and other adverse factors. Microbes are often found in oligotrophic environments. They have evolved strategies to survive in these habitats, including endospore formation and other types of dormancy in growth-arrested states.
- Many microbes form biofilms—aggregations of microbes held together by extracellular polymeric substances (EPS) (**figures 7.23** and **7.24**). Life in a biofilm has several advantages, including protection from harmful agents.
- Bacteria often use chemicals to communicate with one another in a density-dependent way and carry out a particular activity only when a certain population density is reached. This phenomenon is called quorum sensing (**figure 7.25**).

7.7 Laboratory Culture of Microbes Requires Conditions That Mimic Their Normal Habitats

- Culture media can be constructed from chemically defined components (defined media or synthetic media) or constituents, such as peptones and yeast extract, whose precise composition is unknown (complex media). Culture media can be solidified by the addition of agar, a complex polysaccharide from red algae.
- Culture media are classified based on function. Supportive media are used to culture a wide variety of microbes. Enriched media are supportive media that contain additional nutrients needed by fastidious microbes. Selective media contain components that select for the growth of some microbes. Differential media contain components that allow microbes to be differentiated from each other, usually based on some metabolic capability.
- Aerobes are cultivated by supplying oxygen. Anaerobes are cultivated using media and methods that exclude oxygen (**figure 7.28**).
- Enrichment cultures are usually the first step in isolating pure cultures. They use selective conditions (media, environmental conditions) to promote the growth of certain microbes.
- Pure cultures usually are obtained by isolating individual cells with any of three plating techniques: the streak-plate, spread-plate, and pour-plate methods (**figures 7.29–7.31**).
- New approaches are being developed to culture microbes that in the past have not been culturable. These include creating natural habitats within the lab setting and co-culturing microbes with organisms that help them survive.

7.8 Microbial Population Size Can Be Measured Directly or Indirectly

- Microbial populations can be counted directly with counting chambers, flow cytometers, electronic counters, or fluorescence microscopy (**figure 7.32**).
- Indirect measurements include viable counting techniques (standard plate counts) such as the spread-plate, the pour-plate, or the membrane filter method (**figure 7.33**).
- Population size also can be determined indirectly by measuring microbial mass through the measurement of dry weight, turbidity, or the amount of a cell component (**figure 7.34**).

7.9 Chemostats and Turbidostats Are Used for Continuous Culture of Microorganisms

- Microorganisms can be grown in an open system in which nutrients are constantly provided and wastes removed.
- A continuous culture system can maintain a microbial population in a particular phase of the growth curve. There are two types of these systems: chemostats and turbidostats (**figure 7.35**).

Active Learning

1. As an alternative to diffusible signals, discuss another mechanism by which bacteria might quorum sense.
2. If you wished to obtain a pure culture of microbes that could degrade benzene and use it as a carbon and energy source, how would you proceed?
3. How would you determine if a slow-growing microbial culture is in lag phase or exponential phase?
4. Suggest one specific mechanism underlying the observation that the cardinal temperatures of some microbes change depending on other environmental conditions (e.g., pH).
5. Suppose you discovered a new bacterial strain from the permafrost of Alaska. You are able to culture the bacterium, but it grows very slowly. What characteristics of the bacterium would you determine to see if you could optimize its growth rate? Explain your choices.
6. For your new bacterial strain from the previous question, how would you experimentally generate a curve like the one shown in figure 7.18 to determine its cardinal temperatures?
7. What special equipment and techniques would be necessary to culture an organism that is both hyperthermophilic and anaerobic? Or an organism whose growth substrate is a gas, like CO_2?
8. The bacterium *Variovorax* can use *N*-acyl homoserine lactones as a carbon and energy source for growth. How would the presence of *Variovorax* in an ecosystem affect the microbial community?
9. Consider a typical bacterial growth curve, as illustrated in figure 7.10. Assume the organism is growing under its optimal conditions. How would the curve differ if the culture were incubated near its minimum temperature? How would the curve differ if the nutrients were diluted tenfold but the temperature was at its optimum?
10. How might an enrichment culture be used to isolate bacteria capable of growing photosynthetically from a mixed microbial assemblage?
11. Would you expect all organisms present in an environment at 0.5 M NaCl to be halophiles? Explain your answer.
12. Do you think *Campylobacter jejuni* (*see figure 3.2a*) produces FtsZ? What about MreB? Explain your reasoning.
13. Generation times in nature are usually much longer than in a laboratory culture. Why do you think this is the case?

CHAPTER

8

Control of Microorganisms in the Environment

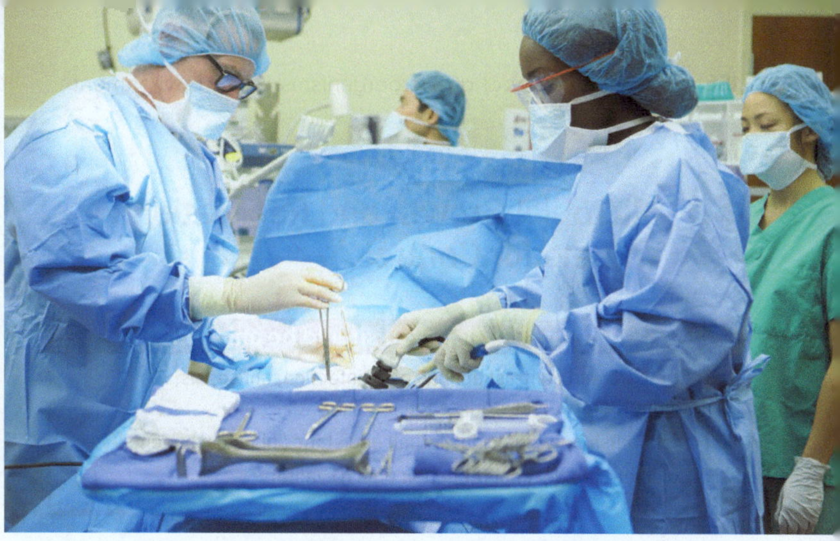

MBI/iStock/iStock/Getty Images

Scalpel, Please

Surgical interventions to address medical conditions have become commonplace in the past century. What were once fearsome and dramatic procedures have now become outpatient treatments in many cases. But while advances in anesthesia and robotics have streamlined surgical practices, concerns over infection remain. Preparing for surgery requires addressing potential pathogens that may enter and exploit a patient's body through skin incisions, via medical implants, from the hospital environment, and from the medical staff.

The human body hosts a rich community of microbes, yet the body's sterile sites should not be compromised by microorganisms once the skin has been breached. Skin protects bodily tissues from a host of external microbes, but the skin itself supports a microbial community. Prior to surgery, patients are instructed to wash with antiseptic soaps and then additional antiseptics are applied to the skin as the procedure begins.

Surgical instruments and tools are made of a variety of materials, all of which must be sterilized, or rendered free from microbes, for use inside the body. Metal instruments are heated or treated with harsh chemicals, but delicate plastics require alternate processes to preserve their integrity.

The operating room itself must be free from airborne microbes. The medical team brings their own microbes to the operating room, so they scrub, then cover up with gowns, masks, caps, and gloves. And the environment—lights, ventilation, and supplies—must all be carefully cleaned and prepared for each surgical event.

What a mix of tasks is essential to a successful surgery! Controlling microbes in the hospital operating room must consider multiple facets: the range of microorganisms that may be present, the variety of materials that must be maintained as sterile, and the choice of cleaning and re-using or disposing of materials. This chapter describes the many methods of controlling microbial growth and the best applications of each technique. We certainly live in a microbial world, but for our own good health, microbes must be kept under control.

Readiness Check:

Based on what you have learned previously, you should be able to:

✓ Identify the cell structures and their functions of bacteria, protists, fungi, viruses, and prions, including their replication processes and energy requirements (sections 3.3–3.8, 5.2–5.7, 6.1–6.7, 7.1–7.6)

8.1 Microbial Growth and Replication: Targets for Control

After reading this section, you should be able to:

a. Compare and contrast actions of disinfection, antisepsis, chemotherapy, and sterilization
b. Distinguish between cidal (lethal) and static (inhibitory) agents
c. Calculate the decimal reduction time (D value)

The principles of microbial control are rooted in microbial nutrition, growth, and development, which are discussed in chapter 7. If you can starve or poison microbes, or inhibit or prevent growth or replication, you can control microorganisms. Of course, it is not as simple as this sounds; a complex vocabulary is applied to describe differences in population density, degree of killing, and the nature of the control mechanism. Terminology is especially important when the control of microorganisms is discussed because words such as *disinfectant* and *antiseptic* often are used loosely, yet they have different meanings. The situation is even more confusing because a particular treatment can either inhibit growth, prevent replication, or kill, depending on the conditions. The types of control methods and their uses are outlined in **figure 8.1**. Antimicrobial agents can be physical, chemical, or biological. To simplify the terminology, we use the term **biocide** to mean all antimicrobial agents that can be used to control microorganisms. It is important to distinguish biocides from antibiotics. **Antibiotics** are used clinically in small quantities and inhibit a specific biochemical target in the cell, like an enzyme, a ribosome, or cell wall synthesis. These are discussed in chapter 9. Biocides have a more general mechanism of killing, like cross-linking, oxidizing, or denaturing biomolecules.

The range of microbial cell types presents a broad target for growth control. The outer layer of a microbe is a permeability barrier that must be overcome for a treatment to be effective. Diderm bacteria are somewhat less susceptible than monoderms as a consequence of their outer membrane. Exterior surfaces on *Mycobacterium* spp., bacterial **endospores,** and protozoan **cysts** render these cell types especially recalcitrant. It is critical to consider the diversity of microbes in a given situation when evaluating control options.

Sterilization (Latin *sterilis,* unable to produce offspring), in our context, is the process by which all living cells, spores, and acellular entities (e.g., viruses, viroids, and prions) are either destroyed or removed from an object or habitat. A sterile object is totally free of viable microorganisms, spores, and other infectious agents. When sterilization is achieved by a chemical agent, the chemical is called a *sterilant.* In contrast, **disinfection** is the killing, inhibition, or removal of microorganisms that may cause disease. Disinfection therefore results in the substantial reduction of the total microbial population and the destruction of potential pathogens. **Disinfectants** are agents, usually chemical,

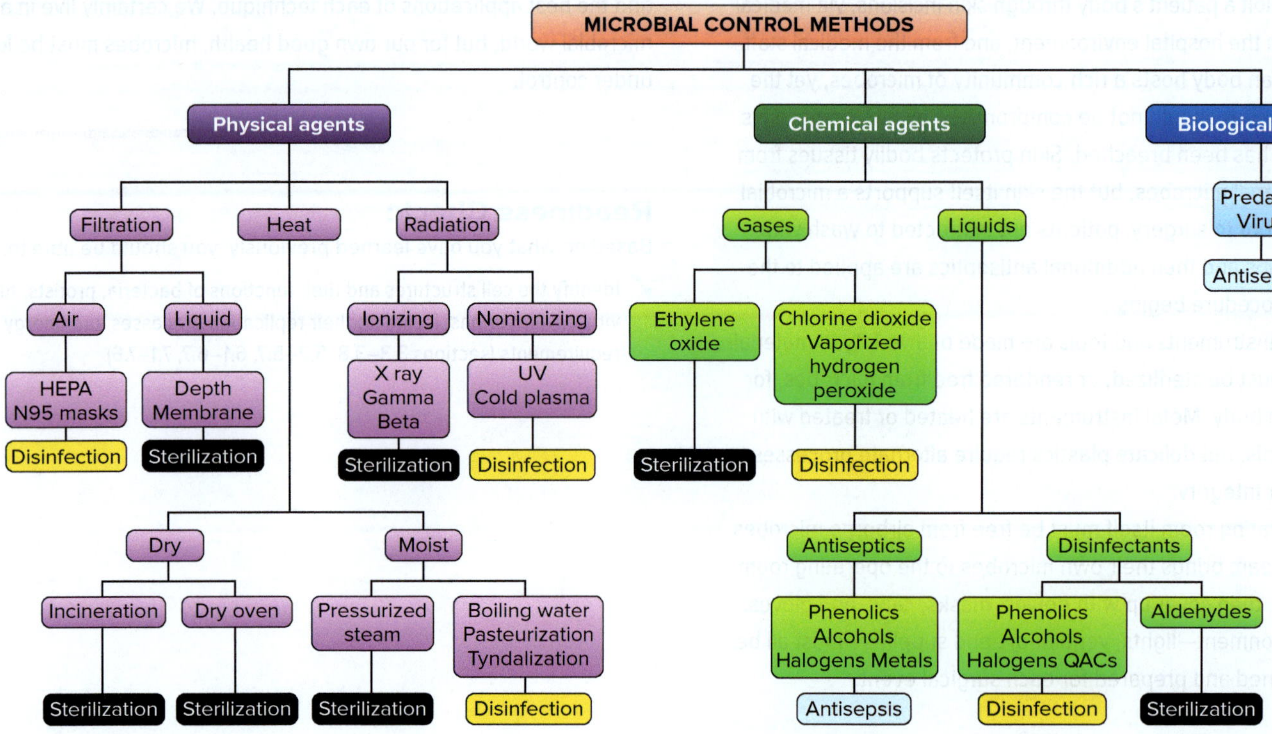

Figure 8.1 Microbial Control Methods.

MICRO INQUIRY *Which types of agents can be used for sterilization? Which can be used for antisepsis? What is the difference?*

used to carry out disinfection and normally used only on inanimate objects. A disinfectant does not necessarily sterilize an object because viable spores and a few microorganisms may remain. In **sanitization,** the microbial population is reduced to levels that are considered safe by public health standards. The inanimate object is usually cleaned as well as partially disinfected. For example, sanitizers are used to clean eating utensils in restaurants. ◀ *Viroids and satellites: nucleic acid–based subviral agents (section 6.6); Prions are composed only of protein (section 6.7)*

Antisepsis (Greek *anti,* against; *sepsis,* putrefaction) is the destruction or inhibition of microorganisms on living tissue; it is the prevention of infection or sepsis. **Antiseptics** are chemical agents applied to tissue to prevent infection by killing or inhibiting pathogen growth; they also reduce the total microbial population. Because they must not cause too much harm to the host, antiseptics are generally not as harsh as disinfectants.

Chemotherapy is the generic term used to describe the application of chemicals to kill or inhibit the growth of microorganisms. Chemotherapeutic agents are often characterized by the outcome of their application. Substances that kill organisms often have the suffix *-cide* (Latin *cida,* to kill); a cidal agent kills microbes but not necessarily endospores. A disinfectant or antiseptic can be particularly effective against a specific group, in which case it may be called a **bactericide, fungicide,** or **viricide.** Other chemicals do not kill but rather prevent growth; if the chemicals are removed, growth will resume. Their names end in *-static* (Greek *statikos,* causing to stand or stopping)—for example, **bacteriostatic** and **fungistatic.**

It is worth noting here that resistance to antimicrobial biocides has been increasing nearly as rapidly as resistance to antibiotics. Biocide resistance mechanisms mimic those of antibiotic resistance in that they alter membrane permeability, modify the target, or activate transporters to expel the control agent. Furthermore, studies have shown that the acquisition of genes that direct both antibiotic and biocide resistance mechanisms are linked. In many cases they are shared on plasmids and upregulated in response to ubiquitously low environmental biocide concentrations. It is clear that microorganisms adapt well to new situations, resulting in the sharing of successful resistance strategies. Notably, in 2016 the broad-spectrum antimicrobial agent triclosan was banned from use in household personal care products due to evidence of widespread resistance. The action was not simply a concern over triclosan resistance alone, but rather because there was demonstrated cross-resistance to other clinically important antibiotics that would impact global health. ▶ *Antimicrobial drug resistance is a public health threat (section 9.8)*

The Pattern of Microbial Death Mirrors the Pattern of Microbial Growth

A microbial population is not killed instantly when exposed to a lethal agent. Population death is generally exponential; that is, the population is reduced by the same fraction at constant intervals (**table 8.1**). If the logarithm of the viable population is plotted against the time of exposure of the microorganism to the agent, a straight-line plot will result (**figure 8.2**). When the population has been greatly reduced, the rate of killing may slow due to the survival of a more resistant strain of the microorganism.

One measure of an agent's killing efficiency is the **decimal reduction time (D)** or **D value.** The decimal reduction time is the time required to kill 90% of the microorganisms or endospores in a sample under specified conditions. For example, in a semilogarithmic plot of the viable microorganisms versus the time of heating, the D value is the time required for the line to drop by one log cycle or 10-fold (figure 8.2*a*). The D value is specific for an organism and a treatment condition. A higher D value indicates that a longer treatment time is required to kill 90% of the viable organisms. A related term, the *Z value,* describes the temperature change at a given D value that decreases the microbial population by one log cycle (90%). The Z value is predicted from a semilogarithmic plot of D values versus temperature (figure 8.2*b*).

Table 8.1 A Theoretical Microbial Heat-Killing Experiment

Minute	Viable Microorganisms at Start of Minute	Microorganisms Killed in 1 Minute (90% of Total)[1]	Viable Microorganisms at End of 1 Minute	Log_{10} of Viable Microorganisms
1	10^6	9×10^5	10^5	5
2	10^5	9×10^4	10^4	4
3	10^4	9×10^3	10^3	3
4	10^3	9×10^2	10^2	2
5	10^2	9×10^1	10	1
6	10^1	9	1	0
7	1	0.9	0.1	−1

[1] Assume that the initial sample contains 10^6 vegetative microorganisms and that 90% of the organisms are killed during each minute of exposure. The temperature is 121°C.

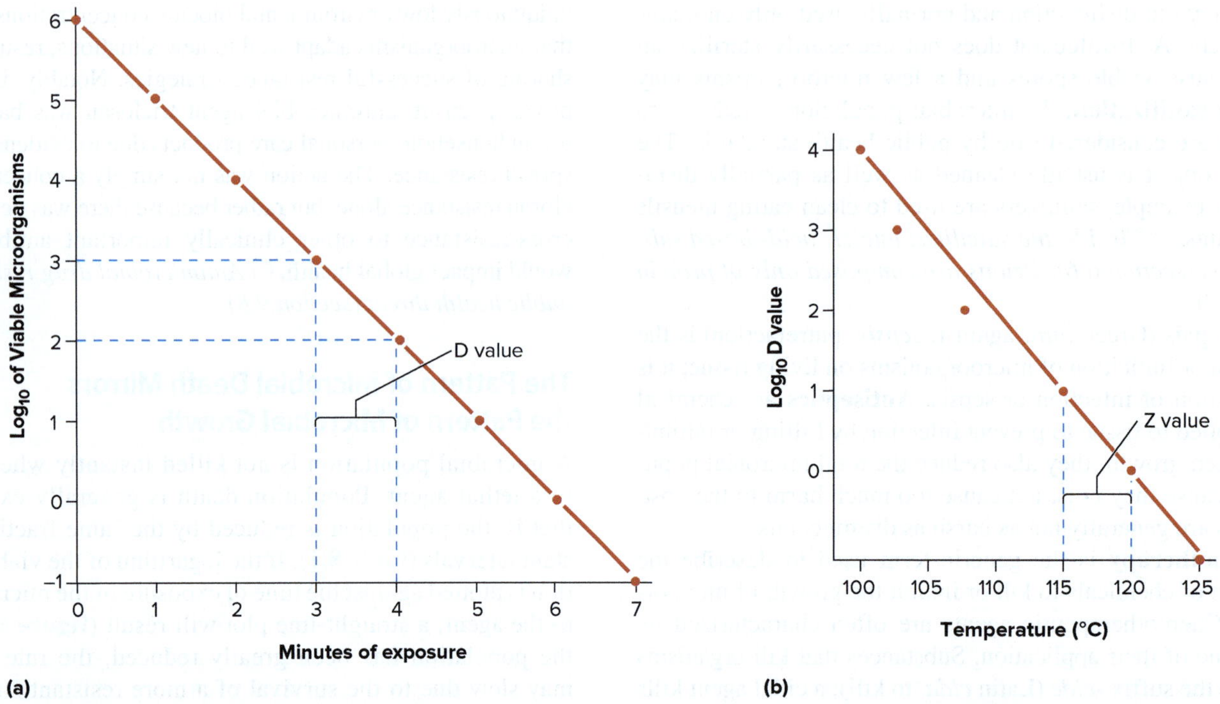

Figure 8.2 The Pattern of Microbial Death. (a) An exponential plot of the survivors versus the minutes of exposure to heating at 121°C. In this example, the D value is 1 minute. The data are from table 8.1. (b) An exponential plot of D values versus temperature. The temperature change at a given D value that reduces the population by one log unit is known as the Z value. In this example, the Z value is 5°C.

MICRO INQUIRY *Examine graph (a). How long would it take to kill one-half of the original population of microorganisms? Keep in mind that the y-axis is exponential.*

To study the effectiveness of a lethal agent, one must be able to decide when microorganisms are dead, which may present some challenges. A microbial cell is often defined as dead if it does not grow when inoculated into culture medium that would normally support its growth. In like manner, an inactive virus cannot infect a suitable host. This definition has flaws, however. It has been demonstrated that when bacteria are exposed to certain conditions, they can remain alive but are temporarily unable to reproduce (*see figure 7.21*). In conventional tests to demonstrate killing by an antimicrobial agent, these **viable but nonculturable** bacteria would be thought to be dead. This is a serious problem because the bacteria may regain their ability to reproduce and cause infection after a period of recovery. ◂ *Most microbes live in growth-arrested states (section 7.6)*

Comprehension Check

1. Define the following terms: *sterilization, sterilant, disinfection, disinfectant, sanitization, antisepsis, antiseptic, chemotherapy, and biocide*.
2. What is the difference between bactericidal and bacteriostatic? To which category do you think most household cleaners belong? Why?
3. Based on what you learned in section 3.8, how would the D value differ between heat treatments to kill vegetative cells vs. endospores for a species of *Bacillus*?

8.2 Microbes Can Be Controlled by Physical Means

After reading this section, you should be able to:

a. Explain the mechanism by which filtration removes microorganisms
b. Differentiate between methods of filter sterilizing liquids and gases
c. Describe the application of heat and radiation to control microorganisms
d. Explain the mechanisms by which heat and radiation control microbes

Physical processes are often used to control viable microbial populations. Filtration imposes a barrier, which isolates the population, while heat or radiation alters the microorganism itself. The processes of desiccation, manipulating pH, and altering osmotic pressure are often used to control microorganisms in food preservation. ▸ *Environmental factors control food spoilage (section 40.2)*

Filtration

Filtration is an excellent way to reduce the microbial population in solutions of heat-sensitive material and can be used to sterilize liquids and gases (including air). Rather than directly destroying contaminating microorganisms, the filter simply acts as a barrier to retain them. There are two types of liquid filtration systems. **Depth filters** consist of fibrous or granular materials that have

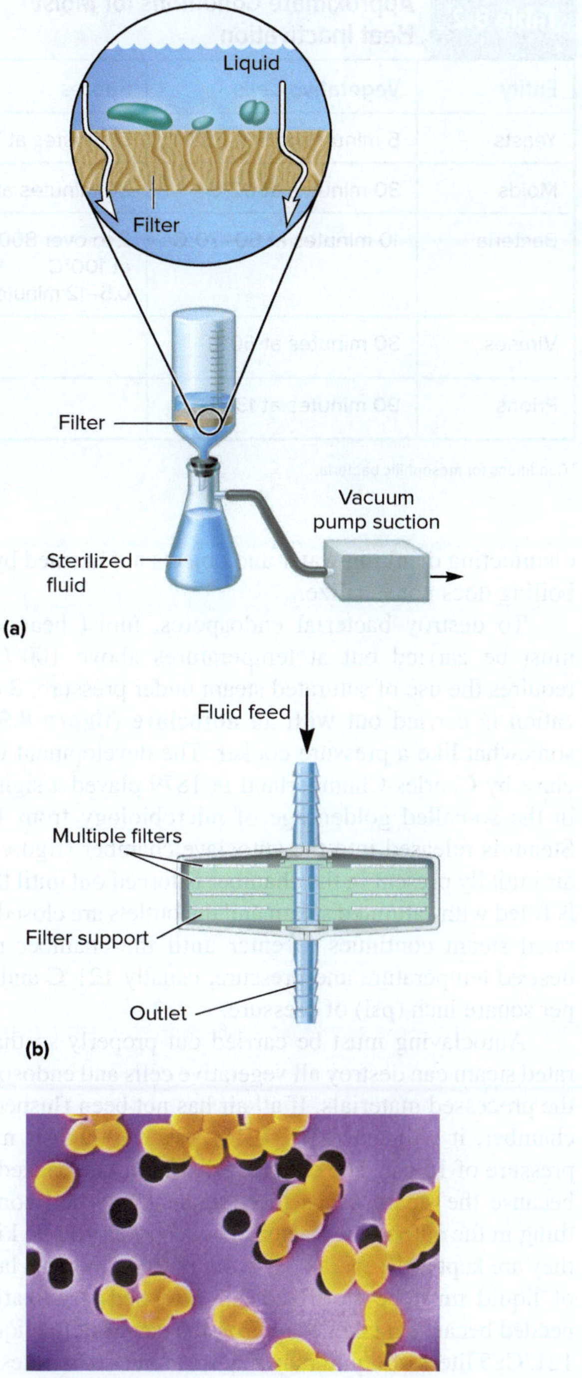

Figure 8.3 Filter Sterilization. (a) Schematic representation of a depth filtration setup that uses a vacuum pump to force liquid through the filter. The inset shows a cross section of the filter and its pores, which are too small for microbes to pass through. (b) Cross section of a membrane filtration unit. Several membranes are used to increase its capacity. (c) *Enterococcus faecalis* resting on a polycarbonate membrane filter with 0.4 µm pores.

(c) Callista Images Cultura/Newscom

MICRO INQUIRY *How might one verify that filtration removed all microorganisms?*

been bonded into a thick layer filled with narrow, twisting channels (**figure 8.3*a***). Depth filters do not have a defined pore size. Microbial cells are removed by entrapment and by adsorption to the filter material.

Membrane filters have replaced depth filters for many purposes. These filters are porous membranes, a little over 0.1 mm thick, made of cellulose acetate, cellulose nitrate, polycarbonate, polyvinylidene fluoride, or other synthetic materials. Membrane filters remove microorganisms by retaining those too large to fit through the pores (figure 8.3*c*). Although a wide variety of pore sizes are available, membranes with pores about 0.2 µm in diameter are used to remove most vegetative cells, but not viruses, from liquids. The membranes are mounted in special holders (figure 8.3*b*), and their use is often preceded by depth filters to remove larger particles that might clog the membrane filter. The liquid is forced through the filter and collected in previously sterilized containers. Membrane filters are used to sterilize pharmaceuticals, ophthalmic solutions, culture media, oils, antibiotics, and other heat-sensitive solutions.

Air also can be filtered to remove microorganisms. Two common examples are N95 disposable masks used in hospitals and labs and **high-efficiency particulate air (HEPA) filters** that let air move freely but restrict microorganisms. N95 masks exclude 95% of particles that are larger than 0.3 µm. HEPA filters (a type of depth filter made from fiberglass) remove 99.97% of particles 0.3 µm or larger by both physical retention and electrostatic interactions. HEPA filters sterilize air by removing viruses that are 0.1 µm and smaller. HEPA-filtered air is critical in clean room environments, such as operating rooms and pharmaceutical manufacturing facilities. In microbiology laboratories, laminar flow biological safety cabinets (commonly called hoods) force air through HEPA filters, then project a vertical curtain of sterile air across the cabinet opening. This protects a worker from microorganisms being handled within the cabinet and prevents contamination of the room (**figure 8.4**). Dangerous agents such as *Mycobacterium tuberculosis,* pathogenic fungi, or tumor viruses must be confined to biological safety cabinets. Additionally, working in a laminar flow biological safety hood maintains the sterility of objects and prevents contamination of cultures.

Comprehension Check

1. What are depth filters and membrane filters, and how are they used to sterilize liquids?
2. Describe the operation of a laminar flow biological safety cabinet/hood.

Heat

Most microorganisms require specific temperatures for normal growth and replication; temperatures that exceed those damage structures and alter chemical reactions (*see figure 7.18*). Moist and dry heat readily destroy viruses, bacteria, and fungi in this way (**table 8.2**).

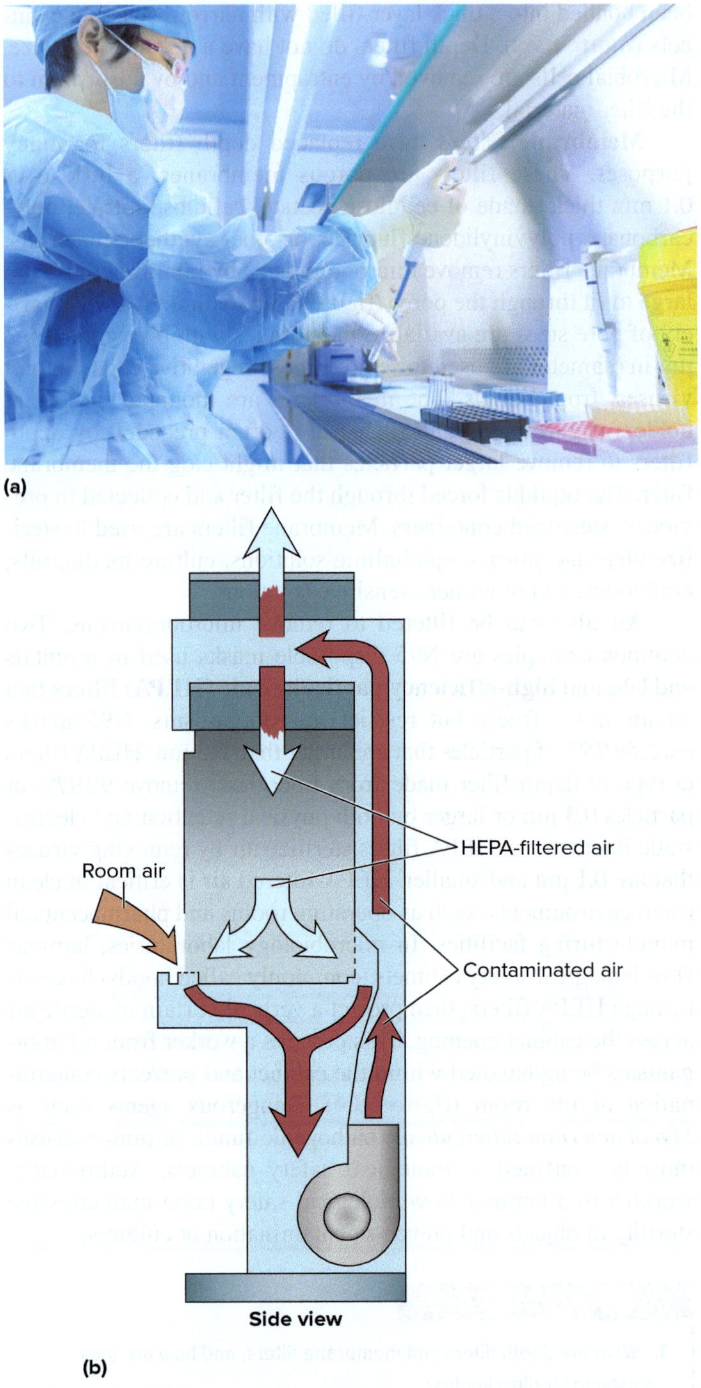

Figure 8.4 A Biological Safety Cabinet. (a) A technician pipetting potentially hazardous material in a safety cabinet. (b) A schematic diagram showing the airflow pattern within a typical biological safety cabinet.

(a) Shutterstock/Eugene Lu

Moist heat destroys cells and viruses by degrading nucleic acids, denaturing proteins, and disrupting cell membranes. Exposure to boiling water for 10 minutes is sufficient to destroy vegetative cells and eukaryotic spores. Unfortunately, the temperature of boiling water (100°C at sea level) is not sufficient to destroy bacterial endospores. Therefore boiling can be used for disinfecting drinking water and objects not harmed by water, but boiling does not sterilize.

Table 8.2 Approximate Conditions for Moist Heat Inactivation

Entity	Vegetative Cells	Spores
Yeasts	5 minutes at 50–60°C	5 minutes at 70–80°C
Molds	30 minutes at 62°C	30 minutes at 80°C
Bacteria[1]	10 minutes at 60–70°C	2 to over 800 minutes at 100°C 0.5–12 minutes at 121°C
Viruses	30 minutes at 60°C	
Prions	90 minutes at 134°C	

[1] Conditions for mesophilic bacteria.

To destroy bacterial endospores, moist heat sterilization must be carried out at temperatures above 100°C, and this requires the use of saturated steam under pressure. Steam sterilization is carried out with an **autoclave** (**figure 8.5**), a device somewhat like a pressure cooker. The development of the autoclave by Charles Chamberland in 1879 played a significant role in the so-called golden age of microbiology from 1850–1915. Steam is released into the autoclave chamber (figure 8.5*b*). The air initially present in the chamber is forced out until the chamber is filled with saturated steam and the outlets are closed. Hot, saturated steam continues to enter until the chamber reaches the desired temperature and pressure, usually 121°C and 15 pounds per square inch (psi) of pressure.

Autoclaving must be carried out properly so that the saturated steam can destroy all vegetative cells and endospores within the processed materials. If all air has not been flushed out of the chamber, it will not reach 121°C, even though it may reach a pressure of 15 psi. The chamber should not be packed too tightly because the steam needs to circulate freely and contact everything in the autoclave. Bacterial endospores will be killed only if they are kept at 121°C for 10 to 12 minutes. When a large volume of liquid must be sterilized, an extended sterilization time is needed because it takes longer for the center of the liquid to reach 121°C; 5 liters of liquid may require about 70 minutes. In view of these potential difficulties, a biological indicator is often included in the autoclave. This indicator commonly consists of a culture tube containing a sterile ampule of medium and a paper strip covered with endospores of *Geobacillus stearothermophilus*. After autoclaving, the ampule is aseptically broken and the culture incubated. If the test bacterium does not grow, the sterilization run has been successful. Sometimes indicator tape or paper that changes color upon heating is autoclaved with a batch of material. These indicate heating has occurred and are convenient but are not as reliable as techniques that directly demonstrate the inactivation of bacterial endospores.

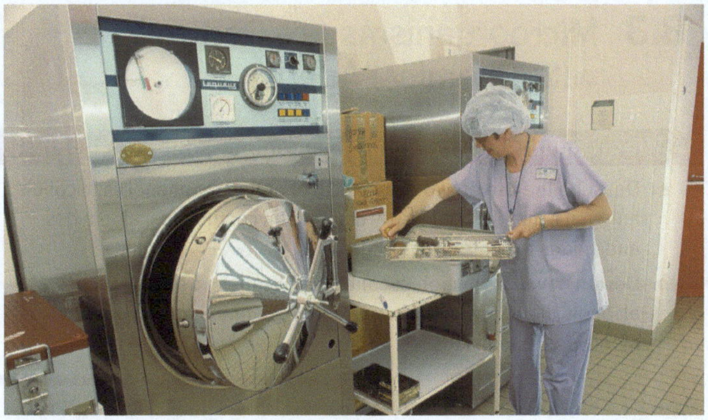

(a)

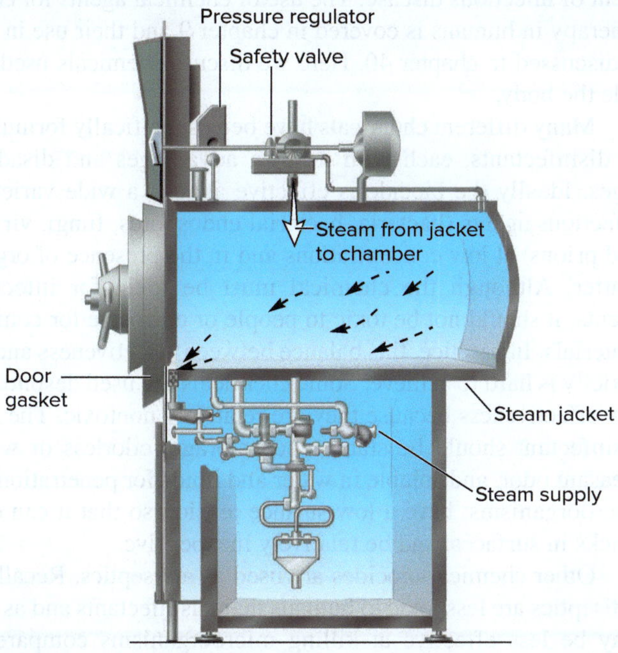

(b)

Figure 8.5 The Autoclave. (a) An automatically controlled autoclave or sterilizer. (b) Longitudinal cross section of a typical autoclave showing some of its parts and the pathway of steam.

(a) Beranger/BSIP SA/Alamy Stock Photo

Many heat-sensitive substances, such as milk, are treated with controlled heating at temperatures well below boiling, a process known as **pasteurization** in honor of its developer, Louis Pasteur (1822–1895). In the 1860s the French wine industry was plagued by wine spoilage, which made wine storage and shipping difficult. Pasteur examined spoiled wine under the microscope and detected microorganisms that looked like the bacteria responsible for lactic acid and acetic acid fermentations; these bacteria do not form endospores. He discovered that a brief heating at 55° to 60°C would destroy these microorganisms and preserve wine for long periods. In 1886 the German chemist F. Soxhlet adapted the technique for preserving milk. Milk pasteurization was introduced in the United States in 1889. Milk, beer, and many other beverages are now pasteurized. Pasteurization does not sterilize, but it does kill pathogens and drastically slows spoilage by reducing the level of nonpathogenic spoilage microorganisms (*see table 40.2*).

Some materials cannot withstand the high temperature of the autoclave, and endospore contamination precludes the use of other methods to sterilize them. For these materials, a process of intermittent sterilization, also known as **tyndallization** (for John Tyndall [1820–1893], the British physicist who used the technique to destroy heat-resistant microorganisms in dust) is used. The process also uses steam (30–60 minutes) to destroy vegetative bacteria. However, steam exposure is repeated for a total of three times with 23- to 24-hour incubations between exposures. The incubations permit remaining endospores to germinate into heat-sensitive vegetative cells that are destroyed upon subsequent steam exposures.

Many objects are best sterilized by dry heat. For instance, inoculating loops, which are used routinely in the laboratory, can be sterilized in a small, bench-top incinerator. Other items are sterilized in an oven at 160° to 170°C for 2 to 3 hours. Microbial death results from the oxidation of cell constituents and protein denaturation. Dry air heat is less effective than moist heat. The endospores of *Clostridium botulinum*, the cause of botulism, are killed in 5 minutes at 121°C by moist heat but only after 2 hours at 160°C by dry heat. However, dry heat has some advantages. It does not corrode glassware and metal instruments as moist heat does, and it can be used to sterilize powders, oils, and similar items. Despite these advantages, dry heat sterilization is slow and not suitable for heat-sensitive materials such as plastic and rubber.

Radiation

Ultraviolet (UV) radiation around 260 nm (*see figure 7.20*) is quite lethal. It causes thymine-thymine dimerization of DNA, preventing replication and transcription (*see figure 16.4*). However, UV radiation does not penetrate glass, dirt films, water, and other substances effectively. Consequently, UV radiation has limited use as a sterilizing agent. UV lamps are sometimes placed on lab ceilings or in biological safety cabinets to sterilize the air and any exposed surfaces. Because UV radiation burns the skin and damages eyes, the UV lamps are off when the areas are in use. Commercial UV units are available for water treatment (**figure 8.6**). Microbes are destroyed when a thin layer of water is passed under the lamps. ▶| *Purification and sanitary analysis ensure safe drinking water (section 42.1)* *UV Irradiation*

Recent studies have shown that **cold plasma,** considered the fourth state of matter, is an effective antimicrobial agent. It is comprised of partially or fully ionized gases where the atoms and/or molecules are stripped of their outer-shell electrons. It emits UV radiation and reactive oxygen and/or nitrogen species, which damage nucleic acids, and oxidizes nucleic acids, proteins, and lipids. It is being considered for use in sterilizing medical devices, inactivating pathogens in agricultural products, and preserving food.

Figure 8.6 Sterilizing with Ultraviolet Radiation. The ultraviolet light system at the Silicon Valley Advanced Water Purification Center.
Sundry Photography/Shutterstock

Ionizing radiation is an excellent sterilizing agent that penetrates deep into objects. Ionizing radiation has sufficient energy to dislodge electrons from atoms or molecules, producing chemically reactive free radicals. The free radicals react with nearby matter to weaken or destroy it. Ionizing radiation destroys bacterial endospores and all microbial cells; however, it is not always effective against viruses.

Gamma radiation (from a cobalt 60 source) and beta radiation (accelerated electrons from high-voltage electricity) are used in the cold sterilization of antibiotics, hormones, sutures, and plastic disposable supplies such as syringes. Gamma radiation and electron beams are also used to treat meat, fruits, vegetables, and spices. Irradiation can eliminate the threat of such pathogens as *E. coli* O157:H7, which causes a life-threatening intestinal disease; *Staphylococcus aureus,* which causes skin and blood infections, and readily colonizes medical devices; and *Campylobacter jejuni,* which contaminates poultry, causing intestinal disease when undercooked meat is eaten. Both the U.S. Food and Drug Administration and the World Health Organization have approved irradiated food and declared it safe for human consumption. ▶| *Human diseases caused by bacteria (chapter 38); Environmental factors control food spoilage (section 40.2)*

Comprehension Check

1. Describe how an autoclave works. What conditions are required for sterilization by moist heat? What three things must one do when operating an autoclave to help ensure success?
2. In the past, spoiled milk was responsible for a significant proportion of infant deaths. Why is untreated milk easily spoiled?
3. List the advantages and disadvantages of ultraviolet light and ionizing radiation as sterilizing agents. Provide a few examples of how each is used for this purpose.

8.3 Microorganisms Are Controlled with Chemical Agents

After reading this section, you should be able to:

a. Describe the use of and mechanism of action for phenolics, alcohols, halogens, metals, quaternary ammonium compounds, aldehydes, and oxides to control microorganisms
b. Design novel antimicrobial control applications using phenolics, alcohols, halogens, metals, quaternary ammonium compounds, aldehydes, and oxides

Chemicals can be employed for sterilization, disinfection, and antisepsis, which encompasses food preservation and the treatment of infectious disease. The use of chemical agents for chemotherapy in humans is covered in chapter 9 and their use in food is discussed in chapter 40. Here we discuss chemicals used outside the body.

Many different chemicals have been specifically formulated as disinfectants, each with its own advantages and disadvantages. Ideally the biocide is effective against a wide variety of infectious agents (bacteria, bacterial endospores, fungi, viruses, and prions) at low concentrations and in the presence of organic matter. Although the chemical must be toxic for infectious agents, it should not be toxic to people or corrosive for common materials. In practice, this balance between effectiveness and low toxicity is hard to achieve. Some chemicals are used despite their low effectiveness because they are relatively nontoxic. The ideal disinfectant should be stable upon storage, odorless or with a pleasant odor, and soluble in water and lipids for penetration into microorganisms; have a low surface tension so that it can enter cracks in surfaces; and be relatively inexpensive.

Other chemical biocides are used as antiseptics. Recall that antiseptics are less toxic to humans than disinfectants and as such may be less effective at killing microorganisms compared to disinfectants. In general, antiseptics should reduce the number of pathogens on human tissue to prevent infection. Examples of antiseptics include hand sanitizers, silver threads woven into clothing, and dilute iodine solutions that can be sprayed onto wounds.

The properties and uses of common disinfectants and antiseptics are surveyed next. Structures of some common agents are shown in **figure 8.7** and many of the characteristics of disinfectants and antiseptics are summarized in **table 8.3.**

Phenolics

Phenol was the first widely used antiseptic and disinfectant. In 1867 Joseph Lister employed it to reduce the risk of infection during surgery. Today phenol and phenol derivatives (phenolics) are used as disinfectants in laboratories and hospitals. The commercial disinfectant Lysol™ is a mixture of phenolics. Phenolics denature proteins and disrupt cell membranes. They have some important advantages as disinfectants: Phenolics are tuberculocidal, are effective in the presence of organic material, and

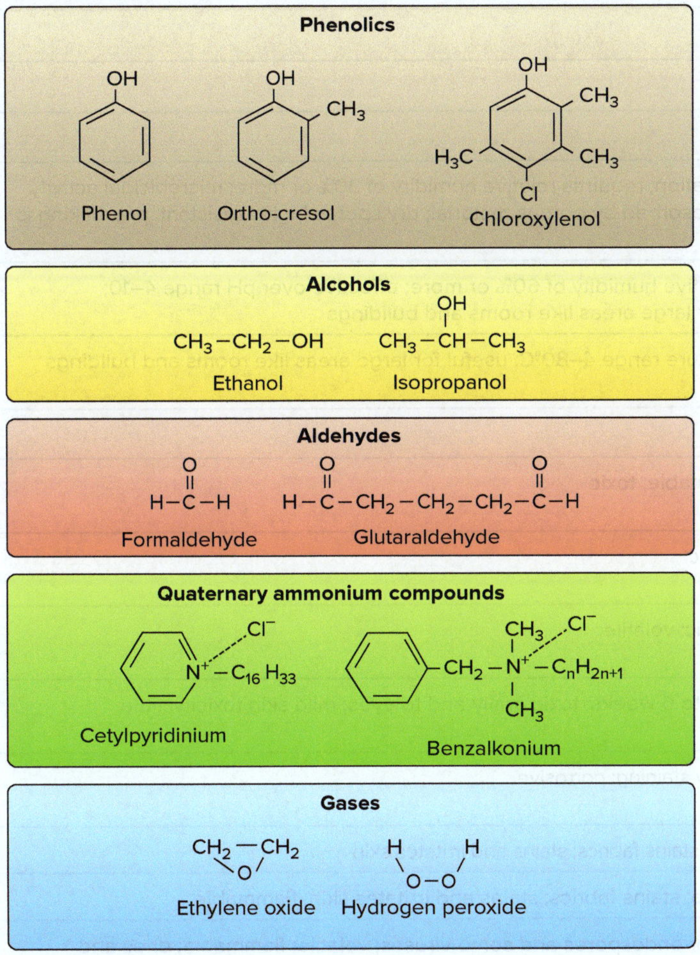

Figure 8.7 Disinfectants and Antiseptics. The chemical structures of some frequently used disinfectants and antiseptics.

MICRO INQUIRY *Why is it important that all of these compounds are relatively hydrophobic?*

remain active on surfaces long after application. Chlorhexidine and chloroxylenol are widely used antiseptics in antimicrobial soaps and preoperative washes.

Alcohols

Alcohols are among the most widely used disinfectants, antiseptics, and sanitizers. They are bactericidal and fungicidal but not sporicidal; some enveloped viruses are also destroyed. The two most popular alcohol germicides are ethanol and isopropanol, usually used in about 60 to 80% concentration. They act by denaturing proteins and permeabilizing membranes. A 10- to 15-minute soaking is sufficient to disinfect small instruments, while rubbing hands with specially formulated alcohol products sanitizes them by killing many pathogens.

Halogens

The halogens iodine and chlorine are important antimicrobial agents. Iodine is used as a skin antiseptic and kills by oxidizing cell constituents and iodinating proteins. At higher concentrations, it may even kill some endospores. Iodine often is applied as tincture of iodine, 2% or more iodine in a water-ethanol solution of potassium iodide. Although it is an effective antiseptic, the skin may be damaged, a stain remains, and iodine allergies can result. Iodine can be complexed with an organic carrier to form an **iodophor.** Iodophors are water soluble, stable, and nonstaining, and release iodine slowly to minimize skin irritation. They are used in hospitals for cleansing preoperative skin and in hospitals and laboratories for disinfecting. Some popular brands are Wescodyne™ for skin and laboratory disinfection and Betadine™ for wounds.

Chlorine is the usual disinfectant for municipal water supplies and swimming pools, and is also employed in the dairy and food industries. It may be applied as chlorine gas (Cl_2), sodium hypochlorite (bleach, NaOCl), or calcium hypochlorite [$Ca(OCl)_2$], all of which yield hypochlorous acid (HOCl):

$$Cl_2 + H_2O \rightarrow HCl + HOCl$$
$$NaOCl + H_2O \rightarrow NaOH + HOCl$$
$$Ca(OCl)_2 + 2H_2O \rightarrow Ca(OH)_2 + 2HOCl$$

The result is oxidation of cellular materials and destruction of vegetative bacteria and fungi. Death of almost all microorganisms usually occurs within 30 minutes. Two important waterborne eukaryotic pathogens, *Cryptosporidium* and *Giardia,* are not killed by chlorine. ▶| *Food and water are vehicles of protozoal diseases (section 39.5); Purification and sanitary analysis ensure safe drinking water (section 42.1)*

Chlorine is also an excellent disinfectant for individual use because it is effective, inexpensive, and easy to employ. Of note is the fact that household bleach (diluted to 10% in water, 10-minute contact time) can be used to disinfect surfaces contaminated by human body fluids and that it is made more effective by the addition of household vinegar.

Metals

For many years metal ions such as those of mercury, silver, arsenic, zinc, and copper were used as germicides. Currently only silver and copper compounds are in use because of toxicity concerns from the others. Silver sulfadiazine is used topically on burns, and silver nanoparticles are incorporated as coatings into products that require slow release over time. Copper sulfate is an effective algicide in lakes and swimming pools. Metals inactivate proteins, often via their sulfhydryl groups.

Quaternary Ammonium Compounds

Quaternary ammonium compounds (QACs) are detergents that have broad-spectrum antimicrobial activity and are effective disinfectants used for decontamination purposes. **Detergents** (Latin *detergere,* to wipe away) are organic cleansing agents that are amphipathic, having both polar hydrophilic and nonpolar hydrophobic components. The hydrophilic portion of a QAC is a positively charged quaternary nitrogen; thus QACs are cationic detergents. They interact with the negatively charged bacterial cell surface and their hydrophobic regions disrupt membranes.

Table 8.3 Activity Level of Selected Biocides

Class	Activity Level[1]	Comment
Gas		
Ethylene oxide	High	Sporicidal; toxic; good penetration; requires relative humidity of 30% or more; microbicidal activity varies with apparatus used; absorbed by porous material; dry spores highly resistant; presoaking is most desirable
Chlorine dioxide (ClO_2)	High	Sporicidal; toxic; requires relative humidity of 60% or more; effective over pH range 4–10; penetrates biofilms; useful for large areas like rooms and buildings
Vaporized hydrogen peroxide (VPH)	High	Toxic; effective over temperature range 4–80°C; useful for large areas like rooms and buildings
Liquid		
Glutaraldehyde, aqueous	High to intermediate	Sporicidal; active solution unstable; toxic
Formaldehyde, aqueous	High to intermediate	Sporicidal; noxious fumes; toxic
Formaldehyde + alcohol	High to intermediate	Sporicidal; noxious fumes; toxic; volatile
Stabilized hydrogen peroxide	High to intermediate	Sporicidal; solution stable up to 6 weeks; toxic orally and to eyes; mild skin toxicity; little inactivation by organic matter
Iodophors	High to intermediate	Somewhat unstable; temporary staining; corrosive
Iodine, aqueous	Intermediate	Rapidly microbicidal; corrosive; stains fabrics; stains and irritates skin
Iodine + alcohol	Intermediate	Rapidly microbicidal; corrosive; stains fabrics; stains and irritates skin; flammable
Alcohols (ethyl, isopropyl)	Intermediate	Rapidly microbicidal except for endospores and some viruses; volatile; flammable; dries and irritates skin
Chlorine compounds	Intermediate	Fast action; inactivation by organic matter; corrosive; irritates skin
Phenolic compounds, aqueous	Intermediate to low	Stable; corrosive; little inactivation by organic matter; irritates skin
Quaternary ammonium compounds	Low	Inactivated by soap and anionics; compounds absorbed by fabrics; old or dilute solution can support growth of Gram-negative bacteria
Chlorhexidine	Low	Soluble in water and alcohol; weakly bactericidal
Hexachlorophene	Low	Insoluble in water, soluble in alcohol; not inactivated by soap; weakly bactericidal

[1] High-level disinfectants destroy vegetative bacterial cells including *M. tuberculosis*, bacterial endospores, fungi, and viruses. Intermediate-level disinfectants destroy all of these except endospores. Low-level agents kill bacterial vegetative cells except for *M. tuberculosis*, fungi, and medium-sized enveloped viruses (but not bacterial endospores or small, naked viruses).

QACs such as benzalkonium chloride and cetylpyridinium chloride kill most bacteria but not *M. tuberculosis* or endospores. They have the advantages of being stable and nontoxic, but they are inactivated by hard water and soap. QACs are often used as disinfectants for food utensils and small instruments, and as antiseptics.

Aldehydes

Both of the commonly used aldehydes, formaldehyde and glutaraldehyde, are highly reactive molecules that inactivate nucleic acids and proteins, by cross-linking and alkylating molecules (**figure 8.8**). They are sporicidal and can be used as chemical sterilants. Formaldehyde is usually dissolved in water or alcohol before use. A 2% buffered solution of glutaraldehyde is an effective disinfectant. It is less irritating than formaldehyde and is used to disinfect hospital and laboratory equipment. Glutaraldehyde usually disinfects objects within about 10 minutes but may require as long as 12 hours to destroy all endospores.

Sterilizing Gases

Many heat-sensitive items such as plastic Petri dishes, heart-lung machine components, sutures, and catheters are sterilized with ethylene oxide (EO) gas. EO is both microbicidal and sporicidal.

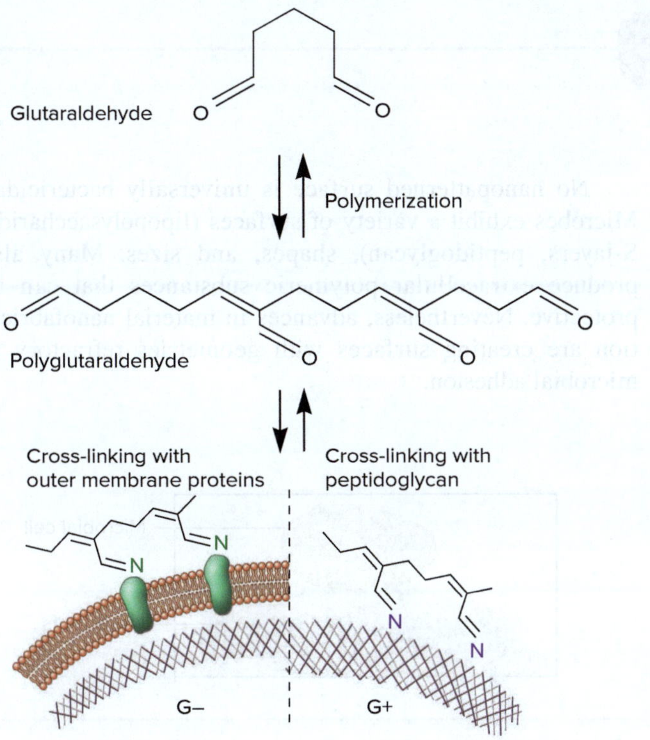

Figure 8.8 Effects of Glutaraldehyde. Glutaraldehyde polymerizes and then interacts with amino groups in outer-membrane proteins of diderm (G–) bacteria (left) or with peptidoglycan in monoderm (G+) bacteria (right). As a result, the molecules are inactivated when they become alkylated and cross-linked to other molecules.

MICRO INQUIRY *Why are cross-linking agents such as glutaraldehyde often called "fixatives" or are said to "fix the cells"?*

It is a strong alkylating agent that kills by reacting with DNA and proteins to block replication and enzymatic activity. It is a particularly effective sterilizing agent because it rapidly penetrates packing materials, even plastic wraps.

Sterilization is carried out in an EO sterilizer, which resembles an autoclave in appearance (**figure 8.9**). Because pure EO is explosive, it is usually supplied in a 10 to 20% concentration mixed with either CO_2 or dichlorodifluoromethane. The EO concentration, humidity, and temperature influence the rate of sterilization. Because it is so toxic to humans, extensive aeration of the sterilized materials is necessary to remove residual EO.

Chlorine dioxide (ClO_2) gas is also used as a disinfectant. Of note is the fact that the chemistry of chlorine dioxide is very different from that of chlorine gas. ClO_2 is typically aerosolized in a humidified environment, 1 mg/liter of air in 60% relative humidity, for at least 4 hours. At this concentration, ClO_2 provides a greater than 6 log reduction of endospores and vegetative bacteria. It has been used to sterilize hospital operating and patient rooms. ClO_2 fumigation is used in the food industry to sanitize fruits and vegetables of contaminating yeasts and molds. ClO_2 has a broad killing spectrum, controlling bacteria, endospores, fungi, and protozoa. ClO_2 appears to have several mechanisms of action: It reacts readily with amino acids cysteine, tryptophan, and tyrosine, denaturing proteins; with free fatty acids, lysing membranes; and with nucleic acids, inhibiting replication. To date, no resistance to ClO_2 has been identified. ClO_2 is highly water soluble, especially in cold water, where it does not hydrolyze but remains as a dissolved gas in solution. Thus a number of municipal water treatment facilities use it for water disinfection.

Vaporized hydrogen peroxide (VHP) can also be used to decontaminate biological safety cabinets, operating rooms, and other large facilities. VHP is produced from a solution of hydrogen peroxide in water, which is then introduced as a sterilizing vapor into the enclosure for some time, depending on the size of the enclosure and the materials within. Hydrogen peroxide and its oxy-radical by-products are toxic and kill a wide variety of microorganisms including *Clostridioides difficile* endospores. During the course of the decontamination process, VHP breaks down to water and oxygen.

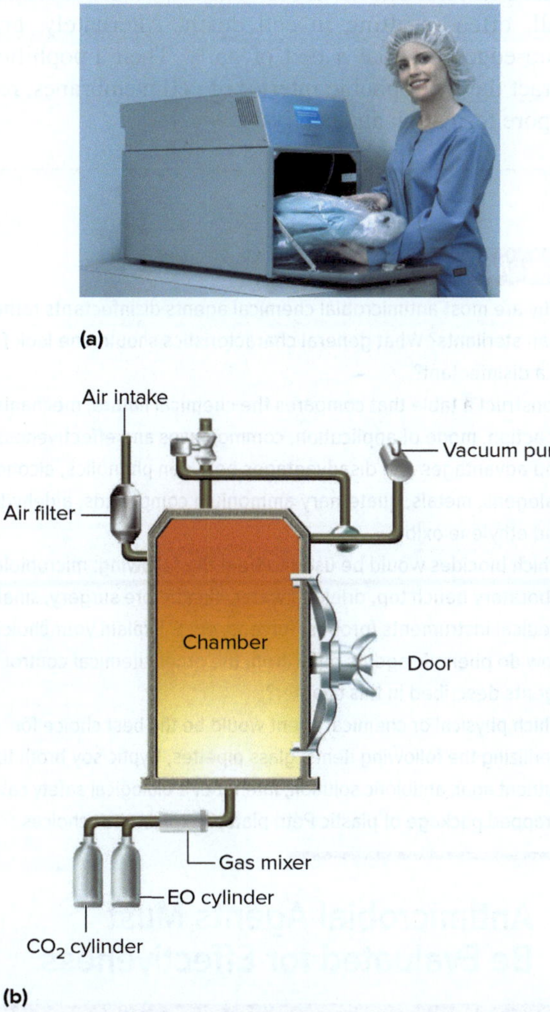

Figure 8.9 An Ethylene Oxide Sterilizer. (a) An automatic ethylene oxide (EO) sterilizer. (b) Diagram of an EO sterilizer. Items to be sterilized are placed in the chamber, and EO and carbon dioxide are introduced. After the sterilization procedure is complete, the EO and carbon dioxide are pumped out of the chamber and air enters.

(a) Anderson Products, www.anpro.com

TECHNIQUES & APPLICATIONS

8.1 Nanosurfaces to Repel Microbes

What if instead of designing biocides to eliminate microbes in unwanted places, we focus on changing the objects in question so that microbes could not colonize them in the first place? Materials scientists and microbiologists are collaborating to develop surfaces to coat catheters and orthopedic and dental implants to prevent the initial step in biofilm formation, microbial attachment.

Bacteria prefer to settle into spaces larger than themselves to maximize surface contact, but nanopatterned surfaces (with spaces measured in nanometers) force microbes to sprawl across a series of nanopillars (**box figure**). Densely spaced features stretch the cell and impose tension on the cell wall, often resulting in cell death. Alternately, graphene nano-edges act like a bed of nails. Their lipophilic edges attract the hydrophobic interior of cell membranes, resulting in pore formation and cell lysis.

No nanopatterned surface is universally bactericidal. Microbes exhibit a variety of surfaces (lipopolysaccharide, S-layers, peptidoglycan), shapes, and sizes. Many also produce extracellular polymeric substances that can be protective. Nevertheless, advances in material nanofabrication are creating surfaces with geometries refractory to microbial adhesion.

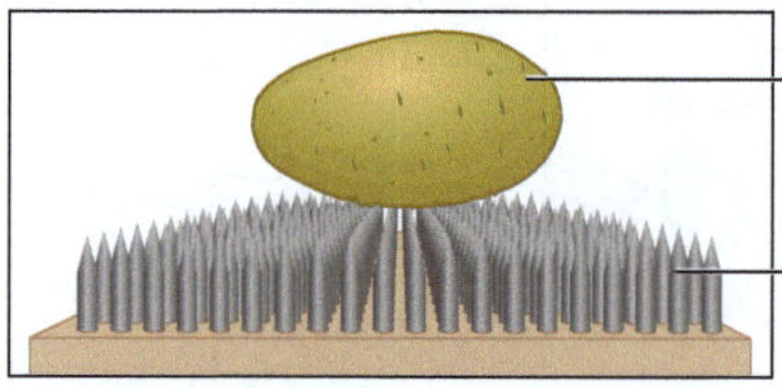

Comprehension Check

1. Why are most antimicrobial chemical agents disinfectants rather than sterilants? What general characteristics should one look for in a disinfectant?
2. Construct a table that compares the chemical nature, mechanism of action, mode of application, common uses and effectiveness, and advantages and disadvantages between phenolics, alcohols, halogens, metals, quaternary ammonium compounds, aldehydes, and ethylene oxide.
3. Which biocides would be used to treat the following: microbiology laboratory bench top, drinking water, skin before surgery, small medical instruments (probes, forceps, etc.)? Explain your choices.
4. How do phenolic agents differ from the other chemical control agents described in this chapter?
5. Which physical or chemical agent would be the best choice for sterilizing the following items: glass pipettes, tryptic soy broth tubes, nutrient agar, antibiotic solution, interior of a biological safety cabinet, wrapped package of plastic Petri plates? Explain your choices.

8.4 Antimicrobial Agents Must Be Evaluated for Effectiveness

After reading this section, you should be able to:

a. Predict the effects of microbial population size and composition, temperature, exposure time, and local environmental conditions on antimicrobial agent effectiveness
b. Describe the processes used to measure microbial killing rates, dilution testing, and in-use testing of antimicrobial agents

The assessment of antimicrobial agent effectiveness is a complex process regulated by two different federal agencies. The Environmental Protection Agency regulates disinfectants, whereas agents used on humans and animals are under the control of the Food and Drug Administration. They establish the guidelines under which these agents are used and agent effectiveness is measured. Importantly, there are a number of variables that must also be considered when evaluating antimicrobial agent effectiveness.

Destruction of microorganisms and inhibition of microbial growth are not simple matters because the effectiveness of an antimicrobial agent is affected by at least six factors.

1. **Population size.** Because an equal fraction of a microbial population is killed during each interval, a larger population requires a longer time to die than does a smaller one (table 8.1 and figure 8.3).
2. **Population composition.** The effectiveness of an agent varies greatly with the nature of the organisms being treated because microorganisms differ markedly in susceptibility. Bacterial endospores are much more resistant to most antimicrobial agents than vegetative forms, and younger cells are usually more readily destroyed than mature organisms. Some species are able to withstand adverse conditions better than others. For instance, *M. tuberculosis,* which causes tuberculosis, is much more resistant to antimicrobial agents than most other bacteria.
3. **Concentration or intensity of an antimicrobial agent.** Often, but not always, the more concentrated a chemical agent or intense a physical agent, the more rapidly microorganisms are destroyed. However, agent effectiveness is not always related to concentration or intensity. Over a short range, a

small increase in concentration leads to an exponential rise in effectiveness; beyond a certain point, increases may not raise the killing rate much at all. Sometimes an agent is more effective at lower concentrations. For example, 70% ethanol is more bactericidal than 95% ethanol because the activity of ethanol is enhanced by the presence of water.

4. **Contact time.** The longer a population is exposed to a microbicidal agent, the more organisms are killed (figure 8.2). To achieve sterilization, contact time should be long enough to reduce the probability of survival by at least 6 logs.
5. **Temperature.** An increase in the temperature at which a chemical is used often enhances its activity. Frequently a lower concentration of disinfectant or sterilizing agent can be used at a higher temperature.
6. **Local environment.** The population to be controlled is not isolated but surrounded by environmental factors that may either offer protection or aid in its destruction. For example, because heat kills more readily at an acidic pH, acidic foods and beverages such as fruits and tomatoes are easier to pasteurize than more alkaline foods such as milk. A second important environmental factor is organic matter, which can protect microorganisms against physical and chemical disinfecting agents. Biofilms are a good example. The organic matter in a biofilm protects the biofilm's microbial residents. Furthermore, bacteria in biofilms are altered physiologically, and this makes them less susceptible to many antimicrobial agents. Because of the impact of organic matter, it may be necessary to clean objects, especially medical and dental equipment, before they are disinfected or sterilized. ◀ *Biofilms are common in nature (section 7.6)*

The actual testing of antimicrobial agents often begins with an initial screening to see if they are effective and at what concentrations. This may be followed by more realistic in-use testing (**figure 8.10**). The best-known disinfectant screening test is the **phenol coefficient test** in which the potency of a disinfectant is compared with that of phenol. A series of dilutions of phenol and the disinfectant being tested are prepared. Standard amounts of *Salmonella enterica* serovar Typhi and *Staphylococcus aureus* are added to each dilution; the dilutions are then placed in a 20° or 37°C water bath. At 5-minute intervals, samples are withdrawn from each dilution and used to inoculate growth medium, which is incubated and examined for growth. Growth indicates that the dilution at that particular time of sampling did not kill the bacteria. The highest dilution (i.e., the lowest concentration) that kills the bacteria after a 10-minute exposure but not after 5 minutes is used to calculate the phenol coefficient. The higher the phenol coefficient value, the more effective the disinfectant under these test conditions. A value greater than 1 means that the disinfectant is more effective than phenol.

The phenol coefficient test is a useful initial screening procedure, but the phenol coefficient can be misleading if taken as a direct indication of disinfectant potency during actual use. This is because the phenol coefficient is determined under carefully controlled conditions with pure bacterial cultures, whereas disinfectants are normally used on complex populations in the presence of organic matter and with significant variations in environmental factors such as pH, temperature, and presence of salts.

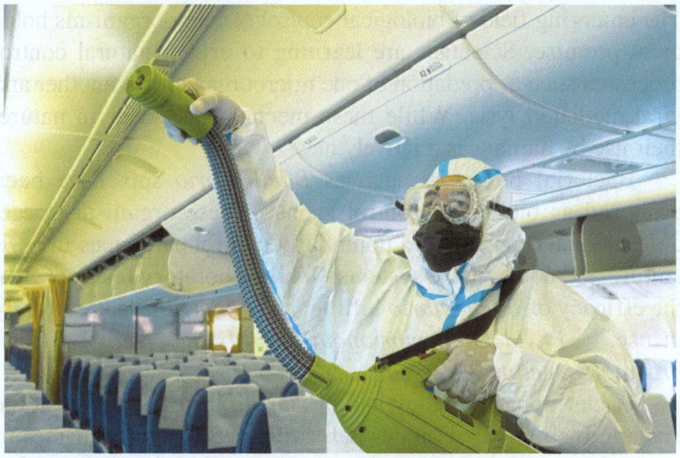

Figure 8.10 Public Disinfection. Effective biocides to combat the spread of COVID-19 were tested for safe use in public places.
Pradpriew/Shutterstock

Disinfectants are also evaluated through **in-use testing,** a process that validates the effectiveness under real-world conditions. Materials used in clinical treatments or food-processing facilities are disinfected as normal, then assessed for the presence of living organisms. Recovery of viable organisms is an indicator that the disinfection procedure is ineffective and must be reevaluated.

Comprehension Check

1. Briefly explain how the effectiveness of antimicrobial agents varies with population size, population composition, concentration or intensity of the agent, contact time, temperature, and local environmental conditions.
2. How does being in a biofilm affect an organism's susceptibility to antimicrobial agents?
3. Suppose hospital custodians have been assigned the task of cleaning all showerheads in patient rooms to prevent the spread of infectious disease. What two factors would have the greatest impact on the effectiveness of the disinfectant the custodians use? Explain what that impact would be.
4. Briefly describe the phenol coefficient test.
5. Why might it be necessary to employ procedures such as in-use tests?

8.5 Microorganisms Can Be Controlled by Biological Methods

After reading this section, you should be able to:

a. Propose predation, competition, and other methods for biological control of microorganisms
b. Suggest alternative decontamination and medical therapies using viruses of bacteria, fungi, and protozoa

The emerging field of biological control of microorganisms holds great promise. Scientists are learning to exploit natural control processes such as predation of one microorganism on another and viral-mediated lysis. While these mechanisms occur in nature, their use by humans is relatively new.

Bacterial predators such as *Bdellovibrio* spp. have been proposed as biological control agents. These organisms target diderm bacteria and could potentially control pathogens in agriculture and aquaculture systems. While laboratory tests confirm the efficacy of *Bdellovibrio* killing, no applications are currently in use. ▶| *Phyla Bdellovibrionota and Myxococcota: bacterial predators (section 20.9)*

Bacteriophage therapy had its start in the early 1900s at the Pasteur Institute in France. Felix d'Herelle isolated bacteriophage from patients recovering from bacillary dysentery. After numerous tests in vitro, d'Herelle concluded that the bacteriophage participated in the destruction of dysentery-causing bacteria. Bacteriophage therapies were under development when penicillin ushered in the age of antibiotics; they are currently used only in Eastern Europe (*see chapter 6, opening story*). However, control of pathogens using bacteriophage is regaining wide support and appears to be effective in the eradication of a number of bacterial species by lysing the pathogenic host. In fact, the U.S. FDA has now approved the use of a bacteriophage spray to eradicate *Listeria, Salmonella,* and *E. coli* in foods. Yet to be approved are proposals for treating human infectious diseases by similar methods. This seems intuitive, knowing that the virus lyses its specific bacterial host, yet unnerving when one thinks about swallowing, injecting, or applying a virus (albeit a bacteriophage) to the human body. |◀ *Viruses and other acellular infectious agents (chapter 6)* ▶| *Environmental factors control food spoilage (section 40.2)*

The distaste of using bacteriophage as a therapeutic can be overcome by the use of *enzybiotics*. These are proteins purified from bacteriophage that cause host cell lysis. They are endolysins whose substrate is peptidoglycan, and the exposed monoderm cell wall renders these cells especially susceptible. Other potentially useful proteins are bacterial depolymerases that hydrolyze a biofilm matrix and microbial toxins such as bacteriocins. ▶| *Bacteriocins (section 31.3)*

Comprehension Check

1. How would you explain to a patient that a virus can be used to eliminate a bone infection caused by bacteria that do not respond to antibiotics?
2. Propose the use of specific bacterial, viral, or fungal products that might be used to kill other, more virulent bacteria, viruses, or fungi.

Key Concepts

8.1 Microbial Growth and Replication: Targets for Control

- Sterilization is the process by which all living cells, viable spores, viruses, viroids, and prions are either destroyed or removed from an object or habitat. Disinfection is the killing, inhibition, or removal of microorganisms (but not necessarily endospores) that can cause disease.
- The main goal of disinfection and antisepsis is the removal, inhibition, or killing of pathogenic microbes. Both processes also reduce the total number of microbes. Disinfectants are chemicals used to disinfect inanimate objects; antiseptics are used on living tissue.
- Antimicrobial agents that kill organisms often have the suffix *-cide,* whereas agents that prevent growth and reproduction have the suffix *-static.*
- Microbial death is usually exponential (**figure 8.2**).
- The decimal reduction time measures an agent's killing efficiency. It represents the time needed to kill 90% of the microbes under specified conditions.

8.2 Microbes Can Be Controlled by Physical Means

- Microorganisms can be efficiently removed by filtration with either depth filters or membrane filters (**figure 8.3**).
- Biological safety cabinets with high-efficiency particulate filters sterilize air by filtration (**figure 8.4**).
- Moist heat kills by degrading nucleic acids, denaturing proteins, and disrupting cell membranes.
- Although treatment with boiling water for 10 minutes kills vegetative forms, an autoclave must be used to destroy endospores by heating at 121°C and 15 pounds of pressure (**figure 8.5**).
- Glassware and other heat-stable items may be sterilized by dry heat at 160° to 170°C for 2 to 3 hours.
- Radiation of short-wavelength or high-energy ultraviolet and ionizing radiation can be used to sterilize objects (**figure 8.6**).

8.3 Microorganisms Are Controlled with Chemical Agents

- Chemical agents usually act as disinfectants or antiseptics because they cannot readily destroy bacterial spores. Disinfectant effectiveness depends on concentration, treatment duration, temperature, and presence of organic material (**table 8.3**).
- Phenolics and alcohols are popular disinfectants and antiseptics that act by denaturing proteins and permeabilizing cell membranes (**figure 8.8**).
- Halogens (iodine and chlorine) kill by oxidizing cellular constituents; cell proteins may also be iodinated. Iodine is applied as a tincture or iodophor. Chlorine may be added to water as a gas, hypochlorite, or organic chlorine derivative.

- Metals tend to be bacteriostatic agents. They are employed in specialized situations such as the use of copper sulfate in lakes and pools.
- Cationic detergents (e.g., quaternary ammonium compounds) are used as disinfectants and antiseptics; they disrupt membranes.
- Aldehydes such as formaldehyde and glutaraldehyde can sterilize as well as disinfect because they kill endospores.
- Ethylene oxide gas penetrates plastic wrapping material and destroys all life forms by reacting with proteins. It is used to sterilize packaged, heat-sensitive materials.
- Chlorine dioxide is aerosolized at low concentrations to kill bacteria, spores, fungi, and protozoa that contaminate living spaces and agriculture. It has multiple killing mechanisms.
- Vaporized hydrogen peroxide is used to decontaminate enclosed spaces (e.g., safety cabinets and small rooms) when the vapor is circulated throughout the space. Peroxide and its oxy-radical by-products are toxic to most microorganisms.

8.4 Antimicrobial Agents Must Be Evaluated for Effectiveness

- The effectiveness of a disinfectant or sterilizing agent is influenced by population size, population composition, concentration or intensity of the agent, exposure duration, temperature, and nature of the local environment.
- The presence of a biofilm can dramatically alter the effectiveness of an antimicrobial agent.
- The phenol coefficient test is frequently used to evaluate the effectiveness of antimicrobial agents. However, it does so using conditions that do not replicate real-life use.
- Other procedures used to determine the effectiveness of disinfectants include measurement of killing rates with germicides and in-use testing.

8.5 Microorganisms Can Be Controlled by Biological Methods

- Control of microorganisms by natural means such as through predation, viral lysis, and toxins is emerging as a promising field.

Active Learning

1. Throughout history, spices have been used as preservatives and to cover up the smell or taste of food that is slightly spoiled. The success of some spices led to ritualized use, and possession of spices was often limited to priests or other powerful members of the community.
   a. Choose a spice and trace its use geographically and historically. What is its common use today?
   b. Spices grow and tend to be used predominantly in warmer climates. Explain.
2. Design an experiment to determine whether an antimicrobial agent is acting as a cidal or static agent. How would you determine whether an agent is suitable for use as an antiseptic rather than as a disinfectant?
3. A scientist is growing mammalian cells in tissue culture using fetal bovine serum as an additive to the growth medium. Discuss the appropriate method of media sterilization to prevent contamination of the cells.
4. Suppose you tested the effectiveness of disinfectants with the phenol coefficient test and obtained the results shown in the "Bacterial Growth After Treatment" table. What disinfectant can you safely say is the most effective? Can you determine its phenol coefficient from these results?

Bacterial Growth After Treatment

Dilution	Disinfectant A	Disinfectant B	Disinfectant C
1/20	–	–	–
1/40	+	–	–
1/80	+	–	+
1/160	+	+	+
1/320	+	–	+

CHAPTER

9

Antimicrobial Chemotherapy

Xinhua News Agency/Getty Images

A Gift from Traditional Chinese Medicine

The 1960s and early 1970s were turbulent. The North Vietnamese, at war with the United States and supported by China, found that drug-resistant *Plasmodium falciparum* was taking an extensive toll on their soldiers and civilians. *P. falciparum* is the most deadly of the five malaria parasites, so it is not surprising that the U.S. found a similar problem in South Vietnam. In response, American drug developers introduced a new drug called mefloquine. The North Vietnamese turned to China for help, but the Chinese research infrastructure had been largely dismantled as part of the Cultural Revolution (1966–1976). Nonetheless, a secret national endeavor called Project 523 commenced, involving over 500 scientists.

One of those scientists, a chemist named Youyou Tu, and her colleagues combed through over 2,000 recipes of traditional Chinese herbal remedies (the figure shows Youyou Tu with her mentor Lou Zhicen in the 1950s). This triggered a massive pharmaceutical bake-off, where they prepared and tested over 500 mixtures in mice infected with a *Plasmodium* species that causes murine malaria. One of the most promising was an extract from the wormwood tree (*Artemisia annu*). To purify this compound, Youyou Tu started with a recipe from *The Handbook of Prescriptions for Emergency Treatments* by Ge Hong (283–343 CE). The instructions roughly translate as "take one bunch of Qinghao, soak in two sheng (~0.4 liters) of water, wring it out to obtain the juice and ingest in its entirety."

After presenting her findings at a meeting of Project 523 investigators in 1972, the active compound, now called artemisinin, was crystallized. And in what amounted to medical diplomacy (or at least détente), Chinese investigators demonstrated the enhanced utility of combining artemisinin with mefloquine. Thus was born artemisinin combined therapy (ACT). Currently, the World Health Organization lists ACT (artemisinin plus another drug from a different class, such as mefloquine) as the treatment of choice for *P. falciparum* infections. In 2015, Youyou Tu was awarded the Nobel Prize for Physiology or Medicine. She titled her Nobel address "Discovery of Artemisinin: A Gift from Traditional Chinese Medicine to the World," and explained "Every scientist dreams of doing something that can help the world."

As this story illustrates, the search for cures to infectious disease is arduous. Nonetheless, very few people living in the industrialized West reach adulthood without having taken an antibiotic. What does the term *antibiotic* really mean? **Antibiotic** (Greek *anti,* against; *bios,* life) refers to natural (not chemically synthesized) products and their derivatives that kill susceptible microorganisms or inhibit their growth. A more inclusive term is **antimicrobial agent,** which includes both natural and chemically synthesized compounds. The rise in antimicrobial resistance among microbes has forced modern medicine to develop new drugs and rethink antimicrobial strategies. This chapter introduces the principles of antimicrobial chemotherapy and resistance.

Readiness Check:

Based on what you have learned previously, you should be able to:

- ✓ Identify the major structures and their functions of bacteria, protists, fungi, and viruses (sections 3.2–3.8, 5.1–5.7, 6.2–6.3, and 7.1–7.7)
- ✓ Compare the various growth and replication processes of viruses and bacteria (sections 6.3, 6.5, 7.1–7.2, and 7.4–7.7)

9.1 Antimicrobial Chemotherapy Evolved from Antisepsis Efforts

After reading this section, you should be able to:

a. Trace the general history of antimicrobial chemotherapy
b. Propose natural sources of new antimicrobial agents

The modern era of chemotherapy began with the work of the German physician Paul Ehrlich (1854–1915). Ehrlich was fascinated with dyes that specifically bind to microbial cells, reasoning that such chemicals might selectively destroy pathogens without harming human cells. Ehrlich and a young Japanese scientist named Sahachiro Hata (1873–1938) developed a treatment for syphilis by testing a variety of arsenic-based chemicals on infected rabbits. They found that arsphenamine was active against the syphilis spirochete. Arsphenamine was made available in 1910 under the trade name Salvarsan.

Ehrlich and Hata's success led to the screening of a vast number of chemicals for similar antimicrobial properties. In 1927 Gerhard Domagk (1895–1964) discovered that Prontosil red, a new dye for staining leather, protected mice completely against pathogenic streptococci and staphylococci without apparent toxicity. Jacques (1897–1977) and Therese (1892–1978) Trefouel later showed that the body metabolized the dye to sulfanilamide. Domagk was awarded the 1939 Nobel Prize in Physiology or Medicine for his discovery, but was prevented from accepting the award until the Nazis were defeated in World War II.

Sulfa drugs are antimicrobial agents, but because they are chemically synthesized, they are not true antibiotics. The first antibiotic was **penicillin,** which was initially discovered in 1896 by a young French medical student named Ernest Duchesne (1874–1912). His work was forgotten until Alexander Fleming (1881–1955) accidentally rediscovered penicillin in 1928. After returning from a vacation, Fleming noticed that a Petri plate of staphylococci with mold growing on it lacked bacterial colonies surrounding the mold (**figure 9.1**). Although the precise events are still unclear, it has been suggested that a *Penicillium chrysogenum* spore had contaminated the Petri dish before it had been inoculated with the staphylococci. Fleming correctly deduced that the mold produced a diffusible substance that inhibited bacterial growth.

Fleming's discovery languished until about a decade later when Howard Florey (1898–1968) and Ernst Chain (1906–1979), professors at Oxford University, obtained the *Penicillium* culture from Fleming and set about purifying the antibiotic. Norman Heatley (1911–2004), a biochemist, was also enlisted to help. When purified penicillin was injected into mice infected with streptococci or staphylococci, almost all the mice survived. This success was reported in 1940, and subsequent human trials were equally successful. Fleming, Florey, and Chain received the Nobel Prize in 1945 for the discovery and production of penicillin.

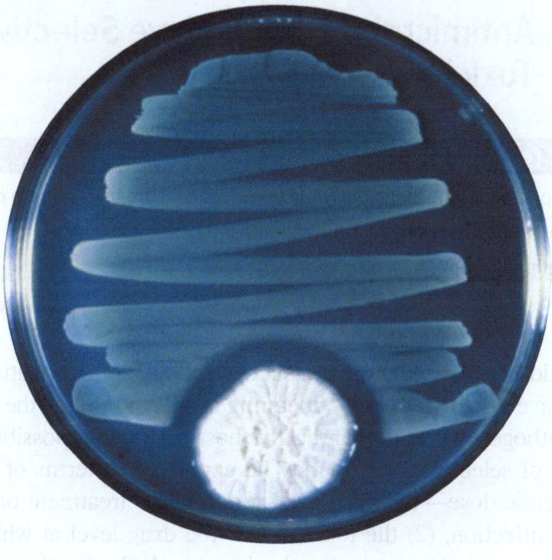

Figure 9.1 Bactericidal Action of Penicillin. The *Penicillium* mold colony secretes penicillin that kills the *Staphylococcus aureus* streaked nearby.
Christine L. Case, Skyline College

The discovery of penicillin stimulated the search for other antibiotics. Selman Waksman (1888–1973), while at Rutgers University, reported in 1944 that he and his associates had found a new antibiotic, streptomycin, produced by *Streptomyces griseus,* after careful screening of about 10,000 strains of soil bacteria and fungi. The importance of streptomycin cannot be overstated, as it was the first drug to successfully treat tuberculosis. Waksman received the Nobel Prize in 1952, and his success led to a worldwide search for other antibiotic-producing soil microorganisms. Neomycin, oxytetracycline, and tetracycline were isolated from other *Streptomyces* spp. by 1953 (*see table 22.1*).

This golden era of antibiotics saw the rapid development of new antimicrobial agents that changed how infectious disease was treated and how medicine was and continues to be practiced. Prior to the advent of antibiotics, elective surgery (e.g., joint replacement and cosmetic surgery) was unthinkable. The use of immune-suppressing chemotherapeutic drugs to treat cancer would likewise be out of the question. Thus many of the tremendous gains in medicine over the last century are due in large part to the dawn of antibiotics and the decline of infectious disease.

Comprehension Check

1. What is the evolutionary benefit to *Penicillium chrysogenum* for producing penicillin?
2. Louis Pasteur is often credited with saying, "Chance favors the prepared mind." How does this apply to Alexander Fleming?

9.2 Antimicrobial Drugs Have Selective Toxicity

After reading this section, you should be able to:

a. Explain the difference between a narrow and broad-spectrum drug
b. Correlate drug action with cidal and static effects
c. Define therapeutic index

It should be self-evident that a successful chemotherapeutic agent must have **selective toxicity,** meaning it kills or inhibits the microbial pathogen while damaging the host as little as possible. The degree of selective toxicity may be expressed in terms of (1) the therapeutic dose—the drug level required for treatment of a particular infection, (2) the toxic dose—the drug level at which the agent becomes too toxic for the host, and (3) the **therapeutic index**—the ratio of the toxic dose to the therapeutic dose. The larger the therapeutic index, the better the chemotherapeutic agent in general.

A drug that disrupts a microbial structure or function not found in host cells often has greater selective toxicity, fewer side effects, and a higher therapeutic index. For example, penicillin inhibits bacterial peptidoglycan synthesis but has little effect on host cells because they lack cell walls; therefore penicillin's therapeutic index is high. A drug may have a low therapeutic index because it inhibits the same process in host cells or damages the host in other ways. This can lead to a diverse range of side effects that may involve almost any organ system. Because side effects can be severe, chemotherapeutic agents must be administered with great care.

Bacteria and fungi naturally produce true antibiotics. In contrast, several important chemotherapeutic agents, such as sulfonamides, trimethoprim, ciprofloxacin, and isoniazid, are synthetic—manufactured by abiotic processes. Some antibiotics are semisynthetic—natural antibiotics that have been structurally modified by the addition of chemical groups to make them less susceptible to stomach acids and inactivation by pathogens (e.g., ampicillin and amoxicillin).

Antimicrobial agents are often classified as either **narrow-spectrum drugs,** meaning that they are effective only against a limited variety of pathogens, or **broad-spectrum drugs** that target many different kinds of bacteria (**figure 9.2**). Drugs may also be classified based on the general microbial group they act against: antibacterial, antifungal, antiprotozoan, and antiviral. A few agents can be used against more than one group; for example, sulfonamides are active against bacteria and some protozoa.

Finally, chemotherapeutic agents can be either **cidal** or **static.** Static agents reversibly inhibit growth; if the agent is removed, the microorganisms will recover and grow again. Although a cidal agent kills the target pathogen, it may be static at low levels. The effect of an agent also varies with the target species; that is, an agent may be cidal for one species and static for another. Because static agents do not directly destroy the

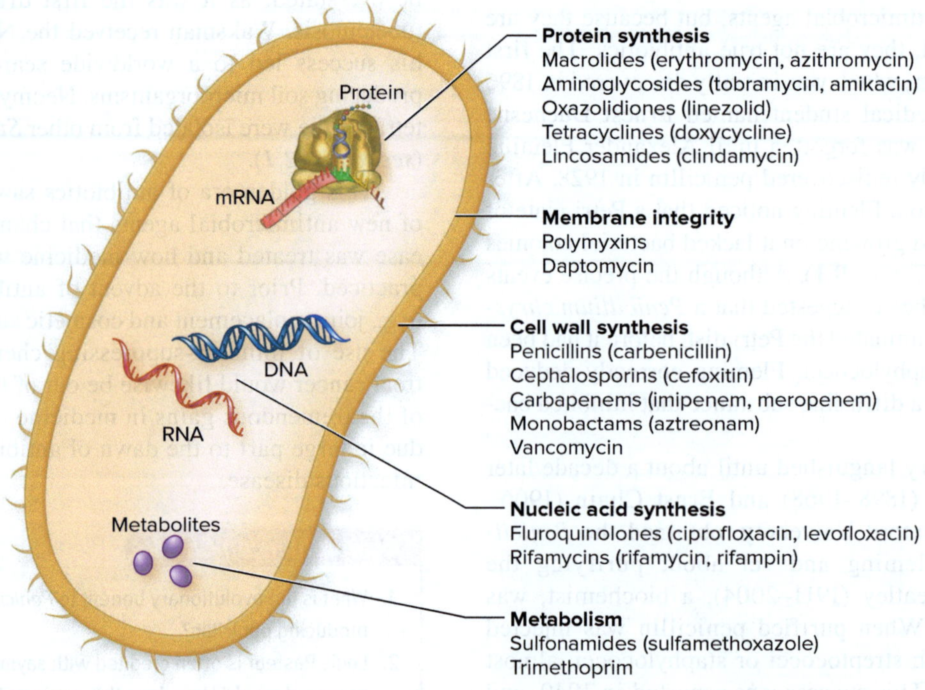

Figure 9.2 Common Targets of Antibiotics. A bacterial cell depicting structures or processes that serve as antibiotic targets.

pathogen, elimination of the infection depends on the host's own immunity mechanisms. A static agent may not be effective if the host is immunosuppressed.

An estimate of the effectiveness of a chemotherapeutic agent against a pathogen can be obtained from the **minimal inhibitory concentration (MIC;** referred to as "mick"). The MIC is the lowest concentration of a drug that prevents growth of a particular pathogen. A cidal drug generally kills pathogens at levels only two to four times more than the MIC, whereas a static agent kills at much higher concentrations, if at all.

Comprehension Check

1. Define the following: *selective toxicity, therapeutic index, side effect, narrow-spectrum drug, broad-spectrum drug, synthetic* and *semisynthetic antibiotics, cidal* and *static agents,* and *minimal inhibitory concentration.*
2. Which drug would have more side effects: a drug with a high or low therapeutic index?
3. How can the MIC concept be applied to distinguish between cidal and static agents?

9.3 Antimicrobial Activity Can Be Measured by Specific Tests

After reading this section, you should be able to:

a. Explain how to determine the level of antibacterial drug activity using the dilution susceptibility test, the disk diffusion test, and the Etest®

Determination of antimicrobial drug effectiveness against specific pathogens is essential for proper therapy. Testing can show which agents are most effective against a pathogen and give an estimate of the proper therapeutic dose.

Dilution Susceptibility Tests

Dilution susceptibility tests can be used to determine MIC values. Antibiotic dilution tests can be done in both agar and broth. Media called Mueller-Hinton broth or agar are used. In the broth dilution test, a series of tubes containing broth with antibiotic concentrations in the range of 0.1 to 128 mg per milliliter (twofold dilutions) is inoculated with a standard density of the test organism. The lowest concentration of the antibiotic resulting in no growth after 16 to 20 hours of incubation is the MIC. The agar dilution test is very similar to the broth dilution test. Plates containing Mueller-Hinton agar and various amounts of antibiotic are inoculated and examined for growth.

A number of automated systems for susceptibility and MIC determination have been commercialized. These products, along with specialized instruments to interpret the tests, miniaturize the broth dilution method to add convenience, decrease time to results, and reduce technical errors. Robotic instruments inoculate the microdilution plates, incubate the cultures, collect data, and interpret the results.

Disk Diffusion Tests

In many situations, such as in analyzing rapidly growing bacteria, a disk diffusion technique is used. This assay is fairly simple. Mueller-Hinton agar is inoculated with the bacterium isolated from the clinical sample (e.g., a urine sample). Small paper disks, each impregnated with a different antibiotic, are placed on the inoculated agar. When the antibiotic diffuses radially outward through the agar, it produces a concentration gradient. The antibiotic is present at high concentrations near the disk and affects even minimally susceptible microorganisms. On the other hand, resistant organisms will grow close to the disk. As the distance from the disk increases, the antibiotic concentration decreases and only more susceptible pathogens are harmed. A clear zone or ring forms around an antibiotic disk after incubation if the agent inhibits bacterial growth (**figure 9.3**). The wider the zone surrounding a disk, the more susceptible the pathogen is. Zone width also is a function of the antibiotic's initial concentration, its solubility, and its diffusion rate through agar. Thus zone width cannot be used to directly compare the effectiveness of different antibiotics.

The **Kirby-Bauer method** is the disk diffusion test most often used. It was developed in the early 1960s by William Kirby, A. W. Bauer, and their colleagues (figure 9.3). Kirby-Bauer test results are interpreted using a table that relates zone diameter to the degree of microbial resistance (**table 9.1**). The values are derived by preparing multiple graphs for drugs tested against specific organisms that represent the MIC as a function of the zone of inhibition diameter measured in millimeters. As the drug diffuses away from the disk into the agar, the concentration diminishes; once it drops below the efficacy level, the organism will grow. Thus, the border of the zone defines the MIC. These data are compiled for combinations of bacteria and drugs, and the resulting table is used as a first step to guide antimicrobial therapy. When a clinical sample is tested, the diameters of inhibition for the pathogen exposed to various drugs are measured and compared to the diameters listed in the table. The designation of resistance or susceptibility to each drug can be established because the data in the table approximately correlate with the expected level of drug in a patient's body when given a therapeutic dose. *Kirby-Bauer*

The Etest®

The Etest® from bioMérieux S.A. is often used in susceptibility testing. In general, each bacterial isolate to be tested for

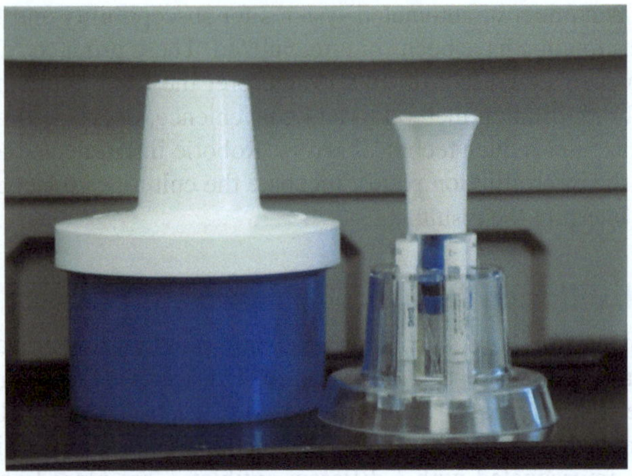
(a)

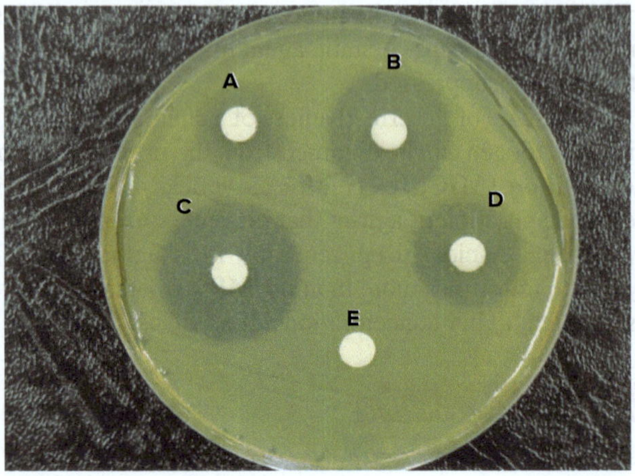

(b)

Figure 9.3 The Kirby-Bauer Method. (a) A multiple antibiotic disk dispenser. (b) Results from a typical disk diffusion experiment. A lawn of bacteria is grown on the surface of an agar plate. Circular zones of growth inhibition appear around the white paper disk.
(a-b) James Redfearn/McGraw Hill

MICRO INQUIRY *To which antibiotic (A, B, C, D, or E) is the plated bacterium fully resistant?*

Table 9.1 Inhibition Zone Diameter of Selected Chemotherapeutic Drugs

		Zone Diameter (millimeter)		
Chemotherapeutic Drug	**Disk Content**	**Resistant**	**Intermediate**	**Susceptible**
Carbenicillin (with *Proteus* spp. and *Escherichia coli*)	100 µg	≤19	20–22	≥23
Carbenicillin (with *Pseudomonas aeruginosa*)	100 µg	≤13	14–16	≥17
Penicillin (with enterococci)	10 U[1]	≤14	—	≥15
Penicillin (with staphylococci)	10 U	≤28	—	≥29
Streptomycin	10 µg	≤11	12–14	≥15
Sulfonamides	250 or 300 µg	≤12	13–16	≥17

[1] One milligram of penicillin G sodium = 1,600 units (U).

antimicrobial sensitivities is inoculated on the surface of an agar medium and then Etest® strips are placed on the surface (**figure 9.4**). Each strip contains a gradient of an antibiotic and is labeled with a scale of MIC values. After 24 to 48 hours of incubation, an elliptical zone of inhibition appears. The MIC is determined from the point of intersection between the inhibition zone and the strip's scale of MIC values.

Comprehension Check

1. Describe the relationship between the diameter of growth inhibition and antibiotic susceptibility in the Kirby-Bauer test.
2. What would you surmise if you examined a Kirby-Bauer assay and found individual bacterial colonies growing within the zone of inhibition?
3. How is the Etest® carried out? When might it be used instead of the Kirby-Bauer test?

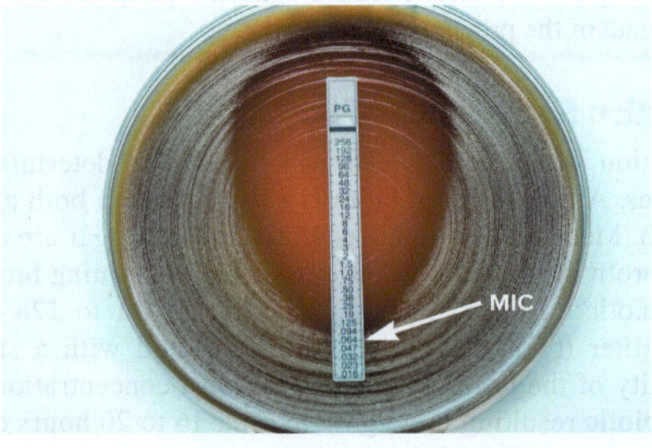

Figure 9.4 Etest®. An example of a bacterial culture plate with an Etest® strip. The MIC value is read at the point where the inhibition ellipse intersects the scale on the strip, as shown by the arrow in this example. Etest® is a registered trademark of bioMérieux S.A. or one of its subsidiaries.
Sirirat/Shutterstock

9.4 Antibacterial Drugs

After reading this section, you should be able to:

a. Compare antibacterial drug mechanisms of action
b. Correlate lack of microbial growth with selective toxicity
c. Explain the relative effectiveness of various antibacterial agents based on drug target

Although there are many different antibiotics, most fall into a limited number of classes based on their molecular structure and mechanism of action (figure 9.2). Part of the challenge in developing new antibiotics is to find bacterial structures or processes not already targeted. These novel targets have the best chance of beating antimicrobial resistance, at least for a while. While antimicrobial drug development remains an active research area, unfortunately most major pharmaceutical companies have pulled back investment in this area. Here we present some of the more common antibiotics according to functional/structural class.

Inhibitors of Cell Wall Synthesis

The most selective antibiotics are those that interfere with bacterial cell wall synthesis. These drugs have a high therapeutic index because they target structures and functions not found in eukaryotic cells. ◀ *Cell walls have many functions (section 3.3)*

Penicillins

Most penicillins (e.g., penicillin G or benzylpenicillin) are derivatives of 6-aminopenicillanic acid and differ from one another with respect to the side chain attached to the amino group (**figure 9.5**). The most crucial feature of the molecule is the **β-lactam ring,** which is essential for bioactivity.

Recall that when new peptidoglycan units are added to a growing cell wall, the peptide side chain on one peptidoglycan subunit must be linked to the peptide side chain on another subunit in a process known as transpeptidation (**figure 9.6*a***; *also see figure 3.17*). The structure of the penicillins resembles the two amino acids at the end of the side chain, called the terminal D-alanyl-D-alanine. This structural similarity blocks the transpeptidase that forms the peptidoglycan cross-links because the enzyme binds penicillin instead of its true substrate (figure 9.6*b*). In fact, these transpeptidases are often called *penicillin-binding proteins (PBPs)*. Because cell wall formation is blocked, osmotic lysis occurs. Penicillins also destroy bacteria by activating their own autolytic enzymes. These enzymes normally break down peptidoglycan at sites of new cell wall synthesis so a cell can enlarge. Thus penicillins act only on growing bacteria actively synthesizing new peptidoglycan. ◀ *Cell walls and osmotic protection (section 3.3)*

Many penicillin-resistant bacteria produce **penicillinase** enzymes that inactivate the antibiotic by hydrolyzing a bond in the β-lactam ring. Penicillinases are therefore members of the **β-lactamase** family of enzymes.

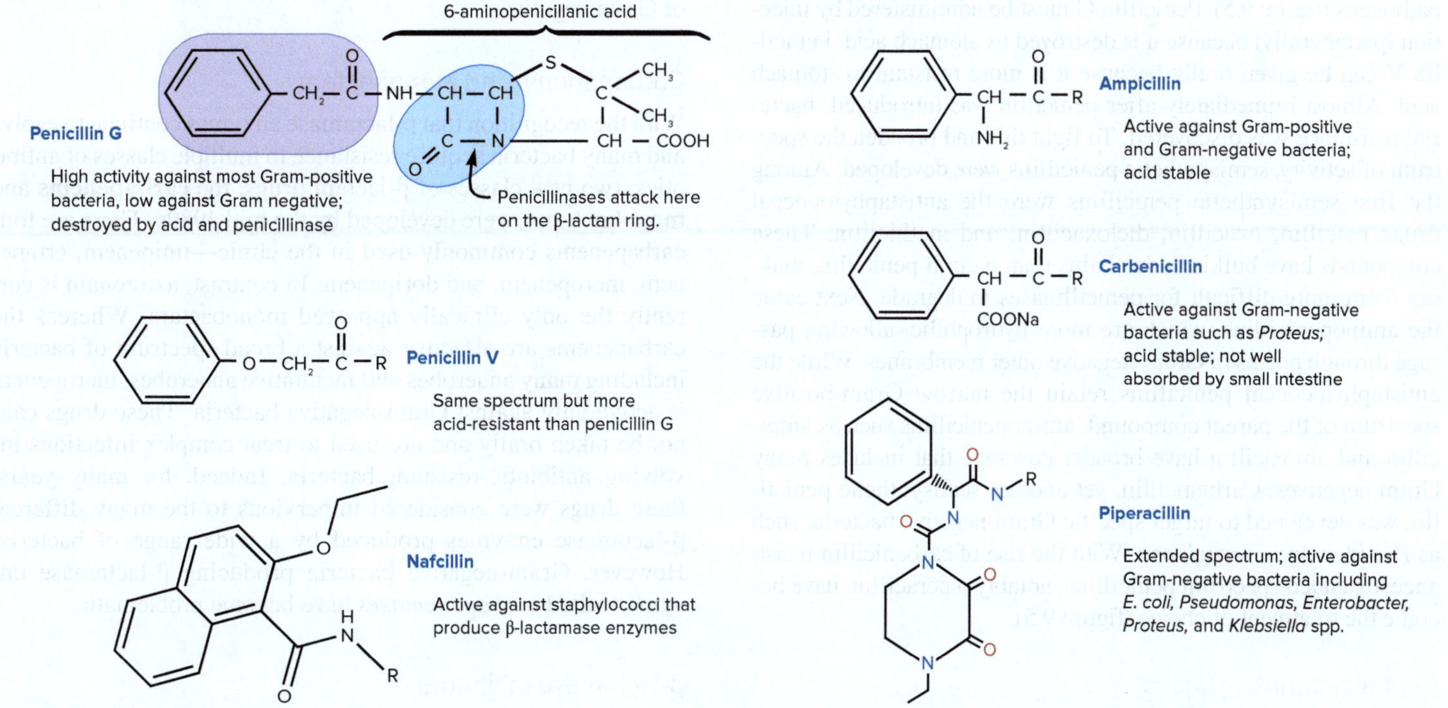

Figure 9.5 Penicillins. The structures and characteristics of representative penicillins. All are derivatives of 6-aminopenicillanic acid (represented by the R group); in each case, the shaded portion of penicillin G (purple) is replaced by the side chain indicated. The β-lactam ring is shaded (blue).
Joanne Willey/McGraw Hill

MICRO INQUIRY *What is the difference between the naturally occurring penicillins (G and V)? How do the remaining (semisynthetic) penicillins differ?*

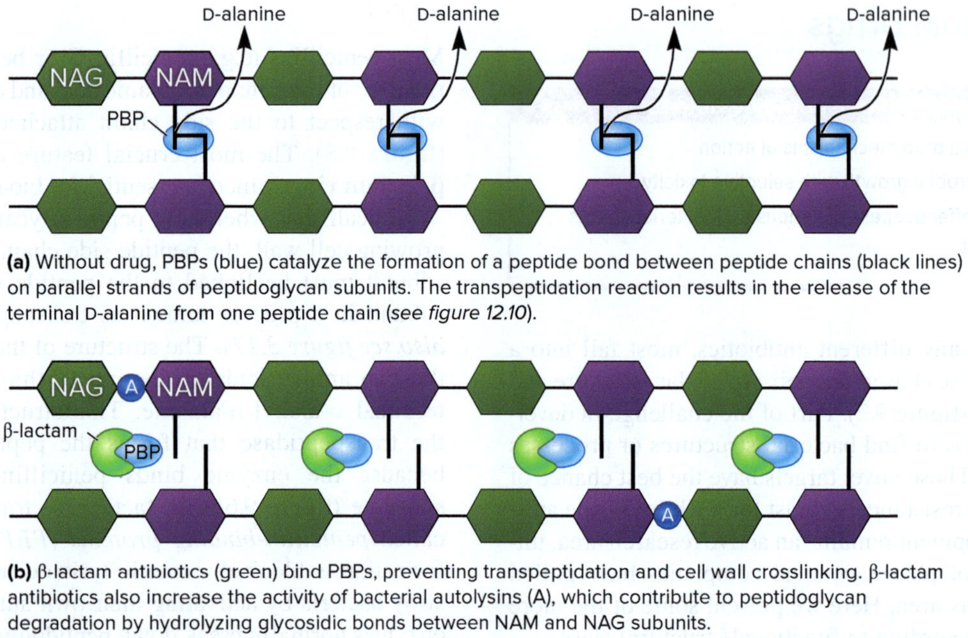

Figure 9.6 β-lactam Antibiotics Block Transpeptidation of Growing Peptidoglycan Chains. *N*-acetylglucosamine (NAG); *N*-acetylmuramic acid (NAM); penicillin-binding protein (PBP)

Penicillins differ from each other in several ways. The two naturally occurring penicillins, penicillin G and penicillin V, are narrow-spectrum drugs, effective against many Gram-positive pathogens (figure 9.5). Penicillin G must be administered by injection (parenterally) because it is destroyed by stomach acid. Penicillin V can be given orally because it is more resistant to stomach acid. Almost immediately after penicillin was introduced, bacterial resistance was discovered. To fight this and broaden the spectrum of activity, semisynthetic penicillins were developed. Among the first semisynthetic penicillins were the antistaphylococcal drugs nafcillin, oxacillin, dicloxacillin, and methicillin. These compounds have bulkier side chains than natural penicillin, making them more difficult for penicillinases to degrade. Next came the aminopenicillins, which are more hydrophilic, allowing passage through porins in Gram-negative outer membranes. While the antistaphylococcal penicillins retain the narrow Gram-positive spectrum of the parent compound, aminopenicillins such as ampicillin and amoxicillin have broader coverage that includes many Gram negatives. Carbenicillin, yet another semisynthetic penicillin, was developed to target specific Gram-negative bacteria, such as *Pseudomonas aeruginosa*. With the rise of carbenicillin resistance, extended spectrum penicillins, notably piperacillin, have become the treatment of choice (figure 9.5).

Cephalosporins

Cephalosporins are a family of antibiotics originally isolated in 1948 from the fungus *Cephalosporium*. They contain a β-lactam structure very similar to that of the penicillins (**figure 9.7**). As might be expected from their structural similarities to penicillins, cephalosporins also inhibit the transpeptidation reaction during peptidoglycan synthesis. The earlier-generation cephalosporins have better activity against a wider range of Gram positives than later generations, which are more effective against a wider range of Gram negatives.

Carbapenems and Monobactam

With the recognition that β-lactamase enzymes continue to evolve and many bacteria acquire resistance to multiple classes of antibiotics, two new classes of β-lactam drugs, the **carbapenems** and **monobactams,** were developed by the mid-1980s. There are four carbapenems commonly used in the clinic—imipenem, ertapenem, meropenem, and doripenem. In contrast, aztreonam is currently the only clinically approved monobactam. Whereas the carbapenems are effective against a broad spectrum of bacteria including many anaerobes and facultative anaerobes, meropenem is active only against Gram-negative bacteria. These drugs cannot be taken orally and are used to treat complex infections involving antibiotic-resistant bacteria. Indeed, for many years, these drugs were considered impervious to the many different β-lactamase enzymes produced by a wide range of bacteria. However, Gram-negative bacteria producing β-lactamase enzymes called *carbapenemases* have become problematic.

β-lactamase Inhibitors

Antibiotic resistance is a major challenge in both the clinic and the research lab. β-lactam antibiotics are among the most safe and effective antibiotics in use. Unfortunately, they are rendered useless when microbes develop resistance, forcing clinicians to use other, less effective antibiotics that may have more toxic side

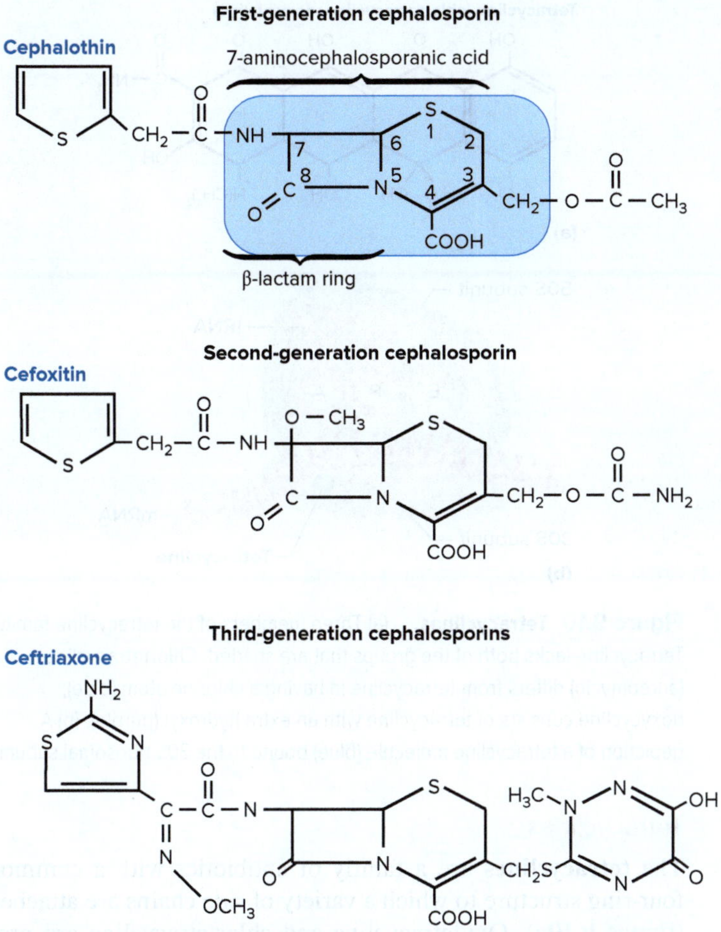

Figure 9.7 Cephalosporin Antibiotics. These drugs are derivatives of 7-aminocephalosporanic acid and contain a β-lactam ring. As newer or next-generation antibiotics are developed, they are numbered first generation, second generation, etc.

MICRO INQUIRY *Do you think cephalosporin antibiotics are susceptible to degradation by β-lactamase enzymes? Explain.*

effects. To give β-lactams new life, researchers have developed a class of molecules that target the principal mechanism of resistance. One prominent example is the β-lactamase inhibitors, of which there are two classes. One class, exemplified by clavulanic acid, inhibits β-lactamases by acting as a competitive inhibitor. This prevents destruction of the β-lactam antibiotic given in combination (e.g., clavulanic acid/amoxicillin). The second class of molecules, like vaborbactam, inhibits more recently evolved β-lactamases (i.e., carbapenemases) by binding noncompetitively to the enzyme's active site. This combination therapy approach gives clinicians a better tool to fight drug-resistant bacterial infections.

Vancomycin

Vancomycin is a glycopeptide antibiotic produced by the bacterium *Amycolatopsis orientalis.* It is a cup-shaped molecule

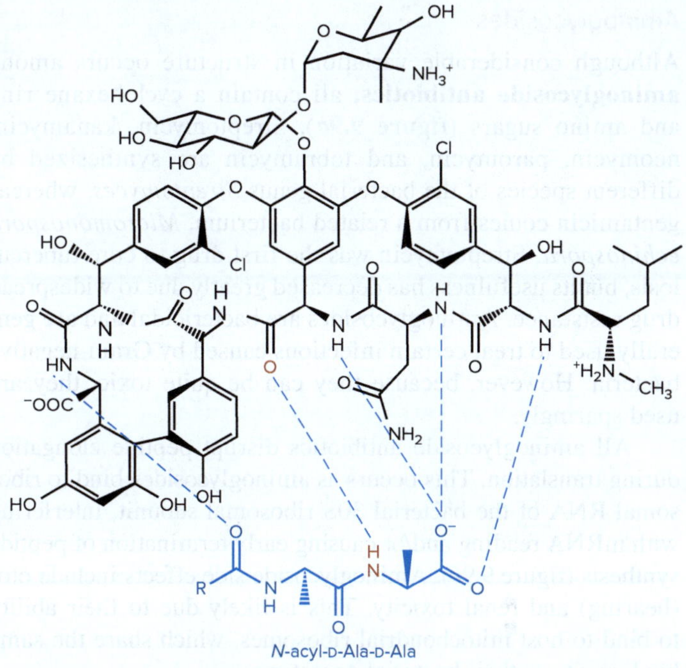

Figure 9.8 Vancomycin. The cup-shaped vancomycin molecule binds to the D-alanyl-D-alanine terminal sequence of peptidoglycan, as shown by the dotted lines.

composed of a peptide linked to a disaccharide (**figure 9.8**). It binds specifically to the D-alanyl-D-alanine terminal sequence on the pentapeptide portion of peptidoglycan. Thus like β-lactam antibiotics, vancomycin blocks transpeptidase activity, but it binds to the enzyme's substrate rather than the enzyme itself.

Vancomycin is bactericidal only for Gram-positive bacteria because it cannot penetrate the Gram-negative outer membrane. Members of the genus *Staphylococcus* and some species in the genera *Bacillus* (food poisoning), *Streptococcus* (strep throat), *Enterococcus* (urinary tract infections), *Clostridium (gangrene),* and *Clostridioides* (debilitating diarrhea) are susceptible. It is usually given intravenously but is taken orally for *Clostridioides difficile* infections. It is an important drug in the treatment of antibiotic-resistant staphylococcal and enterococcal infections. However, vancomycin-resistant strains of *Enterococcus* have emerged, and a few cases of resistant *Staphylococcus aureus* have been reported. Vancomycin resistance poses a serious public health threat since vancomycin is one of the drugs of last resort in cases of antibiotic-resistant *S. aureus.*

Protein Synthesis Inhibitors

Many antibiotics inhibit protein synthesis by binding bacterial ribosomal proteins or rRNA. Because these drugs discriminate between bacterial and eukaryotic ribosomes, their therapeutic index is fairly high, but not as high as that of cell wall synthesis inhibitors. Several different steps in protein synthesis can be affected by drugs in this category. ▶| *Translation in bacteria (section 13.7)*

Aminoglycosides

Although considerable variation in structure occurs among **aminoglycoside antibiotics,** all contain a cyclohexane ring and amino sugars (**figure 9.9*a***). Streptomycin, kanamycin, neomycin, paromycin, and tobramycin are synthesized by different species of the bacterial genus *Streptomyces,* whereas gentamicin comes from a related bacterium, *Micromonospora echinospora.* Streptomycin was the first drug to cure tuberculosis, but its usefulness has decreased greatly due to widespread drug resistance. Aminoglycosides are bactericidal and are generally used to treat certain infections caused by Gram-negative bacteria. However, because they can be quite toxic, they are used sparingly.

All aminoglycoside antibiotics disrupt peptide elongation during translation. This occurs as aminoglycosides bind to ribosomal RNA of the bacterial 30S ribosomal subunit, interfering with mRNA reading and/or causing early termination of peptide synthesis (figure 9.9*b*). Aminoglycoside side effects include oto- (hearing) and renal toxicity. This is likely due to their ability to bind to host mitochondrial ribosomes, which share the same binding site as their bacterial ancestors.

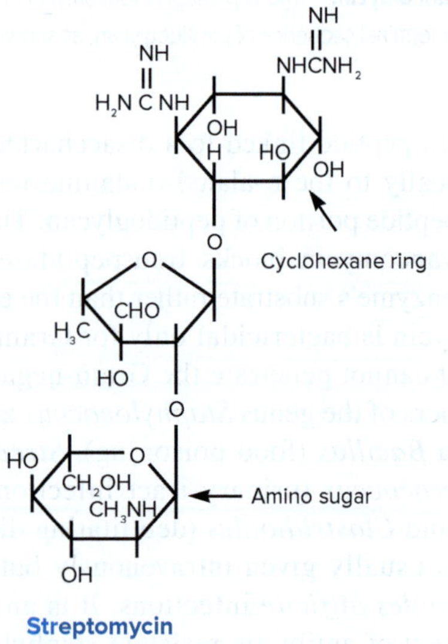

(a)

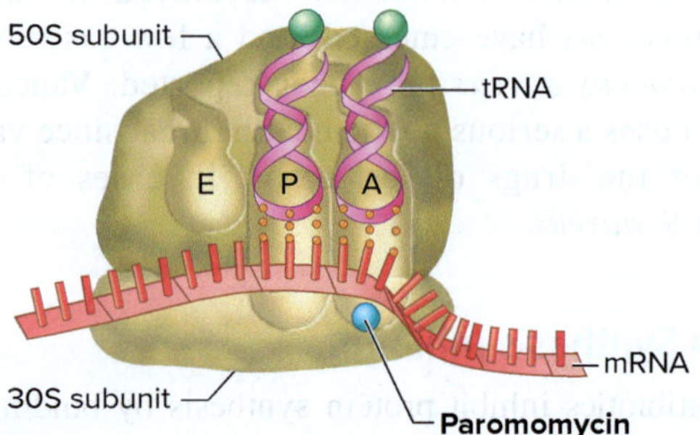

(b)

Figure 9.9 Streptomycin Is an Aminoglycoside Antibiotic. (a) The cyclohexane ring and amino sugar are identified. (b) Aminoglycosides like paromomycin (blue) target the 30S ribosomal subunit.

MICRO INQUIRY *How do aminoglycosides inhibit protein synthesis?*

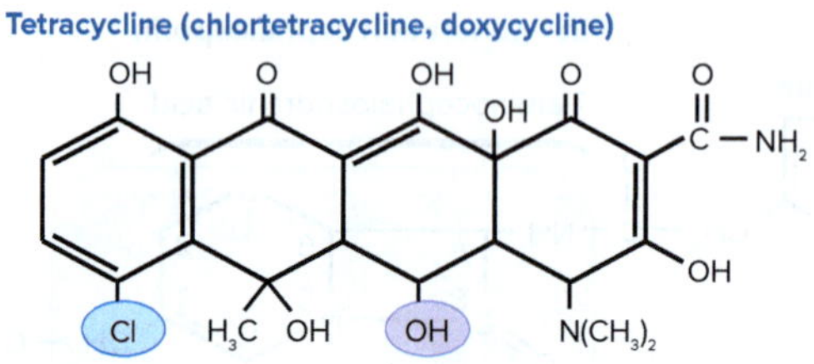

(a)

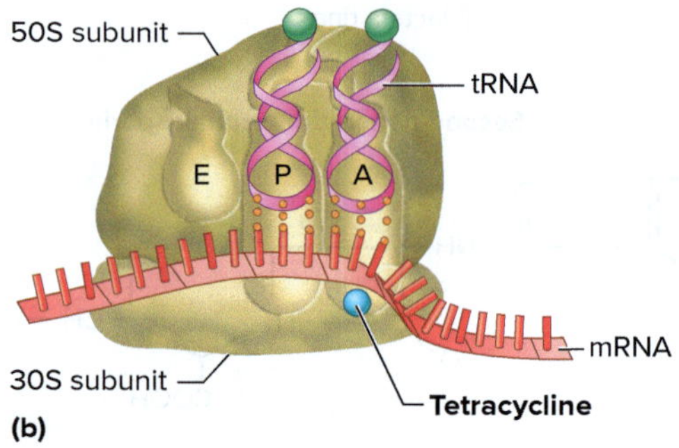

(b)

Figure 9.10 Tetracyclines. (a) Three members of the tetracycline family. Tetracycline lacks both of the groups that are shaded. Chlortetracycline (aureomycin) differs from tetracycline in having a chlorine atom (blue); doxycycline consists of tetracycline with an extra hydroxyl (purple). (b) A depiction of a tetracycline molecule (blue) bound to the 30S ribosomal subunit.

Tetracyclines

The **tetracyclines** are a family of antibiotics with a common four-ring structure to which a variety of side chains are attached (**figure 9.10*a***). Oxytetracycline and chlortetracycline are produced naturally by *Streptomyces* spp., whereas other tetracyclines are semisynthetic. These antibiotics are similar to the aminoglycosides in that they target the 30S subunit of the ribosome, inhibiting protein synthesis (figure 9.10*b*). Tetracyclines are broad-spectrum antibiotics active against most bacteria, including the intracellular pathogens rickettsias, chlamydiae, and mycoplasmas.

Macrolides

The **macrolide antibiotics** contain a ring structure consisting of 12 to 22 carbons called a lactone ring. The lactone ring is linked to one or more sugars (**figure 9.11*a***). Macrolides bind the 50S ribosomal subunit to inhibit bacterial protein elongation (figure 9.11*b*). They are relatively broad-spectrum antibiotics effective against Gram-positive bacteria, mycoplasmas, and some Gram-negative bacteria. They are used with patients who are allergic to penicillins and in the treatment of whooping cough, diphtheria, diarrhea caused by *Campylobacter,* and pneumonia from *Legionella* or *Mycoplasma* infections. Azithromycin has surpassed the older macrolide erythromycin in use largely because it has fewer side effects. Ketolides are semisynthetic macrolide derivatives with a similar mechanism of action. They benefit from better acid stability and oral bioavailability, which is the amount of a drug that reaches the circulation. Telithromycin is a ketolide used to treat community-acquired pneumonia.

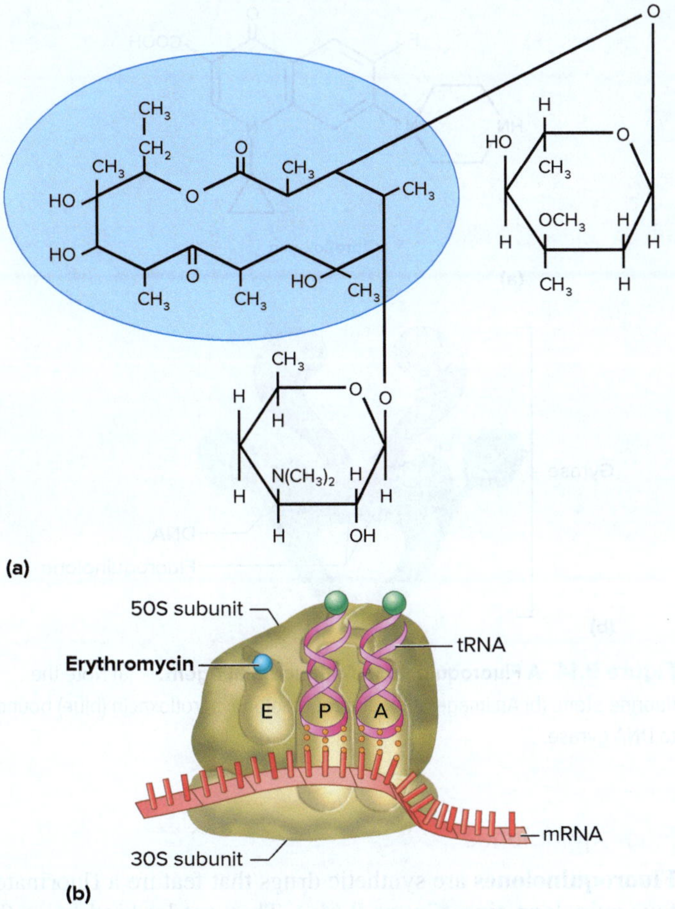

Figure 9.11 Erythromycin, a Macrolide Antibiotic. (a) Erythromycin is composed of a 14-member lactone ring (shaded), so named because it contains a lactone or cyclic ester group that is connected to two sugars. (b) Erythromycin (blue) targets the 50S ribosomal subunit to inhibit translation.

MICRO INQUIRY *How is the mechanism by which macrolides block protein synthesis similar to that of the tetracyclines? How is it different?*

Lincosamides

Lincosamide antibiotics are produced by *Streptomyces* bacteria. They have a broad spectrum of activity against anaerobic bacteria and many Gram-positive cocci are inhibited. Lincosamides are used sparingly, as they can support the overgrowth of *C. difficile,* the cause of a diarrheal infection that can be life-threatening. Clindamycin is used to treat dental infections and some staphylococcal and streptococcal infections.

Oxazolidinones

The **oxazolidinones** are unique synthetic drugs with a heterocyclic five-membered ring structure. The oxazolidinone called linezolid is one of the rare antibiotics to be introduced in the twenty-first century. Oxazolidinones are bacteriostatic, inhibiting protein synthesis by a novel mechanism. They bind to a site on the 23S rRNA within the 50S ribosomal subunit before protein synthesis has begun and prevent the assembly of the 70S initiation complex. Linezolid is active against methicillin susceptible and resistant *Staphylococcus aureus* (MRSA), which is important because MRSA strains are resistant to all β-lactams. It can also be used to treat vancomycin-resistant enterococci, penicillin-resistant *Streptococcus pneumoniae,* and some Gram-positive anaerobes. It can be given by mouth or by injection and has few side effects, but is used primarily in hospitals in an attempt to prevent overuse and the development of resistance.

Metabolic Antagonists

Several valuable drugs act as **antimetabolites,** as they antagonize, or block, the functioning of metabolic pathways. Antimetabolites are structurally similar to substrates of key enzymes and compete with metabolites for the binding site of these enzymes. However, once bound to the enzyme, the antimetabolite drugs are different enough to block enzyme activity and further progression of the pathway. By preventing metabolism, they are broad spectrum but bacteriostatic; their removal reestablishes the metabolic activity. ▶| *Enzymes and ribozymes speed up cellular chemical reactions (section 10.6)*

Sulfonamides or Sulfa Drugs

Sulfonamides, or sulfa drugs, are structural analogues of *p*-aminobenzoic acid, or PABA (**figure 9.12**). PABA is needed for folic acid (folate) synthesis. Folic acid is required for the synthesis of purines. Thus by depleting the pool of folic acid in the cell, sulfonamides prevent synthesis of DNA, RNA, proteins, and other important cell constituents (e.g., ATP). When a sulfa drug enters a bacterial cell, it competes with PABA for the active site of an enzyme (dihydropteroate

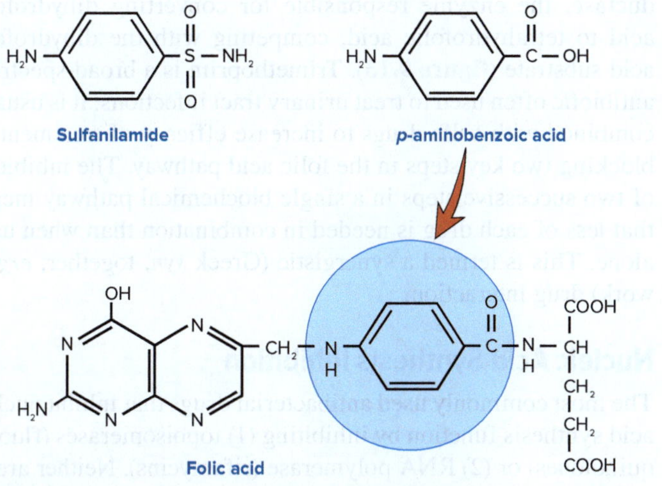

Figure 9.12 Sulfa Drugs. Sulfanilamide competes with *p*-aminobenzoic acid, thereby inhibiting folic acid synthesis.

MICRO INQUIRY *Why do sulfa drugs have a high therapeutic index?*

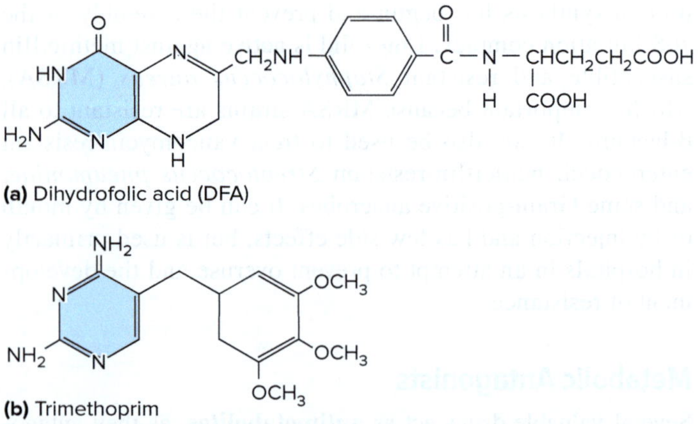

Figure 9.13 Competitive Inhibition of Dihydrofolate Reductase (DHFR) by Trimethoprim. (a) Dihydrofolic acid (DFA) is the natural substrate for the DHFR enzyme of the folic acid pathway. (b) Although the overall structure differs, trimethoprim mimics the structural orientation of DFA and thus competes for the active site of the enzyme. This causes delayed or absent folic acid synthesis because DFA cannot be converted to tetrahydrofolic acid when trimethoprim occupies the DHFR active site.

synthase) involved in one of the first steps in folic acid synthesis *(see figure 10.17)*. Sulfonamides are selectively toxic for many bacteria and protozoa because these microbes manufacture their own folic acid and cannot obtain this cofactor from the environment, whereas humans do not synthesize folate; instead, we must obtain it in our diet. Sulfonamides thus have a high therapeutic index.

Trimethoprim

Trimethoprim is a synthetic antibiotic that also interferes with the production of folic acid but at a later step in the folic acid biosynthetic pathway. Trimethoprim binds to dihydrofolate reductase, the enzyme responsible for converting dihydrofolic acid to tetrahydrofolic acid, competing with the dihydrofolic acid substrate (**figure 9.13**). Trimethoprim is a broad-spectrum antibiotic often used to treat urinary tract infections. It is usually combined with sulfa drugs to increase efficacy of treatment by blocking two key steps in the folic acid pathway. The inhibition of two successive steps in a single biochemical pathway means that less of each drug is needed in combination than when used alone. This is termed a synergistic (Greek *syn,* together; *ergos,* work) drug interaction.

Nucleic Acid Synthesis Inhibition

The most commonly used antibacterial drugs that inhibit nucleic acid synthesis function by inhibiting (1) topoisomerases (fluoroquinolones) or (2) RNA polymerase (rifamycins). Neither are as selectively toxic as other antibiotics because bacteria and eukaryotes do not differ greatly with respect to nucleic acid synthesis. The most commonly used drugs in this category are the fluoroquinolones.

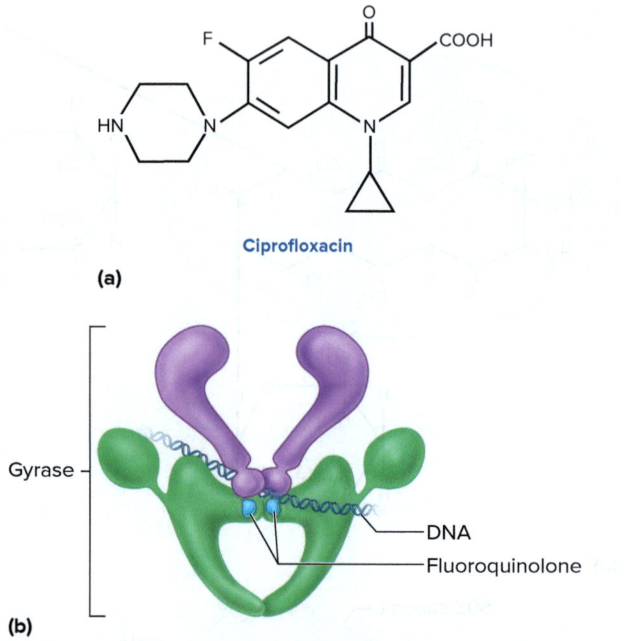

Figure 9.14 A Fluoroquinolone Antimicrobial Agent. (a) Note the fluorine atom. (b) An image of the fluoroquinolone ciprofloxacin (blue) bound to DNA gyrase.

Fluoroquinolones

Fluoroquinolones are synthetic drugs that feature a fluorinated four-quinolone ring (**figure 9.14*a***). They act by binding to the bacterial topoisomerases DNA gyrase and topoisomerase IV (figure 9.14*b*). DNA gyrase introduces negative twist in DNA and helps separate its strands while topoisomerase IV separates linked daughter chromosomes so they can be properly segregated. Inhibition of these enzymes disrupts DNA replication and repair, and bacterial chromosome separation during division. It is therefore not surprising that quinolones are bactericidal. ▶| *DNA replication in bacteria (section 13.3)*

The fluoroquinolones are broad-spectrum antibiotics. They have been used to treat a wide variety of infections. This success has led to their overuse, and fluoroquinolone resistance has become problematic. In addition, because they can cause disabling damage to tendons and nerves, particularly in older adults, the U.S. Food and Drug Administration has deemed them too dangerous to use for all but very severe infections.

Rifamycins

The most commonly used member of the **rifamycin** class is the semisynthetic derivative rifampin. These agents inhibit bacterial transcription by binding to the β-subunit of RNA polymerase. Rifampin is an important member of the multidrug regimen used to treat tuberculosis and other mycobacterial infections. It can also be used to protect those who have come into contact with an individual diagnosed with meningitis caused by *Neisseria meningitidis* or *Haemophilus influenzae.*

Comprehension Check

1. Explain five mechanisms by which chemotherapeutic agents kill or damage bacteria.
2. Why do penicillins and cephalosporins have a higher therapeutic index than most other antibiotics?
3. Why is linezolid not more commonly prescribed?
4. What are antimetabolites? Why are these effective against protozoan pathogens, but the other drugs presented here generally are not?

9.5 Antiviral Drugs

After reading this section, you should be able to:

a. Compare antiviral drug mechanisms of action
b. Provide a rationale for combination drug therapy in the treatment of HIV
c. Explain why there are far fewer antiviral agents than there are antibacterial agents

The discovery of small molecules that inhibit virus-specific enzymes and replication cycle processes has led to the development of antiviral drugs. However, antiviral drugs have had mixed success and, as highlighted by the COVID-19 pandemic, the vast majority of viral infections cannot be cured. Some antiviral drugs simply limit the duration of the illness (e.g., flu) or its severity (e.g., herpes, HIV).

Most antiviral drugs disrupt critical stages in a virus's multiplication cycle. Oseltamivir inhibits the viral enzyme neuraminidase, which is essential for release of newly synthesized influenza A and B virus particles from host cells. However, its widespread use has generated concerns about growing resistance in influenza virus, prompting the Centers for Disease Control and Prevention (CDC) to recommend limiting its use. While it is not a cure, patients who take it within 48 hours of influenza infection are relieved of flu symptoms about a day and a half sooner than patients who do not take oseltamivir. Two other neuraminidase inhibitors, peramivir (for intravenous use) and inhaled zanamivir, can be used to treat influenza infection in adults. In 2018, a novel anti-influenza drug called marboxil baloxavir was approved. It blocks influenza virus transcription by preventing the step known as *cap snatching,* whereby host 5′ caps are added to viral transcripts (*see figure 25.23*). Not surprisingly, resistance to this agent has been reported. ▶| *Viral life cycles have five steps (section 6.3); Influenza (flu) (section 37.1)*

Several drugs are commonly used to treat illnesses caused by herpesviruses, which have DNA genomes (*see table 25.3*). These drugs exploit the fact that the viruses that cause herpes, chickenpox and shingles, and cytomegalovirus use their own (rather than the host's) enzyme to phosphorylate nucleosides to nucleotides. Thus these drugs are nucleoside or mononucleotide analogues that, once phosphorylated, are incorporated into the growing viral genome. Because they are not natural nucleotides, their addition blocks further DNA synthesis, a process called *chain termination.* Examples of these antiviral agents include acyclovir and the ointment vidarabine, which are used to treat herpes infections and shingles. Ganciclovir is used to treat systemic cytomegalovirus illness, and cidofovir, a mononucleotide analog, is used to treat eye infections (retinitis) caused by this virus. In cases of acyclovir or ganciclovir resistance, foscarnet may be used. It is a pyrophosphate analogue that does not require modification by viral enzymes, so it can be used to treat illnesses caused by both herpes simplex viruses and cytomegalovirus. ▶| *Direct contact diseases can be caused by viruses (section 37.3)*

Since the early days of human immunodeficiency virus (HIV) treatment, much effort has been focused on developing new drugs. These drugs target critical steps in the viral replicative processes (**figure 9.15**). There are now six categories of drugs used to manage HIV infection: (1) nucleoside reverse transcriptase inhibitors (NRTIs), which are nucleoside analogues that cause chain termination in much the same way as acyclovir and cidofovir (e.g., azidothymidine or AZT; (2) nonnucleoside reverse transcriptase inhibitors (NNRTIs), which prevent HIV DNA synthesis by selectively binding to and inhibiting the viral reverse transcriptase enzyme; (3) protease inhibitors (PIs) that block the activity of the HIV protease needed for the production of viral proteins, which are translated as a single polypeptide; (4) integrase inhibitors that prevent the incorporation of the HIV genome into the host's chromosomes; (5) fusion inhibitors (FIs), which prevent HIV entry into cells; and (6) capsid inhibitors that prevent nuclear uptake of HIV particles and further capsid assembly. ▶| *Retroviruses: positive-strand viruses that use reverse transcriptase in their life cycles (section 25.7)*

The most successful treatment approach to date in combating HIV/AIDS is to use drug combinations. Most effective is the simultaneous use of three agents to prevent the development of drug resistance. For example, the combination of two NRTIs and an integrase inhibitor reduces HIV concentrations in plasma to undetectable levels. However, no available treatment eliminates proviral HIV DNA residing in certain cells of the immune system (e.g., memory T cells) and possibly other cells. Nonetheless with proper treatment, the virus can disappear from a patient's blood, preventing transmission and the development of drug-resistant strains. But because HIV remains dormant in host cells, it will reactivate if drugs are stopped. Therefore patients cannot be cured and require drug therapy for life, which currently includes an integrase inhibitor/NNRTI combination administered every 6–8 weeks. The CDC recommends a *pre*-HIV *exposure prophylaxis* (prevention) strategy called PrEP to HIV-negative individuals who are at risk of contracting the virus. PrEP involves a daily oral dose of two NRTIs to prevent HIV transmission. In recent years, the newest class of anti-HIV drugs (led by lenacapavir) that target the viral capsid has been introduced. These promising drugs have been shown to drastically reduce the chance of HIV infection, with only two doses needed per year. This is a powerful addition to the global fight against HIV. ▶| *Acquired immunodeficiency syndrome (AIDS) (section 37.3)*

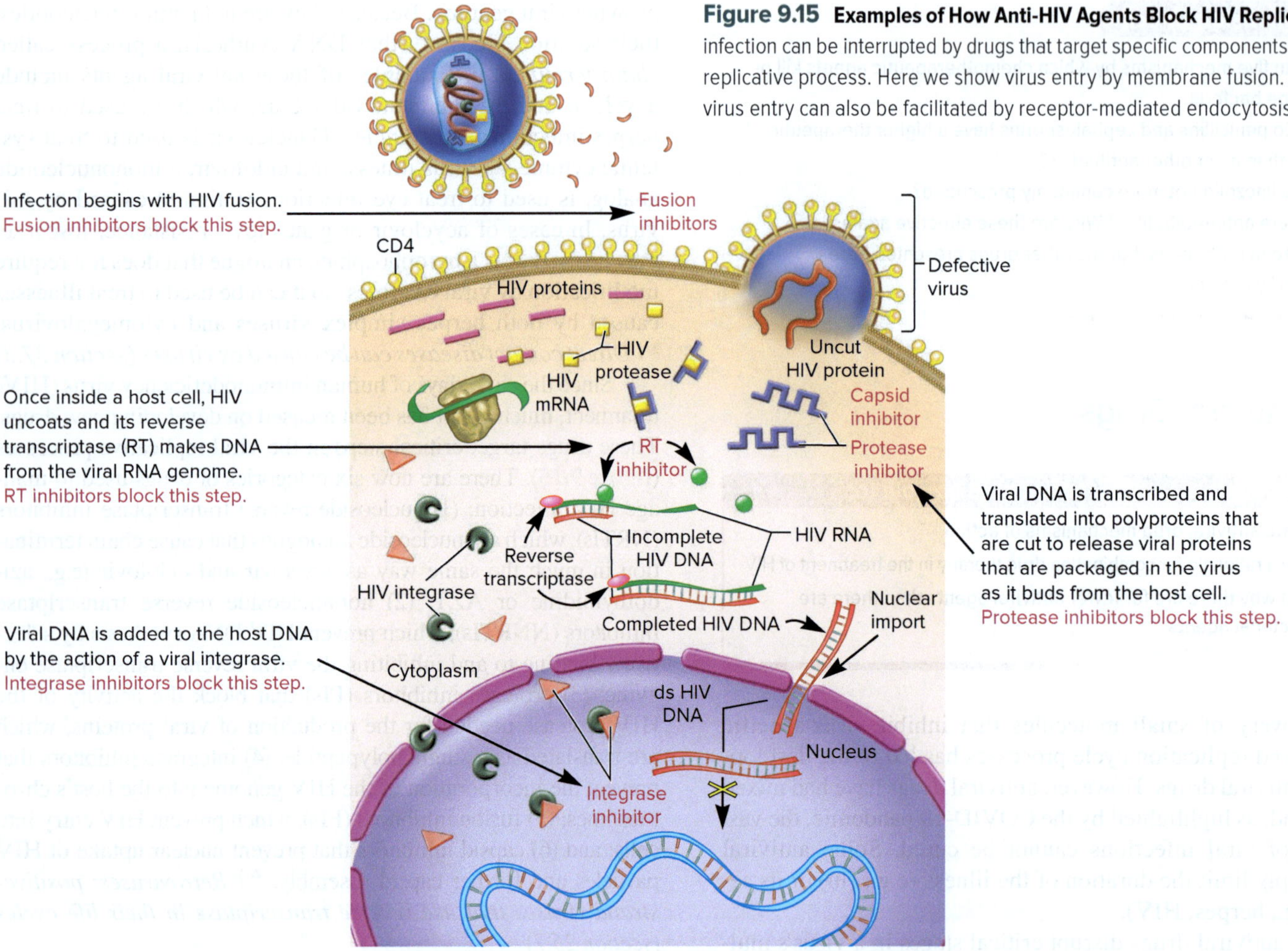

Figure 9.15 Examples of How Anti-HIV Agents Block HIV Replication. Viral infection can be interrupted by drugs that target specific components of the HIV replicative process. Here we show virus entry by membrane fusion. However, virus entry can also be facilitated by receptor-mediated endocytosis.

The ability to cure hepatitis C virus (HCV) infection is an exciting development. The CDC estimates that there are currently at least 2.7 million people with hepatitis C infection in the United States, which is the leading cause of liver transplant. For many years, attempts to limit the damage caused by this RNA virus centered on treatment with the human immune system mediator (cytokine) interferon-α and the antiviral medication ribavirin. This treatment was not particularly effective and patients experienced unpleasant side effects. In 2014 drugs called direct-acting antiviral agents were introduced and changed the prognosis for thousands of patients. These drugs are given in combination. Sofosbuvir is a nucleotide analogue that blocks the activity of the HCV RNA polymerase, and ledipasvir and velpatasvir target a viral enzyme needed for replication. The specific regimen prescribed depends on the genotype of HCV infecting the individual. About 75% of HCV patients in the United States are infected with genotype 1, so are candidates for ledipasvir-sofosbuvir or sofosbuvir-velpatasvir. Almost all others are infected with genotypes 2 and 3 and can be treated with sofosbuvir-velpatasvir. Unfortunately, the high cost of treatment has limited the availability of this (and many other) life-saving treatment.
▶| *Viral hepatitides (section 37.3)*

Comprehension Check

1. Why do drugs used to treat herpes target the virus and spare the host cell?
2. Compare and contrast the activity of a nucleoside reverse transcriptase inhibitor and a nonnucleoside reverse transcriptase inhibitor.
3. Why must individuals with HIV infections take antiretroviral drugs for the rest of their lives?
4. Compare the older and newer drugs used to treat hepatitis C virus. Explain why the newer compounds are called direct-acting antiviral agents.

9.6 Antifungal Drugs

After reading this section, you should be able to:

a. Compare antifungal drug mechanisms of action

b. Explain why there are far fewer antifungal agents than there are antibacterial agents

Treatment of fungal infections generally has been less successful than that of bacterial infections largely because as eukaryotes, fungal cells are much more similar to human cells than are bacterial cells. Many drugs that inhibit or kill fungi are therefore toxic for humans and thus have a low therapeutic index. Most antifungals are fungistatic only as long as repeated application maintains high levels of drug. Nonetheless, a few drugs are useful in treating many major fungal diseases.

There are several major classes of antifungal drugs that can be taken systemically. *Polyenes* and *azoles* interfere with fungal cell membrane biogenesis. This is a good drug target because fungal membranes require the sterol ergosterol, not found in human membranes. Polyenes bind directly to ergosterol, while azoles block the last step in ergosterol biosynthesis. *Echinocandins* inhibit production of cell wall glucans. The *allylamines* (and *benzylamines*) prevent conversion of squalene to lanosterol, an earlier step in ergosterol biosynthesis than targeted by the azoles.

Fungal infections are often subdivided into infections called superficial mycoses, subcutaneous mycoses, systemic mycoses, and opportunistic mycoses. Treatment for each type of mycosis differs. Because superficial and cutaneous mycoses are limited to the skin and hair, these are usually treated with topical forms of allylamines, benzylamines, or azole drugs. Terbinafine, an allylamine, is usually used topically but can be given orally for nail fungus. Nystatin, a polyene antibiotic, is used to control opportunistic *Candida* infections of the skin, vagina, or mouth and esophagus. It is too toxic to be taken systemically; when used as a "swish and swallow" agent for oral thrush, it is not absorbed from the gastrointestinal tract. ▶| *Human diseases caused by fungi and protists (chapter 39)*

Systemic fungal infections are very difficult to control and can be fatal. Drugs commonly used against systemic mycoses include amphotericin B, 5-flucytosine, and azoles such as fluconazole, itraconazole, and voriconazole. Amphotericin B from *Streptomyces* spp. is quite toxic to humans, earning it the nickname "amphoterrible." It is used only for serious, life-threatening infections. The synthetic oral agent 5-flucytosine (5-fluorocytosine) is effective against many systemic fungi, although drug resistance develops rapidly so it is typically used in combination with amphotericin B. The drug is converted to 5-fluorouracil (5FU) only by fungi. 5FU is an analogue of uracil, one of the building blocks of RNA. During transcription, it is incorporated into RNA in place of uracil and disrupts RNA function. It is most commonly used in combination with amphotericin B to treat cryptococcal meningitis. Fluconazole is used in the treatment of candidiasis and coccidioidomycosis (valley fever). Because adverse effects of fluconazole are relatively uncommon, it is used prophylactically to prevent life-threatening fungal infections in individuals who are severely immunosuppressed.

One such life-threatening infection of immunosuppressed individuals is caused by *Pneumocystis jirovecii* (formerly *P. carinii*). Because *P. jirovecii* membranes possess cholesterol rather than ergosterol, polyenes and azoles are ineffective. Rather, infections are treated with the combination of antimetabolite drugs trimethoprim and sulfisoxazole. This is effective because *P. jirovecii* synthesizes folic acid using enzymes that are very similar to those used by bacteria.

Echinocandins are a more recently developed class of antifungal drug. These drugs block the synthesis of the fungal cell wall by inhibiting the enzyme beta (1,3)-D-glucan synthase, which links glucan subunits. Loss of cell wall integrity triggers osmotic lysis, much like that seen with bacterial cell wall inhibitors. Echinocandins are the preferred treatment when fungi are resistant to either polyenes and/or azoles. Most recently, strains of the yeast *Candida auris* have emerged that are resistant to all anti-fungals, including the echinocandins; in effect, these infections are untreatable. This is quite alarming as *C. auris* infection is often life-threatening.

Comprehension Check

1. Compare and contrast the mechanism by which polyenes, azoles, and 5-flucytosine inhibit fungal growth.
2. Go online or visit a drugstore and compare the agents found in triple antibiotic cream with those found in athlete's foot medicine. What can you infer about athlete's foot from this?
3. Why are immunosuppressed individuals sometimes given antifungal agents in the absence of infection?

9.7 Antiprotozoan Drugs

After reading this section, you should be able to:

a. Compare the mechanisms of action of antiprotozoan drugs
b. Explain why there are far fewer antiprotozoan agents than there are antibacterial agents

As with other antimicrobial therapies, antiprotozoan drug effectiveness starts by identifying a unique target to which a drug can bind and thus prevent some vital function. However, because protozoa are eukaryotes, the potential for drug action on host cells and tissues is greater than when targeting bacteria. Most of the drugs used to treat protozoan infections have significant side effects; nonetheless, the side effects are usually acceptable when weighed against the parasitic burden.

The number of antiprotozoan drugs is relatively small, and the mechanism of action for most is not completely understood. The drugs described here have potent antiprotozoan action and appear to act on protozoan nucleic acid or some metabolic process.

Malaria, caused by any of five species in the genus *Plasmodium,* kills about half a million people annually—most of them children. In the 1600s, it was discovered that quinine, from the bark of the cinchona tree, could be used to treat malaria. We now use more purified forms of **quinine,** including chloroquine and quinine sulfate, to treat malaria. These drugs suppress *Plasmodium* reproduction and are effective in eradicating asexual stages of the protozoan's life cycle that occur in red blood cells. An essential part of the *Plasmodium* spp. life cycle is degradation of hemoglobin. During this process, the microbe polymerizes a toxic heme metabolite into a nontoxic form, called hemazoin. Quinine drugs block this process and the toxic metabolite accumulates. During the 2020 COVID-19 pandemic, chloroquine and its derivative hydroxychloroquine garnered much publicity as a potential treatment or prophylaxis (**Disease 9.1**).

Another drug, mefloquine, has been found to swell *P. falciparum* food vacuoles, where it may act by forming toxic complexes that damage membranes and other plasmodial components. Primaquine is active against a dormant form of the protists (hypnozoites) found in the liver, so it prevents relapses of malaria caused by *P. vivax* and *P. ovale,* the two species that produce hypnozoites. Individuals who travel to areas where malaria is endemic should adhere to chemoprophylactic treatment with antimalarial agents. *Malaria (section 39.3)*

DISEASE

9.1 Chloroquine and COVID-19: A Cautionary Tale

In the early days of the COVID-19 pandemic, caused by the coronavirus SARS-CoV-2, the search to repurpose existing drugs as a treatment (or even cure) was intense. As the pandemic grew, the notion of waiting for new drugs to be developed and tested seemed untenable. During the SARS epidemic of 2003, chloroquine and its derivative hydroxychloroquine (**box figure**) were shown to block replication of the causative coronavirus, SARS-CoV, in vitro. Although the antiviral mechanism of these drugs remains debated, one leading theory was that they prevent progression of the viral life cycle by increasing the pH of the endosome in which the virus resides upon entering a host cell.

Unfortunately, the desire for a treatment collided with a lack of understanding of how drug trials must proceed to protect the public from ineffective and unsafe drugs. By rushing clinical trials of dubious quality into print, the scientific community bears some of the blame in the ensuing confusion. Some of the events surrounding chloroquine and hydroxychloroquine in the first half of 2020 include:

February 4: The journal *Cell Research* publishes a letter to the editor by Chinese scientists, including a notable coronavirus expert, suggesting that the antimalarial drug chloroquine might be effective in combating COVID-19.

February 15: A group of French scientists publishes a similar editorial in the *International Journal of Antimicrobial Agents.*

Mid- to late February: Several news outlets report promising results from early Chinese clinical trials using chloroquine and the more bioavailable hydroxychloroquine in treating patients with COVID-19.

March 16: Entrepreneur Elon Musk tweets that chloroquine might be effective in treating COVID-19.

March 20: U.S. President Donald Trump announces chloroquine is a "game changer."

March 23: Patients who take chloroquine and hydroxychloroquine for autoimmune diseases report shortages in getting these medications, which many have been taking for years.

March 24: An Arizona man dies and his wife becomes gravely ill after ingesting chloroquine-containing fish tank cleaner in an effort to prevent contracting COVID-19.

March 25: The World Health Organization announces a large, international clinical trial to test the safety and effectiveness of hydroxychloroquine in treating COVID-19.

March 28: The U.S. Food and Drug Administration (FDA) issues emergency use authorization allowing widespread use of the drug.

April 10: Reports from frontline health-care workers suggest the drug is not effective and may be causing adverse events in some patients.

April 13, April 22: Two clinical trials report that hydroxychloroquine failed to demonstrate any potential benefit in treating patients with COVID-19.

April 24: The FDA issues a warning against using hydroxychloroquine if not hospitalized.

May 18: Donald Trump announces he is taking hydroxychloroquine prophylactically.

May 26: The medical journal *The Lancet* publishes a large clinical trial that concludes hydroxychloroquine is not effective in treating COVID-19 and increases the risk of death.

June 5: *The Lancet* retracts the paper published on May 26 due to concerns about data quality.

June 15: The FDA rescinds its emergency use authorization for chloroquine and hydroxychloroquine.

June 20: The NIH closes a clinical trial due to lack of evidence that the drugs are effective in treating COVID-19.

Since that time, chlroroquine and hydroxychloroquine have largely returned to treating malaria and autoimmune diseases. Let's hope we don't have to wait for the next pandemic to find out if all the stakeholder agencies learned anything from the rise and fall of these drugs.

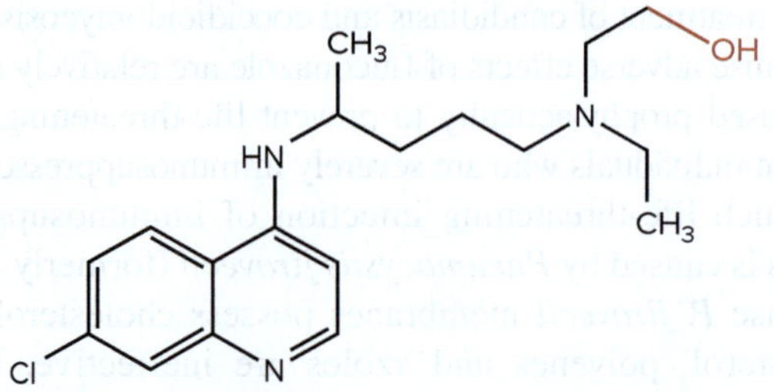

Chloroquine and Hydroxychloroquine. Hydroxychloroquine bears a hydroxyl group, as shown in red.

The World Health Organization recommends a combination of a quinine drug and a derivative of **artemisinin** (see chapter opening story) to treat malaria in endemic areas. Having its origins as a treatment in traditional Chinese medicine, artemisinin is a chemical found in the sweet wormwood plant (*Artemisia annua*). A semisynthetic artemisinin derivative used in combination with at least one other drug is an effective and inexpensive malaria treatment. The mechanism of artemisinin action is not well understood; it appears to form reactive oxygen intermediates inside *Plasmodium*-infected red blood cells, leading to altered hemoglobin catabolism and damage to the protist's electron transport chain. ▶| *Electron transport chains (section 11.4)*

There are many other diseases caused by protozoa, as discussed in chapter 39. Dysentery caused by the amoeba *Entamoeba histolytica* and vaginal infections of the protozoan *Trichomonas vaginalis* can be treated with **metronidazole.** These anaerobic organisms readily reduce metronidazole to a reactive nitrogen intermediate (RNI) within the cytoplasm. RNIs cause oxidative damage to a variety of cellular macromolecules. The drug nitazoxanide has a very similar mechanism of action and is used to treat the waterborne diarrheal diseases caused by the protozoa *Giardia intestinalis* and *Cryptosporidium parvum.* Several antibiotics that inhibit bacterial protein synthesis (e.g., paromycin (an aminoglycoside) and doxycycline (a tetracycline)) are also used to treat protozoan infection (figures 9.9 and 9.10). These agents are used to prevent or treat infections caused by protozoa known as apicomplexans. These microbes possess an organelle derived from plastids, the apicoplast, that is essential for viability. It is thought that these protein synthesis inhibitors bind to the ribosomes in these organelles. ▶| *Metamonada (section 23.2); Amoebozoa clade includes protists with pseudopodia (section 23.3); Alveolata (section 23.4); Food and water are vehicles for fungal and protozoal diseases (section 39.5)*

Toxoplasmosis is a life-threatening infection in individuals who are immunocompromised and can cause severe congenital anomalies in human fetuses. The CDC estimates that 11% of the population over the age of 6 years has been infected with *Toxoplasma gondii,* but the immune system usually keeps the protist from causing symptoms or illness. Toxoplasmosis is typically treated with combinations of pyrimethamine, dapsone, and sulfadiazine, plus folinic acid. Pyrimethamine and dapsone act in the same way as trimethoprim—interfering with folic acid synthesis by inhibition of dihydrofolate reductase. ▶| *Toxoplasmosis (section 39.5)*

Comprehension Check

1. Why are drugs that inhibit bacterial protein synthesis also effective against some protists?
2. Why is malaria, like tuberculosis and HIV, treated with several drugs simultaneously?
3. Synthesis of what metabolite is blocked by the drugs used to treat toxoplasmosis?

9.8 Antimicrobial Drug Resistance Is a Public Health Threat

After reading this section, you should be able to:

a. Report the common reasons for the rise of drug-resistant bacteria
b. Describe common mechanisms by which antimicrobial drug resistance occurs
c. Suggest strategies to slow the emergence of drug-resistant microbes

Antibiotic-resistant bacteria are so widespread that **methicillin-resistant *S. aureus* (MRSA)** is well-known. Fortunately most MRSA strains can still be treated, but the more recent emergence of **vancomycin-resistant enterococci (VRE), carbapenem-resistant Enterobacteriaceae (CRE)** bacteria, and pan-drug resistant *C. auris* has caused great concern among public health officials worldwide. Each year, nearly 3 million antimicrobial-resistant infections are reported in the US. Treating these infections is incredibly expensive, costing $4.6 billion. Unfortunately, many of these infections cannot be cured, resulting in nearly 40,000 deaths. These numbers will continue increasing if the tide of resistance is not stemmed and antibiotic research is not supported.

There are two types of resistance: intrinsic and acquired. An example of intrinsic resistance is that of mycoplasma resistance to β-lactam antibiotics and other cell wall inhibitors simply because these bacteria lack a cell wall. Acquired resistance occurs when there is a change in the genome of a bacterium that converts it from one that is susceptible to an antibiotic to one that is resistant. Another distinction is drug tolerance as opposed to resistance. Drug-tolerant bacteria lack mechanisms for antibiotic resistance. Instead these bacteria "ignore" the presence of antibiotics, usually because they are embedded in biofilms that antibiotics cannot effectively penetrate or are growing too slowly to be inhibited. These bacteria are called persisters and are discussed in chapter 7. When moved to an environment that supports more rapid growth, persister cells demonstrate drug susceptibility. By contrast, antibiotic-resistant bacteria have acquired genome changes, by either mutation or horizontal gene transfer, and now encode proteins that confer true resistance, regardless of growth rate. The mechanisms by which these genetic changes occur is discussed in chapter 16.

There Are Several Mechanisms of Drug Resistance

Bacteria have evolved several resistance mechanisms (**figure 9.16, table 9.2**). One is to modify the target of the antibiotic. For example, MRSA is resistant to all β-lactam antibiotics because it has replaced its transpeptidase (or, as mentioned, penicillin-binding protein [PBP]) with another version of this enzyme (figure 9.6). This altered transpeptidase still functions in cell wall synthesis, but it lacks a binding site for β-lactam antibiotics. This explains why MRSA is resistant to all members of this class of antibiotics but remains susceptible to vancomycin. Recall that vancomycin binds a different target—the terminal D-ala of

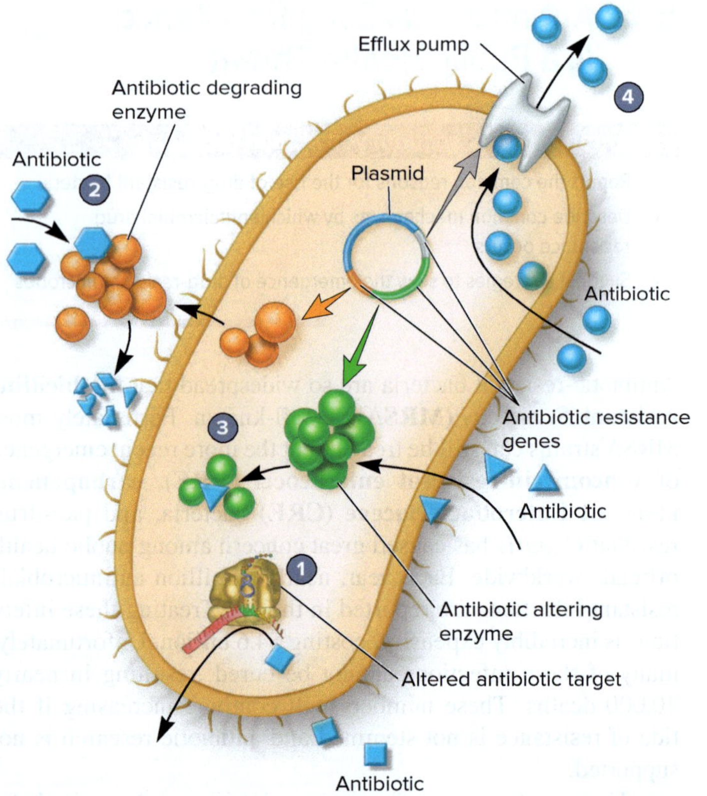

Figure 9.16 Antibiotic Resistance Mechanisms. Bacteria can resist the action of antibiotics by (1) preventing access to (or altering) the target of the antibiotic, (2) degrading the antibiotic, (3) altering the antibiotic, or (4) rapidly extruding the antibiotic.

the peptidoglycan subunit. Resistance to this drug among some enterococci is another example of changing the target. Vancomycin resistance occurs when bacteria gain the capacity to change the last D-ala to D-lactate or D-serine residue to which the drug cannot bind.

A second resistance strategy is drug inactivation. The best-known example is hydrolysis of the β-lactam ring of penicillins by penicillinase and other β-lactamases (**figure 9.17**). Drugs also are inactivated by the addition of chemical groups. For example, aminoglycosides can be modified and inactivated by acetylation of amino groups and phosphorylation or adenylylation of hydroxyl groups present on the drug. Because these resistance strategies involve gaining enzymes that catalyze the inactivating reaction, the new function is most often acquired when bacteria share genes with each other, a process known as horizontal gene transfer.

The third resistance strategy minimizes the concentration of the antibiotic in the cell. This can be accomplished by altering membrane structure, particularly the outer membrane for Gram-negative bacteria, so that less antibiotic enters the cell. Another approach is to pump the drug out of the cell after it has entered, using translocases, called **efflux pumps.** Efflux pumps are relatively nonspecific and expel many different drugs; therefore, they often confer multidrug resistance. Many efflux pumps are drug/proton antiporters; that is, protons enter the cell as the drug leaves.

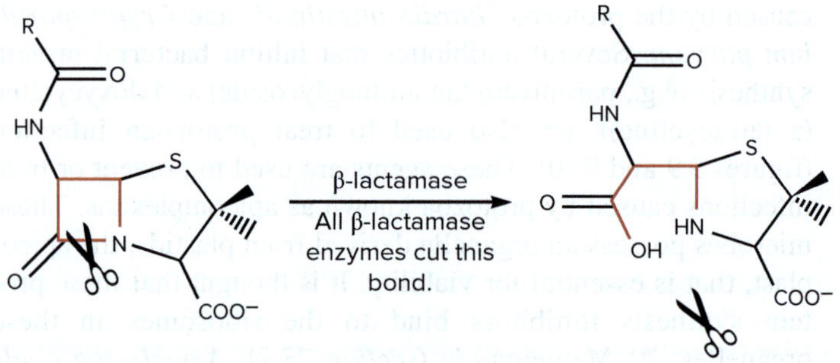

Figure 9.17 β-lactamase Enzymes Cleave the β-lactam Ring. Some β-lactamases degrade only a few different antibiotics, whereas other, more dangerous enzymes have a broader β-lactam specificity. The β-lactam ring is shown in red.

Table 9.2 Mechanisms of Resistance to Antimicrobial Agents

Mechanism	Example
Modify drug target	Methicillin-resistant *Staphylococcus aureus* (MRSA) has alternate transpeptidase, penicillin-binding protein (PBP-2), to which β-lactam drugs cannot bind, thereby blocking activity of all β-lactam drugs. *Enterococcus faecium* replaces terminal D-alanine in peptidoglycan peptide chains with D-lactate blocking vancomycin activity.
Inactivate drug	Degradation: Many Gram-positive and Gram-negative bacteria produce β-lactamase enzymes that degrade β-lactam drugs. Chemical modification: *Mycobacterium tuberculosis* enzymes acetylate aminoglycosides kanamycin and amikacin.
Prevent drug entry into cell	*Neisseria gonorrhoeae* porins block entry of penicillin and tetracycline. *Mycobacterium tuberculosis* efflux pumps block uptake of multiple anti-tuberculosis drugs.
Render drug activity irrelevant	The protozoan pathogen *Leishmania* spp. uses an alternative folic acid biosynthetic pathway that does not require dihydrofolate reductase.

Finally, resistant bacteria may either use an alternate pathway to bypass the biochemical reaction inhibited by the agent or increase the production of the target metabolite. For example, some bacteria are resistant to sulfonamides simply because they use preformed folic acid from their surroundings, rather than synthesize it themselves. Other strains increase their rate of folic acid production and thus counteract sulfonamide inhibition. Again, these changes are most often mediated by horizontal gene transfer.

Detecting Drug Resistance

It is important to know if a clinically isolated microorganism is expressing resistance factors and/or has the genetic potential to do so. Because resistance factors are genetically encoded, it is possible to test for the gene product or for the gene itself. Commercial gene expression systems are designed to identify the production of specific resistance factors, such as a target-modifying enzyme. Several test systems measure the color change induced when a chromophore is acted upon by either a β-lactamase (figure 9.17) or an antibiotic-modifying enzyme. The resulting color changes are measured spectrophotometrically and protein concentration is extrapolated from a standardized curve. Thus direct evidence of induced resistance can be made. Detection systems are also commercially available to identify genes encoding drug resistance factors using polymerase chain reactions (PCRs). ▶| *Polymerase chain reaction amplifies targeted DNA (section 17.2)*

Overcoming Drug Resistance

The World Health Organization cites antibiotic resistance as "one of the biggest threats to global health, food security, and development today." The CDC has classified certain antibiotic-resistant bacteria as "urgent, serious, or concerning threats." These alarms are coupled with calls to action to ensure the effectiveness of antimicrobial agents continues. It has long been understood that the problem has resulted from antibiotic overuse, misuse, and abuse. Put simply, the more drugs are used, the more likely it is that bacteria will become resistant to them. That raises the question: Is there a way to overcome drug resistance?

Several strategies can be employed to limit the emergence of drug resistance. Because about 70% of antibiotics used in the United States are fed to livestock, tighter controls over use of antibiotics in animal feed must be enacted. Whether administered to people or farm animals, drugs should be given in a high enough concentration to destroy susceptible microbes and most spontaneous mutants that might arise during treatment. Patients must be educated to take their antimicrobial agents as prescribed and to finish the course of therapy, which should be as short as possible. If a patient fails to complete treatment, resistant mutants survive and flourish because of their competitive advantage over nonresistant bacteria. Sometimes two or even three different drugs can be administered simultaneously to prevent the emergence of resistance. This approach is used in treating tuberculosis, HIV, and malaria, for example.

Other strategies used to prevent drug resistance include the strict control on use of antimicrobial drugs, especially newly developed drugs to which bacteria have not become widely resistant. The use of broad-spectrum drugs must also be curtailed. If possible, the pathogen should be identified, drug sensitivity tests performed, and the proper narrow-spectrum drug employed. These ideas and others are now incorporated into antibiotic stewardship programs at most health-care facilities.

Despite efforts to control the emergence and spread of drug resistance, the situation continues to worsen. There is an urgent need for new antibiotics that microorganisms have never encountered. Unfortunately, pharmaceutical companies lack incentive to develop new antibiotics (*see chapter 41 opener*). To understand why the economics are so important, consider what we just discussed regarding methods to curb the emergence of antibiotic resistance. The best antibiotics should be used only rarely, and the duration of treatment should be brief. This completely undercuts the demand for a new product! Factor in the potentially multibillion-dollar cost of bringing a new drug to market, it is little wonder why private companies have fled the antimicrobial space (**figure 9.18**). Thankfully there are now several governmental and private organizations working to devise a sustainable framework to encourage investment in drug discovery efforts.

A most interesting response to the current crisis is the renewed interest in an idea first proposed early in the twentieth century by Felix d'Herelle, one of the discoverers of bacterial viruses (bacteriophages). d'Herelle proposed that bacteriophages could be used to treat bacterial diseases. Although most microbiologists did not pursue his proposal due to technical difficulties and the advent of antibiotics, Russian scientists developed the medical use of bacteriophages (*see chapter 6 opener*). The idea has reemerged and trials using bacteriophages to treat certain bacterial infections are underway. |◀ *Viruses and other acellular infectious agents (chapter 6)* ▶| *Microbial genomics (chapter 18); Biotechnology and industrial microbiology (chapter 41)*

Comprehension Check

1. Compare and contrast the four mechanisms bacteria use to block antibiotic activity.
2. Which of the resistance mechanisms described here could be used by viruses? Which cannot? Explain your answer.
3. Why does the use of narrow-spectrum antibiotics help prevent the spread of drug resistance to a greater extent than broad-spectrum antibiotics?
4. What are the primary medical practices that result in antimicrobial drug resistance?

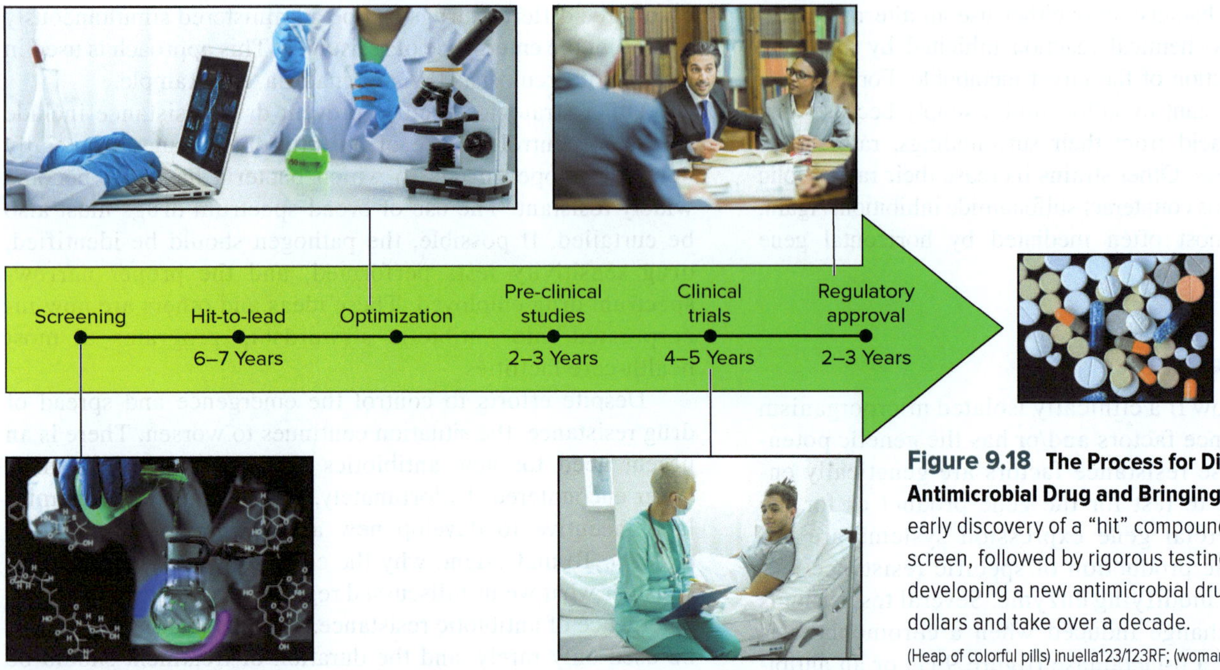

Figure 9.18 The Process for Discovering a New Antimicrobial Drug and Bringing It to Market. From early discovery of a "hit" compound from a high throughput screen, followed by rigorous testing until delivery to market, developing a new antimicrobial drug can cost billions of dollars and take over a decade.

(Heap of colorful pills) inuella123/123RF; (woman scientist) souvikonline2005/123RF; (Male patient) svitlanahulko/123RF; (Laboratory research) pitinan/123RF; (Lawyers meeting) Robert Daly/Caia Image/Glow Images

Key Concepts

9.1 Antimicrobial Chemotherapy Evolved from Antisepsis Efforts

- The modern era of chemotherapy began with Paul Ehrlich's work on drugs against syphilis.
- Other early pioneers were Gerhard Domagk, Alexander Fleming, Howard Florey, Ernst Chain, Norman Heatley, and Selman Waksman.

9.2 Antimicrobial Drugs Have Selective Toxicity

- An effective chemotherapeutic agent must have selective toxicity. A drug with great selective toxicity has a high therapeutic index and usually disrupts a structure or process unique to the pathogen. It has fewer side effects.
- Antibiotics can be classified in terms of the range of target microorganisms (narrow-spectrum versus broad-spectrum); their source (natural, semisynthetic, or synthetic); and their general effect (static versus cidal) (**figure 9.2**).

9.3 Antimicrobial Activity Can Be Measured by Specific Tests

- Antibiotic effectiveness can be estimated through the determination of the minimal inhibitory concentration (MIC) with dilution susceptibility tests.
- Diffusion tests such as the Kirby-Bauer assay are often used to quickly estimate a pathogen's susceptibility to drugs (**figure 9.3**).

9.4 Antibacterial Drugs

- Members of the penicillin family have a β-lactam ring and disrupt bacterial cell wall synthesis, resulting in cell lysis (**figures 9.5 and 9.6**). Natural penicillins are effective against Gram-positive bacteria only. Most semisynthetic penicillins are broader spectrum and are more resistant to penicillinases.
- Cephalosporins are similar to penicillins but are typically broader spectrum and more resistant to β-lactamase attack (**figure 9.7**).
- Vancomycin is a glycopeptide antibiotic that inhibits transpeptidation during peptidoglycan synthesis (**figure 9.8**). It has a narrow spectrum, targeting only Gram positives.
- Aminoglycoside antibiotics such as streptomycin bind to the small ribosomal subunit, inhibit protein synthesis, and are bactericidal (**figure 9.9**).
- Tetracyclines are broad-spectrum antibiotics having a four-ring structure with attached groups (**figure 9.10**). They bind to the small ribosomal subunit and inhibit protein synthesis.
- Erythromycin and azithromycin are bacteriostatic macrolide antibiotics that bind to the large ribosomal subunit and inhibit protein synthesis (**figure 9.11**).
- Linezolid blocks protein synthesis by binding to the large ribosomal subunit and preventing the formation of the 70S initiation complex. Linezolid is used to treat serious infections caused by drug-resistant Gram positives.

- Sulfonamides or sulfa drugs resemble *p*-aminobenzoic acid and competitively inhibit folic acid synthesis (**figure 9.12**).
- Trimethoprim is a synthetic antibiotic that inhibits dihydrofolate reductase, which is required by organisms in the manufacture of folic acid (**figure 9.13**).
- Fluoroquinolones are a family of bactericidal synthetic drugs that inhibit DNA gyrase and topoisomerase IV and thus inhibit DNA replication (**figure 9.14**).
- Rifamycins are synthetic antibiotics that block transcription by binding to bacterial RNA polymerase.

9.5 Antiviral Drugs

- Antiviral drugs interfere with critical stages in the virus life cycle or inhibit the synthesis of virus-specific nucleic acids.
- Drug combinations are more effective than monotherapies when treating HIV and hepatitis C.
- There are five classes of antiretroviral agents used to treat HIV. Drug combinations often include two nucleoside reverse transcriptase inhibitors and an integrase inhibitor (**figure 9.15**).

9.6 Antifungal Drugs

- The two major classes of antifungal agents include the polyenes and the azoles. Both disrupt the fungal plasma membranes. Polyenes, such as amphotericin B, have a low therapeutic index.
- Superficial mycoses can be treated with topical azoles. Amphotericin B, 5-flucytosine, azole, and echinocandin drugs are used for systemic mycoses.

9.7 Antiprotozoan Drugs

- Antiprotozoan drugs interfere with critical steps in nucleic acid synthesis, protein synthesis, electron transport, or folic acid synthesis.

9.8 Antimicrobial Drug Resistance Is a Public Health Threat

- Mechanisms used by microorganisms to resist the toxic effects of antimicrobial agents include changing the drug target so the agent no longer binds, modifying the drug, degrading the drug, or preventing uptake of the drug (**figure 9.16**).
- Drug resistance can often limit drug effectiveness. However, it can be overcome through the judicious use of antimicrobial agents (decreasing selective pressure), use of drug cocktails (attacking more than one microbial structure or function), and use of specific, narrow-spectrum drugs to target infecting microbes.
- New efforts are underway to stabilize the market for antimicrobial drug discovery efforts (**figure 9.18**).

Active Learning

1. Some advocate stockpiling the drug oseltamivir in the event of an influenza pandemic. Others point out that wealthy Western nations would have an unfair advantage because developing nations would not have access to this expensive antiviral. Furthermore, some point out that indiscriminate use of the drug promotes the evolution of resistant flu strains. Given these caveats, discuss the advantages and disadvantages of stockpiling oseltamivir in the event of an influenza pandemic.
2. You are a pediatrician treating a child with an upper respiratory infection that is clearly caused by a virus. The child's parents insist that you prescribe antibiotics. They aren't leaving without them! How do you convince the child's parents that antibiotics will do more harm than good?
3. Antibiotics that target molecules not previously exploited are desperately needed. One such target is the protein FtsZ (*see chapters 3 and 7*). The small molecule 3-methoxybenzamide (3-MBA) is known to inhibit FtsZ in *Bacillus subtilis* but is not bactericidal. Using 3-MBA as a starting point for the synthesis of a molecule that might be a potential drug candidate, how would you go about finding similar molecules that are bactericidal to bacterial pathogens?
4. Newly developed peptide antibiotics are an active area of research. Some of these small proteins target two molecules on bacterial pathogens. For example, a peptide might bind to a cell wall structure and then insert a hole in the underlying plasma membrane, thereby disrupting both structures. What might be the pros and cons of a drug that targets two structures or processes? If you were to design a drug, what two bacterial targets would you pursue?
5. Carbapenem-resistant Enterobacteraceae (CRE) is a growing problem because these Gram-negative pathogens are resistant to almost all β-lactam antibiotics. Most of these drug-resistant strains either make a β-lactamase with broad substrate specificity or have stopped producing a specific porin. Refer to figure 9.17 and table 9.2 and assign these two mechanisms to the categories of mechanisms shown. Strikingly, these two mechanisms do not explain resistance for all CRE isolates. Develop a hypothesis that explains resistance in these remaining cases. What does your hypothesis predict and how would you test this?

CHAPTER

10

Introduction to Metabolism

kittirat roekburi/Shutterstock

Flushed Away

For one day, count the number of times the toilet in your home is flushed. Is your toilet a low-volume toilet that sends no more than 1.6 gallons out of your house with each flush? Or is it an older, water-greedy toilet? And where do the wastes you flush go?

If you live in a city, your wastes travel to a wastewater treatment plant (as shown in the accompanying figure). Now do some math: Multiply the number of gallons your house flushes per day by the number of households in your town. That's about how much wastewater is sent each day to your town's wastewater treatment plant. If you live in a large city, the amount of wastewater generated is astronomical. In New York City, for example, it's about 1.2 billion gallons per day. The drought-prone state of California produces a whopping 4 billion gallons per day!

The cost of treating wastewater so it is safe to release into the environment is also astronomical. New York City spends about $1.5 billion each year. Thus wastewater treatment is a big item in any city's budget. But these wastes can be used to decrease the cost of treatment and, better yet, generate funds for municipalities.

As we describe in chapter 42, wastewater treatment plants are home to a complex microbial community. Working together, these microbes decrease the organic matter (sewage sludge) in wastewater and generate several by-products using their diverse metabolic abilities. One of the most important is methane, which is made when organic matter is degraded under anoxic conditions. Increasingly, wastewater treatment plants are using methane to heat their buildings, and in some countries (including the U.S.), sewage sludge is converted to pipeline-quality natural gas to heat homes.

The treatment of wastewater is only one example of how humans exploit the metabolic prowess of microbes. As we describe in chapter 40, ethanol is used in the production of beer, ale, and wine. Propionic acid flavors Swiss cheese. Carbon dioxide leavens bread, and antibiotics, which are also microbial metabolites, save lives every day.

The impact of microbial metabolites goes beyond their use in industry and food production. It is now recognized that microbes living in and on the human body contribute compounds needed for human homeostasis. Indeed, the metabolic capabilities of both the microbiota and their human hosts are highly integrated as the result of coevolution. ▶| *The microbe-human ecosystem (chapter 33)*

Clearly, microbial metabolism is extremely important to the well-being of humans, and much of the study of microbes has focused on this aspect of their biology. Chapters 10 through 12 focus on the metabolic processes that harness (conserve) energy supplied by an organism's energy source and how it is used to synthesize the building blocks from which an organism is constructed. In this chapter, we lay the foundation for that discussion.

Readiness Check:

Based on what you have learned previously, you should be able to:

✓ Distinguish metabolism, catabolism, and anabolism

✓ Describe the structure of bacterial and archaeal cell envelopes (sections 3.3, 3.4, and 4.2)

✓ Describe mitochondrial structure and function (section 5.6)

10.1 Metabolism: Important Principles and Concepts

After reading this section, you should be able to:

a. List six features common to all types of metabolism
b. Describe the three types of work carried out by cells
c. State the relationship between cellular work and energy
d. Paraphrase the first and second laws of thermodynamics
e. Explain the importance of the equation for free energy change and the equation that links standard free energy change to a reaction's equilibrium constant
f. Infer from a standard free energy change ($\Delta G^{\circ\prime}$) whether a reaction is exergonic or endergonic

It might be convenient to think that many microbes exist for the benefit of humans, but of course they don't. Rather, they use their vast repertoire of chemical reactions to survive and reproduce, just as all organisms do. Indeed, **metabolism,** defined as the sum of all chemical reactions inside a cell, is central to all life. The forces of evolution have shaped the metabolic process used by organisms for billions of years. Despite the diversity of chemical reactions that have evolved, there are several aspects of metabolism common to all organisms. Here we list the common features, noting where they are discussed in this chapter.

- **Life obeys the laws of thermodynamics** (this section). Organisms must obtain energy from their environment —they cannot "make" energy; it must be captured. This capture is often referred to as conservation of energy.
- **The energy cells obtain from their environment is most often conserved as adenosine 5′-triphosphate (ATP)** (section 10.2). ATP supplies the energy needed for many chemical reactions to proceed.
- **Oxidation-reduction (redox) reactions play a critical role in energy conservation** (section 10.3). Many of these redox reactions occur in electron transport chains (section 10.4).
- **The chemical reactions that occur in cells are organized into pathways.** The product of one reaction is the substrate for the next reaction in the pathway (section 10.5).
- **Each reaction of a pathway is catalyzed by an enzyme or a ribozyme.** Enzymes and ribozymes are critical to all life because they speed up the reactions that enable life to exist (section 10.6).
- **The functioning of biochemical pathways is regulated** so that their products are made at the correct place and time (section 10.7).

Cellular Work and Energy Transfers

The common features of metabolism just described contribute to the work cells must do to survive and reproduce. Cells carry out three major types of work. **Chemical work** involves the synthesis of complex biological molecules from much simpler precursors or **anabolism;** energy is needed to increase the molecular complexity of a cell. **Transport work** requires energy to take up nutrients, eliminate wastes, and maintain ion balances. Energy input is needed because molecules and ions often must be transported across cell membranes against concentration and electrochemical (i.e., charge) gradients. The third type of work, **mechanical work,** provides energy for cell motility and the movement of structures within cells, such as partitioning chromosomes during cell division.

As just indicated, cells need energy to do work. Indeed, **energy** may be defined most simply as the capacity to do work because all physical and chemical processes are the result of the application or movement of energy. Organisms obtain the energy they need from an energy source present in their environment. The energy source can be in the form of chemicals or sunlight. They convert the energy from these sources into a useful form.

Cellular energy is most commonly stored in **ATP.** To understand how energy is conserved (harnessed) in ATP and how ATP is used to do cellular work, some knowledge of the basic principles of thermodynamics is required. The science of **thermodynamics** analyzes energy changes in a collection of matter (e.g., a cell or a plant) called a *system.* All other matter in the universe is called the *surroundings.* Thermodynamics focuses on the energy differences between the initial state and the final state of a system. It is not concerned with the rate of the process. For instance, if a pan of water is heated to boiling, only the condition of the water at the start and at boiling is important in thermodynamics, not how fast it is heated.

Laws of Thermodynamics

Two important laws of thermodynamics are critical to understanding energy transfers. The **first law of thermodynamics** says that energy can be neither created nor destroyed. This means the total energy in the universe remains constant, although it can be redistributed. In fact, this redistribution of energy characterizes the energy exchanges that occur during chemical reactions. For example, heat is given off by exothermic reactions and absorbed during endothermic reactions. However, the first law alone cannot explain why heat is released by one chemical reaction and absorbed by another. Explanations for this require the **second law of thermodynamics** and a condition of matter called entropy. **Entropy** is a measure of the randomness or disorder of a system. The greater the disorder of a system, the greater its entropy. The second law states that physical and chemical processes proceed in such a way that the randomness or disorder of the universe (the system and its surroundings) increases. However, even though the entropy of the universe increases, the entropy of any given system within the universe can increase, decrease, or remain unchanged.

Two types of energy units are employed to specify the amount of energy used in or evolving from a particular process. A **calorie** (cal) is the amount of heat energy needed to raise 1 gram of water 1 degree Celsius (specifically from 14.5 to 15.5°C). The amount of energy also may be expressed in terms of **joules** (J), the units of work capable of being done. One cal of heat is equivalent to 4.1840 J of work. One thousand calories, or a kilocalorie (kcal), is enough energy to enable a person weighing 70 kilograms to climb 35 steps.

Free Energy Change Predicts the Nature of a Chemical Reaction

The first and second laws of thermodynamics can be combined in a useful equation, relating the changes in energy that can occur in chemical reactions and other processes:

$$\Delta G = \Delta H - T\,\Delta S$$

where ΔG is the change in free energy, ΔH is the change in enthalpy, T is the temperature in Kelvin (°C + 273), and ΔS is the change in entropy occurring during the reaction. The change in **enthalpy** is the change in heat content. Cellular reactions occur under conditions of constant pressure and volume. Thus the change in enthalpy is about the same as the change in total energy during the reaction. The **free energy change** (ΔG) is the amount of energy a system (e.g., a cell) has to do useful work at constant temperature and pressure. The change in entropy (ΔS) is a measure of the proportion of ΔG unavailable to the system to perform work. Free energy and entropy changes do not depend on how the system gets from start to finish. A reaction will occur spontaneously—that is, without any external intervention—if the free energy of the system decreases during the reaction or, in other words, if ΔG is negative. Following the equation, a reaction with a large positive change in entropy will have a negative ΔG value and therefore occur spontaneously. A decrease in entropy will make ΔG more positive and the reaction less favorable.

The change in free energy has a direct relationship to the direction of chemical reactions. Consider this simple reaction:

$$A + B \rightleftharpoons C + D$$

If molecules A and B are mixed, they will combine to form the products C and D. Eventually C and D will become concentrated enough to combine and produce A and B at the same rate as C and D are formed. The reaction is now at **equilibrium:** The rates in both directions are equal and no further net change occurs in the concentrations of reactants and products. This situation is described by the **equilibrium constant (K_{eq})**, relating the equilibrium concentrations of products and substrates to one another:

$$K_{eq} = \frac{[C]\,[D]}{[A]\,[B]}$$

If the equilibrium constant is greater than one, the products are in greater concentration than the reactants at equilibrium; that is, the reaction tends to go to completion as written.

The equilibrium constant of a reaction is directly related to its change in free energy. When the free energy change for a process is determined at carefully defined standard conditions of concentration, pressure, pH, and temperature, it is called the **standard free energy change** ($\Delta G°$). If the pH is set at 7.0 (which is close to the pH of most living cells), the standard free energy change is indicated by the symbol $\Delta G°'$. The change in standard free energy may be thought of as the maximum amount of energy available from the system for useful work under standard conditions. Using $\Delta G°'$ values allows comparisons of reactions without considering variations in ΔG due to differences in

Exergonic reactions	Endergonic reactions
$A + B \rightleftharpoons C + D$	$A + B \rightleftharpoons C + D$
$K_{eq} = \frac{[C]\,[D]}{[A]\,[B]} > 1.0$	$K_{eq} = \frac{[C]\,[D]}{[A]\,[B]} < 1.0$
$\Delta G°'$ is negative.	**$\Delta G°'$ is positive.**

Figure 10.1 The Relationship of $\Delta G°'$ to the Equilibrium of Reactions. Note the differences between exergonic and endergonic reactions.

MICRO INQUIRY *Which reaction would release heat? Why?*

environmental conditions. The relationship between $\Delta G°'$ and K_{eq} is given by this equation:

$$\Delta G°' = -2.303RT \cdot \log K_{eq}$$

where R is the gas constant (1.9872 cal/mole-degree or 8.3145 J/mole-degree) and T is the absolute temperature. Inspection of this equation shows that when $\Delta G°'$ is negative, the equilibrium constant is greater than one and the reaction goes to completion as written. It is said to be an **exergonic reaction (figure 10.1)**. In an **endergonic reaction,** $\Delta G°'$ is positive and the equilibrium constant is less than one. That is, the reaction is not favorable, and little product will be formed at equilibrium under standard conditions. Keep in mind that the $\Delta G°'$ value shows only where the reaction lies at equilibrium, not how fast the reaction reaches equilibrium.

Comprehension Check

1. Suppose a bacterium was doing the following: synthesizing peptidoglycan, rotating its flagellum to swim, and secreting siderophores. What type of work is the bacterium doing in each case?
2. In your own words, summarize the first and second laws of thermodynamics.
3. Define entropy, enthalpy, and free energy. How do exergonic and endergonic reactions differ?
4. Is a chemical reaction with a large negative $\Delta G°'$ value endergonic or exergonic? What would this indicate about its equilibrium constant? If displaced from equilibrium, would the reaction proceed to completion?

10.2 ATP: The Major Energy Currency of Cells

After reading this section, you should be able to:

a. Draw a simple schematic that illustrates the structure of ATP
b. Describe ATP's role in linking exergonic and endergonic reactions
c. Compare ATP's phosphate transfer potential to that of glucose 6-phosphate and phosphoenolpyruvate (PEP) and relate this to ATP's function in cells
d. Describe the energy cycle observed in all organisms
e. Describe the metabolic functions of three nucleoside triphosphates (other than ATP)

Energy can seem like an abstract concept, so it may be useful to make an analogy to money. Both can be earned, stored, or spent, but not created (at least not legally in the case of money). Energy is earned in exergonic reactions (i.e., those reactions with a negative ΔG). Rather than wasting this energy, much of it is stored in a practical form that allows it to be used to do work. Cells carry out energy consuming endergonic reactions (e.g., anabolism), and stored energy is spent to drive these reactions to completion. In living organisms, the most commonly used practical form of energy is the nucleotide **adenosine 5′-triphosphate (ATP; figure 10.2)**. In a cell's economy, certain processes are performed

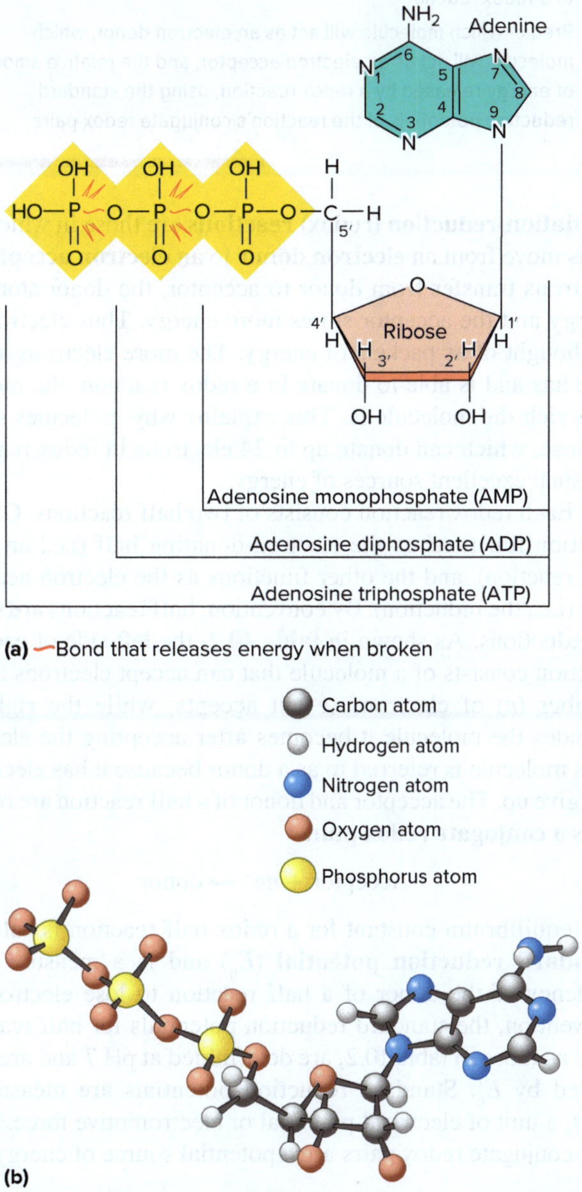

Figure 10.2 Adenosine Triphosphate. (a) Structure of ATP, ADP, and AMP. The two red bonds (~) are more easily broken and release considerable energy that can be used in endergonic reactions. The pyrimidine ring atoms have been numbered, as have the carbon atoms in ribose. (b) A stick-and-ball model of ATP. For simplification, some H atoms are not shown.

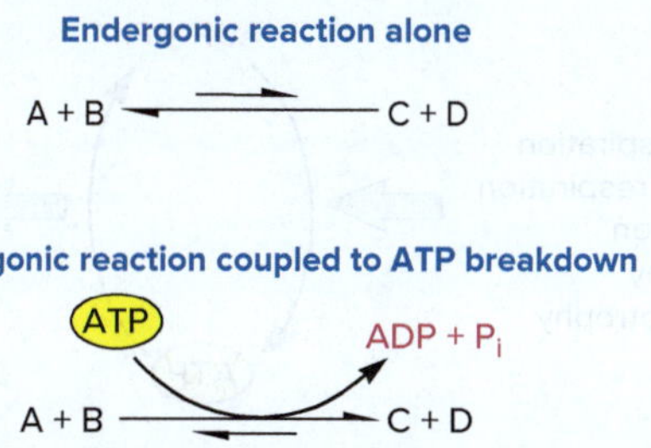

Figure 10.3 ATP as a Coupling Agent. ATP hydrolysis is exergonic and is used to drive endergonic reactions.

so that they can earn ATP, which is spent carrying out other processes. As the cell's energy currency, ATP links exergonic reactions and endergonic reactions (**figure 10.3**).

What makes ATP suited for its role as energy currency? ATP is a high-energy molecule. That is, it is hydrolyzed almost completely to the products **adenosine diphosphate (ADP)** and orthophosphate (P_i), with a strongly exergonic $\Delta G°'$ of −7.3 kcal/mole (−30.5 kJ/mole).

$$\mathbf{ATP + H_2O \rightleftharpoons ADP + P_i + H^+}$$

The energy released is used to power endergonic reactions (figure 10.3).

The very negative $\Delta G°'$ of ATP hydrolysis is related to another important characteristic of ATP: its ability to transfer a phosphoryl group to another molecule. ATP is said to have a high **phosphate transfer potential** because it readily donates a phosphoryl group to other molecules. Note in **table 10.1** that there are

Table 10.1 Phosphate Transfer Potential of Common Phosphorylated Compounds (kJ/mol)[1]

Phosphorylated Molecule	$\Delta G°'$ of Hydrolytic Removal of Phosphate (kJ/mol)	Phosphate Transfer Potential
High-Energy Phosphorylated Compounds		
Phosphoenolpyruvate[2]	−61.9	61.9
1,3-bisphosphoglycerate[2]	−49.3	49.3
ATP (hydrolysis to AMP)	−45.6	45.6
ATP (hydrolysis to ADP)	−30.5	30.5
Low-Energy Phosphorylated Compounds		
Glucose 6-phosphate	−13.8	13.8
Glycerol 1-phosphate	−9.2	9.2

[1] Phosphate transfer potential is defined as the negative of the standard free energy change ($\Delta G°'$) for the hydrolytic removal of phosphate from the phosphorylated molecule.
[2] Phosphoenolpyruvate and 1,3-bisphosphoglycerate are generated during the catabolism of organic molecules such as glucose.

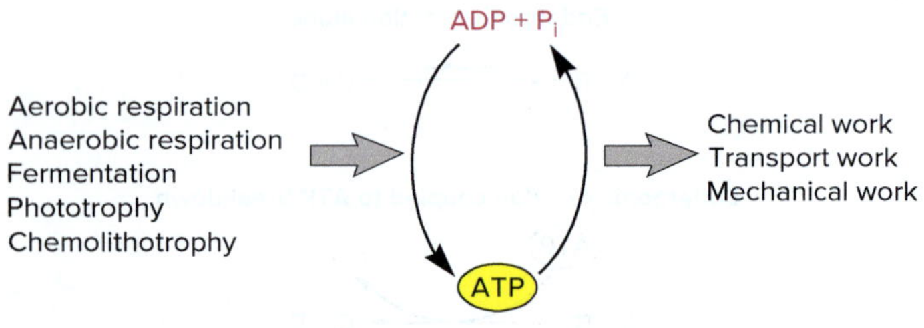

Figure 10.4 The Cell's Energy Cycle. Many processes enable a cell to form ATP, which can be used for chemical, transport, and mechanical work.

other phosphorylated compounds found in cells. These molecules are generated during **catabolism,** or the break down of organic molecules such as glucose. Some of these molecules have even higher phosphate transfer potentials than ATP. The fact that ATP does not have the highest phosphate transfer potential means that it can easily be made by cells from ADP, using molecules such as phosphoenolpyruvate (PEP) as the source of the phosphoryl group. This mechanism for making ATP is called **substrate-level phosphorylation.** ▶| *Aerobic respiration starts with glucose oxidation (section 11.3)*

ATP, ADP, and P_i form an energy cycle. As seen in **figure 10.4**, the energy released from an energy source is used to synthesize ATP from ADP and P_i. When ATP is hydrolyzed, the energy released drives endergonic processes such as anabolism, transport, and mechanical work. The mechanisms for synthesizing ATP are described in more detail in chapter 11.

ATP is the major energy currency for cells, but it is not the only energy currency. Other nucleoside triphosphates (NTPs) have major roles in metabolism. Guanosine 5′-triphosphate (GTP) supplies some of the energy used during protein synthesis. Cytidine 5′-triphosphate (CTP) is used during lipid synthesis, and uridine 5′-triphosphate (UTP) is used for the synthesis of peptidoglycan and other polysaccharides. ▶| *Synthesis of carbohydrates (section 12.4); Lipid synthesis (section 12.7); Translation in bacteria (section 13.7)*

Comprehension Check

1. Why is ATP called a high-energy molecule? How is this related to its phosphate transfer potential?
2. Describe the energy cycle and ATP's role in it. What characteristics of ATP make it suitable for this role?
3. Examine table 10.1. Could glucose 6-phosphate donate its phosphate to ADP to generate ATP? Explain your answer.

10.3 Redox Reactions: Reactions of Central Importance in Metabolism

After reading this section, you should be able to:

a. Describe a redox reaction, noting the role of the two half reactions and identifying the electron donor, electron acceptor, and conjugate redox pairs of the reaction
b. Relate the standard reduction potential of a conjugate redox pair to its tendency to act as an electron-donating half reaction
c. State the equation that links standard free energy change to the difference in standard reduction potentials of the two half reactions of a redox reaction
d. Predict which molecule will act as an electron donor, which molecule will act as an electron acceptor, and the relative amount of energy released by a redox reaction, using the standard reduction potentials of the reaction's conjugate redox pairs

Oxidation-reduction (redox) reactions are those in which electrons move from an **electron donor** to an **electron acceptor.**[1] As electrons transfer from donor to acceptor, the donor stores less energy and the acceptor stores more energy. Thus electrons can be thought of as packets of energy. The more electrons a molecule has and is able to donate in a redox reaction, the more energy rich the molecule is. This explains why molecules such as glucose, which can donate up to 24 electrons in redox reactions, are such excellent sources of energy.

Each redox reaction consists of two half reactions. One half reaction functions as the electron-donating half (i.e., an oxidation reaction), and the other functions as the electron-accepting half (i.e., the reduction). By convention, half reactions are written as reductions. As shown in **table 10.2**, the left side of each half reaction consists of a molecule that can accept electrons and the number (n) of electrons (e^-) it accepts, while the right side includes the molecule it becomes after accepting the electrons. This molecule is referred to as a donor because it has electrons it can give up. The acceptor and donor of a half reaction are referred to as a **conjugate redox pair.**

$$\text{Acceptor} + ne^- \rightleftharpoons \text{donor}$$

The equilibrium constant for a redox half reaction is called the **standard reduction potential** (E_0) and is a measure of the tendency of the donor of a half reaction to lose electrons. By convention, the standard reduction potentials for half reactions, such as those in table 10.2, are determined at pH 7 and are represented by E'_0. Standard reduction potentials are measured in volts, a unit of electrical potential or electromotive force. Therefore conjugate redox pairs are a potential source of energy.

[1] In redox reactions, the electron donor is often called the reducing agent or reductant because it is donating electrons to the acceptor and thus reducing it. The electron acceptor is called the oxidizing agent or oxidant because it is removing electrons from the donor and oxidizing it.

Table 10.2 Selected Biologically Important Half Reactions

Half Reaction (acceptor + ne^- → donor)	E_0' (Volts)[1]
$2H^+ + 2e^- \rightarrow H_2$	−0.42
Ferredoxin (Fe^{3+}) + e^- → ferredoxin (Fe^{2+})	−0.42
$NAD(P)^+ + H^+ + 2e^- \rightarrow NAD(P)H$	−0.32
$S + 2H^+ + 2e^- \rightarrow H_2S$	−0.27
Acetaldehyde + $2H^+ + 2e^-$ → ethanol	−0.20
$Pyruvate^- + 2H^+ + 2e^- \rightarrow lactate^-$	−0.19
$FAD + 2H^+ + 2e^- \rightarrow FADH_2$	−0.18[2]
$Oxaloacetate^{2-} + 2H^+ + 2e^- \rightarrow malate^{2-}$	−0.17
$Fumarate^{2-} + 2H^+ + 2e^- \rightarrow succinate^{2-}$	0.03
Cytochrome *b* (Fe^{3+}) + e^- → cytochrome *b* (Fe^{2+})	0.08
Ubiquinone + $2H^+ + 2e^-$ → ubiquinone H_2	0.10
Cytochrome *c* (Fe^{3+}) + e^- → cytochrome *c* (Fe^{2+})	0.25
Cytochrome *a* (Fe^{3+}) + e^- → cytochrome *a* (Fe^{2+})	0.29
Cytochrome a_3 (Fe^{3+}) + e^- → cytochrome a_3 (Fe^{2+})	0.35
$NO_3^- + 2H^+ + 2e^- \rightarrow NO_2^- + H_2O$	0.42
$NO_2^- + 8H^+ + 6e^- \rightarrow NH_4^+ + 2H_2O$	0.44
$Fe^{3+} + e^- \rightarrow Fe^{2+}$	0.77[3]
$\frac{1}{2}O_2 + 2H^+ + 2e^- \rightarrow H_2O$	0.82

[1] E_0' is the standard reduction potential at pH 7.0.
[2] The value for FAD/$FADH_2$ applies to the free cofactor because it can vary considerably when bound to an apoenzyme.
[3] The value for free Fe, not Fe complexed with proteins (e.g., cytochromes).

MICRO INQUIRY *For the $Fe^{3+} + e^- \rightarrow Fe^{2+}$ half reaction, which form of iron is the acceptor and which is the donor?*

A foundational concept is that conjugate redox pairs with more negative reduction potentials will spontaneously donate electrons to pairs with more positive potentials and greater affinity for electrons. Thus electrons tend to move from donors at the top of the list in table 10.2 to acceptors at the bottom because the latter have more positive potentials. This may be expressed visually in the form of an electron tower in which the most negative reduction potentials are at the top (**figure 10.5**). Electrons "fall down" the tower from donors higher in the tower (i.e., those having more negative potentials) to acceptors lower in the tower (i.e., those having more positive potentials).

Consider the case of the electron acceptor **nicotinamide adenine dinucleotide (NAD⁺)**. The NAD^+/NADH conjugate redox pair has a very negative E_0', so NADH can therefore give electrons to many acceptors, including O_2.

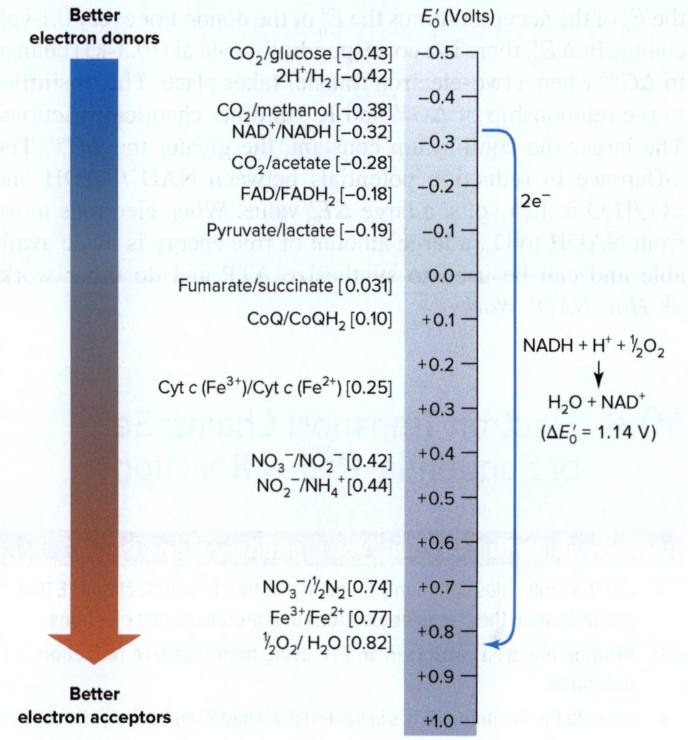

Figure 10.5 Electron Movement and Reduction Potentials. Electrons spontaneously move from donors higher on the tower (more negative potentials) to acceptors lower on the tower (more positive potentials). Some typical conjugate redox pairs are shown on the left and their reduction potentials are given in brackets.

MICRO INQUIRY *Why would energy be required to move electrons "up" the tower, from water to nitrate, for example?*

$$NAD^+ + 2H^+ + 2e^- \rightleftharpoons NADH + H^+ \qquad E_0' = -0.32 \text{ volt}$$
$$\frac{1}{2}O_2 + 2H^+ + 2e^- \rightleftharpoons H_2O \qquad E_0' = +0.82 \text{ volt}$$

Because the reduction potential of NAD^+/NADH is more negative than that of $\frac{1}{2}O_2/H_2O$, electrons flow from NADH (the donor) to O_2 (the acceptor) without added energy, as shown in figure 10.5. The redox reaction is shown here:

$$\mathbf{NADH + H^+ + \frac{1}{2}O_2 \rightarrow H_2O + NAD^+}$$

The relatively negative E_0' of the NAD^+/NADH pair also means that the pair stores more potential energy than redox pairs with less negative (or more positive) E_0' values. Thus when electrons move from a donor to an acceptor with a more positive redox potential, free energy is released. The $\Delta G^{\circ\prime}$ of the reaction is directly related to the magnitude of the difference between the reduction potentials of the two couples ($\Delta E_0'$). The larger the $\Delta E_0'$, the greater the amount of free energy made available, as is evident from the equation

$$\Delta G^{\circ\prime} = -nF \cdot \Delta E_0'$$

in which n is the number of electrons transferred, F is the Faraday constant (23,062 cal/mole-volt; 96,480 J/mole-volt), and $\Delta E_0'$ is the E'_0 of the acceptor minus the E'_0 of the donor. For every 0.1-volt change in $\Delta E'_0$, there is a corresponding 4.6-kcal (19.3-kJ) change in $\Delta G^{\circ\prime}$ when a two-electron transfer takes place. This is similar to the relationship of $\Delta G^{\circ\prime}$ and K_{eq} in other chemical reactions: The larger the equilibrium constant, the greater the $\Delta G^{\circ\prime}$. The difference in reduction potentials between NAD^+/NADH and $\frac{1}{2}O_2/H_2O$ is 1.14 volts, a large $\Delta E'_0$ value. When electrons move from NADH to O_2, a large amount of free energy is made available and can be used to synthesize ATP and do other work. *How NAD^+ Works*

10.4 Electron Transport Chains: Sets of Sequential Redox Reactions

After reading this section, you should be able to:

a. List the molecules commonly found in electron transport chains (ETCs) and indicate if they transfer electrons and protons or just electrons
b. Arrange electron carriers in an ETC using their standard reduction potentials
c. Indicate the location of ETCs in bacterial, archaeal, and eukaryotic cells

We have focused our attention on the reduction of O_2 by NADH because NADH plays a central role in the metabolism of many organisms. For instance, glucose is a common organic energy source. As glucose is catabolized, it is oxidized. Many of the electrons released from glucose are accepted by NAD^+, reducing it to NADH, which in turn donates the electrons to O_2. However, it does not do so directly. Instead, the electrons are transferred to O_2 via a series of additional electron carriers. These electron carriers are membrane-bound and organized into a system called an **electron transport chain (ETC).** An ETC is similar to a bucket brigade that might be used to put out a fire. Each carrier represents a person receiving pails of water (electrons) that are destined for the fire (terminal electron acceptor). Each pail of water is passed down the line, just as electrons are passed from carrier to carrier. As soon as one person hands off a pail of water to the next person in line, a new bucket is received. Likewise, as electrons flow through the ETC, each carrier is sequentially reduced (given electrons) and then reoxidized (passes electrons on to the next carrier in the chain), and is ready to accept more electrons.

The first electron carrier in an ETC has the most negative E'_0, and each successive carrier is slightly less negative (**figure 10.6**). Thus electrons are transferred spontaneously from one carrier to the next. The carriers direct the electrons to the terminal electron acceptor (in this case, O_2). The E'_0 of electron carriers must be relatively close to prevent the cell from losing energy by random reductions of other molecules in the cell or as heat. Returning to our bucket brigade analogy, if people in the brigade are too far apart, water could splash from each bucket and be lost during the transfer.

As we will learn, another fundamental concept is that ETCs are within or associated with plasma membranes and intracytoplasmic membranes of bacterial and archaeal cells (figure 10.6). In eukaryotes they are localized to the internal membranes of mitochondria and chloroplasts. ETCs are important to almost all types of energy-harnessing processes, as we discuss in chapter 11.

The electron carriers associated with ETCs differ in terms of their chemical nature and the way they carry electrons. NADH, which donates electrons to the ETC, and its chemical relative **nicotinamide adenine dinucleotide phosphate (NADPH),** which donates electrons to other molecules during biosynthetic reactions, contain a nicotinamide ring (**figure 10.7**). This ring accepts two electrons and one proton from a donor (e.g., an intermediate formed during the catabolism of glucose), and a second proton is released. **Flavin adenine dinucleotide (FAD)** and **flavin mononucleotide (FMN)** bear two electrons and two protons on the complex ring system shown in **figure 10.8**. Proteins bearing FAD and FMN are often called flavoproteins. **Coenzyme Q (CoQ)** or **ubiquinone** is a quinone that transports two electrons and two protons (**figure 10.9**). **Cytochromes** and several other carriers use iron atoms to transport one electron at a time. In cytochromes, the iron atoms are part of a heme group or other similar iron-porphyrin rings (**figure 10.10**). There are several different cytochromes, each consisting of a protein and an iron-porphyrin ring. Some iron-containing electron-carrying proteins lack a heme group and are called **nonheme iron proteins.** They are often referred to as **iron-sulfur (Fe-S) proteins** because the iron is associated with sulfur atoms (**figure 10.11**). **Ferredoxin** is an Fe-S protein active in photosynthesis-related electron transport and other important redox reactions. Like cytochromes, Fe-S proteins carry only one electron at a time. The differences in the number of electrons and protons transported by electron carriers are of great importance to the operation of ETCs, and we discuss them further in chapter 11. *Proton Pump*

Comprehension Check

1. How is the direction of electron flow between conjugate redox pairs related to the standard reduction potential and the release of free energy?
2. When electrons flow from the NAD^+/NADH conjugate redox pair to the $\frac{1}{2}O_2/H_2O$ redox pair, does the reaction begin with NAD^+ or with NADH? What is produced: O_2 or H_2O?
3. Which among the following would be the best electron donor? Which would be the worst? Ubiquinone/ubiquinone H_2, NAD^+/NADH, FAD/$FADH_2$, NO_3^-/NO_2^-. Explain your answers.
4. In general terms, how is $\Delta G^{\circ\prime}$ related to $\Delta E'_0$? What is the $\Delta E'_0$ when electrons flow from the NAD^+/NADH redox pair to the Fe^{3+}/Fe^{2+} redox pair? How does this compare to the $\Delta E'_0$ when electrons flow from the Fe^{3+}/Fe^{2+} conjugate redox pair to the $\frac{1}{2}O_2/H_2O$ pair? Which will yield the largest amount of free energy to the cell?
5. Name and briefly describe the major electron carriers found in cells. Why is NADH a good electron donor? Why is ferredoxin an even better electron donor? (Hint: Examine table 10.2.)

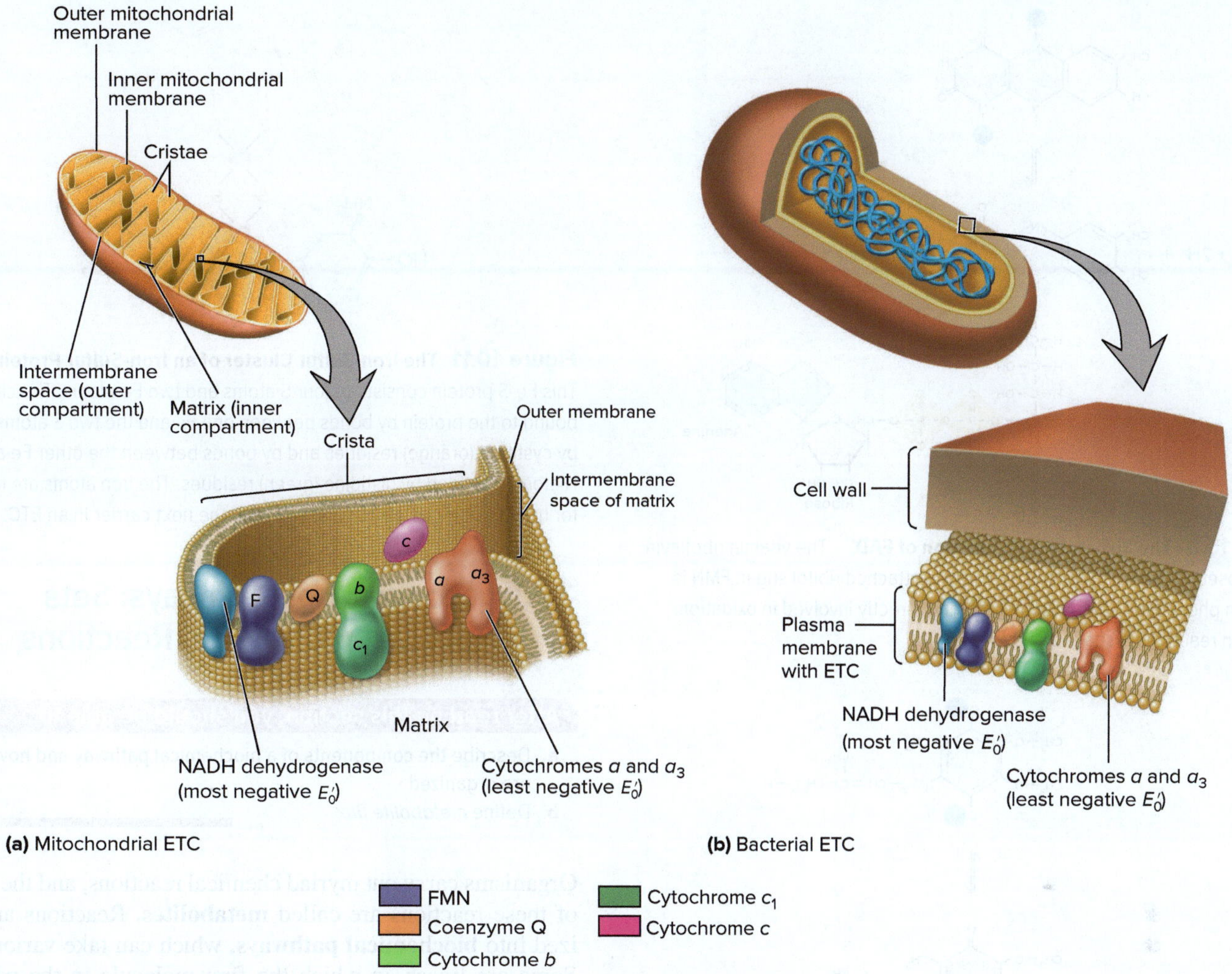

Figure 10.6 Electron Transport Chains. Electron transport chains (ETCs) are associated with membranes. Electrons flow from the electron carrier having the most negative reduction potential to the carrier having the most positive reduction potential. During respiratory processes (e.g., aerobic respiration; see figure 11.2), the source of electrons is a reduced molecule such as glucose and an exogenous molecule such as oxygen serves as the terminal electron acceptor. (a) The mitochondrial ETC. (b) A typical bacterial ETC.

MICRO INQUIRY *Refer to figure 10.5 and determine the E_0' for NAD^+/NADH and coenzyme Q/$CoQH_2$. Suggest a plausible E_0' value for FMN.*

Figure 10.7 The Structure and Function of NAD. (a) The structure of NAD^+ and $NADP^+$. $NADP^+$ differs from NAD^+ in having an extra phosphate on one of its ribose sugar units. (b) NAD^+ can accept two electrons and one proton from a reduced substrate (S). In this example, the electrons and proton are supplied by the hydrogen atoms of the reduced substrate; recall that each hydrogen atom consists of one proton and one electron.

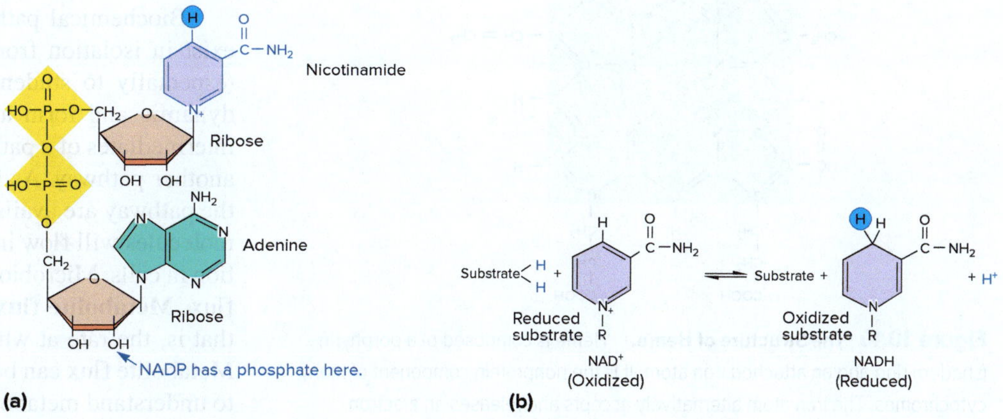

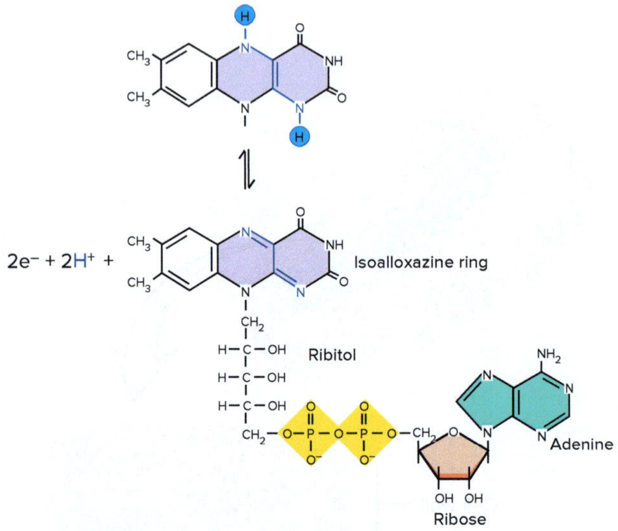

Figure 10.8 The Structure and Function of FAD. The vitamin riboflavin is composed of the isoalloxazine ring and its attached ribitol sugar. FMN is riboflavin phosphate. The portion of the ring directly involved in oxidation-reduction reactions is shaded purple.

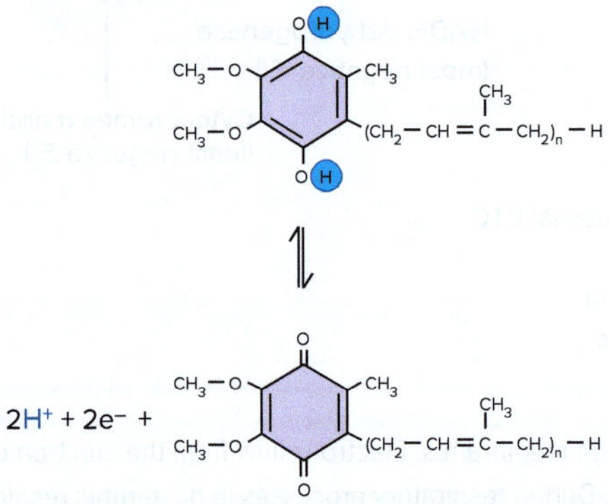

Figure 10.9 The Structure and Function of Coenzyme Q (Ubiquinone). The length of the side chain varies among organisms from n = 6 to n = 10.

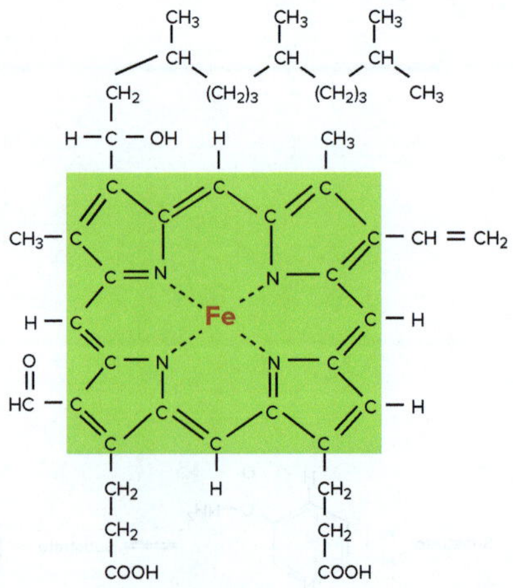

Figure 10.10 The Structure of Heme. Heme is composed of a porphyrin (shaded) ring and an attached iron atom. It is the nonprotein component of many cytochromes. The iron atom alternatively accepts and releases an electron.

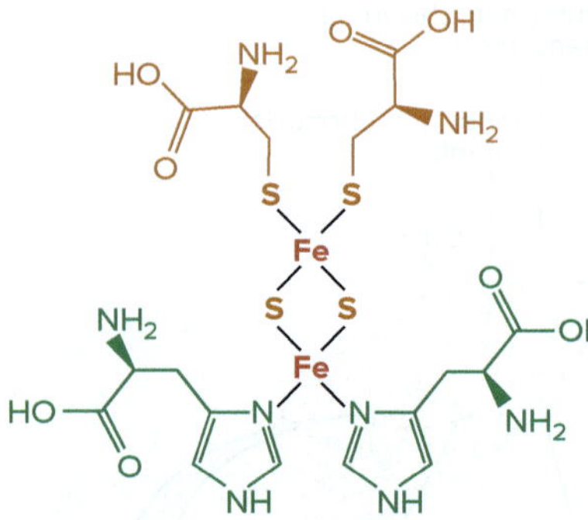

Figure 10.11 The Iron-Sulfur Cluster of an Iron-Sulfur Protein. This Fe-S protein consists of four S atoms and two Fe atoms. The cluster is bound to the protein by bonds between one Fe and the two S atoms provided by cysteine (orange) residues and by bonds between the other Fe and two N atoms provided by histidine (green) residues. The iron atoms are responsible for transferring one electron at a time to the next carrier in an ETC.

10.5 Biochemical Pathways: Sets of Linked Chemical Reactions

After reading this section, you should be able to:

- **a.** Describe the components of a biochemical pathway and how they are organized
- **b.** Define *metabolite flux*

Organisms carry out myriad chemical reactions, and the products of these reactions are called **metabolites.** Reactions are organized into **biochemical pathways,** which can take various forms. Some are linear, in which the first molecule in the pathway is often termed the *starting molecule* or the *substrate* of the pathway (**figure 10.12*a***). The last molecule is called the *end product.* Those metabolites in between are called *intermediates.* As seen in figure 10.12*b,* some linear pathways are branched and can yield more than one product. Biochemical pathways can also be cycles (figure 10.12*c*). In this case, all molecules in the pathway can be thought of as intermediates. If an intermediate is removed, it must be replaced for the cycle to continue running. In both types of pathways, each reaction is represented by an arrow and is catalyzed by an enzyme or ribozyme, as discussed in section 10.6.

Biochemical pathways are often illustrated as though they exist in isolation from each other, but this is very misleading (especially to students). In reality, biochemical pathways are dynamic and form a complex network (figure 10.12*d*). Thus, intermediates of a pathway may be diverted from one pathway to another pathway. As long as the starting molecules or inputs of the pathway are available and some end product is needed, these molecules will flow into and out of the many pathways that function in cells. Microbiologists are often concerned with metabolite flux. **Metabolite flux** is the rate of turnover of a metabolite—that is, the rate at which a metabolite is formed and then used. Metabolite flux can be used as a measure of pathway activity and to understand metabolic networks.

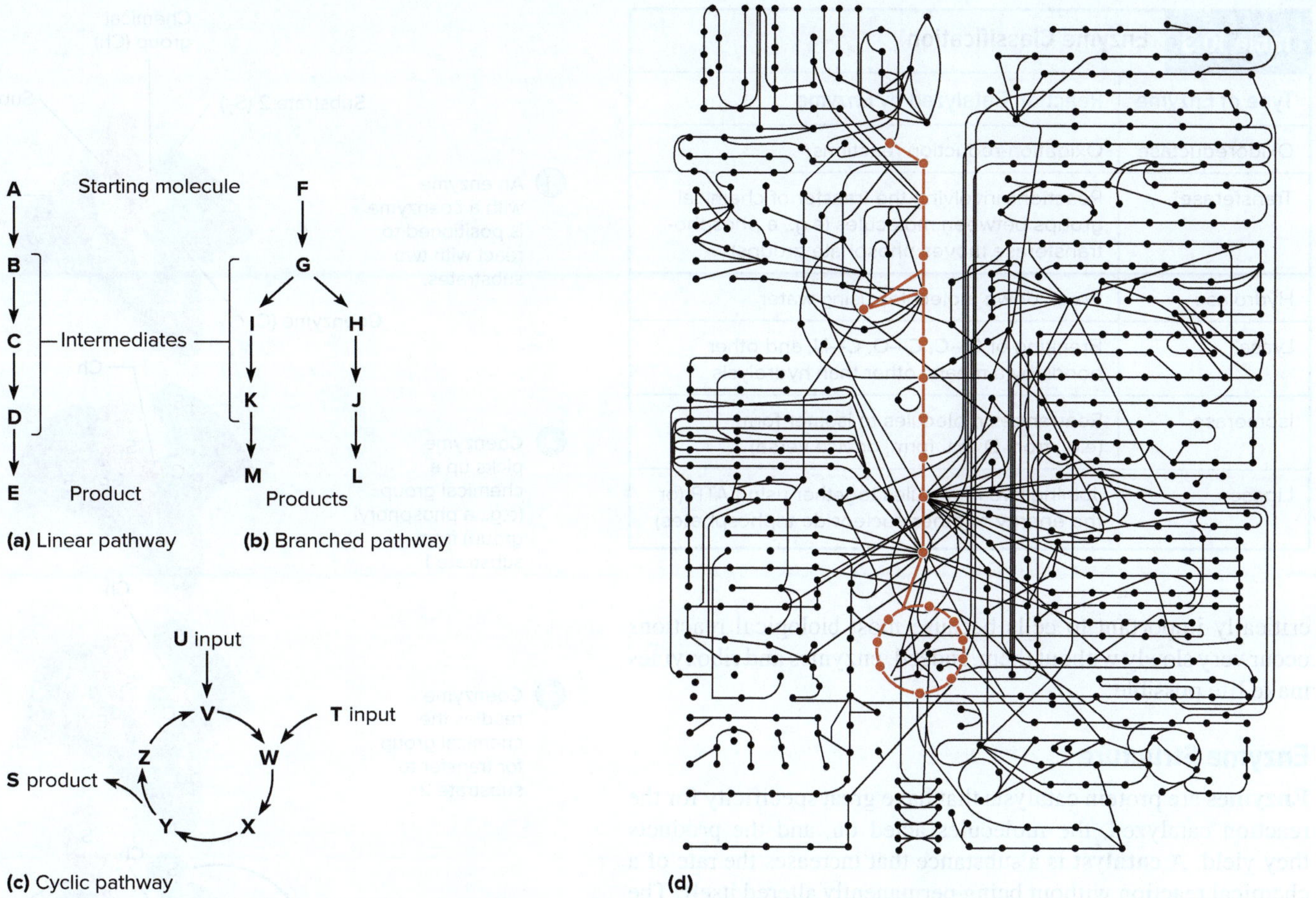

Figure 10.12 Chemical Reactions Carried Out by Cells Are Organized into Biochemical Pathways. Biochemical pathways take two main forms: linear and cyclic. (a) Molecules in the linear forms are distinguished based on their placement in the pathway. (b) Some linear biochemical pathways are branched and thus have more than one end product (molecules M and L). (c) The molecules present in biochemical cycles are intermediates of the pathway (molecules V, W, X, Y, and Z). Biochemical cycles also generate products (molecule S); however, to keep the cycle running, there must be inputs into the cycle (molecules U and T converted to V and W, respectively). (d) Metabolic pathways are interconnected and form a complex network. In this representation of a metabolic network, each point represents a metabolite. Each line represents an enzyme-catalyzed reaction that converts one metabolite into another. Red pathways illustrate the connections between linear, branched, and cyclic pathways.

10.6 Enzymes and Ribozymes Speed Up Cellular Chemical Reactions

After reading this section, you should be able to:

- **a.** Describe the function and chemical makeup of enzymes
- **b.** Distinguish apoenzyme from holoenzyme and prosthetic group from coenzyme
- **c.** Draw a diagram that shows the effect of an enzyme on the activation energy of a chemical reaction
- **d.** Describe the effects of substrate concentration, pH, and temperature on enzyme activity
- **e.** Differentiate competitive and noncompetitive inhibitors of enzymes
- **f.** Compare and contrast ribozymes and enzymes

As we discuss in section 10.1, an exergonic reaction is one with a negative $\Delta G^{\circ\prime}$ and an equilibrium constant greater than one. An exergonic reaction proceeds to completion in the direction written (i.e., toward the right of the equation). Nevertheless, reactants for an exergonic reaction often can be combined with no obvious result. For instance, if a polysaccharide such as starch is mixed in water, hydrolysis of the starch into glucose monomers is exergonic and will occur spontaneously, given enough time. However, the time needed is very long. Even if an organic chemist carried out this reaction in 6 moles/liter (M) HCl and at 100°C, it would still take several hours to go to completion. A cell, on the other hand, can accomplish the same reaction at neutral pH, at a much lower temperature, and in just a fraction of a second. Cells can do this because they manufacture biological catalysts that speed up chemical reactions. Most of these catalysts are proteins called enzymes. Other catalysts are RNA molecules termed ribozymes. Enzymes and ribozymes are

Table 10.3 Enzyme Classification

Type of Enzyme	Reaction Catalyzed by Enzyme
Oxidoreductase	Oxidation-reduction reactions
Transferase	Reactions involving the transfer of chemical groups between molecules (e.g., a phosphotransferase moves phosphate groups)
Hydrolase	Break down molecules using water
Lyase	Breaking of C—C, C—O, C—N, and other bonds by a means other than hydrolysis
Isomerase	Rearranges molecules to isomer form (e.g., from D to L form, or vice versa)
Ligase	Joining two molecules together using ATP (or the energy of other nucleoside triphosphates)

critically important to cells because most biological reactions occur very slowly without them. Indeed, enzymes and ribozymes make life possible.

Enzyme Structure

Enzymes are protein catalysts that have great specificity for the reaction catalyzed, the molecules acted on, and the products they yield. A **catalyst** is a substance that increases the rate of a chemical reaction without being permanently altered itself. The reacting molecules are called **substrates,** and the substances formed are the **products.** Enzymes may be placed in one of six general classes and usually are named in terms of the substrates they act on and the type of reaction catalyzed (**table 10.3**). *Proteins (appendix I)*

Many enzymes are composed only of proteins. However, some are composed of two parts: a protein component called the **apoenzyme** and a nonprotein component called a **cofactor.** Cofactors include metal ions and a variety of organic molecules. The complete enzyme consisting of the apoenzyme and its cofactor is called the **holoenzyme.** If the cofactor is tightly or covalently attached to the apoenzyme, it is a **prosthetic group.** If the cofactor is loosely attached and can dissociate from the apoenzyme after products have been formed, it is called a **coenzyme.** Many vitamins required by humans serve as coenzymes or as their precursors (e.g., riboflavin is incorporated into FAD). Coenzymes often carry one of the products of a chemical reaction to another enzyme or transfer chemical groups from one substrate to another (**figure 10.13**).

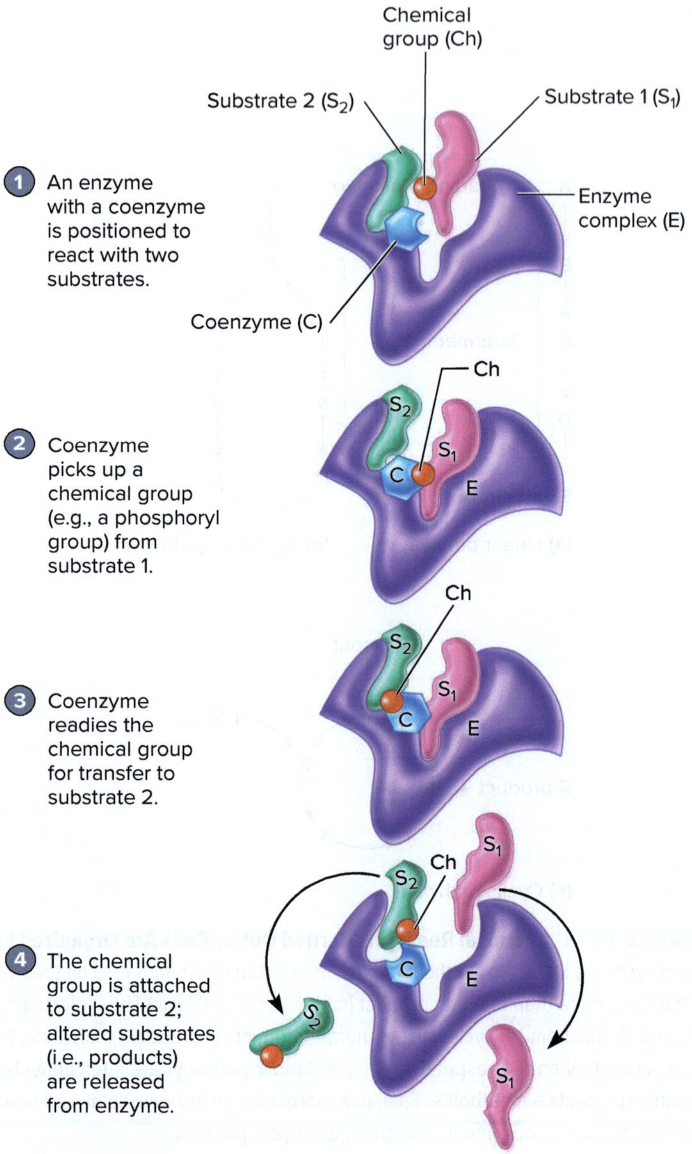

Figure 10.13 Coenzymes as Carriers.

How Enzymes Speed Up Reactions

It is important to keep in mind that enzymes increase the rates of reactions but do not alter their equilibrium constants. This means that if a reaction is endergonic, an enzyme cannot make it exergonic. Enzymes simply speed up the rate at which a reaction proceeds toward its final equilibrium. *How Enzymes Work*

How do enzymes catalyze reactions? Some understanding can be gained by considering the course of a simple exergonic chemical reaction.

$$A + B \rightleftharpoons C + D$$

When molecules A and B approach each other to react, they form a transition-state complex, which resembles both the substrates and the products (**figure 10.14**). **Activation energy** is required to bring the reacting molecules together in the correct way to reach the transition state. The transition-state complex can then resolve to yield the products C and D. The difference in free energy level between reactants and products is $\Delta G^{\circ\prime}$. Thus the equilibrium in our example lies toward the products because $\Delta G^{\circ\prime}$ is negative (i.e., the products are at a lower energy level than the substrates).

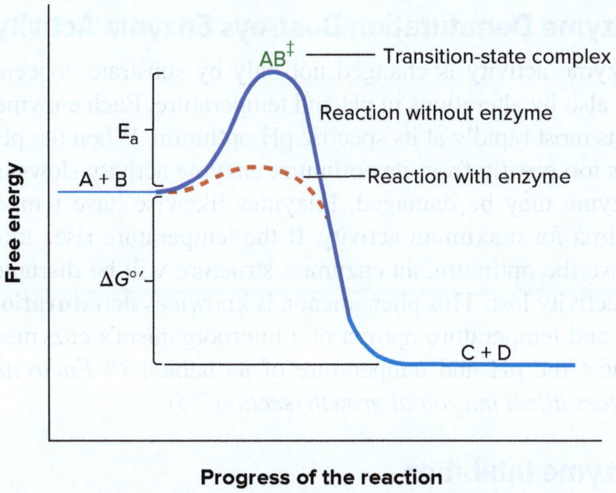

Figure 10.14 Enzymes Lower the Energy of Activation. This figure traces the course of a chemical reaction in which A and B are converted to C and D. The transition-state complex is represented by $AB^{\ddagger}$, and the activation energy required to reach it, by E_a. The red line represents the course of the reaction in the presence of an enzyme. Note that the activation energy is much lower in the enzyme-catalyzed reaction.

As seen in figure 10.14, A and B will not be converted to C and D without energy; the amount of energy is equivalent to the activation energy. Enzymes accelerate reactions by lowering the activation energy; therefore more substrate molecules will have sufficient energy to come together and form products. Even though the equilibrium constant (or $\Delta G^{\circ\prime}$) is unchanged, equilibrium is reached more rapidly in the presence of an enzyme because of this decrease in activation energy.

Researchers have worked hard to discover how enzymes lower the activation energy of reactions. Enzymes bring substrates together at a specific location in the enzyme called the **active** or **catalytic site** to form an enzyme-substrate complex. How the enzyme and substrate interact still has not been fully elucidated. However, the induced fit model describes the interaction for many enzymes. In the induced fit model, the enzyme changes shape when it binds the substrate so that the active site surrounds and precisely fits the substrate. This mechanism is used by hexokinase, as illustrated in **figure 10.15**. The formation of an enzyme-substrate complex can lower the activation energy in many ways. For example, by bringing the substrates together at the active site, the enzyme is, in effect, concentrating them so the reaction can occur more quickly. An enzyme does not simply concentrate its substrates, however. It also binds them so they are correctly oriented with respect to each other. Such an orientation lowers the amount of energy required by the substrates to reach the transition state. These and other catalytic site activities speed up a reaction by hundreds of thousands of times.

Substrate Concentration Affects Enzyme Activity

Enzyme activity varies in response to substrate concentration. Examination of **figure 10.16** shows that the curve describing the change in enzyme activity (usually expressed as the rate of product formation) in response to substrate concentration is a hyperbola. Why is this so? At very low substrate concentrations, an enzyme makes product slowly because it seldom contacts a substrate molecule. As the substrate concentration increases, enzymes bind substrates more often, and the velocity of the reaction is greater than at a lower substrate concentration. Eventually further increases in substrate concentration do not result in a greater reaction velocity because the available enzyme molecules are binding substrate and converting it to product as rapidly as possible. That is, enzymes are saturated with substrate and operating at maximal velocity (V_{max}).

Substrate concentrations in cells are often low. Therefore it is useful to know the substrate concentration an enzyme needs to function adequately. By convention, the substrate concentration required for the enzyme to achieve half-maximal velocity is used as a measure of the apparent affinity of an enzyme for its substrate. This value is called the **Michaelis constant (K_m)**. The lower the K_m value, the lower the substrate concentration at which an enzyme catalyzes its reaction. Enzymes with a low K_m value are said to have a high affinity for their substrates.

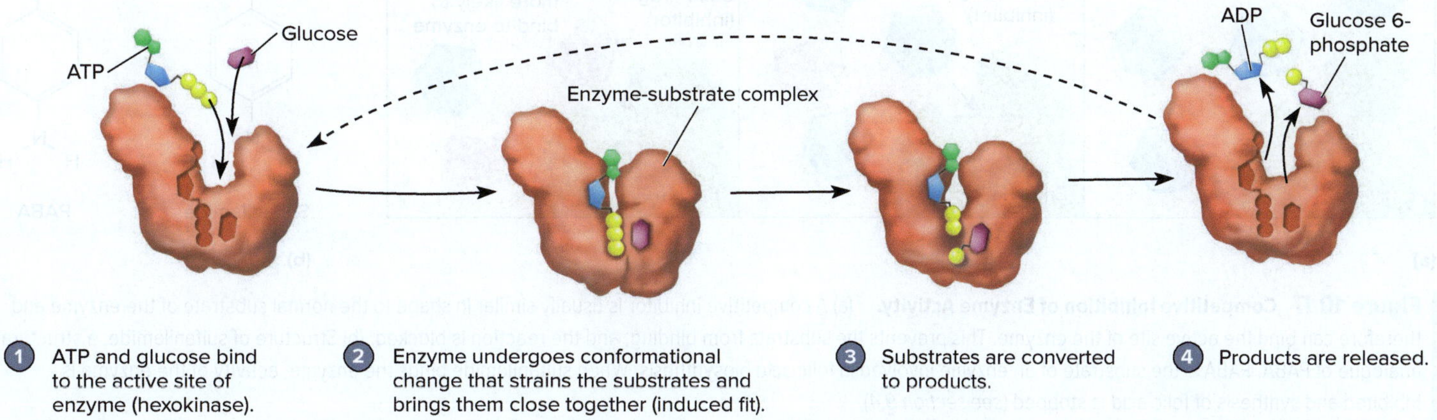

Figure 10.15 The Induced Fit Model of Enzyme Function. Shown is the reaction between glucose and ATP catalyzed by the enzyme hexokinase.

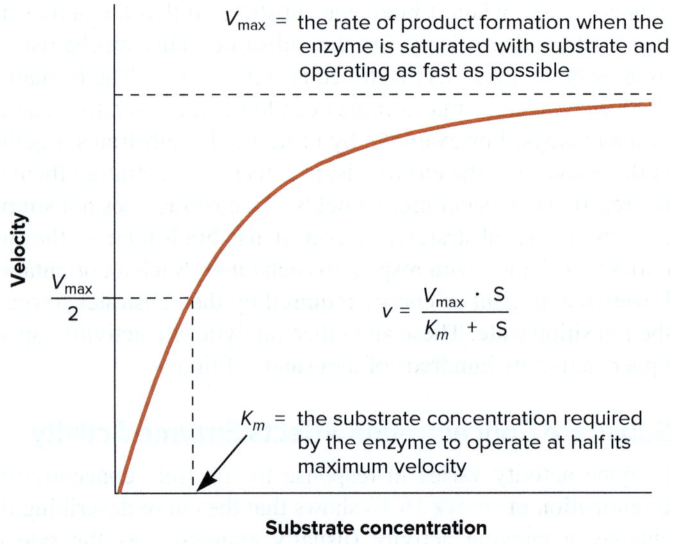

Figure 10.16 The Effect of Substrate Concentration on Enzyme Activity. Enzyme activity increases as substrate concentration increases until a maximum velocity is reached (V_{max}). The resulting curve is described by the Michaelis-Menten equation, which was derived by biochemists Leonor Michaelis (1875–1949) and Maud Menten (1879–1960). It relates reaction velocity (v) to substrate concentration (S) using V_{max} and the Michaelis constant (K_m).

MICRO INQUIRY *Will an enzyme with a relatively high K_m have a high or low affinity for its substrate? Explain.*

Enzyme Denaturation Destroys Enzyme Activity

Enzyme activity is changed not only by substrate concentration but also by alterations in pH and temperature. Each enzyme functions most rapidly at its specific pH optimum. When the pH deviates too greatly from its optimum, enzyme activity slows and the enzyme may be damaged. Enzymes likewise have temperature optima for maximum activity. If the temperature rises too much above the optimum, an enzyme's structure will be disrupted and its activity lost. This phenomenon is known as **denaturation.** The pH and temperature optima of a microorganism's enzymes often reflect the pH and temperature of its habitat. *Environmental factors affect microbial growth (section 7.5)*

Enzyme Inhibition

Microorganisms can be poisoned by a variety of chemicals (e.g., cyanide), and many of the most potent poisons are enzyme inhibitors. A **competitive inhibitor** directly competes with the substrate at an enzyme's catalytic (or active) site and prevents the enzyme from forming product (**figure 10.17**). Competitive inhibitors usually resemble the structure of normal substrates, but they cannot be converted to products.

Competitive inhibitors are important in the treatment of many microbial diseases. Sulfa drugs such as sulfanilamide (figure 10.17*b*) resemble *p*-aminobenzoate (PABA), a molecule used in the formation of the coenzyme folic acid. Sulfa drugs compete with PABA for the catalytic site of an enzyme involved in

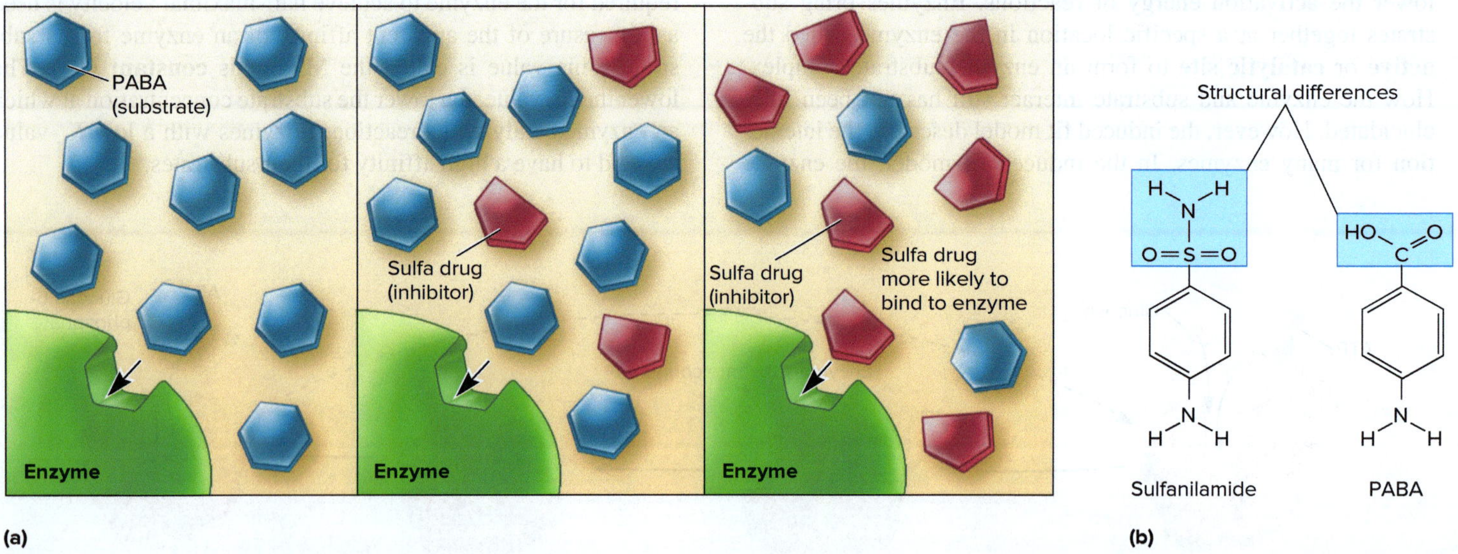

Figure 10.17 Competitive Inhibition of Enzyme Activity. (a) A competitive inhibitor is usually similar in shape to the normal substrate of the enzyme and therefore can bind the active site of the enzyme. This prevents the substrate from binding, and the reaction is blocked. (b) Structure of sulfanilamide, a structural analogue of PABA. PABA is the substrate of an enzyme involved in folic acid biosynthesis. When sulfanilamide binds the enzyme, activity of the enzyme is inhibited and synthesis of folic acid is stopped (*see section 9.4*).

MICRO INQUIRY *Based on this figure, explain the notion of a minimum inhibitory concentration of sulfanilamide.*

Table 10.4 Examples of Self-Splicing Ribozymes

Function	Where Observed
Splicing of pre-rRNA	*Tetrahymena* spp.
Splicing of mitochondrial rRNA and mRNA	Numerous fungi
Splicing of chloroplast tRNA, rRNA, and mRNA	Plants and algae
Splicing of viral mRNA	Viruses (e.g., T4, hepatitis delta virus)

folic acid synthesis. This blocks the production of folic acid and inhibits the growth of organisms that require its synthesis. Humans are not harmed because we do not synthesize folic acid but rather obtain it in our diet. ◀ *Metabolic antagonists (section 9.4)*

Noncompetitive inhibitors affect enzyme activity by binding to the enzyme at a location that is not the catalytic site. This alters the enzyme's shape, rendering it inactive or less active. These inhibitors are called noncompetitive because they do not directly compete with the substrate. Heavy metals such as mercury frequently are noncompetitive inhibitors of enzymes.

Ribozymes: Catalytic RNA Molecules

In the early 1980s Thomas Cech and Sidney Altman discovered that some RNA molecules also catalyze reactions. Catalytic RNA molecules are called **ribozymes.** One important ribozyme is located in ribosomes and is responsible for catalyzing peptide bond formation between amino acids during protein synthesis. Many ribozymes demonstrate self-splicing activity, that is they catalyze their own exisicion from a larger RNA molecule. Examples are listed in **table 10.4**. Just as with enzymes, the shape of a ribozyme is essential to catalytic efficiency. Ribozymes also have Michaelis-Menten kinetics (figure 10.16). ▶ *Translation in bacteria (section 13.7)*

Comprehension Check

1. The enzyme carbonic anhydrase catalyzes the conversion of CO_2 and H_2O into bicarbonate (HCO_3^-) and protons (H^+). In order to function, the enzyme covalently binds a zinc ion. Identify the substrates, product, and components of the carbonic anhydrase holoenzyme. Is zinc a prosthetic group or a coenzyme?
2. Draw a diagram showing the effect enzymes have on the activation energy of the reactions they catalyze and explain why enzymes do not change the equilibria of the reactions they catalyze.
3. Why does the rate of an enzyme catalyzed reaction plateau when V_{max} is reached?
4. Will enzymes need to lower the activation energy more or less in thermophiles and psychrophiles relative to mesophiles? Explain your answer.
5. Compare the binding sites of competitive and noncompetitive inhibitors, and explain how they each inhibit enzymes.
6. Construct a Venn diagram that features enzymes and ribozymes.

10.7 Metabolism Must Be Regulated to Maintain Homeostasis

After reading this section, you should be able to:

a. Describe metabolic channeling and one example of how it is accomplished
b. Distinguish allosteric regulation and covalent modification
c. Create a concept map that illustrates how positive allosteric and negative allosteric effectors regulate the activity of an enzyme
d. List three chemical groups commonly used to covalently modify an enzyme and its activity
e. Explain how feedback inhibition is used to control the functioning of biosynthetic pathways
f. Develop a model illustrating how feedback inhibition can be used to regulate a multiply-branched biosynthetic pathway

Microorganisms must regulate their metabolism to conserve raw materials and energy, and to maintain balance among various cell components. Microbes often live in environments where the nutrients, energy sources, and physical conditions change rapidly. Therefore microbes must continuously monitor internal and external conditions, and respond accordingly. This involves activating or inactivating metabolic pathways as needed. Metabolic pathways can be regulated in three major ways: (1) metabolic channeling, (2) regulation of the synthesis of a particular enzyme (often referred to as regulation of gene expression), and (3) directly controlling the activity of enzymes, referred to as posttranslational regulation. Although we focus here on individual pathways, figure 10.12*d* reminds us that pathways are interconnected. Thus regulating one pathway affects others. Cells often use multiple regulatory approaches to coordinate their complex metabolic activities.

Metabolic Channeling

Metabolic channeling influences pathway activity by localizing metabolites and enzymes to specific parts of a cell. One of the most common metabolic channeling mechanisms is **compartmentation,** the distribution of enzymes and metabolites among separate cell structures or organelles. Compartmentation is particularly important in the organelles of eukaryotic microorganisms. For example, fatty acid catabolism is located within the mitochondrion, whereas fatty acid synthesis occurs in the cytosol. Microcompartments and the periplasm in diderm bacteria are also examples of compartmentation. Compartmentation makes possible the simultaneous but separate operation and regulation of similar pathways. Furthermore, pathway activities can be coordinated through regulation of the transport of metabolites and coenzymes between cell compartments.

Regulation of Gene Expression

The regulation of the synthesis of a particular enzyme involves changing the rate of transcription or translation (or both) to control the amount of an enzyme present in the cell. When the enzyme is not needed, very little if any of the enzyme is synthesized. Conversely, when the enzyme is needed, more enzyme molecules are made. Regulation of gene expression is a relatively slow response to changes in the cell's state, but it saves the cell considerable energy and raw materials. Cells have evolved numerous mechanisms for regulating gene expression. We save the discussion of these mechanisms for chapters 14 and 15.

Posttranslational Regulation of Enzyme Activity

In contrast to regulation of gene expression, the direct stimulation or inhibition of the activity of critical enzymes rapidly alters pathway activity. It is often called **posttranslational regulation** because it occurs after the enzyme has been synthesized. A number of posttranslational regulatory mechanisms are known. Some are irreversible; for instance, cleavage of a protein can either increase or decrease its activity. Others, such as allosteric regulation and covalent modification, are reversible.

Allosteric Regulation

Most enzymes that regulate cell processes are **allosteric enzymes.** The activity of an allosteric enzyme is altered by a small molecule known as an **allosteric effector.** The effector binds reversibly by noncovalent forces to a **regulatory site** separate from the catalytic site and causes a change in the shape (conformation) of the enzyme (**figure 10.18**). The activity of the catalytic site is altered as a result. A positive effector increases enzyme activity, whereas a negative effector decreases activity (i.e., inhibits the enzyme). These changes in activity often result from alterations in the apparent affinity of the enzyme for its substrate, but changes in maximum velocity also can occur.

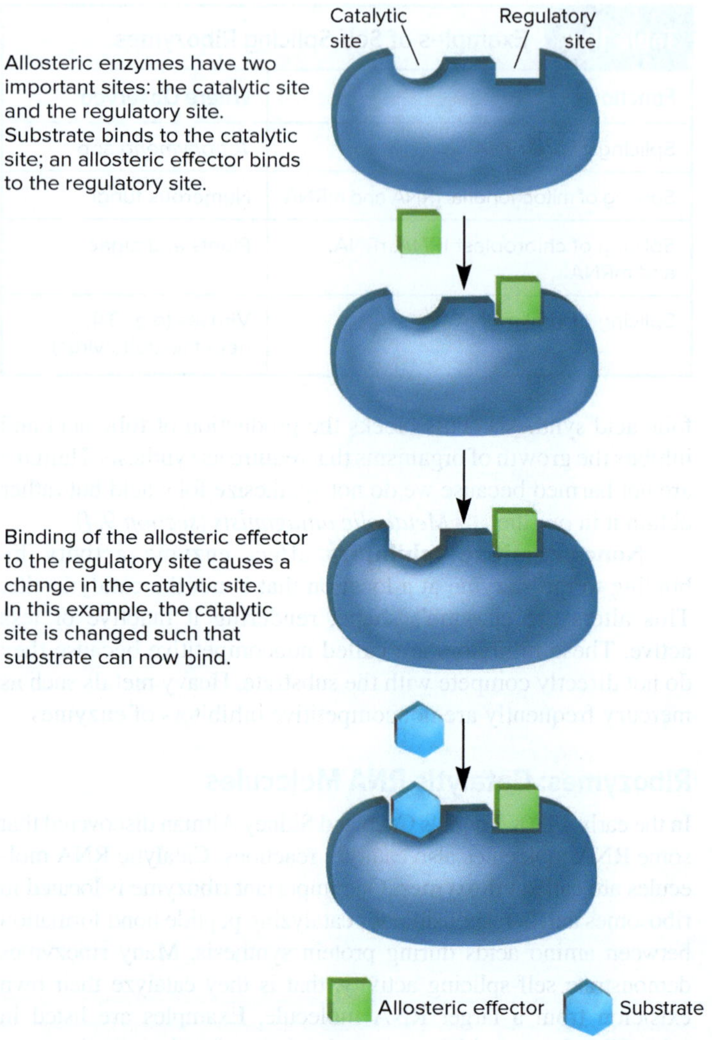

Figure 10.18 Allosteric Regulation. The structure and function of an allosteric enzyme.

MICRO INQUIRY *Is this allosteric effector a positive or negative effector? Is this true of all allosteric effectors?*

Covalent Modification of Enzymes

Whereas allosteric modification involves a small molecule binding to a protein, enzymatic activity also can be switched on and off by reversible covalent modification. Usually this occurs through the addition and removal of a particular chemical group, typically a phosphoryl, methyl, or adenylyl group. The enzymes that catalyze these reactions are themselves tightly regulated.

One of the most intensively studied regulatory enzymes is *Escherichia coli*'s glutamine synthetase, an enzyme involved in nitrogen assimilation. It is a large, complex enzyme consisting of 12 identical subunits, each of which can be covalently modified by an adenylic acid residue (**figure 10.19**). When an adenylic acid residue is attached to all 12 subunits, glutamine synthetase is not very active. Removal of AMP groups produces more active deadenylylated glutamine synthetase, and glutamine is formed. ▶| *Inorganic nitrogen assimilation (section 12.5)*

Covalent modification and allosteric regulation are not mutually exclusive; an enzyme can be subject to both mechanisms. For instance, glutamine synthetase also is regulated allosterically. This enables enzyme activity to be fine-tuned to a variety of conditions.

Feedback Inhibition

The rate of many metabolic pathways is adjusted through control of a pacemaker enzyme, one that catalyzes the slowest or rate-limiting reaction in the pathway. Because other reactions proceed more rapidly than the pacemaker reaction, changes in the activity of this enzyme directly alter the speed with which the entire pathway operates. Usually the first reaction in a pathway is catalyzed by a pacemaker enzyme. The end product of the pathway often inhibits this regulatory enzyme, a process known as **feedback inhibition** or **end product inhibition.** Feedback inhibition ensures balanced production of a pathway end product. If the end product becomes too concentrated, it inhibits the regulatory enzyme and slows its own synthesis. As the concentration of the end product decreases, pathway activity again increases and more product is formed. In this way, feedback inhibition matches

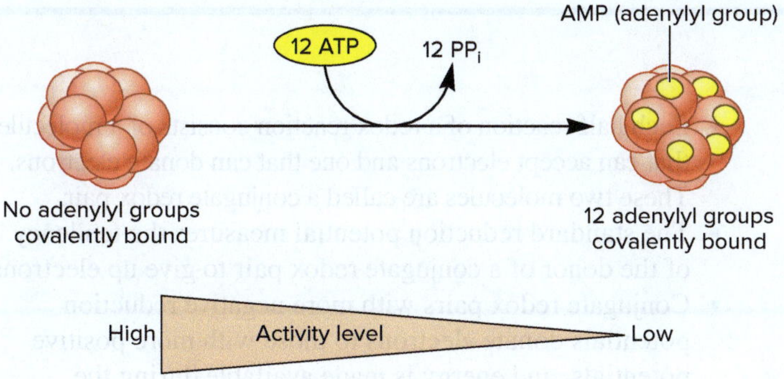

Figure 10.19 Regulation of Glutamine Synthetase Activity by Covalent Modification. Glutamine synthetase consists of 12 identical subunits. Six subunits form a ring, and one ring (darker color) is above the other (lighter color). Not all of the lower ring subunits are shown. Each subunit can be adenylylated. Only the adenylyl groups on the upper ring are shown. As the number of adenylyl groups increases, the activity of the enzyme decreases.

end product supply with demand. *A Biochemical Pathway; Feedback Inhibition of Biochemical Pathways*

Frequently a biosynthetic pathway branches to form more than one end product. In such a situation, the synthesis of all end products must be coordinated precisely. Branching biosynthetic pathways usually achieve a balance among end products by using regulatory enzymes at branch points (**figure 10.20**). If an end product is present in excess, it often inhibits the first enzyme on the branch leading to its formation, thereby regulating its own formation without affecting the synthesis of other products. In figure 10.20, notice that both products also inhibit the initial enzyme in the pathway. An excess of one product slows the flow of carbon (substrate A in figure 10.20) into the whole pathway and also inhibits the appropriate branch-point enzyme. Because less carbon is required when only one branch is functioning, feedback inhibition of the initial pacemaker enzyme helps match the supply with the demand in branching pathways. The regulation of branched pathways is often made even more sophisticated by the presence of **isoenzymes (isozymes),** different forms of an enzyme that catalyze the same reaction. In the pathway shown in figure 10.20, end product P would control one isoenzyme (purple in this case) and end product Q would control a different isoenzyme (blue) of the first enzyme in the pathway. If Q were present in excess, it would affect only its isoenzyme. The P isoenzyme would still function, and the pathway would still operate, but at a decreased level. Thus an initial pacemaker step may be catalyzed by several isoenzymes, each under separate control. In this way, an excess of a single end product reduces pathway activity but does not completely block pathway function because some isoenzymes are still active.

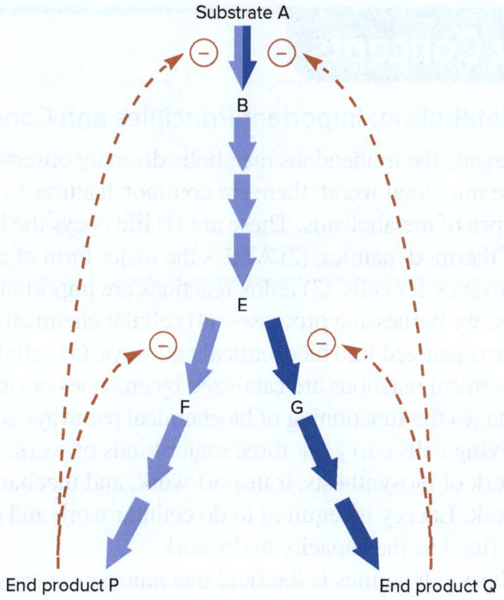

Figure 10.20 Feedback Inhibition. Feedback inhibition in a branching pathway with two end products. The branch-point enzymes, those catalyzing the conversion of intermediate E to F and G, are regulated by feedback inhibition. Products P and Q also inhibit the initial reaction in the pathway, which may be catalyzed by isozymes (indicated by dual colored arrow). A colored line with a minus sign at one end indicates that an end product, P or Q, inhibits the enzyme catalyzing the step next to the minus.

MICRO INQUIRY *Does inhibition of the pacemaker isoenzyme by end product P also shut off production of end product Q?*

Comprehension Check

1. Define the terms *metabolic channeling* and *compartmentation*. How are they involved in the regulation of metabolism?
2. Define *allosteric enzyme* and *allosteric effector*.
3. How can regulatory enzymes be influenced by reversible covalent modification?
4. What is a pacemaker enzyme? How does feedback inhibition automatically adjust the concentration of a pathway end product?
5. Why are regulatory enzymes often located at pathway branch points? What are isoenzymes, and what is their role in pathway regulation?

Key Concepts

10.1 Metabolism: Important Principles and Concepts

- Despite the tremendous metabolic diversity observed in the microbial world, there are common features to all types of metabolisms. These are (1) life obeys the laws of thermodynamics, (2) ATP is the major form of energy currency for cells, (3) redox reactions are important in energy-harnessing processes, (4) cellular chemical reactions are organized into biochemical pathways, (5) cellular chemical reactions are catalyzed by enzymes or ribozymes, and (6) the functioning of biochemical pathways is regulated.
- Living cells carry out three major kinds of work: chemical work of biosynthesis, transport work, and mechanical work. Energy is required to do cellular work and can be defined as the capacity to do work.
- Thermodynamics is the field that analyzes energy changes in systems such as cells. The first law of thermodynamics states that energy is neither created nor destroyed. The second law of thermodynamics states that changes occur in such a way that the randomness or disorder of the universe increases to the maximum possible. That is, entropy always increases during spontaneous processes.
- The first and second laws of thermodynamics can be combined to determine the amount of energy made available for useful work.

$$\Delta G = \Delta H - T \cdot \Delta S$$

In this equation the change in free energy (ΔG) is the energy made available for useful work, the change in enthalpy (ΔH) is the change in heat content, and the change in entropy is ΔS.

- The standard free energy change ($\Delta G^{\circ\prime}$) for a chemical reaction is directly related to the equilibrium constant.
- In exergonic reactions, $\Delta G^{\circ\prime}$ is negative and the equilibrium constant is greater than one; the reaction goes to completion as written. Endergonic reactions have a positive $\Delta G^{\circ\prime}$ and an equilibrium constant less than one (**figure 10.1**).

10.2 ATP: The Major Energy Currency of Cells

- ATP is a high-energy molecule that carries energy in a useful form from one reaction or location in a cell to another (**figure 10.2**).
- ATP is readily synthesized from ADP and P_i using energy released from exergonic reactions; when hydrolyzed back to ADP and P_i, it releases the energy, which is used to drive endergonic reactions. This cycling of ATP with ADP and P_i is called the cell's energy cycle (**figure 10.4**).

10.3 Redox Reactions: Reactions of Central Importance in Metabolism

- In oxidation-reduction (redox) reactions, electrons move from an electron donor to an electron acceptor. Each redox reaction consists of two half reactions: One is the electron-donating half reaction; the other is the electron-accepting half reaction (**table 10.2**).
- Each half reaction of a redox reaction consists of a molecule that can accept electrons and one that can donate electrons. These two molecules are called a conjugate redox pair.
- The standard reduction potential measures the tendency of the donor of a conjugate redox pair to give up electrons.
- Conjugate redox pairs with more negative reduction potentials donate electrons to those with more positive potentials, and energy is made available during the transfer (**figure 10.5**).

10.4 Electron Transport Chains: Sets of Sequential Redox Reactions

- Many metabolic processes use a series of electron carriers to transport electrons from the primary electron donor (e.g., an organism's energy source) to a final electron acceptor (e.g., oxygen). A series of electron carriers is called an electron transport chain (ETC). ETCs are associated with membranes (**figure 10.6**).
- Some of the most important electron carriers in cells are NAD^+, $NADP^+$, FAD, FMN, coenzyme Q, cytochromes, and nonheme iron (Fe-S) proteins (**figures 10.7–10.11**). They differ in several ways, including how many electrons and protons they accept and transfer.

10.5 Biochemical Pathways: Sets of Linked Chemical Reactions

- The chemical reactions carried out by cells are organized to form biochemical pathways. The intermediates and products of pathways are called metabolites (**figure 10.12**).
- Metabolic pathways are interconnected such that the intermediates and end product of one pathway may be the starting molecules of others (figure 10.12*d*). Metabolite flux refers to the turnover of metabolites as they are used in various biochemical pathways.

10.6 Enzymes and Ribozymes Speed Up Cellular Chemical Reactions

- Enzymes are proteins that catalyze specific reactions.
- Many enzymes consist of a protein component, the apoenzyme, and a cofactor. Cofactors may be metals or organic molecules. Some cofactors are prosthetic groups, which are tightly bound to the apoenzyme. Other cofactors are loosely bound and are called coenzymes.
- Enzymes and ribozymes speed chemical reactions by binding substrates at their active sites and lowering the activation energy (**figures 10.14** and **10.15**).
- The rate of an enzyme- or ribozyme-catalyzed reaction increases with substrate concentration at low substrate levels and reaches a plateau (the maximum velocity, V_{max}) at saturating substrate concentrations. The Michaelis constant is the substrate concentration that the enzyme or ribozyme requires to achieve half-maximal velocity (**figure 10.16**).

- Enzymes have pH and temperature optima for activity. Their activity can be slowed by competitive and noncompetitive inhibitors (**figure 10.17**).
- Some ribozymes alter either their own structure or that of other RNAs. An important ribozyme is located in the ribosome, where it links amino acids together during protein synthesis.

10.7 Metabolism Must Be Regulated to Maintain Homeostasis

- The regulation of metabolism keeps cell components in proper balance and conserves metabolic energy and material. There are three major approaches to regulating metabolism: metabolic channeling, regulation of gene expression, and posttranslational regulation of enzyme activity.
- Metabolic channeling localizes metabolites and enzymes in different parts of the cell and influences pathway activity. A common channeling mechanism is compartmentation.
- Regulation of gene expression is brought about by controlling the rate of transcription and translation. This is discussed in chapters 14 and 15.
- Posttranslational regulation is the direct control of enzyme activity. There are two major types of reversible posttranslational regulation: allosteric control and covalent modification. In allosteric control, an allosteric effector binds noncovalently to an enzyme at a regulatory site separate from the catalytic site. This causes a conformational change in the enzyme and alters its activity (**figure 10.18**). In covalent modification, enzyme activity is regulated by reversible attachment of a chemical group, such as a phosphoryl, methyl, or adenylyl group (**figure 10.19**).
- The first enzyme in a pathway and enzymes at branch points often are subject to feedback inhibition by one or more end products. Excess end product slows its own synthesis (**figure 10.20**).

Active Learning

1. Examine the structures of macromolecules in appendix I, figure AI.7. Which type has the most electrons to donate? Suggest why carbohydrates are usually the primary source of energy for bacteria that use organic molecules as energy sources.
2. Examine the branched pathway shown here for the synthesis of the amino acids aspartate, methionine, lysine, threonine, and isoleucine.

   Scenario 1: The microbe is cultured in a medium containing aspartate and lysine but lacking methionine, threonine, and isoleucine.

   Scenario 2: The microbe is cultured in a medium containing a rich supply of all five amino acids.

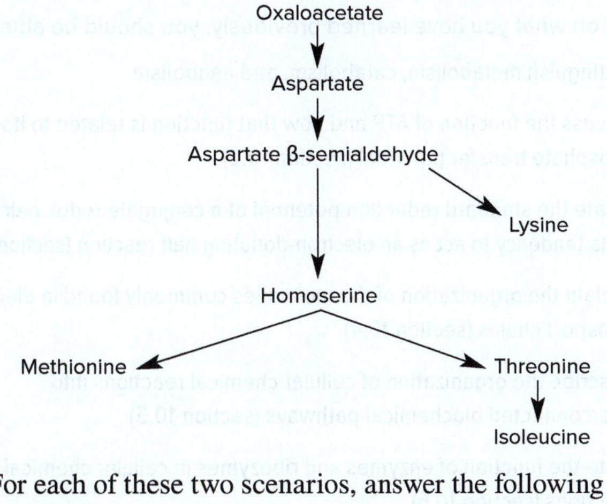

   For each of these two scenarios, answer the following questions:

   a. Which portion(s) of the pathway would need to be shut down in this situation?
   b. How might allosteric control be used to accomplish this?

3. Chlamydia are pathogenic bacteria that must be grown within a eukaryotic host cell. They rely on the host cell for many nutrients, including nucleotides, lipids, and amino acids. It is thus no surprise that when grown in co-culture with a host depleted of any of these metabolites, chlamydial growth declines. However, high levels of certain amino acids also inhibit chlamydial growth. Specifically, the amino acids leucine, isoleucine, methionine, and phenylalanine retard growth by inhibiting an amino acid transporter called BrnQ. On the other hand, if valine is present in the medium, these amino acids have no effect on BrnQ. Discuss how this might be possible.
4. *Pseudomonas aeruginosa* is an important pathogen that shows increasing resistance to antibiotics. In exploring potential new molecules and pathways that antibiotics might target, scientists have focused on dihydrodipicolinate synthase (DHDPS). This enzyme is responsible for producing two cell wall constituents: L-lysine and diaminopimelate (DAP; *see figures 3.15 and 3.18*). DAP and L-lysine are the end products of a branched pathway and *P. aeruginosa* has two distinct versions of DHDPS: one that is allosterically inhibited by L-lysine and one that is not. Explain how inhibition by L-lysine works and hypothesize why this organism evolved to have two different DHDPS isozymes.

CHAPTER

11

Catabolism: Energy Release and Conservation

rkl_foto/Shuttetsock

Solar, Wind, and ... Microbes?

Let's take stock of the current state of the global energy situation: (1) The Earth's population is expected to grow by about 2 billion to peak at about 10 billion in the 2080s; (2) despite efforts to cut oil, gas, and coal production, fossil fuel consumption continues to grow; (3) wind and solar contributed less than 15% to global energy production in 2024 even though their use has grown explosively in recent years. The fact is, the world is designed to run on liquid fuel; rockets may use solar power, but jets don't and it's unlikely they ever will. But (4) it's projected that the world will run out of oil, gas, and coal before the end of this century.

What's the solution? No one can say for sure how coming decades will unfold, but the role biofuels play in the global energy diet will almost certainly expand. Biofuels are liquid forms of energy; they can be produced by algae or microbes from renewable sources like plants. Currently the most common biofuel is bioethanol generated from sugar cane (Brazil), beet sugar (France), and corn (United States). But bioethanol is not the best solution. It's not as energy dense as other potential biofuels and the use of arable land to grow food stuffs that will be converted to biofuel raises ethical concerns considering the number of people who are food insecure globally. Additionally, there is concern that rain forests—an important sink for atmospheric CO_2—are being cleared to plant sugar cane for biofuel production.

Harnessing the catabolic power of microbes, i.e., the ability of microbes to degrade complex organic matter, is a growing and dynamic industry. Microbiologists are discovering new microbes that efficiently degrade cellulose so plant waste can replace corn. Synthetic biologists are even creating microbes with catabolic pathways tailored for production of diverse biofuels. Algae grown in tanks fueled by the sun can be harvested for a variety of biofuels, as well as other products such as cooking oils and even meat alternatives. It is both an urgent and exciting time for microbiology.

In this chapter, we present the most common catabolic pathways used by microbes. Although we cannot cover all the ways in which microbes can capture energy, or "make a living," we hope you develop an appreciation for why microbes are ubiquitous regardless of environmental conditions, and how they may help us save our planet.

Readiness Check:

Based on what you have learned previously, you should be able to:

- ✓ Distinguish metabolism, catabolism, and anabolism
- ✓ Discuss the function of ATP and how that function is related to its phosphate transfer potential (section 10.2)
- ✓ Relate the standard reduction potential of a conjugate redox pair to its tendency to act as an electron-donating half reaction (section 10.3)
- ✓ Explain the organization of the molecules commonly found in electron transport chains (section 10.4)
- ✓ Describe the organization of cellular chemical reactions into interconnected biochemical pathways (section 10.5)
- ✓ State the function of enzymes and ribozymes in cellular chemical reactions (section 10.6)

11.1 Metabolic Diversity and Nutritional Types

After reading this section, you should be able to:

a. Explain why microbes are found in almost all environments on Earth
b. State the carbon, energy, and electron sources of photolithoautotrophs, photoorganoheterotrophs, chemolithoautotrophs, chemolithoheterotrophs, and chemoorganoheterotrophs
c. Describe the products of the fueling reactions
d. Discuss the metabolic flexibility of microorganisms

What would you do without oxygen? What if you had no access to organic carbon (food)? The answer is obvious but illustrates the relatively recent history of mammals, which have evolved for only about 210 million years. Our limited metabolic capacity (must breathe air, must eat food) also explains why mammals are confined to the surface of the Earth.

Let's now compare mammals (like us) with bacteria and archaea, which have been evolving for about 3.5 billion years—about 3.3 billion years longer than mammals. The Great Oxidation Event, when global oxygen levels dramatically increased, occurred about 2.1–2.4 billion years ago. Early microbes, therefore, were all anaerobes. Without oxygen, food as we know it could not exist; instead microbes evolved to "eat rocks"—that is, to extract energy from inorganic molecules. Three and a half billion years have enabled microorganisms to diversify their metabolic capabilities so they can survive—and grow—just about anywhere, as long as they have access to sources of energy, electrons, and carbon.

The central importance of energy, electrons, and carbon has led biologists to describe organisms in terms of how these requirements are met. The terms can be combined to characterize an organism's nutritional type. The energy, electrons, and carbon that an organism obtains from its environment are used in chemical reactions referred to as **fueling reactions.** The nutritional types and fueling reactions are the focus of this section.

Nutritional Types Are Defined by the Source of an Organism's Energy, Electrons, and Carbon

For each requirement—energy, electrons, and carbon—a dichotomy exists for how an organism fulfills that need. Only two sources of energy are available to organisms: light and certain chemical molecules. **Phototrophs** use light as their energy source (**figure 11.1**); **chemotrophs** obtain energy from the oxidation of chemical compounds (either organic or inorganic). Likewise, organisms have only two sources for electrons. **Lithotrophs** (i.e., rock-eaters) use reduced inorganic substances as their electron source, whereas **organotrophs** extract electrons from organic compounds. Finally, organisms may be **heterotrophs**—organisms that use organic molecules as their carbon source for growth—or they may be **autotrophs**—organisms that use carbon dioxide (CO_2) as their source of carbon.

By combining the roots of these terms, most organisms can be placed in one of five nutritional types based on their primary sources of carbon, energy, and electrons (**table 11.1**). For instance, plants are photo (energy from light), litho (electrons from the inorganic molecule water), auto (carbon from CO_2), trophs (Greek *trophe,* nourishment); animals are quite the opposite: chemoorganoheterotrophs.

The majority of microorganisms thus far studied are either chemolithoautotrophic, photolithoautotrophic, or chemoorganoheterotrophic. **Chemolithoautotrophs** oxidize inorganic compounds such as iron-, nitrogen-, or sulfur-containing molecules to derive both energy and electrons for biosynthesis, and use CO_2 as their carbon source. Because they grow using only inorganic compounds, chemolithoautotrophs are most closely related to ancient microbes. They significantly impact global chemical transformations of elements (e.g., the conversion of ammonia to nitrate or sulfur to sulfate). **Photolithoautotrophs,** often called simply **photoautotrophs,** include purple and green sulfur bacteria that, like chemolithotrophs, extract electrons from inorganic donors such as hydrogen, hydrogen sulfide, and elemental sulfur, but gain energy from light. Other photolithoautotrophs, like photosynthetic protists (algae) and cyanobacteria evolved later and are plantlike in that they employ water as the electron

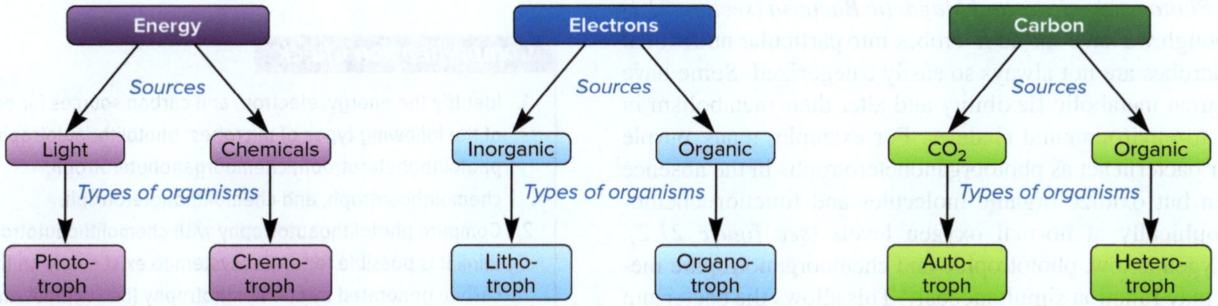

Figure 11.1 A Concept Map of the Sources of Energy, Electrons, and Carbon. For each, there are two different sources, which then defines the nutritional type of a microorganism.

MICRO INQUIRY *Which three nutritional types describe plants? Which describe you?*

Table 11.1 Major Nutritional Types of Microorganisms

Nutritional Type	Energy Source	Electron Source	Carbon Source	Representative Microorganisms
Photolithoautotroph	Light	Inorganic	CO_2	Purple and green sulfur bacteria, cyanobacteria, diatoms
Photoorganoheterotroph	Light	Organic	Organic carbon	Purple nonsulfur bacteria, green nonsulfur bacteria
Chemolithoautotroph	Inorganic chemicals	Inorganic	CO_2	Sulfur-oxidizing bacteria, hydrogen-oxidizing bacteria, methanogens, nitrifying bacteria, iron-oxidizing bacteria
Chemolithoheterotroph	Inorganic chemicals	Inorganic	Organic carbon	Some sulfur-oxidizing bacteria (e.g., *Beggiatoa* spp.)
Chemoorganoheterotroph	Organic chemicals, often same as C source	Organic, often same as C source	Organic carbon	Most nonphotosynthetic microbes, including most pathogens, fungi, and many protists

donor and release oxygen. They are responsible for the initial rise in oxygen during the Great Oxidation Event that in turn enabled the evolution of mammals. Photoautotrophs and chemolithoautotrophs also are important primary producers in ecosystems. That is, they fix CO_2, making reduced organic molecules, such as sugars, that sustain chemoorganoheterotrophs (like us).

Chemoorganoheterotrophs (sometimes called **chemoheterotrophs** or chemoorganotrophs) use reduced organic compounds (e.g., sugars) as sources of energy, electrons, and carbon. Frequently the same organic nutrient will satisfy all these requirements. Many chemoorganotrophs are used industrially to make foods (e.g., yogurt, pickles, cheese), medical products (e.g., antibiotics), and beverages (e.g., beer and wine). Nearly all human pathogenic microorganisms are chemoorganoheterotrophs.

The other nutritional types have fewer known microorganisms but are very important ecologically. Some phototrophic bacteria (e.g., purple nonsulfur bacteria) use organic matter as their electron donor and carbon source. These **photoorganoheterotrophs** are common inhabitants of polluted lakes and streams. **Chemolithoheterotrophs** use reduced inorganic molecules as their energy and electron source but derive their carbon from organic sources. Like chemolithoautotrophs, they contribute to numerous biogeochemical cycles. ▶| *Order Rhodospirillales Includes Photosynthesizers and Magnetic Bacteria (section 21.1)*

Although we have sorted microbes into particular nutritional types, microbes are not always so easily categorized. Some have evolved great metabolic flexibility and alter their metabolism in response to environmental changes. For example, many purple nonsulfur bacteria act as photoorganoheterotrophs in the absence of oxygen but oxidize organic molecules and function chemoorganotrophically at normal oxygen levels (*see figure 21.2*). When oxygen is low, phototrophic and chemoorganotrophic metabolism may function simultaneously. This allows the bacterium to gain energy from both light and organic molecules, and still supply the carbon it needs for biosynthesis. This sort of flexibility gives these microbes a distinct advantage if environmental conditions frequently change.

Fueling Reactions Convert Energy, Electron, and Carbon Sources into ATP, Reducing Power, and Precursor Metabolites

Metabolism is an excellent example of the conservation of highly efficient processes throughout evolution. Despite the diversity of energy, electron, and carbon sources used by organisms, all are used to generate three main products: **ATP,** the primary molecule used to conserve the energy supplied by an energy source; **reducing power,** molecules that serve as a ready supply of electrons for a variety of chemical reactions; and **precursor metabolites,** small organic molecules that provide the carbon skeletons needed for biosynthesis of important chemical building blocks (monomers) such as amino acids (**figure 11.2**). The processes that supply ATP, reducing power, and precursor metabolites are called *fueling reactions.* With the exception of CO_2-fixation, the fueling reactions are part of **catabolism** because molecules are broken down. As seen in figure 11.2, the generation of monomers using the products of the fueling reactions paves the way for anabolic reactions that synthesize macromolecules (e.g., polymers such as proteins). The macromolecules are used to construct cellular structures such as ribosomes and flagella. ▶| *CO_2 fixation: reduction and assimilation of CO_2 carbon (section 12.3)*

Comprehension Check

1. Identify the energy, electron, and carbon sources for each of the following types of microbes: photolithoautotroph, photolithoheterotroph, chemoorganoheterotroph, chemolithoautroph, and chemolithoheterotroph.
2. Compare photolithoautotrophy with chemolithoautotrophy. Do you think it is possible for an ecosystem to exist solely on the organic carbon generated by chemolithotrophy (i.e., without any contribution by photosynthetic organisms)? Explain your reasoning.
3. What are the three major products generated by fueling reactions? Summarize how they are used in anabolic processes.

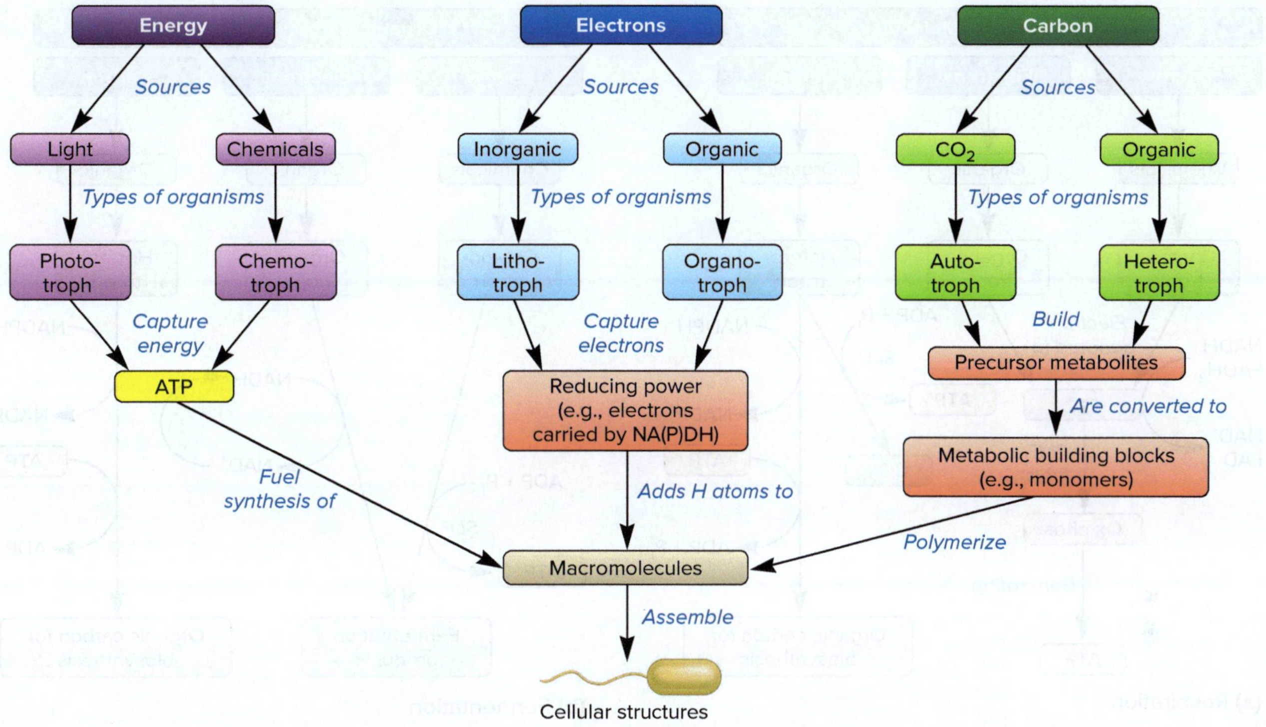

Figure 11.2 A Concept Map of Cellular Synthesis Requirements for Energy, Electron, and Carbon Sources. ATP and reducing power (e.g., NADPH) are used in anabolic reactions that convert precursor metabolites into monomers and other building blocks. Monomers are used to generate macromolecules such as proteins, which are assembled into various cell structures.

11.2 There Are Two Chemoorganotrophic Fueling Processes

After reading this section, you should be able to:

a. List the two types of chemoorganotrophic metabolisms
b. Compare the mechanisms by which ATP is synthesized by microbes using the two different types of chemoorganotrophy

As illustrated in figure 11.2, chemoorganotrophs use reduced organic compounds for catabolic and anabolic processes. When an organic compound is oxidized to release energy (catabolism), it also provides the carbon and electrons needed for **anabolism.** Before we learn about the details of catabolism, it is helpful to consider a brief overview. Chemoorganotrophs use two general types of catabolic strategies: respiration and fermentation. In both cases, when the organic energy source is oxidized, the electrons released must be accepted by electron carriers such as NAD^+ and FAD. When these reduced electron carriers (e.g., NADH, $FADH_2$) in turn donate the electrons to an electron transport chain, the metabolic process is called **respiration** and may be divided into two types (**figure 11.3*a***).

In aerobic respiration, the terminal, or final, electron acceptor is oxygen, whereas the terminal acceptor in anaerobic respiration is a different oxidized molecule such as NO_3^-, SO_4^{2-}, or Fe^{3+}. Sometimes organic acceptors such as fumarate and humic acids may be used. During respiration, electrons pass through the electron transport chain (ETC) to the terminal electron acceptor, generating a type of potential energy called the proton motive force (PMF). The PMF provides the energy for a membrane-bound ATP synthase to convert ADP and phosphate (P_i) to ATP. Respiration typically starts with the oxidation of NADH and $FADH_2$ by enzymes within the ETC. NAD^+ and FAD can now once again accept electrons and donate them to the ETC. In contrast to respiration, the second type of fueling reaction, **fermentation** (Latin *fermentare,* to cause to rise) does not use an ETC (**figure 11.3*b***). Instead, an endogenous (from within the cell) electron acceptor is used. This electron acceptor is usually an intermediate (e.g., pyruvate) of the catabolic pathway used to degrade and oxidize the organic energy source. During fermentation, NADH is oxidized so it can be reused, and ATP is synthesized almost exclusively by substrate-level phosphorylation. ◀ *ATP: The major energy currency of cells (section 10.2); Electron transport chains: sets of sequential redox reactions (section 10.4)*

By convention, the two fueling reactions, respiration and fermentation, are usually described with glucose as the energy source. This is done for several reasons. One is that glucose is used by many chemoorganotrophs as an energy source.

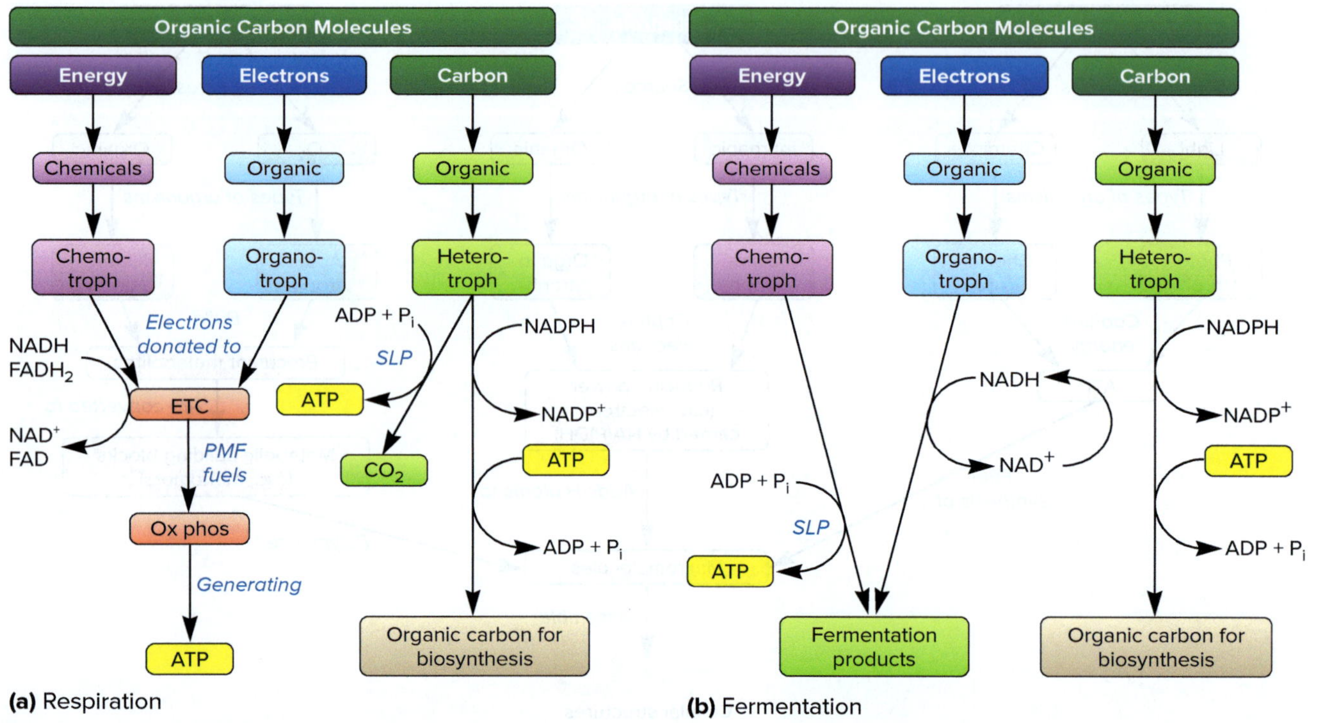

Figure 11.3 Chemoorganotrophic Fueling Processes. (a) In respiration, oxidation of organic molecules provides electrons that pass through an electron transport chain. This generates a proton motive force (PMF), which is used to synthesize most of the cellular ATP by a mechanism called oxidative phosphorylation (Ox phos); a small amount of ATP is made by substrate-level phosphorylation (SLP). In aerobic respiration, O_2 is the terminal electron acceptor, whereas in anaerobic respiration, exogenous molecules other than O_2 serve as electron acceptors (not shown). (b) During fermentation, endogenous organic molecules act as electron acceptors, electron transport chains are not present, and ATP is synthesized by substrate-level phosphorylation (SLP).

But perhaps more important is the way catabolic pathways are organized. Most chemoorganotrophs use a wide variety of organic molecules as energy sources (**figure 11.4**). They are degraded by pathways that generate either glucose or intermediates of the pathways used in glucose catabolism. Thus nutrient molecules are funneled into ever fewer metabolic intermediates. Indeed, a common pathway often degrades many similar molecules (e.g., several different sugars). The existence of a few metabolic pathways that each break down many nutrients greatly increases metabolic efficiency by avoiding the need for a large number of specific pathways.

Our discussion of chemoorganotrophy begins with aerobic respiration and introduces the glycolytic pathways, TCA cycle, and other important processes. Many of these also occur during anaerobic respiration. Fermentation is quite distinct from respiration. It involves only a subset of the reactions that function during respiration and only partially catabolizes the energy source. However, it is widely used by microbes and has important practical applications, including the production of many foods. ▶| *Microbiology of fermented foods: beer, cheese, and much more (section 40.5)*

Comprehension Check

1. Give examples of the types of electron acceptors used in fermentation and respiration. What is the difference between aerobic respiration and anaerobic respiration?
2. Why is it to a cell's advantage to catabolize diverse organic energy sources by funneling them into a few common pathways?

11.3 Aerobic Respiration Starts with Glucose Oxidation

After reading this section, you should be able to:

a. Diagram the major changes made to glucose as it is catabolized by the Embden-Meyerhof, Entner-Doudoroff, and pentose phosphate pathways
b. Calculate the yields of ATP and NAD(P)H by the Embden-Meyerhof, Entner-Doudoroff, and pentose phosphate pathways
c. Explain why amphibolic pathways are needed and give a specific example

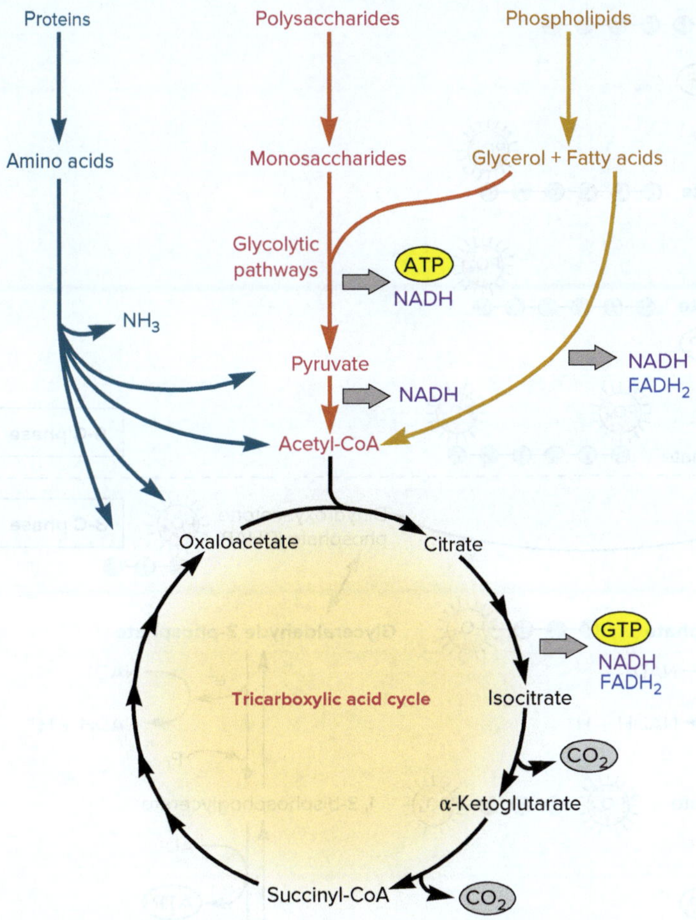

Figure 11.4 Pathways Used by Chemoorganotrophs to Catabolize Organic Energy Sources. These pathways funnel metabolites into the glycolytic pathways and the tricarboxylic acid cycle, increasing metabolic efficiency and flexibility.

d. Diagram the changes made to pyruvate by pyruvate dehydrogenase
e. Identify reactions of the TCA cycle that produce GTP and NADH, generate precursor metabolites, or are redox reactions
f. Calculate the yields of ATP, NADH, and $FADH_2$ by the TCA cycle and explain the fate of each of these molecules

Aerobic respiration is a process that can completely oxidize an organic energy source to CO_2. The catabolism of glucose can be divided into three steps: (1) Pyruvate is formed by using one or more pathways. These pathways also produce NADH, $FADH_2$, or both. (2) Pyruvate is fed into the TCA cycle and oxidized completely to CO_2 with the production of some GTP, NADH, and $FADH_2$. (3) The NADH and $FADH_2$ formed by glycolysis and the TCA cycle are oxidized by an electron transport chain, using O_2 as the terminal electron acceptor. The electron transport chain is responsible for most of the ATP made during aerobic respiration.

Microorganisms employ several pathways to catabolize glucose to pyruvate in the cytosol, including (1) the Embden-Meyerhof pathway, (2) the Entner-Doudoroff pathway, and (3) the pentose phosphate pathway. We refer to these pathways collectively as **glycolytic pathways** or as **glycolysis** (Greek *glyco,* sweet; *lysis,* a loosening). However, in some texts, the term *glycolysis* refers only to the Embden-Meyerhof pathway. For the sake of simplicity, the detailed structures of some metabolic intermediates are not used in pathway diagrams in this chapter. These can be found in appendix II.

Embden-Meyerhof Pathway: The Most Common Route to Pyruvate

The **Embden-Meyerhof pathway (EMP)** is undoubtedly the most common pathway for glucose degradation to pyruvate. It is found in all major groups of microorganisms, as well as in plants and animals, and functions in the presence or absence of O_2. It provides several precursor metabolites, NADH, and ATP for the cell.

The EMP is divided into two parts: a 6-carbon phase and a 3-carbon phase (**figure 11.5** and *appendix II*). In the initial 6-carbon phase, two ATP molecules are used in phosphorylation reactions, yielding fructose 1,6-bisphosphate. This preliminary phase "primes the pump" by adding phosphates to each end of fructose. In essence, the organism invests some of its ATP so that more can be made later in the pathway.

The 3-carbon, energy-conserving phase begins when fructose 1,6-bisphosphate is cleaved into two 3-C molecules: dihydroxyacetone phosphate and glyceraldehyde 3-phosphate. Dihydroxyacetone phosphate is immediately converted to glyceraldehyde 3-phosphate. Thus two molecules of glyceraldehyde 3-phosphate are formed from a single glucose by the cleavage reaction. These are catabolized to pyruvate in a five-step process.

In contrast to the 6-carbon phase when ATP is consumed, in the 3-carbon phase of the pathway NADH and ATP are produced. NADH is formed when glyceraldehyde 3-phosphate is oxidized with NAD^+ as the electron acceptor, and a phosphate is simultaneously incorporated to give a high-energy molecule called 1,3-bisphosphoglycerate. This reaction sets the stage for ATP production. The phosphate on the first carbon of 1,3-bisphosphoglycerate is donated to ADP to produce ATP. This is an example of **substrate-level phosphorylation** because ADP phosphorylation is coupled with the exergonic hydrolysis of a high-energy molecule having a higher phosphate transfer potential than ATP (*see table 10.1*). A second ATP is made by substrate-level phosphorylation when the phosphate on phosphoenolpyruvate (the last intermediate of the pathway) is donated to ADP. This reaction also yields pyruvate, the final product of the pathway. *ATP: the major energy currency of cells (section 10.2)* *How NAD^+ Works*

The yields of ATP and NADH by the EMP may be calculated. In the 6-carbon phase, two ATP are used to form fructose 1,6-bisphosphate. For each glyceraldehyde 3-phosphate transformed into pyruvate, one NADH and two ATP are formed.

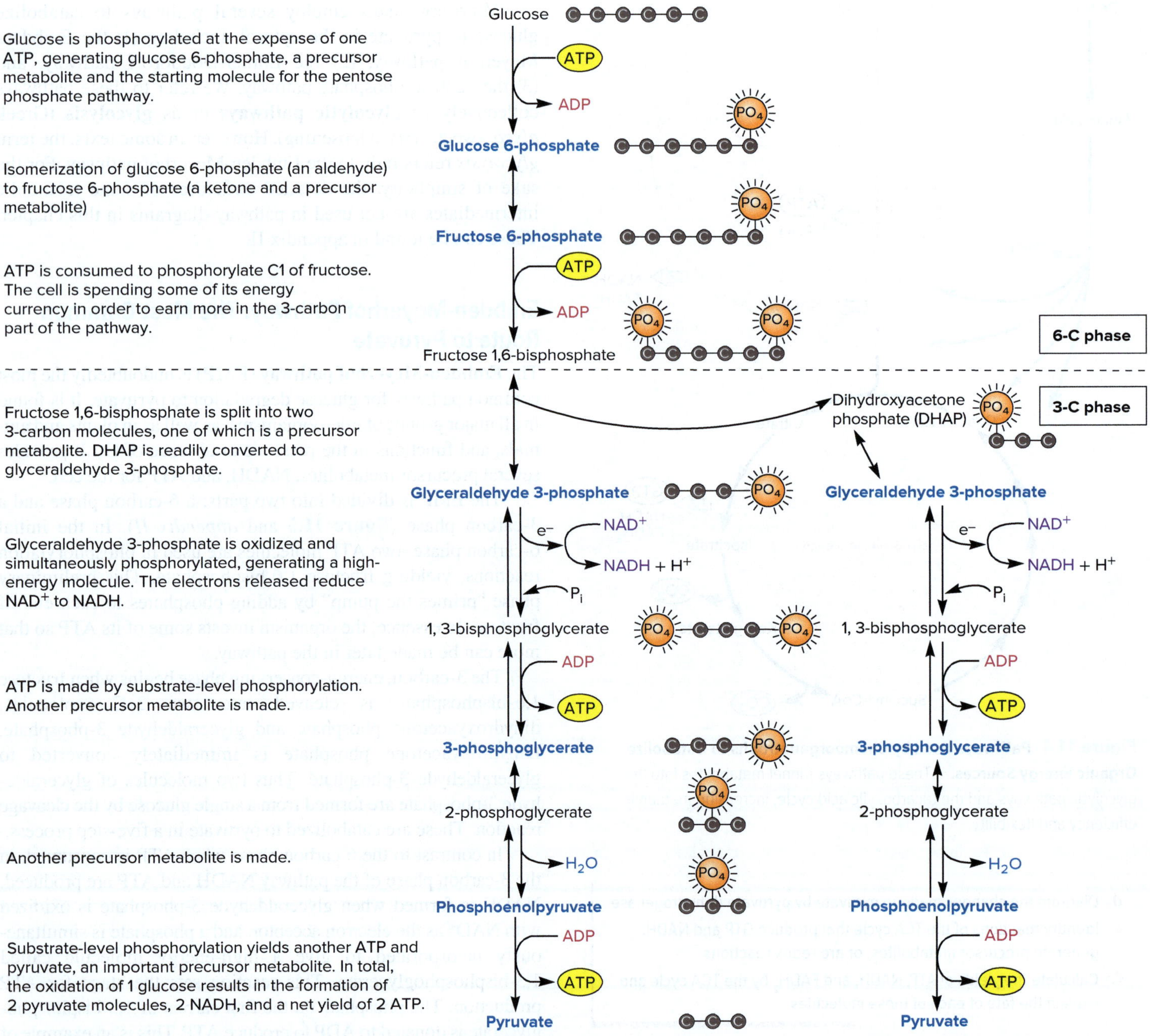

Figure 11.5 Embden-Meyerhof Pathway. This is one of three pathways used to catabolize glucose to pyruvate, and it can function during aerobic respiration, anaerobic respiration, and fermentation. When used during respiration, the electrons accepted by NAD^+ are transferred to an electron transport chain and are ultimately accepted by O_2 or another exogenous electron acceptor. When used during fermentation (section 11.6), electrons accepted by NAD^+ are donated to an endogenous electron acceptor (e.g., pyruvate). Precursor metabolites are shown in blue.

MICRO INQUIRY *Which reactions are examples of substrate-level phosphorylation?*

Because two glyceraldehyde 3-phosphates arise from a single glucose, the 3-carbon phase generates four ATP and two NADH per glucose. Subtraction of the ATP used in the 6-carbon phase from ATP produced by substrate-level phosphorylation in the 3-carbon phase gives a net yield of two ATP per glucose. Thus the catabolism of glucose to pyruvate via the EMP can be represented by the following equation:

$$\text{Glucose} + 2\text{ADP} + 2\text{P}_i + 2\text{NAD}^+ \rightarrow 2\ \text{pyruvate} + 2\text{ATP} + 2\text{NADH} + 2\text{H}^+$$

How Glycolysis Works

The EMP can also run in reverse to convert two pyruvate molecules into a glucose molecule, a process called gluconeogenesis (roughly translated as *making new glucose*). It is therefore an example of an **amphibolic pathway** (Greek, *amphi* on both sides). Amphibolic pathways typically include steps catalyzed by enzymes that work in both directions (both catabolic and anabolic), but also include a few key steps in which catabolic and anabolic reactions are catalyzed by separate, tightly regulated enzymes (*see figure 12.5*). The relative levels of such regulated enzymes help determine which direction the amphibolic pathway proceeds.

Entner-Doudoroff Pathway

The **Entner-Doudoroff pathway (EDP)** is used by some Gram-negative bacteria, especially those found in soil. To date, very few Gram-positive bacteria have been found to use this pathway, with the intestinal bacterium *Enterococcus faecalis* being a rare exception. It is more common in aerobic bacteria, which better tolerate the lower yield of ATP, as compared to the EMP. The loss in ATP is more than offset by the high ATP yield generated during aerobic respiration. In intestinal microbes, such as *Escherichia coli* and *E. faecalis,* this pathway may be important in oxidizing the bile acids these microbes use as a source of carbon. It is not used by eukaryotes.

The EDP replaces the 6-carbon phase of the EMP to yield pyruvate and glyceraldehyde 3-phosphate instead of two molecules of glyceraldehyde 3-phosphate (**figure 11.6** and *appendix II*). A key intermediate of the EDP has the long but descriptive name 2-keto-3-deoxy-6-phosphogluconate (KDPG), which is formed from glucose by three reactions that consume one ATP and produce one NADPH. KDPG is then cleaved to pyruvate and glyceraldehyde 3-phosphate. Bacteria that use this pathway also have the enzymes that function in the 3-carbon phase of the EMP. These enzymes may be used to catabolize glyceraldehyde 3-phosphate to form a second pyruvate molecule. If this occurs, two ATP and one NADH are formed. Thus the catabolism of one glucose molecule to two pyruvates by way of the EDP coupled with the second half of the EMP has a net yield of one ATP, one NADH, and one NADPH. Linking the EDP and EMP in this way can be represented as follows:

EDP:

Glucose + ATP + $NADP^+$ → pyruvate + glyceraldehyde 3-phosphate + ADP + NADPH + H^+

EMP:

Glyceraldehyde 3-phosphate + 2ADP + NAD^+ → pyruvate + 2ATP + NADH + H^+

The NADPH generated by EDP provides electrons for anabolic reactions, whereas during respiration, the NADH is used to donate electrons to the ETC.

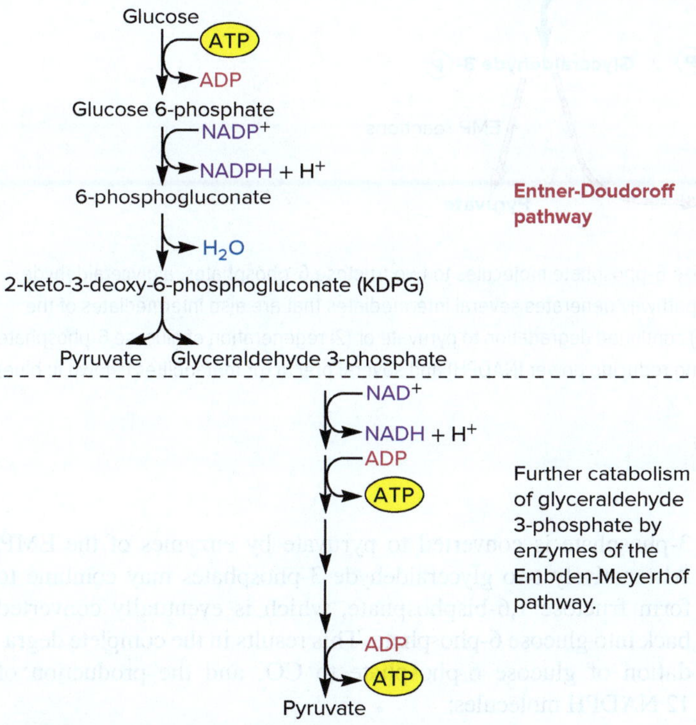

Figure 11.6 The Entner-Doudoroff Pathway.

MICRO INQUIRY *For what kinds of reactions is NADPH used?*

Pentose Phosphate Pathway: A Major Producer of Reducing Power for Anabolic Reactions

The **pentose phosphate pathway** (also called the **hexose monophosphate pathway**) can be used aerobically or anaerobically. It may be used at the same time as either the EMP or the EDP because it is important in biosynthesis as well as catabolism. Both eukaryotes and bacteria use it to provide reducing power (as NADPH) and the 5-carbon sugar (pentose) ribose-5-phosphate, a precursor for nucleotide biosynthesis.

The pentose phosphate pathway begins with a two-step oxidative phase when glucose 6-phosphate is oxidized to 6-phosphogluconate, which in turn is oxidized and decarboxylated to the 5-carbon sugar (i.e., a pentose sugar) ribulose 5-phosphate and CO_2 (**figure 11.7** and *appendix II*). Unlike the EMP where NAD^+ accepts electrons, thereby generating NADH, here $NADP^+$ is the electron acceptor and NADPH is produced. This is another important difference between the two pathways: NADH donates its electrons to the ETC so that energy can be conserved (catabolism), whereas NADPH donates its electrons to biosynthetic reactions, which often consume energy (anabolism).

Following the oxidative phase, ribulose 5-phosphate is converted to a mixture of 3- to 7-carbon sugar phosphates. Two enzymes play a central role in these transformations: (1) transketolase catalyzes the transfer of 2-carbon groups and (2) transaldolase transfers a 3-carbon group from sedoheptulose 7-phosphate (a 7-carbon molecule) to glyceraldehyde 3-phosphate (a 3-carbon molecule). The overall result is that three glucose 6-phosphates

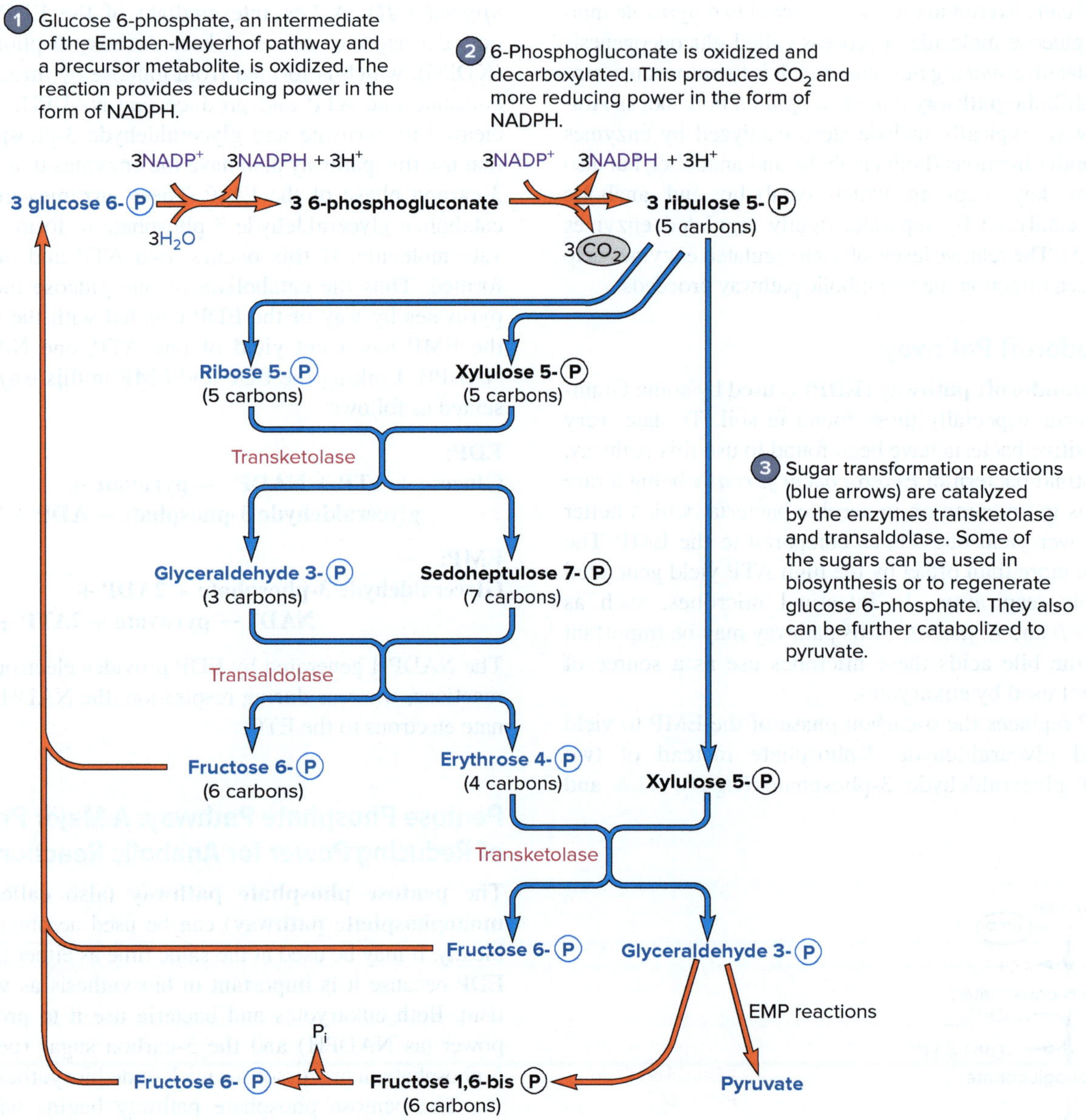

Figure 11.7 The Pentose Phosphate Pathway. The catabolism of three glucose 6-phosphate molecules to two fructose 6-phosphates, a glyceraldehyde 3-phosphate, and three CO_2 molecules is traced. Note that the pentose phosphate pathway generates several intermediates that are also intermediates of the Embden-Meyerhof pathway (EMP). These can be fed into the EMP with two results: (1) continued degradation to pyruvate or (2) regeneration of glucose 6-phosphate by gluconeogenesis. The pentose phosphate pathway plays a major role in producing reducing power (NADPH) and several precursor metabolites (shown in blue) for biosynthesis. The sugar transformations are indicated with blue arrows.

MICRO INQUIRY *For what macromolecule is ribose 5-phosphate a precursor?*

are converted to two fructose 6-phosphates, glyceraldehyde 3-phosphate, and three CO_2 molecules, as shown in the following equation:

$$\textbf{3 glucose 6-phosphate} + 6NADP^+ + 3H_2O \rightarrow \textbf{2 fructose 6-phosphate} + \textbf{glyceraldehyde 3-phosphate} + 3CO_2 + 6NADPH + 6H^+$$

These intermediates are used in two ways. Fructose 6-phosphate can be changed back to glucose 6-phosphate while glyceraldehyde 3-phosphate is converted to pyruvate by enzymes of the EMP. Alternatively two glyceraldehyde 3-phosphates may combine to form fructose 1,6-bisphosphate, which is eventually converted back into glucose 6-phosphate. This results in the complete degradation of glucose 6-phosphate to CO_2 and the production of 12 NADPH molecules:

$$\textbf{Glucose 6-phosphate} + 12NADP^+ + 6H_2O \rightarrow 6CO_2 + 12NADPH + 12H^+ + P_i$$

The pentose phosphate pathway is an important amphibolic pathway because: (1) NADPH serves as a source of electrons for the reduction of molecules during biosynthesis. Indeed, the pentose phosphate pathway is the major source of reducing power for cells, yielding two NADPH molecules for each glucose metabolized to pyruvate. (2) The pathway produces two important precursor metabolites: erythrose 4-phosphate and ribose 5-phosphate. Erythrose 4-phosphate is used to synthesize aromatic amino acids and vitamin B_6; ribose 5-phosphate is a major component of nucleic acids. Furthermore, when a microorganism is growing on a 5-carbon sugar, the pathway can function biosynthetically to supply 6-carbon sugars (hexose sugars, such as the glucose needed for peptidoglycan synthesis). (3) Intermediates in the pathway may be used to produce ATP. For instance, glyceraldehyde 3-phosphate can enter the 3-carbon phase of the EMP. As it is degraded to pyruvate, two ATP are formed by substrate-level phosphorylation. *Synthesis of amino acids consumes many precursor metabolites (section 12.5); Synthesis of purines, pyrimidines, and nucleotides (section 12.6)*

Comprehension Check

1. Summarize the major features of the Embden-Meyerhof, Entner-Doudoroff, and pentose phosphate pathways. Include the starting points, the products of the pathways, the ATP yields, and the metabolic roles of each pathway.
2. What is substrate-level phosphorylation? List an example reaction.
3. Why are the Embden-Meherhof and pentose phosphate pathways considered amphibolic?

Pyruvate Dehydrogenase and the Tricarboxylic Acid Cycle

In the glycolytic pathways, glucose is oxidized to pyruvate. During respiration, the catabolic process continues by oxidizing pyruvate to three CO_2. The first step of this process employs a multienzyme system called the **pyruvate dehydrogenase complex (PDH).** PDH oxidizes and cleaves pyruvate to form one NADH, one CO_2, and the 2-carbon molecule **acetyl-coenzyme A (acetyl-CoA)** (**figure 11.8**, step 1). The oxidization of a molecule to simultaneously release carbon as CO_2 and NADH (or other reduced electron carrier) is called *oxidative decarboxylation.* Recall that one glucose gives rise to two pyruvate molecules, so a single glucose molecule yields:

$$\textbf{2 Pyruvate} + \textbf{2NAD}^+ + \textbf{2CoA} \rightarrow \textbf{2 Acetyl-CoA} + \textbf{2CO}_2 + \textbf{2NADH} + \textbf{2H}^+$$

Acetyl-CoA is an energy-rich molecule because the bond that links the 2-carbon acetyl group to coenzyme A (a thioester bond) has a large negative change in free energy when hydrolyzed. To continue the oxidation of a single glucose, both acetyl-CoA molecules enter the **tricarboxylic acid (TCA) cycle,** also called the **citric acid cycle** or the **Krebs cycle** (figure 11.8 and *appendix II*). Therefore to completely oxidize one molecule of glucose, the TCA cycle must be run twice. For simplicity, we follow a single turn of the cycle.

In the first reaction, acetyl-CoA is condensed with (i.e., added to) the 4-carbon intermediate oxaloacetate to form citrate, a molecule with six carbons. This reaction is fueled by the cleavage of the high-energy thioester bond in acetyl-CoA. Citrate is rearranged to give isocitrate, a more readily oxidized alcohol. Two successive oxidative decarboxylation reactions yield α-ketoglutarate (five carbons) and then succinyl-CoA (four carbons), which is another high-energy molecule containing a thioester bond. Each decarboxylation is accompanied by the production of an NADH, so at this point two NADH and two CO_2 molecules have been produced in the cycle. Succinyl-CoA is then converted to succinate; hydrolysis of the high-energy thioester bond in succinyl-CoA drives the synthesis of one GTP by substrate-level phosphorylation. GTP is a high-energy molecule functionally equivalent to ATP. It is used in protein synthesis and to make other nucleoside triphosphates, including ATP. Two additional oxidation steps follow, yielding one $FADH_2$ and one NADH. The last oxidation step regenerates oxaloacetate, and as long as there is a supply of acetyl-CoA, the cycle can repeat itself. Inspection of figure 11.8 shows that the TCA cycle generates two CO_2 molecules, three NADH molecules, one $FADH_2$, and one GTP for each acetyl-CoA molecule oxidized. *How the Krebs Cycle Works*

TCA cycle enzymes are widely distributed among microorganisms. In bacteria and archaea, they are located in the cytosol. In eukaryotes, they are found in the mitochondrial matrix. The complete cycle is functional in many aerobic bacteria, archaea, free-living protists, and fungi. This is not surprising because the cycle plays an important role in producing NADH and $FADH_2$, which donate electrons to the electron transport chain, as discussed next. Even microorganisms that lack the complete TCA cycle usually have most of its enzymes because the TCA cycle is also a key source of precursor metabolites used in biosynthesis. *Anaplerotic reactions replace the precursor metabolites used for amino acid biosynthesis (section 12.5)*

Comprehension Check

1. Identify the substrate and products of the TCA cycle. Describe its organization in general terms. What are its major functions?
2. What chemical intermediate links pyruvate to the TCA cycle?
3. How many times must the TCA cycle be performed to oxidize one molecule of glucose completely to six molecules of CO_2? Explain.
4. In what eukaryotic organelle are the enzymes that catalyze the TCA cycle found? How does this compare to bacterial and archaeal cells?
5. Why is it desirable for a microbe with the Embden-Meyerhof pathway and the TCA cycle to also have the pentose phosphate pathway?

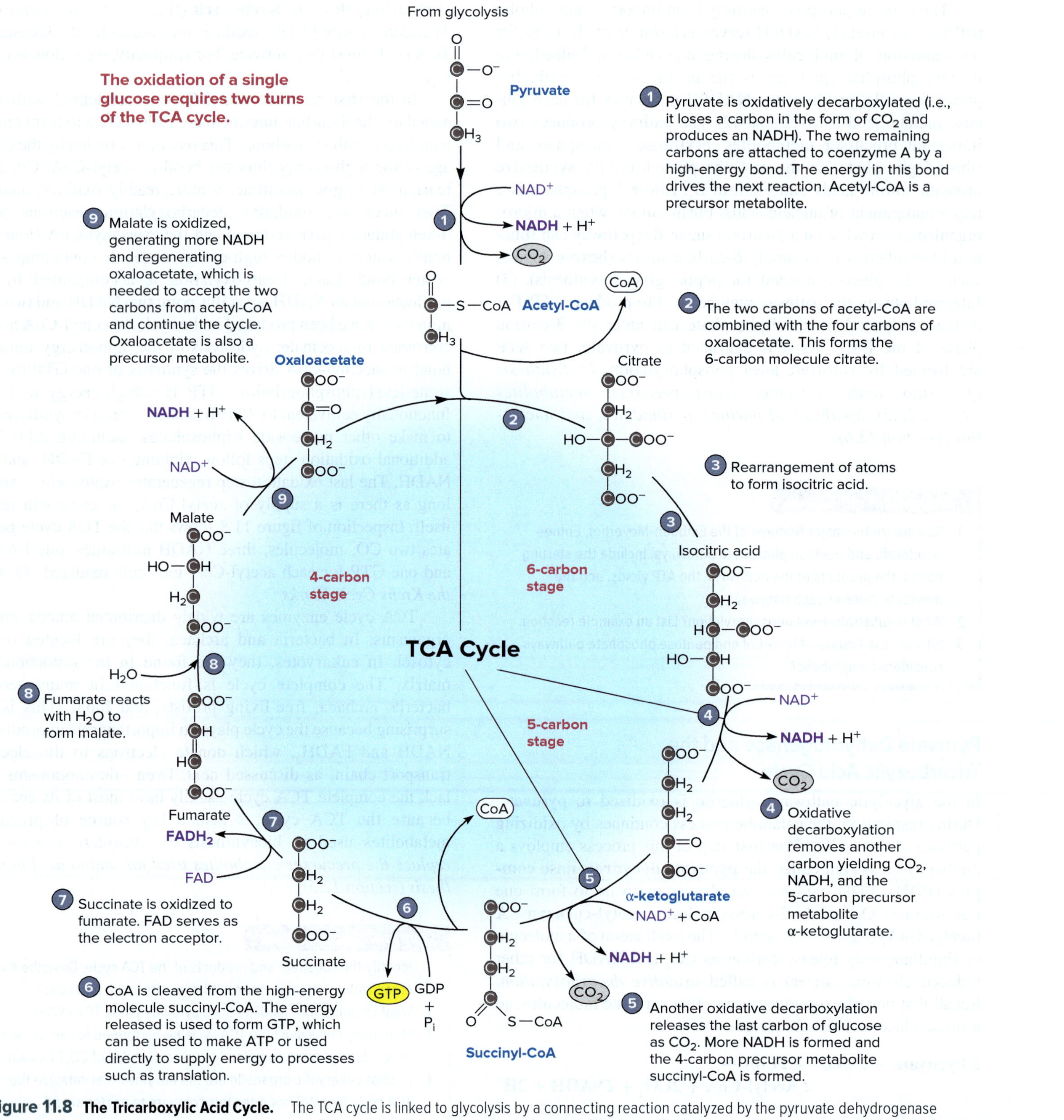

Figure 11.8 The Tricarboxylic Acid Cycle. The TCA cycle is linked to glycolysis by a connecting reaction catalyzed by the pyruvate dehydrogenase complex. The reaction decarboxylates pyruvate and generates acetyl-CoA. The TCA cycle may be divided into three stages based on the number of carbons for each intermediate. The three stages are separated from one another by two oxidative decarboxylation reactions. Precursor metabolites are shown in blue. NADH and $FADH_2$ are shown in purple; they can transfer electrons to an electron transport chain (ETC).

MICRO INQUIRY *What reaction provides the energy to fuel the condensation (joining) of the 4-carbon oxaloacetate with the 2-carbon acetyl-CoA to form citrate?*

11.4 Electron Transport and Oxidative Phosphorylation Generate the Most ATP

After reading this section, you should be able to:

a. Describe the basic structure of bacterial and archaeal electron transport chains (ETCs)
b. Describe the chemiosmotic hypothesis
c. Correlate the length of an ETC and the carriers in it with the magnitude of the proton motive force (PMF) it generates
d. Explain how ATP synthase uses PMF to generate ATP
e. Draw a concept map showing the connections between the glycolytic pathways, TCA cycle, ETC, and ATP synthesis
f. List ways in which the PMF is used by bacterial cells in addition to ATP synthesis
g. Calculate the maximum possible ATP yields when glucose is completely catabolized to six molecules of CO_2 during aerobic respiration

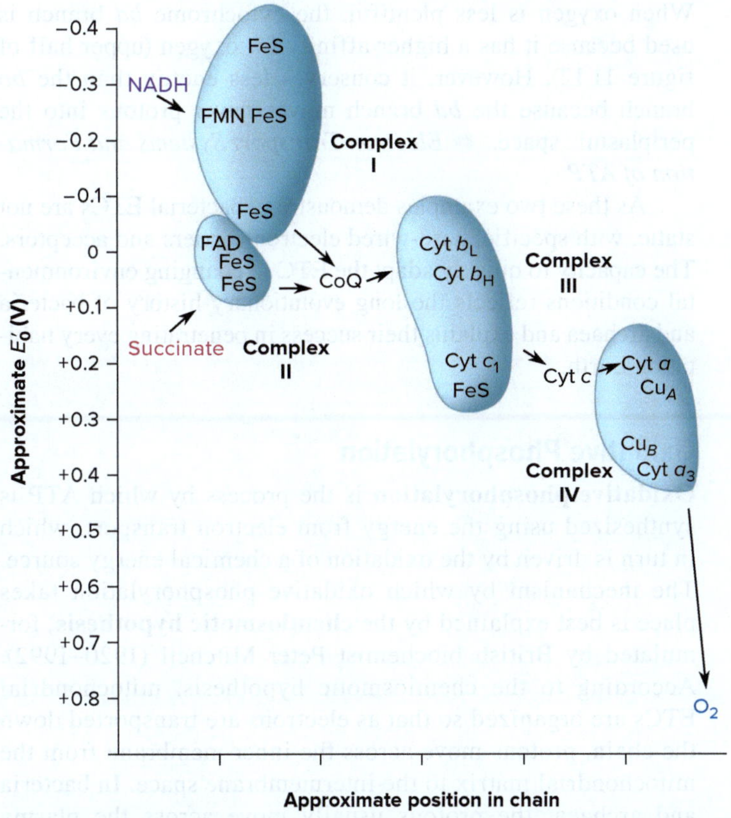

Figure 11.9 The Reduction Potentials of Carriers in the Mitochondrial Electron Transport Chain. In mitochondria, the electron carriers are organized into four complexes linked by the electron carriers coenzyme Q (CoQ) and cytochrome *c* (Cyt *c*, a copper (Cu) containing metalloenzyme). Arrows represent the flow of electrons from NADH and succinate down the reduction potential gradient to oxygen. Bacterial ETCs are arranged similarly but are more variable in their constituents.

During the oxidation of glucose to six CO_2 molecules by glycolysis and the TCA cycle, as many as four ATP molecules are generated by substrate-level phosphorylation. Thus at this point, the work done by the cell has yielded relatively little ATP. However, in oxidizing glucose, the cell has also generated lots of NADH and $FADH_2$. Both of these molecules have a relatively negative E'_0 and can be used to conserve energy. In fact, most of the ATP generated during respiration comes from the energy conserved when these electron carriers donate electrons to an electron transport chain. You may be familiar with the mitochondrial electron transport chain; recall that mitochondria evolved from endosymbiotic bacteria. Bacterial and archaeal ETCs are embedded in the plasma membrane, although some Gram-negative bacteria have ETC carriers in the periplasmic space and even the outer membrane. In the context of the ETC, plasma membranes of bacteria and archaea serve the same function as the inner membranes of mitochondria (*see figure 5.15*).

Electron Transport Chains

The mitochondrial **electron transport chain (ETC)** is composed of a series of electron carriers that operate together to transfer electrons from donors, such as NADH and $FADH_2$, to O_2 (**figure 11.9**). Electrons flow from carriers with more negative reduction potentials to those with more positive potentials and eventually combine with O_2 and H^+ to form water. Each carrier is reduced when it accepts electrons and then reoxidized when the electrons are transferred to the next carrier in the chain (**figure 11.10**). The difference in reduction potentials between O_2 and NADH is large, about 1.14 volts, which makes possible the release of a great deal of energy. The small changes in standard reduction potential from one carrier to the next, as well as their close proximity to one another, enables the ETC to break up the overall energy release (electron transfer) into small steps. This is crucial, as it prevents electrons from reducing cellular components outside the ETC. The energy released is sufficient for ATP production. *Electron transport chains: sets of sequential redox reactions (section 10.4)*

The electron carriers that constitute an ETC can be grouped into those that carry electrons and protons and those that carry only electrons. There are two kinds of molecules that carry both electrons and protons. The first type is a protein that contains the prosthetic group flavin. Flavin is redox active, meaning it is the part of the molecule reduced when it accepts electrons and oxidized when it donates them to the next carrier in the chain. Flavin mononucleotide (FMN) is an example (*see figure 10.8*). The second type are the quinones, such as coenzyme Q (ubiquinone; *see figure 10.9*). Coenzyme Q does not require a prosthetic group like flavin because it can be directly reduced and oxidized. The molecules that carry only electrons contain iron as their redox-active component: Cytochromes have iron bound to heme (*see figure 10.10*), while iron-sulfur proteins have iron that is not bound to heme, so iron-sulfur proteins are

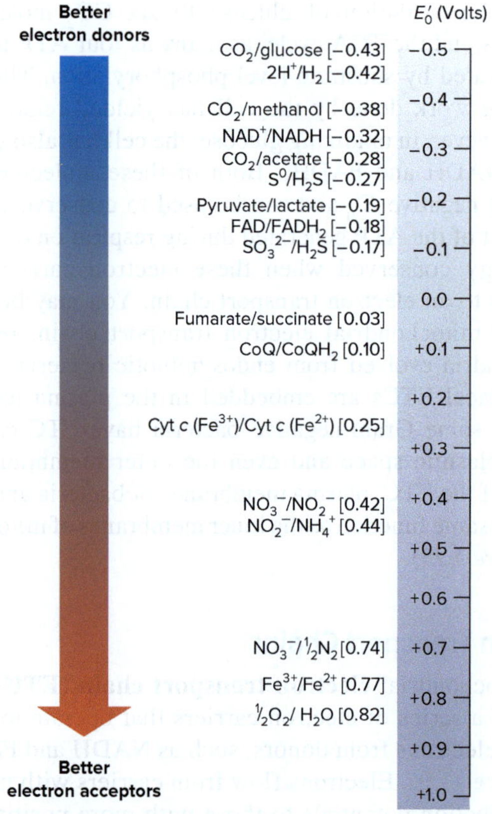

Figure 11.10 The Electron Tower. Electrons move spontaneously from donors with more negative reduction potentials to acceptors with more positive reduction potentials.

sometimes called nonheme proteins (*see figure 10.11*). Because cytochromes and iron-sulfur proteins cannot accept protons being passed down the chain by flavin proteins and coenzyme Q, the protons exit the cell, thereby establishing a PMF. The mechanism by which this occurs in not entirely understood. The last component of an ETC is the terminal electron acceptor (TEA), which resides in a terminal oxidase complex. In aerobic respiration, the TEA is oxygen, which is reduced to water using the two electrons carried down the ETC and two protons from the cytosol.

One of the most amazing aspects of bacterial and archaeal ETCs is their flexibility. Unlike mitochondria, many microbes can change the components of their ETCs to respond to different environmental conditions. For example, a single bacterial or archaeal species can replace electron carriers or employ different terminal oxidases. Additionally, bacterial and archaeal ETCs can be branched. That is, electrons may enter the chain at several points and leave through several different terminal oxidases, not just that which reduces O_2. Bacterial and archaeal ETCs also may be shorter, resulting in the transport of fewer protons across the membrane, and therefore less energy. Although microbial ETCs differ in the details of construction, they operate using the same fundamental principles. *Electron Transport System and ATP Synthesis*

The ETCs of *Paracoccus denitrificans* and *Escherichia coli* serve as examples of bacterial chains. *P. denitrificans* is a Gram-negative, soil bacterium that is a facultative anaerobe that can use several metabolic strategies. Under oxic conditions, it carries out aerobic respiration using the ETC shown in **figure 11.11**. NADH generated by oxidation of organic substrates (during glycolysis and the TCA cycle) is oxidized to NAD^+ by the first component in the ETC, membrane-bound NADH dehydrogenase. The electrons from NADH are transferred to carriers with progressively more positive reduction potentials. As electrons move through the carriers, protons are moved across the plasma membrane to the periplasmic space. The metabolic diversity of the bacterium is also demonstrated in figure 11.11. *P. denitrificans* also can use 1-carbon molecules such as methanol as a source of energy and electrons. In this case, NADH is not produced because methanol is oxidized by the enzyme methanol dehydrogenase, which in turn transfers the electrons directly to cytochrome *c*.

A simplified view of the *E. coli* chain is shown in **figure 11.12**. This ETC has two branches that operate under different oxygen levels. When oxygen is readily available, the cytochrome *bo* branch is used (lower half of figure 11.12). When oxygen is less plentiful, the cytochrome *bd* branch is used because it has a higher affinity for oxygen (upper half of figure 11.12). However, it conserves less energy than the *bo* branch because the *bd* branch moves fewer protons into the periplasmic space. *Electron Transport Systems and Formation of ATP*

As these two examples demonstrate, bacterial ETCs are not static, with specific, hard-wired electron carriers and acceptors. The capacity to quickly adapt the ETC to changing environmental conditions reflects the long evolutionary history of bacteria and archaea and explains their success in penetrating every habitat on Earth.

Oxidative Phosphorylation

Oxidative phosphorylation is the process by which ATP is synthesized using the energy from electron transport, which in turn is driven by the oxidation of a chemical energy source. The mechanism by which oxidative phosphorylation takes place is best explained by the **chemiosmotic hypothesis,** formulated by British biochemist Peter Mitchell (1920–1992). According to the chemiosmotic hypothesis, mitochondrial ETCs are organized so that as electrons are transported down the chain, protons move across the inner membrane from the mitochondrial matrix to the intermembrane space. In bacteria and archaea, the protons usually move across the plasma membrane from the cytoplasm to the periplasmic space (figures 11.11 and 11.12). The movement of protons across the membrane is not completely understood. Some protons are actively pumped across the membrane. In other cases, translocation of protons results from the juxtaposition of carriers that accept both electrons and protons with carriers that accept only electrons.

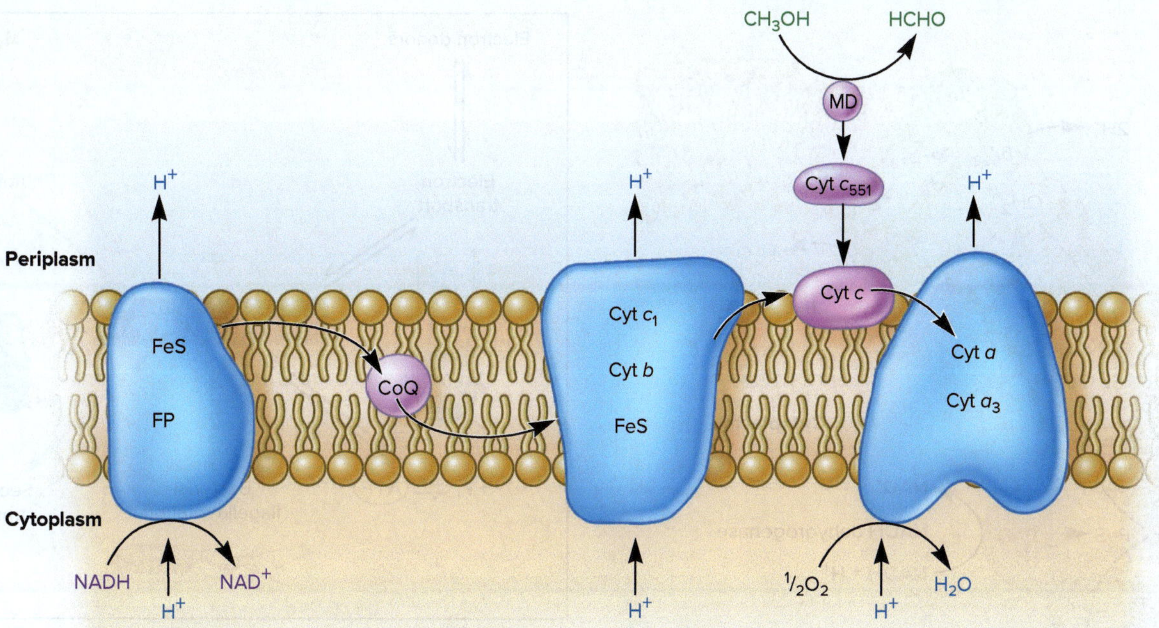

Figure 11.11 ***Paracoccus denitrificans*** **Electron Transport Chain Used During Aerobic Respiration.** *P. denitrificans* can respire both aerobically and anaerobically (figure 11.16). The aerobic ETC uses oxygen as the terminal electron acceptor. This metabolically versatile bacterium can also use methanol and methylamine (not shown) as electron donors. These molecules donate electrons to the chain at the level of cytochrome *c*. Locations of proton movement are shown, but not the number of protons involved. However, 8 to 10 protons are thought to be translocated. FP, Flavoprotein; MD, methanol dehydrogenase.

MICRO INQUIRY *What pathways provide the NADH that is donating electrons to the ETC? Which members of this ETC accept both protons and electrons and which accept only electrons?*

The result of proton expulsion during electron transport is the formation of a proton concentration gradient (ΔpH; chemical potential energy) and a charge gradient (Δψ; electrical potential energy). Because the cell membrane does not permit the diffusion of ions, including protons, the cytoplasm of bacteria and archaea is more alkaline and more negative than the periplasmic space. The combined chemical and electrical potential differences make up the **proton motive force (PMF).** The PMF is used to perform work when protons flow back across the membrane, down the concentration and charge gradients, and into the cytoplasm. This flow is exergonic and is often used to phosphorylate ADP to ATP. PMF is also used by many secondary active transport systems to directly move nutrients into the cell and to rotate the bacterial flagellar motor without the need for ATP hydrolysis (**figure 11.13**). ◀ *Primary and secondary active transport (section 3.2); Bacterial flagella (section 3.7)*

The use of the PMF for ATP synthesis is catalyzed by **ATP synthase.** The best-studied ATP synthases are the F_1F_0 ATP synthases found in bacteria, mitochondria, and chloroplasts. The F_1 component is a globular structure attached to the membrane surface by a stalk (**figure 11.14**). The F_0 component is embedded in the membrane. ATP synthase is on the cytoplasmic surface of the plasma membrane in bacterial cells. F_0 participates in proton movement across the membrane. F_1 is a large complex in which three α subunits alternate with three β subunits. The catalytic sites for ATP synthesis are located on the β subunits. At the center of F_1 is the γ subunit. The γ subunit extends through F_1 and interacts with F_0.

The ATP synthase functions like a rotary engine, much like the rotary motor of bacterial flagella. The flow of protons down the proton gradient through the F_0 subunit causes F_0 and the γ subunit to rotate. As the γ subunit rotates rapidly within the F_1, conformation changes occur in the β subunits (figure 11.14*b*). One conformation change (β_E to β_{HC}) allows entry of ADP and P_i into the catalytic site. Another conformation change (β_{HC} to β_{DP}) loosely binds ADP and P_i in the catalytic site. ATP is synthesized when the β_{DP} conformation is changed to the β_{TP} conformation, and ATP is released when β_{TP} changes to the β_E conformation, to start the synthesis cycle anew.

ATP Yield During Aerobic Respiration

It is possible to estimate the number of ATP molecules synthesized per pair of electrons donated by an NADH or $FADH_2$ to the ETC and ultimately used to reduce an atom of oxygen to H_2O. For historical reasons, this is called the phosphorus to oxygen (P/O) ratio. Prior to the acceptance of the chemiosmotic

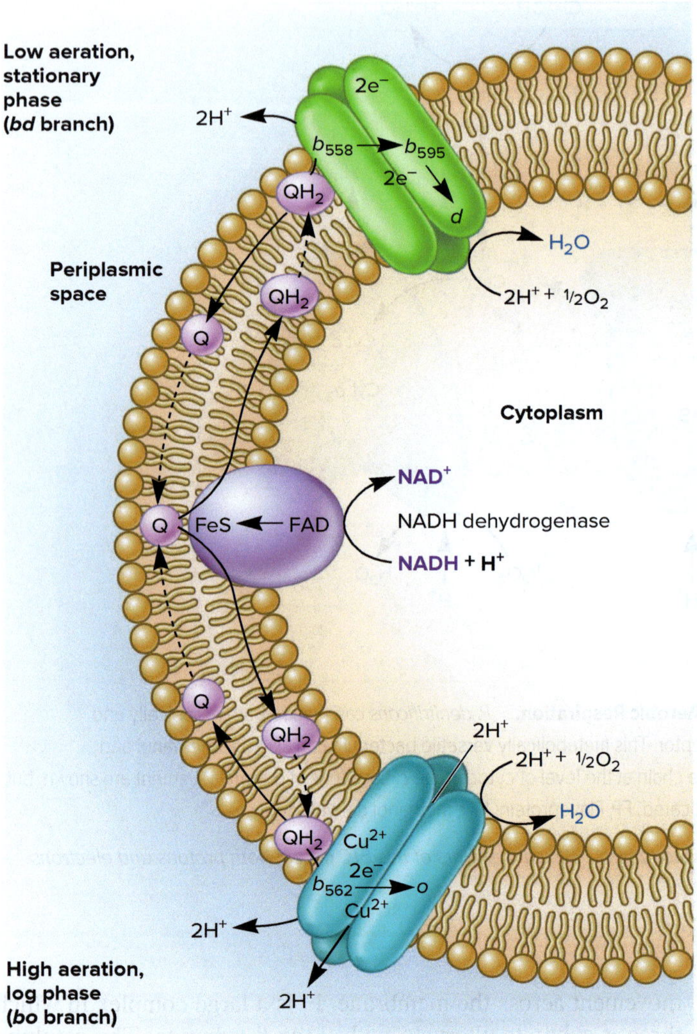

Figure 11.12 The Electron Transport Chain of *E. coli*. NADH transfers electrons from an organic energy and electron source to the ETC. Ubiquinone-8 (Q) connects the NADH dehydrogenase with two terminal oxidase systems. The upper (*bd*) branch operates when the bacterium is in stationary phase and there is little oxygen. It is called the *bd* branch because it uses the cytochromes b_{558}, b_{595}, and *d*. The lower *bo* branch functions when *E. coli* is growing rapidly with good aeration. It is so named because it uses two cytochromes, b_{562} and *o*.

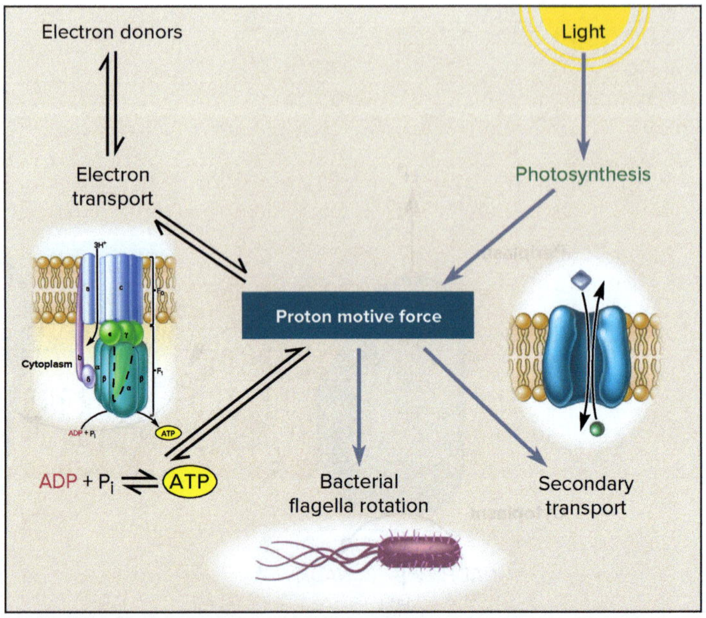

Figure 11.13 The Central Role of Proton Motive Force. Double arrows indicate (1) the ETC operates in both directions in some chemolithotrophs and (2) the ATP synthase can operate in the reverse direction to pump ions out of the cell (e.g., during fermentation).

hypothesis, the concept of the P/O ratio was used as a measure of the number of ATP molecules (phosphorus) generated per oxygen (O) reduced as NADH and $FADH_2$ were oxidized. With the acceptance of the chemiosmotic hypothesis, it was recognized that the important measurement was actually the number of protons transported across the membrane per NADH oxidation and the number of protons needed to drive the synthesis of a single ATP. Nonetheless, the amount of ATP synthesized as the result of the oxidation of one NADH or $FADH_2$ is still often referred to as a P/O ratio. The best estimates of the number of protons moved across the membrane due to NADH oxidation is 10; six electrons are thought to be transferred when $FADH_2$ oxidation initiates electron flow to oxygen. It is currently thought that four protons are consumed during ATP synthesis; three are used by ATP synthase and one is used for transport of ATP, ADP, and P_i. Thus the P/O ratio is estimated to be 2.5 for NADH and 1.5 for $FADH_2$.

Given this information, the maximum ATP yield of aerobic respiration in a eukaryote (i.e., mitochondrion) can be calculated (**figure 11.15**). Substrate-level phosphorylation yields at most two ATP molecules per glucose converted to pyruvate (figure 11.5). Two GTP (ATP equivalent) are generated by substrate-level phosphorylation during the two turns of the TCA cycle needed to oxidize two acetyl-CoA molecules (figure 11.8). Thus during aerobic respiration of glucose, at most four ATP molecules are made by substrate-level phosphorylation. Most of the ATP made during aerobic respiration is generated by oxidative phosphorylation. Up to 10 NADH (2 from glycolysis, 2 from pyruvate conversion to acetyl-CoA, and 6 from the TCA cycle) and 2 $FADH_2$ (from the TCA cycle) are generated when glucose is oxidized completely to 6 CO_2. Assuming a P/O ratio of 2.5 for NADH oxidation and 1.5 for $FADH_2$ oxidation, the 10 NADH could theoretically drive the synthesis of 25 ATP, while oxidation of the 2 $FADH_2$ molecules would add another 3 ATP, for a maximum of 28 ATP generated via oxidative phosphorylation. Thus oxidative phosphorylation accounts for seven times more ATP than does substrate-level phosphorylation.

The maximum total yield of ATP during aerobic respiration by eukaryotes is 32 ATP (including that made by substrate-level

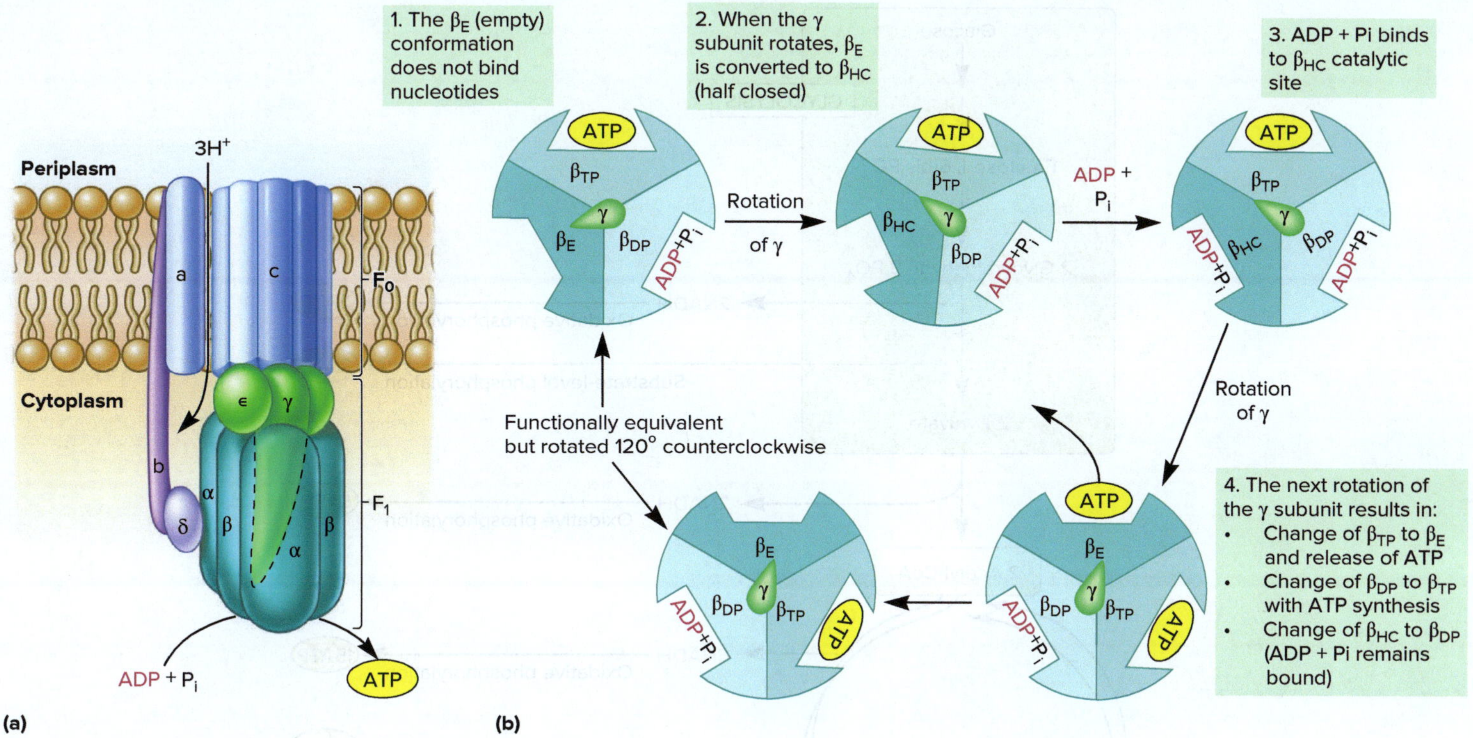

Figure 11.14 ATP Synthase Structure and Function. (a) The major structural features of ATP synthase. F_1 is composed largely of alternating α and β subunits; the three active sites are on the β subunits. The γ subunit (dotted lines) extends through the center and can rotate. The stalk (γ and ε subunits; green) connects the sphere to F_0, the membrane embedded complex that serves as a proton channel. F_0 contains one a subunit, two b subunits, and 9 to 12 c subunits. The stator arm is composed of subunit a, two b subunits, and the δ subunit (purple); it is embedded in the membrane and attached to F_1. A ring of c subunits in F_0 (blue) is connected to the stalk and may act as a rotor and move past the a subunit of the stator. As the c subunit ring turns, it rotates the shaft (γε subunits). (b) The binding change mechanism is a widely accepted model of ATP synthesis. This simplified drawing shows the three catalytic β subunits and the γ subunit, which is located at the center of the F_1 complex. As the γ subunit rotates, it causes conformational changes in each subunit. β_E is the empty conformation; β_{HC} is the half closed conformation; β_{DP} indicates A*DP* is bound; β_{TP} indicates A*TP* is bound.

phosphorylation). Bacterial ETCs are often shorter and therefore transport fewer protons across the plasma membrane. Thus their ETCs have lower P/O ratios, and their ATP yields are smaller. For example, *E. coli*, with its truncated ETC, has a P/O ratio around 1.3 when using the cytochrome *bo* path at high oxygen levels and a ratio of only about 0.67 when employing the cytochrome *bd* branch (figure 11.12) at low oxygen concentrations.

Two other factors prevent the yield of ATP from reaching its theoretical maximum when glucose is catabolized by aerobic respiration. One is that the PMF generated by electron transport is used for functions other than ATP synthesis (e.g., transport and bacterial flagella rotation). The second is related to the amphibolic nature of the pathways used to catabolize glucose. Recall that an important function of these pathways is the generation of precursor metabolites for anabolism. For each molecule of glucose degraded, numerous precursor metabolites are made, and as each is made, a microbe must "decide" if that metabolite is needed for anabolism or if it can continue the catabolic process. If the precursor metabolite is used for biosynthesis, fewer NADH and $FADH_2$ molecules are made and fewer ATP molecules generated. Cells must carefully monitor and regulate the flow of the carbons from glucose such that ATP production is appropriately balanced with biosynthesis. The mechanisms for this regulation are described in chapters 10 and 14.

Comprehension Check

1. Describe the components of an ETC and their relative standard reduction potentials.
2. Describe the current model of oxidative phosphorylation. Briefly describe the structure of ATP synthase and explain how it is thought to function.
3. Compare and contrast substrate-level phosphorylation and oxidative phosphorylation.
4. Suppose a bacterium used the Entner-Doudoroff pathway to degrade glucose to pyruvate and then completed the catabolism of glucose via the TCA cycle. What is the total maximum ATP yield? Explain your reasoning.

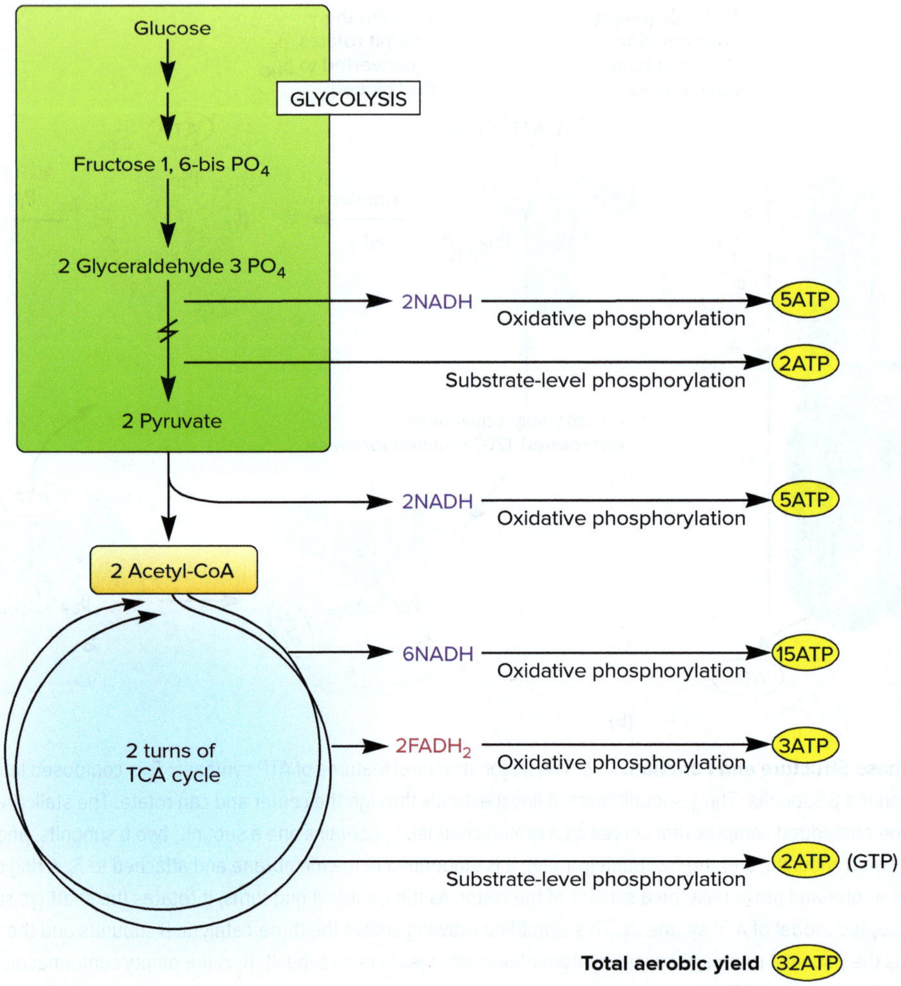

Figure 11.15 Maximum Theoretic ATP Yield from Aerobic Respiration. To calculate the theoretic maximum yield of ATP, P/O ratios of 2.5 for the oxidation of NADH and 1.5 for $FADH_2$ are assumed. These P/O ratios are based on mitochondrial ETCs; values for bacteria and archaea differ and can vary.

11.5 Anaerobic Respiration Uses the Same Steps as Aerobic Respiration

After reading this section, you should be able to:

- **a.** Compare and contrast aerobic respiration and anaerobic respiration using glucose as a carbon source
- **b.** List examples of terminal electron acceptors used during anaerobic respiration
- **c.** Explain why the use of nitrate (NO_3^-) as a terminal electron acceptor is dissimilatory nitrate reduction
- **d.** Explain why less energy is conserved during anaerobic respiration, as compared to aerobic respiration
- **e.** Discuss the evolutionary importance of anaerobic respiration

Recall that early Earth was an oxygen-free planet. It is therefore not surprising that respiration first developed without the use of oxygen as the terminal electron acceptor (TEA). Although aerobic respiration is much more familiar to us, it is a relative late-comer as compared to the many bacteria and archaea that respire anaerobically, either as facultative or obligate anaerobes.

Anaerobic respiration is the chemoorganotrophic process whereby a terminal electron acceptor other than O_2 is used for electron transport. It is carried out by many bacteria and archaea, and a few eukaryotic microbes. The most common terminal electron acceptors used during anaerobic respiration are nitrate, sulfate, and CO_2, but metals and a few organic molecules can also be reduced (**table 11.2**).

In both aerobic and anaerobic respiration processes, sugars and other organic molecules are oxidized and their electrons are transferred to NADH and $FADH_2$, which then donate the electrons to the ETC (figure 11.3). However, for each NADH or $FADH_2$ oxidized, anaerobic respiration yields less ATP. The lower ATP yield is because all TEAs used during anaerobic respiration have less positive reduction potentials than O_2 (figure 11.10). Thus, the difference in standard reduction potentials between NADH and these TEAs is smaller than the difference between NADH and O_2.

Table 11.2 Some Electron Acceptors Used in Respiration

	Electron Acceptor	Reduced Products	Examples of Microorganisms
Aerobic	O_2	H_2O	All aerobic bacteria, archaea, fungi, and protists
Anaerobic	NO_3^-	NO_2^-	*E. coli* and other enteric bacteria
	NO_3^-	NO_2^-, N_2O, N_2	*Pseudomonas, Bacillus,* and *Paracoccus* spp.
	SO_4^{2-}	H_2S	*Desulfovibrio* and *Desulfotomaculum*
	CO_2	CH_4	Methanogens
	CO_2	Acetate	Acetogens
	S^0	H_2S	*Nitratidesulfovibrio* and *Thermoproteus* spp.
	Fe^{3+}	Fe^{2+}	*Pseudomonas, Bacillus,* and *Geobacter* spp.
	$HAsO_4^{2-}$	$HAsO_2$	*Bacillus, Desulfotomaculum, Sulfurospirillum* spp.
	SeO_4^{2-}	Se, $HSeO_3^-$	*Aeromonas, Bacillus, Thauera* spp.
	Fumarate	Succinate	*E. coli, Bacteroides* spp.

This results in a shorter ETC with fewer protons transported to the periplasm per oxidation of NADH or $FADH_2$ as compared to aerobic respiration. The smaller PMF and ATP yield generated by anaerobic respiration explains why facultative anaerobes capable of both aerobic and anaerobic respiration will use aerobic respiration whenever possible.

An example of a facultative anaerobe that can carry out both anaerobic respiration and aerobic respiration is *P. denitrificans.* Figure 11.11 shows the ETC used by this microbe during aerobic respiration. **Figure 11.16** shows that under anoxic conditions, *P. denitrificans* uses NO_3^- as its electron acceptor, reducing it to gaseous dinitrogen (N_2). However, the anaerobic ETC is more complex than the aerobic chain. The anaerobic ETC is branched and uses different electron carriers, some of which are located in the periplasmic space. Not as many protons are pumped across the membrane during anaerobic growth, so a smaller PMF is established.

The anaerobic reduction of NO_3^- makes this nitrogen unavailable for assimilation into the cell. That is, it can't be taken up by the cell to construct N-containing molecules such as amino acids and nucleotides. Therefore, this process is called **dissimilatory nitrate reduction.** When dissimilatory nitrate reduction leads to the production of gaseous compounds such as N_2 that are released into the atmosphere, the process is also called **denitrification.** As can be seen in figure 11.16, denitrification carried out by *P. denitrificans* is a multistep process with four enzymes participating. These enzymes function sequentially to reduce NO_3^- to nitrite (NO_2^-), then nitrite to nitric oxide (NO), followed by NO to nitrous oxide (N_2O), and finally N_2O to N_2. ▶| *Nitrogen cycle (section 28.2)*

Comprehension Check

1. Why does anaerobic respiration yield less ATP than aerobic respiration?
2. What is denitrification?
3. *E. coli* can use O_2, fumarate, or nitrate as a terminal electron acceptor under different conditions. Refer to figure 11.10 to determine the order of energy yield from highest to lowest for these electron acceptors.

11.6 Fermentation Does Not Involve an Electron Transport Chain

After reading this section, you should be able to:

- **a.** Compare and contrast the importance of oxidative phosphorylation and substrate-level phosphorylation in aerobic respiration, anaerobic respiration, and fermentation
- **b.** List the pathways that may function during fermentation if glucose is the organism's carbon and energy source
- **c.** Create a table that lists some of the common fermentation pathways and their products, as well as examples of their uses
- **d.** Compare the use of ATP synthase during respiration and fermentation

Despite the tremendous ATP yield obtained by oxidative phosphorylation, some chemoorganotrophic microbes never respire (i.e., do not use an ETC to generate a PMF) and others may be unable to respire under certain conditions. Microbes completely incapable of respiration lack ETCs. Those that can carry out aerobic respiration may repress the synthesis of ETC components under anoxic conditions, making anaerobic respiration impossible. Finally, some facultative anaerobes can respire both aerobically and anaerobically but may find themselves in an environment that lacks both oxygen and a TEA they can use for anaerobic respiration (e.g., nitrate). The capacity to use such a wide range of catabolic processes demonstrates the extreme flexibility microbes have evolved.

For microbes that ferment, NADH produced by the Embden-Meyerhof pathway reactions during glycolysis (figure 11.5) must still be oxidized back to NAD^+ even though there is no ETC to accept electrons from NADH. If NAD^+ is not regenerated, the oxidation of glyceraldehyde 3-phosphate will cease, glycolysis will stop, and the microbe will not survive. Many microorganisms solve this problem by slowing or

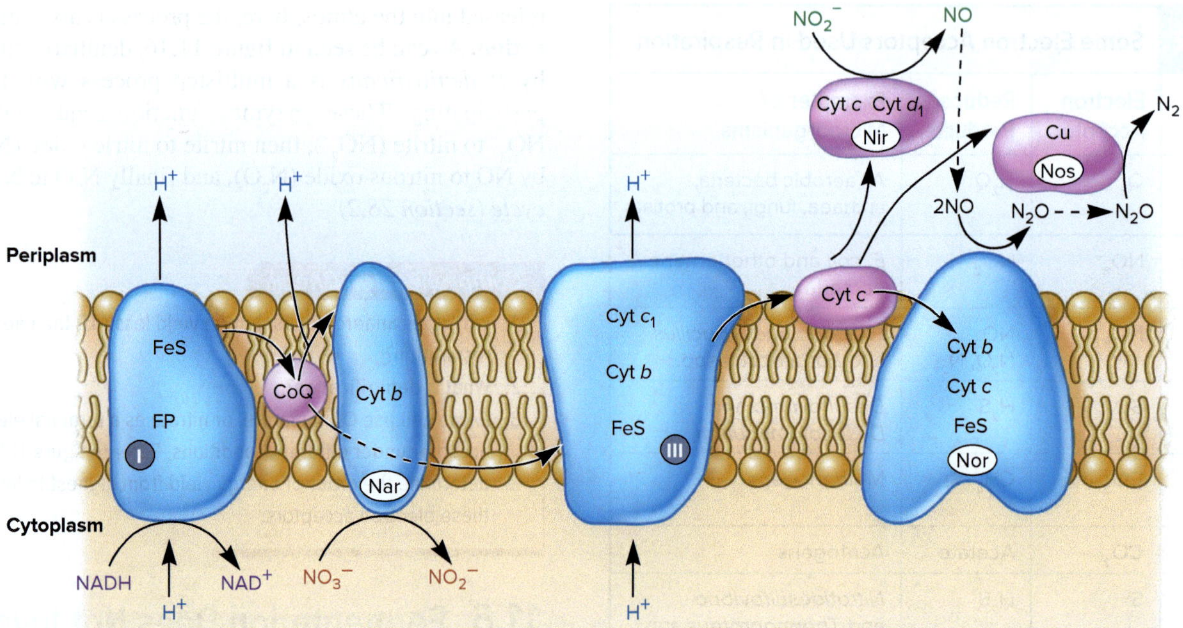

Figure 11.16 ***Paracoccus denitrificans*** **Electron Transport Chain Used During Anaerobic Respiration.** This branched ETC is made of both membrane and periplasmic proteins. Nitrate is the terminal electron acceptor. It is reduced to dinitrogen (N_2) by the collective action of four different reductases that receive electrons from CoQ and cytochrome *c*. Locations of proton movement are indicated. Nitrate (NO_3^-) is transported into the cell and nitrite (NO_2^-) to the periplasm by a NO_3^-/NO_2^- antiporter (not shown for simplicity). Four protons are pumped into the periplasm by complex I upon the oxidation of NADH, two by nitrate reductase (Nar), and two by complex III. However, two protons are used by nitric oxide reductase (Nor) to reduce nitric oxide to nitrous oxide. Thus six protons, net, are used to create a PMF. FP, Flavoprotein; Nir, nitrite reductase; Nos, nitrous oxide reductase, which has an unusual copper (Cu)–sulfur center.

stopping pyruvate dehydrogenase, the enzyme complex that feeds pyruvate into the TCA cycle as acetyl-CoA (figure 11.8). Instead, they use pyruvate or one of its derivatives as an electron acceptor for the reoxidation of NADH in a fermentation process (**figure 11.17**). There are many kinds of fermentations, and they often are characteristic of particular microbial groups (**figure 11.18**). A few of the more common fermentations are introduced here.

Four unifying themes should be kept in mind when considering microbial fermentations: (1) O_2 is not needed; (2) the electron acceptor is usually either pyruvate or a pyruvate derivative; (3) NADH must be oxidized to NAD^+ without an ETC; and (4) the absence of an ETC reduces the ATP yield per glucose significantly. In fermentation, the substrate (e.g., glucose) is only partially catabolized (i.e., not oxidized entirely to CO_2) and ATP is formed in most organisms only by substrate-level phosphorylation. However, not having an ETC creates a problem for fermenting microbes: They still need a PMF to do work, in particular to drive secondary transport (i.e., antiport, symport). To solve this dilemma, they use their ATP synthase in the reverse direction. That is, instead of making ATP from ADP + P_i, the ATP synthase pumps protons out of the cell. Thus the energy to generate a PMF comes from the hydrolysis of ATP to ADP and P_i. In this case, the ATP synthase functions as an ATPase.

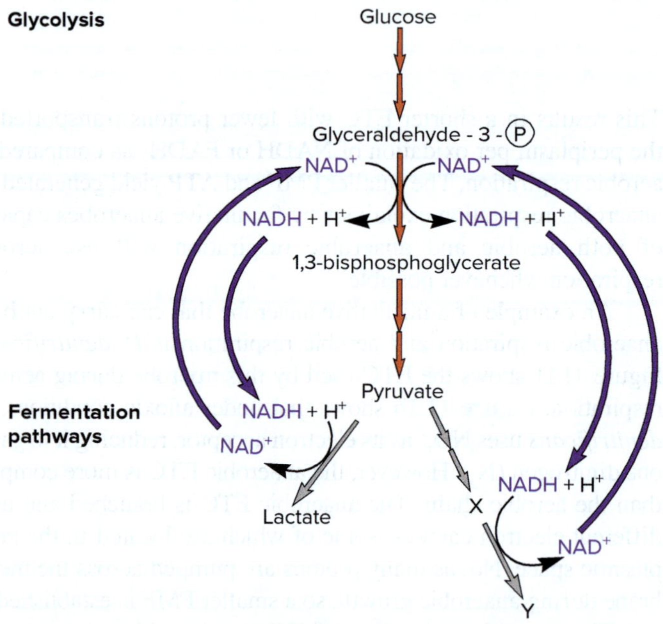

Figure 11.17 Reoxidation of NADH During Fermentation. NADH from glycolysis is reoxidized by donating electrons to pyruvate or a pyruvate derivative (X). Either lactate or reduced product Y results.

MICRO INQUIRY *How many NADH must be reoxidized to NAD^+ for each glucose oxidized to 2 pyruvate molecules?*

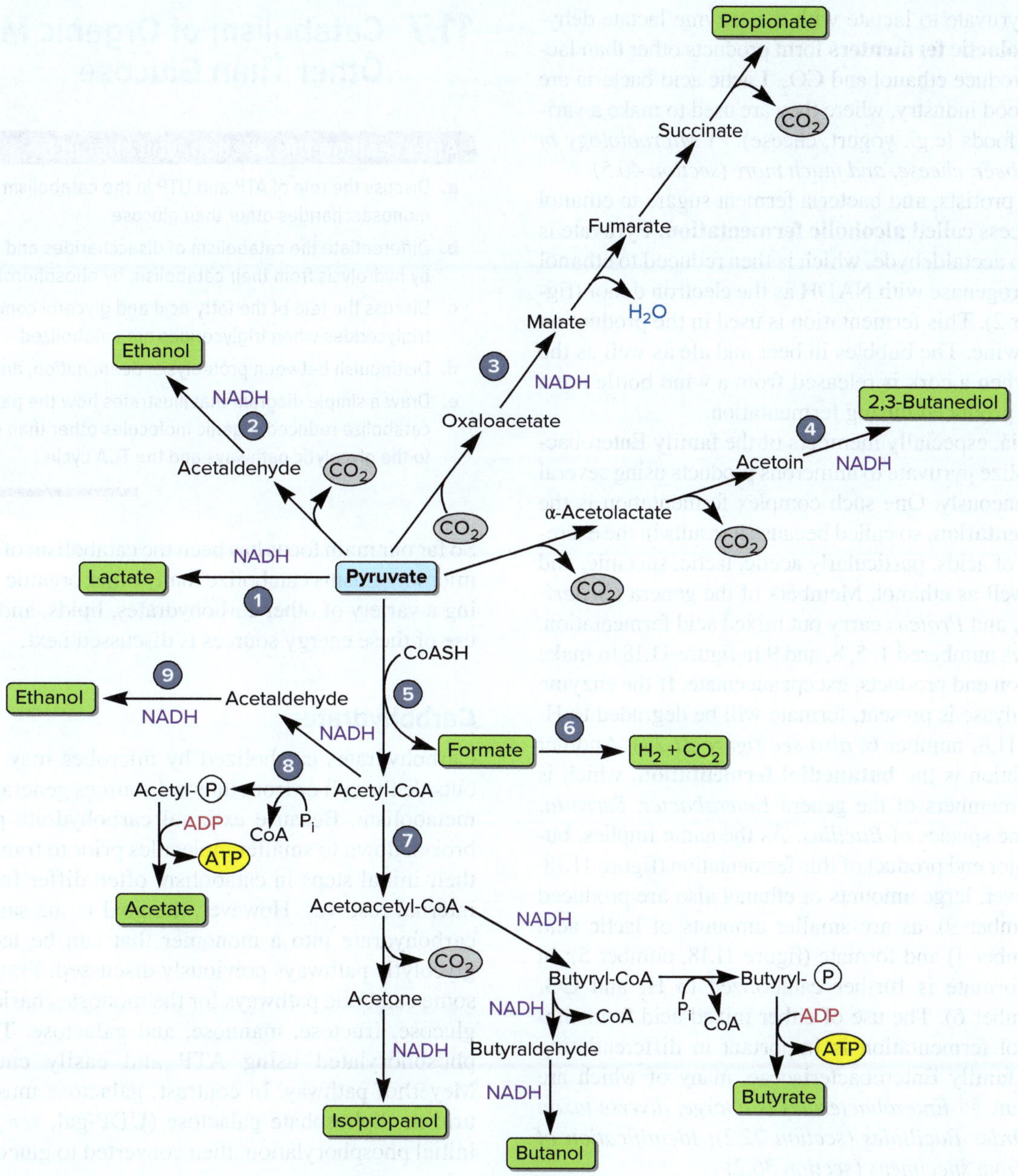

1. Lactic acid bacteria (e.g., *Streptococcus, Lactobacillus* species), *Bacillus* spp., enteric bacteria
2. Yeast, *Zymomonas* spp.
3. Propionic acid bacteria (*Cutibacterium* spp.)
4. *Enterobacter, Serratia, Bacillus* species
5. Enteric bacteria
6. Enteric bacteria
7. *Clostridium* spp.
8. Enteric bacteria
9. Enteric bacteria

Figure 11.18 Some Common Microbial Fermentations. Only pyruvate fermentations are shown for the sake of simplicity; many other organic molecules can be fermented. Each NADH (purple) represents its oxidation to NAD+. Most of these pathways have been simplified by deletion of one or more steps and intermediates. Pyruvate and major end products are shown in color. Enteric bacteria include members of the genera *Escherichia, Enterobacter, Salmonella,* and *Proteus.*

Fermentation pathways are named after the major acid or alcohol produced collectively, called *end products of fermentation.* Lactic acid (lactate) fermentation, the reduction of pyruvate to lactate (figure 11.18, number 1), is most common. It occurs in bacteria (lactic acid bacteria, *Bacillus* spp.), protists (*Chlorella* spp. and some water molds), and animal skeletal muscle. Lactic acid fermenters can be separated into two groups. **Homolactic fermenters** use the Embden-Meyerhof pathway and directly reduce almost all their pyruvate to lactate with the enzyme lactate dehydrogenase. **Heterolactic fermenters** form products other than lactate; many also produce ethanol and CO_2. Lactic acid bacteria are important to the food industry, where they are used to make a variety of fermented foods (e.g., yogurt, cheese). ▶| *Microbiology of fermented foods: beer, cheese, and much more (section 40.5)*

Some fungi, protists, and bacteria ferment sugars to ethanol and CO_2 in a process called **alcoholic fermentation.** Pyruvate is decarboxylated to acetaldehyde, which is then reduced to ethanol by alcohol dehydrogenase with NADH as the electron donor (figure 11.18, number 2). This fermentation is used in the production of beer, ale, and wine. The bubbles in beer and ale as well as the popping sound when a cork is released from a wine bottle are a result of the CO_2 produced during fermentation.

Many bacteria, especially members of the family Enterobacteriaceae, metabolize pyruvate to numerous products using several pathways simultaneously. One such complex fermentation is the **mixed acid fermentation,** so called because it results in the excretion of a mixture of acids, particularly acetic, lactic, succinic, and formic acids as well as ethanol. Members of the genera *Escherichia, Salmonella,* and *Proteus* carry out mixed acid fermentation. They use pathways numbered 1, 5, 8, and 9 in figure 11.18 to make all the fermentation end products, except succinate. If the enzyme formate hydrogenlyase is present, formate will be degraded to H_2 and CO_2 (figure 11.8, number 6; *also see figure 21.24*). Another complex fermentation is the **butanediol fermentation,** which is characteristic of members of the genera *Enterobacter, Serratia, Erwinia,* and some species of *Bacillus.* As the name implies, butanediol is the major end product of this fermentation (figure 11.18, number 4). However, large amounts of ethanol also are produced (figure 11.18, number 9), as are smaller amounts of lactic acid (figure 11.18, number 1) and formate (figure 11.18, number 5); in some bacteria, formate is further catabolized to H_2 and CO_2 (figure 11.18, number 6). The use of either mixed acid fermentation or butanediol fermentation is important in differentiating members of the family Enterobacteriaceae, many of which are clinically important. ▶| *Enterobacterales is a large, diverse taxon (section 21.2); Order Bacillales (section 22.2); Identification of microorganisms from specimens (section 36.2)*

Comprehension Check

1. Why do microorganisms ferment?
2. How do the electron acceptors used in fermentation differ from the terminal electron acceptors used during aerobic respiration and anaerobic respiration?
3. How do homolactic fermenters and heterolactic fermenters differ? How do mixed acid fermenters and butanediol fermenters differ?
4. What is the net yield of ATP during homolactic, acetate, and butyrate fermentations (refer to figures 11.5 and 11.18)? How do these yields compare to aerobic respiration in terms of both quantity and mechanism of phosphorylation?
5. When bacteria carry out fermentation, only a few reactions of the TCA cycle operate. What purpose do you think these reactions might serve? Why are some parts of the cycle shut down?

11.7 Catabolism of Organic Molecules Other Than Glucose

After reading this section, you should be able to:

a. Discuss the role of ATP and UTP in the catabolism of monosaccharides other than glucose
b. Differentiate the catabolism of disaccharides and polysaccharides by hydrolysis from their catabolism by phosphorolysis
c. Discuss the fate of the fatty acid and glycerol components of triglycerides when triglycerides are catabolized
d. Distinguish between proteolysis, deamination, and transamination
e. Draw a simple diagram that illustrates how the pathways used to catabolize reduced organic molecules other than glucose connect to the glycolytic pathways and the TCA cycle

So far our main focus has been the catabolism of glucose. However, microorganisms catabolize many other organic molecules, including a variety of other carbohydrates, lipids, and amino acids. The use of these energy sources is discussed next.

Carbohydrates

Carbohydrates catabolized by microbes may come either from outside the cell or from internal sources generated during normal metabolism. Because external carbohydrate polymers must be broken down to smaller molecules prior to transport into the cell, their initial steps in catabolism often differ from those used for internal reserves. However, the goal is the same: to convert the carbohydrate into a monomer that can be fed into one of the glycolytic pathways previously discussed. **Figure 11.19** outlines some catabolic pathways for the monosaccharides (single sugars) glucose, fructose, mannose, and galactose. The first three are phosphorylated using ATP and easily enter the Embden-Meyerhof pathway. In contrast, galactose must be converted to uridine diphosphate galactose (UDP-gal, *see figure 12.6*) after initial phosphorylation, then converted to glucose 6-phosphate in a three-step process.

The common disaccharides are cleaved to monosaccharides by at least two mechanisms: hydrolysis and phosphorolysis (figure 11.19). Maltose, sucrose, and lactose can be directly hydrolyzed to their constituent sugars. Many disaccharides (e.g., maltose, cellobiose, and sucrose) are split by phosphorolysis, in which a phosphate attacks the bond joining the two sugars. This yields the two constituent monosaccharides, one of which is phosphorylated.

Polysaccharides, like disaccharides, are cleaved by both hydrolysis and phosphorolysis depending on whether they are extracellular or intracellular. Because of their large size, polysaccharides in the environment cannot be transported into the cell. Instead, bacteria, archaea, and fungi degrade external polysaccharides by secreting hydrolytic enzymes. These exoenzymes cleave polysaccharides into smaller molecules that can then be taken up and catabolized. Microbes can catabolize

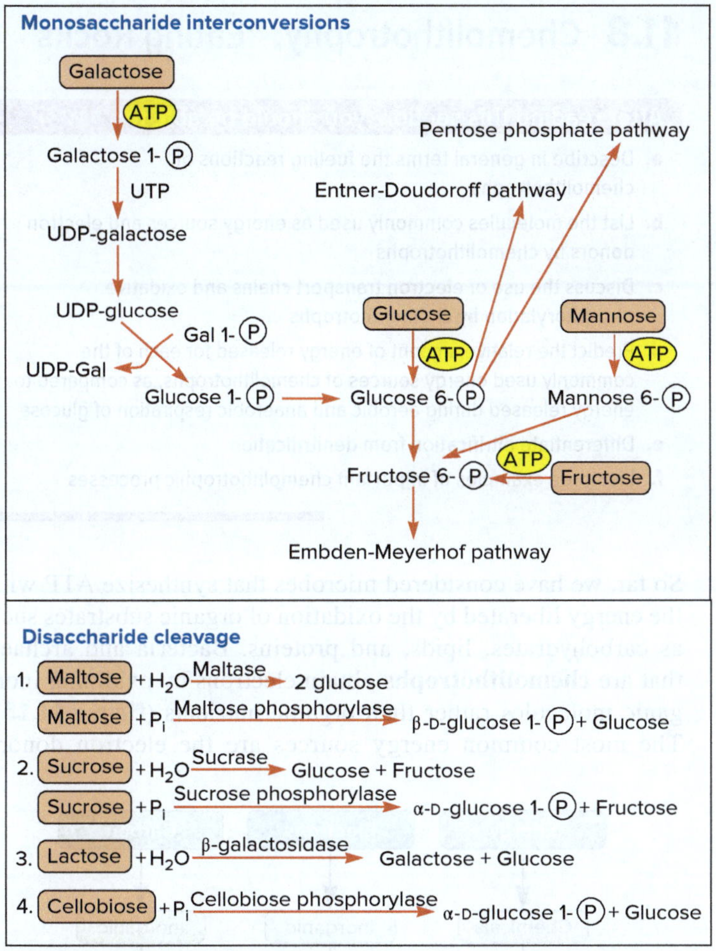

Figure 11.19 Carbohydrate Catabolism. Examples of enzymes and pathways used in disaccharide catabolism. UDP, uridine diphosphate. See figure AI.9 to learn how monosaccharides are named.

MICRO INQUIRY *What is the difference between hydrolysis and phosphorylysis?*

numerous polysaccharides in this way. For instance, starch and glycogen are hydrolyzed by amylases to glucose, maltose, and other products. Cellulose, which is more difficult to digest, is hydrolyzed by the cellulases produced by many fungi and some bacteria, yielding cellobiose and glucose. Many soil bacteria and bacterial plant pathogens degrade pectin, a polymer of galacturonic acid (a galactose derivative) that is an important constituent of plant cell walls and tissues. ▶ *Soils are an important microbial habitat (section 30.1)*

Recall that in times of plenty, microorganisms store excess carbon by synthesizing intracellular reserves such as glycogen and starch. To survive long periods in the absence of exogenous nutrients, they catabolize these intracellular stores. Glycogen and starch are degraded by phosphorylases that shorten the polysaccharide chain by one glucose, yielding glucose 1-phosphate.

$$(\text{Glucose})_n + P_i \rightarrow (\text{glucose})_{n-1} + \text{glucose-1-P}$$

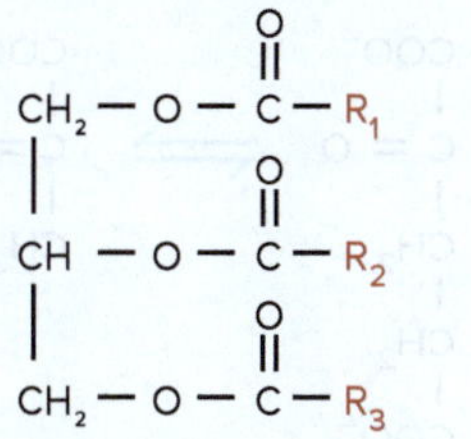

Figure 11.20 A Triglyceride (Triacylglycerol). The R groups represent the fatty acid side chains.

Glucose 1-phosphate can enter glycolytic pathways after conversion to glucose 6-phosphate (figure 11.19).

Lipid Catabolism

Chemoorganotrophic microorganisms frequently use lipids as energy sources. Triglycerides (also called triacylglycerols) are common energy sources. They are composed of a glycerol core attached to three fatty acids (**figure 11.20**). Triglycerides can be hydrolyzed to glycerol and free fatty acids by enzymes called *lipases*. The glycerol is then phosphorylated and oxidized to dihydroxyacetone phosphate, an intermediate of the Embden-Meyerhof pathway, which allows for its continued degradation.

Fatty acids from triglycerides and other lipids can be oxidized in the **β-oxidation pathway** after being linked to coenzyme A (**figure 11.21**). In this pathway, fatty acids are shortened by two carbons with each turn of the cycle. The two carbon units are released as acetyl-CoA, which can be fed into the TCA cycle or used in biosynthesis. One turn of the cycle produces acetyl-CoA, NADH, and $FADH_2$; NADH and $FADH_2$ can

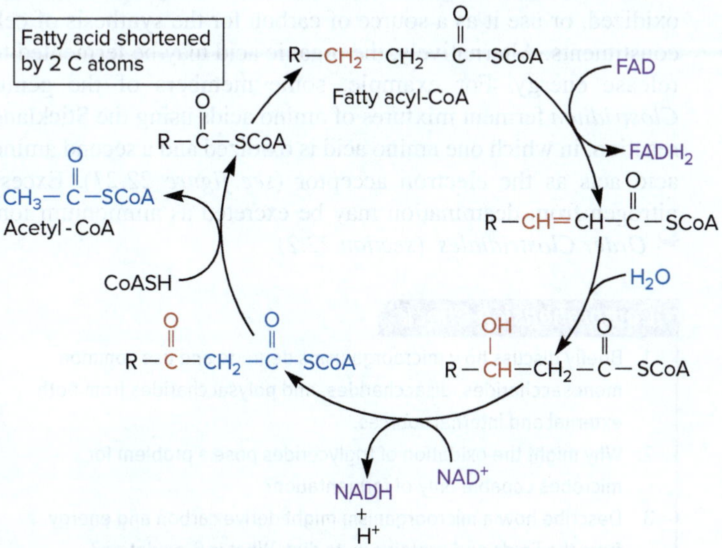

Figure 11.21 Fatty Acid β-Oxidation. The portions of the fatty acid being modified are shown in red.

MICRO INQUIRY *Why are fatty acids a rich source of energy even though no ATP is generated when they are degraded by the β-oxidation pathway?*

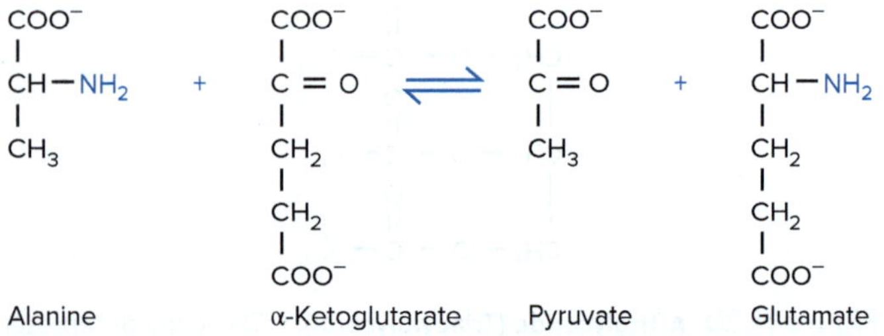

Figure 11.22 An Example of Transamination. The α-amino group (blue) of alanine is transferred to the acceptor α-ketoglutarate, forming pyruvate and glutamate.

be oxidized by an ETC to provide more ATP. Thus, fatty acids are a rich source of energy for organisms that perform respiration.

Protein and Amino Acid Catabolism

Some bacteria and fungi—particularly pathogenic and food spoilage microorganisms—use proteins as their source of carbon and energy. Like polysaccharides, proteins are too large to transport across the plasma membrane, so these microbes secrete enzymes called **proteases** that hydrolyze proteins to amino acids. These are then transported into the cell and catabolized.

The first step in amino acid catabolism is **deamination,** the removal of the amino group from an amino acid. This is often accomplished by **transamination.** The amino group from an amino acid is transferred to an α-keto acid acceptor (**figure 11.22**). The deaminated amino acid is now an organic acid that can be converted to pyruvate, acetyl-CoA, or a TCA cycle intermediate. The resulting organic acid may be used in several ways. Most microbes feed the organic acid into the TCA cycle, where it is oxidized, or use it as a source of carbon for the synthesis of cell constituents. Alternatively, the organic acid may be fermented to release energy. For example, some members of the genus *Clostridium* ferment mixtures of amino acids using the Stickland reaction, in which one amino acid is oxidized and a second amino acid acts as the electron acceptor (*see figure 22.21*). Excess nitrogen from deamination may be excreted as ammonium ion. ▶| *Order Clostridiales (section 22.2)*

Comprehension Check

1. Briefly discuss how microorganisms degrade and use common monosaccharides, disaccharides, and polysaccharides from both external and internal sources.
2. Why might the oxidation of triglycerides pose a problem for microbes capable only of fermentation?
3. Describe how a microorganism might derive carbon and energy from the lipids and proteins in its diet. What is β-oxidation? Deamination? Transamination?

11.8 Chemolithotrophy: "Eating Rocks"

After reading this section, you should be able to:

a. Describe in general terms the fueling reactions of chemolithotrophs
b. List the molecules commonly used as energy sources and electron donors by chemolithotrophs
c. Discuss the use of electron transport chains and oxidative phosphorylation by chemolithotrophs
d. Predict the relative amount of energy released for each of the commonly used energy sources of chemolithotrophs, as compared to energy released during aerobic and anaerobic respiration of glucose
e. Differentiate nitrification from denitrification
f. List three examples of important chemolithotrophic processes

So far, we have considered microbes that synthesize ATP with the energy liberated by the oxidation of organic substrates such as carbohydrates, lipids, and proteins. Bacteria and archaea that are **chemolithotrophs** obtain electrons by oxidizing inorganic molecules rather than organic nutrients (**figure 11.23**). The most common energy sources are the electron donors

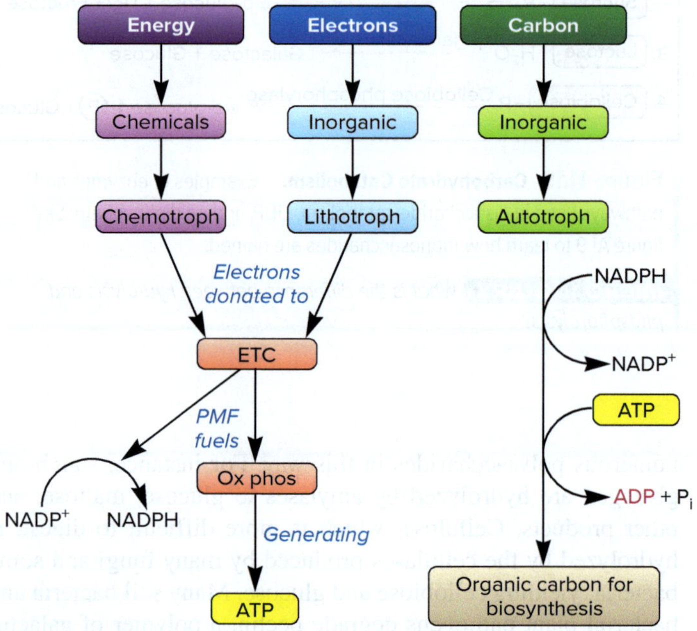

Figure 11.23 Chemolithotrophic Fueling Processes. Chemolithotrophic bacteria and archaea oxidize inorganic molecules (e.g., H_2S and NH_3), which serve as energy and electron sources. The electrons released pass through an electron transport chain, generating a proton motive force (PMF). ATP is synthesized by oxidative phosphorylation (ox phos). Most chemolithotrophs use O_2 as the terminal electron acceptor. Unlike chemoorganoheterotrophy, a molecule other than the energy source provides carbon for biosynthesis. Many chemolithotrophs are autotrophs, as shown here.

Table 11.3 Representative Chemolithotrophs and Their Energy Sources

Bacteria	Electron Donor	Electron Acceptor	Products
Alcaligenes, *Hydrogenophaga*, and *Pseudomonas* spp.	H_2	O_2	H_2O
Nitrobacter spp.	NO_2^-	O_2	NO_3^-, H_2O
Nitrosomonas spp.	NH_4^+	O_2	NO_2^-, H_2O
Thiobacillus denitrificans	S^0, H_2S	NO_3^-	SO_4^{2-}, N_2
Acidithiobacillus ferrooxidans	Fe^{2+}, S^0, H_2S	O_2	Fe^{3+}, H_2O, H_2SO_4

hydrogen, reduced nitrogen compounds, reduced sulfur compounds, and ferrous iron (Fe^{2+}). These inorganic compounds donate electrons directly to the ETC. The terminal electron acceptor is usually O_2, but sulfate and nitrate may also be used (**table 11.3**).

With the exception of H_2, which has a standard reduction potential more negative than the $NAD^+/NADH$ redox pair, much less energy is available from the oxidation of inorganic molecules than from the complete oxidation of glucose to CO_2 (**table 11.4**). This is because unlike the H^+/H_2 redox pair, the reduction potentials of most inorganic substrates used by chemolithotrophs are more positive than the reduction potentials of NADH and $FADH_2$. For example, referring to figure 11.10, it is clear that the $NAD^+/NADH$ couple is more negative than the S^0/H_2S couple. Thus even if O_2 is used as the terminal electron acceptor, when H_2S donates electrons, the ETC is shorter, so fewer protons are expelled from the cell and a smaller PMF

Table 11.4 Energy Yields from Oxidations Used by Chemolithotrophs

Reaction	$\Delta G^{\circ\prime}$ (kcal/mole)[1]
$H_2 + \frac{1}{2}O_2 \rightarrow H_2O$	−56.6
$NO_2^- + \frac{1}{2}O_2 \rightarrow NO_3^-$	−17.4
$NH_4^+ + 1\frac{1}{2}O_2 \rightarrow NO_2^- + H_2O + 2H^+$	−65.0
$S^0 + 1\frac{1}{2}O_2 + H_2O \rightarrow H_2SO_4$	−118.5
$S_2O_3^{2-} + 2O_2 + H_2O \rightarrow 2SO_4^{2-} + 2H^+$	−223.7
$2Fe^{2+} + 2H^+ + \frac{1}{2}O_2 \rightarrow 2Fe^{3+} + H_2O$	−11.2

[1] The $\Delta G^{\circ\prime}$ for complete oxidation of glucose to CO_2 is −686 kcal/mole. A kcal is equivalent to 4.184 kJ.

is generated. Therefore to have enough ATP to grow and reproduce, chemolithotrophs must oxidize a large quantity of inorganic material. This is particularly true of chemolithoautotrophs, which use CO_2 as their source of carbon. CO_2-fixation pathways consume a lot of ATP and reducing power (often NADPH). Because chemolithotrophs consume so much inorganic material, they have significant ecological impacts. They make important contributions to several biogeochemical cycles, including the nitrogen, sulfur, and iron cycles. ▶| *CO_2 fixation: reduction and assimilation of CO_2 carbon (section 12.3); Biogeochemical cycling sustains life on Earth (section 28.1)*

Members of the bacterial and archaeal genera that oxidize hydrogen gas use a hydrogenase enzyme (table 11.3).

$$H_2 \rightarrow 2H^+ + 2e^-$$

The standard reduction potential of the $2H^+/H_2$ conjugate redox pair is −0.42, while that of the $NAD^+/NADH$ pair is −0.32 V (figure 11.10). Thus, electrons from H_2 oxidation can be donated to either an ETC or NAD^+, depending on the hydrogenase. If NADH is produced, it can be used in ATP synthesis by electron transport and oxidative phosphorylation, with O_2, Fe^{3+}, S^0, or carbon monoxide (CO) as the terminal electron acceptor. Often hydrogen-oxidizing microorganisms use organic compounds as energy sources when available.

Some bacteria and archaea use the oxidation of nitrogenous compounds as a source of electrons. Among these chemolithotrophs, **nitrifying bacteria** are best understood. They are soil and aquatic bacteria of considerable ecological significance that carry out **nitrification**—the oxidation of ammonia to nitrate (*see figure 21.9*). Nitrification is a two-step process that until 2015 was thought to always depend on the activity of at least two different microbes. In the first step, ammonia is oxidized to nitrite. In the second step, the nitrite is oxidized to nitrate. When two bacteria are involved, one performs the ammonia oxidation (e.g., *Nitrosomonas* spp.)

while another oxidizes nitrite to yield nitrate (e.g., *Nitrobacter* spp.). Some members of the genus *Nitrospira* can perform both steps, a process called **comammox,** for *com*plete *amm*onia *ox*idation.

Care should be taken not to confuse nitrification and denitrification. Nitrification is a chemolithotrophic process in which ammonia is oxidized in a two-step process to yield nitrate. Electrons from ammonia and nitrite are donated to an ETC that uses oxygen as the terminal electron acceptor. By contrast, denitrification is a form of anaerobic respiration that uses nitrate as the terminal electron acceptor (figure 11.16).

A unique metabolic process that combines aspects of both nitrification and denitrification is carried out by anammox bacteria (phylum Planctomycetota). These bacteria carry out *an*aerobic *amm*onia *ox*idation, or **anammox,** using a membrane-bound compartment called the *anammoxosome.* In the anammoxosome, ammonia oxidation is coupled with the reduction of nitrite, generating N_2 and H_2O (*see figure 20.9*). The anammox reaction differs from comammox, which produces nitrate as the only end product.

Sulfur-oxidizing microbes are the third major group of chemolithotrophs. The metabolism of *Thiobacillus* and *Acidithiobacillus* spp. has been best studied. These bacteria use sulfur (S^0), hydrogen sulfide (H_2S), thiosulfate ($S_2O_3^{2-}$), and other reduced sulfur compounds as electron donors to their ETCs. Interestingly, they generate ATP by both oxidative phosphorylation and substrate-level phosphorylation (*see figure 21.15*). Substrate-level phosphorylation involves adenosine 5′-phosphosulfate (APS), a high-energy molecule formed from sulfite and adenosine monophosphate (**figure 11.24**).

Some sulfur-oxidizing microbes are extraordinarily flexible metabolically. For example, some bacteria and archaea can grow by oxidizing sulfur with oxygen as the electron acceptor. However, in the absence of O_2, they switch from chemolithotrophy to anaerobic respiration and oxidize organic material with sulfur as the electron acceptor. Furthermore, many sulfur-oxidizing chemolithotrophs use CO_2 as their carbon source but will grow heterotrophically if they are supplied with reduced organic carbon sources such as glucose or amino acids.

(a) Direct oxidation of sulfite

$$SO_3^{2-} \xrightarrow{\text{sulfite oxidase}} SO_4^{2-} + 2e^-$$

(b) Formation of adenosine 5′-phosphosulfate

$$2SO_3^{2-} + 2AMP \rightarrow 2APS + 4e^-$$

$$2APS + 2P_i \rightarrow 2ADP + 2SO_4^{2-}$$

$$2ADP \rightarrow AMP + ATP$$

$$2SO_3^{2-} + AMP + 2P_i \rightarrow 2SO_4^{2-} + ATP + 4e^-$$

Figure 11.24 Energy Generation by Sulfur Oxidation. (a) Sulfite can be directly oxidized to provide electrons for electron transport and oxidative phosphorylation. (b) Sulfite can also be oxidized and converted to adenosine 5′-phosphosulfate (APS; *see figure 21.15*). This route produces electrons for use in electron transport and ATP by substrate-level phosphorylation.

Electrons released by the oxidation of ammonia, nitrite, and sulfur-containing compounds are donated to an ETC, thereby generating a PMF that can be used to make ATP by oxidative phosphorylation. As we have already noted, the ATP yields for these chemolithotrophs are low; this reflects their short ETCs. Complicating this already difficult lifestyle is the fact that many of these microbes are also autotrophs, and autotrophs need NAD(P)H (reducing power) as well as ATP to reduce CO_2 and other molecules (figure 11.23). Molecules such as ammonia, nitrite, and H_2S have more positive reduction potentials than the NAD(P)$^+$/NAD(P)H conjugate redox pair (figure 11.10). Therefore, they cannot directly donate their electrons to NAD(P)$^+$. Recall that electrons spontaneously move only from donors with more negative reduction potentials to acceptors with more positive potentials. Thus these chemolithotrophs face a major dilemma: how to make the NAD(P)H they need. They solve this problem by using a process called **reverse electron flow.** During reverse electron flow, electrons derived from oxidation of inorganic substrates (reduced nitrogen or sulfur compounds) are moved up their ETCs to reduce NAD(P)$^+$ to NAD(P)H (**figure 11.25**). Because this is not thermodynamically favorable, energy in the form of the PMF must be diverted from performing other cellular work (e.g., ATP synthesis, transport, motility) to "push" electrons from electron carriers of relatively positive reduction potentials to those that are more negative.

Overall, the use of inorganic molecules by microorganisms can be confusing because some of these molecules can be used as electron donors in chemolithotrophy or terminal electron acceptors in anaerobic respiration. There are two key questions to ask when considering the microbial use of inorganic chemicals for energy conservation. First, "What is the oxidation state of the inorganic molecule?" If it is fully reduced (e.g., H_2S, NH_3), then the molecule can only serve as an electron donor because it simply cannot accept any more electrons. Therefore, fully reduced molecules are used by chemolithotrophs to donate electrons to the ETC. Conversely, if the molecule is fully oxidized (e.g., SO_4^{2-}, NO_3^-), it can accept electrons, but it has none to donate, so it serves as a terminal electron acceptor during anaerobic respiration. The tricky molecules are those that are neither fully oxidized nor reduced (e.g., S^0, NO_2^-), as these can either donate electrons to chemolithotrophic ETCs or serve as terminal electron acceptors in anaerobic respiration. In this case, one must know something about the physiology of the particular microbe.

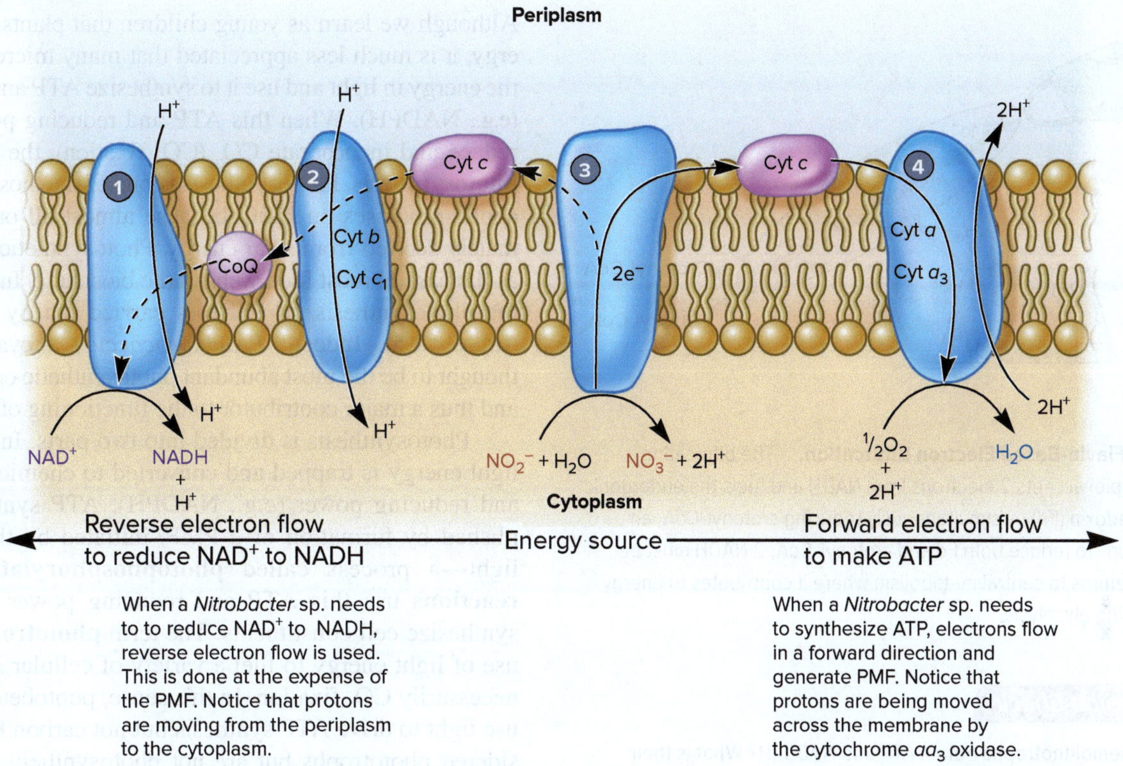

Figure 11.25 Electron Flow in a *Nitrobacter* spp. Electron Transport Chain. *Nitrobacter* spp. oxidize nitrite and carry out normal electron transport to generate proton motive force for ATP synthesis. This is the right-hand branch of the diagram. Some of the proton motive force also is used to force electrons to flow up the reduction potential gradient from nitrite to NAD^+ (left-hand branch). Cytochrome *c* (Cyt *c*), Coenzyme Q (CoQ), and four protein complexes make up this ETC. The four complexes are NADH-ubiquinone oxidoreductase (1), ubiquinol-cytochrome *c* oxidoreductase (2), nitrite oxidase (3), and cytochrome aa_3 oxidase (4).

11.9 Flavin-Based Electron Bifurcation

After reading this section, you should be able to:

- **a.** Explain why electron bifurcation is a fundamental mechanism of energy conservation that differs from ATP hydrolysis or proton motive force
- **b.** Discuss the concept of linking an endergonic reaction with an exergonic one to accomplish an otherwise thermodynamically unfavorable redox reaction

Some anaerobic archaea and bacteria bypass the thermodynamic barrier of unfavorable, endergonic redox reactions by tightly coupling such reactions with favorable, exergonic ones. This process is known as *electron bifurcation* (Latin, *bifurcate,* divided into two forks) because the flow of electrons is branched: One electron (or electron pair) moves in the endergonic direction, while the other moves in the exergonic direction. All electron bifurcation systems reported to date feature a cytoplasmic enzyme with a flavin cofactor, called a *flavoenzyme;* so these systems are commonly called **flavin-based electron bifurcation (FBEB).** All bifurcating flavoenzymes oxidize one electron donor, while donating electrons simultaneously to two different electron acceptors. The reduction of one acceptor has to be exergonic so it can "pull" the thermodynamically less favorable endergonic reduction of the second acceptor. In this way FBEB performs work in the absence of ATP hydrolysis or by harnessing the energy of a proton motive force.

FBEB was first detected in the anaerobic chemoheterotrophic bacterium *Clostridium kluyveri,* which transfers electrons from NADH to ferredoxin (Fd). This reaction is endergonic because NADH has a standard reduction potential (E_0') of –320 mV, which is less negative than that of ferredoxin (E_0' about –450 mV). It is possible for NADH to "push electrons uphill" to reduce Fd only because this is accompanied by an exergonic reaction: The simultaneous flow of electrons to crotonyl-CoA, which is reduced to butyryl-CoA (E_0' = –30 mV). In this reaction electrons are flowing from more negative to less negative E_0'. The role of the cytoplasmic bifurcating flavoenzyme is much like a traffic cop directing the flow of electrons (**figure 11.26**). Since the discovery of FBEB in *C. kluyveri,* other FBEB systems have been discovered, including archaea that use H_2 and CO_2 to produce methane in a pathway called the Wolfe cycle. In these microbes, the last step is an FBEB system (see figure 19.10). FBEB appears to be an important form of energy conservation for anaerobic microbes that live at the edge of their thermodynamic limit. ◀ *Cellular work and energy transfers (section 10.1)* ▶ *Methanogens (section 19.4)*

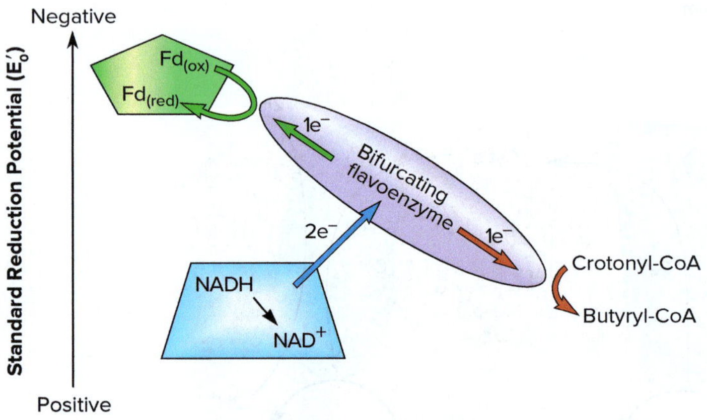

Figure 11.26 Flavin-Based Electron Bifurcation. The bifurcating flavoenzyme (purple) accepts 2 electrons from NADH and fuels the endergonic reduction of ferredoxin (Fd) by simultaneously reducing crotonyl-CoA, an exergonic reaction. To reduce both Fd and crotonyl-CoA, 2 NADH must be oxidized. NAD^+ returns to central metabolism where it contributes to energy conservation during glycolysis.

Comprehension Check

1. How do chemolithotrophs obtain ATP and NAD(P)H? What is their most common source of carbon?
2. Describe energy production by hydrogen-oxidizing bacteria, nitrifying bacteria, and sulfur-oxidizing bacteria.
3. Why can hydrogen-oxidizing bacteria and archaea donate electrons to NAD^+, whereas sulfur- and ammonia-oxidizing bacteria and archaea cannot?
4. What is reverse electron flow and why do many chemolithotrophs perform it?
5. Some H_2-oxidizing chemolithotrophic bacteria use FBEB to oxidize H_2 to $2H^+$ (E'_0 = −420 mV) while reducing Fd and NAD^+ to NADH. Sketch a diagram showing the flow of electrons relative to the E'_0 of these redox pairs.

11.10 Phototrophy

After reading this section, you should be able to:

a. Differentiate phototrophy from photosynthesis
b. Summarize the structure and function of the light-absorbing pigments used by oxygenic and anoxygenic phototrophs
c. Explain why oxidative phosphorylation and photophosphorylation by chlorophyll-based phototrophs differ primarily in the energy source driving the process
d. Distinguish cyclic photophosphorylation from noncyclic photophosphorylation
e. Compare and contrast oxygenic photosynthesis, anoxygenic phototrophy, and rhodopsin-based phototrophy
f. List two examples of the importance of chlorophyll-based phototrophy

Although we learn as young children that plants harvest light energy, it is much less appreciated that many microbes also capture the energy in light and use it to synthesize ATP and reducing power (e.g., NADPH). When this ATP and reducing power are used to reduce and incorporate CO_2 (CO_2 fixation), the process is called **photosynthesis.** Photosynthesis is one of the most significant metabolic processes on Earth because almost all our energy is ultimately derived from solar energy. Photosynthetic organisms serve as the base of most food webs in the biosphere. In fact, at least half the photosynthesis on Earth is carried out by microorganisms (**table 11.5**). Indeed, *Prochlorococcus,* a cyanobacterium, is thought to be the most abundant photosynthetic organism on Earth and thus a major contributor to the functioning of the biosphere.

Photosynthesis is divided into two parts. In **light reactions,** light energy is trapped and converted to chemical energy (ATP) and reducing power (e.g., NADPH). ATP synthesis is accomplished by formation of a PMF, initiated by the absorption of light—a process called **photophosphorylation.** The **dark reactions** use this ATP and reducing power to fix CO_2 and synthesize cell constituents. The term **phototrophy** refers to the use of light energy to fuel a variety of cellular activities but not necessarily CO_2 fixation. In this sense, photoheterotrophs, which use light to drive ATP synthesis but not carbon fixation, are considered phototrophs but are not photosynthetic. In this section, three types of phototrophy are discussed: oxygenic photosynthesis, anoxygenic photosynthesis, and rhodopsin-based phototrophy (**figure 11.27**). The dark reactions of photosynthesis are reviewed in chapter 12. ▶| *CO_2 fixation: reduction and assimilation of CO_2 carbon (section 12.3)*

Light Reactions in Oxygenic Photosynthesis

The Great Oxidation Event 2.4 billion years ago changed Earth forever. The presence of oxygen enabled the evolution of larger and more complex animals fueled by aerobic respiration. However, there would have been no Great Oxidation Event if it were not for the evolution of cyanobacteria, capable of splitting water to hydrogen and oxygen.

Cyanobacteria and the photosynthetic eukaryotes that contain chloroplasts, which arose from endosymbiotic cyanobacteria, carry out **oxygenic photosynthesis,** so named because oxygen is generated when water is reduced (split). Oxygen is released into the environment when light energy is converted to chemical energy.

Table 11.5	Diversity of Phototrophic Microorganisms
Eukaryotes	Multicellular green, brown, and red algae; unicellular protists (e.g., euglenoids, dinoflagellates, diatoms)
Bacteria	Cyanobacteria, green sulfur bacteria, green nonsulfur bacteria, purple sulfur bacteria, purple nonsulfur bacteria, heliobacteria, acidobacteria
Archaea	Halophiles

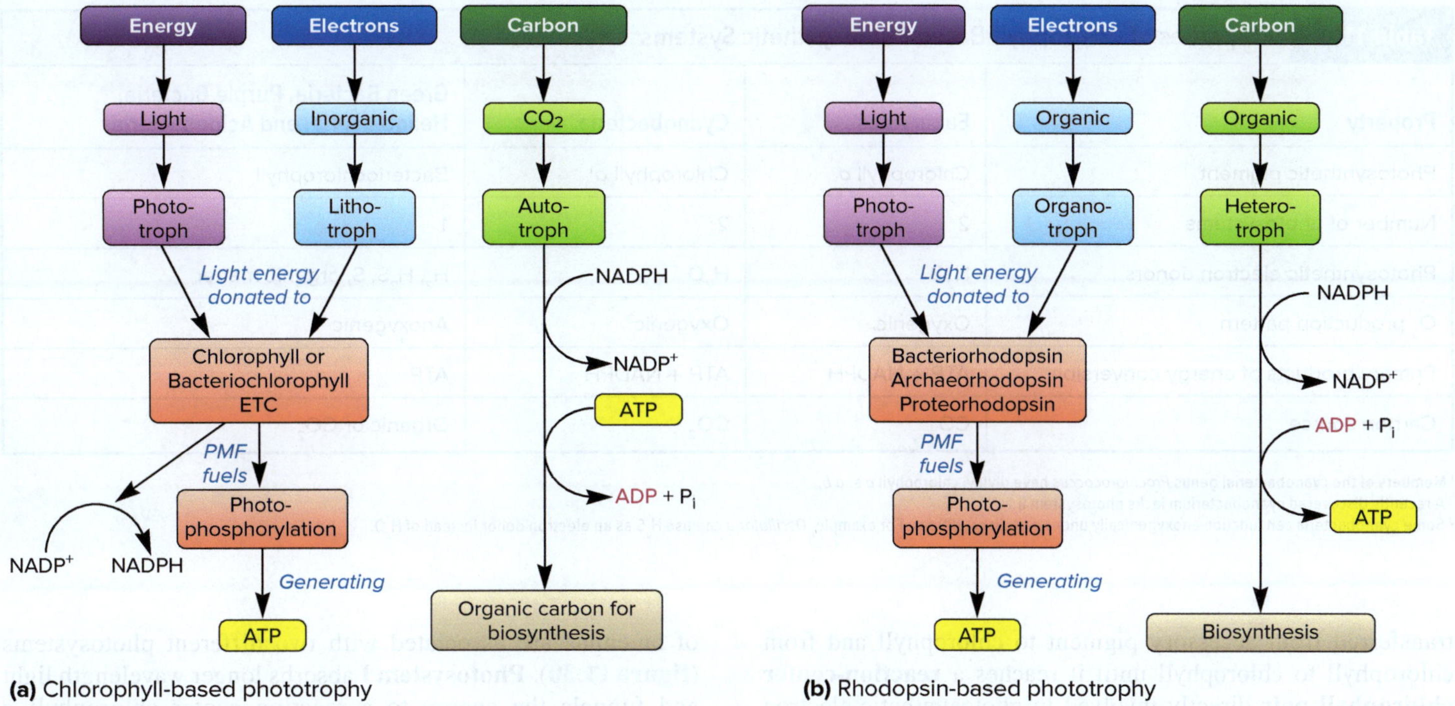

Figure 11.27 Phototrophic Fueling Reactions. Phototrophs use light to generate a proton motive force (PMF), which is then used to synthesize ATP by a process called photophosphorylation. The process requires light-absorbing pigments. (a) When the pigments are chlorophyll or bacteriochlorophyll, the absorption of light triggers electron flow through an electron transport chain, accompanied by the pumping of protons across a membrane. Many of these phototrophs are autotrophs and must use much of the ATP and reducing power they make to fix CO_2. (b) Rhodopsin-based phototrophy differs in that the PMF is formed directly by the light-absorbing pigment, which is a light-driven proton pump.

Light-absorbing pigments are central to all phototrophic processes (**table 11.6**). In oxygenic phototrophs, the most important pigments are **chlorophylls.** Chlorophylls are large planar molecules composed of four pyrrole rings (tetrapyrrole) with a central magnesium atom coordinated to the four nitrogen atoms of the pyrrole rings (**figure 11.28**). The pyrrole rings are decorated with specific chemical groups that distinguish one type of chlorophyll from another. A long hydrophobic tail (the R side chain at the bottom left of figure 11.28) aids in the attachment of chlorophyll to membranes, where light reactions take place. Several chlorophylls are found in eukaryotes; the two most important are chlorophyll *a* and chlorophyll *b*. These two molecules differ slightly in their structure and spectral properties. They both absorb light in the red range of the visible light spectrum (*see figure 7.20*), but chlorophyll *a* has a light absorption peak at a slightly longer wavelength than chlorophyll *b* (665 nm versus 645 nm). In addition to absorbing red light, chlorophylls also absorb blue light strongly. Because chlorophylls absorb primarily in the red and blue ranges, green light is transmitted, and most of these organisms appear green.

Because chlorophylls absorb a narrow range of light wavelengths, other photosynthetic pigments assist in trapping light energy. The most widespread of these are *carotenoids,* long molecules, usually yellowish in color, that possess an extensive conjugated double bond system (**figure 11.29**). β-carotene is present in cyanobacteria belonging to the genus *Prochloron* and most photosynthetic protists; fucoxanthin is found in protists such as diatoms and dinoflagellates. Red algae and cyanobacteria have photosynthetic pigments called **phycobiliproteins,** consisting of a protein with a linear tetrapyrrole attached (see figure 20.6). These pigments give these microbes their characteristic colors.

Carotenoids and phycobiliproteins are often called **accessory pigments** because they help chlorophylls harvest light energy, but they cannot harvest energy directly. They are important because they absorb light in the range not absorbed by chlorophylls (the blue-green through yellow range; about 470 to 630 nm; *see figure 20.3*). This light energy is transferred to chlorophyll, thereby making photosynthesis possible over a broader range of wavelengths. In addition, this allows organisms to use light not used by other phototrophs in their habitat. Finally, by absorbing light in the UV range, some accessory pigments also serve as microbial sunscreen, protecting cellular constituents from intense sunlight.

About 300 chlorophyll molecules and accessory pigments are assembled in highly organized arrays called *antennas* whose purpose is to create a large surface area to trap as many photons as possible. Light energy captured in an antenna is

Table 11.6 Properties of Chlorophyll-Based Photosynthetic Systems

Property	Eukaryotes	Cyanobacteria	Green Bacteria, Purple Bacteria, Heliobacteria, and Acidobacteria
Photosynthetic pigment	Chlorophyll *a*	Chlorophyll *a*[1]	Bacteriochlorophyll
Number of photosystems	2	2[2]	1
Photosynthetic electron donors	H_2O	H_2O	H_2, H_2S, S, organic matter
O_2 production pattern	Oxygenic	Oxygenic[3]	Anoxygenic
Primary products of energy conversion	ATP + NADPH	ATP + NADPH	ATP
Carbon source	CO_2	CO_2	Organic or CO_2

[1] Members of the cyanobacterial genus *Prochlorococcus* have divinyl chlorophyll *a* and *b*.
[2] A recently discovered cyanobacterium lacks photosystem II.
[3] Some cyanobacteria can function anoxygenically under certain conditions. For example, *Oscillatoria* can use H_2S as an electron donor instead of H_2O.

transferred from accessory pigment to chlorophyll and from chlorophyll to chlorophyll until it reaches a **reaction-center chlorophyll pair** directly involved in photosynthetic electron transport (*see figure 20.6*). In oxygenic phototrophs, two kinds of antennas are associated with two different photosystems (**figure 11.30**). **Photosystem I** absorbs longer wavelength light and funnels the energy to a reaction center chlorophyll *a* pair called P700. The term P700 signifies that this molecule most effectively absorbs light at a wavelength of 700 nm. **Photosystem II** traps light at shorter wavelengths and transfers its energy to the reaction center chlorophyll pair P680. Depending on the needs of the organism, photosystem I can function independently, while at other times photosystems I and II are used together. We begin our discussion with the two photosystems working together.

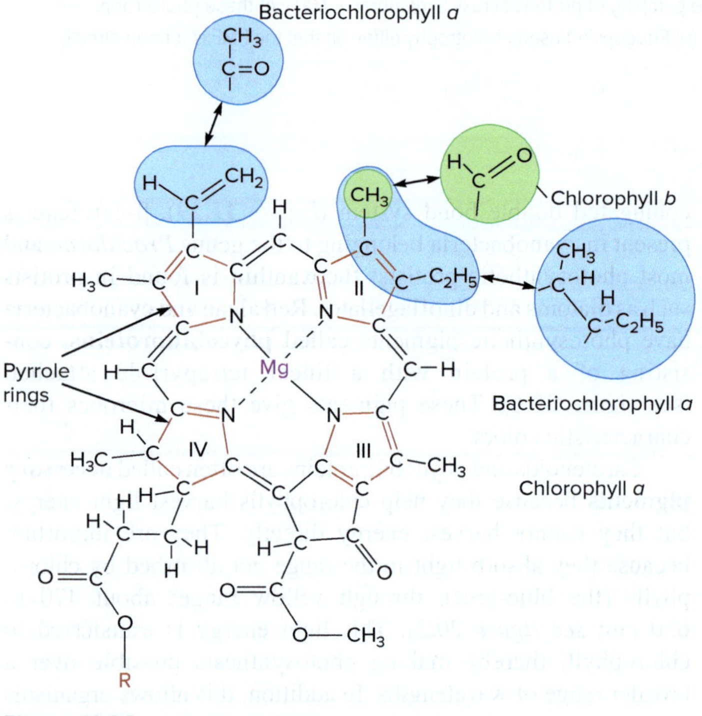

Figure 11.28 Chlorophyll Structure. The structures of chlorophyll *a*, chlorophyll *b*, and bacteriochlorophyll *a*. The complete structure of chlorophyll *a* is shown. Only one group is altered to produce chlorophyll *b* (green), and two modifications in the ring system are found in bacteriochlorophyll *a* (blue). The side chain (R) of bacteriochlorophyll *a* may be either phytyl (a 20-carbon chain also found in chlorophylls *a* and *b*) or geranylgeranyl (a 20-carbon side chain similar to phytyl but with three more double bonds). Pyrroles are shown in red and are numbered.

Two Photosystems Working Together Generate ATP and NADPH; Oxygen Is Released

Close inspection of figure 11.30 reveals that when light is absorbed by a reaction-center chlorophyll pair in a photosystem (i.e., P680 and P700), the chlorophyll molecules change from a very positive standard reduction potential to a very negative standard reduction potential. That is, the absorption of light causes the reaction-center chlorophylls to become excellent electron donors. With this in mind, let's first examine photosystem II. When it absorbs light, the energized P680 (P^*_{680} in figure 11.30) donates electrons to a series of electron carriers that ultimately donate the electrons to P700 of photosystem I. The flow of electrons through the P680-associated ETC generates a PMF, which can be used by ATP synthase to make ATP from ADP and P_i. In the meantime, photosystem I has also absorbed light and its energized reaction-center chlorophylls (P^*_{700} in figure 11.30) donate electrons to a short ETC that ultimately transfers electrons to $NADP^+$, yielding NADPH. Thus when both photosystems work together, both ATP and NADPH are produced. Because the electrons flow in a noncyclic pattern, this is called noncyclic electron flow, and the ATP produced is said to be made by **noncyclic photophosphorylation.**

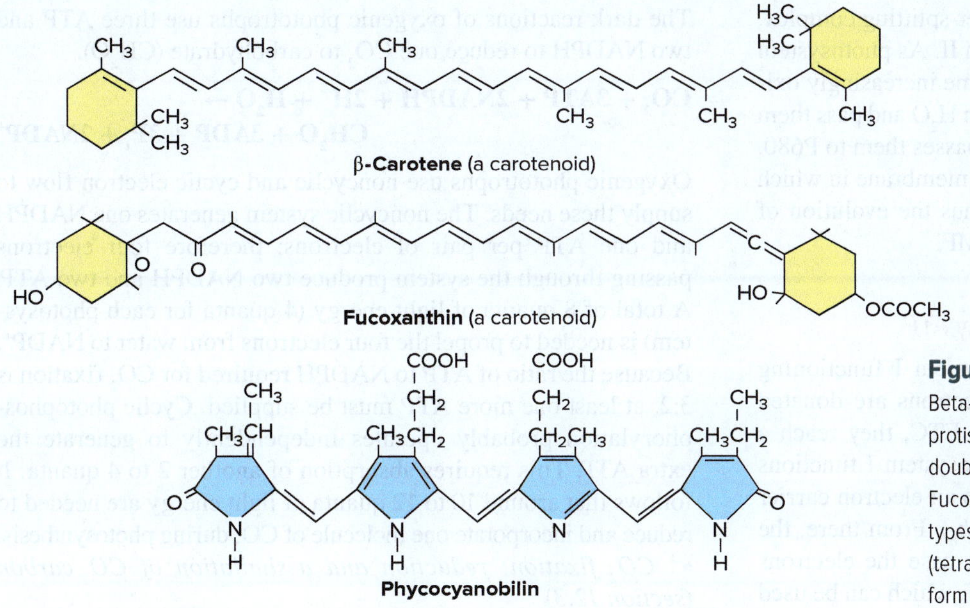

Figure 11.29 Representative Accessory Pigments. Beta-carotene is a carotenoid found in photosynthetic protists and plants. Note that it has a long chain of alternating double and single bonds called conjugated double bonds. Fucoxanthin is a carotenoid accessory pigment in several types of algae. Phycocyanobilin consists of four pyrrole rings (tetrapyrrole; see figure 11.28); it is attached to a protein to form a phycobiliprotein.

At this point the cell has a problem. When P700 was oxidized, its electrons were replaced by electrons arising from the oxidation of P680. But if P680 is to continue to absorb light, its electrons must be replaced. Notice in figure 11.30 that the standard reduction potential of P680 is more positive than that of the ½O_2/H_2O conjugate redox pair. This means that H_2O can donate electrons to P680, resulting in the release of oxygen. This is accomplished by a metal-containing enzyme called the

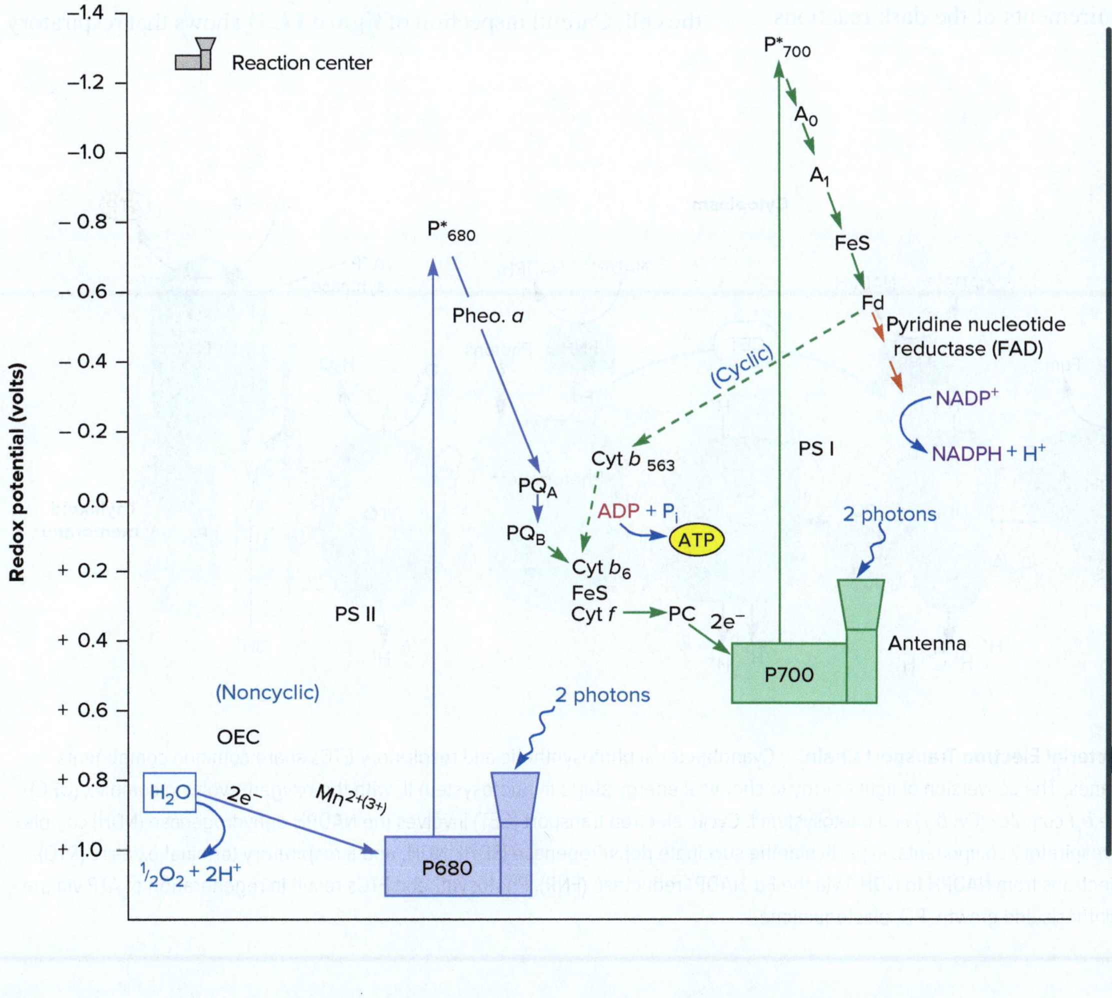

Figure 11.30 Oxygenic Photosynthesis. Cyanobacteria and eukaryotic algae are similar in having two photosystems, although they may differ in some details. The carriers involved in electron transport are ferredoxin (Fd) and other FeS proteins; cytochromes b_6, b_{563}, and f; plastiquinones A and B (PQ_A and PQ_B); copper-containing plastocyanin (PC); pheophytin *a* (Pheo. *a*); a specialized form of chlorophyll (A_0); and phylloquinone (A_1). Both photosystem I (PS I) and photosystem II (PS II) are involved in noncyclic photophosphorylation; only PS I participates in cyclic photophosphorylation. The oxygen-evolving complex (OEC) that extracts electrons from water contains manganese ions that enable electron transfer to the PS II reaction center.

MICRO INQUIRY *When electrons from P700 are used to reduce NADP+, what compound supplies electrons for the re-reduction of P700?*

oxygen-evolving complex (OEC) or the water-splitting complex, which is closely associated with photosystem II. As photosystem II absorbs light, Mn atoms in the OEC become increasingly oxidized, enabling them to extract electrons from H_2O and pass them to a protein in the photosystem, which then passes them to P680. In the process, protons are moved across the membrane in which the photosynthetic machinery is located. Thus the evolution of oxygen from water also contributes to the PMF.

Photosystem I Working Alone Yields Only ATP

Now let us return our attention to photosystem I functioning alone. Notice in figure 11.30 that when electrons are donated from P^*_{700} to the photosystem I–associated ETC, they reach a branch point at ferredoxin (Fd). When photosystem I functions alone, the electrons follow the dashed line to an electron carrier in the photosystem II–associated ETC (cyt b_6). From there, the electrons eventually return to P700. In this case the electrons have flowed in a circle. This generates a PMF, which can be used to synthesize ATP, and is referred to as **cyclic photophosphorylation.** Also note that because the electrons were diverted at Fd, no electrons were available to reduce $NADP^+$ to NADPH. Thus when photosystem I functions alone, only ATP is made available to the cell.

You might be asking why an organism needs to use both photosystems at certain times, and only photosystem I at others. The answer is related to the requirements of the dark reactions. The dark reactions of oxygenic phototrophs use three ATP and two NADPH to reduce one CO_2 to carbohydrate (CH_2O).

$$CO_2 + 3ATP + 2NADPH + 2H^+ + H_2O \rightarrow CH_2O + 3ADP + 3P_i + 2NADP^+$$

Oxygenic phototrophs use noncyclic and cyclic electron flow to supply these needs. The noncyclic system generates one NADPH and one ATP per pair of electrons; therefore four electrons passing through the system produce two NADPH and two ATP. A total of 8 quanta of light energy (4 quanta for each photosystem) is needed to propel the four electrons from water to $NADP^+$. Because the ratio of ATP to NADPH required for CO_2 fixation is 3:2, at least one more ATP must be supplied. Cyclic photophosphorylation probably operates independently to generate the extra ATP. This requires absorption of another 2 to 4 quanta. It follows that around 10 to 12 quanta of light energy are needed to reduce and incorporate one molecule of CO_2 during photosynthesis. ▶| *CO_2 fixation: reduction and assimilation of CO_2 carbon (section 12.3)*

It is worth reemphasizing that although light is the source of energy for chlorophyll-based phototrophy, the process used to make ATP is virtually the same as seen for chemotrophs: Oxidation-reduction reactions occurring in ETCs generate a PMF used by ATP synthase to make ATP. However, rather than occurring across the cell membrane, in cyanobacteria, photosynthetic light reactions are located in thylakoid membranes within the cell. Careful inspection of **figure 11.31** shows that respiratory

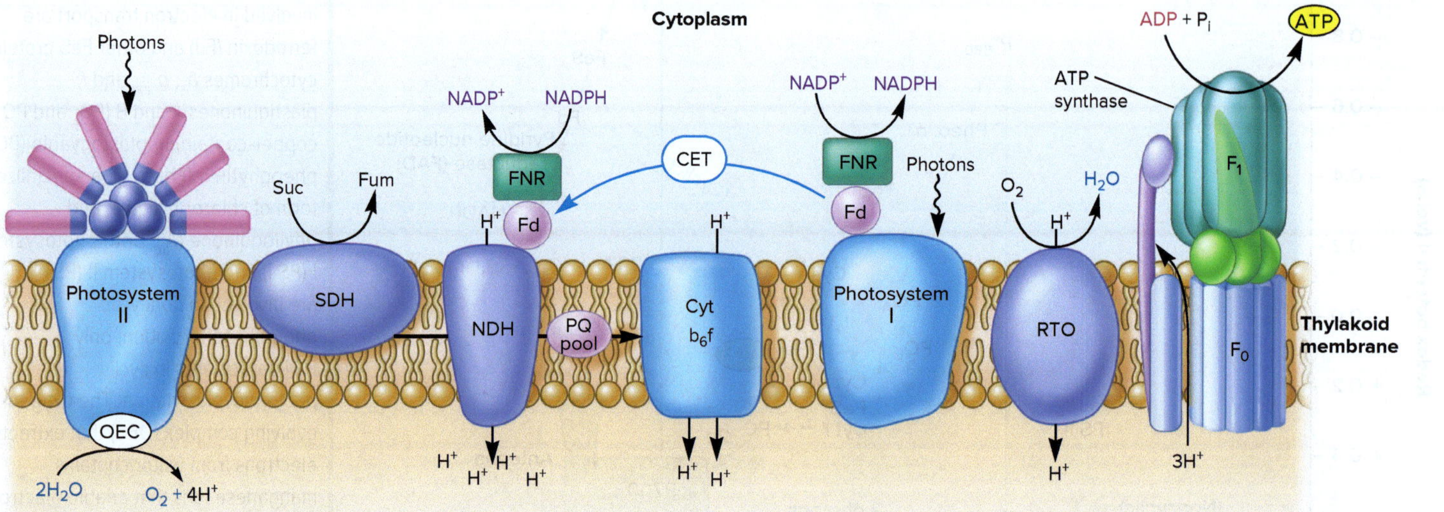

Figure 11.31 Generalized Cyanobacterial Electron Transport Chain. Cyanobacterial photosynthetic and respiratory ETCs share common components and are embedded in thylakoid membranes. The conversion of light energy to chemical energy starts in photosystem II, with the oxygen-evolving complex (OEC). Electrons are donated to the cytochrome b_6f complex (Cyt b_6f) and photosystem I. Cyclic electron transport (CET) involves the NADPH dehydrogenase (NDH) complex. The thylakoid membrane also contains respiratory components, in particular the succinate dehydrogenase (SDH), NDH, and a respiratory terminal oxidase (RTO). Ferredoxin (Fd) is thought to transfer electrons from NADPH to NDH-1 via the Fd-$NADP^+$ reductase (FNR). Photosynthetic ETCs result in regeneration of ATP via the ATP synthase and NADPH for cellular synthesis and growth. PQ, plastoquinione.

ETC proteins are also located within the same membrane. Cyanobacteria, like plants, also respire, especially when it is dark. *Photosynthetic Electron Transport and ATP Synthesis*

Light Reactions in Anoxygenic Photosynthesis

Certain bacteria carry out a second type of photosynthesis called **anoxygenic photosynthesis.** Although the term *photosynthesis* is used, many of these bacteria can function as photoheterotrophs and don't always use the energy gained from light to fix CO_2. *Anoxygenic phototrophy* is a better term; it derives its name from the fact that molecules other than water are used as an electron source and therefore O_2 is not produced. The process also differs in terms of (1) the pigments used, (2) the participation of just one photosystem, and (3) the mechanisms used to generate reducing power. Members of five bacterial phyla carry out anoxygenic phototrophy: Proteobacteria (purple sulfur and purple nonsulfur bacteria), Chlorobi (green sulfur bacteria), Chloroflexota (green nonsulfur bacteria), Firmicutes (heliobacteria), and Acidobacteria. The biology and ecology of these organisms are described in more detail in chapters 20, 21, and 22.

Anoxygenic phototrophs have light-absorbing pigments called **bacteriochlorophylls** (figure 11.28). Although bacteriochlorophylls are structurally very similar to the chlorophylls used in oxygenic photosynthesis, they are sensitive to oxygen. In some bacteria, bacteriochlorophylls are located in membranous vesicles called chlorosomes (*see figure 20.4*). The absorption maxima of bacteriochlorophylls are at longer wavelengths than those of chlorophylls, reaching into the infrared wavelengths. This better adapts anoxygenic phototrophic bacteria to their ecological niches. *Photosynthetic bacteria are diverse (section 20.3)*

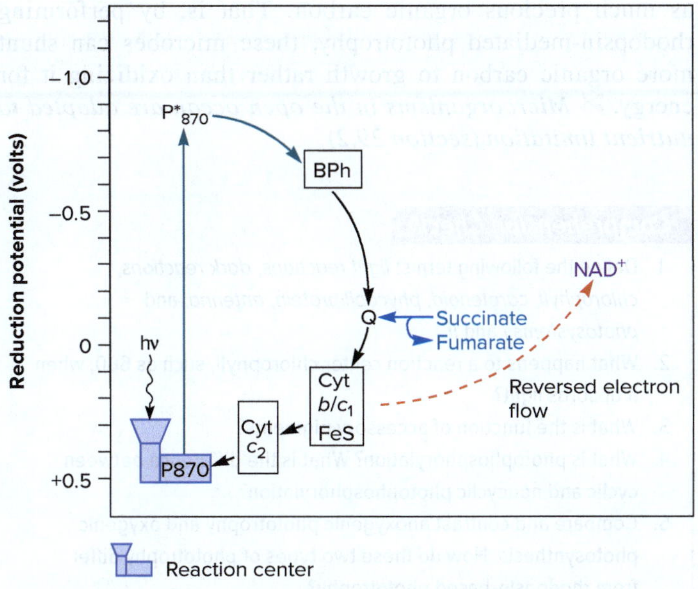

Figure 11.32 Purple Nonsulfur Bacterial Phototrophy. The ETC of the purple nonsulfur bacterium *Cereibacter* (*formerly Rhodobacter*) *sphaeroides.* Ubiquinone (Q) is very similar to coenzyme Q. The electron source succinate is in blue. BPh, Bacteriopheophytin.

The key difference between anoxygenic and oxygenic phototrophs is that anoxygenic phototrophs have only a single photosystem. Because they are restricted to cyclic photophosphorylation, they do not produce O_2 from H_2O. Indeed, almost all anoxygenic phototrophs are strict anaerobes. The schemes for the phototrophic ETCs of two purple nonsulfur bacteria are given in **figures 11.32** and **11.33**. When the reaction-center bacteriochlorophyll P870 is excited, it donates an electron to an ETC that transfers the electron back to P870 while generating sufficient PMF to drive ATP synthesis by ATP synthase. *Green sulfur bacteria (section 20.3); Order Rhodospirillales includes photosynthesizers and magnetic bacteria (section 21.1)*

In addition to making ATP, anoxygenic phototrophs must also generate reducing power (NAD[P]H or reduced ferredoxin) for CO_2 fixation (photoautotrophs) and other biosynthetic processes (photoautotrophs and photoheterotrophs). They are able to generate reducing power in at least three ways, depending on the bacterium. (1) Some have hydrogenases that reduce $NAD(P)^+$ to NAD(P)H directly by donating electrons from hydrogen gas. This is possible because hydrogen gas has a more negative reduction potential than NAD^+ (figure 11.10). (2) Others, such as the photosynthetic purple bacteria, use reverse electron flow to generate NAD(P)H (figure 11.32). In this mechanism, electrons are drawn off the photosynthetic ETC and "pushed" to $NAD(P)^+$ using PMF. This process is similar to that seen for chemolithotrophs having inorganic energy sources with more positive reduction potentials than that of $NAD(P)^+$/NAD(P)H (figure 11.25). (3) Phototrophic green bacteria and heliobacteria also draw electrons from their ETCs. However, because the reduction potential of the carrier where this occurs is more negative than NAD^+, the electrons flow

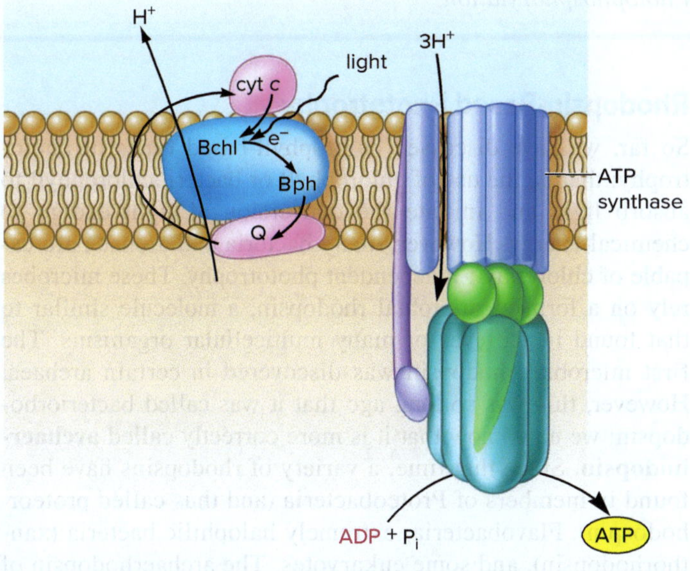

Figure 11.33 Cyclic Electron Flow During Anoxygenic Phototrophy. The photosystem of *Blastochloris viridis* is illustrated. Bchl, bacteriochlorophyll; Bph, bacteriophaeophytin; cyt *c*, cytochrome *c*; Q, quinones.

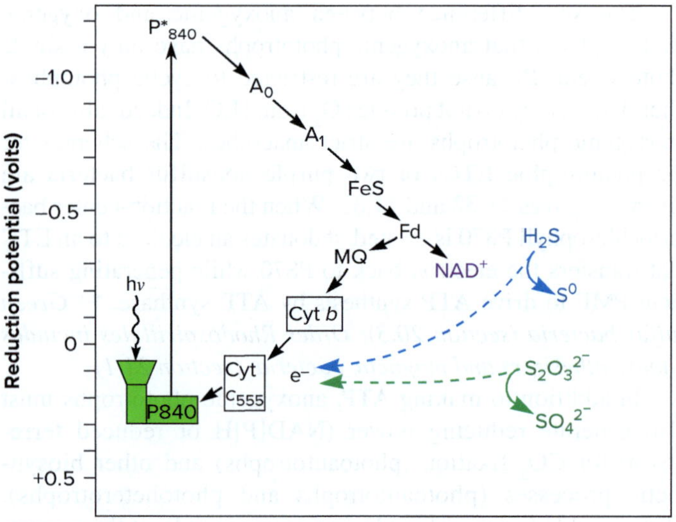

Figure 11.34 Green Sulfur Bacterial Light Reactions. Light energy is used to make ATP by cyclic photophosphorylation and to move electrons from thiosulfate ($S_2O_3^{2-}$) and H_2S (green and blue) to NAD^+. The ETC consists of A_0 (bacteriochlorophyll 663); A_1, a quinonelike molecule; FeS (iron-sulfur molecules within the reaction center); Fd (ferredoxin); MQ (menaquinone); and two different cytochromes (Cyt).

spontaneously to these electron acceptors. Thus these bacteria exhibit a simple form of noncyclic photosynthetic electron flow (**figure 11.34**). The diversion of electrons from the photosynthetic ETC means that continued generation of PMF will cease unless the electrons are replaced. Electrons from environmental electron donors such as hydrogen sulfide, elemental sulfur, and organic compounds replace the electrons removed from the ETC in this way. *Cyclic and Noncyclic Photophosphorylation*

Rhodopsin-Based Phototrophy

So far, we have discussed chlorophyll-based types of phototrophy; that is, the use of chlorophyll or bacteriochlorophyll to absorb light and initiate the conversion of light energy to chemical energy. However, many bacteria and archaea are capable of chlorophyll-independent phototrophy. These microbes rely on a form of microbial rhodopsin, a molecule similar to that found in the eyes of many multicellular organisms. The first microbial rhodopsin was discovered in certain archaea. However, this was so long ago that it was called bacteriorhodopsin; we now know that it is more correctly called **archaerhodopsin.** Since that time, a variety of rhodopsins have been found in members of Proteobacteria (and thus called proteorhodospin), Flavobacteria, extremely halophilic bacteria (xanthorhodopsin), and some eukaryotes. The archaerhodopsin of the halophilic archaeon *Halobacterium salinarum* remains the best studied and is the focus of our discussion here. ▶| *Halobacteriota (section 19.4)*

H. salinarum normally depends on aerobic respiration for the release of energy from its organic energy source. It cannot grow anaerobically by anaerobic respiration or fermentation. However, under conditions of low oxygen and high light intensity, it synthesizes archaerhodopsin, a deep-purple pigment. The color comes from retinal, a carotenoid. Retinal is covalently attached to a protein embedded in the plasma membrane such that retinal is in the center of the membrane.

Archaerhodopsin functions as a light-driven proton pump. When retinal absorbs light, a proton is released and the archaerhodopsin undergoes a sequence of conformation changes that expel the proton from the cell. This light-driven proton pumping generates a pH gradient that can be used to power the synthesis of ATP by chemiosmosis. This phototrophic capacity is particularly useful to *H. salinarum* because oxygen is not very soluble in concentrated salt solutions and may decrease to an extremely low level in its habitat. When the surroundings become temporarily anoxic, the archaeon uses light energy to synthesize sufficient ATP to survive until oxygen levels rise again.

To date, all rhodopsin-containing microbes are chemoorganoheterotrophs, using organic carbon as their source of carbon, electrons, and energy. Note that rhodopsin-based proton pumping generates a PMF without having to oxidize NADH via an ETC. What does this mean for these microorganisms? Many of these microbes live in nutrient-depleted environments where organic carbon is limited. With each organic molecule the organism takes up, it must "decide" to either oxidize it to make NADH and ATP or reduce it at the expense of NADPH and ATP to make more macromolecules for growth. The use of rhodopsin-based proton pumping enables the synthesis of ATP without having to catabolize as much precious organic carbon. That is, by performing rhodopsin-mediated phototrophy, these microbes can shunt more organic carbon to growth rather than oxidizing it for energy. ▶| *Microorganisms in the open ocean are adapted to nutrient limitation (section 29.2)*

Comprehension Check

1. Define the following terms: *light reactions, dark reactions, chlorophyll, carotenoid, phycobiliprotein, antenna,* and *photosystems I* and *II*.
2. What happens to a reaction center chlorophyll, such as 680, when it absorbs light?
3. What is the function of accessory pigments?
4. What is photophosphorylation? What is the difference between cyclic and noncyclic photophosphorylation?
5. Compare and contrast anoxygenic phototrophy and oxygenic photosynthesis. How do these two types of phototrophy differ from rhodopsin-based phototrophy?
6. Suppose you isolated a bacterial strain that carried out oxygenic photosynthesis. What photosystems would it possess, and what group of bacteria would it most likely belong to?

Key Concepts

11.1 Metabolic Diversity and Nutritional Types

- All organisms require a source of energy, electrons, and carbon. Biologists have devised terms to describe how these needs are fulfilled (**figure 11.1**).
- Phototrophs use light energy, and chemotrophs obtain energy from the oxidation of chemical compounds.
- Electrons are extracted from reduced inorganic substances by lithotrophs and from reduced organic compounds by organotrophs.
- Heterotrophs use reduced organic molecules as their source of carbon for biosynthetic purposes. Autotrophs use CO_2 as their primary or sole carbon source.
- Most organisms can be placed into one of five nutritional types: photolithoautotroph, photoorganoheterotroph, chemolithoautotroph, chemolithoheterotroph, and chemoorganoheterotroph (**table 11.1**).

11.2 There Are Two Chemoorganotrophic Fueling Processes

- Chemoorganotrophic microorganisms can use three kinds of electron acceptors during energy metabolism (**figure 11.3**). Electrons from the oxidized nutrient can be accepted by an endogenous electron acceptor (fermentation), by oxygen (aerobic respiration), or by another external electron acceptor (anaerobic respiration).
- Chemoorganotrophs use a variety of molecules as their energy, electron, and carbon source. These molecules are degraded by pathways that funnel into pathways used for glucose catabolism (**figure 11.4**).

11.3 Aerobic Respiration Starts with Glucose Oxidation

- Aerobic respiration of glucose begins with glycolytic pathways, which catabolize glucose to pyruvate.
- Glycolysis, in its broadest sense, refers to all pathways used to break down glucose to pyruvate.
- The Embden-Meyerhof pathway has two stages: a six- and three-carbon stage. The net production is two NADH and two ATP, the latter being produced by substrate-level phosphorylation. It also produces several precursor metabolites (**figure 11.5**).
- In the Entner-Doudoroff pathway, glucose is catabolized to pyruvate and glyceraldehyde 3-phosphate (**figure 11.6**). The latter product can be oxidized by enzymes of the Embden-Meyerhof pathway to provide ATP, NADH, and another molecule of pyruvate.
- In the pentose phosphate pathway, glucose 6-phosphate is oxidized twice and converted to pentoses and other sugars. It is a source of NADPH, ATP, and several precursor metabolites (**figure 11.7**).
- The Embden-Meyerhof and pentose phosphate pathways are amphibolic, functioning in both catabolism and anabolism.
- Pyruvate from the glycolytic pathways is fed into the tricarboxylic acid cycle by a reaction catalyzed by the enzyme pyruvate dehydrogenase. It converts pyruvate to acetyl-CoA. In the process, one of pyruvate's carbons is released in the form of carbon dioxide and one NADH is produced.
- The tricarboxylic acid cycle oxidizes acetyl-CoA to CO_2 and forms one GTP, three NADH, and one $FADH_2$ per acetyl-CoA (**figure 11.8**). It also generates several precursor metabolites.

11.4 Electron Transport and Oxidative Phosphorylation Generate the Most ATP

- The NADH and $FADH_2$ produced from the oxidation of reduced organic molecules can be oxidized by an electron transport chain (ETC). Electrons flow from carriers with more negative reduction potentials to those with more positive potentials (**figures 11.9** and **11.10**).
- As electrons flow through an ETC, protons are transported across the membrane, generating proton motive force (PMF). PMF is used for ATP synthesis by oxidative phosphorylation.
- Bacterial and archaeal ETCs are often different from eukaryotic chains with respect to such aspects as carriers and branching (**figures 11.11** and **11.12**).
- ATP synthase catalyzes the synthesis of ATP (**figure 11.14**). Bacterial and archaeal ATP synthases are on the inner surface of the plasma membrane. ATP synthase uses the PMF for ATP synthesis.
- In eukaryotes, the P/O ratio for NADH is about 2.5 and that for $FADH_2$ is around 1.5; P/O ratios are usually much lower in bacterial and archaeal chains. Aerobic respiration in eukaryotes can theoretically yield a maximum of 32 ATP (**figure 11.15**).

11.5 Anaerobic Respiration Uses the Same Steps as Aerobic Respiration

- Anaerobic respiration uses an exogenous molecule other than O_2 as the terminal electron acceptor for electron transport. The most common acceptors are nitrate, sulfate, and CO_2 (**figure 11.16** and **table 11.2**).
- Less energy is provided by anaerobic respiration than aerobic respiration because the alternate electron acceptors have reduction potentials that are less positive than the reduction potential of oxygen. Thus the ETC is shorter and fewer protons are moved across the membrane.

11.6 Fermentation Does Not Involve an Electron Transport Chain

- During fermentation, an endogenous electron acceptor is used to reoxidize NADH generated by the catabolism of glucose to pyruvate (**figure 11.17**). Flow of electrons from the electron donor to the electron acceptor does not

involve an ETC, and in most organisms, ATP is synthesized only by substrate-level phosphorylation.
- There are many different fermentation pathways. These are of practical importance in clinical and industrial settings (**figure 11.18**).

11.7 Catabolism of Organic Molecules Other Than Glucose

- Microorganisms catabolize many extracellular carbohydrates. Monosaccharides are taken in and phosphorylated; disaccharides may be cleaved to monosaccharides by either hydrolysis or phosphorolysis. External polysaccharides are degraded by hydrolysis and the products are actively transported into the cell. Intracellular glycogen and starch are converted to glucose 1-phosphate by phosphorolysis (**figure 11.19**).
- Triglycerides are hydrolyzed to glycerol and fatty acids by enzymes called lipases. Fatty acids are usually oxidized to acetyl-CoA in the β-oxidation pathway (**figure 11.21**).
- Proteins are hydrolyzed to amino acids that are then deaminated (**figure 11.22**). The carbon skeletons produced by deamination can be fermented or fed into the TCA cycle (**figure 11.4**).

11.8 Chemolithotrophy: "Eating Rocks"

- Chemolithotrophs use inorganic compounds—usually hydrogen, reduced nitrogen and sulfur compounds, or ferrous iron—to donate electrons to an ETC. O_2 is the usual terminal electron acceptor (**figure 11.23** and **table 11.3**). The PMF produced is used by ATP synthase to make ATP.
- Many of the energy sources used by chemolithotrophs have a more positive standard reduction potential than the $NAD^+/NADH$ conjugate redox pair (**figure 11.10**). These chemolithotrophs must expend energy (PMF) to drive reverse electron flow and produce the NADH they need for CO_2 fixation and other processes (**figure 11.25**).

11.9 Flavin-Based Electron Bifurcation

- Some anaerobic bacteria and archaea conserve energy by using a cytoplasmic flavoenzyme that transfers electrons such that an endergonic redox reaction is possible because it is coupled to an exergonic one (**figure 11.26**).

11.10 Phototrophy

- In oxygenic photosynthesis, eukaryotes and cyanobacteria trap light energy with chlorophyll and accessory pigments, and move electrons through photosystems I and II to make ATP and NADPH (the light reactions). The ATP and NADPH are used in the dark reactions to fix CO_2 (**figure 11.30**).
- Cyclic photophosphorylation involves the activity of photosystem I and generates ATP only. In noncyclic photophosphorylation, photosystems I and II operate together to move electrons from water to $NADP^+$, producing ATP, NADPH, and O_2 (**figure 11.31**). In both cases, electron flow generates PMF, which is used by ATP synthase to make ATP.
- Anoxygenic phototrophs differ from oxygenic phototrophs in possessing bacteriochlorophyll and having only one photosystem (**figures 11.32–11.34**). Cyclic electron flow generates a PMF, which is used by ATP synthase to make ATP (i.e., cyclic photophosphorylation). They are anoxygenic because they use molecules other than water as an electron donor for electron flow and the production of reducing power.
- Some bacteria and archaea use a type of phototrophy that involves a rhodopsin, a proton-pumping pigment. This type of phototrophy generates PMF but does not involve an ETC. It supplements the energy needs of heterotrophs.

Active Learning

1. How would you isolate a thermophilic chemolithotroph that uses sulfur compounds as a source of energy and electrons? What changes in the incubation system would be needed to isolate bacteria using sulfur compounds in anaerobic respiration? How would you tell which process is taking place through an analysis of the sulfur molecules present in the spent medium?
2. Certain chemicals block ATP synthesis by allowing protons and other ions to leak across membranes, disrupting the charge and proton gradients established by electron flow through an ETC. Would a bacterium treated with such an agent still be capable of flagellar-mediated swimming motility? Explain your reasoning.
3. Two flasks of *E. coli* are grown in batch culture in the same medium (2% glucose and amino acids; no nitrate) and at the same temperature (37°C). Culture #1 is well aerated. Culture #2 is anoxic. After 16 hours the following observations are made:
   - Culture #1 has a high cell density; the cells appear to be in stationary phase, and the glucose level in the medium is reduced to 1.2%.
   - Culture #2 has a low cell density; the cells appear to be in logarithmic phase, although their doubling time is prolonged. The glucose level is reduced to 0.2%.

   This phenomenon is known as the Pasteur effect. Like Pasteur did, figure out why culture #2 has so little glucose remaining relative to culture #1, even though culture #2 displayed slower growth and has less biomass. (Hint: It may be helpful to think about gas mileage obtained with a sports car versus an old clunker.)

4. For each of the following, suggest the most likely catabolic strategy (e.g., chemoheterotroph, chemolithoautotrophy, etc.) as described in this chapter. Explain your answer for each, referring to figure 11.4 to describe the entry of organic carbon, where relevant, into catabolic pathways.
   - Energy, electrons, and carbon from fatty acids and carbohydrates; oxygen is present
   - Energy and electrons from sunlight; CO_2 is sole carbon source
   - Energy and electrons from hydrogen sulfide (H_2S); oxygen is present; succinate is the carbon source
   - Energy and electrons from amino acids; nitrate (NO_3^-) is present, but oxygen is not
   - Energy, electrons, and carbon from carbohydrates; oxygen is present, but the cell lacks an electron transport chain
   - Energy from sunlight; electrons from thiosulfate ($H_2SO_3^{2-}$); CO_2 is sole carbon source
5. It is speculated that chemolithoautotrophs may have once inhabited Mars. As compared to other types of microbes discussed in this chapter, why might that be the case? What might have been the source of energy, electrons, and carbon? What might have been the terminal electron acceptor(s) in these Martian microbes' ETCs?

CHAPTER 12

Anabolism: The Use of Energy in Biosynthesis

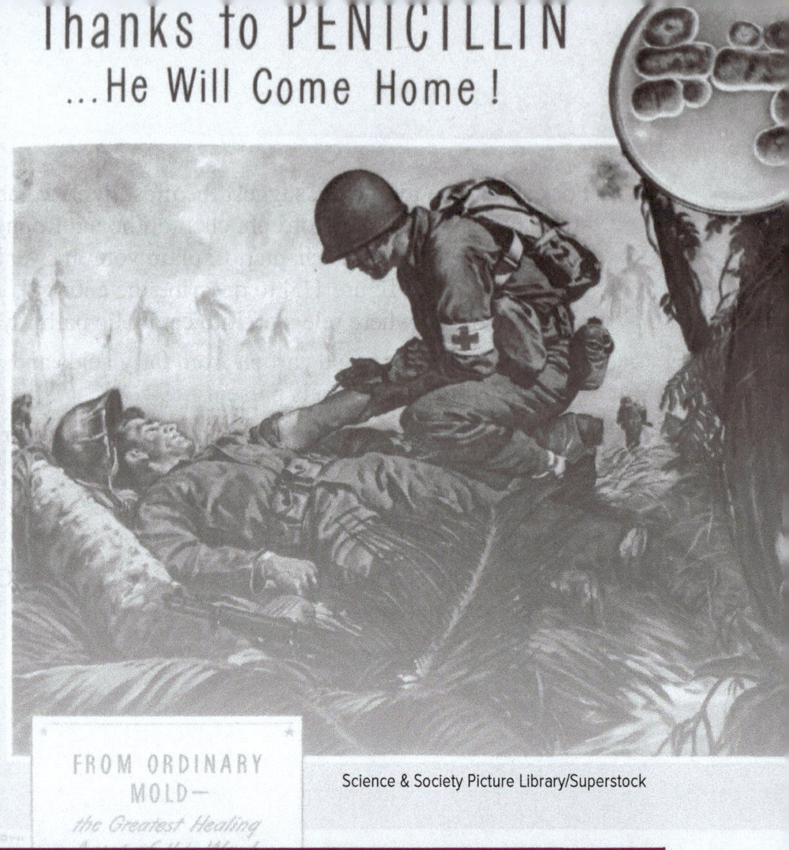

Science & Society Picture Library/Superstock

Building Penicillin

The story of Alexander Fleming's rediscovery of penicillin in 1929 is well-known. However, Fleming had little to do with developing penicillin as a treatment for infectious disease because after some initial experiments, he abandoned the study of its activity. Fortunately, Ernst Chain developed an interest in the chemical and eventually purified it. In 1940, as many nations battled in World War II, Howard Florey and his assistants began clinical trials with penicillin. By that time, Selman Waksman had discovered several other antibacterial agents and had coined the term *antibiotic* to mean an antibacterial agent produced by a microbe. The penicillin trials were successful, and the need for antibiotics during World War II to treat wound infections was the impetus for finding and developing more antibiotics for clinical use. A few years later a ready supply was available. ◀ *Antimicrobial chemotherapy evolved from antisepsis efforts (section 9.1)*

Antibiotics are interesting molecules for several reasons. Why would a microbe expend materials and energy to make an antibiotic? What selective advantage do they provide? How do they contribute to the ecology of the microbe? Do they serve a purpose other than killing other microbes (*see chapter 22 opening story*)? These molecules are also interesting because they are structurally complex but are built from the same precursor metabolites and many of the same biochemical processes used in the anabolic pathways that sustain life.

During anabolism, an organism begins with simple inorganic molecules and a carbon source, and constructs ever more complex molecules until new organelles and cells arise (**figure 12.1**). Thus anabolism is the creation of order. Because a cell is highly ordered and immensely complex, energy is required for biosynthesis (**table 12.1**). Furthermore, even nongrowing cells need energy and reducing power for the biosynthetic processes they carry out. This is because nongrowing cells continuously degrade and resynthesize cellular molecules during a process known as turnover. Although there is considerably less diversity in anabolic processes as compared to catabolic processes, anabolism is amazing in its own right. Using just 12 precursor metabolites, a cell is able to manufacture myriad molecules from which it is constructed. In this chapter, we discuss the synthesis of some of the most important types of cell constituents.

Readiness Check:

Based on what you have learned previously, you should be able to:

- ✓ Describe how carbon, hydrogen, oxygen, nitrogen, phosphorus, and sulfur are used by cells
- ✓ Distinguish metabolism, catabolism, and anabolism
- ✓ Discuss the function of ATP and how that function is related to its phosphate transfer potential (section 10.2)
- ✓ Describe the components of a biochemical pathway and how they are organized (section 10.5)
- ✓ State the function of enzymes and ribozymes (section 10.6)
- ✓ Differentiate between heterotrophy and autotrophy (section 11.1)
- ✓ List the products of the fueling reactions (section 11.1)
- ✓ Distinguish saturated fatty acids from unsaturated fatty acids (appendix I)

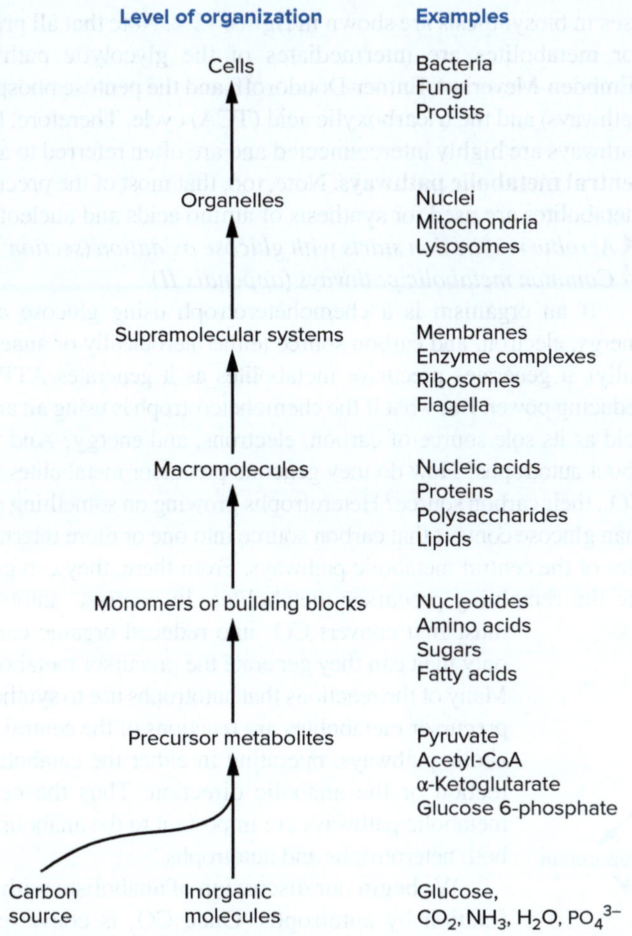

Figure 12.1 The Construction of Cells. The biosynthesis of cells and their constituents is organized in levels of ever greater complexity.

12.1 Principles Governing Biosynthesis

After reading this section, you should be able to:

- **a.** Describe in general terms the steps organisms use to convert a carbon source and inorganic molecules to cells
- **b.** Discuss the principles that govern biosynthesis

The problem faced by all cells is how to make the molecules they need as efficiently as possible. To achieve this efficiency, cells follow a few basic principles, described here.

1. **Large molecules are made from small molecules (figure 12.1).** The construction of **macromolecules** (complex molecules) from a few simple structural units (monomers) linked together by a single type of covalent bond requires fewer genes and saves biosynthetic raw material and energy, making their synthesis highly efficient.
2. **Macromolecules are assembled into larger, more complex structures such as supramolecular systems and organelles.** This process, called *self-assembly*, is possible because many macromolecules contain the necessary information to build multimolecular complexes without enzymatic help.
3. **While many enzymes are bifunctional, some function in one direction only.** As we discuss in chapter 11, many enzymes are used for both catabolic and anabolic processes in pathways that are said to be amphibolic, saving additional materials and energy. Although many steps of amphibolic pathways are catalyzed by enzymes that act reversibly, some are not. Irreversible steps require use of separate enzymes: one to catalyze the catabolic reaction, the other to catalyze the anabolic reaction. The use of two enzymes allows independent regulation of catabolism and anabolism. Thus catabolic and anabolic pathways are never identical, even though they share many enzymes and metabolic intermediates.
4. **Anabolism consumes energy.** Many reactions in anabolic pathways are endergonic and will not proceed in the direction of biosynthesis. Cells solve this problem by connecting endergonic biosynthetic reactions to the hydrolysis of ATP and other nucleoside triphosphates. When these two processes are coupled, the free energy made available during nucleoside triphosphate breakdown drives the biosynthetic reaction to completion. *ATP: the major energy currency of cells (section 10.2)*
5. **Catabolism and anabolism can be physically separated.** Catabolic and anabolic pathways can be localized into distinct subcellular compartments—a process called *compartmentation*. In eukaryotes, this is easily done because membrane-bound organelles carry out specific functions. Compartmentation also occurs in bacterial and archaeal cells. For instance, carboxysomes separate CO_2 fixation from other processes in bacteria. Compartmentation makes it easier for catabolic and anabolic pathways to operate simultaneously yet independently. *The bacterial cytoplasm is complex (section 3.6)*
6. **Catabolism and anabolism often use different cofactors.** Usually catabolic oxidations produce NADH to transport electrons. In contrast, when an electron donor is needed during biosynthesis, NADPH often serves as the donor.

Table 12.1 Biosynthesis of Macromolecules in *E. coli*

Cell Constituent	Number of Molecules per Cell	Rate of Synthesis (sec^{-1})
DNA	2	1,000 nucleotides
RNA	2×10^5	40–80 nucleotides
Polysaccharides	1×10^6	32.5 molecules
Lipids	2×10^7	12,500 molecules
Proteins	3×10^6	10–20 amino acids

12.2 Precursor Metabolites: Starting Molecules for Biosynthesis

After reading this section, you should be able to:

a. List the central metabolic pathways, noting which precursor metabolites are generated by each pathway
b. Draw a simple diagram that lists all the precursor metabolites and illustrates how they are used in biosynthesis

Generation of **precursor metabolites** is critical to anabolism because they give rise to all other molecules made by the cell. Precursor metabolites are carbon skeletons (i.e., carbon chains) used as the starting points for the synthesis of the building blocks needed to make macromolecules (figure 12.1). During biosynthesis, functional moieties such as amino and sulfhydryl groups are added to the carbon chains. The precursor metabolites and their uses in biosynthesis are shown in **figure 12.2**. Note that all precursor metabolites are intermediates of the glycolytic pathways (Embden-Meyerhof, Entner-Doudoroff, and the pentose phosphate pathways) and the tricarboxylic acid (TCA) cycle. Therefore, these pathways are highly interconnected and are often referred to as the **central metabolic pathways.** Note, too, that most of the precursor metabolites are used for synthesis of amino acids and nucleotides.

Aerobic respiration starts with glucose oxidation (section 11.3)
Common metabolic pathways (appendix II)

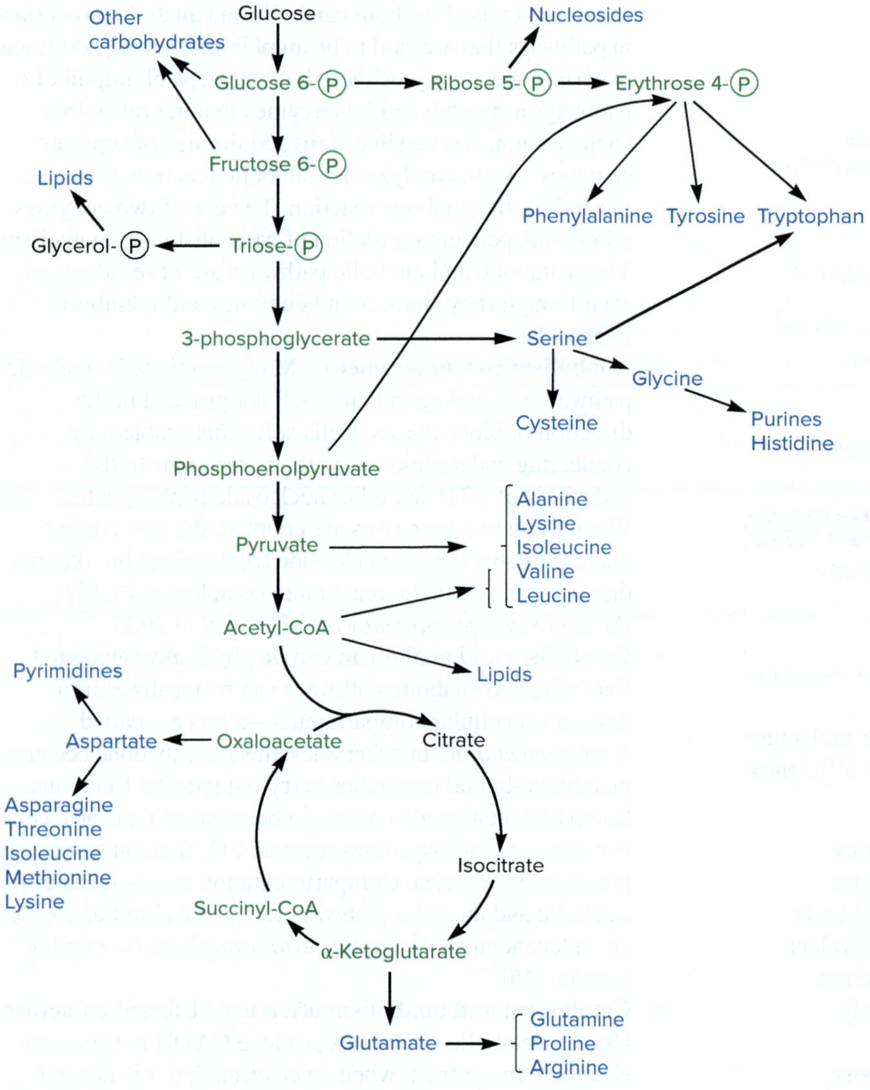

Figure 12.2 The Organization of Anabolism. Biosynthetic products (in blue) are derived from the 12 precursor metabolites (in green).

If an organism is a chemoheterotroph using glucose as its energy, electron, and carbon source (either aerobically or anaerobically), it generates precursor metabolites as it generates ATP and reducing power. But what if the chemoheterotroph is using an amino acid as its sole source of carbon, electrons, and energy? And what about autotrophs? How do they generate precursor metabolites from CO_2, their carbon source? Heterotrophs growing on something other than glucose convert that carbon source into one or more intermediates of the central metabolic pathways. From there, they can generate the remaining precursor metabolites. In contrast, autotrophs must first convert CO_2 into reduced organic carbon; only then can they generate the precursor metabolites. Many of the reactions that autotrophs use to synthesize precursor metabolites are reactions of the central metabolic pathways, operating in either the catabolic direction or the anabolic direction. Thus the central metabolic pathways are important to the anabolism of both heterotrophs and autotrophs.

We begin our discussion of anabolism with CO_2 fixation by autotrophs. Once CO_2 is converted to organic carbon, the synthesis of other precursor metabolites, amino acids, nucleotides, and additional building blocks is similar in both autotrophs and heterotrophs. Recall that the precursor metabolites provide the carbon skeletons for the synthesis of other important organic molecules. In the process of transforming a precursor metabolite into an amino acid or a nucleotide, the carbon skeleton is modified in a number of ways, including the addition of nitrogen, phosphorus, and sulfur. Thus as we discuss the synthesis of monomers from precursor metabolites, we also address the **assimilation** of these elements. Assimilation is the process by which microbes absorb elements from the environment and incorporate them into biomass. *A Biochemical Pathway*

Comprehension Check

1. Explain why most enzymes that operate in amphibolic pathways function in both the anabolic and catabolic direction, but a few function only in either anabolism or catabolism.
2. Provide an example of compartmentation in a eukaryotic cell and a bacterial or archaeal cell.
3. What is self-assembly? What biological structures exhibit self-assembly?

12.3 CO_2 Fixation: Reduction and Assimilation of CO_2 Carbon

After reading this section, you should be able to:

a. List the pathways used by microbes to fix CO_2, noting the types of organisms (eukaryotes, bacteria, archaea) that use each pathway
b. Describe in general terms the three phases of the Calvin-Benson cycle
c. Identify the steps in the Calvin-Benson cycle that consume ATP and NADPH

Autotrophs use CO_2 as their sole or principal carbon source, and the reduction and incorporation of CO_2 require much energy. Autotrophic CO_2 fixation is crucial to life on Earth because it provides the organic matter on which heterotrophs depend. In general, there are two types of autotrophic organisms: photoautotrophs and chemolithoautotrophs. Photoautotrophs obtain energy by trapping light during the light reactions of photosynthesis, while chemolithoautotrophs derive energy from the oxidation of inorganic electron donors. Chemically, the carbon in CO_2 must be reduced and ultimately incorporated into a precursor metabolite. In this manner, the carbon source (CO_2) enters the pathways for biosynthesis. *Chemolithotrophy: "eating rocks" (section 11.8); Phototrophy (section 11.10)* *Carbon cycle (section 28.2)*

Calvin-Benson Cycle

Eukaryotic autotrophs and most aerobic bacterial autotrophs use the **Calvin-Benson cycle** to fix CO_2. These reactions occur in the chloroplast stroma of eukaryotic autotrophs. In cyanobacteria, some nitrifying bacteria, and thiobacilli (sulfur-oxidizing chemolithotrophs), the cycle is associated with inclusions called carboxysomes. These polyhedral structures contain the enzymes critical to the Calvin-Benson cycle. *Inclusions (section 3.7); Pentose phosphate pathway: a major producer of reducing power for anabolic reactions (section 11.3)*

The Calvin-Benson cycle is divided into three phases: carboxylation, reduction, and regeneration (**figure 12.3** and *appendix II*). During the carboxylation phase, the enzyme **ribulose 1,5-bisphosphate carboxylase/oxygenase (RuBisCO)** catalyzes the addition of CO_2 to the 5-carbon molecule ribulose 1,5-bisphosphate (RuBP), forming a 6-carbon intermediate that rapidly and spontaneously splits into two molecules of 3-phosphoglycerate (PGA). PGA is an intermediate of the Embden-Meyerhof pathway (EMP), and in the reduction phase of the Calvin-Benson cycle, PGA is reduced to glyceraldehyde 3-phosphate (G3P) by two reactions that reverse two EMP reactions. The EMP reactions differ from the Calvin-Benson cycle reactions in that the Calvin-Benson cycle enzyme glyceraldehyde 3-phosphate dehydrogenase (GAPDH) uses NADPH rather than NADH (compare figures 12.3 and 11.5). Finally, in the regeneration phase, RuBP is reformed so that the cycle can repeat. In addition, this phase produces the precursor metabolites fructose 6-phosphate and glucose 6-phosphate (figure 12.2). The regeneration phase is essentially the reverse of the pentose phosphate pathway. Not surprisingly, another name for the Calvin-Benson cycle is the reductive pentose phosphate cycle. *The Calvin-Benson Cycle*

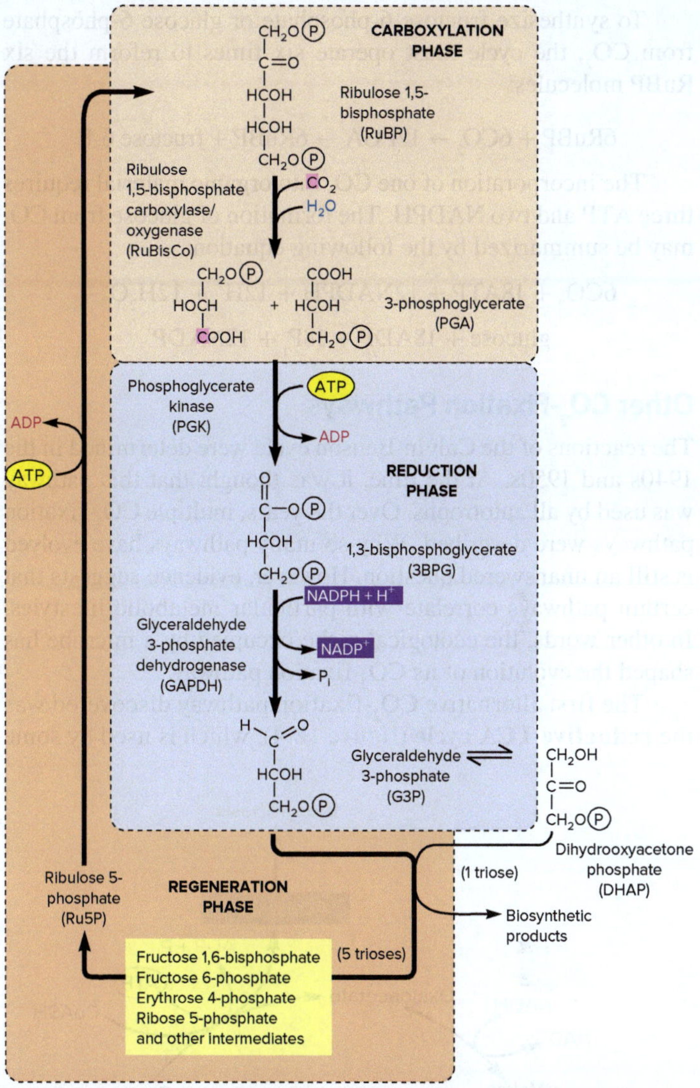

Figure 12.3 The Calvin-Benson Cycle. This overview of the cycle shows only the carboxylation and reduction phases in detail. Three ribulose 1,5-bisphosphates (RuBPs) are carboxylated (the incorporated carbon atom is shaded in pink) to give six 3-phosphoglycerates (PGAs) in the carboxylation phase. In the reduction phase, these are converted to six glyceraldehyde 3-phosphates (G3Ps), which can be converted to dihydroxyacetone phosphate (DHAP). Five of the six trioses (G3P and DHAP) regenerate three RuBPs in the regeneration phase. The remaining triose is used in biosynthesis. 3BPG, 1,3-bisphosphoglycerate; Ru5P, ribulose 5-phosphate; GAPDH, glyceraldehyde 3-phosphate dehydrogenase; PGK, phosphoglycerate kinase; RuBisCo, ribulose 1,5-bisphosphate carboxylase/oxygenase.

To synthesize fructose 6-phosphate or glucose 6-phosphate from CO_2, the cycle must operate six times to reform the six RuBP molecules:

$$6RuBP + 6CO_2 \rightarrow 12PGA \rightarrow 6RuBP + \text{fructose 6-P}$$

The incorporation of one CO_2 into organic material requires three ATP and two NADPH. The formation of glucose from CO_2 may be summarized by the following equation:

$$6CO_2 + 18ATP + 12NADPH + 12H^+ + 12H_2O \rightarrow$$
$$\text{glucose} + 18ADP + 18P_i + 12NADP^+$$

Other CO_2-Fixation Pathways

The reactions of the Calvin-Benson cycle were determined in the 1940s and 1950s. At the time, it was thought that this pathway was used by all autotrophs. Over the years, multiple CO_2-fixation pathways were described. Why so many pathways have evolved is still an unanswered question. However, evidence suggests that certain pathways correlate with particular metabolic lifestyles. In other words, the ecological niche occupied by a microbe has shaped the evolution of its CO_2-fixation pathway.

The first alternative CO_2-fixation pathway discovered was the **reductive TCA cycle** (**figure 12.4**), which is used by some autotrophs in the bacterial phyla Proteobacteria, Nitrospirota, and Bacteroidota. The reductive TCA cycle is so named because it runs in the reverse direction of the normal, oxidative TCA cycle (compare figures 12.4 and 11.8). In the reductive TCA cycle, two molecules of CO_2 are incorporated into carbon skeletons in the reverse of the reactions that release CO_2 in the TCA cycle. CO_2 fixation, like all biosynthetic reactions, requires ATP and reducing power. Operating in the biosynthesis direction, these reactions oxidize cofactors and dephosphorylate ATP. One of the carboxylases relies on a ferredoxin, an oxygen-sensitive redox protein, and thus the reductive TCA cycle is limited to anaerobes. *Thermotogota and Deinococcota live in extreme conditions (section 20.2); Green sulfur bacteria (section 20.3)*

The remaining CO_2-fixation pathways are used by members of the bacterial phylum Chloroflexota (*see figure 20.5*) and the domain Archaea (*see figures 19.1 and 19.2*).

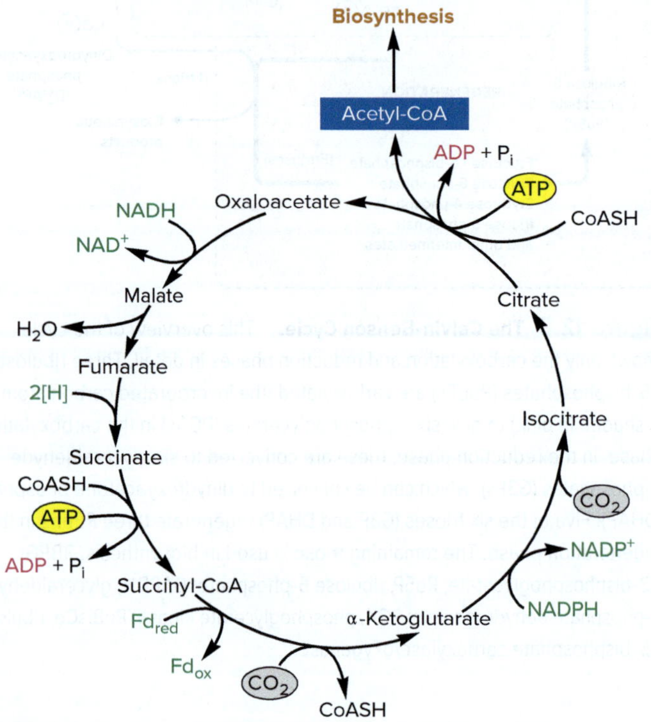

Figure 12.4 The Reductive TCA Cycle. Two CO_2 molecules are incorporated into organic carbon in this pathway. The cycle runs in the opposite direction (i.e., counterclockwise) as the TCA cycle. ATP and reducing power (NADH, NADPH, and [H]) drive the reversal. The product of the pathway is acetyl-CoA.

MICRO INQUIRY *How might acetyl-CoA be used in biosynthesis (see figure 12.2)?*

Comprehension Check

1. Briefly describe the three phases of the Calvin-Benson cycle.
2. Which two enzymes are specific to the Calvin-Benson cycle?
3. Compare figures 12.4 and 11.8. Which enzyme is likely not shared between the TCA cycle and the reductive TCA cycle?
4. Why is it important that the product(s) of a CO_2-fixation pathway be a precursor metabolite?

12.4 Synthesis of Carbohydrates

After reading this section, you should be able to:

a. Compare and contrast gluconeogenesis and the Embden-Meyerhof pathway
b. Describe the role of ATP and UTP in the synthesis of monosaccharides (other than glucose) and polysaccharides
c. Outline the major steps in peptidoglycan synthesis
d. Identify the steps in which peptidoglycan biosynthesis can be targeted by antibiotics

Many microbes need to synthesize glucose 6-phosphate for anabolic purposes (figure 12.2). The synthesis of glucose 6-phosphate from noncarbohydrate precursors is called **gluconeogenesis.** The gluconeogenic pathway shares several enzymes with the Embden-Meyerhof pathway (**figure 12.5**). However, the two pathways are not identical (compare figures 12.5 and 11.5). Three reactions are catalyzed by enzymes that are specific for gluconeogenesis. Two of these enzymes are involved in the conversion of pyruvate to phosphoenolpyruvate, and the third enzyme catalyzes the formation of fructose 6-phosphate from fructose 1,6-bisphosphate. *There are two chemoheterotrophic fueling processes (section 11.2)*

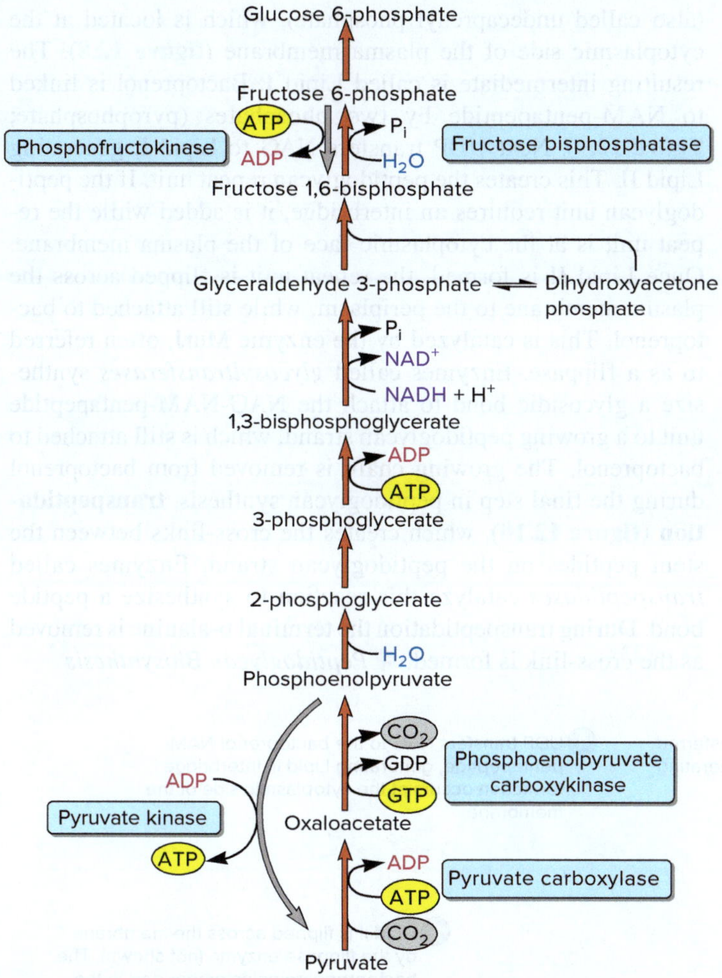

Figure 12.5 Gluconeogenesis Pathway Used by Many Microorganisms. Red arrows indicate the direction of the pathway, and its end product, glucose 6-phosphate, is at the top. Colored boxes show the names of the enzymes catalyzing irreversible reactions that are different from those found in the Embden-Meyerhof pathway (EMP). EMP steps are shown with gray arrows.

MICRO INQUIRY *Why is it important that the enzymes dedicated to gluconeogenesis (rather than those used in both catabolic and anabolic reactions) are found in both the 3-carbon and 6-carbon stages?*

Synthesis of Monosaccharides and Polysaccharides Involves Nucleoside Diphosphate Carriers

As shown in figure 12.5, gluconeogenesis synthesizes fructose 6-phosphate and glucose 6-phosphate. Other common sugars can be synthesized from these precursor metabolites. For example, rearrangement of a hydroxyl group on fructose 6-phosphate yields mannose 6-phosphate. Several sugars are synthesized while attached to a nucleoside diphosphate. One of the most important nucleoside diphosphate sugars is uridine diphosphate glucose (UDPG), which is formed when glucose 1-phosphate reacts with uridine triphosphate (**figure 12.6**). Other important uridine diphosphate (UDP) sugars are UDP-galactose and UDP-glucuronic

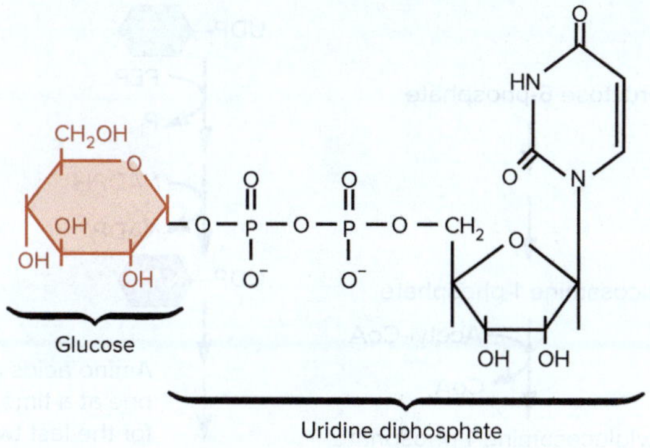

Figure 12.6 Uridine Diphosphate Glucose (UDPG).

acid. UDP carries sugars for participation in enzyme reactions much like ADP bears phosphate in the form of ATP.

Nucleoside diphosphate sugars also play a central role in the synthesis of polysaccharides such as starch and glycogen, both of which are long chains of glucose. Again, biosynthesis is not simply a direct reversal of catabolism. For instance, during glycogen and starch synthesis, adenosine diphosphate glucose (ADP-glucose) is formed from glucose 1-phosphate and ATP. It then donates glucose to the end of growing glycogen and starch chains.

$$\text{ATP} + \text{glucose 1-phosphate} \rightarrow \text{ADP-glucose} + \text{PP}_i$$

$$(\text{Glucose})_n + \text{ADP-glucose} \rightarrow (\text{glucose})_{n+1} + \text{ADP}$$

Synthesis of Peptidoglycan Occurs in the Cytoplasm, at the Plasma Membrane, and in the Periplasmic Space

Nucleoside diphosphate sugars also participate in the synthesis of peptidoglycan. Recall that peptidoglycan is a large, complex molecule consisting of long polysaccharide chains made of alternating *N*-acetylmuramic acid (NAM) and *N*-acetylglucosamine (NAG) residues (*see figure 3.15*). The NAM groups bear stem peptides of five amino acids (a pentapeptide). During peptidoglycan synthesis, adjacent polysaccharide chains are cross-linked by bonds formed between the stem peptides. In diderm bacteria and many monoderm bacteria, the pentapeptide stems are directly linked. In some monoderm bacteria, the stems are connected by an interbridge consisting of one or more amino acids (*see figure 3.19*). ◀ *Cell walls have many functions (section 3.3)*

Not surprisingly, such an intricate structure requires an equally intricate biosynthetic process, especially because some reactions occur in the cytoplasm, others at the plasma membrane, and others in the periplasmic space. Peptidoglycan synthesis involves two carriers. The first, UDP, functions in the cytoplasmic reactions. The second carrier, bactoprenol phosphate, is needed for reactions that occur first on the cytoplasmic and then the periplasmic side of the plasma membrane.

In the first step of peptidoglycan synthesis, UDP derivatives of NAM and NAG are formed (**figure 12.7**). Amino acids are added sequentially to UDP-NAM to form the stem peptide. NAM-pentapeptide is transferred to bactoprenol phosphate

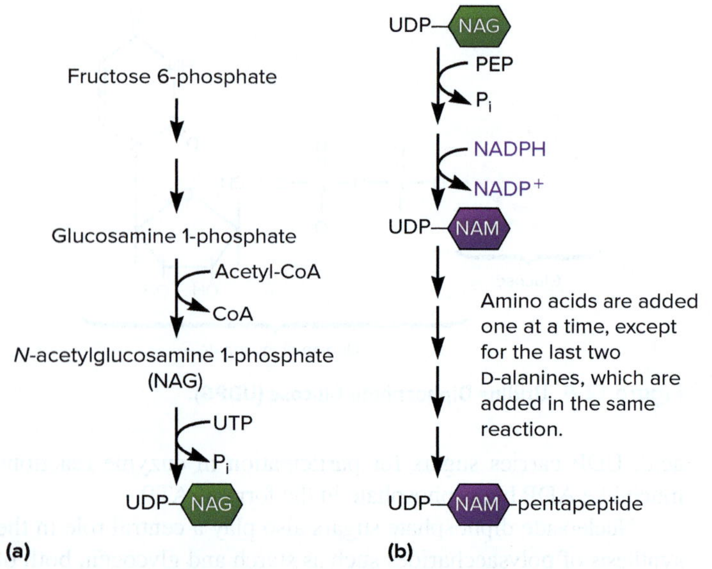

Figure 12.7 Initial steps in peptidoglycan synthesis. (a) Synthesis of UDP-NAG. (b) Synthesis of UDP-NAM-pentapeptide. NAG, *N*-acetylglucosamine; NAM, *N*-acetylmuramic acid; PEP, phosphoenolpyruvate; UDP, uridine diphosphate.

(also called undecaprenyl phosphate), which is located at the cytoplasmic side of the plasma membrane (**figure 12.8**). The resulting intermediate is called Lipid I. Bactoprenol is linked to NAM-pentapeptide by two phosphates (pyrophosphate; **figure 12.9**). Next UDP transfers NAG to Lipid I, generating Lipid II. This creates the peptidoglycan repeat unit. If the peptidoglycan unit requires an interbridge, it is added while the repeat unit is at the cytoplasmic face of the plasma membrane. Once Lipid II is formed, the repeat unit is flipped across the plasma membrane to the periplasm, while still attached to bactoprenol. This is catalyzed by the enzyme MurJ, often referred to as a flippase. Enzymes called *glycosyltransferases* synthesize a glycosidic bond to attach the NAG-NAM-pentapeptide unit to a growing peptidoglycan strand, which is still attached to bactoprenol. The growing chain is removed from bactoprenol during the final step in peptidoglycan synthesis, **transpeptidation** (**figure 12.10**), which creates the cross-links between the stem peptides on the peptidoglycan strand. Enzymes called *transpeptidases* catalyze this reaction to synthesize a peptide bond. During transpeptidation the terminal D-alanine is removed as the cross-link is formed. *Peptidoglycan Biosynthesis*

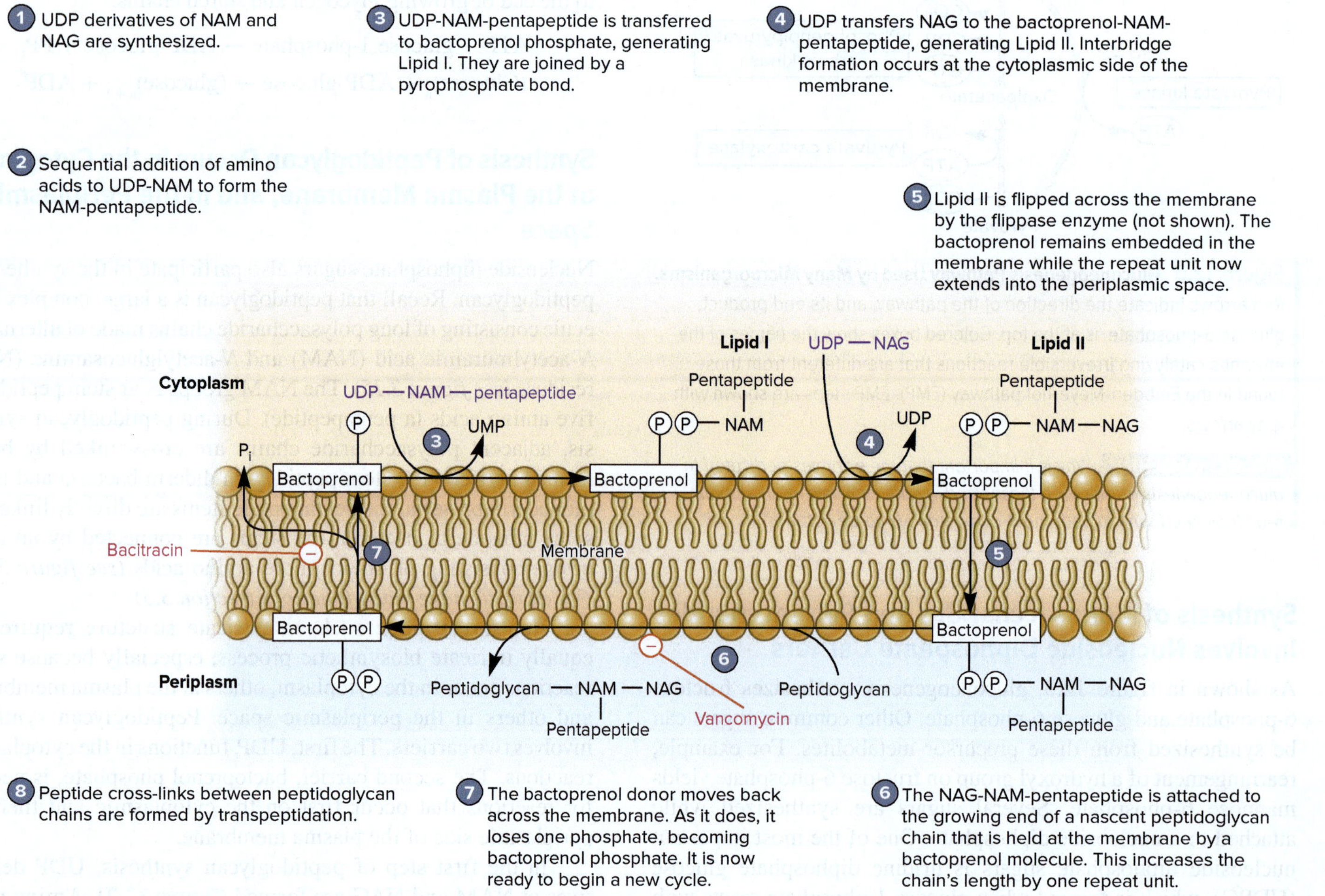

Figure 12.8 Peptidoglycan Synthesis. Points of inhibition by antibiotics are shown in red. NAM, *N*-acetylmuramic acid; NAG, *N*-acetylglucosamine; UDP, uridine diphosphate.

MICRO INQUIRY *How do Lipid I and Lipid II differ?*

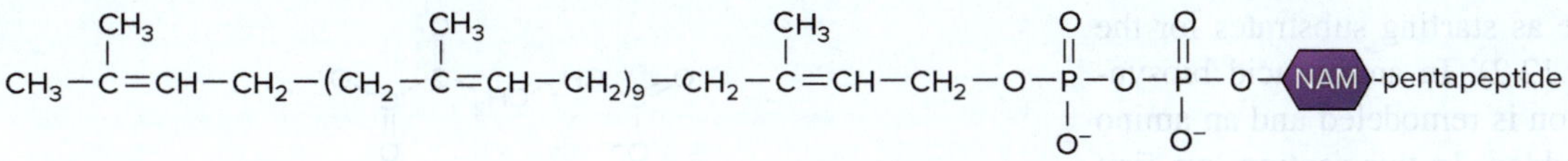

Figure 12.9 Bactoprenol-NAM. Bactoprenol is a 55-carbon, isoprene-derived alcohol (section 12.7) that is connected to *N*-acetylmuramic acid (NAM-pentapeptide) by pyrophosphate.

Peptidoglycan synthesis is a particularly effective target for antimicrobial agents because of its importance to bacterial cell structure and function and because it is absent in animal cells. Inhibition of any stage of synthesis weakens the cell wall and can lead to lysis. For example, penicillin inhibits the transpeptidation reaction (figure 12.10), and bacitracin blocks the dephosphorylation of bactoprenol pyrophosphate (figure 12.8). *Inhibitors of cell wall synthesis (section 9.4)*

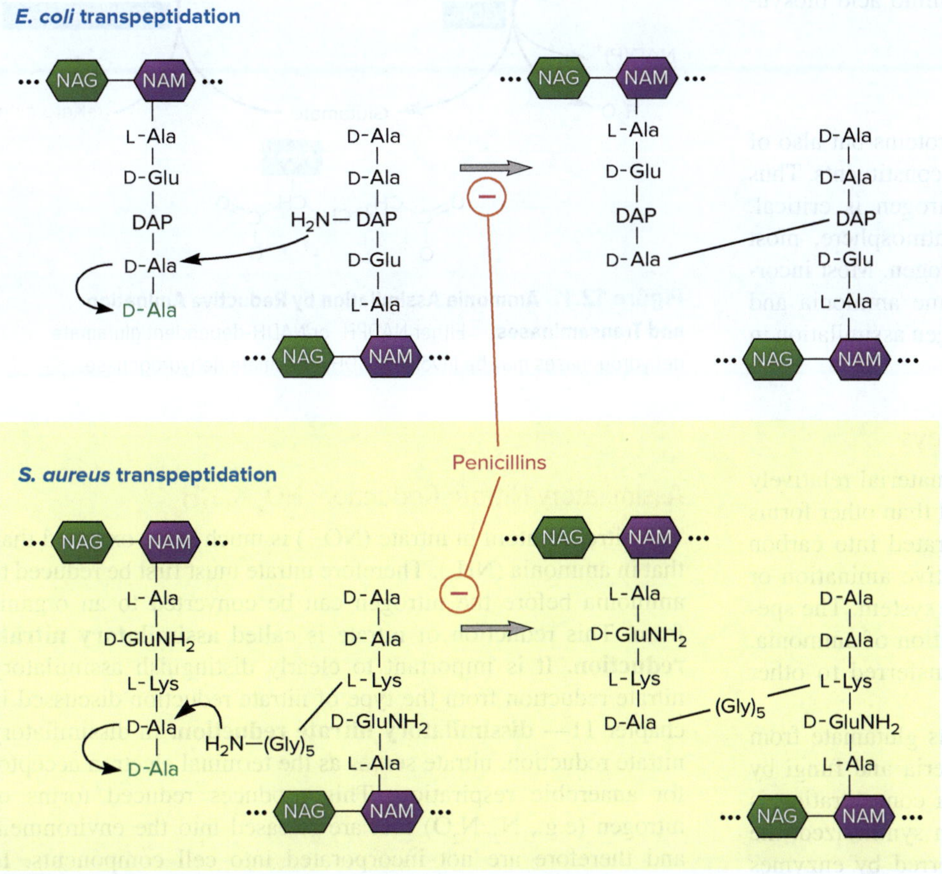

Figure 12.10 Transpeptidation. The transpeptidation reactions in *E. coli* and *Staphylococcus aureus* peptidoglycan synthesis. Note that the terminal D-alanine is removed. The disruption of transpeptidation by penicillins is indicated. DAP, diaminopimelic acid.

Comprehension Check

1. What is gluconeogenesis? Why is it important?
2. Suppose a microorganism is growing on a medium that contains amino acids but no sugars. In general terms, how would it synthesize the pentose and hexose sugars it needs?
3. What are nucleoside diphosphate sugars? Describe the formation of mannose 6-phosphate, starch, and glycogen.
4. Diagram the steps involved in the synthesis of peptidoglycan and show where they occur in the cell. What are the roles of bactoprenol and UDP?
5. Distinguish between glycosyltransferases and transpeptidases in peptidoglycan synthesis. What would happen to a cell if these enzymes were not carefully regulated?

The result of the process just described is a peptidoglycan sacculus that is thicker by one glycan strand. However, the peptidoglycan of any bacterium usually has a typical thickness that does not increase or decrease dramatically. In addition, for cells to grow and divide, the structure of the sacculus must be sculpted and modified to allow elongation and separation during cytokinesis. The glycosyltransferases and transpeptidases mentioned above do so by functioning in a precise and tightly regulated way, all while maintaining wall shape and integrity in the presence of high osmotic pressure (*see figure 7.8*). These enzymes are tightly controlled because unregulated activity would lead to weakening of the peptidoglycan sacculus. The enzymes appear to be synthesized and maintained in an inactive form. They are activated as needed to respond to environmental change, and kept in check by proteolytic degradation. *Cellular growth and determination of cell shape (section 7.2)*

12.5 Synthesis of Amino Acids Consumes Many Precursor Metabolites

After reading this section, you should be able to:

a. Discuss the three mechanisms microorganisms use to assimilate inorganic nitrogen and the role of transaminases in each
b. Describe the three methods microbes use to assimilate sulfur
c. Differentiate assimilatory nitrate reduction from dissimilatory nitrate reduction and assimilatory sulfate reduction from dissimilatory sulfate reduction
d. Evaluate the efficiency of using branched pathways for synthesizing amino acids
e. List the major anaplerotic reactions and explain their importance

Many precursor metabolites serve as starting substrates for the synthesis of amino acids (figure 12.2). In amino acid biosynthetic pathways, the carbon skeleton is remodeled and an amino group, and sometimes sulfur, is added. In this section, we first examine the mechanisms by which nitrogen and sulfur are assimilated and incorporated into amino acids. This is followed by a brief consideration of the organization of amino acid biosynthetic pathways.

Inorganic Nitrogen Assimilation

Nitrogen is a major component not only of proteins but also of nucleic acids, coenzymes, and many other cell constituents. Thus the cell's ability to assimilate inorganic nitrogen is critical. Although nitrogen gas is abundant in the atmosphere, most organisms are reliant on reduced forms of nitrogen. Most incorporate either ammonia or nitrate. We examine ammonia and nitrate assimilation first and then discuss nitrogen assimilation in microbes that fix N_2.

Ammonia Incorporation Occurs in Two Ways

Ammonia can be incorporated into organic material relatively easily and directly because it is more reduced than other forms of inorganic nitrogen. Ammonia is incorporated into carbon skeletons by one of two mechanisms: reductive amination or the glutamine synthetase–glutamate synthase system. The specific pathway used depends on the concentration of ammonia. Once incorporated, the nitrogen can be transferred to other carbon skeletons.

The **reductive amination pathway** forms glutamate from α-ketoglutarate. It is catalyzed in many bacteria and fungi by glutamate dehydrogenase when the ammonia concentration is high (**figure 12.11**). Once glutamate has been synthesized, the newly formed α-amino group can be transferred by enzymes called **transaminases** to other carbon skeletons. Microorganisms have multiple transaminases, each of which catalyzes the formation of several amino acids.

Many bacteria use the **glutamine synthetase–glutamate synthase (GS-GOGAT) system** to assimilate ammonia when its availability is limited (**figure 12.12**). Incorporation of ammonia begins with synthesis of glutamine from glutamate in a reaction catalyzed by glutamine synthetase (**figure 12.13**). In this reaction, the carboxyl group found in the side chain of glutamate reacts with ammonia, and the incorporated nitrogen exists as an amide in glutamine. Glutamate synthase then transfers the amide nitrogen to α-ketoglutarate to generate two glutamate molecules. Because glutamate acts as an amino donor in transaminase reactions, ammonia may be used to synthesize all common amino acids when suitable transaminases are present.

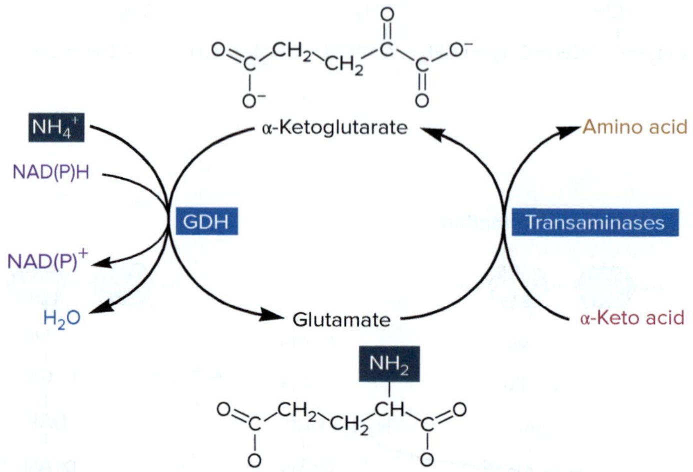

Figure 12.11 Ammonia Assimilation by Reductive Amination and Transaminases. Either NADPH- or NADH-dependent glutamate dehydrogenases may be involved. GDH, glutamate dehydrogenase.

Assimilatory Nitrate Reduction: NO_3 to NH_3

The nitrogen atom in nitrate (NO_3^-) is much more oxidized than that in ammonia (NH_3). Therefore nitrate must first be reduced to ammonia before the nitrogen can be converted to an organic form. This reduction of nitrate is called **assimilatory nitrate reduction.** It is important to clearly distinguish assimilatory nitrate reduction from the type of nitrate reduction discussed in chapter 11— **dissimilatory nitrate reduction.** In dissimilatory nitrate reduction, nitrate serves as the terminal electron acceptor for anaerobic respiration. This produces reduced forms of nitrogen (e.g., N_2, N_2O) that are released into the environment and therefore are not incorporated into cell components. In assimilatory nitrate reduction, nitrate is reduced to ammonia, which is incorporated into cell biomass. The process is widespread among bacteria, fungi, and photosynthetic protists, and it is an important step in the nitrogen cycle. ◀ *Anaerobic respiration uses the same steps as aerobic respiration (section 11.5)* ▶ *Nitrogen cycle (section 28.2)*

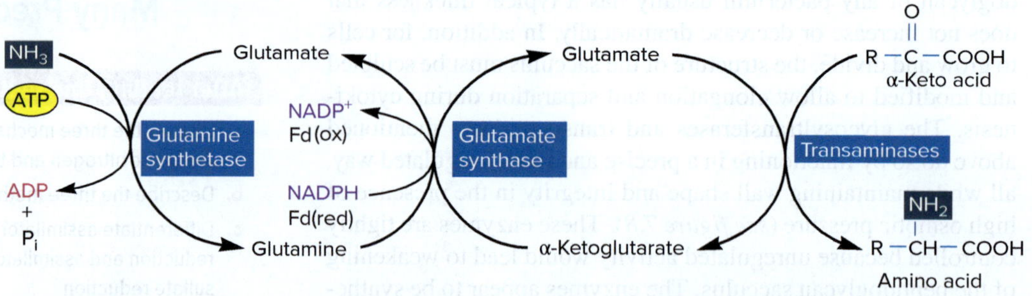

Figure 12.12 Ammonia Incorporation Using Glutamine Synthetase, Glutamate Synthase, and Transaminases. NADPH is the electron donor in the glutamate synthase reaction in archaea and nonphotosynthetic bacteria, while ferredoxin (Fd) is the electron donor in cyanobacteria and photosynthetic eukaryotes.

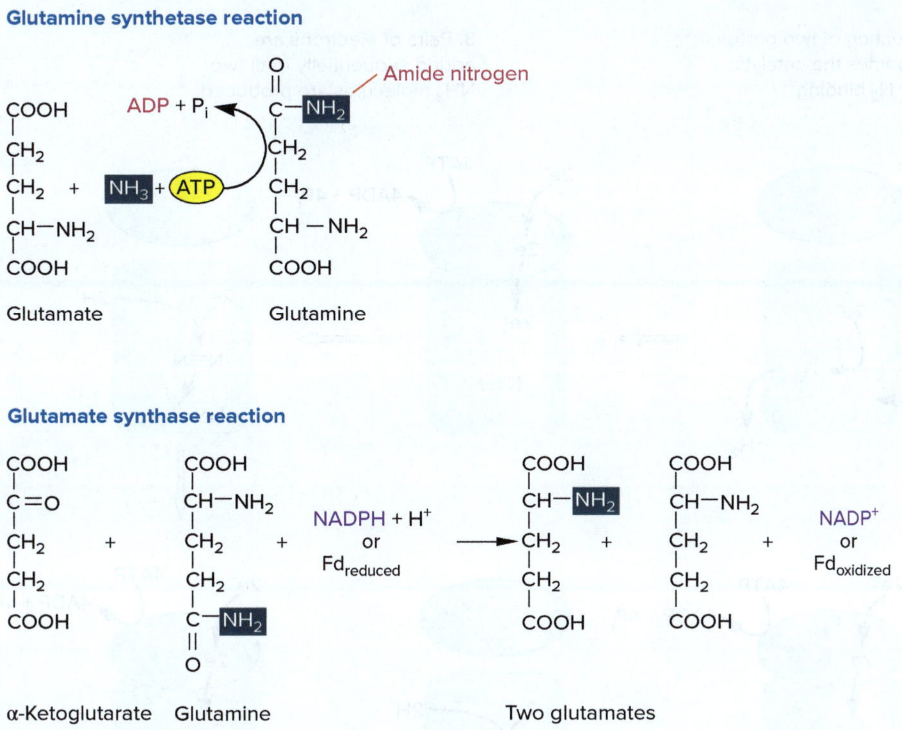

Figure 12.13 Glutamine Synthetase and Glutamate Synthase Reactions Involved in Ammonia Assimilation. Either NADPH or ferredoxin (Fd) is the electron source in the glutamate synthase reaction.

MICRO INQUIRY *What purpose is served for the cell when these reactions are used to produce two glutamates rather than a single glutamine?*

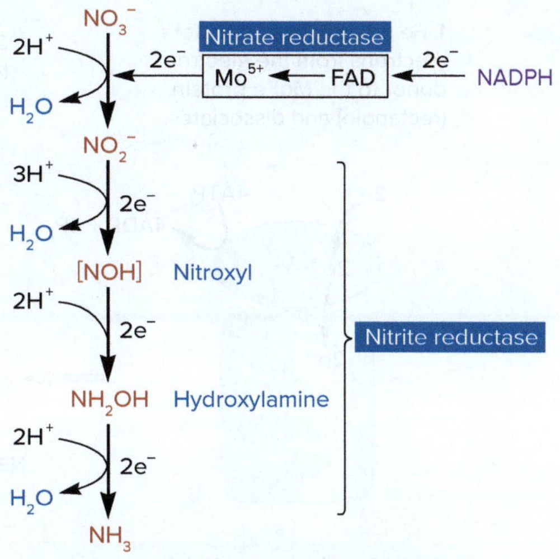

Figure 12.14 Assimilatory Nitrate Reduction. This pathway operates in bacteria that reduce and assimilate nitrate nitrogen. Here we show the electron donor as NADPH. However, some bacteria use other electron donors such as ferredoxin or NADH.

The first step in nitrate assimilation is its reduction to nitrite by nitrate reductase, an enzyme that contains both FAD and molybdenum (**figure 12.14**). Nitrite is next reduced to ammonia through a series of two-electron additions catalyzed by nitrite reductase. The ammonia is then incorporated into amino acids by reductive amination or the GS-GOGAT system.

Nitrogen Fixation: N_2 to NH_3

The reduction of atmospheric gaseous nitrogen to ammonia is called **nitrogen fixation.** Until recently, only a few bacteria and archaea were thought to fix nitrogen. We now know of a eukaryote capable of this feat (*see Microbial Diversity & Ecology 20.1*). These organisms are called **diazotrophs** (Greek *di,* two; French *azo,* containing nitrogen; Greek *troph,* eater), and they play a critical role in the nitrogen cycle. The biological aspects of nitrogen fixation are discussed in chapters 28–30. The biochemistry of nitrogen fixation is the focus of this section.

During nitrogen fixation, nitrogen is reduced by sequential two-electron additions, as illustrated in **figure 12.15**. The reduction of molecular nitrogen (N_2) to ammonia is quite exergonic, but the reaction has a high activation energy because molecular nitrogen is an unreactive gas with a triple bond connecting the two nitrogen atoms. Therefore, nitrogen reduction requires a large energy expenditure—at least 8 electrons and 16 ATP molecules (4 ATP per pair of electrons). Note in figure 12.15 that two of the electrons and four of the ATP molecules are expended to reduce 2 protons to hydrogen gas at the start of the process. This is often thought of as a wasteful expenditure of electrons and energy. However, some bacteria have evolved mechanisms for capturing the hydrogen gas and using it as a source of electrons for nitrogen reduction. As with assimilatory nitrate reduction, the ammonia produced by nitrogen fixation is used for biosynthesis following incorporation into amino acids, as previously described (figures 12.11, 12.12, and 12.13).

The reduction of N_2 to ammonia is catalyzed by the enzyme **nitrogenase** (figure 12.15). The best-studied nitrogenase is a complex enzyme consisting of two major protein components—a MoFe protein joined with a Fe protein. The MoFe protein has an organic cofactor that contains molybdenum, iron, and sulfur atoms, whereas the Fe protein contains an iron-sulfur (Fe-S) cluster. The Fe protein is reduced by ferredoxin and binds ATP. ATP binding changes the conformation of the Fe protein and lowers its reduction potential (−0.29 to −0.40 V), enabling it to reduce the MoFe protein. ATP is hydrolyzed when this electron transfer occurs. Finally, reduced MoFe protein donates electrons to protons (the first two reactions of the process) or sequentially to atomic nitrogen (the series of reactions outlined by number 3 in figure 12.15). Nitrogenase is sensitive to O_2 and must be protected from O_2 inactivation. Microbes use a variety of strategies to protect nitrogenase, as we discuss more fully in chapters 21, 29, and 30.

Sulfur Assimilation: SO_4 to S-Bearing Molecules

Sulfur is needed for the synthesis of amino acids (cysteine and methionine) and several coenzymes (coenzyme A and biotin).

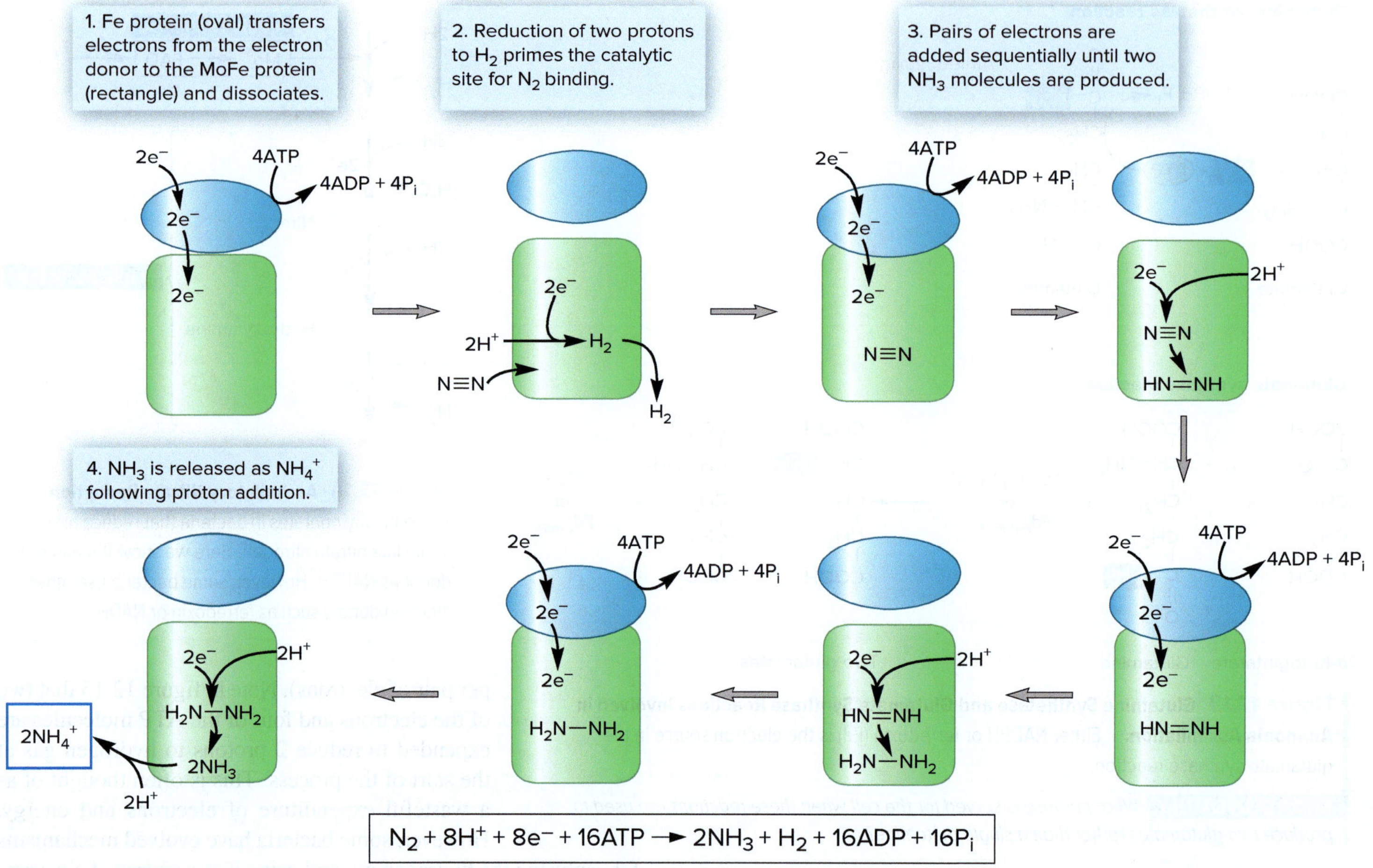

$$N_2 + 8H^+ + 8e^- + 16ATP \rightarrow 2NH_3 + H_2 + 16ADP + 16P_i$$

Figure 12.15 Nitrogen Reduction by Nitrogenase. Ferredoxin is often the electron donor for nitrogen fixation. Fe-S, iron-sulfur; Fe-Mo, iron-molybdenum.

Sulfur atoms can be transferred among organic molecules or obtained from inorganic sulfate. Like nitrate, sulfate is more oxidized than it is in organic molecules; thus it must be reduced before it can be assimilated into biomass. This process is known as **assimilatory sulfate reduction** to distinguish it from the **dissimilatory sulfate reduction** that takes place when sulfate acts as an electron acceptor during anaerobic respiration. *Anaerobic respiration uses the same steps as aerobic respiration (section 11.5)* *Sulfur cycle (section 28.2)*

Assimilatory sulfate reduction involves sulfate activation through the formation of phosphoadenosine 5′-phosphosulfate (PAPS), another modified nucleotide that functions as a carrier (**figure 12.16**). As shown in **figure 12.17**, sulfate (SO_4^{2-}) is first reduced to sulfite (SO_3^{2-}), then to hydrogen sulfide (H_2S). Hydrogen sulfide is used to synthesize cysteine (**figure 12.18**). In the simplest pathway, fungi combine hydrogen sulfide with serine to form cysteine (figure 12.18*a*), whereas many bacteria join hydrogen sulfide with O-acetylserine instead (figure 12.18*b*). Archaea use multiple mechanisms to incorporate H_2S, including the bacterial pathway, and a related pathway via O-phosphoserine (figure 12.18*c*). Once formed, cysteine is used in the synthesis of other sulfur-containing organic compounds, including methionine. Sulfur assimilation is so integral to the cell that PAPS is used by some bacteria as a signal that stimulates growth.

Amino Acid Biosynthetic Pathways

Several amino acids are made directly by transamination of a precursor metabolite. For example, alanine and aspartate are made from pyruvate and oxaloacetate, respectively, using

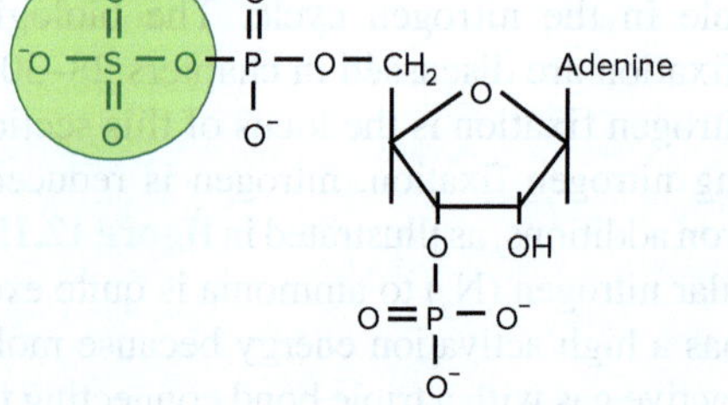

Figure 12.16 Phosphoadenosine 5′-phosphosulfate (PAPS). The sulfate group is shaded green.

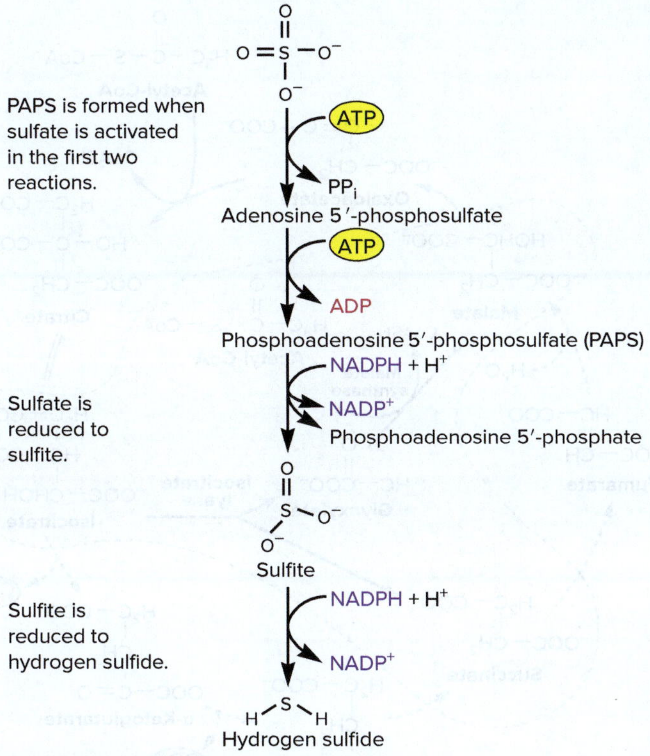

Figure 12.17 The Sulfate Reduction Pathway.

glutamate as the amino group donor. However, most precursor metabolites must be altered by more than just the addition of an amino group. In many cases, the carbon skeleton must be reconfigured, and for cysteine and methionine, the sulfur must be added. These pathways are branched; that is, a single precursor metabolite can be used for the synthesis of a family of related amino acids. For example, lysine, threonine, isoleucine, and methionine are synthesized from oxaloacetate by a branching route (**figure 12.19**). Because of the need to conserve nitrogen, carbon, and energy, amino acid bio synthetic pathways are usually tightly regulated by feedback mechanisms. ◀ *Posttranslational regulation of enzyme activity (section 10.7)*

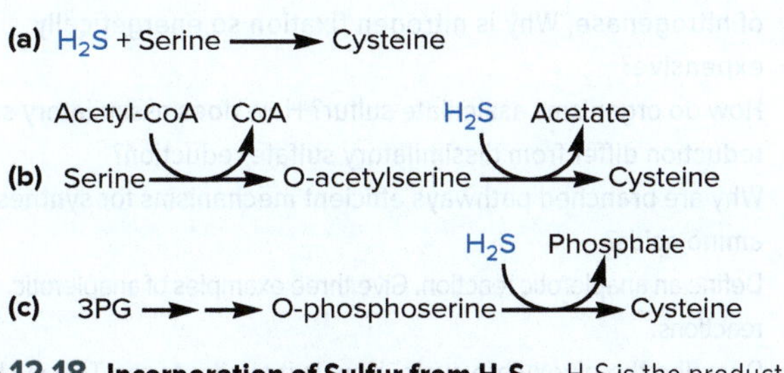

Figure 12.18 Incorporation of Sulfur from H_2S. H_2S is the product of sulfate reduction, as shown in figure 12.17. Pathways used by (a) fungi, (b) many bacteria and some archaea, and (c) some archaea. 3PG, 3-phosphoglycerate.

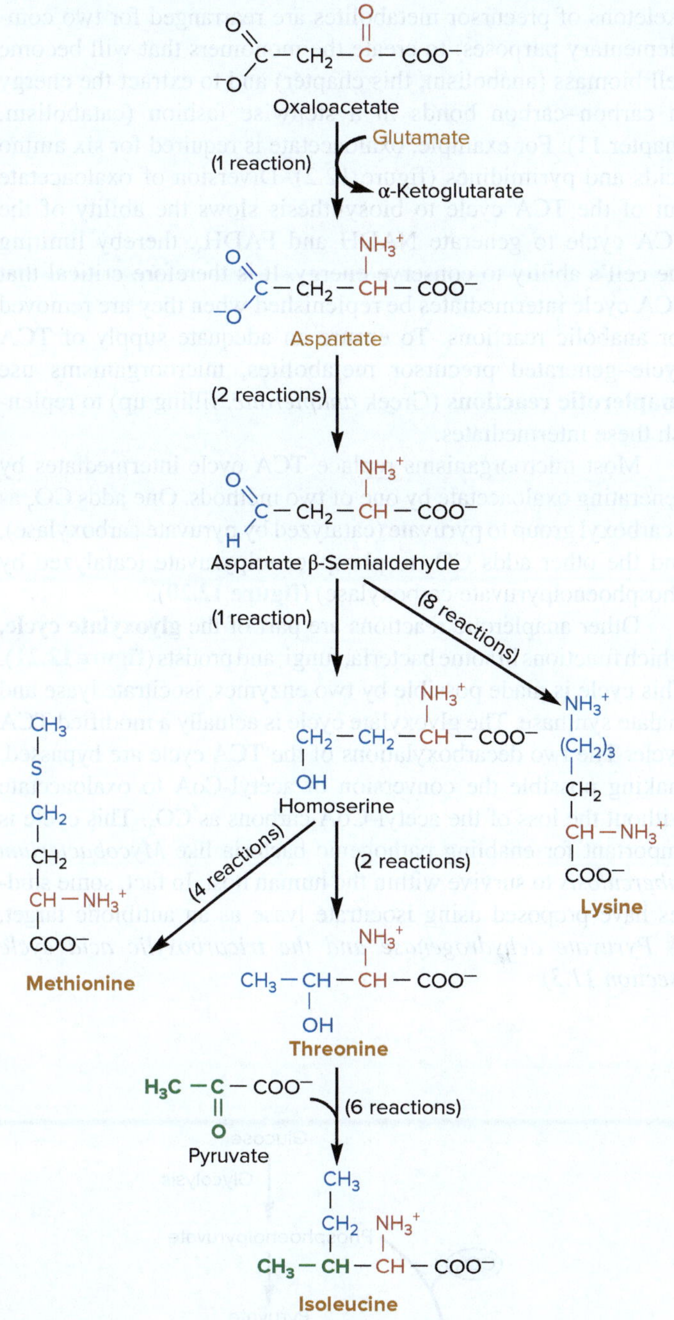

Figure 12.19 The Branching Pathway to Methionine, Threonine, Isoleucine, and Lysine. Most arrows represent numerous enzyme-catalyzed reactions. Not shown is the consumption of reducing power and ATP. Parts of each structure are colored to trace the chemical modifications.

MICRO INQUIRY *Suppose a mutant microbial strain were unable to produce homoserine. What amino acids would it have to obtain from its environment?*

Anaplerotic Reactions Replace the Precursor Metabolites Used for Amino Acid Biosynthesis

When an organism is actively building biomass, there is a heavy demand for precursor metabolites. Recall that the carbon skeletons of precursor metabolites are rearranged for two complementary purposes: to create the monomers that will become cell biomass (anabolism, this chapter) and to extract the energy in carbon–carbon bonds in a stepwise fashion (catabolism, chapter 11). For example, oxaloacetate is required for six amino acids and pyrimidines (figure 12.2). Diversion of oxaloacetate out of the TCA cycle to biosynthesis slows the ability of the TCA cycle to generate NADH and $FADH_2$, thereby limiting the cell's ability to conserve energy. It is therefore critical that TCA cycle intermediates be replenished when they are removed for anabolic reactions. To ensure an adequate supply of TCA cycle–generated precursor metabolites, microorganisms use **anaplerotic reactions** (Greek *anaplerotic,* filling up) to replenish these intermediates.

Most microorganisms replace TCA cycle intermediates by generating oxaloacetate by one of two methods. One adds CO_2 as a carboxyl group to pyruvate (catalyzed by pyruvate carboxylase), and the other adds CO_2 to phosphoenolpyruvate (catalyzed by phosphoenolpyruvate carboxylase) (**figure 12.20**).

Other anaplerotic reactions are part of the **glyoxylate cycle,** which functions in some bacteria, fungi, and protists (**figure 12.21**). This cycle is made possible by two enzymes, isocitrate lyase and malate synthase. The glyoxylate cycle is actually a modified TCA cycle. The two decarboxylations of the TCA cycle are bypassed, making possible the conversion of acetyl-CoA to oxaloacetate without the loss of the acetyl-CoA carbons as CO_2. This cycle is important for enabling pathogenic bacteria like *Mycobacterium tuberculosis* to survive within the human host. In fact, some studies have proposed using isocitrate lyase as an antibiotic target. *Pyruvate dehydrogenase and the tricarboxylic acid cycle (section 11.3)*

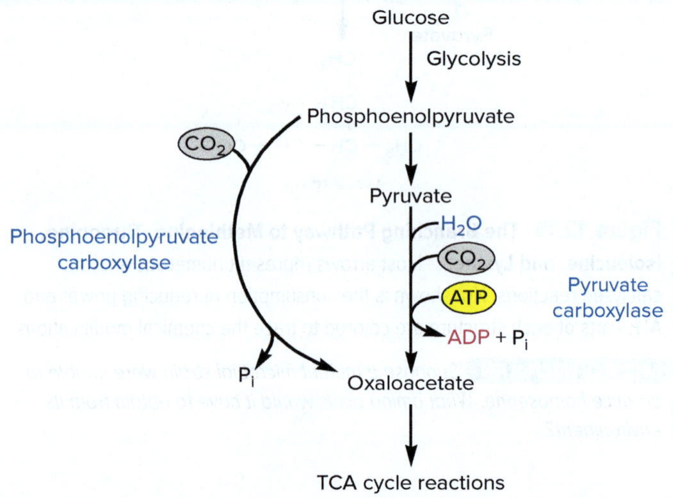

Figure 12.20 Anaplerotic Reactions that Replace Oxaloacetate. Pyruvate and phosphoenolpyruvate are formed by the Embden-Meyerhof pathway.

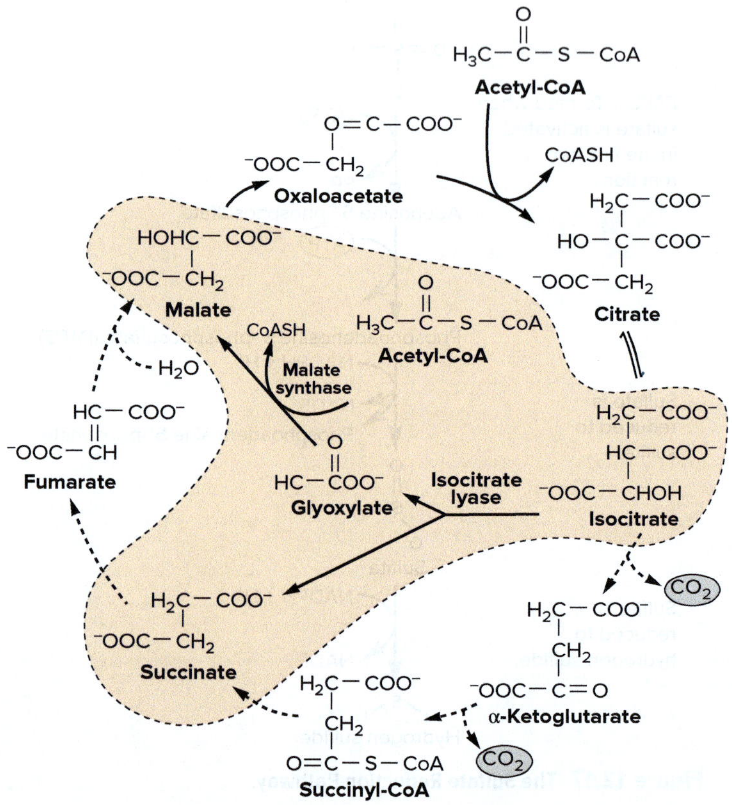

Figure 12.21 The Glyoxylate Cycle. Acetyl-CoA can be converted into succinate or malate by the enzymes isocitrate lyase and malate synthase (shaded reactions). This is a modification of the TCA cycle, bypassing the reactions shown by dashed lines.

Comprehension Check

1. Describe the roles of glutamate dehydrogenase, glutamine synthetase, glutamate synthase, and transaminases in ammonia assimilation.
2. How does assimilatory nitrate reduction differ from dissimilatory nitrate reduction? What is the fate of nitrate in each case?
3. Briefly describe the structure and mechanism of action of nitrogenase. Why is nitrogen fixation so energetically expensive?
4. How do organisms assimilate sulfur? How does assimilatory sulfate reduction differ from dissimilatory sulfate reduction?
5. Why are branched pathways efficient mechanisms for synthesizing amino acids?
6. Define an anaplerotic reaction. Give three examples of anaplerotic reactions.
7. Describe the glyoxylate cycle. How is it similar to the TCA cycle? How does it differ?

12.6 Synthesis of Purines, Pyrimidines, and Nucleotides

After reading this section, you should be able to:

a. Differentiate between purines and pyrimidines
b. Draw a simple diagram that illustrates the chemical moieties found in nucleosides and nucleotides
c. Discuss in general terms how phosphorus is assimilated
d. Compare and contrast purine and pyrimidine biosynthesis
e. Discuss the methods used to convert ribonucleotides to deoxyribonucleotides

Purine and pyrimidine biosynthesis are critical for all cells because these molecules are used to synthesize important cell components like ATP, several cofactors, ribonucleic acid (RNA), and deoxyribonucleic acid (DNA). Nearly all microorganisms can synthesize their own purines and pyrimidines as these are essential for survival. ▶| *DNA replication in bacteria (section 13.3); Transcription in bacteria (section 13.5)*

Purines and **pyrimidines** are cyclic nitrogenous bases with several double bonds. **Adenine** and **guanine** are purines, consisting of two joined rings, whereas pyrimidines (**uracil, cytosine,** and **thymine**) have a single ring (*see figure 13.5*). A purine or pyrimidine base joined with a pentose sugar, either ribose or deoxyribose, is a **nucleoside.** A **nucleotide** is a nucleoside with one or more phosphate groups attached to the sugar.

Amino acids participate in the synthesis of nitrogenous bases and nucleotides in a number of ways, including by providing the organic nitrogen. We begin this section by examining phosphorus assimilation in nucleotides, and then examine the pathways for synthesis of nitrogenous bases and nucleotides.

Phosphorus Assimilation

In addition to nucleic acids, phosphorus is found in phospholipids, in coenzymes such as $NADP^+$, and as a modification to certain proteins. The most common phosphorus sources are inorganic phosphate and organic molecules containing a phosphoryl group. Inorganic phosphate is incorporated through the formation of ATP in one of three ways: (1) photophosphorylation, (2) oxidative phosphorylation, and (3) substrate-level phosphorylation. |◀ *Aerobic respiration starts with glucose oxidation (section 11.3); Electron transport and oxidative phosphorylation generate the most ATP (section 11.4); Phototrophy (section 11.10)*

Microorganisms may obtain organic phosphates from their surroundings in dissolved or particulate form. **Phosphatases** hydrolyze organic phosphate-containing molecules to release inorganic phosphate. Diderms have phosphatases in the periplasmic space, which allows phosphate to be taken up immediately after release. On the other hand, protists can directly use organic phosphates after ingestion or hydrolyze them in lysosomes and then incorporate the phosphate.

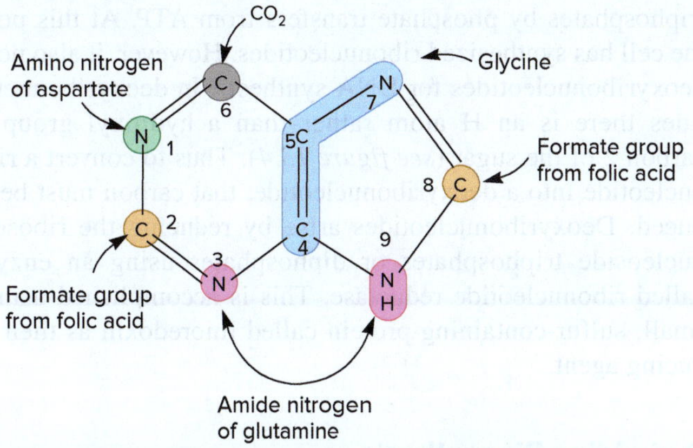

Figure 12.22 Sources of the Nitrogen and Carbon in Purines. The nitrogen numbered 9 is linked to the sugar ribose 5-phosphate.

Purine Biosynthesis

The biosynthetic pathway for purines is a complex, 11-step sequence (*see appendix II*) in which seven different molecules contribute parts to the final purine skeleton (**figure 12.22**). The pathway begins with ribose 5-phosphate, and the purine skeleton is constructed on this sugar. Therefore, the first purine product of the pathway is the nucleotide inosinic acid, not a free purine base. The cofactor folic acid is important in purine biosynthesis because folic acid derivatives contribute two carbons to the purine skeleton.

From inosinic acid, relatively short pathways synthesize adenosine monophosphate and guanosine monophosphate (**figure 12.23**) and produce nucleoside diphosphates and

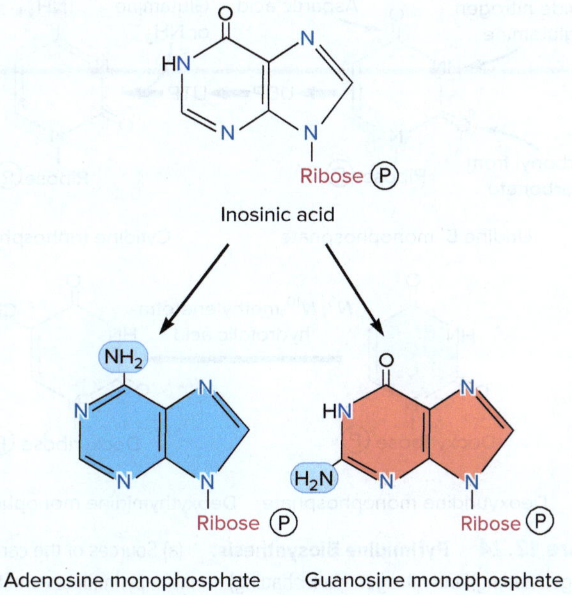

Figure 12.23 Synthesis of Purine Monophosphates. The blue highlighted groups differ from those in inosinic acid.

triphosphates by phosphate transfers from ATP. At this point, the cell has synthesized ribonucleotides. However, it also needs deoxyribonucleotides for DNA synthesis. In deoxyribonucleotides there is an H atom rather than a hydroxyl group on carbon 2 of the sugar (*see figure 13.4*). Thus to convert a ribonucleotide into a deoxyribonucleotide, that carbon must be reduced. Deoxyribonucleotides arise by reducing the ribose in nucleoside triphosphates or diphosphates using an enzyme called ribonucleotide reductase. This is accomplished using a small, sulfur-containing protein called thioredoxin as their reducing agent.

Pyrimidine Biosynthesis

Pyrimidine biosynthesis is equally as complex as purine synthesis. However, unlike purine synthesis, construction of pyrimidine rings is completed before ribose is attached.

Pyrimidine biosynthesis begins with the carbon skeleton of aspartic acid and additonal carbon and nitrogen atoms derived from bicarbonate and glutamine (**figure 12.24*a***). After the six-membered carbon-nitrogen ring is formed, ribose 5-phosphate is attached to form a nucleotide. Phosphate groups are transferred from ATP to form uridine diphosphate, uridine triphosphate, and cytidine triphosphate. These ribonucleotides are reduced to deoxy forms in the same way that purine ribonucleotides are. That leaves the cell with only one deoxynucleotide yet to be made: deoxythimidine monophosphate. It is made when deoxyuridine monophosphate is methylated with a folic acid derivative (figure 12.24*b*).

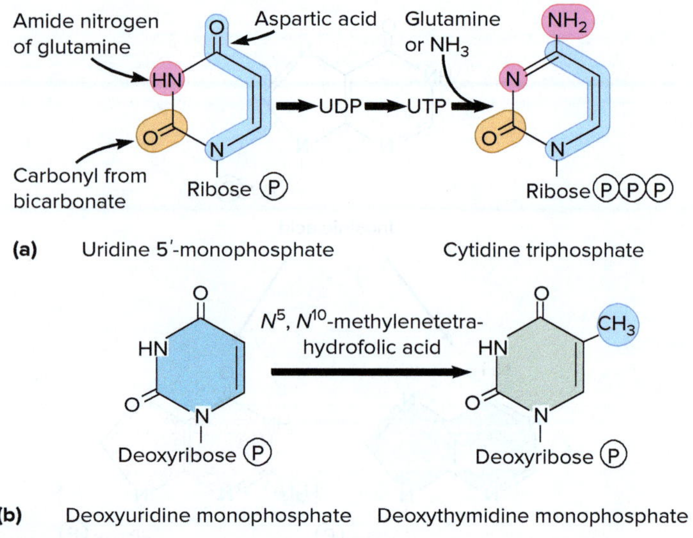

Figure 12.24 Pyrimidine Biosynthesis. (a) Sources of the carbon (orange shading) and nitrogen (pink shading) atoms in pyrimidines and the reactions to synthesize CTP. (b) Deoxythymidine differs from deoxyuridine in the presence of the blue-shaded methyl group.

Comprehension Check

1. How is phosphorus assimilated? What roles do phosphatases play in phosphorus assimilation? Why can phosphate be directly incorporated into cell constituents, whereas nitrate, nitrogen gas, and sulfate cannot?
2. Explain the difference between a purine and a pyrimidine, and between a nucleoside and a nucleotide.
3. Outline the way in which purines and pyrimidines are synthesized. How is the deoxyribose component of deoxyribonucleotides made?

12.7 Lipid Synthesis

After reading this section, you should be able to:

a. Describe the role of acetyl-CoA, acyl carrier protein (ACP), and fatty acid synthase in fatty acid synthesis
b. Compare sterol synthesis to archaeal isoprenoid lipid synthesis
c. Summarize lipopolysaccharide synthesis and insertion into the outer membrane

Lipids are absolutely required by cells, as they are the major components of membranes. Most bacterial and eukaryal lipids contain fatty acids or their derivatives. Archaeal lipids are built using isoprene, an unsaturated 5-carbon unit. In this section we consider the pathways leading to these molecules. We also consider the synthesis of lipopolysaccharides, major components of the outer membrane of typical Gram-negative bacteria.

Fatty Acids and Phospholipids

Fatty acids consist of a carboxyl group attached to a long chain of carbons (alkyl chains); the chains usually have an even number of carbons (the average length is 18 carbons; *see figure 3.7*). Some fatty acids have alkyl chains that are unsaturated; that is, they have one or more double bonds. Most microbial fatty acids are straight chained, but some are branched and some have one or more rings (*see figure 19.8*). Fatty acid synthesis is complex and is the target of some antimicrobial agents used to treat infectious disease caused by eukaryotic pathogens. ◀ *Antifungal drugs (section 9.6)* ▶ *Lipids (appendix I)*

Synthesis of saturated fatty acids is catalyzed by **fatty acid synthase.** In the initiation phase, a molecule of acetyl-CoA is complexed with acyl carrier protein (ACP) with the loss of coenzyme A (CoA) (**figure 12.25**). A second acetyl-CoA molecule is activated by the addition of carbon dioxide to generate a 3-carbon molecule malonyl-CoA. It is then complexed with ACP to form malonyl-ACP. Of the three carbons, the two derived from acetyl-CoA are added to acetyl-ACP, while the remaining carbon (from CO_2) and ACP are released, and a four-carbon molecule proceeds to the elongation phase. In the elongation phase, the β-keto group arising from

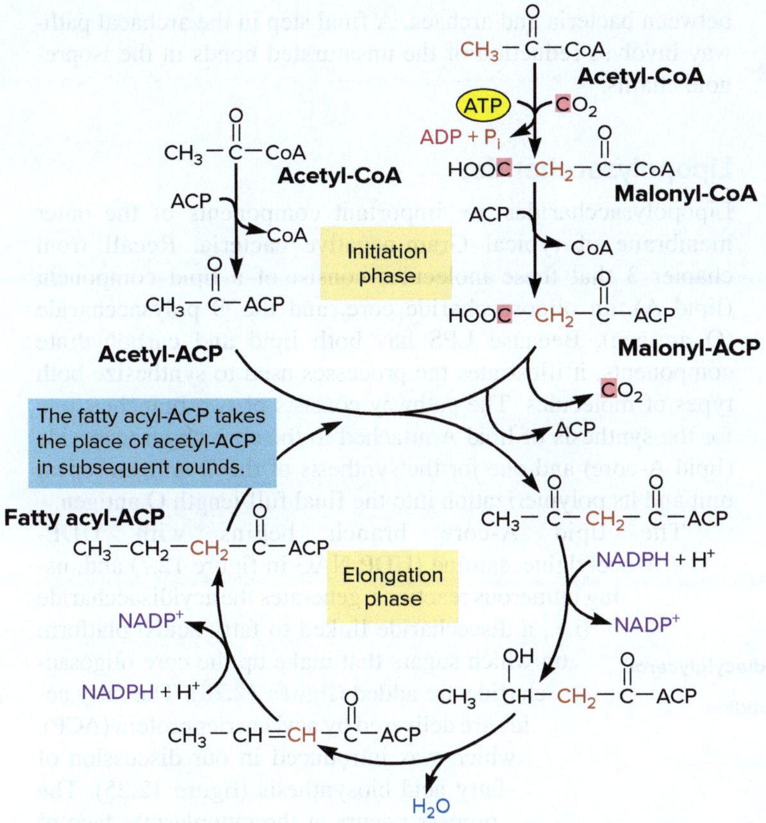

Figure 12.25 Fatty Acid Synthesis. Carbon atoms derived from carbon dioxide are boxed in red; atoms derived from malonyl CoA are in red text. ACP, acyl carrier protein.

the initial condensation reaction is removed in a three-step process involving two reductions and a dehydration. The fatty acid is then ready for the addition of two more carbon atoms by the same mechanism. The cycle repeats until the final length of the hydrocarbon chain is achieved and the fatty acid is released from ACP.

Unsaturated fatty acids are synthesized in two ways. Eukaryotes and many aerobic bacteria use a pathway that introduces double bonds into a saturated fatty acid using both NADPH and O_2, as shown in the following equation:

$$\mathrm{R{-}(CH_2)_9{-}\overset{O}{\overset{\|}{C}}{-}SCoA + NADPH + H^+ + O_2 \rightarrow}$$

$$\mathrm{R{-}CH{=}CH(CH_2)_7{-}\overset{O}{\overset{\|}{C}}{-}SCoA + NADP^+ + 2H_2O}$$

A double bond is formed between carbons 9 and 10, and O_2 is reduced to water with electrons supplied by both the fatty acid and NADPH. Anaerobic bacteria and some aerobes create double bonds during fatty acid synthesis by dehydrating hydroxy fatty acids. Oxygen is not required for double-bond synthesis by this pathway.

Fatty acids are used to make other lipids such as triacylglycerol and phospholipids. Both of these molecules have glycerol linked to fatty acids by ester bonds (*see figure 3.7*) and both are synthesized using the same branched pathway (**figure 12.26**). Glycerol arises from the reduction of dihydroxyacetone phosphate, an intermediate of the Embden-Meyerhof pathway, to glycerol 3-phosphate. Glycerol 3-phosphate is then esterified with two fatty acids to give phosphatidic acid, the branchpoint of the pathway. The branch to triacylglycerol begins with the hydrolysis of phosphate from phosphatidic acid, giving a diacylglycerol; the third fatty acid is then attached to yield a triacylglycerol. The branch to a phospholipid begins when phosphatidic acid is attached to a cytidine diphosphate (CDP) carrier that plays a role analogous to that of uridine and adenosine diphosphate carriers in carbohydrate biosynthesis. The steps in the right branch of figure 12.26 illustrate how the formation of CDP-diacylglycerol initiates the series of reactions to form the polar head of this major cell membrane component. This CDP derivative reacts with serine to form phosphatidylserine, and decarboxylation yields phosphatidylethanolamine. In this way, a complex membrane lipid is constructed from the products of glycolysis, fatty acid biosynthesis, and amino acid biosynthesis.

Sterols and Isoprenoid Lipids

Sterols are important lipids found in eukaryotic membranes, and isoprenoid lipids (also called isoprenes) are found in archaeal membranes. We consider them together because isoprenes are synthesized from two intermediates of the sterol biosynthetic pathway: isopentenyl pyrophosphate (IPP) and its isomer dimethylallyl pyrophosphate (DMAPP) (**figure 12.27*a,b***). These two molecules are most commonly synthesized by a pathway that starts with acetyl-CoA, an important precursor metabolite. Notice that both IPP and DMAPP have two distinct ends. The pyrophosphate end is called the head and the opposite end is called the tail. The two molecules participate in condensation reactions that link molecules head to tail (figure 12.27*c*), head to head, or tail to tail, leading to the production of a variety of molecules, including intermediates that give rise to cholesterol and steroid hormones, bactoprenol, carotenoids, and archaeal lipids (*see figures 3.9, 4.3, and 12.9*). The pathways to the sterols found in eukaryotic membranes are well understood. Although archaeal lipid synthesis pathways have not been fully established, there are themes in common with bacterial lipid synthesis pathways (**table 12.2**). As in the bacterial fatty acid synthesis pathway, building blocks of IPP and DMAPP are added repetitively until the final hydrophobic chain length is reached. Two full-length fatty acids are bound to the glycerolphosphate backbone sequentially, in ether linkages to form **archaeol,** the compound analogous to phosphatidic acid. As a prelude to polar head group addition, both pathways activate the molecule by the addition of a CDP carrier. The polar head group addition reactions are catalyzed by enzymes that are highly conserved

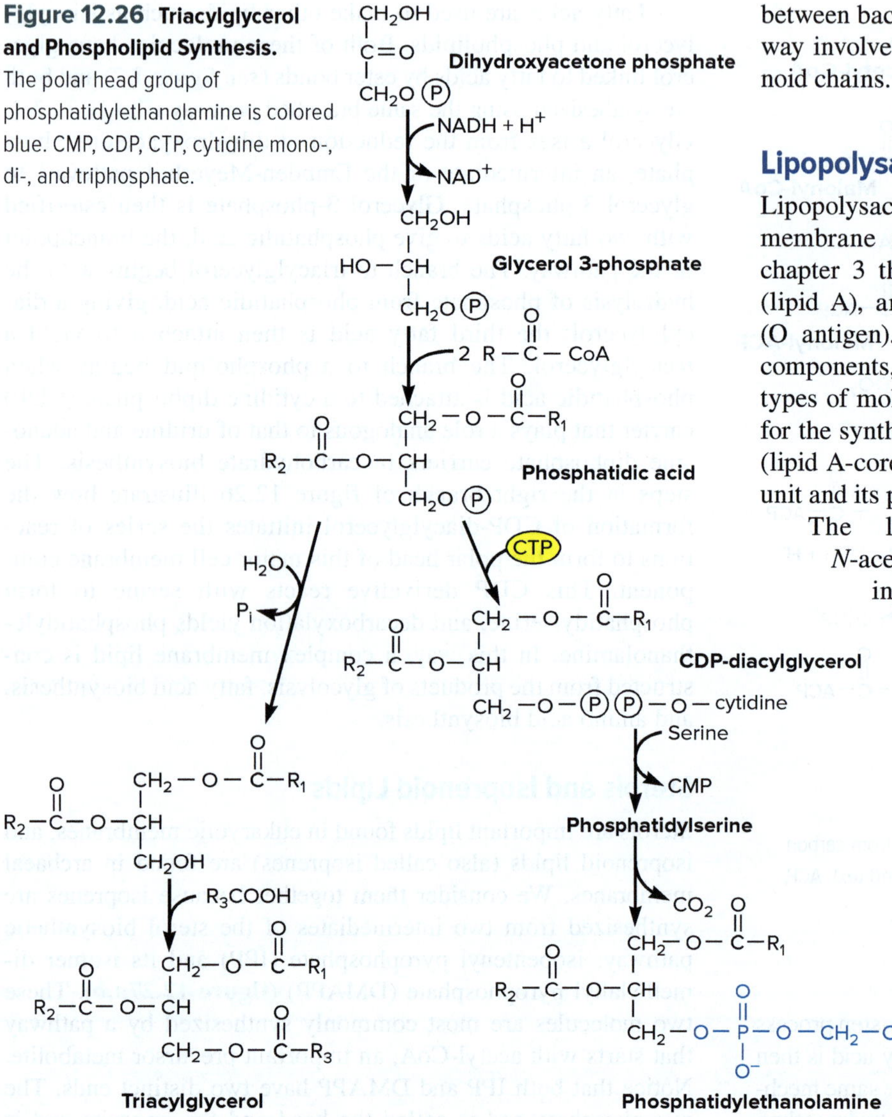

Figure 12.26 Triacylglycerol and Phospholipid Synthesis. The polar head group of phosphatidylethanolamine is colored blue. CMP, CDP, CTP, cytidine mono-, di-, and triphosphate.

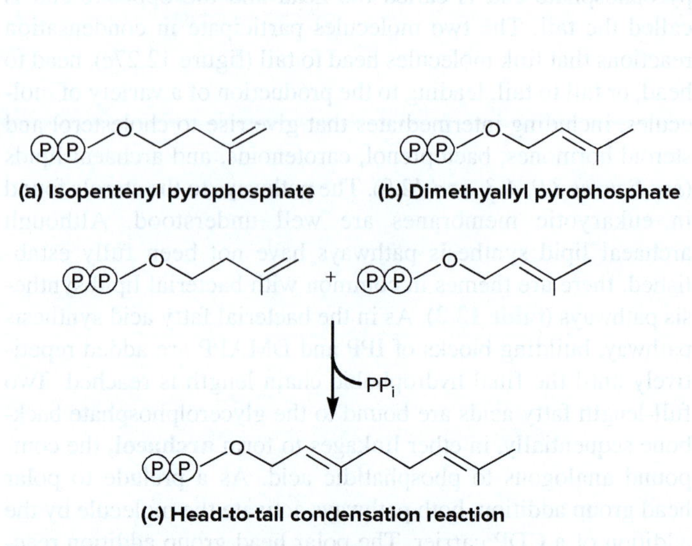

Figure 12.27 Isoprenoid Lipid Building Blocks. (a) Isopentenyl pyrophosphate and (b) dimethylallyl pyrophosphate are shown in (c) undergoing a head-to-tail condensation reaction.

between bacteria and archaea. A final step in the archaeal pathway involves reduction of the unsaturated bonds in the isoprenoid chains.

Lipopolysaccharides

Lipopolysaccharides are important components of the outer membrane of typical Gram-negative bacteria. Recall from chapter 3 that these molecules consist of a lipid component (lipid A), an oligosaccharide core, and the O polysaccharide (O antigen). Because LPS has both lipid and carbohydrate components, it illustrates the processes used to synthesize both types of molecules. The pathway consists of two branches: one for the synthesis of lipid A attached to the core oligosaccharide (lipid A-core) and one for the synthesis of the O antigen repeat unit and its polymerization into the final full-length O antigen.

The lipid A-core branch begins with UDP-*N*-acetylglucosamine (UDP-NAG in figure 12.7) and, using numerous reactions, generates the acyldisaccharide (i.e., a disaccharide linked to fatty acids) platform on which sugars that make up the core oligosaccharide are added (**figure 12.28**). The fatty acids are delivered by acyl carrier protein (ACP), which was introduced in our discussion of fatty acid biosynthesis (figure 12.25). The process occurs at the cytoplasmic face of the plasma membrane, where the fatty acids are inserted into the plasma membrane; the developing core oligosaccharide extends into the cytoplasm (**figure 12.29**). Once synthesis of the lipid A-core portion of LPS is finished, it is flipped to the periplasmic face of the plasma membrane, with the core oligosaccharides in the periplasmic space (figure 12.29).

The O-antigen branch of the pathway begins with the three to five sugars (each on a nucleoside diphosphate carrier) that form the repeat unit of the O antigen. These are added sequentially to bactoprenol on the cytoplasmic face of the plasma membrane; the repeat-unit sugars are in the cytoplasm (recall the use of bactoprenol in peptidoglycan synthesis; figure 12.8). The bactoprenol-repeat unit is next flipped to the periplasmic side of the plasma membrane. The repeat units, which are now extending into the periplasmic space, are polymerized as a block, yielding the O antigen. Finally, the O antigen is ligated to the lipid A-core, generating the final LPS molecule (figure 12.29). The overall structure and chemical composition of the O antigen can vary between different bacteria. This variability is one way to characterize different clinical isolates of the same species. ▶ Escherichia coli *gastroenteritis (section 38.4)*

Figure 12.29 diagrams synthesis of LPS but does not show how the mature molecule is inserted into the outer leaflet of the outer membrane. This is not trivial given the chemical properties of LPS. Since the carbohydrate portion of LPS is hydrophilic, it

Table 12.2 Comparison of Bacterial and Archaeal Lipid Synthesis

	Bacterial Pathway	Archaeal Pathway
Fatty Acid Synthesis		
Carrier for growing fatty acid chain	Malonyl-ACP	None
Number of carbon units added per cycle	2	5
Source of carbon units	Acetyl-CoA	Isopentenyl pyrophosphate
Hydrocarbon length	16–18	20–25
Phospholipid Synthesis		
Glycerophosphate backbone	Glycerol 3-phosphate	Glycerol 1-phosphate
Linkage between backbone and fatty acids	Ester	Ether
Compound with fatty acids and glycerophosphate	Phosphatidic acid	Archaeol
Activated carrier	CDP-diacylglycerol	CDP-archaeol
Common polar head groups	Serine, ethanolamine	Serine, ethanolamine

is difficult to transport across the outer membrane. On the other hand, the lipid A portion of LPS is hydrophobic, impeding its transport across the periplasmic space. Furthermore, there is no power source for this process in the periplasm since it is devoid of ATP. To overcome these challenges, LPS passes through a multi-subunit protein complex that spans the bacterial cell envelope (**figure 12.30**). The proteins are called Lpt proteins (for *l*ipo*p*olysaccharide *t*ransport) and the path they create is called the Lpt pathway. Lpt proteins embedded in the plasma membrane ($LptB_2FC$) engage an LPS molecule for transport. These proteins transfer LPS from the plasma membrane across the aqueous periplasm. The Lpt protein cores are lined with hydrophobic amino acids, which readily interact with lipid A. These hydrophobic grooves of LptF, LptC, LptA, and LptD form a continuous hydrophobic track to allow passage of LPS. Cytoplasmic ATP hydrolysis by LptB provides the energy to push the LPS to the outer membrane. The hydrophilic core oligosaccharides and O antigen extend out from the bridge into the periplasm. This positions the

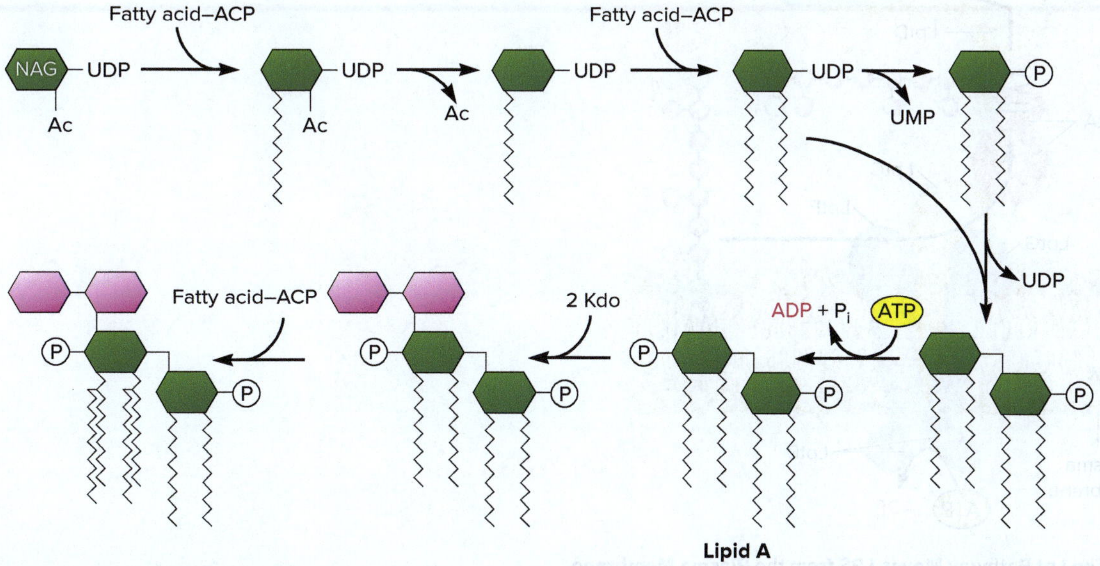

Figure 12.28 Lipid A-Core Biosynthetic Pathway. NAG, *N*-acetylglucosamine; UDP, Uridine diphosphate; Ac, acetyl group; ACP, acyl carrier protein; UMP, Uridine monophosphate; Kdo, 2-keto-3-deoxyoctonate.

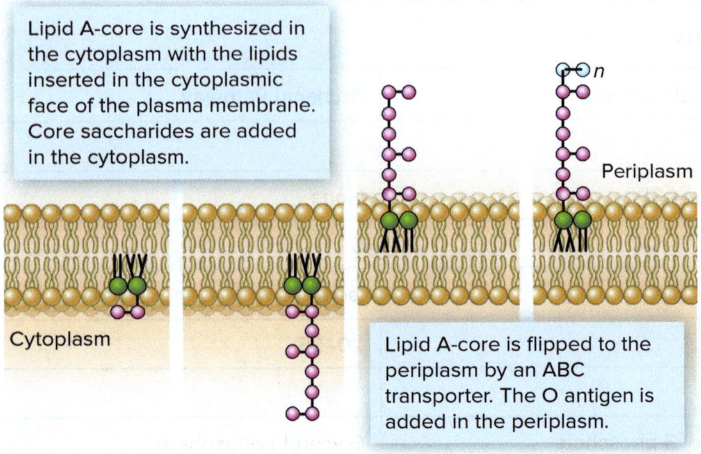

Figure 12.29 Lipopolysaccharide (LPS) Biosynthetic Pathway. Colored portions of LPS correspond to the colors used in figure 12.28.

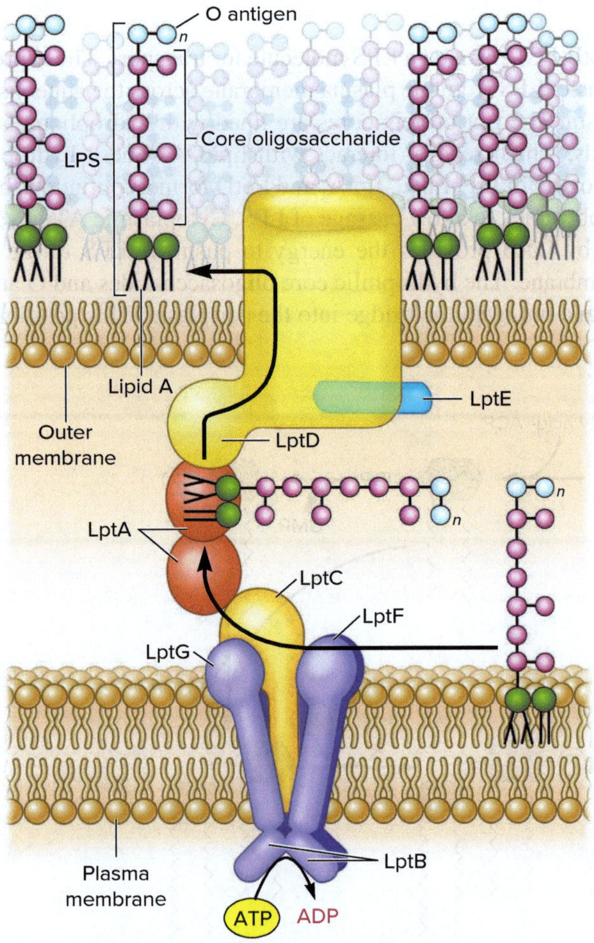

Figure 12.30 The Lpt Pathway Moves LPS from the Plasma Membrane to the Outer Membrane. The arrows track the path of LPS molecules through the proteins of the Lpt pathway.

O antigen near the next part of the pathway, LptD (a β-barrel) and LptE (an outer membrane lipoprotein). Recall from chapter 3 that β-barrels are common in transport proteins (e.g., porin proteins) that translocate molecules across membranes. The second protein, LptE, plugs the LptD β-barrel. LptE directs LPS to enter the LptD lumen, where it is reoriented and inserted into the outer leaflet of the outer membrane.

Comprehension Check

1. What is a fatty acid? Describe in general terms how fatty acid synthase manufactures a fatty acid.
2. How are unsaturated fatty acids made?
3. Briefly describe the pathways for triacylglycerol and phospholipid synthesis. Of what importance are phosphatidic acid and CDP-diacylglycerol?
4. What two building blocks are important for both sterol and isoprene biosynthesis? What pathway is most commonly used to provide them?
5. How does the Lpt complex enable movement of LPS through the periplasm, a compartment that lacks ATP?
6. Diagram how all of the components of the outer membrane fit into the schemes outlined in figures 12.1 and 12.2.

Key Concepts

12.1 Principles Governing Biosynthesis

- Many important cell constituents are macromolecules, large polymers constructed of simple monomers.
- Although many catabolic and anabolic pathways share enzymes, making metabolism more efficient, some enzymes are separate and independently regulated.
- Macromolecular components often undergo self-assembly to form the final supramolecular system.

12.2 Precursor Metabolites: Starting Molecules for Biosynthesis

- Precursor metabolites are carbon skeletons used as the starting substrates for biosynthetic pathways. They are intermediates of glycolytic pathways and the TCA cycle (i.e., the central metabolic pathways) (**figure 12.2**).
- Most precursor metabolites are used for amino acid biosynthesis; others are used for synthesis of purines, pyrimidines, and lipids.

12.3 CO_2 Fixation: Reduction and Assimilation of CO_2 Carbon

- The Calvin-Benson cycle is used by most aerobic autotrophs to fix CO_2. It can be divided into three phases: the carboxylation phase, the reduction phase, and the regeneration phase (**figure 12.3**). Three ATP and two NADPH are required to incorporate one CO_2.
- The reductive TCA cycle is used by a number of autotrophs belonging to the bacterial phyla Bacteroidota, Proteobacteria, and Nitrospirota (**figure 12.4**).

12.4 Synthesis of Carbohydrates

- Gluconeogenesis is the synthesis of glucose 6-phosphate and related sugars from pyruvate.
- Glucose 6-phosphate, fructose 6-phosphate, and mannose 6-phosphate are gluconeogenic intermediates or are made directly from them (**figure 12.5**). Bacteria and protists synthesize glycogen and starch from adenosine diphosphate glucose.
- Peptidoglycan synthesis is a complex process involving both UDP derivatives and the lipid carrier bactoprenol, which helps transport NAG-NAM-pentapeptide units across the plasma membrane. Cross-links are formed by transpeptidation (**figures 12.7–12.10**).

12.5 Synthesis of Amino Acids Consumes Many Precursor Metabolites

- The addition of nitrogen to the carbon chain provided by a precursor metabolite is an important step in amino acid biosynthesis. Ammonia, nitrate, or N_2 can serve as the source of nitrogen.
- Ammonia can be directly assimilated by the activity of transaminases and either glutamate dehydrogenase or the glutamine synthetase–glutamate synthase system (**figures 12.11–12.13**).
- Nitrate is incorporated through assimilatory nitrate reduction catalyzed by the enzymes nitrate reductase and nitrite reductase (**figure 12.14**).
- Nitrogen fixation is catalyzed by nitrogenase. Atmospheric molecular nitrogen is reduced to ammonia, which is then incorporated into amino acids (**figure 12.15**).
- Sulfate must be reduced to sulfide before it is assimilated (**figures 12.17** and **12.18**).
- Some amino acids are made directly by the addition of an amino group to a precursor metabolite, but most amino acids are made by more complex pathways. Many amino acid biosynthetic pathways are branched. Thus a single precursor metabolite can give rise to several amino acids (**figure 12.19**).
- Anaplerotic reactions replace TCA cycle intermediates to keep the cycle in balance while it supplies precursor metabolites. Important anaplerotic reactions include the synthesis of oxaloacetate and the glyoxylate cycle (**figures 12.20** and **12.21**).

12.6 Synthesis of Purines, Pyrimidines, and Nucleotides

- Purines and pyrimidines are nitrogenous bases found in DNA, RNA, and other molecules. The nitrogen is supplied by certain amino acids. Phosphorus is provided by either inorganic or organic phosphate.
- Phosphorus can be assimilated directly by phosphorylation reactions that form ATP from ADP and P_i. Organic phosphorus sources are the substrates of phosphatases that release phosphate from the organic molecule.
- The purine skeleton is synthesized beginning with ribose 5-phosphate and initially produces inosinic acid (**figures 12.22** and **12.23**). Pyrimidine biosynthesis starts with carbamoyl phosphate and aspartate, and ribose is added after the skeleton has been constructed (**figure 12.24**).

12.7 Lipid Synthesis

- Fatty acids are synthesized from acetyl-CoA, malonyl-CoA, and NADPH by fatty acid synthase. During synthesis, the intermediates are attached to the acyl carrier protein (**figure 12.25**). Double bonds can be added in two different ways.
- Triacylglycerols are made from fatty acids and glycerol phosphate. Phosphatidic acid is an important intermediate in this pathway (**figure 12.26**).
- Phospholipids can be synthesized from phosphatidic acid by forming CDP-diacylglycerol, then adding an amino acid.

- Sterols and isoprenes are synthesized from two building blocks, isopentenyl pyrophosphate and dimethylallyl pyrophosphate. Condensation reactions using these molecules in various ways build the ring structures of sterols and the glycerol diethers found in archaeal membranes (**figure 12.27**).
- Lipopolysaccharides are synthesized by a complex branched pathway that uses one branch to make the lipid A-core portion and the other to synthesize the O antigen (**figure 12.28**). These two are joined in the periplasmic space (**figure 12.29**). LPS moves from the plasma membrane to the outer leaflet of the outer membrane via the Lpt pathway (**figure 12.30**).

Active Learning

1. What would happen if a microorganism that depended on the glyoxylate cycle no longer produced the enzyme isocitrate lyase? Could this defect be corrected by nutrients in the media? If so, which nutrients?
2. Nucleoside diphosphates are common carriers in biosynthesis reactions. Identify pathways that use ADP, UDP, and CDP as carriers in this chapter, and review the structures illustrated in chapter 11 to find additional examples.
3. Examine the structure of teichoic acid in figure 3.20. Based on what you learned in this chapter and what you search for online, identify precursors required to synthesize teichoic acid, and their origin in the precursor metabolites listed in figure 12.2.
4. Many nitrogen-fixing microbes are also autotrophs. Recall that oxygenic photosynthesis generates O_2, and that nitrogenase, the enzyme that catalyzes nitrogen fixation, is sensitive to O_2. What mechanisms can you think of for diazotrophic autotrophs to protect nitrogenase from O_2? (Hint: Consider both spatial and temporal mechanisms.)
5. Many antibiotics are naturally synthesized by microbes. This includes vancomycin, produced by the bacterium *Amycolatopsis orientalis*. Review the structure of vancomycin in figure 9.8. What type of precursor metabolites/building blocks are required to build this complex molecule?
6. The road to discovery of the Lpt pathway was long and difficult to travel, not unlike the journey faced by LPS! One major challenge is that the genes encoding the Lpt proteins are all essential for bacterial cells to remain viable. Why is this not surprising? How might you go about studying the Lpt pathway experimentally?

CHAPTER

13

Bacterial Genome Replication and Expression

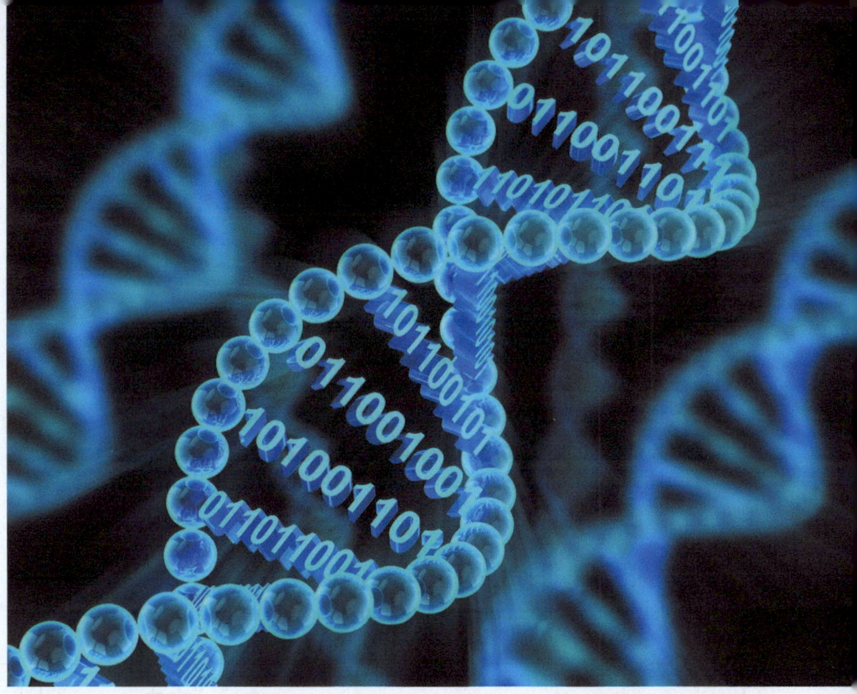

ymgerman/Shutterstock

Making Code

Since ancient times, humans have sent important messages in code. Mary Queen of Scots used code to detail her plans to overthrow Queen Elizabeth I. George Washington regularly communicated to his troops in code, as have commanders in all wars. Today work continues to create codes that are harder to break and more content rich.

The history of communicating in code parallels the history of information storage. Initially messages were coded, or encrypted, in language, song, and dance. Later, coded messages were handwritten on paper; manuscripts and books stored valuable data. As codes became more sophisticated and their content larger, their forms evolved to include radio communications, recordings, and binary digits. Similar methods were developed to store data sets.

In the early years of this century, scientists began exploring a new method of encrypting and storing data called biostorage or bioencryption. The first approaches stored coded data in DNA inserted into a bacterial chromosome. Scientists turned to DNA because they estimated that the information in about 1,000 books having 500 pages of small print is the equivalent of the information stored in one human chromosome. Among the advantages of storing data in DNA is its chemical stability in cold, dry conditions over extended time periods, and its ability to be compacted into a small volume.

To store data in DNA, binary code of zeroes and ones is converted to the four bases of the DNA alphabet, A, G, C, and T. Anything stored digitally, such as music, movies, or this textbook, can be converted to nucleic acid, and later retrieved and decoded. In a stunning demonstration of the principle, a team from Harvard encoded a short GIF of a horse and rider into synthetic DNA pieces. These were introduced into a population of *Escherichia coli,* where they were stably maintained in the chromosome. Rapid, efficient DNA sequencing technology enabled data retrieval.

For millennia organisms have used DNA to store genetic information, and they have devised methods for replicating, recoding, and decoding genetic information. In this chapter, we explore the polymerization reactions that generate DNA, RNA, and protein, reactions central to the inheritance of traits.

Throughout the twentieth century, microorganisms served as model experimental systems for genetic studies and helped elucidate the nature of genetic information, gene structure, the genetic code, and mutation. Thus we begin by considering bacterial genetics in this chapter and the next. Because of the distinctive characteristics of eukaryotes and archaea, the replication and expression of their genomes are discussed in chapter 15.

Readiness Check:

Based on what you have learned previously, you should be able to:

- ✓ Define the terms *genome, chromosome, plasmid, haploid, diploid, genotype,* and *phenotype*
- ✓ Outline the flow of genetic information, identifying the processes used
- ✓ State the base-pairing rules
- ✓ Describe the four levels of protein structure
- ✓ Describe the structure of bacterial genomes (section 3.6)
- ✓ Draw a simple diagram that illustrates the chemical moieties found in nucleotides (section 12.6)

13.1 Experiments Using Bacteria and Viruses Demonstrated That DNA Is the Genetic Material

After reading this section, you should be able to:

a. Summarize Griffith's experiments on transformation
b. Relate how the contributions of Avery, MacLeod, McCarty, Hershey, and Chase confirmed that DNA stores genetic material

Although it is now hard to imagine, it was once thought that DNA was too simple a molecule to store genetic information. With only four different nucleotides, it seemed that a molecule of much greater complexity must house the genetic information of a cell. Since proteins are composed of 20 different amino acids, they were viewed as the better candidate for this important cellular function.

The early work of Frederick Griffith on the transfer of virulence (i.e., the ability to cause disease) in the pathogen *Streptococcus pneumoniae,* commonly called pneumococcus, set the stage for showing that DNA was indeed the genetic material. When exposed to the virulent form of pneumococcus, mice develop pneumonia and die. Griffith found that if he boiled virulent pneumococcal cells and injected them into mice, the mice were unharmed and no bacteria could be recovered from the animals (**figure 13.1**). When he injected a combination of heat-killed virulent bacteria and a living nonvirulent strain, the mice died;

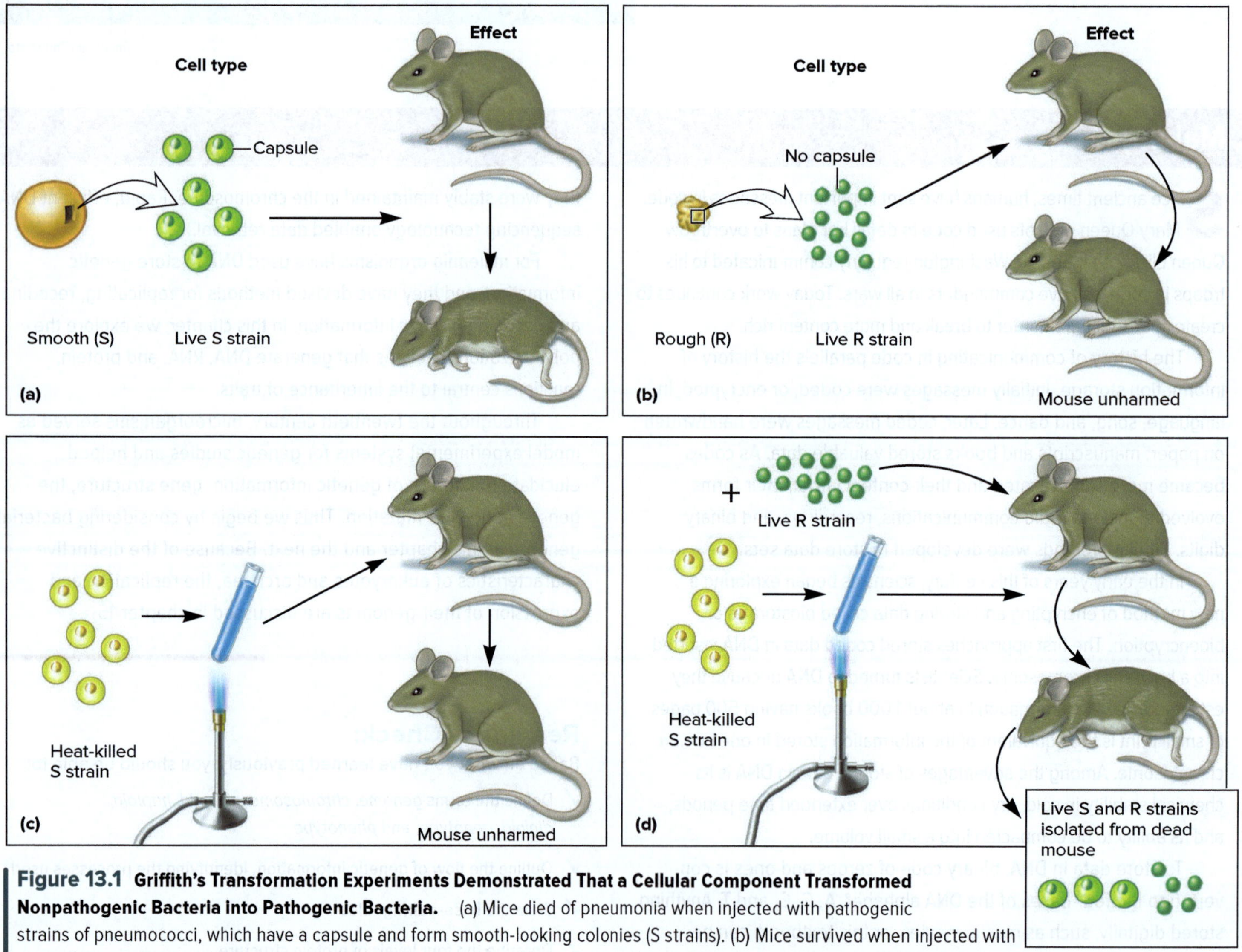

Figure 13.1 Griffith's Transformation Experiments Demonstrated That a Cellular Component Transformed Nonpathogenic Bacteria into Pathogenic Bacteria. (a) Mice died of pneumonia when injected with pathogenic strains of pneumococci, which have a capsule and form smooth-looking colonies (S strains). (b) Mice survived when injected with nonpathogenic strains of pneumococci, which lack a capsule and form rough colonies (R strains). (c) Injection with heat-killed S strains had no effect on the mice. (d) Injection with a live R strain and a heat-killed S strain caused pneumonia, and live S strain pneumococci could be isolated from the dead mice.

MICRO INQUIRY *Based on what we now know about proteins, why can we conclude from this experiment that genetic information was unlikely to be carried by proteins?*

moreover, he could recover living virulent bacteria from the dead mice. Griffith called this change of nonvirulent bacteria into virulent pathogens *transformation*.

Oswald Avery and his colleagues then set out to discover which constituent in the heat-killed virulent pneumococci was responsible for transformation. These investigators used enzymes to selectively destroy DNA, RNA, or protein in purified extracts of virulent pneumococci (S strain). They then exposed nonvirulent pneumococcal strains (R strains) to the treated extracts. Transformation of the nonvirulent bacteria was blocked only if DNA was destroyed, suggesting that DNA was carrying the information required for transformation (**figure 13.2**). The publication of these studies by O. Avery, C. M. MacLeod, and M. J. McCarty provided the first evidence that DNA carried genetic information.

Shortly thereafter Alfred Hershey and Martha Chase wanted to determine if protein or DNA carried the genetic information of a bacterial virus called T2 bacteriophage. In their experiments, radioactivity was used to label the virus's DNA (with ^{32}P) or protein coat (with ^{35}S). They mixed radioactive virions (either ^{32}P-labeled or ^{35}S-labeled) with *E. coli* and incubated the mixture for a few minutes. This allowed the virions to attach to *E. coli* and initiate the infection process (*see figure 6.15*). The culture was then centrifuged. This separated the cells, which formed a pellet at the bottom of the tube, from any unadsorbed phage particles, which remained suspended in the liquid supernatant. The infected cells in the resulting pellet were resuspended and the suspension was agitated violently in a blender. The blender treatment sheared off bacteriophage particles adsorbed to *E. coli* cells (**figure 13.3**). Some of the suspension was used in a plaque assay (*see figure 6.17*) to determine if the blender treatment affected the ability of the phage to multiply within host cells. The remaining suspension was centrifuged to separate cells from the sheared-off phage particles. After centrifugation, type of radioactivity in the supernatant (where the phage particles remained) versus that in the bacterial cells in the pellet was determined.

The results of their experiments showed that blender treatment did not disrupt the infection process because progeny phages were produced. They also demonstrated that when ^{35}S-labeled T2 was used in the experiment, the majority of the radioactive protein was in the supernatant, whereas when ^{32}P-labeled T2 was used, the radioactive DNA was in the bacterial cells that formed the pellet. Because DNA entered the cells and the protein did not, the phage DNA must have been carrying the genetic information needed to complete the infection process. Some luck was involved in their discovery, for the genetic material of many viruses is RNA and the researchers happened to select a

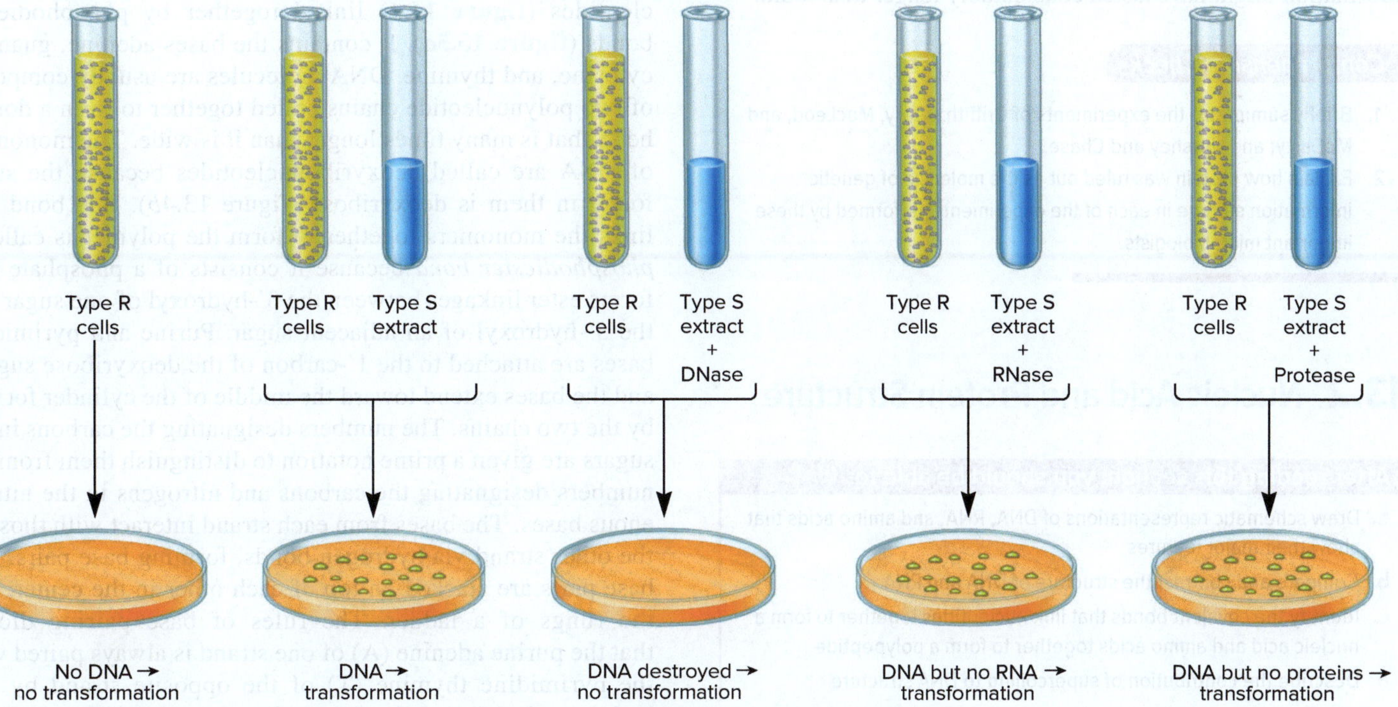

Figure 13.2 Some Experiments on the Transforming Principle. Earlier experiments done by Avery, MacLeod, and McCarty had shown that only DNA extracts from S cells caused transformation of R cells to S cells. To demonstrate that contaminating molecules in the extract were not responsible for transformation, the extract from S cells was treated with RNase, DNase, or protease, and then mixed with R cells. Only treatment of the extract from S cells with DNase destroyed its ability to transform the R cells.

MICRO INQUIRY *Why is the experiment on the far left important to include? How would you interpret the growth of a few colonies on this plate?*

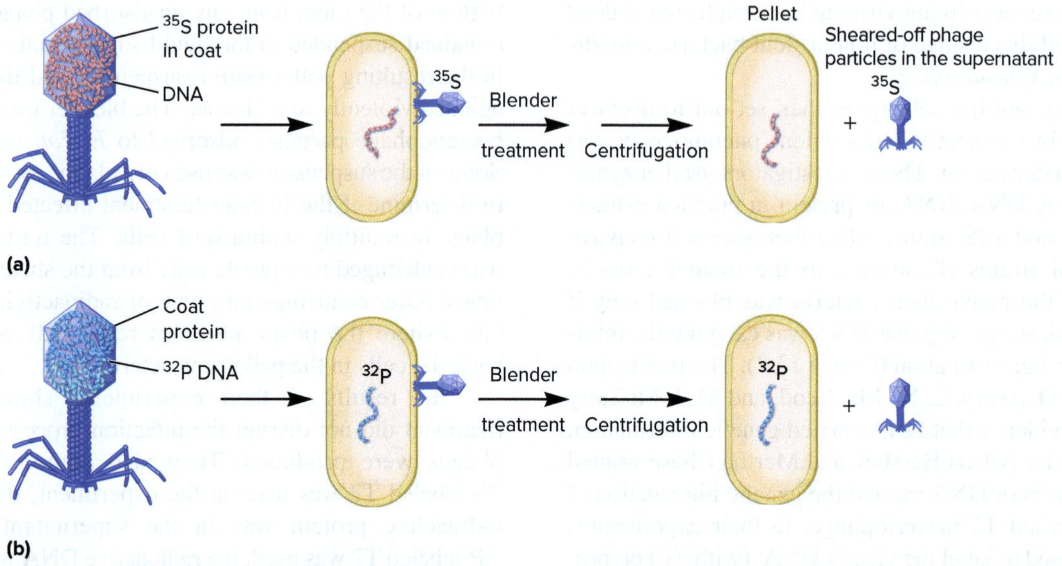

Figure 13.3 The Hershey-Chase Experiment. (a) When *E. coli* was infected with a T2 phage containing ^{35}S protein, after centrifugation most of the radioactivity remained in the supernatant where the sheared-off phage particles were located. (b) When a T2 phage containing ^{32}P DNA was mixed with the host bacterium, after centrifugation the radioactive DNA was in the pellet where the cells were located. Thus DNA was carrying the virus's genetic information.

MICRO INQUIRY *Why is ^{35}S able to selectively label protein but not DNA? Why is ^{32}P able to selectively label DNA but not protein?*

DNA virus for their studies. Imagine the confusion if T2 had been an RNA virus! The controversy surrounding the nature of genetic information might have lasted considerably longer than it did.

Comprehension Check

1. Briefly summarize the experiments of Griffith; Avery, MacLeod, and McCarty; and Hershey and Chase.
2. Explain how protein was ruled out as the molecule of genetic information storage in each of the experiments performed by these important microbiologists.

13.2 Nucleic Acid and Protein Structure

After reading this section, you should be able to:

a. Draw schematic representations of DNA, RNA, and amino acids that show their major features
b. Compare and contrast the structure of DNA and RNA
c. Identify the covalent bonds that link nucleotides together to form a nucleic acid and amino acids together to form a polypeptide
d. Describe the contribution of supercoiling to DNA structure

DNA, RNA, and proteins are often called *informational molecules.* The information exists as the sequence of monomers from which they are built. Here we describe the monomers and how they are linked together to form these important macromolecules.

DNA Is a Polymer of Deoxyribonucleotides

Deoxyribonucleic acid (DNA) is a polymer of deoxyribonucleotides (**figure 13.4**) linked together by phosphodiester bonds (**figure 13.5*a***). It contains the bases adenine, guanine, cytosine, and thymine. DNA molecules are usually composed of two polynucleotide chains coiled together to form a double helix that is many times longer than it is wide. The monomers of DNA are called deoxyribonucleotides because the sugar found in them is deoxyribose (figure 13.4*b*). The bond that links the monomers together to form the polymer is called a *phosphodiester bond* because it consists of a phosphate that forms ester linkages between the 3′-hydroxyl of one sugar and the 5′-hydroxyl of an adjacent sugar. Purine and pyrimidine bases are attached to the 1′-carbon of the deoxyribose sugars, and the bases extend toward the middle of the cylinder formed by the two chains. The numbers designating the carbons in the sugars are given a prime notation to distinguish them from the numbers designating the carbons and nitrogens in the nitrogenous bases. The bases from each strand interact with those of the other strand via hydrogen bonds, forming base pairs. The base pairs are stacked on top of each other in the center, like the rungs of a ladder. The rules of base pairing dictate that the purine adenine (A) of one strand is always paired with the pyrimidine thymine (T) of the opposite strand by two hydrogen bonds, while the purine guanine (G) pairs with cytosine (C) by three hydrogen bonds. This AT and GC base pairing means that the two strands in a DNA double helix are **complementary.** In other words, the bases in one strand match up with those of the other according to specific base-pairing rules. Because the sequences of bases in these strands encode

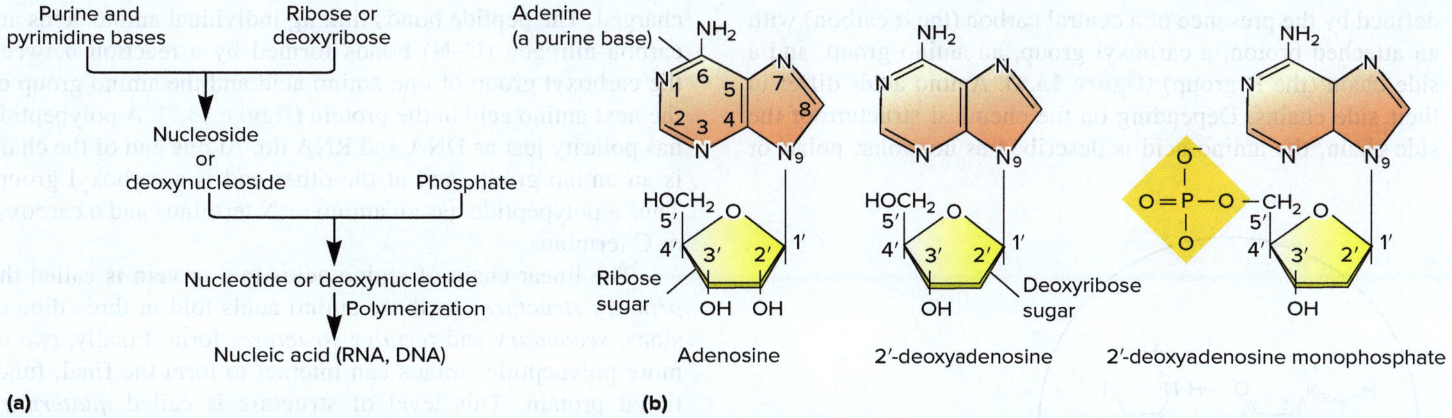

Figure 13.4 The Composition of Nucleic Acids. (a) A diagram showing the relationships of various nucleic acid components. Combination of a purine or pyrimidine base with ribose or deoxyribose gives a nucleoside (a ribonucleoside or deoxyribonucleoside, respectively). A nucleotide contains a nucleoside and one or more phosphates. Nucleic acid polynucleotide chains form when nucleotides are connected by phosphodiester bonds (figure 13.5*a*). (b) Examples of nucleosides—adenosine and 2′-deoxyadenosine—and the nucleotide 2′-deoxyadenosine monophosphate. The carbons of nucleoside and nucleotide sugars are indicated by numbers with primes to distinguish them from the carbons in the bases.

MICRO INQUIRY *To which carbon of ribose (deoxyribose) is each of the following bonded: adenine, the hydroxyl of deoxyadenosine, and phosphate?*

genetic information, considerable effort has been devoted to determining the base sequences of DNA and RNA from many organisms. ▶| *Microbial genomics (chapter 18)*

The two polynucleotide strands of DNA fit together much like the pieces in a jigsaw puzzle. Figure 13.5*b,c* shows that the two strands are not positioned directly opposite one another; rather, they are arranged in an antiparallel orientation. Therefore when the strands twist about one another, a wide major groove and narrower minor groove are formed by the backbone. There are 10.5 base pairs per turn of the helix. The helix is right-handed; that is, the chains turn counterclockwise as they approach a viewer looking down the longitudinal axis. The antiparallel backbones run in opposite directions with respect to the orientation of their sugars. Thus the 5′ end of one strand is paired with the 3′ end of the complementary strand; or stated another way, one strand is oriented 5′ to 3′ and its complement runs 3′ to 5′ (figure 13.5*b*).

The structure of DNA just described is that of the B form, the most common form in cells. The A form, an alternative DNA structure, primarily differs from the B form in that it is slightly wider, with 11 base pairs per helical turn, rather than 10.5. The A form DNA is found in bacterial endospores, and is partially responsible for protecting the DNA from UV damage. In addition, some virus genomes have A form DNA. |◀ *Endospore resistance (section 3.8)* *DNA Structure*

Supercoiling is another property of DNA. DNA is helical; that is, it is a coil. Whenever the rotation of a coil is restrained in some way, it causes it to coil on itself. The coiling of a coil is called *supercoiling* and is readily observed by twisting a rubber band. Recall that most bacterial chromosomes are closed, circular double-stranded DNA molecules. In this state, the two strands are unable to rotate freely relative to each other, and the molecule is said to be constrained. The strain is relieved by supercoiling. There are two types of supercoiling: positive and negative. For DNA, these are defined by the change in number of base pairs per turn in the double helix. Supercoiling that decreases the number of base pairs per turn is said to be negative supercoiling. Likewise, supercoiling that increases the number of base pairs per turn is called positive supercoiling.

Supercoiling is critical for its function. Although bacterial chromosomes are negatively supercoiled overall, smaller regions can vary depending on how the chromosome is used at a given time. Supercoiling loosens up the DNA, making it easier to separate the two strands. Strand separation is an important early step in both DNA replication and transcription, as we discuss in sections 13.3 and 13.5, respectively.

RNA Is a Polymer of Ribonucleotides

Ribonucleic acid (RNA) is a polymer of ribonucleotides containing the sugar ribose and the bases adenine, guanine, cytosine, and uracil. In RNA, uracil (U) replaces thymine. The nucleotides are joined by a phosphodiester bond, just as in DNA. In cells, RNA molecules are single stranded. However, an RNA strand can coil back on itself to form secondary structures such as hairpins with complementary base pairing and helical organization. The formation of double-stranded regions in RNA is often critical to its function. ▶| *Attenuation and riboswitches can stop transcription prematurely (section 14.3); Translational riboswitches (section 14.4)*

Proteins Are Polymers of Amino Acids

Proteins are polymers of amino acids linked by peptide bonds; thus they are also called polypeptides. An amino acid is defined by the presence of a central carbon (the α carbon) with an attached proton, a carboxyl group, an amino group, and a side chain (the R group) (**figure 13.6**). Amino acids differ in their side chains. Depending on the chemical structure of the side chain, the amino acid is described as nonpolar, polar, or charged. The peptide bonds linking individual amino acids are carbon-nitrogen (C–N) bonds formed by a reaction between the carboxyl group of one amino acid and the amino group of the next amino acid in the protein (**figure 13.7**). A polypeptide has polarity just as DNA and RNA do. At one end of the chain is an amino group, and at the other end is a carboxyl group. Thus a polypeptide has an amino or N terminus and a carboxyl or C terminus.

The linear chain of amino acids in a protein is called the *primary structure*. As these amino acids fold in three dimensions, *secondary* and *tertiary structures* form. Finally, two or more polypeptide strands can interact to form the final, functional protein. This level of structure is called *quaternary structure*. These higher levels of structure are stabilized by intra- (and inter-) chain bonds. Protein structure is described in more detail in appendix I.

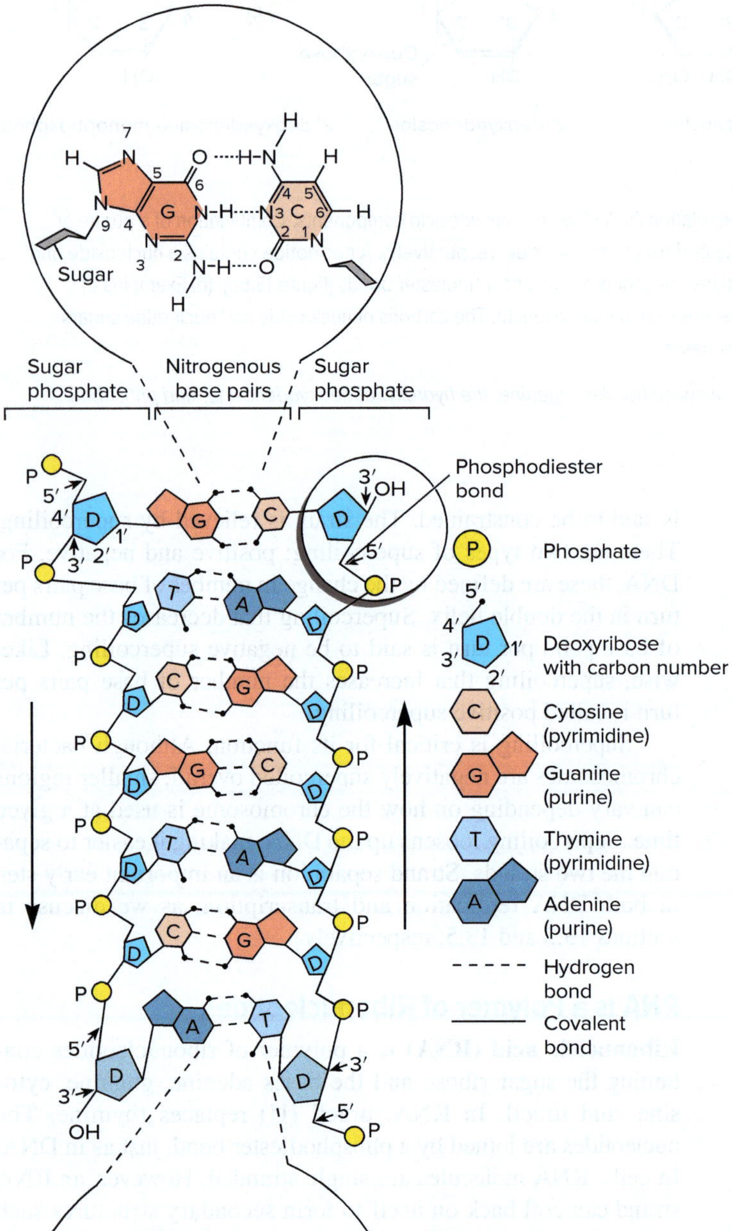

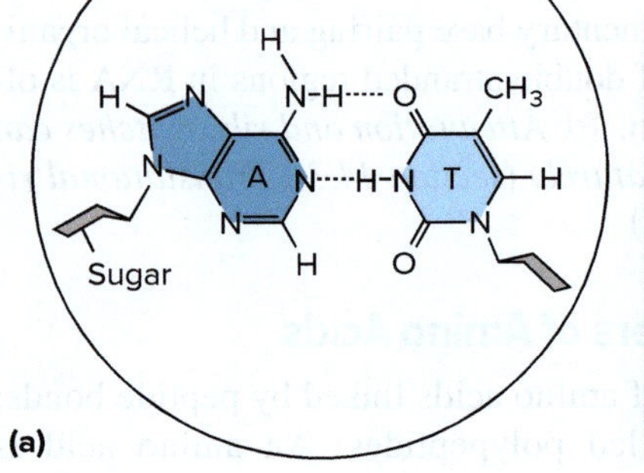

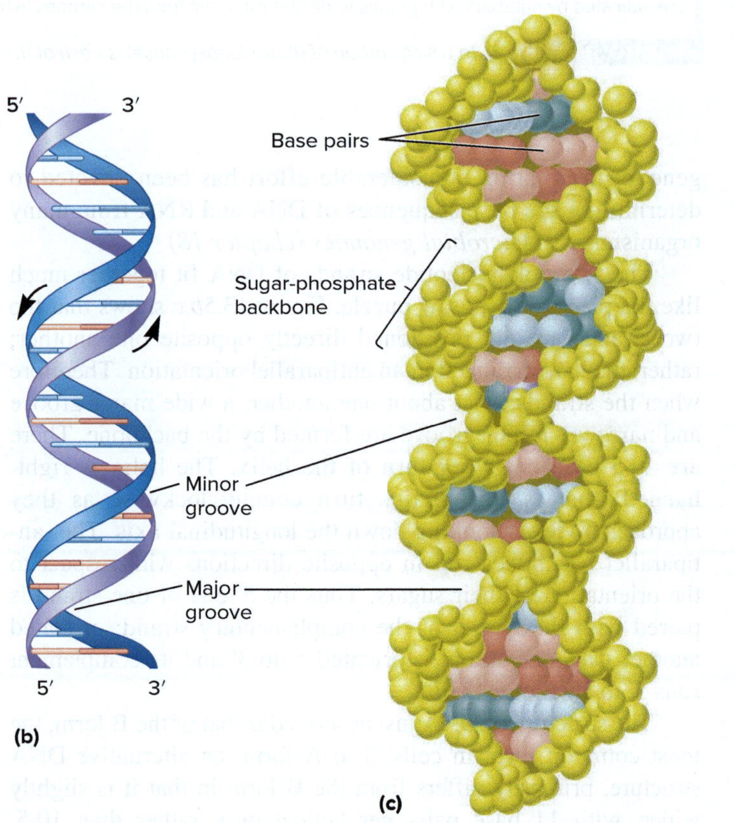

Figure 13.5 DNA Structure—B Form of DNA. DNA Is usually a double-stranded molecule. (a) A schematic, nonhelical model. In each strand, phosphates are esterified to the 3′-carbon of one deoxyribose sugar (blue) and the 5′-carbon of the adjacent sugar. The two strands are held together by hydrogen bonds (dashed lines). Because of the specific base pairing, the base sequence of one strand determines the sequence of the other. The two strands are antiparallel, as indicated by the two arrows, which point in the 5′ to 3′ direction. (b) A simplified model that highlights the antiparallel arrangement and the major and minor grooves. (c) A space-filling model. Note that the sugar-phosphate backbone spirals around the outside of the helix and the base pairs are embedded inside.

MICRO INQUIRY *How many hydrogen bonds are there between adenine and thymine, and between guanine and cytosine?*

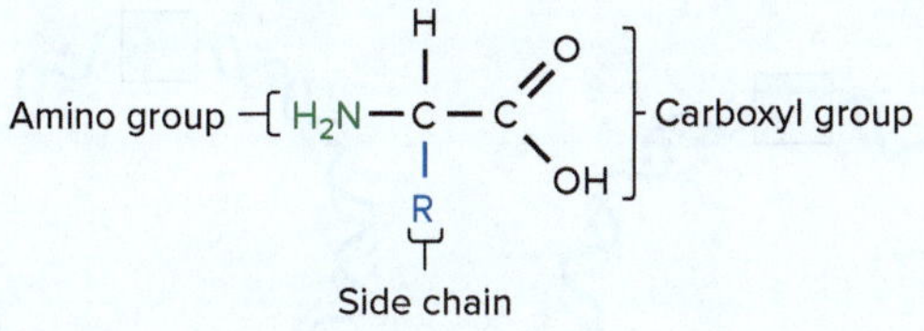

Figure 13.6 The General Structure of an Amino Acid. All amino acids have a central carbon (the α carbon) with an attached proton (H), a carboxyl group (COOH), an amino group (NH_2), and a side chain (R). Amino acids differ in their side chains, which may be nonpolar, polar, negatively charged (acidic), or positively charged (basic).

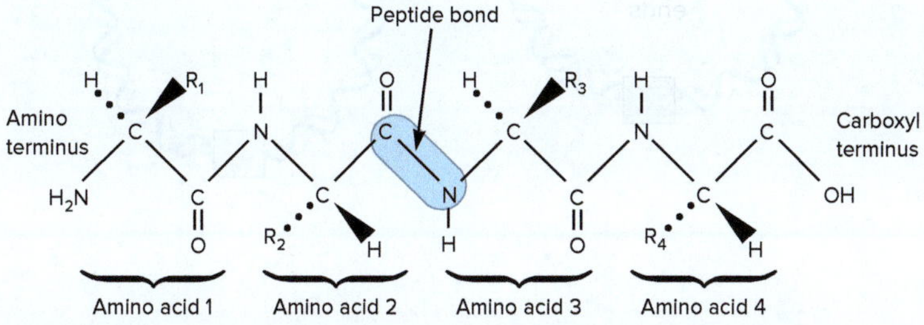

Figure 13.7 Peptide Bonds Link Amino Acids Together in Peptide Chains. A tetrapeptide chain is shown. One of the peptide bonds linking the four amino acids together is highlighted in blue.

MICRO INQUIRY *Identify the two other peptide bonds of the tetrapeptide chain.*

Comprehension Check

1. What are nucleic acids? How do DNA and RNA differ in structure?
2. What does it mean that the two strands of the DNA double helix are complementary and antiparallel? Examine figure 13.5*b* and explain the differences between the minor and major grooves.
3. Amino acids are described as nonpolar, polar, and charged, depending on the molecular makeup of their side chains. Which type of amino acid might be found in the transmembrane portion of a polypeptide located in the plasma membrane? Explain your answer.

13.3 DNA Replication in Bacteria

After reading this section, you should be able to:

a. Describe a bacterial replicon
b. Summarize the events that occur during DNA replication initiation, elongation, and termination
c. Create a concept map that illustrates the function of the major proteins found in a bacterial replisome
d. List the enzymatic and structural elements needed by DNA polymerases for DNA synthesis
e. Outline the major events that occur at the replication fork

DNA replication is an extraordinarily important and complex process upon which all life depends. During DNA replication, the two strands of the double helix are separated; each then serves as a template for the synthesis of a complementary strand according to the base-pairing rules. Each of the two progeny DNA molecules consists of one new strand and one old strand, therefore defining DNA replication as being *semiconservative* (**figure 13.8**). DNA replication is also extremely accurate; *E. coli* makes errors with a frequency of only 10^{-9} or 10^{-10} per base pair replicated (or about one in a million [10^{-6}] per gene per generation). Despite its complexity and accuracy, replication is very rapid. In bacteria, replication rates approach 1,000 base pairs per second. Most of our discussion in this section is based on studies of chromosomal DNA replication in *E. coli.*

Bacterial DNA Replication Initiates from a Single Origin of Replication

The replication of chromosomal DNA begins at a single point, the **origin of replication,** often termed *oriC*. It's important for bacteria to have just a single origin of replication so that replication is coordinated with other cellular events. Recall that chromosome partitioning is carried out by proteins that bind near the

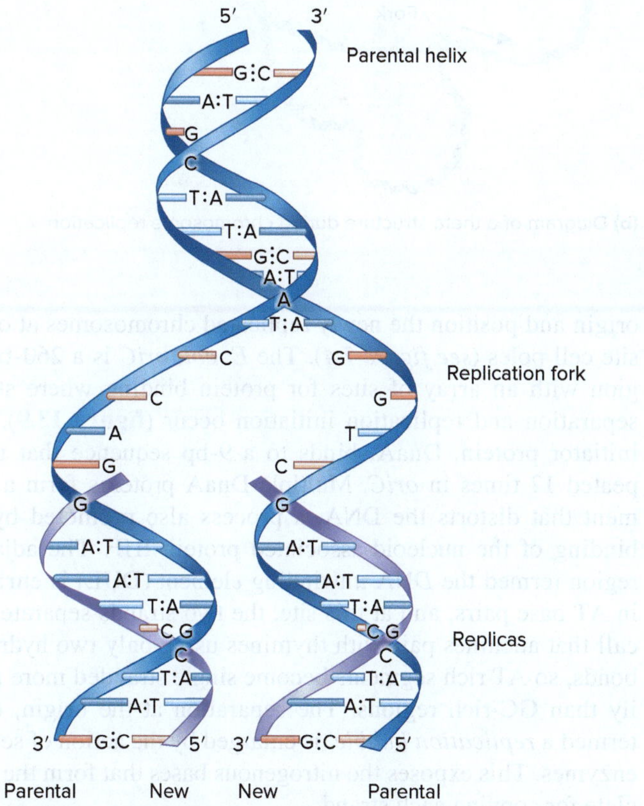

Figure 13.8 Semiconservative DNA Replication. The replication fork of DNA showing the synthesis of two progeny strands. Newly synthesized strands are purple. Black dots indicate hydrogen bonds. Each replica contains one new and one old strand.

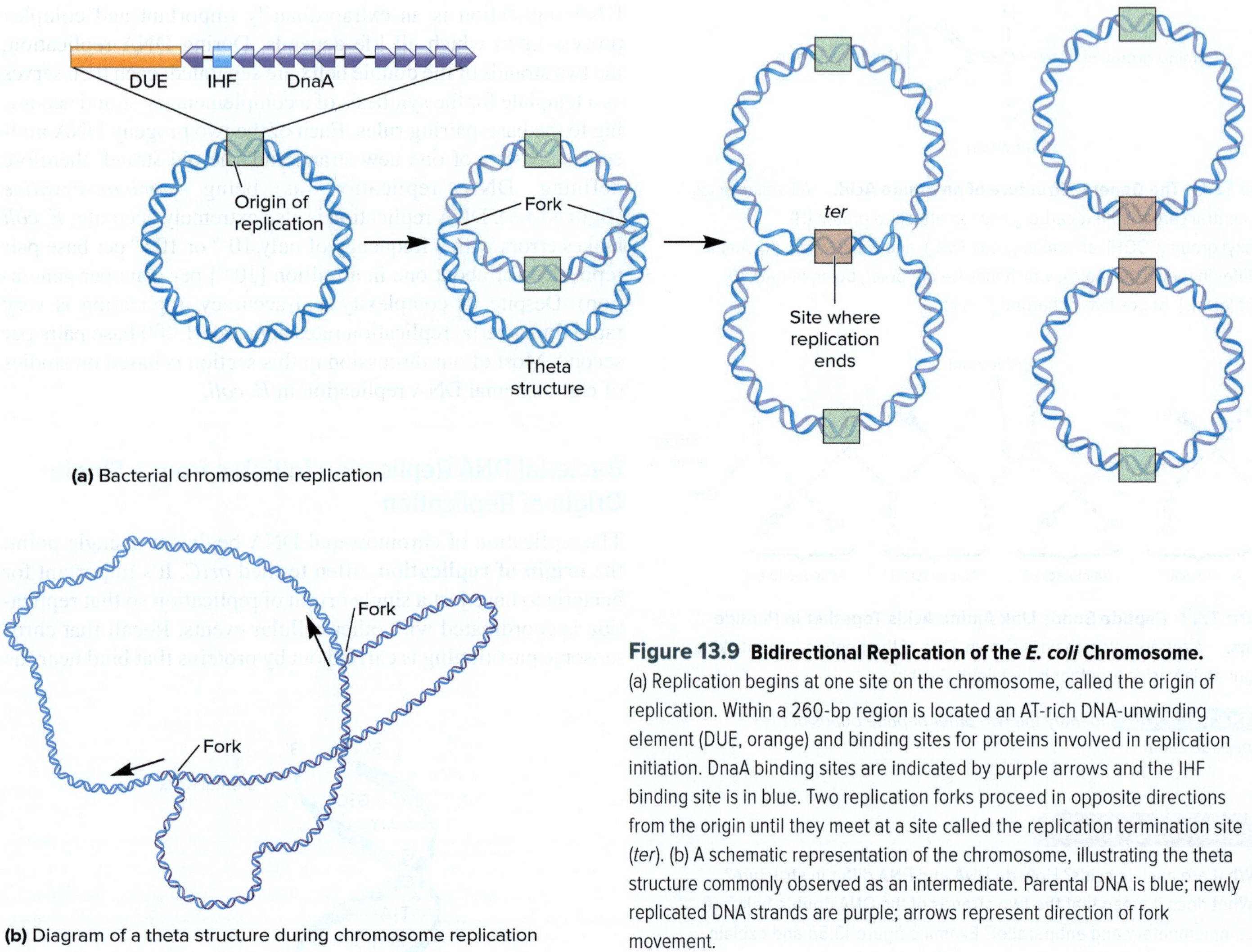

Figure 13.9 Bidirectional Replication of the *E. coli* Chromosome. (a) Replication begins at one site on the chromosome, called the origin of replication. Within a 260-bp region is located an AT-rich DNA-unwinding element (DUE, orange) and binding sites for proteins involved in replication initiation. DnaA binding sites are indicated by purple arrows and the IHF binding site is in blue. Two replication forks proceed in opposite directions from the origin until they meet at a site called the replication termination site (*ter*). (b) A schematic representation of the chromosome, illustrating the theta structure commonly observed as an intermediate. Parental DNA is blue; newly replicated DNA strands are purple; arrows represent direction of fork movement.

origin and position the newly replicated chromosomes at opposite cell poles (*see figure 7.4*). The *E. coli oriC* is a 260-bp region with an array of sites for protein binding where strand separation and replication initiation occur (**figure 13.9**). The initiator protein, DnaA, binds to a 9-bp sequence that is repeated 12 times in *oriC*. Multiple DnaA proteins form a filament that distorts the DNA, a process also promoted by the binding of the nucleoid-associated protein IHF. The adjacent region termed the *D*NA *u*nwinding *e*lement (DUE) is enriched in AT base pairs, and at this site, the two strands separate. Recall that adenines pair with thymines using only two hydrogen bonds, so AT-rich segments become single stranded more readily than GC-rich regions. The separation at the origin, often termed a *replication bubble,* is enlarged by the action of several enzymes. This exposes the nitrogenous bases that form the template for copying each strand.

DNA synthesis occurs at the **replication fork,** the place at which the parental DNA helix is unwound and the two strands are replicated (figure 13.9). Two replication forks move outward from the origin until they have copied the whole **replicon**—the portion of the genome that contains an origin and is replicated as a unit. When the replication forks move around the circular chromosome, a structure shaped like the Greek letter theta (θ) is formed (figure 13.9*b*). Because the bacterial chromosome is a single replicon, the forks meet on the other side and two separate chromosomes are released.
DNA Replication Fork; Bidirectional DNA Replication

Many Enzymes Are Needed for DNA Replication

A great deal of effort has been devoted to understanding the mechanism of DNA replication. In *E. coli,* DNA replication requires at least 12 proteins that form a complex called the **replisome** at the replication fork. Two replisomes move in opposite directions away from the origin. To separate (or denature) the parental DNA strands beyond the DUE, an enzyme termed **helicase** is required (**figure 13.10**). Helicase is a six-subunit ring that encircles one DNA strand. Its movement, powered by ATP, disrupts the hydrogen bonds holding the parental DNA strands together. Thus helicases provide the force to move the replisomes.

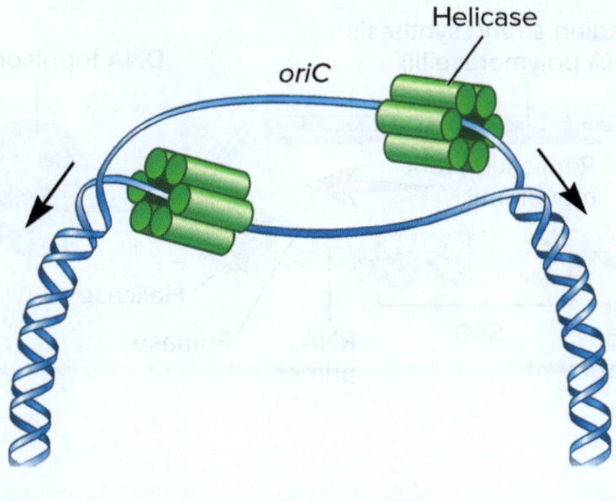

Figure 13.10 DNA Replication Initiation. Following denaturation of the origin region, helicases bind to each strand to open the template.

Within the replisome, enzymes called **DNA polymerases** catalyze DNA synthesis. DNA synthesis occurs in the 5′ to 3′ direction, and the nucleotide to be added is a deoxyribonucleoside triphosphate (dNTP). Deoxynucleotides are linked by phosphodiester bonds formed by a reaction between the hydroxyl group at the 3′ end of the growing DNA strand and the phosphate closest to the 5′ carbon (the α-phosphate, also called the 5′-phosphate) of the incoming deoxynucleotide (**figure 13.11**). The energy needed to form the phosphodiester bond is provided by release of the terminal two phosphates as pyrophosphate (PP_i) from the nucleotide that is added. The PP_i is subsequently hydrolyzed to two separate phosphates (P_i). Thus the deoxynucleoside triphosphates dATP, dTTP, dCTP, and dGTP serve as DNA polymerase substrates while deoxynucleoside monophosphates (dNMPs: dAMP, dTMP, dCMP, dGMP) are incorporated into the growing chain.

Like all enzymes, DNA polymerase has a specific substrate requirement. It needs both a template (the parental DNA strand) and a 3′-OH group from the growing nucleic acid chain to add nucleotides onto. DNA polymerase cannot copy DNA with only the template. To get DNA polymerase started, a nucleotide with a free 3′-OH must be present in the growing strand. To accomplish this, an enzyme termed **primase** synthesizes a short 10-base RNA molecule complementary to the template. This short **primer** is made from RNA rather than DNA because RNA polymerases (such as primase) can initiate RNA synthesis without an existing 3′-OH. The multiprotein complex that includes helicase and primase is termed the *primosome.*

E. coli has five different DNA polymerases (DNA polymerase I–V). DNA polymerase III plays the major role in replication, although it is assisted by DNA polymerase I. Each replisome has two DNA polymerases, and each core polymerase binds one strand of DNA. The two DNA polymerases sandwich the primosome between them, and are oriented such that one replicates the strand that passes through the helicase and the other replicates the strand that passes over it (**figure 13.12**).

A donut-shaped protein termed a clamp attaches to each polymerase and stabilizes the enzyme on the DNA template. Finally a clamp loader complex (CLC) mediates the polymerase attachments to the primosome.

Two additional types of proteins found in the replisome are single-stranded DNA binding protein and topoisomerases (**figure 13.13**). **Single-stranded DNA binding proteins (SSBs),** as their name implies, coat single-stranded DNA to protect it from damage (e.g., enzymatic or chemical modification). **Topoisomerases** relieve the twist generated by the rapid unwinding of the double helix. This is important because rapid unwinding leads to excessive supercoiling in the helix ahead of the replication fork, which can impede helicase if not removed. Topoisomerases transiently break and reseal one or both strands without altering the nucleotide sequence. These are critical enzymes to prevent tangled chromosomes.

As noted, DNA polymerases synthesize DNA in the 5′ to 3′ direction since double-stranded DNA is antiparallel, which means that DNA polymerase reads its template strand in a 3′ to 5′ direction. Therefore one of the DNA polymerase core enzymes moves in the same direction as the replication fork and synthesizes DNA continuously as it is denatured ahead of the replisome. This strand is called the **leading strand** (figure 13.13). The other

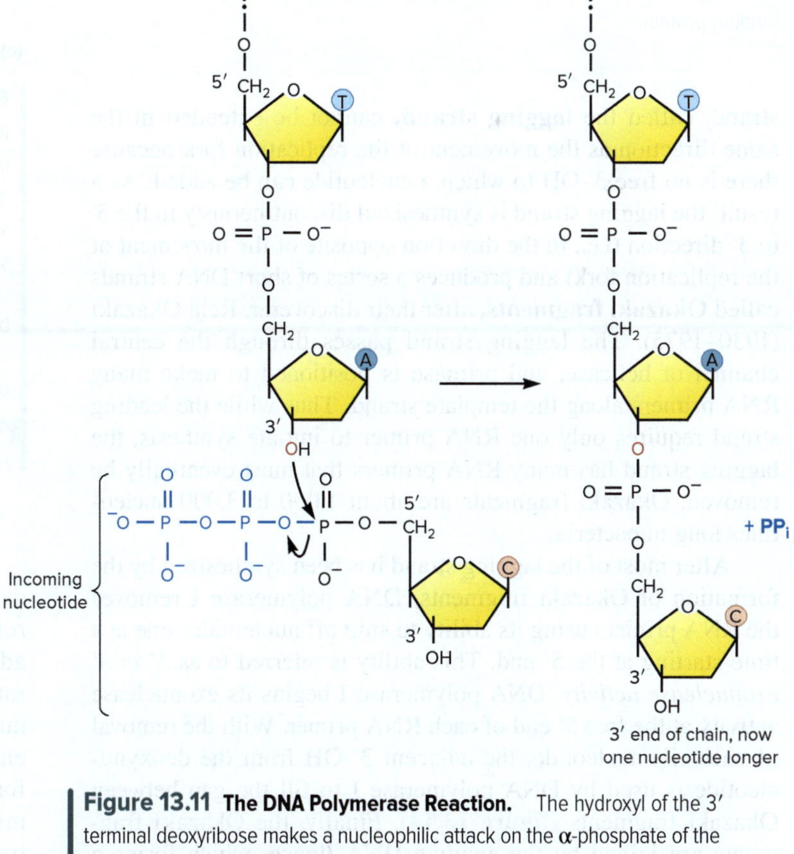

Figure 13.11 The DNA Polymerase Reaction. The hydroxyl of the 3′ terminal deoxyribose makes a nucleophilic attack on the α-phosphate of the nucleotide substrate (in this example, adenosine attacks cytidine triphosphate). Released pyrophosphate is shown in blue.

MICRO INQUIRY *What provides the energy to fuel this reaction?*

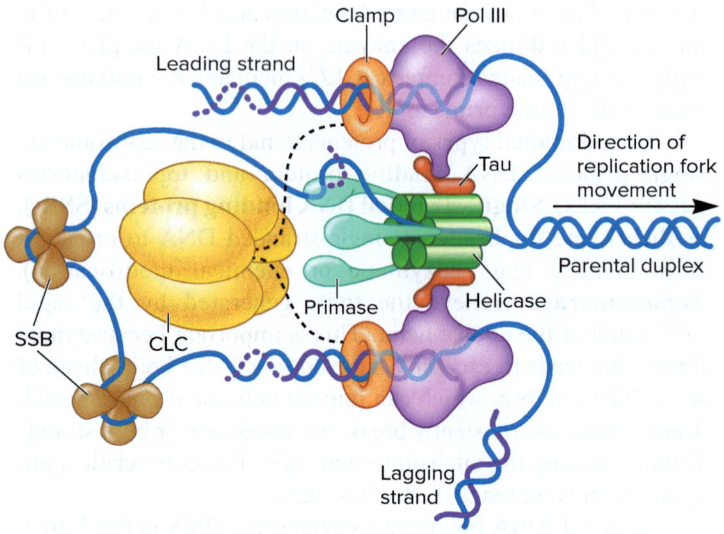

Figure 13.12 Other Replisome Proteins. The parental DNA strands (right) are unwound by helicase (dark green), with the leading strand passing over the enzyme and the lagging strand passing through it. The primosome (light and dark green) is sandwiched between two DNA polymerases. Tau coordinates replication by interacting with DNA polymerase and helicase. In this figure, the clamp loader complex (CLC) is displaced from its normal position; dotted lines indicate where it contacts the two DNA polymerases from atop the primosome. Pol III, DNA polymerase III; SSB, single-stranded binding protein.

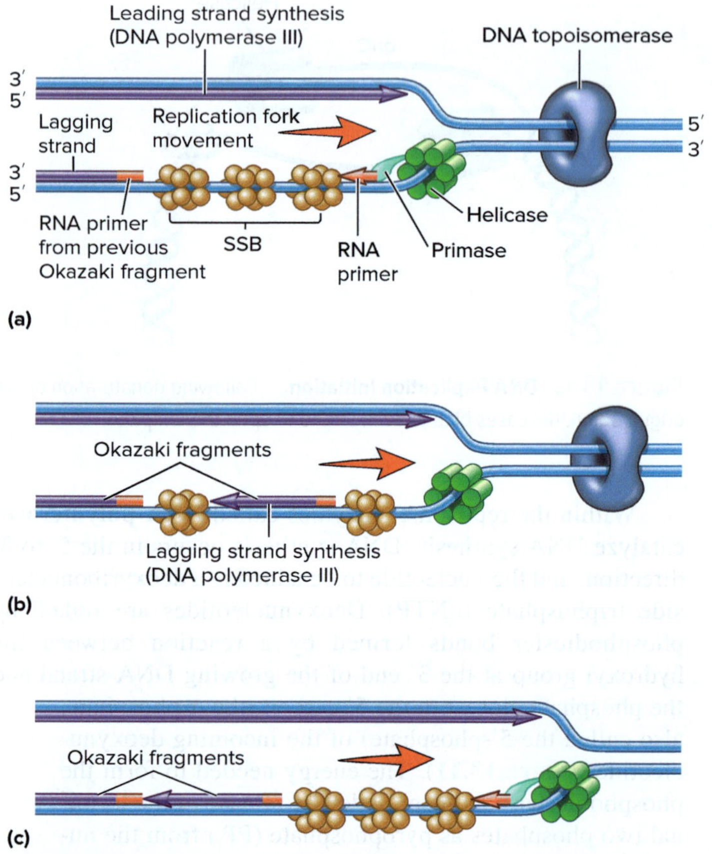

Figure 13.13 Events at the Replication Fork. The three panels illustrate the progress of a single replication fork (from left to right as indicated by red arrow) and replication of both the leading and lagging strands. DNA polymerase III holoenzyme is not shown. Helicase separates the two strands of parental DNA, which are kept apart by single-stranded DNA binding proteins (SSB). The RNA primer is synthesized by primase. DNA topoisomerase eases the strain introduced by helicase activity. The lagging strand is synthesized in short fragments called Okazaki fragments. A new RNA primer is required for the synthesis of each Okazaki fragment.

MICRO INQUIRY *What is the difference between helicase and topoisomerase?*

strand, called the **lagging strand,** cannot be extended in the same direction as the movement of the replication fork because there is no free 3′-OH to which a nucleotide can be added. As a result, the lagging strand is synthesized discontinuously in the 5′ to 3′ direction (i.e., in the direction opposite of the movement of the replication fork) and produces a series of short DNA strands called **Okazaki fragments,** after their discoverer, Reiji Okazaki (1930–1975). The lagging strand passes through the central channel of helicase, and primase is positioned to make many RNA primers along the template strand. Thus while the leading strand requires only one RNA primer to initiate synthesis, the lagging strand has many RNA primers that must eventually be removed. Okazaki fragments are about 1,000 to 3,000 nucleotides long in bacteria.

After most of the lagging strand has been synthesized by the formation of Okazaki fragments, DNA polymerase I removes the RNA primers using its ability to snip off nucleotides one at a time starting at the 5′ end. This ability is referred to as *5′ to 3′ exonuclease activity.* DNA polymerase I begins its exonuclease activity at the free 5′ end of each RNA primer. With the removal of each ribonucleotide, the adjacent 3′-OH from the deoxynucleotide is used by DNA polymerase I to fill the gap between Okazaki fragments (**figure 13.14**). Finally, the Okazaki fragments are joined by the enzyme **DNA ligase,** which forms a phosphodiester bond between the 3′-OH of the growing strand and the 5′-phosphate of an Okazaki fragment (**figure 13.15**).

Amazingly, DNA polymerase III has an additional critically important function: **proofreading.** Proofreading is the removal of a mismatched base immediately after it has been added; its removal must occur before the next base is incorporated. The ε subunit of DNA polymerase III has 3′ to 5′ exonuclease activity. This activity enables the polymerase core enzyme to check each newly incorporated base to see that it forms stable hydrogen bonds with the template. In this way, mismatched bases can be detected. If the wrong base has been mistakenly added, the exonuclease activity removes it, but only as long as it is still at the 3′ end of the growing strand (**figure 13.16**). Once removed, holoenzyme backs up and

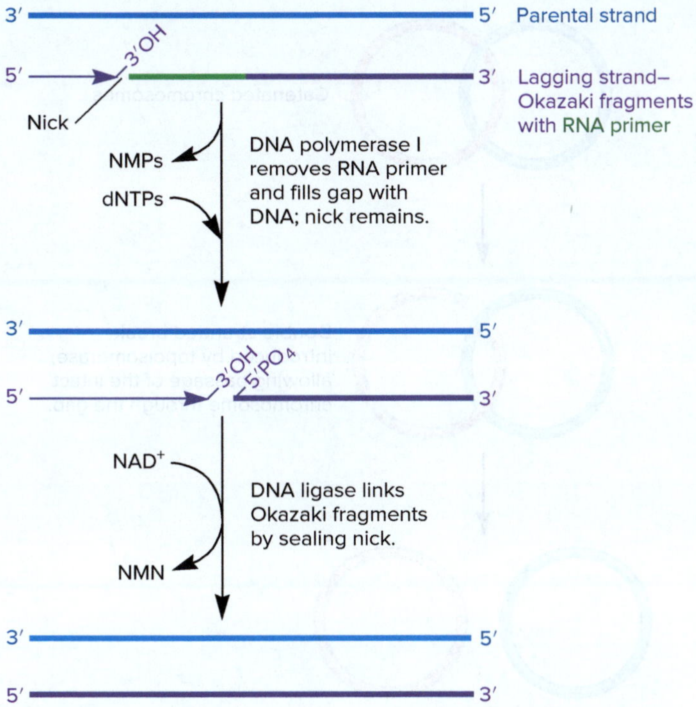

Figure 13.14 DNA Polymerase I Completes Lagging Strand Synthesis. dNTPs, deoxynucleoside triphosphates; NAD^+, nicotinamide adenine dinucleotide; NMN, nicotinamide mononucleotide; NMPs, nucleoside monophosphates.

adds the proper nucleotide in its place. DNA proofreading is not 100% efficient, and as discussed in chapter 16, the mismatch repair system is the cell's second line of defense against the potential harm caused by incorporation of the incorrect nucleotide. *Proofreading Function of DNA Polymerase*

We present replication as a series of discrete steps, but in the cell these events occur quickly and simultaneously on both the leading and lagging strands. Lagging strand synthesis is particularly amazing because of the gymnastic feats performed by the holoenzyme. It must discard old clamps, load new clamps, and tether the template to the core enzyme with each new round of Okazaki fragment synthesis. All of this occurs as DNA polymerase III is synthesizing DNA. *DNA Replication; Structural Basis of DNA Replication*

Termination of Replication Requires Steps to Separate Daughter Chromosomes

In *E. coli,* DNA replication stops when the replisome reaches a termination site (*ter*) on the DNA. A protein called Tus binds to the *ter* sites and halts progression of the forks. In many other bacteria, replication stops spontaneously when the forks meet. Regardless of how fork movement is stopped, there are two problems that must be solved by the replisome. The first is the formation of interlocked chromosomes called **catenanes** (**figure 13.17*a***). Catenanes are produced when topoisomerases break and rejoin DNA strands to ease supercoiling ahead of the replication fork. To resolve this problem, the two daughter DNA molecules are separated by the action of other topoisomerases that break both strands of one molecule, pass the other DNA molecule through the break, and then rejoin the strands. The second problem is a dimerized chromosome—two chromosomes joined together to form a single chromosome twice as long (figure 13.17*b*). Dimerized chromosomes result from DNA recombination that sometimes occurs between two daughter molecules during DNA replication. To separate these dimers, recombinase enzymes (e.g., XerCD in *E. coli*) catalyze an intramolecular crossover that separates the two chromosomes.

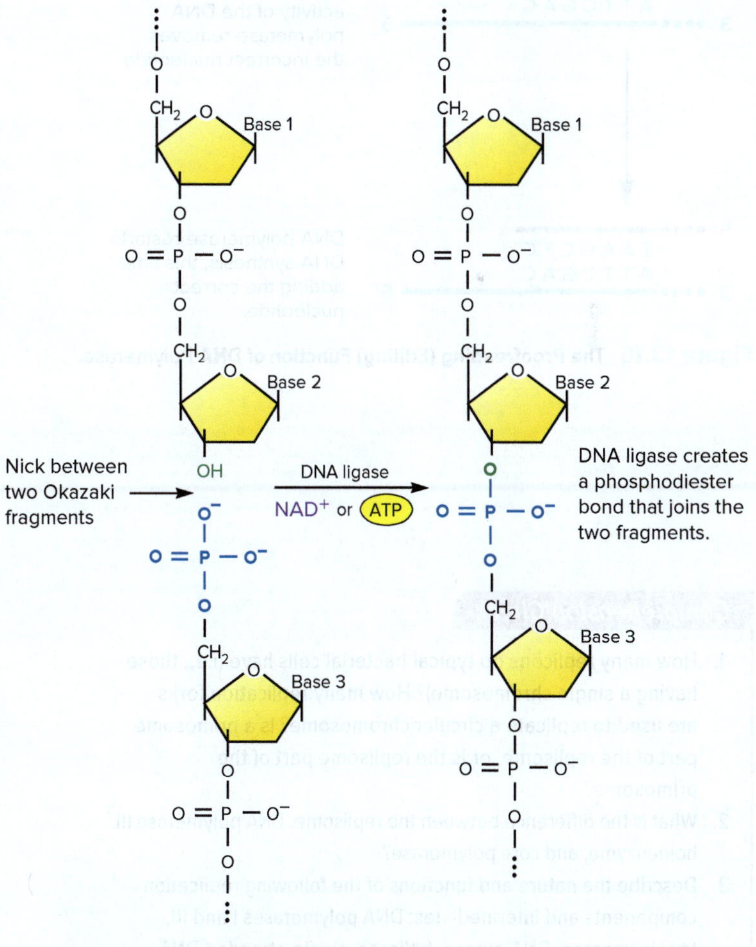

Figure 13.15 The DNA Ligase Reaction. The chemical energy in NAD^+ or ATP fuels this reaction.

MICRO INQUIRY *Why can't DNA polymerase I perform the ligase reaction?*

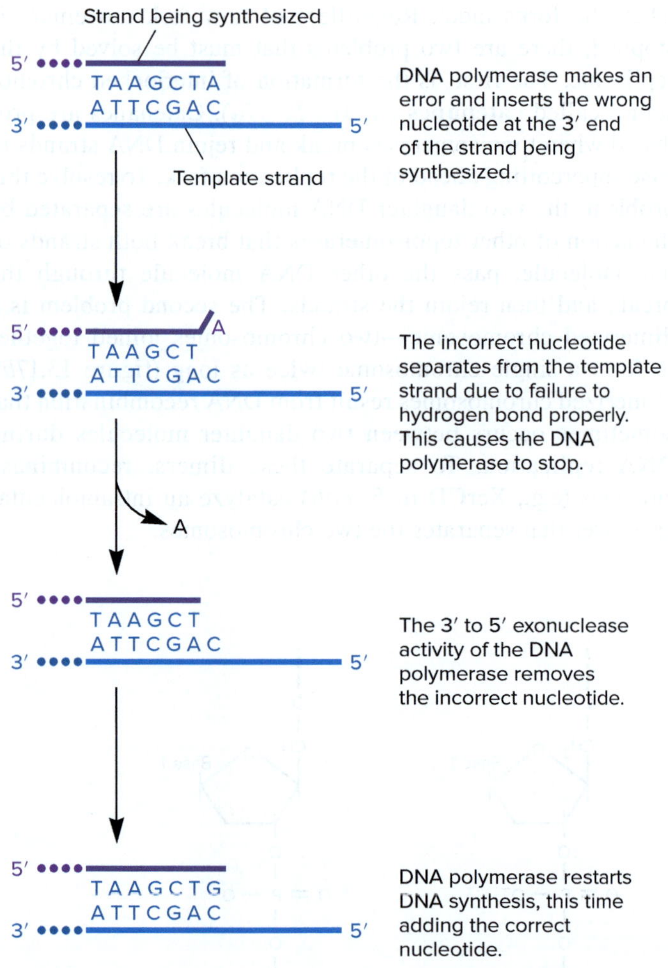

Figure 13.16 The Proofreading (Editing) Function of DNA Polymerase.

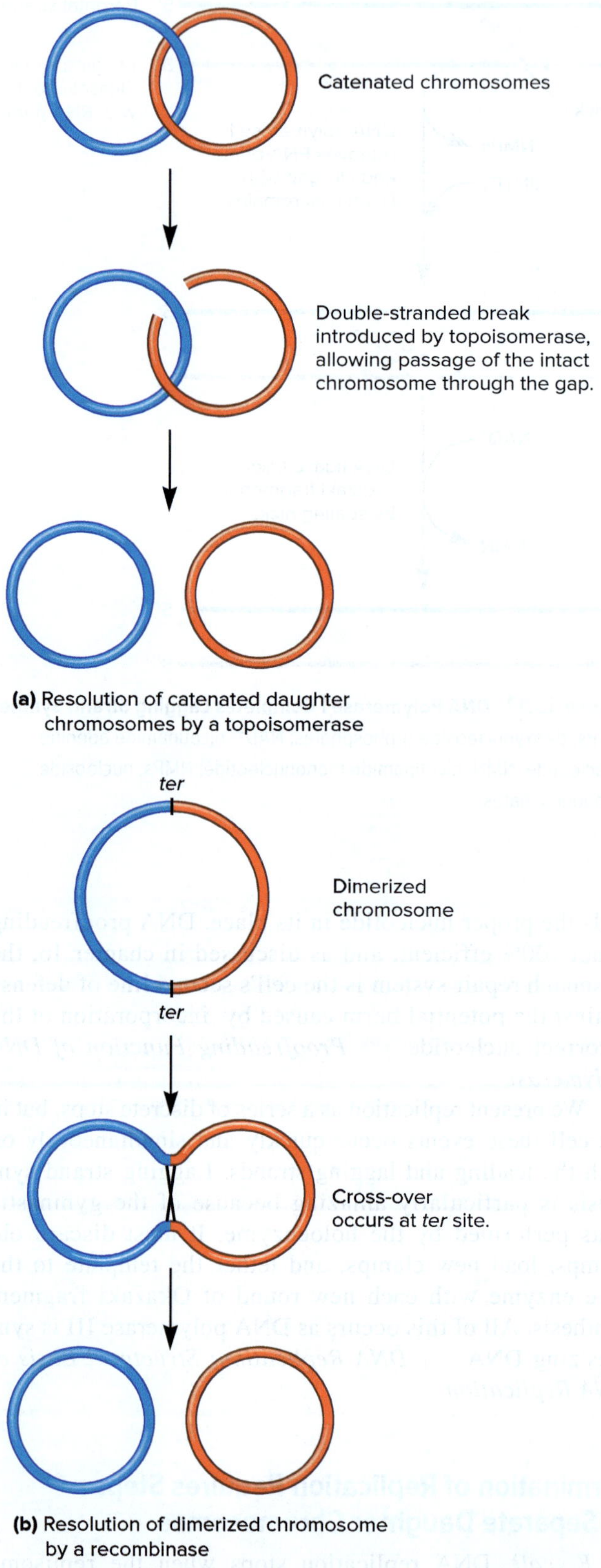

Figure 13.17 Resolving Catenated and Dimerized Chromosomes. (a) Catenated chromosomes are resolved by topoisomerases. (b) Dimerized chromosomes are resolved by recombinase enzymes. These enzymes catalyze a cross-over that separates the two chromosomes.

Comprehension Check

1. How many replicons do typical bacterial cells have (i.e., those having a single chromosome)? How many replication forks are used to replicate a circular chromosome? Is a primosome part of the replisome, or is the replisome part of the primosome?
2. What is the difference between the replisome, DNA polymerase III holoenzyme, and core polymerase?
3. Describe the nature and functions of the following replication components and intermediates: DNA polymerases I and III, topoisomerase, DNA gyrase, helicase, single-stranded DNA binding proteins, Okazaki fragment, DNA ligase, leading strand, lagging strand, and primase.
4. Outline the steps involved in DNA synthesis at the replication fork. How do DNA polymerases correct their mistakes?

13.4 Bacterial Genes Consist of Coding Regions and Other Sequences Important for Gene Function

After reading this section, you should be able to:

a. Draw a typical bacterial protein-coding gene; label the important portions of the gene using the conventions for numbering base pairs

b. Draw typical tRNA- and rRNA-coding genes

DNA replication allows genetic information to be passed from one generation to the next. But how is the information used? To answer that question, we must first look at how genetic information is organized. The basic unit of genetic information is the gene. Genes have been regarded in several ways. At first, it was thought that a gene contained information for the synthesis of one enzyme—the one gene–one enzyme hypothesis. This was modified to the one gene–one polypeptide hypothesis because of the existence of enzymes and other proteins composed of two or more different polypeptide chains (subunits) coded for by separate genes. A segment of DNA that encodes a single polypeptide is sometimes termed a **cistron.** However, not all genes encode proteins; some code instead for **ribosomal RNA (rRNA)** and **transfer RNA (tRNA)** (**figure 13.18**). Synthesis of RNA from a DNA template is called **transcription,** and the RNA product has a sequence complementary to the DNA template directing its synthesis. Transcription generates three major kinds of RNA: tRNA, mRNA, and rRNA. tRNA carries amino acids during protein synthesis, and rRNA molecules are components of ribosomes. **Messenger RNA (mRNA)** bears the message for protein synthesis. Thus a **gene** might be defined as a polynucleotide sequence that codes for one or more functional products (i.e., a polypeptide, tRNA, or rRNA). In this section, we consider the structure of each of these three types of genes.

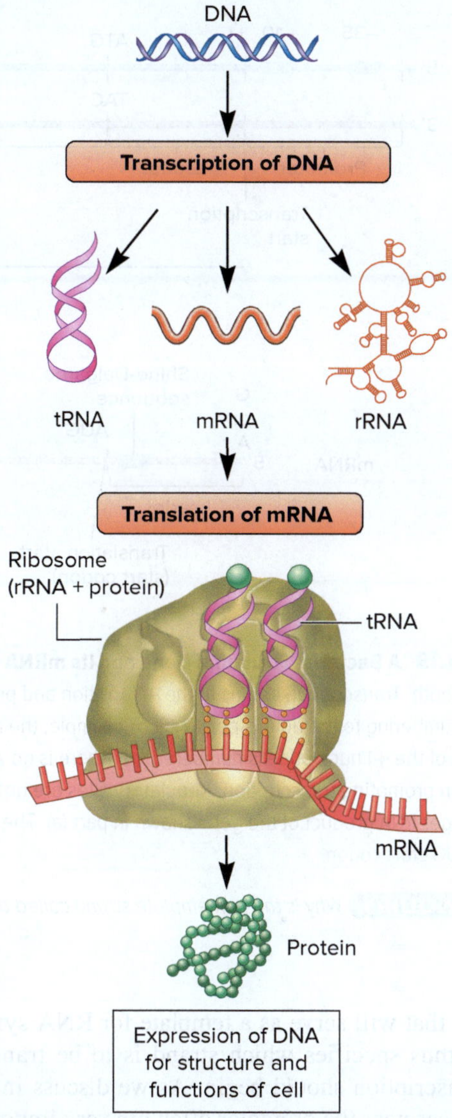

Figure 13.18 Transcription Yields Three Major Types of RNA Molecules. Messenger RNA (mRNA) molecules arise from transcription of protein-coding genes. They are translated into protein with the aid of two other types of RNA: Transfer RNA (tRNA) molecules carry amino acids to the ribosome during translation; ribosomal RNA (rRNA) molecules have several functions, including catalyzing peptide bond formation. The sizes of RNA are enlarged to show structural details.

Protein-Coding Genes

Most of the genes found in bacterial genomes encode proteins; these are called **structural genes.** However, DNA does not serve directly as the template for protein synthesis. Rather, the genetic information in the gene is transcribed (i.e., copied) to give rise to an mRNA, which is translated (i.e., decoded) (section 13.7) into a protein (figure 13.18). For this to occur, protein-coding genes must contain signals that indicate where transcription should start and stop, and signals in the resulting mRNA that indicate where translation should start and stop. As we describe in more detail in section 13.5, during transcription only one strand of a gene directs mRNA synthesis. This strand is called the **template strand,** and the complementary DNA strand is known as the **sense strand** (**figure 13.19**). Messenger RNA is synthesized from the 5′ to the 3′ end in a manner similar to DNA synthesis. Because the template strand and the mRNA are complementary to one another, the sequence of the mRNA is the same as the sense strand, with uracil substituted for thymine. In other words, a human reading the sense strand can read the nucleotide sequence of the gene in the correct order, even though this is not the strand used as the template.

An important site called the **promoter** is located at the start of the gene. The promoter is the binding site for RNA polymerase, the enzyme that synthesizes RNA. The promoter is neither transcribed nor translated; it functions strictly to orient RNA polymerase so it is a specific distance from the first DNA

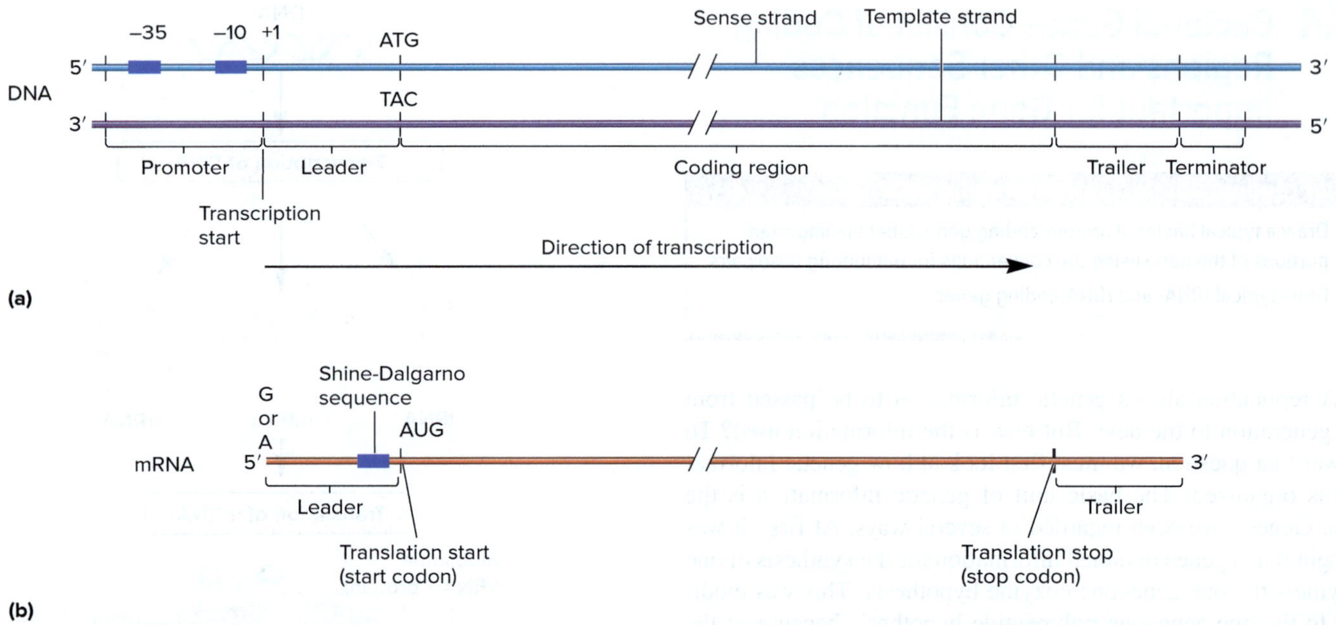

Figure 13.19 A Bacterial Structural Gene and Its mRNA Product. (a) The organization of a typical structural gene in bacteria. Some genes lack leaders or trailers, or both. Transcription begins at the +1 position and proceeds to the right, as shown. The numbering of nucleotides to the left of this spot is negative, while the numbering to the right is positive. For example, the nucleotide that is immediately to the left of the +1 nucleotide is numbered −1, and the nucleotide to the right of the +1 nucleotide is numbered +2. There is no zero nucleotide. In many bacterial promoters, sequence elements at the −35 and −10 regions play a key role in promoting transcription. Regulatory sites are not shown but are usually upstream of the coding region and may overlap with the promoter. (b) Messenger RNA product of the gene shown in part (a). The first nucleotide incorporated into mRNA is usually GMP or AMP. Translation of the mRNA begins with the AUG start codon.

MICRO INQUIRY *Why is the nontemplate strand called the "sense strand"?*

nucleotide that will serve as a template for RNA synthesis. The promoter thus specifies which strand is to be transcribed and where transcription should begin. As we discuss in chapter 14, the sequences near the promoter often are very important in regulating when and at what rate a gene is transcribed. ▶| *Regulation of transcription initiation saves considerable energy and materials (section 14.2)*

The transcription start site (labeled +1 in figure 13.19) represents the first nucleotide in the mRNA synthesized from the gene. However, the initially transcribed portion of the gene does not code for amino acids. Instead, it is a **leader** that is transcribed into mRNA but is not translated into amino acids. In bacteria, the leader includes a region called the **Shine-Dalgarno sequence,** which is important in the initiation of translation. The leader sometimes is also involved in regulation of transcription and translation. ▶| *Attenuation and riboswitches stop transcription prematurely (section 14.3); RNA secondary structures control translation (section 14.4)*

Immediately next to (and downstream of) the leader is the most important part of the gene, the **coding region** (figure 13.19). The coding region typically begins with the template DNA sequence 3′-TAC-5′. This is transcribed into the start codon, 5′-AUG-3′, which codes for the first amino acid of the polypeptide encoded by the gene. The remainder of the coding region specifies the sequence of amino acids for the rest of the protein. The coding region ends with a sequence that signals the end of the protein and stops the ribosome during translation. The stop signal is immediately followed by the **trailer,** which is transcribed but not translated. The trailer contains sequences that prepare RNA polymerase for release from the template strand. Indeed, just beyond the trailer (and sometimes slightly overlapping it) is the **terminator.** The terminator is a sequence that signals RNA polymerase to stop transcription.

tRNA and rRNA Genes

Actively growing cells need a ready supply of tRNA and rRNA molecules for protein synthesis. To ensure this, bacterial cells often have more than one gene for each of these molecules. Furthermore, it is important that the number of each tRNA or rRNA relative to other tRNAs or rRNAs be controlled. This is accomplished in part by having several tRNA or rRNA genes transcribed together from a single promoter.

In bacteria, genes for tRNA consist of a promoter, leader, tRNA coding region, and trailer. This is the same as for protein-coding genes, except that the coding region is not translated. When more than one tRNA is transcribed from the promoter, the coding regions are separated by short spacer sequences (**figure 13.20***a*). Whether the gene encodes a single tRNA or multiple tRNAs, the initial transcript must be processed to remove the noncoding sequences (i.e., leader, trailer, and spacers, if present). This is called *posttranscriptional modification,* and it is accomplished by ribonucleases—enzymes (and in some cases ribozymes) that cut RNA. ◀ *Ribozymes: catalytic RNA molecules (section 10.6)*

Bacterial cells usually also contain more than one rRNA gene. Each gene has a promoter, trailer, and terminator (figure 13.20*b*). As seen for tRNA genes, the transcript from an rRNA gene is a single, large precursor molecule cut by ribonucleases to yield the final rRNA products. Interestingly, in many bacteria, the trailer regions and the spacers often contain tRNA genes. Thus the precursor rRNA encodes for both tRNA and rRNA.

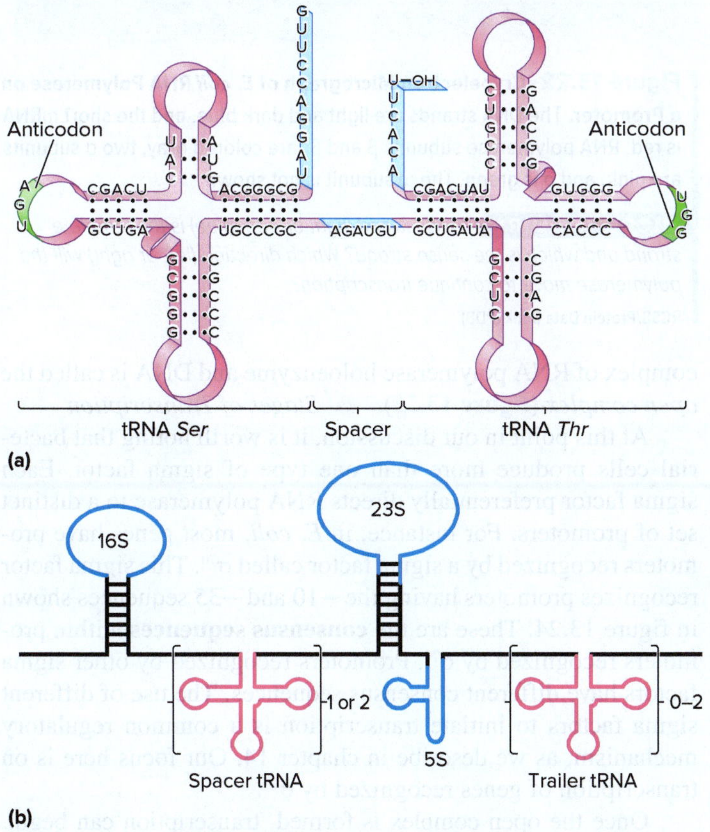

Figure 13.20 tRNA and rRNA Genes. (a) A tRNA precursor from *E. coli* that contains two tRNA molecules. The spacer and extra nucleotides at both ends are removed during processing. (b) The *E. coli* rRNA gene codes for a large transcript that is cleaved into three rRNAs and one to three tRNAs. The 16S, 23S, and 5S rRNA segments are represented by blue lines, and tRNA sequences are placed in brackets. The seven copies of this gene vary in the number and kind of tRNA sequences.

Comprehension Check

1. The coding region of a gene is said to be "downstream" from the leader. Conversely, the leader is said to be "upstream" of the coding region. For each of the following portions of a gene, indicate whether it is downstream or upstream of the coding region: promoter, +1 nucleotide, trailer, terminator. Which portions of a gene are transcribed but not translated?
2. Which strand of a gene has sequences that correspond to the start (i.e., ATG rather than AUG) and stop codons in the mRNA product of the gene?
3. Briefly discuss the general organization of tRNA and rRNA genes. How does their expression differ from that of protein-coding genes with respect to posttranscriptional modification of the gene product?

13.5 Transcription in Bacteria

After reading this section, you should be able to:

- **a.** Illustrate the organization of bacterial genes in a typical operon
- **b.** Describe the structure of a typical bacterial RNA polymerase holoenzyme
- **c.** Outline the events that occur during the three phases of transcription
- **d.** Discuss the role of bacterial promoters and sigma factors in transcription initiation
- **e.** Distinguish factor-independent termination of transcription from rho-dependent termination of transcription

So far we have discussed genes as discrete units bounded by a promoter and terminator. However, not all bacterial genes are organized in this way. Bacterial genes encoding proteins involved in a related process (e.g., encoding enzymes for synthesis of an amino acid) are often located close to each other and are transcribed from a single promoter. Multiple genes controlled by a single promoter comprise an **operon.** Transcription of an operon yields an mRNA consisting of a leader followed by one coding region, which is separated from the second coding region by a spacer, and so on, with the final sequence of nucleotides being the trailer. Such mRNAs are said to be **polycistronic mRNAs** (**figure 13.21***a*). Each coding region in the polycistronic mRNA is defined by a start and stop codon. Thus each coding region is translated separately to give rise to a separate polypeptide. Many archaeal genes are also organized in operons. However, operons are rare in eukaryotes. Instead, their mRNAs are usually **monocistronic** (figure 13.21*b*), containing information for a single gene. ▶| *Transcription in eukaryotes (section 15.3)*

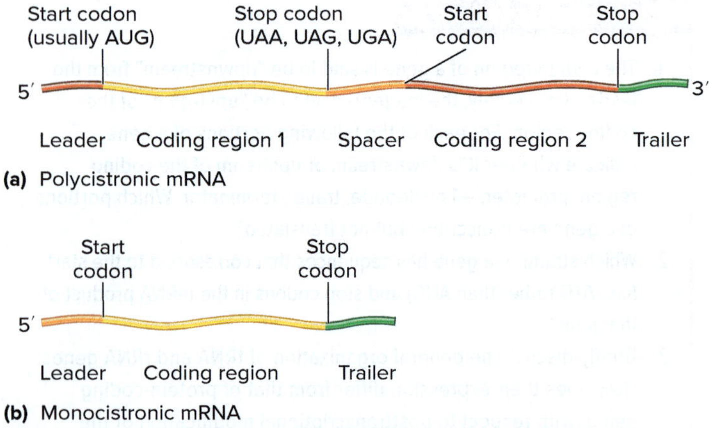

Figure 13.21 Polycistronic and Monocistronic mRNAs.
(a) Polycistronic mRNAs are commonly observed in bacteria and archaea, but not eukaryotes. (b) Monocistronic mRNAs occur in all three domains.

Bacterial RNA Polymerases Consist of Five Different Proteins

RNA is synthesized by enzymes called **RNA polymerases.** Most bacterial RNA polymerases contain five types of polypeptide chains: α, β, β', ω, and σ (**figure 13.22**). The **RNA polymerase core enzyme** is composed of five polypeptides (two α subunits, β, β', and ω) and catalyzes RNA synthesis. The **sigma factor** (σ) has no catalytic activity but instead functions as a transcription factor by helping the core enzyme recognize the promoter. When sigma is bound to the core enzyme, the six-subunit complex is termed **RNA polymerase holoenzyme** (figure 13.22). Only holoenzyme can begin transcription, but the core enzyme completes RNA synthesis once it has been initiated.

Stages of Transcription: Initiation, Elongation, and Termination

Transcription involves three separate processes: initiation, elongation, and termination, which together are referred to as the transcription cycle (**figure 13.23**). The transcription factor sigma is critical to the initiation process. As part of the RNA polymerase holoenzyme, it helps position the core enzyme at the promoter. Bacterial promoters have several characteristic features. Two of particular note are a sequence of six bases (often TTGACA) about 35 base pairs before (upstream of) the transcription starting point and a TATAAT sequence, usually about 10 base pairs upstream of the transcriptional start site (**figure 13.24**; also figure 13.19). These regions are called the -35 and -10 sites, respectively, because these are their approximate distances upstream of the first nucleotide to be transcribed (i.e., the +1 site). Sigma factor first recognizes the -35 sequence, stabilizing the holoenzyme on that region of the promoter. Sigma and the core enzyme undergo conformational changes that cause the DNA strands in the AT-rich -10 region to separate. Sigma then interacts with one of the strands, facilitating the interaction of RNA polymerase with the denatured DNA. The resulting complex of RNA polymerase holoenzyme and DNA is called the *open complex* (**figure 13.25**). *Stages of Transcription*

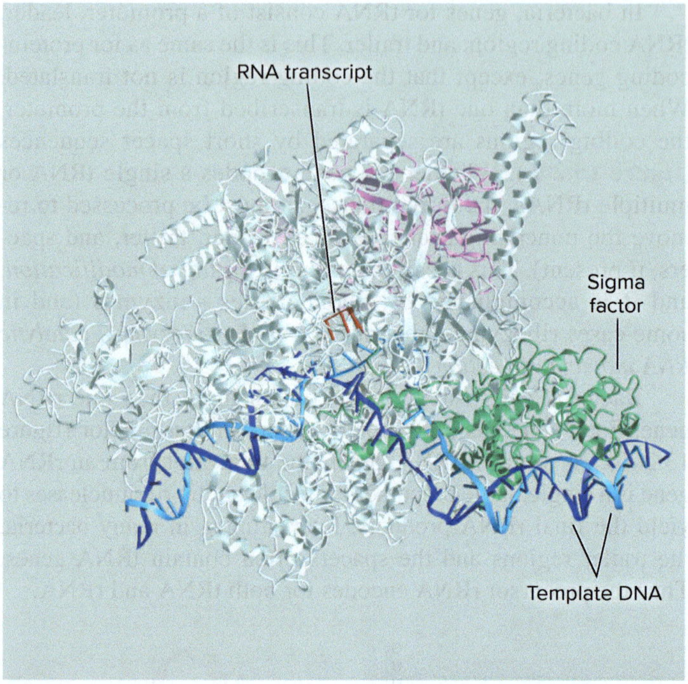

Figure 13.22 Cryoelectron Micrograph of *E. coli* RNA Polymerase on a Promoter. The DNA strands are light and dark blue, and the short mRNA is red. RNA polymerase subunits β and β' are colored gray, two α subunits are pink, and σ is green. The ω subunit is not shown.

MICRO INQUIRY *Which strand (light or dark blue) is the template strand and which is the sense strand? Which direction (left or right) will the polymerase move to continue transcription?*

RCSB/Protein Data Bank (PDB)

At this point in our discussion, it is worth noting that bacterial cells produce more than one type of sigma factor. Each sigma factor preferentially directs RNA polymerase to a distinct set of promoters. For instance, in *E. coli,* most genes have promoters recognized by a sigma factor called σ^{70}. This sigma factor recognizes promoters having the -10 and -35 sequences shown in figure 13.24. These are the **consensus sequences** within promoters recognized by σ^{70}. Promoters recognized by other sigma factors have different consensus sequences. The use of different sigma factors to initiate transcription is a common regulatory mechanism, as we describe in chapter 14. Our focus here is on transcription of genes recognized by σ^{70}.

Once the open complex is formed, transcription can begin. Within the open complex is a region of denatured DNA of about 16 to 20 base pairs. This is the *transcription bubble,* and it moves with the RNA polymerase as it synthesizes mRNA during elongation (figure 13.23). Within the transcription bubble, a temporary RNA:DNA hybrid is formed as RNA is synthesized complementary to the DNA template. As RNA polymerase holoenzyme progresses along the DNA template, the sigma factor dissociates from the open complex and can direct another RNA polymerase core enzyme to initiate transcription (figure 13.23).

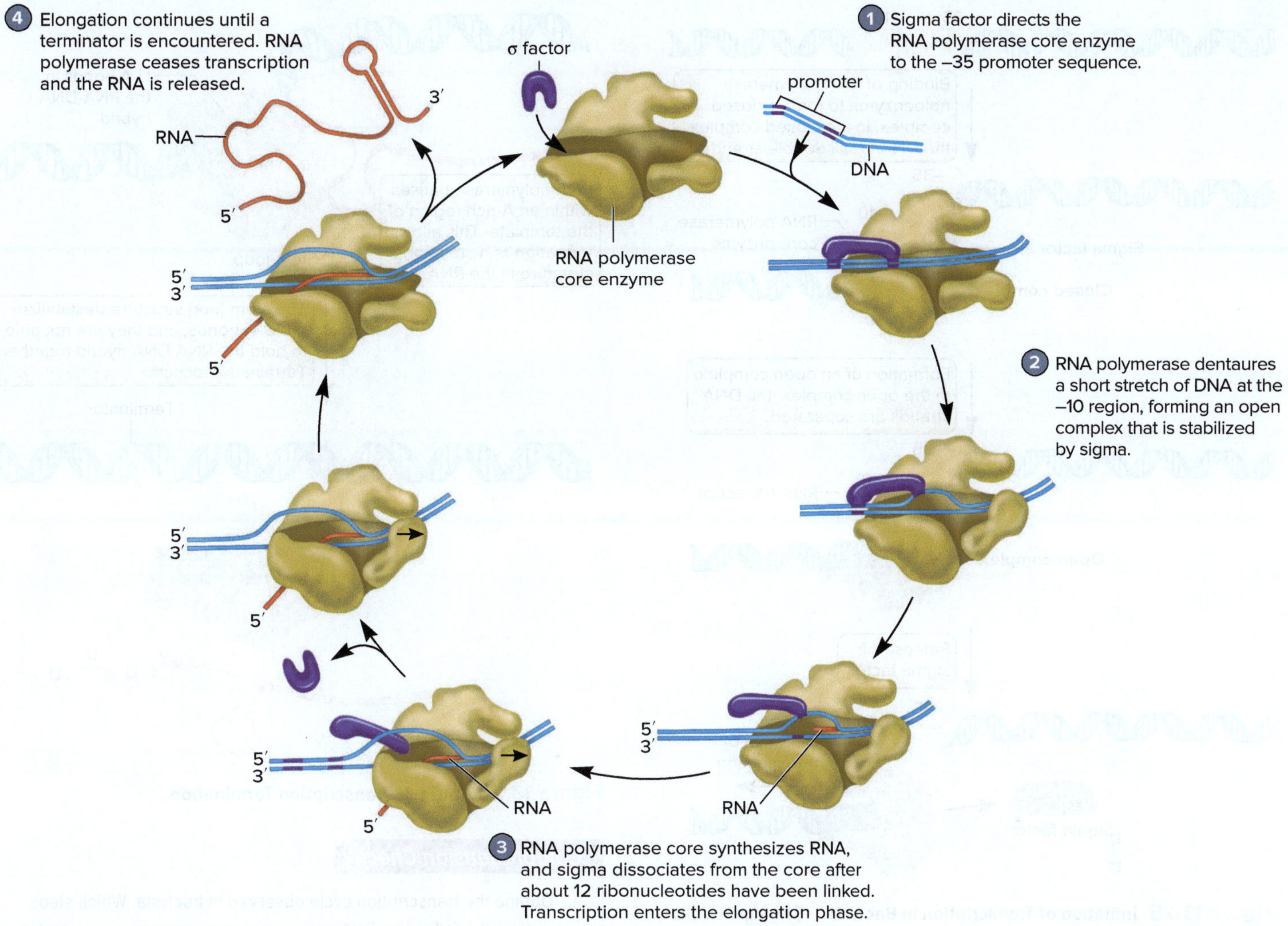

Figure 13.23 The Bacterial Transcription Cycle.

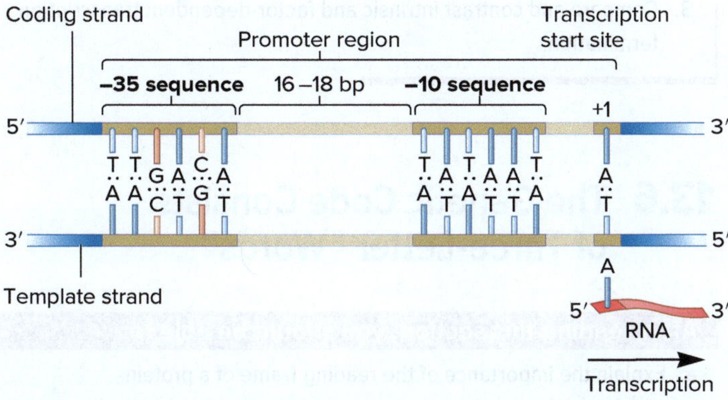

Figure 13.24 A σ^{70} Promoter. Many *E. coli* promoters are recognized by the sigma factor σ^{70}. Its promoters have a characteristic set of nucleotides centered at about –35 and –10. Shown are the consensus sequences for these two important sites.

MICRO INQUIRY *Are the –35 and –10 regions considered upstream or downstream of the +1 nucleotide?*

The reaction catalyzed by RNA polymerase is quite similar to that catalyzed by DNA polymerase (figure 13.11). ATP, GTP, CTP, and UTP are incorporated into the RNA complementary to the DNA template. Pyrophosphate fuels the reaction as ribonucleoside monophosphates are incorporated into the growing RNA chain. RNA synthesis proceeds in a 5′ to 3′ direction with new ribonucleotides added to the 3′ end of the growing chain, making the RNA complementary and antiparallel to the template DNA. As elongation continues, mRNA is released through the enzyme's exit tunnel, and the two strands of DNA behind the transcription bubble reform their double helical structure. As shown in figure 13.23, RNA polymerase is a remarkable enzyme capable of several activities, including separating the DNA strands, moving along the template, and synthesizing phosphodiester bonds in RNA. As is the case with DNA replication, DNA denaturation for transcription introduces supercoiling. A transcribing RNA polymerase requires a topoisomerase to relieve the torsion introduced into the chromosome.

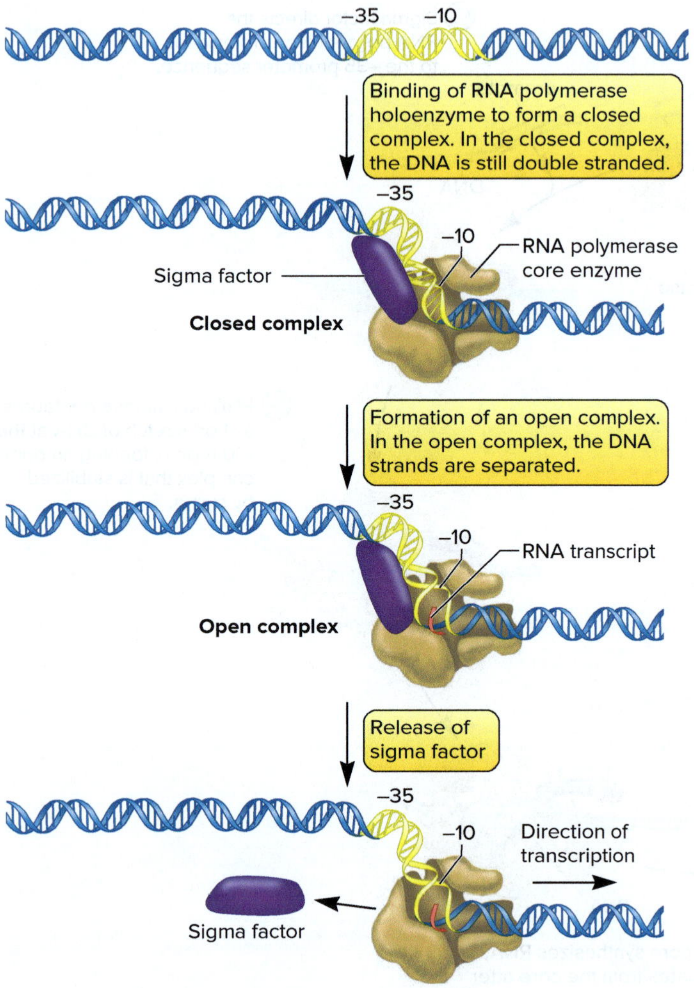

Figure 13.25 Initiation of Transcription in Bacteria. The sigma factor is responsible for positioning the core RNA polymerase at the promoter. Sigma recognizes two promoter regions, one centered at –35 and the other centered at –10. Once positioned properly, the DNA at the –10 region unwinds to form an open complex. The sigma factor dissociates from the core enzyme after transcription is initiated.

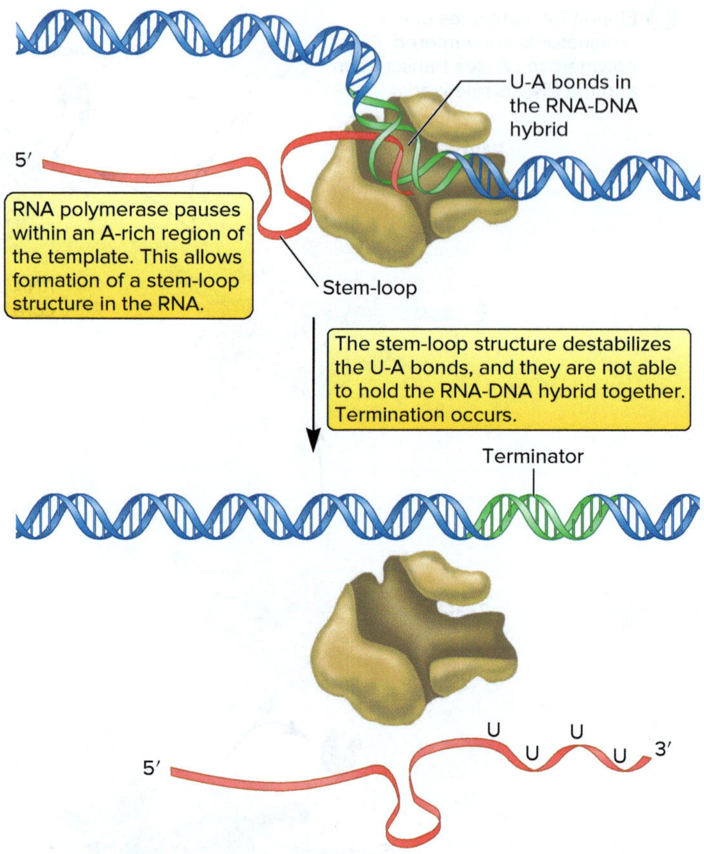

Figure 13.26 Intrinsic Transcription Termination.

An elongating RNA polymerase pauses briefly every 100–200 bases. During a pause, the template base slips in the active site, halting the enzyme. It quickly reengages to continue transcription. Pausing allows the enzyme to interact with sequence-specific regulatory signals described in more detail in Chapter 14.

Transcription termination occurs when the core RNA polymerase dissociates from the template DNA. There are three kinds of termination mechanisms. The first, intrinsic termination, involves signals within the RNA transcript itself and is summarized in **figure 13.26**. The other two types of termination are factor-dependent, meaning that they require the aid of proteins. Termination factor rho (ρ) requires ATP to power its movement along a transcript and physically pull RNA out of the open complex, as shown in **figure 13.27**. The third mechanism is discussed in section 13.8.

Comprehension Check

1. Outline the transcription cycle observed in bacteria. Which steps require the following: RNA polymerase core enzyme, sigma factor, and RNA polymerase holoenzyme?
2. What is a consensus sequence? What happens at the –35 and –10 sequences during transcription initiation?
3. Compare and contrast intrinsic and factor-dependent transcription termination.

13.6 The Genetic Code Consists of Three-Letter "Words"

After reading this section, you should be able to:

- **a.** Explain the importance of the reading frame of a protein-coding gene
- **b.** Describe the genetic code
- **c.** Explain how the wobble hypothesis enables organisms to encode fewer tRNA molecules
- **d.** List deviations from the universal genetic code identified in some microorganisms

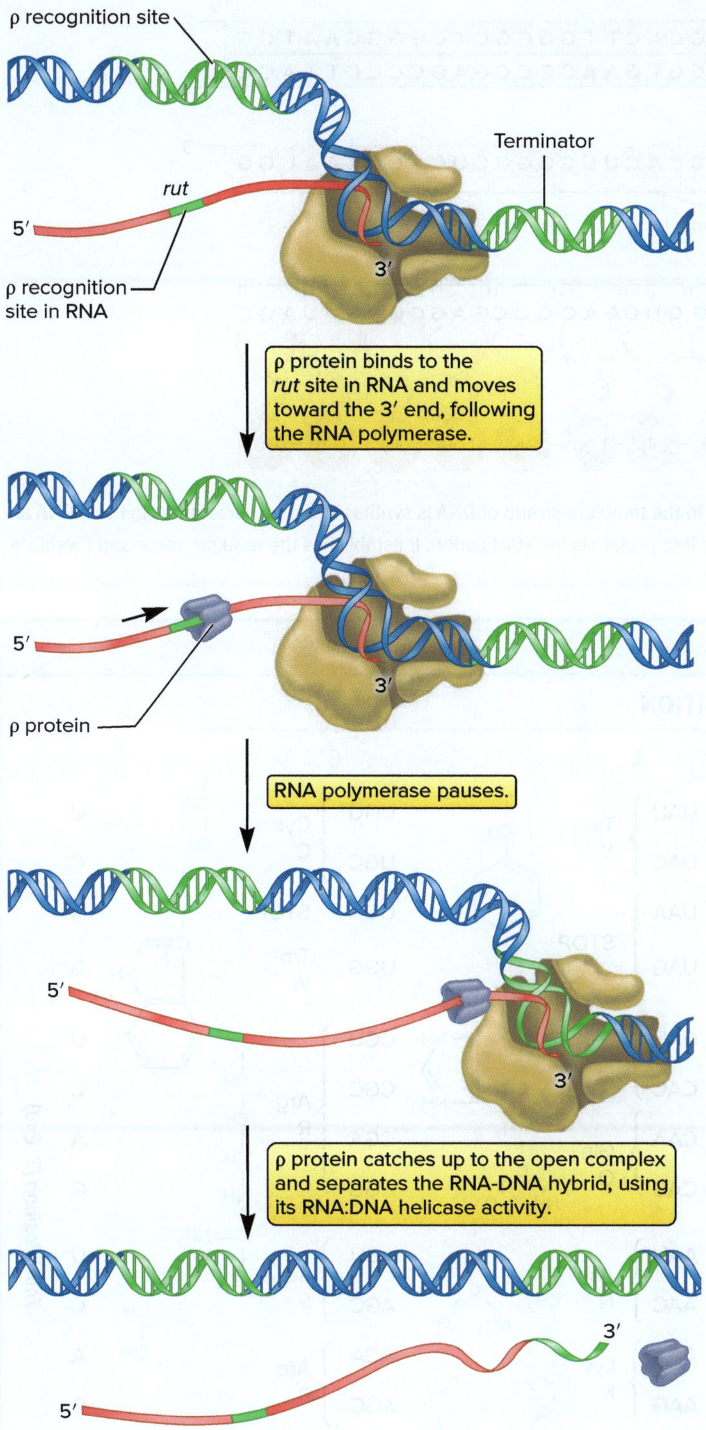

Figure 13.27 Rho-Factor (ρ)–Dependent Transcription Termination. The *rut* site stands for *rho-utilization* site.

The final step in expression of protein-coding genes is translation. Protein synthesis is called translation because the information encoded in the language of nucleic acids must be rewritten in the language of proteins. During translation, the sequence of nucleotides is read in discrete sets of three nucleotides, each set being a **codon.** Each codon specifies a single amino acid. The sequence of codons is read in only one way—the **reading frame** (**figure 13.28**)—to give rise to the amino acid sequence of a polypeptide. Deciphering the genetic code was one of the great achievements of the twentieth century. Here we examine the nature of the genetic code. Translation is considered in section 13.7.

The genetic code, presented in RNA form, is summarized in **table 13.1.** Close inspection of the code reveals several features related not only to the way cells use DNA to store information but also to why it is valuable for storing data, as described in the chapter opening story. One feature is that the code words (codons) are three letters (bases) long; thus one small word conveys a significant amount of information. Each codon is recognized by an anticodon present on a tRNA molecule. Another feature is that the code has punctuation. One codon, AUG, is almost always the first codon in the protein-coding portion of mRNA molecules. It is called the **start codon** because it serves as the start site for translation. Three other codons (UGA, UAG, and UAA) terminate translation and are called **stop** or **nonsense codons.** These codons do not encode an amino acid and therefore do not have a tRNA bearing their anticodon. Thus only 61 of the 64 codons in the code, the **sense codons,** direct amino acid incorporation into protein. Finally, the genetic code exhibits **code degeneracy** (also called *redundancy*); that is, there are up to six different codons for a given amino acid (table 13.1).

Despite the existence of 61 sense codons, there are fewer than 61 different tRNAs. It follows that not all codons have a corresponding tRNA. Cells can successfully translate mRNA using fewer tRNAs because loose pairing between the 5′ base in the anticodon and the 3′ base of the codon is tolerated. Thus as long as the first and second bases in the codon correctly base pair with an anticodon, the tRNA bearing the correct amino acid will bind to the mRNA during translation. This is evident on inspection of the code. Note that the codons for a particular amino acid most often differ at the third position (table 13.1). This somewhat loose base pairing is known as **wobble,** and it relieves cells of the need to synthesize so many tRNAs (**figure 13.29**). Wobble also lessens the impact of some mutations. *Mutations: heritable changes in a genome (section 16.1)*

The description of the genetic code just provided is of the universal genetic code. There are exceptions to the code, although they are limited to organelles and microbes with reduced genomes. The first exceptions discovered were stop codons that encode one of the 20 amino acids. For instance, mycoplasma bacteria use the stop codon UGA to code for glutamine. More dramatic deviations from the code have also been discovered. Members of all three domains of life encode proteins containing the amino acid selenocysteine, the twenty-first amino acid (**figure 13.30**). Most selenocysteine-containing enzymes catalyze redox reactions. Pyrrolysine, the twenty-second amino acid, can be found in the proteins from multiple archaea, including methyltransferases (figure 13.30). Selenocysteine is inserted at certain UGA codons, whereas pyrrolysine is inserted at UAG codons.

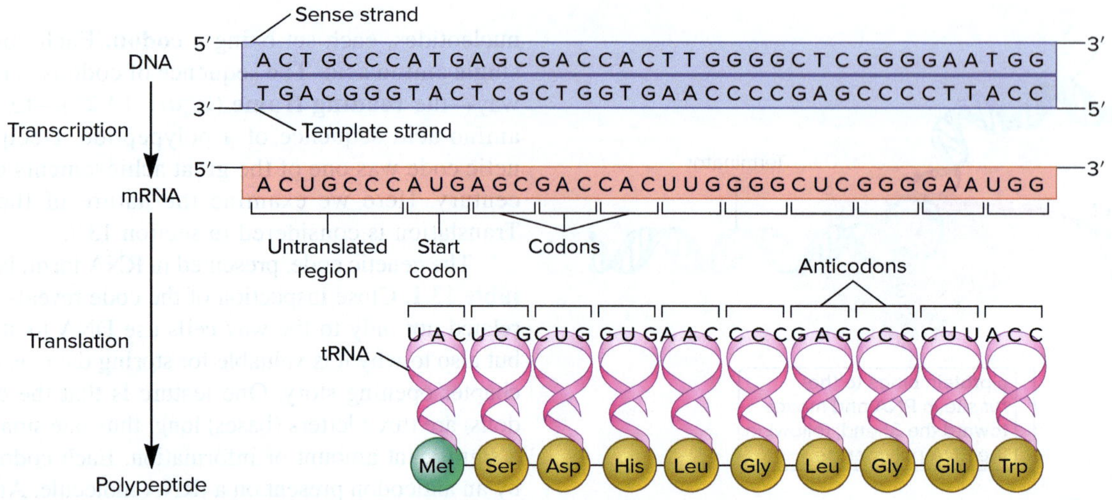

Figure 13.28 Reading Frame. During transcription, an mRNA complementary to the template strand of DNA is synthesized. The nucleotides in the mRNA are organized into groups of three, each group being a codon. The first codon translated into protein is the start codon. It establishes the reading frame and therefore the sequence of amino acids in the polypeptide chain that is made from the mRNA.

Table 13.1 The Genetic Code[1]

First Position (5′ End)[2]	SECOND POSITION: U		C		A		G		Third Position (3′ End)
U	UUU	Phe F	UCU	Ser S	UAU	Tyr Y	UGU	Cys C	U
	UUC	Phe F	UCC	Ser S	UAC	Tyr Y	UGC	Cys C	C
	UUA	Leu L	UCA	Ser S	UAA	STOP	UGA	STOP	A
	UUG	Leu L	UCG	Ser S	UAG	STOP	UGG	Trp W	G
C	CUU	Leu L	CCU	Pro[3] P	CAU	His H	CGU	Arg R	U
	CUC	Leu L	CCC	Pro[3] P	CAC	His H	CGC	Arg R	C
	CUA	Leu L	CCA	Pro[3] P	CAA	Gln Q	CGA	Arg R	A
	CUG	Leu L	CCG	Pro[3] P	CAG	Gln Q	CGG	Arg R	G
A	AUU	Ile I	ACU	Thr T	AAU	Asn N	AGU	Ser S	U
	AUC	Ile I	ACC	Thr T	AAC	Asn N	AGC	Ser S	C
	AUA	Ile I	ACA	Thr T	AAA	Lys K	AGA	Arg R	A
	AUG	Met M	ACG	Thr T	AAG	Lys K	AGG	Arg R	G
G	GUU	Val V	GCU	Ala A	GAU	Asp D	GGU	Gly G	U
	GUC	Val V	GCC	Ala A	GAC	Asp D	GGC	Gly G	C
	GUA	Val V	GCA	Ala A	GAA	Glu E	GGA	Gly G	A
	GUG	Val V	GCG	Ala A	GAG	Glu E	GGG	Gly G	G

[1] Next to each codon or set of codons is the R group characteristic of that amino acid drawn in blue.
[2] The code is presented in RNA form. Codons run in the 5′ to 3′ direction. The 3- and 1-letter abbreviations for the amino acids designated by the codon are provided, as is the structure of the R chain of that amino acid.
[3] Proline is an imino acid, rather than an amino acid.

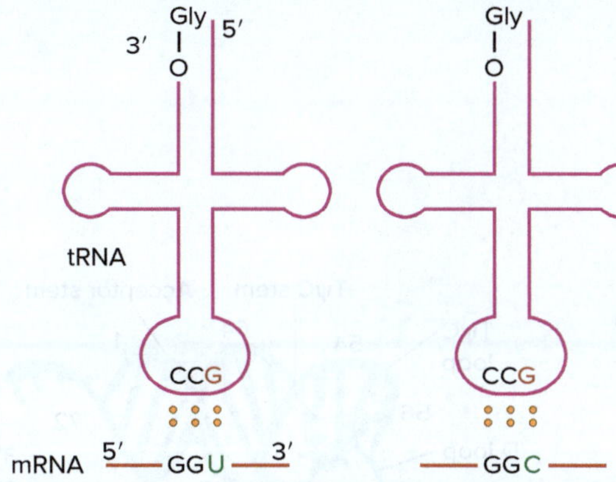

Figure 13.29 Wobble Decreases the Number of tRNAs Needed. Because of wobble, G in the 5′ position of the anticodon can pair with either C or U in the 3′ position of the codon. Thus two codons can be recognized by the same tRNA, in this example, tRNA-*Gly*.

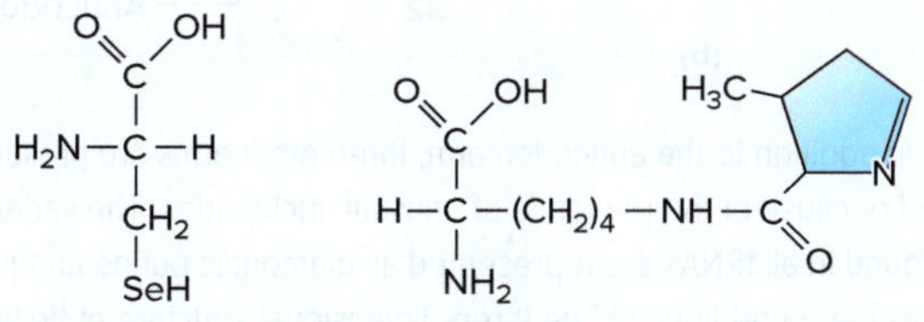

Selenocysteine Pyrrolysine

Figure 13.30 Selenocysteine and Pyrrolysine. These unusual amino acids are not coded for by DNA but are inserted in response to specific stop codons and stem-loop structures in mRNA.

Comprehension Check

1. What is the difference between a codon and an anticodon?
2. List the punctuation functions of the genetic code and explain their importance.
3. What is meant by code degeneracy (redundancy)? How does wobble help alleviate the energy cost exacted on a cell due to code degeneracy?
4. Is the genetic code truly universal? Explain your answer.

13.7 Translation in Bacteria

After reading this section, you should be able to:

a. Relate the general structure of a tRNA molecule to its role in amino acid activation and translation
b. Describe the structure of bacterial ribosomes
c. Summarize the formation of a translation initiation complex
d. Outline the events that occur at the A, P, and E sites of the bacterial ribosome during the elongation phase of translation

Translation involves decoding mRNA and covalently linking amino acids together to form a polypeptide; this occurs within the ribosome. Translation begins when a ribosome binds mRNA and is positioned so that translation will yield the correct amino acid sequence in the polypeptide chain. Transfer RNA molecules carry amino acids to the ribosome so that they can be added to the polypeptide chain as the ribosome moves along the mRNA molecule. Just as DNA and RNA synthesis proceed in one direction, so too does protein synthesis. Polypeptide synthesis begins with the amino acid at the end of the chain with a free amino group (the N-terminal) and moves in the C-terminal direction. Thus translation is said to occur in the amino terminus to carboxyl terminus direction.

Amino Acid Activation: Attachment of an Amino Acid to a Transfer RNA

For translation to occur, a ready supply of tRNA molecules bearing the correct amino acid must be available. Thus a preparatory step for protein synthesis is **amino acid activation,** the process in which amino acids are attached to tRNA molecules. Before we discuss this process, we need to examine the structure of tRNAs.

All tRNAs are about 70 to 95 nucleotides long and possess several characteristic structural features. These features become apparent when the tRNA is folded so that base pairing within the tRNA strand is maximized. When represented two dimensionally, this base pairing causes the tRNA to assume a cloverleaf conformation (**figure 13.31*a***). However, the three-dimensional structure looks like the letter L (figure 13.31*b*). One important feature of tRNAs is the acceptor stem, which holds the activated amino acid. The 3′ end of all tRNAs has the same CCA sequence, and in all cases the amino acid is attached to the 3′-hydroxyl of the adenosine nucleotide. Another important feature of a tRNA is the **anticodon.** The anticodon is complementary to an mRNA codon and is located on the anticodon arm of the tRNA.

Enzymes called **aminoacyl-tRNA synthetases** catalyze amino acid activation (**figure 13.32**). As is true of DNA and RNA synthesis, the reaction is driven to completion when ATP is hydrolyzed to release pyrophosphate. The amino acid is attached to the tRNA by a high-energy bond. The storage of energy in this bond provides the fuel needed to generate the peptide bond when the amino acid is added to the growing peptide chain.

There are at least 20 aminoacyl-tRNA synthetases, each specific for a single amino acid and its tRNAs (cognate tRNAs). It is critical that each tRNA attach the corresponding amino acid because if an incorrect amino acid is attached to a tRNA, it will be incorporated into a polypeptide in place of the correct amino acid. The protein synthetic machinery recognizes only the anticodon of the aminoacyl-tRNA and cannot tell whether the correct amino acid is attached. Some aminoacyl-tRNA synthetases proofread just like DNA polymerases. If the wrong amino acid is attached to tRNA, the enzyme hydrolyzes the amino acid from the tRNA, rather than release the incorrect product. *Aminoacyl-tRNA Structure*

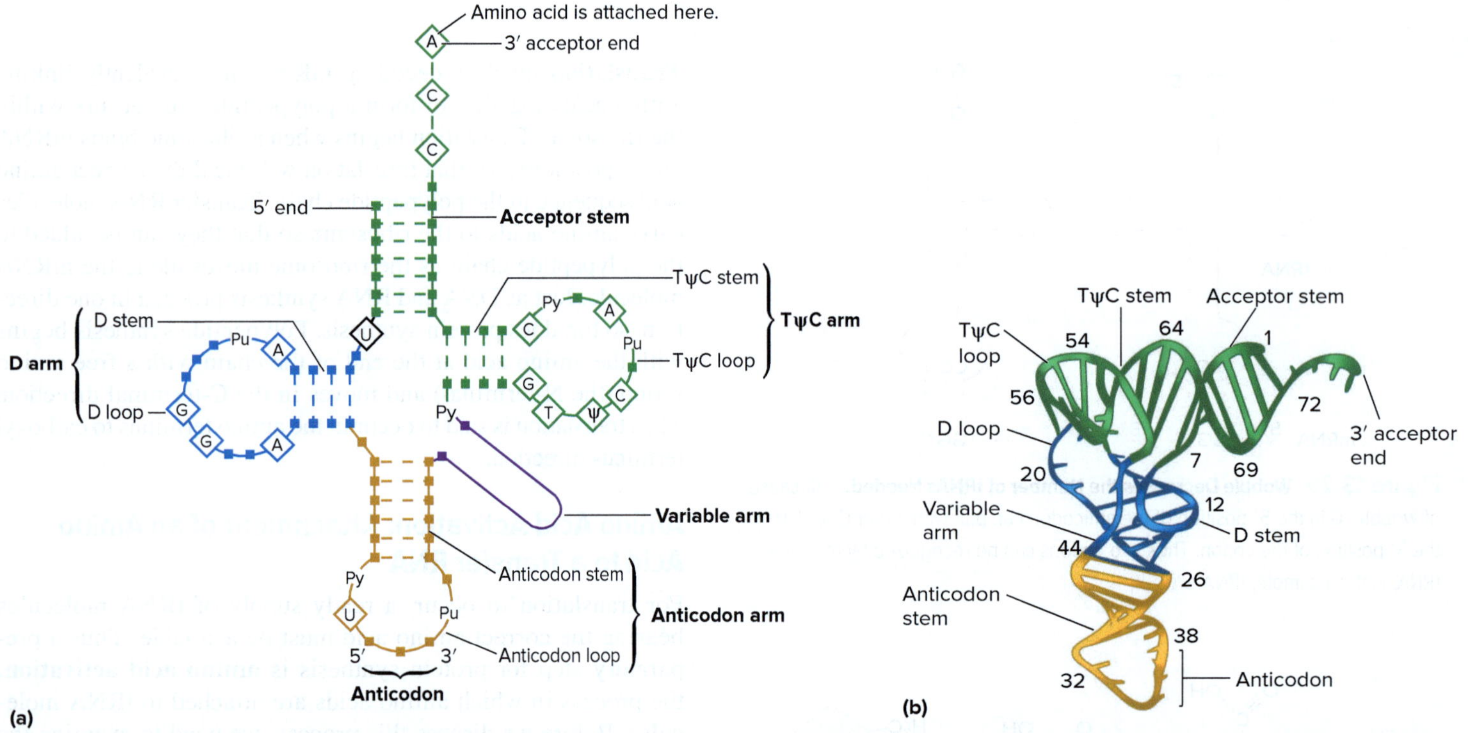

Figure 13.31 tRNA Structure. (a) The two-dimensional cloverleaf structure for tRNA. In addition to the anticodon arm, three other arms are present: the D arm (blue), the T or TψC arm (green), and the variable arm (purple). The D and T arms are named because of the presence of unusual nucleotides. The variable arm is of different lengths depending on the tRNA; the other arms are fairly constant in size. Bases found in all tRNAs are represented as diamonds; purine and pyrimidine positions in all tRNAs are labeled Pu and Py, respectively. ψ represents pseudouridine, an isomer of uridine. (b) The three-dimensional structure of tRNA. The colors of each portion of the molecule are the same as in (a).

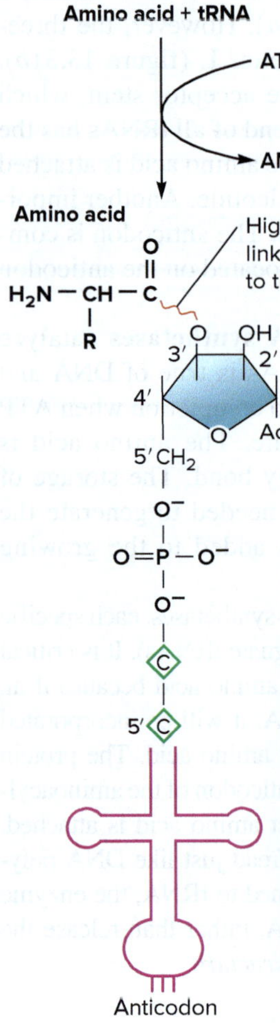

Figure 13.32 Aminoacyl-tRNA Synthetase Attaches an Amino Acid to a tRNA. The amino acid is attached by the appropriate aminoacyl-tRNA synthetase to the 3′-hydroxyl provided by the terminal adenosine nucleoside. The two cytosine nucleotides are indicated in diamonds and the adenine is drawn as atoms.

MICRO INQUIRY *What would be the outcome if an aminoacyl-tRNA synthetase added the wrong amino acid to a tRNA (i.e., the anticodon specified a different amino acid than that added to the 3′ end of the tRNA)?*

Ribosomes Have Three tRNA Binding Sites

Protein synthesis takes place within ribosomes that serve as workbenches, with mRNA acting as the blueprint. Recall that ribosomes are formed from two subunits, the large subunit and the small subunit, and each contains one or more rRNA molecules and numerous proteins. A bacterial ribosome and its components are shown in **figure 13.33.** Three sites are found within the ribosome for binding tRNAs: A, P, and E sites. The **A (aminoacyl** or **acceptor) site** receives tRNAs carrying an amino acid to be added to the protein being synthesized. The **P (peptidyl** or **donor) site** holds a tRNA attached to the growing polypeptide. The **E (exit) site** is the location from which empty tRNAs leave the ribosome.

All three rRNA molecules contribute to ribosome structure. The 16S rRNA of the 30S subunit is needed for initiation of protein synthesis because its 3′ end binds to a site on the leader of the mRNA called the Shine-Dalgarno sequence; thus the Shine-Dalgarno sequence is part of the **ribosome-binding site (RBS).** This positions the mRNA on the ribosome. The 16S rRNA also binds a protein needed to initiate translation and the 3′ CCA end of aminoacyl-tRNA. The 23S rRNA is a ribozyme that catalyzes peptide bond formation.

Stages of Translation: Initiation, Elongation, and Termination

Like transcription and DNA replication, protein synthesis is divided into three stages: initiation, elongation, and termination. Bacteria begin protein synthesis with a modified

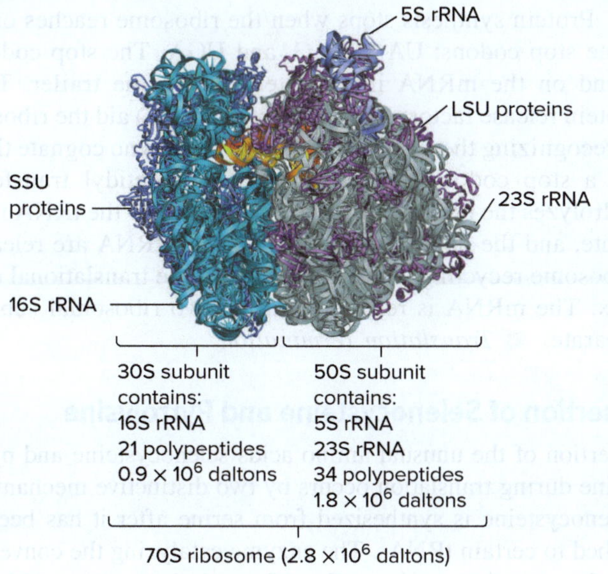

Figure 13.33 Bacterial Ribosome Structure. The *Thermus thermophilus* 70S ribosome viewed with the 30S subunit (SSU) on the left and the 50S subunit (LSU) on the right. The A site tRNA (gold) is visible in the interface cavity. 16S rRNA, cyan; 23S rRNA, gray; 5S rRNA, light blue; 30S proteins, dark blue; 50S proteins, magenta.

aminoacyl-tRNA, *N*-formylmethionyl-tRNA^{fMet} (fMet-tRNA), which is called the **initiator tRNA** and is coded for by the start codon AUG (**figure 13.34**). The amino acid of the initiator tRNA has a formyl group covalently bound to the amino group and can be used only for initiation because the formyl group blocks peptide bond formation. When methionine is to be added to a growing polypeptide chain (i.e., at an AUG codon in the middle of the mRNA), a normal methionyl-tRNA^{Met} is used. Although bacteria start protein synthesis with *N*-formylmethionine, the formyl group is not retained but is hydrolytically removed (section 13.9).

Protein synthesis in bacteria begins with formation of the 30S initiation complex, consisting of the initiator tRNA, the mRNA to be translated, and the 30S ribosomal subunit; two protein **initiation factors** (IF-1 and IF-2) are involved (**figure 13.35**). Positioning of

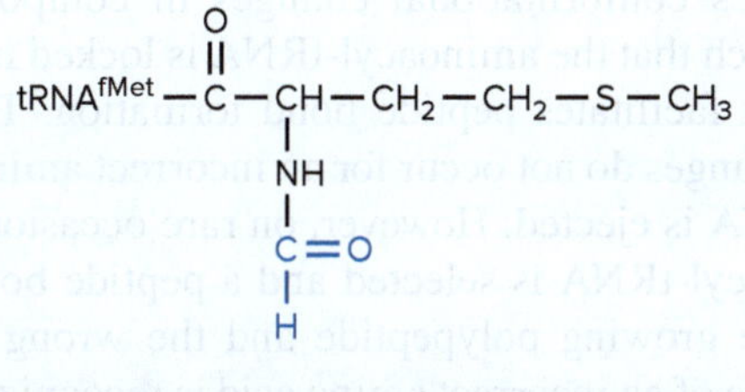

Figure 13.34 *N*-formylmethionyl-tRNA^{fMet} Is the Initiator tRNA Used by Bacteria. The formyl group is in blue.

MICRO INQUIRY *Why would it be impossible for fMet-tRNA to initiate peptide bond formation with another amino acid? (Hint: Examine figure 13.37 closely.)*

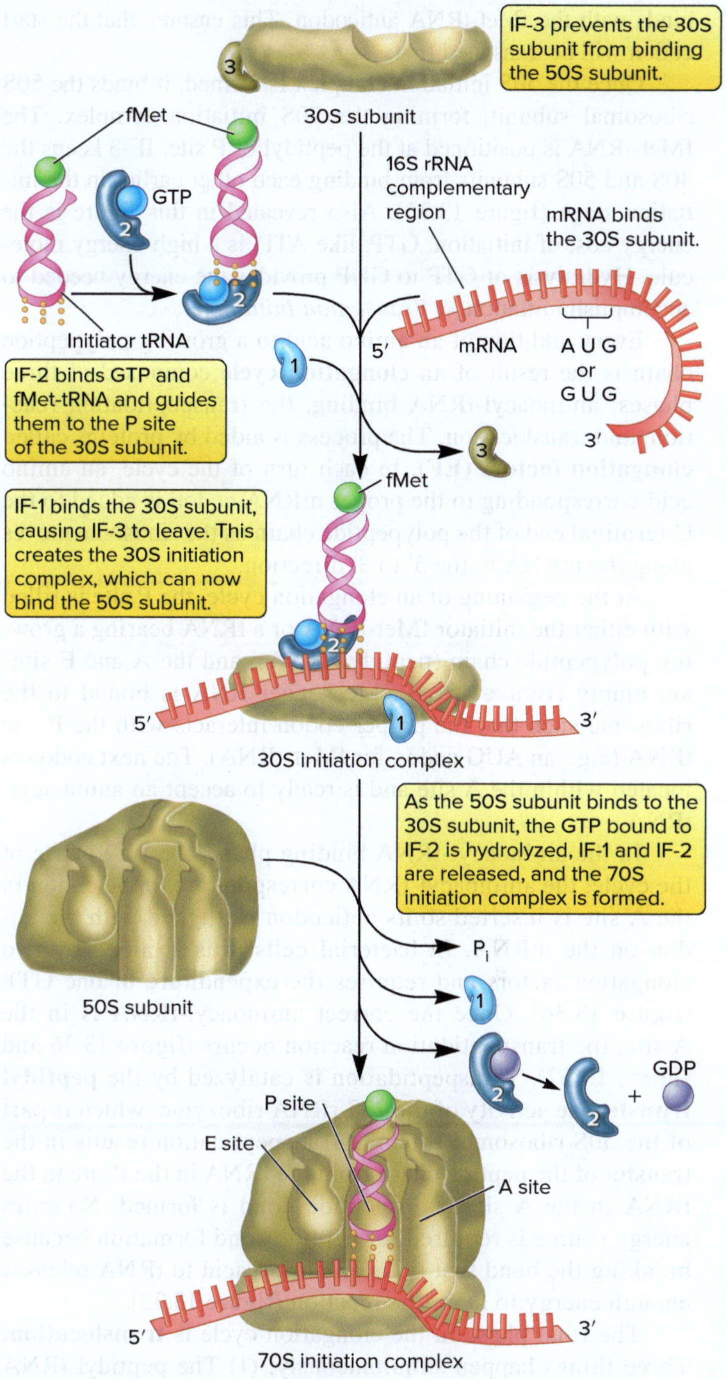

Figure 13.35 Initiation of Protein Synthesis in Bacteria. The proteins labeled 1, 2, and 3 are initiation factors (IF) 1, 2, and 3; initiator tRNA is *N*-formylmethionyl-tRNA^{fMet}. The ribosomal locations of initiation factors are depicted for illustration purposes only. They do not represent the actual initiation factor binding sites.

the initiator fMet-tRNA on the mRNA is crucial for proper translation of the mRNA. This is accomplished with the help of the 16S rRNA within the 30S subunit, which is complementary to and binds the Shine-Dalgarno sequence in the leader sequence of the mRNA. By aligning the Shine-Dalgarno sequence with the 16S rRNA, the start codon (AUG or sometimes GUG) specifically binds with the fMet-tRNA anticodon. This ensures that the start codon will be translated first.

Once the 30S initiation complex is formed, it binds the 50S ribosomal subunit, forming the 70S initiation complex. The fMet-tRNA is positioned at the peptidyl or P site. IF-3 keeps the 30S and 50S subunits from binding each other earlier in the initiation stage (figure 13.35). Also revealed in this figure is the energy cost of initiation. GTP, like ATP, is a high-energy molecule. Hydrolysis of GTP to GDP provides the energy needed to accomplish initiation. *Translation Initiation*

Every addition of an amino acid to a growing polypeptide chain is the result of an **elongation cycle** composed of three phases: aminoacyl-tRNA binding, the transpeptidation reaction, and translocation. The process is aided by proteins called **elongation factors (EF).** In each turn of the cycle, an amino acid corresponding to the proper mRNA codon is added to the C-terminal end of the polypeptide chain as the ribosome moves along the mRNA in the 5′ to 3′ direction.

At the beginning of an elongation cycle, the P site is filled with either the initiator fMet-tRNA or a tRNA bearing a growing polypeptide chain (peptidyl-tRNA), and the A and E sites are empty (**figure 13.36**). Messenger RNA is bound to the ribosome such that the proper codon interacts with the P site tRNA (e.g., an AUG codon for fMet-tRNA). The next codon is located within the A site and is ready to accept an aminoacyl-tRNA.

In the aminoacyl-tRNA binding phase, the first phase of the cycle, the aminoacyl-tRNA corresponding to the codon in the A site is inserted so its anticodon is aligned with the codon on the mRNA. In bacterial cells, this is aided by two elongation factors and requires the expenditure of one GTP (figure 13.36). Once the correct aminoacyl-tRNA is in the A site, the transpeptidation reaction occurs (figure 13.36 and **figure 13.37**). Transpeptidation is catalyzed by the **peptidyl transferase** activity of the 23S rRNA ribozyme, which is part of the 50S ribosomal subunit. Transpeptidation results in the transfer of the peptide chain from the tRNA in the P site to the tRNA in the A site, as a peptide bond is formed. No extra energy source is required for peptide bond formation because breaking the bond that links an amino acid to tRNA releases enough energy to drive the reaction (figure 13.32).

The final phase in the elongation cycle is **translocation.** Three things happen simultaneously: (1) The peptidyl-tRNA moves from the A site to the P site; (2) the ribosome moves one codon along the mRNA so that a new codon is positioned in the A site; and (3) the empty tRNA moves from the P site to the E site and subsequently leaves the ribosome. Translocation involves rotations of the 30S and 50S subunits relative to each other. In addition, the head portion of the 30S subunit swivels. These changes in ribosome structure move the tRNAs into their new locations; the codon-anticodon interactions between the tRNAs and the mRNA move the mRNA as the tRNAs move. One elongation factor participates and one GTP is hydrolyzed during this intricate process. *Translation Elongation*

Protein synthesis stops when the ribosome reaches one of three stop codons: UAA, UAG, and UGA. The stop codon is found on the mRNA immediately before the trailer. Three protein release factors (RF-1, RF-2, and RF-3) aid the ribosome in recognizing these codons. Because there is no cognate tRNA for a stop codon, the ribosome halts. Peptidyl transferase hydrolyzes the bond linking the polypeptide to the tRNA in the P site, and the polypeptide and the empty tRNA are released. Ribosome recycling factor disassembles the translational complex. The mRNA is released and the two ribosomal subunits separate. *Translation Termination*

Insertion of Selenocysteine and Pyrrolysine

Insertion of the unusual amino acids selenocysteine and pyrrolysine during translation occurs by two distinctive mechanisms. Selenocysteine is synthesized from serine after it has been attached to certain tRNAs. The enzyme catalyzing the conversion is selenocysteine synthase. Once formed, the amino acid is recognized by a specific elongation factor and is incorporated when a UGA stop codon is encountered in association with nucleotide sequences called cis-acting *se*leno*c*ysteine *i*nsertion *s*equence elements (SECIS). In bacteria, SECIS are found immediately after the UGA stop codon.

Pyrrolysine insertion differs from that of selenocysteine in several ways. Pyrrolysine is synthesized from lysine before being attached to a tRNA. Organisms that use pyrrolysine make an unusual tRNA with a CUA anticodon; the pyrrolysine is attached by a specific aminoacyl-tRNA synthetase. Pyrrolysine is inserted at UAG stop codons located near a sequence element called *py*rro*l*ysine *i*nsertion *s*equence (PYLIS). Both SECIS and PYLIS form stem-loop structures that prevent translation termination.

Ensuring Accuracy During Translation

Protein synthesis is a very expensive process: Two ATP high-energy bonds are required for amino acid activation, initiation consumes one GTP, two GTP molecules are used during each elongation cycle, and another GTP is hydrolyzed when protein synthesis terminates (figures 13.32, 13.35, 13.36, and 13.38). Presumably this large energy expenditure is required to ensure the fidelity of protein synthesis. Fidelity is assessed both before and after formation of the peptide bond. When an aminoacyl-tRNA enters the A site, correct pairing of the anticodon and codon causes conformational changes in components of the ribosome such that the aminoacyl-tRNA is locked into place in a manner that facilitates peptide bond formation. These conformational changes do not occur for an incorrect aminoacyl-tRNA and the tRNA is ejected. However, on rare occasions, the incorrect aminoacyl-tRNA is selected and a peptide bond is formed between the growing polypeptide and the wrong amino acid. The presence of an incorrect amino acid is recognized by release factors. This leads to hydrolysis of the aberrant polypeptide from the tRNA, its release from the ribosome, and termination of translation.

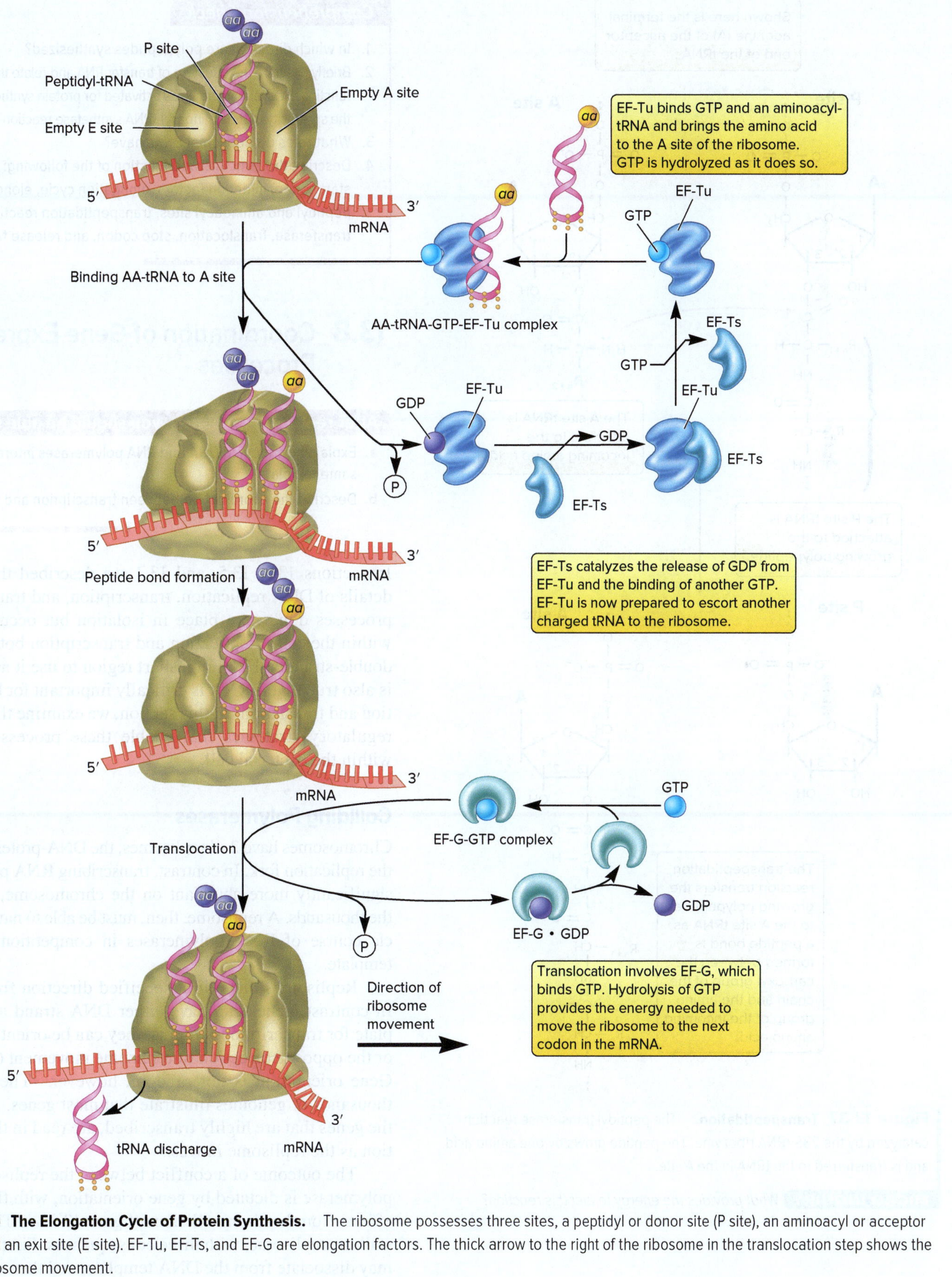

Figure 13.36 The Elongation Cycle of Protein Synthesis. The ribosome possesses three sites, a peptidyl or donor site (P site), an aminoacyl or acceptor site (A site), and an exit site (E site). EF-Tu, EF-Ts, and EF-G are elongation factors. The thick arrow to the right of the ribosome in the translocation step shows the direction of ribosome movement.

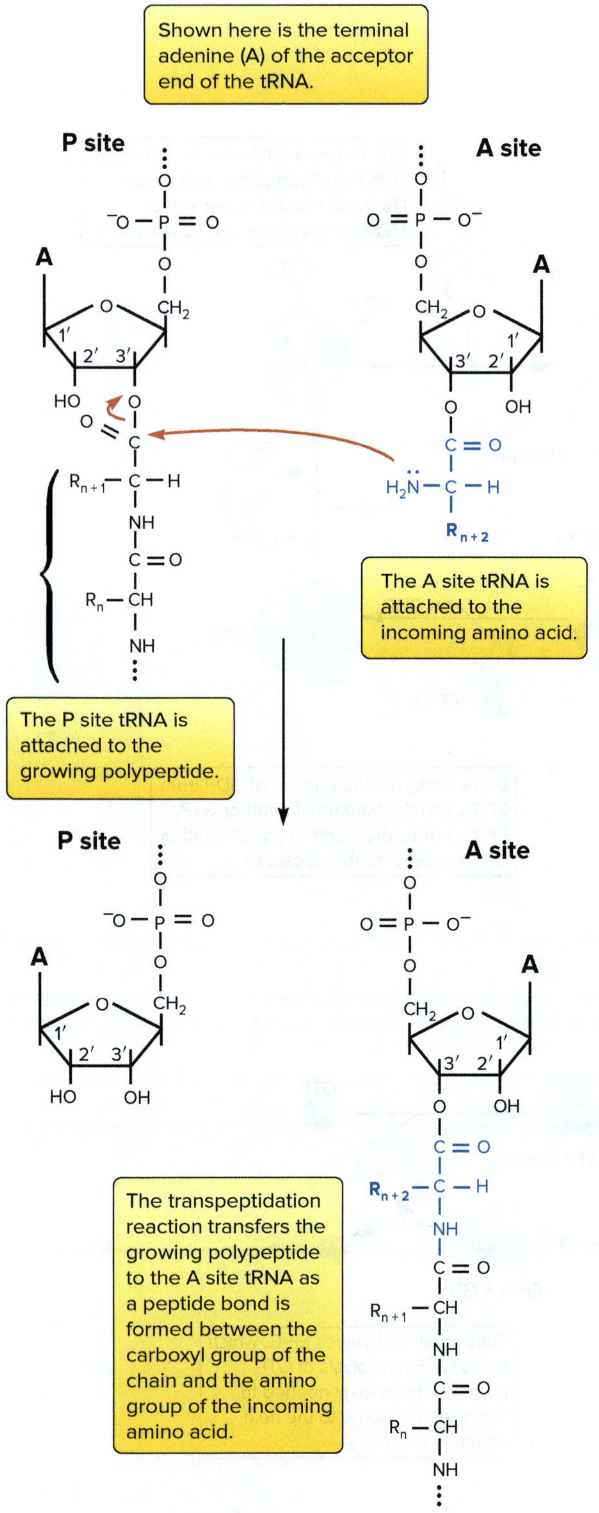

Figure 13.37 Transpeptidation. The peptidyl transferase reaction catalyzed by the 23S rRNA ribozyme. The peptide grows by one amino acid and is transferred to the tRNA in the A site.

MICRO INQUIRY *What provides the energy to fuel this reaction?*

Comprehension Check

1. In which direction are polypeptides synthesized?
2. Briefly describe the structure of transfer RNA and relate this to its function. How are amino acids activated for protein synthesis, and why is the specificity of the aminoacyl-tRNA synthetase reaction so important?
3. What roles do ribosomal RNAs have?
4. Describe the structure and function of the following: fMet-tRNA, start codon, initiation factors, elongation cycle, elongation factors, peptidyl and aminoacyl sites, transpeptidation reaction, peptidyl transferase, translocation, stop codon, and release factors.

13.8 Coordination of Gene Expression Processes

After reading this section, you should be able to:

a. Explain how moving DNA and RNA polymerases interact on the same template
b. Describe the coordination between transcription and translation

In sections 13.3, 13.5, and 13.7, we described the mechanistic details of DNA replication, transcription, and translation. These processes don't take place in isolation but occur coordinately within the cell. Replication and transcription both denature the double-stranded DNA in a short region to use it as a template. It is also true that mRNA is critically important for both transcription and translation. In this section, we examine the physical and regulatory strategies that enable these processes to co-exist within the bacterial cell.

Colliding Polymerases

Chromosomes have 2–4 replisomes, the DNA-protein complexes at the replication fork. In contrast, transcribing RNA polymerases are significantly more abundant on the chromosome, numbering in the thousands. A replisome, then, must be able to navigate an obstacle course of RNA polymerases in competition for the same template.

Replisomes move in a specified direction from the origin. In contrast, genes can have either DNA strand act as the template for transcription—that is, they can be oriented in the same or the opposite direction of replisome movement (**figure 13.38**). Gene orientation is not random, however. The sequences of thousands of genomes illustrate that most genes, and especially the genes that are highly transcribed, are read in the same direction as the replisome moves.

The outcome of a conflict between the replisome and RNA polymerase is dictated by gene orientation, with the most severe effects from head-on collisions (figure 13.38*a*). These collisions stall the replisome, with three consequences. First, the replisome may dissociate from the DNA template, and then requires accessory proteins to reassemble at sites other than *oriC*. Second,

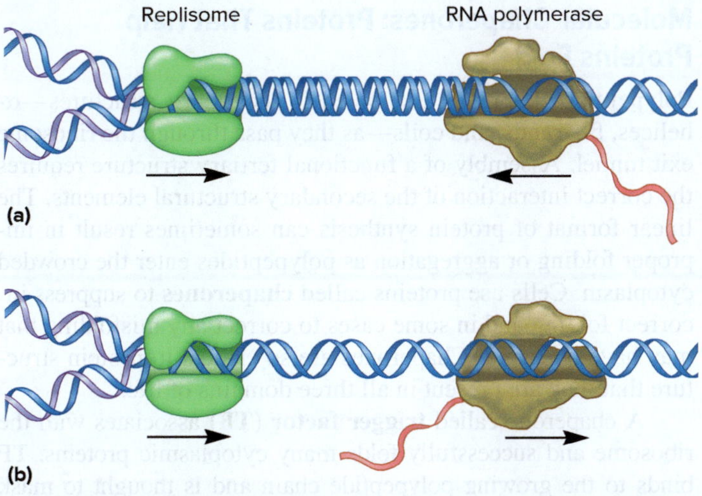

Figure 13.38 Colliding Polymerases. (a) Head-on collision increases supercoiling between the two polymerases, which limits helicase. This destabilizes the replisome and can often result in mutation. (b) Codirectional collision is less severe, generally resulting in RNA polymerase eviction from the template.

replication fidelity is compromised, leading to errors (i.e., mutations). Third, increased supercoiling in front of moving polymerases (both DNA and RNA) builds to an extreme level as the enzymes approach one another (figure 13.38*a*). ▶| *Spontaneous mutations (section 16.1)*

In contrast, codirectional conflicts, or rear-end collisions (figure 13.38*b*), are more easily resolved. In addition to the helicase that propels the replisome, alternative helicases in the cell promote replication fork movement through barriers like RNA polymerases. It is believed that the replisome may use the RNA transcript to re-prime leading strand synthesis. In addition, the protein Mfd is considered a transcriptional terminator because it travels along DNA to evict paused or stalled RNA polymerases.

Coupled Transcription and Translation

To achieve rapid translation rates of protein synthesis, mRNAs often are simultaneously complexed with several ribosomes, each ribosome reading the mRNA message and synthesizing a polypeptide. At maximal translation rates, there may be a ribosome every 80 nucleotides or as many as 20 ribosomes simultaneously reading an mRNA that codes for an average sized protein of 50,000 daltons. A complex of mRNA with several ribosomes is called a **polyribosome** or polysome (**figure 13.39**).

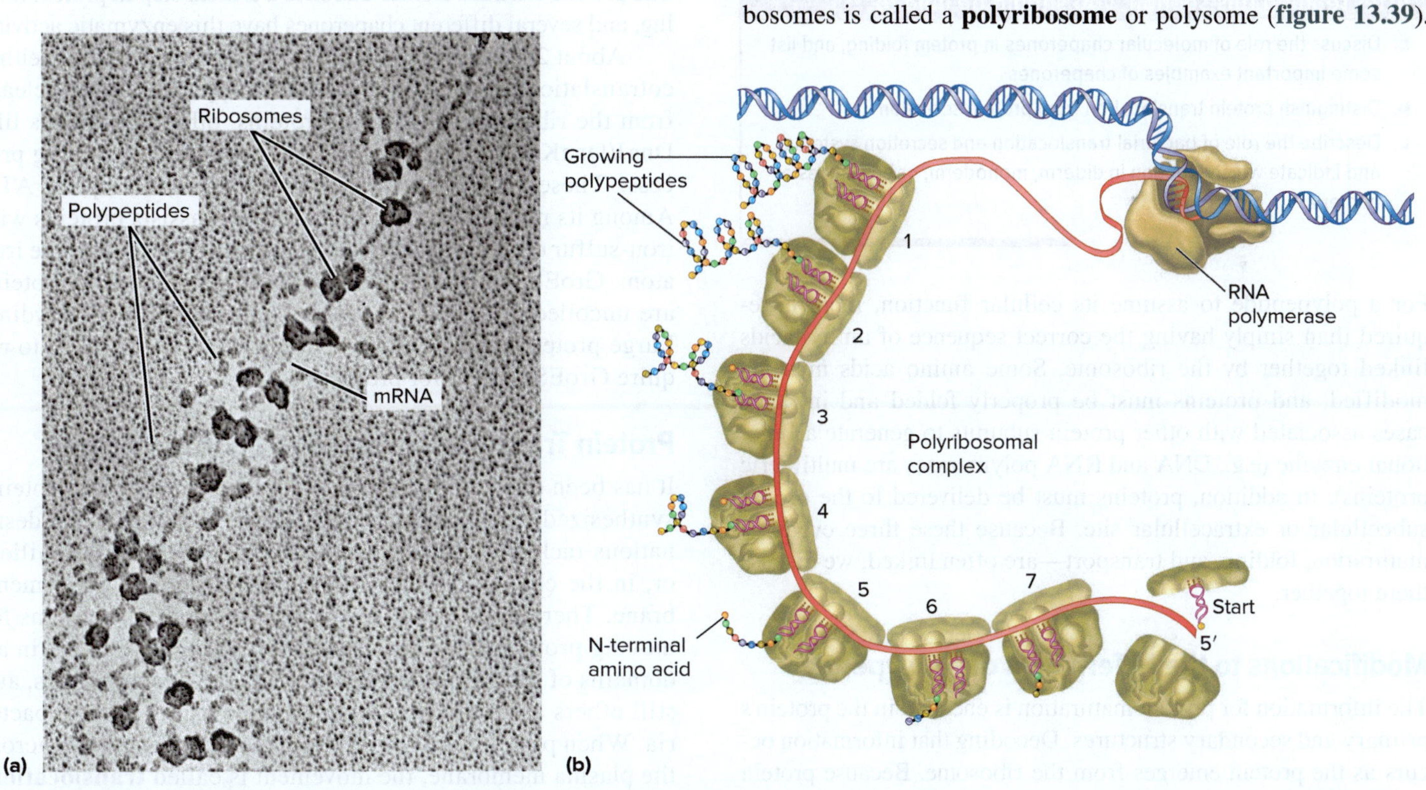

Figure 13.39 Coupled Transcription and Translation in Bacteria. (a) A transmission electron micrograph showing a polyribosome. (b) A schematic representation of coupled transcription and translation. As the DNA is transcribed, ribosomes bind the free 5′ end of the mRNA. Thus translation is started before transcription is completed. Note that there are multiple ribosomes bound to the mRNA, forming a polyribosome. The ribosomes are shown at different points in the translation process. Ribosomes 1, 2, 5, and 7 have completed the transpeptidation reaction, but translocation has not yet occurred. Ribosomes 3 and 4 have an A site containing an incoming aminoacyl-tRNA. Transpeptidation has not occurred. Ribosome 6 shows elongation upon completion of both transpeptidation and translocation. The tRNA bearing the growing polypeptide is in the P site and the empty tRNA is in the E site.

(a) Steven L. McKnight

MICRO INQUIRY *Why is simultaneous transcription and translation impossible in eukaryotes?*

Polysomes are present in all organisms. Bacteria can further increase the efficiency of gene expression by coupling transcription and translation (figure 13.39*b*). While RNA polymerase is synthesizing an mRNA, ribosomes can attach to the mRNA so that transcription and translation occur simultaneously. Coupled transcription and translation is possible in bacterial cells because a nuclear envelope does not separate the translation machinery from DNA, as it does in eukaryotes.

DNA replication and transcription use the same substrate: the chromosome located in the nucleoid. Translation, however, occurs in the cytoplasm. It is believed that a transcribing RNA polymerase moves toward the nucleoid periphery where its transcript interacts with ribosomal subunits. Recent work has demonstrated that the transcription elongation factor NusG bridges RNA polymerase and the 30S ribosomal subunit. It appears to keep the mRNA single stranded for a smooth transition to the ribosome.

13.9 Protein Maturation and Translocation

After reading this section, you should be able to:

a. Discuss the role of molecular chaperones in protein folding, and list some important examples of chaperones
b. Distinguish protein translocation from protein secretion
c. Describe the role of bacterial translocation and secretion systems, and indicate which function in diderm, monoderm, or both types of bacteria

For a polypeptide to assume its cellular function, more is required than simply having the correct sequence of amino acids linked together by the ribosome. Some amino acids must be modified, and proteins must be properly folded and in some cases associated with other protein subunits to generate a functional enzyme (e.g., DNA and RNA polymerases are multimeric proteins). In addition, proteins must be delivered to the proper subcellular or extracellular site. Because these three events—maturation, folding, and transport—are often linked, we discuss them together.

Modifications to the N-Terminus of a Polypeptide

The information for protein maturation is encoded in the protein's primary and secondary structures. Decoding that information occurs as the protein emerges from the ribosome. Because protein synthesis is initiated with *N*-formylmethionine, the N-terminus of a polypeptide matures with the removal of the N-formyl group. Some proteins also require the removal of the methionine, while others remove N-terminal amino acids. This is specified by the identity of the second amino acid in the polypeptide. The enzymes that catalyze these reactions associate with the ribosome near the polypeptide exit tunnel in the large subunit. These reactions occur **cotranslationally,** that is, as protein synthesis is ongoing.

Molecular Chaperones: Proteins That Help Proteins Fold

Polypeptides begin to assume their secondary structures—α-helices, β-strands, and coils—as they pass through the ribosome exit tunnel. Assembly of a functional tertiary structure requires the correct interaction of the secondary structural elements. The linear format of protein synthesis can sometimes result in improper folding or aggregation as polypeptides enter the crowded cytoplasm. Cells use proteins called **chaperones** to suppress incorrect folding and in some cases to correct any misfolding that may have occurred. Chaperones are so critical to protein structure that they are present in all three domains of life.

A chaperone called **trigger factor (TF)** associates with the ribosome and successfully folds many cytoplasmic proteins. TF binds to the growing polypeptide chain and is thought to mask hydrophobic regions so they don't interact with each other prematurely or with other proteins. Recall that hydrophobic regions are generally in the protein interior, so the presence of a hydrophobic surface region indicates that the polypeptide likely needs help to assume its functional shape. TF also acts as an isomerase on the peptide bond that precedes proline residues in the polypeptide. The proline cis/trans isomerization is a critical step in protein folding, and several different chaperones have this enzymatic activity.

About 25% of proteins need additional help folding, either cotranslationally or **posttranslationally** (i.e., after release from the ribosome). For these proteins, other chaperones like DnaJ/DnaK/GrpE or GroES/GroEL complete the folding process. These chaperones are cytoplasmic and require ATP. Among its many roles, DnaJ/DnaK/GrpE assists proteins with iron-sulfur centers to insert and chemically coordinate the iron atom. GroES/GroEL forms a cage in which misfolded proteins are uncoiled and refolded, away from cytoplasmic crowding. Large proteins with complex topologies are most likely to require GroES/GroEL for proper maturation.

Protein Translocation and Secretion in Bacteria

It has been estimated that more than one-third of the proteins synthesized leave the cytoplasm. They can have several destinations including the plasma membrane, the external milieu, or, in the case of diderms, the periplasm or the outer membrane. Therefore it is not surprising that multiple systems for moving proteins have evolved. Some of these are found in all domains of life, while others are unique to bacterial cells, and still others are restricted to only monoderm or diderm bacteria. When proteins are moved from the cytoplasm to or across the plasma membrane, the movement is called **translocation.** Protein **secretion** refers to the movement of proteins from the cytoplasm to the external environment. All protein translocation and secretion systems described here require the expenditure of energy in the form of ATP or the proton motive force. In addition, the protein complexes responsible for movement across the plasma membrane are localized to functional membrane microdomains. ◀ *Plasma membrane structure is dynamic (section 3.2)*

The variations in cell envelope structure pose different challenges for protein secretion. In monoderm bacteria, the proteins must be translocated across the plasma membrane. Once across the plasma membrane, the protein either passes through the relatively porous peptidoglycan into the external environment or becomes attached to the peptidoglycan. Diderm bacteria also transport proteins across the plasma membrane, but then must move them across an additional membrane.

Translocation Systems Are Common to Both Monoderm and Diderm Bacteria

Three translocation systems—the Sec system, the Tat system, and YidC—are observed in both monoderm and diderm bacteria. The **Sec system,** sometimes called the general secretion pathway, is highly conserved, having been identified in all three domains of life (**figure 13.40**). It translocates unfolded proteins across the plasma membrane. YidC is dedicated to folding and translocating plasma membrane proteins, and often coordinates with the Sec system.

Before the Sec system translocates proteins, they must be sorted and targeted to their ultimate destination. The N-terminal polypeptide region, the first part to exit the ribosome, is termed the **signal peptide** or signal sequence. This region of the protein signals factors associated with the ribosome to direct it to one of several routes. **Signal recognition particle (SRP)** is a protein-RNA complex that surveys actively translating ribosomes at the same time as trigger factor (TF). Whereas TF folds and releases cytoplasmic proteins, SRP identifies membrane proteins in part by extremely hydrophobic signal peptides. Membrane proteins in particular have a transmembrane α-helix that SRP recognizes as it probes the ribosome exit tunnel. SRP escorts the entire ribosome to the membrane. Thus transport by this mechanism is defined as cotranslational.

An alternative route to the membrane is via the SecA protein. SecA recognizes signal peptides that are less hydrophobic than those preferred by SRP. SecA transport to the membrane may occur cotranslationally or posttranslationally. SecA does not function alone. During posttranslational translocation, a cytoplasmic chaperone called SecB keeps the newly synthesized preprotein in an unfolded state. Three proteins (SecY, SecE, and SecG) form a channel in the membrane. SecA threads the signal peptide and the preprotein through the channel (figure 13.40). SecA acts as a motor, using the energy from ATP hydrolysis, and two other proteins (SecDF) use the proton motive force to translocate the preprotein. The SecYEG channel is unique in that it opens in two ways, based on the delivery vehicle and the protein destination. SRP-ribosome complexes align the ribosome exit tunnel with the SecYEG channel, and SecYEG opens laterally for membrane proteins to be inserted. Some bacteria have additional Sec proteins that transport a subset of proteins. Examples include the SecA2 system found in mycobacteria and the SecA2-Y2 system found in some streptococci.

The **Tat system** (**figure 13.41**) is also widespread in both monoderm and diderm bacteria, although it translocates only a few dozen proteins. Tat is an acronym for *t*win-*a*rginine *t*ranslocase, referring to two consecutive arginines, a distinctive feature in the signal peptide of proteins translocated by this system. Tat-secreted proteins must completely fold in the cytoplasm; thus unlike the Sec system, the Tat system secretes only folded proteins. Some proteins have dedicated chaperones to insert cofactors and assist in folding. A protein docking complex recognizes the signal peptide and escorts the protein to the pore complex in the membrane. Secretion by Tat relies on the proton motive force for energy.

Following polypeptide translocation, two additional maturation steps occur. Regardless of the secretion system used, the signal peptide must be removed from the preprotein. **Signal**

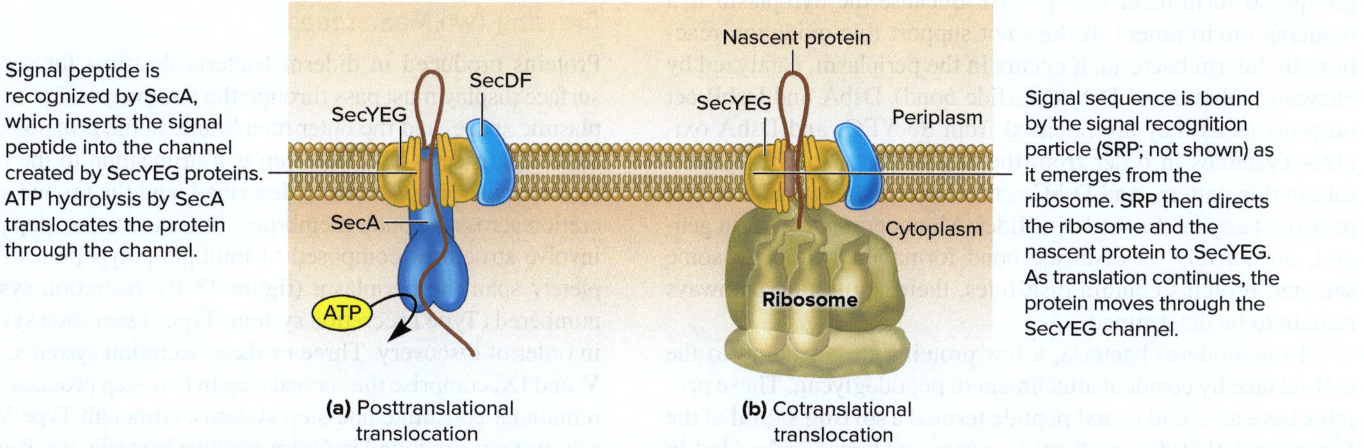

Figure 13.40 Translocation by the Sec System Moves Unfolded Proteins Across or into the Plasma Membrane. (a) Posttranslational translocation uses chaperones to keep the protein unfolded until it is recognized by SecA. Here we show SecA as a dimer. There is also evidence that a monomer of SecA is sufficient for translocation. (b) In cotranslational translocation the protein has not had a chance to fold before it is bound by SRP and delivered, along with the ribosome, to SecYEG.

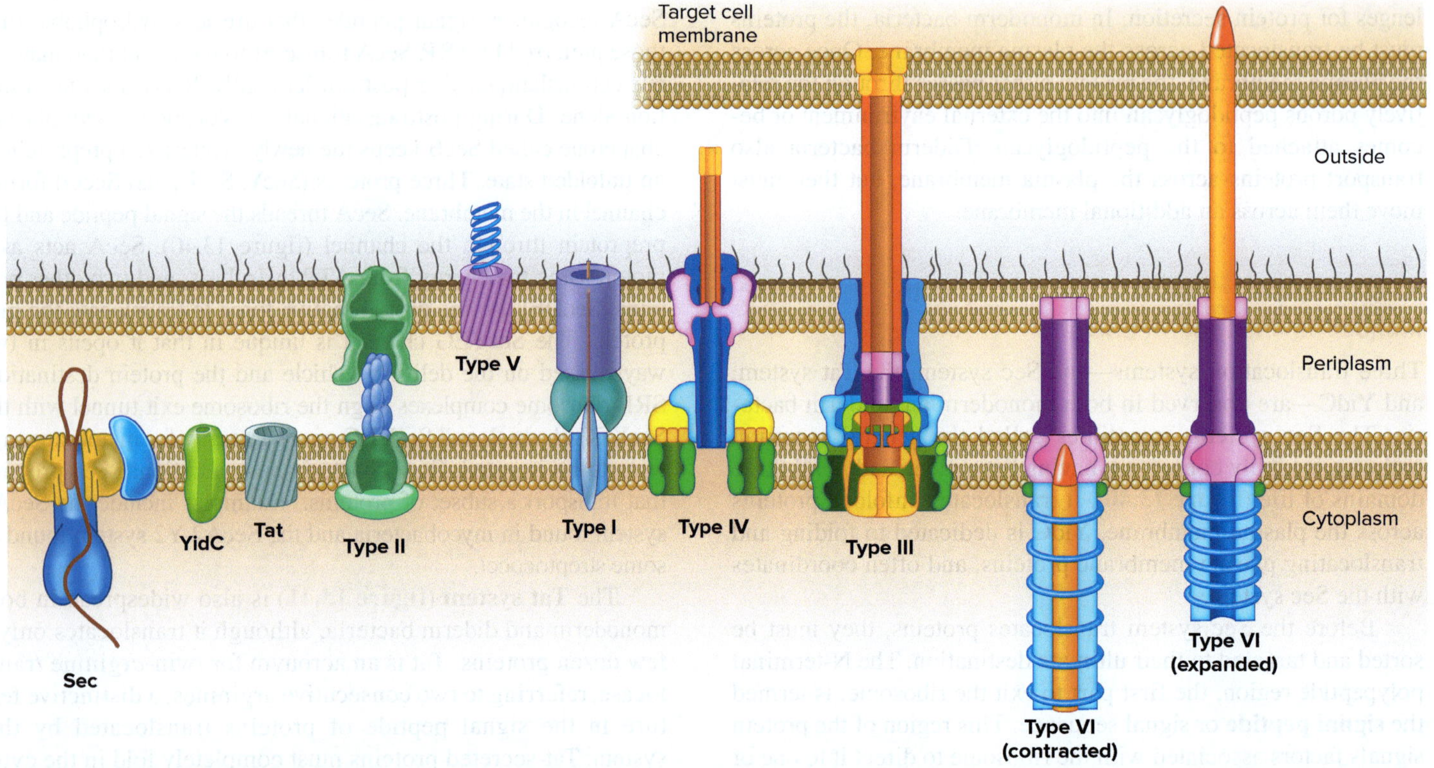

Figure 13.41 Secretion Systems in Bacteria. Translocation systems Sec, YidC, and Tat are shown at the left. These move proteins to the periplasm in diderm bacteria and outside the cell in monoderm bacteria. Types II and V secretion systems export proteins from the periplasm to the outside of a diderm cell. One-step secretion systems move proteins (Type I) and DNA (Type IV) from the cytoplasm directly outside the cell. Types III and VI secretion systems deliver molecules across three membranes, from the bacterial cytoplasm to a target cell cytoplasm. This illustration does not suggest that all translocation and secretion systems exist in any given cell; they are aligned for comparison.

peptidase is a membrane protein that recognizes the signal peptide and hydrolyzes the peptide bond just after it. Finally, for proteins that require disulfide bonds, the enzyme **protein disulfide isomerase (PDI)** joins cysteine residues (with -SH groups) to form disulfides (-S-S-). Because the cytoplasm is a reducing environment, it does not support this oxidation reaction. In diderm bacteria, it occurs in the periplasm, catalyzed by enzymes abbreviated Dsb (*di*sulfide *b*ond). DsbA and DsbB act on proteins as they are released from SecYEG, and DsbA oxidizes cysteines in order from the N-terminus. Not all proteins fold in this pattern, and DsbC acts as both a chaperone and PDI to correct any misformed disulfides. Monoderm bacteria, in general, do not rely on disulfide bond formation. Although some secreted proteins contain disulfides, their maturation pathways remain to be determined.

In monoderm bacteria, a few proteins are anchored to the cell surface by covalent attachment to peptidoglycan. These proteins have a second signal peptide termed a **sorting signal** at the C-terminus that directs them to sortase enzymes embedded in the membrane. **Sortase** catalyzes a reaction between the sorting signal and Lipid II, the peptidoglycan precursor and these surface protein–Lipid II molecules are inserted into the growing cell envelope. ◀ *Synthesis of peptidoglycan occurs in the cytoplasm, at the plasma membrane, and in the periplasmic space (section 12.4)*

Secretion in Diderm Bacteria Requires Crossing Two Membranes

Proteins produced in diderm bacteria destined for secretion or surface display must pass through the plasma membrane, the periplasmic space, and the outer membrane. Some proteins cross in a two-step process. The first step is translocation to the periplasm by the Sec or Tat system, as described, and the second step is secretion across the outer membrane. In contrast, one-step processes involve structures composed of multiple polypeptides that completely span the periplasm (figure 13.41). Secretion systems are numbered (Type I secretion system, Type II secretion system, etc.) in order of discovery. Three of these secretion systems, Types II, V, and IX, comprise the second step in two-step processes; and the remainder constitute one-step systems. Although Type VII secretion systems are found in Gram-positive bacteria, we discuss them here because genera possessing Type VII secretion systems have a diderm cell envelope structure. Several secretion systems are best known for their roles in pathogenicity, where the proteins that they secrete, termed **effectors,** contribute to host injury.

Two-Step Secretion Systems

Type V secretion systems (T5SS) are mechanistically simple in that the protein to be exported passes from the periplasm through a barrel structure in the outer membrane (figure 13.41). The barrel may be either a part of the protein being translocated or a separate polypeptide that is coexpressed and cotranslocated to the periplasm. Chaperones accompany the barrel or barrel domain through the aqueous periplasm to the membrane for insertion and folding. T5SS secrete proteases involved in virulence and adhesins that mediate the attachment of bacterial cells to substrates.

Type II secretion systems (T2SS) are found only in Proteobacteria. A cell may have multiple T2SSs, as each system is dedicated to one or a few proteins. The T2SS is anchored in both membranes by complexes comprised of many proteins, and the two membrane complexes are connected with a pseudopilus (figure 13.41). Proteins for secretion enter from the periplasm and the pseudopilus acts like a piston to push the proteins through the outer membrane. Typical cargo for a T2SS includes degradative enzymes like proteases, lipases, and cellulases. Synthesis of the T2SS is often initiated by quorum sensing. ⏮ *Cell-cell communication within microbial populations (section 7.6)* ⏭ *Proteobacteria (chapter 21)*

Type IX secretion systems (T9SS) also have a narrow phylogenetic distribution. They are limited to the Bacteroidota, where they are responsible for secreting the gliding motility proteins in *Flavobacterium* and the virulence proteases in the dental pathogen *Porphyromonas*. These proteins are translocated by the Sec system and fold in the periplasm. A C-terminal signal peptide tags the proteins for outer membrane localization or surface display. At the outer membrane, the signal peptide is removed and attaches the protein to the lipopolysaccharide using a sortase-like mechanism. ⏭ *Non-flagellar movement—twitching and gliding (section 14.6)*

Type X secretion systems (T10SS) are a recently discovered addition to the array of two-step secretion machines. They are composed of proteins similar to those used by bacteriophage to escape from their bacterial hosts during a lytic infection. Holin proteins act first, puncturing the host cell plasma membrane. These perforations create a portal, which then allows peptidoglycan-degrading enzymes (e.g., endolysins and spanins) to enter the periplasm. The T10SS has been implicated in secreting chitinases in *Serratia marcescens* and toxins in *Yersinia* spp. ⏮ *Double-stranded DNA viruses infect all cell types (section 25.2)*

One-Step Secretion Systems

Type I secretion systems (T1SS) are the simplest of the one-step secretion systems, and a bacterial cell may contain many of these systems for the export of different proteins (figure 13.41). Proteins exported by T1SS include S-layer subunits, degradative enzymes, and proteins important in pathogenesis. Each T1SS has three components: an ABC (*A*TP *b*inding *c*assette) transporter in the plasma membrane (*see figure 3.11*), a membrane fusion protein that bridges the periplasm, and a barrel structure that spans the outer membrane. The ABC component is specific for a single substrate, whereas the outer membrane barrel is common to multiple T1SS. The three components assemble only when the substrate binds to the ABC transporter.

Type III secretion systems (T3SS), also called **injectosomes,** are molecular syringes (*see figure 38.12*) used by pathogens to inject proteins into eukaryotic host cells. The injected proteins are called effector proteins and promote pathogenesis by altering the cell cytoskeleton, signal transduction pathways, or other cellular processes. T3SS have a basal body embedded in the plasma membrane, and an extracellular needle. The basal body includes a channel that directs effector proteins through the periplasm to the hollow needle for delivery to a host cell. The T3SS structure uses energy from both the proton motive force and ATP. T3SS are structurally complex, and the assembly of multiple polypeptides must be carefully regulated. Proteins to be secreted by a T3SS include the distal portions of the T3SS needle, and later the effector proteins targeted to the eukaryotic host. Assembly is initiated in response to both bacterial and host factors, and once assembled, the T3SS remains poised to act until an appropriate target is contacted. The nature of the signal to secrete effectors into the host cell varies with different species. In some cases, chemical signals like a change in pH or calcium concentration initiates secretion, whereas in other cells, the signal appears to be a consequence of mechanical sensing. ⏭ *Plague (section 38.2)*

Type IV secretion systems (T4SS) are most commonly used for DNA transfer, although some are dedicated to the secretion of DNA-protein complexes or proteins. They evolved from the conjugative machinery used to transfer DNA (*see figure 30.13*). Assembly of the T4SS is believed to occur at the cell poles and includes integral membrane proteins in both inner and outer membranes, ATPases in the cytoplasm to provide the energy for secretion, and a pilus. Pili are produced constitutively, but effector secretion requires a host cell for delivery. Understanding how pilus biogenesis and secretion of effector molecules are coupled is an active area of investigation.

Type VI secretion systems (T6SS) are contractile weapons used by bacteria to deliver toxins to neighboring cells, both bacterial and eukaryotic. T6SS are composed of a baseplate in the plasma membrane and a membrane spanning complex in the periplasm. Two concentric cylindrical structures, an outer sheath and an inner tube, form in the cytoplasm and then dock onto the membrane complex (figure 13.41). The sheath and tube assembly may be up to 1 μm long which is about half the length of an average rod-shaped bacterium. In response to a contact signal, the inner tube is ejected, puncturing the target cell and delivering multiple diverse effector proteins to the target cell's cytoplasm. Related bacteria have immunity proteins and are not affected by the secreted toxins. Thus a population of related cells can dominate a niche, as exemplified by *Vibrio cholerae*. In aquatic environments, chitin is a preferred nutrient, and *V. cholerae* competes with other bacteria and predatory amoeba by delivering toxins via the T6SS. Should humans ingest *V. cholerae* via

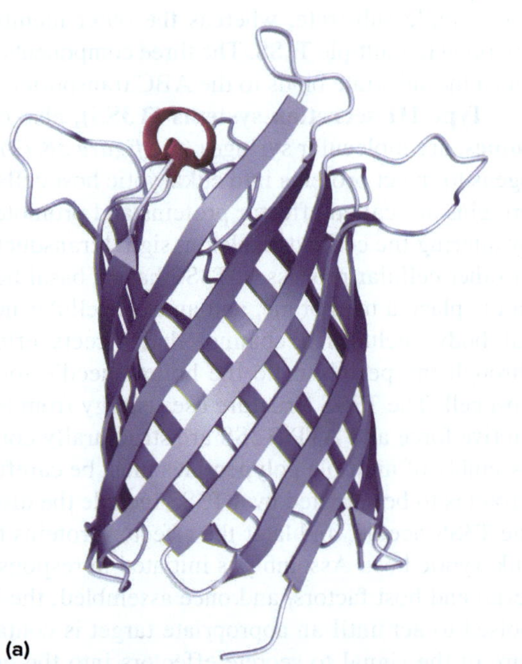

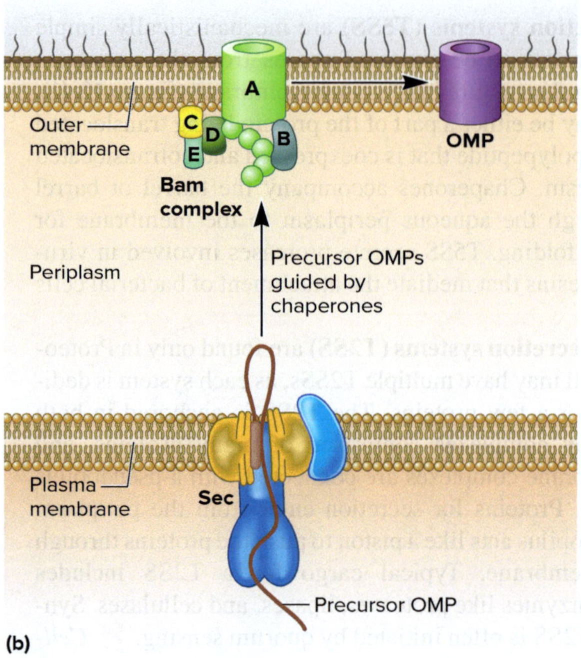

Figure 13.42 The Bam Complex. (a) 3D rendering of OmpG, an outer membrane protein. Note how the beta sheets come together to form a barrel structure. (b) In diderm bacteria, outer membrane proteins (OMPs) are translocated across the plasma membrane as precursors and routed to the Bam complex for assembly into the outer membrane.

(a)VD Image Lab Shutterstock

contaminated water, the same T6SS is used to compete with the gut microbes to establish the infection known as cholera. ▶| *Bacteriophage T4: a virulent bacteriophage (section 25.2); A functional core microbiome is required for host homeostasis (section 33.3); Cholera (section 38.4)*

Type VII secretion systems (T7SS) were initially described in *Mycobacterium* spp. Although these bacteria are classified as Gram-positive, they are considered diderms because they have a cell envelope comprised of a plasma membrane and an outer mycomembrane; the mycomembrane is unique in that it contains mycolic acids, rather than lipopolysaccharides like the outer membrane found in diderms such as *E. coli* (*see figure 22.7*). Protein secretion across both membranes necessitates a secretion system, as found in other diderms. *M. tuberculosis* uses a T7SS to deliver important virulence factors to host immune cells. T7SS have also been characterized in pathogenic species of *Streptococcus*. ▶| Mycobacterium tuberculosis *(section 38.1)*

Assembly of Proteins in Bacterial Outer Membranes

In diderm bacteria, the final destination of some translocated proteins is the outer membrane. Recall how proteins destined for the plasma membrane require the Sec proteins to facilitate their transit. Likewise, proteins localized to the outer membrane also need assistance. Such **outer membrane proteins (OMPs)** have a similar structure: They are barrels composed of beta sheets (**figure 13.42*a***). OMPs are folded and inserted into the outer membrane by a multi-subunit molecular machine called the Bam (*β*-barrel *a*ssembly *m*achine) complex (figure 13.42*b*). Bam is composed of an essential OMP BamA, as well as four associated outer membrane lipoproteins (BamB, BamC, BamD, and BamE). Working together, these proteins receive unfolded OMP precursors—including BamA itself—after they are transported across the plasma membrane. This handoff is assisted by periplasmic chaperone proteins, which prevent the precursor OMPs from aggregating in the aqueous periplasm. It is believed that the Bam complex destabilizes the outer membrane, allowing the emerging OMP to enter the outer membrane. Amazingly, this complex process occurs in the periplasm, an environment devoid of ATP. An analogous system also exists to fold OMPs in mitochondria and chloroplasts, demonstrating the evolutionary history of these organelles (*see figure 1.7*) and the fundamental importance of protein translocation and folding in all living organisms.

Comprehension Check

1. What are molecular chaperones? Describe their functions.
2. What is a signal peptide? Suggest why a protein's signal peptide is not removed until after the protein is translocated across the plasma membrane.
3. How is a signal peptide different from the signal recognition particle?
4. Give the major characteristics and functions of the protein secretion systems described in this section.

Key Concepts

13.1 Experiments Using Bacteria and Viruses Demonstrated That DNA Is the Genetic Material

- DNA is composed of only four different building blocks (nucleotides). Therefore it was originally thought to be too simple to function as an organism's genetic material.
- The knowledge that DNA is the genetic material for cells came from studies on transformation by Griffith and Avery and from experiments on T2 phage by Hershey and Chase (**figures 13.1–13.3**).

13.2 Nucleic Acid and Protein Structure

- DNA and RNA are polymers of deoxyribonucleotides and ribonucleotides, respectively. DNA differs in composition from RNA in having deoxyribose and thymine, rather than ribose and uracil.
- DNA is double stranded, with complementary AT and GC base pairing between the strands. The strands run antiparallel and are twisted into a right-handed double helix (**figure 13.5**).
- RNA is normally single stranded, although it can coil upon itself and base pair to form hairpin structures.
- Proteins are polymers of amino acids (**figure 13.6**) linked by peptide bonds (**figure 13.7**).

13.3 DNA Replication in Bacteria

- Circular bacterial DNAs have a single origin of replication. They are copied by two replication forks to form a theta-shaped (θ) figure (**figures 13.9** and **13.10**).
- The replisome is a complex of proteins responsible for DNA replication (**figures 13.12** and **13.13**).
- DNA polymerase enzymes catalyze the synthesis of DNA in the 5′ to 3′ direction while reading the DNA template in the 3′ to 5′ direction. The double helix is unwound by helicase with the aid of topoisomerases. DNA binding proteins keep the strands separate (**figures 13.11–13.13**).
- DNA polymerase III holoenzyme, the replicative DNA polymerase, synthesizes a complementary DNA strand beginning with a short RNA primer made by the enzyme primase. The leading strand is replicated continuously, whereas DNA synthesis on the lagging strand is discontinuous and forms Okazaki fragments (**figures 13.13** and **13.14**). DNA polymerase I excises the RNA primers and fills in the resulting gap. DNA ligase then joins the fragments together (**figure 13.15**).

13.4 Bacterial Genes Consist of Coding Regions and Other Sequences Important for Gene Function

- A gene is a nucleic acid sequence that codes for a polypeptide, tRNA, or rRNA.
- The template strand of DNA carries genetic information and directs the synthesis of an RNA transcript.
- A gene also contains a promoter, a coding region, and a terminator; it may have a leader and a trailer (**figure 13.19**).
- The genes for tRNA and rRNA often code for a precursor transcript that is subsequently processed to yield several products (**figure 13.20**).

13.5 Transcription in Bacteria

- Genes of related function are often organized into a transcriptional unit called an operon. A single promoter directs transcription of the operon, yielding a polycistronic mRNA. Each coding region in the polycistronic mRNA has its own start and stop codons and gives rise to a single polypeptide (**figure 13.21**).
- RNA synthesis is catalyzed by enzymes called RNA polymerases. Bacterial RNA polymerases are composed of several protein subunits (**figure 13.22**).
- The transcription cycle (**figure 13.23**) includes initiation, elongation, and termination.
- A protein called sigma factor is a transcription factor that directs RNA polymerase to a promoter during transcription initiation. Sigma recognizes certain sequences in the promoter (**figure 13.24**). Once bound, RNA polymerase denatures a short stretch of DNA (**figure 13.25**). It then begins moving down the template strand, synthesizing RNA in a 5′ to 3′ direction (elongation). Transcription termination may be intrinsic (based on mRNA folding) or dependent on protein factors (**figures 13.26** and **13.27**).

13.6 The Genetic Code Consists of Three-Letter "Words"

- Genetic information is carried in the form of 64 nucleotide triplets called codons (**table 13.1**); 61 sense codons direct amino acid incorporation, and three stop codons terminate translation. The code is degenerate (redundant); that is, there is more than one codon for most amino acids.
- Some proteins contain two rare amino acids: selenocysteine and pyrrolysine (**figure 13.30**).

13.7 Translation in Bacteria

- During translation, ribosomes attach to mRNA and synthesize a polypeptide beginning at the N-terminal end.
- Amino acids are activated for protein synthesis by attachment to the 3′ end of transfer RNAs. Activation requires ATP, and the reaction is catalyzed by aminoacyl-tRNA synthetases (**figure 13.32**).
- Ribosomes are large, complex organelles composed of rRNAs and many polypeptides (**figure 13.33**).
- Protein synthesis begins with the formation of the 30S initiation complex made up of the mRNA, initiator

fMet-tRNA, two initiation factors, and the 30S subunit. The initiator tRNA is bound to the start codon on the mRNA. The 50S subunit then binds to form the 70S initiation complex (**figures 13.34** and **13.35**).
- In the elongation cycle, the proper aminoacyl-tRNA binds to the A site (**figure 13.36**). Then the transpeptidation reaction is catalyzed by peptidyl transferase (**figure 13.37**). Finally, during translocation, the peptidyl-tRNA moves to the P site and the ribosome moves down the mRNA by one codon. The empty tRNA leaves the ribosome by way of the exit site.
- Protein synthesis stops when a stop codon is reached. Bacteria require three release factors for stop codon recognition and ribosome dissociation from the mRNA.

13.8 Coordination of Gene Expression Processes

- DNA replication, transcription, and translation occur simultaneously in the cell.
- The replisome and RNA polymerase often collide. Head-on collisions have more severe consequences than rear-end collisions, namely disruption of the replisome, increased likelihood of mutation, and extreme levels of supercoiling (**figure 13.38**).
- A polysome or polyribosome is a complex of mRNA with several ribosomes (**figure 13.39**).

13.9 Protein Maturation and Translocation

- Chaperone proteins help ensure that proteins fold properly or are delivered to their destination site.
- Proteins may require processing to achieve their functional structure. Enzymes may remove the N-formyl group and/or the N-terminal methionine, insert metal cofactors, or oxidize cysteines.
- Many proteins must be transported across the plasma membrane. The most commonly used mechanisms are the Sec system, which is found in all organisms, and the Tat system (**figure 13.40**).
- Protein secretion in diderm bacteria can be a two-step process using the Sec or Tat system and a secretion system located in the outer membrane. Alternately, secretion can occur in a single step when the cargo is passed directly from the cytoplasm through a single complex to the exterior of the cell (**figure 13.41**). In all cases, energy for secretion derives from the cytoplasm, in the form of either ATP or the proton motive force.
- OMPs are assembled in the outer membrane of diderms by the Bam complex (**figure 13.42**).

Active Learning

1. You have isolated several *E. coli* mutants:

   Mutant #1 has a point mutation (a single base-pair change) in the –10 region of the promoter of a gene encoding an enzyme needed for synthesis of the amino acid serine.

   Mutant #2 has a mutation in the –35 region in the promoter of the same gene.

   Mutant #3 is a double mutant with mutations in both the –10 and –35 region of the promoter of the same gene.

   Only Mutant #3 is completely unable to make serine. Why do you think this is so?

2. Compare DNA polymerase and RNA polymerase. How does each enzyme recognize initiation signals in the chromosome DNA sequence? What are their enzymatic activities? What accessory proteins are needed for each enzyme?
3. Review section 3.4 on extracellular vesicles. Should extracellular vesicles be considered a secretion system?

CHAPTER

14

Regulation of Cellular Processes

Mr. Anujak Jaimook/Getty Images

Promoting Expression

If you've ever watched a light show, you know how magnificent and complex a series of electrical switches can be. Technicians use a series of regulators to control each light, making it brighten, dim, or pulse rhythmically. Gene expression, too, can be controlled and coordinated by manipulating promoters—the switches—on individual genes. In fact, this can even result in a bacterial light show.

Scientists have created sophisticated sensors by pairing promoters responsive to a specific stimulus with the gene that encodes green fluorescent protein (GFP). GFP is perfectly named: When made by the cell and excited by ultraviolet (UV) light, it fluoresces green. Fusing the gene that encodes GFP to a promoter effectively creates a switch to turn the green light on and off. The timing and intensity of the green light emitted are a measure of how the promoter is controlled.

In a remarkable demonstration of promoter specificity, scientists built a novel promoter-gene combination. They fused a promoter responsive to the explosive TNT with the gene for GFP and introduced it to *Escherichia coli (E. coli)*. This *E. coli* biosensor glows green after exposure to TNT. In the field, it is sensitive enough to detect the tiny amount of TNT gas that escapes from unexploded land mines buried in sand. This novel application of stimulating a promoter to turn on a light could save life and limb for people living near zones of armed conflict.

Despite the simplicity of this TNT-GFP biosensor, promoters are rarely as simple as an on/off switch. More commonly, cellular components interact with promoters and act as rheostats, or dimmers, to increase or decrease the level of transcription. Gene expression can also be controlled by adjusting the activity of the message, or its translation. In this chapter, we explore mechanisms of gene regulation and how they interact in cellular processes.

Readiness Check:

Based on what you have learned previously, you should be able to:

- ✓ Distinguish secondary and tertiary structures in RNA
- ✓ Describe the phosphoenolpyruvate: sugar phosphotransferase system (section 3.2)
- ✓ Describe the mechanisms often used to regulate enzyme activity (section 10.7)
- ✓ Discuss genome structure and organization of genetic information in bacterial cells (sections 13.3 and 13.4)
- ✓ Summarize the events that occur during the three phases of transcription and translation (sections 13.5 and 13.7)

14.1 Bacteria Use Many Regulatory Strategies

After reading this section, you should be able to:

a. Differentiate constitutive and regulated gene expression
b. List when, during the flow of genetic information, bacterial cells can regulate gene expression

The work of a bacterial cell is performed by its proteins. Many proteins are continuously used for basic cellular functions, like the enzymes of the central metabolic pathways, or the enzymes that build structures like peptidoglycan. The genes that encode these enzymes are regularly transcribed and translated to provide a constant supply of active enzymes. These are termed **constitutive genes,** or **housekeeping genes.** In contrast, other enzymes are required only in specific conditions or environments. For example, at daybreak, a photosynthetic microbe detects light and begins to express the genes that encode the proteins that will capture light energy. To conserve energy and cellular materials, genes encoding these proteins are expressed only when needed, and their expression is *regulated.* **Figure 14.1** diagrams how cells convert environmental signals to gene expression. Bacterial cells must have a sensor to detect an environmental stimulus, and in turn transmit this input signal to molecules that ultimately interact with the machinery that controls gene expression. When gene expression is regulated by controlling mRNA synthesis, it is said to be at the *transcriptional level.* Likewise, when gene expression is governed by controlling protein synthesis, it is said to be at the *translational level.* Translation yields a protein that may or may not be functional. The activity level of proteins may be altered by posttranslational modification, providing yet another potential level of control (*see figure 10.19*).

Transcription can be regulated by governing its initiation or its elongation (*see figure 13.23*). We begin our discussion by introducing two phenomena: induction and repression of enzyme synthesis. Induction and repression provided the first models for gene regulation. These models involve the action of regulatory proteins, and the notion that gene expression is regulated solely by DNA-binding proteins persisted for many years. Eventually it was demonstrated that RNA molecules also have regulatory functions.

14.2 Regulation of Transcription Initiation Saves Considerable Energy and Materials

After reading this section, you should be able to:

a. Compare and contrast constitutive, inducible, and repressible genes
b. Describe the helix-turn-helix motif in DNA-binding proteins
c. Summarize how negative transcriptional control and positive transcriptional control can be used to regulate inducible and repressible genes
d. Contrast the impact of regulating transcription at the level of initiation vs. elongation

Induction and repression were the first regulatory processes for gene expression to be understood in any detail. In this section, we first describe these phenomena and then examine the underlying regulatory mechanisms.

Induction and Repression of Enzyme Synthesis

The appearance of a food source in an organism's environment is one of the most basic stimuli to elicit a change in gene expression. The enzyme β-galactosidase catalyzes hydrolysis of the disaccharide lactose to glucose and galactose (**figure 14.2**). When *E. coli* grows with lactose as its only carbon source, each cell contains about 3,000 β-galactosidase molecules; in the absence of lactose, there are fewer than three. β-galactosidase is an inducible enzyme; this means that the number of β-galactosidase molecules rises in the presence of a particular stimulus. This stimulus is a small effector molecule called an **inducer** (in this case, the lactose derivative allolactose). Likewise, the genes that encode inducible enzymes such as β-galactosidase are referred to as **inducible genes.**

β-galactosidase functions in a catabolic pathway, and many catabolic enzymes are inducible. On the other hand, the genes for enzymes involved in biosynthetic pathways are **repressible genes,** and their products are repressible enzymes. For instance, an amino acid present in the surroundings may inhibit the formation of enzymes responsible for its biosynthesis. This makes sense because the microorganism does not need biosynthetic enzymes for a particular substance if it is already available. In this context, the small effector molecule is termed a **corepressor.** Generally, repressible enzymes are responsible for biosynthesis and are present unless the end product of their pathway is available. Inducible enzymes, in contrast, are required only when their substrate is available; they are missing in the absence of the inducer.

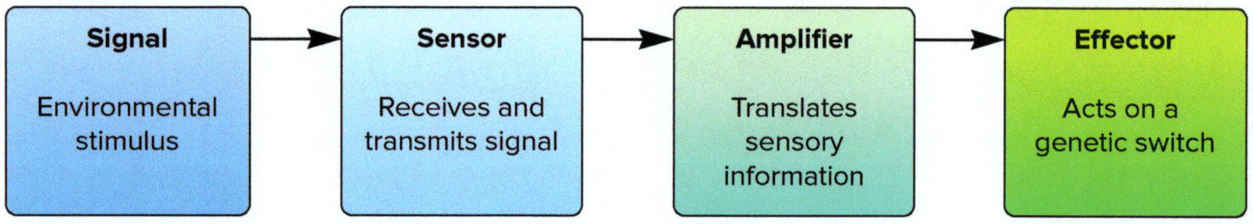

Figure 14.1 Cellular Process Used to Convert an Environmental Signal to Gene Expression.

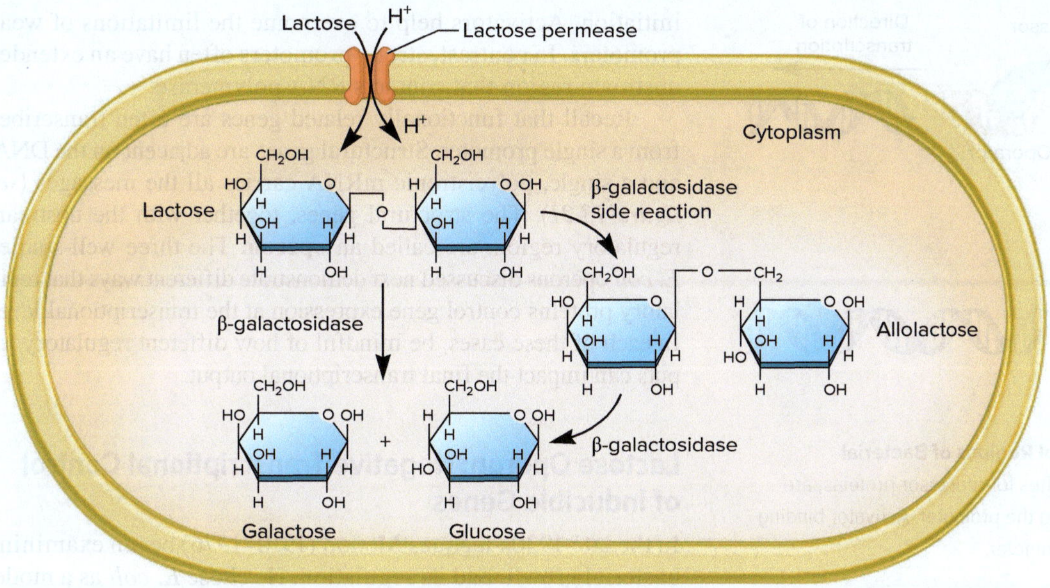

Figure 14.2 The Reactions of β-Galactosidase. The main reaction catalyzed by β-galactosidase is the hydrolysis of lactose, a disaccharide, into the monosaccharides galactose and glucose. The enzyme also catalyzes a minor reaction that converts lactose to allolactose. Allolactose acts as the inducer of β-galactosidase synthesis.

Regulatory Proteins Can Control Transcription Initiation

Bacterial transcription initiation occurs when the sigma subunit of RNA polymerase positions the core enzyme at a promoter. Regulation of transcription initiation, therefore, occurs through the action of **transcription factors.** These regulatory proteins also bind to DNA in the vicinity of a promoter. Many transcription factors are organized into two domains. A **domain** is a portion of a protein with a specific structure and function. In the case of transcription factors, a DNA-binding domain interacts with the promoter, while another domain of the protein interacts with the effector molecule. Effector binding causes a conformational change to the DNA-binding domain, thus affecting its ability to interact with promoter DNA. This is an example of allosteric control. *Allosteric regulation (section 10.7)*

Transcriptional regulatory proteins commonly have a helix-turn-helix structure consisting of two α-helices (*see figure A1.14*) separated by a β-turn. When bound to DNA, one of the helices is positioned in the major groove of the DNA. This *recognition helix* recognizes a specific nucleotide sequence and directly contacts the DNA. Most transcriptional regulatory proteins exist as dimers—that is, two copies of a single polypeptide. When two regulatory polypeptides join to form a dimer, two recognition helices, one from each subunit, interact with the major groove of DNA (**figure 14.3**). Allosteric changes to the effector-binding domain result in the helices shifting such that they are no longer compatible with the major groove, and therefore unable to contact the DNA bases.

Transcriptional regulatory proteins can exert either negative or positive control. **Negative control** occurs when a regulatory protein, termed a **repressor,** binds to DNA to inhibit transcription initiation. **Positive control** occurs when a regulatory protein, termed an **activator,** binds to DNA to boost transcription initiation. *Regulatory Proteins*

Repressor proteins bind to a DNA region called the **operator,** which usually overlaps or is downstream of the promoter (i.e., closer to the coding region) (**figure 14.4**). When bound to the operator, the repressor either blocks binding of RNA polymerase to the promoter or prevents its movement toward the coding sequences. Activator proteins attach to **activator-binding sites** (figure 14.4). These are often upstream of the promoter (i.e., farther away from the coding region). Binding of an activator to its regulatory site enables RNA polymerase binding. Most often, the RNA

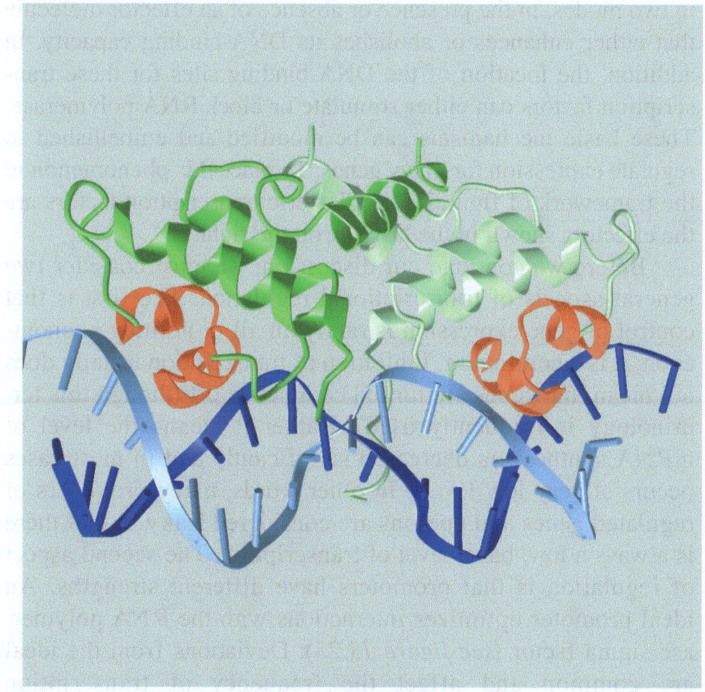

5′ – AACACCGTGCGTGTT – 3′
3′ – TTGTGGCACGCACAA – 5′

Figure 14.3 Helix-Turn-Helix Motif Contacts DNA in the Major Groove. The two identical subunits of the transcriptional regulatory protein are in shades of green; both helix-turn-helix segments are in red. The two DNA strands are in shades of blue. Below the image is an example of a palindromic sequence recognized by a helix-turn-helix regulator. The sequence of bases is palindromic—that is, they are identical when read forward or backward.

Source: Structure 1LMB from RCSB PDB.

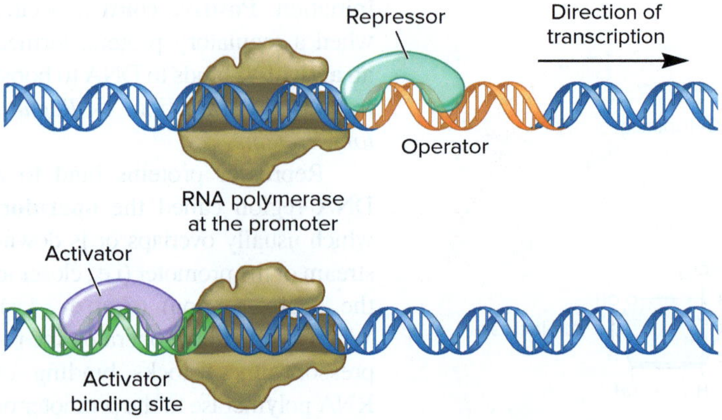

Figure 14.4 The Transcriptional Control Regions of Bacterial Genes. Operators (orange), the binding sites for repressor proteins, are located either downstream of or overlapping the promoter. Activator binding sites (green) are located upstream of the promoter.

polymerase's alpha or sigma subunits contact the activator to facilitate promoter binding.

We have just described how transcription factors can operate in two modes, in the presence or absence of an effector molecule that either enhances or abolishes its DNA-binding capacity. In addition, the location of the DNA-binding sites for these transcription factors can either stimulate or block RNA polymerase. These basic mechanisms can be modified and embellished to regulate expression for many genes. To place this phenomenon in the framework of figure 14.1, allosteric transcription factors are the effectors shown in the last box on the right.

Before we continue our discussion, we must consider two general aspects of transcriptional regulation. The first is that control of gene expression is rarely an all-or-nothing phenomenon; it is a continuum. Inhibition of transcription usually does not mean that genes are turned completely off (though this terminology is frequently used). Rather it means the level of mRNA synthesis is decreased significantly and in most cases occurs at very low levels. In other words, many promoters of regulated genes and operons are considered leaky in that there is always a low, basal level of transcription. The second aspect of regulation is that promoters have different strengths. An ideal promoter optimizes interactions with the RNA polymerase sigma factor (*see figure 13.25*). Deviations from the ideal are common and affect the frequency of transcription initiation. Activators help to overcome the limitations of weak promoters. In contrast, strong promoters often have an extended upstream region that contacts RNA polymerase.

Recall that functionally related genes are often transcribed from a single promoter. Structural genes are adjacent on the DNA, and a single, polycistronic mRNA carries all the messages (*see figure 13.21*). The structural genes, together with the upstream regulatory region, are called an **operon.** The three well-studied *E. coli* operons discussed next demonstrate different ways that regulatory proteins control gene expression at the transcriptional level. In each of these cases, be mindful of how different regulatory inputs can impact the final transcriptional output.

Lactose Operon: Negative Transcriptional Control of Inducible Genes

In the late 1930s Jacques Monod (1910–1976) began examining bacterial growth and its regulation. He chose *E. coli* as a model and focused on genes involved in its growth on lactose. He was joined by François Jacob (1920–2013); together they developed the concept of an operon. Today the lactose (*lac*) operon of *E. coli* is probably the best-studied example of negative transcriptional control.

The *lac* operon contains three structural genes controlled by the *lac* repressor (LacI). LacI is encoded by *lacI,* which is located upstream of the *lac* operon (**figure 14.5**). Note the convention in use here—gene names are written in italics while regular font is used for protein names. The first gene in the operon (*lacZ*) codes for β-galactosidase; the second gene (*lacY*) directs the synthesis of lactose permease, the membrane transporter responsible for lactose uptake. The third gene (*lacA*) encodes β-galactoside transacetylase; while its biochemical activity has been known for decades, the cellular function of LacA remains uncertain. The presence of these genes in the same operon ensures that the rates of lactose uptake and breakdown will vary together.

Lactose is one of many organic molecules *E. coli* uses as a carbon and energy source. It is most efficient to synthesize enzymes of the *lac* operon only when lactose is available and a preferred carbon and energy source is not; the *lac* repressor is responsible for preventing transcription when there is no lactose available.

The *lac* repressor is composed of four identical LacI polypeptides (i.e., it is a tetramer), each with a helix-turn-helix DNA-binding domain. The tetramer is formed when two dimers interact.

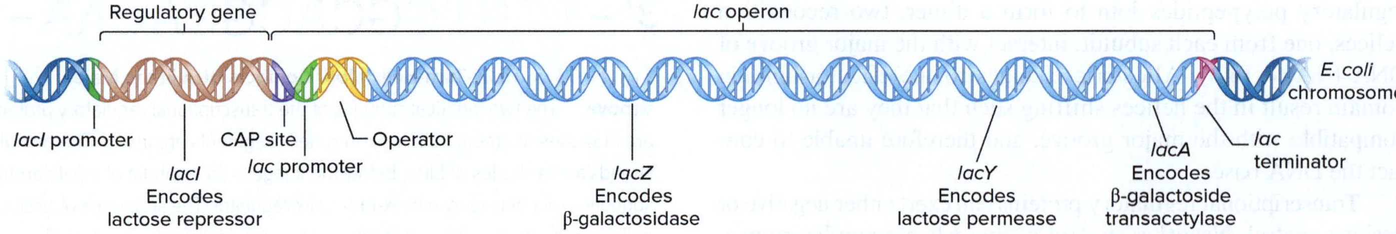

Figure 14.5 The *lac* Operon.

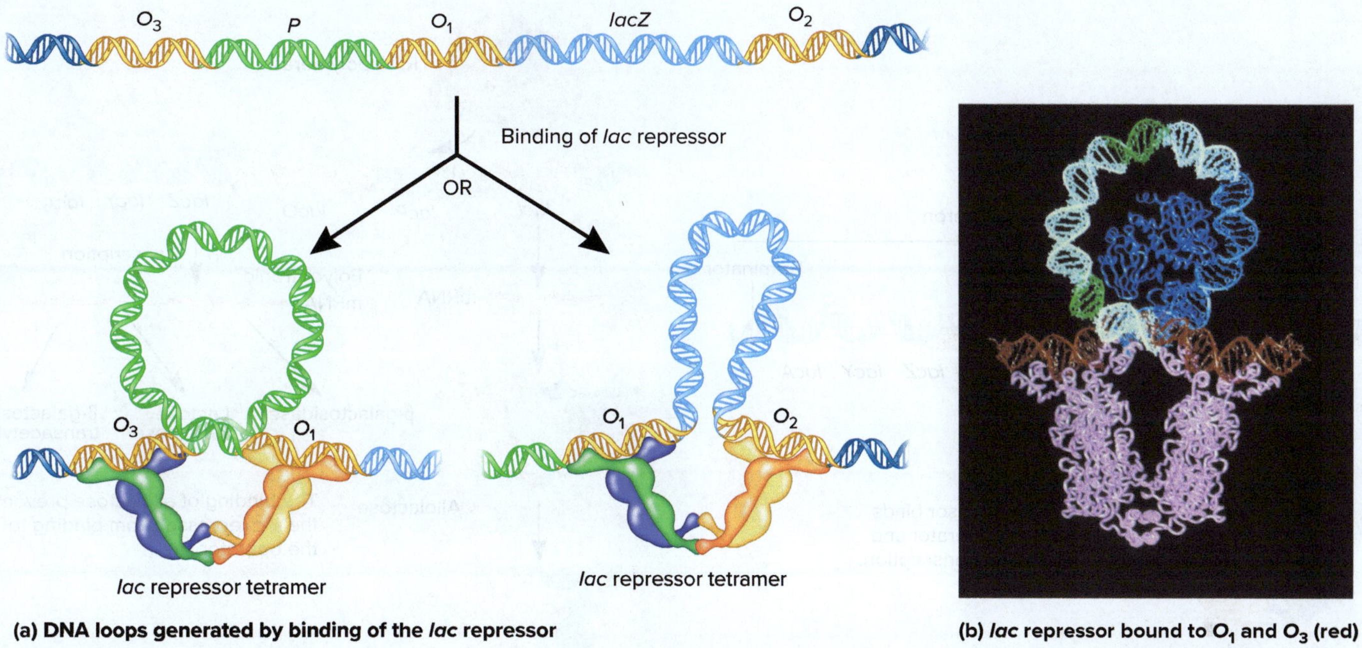

Figure 14.6 The *lac* Operator Sites. (a) The *lac* repressor binds O_1 and one of the other operator sites, forming a DNA loop that contains the −35 and −10 binding sites recognized by RNA polymerase holoenzyme. In this configuration, the operator sites are inaccessible and transcription is blocked. (b) The DNA loop also contains the CAP binding site, and CAP (blue) is shown bound to the DNA.

(b) SPL/Science Source

In the absence of lactose, each dimer of the tetramer recognizes and tightly binds one of three different *lac* operator sites: O_1, O_2, and O_3 (**figure 14.6**). O_1 is the main operator site and must be bound by the repressor if transcription is to be blocked. When one LacI dimer is at O_1 and another is at one of the two other operator sites, the dimers bring the two operator sites close together, with a loop of DNA forming between them. Attachment of the *lac* repressor is a two-step process. First, it binds nonspecifically to DNA. Then it rapidly slides along the major groove until it reaches an operator site. Its helix-turn-helix motifs recognize the DNA and the repressor remains bound to the operator.

How does the repressor inhibit transcription? The *lac* promoter partially overlaps the *lac* operator sites. When there is no lactose, the repressor binds O_1 and one of the other operator sites, bending the DNA in the promoter region. This prevents transcription initiation either because RNA polymerase cannot access the promoter or because it is blocked from moving into the coding region (**figure 14.7*a***). When lactose is available, it is brought into the cell by lactose permease. Once inside the cell, β-galactosidase converts lactose to allolactose, the inducer of the operon (figure 14.2). This occurs because there is always a low level of both enzymes. This illustrates why leaky repression is a feature—not a bug—of transcriptional control! Allolactose noncovalently interacts with the *lac* repressor protein and causes a conformational change, liberating *lac* repressor from the promoter region. RNA polymerase can now proceed and transcription of the operon occurs (figure 14.7*b*).

Close examination of figures 14.5 and 14.6 clearly shows that regulation of the *lac* operon is even more complicated. That is because the *lac* operon is regulated by a second regulatory protein called *c*atabolite *a*ctivator *p*rotein (CAP). CAP functions in a global regulatory network that controls many genes. This network allows *E. coli* to use glucose preferentially over all other carbon and energy sources by a mechanism called *catabolite repression*. The use of two different regulatory proteins to control the expression of an operon illustrates another important point about regulatory processes—there are often layers of regulation. In the case of the *lac* operon, the *lac* repressor regulates gene expression in response to the presence or absence of lactose. CAP regulates the operon (and all others) in response to the presence or absence of glucose. As described in section 14.5, the use of two different regulatory mechanisms generates a continuum of expression levels. The highest levels of transcription occur when lactose is available and glucose is not; the lowest levels occur when lactose is not available and glucose is. For almost all of the examples described in this chapter, regulation is achieved by utilizing more than one mechanism. lac *Operon Induction*

Tryptophan Operon: Negative Transcriptional Control of Repressible Genes

The tryptophan (*trp*) operon of *E. coli* consists of five structural genes that encode enzymes needed for synthesis of the amino acid tryptophan (**figure 14.8**). It is regulated by the *trp* repressor (TrpR), a transcription factor encoded by the *trpR* gene. Because the enzymes encoded by the *trp* operon function in a biosynthetic pathway, it is wasteful to make them when tryptophan is readily available. Therefore this operon is transcribed only when tryptophan must be made de novo from precursor molecules. To accomplish this regulatory goal, TrpR is synthesized in an

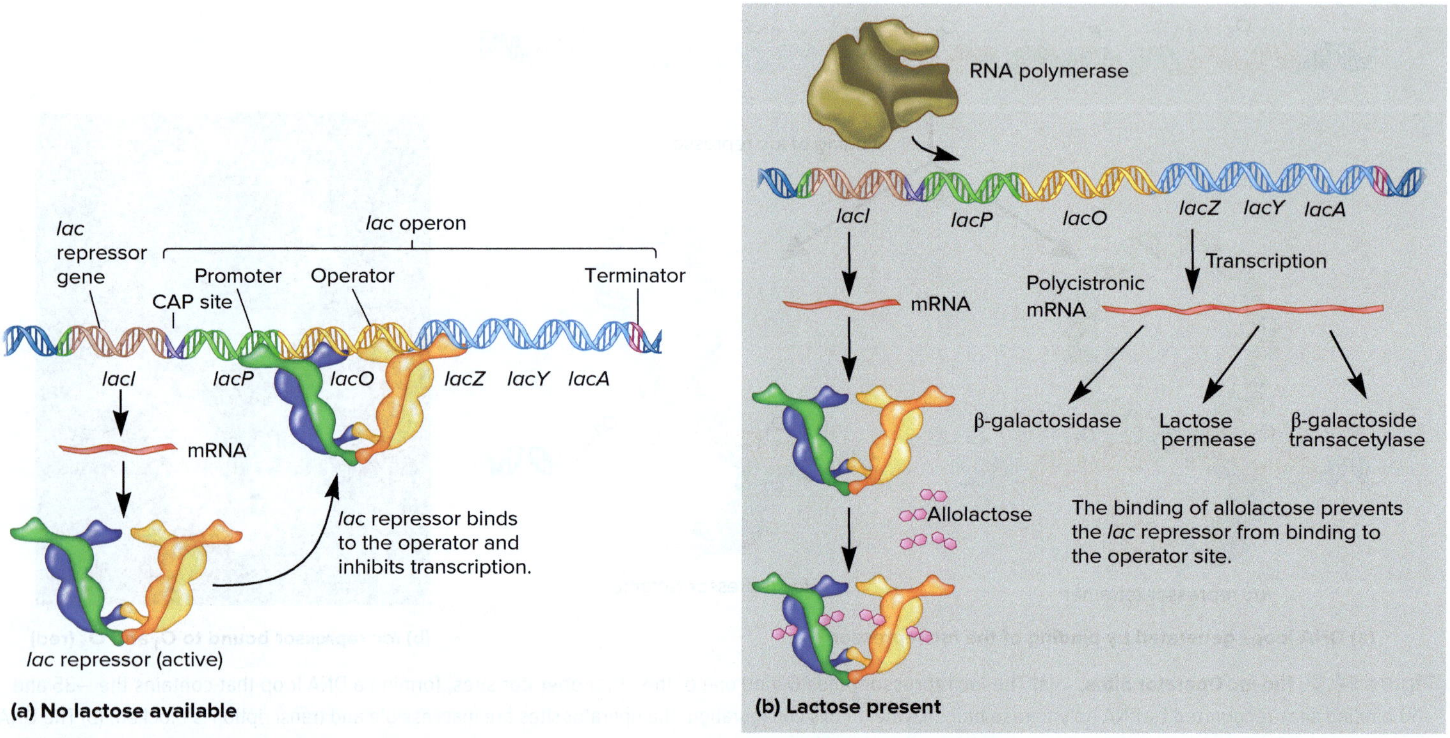

Figure 14.7 Regulation of the *lac* Operon by the *lac* Repressor. (a) The *lac* repressor (encoded by *lacI*) is active and binds the operator when allolactose is not present. Repressor binding to the operator inhibits transcription. (b) When lactose is available, some of it is converted to allolactose by β-galactosidase. When sufficient amounts of allolactose are present, it binds and inactivates the *lac* repressor. The repressor leaves the operator and RNA polymerase is free to initiate transcription.

MICRO INQUIRY *Is allolactose a corepressor or inducer molecule? Explain your answer.*

inactive form that cannot bind the *trp* operator as long as tryptophan levels are low (figure 14.8*a*). When tryptophan is present, it acts as a corepressor, binding one domain of TrpR and thereby altering the orientation of its helix-turn-helix motif. The repressor-corepressor complex then attaches to the operator, blocking transcription initiation (figure 14.8*b*).

Like the *lac* operon, the *trp* operon is subject to another layer of regulation. In addition to being controlled at the level of transcription initiation by TrpR, expression of the *trp* operon is also controlled at the level of transcription elongation by a process called attenuation. This mode of regulation is discussed in section 14.3. *Tryptophan Repressor*

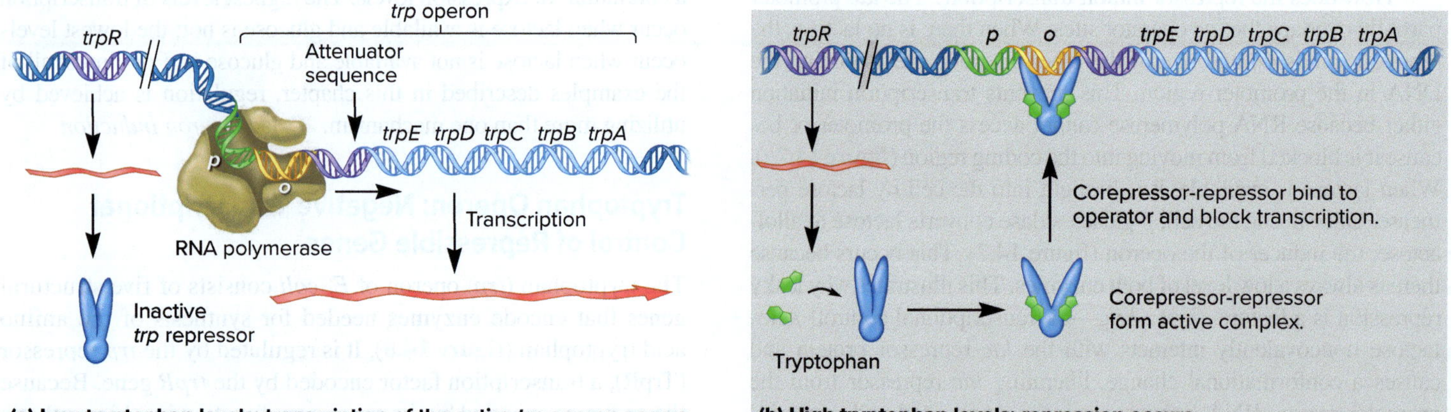

Figure 14.8 Regulation of the *trp* Operon by Tryptophan and TrpR. When synthesized, TrpR is unable to bind the operator. It is activated by the binding of tryptophan, its corepressor. (a) When tryptophan levels are low, TrpR is inactive and transcription occurs. (b) When tryptophan levels are sufficiently high, it binds TrpR. The repressor-corepressor complex binds the operator, and transcription is inhibited.

Arabinose Operon: Transcriptional Control by a Protein That Acts Both Positively and Negatively

Regulation of the *E. coli* arabinose (*ara*) operon illustrates how the same protein can function either positively or negatively, depending on the environmental conditions. The *ara* operon encodes enzymes needed for catabolism of arabinose to xylulose 5-phosphate, an intermediate of the pentose phosphate pathway. The *ara* operon is regulated by the transcription factor AraC, which functions as a dimer and can interact with three different regulatory sequences: *araO*$_2$, *araO*$_1$, and *araI* (**figure 14.9**). When arabinose is present, it binds AraC, inducing a conformational change that promotes DNA binding at *araI* (figure 14.9*a*). In this configuration, AraC functions as an activator that drives transcription of the *araBAD* structural genes. In the absence of arabinose, AraC adopts a different conformation (figure 14.9*b*). In this case, one molecule of AraC binds *araI* while another binds *araO*$_2$. The two AraC proteins interact with each other, causing the DNA between the two sites to form a loop. Genetic analysis shows that in the absence of arabinose, AraC preferentially binds to *araO*$_2$ over *araI* and both AraC subunits must bind *araI* for activation to occur. In this situation, rather than acting as a typical repressor that blocks transcription from occurring, AraC acts as an *anti-activator*. This means that by preventing AraC from functioning as an activator (figure 14.9*a*), when AraC is in the anti-activator conformation (figure 14.9*b*), the overall level of transcription is reduced. The *ara* operon, like the *lac* operon, is also subject to catabolite repression (section 14.5). *Pentose phosphate pathway: a major producer of reducing power for anabolic reactions (section 11.3)*

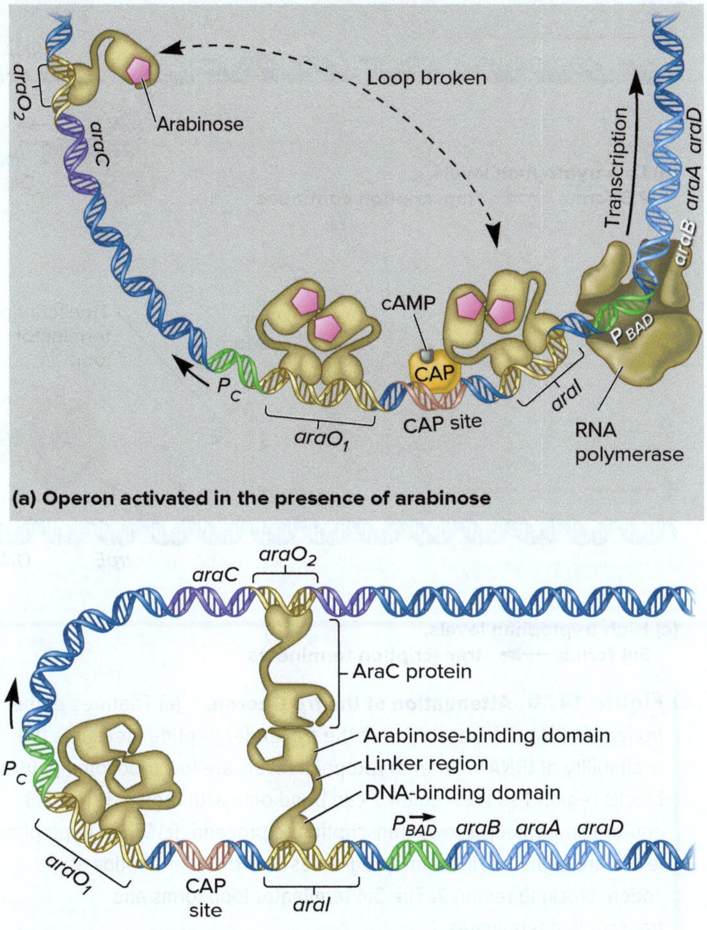

(a) Operon activated in the presence of arabinose

(b) Operon inhibited in the absence of arabinose

Figure 14.9 Regulation of the *ara* Operon by the AraC Transcription Factor. AraC can act both as an activator and as an anti-activator, depending on the presence or absence of arabinose. (a) When arabinose is present, it binds AraC, allosterically disrupting the interaction between the two AraC proteins. Subsequently, two AraC proteins, each bound to arabinose, form a dimer, which binds to *araI*. The AraC dimer functions as an activator. (b) When arabinose is not available, AraC acts as an anti-activator. Two AraC proteins bind at *araI* and *araO*$_2$ and bend the DNA between the two sites, making it inaccessible to RNA polymerase.

Comprehension Check

1. Many genes and operons are regulated at the level of transcription initiation. Why is it advantageous to regulate gene expression at the transcriptional level, rather than during or after translation?
2. What is the difference between induction and repression? How do bacteria use both to respond to changing nutrient supplies?
3. Diagram how negative control would work with both an inducible gene and a repressible gene. Do the same for positive control with both types of genes.

14.3 Attenuation and Riboswitches Stop Transcription Prematurely

After reading this section, you should be able to:

a. Describe in general terms how the *trp* operon of *E. coli* is regulated by transcription attenuation
b. Compare and contrast regulation of transcription elongation by attenuation and riboswitches
c. Explain why the coupling of transcription and translation in bacterial cells is important to the regulatory mechanisms they use

The regulatory circuits described in the previous section control transcription by proteins binding to the promoter region of the DNA. RNA, the product of transcription, also plays a regulatory role in gene expression. Although it is synthesized as a single-stranded molecule, RNA adopts secondary structures (short double-stranded regions) and tertiary structures (three-dimensional configurations) that allow it to interface with proteins and other molecules. In the regulatory mechanisms presented here, mRNAs have a long leader region, and RNA folding patterns in the leader affect transcription, specifically termination. In this type of regulation, transcription is initiated but can be prematurely stopped depending on environmental conditions and the needs of the organism. Attenuation was elucidated by Charles Yanofsky in his studies of the *E. coli trp* operon. More recently riboswitches have been described. These regulatory sequences in the mRNA leader both sense and respond to environmental conditions by either terminating transcription or blocking translation.

Attenuation: Ribosome Behavior Influences Transcription

As noted in section 14.2, the *trp* operon is under the control of a repressor protein with excess tryptophan acting as a corepressor (figure 14.8). Although the operon is regulated mainly by repression, transcription termination is also subject to control. That is, there are two decision points involved in transcriptional control: initiation and elongation into the operon's structural genes. This additional level of control uses a sensor to measure tryptophan availability through a genetic switch. Working together, the two control mechanisms adjust transcription in a subtle fashion. When the repressor protein is not active, RNA polymerase transcribes the leader. However, RNA polymerase often does not progress to the first structural gene in the operon but rather terminates within the leader; this is called **attenuation.** ◀ *Protein-coding genes (section 13.4)*

Transcription attenuation is based on two things: the folding pattern of the leader RNA and the fact that transcription and translation are *coupled* in bacteria (*see figure 13.39*). The RNA leader within the *trp* operon, defined as the region from the transcription start site to the first structural gene of the operon, is unusually long, at 162 base pairs. Recall from chapter 13 that an RNA leader is transcribed but not translated. The *trp* operon leader is unusual because a portion of the leader RNA is translated, giving rise to a small peptide called the *leader peptide*. In addition to encoding the leader peptide, the leader RNA contains **attenuator** sequences (**figure 14.10**). When transcribed, the attenuator forms stem-loop secondary structures in the leader RNA. We define these regions numerically (1, 2, 3, and 4). When regions 1 and 2 pair (1:2; figure 14.10*a*), they form a structure called the pause loop. The terminator loop is made when regions 3 and 4 base pair (3:4; figure 14.10*a*). A poly-U sequence follows the terminator loop, just as it does in intrinsic termination (*see figure 13.26*). However, in this case, the terminator is in the leader RNA, rather than at the end of the operon. Alternatively, a different stem-loop structure can form in the RNA by pairing regions 2 and 3 (2:3, figure 14.10*b*). The formation of this anti-terminator loop prevents the terminator loop from forming, and RNA polymerase continues to transcribe the operon.

Transcription of the *trp* operon only occurs when tryptophan is needed for protein synthesis. The two alternate structures in the RNA leader dictate either transcription termination (when tryptophan supplies are sufficient) or transcription elongation (when tryptophan supplies are low). What determines which of the two alternate structures forms in the leader RNA?

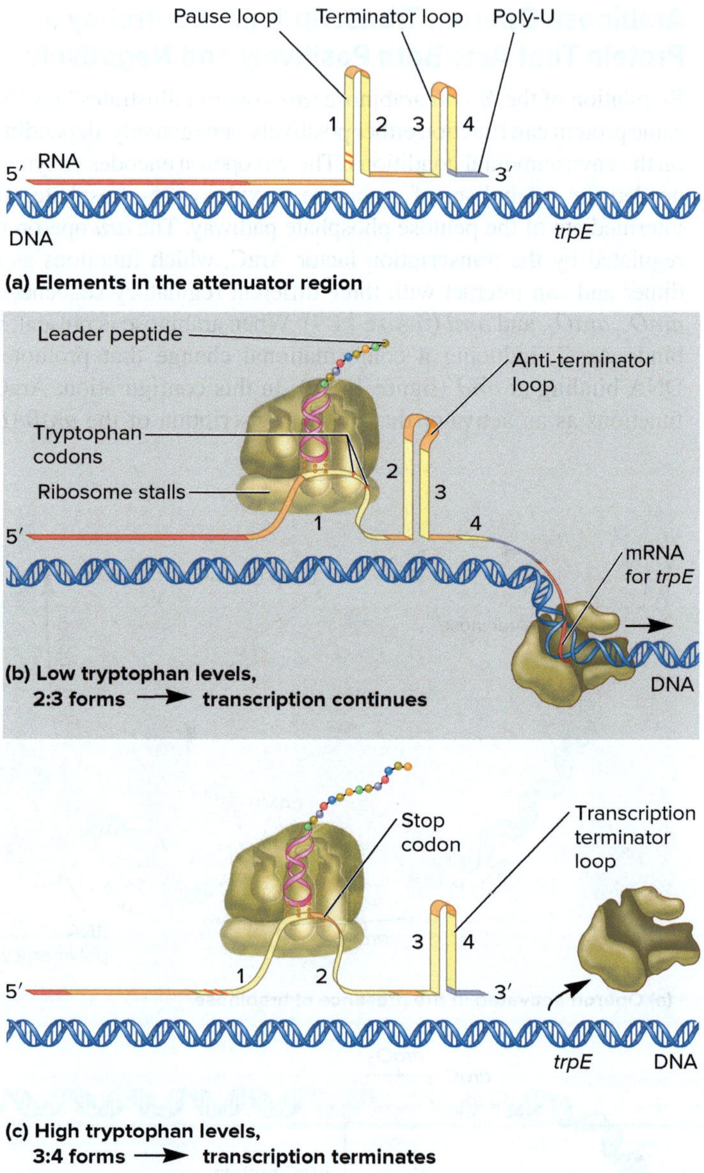

Figure 14.10 Attenuation of the *trp* Operon. (a) Features of the *trp* leader RNA. (b) Translation of the *trp* leader peptide measures the availability of $tRNA^{Trp}$. When tryptophan levels are low, ribosome pausing blocks region 1 so that region 2 can bond only with region 3. The 2:3 anti-terminator loop allows transcription to proceed. (c) When tryptophan levels are high, the ribosome progresses to the leader peptide stop codon, blocking region 2. The 3:4 terminator loop forms and transcription terminates.

MICRO INQUIRY *How does this attenuation respond to rates of protein synthesis in addition to the intracellular level of tryptophan?*

Coordination between RNA polymerase and the ribosome is critical to attenuation. As RNA polymerase moves along the template, a ribosome binds to the Shine-Dalgarno sequence on the RNA to initiate translation of the leader peptide. Among the first several nucleotides of the leader RNA are two tryptophan codons; this is unusual because normally there is only one tryptophan residue per 100 amino acids in *E. coli* proteins. If tryptophan levels are low, there will not be enough charged $tRNA^{trp}$ to fill the A site of the ribosome when it encounters the two tryptophan codons, and it will stall in region 1 (figure 14.10*b*). Meanwhile RNA polymerase continues transcribing, moving away from the stalled ribosome. The presence of the stalled ribosome on region 1 prevents region 1 from base-pairing with region 2. As RNA polymerase continues, region 3 is transcribed, and the 2:3 anti-terminator loop forms. This prevents the formation of the 3:4 terminator loop, and RNA polymerase continues into the *trp* structural genes. If, on the other hand, there is an abundance of charged $tRNA^{trp}$, the ribosome will translate the two tryptophan codons in the leader peptide without hesitation and continue until reaching the UGA stop codon within region 2. Regions 3 and 4 are transcribed, the terminator loop forms, and RNA polymerase is ejected from the DNA template (figure 14.10*c*). *The genetic code consists of three-letter "words"(section 13.6)*

Attenuation measures the level of charged $tRNA^{trp}$ that is available for translation based on the speed with which a ribosome clears the attenuator region. A "too slow" signal releases RNA polymerase to transcribe the operon, ultimately resulting in tryptophan synthesis. In contrast, a fast-moving ribosome forces the RNA polymerase to terminate. Importantly, attenuation occurs on the RNA transcript. It is only after transcription initiates that attenuation further regulates at the level of transcription elongation.

How is attenuation useful to *E. coli?* If the bacterium is deficient in an amino acid other than tryptophan, protein synthesis will slow and $tRNA^{Trp}$ will accumulate. Transcription of the *trp* operon will be inhibited by attenuation. When the bacterium begins to synthesize protein rapidly, tryptophan may be scarce and the concentration of $tRNA^{trp}$ may be low. This stimulates transcription, resulting in larger quantities of the tryptophan biosynthetic enzymes. Acting together, repression (through TrpR, figure 14.8) and attenuation can coordinate the rate of synthesis of amino acid biosynthetic enzymes (based on the availability of amino acid end products) with the overall rate of protein synthesis. When tryptophan is present at high concentrations, RNA polymerases not blocked by the repressor protein probably will not get past the attenuator sequence. Repression decreases transcription about 70-fold and attenuation slows it another 8- to 10-fold; when both mechanisms operate together, transcription can be slowed about 600-fold.

Attenuation is important in regulating other operons that encode amino acid biosynthetic enzymes. Their leader peptide sequences require multiple charged tRNAs for that amino acid, and are followed by an attenuator that is a terminator sequence.

Riboswitches: Effector–mRNA Interaction Regulates Transcription

Like gene regulation by attenuation, regulation by a **riboswitch** is determined by alternative folding patterns in the leader region of RNA. In one configuration, a transcriptional terminator (*see figure 13.26*) forms and the regulated genes located downstream are not transcribed. In the alternative configuration, the terminator cannot form and the operon is transcribed. Which of the two leader folding patterns occurs is determined by the presence or absence of a small effector molecule, which triggers the switch. What makes riboswitches unique is that the RNA folding pattern is determined by effector binding directly to the RNA—a capability previously thought confined to proteins. Other riboswitches function at the translational level, and are considered in section 14.4.

In *Bacillus subtilis,* the production of riboflavin biosynthetic enzymes from the *rfn* operon is regulated by the intracellular concentration of flavin mononucleotide (FMN), which is derived from riboflavin (**figure 14.11**). Upon transcription, sequences in the

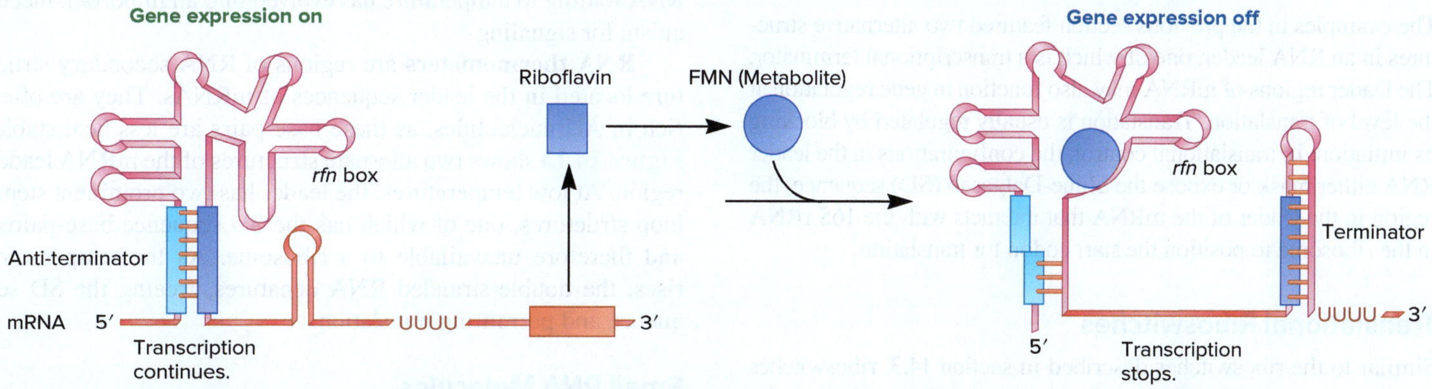

Figure 14.11 Riboswitch Control of the Riboflavin (*rfn*) Operon of *B. subtilis*. The *rfn* operon encodes enzymes that catalyze riboflavin synthesis, a component of flavin mononucleotide (FMN). FMN binding to the *rfn* (*ribo*flavin) box in the mRNA leader causes a transcription terminator to form.

MICRO INQUIRY *In each of the two folded leader RNA structures, a few nucleotides are designated by a light and dark blue box. Why is this termed an anti-terminator (left) when the two sequences base-pair?*

leader RNA fold into a structure called the *rfn* box. When FMN levels are low and the enzymes are needed to synthesize riboflavin, the leader folds so that transcription proceeds. Conversely, when FMN levels are high and the enzymes are not needed, FMN binds to the *rfn* box in the RNA leader such that a terminator loop forms, halting transcription. Transcription is stopped with the aid of rho protein, the RNA helicase that functions in factor-mediated transcription termination (*see figure 13.27*).

Many operons encoding enzymes for amino acid biosynthetic pathways are controlled by a T box riboswitch. The leader RNA in these transcripts adopts two different configurations to cause either transcription termination or continued transcription through the structural genes. The two configurations depend on the binding of a charged tRNA to the leader RNA at a site called the *T box*. Like attenuation in the *trp* operon, the T box riboswitch directly measures the availability of charged tRNA available for translation. High amino acid levels lead to charged tRNA, the T box configuration forms a terminator, and the genes are not transcribed. In contrast, low amino acid levels result in uncharged tRNA that binds to the T box, prevents transcription termination, and leads to gene expression.

Riboswitches are often coordinated with other types of regulatory mechanisms discussed in sections 14.4 and 14.5. For instance, some respond to second messengers, whereas others function in conjunction with small RNAs. This is another example of how bacteria use multiple regulatory mechanisms to control their activities.

14.4 RNA Secondary Structures Control Translation

After reading this section, you should be able to:

a. Distinguish translational riboswitches from transcriptional riboswitches
b. Discuss regulation of translation by small RNAs

The examples in the previous section featured two alternative structures in an RNA leader, one of which is a transcriptional terminator. The leader regions of mRNA may also function in gene regulation at the level of translation. Translation is usually regulated by blocking its initiation. In translational control, the configurations in the leader RNA either mask or expose the Shine-Dalgarno (SD) sequence, the region in the leader of the mRNA that interacts with the 16S rRNA in the ribosome to position the start codon for translation.

Translational Riboswitches

Similar to the riboswitches described in section 14.3, riboswitches that function at the translational level contain effector-binding elements in the long leader RNA. Binding of the effector molecule alters the folding pattern of the leader RNA, which often results in occlusion of the SD sequence. This prevents ribosome binding and, consequently, translation initiation (**figure 14.12**).

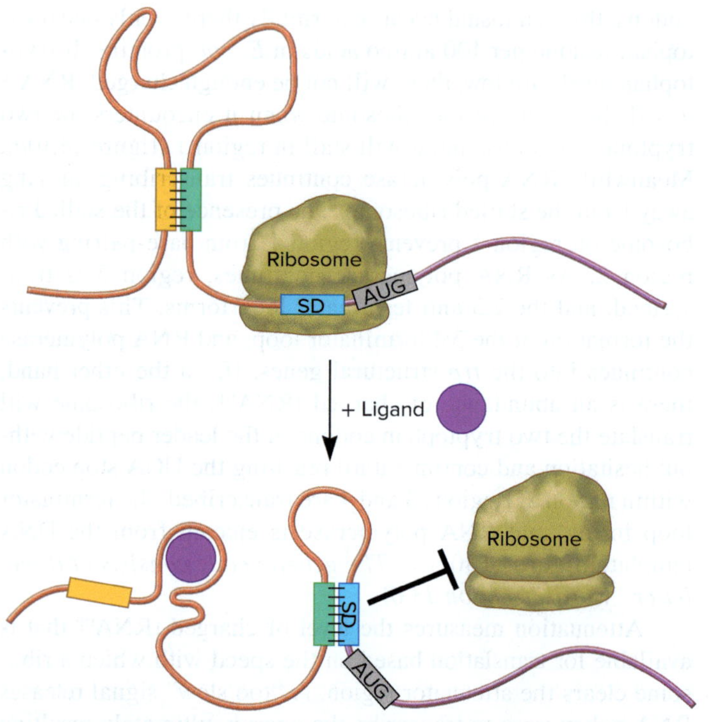

Figure 14.12 Translation Inhibition by a Riboswitch. In the absence of the controlling ligand (purple), the stable configuration of the riboswitch leaves the Shine-Dalgarno (SD) sequence (blue) available to bind a ribosome and translate the mRNA. Abundant ligand binds the leader RNA such that the stable configuration now has an RNA duplex that prevents ribosome binding to the SD sequence, thus inhibiting translation.

RNA Thermometers

One regulatory mechanism widely observed among human pathogens uses a change in temperature to activate gene expression. Microbes detect an opportunity to infect by recognition of constant 37°C body temperature. Genes important in pathogenesis are only expressed in this environment, and the sensitivity of RNA folding to temperature has evolved into an important mechanism for signaling.

RNA thermometers are regions of RNA secondary structure located in the leader sequences of mRNAs. They are often rich in AU nucleotides, as these base pairs are less heat stable. **Figure 14.13** shows two alternate structures of the mRNA leader region. At low temperatures, the leader has two prominent stem-loop structures, one of which has the SD sequence base-paired and therefore unavailable to a ribosome. As the temperature rises, the double-stranded RNA denatures, freeing the SD sequence and permitting translation.

Small RNA Molecules

Many RNA molecules have been discovered that do not function as mRNAs, tRNAs, or rRNAs. Microbiologists often refer to them as **small RNAs (sRNAs)** or noncoding RNAs (ncRNAs); sRNAs range in size from around 50 to 500 nucleotides. Although some

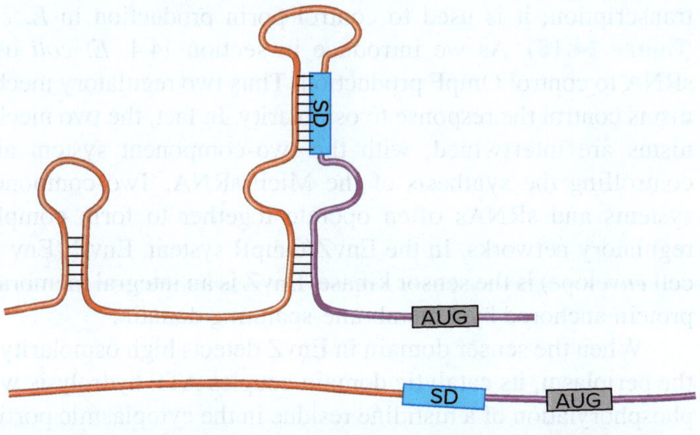

Figure 14.13 Translation Activation by an RNA Thermometer. At low temperatures (top), the leader region forms stem-loops that mask the Shine-Dalgarno (S-D) sequence (blue). Increases in temperature denature the region (bottom) to permit ribosome binding and gene expression.

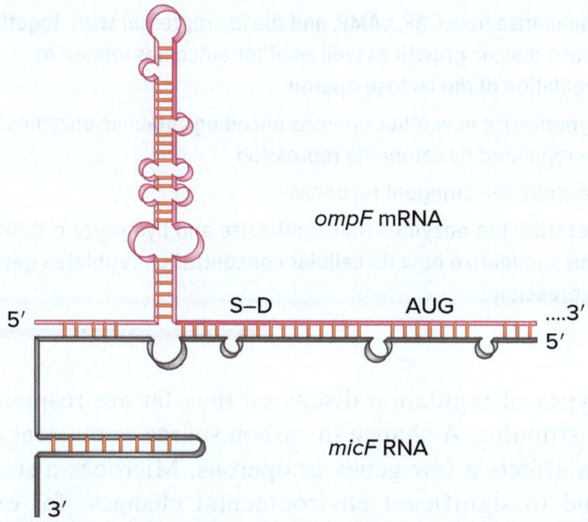

Figure 14.14 Inhibition of Translation by a Trans-Encoded sRNA. The *ompF* mRNA (red) encodes the porin OmpF. Translation of this mRNA is regulated by the trans-encoded sRNA MicF (black), product of the *micF* gene. MicF is partially complementary to the *ompF* mRNA leader and, when bound to it, prevents translation.

MICRO INQUIRY *How does inhibition of translation by sRNA differ from translational inhibition by riboswitches?*

sRNAs have been implicated in the regulation of DNA replication and transcription, many regulate translation by targeting mRNA.

Two main types of mRNA-targeting sRNAs can be described based on where they bind to the genome: cis-acting and trans-acting. Cis-acting sRNAs are synthesized from the nontemplate strand (i.e., the sense strand) of the gene that gives rise to their mRNA target. Thus they are complementary to the target mRNA and are called **antisense RNAs.** Cis-acting sRNAs control expression of a single gene. Trans-acting sRNAs are synthesized from templates located at sites distinct from the genes encoding their target mRNAs. They have short regions of complementarity to their targets and control expression of multiple genes. As you might imagine, binding of an sRNA to an mRNA target can interfere with any aspect of mRNA function. Thus it seems logical that many sRNAs can inhibit translation.

The regulation of the OmpF porin protein provides an example of translational control by a trans-acting sRNA. Recall that the outer membrane of diderm bacteria contains channels made of porins (*see figure 3.23*). Two important *E. coli* porins are OmpF and OmpC (Omp for *o*uter *m*embrane *p*rotein). OmpC pores are smaller; the larger OmpF pores allow solutes to diffuse into the cell more readily. The presence of an antibiotic induces the cells to favor the smaller OmpC pores as a means of excluding the chemical. Expression of the *ompF* gene is regulated by MicF RNA, the product of *micF* (*mic* for *m*RNA-*i*nterfering *c*omplementary RNA). MicF RNA is produced when conditions favor *ompC* expression. MicF RNA is partially complementary to *ompF* mRNA (**figure 14.14**) and when MicF RNA base-pairs with *ompF* mRNA, there are two effects. First, *ompF* expression is rapidly reduced by preventing translation of existing *ompF* mRNA. Second, the *ompF* mRNAs are destroyed because the structure that results from base-pairing between the two RNAs forms the substrate for an RNase. In addition to being regulated by MicF RNA, *ompF* (and *ompC*) is regulated by a two-component signal transduction system, as we describe in section 14.5.

Some sRNAs promote translation upon binding to the mRNA. Whether inhibiting or activating, most trans-acting sRNAs work with the help of an RNA chaperone. Hfq, the RNA chaperone protein in *E. coli,* binds and stabilizes many sRNAs and promotes their interaction with complementary sequences that are the regulatory targets.

Comprehension Check

1. What are the functions of attenuator sequences and the ribosome in attenuation?
2. Sketch the complete *trp* operon in *E. coli*. Include the promoter, operator, attenuator, ribosome binding site, leader, and open reading frame. Which are active on the RNA? Which are on the DNA?
3. How would attenuation in the *E. coli trp* operon change when a culture is shifted from a rich medium to one that is deficient in tryptophan?
4. How are transcriptional and translational riboswitches the same? How do they differ?
5. How are translational riboswitches similar to attenuation as described for the *trp* operon? How do they differ? How do translational riboswitches differ from sRNAs?
6. Suggest a mechanism by which *micF* transcription might be regulated.

14.5 Mechanisms Used for Global Regulation

After reading this section, you should be able to:

a. Explain how alternate sigma factors can initiate transcription of distinct sets of operons

b. List the nucleotides used by bacteria as second messengers

c. Summarize how CAP, cAMP, and the *lac* repressor work together to cause diauxic growth as well as other outcomes related to regulation of the lactose operon
d. Hypothesize how other operons encoding catabolic enzymes might be regulated by catabolite repression
e. Describe the stringent response
f. Describe the enzymes that synthesize and hydrolyze c-di-GMP, and summarize how its cellular concentration regulates gene expression

The types of regulation discussed thus far are responses to a single stimulus. A change in carbon source or nutrient concentration affects a few genes or operons. Microbes also need to respond to significant environmental changes, for example, when a pathogen encounters a susceptible host. In these situations, the microbe's response requires large-scale changes that coordinate sensory input with gene expression and cellular physiology. These responses rely on **global regulatory systems** that affect many genes, operons, and pathways simultaneously. Perhaps the simplest approach to accomplish this is to use a single **global regulatory protein** to alter the expression of multiple operons (or genes). The collection of genes and operons controlled by the global regulator is called a **regulon.** A good example of global regulatory protein is CAP, which enables bacteria to use glucose preferentially regardless of the presence of other potential carbon sources. CAP's regulon includes the *lac* and *ara* operons.

Two-Component Signal Transduction and Phosphorelays

Responses to environmental stimuli are often controlled by regulatory proteins that make up **two-component signal transduction systems.** These systems link events occurring outside the cell to gene expression inside the cell.

Two-component signal transduction systems are named after the two proteins that govern the regulatory pathway. The first is an enzyme called a **sensor kinase** that spans the plasma membrane so that part of it is exposed to the extracellular environment (or the periplasm in diderm bacteria), while another part is exposed to the cytoplasm. In this way, the sensor kinase can detect specific changes in the environment and transmit information to the cell's interior. It responds to environmental stimuli by phosphorylating itself and then transferring the phosphate to the second component, the cytoplasmic **response regulator.** Upon receiving the signal from its partner sensor kinase, the response regulator undergoes a conformational change allowing it to interact with other molecules. Some response regulators bind to DNA to activate or repress transcription initiation. Other response regulators modify enzymes or other proteins in the global network.

The EnvZ/OmpR system is an example of a two-component signal transduction system that functions at the level of transcription; it is used to control porin production in *E. coli* (**figure 14.15**). As we introduce in section 14.4, *E. coli* uses sRNA to control OmpF production. Thus two regulatory mechanisms control the response to osmolarity. In fact, the two mechanisms are intertwined, with the two-component system also controlling the synthesis of the MicF sRNA. Two-component systems and sRNAs often operate together to form complex regulatory networks. In the EnvZ/OmpR system, EnvZ (Env for cell *env*elope) is the sensor kinase. EnvZ is an integral membrane protein anchored by a membrane-spanning domain.

When the sensor domain in EnvZ detects high osmolarity in the periplasm, its catalytic domain couples ATP hydrolysis with phosphorylation of a histidine residue in the cytoplasmic portion of the protein. This is termed *autophosphorylation.* This

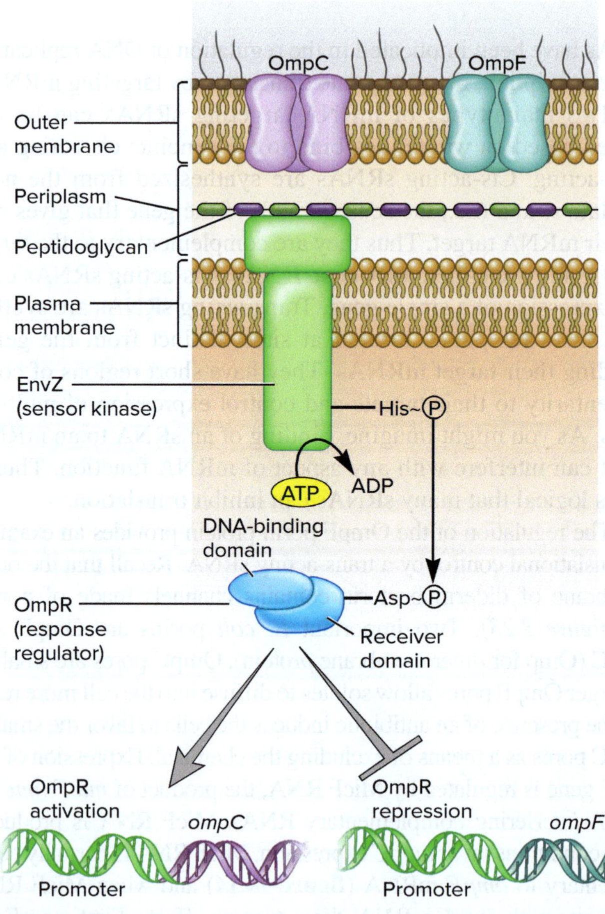

Figure 14.15 Two-Component Signal Transduction and the Regulation of Porins. When EnvZ senses osmolarity change in the periplasm, it autophosphorylates a histidine residue. EnvZ then passes the phosphoryl group to the response regulator OmpR, which accepts it on an aspartic acid residue. This activates OmpR so that it binds DNA to repress *ompF* and induce *ompC* expression.

MICRO INQUIRY *Relative to each promoter, where would you predict phosphorylated OmpR would bind the* ompC *and* ompF *genes?*

phosphoryl group is quickly transferred to an aspartic acid in the receiver domain of OmpR, the second component. Once OmpR is phosphorylated, its DNA-binding domain regulates transcription so that *ompF* is repressed and *ompC* is activated.

Two-component signal transduction systems involve a simple relay where the sensor kinase transfers a phosphoryl group directly to the response regulator. However, there are instances when more proteins participate in the transfer of phosphoryl groups. These more complex pathways are called **phosphorelay systems.** Phosphorelays are important control measures observed in quorum-sensing by *Vibrio harveyi* and endospore formation by *B. subtilis,* both described in detail later in this section.

Sigma Factors

Another way many genes can respond to environmental change in coordinated fashion is through the use of **alternate sigma factors,** which immediately changes expression of many genes as they direct RNA polymerase to specific subsets of bacterial promoters. This is possible because RNA polymerase core enzyme needs the assistance of a sigma factor to initiate transcription (*see figure 13.23*). Each sigma factor recognizes promoters that differ in DNA sequence.

E. coli synthesizes multiple sigma factors (**table 14.1**). Under favorable growth conditions, a sigma factor called σ^{70} directs RNA polymerase activity. (The superscript number indicates the molecular weight of the sigma factor; 70 stands for 70,000 daltons.) When iron acquisition proteins are needed, *E. coli* produces σ^{19}. σ^{19} recognizes its consensus sequences in promoters of genes whose products are needed for survival in iron-limiting conditions. If the temperature rises too high, σ^{32} is produced and stimulates the synthesis of dozens of heat-shock proteins that protect the cell from thermal stress.

Second Messengers

Many bacteria use **second messengers** in their global control systems. Second messengers are small molecules produced in response to a signal (i.e., the first messenger) that is outside the cell. The enzymes that synthesize the second messenger respond to environmental signals, and the second messenger amplifies that information to a larger set of effectors. Three bacterial second messengers were discovered many years ago: cyclic adenosine monophosphate (cAMP), **guanosine tetraphosphate (ppGpp),** and guanosine pentaphosphate (pppGpp) (**figure 14.16**). Other additions to the bacterial second messenger family include **cyclic dinucleotides,** specifically cyclic dimeric GMP (c-di-GMP), c-di-AMP, and cGAMP. Each second messenger responds to a particular type of environmental signal, and redirects cellular physiology and gene expression to respond. We describe three second messenger regulons in detail.

Catabolite Repression in *E. coli*

If *E. coli* grows in a medium containing both glucose and lactose, it uses glucose preferentially until the sugar is exhausted. Then after a short lag, growth resumes at a slower rate with lactose as the carbon source (**figure 14.17**). This allows glucose to be used first, followed by other energy sources that are more difficult to degrade. This biphasic growth pattern is called **diauxic growth** and it is brought about by a global regulatory process called **catabolite repression.** In *E. coli,* two regulatory networks bring about catabolite repression. One network regulates expression of the genes encoding the enzymes that catabolize sugars. The other network regulates the transporters that take up glucose and other sugars. As might be expected, the two networks are linked, as we describe in a moment.

Table 14.1 *E. coli* **Sigma Factors and the Sequences They Recognize**

Sigma Factor	Consensus Promoter Sequences[1]		Genes Transcribed
σ^{70}	TTGACA	TATAAT	Genes needed during exponential growth
σ^{54}	YTGGCAC[2]	TTGC	Genes involved in nitrogen metabolism
σ^{38}	TTGACA	TCTATACTT	Genes needed during stationary phase and the general stress response
σ^{32}	TNTCNCCCTTGAA[2]	CCCCATTTA	Genes needed to protect against heat shock and other stresses
σ^{28}	TAAAG	GYCGATAR[2]	Genes involved in flagellum assembly
σ^{24}	GAACT	TCAAA	Genes needed for the proper folding of membrane proteins
σ^{19}	GAAAAT	TGTCCT	Genes that respond to iron starvation

[1] With the exception of the σ^{54} promoters, all consensus sequences are centered on −35 and −10, respectively. The σ^{54} consensus sequences are located at −24 and −12.
[2] N indicates any nucleotide; Y indicates a pyrimidine, C or T; R indicates a purine, A or G.

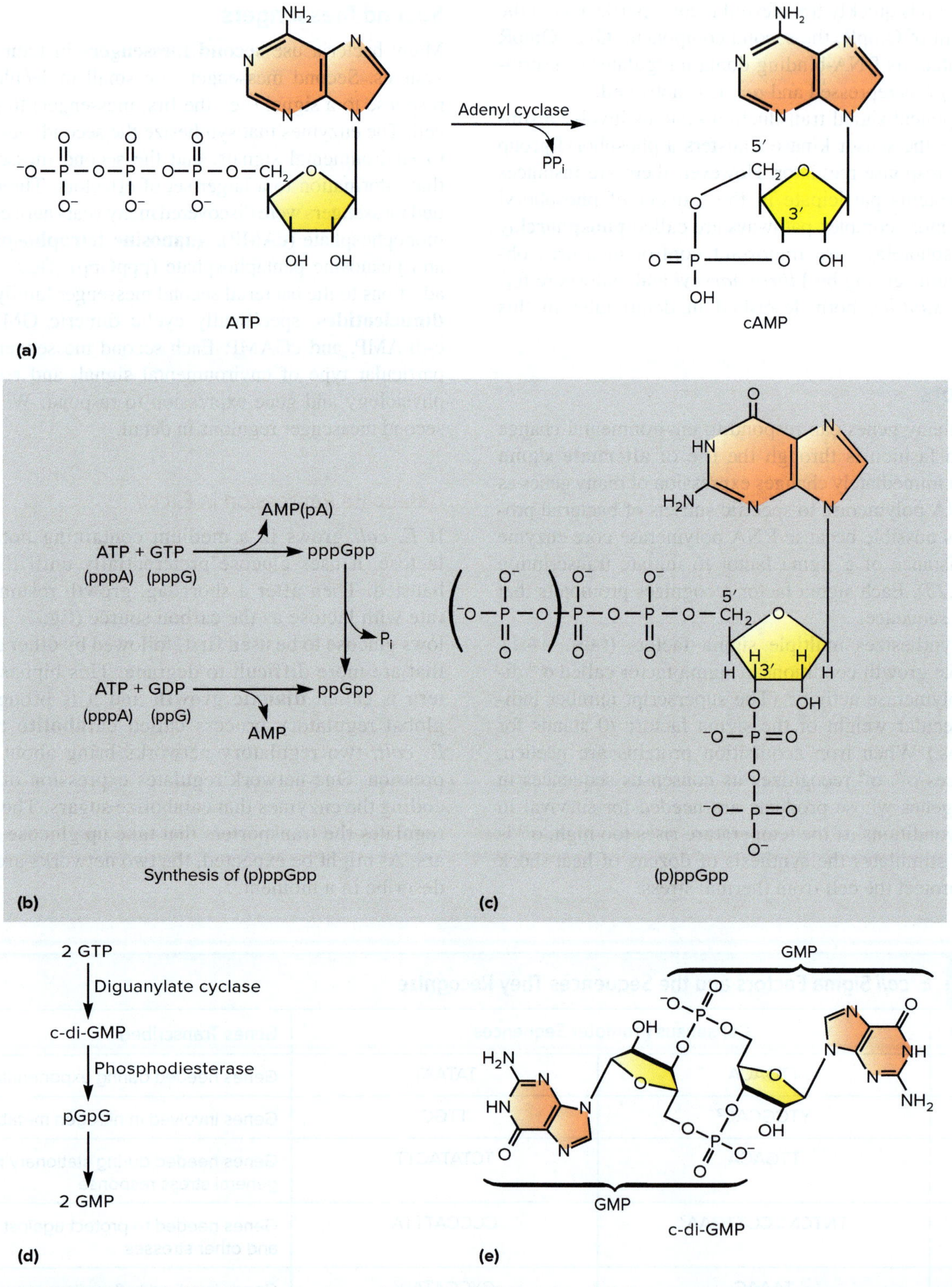

Figure 14.16 Second Messengers. (a) Cyclic adenosine monophosphate (cAMP) forms from ATP. The phosphate extends between the 3′ and 5′ hydroxyls of ribose. (b) Pathways to make guanosine tetraphosphate (ppGpp) and guanosine pentaphosphate (pppGpp). (c) The structure of (p)ppGpp. (d) Two enzymes control levels of cyclic dimeric GMP (c-di-GMP). Diguanylate cyclases form c-di-GMP and phosphodiesterases degrade c-di-GMP. (e) Structure of c-di-GMP, two GMP molecules linked by their phosphates.

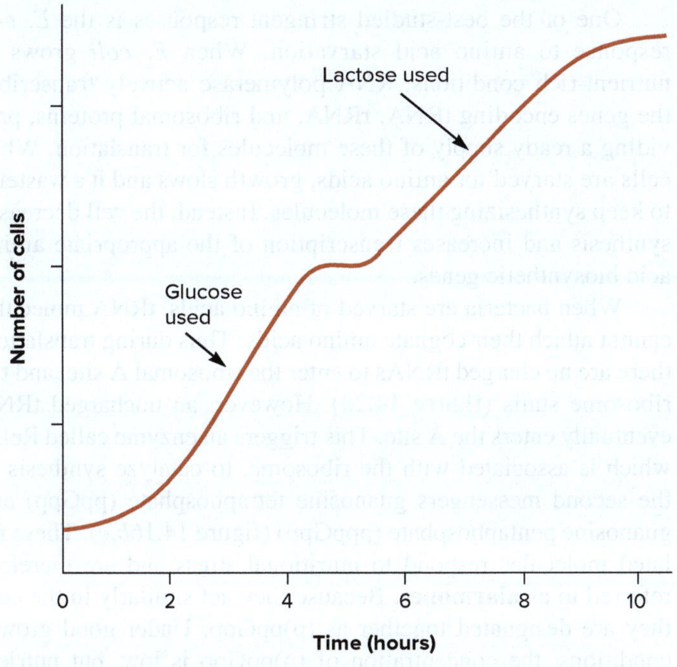

Figure 14.17 Diauxic Growth. The diauxic growth curve of *E. coli* grown with a mixture of glucose and lactose. Glucose is used first, then lactose. A short lag in growth occurs when glucose is exhausted, while the bacteria synthesize the enzymes needed for lactose catabolism.

MICRO INQUIRY *For what other compounds would you also see diauxic growth?*

The enzymes for glucose catabolism are constitutive. However, operons that encode enzymes required for the catabolism of carbon sources that must first be modified before entering glycolysis, such as the *lac, ara, mal* (maltose), and *gal* (galactose) operons (*see figure 11.19*), are called catabolite operons, and their expression is coordinately repressed when glucose is plentiful. This is accomplished by **catabolite activator protein (CAP),** which is also called cyclic AMP receptor protein (CRP). CAP binds DNA through a helix-turn-helix motif that exists in two states determined by the presence or absence of the second messenger **3′, 5′-cyclic adenosine monophosphate (cAMP;** figure 14.16*a*). The enzyme adenyl cyclase synthesizes cAMP from ATP only when little or no glucose is available. Thus the level of cAMP varies inversely with that of glucose. When glucose is unavailable and the catabolism of another sugar might be needed, the amount of cAMP in the cell increases, allowing cAMP to bind to and activate CAP.

All catabolite operons have promoters possessing a CAP recognition site, and cAMP-CAP must bind to this site before RNA polymerase can begin transcription efficiently (figures 14.7 and 14.9). Upon binding, CAP interacts with the α subunits of RNA polymerase and stabilizes the interaction of RNA polymerase with the promoter; this stimulates transcription. Thus all catabolite operons are controlled by two regulatory proteins: CAP and the regulatory protein specific to each operon (e.g., *lac* repressor or AraC protein). In the case of the *lac* operon, if glucose is absent and lactose is present, the inducer allolactose will bind to and inactivate the *lac* repressor protein, CAP will be in the active form (with cAMP bound), and transcription will proceed (**figure 14.18*a***). However, if glucose and lactose are both in short supply, even though CAP binds to its site on the DNA, transcription will be inhibited by the presence of the repressor protein, which remains bound to the operator in the absence of inducer (figure 14.18*c*). Dual control ensures that the *lac* operon is expressed only when lactose catabolic genes are needed. *Combination of Switches: The* lac *Operon*

We have seen how CAP controls catabolite operons; now let's turn our attention to the regulation of cAMP and how glucose

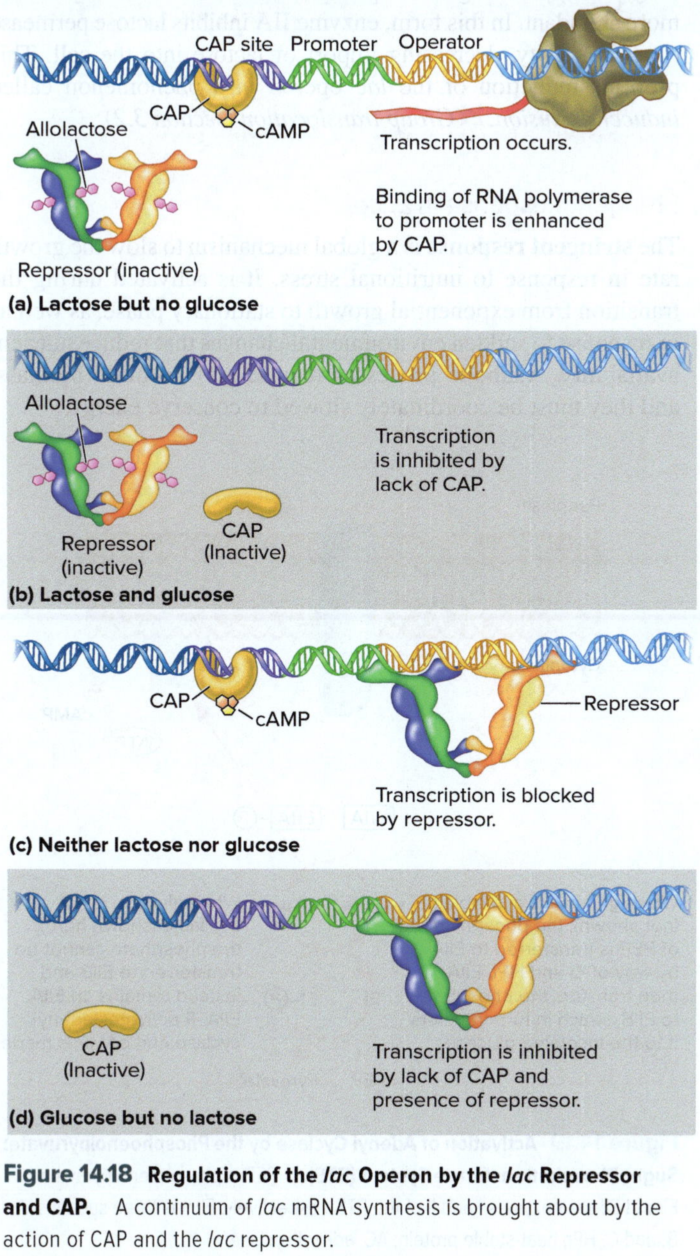

Figure 14.18 Regulation of the *lac* Operon by the *lac* Repressor and CAP. A continuum of *lac* mRNA synthesis is brought about by the action of CAP and the *lac* repressor.

levels are sensed by the cell. Both depend on the state of a protein that functions in the phosphoenolpyruvate: sugar phosphotransferase system (PTS) (**figure 14.19**). Recall from chapter 3 that in the PTS, a phosphoryl group is transferred by a phosphorelay from phosphoenolpyruvate (PEP) to glucose, which then enters the cell as glucose 6-phosphate (*see figure 3.12*). The phosphorylation state of the PTS proteins is an indicator of glucose abundance. When glucose levels are low, enzyme IIA cannot transfer its phosphoryl group to this sugar upon its transport. Instead, the phosphorylated form of enzyme IIA accumulates and activates adenyl cyclase, thus stimulating cAMP production. The cAMP second messenger, then, relays the information about decreasing glucose concentration to molecules that can respond by exploiting alternative carbon sources that may be available. When glucose levels are high, the phosphate is readily transferred from enzyme IIA, making the dephosphorylated form of enzyme IIA more abundant. In this form, enzyme IIA inhibits lactose permease (LacY) activity, decreasing import of lactose into the cell. This prevents induction of the *lac* operon in a phenomenon called *inducer exclusion.* ◀ *Group translocation (section 3.2)*

Stringent Response in *E. coli*

The **stringent response** is a global mechanism to slow the growth rate in response to nutritional stress. It is activated during the transition from exponential growth to stationary phase, as well as in response to sudden environmental changes that reduce nutrient availability. Anabolic processes require ATP to build biomass, and they must be coordinately slowed to conserve energy.

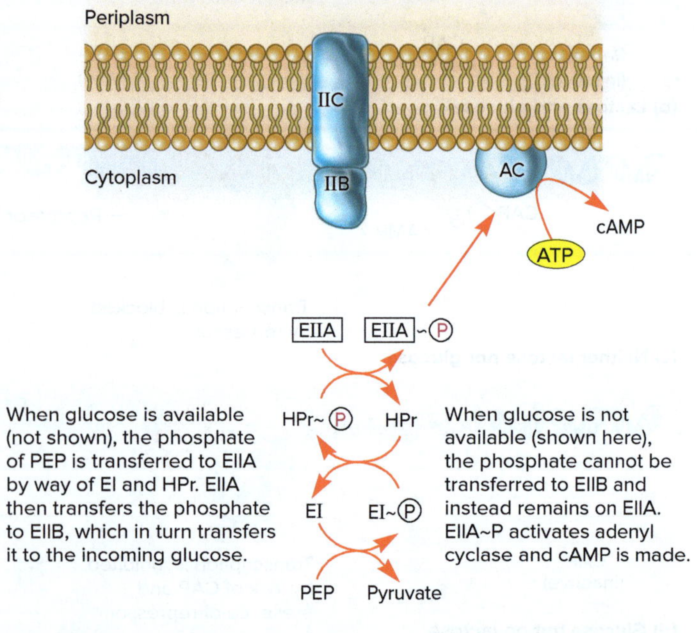

Figure 14.19 Activation of Adenyl Cyclase by the Phosphoenolpyruvate: Sugar Phosphotransferase System (PTS). PEP is phosphoenolpyruvate. EI and EII are enzymes I and II of the PTS, respectively. EII has three subunits: A, B, and C. HPr, heat-stable protein; AC, adenyl cyclase.

One of the best-studied stringent responses is the *E. coli* response to amino acid starvation. When *E. coli* grows in nutrient-rich conditions, RNA polymerase actively transcribes the genes encoding tRNA, rRNA, and ribosomal proteins, providing a ready supply of these molecules for translation. When cells are starved for amino acids, growth slows and it's wasteful to keep synthesizing these molecules. Instead, the cell decreases synthesis and increases transcription of the appropriate amino acid biosynthetic genes.

When bacteria are starved of amino acids, tRNA molecules cannot attach their cognate amino acids. Thus during translation, there are no charged tRNAs to enter the ribosomal A site, and the ribosome stalls (**figure 14.20**). However, an uncharged tRNA eventually enters the A site. This triggers an enzyme called RelA, which is associated with the ribosome, to catalyze synthesis of the second messengers guanosine tetraphosphate (ppGpp) and guanosine pentaphosphate (pppGpp) (figure 14.16*b,c*). These related molecules respond to nutritional stress and are therefore referred to as **alarmones.** Because they act similarly in the cell, they are designated together as (p)ppGpp. Under good growth conditions, the concentration of (p)ppGpp is low, but nutrient limitation increases (p)ppGpp synthesis.

The effects of (p)ppGpp are wide ranging and universally inhibitory. Because (p)ppGpp is structurally related to GTP, it is a competitive inhibitor of translation initiation factor IF2 (*see figure 13.35*). Binding of IF2 to (p)ppGpp instead of GTP inhibits the first step in translation, thereby halting the entire process. It also

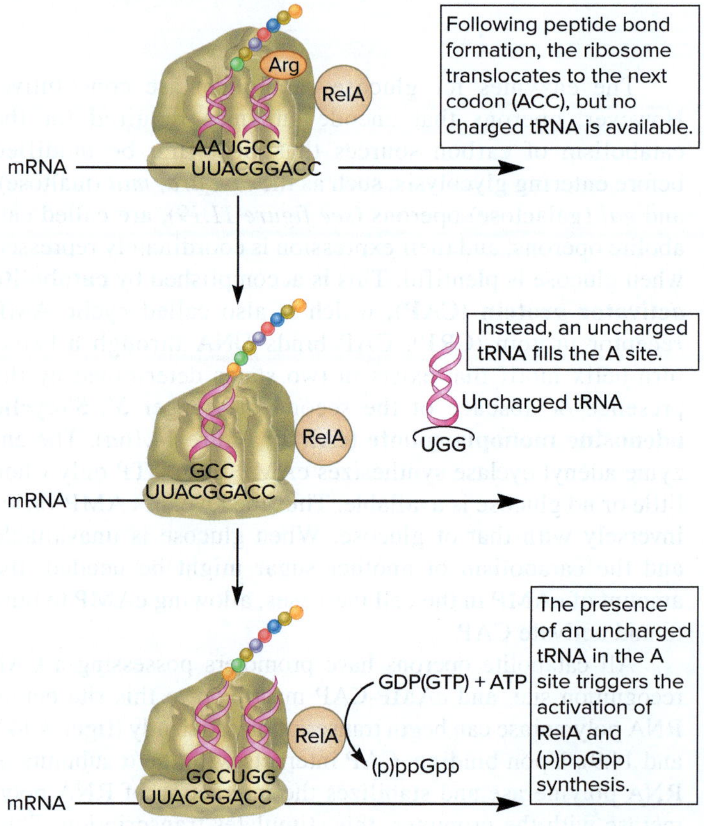

Figure 14.20 Activation of RelA in Response to Amino Acid Starvation.

disrupts transcription by destabilizing RNA polymerase promoter open complexes (*see figure 13.23*), it slows genome replication by inhibiting the replication protein primase (*see figure 13.12*), and it further slows growth by inhibiting several metabolic enzymes. Eventually (p)ppGpp is hydrolyzed, and the cell resets its level of metabolic activity to one better suited to the nutrients available. *DNA replication in bacteria (section 13.3); Transcription in bacteria (section 13.5); Translation in bacteria (section 13.7)*

Although the stringent response functions in many bacteria, the precise triggers differ, reflecting the challenges unique to each microbe's lifestyle. Enzymes called *small alarmone synthetases* and *small alarmone hydrolases* set the cellular concentration of (p)ppGpp and are themselves regulated by global mechanisms like sigma factors. Some cells have these enzymes as domains within a larger protein to directly adjust (p)ppGpp levels upon activation from a sensory domain within the same protein. Recent studies have demonstrated high levels of (p)ppGpp in persister cells, suggesting a possible mechanism for their formation, as an extreme response to slowing the growth rate. *Microbial growth in natural environments (section 7.6)*

Signaling by Cyclic Dinucleotides

Several cyclic purine nucleotides comprise another class of second messengers (figure 14.16*d,e*). The first discovered and best studied is cyclic dimeric guanosine monophosphate (c-di-GMP). Similar messengers are c-di-AMP, a dimer of adenosine monophosphate, and cGAMP, a heterodimer of guanosine and adenosine monophosphates. All three cyclic dinucleotides form in response to specific environmental signals and act on both proteins and riboswitches to regulate gene expression. We focus here on c-di-GMP.

The synthesis of c-di-GMP (figure 14.16*d*) is catalyzed by proteins with diguanylate cyclase (DGC) activity. Conversely, phosphodiesterases (PDE) degrade c-di-GMP. The balance of these two enzymatic activities determines the intracellular concentration, which in turn controls major regulatory decisions. Proteins with DGC and PDE domains are widespread in bacteria and occur as cytoplasmic domains in membrane-located sensors. When a sensor detects a signal, a molecular change results in c-di-GMP synthesis or hydrolysis. Individual microbes may have several environmental inputs that affect c-di-GMP levels, and this permits a cell to integrate multiple signals.

Once formed, c-di-GMP diffuses throughout the cytoplasm, binding to effector molecules. Some effectors respond to a single c-di-GMP molecule, while others require two or four molecules; hence the cellular concentration is important in regulation. In one well-studied example, signaling by c-di-GMP controls the transition from a free-living motile lifestyle to a nonmotile adherent lifestyle and requires coordination of a spectrum of environmental signals to effect morphological and physiological changes to the cell (*see figure 7.23*).

Recent experiments demonstrate the complexity and broad scope of c-di-GMP regulation in *Pseudomonas* spp. As a motile cell nears a solid surface, its flagellum may have its rotation interrupted, generating a mechanical signal in its motor (*see figure 3.35*). A nearby membrane sensor detects the signal and activates its DGC domain. The increasing c-di-GMP concentration initially slows the motor, and then initiates a cascade of gene expression to rapidly shut down motility. It represses transcription of genes encoding flagellum components and activates two-component systems and small RNAs. At the same time, c-di-GMP activates synthesis of adhesin proteins to promote surface attachment (**figure 14.21*a***). The LapA adhesin is displayed on the outer membrane and anchors the cell to the substrate. The LapA C-terminus has two cysteine residues that are oxidized to form a disulfide bond upon reaching the periplasm. The C-terminus plugs the LapE lumen to hold the adhesin to the cell. A second effect of c-di-GMP is on the multidomain protein LapD. LapD spans the plasma membrane with domains in the cytoplasm and the periplasm. When c-di-GMP binds the LapD cytoplasmic domain, the conformational change propagates to the periplasmic domain, where LapD now restrains a protease, LapG. When environmental conditions dictate that a cell should return to a motile state, c-di-GMP levels fall, and cells disperse by a reversal of the LapD-LapG interaction (figure 14.21*b*). As PDE activity eliminates c-di-GMP, the LapD cytoplasmic

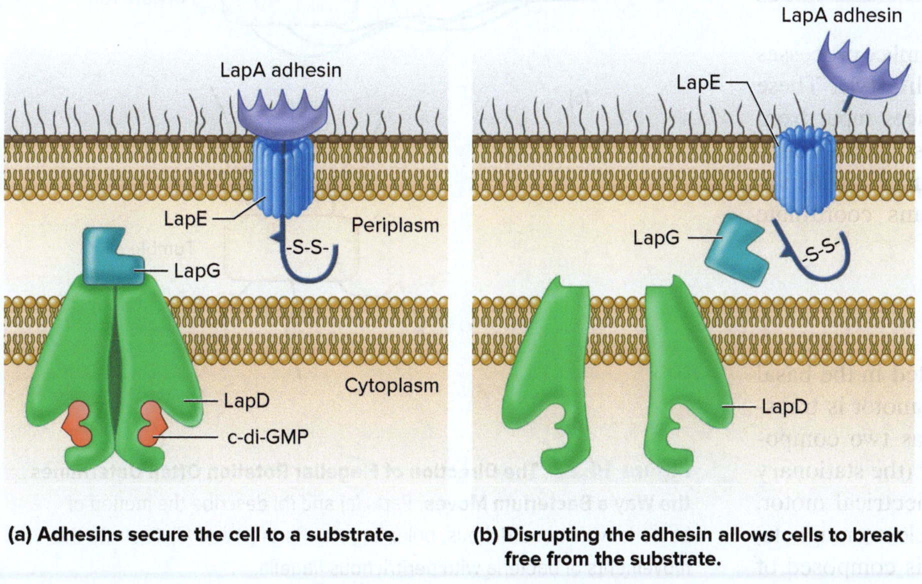

(a) Adhesins secure the cell to a substrate. **(b) Disrupting the adhesin allows cells to break free from the substrate.**

Figure 14.21 Sessile to Motile Transition in a *Pseudomonas* biofilm. (a) Surface adhesins (e.g., LapA) attach cells to a substrate and are anchored to the cell through LapE. c-di-GMP binding to LapD promotes formation of a structure to sequester the LapG protease. (b) When c-di-GMP levels fall, LapG is released from LapD and cleaves the periplasmic domain of LapA. The cell thus breaks free of the substrate and swims away from the biofilm.

domain changes configuration, releasing LapG to the periplasm. LapG cleaves LapA, releasing the cell from the adhesin, thereby allowing it to break free. *Two-step secretion systems (section 13.9)*

Comprehension Check

1. What are global regulatory systems and why are they necessary?
2. Explain how catabolite repression causes diauxic growth.
3. Describe the events that occur with *E. coli* in each of the following growth conditions: in a medium containing glucose but not lactose; in a medium containing both sugars; in a medium containing lactose but no glucose; and in a medium containing neither sugar.
4. Why is the stringent response typically active during nitrogen starvation but not in response to carbon sources?
5. Explain how CDG and PDE domains translate environmental conditions to gene activity and protein activity.

14.6 Bacteria Combine Several Regulatory Mechanisms to Control Complex Cellular Processes

After reading this section, you should be able to:

a. Outline how the proteins of the *E. coli* chemotaxis system control flagellar rotation in the presence of a chemoattractant
b. Predict the events that would occur if *E. coli* cells were in a gradient of a chemorepellent
c. Differentiate quorum sensing observed in *Vibrio fischeri* from that in *V. harveyi*
d. Outline the regulatory mechanisms used by *Bacillus subtilis* to control endospore formation
e. Compare the activities of restriction-modification and CRISPR/Cas in cellular responses to bacteriophage infection

In this section, we present four examples of complex processes that represent a significant change in bacterial lifestyle. These changes are energy-intensive, and the cell balances input from multiple sensors before committing its resources. These examples illustrate how global regulatory systems regulate a wide variety of behaviors, and how individual systems coordinate responses at the molecular level.

Motility and Chemotaxis

The motor that powers flagellar rotation is located in the basal body (*see figure 3.35*). Torque generated by the motor is transmitted to the hook and filament. The motor has two components: the rotor (the moving parts) and the stator (the stationary components). It is thought to function like an electrical motor, where the rotor turns in the center of a ring of electromagnets, the stator. In typical diderm bacteria, the rotor is composed of the central rod and all four rings (*see figure 3.35*). The stator proteins MotA and MotB form a channel across the plasma membrane to power the motor with the proton motive force (PMF). MotB also anchors MotA to cell wall peptidoglycan. Another protein, YcgR, functions as a molecular brake by binding to both the rotor and stator proteins, thereby increasing resistance and slowing down flagellar rotation. YcgR acts as a brake only when bound to c-di-GMP. *Bacterial flagella (section 3.7); Electron transport chains (section 11.4)*

Flagellar Movement—Swimming and Swarming

Chemotaxis is the ability of bacteria to sense chemicals in their environment and move either toward an attractant (i.e., nutrients such as sugars and amino acids) or away from a repellent (e.g., toxic chemicals and bacterial waste products). For simplicity, we concern ourselves here with movement toward an attractant. The best-studied chemotactic system is that of *E. coli,* which, like many other peritrichously flagellated bacteria, exhibits two types of movement: a smooth swimming motion called a **run,** which is interrupted by a **tumble.** A run occurs when the flagellum rotates in a counterclockwise direction (CCW), and a tumble occurs when the flagellum rotates clockwise (CW) (**figure 14.22**). The cell alternates between these movements, with the tumble

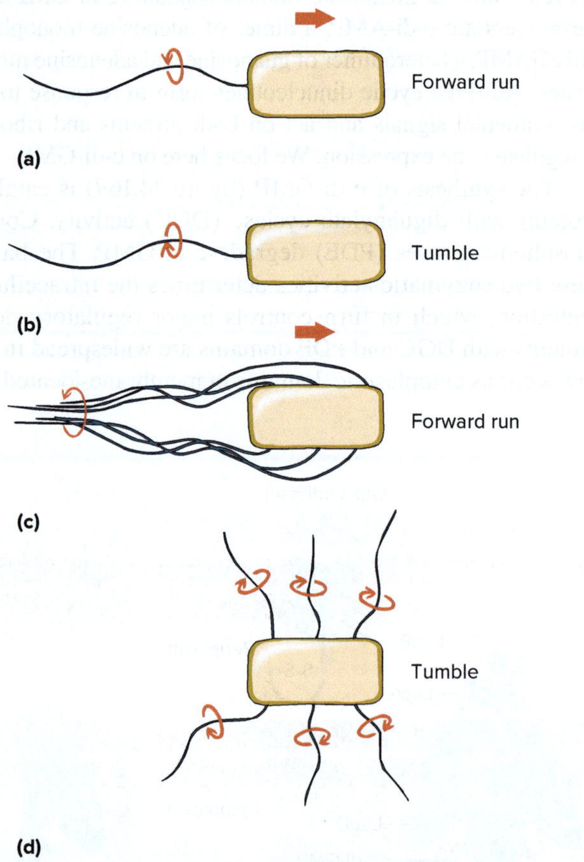

Figure 14.22 The Direction of Flagellar Rotation Often Determines the Way a Bacterium Moves. Parts (a) and (b) describe the motion of bacteria with monotrichous, polar flagella. Parts (c) and (d) illustrate the movements of bacteria with peritrichous flagella.

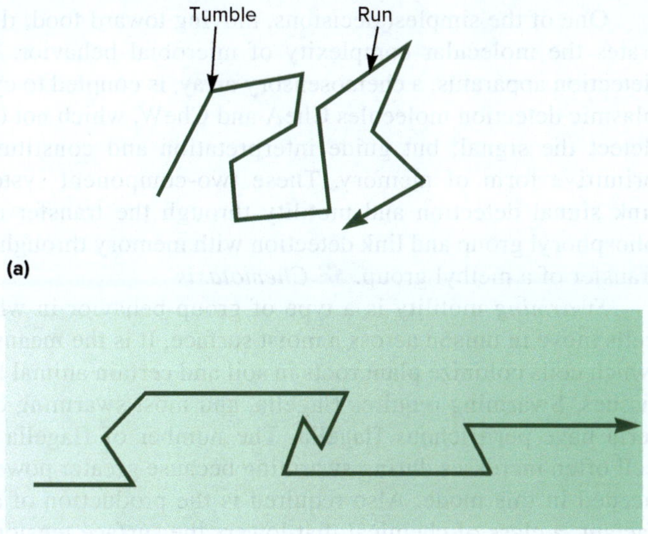

Figure 14.23 Tumbles and Runs Direct the Movement of Bacteria Toward a Chemical Attractant. (a) Random movement of a bacterium in the absence of a concentration gradient. Tumbling frequency is fairly constant. (b) Movement in an attractant gradient. Tumbling frequency is reduced when the bacterium is moving up the gradient. Therefore runs in the direction of increasing attractant are longer.

establishing the direction of movement in the run that follows. Some bacteria display variations to tumbles and runs (e.g., run-stop and run-reverse-flick), but the tumble-run behavior is the most well-studied and is our focus.

When *E. coli* is in a homogeneous environment—that is, the concentration of all chemicals in the environment is the same throughout its habitat—the cell moves randomly, with no apparent direction or purpose; this is called a *random walk* (**figure 14.23*a***). However, if a chemical gradient exists, its path becomes a *biased random walk* (figure 14.23*b*). Runs and tumbles alternate, but runs toward the attractant last longer, and runs away from the attractant are brief. The result is that the cell gradually moves in the direction of the attractant.

For decades, scientists have dissected this complex behavior to understand how *E. coli* senses an attractant, how it switches from a run to a tumble and back again, and how it knows it is heading in the correct direction. These studies reveal that chemotaxis in *E. coli* involves a two-component signal transduction system and covalent modification of chemotaxis receptors. Although most two-component systems regulate transcription initiation, some regulate enzymatic activity. This is the case for chemotaxis.

E. coli senses chemicals in its environment when they bind to **chemoreceptors** (**figure 14.24**). Here we focus on one class of receptors called *m*ethyl-*a*ccepting *c*hemotaxis *p*roteins (MCPs). The signal transduction system that controls the direction of flagellar rotation consists of the sensor kinase CheA and the

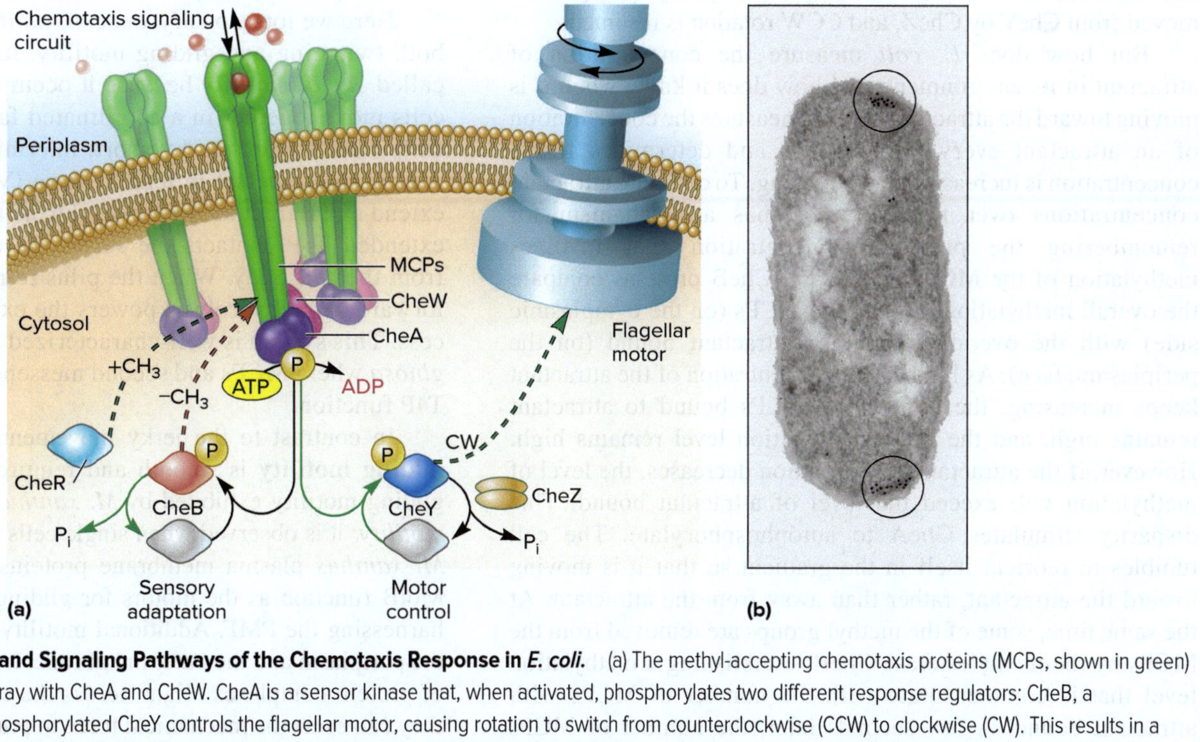

Figure 14.24 Proteins and Signaling Pathways of the Chemotaxis Response in *E. coli*. (a) The methyl-accepting chemotaxis proteins (MCPs, shown in green) cluster in a large sensory array with CheA and CheW. CheA is a sensor kinase that, when activated, phosphorylates two different response regulators: CheB, a methylesterase, or CheY. Phosphorylated CheY controls the flagellar motor, causing rotation to switch from counterclockwise (CCW) to clockwise (CW). This results in a switch from a run to a tumble. The pathway that increases the probability of a run (CCW) is shown in red. CCW rotation is the default. It is periodically interrupted by tumbling (CW). The pathways that lead to CW rotation are shown in green. Molecules in gray are unphosphorylated and inactive. (b) Chemoreceptors cluster at either end of the cell, as shown in this electron micrograph. Gold-tagged antibodies label the receptor clusters, which appear as black dots (encircled).

(b) Prof. Janine Maddock

response regulator CheY. When triggered by an attractant, CheA autophosphorylates using ATP (figure 14.24*a*). This phosphoryl group is quickly transferred to CheY. Phosphorylated CheY diffuses through the cytoplasm to the flagellar motor. Upon interacting with the motor, the direction of rotation is switched from CCW to CW, and a tumble ensues. When CheA is inactive, the flagellum rotates in its default mode (CCW), and the cell moves forward in a smooth run.

As implied by the preceding discussion, the state of the MCPs must be communicated to the CheA/CheY system. How is this accomplished? The MCPs are embedded in the plasma membrane with different domains exposed on each side of the membrane (figure 14.24). The periplasmic side of each MCP has a binding site for attractant molecules. The cytoplasmic side of an MCP is arrayed with two proteins, CheW and CheA, to form large receptor clusters at one or both poles of the cell (figure 14.24*b*).

The MCPs in each receptor cluster work cooperatively to modulate CheA activity. When any one of the MCPs in the cluster binds an attractant, CheA autophosphorylation is inhibited, the flagellum continues rotating CCW, and the cell continues in its run. Because of this cooperation, the cell can respond to very low concentrations of attractant. On the other hand, if attractant levels decrease because the cell is moving in the wrong direction, the level of attractant bound to the MCPs in an array decreases. CheA is stimulated to autophosphorylate, the phosphate is then transferred to CheY, and the cell begins to tumble. However, tumbling does not continue indefinitely. About 10 seconds after the switch to CW rotation occurs, the phosphoryl group is removed from CheY by CheZ, and CCW rotation is resumed.

But how does *E. coli* measure the concentration of attractant in its environment, and how does it know when it is moving toward the attractant? *E. coli* measures the concentration of an attractant every few seconds and determines if the concentration is increasing or decreasing. To compare attractant concentrations over time, *E. coli* has a mechanism for remembering the previous concentration that involves methylation of the MCPs. CheR and CheB proteins compare the overall methylation level of the MCPs (on the cytoplasmic side) with the overall amount of attractant bound (on the periplasmic face). As long as the concentration of the attractant keeps increasing, the number of MCPs bound to attractant remains high, and the MCP methylation level remains high. However, if the attractant concentration decreases, the level of methylation will exceed the level of attractant bound. This disparity stimulates CheA to autophosphorylate. The cell tumbles to reorient itself in the gradient so that it is moving toward the attractant, rather than away from the attractant. At the same time, some of the methyl groups are removed from the MCPs by the methylesterase CheB, establishing a methylation level that is commensurate with the number of MCPs with attractant bound. A few seconds later, the number of MCPs with attractant bound will be compared to this new methylation level. Based on the correspondence of the two, the cell will determine if it is again moving up the gradient. If it is, tumbling will be suppressed and the run will continue.

One of the simplest decisions, moving toward food, illustrates the molecular complexity of microbial behavior. The detection apparatus, a chemosensory array, is coupled to cytoplasmic detection molecules CheA and CheW, which not only detect the signal, but guide interpretation and constitute a primitive form of memory. These two-component systems link signal detection and motility through the transfer of a phosphoryl group and link detection with memory through the transfer of a methyl group. *Chemotaxis*

Swarming motility is a type of group behavior in which cells move in unison across a moist surface. It is the means by which cells colonize plant roots in soil and certain animal host tissues. Swarming requires flagella, and most swarming bacteria have peritrichous flagella. The number of flagella per cell often increases during swarming because greater power is needed in this mode. Also required is the production of surfactant, a class of chemical that lowers the surface tension of the substrate to facilitate movement. Swarming is controlled by multiple mechanisms, including two-component regulators and quorum sensing, which we introduce later in this section.

Nonflagellar Movement—Twitching and Gliding

Twitching and gliding motility occur when cells are on a solid surface. However, unlike swarming, neither involves flagella. We describe the gliding motility of *Flavobacterium* spp. and *Mycoplasma* spp. in detail in chapters 20 and 22. *Phylum Bacteroidota includes important gut microbiota (section 20.6); Order Mycoplasmatales (section 22.2)*

Here we focus on *Myxococcus xanthus* because it exhibits both twitching and gliding motility. Its twitching motility is called *social motility* because it occurs when large groups of cells move together in a coordinated fashion. **Twitching motility** is characterized by short, intermittent motions of up to several micrometers in length. Type IV pili (T4P) alternately extend and retract to move cells during twitching motility. The extended T4P contacts the surface at a point some distance from the cell body. When the pilus retracts, the cell is pulled forward. ATP hydrolysis powers the extension/retraction process. This system is well-characterized in *Pseudomonas aeruginosa* where MCPs and second messengers like cAMP govern T4P function.

In contrast to the jerky movement of twitching motility, **gliding motility** is smooth and requires no appendages. The gliding motility exhibited by *M. xanthus* is called *adventurous motility;* it is observed when single cells move independently. In *M. xanthus* plasma membrane proteins similar to MotA and MotB function as the motors for gliding motility, fueling it by harnessing the PMF. Additional motility proteins are located in the periplasm and the outer membrane. It is only when the full set of protein complexes aligns across the entire cell envelope that they can exert the mechanical force to propel the cell. Thus, gliding motility is regulated by the spatial distribution of these proteins, which depends in part on interactions with the bacterial cytoskeleton. Adhesin proteins on the cell surface are compressed and slide on the substrate, pushing the body of the cell forward. Friction is reduced by slime secretion. *Phyla Bdellovibrionota and Myxococcota: bacterial predators (section 20.9); Predators come in all sizes (section 27.4)*

Quorum Sensing by *Vibrio* spp.

Cell-to-cell communication among bacterial cells often occurs by the exchange of chemicals termed signals or signaling molecules. This exchange is essential to coordinate gene expression in microbial populations. Cell communication plays an essential role in the regulation of genes whose products are needed for the establishment of virulence, symbiosis, biofilm production, and morphological differentiation in a wide range of bacteria. Quorum-sensing mechanisms vary among microbes. Here we focus on two of the best-studied systems.

The marine bioluminescent bacterium *V. fischeri* uses quorum sensing to regulate light production within the light organ of its squid host; cells produce light only at a high density (*see figure 7.26*). Quorum sensing is also used to control genes whose products are needed for maintenance of the symbiotic relationship between *V. fischeri* and its host. As a result, the squid/*V. fischeri* symbiosis has become an important model for understanding animal–bacterial associations. Our focus is on the regulation of a single operon, that involved with bioluminescence. However, it should be kept in mind that **quorum sensing** regulates multiple genes and operons. *Cell-cell communication within microbial populations (section 7.6)*

Quorum sensing in *V. fischeri* and many other diderm bacteria uses an ***N*-acyl homoserine lactone (AHL)** signal (**figure 14.25**). Synthesis of this small molecule is catalyzed by an enzyme called AHL synthase, the product of the *luxI* gene. The *luxI* gene is subject to positive autoregulation; that is, transcription of *luxI* increases as AHL accumulates in the cell. This is accomplished through the transcriptional activator LuxR, which is active only when AHL binds to it. Thus a simple feedback loop is created. Without AHL-activated LuxR, the *luxI* gene is transcribed only at basal levels. When *V. fischeri* cell density within the squid light organ is low, the small amounts of AHL produced by the bacterial cells freely diffuse out of each cell and accumulate in the environment. As cell density increases, the concentration of AHL outside each cell eventually exceeds that inside the cell, and the concentration gradient is reversed (*see figure 7.25*). As the intracellular AHL concentration increases, it binds and activates LuxR. LuxR then increases transcription of *luxI* and the genes whose products are needed for bioluminescence (*luxCDABEG*). Quorum sensing is often called **autoinduction,** and the AHL signal is termed the **autoinducer (AI)** to reflect the autoregulatory nature of this system. *Quorum Sensing*

Another kind of quorum sensing depends on an elaborate phosphorelay signal transduction system. It is found in both neg=diderm and pos=monoderm bacteria, and has been best studied in the bioluminescent bacterium *V. harveyi*.

Unlike *V. fischeri, V. harveyi* responds to three autoinducer molecules. As shown in **figure 14.26**, AI-1, AI-2, and CAI-1 are secreted by cells, which then use separate proteins called LuxN, LuxPQ, and CqsS, respectively, to detect their presence. LuxN, LuxPQ, and CqsS are sensor kinases. At low cell density in the absence of any autoinducer, the three sensor kinases autophosphorylate and converge on a single phosphotransferase protein called LuxU. LuxU accepts phosphates from any of the three sensor kinases and then phosphorylates the

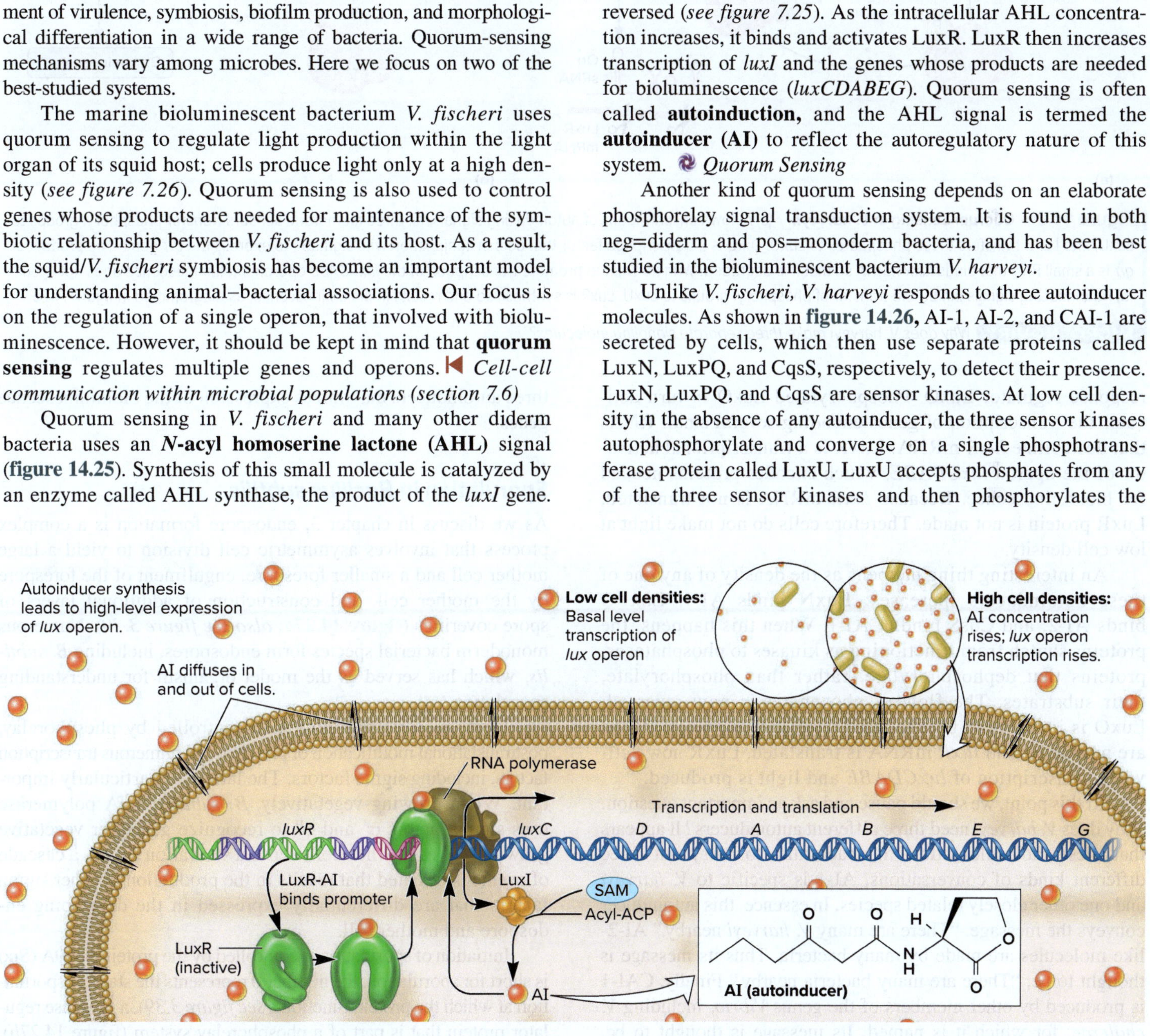

Figure 14.25 Quorum Sensing in *V. fischeri*. The AHL signaling molecule (AI) diffuses out of the cell; when cell density is high, AI diffuses back into the cell, where it binds to and activates the transcriptional regulator LuxR. Active LuxR then stimulates transcription of the gene coding for AHL synthase (*luxI*), as well as the genes encoding proteins needed for light production. SAM, S-adenosylmethionine; ACP, acyl-carrier protein.

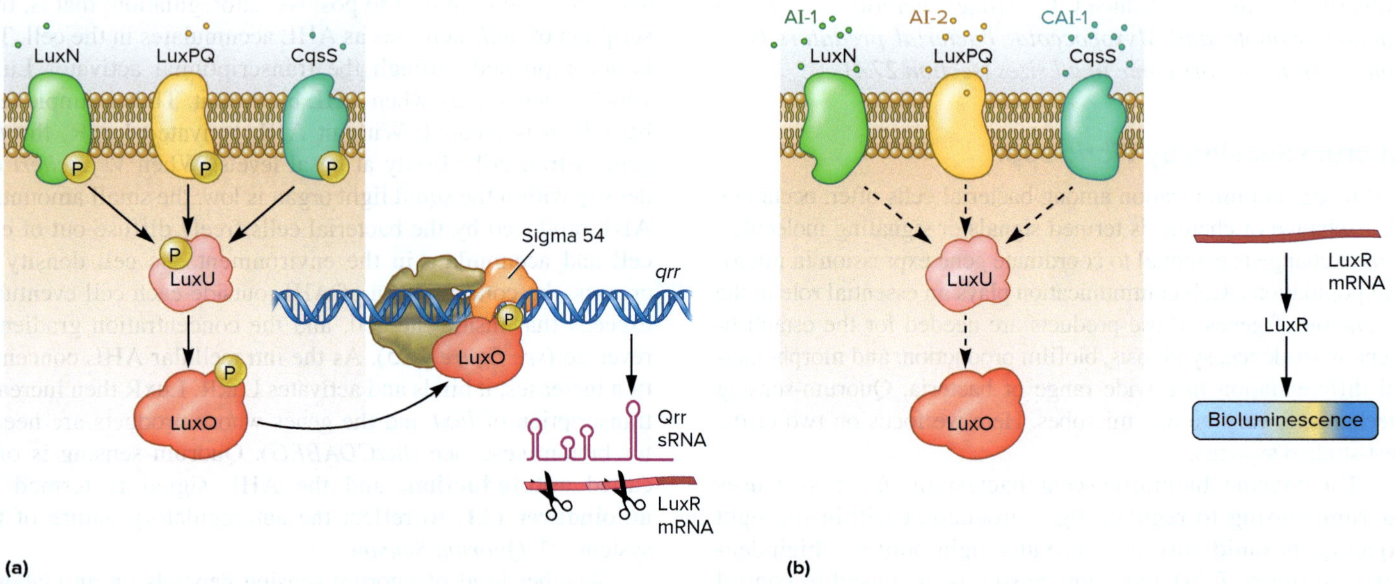

Figure 14.26 Quorum Sensing in *V. harveyi.* (a) At low concentrations of autoinducing signals, all three membrane sensors transfer phosphoryl groups to LuxU, and LuxU transfers these phosphates to LuxO. LuxO is a transcription factor that acts in conjunction with sigma54 to express the *qrr* gene. The product of *qrr* is a small RNA that binds *luxR* mRNA so that it is degraded. With no LuxR to promote transcription of the *luxCDABE* operon, bioluminescence does not occur. (b) Autoinducing signals inhibit the transfer of phosphoryl groups to LuxU. LuxR is synthesized and results in bioluminescence, as diagrammed in figure 14.25.

MICRO INQUIRY *Why does* V. harveyi *make three separate signaling molecules?*

response regulator LuxO. Phosphorylated LuxO in turn activates the transcription of genes encoding several small RNAs that destabilize *luxR* mRNA. LuxR is a transcriptional activator of the operon *luxCDABE,* which encodes proteins needed for bioluminescence. Because *luxR* mRNA is not translated, LuxR protein is not made. Therefore cells do not make light at low cell density.

An interesting thing happens as the density of any one of these autoinducers increases: LuxN binds AI-1, LuxPQ binds AI-2, and CqsS binds CAI-1. When this happens, the proteins switch from functioning as kinases to phosphatases, proteins that dephosphorylate, rather than phosphorylate, their substrates. The flow of phosphates is now reversed; LuxO is inactivated by dephosphorylation, the small RNAs are not made, and *luxR* mRNA is translated. LuxR now activates transcription of *luxCDABE* and light is produced.

At this point, we should pause and ask an important question: Why does *V. harveyi* need three different autoinducers? It appears that these molecules allow the bacterium to carry out three different kinds of conversations. AI-1 is specific to *V. harveyi* and one other closely related species. In essence, this autoinducer conveys the message, "There are many *V. harveyi* nearby." AI-2-like molecules are made by many bacteria. Thus its message is thought to be, "There are many bacteria nearby." Finally, CAI-1 is produced by other members of the genus *Vibrio,* including *V. cholerae,* for which it is named. Its message is thought to be, "There are many *Vibrio* species nearby." As might be expected, each autoinducer has a different signal strength. AI-1 is the strongest and CAI-1 is the weakest. The presence of all three autoinducers maximizes expression of the bioluminescence genes.

Sporulation in *Bacillus subtilis*

As we discuss in chapter 3, endospore formation is a complex process that involves asymmetric cell division to yield a large mother cell and a smaller forespore, engulfment of the forespore by the mother cell, and construction of additional layers of spore coverings (**figure 14.27*a***; *also see figure 3.39*). Numerous monoderm bacterial species form endospores, including *B. subtilis,* which has served as the model organism for understanding sporulation.

In *B. subtilis,* sporulation is controlled by phosphorelay, posttranslational modification of proteins, and numerous transcription factors, including sigma factors. The latter are particularly important. When growing vegetatively, *B. subtilis* RNA polymerase uses sigma factors σ^A and σ^H to recognize genes for vegetative growth. However, when cells sense a starvation signal, a cascade of events is initiated that results in the production of other sigma factors that are differentially expressed in the developing endospore and mother cell.

Initiation of sporulation is controlled by the protein Spo0A (Spo is short for sporulation, the number 0 represents the stage of sporulation at which the protein functions, *see figure 3.39*), a response regulator protein that is part of a phosphorelay system (figure 14.27*b*). Five different sensor kinases associated with this system detect environmental stimuli that trigger sporulation. These five sensors respond to different but overlapping stimuli. Sporulation is a

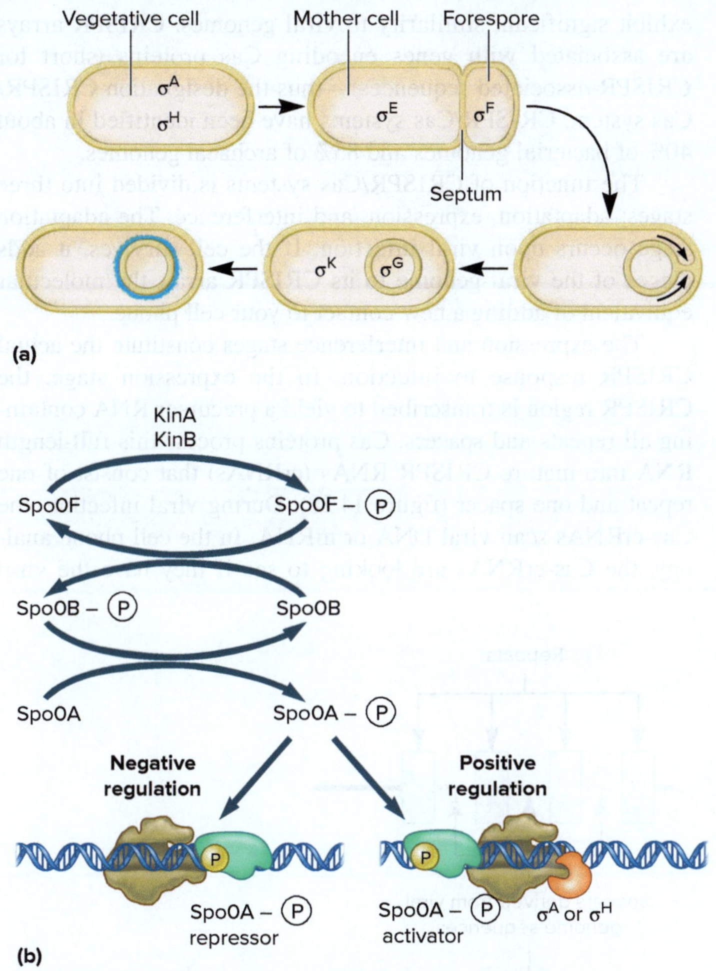

Figure 14.27 Regulation of Sporulation in *B. subtilis*.
(a) Transcriptional regulation in sporulation is governed in part by the activities of spatially separated sigma factors. (b) The phosphorelay system triggered by the activation of the sensor kinases results in phosphorylation of the response regulator SpoOA. SpoOA-P is a transcription factor that represses genes needed in vegetative cells and activates genes needed for endospore formation.

MICRO INQUIRY *Compartmentation—the spatial localization of proteins as a means of regulation—is usually considered in the context of eukaryotic cells. How is the process of sporulation an example of compartmentation?*

significant investment in time and resources, and the cell uses multiple inputs to establish that conditions are ideal before commencing the process. One sensor kinase acting alone is insufficient to initiate sporulation; several acting in concert are required.

One of the most important sensor kinases is KinA. When it senses nutrient depletion, KinA autophosphorylates a specific histidine residue. The phosphoryl group is transferred to an aspartic acid residue on Spo0F, the second member of the phosphorelay. Spo0F donates the phosphoryl group to Spo0B, which in turn relays the phosphoryl group to Spo0A, the response regulator. Phosphorylated Spo0A is a transcription factor that positively controls genes needed for sporulation and negatively controls genes that are not needed. In response to phosphorylated Spo0A, the expression of over 500 genes is altered, earning it the status of master regulator. Among the genes whose expression is stimulated by Spo0A is *sigF*, the gene encoding σ^F, and *spoIIGB*, the gene encoding an inactive form of σ^E (pro-σ^E).

When sporulation starts, the chromosome has replicated, with one copy remaining in the mother cell and another to be partitioned in the forespore. Shortly after the formation of the asymmetric septum, σ^F is found in the forespore, and σ^E is localized to the mother cell (figure 14.27*a*). The two sigma factors bind to the promoters of genes needed in the forespore and mother cell, respectively. There they direct the expression of genes encoding products needed for the early steps of endospore formation, specifically the engulfment process. Another gene regulated by σ^F encodes σ^G, which will replace σ^F in the transcriptional program in the developing endospore. Likewise, σ^E directs the transcription of a mother-cell-specific sigma factor, σ^K. σ^K ensures that genes encoding late-stage sporulation products are transcribed. These include genes for synthesis of the cortex and coat layers. Overall, temporal regulation is achieved because σ^F and σ^E direct transcription of genes that are needed early in the sporulation process, whereas σ^G and σ^K are needed for the transcription of genes whose products function later. In addition, spatial control of gene expression is accomplished because σ^F and σ^G are located in the forespore and developing endospore, whereas σ^E and σ^K are found only in the mother cell.

Responses to Viral Infection

Among the many global regulatory challenges faced by bacteria in dynamic environments is how to protect against infection by a virus. Chapter 6 details the viral infection process from the virus's perspective; here we discuss the phenomenon from the bacterial view. Far from being passive victims, microbial cells have several systems to actively combat viral infection. Interference with the viral life cycle (*see figure 6.10*) at any stage benefits the microbial host. We focus here on bacterial and archaeal responses to viral nucleic acids, as the encounter with the viral genome is the first intracellular event during infection.

The first mechanism discovered was **restriction-modification** in the 1950s by Salvador Luria. He termed the system *restriction* because certain strains of *E. coli* restricted the growth of bacteriophage. We now know that microbes methylate the nitrogenous bases in short stretches of DNA throughout the chromosome. Bacteria and archaea also produce **restriction endonucleases (REs)** that recognize the same DNA sequence as the methylase and hydrolyze the phosphodiester bonds only if the DNA has not been methylated. This enables the microbe to distinguish its own DNA (methylated) from a viral genome (unmethylated). When viral DNA enters a cell, it is destroyed by the RE. However, when large populations are studied, as in Luria's experiments, a few viruses escape restriction and become highly infective. REs are found in most bacteria and archaea, and have been adapted for use in genetic engineering. ▶| *Cloning technologies developed from DNA structure and enzymology (section 17.1)*

In an analogy with the immune system of higher organisms, bacterial responses to infection can be viewed as molecular mechanisms to distinguish self from non-self, and to destroy invaders. Restriction-modification is an innate system, encoded in the genome and expressed constitutively at a low level. Many microbes have another defense mechanism: the **CRISPR/Cas system,** best described as adaptive immunity. The similarity to the immune system is based on the observation that CRISPR/Cas uses part of the invading virus's genome to document an attack and prepare for a possible future attack, just as the adaptive immune system uses parts of an invading microbe to remember and prepare for a future invasion. *Adaptive immunity (chapter 32)*

CRISPR/Cas was discovered when genome analyses identified repeated nucleotide sequences separated by short spacers. These repeats form a CRISPR array, CRISPR being short for *c*lustered *r*egularly *i*nterspersed *s*hort *p*alindromic *r*epeats. The sequence of base pairs in the repeats is identical (or nearly so) within each CRISPR array. The spacers, however, differ considerably (**figure 14.28**). Surprisingly, when nucleic acid databases were queried with the spacer sequences, they were found to exhibit significant similarity to viral genomes. CRISPR arrays are associated with genes encoding Cas proteins (short for *C*RISPR-*as*sociated *s*equences)—thus the designation CRISPR/Cas system. CRISPR/Cas systems have been identified in about 40% of bacterial genomes and 85% of archaeal genomes.

The function of CRISPR/Cas systems is divided into three stages: adaptation, expression, and interference. The adaptation stage occurs upon viral infection. If the cell survives, it adds pieces of the viral genome to its CRISPR array, the molecular equivalent of adding a new contact to your cell phone.

The expression and interference stages constitute the actual CRISPR response to infection. In the expression stage, the CRISPR region is transcribed to yield a precursor RNA containing all repeats and spacers. Cas proteins process this full-length RNA into mature CRISPR RNAs (crRNAs) that consist of one repeat and one spacer (figure 14.28). During viral infection, the Cas-crRNAs scan viral DNA or mRNA. In the cell phone analogy, the Cas-crRNAs are looking to see if they have the viral

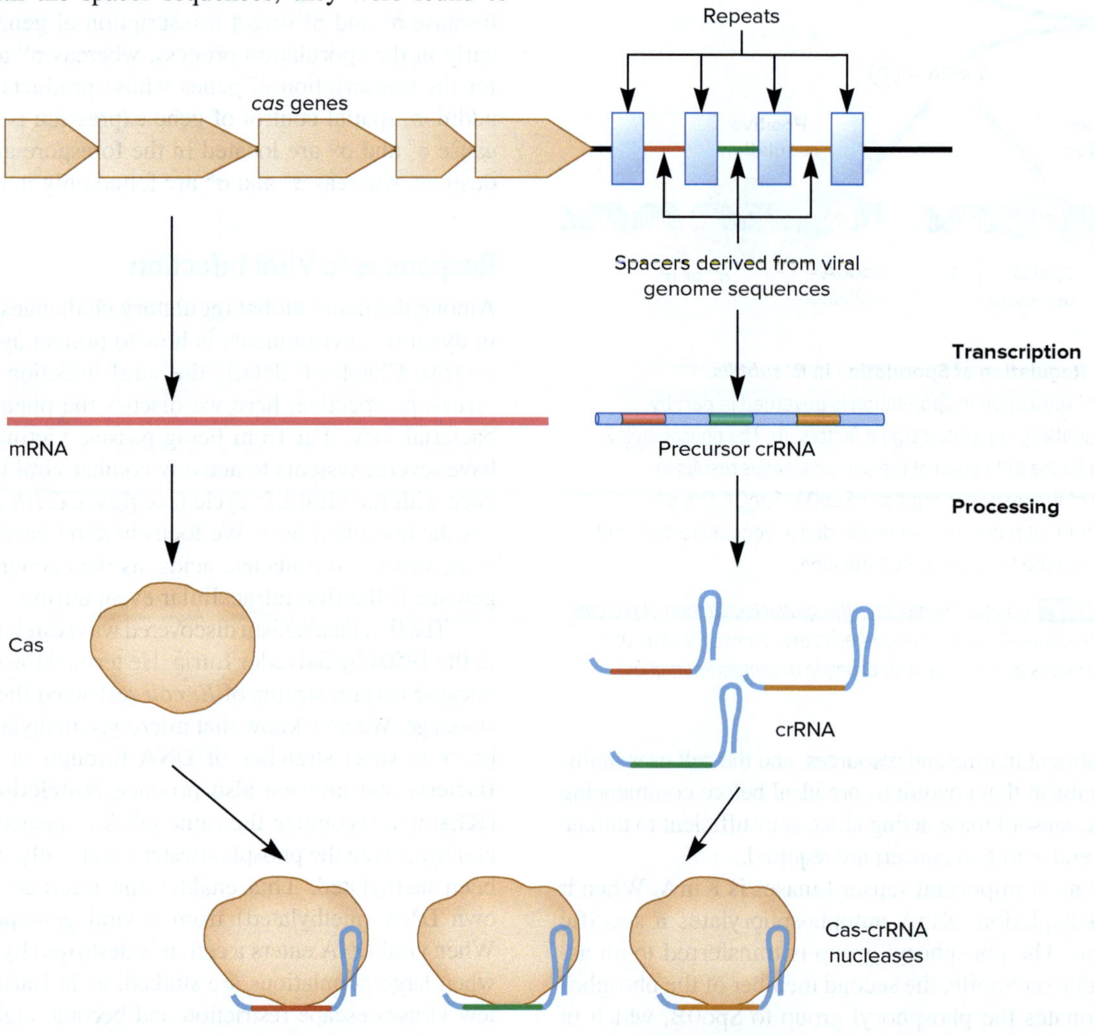

Figure 14.28 CRISPR/Cas System. Bacterial and archaeal genomes often contain a series of repeats and spacer sequences in a CRISPR array that is usually adjacent to *cas* genes. Expression of these loci produces Cas proteins and precursor CRISPR RNA (crRNA). The precursor crRNA is processed to mature crRNAs that each associates with Cas to create a site-specific endonuclease.

contact. Confirmation of a previous encounter triggers destruction of the molecule (interference). This prevents virus multiplication, and infection is thwarted.

CRISPR expression assimilates input from several global regulatory systems. The CRISPR array is believed to be constitutively transcribed at a low level, always primed to respond. Quorum sensing signals stimulate CRISPR/Cas expression, as high cell density renders a population of cells susceptible to infection. Although the detailed molecular mechanisms have not been completely established, one or more two-component systems that sense cell envelope stress can activate CRISPR transcription in response to viral contact. In addition, infection often disrupts host cell metabolism, and the CRISPR array responds to changes in cAMP-CAP levels.

Our focus has been on the use of CRISPR/Cas systems by bacteria to defend against viral attacks. However, there is considerably more to these systems. It is now clear that bacteria use CRISPR/Cas for the more general process of protecting their genomes from foreign nucleic acids that might enter via horizontal gene transfer (*see chapter 16*). Finally, like restriction endonucleases, CRISPR/Cas complexes have been developed into useful tools for genome engineering. ▶| *Cas9 nuclease is a programmable tool for genome editing (section 17.5)*

Comprehension Check

1. Describe the MCP-CheW-CheA receptor complex. What two proteins are phosphorylated by CheA? What is the role of each?
2. How does an MCP regulate the rate of CheA autophosphorylation? How does this mediate chemotaxis?
3. Why might bacteria use quorum sensing to regulate genes needed for virulence? How might this reason be related to the rationale behind using quorum sensing to establish a symbiotic relationship?
4. Briefly describe how a phosphorelay system and sigma factors are used to control sporulation in *B. subtilis.* Give one example of posttranslational modification as a means to regulate this process.
5. How do sigma factors control the timing of gene expression during endospore formation? How do they control the location of gene expression?
6. Briefly describe the three stages of CRISPR/Cas response to viral infection. Why is quorum sensing involved?

Key Concepts

14.1 Bacteria Use Many Regulatory Strategies

- Constitutive or housekeeping genes encode proteins required continuously for basic cellular functions. Genes whose products are required only under specific circumstances are regulated.
- Regulation of gene expression occurs when a stimulus is received and interpreted by the cell for a change in gene expression (**figure 14.1**).

14.2 Regulation of Transcription Initiation Saves Considerable Energy and Materials

- Induction and repression of enzyme levels are two important regulatory phenomena. They usually occur because of the action of regulatory proteins.
- Transcription factors typically have two domains. A sensory domain binds a small effector molecule, which causes a conformational change in its DNA-binding domain (**figure 14.3**). The transcription factor can only bind DNA in one configuration.
- Transcription factors can either inhibit transcription (negative control) or promote transcription (positive control).
- Repressors are responsible for negative control. They block transcription by binding an operator and interfering with the binding of RNA polymerase to its promoter or by blocking the movement of RNA polymerase after it binds DNA (**figure 14.4**).
- Activator proteins are responsible for positive control. They bind DNA sequences called activator-binding sites and, in doing so, promote binding of RNA polymerase to its promoter (**figure 14.4**).
- The *lac* operon of *E. coli* is an example of a negatively controlled inducible operon. When there is no lactose in the surroundings, the *lac* repressor is active and transcription is blocked. When lactose is available, it is converted to allolactose by the enzyme β-galactosidase. Allolactose acts as the inducer of the *lac* operon by binding the repressor and inactivating it. The inactive repressor cannot bind the operator and transcription occurs (**figure 14.7**).
- The *trp* operon of *E. coli* is an example of a negatively controlled repressible operon. When tryptophan is not available, the *trp* repressor is inactive and transcription occurs. When tryptophan levels are high, tryptophan acts as a corepressor and binds the *trp* repressor, activating it. The *trp* repressor binds the operator and blocks transcription (**figure 14.8**).
- The *ara* operon of *E. coli* is an example of an inducible operon that is regulated by the dual-function transcription factor AraC. AraC functions as an anti-activator when arabinose is not available but functions as an activator when arabinose, the inducer, is available (**figure 14.9**).

14.3 Attenuation and Riboswitches Stop Transcription Prematurely

- In the tryptophan operon, a region called the leader lies between the operator and the first structural gene (**figure 14.10**). It codes for synthesis of a leader peptide and contains an attenuator, a factor-independent termination site. Coordination of transcription and translation allows alternative folding patterns in the leader RNA, depending on the availability of charged tRNA. This mechanism of transcription control is called attenuation.
- The 5′ ends (leaders) of some mRNA molecules can bind metabolites and other effectors. Binding of the effector to the mRNA causes a change in the leader structure, which can terminate transcription. This regulatory mechanism is called a riboswitch (**figure 14.11**).

14.4 RNA Secondary Structures Control Translation

- Some riboswitches regulate gene expression at the level of translation. For these riboswitches, the binding of an effector to specific sequences in the leader region of the mRNA alters leader structure and affects ribosome binding either negatively or positively (**figure 14.12**).
- RNA folding patterns in some leader sequences are responsive to temperature changes (**figure 14.13**). These RNA thermometers are often used by pathogens to initiate an infection process.
- Translation can also be controlled by small RNAs. These small RNA molecules are noncoding. They base-pair to the mRNA and usually inhibit translation (**figure 14.14**).

14.5 Mechanisms Used for Global Regulation

- Some cellular processes are so complex they require the action of multiple operons that must be simultaneously controlled. Such operon networks are global regulatory systems.
- Global regulatory systems involve many layers of regulation. Mechanisms such as global regulatory proteins, two-component signal transduction systems, phosphorelay systems, alternate sigma factors, and second messengers are often used.
- Expression of the porin genes *ompF* and *ompC* is controlled by a two-component signal transduction system. These systems have a sensor kinase that detects an environmental change. The sensor kinase transduces the signal to a response-regulator protein by transferring a phosphoryl group. The response regulator then regulates genes to adapt to the new conditions (**figure 14.15**).
- Different sigma subunits control transcription of classes of genes when they bind to core RNA polymerase. Each set of genes controlled by a specific sigma factor has a unique promoter sequence specifically recognized by that sigma.
- Diauxic growth is observed when *E. coli* is cultured in the presence of glucose and another sugar such as lactose (**figure 14.17**). This growth pattern is the result of catabolite repression, where glucose is used preferentially over other sugars. Operons that are part of the catabolite repression system are regulated by the activator protein CAP. CAP activity is modulated by the second messenger cAMP, which is produced only when glucose is not available (**figure 14.19**). Thus when there is no glucose, CAP is active and promotes transcription of operons needed for the catabolism of other sugars (**figure 14.18**).
- The stringent response occurs when bacteria regulate their growth rate. It is mediated by the second messengers guanosine tetraphosphate (ppGpp) and guanosine pentaphosphate (pppGpp). The stringent response results in a decrease in transcription, translation, replication, and metabolism (**figure 14.20**). (p)ppGpp also serves as a general alarm for other stress conditions.
- Cyclic dimeric GMP (c-di-GMP) levels in bacterial cells depend on the activity of enzymes called diguanylate cyclases (DGCs), which synthesize c-di-GMP, and phosphodiesterases (PDEs), which degrade c-di-GMP. The DGCs and PDEs have sensory domains that detect environmental signals. Their response to the signal determines the level of intracellular c-di-GMP. Interaction of c-di-GMP with effector molecules that affect a target brings about the appropriate response to the environmental signal.

14.6 Bacteria Combine Several Regulatory Mechanisms to Control Complex Cellular Processes

- Chemotaxis in *E. coli* is regulated by covalent modification of chemoreceptor proteins and a two-component signal transduction system that controls the direction of flagellar rotation (**figure 14.24**). Methylation is used to measure the amount of a chemoattractant encountered over time. In this way, *E. coli* can determine if it is moving toward or away from an attractant.
- Quorum sensing is a type of cell-to-cell communication mediated by small signaling molecules such as *N*-acylhomoserine lactone. Quorum sensing couples cell density to regulation of transcription. Well-studied quorum sensing systems include the regulation of bioluminescence in *Vibrio* spp. (**figures 14.25** and **14.26**).
- Endospore formation in *B. subtilis* is regulated by a phosphorelay system that is important in initiation of sporulation and the use of alternate sigma factors (**figure 14.27**).
- Bacteria respond to viral infection by deploying enzymes to degrade viral nucleic acids. Restriction endonucleases recognize unmethylated viral DNA and degrade it. CRISPR/Cas systems maintain a record of viruses in a CRISPR array (**figure 14.28**). Following induction by viral contact, expression of the array results in Cas nucleases targeted to viral nucleic acid.

Active Learning

1. Attenuation affects anabolic pathways, whereas repression affects either anabolic or catabolic pathways. Provide an explanation for this.
2. Describe the phenotype of the following *E. coli* mutants when grown in two different media: glucose only and lactose only. Explain the reasoning behind your answer.
   a. A strain with a mutation in the gene encoding the *lac* repressor; the mutant repressor cannot bind allolactose.
   b. A strain with a mutation in the gene encoding CAP; the mutant form of CAP binds but cannot release cAMP.
   c. A strain in which the Shine-Dalgarno sequence has been deleted from the gene encoding adenyl cyclase.
3. What would be the phenotype of an *E. coli* strain in which the tandem tryptophan codons in the *trp* operon leader region were mutated so that they coded for serine instead?
4. What would be the phenotype of a *B. subtilis* strain whose gene for σ^G has been deleted? Consider the ability of the mutant to survive in nutrient-rich versus nutrient-depleted conditions.
5. Propose a mechanism by which a cell might sense and respond to levels of Na^+ in its environment.
6. Some genes whose products confer antibiotic resistance are regulated so that gene expression only occurs in the presence of that antibiotic. Consider an antibiotic that binds to the ribosome to inhibit translation. Diagram a regulatory circuit that would function through a leader peptide.
7. Some genes are regulated by two riboswitches, one that recognizes ppGpp and another that recognizes a specific amino acid. Suggest why two different mechanisms might be useful to control gene expression.
8. Consider the regulatory mechanisms described here. How do they compare in the time elapsed between signal recognition and response?
9. Is a CRISPR/Cas system more likely to target lytic or lysogenic bacteriophages (*see figure 6.15*)? Explain your answer.
10. Magnetotactic bacteria (*see figure 3.30*) build intracellular magnetosomes, particles formed from Fe^{+2} and Fe^{+3} crystals, and biomineralization requires a specific ratio of these ions. Iron import to the cell stimulates production of reactive oxygen species, which triggers oxidative stress. Would you expect a molecular sensor of oxidative stress to be membrane-bound or cytoplasmic?

    When stimulated to activate genes to cope with oxidative stress, the OxyR transcription factor activates many genes. Can you suggest which enzymes might be in the OxyR regulon? *(Review section 7.5)*
11. What would be the phenotype of a *V. fischeri* mutant that could not regulate *luxI* so that it was constantly producing autoinducer at high levels?

CHAPTER

15

Eukaryotic and Archaeal Genome Replication and Expression

Kai4107/Shutterstock

Pharming

Does the above photo look like a vaccine production facility? If you consider each leaf in this field as a bioreactor, this is one giant "farm"aceutical factory. Like all factories, this factory uses raw ingredients (CO_2 from the air and nutrients from soil) and energy (sunlight) to manufacture products (vaccines, antibodies) that have been designed by engineers. Tobacco plants can mass produce biologically active molecules like antibodies and vaccines more rapidly and at a lower cost than traditional cell cultures. Antibodies to treat viral infections like AIDS and Ebola have been produced in tobacco, and additional vaccines made this way are in development.

Decades of research and development have made *Nicotiana tabacum* easy to manipulate in the laboratory, and scientists have optimized methods for gene expression in plant cells. This requires a good understanding of the similarities and differences between the processes that occur in organisms from the different domains of life.

Bacteria, archaea, and eukaryotes share fundamental molecules and cellular structures: lipid-derived boundaries, protein catalysts for cellular reactions, and nucleic acid genomes. Thus far we have compared the three domains in terms of cell structure (chapters 3 through 5), but studied in depth only the bacterial processes of replication, transcription, translation, and regulation (chapters 13 and 14). Here we expand our study to eukaryotes and archaea. The big picture is the same, but differences in these processes reflect the differences in cell structure.

Readiness Check:

Based on what you have learned previously, you should be able to:

- ✓ Trace the movement of materials via the endocytic and secretory pathways of eukaryotic cells (section 5.4)
- ✓ Describe the nucleosomal organization of a eukaryotic chromosome (section 5.5)
- ✓ Discuss bacterial genome structure and the organization of genetic information (sections 13.2, 13.4, and 13.6)
- ✓ Describe the events that occur during the three phases of DNA replication and how the components of the bacterial replisome participate (section 13.3)
- ✓ Describe the structure of a typical bacterial RNA polymerase holoenzyme and how it functions during the three phases of transcription (section 13.5)
- ✓ Describe the events that occur during the three phases of translation and the structure of bacterial ribosomes (section 13.7)
- ✓ Summarize the posttranslational events that result in a properly folded protein in its appropriate location within or outside a bacterial cell (section 13.9)
- ✓ Outline the steps of the eukaryotic cell cycle

15.1 Genetic Processes in the Three Domains

After reading this section, you should be able to:

a. Explain the importance of compartmentation for biochemical reactions in eukaryotic and archaeal cells
b. Distinguish membrane-bound organelles and biomolecular condensates for segregating cellular reactions

To understand cellular processes, biologists have long assumed that molecules diffuse through the cytoplasm and react when they are in close proximity. Given the vast numbers of molecules in a cell, this becomes hard to imagine, especially on the scale and complexity of a eukaryotic cell. Restricting certain reactions to membrane-bound organelles is one way to address this. For example, the carriers in the electron transport chain are ordered within the mitochondrial cristae for efficient electron transfer. Recent studies illustrate that even within organelles, some processes are partitioned into biomolecular condensates**.** These macromolecular complexes segregate and concentrate specific molecules to increase the efficiency of the biochemical reactions in which they participate. The first example discovered was the nucleolus, the site of rRNA synthesis.

Several of the eukaryotic processes described in this chapter (replication, transcription, and RNA processing) occur to some extent within biomolecular condensates in the nucleus. They localize to a specific site on the genome (origin of replication or promoter) and impose a degree of organization on the process. Condensates are dynamic in that they form only when a threshold concentration of molecules is reached, and dissolve when the process is complete.

Biomolecular condensates are not exclusively eukaryotic (*see Molecular Diversity & Ecology 3.1*). For example, FtsZ treadmilling during bacterial and archaeal cell division is a kind of condensate (*see figure 7.6*). However, in comparing replication and gene expression among cells in the three domains, it is interesting to note how the complexity in eukaryotic cells requires the additional organization conferred by biomolecular condensates.

In this chapter we illustrate the similarities of archaea and eukaryotes with respect to how they replicate, express, and regulate the information in their genomes and point out distinct aspects of archaea. Archaea are a mix of bacterial and eukaryotic characteristics. It is this mix that has prompted studies to elucidate the evolutionary history of microorganisms.

15.2 DNA Replication

After reading this section, you should be able to:

a. Create a concept map that illustrates the differences in eukaryotic and archaeal genome structure and organization
b. Distinguish the replicons of eukaryotes and archaeal cells
c. Compare and contrast the replisomes of eukaryotes and archaeal cells
d. Construct a model to demonstrate the role of telomerase in the replication of linear chromosomes

DNA replication is similar in all cellular organisms. It is accomplished by a complex of many different proteins called the **replisome.** Eukaryotic replication is most commonly studied in the yeast *Saccharomyces cerevisiae* (*see figure 24.9*) and cultured cells from insects and mammals.

DNA Replication in Eukaryotes

The distinctive features of eukaryotic DNA replication arise from differences in replication enzymes and genome structure. Eukaryotes use a conserved set of DNA replication proteins (**table 15.1** and **figure 15.1**). Eukaryotic chromosomes are much larger than those of bacteria and archaea. Furthermore, they are linear, which means that a mechanism for replicating chromosome ends is needed. Finally, eukaryotic DNA is organized in nucleosomes (*see figure 5.11*). Nucleosomes must be deconstructed ahead of the replication fork and then formed on the parental and newly synthesized chromosomes behind the replication fork.

DNA Replication Initiates from Multiple Origins

Replication of eukaryotic chromosomes is initiated at multiple origins of replication (**figure 15.2**). This allows these large chromosomes to be replicated much faster than if there were only one origin per chromosome. Two replication forks move outward from each origin until they encounter replication forks at adjacent origins. Thus eukaryotic chromosomes consist of multiple **replicons,** rather than the single replicon (i.e., the entire chromosome) observed in bacteria.

The multiple origins in eukaryotic chromosomes do not all initiate replication simultaneously, or even necessarily within a given cell cycle, and there is not always a specific DNA sequence to mark the origin. In yeast chromosomes, replication originates both from autonomously replicating sequences and from 100 to 200 bp regions that are free of nucleosomes.

Regardless of the mechanism to identify an origin, two steps must occur to initiate DNA replication, termed licensing and firing. *Licensing* refers to replicative helicase binding to origins in the G1 phase of the cell cycle (**figure 15.3**). In yeast, the proteins comprising the **origin recognition complex (ORC)** bind an origin and remain bound throughout much of the cell cycle. The ORC serves as a platform on which additional proteins assemble as the cell progresses through the cell cycle. The first protein to associate with ORC is the helicase loader. In G1, the helicase, a set of proteins called the MCM (*m*ini*c*hromosome *m*aintenance) complex associates with the ORC, thereby forming the *pre-replicative complex* (pre-RC). Each MCM helicase encircles both DNA strands, where it will become part of a larger helicase complex (figure 15.3). As the cell cycle transitions from G1 to S phase, the *firing* step occurs. DNA strand separation occurs when helicase uses the energy from ATP to eject one strand from each complex and propel the complexes in opposite directions (figure 15.3).

Table 15.1 Major Replisome Proteins in Bacteria, Eukaryotes, and Archaea

Function	Bacteria	Eukaryotes	Archaea
Initiator; binds origin[1]	DnaA	ORC	Orc1/Cdc6
Helicase loader	DnaC	Cdc6/Cdt1	Orc1/Cdc6
Helicase	DnaB	CMG complex of Cdc45, MCM, GINS[2]	Complex of RecJ, MCM, GINS[2]
Single-strand DNA binding	SSB proteins	Replication protein A (RPA)	SSB or RPA, depending on the species
Primer synthesis	DnaG	Pol α-primase (subunits PriS and PriL)	PriS, PriL, PriX
Replicative DNA polymerase	DNA polymerase III (C-family polymerase)	DNA polymerase (Pol) δ and DNA polymerase ε (B-family polymerases)	B-family polymerase and/or D-family polymerase
Clamp loader	τ complex	Replication factor C (RFC)-L and RFC-S (1–4)[3]	RFC-L and RFC-S[3]
Clamp	β clamp	Proliferating-cell nuclear antigen (PCNA)	PCNA
Primer removal	Ribonuclease (RNase) H[4]; DNA polymerase I	RNaseH[4]; Dna2; Flap endonuclease (FEN)-1	RNaseH[4]; FEN-1
Ligase	NAD^+-dependent DNA ligase; ATP-dependent DNA ligase[5]	ATP-dependent DNA ligase[5]	ATP-dependent DNA ligase; NAD^+-dependent DNA ligase[5]

[1] Although we have included the initiators in this table, they are involved only in the formation of the replisome and do not remain in the replisome as it advances from the origin of replication.
[2] In eukaryotes, MCM is heteromeric, composed of six different Mcm proteins (Mcm2–7). GINS is short for the first letters of the Japanese numbers representing the proteins composing GINS. In most archaea, MCM is homohexameric. The archaeal MCM is associated with GINS homologues. RecJ is the archaeal version of Cdc45.
[3] The eukaryotic clamp loader is a pentameric structure consisting of one large subunit (RFC-L) and four small subunits encoded by four different genes. The archaeal clamp loader is also pentameric, consisting of RFC-L and four small subunits encoded by one or two genes, depending on the species.
[4] The bacterial RNase H is distinct, whereas the eukaryotic and archaeal enzymes are homologues.
[5] All NAD^+-dependent DNA ligases belong to the same family of ligases and are evolutionarily related. Likewise ATP-dependent DNA ligases are related. NAD^+-dependent DNA ligases are primarily responsible for ligating Okazaki fragments in bacteria, whereas ATP-dependent DNA ligases play this role in eukaryotes and archaea.

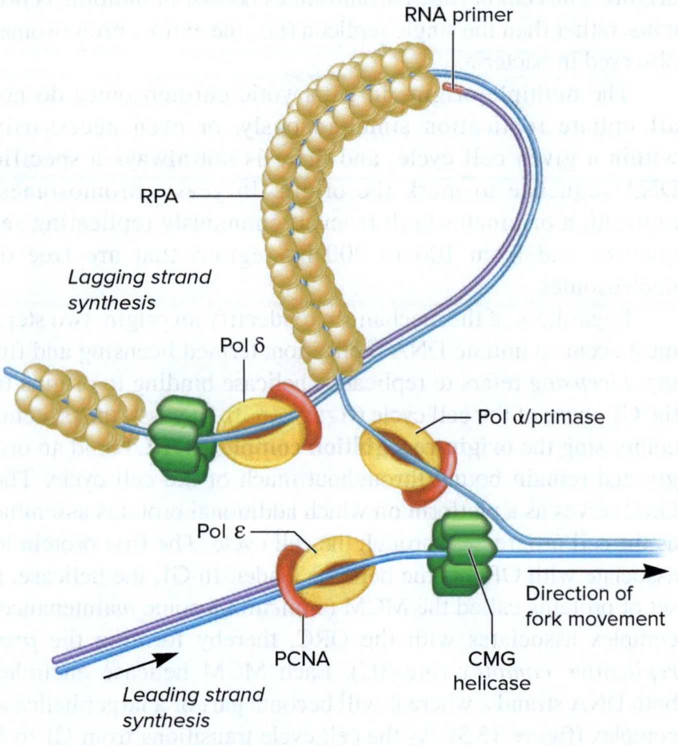

Figure 15.1 A Typical Eukaryotic Replication Fork. The replisome proteins and their functions are defined in table 15.1. Compare this with figure 13.12. Parental DNA strands are blue; newly synthesized DNA strands are purple.

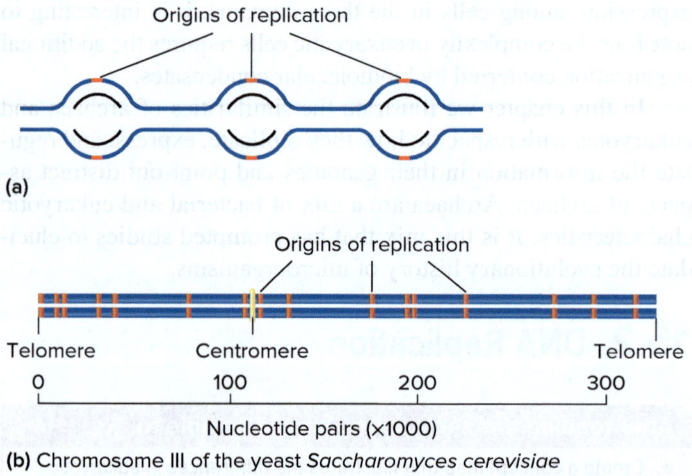

Figure 15.2 Eukaryotic Chromosomes Are Replicated from Multiple Origins of Replication. (a) Drawing of a eukaryotic chromosome being replicated. (b) Schematic of *Saccharomyces cerevisiae* chromosome III. It is 317,000 base pairs and is replicated from multiple origins of replication.

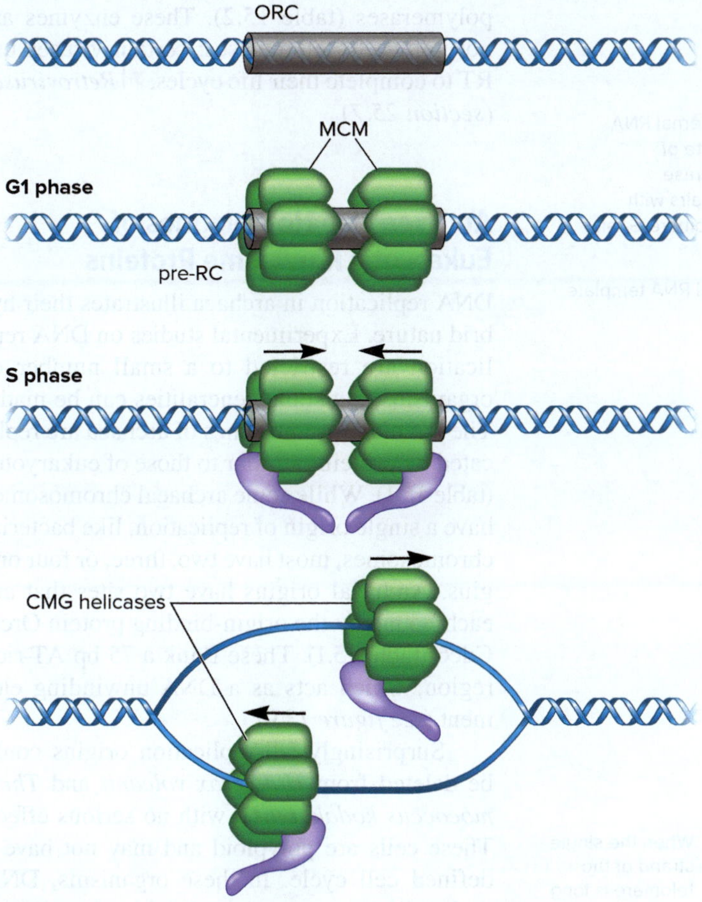

Figure 15.3 Eukaryotic Replication Initiation. The origin recognition complex (ORC) is bound to the origin throughout the cell cycle. In G1 phase, the pre-replicative complex (pre-RC) forms, and in S phase, replication begins when the fully assembled CMG helicases move away from the origin to separate the two DNA strands. MCM, minichromosome maintenance.

Eukaryotic Primases and DNA Polymerases

DNA polymerases (abbreviated Pols) from many organisms have been sorted into seven families based on their amino acid sequence and structure (**table 15.2**). In eukaryotes, three DNA Pols function in replication: DNA Pols α-primase, ε, and δ. Primer synthesis is accomplished with DNA Pol α-primase (often called simply Pol α-primase). Pol α-primase is so named because it has two activities: DNA synthesis (Pol α) and RNA synthesis (primase). This interesting enzyme consists of four subunits: two for DNA synthesis (Pol 1 and Pol 2) and two for RNA synthesis (PriS and PriL). The primer is made in two steps: PriS and PriL make a short RNA strand (~10 nucleotides), which is transferred to the active site of Pol α. Pol α adds an additional 20 or so deoxyribonucleotides to the RNA strand. Thus in eukaryotes, the single-stranded primer for DNA replication is an RNA-DNA hybrid molecule. Once the primer is formed, the other two DNA Pols take over. DNA Pol ε catalyzes leading-strand synthesis, whereas DNA Pol δ carries out lagging-strand synthesis (figure 15.1).

Table 15.2 DNA Polymerase Families

Family	Example(s)	Functions/Comments
A	Bacterial DNA polymerase I	Replacement of RNA primers present in Okazaki fragments
B	Archaeal DNA polymerase B (Pol B) Eukaryotic DNA polymerases α, δ, and ε	Replicative DNA polymerases in eukaryotes and some archaea
C	Bacterial DNA polymerase III	Replicative DNA polymerase
D	Archaeal DNA polymerase D (Pol D)	Replicative DNA polymerase in some archaea; unique to archaea
X	Eukaryotic DNA polymerase β	DNA repair
Y	Bacterial DNA polymerase IV (DinB)	DNA repair
Reverse transcriptase (RT)	Retroviral reverse transcriptases Telomerase RT	RNA-dependent DNA polymerase

Telomeres and Telomerases: Protecting the Ends of Linear DNA Molecules

The linear DNA in eukaryotic chromosomes poses problems not faced by the circular chromosomes in bacteria and archaea. Recall that during replication the lagging strand is comprised of Okazaki fragments. When the primer for the Okazaki fragment at the 5′ end of the daughter strand is removed, the daughter molecule is shorter than the parent molecule. Over numerous rounds of DNA replication and cell division, this leads to a progressively shortened chromosome. Ultimately the chromosome loses critical genetic information. This is called the end replication problem.

Eukaryotic cells have solved this problem by forming structures called **telomeres** at the ends of their chromosomes and by using an enzyme called **telomerase** (**figure 15.4**). Telomeres are protein-DNA complexes that protect the DNA within them from degradation and from accidental fusion with other ends. The protein component of a telomere varies from species to species, as does the length of DNA covered. Telomere DNA contains many copies of a 6–8 base repeated sequence of nucleotides. Importantly, the DNA is single-stranded at the end and is called the *G-tail* because it is rich in guanosine bases; telomerase uses the G-tail to maintain chromosome length.

Telomerase has two important components: an internal RNA template and an enzyme with reverse transcriptase (RT) activity (table 15.2). The internal RNA is complementary to the

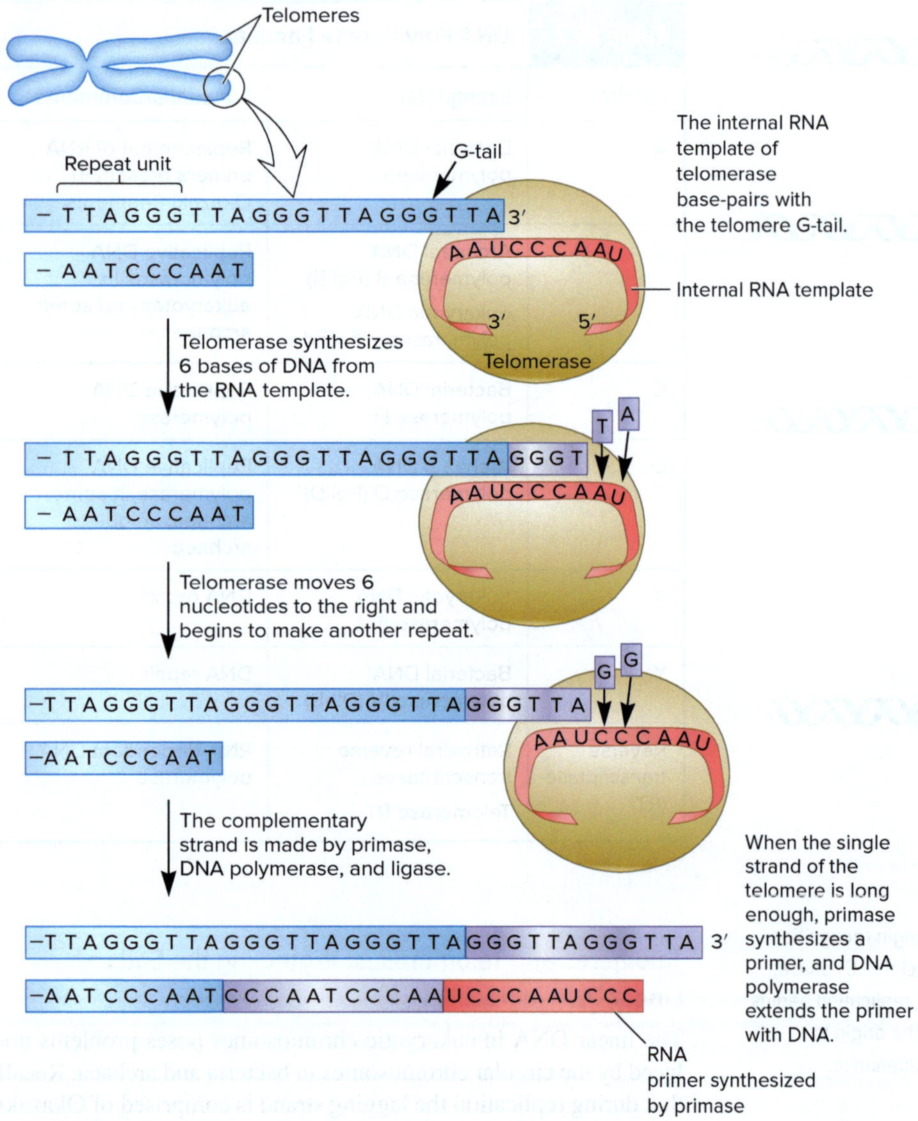

Figure 15.4 Telomerase Replicates the Telomeric DNA of Eukaryotic Chromosomes. Telomerase contains an RNA that base-pairs with the 3′ overhang (the G-tail). The RNA serves as a template for DNA synthesis catalyzed by the reverse transcriptase activity of the enzyme. The 3′-OH of the telomere DNA serves as the primer. The process is repeated until the 3′ overhang is long enough to serve as the template for synthesis of the complementary telomere DNA strand by the replisome.

G-tail and base pairs with it (figure 15.4). This provides a template for DNA synthesis, catalyzed by telomerase RT (i.e., the 3′-OH of the G-tail serves as the primer for DNA synthesis). After being lengthened sufficiently, there is room for synthesis of an RNA primer, and the single strand of telomere DNA serves as the template for synthesis of the complementary strand. Thus the length of the chromosome is maintained. Telomerase is an unusual enzyme that synthesizes DNA using an RNA template. Enzymes with this capability are termed RNA-dependent DNA polymerases (table 15.2). These enzymes are not unique to telomerases; certain viruses use RT to complete their life cycles. ▶| *Retroviruses (section 25.7)*

Archaea Use Homologues of Eukaryotic Replisome Proteins

DNA replication in archaea illustrates their hybrid nature. Experimental studies on DNA replication are restricted to a small number of organisms, but some generalities can be made. The circular chromosomes of archaea are replicated by proteins similar to those of eukaryotes (table 15.1). While some archaeal chromosomes have a single origin of replication, like bacterial chromosomes, most have two, three, or four origins. Archaeal origins have two sites that are each bound by the origin-binding protein Orc1/Cdc6 (table 15.1). These flank a 75 bp AT-rich region, which acts as a DNA unwinding element (*see figure 13.9*).

Surprisingly, all replication origins could be deleted from *Haloferax volcanii* and *Thermococcus kodakarensis* with no serious effect. These cells are polyploid and may not have a defined cell cycle. In these organisms, DNA replication appears to be recombination-driven, which requires at least two genome copies. This replication mechanism requires an archaeal-specific DNA Pol called PolD (table 15.2). In contrast, *Sulfolobus* spp. have a eukaryotic-like cell cycle, multiple origins, and a eukaryotic-like replicative polymerase, PolB. All archaea have eukaryotic-style helicases.

Comprehension Check

1. Compare membrane-bound organelles and biomolecular condensates as compartments for biochemical reactions.
2. What is the advantage of having multiple origins of replication?
3. Refer to figure 15.1 and draw the last replication fork present near the end of a chromosome. Use your drawing to illustrate the end replication problem.
4. Review mitosis and meiosis. Why would having unprotected ends of linear DNA disrupt these two important cellular processes?
5. Outline the process used by telomerase to maintain chromosome length.
6. Why would the discovery of telomerase activity in an archaeon be a surprise?
7. Review section 13.3 and draw a Venn diagram that illustrates the unique and shared properties of chromosome structure and replication among bacteria, eukaryotes, and archaea.

15.3 Transcription

After reading this section, you should be able to:

a. Compare and contrast gene structure of eukaryotic and archaeal cells
b. Create a table to distinguish eukaryotic and archaeal promoters, RNA polymerases, and transcription factors
c. Discuss the modifications made to eukaryotic mRNA molecules during and after synthesis

Transcription is essentially the same in all cellular organisms: The DNA template strand directs RNA synthesis from the 5′ to the 3′ end by RNA polymerases. Although this unifies the process of transcription across all domains of life, several key aspects differ. These arise in part from differences in cell and chromosomal structure, the nature of RNA polymerase, and the organization of protein-coding genes.

Transcription in Eukaryotes

In eukaryotes, transcription occurs in the nucleus, and the RNA products must move to the cytoplasm, where they function in translation. As in DNA replication, chromatin structure must be altered for RNA polymerases to access promoters and to transcribe a gene. This is accomplished by chromatin remodeling described in section 15.5. For the vast majority of eukaryotes, each protein-coding gene has its own promoter, and the transcript synthesized from the gene is **monocistronic.** Furthermore, many eukaryotic protein-coding genes are composed of regions called exons and introns. An **exon** contains sequences that code for part of the polypeptide, whereas an **intron** is a stretch of noncoding sequences. Thus the coding portion of a eukaryotic gene is not continuous. Both exons and introns are transcribed, yielding an exon- and intron-containing transcript called a **primary** or **pre-mRNA.** The introns must be spliced out of the primary transcript before translation.

Eukaryotic RNA Polymerases Include Homologues of the Bacterial RNA Polymerase Core Subunits

Eukaryotes have three major RNA polymerases (RNAP). RNAP I catalyzes rRNA synthesis, RNAP II is responsible for mRNA synthesis, and RNAP III synthesizes all the tRNAs and one rRNA (5S rRNA). All five of the bacterial RNAP core enzyme subunits are conserved in eukaryotes (**figure 15.5**). However, eukaryotic RNAPs have several additional subunits.

Transcription Initiation in Eukaryotes

Most research on transcription in eukaryotes has been on the transcription of protein-coding genes by RNAP II, and that is our focus here. This complex enzyme is at least 500,000 daltons in size, with about 10 subunits.

Eukaryotic promoters are marked by both a depletion of nucleosomes and specific DNA sequences. RNAPII-recognized promoters usually have two regions, the core promoter and a regulatory region. The core promoter is the minimal region needed for transcription to occur. Two conserved sequences found in core promoters are the BRE and TATA boxes (**figure 15.6**). Although it was initially thought that all RNAPII core promoters contain a TATA box, it is now known that many do not. The nucleosome located within the coding sequence and positioned closest to the promoter is termed the **+1 nucleosome.** TATA-less promoters rely on the +1 nucleosome to position the transcription start site, while a TATA box does so when present.

Eukaryotic transcription factors are referred to as extrinsic factors because they are not a subunit of the RNAP, like the sigma (σ) subunits of bacterial RNAP. The TATA-binding protein (TBP) is one of the most important for transcription initiation. Despite its name, TBP first binds to a promoter region regardless of whether

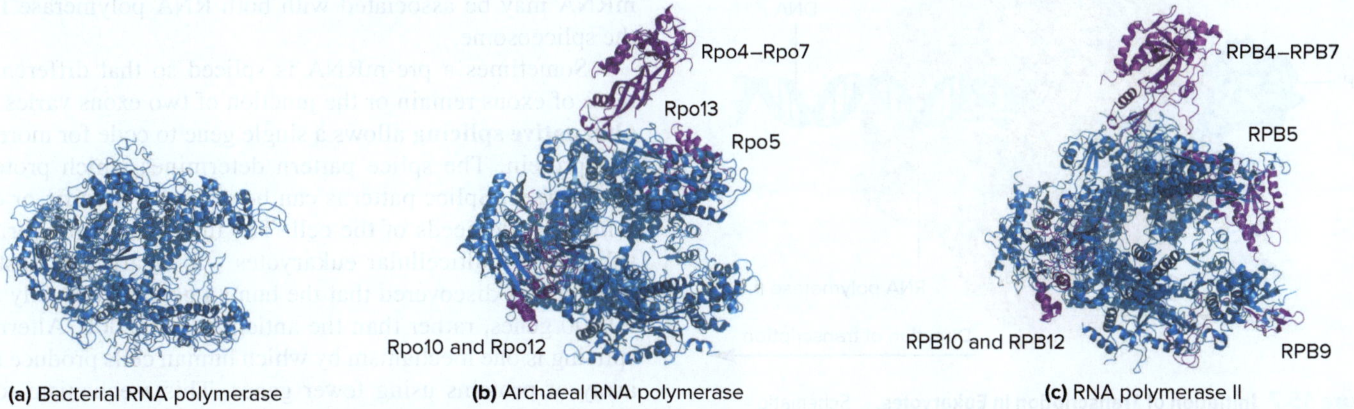

Figure 15.5 Bacterial, Archaeal, and Eukaryotic RNA Polymerases. Ribbon diagrams of the RNA polymerases of (a) the bacterium *Thermus aquaticus,* (b) the archaeon *Saccharolobus shibatae,* and (c) RNA polymerase II of eukaryotes. The subunits found in all three enzymes are shown in blue. Those that are found in archaea (Rpo) and RNA polymerase II (RPB) are shown in magenta.
Finn Werner

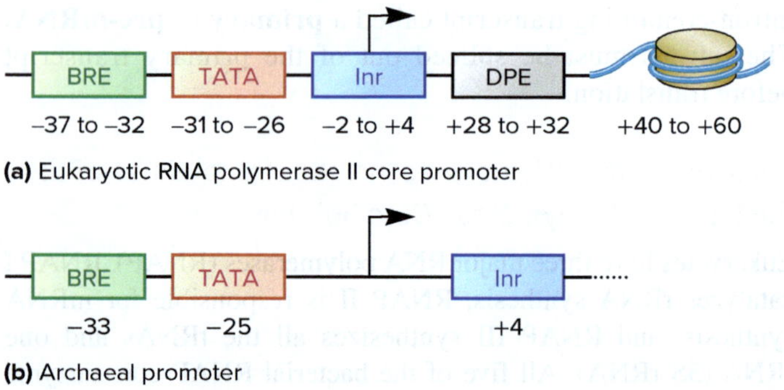

Figure 15.6 Eukaryotic and Archaeal Promoters. (a) Eukaryotic RNA polymerase II core promoters contain one or more of the sequence elements shown. However, it is rare for all of the elements to be present in a single gene. The approximate positions of each element are indicated, relative to the transcriptional start site arrow. (b) Archaeal promoter sequence element designations are the same as those in eukaryotic promoters, although the exact nucleotide sequences in the elements differ. BRE, TFII*B* *r*ecognition *e*lement; Inr, *in*itiato*r* element; DPE, *d*ownstream *p*romoter *e*lement.

a TATA box is present. Two additional TFs enter, TFIIA binding to the TBP and TFIIB binding to the BRE (figure 15.6). RNAPII recognizes this substrate and joins to form the pre-initiation complex (PIC) (**figure 15.7**). TFIIE distorts the DNA and positions TFIIH to denature about 14 base pairs, forming an open complex (*see figure 13.25*). Transcription cannot proceed to the elongation phase until the contacts between TFIIB and RNAPII are broken. This is accomplished by TFIIH, which phosphorylates an RNAPII subunit, and releases RNAPII from the PIC. Elongation proceeds. The transcription factors and co-activators remain on the promoter to recruit additional RNAPIIs.

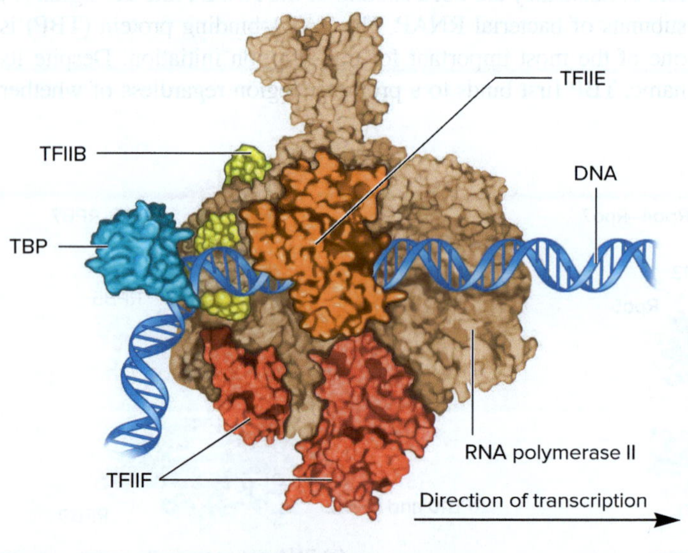

Figure 15.7 Initiation of Transcription in Eukaryotes. Schematic showing the location of some of the transcription factors in the pre-initiation complex (PIC). TFIIH and TFIIA are not visible in this view. Note how closely the transcription factors are associated with RNA polymerase II. The arrow shows the direction of transcription, once it initiates.

Eukaryotic Transcription Elongation Is a Unique Process

Eukaryotic genes are packaged in chromatin and nucleosomes present an impediment to an elongating RNAPII. Recall that promoter regions are depleted in nucleosomes, so RNAPII doesn't encounter nucleosomes until elongation. Each nucleosome must be individually dissociated from the DNA ahead of the RNAPII and reassembled after its passage (**figure 15.8**).

Unlike nearly all bacterial mRNAs, the initial transcript synthesized from a eukaryotic protein-coding gene must be modified before giving rise to a translatable mRNA. Modification occurs in the nucleus and begins during elongation; it is completed after transcription ends. The first modification is the addition of an unusual nucleotide to the 5′ end of the transcript when the transcript is only about 25 ribonucleotides long (**figure 15.9**). The **5′ cap,** as it is called, is 7-methylguanosine. After the 5′ cap is added and a sufficiently long transcript is generated, removal of introns occurs. After transcription is complete, any remaining introns are removed, and the pre-mRNA is modified further by the addition of a **3′ poly-A tail** (figure 15.9). Both the 5′ cap and poly-A tail help protect the mRNA from enzymatic attack. Furthermore, they enable the cell to recognize mRNA ready for transport to the cytoplasm. The 5′ cap also functions as a recognition signal for ribosome binding so that the mRNA can be translated.

Introns are removed from the pre-mRNA by a ribozyme unique to eukaryotic cells, called a **spliceosome.** Like the ribosome, the spliceosome is composed of dozens of proteins and several RNA molecules. Short sequences within each intron are recognized and excised by the spliceosome RNA, and the exons are joined together. The reaction generates two products, the mature mRNA that moves to the ribosome for translation and the excised RNA molecules, each called a *lariat* based on its shape (**figure 15.10**). Lariats may be degraded to ribonucleotides, or a portion may function in gene regulation. Intron removal occurs cotranscriptionally so that an individual mRNA may be associated with both RNA polymerase II and the spliceosome.

Sometimes a pre-mRNA is spliced so that different patterns of exons remain or the junction of two exons varies. This **alternative splicing** allows a single gene to code for more than one protein. The splice pattern determines which protein is synthesized. Splice patterns can be cell-type specific or determined by the needs of the cell. The importance of alternative splicing in multicellular eukaryotes was clearly demonstrated when it was discovered that the human genome has only about 20,000 genes, rather than the anticipated 100,000. Alternative splicing is one mechanism by which human cells produce a vast array of proteins using fewer genes. This mechanism for expanding the coding capacity of a genome is not available to bacteria and archaea because introns in their protein-coding genes are rare. *Processing of Genetic Information: Prokaryotes versus Eukaryotes*

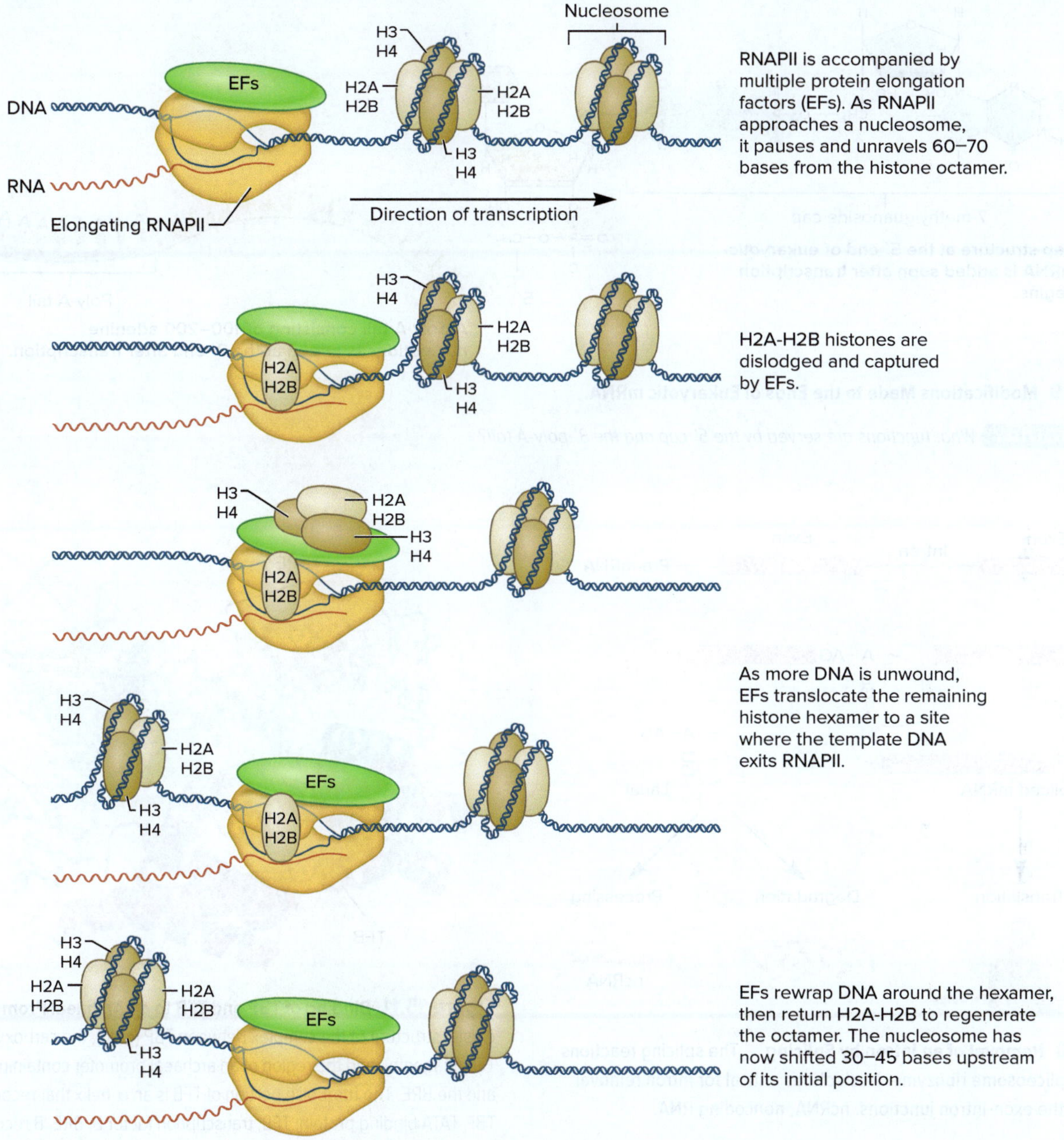

Figure 15.8 Transcription Through a Nucleosome. An elongating RNA polymerase II must dislodge a nucleosome from the DNA template. A nucleosome has eight individual histone proteins. It is dismantled as a dimer (2 proteins) and a hexamer (6 proteins), then relocated to the DNA template behind RNA polymerase II. In this diagram, nucleosomes are oversized relative to RNAPII.

Transcription in Archaea

An article describing archaeal transcription might have the headline: "Eukaryotic-like RNAP functions in bacterial-like environment." This is because archaeal transcription occurs in the nucleoid, as it does for bacteria. Many archaeal mRNAs are polycistronic like bacterial genes. Introns are rare in protein-coding archaeal genes (as they are in bacteria), and the few present are removed by a process different from that used by eukaryotes (again as they are in bacteria). Finally, archaeal cells use a single RNA polymerase to catalyze all transcription.

Despite these many similarities to bacteria, archaeal transcription can be thought of as a simplied version of that in eukaryotes. The archaeal RNAP is large and quite similar to eukaryotic RNAPII (figure 15.5). Likewise, transcription initiation in archaea is very similar to that seen in eukaryotes, as evidenced by the similarity in promoters (figure 15.6) and the use of similar basal transcription factors. The major players, TATA-binding protein (TBP) and transcription factors B (TFB) and E (TFE), are homologues of the eukaryotic transcription factors. As shown in **figure 15.11**, TBP and TFB recruit RNAP to a promoter.

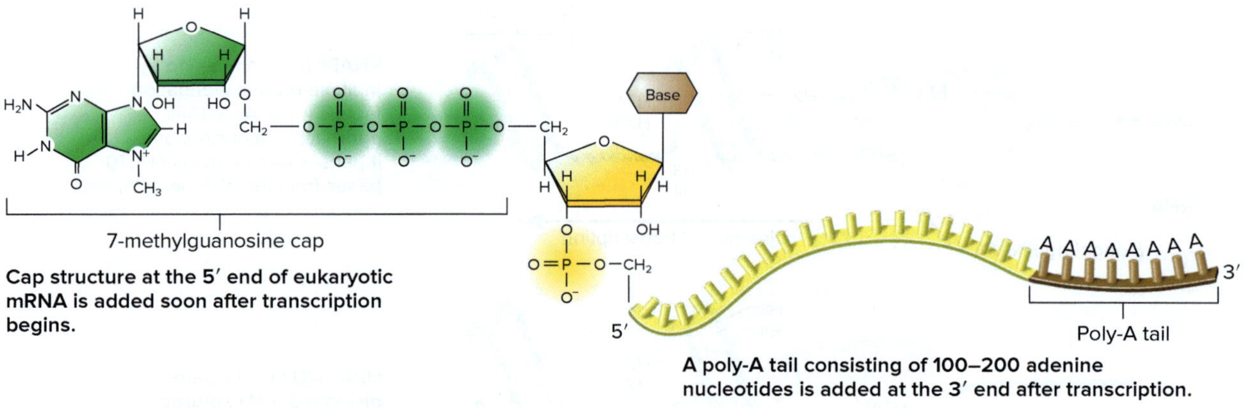

Figure 15.9 Modifications Made to the Ends of Eukaryotic mRNA.

MICRO INQUIRY *What functions are served by the 5′ cap and the 3′ poly-A tail?*

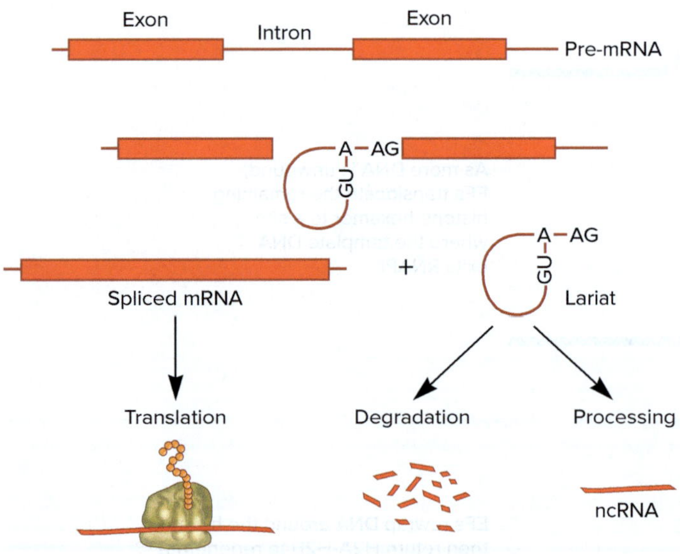

Figure 15.10 Removal of an Intron by Splicing. The splicing reactions occur in the spliceosome ribozyme. Nucleotides critical for intron removal are shown at the exon-intron junctions. ncRNA, noncoding RNA.

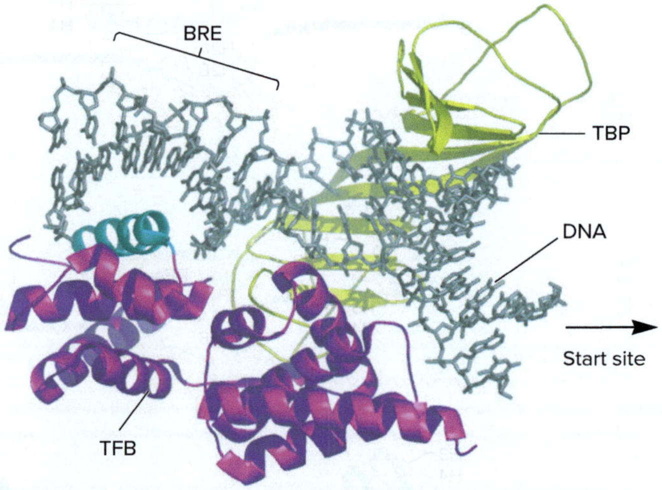

Figure 15.11 Binding of TBP and TFB to an Archaeal Promoter. The crystal structure of the complex between TBP (gold), the carboxyl terminus of TFB (magenta), and the region of an archaeal promoter containing the TATA box and the BRE. The turquoise portion of TFB is an α-helix that recognizes the BRE. TBP, TATA binding protein; TFB, transcription factor B; BRE, B recognition element.
Stephen D. Bell

TFE then associates with RNAP and promotes DNA unwinding to form the open complex (*see figure 13.25*). RNA polymerase then moves along the gene, synthesizing RNA. TFB remains bound to the promoter, ready to initiate another round of transcription. TFE accompanies RNA polymerase until transcription terminates.

While many archaea use histones to organize their DNA, there is not a consistent nucleosomal structure as in eukaryotes. In in vitro experiments, archaeal RNAPs pause at a histone barrier but do not require elongation factors to complete transcription. The diversity among the numerous archaeal histones results in varying nucleosomal forms that play roles beyond organizing the genome and are likely involved in gene regulation.

Comprehension Check

1. What elements in archaeal promoters are also observed in eukaryotic RNA polymerase II (RNAPII) core promoters? What additional elements are observed in RNAPII core promoters?
2. Compare the role of bacterial sigma factors in transcription initiation to that of eukaryotic and archaeal transcription factors. Why are sigma factors referred to as intrinsic transcription factors, whereas eukaryotic and archaeal transcription factors are extrinsic?
3. Outline the events that generate a mature, translatable mRNA in eukaryotes. Where do processing events occur?

15.4 Translation and Protein Maturation and Localization

After reading this section, you should be able to:

a. Differentiate the structure of eukaryotic and archaeal messenger RNAs
b. Identify the initiator tRNA used by eukaryotic and archaeal cells
c. Outline the steps in translation initiation observed in eukaryotes and compare them to those observed in bacteria and archaea
d. Discuss the role of molecular chaperones in eukaryotic and archaeal cells
e. Create a concept map or table that distinguishes the translocation and secretion systems observed in the three domains of life

For organisms to survive, they must be able to synthesize the proteins they need (translation), fold them into a functional conformation, and localize the proteins to the correct site within (or outside) the cell. **Chaperones** help proteins fold, and some chaperones deliver proteins to translocation or secretion systems.

In this section, we focus on translation and protein folding and localization as they occur in eukaryotes and archaea. As with DNA replication and transcription, the overall processes are similar in all organisms. Therefore we focus our attention on those aspects of the eukaryotic and archaeal processes that differ from those observed in bacteria.

Translation in Eukaryotes Initiates in a Unique Manner but Proceeds as in Bacteria

There are many obvious differences between bacterial and eukaryotic translation. Eukaryotic ribosomes are larger than those in bacteria and require more initiation factors to be positioned properly on the mRNA. The 5′ cap and the 3′ poly-A tail of eukaryotic mRNAs are integral to translation initiation. The process begins with mRNA activation, when several eukaryotic initiation factors (eIFs) bind to the 5′ end of the mRNA and poly-A binding proteins (PABPs) attach to the 3′ end. The eIFs and PABPs interact, causing the mRNA to fold back on itself with the eIFs and PABPs forming a bridge between the two ends (**figure 15.12**). Meanwhile, the initiator tRNA (Met-tRNA$_i^{Met}$) is loaded onto the 40S ribosomal subunit (the small subunit, SSU) by several additional eIFs to form a structure termed the 43S complex. The 43S complex then binds the activated mRNA, generating the 48S complex.

As illustrated in figure 15.12, the SSU is not initially positioned at the start codon of the mRNA as it is in bacteria. Rather, it is at the 5′ end of the mRNA. The SSU scans the mRNA for the start codon. Once the start codon is encountered, eIFs are released and the 60S ribosomal subunit (large subunit) joins with the 48S complex, forming the complete 80S ribosome. Translation can now commence.

Translation elongation and termination proceed in a fashion similar to that seen in bacteria. Three elongation factors function in both bacteria and eukaryotes, and they are similar structurally and functionally. Polysomes form during translation of eukaryotic mRNAs. However, because of the bending of the mRNA during translation initiation, the polysomes have a characteristic structure (**figure 15.13**). Another difference is seen in the termination step. In eukaryotes, two release factors (eRFs) function as opposed to three RFs in bacteria.

Protein Folding in Eukaryotes Is Similar to That Seen in Bacteria

Eukaryotes, like bacteria, use chaperones to help many proteins fold properly and restore folding if the protein is denatured. Chaperones were first identified by their role in protecting the proteins of heat-stressed cells and are often called heat-shock proteins (HSPs). There are numerous HSPs, and they are sorted based on their size. For instance, the bacterial chaperones GroEL and GroES have a molecular weight of about 60,000 daltons and are grouped with HSPs of similar size as HSP60s. Representatives of all types of chaperones are found in eukaryotes.

Protein maturation in eukaryotes begins with the N-terminus of the polypeptide, as it emerges from the ribosome. Several enzymes may act on the nascent polypeptide, notably to remove the initiator methionine, and to acetylate the N-terminus. Some chaperones perform their tasks cotranslationally, while others function posttranslationally. About 80% of proteins in yeast are successfully folded cotranslationally by *r*ibosome-*a*ssociated *c*omplex (RAC) and *n*ascent-chain-*a*ssociated *complex* (NAC), acting at the ribosome exit tunnel. Proteins needing additional help to fold are bound by HSP70 and HSP40 cochaperones (similar to the bacterial DnaK and DnaJ proteins). HSP70 is a central folding hub that directs unfolded substrates to a protein called prefoldin, or to other classes of chaperones that may be specific for that substrate. Other proteins are folded by pathways that use HSP60 chaperones.

Protein Localization and Secretion in Eukaryotes

Proteins produced by eukaryotic cells have several possible destinations because these cells are highly compartmentalized, having membrane-bound organelles. Although mitochondria and chloroplasts have genomes that encode some of their proteins, most are encoded in the nucleus and translated in the cytosol. Although some proteins synthesized in the cytosol stay there, many are moved to compartments or are secreted. As described in chapter 5, proteins that are secreted or that move between the endoplasmic reticulum, Golgi apparatus, endosomes, and other membranous components of the endocytic and secretory pathways are delivered to these organelles or the plasma membrane by small vesicles that pinch off from the membrane of one organelle (e.g., the ER) and then fuse with a target membrane. This mechanism for localizing and secreting proteins is often called **vesicular transport,** and it does not involve the translocation of

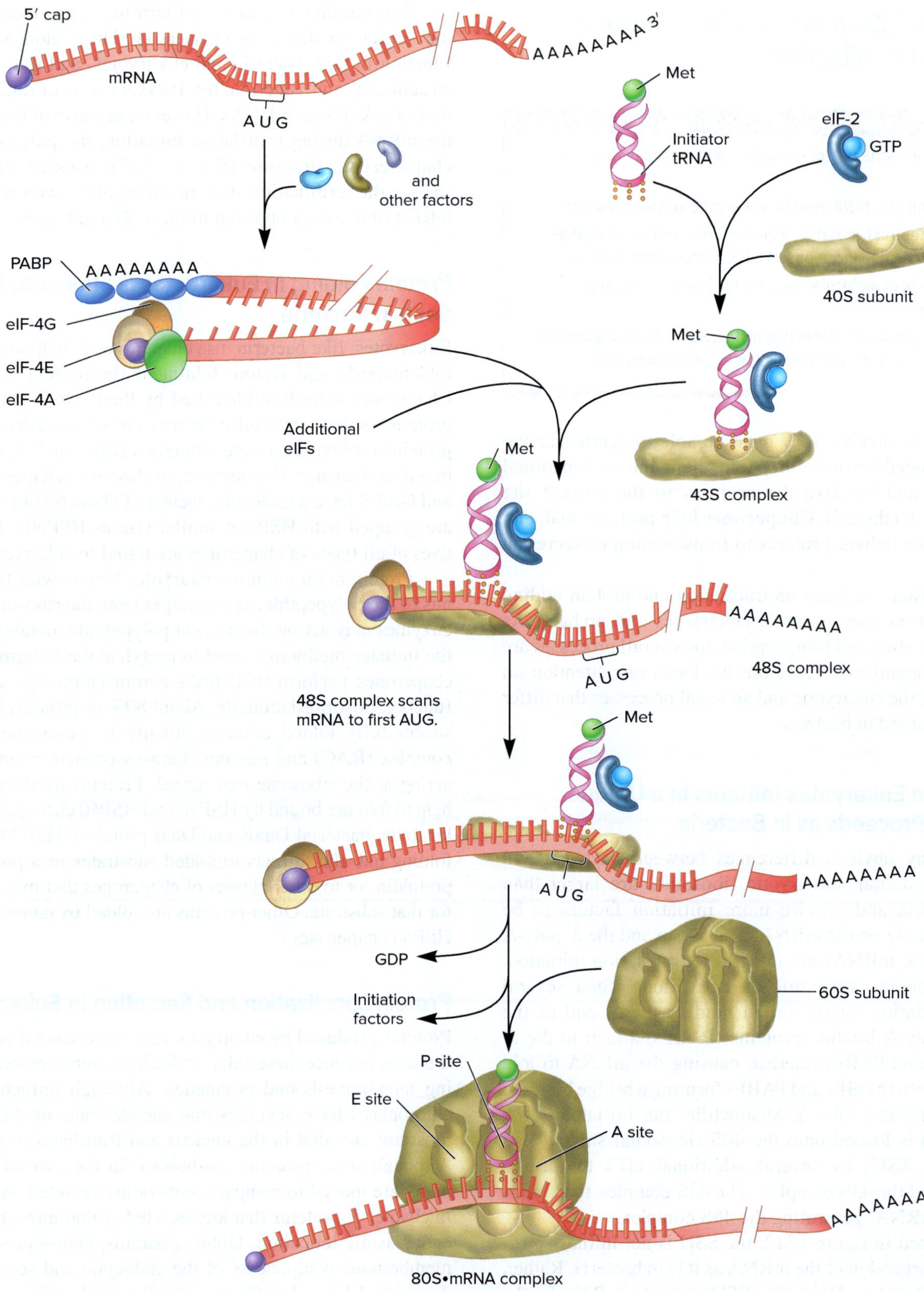

Figure 15.12 Translation Initiation in Eukaryotes. To simplify the drawing, many additional eukaryotic initiation factors (eIFs) are not shown. The upper left portion of the figure shows mRNA activation; the upper right shows the formation of the 43S complex. The 43S complex binds to the 5′ end of the activated mRNA to make the 48S complex, and then moves along the mRNA to find the first AUG. This positions the initiator tRNA carrying methionine in what will become the peptidyl (P) site of the ribosome after the addition of the 60S subunit. The acceptor (A) site is vacant and ready for the arrival of the tRNA carrying the second amino acid encoded by the mRNA. Met, methionine; PABP, poly-A binding protein; E site, exit site.

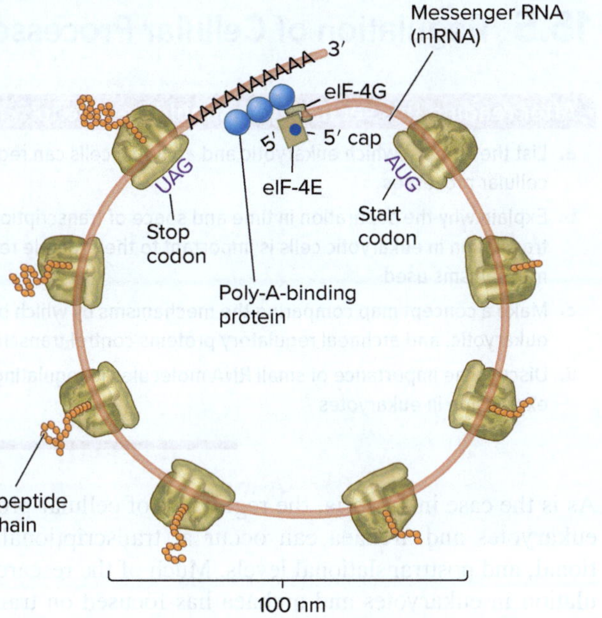

Figure 15.13 Eukaryotic Polysome. Schematic representation of a polysome. The circular form results from the interaction of the two ends of the mRNA brought about by the poly-A-binding proteins, eIF-4E, eIF-4G, and other initiation factors.

proteins across membranes. However, for proteins to move from the cytosol to the ER, mitochondria, and chloroplasts, the proteins must be translocated across membranes. This transmembrane transport involves systems similar to certain translocation and protein secretion systems used by bacteria, as well as other proteins found in bacterial membranes. ◀ *Several organelles function in the secretory and endocytic pathways (section 5.4)*

The best-studied protein translocation system in eukaryotes is the **Sec system** (**figure 15.14**). Central to the functioning of this pathway in eukaryotes is the Sec61αβγ translocon pore. Several proteins that make up the Sec system in eukaryotes are homologous to Sec proteins in bacteria (**table 15.3**). In eukaryotes, the Sec system functions to move unfolded proteins from the cytosol into the membrane or lumen of the ER. It also functions in chloroplasts to move proteins from the stroma into the thylakoid membrane or the thylakoid lumen. As in bacteria, cotranslational translocation involves a signal recognition particle that helps deliver the translating ribosome and nascent polypeptide to the Sec system. Likewise, posttranslationally translocated proteins are targeted to the Sec translocon by a **signal peptide** at their amino-terminus (*see figure 13.40*). In addition to the Sec system, the twin-arginine translocation (Tat) system is thought to function in the chloroplasts and mitochondria of some eukaryotes. ◀ *Protein translocation and secretion in bacteria (section 13.9)*

Translation, Protein Maturation, and Protein Localization in Archaea

Electron micrographs have illustrated that in archaea, transcription and translation are coupled, as in bacteria. Because the transcript is not processed, mRNA can be immediately translated. Some archaeal mRNAs have a Shine-Dalgarno sequence in the leader. Translation initiation of these mRNAs proceeds by a mechanism similar to that in bacteria. However, the initiator tRNA is Met-$tRNA_i^{Met}$, as in eukaryotes. Furthermore, more initiation factors (called archaeal IFs or aIFs) are involved than in bacteria, and these are homologues of eIFs. Many archaeal mRNAs are leaderless—that is, they lack a Shine-Dalgarno sequence and have fewer than five bases before the AUG start codon. Both types are found in almost all archaea. Ribosome structures are more similar in archaea and eukaryotes, and may facilitate translation initiation in leaderless RNAs. For polycistronic mRNAs, each coding-region yields a separate polypeptide, as occurs in bacteria.

Translation elongation and termination have been less well studied than translation initiation. However, several archaeal elongation factors have been characterized. They are similar to those in both bacteria and eukaryotes. Thus elongation factors, as opposed to initiation factors, are highly conserved in all

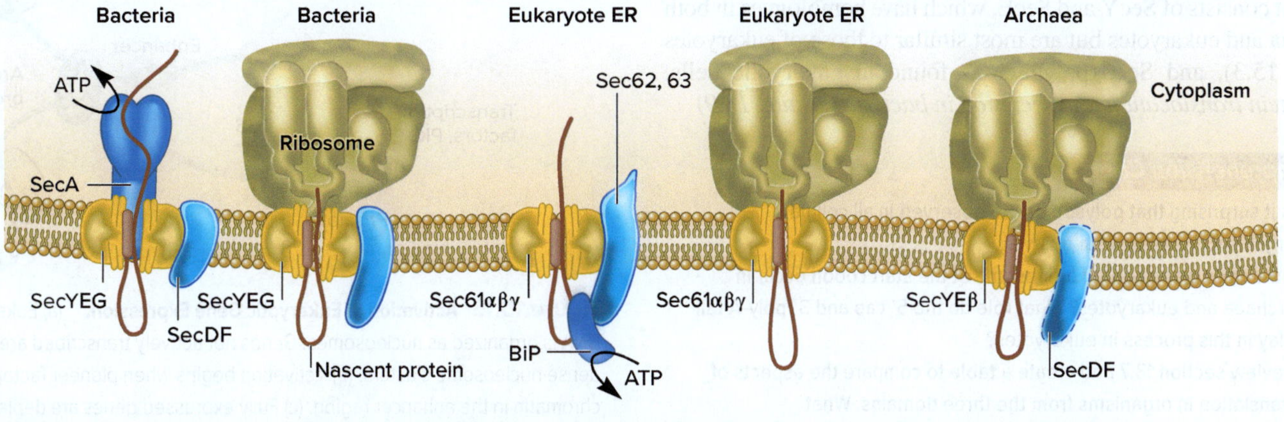

Figure 15.14 Comparison of the Sec Systems Found in Bacteria, Eukaryotes, and Archaea. In bacteria and eukaryotes, translocation can occur posttranslationally or cotranslationally. In archaea, cotranslational translocation is relatively well studied but posttranslational translocation is not. SecDF proteins are found only in some archaea.

Table 15.3 Comparison of Secretion Systems

Bacteria	Eukaryotes	Archaea
SecY	Sec61α	SecY[1]
SecE	Sec61γ	SecE
SecG	—	—
—	Sec61β[2]	Sec61β

[1] Although given the same name as the bacterial protein, this archaeal protein's structure is more like the eukaryotic homologue.

[2] All Sec systems have three proteins that form the pore through which proteins move. However, the Sec61β protein differs from SecG both in structure and in function.

cellular organisms. Finally, archaea use a single release factor during translation termination, as do eukaryotes.

Most archaeal proteins are folded using two chaperones called NAC and prefoldin, homologues of like-named chaperones in eukaryotes. Of the five known HSP families, two (HSP100 and HSP90) are absent from archaea, and a third (HSP70) is absent from thermophiles. Some proteins are folded with the additional help of an HSP60 chaperone called the *thermosome*. As you might imagine, archaeal chaperones have been of particular interest to microbiologists because many archaea are thermophilic. Like other HSP60s, the thermosome hydrolyzes ATP as it helps proteins fold. One well-studied thermosome derives from *Pyrodictium occultum*, an archaeon that grows at temperatures as high as 110°C. The *P. occultum* thermosome hydrolyzes ATP most rapidly at 100°C and makes up almost three-quarters of the cell's soluble protein when *P. occultum* grows at 108°C.

Archaeal protein translocation systems are less well studied than those of bacteria and eukaryotes. Thus far, two protein translocation systems have been identified: the Sec and the Tat systems (*see figure 13.41*). The archaeal Sec system presents a clear example of the hybrid nature of archaea (figure 15.14 and table 15.3). In archaea, the Sec translocon is called the SecYEβ translocon because it consists of SecY and SecE, which have homologues in both bacteria and eukaryotes but are most similar to those of eukaryotes (table 15.3), and Sec61β, which is found in eukaryotic cells. ◀ *Protein translocation and secretion in bacteria (section 13.9)*

Comprehension Check

1. Is it surprising that polysomes are observed in all cellular organisms? Explain your answer.
2. How does placement of a ribosome at the start codon occur in archaea and eukaryotes? What role do the 5′ cap and 3′ poly-A tail play in this process in eukaryotes?
3. Review section 13.7 and create a table to compare the aspects of translation in organisms from the three domains. What generalizations can be made from these comparisons?

15.5 Regulation of Cellular Processes

After reading this section, you should be able to:

a. List the steps at which eukaryotic and archaeal cells can regulate cellular processes
b. Explain why the separation in time and space of transcription and translation in eukaryotic cells is important to the example regulatory mechanisms used
c. Make a concept map comparing the mechanisms by which bacterial, eukaryotic, and archaeal regulatory proteins control transcription
d. Discuss the importance of small RNA molecules in regulating gene expression in eukaryotes

As is the case in bacteria, the regulation of cellular processes in eukaryotes and archaea can occur at transcriptional, translational, and posttranslational levels. Much of the research on regulation in eukaryotes and archaea has focused on transcription initiation. That is our focus here.

Transcription Initiation in Eukaryotes Is Regulated by Chromatin Structure and Activator Proteins

Activating a gene for expression requires modifying its chromatin structure through several processes (**figure 15.15**). First, the

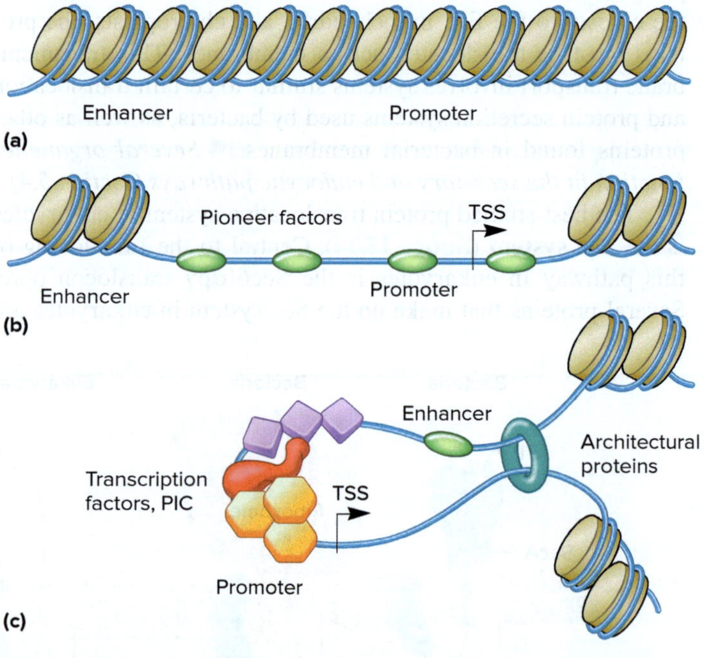

Figure 15.15 Activation of Eukaryotic Gene Expression. (a) Eukaryotic DNA is organized as nucleosomes. Genes not actively transcribed are arrayed in dense nucleosome clusters. (b) Activation begins when pioneer factors loosen chromatin in the enhancer region. (c) Fully expressed genes are depleted of nucleosomes and the enhancer is a platform for multiple transcription factor binding. TSS, transcription start site; PIC, preinitiation complex.

general region of the chromosome must be opened, and ultimately the individual nucleosomes in the promoter region must be loosened for the DNA sequence to become accessible to transcription factors.

In addition to the promoter, a larger region termed the **enhancer** plays a regulatory role in gene regulation. Enhancers are 100–1,000 base-pair platforms for transcriptional activator and repressor binding. The enhancer functions in both positive and negative control of gene expression. While the promoter has recognizable DNA sequence elements like the TATA box and the BRE (figure 15.6), enhancers function within chromatin, meaning that both the DNA sequence and its organization into nucleosomes work in concert to structure the template for transcription.

Recall that a nucleosome is a core of eight histones wrapped by about 146 bp of DNA (*see figure 5.11*). The histones are structured as globular cores with both N- and C-terminal disordered regions termed *tails*. Histone tails protrude from the nucleosome and are platforms for regulatory modifications like methylation and acetylation. The tails convey information about the underlying sequence and moderate its access and stability, while also enabling nucleosome sliding and repositioning.

Surveys of eukaryotic transcription factors (TFs) have revealed that most contain at least one **intrinsically disordered region (IDR).** An IDR is a stretch of at least 250 amino acids that doesn't adopt a single secondary structure. Rather, the IDR occupies a substantial volume of the TF and interacts with a larger set of structures, both ordered and disordered. It is becoming clear that eukaryotic TFs use their IDRs as flexible molecular recognition modules. Eukaryotic transcriptional regulation, then, is based not on binding to a specific DNA sequence, but extensive "fuzzy" protein-protein interactions that ultimately involve histone tails and result in large-scale chromatin restructuring. This contrasts with bacterial TFs, where fewer than 5% possess IDRs.

Gene activation begins with the binding of proteins called *pioneer factors* to the enhancer. Pioneer factors identify regions for activation and engage transcriptional activators like histone modifiers and **chromatin remodeling factors.** Chromatin remodeling factors use ATP to reposition nucleosomes on the DNA template to allow the basal transcription factors to gain access (figure 15.7).

Regulation of the *GAL* genes in the yeast *Saccharomyces cerevisiae* has served as an important model for regulation of transcription initiation in eukaryotes. As outlined in **figure 15.16,** it exemplifies many features of eukaryotic regulation including chromatin remodeling, histone modification, formation of activation

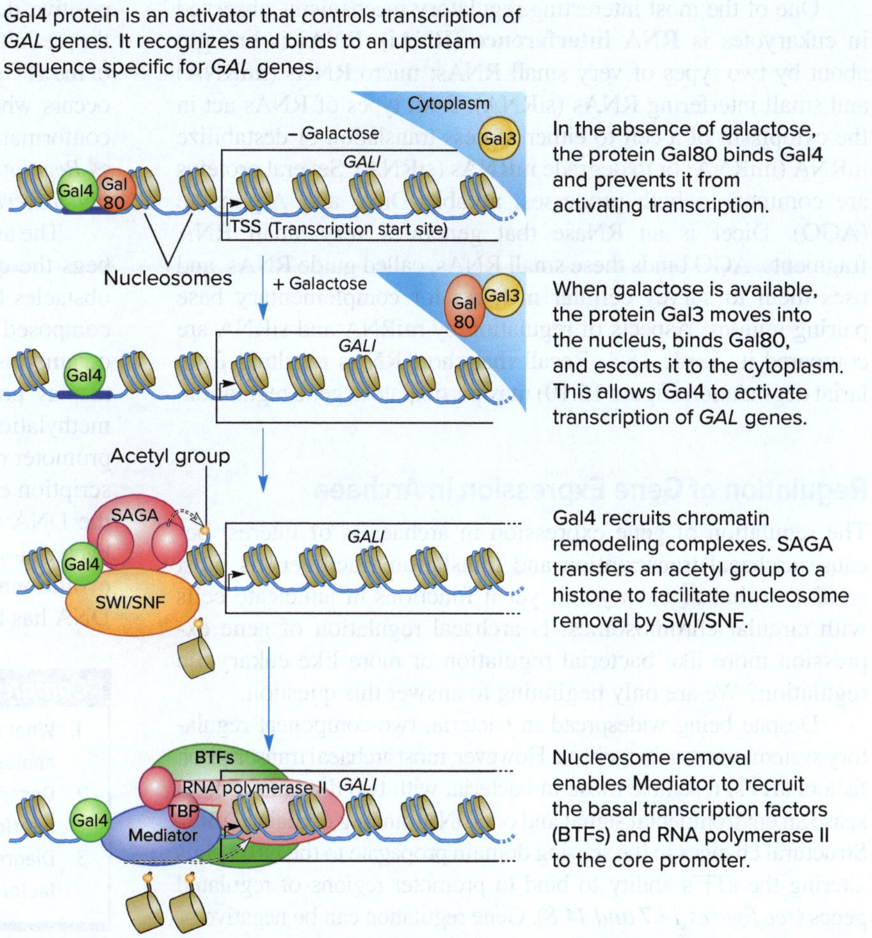

Figure 15.16 Regulation of *GAL* genes in *Saccharomyces cerevisiae*. GAL genes encode enzymes for the catabolism of galactose.

complexes, and the indirect influence of activators and repressors on genes.

Eukaryotes Use RNA Interference Mechanisms for Control

In addition to controlling transcription initiation with activator proteins, eukaryotes use a variety of other methods to regulate gene expression. For example, the nucleotide cAMP serves as a second messenger in eukaryotes, and c-di-GMP signaling functions in fruiting body formation by *Dictyostelium discoideum* and in the innate immune system. Eukaryotes produce cis-encoded small RNA (sRNA), often called antisense RNAs, that function at the level of translation. Eukaryotes also produce RNA molecules that are critical components of the spliceosome, where they contribute to the selection of splice sites used during mRNA processing (alternative splicing). In doing so, different proteins can be made at certain times in the life cycle of the organism by combining different exons.

▶| *Eumycetozoa (section 23.3)*

One of the most interesting regulatory mechanisms observed in eukaryotes is **RNA interference (RNAi).** RNAi is brought about by two types of very small RNAs: micro RNAs (miRNA) and small interfering RNAs (siRNA). Both types of RNAs act in the cytoplasm of a cell to either repress translation or destabilize mRNA (miRNA) or to degrade mRNAs (siRNA). Several proteins are common to both processes, notably Dicer and Argonaute (AGO). Dicer is an RNase that generates very small RNA fragments. AGO binds these small RNAs, called guide RNAs, and uses them to survey cellular mRNAs for complementary base pairing regions. Aspects of regulation by miRNA and siRNA are compared in **Table 15.4.** Recall that short RNAs resulting from lariat degradation (figure 15.10) may participate in gene regulation.

Regulation of Gene Expression in Archaea

The regulation of gene expression in archaea is of interest because archaeal transcription and translation machinery is most similar to that of eukaryotes, yet it functions in anucleate cells with circular chromosomes. Is archaeal regulation of gene expression more like bacterial regulation or more like eukaryotic regulation? We are only beginning to answer this question.

Despite being widespread in bacteria, two-component regulatory systems are rare in archaea. However, most archaeal transcription factors (aTFs) resemble those in bacteria, with two domains, one to sense an environmental signal and one DNA binding domain (DBD). Structural changes in the sensing domain propagate to the DBD, thus altering the aTF's ability to bind to promoter regions of regulated genes (*see figures 14.7 and 14.8*). Gene regulation can be negative or positive, depending on how the aTF interacts with the components of the pre-initiation complex (PIC). Negative regulation is a consequence of the aTF preventing PIC binding to the promoter. Positive regulation occurs when the aTF recruits PIC components or alters the DNA conformation to increase favorable interactions with the PIC. ◀ *Regulation of transcription initiation saves considerable energy and materials (section 14.2)*

The association of histones with some archaeal chromosomes begs the question of whether chromosome architecture imposes obstacles for transcription. Archaeal histone-DNA complexes are composed of a variable number of histones, not a well-defined octamer as in eukaryotes. Archaeal histones lack tails and do not display posttranslational modifications (i.e., the acetylation and methylation used to activate or repress genes). However, archaeal promoter regions have fewer histones than other regions, and transcription elongation in vitro slows upon encountering histones on the DNA template. Clearly the ability of histones to organize a bacterial-sized chromosome located in the nucleoid has important evolutionary implications. This modest platform for wrapping DNA has been expanded and embellished in eukaryotes.

Comprehension Check

1. What regulatory sequences in bacterial genomes are analogous to eukaryotic enhancers?
2. Describe the mechanisms of gene regulation by RNA interference.
3. Diagram an archaeal promoter and indicate how transcription factors could regulate gene expression positively or negatively.

Table 15.4 Comparison of RNA Interference Mechanisms

	MicroRNA (miRNA)	Small Interfering RNA (siRNA)
Origin of guide RNA	Encoded in the genome and transcribed by RNA pol II	dsRNA forms from hybridization of sense and antisense transcripts or from infecting viruses
Base pairing in guide RNA	RNA transcript folds on itself to generate short hairpins with imperfect dsRNA base pairing	Base pairing is fully complementary between the two strands
Dicer cleavage	Dicer cuts hairpins into 20–30 base dsRNAs	Dicer cuts long transcripts into 20–25 base dsRNAs
Strand separation	AGO retains only one strand forming the miRISC[1]	AGO retains only one strand forming the siRISC[1]
Targeting	Imperfect base pairing with the 3′ untranslated region of mRNA	Scans mRNA for perfect base pairing
Mechanism of regulation	Represses translation and destabilizes mRNA	Hydrolyzes mRNA, which is then completely degraded by RNAses
Scope of regulation	Can regulate many genes by imperfect base pairing	Specific for a small number of genes with exact sequence match

[1]RISC, RNA-induced silencing complex

Key Concepts

15.1 Genetic Processes in the Three Domains

- Membraneless compartments termed biomolecular condensates organize gene expression processes in the nucleus of eukaryotic cells.
- The machinery involved in processing and expressing archaeal genomes is more like that of eukaryotes than that of bacteria.

15.2 DNA Replication

- Eukaryotic DNA has many replicons and replication origins (**figure 15.2**). DNA replication is initiated when replicase proteins are recruited to origins of replication by the origin recognition complex (ORC; **table 15.1**).
- In eukaryotes, three different DNA polymerases are part of the replisome (**figure 15.1**). DNA Pol α-primase synthesizes an RNA-DNA hybrid that serves as the primer for DNA synthesis. DNA Pol ε carries out leading strand synthesis, and DNA Pol δ is responsible for lagging strand synthesis.
- Telomerase is responsible for replicating the ends of linear eukaryotic chromosomes (**figure 15.4**).
- All known archaeal cells have circular chromosomes that are thought to be replicated in a manner similar to circular bacterial chromosomes.
- Many archaeal chromosomes have multiple origins of replication, whereas others have been shown to function without an origin. These organisms use a type of recombination to copy their chromosome.
- Archaeal replisomes of all species examined thus far contain a primase and a DNA polymerase (Pol B) similar to those in eukaryotic replisomes. Some archaea have a second DNA polymerase (Pol D) that is unique to archaea (**table 15.2**).

15.3 Transcription

- Most eukaryotic protein-coding genes are not organized into operons. Therefore the mRNAs that arise from them are monocistronic.
- Three RNA polymerases transcribe genes in eukaryotes. RNAPII transcribes protein-coding genes. RNAPI transcribes rRNA genes, and RNAPIII transcribes the 5S rRNA and all tRNA genes. All three eukaryotic RNAPs have more subunits than bacterial RNAPs (**figure 15.5**).
- The RNAPII core promoters of eukaryotes are distinctive, as is the mechanism of transcription initiation (**figure 15.6**). Several basal transcription factors recruit RNAPII to a promoter (**figure 15.7**).
- An elongating RNAP must displace the nucleosomes that cover the template. A series of protein elongation factors dissociate a nucleosome and shift it behind the RNAP (**figure 15.8**).
- RNAPII synthesizes pre-mRNA, which is modified by addition of a 3′ poly-A sequence and a 5′ cap (**figure 15.9**) and by removal of introns (**figure 15.10**).
- Many archaeal genes are organized into operons. The single archaeal RNA polymerase is similar to eukaryotic RNAPII. The archaeal RNA polymerase requires two transcription factors to be positioned properly at a promoter (**figure 15.11**). Archaeal promoters have elements similar to those in the eukaryotic RNAPII core promoters (**figures 15.6** and **15.11**).

15.4 Translation and Protein Maturation and Localization

- Translation initiation involves an interaction of protein factors at the 5′ and 3′ ends of the mRNA, followed by binding of the 43S complex (SSU of the ribosome and initiator tRNA) to the 5′ end of the message. The 48S complex then scans the mRNA until it encounters a start codon (**figure 15.12**).
- Translation elongation is similar in eukaryotes and bacteria. Translation termination requires two release factors, rather than the three used in bacteria.
- Protein folding in eukaryotes is aided by chaperone proteins and occurs cotranslationally or posttranslationally. Homologues of all types of chaperones are found in eukaryotes.
- In eukaryotes, proteins are translocated into the endoplasmic reticulum (ER) membrane by a Sec system (**figure 15.14**). The proteins are then trafficked between the ER, plasma membrane, Golgi, and other components of the endocytic and secretory pathways by vesicular transport.
- In archaea, transcription and translation are coupled. Most archaea lack a Shine-Dalgarno sequence in the mRNA or are leaderless. Other aspects of archaeal translation are similar to those seen in eukaryotes, including the protein factors and initiator tRNA.
- Archaeal chaperones resemble those of eukaryotes.
- Archaea use the Sec and Tat pathways to translocate proteins into or across the plasma membrane. The archaeal Sec proteins are homologous to the eukaryotic Sec proteins (**figure 15.14**).

15.5 Regulation of Cellular Processes

- Eukaryotic gene expression is repressed by chromatin structure. Gene activation requires several levels of loosening the DNA from nucleosomes.

- Enhancers are DNA regions overlapping a promoter and extending up to 1,000 bases away. They are sequences that recruit transcription factors specific for the gene to be activated (**figures 15.15** and **15.16**).
- Nucleosomes are chemically modified by methylation or acetylation, and repositioned by the action of chromatin remodelers. Inherently disordered regions in histone tails and in transcription factors interact spatially to adjust the level of chromatin packing for gene expression.
- RNA interference mechanisms control the accessibility of mRNAs for translation.
- Archaeal gene regulation is similar to bacterial regulation. Transcription factors convert environmental signals to DNA binding in the promoter region of regulated genes. Regulation may be positive or negative. Histones may confer additional levels of regulation to archaeal transcription.

Active Learning

1. Budding yeasts such as *S. cerevisiae* exhibit telomerase activity throughout their life cycles, whereas most human somatic cells do not exhibit telomerase activity. Suggest why this is so.
2. All of the subunits in bacterial RNA polymerases have homologues in both archaeal and eukaryotic RNA polymerases. What does this suggest about the evolution of these enzymes?
3. Compare two ribozymes, the ribosome and the spliceosome, in terms of their substrates and enzymatic activities.
4. Compare the regulation of the *GAL* genes in yeast (figure 15.16) with the regulation of the *lac* operon in *E. coli* (figure 14.7).
5. Review the description of CRISPR/Cas systems for monitoring exposure to mobile genetic elements in section 14.6. How are they similar to and different from RNA interference mechanisms for gene regulation?

CHAPTER

16

Mechanisms of Genetic Variation

Ingram Publishing/SuperStock

Antimicrobial Resistance

Public health victories against infectious diseases in the twentieth century can be attributed to sanitation, vaccination, and antibiotics. While sanitation and vaccination have become cornerstones of good health practices, we are at risk of losing antibiotics as a tool to treat bacterial infections.

It didn't take long after clinical use of antibiotics became common that bacteria became resistant to these drugs. The secret of bacterial success? Evolution. Selective pressure from antibiotic exposure enables the survival of bacteria that develop resistance. Antibiotic resistance can arise quickly thanks to horizontal gene transfer, which allows bacteria to share genes with their neighbors—even if they are a different species. Antimicrobial resistance (AMR) genes can sweep through a bacterial community or an ecosystem at an alarming pace.

Mobile genetic elements, genetic material that moves independently of cell division, often carry AMR genes. Interactions among humans, animals, and the environment promote mingling of different types of bacteria and also promote movement of their mobile genetic elements and, increasingly, AMR genes.

In the twenty-first century, we are more aware of the need for antibiotic stewardship—to prescribe and use antibiotics prudently. The number of antibiotic-resistant infections has doubled in just a decade. The World Health Organization has calculated that unchecked, the number of deaths due to AMR could exceed the number of deaths from cancer by 2050.

Successfully combating AMR requires a multiprong approach to reduce demand for antibiotics. First, many infections can be prevented through stringent sanitation measures and vaccination campaigns. Second, prescribing antibiotics only when necessary reduces exposure and opportunities for resistance to emerge. And third, while antibiotics are effective growth promoters for livestock, the easy transfer of AMR genes from bacteria in industrial farms to human pathogens demands changes to prevent the misuse of medically important antibiotics.

How microbial genomes are altered to increase genetic diversity is the topic of this chapter. Our main focus is on how this is accomplished in the absence of sexual reproduction. Some of the mechanisms can create changes in the genome that are detrimental. Thus we also consider processes that help balance the stability of the genome with the capacity to introduce genetic variation.

Readiness Check:

Based on what you have learned previously, you should be able to:

- ✓ Review cell surface structures in the three domains (sections 3.3, 3.5, 4.2, 5.2)
- ✓ Describe plasmids (section 3.6)
- ✓ Describe the lytic and lysogenic life cycles of a virus (section 6.4)
- ✓ Draw a simple diagram that illustrates the three chemical moieties that define a nucleotide (section 13.2)
- ✓ State the DNA base-pairing rules
- ✓ Explain the importance of the reading frame of a protein-coding gene (section 13.6)
- ✓ Create a concept map or table that distinguishes the secretion systems observed in diderm bacteria (section 13.9)
- ✓ Review meiosis as a means to create genetic variation

16.1 Mutations: Heritable Changes in a Genome

After reading this section, you should be able to:

a. Distinguish spontaneous (endogenous) from induced (exogenous) mutations, and list the common ways each arises
b. Discuss the possible effects of mutations

Genetic diversity can be created in the absence of sexual reproduction by **mutations** (Latin *mutare,* to change), which are heritable changes in DNA sequence. Some mutations arise from the alteration of single pairs of nucleotides or from the addition or deletion of one nucleotide pair. Such small changes in DNA are called **point mutations** because they affect only one base pair. Less commonly, segments of DNA can be inserted, deleted, or duplicated.

Mutations occur in one of two ways: (1) **Spontaneous mutations** arise occasionally in all cells as a consequence of normal cellular activities. (2) **Induced mutations** are the result of environmental exposure to a **mutagen,** which can be either a physical or a chemical agent. These two types of mutations are also termed *endogenous* and *exogenous,* respectively, to emphasize the internal or external origin of the mutation. In this section, we consider the molecular basis of mutations and mutagenesis; then we discuss the phenotypic effects of mutations.

Spontaneous Mutations Are Caused by Cellular Processes

Spontaneous (endogenous) mutations result from reactive oxygen species (ROS) produced during aerobic growth or from errors in DNA replication and repair. A few of the more prevalent mechanisms are described here.

Spontaneous mutations often originate from lesions in DNA caused by reactive forms of oxygen such as oxygen free radicals and peroxides produced during aerobic metabolism. Nucleotides can lose their base by breaking the bond between the base and deoxyribose. The phosphate-sugar backbone remains intact. This results in the formation of an abasic site, which cannot base-pair and may cause a mutation after the next round of replication (**figure 16.1**). *Oxygen concentration (section 7.5).*

DNA replication errors can result in the insertion and deletion of nucleotides. These mutations generally occur where there is a short stretch of repeated nucleotides, especially AT base pairs. The pairing of template and new strand can be displaced, leading to insertions or deletions of bases (**figure 16.2**).

Induced Mutations Are Caused by Environmental Factors

The types of induced (exogenous) mutations depend on an organism's environment. For example, marine microbes are subject to ultraviolet (UV) radiation, while animal pathogens must withstand oxidative bursts from immune cells. Any agent that damages DNA, alters its chemistry, or in some way interferes with its functioning will probably induce mutations and thus can be considered a mutagen.

A common mutation occurs with the loss of an amino group from a nitrogenous base. Cytosine is particularly prone to **deamination,** and the resulting residue is uracil, producing a mismatched GU base pair from a GC base pair (**figure 16.3**). Should DNA replication occur prior to the repair of this mismatch, one daughter DNA molecule will have the original GC base pair, while the other will have an AU base pair.

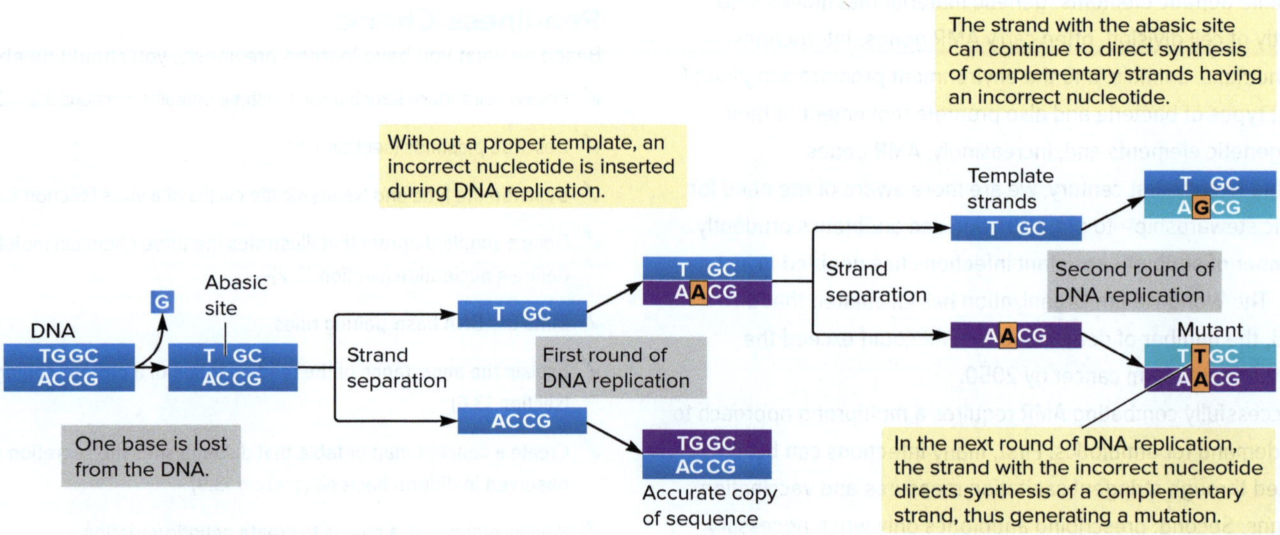

Figure 16.1 Mutation Due to Formation of an Abasic Site.

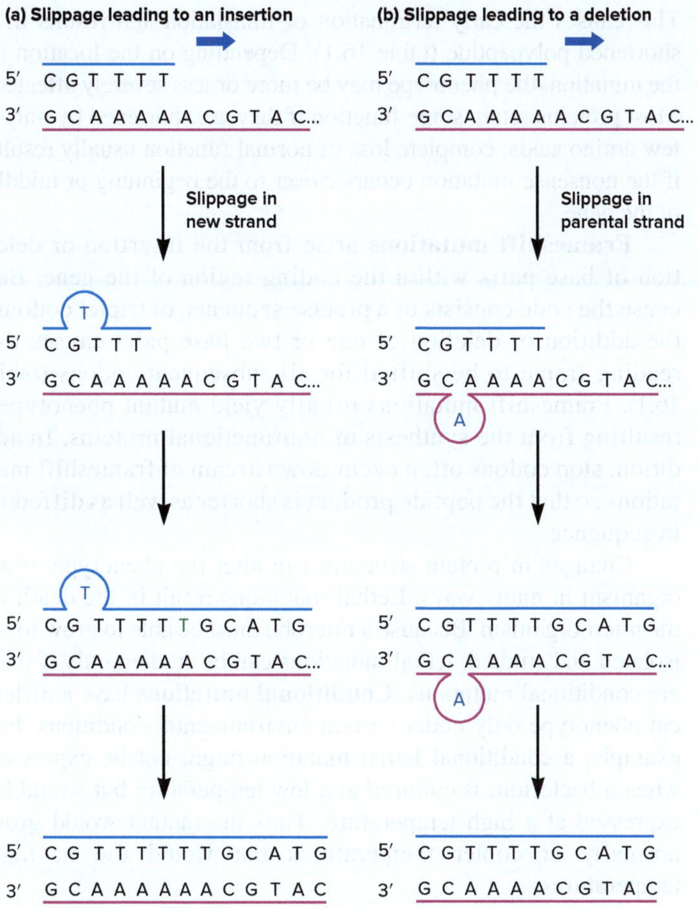

Figure 16.2 A Mechanism for the Generation of Insertions and Deletions During Replication. The direction of replication is indicated by the thick blue arrow. In each case, there is strand slippage resulting in a small loop stabilized by hydrogen bonding in the repetitive AT sequence. (a) If the new strand slips, an addition of one T results. (b) Slippage of the parental strand yields a deletion (in this case, the loss of a T).

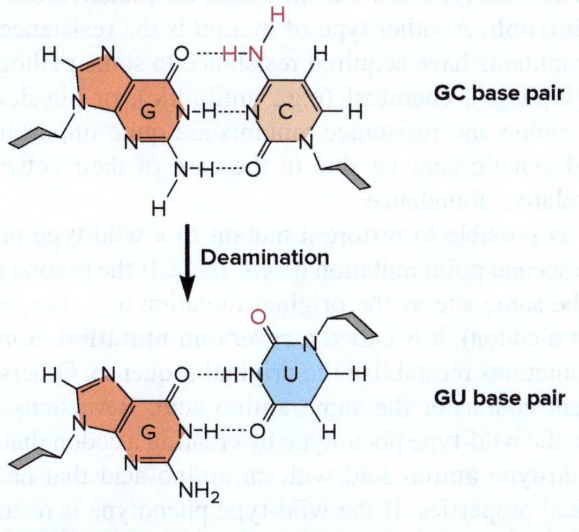

Figure 16.3 Deamination of a GC Base Pair. The amino group shown in red (top) can be lost, resulting in the formation of a GU base pair.

MICRO INQUIRY *Compare this to figure 13.5. How many hydrogen bonds are in a GU base pair?*

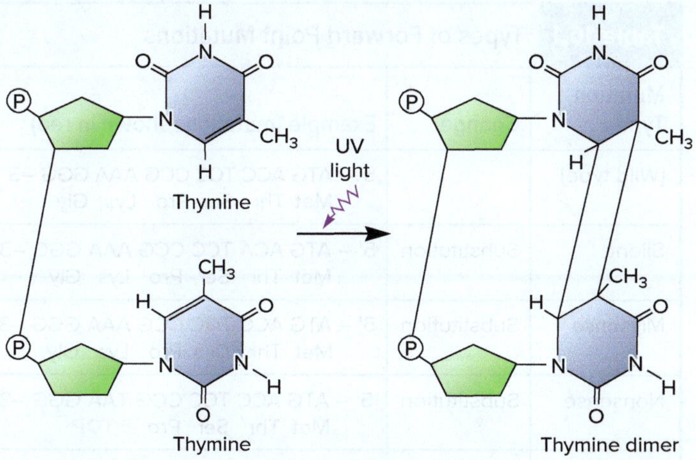

Figure 16.4 Thymine Dimers in the Same DNA Strand Are Formed by UV Radiation.

Abasic sites (figure 16.1) can also result from some environmental factors. Physical agents like radiation induce **cross-links,** that is, the formation of covalent bonds to anchor two molecules together. For example, UV radiation induces thymine dimers in adjacent T residues on the same strand (**figure 16.4**). Interstrand crosslinks form when the two bases in a base pair are linked by a covalent bond in place of the normal hydrogen bonds. Other radiation types, gamma rays, are responsible for single-stranded and double-stranded breaks in the phosphate-sugar backbone of DNA. This severe damage accounts for the effective use of radiation to control microbial growth. *Radiation (section 8.2)* *Thymine Dimer Formation and Repair*

Some common types of chemical mutagens used in the laboratory are base analogues and intercalating agents. Base analogues are structurally similar to normal nitrogenous bases and can be incorporated into the growing polynucleotide chain during replication. Once in place, these compounds exhibit base-pairing properties different from the bases they replace and eventually cause a point mutation. Intercalating agents are planar molecules that insert (intercalate) between the stacked bases of the helix. This results in distortion of the helical structure. An example is ethidium bromide, often used as a stain because it intercalates extensively in the DNA helix. All of these types of DNA damage must be resolved before the DNA can act as a template for replication or transcription.

Effects of Mutations

The effects of mutations can be described at the protein level and in terms of observed phenotypes. In all cases, the impact is noticed only if it produces a change in phenotype. In general, the most prevalent form of a gene and its associated phenotype is called the **wild type.** A change from wild type to a mutant form is a **forward mutation** (**table 16.1**). Because point mutations are the most common types of mutations, their effects are our focus.

Table 16.1 Types of Forward Point Mutations

Mutation Type	Change	Example (mutations shown in red)
[Wild type]		5' – ATG ACC TCC CCG AAA GGG –3' Met Thr Ser Pro Lys Gly
Silent	Substitution	5' – ATG ACA TCC CCG AAA GGG –3' Met Thr Ser Pro Lys Gly
Missense	Substitution	5' – ATG ACC TGC CCG AAA GGG –3' Met Thr Cys Pro Lys Gly
Nonsense	Substitution	5' – ATG ACC TCC CCG TAA GGG –3' Met Thr Ser Pro STOP
Frameshift	Insertion/ Deletion	5' – ATG ACC TCC GCC GAA AGG G –3' Met Thr Ser Ala Glu Arg

Mutations in Protein-Coding Genes

Point mutations in protein-coding genes affect protein structure in a variety of ways. They are named according to if and how they change the encoded protein.

Silent mutations change the nucleotide sequence of a codon but do not change the amino acid encoded by that codon. This is possible because the genetic code exhibits degeneracy. There is more than one codon for some amino acids, so a single base substitution may result in the formation of a different codon for the same amino acid. When there is no change in the protein, there is no change in the phenotype of the organism (table 16.1). ◀ *The genetic code consists of three-letter "words" (section 13.6)*

Missense mutations involve a single base substitution that changes a codon for one amino acid into a codon for a different amino acid (table 16.1). The effects of missense mutations vary. They alter the primary structure of a protein, but the impact of this change may range from complete loss of activity to no change at all. This is because the effect of missense mutations on protein function depends on the type and location of the amino acid substitution. For instance, replacement of a nonpolar amino acid in the protein's interior with a polar amino acid can drastically alter the protein's three-dimensional structure and therefore its function. However, the replacement of one polar amino acid with another at the protein surface may have little or no effect. Such mutations are called **neutral mutations.** In contrast, the replacement of a critical amino acid at the active site of an enzyme often destroys its activity. Missense mutations play an important role in providing new variability to drive evolution because they often are not lethal and therefore remain in the population. ▶ *Proteins (appendix I)* *Mutation by Base Substitution*

Nonsense mutations convert a sense codon (i.e., one that codes for an amino acid) to a nonsense codon (i.e., a stop codon). This causes the early termination of translation and results in a shortened polypeptide (table 16.1). Depending on the location of the mutation, the phenotype may be more or less severely affected. Most proteins retain some function if they are shortened by only a few amino acids; complete loss of normal function usually results if the nonsense mutation occurs closer to the beginning or middle of the gene.

Frameshift mutations arise from the insertion or deletion of base pairs within the coding region of the gene. Because the code consists of a precise sequence of triplet codons, the addition or deletion of one or two base pairs causes the reading frame to be shifted for all subsequent codons (table 16.1). Frameshift mutations usually yield mutant phenotypes resulting from the synthesis of nonfunctional proteins. In addition, stop codons often occur downstream of frameshift mutations so that the peptide product is shorter as well as different in sequence.

Changes in protein structure can alter the phenotype of an organism in many ways. Lethal mutations result in the death of the microorganism. Because a microbe must be able to grow to be isolated and studied, lethal mutations can be studied only if they are conditional mutations. **Conditional mutations** have a different phenotype only under certain environmental conditions. For example, a conditional lethal mutation might not be expressed when a bacterium is cultured at a low temperature but would be expressed at a high temperature. Thus the mutant would grow normally at cooler temperatures but would die at high temperatures.

Other common mutations inactivate a biosynthetic pathway, eliminating the ability of the mutant to make an essential molecule such as an amino acid or nucleotide. A strain bearing such a mutation has a conditional phenotype: It is unable to grow on medium lacking that molecule but grows when the molecule is provided. Such mutants are called **auxotrophs,** and they are said to be auxotrophic for the molecule they cannot synthesize. The wild-type strain from which the mutant arose is called a **prototroph.** Another type of mutant is the resistance mutant. These mutants have acquired resistance to some pathogen (e.g., bacteriophage), chemical (e.g., antibiotic), or physical agent. Auxotrophic and resistance mutants are quite important in microbial genetic analysis due to the ease of their detection and their relative abundance.

It is possible to restore a mutant to a wild-type phenotype with a second point mutation (**table 16.2**). If the second mutation is at the same site as the original mutation (e.g., the same base pair in a codon), it is called a **reversion mutation.** Some reversion mutations reestablish the original sequence. Others create a different codon for the same amino acid. Reversions can also restore the wild-type phenotype by creating a codon that replaces the wild-type amino acid with an amino acid that has similar chemical properties. If the wild-type phenotype is restored by a second mutation at a different site than the original mutation, it is called a **suppressor mutation.** Suppressor mutations may be within the same gene (intragenic suppressor mutation) or in a

Table 16.2 Types of Reversion and Suppressor Mutations

Mutation Type	Example (mutations shown in red)
Reversion to restore the original sequence	5′ – ATG ACC TCC –3′ Met Thr Ser 5′ – ATG **C**CC TCC –3′ Met **Pro** Ser 5′ – ATG **A**CC TCC –3′ Met **Thr** Ser
Reversion and intragenic suppression by base substitution	5′ – ATG ACC TCC –3′ Met Thr Ser 5′ – ATG **C**CC TCC –3′ Met **Pro** Ser 5′ – ATG **T**CC TCC –3′ Met **Ser** Ser
Reversion and intragenic suppression: frameshift generated by 1 base insertion, reading frame restored by 1 base deletion	5′ – ATG ACC TCC CCG AAA GGG –3′ Met Thr Ser Pro Lys Gly 5′ – ATG ACC TCC **G**CC GAA AGG G –3′ Met Thr Ser **Ala** **Glu** **Arg** 5′ – ATG ACC CCG CCG AAA GGG –3′ Met Thr **Pro** Pro Lys Gly

different gene (extragenic suppressor mutation). *Addition and Deletion Mutations*

Two examples of intragenic suppression are shown in table 16.2. An amino acid substitution that changes the protein chemical character (e.g., polar to nonpolar) may be compensated by a second substitution to restore the original character (nonpolar to polar). The second substitution may be a related amino acid, not necessarily the original. Likewise, a second frameshift mutation that occurs near a previous frameshift may restore the proper reading frame but not every amino acid.

Extragenic suppression may take the form of a nonsense suppressor or a physiological suppressor. A nonsense suppressor forms when a point mutation occurs in a tRNA-encoding gene. A change in the tRNA anticodon that allows it to pair with a nonsense codon may rescue a gene with a nonsense mutation. Because the tRNA can decode a stop codon, the effects of this suppressor mutation occur throughout the genome. Physiological suppression occurs when a defect in one pathway is circumvented by a second mutation that induces expression of another pathway to the same product or by a mutation that permits more efficient uptake of a compound affected by the first mutation.

Mutations in Regulatory Sequences

Some of the most interesting and informative mutations are those that occur in the regulatory sequences responsible for controlling gene expression. Constitutive lactose operon mutants in *Escherichia coli* are excellent examples. These mutations cause continuous, unregulated expression of the *lac* operon. They are found in the operator site and produce altered operator sequences that are not recognized by the *lac* repressor protein. Therefore the operon is continuously transcribed, and β-galactosidase is always synthesized. Mutations in promoters also have been identified. If the mutation renders the promoter sequence nonfunctional, the mutant will be unable to synthesize the product, even though the coding region of the structural gene is completely normal. ◀ *Regulation of transcription initiation saves considerable energy and materials (section 14.2)*

Mutations in tRNA and rRNA Genes

Genes encoding tRNA and rRNA exhibit a low frequency of mutations compared to protein-encoding genes. Because of the critical nature of their three-dimensional RNA structure, mutations in these genes are often lethal. When tRNA and rRNA genes are mutated, the phenotype that manifests is impaired translation affecting all proteins. In fact, these mutants often are initially identified because of their slow growth. On the other hand, a suppressor mutation involving tRNA restores normal (or near normal) growth rates.

Comprehension Check

1. List two ways in which spontaneous mutations might arise.
2. Sometimes a point mutation does not change the phenotype. List all the reasons why this is so.
3. Why might a missense mutation at a protein's surface not affect the phenotype of an organism, whereas the substitution of an internal amino acid does?
4. Examine the structure of a porin in figure 3.23. At which locations would an amino acid substitution affect the ability of the porin to regulate passage of solutes? At which location would an amino acid substitution affect the assembly of the three channels? Where might an amino acid substitution have little effect?

16.2 DNA Repair Maintains Genome Stability

After reading this section, you should be able to:

a. Compare and contrast proofreading, mismatch repair, excision repair, and recombinational repair
b. Propose a scenario that would elicit the SOS response and describe the response to those conditions

As we have seen, mutations occur infrequently but consistently and can have disastrous effects for a cell. Therefore it is imperative that a microorganism be able to repair damage to its genome. There are numerous repair mechanisms and these are well conserved across all domains of life. DNA repair is best understood in *E. coli* and is our focus in this section.

Proofreading: The First Line of Defense

As we discuss in chapter 13, replicative DNA polymerases sometimes insert an incorrect nucleotide during DNA replication. However, these DNA polymerases evaluate the hydrogen bonds formed between the newly added nucleotide and the template nucleotide, and correct any errors immediately, before the next nucleotide is added. This is called **proofreading.** When DNA polymerase detects a mistake, it backs up and removes the incorrect nucleotide with its 3′ to 5′ exonuclease activity (*see figure 13.16*). It then restarts DNA replication, this time inserting the correct nucleotide. Proofreading is efficient but does not correct every error; in particular it does not correct induced mutations. *E. coli* thus uses additional repair mechanisms to ensure the stability of its genome. *DNA replication in bacteria (section 13.3)*

DNA Repair Mechanisms: The Second Line of Defense

The importance of protecting the genome by DNA repair was highlighted with the 2015 Nobel Prize in Chemistry to Paul Modrich, Aziz Sancar, and Tomas Lindahl. Each scientist worked out one of the mechanisms that *E. coli* uses to detect and repair lesions in DNA, and we review these mechanisms here.

Mismatched bases that arise, as illustrated in figure 16.1, are usually detected and repaired by the **mismatch repair** system (**figure 16.5**). As established by Modrich, successful mismatch repair depends on the ability of the repair enzymes to distinguish between parental and newly synthesized DNA strands. This distinction is possible because newly synthesized DNA strands are chemically different from the template. They temporarily lack methyl groups on their nitrogenous bases, whereas the parental DNA is methylated. Thus for a short time after the replication fork has passed, the DNA is hemimethylated, that is, methylated only on the parental strand. The repair system cuts out the mismatch from the unmethylated strand. *Mismatch Repair*

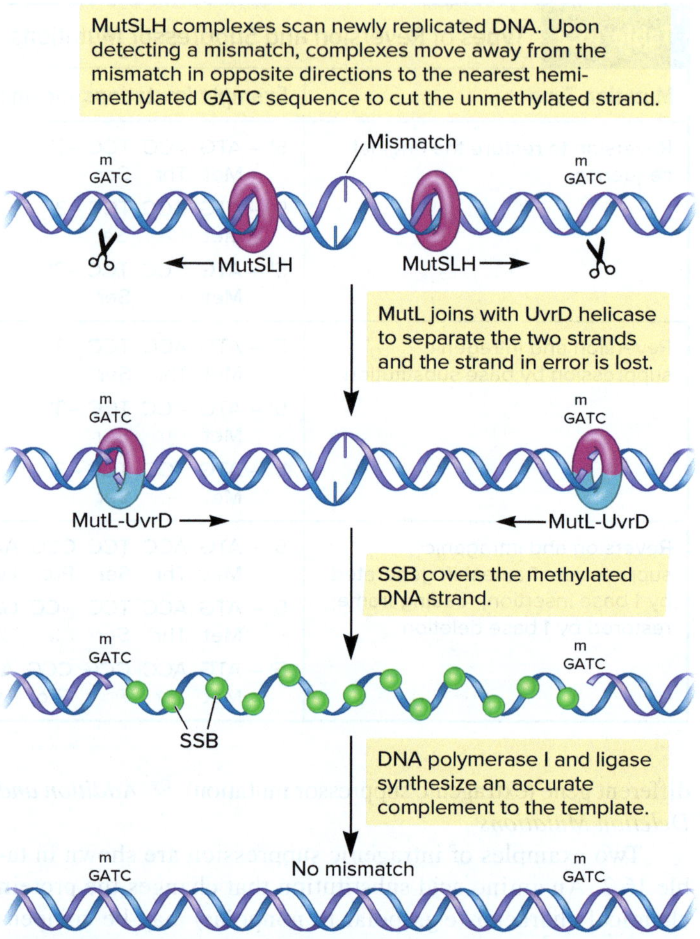

Figure 16.5 Methyl-Directed Mismatch Repair in *E. coli*. The parental template strand is blue and the newly synthesized strand is purple. The methylated adenine is designated by an m. SSB, single-stranded binding protein.

MICRO INQUIRY *How is mismatch repair similar to DNA polymerase proofreading? How is it different?*

Excision repair corrects damage that causes distortions in the DNA double helix. Two types of excision repair systems have been described. The **nucleotide excision repair** system, described by Sancar, can remove thymine dimers (figure 16.4) and repair almost any other injury that produces a detectable distortion in DNA (**figure 16.6**). More recent studies demonstrate that nucleotide excision repair is coupled to transcription. RNA polymerase stalls at DNA damage, thus recruiting the protein components of the repair pathway.

Lindahl explored **base excision repair,** which removes damaged or unnatural bases yielding abasic sites (**figure 16.7**). Both excision repair systems use the same approach: They remove the damaged portion of a DNA strand and use the intact complementary strand as the template for synthesis of new DNA. *Nucleotide Excision Repair*

Recombinational repair corrects damaged DNA in which both bases of a pair are missing or damaged, or where there is a gap opposite a lesion. In this type of repair, the protein **RecA** aligns the damaged DNA with a second copy of the genome and uses the copy as a template to replace the damaged strand (**figure 16.8**). This process is termed **homologous recombination.** It occurs when there are long regions of the same or similar nucleotide sequence in two DNA molecules. Another copy of the damaged segment often is available because either it has recently been replicated or the cell contains more than one copy of its genome. Once a template is in place, the remaining damage can be corrected by another repair system.

SOS Response

Despite having multiple repair systems, sometimes the damage to an organism's DNA is so great that the normal repair

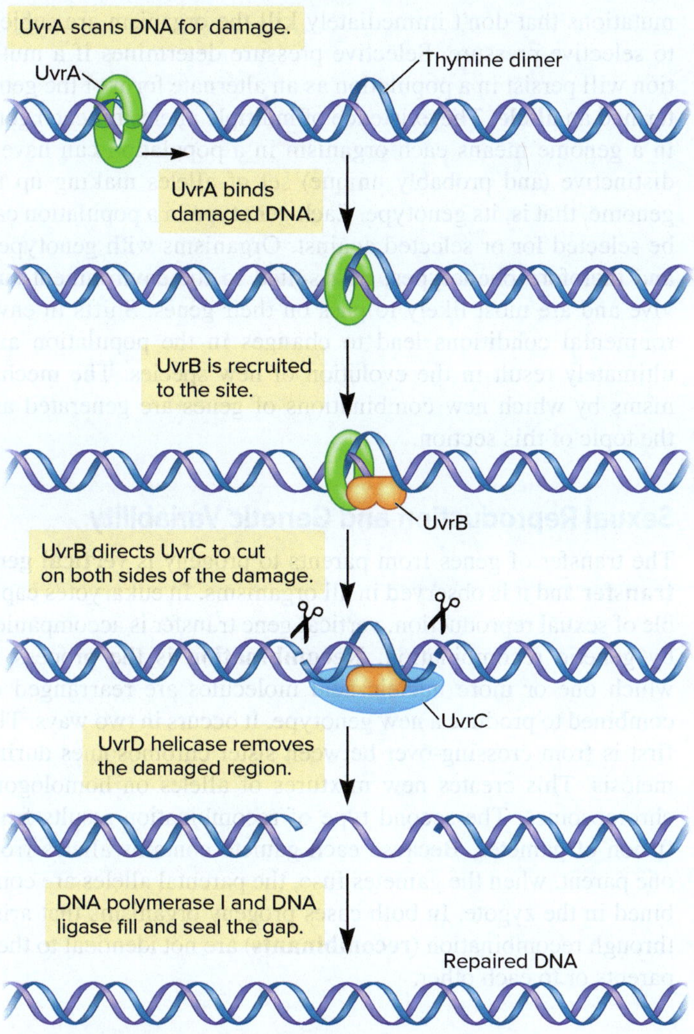

Figure 16.6 Nucleotide Excision Repair in *E. coli*.

MICRO INQUIRY *UvrD helicase acts in both mismatch repair and nucleotide excision repair. How is its activity different in the two mechanisms?*

mechanisms cannot repair all the damage. As a result, DNA synthesis stops completely. In such situations, a global control network called the **SOS response** is activated.

The SOS response, like recombinational repair, depends on the activity of RecA protein. RecA binds to single-stranded DNA and gaps generated by DNA polymerase stalling at lesions. RecA binding to DNA initiates recombinational repair. Simultaneously RecA causes LexA to destroy itself (autoproteolysis). LexA is a transcriptional repressor whose loss activates over 50 genes. The first genes expressed in the SOS response encode the Uvr proteins needed for nucleotide excision repair (figure 16.6). The next class of genes expressed are those whose products are involved in recombinational repair. To give the cell time to repair its DNA, cell division is halted. Finally, if the DNA has not been fully repaired after about 40 minutes, a process called **translesion DNA synthesis** is triggered. In this process, DNA polymerases IV and V synthesize DNA across gaps and other

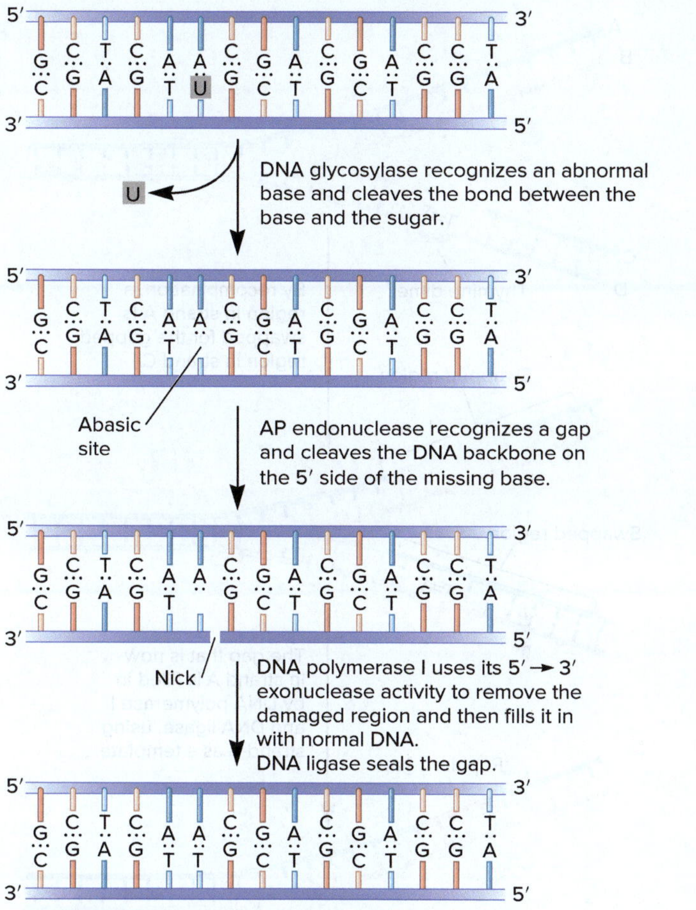

Figure 16.7 Base Excision Repair in *E. coli*.

lesions. However, because an intact template may not exist, these DNA polymerases often insert incorrect bases. Furthermore, they lack proofreading activity. Therefore even though DNA synthesis continues, it is highly error prone and results in numerous mutations.

The SOS response is so named because it is made in a life-or-death situation. The response increases the likelihood that some cells will survive by allowing DNA synthesis to continue. For the cell, the risk of dying because of failure to replicate DNA is greater than the risk posed by the mutations generated by this error-prone process.

Comprehension Check

1. What role does DNA methylation play in mismatch repair?
2. Compare and contrast the two types of excision repair.
3. When *E. coli* cells are growing rapidly, they may contain up to four copies of their chromosomes. Why is this important if *E. coli* needs to carry out recombinational repair?
4. Explain how the following DNA alterations and replication errors would be corrected (there may be more than one way): base addition errors by DNA polymerase III during replication, thymine dimers, deaminated bases, and gaps produced during replication.

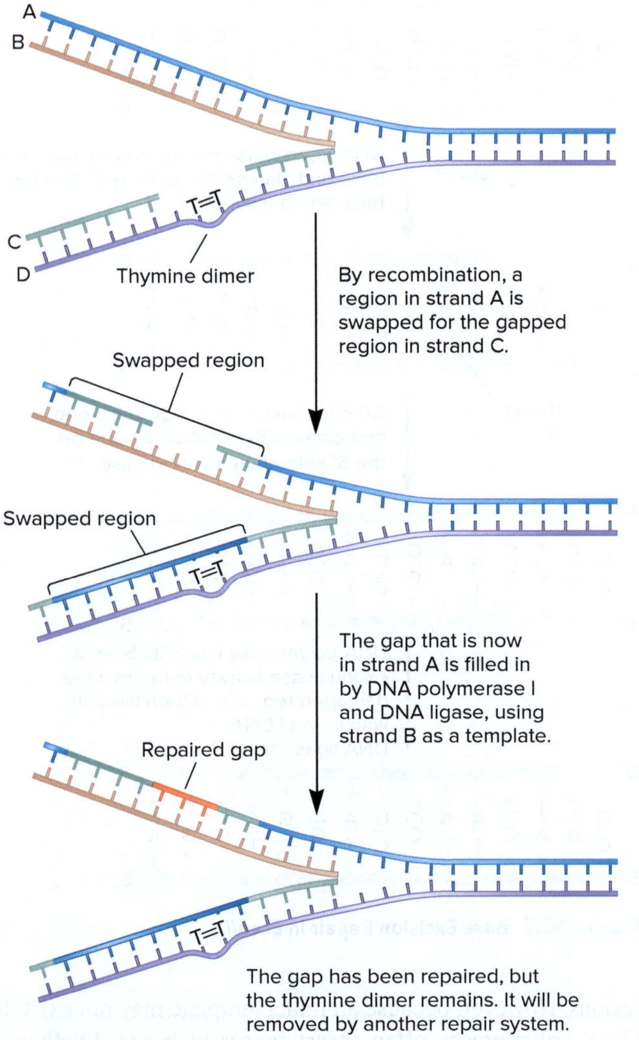

Figure 16.8 Recombinational Repair.

MICRO INQUIRY *Which other repair system will remove the thymine dimer?*

16.3 Microbes Use Mechanisms Other Than Mutation to Create Genetic Variability

After reading this section, you should be able to:

a. Distinguish vertical gene transfer from horizontal gene transfer
b. Summarize the four possible outcomes of horizontal gene transfer
c. Compare and contrast homologous recombination and site-specific recombination

In section 16.1, we discuss the consequences of mutations in terms of their effect on a protein and the phenotype of the organisms bearing the mutation. However, the effect also depends on the environment in which the organism lives. In other words, mutations that don't immediately kill the organism are subject to selective pressure. Selective pressure determines if a mutation will persist in a population as an alternate form of the gene, termed an **allele.** The existence of multiple alleles for each gene in a genome means each organism in a population can have a distinctive (and probably unique) set of alleles making up its genome, that is, its genotype. Each genotype in a population can be selected for or selected against. Organisms with genotypes, and therefore phenotypes, best suited to the environment survive and are most likely to pass on their genes. Shifts in environmental conditions lead to changes in the population and ultimately result in the evolution of new species. The mechanisms by which new combinations of genes are generated are the topic of this section.

Sexual Reproduction and Genetic Variability

The transfer of genes from parents to progeny is **vertical gene transfer** and it is observed in all organisms. In eukaryotes capable of sexual reproduction, vertical gene transfer is accompanied by genetic recombination. **Recombination** is the process in which one or more nucleic acid molecules are rearranged or combined to produce a new genotype. It occurs in two ways. The first is from crossing-over between sister chromosomes during meiosis. This creates new mixtures of alleles on homologous chromosomes. The second type of recombination results from fusion of gametes. Because each gamete contains alleles from one parent, when the gametes fuse, the parental alleles are combined in the zygote. In both cases progeny organisms that arise through recombination (**recombinants**) are not identical to their parents or to each other.

Horizontal Gene Transfer: Creating Variability the Asexual Way

Bacteria and archaea do not reproduce sexually. Nevertheless, microbial populations exhibit variation, generated in ways beyond accumulating mutations. These mechanisms are collectively termed **horizontal (lateral) gene transfer (HGT/LGT)** (**figure 16.9**). HGT is distinct from vertical gene transfer (binary fission) because replication is not required. Instead, genes are transferred from one organism to another, often creating a stable recombinant having characteristics of both donor and recipient. *Horizontal Gene Transfer*

The importance of HGT cannot be overstated. HGT is commonplace and is the main mechanism by which bacteria and archaea evolve. Many examples of DNA transfer between distantly related genera and across domains of life are known, particularly among microbes sharing a habitat (e.g., the human gut). HGT helps bacteria and archaea survive environmental stresses. For instance, some archaea respond to UV radiation by increasing their ability to carry out HGT so that they can obtain and use DNA from other cells to repair UV-damaged DNA in their genomes. Another important example is the transfer of groups of genes encoding virulence factors from

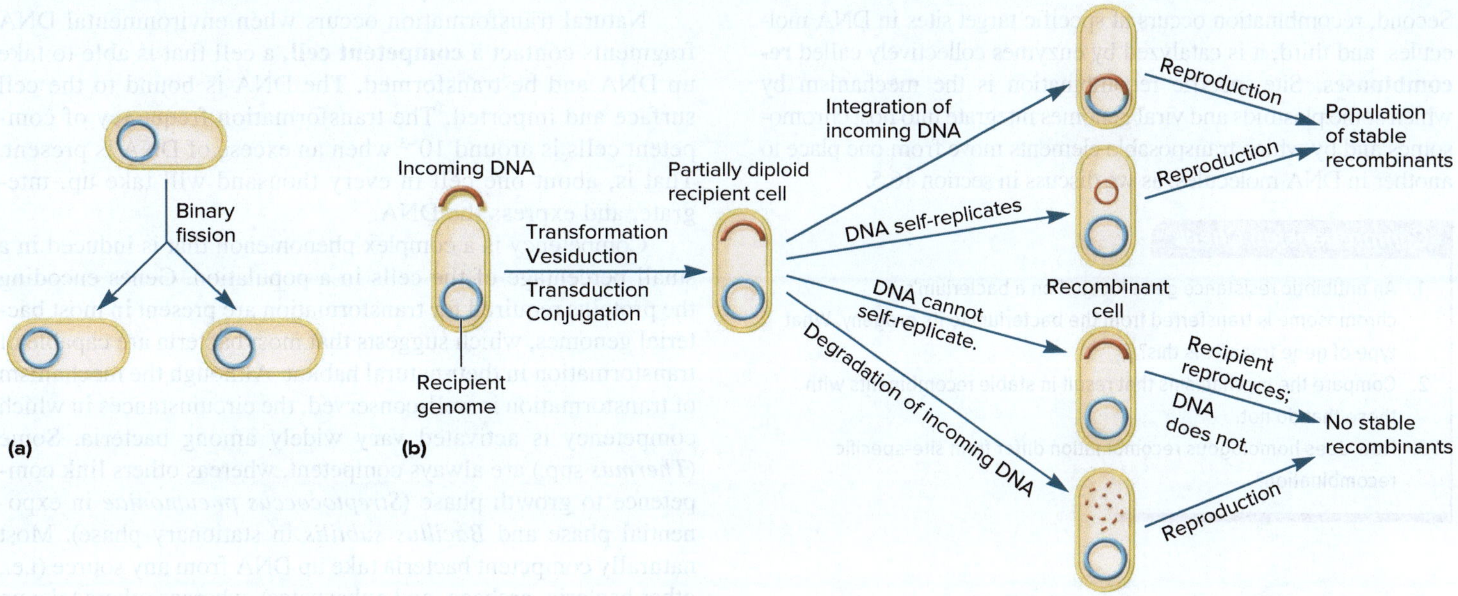

Figure 16.9 Vertical and Horizontal Gene Transfer. Vertical gene transfer (a) is the transfer of genes from a parent to offspring, during binary fission in bacteria and archaea. In horizontal gene transfer (b), DNA moves from one individual to another. Four possible fates of the transferred DNA are shown.

pathogens to other bacteria. These gene clusters are called *pathogenicity islands,* and their discovery has provided valuable insights into the evolution of pathogenic bacteria. Finally, in most cases, the acquisition of new traits by HGT enables microbes to rapidly expand their ecological niche. This is clearly seen when HGT results in the spread of antibiotic resistance genes among bacteria. *Pathogenicity islands encode virulence factors (section 34.4)*

During HGT, a piece of donor DNA enters a recipient cell. The DNA has four possible fates (figure 16.9*b*). First, when the DNA has a sequence similar to the recipient's chromosome, integration may occur by homologous recombination (figure 16.8). The recombinant then reproduces, yielding a population of stable genetic variants. Second, if the donor DNA is able to replicate (e.g., a plasmid), it may persist separate from the recipient's chromosome. When the recipient reproduces, the donor DNA replicates and a population of stable recombinants is formed. Third, the donor DNA remains in the cell but is unable to replicate. After multiple cell divisions, the DNA is eventually lost from the population. Finally, the DNA may be eliminated by host restriction or CRISPR/Cas degradation, thereby preventing the formation of a recombinant cell. *Responses to viral infection (section 14.6)*

HGT in bacteria and archaea occurs through four primary mechanisms (**figure 16.10**). DNA may be acquired from the environment, either directly (transformation) or via extracellular vesicles (vesiduction). Alternately, DNA may be transported from a donor cell via a bacteriophage (transduction) or through direct cell-to-cell contact (conjugation). In all cases, the formation of a stable recombinant requires DNA integration into the recipient genome or the capacity for self-replication in the recipient cell.

Molecular Recombination: Joining DNA Molecules

Donor DNA integration into the recipient's chromosome during HGT in bacteria and archaea occurs by homologous recombination, and many of the enzymes used are similar in all organisms. Homologous recombination results from DNA strand breakage and reunion leading to crossing-over. The enzymes responsible for homologous recombination are also important for DNA repair (e.g., RecA; figure 16.8).

The other major type of recombination is **site-specific recombination.** It differs from homologous recombination in three ways. First, it does not require long regions of sequence homology.

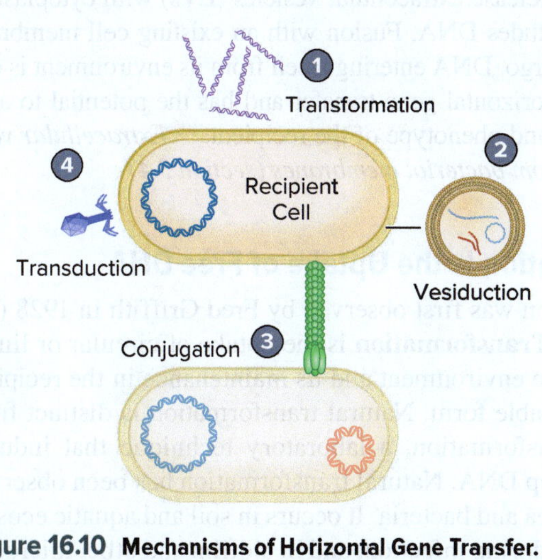

Figure 16.10 Mechanisms of Horizontal Gene Transfer. Four mechanisms are illustrated on a single recipient cell, but they are distinct, unrelated processes.

Second, recombination occurs at specific target sites in DNA molecules, and third, it is catalyzed by enzymes collectively called **recombinases.** Site-specific recombination is the mechanism by which some plasmids and viral genomes integrate into host chromosomes and by which transposable elements move from one place to another in DNA molecules, as we discuss in section 16.5.

Comprehension Check

1. An antibiotic resistance gene located on a bacterium's chromosome is transferred from the bacterium to its progeny. What type of gene transfer is this?
2. Compare the mechanisms that result in stable recombinants with those that do not.
3. How does homologous recombination differ from site-specific recombination?

16.4 Environmental DNA Brings New Genes into Cells

After reading this section, you should be able to:

a. Explain the origin of environmental DNA
b. Describe the factors that contribute to a microbe being naturally competent for transformation
c. Compare transformation and vesiduction in terms of DNA stability and mechanism of entry

Most microbial cells live in diverse communities where individuals are in varied states of growth and senescence. Lysed cells release metabolites and debris, including DNA, into the environment. DNA may be hydrolyzed and consumed as a nutrient, but many microbes have evolved strategies to import stray DNA molecules. Some stressed cells release extracellular vesicles (EVs) with cytoplasmic cargo that includes DNA. Fusion with an existing cell membrane delivers the cargo. DNA entering a cell from its environment is one path toward horizontal gene transfer and has the potential to alter the genotype and phenotype of the recipient. ◀ *Extracellular vesicles emerge from bacterial membranes (section 3.4)*

Transformation Is the Uptake of Free DNA

Transformation was first observed by Fred Griffith in 1928 (*see figure 13.1*). **Transformation** is the uptake of circular or linear DNA from the environment and its maintenance in the recipient cell in a heritable form. Natural transformation is distinct from artificial transformation, a laboratory technique that induces cells to take up DNA. Natural transformation has been observed in both archaea and bacteria. It occurs in soil and aquatic ecosystems, in vivo during infection, and in biofilm and other microbial communities. ▶ *Introducing recombinant DNA into host cells (section 17.1)*

Natural transformation occurs when environmental DNA fragments contact a **competent cell,** a cell that is able to take up DNA and be transformed. The DNA is bound to the cell surface and imported. The transformation frequency of competent cells is around 10^{-3} when an excess of DNA is present. That is, about one cell in every thousand will take up, integrate, and express the DNA.

Competency is a complex phenomenon that is induced in a small percentage of the cells in a population. Genes encoding the proteins required for transformation are present in most bacterial genomes, which suggests that most bacteria are capable of transformation in their natural habitat. Although the mechanism of transformation is well-conserved, the circumstances in which competency is activated vary widely among bacteria. Some (*Thermus* spp.) are always competent, whereas others link competence to growth phase (*Streptococcus pneumoniae* in exponential phase and *Bacillus subtilis* in stationary phase). Most naturally competent bacteria take up DNA from any source (i.e., other bacteria, archaea, and eukaryotes), whereas others take up DNA only from closely related species. For instance, *Haemophilus influenzae* requires a specific 9-base-pair sequence that is common in the *H. influenzae* genome. Archaea also exhibit natural competence, but there are no homologues of bacterial competence proteins encoded in archaeal genomes. This suggests that a novel mechanism operates in these organisms.

Natural transformation is a multistep process, as illustrated for *S. pneumoniae* in **figure 16.11.** Despite the difference in cell envelope architecture, monoderm and diderm bacteria use similar proteins in their DNA uptake machinery. These proteins are related to those used in type II secretion systems and type IV pili *(see figure 13.41)*. (Note that type IV pili are unrelated to type IV secretion systems.) The major differences arise from the need for diderm bacteria to import dsDNA across the outer membrane prior to its translocation as ssDNA across the plasma membrane. *Bacterial Transformation*

Vesiduction Is DNA Delivered in a Vesicle

Extracellular vesicles (EVs) are membrane-encapsulated spheres released from all types of cells. The membrane reflects the cell of origin, typically the plasma membrane of monoderms and archaea, and the outer membrane of diderms. EVs can contain cytoplasmic molecules, notably nucleic acids. **Vesiduction** is a mechanism of horizontal gene transfer mediated by EVs.

An important difference between transformation and vesiduction is that the DNA in an EV is protected from degradation by the vesicle membrane, and likely to remain in the environment longer. In addition, a recipient cell does not require any type of DNA uptake machinery. Mechanistically, little is known about how an EV interacts with a recipient cell to deliver its cargo, but it has been demonstrated that vesicles can fuse with phylogenetically unrelated cells. EVs have been detected in many environments, including cells in both planktonic and biofilm growth states.

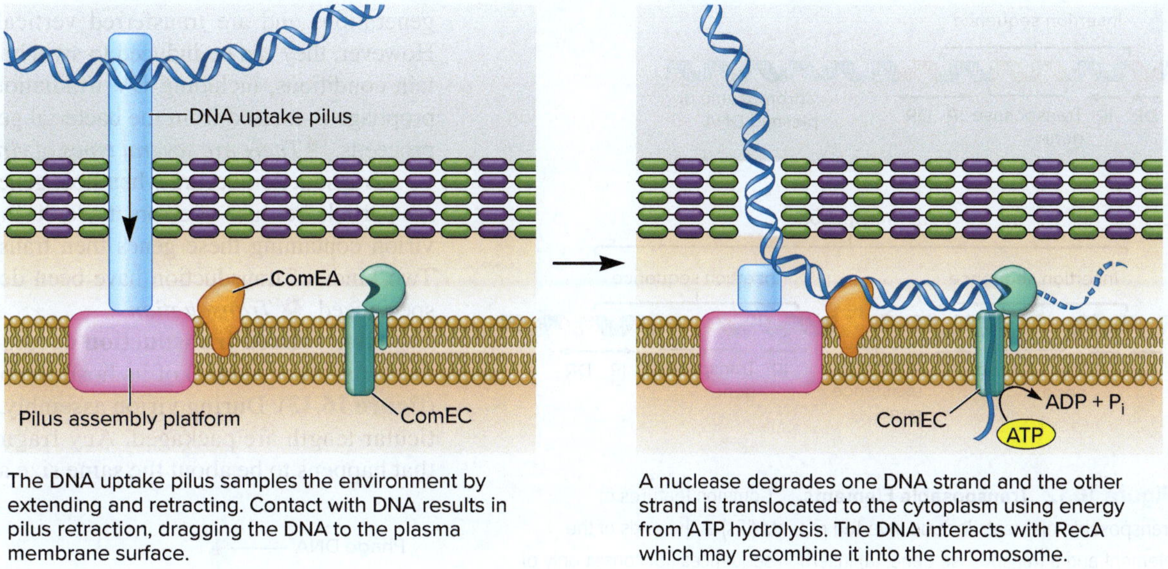

Figure 16.11 Transformation in *Streptococcus pneumoniae*. The contracting DNA uptake pilus brings DNA across the cell wall where it is directed by membrane-bound ComEA to a membrane channel (ComEC) and an endonuclease. One strand passes through the channel to the cytoplasm and the other is degraded.

MICRO INQUIRY *This figure illustrates transformation in the monoderm* S. pneumoniae. *How would it differ in a diderm cell?*

Comprehension Check

1. What is the relationship between transformation and competence?
2. Describe how transformation occurs in *S. pneumoniae*. How does the process differ in *H. influenzae* and *B. subtilis*?
3. How are transformation and vesiduction similar? How do they differ?

16.5 Mobile Genetic Elements Move Genes Within and Between Cells

After reading this section, you should be able to:

a. Differentiate insertion sequences from transposons
b. Distinguish simple transposition from replicative transposition
c. Differentiate generalized transduction from specialized transduction
d. Correlate a virus's life cycle with its capacity to mediate generalized or specialized transduction
e. Describe the features of the F factor that allow it to (1) transfer itself to a new host cell and (2) integrate into a host cell's chromosome
f. Outline the events that occur when an F^+ cell encounters an F^- cell
g. Distinguish F^+, Hfr, and F′ cells from each other
h. Explain how Hfr cells arise
i. Outline the events that occur when an Hfr cell encounters an F^- cell

As genomes and metagenomes are sequenced, annotated, and compared, it is apparent that they are rife with genetic elements that move within and between cells. These elements are referred to as **mobile genetic elements (MGEs)** and include transposable elements, plasmids, and viruses. MGEs are the vectors of horizontal gene transfer, and as such they are critical drivers of evolution. The multiple and varied types of MGE illustrate that the gene pool for bacteria and archaea is both external and vast.

Transposable Elements Move Within a Genome

Transposable elements (TEs) promote DNA mobility within a cell. They were first described in the 1940s by Barbara McClintock during her studies on maize genetics (a discovery for which she was awarded the Nobel Prize in 1983). Movement of a TE, termed **transposition,** can alter gene function either by causing an insertion mutation in the target gene or by providing regulatory sites that change the expression patterns of adjacent genes. The enzymes that catalyze transposition are **transposases,** a type of recombinase that does not require stretches of homologous DNA. As a consequence, TE insertions are random, occurring at any site in a genome. *Transposition: Shifting Segments of the Genome*

The simplest TEs are **insertion sequences,** or IS elements (**figure 16.12*a***). An IS element is a short sequence of DNA (~ 750 to 1,600 base pairs [bp] in length). It typically contains only the gene that encodes transposase, and it is bounded at both ends by inverted repeats—identical or very similar sequences of 15–25 nucleotides in reversed orientation. Each type of IS has a specific nucleotide sequence in its inverted repeats. Transposase recognizes the repeats to catalyze transposition.

Transposons are more complex in structure than IS elements. They consist of a central region containing cargo genes unrelated to transposition (e.g., antimicrobial resistance genes) flanked by IS elements that are identical or very similar in

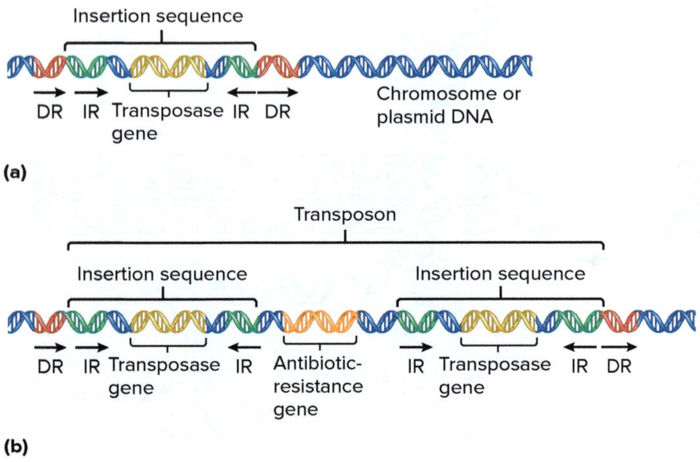

Figure 16.12 Transposable Elements. Common features of transposable elements include inverted repeats (IRs) at the ends of the element and a transposase gene. (a) Insertion sequences (IS) consist only of IRs on either side of the transposase gene. (b) Composite transposons contain additional cargo genes (e.g., antimicrobial resistance genes) in addition to transposases. Direct repeats (DR) in host DNA flank a transposable element.

MICRO INQUIRY *What features are common to all types of transposable elements?*

sequence (figure 16.12*b*). The flanking IS elements encode the transposase used by the transposon to move.

Two major transposition methods have been identified. Simple transposition is also called cut-and-paste transposition. In this method, transposase catalyzes excision of the TE, followed by cleavage of a new genomic insertion site and ligation of the element into this site. In replicative transposition, the MGE remains at the original site and a copy is inserted at the new target. *Simple Transposition*

Transduction Is Virus-Mediated DNA Transfer

Transduction is horizontal gene transfer mediated by viruses and is frequent in nature. Evidence suggests that the number of genes moved by marine viruses from one host cell to another is huge (perhaps 10^{24} per year). Furthermore, viruses in marine environments and hot springs move genes between organisms in all three domains of life. *Marine viruses: mortality at sea (section 29.2)*

Virus particles are structurally simple, often composed of just a nucleic acid genome protected by a protein capsid. Viruses infect and take control of a host cell, forcing it to make many copies of the virus. Viruses that infect bacteria are called bacteriophages, or phages for short. **Virulent bacteriophages** multiply immediately after entry, cause the host to lyse, and are released to infect new host cells (*see figure 6.15*). This process is called the **lytic cycle.** **Temperate bacteriophages,** on the other hand, establish a relationship with their host, called **lysogeny,** by inserting their genome into the bacterial chromosome. The inserted viral genome is called a **prophage** and the host cell is termed a **lysogen.** The host is unharmed and the phage genome is replicated along with the host genome. Temperate phages may remain inactive for many generations and are transferred vertically during cell division. However, they can be induced to switch to a lytic cycle under certain conditions, including UV irradiation. When this occurs, the prophage is excised from the bacterial genome and the lytic cycle proceeds. *There are several types of viral infections (section 6.4)*

Transduction occurs when host genes are packaged in the virus particle because of errors made during the viral life cycle. The virion containing these genes then transfers them to a new cell. Two kinds of transduction have been described: generalized and specialized. *Transduction*

Generalized transduction occurs during a lytic cycle. As the virus takes control of its host, the host genome is degraded (**figure 16.13**). During virion assembly, viral genomes of a particular length are packaged. Any fragment of the host genome that happens to be about the same size as the phage genome can

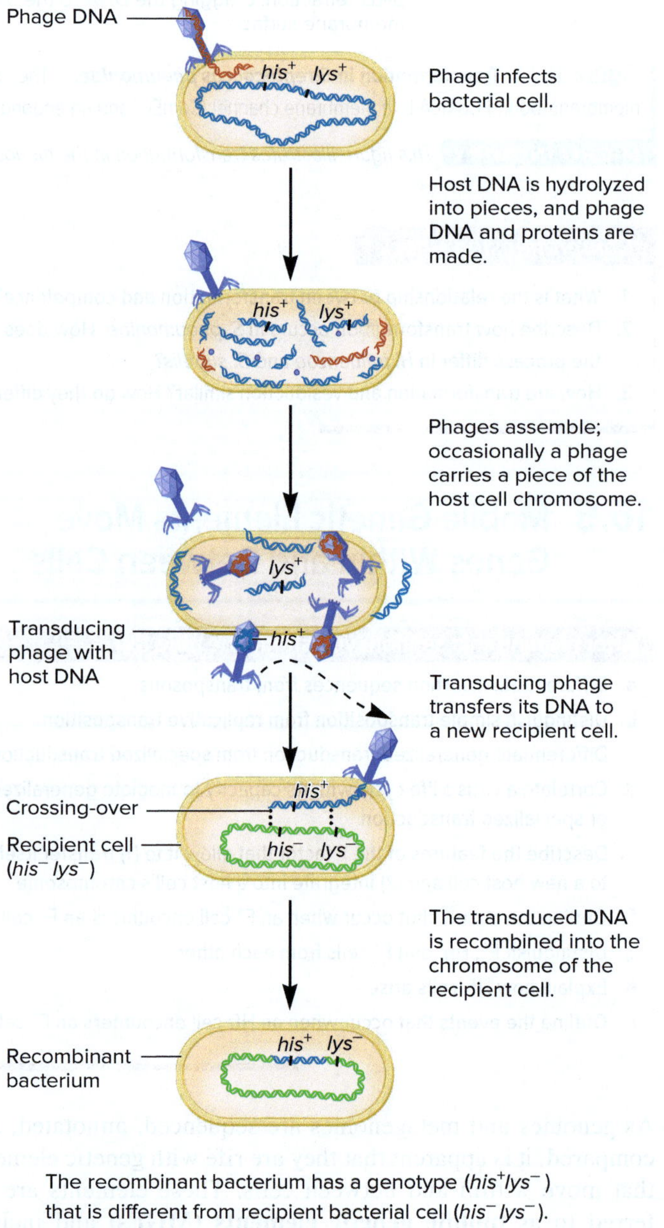

Figure 16.13 Generalized Transduction.

be mistakenly packaged. Such a phage is called a transducing particle because once it is released, it may encounter a susceptible host cell and deliver the bacterial DNA it carries into that cell. However, because it lacks the viral genome, it is not capable of initiating a lytic cycle. Once the DNA fragment is released into the recipient, it must be incorporated into the recipient chromosome by homologous recombination. If the DNA is not integrated, it may remain intact temporarily and be expressed. *Abortive transductants* are bacteria that contain this nonintegrated, transduced DNA and are partial diploids until the transduced DNA is degraded. *Generalized Transduction*

In **specialized transduction,** only specific portions of the bacterial genome are carried by transducing particles. Specialized transduction is made possible by an error in the lysogenic cycle of temperate phages that insert their genomes into a specific site in the host chromosome. When a prophage is induced to leave, excision is sometimes carried out improperly (**figure 16.14**). The resulting phage genome contains portions of the bacterial chromosome next to the integration site. The transducing particle is defective because it lacks some viral genes and cannot produce additional progeny viruses. In spite of this, it will transfer the remaining viral genome and any bacterial genes it carries to another recipient. The bacterial genes may become stably incorporated under the proper circumstances. *Bacteriophage lambda: a temperate bacteriophage (section 25.2)* *Specialized Transduction*

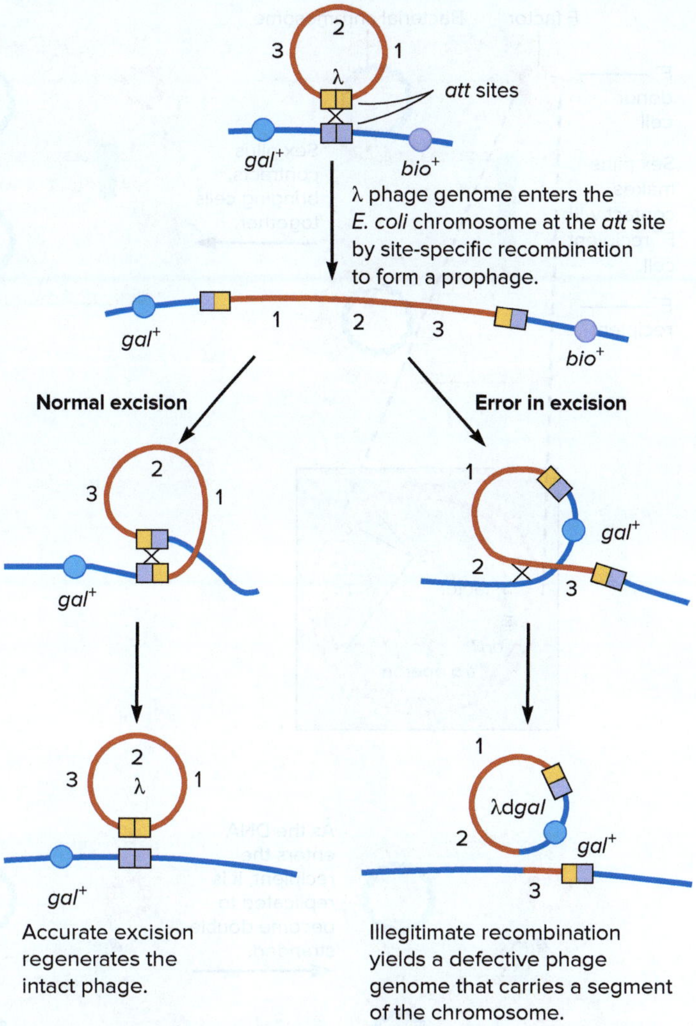

Figure 16.14 Mechanism of Specialized Transduction for Phage Lambda (λ) and *E. coli*. Integrated λ phage lies between the *gal* and *bio* genes. When it excises normally (left), the phage genome is complete. Rarely, excision occurs asymmetrically (right), and either the *gal* or *bio* genes are swapped for an equal amount of phage DNA. (Only aberrant excision involving the *gal* genes is shown.) The result is a defective λ phage (λ*dgal*) that carries bacterial genes and can transfer them to a new recipient.

MICRO INQUIRY *Why can't both the* gal *and* bio *genes be transduced by the same transducing particle?*

Comprehension Check

1. Distinguish mobile genetic elements from transposable elements.
2. How does a transposon differ from an insertion sequence?
3. What is simple (cut-and-paste) transposition? What is replicative transposition? How do the two mechanisms of transposition differ?
4. What effect would you expect the existence of TEs to have on the rate of microbial evolution? Explain your reasoning.
5. Describe generalized transduction and how it occurs. What is an abortive transductant?
6. What is specialized transduction and how does it come about?
7. How might one tell whether horizontal gene transfer was mediated by generalized or specialized transduction?
8. Why doesn't a cell lyse after successful transduction with a lytic phage?

Conjugation Requires Cell-Cell Contact

Conjugation, the transfer of DNA by direct cell-to-cell contact, depends on the presence of a **conjugative plasmid.** Recall from chapter 3 that **plasmids** are small, circular, double-stranded DNA molecules that exist independently of chromosomes. They have their own replication origins, replicate autonomously, and are stably inherited.

Perhaps the best-studied conjugative plasmid is the **F (fertility) factor.** It plays a major role in conjugation in *E. coli,* and it was the first conjugative plasmid to be described. Most of the information required for F factor transfer is encoded on the plasmid in the *tra* operon, which contains at least 35 genes (**figure 16.15**). Many of these direct the formation of the sex pilus that attaches the F^+ cell (the donor cell containing an F plasmid) to an F^- cell (**figure 16.16**).

$F^+ \times F^-$ Mating: Plasmid Transfer

The **sex pilus** is used to establish contact between the F^+ and F^- cells (figure 16.15). Once contact is made, the pilus retracts, bringing the cells into close physical contact. The F^+ cell prepares for DNA transfer by assembling a type IV secretion system, repurposing many of the proteins in the sex pilus membrane-spanning structure (**figure 16.17**). The term *sex pilus* refers to the extracellular structure, while the membrane-bound

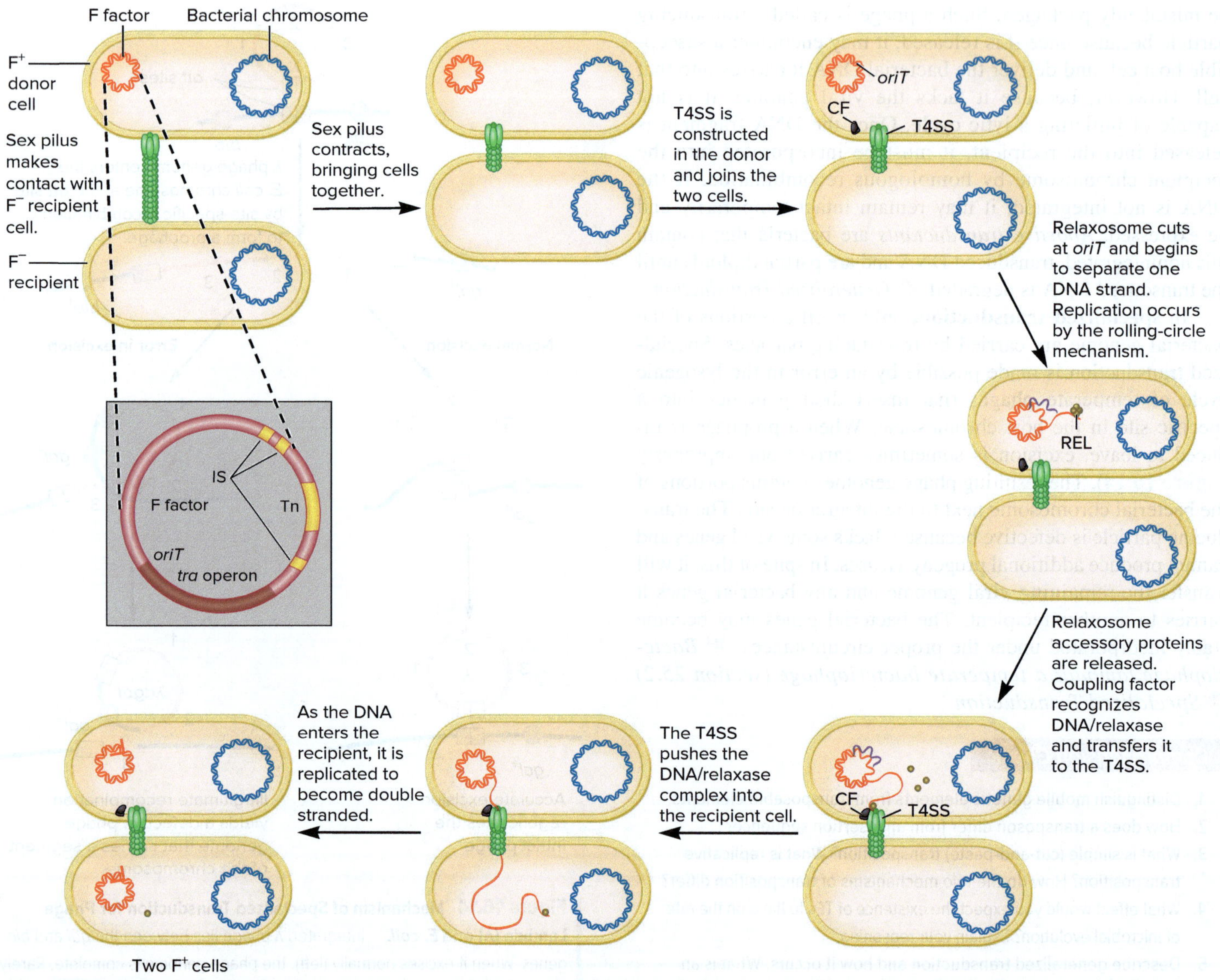

Figure 16.15 F Factor–Mediated Conjugation. During $F^+ \times F^-$ conjugation, only the F factor is transferred because the plasmid is extrachromosomal. The recipient cell becomes F^+. CF, coupling factor. Inset: The F factor transfer (*tra*) operon is solid red. The site for initiation of rolling-circle replication and DNA transfer is *oriT*. IS, insertion sequence; Tn, transposon.

components are remodeled into a **type IV secretion system** (T4SS). Although the two structures share polypeptide components and have complementary roles in conjugation, pilus formation and DNA transfer are independent processes. *Conjugation: Transfer of the F Plasmid*

The F plasmid is copied as it is transferred, using a process called **rolling-circle replication.** It is initiated by the relaxosome, a complex of proteins encoded in the *tra* operon (figure 16.15). The relaxosome nicks one strand of the F factor at a site called *oriT* (for origin of transfer). A protein called TraI attaches to the 5′ end of the nicked strand. As TraI displaces the 5′ end of one strand, replication enzymes extend the 3′ end using the complementary strand as template (**figure 16.18**). TraI collaborates with a protein coupling factor to guide the displaced strand through the T4SS to the recipient cell. It is believed that both ATP hydrolysis and the proton motive force provide the energy for DNA translocation. During plasmid transfer, the entering strand is used as a template to produce the complementary DNA strand. When this is complete, the F^- recipient cell becomes F^+. *Rolling-Circle Replication*

Hfr Conjugation Transfers Chromosomal DNA

By definition, an F^+ cell has the F factor separate from the chromosome, so in an $F^+ \times F^-$ mating, chromosomal DNA is

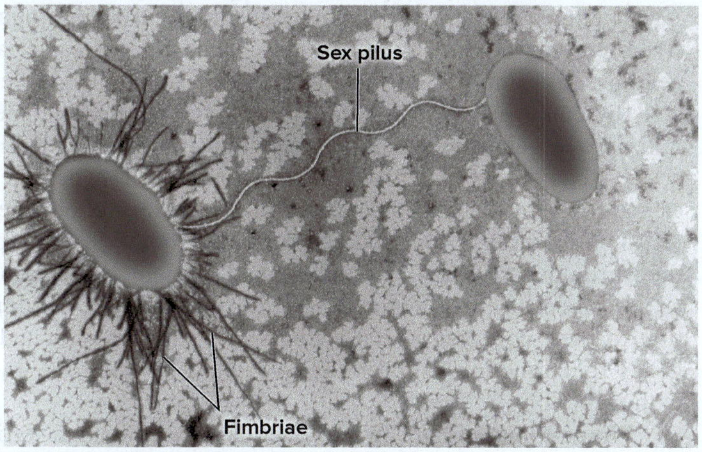

Figure 16.16 Conjugation. An electron micrograph of two *E. coli* cells in an early stage of conjugation. The F^+ cell (left) is covered with fimbriae, and a sex pilus connects the two cells.
Denis Kunkel Microscopy/Science Source

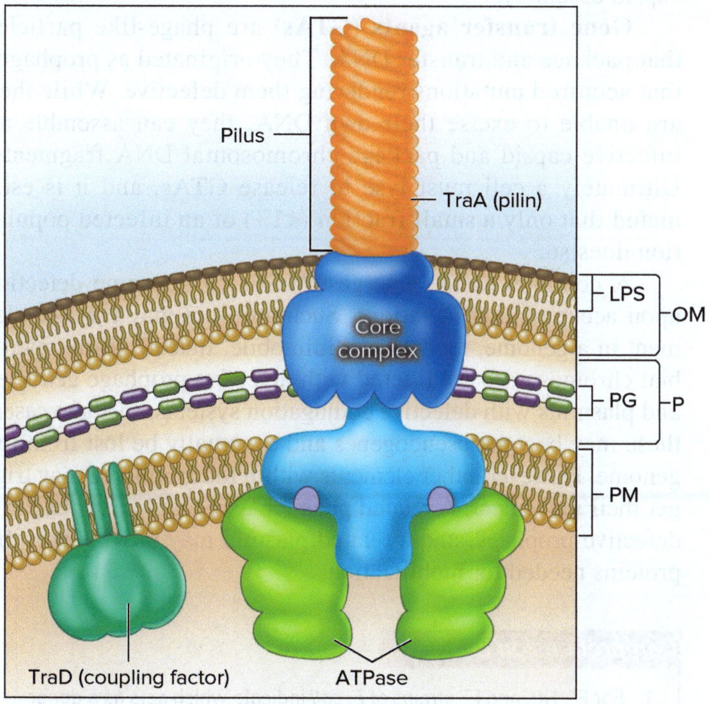

Figure 16.17 The Conjugation Machinery Encoded by the F Factor Is a Type IV Secretion System. The F factor–encoded type IV secretion system is composed of numerous Tra proteins, including TraA, which forms the sex pilus, and TraD, the coupling factor. The core complex is used for both pilus biogenesis and DNA transfer. LPS, lipopolysaccharide; OM, outer membrane; P, periplasm; PG, peptidoglycan; PM, plasma membrane.

not transferred. However, within a population of F^+ cells, a few have the F plasmid integrated (i.e., recombined) into their chromosomes. Notice in figure 16.15 inset that the F factor carries several insertion sequences. Such sequences are not uncommon in the *E. coli* chromosome, and homologous

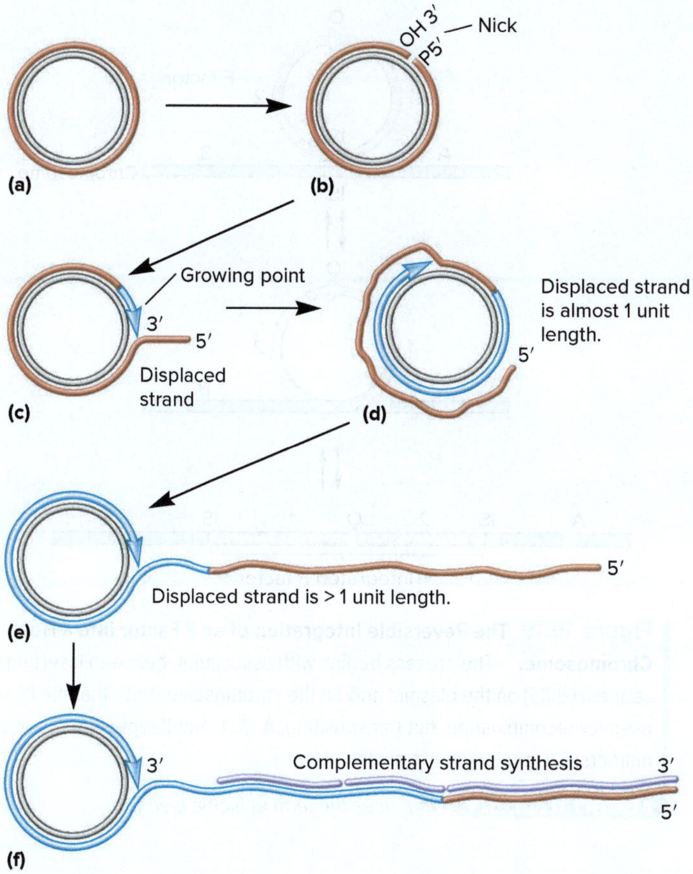

Figure 16.18 Rolling-Circle Replication. OH 3′ is the 3′-hydroxyl and P 5′ is the 5′ phosphate created when the DNA strand is nicked. The free end of the rolling-circle strand is bound to the relaxosome. A single-stranded tail, often longer than one genome copy, is generated and can be converted to the double-stranded form by synthesis of a complementary strand.

recombination between IS in the F factor and IS in the chromosome can occur (**figure 16.19**).

When integrated, the F factor's *tra* operon is still functional; it directs the synthesis of the sex pilus, carries out rolling-circle replication, and transfers genetic material to an F^- recipient. However, rather than transferring just itself, the F factor also directs the transfer of the host chromosome. Because of the *h*igh *f*requency of *r*ecombinants produced by this mating, it is referred to as **Hfr conjugation** and the donor is called an **Hfr strain.** The existence of Hfr strains illustrates that some plasmids, like the F factor, are **episomes,** that is, they function as autonomously replicating plasmids but can also integrate into a host chromosome. *Integration and Excision of a Plasmid*; *Transfer of Chromosomal DNA*

F′ Conjugation

Because the F factor is an episome, it can leave the bacterial chromosome and replicate as an autonomous plasmid. This is accomplished by homologous recombination between IS

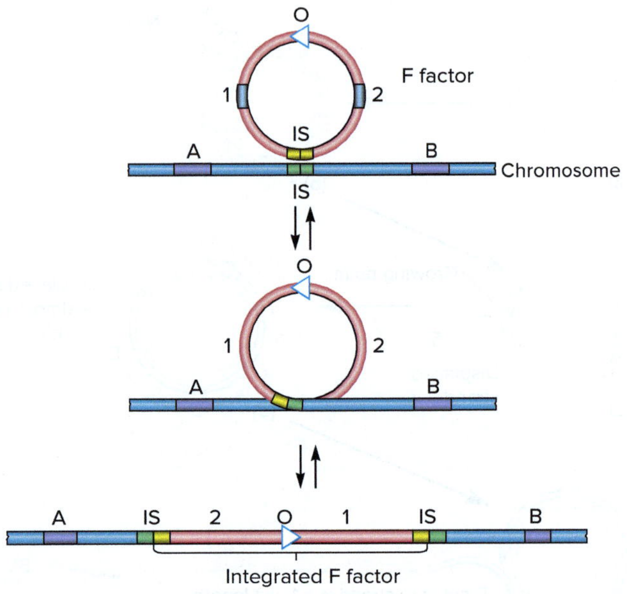

Figure 16.19 The Reversible Integration of an F Factor into a Host Chromosome. The process begins with association between insertion sequences (IS) on the plasmid and on the chromosome. Note that the IS are used for recombination, not transposition. A, B, 1, and 2 represent genetic markers; O is the origin of transfer.

MICRO INQUIRY *What does the term* episome *mean?*

elements and is the reverse of the process shown in figure 16.19. Sometimes during excision an error occurs and a portion of the chromosome is also excised, becoming part of the F plasmid. Because this erroneously excised plasmid is larger and genotypically distinct from the original F factor, it is called an **F′ (F prime) plasmid.** A cell containing an F′ plasmid retains all of its genes, although some of them are on the plasmid. It mates only with an F^- recipient, and $F' \times F^-$ conjugation is similar to an $F^+ \times F^-$ mating. Genes on the chromosome are not transferred, but genes on the F′ plasmid (both F factor genes and chromosomal genes) are transferred. These genes need not be incorporated into the recipient chromosome to be expressed. The recipient becomes F′ and is partially diploid because the same genes present on the F′ plasmid are also found on the recipient's chromosome. In this way, specific genes may spread rapidly throughout a population.

The Mobilome Includes Diverse Mobile Elements

Transposable elements (TEs), phages, and plasmids are the best-known mobile genetic elements (MGEs), but genomic level studies have uncovered additional elements that have a mix of known features and deploy them in unique transfer mechanisms. Collectively, all of the MGEs that act within a microbial population are considered its **mobilome.**

The most widespread elements are **integrative conjugative elements (ICEs),** found in more than 50% of bacterial genomes. As implied by their name, they transmit between cells by conjugation but are integrated into the chromosome. ICEs are large chromosomal regions that transmit vertically at cell division but may also transmit horizontally when they excise from the chromosome. ICEs encode a core set of genes for integration, excision, and conjugation. Upon release from the chromosome, they circularize and initiate conjugation as described for the F factor in *E. coli* (figure 16.15). In addition to their core genes, ICEs carry cargo that may include genes that function in antibiotic resistance, symbiosis, and pathogenicity.

Phage-plasmids are hybrid elements that transfer vertically as plasmids and horizontally as phages, representing about 5–10% of all plasmids. They may be considered a type of lysogen in which the DNA is not integrated but replicates separately from the chromosome. As required for a phage, phage-plasmids encode all functions for viral replication and capsid assembly.

Gene transfer agents (GTAs) are phage-like particles that package and transfer DNA. They originated as prophages that acquired mutations rendering them defective. While they are unable to excise their own DNA, they can assemble an infective capsid and package chromosomal DNA fragments. Ultimately a cell must lyse to release GTAs, and it is estimated that only a small fraction (<1%) of an infected population does so.

A common theme emerges: MGEs can become defective upon accumulating mutations. Such an event may trap the element in a genome, rendering it immobile. Indeed, many microbial chromosomes are littered with defective prophage genomes and plasmids with defective conjugation systems. In some cases, these may become pseudogenes and eventually be lost from the genome. However, other elements within the mobilome may trigger their movement. A related phage may recognize and excise a defective prophage, and a related plasmid may provide specific proteins needed for mobilization.

Comprehension check

1. For F^+, Hfr, and F^- strains of *E. coli* indicate which acts as a donor during conjugation, which acts as a recipient, and which transfers chromosomal DNA.
2. Describe how $F^+ \times F^-$ and Hfr conjugation processes proceed, and distinguish between the two in terms of mechanism and the final results.
3. Compare and contrast $F^+ \times F^-$ and $F' \times F^-$ conjugation.
4. How do ICEs differ from conjugative plasmids? How are they the same?
5. How do phage-plasmids differ from plasmids? How do they differ from phages?
6. How do GTAs differ from phages?

16.6 Evolution in Action: The Development of Antibiotic Resistance in Bacteria

After reading this section, you should be able to:

a. Explain how antibiotic resistance genes evolved from immunity genes in antibiotic producers
b. Describe the selective pressures that maintain antibiotic resistance genes in bacterial populations

In microbes, horizontal gene transfer creates the genetic variability upon which evolution by natural selection acts on a population. We now consider a well-documented example of bacterial evolution: the emergence of populations resistant to antibiotics.

Within 3 years after the widespread use of penicillin began in the 1940s, penicillin-resistant bacteria were found in clinical specimens. To understand the origin of drug resistance, it is important to recall that in nature, antibiotic-producing microbes must also protect themselves from the antibiotics they secrete—otherwise they would succumb to the effects of the antibiotic. In antibiotic producers, the genes that encode resistance proteins are often referred to as immunity genes. Immunity genes are usually coordinately regulated with genes that encode antibiotic biosynthetic enzymes. Many genes encoding antibiotic resistance were likely captured from antibiotic producers and moved by HGT to nonproducers, giving rise to a large pool of resistance-encoding genes outside the producing microorganisms. In an environment with both producers and nonproducers, an organism with a resistance gene has a competitive advantage. What's more competitive than a microbial cell with a resistance gene? A microbial cell with multiple copies of that resistance gene. More copies of the gene result in more of the gene product, with the result being that the cell will be better able to withstand the effects of the antibiotic. Increasing the resistance gene copy number occurs by mobilization of the resistance gene by replicative transposition, especially if it transposes to a multicopy plasmid.

Frequently a bacterial pathogen is drug resistant because it has a plasmid bearing one or more resistance genes. Plasmid-borne resistance genes often code for enzymes that destroy or modify drugs. Plasmid-associated genes have been implicated in resistance to antibiotics in most classes. Once a cell possesses a resistance plasmid, the plasmid (or its genes) may be transferred to other cells quite rapidly through HGT. Because a single plasmid may carry genes for resistance to several drugs, a pathogen population can become resistant to several antibiotics simultaneously, even though a patient may be treated with only one drug.

MGEs carrying a broad mix of cargoes confer flexibility to a population. Shuffling accessory genes for antibiotic resistance or metabolism provides selective advantages under varying pressures. Compared to the relatively stable microbial genome, MGEs are more likely to accumulate mutations, another way to diversify the alleles in a population. While many HGT mechanisms have serious costs (e.g., risk of cell lysis from phage, metabolic burden of plasmid maintenance, risk of insertional mutagenesis from TEs), overall this means of generating population diversity provides an evolutionary advantage.

Our focus here has been on the origin and transmission of resistance genes from antibiotic producers to nonproducers. What is important to recognize is that expression of these genes by nonproducers can be energetically costly. Resistance genes can be lost if the bacteria are not consistently exposed to the antibiotic. Thus exposure to the antibiotic is a selective force that helps maintain the resistance gene in a population. Of concern are the relatively low concentrations of antibiotics needed to maintain resistance genes. Unfortunately, the levels of antibiotic contamination in the environment far exceed this amount. It is therefore imperative that measures be enforced to decrease the use (especially misuse) of antibiotics to help alleviate the antibiotic resistance problem, including a multinational effort to reduce antibiotic use in agriculture.

Key Concepts

16.1 Mutations: Heritable Changes in a Genome

- A mutation is a stable, heritable change in the nucleotide sequence of the genetic material.
- Spontaneous (endogenous) mutations can arise from normal cellular activities and include the formation of abasic sites and insertion and deletion of nucleotides (**figures 16.1** and **16.2**).
- Induced (exogenous) mutations are caused by environmental factors. Mutations may result from mispairing due to alterations of a base caused by deamination, the presence of intercalating agents, crosslinking, and severe damage to the DNA caused by exposure to radiation (**figures 16.3** and **16.4**).
- There are four important types of forward point mutations: silent mutations, missense mutations, nonsense mutations, and frameshift mutations (**table 16.1**).
- Mutations are usually recognized when they cause a change from the more prevalent wild-type phenotype. A mutant phenotype can be restored to wild type by reversion mutations or suppressor mutations (**table 16.2**).
- Mutations can affect phenotype in numerous ways. Some mutations affect colony or cell morphology, viability (i.e., lethal mutations), biochemical pathways, and susceptibility to environmental stresses (e.g.,

resistance mutations). Some mutations are conditional, with effects under some conditions but not others (e.g., high temperature but not low temperatures).

16.2 DNA Repair Maintains Genome Stability

- Proofreading during DNA replication helps correct errors in replication. However, cells have multiple mechanisms for correcting mispaired and damaged DNA when proofreading is not effective.
- Mismatch repair replaces incorrect base pairs (**figure 16.5**).
- Excision repair systems remove damaged portions from a single strand of DNA (e.g., thymine dimers) and use the other strand as a template for filling in the gap (**figures 16.6** and **16.7**).
- Recombinational repair removes damaged DNA by recombination of the damaged DNA with a normal DNA strand elsewhere in the cell (**figure 16.8**).
- When DNA damage is severe, replication is halted, triggering the SOS response. During the SOS response, genes encoding DNA repair enzymes are transcribed at a higher rate. In addition, special DNA polymerases are produced, which replicate damaged DNA. However, they do so without an intact template and therefore can create mutations.

16.3 Microbes Use Mechanisms Other Than Mutation to Create Genetic Variability

- In sexual organisms, the mixing of parental DNA molecules in the progeny, as well as crossing-over that occurs during gamete formation, generates genetic diversity.
- Horizontal gene transfer (HGT) is an important mechanism for creating genetic diversity in bacteria and archaea. It is a one-way process in which DNA is transferred from one cell to another. In many transfers, the donor DNA must be integrated into the recipient's chromosome by homologous recombination to be stably maintained (**figure 16.9**).
- In homologous recombination, DNA molecules pair due to large areas of sequence similarity. Some viral genomes and all known transposable elements integrate into host DNA by site-specific recombination. This type of recombination does not require extensive sequence homology.

16.4 Environmental DNA Brings New Genes into Cells

- Transformation is the uptake of naked DNA by a competent cell and its incorporation into the genome (**figure 16.11**).
- Naturally competent bacteria have uptake machinery composed of proteins similar to those of type II secretion systems and type IV pili. Other species can be made artificially competent.
- In vesiduction, DNA is transferred through membrane vesicles that have pinched off from a donor cell.

16.5 Mobile Genetic Elements Move Genes Within and Between Cells

- Transposable elements are DNA segments that move about the genome to random locations in a process known as transposition.
- There are numerous types of transposable elements, including insertion sequences and several types of transposons (**figure 16.12**).
- Transduction is the transfer of DNA by viruses.
- In generalized transduction, any host DNA fragment can be packaged in a virus capsid and transferred to a recipient (**figure 16.13**). This phenomenon is the result of a DNA packaging error.
- Certain temperate phages carry out specialized transduction by incorporating bacterial genes during prophage induction and then transferring those genes to another bacterium (**figure 16.14**).
- Conjugation is the transfer of genes that depends upon cell-to-cell contact and is mediated by a plasmid. The F factor is one type of conjugative plasmid (**figure 16.15**). F factor conjugation in *E. coli* is accomplished by a sex pilus and a type IV secretion system (**figure 16.16**).
- In $F^+ \times F^-$ mating, the F factor remains independent of the chromosome and a copy is transferred to the F^- recipient; donor chromosomal genes are not transferred (**figure 16.15**).
- Hfr strains transfer bacterial genes to recipients because the F factor is integrated into the host chromosome.
- When the F factor leaves an Hfr chromosome, it occasionally picks up some bacterial genes to become an F′ plasmid, which readily transfers these genes to other bacteria.
- Other types of mobile genetic elements are common in microbial genomes, including integrative conjugative elements (ICEs), phage-plasmids, and gene transfer agents. These exhibit variations of mobility seen in other MGEs.

16.6 Evolution in Action: The Development of Antibiotic Resistance in Bacteria

- Antibiotic resistance genes evolved from the immunity genes in antibiotic producing cells. Horizontal gene transfer and movement of genes to elements with greater copy number have generated vast numbers of antibiotic resistance genes.
- Human overuse and misuse of antibiotics provided the selective pressure to maintain these genes in our environment.

Active Learning

1. Mutations are often considered harmful. Give an example of a mutation that would be beneficial to a microorganism. What gene would bear the mutation? How would the mutation alter the gene's role in the cell, and what conditions would select for this mutant allele?
2. Mistakes made during transcription affect the cell but are not considered mutations. Why not?
3. What would be the evolutionary advantage of having a period of natural competence in a bacterial life cycle? What would be possible disadvantages?
4. Hydroxylamine is a mutagen that attaches a hydroxyl group to cytosine, causing it to base-pair like thymine. Sketch a diagram like figure 16.1 to illustrate how point mutations occur following hydroxylamine treatment.
5. Pseudogenes are found in most genomes. These are genes with mutations that render them completely nonfunctional. Bacteria that are obligate intracellular symbionts contain many pseudogenes, often up to one-third of the genome. It is believed that some genes become dispensable in the intracellular niche, especially for functions that can be obtained from the host cell, like synthesis of biochemical precursors. Genomes of intracellular symbionts usually contain pseudogenes of DNA repair enzymes. Discuss the evolutionary processes that enable pseudogenes to accumulate in obligate intracellular symbionts and how this differs from organisms that do not depend on a host.
6. Hyperthermophiles are especially susceptible to cytosine deamination. What types of DNA repair processes would you expect to find in cells growing at high temperatures?
7. Review section 4.4 and the description of the Ups pilus in *Sulfolobus*. Following exposure to UV light, the Ups pilus promotes cell aggregation. What type(s) of repair might you expect in UV-exposed cells? What mechanism(s) of gene transfer might operate in this situation?
8. Consult section 40.3 and the discussion of toxins produced by the mold *Aspergillus flavus*. In light of what you have learned in this chapter, why are these toxins so deadly?
9. Although the genetic code (table 13.1) can use multiple codons to specify a single amino acid, some organisms have a *codon bias*. This means that their set of tRNAs can decode only a subset of codons. For example, arginine is specified by 6 codons: CGA, CGG, CGC, CGT, AGA, and AGG. Suggest a reason for why a silent mutation, specifically a change from one arginine codon to a different arginine codon, may result in a mutant phenotype.
10. The initial evidence for conjugation came from an elegant experiment performed in 1946 by Joshua Lederberg and Edward Tatum. They mixed two auxotrophic strains of *E. coli*, incubated them for several hours in nutrient medium, and then plated on minimal medium. How could they ensure that they were observing transfer of DNA and not reversion of the auxotrophic mutations?
11. *Synechocystis,* a species of cyanobacteria, is polyploid and may contain up to 100 copies of its genome per cell. As an autotroph, *Synechocystis* inhabits the upper layers of ocean waters and is exposed to UV radiation which can induce mutations. How might polyploidy benefit individual cells? How might polyploidy benefit a population of *Synechocystis*?

CHAPTER

17

Microbial DNA Technologies

Kathleen Sandman

Spinning Stronger Silk

Spider silk is lightweight and elastic, yet strong enough to suspend a dangling spider or capture prey for its next meal. Materials scientists have long sought to replicate spider silk in the laboratory to exploit its properties. Among the potential applications are lightweight bulletproof vests, artificial ligaments, and sutures.

Spider silk fibers are composed of many copies of a single protein, spidroin. In the spider's silk gland, as the protein solution moves through the spinning ducts, it is acidified and dehydrated. Individual polypeptides change conformation, aggregate, and elongate to become silk fiber.

Industrial production of silk fibers requires large quantities of spidroin, more than is feasible to extract from even an army of spiders. By inserting the gene that encodes spidroin into bacteria, yeast, or cultured animal cells, large amounts of recombinant spidroin protein have been produced. But silk from recombinant spidroin is less elastic and weaker than the original spider-made version.

Silkworms also produce silk fibers, as a protective layer for the pupal stage of the insect. However, these fibers don't have the desirable properties found in spider silk. Differences arise from the molecular requirements unique to each organism. While the quaternary structure of silkworm fibroin depends on slow-forming disulfides, spiders spin fibers rapidly on demand, and thus their spidroins must assemble through noncovalent interactions.

Scientists optimized these two silk fiber production systems by replacing part of the silkworm fibroin gene with a part of the spider spidroin. This required cloning the spider spidroin gene and designing a hybrid gene that was synthesized in vitro. Cas9 genome editing in the silkworm introduced the new gene, creating a silkworm that spins spider silk!

In this chapter, we introduce some of the DNA technologies that have made this type of project possible. From cloning a gene in *Escherichia coli (E. coli)* (**figure 17.1**), to assembling a custom gene, to genome editing, biotechnology has been revolutionized by the ease and simplicity of in vitro DNA technologies.

Readiness Check:

Based on what you have learned previously, you should be able to:

- ✓ List differences between a plasmid and a chromosome (section 3.6)
- ✓ Identify the key structural elements of DNA, RNA, and protein (section 13.2)
- ✓ Diagram the flow of cellular information from gene to protein as well as protein localization (e.g., subcellular localization or export) (sections 13.5, 13.7, and 13.9)
- ✓ Explain the general structural differences between a eukaryotic gene and a bacterial gene, including regulatory, coding, and noncoding sequences (sections 13.5 and 15.3)
- ✓ Describe the capacity of proteins to recognize and bind to DNA (section 14.2)
- ✓ Explain the mechanism of CRISPR/Cas9 interference (section 14.6)
- ✓ Describe transformation, transduction, DNA repair, and homologous recombination (sections 16.2, 16.3, 16.4, and 16.5)

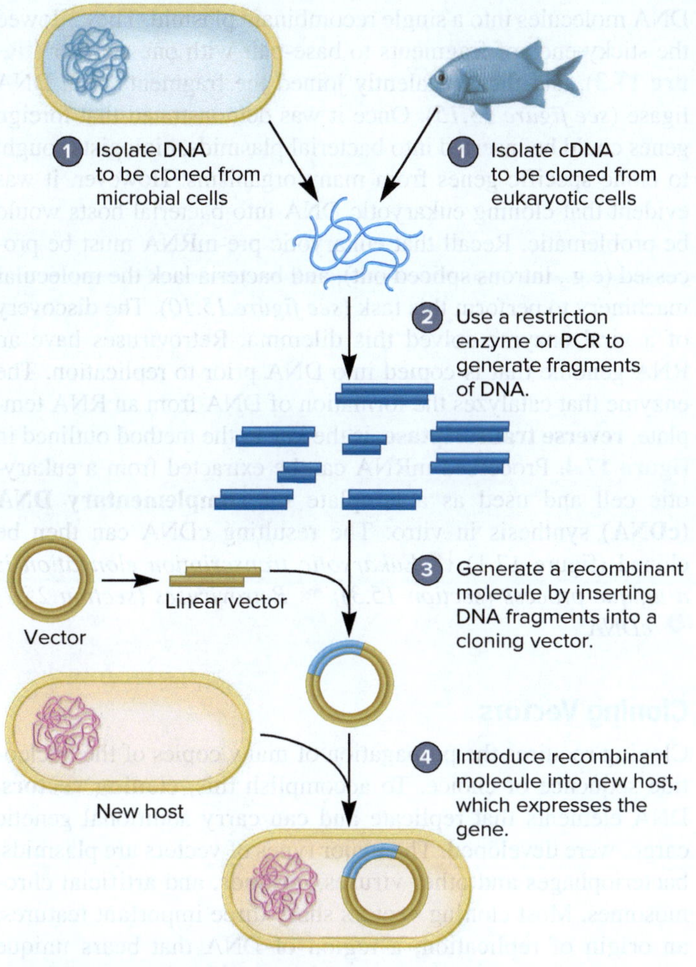

Figure 17.1 Steps in Cloning a Gene. Each step shown in this overview is discussed in more detail in this chapter. Step 1 varies with the type of genome.

MICRO INQUIRY *Which of the DNA molecules shown are recombinant?*

17.1 Cloning Technologies Developed from DNA Structure and Enzymology

After reading this section, you should be able to:

a. Explain how restriction enzymes recognize and hydrolyze DNA to create either blunt or sticky ends

b. Diagram the reaction catalyzed by reverse transcriptase and describe its application to biotechnology

c. List the three features of a cloning vector and why these elements are needed

d. Distinguish plasmids, phage-based cloning vectors, cosmids, and artificial chromosomes in terms of structure and application

e. Explain how DNA is inserted in vitro into a cloning vector

f. Identify host cells commonly used for cloning

g. Compare two common techniques by which DNA constructed in vitro is introduced into host cells

In every living organism, DNA has the same structure (*see figure 13.5*). The central dogma of molecular biology, that information flows from DNA to RNA to protein, applies universally. The genetic code that translates that information from RNA to proteins is the same (*see table 13.1*). This fundamental unity emerged in the years following the discovery of DNA structure and provides strong support for evolution from a common ancestor. This unity also prompted scientists to explore modifying the genome by manipulating DNA, a process termed *genetic engineering*.

For centuries, crop improvement and livestock breeding for desirable traits constituted the only form of genetic engineering. But in the twentieth century, the identification of DNA as the chemical basis for heredity and advances in laboratory methods for manipulating DNA opened new avenues for scientists. Rapid growth and easy laboratory handling made microbes like *E. coli* and the yeast *Saccharomyces cerevisiae* popular model organisms.

The application of DNA technologies to the study of microbes takes two forms. Microbes can be altered to allow scientists to ask direct questions about how they function in the environment and how the basic processes of life work. Additionally, microbes can be treated as factories for the production of biological molecules like insulin (a protein) and small molecules, such as lipids and amino acids. The underlying principle behind these techniques is that the genetic code is universal. Genes can be moved from one organism to another and still encode the same protein. Similarly, enzymes that act on DNA can perform their catalytic role in vitro on DNA from any source. *Microbial cells are customized for specific tasks (section 41.2)*

The earliest methods for DNA manipulation used enzymes and bacterial cells to modify and amplify DNA in a process known as **DNA cloning.** DNA molecules with segments that originate from different organisms are termed *recombinant*. More recent methods bypass the need for cells and use purified enzymes to create recombinant DNA molecules in vitro. As researchers envision more elaborate engineered cells, the components of the biotechnologist's toolbox (e.g., regulatory sequences, structural genes, and enzymes) are growing.

Restriction Enzymes

One of the first breakthroughs leading to DNA cloning was the discovery in the late 1960s of bacterial enzymes that cut double-stranded DNA. These **restriction endonucleases** or restriction enzymes recognize and cleave specific four to eight base pair sequences and can be used to prepare DNA fragments for cloning. For example, the restriction enzyme *Eco*RI, derived from *E. coli,* cleaves between G and A in the sequence 5′-GAATTC-3′ (**figure 17.2**). Because DNA strands are antiparallel, this sequence is also present on the complementary DNA strand and both strands are cut. Following *Eco*RI cleavage, unpaired 5′-AATT-3′ remains at the end of each strand; such overhangs are termed **sticky ends.** In contrast, cleavage by

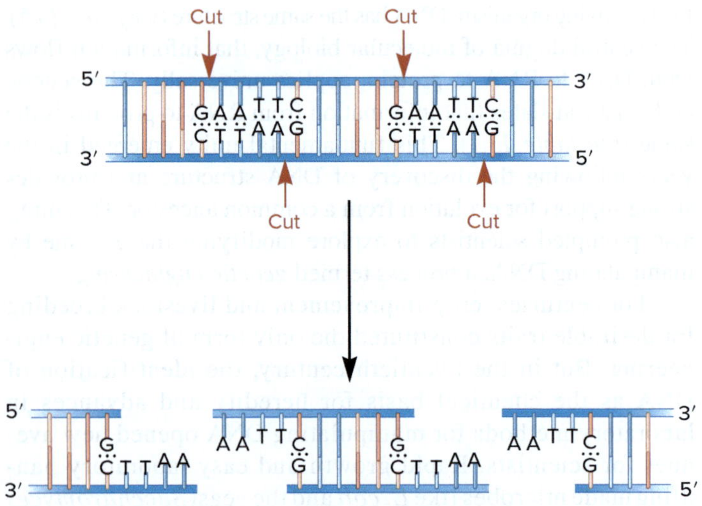

Figure 17.2 Restriction Endonuclease *Eco*RI Action. The enzyme makes staggered cuts on the two DNA strands to form sticky ends.

other restriction enzymes leaves blunt ends or ends with other sequences (**table 17.1**). Because the reaction occurs in the phosphate-sugar backbone, the sequence of nitrogenous bases is preserved. *Responses to viral infection (section 14.6)* *Restriction Endonucleases*

Gene Cloning and cDNA Synthesis

An important advance in cloning DNA came in 1972, when David Jackson, Robert Symons, and Paul Berg reported that they had generated recombinant DNA molecules. By combining the activities of several purified enzymes, they joined individual

Table 17.1 Some Restriction Endonucleases and Their Recognition Sequences

Enzyme	Recognition Sequence[1]	Ends Produced
*Eco*RI	5′ G↓AATTC 3′ 3′ CTTAA↑G 5′	5′ G AATTC 3′ 3′ CTTAA G 5′
*Hin*cII	5′ GTT↓AAC 3′ 3′ CAA↑TTG 5′	5′ GTT AAC 3′ 3′ CAA TTG 5′
*Not*I	5′ GC↓GGCCGC 3′ 3′ CGCCGG↑CG 5′	5′ GC GGCCGC 3′ 3′ CGCCGG CG 5′
*Pst*I	5′ CTGCA↓G 3′ 3′ G↑ACGTC 5′	5′ CTGCA G 3′ 3′ G ACGTC 5′

[1]The arrows indicate the sites of cleavage on each strand.

DNA molecules into a single recombinant plasmid. They allowed the sticky ends of fragments to base-pair with one another (**figure 17.3**), and then covalently joined the fragments with DNA ligase (*see figure 13.15*). Once it was demonstrated that foreign genes could be inserted into bacterial plasmids, biologists sought to clone specific genes from many organisms. However, it was evident that cloning eukaryotic DNA into bacterial hosts would be problematic. Recall that eukaryotic pre-mRNA must be processed (e.g., introns spliced out), and bacteria lack the molecular machinery to perform this task (*see figure 15.10*). The discovery of a viral enzyme solved this dilemma. Retroviruses have an RNA genome that is copied into DNA prior to replication. The enzyme that catalyzes the formation of DNA from an RNA template, **reverse transcriptase,** is the key to the method outlined in **figure 17.4.** Processed mRNA can be extracted from a eukaryotic cell and used as a template for **complementary DNA (cDNA)** synthesis in vitro. The resulting cDNA can then be cloned (figure 17.1). *Eukaryotic transcription elongation is a unique process (section 15.3);* *Retroviruses (section 25.7)* *cDNA*

Cloning Vectors

Cloning requires the propagation of many copies of the nucleotide sequence of choice. To accomplish this, **cloning vectors,** DNA elements that replicate and can carry additional genetic cargo, were developed. The major types of vectors are plasmids, bacteriophages and other viruses, cosmids, and artificial chromosomes. Most cloning vectors share three important features: an origin of replication; a region of DNA that bears unique restriction sites, called a multicloning site; and a selectable marker. *Construction of a Plasmid Vector*

Plasmids make excellent cloning vectors because they replicate independently of the chromosome and are easy to extract from the host and purify. They can be taken up by microbes by conjugation, transformation, or electroporation. Many different plasmids are used in biotechnology, all derived from naturally occurring plasmids that have been genetically engineered (**figure 17.5**). *Plasmids (section 3.6); Environmental DNA brings new genes into cells (section 16.4); Mobile genetic elements move genes within and between cells (section 16.5)*

The **origin of replication** (*ori*) allows the plasmid to replicate in the microbial host, and it dictates the **copy number,** or the number of plasmid molecules present in a cell. High copy number facilitates plasmid purification; an *E. coli* high copy number plasmid like pUC19 is present at hundreds of copies per cell. Some plasmids have two origins of replication, each recognized by different host organisms. These plasmids are called **shuttle vectors** because they can be transferred from one host to another. YEp24 is a shuttle vector that can replicate in yeast (*Saccharomyces cerevisiae*) and in *E. coli* because it has both the 2μ circle yeast replication element and an *E. coli* origin of replication (figure 17.5).

Not all host cells will take up a cloning vector, so it is necessary to discriminate between cells that successfully imported the

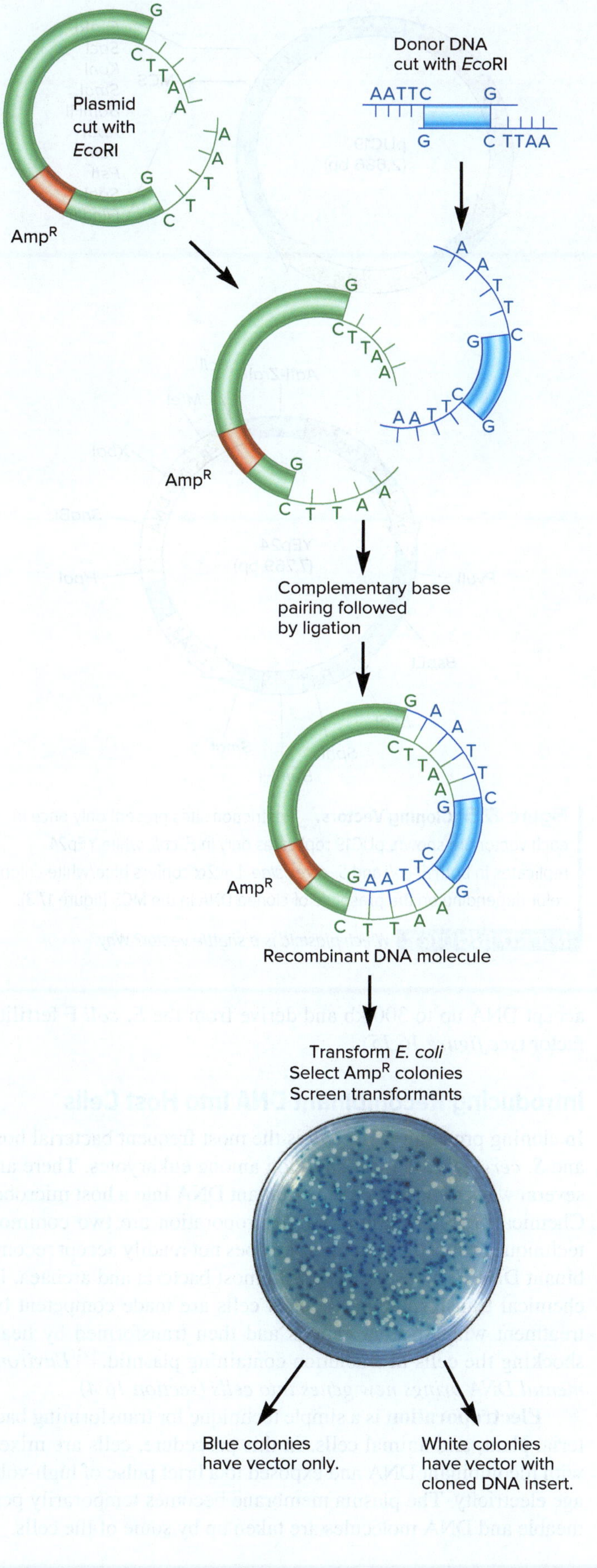

Figure 17.3 Construction of a Recombinant Plasmid. Treatment of both the cloning vector and the donor DNA with the same restriction endonuclease results in complementary sticky ends on both molecules. The scale of the sticky ends has been enlarged to illustrate base pairing. After transformation, *E. coli* cells are plated on medium containing ampicillin and X-Gal. Ampicillin ensures that only ampicillin-resistant transformants grow; X-Gal enables the visualization of colonies that were transformed with recombinant vector (vector + insert, white colonies).
Edvotek, Inc. www.edvotek.com

plasmid (transformants) from those that did not. This is achieved by the presence of a gene on the plasmid encoding a protein needed for the cell to survive under certain conditions. Such a gene is called a **selectable marker.** In the case of pUC19, the selectable marker, Amp^R, encodes the ampicillin-resistance enzyme β-lactamase. Because *E. coli* is normally susceptible to ampicillin, only transformants will grow into colonies when plated on medium containing ampicillin. The shuttle vector YEp24 bears both the Amp^R gene for use in *E. coli* and *URA3,* which encodes a protein essential for uracil biosynthesis in yeast. Therefore this plasmid must be used in strains of *S. cerevisiae* that are uracil auxotrophs; otherwise, it would be impossible to select, or identify, transformants.

Cloning vectors have been engineered with a small region designed to accept the DNA to be cloned. If restriction enzymes are to be used, it is called the **multicloning site (MCS)** (figure 17.5). An MCS contains an array of restriction enzyme recognition sequences. Other cloning techniques use recombination between vector and inserted DNA. In all cases, cloning vectors have been constructed to optimize the success of these techniques.

Phage vectors are phage genomes modified to include useful sites for cloning. When DNA has been inserted, the recombinant phage genome is packaged into viral capsids and used to infect host cells. The resulting phage lysate consists of thousands of phage particles that carry cloned DNA as well as the genes needed for host lysis. Common vectors are derived from the *E. coli* bacteriophage lambda (λ), which has a double-stranded DNA genome. ▶| *Bacteriophage lambda: a temperate bacteriophage (section 25.2)*

Genetic engineering of animal cells is accomplished by the use of viral vectors. These recombinant viruses are shuttle vectors that replicate in *E. coli* as bacterial plasmids, but they also direct the assembly of mature infective virus particles to shuttle the DNA to animal cells. Commonly used viruses for mammalian cells are retroviruses, lentiviruses, and adenoviruses.

DNA inserts larger than about 15 or 25 kb (kilobase pairs, or 1,000 base pairs) cannot be stably maintained in plasmids and phage vectors, respectively. Instead, **cosmids** can be used to clone such large fragments of DNA. These vectors have a *cos* DNA packaging site from phage along with a selectable marker, origin of replication, and MCS from plasmids (thus the term "cos-mid"). These hybrid vectors replicate as plasmids within the host cell, but the *cos* sites permit the vector to be packaged into phage capsids and transferred to new host cells by transduction. |◀ *Mobile genetic elements move genes within and between cells (section 16.5)*

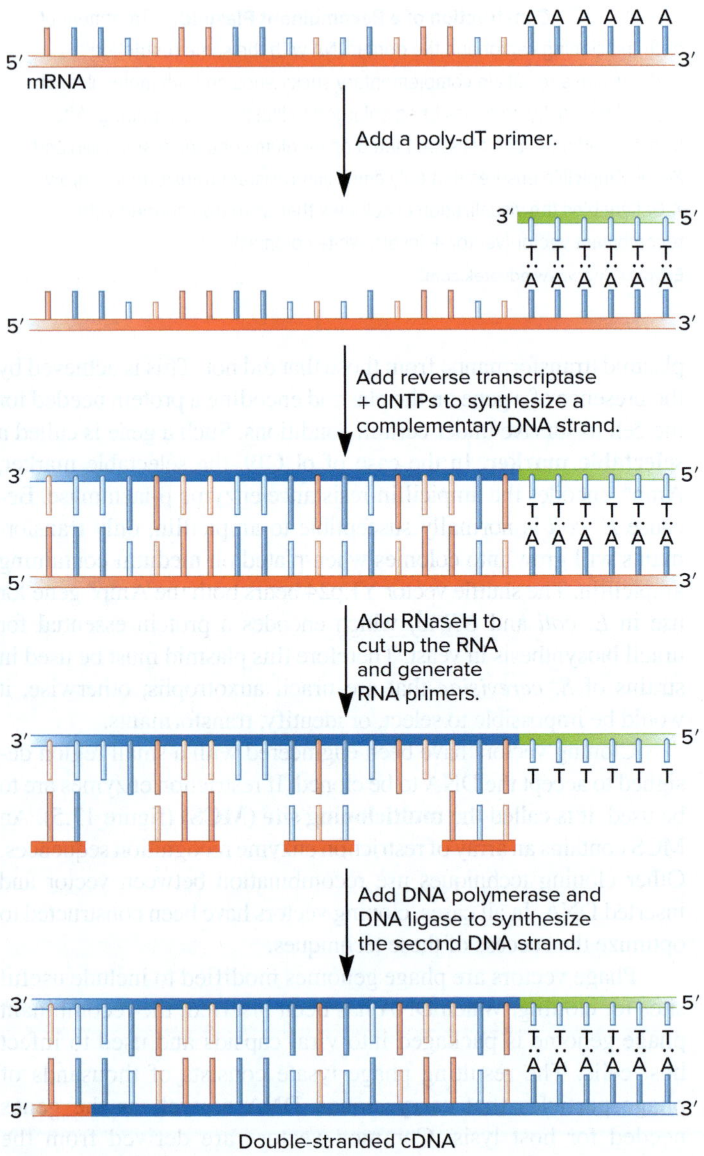

Figure 17.4 Synthesis of cDNA from mRNA. A poly-dT primer anneals to the 3′ poly-A tail of eukaryotic mRNAs. Reverse transcriptase catalyzes the synthesis of a cDNA. RNaseH digests the mRNA into short pieces that are used as primers by DNA polymerase to synthesize the second DNA strand. The DNA polymerase 5′ to 3′ exonuclease activity removes all of the RNA primers except the one at the 5′ end (because there is no primer upstream from this site). After the double-stranded cDNA is made, it can be inserted into vectors, as described in figure 17.3.

Artificial chromosomes are used for large DNA fragments, as when cloning all the genes that comprise a biosynthetic pathway. Like natural chromosomes, artificial chromosomes replicate only once per cell cycle. **Yeast artificial chromosomes (YACs)** consist of a **telomere** at each end, a centromere sequence, an origin of replication, a selectable marker, and an MCS to facilitate DNA insertion (**figure 17.6*a***). YACs are used when extraordinarily large DNA pieces (up to 1,000 kb) are to be cloned. **Bacterial artificial chromosomes (BACs),** figure 17.6*b,* accept DNA up to 300 kb and derive from the *E. coli* F fertility factor (*see figure 16.15*).

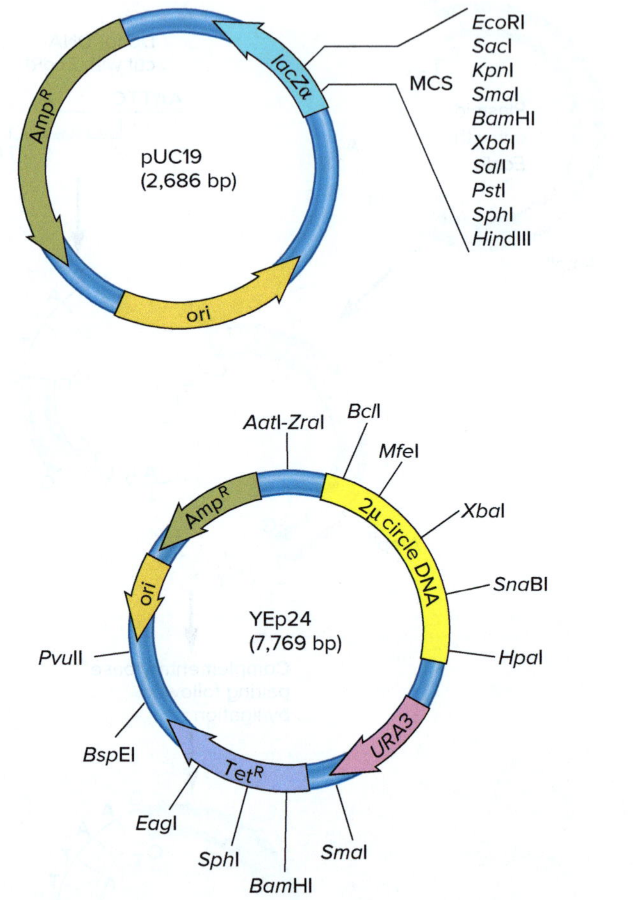

Figure 17.5 Cloning Vectors. Restriction sites present only once in each vector are shown. pUC19 replicates only in *E. coli,* while YEp24 replicates in both *E. coli* and *S. cerevisiae.* LacZα confers blue/white colony color depending on the presence of cloned DNA in the MCS (figure 17.3).

MICRO INQUIRY *Which plasmid is a shuttle vector? Why?*

Introducing Recombinant DNA into Host Cells

In cloning procedures, *E. coli* is the most frequent bacterial host and *S. cerevisiae* is most common among eukaryotes. There are several ways to introduce recombinant DNA into a host microbe. Chemical transformation and electroporation are two common techniques when the host microbe does not readily accept recombinant DNA. This is the case with most bacteria and archaea. In chemical transformation, the host cells are made competent by treatment with divalent cations and then transformed by heat-shocking the cells in a solution containing plasmid. ◀ *Environmental DNA brings new genes into cells (section 16.4)*

Electroporation is a simple technique for transforming bacteria, plant, and animal cells. In this procedure, cells are mixed with recombinant DNA and exposed to a brief pulse of high-voltage electricity. The plasma membrane becomes temporarily permeable and DNA molecules are taken up by some of the cells.

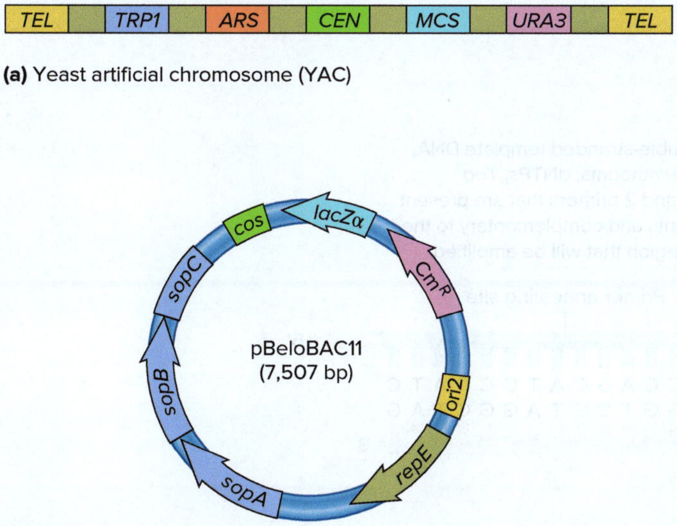

(a) Yeast artificial chromosome (YAC)

(b) Bacterial artificial chromosome (BAC)

Figure 17.6 Artificial Chromosomes. (a) A yeast artificial chromosome. *TRP1* and *URA3* complement tryptophan and uracil auxotrophies. *ARS,* origin of replication; *CEN,* centromere; *TEL,* telomere; MCS, multiple cloning site. (b) The bacterial artificial chromosome pBeloBAC11. Genes *repE* and *sopABC* ensure replication and partitioning in cell division. Cm^R, chloramphenicol resistance.

MICRO INQUIRY *In what ways does the BAC shown here differ from the plasmid pUC19 shown in figure 17.5?*

Comprehension Check

1. Describe restriction enzymes, sticky ends, and blunt ends.
2. What is cDNA? Why is it necessary to generate cDNA before cloning and expressing a eukaryotic gene in a bacterium?
3. You want to clone a 6 kb DNA fragment in *E. coli*. Which cloning vectors would be appropriate? Given the cloning vectors you choose, how will you select transformants?

17.2 Polymerase Chain Reaction Amplifies Targeted DNA

After reading this section, you should be able to:

a. Define the function of each of the three steps in a PCR cycle
b. Explain why PCR results in the amplification of a specific DNA sequence despite many competing sequences
c. Explain why PCR generates billions of products that are all the same size
d. Contrast real-time, quantitative PCR to end-point PCR and identify an application for each
e. Summarize the importance of PCR in biology
f. Explain how 5′ sequences of an oligonucleotide primer become included in a PCR product when they are not homologous to the template
g. Explain how primer design permits seamless cloning

The **polymerase chain reaction (PCR),** invented by Kary Mullis in the 1980s, changed the way genes are cloned, nucleic acids are sequenced, diseases are diagnosed, and crimes are solved. Why is PCR so versatile and important? Quite simply, it has moved the amplification process out of the cell. Rather than using restriction enzymes and a multicopy plasmid to produce a specific DNA, PCR enables the rapid synthesis of billions of copies of a DNA fragment. By adapting DNA replication to be carried out entirely in vitro, tasks involving multiple enzymatic reactions and bacterial growth are bypassed.

Figure 17.7 outlines PCR **amplification,** or repeated rounds of DNA replication. DNA polymerase requires a template to copy, primers to provide a free 3′-OH, and each of the four deoxyribonucleotide triphosphates (dNTPs). **Oligonucleotide primers** are chemically synthesized single-stranded DNA molecules, generally between 15 and 30 nucleotides long. Two oligonucleotide primers are designed so that each primer is complementary to one of the template strands, allowing for both strands to be copied. The primers are positioned so they flank the region to be amplified. Because primers must be complementary to flanking nucleotide sequences, only DNA with a known (or nearly known) sequence can be amplified by PCR.

PCR is a series of DNA replication reactions, called *cycles.* Each cycle precisely executes three steps in a machine called a *thermocycler.* In the first step, the double-stranded template DNA is denatured to single-stranded DNA by raising the temperature to about 95°C. Next the temperature is lowered to about 50°C so the primers can hydrogen bond (anneal) to their complementary regions on the template strands. Primers are included in the reaction at a huge molar excess because they must compete with the other template strand for binding. In the final step, the temperature is raised, usually to 68° to 72°C, so that DNA polymerase can extend the primers and copy the target DNA sequence using dNTPs. Only sturdy DNA polymerases that can tolerate the repeated temperature fluctuations are used in PCR. The most commonly used thermostable enzyme is ***Taq* polymerase** from the thermophilic bacterium *Thermus aquaticus.*

At the end of one PCR cycle, the targeted sequences on both strands have been copied. When the three-step cycle is repeated (figure 17.7), the two strands from the first cycle are copied to produce four DNA molecules. These are amplified in the third cycle to yield eight double-stranded products. Thus each cycle increases the number of target DNA molecules exponentially. Depending on the initial concentration of the template DNA and other parameters such as the G + C content of the DNA to be amplified, it is theoretically possible to produce about 1 million copies of targeted DNA sequence after 20 cycles and over 1 billion after 30 cycles. DNA fragments ranging in size from less than 100 base pairs to several thousand base pairs in length can be amplified, and the initial concentration of target DNA can be as low as 10^{-20} to 10^{-15} M.
DNA replication in bacteria (section 13.3) *Polymerase Chain Reaction*

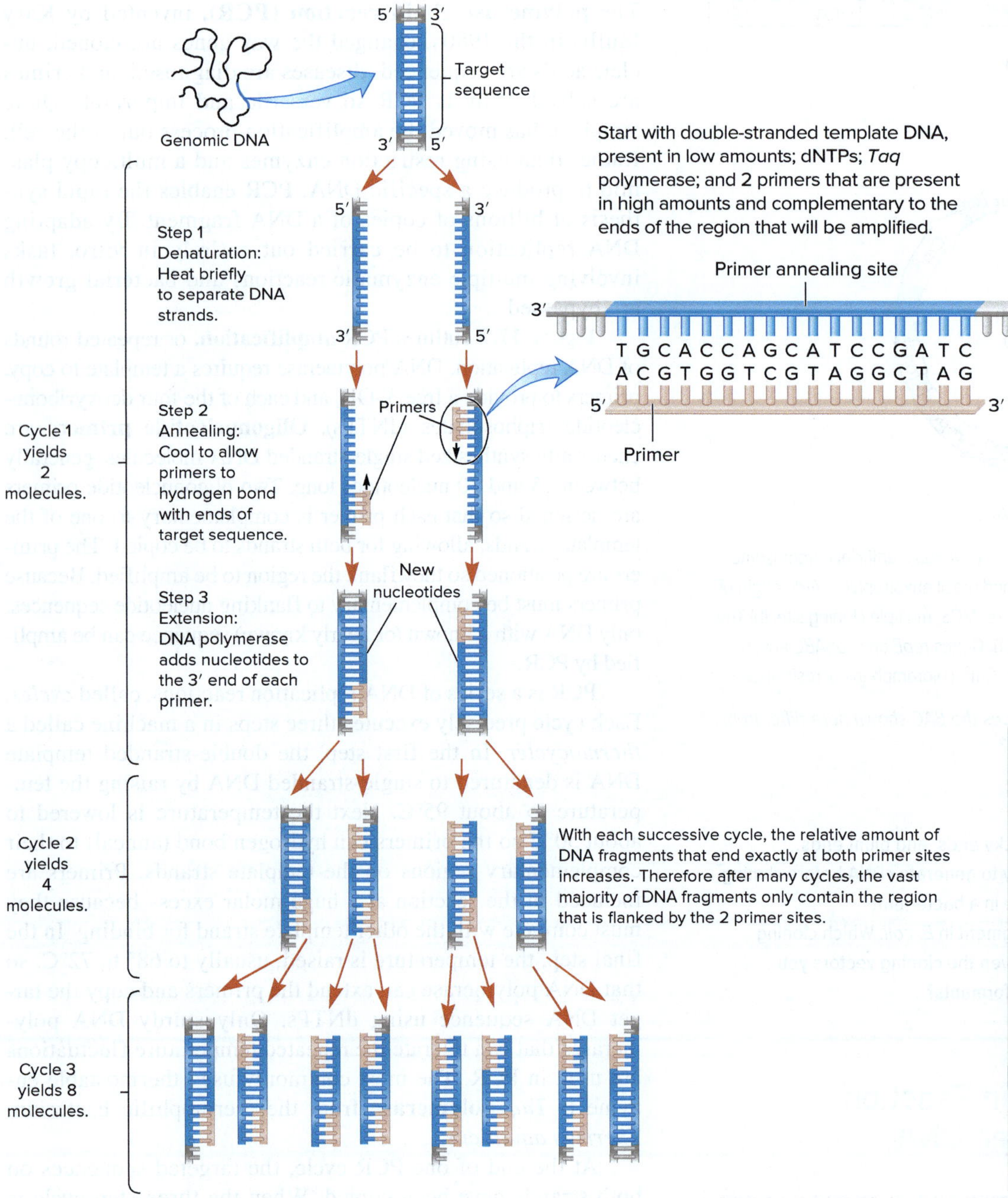

Figure 17.7 The Polymerase Chain Reaction (PCR). During each cycle, oligonucleotides complementary to the ends of the targeted DNA sequence hybridize to the DNA and act as primers for DNA replication. The region between the two primers is typically hundreds of nucleotides in length, not just several nucleotides as shown here. The net result of PCR is many copies of DNA in the region flanked by the two primers.

MICRO INQUIRY *Why, after three cycles, are the majority of amplified DNA molecules (i.e., PCR products) the size defined by the distance between the primers?*

If large quantities of a specific piece of DNA are needed, the PCR products are collected and purified at the end of a designated number of cycles. This is called **end-point PCR,** and the final number of DNA fragments amplified is not quantitative, meaning that the amount of final product does not reflect the amount of template DNA present at the start of the reaction. End-point PCR is useful when looking for the presence or absence of particular DNA, for example, in detecting a pathogen in a clinical laboratory or when synthesizing a DNA fragment to be cloned. In contrast, two variations on PCR are quantitative: **real-time PCR,** also referred to as **qPCR,** and **digital droplet PCR.** The amount of DNA or RNA template (which is converted to DNA with reverse transcriptase prior to starting PCR) present in a given sample can be determined. This is accomplished by adding a fluorescently labeled probe to the reaction mixture and measuring its signal. In qPCR, the product is measured during the initial cycles. This is when the rate of DNA amplification is exponential. However, as the cycles continue, substrates are consumed and polymerase efficiency declines. So although the amount of product increases, its rate of synthesis is no longer exponential (this is why end-point collection of PCR products is not quantitative). Specially designed thermocyclers record the amount of PCR product generated as it occurs, thus the term real-time PCR.

In contrast to qPCR that requires comparison to controls or a standard curve, digital droplet PCR gives absolute quantification of the target DNA. This is important in detecting rare sequences in a mixture or small quantities of a specific nucleic acid. The critical step in digital droplet PCR is partitioning the starting mixture into tens of thousands of individual reactions. Each reaction has zero or one template molecule. The proportion of reactions that give a fluorescent (positive) signal is subject to statistical analysis to give the precise quantitation.

PCR is an essential tool in many areas of molecular biology, medicine, and biotechnology. Because the primers used in PCR target specific DNA, PCR can isolate particular fragments of DNA (e.g., genes) from materials that contain many different genomes, such as soil, water, and blood. This explains why PCR has become an essential part of certain diagnostic tests, including those for COVID-19, *Chlamydia,* human papillomavirus, and other infectious agents. The tests are rapid, sensitive, and specific.

Seamless Cloning

One limitation to conventional cloning is that it relies on the presence of naturally occurring restriction sites in DNA. A modification to PCR allows the addition of custom restriction sites onto the ends of DNA to be cloned, thereby simplifying DNA assembly tasks. This is accomplished by synthesizing primers that in addition to the nucleotide sequences needed to initiate synthesis of the target gene also have a restriction endonuclease recognition sequence on the 5′ end (**figure 17.8*a***). The result is a DNA fragment copied exactly from the template, with the addition of convenient restriction sites at each end.

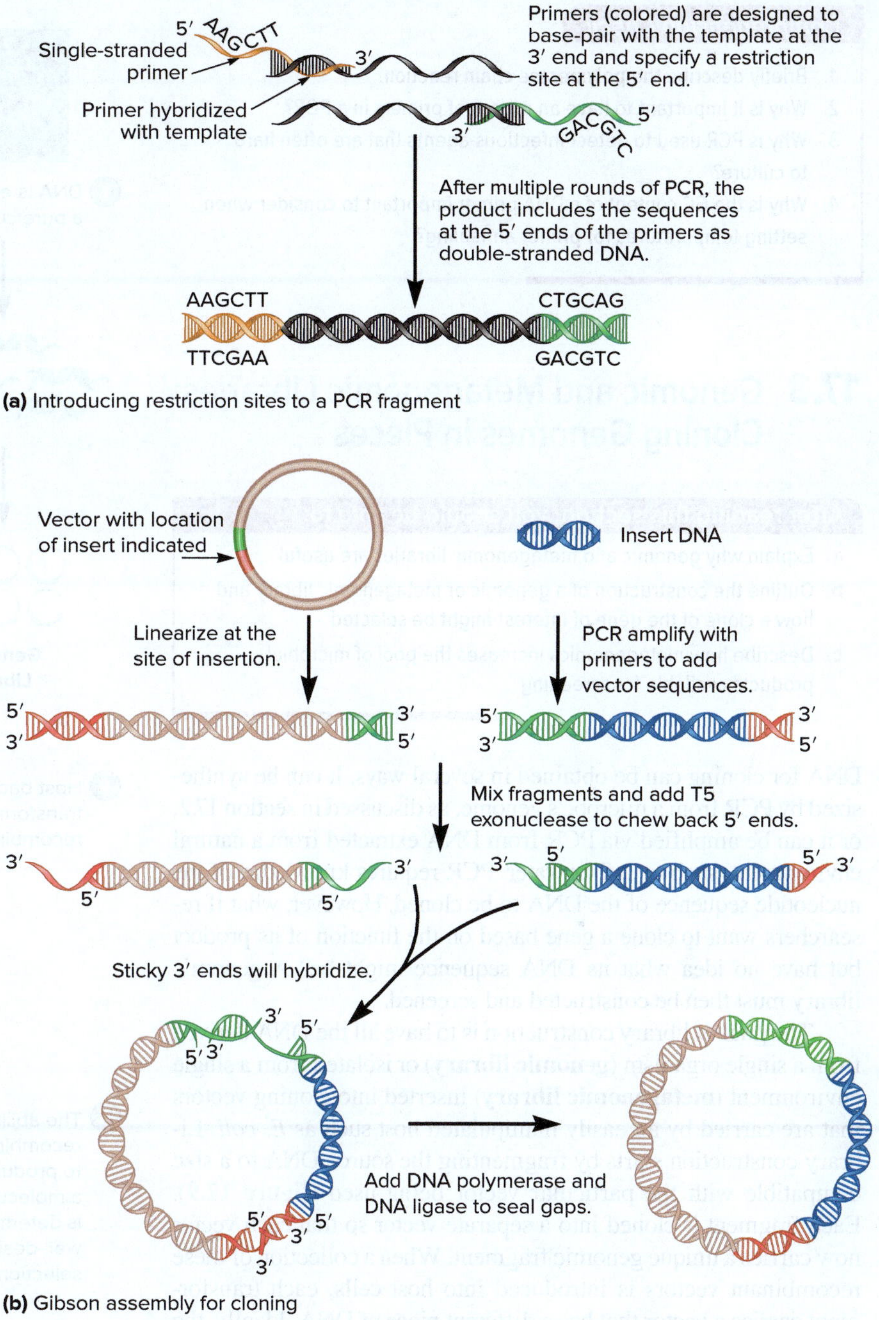

(a) Introducing restriction sites to a PCR fragment

(b) Gibson assembly for cloning

Figure 17.8 Polymerase Chain Reaction Using Primers with a 5′ Extension.

Taking this idea one step further, Daniel Gibson developed a method to eliminate the use of restriction endonucleases entirely. His method, termed Gibson assembly, uses the 5′ ends of oligonucleotide primers to specify the desired junction between the molecules. In a series of reactions diagrammed in figure 17.8*b,* a vector and insert fragment can be joined efficiently. Because the resulting product does not retain any scars or seams from restriction sites in the final product, this approach is sometimes called *seamless cloning.*

With the advent of seamless cloning techniques, scientists are free to envision a DNA molecule and create it in vitro. The procedure diagrammed in figure 17.8*b* can be expanded to assemble multiple fragments by amplifying each fragment with primers that specify the junction between it and its intended neighbor. An entire operon can be synthesized from promoter and regulatory sequences and genes. Indeed, entire genomes have been synthesized by Gibson assembly.

Comprehension Check

1. Briefly describe the polymerase chain reaction.
2. Why is it important to have an excess of primers in a PCR?
3. Why is PCR used to detect infectious agents that are often hard to culture?
4. Why is the GC content of a DNA primer important to consider when setting temperatures for primer annealing?

17.3 Genomic and Metagenomic Libraries: Cloning Genomes in Pieces

After reading this section, you should be able to:

- **a.** Explain why genomic and metagenomic libraries are useful
- **b.** Outline the construction of a genomic or metagenomic library and how a clone of the gene of interest might be selected
- **c.** Describe how metagenomics increases the pool of microbial products available for screening

DNA for cloning can be obtained in several ways. It can be synthesized by PCR from a microbe's genome, as discussed in section 17.2, or it can be amplified via PCR from DNA extracted from a natural environment, such as soil or water. PCR requires knowledge of the nucleotide sequence of the DNA to be cloned. However, what if researchers want to clone a gene based on the function of its product but have no idea what its DNA sequence might be? A genomic library must then be constructed and screened.

The goal of library construction is to have all the DNA derived from a single organism (**genomic library**) or isolated from a single environment (**metagenomic library**) inserted into cloning vectors that are carried by an easily manipulated host such as *E. coli*. Library construction starts by fragmenting the source DNA to a size compatible with the particular vector being used (**figure 17.9**). Each fragment is cloned into a separate vector so that each vector now carries a unique genomic fragment. When a collection of these recombinant vectors is introduced into host cells, each transformant carries a vector that has a different piece of DNA. Ideally, the entire organism's genome or all the DNA from the environmental sample is represented in the collection of recombinant vectors.

Libraries are often constructed to find a gene or genetic element (e.g., small noncoding RNA) with a specific function. In a functional screen, the genes are expressed in microbes such as *E. coli*, which are then screened for the acquisition of a new, specific function. For example, to find genes that encode enzymes needed for the biosynthesis of the amino acid tryptophan from a newly isolated bacterium, the library could be expressed in an *E. coli* tryptophan auxotroph. Colonies that now grow without added tryptophan are good candidates to carry the library fragment with the tryptophan biosynthetic genes. Success with this approach depends on the activity of the cloned gene product being similar in both organisms. The **genetic**

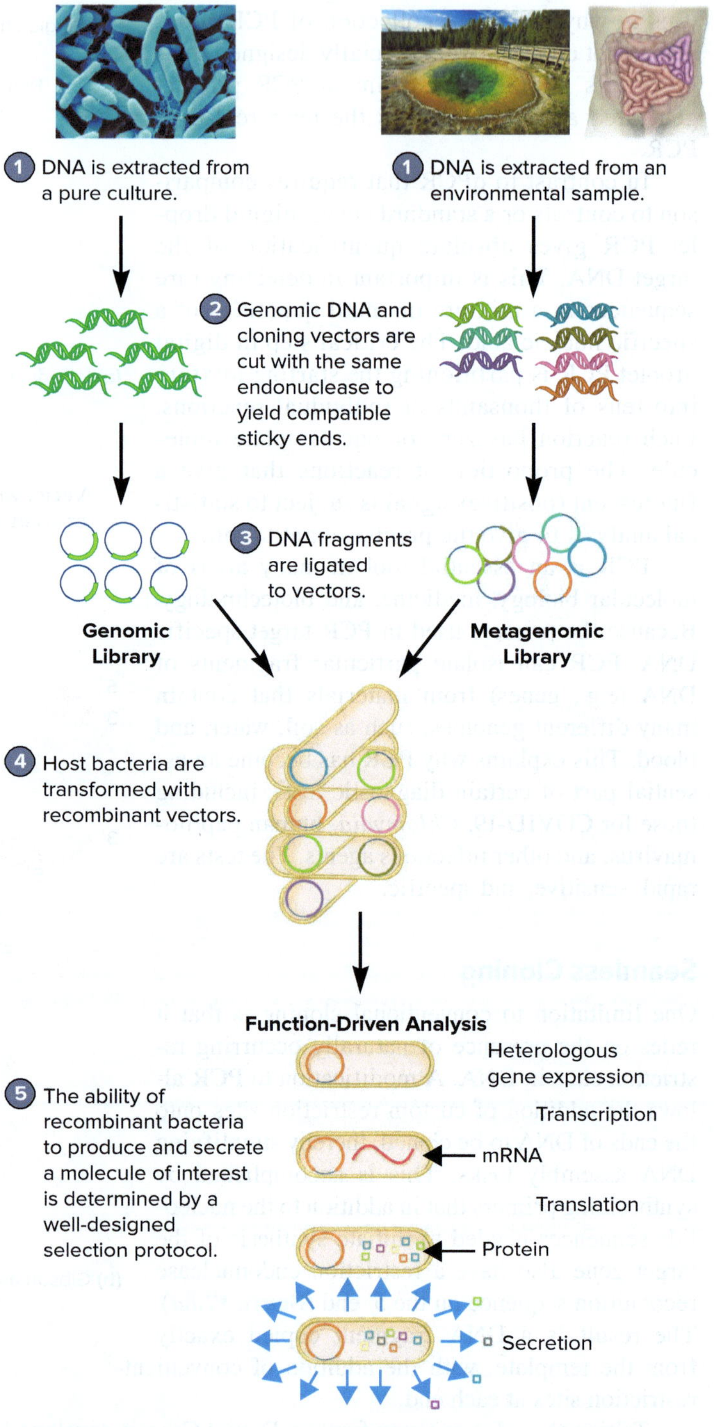

Figure 17.9 Functional Screening of Genomic and Metagenomic Libraries. A genomic library is made by cloning fragments of an entire genome into a vector (top left). In contrast, a metagenomic library begins with fragments derived from total environmental DNA (top right). Both types are constructed and expressed in host bacteria, which are screened for the desired protein function. This is usually performed in a high-throughput fashion.

(top left) Mediscan/Alamy Stock Photo; (middle) Yi Xiang Yeng/iStock/Getty Images

MICRO INQUIRY *Why are long fragments (e.g., 20 kb) of genomic DNA often desired when constructing a genomic library?*

complementation of a deficiency in the host cell is sometimes called **phenotypic rescue** because the mutant phenotype is (at least partially) restored, or rescued, by the cloned DNA.

17.4 Expressing Foreign Genes in Host Cells

After reading this section, you should be able to:

a. Explain the utility of expression vectors
b. Outline the procedure by which a His-tagged protein is generated in vivo and purified in vitro
c. Summarize the role of GFP in protein analysis, and differentiate between a transcriptional and a translational GFP fusion

When genes from one organism are cloned into another and then transcribed and translated into protein, it is called **heterologous gene expression.** An important and early example of heterologous gene expression was the cloning and expression of the human insulin genes in *E. coli.* Heterologous gene expression enables the production of specific proteins and peptides without contamination by other products that might be synthesized in the original organism.

To be expressed, a heterologous gene must have components compatible with host transcription and translation. For instance, if the host is a bacterium and the gene has been cloned from a eukaryote, a bacterial promoter and ribosome binding site must be provided and introns removed.

The problems of expressing recombinant genes in host cells are largely overcome with the help of specially constructed **expression vectors.** These vectors enable high-level transcription because the gene is inserted adjacent to a strong promoter. Often expression vectors contain regulatory regions of the *lac* operon (or another inducible promoter) so that expression of the cloned genes can be controlled by the addition of inducer molecules. This is important because sometimes high-level expression of a heterologous protein can be toxic to the host cell. ◀ *Regulation of transcription initiation saves considerable energy and materials (section 14.2)*

Production of nonnative proteins often disrupts the host cell's balance of redox power (e.g., NADH/NAD^+) and ATP. Therefore it is frequently necessary to optimize the cell's regulatory and metabolic activities so production is maximized. This is called **metabolic engineering.** The application of "omic" data (e.g., genomic, transcriptomic, proteomic, and metabolomic data) to guide genetic manipulation of the host strain for this purpose is called **systems metabolic engineering.**

Purification and Study of Recombinant Proteins

It is often necessary to isolate the protein product of the cloned gene so that its structure and activity can be studied. In addition, it is sometimes desirable to determine the subcellular localization of a protein. Here we discuss some ways in which proteins can be purified and visualized in living cells.

Protein Purification

A common method by which proteins are purified is called **polyhistidine tagging,** or simply **His-tagging.** His-tagging involves adding a series of histidine amino acid residues to either the N- or C-terminus of a protein. Most often six residues are added; this is called a 6×His-tag. Histidine is used because it has a high affinity for metal ions, so a protein bearing multiple consecutive histidine residues will bind preferentially to a solid-phase material, called a resin, which has exposed nickel or cobalt atoms. This permits separation of the His-tagged protein from other cellular constituents. The process of purifying a His-tagged protein is diagrammed in **figure 17.10.** Depending on the purification system used, the His-tag can be cleaved from the protein while it is attached to the resin, or it can be removed following the release of the protein. Often the protein is functional with the His-tag, so it may not require removal. His-tagging is used if the protein is soluble and is found in the cytoplasm. While His-tagging is a popular approach to purifying recombinant proteins, other approaches exist, especially for purifying membrane-associated proteins.

Fluorescence Labeling

What if the goal of cloning a gene is to study the regulation or function of the protein product in vivo? It is possible to visualize the activity of a specific promoter (i.e., when the promoter is on or off) as well as observe the localization of a protein by fluorescence microscopy. The development of fluorescent labeling of living cells began when it was discovered that the jellyfish *Aequorea victoria* produces a protein called **green fluorescent protein (GFP).** GFP is encoded by a single gene that, when translated, undergoes self-catalyzed modification to generate a strong, green fluorescence. This means that the *gfp* gene is easily cloned and expressed in any organism of choice. An entire palette of fluorescent labels is available, including variants of GFP that glow throughout the blue-green-yellow spectrum.

There are two ways in which fluorescent peptides can be used to study gene regulation. A *transcriptional fusion* allows the investigator to observe when a gene is active by replacing the protein coding sequence of the gene of interest with the fluorescence-encoding gene (**figure 17.11*a***). This places the expression of the fluorescence gene under the control of the promoter of the replaced gene. Thus when the promoter is active, it directs transcription of the fluorescence gene and the cells glow. If the transcriptional fusion has been made in a multicellular organism, the specific cell type in which the gene is normally expressed is also determined. On the other hand, a *translational fusion* determines where a particular protein is localized in the cell. Translational fusions add the GFP-coding sequence to the end of the structural gene of interest. This creates a chimeric protein—a protein that consists of two parts: the protein being studied and GFP. Now whenever that protein is made, and wherever it goes in the cell, it will fluoresce. Of course, care must be taken to ensure that the protein fusion still functions like the original protein. Once it has been determined that the chimeric protein is fully functional, the cellular "address" of the protein can be determined (figure 17.11*b*).

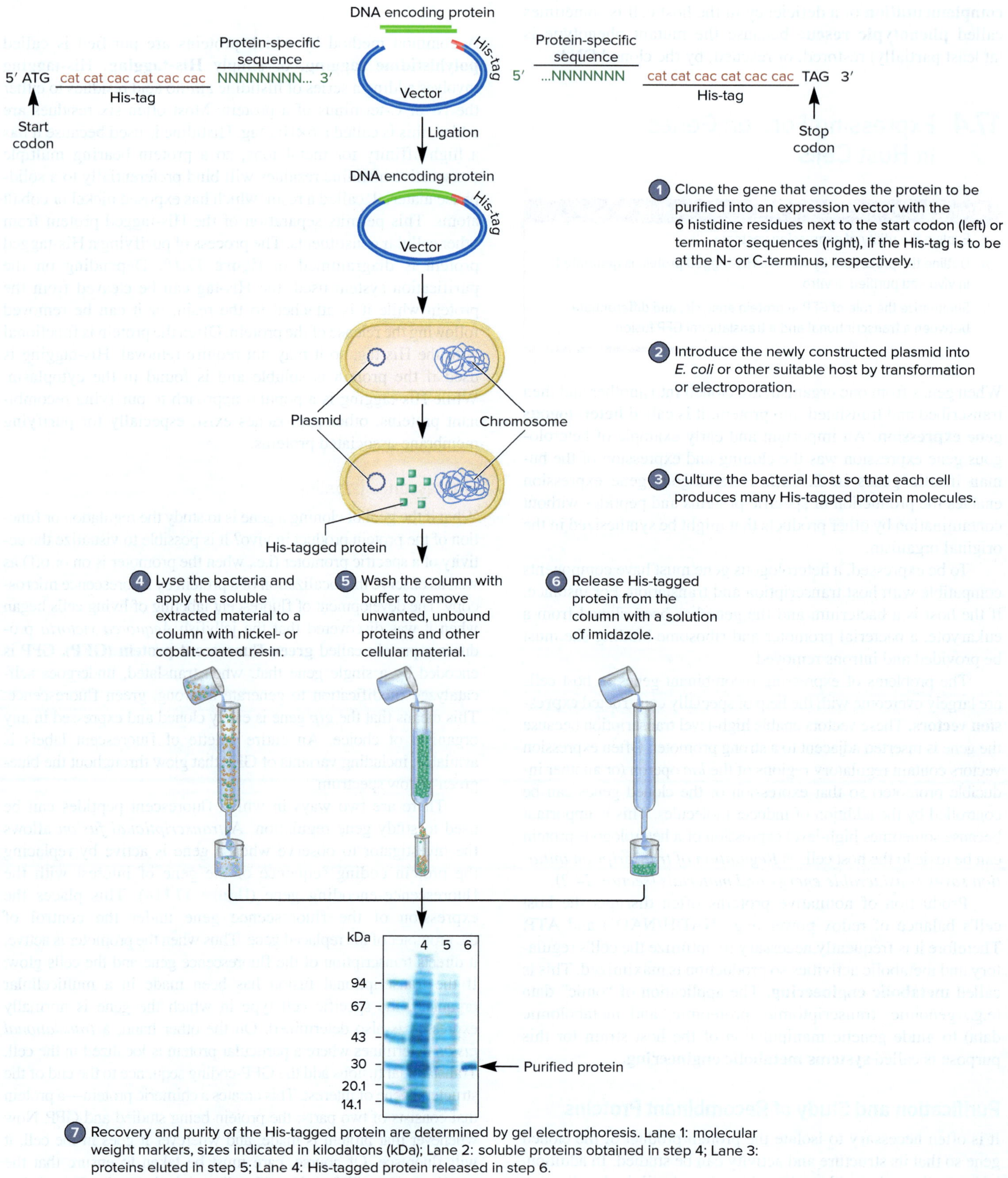

Figure 17.10 Construction and Purification of a Polyhistidine-Tagged Protein.

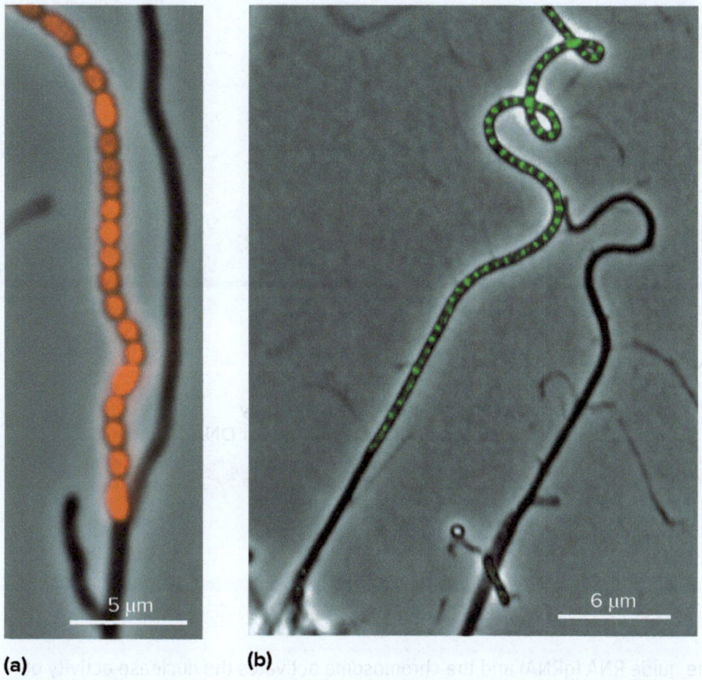

Figure 17.11 Fluorescent Protein Labeling. (a) A transcriptional fusion in which the mCherry gene is controlled by the promoter that normally governs transcription of a gene in *Streptomyces coelicolor* expressed only when aerial hyphae form cross walls (septa) that give rise to a chain of spores. Thus red fluorescence is seen uniformly throughout the chain of spores but not in hyphae without septa. (b) A translational fusion in which the *ftsZ* gene is fused to GFP. This shows that FtsZ localizes at the septa. Notice that the filament to the right is not forming septae, so FtsZ is not present.
(a) Mark Buttner; (b) Klas Flärdh

MICRO INQUIRY *What special considerations are necessary if one is constructing a translational fusion of a protein known to be exported?*

Comprehension Check

1. Describe how a genomic library might be screened for a gene that confers the production of an extracellular enzyme that degrades casein, a protein found in milk. (Hint: Agar with skim milk loses its opacity when casein is degraded.)
2. Why is a genomic library prepared from a eukaryote made with cDNA?
3. List several reasons why a cloned gene might not be expressed in a heterologous host cell.
4. Think of a situation in which you might want to keep the His-tag on a recombinant protein and a situation in which it would be important to remove it.
5. You are studying chemotaxis proteins in a newly described bacterium. You have cloned a gene that encodes the CheA protein. You need to be sure that this protein localizes to the inside face of the plasma membrane (*see figure 14.22*). Will you use a transcriptional or translational fusion? Explain your choice.

17.5 Cas9 Nuclease Is a Programmable Tool for Genome Editing

After reading this section, you should be able to:

a. Distinguish the DNA recognition features of restriction endonucleases and Cas9 nuclease
b. Explain how Cas9 nuclease can be directed to cut at a unique site in a genome
c. Diagram how a specific mutation or a new gene may be inserted into a chromosome by homologous recombination

Cas9 genome editing has rapidly become one of the most widely used tools for altering genomes in vivo. Cas9 genome editing is often referred to as **CRISPR** or **CRISPR/Cas9,** referencing the bacterial genome elements from which it was developed. Cas9 nucleases are encoded in the genomes of most bacteria and archaea, where they are usually adjacent to a CRISPR locus, *c*lustered *r*egularly *i*nterspaced *s*hort *p*alindromic *r*epeats. As the mechanistic details of Cas9 function were discovered, two research groups, one led by Jennifer Doudna and Emmanuelle Charpentier (who were awarded the 2020 Nobel Prize in Chemistry) and the other by Feng Zhang, sought to adapt Cas9 for genome editing. In this process, genomic DNA can be directly modified and the procedures are general enough to be used for any cell into which DNA can be introduced and expressed. ◀ *Responses to viral infection (section 14.6)*

Like restriction enzymes, Cas9 is an endonuclease that cuts both strands of a target DNA. However, unlike restriction enzymes, which recognize four to eight base pairs through contacts between the DNA and the enzyme active site, Cas9 is a ribonucleoprotein consisting of a polypeptide and a **guide RNA (gRNA).** Recognition of target DNA for cleavage occurs by hybridization of about 20 bases between the gRNA and its complementary DNA sequence in the genome (**figure 17.12**). A second element required for recognition is a short series of bases, the **protospacer adjacent motif (PAM),** located next to the hybridizing region on the opposite DNA strand. The extreme specificity conferred by the gRNA is the key to genome editing because each 20-base target sequence almost certainly occurs only once in any given genome. In contrast, a restriction enzyme that recognizes a few nucleotides will cut the genome, on average, every few thousand bases.

In microbes, the CRISPR system evolved to attack DNA from mobile genetic elements (MGEs) that enter the cell. The CRISPR locus encodes the gRNA (*see figure 14.28*), which is based on nucleotide sequences encountered in MGEs. For genome editing in bacteria, the cell's own CRISPR system may need to be inactivated to avoid unintended cross-reactivity. As eukaryotes do not have CRISPR systems, this is not a concern.

Cas9 can be engineered to carry gRNAs with specified nucleotide sequences, thereby programming the recognition sequence for the nuclease. The gRNA directs Cas9 to hybridize

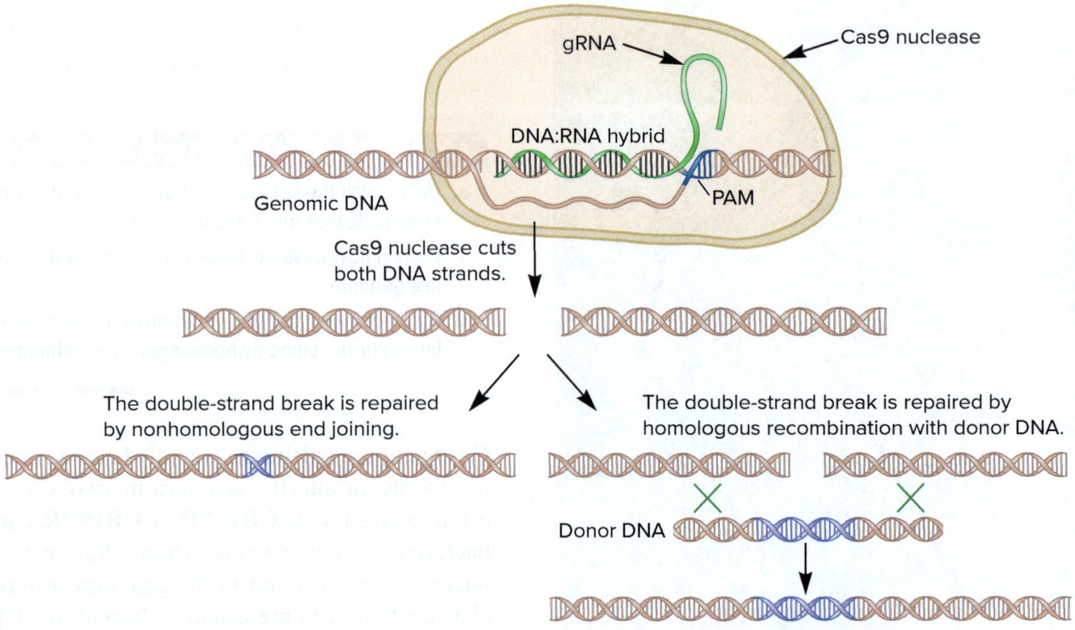

Figure 17.12 Genome Editing with Cas9 Nuclease. Hybridization between the guide RNA (gRNA) and the chromosome activates the nuclease activity of Cas9. PAM, protospacer adjacent motif.

MICRO INQUIRY *How could you assemble the donor DNA molecule for homologous recombination?*

with a defined site in a genome, making it the most precise mechanism available for targeting and cutting DNA.

Genome editing begins by introducing the two components of the mature Cas9 endonuclease, the **apoenzyme** and the gRNA, to the host cells. These molecules may be added directly, or they may be added as cloned DNA regulated by an inducible promoter. In the latter case, upon induction, the Cas9-gRNA complex assembles and performs its nuclease function.

Figure 17.12 illustrates how Cas9 recognizes and hydrolyzes a specific DNA sequence. A portion of the gRNA protrudes from the enzyme, available for hybridization. Upon locating its complement, the gRNA induces a conformational shift in the nuclease (protein) portion of Cas9, which then hydrolyzes phosphodiester bonds in both DNA strands, leaving blunt ends. In the simplest case, a point mutation occurs as the cell attempts to repair the damage. Some microbes and all eukaryotes have a **nonhomologous end joining** system to rejoin the two chromosome pieces. If the repair re-creates the original sequence, it is again susceptible to Cas9 cutting. As a result, imperfect repairs with a deletion or insertion of a few base pairs is the typical outcome. A limitation to this method is that the outcome differs in each cell.

In many microbial cells, the double-strand chromosome break is lethal and the only recourse for survival is **homologous recombination.** Recombination with another wild type chromosome copy merely generates another substrate for Cas9, but introducing a related DNA molecule allows for the introduction of a defined mutation. A donor sequence must be constructed in vitro using cloning techniques. The donor DNA includes chromosomal regions flanking the gRNA binding site to provide regions of homology for recombination. The donor sequence is introduced to the cell at the same time as the Cas9/gRNA molecules. Homologous recombination will repair the double-stranded break and at the same time incorporate the donor molecule into the genome. Selection and screening procedures to identify mutants are not required with Cas9 gene editing, unlike traditional cloning. Cells into which the Cas9-gRNA complex has been introduced survive and reproduce only if the original target sequence is modified. ◀ *Molecular recombination: joining DNA molecules (section 16.3)*

Researchers have developed other applications for Cas9 nuclease that take advantage of its ability to bind chromosomal DNA in a single genomic location. An altered version of Cas9, termed dead Cas9 (dCas9), retains the DNA localization function conferred by the gRNA but no longer cuts DNA. Instead, a dCas9-gRNA molecule bound to DNA can serve as a platform for other enzymes. For example, in the process termed CRISPR interference (CRISPRi), dCas9-gRNA bound at a gene inhibits its expression by blocking the movement of RNA polymerase. Genome base editing has been demonstrated using dCas9-gRNA fused to DNA-modifying enzymes. Because the dCas9 variant is used, there is no nuclease activity to break the phosphate-sugar backbone, but the enzymes directly modify the base—for example, from C to T. The outcome is a precise change to the DNA sequence. Experiments like these have demonstrated the versatility of the dCas9-gRNA module as a homing device to direct a range of effector molecules to a specific chromosomal location. ◀ *Transcription in bacteria (section 13.5); Regulatory proteins often control transcription initiation (section 14.2)*

Cas9 genome editing is a rapidly developing technique that is continually being modified and adapted. Projects as varied as removing allergens from peanuts to curing genetic diseases in humans have been proposed as examples of what might be achievable. Importantly, Cas9 technology opens the prospect of engineering the human genome in germ line cells (e.g., cells that develop into egg cells or sperm cells), thereby inviting discussions on many fronts to address ethical concerns.

Comprehension Check

1. Compare the mechanisms used by restriction endonucleases and Cas9 nucleases to recognize a specific DNA sequence.
2. Commercially grown mushrooms are sensitive to handling during harvest and may rapidly discolor, a reaction catalyzed by a single enzyme. Describe a strategy to extend the shelf life of mushrooms using Cas9 genome editing.
3. A dinucleotide (e.g., GA or CC) appears in DNA with an average frequency of 1 in 16 (4^2) bases, and a trinucleotide appears with an average frequency of 1 in 64 (4^3) bases. Calculate how frequently you would expect a 20-bp sequence to occur. Compare this number to the size of the human genome (3×10^9 bp).

17.6 Biotechnology Develops Custom Microbes for Industrial Use

After reading this section, you should be able to:

a. Compare and contrast three directed evolution technologies
b. Describe the goal of synthetic biology and its advantages over more conventional techniques

With the availability of tools for nucleic acid manipulation and vast numbers of sequences, there are few technical limits to manipulating the genome of living organisms. Here we present approaches for modifying cells to harness the broad range of biological capabilities.

Directed Evolution

Historically, industrial strain development involved repeated rounds of induced mutagenesis and mutations that occur during transfer from culture to culture. **Production strains,** those designed to overproduce a specific product, have a genotype quite different from the original isolate. Until the advent of genomic sequencing, the number and sites of most mutations remained a mystery. **Directed evolution** takes a rational approach to strain development because only genes of interest are targeted for mutagenesis. In many cases, the sequences are altered in vitro, then replaced in the original strain or a different host microbe that may be better suited to produce the desired compound at high levels. Biotechnologists have discovered that making even minor amino acid substitutions can lead to unexpected changes in protein function and stability, resulting in higher yields of existing products or products with new, desirable features. Regulatory sequences can also be altered, resulting in the overproduction of the desired product.

Several technical advancements were key to the development of direct evolution methodologies. The first is **site-directed mutagenesis.** In this technique, the nucleotide sequence of a specific gene is altered. The steps are outlined in **figure 17.13.** Once a desired mutation has been created, it can be moved to the organism of interest through one of the methods of horizontal gene transfer.

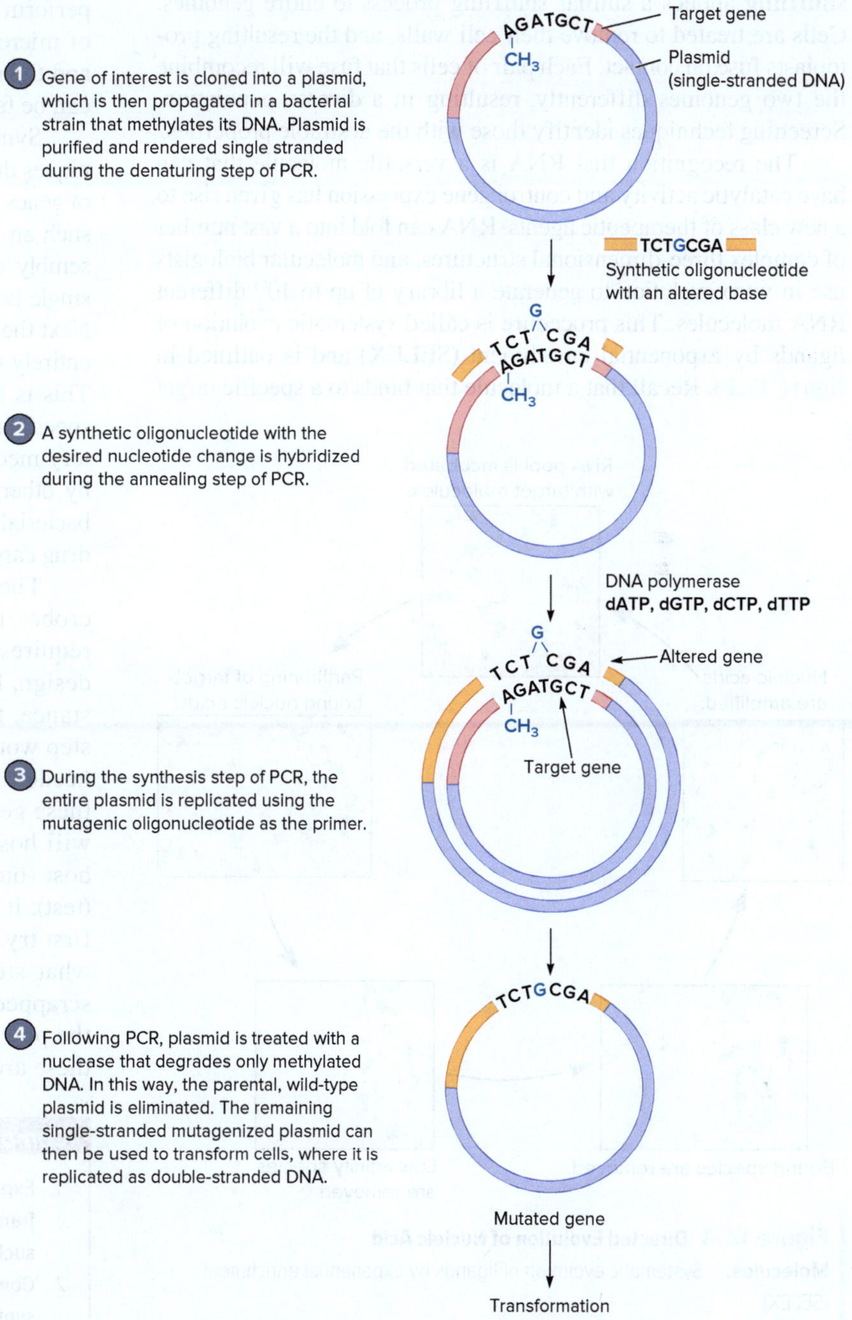

Figure 17.13 Site-Directed Mutagenesis. A synthetic oligonucleotide is used to make a specific change to a gene.

A variety of other directed evolution techniques have been developed that can make larger changes to the gene of interest. **DNA shuffling** takes advantage of homologous genes from multiple sources, breaks them into pieces, and shuffles those pieces to make novel combinations. The pool of genes is randomly fragmented and homologous recombination in vitro generates new alleles. Those encoding desirable traits are identified and the process may be repeated to further increase the diversity.

Many natural products are not proteins and therefore cannot be modified by mutating a single, structural gene. Instead, altering these products requires an understanding of the biosynthetic pathway by which they are catalytically assembled. **Whole-genome shuffling** applies a similar shuffling process to entire genomes. Cells are treated to remove their cell walls, and the resulting protoplasts fuse on contact. Each pair of cells that fuse will recombine the two genomes differently, resulting in a diverse population. Screening techniques identify those with the desirable properties.

The recognition that RNA is a versatile molecule that can have catalytic activity and control gene expression has given rise to a new class of therapeutic agents. RNA can fold into a vast number of complex three-dimensional structures, and molecular biologists use in vitro evolution to generate a library of up to 10^{15} different RNA molecules. This procedure is called *s*ystematic *e*volution of *l*igands by *ex*ponential enrichment (SELEX) and is outlined in **figure 17.14.** Recall that a molecule that binds to a specific target or receptor is called a ligand, so in this case, the goal is to design an RNA ligand that has biological activity because it fits very tightly into a cleft or groove in a specific target. This class of engineered RNAs has been given the name **aptamers** (Latin *aptus*, fitting). ◀ *Ribozymes: catalytic RNA molecules (section 10.6)*

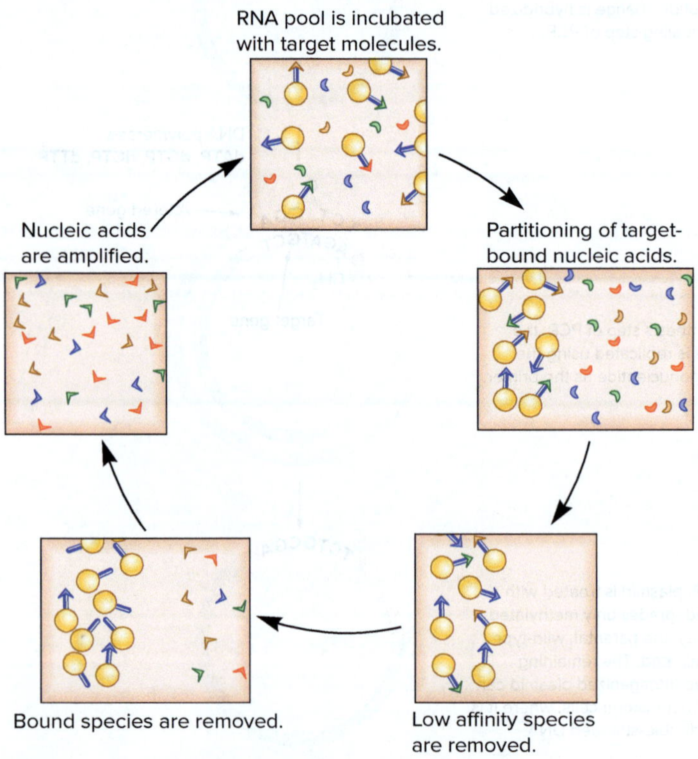

Figure 17.14 Directed Evolution of Nucleic Acid Molecules. Systematic evolution of ligands by exponential enrichment (SELEX).

MICRO INQUIRY *Why is SELEX considered an enrichment technique?*

Synthetic Biology

Synthetic biology looks beyond the constraints of what is already present in an organism and constructs or repurposes entirely new capabilities. Simply put, the field of synthetic biology involves the forward engineering of biological systems (plants, microbes) to perform completely novel functions. Examples include the design of microorganisms that seek out and kill cancerous tumor cells and the redesign of photosynthesis to produce excess energy that can be fed into existing electrical grids.

Synthetic biology offers three chief advantages over the techniques discussed so far. It is more efficient to assemble large arrays of genes encoding enzymes that perform complex tasks than to find such an assembly already existing in nature. For example, the assembly of multiple genes from plants, fungi, and bacteria into a single host microorganism has resulted in a range of new biofuels. Next the need for metabolic engineering can be limited or reduced entirely when new genes are synthesized in vitro for a specific host. This is because the genes can be modified in vitro to meet host specifications. Lastly, synthetic biology can mix-and-match regulatory mechanisms with the expression of genes normally governed by other means. An interesting example is the development of a bacterial strain that seeks out tumor cells and releases its anticancer drug cargo when triggered by a quorum-sensing regulator.

There is, however, no set of instructions for designing microbes to perform completely novel functions. Each project requires careful planning followed by the iterative process of design, build, test, redesign, rebuild, retest, and so on. For instance, to engineer a bacterium to produce a biofuel, the first step would be to review literature to determine what genes are needed (planning). The next step would be to figure out how these genes are going to be cloned and what microbial species will host them (design). Once the genes are inserted into the host (build), the amount of biofuel made would be quantified (test). It is safe to say that perfection is never attained with the first try. However, the results inform the synthetic biologist of what steps need revision (or if the entire design should be scrapped and a fresh approach taken). Synthetic biology has the potential to generate a vast diversity of products. Some of these are illustrated in **Techniques & Applications 17.1.**

Comprehension Check

1. Explain how directed evolution techniques fit into the general framework of having a diverse population under selective pressure such that only the most fit for the environment will survive.
2. Compare and contrast the development of a new product using synthetic biology and functional screening of metagenomic libraries.

TECHNIQUES & APPLICATIONS

17.1 Animal-Free Foods

The meaty flavor of that grilled plant-based burger and the cheesy vegan topping that you melted onto it—all derive from fungal fermentation! Food science meets consumer demand for animal-free foods and the result is ingredients produced by precision fermentation.

Pichia pastoris and *Trichoderma reesei* are workhorses of synthetic biology: They grow to a high cell density in fermentation vats on inexpensive media and are popular platforms for assembling the DNA components for a vast collection of biological molecules. These fungi have few secreted proteins and low toxicity and allergenicity. As eukaryotic cells, they are capable of many post-translational modifications required for recombinant protein activity.

Leghemoglobin (LegH) from soybeans is the not-so-secret ingredient to the burger's meaty flavor. It's a monomeric protein with an iron-containing heme group (*see figure 10.10*). When cooked, LegH denatures and releases its heme, which then reacts with other components to generate the flavor and aroma that we associate with meat. ▶| *Rhizobia are symbionts of leguminous plants (section 30.3)*

It's far more cost-effective to produce LegH in *P. pastoris* than to isolate it from soybeans. This requires cloning the gene encoding the protein portion and the nine enzymes that assemble the heme group into the yeast. All these genes need to be overexpressed to optimize LegH production. Because glycine starts the heme biosynthetic pathway, its levels must be balanced between cell growth and heme synthesis. This complex regulatory system was worked out by **metabolomics,** which identified *P. pastoris* genes to be up- or down-regulated. Genome editing technologies make these changes feasible.

As for the vegan cheese, the milk protein casein is singularly responsible for cheese melting and stretching when heated. *T. reesei* has been engineered to synthesize and secrete recombinant casein. When taken through the cheese-making process, this casein results in a nondairy cheese indistinguishable from the original. Because it is nondairy, the product is a good option for consumers with lactose intolerance. While the details of recombinant casein production in *T. reesei* remain proprietary, multiple phosphorylation and glycosylation reactions occur after casein synthesis, and it is reasonable to expect co-expression of enzymes that modify the recombinant casein substrate.

DENNIS KUNKEL MICROSCOPY/SCIENCE PHOTO LIBRARY/Alamy Stock Photo

RHJPhtotos/Shutterstock

Key Concepts

17.1 Cloning Technologies Developed from DNA Structure and Enzymology

- Genetic engineering developed from essential methods in nucleic acid enzymology.
- Restriction enzymes cut DNA at specific sequences, thereby releasing fragments of DNA that can be cloned or otherwise manipulated (**figure 17.2** and **table 17.1**).
- There are four types of cloning vectors: plasmids, viruses, cosmids, and artificial chromosomes. Cloning vectors have a minimum of three components: an origin of replication, a selectable marker, and a multicloning site (**figures 17.5** and **17.6**).
- A common approach to cloning is to digest both vector and DNA to be inserted with the same restriction enzyme or enzymes so that compatible sticky ends are generated. DNA ligase catalyzes the formation of phosphodiester bonds once the DNA fragment inserts into the vector.
- DNA can be introduced into microbes by transformation or electroporation.
- The bacterium *E. coli* and the yeast *Saccharomyces cerevisiae* are the most common host species.

17.2 Polymerase Chain Reaction Amplifies Targeted DNA

- The polymerase chain reaction (PCR) allows small amounts of specific DNA sequences to be amplified, or increased in concentration thousands of times (**figure 17.7**).
- PCR consists of multiple cycles of three steps each: DNA denaturation, primer annealing, and DNA synthesis.
- PCR can be performed to quantitate the product by two variations with differing levels of precision. Real-time PCR (qPCR) follows synthesis of a specific product in the early cycles when synthesis is linear. Digital droplet PCR partitions the reaction in many different droplets and quantitates based on the fraction that shows amplification.
- PCR has numerous applications. It often is used to obtain genes for cloning and in diagnostic and forensic science.
- Including additional sequences at the 5′ end of an oligonucleotide primer in PCR results in the presence of any desired sequence, for example, a restriction site, in the final product (**figure 17.8**).
- Gibson assembly is a series of reactions that permits seamless cloning—the overlap of specific fragments by enzymatic reactions that do not include restriction enzymes (**figure 17.8**).

17.3 Genomic and Metagenomic Libraries: Cloning Genomes in Pieces

- A genomic library is constructed by cleaving an organism's genome into many fragments, each of which is cloned into a vector to make a unique recombinant plasmid.
- A metagenomic library is constructed similarly, but the starting DNA derives from an environmental sample, not a single organism.
- Genomic and metagenomic libraries are often screened for the gene of interest by phenotypic rescue (genetic complementation) (**figure 17.9**).

17.4 Expressing Foreign Genes in Host Cells

- An expression vector has the necessary features to express in high levels any gene it carries.
- Purification of recombinant proteins is often accomplished by fusing the coding sequence of a protein to six consecutive histidine residue codons found on some expression vectors. When introduced and expressed in bacteria, the His-tagged protein can be selectively purified (**figure 17.10**).
- Fluorescent proteins can be used to study the regulation of gene expression (transcriptional fusions) and protein localization (translational fusions) (**figure 17.11**).

17.5 Cas9 Nuclease Is a Programmable Tool for Genome Editing

- The Cas9-gRNA nuclease is a precise DNA cutting tool. Because the localization component of the enzyme is an RNA molecule, it can be designed to direct the enzyme to any specified location in a genome.
- Repair of the double-stranded DNA break catalyzed by Cas9-gRNA allows for the nearby DNA to be altered by substitution or deletion when the desired change is introduced to the cell as a donor sequence (**figure 17.12**).

17.6 Biotechnology Develops Custom Microbes for Industrial Use

- Site-directed mutagenesis and protein engineering are used to modify genes and their expression. These approaches lead to products with novel properties (**figure 17.13**).
- Directed evolution includes techniques designed to target specific molecules for change with the goal of optimizing their function in a particularly useful way.
- Synthetic biology is the use of molecular methods to design and construct new cell types with phenotypes specified by the user.

Active Learning

1. You have cloned a structural gene that encodes an enzyme required for riboflavin synthesis in *E. coli.* You find that an *E. coli* riboflavin auxotroph carrying the cloned gene on a vector makes less riboflavin than the wild-type strain. Why might this be the case?
2. You are interested in the activity and regulation of a protease made by *Geobacillus stearothermophilus.* What would be the purpose of constructing each of the following: a His-tagged protease, a transcriptional GFP fusion to the protease gene promoter, and a translational GFP fusion to the protease gene coding sequence?
3. The genome sequence of a recently isolated bacterium includes a gene that is predicted to encode a restriction endonuclease. List the steps you could take to express the gene in *E. coli* and purify the enzyme with a His-tag. How could you determine if the His-tag affects the activity of the enzyme? You'd also like to perform some experiments that involve introducing DNA to this new organism, and although it is naturally competent for transformation, the restriction enzyme appears to be active in degrading foreign DNA. Describe a strategy to develop a strain of this isolate that has lost the restriction enzyme activity.
4. Brewer's yeast, *Saccharomyces cerevisiae,* does not metabolize lactose, but an engineered strain of *S. cerevisiae* able to consume lactose is of industrial interest. This is because such a strain can convert whey (an unusable by-product of cheese production) to ethanol. Such a strain has been engineered by cloning the genes that encode β-galactosidase and lactose permease from another yeast, *Kluyveromyces lactis,* into *S. cerevisiae.* However, the recombinant *S. cerevisiae* expresses the *K. lactis* genes only at a low level, so very little lactose is consumed. Discuss a directed evolution technique that could be applied to enhance β-galactosidase production by engineered *S. cerevisiae.*
5. One approach to creating a vaccine against SARS-CoV-2 is to purify the spike protein located on the surface of the virus particle (*see figure 6.7*). Outline the steps required to create such a recombinant vaccine. Start with a sample of the virus, which contains a single-stranded RNA genome whose sequence is available in public databases.

CHAPTER

18

Microbial Genomics

Lophelia II 2010 Expedition/NOAA Ocean Explorer

Life on the Microbial Frontier

One of the joys of microbiology is discovering and describing a new microbe for the first time. Historically, this required obtaining some type of environmental sample or specimen and then cultivating the microbe in the laboratory. As we describe in chapter 7, a plethora of techniques have been developed to enable growth and isolation of microbes in pure culture. These methods are foundational to the science of microbiology. Think of Koch's early experiments showing that *Mycobacterium tuberculosis* is the causative agent of tuberculosis. To some degree, public health officials still employ these methods to study emerging infectious diseases.

As we now appreciate, the number of archaea and bacteria that can be easily manipulated in the lab is extraordinarily low compared to the vast number of diverse microbes found on Earth. Part of this discrepancy is caused by the fact that we can recreate a relatively narrow range of growth conditions experimentally. Think about the thermostat in your microbiology classroom—it is usually set a few degrees on either side of 21°C. The incubators used in your microbiology lab likely range from 20°C to 55°C. This is quite restrictive, given that the air temperature on the Earth's surface can range from −89°C to 56°C. Other conditions like pressure, pH, and salinity also impact growth. Practically speaking, designing growth chambers and culture conditions for all microbes on our planet is impossible. Modern advances in genome analysis provide a powerful workaround to this problem.

One example is the analysis of the microbial community associated with the tube worms that live on the ocean floor. These remarkable creatures exist at a depth of several hundred meters, often forming colonies near hydrocarbon seeps. Obtaining samples from these colonies requires the use of submersible vehicles, which then bring their collections to the surface for analysis. The DNA extracted from these samples includes the genomes of all the microbes living in association with the tube worms. As we describe in this chapter, modern sequence analysis techniques allow rapid and deep analysis of the DNA in these samples. Entire chromosomal maps can be constructed. By comparing these maps with those found in public genome databases, putative genes can be identified and their functions ascribed. This comparative analysis all happens *in silico*—on computers and servers and in the cloud. These remarkable advances are finally allowing scientists to probe the final—microbial—frontier.

Readiness Check:

Based on what you have learned previously, you should be able to:

- ✓ Identify the key structural elements of DNA, RNA, and protein (section 13.2)
- ✓ Discuss what is meant by the term *genome*
- ✓ Explain the process and utility of PCR (section 17.2)
- ✓ Summarize the purpose, construction, and screening of a genomic library (section 17.3)

18.1 DNA Sequencing Methods

After reading this section, you should be able to:

a. Explain how DNA is sequenced by the Sanger chain termination method
b. Contrast and compare the advantages and disadvantages of the Sanger method with next-generation sequencing

The genomic era is a twenty-first-century phenomenon. However, one of the most crucial steps leading to the genomic revolution was figuring out how to sequence DNA. This was accomplished in 1977 by Alan Maxam and Walter Gilbert (collaborating on one technique) and Frederick Sanger, whose method became the most commonly used.

Sanger DNA Sequencing

The Sanger method involves synthesis of a new strand of DNA using the DNA to be sequenced as a template. The reaction begins when single strands of template DNA are mixed with an oligonucleotide primer (DNA about 12 to 20 nucleotides complementary to the region to be sequenced; it primes the initiation of new strand synthesis), DNA polymerase, the four deoxynucleoside triphosphates (dNTPs), and dideoxynucleoside triphosphates (ddNTPs). ddNTPs differ from dNTPs in that the 3′ carbon lacks a hydroxyl group (**figure 18.1**). In such a reaction mixture, DNA synthesis continues until a ddNTP, rather than a dNTP, is added to the growing chain. Without a 3′-OH group to attack the 5′-PO_4 of the next dNTP to be incorporated, synthesis stops (*see figure 13.11*). Indeed, Sanger's technique is frequently referred to as a **chain-termination DNA sequencing method.**

To obtain sequence information, four separate synthesis reactions must be prepared, one for each ddNTP (**figure 18.2**). The ddNTP is mixed with all four normal dNTPs, and as DNA synthesis proceeds, sometimes the ddNTP will be incorporated into the growing DNA strand, rather than its dNTP counterpart. This results in a collection of DNA fragments of varying lengths, each ending in the same ddNTP. For example, a reaction prepared

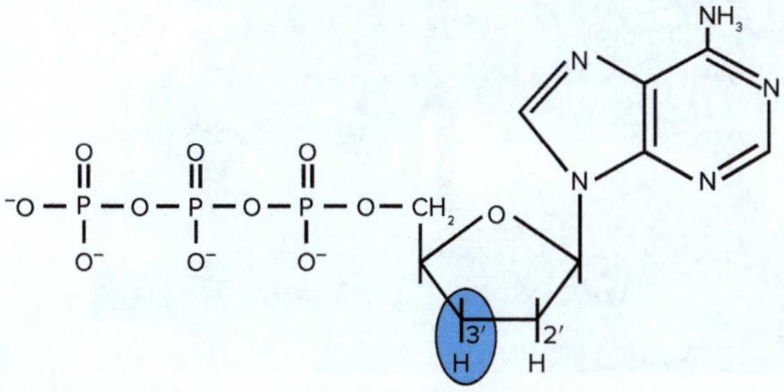

Figure 18.1 Dideoxyadenosine Triphosphate (ddATP). Note the lack of a hydroxyl group on the 3′ carbon (blue), which prevents further chain elongation by DNA polymerase *(compare to figure 13.4)*.

MICRO INQUIRY *What is the function of the 3′-OH during DNA synthesis?*

with ddATP + dATP, dTTP, dGTP, and dCTP produces fragments ending with an A, those with ddTTP produce fragments with T termini, and so forth. After DNA synthesis is complete, the DNA is made single stranded, typically by heating. The resulting fragments are separated by electrophoresis. Recall that each fragment's migration rate is inversely proportional to the log of its molecular weight. The smaller a fragment is, the faster it moves through the gel. Since nucleotides are added to the 3′-OH of the growing DNA strand, the ddNTP at the end of the shortest fragment corresponds to the 5′ end of the DNA sequence, while the largest fragment marks the 3′ end. In this way, the DNA sequence can be read in the 5′ to 3′ direction, which corresponds to the smallest fragment at the bottom of the gel to the largest at the top, respectively. *Gel electrophoresis (Techniques & Applications 17.1)* *Sanger Sequencing*

Today, DNA is sequenced using automated systems (**figure 18.3**). Here each ddNTP is labeled with a different color fluorescent dye. Automated sequencing uses a laser beam to detect DNA fragments as they exit the bottom of the gel from smallest to largest. This is possible because each ddNTP is labeled with a different fluorescent color. The results are recorded on a graph called a chromatogram in which the amplitude of each peak represents the fluorescent intensity of each particular fragment (figure 18.3*b*). The corresponding DNA sequence is listed above the chromatogram. An automated Sanger chain termination sequencing system can accurately read 500 to 800 bp in a single electrophoresis run.

DNA Sequencing by Synthesis

Sanger's chain termination method was used to complete the first human genome sequence in 2001. This cost about $300 million and took about a decade to finish. Although this was an amazing feat at the time, it illustrates two limitations of Sanger sequencing: It is expensive and time consuming. Because scientists want to sequence genomes faster and more cheaply, innovative, newer DNA sequencing techniques have been invented.

These newer technologies are called **next-generation sequencing (NGS).** The setup is quite different from the Sanger approach; rather than long genomic strands in solution, NGS uses short, sheared (cut) pieces of DNA templates attached to a solid substrate. Another major difference is that thousands of identical DNA fragments are sequenced at the same time; therefore these methods are sometimes called **massively parallel sequencing techniques.** In addition to making genomic sequencing faster and cheaper, as we discuss later in this chapter, NGS avoids the need to insert (i.e., clone) individual DNA fragments into vectors. This is important because it is almost impossible to clone every DNA fragment from a genome into a vector. *Genomic and metagenomic libraries: cloning genomes in pieces (section 17.3)*

Although several NGS techniques have been marketed, currently **reversible chain termination sequencing** is most frequently used. This takes place in a flow cell, which is a glass slide into which grooves have been cut. Flow cells allow reagents to be added, flushed out, and new reagents added without the loss of the attached DNA.

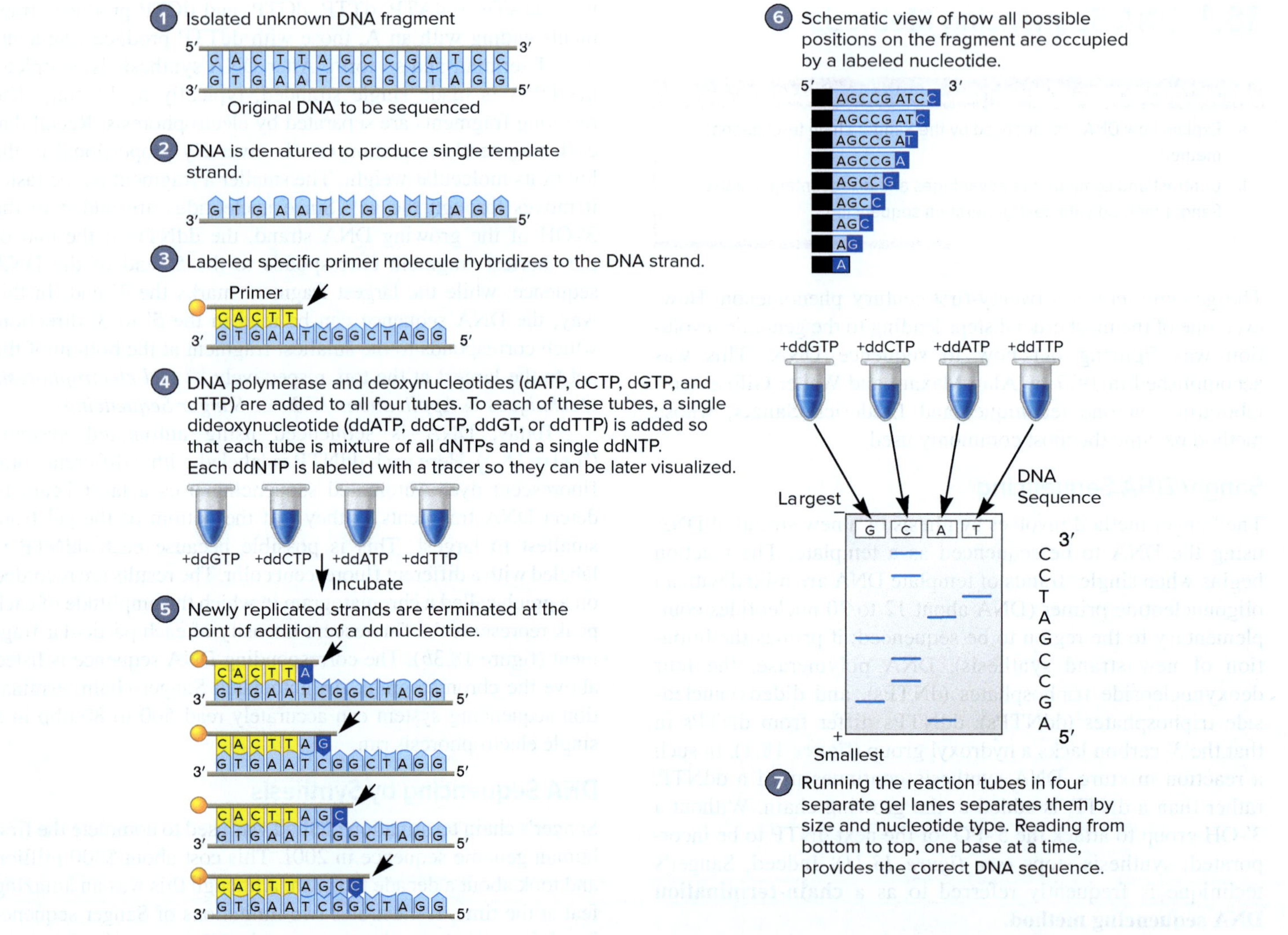

Figure 18.2 The Sanger Method of DNA Sequencing. Steps 1–6 are used for both manual and automated sequencing. Step (7) shows preparation of a gel for manual sequencing in which radiolabeled ddNTPs are used. Although manual sequencing is rarely used, it is instructive in understanding how DNA sequence is determined.

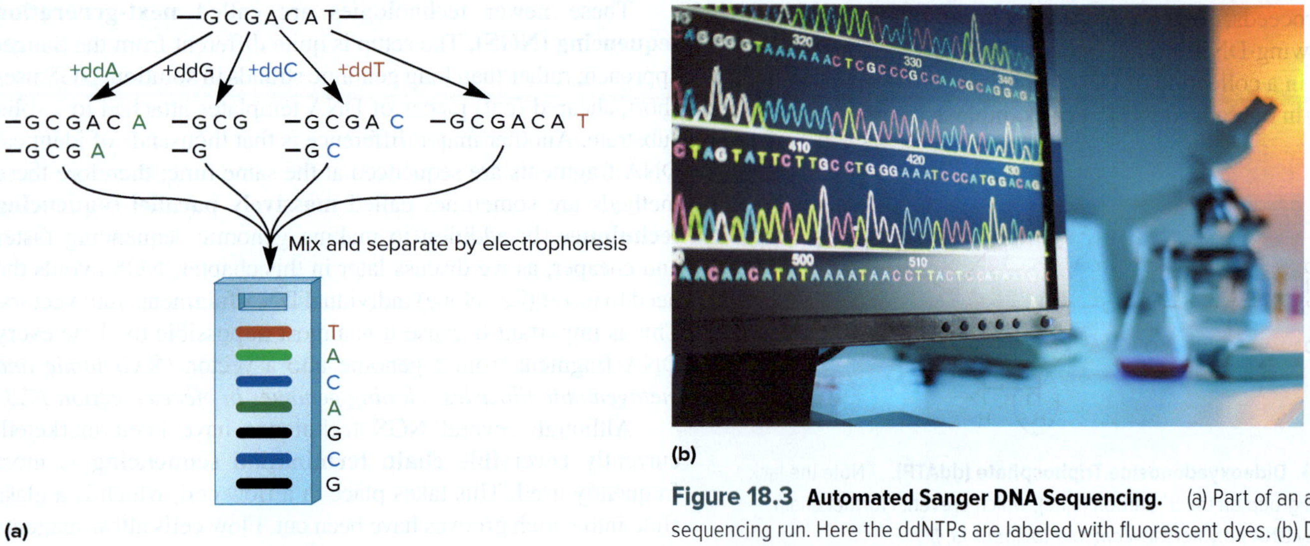

Figure 18.3 Automated Sanger DNA Sequencing. (a) Part of an automated DNA sequencing run. Here the ddNTPs are labeled with fluorescent dyes. (b) Data generated during an automated DNA sequencing run.

(b) Nicolas/E+/Getty Images

The process begins by shearing the DNA template into fragments and adding oligonucleotides called *adapters* to each end (**figure 18.4,** step 1). The adapter at one end of each DNA fragment attaches it to the flow cell. The adapter at the other end also fastens the DNA fragment to the substrate, but does so by annealing it to a short nucleotide fragment (oligonucleotide) already attached to the flow cell (figure 18.4, step 2). Once annealed, the DNA template forms a loop with both ends attached to the flow cell. The oligonucleotide is used as a primer to initiate the polymerase chain reaction (PCR; figure 18.4, step 3). Next in a process called *bridge amplification,* PCR using the DNA templates within the flow cell creates clusters of identical double-stranded fragments of DNA scattered over the surface of the slide (figure 18.4, step 4).

In the next step, the double-stranded fragments are denatured, and the flow cell is flushed. This leaves bundles of single-stranded linear fragments ready for sequencing (figure 18.4, steps 5 and 6). ◀ *Polymerase chain reaction amplifies targeted DNA (section 17.2)*

Now that the DNA attached to the flow cell has been amplified in clusters, it is ready to be sequenced. A technology called **sequencing by synthesis** identifies each nucleotide as it is

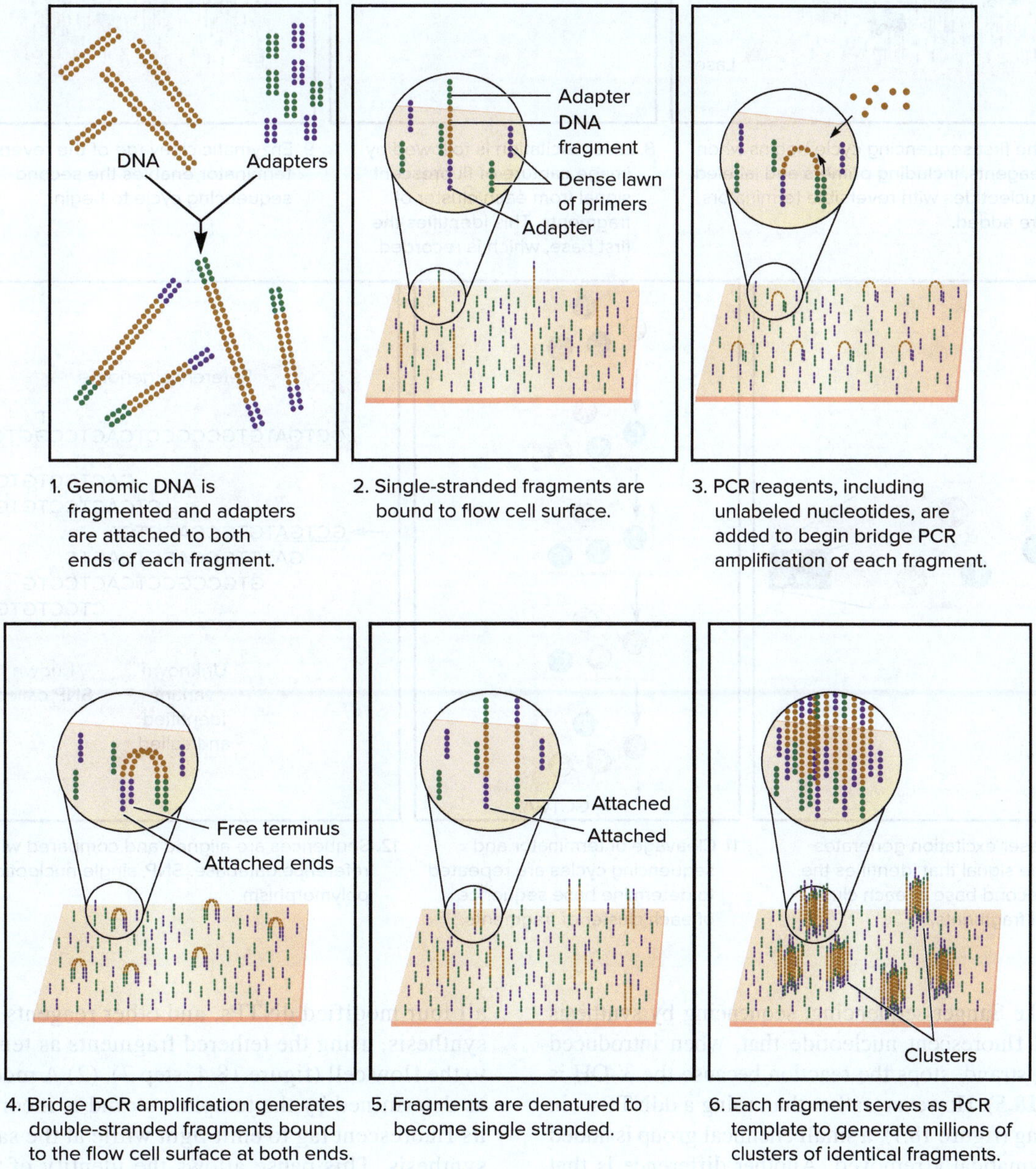

Figure 18.4 **Reversible Chain Termination Sequencing.**

MICRO INQUIRY *Why is it important that identical fragments of DNA to be sequenced are clustered together?*

(Continued)

Figure 18.4 ***(Continued)***

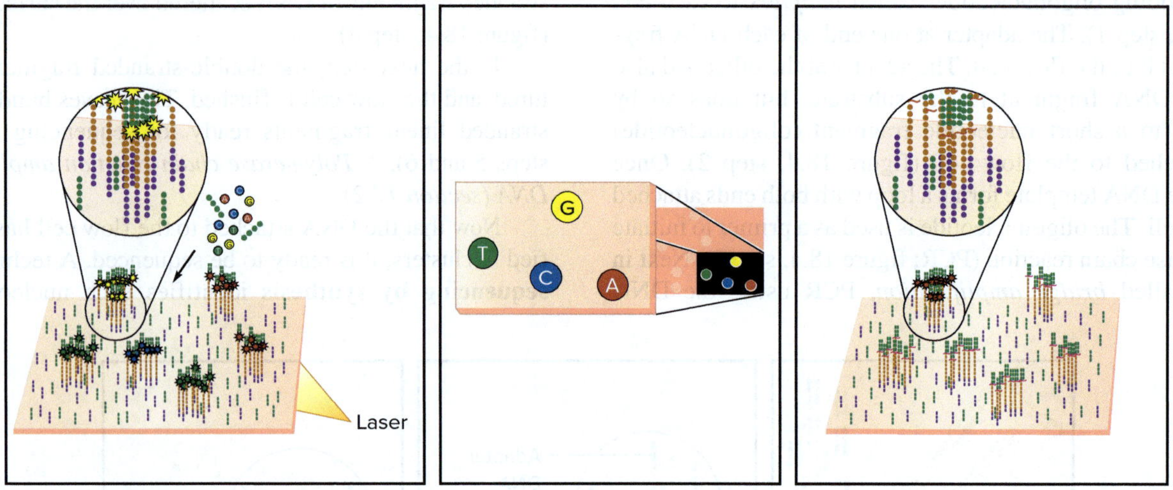

7. The first sequencing cycle begins when reagents, including primers and labeled nucleotides with reversible terminators, are added.

8. Laser excitation is followed by image capture of fluorescent signal from each cluster of fragments. This identifies the first base, which is recorded.

9. Enzymatic cleavage of the reversible terminator enables the second sequencing cycle to begin.

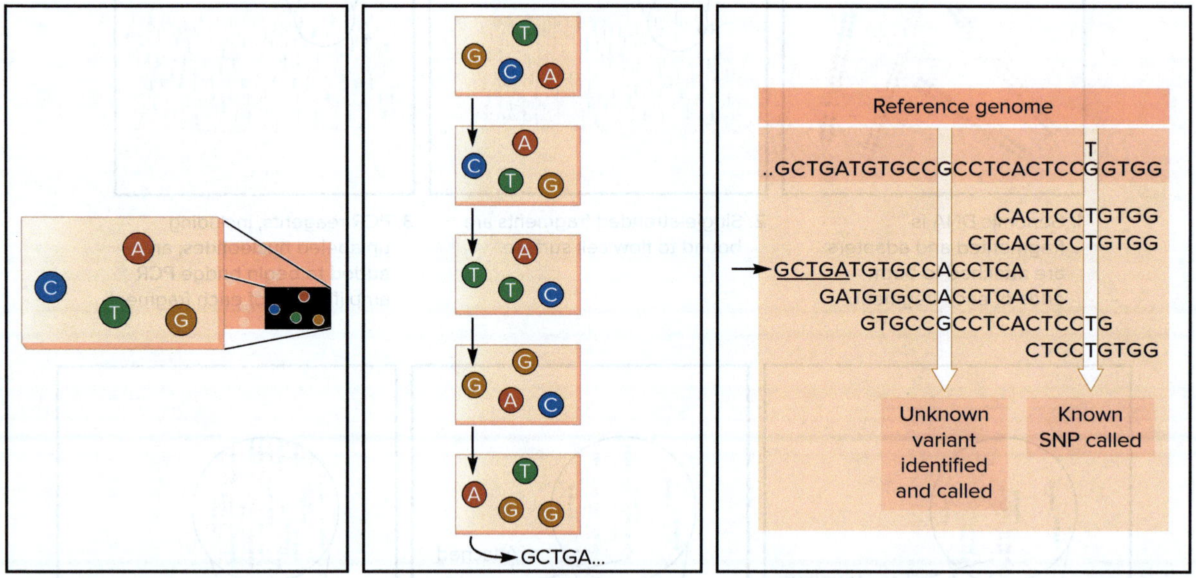

10. Laser excitation generates the signal that identifies the second base in each cluster of fragments.

11. Cleavage of terminator and sequencing cycles are repeated to determine base sequence of each cluster of fragments.

12. Sequences are aligned and compared with reference database. SNP, single nucleotide polymorphism.

incorporated. Like Sanger sequencing, sequencing by synthesis uses a modified fluorescent nucleotide that, when introduced into the growing strand, stops the reaction because the 3′-OH is blocked (**figure 18.5**). However, rather than using a ddNTP as in Sanger sequencing (figure 18.1), a small chemical group is added that can be enzymatically removed. Another difference is that the modified nucleotide does not fluoresce until it is incorporated into the growing DNA strand.

The incorporation of each nucleotide determined by laser optics involves the following events: (1) The DNA polymerase, all four modified dNTPs, and other reagents needed for DNA synthesis, using the tethered fragments as template, are added to the flow cell (figure 18.4, step 7). (2) A modified nucleotide as determined by the template strand is incorporated, causing its fluorescent tag to emit light while at the same time stalling synthesis. This pause allows the identity of the incorporated base to be determined by the color of the fluorescent tag (figure 18.4, step 8). (3) Reagents needed for synthesis are then flushed from the flow cell, and (4) cleavage enzyme reagents are introduced. (5) The fluorescent label and the blocking

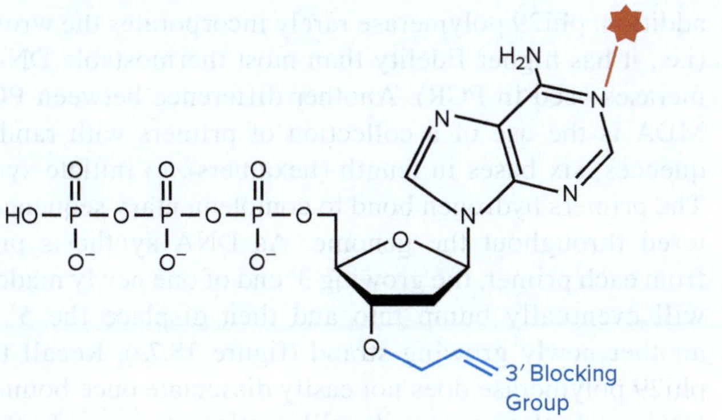

Figure 18.5 Modified Base Used in Reversible Chain Termination Sequencing. This example shows modified dATP, with a 3′-OH blocking group (blue) and a fluorescent tag (red star) attached to the adenine. The base fluoresces only when the PP$_i$ is removed during incorporation into the growing strand. The 3′ blocking group is then enzymatically removed, exposing the 3′-OH needed for continued DNA synthesis.

MICRO INQUIRY *Why is the 3′-OH group so critical for elongation of a DNA strand during replication?*

molecule are removed, exposing the 3′-OH. Finally, (6) the enzyme cocktail is flushed, and synthesis reagents are added once again and the cycle repeats (figure 18.4, step 9). Because the light emitted by any single nucleotide is too dim to accurately record, each cluster of identical fragments must grow synchronously so that a detectable signal is generated. The incorporation of the same nucleotide at the same time into each identical fragment in a cluster generates a signal with sufficient amplitude for detection (figure 18.4, steps 8 and 10). This explains why the length of each read (the number of nucleotides determined in a single run) is much shorter than it is for Sanger sequencing: After about 150 to 300 bases have been incorporated, the synchronicity of nucleotide incorporation deteriorates. That is, the synthesis of complementary strands in a cluster of identical templates becomes "out of sync." Without a strong synchronous signal, the data become ambiguous. Nonetheless, short read length is compensated by the volume of data produced: Several GB (billion bases) are read per flow cell in about a day, with newer high-throughput systems reading hundreds of GB. Compare this with the output of automated Sanger sequencing, which is about 1 million bases per day. With the average bacterial genome measuring about 4 million base pairs, one can begin to appreciate the enormous power of NGS.

In addition to the method described here, various new approaches are being developed. These platforms seek to streamline sample preparation, improve signal detection, optimize surfaces to more rapidly attain high reagent concentrations, and obtain longer read lengths. Although there is room for improvement, next-generation DNA sequencing technologies have already revolutionized genomics by drastically reducing time and cost, while simultaneously increasing accuracy.

18.2 Genome Sequencing

After reading this section, you should be able to:

a. List the steps used in whole-genome shotgun cloning
b. Compare and contrast genome assembly using Sanger and next-generation sequencing (NGS) approaches
c. Describe multiple displacement amplification and how this technique is used

Whole-Genome Shotgun Sequencing

Although revolutionary when it was introduced, sample preparation is now considered one of the biggest disadvantages of **whole-genome shotgun sequencing.** In this technique, each genome to be sequenced must be cut into fragments and cloned into vectors (i.e., the construction of a genomic library; *see figure 17.9*). Once a genomic library has been constructed and sequenced, the fragmented sequences are assembled into increasingly longer, and eventually connected, stretches called **contigs** (**figure 18.6**). Contigs are aligned to form the complete genome sequence, which is called a *scaffold.* Scaffolds may contain gaps between contigs, leaving the genome sequence incomplete; there are several strategies for obtaining the missing sequences. Lastly, the assembled genome undergoes proofreading to resolve any ambiguities or frameshifts.

Next-Generation Genomic Sequencing

Sanger sequencing ushered in the genomic era. However, the advent of next-generation sequencing (NGS) techniques has made genomic sequencing, particularly of microorganisms, much more practical in terms of time, money, and improved outcome. Because genomic fragments are not cloned into

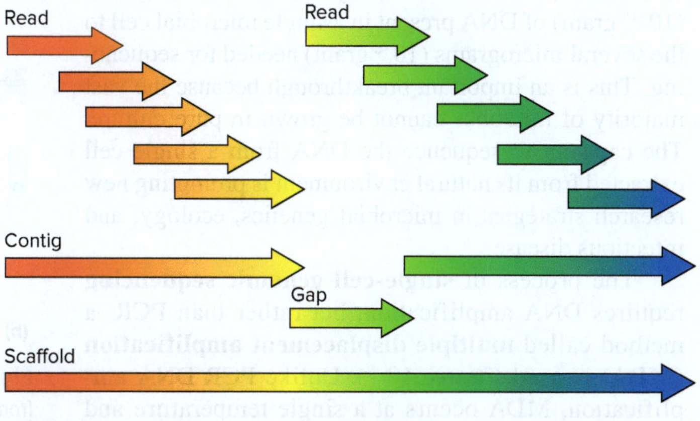

Figure 18.6 Alignment of Sequenced DNA to Construct a Scaffold. DNA fragments (or *reads*) are aligned as determined by overlapping sequences. These fragments form contigs, which are linked together in silico to form a scaffold. Gaps of sequenced DNA in the scaffold must be filled to complete the genome sequence.

vectors as is the case with Sanger sequencing (figure 18.4), NGS is vastly more efficient. Typically close to 100% of the genomic fragments with adapters bind to the solid substrate (e.g., flow cell surface), whereas cloning more than 80% of the genomic fragments into a genomic library requires a lot of skill and luck.

The advent of NGS has significantly impacted the quality of the final genomic sequence, as assessed by two specific factors: depth of coverage and breadth of coverage. *Depth of coverage* refers to the average number of times each nucleotide is sequenced in a genome (or other sequencing project). Depending on the technique used, a single nucleotide might be read (sequenced) 18 to 100 times. The term *deep sequencing* means that each nucleotide in a genome has been sequenced a very high average number of times. *Breadth of coverage* refers to how much of the entire genome was sequenced. A coverage of 100% means the genome sequence is complete, without gaps. Ideally one wants deep sequencing with 100% breadth of coverage. Because NGS does not involve the construction of a genomic library, it almost always yields a higher breadth of coverage. The depth of coverage is also quite different from Sanger sequencing. When Sanger genomic sequencing is used, any given region of the genome is typically sequenced no more than ten times. By contrast, NGS results in the same genomic fragment being read 30 to 100 times, greatly increasing accuracy. This is important because the DNA polymerase used in sequencing reactions lacks proofreading capability, so it cannot correct mismatched bases. Deep NGS overcomes the limitations of the DNA polymerase reaction by re-reading the same stretch of genome many times. A base at a given position that is not in agreement with the majority of the reads is thus recognized as a mistake.

Single-Cell Genomic Sequencing

It is now possible to amplify the few femtograms (10^{-15} gram) of DNA present in a single microbial cell to the several micrograms (10^{-6} gram) needed for sequencing. This is an important breakthrough because the vast majority of microbes cannot be grown in pure culture. The capacity to sequence the DNA from a single cell extracted from its natural environment is prompting new research strategies in microbial genetics, ecology, and infectious disease.

The process of **single-cell genomic sequencing** requires DNA amplification, but rather than PCR, a method called **multiple displacement amplification (MDA)** is used (**figure 18.7**). Unlike PCR DNA amplification, MDA occurs at a single temperature and uses the DNA polymerase from bacteriophage phi29 to synthesize new strands of DNA from the genome template. This polymerase is used because it does not readily dissociate from the template strand; the importance of this feature will be made clear shortly. In addition, phi29 polymerase rarely incorporates the wrong base (i.e., it has higher fidelity than most thermostable DNA polymerases used in PCR). Another difference between PCR and MDA is the use of a collection of primers with random sequences, six bases in length (hexamers), to initiate synthesis. The primers hydrogen bond to complementary sequences scattered throughout the genome. As DNA synthesis proceeds from each primer, the growing 3′ end of one newly made strand will eventually bump into and then displace the 5′ end of another newly growing strand (figure 18.7*a*). Recall that the phi29 polymerase does not easily dissociate once bound to the DNA, so both new strands will continue to grow. In this way, many new strands are rapidly synthesized. The new strands have an average length of about 12,000 bases (12 kb) but can be as long as 100 kb. This makes them suitable for DNA sequencing. NGS is then performed to avoid the requirement of genomic library construction and to ensure higher breadth and depth of coverage.

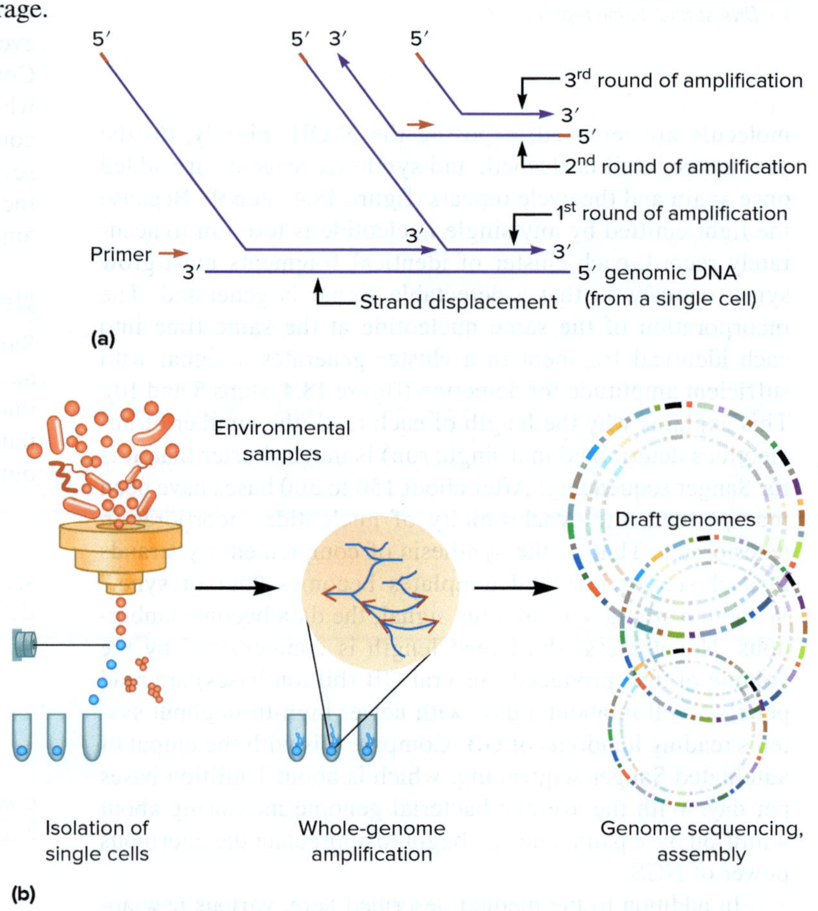

Figure 18.7 Single-Cell Genomic Sequencing. (a) Many copies of DNA extracted from a cell are generated by multiple displacement amplification. Random hexamer primers (red) bind to complementary template sequences, and DNA polymerase from bacteriophage phi29 is used to catalyze synthesis in the 5′ to 3′ direction (purple arrowheads). When the end of a newly synthesized strand meets double-stranded DNA, one strand is displaced by the growing DNA. (b) Single-cell genomic sequencing can be used to discover uncultivated bacteria and archaea from natural samples.

Microbiologists have used single-cell genomics to sequence the genomes of hundreds of uncultivated microbes representing multiple undefined bacterial and archaeal taxa sampled from many different environments (figure 18.7*b*). These microbes constitute what the researchers call *microbial dark matter* because there was no prior knowledge of their existence. Although the average breadth of coverage is typically only about 40% (reflecting a major limitation of single-cell genomics), a variety of discoveries have been made, including at least 20,000 new hypothetical protein families, evidence for new superphyla, and the surprising presence of genes encoding sigma factors in archaea.

Comprehension Check

1. Why is the Sanger technique of DNA sequencing also called the chain-termination method?
2. Explain the difference between a dideoxynucleotide used in Sanger sequencing and the modified bases used in reversible chain termination sequencing.
3. Why does sequencing by synthesis yield short reads?
4. Why is it important that the DNA polymerase used in single-cell genomic sequencing does not easily dissociate from the template DNA?
5. Suggest a medical and an ecological application of single-cell genomic sequencing.

18.3 Metagenomics Provides Access to Uncultured Microbes

After reading this section, you should be able to:

a. Differentiate between the construction and screening of a genomic library and a metagenomic library

b. List two applications of metagenomics in any field of microbiology

DNA sequencing is a powerful tool, but its impact would be diminished if it were applied only to the small minority of microbes currently culturable. Fortunately this is not the case. **Metagenomics**—the study of microbial genomes based on DNA extracted directly from the environment—has emerged as a key technology across a number of biological disciplines, including ecology, environmental microbiology, infectious disease, and immunology. Metagenomics samples the entire pool of nucleic acids found in any given ecosystem (e.g., soil, water, feces) and is most frequently used to determine the members of the microbial community living there. Before the development of NGS, DNA extracted from the environment was typically used as a template to amplify small subunit rRNA genes or other target genes using PCR. This approach continues to be used, but many microbiologists now use shotgun metagenomics instead. The term *shotgun* is used because it involves sequencing all the DNA extracted from an environment, rather than using PCR primers to target a specific gene. In addition to taxonomic information, shotgun metagenomics also catalogs most of the genes present in the sample, providing important clues about microbial activity.

To illustrate the impact of coupling direct DNA extraction with NGS, consider a microbial census based only on the microorganisms that can be cultured from an environmental sample (e.g., soil). Based on the average recovery of microbes in culture, this approach will miss about 98% of the microbial species present. Such an approach would be like taking a census of a big city by randomly sampling a few individuals. A better approach would be to base a microbial census on DNA extracted from soil and cloned into bacterial vectors that are then sequenced by Sanger chain termination, as this would certainly improve representation, but by an unknown amount. Keeping with the city census analogy, this would be like sending out a questionnaire to each household, hoping everyone would respond, but understanding that only an unknown fraction of residents will. Metagenomic analysis using next-generation sequencing, on the other hand, is the equivalent of sending a team of census takers into the community to ensure that each person is counted: The team probably won't reach every single resident, but depth and breadth of coverage will be very high.

Shotgun metagenomics involves two or three general steps (**figure 18.8**). First, the nucleic acids must be extracted from the environment. Because the genome fragments are collected from nature, not from a single organism grown in culture, there is no reliable supply of genome apart from resampling the environment. The researcher may therefore elect to clone the genome fragments into vectors. The resulting metagenomic library is similar to any other genomic library except that each DNA fragment comes from a collection of genomes found in a microbial community, not a single microorganism (*see figure 17.9*). Alternatively, the researcher may elect to sequence the genomic fragments directly, without library construction. Either way, each DNA fragment is then sequenced using an NGS technique.

Once nucleotide sequences are obtained, partial or full genomes can be detected in two ways. First, they can be assembled as previously discussed, by aligning overlapping sequences at the ends of reads (figure 18.6). However, this approach can be difficult with shotgun metagenomics because a collection of 100 to 300 bp reads is generated; thus the number of gapped scaffolds tends to be large. Instead, one can attempt to align these short reads to a previously sequenced genome (called a *reference genome*) in an effort to identify similar sequences from known microorganisms (figure 18.8). This too is limited because many taxa obtained by metagenomics are uncultured and are therefore unrepresented in existing genomic databases; they are microbial dark matter. This problem prompted the establishment of the Genomic Encyclopedia of Bacteria and Archaea Project. The goal of this program is to improve reference databases by sequencing the genomes of a

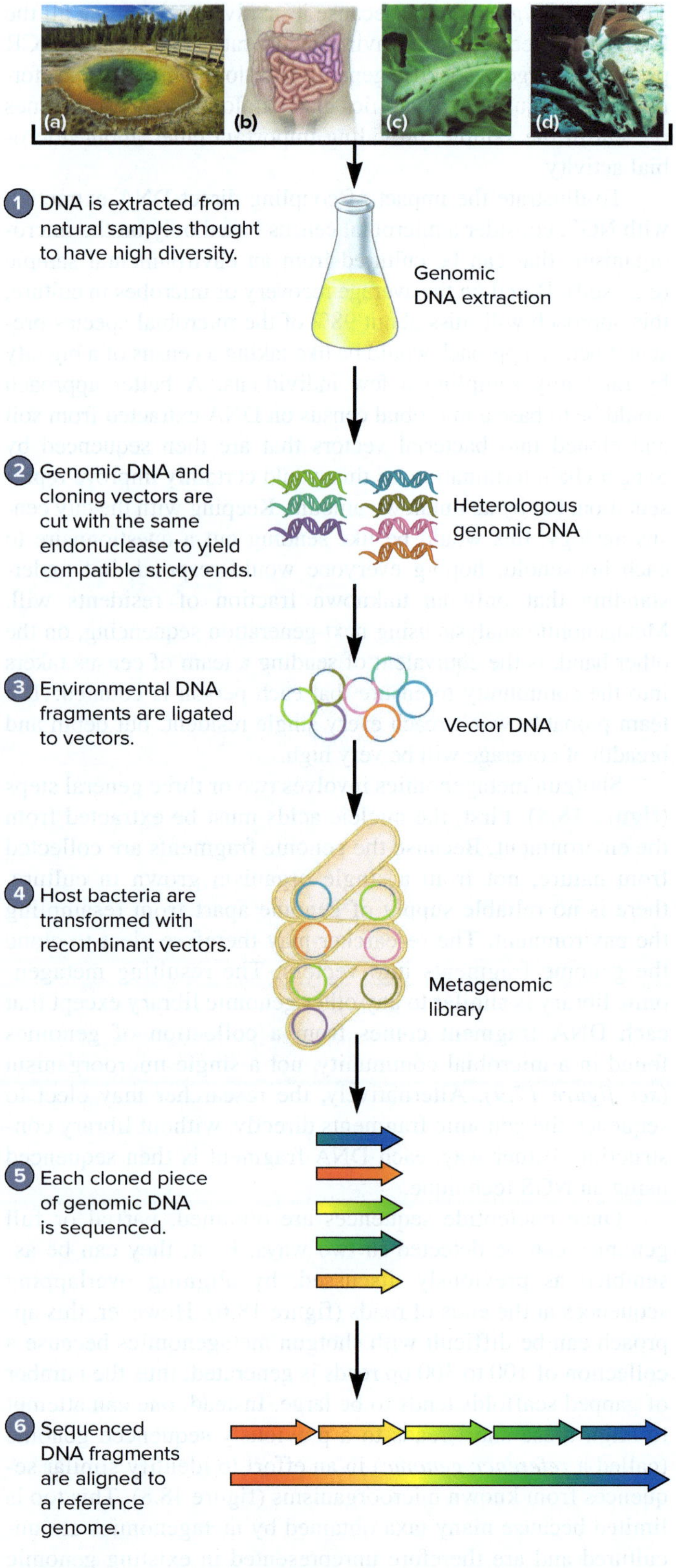

Figure 18.8 Shotgun Metagenomics. DNA has been extracted directly from environments such as (a) bacterial mats at Yellowstone National Park, (b) the human colon, (c) cabbage white butterfly larvae, and (d) tube worms from hydrothermal vents. Following extraction, the DNA is cloned into vectors to create a metagenomic library. Each cloned DNA fragment is sequenced by NGS and aligned to a previously sequenced genome termed a *reference genome*.

(a) Yi Xiang Yeng/iStock/Getty Images; (c) Nigel Cattlin/Science Source; (d) Cindy Van Dover/National Oceanic and Atmospheric Administration

wide diversity of cultured microorganisms. Metagenomic sampling of natural environments has so far led to the discovery of many new bacterial and archaeal phyla. ▶| *Whole-genome comparison (section 26.2)*

Comprehension Check

1. NGS techniques transformed the impact of metagenomic analysis. Why do you think that is the case?
2. Examine figure 18.8. Metagenomics was initially used to take a census of "who is there" but has since become an important tool in deciphering "what they are doing." How can metagenomic data contribute to our understanding of the biochemical activities occurring in a microbial community?

18.4 Bioinformatics: What Does the Sequence Mean?

After reading this section, you should be able to:

a. Explain how a potential protein-coding gene is recognized within a genome sequence
b. Compare the meaning of the terms *orthologue* and *paralogue*
c. Differentiate between a conserved hypothetical protein and a putative protein of unknown function
d. Describe the construction of a physical genome map

As you might imagine, sequencing entire genomes generates a huge volume of information. The field of **bioinformatics** combines biology, mathematics, computer science, and statistics to convert raw nucleotide data into the location and potential function of genes or presumed genes on sequenced genomes using a process called **genome annotation.** Once genes have been identified, bioinformaticists can perform computer or **in silico analysis** to further examine the genome.

Obviously, obtaining nucleotide sequences without understanding the location and function of individual genes would be a pointless exercise. The goal of genome annotation is to identify every potential (putative) protein-coding gene, regulatory element, as well as each rRNA- and tRNA-coding gene. A protein-coding gene is usually recognized as an **open reading frame (ORF);** to find all ORFs, both strands of DNA must be analyzed in all three reading frames (**figure 18.9**). A bacterial or archaeal ORF is generally defined as a sequence of at least 100 codons (300 base pairs) that is not interrupted by a stop codon and has terminator sequences at the 3′ end. The 5′ end of the gene should also bear a ribosome-binding site. Only if these elements are present is an

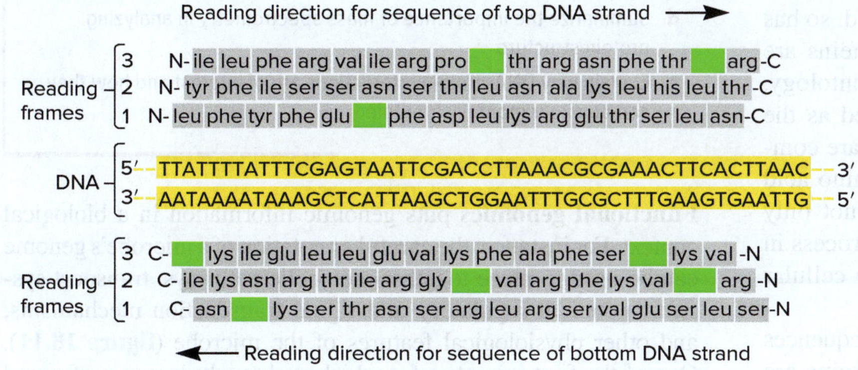

Figure 18.9 Finding Potential Protein-Coding Genes. Annotation of genomic sequence requires that both strands of DNA be translated from the 5′ to 3′ direction in each of three possible reading frames. Stop codons are shown in green.

ORF considered a putative protein-coding gene. This process is performed by gene prediction programs designed to find genes that encode proteins, regulatory sequences, or functional RNA products. Ideally, computer-identified genes are then manually inspected by bioinformaticists to verify the computer-generated gene assignments. This process is called *genome curation.*

ORFs that appear to encode proteins are called ***coding sequences* (CDS).** Bioinformaticists have developed algorithms to compare the sequence of predicted CDS with those in large databases containing nucleotide and amino acid sequences of known proteins. The base-by-base comparison of two or more gene sequences is called **alignment.** Alignments can also be performed by comparing amino acid sequences between two proteins. Scientists most often use **BLAST (*basic local alignment search tool*)** programs to perform this task. These programs compare the nucleotide (or amino acid) sequence of interest, called the *query sequence,* to all other sequences entered in the database. The results ("hits") are ranked in order of decreasing similarity. An E-value is assigned to each alignment; this value measures the possibility of obtaining the alignment by chance; thus highly homologous sequences have very low E-values.

Considerable information can frequently be inferred from translated amino acid sequences of potential genes. Often a short pattern of amino acids, called a domain, will represent a functional unit within a protein, such as the active site of an enzyme. For instance, **figure 18.10** shows the C-terminal domain of the cell division protein MinD from a number of microbes (*see figure 7.5*). Because these amino acids are found in such a wide range of organisms, they are considered phylogenetically well-conserved. In this case, the conserved region is predicted to form a helix needed for proper localization of the protein to the membrane. This motif (α helix) is thought to interact with membrane phospholipids in a way that enables MinD to rapidly attach-release-reattach. Thus this structural motif is also functional domain. Finding a high level of conservation (similarity) allows the genome curator to confidently assign a function to the domain.

Genes from different organisms with highly similar ORFs are called **orthologues.** Sometimes there appear to be duplicated genes on the same genome. This occurs when two or more genes have very similar nucleotide sequences. Such genes are called **paralogues.**

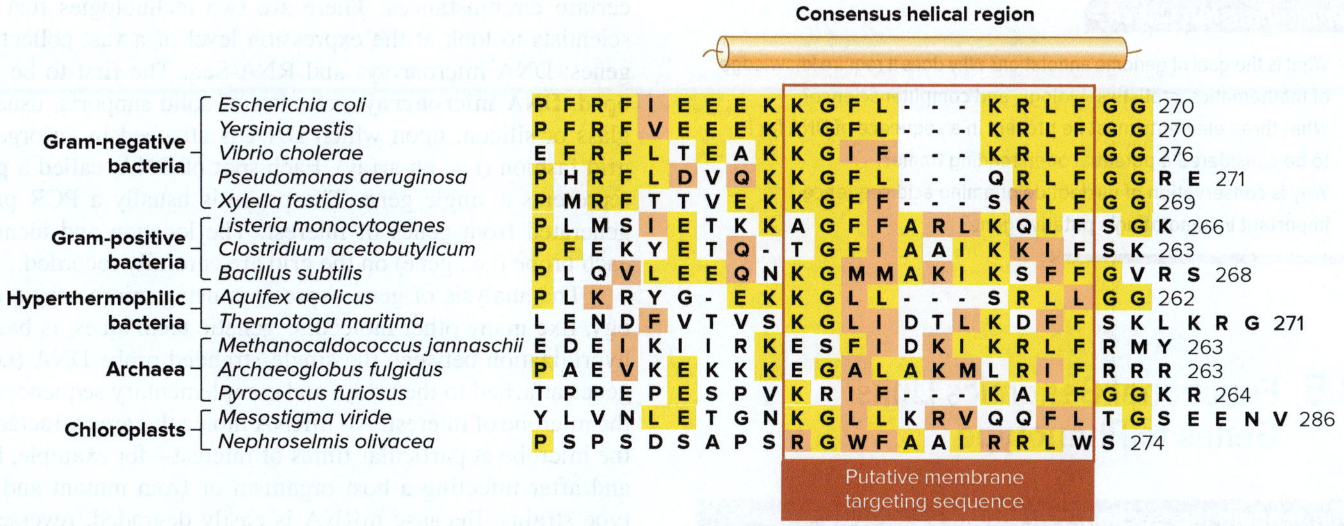

Figure 18.10 Analysis of Conserved Regions of Phylogenetically Well-Conserved Proteins. C-terminal amino acid residues of MinD from 15 organisms and chloroplasts are aligned to show strong similarities. Amino acid residues identical to *Escherichia coli (E. coli)* are boxed in yellow, and conservative substitutions (e.g., one hydrophobic residue for another) are boxed in orange. Dashes indicate the absence of amino acids in those positions; such gaps may be included to maintain an alignment. The number of the last residue shown relative to the entire amino acid sequence is shown at the extreme right of each line.

MICRO INQUIRY *Which amino acids are most highly conserved?*

As the number of sequenced genomes has expanded, so has the need to carefully define how new genes and proteins are named. The use of a structured vocabulary is called ontology, and a standard **gene ontology (GO)** has been adopted as the means by which proteins, or domains within proteins, are commonly named. This is based on the similarities of amino acid sequences among orthologous proteins. A GO term not only reflects protein function but also defines the cellular process in which the protein participates (e.g., motility) and the cellular location of the protein (e.g., flagellum).

Proteins that do not align with known amino acid sequences fall into two classes: (1) **Conserved hypothetical proteins** are encoded by genes that have matches in the database but no function has yet been assigned to any of the sequences. (2) **Proteins of unknown function** are the products of genes unique to that organism. On the one hand, as more genomes are published, the likelihood of finding a match in another organism is increased. But on the other hand, as more metagenomic data become available, so too does the pool of putative proteins of unknown function.

Once all the genes have been annotated, a **physical map** representing the entire genome may be drawn. The map is typically drawn as concentric circles depicting a bacterial or archaeal chromosome with each gene described by functional class (e.g., energy metabolism), which is color coded (figure 18.7*b*). Deviation from the mean % G + C is often indicated by the inner circle. Such deviations are common when DNA has been acquired by horizontal gene transfer. If the microbe has multiple chromosomes or plasmids, a genome map will show each chromosome and plasmid. Recall that the term *genome* includes all of the cell's DNA.

Comprehension Check

1. What is the goal of genome annotation? Why does it require knowledge of mathematics, statistics, biology, and computer science?
2. What three elements must be present in a sequence of DNA for it to be considered a potential open reading frame?
3. Why is conservation of nucleotide or amino acid sequence important in annotating a putative gene?

18.5 Functional Genomics Links Genes to Phenotype

After reading this section, you should be able to:

a. Explain how genome annotation can be used to graphically represent the metabolism, transport, motility, and other key features of a microbe
b. Contrast and compare microarray analysis with RNA-Seq in the study of transcriptomes
c. Explain how 2-D gel electrophoresis separates proteins of identical molecular weight
d. Summarize the importance of mass spectrometry in analyzing protein structure
e. Explain why DNA-protein interactions are of interest and how they can be experimentally identified

Functional genomics puts genomic information in a biological context. For instance, the careful annotation of a microbe's genome can be used to piece together metabolic pathways, transport systems, potential regulatory and signal transduction mechanisms, and other physiological features of the microbe (**figure 18.11**). One of the first microbes for which such analysis was performed was the causative agent of syphilis, *Treponema pallidum.* Because *T. pallidum* was only recently grown in tissue culture, genomics has been key to learning about its metabolism and pathobiology. The sequencing and annotation of the *T. pallidum* genome reveals that it lacks genes encoding key enzymes in the TCA cycle and oxidative phosphorylation. It also lacks many biosynthetic pathways and must rely on molecules supplied by its host. This inference is supported by the observation that about 5% of its genes code for transport proteins. ▶| *Phylum Spirochaetota: bacteria with a corkscrew morphology (section 20.5)*

Transcriptome Analysis

Once the identity and function of the genes that comprise a genome have been analyzed, the key question remains, "Which genes are expressed at any given time?" The entire collection of RNAs produced under specified conditions by an organism is its **transcriptome,** which is studied in the field of *transcriptomics.* Prior to the genomic era, researchers could identify only a limited number of genes whose expression was altered under certain circumstances. There are two technologies that allow scientists to look at the expression level of a vast collection of genes: DNA microarrays and RNA-Seq. The first to be developed, **DNA microarrays,** consists of solid supports, usually of glass or silicon, upon which DNA is attached in an organized grid fashion (i.e., an array). Each spot of DNA, called a **probe,** represents a single gene. The probe is usually a PCR product generated from genes of interest. The location and identity of each probe (i.e., gene) on the grid are carefully recorded.

The analysis of gene expression using microarray technology, like many other molecular genetic techniques, is based on hybridization between the single-stranded probe DNA (i.e., the genes attached to the array) and complementary sequences from the microbe of interest. All mRNA molecules are extracted from the microbe at particular times of interest—for example, before and after infecting a host organism or from mutant and wild-type strains. Because mRNA is easily degraded, reverse transcriptase is used to convert it to more stable complementary DNA (cDNA, *see figure 17.4*). The cDNAs are labeled with fluorescent dyes and incubated with the microarray under conditions that ensure proper binding of cDNA to its complementary probe. Unbound cDNA is washed off and the microarray is then scanned with laser optics. Fluorescence at each spot or probe indicates that cDNA hybridized to its corresponding gene.

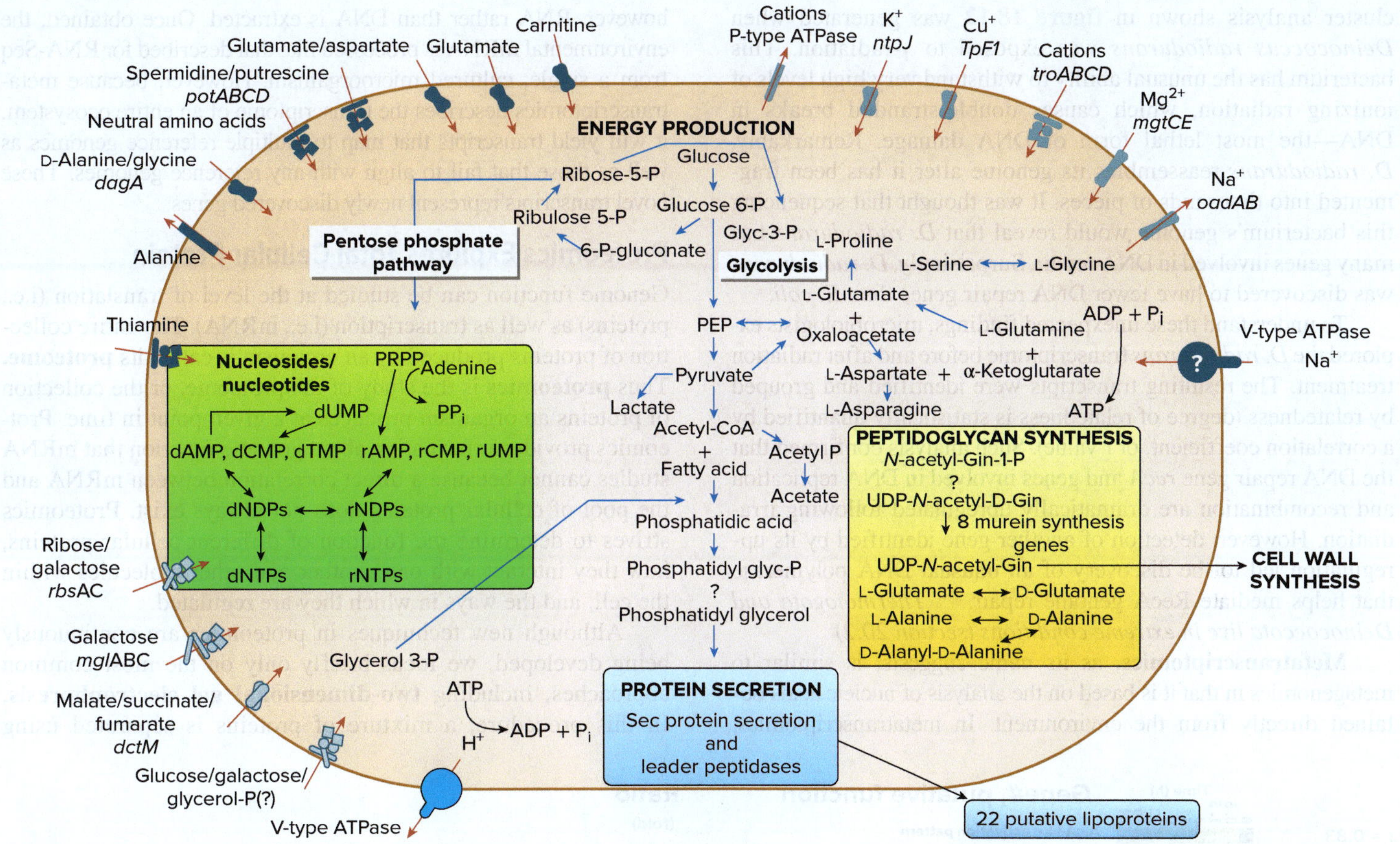

Figure 18.11 Metabolic Pathways and Transport Systems of *Treponema pallidum*. This depicts *T. pallidum* metabolism as deduced from genome annotation. Note the limited biosynthetic capabilities and extensive array of transporters. Although glycolysis is present, the TCA cycle and respiratory electron transport are lacking. Question marks indicate where uncertainties exist or expected activities have not been found.

MICRO INQUIRY *Based on this genomic reconstruction, what are the end products of* T. pallidum *fermentation?*

Recall that each cDNA is derived from an mRNA molecule, so the color and intensity of each probe indicate the relative level of gene expression.

Unfortunately, microarray technology is difficult to standardize and large changes in gene expression cannot be detected. Furthermore, one can only detect changes in transcription level of those genes represented as probes. For these and other reasons, RNA-Seq is now the transcriptomic method of choice. *Cloning technologies developed from DNA structure and enzymology (section 17.1)* *DNA Probe*

The advent of NGS techniques has made it possible to directly sequence total cellular mRNA, a process known as **RNA-Seq.** RNA-Seq has replaced microarray technology for many microbial transcriptomic applications. Quantification of mRNA by RNA-Seq is accomplished by measuring the number of reads (sequenced products) matching each gene.

As with microarray analysis, the process of RNA-Seq begins with extracting all the mRNA from the microbe of interest and converting it to cDNA. Each cDNA, representing an individual mRNA, is then sequenced using NGS (figure 18.4). The results can be analyzed in two ways. Typically, the microorganism being studied has a sequenced genome, so the cDNA sequence is identified by alignment with a reference genome (figure 18.8). If no reference genome is available, the nucleotide sequences are converted to amino acid sequences and compared with databases that house millions of protein sequences.

RNA-Seq generates a wealth of information. Because the number of reads is not limited, there is no upper limit to the level of gene induction that can be detected. For instance, under specific growth conditions, a 9,000-fold increase in *Saccharomyces cerevisiae* gene expression was estimated for several genes when 16 million mapped reads (that is, sequences that could be aligned to the reference genome) were analyzed. Also, because RNA-Seq yields nucleotide sequence, information is revealed about the transcripts themselves. For instance, variations in transcribed sequences of eukaryotic genes can be detected, as can the nucleotides where transcription starts and stops.

Transcriptomic experiments yield a vast amount of information that must be organized in some meaningful fashion. One common approach is to use **hierarchical cluster analysis,** which groups genes according to function or in terms of level of expression: Up-regulated genes (yellow squares) are grouped separately from down-regulated genes (blue squares); genes whose expression remains unaltered are shown in black. The hierarchical cluster analysis shown in **figure 18.12** was generated when *Deinococcus radiodurans* was exposed to γ-radiation. This bacterium has the unusual ability to withstand very high levels of ionizing radiation, which causes double-stranded breaks in DNA—the most lethal form of DNA damage. Remarkably, *D. radiodurans* reassembles its genome after it has been fragmented into thousands of pieces. It was thought that sequencing this bacterium's genome would reveal that *D. radiodurans* has many genes involved in DNA repair. Surprisingly, *D. radiodurans* was discovered to have fewer DNA repair genes than *E. coli*.

To understand these unexpected findings, microbiologists explored the *D. radiodurans* transcriptome before and after radiation treatment. The resulting transcripts were identified and grouped by relatedness (degree of relatedness is statistically quantified by a correlation coefficient, or r value). Such analysis confirmed that the DNA repair gene *recA* and genes involved in DNA replication and recombination are dramatically upregulated following irradiation. However, detection of another gene identified by its upregulation led to the discovery of an unusual DNA polymerase that helps mediate RecA genome repair. ▶| *Thermotogota and Deinococcota live in extreme conditions (section 20.2)*

Metatranscriptomics, as its name suggests, is similar to metagenomics in that it is based on the analysis of nucleic acids obtained directly from the environment. In metatranscriptomics, however, RNA rather than DNA is extracted. Once obtained, the environmental mRNA is processed like that described for RNA-Seq from a single, cultured microorganism. However, because metatranscriptomics describes the transcriptome of an entire ecosystem, it will yield transcripts that map to multiple reference genomes as well as those that fail to align with any reference genomes. These novel transcripts represent newly discovered genes.

Proteomics Explores Total Cellular Protein

Genome function can be studied at the level of translation (i.e., proteins) as well as transcription (i.e., mRNA). The entire collection of proteins produced by an organism is called its **proteome.** Thus **proteomics** is the study of the proteome, or the collection of proteins an organism produces at a given point in time. Proteomics provides information about genome function that mRNA studies cannot because a direct correlation between mRNA and the pool of cellular proteins does not always exist. Proteomics strives to determine the function of different cellular proteins, how they interact with one another and other molecules within the cell, and the ways in which they are regulated.

Although new techniques in proteomics are continuously being developed, we focus briefly only on the most common approaches, including **two-dimensional gel electrophoresis.** In this procedure, a mixture of proteins is separated using

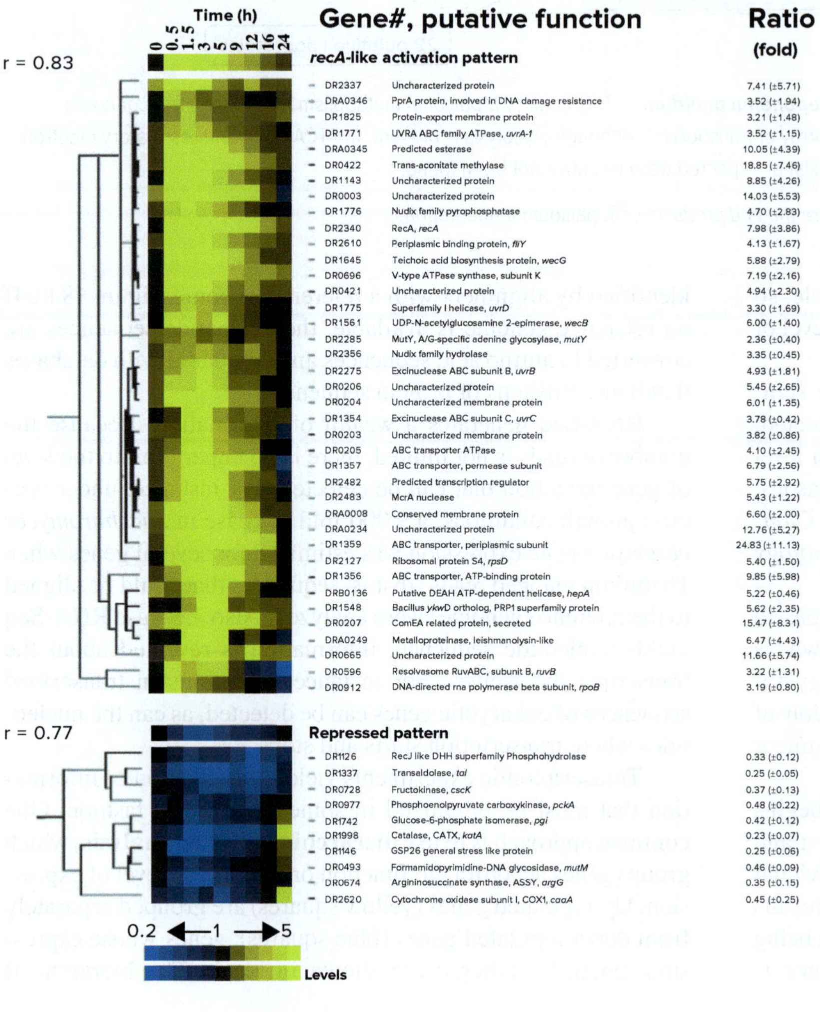

Figure 18.12 Hierarchical Cluster Analysis of Gene Expression of *D. radiodurans* Following Exposure to γ-Radiation. Each row of colored strips represents a single gene, and the color indicates the level of expression over nine time intervals. The far-left column is the control and thus is black (at control levels of expression). The level of induction or repression relative to the control value is indicated as the ratio (fold). Each group of genes has been scored for relatedness, and a tree has been generated on the far left of the clusters, with the indicated correlation coefficient (r value). Many genes encoding DNA repair, synthesis, and recombination proteins are induced upon radiation. These are grouped separately from repressed genes involved in other aspects of metabolism.

Y. Liu et al. "Transciptome dynamics of *Deinococcus radiodurans* recovering from ionizing radiation," *PNAS*, April 2003,Vol. 100 (7): 4191-4196.

MICRO INQUIRY *Based on your interpretation of the data, identify two genes that are strongly induced and two genes that are strongly repressed.*

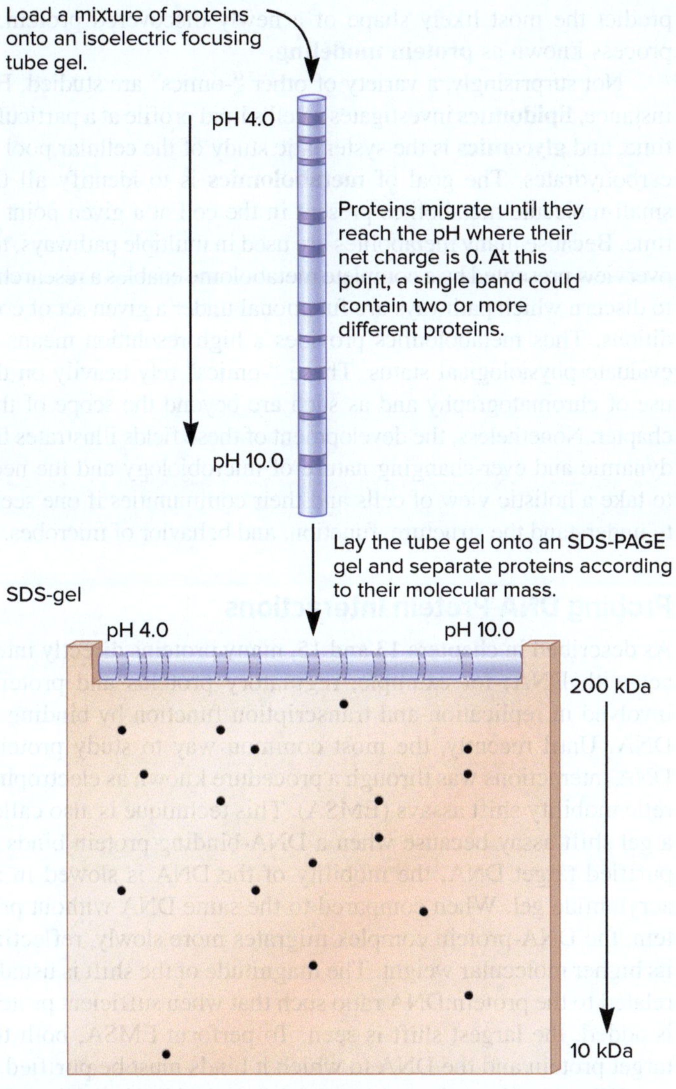

(a) The technique of two-dimensional gel electrophoresis

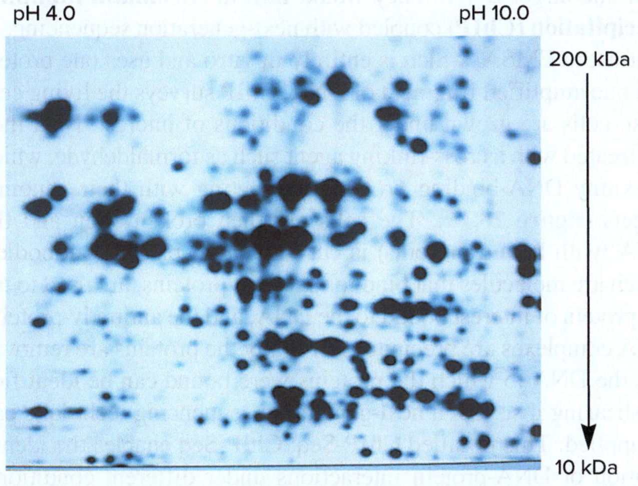

(b) An autoradiograph of a two-dimensional gel. Each protein is a discrete spot.

Figure 18.13 Two-Dimensional Gel Electrophoresis.

(b) Tyne/Simon Fraser/Science Source

two different electrophoretic procedures. This permits the visualization of thousands of proteins that would not otherwise be separated based on molecular weight because many proteins have roughly the same mass. As shown in **figure 18.13*a*,** the first gel involves **isoelectric focusing,** in which proteins move through a pH gradient (e.g., pH 4–10). In isoelectric focusing, a protein mixture is applied to an acrylamide gel in a tube with a pH gradient and electrophoresed. Each protein moves along the pH gradient until the protein's net charge is zero and the protein stops moving. The pH at this point is equal to the protein's **isoelectric point.** Thus this first gel separates proteins based on the content of ionizable amino acids. The second gel is SDS *poly*a*cryl*amide *g*el *e*lectrophoresis (SDS-PAGE). SDS (sodium dodecyl sulfate) is an anionic detergent that denatures proteins and coats them so that now all the proteins have a net negative charge. After isoelectric focusing has been completed, the tube gel is soaked in SDS buffer and placed along the edge of an SDS-PAGE gel. A voltage is then applied. Because all the proteins are negatively charged thanks to SDS treatment, they migrate from the negative pole of the gel to the positive. In this way, polypeptides already separated along the horizontal length of the tube gel by charge are now separated by their molecular weight in the vertical direction; that is, the smallest polypeptide will travel fastest and farthest. Two-dimensional gel electrophoresis can resolve thousands of proteins; each protein is visualized as a spot of varying intensity, depending on its cellular abundance (figure 18.13*b*). Radiolabeled proteins are typically used, enabling greater sensitivity (i.e., proteins present at low concentrations will be detected).

Two-dimensional gel electrophoresis is even more powerful when coupled with **mass spectrometry (MS).** To perform MS, a protein spot is cut from the gel and the extracted protein is cleaved into fragments by treatment with proteolytic enzymes. Then the fragments are analyzed by a mass spectrometer and the mass of the fragments is plotted. This mass fingerprint can be used to estimate the probable amino acid composition of each fragment and tentatively identify the protein. Sometimes proteins or collections of fragments are run through two mass spectrometers in sequence, a process known as **tandem MS** (**figure 18.14**). The first spectrometer separates proteins and fragments, which are further fragmented. The second spectrometer then determines the amino acid sequence (based on molecular weight) of each smaller fragment. The sequence of a whole protein often can be determined by analysis of fragment data. Alternatively, if the genome of the organism has been sequenced, only a partial amino acid sequence is needed. The partial amino acid sequence is aligned to the predicted translated sequences of all the annotated protein-coding genes on the organism's genome. In this way, both the protein and the gene that encodes it can be identified.

In **structural proteomics** the focus is on determining the three-dimensional structures of many proteins and using these to predict the structures of other proteins. The assumption is that proteins fold into a limited number of shapes and can be grouped into families of similar structures. When a number of protein structures are determined for a given family, the patterns of protein structure organization or protein-folding rules will be known. Computational biologists use this information to

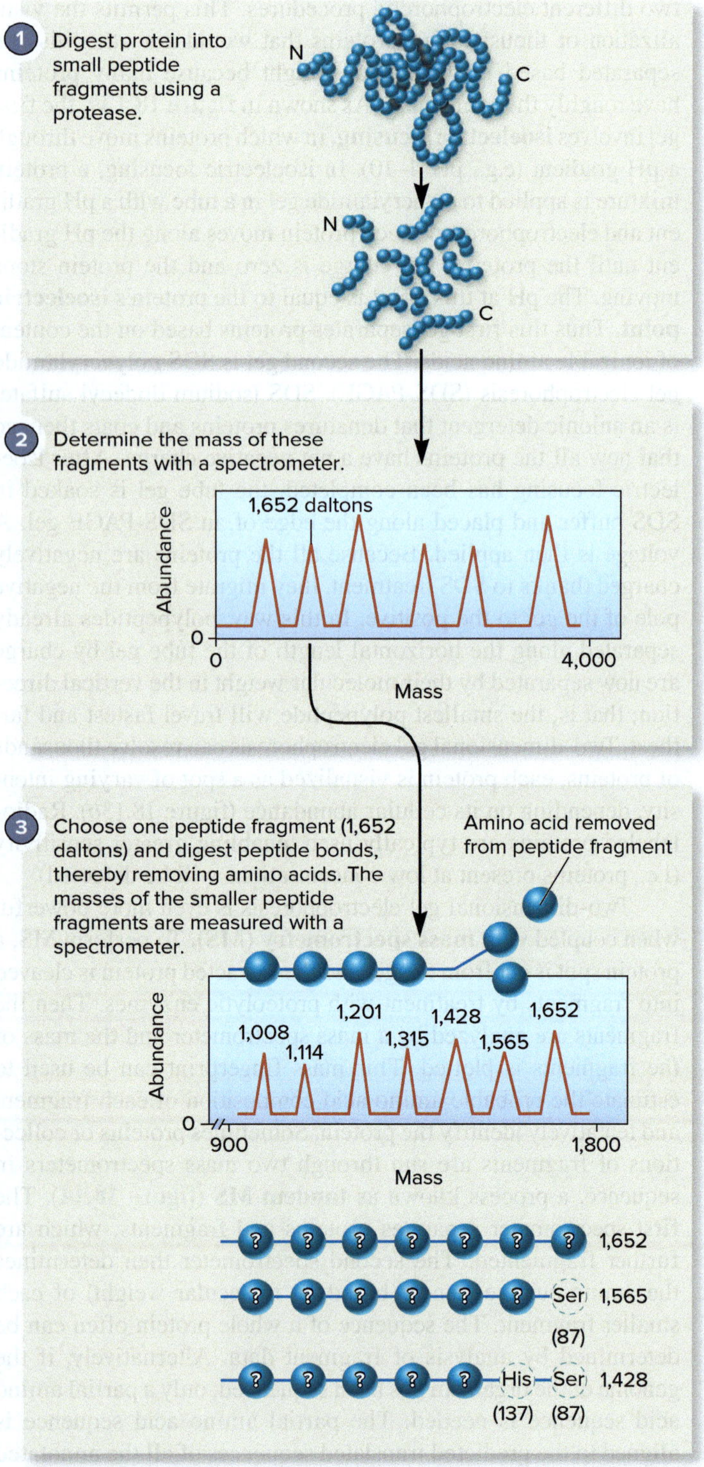

Figure 18.14 The Use of Tandem Mass Spectrometry to Determine the Amino Acid Sequence of a Peptide.

MICRO INQUIRY *How would determining the amino acid sequence of a 30 amino acid residue peptide differ from determining the amino acid sequence of a 200,000 dalton protein?*

predict the most likely shape of a newly discovered protein, a process known as **protein modeling.**

Not surprisingly, a variety of other "-omics" are studied. For instance, **lipidomics** investigates a cell's lipid profile at a particular time, and **glycomics** is the systematic study of the cellular pool of carbohydrates. The goal of **metabolomics** is to identify all the small-molecule metabolites present in the cell at a given point in time. Because many metabolites are used in multiple pathways, the overview presented by a complete metabolome enables a researcher to discern which pathways are functional under a given set of conditions. Thus metabolomics provides a high-resolution means to evaluate physiological status. These "-omics" rely heavily on the use of chromatography and as such are beyond the scope of this chapter. Nonetheless, the development of these fields illustrates the dynamic and ever-changing nature of microbiology and the need to take a holistic view of cells and their communities if one seeks to understand the structure, function, and behavior of microbes.

Probing DNA-Protein Interactions

As described in chapters 13 and 15, many proteins directly interact with DNA; for example, regulatory proteins and proteins involved in replication and transcription function by binding to DNA. Until recently, the most common way to study protein-DNA interactions was through a procedure known as electrophoretic mobility shift assays (EMSA). This technique is also called a gel shift assay because when a DNA-binding protein binds to purified target DNA, the mobility of the DNA is slowed in an acrylamide gel. When compared to the same DNA without protein, the DNA-protein complex migrates more slowly, reflecting its higher molecular weight. The magnitude of the shift is usually related to the protein:DNA ratio such that when sufficient protein is added, the largest shift is seen. To perform EMSA, both the target protein and the DNA to which it binds must be purified.

What if an investigator does not know the identity of the DNA to which a DNA-binding protein binds or suspects it binds to more than one target DNA? They would turn to **chromatin immunoprecipitation (ChIP)** coupled with next-generation sequencing. In contrast to EMSA, which is entirely in vitro and uses one protein and one amplified fragment of DNA, ChIP surveys the living cell. First, cells are grown under the conditions of interest. Next they are treated with a cross-linking agent such as formaldehyde, which fixes any DNA-binding proteins interacting with their genomic targets (**figure 18.15**). The cells are then broken open and the DNA (with bound proteins) is cut into small pieces. Antibodies, which are molecules that bind to specified proteins, are used to tag the protein of interest bound to the DNA, and the antibody-protein-DNA complexes are precipitated. Finally the proteins are removed and the DNA to which the proteins were bound can be identified by shearing it so that a next-generation sequencing technique can be applied. This is called ChIP-Seq. ChIP-Seq enables the identification of DNA-protein interactions under different conditions. The same microbial strain can be treated with cross-linking agents when grown in different situations or at different phases of growth, so the DNA-binding profile of a protein can be followed.

▶| *Antibodies bind specific 3-D antigens (section 32.7)*

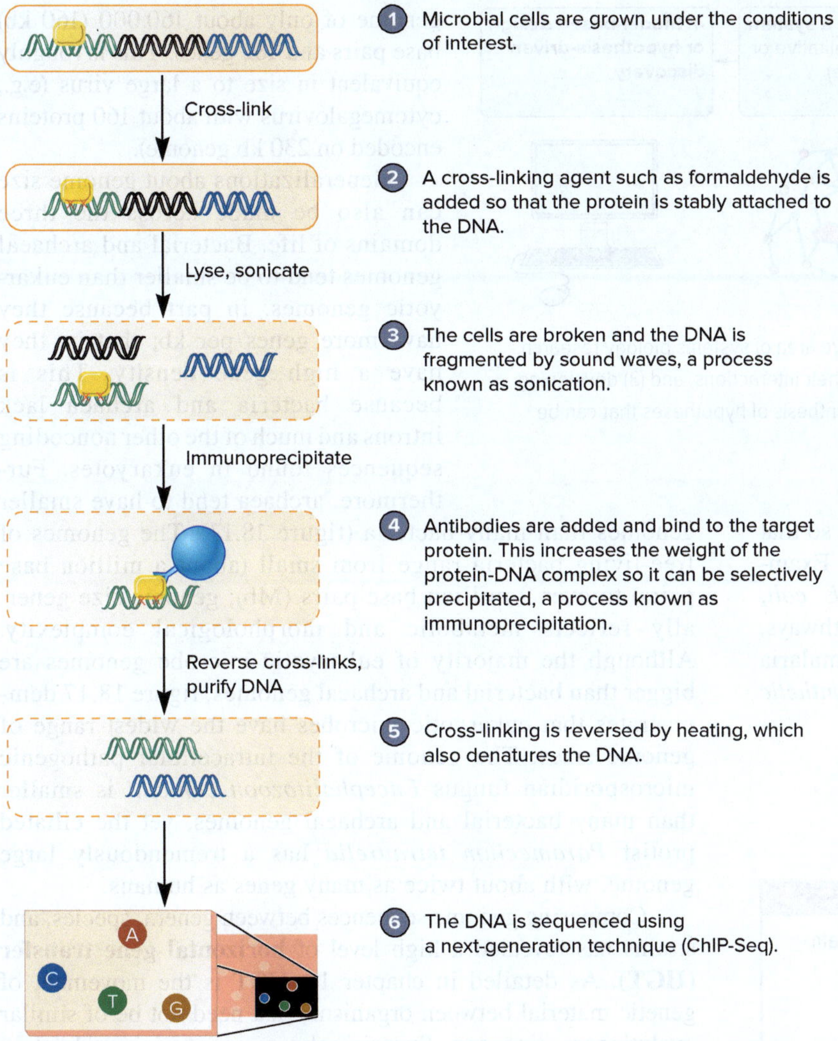

Figure 18.15 ChIP-Seq Analysis. This technology can be used to identify the DNA to which a protein binds.

Comprehension Check

1. Explain the difference between RNA-Seq and metatranscriptomics.
2. How might the following scientists use transcriptomics in their research?
   (a) A microbial ecologist who is interested in how the soil microbe *Rhodopseudomonas palustris* degrades the toxic compound 3-chlorobenzene
   (b) A medical microbiologist who wants to learn how the pathogen *Salmonella* survives within a host cell
3. Why does two-dimensional gel electrophoresis allow the visualization of many more cellular proteins than seen in single-dimension electrophoresis (i.e., SDS-PAGE)?
4. What is the difference between mass spectrometry and tandem mass spectrometry?
5. Describe a ChIP-Seq experiment that will enable the identification of promoters to which the stationary phase sigma factor RpoS binds in *E. coli.* How could you also determine when RpoS binds to DNA?

18.6 Systems Biology: Making and Testing Complex Predictions

After reading this section, you should be able to:

a. Explain the role of genomics in shaping systems biology
b. Compare and contrast systems biology and synthetic biology

Imagine the following experiment: You have created a microbial strain in which you have deleted, or "knocked out," a single gene; it doesn't really matter what the gene encodes. You want to compare the transcriptome and proteome of the knockout mutant with its parent strain, which is identical to it in every way except the presence of this particular gene. To compare the transcriptomes of wild-type and mutant strains, you perform RNA-Seq, and to compare their proteomes, you run two-dimensional gels followed by mass spectroscopy. You find that the loss of just one gene has a huge impact on gene expression and protein synthesis. Not only is the expected pathway altered, but the expression of over a hundred other genes and proteins has also been affected. A single mutation has altered the function of the system, which in this case is a microbial cell.

This scenario has been repeated many times for numerous microorganisms. Life scientists now realize that most genes and the proteins they encode have a vastly broader role in cellular physiology than previously thought. Gone are the days of reductionist biology, when a single metabolic or regulatory pathway could be considered in isolation. **Systems biology** seeks to integrate the "parts list" of cells—mRNA, proteins, small molecules—with the molecular interactions that become pathways for catabolism, anabolism, regulation, behavior, responses to environmental signals, etc. To accomplish this, systems biology is interdisciplinary, requiring the expertise of physiologists, biochemists, bioinformaticists, mathematicians, geneticists, ecologists, and others.

One area where systems biology has been applied to microbiology is the study of metabolism (**figure 18.16**). Here genome sequencing and annotation lead to one set of predictions to be tested by transcriptomics, proteomics, and metabolomics. The interactions between the individual products—RNA, protein, metabolites—can then be linked in a predictive network and used to construct testable hypotheses. Like most scientific endeavors, systems biology is an iterative process involving multiple rounds of data collection, analysis, and hypothesis building. What is unique, however, is the volume of data, the depth of sampling, and the use of sophisticated predictive models to guide the development of new hypotheses.

As systems biology matured, scientists realized that understanding cellular regulatory systems could lead to a new level of metabolic engineering such that a microbe could be constructed with novel genetic networks. The introduction of artificial

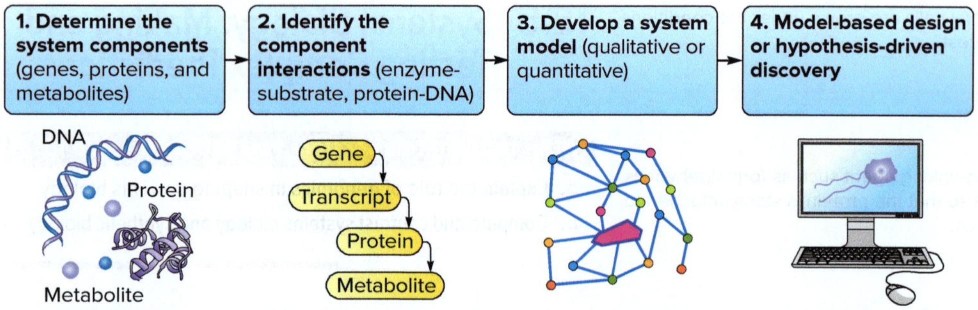

Figure 18.16 Systems Biology. Microbial metabolism is an active area of systems biology research. It can involve (1) identifying the system components, (2) recognizing their interactions, and (3) developing a model that predicts how they interact, which then (4) informs the synthesis of hypotheses that can be tested either by further modeling or by experimentation.

regulatory and metabolic pathways into a microoganism so that it produces specific products is called **synthetic biology.** Examples of synthetic biology include biofuel production in *E. coli,* accomplished by rerouting amino acid biosynthetic pathways, and the commercial production of the important antimalaria drug artemisinin from *Saccharomyces cerevisiae.* ◀ *Synthetic biology (section 17.6)*

18.7 Comparative Genomics

After reading this section, you should be able to:

a. Predict the relative genome size of an organism given its domain and its morphologic and metabolic complexity
b. Describe the genomic differences that distinguish intracellular parasites from free-living microbes

A natural consequence of the tremendous number of sequenced genomes is the ability to make meaningful genome comparisons, which can reveal much information about the evolution of microbes. **Comparative genomics** focuses on similar nucleotide and amino acid sequences among organisms to infer gene function and evolutionary relationships. Perhaps the most fundamental comparison is the relative sizes of microbial genomes. Several generalizations can be deduced from such analysis. Inspection of **figure 18.17** reveals that the smallest genomes published to date, across all three domains, belong to parasitic microbes. These organisms depend on metabolites from their hosts, with whom they have coevolved. Parasitic microbes have therefore lost the genes encoding enzymes needed for the assembly of these molecules. In addition to losing genes, these microbes also possess genes that are no longer functional. These genes, which have degraded through lack of selection (their products no longer needed), are called **pseudogenes.** For example, one of the smallest genomes belongs to *Candidatus* Carsonella ruddii, which lives within the sap-sucking insect *Pachypsylla venusta.* This bacterium, which cannot be grown without its host, has a genome of only about 160,000 (160 kb) base pairs and 182 genes. This is roughly equivalent in size to a large virus (e.g., cytomegalovirus with about 160 proteins encoded on 230 kb genome).

Generalizations about genome size can also be made across the three domains of life. Bacterial and archaeal genomes tend to be smaller than eukaryotic genomes, in part because they have more genes per kb; that is, they have a high gene density. This is because bacteria and archaea lack introns and much of the other noncoding sequences found in eukaryotes. Furthermore, archaea tend to have smaller genomes than many bacteria (figure 18.17). The genomes of free-living bacteria range from small (about a million base pairs) to over 9 million base pairs (Mb); genome size generally reflects metabolic and morphological complexity. Although the majority of eukaryotic microbe genomes are bigger than bacterial and archaeal genomes, figure 18.17 demonstrates that eukaryotic microbes have the widest range of genome sizes. The genome of the intracellular pathogenic microsporidian fungus *Encephalitozoon cuniculi* is smaller than many bacterial and archaeal genomes, yet the ciliated protist *Paramecium tetraurelia* has a tremendously large genome, with about twice as many genes as humans.

Comparing genome sequences between genera, species, and strains has revealed a high level of **horizontal gene transfer (HGT).** As detailed in chapter 16, HGT is the movement of genetic material between organisms that need not be of similar evolutionary lineages. Surprisingly, many genes involved in myriad cellular functions reside on mobile genetic elements, including transposons, plasmids, and phages. Genome analysis has revealed that HGT is frequently mediated by viruses and that lysogeny may be the rule, rather than the exception. In fact, some bacteria and archaea carry multiple proviruses. Some temperate phages carry lysogenic conversion genes, so called because they change the phenotype of their hosts. For instance, the virulence genes for some important pathogens, including the bacteria that cause diphtheria and cholera, are encoded by phages. ◀ *Lytic and lysogenic infections are common for bacterial and archaeal cells (section 6.4)*

Mobile genetic elements permanently integrated into a microbial genome are called **genomic islands.** If these new genes encode proteins that confer virulence, they are known as **pathogenicity islands.** The term *island* reflects the fact that these imported DNA sequences are different than the rest of the organism's genome. For example, genomic and pathogenicity islands often have a G + C content (measured as percentage) unlike the remainder of the genome. Genomic islands also sometimes retain inverted repeat sequences or other elements that mark their transposon or viral origin. It has become clear that HGT is a major evolutionary force in short-term microbial

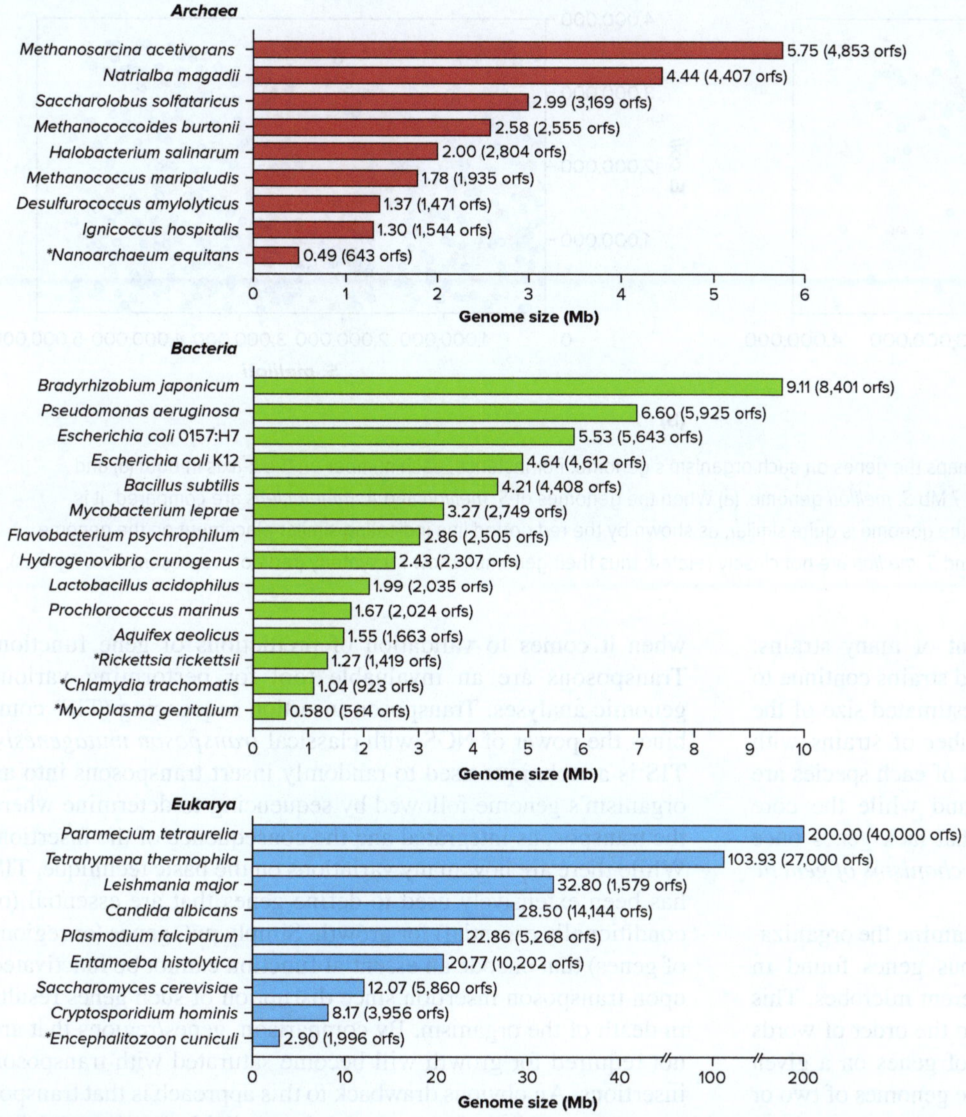

Figure 18.17 Microbial Genome Sizes. The distribution of genome sizes in millions of base pairs (Mb) among some archaea, bacteria, and eukaryotic microbes. Asterisk denotes that the microbe is parasitic. Note that the x-axis in each panel has a different scale.

MICRO INQUIRY *Two E. coli strains are shown in this figure. One is a pathogen and one is not. Which do you think is pathogenic and why?*

evolution and long-term speciation. *Microbes use mechanisms other than mutation to create genetic variability (section 16.3)*

The impact of HGT on microbial evolution has been studied by comparative genomics. Genome comparisons of strains within a species and among species within a genus have revealed that microbial genomes consist of an older core genome and a more recently acquired pan-genome (**figure 18.18**). The **core genome** is the set of genes found in all members of a species (or other monophyletic group). Thus it is thought to represent the minimal number of genes needed for the group of microbes to survive. In general, these genes encode informational proteins involved in DNA replication, transcription, and translation (e.g., rRNA genes). Because these genes are so highly conserved, they were likely present in the group's common ancestor. By contrast, the **pan-genome** consists of every gene in all strains of a species (or other taxonomic unit), so it includes the core genome plus every additional gene found in at least one strain. Genes outside the core genome were more recently acquired and enable microbial colonization of new niches. In general, genes unique to the pan-genome are considered to have been acquired by HGT. A comparison of the core genome size with the actual genome size of a particular strain thus indicates the evolution of new traits. For instance, current values of the core genome and pan-genome of *Bacillus anthracis* differ by only roughly 200 genes (about 3,600 versus 3,800, respectively), reflecting the limited genetic diversity within this species. By contrast, the *E. coli* core genome consists of about 2,800 genes, whereas some estimate that the pan-genome consists of roughly 25,000 genes. The broad genetic diversity among *E. coli* strains is illustrated by comparing the genomes of the nonpathogenic *E. coli* strain K12 and the pathogenic strain O157:H7 (figure 18.17). Their last common ancestor is estimated to have lived about 4.5 million years ago. During this time *E. coli* has mutated and

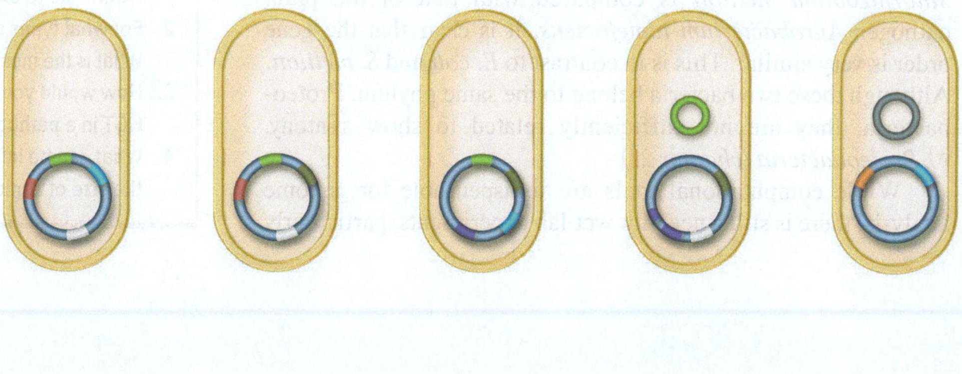

Figure 18.18 Core and Pan-genomes. The core and pan-genomes in cells comprising strains within a species. The core genome is common to all members (blue), whereas the pan-genome includes all sequences present in any member of the set.

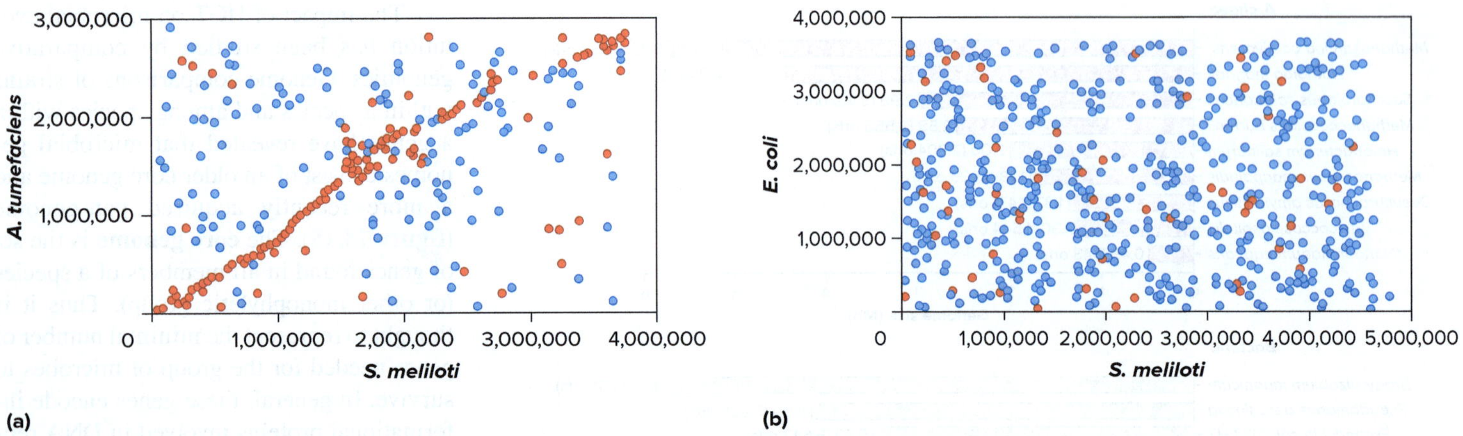

Figure 18.19 Synteny. Each axis on the graphs maps the genes on each organism's genome. For instance, each number on the x-axis in both (a) and (b) indicates the position of genes (blue dots) on the 3.7 Mb *S. meliloti* genome. (a) When the genomes of *S. meliloti* and *A. tumefaciens* are compared, it is evident that the distribution of orthologous genes on the genome is quite similar, as shown by the red dotted line indicating similar placement on the genome (red dots represent *A. tumefaciens* genes). (b) *E. coli* and *S. meliloti* are not closely related; thus their genomes show no synteny (red dots represent *E. coli* genes).

exchanged genes, allowing the development of many strains. These have radiated to numerous habitats and strains continue to adapt. Obviously for any given species the estimated size of the core and pan-genome depends on the number of strains with sequenced genomes. Indeed, as more strains of each species are sequenced, the pan-genome tends to expand while the core genome shinks, as strains are discovered that lack genes once thought to be common to all genomes. ◀ *Mechanisms of genetic variation (chapter 16)*

Comparative genomics is also used to examine the organization of orthologous genes (i.e., homologous genes found in different organisms) in the genomes of different microbes. This is called **synteny,** and just as syntax refers to the order of words in a sentence, synteny describes the order of genes on a given genome and compares this order between the genomes of two or more organisms. Closely related microbes share a more recent common ancestor, so they generally have a higher degree of synteny than more distantly related organisms. It may be easiest to understand synteny by first considering the order of genes in similar operons found in two different microbes. For instance, one would not be surprised to learn that the order of tryptophan biosynthetic genes is the same (conserved) in the *E. coli* and *Salmonella* spp. genomes because these two bacteria are related (both are Enterobacteriacea). Analysis of synteny takes this comparison to the level of whole genomes. As shown in **figure 18.19,** when the order of genes in the genome of the plant symbiont *Sinorhizobium meliloti* is compared with that of the plant pathogen *Agrobacterium tumefaciens,* it is clear that the gene order is very similar. This is in contrast to *E. coli* and *S. meliloti.* Although these two bacteria belong to the same phylum, Proteobacteria, they are not sufficiently related to show synteny. ▶ *Proteobacteria (chapter 21)*

While computational tools are indispensable for genome analysis, there is still a need for wet-lab experiments, particularly when it comes to validation of predictions of gene function. Transposons are an invaluable tool for performing various genomic analyses. Transposon insertion sequencing (TIS) combines the power of NGS with classical *transposon mutagenesis.* TIS is a technique used to randomly insert transposons into an organism's genome followed by sequencing to determine where the transposons integrated and the consequence of the insertion. While there are now many variations on the basic technique, TIS has been extensively used to define genes that are essential (or conditionally essential) for growth. Simply put, genes (or regions of genes) that encode an essential function cannot be inactivated upon transposon insertion since disruption of such genes results in death of the organism. By comparison, genes/regions that are not required for growth will become saturated with transposon insertions. An obvious drawback to this approach is that transposon mutagenesis techniques must be available for the organism under study, and that organism also must be culturable in the laboratory. However, performing TIS with a larger, more diverse population of culturable organisms will strengthen the bioinformatic predictions on which we often must rely.

Comprehension Check

1. Cite an infectious disease for which you think a systems biology approach would be appropriate. List three specific questions you would like to explore regarding this host-pathogen interaction.
2. For what types of microorganisms is extensive gene loss common? What is the most likely explanation for this phenomenon?
3. How would you infer that a toxin gene was recently acquired by HGT in a pathogen such as *E. coli* O157:H7?
4. What can be inferred about evolution by the difference between the size of a microbial species' core and pan-genome?

Key Concepts

18.1 DNA Sequencing Methods

- Genomics is the study of the molecular organization of genomes, their information content, and the gene products they encode.
- DNA fragments have traditionally been sequenced using dideoxynucleotides and the Sanger chain termination technique (**figures 18.2** and **18.3**).
- Next-generation sequencing (NGS) technologies are faster and cheaper than the Sanger method and are based on sequencing by synthesis, employing reversible chain termination (**figures 18.4** and **18.5**).

18.2 Genome Sequencing

- The Sanger sequencing technique uses the whole-genome shotgun technique. Four stages are involved: library construction, sequencing of randomly produced fragments, fragment alignment and gap closure, and editing the final sequence.
- Genome sequencing using NGS techniques has accelerated the volume and pace of microbial genome sequencing.
- It is now possible to use multiple displacement amplification to copy the genome of a single microbial cell so there is sufficient DNA for sequencing. In this way, the genomes of uncultivated microbes can be studied (**figure 18.7**).

18.3 Metagenomics Provides Access to Uncultured Microbes

- Metagenomics enables the study of the biodiversity and metabolic potential of microbial communities without culturing individual microbes.
- Metagenomics involves sequencing DNA extracted directly from the environment. NGS approaches are used to avoid genomic library construction, and genome sequences are assembled using reference genomes (**figure 18.8**).
- Metagenomics has been applied to many natural microbial habitats, including the human body.

18.4 Bioinformatics: What Does the Sequence Mean?

- Analysis of vast amounts of genome data requires sophisticated analytical procedures; this is the basis of the discipline of bioinformatics.
- An open reading frame is a putative protein-coding gene; the identity of the gene and its product are inferred by aligning the translated amino acid sequence with similar sequences in the database (**figures 18.9** and **18.10**).
- Bioinformatics enables the comparison of genes within genomes to identify paralogues and the comparison of genes between different organisms to identify orthologues.
- The physical map of a microbe's genome is often displayed so the location of genes with related functions are easily identified.

18.5 Functional Genomics Links Genes to Phenotype

- The presence of genes whose products are known to function in specific metabolic processes such as catabolic pathways and transport enables the genomic reconstruction of the organism—a type of map that illustrates these processes within the cell (**figure 18.11**).
- Transcriptomics is the study of all the RNA an organism makes at a given time. It is best studied with RNA-Seq, which assesses gene expression as a measure of individual gene transcripts (mRNA). Gene expression can be compared between mutant and wild-type strains or for a given organism under different environmental conditions.
- The entire collection of proteins that an organism can produce is its proteome, and its study is called proteomics.
- The proteome is often analyzed by two-dimensional gel electrophoresis, in which the total cellular protein pool can be visualized. In many cases, the amino acid sequence of individual proteins is determined by mass spectrometry; if this is coupled to genomics, both a protein of interest and the gene that encodes it can be identified (**figures 18.13** and **18.14**).
- Structural proteomics seeks to model the three-dimensional structure of proteins based on computer analysis of amino acid sequence data.
- Protein-DNA interactions can be detected by electrophoretic mobility shift assay (EMSA) or more precisely by chromatin immunoprecipitation (**figure 18.15**).

18.6 Systems Biology: Making and Testing Complex Predictions

- Systems biology combines experimentally derived data sets with predictive modeling to study an organismal process (**figure 18.16**).
- Systems biology is the opposite of reductionist biology because it integrates molecular networks with the goal of understanding the function of the cell as a whole.

18.7 Comparative Genomics

- Comparing genomes reveals information about genome structure and evolution, including the importance of horizontal gene transfer (HGT).
- The core genome includes genes common to all strains within a microbial species, whereas the pan-genome represents every gene found in all strains. Genes on the pan-genome are thought to have been acquired by HGT and enable strains to rapidly adapt to new environments (**figure 18.18**).
- Comparative genomics is an important tool in discerning how microbes have adapted to particular ecological niches and in developing new therapeutic agents.

Active Learning

1. You are annotating the genomic sequence of a new soil bacterium. You believe you have found a gene that encodes nitrate reductase. How did you arrive at this conclusion, and how will you strengthen this annotation using only in silico analysis?
2. You have discovered a new DNA-binding protein. You show by EMSA that it binds near the promoter of an operon that encodes enzymes required for carotenoid biosynthesis. You wonder if any other genes are controlled by this new protein. How will you find other regions of DNA to which the protein might bind? How will you determine if the new protein increases or decreases transcription of the genes it regulates?
3. You are developing a new vaccine for a pathogen. You want your vaccine to recognize specific cell-surface proteins. Explain how you will use genome analysis to identify potential protein targets. What functional genomics approach will you use to determine which of these proteins is produced when the pathogen is in its host?

CHAPTER

19

Archaea

Chatrawee Wiratgasem/Shutterstock

Taking the Pledge

Since 2021, more than 155 nations have committed to lowering human-caused methane emissions by taking the Global Methane Pledge. As one of the three gases (carbon dioxide, CO_2; methane, CH_4; nitrous oxide, N_2O) most responsible for climate change, methane is targeted for a 30% reduction by 2030. While long-term changes are critical to control atmospheric concentrations of greenhouse gases, the Global Methane Pledge addresses actions with a significant effect in the short term, namely agricultural practices.

Rice is widely cultivated as a food staple. Traditionally, rice has been grown in flooded, terraced paddies—conditions that promote the growth of microbes that generate significant quantities of both CH_4 and N_2O. It's not surprising, then, that agricultural practices in rice fields are under critical review.

The microbial communities that thrive in rice paddies generate N_2O through denitrification, and methane is produced by archaeal **methanogens** as the end product of organic matter decomposition. While this greenhouse gas production can be viewed as microbially induced, it's clear that there is also a role for microbes in strategies to reduce those emissions.

Several methods are being adopted to mitigate CH_4 emission. Application of sulfate to rice paddies promotes the growth of sulfate-reducing bacteria that compete with methanogens for carbon and electron sources. Alternate wetting and drying practices reduce water use and lessen the anoxic environment where methanogens thrive, and new cultivars of rice have been developed for these conditions. In addition, these oxic niches promote methane consumption by aerobic methanotrophic bacteria.

In this chapter, we review archaea, microorganisms that are metabolically and ecologically diverse. Although they rarely dominate an environment, their metabolic capabilities make them critical components of the global carbon and nitrogen cycles. Exploring archaeal biology and ecology is critical for addressing the role of these organisms in climate change.

Readiness Check:

Based on what you have learned previously, you should be able to:

- ✓ Draw the basic structure of the phylogenetic tree of life based on SSU rRNA sequences and explain the placement of Archaea (section 1.2)
- ✓ Describe an archaeal cell wall and membrane, including component parts (section 4.2)
- ✓ List environmental factors that influence microbial growth, including oxygen (section 7.5)
- ✓ Explain the difference between heterotrophy and autotrophy (section 11.1)
- ✓ Define chemolithotrophy and identify an example of an electron donor and an acceptor (section 11.8)
- ✓ Review archaeal DNA replication, transcription, and protein synthesis and secretion (sections 15.2, 15.3, and 15.4)

19.1 Overview of Archaea

After reading this section, you should be able to:

a. List some common habitats in which archaea reside
b. Describe the challenges in archaeal taxonomy
c. Compare at least three key pathways that are central to archaeal metabolism with those used by bacteria

Archaea have many features in common with eukaryotes, others in common with bacteria, and still other elements that are uniquely archaeal (*see table 1.1*). In general, archaeal informational processes—that is, DNA replication, transcription, and translation—are most like those of eukaryotes, whereas metabolic processes are most similar to bacteria. Some unique archaeal features include their membrane lipid structure and the production of methane. Like bacteria, archaea are morphologically diverse. They include the full range of cell shapes observed in bacteria. Multiplication is usually by binary fission, but some undergo budding, fragmentation, or other mechanisms. Archaea are just as diverse metabolically. They perform important roles in global carbon and nitrogen cycles. Nutritionally, they range from chemolithoautotrophs to chemoheterotrophs.

Archaea inhabit a wide variety of habitats, but for many years they were considered microbes of extreme environments, or *extremophiles*. Indeed, many, but by no means all, archaea inhabit niches that have very high temperatures, low pH, or concentrated salts, or are completely **anoxic,** without oxygen. For instance, in some hypersaline environments, their populations become so dense that the brine is red with archaeal pigments. However, archaea are also important members of marine and soil communities—environments that cannot be considered extreme. In addition, some archaea are members of the human microbiome, residing in the digestive tract or oral cavity. Thus the notion that archaea are exclusively extremophiles is no longer valid.

Archaeal Taxonomy

In the current taxonomic framework derived from comparison of sequenced genomes and metagenomes, the domain Archaea contains 21 phyla, five of which are discussed here. The evolutionary relationships among archaeal lineages is an area of active research that has not yet reached consensus. Several archaeal phyla are composed largely of **microbial dark matter**—that is, the organisms have never been grown in laboratory culture—so what we know about them is inferred from their genome sequences. Development of single-cell genomic sequencing, and metagenome assembly methods have massively expanded the known members of archaea. The importance of identifying and classifying microbes by culture-free methods cannot be understated when discussing archaea. ◀ *Single-cell genomic sequencing (section 18.2); Metagenomics provides access to uncultured microbes (section 18.3)*

Metabolism

In view of the diversity of archaeal lifestyles, it is not surprising that archaeal metabolism varies greatly. Some archaea are autotrophic; others are heterotrophs. Here we review the pathways by which archaea fix carbon and catabolize carbohydrates.

CO_2-Fixation Pathways

Like all anabolic processes, carbon fixation requires reducing power and energy. Because inorganic carbon (CO_2 or HCO_3^-) has an oxidation state of +4, it requires four reducing equivalents for its assimilation into cellular carbon (oxidation state of 0). Most known autotrophic archaea either are anaerobic or live in low oxygen conditions. These microbes tend to use ferredoxin (Fd) instead of NADPH as their electron carrier. Because reduced Fd has a standard reduction potential (E_0') of −400 mV, it bears more energy than NADPH (E_0' of −320 mV), and most anaerobic pathways require less ATP per CO_2 fixed than aerobic carbon fixation pathways (e.g., the Calvin-Benson cycle). This is important as some autotrophic archaea, particularly methanogens, live close to their thermodynamic limit and therefore have very little (if any) ATP to spare. ◀ *Redox reactions: reactions of central importance in metabolism (section 10.3); CO_2-fixation: reduction and assimilation of CO_2 carbon (section 12.3)*

Three different carbon fixation pathways have been characterized in archaea: the Wood-Ljungdahl pathway (also known as the reductive acetyl-CoA pathway), the 3-hydroxypropionate/4-hydroxybutyrate (HP/HB) cycle, and the dicarboxylate/4-hydroxybutyrate (DC/HB) cycle (**table 19.1**).

The Wood-Ljungdahl pathway is the most energetically favorable route and is widespread in archaea. In this branched pathway, two CO_2 molecules are reduced and incorporated into one molecule of acetyl-CoA (**figure 19.1**). One CO_2 is reduced to carbon monoxide (CO) and the other CO_2 is reduced to the level of CH_3 while bound to a tetrahydromethanopterin coenzyme. A single enzyme complex catalyzes both pathway branches and the pathway end-product, acetyl-CoA, is a precursor metabolite used in biosynthesis. ◀ *Precursor metabolites (section 12.2)*

The HP/HB cycle (**figure 19.2**) requires more energy than any other CO_2-fixation pathway (9 ATP/pyruvate synthesized)

Table 19.1 Comparison of Archaeal CO_2-Fixation Pathways to the Calvin-Benson Cycle

CO_2-Fixation Pathway	Oxygen Tolerance	ATP Consumed per Pyruvate Produced
Wood-Ljungdahl (Reductive Acetyl-CoA)	Anaerobic	1
3-Hydroxypropionate/ 4-Hydroxybutyrate	Aerobic	9
Dicarboxylate/ 4-Hydroxybutyrate	Anaerobic	5
Calvin-Benson	Aerobic	7

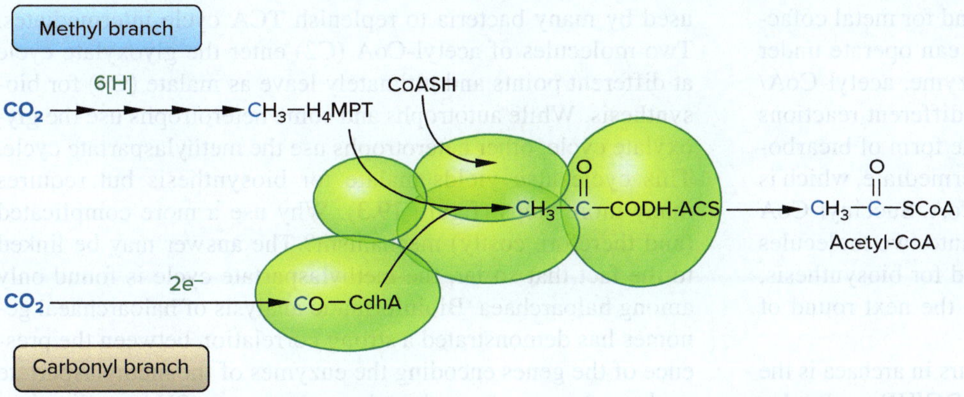

Figure 19.1 The Wood-Ljungdahl Pathway. Two molecules of CO_2 are reduced, each by a different mechanism, and combined to form an acetyl group by the enzyme carbon monoxide dehydrogenase/acetyl-CoA synthase complex (CODH/ACS; green). In the methyl branch, the carbon is attached to a tetrahydromethanopterin (H_4MPT) carrier (figure 19.10). In the carbonyl branch, two electrons are transferred to CO_2 to generate CO. Finally, CO is combined with the methyl group and CoA to form acetyl-CoA.

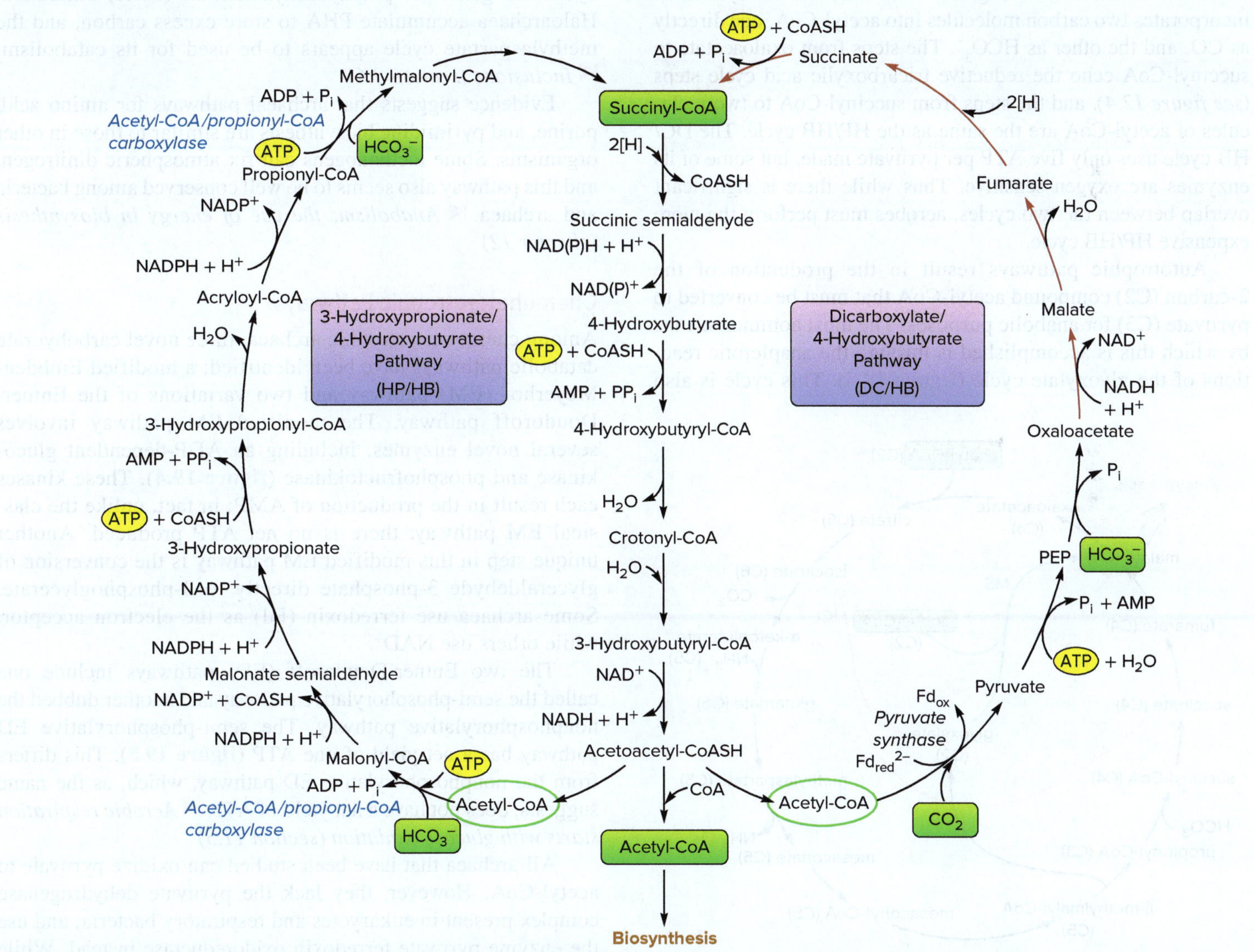

Figure 19.2 The 3-Hydroxypropionate/4-Hydroxybutyrate (HP/HB) and Dicarboxylate/4-Hydroxybutyrate (DC/HB) Cycles. Follow each cycle beginning with the acetyl-CoA molecules at the bottom center (circled in green). In the HP/HB cycle (left), acetyl-CoA/propionyl-CoA carboxylase (blue) catalyzes two related reactions to incorporate bicarbonate into an organic molecule. In the anaerobic DC/HB cycle (right), one molecule of CO_2 and one bicarbonate ion are fixed. Both cycles use the same pathway and enzymes to convert succinyl-CoA to acetyl-CoA (center). The reactions shown with red arrows are also in the reductive tricarboxylic acid cycle (*see figure 12.4*).

MICRO INQUIRY *Given that the DC/HB pathway uses far less ATP per pyruvate synthesized than the HP/HB pathway, why do some archaea use the HP/HB pathway instead?*

but has other advantages. It has a low demand for metal cofactors, and unlike the other two pathways, it can operate under aerobic conditions (table 19.1). The key enzyme, acetyl-CoA/propionyl-CoA carboxylase, catalyzes two different reactions in the pathway. In both reactions, CO_2 in the form of bicarbonate ion (HCO_3^-) is added to an organic intermediate, which is rearranged and ultimately yields succinyl-CoA. Succinyl-CoA is transformed through 4-hydroxybutyrate into two molecules of acetyl-CoA. One acetyl-CoA is then used for biosynthesis, and the other is used as a CO_2 acceptor in the next round of the cycle.

An additional autotrophic cycle that occurs in archaea is the anaerobic DC/HB cycle (figure 19.2). The DC/HB cycle also incorporates two carbon molecules into acetyl-CoA, one directly as CO_2 and the other as HCO_3^-. The steps from oxaloacetate to succinyl-CoA echo the reductive tricarboxylic acid cycle steps (*see figure 12.4*), and the steps from succinyl-CoA to two molecules of acetyl-CoA are the same as the HP/HB cycle. The DC/HB cycle uses only five ATP per pyruvate made, but some of its enzymes are oxygen sensitive. Thus while there is significant overlap between the two cycles, aerobes must perform the more expensive HP/HB cycle.

Autotrophic pathways result in the production of the 2-carbon (C2) compound acetyl-CoA that must be converted to pyruvate (C3) for anabolic purposes. The most common means by which this is accomplished is through the anaplerotic reactions of the glyoxylate cycle (**figure 19.3**). This cycle is also used by many bacteria to replenish TCA cycle intermediates. Two molecules of acetyl-CoA (C2) enter the glyoxylate cycle at different points and ultimately leave as malate (C4) for biosynthesis. While autotrophs and some heterotrophs use the glyoxylate cycle, other heterotrophs use the methylaspartate cycle. This cycle also yields malate for biosynthesis but requires many more steps (figure 19.3). Why use a more complicated (and therefore costly) mechanism? The answer may be linked to the fact that so far, the methylaspartate cycle is found only among haloarchaea. Bioinformatic analysis of haloarchaeal genomes has demonstrated a strong correlation between the presence of the genes encoding the enzymes of the methylaspartate cycle and genes for polyhydroxyalkanoate (PHA) utilization. Haloarchaea accumulate PHA to store excess carbon, and the methylaspartate cycle appears to be used for its catabolism. ◀ *Inclusions (section 3.6)*

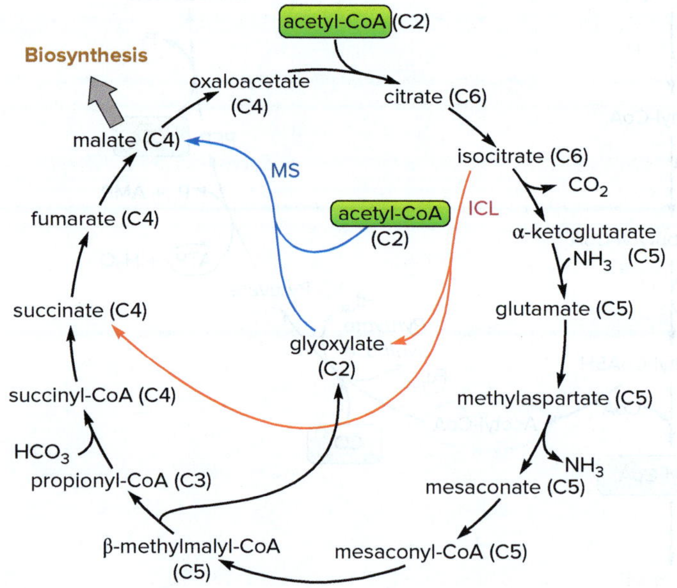

Figure 19.3 Glyoxylate and Methylaspartate Cycles Are Used for Acetyl-CoA Assimilation. Autotrophic archaea and some halophiles use the glyoxylate cycle (red and blue arrows) to incorporate acetyl-CoA. Key enzymes are isocitrate lyase (ICL; red) and malate synthase (MS; blue). Other halophiles lack ICL and assimilate acetyl-CoA via the longer methylaspartate cycle instead (black and blue arrows).

Evidence suggests that archaeal pathways for amino acid, purine, and pyrimidine biosynthesis are similar to those in other organisms. Some methanogens can fix atmospheric dinitrogen, and this pathway also seems to be well conserved among bacteria and archaea. ◀ *Anabolism: the use of energy in biosynthesis (chapter 12)*

Chemoheterotrophic Pathways

Among chemoheterotrophic archaea, three novel carbohydrate catabolic pathways have been identified: a modified Embden-Meyerhof (EM) pathway and two variations of the Entner-Doudoroff pathway. The modified EM pathway involves several novel enzymes, including an ADP-dependent glucokinase and phosphofructokinase (**figure 19.4**). These kinases each result in the production of AMP; in fact, unlike the classical EM pathway, there is no net ATP produced. Another unique step in this modified EM pathway is the conversion of glyceraldehyde 3-phosphate directly to 3-phosphoglycerate. Some archaea use ferredoxin (Fd) as the electron acceptor, while others use NAD^+.

The two Entner-Doudoroff (ED) pathways include one called the semi-phosphorylative pathway and another dubbed the nonphosphorylative pathway. The semi-phosphorylative ED pathway has a net yield of one ATP (**figure 19.5**). This differs from the nonphosphorylative ED pathway, which, as the name suggests, does not have a net yield of ATP. ◀ *Aerobic respiration starts with glucose oxidation (section 11.3)*

All archaea that have been studied can oxidize pyruvate to acetyl-CoA. However, they lack the pyruvate dehydrogenase complex present in eukaryotes and respiratory bacteria, and use the enzyme pyruvate ferredoxin oxidoreductase instead. While many archaeal species appear to have a functional tricarboxylic acid (TCA) cycle, those that do not catabolize glucose to any significant extent (e.g., methanogens) lack a complete TCA cycle. However, some store glycogen, which is catabolized via a modified EM pathway (figure 19.4). ◀ *Pyruvate dehydrogenase and the tricarboxylic acid cycle (section 11.3)*

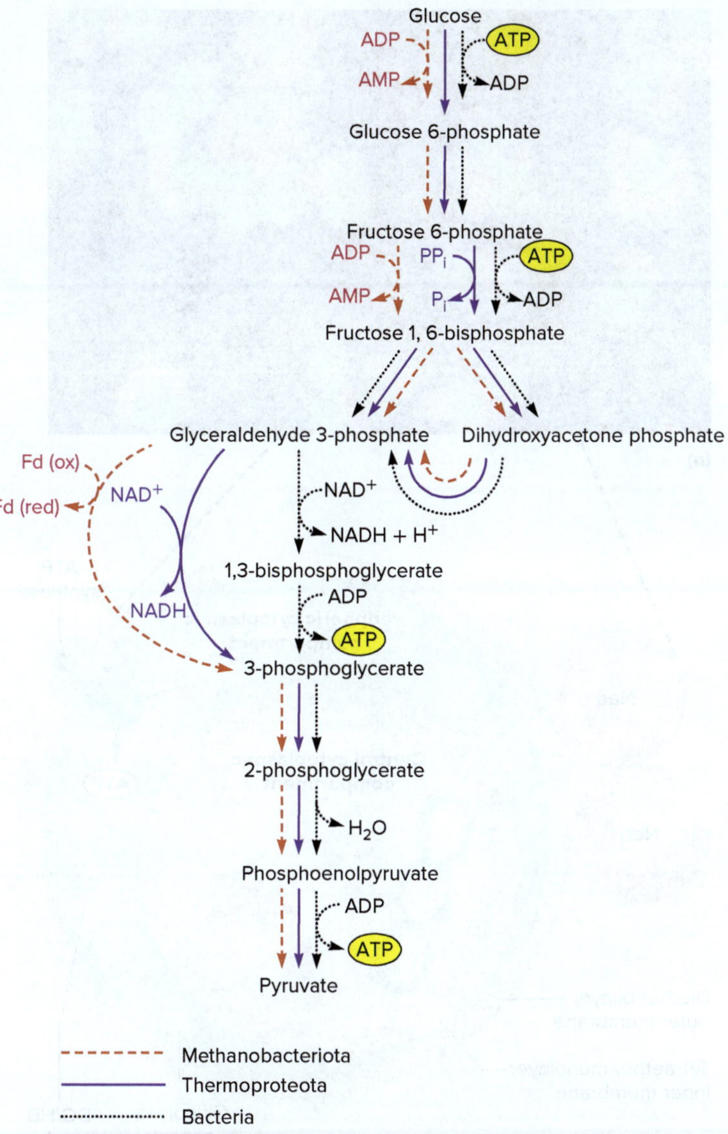

Figure 19.4 Modified Embden-Meyerhof Pathways. The pathways used by some archaea compared to that of bacteria. There is no net ATP produced in the archaeal pathways. Fd (ox), oxidized ferredoxin; Fd (red), reduced ferredoxin.

Comprehension Check

1. Why have genomic and metagenomic sequencing made a tremendous impact on archaeal taxonomy?
2. Explain how the two carbon atoms in the two molecules of CO_2 entering the Wood-Ljungdahl pathway are converted to the two-carbon molecule acetyl-CoA.
3. Why do some archaea use the methylaspartate rather than the glyoxylate cycle for the incorporation of acetate?
4. Compare the ATP and NADH yield of the glycolytic pathways used by archaea and bacteria.

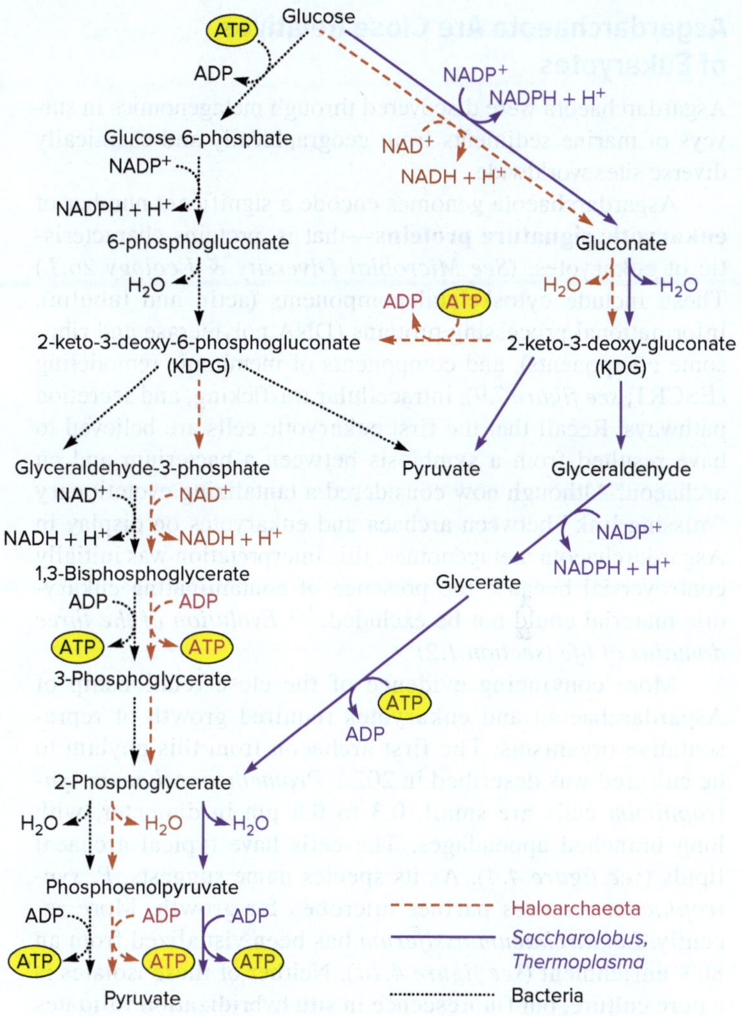

Figure 19.5 Two Alternate Versions of the Entner-Doudoroff Pathway. The semi-phosphorylative pathway (red) involves the ATP-dependent phosphorylation of KDG to KDPG; the remainder of the pathway is like that of bacteria and an ATP is gained. The nonphosphorylative pathway (purple) joins the bacterial ED pathway when glycerate is converted to 2-phosphoglycerate, so no net ATP is produced.

19.2 Many Archaea Are Known Primarily from Metagenomics

After reading this section, you should be able to:

- **a.** Discuss the relationship of Asgardarchaeota to eukaryotes
- **b.** Describe the obligate symbionts Nanoarchaeota
- **c.** Explain how the metabolism of uncultured organisms can be inferred from genome sequence data

Asgardarchaeota Are Close Relatives of Eukaryotes

Asgardarchaeota were discovered through metagenomics in surveys of marine sediments from geographically and chemically diverse sites worldwide.

Asgardarchaeota genomes encode a significant number of **eukaryotic signature proteins**—that is, proteins characteristic of eukaryotes. (*See Microbial Diversity & Ecology 26.1.*) These include cytoskeletal components (actin and tubulin), informational processing proteins (DNA polymerase and ribosome components), and components of membrane remodeling (ESCRT; *see figure 7.9*), intracellular trafficking, and secretion pathways. Recall that the first eukaryotic cells are believed to have resulted from a symbiosis between a bacterium and an archaeon. Although now considered a tantalizing evolutionary "missing link" between archaea and eukaryotes on display in Asgardarchaeota metagenomes, this interpretation was initially controversial because the presence of contaminating eukaryotic material could not be excluded. ◀ *Evolution of the three domains of life (section 1.2)*

More convincing evidence of the close relationship of Asgardarchaeota and eukaryotes required growth of representative organisms. The first archaeon from this phylum to be cultured was described in 2020. *Prometheoarchaeum syntrophicum* cells are small, 0.3 to 0.8 μm in diameter, with long branched appendages. The cells have typical archaeal lipids (*see figure 4.4*). As its species name suggests, *P. syntrophicum* requires partner microbes for growth. More recently, *Lokiarchaeum ossiferum* has been visualized from an 80% enrichment (*see figure 4.1d*). Neither of these isolates is a pure culture, but fluorescence in situ hybridization validates that the images of *L. ossiferum* correspond to the genome in question. Both cultured representatives of Asgardarchaeota have small cells with appendages and typical archaeal cellular components. Cryo-electron tomography of *L. ossiferum* illustrates the presence of actin-related filaments. The absence of a cell wall and the abundant filaments arrayed around a central cellular core support the proposal that a similar ancestral cell type might have been capable of engulfing a bacterial cell, perhaps even a syntrophic partner, in the early stages of eukaryogenesis. ◀ *Archaeal cell envelopes are structurally diverse (section 4.2)* ▶ *Mapping sequence data to the community (section 26.4)*

Nanoarchaeota Are Obligate Symbionts

Nanoarchaeum equitans (**figure 19.6**) was the first isolate of Nanoarchaeota, and it was immediately recognized to be phylogenetically distinct from other archaea. Related organisms were subsequently described from culture-independent single-cell and metagenomic studies from a wide variety of environments. Nanoarchaeota genomes are small (0.5–1.5 Mb) and feature limited biosynthetic potential. Only a few Nanoarchaeota strains have been cultured, and these have extremely small cells obligately attached to a host (*see figure 4.1e*). The small sizes and reduced genomes together point to a dependence on a larger microbial community. The cell biology of the symbiotic interactions, that is, the compatibility of S-layers and plasma membranes, remains an active area of investigation.

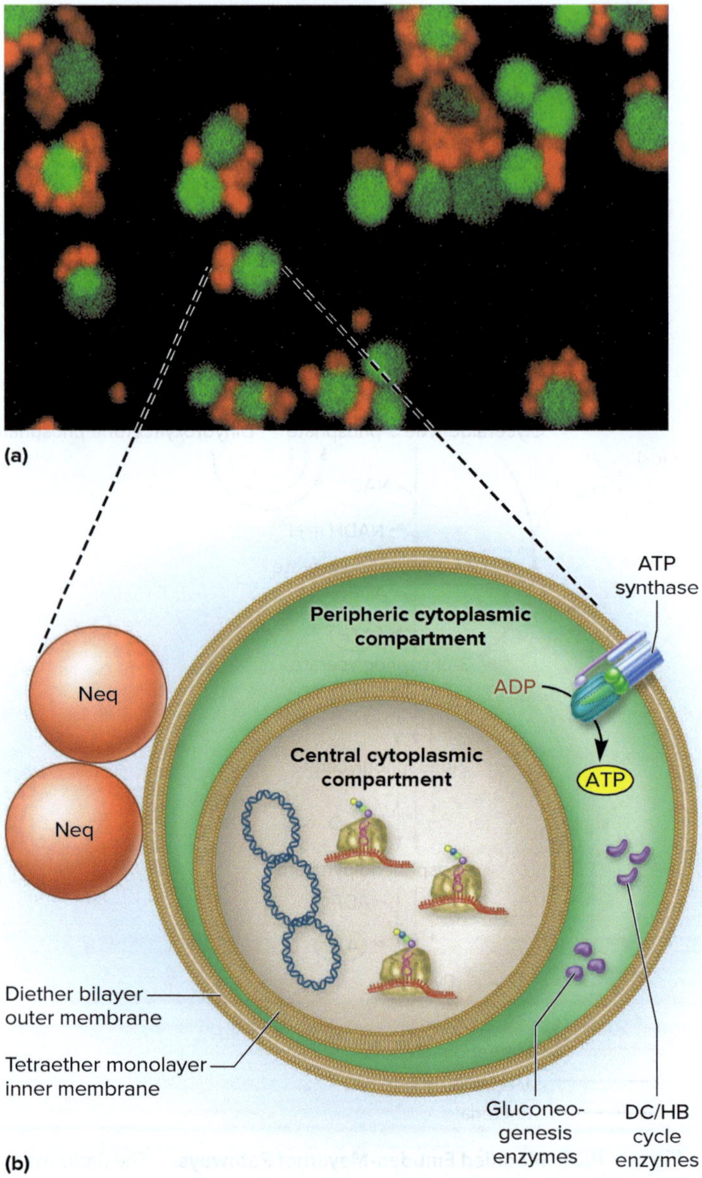

Figure 19.6 Symbiotic Cells of *Nanoarchaeum equitans* Attached to the Surface of Their Host, *Ignicoccus hospitalis*. (a) In this confocal laser scanning micrograph, *I. hospitalis* is stained green and *N. equitans* is red. (b) Diagram of the two compartments in an *I. hospitalis* cell.

Karl O. Stetter

MICRO INQUIRY *Compare the double-membrane structure of* Ignicoccus *to a typical diderm bacterium. How is the peripheric cytoplasmic compartment different from the periplasm?*

Comprehension Check

1. Which properties do Asgardarchaeota share with eukaryotic cells? Which properties are uniquely archaeal?
2. Nanoarchaeota cells clearly require hosts, but what advantages might they confer on those hosts?

19.3 Phylum Thermoproteota Features Temperature Adaptations

After reading this section, you should be able to:

a. Describe the environments in which Thermoproteota thrive
b. Describe the symbiotic relationship between *Ignicoccus* and *Nanoarchaeum*
c. Describe the metabolism of ammonia oxidizing archaea
d. List molecular adaptations to extreme growth temperatures

The first-described members of Thermoproteota were thermophilic and often sulfur-dependent. Taxonomic reassessments now include other clades in this phylum, but all appear to descend from thermophilic ancestors. Some groups remain adapted to extreme heat, while others cooled off.

Sulfur-Dependent Thermophiles

In these organisms, sulfur may be used either as an electron acceptor in anaerobic respiration or as an electron donor by lithotrophs. They are often found in geothermally heated water or soils that contain elemental sulfur, which imparts a milky white appearance to water. These environments are sometimes called solfataras and are found worldwide (**figure 19.7**). Many other Thermoproteota have been isolated from the waters surrounding submarine volcanoes. These archaea are classified as **hyperthermophiles** if their optimum growth temperature exceeds 85°C. The most extreme example is *Pyrolobus fumarii,* which was isolated from an active hydrothermal vent in the northeast Pacific Ocean. It grows best at about 105°C but can survive autoclaving at 121°C for 1 hour (*see figure 7.18*). It is strictly anaerobic, using Fe(III) as a terminal electron acceptor and H_2 or formate as electron donors and energy sources.

Figure 19.7 A Solfatara on Mt. Krafla, Iceland, Supports the Growth of Thermophilic Archaea.
Joanne M. Willey, Ph.D.

Hydrothermal vents are also home to the symbiotic pair *Ignicoccus hospitalis* and *N. equitans* (figure 19.6). While *N. equitans* is a member of Nanoarchaeota, its host *I. hospitalis* is a sulfur-dependent member of the phylum Thermoproteota. *N. equitans* requires its host for growth, but *Ignicoccus* species readily grow as axenic cultures. *I. hospitalis* is an anaerobic, chemolithoautotrophic hyperthermophile that uses H_2 as an electron donor and elemental sulfur as its electron acceptor. Its cell envelope lacks an S-layer but possesses a double membrane. The inner membrane is a tetraether monolayer (*see figures 4.4 and 4.5*), while the outer membrane is an archaeol bilayer (*see figure 4.3*). This double membrane divides the *Ignicoccus* cell into two functional compartments. The central cytoplasmic compartment contains the nucleoid and ribosomes, while the *peripheric cytoplasmic compartment* holds metabolic enzymes. The outer membrane supports ATP synthase and is thereby energized. Enzymes in metabolic cycles (figure 19.2) are localized to the peripheric cytoplasmic compartment. It has been proposed that the DC/HB enzymes are segregated to protect the nucleic acids from the reactive aldehyde intermediates that form at high temperatures. *N. equitans,* which attaches to the *I. hospitalis* outer membrane, obtains its ATP directly from the peripheric cytoplasmic compartment, thus explaining its obligate association.

Saccharolobus and *Sulfolobus* spp. are aerobic, spherical **thermoacidophiles,** with a temperature optimum of 75° to 85°C and a pH optimum of 2 to 4. *Saccharolobus solfataricus* is a model archaeon due to its relative ease in culturing. Its cell envelope is a glycosylated S-layer and its membrane contains primarily tetraether lipids. It grows chemoheterotrophically under oxic conditions, but can also grow chemolithoautotrophically, using H_2, H_2S, and FeS_2 as electron donors and with oxygen as the terminal electron acceptor. When growing chemolithoautotrophically, it uses the HP/HB cycle for carbon fixation, and when growing chemoheterotrophically, it employs the nonphosphorylative Entner-Doudoroff pathway (figure 19.5) and a complete TCA cycle (*see figure 11.8*). Unlike most organisms, *S. solfataricus* rarely uses NAD^+ as an electron acceptor; instead, it uses $NADP^+$ and ferredoxin-dependent oxidoreductases.

Although *S. solfataricus* grows at pH 2 to 4, it maintains a cytoplasmic pH of about 6.5, thereby generating a large pH gradient across the plasma membrane. This energy is conserved by membrane-bound ATP synthases and at least 15 secondary transport systems that couple the transport of organic solutes (e.g., sugars, amino acids) with the movement

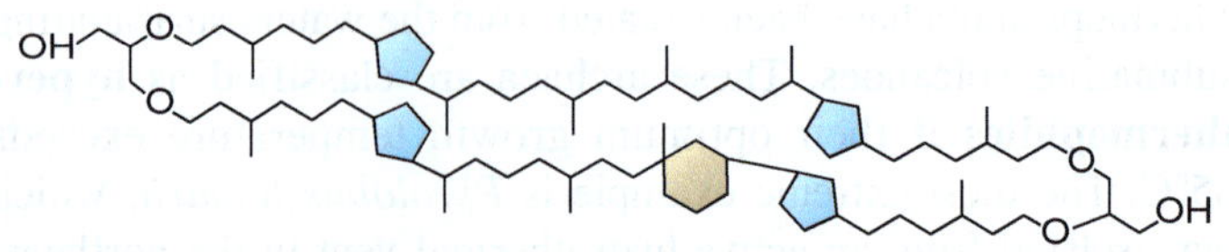

Figure 19.8 Thaumarchaeol. The presence of a cyclohexane ring (orange) is the molecular signature of thaumarchaeol.

of protons. The ABC transporters also used for nutrient uptake have high substrate affinity, an important adaptation in low nutrient environments. *Primary and secondary active transport (section 3.2); Electron transport and oxidative phosphorylation generate the most ATP (section 11.4)*

Mesophilic Ammonia Oxidizers

A prominent class within the phylum Thermoproteota is the Nitrososphaeria, mesophiles that are predominantly ammonia oxidizers. They are referred to as ammonia-oxidizing archaea (AOA), collaborating with ammonia-oxidizing bacteria (AOB) and comammox bacteria in nitrification (*see figure 21.9*). AOA comprise a significant fraction of the planetary microbial biomass and play critical roles in the carbon and nitrogen cycles. These archaea are considered **mixotrophic** rather than heterotrophic because they capture energy through the oxidation of ammonia to nitrite using oxygen as the terminal electron acceptor while using organic carbon for anabolic purposes. AOA predominate in oxic surface waters and have adapted to thrive at a lower ammonia concentration than AOB. The S-layer of *Nitrosopumilus maritimus* has a net negative charge, which attracts ammonium ions that pass through pores in the S-layer. The ions are then channeled to the enzyme ammonia monooxygenase (AMO) that catalyzes the first step in ammonia oxidation. *Nitrifying bacteria oxidize ammonium or nitrite to gain energy and electrons (section 21.1)*

The lipid **thaumarchaeol** (**figure 19.8**) serves as a molecular signature for AOA. A single cyclohexane ring differentiates thaumarchaeol from other diglycerol tetraether lipids (*see figure 4.4*). The lipids found in thermophilic Thermoproteota contain cyclopentane rings that confer membrane stability at high temperature. The addition of a cyclohexane ring promotes greater membrane fluidity, as required by archaea growing at lower temperatures. Phylogenetic analysis points to a thermophilic origin of AOA that dispersed to soil and marine environments. *Plasma membrane structure is dynamic (section 3.2)*

Comprehension Check

1. In what types of metabolic reactions do sulfur-dependent thermophiles use sulfur?
2. Discuss the role of external pH on the magnitude of the proton motive force generated by *Saccharolobus* spp.
3. What types of molecular adaptations to growth temperature are found in Thermoproteota?

19.4 Phyla Methanobacteriota and Halobacteriota: Methanogens, Halophiles, and Others

After reading this section, you should be able to:

a. Outline the process of methanogenesis and discuss its importance in the flow of carbon through the biosphere as well as in the production of energy for human needs
b. Discuss the physiology and ecology of anaerobic methane oxidation
c. Explain the strategies halophiles have evolved to cope with osmotic stress and why these strategies are needed
d. Outline rhodopsin-based phototrophy as used by halophiles
e. Describe the habitats in which methanogens and halophiles reside

The vast majority of cultured archaea are found in these phyla. Their common names, methanogens and halophiles, describe phenotypes. Exploring their phylogenetic relationships establishes that these are not monophyletic groups, and their detailed taxonomic relationships are under revision.

Methanogens

Late in the eighteenth century, Italian physicist Alessandro Volta discovered that he could ignite gas released from anoxic marshes and in this way Volta discovered the biological production of methane. The production of methane—**methanogenesis**—is the last step in the anaerobic degradation of organic compounds. Because the ΔG of methanogenesis compares unfavorably to other forms of respiration, methanogenesis occurs only when O_2 and most other electron acceptors are unavailable, making it a strictly anaerobic process (*see figure 28.1*). Archaea in multiple phyla perform methanogenesis (**Table 19.2**). Genome analysis indicates a widespread distribution of genes that encode the enzymes for methanogenesis and the biosynthesis of methanogenic cofactors. The capability to perform methanogenesis in these divergent lineages demonstrates the antiquity of this metabolism and it has been proposed that methanogenesis was the earliest energy-generating pathway to evolve, perhaps even before cells. Our discussion focuses on the well-established pathways in cultured methanogens.

Methanogens generate methane (CH_4) from H_2 and CO_2 or from short-chain organic compounds, such as acetate and methanol. These compounds are usually the fermentation products of other microbes that live in the same community. When using H_2 and CO_2, methanogen growth is autotrophic and uses the Wood-Ljungdahl pathway to incorporate CO_2 (figure 19.1).

Methanogens use one of three pathways to make methane, depending on the carbon substrates available. Hydrogenotrophic methanogens use electrons from H_2 to reduce CO_2 to methane (CH_4). The production of CH_4 requires that a carbon atom be

Table 19.2 Selected Characteristics of Methanogens

Phylum Class	Habitat	Methanogenic Substrates[1]	Temperature Range	Other Features
Methanobacteriota Methanobacteria	Marine and freshwater sediments, permafrost, hot springs, human GI tracts	CO_2, H_2 (formate, CO, secondary alcohols[2])	Mesophiles, thermophiles	Some autotrophs, many heterotrophs
Methanobacteriota Methanopyri	Hydrothermal marine sediments	CO_2, H_2	Hyperthermophiles	
Methanobacteriota Methanococci	Marine sediments, hydrothermal vents	CO_2, H_2 (formate)	Mesophiles, thermophiles, hyperthermophiles	Highest growth rates
Halobacteriota Methanocellia	Rice paddies	CO_2, H_2 (formate)	Mesophiles	Require acetate for biosynthesis
Halobacteriota Methanonatronarchaeia	Alkaline lake sediment	CO_2, H_2, formate, methanol, methylamines	Thermophiles	Alkaliphiles, halophiles
Halobacteriota Methanosarcinia	Marine and freshwater sediments, human GI tracts, anaerobic sewage digesters	(CO_2, H_2, CO, acetate, methanol, methyl sulfide, methylamines)	Mesophiles, thermophiles	Contain cytochromes, methanophenazine
Halobacteriota Methanomicrobia	Marine sediments, anaerobic sewage digesters, ciliate symbionts	CO_2, H_2, formate (secondary alcohols[2])	Mesophiles	

[1] Compounds in parentheses are used by some but not all members.
[2] Secondary alcohols are 2-propanol and 2-butanol.

converted from its most oxidized state in CO_2 (+4) to its most reduced state (−4). Recall that carbon fixation consumes ATP when it takes a carbon atom from +4 to 0. By contrast, methanogenesis is an energy-generating mechanism that requires several unique cofactors (**figure 19.9**). Some are carriers for the carbon atom, whereas others are electron donors. These are used in the following series of reactions termed the **Wolfe cycle** (**figure 19.10**): (1) CO_2 enters an enzyme complex that reduces it to formate at one active site. Formate is directed to a second active site, where it is further reduced by binding to the coenzyme methanofuran. This is an endergonic reaction, coupled to ferredoxin oxidation. The carbon atom is now activated and part of a formyl group (—HC═O). (2) The formyl group is transferred to tetrahydromethanopterin and dehydrated. (3) Using coenzyme F_{420} as the electron donor, the dehydrated formyl group (═HC—) is reduced to the methyl level (—CH_3) in a multistep process. (4) The methyl group is transferred to coenzyme M (CoM) by a membrane-bound enzyme that couples the reaction to Na^+ translocation across the membrane. This results in a sodium motive force to drive an ATP synthase. (5) The methyl group is now ready for final reduction to methane (CH_4) by coenzyme B (CoB). This is catalyzed by the coenzyme F_{430}-containing methyl-coenzyme M reductase. A second reaction product is a disulfide of the two cofactors, CoM-S-S-CoB. The final step in the cycle (6) regenerates the two coenyzmes by reducing this disulfide in an exergonic reaction. The enzyme that catalyzes this reaction, heterodisulfide reductase, uses a flavin-based electron bifurcation mechanism to reduce CoM-S-S-CoB and reduce the ferredoxin required for reaction (1). ◀ *Flavin-based electron bifurcation (section 11.9)*

There are two other methanogenesis pathways by which archaea make methane: aceticlastic and methylotrophic (table 19.2). Aceticlastic methanogens import acetate and convert it to acetyl CoA, and the C-C bond is broken. The carbonyl carbon is oxidized to CO_2 and the methyl group is reduced to CH_4 using the enzymes in the Wolfe cycle. Methylotrophic methanogens reduce the carbon atoms in compounds like methanol and methylamines. In these organisms, the substrate is partitioned into ~25% that is oxidized to CO_2 and ~75% that is reduced to CH_4. In all cases, carbon for anabolism is fixed using the Wood-Ljungdahl pathway (figure 19.1).

Aceticlastic methanogens have the highest growth yield, typically two to three times greater than that of hydrogenotrophic methanogens. Given their better growth yields, it is not surprising that these archaea account for about two-thirds of methanogenesis. It has been noted that the evolution of this methane-generating pathway about 475 million years ago corresponds with increased planetary temperatures that led to mass extinctions, a fact that underscores the importance of methane in climate change.

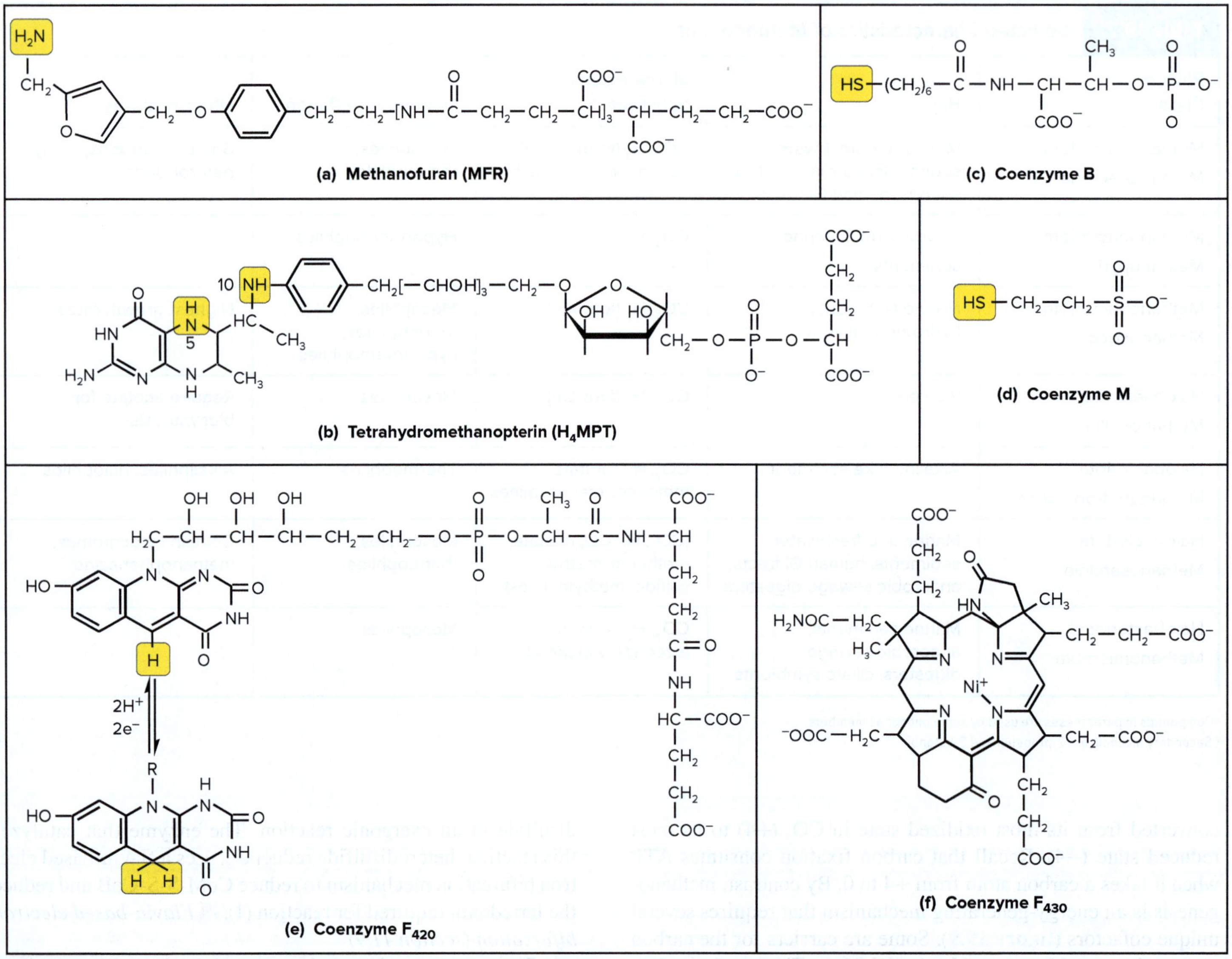

Figure 19.9 Methanogen Coenzymes. (a) Coenzymes MFR, (b) H_4MPT, and (d) M carry 1-carbon units during methanogenesis. The portions of the coenzymes that carry the 1-carbon units are highlighted. (c) Coenzyme B and (e) Coenzyme F_{420} participate in redox reactions. The part of the molecule that is reversibly oxidized and reduced is highlighted. (f) Coenzyme F_{430} participates in reactions catalyzed by methyl-CoM reductase.

Methanogenic archaea have great practical importance, as discussed in the chapter opening story. Industrial methane production illustrates that methanogenesis from organic waste is an important source of energy. Anaerobic digesters use fermentative bacteria to degrade particulate wastes such as sewage sludge to H_2, CO_2, and acetate (*see figure 42.6*), all substrates for methanogenesis. The tight thermodynamic relationship between the fermenting bacteria that produce H_2 and methanogens is called *interspecies hydrogen transfer* and is discussed in chapter 27. ▶| *Syntrophy (section 27.2)*

On a global scale, methanogenic archaea are estimated to produce about 1 billion tons of methane annually. In freshwater ecosystems, rates of methane production can be so great that bubbles of methane rise to the surface, as Volta observed over 200 years ago. Rumen methanogens are so active that a cow can release 200 liters of methane a day. ▶| *The rumen ecosystem (section 27.2)*

Although they do not perform methanogenesis for energy conservation, the extremely thermophilic sulfur reducers in the class Thermococci are closely related to methanogens. These archaea are strictly anaerobic chemoheterotrophs that use a variety of carbon substrates (e.g., peptides and carbohydrates) and reduce sulfur to sulfide. They are motile by archaella and have optimum growth temperatures around 88° to 100°C. Thermococci are an important source of thermostable enzymes, including several DNA polymerases used for PCR.

Despite being thermophilic anaerobes, thermococci are among the easiest archaea to grow in the laboratory, as they tolerate brief room temperature handling. *Thermococcus kodakarensis* has become a model organism due to its ease of genetic

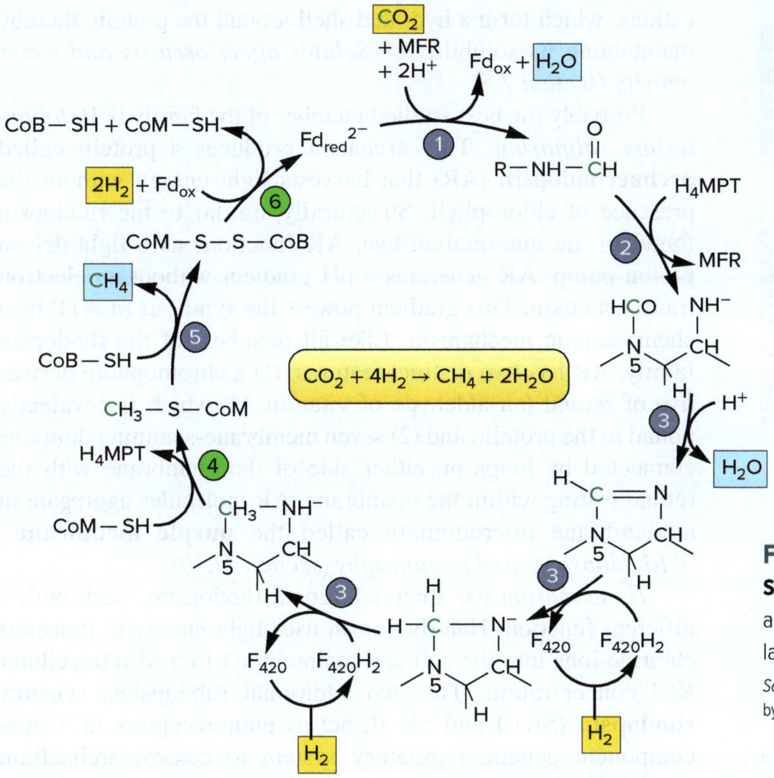

Figure 19.10 The Wolfe Cycle. Cycle for CH_4 synthesis from CO_2 and H_2 in hydrogenotrophic methanogens. The carbon atom derived from CO_2 is in green in each chemical structure. The circled numbers correspond to steps described in the text. Blue circles are endergonic reactions; green circles are exergonic reactions. CoB, Coenzyme B; CoM, coenzyme M; Fd, ferredoxin; MFR, methanofuran; H_4MPT, tetrahydromethanopterin.

MICRO INQUIRY *What are the mechanisms by which methanogens are thought to couple CO_2 reduction to ATP generation?*

transformation and the presence of plasmids to create shuttle vectors. The development of genetic techniques for an archaeal hyperthermophile has opened new avenues to explore the biology of these organisms. *Cloning vectors (section 17.1)*

Methanotrophs

No doubt the levels of archaeal-generated methane would render this planet uninhabitable were it not for the oxidation of methane by **methanotrophs,** microbes with the capacity to oxidize methane and use it as a carbon source. About 90% of methane produced in marine sediments is oxidized anaerobically by these archaea. The use of fluorescent probes for specific DNA sequences has revealed assemblages of archaea and sulfate-reducing bacteria inhabiting anoxic, methane-rich sediments (**figure 19.11**).

Anaerobic methane-oxidizing archaea (ANME) have thus far resisted cultivation, and the mechanism of methane oxidation has been deciphered by metagenomic analysis. ANME possess all the genes needed for methanogenesis, implying that anaerobic methane oxidation is accomplished by running methanogenesis

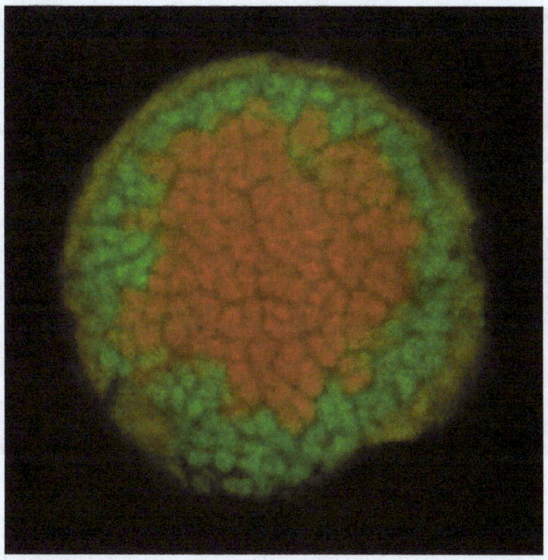

Figure 19.11 Methanotrophic Archaea Grow in Association with Sulfate-Reducing Bacteria. Methanotrophic archaea, stained red by an archaeal 16S rRNA probe, surrounded by a layer of sulfate-reducing bacteria labeled by a green bacterial 16S rRNA probe.

Source: V. Orphan. "Multiple archaeal groups mediate methane oxidation in anoxic cold seep sediments." by *PNAS*, Vol. 99: 7663–7668. Copyright (2002) National Academy of Sciences, U.S.A.

backward. In the ANME consortia with sulfate-reducing bacteria, sulfate is the terminal electron acceptor:

$$CH_4 + SO_4^{2-} \rightarrow HCO_3^- + HS^- + H_2O \; (\Delta G'^\circ = -17 \text{ kJ/mol})$$

Electron transfer within these cell clusters appears to be mediated through nanowires—that is, conductive pili (*see figure 20.16*). The archaea provide the bacteria with much of their organic carbon. In addition, ANME archaea also fix N_2, thereby providing the bacteria with organic nitrogen. Sulfate is not the only electron acceptor used in the anaerobic oxidation of methane. One group of ANME couples nitrate reduction by anammox bacteria to methane oxidation. Other ANME archaea use Fe(III) and Mn(IV) minerals as electron acceptors, representing new aspects of biogeochemical cycling (*see figure 28.3*).

Halobacteriota

The **extreme halophiles** include some bacteria, eukaryotic microorganisms, and **haloarchaea** that belong to the phylum Halobacteriota. This phylum also includes some methanogens (table 19.2) and some methanotrophs. Our discussion focuses on the halophilic archaea. Most are aerobic chemoheterotrophs with respiratory metabolism. Extreme halophiles demonstrate a wide variety of nutritional capabilities, using carbohydrates, glycerol, and organic or amino acids as their carbon source. Halophiles are found in a variety of cell shapes, including cubes and pyramids in addition to rods and cocci (**figure 19.12*a***). Some are motile by archaella.

The most obvious distinguishing trait of the haloarchaea is their absolute dependence on a high concentration of NaCl,

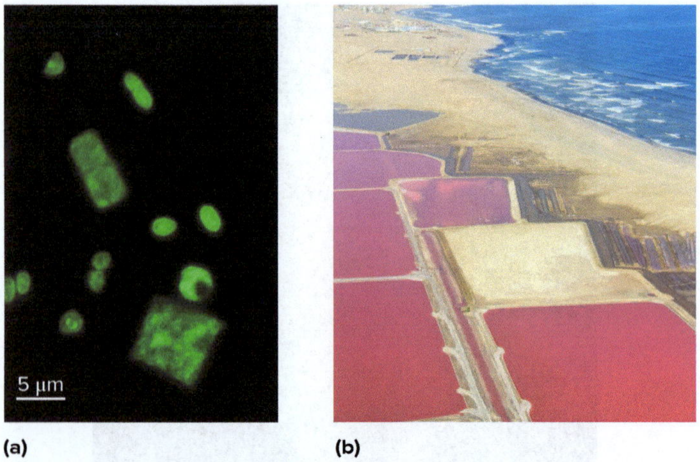

Figure 19.12 Halophilic Archaea. (a) A sample taken from a saltern viewed by acridine orange staining and fluorescence microscopy. Acridine orange fluoresces green when bound to double-stranded DNA. Cell shapes include cocci, rods, and cubes. (b) This solar evaporation pond is extremely high in salt and mineral content. The archaea that dominate this hot, saline habitat produce a brilliant red pigment, a carotenoid termed bacterioruberin.

(a) Dr. Mike Dyall-Smith; (b) Chris Sattlberger/Blend Images

requiring at least 1.5 M NaCl (about 8% wt/vol). Growth optima are usually 3 to 4 M NaCl (17 to 23%), but some will grow at salt concentrations approaching saturation (about 36%). The cell walls of most halophiles are so dependent on the presence of NaCl that they disintegrate when the NaCl concentration drops below 1.5 M. Thus they are restricted to high-salinity habitats such as marine salterns and salt lakes. Haloarchaea often have red-to-yellow pigmentation from carotenoids present in their membranes. They can reach such high population levels that salt lakes, called salterns, turn red (figure 19.12*b*).

Haloarchaea take a "salt-in" approach to cope with osmotic stress. They use Na^+/H^+ antiporters and K^+ symporters to concentrate KCl and NaCl to levels equivalent to the external environment. Thus proteins need to be protected from salt denaturation and dehydration. Proteins in these microbes have evolved to possess a limited number of hydrophobic amino acids and a greater number of acidic residues. The acidic amino acids tend to be located on the surface of the folded protein, where they attract cations, which form a hydrated shell around the protein, thereby maintaining its solubility. ◀ *Solutes affect osmosis and water activity (section 7.5)*

Probably the best-studied member of the family is *Halobacterium salinarum*. This archaeon produces a protein called **archaerhodopsin** (AR) that harvests light energy without the presence of chlorophyll. Structurally similar to the rhodopsin found in the mammalian eye, AR functions as a light-driven proton pump. AR generates a pH gradient without an electron transport chain. This gradient powers the synthesis of ATP by a chemiosmotic mechanism. Like all members of the rhodopsin family, AR has two distinct features: (1) a chromophore derivative of retinal (an aldehyde of vitamin A), which is covalently bound to the protein; and (2) seven membrane-spanning domains connected by loops on either side of the membrane with the retinal resting within the membrane. AR molecules aggregate in a membrane microdomain called the **purple membrane.** ◀ *Rhodopsin-based phototrophy (section 11.10)*

H. salinarum has three additional rhodopsins, each with a different function. Halorhodopsin uses light energy to transport chloride ions into the cell and maintain a 4 to 5 M intracellular KCl concentration. The two additional rhodopsins, **sensory rhodopsin** (SR) I and SR II, act as photoreceptors in a two-component genetic regulatory system to control archaellum activity. *Halobacterium* cells move to a location of high light intensity that optimizes AR proton pumping but limits the lethal effects of ultraviolet light.

Comprehension Check

1. Why are the growth yields of most methanogens much lower than those of microbes that grow aerobically?
2. Why do you think methanogenesis requires so many cofactors? (Hint: What would happen to CO_2 and CH_4 if they were not bound to larger molecules?)
3. What are the ecological and practical importance of methanogens?
4. Where are extreme halophiles found, and what is unusual about their growth requirements?
5. What is the difference between sensory rhodopsin and archaerhodopsin?

Key Concepts

19.1 Overview of Archaea

- Archaea contains a highly diverse group of microbes with respect to morphology, reproduction, metabolism, and ecology. Although they were first isolated from anoxic, hypersaline, and high-temperature habitats, they also inhabit marine Arctic, temperate, and tropical waters, as well as the human body.
- Many archaea are known only from culture-independent methods. Information about their metabolism is inferred from genome sequences and metagenome assembled genomes.
- Autotrophic archaea generally use one of three pathways for CO_2 assimilation: the Wood-Ljungdahl pathway, the

3-hydroxypropionate/4-hydroxybutyrate (HP/HB) cycle, or the dicarboxylate/4-hydroxybutyrate (DC/HB) cycle. Acetate is incorporated either by the glyoxylate or methylaspartate cycle. Most other anabolic pathways appear to be similar to those found in bacteria (**table 19.1; figures 19.1–19.3**).

- Although much of archaeal catabolism appears similar to that of bacteria, archaea differ in glucose catabolism, using modified versions of the Embden-Meyerhof and Entner-Doudoroff pathways (**figures 19.4** and **19.5**).

19.2 Many Archaea Are Known Primarily from Metagenomics

- Microbes in these phyla are microbial dark matter, meaning that few strains have been cultured in the laboratory.
- Asgardarchaeota encode multiple eukaryotic signature proteins, and are believed to resemble the earliest eukaryotes.
- Nanoarchaeota are small cells with small genomes and probably comprise symbionts (**figure 19.6**).

19.3 Phylum Thermoproteota Features Temperature Adaptations

- The extremely thermophilic S^0-metabolizers in the phylum Thermoproteota depend on sulfur for growth and are frequently acidophiles. The sulfur may be used as an electron acceptor in anaerobic respiration or as an electron donor by chemolithotrophs. Many are strict anaerobes and grow in geothermally heated soil and water that are rich in sulfur.
- *Ignicoccus* has a compartmentalized cell with distinct membrane types and functional separation of genetic and metabolic processes (**figure 19.6**).
- Nitrososphaeria are primarily mesophilic ammonia oxidizers. They are found in many marine and terrestrial habitats, where they can be important contributors to nitrification.

19.4 Phyla Methanobacteriota and Halobacteriota: Methanogens, Halophiles, and Others

- Methanogenic archaea are strict anaerobes that obtain energy through the synthesis of methane from CO_2, acetate, methanol, or methylamines. They have several unusual cofactors involved in methanogenesis (**figures 19.9** and **19.10**).
- The class Thermococci contains extremely thermophilic organisms that can reduce sulfur to sulfide.
- Anaerobic methane-oxidizing archaea commonly use sulfate or nitrate as electron acceptors in symbiotic relationships with other microbes (**figure 19.11**). These methanotrophs oxidize methane by reversing the methanogenic pathway.
- Extreme halophilic archaea are aerobic chemoheterotrophs that require at least 1.5 M NaCl for growth. They are found in habitats such as salterns (**figure 19.12**).
- *Halobacterium salinarum* carries out phototrophy without chlorophyll or bacteriochlorophyll by using archaerhodopsin, which employs retinal to pump protons across the plasma membrane.

Active Learning

1. When the temperature increases, some archaea change their shapes from elongated rods into spheres. Suggest one reason for this change.
2. Suppose you wished to isolate an archaeon from a hot spring in Yellowstone National Park. How would you go about it?
3. Do you think there will be a discovery of an archaeal pathogen of humans? Explain your reasoning.
4. Currently Asgardarchaeota are considered members of the domain Archaea. Most have yet to be cultured and their genomes encode eukaryotic signature proteins related to membrane movement. If a member of Asgardarchaeota is grown in pure culture and is shown to have internal membranes, can you foresee classifying it as eukaryotic on the basis of cellular structure?

CHAPTER

20

Nonproteobacterial Gram-Negative Bacteria

Evan Lorne/Shutterstock

From Food Waste to Fuel

After finishing a meal, many people collect their food scraps and feed them to microbes. Home composters degrade organic waste, and the resulting material is recycled in gardens. But what if, instead of tilling the compost into soil, it could be fed to the electric grid? Microbes such as *Thermotoga* perform dark fermentation, an anaerobic process that converts the complex carbohydrates found in food waste to acetate, CO_2, and, most importantly, hydrogen gas (H_2). Rather than a composter, consider a bioreactor that captures H_2 and converts it to electricity to power your kitchen appliances.

Thermotoga has distinguished itself by producing so much hydrogen that it almost reaches the theoretical maximum yield. It does so by secreting abundant hydrolytic enzymes that degrade plant biomass to release smaller sugar monomers. These sugars are fermented following glycolysis and produce H_2. H_2 is an attractive alternative fuel because its combustion generates water instead of greenhouse gases. It has a high energy density, meaning that a given weight of H_2 has more energy than a comparable weight of fuels such as methane, ethanol, or gasoline.

Scientists in Tunisia studied dark fermentation in *Thermotoga maritima* using an unusual medium—the fruit and vegetable waste from the city of Tunis and seawater from a nearby bay. The food waste provided the carbon and energy sources with additional micronutrients provided in the seawater. With only nitrogen and sulfur sources added for growth, the cultures produced abundant H_2. In a win-win situation, organic waste was removed from the city and converted to a clean fuel. H_2 generated from dark fermentation can be used to generate electricity or to power vehicles. While many organisms can perform dark fermentation, an important advantage to using *Thermotoga* is that at its growth temperature of 80°C, few contaminants survive.

In this chapter, we discuss these microbes as well as other phyla of diderm (Gram-negative) bacteria. From deinococci, which can withstand levels of ionizing radiation that would kill a human, to the spirochetes that cause diseases such as Lyme disease and syphilis, this chapter truly illustrates the depth of microbial diversity.

Readiness Check:

Based on what you have learned previously, you should be able to:

- ✓ Draw a diderm bacterial cell envelope and list its constituent parts (section 3.3)
- ✓ List environmental factors that influence microbial growth (section 7.5)
- ✓ Compare and contrast the growth and physiology of obligate aerobes, facultative anaerobes, and strict anaerobes (section 7.5)
- ✓ Explain the difference between heterotrophy and autotrophy (section 11.1)
- ✓ Define chemolithotrophy and identify an example of an electron donor and an acceptor (section 11.8)
- ✓ Compare and contrast oxygenic and anoxygenic phototrophy (section 11.10)
- ✓ Differentiate between swimming and gliding motility (section 14.6)

20.1 Bacterial Taxonomy and Cell Envelopes

After reading this section, you should be able to:

a. Describe the challenges in bacterial taxonomy
b. Explain how genomic analysis contributes to the identification of cell envelope components

Since Christian Gram's staining technique resolved two bacterial cell types—Gram positive and Gram negative—in the late nineteenth century, microbiologists have used this scheme to organize bacteria. Structural studies from microscopy complemented metabolic and ecological data to name and group similar isolates. For about 100 years, *Bergey's Manual of Systematic Bacteriology* was the dominant resource in bacterial taxonomy, tabulating the properties of described bacterial species. Regular updates were made to keep pace with research discoveries, and in fact, the format of these bacterial diversity chapters in *Prescott's Microbiology* originated with the organizational scheme from *Bergey's Manual.*

With Carl Woese's use of 16S rRNA as a tool to establish relatedness among bacteria, bacterial taxonomy became quantitative. As databases of nucleic acid sequences grew, debate ensued about what degree of identity or similarity between organisms is needed to constitute a new species. Nucleic acid and genome sequencing continue to advance, with metagenomic sequences of uncultivated microbes adding significantly to our appreciation of bacterial diversity. This astounding quantity of data has driven the need for standardizing the taxonomic ranks: phylum, class, order, family, and genus.

Taxonomy reflects phylogeny, that is, taxonomy describes evolutionary history. The more similar two organisms are at the DNA level, the more recently they diverged from a common ancestor. By quantitating differences in genome structures and normalizing the taxonomic ranks, the Genome Taxonomy Database is a major step forward in resolving relationships among bacterial taxa. This resource is the basis for the taxonomic classifications presented here.

The ongoing revision of bacterial phylogeny has uncovered some surprises. What has long been considered the fundamental division of bacterial cell types—their response to the Gram stain—does not hold under current phylogenetic analysis. The two cell types defined by the Gram stain have more recently been termed *monoderm* (having a single membrane, the plasma membrane) and *diderm* (having two membranes, the plasma membrane and the outer membrane). The common bacterial ancestor appears to have been a diderm, and monoderms arose not once, but several times during evolution. Yet these terms (Gram-positive/negative; monoderm/diderm) imply that there are two types of cell envelopes, when, in fact, significant variation exists among bacteria. ◀ *Overview of bacterial cell wall structure (section 3.3)*

Genomic data complement cell envelope structural information gleaned by electron microscopy and biochemical analyses. For example, the presence of genes encoding proteins that synthesize and localize lipopolysaccharide to the outer membrane (*see figures 12.28 and 12.29*) indicate that such a structure should be present on the cell, even if visualization of such structures is lacking. As more genomes become available, especially for organisms that have proved challenging to grow in laboratory culture, we can get a greater understanding of the diversity of cell envelope structures.

In chapters 20 through 22, we profile many types of bacteria, using the traditional Gram-stain-based classification. This scheme is widely recognized and deeply entrenched in all aspects of microbiology. We note variations in cell envelopes for certain taxa, and, in some places, bacteria are grouped by metabolic type.

20.2 Thermotogota and Deinococcota Live in Extreme Conditions

After reading this section, you should be able to:

a. Describe the cell envelope structure of *Thermotoga* spp.
b. Describe the cellular and genome structure of deinococci
c. Discuss the unique capacity of deinococci to tolerate desiccation and high doses of radiation
d. List an important commercial use of *Thermus* spp.

Members of the genus *Thermotoga* (Greek *therme,* heat; Latin *toga,* outer garment) are thermophiles with a growth optimum of 80°C and a maximum of 90°C. They are rods with a cell envelope that balloons out from the ends of the cell enclosing a large periplasmic space (**figure 20.1**). The toga is an outer surface layer composed of arrays of proteins interspersed with regions of outer membrane. This unusual structure may be a contemporary intermediate between monoderms and diderms. *Thermotoga* membranes (both the plasma membrane and the toga) contain some

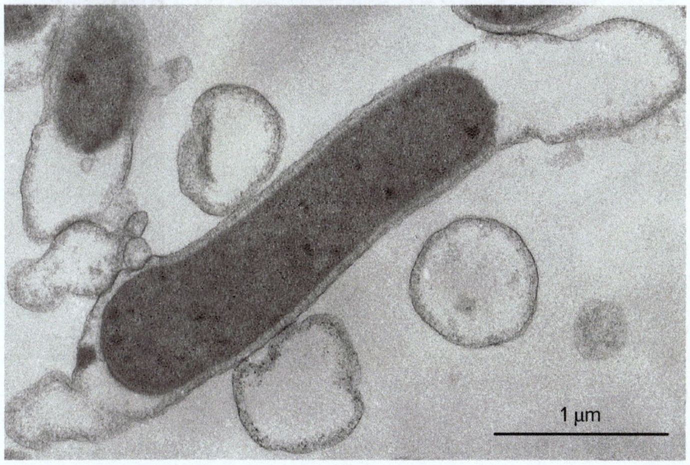

Figure 20.1 ***Thermotoga maritima.***
Karl O. Stetter

ether-linked membrane-spanning lipids more commonly found in archaea. *Thermotoga* grows in active geothermal areas in marine hydrothermal systems and terrestrial solfataric springs, environments often shared with archaea. It has been proposed that the capacity to build atypical lipids may have been horizontally transferred from archaea. *Thermotoga* species are chemoheterotrophs with a functional glycolytic pathway that can grow anaerobically using protons, elemental sulfur, or ferric iron as terminal electron acceptors (*see chapter opening story*). *Archaeal plasma membranes are composed of unique lipids but function like bacterial membranes (section 4.2)*

The phylum Deinococcota contains the families Deinococcaceae and Thermaceae. *Deinococcus* spp. are spherical or rod-shaped and nonmotile, and they are often seen in pairs or tetrads (**figure 20.2*a***). They are aerobic heterotrophs producing acid from only a few sugars. The *Deinococcus* cell envelope includes an outer membrane with a lipid composition resembling the plasma membrane and lacks lipopolysaccharide (figure 20.2*b*). The thick peptidoglycan layer is more typical of monoderm bacteria, which probably accounts for its Gram-positive staining behavior. Deinococci are ubiquitous but not abundant.

Deinococci are extraordinarily resistant to desiccation and radiation. The D value for radiation-induced cell death in deinococci is more than 10-fold higher than the D value for most other bacteria. Ionizing radiation generates radical oxygen species (ROS) that damage all types of macromolecules, notably fragmenting the genome into hundreds of pieces. Early experiments focused on the DNA repair capabilities of deinococci, but genomic surveys document that these organisms have the same set of repair enzymes as non–radiation-resistant strains (*see figure 18.12*). The mechanism of radiation resistance lies in the

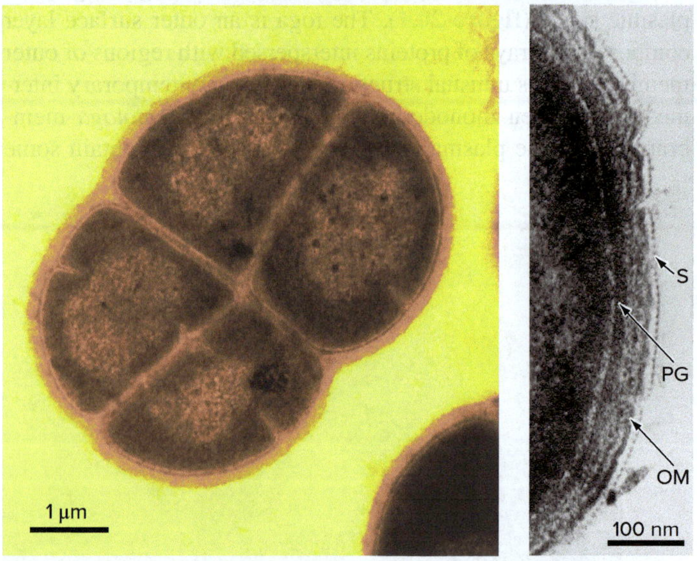

Figure 20.2 Deinococci. (a) A *Deinococcus radiodurans* tetrad. (b) The cell envelope of *D. radiodurans* with an S-layer (S), a peptidoglycan layer (PG), and an outer membrane (OM).

(a) Michael J. Daly/Science Source; (b) Dr. Robert G.E. Murray/University of Western Ontario

ability of deinococci to protect proteins from ROS-induced oxidation. This chemical protection is conferred by Mn^{+2} complexes with amino acids, peptides, small organic acids, and inorganic phosphate. Following ROS detoxification, chromosomes are reassembled by recombination (*see figure 16.8*). Deinococci are polyploid, which provides overlapping fragments, and within 12 to 24 hours, the chromosome is pieced back together. *Oxygen concentration (section 7.5); The pattern of microbial death mirrors the pattern of microbial growth (section 8.1)*

Thermus spp. are readily cultivated in the laboratory and are the source of many thermostable enzymes used in industrial applications. *Thermus aquaticus* was one of the first thermophilic organisms cultured. It grows optimally at 70°C, and its DNA polymerase, *Taq,* is commonly used in polymerase chain reactions. *Polymerase chain reaction amplifies targeted DNA (section 17.2)*

Comprehension Check

1. While inspecting the predicted proteins from a genome, which proteins would indicate that a cell is monoderm or diderm?
2. How do the cell envelopes of *Thermotoga* and *Deinococcus* differ from those of a typical diderm?
3. How do deinococci survive high doses of radiation?

20.3 Photosynthetic Bacteria Are Diverse

After reading this section, you should be able to:

a. Assess the importance of photosynthetic pigments in the distribution of photosynthetic bacteria in nature
b. Draw a generic chlorosome and identify the function of its structural elements
c. Predict the habitat closest to your home where you might find green sulfur bacteria, green nonsulfur bacteria, and cyanobacteria
d. Draw a generic cyanobacterial cell and label its intracellular structures
e. List three types of specialized cells made by cyanobacteria and describe the function of each
f. Compare and contrast prochlorophytes with other cyanobacteria
g. Draw a concept map describing the structure and metabolism of the six types of photosynthetic bacteria listed in table 20.1

Historically, the term *photosynthetic* has been applied to both photoheterotrophic and photoautotrophic microbes. We continue to use the term in this way, but care should be taken not to assume that the capture of light energy (phototrophy) is always linked with CO_2 fixation (autotrophy). As we shall see, unlike plants, photosynthetic microbes often have more than one metabolic strategy and may perform phototrophy but not autotrophy and grow photoheterotrophically (*see table 11.1*).

Bacteria capable of chlorophyll-based photosynthesis have been identified in multiple phyla. Properties of the major groups are summarized in **table 20.1**. There appears to have been

Table 20.1 Characteristics of the Major Groups of Gram-Negative Photosynthetic Bacteria

Characteristic	Green Sulfur	Green Nonsulfur	Purple Sulfur	Purple Nonsulfur	Aerobic Anoxygenic Phototrophs	Cyanobacteria
Major photosynthetic pigments	Bacteriochlorophylls *a* plus *c, d,* or *e*	Bacteriochlorophylls *a* and *c*	Bacteriochlorophyll *a* or *b*	Bacteriochlorophyll *a* or *b*	Bacteriochlorophyll *a*	Chlorophyll *a* plus phycobiliproteins. *Prochlorococcus* has divinyl derivatives of chlorophyll *a* and *b*.
Morphology of photosynthetic membranes	Photosynthetic system partly in chlorosomes	Chlorosomes present when grown anaerobically	Photosynthetic system contained in spherical or lamellar membrane complexes that are continuous with the plasma membrane	Photosynthetic system contained in spherical or lamellar membrane complexes that are continuous with the plasma membrane	Few, if any, intracytoplasmic membranes	Thylakoid membranes lined with phycobilisomes
Photosynthetic electron donors	H_2, H_2S, S^0	Photoheterotrophic donors—a variety of sugars, amino acids, and organic acids; photoautotrophic donors—H_2S, H_2	H_2, H_2S, S^0	Usually organic molecules: sometimes reduced sulfur compounds or H_2	A variety of sugars, amino acids, and organic acids	H_2O
Sulfur deposition	Outside of the cell	N/A[1]	Inside the cell[2]	Outside of the cell in a few cases	N/A	N/A
Nature of phototrophy	Anoxygenic	Anoxygenic	Anoxygenic	Anoxygenic	Anoxygenic	Oxygenic (some are also facultatively anoxygenic)
General metabolic type	Most are anaerobic photolithoautotrophs; some aerobic photoheterotrophs	Usually photoheterotrophic; sometimes photoautotrophic or chemoheterotrophic (when aerobic and in the dark)	Obligate anaerobic photolithoautotrophs	Usually anaerobic photoorgano-heterotrophs; some facultative photolithoautotrophs; some are chemo-organoheterotrophs in the dark	Aerobic photoorgano-heterotrophs	Aerobic photo-lithoautotrophs
Motility	Many nonmotile; some glide; some have gas vesicles	Gliding	Motile with polar or peritrichous flagella	Motile with polar flagella or nonmotile; some have gas vesicles.	Some are motile with a single or a few polar or subpolar flagella.	Nonmotile, swimming motility without flagella; some glide; some have gas vesicles.
Taxonomic distribution	Bacteroidota	Chloroflexota	γ-proteobacteria	α-proteobacteria	α-proteobacteria, γ-proteobacteria	Cyanobacteriota

[1] N/A: Not applicable.
[2] With the exception of *Ectothiorhodospira*.

considerable horizontal transfer of photosynthesis genes among the phyla, with at least 50 genes related to photosynthesis common to all. In this chapter, we describe cyanobacteria and green bacteria; purple bacteria are discussed in chapter 21, while heliobacteria are discussed in chapter 22.

Cyanobacteria differ fundamentally from the other photosynthetic bacteria because they perform **oxygenic photosynthesis:** They have photosystems I and II, use water as an electron donor, and generate oxygen during photosynthesis. In contrast, purple bacteria, green bacteria, and aerobic anoxygenic phototrophs (AAnPs) have only one photosystem and use **anoxygenic photosynthesis.** Because they are unable to use water as an electron source, they employ other reduced molecules as electron donors. AAnPs and purple nonsulfur bacteria are photoheterotrophs, so they use electrons obtained during the oxidation of organic substrates but make ATP from light energy. The purple and green sulfur bacteria use inorganic electron donors such as hydrogen sulfide (H_2S), sulfur, and hydrogen, as well as organic matter, as the source of electrons for the reduction of $NAD(P)^+$ to NAD(P)H. ◀ *Phototrophy (section 11.10)*

Photosynthetic microbes are major contributors of fixed carbon, termed *primary productivity,* in a variety of habitats. The differences in photosynthetic pigments and oxygen requirements among photosynthetic bacteria have important ecological implications. As shown in **figure 20.3,** chlorophylls (Chl), bacteriochlorophylls (Bchl), and their associated accessory pigments have narrow absorption spectra. Microbes therefore use accessory pigments to broaden the spectrum of their energy input (*see figure 11.29*). These pigments collect light of different wavelengths and transfer it to Chl or Bchl, where it is converted to chemical energy. Oxygenic cyanobacteria and diatoms dominate the aerated upper layers of freshwater and marine microbial communities, where they absorb large amounts of red and blue light. Anoxygenic purple and green photosynthetic bacteria inhabit the deeper anoxic zones rich in hydrogen sulfide and other reduced compounds that can be used as electron donors. Their Bchl and accessory pigments enable their use of light in the far-red spectrum (i.e., >750 nm) not used by other photosynthetic organisms (compare the Bchl *a* and Bchl *b* with Chl *a* in figure 20.3). Bchl absorption has additional peaks between 350 and 550 nm, allowing growth at greater depths because shorter-wavelength light can penetrate water farther.

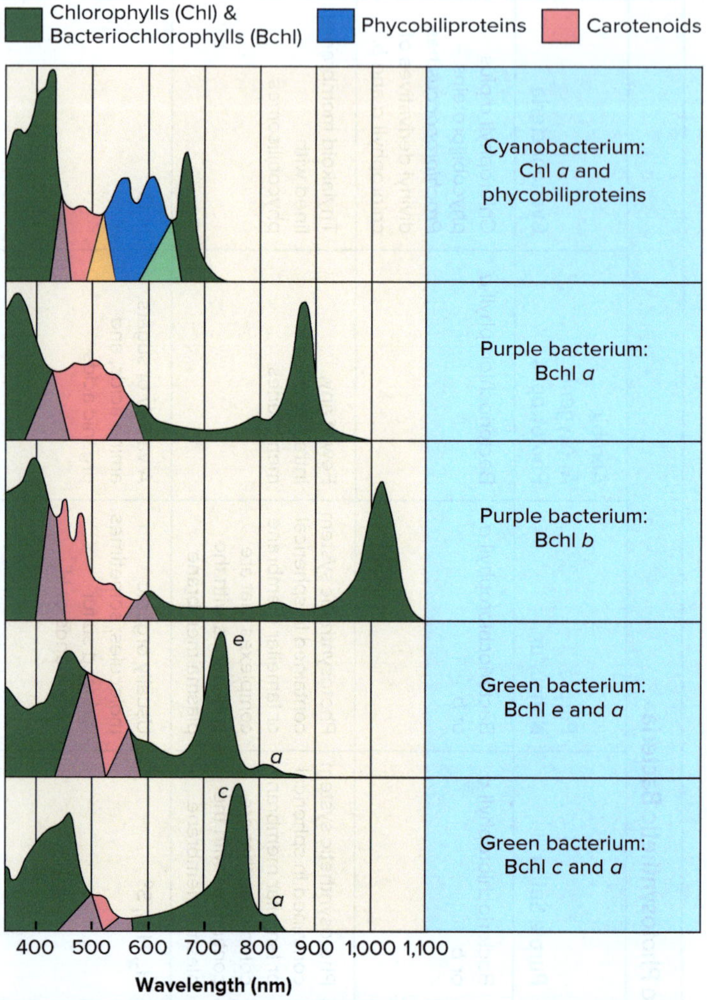

Figure 20.3 Photosynthetic Pigments. Absorption spectra of five photosynthetic bacteria showing the differences in absorption maxima and the contributions of various accessory pigments.

Green Sulfur Bacteria

The **green sulfur bacteria** belong to the phylum Bacteroidota, discussed in section 20.6. They are presented here to illustrate their similarity with other green bacteria. This is a small group of obligately anaerobic photolithoautotrophs that use H_2S, elemental sulfur, and hydrogen as electron sources. When sulfide is oxidized, elemental sulfur is deposited outside the cell. Photosynthetic pigments are located in ellipsoidal vesicles called **chlorosomes** attached to the plasma membrane by a proteinaceous baseplate that contains Bchl *a* molecules (**figure 20.4**). The chlorosome membrane is a lipid monolayer, rather than bilayer, with a hydrophobic interior. Within chlorosomes, about 200,000 molecules of Bchl aggregate into rodlike structures held together by carotenoids and lipids. The light energy harvested by these pigments is transferred to reaction centers in the plasma membrane. A **reaction center** is a complex of proteins and cofactors that capture light energy from photons and transfer it to electrons. The movement of electrons through an electron transport chain contributes to the proton gradient across the membrane that powers an ATP synthase (*see figure 11.34*). The ATP produced by photophosphorylation is used to fuel CO_2 fixation by the reductive TCA cycle (*see figure 12.4*); the more familiar Calvin-Benson cycle is used only by aerobes.

These bacteria flourish in anoxic, sulfide-rich zones of lakes. Although they lack flagella and most are nonmotile, members of at least one genus glide. Some have gas vesicles to adjust their depth for optimal light and H_2S (*see figure 3.29d*). Those without gas vesicles are found in sulfide-rich sediment at the bottom of lakes and ponds.

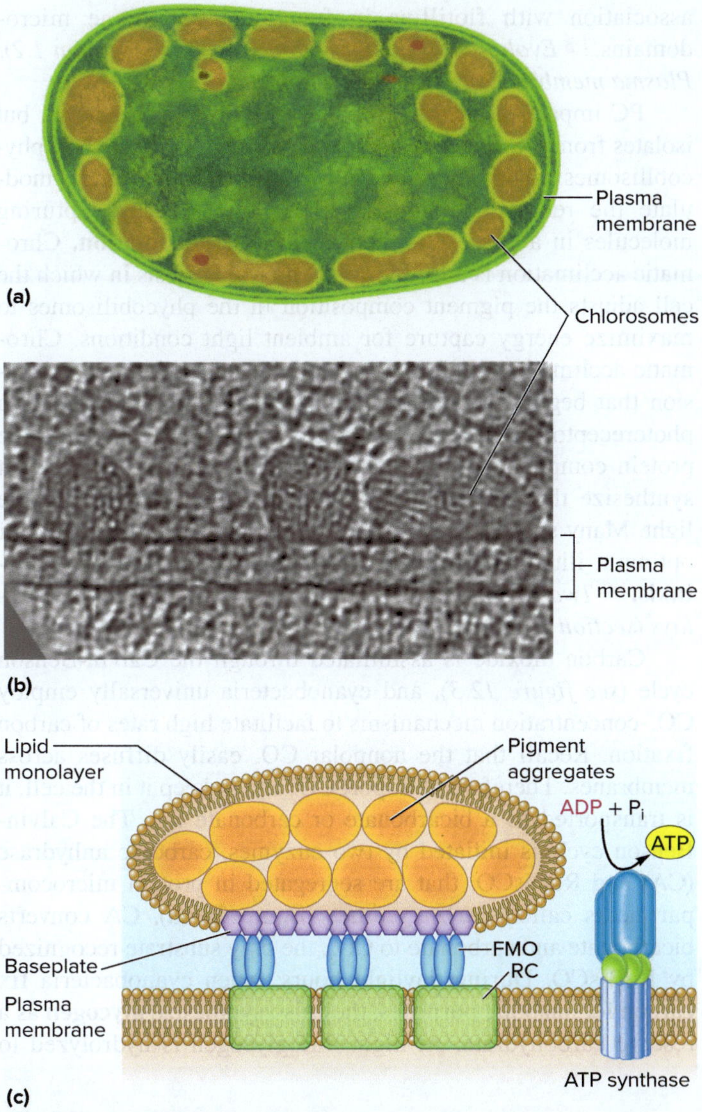

Figure 20.4 Light-Harvesting Chlorosomes Surround the Inside Perimeter of the Plasma Membrane. (a) Representation of chlorosomes associated with the membrane in a green sulfur bacterium. (b) Cryoelectron tomogram of chlorosomes. (c) Diagram of a single chlorosome, packed with bacteriochlorophyll molecules and other pigments. Light energy captured by photosynthetic pigments is transferred to the reaction centers (RC) via FMO proteins. ATP synthase is also located in the plasma membrane.

(b) Dr. Vasil Gaisin, Professor Martin Pilhofer, Dr. Romain Kooger

Green sulfur bacteria have a typical diderm cell envelope and are diverse morphologically. They may be rods, cocci, or vibrios; some grow singly, and others form chains and clusters. They are either grass-green or chocolate-brown in color. Representative genera are *Chlorobium, Prosthecochloris,* and *Chlorobaculum.*

Phylum Chloroflexota: Green Nonsulfur Bacteria

Members of the genus *Chloroflexus* make up the majority of photosynthetic **green nonsulfur bacteria.** However, the term *green nonsulfur* is a misnomer because not all members of this group are green, and some can use sulfide. *Chloroflexus* spp. are filamentous, gliding bacteria often isolated from neutral to alkaline hot springs, where they grow in the form of orange-reddish mats, usually in association with cyanobacteria. Like the green sulfur bacteria, they possess small chlorosomes associated with the plasma membrane.

The cell envelope in organisms in this phylum is a typical monoderm structure. While some members have only an S-layer surrounding the plasma membrane, others have a thin peptidoglycan layer.

Chloroflexus spp. are metabolically diverse. Some are facultative anaerobes, others are anaerobic, and while some are chemoheterotrophic, most are photoheterotrophic. They can carry out anoxygenic photosynthesis using sulfide or hydrogen as a source of electrons. Heterotrophic growth is supported by pyruvate, acetate, glycerol, and glucose oxidation. When growing autotrophically, they use an unusual pathway, the 3-hydroxypropionate bi-cycle to fix CO_2 (**figure 20.5**).

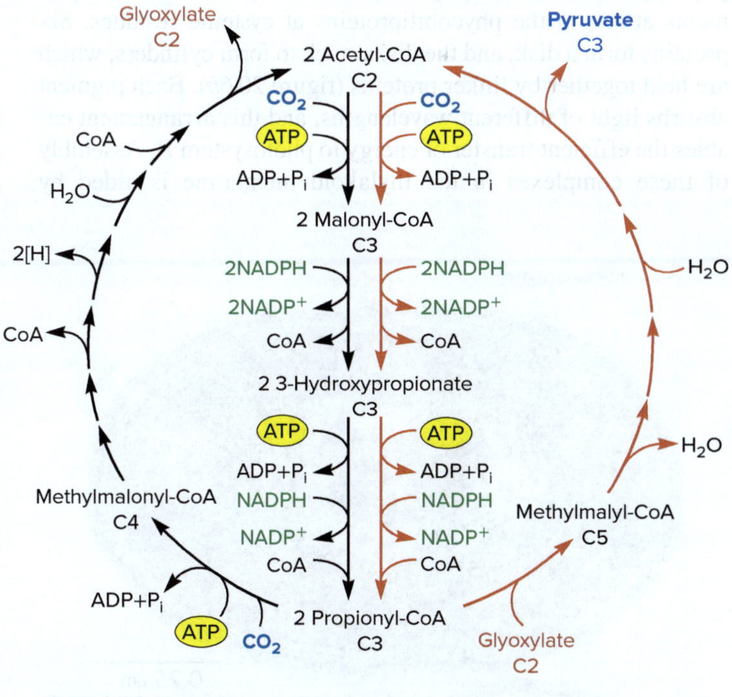

Figure 20.5 The 3-Hydroxypropionate Bi-Cycle. This cycle functions during autotrophic growth; CO_2 exists as HCO_3^- when it is added to intermediates of the pathway. The left half of the cycle produces glyoxylate, which is then consumed in the right half to make pyruvate. Overall, three carbons from CO_2 are fixed, resulting in the synthesis of the three-carbon compound pyruvate.

Comprehension Check

1. Which of the phototrophs discussed perform anoxygenic photosynthesis? Which are also capable of chemoheterotrophy?
2. What is the function of chlorosomes? How does their unique structure contribute to this function?
3. Compare the capacity of green nonsulfur and green sulfur bacteria to position themselves in light of optimum wavelength.

Phylum Cyanobacteriota: Oxygenic Photosynthetic Bacteria

Cyanobacteria are the largest and most diverse group of photosynthetic bacteria. Originally studied by botanists, they were erroneously known as blue-green algae.

Recall that endosymbiotic cyanobacteria are thought to have evolved into chloroplasts. It follows that the cyanobacterial photosynthetic apparatus closely resembles that of eukaryotes; cyanobacteria have chlorophyll *a* and photosystems I and II, and thereby perform oxygenic photosynthesis (*see figure 11.30*). Light-harvesting and electron transport chain components are located in thylakoid membranes to which particles called **phycobilisomes** are attached (**figure 20.6**). Phycobilisomes are protein complexes made up of **phycobiliprotein** pigments, particularly **phycocyanin** (PC) and **phycoerythrin** (PE). Chromophore pigments attach to the phycobiliproteins at cysteine residues. Six proteins form a disk, and the disks stack to form cylinders, which are held together by linker proteins (figure 20.6*b*). Each pigment absorbs light of different wavelengths, and this arrangement enables the efficient transfer of energy to photosystem II. Assembly of these complexes in the thylakoid membrane is aided by association with flotillins in functional membrane microdomains. *Evolution of the three domains of life (section 1.2); Plasma membrane structure is dynamic (section 3.2)*

PC imparts a blue-green color to many cyanobacteria, but isolates from the open ocean are red or brown because their phycobilisomes contain more PE (figure 20.6*b*). Cyanobacteria modulate the relative amounts of these light energy–capturing molecules in a process termed **chromatic acclimation.** Chromatic acclimation is a form of photomorphogenesis in which the cell adjusts the pigment composition in the phycobilisomes to maximize energy capture for ambient light conditions. Chromatic acclimation mechanisms involve changes in gene expression that begin with a two-component system comprised of a photoreceptor and a DNA-binding protein. Genes encoding the protein components of the phycobilisome or the enzymes that synthesize the pigments are regulated based on the available light. Many species use gas vacuoles to position themselves for optimum illumination in the water column, a form of **phototaxis.** *Two-component signal transduction and phosphorelays (section 14.5)*

Carbon dioxide is assimilated through the Calvin-Benson cycle (*see figure 12.3*), and cyanobacteria universally employ CO_2-concentration mechanisms to facilitate high rates of carbon fixation. Recall that the nonpolar CO_2 easily diffuses across membranes. Therefore to import CO_2 and to keep it in the cell, it is transported as a bicarbonate or carbonate ion. The Calvin-Benson cycle is initiated by two enzymes, carbonic anhydrase (CA) and RuBisCO, that are segregated in protein microcompartments called **carboxysomes** (figure 20.6*a*). CA converts bicarbonate and carbonate to CO_2, the only substrate recognized by RuBisCO. During daylight hours, when cyanobacteria fix carbon for anabolic purposes, they also synthesize **glycogen** as a reserve carbohydrate. At night, the glycogen is hydrolyzed to

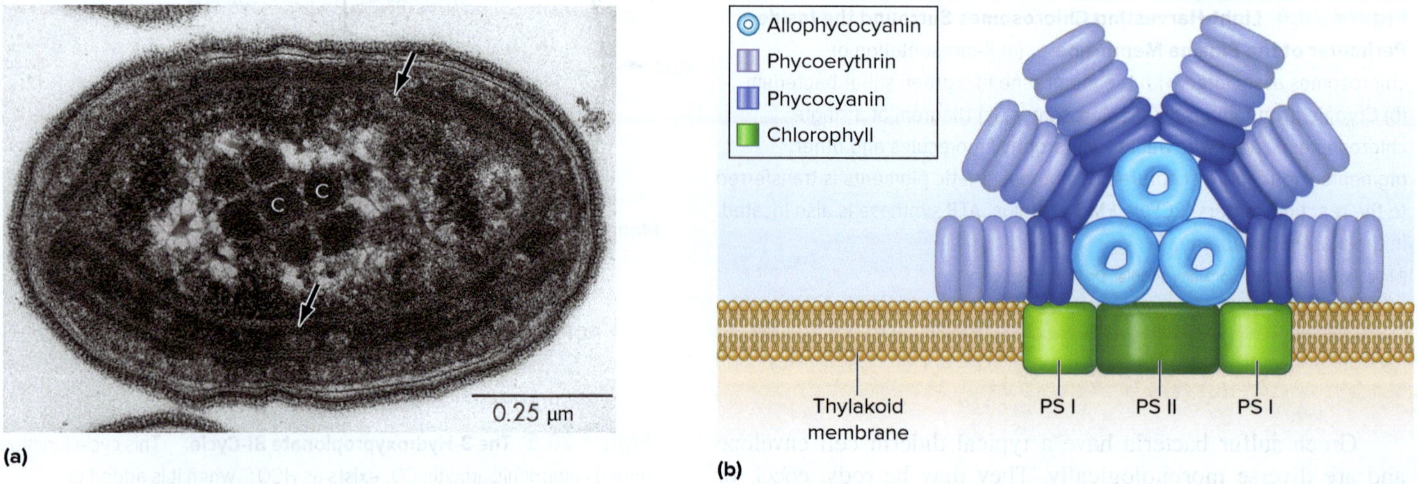

Figure 20.6 Cyanobacterial Thylakoids and Phycobilisomes. (a) Marine *Synechococcus* with thylakoids (arrows). C, Carboxysomes. (b) Phycobilisomes, located on the thylakoid membrane, have an allophycocyanin core that transfers energy from the radially arranged pigment stacks to photosystem II. Phycobilisomes are found in all cyanobacteria except prochlorophytes. PS, photosystem.

(a) Joanne M. Willey, Ph.D.

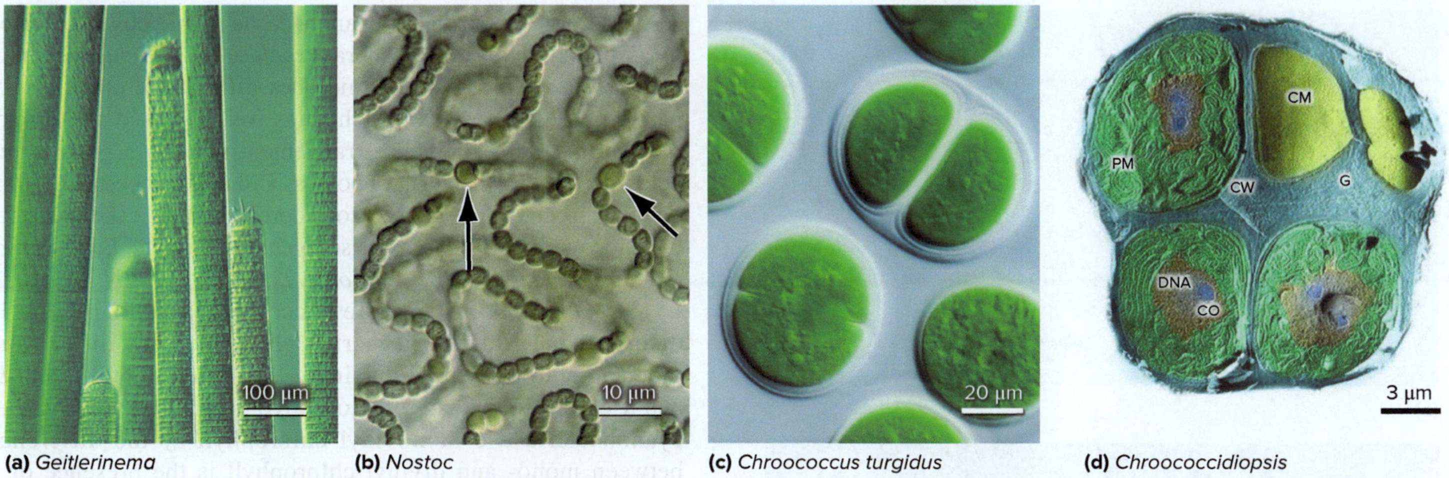

Figure 20.7 Cyanobacteria Are Morphologically Diverse. (a) *Geitlerinema* trichomes. (b) *Nostoc* sp. produces heterocysts (arrows) in the absence of combined nitrogen. (c) *Chroococcus turgidus* produces a gelatinous sheath that can surround one, two, or four cells. (d) Colorized electron micrograph taken after freeze fracture, illustrating baeocytes in *Chroococcidiopsis*. PM, photosynthetic membranes; CO, carboxysomes; CM, plasma membrane; DNA, nucleoid; G, capsule; CW, cell wall.

(a) M.I. Walker/Science Source; (b) Don Rubbelke/McGraw Hill; (c) Gerd Guenther/SPL/Science Source; (d) Dr. Burkhard Büdel

glucose monomers, which are then oxidized. Some cyanobacteria also store extra nitrogen as polymers of arginine and aspartic acid called **cyanophycin** granules. Phosphate is stored in polyphosphate granules. Acetyl-CoA is assimilated through the glyoxylate cycle (*see figure 19.3*). *Inclusions (section 3.6)*

Although most cyanobacteria perform only oxygenic photosynthesis, *Geitlerinema* switches to anaerobic photosynthesis under anoxic conditions, where it oxidizes H_2S instead of water much like green photosynthetic bacteria. *Geitlerinema* is among the cyanobacteria that produce geosmins, volatile organic compounds that have an earthy odor. When populations of these cyanobacteria become dominant in lakes and rivers, this can be a nuisance, even changing the taste of drinking water. *Microorganisms in marine and freshwater ecosystems (chapter 29)*

Cyanobacteria vary greatly in shape and appearance. They range in diameter from about 1 to 10 μm and may be unicellular, exist as colonies of many shapes, or form filaments called trichomes (**figure 20.7*a,b,c***). A **trichome** is a row of bacterial cells in close contact with one another over a large area. Cells in a trichome have individual plasma membranes and peptidoglycan but share a common periplasmic space and outer membrane that contains lipopolysaccharide. The peptidoglycan layer in cyanobacteria is significantly thicker and exhibits more extensive crosslinking than in other diderms.

Gliding motility is used by many filamentous cyanobacteria, including those found in dense microbial assemblages in sediments called *microbial mats*. Although cyanobacteria lack flagella, about one-third of marine *Synechococcus* strains swim at rates up to 25 μm/second. Swimming motility is used for chemotaxis toward simple nitrogenous compounds such as urea. The mechanisms for both gliding and swimming motility in cyanobacteria involve motor proteins attached to the peptidoglycan that move in a helical path in the periplasm. Such a movement creates local ridges in the S-layer to provide thrust for the cell. *Motility and Chemotaxis (section 14.6)*

Cyanobacteria show great diversity with respect to reproduction and employ a variety of mechanisms: binary fission, budding, fragmentation, and multiple fission. In the last process, a cell enlarges and then divides several times to produce many smaller progeny called **baeocytes** (figure 20.7*d*), which are released upon rupture of the parental cell. Some baeocytes disperse through gliding motility. Fragmentation of filamentous cyanobacteria can generate small, motile filaments called **hormogonia.** Some filamentous species develop **akinetes,** specialized, dormant, thick-walled resting cells resistant to desiccation (**figure 20.8*a***). These later germinate to form new filaments.

Many filamentous cyanobacteria fix atmospheric nitrogen by means of special cells called **heterocysts** (figure 20.8). Around 5 to 10% of cells in a filament develop into heterocysts when deprived of both nitrate and ammonia, their preferred nitrogen sources. Heterocyst structure and physiology ensure that these specialized cells remain anaerobic; the heterocyst is dedicated to nitrogen fixation, not reproduction. Within heterocysts, photosynthetic membranes are reorganized and proteins that make up photosystem II and phycobilisomes are degraded. Photosystem I remains functional to produce ATP, but no oxygen is generated. Lack of O_2 production is critical because the N_2-fixing enzyme nitrogenase is extremely oxygen sensitive. Heterocysts develop thick cell walls, which slow or prevent O_2 diffusion into the cell, and any O_2 present is consumed in respiration. Nutrients are obtained from adjacent vegetative cells, while the heterocysts contribute fixed nitrogen in the form of amino acids. Nitrogen fixation also is carried out by some cyanobacteria that lack heterocysts. Some fix nitrogen under dark, anoxic conditions in

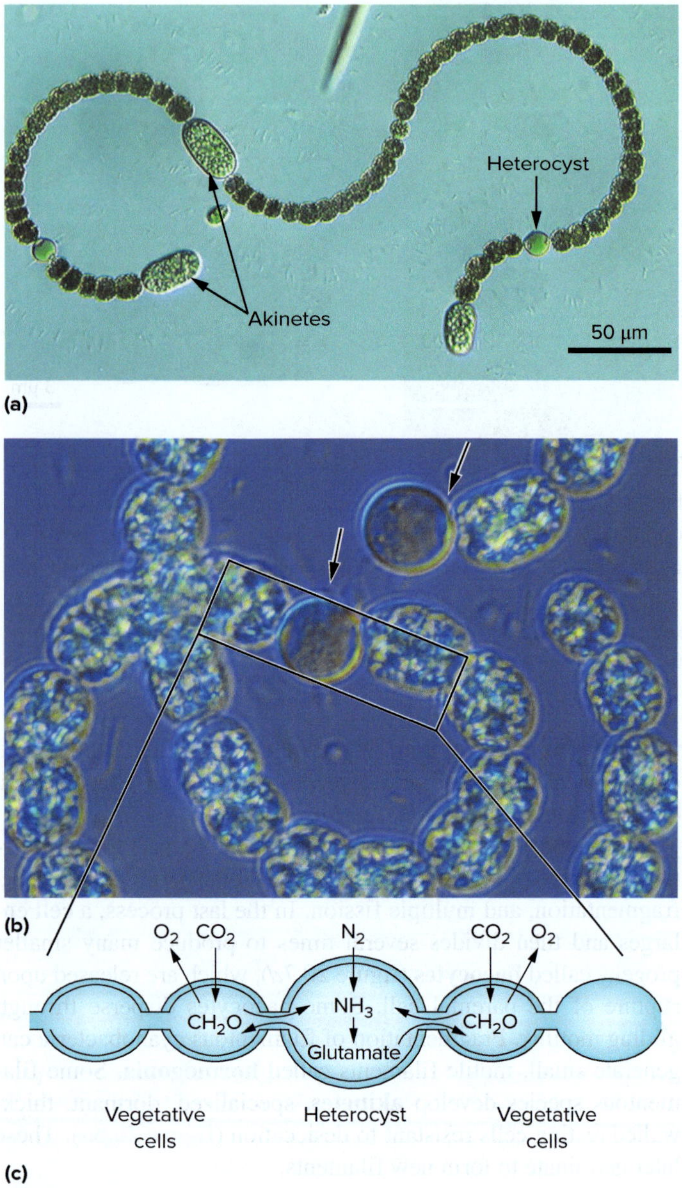

Figure 20.8 Cyanobacteria with Akinetes and Heterocysts.
(a) The filamentous cyanobacterium *Anabaena*. (b) *Anabaena* heterocysts (arrows) and vegetative cells. (c) Heterocysts fix dinitrogen to form glutamate, which is exchanged with adjacent vegetative cells for carbohydrate. Heterocysts use only photosystem I, so photosynthesis is anoxygenic in these differentiated cells. When fixing nitrogen, adjacent cells form proteinaceous septal junctions, and nanopores in these junctions promote diffusion between cells.

(a) Jason K. Oyadomari; (b) Biophoto Associates/Science Source

MICRO INQUIRY *Why do you think NH_3 is converted to the amino acid glutamate in heterocysts before its transfer to neighboring vegetative cells?*

microbial mats. Planktonic forms such as those belonging to the genus *Trichodesmium* fix nitrogen and contribute significantly to the marine nitrogen budget (*see figure 29.8*). *Nitrogen fixation: N_2 to NH_3 (section 12.5)* *Nitrogen cycle (section 28.2)*

An important group of cyanobacteria is collectively referred to as **prochlorophytes** (genera *Prochloron, Prochlorococcus,* and *Prochlorothrix*). These microbes differ from other cyanobacteria by the presence of Chl *b* as well as *a* and the lack of phycobilins. This pigment arrangement imparts a grass-green color. As the only bacteria to possess Chl *b,* the ancestors of unicellular prochlorophytes are considered the best candidates for the endosymbionts that gave rise to chloroplasts. *Evolution of the three domains of life (section 1.2)*

Prochlorococcus spp. are coccobacilli less than 1 μm in diameter that flourish in marine plankton. They differ from other prochlorophytes in having divinyl chlorophyll *a* and *b,* and α-carotene, instead of chlorophyll *a* and β-carotene. Other cyanobacteria use monovinyl chlorophyll *a;* the difference between mono- and divinyl chlorophyll is the presence of a double bond between carbons attached to pyrrole ring II (*see figure 11.28*). *Prochlorococcus marinus* is globally distributed in the ocean between the latitudes 40°N and 40°S and to 200 m in depth, making it the most abundant photosynthetic organism on Earth.

Unicellular cyanobacteria have a profound effect on the global carbon cycle. *Synechococcus* and *Prochlorococcus* account for at least one-fourth of global CO_2 fixation. These microbes have coevolved to take maximum advantage of incident light. *Synechococcus* spp. occupy the upper 25 m of the open ocean, where white and blue-green light can be harvested. In deeper waters, two ecotypes of *Prochlorococcus* species exist: The high-light ecotype grows between about 25 and 100 m in depth, while the low-light ecotype lives from 80 to 200 m, where the waters are penetrated by blue-violet light. This prevents the two genera of cyanobacteria from competing for the essential resource of light. *Microorganisms in the open ocean are adapted to nutrient limitation (section 29.2)*

In addition to those in coastal and open ocean waters, many other cyanobacteria inhabit a wide range of other habitats. Many cyanobacteria live in symbiotic association with protists and fungi (**Microbial Diversity and Ecology 20.1**). Thermophilic species may grow at temperatures up to 75°C in neutral to alkaline hot springs. Because these photoautotrophs are so hardy, they are primary colonizers of soils and surfaces devoid of plant growth. Some unicellular forms even grow in the fissures of desert rocks. In nutrient-rich warm ponds and lakes, filamentous cyanobacteria such as *Microcystis* and *Anabaena* spp. can reproduce rapidly to form blooms. *Microorganisms in coastal ecosystems are adapted to a changeable environment (section 29.2)*

Comprehension Check

1. What are the major characteristics that distinguish cyanobacteria from other photosynthetic organisms?
2. What is a trichome and how does it differ from a simple chain of cells?
3. Briefly discuss the ways in which cyanobacteria reproduce.

MICROBIAL DIVERSITY & ECOLOGY

20.1 UCYN-A: From PCR Fragment to Genome to Organelle

Diazotrophs, organisms that fix atmospheric nitrogen to ammonia, are critical components of marine ecosystems. *Atelocyanobacterium thalassa* (originally designated UCYN-A) was first detected as a PCR fragment of a gene encoding a component of nitrogenase. This evidence of diazotrophy was later linked to a cyanobacterial genome affiliated with a single-celled algae, *Braarudosphaera bigelowii.* Challenges associated with culturing this duo intact left researchers reliant on genome reconstruction to assess the metabolic capabilities of *A. thalassa.* Neither its carbon fixation nor photosynthetic pathways were complete, indicating dependence on its *B. bigelowii* host. *Nitrogen fixation:* N_2 *to* NH_3 *(section 12.5)*; *Microorganisms in the open ocean are adapted to nutrient limitation (section 29.2)*

Finally obtaining a stable co-culture of the two organisms allowed researchers to probe their association. Surprisingly, it appears that *A. thalassa* is not a symbiont, but the first (and, so far, the only) nitrogen-fixing organelle, termed a **nitroplast.** But how was the line drawn between organism vs. organelle? The nitroplast genome (no longer known as *A. thalassa*) is much smaller than needed to be free-living, and the nitroplast proteome contains multiple host-derived proteins that complement missing functions. The nitroplast divides by binary fission and coordinates its division with the host cell. *B. bigelowii* is now unique among eukaryotes in its capacity to fix nitrogen, courtesy of its novel organelle, the nitroplast (**box figure**).

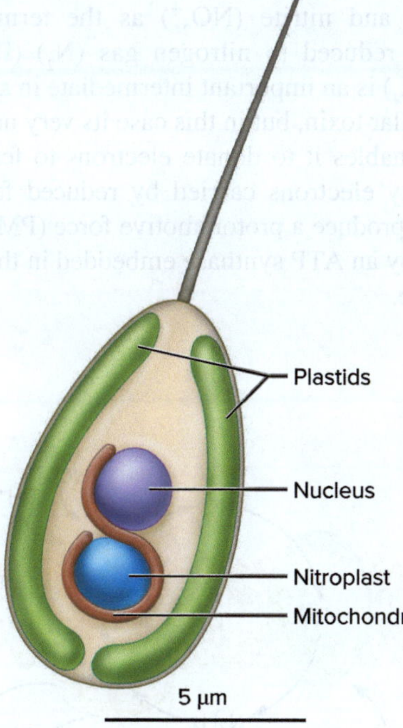

4. Describe how a vegetative cell, a heterocyst, and an akinete are different. How are heterocysts modified to carry out nitrogen fixation?
5. Compare prochlorophytes with other cyanobacteria. Why do they occupy stratified layers in the ocean?
6. Describe an important positive and negative impact cyanobacteria have on humans and the environment.

20.4 PVC Superphylum (Planctomycetota and Verrucomicrobiota): Atypical Cell Division

After reading this section, you should be able to:

a. List unusual features that unite some members of Planctomycetota and Verrucomicrobiota
b. Identify the electron donor and acceptor used in the anammox reaction
c. Explain why the anammox reaction is important to the global flux of nitrogen
d. Explain the term *obligate intracellular parasite*
e. Diagram the chlamydial life cycle

This superphylum includes multiple phyla, most notably Planctomycetota and Verrucomicrobiota. The grouping is supported by multiple phylogenetic markers and the observation that some genera within each of these phyla synthesize sterols and lack the cell division protein FtsZ. Cell division in these bacteria occurs by budding or binary fission that uses other cytoskeletal proteins. For many years the presence of peptidoglycan was debated, but genomic studies document the capacity to synthesize peptidoglycan. Because these features are not uniformly present, it is believed that the last common ancestor of these microbes had all these characteristics, which were subsequently lost by some. *Cytokinesis (section 7.2)*

Phylum Planctomycetota: Anammox Bacteria

Members of this phylum are cocci with a membrane-bound organelle called the **anammoxosome** that occupies much of the cell volume. This is the site of *an*aerobic *amm*onia *ox*idation, called the **anammox reaction,** or simply anammox. This unique form of chemolithoautotrophy employs ammonium ion (NH_4^+) as the electron donor and nitrite (NO_2^-) as the terminal electron acceptor—it is reduced to nitrogen gas (N_2) (**figure 20.9*a***). Hydrazine (N_2H_4) is an important intermediate in anammox. It is normally a cellular toxin, but in this case its very negative reduction potential enables it to donate electrons to ferredoxin. The four high-energy electrons carried by reduced ferredoxin can then be used to produce a proton motive force (PMF) that drives ATP synthesis by an ATP synthase embedded in the anammoxosome membrane.

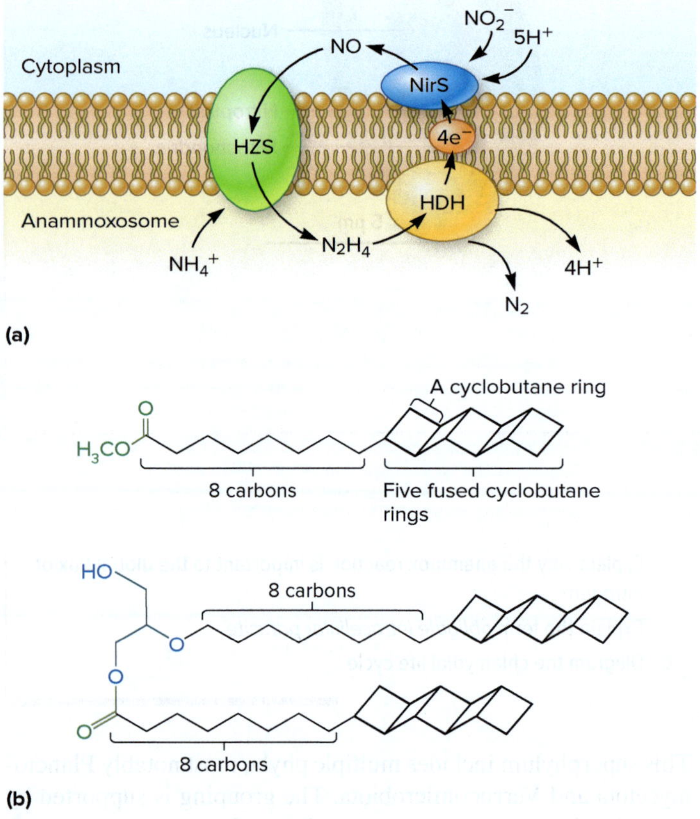

Figure 20.9 Features of Anammox Bacteria. (a) In the anammox reaction, ammonium (NH_4^+) combines with nitric oxide (NO) to form hydrazine (N_2H_4) in a reaction catalyzed by hydrazine synthase (HZS). Oxidation of hydrazine by hydrazine dehydrogenase (HDH) results in the formation of N_2 and four protons. The electrons are passed to NO_2^-, which is reduced by nitrite:nitric oxide reductase (NirS) to form another molecule of NO and the cycle continues. (b) Ladderane lipids are found in anammoxosomes.

MICRO INQUIRY *How does a planctomycete use the anammox reaction to generate a proton motive force?*

One interesting feature of anammox bacteria is that they use nitrite not only as the electron acceptor for ammonium oxidation and generation of a PMF but also as an electron donor for CO_2 fixation, which is accomplished using the Wood-Ljungdahl pathway (*see figure 19.1*). Therefore nitrite acts in both anabolic and catabolic processes, as shown in the following reactions:

An anabolic reaction:

$$2NO_2^- + CO_2 + H_2O \rightarrow CH_2O + 2NO_3^-$$

And a catabolic reaction:

$$NO_2^- + NH_4^+ \rightarrow N_2 + 2H_2O$$

The anammoxosome is bounded by a single bilayer membrane that contains ladderane lipids (figure 20.9*b*). These lipids are unique to planctomycetes. They have a low permeability to protons, and therefore maintain the proton motive force across the anammoxosome membrane. Anammox is estimated to contribute as much as 70% to the cycling of nitrogen in the world's oceans. ▶| *Nitrogen cycle (section 28.2)*

Members of the genus *Planctomyces* attach to surfaces through a stalk and holdfast. Most of these bacteria reproduce by budding. Sessile cells bud to produce motile swarmer cells. The swarmer cells are flagellated and swim for a while before attaching to a substrate.

Phylum Verrucomicrobiota Includes Human Symbionts and Obligate Intracellular Parasites

Verrucomicrobiota comprise a significant fraction of soil microbes and are found in a variety of aquatic and terrestrial environments. Their broad distribution may reflect their ability to degrade complex carbohydrates.

Akkermansia muciniphila is a strict anaerobe with a fermentative metabolism and is a prominent member of the human gut microbial community. As its name suggests, *A. muciniphila* consumes mucus and was initially grown in pure culture by providing gastric mucin as the sole carbon source. Although an organism that grows by degrading mucus might appear to be a pathogen, it seems that *A. muciniphila* helps to turn over and replenish the intestinal mucus layer. Greater numbers of *A. muciniphila* correlate with a healthy gut. Its fermentative end-products are short-chain fatty acids that protect against host metabolic diseases like obesity. Although the role of *A. muciniphila* in maintaining gut health remains under investigation, it is available commercially as a purported probiotic. ▶| *A functional core microbiome is required for host homeostasis (section 33.3)*

Chlamydiae are obligate intracellular parasites, that is, they must grow and reproduce within host cells. Although their ability to cause disease is widely recognized, many species grow within protists and animal cells without adverse effects. These hosts represent a natural reservoir for chlamydiae. Many uncultured chlamydiae have been identified through metagenomic sequencing and single amplified genomes. This has expanded their known environmental distribution to a broad range of ecosystems, including freshwater, seawater, wastewater, soil, and sediments. It has been proposed that their ubiquity may be a consequence of their residence in protists.

Human pathogens include *Chlamydia trachomatis* (which causes a sexually transmitted infection) and *Chlamydophila pneumoniae* (a cause of pneumonia); *Chlamydophila psittaci* is a bird pathogen that causes a zoonotic infection in humans. Members of these genera are nonmotile, coccoid bacteria, ranging in size from 0.2 to 1.5 μm. Like other intracellular parasites, their genomes are small at 1.0 to 1.3 Mb (*see figure 18.17*). ▶| *Chlamydial diseases (section 38.3)*

Chlamydia spp. have an unusual life cycle that involves two cell types. The **elementary body** (EB) is the infectious form, featuring small cells with reduced metabolic activity; this cell type is adapted to survive harsh extracellular conditions. The **reticulate body** (RB) develops from the EB upon host epithelial cell infection and is the metabolically active and replicative form.

In the *C. trachomatis* infection cycle, the EB attaches to a host cell membrane, where it is activated by the host enzyme protein disulfide isomerase (**figure 20.10**). EB surface proteins are cross-linked with disulfide bonds, and reducing these bonds triggers events within the EB to initiate infection. A type III secretion system (T3SS) delivers effectors from the EB to the host cytoplasm. These induce a reorganization of the host cytoskeleton to surround the EB and bring it into the cell as a membrane-bound inclusion. Once established within the host cell, a T3SS delivers inclusion membrane proteins to accomplish three tasks: avoid fusion with a lysosome, protect against intracellular immune surveillance, and import nutrients. |◀ *One-step secretion systems (section 13.9)*

Because of their reduced genomes, chlamydiae rely on their host cells for key metabolites. Metabolism in the EB is based on glycogen, derived from storage in the EB and acquisition from the host. Several TCA cycle enzymes are absent, and glutamine and malate must be imported. ATP is generated from oxidative phosphorylation.

Glutamine is a signal for the EB to intiate the transition to the RB form within the inclusion. RBs are larger cells than EBs, but as their cell division proceeds by budding, the cells are reduced to the size of EBs after a few divisions. Recall that many bacteria in the PVC phylum lack FtsZ. Cell division is accomplished by the formation of a contractile ring formed from peptidoglycan and MreB, the cell-shape-determining protein. Chlamydiae synthesize peptidoglycan only transiently during cell division, using the imported glutamine as a precursor. As the dividing RBs decrease in size, they transition to the EB form.

Nucleoids condense, and disulfide-linked proteins assemble on the cell surface. During the RB replicative phase, chlamydiae suppress apoptosis of the host cell, but once the EBs form, apoptosis is induced, thus releasing the EBs. |◀ *Bacterial cell cycles are divided into three phases (section 7.2)*

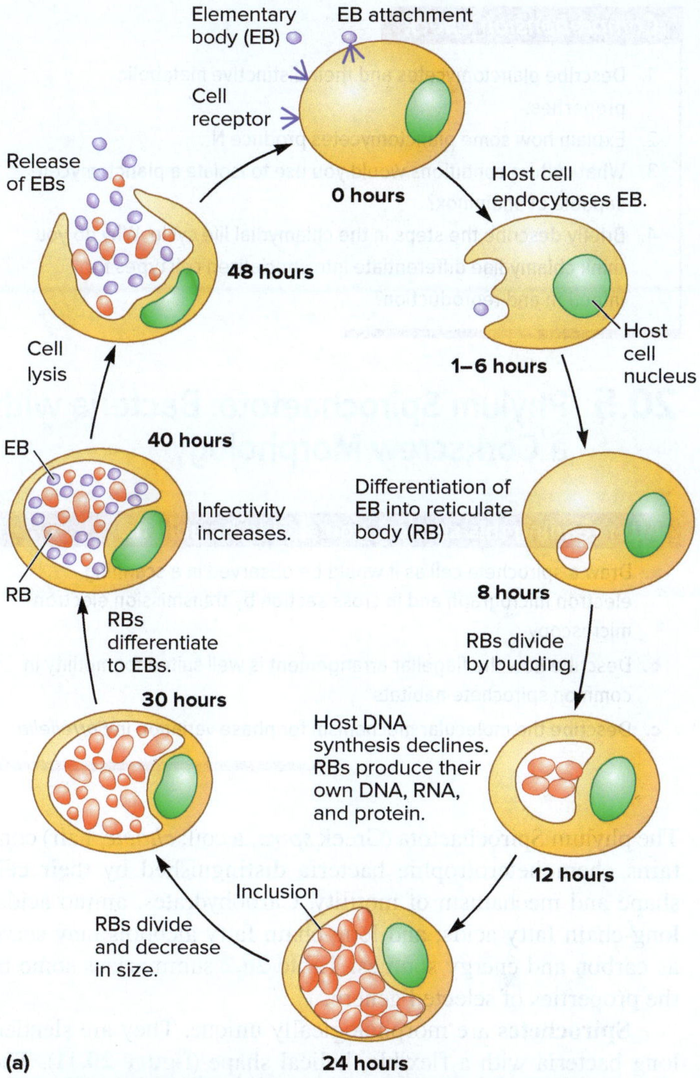

Figure 20.10 The Chlamydial Life Cycle. (a) Generalized life cycle of chlamydiae. (b) Colorized scanning electron micrograph of a human cell infected by *Chlamydia trachomatis* (green spheres).

(b) SPL/Science Source

Comprehension Check

1. Describe planctomycetes and their distinctive metabolic properties.
2. Explain how some planctomycetes produce N_2.
3. What culture conditions would you use to isolate a planctomycete capable of anammox?
4. Briefly describe the steps in the chlamydial life cycle. Why do you think chlamydiae differentiate into specialized cell types for infection and reproduction?

20.5 Phylum Spirochaetota: Bacteria with a Corkscrew Morphology

After reading this section, you should be able to:

a. Draw a spirochete cell as it would be observed in a scanning electron micrograph and in cross section by transmission electron microscopy

b. Describe how the flagellar arrangement is well suited for motility in common spirochete habitats

c. Describe the molecular mechanism for phase variation in *Borreliella*

The phylum Spirochaetota (Greek *spira,* a coil; *chaete,* hair) contains chemoheterotrophic bacteria distinguished by their cell shape and mechanism of motility. Carbohydrates, amino acids, long-chain fatty acids, and long-chain fatty alcohols may serve as carbon and energy sources. **Table 20.2** summarizes some of the properties of selected genera.

Spirochetes are morphologically unique. They are slender, long bacteria with a flexible, helical shape (**figure 20.11**). The central protoplasmic cylinder, which contains cytoplasm and the nucleoid, is bounded by a plasma membrane, peptidoglycan, and an outer membrane that lacks lipopolysaccharide. Two or more

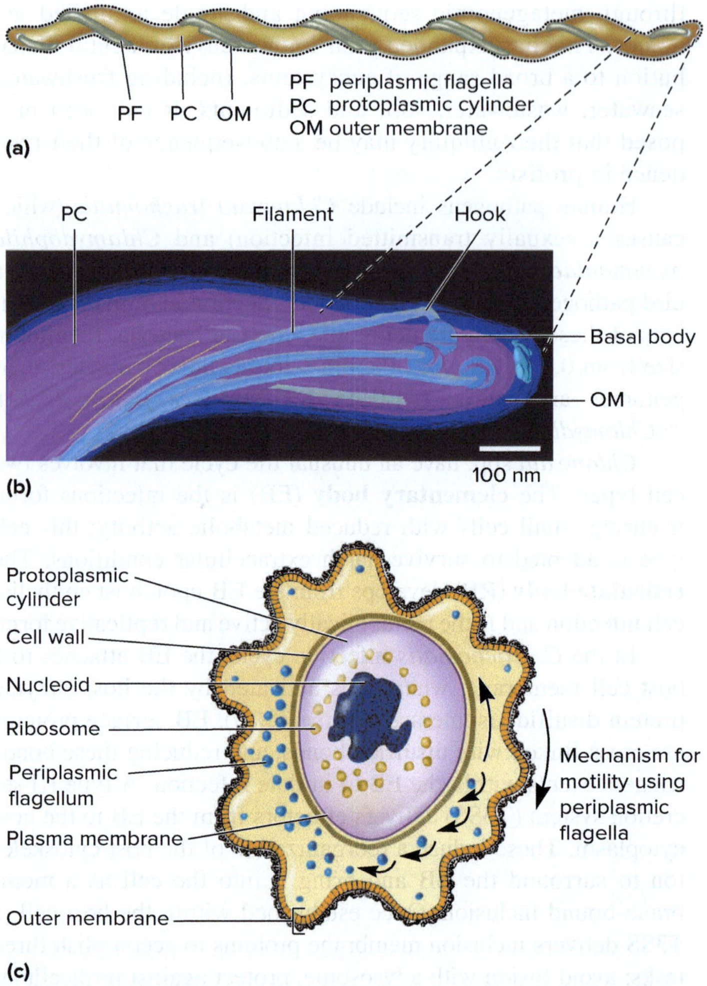

Figure 20.11 Spirochete Morphology. (a) A surface view of spirochete structure. (b) Electrocryotomographic image of *Treponema denticola* showing three flagella attached at the tip of the cell. (c) A cross section of a typical spirochete.

(b) Jacques Izard

Table 20.2 Characteristics of Some Spirochete Genera

Genus	Dimensions (μm) and Flagella	Oxygen Relationship	Carbon + Energy Source	Habitats (and Diseases)
Spirochaeta	0.2–0.75 × 5–250; usually 2 flagella	Facultatively anaerobic or anaerobic	Carbohydrates	Aquatic and free-living
Treponema	0.1–0.7 × 1–20; 2–16 flagella	Anaerobic or microaerophilic	Carbohydrates or amino acids	Mouth, intestinal tract, and genital areas of animals; some are pathogenic (syphilis, yaws)
Borreliella	0.2–0.5 × 3–30; 15–20 flagella	Microaerophilic	Carbohydrates	Arthropods; pathogens (relapsing fever, Lyme disease)
Leptospira	0.1 × 6–12; 2 flagella	Aerobic or microaerophilic	Fatty acids and alcohols	Free-living or pathogens of mammals, usually located in the kidney (leptospirosis)

periplasmic flagella extend from both ends of the cylinder and often overlap one another in the center third of the cell. The periplasmic flagella insert into the membrane with basal bodies located at the cell poles. A short hook connects to a filament that wraps around the peptidoglycan (figure 20.11*b*). Mutant spirochetes lacking flagella have a cylindrical shape, thus establishing the importance of the flagellum for cell shape.

Spirochete motility occurs by the rotation of the periplasmic flagella, like the external flagella of other bacteria. This results in the rotation of the outer membrane in the opposite direction relative to the protoplasmic cylinder, and the cell moves in a corkscrewlike movement (figure 20.11*c*). Flagellar rotation also enables flexing and crawling on solid substrates. ◀ *Bacterial flagella (section 3.7)*

Spirochetes are exceptionally diverse ecologically and grow in habitats ranging from mud to the human mouth (table 20.2). Several spirochetes are pathogenic to humans. *Leptospira* spp. are pathogens that infect the kidneys of many mammals; they are transmitted through exposure to urine. *Treponema pallidum* is the cause of the sexually transmitted infection syphilis. Some spirochetes form symbiotic associations with other organisms and are found in a variety of locations: the hindguts of termites and wood-eating roaches (*see figure 27.2c*), the digestive tracts of mammals, and the oral cavities of animals (*Treponema denticola*). Spirochetes coat the surfaces and drive the motility of many protists. ▶ *Syphilis (section 38.3)*

Borreliella burgdorferi (*see figure 3.2b*) is one of several *Borreliella* species that cause Lyme disease (Lyme borreliosis), which was originally documented among residents of Old Lyme, Connecticut. It is an unusual bacterium in several respects. Its genome consists of a 1 Mb linear chromosome and at least 20 linear and circular plasmids, making it the most segmented bacterial genome yet described. Metabolically, it is also unusual. For instance, *N*-acetylglucosamine (NAG), a component of peptidoglycan, is required for growth. *B. burgdorferi* may also use it as an energy source. The Embden-Meyerhof pathway is used to oxidize glucose monomers to pyruvate. Consistent with the microaerophilic nature of the bacterium, *B. burgdorferi* lacks a TCA cycle and oxidative phosphorylation. Its genome does not encode many biosynthetic enzymes, but transport systems are common.

Despite being a diderm, *B. burgdorferi* does not synthesize lipopolysaccharide. Rather, its cell envelope has lipoprotein as the external layer. The abundant outer membrane protein VlsE undergoes regular changes during an infection, a phenomenon called **phase variation.** Phase variation (also called antigenic variation) allows the microbe to avoid immune clearance, and in this case involves regular modification of the amino acid sequence in VlsE through changes to the *vlsE* gene. The protein has a conserved structure of long α-helices embedded in the outer membrane through hydrophobic interactions (**figure 20.12*a***). The surface loops vary significantly in sequence, and this variation becomes an endlessly changing disguise that allows *Borreliella* to escape the host immune system. The detailed mechanism for this remains to be discovered, but the *vlsE* gene is termed an *expression cassette* because it is the platform on which the *vlsE* gene is assembled. Adjacent to *vlsE* is a series of short open reading frames that are the raw material for insertion into the *vlsE* expression cassette (figure 20.12*b*). In addition to this modular genetic organization, the region has a higher rate of spontaneous mutation than the rest of the genome. Both mechanisms contribute to expanding the variability in the surface-exposed loops in the VlsE protein.

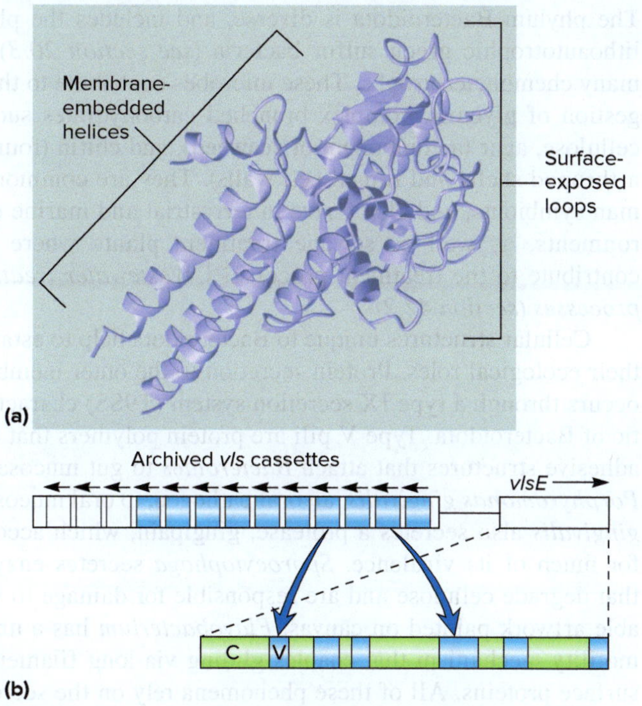

Figure 20.12 Phase Variation in *Borreliella*. (a) The three-dimensional structure of VlsE has a conserved region for membrane attachment, and variable surface-exposed loops. (b) The *vlsE* region of the *Borreliella* genome is at the telomere of a linear plasmid, and adjacent to an archive of short gene segments (top line). The *vlsE* expression cassette has constant (C) and variable (V) regions. The variable regions comprise the surface-exposed loops.

20.6 Phylum Bacteroidota Includes Important Gut Microbiota

After reading this section, you should be able to:

a. Explain how humans host members of the phylum Bacteroidota
b. Describe unusual carbon substrates used by some Bacteroidota
c. Relate the type IX secretion system to outer membrane functions in Bacteroidota
d. Describe *Flavobacterium johnsoniae* gliding motility, and relate it to the microbe's capacity to degrade complex organic substrates

The phylum Bacteroidota is diverse, and includes the photolithoautotrophic green sulfur bacteria (*see section 20.3*) and many chemoheterotrophs. These microbes contribute to the digestion of **glycans**, complex branched carbohydrates such as cellulose, agar (a component of seaweed), and chitin (found in arthropod shells and fungal cell walls). They are common human symbionts, and also thrive in terrestrial and marine environments, as well as sewage treatment plants where they contribute to the treatment process. ▶| *Wastewater treatment processes (section 42.2)*

Cellular structures unique to Bacteroidota help to establish their ecological roles. Protein secretion to the outer membrane occurs through a type IX secretion system (T9SS) characteristic of Bacteroidota. Type V pili are protein polymers that form adhesive structures that attach *Bacteroides* to gut mucosa and *Porphyromonas gingivalis,* an oral pathogen, to oral mucosa. *P. gingivalis* also secretes a protease, gingipain, which accounts for much of its virulence. *Sporocytophaga* secretes enzymes that degrade cellulose and are responsible for damage to valuable artwork painted on canvas. *Flavobacterium* has a unique motility mechanism that enables gliding via long filamentous surface proteins. All of these phenomena rely on the secretion and anchoring of proteins to the outer membrane. |◀ *Bacterial pili and fimbriae (section 3.7); Two-step secretion systems (section 13.9)*

The genus *Bacteroides* contains anaerobic, diderm, nonspore-forming rods; motility is rare. They are important constituents of animal rumens and the human intestinal tract. About 30% of the bacteria cultured from human feces are members of the genus *Bacteroides,* and these organisms provide nutrients by degrading glycans. Although they often benefit their hosts, they are also involved in human disease. For example, *B. fragilis* is a particularly common anaerobic pathogen found in abdominal, pelvic, pulmonary, and blood infections. ▶| *The rumen ecosystem (section 27.2); The microbe-human ecosystem (chapter 33)*

Bacteroides genomes are dominated by polysaccharide *u*tilization *l*oci (PULs), clusters of genes encoding the enzymes required for glycan capture, processing, transport, and hydrolysis. Glycans are diverse polysaccharides, and each has a dedicated PUL; overall genes encoding PULs constitute about 20% of the genome. While the final steps of glycan hydrolysis occur in the cytoplasm, the initial binding and processing are extracellular. Outer membrane complexes termed *utilisomes* consist of four types of proteins (**figure 20.13**). Surface glycan binding proteins are displayed on the outer membrane to capture the substrate. An attached glucosyl hydrolase (GH) generates oligosaccharides, and an outer membrane barrel protein with a lid directs the oligosaccharide to the periplasm. In addition to this import mechanism, *Bacteroides* has also been observed to package GH enzymes in outer membrane vesicles for extracellular processing. |◀ *Extracellular vesicles emerge from bacterial membranes (section 3.4)*

The **gliding motility** characteristic of *Flavobacterium* can be rapid; cells typically travel 2 μm/second, but some are considerably faster. Low nutrient levels usually stimulate gliding. Unlike flagellar-mediated swimming, the mechanisms that enable gliding motility are not well understood. *F. johnsoniae* depends on motility proteins called glide (Gld) proteins that comprise the T9SS. Like the motor proteins in flagellar basal bodies, the Gld proteins harvest the energy of the proton motive force. Gld proteins propel large, filamentous surface proteins, resulting in cell movement along helical tracks. |◀ *Non-flagellar movement—twitching and gliding (section 14.6)*

Gliding motility gives a bacterium many advantages. Many aerobic chemoheterotrophic gliding bacteria actively digest insoluble macromolecular substrates such as cellulose and chitin, and

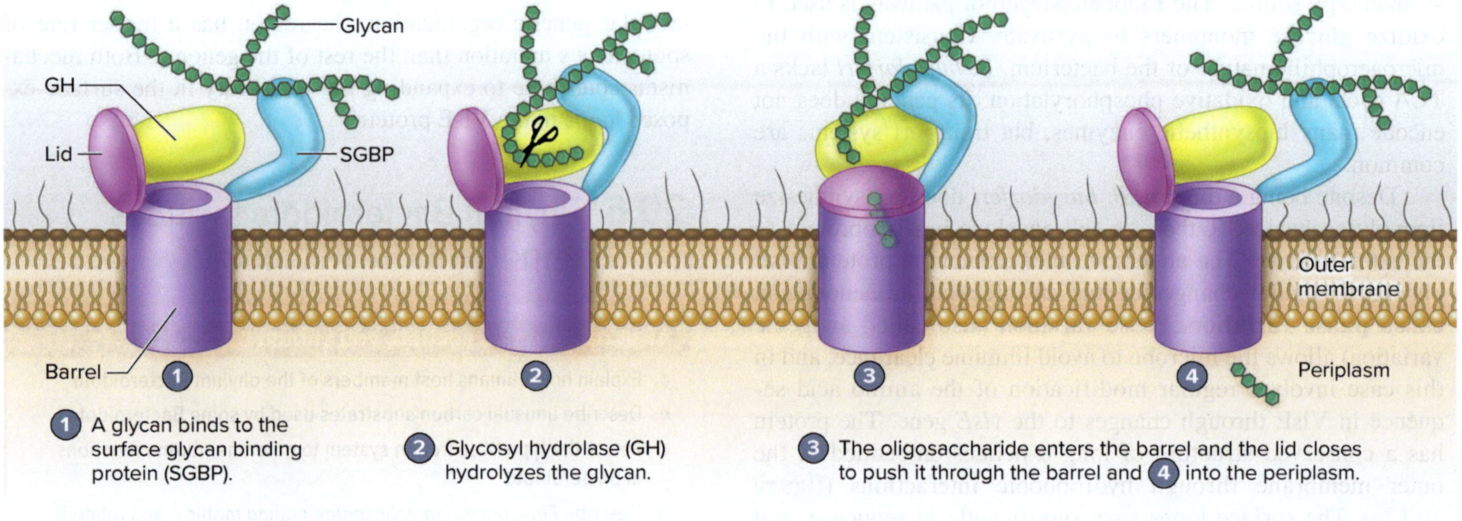

Figure 20.13 Utilisomes on the Outer Membrane of *Bacteroides*.

gliding motility is ideal for searching these out. Because the digestive enzymes are outer membrane associated, the bacteria must be in contact with insoluble nutrient sources; gliding motility makes this possible. Gliding movement is well adapted to drier habitats and to movement within solid masses such as soil, sediments, and rotting wood that are permeated by small channels. Finally, gliding bacteria, like flagellated bacteria, can position themselves at optimal levels of oxygen, H_2S, light intensity, temperature, and other factors that influence growth and survival.

Comprehension Check

1. Draw and label a diagram of spirochete morphology, locating these structures: protoplasmic cylinder, periplasmic flagella, and outer membrane.
2. List the major metabolic activities of the genus *Bacteroides*.
3. List three advantages of gliding motility.
4. What feature of Fusobacteriota growth requirements is responsible for its ability to colonize tumors?

20.7 Phylum Fusobacteriota: Commensal Anaerobes

After reading this section, you should be able to:

a. Describe metabolism in Fusobacteriota
b. Describe habitats in the human body where Fusobacteriota are found

Bacteria in the phylum Fusobacteriota (Latin *fuso,* spindle) exhibit a characteristic spindle, or football, shape (**figure 20.14**). They are obligate or facultative anaerobes that inhabit both anoxic sediments and oral and intestinal habitats of mammals and insects. Amino acids and proteins are the primary energy source. The *Fusobacterium nucleatum* genome encodes biosynthetic pathways for only three amino acids. Amino acid fermentation results in end products that include butyric acid, a metabolite produced by gut microbes that is important for host homeostasis, and H_2S, which promotes pathogenesis through damage to epithelial tissue. ▶| *A functional core microbiome is required for host homeostasis (section 33.3)*

As commensals, Fusobacteriota are widespread in anaerobic niches in the human body and can cause opportunistic infections. Oral biofilms feature Fusobacteriota as bridge bacteria—that is, they physically interact with other species to support the overall biofilm architecture (*see figure 38.36*). Surveys of the human microbiome associated with specific tumors, notably in colorectal cancer, have revealed an overabundance of *F. nucleatum.* It is believed that oral bacteria enter the bloodstream through periodontal lesions and are transported throughout the body. An *F. nucleatum* surface adhesin binds carbohydrate receptors displayed on some cancerous cells and enters those cells. *F. nucleatum* grows intracellularly, where it induces the production of cytokines that promote tumor growth. ▶| *Cancer (section 33.4)*

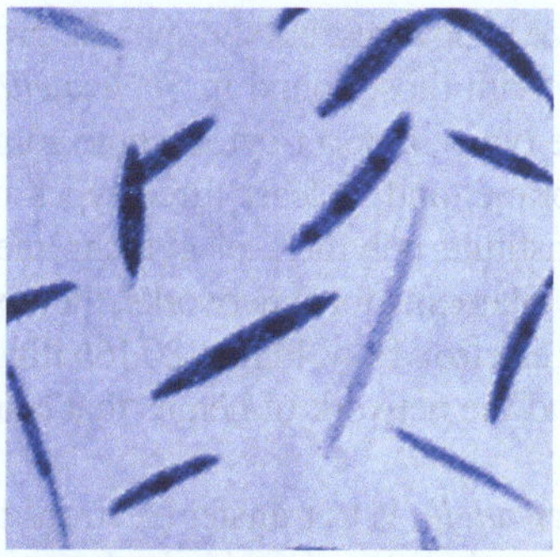

Figure 20.14 ***Fusobacterium* Species Are Spindle-Shaped.**
Centers for Disease Control and Prevention

20.8 Phylum Desulfobacterota: Anaerobic Sulfate/Sulfur or Metal Reducers

After reading this section, you should be able to:

a. Characterize the common metabolic features and identify carbon substrates used by sulfate-reducing bacteria (SRB)
b. Draw an electron transport chain capable of dissimilatory sulfate reduction
c. Relate the metabolism of SRB to their common environmental niches
d. List metal acceptors that can be used and explain the term *dissimilatory metal reduction*
e. Describe strategies used by dissimilatory metal reducing bacteria that enable their use of insoluble compounds as electron acceptors
f. Explain the redox reactions involved in the long-distance electron transport in cable bacteria

Bacteria in the phylum Desulfobacterota are a diverse group of bacteria united by their ability to transfer electrons to sulfur compounds or metals during anaerobic respiration.

Dissimilatory Sulfate Reduction

Sulfate- or **sulfur-reducing bacteria (SRB)** reduce elemental sulfur or sulfate and other oxidized sulfur compounds to H_2S. These sulfur molecules are used as terminal electron acceptors and are not incorporated into the cell's biomass; hence this is dissimilatory sulfate reduction. Recall that anaerobic respiration is coupled with the oxidation of reduced organic compounds and in general the SRB can be grouped according to carbon metabolism. Many SRB oxidize lactate, formate, butyrate, propionate, pyruvate, and aromatic compounds to acetate, which is not further metabolized. Other SRBs oxidize acetate completely to CO_2. |◀ *Anaerobic respiration uses the same steps as aerobic respiration (section 11.5)*

The use of sulfate as a terminal electron acceptor is a multistep process (**figure 20.15**). Sulfate cannot be directly reduced. Because sulfate is a stable molecule, it must first be activated by reacting with ATP to form adenosine 5′-phosphosulfate, or APS (*see figure 21.15*). APS accepts two electrons, which results in

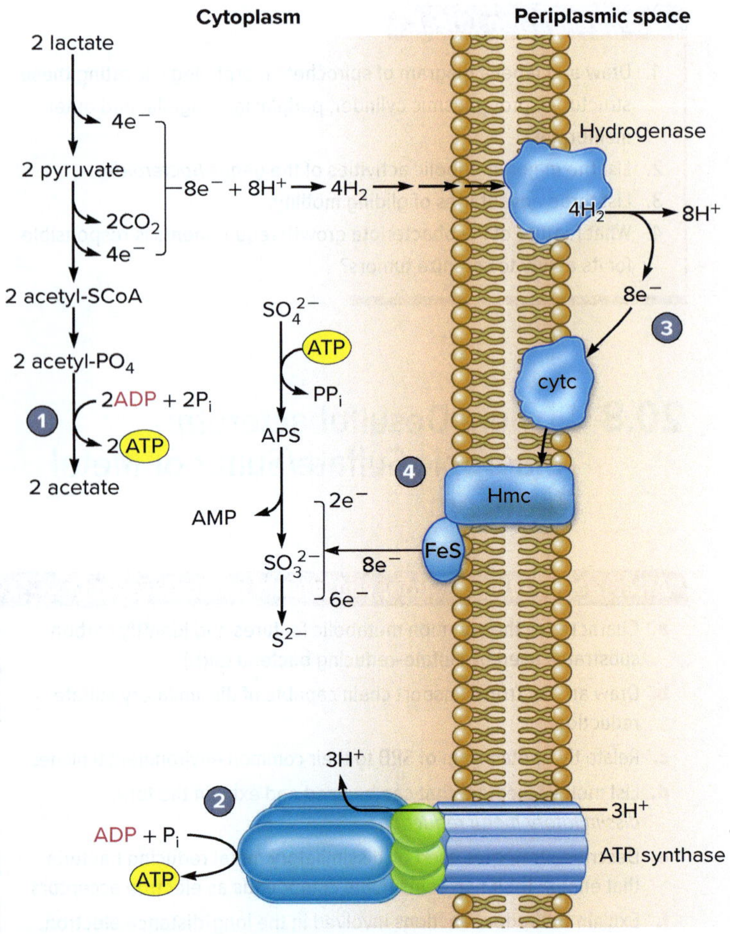

Figure 20.15 Dissimilatory Sulfate Reduction by *Desulfovibrio*. These microbes make ATP by (1) substrate-level phosphorylation during acetate fermentation and (2) ATP synthase. ATP synthase is driven by a proton motive force generated by a periplasmic hydrogenase (3), which oxidizes H_2 to protons and electrons. The electrons are passed back into the cell, (4) where they are used to reduce sulfate completely to sulfide. Cytc, cytochrome c_3; Hmc, high molecular mass cytochrome complex.

the production of sulfite (SO_3^{2-}) and adenosine monophosphate (AMP). Six additional electrons are needed to reduce sulfite completely to H_2S. These are captured by a hydrogenase enzyme located on the periplasmic face of the plasma membrane. The hydrogenase donates the electrons from molecular hydrogen to cytochrome c_3, which in turn transfers them to the integral membrane protein Hmc. Finally, an iron-sulfur protein delivers the electrons to sulfite reductase, and H_2S is produced. Note that hydrogenase releases protons in the periplasm; the resulting proton motive force is sufficient to drive the synthesis of ATP.

SRB are important in sulfur cycling within ecosystems. Because significant amounts of sulfate are present in almost all aquatic and terrestrial habitats, SRB are widespread and active in locations made anoxic by microbial digestion of organic materials (*see figure 28.6*). The SRB thrive in habitats such as muds and sediments of polluted lakes and streams, sewage lagoons and digesters, and waterlogged soils. *Desulfuromonas* spp. are most prevalent in anoxic marine and estuarine sediments. They also can be isolated from methane digesters and anoxic hydrogen sulfide–rich muds of freshwater habitats. Often sulfate and sulfur reduction are apparent from the smell of H_2S (a rotten-egg odor) and the blackening of water and sediment by iron sulfide. H_2S production in waterlogged soils can kill animals, plants, and microorganisms. SRB negatively impact industry because of their primary role in the anaerobic corrosion of iron in pipelines, heating systems, and other structures. ▶| *Sulfur cycle (section 28.2)*

Dissimilatory Metal Reduction

Bacteria in the family Geobacteraceae are strict anaerobes with respiratory or fermentative metabolism. They can be chemolithoheterotrophs, obtaining electrons from reduced inorganic compounds, or chemoheterotrophs. They are mesophilic and have been isolated from anoxic marine and freshwater environments.

Geobacter metallireducens was the first microbe described to couple the oxidation of organic substrates such as acetate with the use of the solid metal Fe(III) as a terminal electron acceptor. The use of metals as terminal electron acceptors is considered dissimilatory metal reduction because the metals are not incorporated into biomass. Dissimilatory metal reduction presents a problem for the microbe because many metals are insoluble in their oxidized forms, which is the form normally needed by microbes for use as an electron acceptor.

Geobacter spp. and *Shewanella* spp. (in the class γ-proteobacteria) have been extensively studied as models of extracellular electron transfer (EET). EET requires electrons to be transported through the layers of the cell envelope to the metal surface. In both organisms, electrons derived from oxidation of organic molecules are moved to quinones in the plasma membrane and periplasmic cytochromes. In *Shewanella* the polypeptides MtrCAB form a complex that spans the outer membrane (**figure 20.16*a***). MtrA and MtrC constitute a **nanowire;** that is, the proteins establish an electron transport pathway from the periplasm to the cell exterior by each supporting 10 heme molecules. MtrB is a porin that functions as an insulator and a membrane anchor. |◀ *Electron transport chains: sets of sequential redox reactions (section 10.4)*

Geobacter spp. also synthesize nanowires, but the specific proteins involved share no similarity with those found in *Shewanella. Geobacter* appears to use multiple types of porin-cytochrome complexes that span the outer membrane, and at least three types of nanowires are known. These constitute polymers of cytochromes that form long filaments; multiple filaments form networks, thereby providing many possible paths for electron flow. Figure 20.16*b* illustrates one *Geobacter* nanowire and the network of hemes that runs through it.

Another strategy used by *Shewanella* is to transfer electrons to the metal via external compounds such as humic acids in the environment, or its own secreted metabolites like flavins. These compounds are collectively known as **electron shuttles** because they pass electrons from the last point of the electron transport chain through the growth substrate to the metal surface. Interestingly,

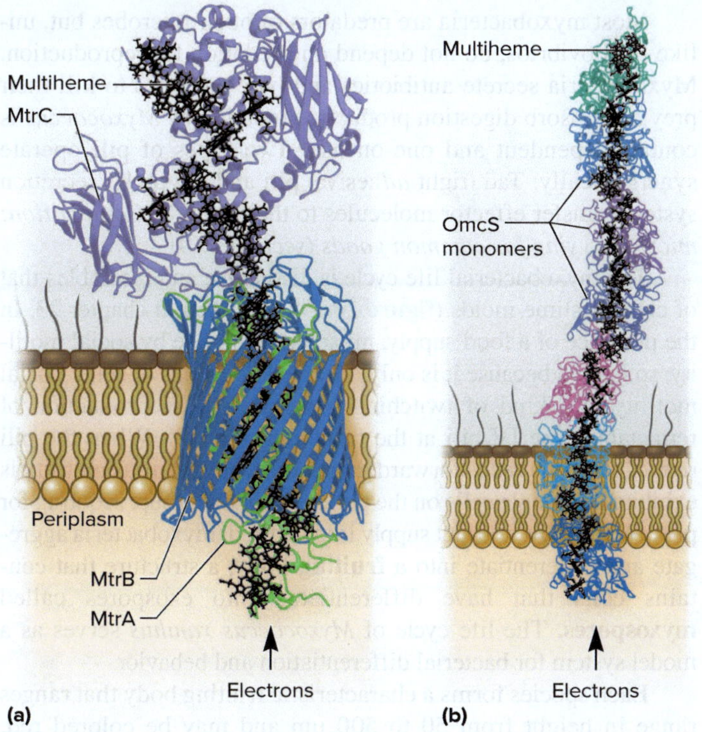

figure 20.16 Nanowires Direct Electrons Across the Outer Membrane. (a) The protein complex from *Shewanella* spp. and (b) a protein nanowire from *Geobacter* spp. Both structures have the hemes highlighted in black to illustrate the path for electron flow.

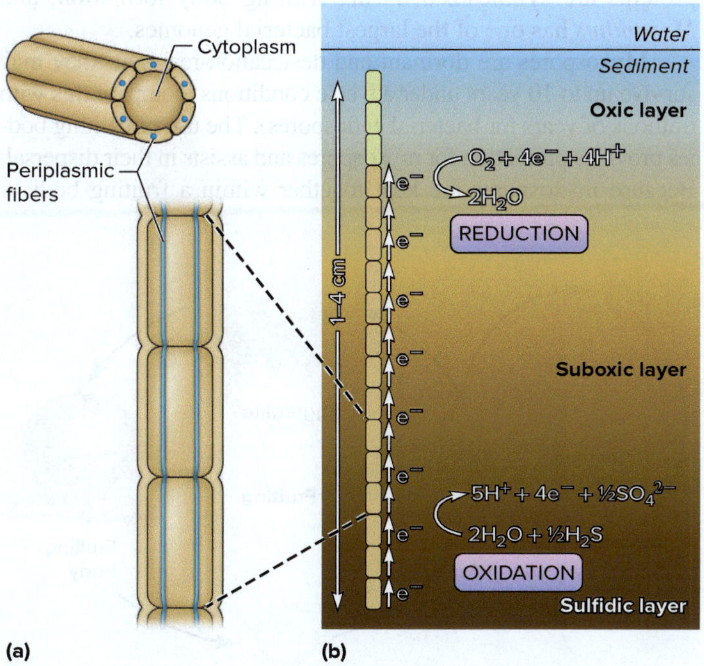

Figure 20.17 Cable Bacteria Transfer Electrons over Long Distances. Electrons resulting from sulfide oxidation at the base of the cable move through the periplasm of cable bacteria to ultimately reduce oxygen near the surface. The cell shown here is not to scale; several thousand bacterial cells comprise a cable. (a) An expanded illustration of the cells in (b) to highlight the periplasmic nanowires.

when near a redox active metal, *Shewanella* cells display electrolocation—behavior analogous to chemotaxis.

Two genera, *Electrothrix* and *Electronema* take electrically conductive nanowires to new lengths, literally. Also termed **cable bacteria,** these microbes are ubiquitous in aquatic sediments where long chains of bacteria extend from the deep anaerobic sulfide-dominant sediment upward to the oxic layer (**figure 20.17**). These e-cables are a multicellular system with two different cell types. Cells in the anoxic zone oxidize sulfide by reversing the pathway diagrammed in figure 20.15. Cytochromes transfer the electrons to fibers located in the periplasm, where they travel distances of several centimeters to the surface. Surface-located cells use the electrons to reduce oxygen but without conserving energy. Overall, the cable bacteria function as aerobic sulfide oxidizers, but each cell type has a distinct metabolism. The electrically conductive nanowire is not external, but rather it is located in the periplasm, and the connection is maintained across cell-cell interfaces. Cell division in cable bacteria is synchronous within a filament.

20.9 Phyla Bdellovibrionota and Myxococcota: Bacterial Predators

After reading this section, you should be able to:

a. Draw the life cycle of a bdellovibrio and explain why it is considered a bacterial predator
b. Relate the life cycle of myxobacteria to their habitat and their means of acquiring nutrients

The phylum Bdellovibriota is best exemplified by the genus *Bdellovibrio* (Greek *bdella,* leech), which contains aerobic, curved rods with polar flagella. The flagellum is unusually thick due to the presence of a sheath that is continuous with the cell envelope. Bdellovibrios have a distinctive lifestyle: They prey on other diderm bacteria and alternate between a nonreplicating predatory phase and an intracellular reproductive phase.

The life cycle of a bdellovibrio is complex, requiring 3 to 4 hours for completion (**figure 20.18**). The free bacterium swims rapidly and collides violently with its prey. Using a type IV pilus to make the initial connection, it attaches to the prey bacterium's surface, rotates as fast as 100 revolutions/second, and bores a hole through the host cell envelope. These bacteria can also scout for bacterial prey by gliding over solid substrates. ▶| *Predators come in all sizes (section 27.4)*

Secreted enzymes cleave the prey bacterium's Braun's lipoprotein, thereby detaching the outer membrane from the peptidoglycan. This enables the bdellovibrio to gain entry to the prey's periplasm. The entry pore is then sealed, providing a refuge for the predator, now termed a *bdelloplast,* to exploit all available resources. The predator assesses the quality of the prey to adjust the number of progeny that it can produce. The growing bdellovibrio uses prey amino acids as its carbon, nitrogen, and energy sources.

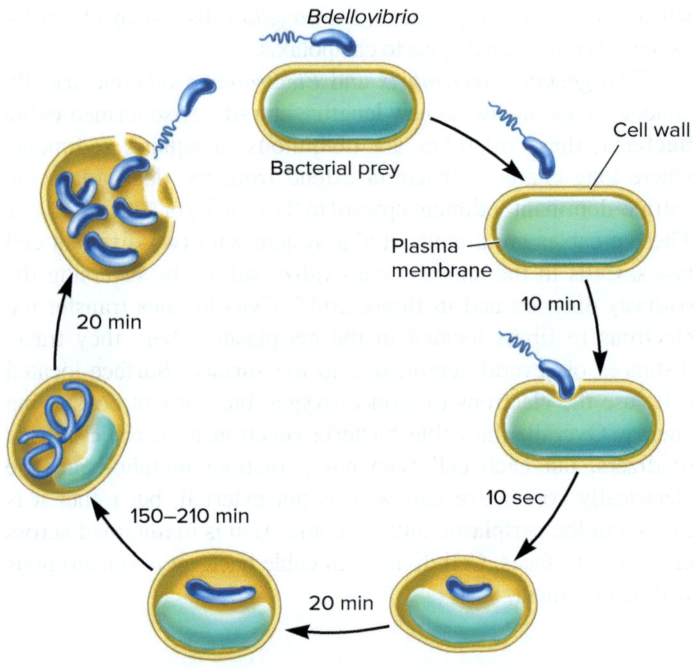

Figure 20.18 The Life Cycle of *Bdellovibrio*.

It employs prey fatty acids and nucleotides directly in biosynthesis. *Bdellovibrio* progeny result from filamentation of the invading bacterium; they are released upon host cell lysis. *Bdellovibrio*

Myxobacteria are aerobic soil bacteria characterized by gliding motility and a life cycle that includes production of multicellular fruiting bodies and the formation of myxospores. Myxobacteria are most commonly isolated from neutral soils or decaying plant material such as leaves and tree bark, and from animal dung. Most myxobacteria use amino acids and lipids as their major sources of carbon, nitrogen, and energy. All are aerobic chemoheterotrophs with respiratory metabolism.

Most myxobacteria are predators of other microbes but, unlike bdellovibrios, do not depend on predation for reproduction. Myxobacteria secrete antibiotics and lytic enzymes to kill their prey and absorb digestion products. Predation by *Myxococcus* is contact-dependent and one-on-one. Two types of pili operate synergistically: Tad (*t*ight *ad*hesive) pili and a type III secretion system transfer effector molecules to their prey. *Competition: microbes vying for common goods (section 27.4)*

The myxobacterial life cycle is distinctive and resembles that of cellular slime molds (**figure 20.19**) discussed in chapter 23. In the presence of a food supply, myxobacteria glide by social motility, so named because it is only observed in groups of cells. Social motility is a kind of twitching governed by the production of retractable type IV pili at the front end of a cell. When the pili retract, the cell creeps forward. In contrast, adventurous motility is exhibited by single cells on the periphery of the group, scouting for prey. When their nutrient supply is exhausted, myxobacteria aggregate and differentiate into a **fruiting body,** a structure that contains cells that have differentiated into exospores called **myxospores.** The life cycle of *Myxococcus xanthus* serves as a model system for bacterial differentiation and behavior.

Each species forms a characteristic fruiting body that ranges range in height from 50 to 500 μm and may be colored red, yellow, or brown by carotenoids. Fruiting bodies vary in complexity from simple globular objects (*Myxococcus;* figure 20.19) to the elaborate, branching structures formed by members of the genera *Stigmatella* and *Chondromyces* (*see figure 3.2d*). As might be expected for such complex structures, a variety of new proteins are synthesized during fruiting body formation, and *M. xanthus* has one of the largest bacterial genomes.

Myxospores are dormant and desiccation-resistant; they may survive up to 10 years under adverse conditions (compare this with millions of years for bacterial endospores). The use of fruiting bodies provides protection for myxospores and assists in their dispersal. Because myxospores are kept together within a fruiting body, a

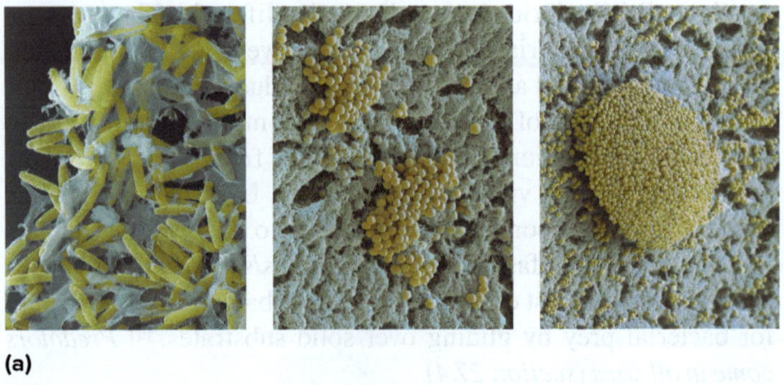

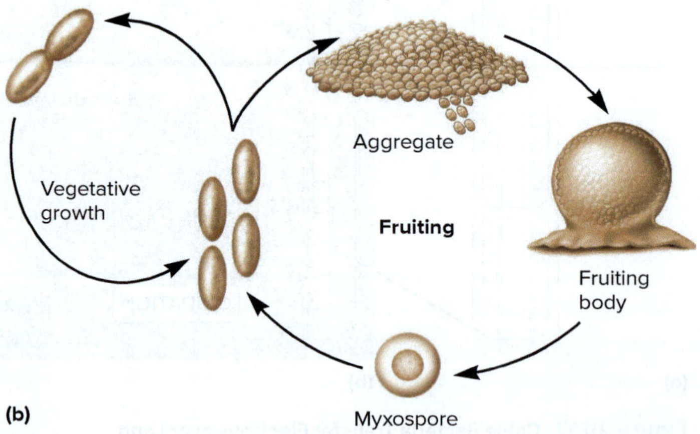

Figure 20.19 Life Cycle of *Myxococcus xanthus*. (a) Scanning electron micrographs taken (from left to right) during vegetative growth, aggregation, and fruiting body formation. (b) When nutrients are plentiful, *M. xanthus* grows vegetatively. However, when nutrients are depleted, the exchange of extracellular signaling molecules triggers cells to aggregate and form fruiting bodies. Most of the cells within a fruiting body become myxospores that will germinate when nutrients are available.

(a) Eye of Science/Science Source

colony of myxobacteria automatically develops when myxospores germinate. This communal organization is advantageous because a mass of myxobacteria produces enzyme concentrations sufficient to digest prey more easily than an individual cell.

20.10 Phylum Campylobacterota: Human and Animal Commensals

After reading this section, you should be able to:

a. List two common human illnesses caused by Campylobacterota
b. Describe the unusual metabolism of *Helicobacter* spp. and explain how *H. pylori* can tolerate the acidic conditions within the stomach

The order Campylobacterales contains two families of interest, Campylobacteraceae and Helicobacteraceae, in which are placed, respectively, two genera, *Campylobacter* and *Helicobacter.* Both contain members that are microaerophilic, motile, helical or vibrioid rods (*see figure 3.2a*).

The genus *Campylobacter* contains both nonpathogens and species pathogenic for humans and other animals. *C. fetus* causes reproductive disease and abortions in cattle and sheep. Other campylobacters are associated with conditions in humans ranging from septicemia (pathogens or their toxins in the blood) to enteritis (inflammation of the intestinal tract). Infection with some species can precede the development of Guillain-Barré syndrome (GBS), the most common form of flaccid paralysis. It is thought that these bacteria trigger GBS by a phenomenon known as *molecular mimicry* (*see figure 38.29*). In this case, the structure of *Campylobacter* spp. lipooligosaccharide (lipopolysaccharide without an O-antigen) resembles components found on nerve cells. This causes the immune system to mistakenly attack the host's nervous system even after infection has cleared.

C. jejuni is found in the intestinal tract of animals. It causes an estimated 1.5 million human cases of *Campylobacter* gastroenteritis—inflammation of the stomach and intestine with subsequent diarrhea—in the United States each year. Studies with chickens, turkeys, and cattle have shown that as much as 50 to 100% of a flock or herd of these birds or other animals excrete *C. jejuni*. Campylobacteria are transmitted to humans by contaminated food and water, contact with infected animals, or anal-oral sexual activity.

There are several dozen species of *Helicobacter,* all isolated from the stomachs and upper intestines of humans and other mammals. About 50% of the global population is infected. The major human pathogen is *Helicobacter pylori,* which causes gastritis, peptic ulcer disease, and gastric cancer. *H. pylori* is an obligate microaerophile. The only carbohydrate the bacterium can use is glucose, which it metabolizes by both respiratory and fermentative pathways. It uses the pentose phosphate pathway, which provides NADPH for biosynthesis, and the Entner-Doudoroff pathway to produce pyruvate. The microbe does not have a complete TCA cycle and appears to run the cycle forward (i.e., oxidatively) from oxaloacetate to α-ketoglutarate and backward (i.e., in the reductive direction) from oxaloacetate to succinate. Fumarate may act as the terminal electron acceptor during anaerobic respiration. *H. pylori* can also carry out aerobic respiration. Pyruvate is fermented to lactate, acetate, and formate during fermentation. As a resident of the gastric mucosa, it would seem that *H. pylori* is exposed to very acidic conditions, yet in the laboratory, it cannot be cultured below pH 4.5. *H. pylori* burrows within the gastric mucosa using pili as adhesins, thereby avoiding the acidity of the stomach lumen. It also produces the enzyme urease, which converts urea to CO_2 and ammonia (NH_3). The release of NH_3 drives up the local pH while the microbe uses chemotaxis to maintain its position within the mucous membranes. Indeed, motility is required for colonization of the host. ▶| *Peptic ulcer disease and gastritis (section 38.4)*

Comprehension Check

1. What is the role of fermentation in dissimilatory sulfate reduction? How do SRB make ATP?
2. What mechanisms do *Geobacter* spp. and *Shewanella* spp. use to transfer electrons to extracellular insoluble metals? How does this compare with cable bacteria?
3. Characterize *Bdellovibrio* spp. and outline the life cycle in detail. Why do they prey only on diderm bacteria?
4. Compare the growth of a myxobacterium in nutrient-rich and nutrient-poor conditions.
5. Why is it important to control *Campylobacter* in animals in the human food supply?
6. How does *H. pylori* survive the acidic conditions of the stomach?

Key Concepts

20.1 Bacterial Taxonomy and Cell Envelopes

- Current bacterial taxonomy relies on nucleic acid and genome sequencing. Standardizing taxonomic ranks based on sequence data has upended some traditional taxa, namely Gram-positive bacteria.
- Genomic analysis allows for the identification of genes encoding the components of cell envelope structures or the enzymes that synthesize those components. Taxa within a phylum tend to be uniform in cell envelope components.

- Although the traditional division of bacteria into Gram positives and Gram negatives (or monoderms and diderms) does not reflect their evolutionary history, these designations are widely used and entrenched in the common language of microbiologists.

20.2 Thermotogota and Deinococcota Live in Extreme Conditions

- *Thermotoga* contains hyperthermophilic, chemoheterotrophic rods. Their cell envelopes feature both protein and membrane surfaces, as well as some lipids reminiscent of archaeal membrane-spanning monolayer lipids.
- Deinococcota are aerobic cocci and rods that have an unusual cell wall.
- *Deinococcus radiodurans* is distinctive because of its tremendous resistance to desiccation and radiation.
- *Thermus* spp. are readily culturable thermophiles, and are the source of the *Taq* DNA polymerase used in PCR.

20.3 Photosynthetic Bacteria Are Diverse

- Cyanobacteria carry out oxygenic photosynthesis; purple and green bacteria use anoxygenic photosynthesis.
- The four most important groups of purple and green photosynthetic bacteria are the purple sulfur bacteria, purple nonsulfur bacteria, green sulfur bacteria, and green nonsulfur bacteria (**table 20.1**).
- The bacteriochlorophyll pigments of purple and green bacteria enable them to live in deeper, anoxic zones of aquatic habitats (**figure 20.3**).
- Green sulfur bacteria are obligately anaerobic photolithoautotrophs that use hydrogen sulfide, elemental sulfur, and hydrogen as electron sources.
- Green nonsulfur bacteria include members of the genus *Chloroflexus,* gliding thermophilic filaments metabolically similar to purple nonsulfur bacteria. Both green sulfur and green nonsulfur bacteria have chlorosomes, which function to harvest light and transfer it to the reaction center in the plasma membrane (**figure 20.4**).
- Cyanobacteria carry out oxygenic photosynthesis by means of a photosynthetic apparatus similar to that of the eukaryotes. Phycobilisomes contain the light-harvesting pigments phycocyanin and phycoerythrin (**figure 20.6**).
- Cyanobacteria reproduce by binary fission, budding, multiple fission, and fragmentation by filaments to form hormogonia. Some produce dormant akinetes.
- Many nitrogen-fixing cyanobacteria form heterocysts, specialized cells in which nitrogen fixation occurs (**figure 20.8**).
- *Prochlorococcus* spp. are the most abundant photosynthetic organism on Earth. Together with other unicellular cyanobacteria in the genus *Synechococcus,* they account for about one-third of global carbon fixation.

20.4 PVC Superphylum (Planctomycetota and Verrucomicrobiota): Atypical Cell Division

- Members of the PVC superphylum undergo cell division by budding and other atypical processes in the absence of cytoskeletal protein FtsZ.
- Some planctomycetes perform the anammox reaction (**figure 20.9**).
- Verrucomicrobiota include human symbionts and bacteria that degrade plant carbohydrate polymers.
- Chlamydiae are nonmotile, coccoid bacteria that lack peptidoglycan and must reproduce within cytoplasmic vacuoles of host cells. Their life cycle involves elementary bodies (EBs) and reticulate bodies (RBs) (**figure 20.10**).

20.5 Phylum Spirochaetota: Bacteria with a Corkscrew Morphology

- Spirochetes are slender, long, helical bacteria. They are motile by means of periplasmic flagella underlying an outer membrane (**figure 20.11**).
- *Borreliella burgdorferi,* the causative agent of Lyme disease, changes its surface proteins through phase variation. The mechanism for phase variation includes frequent exchanges of short DNA segments.

20.6 Phylum Bacteroidota Includes Important Gut Microbiota

- Members of the phylum Bacteroidota are metabolically diverse motile or nonmotile rods.
- Some of these bacteria are important rumen and intestinal symbionts; others can cause disease. Many are important in degrading cellulose, chitin, and other glycans in terrestrial and aquatic environments.
- Utilisomes are protein complexes in the outer membrane of *Bacteroides* cells. Glycan degradation occurs through the capture, transport, and hydrolysis of macromolecular carbohydrates.
- Gliding motility is present in a diverse range of bacteria; in members of Bacteroidetes, gliding is mediated by a unique group of transmembranous and surface proteins.

20.7 Phylum Fusobacteriota: Commensal Anaerobes

- Spindle-shaped Fusobacteriota are anaerobes found in multiple environments including the human body. They are components of oral biofilms and have been associated with tumor growth.

20.8 Phylum Desulfobacterota: Anaerobic Sulfate/Sulfur or Metal Reducers

- Desulfurobacterota are aerobic chemoheterotrophic bacteria that can use elemental sulfur and oxidized sulfur compounds or metals as electron acceptors in anaerobic respiration.

- Sulfate-reducing bacteria are important in sulfur cycling in ecosystems. *Desulfovibrio* spp. conserve energy by fermentation and a proton motive force (**figure 20.15**).
- *Geobacter* spp., *Shewanella* spp., and cable bacteria produce extracellular nanowires to transfer electrons to terminal electron acceptors outside the cell (**figures 20.16** and **20.17**).

20.9 Phyla Bdellovibrionota and Myxococcota: Bacterial Predators

- *Bdellovibrio* spp. are aerobic curved rods with a sheathed polar flagellum that consume other diderm bacteria and reproduce within the periplasmic space of their prey (**figure 20.18**).
- *Myxobacteria* spp. are aerobic soil bacteria with gliding motility and a complex life cycle that leads to the production of dormant myxospores held within fruiting bodies (**figure 20.19**).

20.10 Phylum Campylobacterota: Human and Animal Commensals

- Campylobacterota includes two important genera with pathogenic species: *Campylobacter* and *Helicobacter*. These bacteria are microaerophilic, motile, helical or vibroid, diderm rods.
- *Campylobacter* spp. are a common cause of foodborne illness, which in rare cases leads to Guillain Barré syndrome as a result of molecular mimicry.
- *Helicobacter* spp. are resident in the stomach and upper intestines of mammals. *H. pylori* escapes the effects of acidic conditions to colonize the stomach, resulting in gastritis, ulcers, or cancer.

Active Learning

1. The cyanobacterium *Anabaena* grows well in liquid medium that contains nitrate as the sole nitrogen source. Suppose you transfer some of these filaments to the same medium but lacking nitrate. Describe the morphological and physiological changes you would observe.
2. Compare the structural and functional differences between chlorosomes and thylakoid membranes. Do you think this is an example of convergent evolution—two structures evolving separately to fulfill the same function, or divergent evolution—two structures that arose from a single ancestor but adapted to meet the particular needs of the organism that bears them? What other data would you need (apart from structure and function) to answer this question?
3. Many types of movement are employed by bacteria discussed in this chapter. Review them and propose mechanisms by which energy (ATP or proton gradients) might drive the locomotion.
4. Propose an evolutionary scenario to explain how bacteria in the phylum Chloroflexota came to lack an outer membrane.
5. Compare and contrast the life cycle of *Chlamydia* with that of a virus (*see figure 6.9*) and with the process of endospore formation (*see figure 3.39*).
6. *Bdellovibrio* preys on diderm bacteria, including the well-studied *E. coli*. Since the morphology and biosynthesis of *E. coli* surface structures (lipopolysaccharide, capsule, flagella, pili) are well-known, propose experiments to dissect the process of *Bdellovibrio* recognition and infection of *E. coli*. How would you determine which surface structures are recognized by *Bdellovibrio?*
7. Compare and contrast the organelles/compartments discussed in this chapter: chlorosomes, carboxysomes, and anammoxosomes. What purpose is served by segregating activities to these compartments?
8. How might electron shuttles facilitate growth of bacteria in a biofilm that has become anoxic in the interior?

CHAPTER

21

Proteobacteria

Marnie Burkhart/Corbis/Glow Images

Welcome back!

When the SARS-CoV-2 virus began spreading in the United States in March 2020, lockdowns went into effect in many communities. Schools, office buildings, gyms, and shopping centers closed down, with the intent of reopening in weeks or months. In reality, it was well over a year later, after widespread vaccination and increased public awareness of how to control respiratory pathogens, that many such buildings were reopened.

So unlock the doors and flip on the light switch and we're back in business, right? It's not that easy, as we are learning after the fact. Respiratory illnesses in general declined during and immediately after the pandemic as we took precautions like masking and distancing. But about two years after the pandemic, epidemiologists worldwide saw a spike in cases of legionellosis, a type of pneumonia caused by *Legionella* infection.

Legionella lurks in plumbing systems, including hot water heaters, cooling towers, and air-conditioning units. During the months of underutilization, stagnation in building water systems promoted the growth of biofilms and *Legionella*. *Legionella* enters the lungs by inhalation of aerosolized water droplets, as produced in showers, in aerated sinks, or from air-conditioning ducts. Older people and those with weakened immune systems are especially susceptible to *Legionella* infection.

Several years later, building managers are still encountering *Legionella* in water systems, and government authorities have developed detailed plans for the testing and management of building water systems. From a federal courthouse in Manhattan to even the offices of the Centers for Disease Control and Prevention (CDC) in Atlanta, our built environment infrastructure requires a high level of air quality.

Legionella is a Gram-negative diderm classified in the phylum Pseudomonadota, a large taxon that is divided into two classes of proteobacteria, alpha and gamma. Many of these organisms are among the most commonly encountered human pathogens, while others play important roles in environmental carbon, nitrogen, and sulfur cycles.

Readiness Check:

Based on what you have learned previously, you should be able to:

- ✓ Draw a diderm cell envelope and list its constituent parts (section 3.3)
- ✓ List environmental factors that influence microbial growth with a particular emphasis on oxygen (section 7.5)
- ✓ Draw an electron transport chain indicating the relative standard reduction potentials of the electron donor, acceptor, and electron transporters (sections 10.4 and 11.4)
- ✓ Explain the difference between heterotrophy and autotrophy (section 11.1)
- ✓ Compare anaerobic and aerobic respiration and fermentation (sections 11.5 and 11.6)
- ✓ Define chemolithotrophy and identify an example electron donor and acceptor (section 11.8)
- ✓ Compare and contrast oxygenic and anoxygenic phototrophy (section 11.10)
- ✓ Summarize the function of the Calvin-Benson cycle and the enzyme nitrogenase (sections 12.3 and 12.5)
- ✓ Differentiate between swimming and gliding motility (section 14.6)

21.1 Class Alphaproteobacteria Includes Many Oligotrophs

After reading this section, you should be able to:

a. Describe oligotrophic environments inhabited by some α-proteobacteria
b. Outline the metabolic flexibility of purple nonsulfur bacteria
c. Draw the photosynthetic apparatus found in purple nonsulfur bacteria and label its constituent parts
d. Diagram the life cycle of *Caulobacter* spp.
e. Compare and contrast the relationship between *Rhizobium* and *Agrobacterium* spp. and plants
f. Identify the electron donors and the electron acceptor used in nitrification
g. List the physiological and taxonomic groups of microbes capable of nitrification
h. Describe the β-hydroxyaspartate cycle and its importance in the global carbon cycle
i. List two unusual metabolic features of rickettsias and relate these to their intracellular lifestyle
j. Explain the relationship between mitochondria and α-proteobacteria

A wide variety of bacteria belong to the class Alphaproteobacteria, many of which are oligotrophs, capable of growing at low nutrient levels. Some α-proteobacteria have unusual metabolic modes such as anoxygenic photosynthesis (purple nonsulfur bacteria), chemolithotrophy (*Nitrobacter*), and the ability to fix nitrogen (*Rhizobium*). **Table 21.1** summarizes the general characteristics of many of the bacteria discussed in this section.

Order Rhodospirillales Includes Photosynthesizers and Magnetic Bacteria

Purple nonsulfur bacteria vary considerably in morphology. They may be spirals (*Rhodospirillum*) or rods (*Rhodopseudomonas*), or they may even form prosthecae (stalks) and buds (*Rhodomicrobium*). Most are motile by polar flagella. They are prevalent in the mud and water of lakes and ponds with abundant organic matter and low sulfide levels.

Purple nonsulfur bacteria use anoxygenic photosynthesis and possess bacteriochlorophylls *a* or *b* (*see table 20.1*). Their photosynthetic apparatus consists of invaginated plasma membrane, which makes many intracellular folds (**figure 21.1*a***). These *i*ntra*c*ytoplasmic *m*embranes (ICMs) increase the membrane surface area, thereby providing more space for chromatophores, the sites where light energy is captured. Each chromatophore includes a *r*eaction *c*enter (RC), consisting of a protein-bacteriochlorophyll (Bchl) complex that absorbs light in the infrared range (*see figure 20.3*). The RC-Bchl is responsible for donating electrons to the electron transport chain that results in a proton motive force sufficient to fuel ATP synthesis. The energy required for charge separation is supplied by the photons directly absorbed by the RC-Bchl or transferred from a light-harvesting complex. All purple bacteria (nonsulfur and sulfur) have a primary light-harvesting complex called LH-I closely associated with the RC (figure 21.1*b*). Most also have additional light-harvesting

Table 21.1 Characteristics of Selected α-Proteobacteria

Genus	Dimensions (μm) and Morphology	Oxygen Requirement	Other Distinctive Characteristics
Agrobacterium	0.6–1.0 × 1.5–3.0; motile rods with peritrichous flagella	Aerobic; microaerophilic	Chemoheterotrophs that can invade plants and cause tumors
Caulobacter	0.4–0.6 × 1–2; comma-shaped with polar flagellum and prostheca with holdfast	Aerobic	Chemoheterotrophic and oligotrophic; asymmetric cell division
Magnetospirillum	0.2–0.7 × 3–4; spiral cells with a single polar flagellum; single chain of cytoplasmic magnetite crystals	Microaerophilic	Chemoheterotrophic; magnetotactic
Nitrobacter	0.5–0.9 × 1.0–2.0; rod- or pear-shaped, sometimes motile by flagella	Aerobic; microaerophilic	Chemolithotrophs, oxidize nitrite to nitrate
Pelagibacter	0.2 × 0.5; rod-shaped	Aerobic	Chemoheterotrophic; proteorhodopsin present
Rhizobium	0.5–1.0 × 1.2–3.0; motile rods with flagella	Aerobic	Symbiotic with leguminous plants in nitrogen-fixing root nodules
Rhodospirillum	0.8–1.0 × 7–10; spiral cells with polar flagella	Aerobic; anaerobic; microaerophilic	Anoxygenic photoheterotrophs under anoxic conditions
Rickettsia	0.3–0.5 × 0.8–2.0; short nonmotile rods	Aerobic	Obligate intracellular parasites

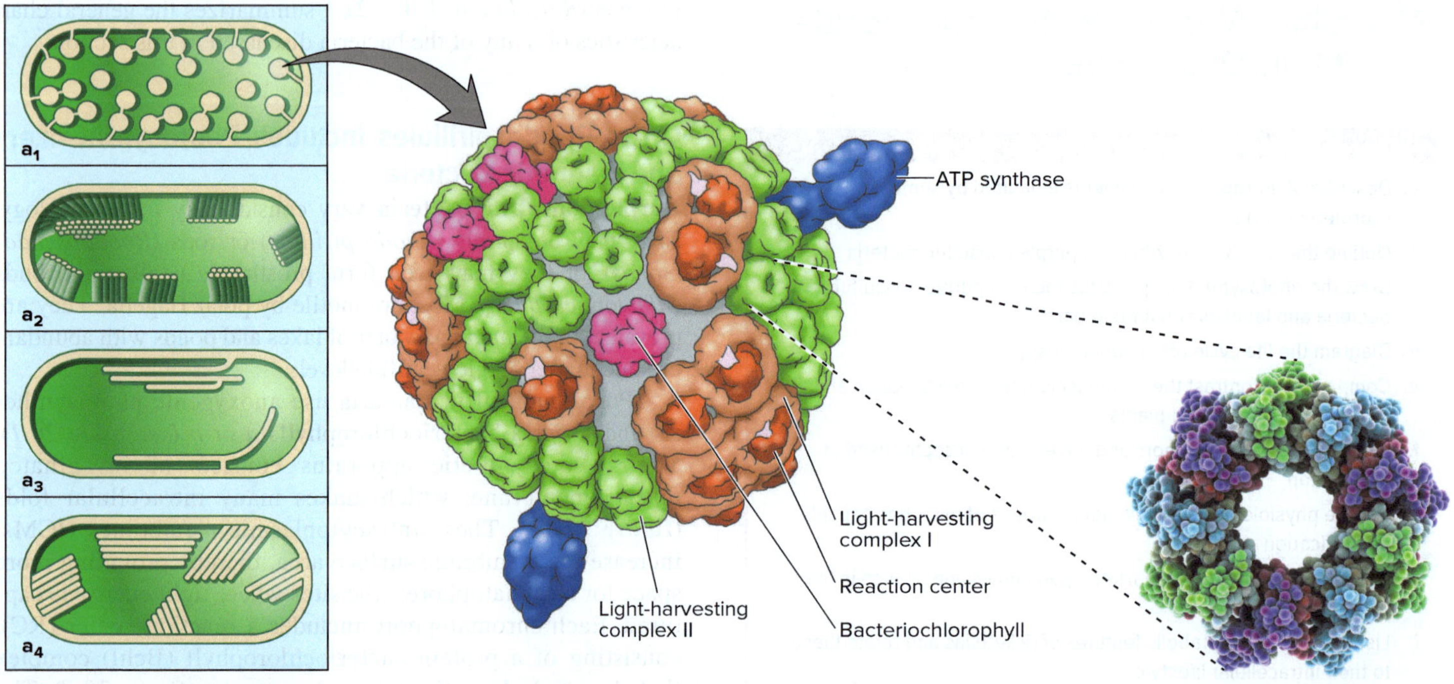

Figure 21.1 The Photosynthetic Apparatus of Purple Bacteria. (a) The arrangement of intracytoplasmic membranes (ICMs) associated with the plasma membrane. (a_1) Vesicular ICMs (chromatophores). (a_2) Lamellar ICMs. (a_3) ICMs arranged in small stacks. (a_4) Larger stacked ICMs. (b) Each chromatophore carries photosynthetic pigments organized in light-harvesting (LH) complexes I and II. These transfer energy to the centrally located reaction center. (c) Molecular model of LH-II from *Rhodoblastus acidophilus*. LH-II has nine identical subunits, each of which contains three bacteriochlorophyll *a* molecules that transfer electrons to the reaction center and LH-I.

(c) Laguna Design/Science Source

MICRO INQUIRY *How do chromatophores compare to chlorosomes in green sulfur bacteria (see figure 20.4)?*

complexes, called LH-II (figure 21.1*c*). Light-harvesting complexes are composed of Bchl *a* and carotenoid pigments. Their proximity to the RC/LH-I complex within the membrane maximizes the efficiency of energy transfer. *Light reactions in anoxygenic photosynthesis (section 11.10); Photosynthetic bacteria are diverse (section 20.3)*

Purple nonsulfur bacteria show exceptional metabolic flexibility. Normally they grow anaerobically as photoorganoheterotrophs, trapping light energy and employing organic molecules as both electron and carbon sources (*see table 20.1*). Although they are called nonsulfur bacteria, some species can oxidize very low, nontoxic levels of sulfide to sulfate, but they do not oxidize elemental sulfur. Some can also use hydrogen as an electron source. In the absence of light, most purple nonsulfur bacteria grow aerobically as chemoorganoheterotrophs, but some species carry out fermentation anaerobically. Oxygen inhibits bacteriochlorophyll and carotenoid synthesis so that cultures growing aerobically in the dark are colorless.

Rhodospirillum rubrum exemplifies the metabolic flexibility of this group. In a strategy analogous to *E. coli* catabolite repression, *R. rubrum* must assess its oxygen and light conditions and the available carbon sources, and adjust gene expression to optimize its metabolism for its environment. The "decision-making" process this microbe uses is displayed in **figure 21.2**. *R. rubrum* has sparked the interest of industrial microbiologists because it produces novel biodegradable plastics when grown on β-hydroxycarboxylic and *n*-alkanoic acids. *Catabolite repression in* E. coli *(section 14.5)*

In addition to the anaerobic anoxygenic photosynthesis displayed by the purple nonsulfur bacteria, some α-proteobacteria are capable of **aerobic anoxygenic phototrophy (AAnP).** These aerobic bacteria are photoheterotrophs that catabolize carbon and other organic materials using O_2-dependent respiration. They supplement this chemoheterotrophic source of energy using Bchl *a*–dependent phototrophy. This allows them to shunt more organic carbon to anabolic pathways, thereby enhancing their growth. They are found in marine and freshwater environments and are often present in microbial mat communities. Metagenomic analysis reveals that a wide diversity of bacteria are capable of AAnP (*see table 20.1*). Their contribution to the global carbon cycle is an area of active investigation. *Microorganisms in the open ocean are adapted to nutrient limitation (section 29.2)*

Rhodospirillum and *Azospirillum* are among several bacterial genera with species capable of forming **cysts** in response to nutrient limitation (**figure 21.3**). In addition to nutritional cues, cyst

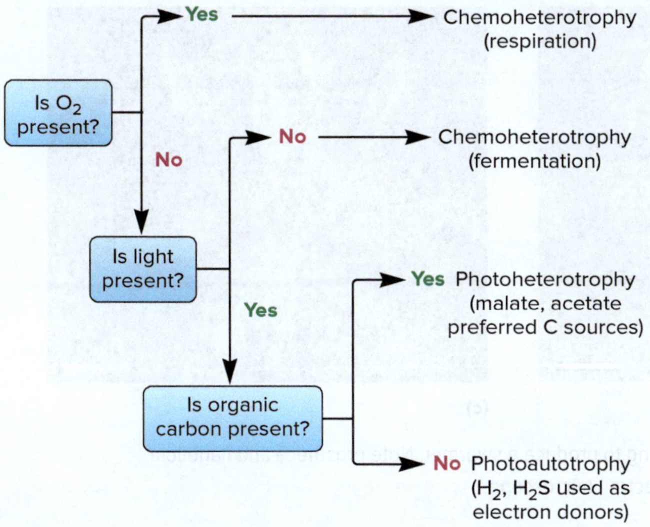

Figure 21.2 The Metabolic Flexibility of *Rhodospirillum rubrum*. This microbe can switch from one type of metabolism to another, depending on the environmental conditions.

MICRO INQUIRY *What would* R. rubrum *use for carbon and energy sources if grown (a) anaerobically in the dark in the presence of malate, (b) anaerobically in the light without organic carbon, and (c) aerobically in the light in the presence of malate?*

formation is regulated by the secretion and reuptake of cyclic GMP, a molecule better known for its role as a second messenger. These resting cells differ from the well-characterized endospores made by monoderm bacteria. Although cysts are also resistant to desiccation,

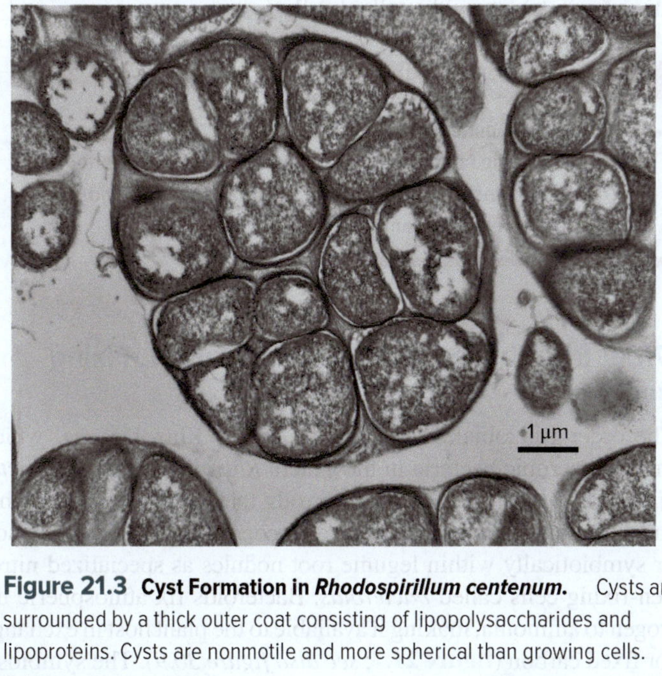

Figure 21.3 Cyst Formation in *Rhodospirillum centenum*. Cysts are surrounded by a thick outer coat consisting of lipopolysaccharides and lipoproteins. Cysts are nonmotile and more spherical than growing cells.
Carl Bauer

MICRO INQUIRY *How are bacterial cysts different from the endospores of certain monoderm bacteria?*

they are less tolerant of other environmental stresses such as heat and ultraviolet light. They have a thick outer coat and store an abundance of poly-β–hydroxybutyrate. Cyst-forming bacteria are not limited to α-proteobacteria; for instance, members of *Azotobacter,* which are γ-proteobacteria, also form cysts. *Inclusions (section 3.6); Bacterial endospores are a survival strategy (section 3.8)*

A fascinating group of bacteria are magnetotactic—that is, they use intracellular magnets called *magnetosomes* to align themselves with Earth's magnetic field. Magnetotactic bacteria are found in several phylogenetic groups, but the best characterized genus is *Phaeospirillum* from the order Rhodospirillales (**figure 21.4**). They occupy freshwater or marine sediments and are highly motile, using their magnetosomes to orient in the planet's magnetic field. Studies of *Phaeospirillum* have identified enzymes and structural proteins involved in magnetosome formation. The magnetosome intracytoplasmic membrane forms from plasma membrane invagination but displays unique *ma*gnetosome *m*embrane (Mam) proteins. Iron is transported from the cytoplasm in an ATP-dependent process, and its nucleation to the growing crystal is driven by other Mam proteins. Each magnetosome contains a single magnetite crystal, and its size and structure are characteristic of a species. *Phaeospirillum* cells contain up to 60 magnetosomes, aligned across the center of the cell by their interaction with the cytoskeletal protein MamK (*see figure 3.30*). Genes encoding the magnetosome structure and assembly proteins are clustered on a genomic magnetosome island (GMI) of about 100 kilobases. Interestingly, the GMI can be transferred to *R. rubrum,* which then forms ordered magnetosome chains in its cells.

Caulobacterales Reproduce in Unusual Ways

Members of the Caulobacterales have life cycles that feature a prostheca. A **prostheca** (pl., prosthecae), also called a **stalk,** is an extension of the cell, including the plasma membrane and cell wall, that is narrower than the mature cell. The prostheca terminates in a **holdfast,** by which the cells attach to solid substrates (**figure 21.5**).

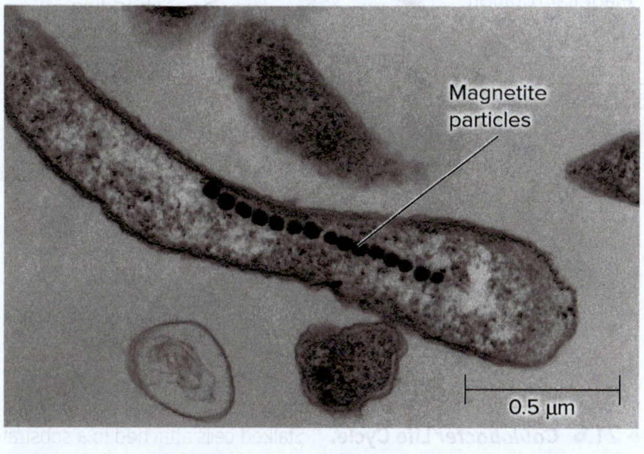

Figure 21.4 Magnetosomes. Transmission electron micrograph of the magnetotactic bacterium *Phaeospirillum magnetotacticum.*
Dennis Kunkel Microscopy/Science Source

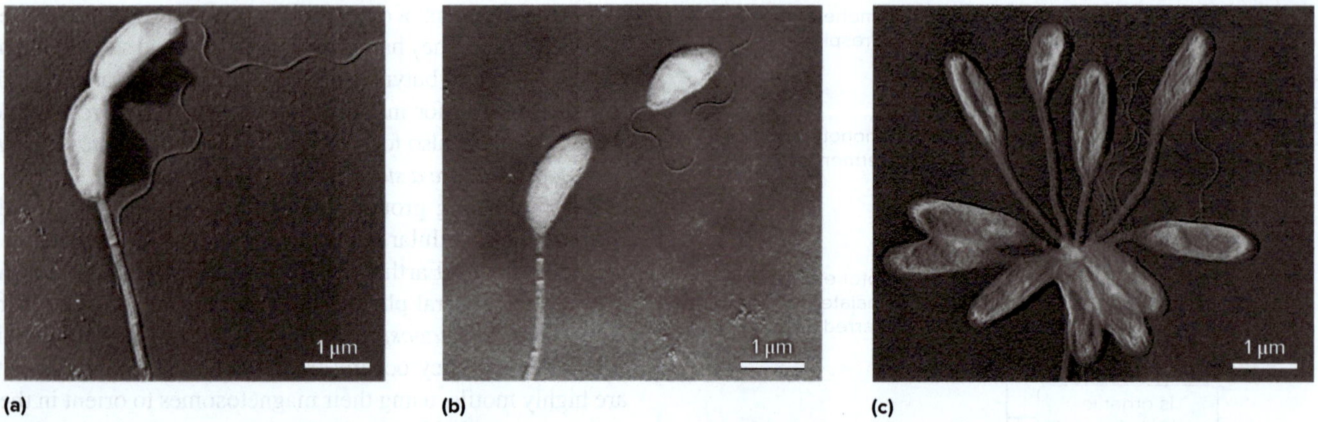

Figure 21.5 ***Caulobacter* spp. Morphology and Reproduction.** (a) A cell dividing to produce a swarmer. Note prostheca and flagellum. (b) A stalked cell and a flagellated swarmer cell. (c) A rosette of cells as seen in the electron microscope.

(a, b) Jeanne S. Poindexter/Barnard College; (c) James T. Staley

MICRO INQUIRY *If* Caulobacter *were grown in a phosphate-limited broth, what would happen to the length of the stalks?*

Incredibly, the material secreted at the end of the *Caulobacter vibrioides* holdfast is the strongest biological adhesion molecule known—bacterial superglue. These bacteria often adhere to other microorganisms and may absorb nutrients released by those microbes. The prostheca lacks cytoplasmic components and is composed almost totally of the plasma membrane and cell wall. It grows longer in nutrient-poor media and can reach more than 10 times the length of the cell body. The prostheca improves the efficiency of nutrient uptake from dilute habitats by increasing surface area.

Bacteria in the genus *Caulobacter* form vibroid-shaped cells (*see figure 7.8*), which alternate between polarly flagellated rods and stalked cells. The life cycle of *C. vibrioides* (formerly *C. crescentus*) is a model for the investigation of the bacterial cell cycle (**figure 21.6**). When ready to reproduce, the stalked cell

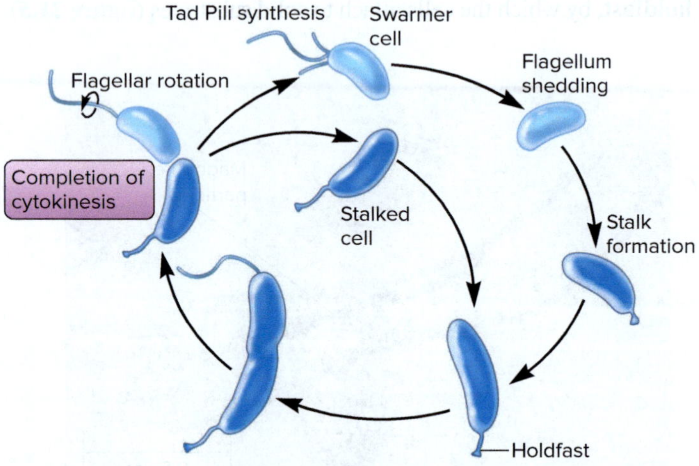

Figure 21.6 ***Caulobacter* Life Cycle.** Stalked cells attached to a substrate undergo asymmetric binary fission, producing a stalked and a flagellated swarmer cell. The swarmer cell swims freely and synthesizes pili until it settles, ejects its flagella, and forms a stalk. Only stalked cells can divide.

elongates and a single polar flagellum forms at the end opposite the prostheca. The cell undergoes asymmetric binary fission to produce a flagellated swarmer cell that swims away. At the same pole as the flagellum, multiple *t*ight *ad*hesive pili (tad pili) form. These pili extend and retract as a means of detecting a solid surface for attachment. When tad pili cannot retract because they have bonded to a surface, a developmental sequence is initiated. The swarmer, which cannot reproduce, ejects its flagellum and forms a new prostheca in that location. The new stalked cell starts the cycle anew. This process allows *Caulobacter* to sample new environments via swarmer cells, thereby decreasing competition for nutrients with stalked cells.

Comprehension Check

1. What feature unites α-proteobacteria?
2. Why are purple nonsulfur bacteria called nonsulfur bacteria?
3. What is the function of the prostheca in *Caulobacter vibrioides* growth and reproduction?

Order Rhizobiales Includes Important N_2-Fixing Bacteria

The order Rhizobiales contains the family Rhizobiaceae, which includes aerobic bacteria in the genera *Rhizobium* and *Agrobacterium*. *Rhizobium* spp. are motile rods that become pleomorphic (having many shapes) under adverse conditions. They live in soil or symbiotically within legume root nodules as specialized nitrogen-fixing cells called *bacteroids*. Bacteroids fix atmospheric nitrogen to ammonia, making it available to the plant host in exchange for fixed carbon (**figure 21.7**; *see also figure 30.9*). The symbiosis has led to the success of leguminous plants like soybeans. Indeed, Fabaceae, formerly called Leguminosae, is among the most successful plant families on Earth, with over 19,000 species.

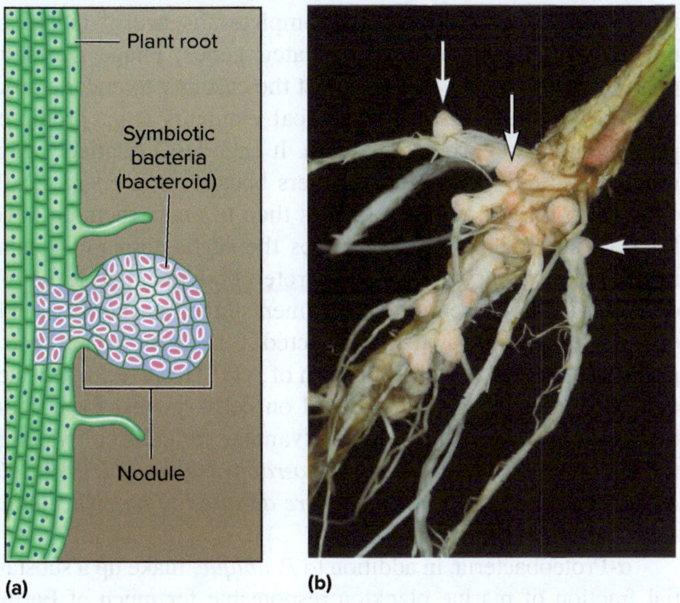

Figure 21.7 ***Rhizobium.*** (a) Diagram of differentiated bacteria, called bacteroids, in a legume root nodule. (b) Nodules (arrows) on a legume root have a faint pink color.

(b) Shutterstock / Worachat Tokaew

The process by which bacteria perform this fascinating and important symbiosis is discussed in chapter 30. *Nitrogen fixation: N_2 to NH_3 (section 12.5)* *Rhizobia are symbionts of leguminous plants (section 30.3)*

The genus *Agrobacterium* contains plant pathogens. Agrobacteria invade the crown, roots, and stems of many plants and transform plant cells into autonomously proliferating tumor cells (*see figures 30.12 and 30.13*). The best-studied species is *A. tumefaciens,* which enters many broad-leaved plants through wounds and causes crown gall disease (**figure 21.8**). The ability to produce tumors depends on a large Ti (for *t*umor *i*nducing) plasmid. Tumor production by *A. tumefaciens* is discussed in greater detail in chapter 30, while the use of the Ti plasmid in plant genetic engineering is covered in chapter 41.

Figure 21.8. Crown Gall Tumor of a Plant Caused by *Agrobacterium tumefaciens.*

Custom Life Science Images/Alamy Stock Photo

Nitrifying Bacteria Oxidize Ammonium or Nitrite to Gain Energy and Electrons

Nitrifying bacteria fall into the α- and γ-proteobacterial classes. They vary widely in morphology and may be rod-shaped, ellipsoidal, spherical, spiral, or lobate (**figure 21.9**). They may possess either polar or peritrichous flagella. Often they have extensive intracytoplasmic membrane complexes that increase the surface area available for electron transport (*see figure 3.28*).

Nitrifiers are chemolithoautotrophs that use either ammonium or nitrite as electron donors in a process called **nitrification.** Electrons are donated to an electron transport chain from ammonia, which is oxidized to nitrite, or from nitrite, which is then oxidized to nitrate (figure 21.9; *see also figure 11.25*). Nitrification is usually performed as a two-step process involving

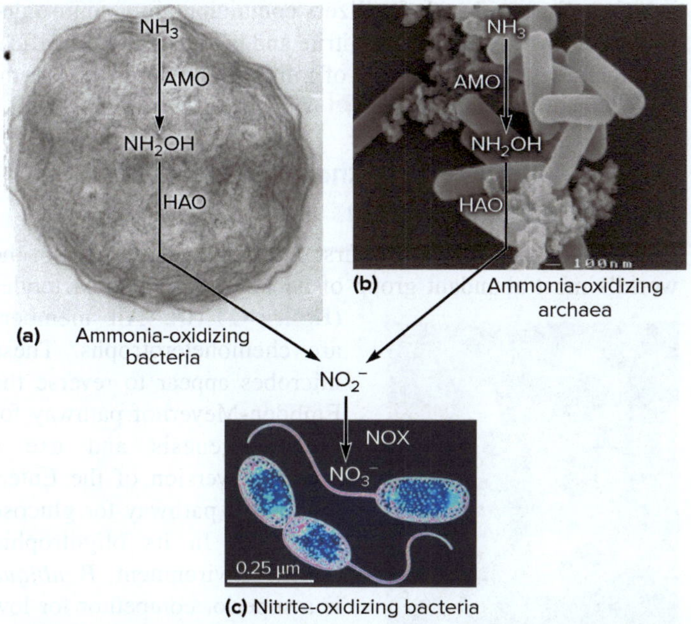

Figure 21.9 Nitrification. (a, b) Ammonia-oxidizing bacteria or archaea use ammonia monooxygenase (AMO) and hydroxylamine oxidoreductase (HAO) to convert ammonia to nitrite. (c) Nitrite-oxidizing bacteria use nitrite oxidoreductase (NOX) to catalyze the oxidation of nitrite to nitrate.

(a) Courtesy of Luis Sayavedra-Soto and the Electron Microscopy Facility, Oregon State University; (b) Dr. Martin Konneke; (c) Alfred Pasieka/Science Source

MICRO INQUIRY *Why do nitrifying microbes possess intracellular membranes?*

consortia of bacteria and archaea. Ammonia monooxygenase (AMO) and hydroxylamine oxidoreductase (HAO) oxidize ammonia to nitrite in the reactions:

$$NH_3 + O_2 + 2H^+ + 2e^- \rightarrow NH_2OH + H_2O \rightarrow NO_2 + 5H^+ + 4e^-$$

Of the four electrons released, two must return to AMO to continue ammonia oxidation, while the other two can be used either as reductants for biosynthesis or donated to the electron transport chain to reduce the terminal electron acceptor, oxygen.

$$NO_2^- + H_2O \rightarrow NO_3^- + 2H^+ + 2e^-$$

Alternately, some bacteria in the phylum Nitrospirota perform both steps, oxidizing ammonia to nitrate in a process called **comammox,** for *com*plete *amm*onia *ox*idation. Although most nitrifying bacteria are chemolithoautotrophs, some can also use reduced organic carbon sources and are considered chemolithoheterotrophs. ⏮ *Chemolithotrophy: "eating rocks" (section 11.8); Mesophilic ammonia oxidizers (section 19.3)*

As the need for intracellular membranes indicates, chemolithotrophy captures little free energy, so reduced inorganic compounds such as ammonia or nitrite must be oxidized at a great rate to conserve enough energy to support growth. This is analogous to an old car that gets bad gas mileage: It must burn more gas (electrons) to cover the same distance (do work) than an efficient car (cell).

Nitrifying microorganisms make important contributions to the nitrogen cycle (*see figure 28.4*). Nitrification occurs rapidly in oxic soil treated with fertilizers containing ammonium salts. By converting ammonium to nitrite and then to nitrate, nitrification increases the availability of nitrogen to plants because nitrate is readily used by plants. ⏭ *Nitrogen cycle (section 28.2)*

Order Pelagibacterales, the Most Abundant Microorganisms on Earth

Pelagibacter ubique was the first cultured representative of the world's most abundant group of bacteria, the Pelagibacterales (**figure 21.10**). All members are chemoheterotrophs. These microbes appear to reverse the Embden-Meyerhof pathway for gluconeogenesis and use a modified version of the Enter-Douderoff pathway for glucose oxidation. In its oligotrophic marine environment, *P. ubique* is a superior competitor for low molecular weight dissolved organic matter, specifically compounds derived from eukaryotic algae and marine cyanobacteria.

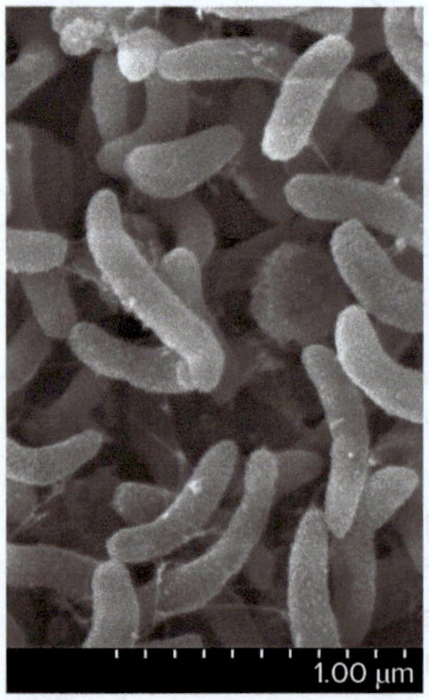

Figure 21.10 ***Pelagibacter ubique.***

Michael Rappé, Hawaii Institute of Marine Biology, University of Hawaii

The genome of *P. ubique* is 1.31 Mb, quite small for an independently growing microbe. As this bacterium evolved, it became optimized for a life with scant nutrients. The small size of its genome reflects the almost complete absence of intergenic nucleotides. There are no duplicated genes, phage genes, or **pseudogenes**—genes that have lost the capacity to encode products due to the mutation of critical elements (e.g., promoter, placement of an early stop codon). It encodes a relatively large number of high-affinity transporters because it is energetically cheaper to import macromolecules than to make them de novo. Even its small cell size maximizes the surface area-to-volume ratio. The discovery of genes for proteorhodopsin perplexed microbiologists because initial experiments failed to show that light enhanced rates of growth, as predicted if this light-driven proton pump contributed to the production of ATP. However, when such experiments were later performed on cells starved for carbon, there was an observable growth advantage for cells cultivated in the light. ⏮ *Rhodopsin-based phototrophy (section 11.10)* ⏭ *Microorganisms in the open ocean are adapted to nutrient limitation (section 29.2)*

α-Proteobacteria, in addition to *P. ubique,* make up a substantial fraction of marine plankton responsible for much of Earth's primary production. Following autotrophic carbon fixation, a predominant waste product is glycolate. α-Proteobacterial heterotrophs in the Rhizobiales and Rhodobacterales orders capitalize on glycolate availability and metabolize it via the β-hydroxyaspartate cycle (**figure 21.11**). Glycolate is first oxidized to glyoxylate, and in only four steps, this cycle converts two molecules of gloxylate to the TCA cycle intermediate oxaloacetate at the expense of a single NADH. Importantly, the cycle does not release any carbon as CO_2. This trophic linking of autotrophs and heterotrophs in the open ocean via glycolate is a mechanism to remove the greenhouse gas CO_2 from the atmosphere.

Order Rickettsiales Includes Obligate Intracellular Bacteria

The genera *Rickettsia* and *Wolbachia* contain rod-shaped, coccoid, or pleomorphic cells with typical diderm cell envelopes.

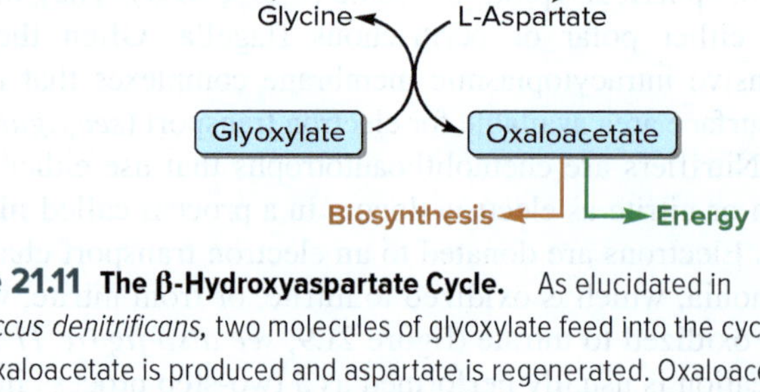

Figure 21.11 The β-Hydroxyaspartate Cycle. As elucidated in *Paracoccus denitrificans,* two molecules of glyoxylate feed into the cycle, while oxaloacetate is produced and aspartate is regenerated. Oxaloacetate is used both in energy conservation via the TCA cycle and as a precursor metabolite for biosynthesis.

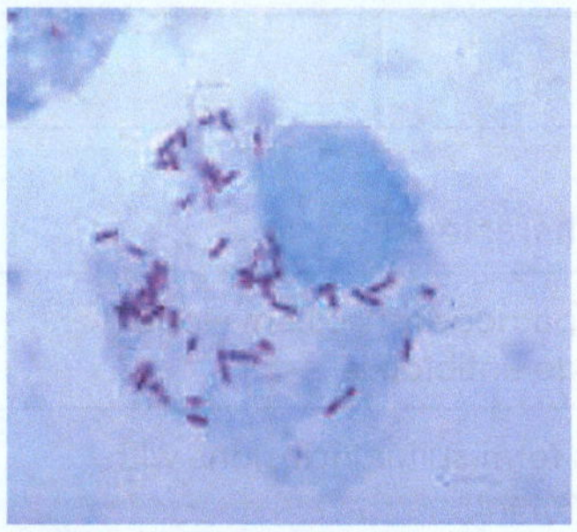

Figure 21.12 ***Rickettsia rickettsii.*** A tick hemolymph cell filled with *Rickettsia rickettsii,* the causative agent of Rocky Mountain spotted fever.

Centers for Disease Control and Prevention

Although their size varies, they tend to be smaller than most bacteria. All species are parasitic or mutualistic, living in association with eukaryotes from protozoa to vertebrates. The parasitic forms grow in erythrocytes, macrophages, and endothelial cells. Others live in blood-sucking arthropods such as fleas, ticks, mites, or lice, which serve as vectors or primary hosts (**figure 21.12**). ▶| Wolbachia: *the world's most infectious microbe? (Microbial Diversity & Ecology 27.1)*

Rickettsias are believed to have descended from a free-living, aerobic bacterium that became an intracellular parasite of an ancestral eukaryotic cell lacking organelles. In this way, the bacterium was able to assimilate many host substrates, making genes for the biosynthesis of these substrates unnecessary. As these genes mutated, there was no selection to prevent their loss and conversion to pseudogenes. To decrease cellular energy burden, many pseudogenes were eventually lost. This gene reduction gave rise to a small genome that prevented the bacterium from living outside the host cell (*see figure 18.17*).

Rickettsias are therefore different from most other bacteria in physiology and metabolism. They have lost the genes encoding glycolytic pathway enzymes, so they cannot use glucose as an energy source. Instead, they oxidize glutamate and tricarboxylic acid cycle intermediates such as succinate. The rickettsial plasma membrane has carrier-mediated transport systems for the assimilation of host cell nutrients and coenzymes. Its membrane also has a carrier that exchanges ADP for host cell ATP. An exception is *R. prowazekii,* which possesses a complete TCA cycle and a membrane-bound ATP synthase.

This genus includes important human pathogens such as *R. prowazekii* and *R. typhi,* the causative agents of louse-borne and murine typhus, respectively, as well as *R. rickettsii,* which causes Rocky Mountain spotted fever. These microbes enter host cells by inducing phagocytosis but immediately escape the phagocytic vacuole to reproduce by binary fission in the cytoplasm. Eventually the host cell bursts, releasing the organisms. ▶| *Phagocytosis destroys invaders (section 31.6); Rocky Mountain spotted fever (section 38.2)*

The Origin of Mitochondria

It has long been held that mitochondria originated from within the class α-proteobacteria. However, recent evidence based on large metagenomic data sets suggests that mitochondria may have evolved from a group of bacteria related to, but more ancient than, the α-proteobacteria.

Mitochondrial origin is like the chicken-or-egg dilemma: Did the protomitochondrion first become an intracellular entity (like rickettsiae) or did the protomitochondrion first develop extensive intracytoplasmic membranes (ICMs) as architectural supports for its energy conservation system, like purple nonsulfur bacteria, nitrifiers, and magnetobacteria?

Evidence in favor of the intracellular-first proposal includes phylogenetic analysis and the diversity of rickettsiae. Many rickettsiae are symbionts of protozoa, associations that may closely resemble ancestral engulfment. Alternately, comparison of figures 21.1*a* and 5.14 illustrates the structural similarity between mitochondrial cristae and the ICMs of α-proteobacteria. These extended plasma membranes provide additional surface area for the membrane-embedded bioenergetic machinery. Beyond the visual similarities, the protein complex responsible for the growth and curvature of the ICMs appears to be conserved. The *mi*tochondrial *c*ontact site and cristae *o*rganizing *s*ystem (MICOS) proteins are present in all eukaryotes, and the gene encoding the MICOS core subunit is widespread among α-proteobacteria, but not other bacteria. Assuming the proteins play the same role in these organisms, this strongly supports the proposal that ICMs are the functional counterparts to mitochondrial cristae.

Comprehension Check

1. How do *Agrobacterium* and *Rhizobium* spp. differ in lifestyle?
2. How is *P. ubique* adapted for life with limited nutrients?
3. Why do rickettsias have very small genomes?

21.2 Gammaproteobacteria Is the Largest Bacterial Class

After reading this section, you should be able to:

a. List human diseases caused by *Neisseria* and *Bordetella* spp.
b. Discuss the function of the sheath in *Leptothrix* spp.
c. Explain how sulfur-oxidizing bacteria obtain ATP from reduced sulfur species via substrate-level phosphorylation
d. Relate the electron donors used by γ-proteobacteria to the habitats in which they thrive
e. Describe the multicellular arrangements and unusual cellular morphologies displayed by some γ-proteobacteria
f. Outline the oxidation of methane and related C1 compounds
g. Compare the intracellular life cycle of *Legionella* spp. to that of *Coxiella* spp.
h. Describe practical aspects of pseudomonads
i. Relate the multidrug resistance of *Acinetobacter* to its cell envelope structures
j. Compare and contrast the physiology and importance of members of Vibrionales with those in Enterobacterales
k. Summarize mixed acid fermentation

The γ-proteobacteria impact humans in many ways, as pathogens and as drivers of carbon, nitrogen, and sulfur cycles. **Table 21.2** summarizes general characteristics of many of the bacteria discussed in this section.

Table 21.2 Characteristics of Selected γ-Proteobacteria

Genus	Dimensions (μm) and Morphology	Oxygen Requirement	Other Distinctive Characteristics
Acinetobacter	1.0–1.5 × 1.5–2.5; coccobacilli in pairs or short chains, nonmotile	Aerobic	Multidrug-resistant pathogens; cell envelope modifications exclude antibiotics
Beggiatoa	1–200 × 2–10; colorless cells form filaments, either single or in bundles	Aerobic or microaerophilic	Gliding motility; can form sulfur inclusions with hydrogen sulfide present
Burkholderia	0.5–1.0 × 1.5–4; straight rods with single flagellum or a tuft at the pole	Aerobic, some capable of anaerobic respiration with NO_3^-	Poly-β-hydroxybutyrate as reserve; can be pathogenic to humans and animals
Chromatium	1–6 × 1.5–16; rod-shaped or ovoid, straight or slightly curved, polar flagella	Anaerobic	Anoxygenic photolithoautotrophs that can use sulfide; sulfur stored within the cell
Escherichia	1.1–1.5 × 2–6; straight rods, peritrichous flagella or nonmotile	Facultatively anaerobic	Mixed acid fermenters; formic acid converted to H_2 and CO_2, lactose fermented
Leptothrix	0.6–1.5 × 2.5–15; straight rods in chains with sheath, free cells flagellated	Aerobic	Sheaths encrusted with iron and manganese oxides
Leucothrix	Long filaments of short cylindrical cells, usually holdfast is present	Aerobic	Dispersal by gonidia, filaments don't glide; rosettes formed; chemoheterotrophic
Methylococcus	0.8–1.5 × 1.0–1.5; cocci with capsules, nonmotile	Aerobic	Can form cysts; uses methane, methanol, and formaldehyde as sole carbon and energy sources; fixes nitrogen nonsymbiotically
Neisseria	0.6–1.9; cocci in pairs with flattened adjacent sides	Aerobic	Inhabitants of mammalian mucous membranes
Nitrosomonas	Size varies with strain; spherical to ellipsoidal cells with intracytoplasmic membranes	Aerobic	Chemolithotrophs that oxidize ammonia to nitrite
Pseudomonas	0.5–1.0 × 1.5–5.0; straight or slightly curved rods, polar flagella	Aerobic or facultatively anaerobic	Respiratory metabolism with oxygen or nitrate as acceptor; degrade a broad range of organic compounds
Thiobacillus	0.3–0.5 × 0.9–4; rods, often with polar flagella	Aerobic	All chemolithotrophic; oxidize reduced sulfur compounds to sulfate; some also capable of chemoheterotrophy
Vibrio	0.5–0.8 × 1.4–2.6; straight or curved rods with sheathed polar flagella	Facultatively anaerobic	Fermentative or respiratory metabolism; sodium ions stimulate or are needed for growth; oxidase positive

Order Burkholderiales Includes Chemoheterotrophs and Chemolithotrophs

Microbes in the order Burkholderiales are metabolically similar to α-proteobacteria but tend to use substances released from organic decomposition in anoxic habitats. Some of these bacteria use hydrogen, ammonia, volatile fatty acids, and similar substances. As with α-proteobacteria, there is considerable metabolic diversity; Burkholderiales may be chemoheterotrophs, photolithotrophs, and chemolithotrophs.

Burkholderia, the namesake of the family Burkholderiaceae, is an important genus with widespread environmental distribution. These bacteria are known to degrade over 100 different organic molecules and are active in recycling organic materials in nature. In the twenty-first century, *Burkholderia* has emerged as an important hospital-acquired pathogen. Because of its expansive metabolic capabilities, it is a common contaminant of pharmaceuticals. *Burkholderia* spp. are a particular problem for patients with cystic fibrosis, where they form biofilms in the lungs. *B. mallei,* is a pathogen that can be misused as a bioterrorism agent. ▶| *Bioterrorism readiness is an integral component of public health microbiology (section 35.7)*

Bacteria in the genera *Sphaerotilus* and *Leptothrix* have a **sheath**—a hollow, tubelike structure surrounding a chain of cells. *Sphaerotilus* bacteria attach to submerged plants, rocks, and other objects. These chains release flagellated swarmer cells that attach by means of a holdfast and form a new chain. They grow best in slowly running, polluted freshwater. In fact, they thrive in activated sewage sludge and sometimes form tangled masses of filaments that interfere with the proper settling of sludge. ▶| *Wastewater treatment processes (section 42.2)*

Leptothrix spp. sheaths are an aggregation of nanofibrils, polysaccharide structures 10 nm wide and up to 10 µm long. The carbohydrates comprising the sheath are decorated with chemical functional groups that mediate attachment to a substrate and to other nanofibrils. As nanofibrils aggregate, separation of daughter cells following cell division is constrained so that new cells are pushed away from the attachment point (**figure 21.13**). As the chain lengthens, it may fragment to release single cells to start the cycle anew. The sheath holds cells in a location with a strong current and also protects against protozoan predators. As the sheath matures, iron and manganese oxides accumulate and rigidify the structure.

The genus *Bordetella* is composed of aerobic coccobacilli. They are chemoheterotrophs with respiratory metabolism and require organic sulfur and nitrogen (amino acids) for growth. As mammalian parasites, they multiply in respiratory epithelial cells. *B. pertussis* is a nonmotile, encapsulated species that causes pertussis (whooping cough); related strains of the same species cause the canine condition kennel cough. ▶| *Pertussis (section 38.1)*

The family Neisseriaceae is notable for the genus *Neisseria,* nonmotile, aerobic cocci that most often occur in pairs with adjacent sides flattened (**figure 21.14**). They may have capsules and pili. These chemoheterotrophs inhabit mammalian mucous membranes, and some are human pathogens. *Neisseria gonorrhoeae* is the causative agent of gonorrhea; *Neisseria meningitidis* can cause some cases of bacterial meningitis. These two species are believed to have diverged from a common ancestor only in the last few hundred years, and each has adapted to a distinct ecological niche within the human body. ▶| *Meningitis (section 38.1); Gonorrhea (section 38.3)*

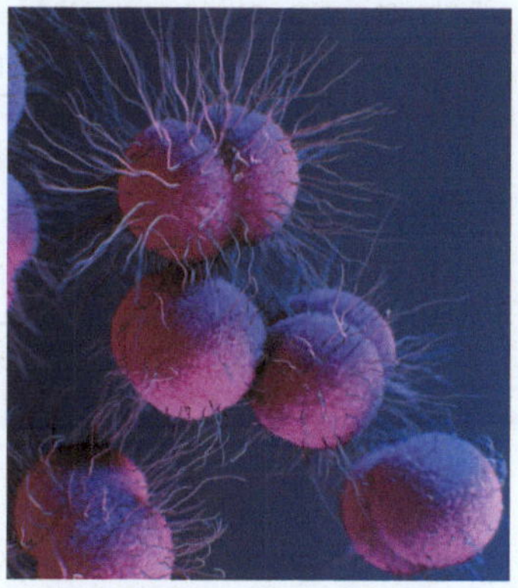

Figure 21.14 ***Neisseria gonorrhoeae.***

Antibiotic Resistance Coordination and Strategy Unit/CDC

The family Thiobacillaceae contains the genus *Thiobacillus,* which are the best-studied **colorless sulfur bacteria,** so named because they use sulfur-containing compounds as energy sources but are not photosynthetic. Many are unicellular, rod-shaped or spiral, sulfur-oxidizing bacteria that can be either nonmotile or flagellated.

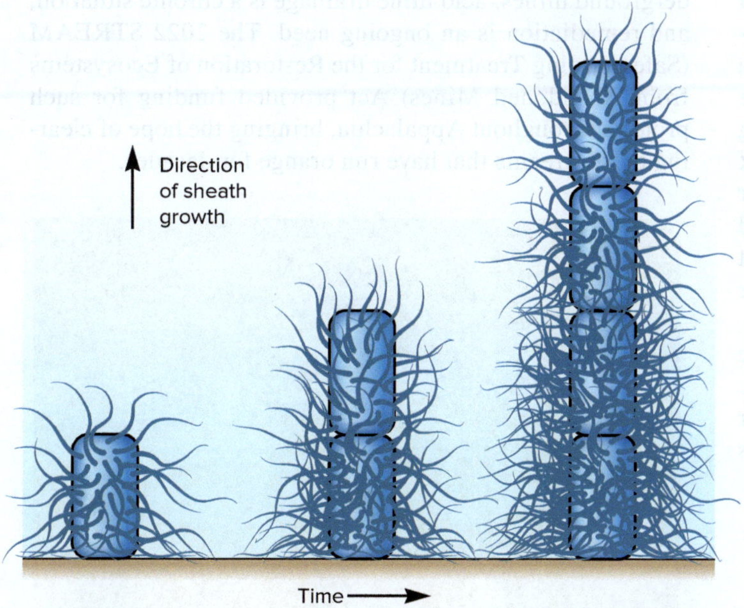

Figure 21.13 Growth of a *Leptothrix* Sheath. Following attachment to a solid substrate, *Leptothrix* cells secrete nanofibrils that aggregate to form the sheath.

Thiobacillus spp. metabolism has been intensely studied. The genus includes obligate aerobes and facultative denitrifiers. They use a variety of reduced, inorganic sulfur compounds as energy and electron sources and oxygen as electron acceptor. ATP is generated in two ways, both of which require that the cell first oxidize sulfide or thiosulfate to sulfite (SO_3^{2-}) as described in **figure 21.15**. Although all *Thiobacillus* spp. are autotrophic, fixing CO_2 via the Calvin-Benson cycle (*see figure 12.3*), a few species can also grow heterotrophically. |◀ *Chemolithotrophy: "eating rocks" (section 11.8)*

The metabolic flexibility of *Thiobacillus denitrificans* has made it the focus of much investigation. Remarkably it can use either oxygen or nitrate as its terminal electron acceptor when growing chemolithotrophically, using hydrogen sulfide or thiosulfate to donate electrons to its electron transport chain. When using nitrate as a terminal electron acceptor, *T. denitrificans* can sequentially reduce nitrate to nitrogen gas—a process called **denitrification.**

Thiobacillus spp. are ubiquitous in soil, freshwater, and marine habitats. Many demonstrate great acid tolerance and prosper in habitats they have acidified by sulfuric acid production. Acid mine drainage is the most important high-sulfide, low-pH environment in the United States (**Microbial Diversity & Ecology 21.1**). Conversely, sulfur-oxidizing bacteria can be beneficial. Neutrophilic species can increase soil fertility by oxidizing elemental sulfur to sulfate. Thiobacilli are used in processing low-grade metal ores because of their ability to leach metals from ore. ▶| *Sulfur cycle (section 28.2)*

Comprehension Check

1. Members of which genera within Burkholderiales make important contributions to the environment? Which are human or animal pathogens?
2. How do colorless sulfur bacteria obtain energy by oxidizing sulfur compounds?
3. What role does iron play in the redox metabolism of *Thiobacillus?*

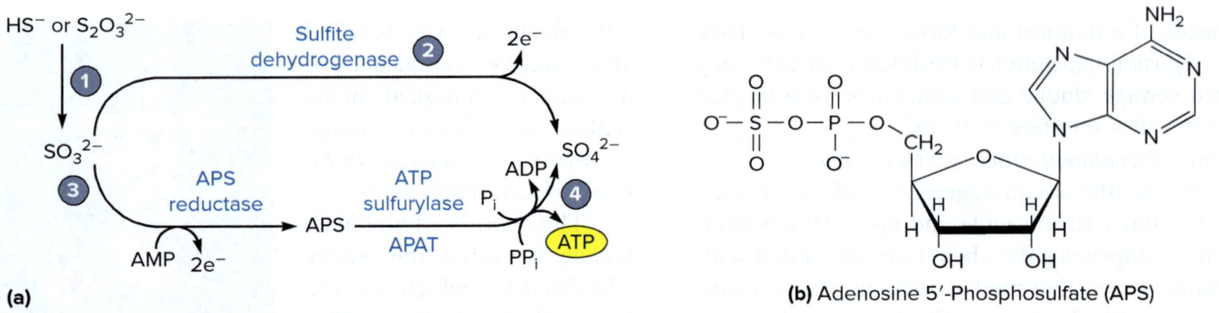

Figure 21.15 Sulfur Oxidation in *Thiobacillus*. (a) (1) Reduced sulfur compounds such as sulfide (HS^-) or thiosulfate ($S_2O_3^{2-}$) are first oxidized to sulfite (SO_3^{2-}). (2) Electrons from sulfite can be donated to an electron transport chain via sulfite dehydrogenase and used to generate a proton motive force. Alternatively, (3) APS reductase catalyzes the oxidation of sulfite to sulfate, which is used to convert adenosine monophosphate (AMP) to adenosine 5′-phosphosulfate (APS). (4) ATP sulfurylase and APS:phosphate adenylyltransferase (APAT) then catalyze the exchange of the sulfate on APS with PP_i to generate ATP via substrate-level phosphorylation. (b) The structure of APS.

MICRO INQUIRY *Identify the reactions that contribute to substrate-level phosphorylation and those that might contribute to ATP synthesis by a membrane-bound ATP synthase.*

MICROBIAL DIVERSITY & ECOLOGY

21.1 The Microbial Legacy of Coal Mining

You don't have to travel far in the central Appalachian Mountains to come across surface waters that appear bright orange (**box figure**). Over 6,000 miles of orange streams are one legacy of the coal mining industry. This water is toxic for most aquatic life and unfit for human consumption.

The culprit is *Acidithiobacillus ferrooxidans,* a chemolithotrophic bacterium that derives its energy from oxidizing ferrous ion (Fe^{2+}) to ferric ion (Fe^{3+}) and sulfide ion to sulfate ion. The combination of these two energy sources is important because of the solubility properties of iron. When the pH is greater than 4.0 to 5.0, ferrous ion is spontaneously oxidized to ferric ion by O_2 in the water and precipitates as a hydroxide. Ferrous ion is somewhat soluble and can be formed at pH values of 3.0 or less in moderately reducing environments. If the pH drops below 2.0 to 3.0 because of sulfuric acid production by spontaneous oxidation of sulfur or sulfur oxidation, the ferrous ion remains reduced, soluble, and available as an energy source. *A. ferrooxidans* grows well at such acidic pHs and oxidizes ferrous ion to an insoluble ferric precipitate.

The ecological consequences of this metabolic lifestyle arise from the presence of pyrite (FeS_2) in coal mines. The bacteria oxidize both elemental components of pyrite for their growth and in the process form sulfuric acid, which leaches the remaining minerals.

Autoxidation or bacterial action

$$2FeS_2 + 7O_2 + 2H_2O \rightarrow 2Fe^{2+} + 4SO_4^{2-} + 4H^+$$

A. ferrooxidans

$$2Fe^{2+} + 1/2O_2 + 2H^+ \rightarrow 2Fe^{3+} + H_2O$$

Pyrite oxidation is further accelerated because the ferric ion generated by bacterial activity readily oxidizes more pyrite to sulfuric acid and ferrous ion. In turn, the ferrous ion supports further bacterial growth. It is difficult to prevent *A. ferrooxidans* growth as it requires only pyrite and common inorganic salts.

Environmental remediation efforts range from passive efforts like adding limestone to raise the water pH, to active systems that divert a waterway for treatment and subsequent release of clean water. Because of the extensive underground mines, acid mine drainage is a chronic situation, and remediation is an ongoing need. The 2022 STREAM (Safeguarding Treatment for the Restoration of Ecosystems from Abandoned Mines) Act provided funding for such projects throughout Appalachia, bringing the hope of clearing these streams that have run orange for decades.

K Steve Cope/Shutterstock

Order Chromatiales: Purple Sulfur Bacteria Perform Anoxygenic Photosynthesis

Purple sulfur bacteria are strict anaerobes that are usually photolithoautotrophic. Recall that during anoxygenic photosynthesis, electrons are required for the reduction of $NADP^+$. These microbes use hydrogen sulfide as an electron donor and deposit the resulting oxidized sulfur globules internally (*see figure 3.29*); often they further oxidize the sulfur to sulfate. Hydrogen may also serve as an electron donor. *Thiocapsa* and *Chromatium* are typical purple sulfur bacterial genera found in anoxic, sulfide-rich zones of lakes, marshes, and lagoons where large blooms can occur under certain conditions (**figure 21.16*a***). Okenone, a carotenoid pigment produced only in the family Chromatiaceae, is a useful geochemical biomarker (figure 21.16*b*). Its occurrence in sedimentary rock formations provides geochemists with an indicator of the environmental conditions at a specific historic time interval, thus aiding in climate reconstruction. *Light reactions in anoxygenic photosynthesis (section 11.10)*

Orders Thiotrichales and Beggiatoales Include Large Filamentous Bacteria

These orders have several members that oxidize sulfur compounds. Morphologically, both rods and filamentous forms are present, but the most striking members are filamentous.

Beggiatoa spp. are microaerophilic and grow in sulfide-rich habitats such as sulfur springs, freshwater with decaying plant material, rice paddies, salt marshes, and marine sediments. Their filaments contain short, disklike cells (**figure 21.17**). These bacteria oxidize hydrogen sulfide to form large sulfur grains located in pockets formed by invaginations of the plasma membrane. *Beggiatoa* spp. also oxidize the sulfur to sulfate when donating electrons to the electron transport chain for energy production. Many strains grow heterotrophically with acetate as a carbon source, and some incorporate CO_2 autotrophically. Carbon fixation typically occurs using the Calvin-Benson cycle, but there is metagenomic evidence that some *Beggiatoa* from marine sediments perform the reductive TCA cycle as well (*see figure 12.4*). Freshwater species are much smaller than marine.

(a)

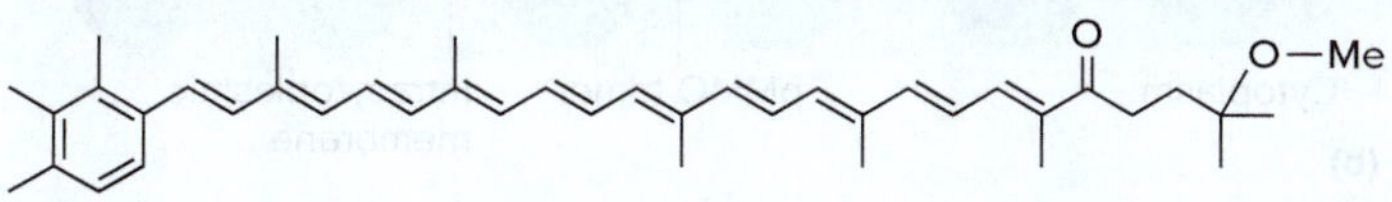

(b)

Figure 21.16 Photosynthetic Purple Sulfur Bacteria. (a) Photosynthetic purple sulfur bacteria growing in a salt marsh. (b) The chemical structure of okenone.

Johnny Madsen/Alamy Stock Photo

Figure 21.17 ***Beggiatoa alba.*** A light micrograph showing part of a colony. Note the dark sulfur granules within many of the filaments.

Roger Burks/University of California at Riverside and Mark Schneegurt/Wichita State University, and Cyanosite (www-cyanosite.bio.purdue.edu)

Members of three genera—*Beggiatoa, Thioploca,* and *Thiomargarita*—are notable because they form large mats on the seafloor near hydrothermal vents and adjacent to nutrient-rich coasts where methane seeps out of Earth's crust. These filamentous microbes are among the largest known bacteria. For example, *Thiomargarita* spp. can be over 100 μm in diameter and hundreds of centimeters long (**figure 21.18**). All three groups grow in bundles, making them even more conspicuous. These bacteria appear hollow because they form large intracellular vacuoles where high concentrations of nitrate (up to 500 mM) are stored.

Leucothrix spp. are aerobic chemoheterotrophs that form filaments or trichomes up to 400 μm long. Most are marine and attach to solid substrates by a holdfast. These filamentous bacteria have a complex life cycle in which they are dispersed by the formation of small spherical cells called gonidia (**figure 21.19**). *Thiothrix* is a related genus whose members form sheathed filaments and release gonidia from the open end of the sheath. *Thiothrix* includes species that are facultatively autotrophic, chemoheterotrophic, or mixotrophic, oxidizing hydrogen sulfide and depositing sulfur granules internally. *Thiothrix* spp. grow in sulfide-rich flowing water and activated sludge sewage systems.

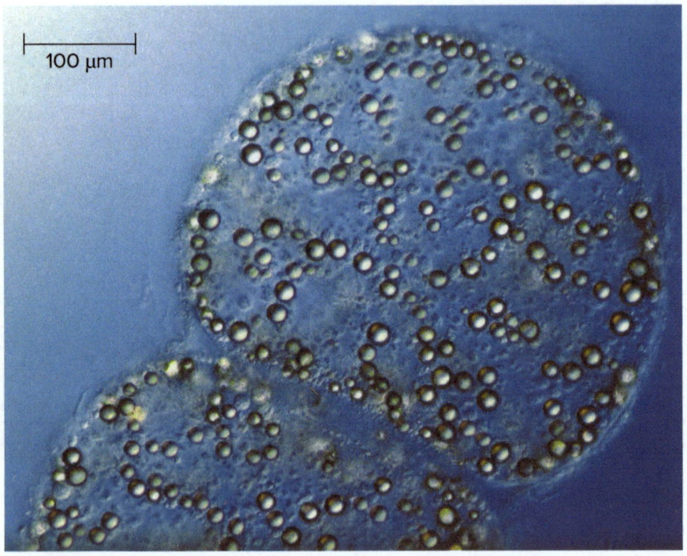

Figure 21.18 ***Thiomargarita*, the World's Largest Known Bacterium.** This bacterium, usually 100–300 μm in diameter, occasionally reaches a size of 750 μm (larger than a period on this page). It uses sulfide from bottom sediments as an energy source and nitrate from overlying waters as an electron acceptor. Sulfur granules appear beadlike in this image.
Dr. Heide Schulz-Vogt

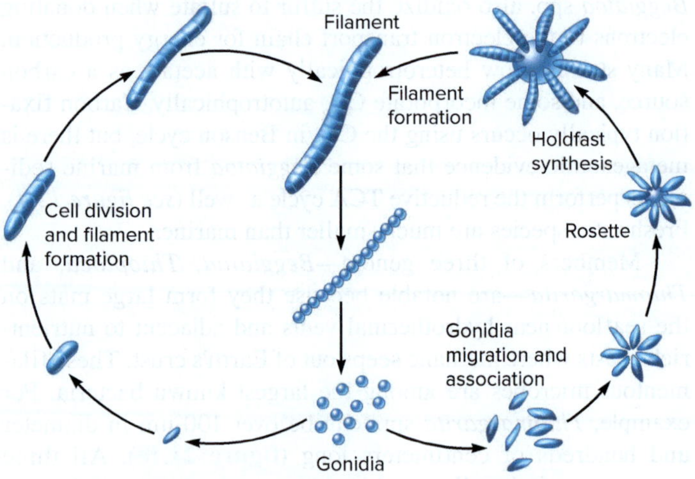

Figure 21.19 Life Cycle and Morphology of *Leucothrix mucor.*

Order Methylococcales Consists of C1-Metabolizing Bacteria

The genera *Methylomonas, Methylobacter,* and *Methylococcus* are obligate **methylotrophs**—bacteria whose growth is supported only by the assimilation of one-carbon (C1) compounds. They oxidize methane, methanol, formate, or methylamine as their carbon and energy source. Those that grow on methane are termed **methanotrophs.** Aerobic methanotrophs are typically found in environments where methanogenic archaea produce methane in close proximity to air. In contrast, anaerobic methanotrophs (ANME archaea) reside in deep-sea marine sediments near methane seeps. ◀ *Methanotrophs (section 19.4)*

Prominent in the cells of all methanotrophs are *i*ntra*c*ytoplasmic *m*embranes (ICMs) that support the enzyme critical for methane oxidation, *p*articulate *m*ethane *mo*nooxygenase (pMMO) (**figure 21.20**). ICMs are contiguous with the plasma membrane and occupy more than half of the cell volume. pMMO assembles as trimers and packs tightly into the ICMs, forming extensive enzyme arrays. Oxidation proceeds through several steps, ultimately generating formaldehyde (**figure 21.21*a***). Formaldehyde is either assimilated to produce biomass or oxidized to CO_2, and its electrons used to generate a proton motive force. The use of formaldehyde as the central intermediate is paradoxical because it is poisonous. The key to survival is the speed of these reactions. It is incorporated into cell material by the activity of one of two cycles. γ-Proteobacterial methylotrophs use the ribulose monophosphate cycle (figure 21.21*b*). By contrast, α-proteobacterial methylotrophs use the serine cycle (figure 21.21*c*), which features the deamination of serine to generate acetyl-CoA.

Methane is one of the greenhouse gases of concern for climate change (*see figure 28.10*). Methane removal technologies seek to exploit methanotrophs in major methane emission sites like rice paddies and anaerobic digesters (*see chapter 19 opening story*). ▶ *Global climate change: biogeochemical cycling out of balance (section 28.3)*

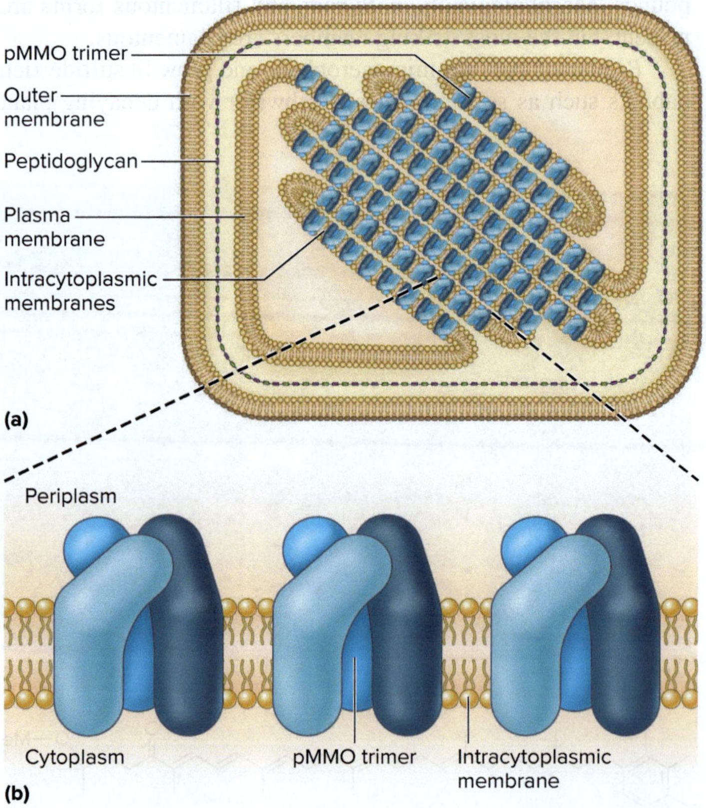

Figure 21.20 Methanotrophic Bacteria Are Packed with Particulate Methane Monooxygenase (pMMO). (a) Intracytoplasmic membranes supporting pMMO almost completely fill the cell. (b) pMMO is a trimer of identical polypeptides and constitutes about 10% of the cellular protein.

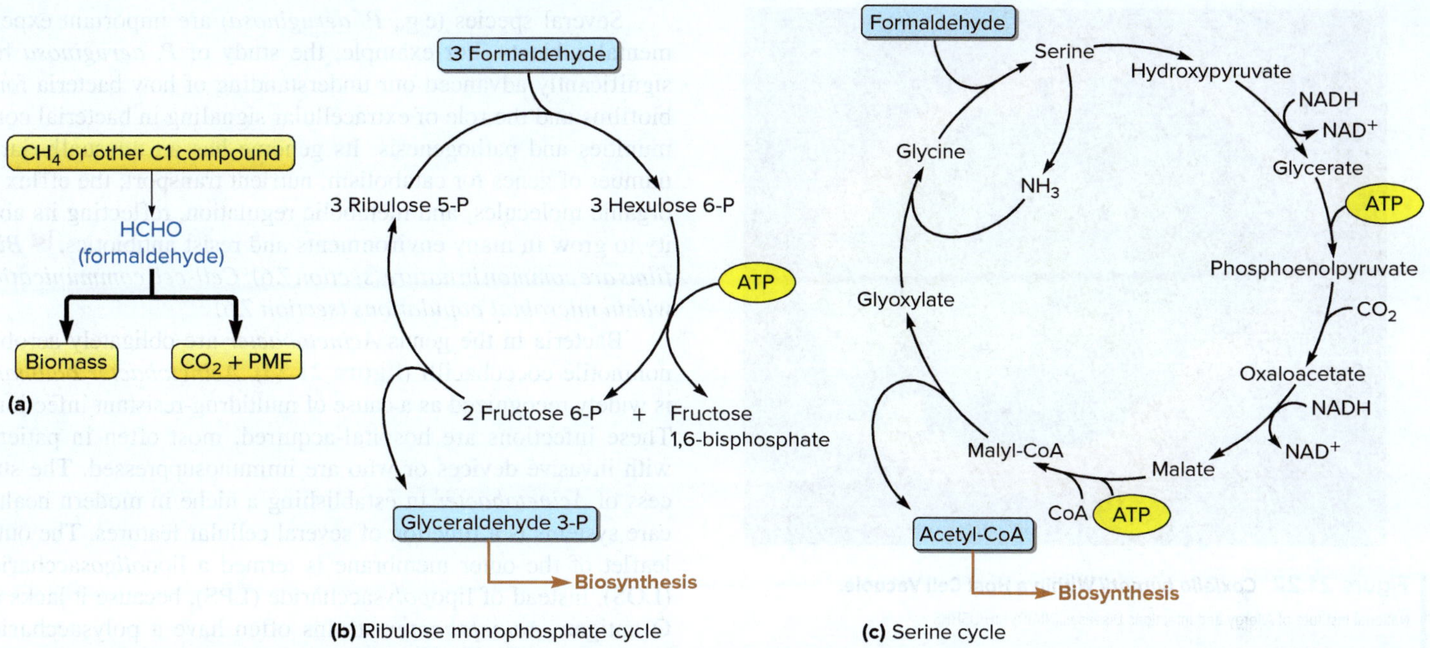

Figure 21.21 Methanotrophy and Methylotrophy. (a) Methane or another reduced C1 compound (i.e., not CO_2 or CO) is converted to the toxic intermediate formaldehyde, which is rapidly incorporated into biomass or oxidized to generate a proton motive force (PMF). Thick arrows indicate that this must happen faster than it can accumulate in the cell. (b) The net yield of the ribulose monophosphate cycle is a molecule of glyceraldehyde 3-phosphate that can be used in central metabolism. (c) The net yield of the serine cycle is an acetyl-CoA molecule that can be used in central metabolism.

MICRO INQUIRY *How do the cycles compare in terms of energy and reducing power used? Which might be the ecological consequence of the difference?*

Orders Legionellales and Coxiellales Include Intracellular Pathogens

We discuss the genus *Legionella* in the order Legionellales and the genus *Coxiella* in the Coxiellales. These microbes are intracellular pathogens that display a dimorphic (i.e., two cell types) lifestyle reminiscent of the chlamydial life cycle. ◀ *Phylum Verrucomicrobiota includes human symbionts and obligate intracellular parasites (section 20.5)*

Legionella pneumophila has been intensely studied because it causes a pneumonia in humans called Legionnaires' disease. In nature, *L. pneumophila* is an intracellular parasite of protozoa that live in moist soil and aquatic environments. It is believed that the similarities between *Legionella*'s protozoan hosts and macrophages have made *L. pneumophila* an accidental pathogen of humans. Within a host cell (either protozoan or human), the bacteria are contained within a *Legionella*-containing vacuole (LCV). The bacteria use a type IV secretion system (T4SS) to transfer over 330 types of effector proteins to the host cytoplasm. Some effector molecules modify the host phagolysosome pathway, thereby allowing the LCV to avoid destruction. Others initiate proteolysis to provide the amino acids required as the carbon and energy source for bacterial growth within the LCV. As amino acid nutrients become limited, *Legionella* initiates a stringent response by synthesizing (p)ppGpp. Accumulation of this secondary messenger results in the synthesis of the stress sigma factor RpoS, which in turn activates hundreds of genes. The result is a metabolic shift and differentiation of *Legionella* cells to a motile nonreplicating transmissible cell. ◀ *Stringent response in* E. coli *(section 14.5)* ▶ *Monocytes, macrophages, and dendritic cells (section 31.4); Legionnaires' disease (section 38.1)*

Coxiella spp. also display two cell types within the host, although they have a much broader range of hosts, including insects, fish, birds, rodents, sheep, goats, and humans. As with *L. pneumophila,* humans become infected by inhaling contaminated aerosols, although in the case of the pathogen *C. burnetii,* the resulting Q fever is a flulike illness. The developmental life cycle of *C. burnetii* is similar to that of *L. pneumophila.* A small, infectious form called the *small cell variant* (SCV) enters host cells by phagocytosis. Once the phagosome has fused with a lysosome, the low pH triggers SCVs to become metabolically active. Within the resulting vacuole, *C. burnetii* SCVs differentiate into the replicative form known as the *large cell variant* (LCV) (**figure 21.22**). LCVs undergo binary fission and secrete pathogenesis effectors through a T4SS. After several days they differentiate back to SCVs, the cellular form responsible for long-term survival outside the host. In both *L. pneumophila* and *C. burnetii,* an important aspect of the differentiation of LCVs to SCVs involves synthesis and modification of peptidoglycan, resulting in a thicker, less porous cell envelope in the SCV. The alternation of two cell types makes these pathogens distinct from the rickettsiae, which cannot

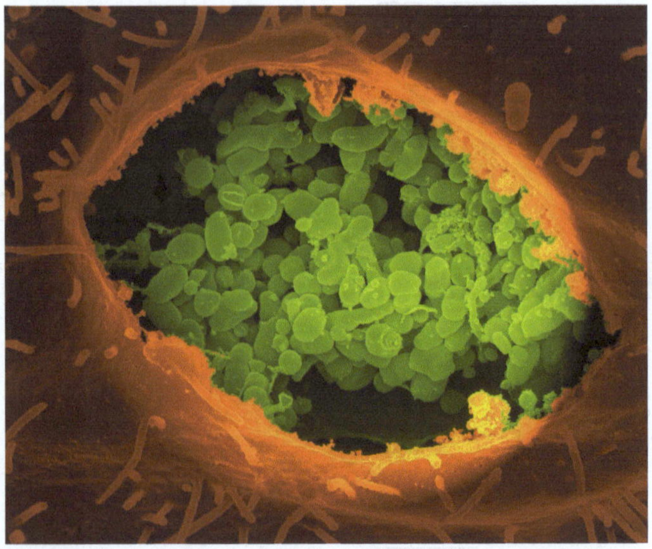

Figure 21.22 ***Coxiella burnetii*** **Within a Host Cell Vacuole.**
National Institute of Allergy and Infectious Diseases (NIAID)/NIH/USHHS

MICRO INQUIRY *Compare this figure with figure 20.10b. How are these two bacterial pathogens similar?*

survive outside a host cell. *Phylum Verrucomicrobiota includes human symbionts and obligate intracellular parasites (section 20.4)* *Phagocytosis destroys invaders (section 31.6); Q fever (section 38.5)*

Order Pseudomonadales Includes Opportunistic Pathogens

Pseudomonas is the most important genus in the family Pseudomonaceae. These bacteria are straight or slightly curved rods, and are motile by one or several polar flagella. They are chemoheterotrophs that usually carry out aerobic respiration, although some use nitrate as the terminal electron acceptor in anaerobic respiration. Most hexoses are degraded by the Entner-Doudoroff pathway; all pseudomonads have a functional tricarboxylic acid cycle. *Aerobic respiration starts with glucose oxidation (section 11.3)*

The genus *Pseudomonas* is an exceptionally heterogeneous taxon. Four important species are *P. aeruginosa,* an opportunistic pathogen associated with cystic fibrosis, burns, and sepsis in patients who are immunocompromised; *P. putida,* a ubiquitous soil resident with broad degradative capacity; *P. fluorescens,* a fluorescent pigment producer associated with food spoilage; and *P. syringae,* a widespread plant pathogen.

A distinctive attribute of *Pseudomonas* species is their capacity to degrade a wide variety of organic molecules. Thus they are important in the microbial breakdown of organic materials to inorganic substances (mineralization) in nature and in sewage treatment. Likewise, secreted proteases and lipases make *Pseudomonas* spp. common food spoilage organisms.

Several species (e.g., *P. aeruginosa*) are important experimental subjects. For example, the study of *P. aeruginosa* has significantly advanced our understanding of how bacteria form biofilms and the role of extracellular signaling in bacterial communities and pathogenesis. Its genome has an unusually large number of genes for catabolism, nutrient transport, the efflux of organic molecules, and metabolic regulation, reflecting its ability to grow in many environments and resist antibiotics. *Biofilms are common in nature (section 7.6); Cell-cell communication within microbial populations (section 7.6)*

Bacteria in the genus *Acinetobacter* are obligately aerobic, nonmotile coccobacilli (**figure 21.23**). *Acinetobacter baumanii* is widely recognized as a cause of multidrug-resistant infections. These infections are hospital-acquired, most often in patients with invasive devices or who are immunosuppressed. The success of *Acinetobacter* in establishing a niche in modern health-care systems is a function of several cellular features. The outer leaflet of the outer membrane is termed a lipo*oligo*saccharide (LOS), instead of lipo*poly*saccharide (LPS), because it lacks an O antigen. *Acinetobacter* strains often have a polysaccharide capsule covering the LOS. Like many capsules on pathogens, this blocks complement activity by binding factor H (*see table 31.1*). The OmpA porin has a low rate of solute passage, and multiple drug efflux systems are present. All of these factors contribute to its resistance to disinfectants and most clinically

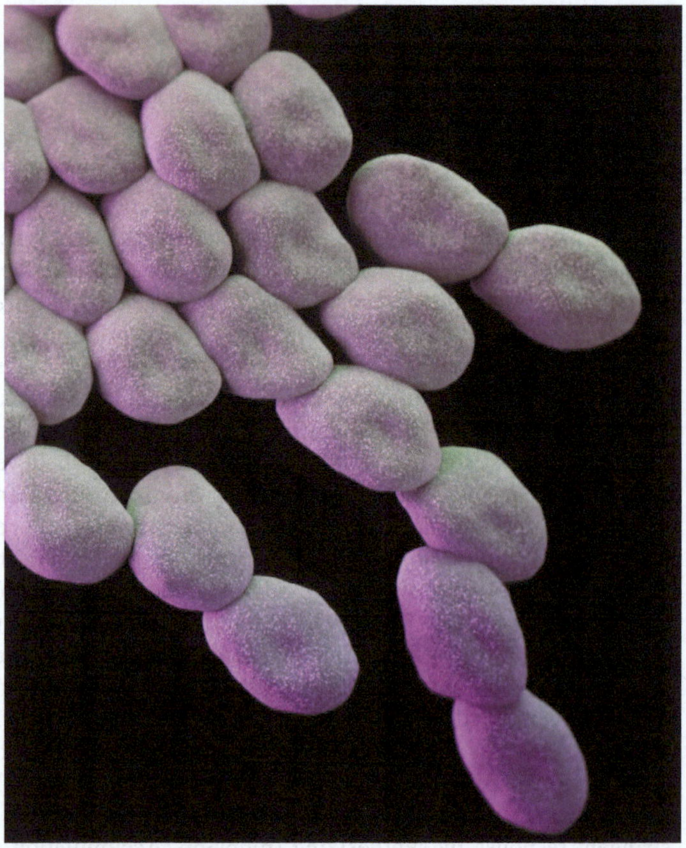

Figure 21.23 ***Acinetobacter baumanii.***
Melissa Brower/CDC

useful antibiotics, and also offer protection from the host immune response. *Cell walls have many functions (section 3.3)*

Order Enterobacterales Is a Large, Diverse Taxon

Bacteria in the order Enterobacterales are rod-shaped facultative anaerobes. They inhabit many ecosystems and include common human pathogens. We discuss two notable families within the order.

The family Vibrionaceae contains the genus *Vibrio*, which includes several important pathogens. *Vibrio cholerae* causes cholera, and *V. parahaemolyticus* causes gastroenteritis and is transmitted by consumption of contaminated seafood. *V. vulnificus* is a marine bacterium that causes severe skin and soft tissue infections that can be life-threatening. As ocean temperatures warm, the incidence of *V. vulnificus* infections is on the rise. *V. anguillarum* and others are responsible for fish diseases, which can be especially problematic in aquatic farms. These bacteria prefer warm water, and climate change has allowed them to expand their range into ecosystems beyond the tropics. *Cholera (section 38.4)*

Some members of Vibrionaceae are unusual in being bioluminescent—that is, they can emit light of a blue-green color. The enzyme luciferase catalyzes the reaction and uses reduced flavin mononucleotide (FMN), molecular oxygen, and a long-chain aldehyde as substrates.

$$FMNH_2 + O_2 + RCHO \xrightarrow{\text{luciferase}} FMN + H_2O + RCOOH + \text{light}$$

An enzyme-bound, excited flavin intermediate is the direct source of luminescence. Because the electrons used in light generation are diverted from the electron transport chain and ATP synthesis, the bacteria expend considerable energy on luminescence.

In fact, there is much speculation about the role of bacterial luminescence and its value to bacteria, particularly because it is such an energetically expensive process. Luminescent bacteria do not emit light when they grow as free-living organisms in seawater; the light emitted by a single cell would not be visible. When free-living luminescent bacteria infect the light organ of young fish or squid (*see figure 7.26*), they reproduce and reach densities where light emission is visible. Bacterial luminescence was the first example of density-dependent gene regulation by a quorum-sensing molecule autoinducer and it remains an important model for understanding quorum sensing in many bacteria, including a number of pathogens. *Cell-cell communication within microbial populations (section 7.6); Quorum sensing by* Vibrio *spp. (section 14.6)*

The family Enterobacteriaceae contains peritrichously flagellated or nonmotile, facultatively anaerobic, straight rods with simple nutritional requirements. The metabolic properties (particularly during anaerobic growth) of members of Enterobacteriaceae are useful in characterizing its constituent genera. These bacteria, often called **enterobacteria** or **enteric bacteria** (Greek *enterikos*, pertaining to the intestine), degrade sugars by means of the Embden-Meyerhof pathway. Under microaerobic or anoxic conditions, the enzyme **pyruvate formate-lyase (PFL)** catalyzes the cleavage of pyruvate to formate and acetyl-CoA. Those enteric bacteria that produce large amounts of gas during sugar fermentation, such as *Escherichia* spp., also have the enzyme **formate dehydrogenase (FDH),** which splits formic acid to H_2 and CO_2 (**figure 21.24**). This enzyme complex is crucial to sur-

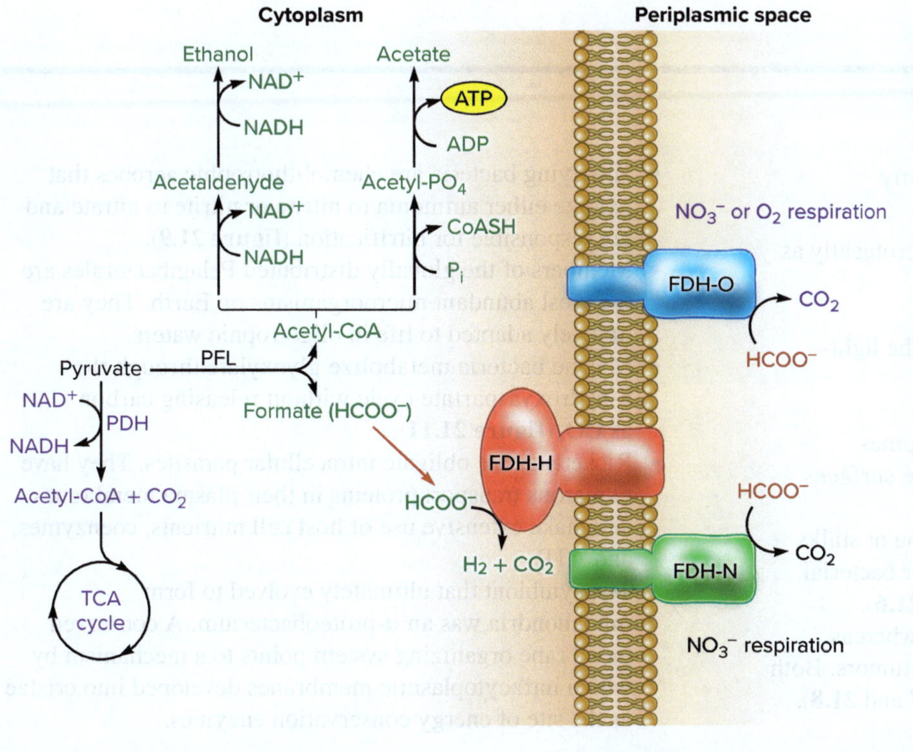

Figure 21.24 Metabolism of Pyruvate and Formate in *E. coli*. During aerobic growth, pyruvate is converted to CO_2 and acetyl-CoA by the pyruvate dehydrogenase complex (PDH) and acetyl-CoA enters the TCA cycle. Under low oxygen conditions, some pyruvate is also cleaved to formate and acetyl-CoA by pyruvate formate-lyase (PFL). Formate must then be converted into CO_2 and H_2 by one of three formate dehydrogenase (FDH) enzymes. FDH-O is active during aerobic respiration when oxygen levels are low. If respiration continues under anoxic conditions with nitrate as the terminal electron acceptor, both FDH-N and FDH-O split formate. These enzymes are located on the periplasmic face of the plasma membrane. During fermentative conditions, FDH-H is made, resulting in mixed acid fermentation: Formate is cleaved to yield CO_2 and H_2, whereas acetyl-CoA is converted to ethanol and acetate. FDH-H is a peripheral membrane protein, with its active site exposed to the cytoplasm. Aerobic metabolism is shown in purple, fermentative metabolism in green.

MICRO INQUIRY *Why do you think FDH-O and FDH-N are on the periplasmic face of the plasma membrane while FDH-H is on the cytoplasmic face?*

vival during fermentation, when one-third of the carbon derived from glucose is converted to formate. If this intermediate were allowed to accumulate within the cell, the internal pH would become too acidic to support viability. In fact, *E. coli* has three different FDH enzymes. FDH-H (H for hydrogen production) is synthesized only during fermentative conditions. FDH-N is produced under anoxic conditions when nitrate serves as the terminal electron acceptor. FDH-O is produced under microaerobic conditions as well as when nitrate is the terminal electron acceptor.

The family can be divided into two groups based on their fermentation products. The majority (e.g., the genera *Escherichia, Proteus,* and *Salmonella*) carry out mixed acid fermentation and produce mainly lactate, acetate, succinate, ethanol, formate, H_2, and CO_2. In contrast, *Enterobacter, Serratia, Pectobacterium,* and *Klebsiella* spp. are butanediol fermenters, producing butanediol, ethanol, and CO_2 (*see figure 11.18*). ◀ *Fermentation does not involve an electron transport chain (section 11.6)*

Because enteric bacteria are so similar in morphology, biochemical tests have traditionally been used to identify them after a preliminary examination of their morphology, motility, and growth responses. Some more commonly used tests are those for the type of fermentation, lactose and citrate utilization, indole production from tryptophan, urea hydrolysis, and hydrogen sulfide production. For example, lactose fermentation occurs in some genera (e.g., *Escherichia, Klebsiella,* and *Enterobacter*) but not in others (e.g., *Salmonella* and *Proteus*). ▶ *Identification of microorganisms from specimens (section 36.2)*

Members of Enterobacteriaceae are so widespread and important that they are probably seen more often in laboratories than any other bacteria. *E. coli* is undoubtedly the best-studied bacterium and the experimental organism of choice for many microbiologists. It is an inhabitant of the colon of humans and other warm-blooded animals, and it is useful in detecting fecal contamination in water. Some strains are the causative agents of food-borne diarrhea and urinary tract infections; however, it should be emphasized that the vast majority of *E. coli* strains are harmless and help metabolize the food we eat. Several other genera contain important human pathogens responsible for a variety of diseases: *Salmonella,* typhoid fever and gastroenteritis; *Klebsiella,* pneumonia; and *Yersinia,* plague. Members of the genus *Pectobacterium* are major pathogens of crop plants and cause blights, wilts, and several other plant diseases. ▶ Escherichia coli *gastroenteritis (section 38.4)*

Comprehension Check

1. What are the major characteristics of purple sulfur bacteria?
2. Describe the metabolism of *Beggiatoa, Leucothrix,* and *Thiothrix* spp.
3. Describe the oxidation of methane and other 1-carbon compounds.
4. List three attributes that account for the importance of pseudomonads.
5. List the major distinguishing traits of bacteria in the families Vibrionaceae and Enterobacteriaceae.
6. What two cofactors are required for bioluminescence?
7. Into what two groups can enteric bacteria be placed based on their fermentation patterns?

Key Concepts

21.1 Class Alphaproteobacteria Includes Many Oligotrophs

- The purple nonsulfur bacteria can grow anaerobically as photoorganoheterotrophs and aerobically as chemoorganoheterotrophs. When growing photosynthetically, purple bacteria localize the light-harvesting proteins and reaction centers on intracytoplasmic membranes (**figure 21.1**).
- *Magnetospirillum* spp. assemble magnetosomes from iron on invaginated plasma membrane surfaces (**figure 21.4**).
- *Caulobacter* spp. are bacteria with prosthecae or stalks and holdfasts that are an important model for bacterial growth and reproduction (**figures 21.5** and **21.6**).
- *Rhizobium* spp. carry out nitrogen fixation, whereas agrobacteria cause the development of plant tumors. Both are in the family Rhizobiaceae (**figures 21.7** and **21.8**).
- Nitrifying bacteria are chemolithotrophic aerobes that oxidize either ammonia to nitrite or nitrite to nitrate and are responsible for nitrification (**figure 21.9**).
- Members of the globally distributed Pelagibacterales are the most abundant microorganisms on Earth. They are uniquely adapted to life in oligotrophic waters.
- Marine bacteria metabolize glyoxylate through the β-hydroxyaspartate cycle without releasing carbon as CO_2 (**figure 21.11**).
- Rickettsias are obligate intracellular parasites. They have numerous transport proteins in their plasma membranes and make extensive use of host cell nutrients, coenzymes, and ATP.
- The symbiont that ultimately evolved to form mitochondria was an α-proteobacterium. A conserved membrane organizing system points to a mechanism by which intracytoplasmic membranes developed into cristae as the site of energy conservation enzymes.

21.2 Gammaproteobacteria Is the Largest Bacterial Class

- γ-Proteobacteria are the largest subgroup of bacteria and have great variety in metabolic types (**table 21.2**).
- *Sphaerotilus* and *Leptothrix* contain species with sheaths, tubelike structures made of nanofibers that surround chains of cells without being in intimate contact with the cells (**figure 21.13**).
- The genus *Neisseria* contains nonmotile, aerobic cocci that usually occur in pairs (**figure 21.14**). They colonize mucous membranes and cause several human diseases, including meningitis and gonorrhea.
- Colorless sulfur bacteria such as *Thiobacillus* spp. oxidize elemental sulfur, hydrogen sulfide, and thiosulfate to sulfate while generating energy chemolithotrophically (**figure 21.15**).
- Purple sulfur bacteria are anaerobes and usually photolithoautotrophs. They oxidize hydrogen sulfide to sulfur and deposit the granules internally (**figure 21.16**).
- Bacteria such as *Beggiatoa* and *Leucothrix* spp. grow in long filaments or trichomes. They have gliding motility. *Beggiatoa* spp. are primarily chemolithotrophs and *Leucothrix* spp. are chemoheterotrophs.
- Marine *Beggiatoa, Thioploca,* and *Thiomargarita* spp. are among the largest microbes. Members of the genus *Thiomargarita* form large intracellular vacuoles in which nitrate is stored (**figure 21.18**).
- Proteobacteria that use 1-carbon compounds for carbon and energy are called methylotrophs; some of these microbes can use methane and are called methanotrophs. In bacteria, this is an aerobic process that produces the intermediate formaldehyde, which is either oxidized for energy or assimilated for anabolic purposes (**figure 21.21**).
- Members of the genus *Pseudomonas* are straight or slightly curved, aerobic rods that are motile by one or several polar flagella.
- Pseudomonads participate in natural mineralization processes, are major experimental subjects, cause many diseases, and often spoil refrigerated food.
- *Acinetobacter* is responsible for hospital-acquired infections. Several features of their cell envelope structure contribute to their persistence in the health-care environment.
- Enterobacteriaceae, often called enterobacteria or enteric bacteria, are peritrichously flagellated or nonmotile, facultatively anaerobic, straight rods with simple nutritional requirements.
- Enteric bacteria are usually identified by a variety of physiological tests. *E. coli* performs a mixed acid fermentation, which depends on the enzymes pyruvate formate-lyase and formate dehydrogenase (**figure 21.24**).

Active Learning

1. The advantages of metabolic flexibility in *Rhodospirillum rubrum* must offset the energetic burden of carrying all the structural and regulatory genes involved in these processes. What makes this trade-off possible?
2. Compare nitrification with denitrification: Which is found in heterotrophs and which is found in autotrophs? Are they both dissimilatory processes?
3. Why might the ability to form dormant cysts be of great advantage to an agrobacterium but not as much to a rhizobium?
4. Obligate intracellular pathogens like *Rickettsia* are transmitted when an infected insect feeds on a human. In their intracellular niche in both hosts, the microbes are an isolated population and there is little opportunity for horizontal gene transfer. Relate this lack of community to the genome reduction that has been observed for obligate intracellular parasites.

CHAPTER

22

Gram-Positive Bacteria

Mateusz Kropiwnicki/Shutterstock

Antibiotic Production: Is It Actually Bacterial Chitchat?

Among bacteria, it is hard to find a genus whose constituents have given more to humankind than *Streptomyces.* When growing in their natural soil habitat, streptomycetes are key players in a microbial community that eats decaying plant and animal material. And like diners in a crowded restaurant, these microbes are "talking" up a storm, using a vocabulary of small molecules.

In fact, microbiologists hypothesize that antibiotics are the language of inter- and intraspecies communication. If this is the case, then streptomycetes are very talkative indeed. About two-thirds of all antibiotics prescribed are made by these bacteria. Noting that streptomycetes produce and secrete such a wide range of natural products, pharmaceutical companies have screened countless isolates, discovering that they also make anticancer compounds and antifungals. All told, *Streptomyces* species produce about 10,000 bioactive products, but models suggest there may be a million yet undiscovered!

How can it be that so few streptomycete compounds are known? And how do we know this? The genome sequence of the model streptomycete, *S. coelicolor,* encodes over 20 potential antimicrobial compounds, yet this bacterium produces only four when grown in culture. To tap this reserve, microbiologists must figure out how to turn on the genetic switch that represses expression of these genes. This is where the ability of streptomycetes to communicate gets interesting. About 80 years ago, streptomycin, the first drug to treat tuberculosis, was purified from *S. griseus*. Ten years later, researchers discovered that streptomycin production depended on the release of tiny amounts of a small molecule called A-factor, a gamma-butyrolactone quorum-sensing molecule. If *S. griseus* cells fail to exchange A-factor, no streptomycin is produced. Much later it was discovered that a different group of chemicals controls the release of other antibiotics. When cultured alone, many *Streptomyces* spp. lack discernible antibiotic activity, but when grown in the company of other microbes, and allowed to exchange secreted molecules, antibiotics are released.

What does all this bacterial chitchat mean? To the microbes it probably allows them to assess their local population size. To the world of drug discovery, it means that screening for new natural products is more complicated than simply growing a *Streptomyces* isolate in pure culture. New and creative methods of culturing and screening must be developed.

In this chapter, we survey Gram-positive bacteria. Most Gram-positive bacteria are **monoderms,** with a cell envelope comprised of the plasma membrane and a thick peptidoglycan exoskeleton. Among the Gram positives are actinomycetes such as *Streptomyces,* as well as endospore formers, notable pathogens, and bacteria important for food production.

Readiness Check:

Based on what you have learned previously, you should be able to:

- ✓ Draw a monoderm cell envelope and list its constituent parts (section 3.3)
- ✓ Outline the process of endospore formation and germination (section 3.8)
- ✓ Compare and contrast the growth and physiology of obligate aerobes, facultative anaerobes, microaerophiles, and strict anaerobes (section 7.5)
- ✓ Explain how biofilms form and why they are important (section 7.6)
- ✓ Draw an electron transport chain indicating the relative standard reduction potentials of the electron donor, acceptor, and electron transporters (sections 10.4 and 11.4)
- ✓ Distinguish heterotrophy from autotrophy (section 11.1)
- ✓ Compare and contrast anaerobic and aerobic respiration, oxygenic and anoxygenic phototrophy, and substrate-level phosphorylation and oxidative phosphorylation (sections 11.4, 11.5, and 11.10)
- ✓ Summarize the generation of ATP and the fate of pyruvate in fermentation (section 11.6)
- ✓ Compare monoderm and diderm cell envelope structures and explain their correlation with the Gram stain (section 20.1)

22.1 Phylum Actinomycetota

After reading this section, you should be able to:

a. Draw the life cycle of at least two different actinomycetes
b. Distinguish an exospore from an endospore
c. Explain how streptomycete morphology helps this group thrive in soils
d. Discuss the origin of the antibiotic streptomycin and list five other medicinal agents made by streptomycetes
e. List common environments in which actinobacteria can be found
f. Describe the morphology of *Bifidobacterium*
g. Relate the structure of mycolic acids to the properties of mycobacterial cell envelopes
h. Explain snapping division and how this results in the palisade arrangement of cells
i. Describe the unusual morphology of actinoplanete spores
j. Describe the morphology of *Frankia* spp., relate this to their colonization of plants, and explain the advantage these bacterial symbionts confer to plants such as alder trees

Actinomycetes (Greek, *actino,* ray or beam; *myces,* fungus) are fascinating bacteria. Metabolically, all are chemoheterotrophs. As described in the chapter opening story, they also produce metabolites used as antibiotics, as well as anticancer drugs, antihelminthics, and drugs that suppress the immune system. This practical aspect of actinomycetes is linked closely to their mode of growth, which in many cases involves a complex life cycle. Many actinomycetes exist as filaments, called **hyphae,** instead of more familiar rods or cocci. Septae (cross walls) usually divide the hyphae into long cells (20 μm and longer), each containing several nucleoids. When growing on a solid substratum such as soil or agar, filamentous actinomycetes develop a branching network of hyphae that grows on and into the substratum to form a dense mat called a **substrate mycelium** (**figure 22.1**). In many actinomycetes, substrate hyphae go on to differentiate into upwardly growing hyphae to form an **aerial mycelium** that gives the colony a fuzzy appearance. It is at this time that medically useful compounds form. Because the actinomycete has switched from actively growing vegetative cells into this special, secondary cell type, these compounds are often called **secondary metabolites.** ▶| *Growing microbes in industrial setting presents challenges (section 41.3)*

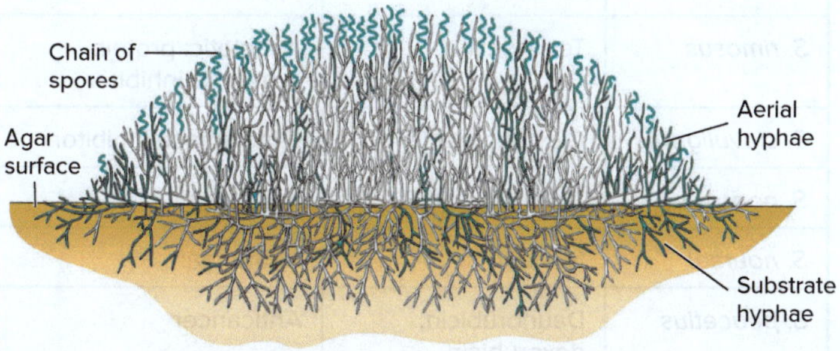

Figure 22.1 Diagram of an Actinomycete Colony. The colony consists of actively growing aerial hyphae that cannibalize the substrate hyphae to obtain nutrients. Live hyphae are blue; dead are white.

Over time, aerial hyphae septate to form chains of spores (**figure 22.2;** *see also figure 7.2*). When the septum completes constriction, individual exospores are released into the environment. These spores are considered **exospores** because they do not develop within a mother cell as do endospores. If the spores are located in a sporangium (a sheathlike structure), they may be called **sporangiospores.** Like spore formation in other bacteria, actinomycete sporulation is in response to nutrient deprivation and chemical signaling between microbes. In general, actinomycete spores are not particularly heat resistant but withstand desiccation well, so they are well suited to the sediment environment. Most actinomycetes are nonmotile, and spores dispersed by wind or animals may find a new habitat to provide needed nutrients. Interestingly, among the few motile actinomycetes, movement is confined to flagellated spores. |◀ *Bacterial endospores are a survival strategy (section 3.8)*

Order Streptomycetales: An Important Source of Antibiotics

The most important genus in the order Streptomycetales is *Streptomyces.* This is a large genus with hundreds of species. All are strict aerobes. These bacteria are often called **streptomycetes** (Greek *streptos,* bent or twisted; *myces,* fungus).

Like diderm myxobacteria, streptomycetes exhibit true multicellularity among bacteria. This is illustrated by their life cycle (**figure 22.3*a***). Single, nonmotile exospores germinate to produce long substrate hyphae. These form a dense substrate mycelium. Upon receipt of intercellular signals that indicate nutrient limitation, the hyphae initiate a coordinated division of

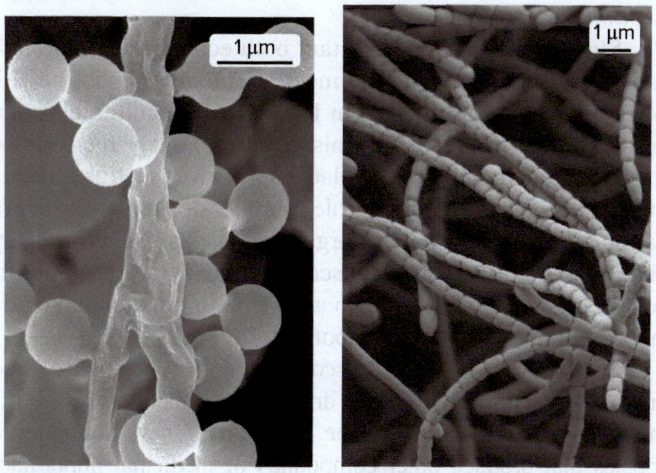

(a) *Pseudonocardia* (b) *Streptomyces venezuelae*

Figure 22.2 Examples of Actinomycete Spores as Seen in the Scanning Electron Microscope.

(a) C. L. Jiang, L. H. Xu & S. Suzuki/Society for Actinomycetes Japan; (b) Kim Findlay and Mark Buttner

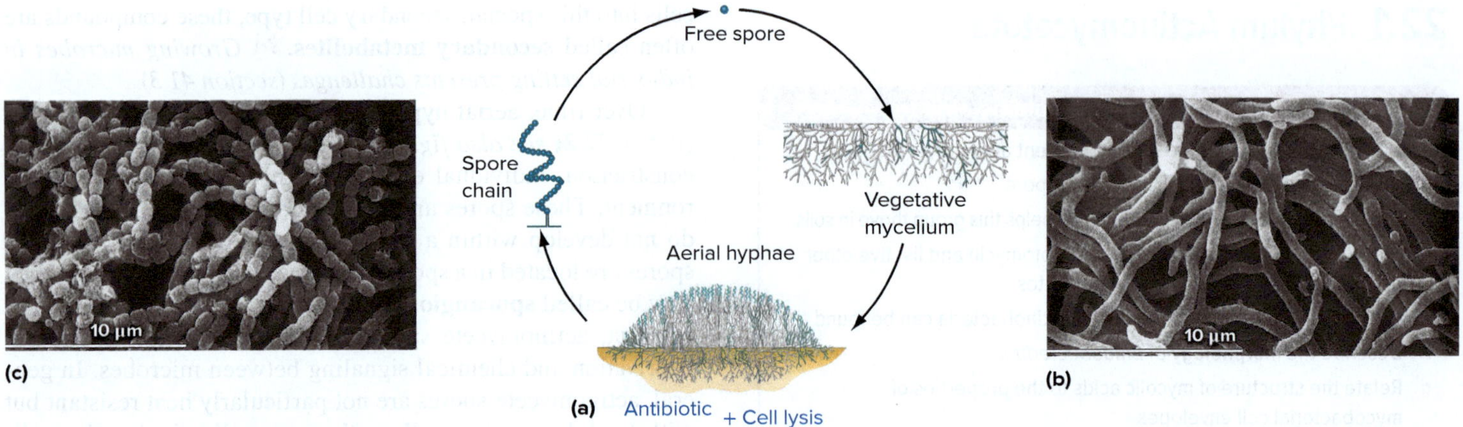

Figure 22.3 ***Streptomyces* Life Cycle.** (a) Streptomycetes form three types of cells: Exospores germinate to produce vegetative (also called substrate) hyphae, which then differentiate into aerial hyphae. These give rise to chains of exospores. Scanning electron micrographs are shown of (b) *S. coelicolor* vegetative hyphae and (c) chains of exospores.

(b-c) Joanne M. Willey, Ph.D.

labor. Cells in the substrate hyphae produce and secrete secondary metabolites to compete with other soil microbes. Other hyphae grow up into the air to form aerial hyphae and ultimately spores. Aerial hyphae obtain nutrients by cannibalizing the filaments within the substrate mycelium. As aerial hyphae continue to grow, they curl and septate in regular intervals, ultimately becoming chains of spores to begin the cycle anew. Interestingly, exospore formation is dependent on the cytoskeletal proteins FtsZ and MreB, but vegetative growth is not because substrate hyphae grow by extending their tips rather than by binary fission. This results in long filaments with only an occasional septa (*see figure 17.11*). Multiple nucleoids are enclosed in the same cell; this is called **syncytial growth.** ◀ *Cytokinesis (section 7.2); Phyla Bdellovibrionota and Myxococcota: bacterial predators (section 20.9)* ▶ *Antibiotics (section 41.1)*

Streptomycetes are important both ecologically and medically. The natural habitat of most streptomycetes is the soil, where they may constitute from 1 to 20% of the culturable population. In fact, the odor of moist earth is largely the result of streptomycete production of volatile substances such as geosmin. Streptomycetes play a major role in mineralization—the degradation of organic matter to inorganic elements. As substrate hyphae grow into the soil, they secrete exoenzymes enabling the degradation of substances such as pectin, lignin, chitin, keratin, latex, agar, and aromatic compounds.

Selman Waksman's 1944 discovery that *S. griseus* (**figure 22.4a**) produces streptomycin, the first drug to cure tuberculosis, set off a massive search resulting in the isolation of new *Streptomyces* species that produce other compounds of medicinal importance. Since that time, streptomycetes have been found to produce over 10,000 bioactive compounds; a few are listed in **table 22.1.** About two-thirds of the antimicrobial agents used in human and veterinary medicine are derived from streptomycetes. Antibiotic-producing bacteria have immunity genes that encode proteins (or small

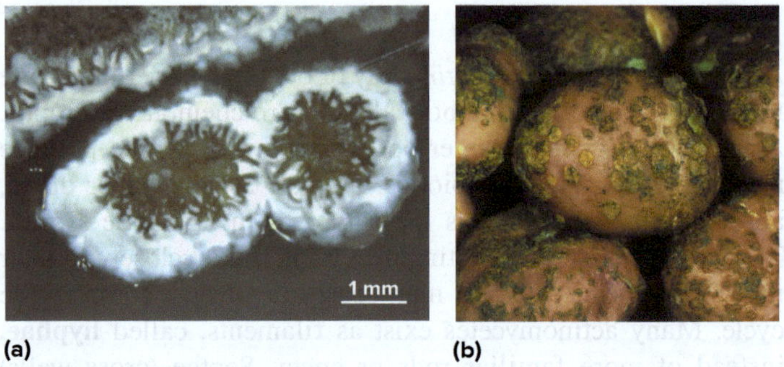

Figure 22.4 Streptomycetes of Practical Importance. (a) Colonies of *Streptomyces griseus,* which produces streptomycin. (b) *S. scabiei* growing on potatoes.

(a) Roger Greenwell; (b) Nigel Cattlin/Alamy Stock Photo

Table 22.1 Examples of Natural Products Made by Streptomycetes

Microbe	Natural Product	Application
S. rimosus	Tetracycline	Antibiotic: protein synthesis inhibitor
S. clavuligerus	Clavulanic acid	β-lactamase inhibitor
S. nodosus	Amphotericin B	Antifungal
S. noursei	Nystatin	Antifungal
S. peucetius	Daunorubicin, doxorubicin epirubicin	Anticancer
S. mobaraensis	Bleomycin	Anticancer

noncoding RNA) that make them resistant to the antibiotics they synthesize. *Antimicrobial chemotherapy (chapter 9)*

Secondary metabolites like antibiotics are synthesized by biochemical pathways whose enzymes are encoded in large operons termed **biosynthetic gene clusters (BGCs).** These genes are coregulated so they are expressed only under conditions when the pathway end product is needed. For example, as sporulation begins, dying cells from the substrate hyphae release peptidoglycan fragments whose monomers enter the surviving cells and act as transcriptional inducers for the antibiotic biosynthesis operon. In this way, the cell coordinates BGC expression with spore formation. *Induction and repression of enzyme synthesis (section 14.2)*

Genomic studies document multiple BGCs in many *Streptomyces* spp. Two current challenges are identifying the pathway end product and determining conditions that trigger expression. Few BGCs are expressed in laboratory conditions, and often the identity of the natural product cannot be inferred from the genome sequence. Experiments like co-culturing the streptomycete with other microbes have brought some success. Given the vast pool of natural products already known from *Streptomyces* (table 22.1), these BGCs are tantalizing.

Streptomyces spp. have linear chromosomes, and replication proceeds bidirectionally from a central origin. Their genomes are among the largest bacterial genomes, often exceeding 10 Mb. The large number of genes no doubt reflects the number of proteins required to undergo a complex life cycle. Likewise, synthesis of natural products requires significant genetic capacity. *Streptomyces* spores have a single genome copy, but, as noted, hyphae have multiple nucleoids. This appears to correlate with the instability that has been observed in these linear chromosomes. Essential genes are located in a core region in the genome center, with BGCs near the two termini. Large deletions of up to 1 Mb have been documented at the chromosome ends in some regions of the substrate hyphae. Multiple chromosome copies allow these hyphae to synthesize secondary metabolites, while horizontal transfer of integrative conjugative elements contributes to population diversity. *The mobilome includes diverse mobile elements (section 16.5)*

Although most streptomycetes are almost exclusively nonpathogenic saprophytes, a few are associated with plant and animal diseases. *Streptomyces scabiei* causes scab disease in potatoes and beets (figure 22.4*b*). *S. somaliensis* is associated with actinomycetoma, a rare infection of subcutaneous tissues.

Order Actinomycetales

The order Actinomycetales includes four important genera: *Actinomyces, Bifidobacteria, Arthrobacter,* and *Micrococcus.* Actinomycetales make a significant contribution to degrading complex polysaccharides. They synthesize and secrete enzymes to hydrolyze molecules like chitin, pectin, and cellulose.

Members of the genus *Actinomyces* are either straight or slightly curved rods, or slender filaments with true branching (**figure 22.5*a,c*;** *see also figure 3.4a*). The rods and filaments may have swollen or clubbed ends. They are either facultative or strict anaerobes that require CO_2 for optimal growth. *Actinomyces* spp. are normal inhabitants of mucosal surfaces of humans and other warm-blooded animals; the oral cavity is their preferred habitat. *A. oris* is a significant component of dental plaque, where its pili mediate attachment both to the tooth surface and the cell envelope of other oral bacterial species. *Actinomyces* spp. are associated with severe dental infections and osteonecrosis in the jaw, commonly termed lumpy jaw.

Members of the genus *Bifidobacterium* are nonmotile, nonsporing, anaerobic rods that are slightly curved and clubbed; often they are branched (figure 22.5*d*). The rods are single or in clusters and V-shaped pairs. *Bifidobacterium longum* is a pioneer colonizer of the human intestinal tract, dominating the gut microbiome of infants fed breast milk. Following the dietary shift of weaning or the introduction of solid foods, *Bifidobacterium* decreases in abundance. *Bifidobacterium* spp. are sold as probiotic agents (particularly in yogurt) because they are thought to impart a health benefit. *The microbiome develops from birth to adulthood (section 33.2); Microbiome manipulation can be therapeutic (section 33.5)*

The genus *Micrococcus* contains aerobic, catalase-positive cocci that occur mainly in pairs, tetrads, or irregular clusters and are usually nonmotile (figure 22.5*b*). Unlike other actinomycetes, micrococci do not undergo morphological differentiation. On solid media, their colonies often appear yellow, orange, or red. They are widespread in soil, in water, and on mammalian skin.

The genus *Arthrobacter* contains aerobic rods with respiratory metabolism. Their most distinctive feature is a rod-coccus growth cycle (**figure 22.6**). Like many other actinomycetes, their most important habitat is soil, where they constitute a significant component of the culturable microbial community. They are well adapted to this niche because they are resistant to desiccation and nutrient deprivation, even though they do not form spores. Arthrobacters are nutritionally flexible and can degrade some herbicides and pesticides.

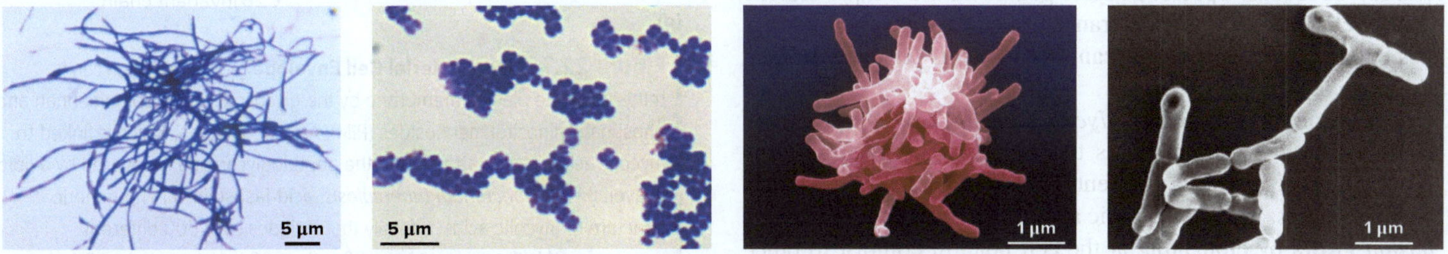

(a) *Actinomyces oris* **(b)** *Micrococcus* spp. **(c)** *Actinomyces* spp. **(d)** *Bifidobacterium bifidum*

Figure 22.5 Light (a,b) and Scanning Electron (c,d) Micrographs of Bacteria in the Order Actinomycetales.

(a) Dr. Lucille K. Georg/CDC; (b) Dr. Richard Facklam/CDC; (c) David M. Phillips/Science Source; (d) S. Amano & S. Miyadoh/Society for Actinomycetes Japan

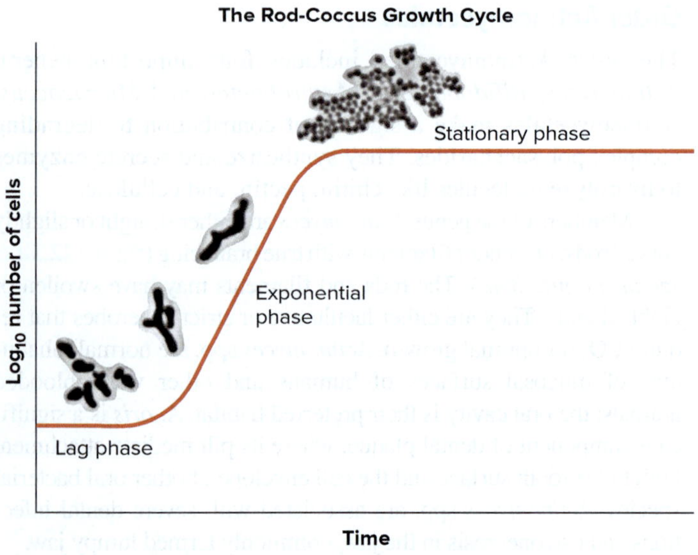

Figure 22.6 The Rod-Coccus Growth Cycle of *Arthrobacter* spp. When stationary phase cells are inoculated into fresh medium at 25°C, the coccoid cells of about 1 µm in diameter swell and elongate to form branching rods of about 7–10 µm in exponential phase. As nutrients are depleted and cells enter stationary phase, the cells cluster and become spherical.

Comprehension Check

1. Describe the function of the substrate mycelium, aerial mycelium, and exospore. Explain how these structures confer a survival advantage.
2. List three types of natural products made by *Streptomyces*.
3. Describe three ways that *Streptomyces* spp. are of ecological importance.
4. List two genera that are commonly associated with humans.
5. Examine the ingredients list on a commercial yogurt cup. Are there Actinomycetales present?

Order Mycobacteriales Includes Important Human Pathogens

Within this order, the family Mycobacteriaceae includes four important genera: *Mycobacterium, Corynebacterium, Nocardia,* and *Rhodococcus.* Species in these genera have a unique diderm cell envelope. The outer membrane, termed the **mycomembrane,** lies external to the peptidoglycan. ◀ *Cell walls have many functions (section 3.3)*

Bacteria in the genus *Mycobacterium* are composed of slightly curved or straight rods that sometimes branch or form filaments. Mycobacterial filaments readily fragment into rods and coccoid bodies. They are aerobic and catalase positive. *Mycobacterium* grows by elongating at the cell pole, in contrast to other bacterial growth strategies (*see figure 7.8b*), resulting in asymmetric cells. On solid complex media, *Mycobacterium* must be incubated for 2 to 40 days after inoculation to form a visible colony. ◀ *Cellular growth and determination of shape (section 7.2)*

The mycobacterial cell envelope is diagrammed in **figure 22.7a.** The 7-nm-thick mycomembrane is constructed of tightly packed, covalently linked **mycolic acids** on the inner leaflet and noncovalently attached acyl lipids on the outer leaflet.

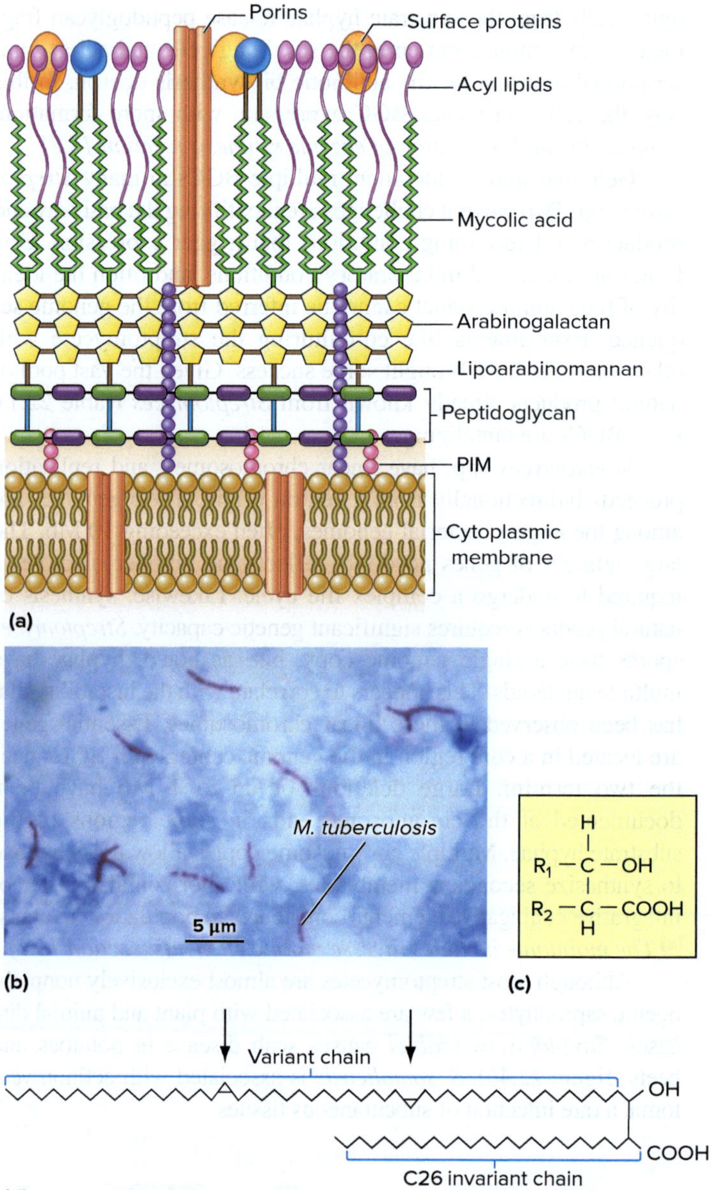

Figure 22.7 Mycobacterial Cell Envelope. (a) Peptidoglycan is tethered to the plasma membrane by the glycolipids lipoarabinomannan and phosphatidylinositol mannosides (PIM). Arabinogalactan, which is linked to mycolic acids, is also attached to the peptidoglycan and is covered by a lipid bilayer. (b) *Mycobacterium tuberculosis,* acid-fast stain. (c) The generic structure of mycolic acids, a family that includes over 500 different molecules. R1 is the variant chain; R2 is the C26 invariant chain. (d) An example of mycolic acid with two cyclopropane rings (arrows).

(b) Dr. George P. Kubica/CDC

MICRO INQUIRY *Why do* Mycobacterium *spp. have porins?*

Mycolic acids are complex fatty acids with an invariant 26-carbon (C26) fatty acid that is attached to a longer, C18 to C76 fatty acid (figure 22.7*c,d*). Chemical modifications to the variable fatty acids include the formation of cyclopropane units, double bonds, and methylation; these modifications create diversity within the fatty acid. Mycolic acids are covalently linked to peptidoglycan by arabinogalactan, a polysaccharide of arabinose and galactose subunits. The lipid content makes the cell envelope **acid-fast** (figure 22.7*b*; carbol fuchsin dye cannot be removed from the cell by acid alcohol treatment). *Differential staining (section 2.3)*

The cell envelope is extremely hydrophobic and impenetrable to most organic molecules, including many antibiotics. In fact, mycobacteria require porins to enable the entry of small molecules (*see figure 3.23*). In addition, the mycomembrane possesses a type VII secretion system through which proteins are exported. *Mycobacterium* spp. are not naturally competent, probably due to the hydrophobic cell envelope. These microbes use a novel genetic transfer mechanism, distributive conjugal transfer. Donor and recipient strains are in direct contact. The donor chromosome is entirely fragmented, while the recipient synthesizes proteins to import donor DNA fragments. Multiple fragments are transferred into a single recipient cell and the resulting transconjugant chromosome is a mosaic of the two original strains. *Protein maturation and secretion (section 13.9)*

Although some mycobacteria are free-living saprophytes, they are best known as animal pathogens. *Mycobacterium tuberculosis* complex (MTBC) describes closely related strains that cause tuberculosis in primates (including humans) and ruminants (including cattle). It is estimated that about one-third of the world's population is infected with MTBC. Leprosy is also a mycobacterial disease; it is caused by *M. leprae*. Finally, *M. avium* and the related species *M. intracellulare* comprise *M. avium* complex, which has become a common mycobacterial disease, particularly among individuals with compromised immunity. *Mycobacterium infections (section 38.1); Mycobacterial skin infections (section 38.3)*

The genus *Corynebacterium* includes aerobic and facultative, straight to slightly curved rods. Club-shaped cells are common. Both corynebacteria and arthrobacters have a two-layered cell wall that leads to **snapping division** (**figure 22.8*a***). This occurs because during cell division, only the inner layer grows inward to generate a septum dividing new cells. As the septum thickens, tension rises on the outer wall layer, which still holds the cells together. Eventually increasing tension ruptures the outer layer at its weakest point, and a snapping movement tears the outer layer apart so that the new cells now rest at an angle to each other. This results in the cellular arrangement termed a **palisade arrangement** in which cells are lined up side by side (figure 22.8*b*). Although some species are harmless saprophytes, many corynebacteria are plant or animal pathogens. For example, *C. diphtheriae* is the causative agent of diphtheria in humans. In contrast, *C. glutamicum* is widely used in industrial settings as it naturally produces large quantities of several amino acids. *Diphtheria (section 38.1)*

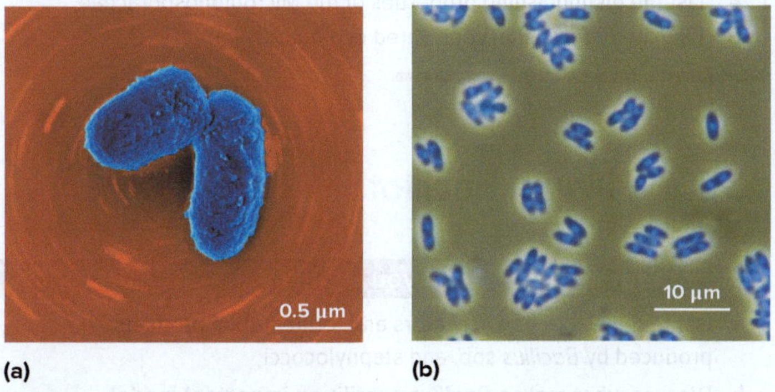

Figure 22.8 ***Corynebacterium diphtheriae* Undergoes Snapping Division, Giving Rise to Cells in Palisades.** (a) Following snapping division, daughter cells remain attached and at an angle. (b) This angular association of pairs of cells results in a palisade arrangement.

(a) Dr. Gary Gaugler/Science Source; (b) M. Wachi/Society for Actinomycetes Japan

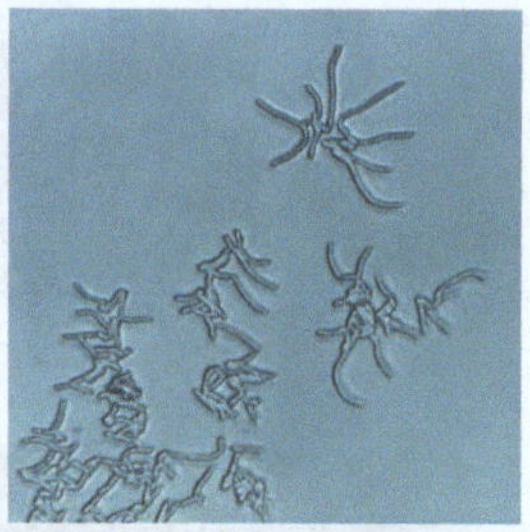

Figure 22.9 ***Rhodococcus.***

Dr. David Berd/CDC

Nocardia spp. and *Rhodococcus* spp. develop a substrate mycelium with hyphae of 0.5 to 1.2 μm in diameter that readily breaks into rods and coccoid fragments (**figure 22.9**). Almost all are strict aerobes.

Nocardia spp. are distributed worldwide in soil and aquatic habitats. They are involved in the degradation of hydrocarbons and waxes and contribute to the biodeterioration of rubber joints in water and sewage pipes. Although most are free-living saprophytes, some species, particularly *N. asteroides*, are opportunistic pathogens of humans and other animals. People with impaired immune responses are most at risk. The lungs are typically infected; other organs and the central nervous system may be invaded as well.

Rhodococcus contains species widely distributed in soils and aquatic habitats. They are of considerable interest because they degrade an enormous variety of molecules, such as petroleum hydrocarbons, detergents, benzene, polychlorinated biphenyls (PCBs), and various pesticides.

Also within the Order Mycobacteriales, the family Micromonosporaceae contains the genera *Micromonospora, Dactylosporangium,* and *Actinoplanes*. Together, these bacteria illustrate the variation in actinomycete life cycles and spore dispersal strategies. They have an extensive substrate mycelium but usually lack all but a rudimentary aerial mycelium. Spores are usually formed within a sporangium raised above the surface of the substratum at the end of a special hypha called a sporangiophore (**figure 22.10*a,b***). Compared with *Streptomyces* spp., which have up to 30 spores on each aerial hypha, *Micromonospora,* as its name describes, bears a single spore on a short vegetative hypha. Increasing in complexity, *Dactylosporangium* generates two short, fingerlike aerial hyphae, each of which forms 1 to 3 spores (figure 22.10*c*). Finally, *Actinoplanes* spp. may produce several to thousands of spores, also on short sporangia (figure 22.10*a,b*). These differences reflect the variation in ecological niche for each species. Two important aspects of spore formation are dispersal and persistence. Spore motility permits

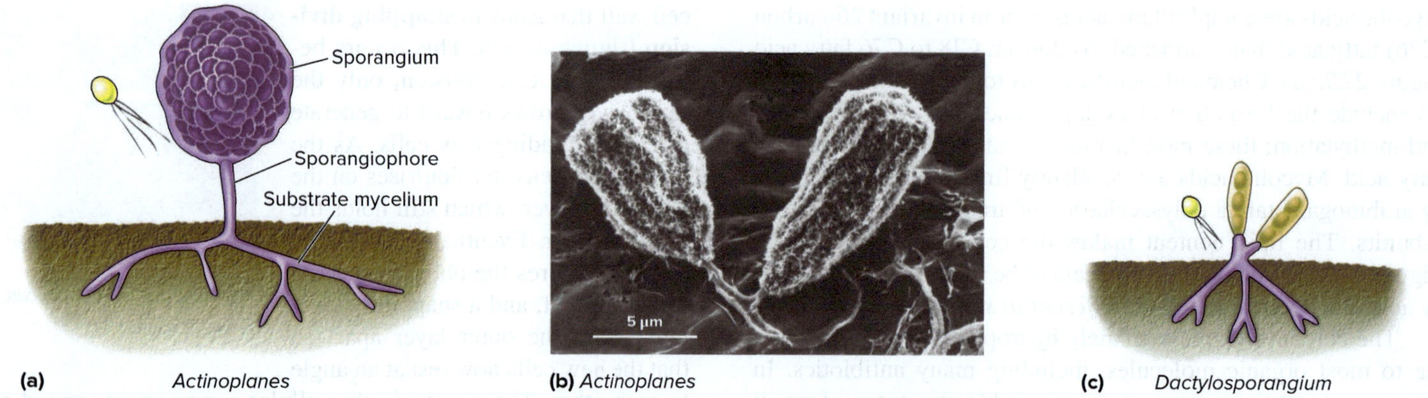

(a) *Actinoplanes* (b) *Actinoplanes* (c) *Dactylosporangium*

Figure 22.10 Micromonosporaceae. Actinoplanete morphology. (a) *Actinoplanes* structure. (b) A scanning electron micrograph of mature *Actinoplanes* sporangia. (c) *Dactylosporangium* structure.

(b) S. Amano & S. Miyadoh/Society for Actinomycetes Japan

local dispersal, while aerial hyphae promote more widespread dispersal by wind, water, or insects. Persistence refers to the ability of a spore to germinate and begin another life cycle; the more cosmopolitan species gamble on an abundance of spores with broad dispersal.

Bacteria in the family Micromonosporaceae grow in almost all soil habitats, ranging from forest litter to beach sand; they play an important role in the decomposition of plant and animal material. *Micromonospora* spp. actively degrade chitin and cellulose and produce antibiotics such as gentamicin.

In the family Frankiaceae, bacteria in the genus *Frankia* form multilocular sporangia (having many cells or compartments). These are, characterized by clusters of nonmotile spores when hyphae divide both transversely and longitudinally (**figure 22.11*a*; *see also figure 30.11***). *Frankia* grows in symbiotic association

(a) (b)

Figure 22.11 *Frankia*. (a) A phase contrast micrograph showing hyphae and multilocular spores. (b) Nodules of the alder *Alnus rubra*.

David Benson/University of Connecticut

with the roots of at least eight families of higher nonleguminous plants (e.g., alder trees). They are microaerophilic and can fix atmospheric nitrogen. In fact, *Frankia* spp. fix nitrogen so efficiently that a tree such as an alder can grow in the absence of combined nitrogen (e.g., NO_3^-). Within the root nodule, these bacteria form branching hyphae with globular vesicles at their ends, which may be the sites of nitrogen fixation (figure 22.11*b*). Their nitrogenase resembles that of rhizobia in that it is oxygen sensitive and requires molybdenum and cobalt cofactors. ◀ *Rhizobiales includes important N_2-fixing bacteria (section 21.1)*; ▶ Frankia *form actinorhizae (section 30.3)*

Comprehension Check

1. What implications does the complex hydrophobic mycobacterial cell envelope have in terms of treating diseases caused by these bacteria?
2. What is snapping division? How does it lead to palisade formation?
3. *Nocardia* and *Rhodococcus* have cell morphologies reminiscent of Actinobacteria. What feature (besides genomic sequence) groups them with the Mycobacteriales?
4. List the distinguishing properties of the Micromonosporaceae.
5. Why are *Frankia* spp. considered ecologically important?

22.2 Phylum Bacillota

After reading this section, you should be able to:

a. List terminal electron acceptors and fermentation products produced by *Bacillus* spp. and staphylococci
b. Discuss what makes *Bacillus subtilis* an important model organism
c. List three reasons *Bacillus* spp. are of practical importance
d. Explain why *Thermoactinomyces* is not considered an actinomycete

e. Identify the genera considered lactic acid bacteria, and discuss their importance in the food industry
f. Distinguish between pathogenic and commensal species of *Staphylococcus*
g. Name bacterial genera with species capable of hemolysis, and differentiate between α-hemolysis and β-hemolysis
h. Identify several *Clostridium* spp. that form endospores
i. Outline the reaction by which some clostridia ferment amino acids, and relate this carbon substrate to their environmental distribution and pathogenic potential
j. Explain how *Clostridioides difficile* causes disease
k. Distinguish phototrophy in heliobacteria from that of other anoxygenic phototrophic bacteria
l. Relate the structure of the mycoplasma cell envelope to its morphology
m. Diagram the flux of protons that enables *Ureaplasma urealyticum* to generate a proton motive force
n. Illustrate how gliding motility is accomplished by mycoplasmas
o. List several habitats in which mycoplasmas reside and relate this to their capacity to cause disease

There are several important orders in the phylum Bacillota, including pathogens and bacteria essential for fermented food production. We start our discussion with the two orders that form endospores: Bacillales and Clostridiales. Recall that **endospores** are made within a mother cell upon nutrient depletion. Once released, endospores are remarkably resistant to environmental challenges, sometimes surviving millions of years. ◀ *Bacterial endospores are a survival strategy (section 3.8)*

Order Bacillales

The genus *Bacillus* contains endospore-forming, chemoheterotrophic rods that are usually motile with peritrichous flagella (**figure 22.12*a***). *B. subtilis* is a facultative anaerobe that can use nitrate as a terminal electron acceptor or perform mixed acid fermentation with lactate, acetate, and acetoin as major end products. It is nonpathogenic and a popular model organism for the study of gene regulation, cell division, quorum sensing, and cellular differentiation. Its 4.2 Mb genome reveals a number of interesting elements. For instance, several families of genes have been expanded by gene duplication; the largest such family encodes ABC transporters (*see figure 3.11*)—the most abundant class of protein in *B. subtilis*. There are 18 genes that encode sigma factors. Recall that the use of alternative sigma subunits of RNA polymerase is one way in which bacteria regulate gene expression. In this case, many of the sigma factors govern endospore formation and other responses to stressful conditions. ◀ *Sporulation in* B. subtilis *(section 14.6)*

B. subtilis has a remarkable diversity of specialized cell types, including actively dividing vegetative cells, endospores, and biofilm cells. Like the actinomycetes described in section 22.1, an entire population doesn't follow a single developmental path, but rather a division of labor operates. This strategy to adapt to environmental changes provides different opportunities to subpopulations, thereby ensuring that some fraction survives. In figure 22.12*c* a pellicle biofilm establishes at the air-liquid interface: Cells at the bottom are motile, cells in the middle secrete a protective matrix of extracellular polysaccharide and protein, while cells on the top form endospores. Even within the endospore population, development is not synchronous, and within a germinating population, recovery to vegetative growth is not synchronous. *B. subtilis* has mastered the survival gamble by maintaining a repertoire of cell types. ◀ *Biofilms are common in nature (section 7.6)*

Many species of *Bacillus* are of considerable importance. They produce the antibiotics bacitracin, gramicidin, and polymyxin. *B. cereus* (figure 22.12*b*) can cause food poisoning. *B. anthracis* is the causative agent of the disease anthrax, which can kill both hoofed animals and humans. *B. thuringiensis* produces an insecticide called Bt. During endospore formation, a solid protein crystal, the **parasporal body,** forms in the mother cell (**figure 22.13**). As we discuss in chapter 41, the Bt crystal contains protein toxins that can kill over 100 insect species. When activated in the alkaline gut of caterpillars, the toxins destroy the epithelium. ▶ *Anthrax (section 38.5); Agricultural biotechnology repurposes plant and insect pathogens (section 41.4)*

One other genus within the order Bacillales bears mentioning: *Sporosarcina*. This genus contains the only known endospore-forming bacterium that has a coccoid rather than rod shape. *S. ureae* is particularly interesting because it is tolerant of

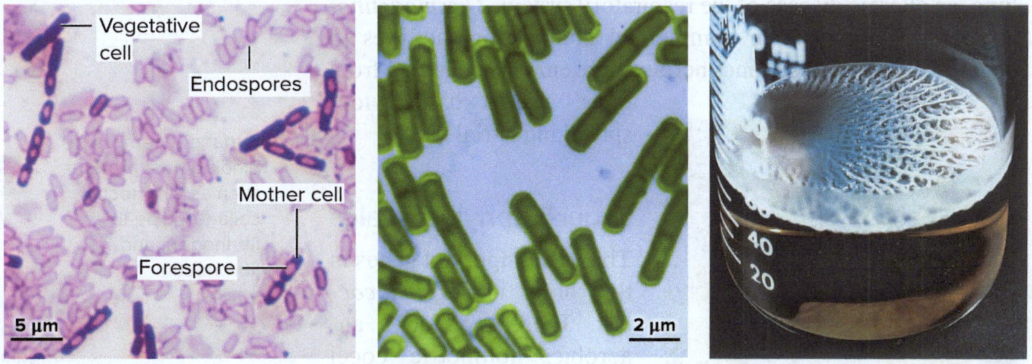

(a) *B. megaterium* (b) *B. cereus* (c) *B. subtilis*

Figure 22.12 ***Bacillus.*** (a) *B. megaterium* endospores are elliptical and central. (b) False colorized phase contrast image of *B. cereus.* (c) When grown for several days without agitation in nutrient-limiting medium, *B. subtilis* forms a hydrophobic pellicle at the air-liquid interface.

(a) Lisa Burgess/McGraw Hill; (b) Science History Images/Alamy Stock Photo; (c) Branda et al., PNAS 25 Dept. 2001, vol. 98, fig 1A, p. 11623 Copyright (2001) National Academy of Sciences, U.S.A.

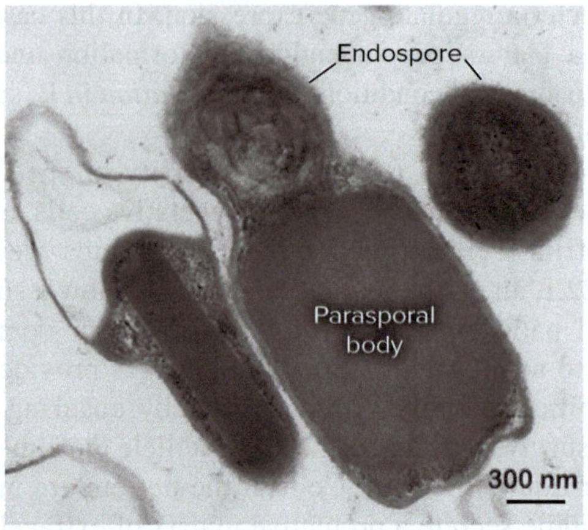

Figure 22.13 The Parasporal Body. An electron micrograph of a sporulating *B. thuringiensis* cell containing a parasporal crystal. A free endospore is also shown.

From: "Protein crystal structure obtained at 2.9 Å resolution from injecting bacterial cells into an X-ray free-electron laser beam," Michael R. Sawaya et al. *PNAS* vol. 111, no. 35, fig 1, pp. 12769–12774. Copyright (2014) National Academy of Sciences, U.S.A.

alkaline conditions (up to pH 10), which occur when it degrades urea to CO_2 and NH_3. It is easily isolated from livestock fields on which animals urinate.

Order Thermoactinomycetales

Bacteria in the genus *Thermoactinomyces* were originally classified as actinomycetes because, like these microbes, their soil-associated substrate hyphae (filamentous cells) differentiate into upwardly growing aerial hyphae. However, genomic analysis places them in the phylum Bacillota. Unlike actinomycetes, *Thermoactinomyces* species form true endospores within both aerial and substrate hyphae. These bacteria are thermophilic, growing between 45° and 60°C, and are commonly found in damp haystacks and compost piles. In fact, *T. vulgaris* from grain storage silos and compost piles is a causative agent of farmer's lung, an allergic disease of the respiratory system. *Thermoactinomyces* spp. are used in commercial composting facilities to degrade organic matter, and thermostable enzymes purified from these strains have widespread industrial uses.

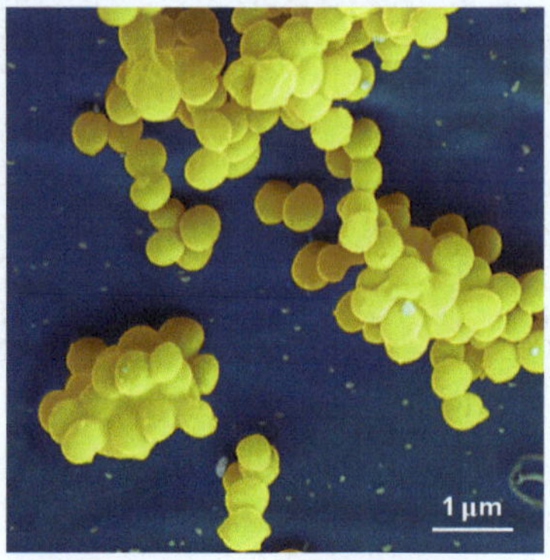

Figure 22.14 *Staphylococcus*.

Eye of Science/Science Source

Order Staphylococcales

The most important genus in this order is *Staphylococcus*. These facultative anaerobic, nonmotile cocci divide in more than one plane to form irregular clusters (**figure 22.14**). They are usually catalase positive, oxidase negative, and salt-tolerant. They respire using oxygen as the terminal electron acceptor, although some can reduce nitrate to nitrite. They are also capable of fermentation and convert glucose principally to lactate.

Staphylococci are associated with the skin and mucous membranes of warm-blooded animals. On human skin, *S. aureus* is responsible for many skin and soft tissue infections, while *S. epidermidis* is a commensal that plays a protective role and antagonizes pathogens (*see figure 33.3*). *S. aureus* has an impressive collection of virulence factors to evade the immune system and subvert the host (**figure 22.15**). These include coagulase that enables bacteria to attach to platelets and wounded tissue, thereby promoting initial colonization, and staphylokinase that dissolves fibrin clots that would keep the bacterium from invading beyond the site of colonization. Once *S. aureus* has penetrated and colonized tissue, Protein A binds host antibodies by the conserved Fc region to the cell surface, rendering them ineffective while at the same time concealing the bacteria from the immune system. Exotoxin production induces severe tissue damage and its superantigen causes toxic shock syndrome. *Antibody structure (section 32.7); Toxins are biological poisons (section 34.4)*

In contrast to the pathogenicity of *S. aureus, S. epidermidis collaborates* with the immune system to establish effective barriers and limits the colonization by pathogens. Metagenomic surveys have correlated higher levels of *S. epidermidis* with lower levels of opportunistic pathogens on skin. *S. epidermidis* surface proteins promote attachment to skin cells and tissue matrix surfaces. To maintain their ecological niche in the skin, *S. epidermidis* cells secrete antimicrobial peptides and proteases that restrict attachment by other microbes. *Skin (section 33.2)*

Staphylococcal species engage in frequent horizontal gene transfer. Exotoxin genes and antibiotic resistance genes have been observed moving through populations. *Staphylococcal diseases (section 38.3); Staphylococcal food poisoning (section 38.4)*

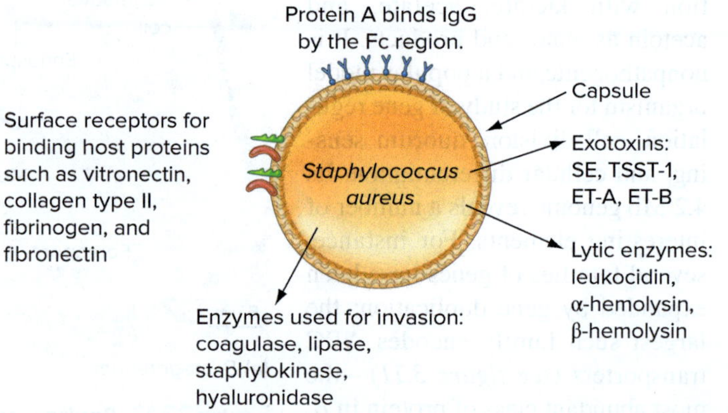

Figure 22.15 *Staphylococcus* Virulence Factors. *S. aureus* produces many virulence factors that contribute to its success as a pathogen. SE, staphylococcal enterotoxin; TSST-1, toxic shock syndrome toxin; ET, exfoliative toxin.

Comprehension Check

1. How do the different types of cells produced by *B. subtilis* allow the organism to persist in different environments?
2. What makes the genus *Thermoactinomyces* unique? What disease does *T. vulgaris* cause?
3. List environments where you would find *Staphylococcus* spp.

Order Lactobacillales

Streptococcus, Enterococcus, Lactococcus, Lactobacillus, and *Leuconostoc* are all genera in the order Lactobacillales. These bacteria do not form endospores and are usually nonmotile. They are strictly fermentative, depending on sugars for carbon and energy. They lack cytochromes and an electron transport chain so they obtain ATP only by substrate-level phosphorylation. Those that produce lactate as their major or sole fermentation product are sometimes collectively called **lactic acid bacteria.** Nutritionally, they are fastidious, and many vitamins, amino acids, purines, and pyrimidines must be supplied during culture because of their limited biosynthetic capabilities. Members of this order are usually categorized as facultative anaerobes but can be considered aerotolerant. ◀ *Oxygen concentration (section 7.5); Fermentation does not involve an electron transport chain (section 11.6)*

The largest genus in this order is *Lactobacillus,* with dozens of species. Lactobacilli include rods and some coccobacilli. All lack catalase and cytochromes and produce lactate as their main or sole fermentation product (*see figure 11.18*). This process, termed **homolactic fermentation,** uses the Embden-Meyerhof pathway. Alternately, **heterolactic fermentation** uses the phosphoketolase pathway producing lactate and ethanol (**figure 22.16**). They grow optimally under slightly acidic conditions, at a pH between 4.5 and 6.4, and are found in dairy products, meat, water, sewage, beer, fruits, and many other materials. Lactobacilli also are part of the normal flora of the human body; in particular, they are important normal vaginal flora. ▶ *Genitourinary tract (section 31.2)*

Lactobacillus species are indispensable to the food and dairy industry. They are used in the production of fermented vegetable foods (e.g., sauerkraut, pickles), sourdough bread, Swiss and other hard cheeses, yogurt, and sausage. Yogurt is probably the most popular fermented milk product in the United States, and *Streptococcus thermophilus* and *Lactobacillus delbrueckii* subspecies *bulgaricus* are used in its production. At least one species, *L. acidophilus,* is sold commercially as a probiotic agent that may provide some health benefits for the consumer. On the other hand, some lactobacilli are problematic. They can be responsible for the spoilage of beer, wine, milk, and meat. ▶ *Microbiome manipulation can be therapeutic (section 33.5); Fermented milks (section 40.5)*

Leuconostoc are facultative anaerobic cocci, which may be elongated or elliptical and arranged in pairs or chains (**figure 22.17**). Leuconostocs carry out heterolactic fermentation (figure 22.16) and can be isolated from plants, silage, and milk. They are used in wine production, in the fermentation of vegetables, and in the

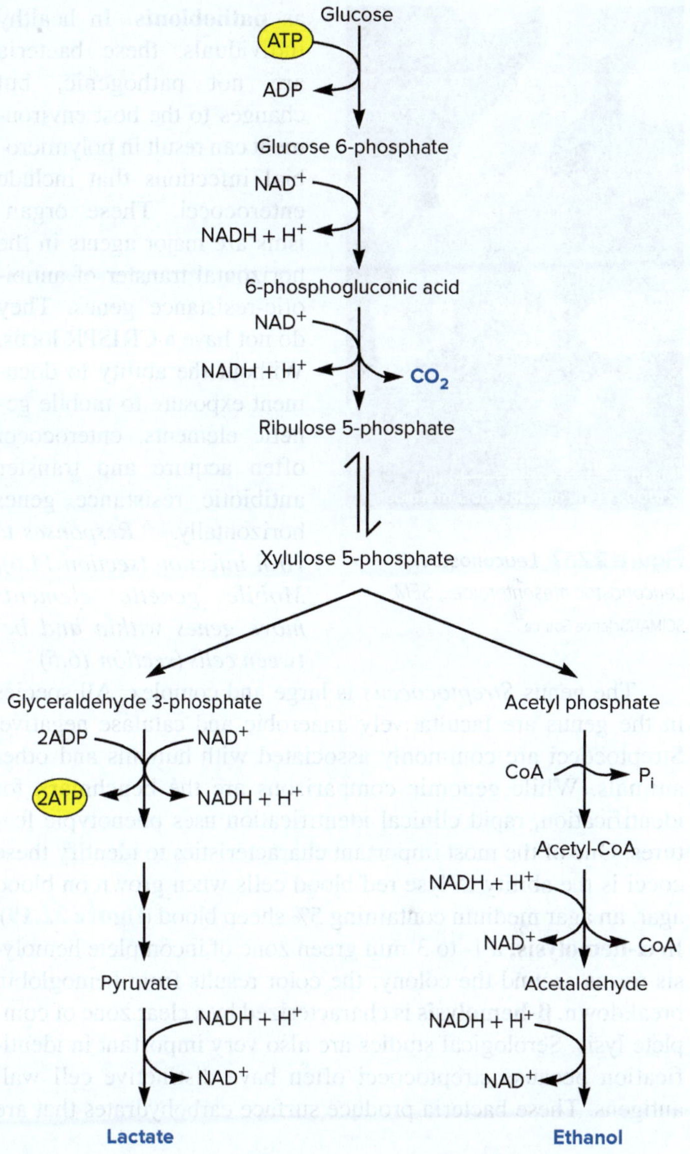

Figure 22.16 Heterolactic Fermentation and the Phosphoketolase Pathway. The phosphoketolase pathway converts glucose to lactate, ethanol, and CO_2. In the branch to form lactate, the steps from glyceraldehyde 3-phosphate to pyruvate is the Embden-Meyerhof pathway (*see figure 11.5*).

MICRO INQUIRY *What is the net yield of ATP for each glucose that enters the phosphoketolase pathway?*

manufacture of buttermilk, butter, and cheese. *L. mesenteroides* synthesizes dextran from sucrose. These are important in industrial dextran production and when ingested from fermented foods, they maintain homeostasis in the human gut.

Bacteria in the important genera *Enterococcus* and *Streptococcus* occur in pairs or chains when grown in liquid media (**figure 22.18**). They are unable to respire and carry out homolactic fermentation. Enterococci are normal residents of the intestinal tracts of humans and many other animals. *E. faecalis* and *E. faecium* are opportunistic pathogens that may also be described

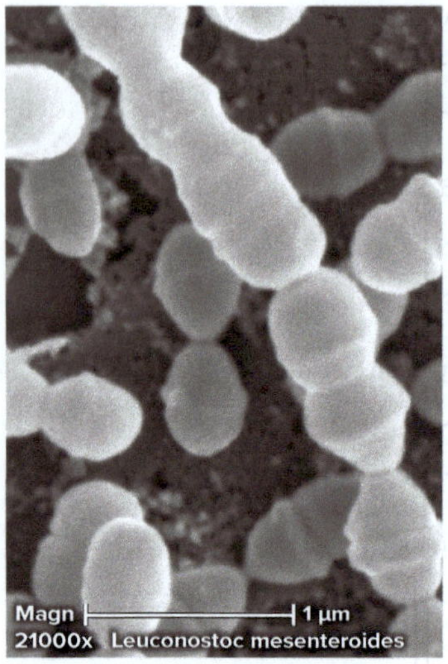

Figure 22.17 ***Leuconostoc.*** *Leuconostoc mesenteroides; SEM.*
SCIMAT/Science Source

as **pathobionts.** In healthy individuals, these bacteria are not pathogenic, but changes to the host environment can result in polymicrobial infections that include enterococci. These organisms are major agents in the horizontal transfer of antibiotic-resistance genes. They do not have a CRISPR locus. Without the ability to document exposure to mobile genetic elements, enterococci often acquire and transfer antibiotic resistance genes horizontally. *Responses to viral infection (section 14.6); Mobile genetic elements move genes within and between cells (section 16.5)*

The genus *Streptococcus* is large and complex. All species in the genus are facultatively anaerobic and catalase negative. Streptococci are commonly associated with humans and other animals. While genomic comparisons are the benchmark for identification, rapid clinical identification uses phenotypic features. One of the most important characteristics to identify these cocci is the ability to lyse red blood cells when grown on blood agar, an agar medium containing 5% sheep blood (**figure 22.19**). In **α-hemolysis,** a 1- to 3-mm green zone of incomplete hemolysis forms around the colony; the color results from hemoglobin breakdown. **β-hemolysis** is characterized by a clear zone of complete lysis. Serological studies are also very important in identification because streptococci often have distinctive cell wall antigens. These bacteria produce surface carbohydrates that are important in species identification through the **Lancefield grouping system,** which divides the streptococci into 20 groups. For example, *S. pyogenes* is often referred to as GAS, for group A streptococcus. Biochemical and physiological tests, and sensitivity to bacitracin, sulfa drugs, and optochin (a quinine analogue), are also used to identify particular species. Blood Agar

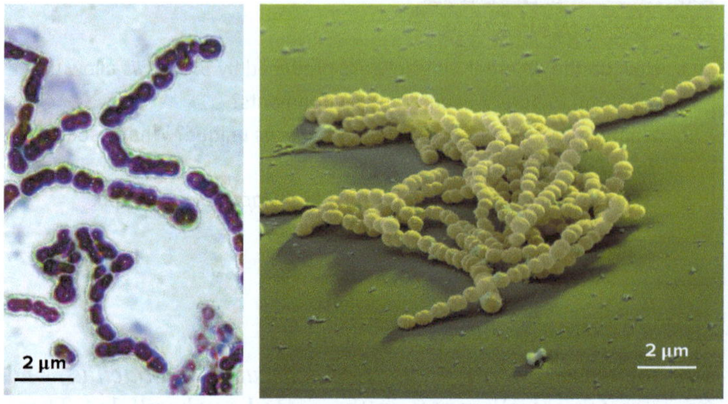

(a) *S. pyogenes* **(b)** *S. agalactiae*

Figure 22.18 ***Streptococcus.*** (a) *S. pyogenes*. (b) *S. agalactiae,* the cause of group B streptococcal infections. Note the long chains of cells; color-enhanced SEM.
(a) James Cavallini/Science Source; (b) Eye of Science/Science Source

The streptococci are divided into three groups based on pathogenicity: pyogenic streptococci, oral streptococci, and other streptococci. Pyogenic streptococci are pathogens associated with pus formation. Most species are β-hemolytic and form chains of cells. The major human pathogen in this group is *S. pyogenes,* which causes streptococcal sore throat. If untreated, it can lead to rheumatic fever. *S. pyogenes* also causes a variety of skin infections that may give rise to glomerulonephritis.

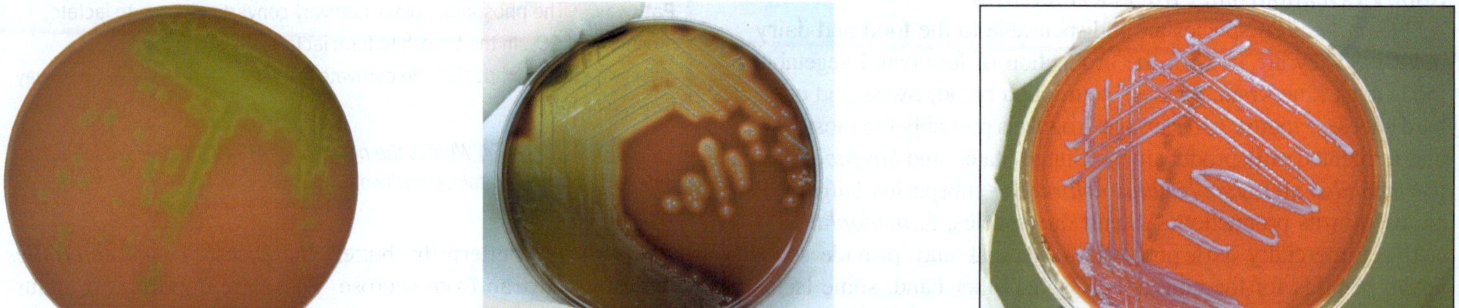

(a) α-Hemolysis **(b)** β-Hemolysis **(c)** No hemolysis

Figure 22.19 Staphylococcal and Streptococcal Hemolytic Patterns. (a) *Staphylococcus aureus* on blood agar, illustrating α-hemolysis. (b) *Streptococcus pyogenes* on blood agar, illustrating β-hemolysis. (c) *Staphylococcus epidermidis* on blood agar with no hemolysis, also referred to as γ-hemolysis.
(a) Matt Mezydlo; (b) Linde Stewart/iStock/Getty Images; (c) James Redfearn/McGraw Hill

MICRO INQUIRY *Why is there a green zone around α-hemolytic colonies?*

Oral streptococci are part of the normal microbiota of the mouth and upper respiratory tract of humans and other animals. *S. mutans* is associated with the formation of dental caries. *S. pneumoniae* is α-hemolytic and grows as pairs of cocci. It is associated with lobar pneumonia and otitis media (inflammation of the middle ear) in young children. *Streptococcal diseases (sections 38.1 and 38.3; Streptococcal pneumonia (section 38.6))*

Order Clostridiales

The order Clostridiales contains two important families, Clostridiaceae and Peptostreptococcaceae, which feature, respectively, the genera *Clostridium* and *Clostridioides*. Both genera contain obligately anaerobic, fermentative endospore formers (**figure 22.20**).

Clostridium spp. have a great practical impact. Because they are anaerobic and form heat-resistant endospores, they are responsible for many cases of food spoilage, even in canned foods. During anaerobic protein decomposition, clostridia often ferment amino acids via the **Stickland reaction** (**figure 22.21**). In this process, one amino acid is oxidized to produce pyruvate, which is then fermented to yield acetate and ATP. A different amino acid serves as an electron acceptor, so it is reduced, thereby reoxidizing NADH to NAD^+ and forming two additional acetate molecules. Glycine and alanine are particularly important to *Clostridium* spp. because they act as germinants, signals that indicate a favorable local nutritional landscape to trigger spore germination. Both amino acids are also abundant in peptidoglycan. The Stickland reaction also generates ammonia, hydrogen sulfide, and fatty acids. These products are responsible for unpleasant odors arising during putrefaction.

Although some clostridia are industrially valuable (e.g., *C. acetobutylicum* is used to manufacture butanol), the pathogenic species produce well-known toxins. For instance, *C. botulinum* and *C. tetani* produce the botulism and tetanus toxins, respectively. These diseases are toxin-mediated, meaning that the disease results from the presence of the toxin, not from the presence of the bacteria. *Botulism (section 38.4); Tetanus (section 38.3)*

C. tetani ferments a number of amino acids to lactate, butyrate, and H_2 (**figure 22.22**). It uses an ATP-binding cassette transporter and a V-type ATPase to couple ATP hydrolysis to the flow of sodium ions out of the cell. V-type ATPases pump ions out of the cell at the expense of ATP made by substrate-level phosphorylation during fermentation. This is not a futile exercise because it creates a sodium motive force, which in turn drives amino acid and sugar uptake systems. *Primary and secondary active transport (section 3.2)*

Clostridioides difficile, formerly *Clostridium difficile,* is often abbreviated C. diff. It is a human pathogen responsible for

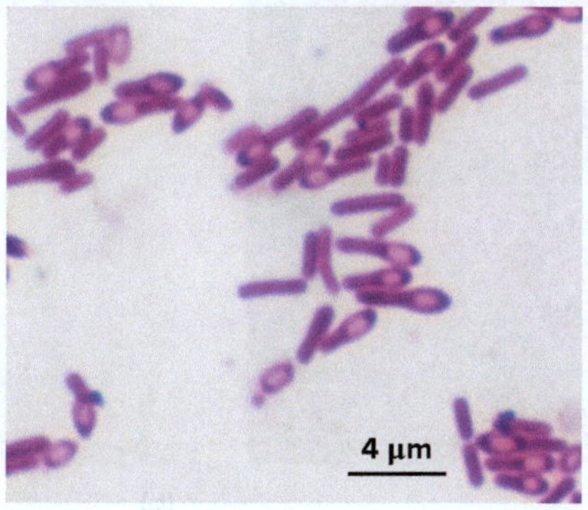

Figure 22.20 ***Clostridium tetani.*** This pathogen makes endospores that are round and terminal.

Lisa Burgess/McGraw Hill

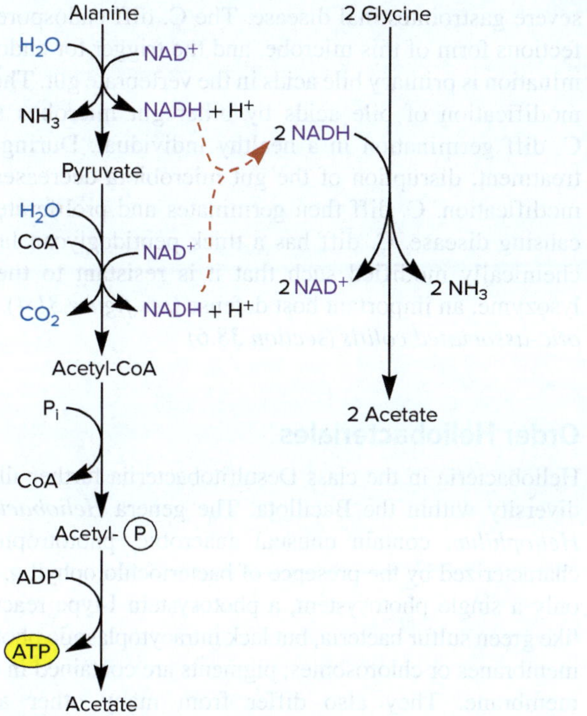

Figure 22.21 The Stickland Reaction. Alanine is oxidized to acetate, and glycine is deaminated to acetate to reoxidize the NADH generated during alanine fermentation. The fermentation also produces ATP.

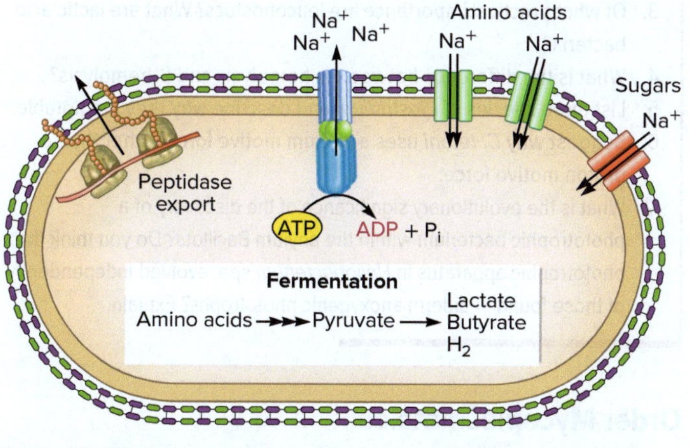

Figure 22.22 Amino Acid Fermentation and Na^+ Flux in *Clostridium tetani.* The cell exports peptidases to hydrolyze proteins. A V-type ATPase pumps Na^+ out of the cell at the expense of ATP hydrolysis. Amino acids are imported by Na^+-dependent uptake systems and fermentation yields lactate, butyrate, and H_2. Sugars are imported but used only in anabolism.

severe gastrointestinal disease. The C. diff endospore is the infectious form of this microbe, and the trigger for endospore germination is primary bile acids in the vertebrate gut. The chemical modification of bile acids by other gut microbes suppresses C. diff germination in a healthy individual. During antibiotic treatment, disruption of the gut microbiota decreases bile acid modification. C. diff then germinates and proliferates, thereby causing disease. C. diff has a thick peptidoglycan layer that is chemically modified such that it is resistant to the effect of lysozyme, an important host defense *(see figure 31.4)*. ▶| *Antibiotic-associated colitis (section 38.6)*

Order Heliobacteriales

Heliobacteria in the class Desulfitobacteriia further illustrate the diversity within the Bacillota. The genera *Heliobacterium* and *Heliophilum* contain unusual anaerobic, phototrophic species characterized by the presence of bacteriochlorophyll *g*. They have only a single photosystem, a photosystem I-type reaction center like green sulfur bacteria, but lack intracytoplasmic photosynthetic membranes or chlorosomes; pigments are contained in the plasma membrane. They also differ from many other anoxygenic phototrophic bacteria because they cannot grow autotrophically. Pyruvate is used during photoheterotrophic growth in light; it is fermented in dark, anoxic conditions. Heliobacteria are free-living in soil, and are capable of nitrogen fixation. Although they have a monoderm cell wall, they have low peptidoglycan content, and stain Gram negative. Some heliobacteria form endospores. |◀ *Light reactions in anoxygenic photosynthesis (section 11.10); Photosynthetic bacteria are diverse (section 20.3)*

Comprehension Check

1. List the major properties that define the genus *Lactobacillus*. In what ways are species in this genus important in the food and dairy industries?
2. How are various strains of *Streptococcus* identified clinically?
3. Of what practical importance are leuconostocs? What are lactic acid bacteria?
4. What is the difference between α-hemolysis and β-hemolysis?
5. List three species of *Clostridium* and describe why they are notable.
6. Suggest why *C. tetani* uses a sodium motive force rather than a proton motive force.
7. What is the evolutionary significance of the discovery of a phototrophic bacterium within the phylum Bacillota? Do you think the phototrophic apparatus in *Heliobacterium* spp. evolved independently of those found in diderm anoxygenic phototrophs? Explain.

Order Mycoplasmatales

Taxonomic placement of the bacteria in the order Mycoplasmatales, commonly called **mycoplasmas,** has been controversial. At times they have been considered Gram negative and at other times, Gram positive. They are neither monoderm nor diderm, as they do not have a cell wall. Genomic analysis places them in the phylum Bacillota. In addition to their lack of a cell wall, these bacteria are characterized by small genomes and simplified metabolic pathways. Because mycoplasmas are bound only by a plasma membrane, they are pleomorphic, varying in shape from spherical or pear-shaped microbes of 0.3 to 0.8 μm in diameter, to slender, branched, or helical filaments. Some mycoplasmas (e.g., *Mycoplasmoides genitalium*) have a specialized terminal structure that projects from the cell and gives them a flask or pear shape (**figure 22.23*a***). This structure aids in attachment to eukaryotic cells. They are among the smallest bacteria capable of self-reproduction. Most species are facultative anaerobes, but a few are obligate anaerobes. When grown on agar, most form tiny colonies with a fried-egg appearance because they grow into the agar surface at the center while spreading outward on the surface at the colony edges (figure 22.23*b*).

Mycoplasmas possess among the smallest bacterial genomes, fewer than 1,000 genes, suggesting a minimal genome size for free-living existence. Their small genome reflects their inability to synthesize a number of macromolecules. For instance, because *M. genitalium* relies so heavily on its host's biosynthetic capacity, it has lost genes that encode enzymes needed to synthesize amino acids, purines, pyrimidines, and fatty acids. In addition, most species require sterols for growth, which they obtain from the host as cholesterol. Sterols are an essential component of the mycoplasma plasma membrane, where they facilitate osmotic stability in the absence of a cell wall. Some mycoplasmas produce ATP by the Embden-Meyerhof pathway and lactic acid fermentation. The pentose phosphate pathway seems to be functional in at least some mycoplasmas; none appears to have the complete tricarboxylic acid cycle. Arginine can be catabolized to generate ATP in the following series of reactions:

$$\text{Arginine} + H_2O \rightarrow \text{citrulline} + NH_3$$
$$\text{Citrulline} + P_i \rightarrow \text{ornithine} + \text{carbamyl-P}$$
$$\text{Carbamyl-P} + \text{ADP} \rightarrow \text{ATP} + CO_2 + NH_3$$

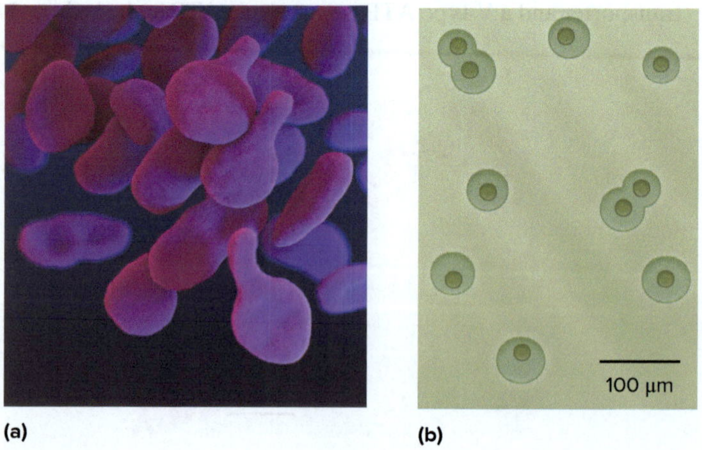

Figure 22.23 Mycoplasma Cells and Colonies. (a) *Mycoplasmoides genitalium* cells are pleomorphic. (b) Note the fried-egg appearance of mycoplasma colonies.

(a) Jennifer Oosthuizen/CDC

Ureaplasma urealyticum, a causative agent of some urinary tract infections, has an interesting way of conserving energy: It hydrolyzes urea (found in urine) to generate an electrochemical gradient via the accumulation of ammonia/ammonium (NH_3/NH_4^+). This gradient is responsible for the chemiosmotic potential that drives ATP synthesis (**figure 22.24**).

Some mycoplasmas are capable of gliding motility that propels them across host cell surfaces. *Mycoplasma mobile* moves at a steady pace of 2 to 5 μm/second. This is mediated by cell surface adhesins, termed *legs,* that surround the neck of the cell (**figure 22.25**). In *M. mobile,* these proteins attach to a contractile protein just inside the plasma membrane. Unlike bacterial flagella that derive energy from the proton motive force, movement in mycoplasmas is powered by ATP hydrolysis. By contrast, most other motile mycoplasmas move more like inchworms, at about 1 μm per second. Motility in *Mycoplasmoides pneumoniae,* the causative agent of walking pneumonia, accounts for its ability to move from bronchial cilia to the host cell surface.

Mycoplasmas are remarkably widespread and can be isolated from animals, plants, and soil. Although their complex growth requirements can make their growth in pure (axenic) cultures difficult, at least 10% of the mammalian cell cultures in use are probably contaminated with mycoplasmas. This seriously interferes with experiments and has prompted the development of rapid test kits for mycoplasma detection in cell culture.

In animals, mycoplasmas colonize mucous membranes and joints, and often are associated with diseases of the respiratory and urogenital tracts. *Spiroplasma* spp. have been isolated from insects, ticks, and a variety of agriculturally important plants. Arthropods often act as vectors and carry the spiroplasmas between plants. In humans, *U. urealyticum* and *Metamycoplasma hominis* are sexually transmitted bacteria that cause inflammation of the reproductive organs.

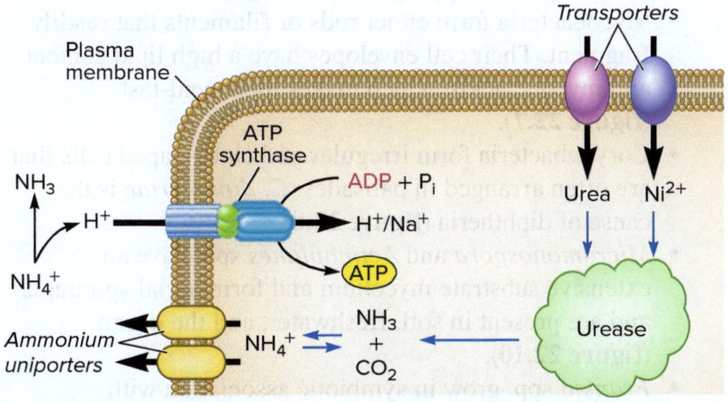

Figure 22.24 Energy Conservation in *Ureaplasma urealyticum*. To generate an electrochemical gradient through the accumulation of NH_3/NH_4^+, *U. urealyticum* imports urea through specific transporters. Once inside the cell, urease (a nickel-binding enzyme) catalyzes the hydrolysis of urea to NH_3 and CO_2. Ammonia accepts a proton to become NH_4^+, which is exported through NH_4^+ uniporters. Outside the cell, NH_4^+ is converted to NH_3 and H^+, providing protons to drive ATP synthesis via the membrane-bound ATP synthase.

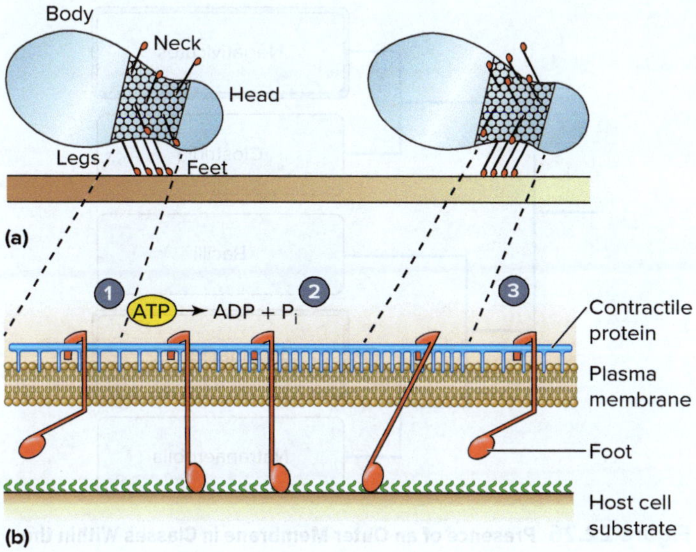

Figure 22.25 Motility in *Mycoplasma mobile*. (a) Leg proteins confer a type of walking motility to the cell. (b) Each leg undergoes a cycle of ATP hydrolysis as it takes a step. (1) ATP binding stiffens the leg and it contacts the oligosaccharide surface of a host epithelial cell. (2) ATP hydrolysis contracts a protein (blue) inside the plasma membrane to move the cell body forward while the foot remains bound to the substrate. (3) Release of ADP and P_i also releases the foot protein from the substrate.

22.3 Gram-Positive Taxonomic Groups Include Some Bacteria with Outer Membranes

After reading this section, you should be able to:

a. Describe the variations in cell envelope structures found in Gram-positive bacteria

b. Discuss evolutionary scenarios to explain the presence of monoderms and diderms within the phylum Bacillota

In Chapter 20, we distinguish Gram-staining behavior from cell envelope structure. The bacteria described in this chapter have been traditionally known as Gram positive and most, but not all, are monoderms. The current genome-based taxonomy places bacteria with variant cell envelopes in the same taxon as those traditionally termed Gram positive. For example, mycobacteria have long been known to synthesize an outer membrane (OM) based on mycolic acid lipids (figure 22.7) and mycoplasmas do not have any envelope layers beyond a sterol-reinforced plasma membrane (figure 22.23). ◀ *Bacterial taxonomy and cell envelopes (section 20.1)*

This organizational scheme that includes a few OM-containing bacteria with the Gram positives (**figure 22.26**) has prompted detailed structural studies of the cells and assessment of possible evolutionary trajectories to explain the implied

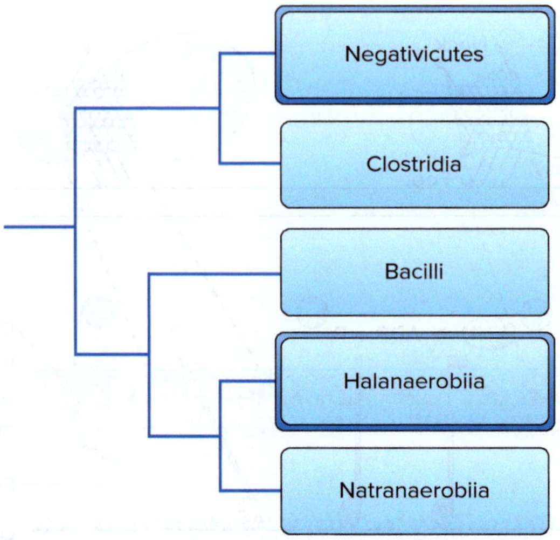

Figure 22.26 Presence of an Outer Membrane in Classes Within the Phylum Bacillota. Classes are organized based on similarities of ribosomal proteins. A thick border indicates diderms.

relationships. The genomes of model bacteria *Veillonella parvula* (Negativicutes) and *Halobacteroides halobius* (Halanaerobiia) contain dozens of genes whose products are required for OM synthesis. Detailed comparisons of these genes establish that they are related, thus supporting the proposal that the common bacterial ancestor was diderm. *Lipopolysaccharides (section 12.7)*

How might the loss of the OM be tolerated in bacteria? A mutant that deposited excess peptidoglycan on its surface, thereby disrupting the OM, might survive and replicate. Alternately, it has been proposed that the endospore-forming members of the genera *Bacillus* and *Clostridium* do not have an OM because the intricate membrane folding processes during endospore engulfment (*see figure 3.39*) require only a single membrane. The evolutionary success of endospore formation may have driven the stabilization of microbes lacking an OM. *Sporulation: making endospores (section 3.8)*

Comprehension Check

1. Design an experiment to explore the roles of sterols in mycoplasma structure and/or physiology.
2. Why is genome reduction common in intracellular microbes?
3. How do you think *Mycoplasma* uses gliding motility in the process of infection?
4. What types of enzymes and structural proteins would indicate the presence of an outer membrane?
5. How might the cell envelope structure of *Mycoplasma* have evolved?

Key Concepts

22.1 Phylum Actinomycetota

- Members of this phylum that form branching, usually nonfragmenting hyphae and asexual spores are collectively called actinomycetes (**figure 22.1**).
- The asexual spores borne on aerial hyphae are called exospores because they do not develop within a mother cell. Some produce sporangiospores.
- Members of the genus *Streptomyces* produce substrate hyphae that differentiate into aerial hyphae-bearing chains of 3 to 50 or more exospores (**figure 22.3**).
- Streptomycetes are important in the degradation of organic material in the soil and produce many useful antibiotics. A few species cause diseases in plants and animals. However, they are most notable for the variety of medically and industrially important compounds they produce (**table 22.1**).
- *Streptomyces* genomes feature biosynthetic gene clusters, large operons that encode the enzymes that catalyze the synthesis of natural products such as antibiotics.
- Members of the genus *Actinomyces* are irregularly shaped, nonsporing rods that can cause disease in animals and humans.
- Members of the genus *Bifidobacterium* are irregular, anaerobic rods and one of the first colonizers of the intestinal tract in nursing babies. These bacteria are often commercialized for their probiotic activity.
- Members of the genus *Micrococcus* form simple coccoid cells. Some are normal inhabitants on human skin.
- Arthrobacters have an unusual rod-coccus growth cycle and carry out snapping division (**figure 22.6**).
- Mycobacteria form either rods or filaments that readily fragment. Their cell envelopes have a high lipid content and mycolic acids, which makes them acid-fast (**figure 22.7**).
- Corynebacteria form irregular and club-shaped cells that are often arranged in palisades. *C. diphtheriae* is the cause of diphtheria (**figure 22.8**).
- *Micromonospora* and *Actinoplanes* spp. have an extensive substrate mycelium and form aerial sporangia and are present in soil, freshwater, and the ocean (**figure 22.10**).
- *Frankia* spp. grow in symbiotic association with nonleguminous plants and fix nitrogen (**figure 22.11**).

22.2 Phylum Bacillota

- The genus *Bacillus* contains aerobic and facultative, endospore-forming, chemoheterotrophic rods that are usually motile with peritrichous flagella. *B. subtilis* is a model organism that forms multiple cell types, including endospores, biofilms, and fruiting bodies (**figure 22.12**).

- *Thermoactinomyces* spp. are thermophilic and form a mycelium and true endospores. *T. vulgaris* causes allergic reactions that lead to farmer's lung.
- Members of the genus *Staphylococcus* are facultatively anaerobic, nonmotile cocci that form irregular clusters. They carry out aerobic respiration; some also ferment and use nitrate as a terminal electron acceptor. *S. aureus* is a pathogen that possesses an array of virulence factors (**figure 22.16**). *S. epidermidis* resides on human skin and establishes barriers and limits growth of pathogens.
- Members of the genus *Lactobacillus* are regular, nonsporing rods that carry out lactic acid fermentation and are extensively used in the food and dairy industries. *Leuconostoc* spp. carry out heterolactic fermentation using the phosphoketolase pathway (**figure 22.16**); they are involved in the production of fermented vegetable and dairy products.
- The genera *Streptococcus* and *Enterococcus* contain cocci arranged in pairs and chains. These are usually facultative anaerobes that carry out fermentation. Some important species are the pyogenic coccus *S. pyogenes, S. pneumoniae* and enterococci *E. faecalis* and *E. faecium*.
- Members of the genus *Clostridium* are anaerobic rods that form endospores. They are responsible for food spoilage, botulism, tetanus, and gas gangrene. They have a fermentative metabolism (**figures 22.20–22.22**).
- *Clostridioides difficile* is a human pathogen responsible for gastrointestinal disease. Its modified peptidoglycan that is not susceptible to lysozyme aids in avoiding the immune system.
- Heliobacteria are anaerobic, phototrophic bacteria with bacteriochlorophyll *g*. Some form endospores.
- Mycoplasmas lack cell walls and cannot synthesize peptidoglycan but require sterols for growth.
- Many mycoplasmas are plant and animal pathogens. They are some of the smallest bacteria capable of self-reproduction.

22.3 Gram-Positive Taxonomic Groups Include Some Bacteria with Outer Membranes

- Genomic taxonomy has changed the traditional groupings, and the Gram positives now include mycoplasmas (no cell wall) and organisms that have an outer membrane. It has been proposed that the ancestral Bacillota had an outer membrane and that it may have been lost to facilitate endospore formation.

Active Learning

1. Given that mycolic acids are essential for mycobacterial viability, how would you test their importance in mycobacterial infection?
2. *Streptomyces coelicolor* is studied as a model system for cellular differentiation. Some of the genes involved in sporulation contain an AT-rich leucine codon not used in vegetative genes. Suggest how *Streptomyces* might use this rare codon to regulate sporulation.
3. Suppose that you discovered a nodulated plant that could fix atmospheric nitrogen. How might you show that a bacterial symbiont was involved and that a *Frankia* sp. rather than a *Rhizobium* sp. was responsible?
4. Some species of *Clostridium* demonstrate much faster doubling times when in a mammalian host than when grown in vitro. Design a demonstration that would explore this phenomenon.
5. What physiological properties might account for the ease with which anaerobic clostridia can be isolated from soil and other generally aerobic niches?
6. *S. aureus* strains resistant to methicillin (MRSA) are also resistant to many other antibiotics and are more virulent than non-MRSA strains. Refer to chapters 16 and 18, and develop a testable hypothesis that predicts how these many traits came to be in a single group of strains.
7. How does a mycelial growth form promote degradation of complex organic substrates?
8. Compare and contrast the division of labor in *Streptomyces* cell types that produce antibiotics with that observed in *Anabaena* cell types that fix nitrogen (*see figure 20.8*).
9. How might you use CRISPR technology (section 17.5) to investigate BCG clusters in *Streptomyces?*

CHAPTER

23

Protists

Filipe Frazao/Shutterstock

Human Evolution Shaped by a Protist

Chagas disease (CD) is transmitted by insects ironically nicknamed "kissing bugs." More accurately, several species of triatomine bugs transmit the causative pathogen, a protist called *Trypanosoma cruzi,* while taking a blood meal from their host. Dubbed kissing bugs because they often alight on and bite a sleeping individual's face, the protist is transmitted when the bugs defecate. This stimulates the victim to scratch the site, and the protist enters through microabrasions in the skin.

CD is also called American trypanosomiasis because it is endemic to 21 countries in the Americas where triatomine bugs reside. It currently infects about 6 million people, with about 12,000 deaths per year. If diagnosed early, CD can be cured. However, early infection is often asymptomatic as *T. cruzi* enters muscle tissue, particularly in the heart and gut. Up to 20 years later, a person may have developed irreversible heart and gastrointestinal tract damage—both of which can be deadly.

CD is an ancient disease, as shown by evidence of its presence in 7,000-year-old South American mummies. Indigenous peoples in the region have coexisted with CD for so long that natural selection appears to offer some protection. Genome data from over 100 individuals belonging to 19 indigenous groups throughout the southern Amazon stored in the Human Genome Diversity Project database illustrate their coevolution with *T. cruzi*. Variants for several genes whose products are related to cardiovascular function and metabolism appear more frequently in these populations as compared to others, suggesting that living in a CD endemic area selected for these alleles. Two genes in particular stand out: *PPP3CA* and *DYNC1I1* promote a host's immune response to CD, with the *PPP3CA* variant thought to protect heart cells during *T. cruzi* infection.

CD and its role in human natural selection exemplify the importance of protists. Despite this, protists suffer from a confused taxonomic history. The kingdom Protista, as presented in many biology texts, is an artificial grouping of over 64,000 different single-celled life forms that lack common evolutionary heritage; that is, they are not monophyletic (*see figure 1.3*). In this text, we refer to these organisms simply as **protists.** The term **protozoa** (s., protozoan; Greek *protos,* first; *zoon,* animal) refers to chemoorganotrophic protists. The term **algae** describes photosynthetic protists that, like plant cells, possess a cell wall. The study of algae is often referred to as **phycology** and is the realm of both botanists and protistologists (also called protozoologists). The study of all protists, regardless of their metabolic type, is called **protistology,** although the term **protozoology** is more commonly used.

Readiness Check:

Based on what you have learned previously, you should be able to:

- ✓ Draw a eukaryotic cell envelope and identify the structure and function of its component layers (section 5.2)
- ✓ Identify the structure and function of eukaryotic cellular organelles and cytoskeleton filaments (sections 5.3 through 5.6)
- ✓ Describe the endocytic pathway in general terms (section 5.4)
- ✓ Distinguish a mitochondrion from a hydrogenosome (section 5.6)
- ✓ Describe the structure and function of eukaryotic flagella and cilia (section 5.7)

23.1 Protist Diversity Reflects Broad Phylogeny

After reading this section, you should be able to:

a. Describe three nutritional strategies used by protists
b. Draw a typical protist cell membrane and underlying features, labeling the plasmalemma, ecto- and endoplasm, and pellicle
c. Explain how contractile vacuoles are used to maintain osmotic stability
d. Trace the path of a food particle as it is consumed by a protist
e. Compare binary fission as it occurs in a protist with that in a bacterium or archaeon
f. Outline the process of sexual reproduction and how it can vary in protists

Most eukaryotes are microbes, and most eukaryotic microbes are protists. Their abundance is so great that global protist biomass is thought to exceed that of all animals. Their vast diversity is a function of their capacity to thrive in a wide variety of habitats. One common requirement among protists is moisture because all are susceptible to desiccation. Most known protists are free living and inhabit freshwater or marine environments. Many terrestrial chemoorganotrophic forms can be found in decaying organic matter and soil. Whether terrestrial or aquatic, protists play an important role in nutrient cycling. ▶| *Biogeochemical cycling sustains life on Earth (section 28.1)*

Some protozoa, or chemoorganoheterotrophic protists, are **saprophytes,** securing nutrients from dead organic material by releasing degradative enzymes into the environment. They then absorb the soluble products—a process sometimes called **osmotrophy.** Many protozoa employ **holozoic nutrition,** in which solid nutrients are acquired by phagocytosis. Photoautotrophic protists are strict aerobes and perform oxygenic photosynthesis (*see figure 11.30*). Symbiotic protists include both protozoan and algal forms; they depend on larger, multicellular hosts for nutrients. Finally, it is difficult to classify the nutritional strategies of some protists because they simultaneously use reduced organic molecules and CO_2 as carbon sources. This strategy is called **mixotrophy.**

Protist Morphology

Protists can be smaller than bacteria or, in the case of giant seaweeds, taller than a 10-story building. Despite their diversity, protists share some common features. In many respects, their morphology and physiology are the same as the cells of multicellular plants and animals. However, because many protists are unicellular, all of life's various functions must be performed within a single cell. Those that are multicellular, like giant kelp (seaweed), lack highly differentiated tissues. The structural complexity observed in protists arises at the level of specialized organelles.

The protist cell membrane is called the **plasmalemma** and is identical to that of multicellular organisms (**figure 23.1**).

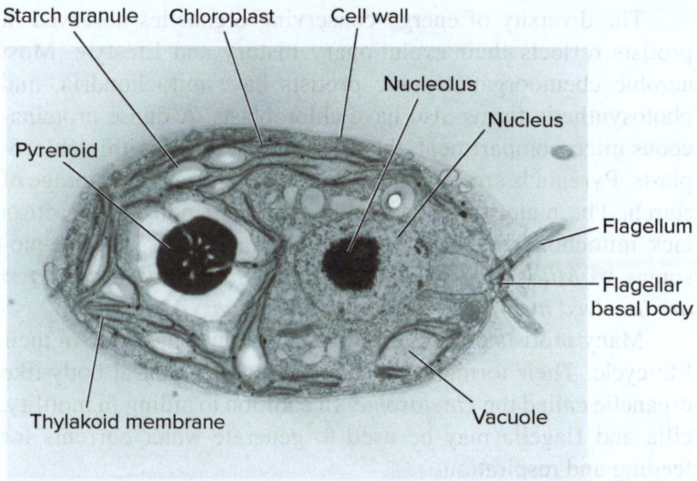

Figure 23.1 Protist Structure. A transmission electron micrograph of the photosynthetic protist *Chlamydomonas reinhardtii.*
Biophoto Associates/Science Source

In some protists, the cytoplasm immediately under the plasmalemma is divided into an outer gelatinous region called the **ectoplasm** and an inner fluid region, the **endoplasm.** The ectoplasm imparts rigidity to the cell body. Many protists also have a supportive layer or cuticle called the **pellicle.** The pellicle consists of the plasmalemma and a relatively rigid layer just beneath it. The pellicle may be simple in structure. For example, *Euglena* spp. are protists with a series of overlapping proteinaceous strips with a ridge at the edge of each strip fitting into a groove on the adjacent one, much like the tongue-and-groove boards of a hardwood floor. In contrast, the pellicles of ciliate protists are protein complexes with two membranes and a variety of associated structures. Although pellicles are not as strong and rigid as cell walls, for those that possess them, pellicles serve a protective function and impart the characteristic shape associated with that particular species.

One or more vacuoles are usually present in the cytoplasm of protists. These include contractile, secretory, and food or phagocytic vacuoles. **Contractile vacuoles** are osmoregulatory organelles in protists that live in hypotonic environments, such as freshwater lakes. They maintain osmotic balance by continuous water expulsion. Secretory vacuoles, or vesicles, are discussed in chapter 5 (section 5.4). **Phagocytic vacuoles** are conspicuous in protists that ingest food by phagocytosis (holozoic protists). Phagocytic vacuoles are the sites of food digestion. In some organisms, they may occur anywhere on the cell surface, while others have a specialized structure for phagocytosis called the **cytostome** (cell mouth). Once the cytostome has engulfed a food particle, digestion commences when the phagocytic vacuole becomes acidic. As digestion proceeds, the vacuole membrane forms small blebs that pinch off and carry nutrients throughout the cytoplasm. The undigested contents of the original phagocytic vacuole are expelled from the cell either at a random site on the cell membrane or at a designated position called the **cytoproct** (Greek *proktos,* anus).

The diversity of energy-conserving organelles observed in protists reflects their evolutionary history and lifestyle. Most aerobic chemoorganotrophic protists have mitochondria, and photosynthetic forms also have chloroplasts. A dense proteinaceous microcompartment, the **pyrenoid,** is found within chloroplasts. Pyrenoids are associated with the synthesis and storage of starch. The majority of anaerobic chemoorganotrophic protists lack mitochondria; some of these organisms have hydrogenosomes. ◀ *Mitochondria, related organelles, and chloroplasts are involved in energy conservation (section 5.6)*

Many protists possess cilia or flagella at some point in their life cycle. Their formation is associated with a basal body-like organelle called the *kinetosome.* In addition to aiding in motility, cilia and flagella may be used to generate water currents for feeding and respiration.

Encystment and Excystment

Actively growing and replicating protists are called **trophozoites,** which roughly means "little animals that eat." Many are also capable of forming dormant cysts, in a process called **encystment.** The cyst is morphologically simpler than the trophozoite and has a cell wall even if the trophozoite form of the protist lacks one. Cyst formation is particularly common among aquatic protists and parasitic forms. Cysts serve three major functions: (1) They protect against adverse changes in the environment, such as nutrient deficiency, desiccation, adverse pH, and low levels of O_2; (2) they are sites for nuclear reorganization and cell division (reproductive cysts); and (3) in parasitic species they are the infectious stage between hosts. Protists escape from cysts by a process called **excystment** (**figure 23.2**). Although the exact stimulus for excystment is unknown for most protists, it is generally triggered by a return to favorable environmental conditions. For example, cysts of parasitic species excyst after ingestion by the host.

Protist Reproductive Cells and Structures

Most protists have both asexual and sexual reproductive phases in their life cycles. The most common method of asexual reproduction is **binary fission.** During this process, the nucleus first undergoes **mitosis** and then the cytoplasm divides by cytokinesis to form two identical individuals (**figure 23.3**). Multiple fission is also common, as is budding. Some filamentous, photosynthetic protists undergo fragmentation so that each piece of the broken filament contains a nucleus and grows independently.

Sexual reproduction involves the formation of gametes. Protist cells that produce gametes are termed **gamonts.** The fusion of haploid gametes is called **syngamy.** Among protists, syngamy can involve the fusion of two morphologically similar gametes (isogamy) or two morphologically different types (anisogamy). **Meiosis** may occur before the formation and union of gametes, as in most animals, or just after fertilization, as is the case with lower plants. Furthermore, the exchange of nuclear material may occur in the familiar fashion—between two different individuals (**conjugation**)—or by the development of a genetically distinct nucleus within a single individual (autogamy).

With this level of reproductive complexity, perhaps it is not surprising that the nuclei among protists show considerable diversity. Most commonly, a vesicular nucleus is present. This is 1 to 10 μm in diameter and spherical, and has a distinct nucleolus and uncondensed chromosomes (figure 23.1). The uncondensed chromosomes give these nuclei a lacy or vesicular appearance when viewed microscopically. Ovular nuclei are up to 10 times this size and possess many peripheral nucleoli. Still others have chromosomal nuclei, in which the chromosomes remain condensed throughout the cell cycle. Finally, many ciliated forms

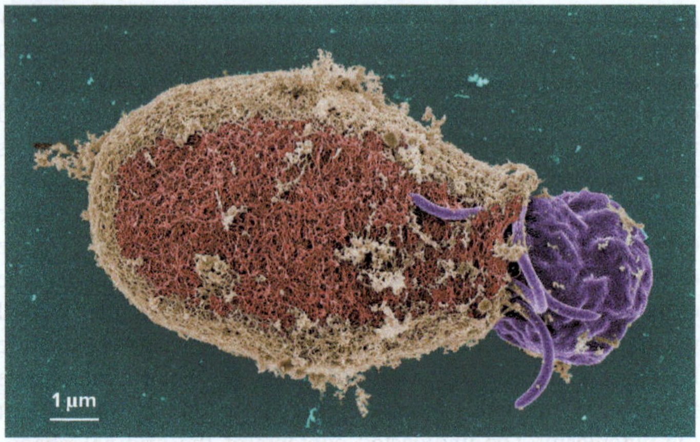

Figure 23.2 Excystment. This colorized scanning electron micrograph shows the protist *Giardia* sp. undergoing excystment. The flagellated vegetative cell, called a trophozoite, is emerging from the right of the cyst.

Dr. Stan Erlandsen/CDC

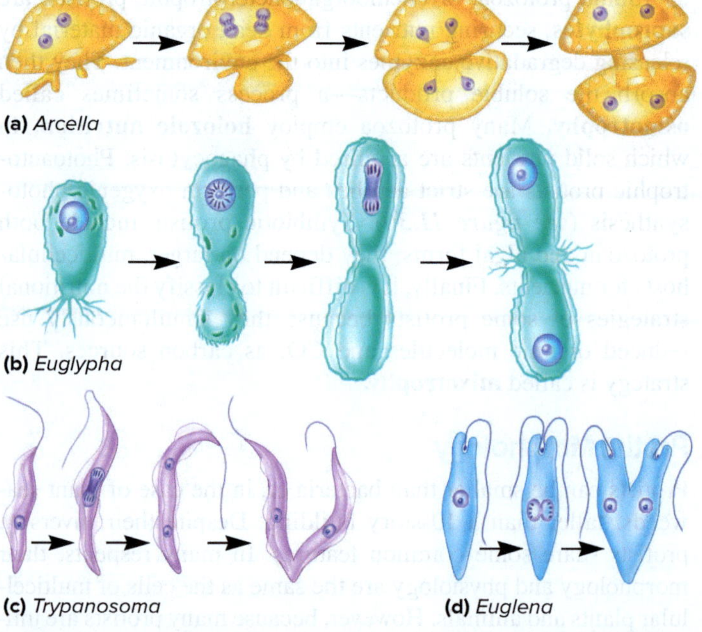

Figure 23.3 Binary Fission in Protists. (a) The two nuclei of a testate (shelled) amoeba in the genus *Arcella* divide as some of its cytoplasm is extruded and a new test for the daughter cell is secreted. (b) For members of another testate amoeba genus, *Euglypha,* the nucleus divides while platelets (siliceous scales) are used to construct the test for the daughter cell. For many protists, all organelles must be replicated and segregated before the cell divides. This is the case with species of (c) *Trypanosoma* and (d) *Euglena*.

have two types of nuclei: a large macronucleus with distinct nucleoli and condensed chromatin, and a smaller, diploid micronucleus with dispersed chromatin but lacking nucleoli. Genes in the macronucleus encode proteins involved in housekeeping functions and binary fission, whereas genes on the micronucleus function only in sexual reproduction.

Protist Taxonomy Is Evolving

Ever since Antony van Leeuwenhoek described the first "animalcule" in 1674, the taxonomic classification of protists has remained in flux. During the twentieth century, classification schemes were based on morphology and protists were classified into four major groups based on their means of locomotion: flagellates (*Mastigophora*), ciliates (*Infusoria* or *Ciliophora*), amoebae (*Sarcodina*), and stationary forms (*Sporozoa*). Although these terms are still used, they describe morphology and are without evolutionary context. While it is now agreed that the old classification system is best abandoned, little agreement exists on what should take its place. In the early 2000s, a system of supergroups was developed based on shared biological features. As more protist genomes were sequenced, phylogenomic analysis was possible. **Phylogenomics** involves aligning as many genes as possible across as many species as possible to assess evolutionary relationships. This analysis has reshaped our understanding of protist diversity and evolution. Here we present well-known **clades** of protists of medical and/or environmental importance.

Comprehension Check

1. Describe the pellicle. What is its function?
2. Describe the difference between a cytostome to cytoproct.
3. What function does a cyst serve for a typical protist? What causes excystment to occur?
4. Construct a concept map that summarizes sexual reproduction in protists.
5. Describe vesicular, ovular, and chromosomal nuclei. Which is most like the nuclei found in higher eukaryotes?
6. How has phylogenomics influenced the classification of protists?

23.2 Discoba–Metamonada Clade

After reading this section, you should be able to:

a. Identify a location in your community where you might find protists belonging to the Discoba–Metamonada clade
b. List one human and one veterinary disease caused by protists belonging to the Discoba–Metamonada clade
c. Draw a typical euglenoid cell and explain the function of the stigma and pyrenoid
d. Discuss the etiology of Chagas disease and African sleeping sickness
e. Describe two pathogenic metamonads

Members of the Discoba–Metamonada clade include some of the most primitive eukaryotes. Most possess a cytostome characterized by a suspension-feeding groove with a posteriorly directed flagellum used to generate a feeding current. This enables the capture of small particles. Those that lack this morphological feature are presumed to have had it at one time during their evolution, but it was secondarily lost.

Discoba

Members of the Discoba include a variety of free-living and parasitic flagellated protists as well as some amoebae. Euglenozoa are important members of this clade. They are commonly found in freshwater, although a few species are marine. About one-third are photoautotrophic; the remaining are free-living chemoorganotrophs. Most chemoorganotrophic forms are saprophytic, although a few parasitic species have been described. Members of the genus *Euglena* are photoautotrophic. A typical *Euglena* spp. cell (**figure 23.4**) is elongated and bounded by a plasmalemma. The pellicle consists of proteinaceous strips and microtubules; it is elastic enough to enable turning and flexing of the cell, yet rigid enough to prevent excessive alterations in shape. Chloroplasts contain chlorophylls *a* and *b* together with carotenoids. The large nucleus contains a prominent nucleolus. A red eyespot called a **stigma** helps the organism orient to light and is located near an anterior reservoir. A contractile vacuole near the reservoir continuously collects water from the cell and empties it into the reservoir, thus regulating the osmotic pressure within the organism. Two flagella arise from the base of the reservoir—although only one may fully emerge—and actively beat to move the cell. Reproduction in euglenids is by longitudinal binary fission (figure 23.3).

Trypanosomes are important members of Euglenozoa. These microbes exist only as parasites of plants and animals and have global significance. For example, members of the genus *Leishmania,* which cause a group of diseases collectively termed leishmaniasis, are typanosomes that affect between 700,000 and 1 million people annually. Chagas disease is caused by *Trypanosoma cruzi,* which is transmitted by "kissing bugs" (*Triatominae*), so called because they bite the face of sleeping

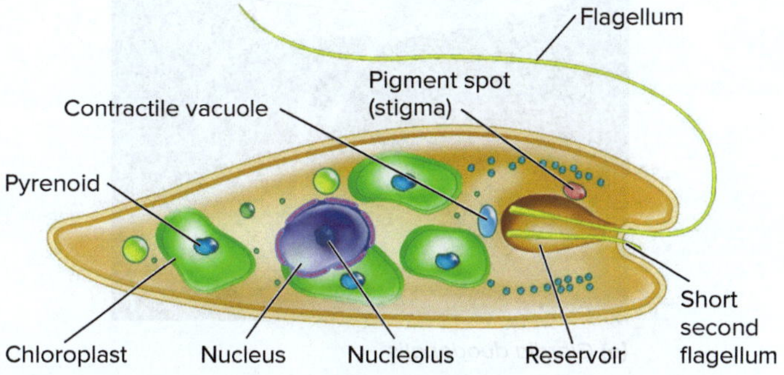

Figure 23.4 ***Euglena.*** Cells range from about 35–75 μm in length. In this image, the short second flagellum does not emerge from the anterior reservoir. In some euglenids, both flagella are emergent.

MICRO INQUIRY *What are the functions of the stigma and the reservoir?*

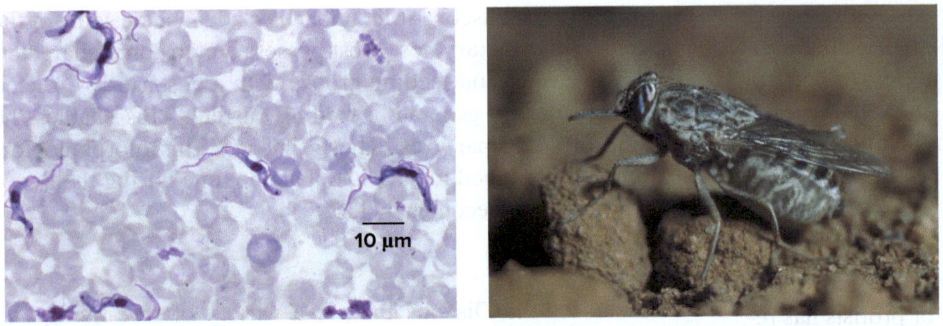

(a) (b)

Figure 23.5 ***Trypanosoma* and Its Insect Host.** (a) *Trypanosoma* among red blood cells. Note the dark-staining nuclei, anterior flagella, and undulating, changeable shape. (b) The tsetse fly is an important vector of the *Trypanosoma* species that causes African sleeping sickness.

(a) Dr. Mae Melvin/CDC; (b) Oxford Scientific/Getty Images

victims (chapter opener; *see figures 39.7 to 39.10*). ▶| *Leishmaniasis (section 39.3); Trypanosomiasis (section 39.3)*

The trypanosomes *T. brucei* subspecies *rhodesiense* and subspecies *gambiense* cause African sleeping sickness (**figure 23.5**). Ingestion of these parasites by the blood-sucking tsetse fly triggers a complex cycle of development and reproduction, first in the fly's gut and then in its salivary glands (*see figure 39.9*). From there, the parasite is transferred to a vertebrate host, where it often causes a fatal infection. Efforts to control the tsetse fly have dramatically reduced the disease burden. However, the presence of this dangerous parasite prevents the use of about 11 million square kilometers of African grazing land.

Metamonada

Ever curious, van Leeuwenhoek described *Giardia lamblia* from his own diarrheic feces. Over 300 years later, this species, now known as *G. duodenalis* (**figure 23.6*a***), continues to be a public health concern. Today this flagellated protist infects campers and other individuals who unwittingly consume contaminated water. It is the most common cause of epidemic waterborne diarrheal disease in the United States with about 200 million cases globally each year.

Like all members of this clade, *G. duodenalis* is anaerobic. It lacks mitochondria and conserves energy by substrate-level phosphorylation in hydrogenosomes (*see figures 5.14 and 5.16*). It also has double-bounded membrane organelles called **mitosomes.** Although these primitive organelles share ancestry with mitochondria and hydrogenosomes, they are not involved in ATP synthesis. Instead the mitosomes of *G. duodenalis* are involved in producing FeS proteins. |◀ *The aerobic mitochondrion belongs to the family of mitochondrial-related organelles (section 5.6)*

Cyst formation is key to *G. duodenalis* survival. Ingestion of only about 10 cysts is needed for infection. Within the small intestine, bile and slight alkalinity trigger excystment (figure 23.2). Trophozoites then attach to gut cells called enterocytes. When these enterocytes die and are sloughed from the gut, the microbe must "decide" whether to attach to another enterocyte or encyst and exit to find a new host. ▶| *Giardiasis (section 39.5)*

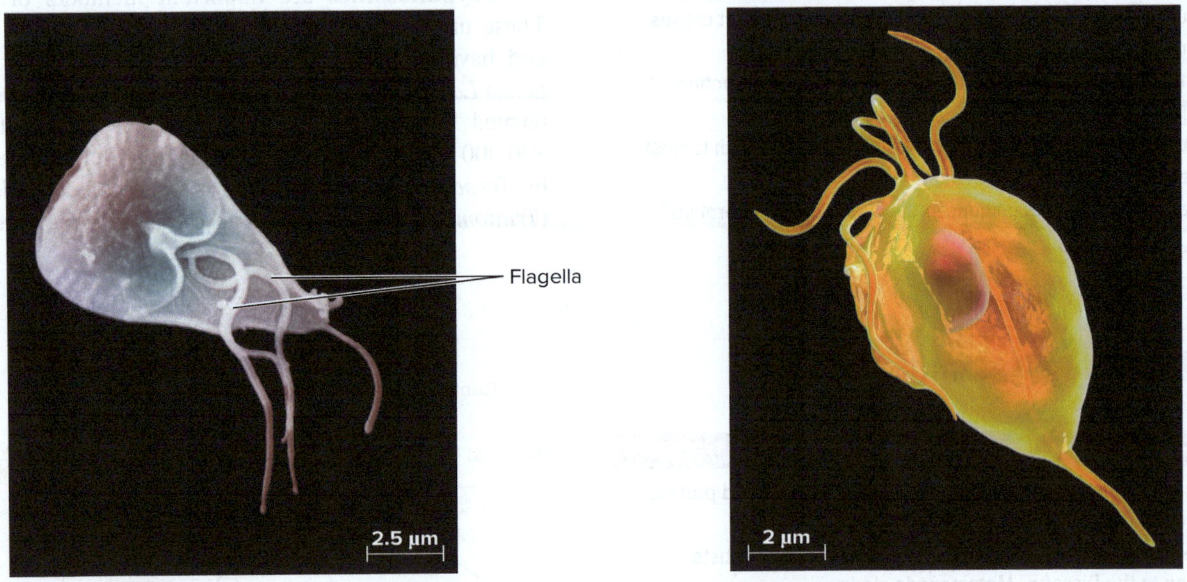

(a) *Giardia duodenalis* (b) *Trichomonas vaginalis*

Figure 23.6 Parasitic Metamonada. These specialized parasitic flagellates absorb nutrients from their hosts.

(a) Science Source; (b) Kateryna Kon/Shutterstock

MICRO INQUIRY *How are these two protists morphologically similar but physiologically different?*

The Metamonoda also includes two subgroups: Trichonymphida and Trichomonadida. Trichonymphids are obligate mutualists in the digestive tracts of wood-eating insects such as termites and wood roaches. They secrete the enzyme cellulase needed for the digestion of wood particles, which they entrap with pseudopodia. One species, *Trichonympha campanula,* can account for up to one-third of the biomass of an individual termite (*see figure 27.2b*). Although asexual reproduction is the norm, a hormone called ecdysone produced by the host when molting triggers sexual reproduction. ▶| *Microorganism-insect mutualisms (section 27.2)*

Trichomonadida spp., or simply trichomonads, are symbionts of the digestive, reproductive, and respiratory tracts of many vertebrates. *Tritrichomonas foetus* is a cattle parasite and an important cause of spontaneous abortion in these animals. Four species infect humans: *Dientamoeba fragilis, Pentatrichomonas hominis, Trichomonas tenax,* and *T. vaginalis. D. fragilis* infection in the human gut can be asymptomatic or can cause a range of gastrointestinal problems. *T. vaginalis* is found in the genitourinary tract of both males and females, and causes the sexually transmitted infection trichomoniasis. Because this protist never lives outside the host, it does not have a cyst stage in its life cycle. ▶| *Trichomoniasis (section 39.4)*

Unlike *G. intestinalis* and *T. vaginalis,* most metamonads are harmless symbionts. The few free-living forms are most often found in waters that are heavily polluted with organic nutrients (a condition known as eutrophication). Other pathogenic species include *Hexamida salmonis,* a fish parasite found in hatcheries and fish farms, and *H. meleagridis*. The latter along with the unrelated protist *Histomonas meleagridas* are turkey pathogens responsible for the annual loss of millions of poultry.

Comprehension Check

1. List at least two features that distinguish the Metamonada–Discoba clade.
2. What is the function of the plasmalemma and pellicle in euglenid cells? How are these structures similar to the bacterial cell wall?
3. What is the difference between a hydrogenosome and a mitosome?

23.3 Amoebozoa Clade Includes Protists with Pseudopodia

After reading this section, you should be able to:

- **a.** Draw three different types of pseudopodia
- **b.** Differentiate between a naked and a testate amoeba
- **c.** Compare and contrast the structure, motility, and life cycles of acellular and cellular slime molds
- **d.** Explain the importance of molecular signaling in the morphological development of *Dictyostelium discoideum*
- **e.** Identify a nearby location where you might find examples of *Tubulinea, Entamoebida,* and *Eumycetozoa*

Because the amoeboid form arose independently numerous times from various flagellated ancestors, amoebae are placed in two clades: Amoebozoa and Rhizaria. One of the morphological hallmarks of amoeboid motility is the use of **pseudopodia** (Latin *pseudo,* false, and *podium,* feet) for both locomotion and feeding (**figure 23.7**). Pseudopodia can be rounded (**lobopodia**) or long and narrow (**filopodia**), or form a netlike mesh (**reticulopodia**). Amoebae that lack a cell wall and are surrounded only by a plasma membrane are called **naked amoebae.** In contrast, the plasma membrane of a shelled or **testate amoeba** is covered by material that is either made by the protist itself or collected by the organism from the environment. Binary fission is the usual means of asexual division, although some amoebae form cysts that undergo multiple fission (figure 23.3).

Tubulinea

Members of Tubulinea inhabit almost any environment where they can remain moist; this includes glacial meltwater, marine environments, tide pools, lakes, and streams. Free-living forms are known to dwell in ventilation ducts and cooling towers, where they feed on microbial biofilms. Others are endosymbionts, commensals, or parasites of invertebrates, fishes, and mammals. Some harbor intracellular symbionts, including algae, bacteria, and viruses, but the nature of these relationships is not well understood. *Amoeba proteus,* a favorite among introductory biology laboratory instructors, is included in this group.

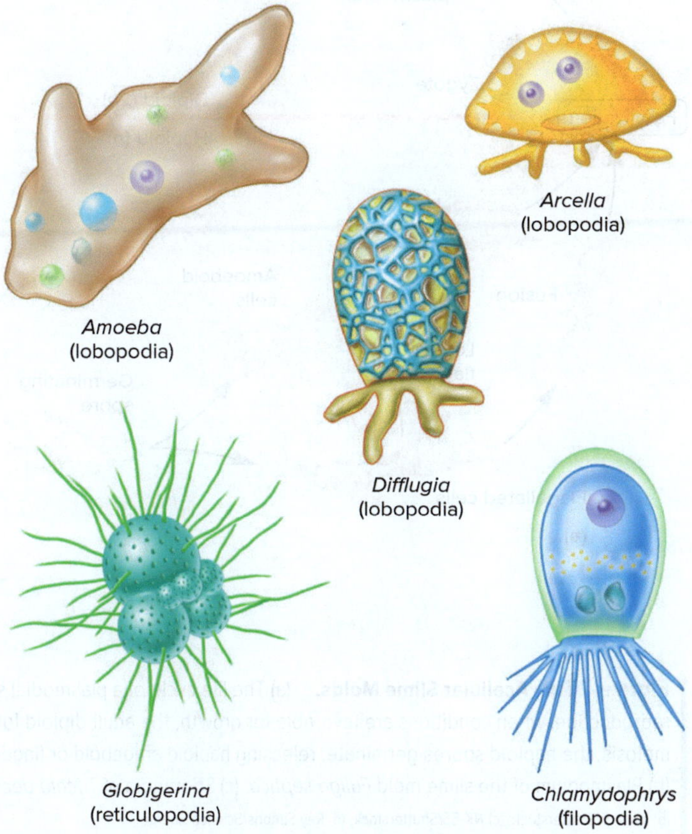

Figure 23.7 Types of Pseudopodia Found on Amoebozoa.

Entamoebida

Like *G. duodenalis,* entamoebid protists lack mitochondria and contain mitosomes. However, the identification of about 20 putative mitochondrial proteins in the free-living amoeba *Mastigamoeba balamuthi* suggests that these mitosomes are more advanced than those found in *Giardia* spp. *M. balamuthi* is a relative of *Entamoeba histolytica,* the cause of amoebic dysentery—the third-leading cause of parasitic death worldwide. In fact, all *Entamoeba* spp. are either parasites like *E. histolytica* or commensals—that is, living in, but causing no harm to, their hosts. ▶| *Amebiasis (section 39.5)*

Eumycetozoa

First described in the 1880s, members of Eumycetozoa or slime molds have been classified as plants, animals, and fungi. As we examine their morphology and behavior, the source of this confusion should become apparent. Phylogenomic analysis as well as physiological, behavioral, biochemical, and developmental data point to a monophyletic group of protists.

The life cycle of **acellular slime molds,** exemplified by those in the genus *Myxogastria,* includes a distinctive stage when the organism exists as streaming masses of colorful protoplasm-cytosol and organelles without a cell membrane. The protoplasm creeps along in amoeboid fashion over moist, rotting logs, leaves, and other organic matter, which it degrades (**figure 23.8*a***). Acellular slime molds derive their name from the stage in their life cycle when a large, multinucleate mass called a *plasmodium* is formed; there can be as many as 10,000 synchronously dividing nuclei within a single plasmodium (figure 23.8*b*). When starved or dried, the plasmodium develops ornate fruiting bodies. As these mature, they form stalks with cellulose walls that are resistant to environmental stressors (figure 23.8*c,d*). When conditions improve, spores germinate and release haploid flagellated cells. These fuse, and as the resulting zygotes feed by endocytosis, nuclear divisions give rise to the multinucleate plasmodium.

Cellular slime molds (Dictyostelia) use endocytosis to feed on bacteria and yeasts. The species *Dictyostelium discoideum* is an attractive model organism. During its life cycle individual cells cooperate to form different morphologies.

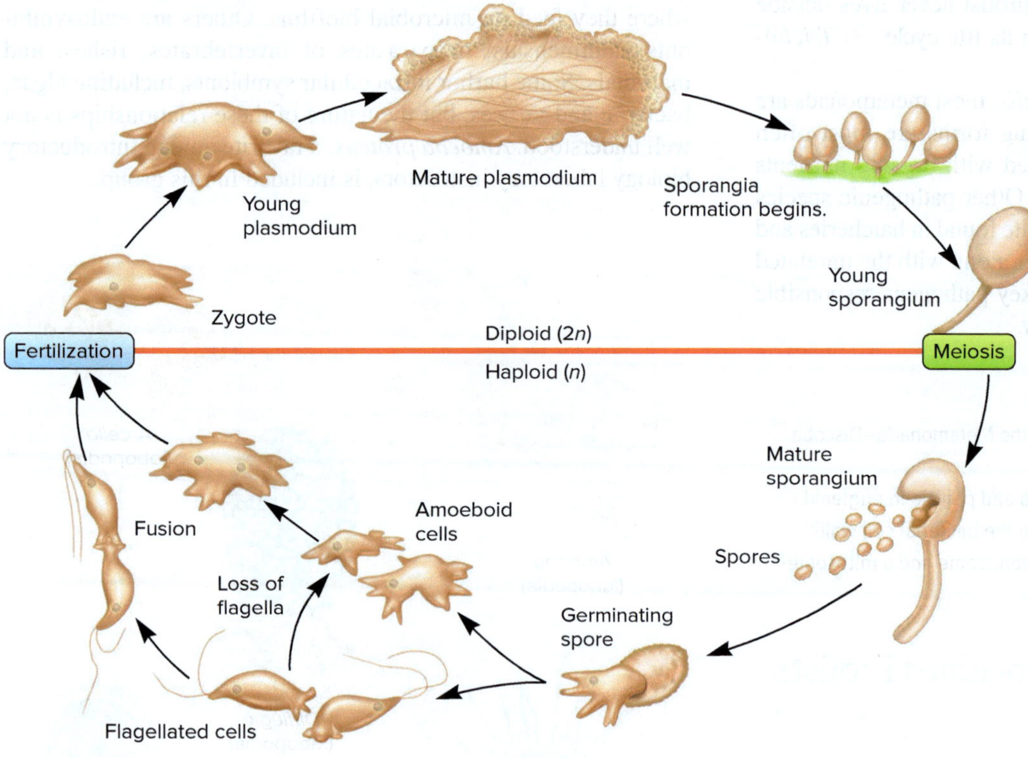

(a)

(b) *Fuligo septica*

(c) *Trichia decipiens*

(d) *Stemonitis*

Figure 23.8 Acellular Slime Molds. (a) The life cycle of a plasmodial slime mold includes sexual reproduction; when conditions are favorable for growth, the adult diploid forms sporangia. Following meiosis, the haploid spores germinate, releasing haploid amoeboid or flagellated cells that fuse. (b) Plasmodium of the slime mold *Fuligo septica.* (c) Sporangia of *Trichia decipiens* and (d) *Stemonitis* sp.

(b) iwciagr/Shutterstock; (c) NK-55/Shutterstock; (d) Ray Simons/Science Source

MICRO INQUIRY *How does a plasmodium differ from a single cell?*

Multicellularity is demonstrated in the pseudoplasmodium, which consists of thousands of individual vegetative cells moving together as a mass (**figure 23.9**, step 2). The pseudoplasmodium forms when starved cells release the signaling molecules cyclic AMP (cAMP) and a specific glycoprotein. Other *D. discoideum* cells sense these compounds and respond by forming an aggregate around the signal-producing cells. In this way, large, motile, multicellular slugs develop and serve as precursors to fruiting body formation (figure 23.9, step 3). Fruiting body morphogenesis commences when the slug stops and cells pile on top of each other. Cells at the bottom of this vertically oriented structure form a stalk by secreting cellulose, while cells at the tip differentiate into spores (figure 23.9, steps 4 and 5). Germinated spores become vegetative amoebae to start this asexual cycle a new (figure 23.9, step 6).

Dictyostelium spp. display complex behaviors. For example, they differentiate to resemble primitive immune cells. During slug formation, some cells become sentinel cells and roam within the slug vanquishing pathogenic bacteria such as *Legionella pneumophila,* which infect *Dictyostelium* spp. Sentinel cells accomplish this by producing proteins similar to those involved in immune responses in higher organisms.

Sexual reproduction in *D. discoideum* involves the formation of special spores called *macrocysts*. These arise by a form of conjugation that has some unusual features. First, a group of amoebae become enclosed within a wall of cellulose. This brings cells into close contact so that conjugation can occur between members of different mating types. There are three mating types determined by the sequence of a single gene, as well as those capable of self-fertilization. Following conjugation, a large amoeba forms and cannibalizes the remaining amoebae. The now-giant amoeba matures into a macrocyst. Macrocysts can remain dormant within their cellulose walls for extended periods. Vegetative growth resumes after the diploid nucleus undergoes meiosis to generate haploid amoebae.

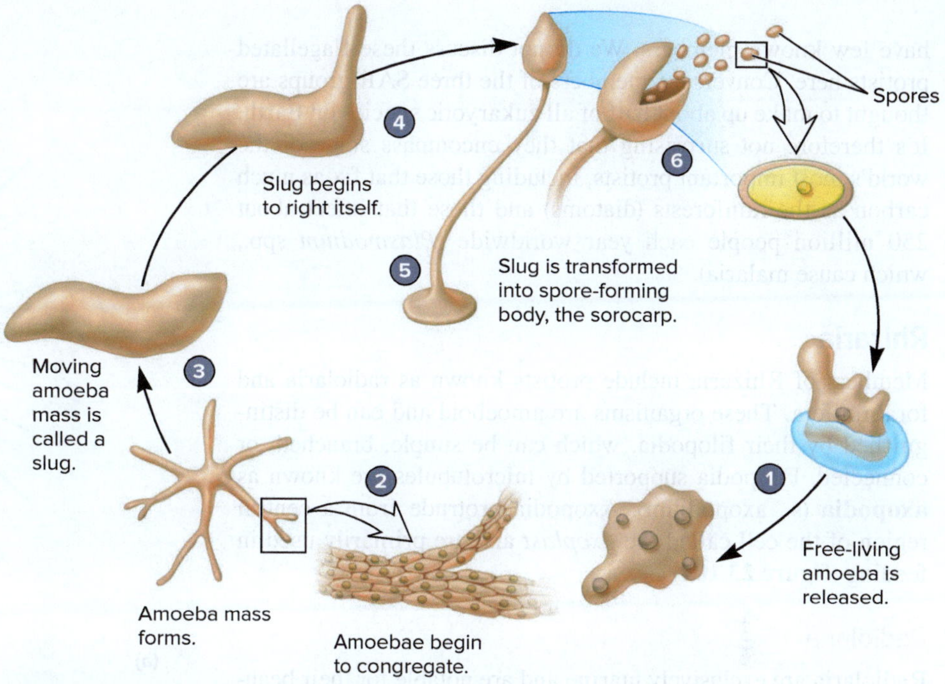

Figure 23.9 Life Cycle of *Dictyostelium discoideum*, a Cellular Slime Mold.

MICRO INQUIRY *How would a mutant strain unable to secrete cAMP behave if mixed with a wild-type strain? What if it were mixed with another cAMP mutant?*

Comprehension Check

1. Describe filopodia, lobopodia, and reticulopodia form and function.
2. Why have slime molds been so hard to classify?
3. What is a plasmodium?
4. Describe the life cycle of *Dictyostelium discoideum*.

23.4 TSAR Clade: Protists of Global Importance

After reading this section, you should be able to:

- **a.** Describe the morphology of the filopodia and axopodia found in members of Rhizaria
- **b.** Explain the morphology, feeding mechanisms, and reproduction of members of Radiolaria and Foraminifera
- **c.** Describe the morphology, motility, and feeding behaviors of dinoflagellates
- **d.** Explain the importance of dinoflagellates to coral biology and the generation of red tides
- **e.** Diagram how a ciliated protist captures and consumes a food particle
- **f.** Compare and contrast the morphology and function of the micro- and macronucleus and trace their fate during conjugation
- **g.** Identify the unique structures found in an apicomplexan cell and explain their function
- **h.** List at least two human diseases caused by apicomplexans
- **i.** Explain the importance of heterokont flagella in the taxonomy of the Stramenopila
- **j.** Review the structure, morphology, reproduction, and ecological importance of diatoms
- **k.** Discuss the agricultural and historical significance of peronosporomycetes

TSAR stands for telonemids, stramenopiles, alveolates, and Rhizaria. Telonemids have only recently been recognized and have few known members. We do not discuss these flagellated protists here. Conversely, members of the three SAR groups are thought to make up about half of all eukaryotic species on Earth. It's therefore not surprising that they encompass some of the world's most important protists, including those that fix as much carbon as the rainforests (diatoms) and those that infect about 250 million people each year worldwide (*Plasmodium* spp., which cause malaria).

Rhizaria

Members of Rhizaria include protists known as radiolaria and foraminifera. These organisms are amoeboid and can be distinguished by their filopodia, which can be simple, branched, or connected. Filopodia supported by microtubules are known as **axopodia** (s., axopodium). Axopodia protrude from a central region of the cell called the *axoplast* and are primarily used in feeding (**figure 23.10*a***).

Radiolaria

Radiolaria are exclusively marine and are notable for their beautifully intricate skeletons (figure 23.10*b*). Most have an internal skeleton made of siliceous material; however, members of the subgroup Acantharia have endoskeletons consisting of strontium sulfate. A few genera contain species with an exoskeleton of siliceous spines or scales, while some lack a skeleton completely. Skeletal morphology is highly variable and often includes radiating spines that help the organisms float, as does the storage of oils and other low-density fluids.

Radiolaria feed by endocytosis, using mucus-coated filopodia to entrap prey including bacteria, other protists, and even small invertebrates. Large prey items are partially digested extracellularly by the secretion of degradative enzymes before becoming encased in a food vacuole. Many surface-dwelling radiolaria have algal symbionts that enhance their net carbon assimilation. Asexual reproduction (binary or multiple fission, or budding) is most common in radiolaria, but sexual reproduction can be triggered by either a heavy feeding or nutrient limitation. In this case, two haploid nuclei fuse to form a diploid zygote encased in a cyst from which it is released when survival conditions improve.

Foraminifera

Members of Foraminifera (called forams for short) range in size from roughly 20 μm to several centimeters. Their filopodia are arranged in a branching network called reticulopodia (Latin *reticulum*, net or mesh; **figure 23.11**). These bear vesicles at their tips that secrete a sticky substance used to trap prey. Many species harbor endosymbiotic algae that migrate out of reticulopodia (without being eaten) to expose themselves to more sunlight. These algae contribute significantly to foram nutrition. Forams have characteristic tests (plates), usually made of calcite ($CaCO_3$), arranged in multiple chambers that are sequentially added as the protist grows. Their ability to sequester CO_2 as calcite contributes to the removal of this greenhouse gas from the atmosphere.

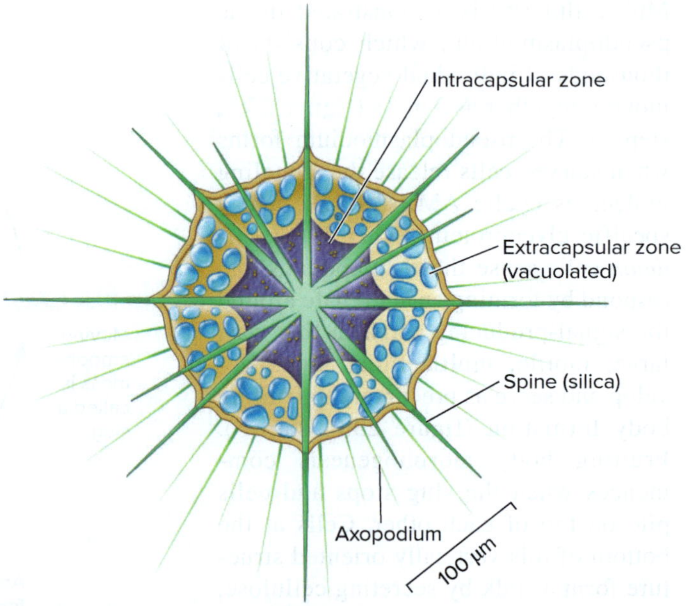

(a)

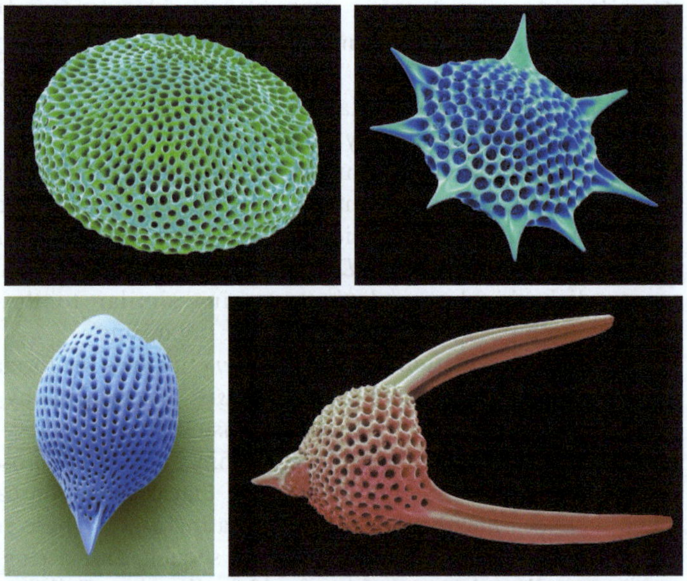

(b)

Figure 23.10 Radiolaria. (a) The radiolarian *Acanthometra elastica* demonstrates the internal skeleton. Endoplasm is within the intracapsular zone; the vacuolated extracapsular zone contains ectoplasm. (b) Radiolarian shells made of silica. Radiolaria range in size from 100 to 200 μm (excluding spines).

(b) (left) Science Photo Library/Alamy Stock Photo; (right) Steve Gschmeissner/Science Photo Library/Alamy Stock Photo

Some forams are free-floating planktonic forms, but most are benthic, living within the sediments. A few benthic forams inhabiting anoxic sediments perform denitrification, reducing nitrate completely to dinitrogen gas. Foraminifera were the first example of a denitrifying eukaryote; it is now known that some diatoms and fungi are also capable of this process. ▶| *Nitrogen cycle (section 28.2)*

Figure 23.11 A Foraminiferan. The reticulopodia project through pores in the calcareous tests of this protist.
David Caron/Science Source

Foraminiferan life cycles can be complex. While some smaller species reproduce only asexually by budding or multiple fission, larger forms frequently alternate between sexual and asexual phases. During the sexual phase, flagellated gametes pair, fuse, and generate asexual individuals (agamonts). Meiotic division of agamonts gives rise to haploid gamonts. There are several mechanisms by which gamonts become diploid again. For instance, a variety of forams release flagellated gametes that become fertilized in the open water. In others, two or more gamonts attach to one another, enabling gametes to fuse within the chambers of the paired tests. When the shells separate, newly formed agamonts are released.

Forams are found in marine and estuarine habitats. Their tests accumulate on the seafloor, where they create a fossil record dating back to the Early Cambrian (543 million years ago). Their remains can be up to hundreds of meters deep in some tropical regions. Foram tests make up most modern-day chalk, limestone, and marble, and formed the famous White Cliffs of Dover in England (**figure 23.12**). The stones used to build the great pyramids of Egypt are also made of forams.

Alveolata

Alveolata is a large group that includes three important taxa: Dinoflagellata (dinoflagellates), Ciliophora (ciliates), and Apicomplexa (apicomplexans). We begin our discussion with

Figure 23.12 White Cliffs of Dover. The limestone that forms these cliffs is composed almost entirely of fossil shells of protists, including foraminifera.
elena1977/123RF

dinoflagellates—a large group most commonly found in marine plankton, where some species are responsible for the phosphorescence sometimes seen in seawater. Their nutrition is complex; photoautotrophy, heterotrophy, and mixotrophy are all observed. Most of the heterotrophic forms are saprophytic (using osmotrophy to assimilate nutrients), but some also use endocytosis. All dinoflagellates have two distinctively placed flagella: One is wrapped around a transverse groove (the girdle), and the other is draped in a longitudinal groove (the sulcus; **figure 23.13**). The orientation and beating patterns of these flagella cause the cell to spin as it is propelled forward; the name dinoflagellate is derived from the Greek *dinein,* to whirl. Many dinoflagellates are covered with cellulose plates and are said to be *thecate* or armored. Those without plates are called *athecate* or naked. Like some other protists, dinoflagellates have specialized compressed thread-like structures called **trichocysts.** When attacked, trichocysts shoot out from the cell as a means of defense.

Most dinoflagellates are free living, although some form important associations with other organisms. Endosymbiotic dinoflagellates that live as undifferentiated cells occasionally send out motile cells called **zooxanthellae.** The most well-known zooxanthellae belong to the genus *Symbiodinium.* These are photosynthetic endosymbionts of reef-building coral (*see figure 27.3*). They provide fixed carbon to the coral animal and help maintain the internal chemical environment needed for the coral to secrete its calcium carbonate exoskeleton. Dinoflagellates are also responsible for toxic red tides that harm other organisms, including humans. Common red tide dinoflagellates include members of the genera *Alexandrium* and *Lingulodinium* (*see figure 29.3*). *The coral holobiont (section 27.2); Microorganisms in coastal ecosystems are adapted to a changeable environment (section 29.2)*

Ciliates (Ciliophora) include about 8,000 known species. All are chemoorganotrophic and range from about 10 μm to 4.5 mm long. They inhabit both benthic and planktonic

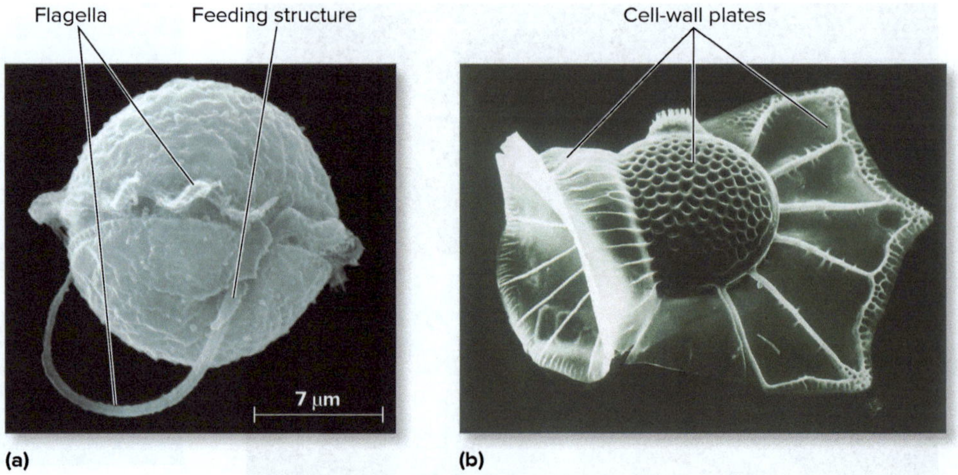

Figure 23.13 Dinoflagellates. (a) The freshwater dinoflagellate *Peridiniopsis berolinensis* displays both types of flagella. One flagellum coils around the girdle; as it moves, this flagellum causes the cell to spin. The straight flagellum that extends from the cell acts as a rudder to determine the direction of movement. (b) Marine dinoflagellates in the genus *Ornithocercus* contain cellulose cell-wall plates. The cell wall extends to form sail-like structures.

(a) Lee W. Wilcox, Ph.D.; (b) Claude Taylor, III

MICRO INQUIRY *What do you think is the function of the "sails" in* Ornithocercus *sp.?*

communities in marine and freshwater systems, as well as moist soils. As their name implies, they employ many cilia for locomotion and feeding. The cilia are generally arranged either in longitudinal rows (**figure 23.14**) or in spirals around the body of the organism. They beat with an oblique stroke, causing the protist to revolve as it swims. Ciliary beating is so precisely coordinated that ciliates can go both forward and backward. There is great variation in shape, and most ciliates do not look like the slipper-shaped members of the genus *Paramecium.* Some species, including *Vorticella* spp., attach to substrates by a long stalk. *Stentor* spp. attach to substrates and stretch out in a trumpet shape to feed. Some species have trichocysts, while others possess toxicysts. Toxicysts are similar to trichocysts but discharge toxic substances to capture prey. A striking feature of ciliates is their ability to quickly entrap many particles by the action of the cilia. Food enters the cytostome and passes into food vacuoles that fuse with lysosomes after detachment from the cytostome and the vacuole's contents are digested. After the

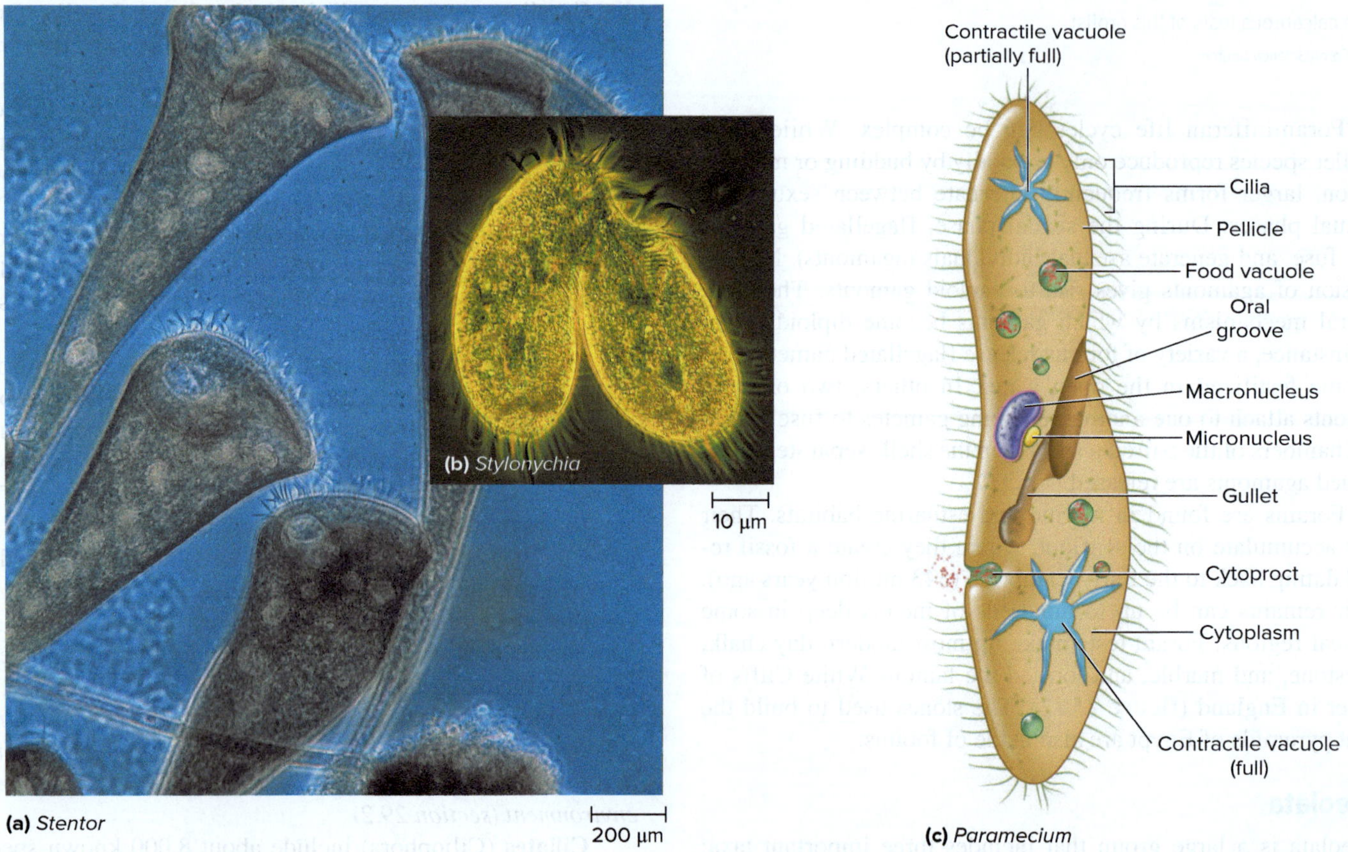

Figure 23.14 ***Ciliophora.*** (a) *Stentor,* a large, vase-shaped, freshwater protozoan. (b) Two *Stylonychia* cells conjugating. (c) Structure of *Paramecium* spp.

(a,b) Eric Grave/Science Source

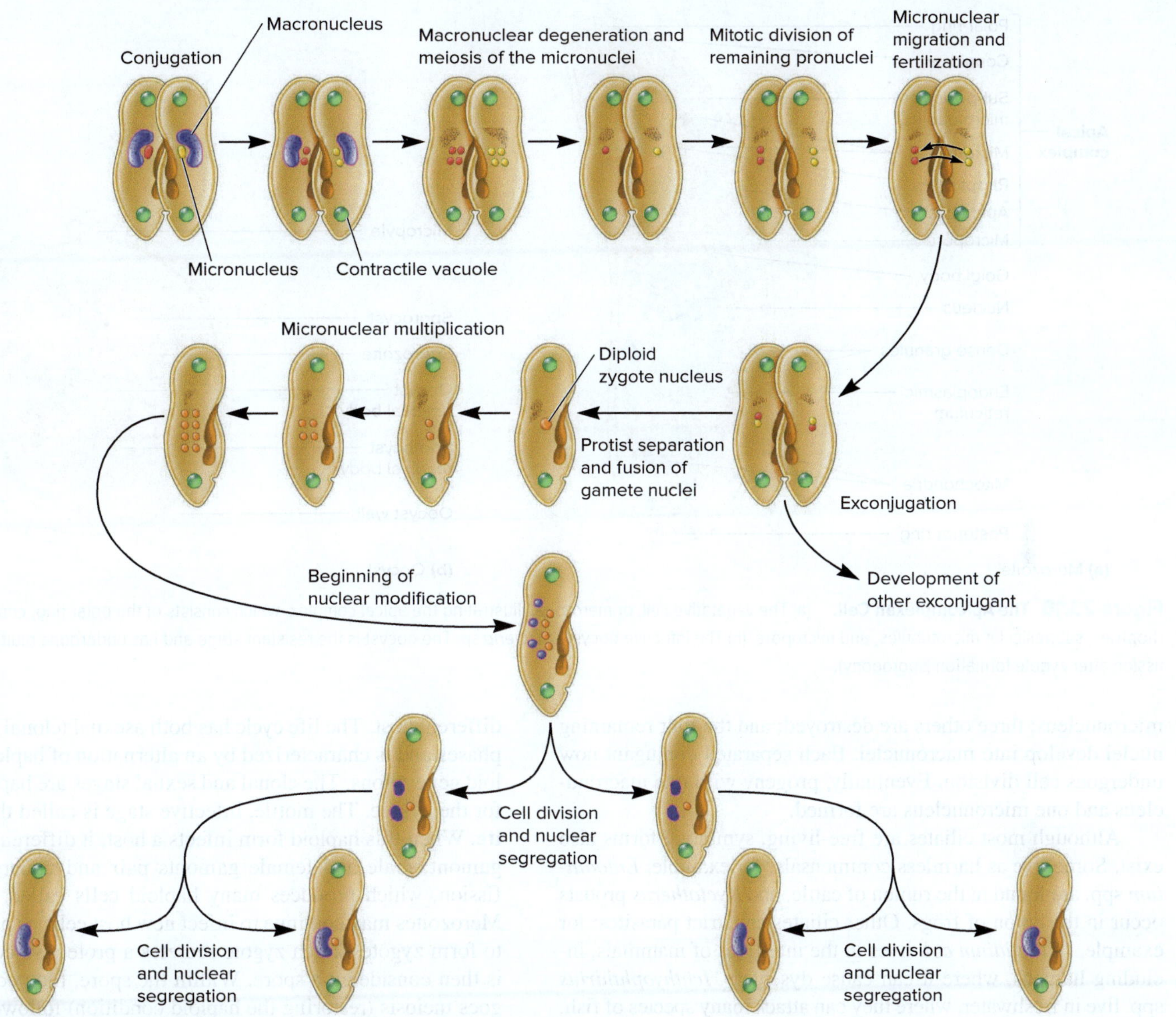

Figure 23.15 Conjugation in *Paramecium caudatum*. After the conjugants separate, only one of the exconjugants is shown; however, a total of eight new protists result from each conjugation.

MICRO INQUIRY *What are the functions of the micronucleus and macronucleus?*

digested material has been absorbed into the cytoplasm, the vacuole fuses with the cytoproct and waste material is expelled.

Most ciliates have two types of nuclei: a large **macronucleus** and a smaller **micronucleus.** The micronucleus is diploid and contains the normal somatic chromosomes. It divides by mitosis and transmits genetic information through meiosis and sexual reproduction. Macronucleus gene expression is needed to maintain routine cellular functions and control normal cell metabolism. Macronuclei are derived from micronuclei by a complex series of steps. Within the macronucleus are many chromatin bodies, each containing many copies of only one or two genes. Macronuclei are thus polyploid and divide by elongating and then constricting.

The fate of the macro- and micronuclei can be followed during conjugation of *Paramecium caudatum* (**figure 23.15**). In this process, there is an exchange of gametes between paired cells of complementary mating types (conjugants). At the beginning of conjugation, two ciliates fuse their pellicles at the site of contact. The macronucleus is no longer needed, so it is degraded in each cell. The individual micronuclei undergo meiosis to form four haploid pronuclei, three of which disintegrate. The remaining pronucleus divides again mitotically to form two gametic nuclei—a stationary one and a migratory one. The migratory nuclei pass into the respective conjugants. Then the ciliates separate, the gametic nuclei fuse, and the resulting diploid zygote nucleus undergoes three rounds of mitosis. The eight resulting nuclei have different fates: One nucleus is retained as a

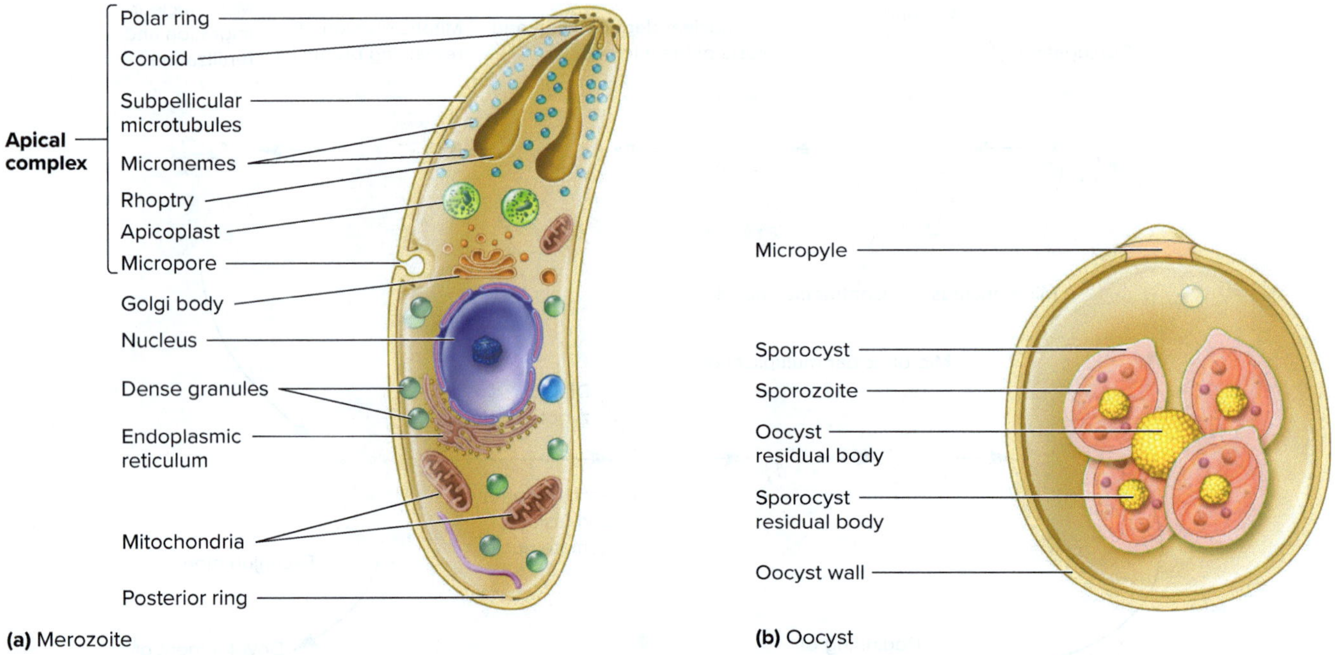

Figure 23.16 The Apicomplexan Cell. (a) The vegetative cell, or merozoite, illustrating the apical complex, which consists of the polar ring, conoid, rhoptries, subpellicular microtubules, and micropore. (b) The infective oocyst of *Eimeria* sp. The oocyst is the resistant stage and has undergone multiple fission after zygote formation (sporogony).

micronucleus; three others are destroyed; and the four remaining nuclei develop into macronuclei. Each separated conjugant now undergoes cell division. Eventually, progeny with one macronucleus and one micronucleus are formed.

Although most ciliates are free-living, symbiotic forms also exist. Some live as harmless commensals; for example, *Entodinium* spp. are found in the rumen of cattle, and *Nyctotherus* protists occur in the colon of frogs. Other ciliates are strict parasites; for example, *Balantidium coli* lives in the intestines of mammals, including humans, where it can cause dysentery. *Ichthyophthirius* spp. live in freshwater, where they can attack many species of fish, producing a disease known as "ick."

All **apicomplexans** are parasites of animals and are distinguished by a unique arrangement of fibrils, microtubules, vacuoles, and other organelles, collectively called the **apical complex,** which is located at one end of the cell (**figure 23.16*a***). There are three secretory organelles—the rhoptries (pronounced "rope-trees"), micronemes, and dense granules—that release enzymes and calcium needed to penetrate host cells. In addition, apicomplexans (with the exception of *Cryptosporidium* spp.) contain plastids, called **apicoplasts.** Although these arose by endosymbiosis of an ancient cyanobacterium, unlike chloroplasts, apicoplasts are not the site of photosynthesis. In the malaria-causing pathogen *Plasmodium* spp., apicoplasts are known to be the site of acetyl-CoA and isoprenoid biosynthesis. In all apicomplexans studied to date, the apicoplast is essential for viability. This makes them attractive targets for the development of drugs to inhibit the growth of pathogenic apicomplexans. *Sterols and isoprenoid lipids (section 12.7)*

Apicomplexans have complex life cycles in which certain stages sometimes occur in one host and other stages occur in a different host. The life cycle has both asexual (clonal) and sexual phases and is characterized by an alternation of haploid and diploid generations. The clonal and sexual stages are haploid, except for the zygote. The motile, infective stage is called the **sporozoite.** When this haploid form infects a host, it differentiates into a gamont; male and female gamonts pair and undergo multiple fission, which produces many haploid cells called *merozoites.* Merozoites may continue to infect new host cells or pair and fuse to form zygotes. Each zygote secretes a protective covering and is then considered a spore. Within the spore, the nucleus undergoes meiosis (restoring the haploid condition) followed by mitosis to generate eight sporozoites ready to infect a new host (figure 23.16*b*). Motility (flagellated or amoeboid) is confined to the gametes and zygotes of a few species.

Among the most important human pathogens are the five species in the apicomplexan genus *Plasmodium,* which cause malaria: *P. falciparum, P. malariae, P. vivax, P. ovale,* and *P. knowlesi.* Other *Plasmodium* spp. cause malaria in animals such as monkeys, birds, bats, lizards, and antelopes. The life cycle of the human pathogen *P. vivax* is shown in figure 39.6; it infects both liver and red blood cells. Prevention measures have resulted in over a 50% decline in malaria since 2000; however, the disease still causes over 600,000 deaths per year. *P. falciparum* is the most deadly of the five species, accounting for at least one-third of malaria deaths.

The energy metabolism of *Plasmodium* spp. is unusual. Despite possessing genes for a mitochondrial ATP synthase, it is largely fermentative when in host blood cells. Like all apicomplexans, *P. falciparum* cannot convert pyruvate to acetyl-CoA in the mitochondrion. Instead, phosphoenolpyruvate generated by the Embden-Meyerhof pathway in the cytosol is transported to

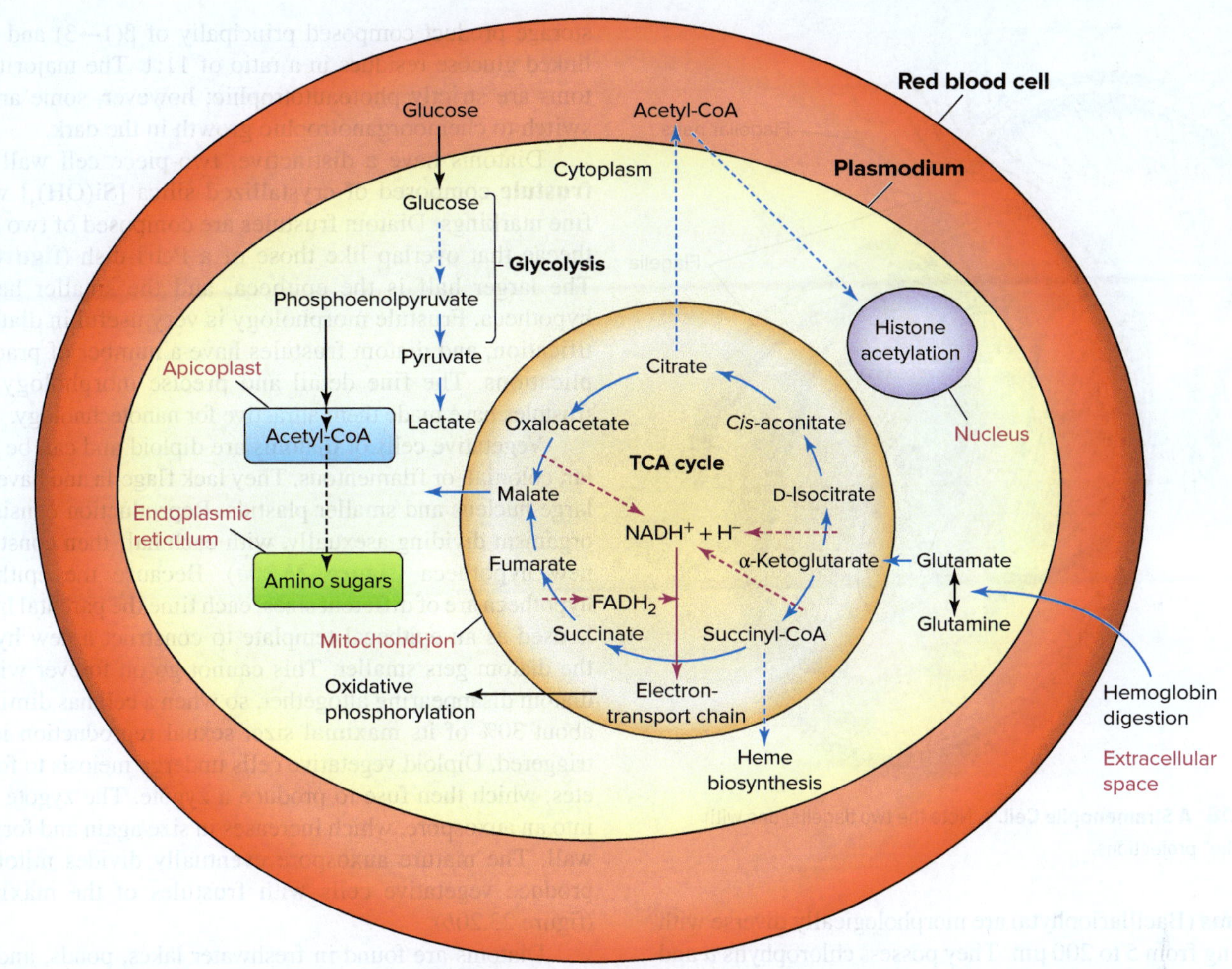

Figure 23.17 ***Plasmodium falciparum*** **Drives the TCA Cycle in Two Directions.** Glutamate is deaminated to yield α-ketoglutarate, from which reactions proceed in both the oxidative and reductive directions. Pyruvate is fermented in the parasite's cytosol to lactate rather than oxidized to acetyl-CoA. Phosphoenolpyruvate is converted to acetyl-CoA in the apicoplast and used to synthesize amino sugars. The host may convert citrate produced by the protist to acetyl-CoA, which then re-enters the microbe, where it is thought to be used to acetylate histones. The uncertainty of these steps is denoted by a dashed line.

the apicoplast, where it is converted to acetyl-CoA. Without acetyl-CoA in the mitochondrion, the TCA cycle begins with the deamination of glutamate to α-ketoglutarate. Reactions then proceed in both the reductive and oxidative directions. Reductive TCA steps result in citrate and malate, which are exported from the mitochondrion, while in the oxidative direction, succinyl-CoA, a key precursor for heme biosynthesis, is generated (**figure 23.17**). ▶| *Malaria (section 39.3)*

A number of other apicomplexans are important infectious agents. Several species of *Eimeria* are the causative agents of cecal coccidiosis in chickens, a condition that costs over a billion dollars in lost animals each year in the United States. *Theileria parva* and *T. annulata* are tick-borne parasites that cause a fatal disease in cattle called East Coast fever. This kills over a million animals annually in sub-Saharan Africa. Toxoplasmosis, caused by members of the genus *Toxoplasma,* is transmitted either by consumption of undercooked meat or by fecal contamination from a cat's litter box. *Cryptosporidium* spp. are responsible for cryptosporidiosis, which causes watery diarrhea. *Cryptosporidium* and members of another apicomplexan genus, *Cyclospora,* most commonly infect individuals who are immunocompromised. ▶| *Toxoplasmosis (section 39.5); Cryptosporidiosis (section 39.5); Cyclosporiasis (section 39.5)*

Stramenopila

The large and diverse Stramenopila group (informally known as stramenopiles) includes photosynthetic protists such as diatoms, brown and golden algae (Chrysophyceae), as well as chemoorganotrophic (saprophytic) protists such as öomycetes (Peronosporomycetes), labyrinthulids (slime nets), and members of Hyphochytriales. Stramenopiles also include brown seaweeds and kelp that were once considered fungi and plants. One unifying feature of these very diverse protists is the presence of tubular cristae in their mitochondria. Another is the possession of **heterokont flagella** at some point in the life cycle. This is characterized by two flagella—one extending anteriorly and the other posteriorly. These flagella bear small hairs with a unique, three-part morphology; stramenopila means "straw hair" (**figure 23.18**).

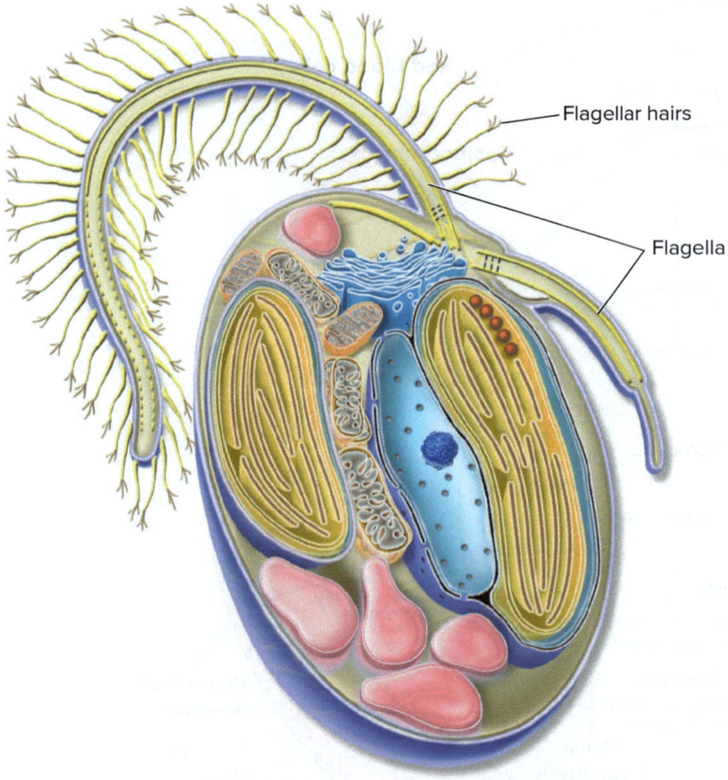

Figure 23.18 A Stramenopile Cell. Note the two flagella, one with "straw hair-like" projections.

Diatoms (Bacillariophyta) are morphologically diverse with sizes ranging from 5 to 200 μm. They possess chlorophylls *a* and c_1/c_2 and the carotenoid fucoxanthin. When fucoxanthin is the dominant pigment, cells have a golden-brown color. Their major carbohydrate reserve is chrysolaminarin, a polysaccharide storage product composed principally of β(1→3) and β(1→6)–linked glucose residues in a ratio of 11:1. The majority of diatoms are strictly photoautotrophic; however, some are able to switch to chemoorganotrophic growth in the dark.

Diatoms have a distinctive, two-piece cell wall called a **frustule** composed of crystallized silica [$Si(OH)_4$] with very fine markings. Diatom frustules are composed of two halves or thecae that overlap like those of a Petri dish (**figure 23.19**). The larger half is the epitheca, and the smaller half is the hypotheca. Frustule morphology is very useful in diatom identification, and diatom frustules have a number of practical applications. The fine detail and precise morphology of these frustules have made them attractive for nanotechnology.

Vegetative cells of diatoms are diploid and can be unicellular, colonial, or filamentous. They lack flagella and have a single, large nucleus and smaller plastids. Reproduction consists of the organism dividing asexually, with each half then constructing a new hypotheca (**figure 23.20*a***). Because the epitheca and hypotheca are of different sizes, each time the parental hypotheca is used as an epithecal template to construct a new hypotheca, the diatom gets smaller. This cannot go on forever without the diatom disappearing altogether, so when a cell has diminished to about 30% of its maximal size, sexual reproduction is usually triggered. Diploid vegetative cells undergo meiosis to form gametes, which then fuse to produce a zygote. The zygote develops into an auxospore, which increases in size again and forms a new wall. The mature auxospore eventually divides mitotically to produce vegetative cells with frustules of the maximal size (figure 23.20*b*).

Diatoms are found in freshwater lakes, ponds, and streams and throughout the world's oceans. Marine planktonic diatoms produce about 50% of the organic carbon in the ocean and are responsible for about 20% of annual carbon fixation globally.

(a) *Endictya oceanica*

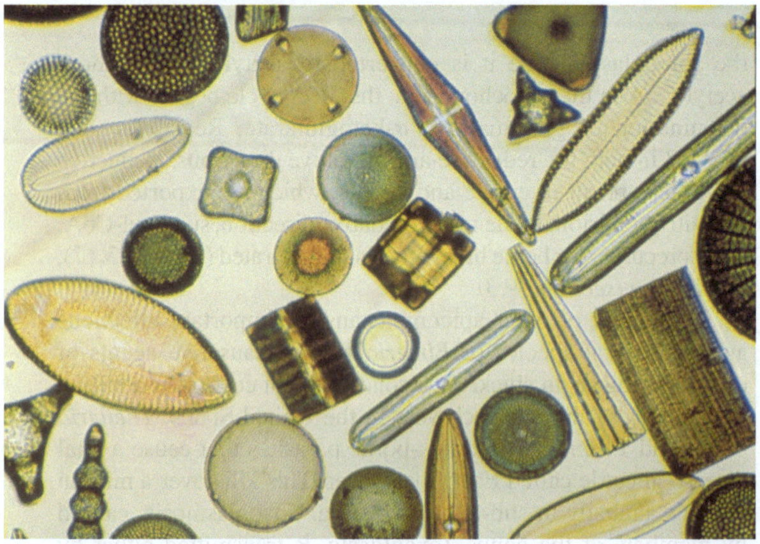

(b)

Figure 23.19 Diatoms. (a) The siliceous epitheca and hypotheca of the marine diatom *Endictya oceanica* fit together like halves of a Petri dish. (b) A variety of diatoms show the intricate structure of the silica cell wall.

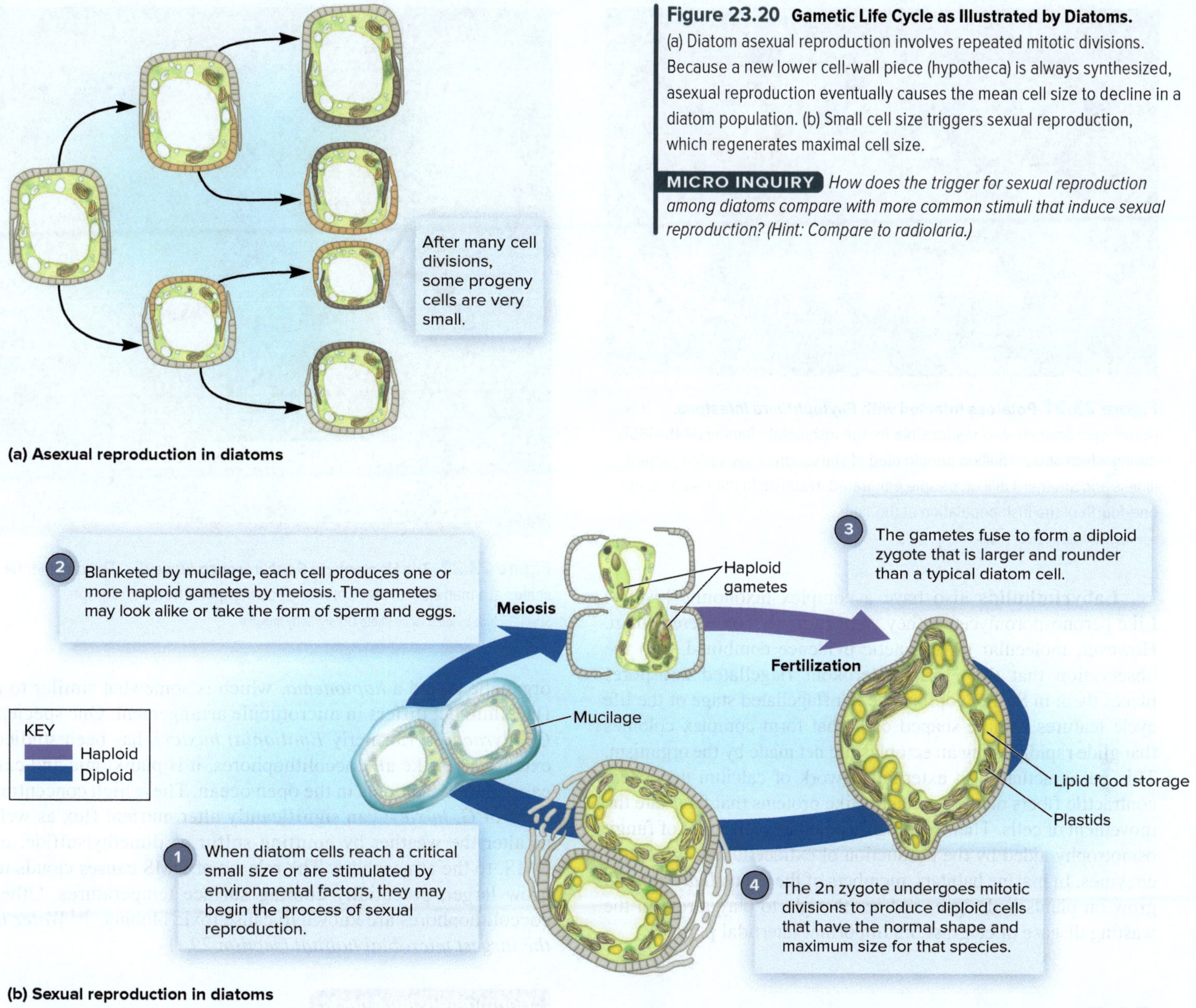

Figure 23.20 Gametic Life Cycle as Illustrated by Diatoms. (a) Diatom asexual reproduction involves repeated mitotic divisions. Because a new lower cell-wall piece (hypotheca) is always synthesized, asexual reproduction eventually causes the mean cell size to decline in a diatom population. (b) Small cell size triggers sexual reproduction, which regenerates maximal cell size.

MICRO INQUIRY *How does the trigger for sexual reproduction among diatoms compare with more common stimuli that induce sexual reproduction? (Hint: Compare to radiolaria.)*

They are therefore critical in global carbon cycling and are sometimes called the rainforests of the oceans. *Carbon cycle (section 28.2); Microorganisms in the open ocean are adapted to nutrient limitation (section 29.2)*

A group of protists once called **öomycetes,** meaning egg fungi, are now called **peronosporomycetes.** Historically classified as fungi because of their filamentous growth and osmotrophic lifestyle, they differ from fungi in a number of ways, including cell walls made of cellulose and the fact that they are diploid throughout their life cycle, except for the short period of time when they are haploid gametes. When undergoing sexual reproduction, they form a relatively large egg cell (öogonium) that is fertilized by either a sperm cell or a smaller gametic cell (called an antheridium) to produce a zygote. When the zygote germinates, asexual zoospores display heterokont flagellation.

Some peronosporomycetes help degrade plant and animal detritus, while others are pathogenic. For instance, species of *Saprolegnia* and *Achlya* are saprophytes that form cottony masses on dead algae and animals, mainly in freshwater environments. Certainly the most famous peronosporomycete is *Phytophthora infestans,* which attacked the European potato crop in the mid-1840s, spawning the Irish famine (**figure 23.21**). The original classification of *P. infestans* as a fungus was misleading, and for decades farmers attempted to control its growth with fungicides, to which it is (of course) resistant. This protist continues to take its toll; crop losses due to *P. infestans* cost an estimated \$10 to \$14 billion annually worldwide. *P. sojae* is another agriculturally important species; it infects soybeans and costs this industry millions of dollars per year. Finally, *P. ramorum* is a plant pathogen that causes a disease called sudden oak death. It kills oak trees and a variety of woody shrubs.

Figure 23.21 Potatoes Infected with *Phytophthora infestans*. This peronosporomycete was responsible for the Irish potato famine (1845–1851), during which about 1 million people died of starvation or starvation-related illness and another 1 million people emigrated, resulting in the loss of about one-fourth of the Irish population at the time.
agefotostock /Alamy Stock Photo

Labyrinthulids also have a complex taxonomic history: Like peronosporomycetes, they were formerly considered fungi. However, molecular phylogenetic evidence combined with the observation that they form heterokont flagellated zoospores places them in Stramenopila. The nonflagellated stage of the life cycle features spindle-shaped cells that form complex colonies that glide rapidly along an ectoplasmic net made by the organism. This net is actually an external network of calcium-dependent contractile fibers made up of actin-like proteins that facilitate the movement of cells. Their feeding mechanism is like that of fungi: osmotrophy aided by the production of extracellular degradative enzymes. In marine habitats, members of the genus *Labyrinthula* grow on plants and algae, and are thought to play a role in the wasting disease of eelgrass, an important intertidal plant.

23.5 Haptista Clade

After reading this section, you should be able to:

a. Describe the structure of coccolithophores
b. Explain the global importance of coccolithophores

Haptophyta were previously thought to be a sister group to TSAR; however, phylogenomic analysis places the haptophytes in a separate clade. These include Coccolithales, photosynthetic protists that bear ornate calcite scales called *coccoliths* (**figure 23.22**). Together with forams, **coccolithophores** precipitate calcium carbonate ($CaCO_3$) in the ocean, thereby removing soluble CO_2 from the ocean and influencing Earth's carbon budget. Cells are usually biflagellated and possess a unique organelle called a *haptonema,* which is somewhat similar to a flagellum but differs in microtubule arrangement. One species, *Gephyrocapsa* (formerly *Emiliania) huxleyi,* has been studied extensively. Like all coccolithophores, it is planktonic and can cause massive blooms in the open ocean. These high concentrations of *G. huxleyi* can significantly alter nutrient flux as well as alter the weather by emitting sulfur as dimethylsulfide, or DMS, to the atmosphere. The release of DMS causes clouds to grow larger, potentially cooling surface temperatures. Other coccolithophores are known to cause toxic blooms. ▶| *Water is the largest microbial habitat (section 29.1)*

Figure 23.22 The Haptophyte *Gephyrocapsa huxleyi*. The ornamental scales are made of calcite. Colorized scanning electron micrograph.
Steve Gschmeissner/Science Photo Library/Getty Images

Comprehension Check

1. What adaptations do planktonic radiolaria have to help them float?
2. Compare how radiolaria use axopodia with the way forams use reticulopodia.
3. Describe the difference between a naked and an armored dinoflagellate.
4. Describe conjugation as it occurs in ciliated protists. What is the fate of the micronucleus and the macronucleus during this process?
5. What function does the apical complex serve in apicomplexans?
6. Explain the unique structural features of diatoms. How does their morphology play a role in the alternation between asexual and sexual reproduction?
7. Why were peronosporomycetes and labyrinthulids formerly considered fungi?
8. What is the ecological importance of coccolithophores?

23.6 Archaeplastida Clade Includes Green and Red Algae

After reading this section, you should be able to:

a. List at least three different morphologies found among the Chloroplastida
b. Describe the structure and life cycle of the model organism, *Chlamydomonas*

Archaeplastida includes all organisms with a photosynthetic plastid that arose through an ancient endosymbiosis with a cyanobacterium. It thus includes all higher plants as well as three types of algae: glaucophyte, red, and green. Glaucophyte algae consist of fewer than 20 known species. Red algae get their color from the pigment phycoerythrin. Some species contribute to the formation of tropical reefs; these are called coralline algae. Here we focus on Chloroplastida, also known as green algae. *Microbial diversity and ecology 5.1*

Chloroplastida

Members of Chloroplastida (Greek *chloros,* green) grow just about anywhere: in freshwater and saltwater, in soil, and on and within other organisms. They are photoautotrophic with chlorophylls *a* and *b* along with specific carotenoids, and they store carbohydrates such as starch. Many have cell walls made of cellulose. They exhibit a wide diversity of body forms, ranging from unicellular to colonial, filamentous, membranous or sheet-like, and tubular types (**figure 23.23**). Some species have a holdfast structure that anchors them to a substratum. Both asexual and sexual reproduction are observed.

Chlamydomonas spp. are members of the subgroup Chlorophyta (**figures 23.24** and 23.1). Individuals have two flagella of equal length at the anterior end by which they move rapidly in water. Each cell has a single haploid nucleus, a large chloroplast, a conspicuous pyrenoid, and a stigma that aids the cell in phototactic responses. Two small contractile vacuoles at the base of the flagella function as osmoregulatory organelles. *Chlamydomonas* spp. reproduce asexually by producing zoospores through mitotic division. Sexual reproduction occurs when some cells act as gametes and fuse to form a four-flagellated, diploid zygote that ultimately loses the flagella and enters a resting phase. Meiosis occurs at the end of this resting phase and produces four haploid cells that give rise to adults.

Picozoa

Evidence for the smallest eukaryotes, the Picozoa (Greek *pico,* tiny; *zoa,* animal), emerged in the early 2000s when 18S rRNA metagenomic data revealed the presence of eukaryotic organisms small enough to fit through filters of 2 to 3 μm. Phylogenomic analysis places these microbes in the Archaeplastida clade most closely related to red algae. However, this finding is perplexing as picozoans lack plastids and are heterotrophic. They are broadly distributed in high numbers throughout the world's oceans.

The model species *Picomonas judraskeda* has been isolated from marine coastal waters. It consists of two roughly spherical halves divided by a deep cleft. Total cell size ranges from 2×2.5 μm to 2.5×3.8 μm. They have two flagella and demonstrate an unusual kind of motility that has been described as cycles of "jump-drag-skedaddle." They consume colloidal organic materials by endocytosis, and it is thought that their unusual motility assists in feeding.

Comprehension Check

1. Compare the morphology of members of the genus *Chlamydomonas* with those belonging to *Euglena* (figures 23.4 and 23.24).
2. Like many protists, sexual reproduction in *C. reinhardtii* involves individuals from different mating types. From an evolutionary point of view, why is this the case?
3. What does the placement of Picozoa as relatives of red algae despite their lack of plastids (e.g., chloroplasts) suggest about the evolution of these organelles?

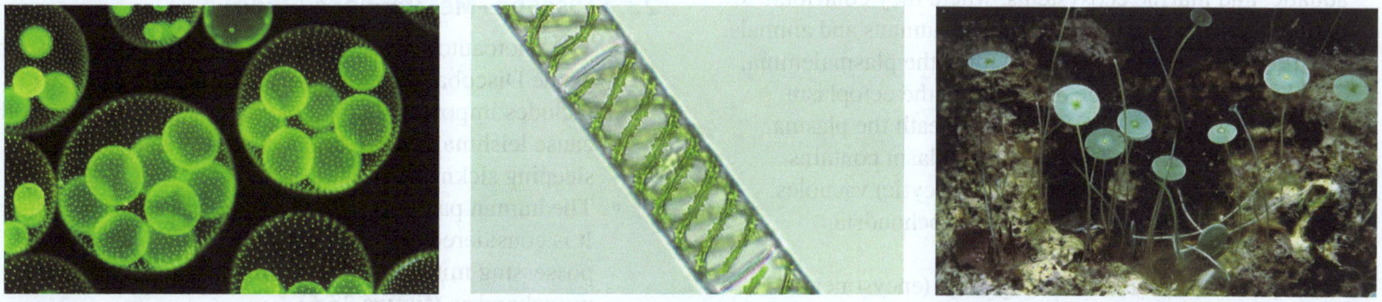

(a) *Volvox* (b) *Spirogyra* (c) *Acetabularia*

Figure 23.23 Chlorophyta (Green Algae); Light Micrographs. (a) *Volvox,* which demonstrates colonial growth. (b) *Spirogyra;* each filament contains ribbon-like, spiral chloroplasts. (c) *Acetabularia,* the mermaid's wine goblet.

(a) Micro photo/iStock/Getty Images; (b) Elif Bayraktar/Shutterstock; (c) De Agostini Picture Library/Science Source

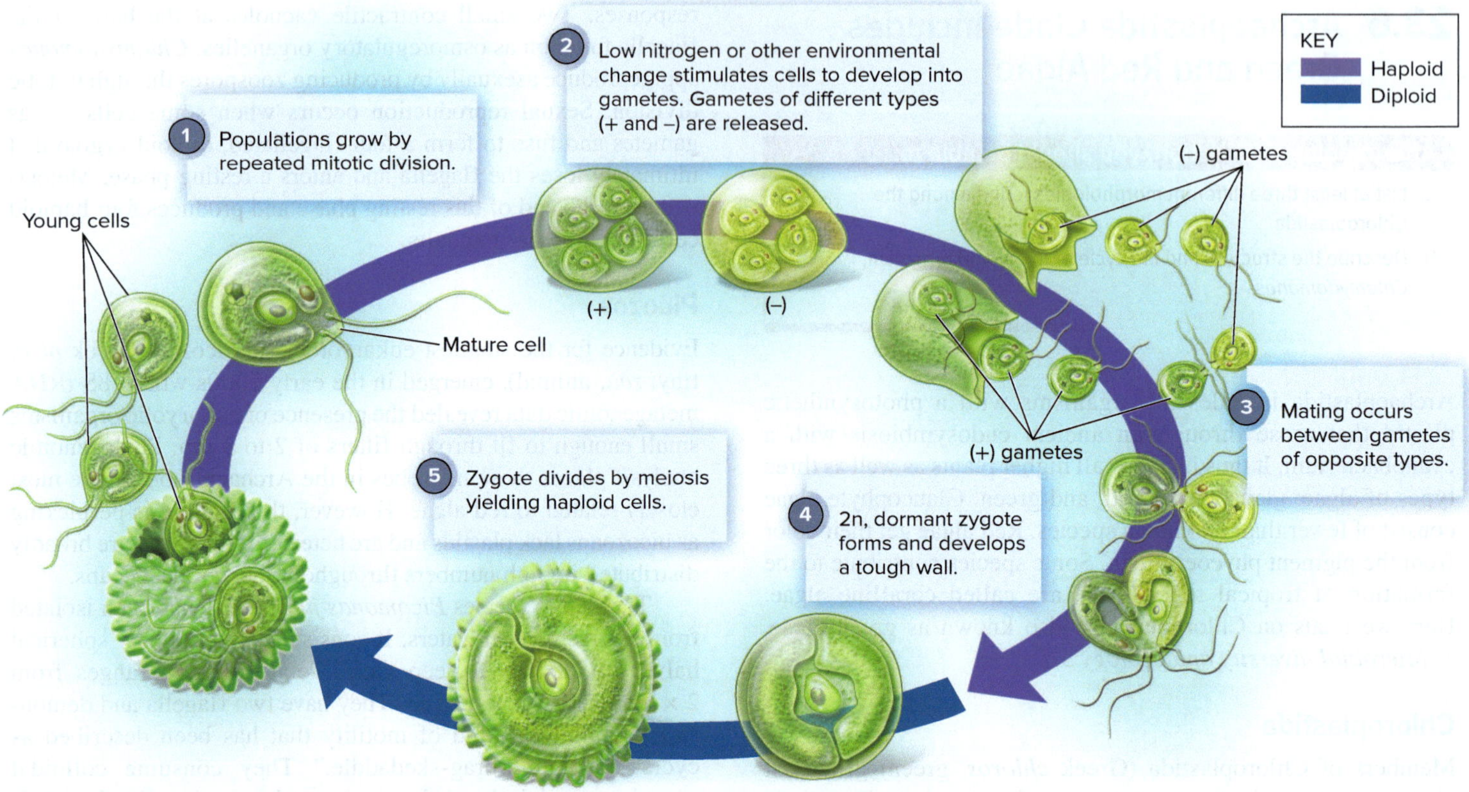

Figure 23.24 The Structure and Life Cycle of *Chlamydomonas reinhardtii*. During asexual reproduction, all structures are haploid; during reproduction, only the zygote is diploid.

MICRO INQUIRY *Why do you think environmental stress stimulates gamete development?*

Key Concepts

23.1 Protist Diversity Reflects Broad Phylogeny

- Protists are widespread in the environment, found wherever water, suitable nutrients, and an appropriate temperature occur.
- Protists are important components of many terrestrial, aquatic, and marine ecosystems, where they contribute to nutrient cycling. Many are parasitic in humans and animals.
- The protistan cell membrane is called the plasmalemma, and the cytoplasm can be divided into the ectoplasm and the endoplasm. A pellicle lies beneath the plasma membrane of some protists. The cytoplasm contains contractile, secretory, and food (phagocytic) vacuoles.
- Energy metabolism occurs within mitochondria, hydrogenosomes, or chloroplasts.
- Some protists can enter a resting stage (encystment) called a cyst. Cysts protect the organism against adverse environments, function as a site for nuclear reorganization, and serve as a means of transmission in parasitic species (**figure 23.2**).
- Most protists reproduce asexually by binary or multiple fission or budding (**figure 23.3**).
- Protists also use a variety of sexual reproductive strategies, including syngamy and autogamy.
- Protist phylogeny is the subject of active research and debate.

23.2 Discoba–Metamonada Clade

- The photoautotrophic flagellates Euglenozoa are members of the Discoba lineage (**figure 23.4**). This group also includes important human pathogens such as protists that cause leishmaniasis, Chagas disease, and African sleeping sickness.
- The human pathogen *Giardia duodenalis* is a metamonad. It is considered one of the most primitive eukaryotes, possessing mitosomes and hydrogenosomes but not mitochondria (**figure 23.6**).
- Metamonads also include some flagellated endosymbionts of animals, such as the obligate mutualists of wood-eating insects such as *Trichonympha* spp. and the human pathogen *Trichomonas vaginalis*.

23.3 Amoebozoa Clade Includes Protists with Pseudopodia

- Amoeboid forms use pseudopodia, which can be lobopodia, filopodia, or reticulopodia (**figure 23.7**).
- Amoebae that bear external plates are called testate; those without plates are called naked or atestate amoebae.
- Eumycetozoa includes acellular and cellular slime molds. Acellular slime molds form a large, multinucleate mass of protoplasm (**figure 23.8**).
- Cellular slime molds produce a pseudoplasmodium, consisting of many individual cells moving together as a mass. *Dictyostelium discoideum* is a cellular slime mold that is used as a model organism in the study of chemotaxis, cellular development, and cellular behavior (**figure 23.9**).

23.4 TSAR Clade: Protists of Global Importance

- Rhizaria encompasses amoeboid forms, including members of *Radiolaria,* which have a siliceous internal skeleton and filopodia that are used to entrap prey (**figure 23.10**).
- Foraminifera bear netlike reticulopodia and tests. Most foraminifera are benthic, and their tests accumulate on the ocean floor (**figure 23.11**).
- Alveolata consists of dinoflagellates, ciliates, and apicomplexans.
- Dinoflagellates are a large group of nutritionally complex protists. Most are marine and planktonic. They are known for their phosphorescence and for causing toxic blooms (**figure 23.13**).
- Ciliates are chemoorganotrophic protists that use cilia for locomotion and feeding. In addition to asexual reproduction, conjugation is used in sexual reproduction (**figures 23.14** and **23.15**).
- Apicomplexans are parasitic with complex life cycles. The motile, infective stage is called the sporozoite. This group includes a number of pathogens, including five species of *Plasmodium,* which cause malaria in humans (**figure 23.16**).
- Stramenopiles are extremely diverse and include diatoms, golden and brown algae, peronosporomycetes, and labyrinthulids. Diatoms are found in freshwater and saltwater and are important components of marine plankton (**figures 23.18 through 23.20**).

23.5 Haptista Clade

- Coccolithophores are marine photosynthetic protists that fix CO_2 and precipitate $CaCO_3$ to construct their ornate scales (**figure 23.22**).
- Because of their abundance, coccolithophores help regulate the global carbon budget.

23.6 Archaeplastida Clade Includes Green and Red Algae

- Archaeplastida includes Chlorophyta, also known as green algae. All are photosynthetic with chlorophylls *a* and *b* along with specific carotenoids. They exhibit a wide range of morphologies (**figure 23.23**).
- *Chlamydomonas* spp. are model protists that undergo sexual and asexual reproduction (**figure 23.24**).
- Picozoa include very small, heterotrophic, marine protists with a global distribution.

Active Learning

1. Protist encystment is usually triggered by changes in the environment. Compare and contrast this with endospore formation in monoderm bacteria.
2. Vaccine development for diseases caused by protists (e.g., malaria, Chagas disease) has been much less successful than for bacterial or viral diseases. Discuss one biological reason and one geopolitical reason for this fact.
3. Most apicomplexans bear the unpigmented chloroplast remnant called the apicoplast. Only the recently described apicomplexan *Chromera velia* has been shown to use the apicoplast as the site of photosynthesis. What genetic and physiological features do you think were characterized to show that this organism is truly photosynthetic? How would you determine if this protist is photoautotrophic, photoheterotrophic, or mixotrophic?
4. Benthic foraminifera inhabit marine sediments. It was discovered that the benthic foram *Globobulimina pseudospinescens* stores intracellular nitrate, which it reduces to nitrogen gas. How would you go about designing a comparative analysis of eukaryotic and bacterial denitrification? What physiological and enzymatic features would you characterize?
5. Protist taxonomy was once based entirely on morphology and ecology. This has been replaced by molecular characterization. In an effort to unite classical and molecular taxonomic approaches, an international group of scientists has started an open-access platform called *UniEuk* to build a common community of expert-driven taxonomy. Select a protist discussed in this chapter and list the kinds of information that would be helpful in exploring its taxonomy. You can then compare your proposed inputs with those described on the *UniEuk* website.

CHAPTER

24

Fungi

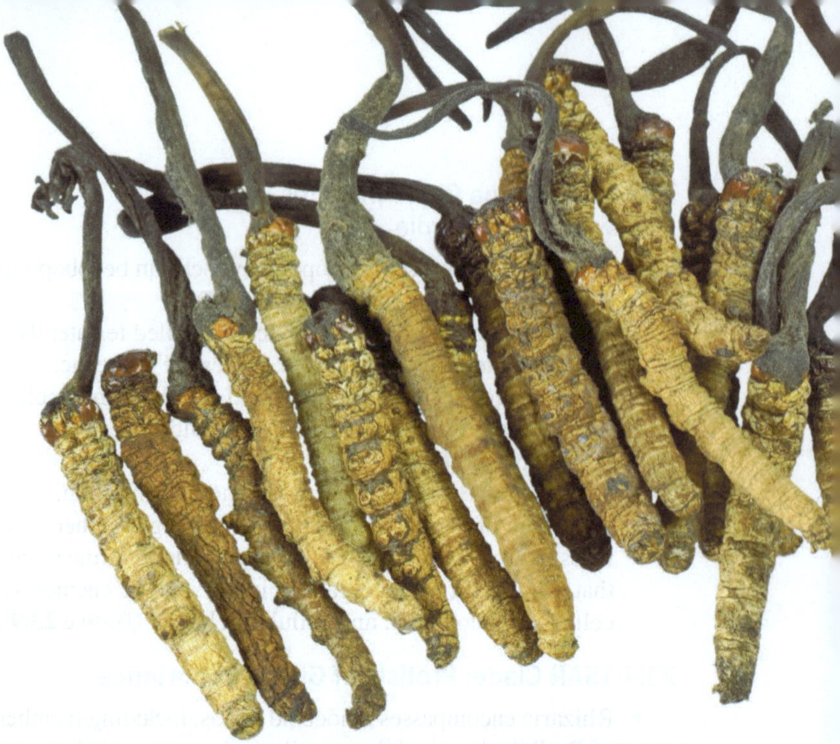

Y H Lim/Alamy Stock Photo

The Complex Story of Caterpillar Fungus

Question: What do caterpillars, the economy of the Tibetan Autonomous Region (TAR), and global climate change have in common? *Answer:* The fungus *Ophiocordyceps sinensis,* more commonly known as cordyceps. This fungus parasitizes caterpillars of ghost moths (*Thitarodes* spp.). Each spring, when the fungal spores germinate, their hyphae (filaments) infect these soil-dwelling caterpillars. By the following spring, fruiting bodies called stoma have formed within the heads of the caterpillars, which poke through the soil surface. Residents of the TAR comb hillsides looking for fungal stomas poking through the earth. Once found, the mummified caterpillar with its attached fungus, as shown, is harvested, dried, and sold for its purported medical qualities.

What are these presumed medical benefits? Since the fifteenth century, cordyceps has been considered an aphrodisiac with the ability to improve myriad medical conditions, from asthma to cancer. More recently, it was credited with the outstanding performance of Chinese track athletes. This triggered a run on the cordyceps market, and between 2000 and 2011, the value of caterpillar fungus increased about 1,000%. In 2024, the highest-quality cordyceps fetched about $110,000 per kilogram.

The tremendous popularity of cordyceps has had a profound effect on the TAR economy as subsistence farmers turned to caterpillar fungus as a means of income. Predictably as the number of people harvesting cordyceps grew, the yield per person declined. The reasons for this extend beyond supply and demand. First, there was a lack of understanding of the fungal life cycle. Harvesting late-season caterpillar fungus removes spores from the environment, eliminating spore dissemination, which ensures sustainable harvests. Another reason is global climate change. Rising temperatures in the TAR are correlated with reduced harvest size. Hopefully, the implementation of sustainable harvesting techniques will allow economically marginalized people to continue to profit from cordyceps.

In this chapter, we introduce **Fungi,** sometimes referred to as Eumycota (Greek *eu,* true; *mykes,* fungus). The term *fungus* (pl., fungi; Latin *fungus,* mushroom) describes eukaryotic organisms that produce spores, have absorptive nutrition, lack chlorophyll, and reproduce both sexually and asexually. Scientists who study fungi are **mycologists,** and the scientific discipline devoted to fungi is called **mycology.** The study of fungal toxins is called **mycotoxicology,** and the diseases caused by fungi in animals are known as **mycoses** (s., mycosis).

Fungi is an enormous group of organisms, with about 150,000 known species but an estimated additional 2 to 11 million species yet to be isolated. Like protists, the taxonomy of these organisms has been revised numerous times and remains uncertain. Recently, the application of phylogenomics has led to a better understanding of the evolutionary relationships between various fungal groups. Here we present the six major fungal groups Chytridiomycota, Microsporidia, Mucoromycota, Glomeromycota, Ascomycota, and Basidiomycota, shown in **figure 24.1** and summarized in **table 24.1**.

Readiness Check:

Based on what you have learned previously, you should be able to:

- ✓ Draw a eukaryotic cell envelope and identify the structure and function of its component layers (section 5.2)
- ✓ Identify the structure and function of eukaryotic cellular organelles (sections 5.4–5.6)
- ✓ Describe the structure and function of eukaryotic flagella and cilia (section 5.7)

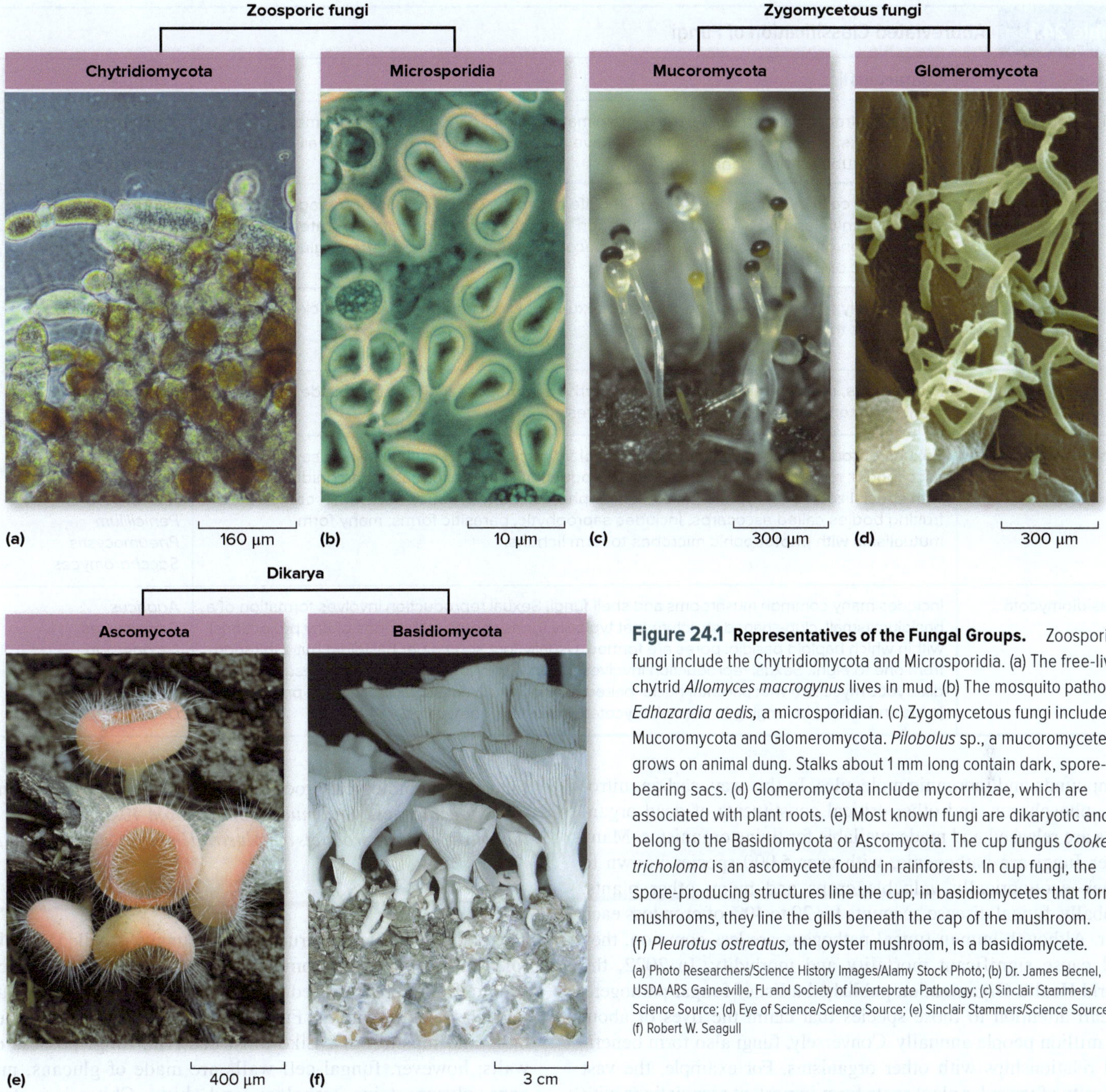

Figure 24.1 Representatives of the Fungal Groups. Zoosporic fungi include the Chytridiomycota and Microsporidia. (a) The free-living chytrid *Allomyces macrogynus* lives in mud. (b) The mosquito pathogen *Edhazardia aedis,* a microsporidian. (c) Zygomycetous fungi include Mucoromycota and Glomeromycota. *Pilobolus* sp., a mucoromycete, grows on animal dung. Stalks about 1 mm long contain dark, spore-bearing sacs. (d) Glomeromycota include mycorrhizae, which are associated with plant roots. (e) Most known fungi are dikaryotic and belong to the Basidiomycota or Ascomycota. The cup fungus *Cookeina tricholoma* is an ascomycete found in rainforests. In cup fungi, the spore-producing structures line the cup; in basidiomycetes that form mushrooms, they line the gills beneath the cap of the mushroom. (f) *Pleurotus ostreatus,* the oyster mushroom, is a basidiomycete.

(a) Photo Researchers/Science History Images/Alamy Stock Photo; (b) Dr. James Becnel, USDA ARS Gainesville, FL and Society of Invertebrate Pathology; (c) Sinclair Stammers/Science Source; (d) Eye of Science/Science Source; (e) Sinclair Stammers/Science Source; (f) Robert W. Seagull

24.1 Fungal Biology Reflects Vast Diversity

After reading this section, you should be able to:

- **a.** Discuss the importance of fungi in the environment as well as their practical importance
- **b.** Differentiate between a yeast and a mold, and between hypha and mycelium
- **c.** Compare and contrast different sporangia morphologies
- **d.** Explain how mating types govern sexual reproduction

Fungal Distribution and Importance

Fungi and animals diverged from a common ancestor about a billion years ago. Since that time, fungi have gained a global distribution from polar to tropical regions. Many are **saprophytes,** securing nutrients from dead organic material by releasing degradative enzymes into the environment. This enables their absorption of the soluble products—a process sometimes called **osmotrophy** (Greek *osmos,* pushing; perhaps a misnomer, as nutrients are transported rather than pushed into the cell). Saprophytic fungi are important decomposers. They degrade complex organic materials in the environment to simple organic

Table 24.1 Abbreviated Classification of Fungi

Taxon	Characteristics	Examples
Microsporidia	Obligate intracellular parasites usually of animals. Lack mitochondria, peroxisomes, kinetosomes, cilia, and centrioles; spores have an inner chitin wall and outer wall of protein; produce a tube for host penetration.	*Amblyospora* *Encephalitozoon* *Enterocytozoon*
Chytridiomycota	Flagellated cells in at least one stage of the life cycle; may have one or more flagella. Cell walls with chitin and β-1,3/1,6-glucan; glycogen is used as a storage carbohydrate. Sexual reproduction often results in a zygote that becomes a resting spore or sporangium; saprophytic or parasitic.	*Allomyces* *Blastocladiella* *Coelomomyces* *Synchytrium*
Mucoromycota	Thalli usually filamentous and nonseptate; sexual reproduction gives rise to thick-walled zygospores that are often ornamented.	*Mucor* *Phycomyces* *Rhizopus*
Glomeromycota	Filamentous, most are arbuscular endomycorrhizae; form asexual spores outside of host plant; lack centrioles, conidia, and aerial spores.	*Acaulospora* *Glomus*
Ascomycota	Sexual reproduction involves meiosis of a diploid nucleus in an ascus, giving rise to haploid ascospores; most also undergo asexual reproduction with the formation of conidiospores with specialized aerial hyphae called conidiophores. Many produce asci within complex fruiting bodies called ascocarps. Includes saprophytic, parasitic forms; many form mutualisms with phototrophic microbes to form lichens.	*Aspergillus* *Candida* *Neurospora* *Penicillium* *Pneumocystis* *Saccharomyces*
Basidiomycota	Includes many common mushrooms and shelf fungi. Sexual reproduction involves formation of a basidium (small, club-shaped structure that typically forms spores at the ends of tiny projections) within which haploid basidiospores are formed. Usually four spores per basidium but can range from one to eight. Sexual reproduction involves fusion with opposite mating type, resulting in a dikaryotic mycelium with parental nuclei paired but not initially fused. Includes plant pathogens (rusts and smuts) belonging to urediniomycetes and ustilaginomycetes.	*Agaricus* *Dacrymyces* *Lycoperdon* *Polyporus* *Uromyces* *Ustilago*

compounds and inorganic molecules. In this way, carbon, nitrogen, phosphorus, and other critical constituents of dead organisms are released and made available for living organisms. Many other fungi are pathogenic, with over 5,000 species known to attack economically valuable crops and many other plants. Globally, fungal disease accounts for 20 to 40% of crop loss each year. Although human fungal pathogens are less common, they still cause significant mortality and morbidity. In 2022, the World Health Organization published a list of fungal pathogens to call attention to those species that claim the lives of about 1.5 million people annually. Conversely, fungi also form beneficial relationships with other organisms. For example, the vast majority of vascular plant roots form important associations with fungi called mycorrhizae. It is estimated that annually about 13 gigatons of carbon fixed by plants (equivalent to roughly a third of all CO_2 emitted by burning fossil fuels) flow through underground fungi. ▶| *Mycorrhizae facilitate the growth of most plants (section 30.3)*

Fungi, especially yeasts (single-celled fungi), are essential to many industrial processes involving fermentation. Examples include the making of bread, wine, beer, cheeses, and soy sauce. They are also important in the commercial production of many organic acids and certain drugs, and in the manufacture of many antibiotics (e.g., citric acid, cortisone, and penicillin, respectively). In addition, fungi are important research tools in the study of fundamental biological processes. ▶| *Microbiology of fermented foods: beer, cheese, and much more (section 40.5); Microbes are the source of many products of industrial importance (section 41.1)*

Fungal Structure

The body or vegetative structure of a fungus is called a **thallus** (pl., thalli). It varies in complexity and size. Single-cell microscopic fungi are referred to as yeasts, while multicellular masses are called molds. Fungi also include macroscopic puffballs and mushrooms. Like most bacteria, fungi possess cell walls; however, fungal cell walls are made of glucans, mannans, glycoproteins, trehalose, and chitin. Chitin is a strong but flexible nitrogen-containing polysaccharide consisting of *N*-acetylglucosamine residues.

A **yeast** is a unicellular fungus with a single nucleus that reproduces either asexually by **budding** or transverse division or sexually through spore formation. Generally, yeast cells are larger than bacteria and are commonly spherical to egg-shaped. They lack flagella and cilia but have most other eukaryotic organelles (*see figures 5.2 and 5.3a*). Unlike molds, yeasts grow as smooth colonies on agar plates much like bacteria.

The thallus of a **mold** consists of long, branched, thread-like filaments called **hyphae** (s., hypha; Greek *hyphe,* web) that form a tangled mass called a **mycelium** (pl., mycelia). In some fungi,

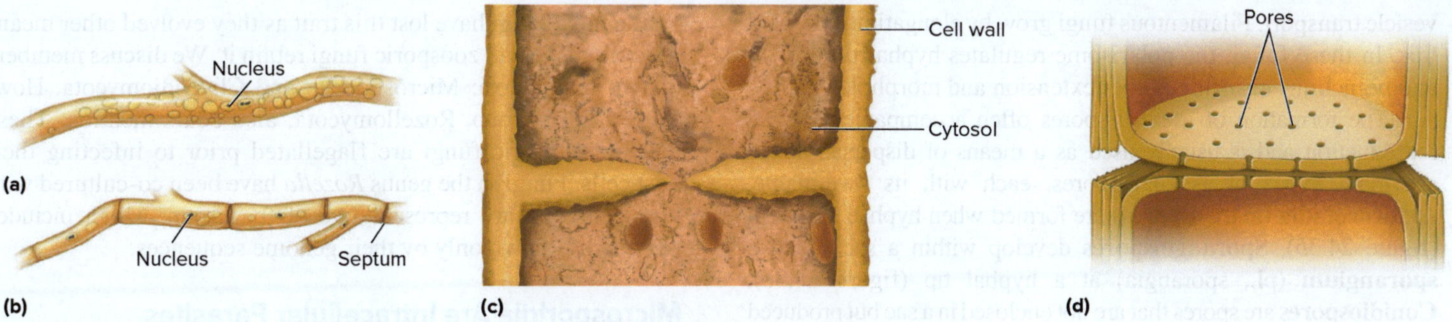

Figure 24.2 Hyphae. Drawings of (a) coenocytic hyphae (aseptate) and (b) hyphae divided into cells by septa. (c) A section of a representative fungal filament demonstrates that in some species, cellular compartments communicate by a single pore. (d) Drawing of a multiperforate septal structure.

protoplasm streams lengthwise through the hyphae, uninterrupted by cross walls. These hyphae are called **coenocytic** or **aseptate hyphae** (**figure 24.2*a***). The hyphae of other fungi (figure 24.2*b*) have cross walls called **septa** (s., septum) with either a single pore (figure 24.2*c*) or multiple pores (figure 24.2*d*) that enable cytoplasmic streaming. These hyphae are termed **septate hyphae.** When grown in the laboratory, molds can be distinguished from yeast by their fuzzy colony appearance.

Like yeast, hyphae are composed of an outer cell wall that surrounds the plasma membrane, which encloses the cytoplasm. Unlike yeast, the filamentous nature of hyphae results in a large surface area relative to the volume of cytoplasm. This optimizes nutrient absorption by osmotrophy.

Fungal Growth and Reproduction

Reproduction in fungi can be either asexual or sexual. Asexual reproduction is accomplished in two ways: (1) A parent cell undergoes mitosis and divides into two daughter cells by a central constriction and formation of a new cell wall (**figure 24.3*a***) or (2) mitosis in vegetative cells may be concurrent with budding to produce a smaller daughter cell. This is very common in yeasts. Both budding yeasts and filamentous fungi orchestrate growth and cell polarity through the development of a multiprotein complex called the *polarisome*. In budding yeast, the polarisome directs localized growth and determines the site at which budding will occur by organizing the actin cytoskeleton and guiding

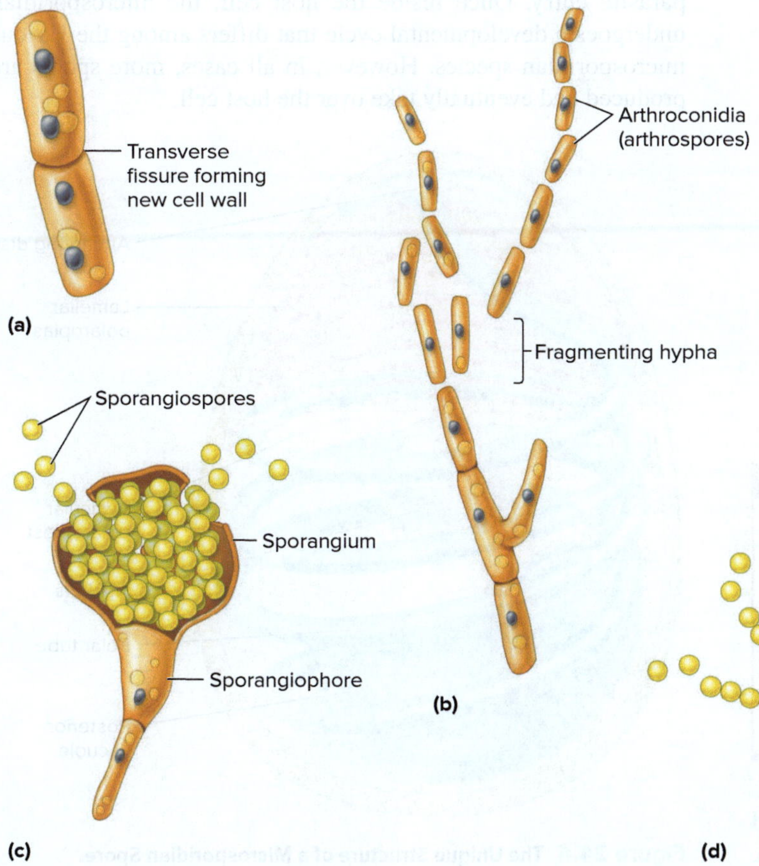

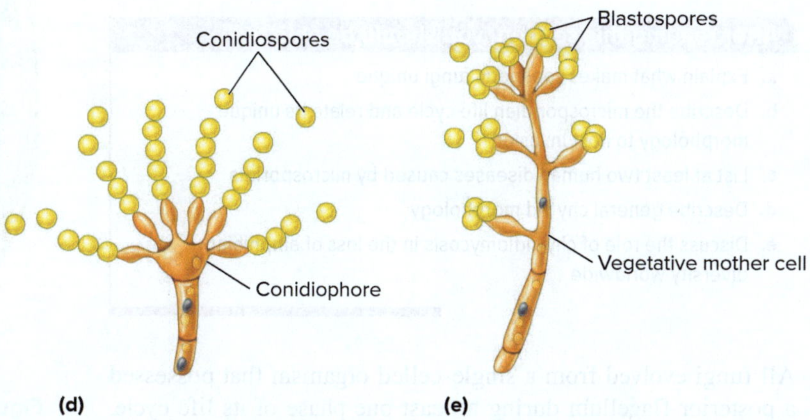

Figure 24.3 Asexual Reproduction in Fungi and Some Representative Asexual Spores. (a) Transverse fission. (b) Hyphal fragmentation results in arthroconidia (arthrospores). (c) Sporangiospores in a sporangium. (d) Conidiospores arranged in chains at the end of a conidiophore. (e) Blastospores are formed by budding off of the parent cell.

vesicle transport. Filamentous fungi grow by elongation of hyphal tips. In these fungi, the polarisome regulates hyphal tip growth and branching, ensuring proper extension and morphology.

The formation of asexual spores often accompanies asexual reproduction and is usually used as a means of dispersal. There are many types of asexual spores, each with its own name. **Arthroconidia (arthrospores)** are formed when hyphae fragment (figure 24.3*b*). **Sporangiospores** develop within a sac called a **sporangium** (pl., sporangia) at a hyphal tip (figure 24.3*c*). **Conidiospores** are spores that are not enclosed in a sac but produced at the tips or sides of the hypha (figure 24.3*d*). **Blastospores** are produced from a vegetative mother cell by budding (figure 24.3*e*).

Sexual reproduction in fungi involves the fusion of compatible nuclei. *Homothallic* fungal species are self-fertilizing and produce sexually compatible gametes on the same mycelium. *Heterothallic* species require crossing between different but sexually compatible mycelia called mating types. Depending on the species, sexual fusion may occur between haploid gametes, gamete-producing bodies called **gametangia,** or hyphae.

Fungal spores are important for several reasons. They enable fungi to survive environmental stresses such as desiccation, nutrient limitation, and extreme temperatures, although they are not as stress resistant as bacterial endospores. They aid in fungal dissemination because spores are small and light, and thus remain suspended in air for long periods. Spores are also spread by adhering to the bodies of insects and other animals. Finally, the size, shape, color, and number of spores are useful in the identification of fungal species.

Comprehension Check

1. How can fungi be defined? What is the difference between a yeast and a mold?
2. List one important fact about fungal pathogens and describe one practical use of fungi.
3. Compare and contrast the following types of asexual fungal spores: sporangiospore, arthrospore, conidiospore, and blastospore.

24.2 Zoosporic Fungi Form Motile Spores

After reading this section, you should be able to:

a. Explain what makes zoosporic fungi unique
b. Describe the microsporidian life cycle and relate its unique morphology to host invasion
c. List at least two human diseases caused by microsporidia
d. Describe general chytrid morphology
e. Discuss the role of chytridiomycosis in the loss of amphibian diversity worldwide

All fungi evolved from a single-celled organism that possessed a posterior flagellum during at least one phase of its life cycle. While most fungi have lost this trait as they evolved other means of spore dispersal, zoosporic fungi retain it. We discuss members of two major taxa: Microsporidia and Chytridiomycota. However, a third group, Rozellomycota, also bears mention. These aquatic, parasitic fungi are flagellated prior to infecting their host cells. Fungi in the genus *Rozella* have been co-cultured with host cells and are representative of the group, which includes organisms known only by their genome sequences.

Microsporidia Are Intracellular Parasites

Microsporidia are obligately endoparasitic fungi, meaning these fungi must infect host cells to complete their life cycle. As parasites, their genomes are markedly reduced in size and complexity. Microsporidia were once considered protists. The confusion stems from their lack of mitochondria, peroxisomes, and centrioles, and the presence of mitosomes, an organelle found in some protists (*see figure 5.14*). Like most fungi, microsporidian cell walls contain chitin and trehalose.

Microsporidial morphology is also unique among eukaryotes. It is distinct in the possession of an organelle called the *polar tube,* which is essential for host invasion. Spores of 1 to 40 µm may be spherical, rodlike, or egg- or crescent-shaped, depending on the species. Spore germination occurs only in the host, where the spore uses its anchoring disk to attach to the host cell and expel the tightly packed polar tube (**figures 24.4** and 24.1*b*). Aided by the lamellar polaroplast, the polar tube is ejected with enough force to pierce the host cell membrane, permitting parasite entry. Once inside the host cell, the microsporidian undergoes a developmental cycle that differs among the various microsporidian species. However, in all cases, more spores are produced and eventually take over the host cell.

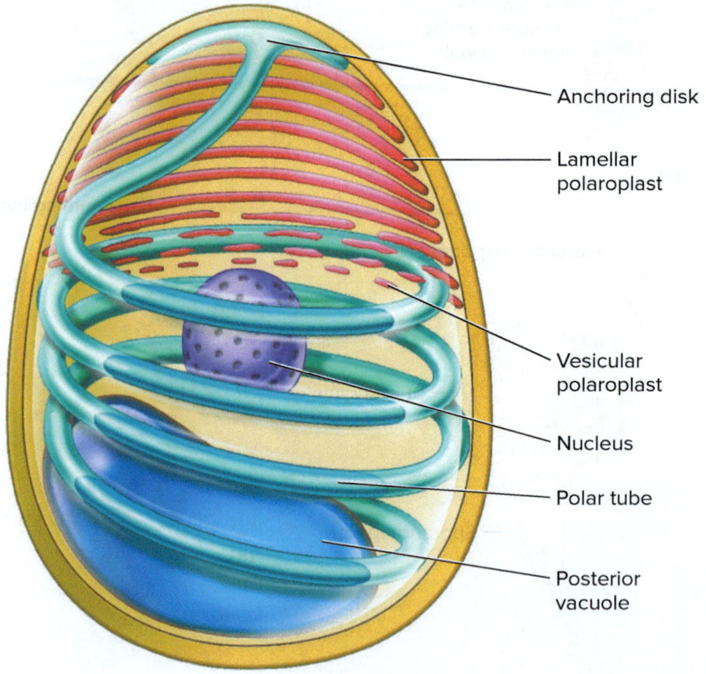

Figure 24.4 The Unique Structure of a Microsporidian Spore.

Microsporidia are important obligate intracellular parasites that infect insects, fish, and humans. They are particularly problematic for immunocompromised individuals. Human pathogens include *Enterocytozoon bieneusi,* which causes diarrhea and pneumonia (depending on whether it was acquired through ingestion or inhalation, respectively), and *Encephalitozoon cuniculi,* which causes encephalitis (infection of the brain) and nephritis (kidney disease). Aquatic birds are an important source of pathogenic microsporidia in nature; indeed, a single flock of waterfowl can deposit over 10^8 microsporidian spores in surface waters after one visit. ▶| *Microsporidiosis (section 39.6)*

Chytridiomycota Are Killing Amphibians

Members of Chytridiomycota are commonly called **chytrids** and include free-living and parasitic genera. Free-living chytrids are saprophytic, living on plant or animal matter in freshwater, mud, or soil. Parasitic forms infect aquatic plants and animals, including insects. A few are found in the anoxic rumen of herbivores. Chytrids display a variety of life cycles involving both asexual and sexual reproduction. Members of this group are microscopic in size and may consist of a single cell, a small multinucleate mass, or a true mycelium. As zoosporic fungi, chytrid spores have a single, posterior, whiplash flagellum.

Two chytrids that are having a profound impact are *Batrachochytrium dendrobatidis,* or Bd for short, and its close relative *B. salamandrivorans* (Bsal). Bd and Bsal cause chytridiomycosis, which has decimated amphibian populations worldwide (**figure 24.5**). Whereas Bd emerged from East Asia about 100 years ago, it has since evolved into several genetically distinct lineages that are now found across the globe. Bsal was introduced to Europe from Asia in 2010 via the pet trade, and large die-offs of salamander populations have been reported. Bsal is particularly hardy in the environment because it produces an encysted spore covered by a thick, water-resistant cell wall. Over 500 species of amphibians have been decimated by Bd and Bsal infections and at least 90 species have been driven to extinction. In fact, Bd and Bsal appear to be causing the greatest disease-induced loss of biodiversity in recorded history. It is unclear why chytridiomycosis has established such a firm grip on some species but not others. Those that are unaffected often serve as carriers. Bd and Bsal affect the skin of infected amphibians, which becomes overly thick (hyperkeratinized). Because gas exchange occurs through amphibian skin, infected animals are unable to regulate the concentration of blood electrolytes and die in cardiac arrest.

Figure 24.5 A Chytridiomycosis Susceptible Host. Chytridiomycosis is killing amphibians. Shown here is the red-eyed tree frog (*Agalychnis callidryas*), whose population has succumbed to infection.

Irina Kozorog/Shutterstock

Comprehension Check

1. Why were microsporidia formerly considered protists?
2. What structure is essential for the germination of a microsporidian spore? Describe its function.
3. Which fungi are responsible for the current destruction of amphibians? What is the mechanism by which these fungi cause disease?
4. Compare and contrast habitats of microsporidia and chytrids.

24.3 Zygomycetous Fungi Have Coenocytic Hyphae

After reading this section, you should be able to:

a. Recognize zygomycetous fungi in nature
b. Outline the life cycle of a *Rhizopus* species
c. Discuss the benefits and harm attributed to mucoromycetes
d. Describe the functional importance of Glomeromycota
e. Discuss the contributions of each partner in a mycorrhizal fungus-plant symbiosis

With the loss of flagella, fungi evolved alternative means of spore dispersal. An important mechanism is the ability to form hyphae and spores that can be blown in the wind or cling to animals. Sexual reproduction in zygomycetous fungi produces tough, thick-walled zygotes called **zygospores** that can remain dormant in unfavorable environments. Zygomycetous fungi form hyphae that lack septae (cross walls) and are therefore coenocytic. Although formerly considered a single group (Zygomycota), phylogenomic analysis places these fungi in two phyla: Zoopagomycota, which often are animal-associated, and Mucoromycota. Here we discuss the saprophytic Mucoromycota and the plant symbiont Glomeromycota, currently classified as a subphylum of Mucoromycota.

Mucoromycota Include Fungi of Industrial Importance

Most members of Mucoromycota, also called mucoromycetes, are free-living saprophytes, but some are parasitic or mutualistic (figure 24.1*c, d*). Saprophytic mucoromycetes include fungi used in the food industry, as well as those that cause food spoilage. Ectomycorrhizae, which form mutualistic relationships with plants, are also members of this group. *Mycorrhizae facilitate the growth of most plants (section 30.3); Microbiology of food (chapter 40)*

The mold *Rhizopus stolonifer* is a common mucoromycete. This fungus grows on the surface of moist, carbohydrate-rich foods, such as breads, fruits, and vegetables. Hyphae called rhizoids extend into the substrate and absorb nutrients. Other hyphae become erect, then arch back into the substratum, forming new rhizoids. Still others remain erect and their tips develop asexual sporangia filled with hundreds of black spores, giving the mold its characteristic color. Each spore, when liberated, can germinate to start a new mycelium.

Rhizopus spp. usually reproduce asexually, but if food becomes scarce or environmental conditions unfavorable, sexual reproduction occurs (**figure 24.6**). Sexual reproduction requires compatible strains of opposite mating types, called + and −. When + and − hyphae are close, each produces a different hormone, called a pheromone, that causes their hyphae to form projections called *progametangia;* these mature into *gametangia.* First, the gametangia fuse, a process called *plasmogamy.* Then, the nuclei of the two gametes fuse (*karyogamy*), forming a zygote. The zygote develops a thick, rough, black coat and becomes a dormant zygospore. Meiosis often occurs at the time of germination; the zygospore then splits open and produces a hypha that bears an asexual sporangium filled with haploid spores that begin the cycle again.

One important mucoromycete is *Rhizopus chinensis,* the cause of the rice disease known as seedling blight. If one considers that about 3.5 billion people depend on rice for food security, the implications of this disease are obvious. It was thought that the fungus secreted a toxin that kills rice seedlings, so scientists set about isolating the toxin and the genes that produce it. Much to everyone's surprise, the toxin is produced by a γ-proteobacterium, *Burkholderia* sp., growing within the fungus.

Mucoromycetes also contribute to human welfare. For example, *Rhizopus oligosporus* is used in Indonesia to produce tempeh from boiled soybeans. Other mucoromycetes (*Mucor* spp.) are used with soybeans in Asia to make a curd called sufu. Others are employed in the commercial preparation of some anesthetics, birth control agents, industrial alcohols, meat tenderizers, and the yellow coloring used in margarine and butter substitutes. *Mucor* and *Rhizopus* spp. also cause a rare but potentially fatal infection in humans called *mucormycosis.* Although rare, it most commonly infects people with diabetes whose blood sugar levels are poorly controlled; the fungi thrive on the patient's extracellular glucose. *Microbiology of fermented foods: beer, cheese, and much more (section 40.5)*

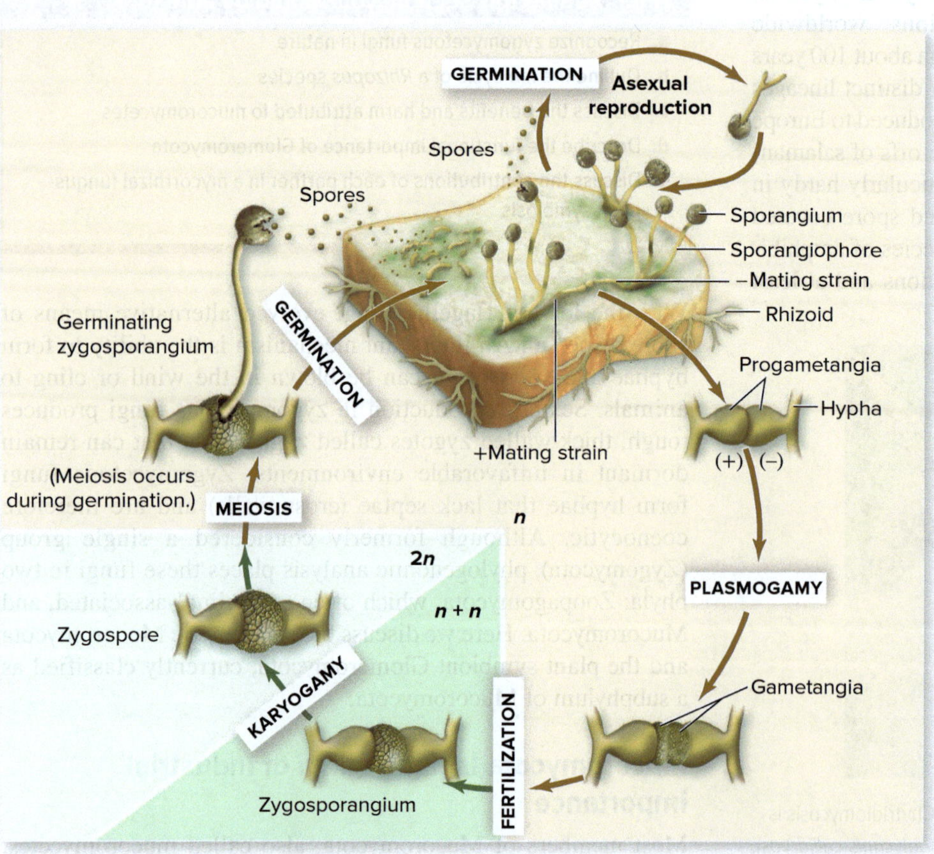

Figure 24.6 ***Rhizopus stolonifer*, a Mucoromycete That Grows on Simple Sugars.** This fungus is often found on moist bread or fruit.

Glomeromycota Are Mycorrhizal Symbionts

Members of Glomeromycota, or glomeromycetes, are of critical ecological importance because they are symbionts of vascular plants. All glomeromycetes are **mycorrhizal fungi** that grow in association with the roots of their plant hosts, which include herbaceous plants and tropical trees. This is a mutualistic relationship because both the host plant and the fungus benefit: The fungus delivers soil nutrients to the plant, and the plant in turn provides carbohydrate to the fungus. Among the types of mycorrhizal fungi discussed in chapter 30, only the arbuscular mycorrhizal hyphae are glomeromycetes. Arbuscular mycorrhizae are important because of the number and diversity of plants they colonize. *Mycorrhizae facilitate the growth of most plants (section 30.3)*

Like mucoromycetous fungi, glomeromycetes have aseptate hyphae and produce large, multinucleate spores. Only asexual reproduction has been documented, but there is evidence sexual reproduction may occur as well. Spores germinate when in contact with the roots of a suitable host plant. Glomeromycetes have evolved specialized flat hyphae called **appressoria** (s., appressorium) to penetrate their host plants.

Comprehension Check

1. What features unite the zygomycetous fungi?
2. How do different mating types of *Rhizopus stolonifer* recognize each other?
3. Given that mycorrhizal fungi obtain carbohydrates from their plant partners, what specific nutrients are the fungi likely to provide to plants?

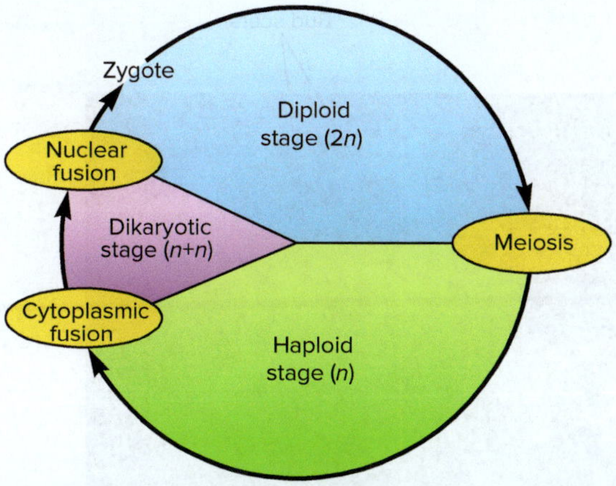

Figure 24.7 Alternation of Haploid and Diploid Reproduction in Dikaryotic Fungi. Following cytoplasmic fusion, the two parental nuclei exist in a single cell for some variable period of time prior to fusing. The asexual (haploid) stage is used to produce spores that aid in the dissemination of the species. The sexual (diploid) stage involves the formation of spores that survive adverse environmental conditions (e.g., desiccation and temperature extremes).

24.4 Dikarya Is the Most Diverse Fungal Group

After reading this section, you should be able to:

- **a.** Describe the ecological, economic, and medical importance of ascomycetes
- **b.** Diagram the *Saccharomyces cerevisiae* life cycle
- **c.** Outline the mechanisms used by typical filamentous ascomycetes to reproduce, survive unfavorable conditions, and disperse
- **d.** Explain the function of sclerotia
- **e.** Identify the origin of St. Anthony's fire, LSD, and sick building syndrome
- **f.** Sketch the life cycle of a typical basidiomycete
- **g.** Identify the fungi that cause plant rusts
- **h.** Describe a plant disease caused by a basidiomycete

Dikarya, or dikaryotic fungi, is the largest group of fungi and includes two large taxa, Ascomycota and Basidiomycota. Dikaryotic fungi have septate hyphae and cell membranes with the sterol ergosterol (an important drug target for those that cause human disease). The term dikaryotic, which roughly translates as "two nuclei," describes the nature of reproduction in these fungi. In other fungi (as well as higher eukaryotes), when two haploid cells come together, their cytoplasms mix and the two nuclei fuse immediately to produce a diploid zygote (2*n*). By contrast, cytoplasmic and nuclear fusion do not occur in rapid succession in dikaryotic fungi. This produces a stage in which a single cell contains two separate haploid nuclei ($n + n$), one from each parent (**figure 24.7**). After a period of time, the two nuclei fuse and undergo meiosis to yield haploid spores.

Ascomycota Includes Yeasts and Molds

Members of Ascomycota, or **ascomycetes,** commonly known as **sac fungi,** are named for their characteristic reproductive structure, the **ascus** (pl., asci; Greek *askos,* sac), which holds **ascospores** (figure 24.1*e*). Ascomycetes are ecologically important in freshwater, marine, and terrestrial habitats because they degrade many chemically stable organic compounds, including lignin, cellulose, and collagen. Many species are quite familiar and economically important. For example, most of the red, brown, and blue-green molds that cause food spoilage are ascomycetes. The powdery mildews that attack plant leaves and the fungi that cause chestnut blight and Dutch elm disease are ascomycetes. Most fungi that infect humans are ascomycetes. On the other hand, many yeasts as well as edible morels and truffles are also ascomycetes (**figure 24.8**). The pink bread mold *Neurospora crassa* is an important research tool in genetics and biochemistry.

Some ascomycetes are yeasts, whereas others have a life cycle that alternates between yeast and mold forms. The life cycle of the yeast *Saccharomyces cerevisiae,* commonly known as brewer's or baker's yeast, is well understood (**figure 24.9**). *S. cerevisiae* alternates between haploid and diploid states. As long as nutrients remain plentiful, haploid and diploid cells

(a) *Morchella esculenta*

(b) *Tuber brumale*

Figure 24.8 ***Ascomycota.*** (a) The common morel, *Morchella esculenta,* is one of the choicest edible fungi. It fruits in the spring. (b) The black truffle, *Tuber brumale,* is highly prized for its flavor.

(a) Juniors Bildarchiv GmbH/Alamy Stock Photo; (b) Kelly Cline/E+/Getty Images

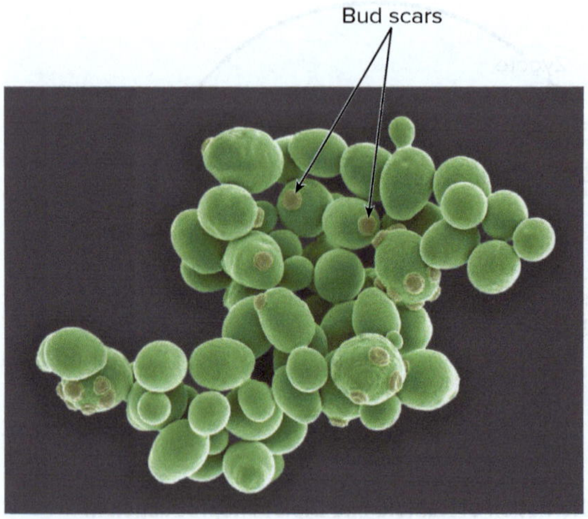

(a) *Saccharomyces cerevisiae:* budding division

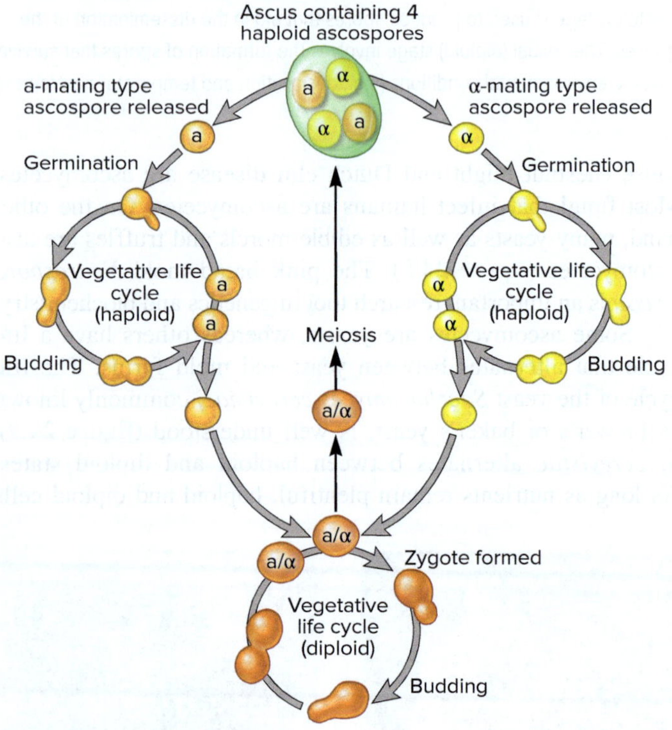

(b) *S. cerevisiae* life cycle

Figure 24.9 The Life Cycle of the Yeast *Saccharomyces cerevisiae*. (a) Budding division; in this colorized scanning electron micrograph, bud scars are tan. Cells are 5–10 μm in diameter. (b) When nutrients are abundant, haploid and diploid cells undergo mitosis and grow vegetatively. When nutrients are limited, diploid *S. cerevisiae* cells undergo meiosis. Upon the addition of nutrients, two haploid cells of opposite mating types (a and α) fuse to create a diploid cell.

(a) Steve Gschmeissner/Science Photo Library/Alamy Stock Photo

MICRO INQUIRY *Which cell in panel (a) appears to be closest to senescence?*

undergo mitosis to produce haploid and diploid daughter cells, respectively. Each daughter cell leaves a scar on the mother cell as it separates, and daughter cells bud only from unscarred regions of the cell wall. When a mother cell has no more unscarred

Figure 24.10 Asexual Reproduction in Filamentous Ascomycetes. Characteristic conidiospores of *Aspergillus* sp. as viewed with the scanning electron microscope.

Microfield Scientific Ltd./SPL/Science Source

cell wall remaining, it can no longer reproduce and will senesce (die). When nutrients are limited, diploid *S. cerevisiae* cells undergo meiosis to produce four haploid cells that remain bound within a common cell wall, the ascus. Upon the addition of nutrients, two haploid cells of opposite mating types (a and α) come into contact and fuse to create a diploid. Typically only cells of opposite mating types can fuse; this process is tightly regulated by the action of pheromones.

S. cerevisiae is a valuable model organism. Research on this organism has revealed the importance of many cellular processes. For instance, it is a favorite model system for studying the eukaryotic cell cycle. This research is critical not only for our understanding of normal cell division, but also for exploring the loss of cell cycle control that occurs in cancerous cells.

Among filamentous ascomycetes, asexual reproduction is common and is associated with the production of asexual spores called **conidia** (**figure 24.10**). Sexual reproduction involves two strains of opposite mating types. Hyphae from one mating type form a specialized structure called an *ascogonium,* and hyphae of the other mating type form an *antheridium* (**figure 24.11**). The nuclei from the antheridium migrate to the ascogonium during the exchange of cytoplasm (plasmogamy). The nuclei from each mating type pair and replicate in unison during the ensuing dikaryotic phase. The ascogonium is said to be fertilized and the hyphae are called **ascogenous hyphae.** Ascogenous hyphae next develop into a cuplike structure called an **ascocarp.** This is where ascogenous hyphae mature and nuclear fusion (karyogamy) takes place. The diploid zygote nucleus then undergoes meiosis, and the resulting four haploid nuclei divide mitotically again to produce a row of eight nuclei in each developing ascus. These nuclei are walled off from one another and each will develop into an **ascospore.** Thousands of asci may be packed together in an ascocarp. When ascospores mature, they often are released from the asci with great force. If the mature ascocarp is jarred, it may appear to belch puffs of smoke consisting of thousands of ascospores. Upon reaching a suitable environment, the ascospores germinate and start the cycle anew.

Although conidia are the major form of dissemination, structures called **sclerotia** are compact masses of dormant hyphae that

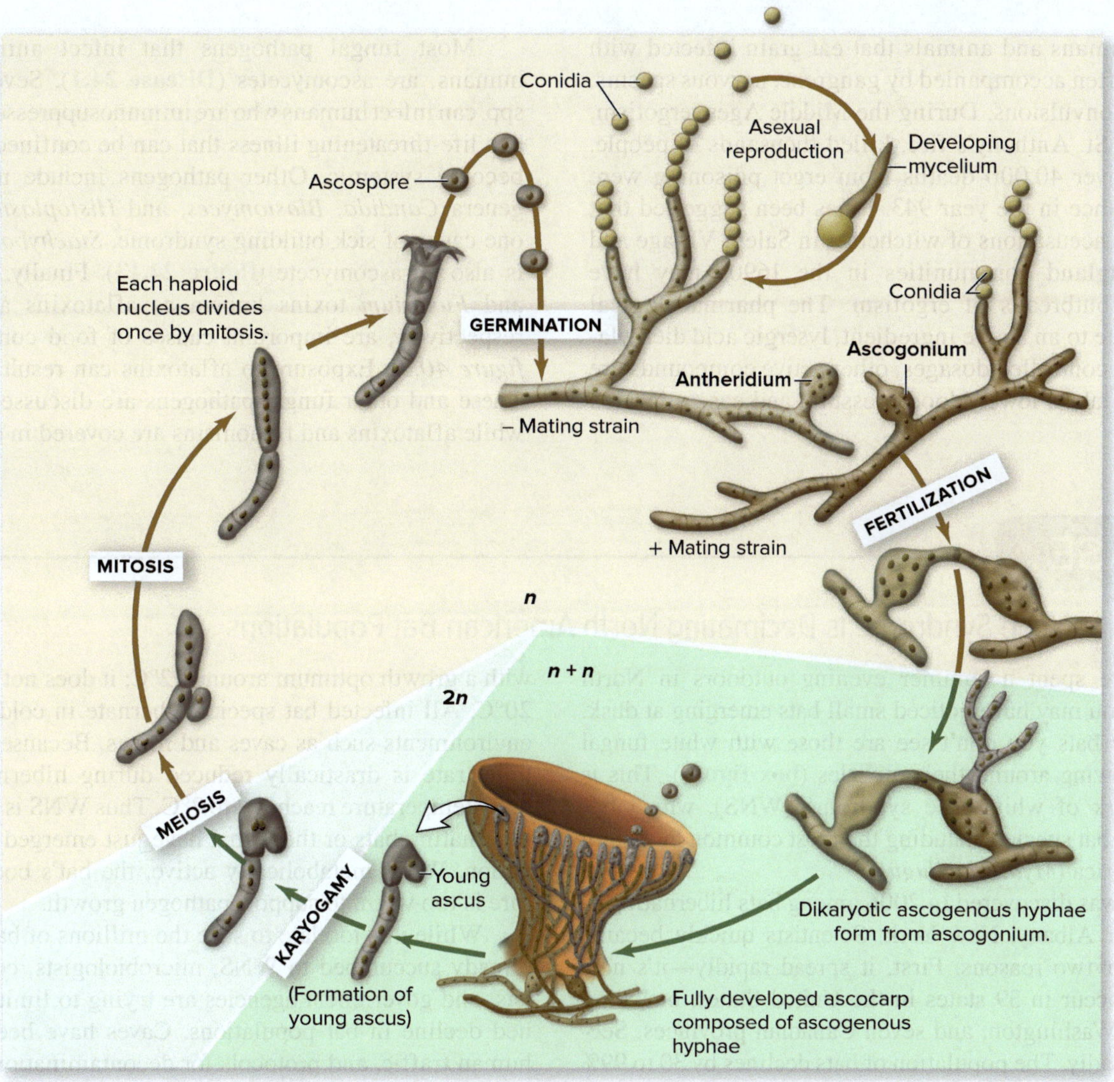

Figure 24.11 The Typical Life Cycle of a Filamentous Ascomycete. Sexual reproduction involves the formation of asci and ascospores. The events pictured on the left side of the figure (karyogamy, meiosis, mitosis, and release of ascospores) occur within the ascocarp.

MICRO INQUIRY *In what ways is the life cycle of filamentous ascomycetes similar to that of the mucoromycete shown in figure 24.6? How do the two life cycles differ?*

permit survival in stressful conditions. Sclerotia confer a competitive advantage to the fungi that produce them. For instance, some species of *Aspergillus* that infect plants form sclerotia to remain viable in the soil during cool months. They then take advantage of nutrient resources when the temperature rises.

Several *Aspergillus* species are noteworthy. *A. fumigatus* is ubiquitous, commonly found in homes and the workplace. It is known to trigger allergic responses and is implicated in the increased incidence of severe asthma and sinusitis. *A. nidulans* is a model organism used to study questions of eukaryotic cell and developmental biology. *A. oryzae* is used in the production of traditional fermented foods and beverages in Japan, including sake and soy sauce. Because *A. oryzae* secretes many industrially useful proteins and can be genetically manipulated, it has become an important organism in biotechnology.

Many ascomycetes are parasites of higher plants. *Claviceps purpurea* parasitizes rye and other grasses, causing the plant disease **ergot** (**figure 24.12**). Ergotism, the toxic, delusional

Figure 24.12 Ergot of Rye. The ascomycete *Claviceps purpurea* infects rye and other grasses, producing hard masses of hyphae known as ergots in place of some of the grains (fruits).

Carmen Rieb/Shutterstock

condition in humans and animals that eat grain infected with the fungus, is often accompanied by gangrene, nervous spasms, abortion, and convulsions. During the Middle Ages, ergotism, then known as St. Anthony's fire, killed thousands of people. For example, over 40,000 deaths from ergot poisoning were recorded in France in the year 943. It has been suggested that the widespread accusations of witchcraft in Salem Village and other New England communities in the 1690s may have resulted from outbreaks of ergotism. The pharmacological activities are due to an active ingredient, lysergic acid diethylamide (LSD). In controlled dosages, other active compounds are used to induce labor, lower blood pressure, and ease migraine headaches.

Most fungal pathogens that infect animals, including humans, are ascomycetes (**Disease 24.1**). Several *Aspergillus* spp. can infect humans who are immunosuppressed. Aspergillosis is a life-threatening illness that can be confined to the lungs or become systemic. Other pathogens include members of the genera *Candida, Blastomyces,* and *Histoplasma.* In addition, one cause of sick building syndrome, *Stachybotrys chartarum,* is also an ascomycete (**figure 24.13**). Finally, the *Aspergillus* and *Fusarium* toxins known as aflatoxins and fumonisins, respectively, are important causes of food contamination (*see figure 40.3*). Exposure to aflatoxins can result in liver cancer. These and other fungal pathogens are discussed in chapter 39, while aflatoxins and fumonisins are covered in chapter 40.

DISEASE

24.1 White-Nose Syndrome Is Decimating North American Bat Populations

If you have spent a summer evening outdoors in North America, you may have noticed small bats emerging at dusk. Among the bats you don't see are those with white fungal hyphae growing around their muzzles (**box figure**). This is the hallmark of white-nose syndrome (WNS), which has infected 13 bat species including the most common in eastern North America (*Myotis lucifugus*).

WNS was discovered in 2006 among bats hibernating in a cave near Albany, New York. Scientists quickly became alarmed for two reasons. First, it spread rapidly—it's now known to occur in 39 states in the United States, including Texas and Washington, and seven Canadian provinces. Second, it is deadly. The population of bats declines by 30 to 99% in any given infected hibernacula (the place where bats hibernate; rhymes with Dracula).

WNS is caused by the ascomycete *Pseudogymnoascus destructans.* It colonizes a bat's wings, muzzle, and ears, where it erodes the epidermis before invading the underlying skin and connective tissue. Despite the name WNS, the primary site of infection is the wing. Wings provide a large surface area for colonization, and once infected, the thin layer of skin is easily damaged, leading to adverse physiological changes during hibernation. These in turn result in premature awakening, loss of essential fat reserves, and strange behavior.

Where did this pathogen come from and why does it infect bats? The best hypothesis is that in the early 2000s humans inadvertently brought it from Europe, where it causes mild infection in at least one hibernating bat species. This makes *P. destructans* an apparent case of pathogen pollution—the human introduction of invasive pathogens of wildlife and domestic animal populations that threaten biodiversity and ecosystem function.

The capacity of *P. destructans* to sweep through bat populations results from a perfect storm of host- and pathogen-associated factors. *P. destructans* is psychrophilic, with a growth optimum around 12°C; it does not grow above 20°C. All infected bat species hibernate in cold and humid environments such as caves and mines. Because their metabolic rate is drastically reduced during hibernation, their body temperature reaches 2 to 7°C. Thus WNS is only seen in hibernating bats or those that have just emerged from hibernation. When metabolically active, the bat's body temperature is too warm to support pathogen growth.

While it is too late to save the millions of bats that have already succumbed to WNS, microbiologists, conservationists, and government agencies are trying to limit the continued decline in bat populations. Caves have been closed to human traffic, and protocols for decontamination after visiting hibernacula have been developed to limit the spread from cave to cave. Although we cannot cure sick bats, it is our responsibility to stop the continued spread of this pathogen.

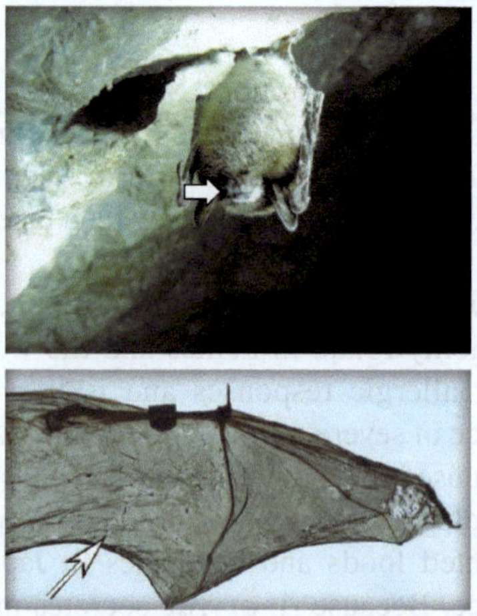

Pseudogymnoascus destructans **Causes WNS.** A little brown bat (*Myotis lucifugus*) with the white fungal hyphae (arrow) for which WNS is named. Lower figure shows damage to wing as a result of fungal infection.

Figure 24.13 Fungal "Sick Building Syndrome" Can Be Caused by an Ascomycete. Scanning electron micrograph of dense mycelia and conidiophores characteristic of *Stachybotrys chartarum*.

Scimat/Science Source

Basidiomycota Includes Mushrooms and Plant Pathogens

Fungi in Basidiomycota are commonly known as **basidiomycetes,** or **club fungi.** Examples include jelly fungi, rusts, shelf fungi, stinkhorns, puffballs, toadstools, mushrooms, and bird's nest fungi. Most are saprophytes that decompose plant debris, especially cellulose and lignin. For example, the common fungus *Polyporus squamosus* forms large, shelf-like structures that project from the lower portion of dead trees, which they help decompose. Many mushrooms are used as food, and the cultivation of the mushroom *Agaricus campestris* is a multimillion-dollar business. Of course, not all mushrooms are edible. Many mushrooms produce alkaloids that act as either poisons or hallucinogens. One such example is the death cap mushroom, *Amanita phalloides* (**figure 24.14*a***). Two toxins isolated from this species are phalloidin and α-amanitin. Phalloidin primarily attacks liver cells, where it binds to and ruptures plasma membranes. Alpha-amanitin attacks the cells lining the stomach and small intestine, causing severe gastrointestinal symptoms associated with mushroom poisoning.

(a)

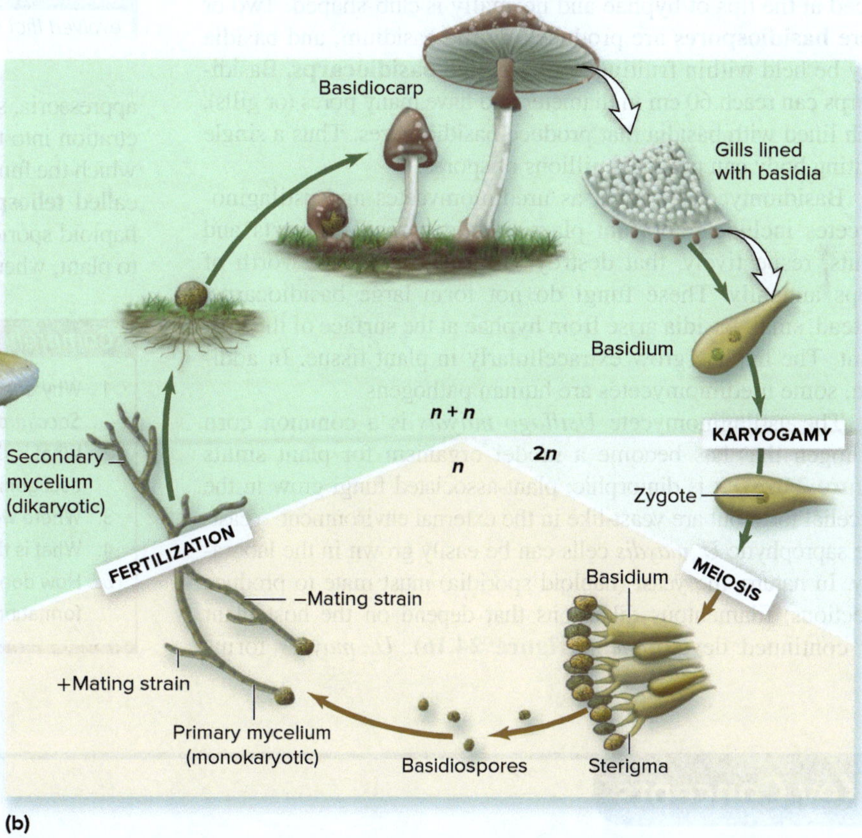

(b)

Figure 24.14 Basidiomycota. (a) The death cap mushroom *Amanita phalloides* is usually fatal when eaten. (b) The life cycle of a typical soil basidiomycete starts with a basidiospore germinating to produce a monokaryotic mycelium (a single nucleus in each septate cell; bottom center). The mycelium grows and spreads throughout the soil. When this primary mycelium meets another monokaryotic mycelium of a different mating type, the two fuse to initiate a dikaryotic secondary mycelium. Hyphae in the secondary mycelium are divided by septa into cells, each of which contains two nuclei, one of each mating type. This dikaryotic mycelium is eventually stimulated to produce basidiocarps. A solid mass of hyphae forms a button that pushes through the soil, elongates, and develops a cap. The cap contains many plate-like gills, each of which is coated with basidia. The two nuclei in the tip of each basidium fuse to form a diploid zygote nucleus, which immediately undergoes meiosis to form four haploid nuclei. These nuclei push their way into the developing basidiospores, which are then released at maturity.

(a) Christine Whitehead/Alamy Stock Photo

Figure 24.15 ***Ustilago maydis.*** This pathogen causes tumor formation in corn. Note the enlarged tumors in place of normal corn kernels.

Marina Mariya/iStock/Getty Images

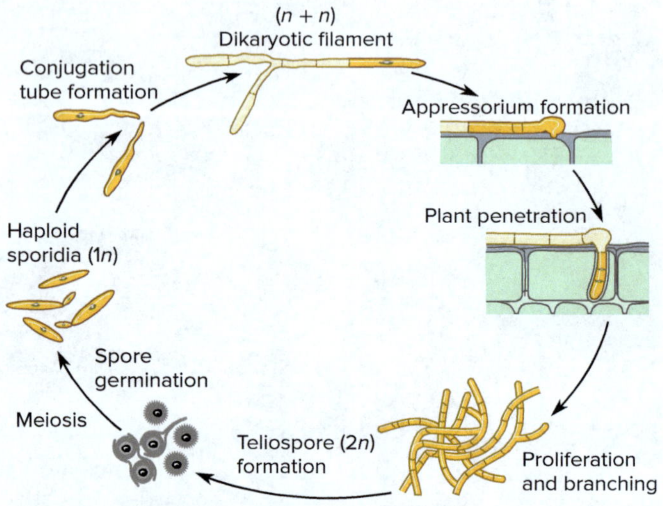

Figure 24.16 Life Cycle of the Plant Pathogen *Ustilago maydis*. Haploid sporidia (left) mate to form infectious dikaryons, which depend on the host plant for continued development. The formation of fungal appressoria within the host induces tumor formation.

MICRO INQUIRY *What adaptations do you think* U. maydis *has evolved that has led to its pathogenicity? Contrast this with mycorrhizae.*

Basidiomycetes are named for their characteristic structure or cell, the **basidium,** which is involved in sexual reproduction (figure 24.14*b*). A basidium (Greek *basidion,* small base) is produced at the tips of hyphae and normally is club-shaped. Two or more **basidiospores** are produced by the basidium, and basidia may be held within fruiting bodies called **basidiocarps.** Basidiocarps can reach 60 cm in diameter and have many pores (or gills), each lined with basidia that produce basidiospores. Thus a single fruiting body can produce millions of spores.

Basidiomycetes known as urediniomycetes and ustilaginomycetes include important plant pathogens causing rusts and smuts, respectively, that destroy millions of dollars' worth of crops annually. These fungi do not form large basidiocarps. Instead, small basidia arise from hyphae at the surface of the host plant. The hyphae grow extracellularly in plant tissue. In addition, some urediniomycetes are human pathogens.

The ustilaginomycete *Ustilago maydis* is a common corn pathogen that has become a model organism for plant smuts (**figure 24.15**). It is dimorphic; plant-associated fungi grow in the mycelial form but are yeast-like in the external environment. Yeast-like saprophytic *U. maydis* cells can be easily grown in the laboratory. In nature, the yeast (haploid sporidia) must mate to produce infectious, filamentous dikaryons that depend on the host plant for continued development (**figure 24.16**). *U. maydis* forms appressoria, specialized filaments that mediate attachment and penetration into the host cell. This triggers the plant to form tumors in which the fungus proliferates and eventually produces diploid spores called **teliospores.** Upon germination, cells undergo meiosis and haploid sporidia are released. These sporidia can spread from plant to plant, where they start the growth and infection cycle again.

Comprehension Check

1. Why does nutrient deprivation trigger sexual reproduction in *Saccharomyces cerevisiae?*
2. What is the difference between an antheridium and an ascogonium? What is an ascocarp?
3. Where would you most likely find ascomycetes that form sclerotia?
4. What is the difference between a basidiocarp and a basidium?
5. How does *Ustilago maydis* trigger tumor formation? How is tumor formation advantageous to the fungi?

Key Concepts

24.1 Fungal Biology Reflects Vast Diversity

- Fungi are important decomposers that break down organic matter; live as parasites on animals, humans, and plants; play a role in many industrial processes; and are used as research tools in the study of fundamental biological processes. Fungi secrete enzymes and absorb the digested food.
- A fungus is a eukaryotic, spore-bearing organism that has absorptive nutrition and lacks chlorophyll; reproduces asexually, sexually, or by both methods; and normally has a cell wall containing glucans and chitin.
- Hyphae are long, branched, thread-like filaments that form a tangled mass called a mycelium. Hyphae may be either

septate or coenocytic (nonseptate). The mycelium can produce reproductive structures (**figure 24.2**).

- Asexual reproduction in fungi often leads to the production of specific types of spores that are easily dispersed (**figure 24.3**).
- Sexual reproduction is initiated in fungi by the fusion of hyphae (or cells) of different mating types within a species.

24.2 Zoosporic Fungi Form Motile Spores

- Zoosporic fungi display a single, posterior flagellum at some point in their life cycle.
- Members of Microsporidia have a unique morphology and are obligate intracellular pathogens (**figure 24.4**).
- Fungi belonging to Chytridiomycota produce motile spores. Most are saprophytic; chytridiomycosis is causing massive, global amphibian die-offs (**figure 24.5**).

24.3 Zygomycetous Fungi Have Coenocytic Hyphae

- Zygomycetous fungi form hyphae that lack septa and are therefore coenocytic. In most cases, sexual reproduction produces tough, thick-walled zygotes called zygospores.
- Most members of the Mucoromycota are saprophytic; the group also includes ectomycorrhizae.
- One example of mucoromycete is the common bread mold, *Rhizopus stolonifer.* Sexual reproduction occurs through fusion of strains of opposite mating type (**figure 24.6**).
- Mucoromycetes are used in food and pharmaceutical manufacturing.
- Glomeromycota includes arbuscular mycorrhizal fungi that grow in association with plant roots. They provide soil nutrients to the plant, which in turn feeds the fungus organic carbon.

24.4 Dikarya Is the Most Diverse Fungal Group

- Dikaryotic fungi are so named because during sexual reproduction, cytoplasmic fusion precedes nuclear fusion, so there is a period of time during which the two haploid nuclei co-exist in the same cell prior to nuclear fusion (**figure 24.7**).
- Members of Ascomycota are known as sac fungi because they form a sac-shaped reproductive structure called an ascus. Ascomycetes can have either a yeast morphology (**figure 24.9**) or a mold morphology (**figure 24.11**).
- Important ascomycetes include the yeast *Saccharomyces cerevisiae, Aspergillus* spp., *Claviceps purpurea* (the cause of ergotism), and the indoor mold *Stachybotrys chartarum.*
- Members of Basidiomycota are called club fungi. They are named after their basidium that produces basidiospores (**figure 24.14**).
- Genera from urediniomycetes and ustilaginomycetes include important plant pathogens called smuts and rusts (**figures 24.15** and **24.16**).

Active Learning

1. Some fungi can be viewed as coenocytic organisms that exhibit little differentiation. When differentiation does occur, such as in the formation of reproductive structures, it is preceded by septum formation. Why is the development of cross-walls necessary at this time?
2. Both bacteria and fungi are major environmental decomposers. Obviously competition exists in any given environment, but fungi usually have an advantage. Discuss characteristics specific to fungi that provide this advantage.
3. *Neurospora crassa* and several other filamentous fungi demonstrate a circadian rhythm such that conidiation (sporulation) occurs in roughly 21.5-hour cycles, depending on temperature and the light/dark regime. This can be observed in nature as bands of conidia alternating with bands of undifferentiated fungi. Suggest why these fungi evolved a circadian rhythm. If you were to study this phenomenon, what questions would you want to explore?
4. Analysis of sclerotia and conidia formation in the filamentous ascomycete *Aspergillus flavus* suggests that formation of these structures is cell density dependent. Specifically, high cell density cultures yield conidia, whereas low cell density triggers sclerotia formation. Review the structure and function of conidia and sclerotia, and formulate a hypothesis to explain why their development would be regulated in a density-dependent manner.
5. *Ustilago maydis* causes corn smut disease (figure 24.15). Little is known about what triggers the yeast form of the fungus to switch to the hyphal form that invades host plants. This dimorphic transition is governed by the two mating loci *a* and *b*. If you were to explore this morphological switch, what environmental factors would you test? Describe an experiment, with controls, to test hypotheses about specific conditions that might trigger the morphological switch. How would you monitor the changes in gene expression? (Refer to chapter 18.)

CHAPTER

25

Viruses

K. Sandman/McGraw Hill

Disrupting the Viral Life Cycle

Cats and humans have cohabited for thousands of years, and as every cat owner knows, they live in intimate contact. From the cat's perspective, all household goods, especially beds, are communal. As with all close relationships, partners also share microbiomes. While we know that you can catch a virus from another person in a household, we also know that it is uncommon to get a virus from a pet. Why? As we discuss in chapter 6, for a virus to complete its life cycle, it must bind to a specific receptor on the surface of a host cell, enter, and subvert cellular mechanisms to ensure its replication and release. That first step in infection is critical—virus binding to a host cell receptor. All tissues have a unique and characteristic set of surface structures. A virus typically infects only a single tissue type because that's where its compatible receptor is located. This tissue tropism explains why we avoid touching our face to protect ourselves from respiratory viruses like influenza or SARS-CoV-2—while the virus on our hands can't cause infection directly, it can be transferred to the eyes or nose, introducing it to the susceptible respiratory epithelium.

Feline leukemia virus (FeLV) causes an immune deficiency in cats that may make them vulnerable to other potentially fatal infections and cancers. The virus is a retrovirus that infects several cell types in the animal. Transmission of FeLV between cats occurs by contact with an infected animal's secretions (saliva, nasal secretions, and sometimes urine). As cats groom with a coating of saliva, simply touching the animal exposes humans to the virus. How does the FeLV life cycle work at a molecular level, and what might this reveal about how humans avoid infection, despite repeated exposure?

FeLV enters cat cells through a membrane protein that normally transports phosphate. This common and widespread protein is highly conserved between cats and humans, and indeed, phosphate transporters in human plasma membranes transport FeLV. Most human cells are susceptible to FeLV infection. Importantly, though, the virus and host cell must have a prolonged molecular conversation for the virus to succeed in infecting the host. Any breakdown in communication can stall viral genome replication, viral gene expression, virion assembly, or virion release. It is in these later steps in the viral life cycle that FeLV replication is halted in human cells.

In this chapter, we examine the varied strategies used by viruses. In many ways, this discussion is a continuation of topics introduced in chapters 14 and 15, where we consider the regulation of cellular processes, and is continued in chapter 37, where we discuss diseases caused by human viruses.

Readiness Check:

Based on what you have learned previously, you should be able to:

- ✓ Describe common virion structures and morphology (section 6.2)
- ✓ Summarize the five steps common to the life cycle of most viruses (section 6.3)
- ✓ Describe the events that occur during the three phases of DNA replication, transcription, and translation (sections 13.3, 13.5, 13.7, 15.2, 15.3, and 15.4)
- ✓ Draw typical bacterial, archaeal, and eukaryotic protein-coding genes (sections 13.4, 13.5, and 15.3)

25.1 Virus Phylogeny Relies on Genomics

After reading this section, you should be able to:

a. Explain the challenge to establishing viral phylogeny
b. Distinguish between the Baltimore system of grouping viruses and the official taxonomy of viruses curated by the International Committee on Taxonomy of Viruses
c. Determine if an RNA virus has a positive- or negative-strand genome

Virus classification is overseen by the International Committee on Taxonomy of Viruses (ICTV). For decades, the ICTV used a system that considered features like nucleic acid type, the presence or absence of an envelope, capsid symmetry, and capsid dimensions. More recently, metagenomic sequencing has revealed viruses known only by their nucleotide sequence. Because neither the host organism nor the virion morphology can be discerned from metagenomic sequence, these uncultivated viruses were previously not included in the ICTV classification. ◀ *Virion structure is defined by capsid symmetry and presence or absence of an envelope (section 6.2)*

To piece together the phylogenetic relationships of all viruses regardless of whether they are cultivated, two methods that rely on comparative sequence analysis are used by virologists and the ICTV. The first method identifies gene sharing networks. The viral gene pool is vast, but a virus species carries only the minimal genome needed to carry out its life cycle. Horizontal gene transfer between viruses and between viruses and their hosts is common. Genome network analysis tracks the evolutionary history of viral clades. The second method relies on proteins that are not universal, but widespread within viral clades. These virus hallmark proteins specify capsid structure and genome replication enzymes.

In 2017, the ICTV expanded the taxonomic criteria to include data from uncultivated viruses, thereby modernizing its classification system. Viral taxa echo the familiar scheme for cellular organisms, but with the highest taxon, **realm,** equivalent to domain (**table 25.1**). In 2021, the ICTV adopted a standard binomial format for virus species names that includes a genus and a species epithet.

Although the ICTV classification is the official authority on viral taxonomy, many virologists find it useful to group viruses using a scheme devised by Nobel laureate David Baltimore. The Baltimore system complements the ICTV phylogenetic system but focuses on the viral genome and the process used to synthesize viral mRNA. Recall from chapter 6 that all four nucleic acid types can be found in viruses: dsDNA, ssDNA, dsRNA, and ssRNA. The characterization of the genome of a ssRNA virus is further differentiated by the sense of the ssRNA—that is, whether the genome is identical to or complementary to the mRNA produced by the virus. ssRNA viruses with an RNA genome that is identical in base sequence to that of the mRNA it produces are said to have **positive-strand** or **plus-strand RNA** genomes (**figure 25.1**). Other ssRNA viruses have genomes that are complementary to the mRNA they produce. These viruses are said to have **negative-strand** or **minus-strand RNA.** Baltimore's system organizes viruses into seven groups (**table 25.2**). This system helps virologists (and microbiology students) simplify the vast array of viral life cycles into a relatively small number of basic types; thus we use it here while acknowledging that it does not reflect evolutionary relationships.

Table 25.1 Taxonomic Ranks for Viruses

Rank	Suffix	Classification of Selected Viruses	
		Measles Virus (MeV)	**Bacteriophage Lambda**
Realm	-viria	Riboviria	Duplodnaviria
Phylum	-viricota	Negarnaviricota	Uroviricota
Class	-viricetes	Monjiviricetes	Caudoviricetes
Order	-virales	Mononegavirales	Caudovirales
Family	-viridae	Paramyxoviridae	Siphoviridae
Genus	-virus	*Morbillivirus*	*Lambdavirus*
Species	(irregular)	*Morbillivirus hominis*	*Lambdavirus lambda*

Comprehension Check

1. List some characteristics used in classifying viruses. Which seem to be the most important?
2. Consider these terms: *Feline leukemia virus,* Artverviricota, Riboviria, Retroviridae, *Gammaretrovirus,* Retriviricetes, Ortervirales. Place each in its proper rank as described in table 25.1.

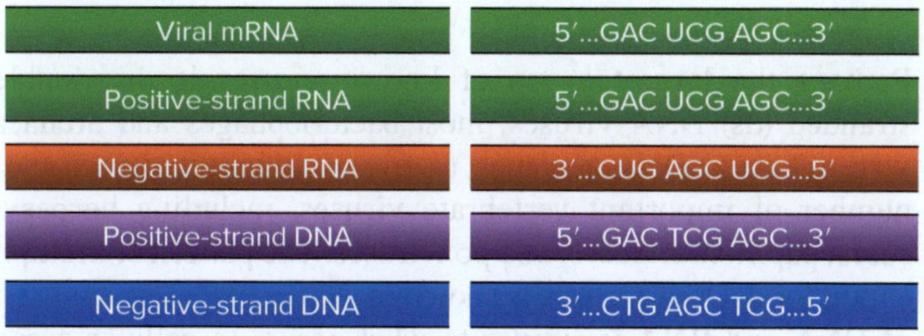

Figure 25.1 Positive-Strand and Negative-Strand Viral Genomes. The genomes and replication intermediates of viral genomes can be either positive or negative strand. This designation is relative to the sequence of nucleotides in the viral mRNA. Positive-strand genomes have the same sequence as the mRNA, either using DNA nucleotides if a DNA genome or RNA nucleotides if an RNA genome. Negative-strand genomes are complementary to the viral mRNA.

Table 25.2	The Baltimore System
Group	**Description**
Double-stranded (ds) DNA viruses	Genome replication: dsDNA → dsDNA Protein synthesis: dsDNA → mRNA → Protein
Single-stranded (ss) DNA viruses	Genome replication: ssDNA → dsDNA → ssDNA Protein synthesis: ssDNA →dsDNA → mRNA → Protein
Double-stranded RNA viruses	Genome replication: dsRNA → ssRNA → dsRNA Protein synthesis: dsRNA → mRNA → Protein
Positive-strand RNA (+RNA) viruses	Genome replication: +RNA → −RNA → +RNA Protein synthesis: +RNA → −RNA → mRNA → Protein
Negative-strand RNA (−RNA) viruses	Genome replication: −RNA → +RNA → −RNA Protein synthesis: −RNA → mRNA → Protein
Retroviruses	Genome replication: ssRNA → dsDNA → ssRNA Protein synthesis: ssRNA → dsDNA → mRNA → Protein
Reverse transcribing DNA viruses	Genome replication: dsDNA → ssRNA → dsDNA Protein synthesis: dsDNA → mRNA → Protein

25.2 Double-Stranded DNA Viruses Infect All Cell Types

After reading this section, you should be able to:

a. Describe in general terms the strategy used by double-stranded (ds) DNA viruses to synthesize their nucleic acids and proteins

b. Describe in general terms how bacteriophage lambda regulates the switch between lytic and lysogenic cycles

c. Choose one specific dsDNA virus and outline the major events in its life cycle, noting, when possible, the specific mechanisms used to accomplish each step

Perhaps the largest group of known viruses is the double-stranded (ds) DNA viruses; most bacteriophages and archaeal viruses have dsDNA genomes, as do several insect viruses and a number of important vertebrate viruses, including herpesviruses, papillomaviruses, and poxviruses. The pattern of multiplication for dsDNA viruses is shown in **figure 25.2**. The synthesis of DNA and RNA is similar to what occurs in cellular organisms; therefore many dsDNA viruses can rely entirely on their host's DNA and RNA polymerases.

Escherichia virus T4: A Virulent Bacteriophage

The life cycle of *Escherichia virus* T4 (family Myoviridae) serves as our example of a virulent (lytic) dsDNA phage. Its

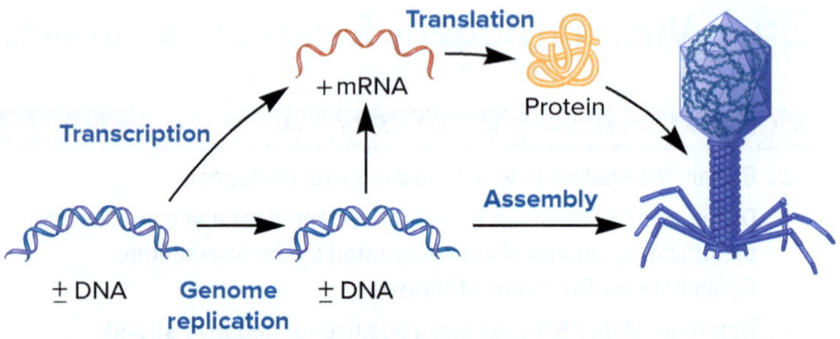

Figure 25.2 Multiplication Strategy of Double-Stranded DNA Viruses.

structure has become the iconic image of bacterial viruses (*see figure 6.6*). Cell lysis is the outcome of an infection with this type of bacteriophage. T4 attachment begins when a long tail fiber contacts either the lipopolysaccharide or certain proteins in the outer membrane of its *Escherichia coli* host. A sequence of events leads to piercing of the cell envelope by a puncturing device composed of viral gene products gp5 (also called *tail lysozyme*), gp5.4, and gp27. Phage DNA is ejected into the host cell's cytoplasm, where gene expression begins (**figure 25.3**). Genes and proteins are designated early or late, depending on when they are required in the phage life cycle. Within 2 minutes after entry of T4 DNA into an *E. coli* cell, the bacterial RNA polymerase starts synthesizing T4 early mRNA (**figure 25.4**). One early T4 gene encodes a protein that binds to the host enzyme RNaseE and directs it to degrade host mRNA. This frees host ribonucleotides and ribosomes for transcription and translation of T4 genes. Within 5 minutes, viral DNA synthesis commences, catalyzed by a virus-encoded **DNA-dependent DNA polymerase** (DdDp). DNA replication is initiated from several origins of replication and proceeds bidirectionally from each. This allows for rapid production of many copies of viral DNA within a short time frame. Viral DNA replication is followed by the synthesis of late mRNAs, which encode proteins important in later stages of the infection.

T4 gene expression is tightly regulated and this is aided by its genome organization. Genes coding for products with related functions—such as those for phage head or tail fiber construction—are clustered together. Early and late genes are clustered separately on the genome; they are even transcribed from different DNA strands. A considerable portion of the T4 genome encodes products needed for replication, including the protein subunits of its replisome and enzymes needed to prepare for DNA synthesis. Some of these enzymes synthesize hydroxymethylcytosine (HMC), a modified nucleotide that replaces cytosine in T4 DNA (**figure 25.5**). Once HMC is synthesized, replication ensues by a mechanism similar to that seen in bacteria. After T4 DNA synthesis, it is modified by the addition of glucose to HMC residues, a process called *glycosylation*. Glycosylated HMC residues protect T4 DNA from attack by *E. coli* endonucleases called **restriction enzymes,** which would otherwise destroy the viral DNA. This bacterial defense mechanism is called **restriction.** Other phages also chemically modify their

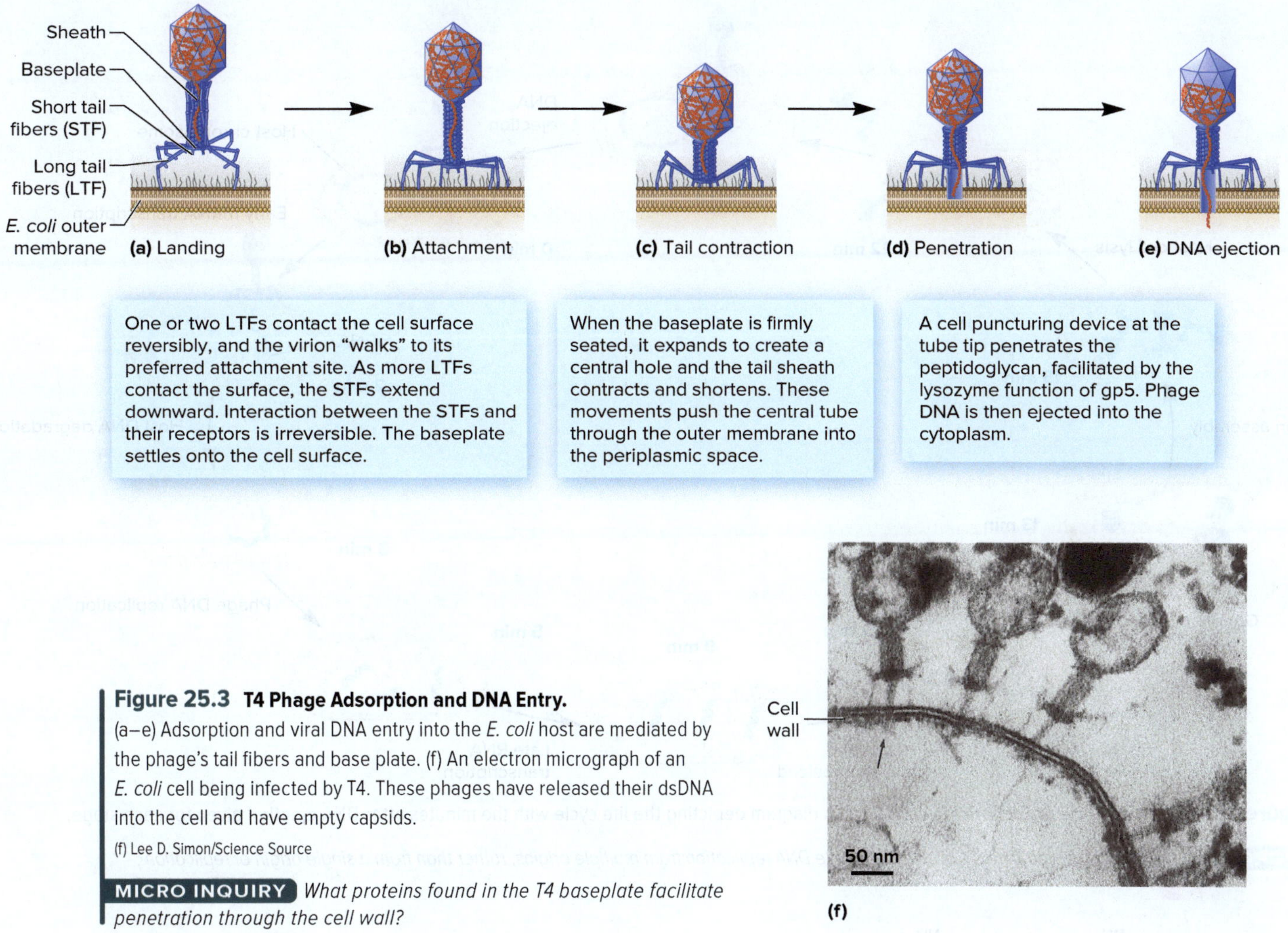

Figure 25.3 T4 Phage Adsorption and DNA Entry. (a–e) Adsorption and viral DNA entry into the *E. coli* host are mediated by the phage's tail fibers and base plate. (f) An electron micrograph of an *E. coli* cell being infected by T4. These phages have released their dsDNA into the cell and have empty capsids.

(f) Lee D. Simon/Science Source

MICRO INQUIRY *What proteins found in the T4 baseplate facilitate penetration through the cell wall?*

DNA to protect against host restriction. ◀ *Responses to viral infection (section 14.6); Restriction enzymes (section 17.1)*

Synthesis of the linear dsDNA genome of T4 involves the formation of long DNA molecules called **concatemers,** which are composed of several genome units linked together in the same orientation (**figure 25.6**). Concatamer formation is made possible because each progeny viral DNA molecule has single-stranded 3′ ends. These ends participate in homologous recombination with double-stranded regions of other progeny DNA molecules, generating concatemers. During assembly, concatemers are cleaved such that the genome packaged in the capsid is slightly longer than the T4 gene set. Thus each progeny virus has a genome unit that begins with a different gene and ends with the same set of genes. If each viral genome were circularized, the sequence of genes in each virion would be the same. Therefore the T4 genome is said to be *terminally redundant* and *circularly permuted,* and the genetic map of T4 is drawn as a circular molecule.

The formation of new T4 phage particles is a complex self-assembly process that involves viral proteins and some host cell factors (*see figure 6.12*). A critical step in T4 virion construction is filling the head portion of the virion with the T4 genome. This is no simple matter—the dsDNA genome is somewhat rigid and has many negatively charged moieties. Therefore the dsDNA must be crammed into the capsid. This is accomplished by a complex of proteins sometimes called the *packasome.* The T4 packasome can package DNA at a rate of 2,000 base pairs per second.

The packasome includes a protein called *terminase,* which has two functions: to cut the concatemers formed during T4 genome replication and to coordinate the insertion of DNA into the T4 head. Terminase threads the T4 genome through a portal, where it enters the phage head using energy supplied by ATP hydrolysis. The packasome has been proposed to move the DNA by causing a transition from B-form DNA to A-form DNA, and then reversing back to B-form DNA. This compresses the helix like a spring. When the phage head is filled with a DNA molecule roughly 3% longer than the length of one set of T4 genes, terminase makes a second cut, and the packaging process for that head is complete. Terminase then leaves the head, and other viral proteins bind at the portal. This seals the head and prepares it for addition of the tail and tail fibers.

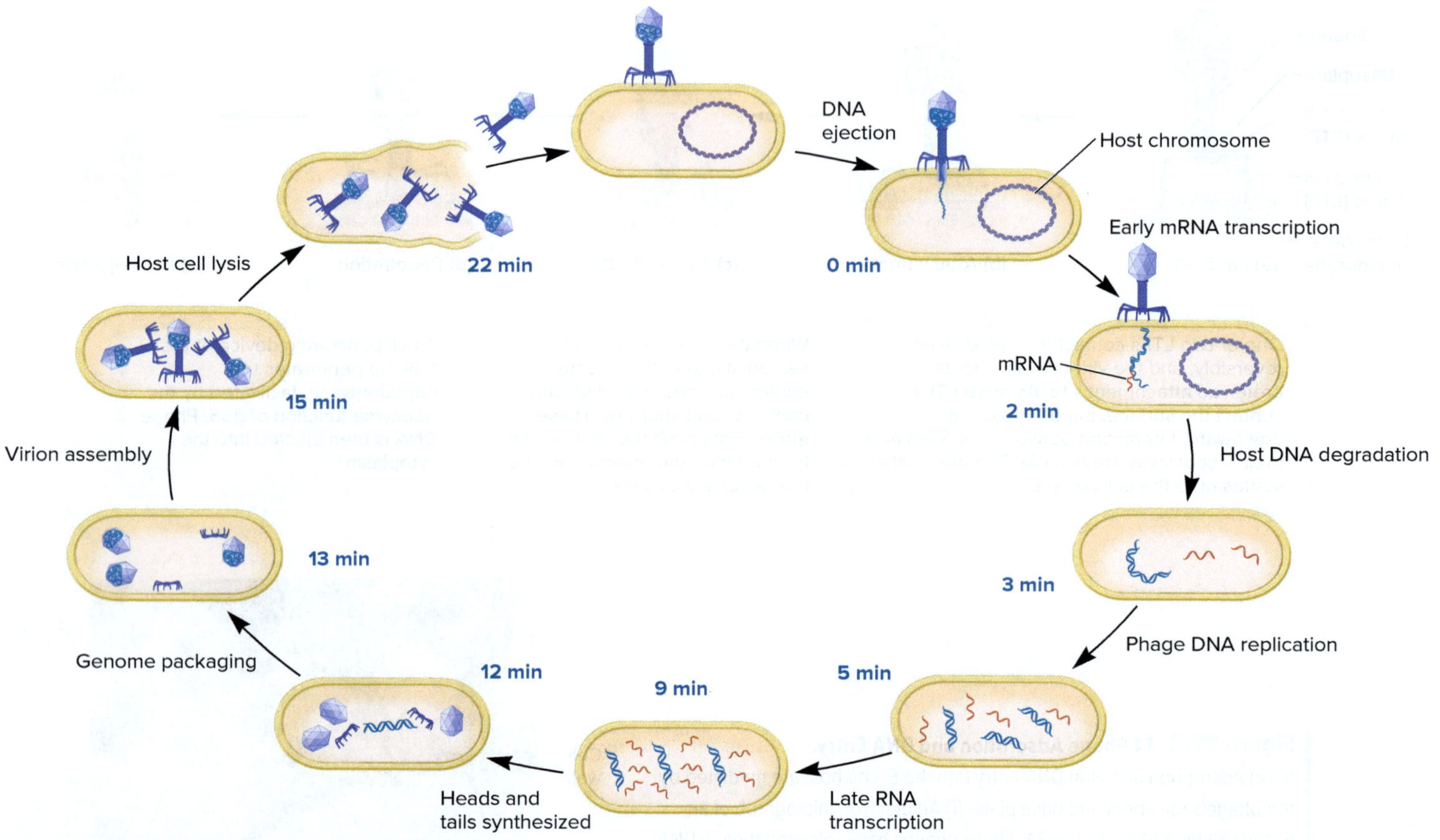

Figure 25.4 The Life Cycle of *Escherichia virus* T4. A diagram depicting the life cycle with the minutes after DNA ejection given for each stage.

MICRO INQUIRY *Why do you think T4 evolved to initiate DNA replication from multiple origins, rather than from a single origin of replication?*

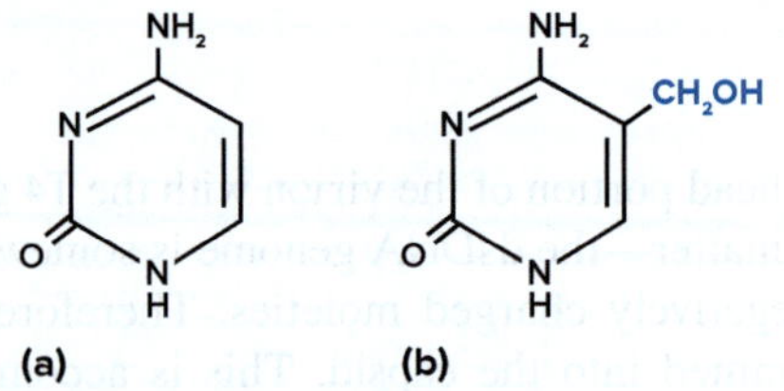

Figure 25.5 5-Hydroxymethylcytosine (HMC). Cytosine (a) is a regular DNA base. HMC (b) has a hydroxymethyl group (blue). In T4 DNA, HMC often has glucose attached to its hydroxyl.

MICRO INQUIRY *What function does HMC glycosylation serve?*

When about 150 virus particles have accumulated in the host cell, T4 lyses *E. coli* through the action of two proteins. The first, called *holin,* creates holes in the plasma membrane. The second, an endolysin called *T4 lysozyme* (distinct from the gp5 tail lysozyme used for bacteriophage entry), degrades peptidoglycan in the host's cell wall. Thus the activity of holin enables T4 lysozyme to move from the cytoplasm to the periplasm so that both the plasma membrane and the peptidoglycan cell wall are destroyed. Finally, virions are released so they can infect other cells and begin the cycle anew. *Steps in the Replication of T4 Phage in E. coli*

Escherichia virus Lambda: A Temperate Bacteriophage

Temperate bacteriophages, such as *Escherichia virus Lambda* (family Siphoviridae), can enter either the lytic or lysogenic cycle upon infecting a host cell. If a temperate virus enters the lysogenic cycle, its dsDNA genome often is integrated into the host's chromosome, where it resides as a prophage until conditions for induction occur (*see figure 6.15*). Upon induction, the viral genome is excised from the host genome and the lytic cycle is initiated. *Lytic and lysogenic infections are common for bacterial and archaeal cells (section 6.4)*

Events that determine which life cycle is followed are best illustrated by bacteriophage lambda (λ), which infects *E. coli.* As shown in **figure 25.7**, λ has an icosahedral head 55 nm in diameter and a noncontractile tail with a thin tail fiber at its end. Its DNA genome is a linear molecule with cohesive ends—single-stranded stretches, 12 nucleotides long, that are complementary to each other and can base-pair.

Like most bacteriophages, λ attaches to its host and releases its genome into the cytoplasm, leaving the capsid outside. Once inside the cell, the linear genome is circularized when the two cohesive ends base-pair with each other; the breaks in the strands

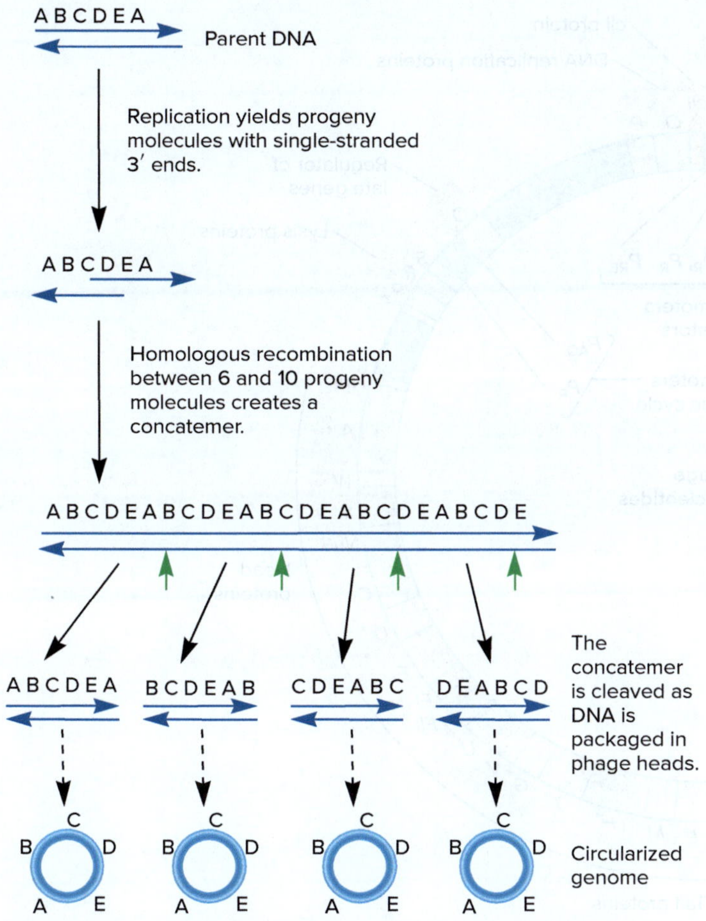

Figure 25.6 The Terminally Redundant, Circularly Permuted Genome of T4. Concatamer formation during T4 genome replication is an important step in phage multiplication. During virion assembly, the phage head fills with DNA cleaved from the concatemer. Because slightly more than one set of T4 genes is packaged in each head, every virion contains a different DNA fragment (note that the ends of the fragments are different). However, if each genome were circularized, the sequence of genes would be the same.

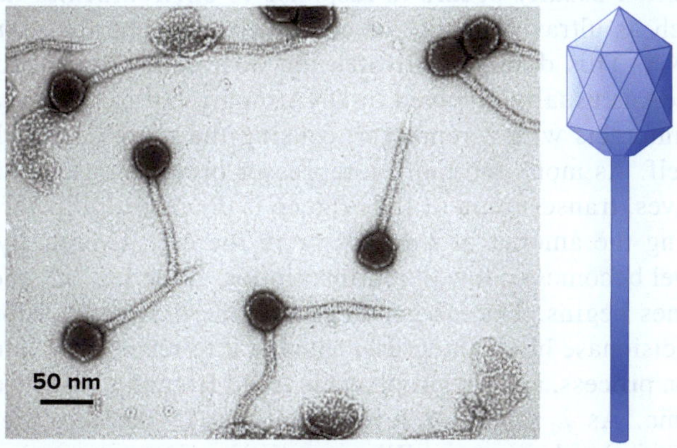

Figure 25.7 *Escherichia virus Lambda* (λ). The icosahedral head illustrated in the drawing is not obvious in the photomicrograph at this resolution.
M. Wurtz/Science Source

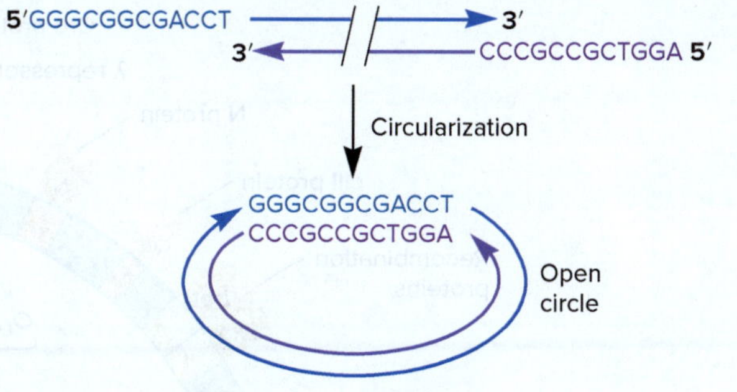

Figure 25.8 *Escherichia virus Lambda* DNA. A diagram of λ DNA showing its 12-base, single-stranded cohesive ends.

are sealed by the host cell's DNA ligase (**figure 25.8**). The λ genome has been carefully mapped, and over 70 genes have been identified (**figure 25.9**). Most genes are clustered according to their function, with separate groups involved in head synthesis, tail synthesis, lysogeny, DNA replication, and cell lysis. This organization is important because once the genome is circularized, a cascade of regulatory events occurs that determine if the phage pursues a lytic cycle or establishes lysogeny. Regulation of appropriate genes is facilitated by clustering and coordinated transcription from the same promoters.

The cascade of events leading to either lysogeny or the lytic cycle serves as a model for complex regulatory processes. It involves the action of several proteins that function as transcriptional repressors or activators or both, proteins that regulate transcription termination, and antisense RNA molecules (**figure 25.10**). The protein cII is an activator that plays a pivotal role in determining if λ will establish lysogeny or follow a lytic pathway.

Transcription of the λ genome is catalyzed by the host cell's **DNA-dependent RNA polymerase** (DdRp; an RNA polymerase that uses DNA as the template for RNA synthesis), and the cII protein is synthesized early in the infection. If cII protein levels are high enough, it will increase transcription of the *int* gene, which encodes the enzyme integrase. **Integrase** catalyzes integration of the λ genome into the host cell's chromosome, thus establishing lysogeny. The cII protein also increases transcription of the *cI* gene. This gene encodes a regulatory protein often called *λ repressor* because it represses transcription of all bacteriophage genes (except its own). This repression maintains the lysogenic state. As just noted, the cI protein (λ repressor) allows transcription of its own gene. This is because cI functions as an activator when it binds to the P_{RM} promoter from which the *cI* gene is transcribed.

Integration of the λ genome into the host chromosome takes place at a site called the attachment site (*att*). A corresponding site is present on the phage genome, and the λ integrase catalyzes site-specific recombination between the two sites. The bacterial site is located between the galactose (*gal*) and biotin (*bio*) operons, and upon integration, the circular λ genome becomes a linear stretch of DNA located between these two host operons (*see figure 16.14*). The prophage can remain integrated

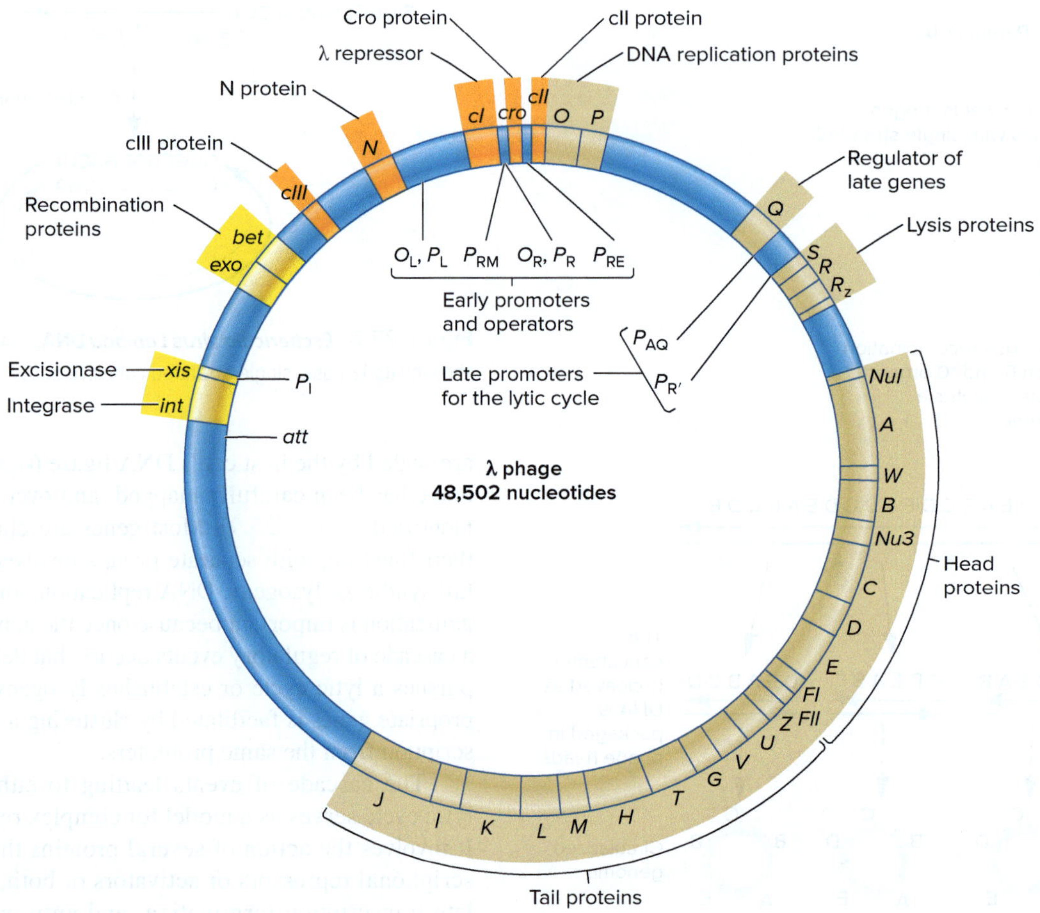

Figure 25.9 The Genome of *Escherichia virus* λ. The genes are color-coded according to function. Genes shaded in orange encode regulatory proteins that determine if the lytic or lysogenic cycle will be followed. Genes involved in establishing lysogeny are shaded in yellow. Those required for the lytic cycle are shaded in tan. Important promoters and operators are noted inside the circle.

MICRO INQUIRY *Why are so many more genes needed for the lytic cycle as compared with lysogeny?*

for an indefinite period of time, being replicated as the bacterial genome is replicated. *Molecular recombination: joining DNA molecules (section 16.3)*

Because the cII protein is made early in the infection, it might seem that it would accumulate quickly and ensure that lysogeny always occurs. However, cII is degraded by a host protease unless it is protected by a viral protein called cIII. The cIII protein is synthesized at the same time as cII, and as long as its levels remain high enough, cII will be protected. However, if cII is not protected from proteolysis, the level of a protein called Cro (product of the *cro* gene) will increase. Cro is both a repressor and an activator. It inhibits transcription of the *cIII* and *cI* genes, further decreasing the amount of cII and λ repressor. However, it increases its own synthesis as well as the synthesis of another regulatory protein called Q. When Q accumulates at a high enough level, it activates transcription of genes required for the lytic cycle. When that occurs, the infection process is committed to the lytic cycle (figure 25.10).

Now we consider how induction reverses lysogeny. Induction usually occurs in response to environmental factors such as ultraviolet light or chemical mutagens that damage DNA. This damage activates the host cell's **RecA** protein. RecA is usually involved in DNA repair, but in this situation it interacts with λ repressor, causing the repressor to cleave itself. As more and more λ repressor proteins destroy themselves, transcription of the *cI* gene is decreased, further lowering the amount of λ repressor in the cell. Eventually, the level becomes so low that transcription of the *xis, int,* and *cro* genes begins. The *xis* gene encodes the protein excisionase. Excisionase binds integrase, causing it to reverse the integration process, and the prophage is freed from the host chromosome. As λ repressor levels continue to decline, the Cro protein levels increase. When synthesis of λ repressor is completely blocked, protein Q levels increase, and the lytic cycle proceeds to completion. *DNA repair maintains genetic stability (section 16.2)*

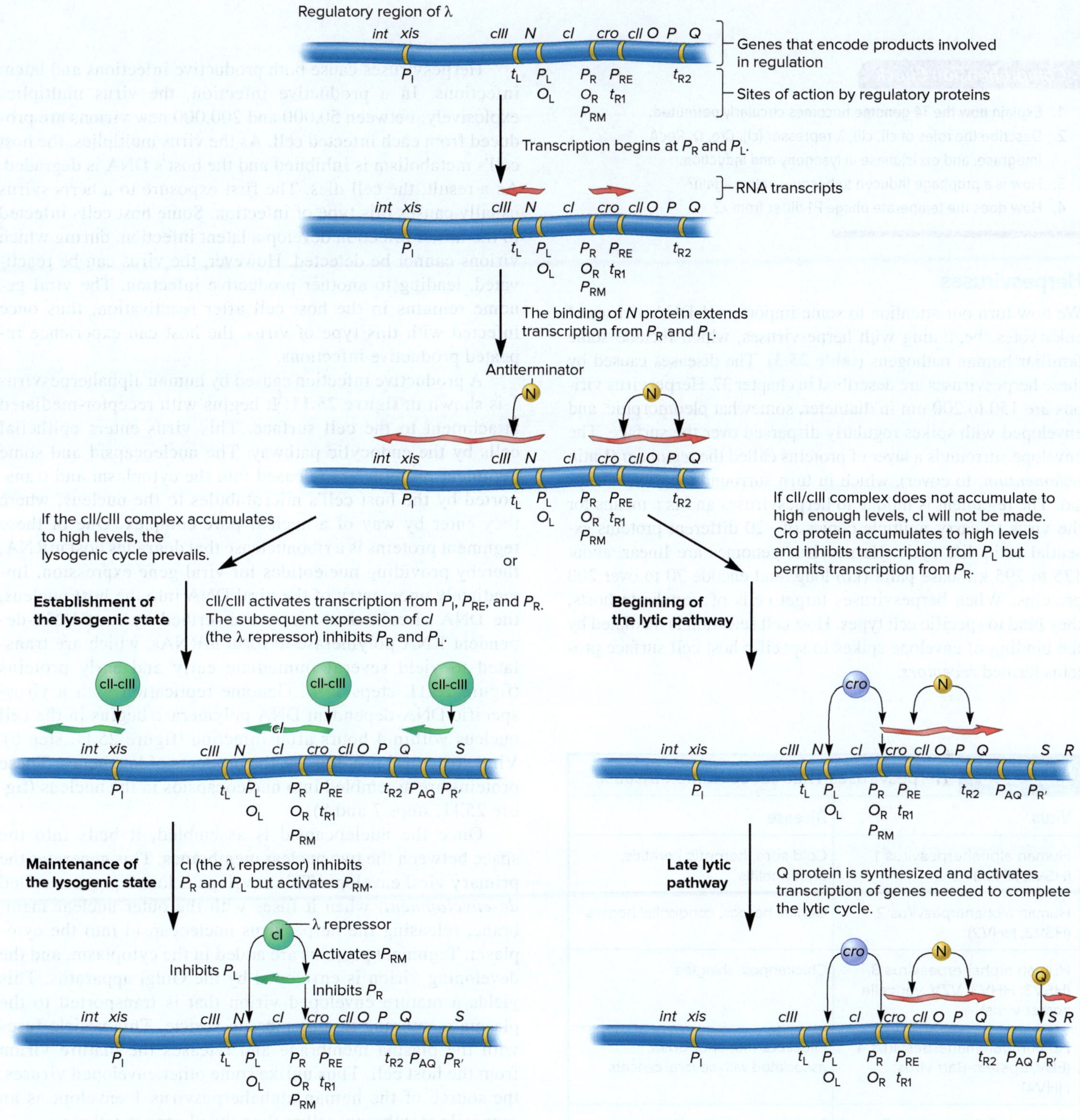

Figure 25.10 The *Escherichia virus Lambda* Decision-Making Process for Establishing Lysogeny or the Lytic Pathway. Initial transcripts synthesized by host RNA polymerase encode the N and Cro proteins. The N protein is an antiterminator that allows transcription to proceed past the terminator sequences t_L, t_{R1}, and t_{R2}. This allows transcription of other regulatory genes, as well as the *xis* and *int* genes. The latter genes encode excisionase and integrase, respectively. The left side illustrates what occurs if lysogeny is established. The right side shows the lytic pathway.

Our attention has been on λ, but there are many other temperate phages. Most, like λ, exist as integrated prophages in the lysogen. However, integration is not an absolute requirement for lysogeny. *Escherichia virus* P1 is similar to λ in that it circularizes after infection and begins to manufacture repressor. However, it remains as an independent circular DNA molecule in the lysogen and is replicated at the same time as the host chromosome. When *E. coli* divides, P1 DNA is apportioned between the daughter cells so that all lysogens contain one or two copies of the phage genome. *Lambda Phage Replication Cycle*

Comprehension Check

1. Explain how the T4 genome becomes circularly permuted.
2. Describe the roles of cII, cIII, λ repressor (cI), Cro, Q, RecA, integrase, and excisionase in lysogeny and induction.
3. How is a prophage induced to become active again?
4. How does the temperate phage P1 differ from λ?

Herpesviruses

We now turn our attention to some important dsDNA viruses of eukaryotes, beginning with herpesviruses, which include some familiar human pathogens (**table 25.3**). The diseases caused by these herpesviruses are described in chapter 37. Herpesvirus virions are 150 to 200 nm in diameter, somewhat pleomorphic, and enveloped with spikes regularly dispersed over the surface. The envelope surrounds a layer of proteins called the *tegument* (Latin *tegumentum,* to cover), which in turn surrounds the nucleocapsid. The tegument is unique to herpesviruses and is a means for the virus to carry multiple copies of ~20 different proteins essential to its life cycle. Herpesvirus genomes are linear, about 125 to 295 kilobase pairs (kb) long, and encode 70 to over 200 proteins. When herpesviruses target cells of vertebrate hosts, they bind to specific cell types. Host cell selection is mediated by the binding of envelope spikes to specific host cell surface proteins termed *receptors.*

Table 25.3 Herpesviruses That Cause Human Disease

Virus	Disease
Human alphaherpesvirus 1 (HSV1; HHV1)[1]	Cold sore, herpetic keratitis, encephalitis
Human alphaherpesvirus 2 (HSV2; HHV2)	Genital herpes, congenital herpes
Human alphaherpesvirus 3 (HSV3; HHV3; VZV, varicella zoster virus)	Chickenpox, shingles
Human gammaherpesvirus 4 (EBV, Epstein-Barr virus; HHV4)	Infectious mononucleosis; associated with several cancers
Human betaherpesvirus 5 (CMV, cytomegalovirus; HHV5)	Mononucleosis-like illness, congenital cytomegalovirus infection
Human betaherpesvirus 6A, 6B, 7 (HHV6; HHV7)	Roseola infantum, also called exanthem subitum
Human gammaherpesvirus 8 (KSHV, Kaposi's sarcoma herpesvirus; HHV8)	Kaposi's sarcoma

[1] Abbreviations in parentheses are alternate designations in common use. HSV (Herpes Simplex Virus); HHV (Human Herpes Virus).

Herpesviruses cause both productive infections and latent infections. In a productive infection, the virus multiplies explosively; between 50,000 and 200,000 new virions are produced from each infected cell. As the virus multiplies, the host cell's metabolism is inhibited and the host's DNA is degraded. As a result, the cell dies. The first exposure to a herpesvirus usually causes this type of infection. Some host cells infected in the initial infection develop a latent infection, during which virions cannot be detected. However, the virus can be reactivated, leading to another productive infection. The viral genome remains in the host cell after reactivation; thus once infected with this type of virus, the host can experience repeated productive infections.

A productive infection caused by human alphaherpesvirus 1 is shown in **figure 25.11**. It begins with receptor-mediated attachment to the cell surface. This virus enters epithelial cells by the endocytic pathway. The nucleocapsid and some tegument proteins are released into the cytoplasm and transported by the host cell's microtubules to the nucleus, where they enter by way of a nuclear pore complex. One of these tegument proteins is a ribonuclease that degrades host mRNA, thereby providing nucleotides for viral gene expression. Immediately upon entry of the viral DNA into the host nucleus, the DNA circularizes and is transcribed by host DNA-dependent RNA polymerase to form mRNAs, which are translated to yield several immediate early and early proteins (figure 25.11, steps 1–4). Genome replication with a virus-specific DNA-dependent DNA polymerase begins in the cell nucleus within 4 hours after infection (figure 25.11, step 6). Viral structural proteins are the products of late genes. These proteins are assembled into nucleocapsids in the nucleus (figure 25.11, steps 7 and 8).

Once the nucleocapsid is assembled, it buds into the space between the two nuclear membranes. This generates the primary viral envelope. This envelope is lost, a process called *de-envelopment,* when it fuses with the outer nuclear membrane, releasing the herpesvirus nucleocapsid into the cytoplasm. Tegument proteins are added in the cytoplasm, and the developing virion is enveloped by the Golgi apparatus. This yields a mature enveloped virion that is transported to the plasma membrane in a membrane vesicle. This vesicle fuses with the plasma membrane and releases the mature virion from the host cell. Thus unlike some other enveloped viruses, the source of the human alphaherpesvirus 1 envelope is an organelle membrane, rather than the plasma membrane.

Just as the life cycle of λ has been of interest as a model of regulatory processes, so too has the establishment of latency by herpesviruses. Both viral and host proteins play a role in establishing a latent human alphaherpesvirus 1 infection. In epithelial cells, where productive infections occur, a protein called VP16 and a host protein called host cell factor 1 (HCF-1) are both required for full expression of the virus's immediate early genes. HCF-1 is a nuclear protein that establishes the chromatin structure in the human alphaherpesvirus 1 genome for gene expression. In infected neurons, however, where the virus is latent,

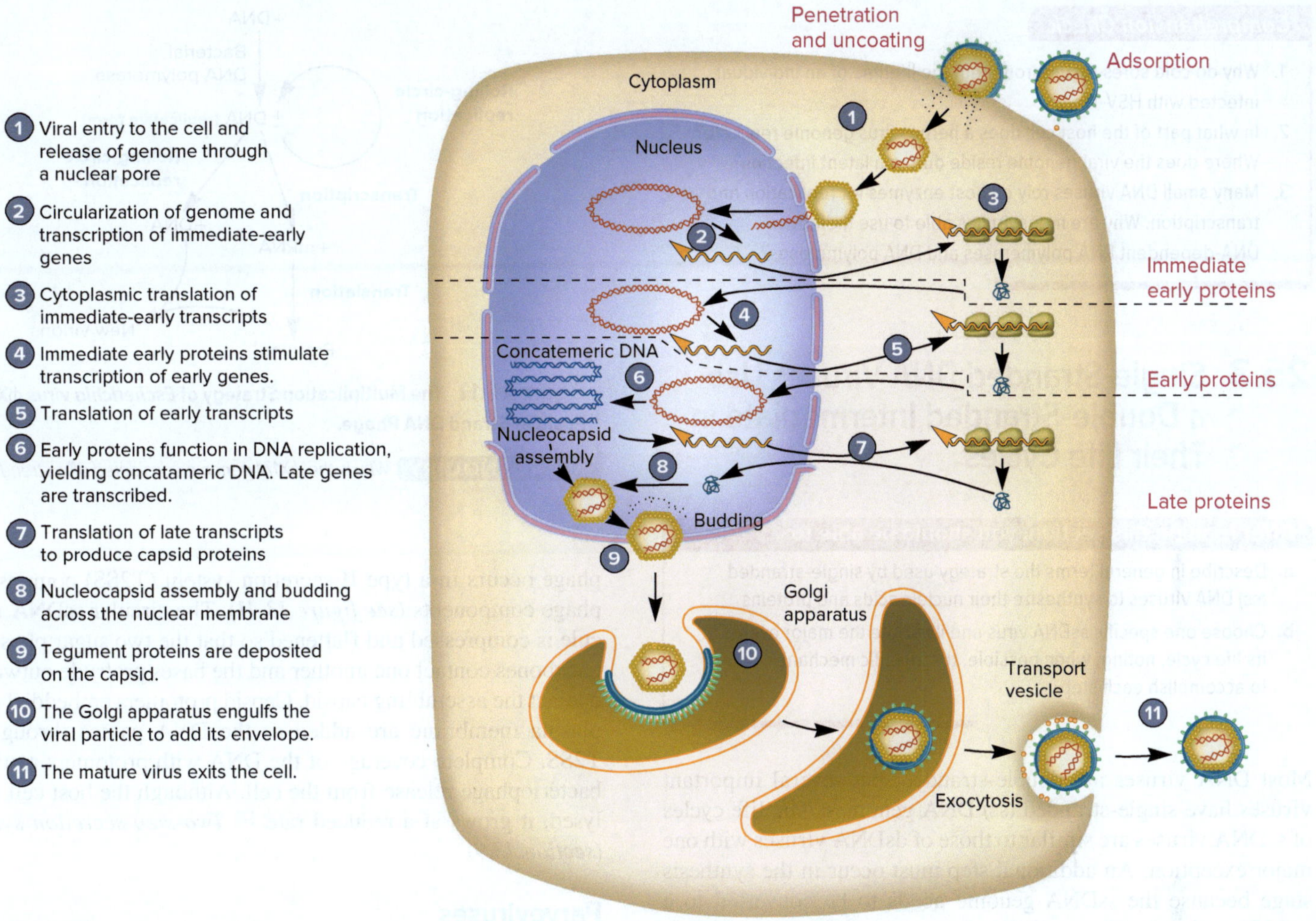

Figure 25.11 Life Cycle of Human Alphaherpesvirus 1.

MICRO INQUIRY *How is the envelope of this virus formed? How does this differ from most enveloped viruses?*

HCF-1 is cytoplasmic. In addition, small noncoding RNAs produced only in the neuron act to repress the viral lytic cycle. The inhibition of early gene expression helps establish latency. Interestingly, during viral reactivation in neurons, one of the earliest events is the movement of HCF-1 to the nucleus and the production of VP16.

Nucleocytoplasmic Large DNA Viruses (Megaviruses)

Nucleocytoplasmic large DNA (NCLD) viruses, often referred to as megaviruses, are large enough to be visible by light microscopy (*see figure 6.1f*). Most have icosahedral capsids that enclose a lipid membrane and all have a large dsDNA genome of 500 to over 2,500 kilobases. Thus both the virion itself and the genome often rival the size of small bacteria and their genomes. Megaviruses infect a variety of eukaryotic hosts, primarily protists, but also insects and humans. This extraordinarily broad host range is a consequence of the viral uptake mechanism by the host cell—phagocytosis. There is no viral envelope–cell surface receptor pairing to confer specificity.

The large genomes of megaviruses enable them to encode many proteins. Thus they are self-sufficient and rely on their hosts much less than other viruses. They generally encode proteins needed for DNA replication, recombination, transcription, and protein maturation. In addition, some NCLD viruses encode components of the translational machinery (tRNAs and aminoacyl-tRNA synthetases). However, their lack of ribosomes or a mechanism to generate ATP, as well as their method of reproduction, distinguishes NCLD viruses from cells. Although it has been suggested that these viruses may represent a primitive cell type, it is more plausible that a large virion size has been selected by the particle size requirements for phagocytosis.

Comprehension Check

1. Why do cold sores recur throughout the lifetime of an individual infected with HSV-1?
2. In what part of the host cell does a herpesvirus genome replicate? Where does the viral genome reside during a latent infection?
3. Many small DNA viruses rely on host enzymes for replication and transcription. Why are megaviruses able to use their own DNA-dependent RNA polymerases and DNA polymerases?

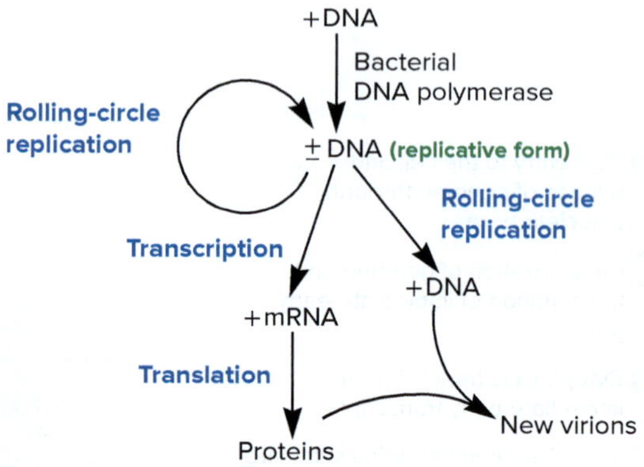

Figure 25.12 The Multiplication Strategy of *Escherichia virus* ϕX174, a Positive-Strand DNA Phage.

MICRO INQUIRY *Why is the ϕX174 genome considered plus stranded?*

25.3 Single-Stranded DNA Viruses Use a Double-Stranded Intermediate in Their Life Cycles

After reading this section, you should be able to:

a. Describe in general terms the strategy used by single-stranded (ss) DNA viruses to synthesize their nucleic acids and proteins
b. Choose one specific ssDNA virus and illustrate the major events in its life cycle, noting, when possible, the specific mechanisms used to accomplish each step

Most DNA viruses are double-stranded, but several important viruses have single-stranded (ss) DNA genomes. The life cycles of ssDNA viruses are similar to those of dsDNA viruses with one major exception: An additional step must occur in the synthesis stage because the ssDNA genome needs to be converted to a dsDNA molecule. Three ssDNA viruses are discussed next.

Escherichia virus ϕX174 and *Pseudomonas virus Pf1*

The life cycle of *Escherichia virus phiX174* (ϕX174; family Microviridae) begins with attachment to the cell envelope of its *E. coli* host. The circular ssDNA genome enters the cell via a tubelike structure formed from a viral spike protein. The protein capsid remains outside the cell. The ϕX174 ssDNA genome has the same base sequence as viral mRNA and is therefore **positive-strand DNA.** For either transcription or genome replication to occur, the phage DNA must be converted to a double-stranded **replicative form (RF)** (**figure 25.12**). This is catalyzed by the host's DNA polymerase. The RF directs the synthesis of more RF copies and positive-strand DNA, both by rolling-circle replication (*see figure 16.18*). Virion assembly begins when scaffold proteins interact with capsid proteins, causing them to undergo conformational changes to form a procapsid. The ssDNA genome is inserted into the procapsid, creating the mature virion. The host is lysed by a viral enzyme that blocks peptidoglycan synthesis. This weakens the host cell wall, causing it to lyse and release the progeny virions.

We include *Pseudomonas virus Pf1* (family Inoviridae) in our discussion because of one striking feature: its release from its host by a secretory process, without lysis. Assembly of the mature phage occurs in a type II secretion system (T2SS) composed of phage components (*see figure 13.41*). The circular ssDNA molecule is compressed and flattened so that the two sugar phosphate backbones contact one another and the bases protrude outward to contact the assembling capsid. Capsid protomers embedded in the plasma membrane are added as the DNA passes through the T2SS. Complete coverage of the DNA with protomers results in bacteriophage release from the cell. Although the host cell is not lysed, it grows at a reduced rate. ◀ *Two-step secretion systems (section 13.9)*

Parvoviruses

Parvoviruses (family Parvoviridae) infect numerous animal hosts, including crustaceans, dogs, cats, mice, and humans. Since its discovery in 1974, primate erythroparvovirus 1 (parvovirus B19) has emerged as a significant human pathogen, responsible for a common childhood illness called fifth disease. Parvovirus virions are icosahedral, nonenveloped particles approximately 26 nm in diameter. Their genomes are one ssDNA molecule of about 5,600 bases. Most genomes are **negative-strand DNA** molecules—that is, their sequence of nucleotides is complementary to that of the viral mRNA (figure 25.1). Parvovirus genomes are so small that they direct the synthesis of only three proteins and some smaller polypeptides, none of which have enzymatic activity. ▶ *Human parvovirus B19 infection (section 37.3)*

A productive parvovirus infection requires attachment to a particular cell type. Virions enter host cells by receptor-mediated endocytosis. They escape the endosome and are transported to the nucleus by the host cell's microtubules, directed by a nuclear localization signal on the capsid. The nucleocapsid enters the nucleus and viral DNA is released.

Because the parvovirus genome does not code for any enzymes, the virus must use host cell enzymes for all biosynthetic processes. Viral infection stops the cell cycle in S phase, when the host cell can provide all of the components needed for viral replication. Because the genome is negative-strand DNA, it serves as

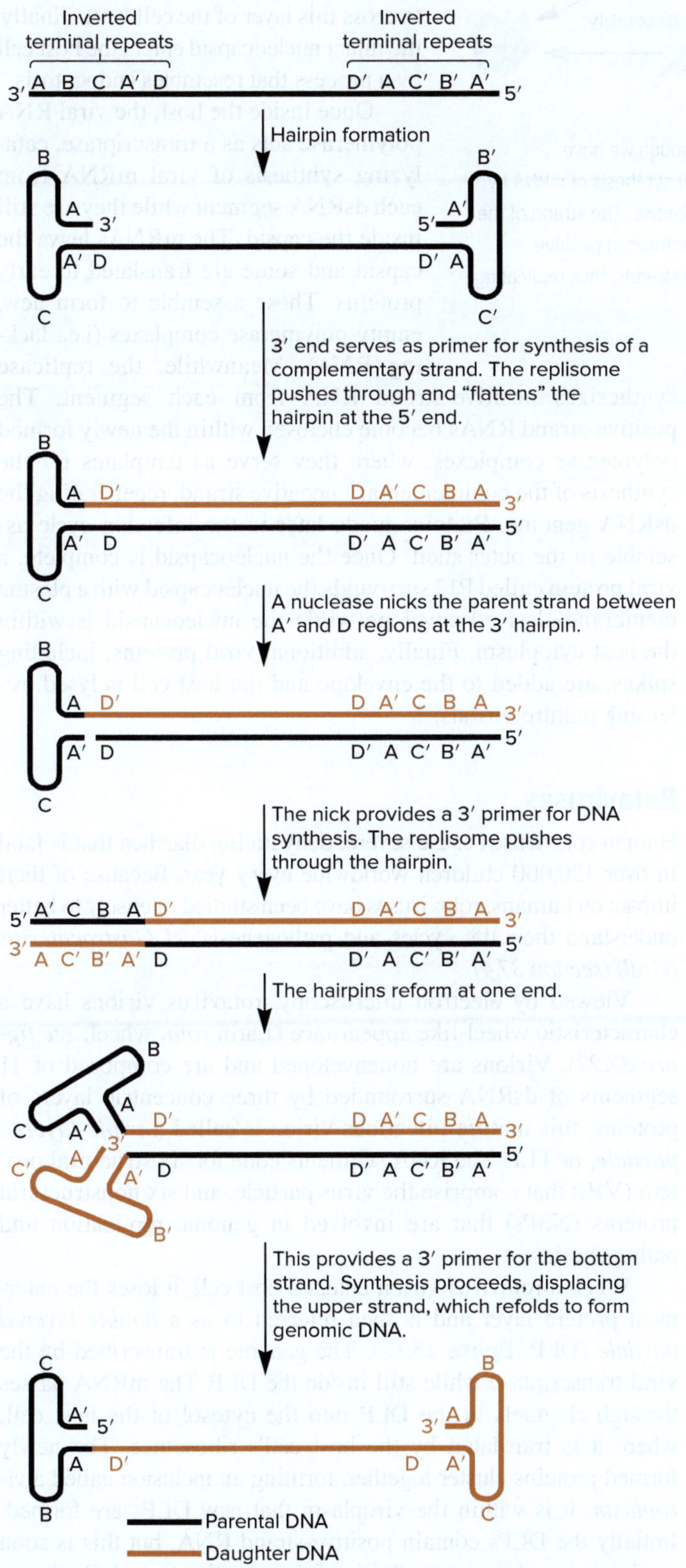

Figure 25.13 The Rolling-Hairpin Replication Used by Parvoviruses to Replicate Their DNA.

the template for mRNA synthesis. Some of the RNA products encode polypeptides required for the virus's novel genome replication mechanism. The genome ends are palindromic sequences that fold back on themselves. Formation of a complex composed of a viral protein and a hairpin at the 3′ end of the genome provides the primer needed for replication (**figure 25.13**). This is recognized as a substrate by the host DNA polymerase, and DNA replication ensues by a process that is somewhat similar to rolling-circle replication. The parvovirus version of this replication method is often called *rolling-hairpin replication* because DNA polymerase seems to shuttle back and forth as it synthesizes genomes. The process involves a dsDNA intermediate, much like ϕX174.

Comprehension Check

1. Why is it necessary for some ssDNA viruses to manufacture a replicative form?
2. From the point of view of the virus, compare the advantages and disadvantages of host cell lysis, as seen in ϕX174, with the continuous release of phage without lysis, as with Pf1 phage.
3. Why does parvovirus replication occur in the nucleus only during the S phase of the host's cell cycle?

25.4 Double-Stranded RNA Viruses: RNA-Dependent RNA Polymerase Replicates the Genome and Synthesizes mRNA

After reading this section, you should be able to:

a. Distinguish RNA-dependent RNA polymerases from DNA-dependent RNA polymerases
b. Describe the enzymatic activity of an RNA-dependent RNA polymerase when functioning as a replicase and when functioning as a transcriptase
c. Describe in general terms the strategy used by dsRNA viruses to synthesize their nucleic acids and proteins
d. Describe the major events in the life cycles of ϕ6 and rotaviruses, noting, when possible, the specific mechanisms used to accomplish each step

The Baltimore system divides viruses with RNA genomes into four groups (table 25.2). However, they all share the same dilemma: Their host cells have dsDNA genomes and thus lack a polymerase that can make RNA from an RNA template. Therefore double-stranded, positive-strand, and negative-strand RNA viruses must all produce an enzyme called **RNA-dependent RNA polymerase** (RdRp; an RNA polymerase that uses an RNA molecule as the template for RNA synthesis). When the

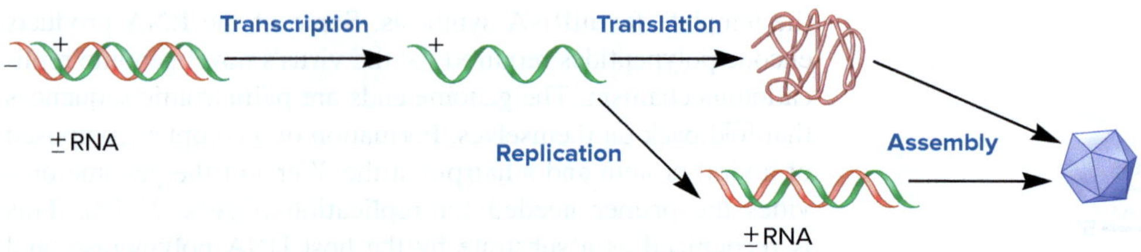

Figure 25.14 Multiplication Strategy of Double-Stranded RNA Viruses. Although we have designated positive and negative RNA strands, both strands may serve as templates for synthesis of mRNA by transcriptase, depending on the virus. For those viruses, each strand is said to be ambisense. The strand of the viral genome we have designated as the negative strand serves as the template for synthesis of positive strands by replicase. The positive strands are used by replicase to synthesize negative strands, thus replicating the dsRNA genome.

RdRp is used to replicate the viral RNA genome, it is often referred to as a **replicase.** When it is used to synthesize mRNA, the RdRp is often said to have **transcriptase** activity. In most cases, the same enzyme carries out both functions. Viruses in the fourth group, retroviruses, use a very different approach: They use the multifunctional enzyme **reverse transcriptase** (RT) to synthesize a dsDNA copy of their ssRNA genomes, a process called *reverse transcription.* The dsDNA is then recognized by the host cell's DNA-dependent RNA polymerase during transcription.

Among viruses with RNA genomes, those with dsRNA are uncommon. These viruses share a multiplication strategy (**figure 25.14**). Here we discuss two representative dsRNA viruses: a bacteriophage and a vertebrate virus.

Pseudomonas virus ϕ6

Several dsRNA phages have been discovered and all are virulent. The best studied, *Pseudomonas virus phi6* (ϕ6), is unusual for several reasons. First, it is an enveloped phage. The envelope encloses the nucleocapsid, which consists of an outer protein shell and an inner core, sometimes called the *polymerase complex.* It is so named because it consists of the virus's RNA-dependent RNA polymerase (RdRp) and three other proteins, and encloses the genome. The ϕ6 genome illustrates one of the methods used by RNA viruses to ensure synthesis of all viral proteins: It is a **segmented genome,** composed of more than one distinct RNA molecule. For most viruses, each segment encodes a single protein. However, in the case of ϕ6, each of its three dsRNA molecules generates a polycistronic mRNA and several proteins. Many other viruses also generate polycistronic mRNAs, much like the polycistronic mRNAs observed in bacteria and archaea (*see figure 13.21*).

The life cycle of the virus also has unusual features. ϕ6 attaches to a pilus on the *Pseudomonas* host cell via an envelope spike. Pilus retraction brings the phage into contact with the outer membrane of its diderm host. The viral envelope then fuses with the cell's outer membrane, a process mediated by another envelope spike. Fusion of the two membranes delivers the nucleocapsid to the periplasmic space. A protein associated with the nucleocapsid digests the peptidoglycan, allowing the nucleocapsid to cross this layer of the cell wall. Finally, the intact nucleocapsid enters the host cell by a process that resembles endocytosis.

Once inside the host, the viral RNA polymerase acts as a transcriptase, catalyzing synthesis of viral mRNA from each dsRNA segment while they are still inside the capsid. The mRNAs leave the capsid and some are translated to early proteins. These assemble to form new, empty polymerase complexes (i.e., lacking RNA). Meanwhile, the replicase synthesizes positive-strand RNA from each segment. The positive-strand RNAs become enclosed within the newly formed polymerase complexes, where they serve as templates for the synthesis of the complementary negative strand, regenerating the dsRNA genome. Proteins made later in the infection cycle assemble in the outer shell. Once the nucleocapsid is complete, a viral protein called P12 surrounds the nucleocapsid with a plasma membrane–derived envelope while the nucleocapsid is within the host cytoplasm. Finally, additional viral proteins, including spikes, are added to the envelope and the host cell is lysed, releasing mature virions.

Rotaviruses

Human rotaviruses cause severe, dehydrating diarrhea that is fatal in over 120,000 children worldwide every year. Because of their impact on humans, rotaviruses have been studied intensely to better understand their life cycles and pathogenesis. ▶| *Gastroenteritis (viral) (section 37.4)*

Viewed by electron microscopy, rotavirus virions have a characteristic wheel-like appearance (Latin *rota,* wheel; *see figure 37.22*). Virions are nonenveloped and are composed of 11 segments of dsRNA surrounded by three concentric layers of proteins; this mature infectious virion is called a *triple-layered particle,* or TLP. The RNA segments code for six structural protein (VPs) that comprise the virus particle, and six nonstructural proteins (NSPs) that are involved in genome replication and pathogenesis.

When a rotavirus virion enters a host cell, it loses the outermost protein layer and is then referred to as a *double-layered particle* (DLP; **figure 25.15**). The genome is transcribed by the viral transcriptase while still inside the DLP. The mRNA passes through channels in the DLP into the cytosol of the host cell, where it is translated by the host cell's ribosomes. The newly formed proteins cluster together, forming an inclusion called a *viroplasm.* It is within the viroplasm that new DLPs are formed. Initially the DLPs contain positive-strand RNA, but this is soon used as a template for synthesis of the negative strand. Replicase activity of the RdRp requires that the enzyme be in complex with the core shell protein. Thus the dsRNA molecules are synthesized

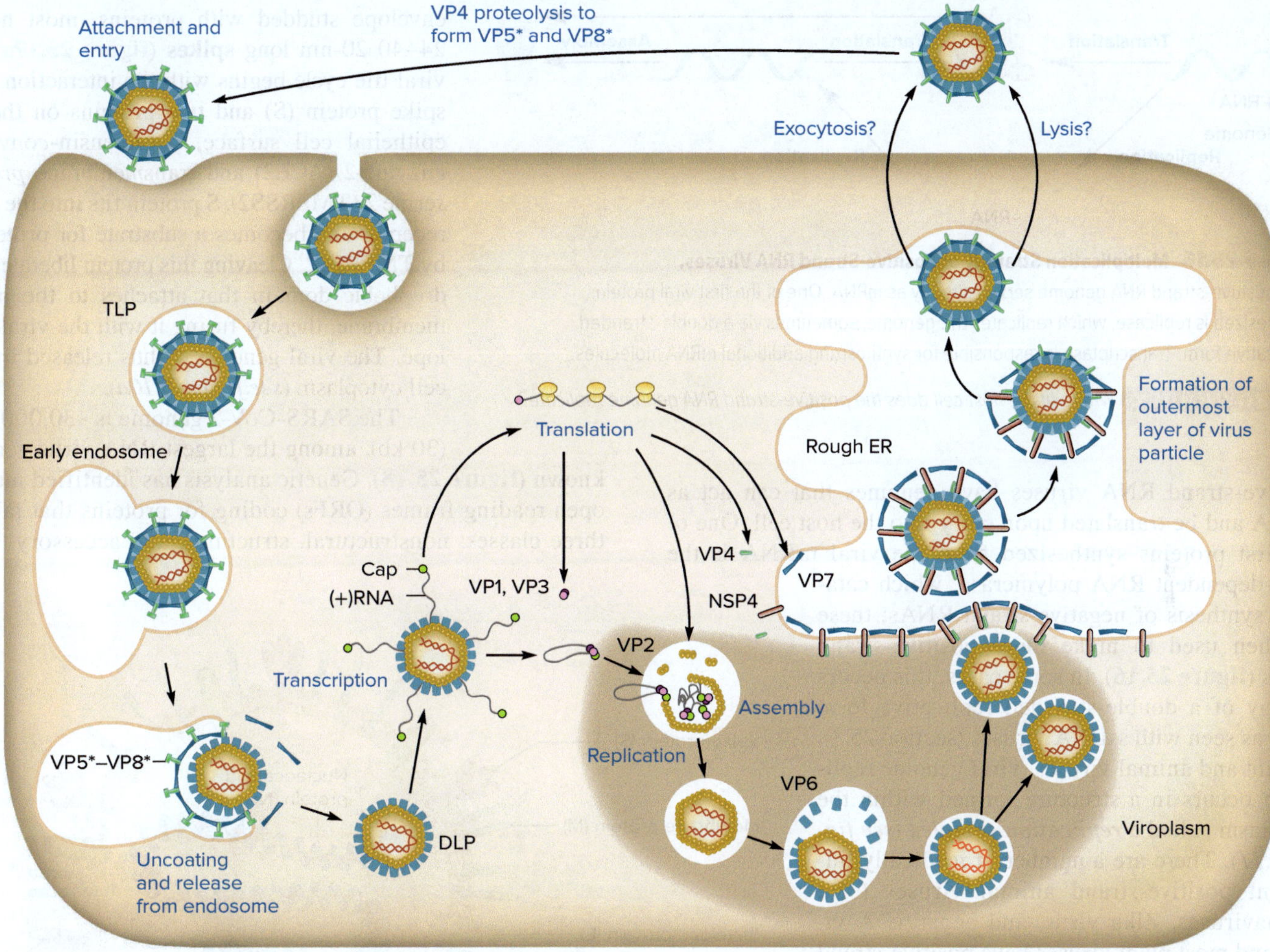

Figure 25.15 **Life Cycle of a Rotavirus.** Proteases in the intestinal tract act on the outer shell protein (VP4) to generate the infective form (triple-layered particle or TLP) of the virus. VP5* and VP8* are subunits of VP4 that bind cell surface receptors. Upon viral uncoating and release of the double-layered particle (DLP) into the host cytoplasm, RdRp (VP1) transcribes the genome and VP3 caps the mRNA. Genome replication and virion assembly are coupled in the viroplasm. NSP4 is a viroporin that creates a membrane channel through which ions leak into the cytoplasm from the endoplasmic reticulum (ER). This leakage signals outer shell proteins VP7 and VP4 to coat the virion as it enters the ER. The mechanism of virion release has not been definitively determined.

within the developing DLP. DLPs containing dsRNA are enveloped by membranes of the endoplasmic reticulum, thereby converting it to the mature TLP. The mature virion is released by an unknown mechanism.

Comprehension Check

1. Describe the life cycle of ϕ6. What makes this phage unusual when compared with other bacteriophages?
2. How does the host contribute to generating infectious rotavirus virions?
3. In what ways are the life cycles of ϕ6 and rotaviruses similar? How do they differ?

25.5 Positive-Strand RNA Viruses: Genomes That Are Translated upon Entry

After reading this section, you should be able to:

a. Describe in general terms the strategy used by positive-strand RNA viruses to synthesize their nucleic acids and proteins

b. Outline the major events in the life cycles of coronavirus and tobacco mosaic virus, noting, when possible, the specific mechanisms used to accomplish each step

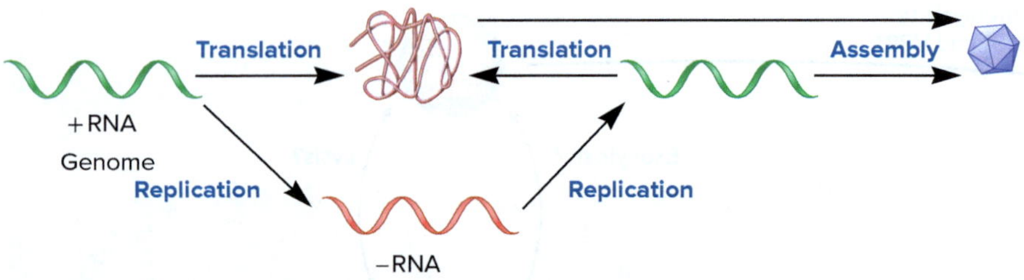

Figure 25.16 Multiplication Strategy of Positive-Strand RNA Viruses. The positive-strand RNA genome serves directly as mRNA. One of the first viral proteins synthesized is replicase, which replicates the genome, sometimes via a double-stranded replicative form. Transcriptase is responsible for synthesizing additional mRNA molecules.

MICRO INQUIRY *Where in the host cell does the positive-strand RNA genome replicate?*

Positive-strand RNA viruses have genomes that can act as mRNA and be translated upon entry into the host cell. One of the first proteins synthesized from the viral mRNA is the RNA-dependent RNA polymerase, which catalyzes synthesis of negative-strand RNAs; these are then used to make more positive-strand RNAs (**figure 25.16**). In some cases, this occurs by way of a double-stranded replicative form (RF), as seen with ssDNA viruses (section 25.3). In plant and animal viruses, viral genome replication occurs in a structure formed within the cytoplasm called a *replication complex* (*see figure 6.11*). There are a number of medically important positive-strand animal viruses (e.g., coronaviruses, Zika virus, and hepatitis A virus), and most plant viruses have positive-strand RNA genomes.

Coronavirus

Coronaviruses (Latin *corona,* crown) were named for their distinctive profile in electron micrographs (*see figure 6.7b*). Virologists have known for decades that coronaviruses are one causative agent of the common cold. In the twenty-first century, several animal coronaviruses developed the capacity to cause *s*evere *a*cute *r*espiratory *s*yndrome (SARS) in humans. While outbreaks of SARS and Middle East respiratory syndrome (MERS) caused regional epidemics, the rapid spread of SARS-CoV-2 resulted in the COVID-19 pandemic. We describe here the SARS-CoV-2 viral life cycle, typical of positive-strand RNA viruses. The disease manifestations of SARS-CoV-2 infection are discussed in Chapter 37. ▶| *Epidemiology is an evidence-based science (section 35.1);* ▶| *COVID-19 (section 37.1)*

SARS-CoV-2 is an enveloped virus. Surrounding the 120-nm capsid is a lipid bilayer envelope studded with proteins, most notably 24–40 20-nm long **spikes** (**figure 25.17*a***). The viral life cycle begins with the interaction of the spike protein (S) and two proteins on the host epithelial cell surface, *a*ngiotensin-*c*onverting *e*nzyme-2 (ACE2) and *t*rans*m*embrane *p*rotease *s*erine 2 (TMPRSS2). S protein fits into the ACE2 receptor and becomes a substrate for proteolysis by TMPRSS2. Cleaving this protein liberates a hydrophobic domain that attaches to the plasma membrane, thereby fusing it with the viral envelope. The viral genome is thus released into the cell cytoplasm (*see figure 6.10a*).

The SARS-CoV-2 genome is ~30,000 bases (30 kb), among the largest RNA virus genomes known (**figure 25.18**). Genetic analysis has identified multiple open reading frames (ORFs) coding for proteins that fall into three classes: nonstructural, structural, and accessory. These

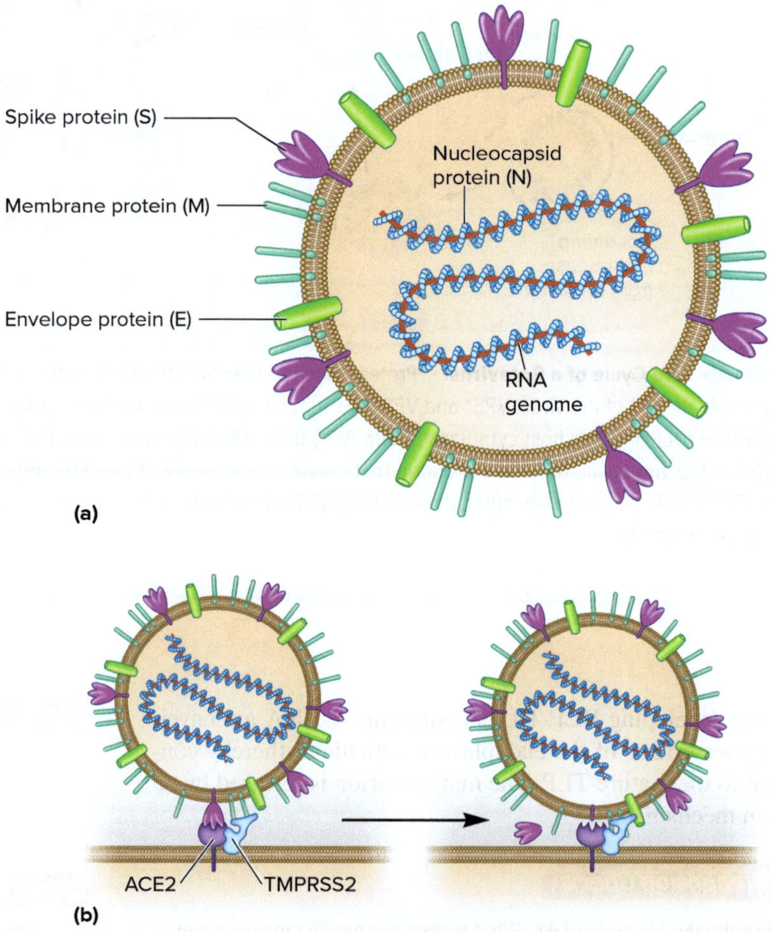

Figure 25.17 Attachment and Entry of SARS-CoV-2. (a) Schematic drawing of the SARS-CoV-2 virion. (b) The virus gains entry to a cell through the interaction of S with the ACE2 cellular receptor (left). Proteolysis by TMPRSS2 exposes a spike domain that promotes membrane fusion (right).

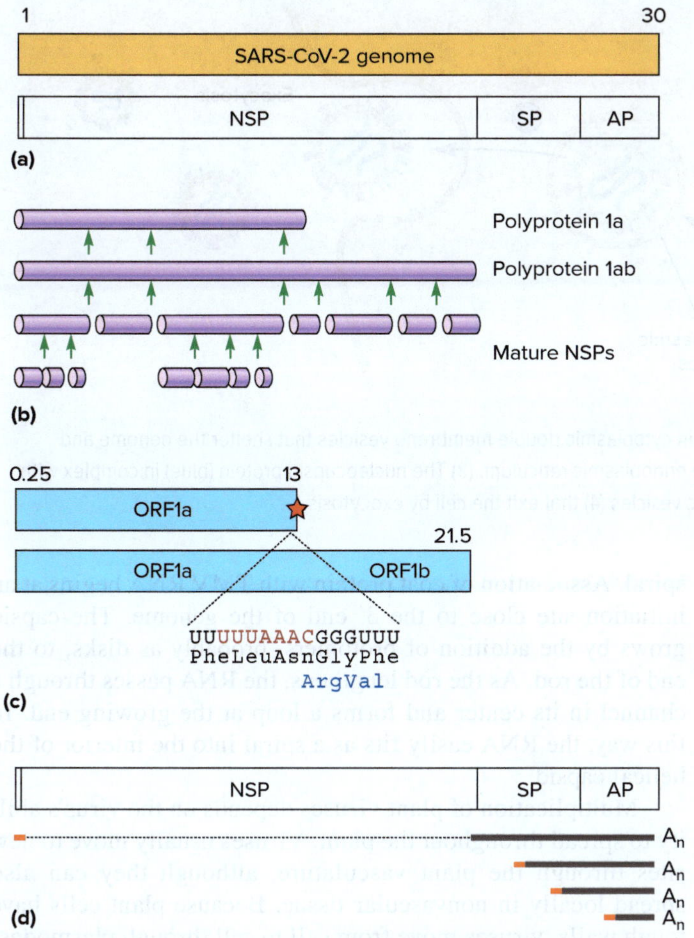

Figure 25.18 Translation and Transcription of the SARS-CoV-2 Genome. (a) Genome structure. Numbers indicate the genome length in kilobases. (b) Proteolysis of the polyproteins generates nonstructural proteins. Not all cleavages are illustrated here. (c) Nonstructural proteins are synthesized from ORF1, which is broken into two separate reading frames. The star indicates the position of ribosomal frameshifting. (d) Subgenomic RNAs are formed by splicing translational signals from the 5′ genome end (red) to the individual genes. NSP - nonstructural proteins; SP - structural proteins; AP - accessory proteins.

are grouped in the genome and expressed as units (figure 25.18*a*). Nonstructural proteins (NSP) include the enzymes required for uniquely viral processes—for example, the RNA-dependent RNA polymerase (RdRp) that synthesizes RNA from an RNA template. Structural proteins (SP) comprise the virion, including the distinctive surface spike. Accessory proteins (AP) aid the virus in subverting normal cellular processes and antagonizing the innate immune response.

The positive-strand genome is packaged with a 5′-cap so it is ready for immediate translation by host ribosomes. NSPs are translated first as a consequence of their position at the 5′ end. Two overlapping ORFs (ORF1a and ORF1b) encode polyproteins (figure 25.18*b*). A **polyprotein** is a large protein that is cleaved into smaller functional proteins by proteases. Translation of ORF1a yields polyprotein 1a encoding 11 NSPs, and translation of both ORF1a and ORF1b (ORF1ab) yields polyprotein 1ab encoding 16 NSPs. These polyproteins are cleaved by viral proteases NSP3 and NSP5, which undergo self-proteolysis and then generate the other NSPs.

Translation of ORF1a to generate polyprotein 1a is straightforward, but the origin of polyprotein 1ab is less obvious. ORF1a and ORF1b encode polypeptides in different reading frames, and the virus uses ribosomal frameshifting to synthesize polyprotein 1ab. In **ribosomal frameshifting,** a translating ribosome encounters RNA secondary structure that causes it to pause with a string of identical nucleotides in the A and P sites (figure 25.18*c; see also figure 13.36*). As the secondary structure unfolds to restart translation, the tRNAs slip by one nucleotide and continue in an alternate reading frame; translation of ORF1ab occurs because the ribosomal frameshift allows the process to proceed beyond the stop codon of ORF1a.

The proteins synthesized from polyprotein 1a (NSP1 through NSP11) enable the virus to exploit the host cell resources, while proteins synthesized from polyprotein 1ab (NSP12 through NSP16) act during genome replication. NSP1 inhibits translation of cellular mRNA by two mechanisms. It plugs the 40S ribosomal subunit entry portal preventing initiation, and it stimulates cellular RNases to cleave mRNAs in association with ribosomes. NSP4 and NSP6 promote rearrangement of endoplasmic reticulum (ER) membranes to form **double membrane vesicles, DMVs** (**figure 25.19**). These are the location of the replication complex (*see figure 6.11*). It is here that the RdRp (NSP12), acting as a replicase, catalyzes formation of the negative-strand RNA. Using that as a template and acting as a transcriptase, RdRp synthesizes positive-strand RNA. SARS-CoV-2 is unusual among RNA viruses in that one of its proteins (NSP14) proofreads the newly synthesized RNA: A domain of NSP14 called ExoN complexes with NSP10 and excises misincorporated nucleotides.

Negative-strand synthesis forms two types of positive-strand RNA. Full-length positive-strand genomic RNA will be packaged into mature particles. Subgenomic RNAs (sgRNAs) serve as transcripts for structural and accessory proteins (figure 25.18*d*). To ensure efficient translation by host ribosomes, the 5′ end of each sgRNA is fused with short sequences from the 5′ end of the genome. At the 3′ end, sgRNAs are polyadenylated. Because sgRNAs are copied from the negative strand, they are more abundant than the original positive-strand genome, and can therefore synthesize the quantities of proteins needed for the final stages of the life cycle.

Accessory proteins distract the host cell from the viral reactions. Double-stranded RNA is a hallmark of infection and triggers the innate immune response. Viral accessory proteins disrupt the cell's signaling networks, thereby delaying the response. *Recognition of foreignness is key for survival (section 31.6)*

Assembly of mature SARS-CoV-2 virions begins when the positive-strand RNA genome joins with nucleocapsid (N) proteins

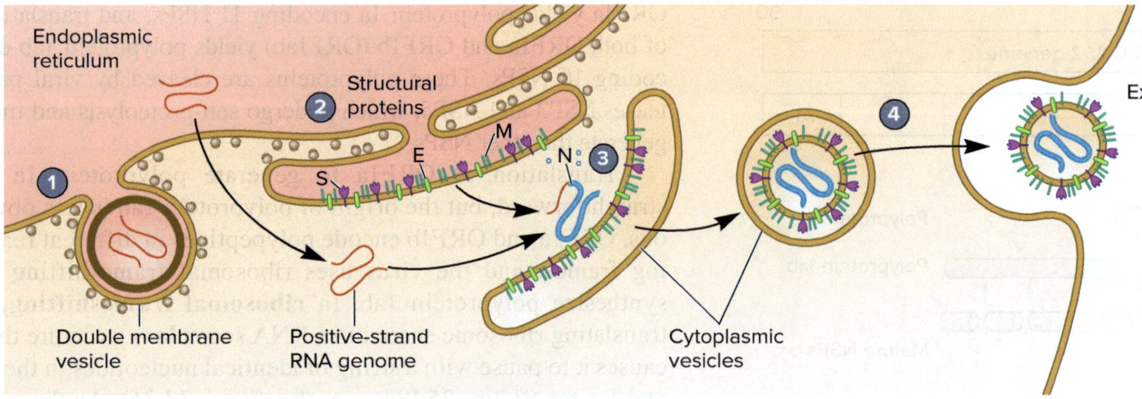

Figure 25.19 Replication and Exit of SARS-CoV-2. (1) Viral replication occurs in cytoplasmic double membrane vesicles that shelter the genome and concentrate ribonucleotide precursors. (2) Structural proteins are synthesized on the endoplasmic reticulum. (3) The nucleocapsid protein (blue) in complex with the positive-strand RNA genome (red) is engulfed by membranes to form cytoplasmic vesicles (4) that exit the cell by exocytosis.

to form ribonucleoprotein complexes (figure 25.19). The spike (S), envelope (E), and membrane (M) proteins are retained in the endoplasmic reticulum (ER) membranes. The nucleocapsid transits from the ER, and the M protein coordinates envelope assembly. E and S proteins are inserted in the membrane, and E mediates membrane curvature around the nucleocapsid. A host cell protease, furin, cleaves the S protein as it is inserted in the membrane. This promotes more rapid entry into the next host cell. Mature virions accumulate in cytoplasmic vesicles and exit by exocytosis.

Tobacco Mosaic Virus

Most plant viruses are RNA viruses, and of these, positive-strand RNA viruses are most common. Tobacco mosaic virus (TMV; family Virgaviridae) is the best-studied positive-strand RNA plant virus. TMV virions are filamentous with coat proteins arranged in a helical pattern. The TMV life cycle illustrates another strategy to produce multiple proteins from a single mRNA transcript: readthrough. **Readthrough** occurs during translation when the ribosome ignores a stop signal and continues translation.

Plant viruses, including TMV, usually enter the host through an abrasion on the plant, or via sap-sucking insects such as aphids. After entering its host, the TMV RNA genome is translated. Two proteins are produced, one ~125 kilodaltons (kD) and one ~185 kD. The larger protein is produced by readthrough of the stop codon at the end of the coding region for the 125-kD protein. The 185-kD protein has RdRp activity and functions as both a transcriptase and a replicase. TMV uses membranes of the endoplasmic reticulum to form a replication complex. There, the RdRp synthesizes negative-strand RNA using the positive-strand genome as template. The negative-strand RNA serves as a template both for synthesis of new genomes and for synthesis of subgenomic mRNAs that are translated into viral proteins.

After the coat protein and RNA genome of TMV have been synthesized, they spontaneously assemble into complete virions in a highly organized process (**figure 25.20**). The protomers form disks composed of two layers arranged in a helical spiral. Association of coat protein with TMV RNA begins at an initiation site close to the 3′ end of the genome. The capsid grows by the addition of protomers, probably as disks, to the end of the rod. As the rod lengthens, the RNA passes through a channel in its center and forms a loop at the growing end. In this way, the RNA easily fits as a spiral into the interior of the helical capsid.

Multiplication of plant viruses depends on the virus's ability to spread throughout the plant. Viruses usually move to new sites through the plant vasculature, although they can also spread locally in nonvascular tissue. Because plant cells have tough walls, viruses move from cell to cell through plasmodesmata. These are slender bridges of cytoplasmic material, including extensions of the ER, that extend through holes in cell walls to join adjacent cells. Viral movement proteins are required for transfer through the plasmodesmata. Movement proteins are associated with ER membranes and so are near the replication complex. They facilitate the movement of viral RNA from the replication complex to the plasmodesmata.

Several cytological changes take place in TMV-infected cells. These include intracellular inclusions similar to the replication complexes observed in animal cells infected with positive-strand viruses. Hexagonal crystals of almost pure TMV virions sometimes develop. In addition, chloroplasts become abnormal and often degenerate, while new chloroplast synthesis is inhibited.

Comprehension Check

1. How do some positive-strand viruses use polyproteins to complete their life cycles?
2. What are the three categories of proteins synthesized by SARS-CoV-2? What are their functions?
3. How does the SARS-CoV-2 coronavirus control the amount of each protein synthesized?
4. How does TMV spread from one host cell to another?

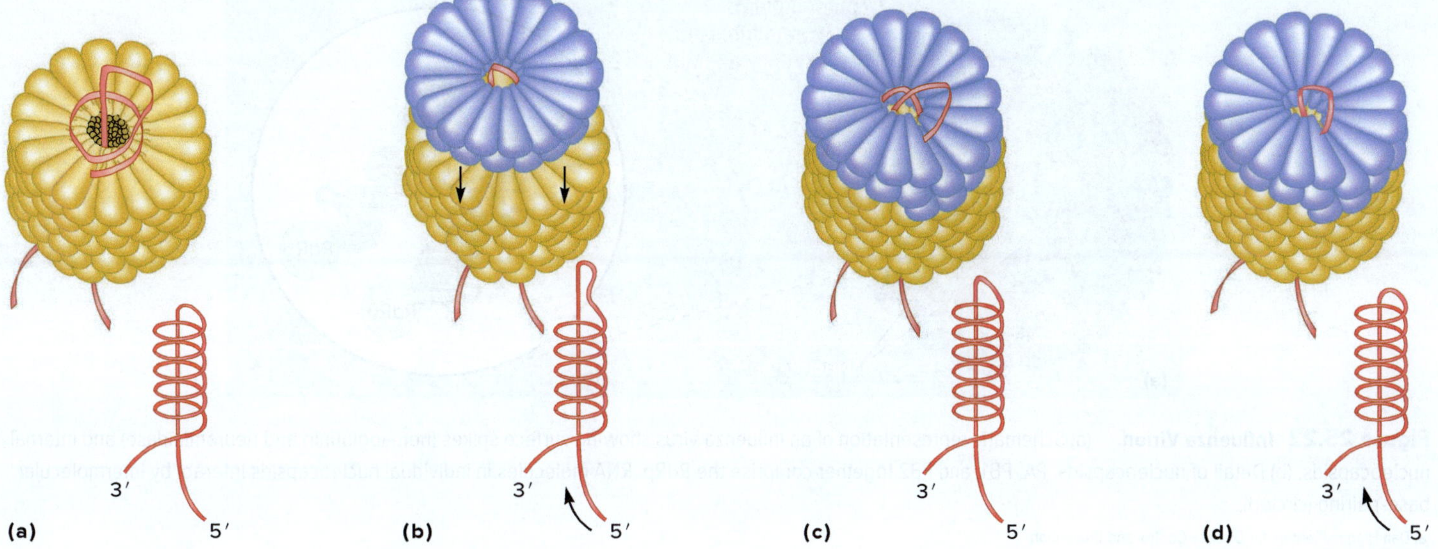

Figure 25.20 TMV Assembly. The elongation phase of tobacco mosaic virus nucleocapsid construction is illustrated. The lengthening of the helical capsid through the addition of a protein disk to its end is shown; line drawings depict RNA behavior. The RNA genome inserts itself through the hole of an approaching disk and then binds to the groove in the disk as it locks into place at the end of the cylinder.

25.6 Negative-Strand RNA Viruses: RNA-Dependent RNA Polymerase Is Part of the Virion

After reading this section, you should be able to:

a. Describe in general terms the strategy used by negative-strand RNA viruses to synthesize their nucleic acids and proteins
b. Explain how having a segmented genome impacts synthesis of viral mRNA and proteins and the generation of new strains of a virus
c. Create a flow chart that summarizes the life cycle of influenza virus, noting the specific mechanisms it uses to accomplish each step of its life cycle

Negative-strand viruses are found in many families, including Rhabdoviridae (e.g., rabies virus), Filoviridae (e.g., Ebola virus), Paramyxoviridae (e.g., measles and mumps viruses), and Orthomyxoviridae (e.g., influenza viruses). Most negative-strand RNA viruses are enveloped viruses that vary in morphology from spherical, to filamentous, rod-shaped, bullet-shaped, and pleomorphic.

The genomes of negative-strand RNA viruses cannot function as mRNA. Therefore these viruses must bring at least one RNA-dependent RNA polymerase (RdRp) into the host cell during entry. Initially the viral genome serves as the template for mRNA synthesis (**figure 25.21**). Later the virus switches from mRNA synthesis to genome replication, as the RdRp synthesizes a positive-strand RNA. During this phase of the life cycle, the positive-strand RNA molecules serve as templates for the manufacture of new negative-strand RNA genomes.

Our focus in this section is on influenza A and B viruses. Influenza virus virions are composed of an envelope containing eight segments of RNA, each enclosed in a separate nucleocapsid (**figure 25.22**). Each RNA molecule is configured in a double helical hairpin structure (figure 25.22*b*). Formation and maintenance of the double helix do not involve base pairing. Rather, this is accomplished by contact between numerous copies of the nucleocapsid protein (NP), which coats the RNA and allows it to fold back on itself. The three polypeptides of the influenza RdRp (PA, PB1, PB2) are attached to the open end of the hairpin. The life cycle of the influenza virus is outlined in **figure 25.23**.

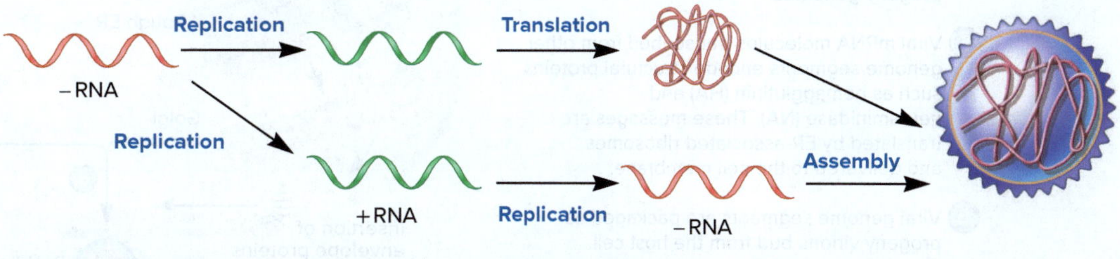

Figure 25.21 Multiplication Strategy of Negative-Strand RNA Viruses. An RNA-dependent RNA polymerase enters the host cell along with the negative-strand RNA genome. The genome serves as a template for synthesis of mRNA. Later in the infection, the negative-strand genome is used for positive-strand synthesis. These positive strands then act as templates for replication of the negative-strand genomes.

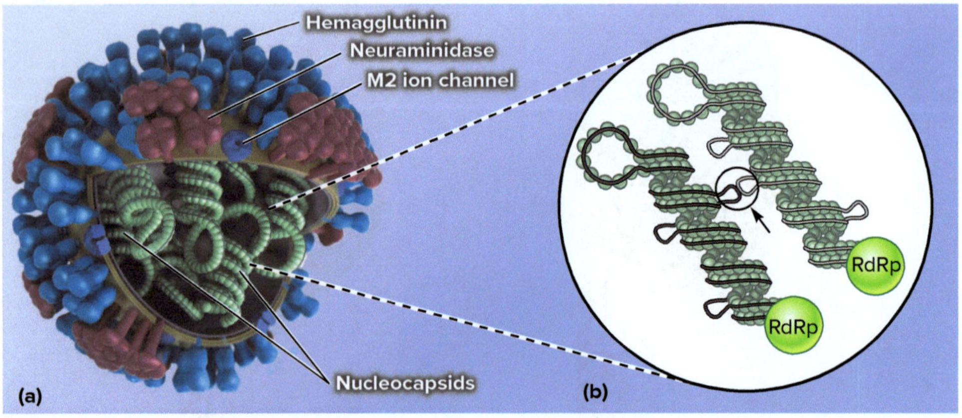

Figure 25.22 Influenza Virion. (a) Schematic representation of an influenza virus showing surface spikes (hemagglutinin and neuraminidase) and internal nucleocapsids. (b) Detail of nucleocapsids. PA, PB1, and PB2 together comprise the RdRp. RNA molecules in individual nucleocapsids interact by intermolecular base-pairing (circled).

(a) Dan Higgins/Centers for Disease Control and Prevention

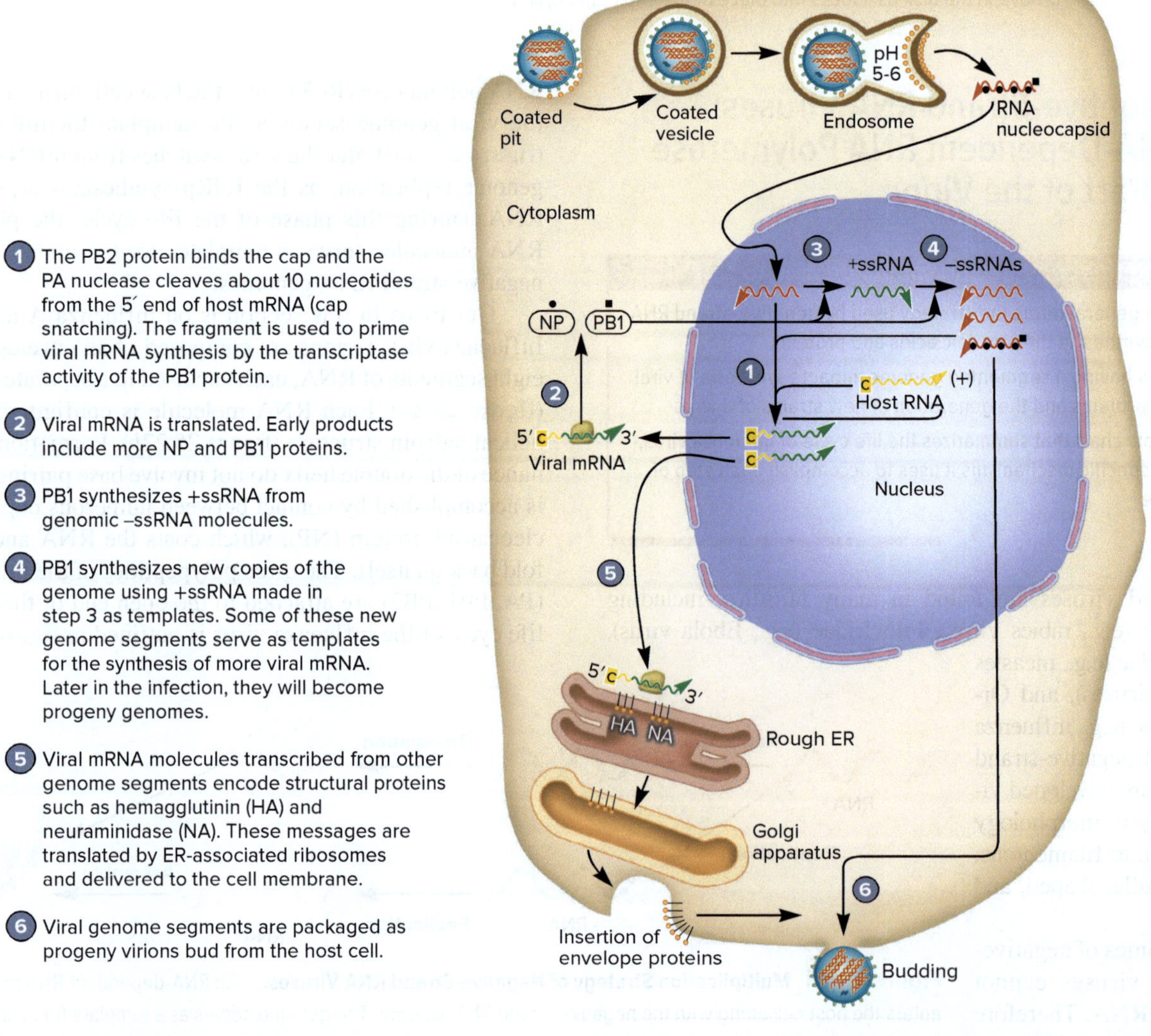

Figure 25.23 The Influenza Virus Life Cycle. This simplified version of the influenza life cycle omits several steps and represents the viral RNA molecules as simple lines. Negative-strand RNA is red, positive-strand RNA is green, and host-derived RNA is yellow. Steps 1 through 6 illustrate the steps following endocytosis.

An influenza virion enters a host cell by receptor-mediated endocytosis. This encloses the virus in an endosome (figure 25.23). When the endosomal pH decreases, a conformational change in the virion facilitates contact with the endosome membrane, allowing fusion of the endosome membrane and viral envelope to occur, releasing the nucleocapsids into the cytosol. ▶| *Influenza (Flu) (section 37.1)*

To carry out its life cycle, influenza components shuttle between the nucleus and cytoplasm using nuclear localization signals on proteins and nuclear export signals on RNA. Uncoated nucleocapsids are transported to the nucleus for mRNA synthesis, mRNAs move to the cytoplasm for translation, and the newly synthesized viral RdRps return to the nucleus for replication and capsid assembly. When RNA segments and associated RdRp enter the host cell nucleus, genome segments serve as templates for mRNA synthesis (figure 25.23, step 1). Copies of the negative-strand genome destined for translation are distinguished from copies destined for replication by the presence of a 5′ cap structure. Both termini of each genome segment form a panhandle (double-stranded) structure recognized by RdRp as the substrate for transcriptase activity. Once bound to the panhandle, the RdRp acquires a 5′ cap by a process termed **cap snatching.** Host mRNAs with a 5′ cap are cut by the RdRp about 8 to 14 bases after the cap, and that short primer is used for generating viral mRNA. Later the virus switches from mRNA synthesis to genome replication. During this phase of the life cycle, the positive-strand RNA molecules synthesized from the negative-strand genome segments serve as templates for new negative-strand RNA genomes (figure 25.23, step 4). Virions exit the host cell by budding and thus acquire their envelope (figure 25.23, step 6; *see also figure 6.14*).

Comprehension Check

1. How does the use of a segmented genome by influenza viruses differ from that used by *Pseudomonas virus* ϕ6?
2. Trace the multiplication of an influenza virus starting with host cell attachment and ending with the exit of virions.

25.7 Retroviruses: Positive-Strand Viruses That Use Reverse Transcriptase in Their Life Cycles

After reading this section, you should be able to:

a. Describe in general terms the strategy used by retroviruses to synthesize their nucleic acids and proteins
b. Differentiate a segmented genome from the genome of a retrovirus
c. Describe the life cycle of HIV, noting the specific mechanisms it uses to accomplish each step
d. Defend the statement "all three enzymatic activities of reverse transcriptase are required for dsDNA to be synthesized from RNA"
e. Distinguish DNA-dependent DNA polymerases, DNA-dependent RNA polymerases, RNA-dependent RNA polymerases, and RNA-dependent DNA polymerases in terms of their templates, products synthesized, and proofreading activity, as discussed in this and previous sections

Retroviruses have positive-strand RNA genomes. However, their genomes do not function as mRNA (**figure 25.24**). Instead, retroviruses first convert their ssRNA genomes into dsDNA using reverse transcriptase (RT). The dsDNA is then integrated into the host's DNA, where it can serve as a template for mRNA synthesis and synthesis of the positive-strand RNA genome. The host cell's DNA-dependent RNA polymerase catalyzes both of these processes.

Numerous retroviruses have been identified and studied. However, human immunodeficiency virus (HIV), the cause of acquired immune deficiency syndrome (AIDS), is of particular interest. Because of its global importance, we focus exclusively on HIV in this section. ▶| *Acquired immune deficiency syndrome (AIDS) (section 37.3)*

HIV is a member of the genus *Lentivirus* within the family Retroviridae. In the United States, AIDS is primarily caused by HIV-1, while in many developing nations HIV-2, a different subtype, is prevalent. HIV is an enveloped virus. The envelope

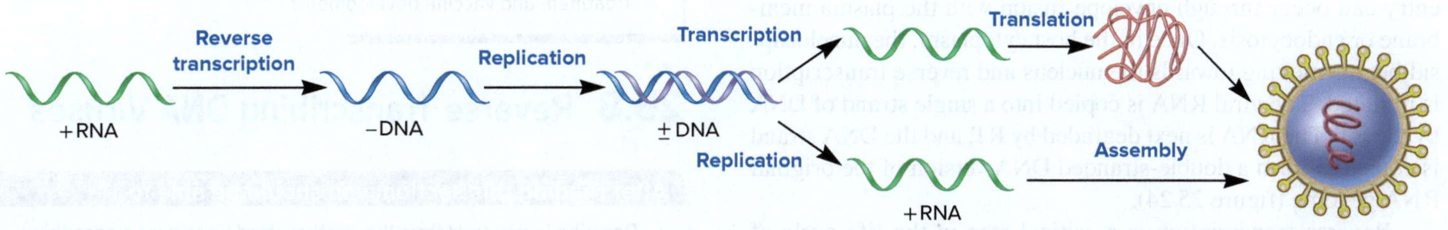

Figure 25.24 Multiplication Strategy of Retroviruses. Retroviruses have a positive-strand RNA genome that is first converted into dsDNA by reverse transcriptase. The viral dsDNA integrates into the host chromosome, where it serves as the template for synthesis of viral mRNA and viral genomes. Both are synthesized using the host cell's DNA-dependent RNA polymerase.

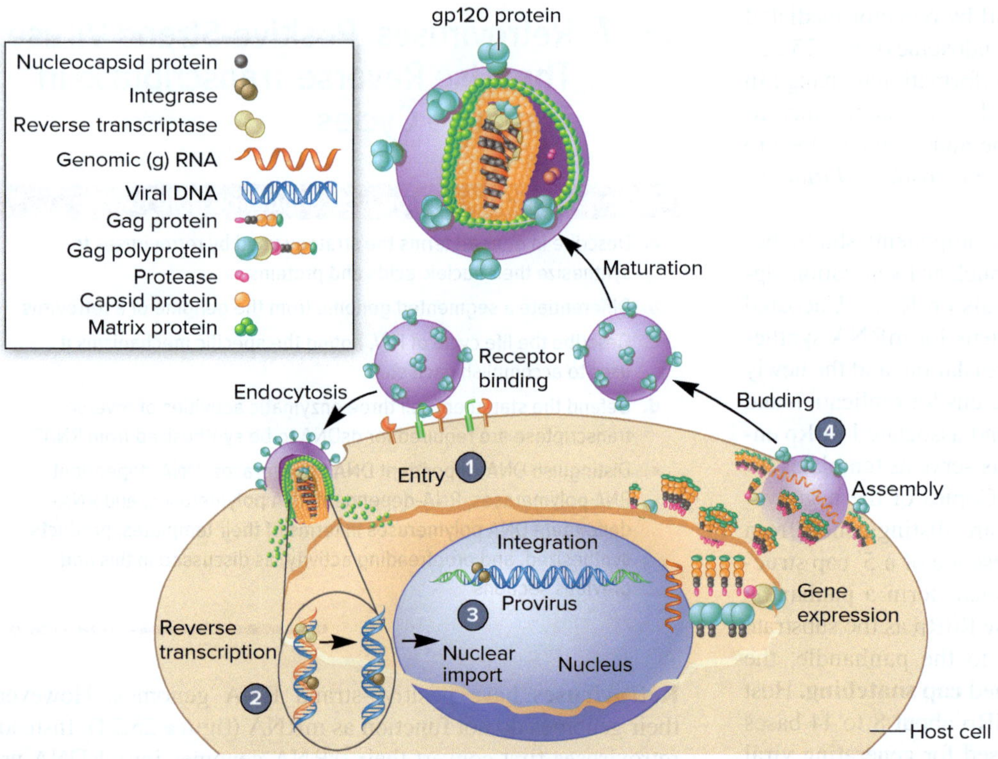

Figure 25.25 The HIV Life Cycle. (1) HIV nucleocapsid enters the cell following binding of viral surface protein gp120 to cell surface CD4 glycoprotein. (2) Within the nucleocapsid, reverse transcriptase copies viral RNA to ssDNA and that strand is copied to form dsDNA. (3) Following uncoating, the dsDNA viral genome is integrated into the host chromosome. Viral genes are expressed via the normal cellular processes. (4) Viral proteins and the RNA genome are assembled into nucleocapsids that bud from the infected host cell. Viral particles require maturation by proteases.

MICRO INQUIRY *What three enzymatic activities does reverse transcriptase display? When during the HIV life cycle is each activity used?*

surrounds an outer shell, which encloses a cone-shaped core (**figure 25.25**). The core contains two copies of the 10,000-base HIV ssRNA genome and several enzymes, including RT, protease, and integrase. The genome has nine genes that encode 15 virus-specific proteins.

After HIV enters the body, the gp120 viral envelope protein binds to host cells that have a surface glycoprotein called CD4. The virus requires a coreceptor in addition to CD4, and this varies depending on the host cell infected (*see figure 37.12*). Viral entry can occur through envelope fusion with the plasma membrane or endocytosis. Once in the host cytoplasm, the nucelocapsid begins moving towards the nucleus and reverse transcription is initiated. The viral RNA is copied into a single strand of DNA by the RT. The RNA is next degraded by RT, and the DNA strand is copied to form a double-stranded DNA version of the original RNA genome (figure 25.24).

Reverse transcription is a critical step in the life cycle of HIV, and RT is a remarkable enzyme with multiple activities. It is an **RNA-dependent DNA polymerase,** a DNA-dependent DNA polymerase, and a ribonuclease. Despite its versatility, RT lacks an important function observed in other DNA polymerases: proofreading. Thus it makes errors as it synthesizes DNA.

Uncoating begins soon after the nucleocapsid enters the cytoplasm, but evidence suggests that the process is completed after entry into the nucleus. Once the HIV dsDNA is released, it is inserted into the host cell's genome by viral integrase and some host proteins. HIV has no specific integration site and may recombine anywhere in the genome. Once integrated, the provirus forces the cell to synthesize viral mRNA (figure 25.25). As described in other ssRNA life cycles, gene expression strategies for HIV include splicing to generate subgenomic RNAs and ribosomal frameshifting.

Viral proteins and the RNA genome are assembled into new nucleocapsids that bud from the infected host cell (figure 25.25). The viral particles released from the host cell are not infective and require a maturation process. The viral Gag protein, a structural component of the virion, is cleaved by proteases. The result is a change in capsid shape from a sphere to a cone and RNA genome repackaging. Drugs for treating HIV target multiple steps in the life cycle, including maturation. ◄ *Antiviral drugs (section 9.5)*

Comprehension Check

1. What is the function of each of the following HIV proteins: gp120, reverse transcriptase, and integrase?
2. CD4 is found on several different immune system cells. Consult chapter 32. Why is the presence of this molecule on the surface of many immune system cells important to the development of AIDS?
3. Why does HIV have a high mutation rate? How does this influence treatment and vaccine development?

25.8 Reverse Transcribing DNA Viruses

After reading this section, you should be able to:

a. Describe in general terms the strategy used by reverse transcribing DNA viruses to synthesize their nucleic acids and proteins
b. Compare the role of reverse transcriptase in the life cycle of a retrovirus to that in the life cycle of a hepadnavirus

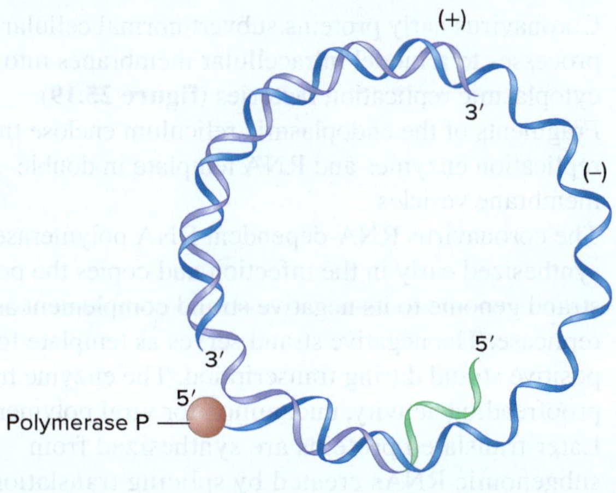

Figure 25.26 The Gapped Genome of Hepadnaviruses. The negative strand of the dsDNA molecule is complete but nicked (blue). Viral DNA polymerase P is attached to its 5′ end. The positive strand is gapped; that is, it is incomplete (purple). A short stretch of RNA is attached to its 5′ end (green).

Reverse transcribing DNA viruses use reverse transcriptase (RT) to replicate their genomes. Hepatitis B virus (HBV) is one of the best-studied reverse transcribing DNA viruses and is our focus here.

HBV virions are 42 nm spherical particles that contain the viral genome. The HBV genome is a 3.2 kb circular dsDNA molecule that consists of one complete but nicked strand and a complementary strand that has a large gap; that is, it is incomplete (**figure 25.26**). In addition, the viral polymerase P is covalently bound to the genome. After infecting a host cell, the viral genome is released into the nucleus. Before the virus can initiate its life cycle in cells of the liver (hepatocytes), its genome must be converted to a covalently *c*losed *c*ircular DNA (cccDNA) form. First, host proteases cleave the polymerase P from the genome. But rather than using polymerase P, host DNA repair enzymes synthesize the remainder of the positive strand, cleave the RNA primer, and seal the nicks. Note that these processes are the same as are required to join Okazaki fragments following DNA replication. ◀ *Many enzymes are needed for DNA replication (section 13.3)*

The HBV genome has four partially overlapping open reading frames that encode viral proteins. Transcription of viral genes occurs in the nucleus using host RNA polymerase and yields several mRNAs, including a large 3.4-kb RNA known as the *pregenome* (a positive-strand RNA). The RNAs move to the cytoplasm, and the mRNAs are translated to produce viral proteins, including core protein and polymerase P, which also has RT activity. Polymerase P then associates with the positive-strand RNA pregenome and core protein to form an immature core particle. Within the immature core particle, polymerase P uses its RT activity to generate negative-strand DNA. This process is unusual because a protein, rather than RNA, is used to prime synthesis. After almost all the pregenome RNA has been degraded by the RNaseH activity of RT, the remaining RNA fragment serves as a primer for synthesis of the gapped dsDNA genome, using the negative-strand DNA as template. When the nucleocapsid is complete, progeny virions are released.

Comprehension Check

1. Describe the HBV genome. How is it converted to cccDNA in the host?
2. Trace the HBV multiplication cycle, paying particular attention to localization within the host cell during the biosynthetic stage.

Key Concepts

25.1 Virus Phylogeny Relies on Genomics

- In the absence of a universal genetic element, viruses are classified by nucleic acid type. The highest level taxon of realm divides viruses by their genome form: single- or double-stranded DNA or RNA (**table 25.1**).
- The Baltimore system is used by many virologists to organize viruses based on their genome type and the mechanisms used to synthesize mRNA and replicate their genomes (**table 25.2**).

25.2 Double-Stranded DNA Viruses Infect All Cell Types

- *Escherichia virus* T4 is a virulent bacteriophage. After attachment to a specific surface receptor, T4 releases its dsDNA into the cell (**figures 25.3** and **25.4**). T4 DNA contains hydroxymethylcytosine (HMC) in place of cytosine, and glucose is often added to the HMC to protect the phage DNA from attack by host restriction enzymes (**figure 25.5**). T4 DNA replication produces concatemers, long strands of several genome copies linked together (**figure 25.6**).
- *Escherichia virus Lambda* (λ) is a temperate bacteriophage. It can establish lysogeny, rather than pursuing a lytic infection. During lysogeny, the viral DNA, called a prophage, is replicated along with the cell's genome. Lysogeny is reversible, and the prophage can be induced to become active and lyse its host. This highly regulated process is an important model system for gene regulation.
- The protein cII plays a central role in regulating the switch between lysogeny and a lytic cycle. If cII protein levels are high enough, lysogeny is established. If not, the lytic cycle is initiated (**figures 25.9** and **25.10**).
- Herpesviruses are a large group of dsDNA viruses. In humans, they cause acute infections such as cold sores, genital herpes, chickenpox, and mononucleosis. This is

followed by a lifelong latent infection in which the virus genome resides within host cells. The virus can be reactivated at later times to cause a productive infection.

- Nucleocytoplasmic large DNA (NCLD) viruses, also called megaviruses, are members of several virus families. They are thought to share an evolutionary history and are notable for their large genome and virion size. Furthermore, their genomes contain genes for translation-related functions not usually observed in viral genomes.

25.3 Single-Stranded DNA Viruses Use a Double-Stranded Intermediate in Their Life Cycles

- *Escherichia virus* ϕX174 is an example of a ssDNA bacteriophage. Its replication involves the formation of a dsDNA replicative form (RF) (**figure 25.12**).
- *Pseudomonas virus Pf1* is a filamentous phage that upon infection is continuously released by the host without causing lysis.
- Parvoviruses cause a spectrum of diseases in a wide variety of animals. The parvovirus genome is replicated by host DNA polymerase in the host cell's nucleus using a process that is similar to rolling-circle replication (**figure 25.13**).

25.4 Double-Stranded RNA Viruses: RNA-Dependent RNA Polymerase Replicates the Genome and Synthesizes mRNA

- Double-stranded RNA viruses use a viral enzyme called RNA-dependent RNA polymerase to synthesize mRNA (transcriptase activity) and replicate their genomes (replicase activity) (**figure 25.14**).
- *Pseudomonas virus* ϕ6 is unusual in that it is enveloped and enters the host bacterium through a process that resembles endocytosis. It has a segmented genome, in which each segment encodes a polycistronic mRNA that yields several proteins.
- Rotaviruses have segmented genomes, in which each segment codes for one or two proteins. They multiply in an inclusion called the viroplasm (**figure 25.15**).

25.5 Positive-Strand RNA Viruses: Genomes That Are Translated upon Entry

- The genomes of positive-strand RNA viruses serve directly as mRNA molecules. Among the first viral proteins synthesized is an RNA-dependent RNA polymerase that replicates the positive-strand RNA genome, sometimes by forming a dsRNA replicative intermediate. The negative RNA strands produced by the RNA-dependent RNA polymerase are used to make more genomes and mRNA (**figure 25.16**).
- Coronaviruses have large genomes. The largest open reading frames encode polyproteins that are translated immediately upon entry to the cell (**figure 25.18**). The extent of translation is controlled by ribosomal frameshifting.
- Coronavirus early proteins subvert normal cellular processes to remodel intracellular membranes into cytoplasmic replication factories (**figure 25.19**). Fragments of the endoplasmic reticulum enclose the viral replication enzymes and RNA template in double-membrane vesicles.
- The coronavirus RNA-dependent RNA polymerase is synthesized early in the infection and copies the positive strand genome to its negative strand complement as a replicase. The negative strand serves as template for the positive strand during transcription. The enzyme has proofreading activity, uncommon for viral polymerases.
- Later-translated proteins are synthesized from subgenomic RNAs created by splicing translational signals to each open reading frame. These proteins include viral structural proteins and accessory proteins that promote viral release and antagonize the innate immune system.
- Tobacco mosaic virus (TMV), like most known plant viruses, has a positive-strand RNA genome. Translation of TMV genomic RNA yields two proteins, the larger of which is produced by a readthrough mechanism. The RNA-dependent RNA polymerase encoded by the genome synthesizes negative-strand RNAs that serve as templates for synthesis of subgenomic mRNAs that are translated to yield other viral proteins. The negative-strand RNAs also serve as templates for synthesis of new positive-strand RNA genomes. The TMV nucleocapsid forms spontaneously when disks of coat protein protomers complex with the RNA (**figure 25.20**).

25.6 Negative-Strand RNA Viruses: RNA-Dependent RNA Polymerase Is Part of the Virion

- For negative-strand RNA viruses to synthesize mRNA and replicate their genomes, their virions must carry an RNA-dependent RNA polymerase that enters host cells with the genome. The polymerase first synthesizes mRNA from the negative-strand genome. Later it is used to replicate the genome by way of a positive-strand intermediate (**figure 25.21**).
- Influenza viruses have segmented genomes; most segments encode a single protein. The viruses exit by budding (**figure 25.23**).

25.7 Retroviruses: Positive-Strand Viruses That Use Reverse Transcriptase in Their Life Cycles

- Retroviruses replicate their genomes and synthesize mRNA via a dsDNA intermediate. The dsDNA is formed by a multifunctional enzyme called reverse transcriptase (**figure 25.24**).
- HIV is an enveloped retrovirus with a cone-shaped core that contains two copies of its genome and several enzymes, including reverse transcriptase. Upon infection, its ssRNA genome is converted to dsDNA, which is then integrated into the host genome (**figure 25.25**). HIV uses

alternative splicing to generate different mRNA molecules. One polyprotein arises by a combined ribosomal frameshifting, readthrough mechanism.

25.8 Reverse Transcribing DNA Viruses

- This class of viruses is exemplified by hepatitis B virus, which has a dsDNA genome that consists of one complete but nicked strand and an incomplete (i.e., gapped) complementary strand (**figure 25.26**).
- Upon infection with hepatitis B virus, host cell enzymes repair the gap and seal the nick to generate a covalently closed, circular viral genome. This serves as the template for the synthesis of pregenome RNA, which is the template for reverse transcription. Reverse transcription produces the dsDNA, gapped genome.

Active Learning

1. No temperate RNA phages have yet been discovered. How might this be explained?
2. The switch between lysogeny and lysis is influenced by many factors. How would external conditions such as starvation or crowding be sensed and communicated to the transcriptional machinery to influence this choice?
3. The most straightforward explanation as to why the T4 endolysin (i.e., the T4 lysozyme) is expressed so late in infection is that its promoter is recognized by the virus-encoded gp55 alternative sigma factor. Propose a different explanation.
4. You are studying RNA viruses and have discovered one that grows well in a culture of eukaryotic cells. You know that the virus is a single-stranded RNA virus, but you don't know if it is positive or negative stranded. Your lab-mate says, "Well, just treat your cell culture with cyclohexamide and see if the virus replicates its genome." You know that cyclohexamide inhibits protein elongation by binding to eukaryotic ribosomes. What is the basis of your lab-mate's suggestion?
5. Although they do not contain ribosomes, and therefore genes to encode small subunit rRNA, nucleocytoplasmic large DNA (NCLD) viruses do contain conserved genes for functions related to translation. When phylogenetic trees are created using these conserved genes, the NCLD viruses group as a fourth domain, beside Bacteria, Archaea, and Eukaryotes. Defend or refute the creation of a fourth domain to classify these viruses.
6. Compare and contrast an enveloped virus and an extracellular vesicle (*see figure 3.24*). What cellular components can be mobilized in each entity?
7. Since the beginning of the pandemic, multiple SARS-CoV-2 variants have been discovered that are more transmissible than the strain of virus that had initially spread worldwide. These variants were found to carry several genomic changes, including in the S gene that encodes the spike protein. How would you expect a change to the spike-ACE2 interaction to manifest as a more transmissible virus?

CHAPTER

26

Exploring Microbes in Ecosystems

malshkoff/Shutterstock

Can Soil Have Memory?

Soybean farmers, when purchasing seeds, have a menu of options: *Trichoderma, Bradyrhizobium japonicum, Bacillus* spp. Which microbial soil inoculant(s) would you like with your soybean seeds? If you haven't grown soybeans in a field for a few years, you should choose them all, as the soil has lost its memory for how to grow soybeans.

The menu of microbial choices is expanding as we learn that agricultural productivity encompasses more than seeds and soil. Historically, farmers transferred productive soils between fields to positively affect crop yields. The soil was said to have a memory, but we now know that this "memory" is the soil microbiota.

Exploration of the microbial communities in soils reveals an astounding diversity of microorganisms. Microbial ecologists can now correlate the presence or absence of certain species with good soil health and productivity: *Bacillus* spp. promote seed germination and root growth; *B. japonicum* forms root nodules for nitrogen fixation; *Trichoderma,* a fungus, controls fungal pathogens and promotes phosphate uptake. Establishment of these organisms in a soybean field reduces the need for fertilizer.

The green revolution of the mid-twentieth century brought vast expanses of high-yield corn, soybeans, and wheat. The next green revolution may well see microbe engineering that pairs growth-promoting microbes with specific food crops designed for local soil types.

Exploring the soil's microbial memory requires cataloging the organisms present and exploring their interactions within the soil ecosystem. These communities differ between soil, the rhizosphere, and the phyllosphere (plant tissues below and above the ground, respectively). And critically, these communities differ between healthy and diseased plants.

Microbial ecology encompasses traditional, laboratory-based analyses; genomics and metagenomics; and in situ (Latin, in place) biogeochemical assessments. The application of these techniques has brought an explosion of recent advances, generating new appreciation of the vast diversity of microbes and their roles in ecosystems.

Readiness Check:

Based on what you have learned previously, you should be able to:

- ✓ Explain why SSU rRNA is used as a taxonomic marker (section 1.2)
- ✓ List environmental factors that influence microbial growth (section 7.5)
- ✓ Describe conventional techniques used to isolate single colonies and establish a pure culture (section 7.7)
- ✓ Explain why PCR results in the amplification of a specific DNA sequence despite many competing sequences (section 17.2)
- ✓ Review the theory and practice of metagenomics, bioinformatics, and functional genomics (sections 18.3, 18.4, and 18.5)

26.1 The Vocabulary of Ecology Applies to Microbial Environments

After reading this section, you should be able to:

a. Define terms used by ecologists to describe relationships among living organisms
b. Explain why it is challenging to define a bacterial or archaeal species

Ecology is the study of living organisms and their relationships with other organisms and the environment. *Microbial* ecology acknowledges the ubiquity of microorganisms and their impacts on all levels of living organisms. An individual is a member of a species, while multiple organisms of the same species are a **population** (e.g., all the *B. subtilis* in a soybean field). Populations are also defined geographically and at a specific time. A **community** includes multiple populations (and therefore multiple species) that occupy the same region (e.g., soil microbes, plants, and insects in a soybean field). Communities exist in environments with nonliving components, and together these define an ecosystem (e.g., all microbes and soil components like sand, clay, and silt; *see figure 30.1*). An ecologist focuses on communities and ecosystems: the numbers and types of organisms present, their roles and interactions, and how the organisms respond to change.

The greatest challenge in applying these terms to a microbial ecosystem is the definition of a **species.** The biological species definition is a population whose members can interbreed and produce fertile offspring. This definition readily applies to sexually reproducing species that transmit DNA only vertically, from parent to offspring. However, such a definition cannot apply to organisms that reproduce asexually like bacteria and archaea. They transmit DNA both vertically and horizontally: Cell division faithfully passes the genome to daughter cells, but horizontal gene transfer creates substantial genetic variation within a microbial population that is not inherited. As we discuss in chapter 1, bacterial and archaeal species are now defined based on genome similarity. DNA sequence identity of greater than 97% between two genomes warrants classifying the source organisms as the same species. ◀ *Microbial taxonomy (section 1.2); Microbes use mechanisms other than mutation to create genetic variability (section 16.3)*

26.2 Microbial Identification Is Largely Based on Molecular Characterization

After reading this section, you should be able to:

a. Construct a concept map describing the approaches commonly used to determine taxonomic classification
b. Select an appropriate technique to identify a microbial genus, species, and strain
c. Predict the basic biological as well as public health implications of microbial taxonomic identification

Microbiologists use a variety of approaches to identify and classify a microbe. For clarity, we divide these approaches into two groups: classical and molecular. The most durable identifications are based on a combination of approaches.

Classical Characteristics

Classical approaches to taxonomy make use of morphology, physiology, and ecology. This approach has been employed in microbial taxonomy for many years and forms the basis for phenetic (phenotypic) classification. When used in combination, they are quite useful in routine identification of well-characterized microbes, for example, those typically encountered in clinical settings.

Morphological features such as cell shape, size, and staining behavior (e.g., Gram positive or negative) are important in microbial taxonomy for many reasons (**figure 26.1**). Morphology is easy to study, particularly in eukaryotic microorganisms. In addition, morphological comparisons are valuable because structural features depend on the expression of many genes and are usually genetically stable.

Physiological and metabolic characteristics are useful because they are directly related to the activity of microbial proteins. For instance, the detection of fermentation end products in a newly discovered microorganism reveals the presence of specific catabolic enzymes and the genes that encode them. Therefore analysis of characteristics such as energy metabolism and nutrient transport provides an indirect comparison of microbial genomes.

The ability of a microorganism to colonize a specific environment is of taxonomic value. Some microbes may be similar in a number of respects but inhabit different ecological niches,

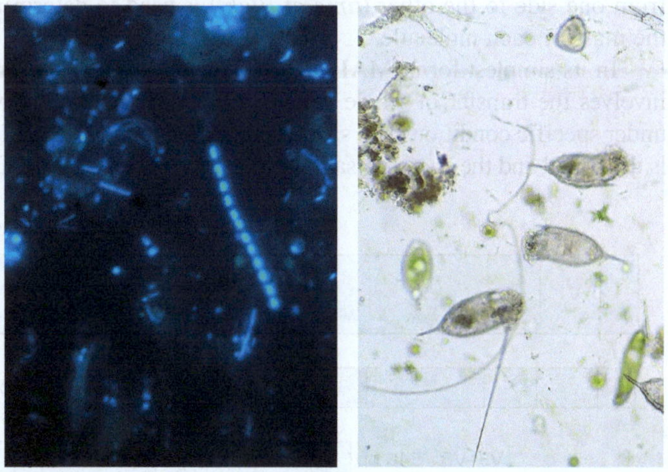

(a) (b)

Figure 26.1 Microscopic morphology. (a) A microbial assemblage from the North Sea is visualized following DAPI staining. The fluorescent stain DAPI (4′,6-diamido-2-phenylindole) specifically labels nucleic acids and is commonly used to visualize microbes in environmental, food, and clinical samples. (b) Light microscopy of freshwater aquatic plankton.

(a) David C. Gillan/Mons University, Belgium; (b) Choksawatdikorn/Shutterstock suggesting they may not be as closely related as first suspected. Some examples of taxonomically important ecological properties are life-cycle patterns: the nature of symbiotic relationships; the ability to cause disease in a particular host; and habitat preferences such as requirements for temperature, pH, oxygen, and osmotic concentration. ◀ *Environmental factors affect microbial growth (section 7.5)* ▶ *Many types of microbial interactions exist (section 27.1)*

Molecular Characteristics

It is hard to overstate how the study of DNA, RNA, and proteins has advanced our understanding of microbial evolution and taxonomy. Although much emphasis has been placed on nucleotide sequence analysis, we start our discussion with a biochemical approach.

Biochemical Characteristics

Advances in mass spectrometry (MS) have resulted in the fast and accurate identification of bacteria based on the presence of specific, highly abundant proteins. The specific type of MS used is called *m*atrix-*a*ssisted *l*aser *d*esorption/*i*onization-*t*ime *o*f *f*light (MALDI-TOF). MALDI-TOF enables the analysis of complex biomolecules that could not previously be studied by MS. The material to be analyzed is dried on a sample holder (called a *target*) and mixed with a molecular film called a *matrix*. When an ultraviolet (UV) laser beam strikes the sample target, the matrix helps stimulate release of the sample from the surface; this is *matrix-assisted desorption*. Once the matrix-sample is released (desorbed), the matrix transfers protons to the sample, which then becomes ionized; this is *matrix-assisted ionization*. The ionized biomolecules are then "flown" across a space within the instrument; the time taken for a molecule to fly from one side to the other (*time of flight*) is used to determine the mass of each molecule.

In its simplest form, MALDI-TOF identification of bacteria involves the transfer of whole cells from a single colony grown under specific conditions to a sample target. Once dried, a matrix is deposited and the bacterial samples are analyzed. MALDI-TOF yields the masses of highly abundant bacterial proteins. Each experimentally derived protein profile must be matched to a protein profile from a known bacterium. MALDI-TOF is commonly used in medical microbiology laboratories where the same strains of organisms are regularly encountered. This method is limited by the need to grow microbes under specific conditions and it cannot be used to identify newly discovered microbes. ◀ *Proteomics explores total cellular protein (section 18.5)*

Signature Sequences

As we introduce in chapter 1, **small subunit ribosomal RNAs (SSU rRNAs)** are commonly used to infer microbial phylogenies and make taxonomic assignments at the genus level. The ability to amplify regions of rRNA genes (rDNA) by the polymerase chain reaction (PCR) and sequence the DNA using next-generation sequencing techniques has greatly increased the efficiency by which SSU rRNA sequences are obtained. While the complete rDNA gene yields the most information for taxonomic classification, it is impractical to sequence the entire ~1,500 base pairs (bp) on the most commonly used DNA sequencing platforms. Therefore shorter, information-rich regions of the rDNA gene that possess **oligonucleotide signature sequences** or **barcodes** are used. These are short, conserved regions of a gene that are specific to phylogenetically defined groups of organisms (*see figure 1.8*). Much like the distinctive black and white barcodes on consumer goods, these sequences identify the source microorganism. As illustrated in **figure 26.2**, some regions of the bacterial 16S rRNA gene are invariant among isolates, while variable regions contain the taxonomically valuable information. PCR primers are designed to hybridize to the conserved regions while amplifying the internal variable regions. Archaeal-specific 16S rRNA primers (distinct from bacterial 16S rRNA primers) and eukaryotic 18S rRNA primers are used to amplify rDNA genes from these groups. ◀ *Microbial taxonomy (section 1.2)*; *Polymerase chain reaction amplifies targeted DNA (section 17.2).*

Genes other than those encoding rRNA can be used as barcodes. In bacteria, two universal proteins, the chaperone GroEL

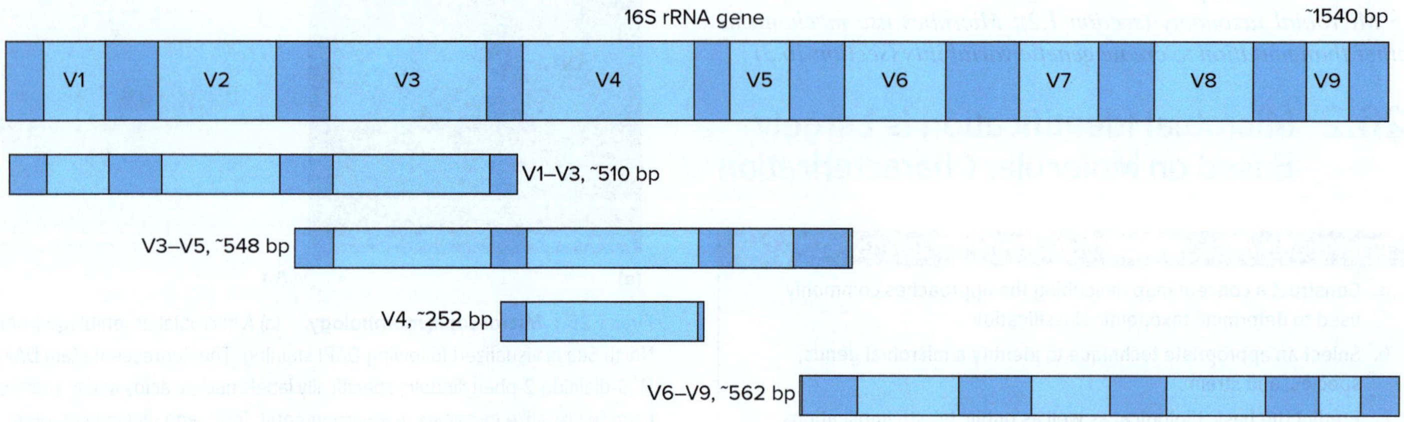

Figure 26.2 Barcode Amplification of the Small Ribosomal Subunit Gene. Variable regions (V1 through V9) of the bacterial 16S rRNA gene are amplified by PCR and sequenced.

and RNA polymerase, are encoded by genes (*cpn60* and *rpoB*, respectively) that can be amplified and compared to databases for identification, often to the level of species and subspecies.

A different type of signature sequence is the **indel.** Many genes have nucleotide *in*sertions or *del*etions of specific lengths and sequences at fixed positions. A particular indel may be found exclusively among all members of a taxon. Indels are particularly useful in phylogenetic studies when they are flanked by conserved regions. In such cases, changes in the signature sequence cannot be due to sequence misalignments. Alignment of the inferred amino acid sequences reveals indels useful for phylogenetic analysis.

Whole-Genome Comparison

As the field of taxonomy shifts to whole-genome sequencing (WGS) for species and strain identification and taxonomic classification, new quantitative measures of relatedness are being developed and evaluated. As with any metric, standardization of methods and interpretations are required. These processes are ongoing. An initial step to identify a newly discovered microbe is SSU rRNA comparison, which provides confident identification to the genus level. ◀ *Genome sequencing (section 18.2)*

G + C content is a simple percentage of the bases in DNA that are G + C, and it is readily determined computationally from WGS data. Organisms range from around 30 to 80% G + C content. Despite the wide range of variation, the G + C content of strains within a species is constant and varies little within a genus.

Identification to the species level requires detailed gene-by-gene comparison of the new isolate and type strains of its closest relatives. **Average nucleotide identity (ANI)** uses pairwise alignments between all sequences shared between two genomes and calculates the fraction of identical nucleotides. ANI values for two genomes of the same species should be at least 97%.

Comprehension Check

1. Distinguish the terms *organism, population, community,* and *ecosystem.*
2. Why is it a challenge to define *species* for bacteria and archaea?
3. What are the advantages of using each major group of characteristics (morphological, physiological and metabolic, ecological, and molecular) in classification and identification? Give examples of each type of characteristic.
4. List the steps you would take to identify an organism just isolated in pure culture and for which you have a genome sequence. What criteria would tell you if it is a new species?

26.3 Microbial Ecology Also Relies on Cultures

After reading this section, you should be able to:

a. Explain why, to the best of our knowledge, most microorganisms resist culturing in the laboratory
b. Describe the use of enrichment cultures
c. Describe at least two innovative approaches used to coax growth in the laboratory of previously uncultured microbes
d. Explain the fundamentals of flow cytometry

To thoroughly assess microbial communities, it is often desirable to have a steady supply of the microorganisms of interest, ideally in **axenic (pure) culture.** However, it is well understood that less than 5% of microbes have been grown under laboratory conditions. There are many reasons a particular microbe may resist culturing; a few are listed in **table 26.1.** The discrepancy between

Table 26.1 Example Reasons Some Microbes Have Not Yet Been Cultured

Potential Problem	Example Method Designed to Overcome
Microbe is slow growing.	Incubation times of weeks to months
Microbe is present in very low abundance.	Extinction cultures, using many replicates
Different microbes in same habitat are physiologically very similar. OR Inhibition by other microbes in a mixed culture	Remove other microbes from the sample by physical methods such as filtration or density-gradient centrifugation, or use extinction cultures.
Fastidious growth requirement	Assess growth requirements of similar microbes, if known; use annotation of metagenomic sequences to infer nutritional capabilities and requirements; grow in diffusion chambers that allow influx of small molecules from natural environmental samples without microbial contamination.
Cross-feeding or communication signals from other microbes are needed.	Co-cultivate strains; use diffusion chambers; grow in conditioned spent medium (medium in which microbes have grown but have been removed) of helper microbe.
Triggers for growth or exit from a dormant state are not present.	Add known growth triggers (e.g., *N*-acetylmuramic acid, bile acids).

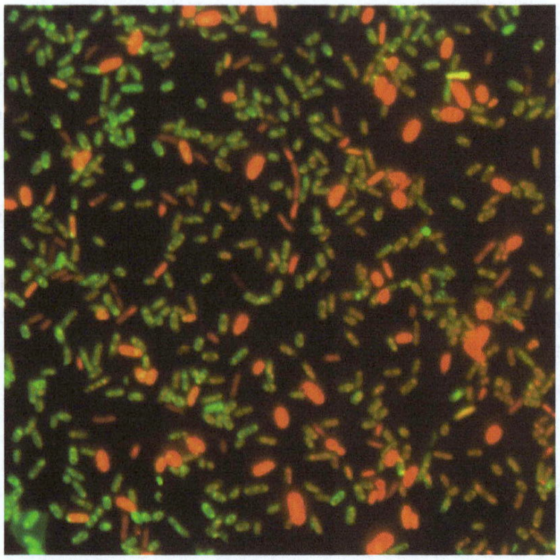

Figure 26.3 Assessment of Microbial Viability by Use of Direct Staining. In this procedure, two stains are used: a membrane-permeable green fluorescent nucleic acid stain and propidium iodide (red), which penetrates only cells with damaged membranes. Living cells stain green, while dead and dying cells stain red.

Dr. Rita B. Moyes

the number of microbial cells observed by microscopic examination and the number of colonies cultivated from the same natural sample has been called the **great plate count anomaly.** This has led to the description of potentially viable microbes as being "nonculturable." A microbe is deemed **viable but nonculturable (VBNC)** if it fails to be cultivated but is determined to be viable by one of several methods. A common microscopy technique uses dyes that enable the discrimination between live and dead cells (**figure 26.3**). A molecular approach, called *viability polymerase chain reaction* (vPCR), also uses a dye to differentiate between live and dead cells. In this method, the dye enters dead cells whose membranes have become damaged. Upon photoactivation, the dye cross-links the DNA, thereby preventing subsequent PCR amplification of target genes (e.g., small subunit rRNA). In this way, the DNA of only viable cells is present in the pool of PCR templates and is therefore detected as amplified DNA product. ◂ *Most microbes live in growth-arrested states (section 7.6)*

Although the gold standard is an axenic culture of the microbe of interest, one typically has to purify the microbe from a natural microbial assemblage. In such cases, it is often best to increase the number of microbes of interest relative to others in the same sample. **Enrichment culture** techniques are based on creating a microenvironment in the laboratory to promote the growth of a microorganism formerly restricted to a small ecological niche. At the same time, growth of other microbial types is inhibited. This approach often plays a central role in finding new microbes. As we describe in chapter 7, the success of an enrichment culture depends on a thorough understanding of the specific niche the microbe of interest inhabits and the physiological features that set that microbe apart from others. ◂ *Enrichment cultures (section 7.7)*

Although most microorganisms found in nature have not yet been grown in the laboratory, it would be premature to claim that microbes currently unculturable will never be cultivated. Traditional growth media are now giving way to novel media and incubation conditions. One approach is to use diffusion chambers, which are enrichment vessels incubated within porous membranes placed in a replica of the original environment. For example, marine sediment can be mixed with an artificial seawater medium and incubated in an aquarium filled with the same sediment but at a higher density. Nutrients, metabolites, and signaling molecules from the sediment diffuse through the porous membrane and promote the growth of significantly more isolates than traditional plating methods.

It has also become clear that growing novel microbes requires patience. Microbiologists are most familiar with doubling times in the order of hours or, at the most, a few days, when in fact this is not always the case, as noted in **Microbial Diversity & Ecology 26.1**. Furthermore, microbiologists have been trained to equate turbidity in liquid culture with growth. This is not always true, as some bacteria stop dividing (i.e., reach stationary phase) at cell densities too low for the medium to appear turbid.

A variety of new culture techniques have been applied that address specific needs of microorganisms growing in diverse environments. Efficiency can be greatly improved by using high-throughput methods. One popular approach is the **extinction culture technique,** so named because natural samples are first microscopically examined and then diluted to a density of 1 to 10 cells—that is, diluted to extinction. Multiple 1-milliliter cultures are established in 48-well microtiter dishes. After incubating for the desired time under appropriate conditions, the samples are stained and examined microscopically to screen for growth. In many cases, cultures that do not appear turbid will nonetheless show signs of viable microbial growth and can be transferred to fresh medium. It is important that the initial culture has fewer than 10 cells; additional transfers and dilutions can ensure that a culture arising from a single cell (i.e., clonal growth) eventually is obtained.

Recent progress in cataloging microbes that make up the human microbiome has impelled the development of the high-throughput cultivation technique called culturomics. **Culturomics** involves at least three steps: First, samples of a mixed assemblage of microbes are used as inoculum for hundreds of different culture media incubated under a variety of physical conditions (e.g., temperature, pH, and oxygen concentration). Microorganisms from colonies that grow are then identified by barcoding. From these initial cultures, larger cultures can be started—that is, cultures are scaled up and, in some cases, organisms are grown in pure culture. Even if pure culture is not possible, these microbes can be the source for genome sequencing by single-cell genomics. ◂ *Single-cell genomic sequencing (section 18.2); Proteomics explores total cellular protein (section 18.5)*

Flow cytometry has become common to isolate single cells from a mixed population. This can provide isolated cells to

MICROBIAL DIVERSITY & ECOLOGY

26.1 The Case of the Endosymbiotic Hypothesis

One of the most vexing questions in evolution concerns the origin of eukaryotes. One hypothesis posits that eukaryotes evolved from ancestors of so-called Asgard archaea (Asgardarchaeota). It is suggested that these archaeal ancestors and their endosymbiotic bacteria gave rise to the earliest eukaryote with mitochondria of bacterial origin (*see sections 1.2 and 21.1*). But piecing together evolutionary events is like developing a case in a court of law. There are two types of evidence. The first is circumstantial evidence such as a person who resembles the suspect was seen carrying stolen items. The microbial equivalent might be the genome of a living archaeon that encodes proteins with functions similar to eukaryotes (e.g., genes involved in replication and transcription). The second is direct evidence: the suspect was caught engaging in illegal activity. Here the microbial parallel would be a living archaeon that encodes signature eukaryotic genes—genes that were thought to be uniquely eukaryotic, such as those involved in membrane trafficking vesicle formation, ubiquitin, and cytoskeleton formation.

In 2010, tantalizing circumstantial evidence was found. Metagenomic analysis of samples taken from a deep-sea hydrothermal vent (**box figure**) called Loki's castle revealed most of the genome of a novel archaeon dubbed Lokiarchaeum. In addition to archaeal genes, it encoded eukaryotic signature proteins (ESPs). However, this evidence was considered inadmissible in the court of evolutionary science because the risk of contamination by eukaryotic genes was too high. After all, ESPs had never been seen in bacteria or archaea before.

As the evolutionary implications of the Lokiarcheaum metagenome were being debated, Hiroyuki Imachi's laboratory in Japan had placed some hydrothermal vent sediment in a bioreactor that mimicked the vent environment. They waited 5 years to take a sample from the bioreactor, which they incubated for yet another year in separate enrichment cultures. Finally, in 2019 they had a co-culture of Lokiarchaeum and a methanogenic archaeon. They named the former *Prometheoarchaeum syntrophicum*. Turns out *P. syntrophicum* has a doubling time of 2 to 3 weeks. Dr. Imachi's group sequenced the genome of *P. syntrophicum,* thereby providing direct evidence that it does encode ESPs.

But the case was not yet closed. The challenges to culturing *P. syntrophicum* hindered studies of the cells and their ESPs. Christa Schleper and colleagues also spent years developing enrichments of Asgardarchaeota, from estuary sediments on the coast of Slovenia. In 2023, they published a genome sequence from the organism that they called *Lokiarcheum ossiferum*. Like the earlier isolate, *L. ossiferum* encodes multiple ESPs. The faster-growing cells were also amenable to microscopy, and those cells exhibit an extensive cytoskeleton with multiple protrusions (*see figure 4.1d*). Lokiactin, a relative of eukaryotic actin, is an abundant ESP in *L. ossiferum* cells.

Why is the cytoskeleton so important in this Case of the Endosymbiotic Hypothesis? Recall that the Asgardarchaeota host cell is proposed to have engulfed a bacterial cell, as the precursor to mitochondria. Extensive cell-cell contacts are required for such an interaction, and one proposed mechanism for this event is termed E3: entanglement, engulfment, and endogenization (growth within a cell). Those first two Es require flexible cellular movements that would be enabled by the cytoskeleton. This stable *L. ossiferum* culture now enables further studies on Asgardarchaeota cells. ◀ *Asgardarchaeota are close relatives of eukaryotes (section 19.2)*

P. Rona/AR/National Undersea Research Program (NURP)/NOAA

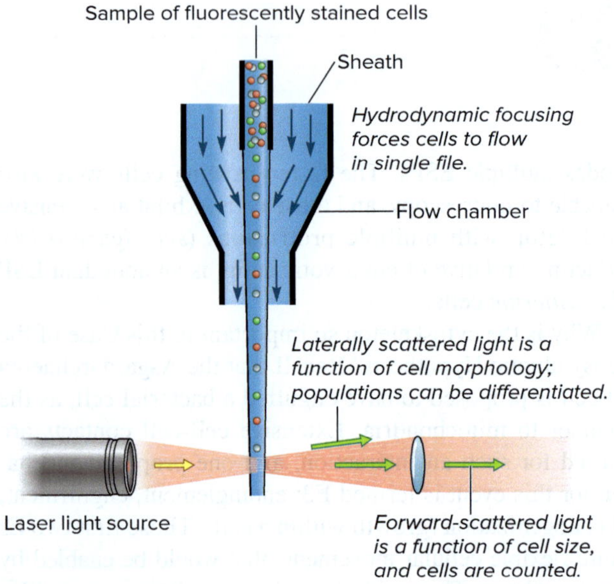

Figure 26.4 **Flow Cytometry.**

MICRO INQUIRY *What happens to the flow stream in fluorescence activated cell sorting (FACS)?*

start cultures. In **flow cytometry,** cells are tagged with a fluorescent dye and injected into a fluid stream. As shown in **figure 26.4,** as the diameter of the stream narrows, one cell at a time is forced through a thin tube. Each cell is detected by a laser beam. The laser excites the fluorescent label and light emitted by each cell is collected by a detector that converts the signal made by the cell into a voltage pulse proportional to the size of the cell. Some of the cell's emitted light scatters to the side; this is a function of morphological complexity of the cell. This is also detected and quantified. Thus based on cell size and morphology, flow cytometry can detect different populations of cells. This enables an additional function, **fluorescence activated cell sorting (FACS).** Here, as the stream leaves the detector, vibration causes the fluid to break into single droplets with one cell per drop. Cells become electrostatically charged as they pass the detector. The charge is used to guide each cell into a collection vessel. Thousands of cells per second pass through the laser beam, making "flow" an extremely efficient means to count, sort, and collect cells.

Comprehension Check

1. When might a microbe be considered VBNC?
2. How would you determine the ratio of dead to living microbial cells in a soil sample?
3. Explain how viability PCR is a critical tool in determining that food samples are free of pathogens.
4. Describe the rationale upon which the extinction culture technique is based.
5. How might you use FACS to obtain bacterial and archaeal cells for analysis from lake water?

26.4 Assessing Microbial Communities

After reading this section, you should be able to:

a. Distinguish marker gene analysis from metagenomic analysis
b. Explain the metrics that define phylogenetic diversity
c. Describe the rare biosphere and its implications for community diversity
d. Describe why, when, and how FISH and CARD-FISH are used
e. Differentiate deterministic and stochastic factors that control communities

A census relies on counting the individuals in a community as accurately as possible. The great plate count anomaly established that laboratory cultivation gives only a glimpse of the true numbers of microorganisms in the environment. In contrast, culture-independent techniques allow a greater proportion of microbial community members to be recognized and identified. Extracting and analyzing total nucleic acids from an ecosystem has provided microbial ecologists with the first real opportunity to explore, document, and characterize microbial communities, in part by evaluating their diversity.

Microbial diversity can be described in two ways. *Phylogenetic* diversity enumerates and classifies organisms in a taxonomic system. *Functional* diversity evaluates the metabolic processes in an ecosystem. We focus first on phylogenetic diversity.

Culture-Independent Surveys of Microbial Communities

Microbial ecologists investigating a natural community in the soil or water, or in association with other organisms, like humans, begin by extracting total DNA from the community. Depending on the type of information desired, the microbial ecologist may choose to get a broad sense of the microbes present or an in-depth picture. The former can be approached by marker gene analysis. This is less costly in terms of time and materials invested. The latter uses metagenomics to reveal a wealth of sequence data from the community.

Marker gene analysis focuses on one or more genes present in most organisms; often this is the 16S or 18S rRNA gene. Alternately, genes specific to the community in question may be surveyed. For example, the genes encoding methyl CoM reductase may be used to survey a community performing methanogenesis. PCR amplification is performed as diagrammed in figure 26.2, but because the template derives from a community and therefore contains multiple templates, the products are heterogeneous. DNA sequencing is performed on all products, yielding a set of sequences. Signature sequences derived in this manner are termed **amplified sequence variants (ASVs).** Because the ASVs are only a portion of the marker gene, limited taxonomic information can be gleaned from these sequences, but a collection of ASVs can provide a broad indication of the community's phylogenetic diversity. ◀ *Methanogens (section 19.4)*

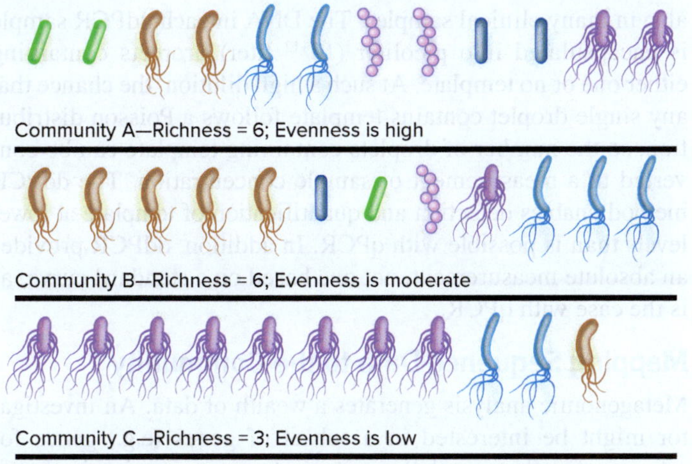

Figure 26.5 Richness and Evenness in Microbial Communities.

Libraries of ASVs from all types of ecosystems have revealed numbers of distinct sequences in the thousands and tens of thousands. Each unique ASV is assumed to represent an individual taxon present in the community. Such demonstrations validate culture-independent methods as more thorough and complete in describing the phylogenetic diversity of microbes. For instance, *Bacillus* and *Streptomyces* are the most commonly cultivated microorganisms when soil samples are plated, yet they are not abundant in soil communities surveyed by culture-independent approaches.

Marker gene analysis has some important limitations. First, the primers used for PCR are not universal and introduce some bias into which sequences are amplified. Second, genomes from closely related species may have identical sequences within the region that has been targeted for amplification, thus underestimating diversity. Third, many bacteria and archaea have multiple copies of their rDNA gene; bacterial genomes average five rDNA genes. If, for example, all five are distinct sequences, a single genome could generate five ASVs, overestimating the diversity. Fourth, because marker gene analysis depends on databases both for primer design and for identification, novel taxa may be underrepresented or absent.

A second culture-independent approach, **metagenomics,** moves beyond some of the limitations of marker gene analysis. As we describe in chapter 18, metagenomic analysis is the most comprehensive approach to survey genomes in a given environment (*see figure 18.8*). DNA from samples of a microbial community is extracted and sequenced in shotgun metagenomics. This generates millions of sequences, termed reads. Using computer software to detect overlaps, complete or almost complete bacterial or archaeal genome sequences are reconstructed. These are termed **metagenome-assembled genomes (MAGs).** Because MAGs require reference genomes for comparison, identification is restricted to microbes closely related to those with previously determined genome sequences. It should be stressed that novel microorganisms can be (and have been) discovered using metagenomics. It requires the detection of enough overlaps so that most (or ideally all) of the genome of the new microbe can be assembled. ◀ *Metagenomics provides access to uncultured microbes (section 18.3)*

Phylogenetic Diversity in Microbial Communities

Ecologists use two terms to describe community structure: richness and evenness. **Richness** quantifies the number of taxa that have been detected, while **evenness** evaluates the relative abundance of those taxa in the community (**Figure 26.5**).

Two types of graphs quantify phylogenetic diversity. **Figure 26.6*a*** is a rarefaction curve, where sample richness is plotted as a function of the total number of sequences analyzed. As increasing numbers of unique ASVs are added to the dataset, the plot levels off, indicating that most organisms have been

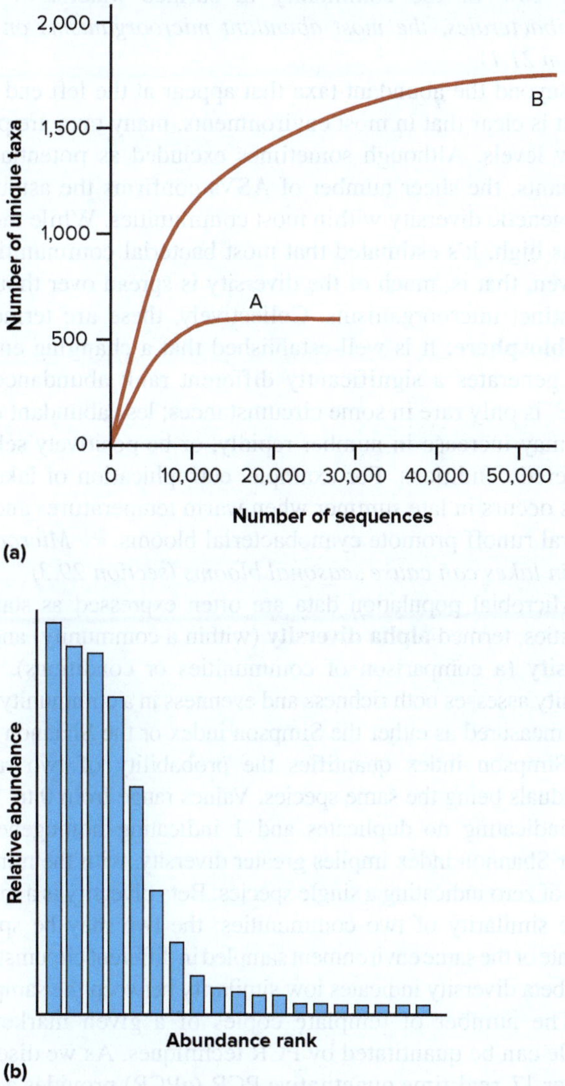

Figure 26.6 Graphic Illustrations of Microbial Diversity. (a) Rarefaction analysis of two communities (A and B) illustrates that the number of unique taxa increases with increasing sequence reads until fewer distinct taxa are identified. Community B is the more phylogenetically diverse. (b) Rank abundance plot tails off slowly, highlighting the rare sequences.

detected from that community. A plot that does not level off implies that the community is not fully described by the sequences obtained thus far.

The same data displayed as a rank abundance plot (figure 26.6b) assesses the community evenness. ASVs are grouped into sets of related sequences and ranked by the frequency with which each set appears in the data. For example, the most abundant ASV is rank 1, the next most abundant is rank 2, and so on. Some environments are so restrictive that the rank abundance plot shows a limited number of ASVs, implying low richness. For example, extreme halophiles are the only organisms that can tolerate high salt, and just a few taxa are present in a saltern (*see figure 19.12*). In other environments, certain organisms dominate by exploiting the available resources, as *Pelagibacter* does in the open ocean worldwide. It is often the most abundant taxon and can comprise up to 40% of the community in surface waters. ⏮ *Order Pelagibacterales, the most abundant microorganisms on earth (section 21.1)*

Beyond the abundant taxa that appear at the left end of the plot, it is clear that in most environments, many taxa are present at low levels. Although sometimes excluded as potential contaminants, the sheer number of ASVs confirms the astounding phylogenetic diversity within most communities. While the richness is high, it's estimated that most bacterial communities are not even, that is, much of the diversity is spread over thousands of distinct microorganisms. Collectively, these are termed the **rare biosphere.** It is well-established that a changing environment generates a significantly different rank abundance plot. "Rare" is only rare in some circumstances; less abundant organisms may increase in number rapidly, or be positively selected, in altered conditions. For example, eutrophication of lakes and ponds occurs in late summer when warm temperatures and agricultural runoff promote cyanobacterial blooms. ⏭ *Microorganisms in lakes can cause seasonal blooms (section 29.3)*

Microbial population data are often expressed as statistical quantities, termed **alpha diversity** (within a community) and **beta diversity** (a comparison of communities or conditions). Alpha diversity assesses both richness and evenness in a community and is often measured as either the Simpson index or the Shannon index. The Simpson index quantifies the probability of two random individuals being the same species. Values range from 0 to 1, with zero indicating no duplicates and 1 indicating homogeneity. A higher Shannon index implies greater diversity, with the minimum value of zero indicating a single species. Beta diversity is a measure of the similarity of two communities; the two may be spatially separate or the same environment sampled in different circumstances. High beta diversity indicates low similarity between the samples.

The number of template copies of a given marker in a sample can be quantitated by PCR techniques. As we discuss in chapter 17, real-time quantitative PCR (qPCR) provides a measure of template number and has been commonly used. More recently, **digital droplet PCR (ddPCR)** has become the method of choice when samples contain complex mixtures of molecules, including contaminating proteins that might otherwise inhibit PCR. This is the case not only in environmental samples, but also in many clinical samples. The DNA in each ddPCR sample is encapsulated into picoliter (10^{-12} liter) droplets containing either one or no template. At such a high dilution, the chance that any single droplet contains template follows a Poisson distribution, so the number of droplets containing template can be converted to a measurement of sample concentration. The ddPCR method enables detection and quantification of template at lower levels than is possible with qPCR. In addition, ddPCR provides an absolute measurement, not one based on a standard curve, as is the case with qPCR.

Mapping Sequence Data to the Community

Metagenomic analysis generates a wealth of data. An investigator might be interested in a subset of genes (e.g., genes for nitrogen fixation), and the power of metagenomics is that it reveals the context in which these genes of interest exist. Generally speaking, the genetic diversity (microdiversity) includes genes in high, medium, and low abundance. High-frequency genes typically represent the **core genome** and include housekeeping genes (encoding metabolic enzymes) and genes whose expression is specific to a growth substrate. Medium-frequency genes encode products that enable local biological interactions (e.g., siderophores, antibiotics, and quorum-sensing molecules). Low-frequency genes often encode surface structures that are species- or strain-specific (e.g., phage receptors, enzymes that assemble lipid A and O-antigen). These genes may reside on genomic islands or plasmids near transposases that have been acquired by horizontal gene transfer, thereby making them part of a strain's **pan-genome.** Phylogenetic diversity benefits occur at the population level, not the individual level for a number of reasons, such as increasing potential interactions and expanding the number of niches available for growth. ⏮ *The mobilome includes diverse mobile elements (section 16.5); Comparative genomics (section 18.7)*

Culture-independent methods allow one to count the total number of microbes, but microbiologists often want to enumerate a specific genus or type of microbe. In these cases, a technique called ***f*luorescent *i*n *s*itu *h*ybridization (FISH)** can be used. Here fluorescently labeled, single-stranded oligonucleotides, called **probes,** known to be specific to the microbe of interest are used to label natural samples (**figure 26.7**). SSU rRNA signature sequences are commonly used probes. To perform FISH, microbial cell membranes are made permeable so the probe can be internalized. Once inside the cell, it hybridizes to complementary nucleotide sequences. A probe that does not bind is washed off, and the fluorescence of a bound probe reveals the organisms of interest. Probe fluorescence is visualized by **epifluorescence microscopy** (**figure 26.8**). Alternatively, individual cells can be separated and counted (but not visualized) by flow cytometry in the technique termed flow-FISH (figure 26.4). ⏮ *Fluorescence microscopes use emitted light to create images (section 2.2)*

Depending on the probe used, FISH can identify a particular species or strain. This has made it a popular tool in clinical diagnostics and food microbiology, as well as microbial ecology.

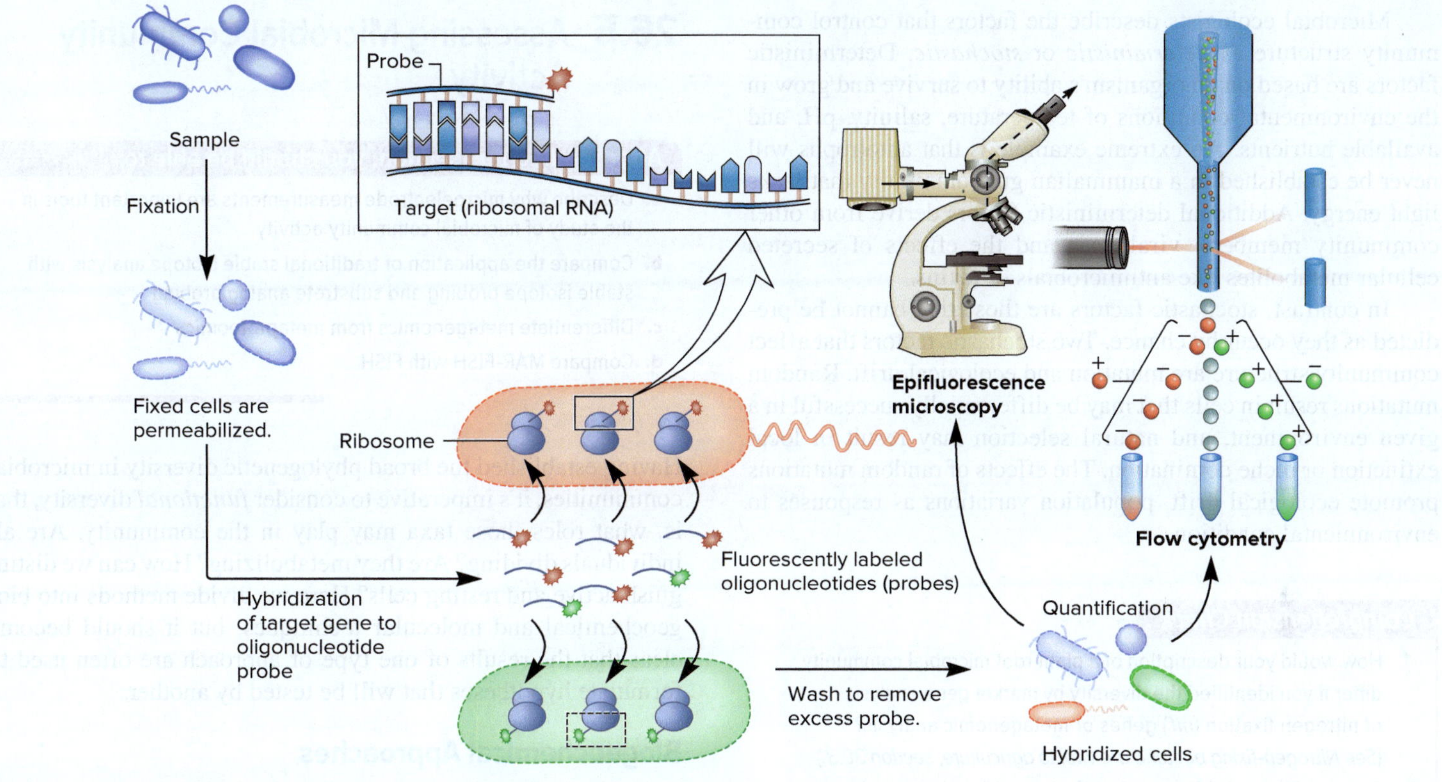

Figure 26.7 Fluorescence In Situ Hybridization (FISH).

For example, FISH was used to discover the association between the archaea *Nanoarchaeum equitans* and *Ignicoccus hospitalis* (*see figure 19.6*).

Figure 26.8 The Use of FISH to Study Microbial Populations. In this study, 16S rRNA probes tagged with different color fluorescent labels were used to study the microbial diversity within an earthworm gut.

Seana Davidson

MICRO INQUIRY *Under what circumstances would one use epifluorescence microscopy, rather than flow cytometry, and vice versa?*

Sometimes when FISH is used in natural samples, the fluorescent signal is not bright enough to be detected microscopically. This is because microbes living in these environments often grow quite slowly. Thus rather than having tens of thousands of ribosomes (and therefore many rRNA targets), each cell has relatively few ribosomes and the amount of probe that binds is below detectable limits. A clever modification of the FISH technique amplifies the signal produced by each cell. The fluorescent probe is labeled with an enzyme that makes lots of fluorescent product when exposed to substrate. This is called *c*atalyzed *r*eported *d*eposition-FISH, or simply CARD-FISH. The enzyme attached to the oligonucleotide probe is usually horseradish peroxidase (HRP). After the oligonucleotides are hybridized to the sample, the HRP substrate is added and the level of fluorescence is amplified. In this way, a single oligonucleotide probe bound to just one molecule of rRNA can generate an "enzyme boosted" fluorescent signal.

Multiple Factors Affect Community Structure

Few environments are static. The existence of the rare biosphere demonstrates that microbial communities preserve species richness as a buffer against change. While many species can coexist in an ecosystem, their activity levels vary according to the immediate conditions and their numbers vary according to several factors.

Microbial ecologists describe the factors that control community structure as *deterministic* or *stochastic*. Deterministic factors are based on an organism's ability to survive and grow in the environmental conditions of temperature, salinity, pH, and available nutrients. An extreme example is that autotrophs will never be established in a mammalian gut community that lacks light energy. Additional deterministic factors derive from other community members: viral lysis and the effects of secreted cellular metabolites like antimicrobials or toxins.

In contrast, stochastic factors are those that cannot be predicted as they occur by chance. Two stochastic factors that affect community structure are mutation and ecological drift. Random mutations result in cells that may be differentially successful in a given environment, and natural selection may result in local extinction or niche domination. The effects of random mutations promote ecological drift, population variations as responses to environmental conditions.

Comprehension Check

1. How would your description of a plant root microbial community differ if you identified the diversity by marker gene analysis of nitrogen fixation (*nif*) genes or metagenomic analysis? (*See Nitrogen-fixing bacteria are vital to agriculture, section 30.3.*)
2. Consider the establishment of the microbiota in a newborn baby. How does species richness change in the child's first year of life? (*See The microbiome develops from birth to adulthood, section 33.2.*)
3. Consider a community of microorganisms at an acid mine drainage site. What would you expect for a Simpson diversity value? How would that value change upon bioremediation of the site? (*See The microbial legacy of coal mining, Microbial Diversity & Ecology box 21.1.*)
4. What is FISH? Describe an example in which the application of CARD-FISH might yield more informative results than traditional FISH.
5. List some deterministic factors that affect the bacterial community in yogurt. (*See Fermented milks, section 40.5.*)

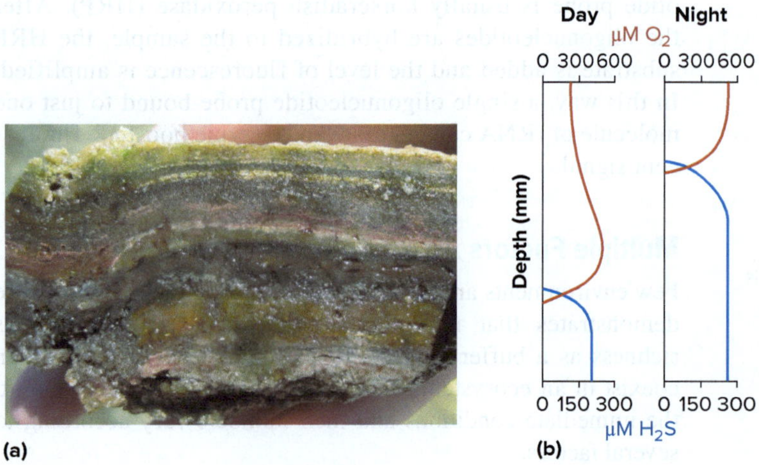

26.5 Assessing Microbial Community Activity

After reading this section, you should be able to:

a. Describe why microelectrode measurements are important tools in the study of microbial community activity
b. Compare the application of traditional stable isotope analysis with stable isotope probing and substrate analog probing
c. Differentiate metagenomics from metaproteomics
d. Compare MAR-FISH with FISH

Having established the broad phylogenetic diversity in microbial communities, it's imperative to consider *functional* diversity, that is, what roles those taxa may play in the community. Are all individuals dividing? Are they metabolizing? How can we distinguish active and resting cells? Here we divide methods into biogeochemical and molecular techniques, but it should become clear that the results of one type of approach are often used to formulate hypotheses that will be tested by another.

Biogeochemical Approaches

To determine community activity, it is necessary to have a good understanding of the microenvironment inhabited by the microbial community in question. In many ecosystems, these microhabitats are studied by **microelectrodes**—electrodes capable of measuring pH, oxygen tension, H_2, H_2S, nitrogen species, or other ions. The tips of these electrodes range from 2 to 5 µm wide and 100 to 200 µm long, enabling measurements to be taken as the tip penetrates the sample in increments of a millimeter or less. Because the concentration and distribution of electron donors (e.g., sulfide, ammonium) and acceptors (e.g., oxygen, nitrate, sulfate) are measured, microbial activities can be inferred. Microelectrodes have widely been used in the study of **microbial mats** (**figure 26.9**). These highly striated communities of photoautotrophs and chemotrophs develop in saline lakes, marine intertidal zones, and hot springs.

Figure 26.9 Microelectrode Analysis of Microbial Mats. (a) Microorganisms, through their metabolic activities, can create environmental gradients resulting in layered ecosystems. A vertical section of a hot spring (55°C) microbial mat, showing the layers of microorganisms. (b) The measurement of oxygen (red) and H_2S (blue) with microelectrodes demonstrates the diurnal shift in gas production: During the day, oxygen produced by cyanobacteria in the upper layers inhibits anaerobic respiration, so sulfate reduction is confined to the lower (dark) strata. At night, when oxygenic photosynthesis is diminished, atmospheric oxygen has a limited range of diffusion into the mat, and H_2S diffuses upward, where it can be used as an electron donor by purple photosynthetic bacteria.
(a) Reut S. Abramovich

Perhaps the most fundamental microbial activity is growth. So the first question that may be asked is, "Are the microbes in this environment growing, and if so, at what rate?" Microbial growth rates in complex systems can be measured directly. Changes in microbial numbers are followed over time, and the frequency of dividing cells is also used to estimate production. Alternatively, the incorporation of radiolabeled thymidine (usually labeled with ^3H) into microbial biomass provides information about growth rates and microbial turnover. Because the nucleoside thymidine is found in DNA and not in RNA, its incorporation measures growth. This approach is most easily applied to aquatic microorganisms; it is more difficult to directly observe terrestrial microbial communities. Radiolabeled carbon compounds in the form of $^{14}CO_2$ or glucose are used to track carbon cycling in autotrophs and heterotrophs, respectively.

In addition to radioactive isotope incorporation, **stable isotope analysis** may be used. Isotopes are those forms of elements that differ in atomic weight because they bear different numbers of neutrons. Isotopes can be stable or unstable; only radioactive isotopes are unstable and decay. Stable isotopes are rare when compared to their lighter counterparts. For example, both ^{14}N and ^{15}N occur in nature and are stable, but ^{14}N is found in much greater abundance. Organisms discriminate between stable isotopes, and for any given element, the lighter (in this case, ^{14}N) is preferentially incorporated into biomass, a phenomenon known as **isotope fractionation.** This discrimination can be used to follow the fate of a specific element by labeling the compound of interest with the heavier isotope. For instance, to explore the fate of NO_3^- in soil, samples of the soil would be incubated in the presence of $^{15}NO_3^-$. One would then measure the amount of ^{15}N incorporated into biomass, released as $^{15}N_2$, and reduced to $^{15}NH_4^+$. Unfortunately, the absolute value of ^{15}N is so small, it cannot be directly measured. Instead, the ratio of ^{15}N to ^{14}N in the sample is compared to the standard ratio found in inorganic material. An organism that assimilates $^{15}NO_3^-$ would contain more ^{15}N when compared to the standard. This difference, called delta (δ), is calculated as follows:

$$[(R_{sample} - R_{standard})/(R_{standard})] \times 1{,}000 = \delta_{sample\text{-}standard}$$

where R_{sample} is the ratio of heavy to light isotope in the sample and $R_{standard}$ is the ratio of heavy to light isotope in the standard. For instance, if our hypothetical soil community is found to have a $^{15}N/^{14}N$ ratio greater than the standard by 3 parts per thousand (ppt), this value is reported as $\delta\ ^{15}N = +3\ \delta$ ‰ (the symbol ‰ represents ppt). The true value of stable isotope fractionation rests in the fact that different microbes discriminate between heavy and light isotopes differently. For instance, nitrate production can be discerned from N_2 formation because the isotopic enrichment factor for denitrification is less than that for nitrification (about −26 ‰ vs. −34 ‰, respectively).

Another technique, called **stable isotope probing,** can be used to examine nutrient cycling in microbial communities as well as identify the microbes taking up the element of interest. In these experiments, labeled precursors are added to an ecosystem and, following incubation, the target macromolecule is purified and analyzed. As an example, this technique was employed to explore methanogenesis in a rice paddy soil ecosystem. Researchers built **microcosms**—small incubation chambers with conditions that mimic the natural setting. In this case, they simulated natural rice paddies and $^{13}CO_2$ was introduced. Gaseous $^{13}CH_4$ was collected and RNA was isolated from the soil. The ^{12}C-containing RNA was separated from the ^{13}C-RNA on the basis of differing densities. The ^{13}C-containing 16S rRNA, which could only be synthesized by those microbes that assimilated $^{13}CO_2$, was then used to identify these archaea; recall that methanogenic microbes are archaea. ◀ *Methanogens (section 19.4)*

Substrate analog probing uses a similar approach as stable isotope probing, but the chemical probes are synthetic organic compounds with a structural similarity to naturally occurring biomolecules. They are incorporated into cellular macromolecules, and many carry a tag for detection. For example, 5-bromo-2′-deoxyuridine (BrdU) is a thymidine analog and labels DNA as a proxy for growth. BrdU detection is through a conjugated fluorescent dye or through the use of an antibody specific for the analog. Although ^3H-thymidine and BrdU are both used to assess growth, BrdU is often preferred because of the challenges involved in the use and disposal of radioactive materials.

Molecular Approaches

Recall that shotgun metagenomics involves the nucleotide sequencing of community DNA fragments. So far we have focused on using these nucleotide sequences to identify community members, but metagenomic sequence can also be used to infer community function. Metagenomes from a variety of biomes (e.g., mines, hypersaline ponds, estuaries) reflect the biogeochemical conditions of a habitat. For instance, community metagenomes from a well-oxygenated coral reef are rich in genes encoding respiration proteins, while the metagenomes from the colons of terrestrial animals have a high number of genes encoding enzymes involved in fermentation.

mRNA present in the environment can also be monitored, a technique called **metatranscriptomics.** Here mRNA extracted from the natural environment is reverse transcribed to cDNA (*see figure 17.4*) and directly analyzed by next-generation sequencing. The presence of mRNA demonstrates that a gene is actively transcribed and the gene product is probably active. Thus one can assess community activity at the time the sample was collected. However, rather than surveying the entire genetic landscape, a researcher may be interested in a particular metabolic function. In this case, a specific gene or mRNA encoding enzymes of interest would be targeted for PCR amplification by specific primers. Example genes are listed in **table 26.2.** Such studies have shown that both dissimilatory nitrate reduction and the anammox reaction account for loss of nitrogen in specific habitats. ◀ *Transcriptome analysis (section 18.5)* ▶ *Nitrogen cycle (section 28.2)*

Unfortunately, most mRNAs are unstable, making recovery technically challenging. Instead of measuring mRNA, **metaproteomics** identifies each protein present at the time of sampling.

Table 26.2 Example Genes Used to Assess Functional Diversity

Process	Gene	Gene Product
Ammonium oxidation	*amoA, amoB, amoC*	Ammonia monooxygenase subunits
Anammox	*hdh*	Hydrazine dehydrogenase
Dissimilatory nitrate reduction	*narG*	Nitrate reductase
	nosZ	Nitrous oxide reductase
	nir	Nitrite reductase
Methane oxidation	*mmoA*	Methane monooxygenase
Methanogenesis	*mcrA*	Methyl coenzyme M reductase
Nitrogen fixation	*nifH, nifD, nifK*	Nitrogenase subunits
Sulfate reduction	*apsA*	Adenosine phosphosulfate reductase

There are two general approaches by which all the proteins in a community can be sampled and identified. The more labor-intensive technique uses 2D-polyacrylamide gel electrophoresis (2D-PAGE) so that each protein recovered from the environment is visualized (*see figure 18.13*). Individual proteins can be extracted from the gel and analyzed by mass spectroscopy (MS). Total peptide mass and amino acid sequence information is then used to search databases to identify the protein. This approach is best if the relative concentration of individual proteins is considered important. If a census of proteins is the goal, a more automated, high-throughput approach is taken. Here community proteins are extracted and digested with a protease to generate protein fragments called peptides. This mixture of peptides is separated by a technique known as 2D nano-liquid chromatography (nano-LC). Just as 2D gels separate proteins first by charge and then by molecular weight, 2D nano-LC separates peptides by two different properties: Peptides are first separated by charge and then by hydrophobicity. The peptide masses are determined by mass spectrometry, and, finally, tandem MS is used to obtain amino acid sequence information (*see figure 18.14*). In both cases, the protein sequences obtained are compared to metagenomic data obtained from the same environment. Metaproteomics is limited by the complexity of the environment being assessed. Complex environments like soil may have as many as a billion or more different proteins per gram. Even the most advanced and automated MS approaches are not yet able to resolve such a large and diverse group of proteins. ◀ *Proteomics explores total cellular protein (section 18.5)*

The technique termed **BONCAT** (*bio*orthogonal *n*on-*c*anonical *a*mino acid *t*agging) is a popular mechanism to identify translationally active cells. BONCAT uses a modified amino acid that is not one of the (canonical) 20 amino acids normally found in proteins (**figure 26.10**). In cells, these non-canonical amino acids (NCAAs) are attached to a methionine-tRNA by methionine-tRNA synthetase with a relaxed specificity. During translation, the ribosome incorporates the NCAA in place of methionine with minimal impact on protein synthesis. Detection in a BONCAT experiment involves fluorescence from the tag on the NCAA when its complementary reagent is added.

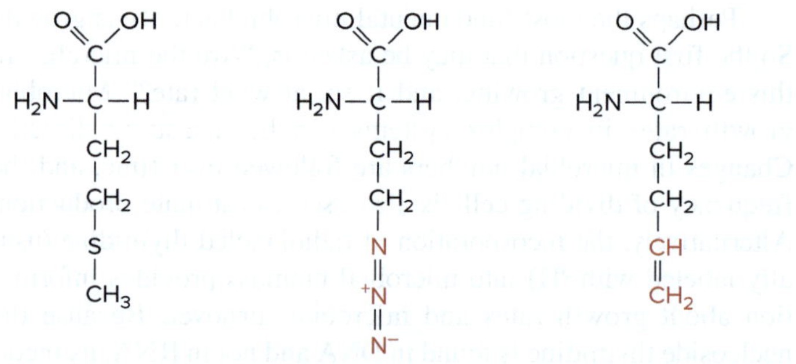

Figure 26.10 L-azidohomoalanine (AHA) and L-homopropargylglycine (HPG). Structures of methionine and two analogs that carry detection tags. All three compounds can be bound to methionine-tRNA synthetase for addition to a polypeptide.

To simultaneously assess metabolic activity and phylogenetic identity, a technique that combines an older method, **microautoradiography** (**MAR; figure 26.11**), with FISH can be employed. This method is thus called MAR-FISH. In MAR, a radioactive substrate (e.g., thymidine labeled with ^3H) is mixed with the sample and incubated under specific conditions, usually closely resembling those in the field. At the completion of the incubation period, the sample is washed to remove any unincorporated radioactive substrate. The sample is viewed in two ways: by epifluorescence microscopy and by radiolabeling. To accomplish the latter, the sample is attached to a microscope slide and treated with a photographic emulsion. As the radioactive material decays, the silver grains in the photographic emulsion are exposed. When developed, this leaves a series of black dots around the cells that assimilated the radioactive substrate. MAR-FISH can be used to analyze genetic composition and specific substrate uptake patterns of microbes at the level of a single cell in complex microbial communities.

Several of the techniques just described, along with those described in section 26.3, can be used in combination, and have become powerful tools to dissect microbial communities. Microscopy in its various forms (light, fluorescence, and confocal) along with microfluidic techniques (flow cytometry and cell sorting) allow a microbial ecologist to identify individual microbes and spatially manipulate them within or away from the community. For example, FISH-BONCAT involves labeling with both a nucleic acid probe and a substrate analog probe. The overall community, then, can be visually categorized both phylogenetically and functionally in a single experiment. In addition, fluorescent populations can be physically separated for further studies.

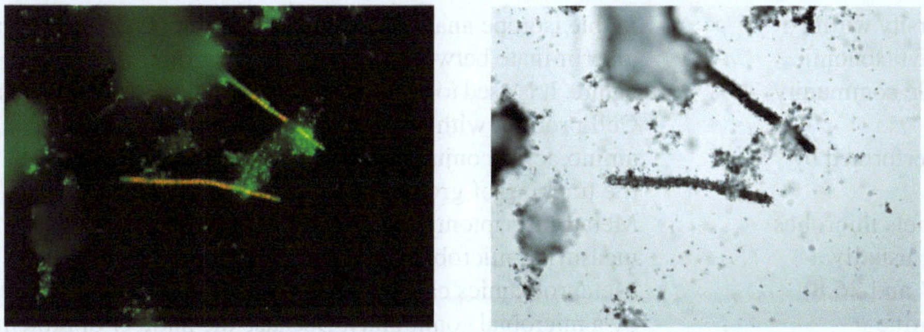

(a) (b)

Figure 26.11 MAR-FISH Technique Combines Microautoradiography with FISH in a Single Sample. Cells of interest are triple-labeled for identification and measurement of metabolic activity, even if present in a complex microbial community. (a) The filamentous microbe of interest identified using a FISH probe that generates an orange signal under epifluorescence microscopy. The green fluorescence is the result of a dye (YO-PRO-1) that binds to double-stranded nucleic acid such as DNA. (b) MAR resulted in black staining of the filamentous bacteria when viewed under bright-field microscopy, indicating that the microbes took up (assimilated) ^{3}H-labeled amino acids.

Cleber Ouverney

Comprehension Check

1. How do radioactive and stable isotopes differ? Design an experiment to determine whether denitrification is taking place in a soil ecosystem. (*See Nitrogen cycle, section 28.2.*)
2. How might microelectrodes be used to gather data in preparation for a stable isotope study?
3. Develop a question that could be answered by metatranscriptome analysis.
4. What is metaproteomics? How is it similar to metagenomics? How does it differ?
5. What is the difference between MAR and MAR-FISH? Suggest a microbial ecology question that could be answered by MAR and one that would require MAR-FISH.

Key Concepts

26.1 The Vocabulary of Ecology Applies to Microbial Environments

- Individual organisms belong to a species, and multiple organisms constitute a population. A community has many populations that occupy the same space, and communities together with nonliving components define an ecosystem.
- The standard biological species definition does not apply to bacteria and archaea because of the prevalence of horizontal gene transfer. For these organisms, a species is defined at the genome level, that is, organisms with greater than 97% sequence identity.

26.2 Microbial Identification Is Largely Based on Molecular Characterization

- Historically, microbial taxonomic and phylogenetic analysis used morphological, physiological, and ecological characteristics. These remain important in building a complete picture that also includes molecular information.
- Nucleic acid sequencing is the most powerful and direct method for comparing genomes. Whole-genome sequencing is rapidly replacing other methods for genome comparison.
- The sequences of SSU rRNA are used in phylogenetic studies of microbes and can identify an organism to the genus level (**figure 26.2**).

26.3 Microbial Ecology Also Relies on Cultures

- Less than 5% of microbial taxa have been cultured. The discrepancy between the number of microbial taxa observed and that grown is termed the great plate count anomaly and reflects the large proportion of microbes that are viable but nonculturable.
- Although most microbes cannot be grown in pure culture, novel approaches such as extinction culture and culturomics have resulted in the isolation of previously uncultured microorganisms (**table 26.1**).
- Flow cytometry is a fast and efficient method of counting and sorting cells (**figure 26.4**).

26.4 Assessing Microbial Communities

- Phylogenetic diversity describes the taxonomic classification of organisms in a community and is often assessed by culture-independent surveys that use marker gene analysis and metagenomics.
- Shotgun metagenomics provides the most complete information about the microbial population because all genes present in DNA extracted from natural samples are sequenced and analyzed.
- Ecologists describe microbial communities in terms of richness (the number of taxa detected) and evenness (the relative abudance of those taxa in the community) (**figures 26.5** and **26.6**).
- The term *rare biosphere* describes organisms that are in low abundance. While an organism may be rare in a community under certain conditions, environmental perturbations likely change the abundance of many organisms.

- Alpha diversity describes the taxonomic diversity within a community, while beta diversity compares the taxonomic diversity between two communities, or a single community in different environmental conditions.
- Quantitation of certain genes in a sample is performed by digital droplet PCR (ddPCR).
- Fluorescent in situ hybridization (FISH) labels microbes that possess a specific nucleotide sequence, usually a region of the SSU rRNA gene (**figures 26.7** and **26.8**).
- Deterministic and stochastic factors determine community structure. Deterministic factors are the environmental conditions present in the ecosystem, while stochastic factors include mutations and ecological drift.

26.5 Assessing Microbial Community Activity

- Microelectrodes can be used to determine parameters such as pH, O_2, and H_2S concentrations in microniches (**figure 26.9**).
- Stable isotope analysis is based on the fact that organisms discriminate between heavy and light isotopes found in nature. It is used to monitor nutrient flux through ecosystems.
- Cell probing with synthetic analogs of nucleotides or amino acids conjugated to fluorescent reporters allows the tracking of growth and translation.
- Metatranscriptomics assesses community activity by measuring microbial gene expression.
- Metaproteomics can be used to identify proteins expressed by a microbial community. Because the number of proteins is so large, traditional 2D gel analysis is usually not practical and high-throughput chromatographic techniques are used instead.
- Techniques can be used in tandem to label the genome and other aspects of the cell, and subsequent applications of cell sorting can separate communities for further analysis. Microautoradiography (MAR) combined with FISH simultaneously assesses metabolic activity and genetic composition (**figure 26.11**).

Active Learning

1. How might you attempt to culture a chemolithoautotroph that uses ferrous iron as an electron donor and oxygen as an electron acceptor?
2. Both acetate and CO_2 can be used by methanogenic archaea to generate methane (CH_4). How would you determine the source of carbon for methanogenesis in a waterlogged peat?
3. There are two types of microbial mats in Shark Bay, Australia: smooth and pustular (bumpy). Smooth mats possess a diverse collection of methanogenic and methylotrophic archaea, while halobacteria dominate the archaeal community in pustular mats. How would you go about identifying and enumerating archaeal genera found in these mats? How might you determine their metabolic strategies (e.g., methanogenesis vs. methylotrophy) and determine their spatial distribution relative to the physical features of the mats, such as oxygen penetration? Discuss techniques that might be used to answer these questions.
4. Ecologists have defined *keystone species* in macroecology as species that have an outsized impact on the ecosystem relative to their own abundance. Does this term apply to microbial communities?
5. After a metagenomic survey of lake sediment, you identify several MAGs from a phylum that has been described, but for which no cultured representatives exist. You set up enrichment cultures based on metabolism inferred from genes present in the MAG dataset. How can you follow the progress of the enrichment over time to determine if your target microbes are becoming more abundant?
6. Ecologists describe animal species as generalists or specialists based on their dietary needs. For example, scavengers like vultures or rats are generalists, but pandas that eat only bamboo are specialists. Review the microorganisms described in chapters 19 through 24 and identify some from each category. How would you expect the proteomes of generalists and specialists to differ?
7. How might culture-independent surveys be biased for or against some types of microorganisms?
8. Suggest some ways that you could explore the virome, the viruses present in an ecosystem.

CHAPTER 27

Microbial Interactions

Kevin Wells Photography/Shutterstock

Microbes in Community

How is a Central American ant like a contemporary farmer? They both cultivate crops important for feeding their community and rely on chemicals to tackle pests that attack their crops. How are they different? Ants grow a crop of fungi, but with the exception of mushrooms, humans grow crops of plants.

Leaf-cutter ants do exactly that—cut leaves. They transport the leaves to their nests where they are fed to the fungal crop. These fungi have been domesticated over millennia and are cultivated as food for ant larvae. Unfortunately, the fungal crop is susceptible to infection by a parasitic fungus: *Escovopsis*. As it turns out, *Escovopsis* can be held in check by actinobacteria in the genus *Pseudonocardia*. These bacteria live on the ants and produce inhibitory compounds that protect the fungal crop from *Escovopsis*. The community of interacting members includes cooperative relationships (actinobacteria protecting the fungal crop) and antagonistic relationships (*Escovopsis* infecting the fungal crop).

Surprisingly, there's another player, a quality control manager checking to ensure the antifungal compounds produced by actinobacteria are up to standards. The black yeast *Phialophora* grows in association with *Pseudonocardia* on which it preys. It seems odd that *Phialophora* is tolerated when it antagonizes the bacteria protecting the fungal garden. The explanation has to do with cheating. It is energetically expensive for *Pseudonocardia* cells to produce an antimicrobial compound, but its production ensures a stable existence for the bacteria. However, some *Pseudonocardia* could cheat by letting their neighbors produce the antifungal. *Phialophora* appears to contribute by ensuring full participation in pest control by preventing the rise of cheating actinobacteria.

This six-part symbiosis (leaves, ants, cultivated fungus, *Escovopsis* parasite, *Pseudonocardia,* and *Phialophora*) illustrates the complexity of microbial communities in multicellular organisms. These associations have been optimized over years of coevolution. Symbionts affect many aspects of their hosts: health, behavior, evolution, and reproductive success. In this chapter, we define types of microbial interactions and present illustrative examples.

Readiness Check:

Based on what you have learned previously, you should be able to:

- ✓ Explain the general morphology of protists and the structure and function of hydrogenosomes (sections 5.6 and 23.1)
- ✓ List environmental factors that influence microbial growth (section 7.5)
- ✓ Diagram standard reduction potentials of electron donors compared to electron acceptors; relate this to standard free energy change (section 10.3)
- ✓ Describe the diversity of catabolic strategies used by bacteria and archaea (chapter 11)
- ✓ Explain carbon and nitrogen fixation (sections 12.3 and 12.5)
- ✓ Relate genome reduction to obligate intracellular lifestyles (section 18.7)
- ✓ Describe the application of metagenomics (sections 18.3 and 26.4)
- ✓ Identify substrates used in methanogenesis (section 19.4)

27.1 Many Types of Microbial Interactions Exist

After reading this section, you should be able to:

a. Compare and contrast different types of microbial interactions, including mutualism, cooperation, and antagonism
b. Differentiate obligatory and facultative partnerships

You need look no further than in a mirror to grasp the extent of microbial interactions. Your body possesses at least as many bacterial cells as it does somatic cells (and that doesn't include the archaea, fungi, viruses, bacteriophages, and protists you host). Looking deeper, the mitochondria in every cell of your body originated from an ancient bacterial symbiosis based on energy transfer. The vast majority of microbes live in complex communities that include close relatives and organisms from other domains of life. The stable association of one organism with another is called **symbiosis,** regardless of whether the association is beneficial or neutral, or has a negative impact **(figure 27.1)**. Symbiotic interactions fall on a continuum, and the benefit or harm to an individual may shift with environmental conditions and community structure.

Consider a bacterial cell and a temperate bacteriophage (*see figure 6.15*). Environmental factors determine whether the bacteriophage will enter a lytic or lysogenic cycle. The outcome of a lytic infection always harms the host cell, but infection by a temperate bacteriophage may have a neutral or beneficial outcome. The host bacterium may not be significantly impacted by replicating a few extra kilobases of prophage DNA (neutral outcome). In the case of bacteriophages that carry extra metabolic genes, the host may now have expanded capabilities to exploit its niche (beneficial outcome). ◀ *There are several types of viral infections (section 6.4)*

An important aspect of bacterial interaction is whether it is facultative or obligatory. **Facultative** interaction means that the microbe has alternative lifestyles. This is illustrated by *Rhizobium* spp. that inhabit soil as free-living bacteria, but also thrive within the nodules of leguminous plants (*see figures 21.7 and 30.9*). Viruses exemplify **obligatory** interactions—that is, a host cell is an absolute requirement for viral replication. Likewise, many bacteria that live within eukaryotic cells are obligate intracellular symbionts: The relationship is cemented in place such that the microbe cannot live outside the host cell. ◀ *Order Ricketsiales includes obligate intracellular bacteria (section 21.1)* ▶ *Rhizobia are symbionts of leguminous plants (section 30.3)*

For many years bacterial interactions were viewed as the relationship between one host and one **symbiont,** distinguished by their relative sizes. However, the wide application of metagenomics, single-cell sequencing, next-generation sequencing, and new approaches to culturing microbes have revealed that microbial interactions range from a single species of microorganism and its host to systems with thousands of different kinds of microbes. In fact, there have been so many discoveries of new and never imagined relationships that the vocabulary of microbial interactions is changing. What was once considered a microbe-host relationship is now sometimes called a *metaorganism, holobiont,* or *superorganism* to reflect the critical nature of the interaction. ◀ *Metagenomics provides access to uncultured microbes (section 18.3); Microbial ecology also relies on cultures (section 26.1)* ▶ *Humans are holobionts (section 33.1)*

We use three broad categories to describe microbial interactions. **Mutualism** defines the relationship in which some reciprocal benefit accrues to both partners. This is an obligatory relationship in which the organisms are dependent on each other. In many cases, the individuals cannot survive when separated. **Cooperation** is differentiated from mutualism by its nonobligatory aspect. Unfortunately, it is often difficult to distinguish the two because what is obligatory in one habitat may

Figure 27.1 Microbial Interactions. Microbial interactions form a continuum from cooperation to conflict. (a) Methanotrophs and sulfate-reducing bacteria are linked by metabolism. (b) *Aliivibrio fischeri* colonizes the bobtail squid light organ. (c) Tree bark supporting a lichen, a symbiosis between fungus and either cyanobacteria or algae. (d) *Penicillium* secretes an antibiotic that inhibits bacterial growth. (e) T4 bacteriophage infect *E. coli.*

MICRO INQUIRY *Review the microbial interactions illustrated here and identify benefits or harms to each partner.*

not be in another (e.g., the laboratory). **Antagonism** describes relationships in which one organism negatively impacts another. Defensive maneuvering is an important component of antagonistic relationships. Keep in mind that while these terms can define the relationship between two organisms, within a community there are multiple relationships, as illustrated in the chapter opening story.

27.2 Mutualism: Obligatory Positive Interaction

After reading this section, you should be able to:

a. List the benefits to each member of a mutualistic interaction
b. Discuss the metabolic relationships between symbiotic partners
c. Relate coral bleaching to a symbiotic relationship
d. Describe the trophic links between microbes and complex organisms as illustrated in corals, tube worms, and the mammalian rumen

Many mutualisms typically have their basis in metabolism, in which organisms provide growth substrates or electron donors to others. These relationships are mutually beneficial, and in several cases, the microbial community provides essential nutrients to an animal. Two agriculturally important mutualisms—mycorrhizae/plants and nitrogen fixers/legumes—are discussed in chapter 30. ▶| *The plant microbiome (section 30.3)*

Syntrophy

Syntrophy (Greek *syn,* together; *trophe,* nourishment) occurs when one species benefits from the metabolic products of another. In most marine, freshwater, and soil habitats, syntrophy accounts for the majority of microbial interactions between free-living organisms. For example, marine planktonic communities are structured for the efficient transfer of nutrients from primary producers to heterotrophs (*see figure 29.5*).

Methanogenic archaea are often found in syntrophic associations based on **interspecies hydrogen transfer.** This involves the removal of H_2 from the environment, thereby enabling fermentations that would otherwise be thermodynamically unfavorable.

Anoxic methanogenic ecosystems such as sludge digesters, anaerobic freshwater sediments, and flooded soils host communities that generate H_2, CO_2, and acetate as fermentation end products (*see figure 42.6*). Anaerobes such as *Syntrophobacter* spp. produce H_2 when they use protons ($2H^+ + 2e^- \rightarrow H_2$) as the terminal electron acceptor. Under standard conditions, this reaction is endergonic (note the positive ΔG) and yields H_2 as well as HCO_3^- (which is oxidized to CO_2):

$$\text{Propionate}^- + 3H_2O \rightarrow \text{acetate}^- + HCO_3^- + H^+ + 3H_2 \quad (\Delta G^\circ = +76.1 \text{ kJ/mol})$$

However, these products provide methanogenic archaea with substrates for the following:

$$4H_2 + CO_2 \rightarrow CH_4 + 2H_2O \ (\Delta G^\circ = -25.6 \text{ kJ/mol})$$

By synthesizing methane, methanogens consume H_2, keeping it at a low concentration in the immediate environment of both microbes. As a consequence, the reaction becomes exergonic, with a ΔG between –20 and –25 kJ/mol. Continuous removal of H_2 promotes further fatty acid fermentation and H_2 production. Because increased H_2 production and consumption stimulate the growth of both bacteria, all participants in the relationship benefit. |◀ *Methanogens (section 19.4)* ▶| *Wastewater treatment processes (section 42.2)*

Microorganism-Insect Mutualisms

Mutualistic associations are common between microbes and insects partly because insects often consume plant sap or animal fluids lacking essential vitamins and amino acids. Aphids and their gamma-proteobacterium *Buchnera aphidicola* partner are a model system for the study of mutualistic symbioses (**figure 27.2*a***). *B. aphidicola* is an **endosymbiont**—that is, it lives within its host's cells. A mature aphid contains millions of these bacteria within specialized cells called **bacteriocytes.** In exchange for a secure habitat and ample nutrients, strains of *B. aphidicola* provide their hosts with amino acids absent in sap. If the insect is treated with antibiotics, it dies. Likewise, *B. aphidicola* is an obligate symbiont. The inability of either partner to grow without the other indicates that the two organisms underwent **coevolution;** that is, they have evolved together. In fact, the pea aphid *Acyrthosiphon pisum* and its *B. aphidicola* symbionts share the pathway for the biosynthesis of several amino acids such that certain steps occur only in one of the organisms.

Termites, whose diet is mostly wood, are another good model system for the study of endosymbiosis. The main structural polysaccharides of wood are cellulose and hemicellulose (glucose polymers), which combine with lignin (*see figure 30.2*) to form lignocellulose. A diet of wood poses two problems for the termite: how to degrade polysaccharides that may possess as many as 15,000 glucose monomers, and where to get organic nitrogen needed for nucleotide and protein synthesis. The solutions to both problems are found in the protist, bacterial, and archaeal endosymbionts that live within the insects. Although termites produce a cellulolytic enzyme, their mutualistic protists complete lignocellulose degradation, solving the first problem. Nitrogen-fixing bacteria that live in the termite gut solve the second problem. |◀ *Nitrogen fixation (section 12.5)*

Most termite-associated protists are members of the Metamonada clade and possess hydrogenosomes rather than mitochondria (*see figure 5.16*). In the anaerobic termite gut, these protists ferment cellulose to acetate, CO_2, and H_2. Acetate is the termite's preferred carbon source, and termites rely on bacterial symbionts to convert the CO_2 and H_2 released by the protists to acetate by the Wood-Ljungdahl pathway (*see figure 19.1*).

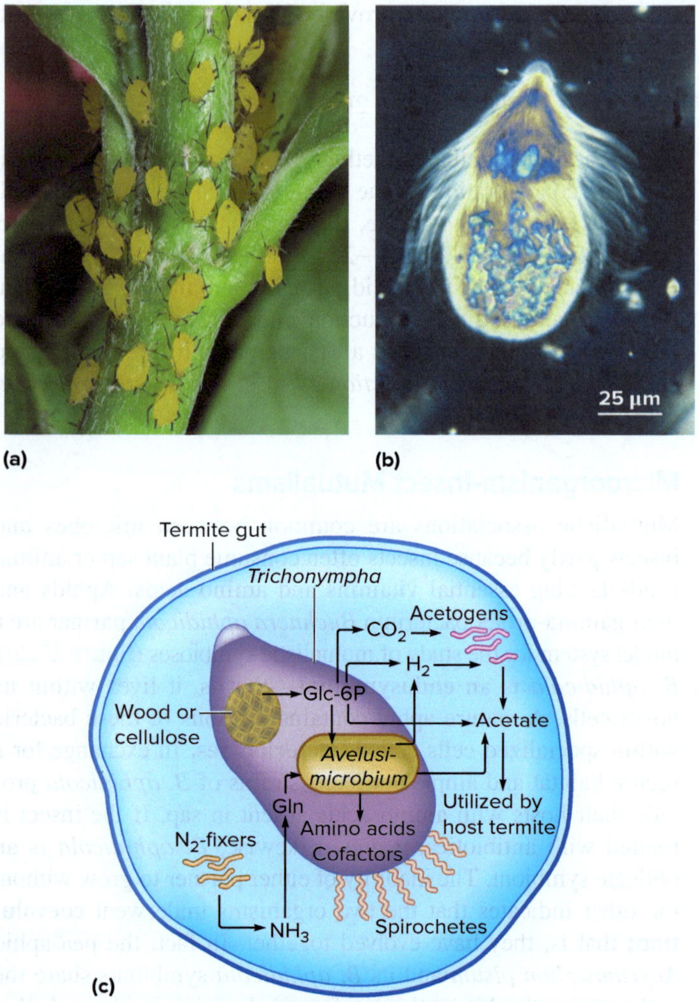

Figure 27.2 Microorganism-Insect Mutualisms. (a) *Aphis nerii.* A mature aphid hosts millions of *Buchnera* symbionts. (b) Light micrograph of *Trichonympha* sp., a multiflagellated protozoan from the termite gut. (c) Interaction between *Trichonympha* and its bacterial endosymbiont *Avelusimicrobium.* Within the termite gut, bacteria fix N_2 to ammonia, which is taken up and converted to glutamine (Gln) by *Trichonympha.* The protist relies on *Avelusimicrobium* to convert Gln to other amino acids, which are released with other cofactors that the protist cannot synthesize. Both organisms ferment glucose to acetate, H_2, and CO_2. Glc-6P, glucose 6-phosphate.

(a) Steven P. Lynch; (b) Alamy Stock Photo

MICRO INQUIRY *What methanogenic substrates are available in the termite gut?*

In addition, termites harbor methanogens that thrive based on interspecies hydrogen transfer. *Mitochondria, related organelles, and chloroplasts are involved in energy conservation (section 5.6); Discoba–Metamonada clade (section 23.2)*

Some termite protists also harbor endosymbionts (i.e., endosymbionts of endosymbionts). For instance, the protist *Trichonympha* sp. relies on a bacterial endosymbiont in the genus *Avelusimicrobium* to convert glutamine to other amino acids and nitrogenous compounds (figure 27.2*b,c*). In exchange, the protist supplies *Avelusimicrobium* with glucose 6-phosphate, which can directly enter glycolysis. In addition, motility is often conferred by spirochetes that cover the protist surface. Motility is essential to prevent protist expulsion from the termite gut and to acquire food.

Mutualisms with Primary Producers

The **primary producers** in an ecosystem are the organisms that capture energy from light or inorganic chemicals. All primary producers are autotrophs—that is, their sole source of carbon is CO_2, which they fix to build biomass. Autotrophs fuel these reactions either by phototrophy or lithotrophy. We are all familiar with **phototrophy** in the plants that surround us, so it is easier to understand microbial photosynthetic processes. **Lithotrophy** is uniquely microbial and involves the use of reduced, inorganic chemicals as sources of electrons. We discuss two examples of symbioses in which primary producers support an animal that can be considered a holobiont. *Metabolic diversity and nutritional types (section 11.1)*

The Coral Holobiont

Many marine invertebrates (sponges, jellyfish, sea anemones, corals) harbor endosymbiotic photosynthetic dinoflagellates (formerly called *zooxanthellae*) within their tissues. In the oligotrophic marine environment, photosynthesizers are the basis for all food chains. Because the degree of host dependence on the dinoflagellate is variable, one well-known example is presented here. *Alveolata (section 23.4)*

The hermatypic (reef-building) corals satisfy most of their energy requirements using their photosynthetic symbionts, which include members of the dinoflagellate family Symbiodiniaceae, exemplified here as the genus *Symbiodinium* (**figure 27.3*a***). These protists line the coral gastrodermis at densities greater than 10^6 cells per square centimeter of coral animal. *Symbiodinium* cells transfer up to 95% of their fixed carbon in exchange for nitrogenous compounds, phosphates, and protection from ultraviolet light from their coral hosts. In addition to dinoflagellates, corals host distinct microbial communities in microhabitats (figure 27.3*b*). The nutrient-rich mucus layer supports the growth of nitrogen-fixing and chitin-degrading bacteria. The coral skeleton hosts diverse bacteria, ammonium-oxidizing archaea, and fungal species that reduce nitrate and nitrite to ammonium. These microbial communities are responsible for extremely efficient nutrient cycling and tight coupling of trophic levels, which accounts for the stunning success of reef-building corals in developing vibrant ecosystems.

During the past several decades, the number of coral bleaching events has increased dramatically. **Coral bleaching** is defined as a loss of either the photosynthetic pigments from the dinoflagellates or the complete expulsion of the dinoflagellates from the coral (figure 27.3*d*). Damage to photosystem II generates reactive oxygen species (ROS); these ROS appear to directly damage many corals (recall that photosystem II uses water as the electron source, resulting in the evolution of oxygen). Coral bleaching results from a variety of stressors, but one important

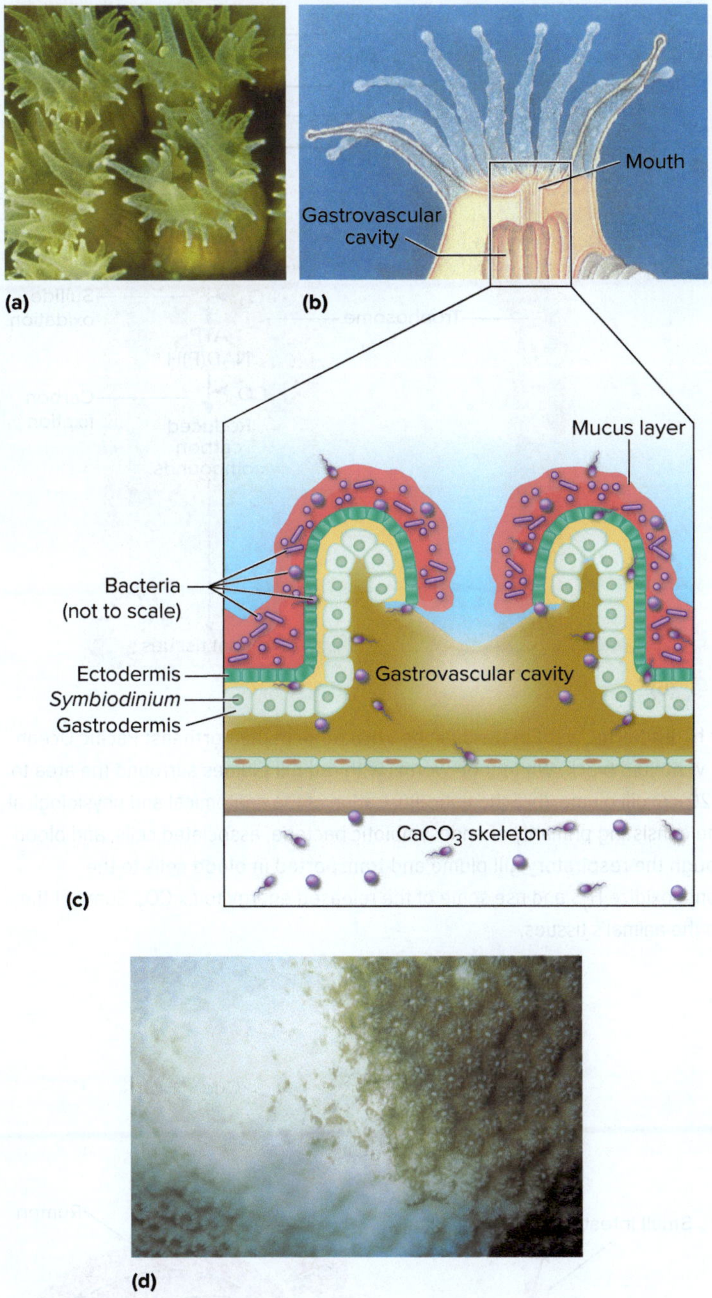

Figure 27.3 Coral Holobiont. (a) The coral holobiont is composed of polyps. (b) Cross section of a single polyp. (c) The gastrodermis layer of the polyp hosts photosynthetic dinoflagellates, while the external mucus layer supports diverse bacterial and fungal species. (d) Coral bleaching: Golden-brown color indicates healthy coral tissue; white (bleached) area has lost dinoflagellates.

(a) NOAA/NOS/NMS/FKNMS; (d) Diane Nelson

factor is temperature. Temperature increases as small as 2°C above the average summer maxima can trigger coral bleaching. Increased temperature has also been correlated with coral infectious diseases. Some mucus residents are opportunistic pathogens. During and following bleaching, the microbial community is in disequilibrium, and stressed corals are more susceptible to pathogenic attack. Finally, ocean acidification caused by increased CO_2 dissolution in seawater also threatens coral health (*see figures 29.1 and 29.2*). Sadly, evidence suggests that the coral holobiont populations will be unable to evolve quickly enough to keep pace with ocean acidification and warming if global climate change continues unchecked. ◀ *Light reactions in oxygenic photosynthesis (section 11.10)* ▶ *Global climate change: biogeochemical cycling out of balance (section 28.3); Water is the largest microbial habitat (section 29.1)*

The Tube Worm Holobiont

Several thousand meters below the surface of the ocean, hydrothermal vents occur where the Earth's crustal plates are spreading apart (**figure 27.4*a***). Vent fluids are anoxic, contain high concentrations of hydrogen sulfide (H_2S), and can reach a temperature of 350°C. The seawater surrounding these vents has sulfide concentrations around 250 μM and temperatures 10° to 20°C above the ambient seawater temperature of about 2°C. Although the high pressure at such depths prevents the water from boiling, one might think these conditions are inhospitable to life. But nothing could be further from the truth.

All animals in the hydrothermal vent community harbor bacterial endosymbionts, chemolithoautotrophs that are the primary producers in this ecosystem. These bacteria are the sole organisms capable of conserving energy from reduced sulfur. Without sunlight, the environment lacks fixed carbon to promote the growth of heterotrophs, leaving animals entirely dependent upon their endosymbionts.

Giant (>1 m in length) tube worms in the genus *Riftia* (figure 27.4) growing near hydrothermal vents exemplify a remarkably successful mutualism in which bacterial endosymbionts are maintained within bacteriocytes of the tube worm. *Riftia* lives at the interface between hot, anoxic, sulfide-containing vent fluids and cold, oxygenated seawater. The worm lacks a digestive tract and relies exclusively on endosymbiotic bacteria to provide organic carbon. The bacteria, *Endoriftia persephone,* are chemolithotrophic, using sulfide as an electron donor and oxygen as a terminal electron acceptor. *Riftia* worms have unique hemoglobin that transports both reduced sulfur and oxygen to the endosymbionts. H_2S and O_2 are removed from seawater by the worm's hemoglobin and delivered to bacteriocytes that contain *E. persephone.* These bacteria reach densities of up to 10^{11} cells per gram of worm tissue. The endosymbionts comprise two subpopulations representing different stages of the symbiotic relationship. Smaller cells divide frequently to maintain their population, while larger cells are dedicated to producing biomass to feed the tube worm tissues. This mutualism enables *Riftia* worms to grow to an astounding size in densely packed communities.

What advantage does the mutualism provide to each partner? The advantage to *Riftia* is clear: It could not survive without fixed carbon provided by its endosymbionts. The advantage to *E. persephone* is a stable niche and the convenient delivery of its carbon source, electron donor, and acceptor.

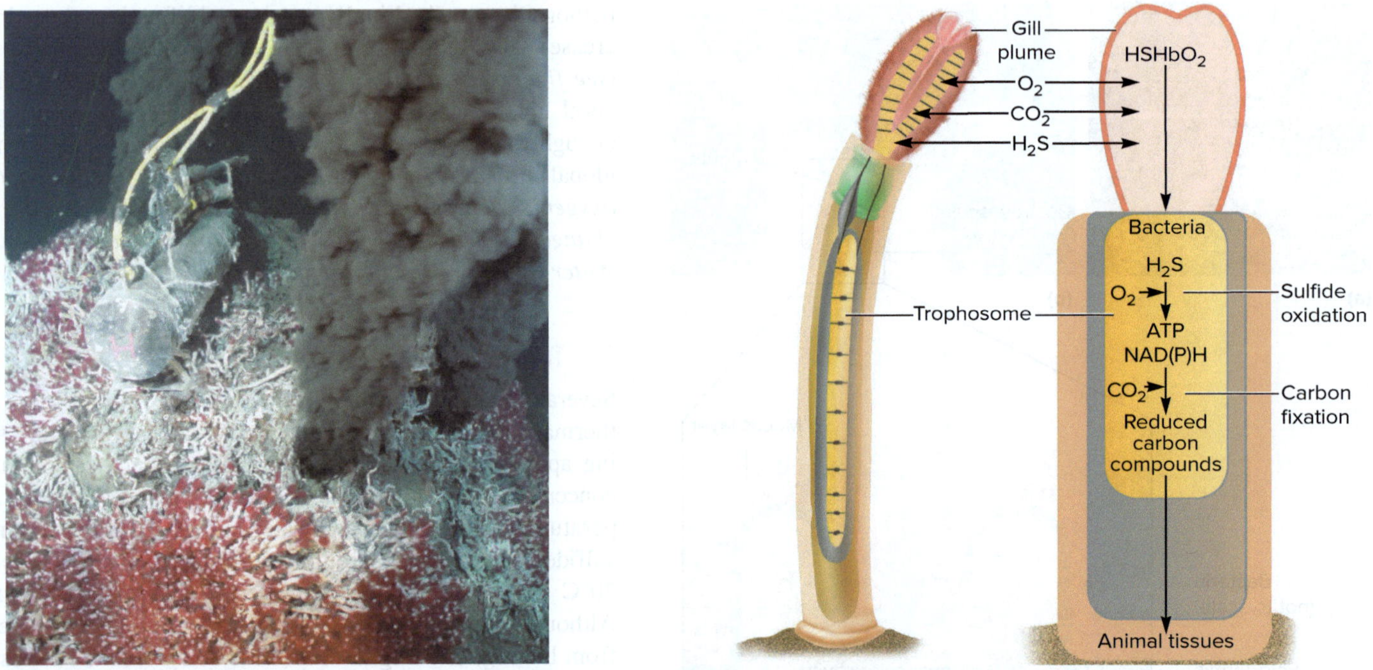

(a) (b)

Figure 27.4 The Tube Worm–Bacterial Relationship. (a) A black smoker hydrothermal vent at the Juan de Fuca ridge in the northeast Pacific Ocean. Dissolved sulfides precipitate upon reaching the cold ocean water, coloring the vent fluid black. White tube worms with red gill plumes surround the area to exploit the chemical energy. Each worm is more than a meter in length and has a 20-cm gill plume. (b) Schematic illustration of the anatomical and physiological organization of the adult tube worm. Inside the trunk of the worm is a trophosome consisting primarily of endosymbiotic bacteria, associated cells, and blood vessels. Oxygen, carbon dioxide, and hydrogen sulfide (H_2S) are absorbed through the respiratory gill plume and transported in blood cells to the trophosome. H_2S is bound to the worm's hemoglobin ($HSHbO_2$). The endosymbionts oxidize H_2S and use some of the released energy to fix CO_2. Some of the reduced carbon compounds synthesized by the endosymbiont are translocated to the animal's tissues.

(a) PMEL Vents Program/NOAA

MICRO INQUIRY *What is the nutritional type of the endosymbiotic bacteria?*

The Rumen Ecosystem

Ruminants include cattle, deer, bison, camels, sheep, goats, and giraffes. These animals spend vast amounts of time chewing their cud—a small ball of partially digested grasses that the animal has consumed but not yet completely digested. It is thought that ruminants evolved an "eat now, digest later" strategy because their grazing can be interrupted by predator attacks.

These herbivorous animals have stomachs divided into four chambers (**figure 27.5**). The upper part of the stomach is expanded to form a large pouch called the rumen. The microbial community in the rumen is dynamic and, as a whole, forms a mutualistic relationship with the animal host. The rumen is slightly warmer than the rest of the animal, and with a redox potential of about –30 mV, all resident microorganisms must carry out anaerobic metabolism. Because of the chemical complexity of the plant material consumed by the animal, hydrolysis to simple carbohydrates requires the action of multiple enzymes. No single microbe encodes the complete set of enzymes; rather the community acts together to degrade the plant polymers cellulose, hemicellulose, lignin, and pectin. One group of bacteria secretes cellulases that cleave the $\beta(1\rightarrow4)$ linkages between successive D-glucose molecules that form plant cellulose.

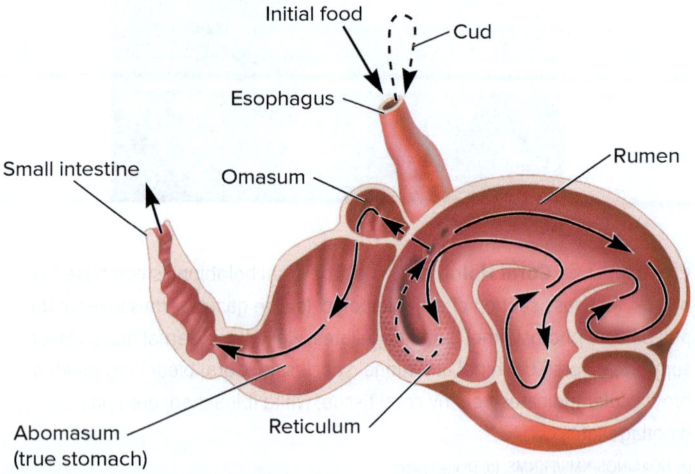

Figure 27.5 Ruminant Stomach. The stomach compartments of a cow. Grasses are swallowed without chewing and enter the rumen, where they are covered with microbes and partially digested. The food mass moves to the reticulum and is regurgitated, chewed, and reswallowed. As the mass is hydrolyzed by the microbial community, it becomes more liquefied and moves into the omasum, and finally to the abomasum. Here the food encounters digestive enzymes, and soluble nutrients are absorbed into the animal's bloodstream. Arrows indicate direction of food movement.

Other community members ferment glucose to organic acids such as acetate, butyrate, and propionate. These organic acids, as well as fatty acids, are the true carbon and energy sources for the animal. *Fermentation does not involve an electron transport chain (section 11.6)*

Comprehension Check

1. Review the bacterial and archaeal community responsible for nitrification (*see figure 21.9*). What terms best describe these microbial interactions?
2. How could one test to see if an insect-microbe relationship is mutualistic?
3. What is the role of the *Riftia* tube worm's hemoglobin in the success of the tube worm–endosymbiont mutualistic relationship?
4. What is meant by "tight trophic coupling" between the coral animal and its microbial symbionts?
5. Why is it important that the rumen is a reducing environment?

27.3 Cooperation: Nonobligatory Positive Interaction

After reading this section, you should be able to:

a. Distinguish cooperation from mutualism
b. Describe benefits to each partner in examples of cooperation

Cooperation is distinct from mutualism in that both organisms are capable of growth outside the partnership. This has enabled laboratory manipulation of cooperative partnerships to explore the features that drive these associations.

The interaction between the diderm bacterium *Xenorhabdus nematophila* and its nematode (worm) host *Steinernema carpocapsae* is remarkable because the bacterium contributes directly to the reproductive success of its host. Juvenile nematodes harboring *X. nematophila* within their guts live in the soil. To mature, the juvenile worms must find an insect to infect and consume; this is when things get really interesting (**figure 27.6**). As the hungry juvenile consumes insect blood (haemolymph), *X. nematophila* is excreted in the worms' feces. *X. nematophila* is now free-living, and as the bacteria replicate, they use a type III secretion system to deliver enzymes that kill the insect. Once the insect is dead, the bacteria switch to the production of compounds that protect the insect from degradation by other bacteria and from attack by ants, thereby protecting the home of their host nematode. Amazingly, as the process continues, *X. nematophila* produces yet a different set of molecular signals that trigger *S. carpocapsae* worms development to adulthood. A single insect cadaver may host many adult nematodes that mate, yielding a cadaver that ultimately teems with *S. carpocapsae* eggs. As these eggs mature to the juvenile stage with *X. nematophila* symbionts, they leave to find new insect homes and start the cycle anew.

Cooperation is also documented in several types of interactions between fungi and motile bacteria. In solid dry niches like soil or a cheese rind, the physical network of fungal hyphae (filaments) comprises a **fungal highway** to promote bacterial dispersal. Hyphae spread on solid surfaces, growing toward nutrients. Bacterial motility is restricted to the liquid phase, but surface tension maintains a water layer around hyphae. This permits bacteria to move along the hyphal surface to explore new environments, perhaps attracted by the same chemical signals as

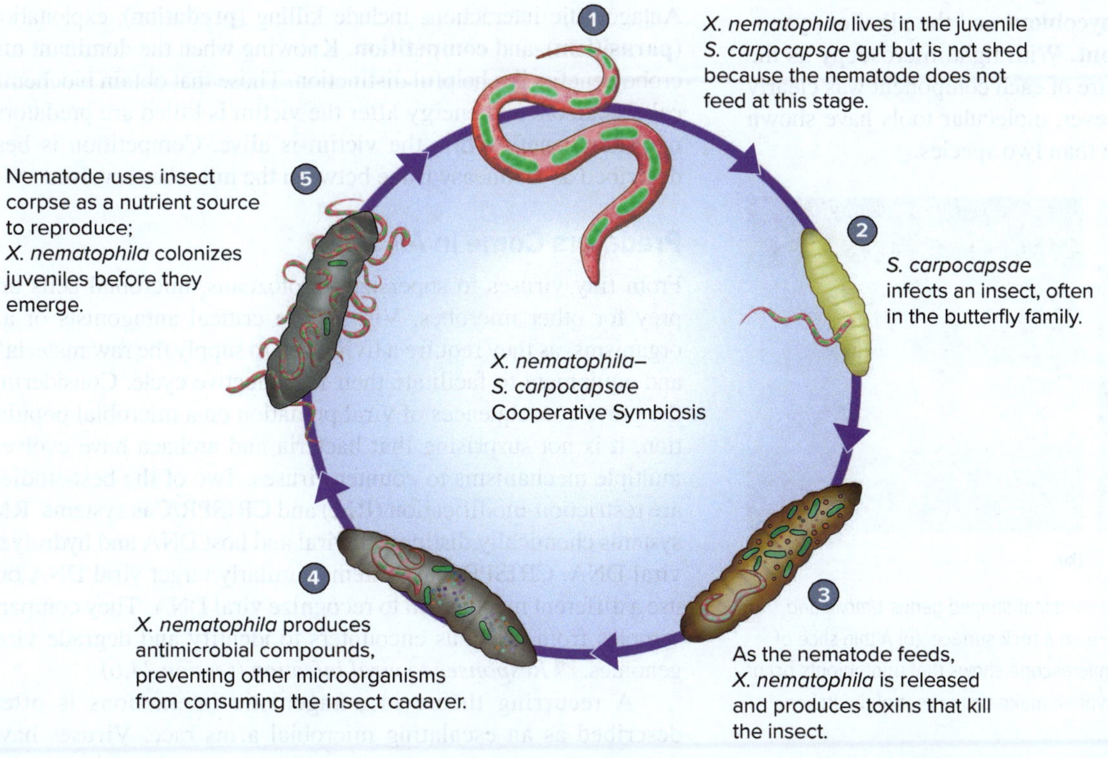

Figure 27.6 The *Xenorhabdus nematophila–Steinernema carpocapsae* Symbiosis.

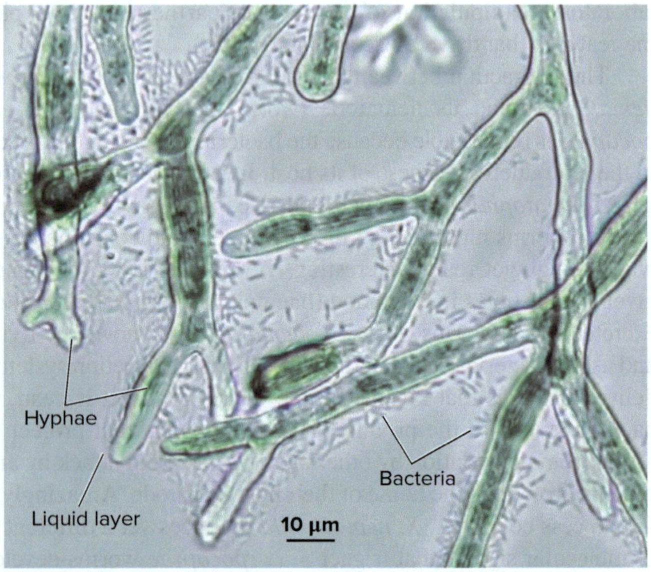

Figure 27.7 Fungal Highways Promote Bacterial Dispersal. A liquid layer (faint pink) surrounding *Mucor* hyphae provides a means for motile *Serratia* to explore new environments in a cheese rind.
Benjamin Wolfe

the fungus (**figure 27.7**). Transcriptomic studies of fungus/bacterial pairs reveal molecular details of the cooperative interaction. In a model system exemplified by *Aspergillus* and *Bacillus* spp., the fungi downregulates its thiamine biosynthesis genes, while the bacteria upregulate theirs, thereby establishing nutrient transfer as a benefit to the fungus.

For over a century, lichens (**figure 27.8**) were defined as the association between one species of ascomycetes (fungi) and another species of either green algae or cyanobacteria. The fungal partner is termed the **mycobiont** and the algal or cyanobacterial partner, the **photobiont.** With light microscopy as the primary tool, the spatial structure of each component was clearly delineated (figure 27.8*b*). However, molecular tools have shown that lichens often contain more than two species.

(a) (b)

Figure 27.8 Lichens. (a) The flattened leaf-shaped genus *Umbilicaria* is a common foliose lichen, shown here on a rock surface. (b) A thin slice of *Umbilicaria* viewed with a compound microscope shows that photobionts occur in a thin upper layer and mycobiont hyphae make up the rest of the lichen.
(a) Jeff Holcombe/Shutterstock; (b) Lee W. Wilcox

The photobionts are photosynthetic, secreting fixed carbon. If cyanobacteria are present, they fix nitrogen in specialized compartments. The mycobionts obtain nutrients by hyphal projections called *haustoria* that penetrate the photobiont cell wall. The mycobionts also use O_2 produced during photosynthesis for respiration. In turn, the mycobionts protect the photobionts from high light intensities, provide water and minerals, and create a firm substratum within which the photobiont can grow. This is an ancient association, having developed prior to the evolution of vascular plants.

Comprehension Check

1. How does cooperation differ from mutualism? What might be some of the evolutionary implications of both types of symbioses?
2. Why is the *X. nematophila–S. carpocapsae* symbiosis considered cooperation? What other types of microbial interactions are in play during the nematode life cycle?
3. How do the individuals in a lichen benefit from the physical association?

27.4 Antagonistic Interactions Prompt Microbial Responses

After reading this section, you should be able to:

a. Distinguish predation from parasitism
b. Describe contact-dependent and contact-independent growth inhibition

Antagonistic interactions include killing (**predation**), exploitation (**parasitism**), and **competition.** Knowing when the dominant microbe benefits is a helpful distinction. Those that obtain biochemical precursors and energy after the victim is killed are predators; parasites benefit while the victim is alive. Competition is best described as an uneasy truce between the microbes involved.

Predators Come in All Sizes

From tiny viruses to supersized protozoans, microbial cells are prey for other microbes. Viruses are critical antagonists of all organisms, as they require a living cell to supply the raw materials and equipment to facilitate their reproductive cycle. Considering the grave consequences of viral predation on a microbial population, it is not surprising that bacteria and archaea have evolved multiple mechanisms to counter viruses. Two of the best-studied are restriction-modification (RM) and CRISPR/Cas systems. RM systems chemically distinguish viral and host DNA and hydrolyze viral DNA. CRISPR/Cas systems similarly target viral DNA but use a different mechanism to recognize viral DNA. They compare samples from previous encounters to identify and degrade viral genomes. *Responses to viral infection (section 14.6)*

A recurring theme in antagonistic interactions is often described as an escalating microbial arms race. Viruses have

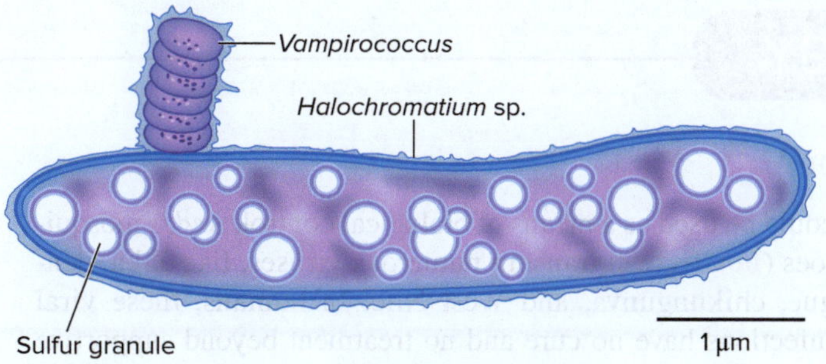

Figure 27.9 ***Vampirococcus* is an epibiotic predator.** *Vampirococcus* is an obligate parasite that grows while attached to a host cell. After killing the host, only the sulfur granules remain.

developed strategies to get around host defenses. Some bacteriophages synthesize small proteins that inhibit host RM and CRISPR/Cas systems. Similarly, animal viruses often synthesize proteins that interfere with the host immune response. *Coronavirus (section 25.5)*

Bacteria are also susceptible to predation by protozoa, particularly heterotrophic flagellates and ciliates. In soil, freshwater, and marine environments, **grazing** is the process by which protozoa consume bacteria. This is an important component of food webs, as it is the primary mechanism to transfer carbon and energy from bacterial biomass to more complex organisms (*see figure 29.5*).

Defensive mechanisms to counter grazing prominently feature an increase in growth form. Grazing protozoa generally choose midsize bacteria, so bacterial cells in larger assemblages such as filaments or microcolonies are more likely to avoid predation. The extracellular matrix of a biofilm offers some degree of protection, but some predators simply consume the entire biofilm.

Bacterial predators of other bacteria are classified according to their predation strategy. Epibiotic predators attach to their prey organism's surface. *Vampirococcus* cells (**figure 27.9**) attach to the outer membrane and secrete degradative enzymes that result in lysis and the release of the prey's cytoplasmic contents. Endobiotic predators invade the victim's cytoplasm or periplasm where they consume the contents to obtain the energy and precursors needed for cell division. *Bdellovibrio* (*see figure 20.18*) uses this predation strategy. *Bdellovibrio*

Myxococcus is a facultative predator that consumes organic matter released from dead organisms, but at other times actively preys on other microbes. *Myxococcus* displays predation whereby populations of cells use gliding motility to creep toward and cover their prey while releasing an arsenal of degradative enzymes and antibiotics (**figure 27.10**). In an illustration of the interplay between antagonistic and cooperative interactions within a larger community, *Myxococcus* is known to aggressively prey on plant pathogenic fungi in the soil, prompted by chemical signals from plants. *Phyla Bdellovibrionota and Myxococcota: bacterial predators (section 20.9)*

Importantly, predation has many beneficial effects, especially when one considers predators and prey at the population and community levels. Ingestion and assimilation of a prey bacterium can lead to increased rates of nutrient cycling, critical for the functioning of the microbial loop (*see figure 29.5*). The impact of viral predation on communities, ecosystems, and even the global carbon budget is vast. As such, it is an active area of research. *Marine viruses: mortality at sea (section 29.2)*

Parasitism: Microbes Harm the Host

Parasitism is one of the most complex microbial interactions; **parasite** and host must coexist at least temporarily, so the microbe has enough time to reproduce and colonize a new host. Coexistence may involve nutrient acquisition, physical maintenance in or on the host, or both. But what happens if the host-parasite equilibrium is upset? If the balance favors the host (perhaps by a strong immune defense or antimicrobial therapy), the parasite loses its habitat and may be unable to survive. On the other hand, if the equilibrium is shifted to favor the parasite, the host becomes ill and, depending on the specific host-parasite relationship, may die.

A controlled parasite-host relationship can be maintained for long periods of time and the nature of the relationship may not be obvious. For example, the *Wolbachia*-insect symbiosis has been described as both parasitic and mutualistic, depending on whether it is viewed through the lens of an insect cell and an endosymbiont, or through insect population genetics (**Microbial Diversity & Ecology 27.1**).

An important aspect of many symbiotic relationships, including parasitism, is that over time, the symbiont will discard excess, unused genomic information, a process called **genomic reduction** (*see figure 18.17*). This occurs when the symbiont has become dependent on the host for specific functions, such as synthesis of key metabolites. It is observed in the aphid endosymbiont *Buchnera aphidicola;* it has also occurred in the human pathogens *Mycobacterium leprae, Mycoplasmoides genitalium,*

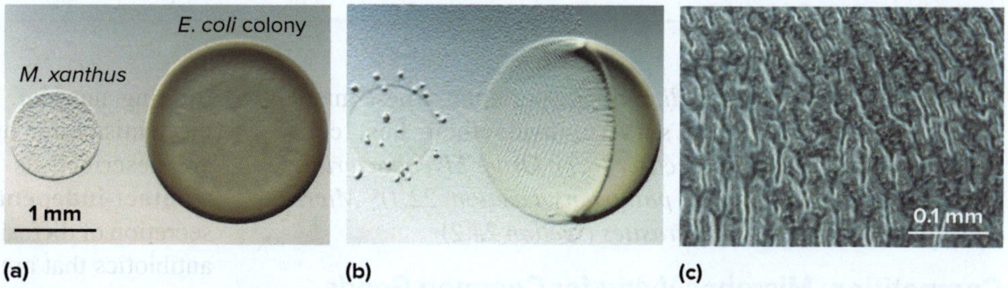

Figure 27.10 Predation by *Myxococcus xanthus*. (a) In this demonstration, a drop of *M. xanthus* cells is placed to the left of a colony of prey bacteria (*Escherichia coli*). (b) The *E. coli* cells lyse as *M. xanthus* glides over and consumes them; thus the colony becomes clear. (c) *M. xanthus* cells swarm in waves over the prey bacteria and will go on to form fruiting bodies (*see figure 20.19*).
James Berleman

MICRO INQUIRY *How does myxobacterial predation differ from that of* Vampirococcus?

MICROBIAL DIVERSITY & ECOLOGY

27.1 *Wolbachia:* The World's Most Infectious Microbe?

Most people have never heard of the bacterium *Wolbachia,* but this rickettsia infects more organisms than any other microbe. It infects a broad range of crustaceans, spiders, mosquitoes, millipedes, and nematodes, and may infect more than 2 million insect species worldwide. To what does *Wolbachia* owe its extraordinary success? Quite simply, this endosymbiont is a master at manipulating its hosts' reproductive biology.

Wolbachia inhabits the cytoplasm of its host's cells and is transferred from one generation to the next through the eggs of infected females. To survive, *Wolbachia* must ensure the fertilization and viability of infected eggs while decreasing the viability of uninfected eggs. The mechanism by which this is accomplished depends on the host. In wasps and mosquitoes, *Wolbachia* causes *cytoplasmic incompatibility,* which means that embryonic development will be abnormal if only the male is infected. For instance, when infected sperm of the wasp *Nasonia vitripennis* fertilize uninfected eggs, chromosomes from *Wolbachia*-laden sperm prematurely try to align with the egg's chromosomes. These eggs then divide as if never fertilized. However, chromosomes behave normally when an infected female mates with an uninfected male. This yields a normal sex distribution, and all progeny are infected with the rickettsia.

In other infected insects, *Wolbachia* may simply kill all the male offspring and induce parthenogenesis in infected females; that is, the females simply clone themselves. This limits genetic diversity but allows 100% transmission of *Wolbachia* to the next generation. In still other hosts, the microbe modifies male hormones so the males become feminized and produce eggs.

Another effect of *Wolbachia* infection of insect cells is interference with viral replication. Viruses normally spread by insects may not be transmissible if the insect is infected with *Wolbachia.* This has led to the notion that this infection could be used as a means of biological control. *Aedes* mosquitoes (**box figure**) transmit numerous viruses, like Zika, dengue, chikungunya, and West Nile. In humans, these viral infections have no cure and no treatment beyond supportive care, so insect control is the best prevention.

Over the past decade, researchers have released *Wolbachia*-infected *Aedes* mosquitoes to control the spread of dengue in several regions. Depending on the local mosquito population, the released *Aedes* are either all male or both sexes. In the first case, the goal is to reduce the overall insect population, but in the second case, *Wolbachia* is spread through the population to reduce virus transmission. Promising results from Australia, Indonesia, and Brazil show dramatic reductions in the disease among residents of the test area compared to populations in nearby control sites.

▶| *Arthropods can transmit viral diseases (section 37.2)*

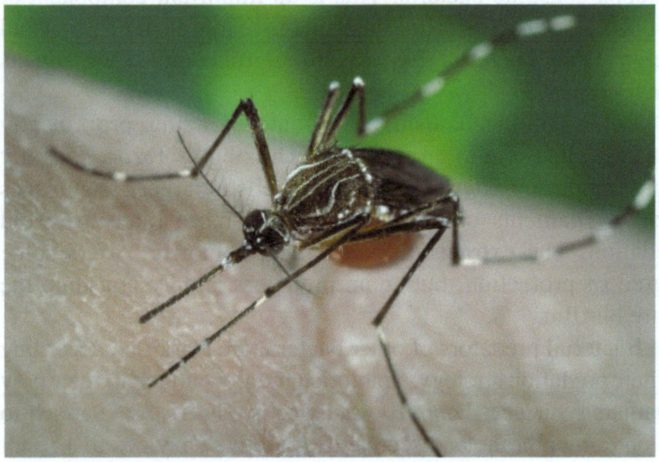

Female *Aedes aegypti* mosquito. Infection with *Wolbachia* renders these insects incapable of spreading arboviruses to humans.
James Gathany/CDC

and *Encephalitozoon cuniculi,* a microsporidium. These microorganisms now can only survive inside their host cells. |◀ *Comparative genomics (section 18.7); Order Mycobacteriales includes important human pathogens (section 22.1); Microsporidia are intracellular parasites (section 24.2)*

Competition: Microbes Vying for Common Goods

Competition arises when different organisms within a population or community try to acquire the same resource, whether this is a physical location or a particular limiting nutrient. If one of the two competing organisms can dominate the environment, whether by occupying the physical habitat or by consuming a limiting nutrient, it will overtake the other. The two general mechanisms for offensive action against neighboring microbes are described as contact-independent and contact-dependent. **Contact-independent growth inhibition** is carried out by secretion of diffusible compounds (**figure 27.11*a***). These include antibiotics that are active against a range of target microbes and bacteriocins that act on closely related strains. Bacteriocins are often named for the particular type of microbe that produces them—for example, halocins are produced by and are active against halophilic archaea. Another contact-independent mechanism involves secretion of toxin-bearing outer membrane vesicles. |◀ *Extracellular vesicles form from bacterial cells*

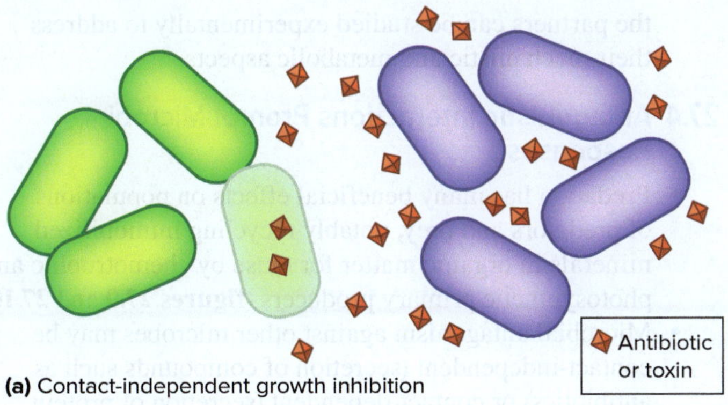

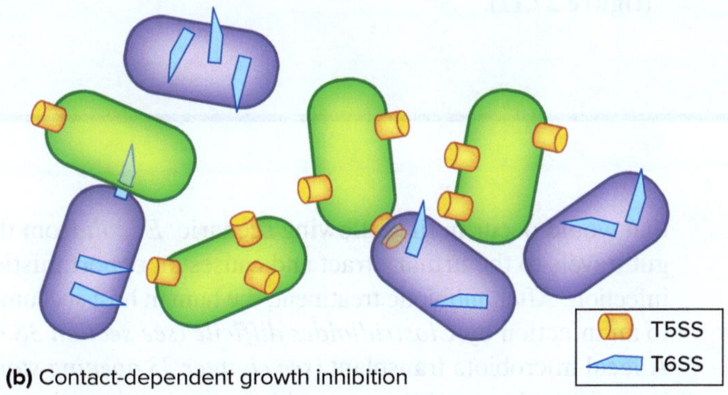

Figure 27.11 Competition Between Microbes. (a) Contact-independent growth inhibition involves the secretion of soluble antagonist molecules. Here, the antagonist is secreted by the purple cells and acts on the green cells. (b) Contact-dependent growth inhibition is mediated by molecules transferred through type V and VI secretion systems.

(section 3.4); Order Streptomycetales: an important source of antibiotics (section 22.1); ▶▌ *Bacteriocins (section 31.3)*

A bacterial population synthesizing and secreting a compound that benefits the entire population, such as an antibiotic, includes some members best described as **cheaters.** This subpopulation accrues the benefit of the compound (inhibition of a competitor) but avoids the energetic cost of its production. Cheaters must be in the minority, or the population as a whole would be less competitive. Nature has evolved multiple ways in which populations police themselves, or are policed by others, as described in the chapter opening story.

Contact-dependent growth inhibition (CDI) requires that the cells be in physical proximity for efficient toxin delivery, often mediated by type V and VI secretion systems (T5SS, T6SS) in diderm bacteria (figure 27.11*b; see figure 13.41*). T5SS, also termed *CDI systems,* are a two-partner mechanism. One protein, CdiB, is an outer membrane beta barrel protein through which its partner CdiA is secreted. A portion of the CdiA protein is a toxin that is cleaved and delivered to the cytoplasm of the target cell. Toxins delivered via the T5SS include nucleases to damage the chromosome and/or ribosomes and pore-forming molecules that result in the release of cytoplasmic contents. ▌◀ *Two-step secretion systems (section 13.9)*

The T6SS likewise secretes toxic proteins, but via a projectile. T6SS synthesis requires an energy commitment, as significant cellular resources are dedicated to synthesizing the T6SS and its toxins. The T6SS assembles in the cytoplasm and is poised to penetrate three membranes, the two membranes of the producing cell and a third membrane. When the target is another diderm bacterium, the toxins are released to its periplasm; however, they are released to the cytoplasm of a monoderm bacterium or eukaryotic cell. It is believed the T6SS originated as a weapon for bacterial competition, but it confers an advantage in promoting infection and pathogenesis. Human gut pathogens *Vibrio* and *Salmonella* employ a T6SS against the normal microbiota to establish themselves in the niche, and also against the host to cause disease. ▌◀ *One-step secretion systems (section 13.9)*

In contrast to the genomic reduction observed in symbionts, free-living bacteria that engage in competition have large genomes. The more competitive an organism, the more of its genome is dedicated to offensive and defensive systems. Antibiotics and bacteriocins require complex pathways for biosynthesis from precursor metabolites, and secretion systems require synthesis, assembly, and maintenance. All of these place an energetic burden on the cell.

Comprehension Check

1. Define predation and parasitism. How are these similar and different?
2. Summarize the types of predation that may affect bacterial cells.
3. What is a cheater? Why are cheaters only a small fraction of a population?

Key Concepts

27.1 Many Types of Microbial Interactions Exist

- Symbiotic interactions range over a continuum from cooperation to conflict (**figure 27.1**). Interactions between individuals occur in a community context.

27.2 Mutualism: Obligatory Positive Interaction

- Mutual advantage is central to many organism-organism interactions. These interactions can be based on material transfers related to energetics or the creation of physical environmental changes that offer protection. With several important mutualistic interactions, microbial primary producers play a critical role in making organic matter available for use by a larger community (**figures 27.3** and **27.4**).

- The rumen is an example of a mutualistic interaction between an animal and a complex microbial community. In this microbial community, complex plant materials are broken down to simple organic compounds that can be absorbed by the ruminant, as well as forming waste gases such as methane that are released to the environment.

27.3 Cooperation: Nonobligatory Positive Interaction

- Cooperation, like mutualism, benefits both partners. Because cooperative interactions are not obligatory, the partners can be studied experimentally to address their mechanistic and metabolic aspects.

27.4 Antagonistic Interactions Prompt Microbial Responses

- Predation has many beneficial effects on populations of predators and prey, notably recycling immobilized minerals in organic matter for reuse by chemotrophic and photosynthetic primary producers (**figures 27.9** and **27.10**).
- Microbial antagonism against other microbes may be contact-independent (secretion of compounds such as antibiotics) or contact-dependent (secretion of protein toxins through type V or VI secretion systems (**figure 27.11**).

Active Learning

1. Describe an experimental approach to determine if a microbe and its associated plant interact through cooperation or mutualism. How could you compare microbial growth in both conditions quantitatively?
2. Some patients who take antibiotics for acne develop yeast infections of the mouth or genitourinary tract. Explain.
3. *Variovorax paradoxus* is a soil organism that has adapted to use *N*-acyl homoserine lactones (*see figure 14.23*) as a carbon source. In so doing, *V. paradoxus* interferes with quorum sensing in other microbes in its habitat. Classify this interspecies interaction.
4. Consider the human holobiont and its gut microbiome. Identify and classify the types of microbial interactions that would occur in the following scenario. *E. coli* from the gut travels to the urinary tract and causes an opportunistic infection. After antibiotic treatment, the human host succumbs to an infection by *Clostridioides difficile* (*see section 38.6*). A fecal microbiota transplant (*see chapter 33 opening story*) is performed to treat this second infection, with success.
5. How might a bacterial cell defend against contact-dependent growth inhibition?
6. Compare and contrast *Vampirococcus* with a lytic bacteriophage.
7. It has been suggested that horizontal gene transfer (*see section 16.3*) is an example of microbial cooperation. Support or refute this idea.

CHAPTER

28

Biogeochemical Cycling and Global Climate Change

Voishmel/AFP/Getty Images

Global Climate Change; Infectious Disease Change

Warnings that global climate change could impact the transmission of infectious disease are no longer theoretical. Evidence now demonstrates that rising temperatures, floods, droughts, and fires—attributable at least in part to climate change—have resulted in the exacerbation of over 200 infectious diseases. Vector-borne diseases lead this increase.

Insect vectors carry and transmit the microbes that cause plague, malaria, West Nile fever, Lyme disease, and yellow and dengue fevers (just to name a few). Warmer temperatures not only expand the geographical range of many insects, but also lengthen their breeding season, boost their reproductive rates, and drive them to feed (i.e., bite) more frequently. In addition, many infectious agents, such as the viruses causing Ebola and COVID-19, are shared between humans and animals. Historically, changes in the environment have been associated with the emergence of new diseases or the reemergence of old ones, particularly those transmitted by animals and insects (i.e., zoonoses and arthropod-borne, respectively).

The effects of global climate change on infectious disease can be broadly divided into three categories: altered disease patterns already happening, those that can be predicted with confidence, and those that will surprise us. Let's start with a disease whose incidence has seen a dramatic increase—dengue. Its mosquito vector, *Aedes aegypti* thrives in warmer temperatures. In 2024, the incidence of dengue broke records in South America and southern Europe, where it had not been widely seen before. In the United States, dengue had only been diagnosed in returning travelers, but mosquitoes in Florida, Texas, Arizona, and California now transmit the disease.

Malaria is another mosquito-borne disease, but it is transmitted by a different genus, *Anopheles*. Modeling changes in climate and climate-related impacts such as flooding, as well as the reproductive rates of *Anopheles* mosquitos, have enabled scientists to predict that while cases of malaria in Sub-Saharan Africa are currently increasing, if temperatures continue to rise, it may become so hot that the transmission rate will decline in that region but increase in others.

What about diseases that could surprise us? These could be completely new diseases, like COVID-19, or diseases that were thought to be well controlled. But whether the infectious diseases are new or old, their prevention and treatment will require the expertise of a new generation of epidemiologists, health-care professionals, and (of course) microbiologists.

It is obvious why microbiologists are involved in the study of the link between climate and infectious disease. What may be less obvious is the role microbes play in climate change. However, the cycling of all elements is mediated by microorganisms. The sum of the microbial, physical, and chemical processes that drive the flow of elements between sediments, waters, and the atmosphere is known as **biogeochemical cycling.** In this chapter we focus on the flux of the major elements: carbon, nitrogen, sulfur, phosphorus, iron, and manganese. We then turn our attention to global climate change, as it is driven by changes in the carbon and nitrogen cycles.

Readiness Check:

Based on what you learned previously, you should be able to:

- ✓ Explain how nutrient availability and environmental factors influence microbial growth (sections 3.2, 7.5, and 7.6)
- ✓ Explain oxidation-reduction (redox) reactions (sections 10.3 and 10.4)
- ✓ Describe the difference between heterotrophy and autotrophy (section 11.1)
- ✓ Compare aerobic respiration, anaerobic respiration, and fermentation, as well as chemolithoautotrophy and photoautotrophy (sections 11.3, 11.5, 11.6, 11.8, and 11.10)

28.1 Biogeochemical Cycling Sustains Life on Earth

After reading this section, you should be able to:

a. Define the term *biogeochemical cycling*
b. Defend the statement "Biogeochemical cycling sustains life on Earth."
c. Explain the term *redox potential* and how it influences the flux of elements found in specific habitats
d. Contrast and compare mineralization and immobilization

Biogeochemical cycling involves both abiotic, or nonliving, processes, such as erosion, and biotic activities. The biotic processes are sometimes called *nutrient cycling.* The term *nutrient* refers to chemicals, both organic and inorganic, that support life. The term *cycle* indicates that for each nutrient, there is movement from one pool (or stockpile) to another. In one pool, a nutrient exists in a certain state (e.g., highly oxidized), while in another pool it exists in a different state (e.g., more reduced). The process of moving from one pool to another changes the element from one state to another and is called *flux.* An important example of flux is the conversion of atmospheric CO_2 (oxidized; gaseous) by autotrophic microbes in the ocean to carbohydrate (reduced; solid). Although some changes are driven by abiotic factors, the majority are performed by bacteria, archaea, and eukaryotic microbes. Without elemental cycling, all life on Earth would cease. ◀ *CO_2 fixation: reduction and assimilation of CO_2 carbon (section 12.3)*

The notion of a cycle also implies mass balance: The number of atoms of a given element like carbon or nitrogen that revolve around the cycle remains constant because any input to the cycle is balanced by an equal loss. Indeed, that was the case for billions of years. But around the turn of the twentieth century, humans started moving carbon and nitrogen from one pool (Earth) into another pool (the atmosphere) faster than they could be returned. The burning of fossil fuel releases tons of carbon that had been sequestered for millions of years. The application of nitrogenous fertilizer to support agriculture for the ever-increasing human population also releases nitrogenous compounds faster than they can be reintroduced to the cycle. The release of more heat-trapping carbon and nitrogen-containing compounds into the atmosphere than can be removed by microbes and other organisms generates an imbalance that is the crux of global climate change. Global climate change is a mass balance problem.

Our understanding that global climate change depends on the flow of carbon and nitrogen between sources and sinks has coincided with the recent discovery that much microbial activity occurs deep beneath Earth's surface (*see figure 1.2*). As we explore the biogeochemical cycles that make life possible, keep in mind that microbes on the surface of the Earth contribute to the rapid cycling of nutrients, while the very slow-growing and largely unexplored deep subsurface microbes impact elemental cycling over geological time periods and represent a very large and under-explored carbon sink. ▶ *Microbial communities in benthic marine sediments are enormous and mysterious (section 29.2); The subsurface biosphere is vast (section 30.4)*

Throughout the text, we consider how differences in temperature, pressure, salinity, pH, and nutrient availability impact microbial growth and diversity. We now introduce another factor critical in determining microbial activity: redox potential. **Redox potential,** or, more formally, oxidation-reduction potential, is a measure of the tendency of molecules in a system to accept or donate electrons. In chapter 10, we discuss redox potential as it relates to specific conjugate redox pairs (*see table 10.2*). Here our focus is on redox potential in the context of natural microbial environments. The redox potential of each environment can be quantified in volts or millivolts as compared to a standard. However, no "standard environment" exists, so a hydrogen electrode set to zero volts is used. Thus the redox potential of an environment is determined by the electrical potential difference between the environment and a standard hydrogen electrode. The value is denoted as E_h. Environments composed principally of compounds with high redox potentials (more positive) are more likely to accept electrons (i.e., become reduced) when a new compound is added. The newly added compound that donates electrons then becomes oxidized (**table 28.1**).

Why do microbiologists care about redox potentials? The redox state of the environment plays a critical role in determining the types of microbes present because it determines (1) which oxidized compounds are available as terminal electron acceptors for anaerobic respiration and (2) what reduced molecules are present for use as electron donors in lithotrophy. In general, electron acceptors used in anaerobic respiration follow a continuum defined by the free energy yield of each half reaction (*see figure 10.5*). The terminal electron acceptors and their vertical zonation are shown in **figure 28.1;** however, microbial metabolic versatility is far more amazing than illustrated here. For example, the discovery of cable bacteria—multicellular sulfide-oxidizing bacteria in the Desulfobulbaceae family—demonstrates nature's ingenuity. Inspection of figure 28.1 reveals that sulfide and oxygen are spatially separated in the sediment column. Cable bacteria position themselves vertically in marine sediments (*see figure 20.17*). This enables cells at the anoxic end of the filament to use sulfide as an electron donor. The electrons are passed from cell to cell until they reach those in oxic sediments, where O_2 is used as the terminal electron acceptor. ◀ *Redox reactions: reactions of central importance in metabolism (section 10.3)*

Finally, two terms are helpful in considering the fate and availability of organic materials in nature: mineralization and immobilization. **Mineralization** describes the decomposition of organic matter to simpler, inorganic compounds (e.g., CO_2, NH_3, H_2). These compounds may or may not be recycled within

Table 28.1 Elements Important in Biogeochemical Cycling and Their Redox States

		MAJOR FORMS AND VALENCES				
Element	**State of matter**	**Reduced Form**	**Intermediate Oxidation State Forms**			**Oxidized Form**
C	Gaseous and solid	Methane: CH_4 (−4)	Carbon monoxide: CO (+2)			Carbon dioxide: CO_2 (+4)
N	Gaseous and solid	Ammonium: NH_4^+; organic N (−3)	Nitrogen gas: N_2 (0)	Nitrous oxide: N_2O (+1)	Nitrite: NO_2^- (+3)	Nitrate: NO_3^- (+5)
S	Gaseous and solid	Hydrogen sulfide: H_2S; SH groups in organic matter (−2)	Elemental sulfur: S^0 (0)	Thiosulfate: $S_2O_3^{2-}$ (+2)	Sulfite: SO_3^{2-} (+4)	Sulfate: SO_4^{2-} (+6)
Fe	Solid	Ferrous iron: Fe^{2+} (+2)				Ferric iron: Fe^{3+} (+3)
P	Solid	Phosphonates and phosphite: (PO^{3-}) (+3)				Polyphosphate, phosphate: PO_4^- (+5)
Mn	Solid	Manganese ion: Mn^{2+} Manganous oxide: MnO (+2)				Manganese dioxide: MnO_2 (+4)

the same ecosystem. By contrast, nutrients (including carbon) converted into biomass become temporarily unavailable for nutrient cycling; this is called nutrient **immobilization.** Saprophytes, which degrade dead organic material, as well as viruses, protists, and other predators are important in mineralizing immobilized organic compounds.

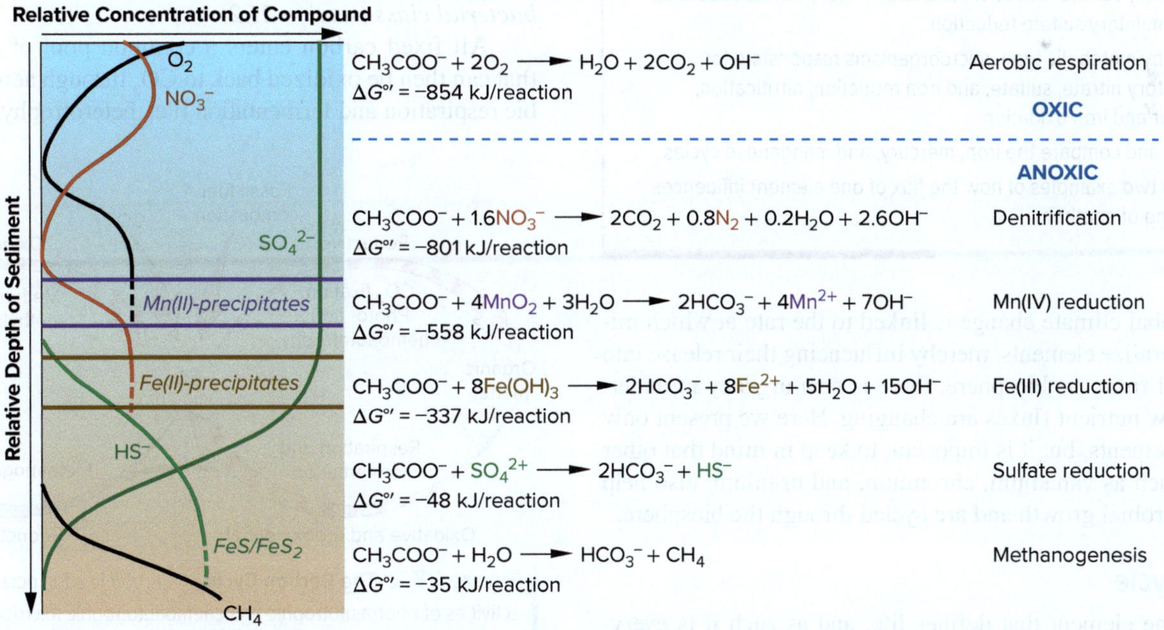

Figure 28.1 The Redox Potential, pH, and Oxygen Concentration Influence the Availability of Terminal Electron Acceptors. As oxygen concentration decreases with depth, other terminal electron acceptors are used in anaerobic respiration. The distribution of microbes in sediments or soils is dictated by the free energy yield for each respiratory process and the availability of the oxidized element. The organic compound acetate is used as the substrate in all equations to provide comparable $\Delta G^{o\prime}$ values. Note that fermentation takes place throughout the anoxic region of the column. The relative concentration of elements with depth is shown for marine sediments; freshwater sediments and soil have less sulfate; thus methanogenic archaea may be more abundant.

MICRO INQUIRY *Why do sulfate reduction and methanogenesis rarely occur at the same depth or in the same microhabitat?*

Comprehension Check

1. Explain in general terms the relationship between nutrient cycling and global climate change.
2. In an anoxic environment containing equal concentrations of nitrate and sulfate, which terminal electron acceptor will support the growth of most anaerobic microbes? Explain your answer.
3. What is the difference between mineralization and immobilization?

28.2 Microbes Mediate Nutrient Cycling

After reading this section, you should be able to:

a. Sketch generalized carbon, nitrogen, phosphorus, sulfur, iron, manganese, and mercury cycles
b. Identify, by metabolic type, the microorganisms responsible for carbon and nitrogen fixation, and the mineralization of organic carbon and nitrogen under oxic and anoxic conditions
c. List environments in which methanogenesis and methane oxidation occur
d. Trace the fate of an N atom during dissimilatory nitrate reduction, nitrification, denitrification, and the anammox reaction, and indicate whether each must occur in an oxic or anoxic environment
e. Contrast and compare dissimilatory nitrate reduction with dissimilatory sulfate reduction, and assimilatory nitrate reduction with assimilatory sulfate reduction
f. Identify, by metabolic type, microorganisms responsible for dissimilatory nitrate, sulfate, and iron reduction; nitrification; and sulfur and iron oxidation
g. Contrast and compare the iron, mercury, and manganese cycles
h. Describe two examples of how the flux of one element influences the cycling of another

Because global climate change is linked to the rate at which microbes mineralize elements, thereby influencing their release into and removal from the biosphere, there is new urgency in understanding how nutrient fluxes are changing. Here we present only the major elements, but it is important to keep in mind that other elements, such as vanadium, chromium, and uranium, also help support microbial growth and are cycled through the biosphere.

Carbon Cycle

Carbon is the element that defines life, and as such it is everywhere. To understand the carbon cycle and global climate change, we must identify the sources, sinks, and reservoirs of CO_2. *Sources,* such as heterotrophic microbes and the burning of fossil fuels, release carbon as CO_2, whereas *sinks* include plants, phytoplankton, and other autotrophic microbes that take up CO_2 from the atmosphere. *Reservoirs* store carbon for geological periods of time. Carbon is present in reduced forms, such as methane (CH_4) and other, more complex organic matter, and in the oxidized, inorganic forms, carbon monoxide (CO) and carbon dioxide (CO_2). Although carbon is continuously transformed from one form to another, for the sake of clarity, we shall say that the cycle "begins" with carbon fixation—the conversion of CO_2 into organic matter (**figure 28.2**). Plants are often regarded as the principal CO_2-fixing organisms, but at least half the carbon on Earth is fixed by microbes, particularly by diatoms and marine cyanobacteria in the genera *Prochlorococcus* and *Synechococcus*. Importantly, microbes also fix carbon in anoxic environments using anoxygenic photosynthesis as well as by chemolithoautotrophy in the absence of light. In fact, bacterial and archaeal chemolithoautotrophy in deep, dark subsurface sediments may constitute a significant fraction of global carbon fixation. ◀ *Calvin-Benson cycle (section 12.3); Phylum Cyanobacteriota: oxygenic photosynthetic bacteria (section 20.3); Stramenopila (section 23.4)*

Alternatively, CO_2 can be reduced anaerobically to methane (CH_4). Recall that only archaea form methane, and they do so using primarily $H_2 + CO_2$ or H_2 + acetate. As discussed in chapters 19 and 21, methane produced in sediments is then oxidized either aerobically by bacteria or anaerobically by archaea and denitrifying bacteria (**figure 28.3**). Globally, sediments found in rice paddies, coal mines, sewage treatment plants, landfills, marshes, and mangrove swamps and archaea found in the guts of ruminant animals and even termites are important sources of methane. ◀ *Methanogens and methanotrophs (section 19.4); Class* Alphaproteobacteria *includes many oligotrophs (section 21.1);* Gammaproteobacteria *is the largest bacterial class (section 21.2)*

All fixed carbon enters a common pool of organic matter that can then be oxidized back to CO_2 through aerobic or anaerobic respiration and fermentation (i.e., heterotrophy). In the carbon

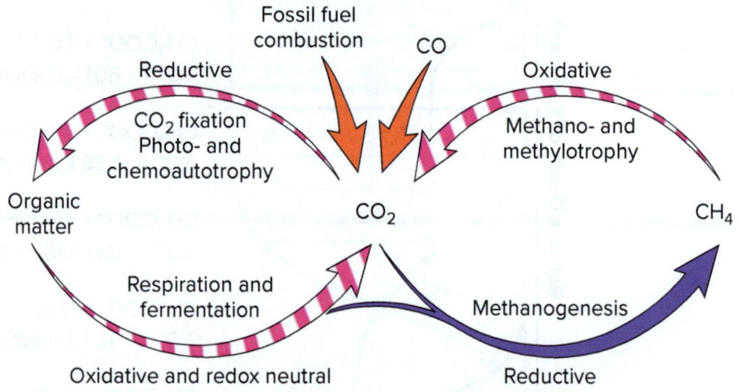

Figure 28.2 The Carbon Cycle. Carbon fixation occurs through the activities of photoautotrophic and chemoautotrophic microorganisms. Methane is produced from inorganic substrates ($CO_2 + H_2$) or from organic matter. Carbon monoxide (CO), produced by sources such as automobiles and industry, is returned to the carbon cycle as CO_2 by CO-oxidizing bacteria. Aerobic processes are indicated by red arrows, those that occur under both oxic and anoxic conditions are noted with striped arrows, and the purple arrow leads to methanogenesis, an anaerobic process.

MICRO INQUIRY *What microbes are capable of methanogenesis and methane oxidation?*

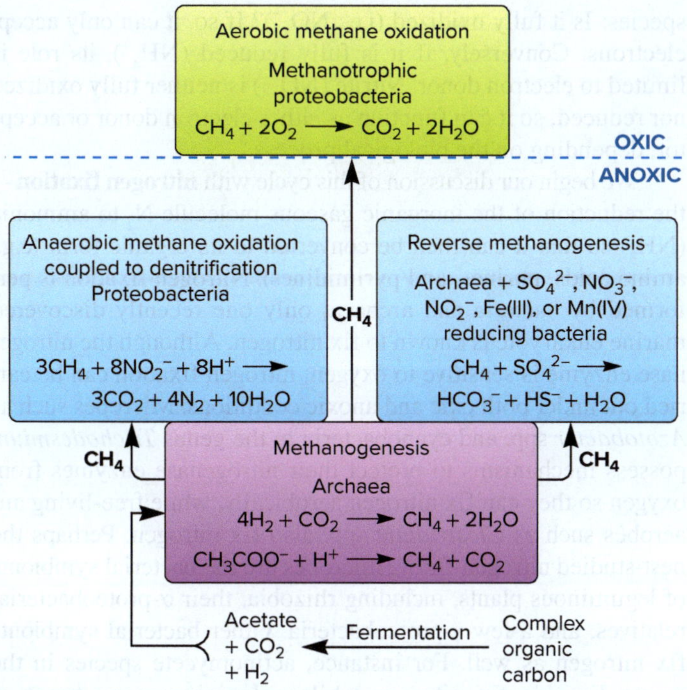

Figure 28.3 Methanogenesis and Methanotrophy. Starting at the bottom of the figure, bacterial fermentation products provide substrates for methanogenic archaea (purple box). As methane is produced, it diffuses upward, where it is oxidized by one of three general types of methanotrophs: proteobacteria that couple methane oxidation with denitrification (left blue box); anaerobic archaea that reverse the methanogenic pathway in association with sulfate (shown), nitrate, nitrite Fe(III), or manganese reduction (right blue box); and aerobic proteobacteria (yellow box).

cycle depicted in figure 28.2, no distinction is made between the different types of organic matter formed and degraded. This is a marked oversimplification because organic matter varies widely in terms of elemental composition, structure of repeating units, and linkages between repeating units. Its bioavailability and degradation are also influenced by other factors, including (1) oxidation-reduction potential; (2) availability of competing nutrients; (3) abiotic conditions such as pH, temperature, O_2, and osmotic conditions; and (4) the microbial community present.

Many of the complex organic substrates used by microorganisms are summarized in **table 28.2.** Lignin, an important structural component in mature plant materials, is notoriously stable. Lignin is actually a family of complex polymers linked by carbon–carbon and carbon–oxygen (ether) bonds (*see figure 30.2*). Filamentous microorganisms—fungi and the streptomycetes—secrete hydrolytic enzymes that degrade lignin by a depolymerization process that requires oxygen. Their filamentous form provides a large surface area for exoenzyme release and nutrient uptake. A few unicellular microbes, such as the purple nonsulfur bacterium *Rhodopseudomonas palustris,* can degrade lignin anaerobically but very slowly. Diminished biodegradability under anoxic conditions results in accumulation of lignified materials, including the formation of peat bogs, which serve as significant carbon reservoirs.

Many complex substrates, including lignin, contain only carbon, hydrogen, and oxygen (table 28.2). Growth using these substrates demands that microbes acquire the remaining nutrients (e.g., N, P, S, Fe) elsewhere in the environment. This is often very difficult, as the concentration of these other elements may be very low. When the supply of a nutrient is insufficient to support maximal growth, that nutrient is said to be limiting. For instance, in open-ocean microbial communities, growth of many microbes is often nitrogen limited. In other words, if higher concentrations of usable nitrogen (e.g., NO_3^-, NH_4^+) were available, the rate of microbial growth would increase. This phenomenon is called *Leibig's law of the minimum,* named after the nineteenth-century chemist Justus van Leibig.

Table 28.2 Complex Organic Substrate Characteristics That Influence Decomposition and Degradability

			ELEMENTS PRESENT IN LARGE QUANTITY					DEGRADATION	
Substrate	**Basic Subunit**	**Linkages**	**C**	**H**	**O**	**N**	**P**	**With O_2**	**Without O_2**
Starch	Glucose	α(1→4) α(1→6)	+	+	+	−	−	+	+
Cellulose	Glucose	β(1→4)	+	+	+	−	−	+	+
Hemicellulose	C6 and C5 monosaccharides	β(1→4), β(1→3), β(1→6)	+	+	+	−	−	+	+
Lignin	Phenylpropene	C—C, C—O bonds	+	+	+	−	−	+	+/−
Chitin	*N*-acetylglucosamine	β(1→4)	+	+	+	+	−	+	+
Hydrocarbon	Aliphatic, cyclic, aromatic	C—C, C═O bonds	+	+	−	−	−	+	+/−

MICRO INQUIRY *What nutrient(s) would be limiting if a microbe consumed chitin, as is the case with* Vibrio cholerae?

Comprehension Check

1. What are the microbial sources and sinks of CO_2? In what form is carbon stored?
2. What is the microbial contribution to global carbon fixation?
3. Where will lignin degrade faster: in a shallow riverbed or in a rice paddy? Explain.
4. Compare and contrast the processes that lead to mineralization and immobilization of carbon.

Nitrogen Cycle

Nitrogenous compounds also contribute to global climate change. The nitrogen cycle, however, is more complex than the carbon cycle because nitrogen exists in redox states from −3 to +5, and this is reflected in the transformations that drive nitrogen cycling (table 28.1). As shown in **figure 28.4,** depending on its level of oxidation, nitrogen species can serve as electron acceptors in anaerobic respiration (e.g., NO_3^- and NO_2^-) or as electron donors (e.g., NO_2^- and NH_4^+) in chemolithotrophy. This can lead to some confusion, so keep in mind the oxidation state of the nitrogen species: Is it fully oxidized (i.e., NO_3^-)? If so, it can only accept electrons. Conversely, if it is fully reduced (NH_4^+), its role is limited to electron donor. Nitrite (NO_2^-) is neither fully oxidized nor reduced, so it can function as either electron donor or acceptor, depending on the biological process.

We begin our discussion of this cycle with **nitrogen fixation**—the reduction of the inorganic gaseous molecule N_2 to ammonia (NH_3) so that it can then be converted to its organic form (e.g., amino acids, purines, and pyrimidines). Nitrogen fixation is performed by bacteria and archaea; only one recently discovered marine eukaryote is known to fix nitrogen. Although the nitrogenase enzyme is sensitive to oxygen, nitrogen fixation can be carried out under both oxic and anoxic conditions. Microbes such as *Azotobacter* spp. and cyanobacteria in the genus *Trichodesmium* possess mechanisms to protect their nitrogenase enzymes from oxygen so they can fix nitrogen aerobically, while free-living anaerobes such as *Clostridium* spp. also fix nitrogen. Perhaps the best-studied nitrogen-fixing microbes are the bacterial symbionts of leguminous plants, including rhizobia, their α-proteobacterial relatives, and a few γ-proteobacteria. Other bacterial symbionts fix nitrogen as well. For instance, actinomycete species in the genus *Frankia* fix nitrogen while colonizing many types of

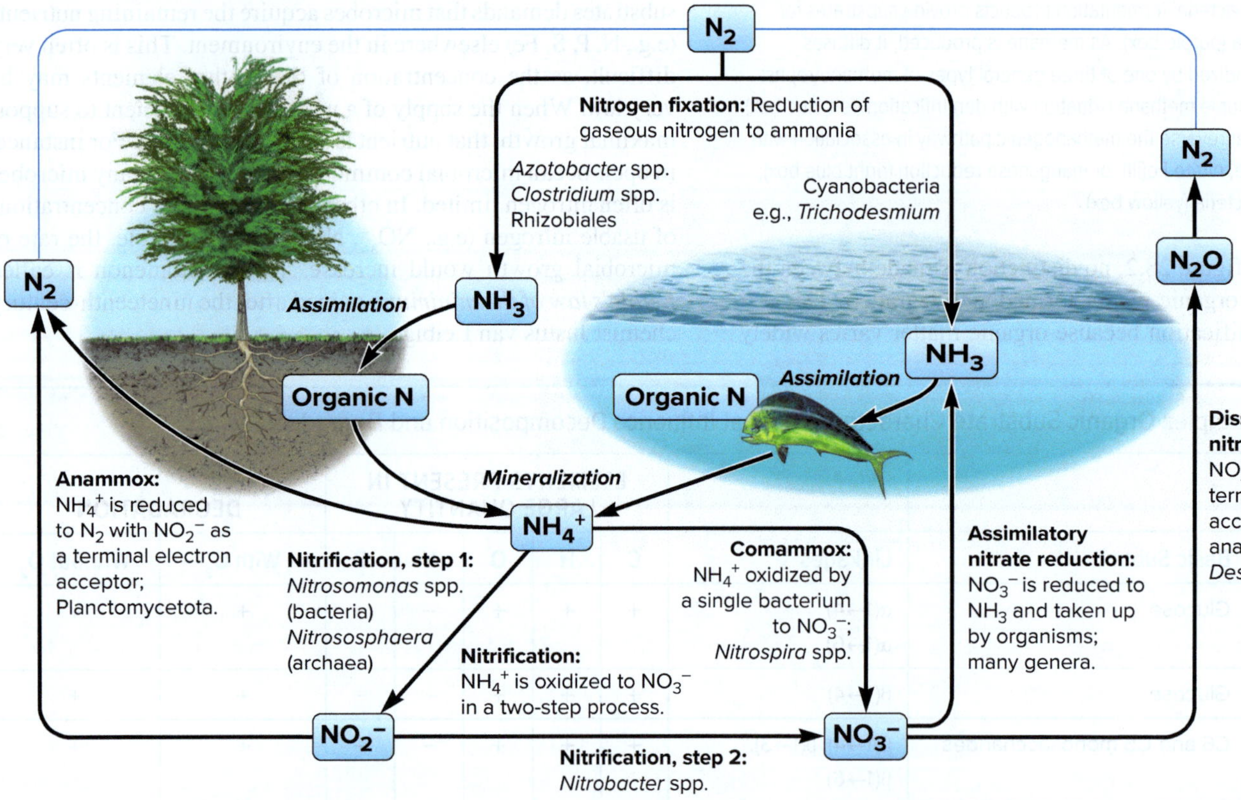

Figure 28.4 A Simplified Nitrogen Cycle. Depending on their oxidative state, some nitrogen species can function as either electron donors or electron acceptors and therefore play very different roles in the environment. NH_4^+ is fully reduced; therefore it can serve only as an electron donor. The fully oxidized species NO_3^- can function as a terminal electron acceptor, or plants and microbes can reduce it and incorporate it into biomass during assimilatory reduction. All parts of the nitrogen cycle occur in both terrestrial and marine ecosystems; we differentiate nitrogen fixation because this essential process is performed by different genera in these environments.

MICRO INQUIRY *What are the products of denitrification, nitrification, dissimilatory nitrate reduction, and assimilatory nitrate reduction?*

woody shrubs, and heterocyst-forming cyanobacteria *Anabaena* spp. fix nitrogen when in association with the water fern *Azolla* spp. ◀ *Nitrogen fixation: N_2 to NH_3 (section 12.5); Phylum Cyanobacteriota: oxygenic photosynthetic bacteria (section 20.3); Phylum Actinomycetota (section 22.1)* ▶ *Microorganisms in the open ocean are adapted to nutrient limitation (section 29.2); Nitrogen-fixing bacteria are vital to agriculture (section 30.3)*

The product of N_2 fixation is ammonia (NH_3), which is immediately incorporated into organic matter as an amine. These amine N atoms are immobilized through their incorporation into proteins, nucleic acids, and other biomolecules. When these organic molecules are degraded (dissimilated) and mineralized, nitrogen is released as ammonium (NH_4^+). Many microbial genera are capable of dissimilation of organic nitrogen substrates, but complete mineralization requires an assemblage of microbes.

Because ammonium is fully reduced, it can only donate electrons, and this is what happens during **nitrification** (figure 28.4). In the first step of this chemolithotrophic process, ammonium is oxidized to nitrite (NO_2^-), and in the second step, NO_2^- is oxidized to nitrate (NO_3^-). For many years, it was thought that nitrification had to be performed by two different groups of bacteria. For example, some archaea and bacteria in the genera *Nitrosomonas* and *Nitrosococcus* play important roles in ammonia oxidation, while *Nitrobacter* spp. and related bacteria carry out nitrite oxidation. Both steps of nitrification are aerobic, with O_2 as the terminal electron acceptor *(see figure 21.9)*. There are now two known exceptions. First is the γ-proteobacterium *Nitrosomonas eutropha,* which oxidizes ammonium anaerobically to nitrite and nitric oxide (NO) using nitrogen dioxide (NO_2) as an acceptor in a denitrification-related reaction. The second exception involves several *Nitrospira* species that perform both nitrification steps (ammonium to nitrite and nitrite to nitrate). This ability has been dubbed **comammox,** for *com*plete *amm*onia *ox*idation (figure 28.4). ◀ *Nitrifying bacteria oxidize ammonium or nitrite to gain energy and electrons (section 21.1)*

The production of nitrate is important because microbes and plants can reduce it and incorporate it into biomass in a process known as **assimilatory nitrate reduction.** Alternatively, some microorganisms use nitrate as a terminal electron acceptor during anaerobic respiration. When respired, the nitrogen is not incorporated into cellular material, so this is called **dissimilatory nitrate reduction.** A variety of microbes, including *Geobacter* spp. and *Desulfovibrio* spp., are capable of dissimilatory nitrate reduction. When nitrate is fully reduced to dinitrogen gas (N_2), nitrogen is removed from the ecosystem and returned to the atmosphere through a series of reactions collectively known as **denitrification.** This dissimilatory process is performed by a variety of heterotrophic bacteria, such as *Pseudomonas* spp. The major products of dissimilatory nitrate reduction include nitrogen gas (N_2) and nitrous oxide (N_2O), although nitrite (NO_2^-) can also accumulate (*see figure 11.16*). N_2O is an important greenhouse gas, as discussed in section 28.3. ◀ *Anaerobic respiration uses the same steps as aerobic respiration (section 11.5); Inorganic nitrogen assimilation (section 12.5)*

Lastly, the **anammox reaction** (*an*oxic *amm*onium *ox*idation) is an anaerobic reaction performed by chemolithotrophs in the phylum Planctomycetota. Here ammonium ion (NH_4^+) serves as the electron donor and nitrite (NO_2^-) as the terminal electron acceptor; it is reduced to nitrogen gas (N_2) (*see figure 20.9*). In effect, the anammox reaction is a shortcut to N_2, proceeding directly from ammonium and nitrite, without having to cycle first through nitrate (figure 28.4). The discovery that planctomycete bacteria oxidize considerable amounts of NH_4^+ to N_2, thereby removing ammonia from the environment, has sparked keen interest in the wastewater management field, where high levels of ammonia are undesirable. ◀ *Phylum Planctomycetota: anammox bacteria (section 20.4)* ▶ *Wastewater treatment processes (section 42.2)*

Comprehension Check

1. Organic nitrogen introduced into an ecosystem directly by nitrogen fixation is sometimes called "new" nitrogen. Explain why.
2. Describe the two-step process that constitutes nitrification. Could microbes performing nitrification and anammox live in the same environment? Why or why not?
3. What is the difference between assimilatory nitrate reduction and denitrification? Which reaction is performed by many microbes, and which is a more specialized metabolic capability?

Phosphorus Cycle

Biogeochemical cycling of phosphorus is important for a number of reasons. All living cells require phosphorus for ATP, nucleic acids, and some lipids and polysaccharides. For many years, it was believed that phosphorus was derived solely from the weathering of phosphate-containing rocks and assimilated as phosphate (**figure 28.5**). However, phosphonates, which bear a C—P bond and are thus organic, appear to be significant sources of phosphorus for some marine microorganisms.

In soil, phosphorus exists in both inorganic and organic forms. Organic phosphorus includes not only that found in biomass but also that in materials such as humus and other organic compounds. The phosphorus in such organic materials is readily recycled by microbial activity. Inorganic phosphorus is negatively charged, so it complexes with positively charged elements in the environment, such as iron, aluminum, and calcium. These compounds are relatively insoluble, and their dissolution is pH dependent such that phosphate is most available to plants and microbes between pH 6 and 7. Microbes transform simple orthophosphate (PO_4^{3-}) to more complex forms. These include the polyphosphates present as inclusions, as well as more familiar macromolecules such as nucleotides and phospholipids. ◀ *Inclusions (section 3.6); Phosphorus assimilation (section 12.6)*

Sulfur Cycle

Microorganisms contribute greatly to the sulfur cycle; a simplified version is shown in **figure 28.6.** The sulfur cycle is similar

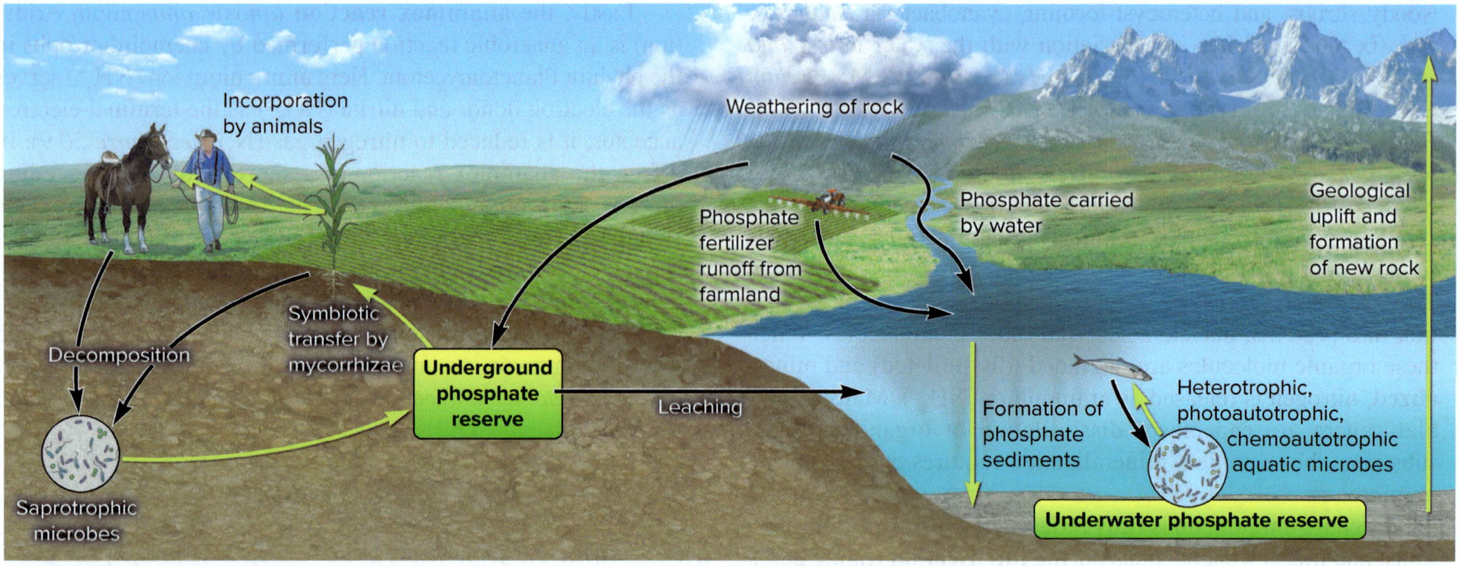

Figure 28.5 A Simplified Phosphorus Cycle. Phosphorus enters soil and water through the degradation of plants and animals, weathering of rocks, and fertilizer application. Phosphate in soil and phosphate, phosphite, and phosphonates in waters are consumed by terrestrial and aquatic microbes and passed on to larger organisms. However, much of the soil phosphorus can leach great distances or complex with cations to form relatively insoluble compounds. Terrestrial chemolithotrophic microbes also take up soil phosphate but are not shown.

to the nitrogen cycle in that, depending on the oxidation state of the sulfur species, it may serve as an electron acceptor, an electron donor, or both. Sulfate, the fully oxidized form, is reduced by plants and microbes for use in amino acid and protein biosynthesis during **assimilatory sulfate reduction.** By contrast, when sulfate diffuses into anoxic habitats, it provides an opportunity for microbial **dissimilatory sulfate reduction.** Here sulfate serves as a terminal electron acceptor during anaerobic respiration by a variety of bacteria, including *Desulfovibrio* and *Desulfonema* spp. and archaea belonging to the genus *Archaeoglobus*.

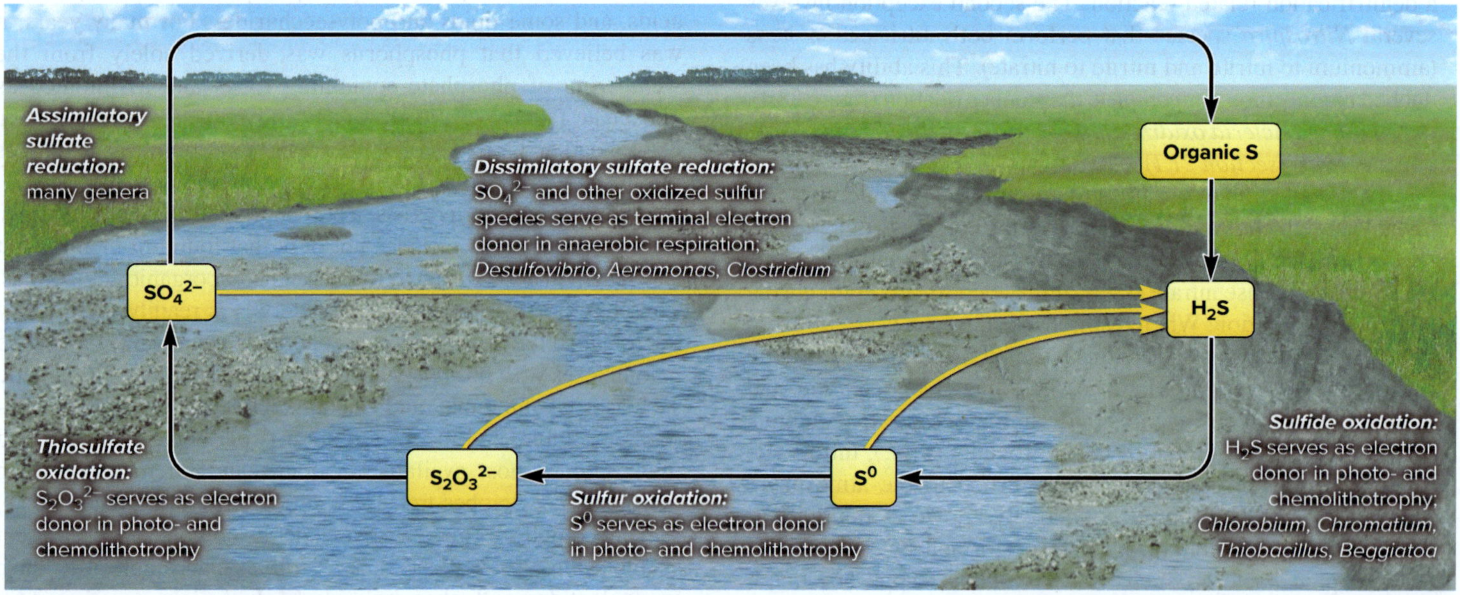

Figure 28.6 A Simplified Sulfur Cycle. Sulfate is fully oxidized; thus it serves as the terminal electron acceptor in anaerobic respiration, termed dissimilatory sulfate reduction. Alternatively, plants and microbes can donate the S atom to organic compounds during assimilatory sulfate reduction. H_2S is fully reduced, so it can serve as an electron donor during chemo- and photolithotrophy. Because elemental sulfur and thiosulfate are neither fully oxidized nor reduced, they can serve as either electron donors or acceptors.

MICRO INQUIRY *What is the role of elemental sulfur (S^0) when it is being reduced? What is its role when it is being oxidized?*

This results in sulfide accumulation and explains the odor of rotten eggs at a salt marsh during low tide. Sulfide, which is fully reduced, can then serve as an electron source for anoxygenic photosynthetic microorganisms such as green sulfur bacteria and chemolithoautotrophs including *Thiobacillus* spp. These microbes convert sulfide to elemental sulfur and sulfate.

Other microorganisms have been found to carry out dissimilatory elemental sulfur (S^0) reduction. These include members of the genus *Desulfuromonas,* thermophilic archaea, and cyanobacteria in hypersaline sediments. Sulfite (SO_3^{2-}) is another critical intermediate that can be reduced to sulfide by a wide variety of microorganisms, including members of the genera *Alteromonas, Clostridium, Desulfovibrio,* and *Desulfotomaculum.*

Dimethylsulfoniopropionate (DMSP) is an important organic sulfur compound. It is produced by marine phytoplankton as a compatible solute. When these cells die and release DMSP, bacterioplankton (floating bacteria) use it as a sulfur and carbon source. In the process, it is metabolized to dimethylsulfide (DMS) and released into the atmosphere. There DMS is rapidly converted into a variety of sulfur compounds that serve as nuclei for water droplet formation, contributing to the formation of clouds. Because clouds help keep the Earth's surface cool, it is possible that seeding clouds with DMS could help mitigate the effects of global climate change.

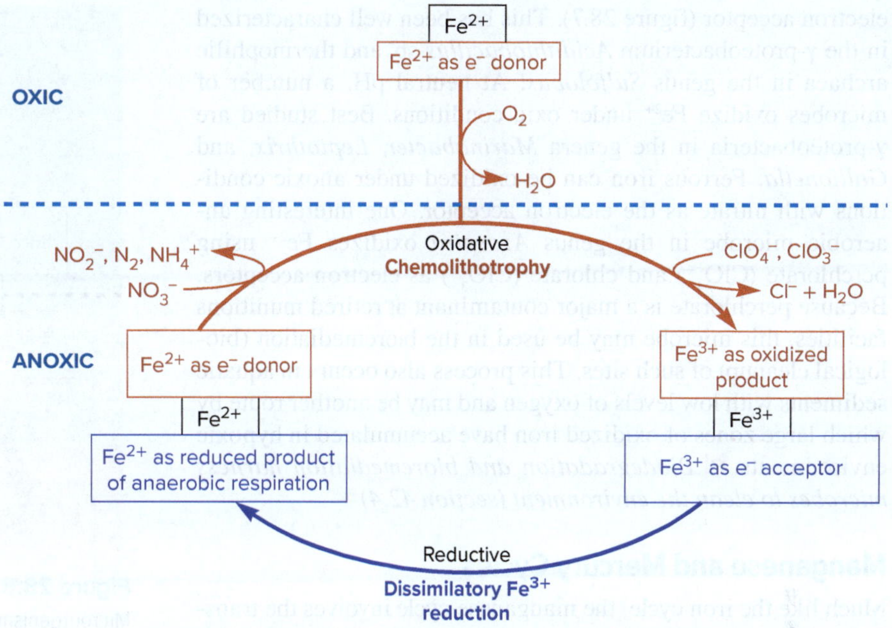

Figure 28.7 The Iron Cycle. Dissimilatory reduction of iron takes place when Fe^{3+} is used as a terminal electron acceptor during anaerobic respiration. This occurs only in anoxic environments. By contrast, oxidation of Fe^{2+} can occur under both oxic and anoxic conditions. Fe^{2+} can be used as an electron donor with O_2 as electron acceptor, although some lithotrophs use NO_3^- or even the environmental contaminants chlorate (ClO_3^-) and perchlorate (ClO_4^-) as electron acceptors.

MICRO INQUIRY *Why is Fe^{2+} used as an electron donor chiefly in acidic environments?*

Iron Cycle

The iron cycle features the interchange of ferrous iron (Fe^{2+}) to ferric iron (Fe^{3+}) (**figure 28.7**). In environments that are fully aerated and have a neutral pH, iron is present primarily as insoluble minerals of either oxidation state. The solubility of both reduced, ferrous iron (Fe^{2+}) and oxidized, ferric iron (Fe^{3+}) increases with acidification so that below pH 4, iron is found in the aqueous form. Its assimilation presents two challenges for most aerobic microorganisms. First, while iron is an essential element, free iron is usually present in very small quantities. Second, iron typically exists in the wrong oxidation state for uptake: Fe^{3+} dominates in oxic environments, but microbes generally incorporate Fe^{2+}. The use of siderophores solves both problems. **Siderophores** are low molecular weight organic molecules that bind Fe^{3+}, facilitating its transport into the cell, where it is reduced to Fe^{2+} (*see figure 3.13*). ◀ *Iron uptake (section 3.2)*

Once again, we see that dissimilatory reduction involves anaerobic respiration. In the case of dissimilatory iron reduction, ferric iron (Fe^{3+}) serves as a terminal electron acceptor during anaerobic respiration. In most environments, Fe^{3+} is found chiefly in a crystalline phase (e.g., hematite and magnetite) and as a component of sediment clays. This helps explain why some microbes donate electrons from the electron transport chain to these solid forms of Fe^{3+} outside the cell. ◀ *Anaerobic respiration uses the same steps as aerobic respiration (section 11.5)*

Both archaea and bacteria are capable of dissimilatory Fe^{3+} reduction. This phylogenetic diversity may reflect the antiquity of Fe^{3+} reduction. It is believed that life began 3.5 to 3.8 billion years ago in an environment rich in Fe^{2+}. The photooxidation of Fe^{2+} to Fe^{3+} and H_2 would have provided an electron acceptor and energy source, respectively, to early cellular forms. Banded iron formation, which feature alternating layers of iron oxides and iron-poor chert, occurred when atmospheric oxygen levels were beginning to increase at the end of the Precambrian era. These may be evidence of increased bacterial iron metabolism. ◀ *Microbes have evolved and diversified for billions of years (section 1.2)*

In addition to ferric iron (Fe^{3+}) as a terminal electron acceptor, some magnetotactic bacteria transform extracellular iron to the mixed valence iron oxide mineral magnetite (Fe_3O_4) and construct intracellular magnetic compasses (*see figure 3.30*). These bacteria are more precisely magnetoaerotactic because they are thought to use magnetic fields to migrate to the position in a bog or swamp where the oxygen level best meets their needs. Furthermore, some dissimilatory iron-reducing bacteria accumulate magnetite as an extracellular product. ◀ *Inclusions (section 3.6)*

In the reduced state, ferrous iron (Fe^{2+}) can be used as an electron donor by lithotrophic microbes in acidic, oxic environments where it is soluble and oxygen can serve as the terminal electron acceptor (figure 28.7). This has been well characterized in the γ-proteobacterium *Acidithiobacillus* sp. and thermophilic archaea in the genus *Sulfolobus*. At neutral pH, a number of microbes oxidize Fe^{2+} under oxic conditions. Best studied are γ-proteobacteria in the genera *Marinobacter, Leptothrix,* and *Gallionella*. Ferrous iron can be oxidized under anoxic conditions with nitrate as the electron acceptor. One interesting anaerobic microbe in the genus *Azospira* oxidizes Fe^{2+} using perchlorate (ClO_4^-) and chlorate (ClO_3^-) as electron acceptors. Because perchlorate is a major contaminant at retired munitions facilities, this microbe may be used in the bioremediation (biological cleanup) of such sites. This process also occurs in aquatic sediments with low levels of oxygen and may be another route by which large zones of oxidized iron have accumulated in hypoxic environments. ▶| *Biodegradation and bioremediation harness microbes to clean the environment (section 42.4)*

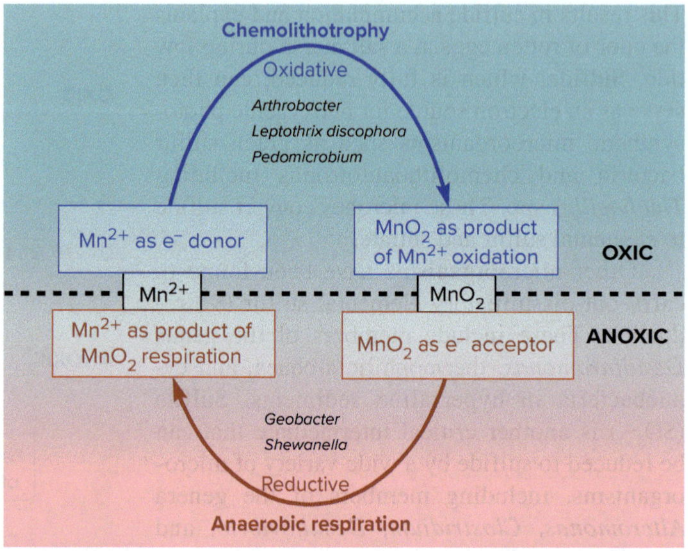

Figure 28.8 The Manganese Cycle, Illustrated in a Stratified Lake. Microorganisms make many important contributions to the manganese cycle. After diffusing from anoxic (pink) to oxic (blue) zones, manganous ion (Mn^{2+}) is oxidized chemically and by many morphologically distinct microorganisms in the oxic water column to manganic oxide—MnO_2, valence equivalent to 4+. When the MnO_2 diffuses into the anoxic zone, bacteria such as *Geobacter* spp. and *Shewanella* spp. carry out the complementary reduction process. Similar processes occur across oxic/anoxic transitions in soils, muds, and other environments.

Manganese and Mercury Cycles

Much like the iron cycle, the manganese cycle involves the transformation of the reduced form, manganous ion (Mn^{2+}), to the oxidized species, MnO_2 (equivalent to manganic ion [Mn^{4+}]). This occurs at the oxic-anoxic interface in hydrothermal vents, bogs, and stratified lakes (**figure 28.8**). Phylogenetically diverse bacteria use Mn^{2+} as an electron donor with oxygen or nitrate at the terminal electron acceptor. *Shewanella* spp., *Geobacter* spp., and other chemoorganotrophs can carry out the complementary manganese reduction process, using MnO_2 as an electron acceptor.

The mercury cycle illustrates characteristics of metals that can be methylated. Mercury compounds were once widely used in industrial processes. Inorganic mercury that accumulates in anoxic sediments is methylated by anaerobic bacteria of the genus *Desulfovibrio* as well as iron-reducing bacteria and methanogens (**figure 28.9**). Methylated mercury forms are volatile and

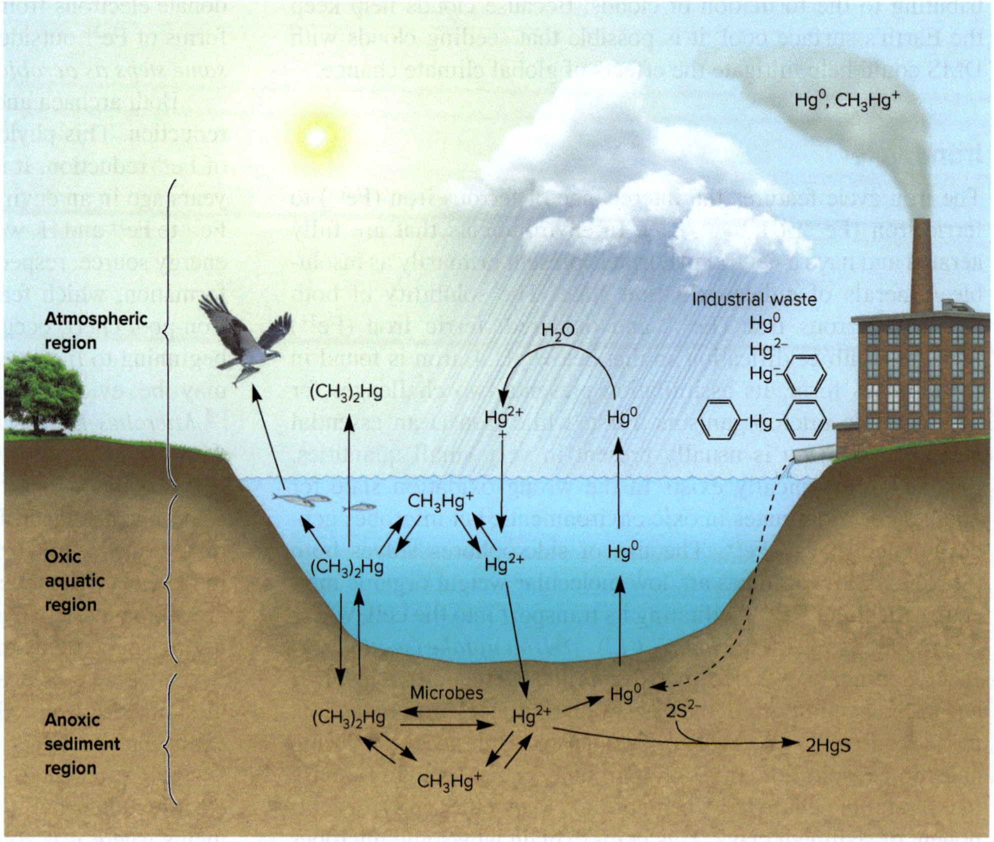

Figure 28.9 The Mercury Cycle. Interactions between the atmosphere, oxic water, and anoxic sediment are critical. Microorganisms in anoxic sediments, primarily *Desulfovibrio* spp., can transform mercury to methylated forms that can be transported to water and the atmosphere. These methylated forms also undergo biomagnification. The production of volatile elemental mercury (Hg^0) releases this metal to waters and the atmosphere. Sulfide, if present in the anoxic sediment, can react with ionic mercury to produce less soluble HgS.

lipid soluble, and mercury concentrations increase with each link of the food web in a process known as **biomagnification.** When fish and shellfish containing high levels of methylated mercury are ingested by humans, severe neurological disorders, particularly in children, can occur. This danger was tragically demonstrated during the 1950s in Japan when large-scale mercury poisoning occurred in the Minamata Bay region because industrial mercury was released into the marine environment.

Interaction Between Elemental Cycles

So far, we have presented biogeochemical cycling as discrete nutrient fluxes. However, it is important to understand that biogeochemical cycling involves dynamic, interconnected processes that over geologic time keep the biosphere in a self-sustaining steady state. Although links between cycles occur at a global scale, it is easiest to consider such links at the level of a single type of microbe. Here we present two illustrative examples: chemolithotrophy and nitrogen fixation.

When a community of chemolithotrophic microbes uses ammonium (NH_4^+) and nitrite (NO_2^-) as its source of electrons and CO_2 as its sole source of carbon, it contributes to the flux of both carbon and nitrogen by producing organic carbon and nitrate, which can be taken up by assimilatory nitrate reduction or returned to the atmosphere following denitrification. Because the use of reduced inorganic compounds such as ammonium as a source of electrons generates a smaller proton motive force than the use of NADH, chemolithoautotrophs must oxidize large amounts of these compounds to generate enough ATP and NADPH for carbon fixation. In this way, chemolithoautotrophs significantly impact nitrogen, sulfur, or iron (or other metal) cycles while removing, or "drawing down," CO_2 from the atmosphere.

In the case of nitrogen fixation, bacteria and archaea convert gaseous N_2 to ammonia, which is then incorporated into organic nitrogen molecules such as amino and nucleic acids. In many environments, the amount of carbon taken up by plants and microbes (either as CO_2 or as dissolved organic carbon) is dictated by how much nitrogen is available to support growth. Thus by introducing "new" nitrogen into the system, nitrogen-fixing microbes can also increase the rate of carbon flux. These examples present important concepts when considering the removal of CO_2 from the atmosphere.

Comprehension Check

1. Compare the products of dissimilatory sulfate reduction to those of dissimilatory nitrate reduction.
2. How might iron have been important in the evolution of early life?
3. What are some important microbial genera that contribute to manganese cycling?
4. How can microbial activity render some metals more or less toxic to animals?
5. Suggest how a chemolithoautotroph might link the carbon and sulfur cycles.

28.3 Global Climate Change: Biogeochemical Cycling Out of Balance

After reading this section, you should be able to:

a. Differentiate between the terms *global warming* and *global climate change* and identify greenhouse gases
b. Explain the link between greenhouse gas accumulation and microbial cycling of carbon and nitrogen
c. Discuss the origin of the greenhouse gases CO_2, methane, and nitrogen oxide species

Microbial activity is critical in maintaining the dynamic equilibrium that defines our biosphere. The capacity of microbes to thrive in seemingly every niche on Earth is the result of 3.5 billion years of microbial evolution. Importantly, changes in the physical and chemical environment to which microbes have adapted generally occurred over geological time scales. However, since the beginning of the twentieth century, the rate at which CO_2 and other so-called greenhouse gases have entered the atmosphere has been much faster than at any other time in the known history of life on Earth. Atmospheric gases such as CO_2, CH_4, and nitrogen oxides are called **greenhouse gases** because they trap heat reflected from Earth's surface in the atmosphere, rather than allowing it to radiate into space. The rate at which these gases are entering the atmosphere exceeds the rate by which the natural carbon and nitrogen cycles can remove them. It is the ever-increasing atmospheric concentration of these gases that has resulted in what is sometimes called global warming. However, the term *global climate change* more accurately reflects the current changes in patterns of wind, precipitation, and ocean and atmospheric currents and temperatures.

The most abundant greenhouse gas is CO_2. Since the onset of the industrial era, about 150 years ago, global CO_2 levels have risen from 278 ppm (parts per million) to over 420 ppm (**figure 28.10**). Most of this can be attributed to the burning of fossil fuel, which in the 1960s released about 10 billion tons of CO_2 each year. By 2024, this increased to over 35 billion tons annually. Changes in land use management, principally through deforestation, account for another 2.5 billion tons of CO_2 annually. To understand why these activities result in the accelerated release of CO_2, consider that fossil fuel (e.g., oil) is formed from decaying organic matter. This organic matter arose through millions of years of CO_2 fixation, slowly removing CO_2 from the atmosphere. It has been locked deep within the Earth for millennia. When fossil fuels are burned, the energy of combustion oxidizes this material back to CO_2 in a tiny fraction of the time it took to accumulate. The equilibrium of carbon flux is lost because the rate of atmospheric inputs (CO_2 release) greatly exceeds the rate of removal (CO_2 fixation).

The situation is different with deforestation. Forests are generally regarded as CO_2 sinks because the abundance of

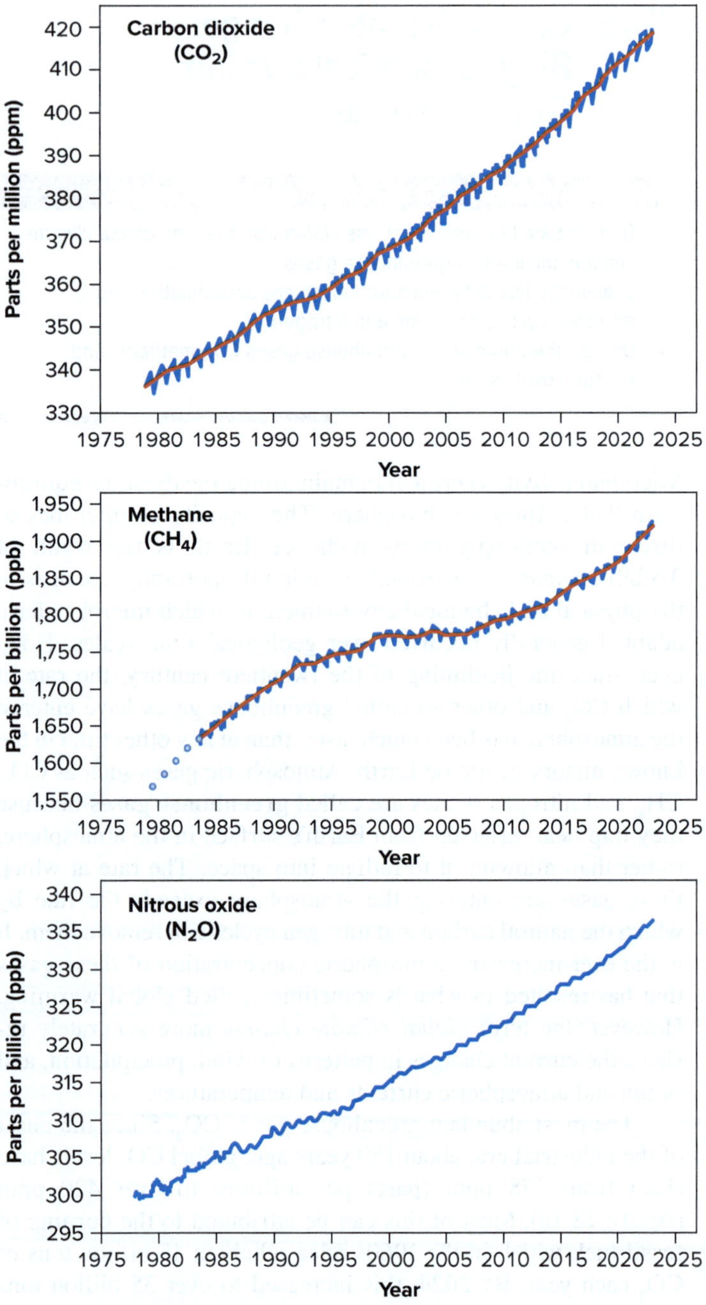

Figure 28.10 Global Atmospheric Concentrations of Greenhouse Gases.
Source: Butler, James, & Montzka, Stephen. "The NOAA Annual Greenhouse Gas Index (AGGI)." 2024.

photoautotrophic plants pulls a great deal of CO_2 out of the atmosphere, converting it to biomass. Indeed, during the 1990s terrestrial ecosystems sequestered about 3 gigatons of carbon per year (a gigaton is 1 billion metric tons). Thus when forests are cut down for firewood or agriculture, the Earth loses a carbon sink. While some speculate that other, remaining plants will assimilate the extra CO_2 and simply grow faster, accelerated rates of soil microbial respiration complicate matters. Recall that as respiration rates increase, so does the quantity of CO_2, CH_4, and N_2O released. In fact, warming permafrost in subarctic regions has dramatically increased soil microbial respiration, and the release of the 1,700 billion tons of carbon buried in permafrost for thousands of years has begun.

In addition to CO_2, methane is a greenhouse gas of increasing concern because it has about 30 times the global warming potential of CO_2. Based on analyses of gas bubbles in glacier ice cores, the levels of methane in the atmosphere remained essentially constant until about 150 years ago, reflecting the balance of abiotic and archaeal methanogensis with microbial methane oxidation (figure 28.3). Since then, the release of methane has increased 2.5-fold to about 1,930 parts per billion (ppb) (figure 28.10). Interestingly, agriculture (e.g., methanogenesis in cattle rumen and rice paddies) is responsible for over half of all anthropogenic CH_4 (*see chapter 19 chapter opener*). Considering these trends, there is worldwide interest in understanding the factors that control methane production and oxidation by microorganisms.

The nitrogen oxide greenhouse gases NO and N_2O, collectively referred to as "Nox," have about 280 times the global warming potential of CO_2. To understand the origin of rising Nox, we need to examine food production. The "green revolution" of the mid-twentieth century introduced human-made fertilizer as a key means of ensuring that agricultural output could keep pace with population growth. However, manufacturing fertilizer is an energy-intensive process that uses hydrogen gas to reduce N_2 to NH_4^+ at high temperature and pressure. This process was first described by Fritz Haber and Karl Bosch in 1913, but it was not until the middle of the twentieth century that the Haber-Bosch reaction was used at an industrial level. Currently, about 200 million metric tons of fertilizer is used per year, which is about 500 times more than in 1940. In contrast to the positive outcome of increased agricultural output, the addition of so much ammonium to the soil has drastically altered the balance of the nitrogen cycle.

The problem is that not all the NH_4^+ is taken up by plants. The extra NH_4^+ generally has one of two fates: runoff or nitrification followed by denitrification (**figure 28.11**). Ammonium runoff leaches into lakes and streams, frequently causing **eutrophication**—an increase in nutrient levels that stimulates the growth of a limited number of organisms, thereby disturbing the ecology of these aquatic ecosystems (*see figure 29.3*). By contrast, microbial nitrification results in the oxidation of ammonium to more nitrate than can be immobilized by plants and microbes because organisms need a specific ratio of C:N:P. The extra nitrate is reduced to N_2 and reactive greenhouse Nox gases through the process of denitrification. Cycles of nitrification/denitrification fueled by NH_4^+ fertilizer are responsible for the highest N_2O levels in 650,000 years.

What are the consequences of disrupting the carbon and nitrogen cycles? Global climate change is measured over decades and includes many parameters such as surface temperature on land and sea, and in the atmosphere; rates of precipitation; and frequency of extreme weather. Based on these analyses, the average global temperature has increased about 1.2°C since 1880, with two-thirds of the increase occurring since 1975 (**figure 28.12**).

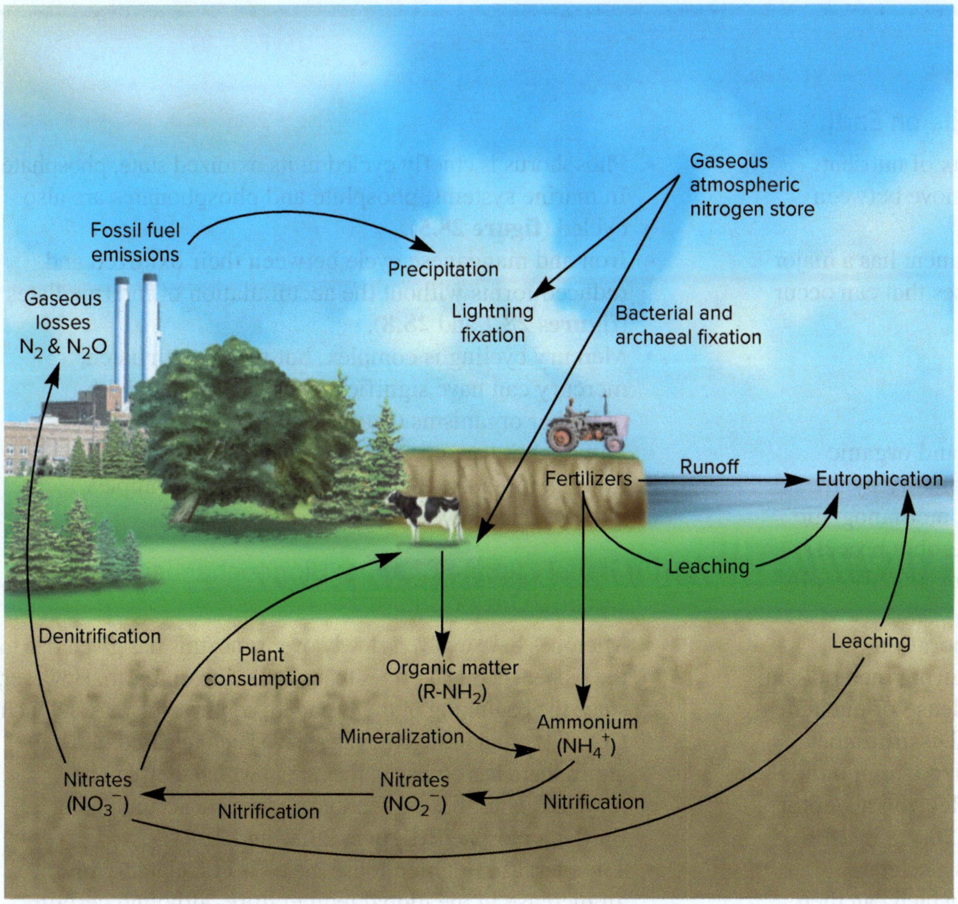

Figure 28.11 **Natural and Human-Made Influences on the Nitrogen Cycle.**

MICRO INQUIRY *What organisms benefit from nitrification?*

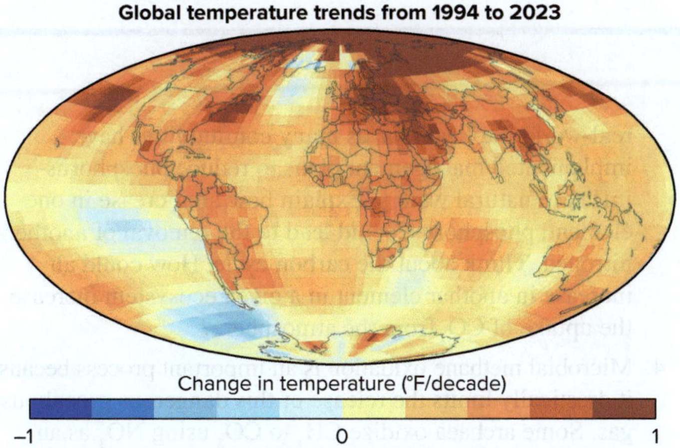

Figure 28.12 **Global Temperature Trends in the Past 30 Years.**

Source: NOAA

However, the rate of warming has not been uniform globally, with certain areas getting hotter faster. For example, the arctic is warming at twice the rate of the rest of the globe. Melting glaciers are contributing to sea level rise, which has made some communities and islands uninhabitable. Depending on the rate of continued increase in greenhouse gases, the average global surface temperature is predicted to rise, as much as 4.5°C by 2100. This would be accompanied by an average 1-meter increase in sea level and the displacement of about 680 million people (about twice the current U.S. population) from impacted coastal regions.

An important question is how will microbes respond to a changing world. For the vast majority of Earth's history, microorganisms have been the drivers of balanced elemental cycling, so changes in microbial activities will have a major impact on the rate and magnitude of greenhouse gas accumulation and global climate change. As introduced in chapters 29 and 30, the role microbes play in balancing carbon and nitrogen fluxes has opened new avenues of research and discovery in microbial ecology.

Comprehension Check

1. List three greenhouse gases. Discuss their origins.
2. Discuss the possible role of forests in controlling atmospheric concentrations of CO_2.
3. How do changes in the nitrogen cycle caused by fertilization influence the carbon cycle?
4. Given that each microbial group has an optimum temperature range for growth, how might you predict changes to a soil microbial community living in your geographic area?

Key Concepts

28.1 Biogeochemical Cycling Sustains Life on Earth

- Biogeochemical cycling involves the flux of nutrients from one pool to another. As nutrients move between pools, they are chemically modified.
- The redox potential of the local environment has a major impact on the type of metabolic processes that can occur (**figure 28.1**).

28.2 Microbes Mediate Nutrient Cycling

- The cycling between inorganic (CO_2) and organic carbon occurs aerobically and anaerobically by all types of microbes (**figure 28.2**). Only methanogenic archaea can produce methane, an anaerobic process. Methane can be oxidized aerobically or anaerobically (**figure 28.3**).
- Assimilatory processes involve incorporation of nutrients into the organism's biomass during metabolism; dissimilatory processes involve the release of nutrients to the environment after metabolism. Dissimilatory reduction involves the use of the oxidized compound (e.g., NO_3^- or SO_4^{2-}) as the terminal electron acceptor during anaerobic respiration (**figures 28.4** and **28.6**).
- Chemolithotrophs use reduced elements as electron donors and generate the oxidized form, which can then be used for assimilatory or dissimilatory processes (**figures 28.4** and **28.6**).
- Phosphorus is chiefly cycled in its oxidized state, phosphate. In marine systems, phosphite and phosphonates are also cycled (**figure 28.5**).
- Iron and manganese cycle between their oxidized and reduced forms without the accumulation of intermediates (**figures 28.7** and **28.8**).
- Mercury cycling is complex, but when methylated, mercury can have significant effects on the health of higher organisms (**figure 28.9**).
- Although it is convenient to consider a single element cycling, in nature there are extensive and complex interactions between nutrient cycles.

28.3 Global Climate Change: Biogeochemical Cycling Out of Balance

- Microbes have evolved slowly over time; this has given rise to the biogeochemical cycles that sustain life on Earth. However, since the beginning of the twentieth century, the rate at which CO_2, CH_4, and nitrogen oxides have been released into the atmosphere has outpaced the rate at which they can be recycled. Thus these greenhouse gases are accumulating in Earth's atmosphere (**figures 28.10** and **28.11**).
- The increase in greenhouse gases is correlated with an increase in the global temperature, although certain regions, such as the arctic, are warming faster than others (**figure 28.12**).

Active Learning

1. Examine figure 28.11. Farmers in Argentina use a crop/cattle rotation system in which cattle graze on pastures for about 5 years. Cattle are then moved to a different area and crops are planted where they had grazed. Crops are grown for 3 years, and cattle are rotated back and the field reverts to a pasture. This dramatically reduces the amount of nitrogen fertilizer that must be added when crops are grown. Discuss why this practice minimizes the release of greenhouse gases.
2. A bacterium isolated from sewage sludge was recently found to be capable of anoxygenic photosynthesis using nitrite as the electron donor, converting it to nitrate. Compare this form of nitrification to that which is well characterized. By removing nitrite from the environment, what other processes within the nitrogen cycle might this bacterium be influencing?
3. It has been shown that the addition of phosphorus to lake ecosystems promotes excess nitrogen removal. This has real-world implications, as many communities have implemented management plans to reduce phosphorus influx to natural waters. Explain how an increase in one element, phosphorus, could lead to the removal of another, nitrogen. Think about the carbon cycle. How could an increase in another element in a given ecosystem increase the uptake of CO_2 from the atmosphere?
4. Microbial methane oxidation is an important process because it drastically limits the release of this dangerous greenhouse gas. Some archaea oxidize CH_4 to CO_2 using NO_3^- as an electron acceptor to generate N_2. Others reduce Fe(III) to Fe(II) during methane oxidation. Refer to figures 28.2, 28.3, 28.4, and 28.7 to review the parts of the carbon, nitrogen, and iron cycles involved. If you mixed the two isolates with equal amounts of nitrate and Fe(III), which methane-oxidizing archaeon do you think would grow fastest? Why?

CHAPTER

29

Microorganisms in Marine and Freshwater Ecosystems

NASA Photo/Alamy Stock Photo

Ocean Death Coming Soon to a Coast Near You

If you live near a coast, you're probably not far from an ocean dead zone—a region where all plant and animal life dwelling below the surface has suffocated. These regions have been increasing in size and frequency since the 1960s, and could occupy as much as 20% of the world's oceans by 2100. Dead zones result when phytoplankton respond to increases in nutrients and temperature by growing (and growing and growing) and then sinking. Initially, phytoplankton on the ocean floor means more food for heterotrophic organisms. But as the number of oxygen-consuming animals and microbes increases, available oxygen dwindles. Invertebrates and fish die. Plants cannot grow because there is not enough light. The only organisms left are anaerobic microbes.

Of the more than 500 ocean dead zones globally covering over 245,000 square kilometers, the Baltic Sea is the largest, while the second largest is the mouth of the Mississippi River. The addition of literally tons of extra nitrogen and phosphorus to the Gulf of America (also called Gulf of Mexico) has resulted in a dead zone that is slightly bigger than New Jersey as shown in the sattelite image, above. The presence of such a large dead zone has huge and obvious implications for fisheries, seabirds, tourism, and the overall health of the Gulf.

Are dead zones reversible? In some cases, yes. But like most problems, prevention is best. Prudent use of fertilizer and limiting surface runoff are good places to start. Another area of concern is the deposition of animal waste into local waterways. Finally, proper sewage treatment (*section 42.2*) and the restoration of natural wetlands to filter and consume much of the nitrogen and phosphorus before it flows to the coast are projects worth pursuing.

In this chapter, we turn our attention to the microbial ecology of marine and freshwater environments. **Oceanography** is the study of marine systems and the biological, physical, geological, and chemical factors that impact biogeochemical cycling, water circulation, and climate. Microbial oceanography seeks to understand marine microorganisms at the level of the individual taxon and at the same time elucidate how they influence biogeochemical cycling at a global level. **Limnology** is the investigation of aquatic systems within continental boundaries, including glaciers, groundwater, rivers, streams, and wetlands. All terrestrial life depends on the sustained provision of clean freshwater from these ecosystems.

Readiness Check:

Based on what you have learned previously, you should be able to:

- ✓ Diagram the viral life cycle (section 6.3)
- ✓ List environmental factors that influence microbial growth (section 7.5)
- ✓ Compare and contrast the growth and physiology of obligate aerobes, facultative anaerobes, and strict anaerobes (section 7.5)
- ✓ Explain why and how microbes frequently grow in biofilms (section 7.6)
- ✓ Review innovative methods for growing microbes isolated from nature in the laboratory (section 26.2)
- ✓ Explain how the redox state of an environment impacts the types of microbial metabolic processes found there (section 28.1)
- ✓ Explain the global impact that microbes have on carbon and nitrogen cycling (section 28.2)
- ✓ Identify, by nutritional type, microorganisms responsible for carbon fixation, nitrogen fixation, nitrification, dissimilatory nitrate reduction, denitrification, nitrification, the anammox reaction, and sulfur reduction (sections 19.1, 20.3, 20.4, 20.8, 21.1, and 21.2)

29.1 Water Is the Largest Microbial Habitat

After reading this section, you should be able to:

a. Explain the carbonate equilibrium system
b. Discuss the cause and consequences of ocean acidification

Microbes drive marine and freshwater ecosystems. The importance of microbial activities in oceans, rivers, and lakes has been brought into sharp focus by global climate change. To understand the capacity of microbes to influence the worldwide flux of nutrients and greenhouse gases, we must first consider marine and freshwater habitats.

The nature of water as a microbial habitat depends on a number of physical factors such as temperature, pH, and light penetration. One of the most important factors is dissolved oxygen. The flux rate of oxygen through water is about 10,000 times less than its rate through air. However, in some habitats, the limits to oxygen diffusion can be offset by a combination of global seawater circulation, microbial consumption patterns, and increased solubility of oxygen at colder temperatures and higher pressures. For instance, in the very deep ocean, the dissolved oxygen concentration actually increases with depth, even though the air-water interface may be literally miles away. On the other hand, tropical lakes, summertime temperate lakes, and coastal regions may become oxygen limited only meters below the surface.

As seen in the development of dead zones (chapter opening story), light is also critical for the health of marine and freshwater ecosystems. Like all life on Earth, all organisms in these environments depend on **primary producers:** autotrophic organisms that fix CO_2, thereby providing organic carbon. In streams, lakes, and estuary systems, macroscopic algae and plants are the chief primary producers. Organic carbon also enters these systems in terrestrial runoff. The situation is very different in the open ocean, where all organic carbon is the product of microbial autotrophy. In fact, about half of all **primary production** (carbon fixation) on Earth is the result of marine microbial carbon fixation.

Solar radiation warms the water, and this can lead to thermal stratification. Warm water is less dense than cool water, so as the sun heats the surface in tropical and temperate waters, a **thermocline** develops. A thermocline can be thought of as a mass of warmer water floating on top of cooler water. These two water masses remain separate until there is either a substantial mixing event, such as a severe storm, or in temperate climates, the onset of autumn. As the weather cools, the upper layer of warm water becomes cooled and the two water masses mix. This is often associated with a pulse of nutrients from the lower, darker waters to the surface. As discussed in section 29.3, this pulse of nutrients can trigger a sudden and rapid increase in the population of certain microbes (i.e., a bloom).

Because CO_2 is critical in many chemical and biological processes, it is important to understand its fate in freshwater and marine ecosystems. The pH of unbuffered distilled water is determined by dissolved CO_2 in equilibrium with the air and is typically acidic, at pH 5.0 to 5.5. The pH of freshwater systems such as lakes and streams, which are usually only weakly buffered, is therefore controlled by terrestrial input (e.g., minerals that may be either acidic or alkaline) and the rate at which CO_2 is removed by plants and photoautotrophic microorganisms. The loss of CO_2 by its fixation can increase the pH of freshwater.

In contrast, seawater is strongly buffered by the balance of CO_2, bicarbonate (HCO_3^-), and carbonate (CO_3^{2-}). Atmospheric CO_2 enters the oceans and either is converted to organic carbon by photosynthesis or reacts with seawater to form carbonic acid (H_2CO_3), which quickly dissociates to form bicarbonate and carbonate (**figure 29.1**):

$$CO_2 + H_2O \rightleftharpoons H_2CO_3 \rightleftharpoons H^+ + HCO_3^- \rightleftharpoons 2H^+ + CO_3^{2-}$$

The oceans are therefore alkaline, buffered between pH 7.6 and 8.2 by this **carbonate equilibrium system.** Much like the buffer one might use in a chemistry experiment, the pH of seawater is determined by the relative concentrations of bicarbonate and carbonate.

The reactions of the carbonate equilibrium system have taken on new importance as increases in atmospheric CO_2 result in higher rates of CO_2 dissolution in seawater. In fact, the ocean contains 50 times as much carbon as the atmosphere and is already absorbing about 30% of the anthropogenically produced CO_2. Much of this carbon is converted to the exoskeletons of microscopic plankton and is deposited on the seafloor in a *carbonate ooze* (figure 29.1).

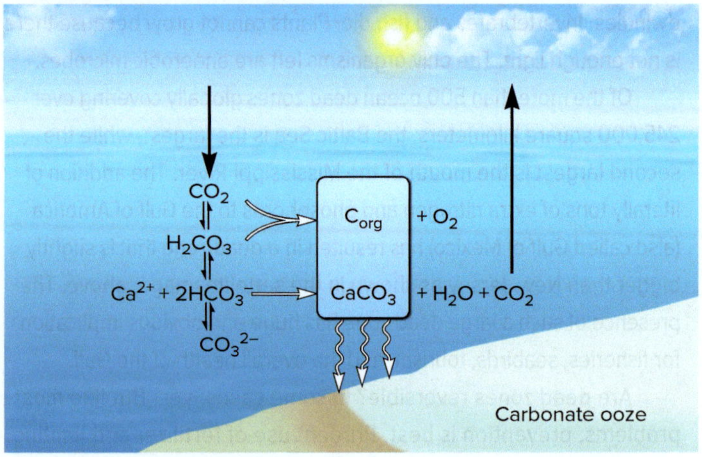

Figure 29.1 The Carbonate Equilibrium System. Atmospheric CO_2 enters seawater and is converted to organic carbon (C_{org}) or to carbonic acid (H_2CO_3) that spontaneously dissociates into the weak acids bicarbonate (HCO_3^-) and carbonate (CO_3^{2-}). Calcium carbonate ($CaCO_3$), a solid, precipitates to the seafloor, where it helps form a carbonate ooze. This system removes CO_2 from the atmosphere and keeps seawater buffered at about pH 8.0.

MICRO INQUIRY *Why does the production of $CaCO_3$ serve to reduce atmospheric CO_2 concentrations?*

Figure 29.2 Coccolithophores Respond to Seawater Acidification. *Gephyrocapsa huxleyi*, an abundant species in the open ocean.
Steve Gschmeissner/Science Photo Library/Getty Images

However, the current input of carbon (as CO_2) is more than the oceans can sequester and has led to **ocean acidification.** The average pH of the ocean has dropped by 0.1 unit (to 8.1) since the preindustrial age. It is expected to decline another 0.35 to 0.50 by the year 2100, unless effective means of limiting CO_2 emissions are implemented. This is significant because as marine CO_2 increases, the carbonate equilibrium shifts away from $CaCO_3$, potentially limiting its availability for the formation of shells and skeletons of marine organisms. Because protists known as coccolithophores produce about a third of the marine $CaCO_3$, they have been the focus of investigations seeking to predict the effects of ocean acidification. While responses are species specific, it is clear that even slight shifts in pH have an impact on their ability to form their shells, called *coccoliths* (**figure 29.2**). Ocean acidification is also having a negative impact on corals. Although it is difficult to predict, some models estimate that when atmospheric CO_2 reaches about 550 ppm, many corals will begin to dissolve faster rather than they grow. If left unchecked, this threshold will be crossed in about 40 years. ◀ *Haptista clade (section 23.5); Global climate change: biogeochemical cycling out of balance (section 28.3)*

Comprehension Check

1. What factors influence oxygen solubility? Why is this important in considering aquatic environments?
2. Describe the buffering system that regulates the pH of seawater. What might be the implications of this stable buffering on microbial evolution?
3. How does primary production in lakes and coastal ecosystems differ from that in the open ocean?
4. What features of a thermocline make it similar to hot and cool masses of air?

29.2 Microorganisms in Marine Ecosystems

After reading this section, you should be able to:

a. Explain the ecological and economic importance of estuaries and harmful algal blooms
b. Diagram carbon and sulfur flux in a Winogradsky column
c. Assess the importance of marine microbial primary productivity in the global carbon budget
d. Illustrate the microbial loop and evaluate its importance in the open ocean
e. Evaluate the importance of members of the order Pelagibacterales in marine ecosystems
f. Describe at least three mechanisms microbes have evolved to adapt to the low nutrient conditions found in the open ocean
g. Explain the role of viruses in maintaining and influencing marine microbial populations, as well as their contribution to nutrient cycling
h. Explain the distribution of microbial metabolic types that inhabit marine subsurface sediment

As terrestrial organisms, we must remind ourselves that about 96% of Earth's water is in marine environments. From a microbiological perspective, surface ocean waters have been most intensely studied; only recently have scientists begun to probe the benthos—deep-sea sediments and the subsurface. We begin our discussion of marine ecosystems with estuaries, then consider the microbial communities that inhabit the open ocean and, finally, the dark, cold, high-pressure benthos.

Microorganisms in Coastal Ecosystems Are Adapted to a Changeable Environment

An estuary is a semi-enclosed coastal region where a river meets the sea. Estuaries are defined by tidal mixing between freshwater and saltwater. Many estuaries undergo large-scale tidal flushing; this forces organisms to adapt to changing salt concentrations on a twice-daily basis. Microbes that live under such conditions combat the resulting osmotic stress by adjusting their intracellular osmolarity to limit the difference with that of the surrounding water. Most protists and fungi produce compatible solutes such as dimethyl sulfonyl propionate for this purpose. Bacteria and archaea regulate internal concentrations of potassium or compounds such as proline and betaine. Salt is another stressor, and most microbes that inhabit estuaries are halotolerant, which is distinct from halophilic. **Halotolerant** microbes can withstand significant changes in salinity; halophilic microorganisms have an absolute requirement for high salt concentrations. ◀ *Solutes affect osmosis and water activity (section 7.5)*

Estuaries are unique in many respects. Their calm, nutrient-rich waters serve as nurseries for juvenile forms of many commercially important fish and invertebrates. However, despite their importance to the commercial fishing industry, estuaries are among the most polluted marine environments. Estuaries and the rivers that feed them are the receptacles of toxins discharged during industrial processes as well as agricultural and street runoff. These pollutants include inorganic materials, particularly fertilizer, that can be used as nutrients by resident microbes. If too many nitrates and phosphates are consistently added, the system is said to be **eutrophic,** and the rate of respiration greatly exceeds that of photosynthesis (*see figure 28.11*). Thus the water may become anoxic (chapter opening story). By contrast, if nutrients are introduced in a pulse-like fashion, a single photoautotrophic species, either algal or cyanobacterial, grows at the expense of all other organisms in the community. This phenomenon, called a **bloom,** often results from the introduction of nutrients combined with changes in light, temperature, and mixing of water masses. If the microbes produce a toxic product or are in themselves poisonous to other organisms such as shellfish or fish, the term **harmful algal bloom (HAB)** is used. Some HABs are called **red tides** because the microbial density is so great that the water becomes red or pink, the color of the algae (**figure 29.3**).

Climate change contributes to the recent increase in HABs in several ways. These include warming water temperatures, changing local salinity and nutrient runoff as a result of both flooding and drought, and altering the timing and intensity of coastal upwelling of deeper, nutrient-rich water. HABs are sometimes responsible for killing large numbers of fish or marine mammals. For instance, off the coast of California, periodic HABs are responsible for sudden, large-scale deaths of sea lions. These blooms are caused by diatoms in the genus *Pseudo-nitzschia*. Anchovies consume the diatoms, and a potent neurotoxin, domoic acid, accumulates in the fish. The mammals are poisoned after they ingest large quantities of these toxin-laden fish. ◀ *Stramenopila (section 23.4)*

Figure 29.3 Harmful Algal Blooms Are Often Called Red Tides. Blooms may color the water with the pigment of the microbe responsible. This HAB occurred in Puget Sound, Washington, the United States.

Don Paulson Photography/Purestock/Alamy Stock Photo

HABs are also often caused by dinoflagellates. Some bloom-causing dinoflagellates also produce neurotoxins. Dinoflagellates of the genus *Alexandrium* produce a toxin responsible for paralytic shellfish poisoning, which affects humans as well as other animals in coastal, temperate North America. The neurotoxin produced by an HAB of the dinoflagellate *Karenia brevis* killed a number of endangered manatees, sea turtles, and dolphins in 2019 in Florida. ◀ *Alveolata (section 23.4)*

Salt marshes generally differ from estuaries because their freshwater input is from multiple sources, rather than a single river. The microbial communities within salt marsh sediments are very dynamic. These ecosystems can be modeled in **Winogradsky columns** (**figure 29.4**), named after the pioneering microbial ecologist Sergei Winogradsky (1856–1953).

A Winogradsky column is easily constructed using a glass cylinder into which sediment is placed and then overlaid with seawater. When the top of the column is sealed, the cylinder becomes a closed system, supplied only with light energy. The top of the column is populated with cyanobacteria and photosynthetic protists (mostly diatoms) that perform oxygenic photosynthesis. The addition of shredded paper introduces a source of organic carbon for aerobic heterotrophs, which consume the oxygen produced by the cyanobacteria and diatoms. Just below

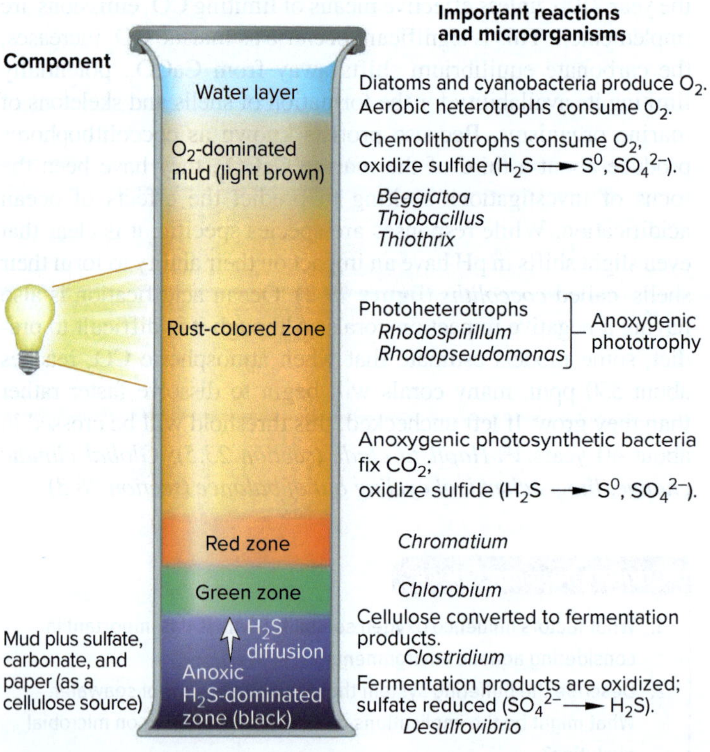

Figure 29.4 The Winogradsky Column. Microorganisms and nutrients interact over a vertical gradient in these microcosms.

MICRO INQUIRY *How is sulfur cycled between the anoxygenic photosynthetic microbes belonging to the genera* Chlorobium *and* Chromatium, *and anaerobic heterotrophic* Desulfovibrio *spp.?*

these heterotrophs, chemolithotrophs in the genera *Beggiatoa* and *Thiothrix* fix CO_2 to organic carbon, use sulfide as a source of electrons, and use oxygen as their terminal electron acceptor. The growth of these chemolithotrophs illustrates that in addition to cycling carbon, Winogradsky columns, like estuaries and salt marshes, have an active sulfur cycle. The consumption of oxygen by heterotrophs and chemolithotrophs introduces an oxygen gradient resulting in the growth of purple nonsulfur bacteria of the genera *Rhodospirillum* and *Rhodopseudomonas* where the sediment becomes anoxic. These photoheterotrophs use organic matter as both a carbon source and an electron donor (*see figure 21.2*). Below these microbes, anoxygenic photoautotrophs in the genera *Chlorobium* and *Chromatium* develop as visible olive green and purple zones, respectively. These green and purple sulfur photosynthetic bacteria fix CO_2 and use H_2S as an electron source (*see figure 11.34*). Turning our attention to the very bottom of the column, heterotrophic fermenters, including members of the genus *Clostridium,* generate fermentation products that support the growth of bacteria that perform anaerobic respiration. For example, *Desulfovibrio* spp. use sulfate (found in seawater) as a terminal electron acceptor, thereby producing hydrogen sulfide (H_2S). The H_2S diffuses upward, creating a stable H_2S gradient where anoxygenic photoautotrophs oxidize it, thereby providing the sulfate needed by *Desulfovibrio* spp. and others performing anaerobic respiration. ◀ *Anaerobic respiration uses the same steps as aerobic respiration (section 11.5); Phototrophy (section 11.10); Photosynthetic bacteria are diverse (section 20.3); Order Rhodospirillales includes photosynthesizers and magnetic bacteria (section 21.1); Gammaproteobacteria is the largest bacterial class (section 21.2); Order Clostridiales (section 22.2)*

Comprehension Check

1. Name two groups of protists known to cause HABs in marine ecosystems. What hazards do HABs pose for other marine organisms and humans?
2. Describe the sulfur cycle within a Winogradsky column. What would happen if the column were incubated in the dark for an extended period of time?

Microorganisms in the Open Ocean Are Adapted to Nutrient Limitation

Sometimes called "the invisible rain forest," the illuminated upper 100 m of the open ocean, termed the **photic zone,** is home to a diverse collection of microbes. **Phytoplankton** (Greek *phyto,* plant; *planktos,* wandering) are the source of all organic matter in the open ocean, with the exception of the hydrothermal vent and subsurface chemolithoautotrophic communities (*see figure 27.4*). Roughly half the carbon fixation in the open ocean is performed by cyanobacteria in the genera *Prochlorococcus* and *Synechococcus*. These microbes help constitute the **picoplankton** (planktonic microbes between 0.2 and 2.0 μm in size) and can represent 20 to 80% of the total phytoplankton biomass. Larger unicellular cyanobacteria (*Crocosphaera watsonii*) and eukaryotic autotrophs, especially diatoms, contribute the other 50% of fixed carbon. ◀ *Phylum Cyanobacteria: oxygenic photosynthetic bacteria (section 20.3); Stramenopila (section 23.4); Mutualisms with primary producers (section 27.2)*

In contrast to coastal regions, which are heavily impacted by rivers, streams, and terrestrial runoff, nutrient levels are extremely low in the open ocean. Therefore the key to survival in the photic zone is tight recycling of nutrients, rather than allowing them to sink to the seafloor; this is the essence of the **microbial loop** (**figure 29.5**). Within the microbial loop, heterotrophic microbes consume **dissolved organic matter (DOM)** released as exudates by other microbes as well as larger organisms. The DOM released by photoautotrophic microbes is sometimes called **photosynthate.** Marine viruses, present at concentrations up to 10^8 per milliliter, are also a source of DOM as the lysis of their host cells contributes significantly to the return of nutrients into the microbial loop. This generates a common pool of DOM that serves as a nutrient source for all osmotrophic organisms—that is, microbes that gain all their nutrition by the transport of DOM. Heterotrophic protists, including flagellates and ciliates, consume smaller microbes, which may include other protists (figure 29.5*b,c*). Microbial cells, insoluble detritus, and other solid organic materials are called **particulate organic matter (POM).** Because protists are consumed by zooplankton, both DOM and POM are recycled at a number of trophic levels. The net effect of the microbial loop is the continuous recycling of essential nutrients such as organic carbon, nitrate, and phosphate within the photic zone.

The fate of all the carbon fixed by marine phytoplankton has practical importance. Global climate change has focused intense scrutiny on determining how much CO_2 phytoplankton can remove from the atmosphere, fall through the water column, and be sequestered in the benthos. This is sometimes called the *biological carbon pump,* and it describes the export of carbon to the seafloor.

To understand how the microbial loop prevents most carbon from escaping to the seafloor, we need to consider the fate of organic matter that leaves the photic zone. Much of this material falls through the depths as **marine snow** (**figure 29.6*a***). This material gets its name from its appearance, sometimes seen in video images of mid- and deep-ocean water, where drifting flocculent particles resemble snow. Marine snow consists primarily of fecal pellets, diatom frustules, and other materials not easily degraded. During its fall to the bottom, marine snow is colonized by microbes so that nutrient recycling and mineralization continue. Bacterial population densities associated with marine snow can reach 10^8 to 10^9 per milliliter, which is orders of magnitude higher than the surrounding water. By forming a biofilm, these microbes work together, secreting a variety of exoenzymes that break down organic polymers so the resulting monomers can be assimilated. As marine snow sinks, it leaves a plume of

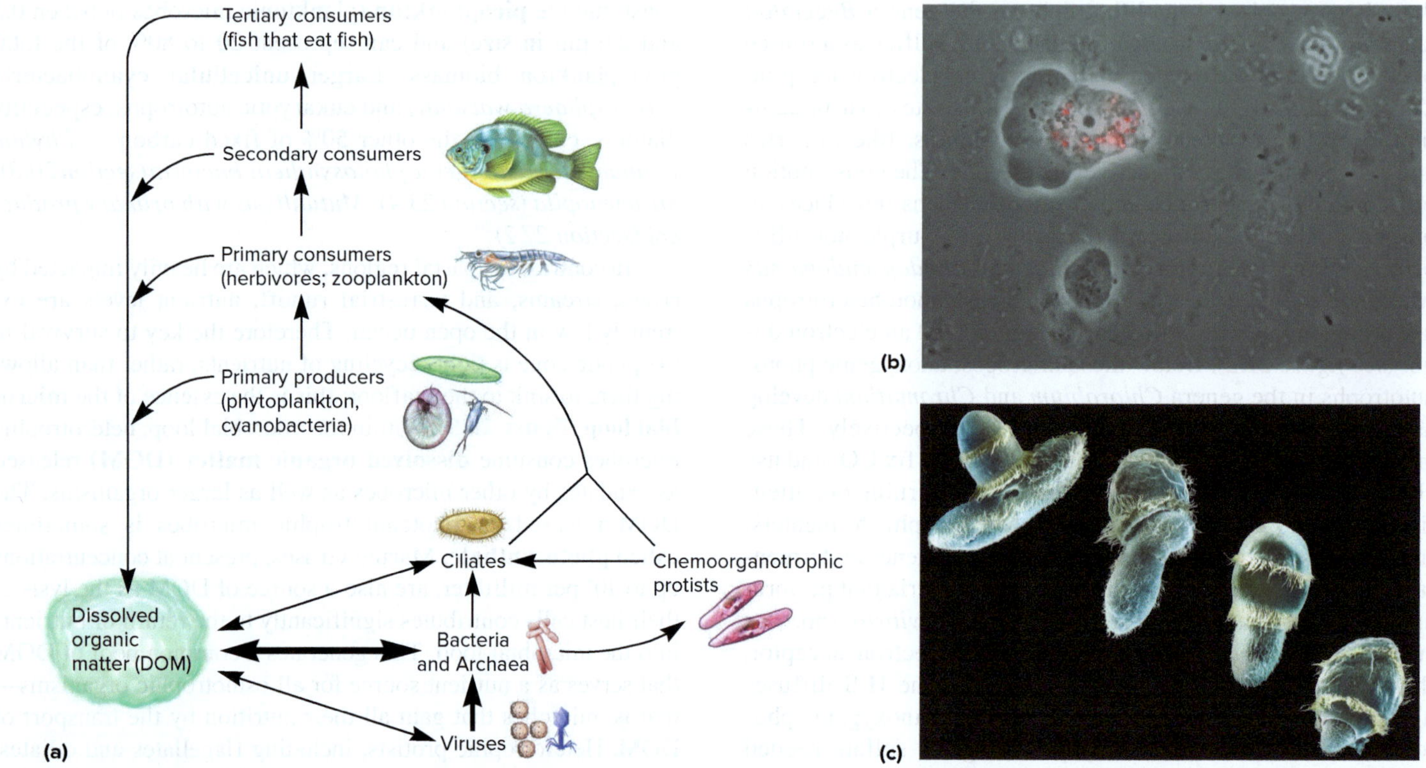

Figure 29.5 The Microbial Loop. (a) Microorganisms play vital roles in the open ocean photic zone as primary producers, decomposers, and primary consumers. All organisms contribute to a common pool of dissolved organic matter (DOM) consumed by microbes. Viruses contribute to DOM by lysing their hosts, and bacterial and archaeal cells are consumed by protists, which also consume other protists. These microbes are then consumed by herbivores and zooplankton that often select food items by size, thereby ingesting both heterotrophic and autotrophic microbes. (b) Protists consume bacteria; in this case, an amoeba has consumed the cyanobacterium *Synechococcus* sp., which fluoresces red. (c) Protists consume protists; in this time-lapse series, the ciliate *Didinium* sp. (rounded organism with two rows of cilia) preys upon another ciliate, *Paramecium* sp.

(b) Jackie Parry; (c) Greg Antipa/Biophoto Associates/Science Source

MICRO INQUIRY *How do heterotrophic microbes contribute to the pool of DOM? How does this compare with the manner by which microbial autotrophs contribute to the DOM pool?*

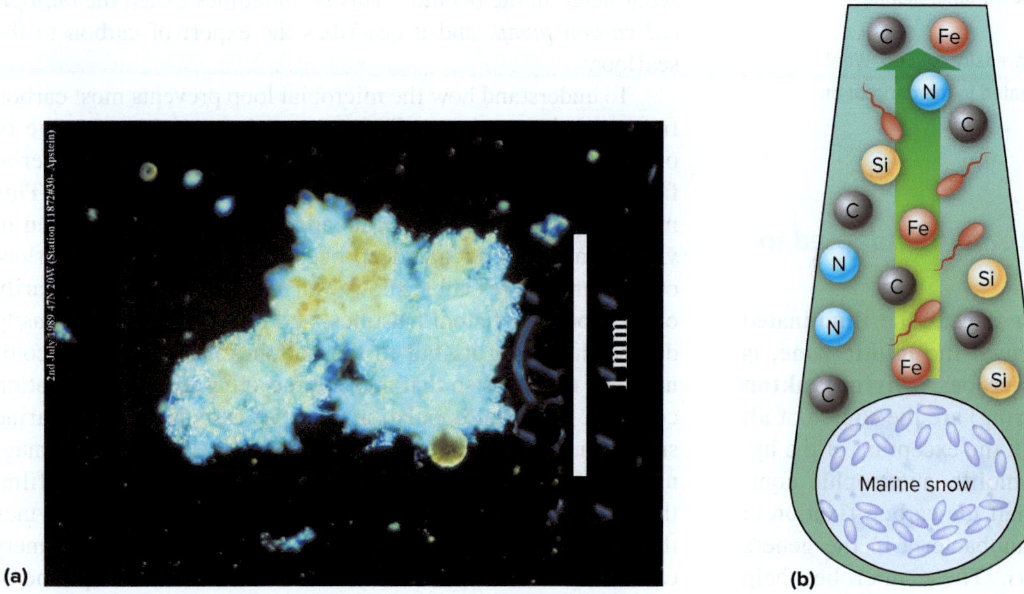

Figure 29.6 Marine Snow. The export of organic matter out of the photic zone to deeper water occurs in part as marine snow. (a) Marine snow includes fecal pellets, planktonic tests, transparent snail-like pteropod shells, radiolarians, and diatoms. (b) The hydrolytic activity of the exoenzymes secreted by the microbes that colonize marine snow generates a plume of nutrients during sinking. This in turn attracts more microbes.

(a) Prof. Richard Lampitt/National Oceanography Centre

DOM, nitrogen, phosphorus, and iron. These nutrients attract chemotactic free-swimming microbes, which help to deplete the nutrient pool even further (figure 29.6*b*). By the time a piece of marine snow reaches the seafloor, less than 1% of photosynthetically derived organic carbon remains unaltered. Once there, only a tiny fraction of the carbon will be buried in the sediments; most will return to the surface in upwelling regions (**figure 29.7**).

An urgent question is whether or not the oceans can increase CO_2 drawdown (removal and burial) as the atmospheric concentration continues to climb. Proposed and experimental ocean-based solutions to mitigating climate change are based on the premise that CO_2 sequestration deep beneath the ocean floor is a viable strategy (**table 29.1**).

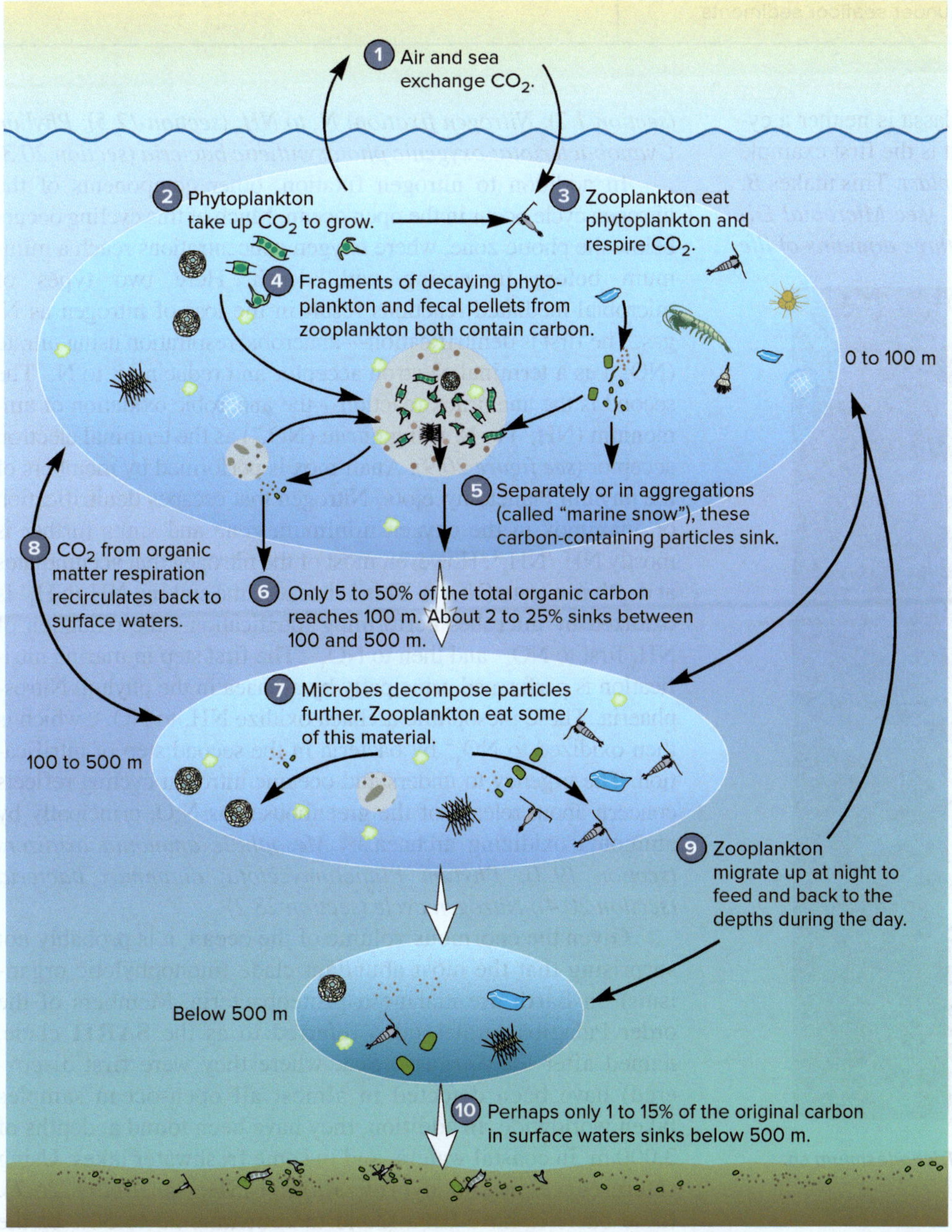

Figure 29.7 Fate of Carbon from the Photic Zone to the Deep Benthos. The vast majority of the carbon fixed by microbial autotrophs at the surface of the open ocean remains in the photic zone (upper 100 m). Although a small fraction reaches the seafloor, much of it returns to the surface in regions where deep and surface waters mix (upwelling regions; not shown).

Marine microbiologists have been documenting the biology of microbes in the open oceans for decades. They have learned that pelagic (open ocean) microbes have evolved diverse strategies to survive in ecosystems where nutrients are in very short supply, that is, oligotrophic environments.

Importantly, microbial growth in the open ocean is often limited by nitrogen. Because there is no terrestrial runoff, the only source of "new" organic nitrogen at sea is biological nitrogen fixation. There appear to be at least four significant sources of N_2 fixation. (1) Many years ago, it was thought that the filamentous cyanobacterium *Richelia intracellularis* was the only marine microbe that fixed significant amounts of N_2. This cyanobacterium lives in association with certain diatoms. (2) We now know that filamentous cyanobacteria in the genus *Trichodesmium* are much more important in terms of the amount of N_2 fixed (**figure 29.8**). This microbe drifts on the surface in rafts that cover many square meters in tropical and subtropical oceans. *Trichodesmium* protects its oxygen-sensitive nitrogenase enzyme by partitioning it to regions of trichomes (filaments) called *diazocytes.* Because the region of a filament that functions as a diazocyte is not separated from the rest of the trichome by a septum, these structures differ from heterocysts seen in other filamentous cyanobacteria (*see figure 20.8*). (3) Another source of new nitrogen is the N_2-fixing, unicellular cyanobacterium *Crocosphaera watsonii.* It measures 3 to 8 μm in diameter, making it larger than the other unicellular cyanobacteria in the genera *Synechococcus* or *Prochlorococcus. C. watsonii* is found in tropical surface waters.

The fourth source of new nitrogen was first reported in 2008 and described as a cyanobacterium originally dubbed UCYN-A. It was later known as *Candidatus* Atelocyanobacterium thalassa and thought to be an endosymbiont of the alga *Braarudosphaera bigelowii,* which is found in temperate coastal and open ocean waters. Like an endosymbiont, it lacks key metabolic enzymes but is unique in its ability to fix N_2. Most recently, genetic and structural

Table 29.1 Potential Ocean-Based Climate Change Solutions

Proposed Solution	How It Would Work	Potential Ecosystem Impacts
Ocean fertilization	Addition of a limiting nutrient (e.g., iron) to induce a diatom bloom; diatoms and other POM sinks and is sequestered at depth	Unknown impact on cycling of other nutrients and growth of other organisms; most carbon may be cycled within water column
Afforestation	Growth of vast kelp gardens; the kelp is then harvested and buried at great depths	Release of POM and DOM while sinking impacts microbial growth and O_2 consumption, possibly leading to dead zones
Alkalinity enhancement	Add chemicals (e.g., calcium carbonate or calcium silicate) to increase ocean pH, thereby shifting carbonate equilibrium to carbonate and bicarbonate, increasing capacity for additional CO_2 drawdown (figure 29.1)	Increased turbidity and light distribution through water column would impact photosynthetic microbes
Direct CO_2 injection	Capture atmospheric CO_2 and pump it directly into crustal ocean rock under seafloor sediments	Uncertain impacts and technologically challenging

analysis demonstrate that *Candidatus* A. thalassa is neither a cyanobacterium nor an endosymbiont. Rather, it is the first example of a nitrogen-fixing organelle, dubbed a *nitroplast*. This makes *B. bigelowii* the first eukaryote known to fix N_2 (*see Microbial Diversity & Ecology 20.1*) ◀ *Evolution of the three domains of life*

Figure 29.8 The Filamentous Cyanobacterium *Trichodesmium* sp. Fixes N_2 in the Open Ocean.

Image Quest 3-D/NHPA/Photoshot/Newscom

(section 1.2); Nitrogen fixation: N_2 to NH_3 (section 12.5); Phylum Cyanobacteriota: oxygenic photosynthetic bacteria (section 20.3)

In addition to nitrogen fixation, other components of the nitrogen cycle occur in the open ocean. Much of this cycling occurs below the photic zone, where oxygen concentrations reach a minimum before increasing with depth. Here two types of microbial-mediated reactions result in the loss of nitrogen as N_2 gas. The first is denitrification—anaerobic respiration using nitrate (NO_3^-) as a terminal electron acceptor and reducing it to N_2. The second is the anammox reaction—the anaerobic oxidation of ammonium (NH_4^+) to N_2 using nitrite (NO_2^-) as the terminal electron acceptor (*see figure 20.9*). Anammox is performed by members of the phylum Planctomycetota. Nitrogen that escapes denitrification or anammox in the oxygen minimum zone and sinks further is mostly NH_3/NH_4^+. However, most of the nitrogen that accumulates at depth is nitrate (NO_3^-). This is because the sinking NH_3/NH_4^+ is oxidized by microbes performing nitrification—the oxidation of NH_3 first to NO_2^- and then to NO_3^-. The first step in marine nitrification is performed principally by archaea in the phylum Nitrosphaeria. These mesophilic archaea oxidize NH_3 to NO_2^-, which is then oxidized to NO_3^- by bacteria in the second step of nitrification. The urgency to understand oceanic nitrogen cycling reflects concern about release of the greenhouse gas N_2O, principally by ammonia oxidizing archaea. ◀ *Mesophilic ammonia oxidizers (section 19.3); Phylum Planctomycetota: anammox bacteria (section 20.4); Nitrogen cycle (section 28.2)*

Given the enormous volume of the ocean, it is probably not surprising that the most abundant clade (monophyletic organisms) on Earth are marine α-proteobacteria. Members of the order Pelagibacterales (also referred to as the **SAR11** clade, named after the *Sargasso* Sea, where they were first discovered) have been detected in almost all open-ocean samples taken worldwide. In addition, they have been found at depths of 3,000 m, in coastal waters, and in some freshwater lakes. Using fluorescence in situ hybridization (FISH; *see figure 26.7*), these bacteria have been found to constitute 25 to 50% of the microbial cells in surface waters of both coastal and open ocean samples. Indeed, some scientists estimate that bacteria in the order Pelagibacterales constitute about 25% of all microbial life on the planet. *Order Pelagibacterales, the most abundant microorganisms on Earth (section 21.1)*

The physiology of bacteria in the order Pelagibacterales illustrates a recurring metabolic theme that has emerged in the study of marine microorganisms: Many have evolved strategies to supplement their energy reserves. This metabolic diversity blurs the trophic divisions between marine microbes and further contributes to the microbial loop (figure 29.5). For example, genes for rhodopsin, first identified in halophilic archaea, are found in α-proteobacteria (i.e., Pelagibacterales), as well as γ-proteobacteria, bacteroidetes, and marine dinoflagellates. Rhodopsin found in proteobacteria is termed *proteorhodopsin* (PR). None of these microorganisms is autotrophic; rather they use the light-driven rhodopsin pump to supplement their proton motive force to drive ATP synthesis as well as other forms of cellular work (*see figure 11.13*). They are thus considered photochemotrophs (energy from both light and chemicals). Given that the majority of heterotrophic marine bacteria and archaea in the photic zone have the genes needed for rhodopsin biosynthesis, we can infer that this form of phototrophy is important for survival in oligotrophic waters. Indeed, with bacterial and archaeal rhodopsins as the rule rather than the exception, the importance of rhodopsins cannot be overstated. In the vast pelagic photic zone, all energy originates from light, and the amount of light energy captured by microbial rhodopsins rivals that of chlorophyll and bacteriochlorophyll combined. *Rhodopsin-based phototrophy (section 11.10); Halobacteriota (section 19.4)*

Another group of marine microbes that supplement their energy reserves are aerobic anoxygenic phototrophs (AAnPs). Bacteria capable of **aerobic anoxygenic phototrophy** are known to possess bacteriochlorophyll *a* (Bchl *a*). Some α-proteobacteria (the *Roseobacter* clade), as well as γ-proteobacteria, are capable of this form of phototrophy (*see table 20.1*). Like rhodopsin-containing microbes, AAnPs are photoheterotrophs. That is, they use light to drive the synthesis of ATP (photophosphorylation) but meet their carbon needs by the assimilation of organic carbon. By meeting some of their ATP demands phototrophically, photoheterotrophs divert more organic carbon to biosynthesis and less to catabolic processes. AAnPs constitute about 10% of the open ocean microbial community and are therefore less abundant than those that produce proteorhodopsin. This may reflect the fact that genes for proteorhodopsin have widely been subject to horizontal gene transfer, as only six genes are required for a functional light-driven proton pump. This is in contrast to the roughly 50 genes needed to assemble the anoxygenic light-harvesting complex, reaction center, Bchl *a*, and carotenoids found in AAnPs.

The α-proteobacterium *Ruegeria pomeroyi* appears to be capable of a third interesting metabolic strategy to increase its ATP yield: **lithoheterotrophy** (or chemolithoheterotrophy). Lithoheterotrophs use inorganic chemicals as a source of energy and electrons but require organic carbon for anabolic purposes. *R. pomeroyi* is one of a number of marine carbon monoxide (CO) oxidizing heterotrophs. These bacteria produce CO dehydrogenase, thereby oxidizing CO to CO_2, but unlike other CO-oxidizing bacteria, they do not have the enzymes necessary for converting CO_2 into biomass. Instead, they use CO as a supplemental electron donor for their electron transport chain. Thus they need not oxidize as much organic carbon to generate NADH for this purpose. Instead, more organic carbon can be used for biosynthesis. This is supported by thermodynamic considerations: The oxidation of one molecule of CO releases enough energy to drive the production of one ATP. Although the actual contribution of CO oxidation to growth in the sea remains to be determined, we highlight it here to illustrate the metabolic diversity that has evolved within the oligotrophic open ocean.

Marine Viruses: Mortality at Sea

As mentioned in our discussion of the microbial loop (figure 29.5), viruses are important members of marine and freshwater microbial communities. In fact, **virioplankton** are the most numerous members of marine ecosystems. However, quantifying viruses is tricky: The traditional method of plaque formation requires knowledge of both the virus and its host, and the ability to grow the host in the laboratory. Because so few microbes have been cultured, measurement of virus diversity by demonstrating virus infection is often impossible. Instead, virus particles may be visualized directly. There are at least two approaches for this. Electron microscopy is the most rigorous, but fluorescence microscopy is a common and convenient method used to quantify viruses. When seawater is filtered, viruses can be collected on filters with a pore size of 0.02 μm. Viral nucleic acids are stained with a fluorescent dye and samples can be visualized by fluorescence microscopy. Because microscopy reveals the presence of virions outside the host cell, it does not prove that the observed viral particles are capable of infecting a particular host species. Therefore viruses enumerated in this way are called **virus-like particles (VLPs).** Using this approach, the average VLP density in seawater is typically between 10^6 and 10^7 per milliliter; their numbers decline to roughly 10^6 below about 250 m. Marine viruses are so abundant that VLPs are now recognized as the most abundant microbes on Earth (**figure 29.9**).

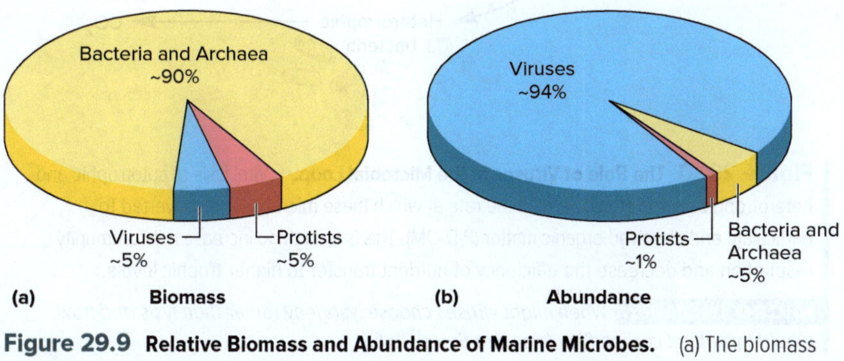

Figure 29.9 Relative Biomass and Abundance of Marine Microbes. (a) The biomass of bacteria and archaea far exceeds that of both protists and viruses. (b) When individuals are counted, however, the vast majority of marine microbes are viruses.

Viral diversity is vast, including single- and double-stranded RNA and DNA viruses that infect archaea, bacteria, and protists. Metagenomic analysis of viral genomes has revealed that as many as 90% of the genomes have no recognizable match on any database. Thus not only is marine viral genetic diversity immense, it is also largely unexplored. ◀ *Viruses (chapter 25)*

The abundance of marine viruses indicates that they are major agents of mortality in the sea. Although highly variable, viruses are thought to kill on average about 20% of the marine microbial biomass every day. This mass lysis accelerates the rate at which large populations of microbes are converted to POM and DOM. In effect, viruses short-circuit the microbial loop (**figure 29.10**).

Because viruses are critical in determining the abundance of their hosts, several theories have been proposed to describe virus-host dynamics. One model predicts that as one microbial species (or strain) becomes numerically dominant, it soon will be targeted by lytic viruses and its population will decline. This permits another microbial species (or strain) to flourish, which then becomes subject to intense viral lysis, and so on. This *kill the winner* model has been challenged by the *piggyback the winner* model based on observations of higher-than-expected rates of lysogeny in marine bacterial populations. In this model, virus infection ends in insertion of the viral genome in the host chromosome (lysogeny) rather than lysis. Finally, some evidence supports a *Red Queen* model, whereby viruses and their hosts continually evolve to remain in opposition to one another; a sort of viral-host arms race. ◀ *Bacteriophage lambda: a temperate bacteriophage (section 25.2)*

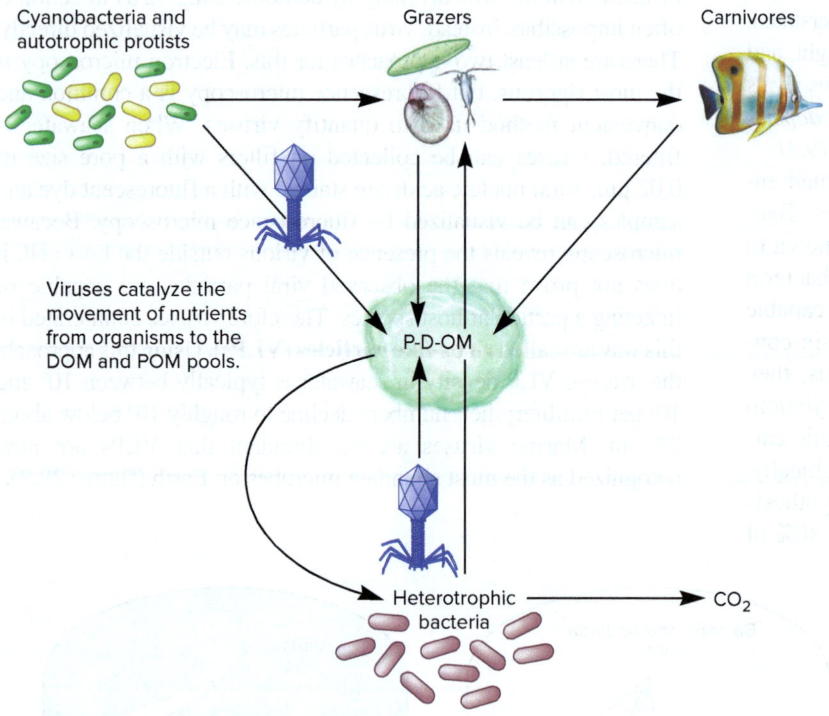

Figure 29.10 The Role of Viruses in the Microbial Loop. Viral lysis of autotrophic and heterotrophic microbes accelerates the rate at which these microbes are converted to particulate and dissolved organic matter (P-D-OM). This is thought to increase net community respiration and decrease the efficiency of nutrient transfer to higher trophic levels.

MICRO INQUIRY *When might viruses choose lysogeny rather than lysis and how would this affect carbon flow (also see figure 29.5)?*

Microbial Communities in Benthic Marine Sediments Are Enormous and Mysterious

Sixty-seven percent of the Earth's surface lies under the sea, which means that of all the world's microbial ecosystems, we know the least about the largest. However, the combination of deep-ocean drilling projects and the exploration of geologically active sites, such as submarine volcanoes and hydrothermal vents, has revealed that the study of life within ocean sediments, or *benthos,* is both rewarding and surprising. Marine sediments range from the very shallow to the deepest trenches, from dimly illuminated to completely dark, and from the newest sediment on Earth to material that is millions of years old. The temperature and age of marine sediments depend on their proximity to geologically active areas. Hydrothermal vent communities with large and diverse invertebrates, some of which depend on endosymbiotic chemolithotrophic bacteria, have been intensely investigated since their exciting discovery in the late 1970s. These microbes are discussed in chapter 27. Because the vast majority of Earth's crust lies at great depths far from geothermally active regions, most benthic marine microbes live under high pressure, without light, and at temperatures between 1° and 4°C.

Deep-ocean sediments were once thought to be devoid of all life, so imagine how surprising it has been to discover that ocean sediments are the Earth's largest microbial habitat (*see figure 1.2*). Indeed, sediment samples from water depths up to 10,840 m (at its deepest, the ocean is about 11,000 m) reveal the presence of a vast *piezosphere* (Greek *piezein,* to squeeze or press). The piezosphere describes the biome at depths with pressures exceeding 100 atm (pressure increases about 1 atm per 10 m depth). Microbes that grow at such great pressures are termed **piezophilic** and their abundance is surprisingly high, typically exceeding that of surface waters. Reasons for this difference are complicated but include energy availability and limited predation.

In this this relatively new realm of ocean exploration, two benthic habitats are under active investigation: surface sediments (0–15 cm deep) within ocean trenches and deep-subsurface sediments collected from hundreds of meters below the seafloor. Surface sediments from the Mariana Trench system (the deepest habitat on Earth with a maximum depth of 10,984 meters) collected by the Challenger Deep expedition in the Western Pacific reveal the presence of a wide diversity of bacteria, archaea, fungi, and viruses. Importantly, metagenomic analysis of assembled microbial genomes provides an assessment of the metabolic activities that drive these communities. As listed in **table 29.2,** these microbes engage in complex nutrient cycling thousands of meters below the sea surface. These communities include diverse bacteria, archaea, six phyla of fungi, protists in the SAR supergroup, and a variety of viruses. Interestingly, genes encoding enzymes for dissimilatory nitrate and sulfate reduction were found in several fungal groups.

Table 29.2 Metagenomic Evidence for Metabolic Activities in Mariana Trench System Surface Sediments

Metabolic Activity
Aerobic respiration
Mixotrophy
Anaerobic respiration
• Dissimilatory nitrate reduction
• Dissimilatory sulfate reduction
Fermentation
Autotrophy (6 different pathways; dominated by the 3-HP bi-cycle, *see figure 20.5*)
Assimilatory nitrate/sulfate reduction
Nitrogen fixation
Denitrification
Methanogenesis
Chemolithotrophy
• Nitrification
• H_2 oxidation
• CO oxidation
• Sulfide oxidation

Global surveys show that deep subsurface sediments within the continental margin differ from those in the distant abyss. Seawater is rich in sulfate, which is used as an electron acceptor in anaerobic respiration. In the continental margin, sulfate in sediments is depleted within 100 m depth, at which point methanogenesis becomes dominant. However, in the abyss, sulfate penetrates much more deeply, and distinct zonation between sulfate-reducing bacteria and methanogenic archaea is absent. Furthermore, the sediment surface and subsurface appear to be teeming with viruses; it is estimated that each year viral lysis in the benthos releases up to 630 million tons of carbon that had been sequestered by marine snow and other falling particulates (figure 29.6). These exciting findings suggest that a significant fraction of Earth's biomass is amazingly active yet largely uncharacterized.

Exploration of deep-sea subsurface microbiology has turned our understanding of bacterial energetics literally upside down. As discussed in chapters 11 and 28, it is generally understood that anaerobic respiration occurs such that there is preferential use of available terminal electron acceptors (*see figure 28.1*). Acceptors that yield the most energy (more negative ΔG) from the oxidation of NADH or reduced inorganic compounds (e.g., H_2, H_2S) will be used before those that produce a less negative ΔG. Thus following oxygen depletion, nitrate will be reduced, then manganese, iron, sulfate, and, finally, carbon dioxide. This is demonstrated in cores collected in sediments 420 m deep off the coast of Peru (**figure 29.11**). However, the profile of these signature chemical compounds is reversed at depths greater than about 400 m. This suggests the presence of unknown sources of these electron acceptors at subsurface depths of more than 420 m. In addition, contrary to our long-held notion of thermodynamic limits, methanogenesis and iron and manganese reduction co-occur in sediments with high sulfate concentrations. Could this be due to highly localized microenvironments or metabolic activities that occur at different times but in the same place? Although these and other questions await further study, it is clear that with densities of 10^8 cells per gram of sediment at the seafloor surface and 10^4 cells per gram in deep subsurface sediments, these communities are important. ◀ *Redox reactions: reactions of central importance in metabolism (section 10.3); Anaerobic respiration uses the same steps as aerobic respiration (section 11.5); Chemolithotrophy: "eating rocks" (section 11.8)*

Another outcome of deep-ocean sediment exploration has been the discovery of microbial communities on continental margins fueled by the release of hydrocarbons at depths between 200 and 3,500 m. Depending on the surface topology and the rate at which hydrocarbons are emitted, they are called pockmarks, gas chimneys, mud volcanoes, brine ponds, and oil and asphalt seeps. A variety of microbes use hydrocarbons as a sole carbon source, oxidizing them with oxygen or sulfate as the

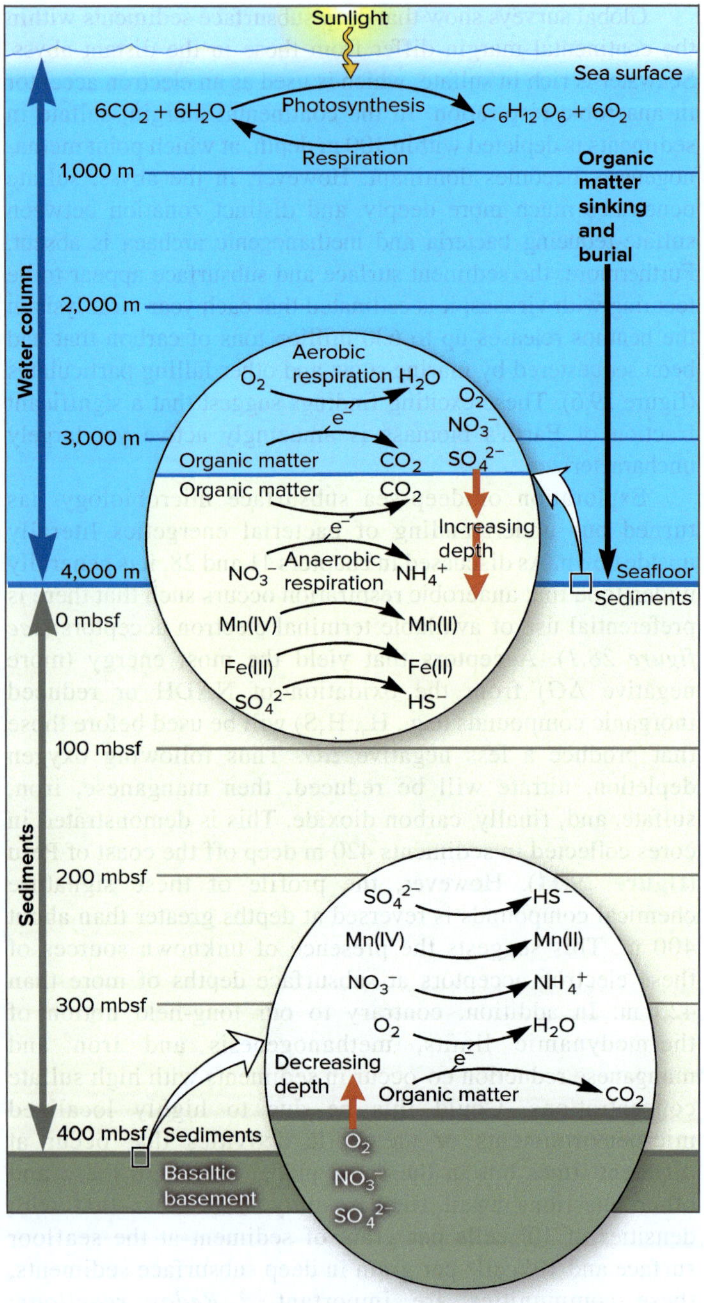

Figure 29.11 Microbial Activity in Deep-Ocean Sediments. At the surface of the seafloor, reduction of oxidized substrates that serve as electron acceptors in anaerobic respiration follows a predictable stratification based on thermodynamic considerations (*see also figure 28.1*). This sequence reverses in very deep subsurface sediments, suggesting a source of electron acceptors from deep within Earth's crust. mbsf, meters below seafloor.

electron acceptor. Some regions are the sites of microbial mats, which can sometimes cover several hundred meters. These mats are dominated by the giant, vacuolated sulfur-oxidizing bacteria in the genera *Beggiatoa* and *Thiomargarita (see figures 21.17 and 21.18).* ◀ *Gammaproteobacteria is the largest bacterial class (section 21.2)*

Another fascinating community is found at **methane hydrates.** Methane hydrates are lattice-like ice structures in which methane is trapped. The methane is produced by both geothermal activity and methanogenic archaea. The lack of associated methane-oxidizing microbes has enabled pools of natural gas to accumulate in lattice-like cages of crystalline water 500 m or more below the sediment surface in many regions of the world's oceans. In fact, there may be up to 10^{13} metric tons of methane hydrate worldwide. ◀ *Methanogens (section 19.4)*

Comprehension Check

1. What is marine snow? Why is it important in CO_2 drawdown?
2. Explain how new organic nitrogen is introduced into the open ocean.
3. Explain how proteorhodopsin contributes to the success of bacteria in the Pelagibacterales.
4. Compare and contrast the use of lithoheterotrophy and photoheterotrophy by bacteria in oligotrophic waters.
5. How might community dynamics change in the absence of viral lysis?
6. Explain what is meant by "upside-down microbial energetics," as described for some deep-subsurface ocean sediments.

29.3 Microorganisms in Freshwater Ecosystems

After reading this section, you should be able to:

a. Appreciate the scarcity of freshwater on Earth
b. Contrast and compare the microbial communities found in lotic systems with those in lentic systems
c. Describe the differences between the microbial communities found in a deep, stratified lake with those of a shallow lake that lacks stratification
d. Describe how seasonal changes influence lake microbial communities and how these changes can lead to blooms

While the vast majority of water on Earth is in marine environments, freshwater is crucial to our terrestrial existence. As shown in **figure 29.12,** of the 3.9% of Earth's water that is not saline, only a tiny fraction represents the familiar streams, rivers, and wetlands that we associate with the term *freshwater.* We begin our discussion with the largest fraction of freshwater—that which is frozen at the Earth's polar regions. We then move to streams, rivers, and lakes.

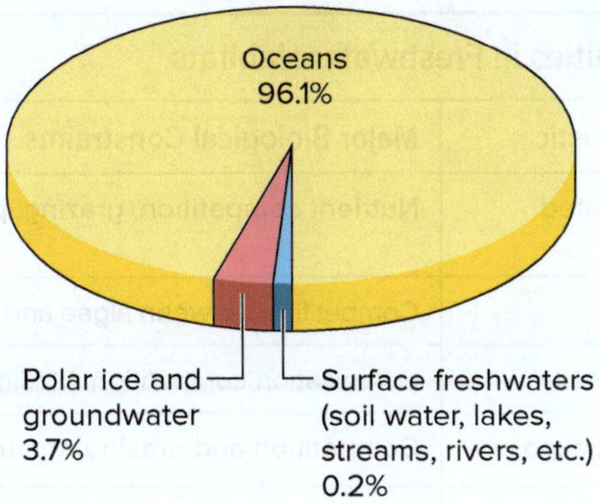

Figure 29.12 Relative Distribution of Freshwater on Earth.

Glaciers and Permanently Frozen Lakes Support Active Microbial Communities

Much of the ice on Earth has remained frozen for millions of years. Indeed, it is important to note that a majority of the Earth's surface never exceeds 5°C. This includes polar regions, the deep ocean, and high-altitude terrestrial locations. Surprisingly, microbes within glaciers are not dormant. Rather, evidence that active microbial communities exist in these environments has been growing over the last few decades. In fact, this is an interesting time in glacial microbiology; determining the diversity in these systems and assessing the role of these microorganisms in biogeochemical cycling as glaciers melt can involve novel and creative techniques. In addition, since the discovery of ice on Mars and icy moons within and beyond our solar system, astrobiologists have become very interested in ice-dwelling psychrophilic microbes. ◂ *Environmental Factors That Affect Microbial Growth (section 7.5).*

The biology of lakes associated with glaciers has gained importance as it has been recognized that these ecosystems function as sentinels of climate change. Glacial lakes are constrained by low temperature and reduced solar radiation, and in the Antarctic, lakes are separated from macrofauna as well. Here phytoplankton, bacteria, and protists form simple food webs almost completely devoid of multicellular organisms (**figure 29.13**). The deposition of organic carbon from the surrounding environment supplements the DOM pool generated by photosynthetic microbes. Furthermore, phototrophic flagellated and ciliated protists are also capable of consuming bacteria and dissolved organic carbon. These **mixotrophs** are not the only consumers of bacteria. Viral lysis of bacteria in Antarctic lakes is also significant. Note that archaea have been found in glacial lakes but in lower numbers and diversity than other microbes. Their contribution to these ecosystems awaits discovery.

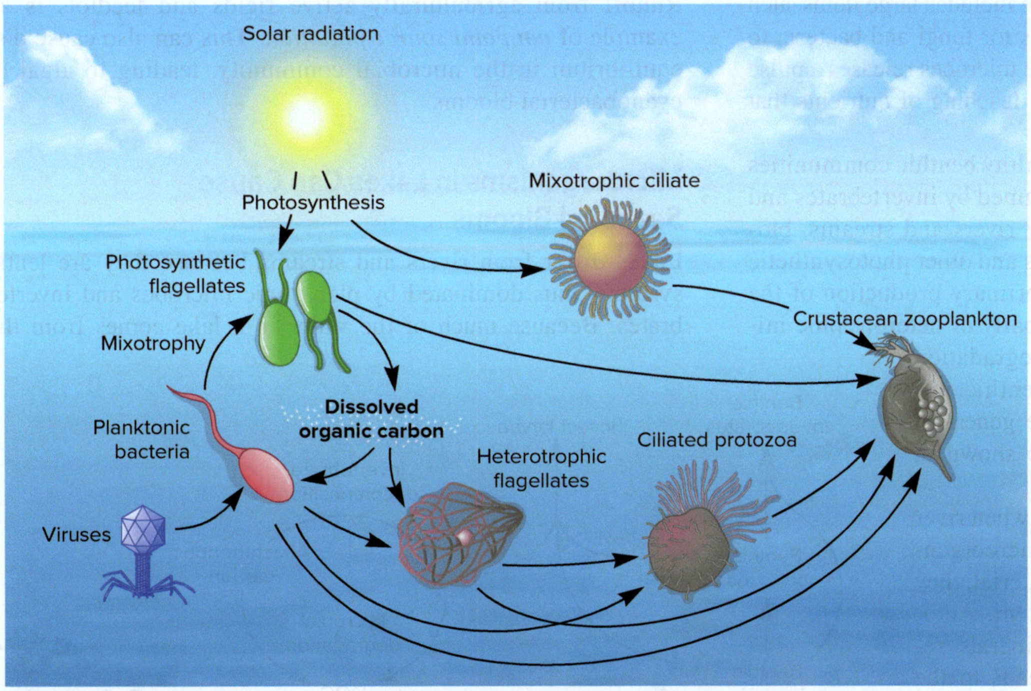

Figure 29.13 Nutrient Cycling in Antarctic Lakes Is Performed Entirely by Microbes. Antarctic glacial lakes lack multicellular organisms, so all nutrient cycling is between various types of microbes, including bacteriophages and viruses that infect protists (not shown). The surrounding environment provides some organic carbon.

MICRO INQUIRY *List three major differences between this microbial loop and that found in the open ocean (figure 29.5).*

Microbial Communities in Streams and Rivers Differ

Streams and rivers are much more dynamic than glacial lakes. Some of the physical and biological factors that help regulate microbial communities in these freshwater systems are listed in **table 29.3.** Lakes, rivers, and streams can be broadly classified as either **lotic** (Latin *lotus,* washing) systems with free-running waters (streams, most rivers) or **lentic** (Latin *lentus,* calm or slow) systems with free-standing waters (e.g., ponds, lakes, marshes, slow-moving rivers). In large, lentic rivers and lakes, phytoplankton thrive and are responsible for much of the fixed carbon. The biologically available carbon produced within the system is called **autochthonous.** By contrast, the constant flow of water in lotic streams and all but the largest rivers prevents the development of significant planktonic communities. In addition, phytoplankton growth can be further limited by the amount of available light due to overhanging foliage, turbidity, and rapid mixing of the water. In most streams and rivers, the source of nutrients therefore comes from the surrounding land. Such nutrients are called

Table 29.3 Physical and Biological Factors Influencing Microbial Communities in Freshwater Habitats

Habitat	Major Microbial Community	Major Physical Characteristic	Major Biological Constraints
Lake	Plankton	Stratification, wind-generated turbulence	Nutrient competition, grazing, parasitism
Floodplain	Plankton, biofilms attached to plants	Periodic desiccation	Competition between algae and plants
Rivers and streams	Benthic microbes, biofilms	Flow-generated turbulence	Colonization competition, biofilm grazing
Estuary—mudflats	Biofilms	Desiccation, high light, salt exposure	Competition and grazing at mud surface
Estuary—outflow	Plankton	Mixing with saltwater, turbidity	Grazing in water column

allochthonous. Thus lentic systems often have a net autotrophic metabolism, whereas lotic systems are generally heterotrophic.

The allochthonous carbon that enters rivers and streams includes both dissolved and particulate organic carbon (DOC and POC, respectively). The amount of carbon entering freshwater systems depends on surrounding vegetation, proximity to human activity, and local geology. In addition, hydrological factors such as how long water stays in the soil before entering the river or stream also influence the nature of the organic carbon. Thus DOC ranges from material that can be readily assimilated to refractory carbon that only a few microbial groups can slowly degrade. Much of the particulate organic carbon includes large items such as leaf litter, which provide substrates for fungi and bacteria to colonize. Initially, the activity of these microbes releases a pulse of DOC; this is followed by the slow leaching of nutrients that may not be as easy to assimilate.

Most rivers and streams have healthy benthic communities of microbes that form biofilms consumed by invertebrates and other grazing organisms. In shallow rivers and streams, biofilms are often dominated by diatoms and other photosynthetic protists that contribute to the total primary production of the system. Biofilms that consist principally of heterotrophic microbes are important sources of degradation and mineralization. The role of benthic microbes in rivers and streams can be generally characterized by the microbial loop shown in **figure 29.14.**

As in any microbial community, when river and stream chemoorganotrophic microorganisms metabolize available organic material, they recycle nutrients within the ecosystem. Autotrophic microorganisms require minerals released from organic matter. This leads to the production of O_2 during daylight hours, when the rates of photosynthesis and respiration are in dynamic equilibrium. When the amount of organic matter added to streams and rivers does not upset this balance, so that heterotrophy and autotrophy are roughly equal, productive streams and rivers are maintained. However, when the supply of organic matter supports a vast heterotrophic population whose respiration rate outpaces oxygenic photosynthesis, the water may become anoxic. This can occur in streams and rivers adjacent to urban and agricultural areas. The release of inadequately treated municipal wastes and other materials from a specific location along a river or stream represents *point source pollution.* Such additions of organic matter can produce distinct and predictable changes in the microbial community and oxygen concentration, creating an *oxygen sag curve* in which oxygen is depleted just downstream of the pollution source (**figure 29.15**). Runoff from agriculturally active fields and feedlots is an example of *nonpoint source pollution.* This can also cause disequilibrium in the microbial community, leading to algal or cyanobacterial blooms.

Microorganisms in Lakes Can Cause Seasonal Blooms

Lakes differ from rivers and streams because they are lentic systems, thus dominated by planktonic microbes and invertebrates. Because much of the water in a lake comes from the

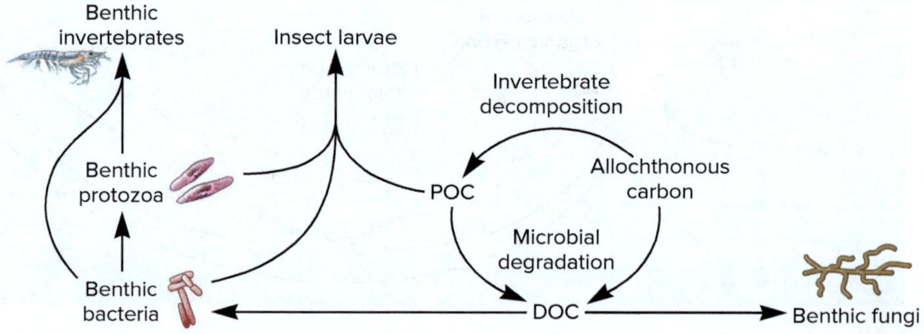

Figure 29.14 A Benthic Microbial Loop in Lotic Systems. In contrast to the microbial loop in planktonic systems (figure 29.5), the benthos of a lotic system, such as a fast-moving river or stream, features small eukaryotes such as invertebrates and insect larvae. These organisms play an important part in nutrient cycling. At the same time, macrofauna such as fish and other top carnivores may play minor roles.

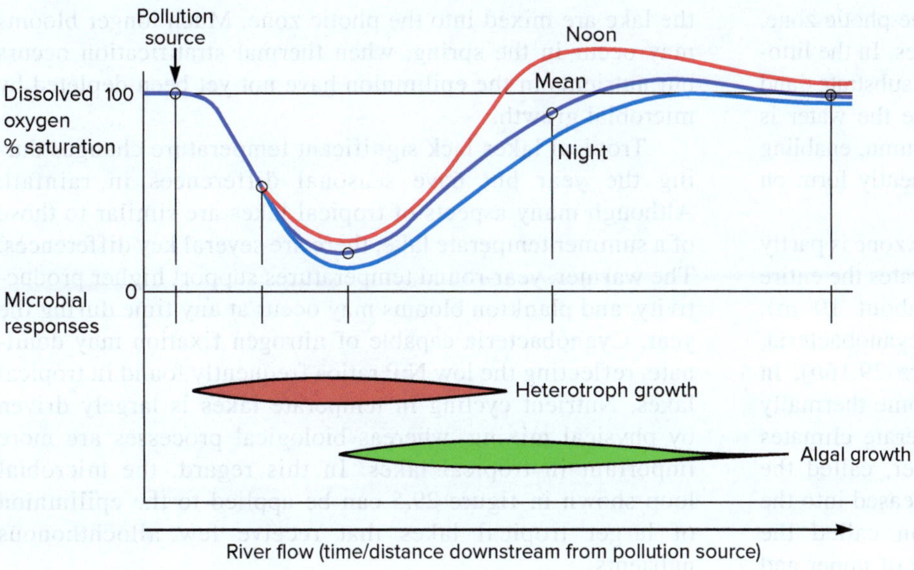

Figure 29.15 The Dissolved Oxygen Sag Curve. Microorganisms can create gradients over distance and time when nutrients are added to rivers. When organic wastes are added at a single site to a clean river system, the rate of O_2 consumption (depletion) directly downstream increases, resulting in a dissolved O_2 sag curve. Further downstream, the phototrophic community will again become dominant, resulting in diurnal changes in river oxygen levels.

surrounding land, the geology and use of the land that feeds the lake largely determine the chemistry of the inflow water. For instance, mountain lakes are fed water from relatively infertile land, resulting in oligotrophic conditions within the lake. This explains why mountain lakes tend to be so clear: Microbial growth is limited to those species that tolerate low nutrient levels. On the other hand, low-lying lakes generally catch water from fertile, cultivated soils, so they are nutrient rich. This supports a high level of planktonic growth, and the lakes may appear turbid. The picoplankton often include freshwater species of *Synechococcus*, whereas filamentous cyanobacteria such as those in the genus *Anabaena* and freshwater protists constitute the macroplankton (microbes larger than 200 µm).

Deep lakes can be divided into two main regions: the **littoral zone,** or shoreline, and the central, **pelagic zone** (these terms also apply to marine systems). As shown in **figure 29.16*a*,**

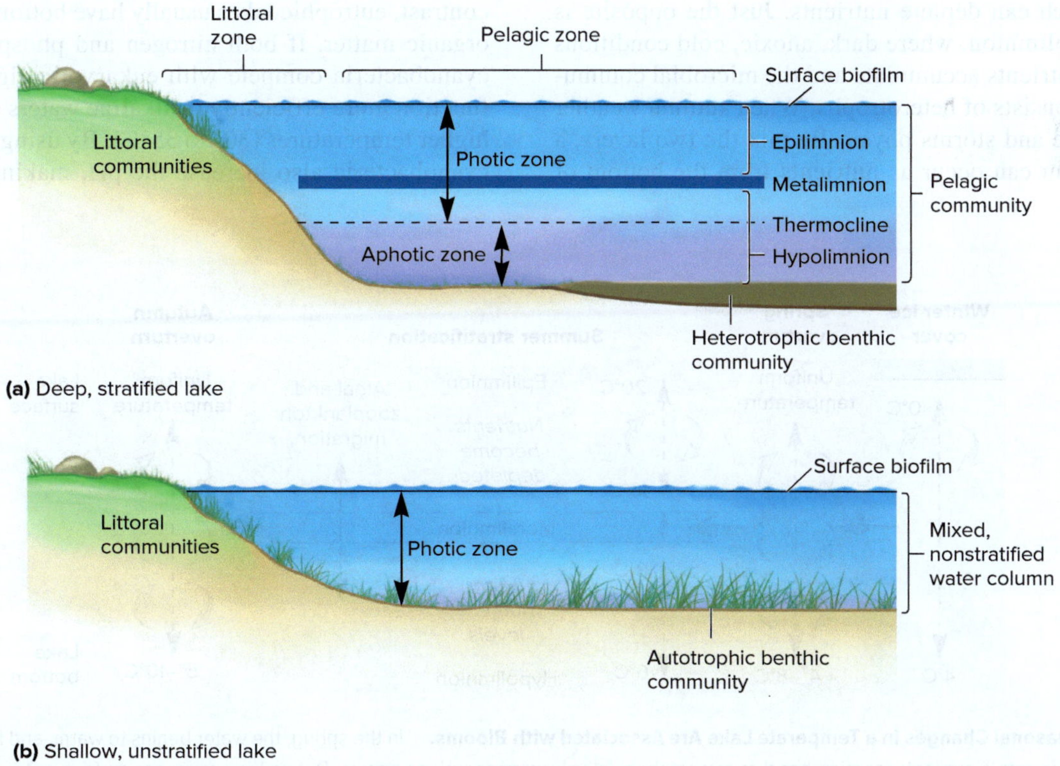

Figure 29.16 The Depth of a Lake Is a Key Factor in Shaping the Microbial Community. (a) Deep lakes have autotroph-dominated littoral and epilimnion microbial communities. Waters below the thermocline and photic zone form the hypolimnion, which can become oxygen depleted. Here a heterotrophic benthic community thrives. (b) Shallow lakes have little mixing, and the photic zone reaches the lake bottom. Thus autotrophs dominate the water column and benthic regions throughout the lake.

MICRO INQUIRY *How does the contribution of benthic autotrophs differ in deep lakes compared to that in shallow lakes? How does this affect the relative contribution of phytoplankton in these two aquatic systems?*

the littoral zone is also defined by the depth of the photic zone. The two zones have distinct microbial communities. In the littoral zone, allochthonous DOC introduces organic substrates and nutrients such as phosphate and nitrate. Because the water is shallow, light penetrates throughout the water column, enabling the development of larger plants. Biofilms frequently form on these plants and in the sediments.

The microbial community within the pelagic zone is partly determined by the depth of the lake. Light penetrates the entire water column in shallow lakes (less than about 10 m), so benthic autotrophs, including diatoms and cyanobacteria, may be the principal primary producers (figure 29.16*b*). In deeper lakes, the central, pelagic zone can become thermally stratified. This is the case in areas with temperate climates (**figure 29.17**). In the summer, the upper layer, called the **epilimnion,** becomes warmed and oxygen is released into the atmosphere. The thermocline spans a region called the **metalimnion**, which acts as a barrier to mixing of upper and lower water masses. The lower, colder, dark zone, known as the **hypolimnion,** can become anoxic as a result of two factors at play simultaneously: (1) The oxygen penetration rate by diffusion is slower than its loss by respiration and (2) oxygenic photosynthesis is confined to the epilimnion. Because the epilimnion and hypolimnion do not mix (or mix very little), the upper layer supports a rich diversity of primary producers and consumers, which can deplete nutrients. Just the opposite is true for the hypolimnion, where dark, anoxic, cold conditions limit growth. Nutrients accumulate and the microbial community primarily consists of heterotrophs. When autumn weather cools the surface and storms physically mix the two layers, a short-lived bloom can occur as nutrients from the bottom of the lake are mixed into the photic zone. Much longer blooms may occur in the spring, when thermal stratification occurs but nutrients in the epilimnion have not yet been depleted by microbial growth.

Tropical lakes lack significant temperature changes during the year but have seasonal differences in rainfall. Although many aspects of tropical lakes are similar to those of a summer temperate lake, there are several key differences. The warmer, year-round temperatures support higher productivity, and plankton blooms may occur at any time during the year. Cyanobacteria capable of nitrogen fixation may dominate, reflecting the low N:P ratios frequently found in tropical lakes. Nutrient cycling in temperate lakes is largely driven by physical mixing, whereas biological processes are more important in tropical lakes. In this regard, the microbial loop shown in figure 29.5 can be applied to the epilimnion of larger tropical lakes that receive few allochthonous nutrients.

Lakes vary in nutrient status. Some are oligotrophic and others are eutrophic. Nutrient-poor lakes remain oxic throughout the year, and seasonal temperature shifts usually do not result in distinct oxygen stratification. In such oligotrophic lakes, cyanobacteria capable of nitrogen fixation may bloom. Members of several genera, notably *Anabaena, Nostoc,* and *Cylindrospermum,* can fix nitrogen under oxic conditions. In contrast, eutrophic lakes usually have bottom sediments rich in organic matter. If both nitrogen and phosphorus are present, cyanobacteria compete with eukaryotic algae. Cyanobacteria function more efficiently in alkaline waters (pH 8.5 to 9.5) and higher temperatures (30° to 35°C). By using CO_2 at rapid rates, cyanobacteria also increase the pH, making the environment

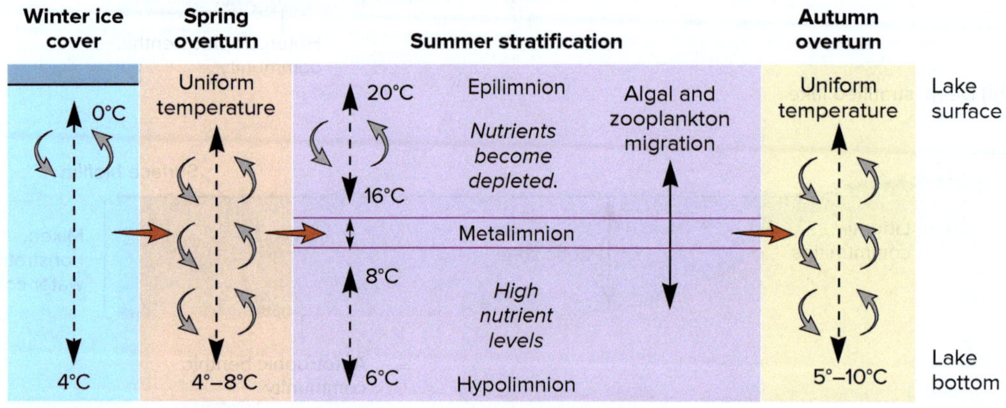

Figure 29.17 Seasonal Changes in a Temperate Lake Are Associated with Blooms. In the spring, the water begins to warm, and lack of thermal stratification ensures ample nutrients for microbes that grow in the relatively warmer surface waters. During the summer as surface waters warm, thermal stratification separates the cool, deep, nutrient-rich waters from the warm, illuminated waters. Because microbial growth is favored in the warmer surface water, this water becomes depleted of nutrients. During the autumn, waters cool and storms mix the layers, bringing the bottom nutrients to the surface. This pulse of nutrients can support a bloom of microbes or algae. Water turbulence is indicated by gray arrows.

MICRO INQUIRY *Why does water turbulence play only a minor role in mixing the water column in the summer?*

MICROBIAL DIVERSITY & ECOLOGY

29.1 Attention All Dog Owners!

Freshwater harmful algal blooms (HABs) really hit home every summer when dog owners across the United States report the deaths of their beloved pets after swimming in lakes and ponds with cyanobacterial blooms. The odor and sight of HABs typically repel humans, but dogs will be dogs. They drink and swim in water we would not set foot in and they ingest the cyanobacteria when licking their wet fur. In one instance, a Texas woman's dog died within an hour of swimming in a HAB-contaminated lake. Another family in North Carolina lost three dogs this way.

Although a necropsy is needed to identify the exact cause of death of any animal, it is believed two cyanobacterial toxins are the culprits based on the symptoms described by dog owners. One toxin that targets liver cells is called microcystin. This toxin, produced by the unicellular cyanobacterium *Microcystis* spp., blocks eukaryotic protein phosphatases involved in crucial cell signaling events. Neurotoxins are the second type of cyanobacterial toxin; dogs have been reported as having seizures just prior to succumbing. There are two such neurotoxins, anatoxin-a and saxitoxin, both of which block voltage-gated sodium channels in neurons resulting in neuromuscular weakness, respiratory arrest, and death. Anatoxin-a is produced by over 40 different cyanobacterial species and saxitoxin by over 15 species.

Signs that a dog (or any animal, including humans) has ingested cyanobacterial toxins include nausea, vomiting, twitching, weakness, seizures, and trouble breathing. These symptoms can occur minutes to hours after exposure, or even days later. Because there is no antidote, it is best to keep pets out of scummy, smelly ponds and lakes. Remember, dogs just want to have fun (**Box figure**).

Andraž Cerar/Shutterstock

less suitable for protists. Cyanobacteria have additional competitive advantages. Many produce siderophores to bind iron, making this important trace nutrient less available for protists. Some cyanobacteria also resist predation because they produce toxins (**Microbial Diversity & Ecology 29.1**). In addition, some synthesize odor-producing and toxic compounds that affect the quality of drinking water. These problems become magnified when cyanobacteria or protists (algae) produce massive blooms in strongly eutrophied lakes. ◀ *Phylum Cyanobacteriota: oxygenic photosynthetic bacteria (section 20.3)*

Comprehension Check

1. Why is mixotrophy suited for survival in Antarctic glacial lakes?
2. What is an oxygen sag curve? What changes in a river cause this effect?
3. What are point and nonpoint source pollution? Can you think of examples in your community?
4. Compare and contrast the types of microbes found in the epilimnion and hypolimnion of a deep lake.
5. What are some important effects of eutrophication on lakes?
6. Why do cyanobacteria often dominate waters that have been polluted by the addition of phosphorus?

Key Concepts

29.1 Water Is the Largest Microbial Habitat

- Most water on the Earth is marine (97%). The majority of this is cold (2°–5°C) and at high pressure. Freshwater is a minor but important part of Earth's biosphere.
- The carbonate equilibrium system keeps the oceans buffered at pH 7.6 to 8.2; however, ocean acidification threatens the health of marine ecosystems (**figure 29.1**).
- The penetration of light into surface waters determines the depth of the photic zone. Warming the surface waters can lead to the development of a thermocline.

29.2 Microorganisms in Marine Ecosystems

- Tidal mixing in estuaries causes daily swings in osmolarity. Halotolerant microbes within estuaries have evolved mechanisms to cope with rapid changes in salinity.
- Coastal regions such as estuaries and salt marshes can be the sites of harmful algal blooms (**figure 29.3**).
- Autotrophic microbes in the photic zone within the open ocean account for about one-half of all the carbon fixation on Earth.
- The microbial loop describes the transfer of nutrients between trophic levels while taking into account the multiple contributions of microbes to recycling nutrients. Nutrients are recycled so efficiently, very little organic carbon is buried in the seabed (**figures 29.5–29.7**).
- The carbon and nitrogen budgets of the open ocean are intensely studied because of their implications for the future of climate change.
- Bacteria in the α-proteobacterial order Pelagibacterales are the most abundant organisms on Earth and like many marine bacteria demonstrate unique adaptations to life in the oligotrophic open ocean.
- Viruses are present at high concentrations in many waters and occur at 10-fold higher levels than bacteria. In marine systems, they play a major role in nutrient turnover (**figures 29.9** and **29.10**).
- Sediments deep beneath the ocean's surface are home to about as many bacteria and archaea as found in the world's ocean waters.
- Methane hydrates, the result of psychrophilic archaeal methanogenesis under extreme atmospheric pressure, contain abundant natural gas.

29.3 Microorganisms in Freshwater Ecosystems

- Glaciers and permanently frozen lakes are sites of active microbial communities. Antarctic lakes lack multicellular organisms and have truncated food webs (**figure 29.13**).
- Nutrient sources for streams and rivers may be autochthonous or allochthonous. Often allochthonous inputs include urban, industrial, and agricultural runoff.
- The depth of a lake determines the microbial communities it will support. Deep lakes have a planktonic, autotrophic community in the epilimnion, and shallow lakes support large autotrophic communities (**figure 29.16**).
- Seasonal mixing of temperate lakes can introduce conditions that support microbial blooms in the spring and fall (**figure 29.17**).

Active Learning

1. The unicellular cyanobacterium *Prochlorococcus* sp. is the most abundant photosynthetic microbe in tropical and subtropical oceans. At least two ecotypes exist: One is adapted to high light and the other to lower light intensities. How does the presence of these two ecotypes contribute to their physiological success and their numerical success? How would you determine the amount of fixed carbon each ecotype contributes to open-ocean ecosystems?
2. How might it be possible to cleanse a eutrophic lake? Consider chemical, biological, and physical approaches as you formulate your plan.
3. It is well known that bacterivory (the consumption of bacteria) supports the growth of marine protists that lack plastids. However, measurements of bacterivory among plastid-containing planktonic algae show that small (less than 5 μm) algae carry out between about 40 and 95% of the bacterivory in the photic zone of the North Atlantic. Discuss how this level of mixotrophy impacts the concept of the microbial loop (figure 29.5).
4. Discuss three specific questions you would ask if you belonged to a research team seeking to investigate how microbial community dynamics within the photic zone impact the export of carbon to the seafloor (i.e., the biological pump). Assume you have access to a variety of sampling and genomic techniques. Refer to chapters 18 and 26 to suggest methods you might use to answer these questions.

CHAPTER

30

Microorganisms in Terrestrial Ecosystems

Yue Jin/USDA-ARS

Bread for a Hungry World

Aristotle wrote about them. The Romans had a ceremony to prevent them. Over 2 million South Asians starved in the 1940s because of them. They are responsible for the loss of more than a million metric tons of wheat in the United States every year. Norman Borlaug won the 1970 Nobel Peace Prize fighting them. What are they? Plant pathogenic fungi: the cause of more than 70% of all major crop diseases.

The worst offenders among fungal plant pathogens are the "rusts," which are basidiomycetes (*see section 24.4*). In North America, probably the most important rust is *Puccinia graminis* subsp. *tritici,* the causative agent of wheat stem rust, shown in the above image. Other important rusts include barley stem rust, soybean rust, cedar-apple rust, and poplar leaf rust. The introduction of fungi from Europe triggered chestnut blight (*Cryphonectria parasitica*), Dutch elm disease (*Ophiostoma ulmi*), and white pine blister rust (*Cronartium ribicola*). With over 6,000 different rust species known to cause plant diseases, it would seem the only approach to saving millions of dollars of valuable agricultural products would be to spray heavily with a fungicide and hope for the best.

Borlaug knew that fungicides were not an option for inhabitants of developing countries. Shortly after earning his Ph.D. in plant pathology at the University of Minnesota, he moved to Mexico to breed new, disease-resistant wheat strains. He later turned his attention to South Asia, where he helped develop hardy strains of rice. In 1970 the Nobel Committee noted, "More than any other single person of this age, he has helped provide bread for a hungry world." It is estimated that each day at least half the global population consumes grain that arose through Borlaug's plant breeding efforts. Borlaug was an active scientist until his death at age 91 in 2009.

Most microbes that associate with plants, whether that association is good, bad, or neutral, live in soil. Soil is a hugely underappreciated resource. It is a vast reservoir of nutrients, wastes, energy, and even pharmaceuticals. Although much is known about the geology and chemistry of soil, our understanding of the microbial communities that drive the health of soils and biogeochemical cycling is far from complete. However, the introduction of new, innovative culture and genomic techniques has made the study of terrestrial microbiology an exciting and dynamic field.

Readiness Check:

Based on what you learned previously, you should be able to:

- ✓ Explain why SSU rRNA is used as a taxonomic marker (section 1.2)
- ✓ List environmental factors that influence microbial growth (section 7.5)
- ✓ Compare and contrast the growth and physiology of obligate aerobes, facultative anaerobes, and strict anaerobes (section 7.5)
- ✓ Evaluate the importance of intercellular molecular signaling to microbial growth in nature (section 7.6)
- ✓ Recall the basic enzymology of nitrogenase (section 12.5)
- ✓ Identify microorganisms responsible for nitrogen fixation (sections 21.1 and 22.1)
- ✓ Explain the global impact that microbes have on carbon and nitrogen cycling (section 28.2)

30.1 Soils Are an Important Microbial Habitat

After reading this section, you should be able to:

a. Differentiate between a mineral and an organic soil, and describe the role of microbes in generating soil organic matter
b. Explain the importance of extracellular enzymes in cellulose and lignin degradation
c. Predict how a specific carbon-to-nitrogen ratio will influence the microbial community within a given soil

The level of microbial diversity in soil is astonishing, exceeding that of any other habitat on Earth, with up to 50,000 species at densities of 10^7 to 10^{10} cells per gram. The soil environment supports active microbial growth that contributes to global elemental cycling and makes nutrients available to plants. Simply put, without soil microbes, there would be no plants, which would eliminate herbivores, and so on up the food chain.

Although many of us have not given soil much thought, it is a vast array of microhabitats created by interdependent biological, physical, and chemical factors. These include soil particles and the pore space between them, critical for the movement of water and gases (**figure 30.1**). Total pore space, and thus gas diffusion, is determined by soil texture. For instance, sandy soils have larger pore spaces than clay soils, so sandy soils tend to drain more quickly. Pores are also critical because they provide the optimum environment for microbial growth. Here microbes are within thin water films on particle surfaces where oxygen may be present at high levels and can be easily replenished by diffusion.

Soil environments can change rapidly. For instance, rainfall or irrigation may quickly change a well-aerated soil to one with isolated pockets of water, generating miniaquatic habitats. If the rate of oxygen consumption exceeds that of oxygen diffusion, waterlogged soils can become anoxic. The roots of plants growing in aerated soils also consume oxygen and release CO_2, influencing the concentrations of these gases in the root environment.

As any farmer knows, not all soils are the same. Soils can be divided into two general categories: A **mineral soil** contains less than 20% organic carbon, whereas an **organic soil** possesses at least this amount. By this definition, the vast majority of Earth's soils are mineral. The importance of organic matter within soils cannot be overstated. **Soil organic matter (SOM)** helps retain nutrients, maintains soil structure, and holds water for plant use. SOM is subject to gains and losses, depending on changes in environmental conditions and agricultural practices. Human activities like plowing expose SOM to more oxygen, leading to extensive microbiological degradation of organic matter. Irrigation causes periodic wetting and drying, which can also lead to increased degradation of SOM, especially at higher temperatures.

Emilio Ereza/Pixtal/age fotostock

Microbial degradation of plant material results in the release of CO_2 and the incorporation of plant organic carbon into microbial biomass. However, a small fraction of the decomposed plant material remains in the soil as SOM. This helps explain why forests are considered a sink for CO_2 (i.e., removing more CO_2 than produced), thereby slowing global climate change. For example, between 2001 and 2021 the Amazon rainforest was a carbon sink, removing about 460 million tons of CO_2 out of the atmosphere annually. However, the increasing rate of fires, drought, deforestation, and the release of CO_2 by soil microbes indicate that without intervention this rain forest could flip from sink to source by 2035. In fact, some regions of the Amazon rainforest already have.

Microbial degradation of plant material and the development of SOM can be thought of as a three-step process. First, easily degraded compounds such as soluble carbohydrates and proteins are broken down by a variety of heterotrophic microorganisms. About half the carbon is respired as CO_2, and the remainder is rapidly incorporated into new biomass. During the second stage, complex carbohydrates, such as cellulose, are degraded. Cellulose, an important plant cell wall structural polysaccharide, represents a vast store of carbon and is the most abundant organic compound on Earth. Microbes secrete extracellular enzymes to break down cellulose. Brown rot fungi and filmentous bacteria (e.g., *Coniophora* and *Streptomyces* spp., respectively) degrade cellulose into cellobiose, while nonfilamentous bacteria (e.g., *Cytophaga* and *Bacillus* spp.) degrade cellulose into cellotriose. These smaller compounds are readily degraded and assimilated as glucose monomers.

Lignin, an important structural component of woody plants, is the soil component most resistant to degradation. While its exact structure differs among plant species, the common building block is the phenylpropene unit. This consists of a hydroxylated benzene ring and a three-carbon linear side chain (**figure 30.2**). A single lignin molecule can consist of up to 600 cross-linked phenylpropene units. It is therefore not surprising that degradation of lignin is much slower than that of cellulose. Basidiomycete fungi commonly termed white rot fungi (e.g., *Phanerochaete* and *Phlebia* spp.) and actinobacteria (e.g., *Streptomyces* spp.) secrete extracellular phenoloxidase enzymes that degrade lignin aerobically. Lignin can be degraded anaerobically, but this process is very slow, enabling lignin to accumulate in wet, poorly oxygenated soils, such as peat bogs.

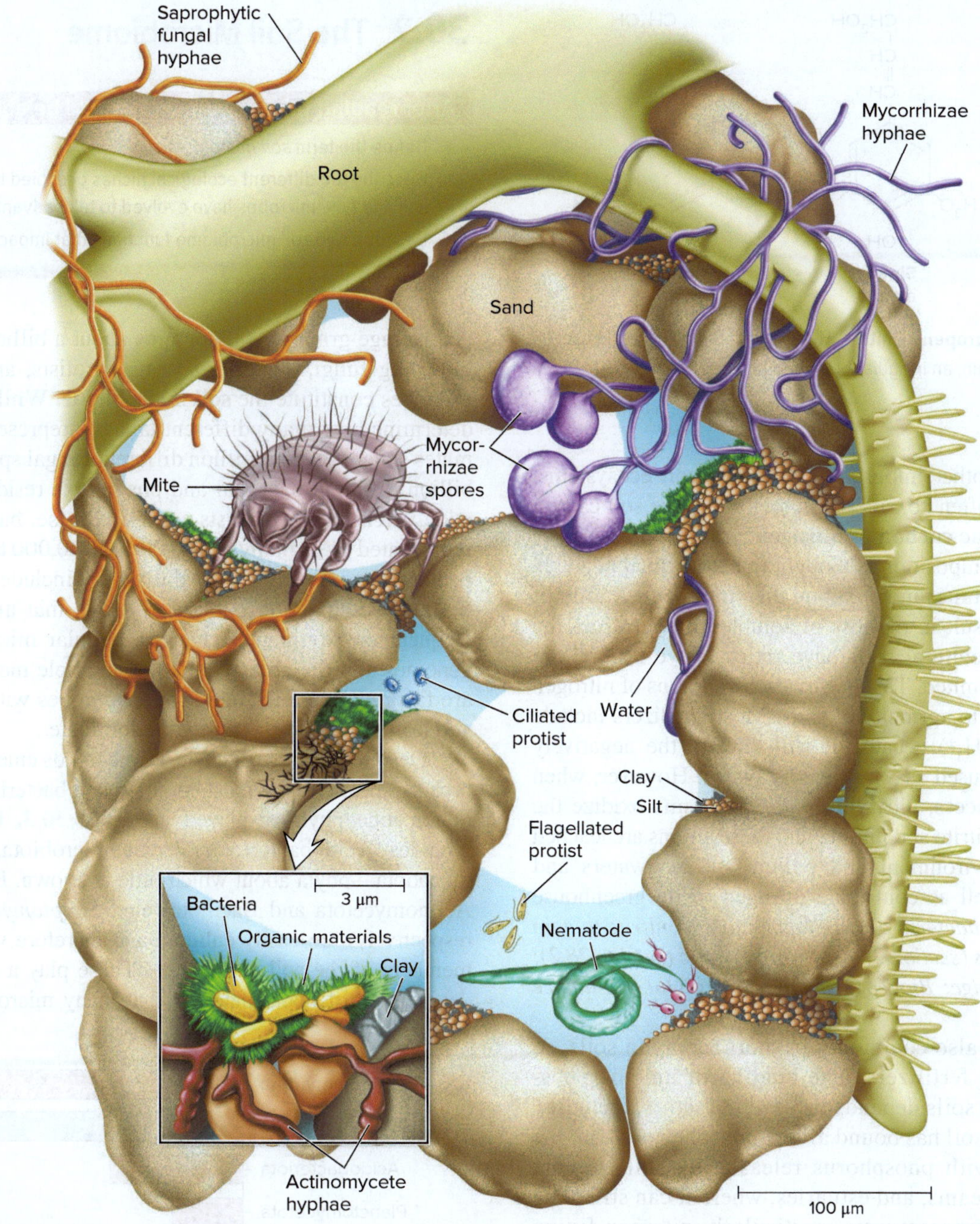

Figure 30.1 The Soil Habitat. A typical soil habitat contains a mixture of clay, silt, and sand along with soil organic matter. Roots, animals (e.g., nematodes and mites), and chemoheterotrophic bacteria consume oxygen, which is replaced by diffusion within the soil pores where microbes live. Two types of fungi are present: mycorrhizal fungi, which derive their organic carbon from their symbiotic partners, plant roots; and saprophytic fungi, which contribute to the degradation of organic material. In addition to heterotrophs, chemolithotrophic microorganisms are important in elemental cycling within the soil environment.

MICRO INQUIRY *How are filamentous microbes especially well suited to the soil ecosystem?*

Regardless of the presence or absence of oxygen, no more than 10% of lignin is converted to biomass.

Lignocellulose is another soil biopolymer, made up of cellulose and hemicellulose with lignin serving as the molecular glue that holds these polysaccharides together. Research is currently focused on optimizing microbial degradation of lignin to generate the building blocks for biofuel and bioplastics.

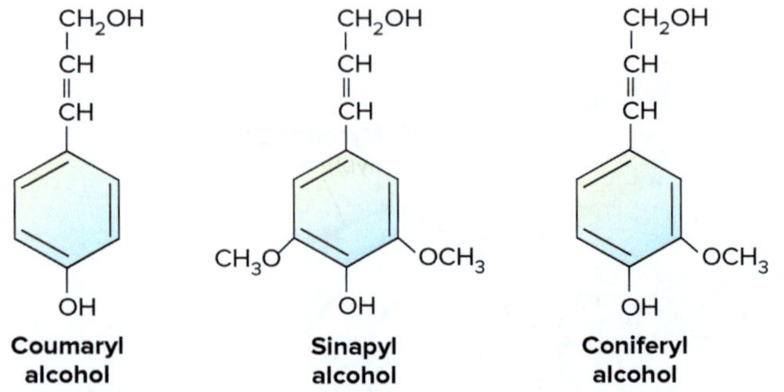

Figure 30.2 Phenylpropene Units. Hundreds of these molecules are polymerized to form lignin, an irregular branched polymer.

Nitrogen is another important element in soil ecosystems. Nitrogen in soil is often considered in relation to the soil carbon content as the organic **carbon to nitrogen ratio (C:N ratio).** As a general rule, decomposition is thought to be maximal at a C:N ratio of 30 (i.e., 30 times more carbon than nitrogen). A ratio of less than 30 can result in the loss of soluble nitrogen from the system. Many agricultural soils have ratios above 30 and are therefore nitrogen limited. This is why each year tons of nitrogen fertilizer are added to agricultural soils. Most fertilizers include ammonium ion (NH_4^+) because it will bind to the negatively charged clays until used as a nutrient by plants. However, when ammonium is in excess, nitrifying microorganisms oxidize the ammonium ion to nitrite and nitrate, and these anions are leached from the plant environment and pollute surface waters and groundwater, as well as contribute nitrogen oxide greenhouse gases. *Nitrifying bacteria oxidize ammonium or nitrite to gain energy and electrons (section 21.1); Nitrogen cycle (section 28.2); Global climate change: Biogeochemical Cycling Out of Balance (section 28.3)*

Phosphorus is also commonly in short supply in soils, so it is also added to fertilizers. The binding of anionic phosphate molecules to soils depends on several factors, including soil pH. When the soil has bound all the phosphorus it can, the excess, together with phosphorus released by soil erosion, moves to lakes, streams, and estuaries, where it can stimulate the growth of microorganisms, particularly nitrogen-fixing cyanobacteria, resulting in eutrophication. *Microorganisms in lakes can cause seasonal blooms (section 29.3)*

Comprehension Check

1. List three reasons why soil organic matter (SOM) is important.
2. In what order will these compounds be degraded in soils: peptides, lignin, cellulose, fructose?
3. Which of the following soils would a farmer be most likely to fertilize: a soil with a C:N ratio of 23, 28, or 37?
4. Why is most nitrogen fertilizer added as ammonium ion?

30.2 The Soil Microbiome

After reading this section, you should be able to:

a. Define the term *soil microbiome*
b. Describe the different ecological niches occupied by microbes in soil and how microbes have evolved to take advantage of the niches
c. Describe two soil microbiome functions that impact humans

An average gram of soil contains about a billion (10^9) microbes including fungi, bacteria, archaea, protists, and viruses; these microbes constitute the **soil microbiome.** While it is difficult to determine how many different taxa are represented, recent estimates suggest over a million different fungal species and about a billion different bacterial and phage taxa reside in soils worldwide. Archaea and protists are less diverse, but nonetheless are represented by as many as 100,000 and 10,000 taxa, respectively. Factors that impact microbial diversity include soil pH, C:N ratio, and precipitation patterns. In soils that undergo rapid and frequent environmental changes, cellular microbes with large genomes and the ability to deploy multiple metabolic strategies are favored, while in stable soils, microbes with small genomes and less metabolic flexibility predominate.

Despite this wide diversity, some trends emerge. For example, broadly speaking, the most common soil bacteria can be assigned to the nine phyla displayed in **figure 30.3.** Interestingly, this includes Acidobacteriota, Verrucomicrobiota, and Gemmatimonadota—phyla about which little is known. By contrast, many Actinomycetota and Bacillota (e.g., *Streptomyces* and *Bacillus,* respectively) are easily cultured and therefore well-characterized members of the soil community. These play a major role in the degradation of plant materials that many microbes cannot break

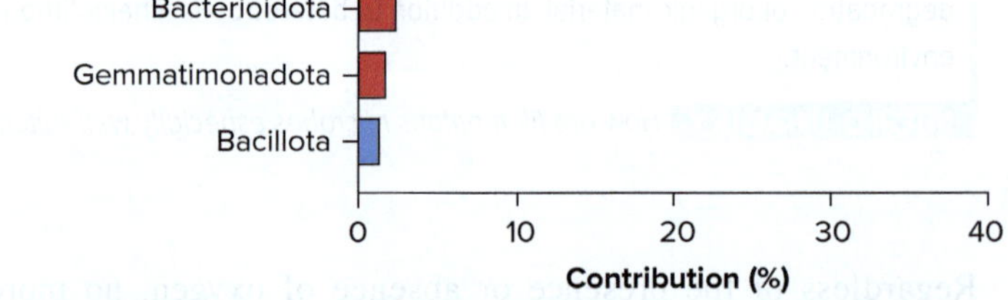

Figure 30.3 Known Bacterial Phyla Identified from 16S rRNA and 16S rRNA Genes Extracted from Soil. Gram-negative phyla are shown in red, Gram-positive in violet.

down. In addition, some members of these groups actively degrade pesticides. Filamentous bacteria, primarily of the genus *Streptomyces,* produce a volatile compound called geosmin, which gives soils their characteristic earthy odor. *Phylum Actinomycetota (section 22.1)*

Among archaea, ammonia-oxidizing members of the phylum Thermoproteota often prevail and may constitute up to 10% of the microorganisms. The abundance of archaea appears to be related to the C:N ratio, with larger archaeal populations occurring in soils with higher concentrations of nitrogen (lower C:N ratios). The relative contribution of bacterial and archaeal nitrification in soils is an active area of research (*see figure 21.9*). *Nitrogen cycle (section 28.2)*

Fungi are also very important members of the soil microbial community. Although we tend to think of soil fungi as the mushrooms sprouting on our lawns, the vast majority of fungal biomass is below ground. Their filaments can extend hundreds of meters per gram of soil, bridging open areas between soil particles. These fungi form darkly pigmented resting structures called sclerotia and hyphal cords. *Sclerotia* are compact masses of fungal hyphae designed for survival in harsh conditions. *Hyphal cords* consist of parallel fungal filaments that move nutrients and water over great distances, including across air spaces, a unique part of their functional strategy. Large, well-organized hyphal cords look similar to plant roots, so are called **rhizomorphs** (Greek *rhiza,* root; *morphe,* form).

Fungal biomass can be enormous. In fact the largest organism on Earth is found mostly underground in the mountains of Oregon. There a clone of the mushroom *Armillaria ostoyae,* whose rhizomorphs infect the roots of many conifers, has been determined to cover over 4 square miles. Estimated to be over 2,000 years old, it is one of the oldest living organisms as well.

Viruses in soils are important, but as predators of bacteria, archaea, fungi, protozoa, and nematodes (worms), they are difficult to quantify. A surprising additional function may be metabolic: Soil virome analysis reveals that many viruses carry genes that help degrade carbohydrates, such as glycoside hydrolases and carbohydrate esterases that may be expressed following lysogeny. This suggests that the means by which viruses contribute to biogeochemical cycling in soils is more complex than helping to regulate the population size of their prey.

Taken together, the abundance and diversity of microorganisms that constitute the soil microbiome underscore their importance as drivers of biogeochemical cycling, as discussed in chapter 28. The soil microbiome also represents a rich source and reservoir of plant and animal (including human) pathogens as well as microbes that produce useful natural products, such as antibiotics. Finally, the soil microbiome is of immense importance to agriculture as it directly determines soil fertility and crop productivity. As we discuss next, some microbes within the soil microbiome are also part of the plant microbiome.

Comprehension Check

1. What are the differences in preferred soil microhabitats between bacteria and filamentous fungi?
2. What types of archaea have been detected in soils?
3. Outline the contribution of plants, bacteria, fungi, protists, and viruses to the flux of carbon in soil.

30.3 The Plant Microbiome

After reading this section, you should be able to:

a. Define the term *plant microbiome*
b. Differentiate between the phyllosphere, rhizosphere, and rhizoplane environments and the microbes that dwell in each
c. Explain the importance of mycorrhizae to the growth of vascular plants
d. Contrast and compare the ectomycorrhizal life cycle and that of arbuscular fungi
e. Diagram the steps that result in the formation of rhizobium nodules on leguminous plants
f. Compare the process of stem nodulation with root nodulation
g. Discuss other microbes, in addition to rhizobia, that fix nitrogen
h. Explain how *Agrobacterium tumefaciens* causes tumors in plants
i. List at least five plant pathogens other than *Agrobacterium tumefaciens* and the diseases they cause
j. Identify a plant disease that is likely to be found near your home or school

All plants support an active assemblage of microorganisms that make up the **plant microbiome.** The plant microbiome can be broadly divided into two classes: microbes that live on the surface of plants, called **epiphytes,** and those that colonize internal plant tissues, called **endophytes.** Some of these microbes are commensals—no harm is done to the plant, but the microbe gains some advantage. Many others are beneficial to both the microorganism and the plant (i.e., they are mutualistic). Finally, other microbes are plant pathogens that parasitize their plant hosts. In all cases, the microbe and the plant have established the capacity to communicate. The microbe detects and responds to plant-produced chemical signaling molecules. This generally triggers the release of microbial compounds recognized by the plant, thereby beginning a two-way chemical conversation. Once a microbe-plant relationship is initiated, microbes and plants continue to monitor the physiology of their partner and adjust their own activities accordingly. The nature of the signaling molecules and the mechanisms by which both plants and microbes respond stimulate exciting multidisciplinary research in soil microbiology, ecology, molecular biology, genetics, and biochemistry. *Cell-cell communication within microbial populations (section 7.6); Many types of microbial interactions exist (section 27.1)*

Just as the human microbiome is differentiated by its anatomical site (e.g., the microbes in your gut are not the same as those on your skin), so too does the plant microbiome vary across different regions of any given plant. We therefore further classify the plant microbiome into microbes that live in the aboveground, or aerial, surfaces of plants and those that inhabit belowground plant tissues. We begin our discussion by introducing the microbiome within the aerial regions of plants.

Phyllosphere Microorganisms Are Stress Resistant

The environment of the aerial portion of a plant (e.g., stems, leaves, fruit, and flowers), called the **phyllosphere,** undergoes frequent and rapid changes in humidity, ultraviolet (UV) light exposure, and temperature. This causes fluctuations in the leaching of organic material (primarily simple sugars) that supports a diverse assortment of microbes, including bacteria, fungi, photosynthetic and heterotrophic protists, and viruses. Bacterial densities are estimated to be 10^6 to 10^7 cells per cm^2 of plant tissue. Culturable bacteria include the γ-proteobacteria *Pseudomonas syringae, Erwinia* spp., and *Pantoea* spp. In addition, methylotrophic α-proteobacteria such as *Methylobacterium* spp. are supported by plant release of methanol. Members of another abundant bacterial genus, *Sphingomonas,* produce pigments that function like sunscreen so they can survive the UV radiation striking plant surfaces. Importantly, microbial diversity in the phyllosphere is positively correlated with the productivity of the entire ecosystem. Furthermore, the phyllosphere microbial community contributes to global carbon and nitrogen cycling, the removal of airborne pollutants, and the decomposition of leaf litter.

The Rhizoplane and Rhizosphere Support Complex Microbial Communities

Plant roots receive between 30 and 60% of the net carbon fixed by photosynthesis. Of this, an estimated 40 to 90% enters soils as a wide variety of materials, including alcohols, ethylene, sugars, amino and organic acids, vitamins, nucleotides, polysaccharides, and enzymes. The plant root surface, termed the **rhizoplane** (Greek *rhiza,* root; Latin *planum,* flat surface), provides an exceptional environment for microorganisms, as gaseous, soluble, and particulate materials move from plant to soil.

The **rhizosphere** is the region surrounding a plant root rich in plant exudates more easily degraded than lignin and cellulose. This in turn supports microbial populations that can be as much as 100 times greater than surrounding bulk soil. Despite the seeming abundance of nutrients, microbes in rhizosphere communities must compete with one another, and they do so by the secretion of antimicrobial agents, lytic enzymes, and even hydrogen cyanide. Other microbes counteract these inhibitors by efflux pumps, detoxification enzymes, and secretion of molecules that alter competitor gene expression. Conversely, rhizosphere and rhizoplane microorganisms provide nutrients for other organisms, including plants and other microbes, thereby playing critical roles in organic matter synthesis and degradation.

Microbes in the rhizoplane and rhizosphere enhance plant growth by producing a variety of compounds. These include molecules that regulate plant growth called *phytohormones* (e.g., auxin and gibberellin) and compounds that stimulate host plant production of phytohormones. The production of volatile organic compounds including acetoin and phenazines—a diverse class of heterocyclic compounds—stimulates plant responses to stress. In addition, some bacteria secrete compounds that suppress the growth of plant fungal pathogens. The use of these compounds in agriculture is an active area of research.

Associative nitrogen fixation—N_2 fixation by epiphytic bacteria—is a critical process performed in the rhizoplane as well as in the rhizosphere. It is carried out by members of a number of genera, including *Azotobacter, Azospirillum,* and *Acetobacter.* These bacteria enhance the growth of agriculturally important crops such as wheat, potato, and barley. In addition to fixing N_2, many of these bacteria produce growth-promoting hormones that increase root hair development, thereby enhancing plant nutrient uptake.

Comprehension Check

1. Define *rhizosphere, rhizoplane,* and *associative nitrogen fixation.*
2. What stresses does a microorganism on a leaf contend with that a microbe in the soil does not?
3. List two ways in which compounds produced by rhizosphere and rhizoplane microbes promote plant growth.
4. What important genera are involved in associative nitrogen fixation?

Mycorrhizae Facilitate the Growth of Most Plants

Mycorrhizae (Greek *myco,* fungus; *rhizo,* root) are mutualistic relationships that develop between about 80% of all land plants and a limited number of filamentous fungal species. Mycorrhizae are so abundant that up to 100 m of mycorrhizal filaments can be found in a gram of soil. **Mycorrhizal fungi** use photosynthetically derived carbohydrate from their host. In return, they provide a number of services for their plant hosts. In moist environments, they increase the availability of nutrients, especially phosphorus and nitrogen. In arid environments, mycorrhizae aid in water uptake, increasing transpiration rates in comparison with nonmycorrhizal plants. Mycorrhizae also mediate plant interactions with the soil microbiome, helping to defend against pathogens and promote colonization by nitrogen-fixing bacteria. Indeed, the impact mycorrhizae have on the terrestrial biosphere cannot be understated. ▶| *Mutualism: obligatory positive interactions (section 27.2)*

Mycorrhizae can be broadly classified as **endomycorrhizae**—those with fungi that enter root cells, or as **ectomycorrhizae**—those that remain extracellular, forming a sheath of interconnecting filaments (hyphae) around roots. Although six types of mycorrhizae are detailed in **table 30.1** and **figure 30.4,** we focus primarily on the two most important types: ectomycorrhizae and arbuscular mycorrhizae.

Table 30.1 Mycorrhizal Associations

Mycorrhizal Classification	Fungi Involved	Plants Colonized	Fungal Structural Features	Fungal Function
Ectomycorrhizae	Most are basidiomycetes, including those with large fruiting bodies (e.g., toadstools); some ascomycetes	~90% of trees and woody plants in temperate regions; fungal-plant colonization is often species specific.	Hartig net, mantle or sheath; rhizomorphs; root hair development is usually limited.	N and P uptake and transfer; promote growth of plants of the same species the fungus has colonized; protect host plants from pathogens by ensheathing roots and acidifying soil
Arbuscular	Glomeromycota, in particular six genera of glomeromycetes	~80% wild and crop plants, tropical trees; fungal-plant colonization is not highly specific.	Arbuscules: hyphae-filled invaginations of cortical root cells	P > N uptake and transfer; promote seed production; reduce pest and nematode infection; increase drought and disease resistance
Ericaceous	Ascomycetes Basidiomycetes	Low evergreen shrubs, heathers	Intracellular coils	N and P uptake and transfer; mineralization of organic matter
Orchidaceous	Basidiomycetes	Orchids	Hyphal coils called pelotons within host tissue	Some orchids are non-photosynthetic and others produce chlorophyll only when mature; these organisms depend on mycorrhizae for organic carbon and nutrients.
Ectendomycorrhizae	Ascomycetes	Conifers	Hartig net with some intracellular hyphae	Nutrient uptake and mineralization of organic matter
Monotropoid mycorrhizae	Ascomycetes Basidiomycetes	Flowering plants that lack chlorophyll (*Monotropaceae;* e.g., Indian pipe)	Hartig net one cell deep in the root cortex	Nutrient uptake and transfer

Ectomycorrhizae (ECM) are formed primarily by basidiomycete fungi, although some ECM are ascomycetes. ECM colonize almost all trees in temperate climates. Their importance arises from their ability to transfer essential nutrients, especially phosphorus and nitrogen, to the root. The development of ECM starts with the growth of a fungal mycelium around the root. As the mycelium thickens, it forms a sheath or mantle so that the entire root may be covered by the fungal mycelium (**figure 30.5**). This increases the surface area in contact between the fungus and plant, thereby optimizing nutrient transfer. Most ECM produce signaling molecules that limit the growth of root hairs; thus ECM-colonized roots often appear blunt. From the root surface, the fungi extend hyphae into the soil; these filaments may aggregate to form rhizomorphs visible to the naked eye. Rhizomorphs consist of hyphae enclosed in a sheath; hyphae in the sheath penetrate between (but not within) root cells, forming a characteristic meshwork of hyphae called the **Hartig net** (figure 30.4*d*). Soil nutrients taken up by rhizomorphs pass through the hyphal sheath and into the Hartig net filaments, which form numerous contacts with root cells. This results in efficient exchange of soil nutrients from fungus to plant and carbohydrates from plant to fungus. This relationship has co-evolved to the point that some plants synthesize sugars solely for the use of their fungal partners, such as mannitol and trehalose.

Arbuscular mycorrhizae (AM) are the most common type of mycorrhizae and make up about half the fungal biomass in most soils. They are found in association with many tropical plants and, importantly, with most crop plants. AM fungi are Glomeromycota that can be cultured only with roots of their plant hosts. These microbes enter root cells between the plant cell wall and invaginations in the plasma membrane (figure 30.4*a*). So, although AM are endomycorrhizae, they do not breach the root cell membrane. Instead, tree-like hyphal networks called **arbuscules** develop within folds of the plasma membrane. Individual arbuscules are transient, lasting at most 2 weeks. Spores released by nearby root-associated AM germinate and colonize roots. This requires a number of specific steps, as shown in **figure 30.6.** This includes the development of *appressoria* (plural; appressorium, *s*), flattened and thickened tips of specialized hyphae used to attach and penetrate host plant cells. AM can be vigorous colonizers: A 5-cm segment of root can support the growth of as many as eight species, and hyphae from a single germinated spore can simultaneously colonize multiple roots of unrelated plant species, forming a complex underground highway for nutrient transport.

◀ *Glomeromycota are mycorrhizal symbionts (section 24.3)*

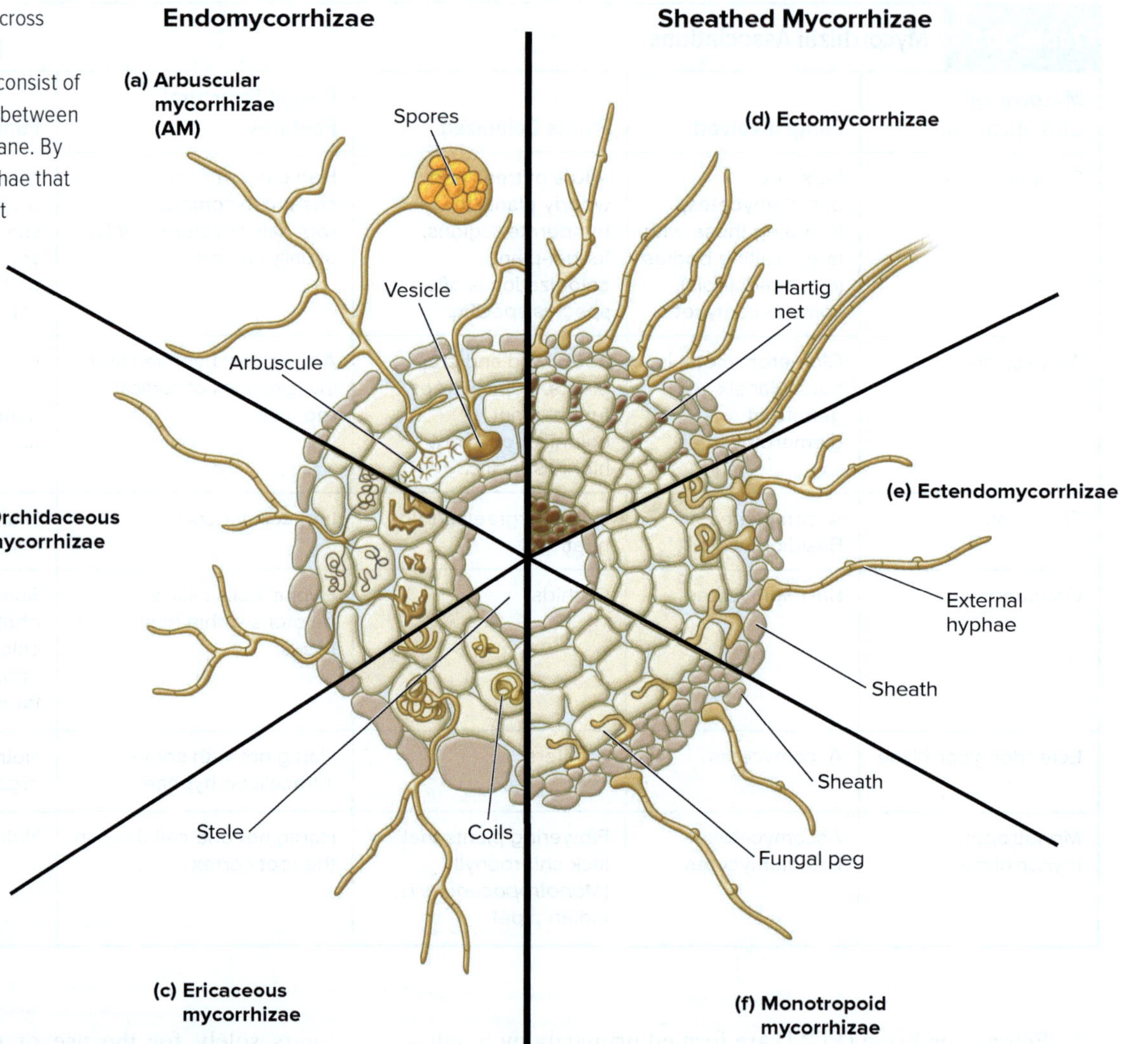

Figure 30.4 Mycorrhizae. Root cross sections illustrate different mycorrhizal relationships. Note that (a) abuscules consist of fungal filaments growing in networks between the plant cell wall and plasma membrane. By contrast (d) Hartig nets consist of hyphae that remain between plant cells and do not penetrate plant cell walls.

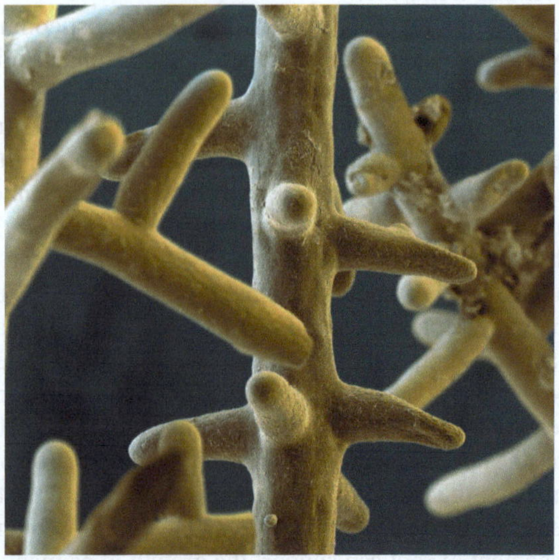

Figure 30.5 Ectomycorrhizae as Found on Roots of a Pine Tree. Note that the fungal hyphae grow as a sheath, enclosing the root surface and blunting root development.

Eye of Science/Science Source

Like the α-proteobacteria belonging to the rhizobium family (discussed next), AM establish symbioses with legumes (among other plants). The secretion of **Nod factors** by rhizobia has been well established, and analogous **Myc factors** have been purified from AM fungi (**figure 30.7**). Both Nod and Myc factors are lipochitooligosaccharides that prompt legume roots to accept symbionts. Whereas AM were present with the development of land plants 407 Mya (MYA), the rhizobium-legume symbiosis is believed to have arisen only about 60 Mya. This suggests that this signaling mechanism was adopted by rhizobia, which then further modified it.

AM are believed to provide a number of services to their plant hosts, including protection from disease, drought, nematodes, and other pests. AM use molecular signals to alert their host plants of impending pathogen and herbivore attack. These signals induce plant genes that encode defense- and stress-related products.

AM are probably most well known as providers of phosphorus and, to a lesser extent, nitrogen for their plant partners. Interestingly, AM assimilate NO_3^- and NH_4^+ through the glutamine synthetase–glutamate synthase (GOGAT) pathway (*see figure 12.12*), and therefore incorporate nitrogen into glutamine, which they then convert to arginine. But rather than transferring arginine to host cells, intracellular fungal hyphae degrade the amino acid, transferring only the NH_4^+ (**figure 30.8**). Thus while the fungus provides its host with much-needed ammonium, it retains the carbon skeleton required for nitrogen uptake. In contrast, less is known about how mycorrhizal fungi take up carbon from their hosts. Plants can detect, discriminate, and reward the best fungal partners with more carbohydrates and, in at least one case, fatty acids as well. In return, the fungi transfer more nutrients to the plants that "feed" them best. This tight plant-microbe coupling helps to account for the evolutionary stability of this partnership. ◀ *Inorganic nitrogen assimilation (section 12.5)*

In addition to these two most abundant mycorrhizal types, the orchidaceous mycorrhizae are of particular interest (figure 30.4*b*). Orchids are unusual plants because many never produce chlorophyll, while others only do so after they have matured past the seedling stage. Therefore all orchids have an absolute dependence on their endomycorrhizal partners for at least part of their lives. Indeed, orchid seeds will not germinate unless first colonized by a basidiomycete orchid mycorrhiza. Because orchids produce very little or no photosynthetically derived organic carbon, orchid mycorrhizal fungi are saprophytic. They must degrade soil organic matter to obtain carbon, which the orchids then also consume. In this case, the orchid functions as a parasite.

Regardless of the type of mycorrhizae, as the external hyphal network radiates out into the soil, a *mycorrhizosphere* is formed characterized by the flow of carbon from the plant into the mycorrhizal hyphal network and then into the surrounding soil. Many bacteria in this zone are called mycorrhization

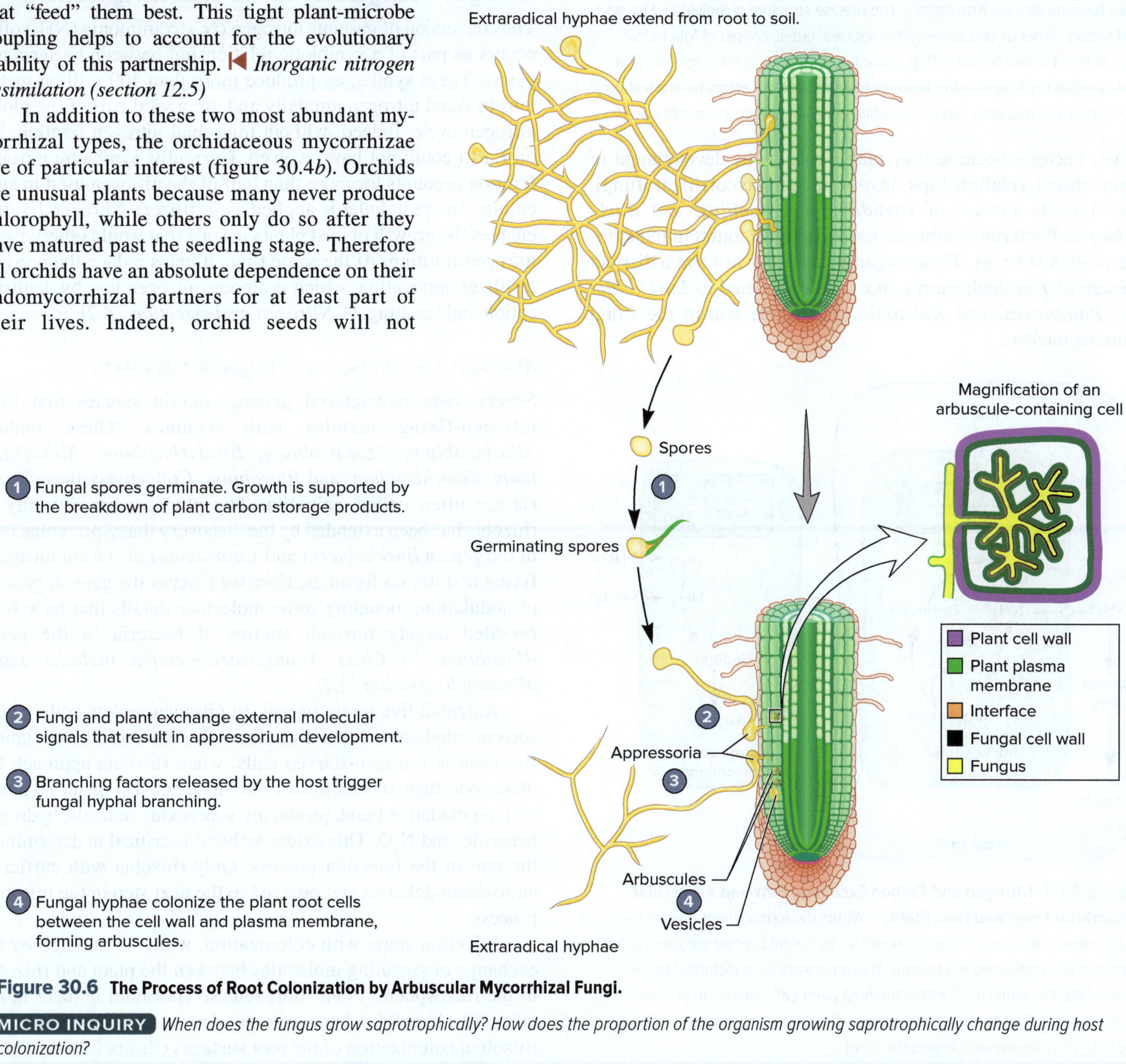

Figure 30.6 The Process of Root Colonization by Arbuscular Mycorrhizal Fungi.

MICRO INQUIRY *When does the fungus grow saprotrophically? How does the proportion of the organism growing saprotrophically change during host colonization?*

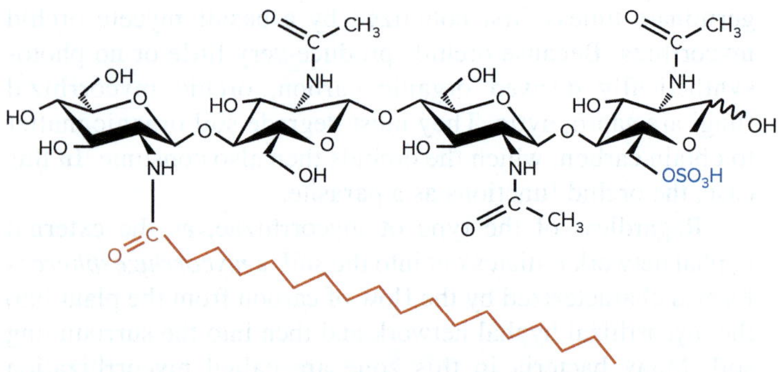

Myc factor produced by *Glomus intraradices.*

Figure 30.7 Myc Factor. Both arbuscular mycorrhizae and rhizobia secrete lipochitooligosaccharide signaling molecules. AM produce Myc factors, while bacteria secrete Nod factors. The precise structure of individual Myc and Nod factors depends on the microbial species, but all consist of four to five units of β–1,4 linked *N*-acetyl-*D*-glucosamine bearing a species-specific acyl chain (red) at the nonreducing terminal residue. All Nod factors have a sulfate attached to the reducing end (blue), while not all Myc factors are sulfonated.

helper bacteria because they play a role in the development of mycorrhizal relationships involving ectomycorrhizal fungi; they include species of *Pseudomonas, Bacillus,* and *Burkholderia.* Bacterial symbionts have also been found in the cytoplasm of AM fungi. These organisms appear to form a distinct lineage of γ-proteobacteria, most closely related to *Burkholderia, Pandoraea,* and *Ralstonia.* Their role within the fungi remains unclear.

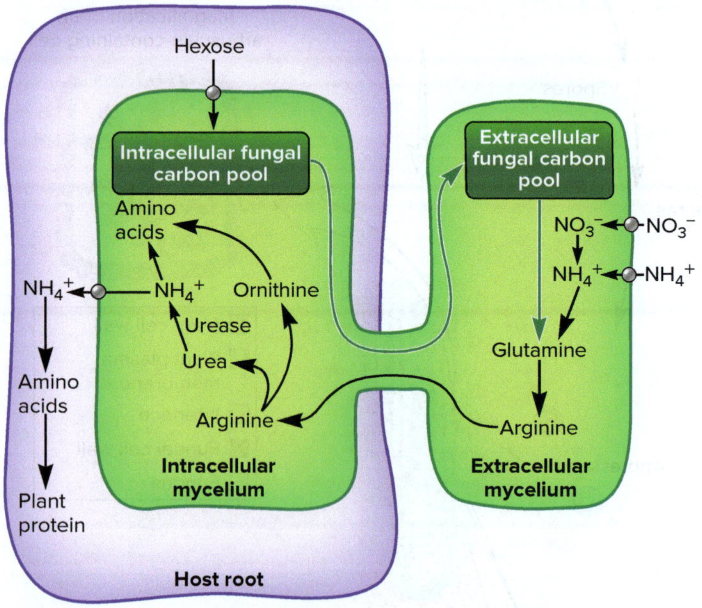

Figure 30.8 Nitrogen and Carbon Exchange Between Arbuscular Mycorrhizal Fungi and Host Plant. Within the extracellular mycelium (right), nitrate and ammonium are taken up by fungal hyphae outside the host plant cell and converted to arginine. This amino acid is transferred to the intracellular mycelium (left) within the host plant cell, where the enzyme arginase converts it to urea and ornithine. Urease then cleaves urea to CO_2 and NH_3. Only ammonium enters the plant.

Comprehension Check

1. Describe the two-way relationship between mycorrhizal fungi and the plant host.
2. List three major differences between AM and EM. What plants in your neighborhood do you think rely on either AM or EM?
3. What is the function of the rhizomorph and the Hartig net?
4. Describe the uptake and transfer of ammonium by AM to the plant host. Why is only the ammonium transferred?
5. Propose two potential functions for mycorrhization helper bacteria.

Nitrogen-Fixing Bacteria Are Vital to Agriculture

The conversion of gaseous nitrogen (N_2) to ammonia (NH_3) often occurs as part of a symbiotic relationship between bacteria and plants. These symbioses produce more than 100 million metric tons of fixed nitrogen annually and are a vital part of the global nitrogen cycle. Indeed, without microbial nitrogen fixation, life on Earth could not have evolved. Currently, symbiotic nitrogen fixation accounts for more than half of the nitrogen used in agriculture (the remainder is applied as fertilizer). Nitrogen fixation enables the growth of host plants in soils that would otherwise be nitrogen limiting. At the same time, it helps reduce the need for fertilizer application, which promotes nitrogen loss by denitrification and leaching. ◀ *Nitrogen cycle (section 28.2)*

Rhizobia Are Symbionts of Leguminous Plants

Several α-proteobacterial genera contain species that form nitrogen-fixing nodules with legumes. These include *Allorhizobium, Azorhizobium, Bradyrhizobium, Mesorhizobium, Sinorhizobium,* and *Rhizobium.* Collectively these bacteria are often called **rhizobia.** The phylogenetic diversity of rhizobia has been extended by the discovery that γ-proteobacteria in the genera *Burkholderia* and *Cupriavidus* also form nitrogen-fixing nodules on legumes. Here we discuss the general process of nodulation, including some molecular details that have been revealed largely through studies of bacteria in the genus *Rhizobium.* ◀ *Class Alphaproteobacteria includes many oligotrophs (section 21.1)*

Rhizobia live freely in soil. In nitrogen-replete soils, plants secrete ethylene-mediated compounds that inhibit nodulation. But even in nitrogen-starved soils, when rhizobia approach the plant root, they are assumed to be invaders. The plant responds with an oxidative burst, producing superoxide radicals, hydrogen peroxide, and N_2O. This oxidative burst is critical in determining the fate of the infection process. Only rhizobia with sufficient antioxidant defenses can proceed to the next step in the infection process.

Infection starts with colonization, which is initiated by the exchange of signaling molecules between the plant and rhizobia in the rhizosphere. Plant roots release flavonoid inducer molecules (2-phenyl-1,4-benzopyrone derivatives) that stimulate rhizobial colonization of the root surfaces (**figure 30.9*a*,** step 1).

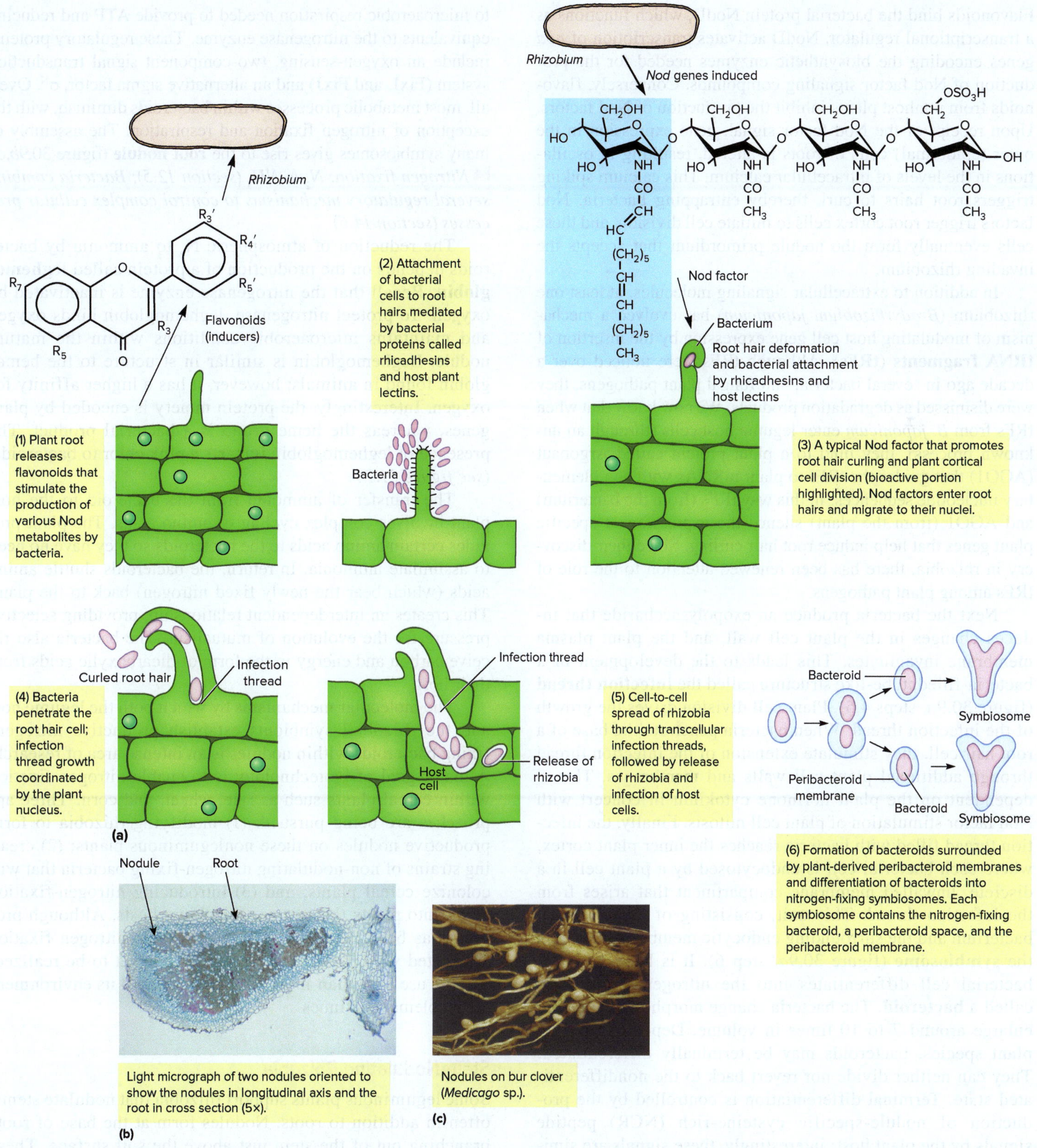

Figure 30.9 Root Nodule Formation by a *Rhizobium* Species. (a) Initiation of bacterial penetration into the root hair cell, infection thread development, and differentiation of bacterial cells into bacteroids. (b) Light micrograph of two nodules that develop by cell division (×5). This section is oriented to show the nodules in longitudinal axis and the root in cross section. (c) Nodules on bur clover (*Medicago* sp.). (b) Biology Pics/Science Source; (c) Hugh Spencer/Science Source

Flavonoids bind the bacterial protein NodD, which functions as a transcriptional regulator. NodD activates transcription of *nod* genes encoding the biosynthetic enzymes needed for the production of Nod factor signaling compounds. Conversely, flavonoids from nonhost plants inhibit the production of Nod factors. Upon receipt of the Nod factor signal, gene expression in the outer (epidermal) cells of roots is altered, resulting in oscillations in the levels of intracellular calcium. This calcium spiking triggers root hairs to curl, thereby entrapping bacteria. Nod factors trigger root cortex cells to initiate cell division, and these cells eventually form the nodule primordium that accepts the invading rhizobium.

In addition to extracellular signaling molecules, at least one rhizobium (*Bradyrhizobium japonicum*) has evolved a mechanism of modulating host cell gene expression by the insertion of **tRNA fragments (tRFs).** Although tRFs were noticed over a decade ago in several bacterial and fungal plant pathogens, they were dismissed as degradation products. We now know that when tRFs from *B. japonicum* enter legume host cells (through an unknown process), they bind to a plant protein called Argonaut (AGO1) that guides the tRFs to plant mRNAs with complementary nucleotide sequences. In this way tRFs (from the bacterium) and AGO1 (from the plant) silence the expression of specific plant genes that help induce root hair curling. Since their discovery in rhizobia, there has been renewed attention to the role of tRFs among plant pathogens.

Next the bacteria produce an exopolysaccharide that induces changes in the plant cell wall, and the plant plasma membrane invaginates. This leads to the development of a bacteria-filled, tube-like structure called the **infection thread** (figure 30.9*a,* steps 4,5). Plant cell division drives the growth of the infection thread. When bacteria penetrate the base of a root hair cell, they stimulate extension of the infection thread through additional plant cell walls and membranes. This is dependent on the plant hormone cytokinin in concert with Nod factor stimulation of plant cell mitosis. Finally, the infection thread filled with bacteria reaches the inner plant cortex, where each bacterial cell is endocytosed by a plant cell in a discrete, unwalled membrane compartment that arises from the infection thread. This unit, consisting of an individual bacterium and the surrounding endocytic membrane, is called the **symbiosome** (figure 30.9*a,* step 6). It is here that each bacterial cell differentiates into the nitrogen-fixing form called a **bacteroid.** The bacteria change morphologically and enlarge around 7 to 10 times in volume. Depending on the plant species, bacteroids may be terminally differentiated: They can neither divide nor revert back to the nondifferentiated state. Terminal differentiation is controlled by the production of nodule-specific cysteine-rich (NCR) peptide signals by the plant host; interestingly, these signals are similar in structure to antimicrobial peptides produced by humans called defensins.

The low levels of oxygen within the symbiosome trigger a regulatory cascade that controls nitrogen fixation and the switch to microaerobic respiration needed to provide ATP and reducing equivalents to the nitrogenase enzyme. These regulatory proteins include an oxygen-sensing, two-component signal transduction system (FixL and FixJ) and an alternative sigma factor, σ^N. Overall, most metabolic processes within bacteroids diminish, with the exception of nitrogen fixation and respiration. The assembly of many symbiosomes gives rise to the **root nodule** (figure 30.9*b,c*).

◀ *Nitrogen fixation: N_2 to NH_3 (section 12.5); Bacteria combine several regulatory mechanisms to control complex cellular processes (section 14.6)*

The reduction of atmospheric N_2 to ammonia by bacteroids depends on the production of a protein called **leghemoglobin.** Recall that the nitrogenase enzyme is inactivated by oxygen. To protect nitrogenase, leghemoglobin binds oxygen and maintains microaerobic conditions within the mature nodule. Leghemoglobin is similar in structure to the hemoglobin found in animals; however, it has a higher affinity for oxygen. Interestingly, the protein moiety is encoded by plant genes, whereas the heme group is a bacterial product. The presence of leghemoglobin imparts a pink color to bacteroids. (*see figure 21.7*)

The transfer of ammonia from the bacteroid to the host plant involves complex cycling of amino acids. The plant provides certain amino acids to the bacteroids so they have no need to assimilate ammonia. In return, the bacteroids shuttle amino acids (which bear the newly fixed nitrogen) back to the plant. This creates an interdependent relationship, providing selective pressure for the evolution of mutualism. The bacteria also receive carbon and energy in the form of dicarboxylic acids from their host.

The molecular mechanisms by which both the legume host and the rhizobial symbionts establish productive nitrogen-fixing bacteroids within nodules is an intense area of research. A major goal of biotechnology is to enable nitrogen fixation within cereal plants such as rice, wheat, and corn. Three approaches are being pursued: (1) modifying rhizobia to form productive nodules on these nonleguminous plants; (2) creating strains of non-nodulating nitrogen-fixing bacteria that will colonize cereal plants; and (3) introducing nitrogen-fixation genes into plants that are not rhizobium hosts. Although progress has been made using each approach, nitrogen fixation associated with these important crops has yet to be realized, and the use of human-made fertilizer, with all its environmental problems, continues.

Stem-Nodulating Rhizobia

Some leguminous plants support rhizobia that nodulate stems, often in addition to roots. Nodules form at the base of roots branching out of the stem just above the soil surface. These plants are generally found in waterlogged soils and riverbanks.

Stem nodulation has not been as well studied as root nodulation, but it is clear that the two processes differ. Stem nodulation starts with the entry of bacteria through ruptures in stem epidermal cells that occur during emergence of the root tip. Unlike root nodulation, stem nodulation does not appear to involve the production of Nod factors or plant calcium spikes (**figure 30.10**). Nonetheless, as with root nodulation, bacteria within stem nodules are protected from the external environment and can obtain plant-derived energy substrates.

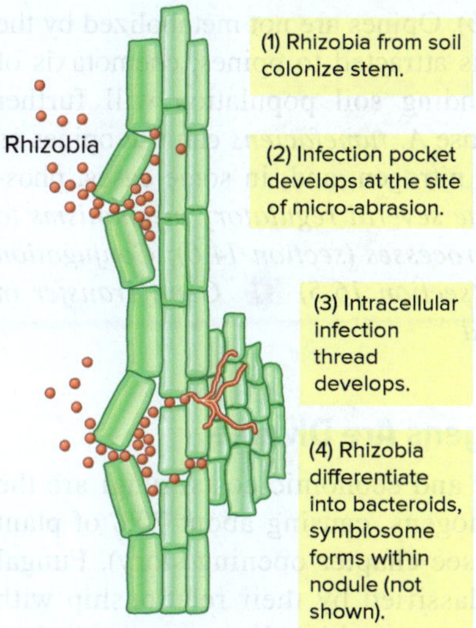

Figure 30.10 Stem Nodulation.

Frankia form Actinorhizae

Another example of symbiotic nitrogen fixation occurs between actinobacteria of the genus *Frankia* and eight nonleguminous host plant families. These bacterial associations with plant roots are called **actinorhizae** (**figure 30.11**). *Frankia* spp. are important particularly in trees and shrubs. The nodules of some plants (*Alnus, Ceanothus* spp.) can be as large as baseballs. The nodules of *Casuarina* spp. (Australian pine) approach soccer ball size.

Depending on the host plant, the infection process proceeds either by root hair infection or by direct penetration of the matrix between root cells. When infecting through root hairs, the bacteria induce plant cell division, leading to the development of a prenodule. *Frankia* cells then enter the cortex of the root to form a nodule that resembles a lateral root. When infection is independent of root hair invasion, the bacteria are encapsulated in a shell made of pectin that then extends through the plant's intercellular spaces. Although *Frankia* nodules vary in morphology, they all bear the general appearance of lateral root clusters that grow by repeated branching of the tips (*see figure 22.11*). Some nodules are found just below the surface of the soil, whereas others—particularly those forming on plants that grow in dry, rocky soil—develop very deep below the surface. ◀ *Order Actinomycetales (section 22.1)*

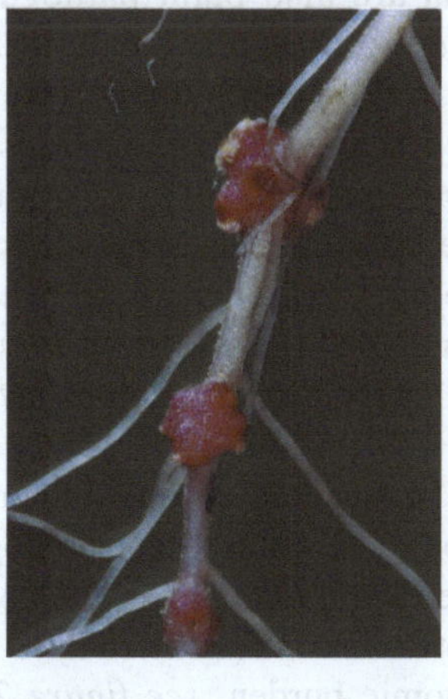

Figure 30.11 Actinorhizae. *Frankia*-induced actinorhizal nodules. Leghemoglobin makes the nodules appear red.

Biophoto Associates/Science Source

MICRO INQUIRY *How do you think* Frankia *spp. protect nitrogenase from oxygen damage?*

Comprehension Check

1. List several bacteria that are considered rhizobia.
2. What is the function of the infection thread?
3. What does the term *terminally differentiated* mean?
4. How does nitrogen transfer between a rhizobium and its host differ from that between arbuscular mycorrhizae and host?
5. Describe two examples that demonstrate coevolution between rhizobia and legumes.
6. What are the two general mechanisms by which *Frankia* spp. nodulate their plant hosts? Which types of plants are infected?

Agrobacterium Is a Model Plant Pathogen

Not all microbe-plant interactions are beneficial for both partners as many involve microbial pathogens that harm or even kill their host. *Agrobacterium tumefaciens* is an α-proteobacterium that has been studied intensely. Initially research focused on how this microbe causes crown gall disease, which results in the formation of tumor-like growths in a wide variety of plants (**figure 30.12**). A better understanding of this mechanism led to *A. tumefaciens* becoming one of biotechnology's most important tools because the molecular genetics by which this pathogen infects its host are the basis for plant genetic engineering. ▶ *Agricultural biotechnology repurposes plant and insect pathogens (section 41.4)*

The genes for plant infection and virulence are encoded on an *A. tumefaciens* plasmid called the **Ti (tumor-inducing) plasmid.** These genes include 21 *vir* genes (*vir* stands for virulence), found in six separate operons. Two of these genes, *virD1* and *virD2,* encode proteins that excise a specific region of the Ti plasmid, called **T-DNA.** After excision, the T-DNA fragment is integrated into the host plant's genome. Once incorporated into a plant cell's genome, T-DNA directs the overproduction of phytohormones that cause unregulated growth and reproduction of plant cells, thereby generating a tumor or gall in the plant.

The *vir* genes are not expressed when *A. tumefaciens* is living saprophytically in the soil. Instead, they are induced by the presence of plant phenolics and monosaccharides in an

Figure 30.12 ***Agrobacterium.*** *Agrobacterium tumefaciens*–caused tumor on a maple tree.
Joanne M. Willey/McGraw Hill

acidic (pH 5.2–5.7) and cool (below 30°C) environment (**figure 30.13*a***). The microbe infects its host through a wound. Upon reception of the plant signal, a two-component signal transduction system is activated: VirA is a sensor kinase that, in the presence of a phenolic signal, phosphorylates the response regulator VirG. Activated VirG then induces transcription of the other *vir* genes. This enables the bacterial cell to become adequately positioned relative to the plant cell, at which point the *virB* operon expresses the apparatus that will transfer the T-DNA. This transfer is similar to bacterial conjugation and involves a type IV secretion system. After VirD1 and VirD2 excise the T-DNA from the Ti plasmid, the T-DNA, with the VirD2 protein attached to the 5′ end, is delivered to the plant cell cytoplasm. The protein VirE2 is also transferred, and together with VirD2, the T-DNA is transported to the plant cell nucleus, where it is integrated into the host's genome.

Once it is part of the host cell genome, the T-DNA has two specific functions. First, it directs the host cell to overproduce phytohormones that cause tumor formation. Second, it stimulates the plant to produce special amino acid and sugar derivatives called opines (figure 30.13*b*). Opines are not metabolized by the plant, but *A. tumefaciens* is attracted to opines; chemotaxis of bacteria from the surrounding soil population will further advance the infection because *A. tumefaciens* can use opines as sources of carbon, energy, nitrogen, and, in some cases, phosphorus. *Bacteria combine several regulatory mechanisms to control complex cellular processes (section 14.6); Conjugation requires cell-cell contact (section 16.5)* *Gene Transfer in Plants Using the Ti Plasmid*

Microbial Plant Pathogens Are Diverse

In terms of both diversity and economic costs, fungi are the most important plant pathogens, causing about 70% of plant disease (**table 30.2;** also see chapter opening story). Fungal plant pathogens can be classified by their relationship with their plant hosts: They consume either living or dead plant material. **Biotrophic fungi** assimilate living plant material but do not kill their hosts. To do so, they must overcome the plant immune hormone salicylic acid. Parasitic biotrophic fungi include powdery mildew and rust fungi; these microorganisms have evolved a special feeding apparatus called the *haustorium* to take up nutrients. Haustoria develop by the expansion of fungal hyphae that penetrate plant host cells. By contrast, **necrotrophic fungi** kill their host plant cells by releasing toxins or enzymes prior to consuming plant nutrients. These pathogens have evolved to evade the plant immunity hormones jasmonic acid and ethylene. Necrotrophic fungi infect (and kill) a broader range of hosts. *Sclerotinia sclerotiorum* (white mold fungus) and *Botrytis cinerea* (grey mold fungus) are two notorious necrotrophic fungi that each parasitize over 200 different plant species and are responsible for the loss of millions of pounds of produce both pre- and post-harvest annually.

Many bacteria are also plant pathogens (**Disease 30.1**). Like *A. tumefaciens,* some plant pathogenic bacteria infect through a wound in the plant. Others, like the soft-rot bacteria *Dickeya dadantii* and *Pectobacterium carotovorum* (formerly *Erwinia chrysanthemi* and *E. carotovora,* respectively), digest plant tissue by secreting enzymes that degrade pectin, cellulose, and proteins. Another group of important bacterial plant pathogens includes the wall-less *Phytoplasma* spp. that infect vegetable and fruit crops such as sweet potatoes, corn, and citrus.

As discussed in chapters 23 and 25, protists, viruses, and viroids can also be devastating plant pathogens. Examples include the öomycete *Phytophthora infestans,* which was responsible for the Irish potato famine (from 1845 to 1852, but continues to extract an economic burden, *see figure 23.21*), and tobacco mosaic virus (TMV), the first virus to be characterized. A virus of particular interest in terms of plant-pathogen interactions is a hypovirus (family *Hypoviridae*) that infects the fungus *Cryphonectria parasitica,* the cause of chestnut blight. Infection of trees with less lethal strains of this virus is the basis of biological control efforts for chestnut blight in Europe and North America.

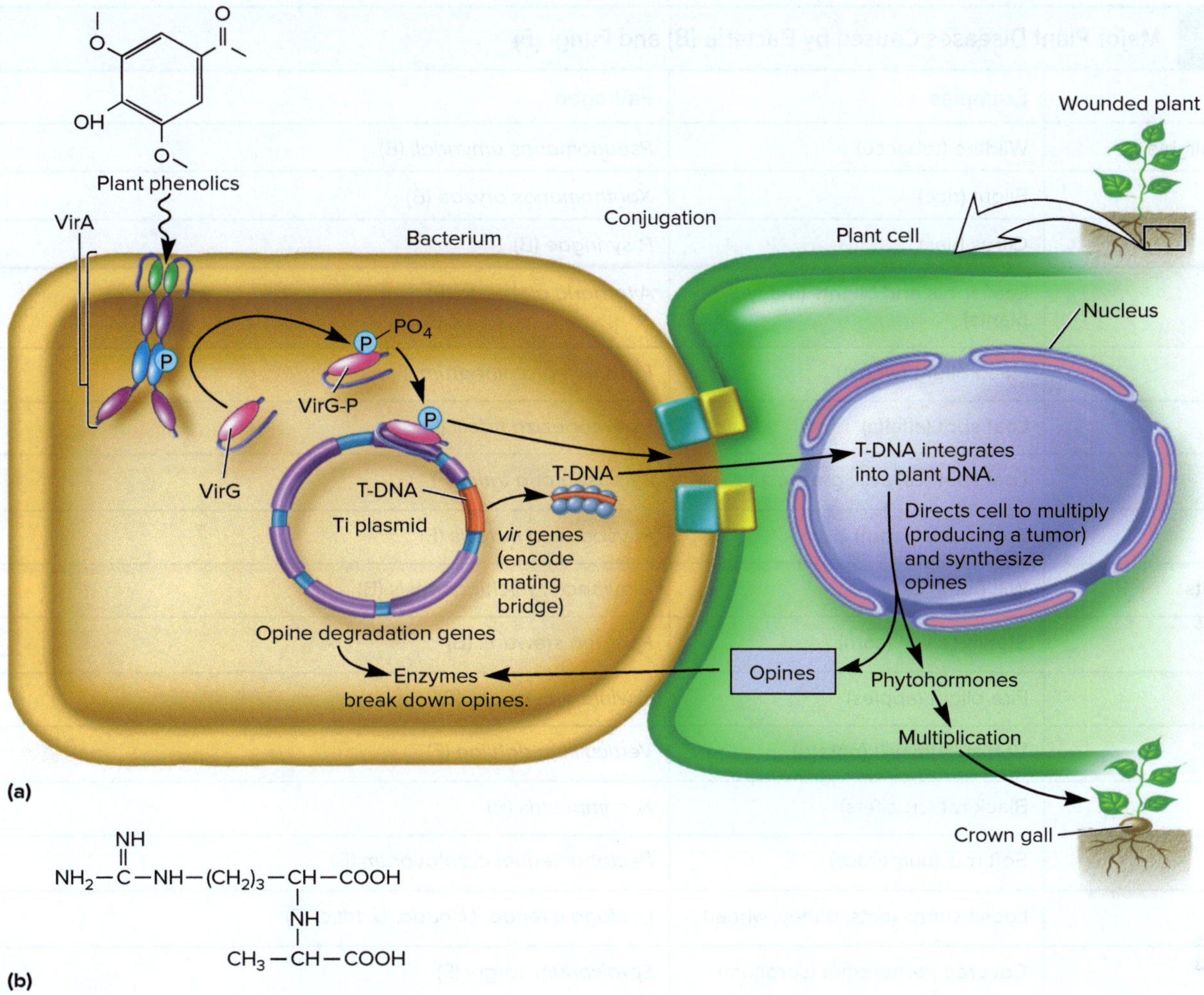

Figure 30.13 Functions of Genes Carried on the *Agrobacterium tumefaciens* Ti Plasmid. (a) Genes carried on the Ti plasmid of *A. tumefaciens* control tumor formation by a two-component regulatory system that stimulates formation of the mating bridge (a type IV secretion system) and excision of the T-DNA. The T-DNA is moved into the plant cell, where it integrates into the plant DNA. T-DNA induces the production of plant hormones that cause the plant cells to divide, producing a tumor. Note the plant and bacterial cells are not drawn to scale for illustrative purposes. (b) An example opine produced by tumor cells. Opines serve as a carbon source for the infecting *A. tumefaciens.* Ultimately a crown gall is formed on the stem of the wounded plant above the soil surface.

MICRO INQUIRY *How does the production of opines by the plant accelerate and amplify the* A. tumefaciens *infection process?*

Comprehension Check

1. What is the difference between the Ti plasmid and T-DNA?
2. What functions do the members of the two-component system play in infection of a plant by *A. tumefaciens?* What are the roles of phenolics and opines in this infection process?
3. What are the two general ways by which plant pathogens gain entry into the host plant tissue? What are the advantages and disadvantages of each mechanism?
4. What is the difference between biotrophic and necrotrophic fungi?

30.4 The Subsurface Biosphere Is Vast

After reading this section, you should be able to:

a. Describe four subsurface zones and the types of microbes that inhabit each

b. Explain at least one new discovery that has come from exploring the "deep hot biosphere"

For many years, it was thought that life could exist only in the thin veneer of Earth's surface and that any microbes recovered from sediments hundreds of meters deep were contaminants

Table 30.2 Major Plant Diseases Caused by Bacteria (B) and Fungi (F)

Symptoms	Examples	Pathogen
Spots and blights	Wildfire (tobacco)	*Pseudomonas amygdali* (B)
	Blight (rice)	*Xanthomonas oryzae* (B)
	Citrus blast	*P. syringae* (B)
	Leaf spots and blights (many plants)	*Alternaria alternata* (F)
	Head blight (wheat)	*Fusarium graminearum* (F)
	Leaf spot (alfalfa)	*Pseudopeziza trifolii* (F)
	Leaf spot (cereals, grasses)	*Zymoseptoria tritici* (F)
	Brown spot (corn)	*Physoderma maydis* (F)
Vascular wilts	Wilt (tomato)	*Clavibacter michiganensis* (B)
	Stewart's wilt (corn)	*Pantoea stewartii* (B)
	Fire blight (apples)	*Erwinia amylovora* (B)
	Verticullium wilt (potato)	*Verticullium dahliae* (F)
Soft rots	Black rot (crucifers)	*X. campestris* (B)
	Soft rots (numerous)	*Pectobacterium carotovorum* (B)
Smuts	Loose smuts (oats, barley, wheat)	*Ustilago avenae, U. nuda, U. tritici* (F)
	Covered kernel smut (sorghum)	*Sproisorium sorghi* (F)
Cankers	Canker (stone fruit)	*P. syringae* (B)
	Hyoxylon canker (oak trees)	*Biscogniauxia atropunctatum* (F)
Rusts	Asian soybean rust	*Phakopsora pachyrrhizi* (F)
	Leaf rust (cereals)	*Puccinia recondita* (F)
	White pine blister rust	*Cronartium ribicola* (F)

Source: From Lengler, J. W., et al. 1999. Biology of the prokaryotes. *Malden, MA: Blackwell Science, table 34.4.*

introduced during sampling. This view was drastically altered in the 1980s when the U.S. Department of Energy started looking for novel ways to store toxic waste. The agency began funding studies that applied modern technologies to sample the deep subsurface. Subsequent reports of microbes at great depth gained credibility, and international teams of geologists and microbiologists have since recovered microbes from thousands of meters below Earth's surface. The application of culture-independent techniques to quantify the density, diversity, and activity of microbes has revealed that subsurface microbes constitute about one-third of Earth's living biomass. This realization has made deep subsurface microbiology an exciting and active field.

Microbial processes take place in different subsurface regions, including (1) the shallow subsurface where water flowing from the surface moves below the plant root zone; (2) regions where organic matter, originally from the Earth's surface, has been transformed by chemical and biological processes to yield coal (from land plants), kerogens that yield fossil fuels upon distillation (from marine and freshwater microorganisms), and oil and gas; (3) zones where methane is synthesized as a result of microbial activity; and (4) the energy-starved deep biosphere.

DISEASE

30.1 Citrus Greening and the Power of "Why?"

The disease citrus greening, also known as huanglongbing (HLB), is devastating the citrus industry. First seen in the early 1900s in China, HLB is now also found in Africa and the Americas. Since an HLB outbreak in Florida began in 2005, it has reduced the state's citrus production by 75% while more than doubling the cost. And it's spreading: The disease now infects citrus trees in Louisiana, Georgia, South Carolina, Texas, California, Puerto Rico, and the U.S. Virgin Islands, as well as Cuba, Belize, Mexico, and Brazil.

HLB is a bacterial disease, caused by an unculturable Gram-negative bacterium in the Rhizobiaceae family. There are three species of this plant pathogen: In Africa it is caused by *Candidatus* Liberibacter africanus, in Asia and the United States by *Ca.* L. asiaticus, and in Brazil by *Ca.* L. americanus. HLB is transmitted by an insect called the Asian citrus psyllid, which is about the size of a grain of rice. When the insect takes a meal of plant sap, bacteria move from its salivary glands to the plant's phloem. The bacteria attack the plant's vasculature, blocking the flow of nutrients. Leaves become mottled, the fruit fails to ripen, and the juice is acidic, unpalatable, and unmarketable (**box figure**). Unfortunately, this isn't just impacting your morning glass of OJ; lots of citrus trees are susceptible, including grapefruit, lime, lemon, tangerine, and many more. Once a tree is infected, there is no cure and most infected trees die within a few years. This has triggered quarantining trees and enhanced cleaning of citrus fruit before shipping.

The obvious question is, what can be done to slow or stop citrus greening? Efforts to control the insect vector through pesticides and biocontrol measures help, but are not enough. In 2019 the dire situation prompted the U.S. Environmental Protection Agency to issue emergency authorization for the spraying of streptomycin and oxytetracycline antibiotics on citrus orchards. This was despite loud objections from the Centers for Disease Control and Prevention and the Food and Drug Administration, who point out that releasing tons of antibiotics meant for clinical use into the environment is a sure way to increase our already grim antimicrobial resistance problem. Surely, there must be a better strategy.

This brings us to the Australian finger lime tree, which is naturally resistant to HLB. Scientists at University of California, Riverside (UCR), took notice and asked the simple question: Why? They discovered that finger lime fruit produce a peptide toxic to *Ca.* L. asiaticus, harmless to people, and not part of our antibiotic armamentarium. And the peptide is practical: It's heat stable—so it will remain active in the hot sun, can be applied by spraying or injection, and has lasting activity, so it only needs to be applied a few times each year. The UCR team is currently field testing peptide application to treat HLB and prevent further spread of disease. Such is the power of "Why?"

U.S. Department of Agriculture (USDA)

In the shallow subsurface, surface waters often move through aquifers—porous geological structures below the plant root zone. In a pristine system with an oxic surface zone, the electron acceptors used in catabolism are distributed from the most oxidized and energetically favorable (oxygen) near the surface to the least favorable (in which CO_2 is used in methanogenesis) in lower zones (*see figures 28.1 and 29.12*).

In subsurface regions where organic matter from the Earth's surface has been buried and processed by thermal and biological processes, insoluble kerogens and coals break down to yield gas and oil. These mobile products move upward into more porous geological structures where microorganisms are active. Chemical signature molecules from plant and microbial biomass are present in these petroleum hydrocarbons. Microorganisms also appear to be growing in intermediate-depth, oil-bearing structures. Active bacterial and archaeal assemblages are present in high-temperature (60° to 90°C) oil reservoirs, including members of such genera as *Thermotoga, Thermoanaerobacter,* and *Thermococcus*. Archaeal species are dominated by methanogens.

The "deep hot biosphere" is a term suggested by Thomas Gold to describe the largest and most mysterious biosphere on Earth: Bacteria and archaea living in biofilms kilometers beneath the Earth's crust. The lack of apparent substrates to support growth and the calculated generation times in centuries to even 1,000 years seem alien to our understanding of microbial life. Anaerobic chemolithoautotrophy, independent of photosynthetically derived compounds, fuels these microbes; geologically produced H_2 is believed to be the most important energy and electron source. Genomic studies suggest the presence of N_2 fixation and denitrification form a subterranean nitrogen cycle. Bacterial and archaeal cells in this ecosystem number roughly 3×10^{29} (*see figure 1.2*) and are important not just because of their abundance. Like microbial communities deep below the seafloor, these terrestrial microorganisms contribute to all biogeochemical cycles. In fact, some argue they are the main drivers of these cycles. ◂ *Biogeochemical cycling sustains life on Earth (section 28.1)*

The discovery of deep subsurface microbes has a variety of implications. For example, the presence of primary production in the absence of sunlight suggests that autotrophy might also be possible in subsurface sediments of Mars. Exploring this possibility, along with investigating the level of metabolic activity among Earth's deep subsurface microbes, remains the next challenge. ◂ *Microbial communities in benthic marine sediments are enormous and mysterious (section 29.2)*

Comprehension Check

1. Compare and contrast the metabolism (specifically, electron donors and acceptors) of microbes in the shallow subsurface with that of microbes in the very deep subsurface.
2. What happens in terms of microbiological processes when organic matter leaches from the surface into the subsurface?
3. What microbial genera have been observed in oil field materials?

Key Concepts

30.1 Soils Are an Important Microbial Habitat

- Soil is a complex mixture of clay, sand, silt, organic matter, and water. This provides many microenvironments that can support a diverse microbial community (**figure 30.1**).
- Soils can be classified as mineral or organic. Soil organic matter (SOM) helps retain nutrients and water, and maintain soil structure. A C:N ratio of 30 is typically ideal.

30.2 The Soil Microbiome

- Soils support the highest levels of microbial diversity measured.
- Archaea are not as abundant as bacteria in soils; ammonia-oxidizing archaea contribute to nitrification in soils.
- Saprophytic and mycorrhizal fungi constitute about half the soil microbial biomass in most soils.

30.3 The Plant Microbiome

- The phyllosphere includes all parts of a plant that are aboveground. These structures offer a unique but changeable environment for diverse microbial growth.
- The rhizosphere is the region around plant roots into which plant exudates are released. The rhizoplane is the plant root surface. A variety of microbes growing in these regions promote plant growth.
- Mycorrhizal relationships (plant-fungal associations) are varied and complex. Six basic types are observed, including endomycorrhizal and sheathed/ectomycorrhizal types. The hyphal network of the mycobiont can lead to the formation of a mycorrhizosphere (**figures 30.4** through **30.8; table 30.1**).
- The Rhizobium-legume symbiosis is one of the best-studied examples of plant-microorganism interactions. Rhizobia form nodules on the roots and stems of host plants in which specialized bacterial cells (bacteroids) fix nitrogen (**figure 30.9**).
- The interaction between rhizobia and host plants is mediated by complex chemicals that serve as communication signals (**figures 30.9** and **30.10**).
- Actinobacteria in the genus *Frankia* form nitrogen-fixing symbioses with some trees and shrubs.
- *Agrobacterium tumefaciens* establishes a complex communication system with plant hosts into which they transfer a fragment of DNA. Genes on this DNA encode proteins that promote bacterial growth and result in the formation of plant tumors or galls (**figures 30.12** and **30.13**).
- Most plant pathogens are fungi, which can be broadly divided into biotrophic and necrotrophic fungi. Fungal plant diseases have a variety of names such as spots, blights, smuts, and rusts (**table 30.2**).

30.4 The Subsurface Biosphere Is Vast

- The subsurface includes at least four zones: the shallow subsurface; the zone where gas, oil, and coal have accumulated; intermediate depths where oil and methane are found; and the very deep, hot subsurface.
- Microbial activity in the deep subsurface suggests extraordinarily slow growth fueled by substrates of geological, rather than photosynthetic, origin. These compounds support anaerobic chemolithoautotrophy.

Active Learning

1. Considerable research is currently devoted to developing a better understanding of the rhizobiome, defined as all microbes within the rhizosphere, particularly of commercial crops. The goal is to increase the use of biocontrol microbes that will reduce the need for pesticides and fertilizers, limit plant pathogens, and provide larger yields. Select a crop and suggest microbes that might be candidates to be used in this way. Be sure to explain what each microbe would contribute to the microbe-plant relationship.
2. The Nod factor signals produced by the root nodulating bacterium *Sinorhizobium meliloti* are essential for biofilm formation as well as N_2 fixation. In what ways are nodulation and biofilm formation similar, and why do you think this microbe evolved to use the same signal for the development of both?
3. The ectomycorrhizal basidiomycete *Laccaria bicolor* produces a large number of small, secreted proteins (SSPs), some of which are only expressed in hyphae that colonize the host root. Discuss the potential role these proteins might have in the life cycle of *L. bicolor* as it colonizes host plants and moves from one host to another.
4. Although the role of viruses and bacteriophages in marine and freshwater systems has been studied for a number of years, little is known about viruses in soils. Stable isotope probing (*see section 26.5*) shows that carbon in rice plants is incorporated into the capsid of a soil bacteriophage. What does this imply about the trophic transfer of carbon in this soil? Formulate two questions that might increase our understanding of viruses in soil ecosystems.
5. Many have predicted that rising atmospheric CO_2 levels will drive increased carbon fixation by plants. However, others argue that this CO_2 fertilization effect will be limited by nitrogen availability. Discuss the role fungal and bacterial microbes may have in providing nitrogen to plants. Refer to chapter 26 and suggest a method that might allow you to determine changes in microbial communities in response to changes in atmospheric CO_2.
6. In soils, mycorrhizae compete with bacteria for nitrogen; the result of this competition helps determine the amount of soil organic matter. Consider the interaction between mycorrhizal fungi and denitrifying and nitrifying bacteria. List the metabolic strategies used by each of these microbes and hypothesize how their activities influence the C:N ratio in soils. Predict the outcome in the relative abundance of these microbes if nitrogen fertilizer were added. How would you test your hypothesis?

CHAPTER

31

Innate Host Resistance

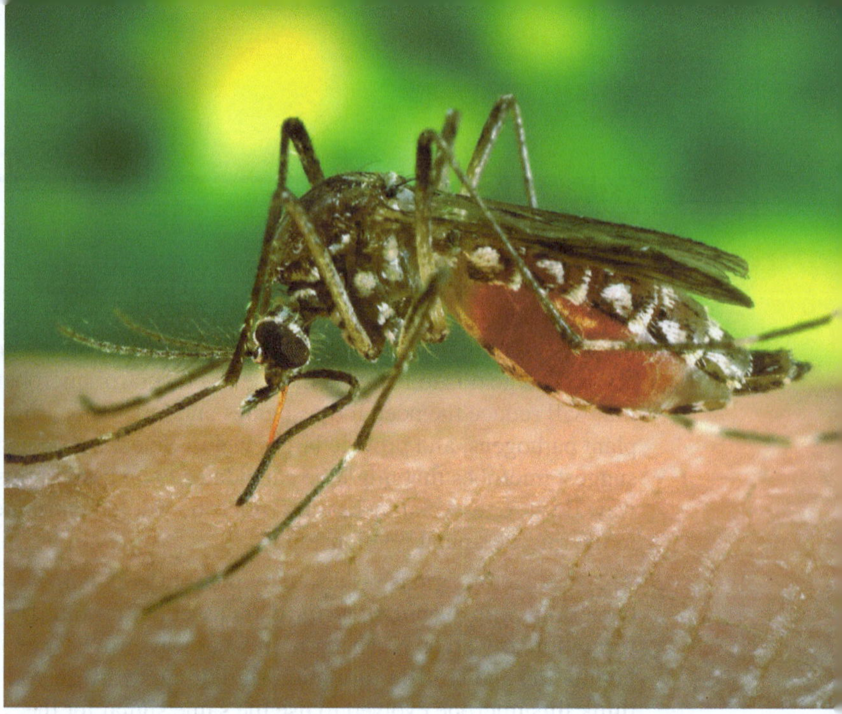

James Gathany/CDC

Vaccines for a Changing Planet

Think about a mosquito proboscis slicing into your skin. Next the mosquito injects its saliva, which contains a complex mixture of over 100 different molecules that promote feeding by suppressing blood clotting and dilating blood vessels. Although you may think your only defense is to swat the mosquito, all animal hosts (including humans) possess innate immunity: a set of immune chemicals, cells, and tissues at the ready to shield against the invasion of all foreign molecules, including those from blood-sucking insects. However, because such arthropods coevolved with their hosts, their saliva also includes proteins that modulate the animal's innate immune response, setting up a host-mosquito interaction that leaves that familiar, itchy welt.

What if that mosquito's saliva also carries protozoa that cause malaria or viruses that cause diseases like yellow fever, Zika virus disease, or dengue (like the mosquito *Aedes aegypti* shown above)? As you may be aware, or even have experienced, the last decade has seen an increase in mosquito-borne diseases. As the climate warms, the breeding season and geographical distribution of mosquitos that transmit these diseases are growing, thereby putting ever larger numbers of human populations at risk. New approaches to controlling disease transmission are needed.

One innovative strategy targets mosquito saliva, aiming to block transmission at its source. The specific goal of saliva vaccines is to train the host's immune system to block the activity of salivary proteins, which in turn would prevent the mosquito-borne pathogen from gaining a foothold in the human host. Such vaccines are in development and offer broad-spectrum defense by targeting proteins found in multiple mosquito species. Thus a single vaccine could protect against many mosquito-borne diseases.

Early trials of one such anti–mosquito saliva vaccine illustrate the importance of innate immunity. An experimental vaccine targeting four different salivary proteins was prepared in two ways: Some vaccines stimulated both innate immunity as well the more specific, adaptive arm of the immune system, while other vaccine triggered just an adaptive response. When tested on human volunteers, only those who received the vaccine that stimulated both arms of the immune system showed a clear immune response against salivary proteins, demonstrating that, in addition to innate immunity serving as our first line of defense, it is critical in calling up the responses of adaptive immunity. Although saliva vaccines are still in the early stage of development, they represent the kind of groundbreaking approaches needed in a world of rapidly evolving infectious diseases. To better understand how infectious agents cause disease, we first explore the components of innate immunity in this chapter and learn about adaptive immunity in chapter 32.

Readiness Check:

Based on what you have learned previously, you should be able to:

✓ Explain basic eukaryotic cell biology (sections 5.1 through 5.6)

✓ Diagram the flow of information in eukaryotes from DNA to protein and regulation of this flow (chapter 15)

31.1 Immunity Arises from Innate Resistance and Adaptive Defenses

After reading this section, you should be able to:

a. In general terms, differentiate between the innate and adaptive arms of the immune system
b. Define the term *antigen*

Every day, we encounter thousands of microbes without getting sick. However, we cannot credit our success entirely to our exquisite immune systems. Most microorganisms cannot invade and establish an infection. To do so, a microorganism must possess mechanisms to overcome many surface barriers, such as skin and mucous membranes. While the vast majority of microbes cannot cross these barriers, most **pathogens,** or disease-causing microorganisms, can. Fortunately animals have evolved an **immune system** composed of widely distributed proteins, cells, tissues, and organs. Together they act to neutralize or destroy microbes and other substances recognized as foreign, or non-self.

Immunity

The term **immunity** (Latin *immunis,* free of burden) refers to the general ability of a host to resist infection or disease. **Immunology** is the study of immune responses and how they protect the host. It includes the distinction between "self" and "non-self" and all the biological, chemical, physiological, metabolic, and physical aspects of the immune response.

There are two different yet complementary components of the mammalian immune response. **Innate immunity** is the first line of defense against any foreign material, including microorganisms, encountered by the host. It includes general mechanisms such as skin, mucus, and constitutively produced antimicrobial chemicals. Innate resistance mechanisms are always prepared to defend against foreign invaders to the same maximal extent each time a foreign invader is encountered. In contrast, the **adaptive immune response** must be activated (turned on) by cells and chemicals of innate immunity. Unlike innate immunity, each adaptive immune response is tailored to a particular foreign agent (e.g., virus, bacterium, protozoan, worm, fungus, abnormal host cell, or toxin). Adaptive immune responses coordinate and amplify a variety of cellular and chemical responses that evolved to meet the challenge of each specific invader. The effectiveness of adaptive immune responses increases if the host encounters the specific foreign agent again; that is to say, adaptive immunity has a memory. We learn about adaptive immunity in chapter 32.

Any substance recognized by the immune system—either the innate or adaptive arm—is called an **antigen.** An absolutely critical feature of the immune system is that it can differentiate between self antigens and non-self (foreign) antigens. As discussed later in this chapter and in the next, discriminating self from non-self enables the immune system to rid the host of foreign invaders, as well as eliminate cancer cells. However, when this discrimination goes awry, it is the basis for allergic responses and autoimmune diseases such as rheumatoid arthritis and multiple sclerosis.

There are many bridges that interconnect innate and adaptive immunity. Although this may present challenges for students of immunology, the complexity of the mammalian immune system and its capacity to prevent infection from the billions of microorganisms encountered every day is truly remarkable. We begin our discussion with an introduction to the physical and chemical barriers that microbes face when encountering a host. We then present the various cells, tissues, and organs of innate defenses, discussing the mechanisms by which they detect and respond to foreign invaders. In this discussion, reference is made to processes and chemicals described in detail either later in this chapter or in chapter 32 (adaptive immunity). To help you navigate this information, cross-references to further discussion are provided.

Comprehension Check

1. List two key differences between innate and adaptive immunity.
2. What is an antigen?

31.2 Innate Resistance Starts with Barriers

After reading this section, you should be able to:

a. Identify the barriers that help prevent microbial invasion of the host
b. Explain how physical and mechanical barriers function to prevent microbial invasion of the host

With few exceptions, a microbial pathogen attempting to invade a human host immediately confronts a vast array of innate defense mechanisms (**figure 31.1**). All vertebrate hosts have specific physical and mechanical barriers. These barriers, along with the host's flushing mechanisms (coughing, urinating, etc.), are the first line of defense against chemical, microbiological, and physical assaults. Protection of body surfaces and the mucous membrane interfaces between the host and its environment is essential in preventing microbial access to the host.

Skin

Skin consists of stratified, cornified (converted into keratin) epithelial cells (**figure 31.2**). Intact skin is a very effective mechanical barrier, and microbes living on the skin (i.e., skin-associated microbiota) defend against invaders. Its outer layer consists of thick, closely packed cells called **keratinocytes.** These produce keratins, insoluble proteins that are the main components of hair, nails, and the outer skin cells. These outer

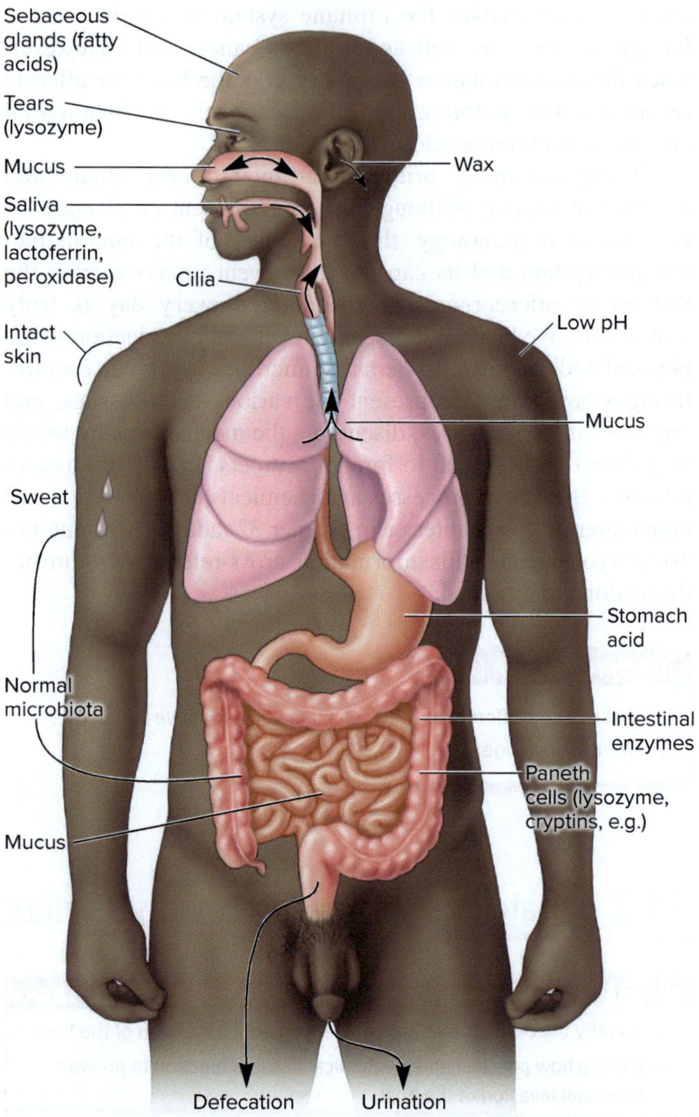

Figure 31.1 Host Defenses. Normal microbiota are found in all body sites exposed to the environment; skin and gut are shown for simplicity.

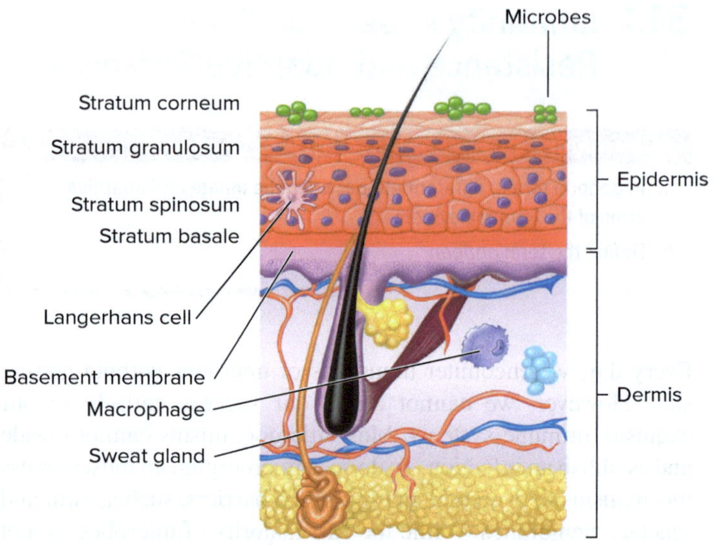

Figure 31.2 Human skin. Skin is composed of stratified cell layers, epidermis and dermis. Macrophages and Langerhans cells are phagocytic immune cells.

skin cells shed continuously. Select areas of the skin are bathed in sebum, a lubricating oil produced by the sebaceous glands. The skin surface is topographically, ecologically, and physiologically diverse; this yields numerous microenvironments, each with distinct moisture, pH, temperature, and sebum content. These niches are colonized by resident microbial populations that evolved to thrive in the specific skin environment and help prevent colonization of pathogens (*see figure 33.2*). In addition, immune cells embedded within the skin called *Langerhans cells* detect and defend against invading microbes.

Mucous Membranes

Mucous membranes line internal structures connected to the external environment, such as the respiratory, gastrointestinal, and genitourinary tracts. Also called *mucosa,* mucous membranes form a protective covering that resists penetration and traps microbial invaders. These membranes consist of a layer of epithelial cells and several types of specialized cells. For example, cells called *goblet cells* bathe membrane surfaces with mucus (a glycoprotein-rich watery fluid), thereby increasing barrier defense. Within the gastrointestinal (GI) tract, specialized cells called *Paneth cells* secrete antimicrobial peptides and the antibacterial enzyme lysozyme.

Respiratory System

The average person inhales at least eight microorganisms a minute, or 10,000 each day. Once inhaled, a microorganism must first survive the defenses of mucous membranes that line the upper and lower respiratory tracts. Because the airflow in these tracts is turbulent, microorganisms are deposited on the moist, sticky mucosal surfaces. Microbes larger than 10 µm usually are trapped by hairs and cilia lining the nasal cavity. The cilia in the nasal cavity beat downward, toward the pharynx, moving mucus with trapped microorganisms toward the mouth to be expelled (**figure 31.3**). Microbes smaller than 10 µm (i.e., most bacterial cells) pass through the nasal cavity and are trapped by the **mucociliary escalator** that coats mucosal surfaces of lower portions of the respiratory system. The mucociliary escalator moves particles upward, away from the lungs toward the mouth. Coughing and sneezing reflexes also clear the respiratory system of microorganisms. Microorganisms that succeed in reaching the alveoli of the lungs (where gas exchange occurs) encounter a population of specialized phagocytic cells called **alveolar macrophages.** These cells ingest and kill most inhaled microorganisms by phagocytosis (sections 31.4 and 31.6).

Gastrointestinal (GI) Tract

With every snack, meal, and drink, we ingest thousands of microorganisms that reach the stomach. There, most are killed by acidic gastric secretions (pH 2–3), containing a mixture of hydrochloric

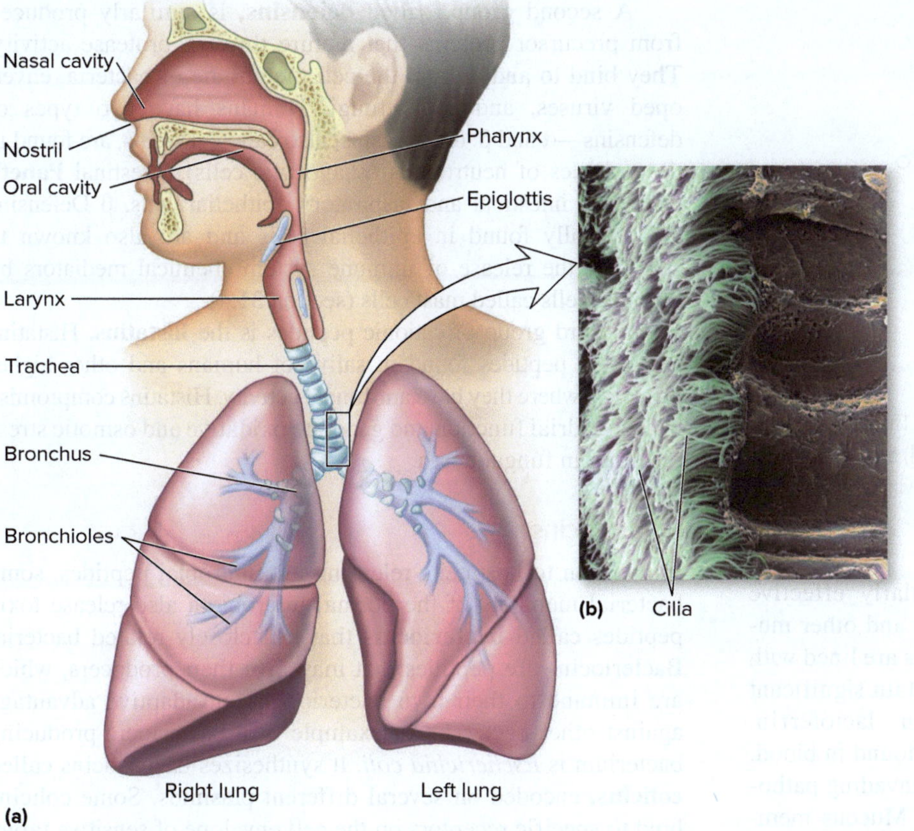

Figure 31.3 Respiratory Tract. (a) A schematic of the respiratory tract showing its components. (b) The respiratory tract is lined with a mucous membrane of ciliated epithelial cells. SEM of cilia; these specialized structures sweep particles toward the mouth to be expectorated or swallowed (5,000×).

(b) SPL/Science Source

acid, proteolytic enzymes, and mucus. However, some microorganisms and their products (e.g., protozoan cysts, *Helicobacter pylori, Clostridium* spp., and staphylococcal toxins) can survive stomach acidity. Furthermore, organisms embedded in food particles may be protected and reach the small intestine. There, these microorganisms may be damaged by various pancreatic enzymes, bile, enzymes in intestinal secretions, and cells and molecules within the gut-associated lymphoid tissue (GALT) (section 31.5). **Peristalsis** (Greek *peri,* around; *stalsis,* contraction) and the normal shedding of columnar epithelial cells are important mechanisms that purge intestinal microorganisms. Despite these mechanisms, the GI tract is home to an enormous microbial community that is essential for human health and homeostasis. These microbes are required for maturation of the immune system and proper immune function. In addition, gut microbiota help prevent the growth of potential pathogens. ▶| *A functional core microbiome is required for host homeostasis (section 33.3)*

Genitourinary Tract

The normal sterility of the kidney, ureters, and urinary bladder is maintained by a complex set of factors. In addition to removing microbes by flushing action, urine kills some bacteria due to its low pH and the presence of urea and other metabolic end products (e.g., uric acid, fatty acids, mucin, enzymes). In males, the anatomical length of the urethra (20 cm) provides a distance barrier that excludes microorganisms from the urinary bladder. Conversely, the short urethra (5 cm) in females is more readily traversed by microorganisms; this explains why females are about four times more likely than males to develop a urinary tract infection.

The innate defense of the vagina consists primarily of the normal microbiota combined with the protective measures provided by the mucous membrane. The vaginal microbiota is dominated by *Lactobacillus* spp. These lactic-acid-producing bacteria generate an acidic environment inhospitable to pathogenic organisms. Cervical mucus also has antibacterial activity from lysozyme, lactoferrin, and natural antimicrobial peptides.

Comprehension Check

1. Why is skin such a good first line of defense against pathogenic microorganisms?
2. Describe one specific antimicrobial defense mechanism that operates within each of the following: the respiratory, gastrointestinal, and genitourinary systems of mammals. For each, explain how it protects against pathogenic microorganisms.

31.3 Innate Resistance Relies on Chemical Mediators

After reading this section, you should be able to:

a. List host proteins with antimicrobial activity and compare their mechanisms of action
b. Describe in general terms the activation of the host complement system and its three outcomes
c. List three categories of cytokines and discuss their major functions
d. Explain the function of acute-phase proteins

Mammalian hosts have a chemical arsenal with which to combat the continuous onslaught of microorganisms. Importantly, the normal microbiota also assist in host resistance by contributing their own chemicals to fend off competitors, as we discuss in chapter 33. How chemical defenses protect against microbial invasion is the next topic.

Lysozyme (muramidase) is an enzyme that lyses bacteria by hydrolyzing the bond connecting *N*-acetylmuramic acid and *N*-acetylglucosamine of the bacterial cell wall peptidoglycan

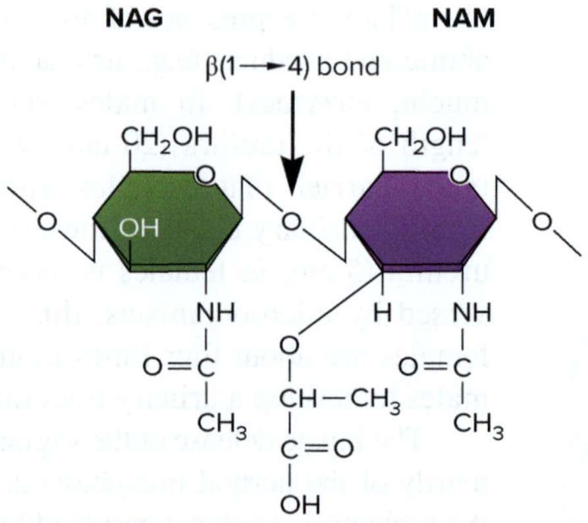

Figure 31.4 Action of Lysozyme on Peptidoglycan. This enzyme degrades cell wall peptidoglycan by hydrolyzing the β(1⟶4) bonds that connect alternating *N*-acetylglucosamine (NAG) with *N*-acetylmuramic acid (NAM) residues.

(**figure 31.4**). This makes lysozyme particularly effective against monoderm bacteria. It is found in tears and other mucous secretions, including breast milk (milk ducts are lined with mucous membranes). These secretions also contain significant amounts of the iron-binding glycoprotein lactoferrin. **Lactoferrin** is similar to **transferrin,** which is found in blood. Both sequester iron, reducing its availability to invading pathogens, thereby limiting their ability to multiply. Mucous membranes also produce lactoperoxidase, an enzyme that catalyzes the production of superoxide radicals, a reactive oxygen species toxic to many microorganisms. ◀ *Peptidoglycan structure (section 3.3); Oxygen concentration (section 7.5)*

Antimicrobial Peptides

Peptides that exhibit broad-spectrum antimicrobial activity are generally considered to be the most ancient primary defense mechanism of animals. Many are positively charged, and therefore electrostatically attracted to microbial cell surfaces. Antimicrobial peptides are amphipathic; that is, they have hydrophobic and hydrophilic components so they are soluble in both aqueous and lipid-rich environments. They kill target cells through diverse mechanisms after interacting with their membranes. Two major types of antimicrobial peptides are cationic peptides produced by host cells and bacteriocins secreted by our microbiota.

Cationic Peptides

Humans produce three generic classes of cationic peptides. The first group includes linear, α-helical peptides. Important examples are the **cathelicidins.** There are about 30 members of this family found in many vertebrates, including humans. They are secreted as large proteins that become activated by proteolytic cleavage. Cathelicidins have broad antimicrobial activity against bacteria, enveloped viruses, and fungi. In humans, cathelicidins are produced by a variety of cells, including respiratory and urogenital epithelial cells, neutrophils, and alveolar macrophages (section 31.4).

A second group, called **defensins,** is similarly produced from precursor proteins that mature through protease activity. They bind to and disrupt the cell membrane of bacteria, enveloped viruses, and some fungi. Humans have two types of defensins—α and β. α Defensins, also called cryptins, are found in the granules of neutrophils (phagocytic cells), intestinal **Paneth cells,** and intestinal and respiratory epithelial cells. β Defensins are typically found in epithelial cells and are also known to stimulate the release of immune system chemical mediators by immune cells called mast cells (section 31.4).

A third group of cationic peptides is the **histatins.** Histatins are larger peptides found in saliva of humans and other higher primates, where they have antifungal activity. Histatins compromise mitochondrial function and generate oxidative and osmotic stress resulting in fungal death.

Bacteriocins

In addition to host cells releasing antimicrobial peptides, some bacterial members of the normal microbiota also release toxic peptides called **bacteriocins** that kill closely related bacteria. Bacteriocins are peptides that may give their producers, which are immune to their own bacteriocins, an adaptive advantage against other bacteria. An example of a bacteriocin-producing bacterium is *Escherichia coli.* It synthesizes bacteriocins called **colicins,** encoded on several different plasmids. Some colicins bind to specific receptors on the cell envelope of sensitive target bacteria and cause cell lysis; others attack intracellular sites such as ribosomes or disrupt energy production. A different group of bacteriocins include peptides called *lantibiotics* produced by monoderm genera such as *Streptococcus, Bacillus, Lactococcus,* and *Staphylococcus.* Lantiobiotic peptides are active against other monoderm bacteria. ▶ *A functional core microbiome is required for host homeostasis (section 33.3)*

Complement Is a Cascading Enzyme System

The **complement system** is an ancient form of innate immunity found in primitive invertebrates as well as mammals. It is so named because when it was discovered, this collection of over 30 heat-labile proteins was found to "complement" the antimicrobial effects of other components of the immune system. The purpose of the complement system is to quickly clear the host of invading pathogens through three distinct mechanisms. These include: (1) stimulating an inflammatory response, (2) lysing microbial cells, and (3) promoting phagocytosis (engulfment and killing) of microbial invaders through a process called opsonization. **Opsonization** (Greek *opson,* to prepare for eating) is the term used when microorganisms or other non-self antigens are coated with complement proteins, antibodies, or both. In this context, these proteins are called **opsonins** and the microorganism is said to be *opsonized.* The presence of opsonins on the surface of antigens acts like a beacon that promotes recognition by phagocytic cells (also called *phagocytes*). This is because phagocytes have surface receptors that specifically recognize opsonins (i.e., complement proteins and antibodies). Therefore

Phagocytic cell–microbe interaction	Degree of binding	Opsonin
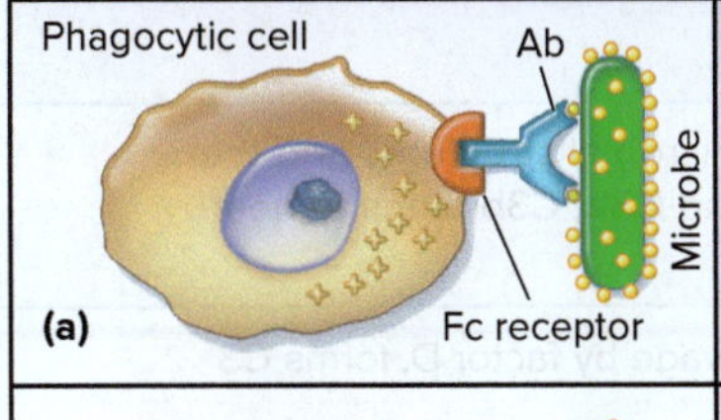	+	Antibody
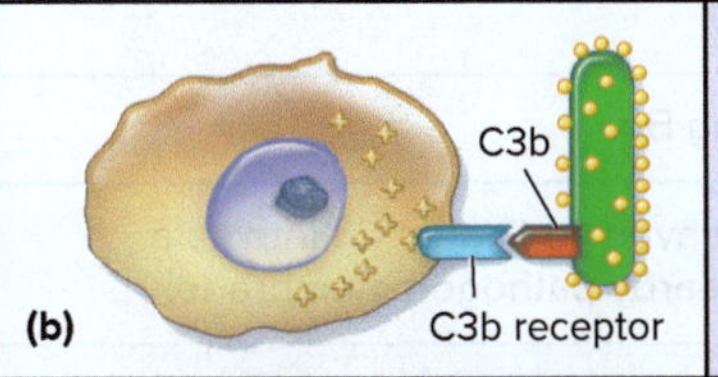	++	Complement C3b
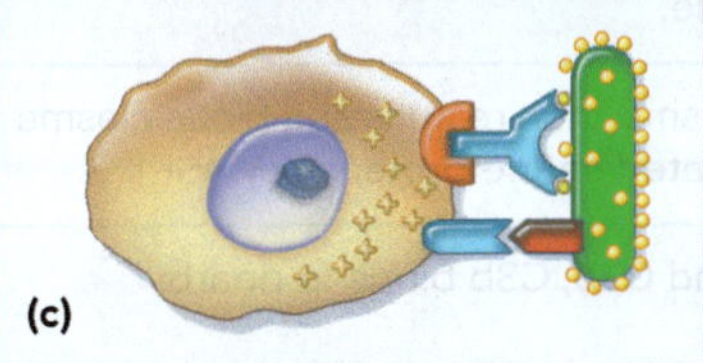	++++	Antibody and complement C3b

Figure 31.5 Opsonization. (a) The ability of a phagocyte to bind to a microorganism is enhanced if the microorganism has antibodies (Ab) bound to it. Antibodies act as bridges by attaching the antibody-bound microbe to the antibody receptor (Fc) on the phagocytic cell. (b) If the activated complement protein C3b is bound to the microbe, the degree of binding to the phagocyte is further increased by attachment to the C3b receptor. (c) If both antibody and C3b opsonize, binding is maximal.

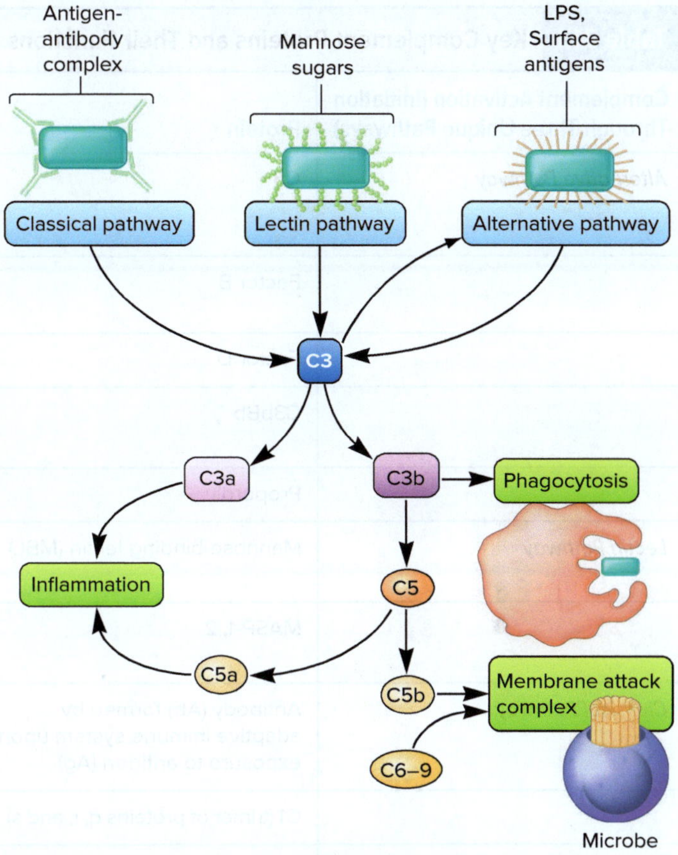

Figure 31.6 Complement Activation and Action. Complement activation involves a series of enzymatic reactions that lead to three outcomes (green boxes). Each requires the formation of a C3 convertase, which cleaves C3 into C3a and C3b. C3a is a peptide mediator of local inflammation. C3b is an opsonin that binds covalently to microbial cells, thereby enhancing phagocytosis. C5a and C5b are generated by the cleavage of C5 by a C5 convertase. C5a functions together with C3a to mediate inflammation. C5b promotes assembly of a membrane attack complex from the terminal C6–9 complement components.

pathogen clearance by phagocytes increases dramatically when they are coated with opsonins (**figure 31.5**).

Complement proteins are produced in the liver and released into the serum (the liquid component of blood) and extracellular fluids (fluid found in tissues, outside of cells). In the absence of pathogens, they are in an inactive form. When pathogens are present, complement molecules become activated by a cascade of proteolytic cleavage, whereby one protease cleaves another and so on. Because activation pathways are complex, we provide a simplified overview. The three pathways of complement activation include the alternative, lectin, and classical pathways (**figure 31.6**). The major differences between them are (1) the types of molecules that activate each pathway, (2) the means by which key proteases are activated, and (3) the identity of these proteases. *Activation of Complement*

Recall that as part of the innate immune system, complement proteins are always present at low, inactive levels in blood serum and tissues. When either is invaded by pathogens with repetitive surface molecules like lipopolysaccharide (LPS) that possess hydroxyls (–OH) and/or amino groups ($-NH_2$), the **alternative complement pathway** (figure 31.6) is initiated. These repetitive molecules activate complement (C) protein 3 (C3) to cleave itself into two fragments: C3a and C3b (**table 31.1**). We first focus on C3b, which is unstable when in the bloodstream and tissue fluid but becomes stabilized upon binding to microbial cell surfaces. Membrane-bound C3b acts as an opsonin, enhancing phagocytosis (figure 31.5). To bring about the other two outcomes (inflammation and cell lysis), C3b must interact with another complement protein called factor B. After binding to C3b, factor B is cleaved by the complement protease factor D (**figure 31.7**). The cleavage product factor Bb remains bound to C3b, thereby forming a complex called C3bBb. The C3bBb functions as a protease called the *C3 convertase* of the alternative complement pathway. This C3 convertase amplifies the pathway by cleaving more C3 to C3a and C3b, thereby increasing the amount of C3b coating microbial structures. Thus brief exposure to a microbial surface structure like LPS triggers a proteolytic cascade that starts with C3 generating C3b and ends with C3bBb. This in turn starts a chain reaction that is propagated because C3bBb is a protease that cleaves more C3, ultimately leading to an abundance of C3b on pathogen cell surfaces.

Table 31.1 Key Complement Proteins and Their Functions

Complement Activation (Initiation Through Three Unique Pathways)	Protein	Function
Alternative Pathway	C3	C3 binds repeating patterns within microbial structures and then is spontaneously cleaved into C3a and C3b; C3b binds to nearby pathogen membranes.
	Factor B	Factor B binds to C3b, and, upon cleavage by factor D, forms C3 convertase (protease) C3bBb.
	Factor D	Factor D cleaves Factor B, generating Bb.
	C3bBb	C3bBb (the active C3 convertase) cleaves additional C3 into its a and b fragments; C3b binds to the nearby pathogen membrane.
	Properdin	Properdin stabilizes the C3 convertase.
Lectin Pathway	Mannose-binding lectin (MBL)	MBL binds to mannose on microorganisms, then recruits and binds plasma esterase to become mannose-associated serine protease (MASP).
	MASP-1, 2	MASP proteins cleave C3 into C3a and C3b; C3b binds to nearby microbial membranes.
Classical Pathway	Antibody (Ab) formed by adaptive immune system upon exposure to antigen (Ag)	Ab binds to Ag on microorganism, causing three-dimensional shift in Ab structure, which exposes amino acids near the Fc region; the newly revealed amino acids attract plasma C1 protein.
	C1 (trimer of proteins q, r, and s)	C1q binds to Fc region of Ab-Ag complex; C1r,s binds plasma calcium.
	Ag-Ab-C1q,r,s complex	Ag-Ab-C1 complex is an activated protease (called the C2/C4 esterase) that cleaves plasma C2 and C4 into their a and b fragments, respectively.
	C2a	C2a binds the Ag-Ab-C1 complex.
	C4b	C4b binds the Ag-Ab-C1-C2a complex, forming a C3 convertase (protease).
	Ag-Ab-C1-C2a-C4b	Ag-Ab-C1-C2a-C4b (the active C3 convertase) cleaves plasma C3 into its a and b fragments; C3b binds to the nearby pathogen membranes.

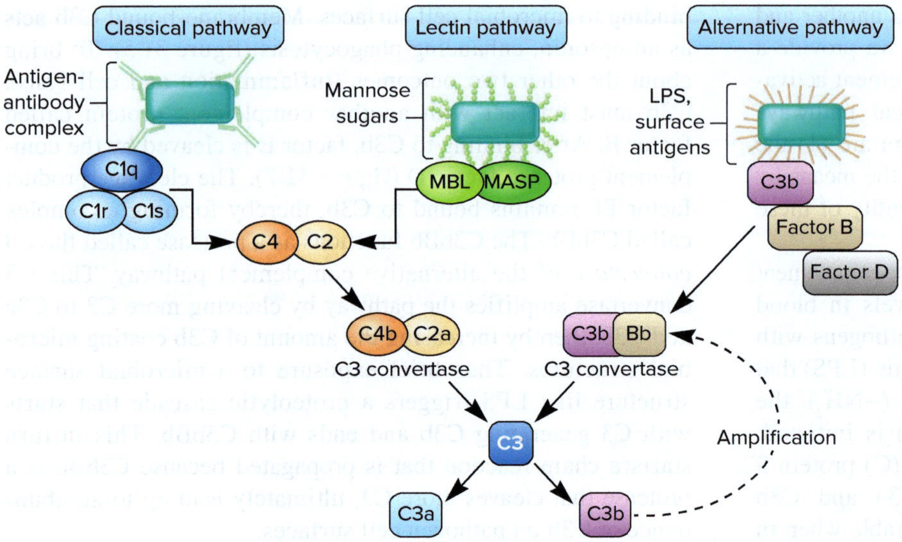

Figure 31.7 Activation of C3 Convertases by the Three Complement Pathways. C3 convertases are activated by a cascade of proteases specific to each pathway. Note that the lectin and classical pathways use the same C3 convertase, C4bC2a, but its activation differs in each pathway. There are two MBL-associated serine proteases (MASPs)—MASP-1 and MASP-2—involved in activation of the lectin pathway, but only one is shown for simplicity. Whereas the C3 convertases of the alternative and lectin pathways are bound to microbial surfaces, the C3 convertase of the classical pathway is bound to the antigen-antibody-C1 complex. MBL, mannose binding lectin.

MICRO INQUIRY *Which complement components are bound to microbial surfaces?*

Not only does C3bBb serve to amplify C3b, but it also proteolytically activates another complement protein, C5. Because C3bBb cleaves both C3 and C5, it is both a C3 and C5 convertase (protease). As a *C5 convertase,* it leads to the production of the cleavage products C5a and C5b. C5b rapidly complexes with two additional complement proteins, C6 and C7, forming a C5b67 complex that binds to microbial membranes. C8 and C9 then join C5b67, forming the **membrane attack complex (MAC),** which creates a pore in the plasma membrane or outer membrane of the target cell (**figure 31.8**). When multiple pore-forming complexes cover a microbial membrane, MAC is a very effective killing mechanism. MAC is particularly good at puncturing diderm outer membranes, but less so against monoderm bacteria and fungi, which are protected by thick cell walls. Because MAC cannot discriminate between pathogen and host cell membranes, host cells have several mechanisms to protect their membranes from complement attack. Chief among these is the protein factor H and *d*ecay *a*ccelerating *f*actor (DAF) on the surface of host cells. Factor H inactivates Bb while DAF binds to other complement factors.

We now turn our attention to the cleavage products C3a and C5a, which have important pro-inflammatory effects. Binding of C3a and C5a to receptors on specific host cells induces the release of other biological signals that dilate vessels, stimulate nerves, and increase permeability of local blood vessels so that phagocytic cells, particularly neutrophils, can be recruited to the site of infection. Neutrophils are voracious phagocytes critical to the removal of extracellular pathogens. The inflammatory signals released as a result of C3a and C5a activation also stimulate other innate immune cells (dendritic cells and macrophages), which in turn help activate an adaptive immune response (section 31.4).

The **lectin complement pathway** is activated in response to bacteria, fungi, and some enveloped viruses. These microbes are detected by the presence of mannose, a carbohydrate not found on human cells. Mannose is detected by a protein called the **mannose-binding lectin** (**MBL**; figure 31.7, table 31.1). Activation of the lectin complement pathway also requires the activation of a C3 convertase. When MBL binds to mannose on pathogens, it forms a complex with other blood proteins known as *m*annose-binding lectin-*a*ssociated *s*erine *p*rotease-1 (MASP-1) and MASP-2. The MASPs become activated to serve as proteases. MASPs first cleave the complement protein C4 to generate C4a and C4b. C4b is much like C3b because it too binds to microbial surfaces. Once C4b is bound to the target microbe, another complement protein called C2 complexes with it. C2 is then cleaved by a MASP to release C2a and C2b. C2a remains bound to C4b, forming a C3 convertase. Like the C3 convertase of the alternative pathway, C4bC2a cuts C3 to generate C3a and C3b. In addition to acting as an opsonin, some of the C3b joins forces with C4b and C2a to form a C5 convertase (C3b plus C4bC2a) so that MAC components can be assembled and C5a and C3a can help trigger inflammation. Thus although different C3 and C5 convertases are used, the three outcomes are achieved, as seen in the alternative pathway (figure 31.7).

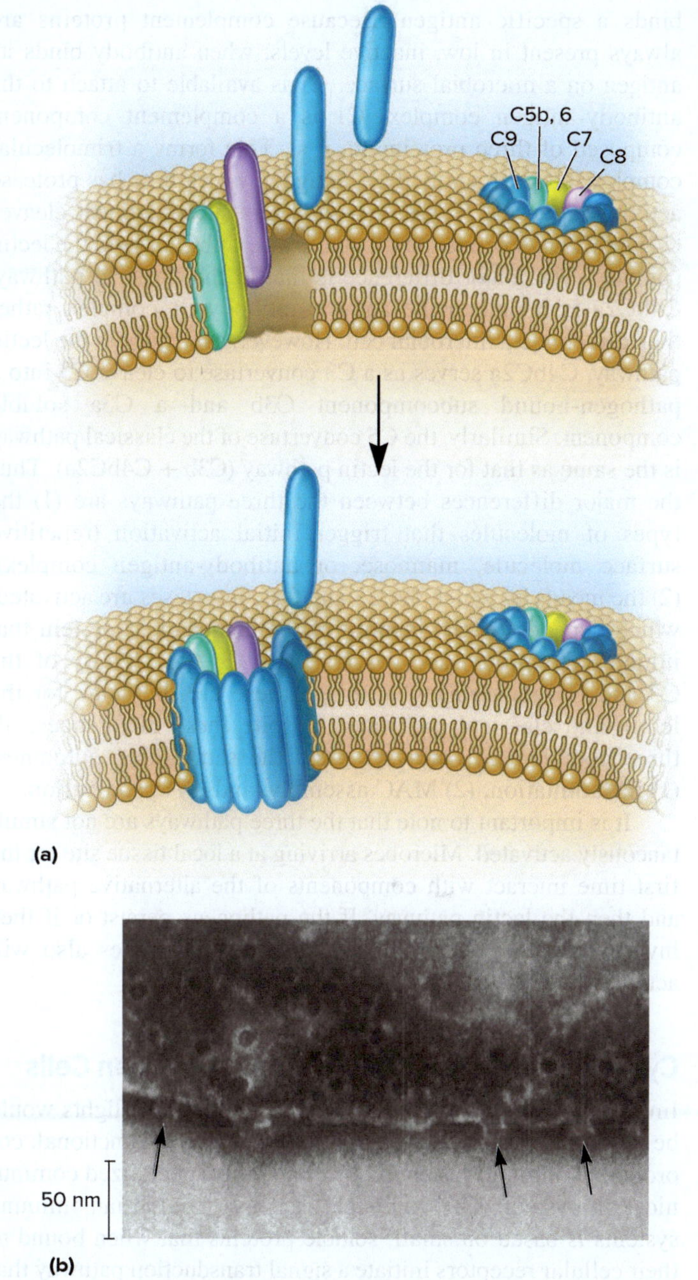

Figure 31.8 The Membrane Attack Complex. The membrane attack complex (MAC) is a barrel-shaped structure that forms a transmembrane pore in the pathogen's membrane. (a) The transmembrane pore is formed by a C5b678 complex and 10 to 16 polymerized molecules of C9. (b) MAC pores appear as craters or donuts by electron microscopy.

(b) Sucharit Bhakdi

MICRO INQUIRY *What strategies could a diderm bacterium use to protect itself against lysis by complement?*

Activation of the **classical complement pathway** typically depends on the presence of antibody specific to the pathogen (table 31.1). Antibodies are glycoproteins produced by activated B cells as part of an adaptive response (*section 32.7*). Each antibody binds a specific antigen. Because complement proteins are always present in low, inactive levels, when antibody binds its antigen on a microbial surface, C1 is available to attach to the antibody-antigen complex. C1 is a complement component composed of three proteins (q, r, s). This forms a trimolecular complex (C1qrs + antibody + antigen), which now has protease activity (figure 31.7). The activated C1s subcomponent cleaves C4 and C2 to generate the C3 convertase C4bC2a, as in the lectin pathway. The major difference is that in the classical pathway, C4bC2a is bound to the antigen-antibody-C1 complex rather than directly to a microbial cell. However, as we saw in the lectin pathway, C4bC2a serves as a C3 convertase to cleave C3 into a pathogen-bound subcomponent C3b and a C3a soluble component. Similarly, the C5 convertase of the classical pathway is the same as that for the lectin pathway (C3b + C4bC2a). Thus the major differences between the three pathways are (1) the types of molecules that trigger initial activation (repetitive surface molecule, mannose, or antibody-antigen complex); (2) the means by which the C3 and C5 convertases are activated, which depends on the identity of the complement protein that initiates the proteolytic cascade; and (3) the identity of the C3 and C5 convertases, although these are the same for the lectin and classical pathways. Despite these differences, all three complement pathways yield the same three outcomes: (1) inflammation, (2) MAC assembly, and (3) opsonization.

It is important to note that the three pathways are not simultaneously activated. Microbes arriving at a local tissue site for the first time interact with components of the alternative pathway and then the lectin pathway. If the pathogens persist or if they invade the host a second time, antibody responses also will activate the classical complement pathway.

Cytokines Are Chemical Messages Between Cells

Imagine an airport without air traffic control: All flights would be grounded and chaos would ensue. Similarly, a functional, coordinated immune system requires a highly specialized communication system. The language used by mammalian immune systems is based on small, soluble proteins that when bound to their cellular receptors initiate a signal transduction pathway that regulates the transcription of genes and the activation of proteins needed for an appropriate immune response. Because each cytokine triggers a specific cellular response or set of responses, the resulting immune activity depends on which signals are received, how many receptors become occupied, and interference from either external or internal sources.

Cytokines are soluble, low molecular weight proteins or glycoproteins released by one cell population that, upon binding to specific receptors on target cells, act as intercellular (between cells) signaling molecules. Cytokines are produced by and impact a broad range of cells, resulting in a variety of cellular responses. Cytokine synthesis is induced by stimuli such as a viral, bacterial, or parasitic infection, malignancy, or inflammation. Their synthesis must be tightly regulated. This is accomplished at the transcriptional level as well as posttranslationally. For example, the release of certain cytokines by one kind of immune cell can induce the production of other cytokines by other types of cells.

Although there are many ways to group cytokines, we can place them in three simplified categories: (1) cytokines that principally regulate innate immunity, (2) cytokines that principally regulate adaptive immunity, and (3) cytokines that stimulate hematopoiesis—the development of blood cells (**table 31.2**). Cytokines are also divided into those that stimulate or those that inhibit host inflammation, that is, *pro-inflammatory* and *anti-inflammatory,* respectively.

The naming of cytokines has been one of those unfortunate stories of the left hand not knowing what the right has done. Historically, each newly discovered protein would be named based on its function. As protein biochemistry became more exact, scientists grouped proteins based on structure. This resulted in a list of proteins with different names yet the same function, and proteins with identical structures but different names. For simplicity, cytokines can be divided into four functional groups: (1) Cytokines that stimulate cell migration (i.e., chemotaxis) are called **chemokines,** (2) cytokines produced by one **leukocyte** (**white blood cell**) acting on another leukocyte are called **interleukins,** (3) regulatory cytokines produced in response to infection are called **interferons,** and (4) cytokines that stimulate the growth and differentiation of immature leukocytes in the bone marrow are called **colony-stimulating factors (CSFs).** The last important class of cytokines is called **tumor necrosis factors (TNFs),** based on early observations of their in vitro activity; their immune function is to stimulate an inflammatory response. Although these terms are a helpful shorthand method of recognizing function, the groups are not mutually exclusive. For instance, interferons produced by one type of leukocyte will influence the behavior of others—that is, interferons also serve as interleukins.

Acute-Phase Proteins

In addition to cytokines, a number of biologically active molecules can cause dramatic physiological changes in the host. An important example is **acute-phase proteins.** These proteins (also called acute-phase reactants) are produced by liver cells in response to pro-inflammatory cytokines released by macrophages and other cells following tissue injury or infection. Acute-phase proteins assist in the prevention of blood loss and ready the host for microbial invasion. They include the complement protein MBL, C-reactive protein (CRP), and surfactant proteins A (SP-A) and D (SP-D), all of which bind bacterial surfaces and act as opsonins. After binding to a microbial surface, CRP interacts with C1q. This interaction mimics the binding of C1q to antibody to activate the classical complement pathway. SP-A, SP-D, MBL, and C1q are also known as **collectins**—proteins composed of a collagen-like motif connected by α-helices to globular binding sites. These proteins (along with others) bind to foreign sugars and lipids, thereby assisting in the removal of microorganisms and their products through aggregation, complement activation, and opsonization, as well as other mechanisms (section 31.6).

Table 31.2 Example Cytokines Grouped by Function

	Cytokine[1]	Source	Role
Innate Immunity	IL-1	Macrophages, endothelial and epithelial cells	Upregulates inflammatory response, including fever; stimulates release of acute-phase proteins
	IL-6	Macrophages, T cells, endothelial cells	Upregulates acute-phase response, including fever; stimulates neutrophil maturation; induces differentiation of T cells into T helper cells (T_H17)
	IL-8	Neutrophils, macrophages, dendritic cells, endothelial and epithelial cells	Activates neutrophils, induces chemotaxis, enhances inflammation
	TGF-β	Dendritic cells, T helper cells	Induces differentiation of T cells into T helper cells (T_H17, Treg); stimulates antibody production
	TNF-α	Macrophages, NK cells, and T cells	Upregulates inflammatory response, including fever; stimulates acute-phase protein synthesis; stimulates neutrophil activity
	Type I IFN (IFN α/β)	Virus-infected host cells; plasmacytoid dendritic cells, macrophages, and NK cells	Induces broad-spectrum antiviral and antitumor activity; enhances other cytolytic function; promotes inflammation
Adaptive Immunity	IL-2	T cells	Stimulates growth and differentiation of T cells and NK cells
	IL-4	T helper cells (T_H2), dendritic cells	Induces differentiation of T cells into T helper cells (T_H2 cells); stimulates antibody production
	IL-5	T-cell subset and mast cells	Stimulates growth of B cells; enhances antibody secretion; activates eosinophils
	IL-10	T regulatory cells, macrophages	Anti-inflammatory effects largely mediated by inhibiting release of pro-inflammatory cytokines
	IL-12	T helper cells (T_H1), dendritic cells, macrophages	Induces differentiation of T cells into T helper cells (T_H1 cells); stimulates activity of NK cells
	IL-17	T helper cells (T_H17)	Monocyte and neutrophil chemokine; induces pro-inflammatory cytokines IL-6, TNF-α, IL-1, chemokines, and prostaglandins from various cells
	IFN γ	T helper cells (T_H1), dendritic cells, NK cells	Induces differentiation of T cells into T helper cells (T_H1 cells); stimulates antibody production; upregulates macrophage and NK cell activity
Hematopoiesis	IL-3	Basophils and activated T cells	Stimulates pluripotent hematopoietic stem cells to become myeloid progenitor cells; stimulates myeloid cell proliferation
	IL-7	Bone marrow and thymic stromal cells, dendritic and epithelial cells, and hepatocytes	Stimulates pluripotent hematopoietic stem cells to become lymphoid progenitor cells; stimulates lymphoid cell proliferation
	CSF-2	Macrophages, T cells, endothelial and mast cells, and fibroblasts	Induces hematopoietic stem cells to proliferate and differentiate into granulocytes and monocytes; also called granulocyte-macrophage (GM)-CSF

[1]CSF, colony-stimulating factor; IFN, interferon; IL, interleukin; TNF, tumor necrosis factor

Comprehension Check

1. How do lysozyme and transferrin limit pathogen growth?
2. Describe the function of cationic peptides and bacteriocins. Which are host products and which are bacterial?
3. Create a diagram showing how the alternate, lectin, and classical pathways activate their respective C3 and C5 convertases.
4. Describe the three outcomes of complement activation.
5. Based on the information in table 31.2, which cytokines are pro-inflammatory and which are anti-inflammatory? Can hematopoietic cytokines be classified this way?
6. How do acute-phase proteins assist in pathogen removal?

31.4 Each Type of Innate Immune Cell Has a Specific Function

After reading this section, you should be able to:

a. Recognize the different types of leukocytes involved in innate immunity
b. Outline the response of professional phagocytes to microbial invasion
c. Relate leukocyte distribution within the host to host immunity

Unlike organ systems (e.g., respiratory, cardiac), the immune system consists of both organs, such as the thymus discussed later, and white blood cells or *leukocytes* (Greek *leukos,* white; *kytos,* cell). Therefore an understanding of the immune system requires an introduction to the origin and function of each type of leukocyte. Blood cell development occurs in the bone marrow of mammals during the process of **hematopoiesis** (Greek *haima,* blood; *poiesis,* to produce something). All leukocytes originate from hematopoietic precursor cells in the fetal liver and after birth, in the bone marrow (**figure 31.9**). Some leukocytes reside in tissues where they respond to local trauma and infection. These cells sound the alarm that signals invasion by foreign organisms. Other leukocytes circulate in body fluids and are recruited to the sites of infection after the alarm has been raised. Finally, some leukocytes migrate from the site of harm to the nearest lymph nodes to communicate the presence of a microbial invader.

The average adult has between 4,500 and 10,000 leukocytes /mL of blood. This value shifts substantially during an immune response. For example, in most bacterial and fungal infections, the white blood cell count increases as leukocytes migrate from the bone marrow to the blood, then to the site of infection. This transient increase in circulating leukocytes (leukocytosis) is a helpful biomarker for clinicians in diagnosing an infectious process. We now introduce these cells and their specialized functions.

Mast Cells

Mast cells are bone marrow–derived cells that travel to a variety of tissues, where they differentiate into mature cells. Mast cells (figure 31.9) are filled with specialized organelles called *granules.* When stimulated, they rapidly release the contents of their granules into the extracellular environment, a process called *degranulation.* Mast cell granules contain **vasoactive mediators**—compounds that influence the tone and diameter of blood vessels. These include histamine, prostaglandins, serotonin, heparin, dopamine, platelet-activating factor, and leukotrienes. Additionally, the surface of mast cells has high-affinity receptors for the type of antibody associated with worm infections and allergic responses (IgE). When IgE is bound in sufficient quantity to these receptors, the release of vasoactive mediators occurs. As discussed in chapter 32, this plays a major role in certain allergic responses such as eczema, hay fever, and asthma. ▶| *Antibody function (section 32.7); Hypersensitivities (section 32.10)*

Granulocytes

Granulocytes have irregularly shaped nuclei with two to five lobes (figure 31.9). Their cytoplasm has granules that contain reactive substances that kill microorganisms and enhance inflammation. Three types of granulocytes exist: basophils, eosinophils, and neutrophils.

Basophils (Greek *basis,* base; *philein,* to love) have irregularly shaped nuclei with two lobes (figure 31.9). The granules contain histamine and other substances similar to those in mast cells. Like mast cells, basophils are important in the development of allergies because they also possess high-affinity receptors for IgE and release preformed vasoactive mediators.

Eosinophils (Greek *eos,* dawn; *philein*) have a two-lobed nucleus connected by a slender thread of chromatin (figure 31.9). Eosinophil granules contain hydrolytic enzymes (e.g., nucleases), in addition to peroxidases and a caustic protein called major basic protein. Eosinophils circulate in low numbers and migrate from the bloodstream into tissue spaces, especially mucous membranes, where they defend against worm parasites. These organisms are too large to be phagocytosed, so when eosinophils release hydrolytic enzymes, cationic peptides, and reactive oxygen species into extracellular fluid, the parasite's plasma membrane is damaged and the organism is (hopefully) killed. Eosinophils also play a role in allergic reactions. Thus their numbers in the bloodstream often increase during both parasite infections and allergic reactions. ▶| *Type I hypersensitivity (section 32.10)*

Neutrophils (Latin *neuter,* neither; *philein*) have nuclei with three to five lobes connected by slender threads of chromatin (figure 31.9). They are also called **polymorphonuclear cells (PMNs).** Neutrophils can be described as "phagocytic machines" that harbor inconspicuous organelles known as primary and secondary granules. *Primary granules* contain peroxidase, lysozyme, defensins, and various hydrolytic enzymes, whereas the smaller *secondary granules* have collagenase, lactoferrin, cathelicidin, and lysozyme. These enzymes and other molecules help digest foreign material after it is phagocytosed (section 31.6).

Mature neutrophils leave the bone marrow and circulate in blood so they can rapidly migrate to a site of tissue damage or infection, where they become the principal phagocytic responders. Neutrophils migrate to sites where complement C3a and C5a have been released. There they bind to opsonized particles, which they readily phagocytose. Neutrophils have a limited lifespan of about a week. Neutrophils and their antimicrobial compounds are described in more detail in the context of phagocytosis (section 31.6) and the inflammatory response (section 31.7).

Monocytes, Macrophages, and Dendritic Cells

Monocytes (Greek *monos,* single; *cyte,* cell) are leukocytes with an ovoid- or kidney-shaped nucleus and granules in the cytoplasm (figure 31.9). They are produced in the bone marrow and enter the blood, circulate for about 8 hours, enlarge, migrate into tissues, and mature into macrophages.

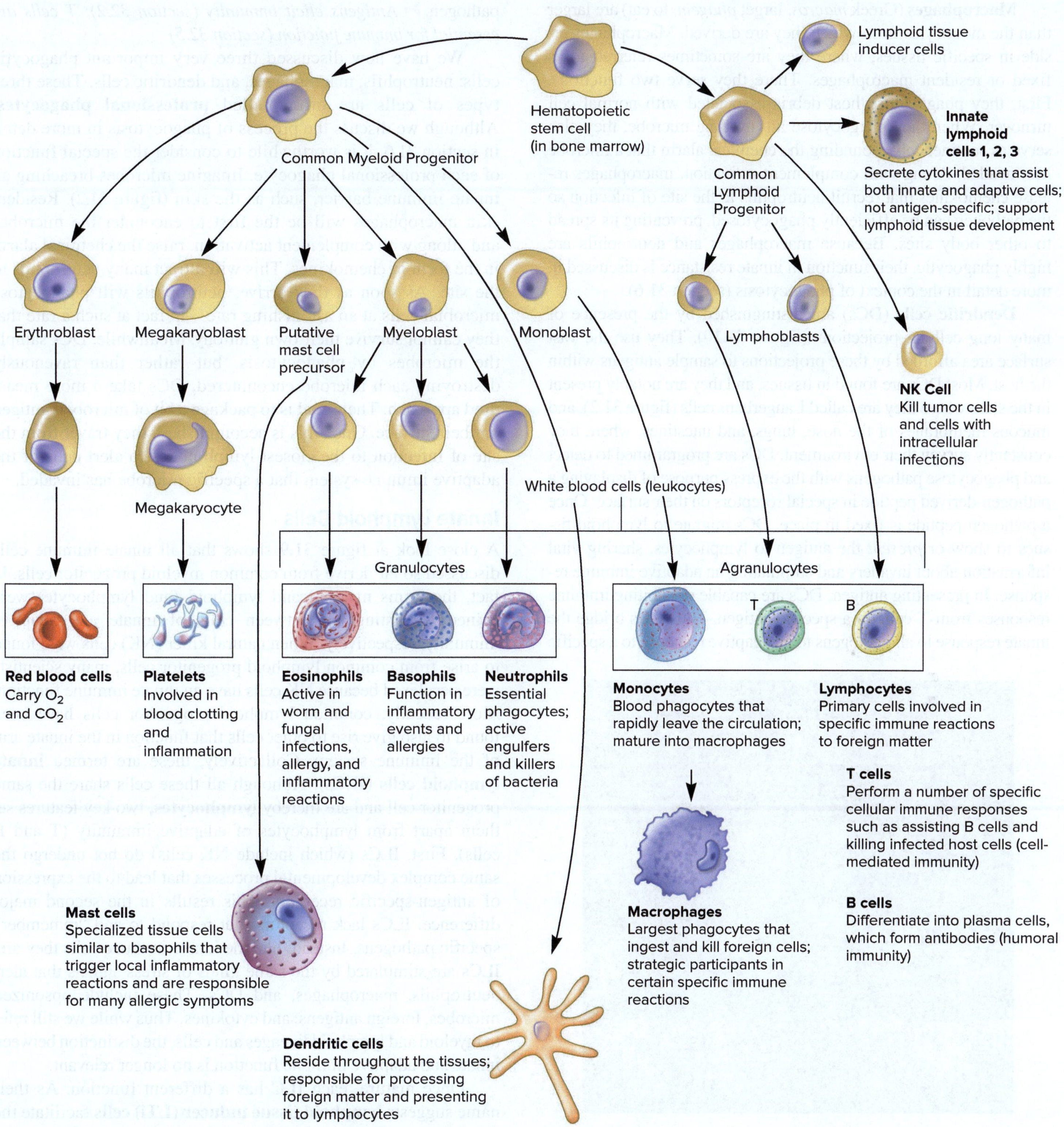

Figure 31.9 The Different Types of Human Blood Cells. Hematopoietic stem cells in the bone marrow divide to form two blood cell lineages: (1) The common myeloid progenitor cell gives rise to erythroblasts that produce erythrocytes (red blood cells), megakaryocytes that produce platelets, and an unknown precursor that gives rise to mast cells. Myeloblasts mature into granulocytes (eosinophils, basophils, and neutrophils) and monocytes give rise to macrophages. A separate lineage generates dendritic cells. (2) The common lymphoid progenitor cell gives rise to B cells; T cells; innate lymphoid cells (ILCs) 1, 2, and 3; natural killer (NK) cells; and lymphoid tissue inducer cells.

Macrophages (Greek *macros,* large; *phagein,* to eat) are larger than the monocytes from which they are derived. Macrophages reside in specific tissues, where they are sometimes referred to as fixed or resident macrophages. There they serve two functions. First, they phagocytose host debris associated with normal cell turnover. When they phagocytose an invading microbe, they also serve as sentinel cells, sounding the chemical alarm that a microbe has invaded. Along with complement activation, macrophages release chemokines that recruit neutrophils to the site of infection so the pathogen can be efficiently phagocytosed, preventing its spread to other body sites. Because macrophages and neutrophils are highly phagocytic, their function in innate resistance is discussed in more detail in the context of phagocytosis (section 31.6).

Dendritic cells (DCs) are distinguished by the presence of many long cellular projections (**figure 31.10**). They use the vast surface area afforded by these projections to sample antigens within the host. Most DCs are found in tissues, and they are notably present in the skin, where they are called Langerhans cells (figure 31.2), and mucous membranes of the nose, lungs, and intestines, where they constantly survey their environment. DCs are programmed to detect and phagocytose pathogens with the express purpose of displaying a pathogen-derived peptide in special receptors on their surface. Once a pathogen peptide is fixed in place, DCs migrate to lymphoid tissues to show or *present* the antigen to lymphocytes, sharing vital information about invaders and stimulating an adaptive immune response. In presenting antigen, DCs are capable of eliciting immune responses from T cells to a specific antigen. Thus DCs bridge the innate response to all pathogens to an adaptive response to a specific pathogen. ▶| *Antigens elicit immunity (section 32.2); T cells are essential for immune function (section 32.5)*

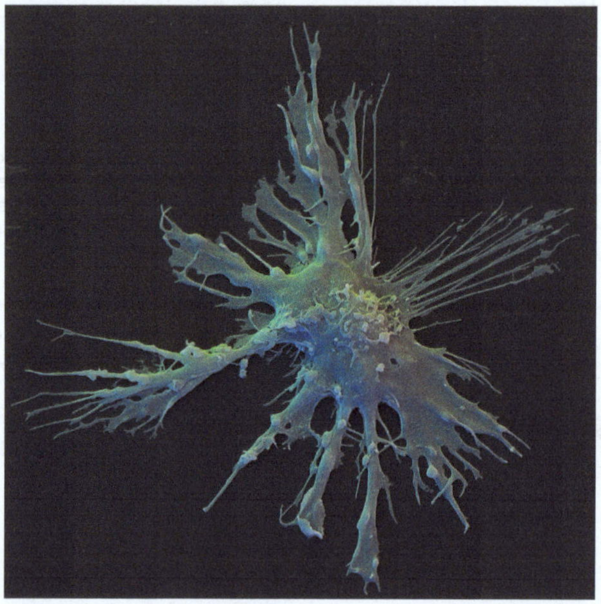

Figure 31.10 The Dendritic Cell. The dendritic cell was named for its cellular extensions, which resemble the dendrites of nerve cells. Dendritic cells migrate to lymphoid tissue, where they present antigen to lymphocytes to trigger an adaptive immune response.
David Scharf/Science Source

MICRO INQUIRY *What is the purpose of the cellular extensions?*

We have now discussed three very important phagocytic cells: neutrophils, macrophages, and dendritic cells. These three types of cells are often called **professional phagocytes.** Although we discuss the process of phagocytosis in more detail in section 31.6, it is worthwhile to consider the special function of each professional phagocyte. Imagine microbes breaching an innate immune barrier, such as the skin (figure 31.2). Resident skin macrophages will be the first to encounter the microbes and, along with complement activation, raise the chemical alarm in the form of chemokines. This will attract many neutrophils to the site. As soon as they arrive, neutrophils will phagocytose microbial cells at an astonishing rate—in fact at such a rate that they cannot survive their own gluttony. Meanwhile, DCs sample the microbes by phagocytosis, but rather than ravenously destroying each microbe encountered, DCs take a more measured approach. Their goal is to package a bit of microbial antigen on their surface. Once this is accomplished, they travel from the site of infection to the closest lymph nodes to alert cells of the adaptive immune system that a specific microbe has invaded.

Innate Lymphoid Cells

A close look at figure 31.9 shows that all innate immune cells discussed so far derive from common myeloid progenitor cells. In fact, the terms myeloid and lymphoid (and lymphocyte) were coined to distinguish between cells of innate and adaptive immunity, respectively. When natural killer (NK) cells were found to arise from common lymphoid progenitor cells, many scientists were perplexed because NK cells have an innate immune function. More recently, common lymphoid progenitor cells have been found to also give rise to other cells that function in the innate arm of the immune system. Collectively, these are termed **innate lymphoid cells (ILCs).** Although all these cells share the same progenitor cell and are thereby **lymphocytes,** two key features set them apart from lymphocytes of adaptive immunity (T and B cells). First, ILCs (which include NK cells) do not undergo the same complex developmental processes that lead to the expression of antigen-specific receptors. This results in the second major difference. ILCs lack receptors that respond to and "remember" specific pathogens. Instead, like the innate immune cells they are, ILCs are stimulated by the same kinds of stress signals that alert neutrophils, macrophages, and DCs. These include opsonized microbes, foreign antigens, and cytokines. Thus while we still refer to myeloid and lymphoid lineages and cells, the distinction between innate and adaptive immune function is no longer relevant.

Once mature, each ILC has a different function. As their name suggests **lymphoid tissue inducer (LTi) cells** facilitate the development of tissue found in lymph nodes and Peyer's patches (specialized lymphoid tissue in the small intestine), as discussed in section 31.5. Other ILCs can be divided into three functional types, *ILCs 1, 2, and 3,* with each type responding to a different kind of antigenic trigger (**figure 31.11**). When stimulated by intracellular pathogens (e.g., viruses), ILC-1s help activate macrophages. In response to extracellular pathogens such as

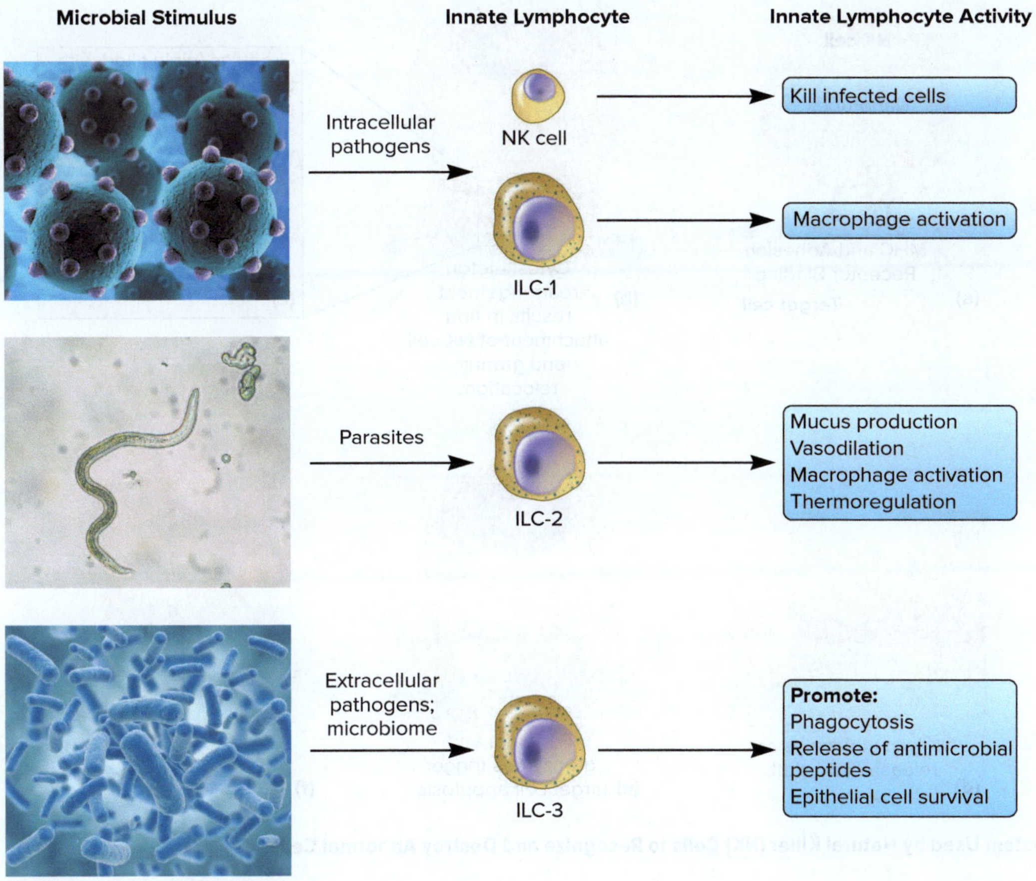

Figure 31.11 Innate Lymphocyte Stimuli and Responses.

(Left top) Nixxphotography/Getty Images; (Left middle) Olgaru79/Shutterstock; (Left bottom) Jezperklauzen/iStock/Getty Images

worms (particularly those that inhabit the bowel), ILC-2s trigger vasodilation, the release of more mucus, and macrophage activity. ILC-2s also have a role in thermoregulation that is not well understood. ILC-3s have proven to be very important in immune homeostasis because they respond to our normal microbiota in addition to microbial pathogens. ILC-3s help maintain the barrier function along mucous membranes by stimulating the release of antimicrobial peptides, preventing an immune response to normal gut microbiota, and promoting phagocytosis and the survival of epithelial cells that line the mucosa. ▶| A *functional core microbiome is required for host homeostasis (section 33.3)*

Natural killer (NK) cells are sometimes also referred to as cytotoxic ILCs (figure 31.11). These large, granular cells detect and destroy stressed, malignant, or cells infected with an intracellular pathogen, such as a virus (**figure 31.12**). NK cells survey adjacent cells by contacting them via several surface receptors (figure 31.12*a*). These interactions provide both "kill" and "don't kill" signals to the NK cell. It is the balance of these signals that determines the fate of the target cell. If the target cell is found to be defective (e.g., malignant) or infected, the kill signals outweigh the survival signals and the NK cell becomes deadly.

Once activated to kill the target cell, the NK cell mobilizes its cytoskeletal proteins to closely adhere to the target cell and releases a cargo of lethal molecules (figure 31.12). These include proteins called **perforins** and enzymes called **granzymes.** Perforins form pores in the target cell membrane to facilitate entry of granzyme enzymes, which are proteases that trigger the target cell to undergo programmed cell death or **apoptosis.**

In addition to responding to the balance of kill versus survival signals transmitted by the target cell, NK cells have another way to detect cells that need to be cleared. NK cells have receptors for antibodies, and therefore can attack cells that are opsonized by antibodies. Recall that self cells should not be antigenic, and should therefore not be coated with antibodies. However, cancerous cells may produce aberrant proteins perceived to be non-self by the adaptive immune system. In addition, host cells infected with viruses or other intracellular pathogens may attach microbial proteins to their surfaces. Both cases lead to antibody attachment to host cells. The process by which NK cells recognize and kill such cells is therefore called **antibody-dependent cell-mediated cytotoxicity (ADCC)** (**figure 31.13**).

Comprehension Check

1. Describe the structure and function of each of the following blood cells: basophil, eosinophil, neutrophil, monocyte, macrophage, and dendritic cell. Which cells are phagocytic?
2. How are macrophages and dendritic cells different? How are they similar?

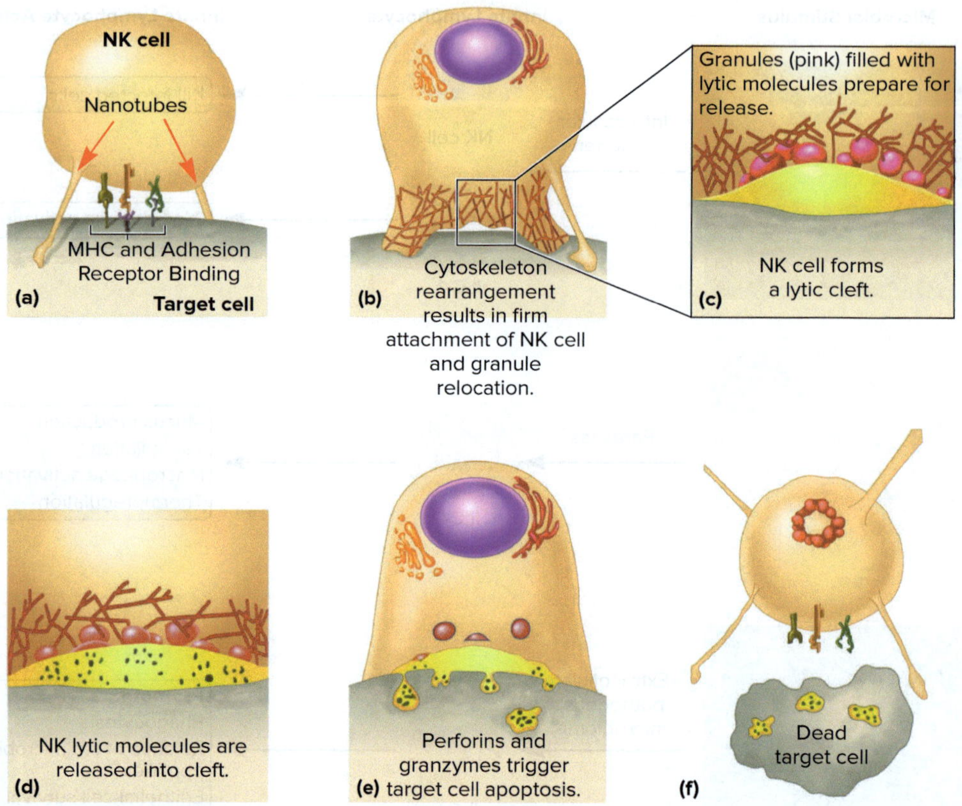

Figure 31.12 The System Used by Natural Killer (NK) Cells to Recognize and Destroy Abnormal Cells.

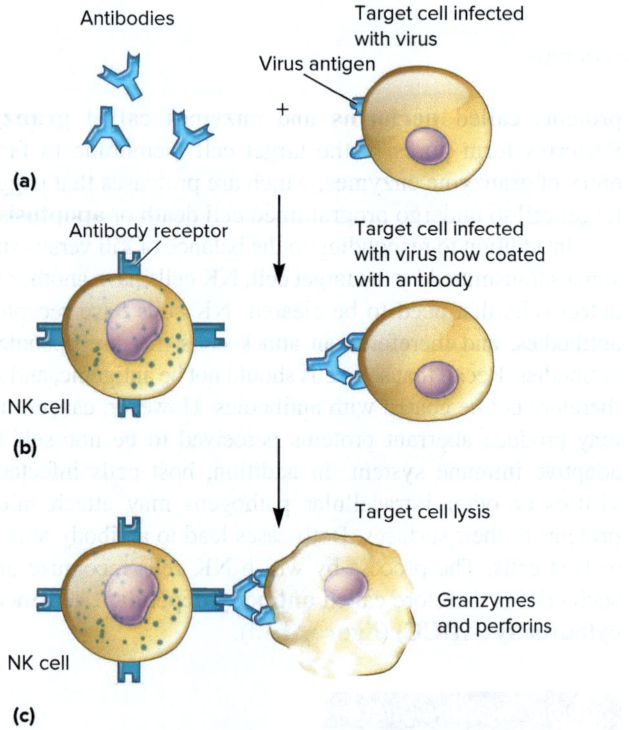

Figure 31.13 Antibody-Dependent Cell-Mediated Cytotoxicity. (a) In this mechanism, antibodies bind to a malignant cell or a host cell infected with a virus. (b) NK cells have antibody receptors on their surface. (c) The antibody bridges the infected cell with the NK cell so that the target is close enough for enzymatic attack.

3. How does a dendritic cell bridge the innate and adaptive arms of immunity?
4. Compare and contrast the function of ILCs –1, –2, and –3.
5. Discuss the role of NK cells in protecting the host. Describe the two mechanisms NK cells use to detect altered cells.

31.5 Organs and Tissues of the Immune System Are Sites of Host Defense

After reading this section, you should be able to:

a. Differentiate between primary and secondary lymphoid tissues and organs in terms of structure and function
b. Explain the role of lymph nodes in host defense
c. Explain how the cellular architecture of skin- and mucosal-associated lymphoid tissues is optimized to defend the host

The organs and tissues of the immune system can be divided into primary or secondary lymphoid organs and tissues based on function (**figure 31.14**). The primary organs are where hematopoietic stem cells differentiate into innate immune cells and the adaptive immune cells B and T lymphocytes. The thymus is the primary lymphoid organ for T cells, and the bone marrow is

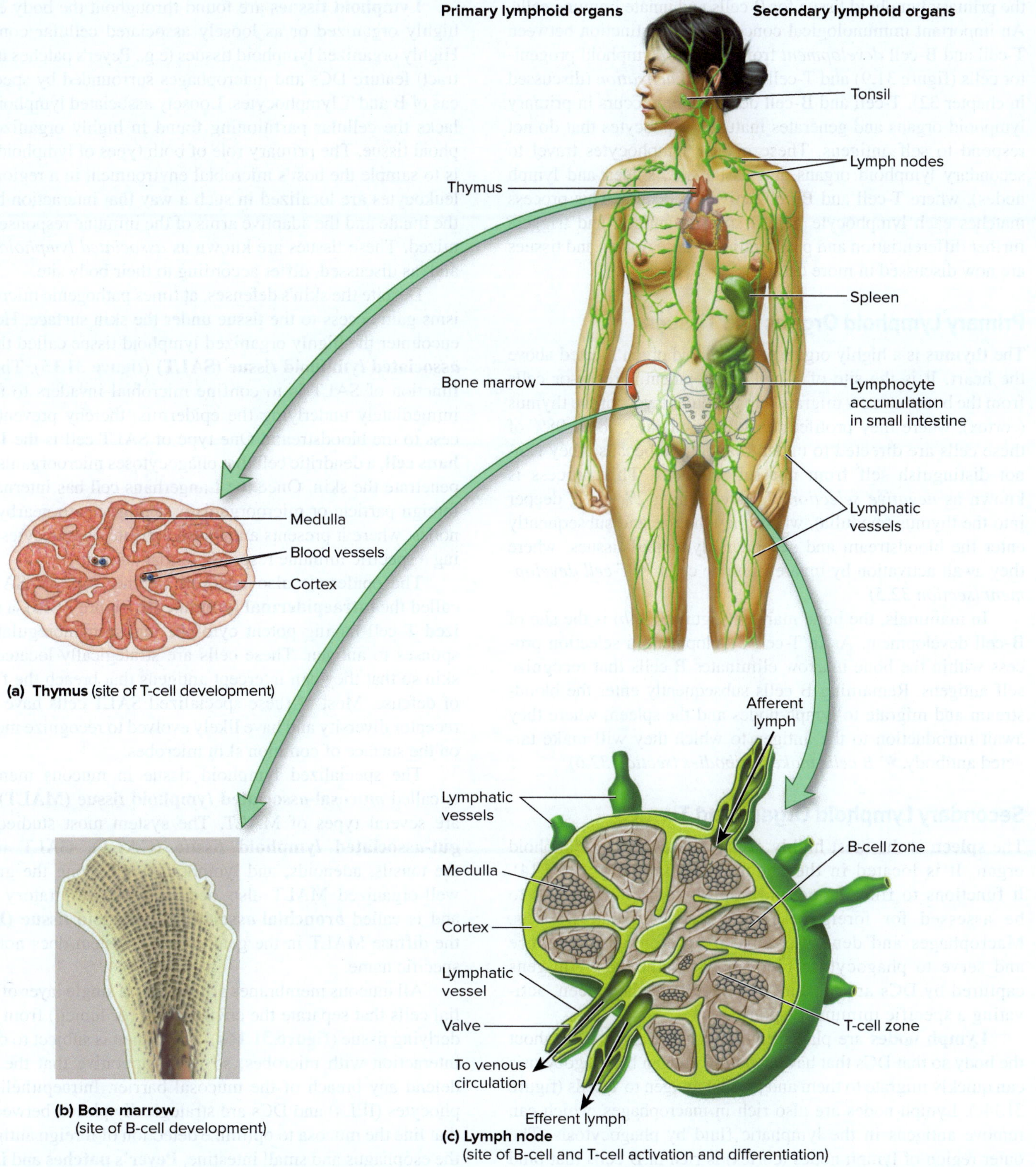

Figure 31.14 Anatomy of the Lymphoid System. Primary lymphoid organs include (a) the thymus and (b) bone marrow. (c) Lymph nodes are secondary lymphoid tissues. Lymphatic fluid (lymph) enters the lymph node through afferent ("arriving") lymphatic vessels; percolates through and around the follicles, made up of B-cell and T-cell zones in the node; and leaves through the efferent ("exiting") lymphatic vessels. The lymphoid follicles are the sites of cellular interactions and extensive immunologic activity.

MICRO INQUIRY *In which of these organs and tissues are macrophages and dendritic cells found in abundance?*

the primary lymphoid tissue for B cells and innate immune cells. An important immunological concept is the distinction between T-cell and B-cell *development* from common lymphoid progenitor cells (figure 31.9) and T-cell and B-cell *activation* (discussed in chapter 32). T-cell and B-cell development occurs in primary lymphoid organs and generates mature lymphocytes that do not respond to self antigens. These mature lymphocytes travel to secondary lymphoid organs and tissue (i.e., spleen and lymph nodes), where T-cell and B-cell activation occurs. This process matches each lymphocyte with a specific antigen and triggers further differentiation and proliferation. These organs and tissues are now discussed in more detail.

Primary Lymphoid Organs and Tissues

The **thymus** is a highly organized lymphoid organ located above the heart. It is the site of T-cell development. Precursor cells from the bone marrow migrate into the outer layers of the thymus (cortex), where they proliferate. As they mature, about 98% of these cells are directed to undergo apoptosis because they cannot distinguish self from non-self antigens. This process is known as *negative selection*. The remaining 2% move deeper into the thymus (medulla), where they mature and subsequently enter the bloodstream and secondary lymphoid tissues, where they await activation by innate immune cells. ▶| *T-cell development (section 32.5)*

In mammals, the bone marrow (figure 31.14*b*) is the site of B-cell development. As in T-cell development, a selection process within the bone marrow eliminates B cells that recognize self antigens. Remaining B cells subsequently enter the bloodstream and migrate to lymph nodes and the spleen, where they await introduction to the antigen to which they will make targeted antibody. ▶| *B cells make antibodies (section 32.6)*

Secondary Lymphoid Organs and Tissues

The **spleen** is the most highly organized secondary lymphoid organ. It is located in the abdominal cavity (figure 31.14). It functions to filter blood and trap blood-borne particles to be assessed for foreignness by phagocytes and B cells. Macrophages and dendritic cells are present in abundance and serve to phagocytose blood-borne pathogens. Antigens captured by DCs are presented to T cells in the spleen, activating a specific immune response.

Lymph nodes are placed in strategic locations throughout the body so that DCs that have captured antigen by phagocytosis can quickly migrate to them and present antigen to T cells (figure 31.14*c*). Lymph nodes are also rich in macrophages, which can remove antigens in the lymphatic fluid by phagocytosis. The outer region of lymph nodes (cortex) is rich in B cells that bind antigen directly from the blood or lymph. The inner region of lymph nodes (medulla) contains regions called *follicles,* which are densely packed with B cells surrounded by the T-cell zone. There T cells await interaction with DCs or B cells. ▶| *T-cell activation (section 32.5)*

Lymphoid tissues are found throughout the body either as highly organized or as loosely associated cellular complexes. Highly organized lymphoid tissues (e.g., Peyer's patches in the GI tract) feature DCs and macrophages surrounded by specific areas of B and T lymphocytes. Loosely associated lymphoid tissue lacks the cellular partitioning found in highly organized lymphoid tissue. The primary role of both types of lymphoid tissues is to sample the host's microbial environment in a region where leukocytes are localized in such a way that interaction between the innate and the adaptive arms of the immune response is optimized. These tissues are known as *associated lymphoid tissues* and, as discussed, differ according to their body site.

Despite the skin's defenses, at times pathogenic microorganisms gain access to the tissue under the skin surface. Here they encounter the highly organized lymphoid tissue called the **skin-*a*ssociated *l*ymphoid *t*issue (SALT)** (**figure 31.15**). The major function of SALT is to confine microbial invaders to the area immediately underlying the epidermis, thereby preventing access to the bloodstream. One type of SALT cell is the **Langerhans cell,** a dendritic cell that phagocytoses microorganisms that penetrate the skin. Once the Langerhans cell has internalized a foreign particle or microorganism, it migrates to nearby lymph nodes, where it presents antigen to activate lymphocytes, inducing a specific immune response to that antigen.

The epidermis also contains another type of SALT cell called the **intraepidermal lymphocyte** (figure 31.15), a specialized T cell having potent cytolytic and immunoregulatory responses to antigen. These cells are strategically located in the skin so that they can intercept antigens that breach the first line of defense. Most of these specialized SALT cells have limited receptor diversity and have likely evolved to recognize molecules on the surface of common skin microbes.

The specialized lymphoid tissue in mucous membranes is called ***m*ucosal-*a*ssociated *l*ymphoid *t*issue (MALT).** There are several types of MALT. The system most studied is the ***g*ut-*a*ssociated *l*ymphoid *t*issue (GALT).** GALT includes the tonsils, adenoids, and lymphoid tissue along the gut. Less well-organized MALT also occurs in the respiratory system and is called ***b*ronchial-*a*ssociated *l*ymphoid *t*issue (BALT);** the diffuse MALT in the genitourinary system does not have a specific name.

All mucous membranes are lined by a single layer of epithelial cells that separate the environment (or lumen) from the underlying tissue (**figure 31.16*a***). The lumen is subject to constant interaction with microbes, so it is imperative that the MALT defend any breach of the mucosal barrier. Intraepithelial lymphocytes (IELs) and DCs are strategically placed between cells that line the mucosa to optimize detection of foreign antigens. In the esophagus and small intestine, **Peyer's patches** and **isolated lymphoid follicles** (**ILFs;** throughout the gut) have specialized cells called **M cells,** in between epithelial cells. M cells are flattened on the lumen side of the gut and have a large pocket facing the interior (figure 31.16*b*). When a foreign particle contacts an M cell, it is endocytosed on the lumenal side and released into

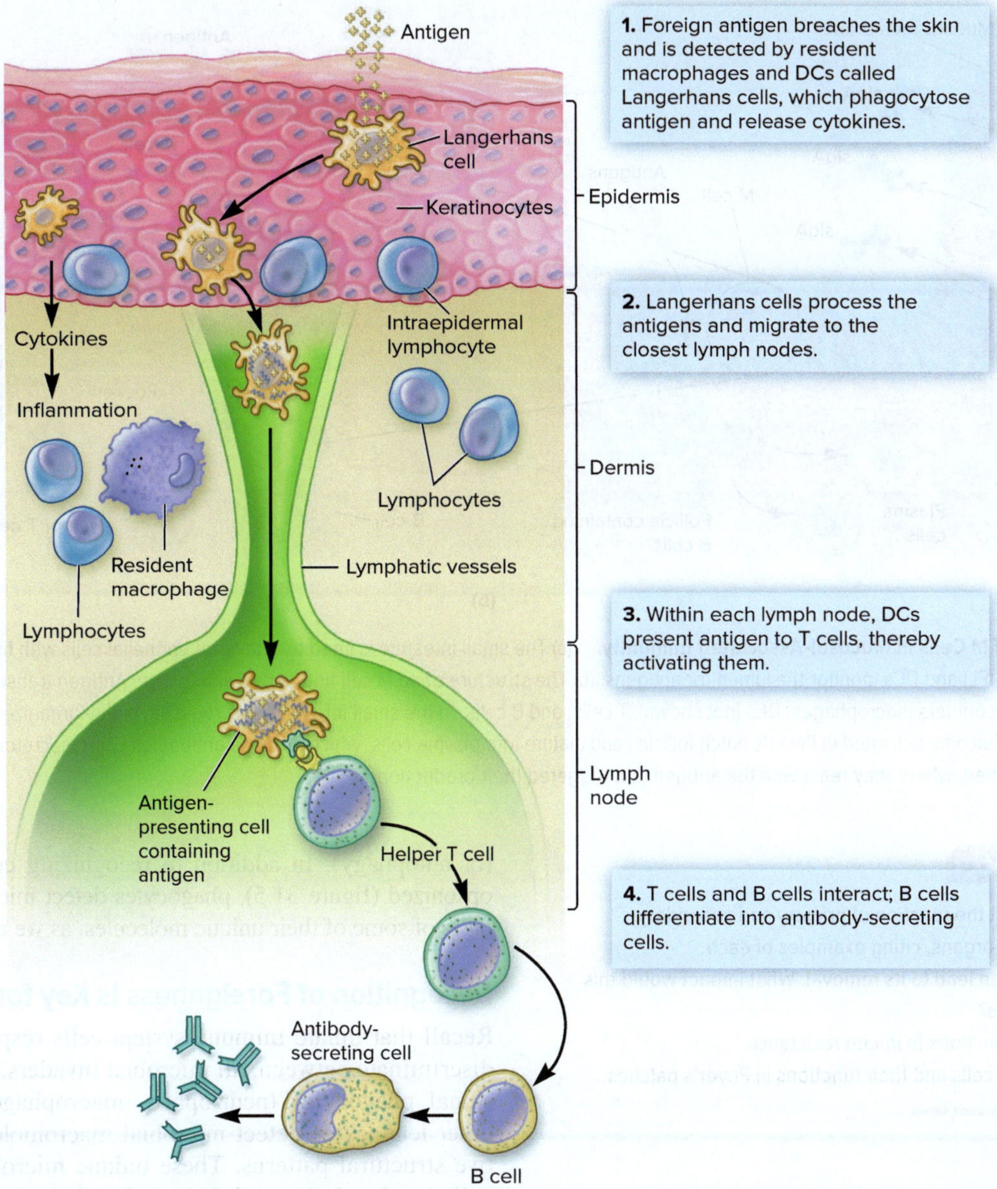

Figure 31.15 Skin-Associated Lymphoid Tissue (SALT). Keratinocytes make up 90% of the epidermis. They can secrete cytokines that cause an inflammatory response to invading pathogens.

the pocket, a process called *transcytosis*. There macrophages and DCs engulf the particle or pathogen and degrade it. In addition to macrophages, T and B cells are clustered in follicles within Peyer's patches and ILFs. As microbes are introduced into the M-cell pocket, they can be captured by B cells and stimulate the production of antibody that can be secreted into the gut lumen (secretory IgA; *section 32.7*). DCs activate T cells within the follicles or migrate to local lymph nodes (called mesenteric lymph nodes) to activate T cells there. In addition to Peyer's patches and ILFs, diffuse lymphoid tissue is widely distributed in the large intestine, as well as the respiratory and genitourinary tracts, breast, eye, salivary glands, and skin. These areas typically have a high density of immune cells but do not have T and B cells in organized follicles. ▶| *Antibodies bind specific 3-D antigens (section 32.7)*

As seen in our discussion thus far, a microbe attempting to invade a potential host encounters innate mechanisms including physical, chemical, and cellular barriers designed to kill the invader, digest the carcass into small antigens, and assist lymphocytes in developing long-term protection against the next invasion. We now examine the phagocytic processes in more detail and consider how the host integrates many of the innate resistance activities into the coordinated activities that define an inflammatory response.

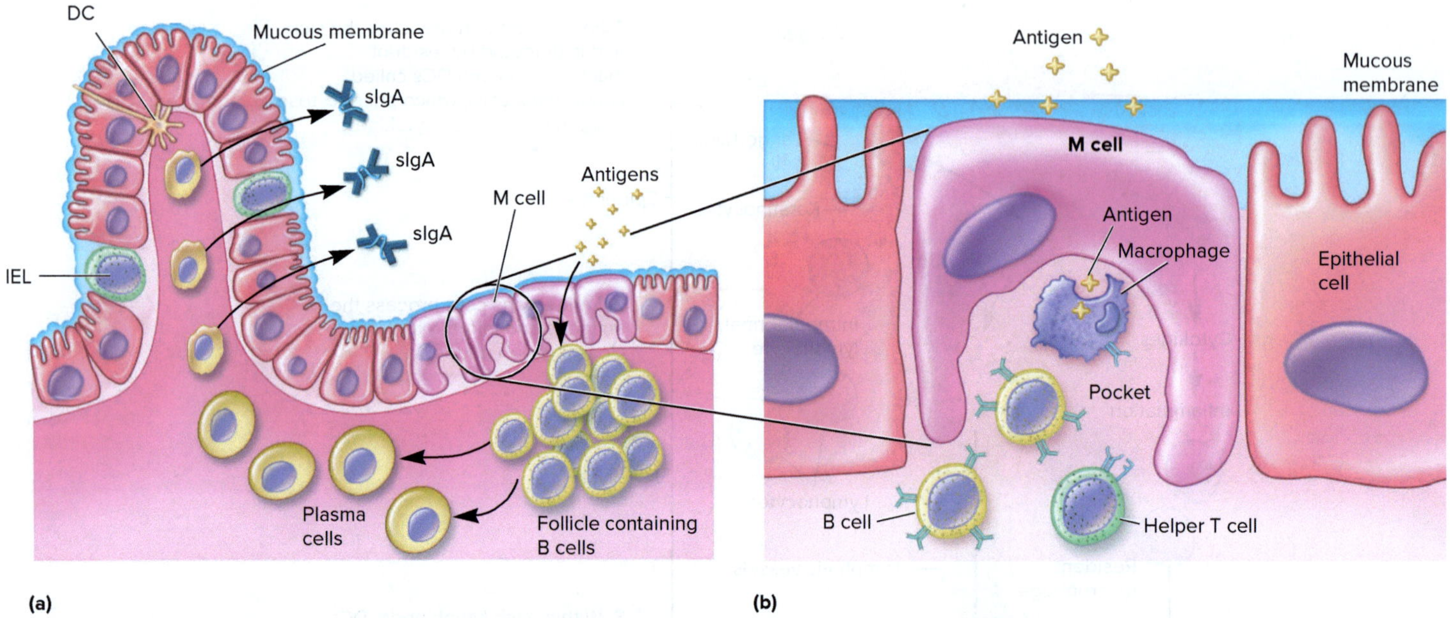

Figure 31.16 Function of M Cells in Mucosal-Associated Immunity. (a) The small intestine is lined by a layer of epithelial cells with intervals of M cells. Intraepithelial lymphocytes (IEL) and DCs monitor the lumen for antigens. (b) The structure of an M cell and underlying follicle. Antigen transcytosed through the M-cell and into its pocket encounters macrophages, DCs (not shown), T cells, and B cells. In the small intestine, this region is highly organized into tissue called Peyer's patches. B cells can become activated in Peyer's patch follicles and mature into plasma cells, which produce antibodies (SIgA; secretory Immunoglobulin A) that are released into the lumen, where they react with the antigen that triggered their production.

Comprehension Check

1. Compare and contrast the function of primary and secondary lymphoid tissues and organs, citing examples of each.
2. Injury to the spleen can lead to its removal. What impact would this have on host defenses?
3. Describe how SALT functions in innate resistance.
4. Describe the immune cells and their functions in Peyer's patches.

31.6 Phagocytosis Destroys Invaders

After reading this section, you should be able to:

a. Explain the methods by which pathogens are recognized by phagocytes
b. Compare and contrast the processes of autophagy and phagocytosis
c. Correlate biochemical activities within the phagolysosome with pathogen destruction

Phagocytic cells (macrophages, DCs, and neutrophils) are an important early defense against invading microorganisms. These cells recognize, ingest, and kill cellular microbes by the process of **phagocytosis** (Greek *phagein,* to eat; *cyte,* cell; *osis,* a process) and intracellular pathogens by the process of **autophagocytosis** (or autophagy). In addition to recognizing cells that have been opsonized (figure 31.5), phagocytes detect microorganisms on the basis of some of their unique molecules, as we now discuss.

Recognition of Foreignness Is Key for Survival

Recall that innate immune system cells respond to, but do not discriminate between, all microbial invaders. To do this, professional phagocytes (neutrophils, macrophages, and DCs) and other leukocytes detect microbial macromolecules with repetitive structural patterns. These unique microbial signatures are called ***m**icrobe-**a**ssociated **m**olecular **p**atterns* **(MAMPs),** although they are referred to as pathogen-associated molecular patterns (PAMPs), too. We use the more inclusive term, MAMP, because the immune system responds to many microorganisms that are not pathogens. MAMPs are specific regions within common microbial macromolecules, such as lipopolysaccharide (LPS), peptidoglycan, fungal cell wall components, viral nucleic acids, and other microbial structures. MAMPs do not identify a specific microorganism; rather they inform the host that a microbe is present, sounding the alarm of potential infection.

Microbial invaders can be either extracellular, like most fungi and many bacteria, or intracellular, like viruses, some bacteria, and a few fungi. It therefore follows that cells of the innate immune system recognize extracellular infectious agents with receptors on the surface of host immune cells, while intracellular pathogens are detected by host receptors in the cytosol. All receptors that recognize MAMPs, regardless of cellular location, are called ***p**attern **r**ecognition **r**eceptors* **(PRRs).** PRR stimulation results in the production of cytokines that function to promote immune defenses. As we take a closer look at these immune cell receptors, keep in mind they are critical in distinguishing between self and non-self.

C-Type Lectin Receptors

C-type lectin receptors (CLRs) are a large group of calcium-dependent membrane-bound proteins having one or more lectin-binding domains specific for one type of carbohydrate (lectin) found on MAMPs (**table 31.3**). Ligands include mannose, fucose, or glucan carbohydrates found on some bacteria (e.g., *Mycobacterium tuberculosis* and *Helicobacter pylori*), fungi (e.g., *Candida albicans* and *Aspergillus fumigatus*), helminths (worms), and some viruses. When a CLR on the surface of a phagocyte binds a lectin MAMP, a signaling cascade is triggered that induces cytokine gene expression by the phagocyte. Binding also results in receptor-MAMP internalization, its degradation, and subsequent antigen presentation to lymphocytes (section 32.4).

Toll-Like Receptors

Toll-like receptors (TLRs) are an important class of transmembrane PRRs expressed by a variety of cells that participate in innate immunity, especially macrophages and DCs (table 31.3). They span the plasma membrane as well as membranes of endosomes, lysosomes, endolysosomes, and the endoplasmic reticulum, where they detect MAMPs that have gained access to the host cell cytosol. TLRs have an extracellular (or extra-organelle) domain that binds MAMPs and a cytoplasmic Toll/interleukin-1 receptor (TIR) domain that activates downstream signaling (**figure 31.17**). Binding of a MAMP to a TLR causes the TIR to recruit a specific adapter protein, which in turn activates a signal transduction cascade that ends in the activation of transcription factors required for expression of genes encoding cytokines. There are at least 10 human TLRs that recognize distinct but sometimes overlapping MAMPs and subsequently initiate an immune response.

NOD-Like Receptors

Unlike the transmembranous TLRs, **nucleotide-binding and oligomerization domain (NOD)–like receptors** are located only in the host cell cytosol (**figure 31.18**). There NOD-like receptors (NLRs) can sense intracellular MAMPs (table 31.3). In addition, NLRs also respond to host-derived cellular stress. This includes reactive oxygen species, loss of cellular K^+, and host molecules called *damage-associated molecular patterns* (DAMPs), such as uric acid and heat-shock proteins. Because eukaryotic cytosol is normally a DNA-free zone, intracellular receptors often detect DNA as evidence of invasion by intracellular pathogens or a sign of host nuclear stress. Upon detection of one of these foreign or native signals, several NLRs oligomerize cofactors and the protease caspase-1 is recruited to the oligomerized unit. This NLR-caspase complex is called the **inflammasome** because once assembled, it triggers the release of pro-inflammatory cytokines.

RIG-I-Like Receptors

Retinoic-acid-inducible gene I (RIG-I)–like receptors (RLRs) are another family of cytoplasmic receptors. These receptors evolved to sense viral RNA. Replication of viruses with RNA genomes within the cytosol leads to the production of dsRNA

Table 31.3 Pattern Recognition Receptors

Receptor	Cellular Location	Cell Type	Ligands
C-type lectin	Cell surface	Macrophages, dendritic cells, NK cells	Carbohydrate MAMPs (e.g., mannose, fucose, glucan, *N*-acetylglucosamine)
Nucleotide-binding oligomerization domain (NOD)–like receptors (NLRs)	Cytosol	Macrophages, dendritic cells, lymphocytes, some epithelial cells (e.g., those lining mucous membranes)	Viral RNA Intracellular pathogen MAMPs Host damage-associated molecular pattern molecules (e.g., uric acid, heat shock proteins)
Retinoic-acid-inducible gene 1 (RIG-1)	Cytosol	All cell types	Viral RNA
Stimulator of interferon genes (STING)	ER membrane	All cell types	Bacterial and host-generated cyclic dinucleotides
Toll-like receptors (TLRs)	Cell surface and ER membrane, surface of endosomes, lysosomes, endolysosomes	Macrophages, dendritic cells, some epithelial cells (e.g., those lining mucous membranes)	Lipopolysaccharide, flagellin, teichoic acid, lipopeptides, ssRNA,[1] dsRNA,[2] DNA with the CpG motif[3]

[1] Single-stranded RNA
[2] Double-stranded RNA
[3] Cytosine triphosphate deoxynucleotide (C)-phosphodiester bond (p)-guanine triphosphate deoxynucleotide(G); unmethylated CpG is a MAMP

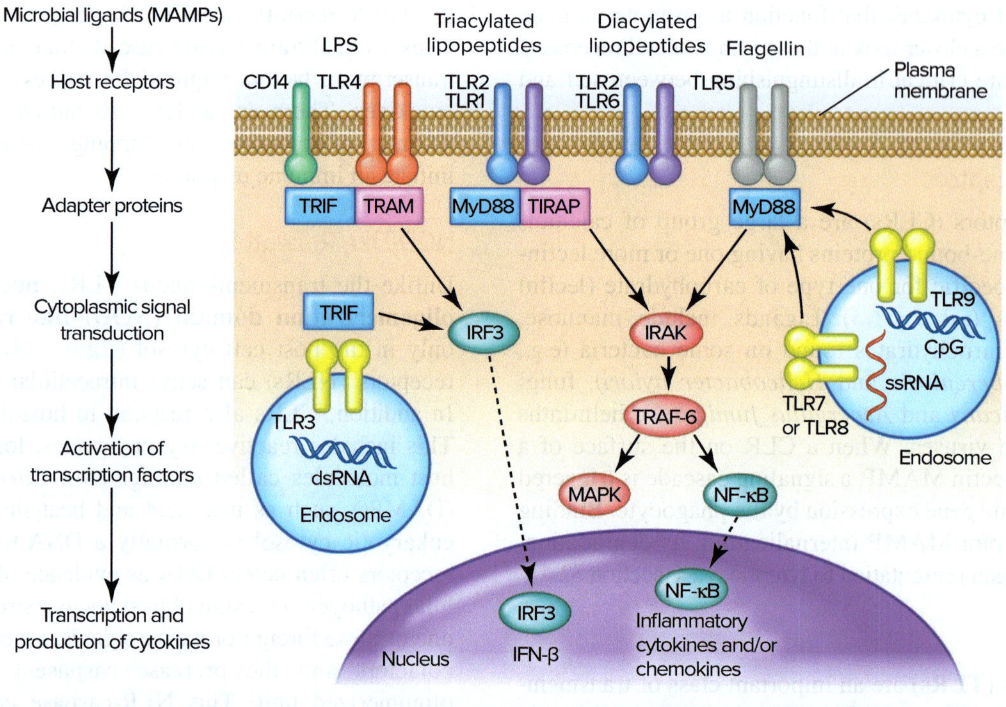

Figure 31.17 Recognition of Microbe-Associated Molecular Patterns (MAMPs) by Pattern Recognition Receptors (PRRs). MAMPs bind toll-like receptors (TLRs) found on both the plasma membrane and endosomal membranes. PRR binding results in signaling that upregulates cytokine gene expression through common signal transduction pathways, like the transcription factors NF-κB and IRF3. Interferon-β (IFN-β) is an antiviral cytokine. TRIF, TRAM, MyD88 and TIRAP are proteins that help mediate intracellular signaling. IRAK, TRAF-6, and MAPK proteins contribute to the regulation of NF-kB In addition to TLR4, CD14 also binds lipopolysaccharide (LPS).

Source: "Introduction to the Immune System." What-When-How.

***K s.b kappa as in figure

and triphosphate-capped ssRNA. These forms of RNA are not found in the cytosol of healthy cells. RLRs recognize these forms of RNA and stimulate an antiviral cytokine response. *Viruses and other acellular infectious agents (chapter 6)*

STING Receptors

*S*timulator of *in*terferon *g*enes (STING) receptors are localized to the surface of the endoplasmic reticulum of all cell types, where they serve to alert the cell of intracellular bacteria and

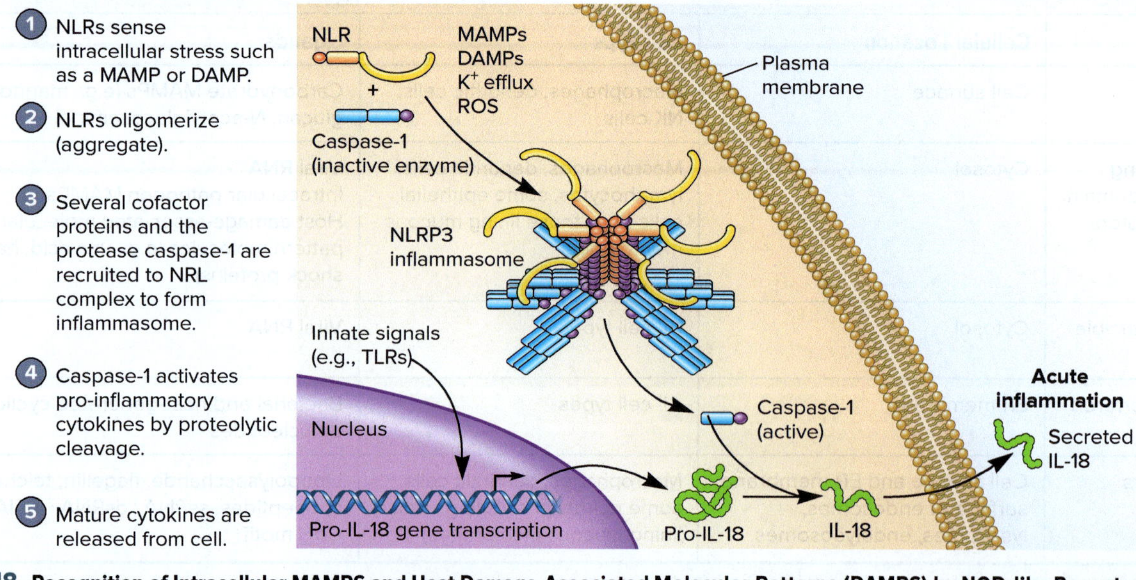

Figure 31.18 Recognition of Intracellular MAMPS and Host Damage-Associated Molecular Patterns (DAMPS) by NOD-like Receptors (NLRs) Results in Inflammasome Formation. The oligomerization of NLRs upon sensing MAMPs or DAMPs results in the recruitment and activation of caspase-1, a protease, which cleaves, thereby activating immature cytokines (e.g., Pro-interleukin-18, Pro-IL-18). IL-18 is shown; IL-1 is activated in a similar manner. ROS, reactive oxygen species.

host nuclear stress (table 31.3). This is accomplished by detecting cyclic dinucleotides such as cyclic di-GMP and cyclic di-AMP produced by intracellular pathogens such as *Listeria monocytogenes*. Host nuclear stress is also detected if DNA is present in the cytosol rather than confined to the nucleus. If dsDNA larger than about 70 bp is found in the cytosol, the host enzyme cyclic GMP-AMP synthase (cGAS) converts it to cyclic dinucleotides so that the alarm can be raised by STING receptors. As the name suggests, activation of these receptors results in the induction of the cytokine genes for type I interferons, which defend against intracellular (e.g., viral) infections. Additionally, STING activation can induce autophagy or apoptosis as a means of protecting the host from nuclear stress.

Intracellular Digestion

Once an invader has been detected, it must be removed. Two parallel processes, phagocytosis and autophagy, accomplish ingestion and digestion of extracellular and intracellular foreign materials, respectively. We focus first on phagocytosis, whereby extracellular particles are engulfed by pseudopodia (Latin *pseudo*, false; *podium*, foot) formed by the extension of the phagocyte cell membrane. Once pseudopodia surround the particle, an internal *phagosome* is formed. This is often opsonin and receptor-mediated (**figure 31.19**, steps 1 and 2). *There are several endocytic pathways (section 5.4)*

Next lysosomes fuse to the phagosome (figure 31.19, step 3) creating a **phagolysosome. Lysosomes** deliver a variety of

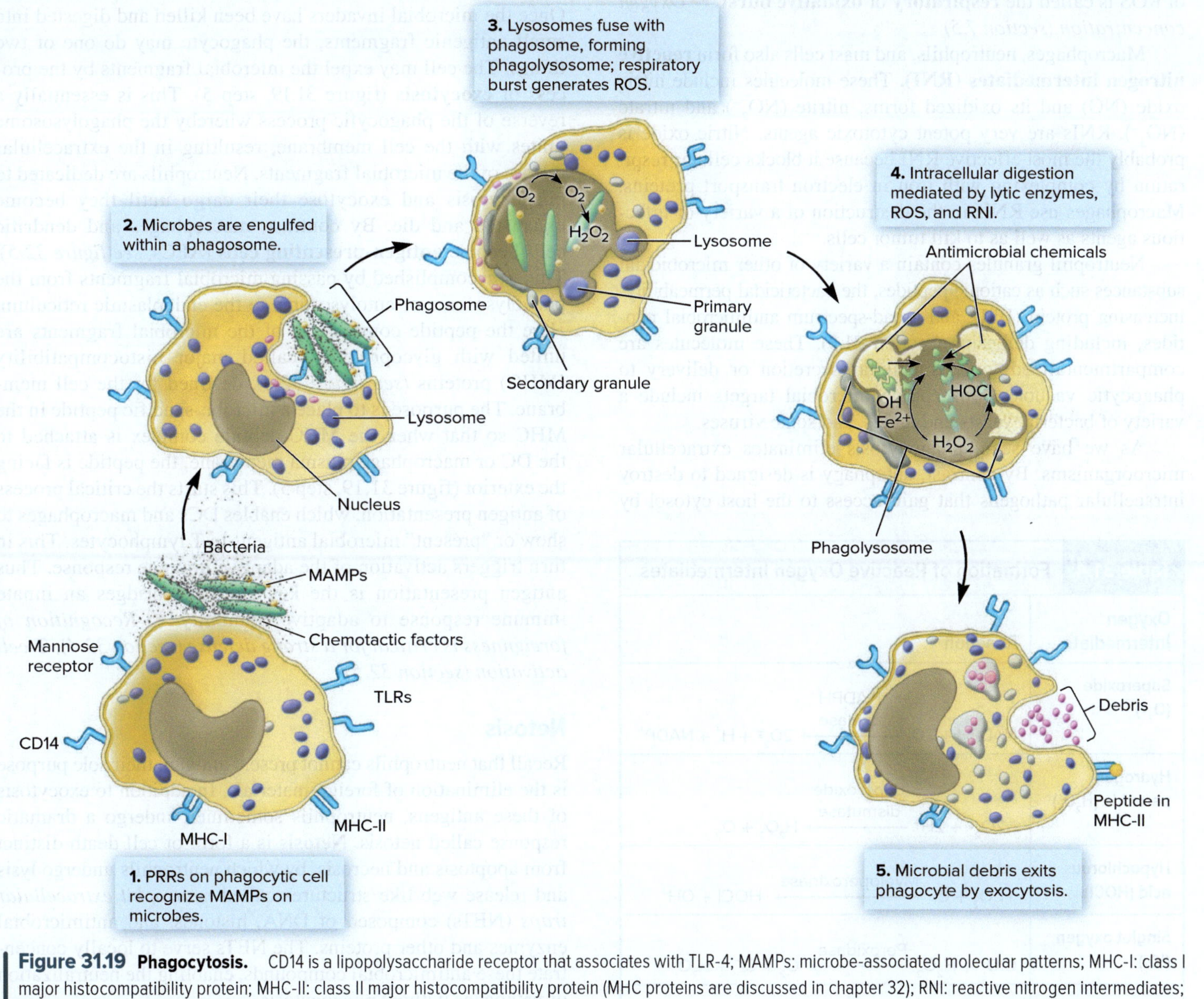

Figure 31.19 Phagocytosis. CD14 is a lipopolysaccharide receptor that associates with TLR-4; MAMPs: microbe-associated molecular patterns; MHC-I: class I major histocompatibility protein; MHC-II: class II major histocompatibility protein (MHC proteins are discussed in chapter 32); RNI: reactive nitrogen intermediates; ROS: radical oxygen species; TLRs: toll-like receptors.

MICRO INQUIRY *How do the enzymes delivered by lysosomes differ from the NADPH oxidase that is also recruited to the phagosome?*

hydrolases such as lysozyme, phospholipase A2, ribonuclease, deoxyribonuclease, and proteases. The activity of these degradative enzymes is enhanced by the acidic vacuolar pH. Collectively these enzymes participate in the destruction of entrapped microorganisms (figure 31.19, step 4). In addition, toxic **reactive oxygen species (ROS)** such as superoxide radical ($O_2^{\bar{\cdot}}$), hydrogen peroxide (H_2O_2), singlet oxygen ($^1O_2$), and hydroxyl radical (·OH) are produced. Neutrophils also contain the heme-protein enzyme *myeloperoxidase,* which catalyzes the production of hypochlorous acid (bleach). The generation of ROS begins with the recruitment of an NADPH oxidase to the phagosome or autosome membrane. This enzyme uses molecular oxygen to oxidize NADPH to $NADP^+$ and generate superoxide radicals (**table 31.4**). The rapid consumption of O_2 and release of ROS is called the **respiratory** or **oxidative burst.** ◀ *Oxygen concentration (section 7.5)*

Macrophages, neutrophils, and mast cells also form **reactive nitrogen intermediates (RNI).** These molecules include nitric oxide (NO) and its oxidized forms, nitrite (NO_2^-) and nitrate (NO_3^-). RNIs are very potent cytotoxic agents. Nitric oxide is probably the most effective RNI because it blocks cellular respiration by complexing with iron in electron transport proteins. Macrophages use RNIs in the destruction of a variety of infectious agents as well as to kill tumor cells.

Neutrophil granules contain a variety of other microbicidal substances such as cationic peptides, the bactericidal permeability-increasing protein (BPI), and broad-spectrum antimicrobial peptides, including defensins (section 31.3). These molecules are compartmentalized for extracellular secretion or delivery to phagocytic vacuoles. Susceptible microbial targets include a variety of bacteria, yeasts and molds, and some viruses.

As we have seen, phagocytosis eliminates extracellular microorganisms. By contrast, autophagy is designed to destroy intracellular pathogens that gain access to the host cytosol by either membrane fusion or endocytosis (**figure 31.20;** *also see figure 6.10*). Once within a host cell, autophagy begins when microorganisms, or their products, are recognized by intracellular membranes through PRRs, or are coated with ubiquitin, the molecule used to tag proteins for recycling by proteasomes (*see figure 5.8*). In this case, rather than targeting proteins, ubiquitin labels microorganisms for capture by a *phagophore,* which is a free-floating, open membrane in the cytosol. The phagophore encircles the intracellular microorganism, creating a double-membrane *autophagosome.* Like a phagosome, the autophagosome fuses with lysosomes, which deliver their toxic cargo to neutralize the pathogen entrapped within the *autolysosome.*

Table 31.4 Formation of Reactive Oxygen Intermediates

Oxygen Intermediate	Reaction
Superoxide ($O_2^{\bar{\cdot}}$)	$NADPH + 2O_2 \xrightarrow{\text{NADPH oxidase}} 2O_2^{\bar{\cdot}} + H^+ + NADP^+$
Hydrogen peroxide (H_2O_2)	$2O_2^{\bar{\cdot}} + 2H^+ \xrightarrow{\text{Superoxide dismutase}} H_2O_2 + O_2$
Hypochlorous acid (HOCl)	$H_2O_2 + Cl^- \xrightarrow{\text{Myeloperoxidase}} HOCl + OH^+$
Singlet oxygen ($^1O_2$)	$ClO^- + H_2O_2 \xrightarrow{\text{Peroxidase}} {}^1O_2 + Cl^- + H_2O$
Hydroxyl radical (•OH⁻)	$O_2^{\bar{\cdot}} + H_2O_2 \xrightarrow{\text{Peroxidase}} 2{\cdot}OH^- + O_2$

Exocytosis vs. Antigen Presentation

Once the microbial invaders have been killed and digested into small antigenic fragments, the phagocyte may do one of two things. The cell may expel the microbial fragments by the process of **exocytosis** (figure 31.19, step 5). This is essentially a reverse of the phagocytic process whereby the phagolysosome unites with the cell membrane, resulting in the extracellular release of the microbial fragments. Neutrophils are dedicated to phagocytosis and exocytose their cargo until they become exhausted and die. By contrast, macrophages and dendritic cells become **antigen-presenting cells** (APCs; *see figure 32.5*). This is accomplished by passing microbial fragments from the phagolysosome or autolysosome to the endoplasmic reticulum. Here the peptide components of the microbial fragments are united with glycoproteins, called major histocompatibility (MHC) proteins *(see figure 32.4),* destined for the cell membrane. The purpose is to place a microbe-specific peptide in the MHC so that when the MHC-peptide complex is attached to the DC or macrophage plasma membrane, the peptide is facing the exterior (figure 31.19, step 5). This starts the critical process of antigen presentation, which enables DCs and macrophages to show or "present" microbial antigens to T lymphocytes. This in turn triggers activation of the adaptive immune response. Thus antigen presentation is the key step that bridges an innate immune response to adaptive immunity. ▶ *Recognition of foreignness is critical for a strong defense (section 32.4); T-cell activation (section 32.5)*

Netosis

Recall that neutrophils cannot present antigen; their sole purpose is the elimination of foreign materials. In addition to exocytosis of these antigens, neutrophils sometimes undergo a dramatic response called netosis. **Netosis** is a form of cell death distinct from apoptosis and necrosis, by which neutrophils undergo lysis and release web-like structures called *neutrophil extracellular traps* (NETs) composed of DNA, histones, and antimicrobial enzymes and other proteins. The NETs serve to locally concentrate these antimicrobial compounds, enabling the neutralization of pathogens without phagocytosis.

The means by which netosis is induced is still being discovered, but we know that it begins with the activation of neutrophils by

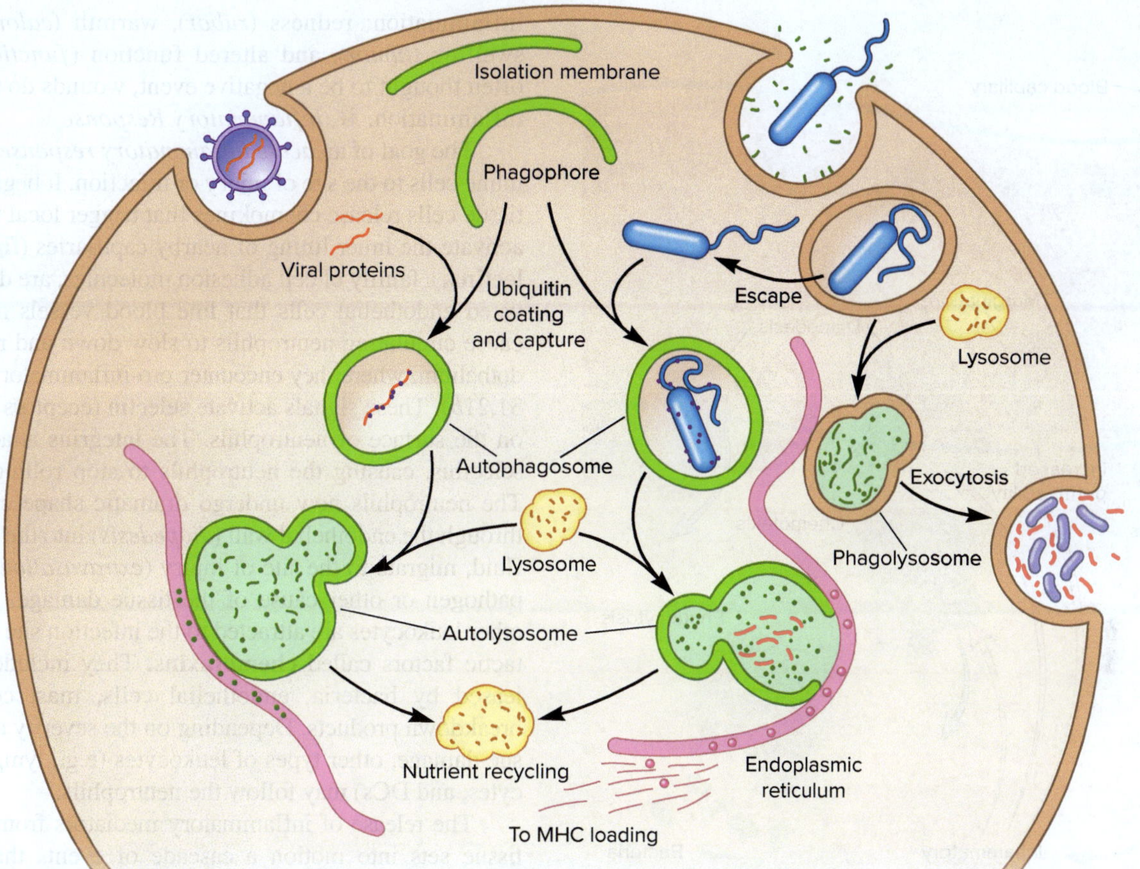

Figure 31.20 Autophagy Is Similar to Phagocytosis. Intracellular infectious agents within the cytosol are captured by free-floating membranes that form a phagophore, often through membrane-bound toll-like receptors. Cytosolic ubiquitin molecules coat the intracellular invaders and their macromolecules (such as viral proteins). The ubiquitinated particles are then surrounded by the phagophore to form an autophagosome. Autophagosome fuses to lysosomes, forming an autolysosome, and its contents are degraded. Both phagolysosomes and autolysosomes can donate some of their debris to the endoplasmic reticulum, where antigenic peptides are further processed to bind MHC molecules destined for the phagocyte's cell membrane (section 32.4). The remaining debris is recycled or released to the environment.

signals such as MAMPs and pro-inflammatory cytokines. One pathway involves NADPH oxidase generation of ROS, which leads to the decondensation of chromatin and the expulsion of NETs. While netosis is crucial for controlling infections, excessive or dysregulated NET formation can contribute to tissue damage and inflammatory diseases, including autoimmune disorders and sepsis. Therefore netosis plays a dual role in host defense and immunopathology.

Comprehension Check

1. How are intracellular and extracellular microbes recognized by leukocytes? Consider both the microbial and host components.
2. What is the purpose of the respiratory burst that occurs within phagocytic cells? Describe the nature and function of reactive oxygen and nitrogen species.
3. Of the three professional phagocytes, which are capable of antigen presentation and what is the purpose of this process?

31.7 Inflammation Unites All Components of Immunity

After reading this section, you should be able to:

a. Outline the sequence of innate host responses that result in inflammation
b. Distinguish between acute and chronic inflammation in terms of the host responses involved in each
c. Construct a concept map relating host cells and processes that remove pathogens

Inflammation (Latin *inflammatio,* to set on fire) is an important defense reaction to tissue injury, such as that caused by a pathogen or wound. Acute inflammation is the immediate response of the body to injury or cell death. The gross features were described over 2,000 years ago and are still known as the cardinal signs of

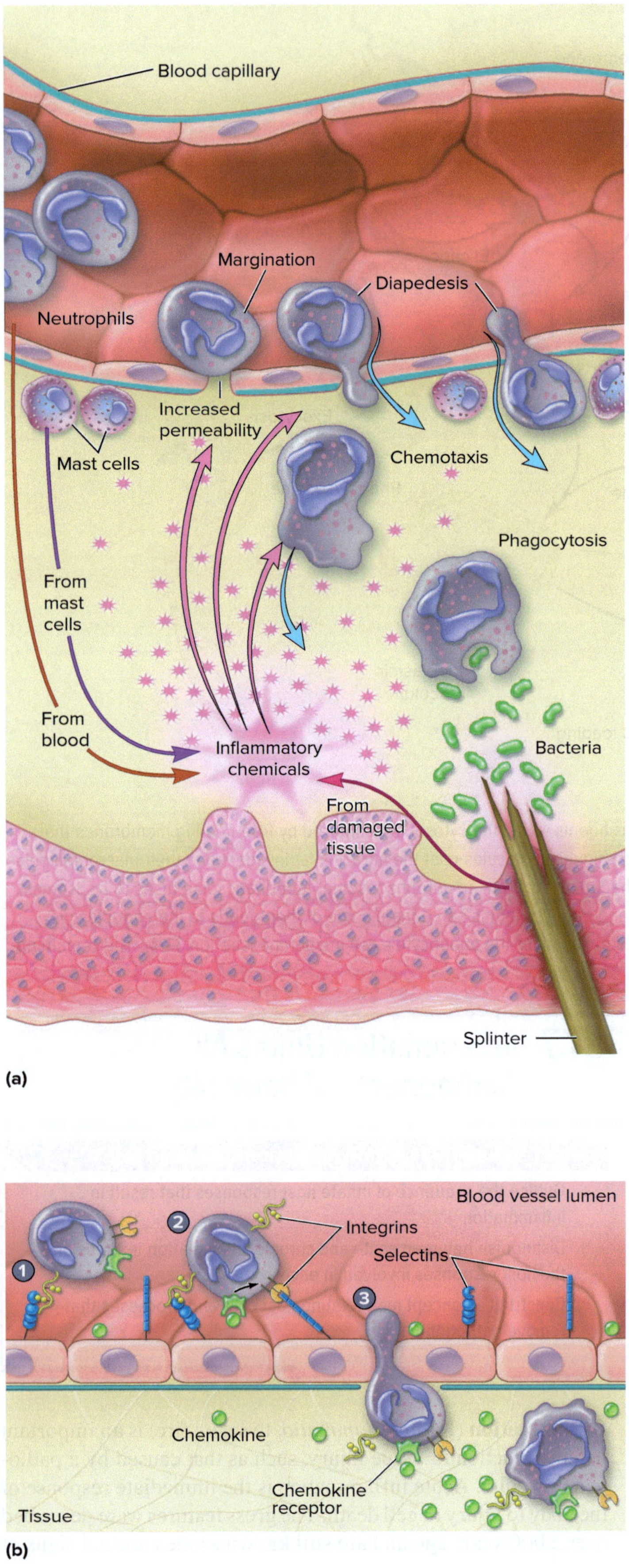

inflammation: redness (*rubor*), warmth (*calor*), pain (*dolor*), swelling (*tumor*), and altered function (*functio laesa*). While often thought to be a negative event, wounds do not heal without inflammation. *Inflammatory Response*

The goal of an *acute inflammatory response* is to deliver immune cells to the site of injury or infection. It begins when injured tissue cells release chemokines that trigger local vasodilation and activate the inner lining of nearby capillaries (**figure 31.21**). **Selectins,** a family of cell adhesion molecules, are displayed on activated endothelial cells that line blood vessels nearby. Selectins cause circulating neutrophils to slow down and roll along the endothelium, where they encounter pro-inflammatory signals (figure 31.21*b*). These signals activate selectin receptors called **integrins** on the surface of neutrophils. The integrins attach tightly to the selectins, causing the neutrophils to stop rolling (*margination*). The neutrophils now undergo dramatic shape changes, squeeze through the endothelial wall (*diapedesis*) into the interstitial tissue fluid, migrate to the site of injury (*extravasation*), and attack the pathogen or other cause of the tissue damage. Neutrophils and other leukocytes are attracted to the infection site by other chemotactic factors called **chemotaxins.** They include substances released by bacteria, endothelial cells, mast cells, and tissue breakdown products. Depending on the severity and nature of tissue damage, other types of leukocytes (e.g., lymphocytes, monocytes, and DCs) may follow the neutrophils.

The release of inflammatory mediators from cells in injured tissue sets into motion a cascade of events that results in the development of the signs of inflammation. One response is the stimulation of resident macrophages and distant liver cells to release antimicrobial and acute-phase proteins, respectively. In the local response, these mediators increase the acidity in the surrounding extracellular fluid, which activates the extracellular enzyme **kallikrein** (**figure 31.22**). Cleavage of kallikrein releases the smaller peptide bradykinin. Bradykinin then binds to receptors on capillary walls, relaxing the tight junctions between cells, which enables fluid, red blood cells, and infection-fighting leukocytes to leave the capillaries and enter infected tissue. Simultaneously, bradykinin binds to mast cells in connective tissue. This activates mast cells by causing an influx of calcium ions, which leads to degranulation and release of preformed mediators such as histamine. Histamine in turn also makes intercellular junctions in the capillary wall wider so that more fluid, leukocytes, kallikrein, and bradykinin move out, causing swelling or edema. Bradykinin also stimulates the production of prostaglandins (PGE_2 and PGF_2) by endothelial

Figure 31.21 Physiological Events of the Acute Inflammatory Response. (a) At the site of injury (splinter), chemical messengers are released from the damaged tissue, mast cells, and macrophages. These inflammatory mediators stimulate neutrophil migration, diapedesis, chemotaxis, and phagocytosis. (b) Neutrophil integrins interact with endothelial selectins (1) to facilitate margination (2) and diapedesis (3).

MICRO INQUIRY *How does the movement of leukocytes in response to chemokines contribute to the cardinal signs of inflammation?*

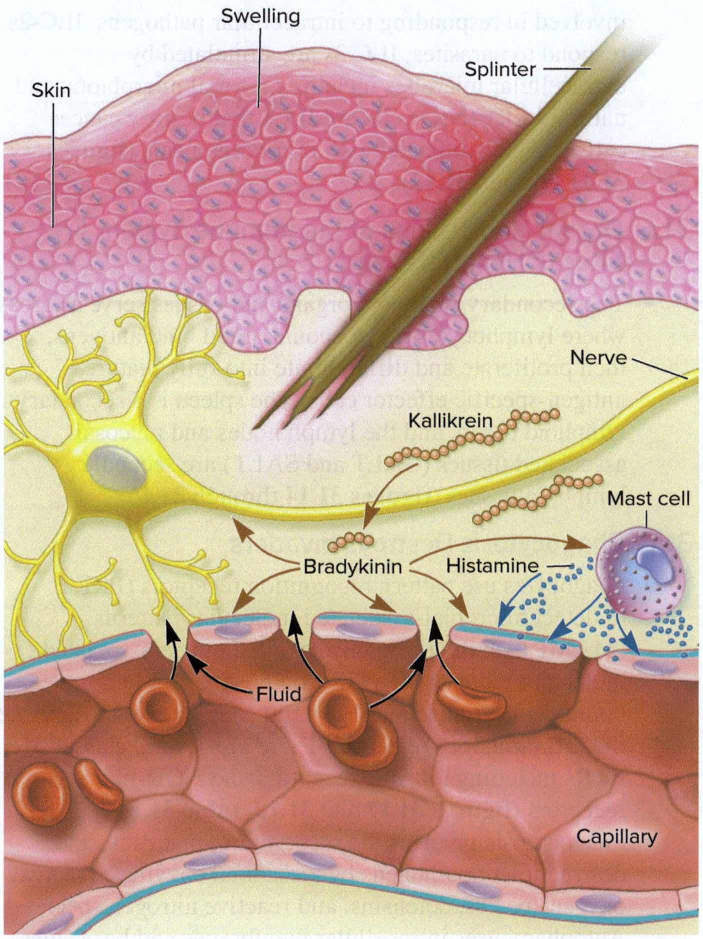

Figure 31.22 Tissue Injury Results in the Recruitment of Kallikrein, from Which Bradykinin Is Released. Bradykinin acts on endothelial and nerve cells, resulting in edema and pain, respectively. It also stimulates mast cells to release histamine. Histamine and bradykinin act on endothelial cells, increasing blood leakage into injured tissue sites.

cells to promote tissue swelling in the infected area. Prostaglandins also bind free nerve endings, making them fire and start a pain impulse. At the same time, liver cells release additional complement proteins, the iron-binding glycoprotein lactoferrin, and collectins.

As the inflammatory response continues, the combination of vasodilation and chemokines results in the influx of more phagocytes from the bloodstream. To keep up with this demand, chemicals stimulate the bone marrow to release neutrophils and increase the rate of granulocyte production. Another important step in resolving a local infection is the formation of a fibrin clot to limit the spread of the infectious agents to the bloodstream or other parts of the body.

An acute inflammatory response resolves in days to a few weeks. Prolonged stimulation is known as **chronic inflammation.** Whereas tissue healing and repair mark the end of acute inflammation, chronic inflammation is characterized by infiltration of lymphocytes and macrophages into the affected site and the formation of new connective tissue. This can cause permanent tissue damage. As cycles of cellular infiltration and resolution continue, degradative enzymes from macrophages destroy more underlying tissue than is replaced. Furthermore, if innate responses are unable to protect the host from ongoing infection, chronic tissue injury, or the persistence of poorly degradable materials (sutures, implants, etc.), the body attempts to wall off and isolate the site by forming a **granuloma** (Latin *granulum,* a small particle; Greek *oma,* to form). A granuloma is a well-organized mass of macrophages, eosinophils, multinucleated giant cells (two or more cells fused into one large cell), T lymphocytes, fibroblasts, and collagen. Together, the cells and extracellular matrix proteins form a spherical mass that can bind calcium and form a recalcitrant nodule, making the granuloma harder to break down.

Importantly, chronic inflammation can occur as a distinct process without much acute inflammation. The persistence of bacteria can also stimulate chronic inflammation. For example, mycobacteria, some of which cause tuberculosis and leprosy, have evolved to survive phagocytosis and reside within macrophages. Granulomas formed around mycobacteria-containing macrophages often calcify and appear on chest X rays, indicating latent (inactive) tuberculosis. Importantly, mycobacterial granulomas keep the bacteria from spreading throughout the body. However, if immune function declines, granuloma integrity will fail and bacteria will be released into the alveoli, causing active tuberculosis. ◀ *Order Mycobacteriales includes important human pathogens (section 22.1)* ▶ *Mycobacterium infections (section 38.1)*

Comprehension Check

1. Explain how the events that cause inflammation contribute to the development of each of the five cardinal signs of inflammation.
2. Compare and contrast acute and chronic inflammation.

Key Concepts

31.1 Immunity Arises from Innate Resistance and Adaptive Defenses

- There are two interdependent components of the immune response to invading microorganisms and foreign material: innate resistance and adaptive immune responses.
- Innate resistance mechanisms offer substantial host defense against any microorganism or foreign material. It has no immunological memory and innate responses are always ready for host defense.
- Cells and molecules of innate immunity activate an adaptive immune response, which defends against a particular foreign agent. Each adaptive immune response improves on repeated exposure to the agent.

31.2 Innate Resistance Starts with Barriers

- Physical and mechanical barriers along with host secretions are the host's first line of defense against pathogens. Examples include the skin and mucous membranes and the epithelia of the respiratory, gastrointestinal, and genitourinary systems (**figures 31.1** through **31.3**).

31.3 Innate Resistance Relies on Chemical Mediators

- Mammalian hosts have chemical barriers that help combat the continuous onslaught of pathogens. Examples include lysozyme, cationic peptides, bacteriocins, cytokines, acute-phase proteins, and complement.
- The complement system is composed of a large number of serum proteins that play a major role in an animal's immune response. There are three pathways of complement activation: the alternative, lectin, and classical pathways (**figures 31.6** and **31.7, table 31.1**). All pathways lead to opsonization, enhanced inflammation, and development of the membrane attack complex (**figure 31.8**).
- Cytokines are required for regulation of innate and adaptive immune responses. Cytokines have a broad range of actions on eukaryotic cells (**table 31.2**).
- Acute-phase proteins are released from liver cells to bind microbial surfaces. They act as opsonins.

31.4 Each Type of Innate Immune Cell Has a Specific Function

- The cells responsible for both nonspecific and adaptive immunity are white blood cells, also called leukocytes (**figure 31.9**). Examples include monocytes, macrophages, dendritic cells (DCs), granulocytes, and mast cells.
- Neutrophils, macrophages, and DCs are professional phagocytes; macrophages and DCs present antigen to lymphocytes.
- There are five kinds of innate lymphoid cells (ILCs): Lymphoid tissue inducer (LTi) cells stimulate the production of secondary lymphoid tissue; ILC-1s are involved in responding to intracellular pathogens; ILC-2s respond to parasites; ILC-3s are stimulated by extracellular microbes including normal microbiota; and natural killer (NK) cells are ILCs that destroy cancer cells and cells infected with microorganisms (**figures 31.11** through **31.13**).

31.5 Organs and Tissues of the Immune System Are Sites of Host Defense

- The secondary lymphoid organs and tissues serve as areas where lymphocytes may encounter and bind antigens, then proliferate and differentiate into fully mature, antigen-specific effector cells. The spleen is a secondary lymphoid organ, and the lymph nodes and mucosal-associated tissues (GALT and SALT) are secondary lymphoid tissues (**figures 31.14** through **31.16**).

31.6 Phagocytosis Destroys Invaders

- Phagocytes use pattern recognition receptors (PRRs) to detect microbial-associated molecular patterns (MAMPs) on microorganisms and damage-associated molecular patterns (DAMPs) on host cells. Membrane-bound and cytoplasmic PRRs transduce signals to activate the inflammatory response. There are several classes of PRRs including toll-like receptors and NOD-like receptors (**figures 31.17** and **31.18, table 31.3**).
- Phagocytosis involves the recognition, ingestion, and destruction of pathogens by lysosomal enzymes, reactive oxygen species, defensins, and reactive nitrogen species. Autophagy uses intracellular membranes and lysosomes to kill and degrade intracellular pathogens (**figures 31.19** and **31.20**).

31.7 Inflammation Unites All Components of Immunity

- Inflammation is a defense mechanism to tissue injury and infection.
- Inflammation can be either acute or chronic (**figures 31.21** and **31.22**).

Active Learning

1. Develop a list of four attributes a bacterium would need to evade or subvert innate immune responses. For each attribute, suggest a mechanism that might have evolved to accomplish this trait.
2. Patients who become very ill with COVID-19 have an over stimulation of the immune system. Early drug trials sought to limit the amount of IL-6 produced by these patients. Other trials looked at the efficacy of type 1 interferons delivered through an inhaler used early in infection. Refer to table 31.2 and develop a rationale why each of these drug trials was deemed worthy of pursuing.
3. Think about the last time you had a cold, flu, skin infection, or a stomach "bug." Which innate barrier was breached by the infectious agent? Refer to the cytokines listed in table 31.2. Which cytokines were part of your immune response? What immune cells do you think were activated?
4. TLR-10 is the only human toll-like receptor for which the ligand is unknown. Evidence now suggests that it functions very differently from other TLRs. TLR-10 receptor activity appears to suppress pro-inflammatory responses and monocyte differentiation into fully functional macrophages. Discuss why the host cell may have evolved a strategy to down-regulate innate immunity. Refer to figure 31.17 and suggest how TLR-10 receptor activity could diminish a pro-inflammatory response.

CHAPTER

32

Adaptive Immunity

Blend Images/Alamy

Killing Cancer, Immunologically

This year, about 150,000 people in the United States will be diagnosed with melanoma. About 8,000 people will die of this cancer, which has been notoriously hard to treat. It is one of the most common cancers in people under 30 years of age. But these grim truths are beginning to be countered by reasons for optimism. The field of immunology is changing the way we treat melanoma and other cancers.

Since the beginning of the twenty-first century, scientists and clinicians have come to understand much more about how cancer cells and other types of cells in the tumor microenvironment interact with the immune system. This in turn has led to the development of new approaches with real promise for how cancer is treated. The goal is to use the patient's adaptive immune response to specifically target, and kill, cancer cells.

One approach is termed immunomodulation, and it changes how the immune system responds to cancer cells. Cancer cells display unique proteins on their surfaces called *neoantigens,* which simply means new antigen. Because neoantigens have not previously been detected by the patient's immune cells, they trigger an immune response. If the response is vigorous enough, cancer cells bearing the neoantigen are killed. The goal of immunomodulatory cancer therapy is to ensure this kind of robust immune response occurs. It has proven successful in prolonging the lives in many but not all patients, and some with late-stage disease have even been cured.

Another approach uses a patient's T cells, which are engineered to express chimeric *a*ntibody *r*eceptors (CAR T-cells). As we learn in this chapter, antibodies bind antigen with high specificity and one group of T cells, called cytotoxic T lymphocytes (CTLs), kills host cells expressing neoantigens. The idea behind CAR T-cells is to construct superkiller-CTLs. To do this, the domain of the T-cell receptor (TCR) that normally binds antigen is replaced with an antibody that targets a cancer neoantigen, while the part of the TCR that spans the membrane and instructs the T cell to kill remains intact. The resulting CAR T-cells are able to seek out cancer cells with laser focus and kill them. This approach has been tremendously successful in treating B-cell acute lymphocytic lymphoma and leukemia. Now the challenge is to modify these immunotherapies to treat other cancers, including melanoma.

In chapter 31 we learned that if the innate immune response is not sufficient to eliminate a foreign invader, it activates an adaptive immune response. We now continue our discussion of the immune response by describing adaptive immunity.

Readiness Check:

Based on what you have learned previously, you should be able to:

- ✓ Discuss the importance of the host complement response (section 31.3)
- ✓ Describe the function of host leukocytes (section 31.4)
- ✓ Outline the events that occur during phagocytosis and inflammation (sections 31.6 and 31.7, respectively)

32.1 Adaptive Immunity Relies on Recognition and Memory

After reading this section, you should be able to:

a. Contrast host innate resistance with adaptive immunity
b. Define the two branches of adaptive immunity by their cell types and general function

Although the separation of innate and adaptive functions of immunity is artificial, it allows us to distinguish the preprogrammed, innate responses from the more specific adaptive responses. In reality the two sides of this coin we call immunity represent a single, fully integrated host defense system.

The adaptive immune system of mammals has three major functions: (1) recognize anything foreign to the body ("non-self"); (2) defend the host from this foreign material; and (3) remember the foreign invader in case of future infection. Like innate immunity, adaptive immunity must differentiate between self and non-self, but in contrast to innate immunity, recognition of non-self antigens by adaptive cells is highly specific. Adaptive immunity involves the activation and expansion of populations of **T and B lymphocytes** that recognize a single, specific non-self antigen.

In response to activation by antigen detection, most B and some types of T lymphocytes elicit an **effector response** and are sometimes called *effector cells* (**figure 32.1**). A successful effector response either eliminates the foreign material or renders it harmless to the host. In addition to effector cells, some activated B and T lymphocytes enter a state of limited activity and are referred to as **memory cells** because they remember the foreign material that initially activated them. If the same substance is encountered at a later time, B and T memory cells are already programmed to mount a more intense and rapid response to eliminate that specific non-self antigen.

Thus the following five characteristics distinguish adaptive immunity:

1. **Discrimination.** While both innate and adaptive immune cells must discriminate between self and non-self, the adaptive immune system responds much more selectively to non-self.
2. **Specificity.** Activated T and B lymphocytes respond to each specific, non-self antigen (among trillions); the immunity to one pathogen or substance usually does not confer immunity to others.
3. **Diversity.** To mount these very specific responses, the system generates an enormous diversity of cellular receptors and antibodies that recognize trillions of unique foreign substances.
4. **Timing.** The adaptive immune system must be stimulated to recognize a new antigen during the process of *activation*. By contrast, the innate immune system is always at the ready to defend the host.
5. **Memory.** When reexposed to the same pathogen or substance, adaptive immunity reacts so quickly there is usually no noticeable illness. By contrast, the reaction time for innate defenses remains constant.

Two branches or arms of adaptive immunity are recognized: antibody-mediated immunity and cell-mediated immunity. B cells constitute **antibody-mediated (humoral) immunity** (figure 32.1). The older term *humoral* reflects the fact that antibodies were discovered in the blood plasma (serum), so the term refers to the fluids or "humors" of the body. Circulating antibodies bind to microorganisms, toxins, and extracellular viruses, neutralizing them or tagging them for destruction by phagocytes and other mechanisms described in section 32.8. T cells constitute **cell-mediated (cellular) immunity.** Broadly speaking, there are two types of T cells. T lymphocytes that differentiate into cytotoxic T cells (CTLs) directly attack target cells infected with intracellular pathogens such as viruses as well as cancer cells. The other type of T cells, T helper cells, are so named because they can direct CTLs to target cell lysis, interact with B cells to stimulate antibody production, as well as up- or downregulate innate defenses, such as phagocytosis and inflammation. Because the activity of the adaptive immune response is so potent, it is imperative that T and B cells consistently discriminate between self and non-self with great accuracy.

32.2 Antigens Elicit Immunity

After reading this section, you should be able to:

a. Predict the types of molecules that can serve as antigens
b. Describe the relationship between haptens and antigens

While innate immune cells (e.g., macrophages, dendritic cells) are limited to detecting molecular patterns as non-self, the adaptive arm of immunity uses a more sophisticated mechanism

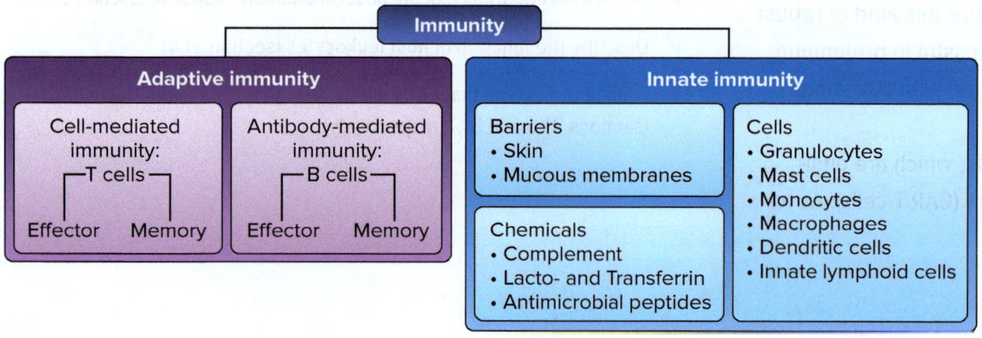

Figure 32.1 The Immune System Consists of Two Arms. Innate immunity is the subject of chapter 31.

to recognize foreignness. The term **antigen** refers to any self or non-self substance that elicits an immune response and is therefore also termed *immunogen* (*immunity generating*). Antigens include microbe-associated molecular patterns (MAMPS, described in section 31.6) as well as other molecules, including proteins, nucleoproteins, polysaccharides, and some glycolipids (**figure 32.2**). The ability of a molecule to function as an antigen depends on its size, structural complexity, chemical nature, and degree of foreignness to the host. Most antigens are complex molecules with a molecular mass generally greater than 10,000 daltons (Da).

Each antigen can have several **antigenic determinant sites,** or **epitopes** (figure 32.2*b*). An epitope is the precise site of the antigen to which a specific antibody binds. Chemically, epitopes include sugars, organic acids and bases, amino acids, hydrocarbons, and aromatic groups. The number of epitopes on the surface of an antigen is its **valence.** The valence determines the number of antibody molecules that can combine with the antigen at one time. If one epitope is present, the antigen is monovalent. Most antigens, however, have more than one copy of the same epitope and so are multivalent. Multivalent antigens generally elicit a stronger immune response than monovalent antigens. Thus structurally repetitive molecules are multivalent and said to be strongly immunogenic. **Antibody affinity** measures the strength of the interaction between a single antigen-binding site and its epitope. By contrast, the **avidity** of an antibody defines the overall strength of antibody-antigen interactions at all antigen-binding sites present on the antigen. Thus avidity depends, in part, on an antigen's valence.

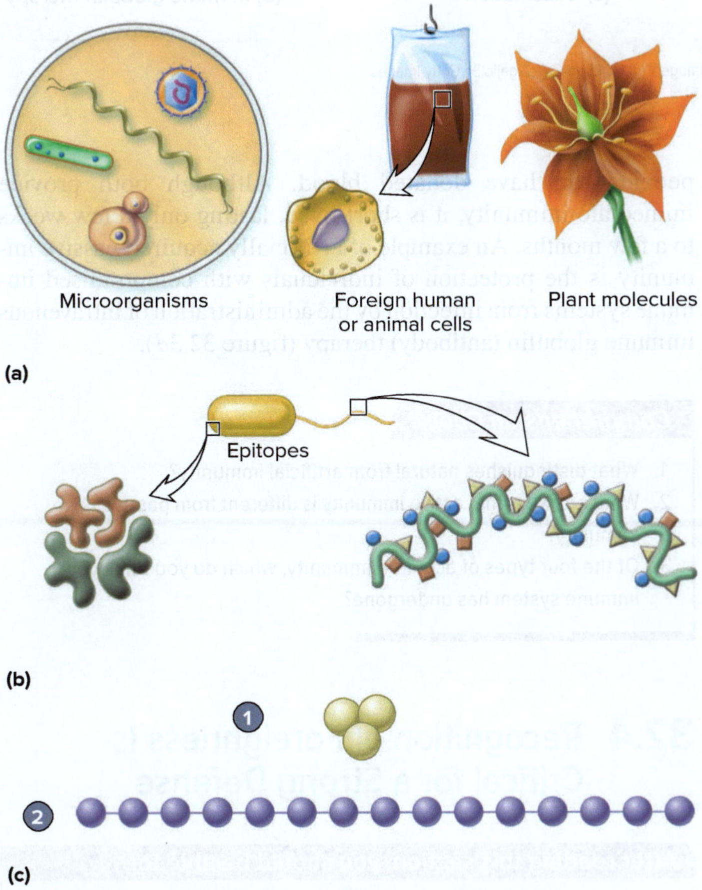

Figure 32.2 Antigen Characteristics Are Numerous and Diverse. (a) Whole cells, viruses, and some plant molecules are immunogenic. (b) The specific regions of an antigen recognized by antibodies are called epitopes. (c) Small molecules may be either (1) poorly immunogenic, such as small molecules (haptens) not attached to a carrier molecule, or (2) very immunogenic if they are structurally repetitive such as flagella and peptidoglycan.

MICRO INQUIRY *What does the term* valence *mean and how might an antigen's valence influence an immune response?*

Haptens

As shown in figure 32.2*c,* some small antigens may not trigger an immune response—that is, they are not immunogenic. However, such molecules may become immunogenic if they become attached to a larger host protein, called a carrier protein. Small molecules that become immunogenic when combined with a carrier protein are called **haptens** (Latin *haptein,* to grasp). One example of a hapten is penicillin. Penicillin is a very small molecule that becomes immunogenic only when it combines with certain serum proteins, thereby triggering an allergic immune reaction in sensitive individuals (section 32.10).

Comprehension Check

1. What structural features make an antigen immunogenic?
2. Define the terms *epitope, affinity,* and *avidity.*
3. What are the distinguishing features of a hapten? How does it become immunogenic?

32.3 Adaptive Immunity Can Be Earned or Borrowed

After reading this section, you should be able to:

a. Compare and contrast methods by which immunity occurs by natural and artificial means
b. Differentiate between the active and passive forms of adaptive immunity

Humans evolved the natural ability to mount an adaptive immune response, but we now also use technology to artificially stimulate such a response to protect the host. Here we discuss these two types of adaptive immunity—natural and artificial—and the means by which they can be either passively or actively developed (**figure 32.3**).

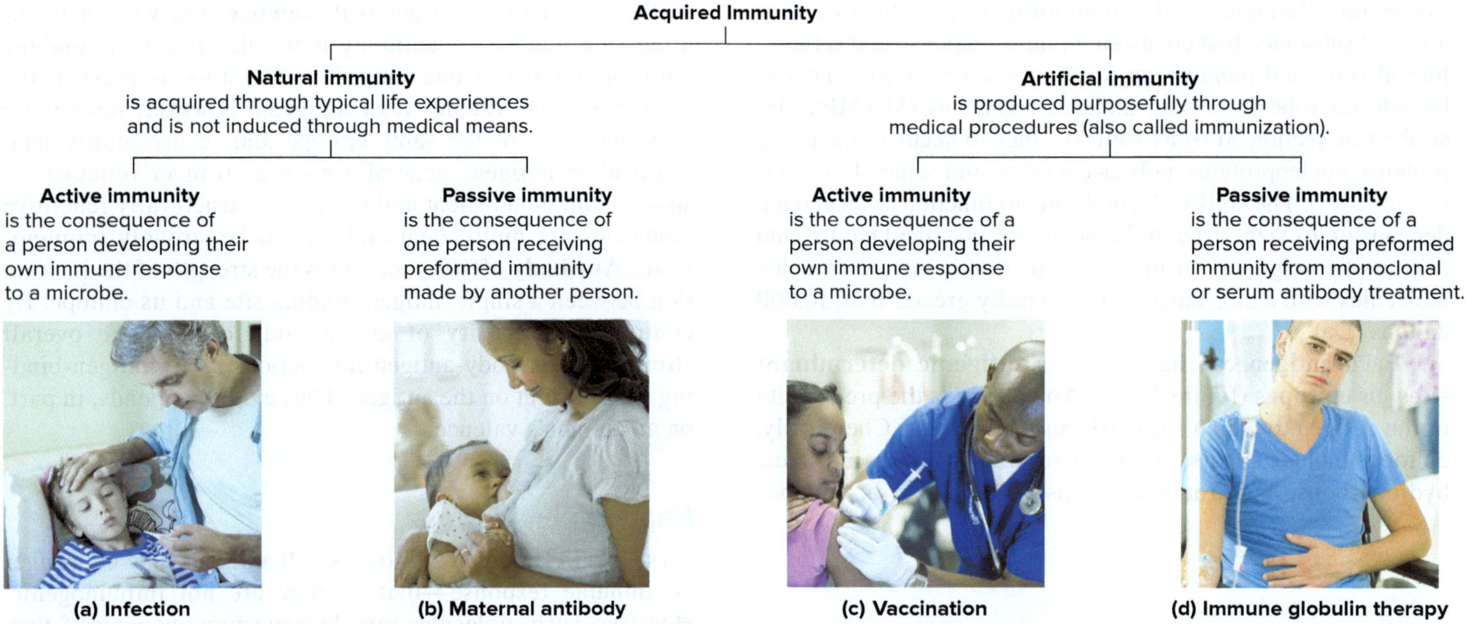

Figure 32.3 Immunity Can Be Acquired by Various Means.
(a) Paul Bradbury/Caiaimage/Getty Images; (b) JGI/Blend Images; (c) ER Productions Limited/DigitalVision/Getty Images; (d) Miodrag Gajic/vgajic/E+/Getty Images

Naturally Acquired Immunity

Naturally acquired active immunity occurs when an individual's immune system contacts an antigen such as a pathogen that causes an infection. The adaptive immune system responds by producing antibodies and activated T lymphocytes that neutralize or destroy the pathogen. The immunity produced can either be lifelong, as with measles, or last for only a few years, as with influenza. **Naturally acquired passive immunity** involves the transfer of premade antibodies from parent to child. This occurs in two ways. First, a pregnant person's antibodies cross the placenta to the fetus. If the parent is immune to diseases such as polio or diphtheria, this placental transfer gives the fetus and newborn temporary immunity to these diseases. Second, antibodies pass from a parent to their offspring when the child is breastfed. Maternal antibodies are essential for providing immunity to the newborn for the first 6 months of life (or for the duration of breastfeeding), as the child's own immune system matures.

Artificially Acquired Immunity

Vaccination stimulates the recipient's adaptive immune system to respond, and therefore generates **artificially acquired active immunity.** A **vaccine** consists of antigens that can be prepared in several different ways: (1) killed microorganisms; (2) living, weakened (attenuated) microorganisms; (3) genetically engineered organisms, their products, or nucleic acids; or (4) inactivated bacterial toxins (toxoids). Vaccines and immunizations are discussed in detail in chapter 35.

Artificially acquired passive immunity results when premade antibodies are given to a person therapeutically. This can be in the form of monoclonal antibody treatment or by the transfer of all the antibodies in serum collected and pooled from many people who have donated blood. Although both provide immediate immunity, it is short-lived, lasting only a few weeks to a few months. An example of artificially acquired passive immunity is the protection of individuals with compromised immune systems from infection by the administration of intravenous immune globulin (antibody) therapy (figure 32.3*d*).

Comprehension Check

1. What distinguishes natural from artificial immunity?
2. What are ways that active immunity is different from passive immunity?
3. Of the four types of acquired immunity, which do you think your immune system has undergone?

32.4 Recognition of Foreignness Is Critical for a Strong Defense

After reading this section, you should be able to:

a. Compare the processes by which MHC class I and class II receptors process and display peptides
b. Identify cells that function as antigen-presenting cells (APCs)

We have stressed that distinguishing between self and non-self is essential in maintaining host integrity. This distinction must be highly specific and selective so that as invading pathogens are eliminated, host tissue is not also damaged or destroyed.

Therefore it is not enough for the immune system to recognize foreign antigens as non-self; it must also be able to recognize host cells as self. To accomplish this, each host cell expresses surface proteins that mark it as a resident of that host. This self signal must be broadcast to all cells of the immune system so that only non-self antigens are targeted for destruction. This system has evolved in mammals and is called the major histocompatibility complex.

The Major Histocompatibility Complex

The ***m**ajor **h**istocompatibility **c**omplex* **(MHC)** is a collection of genes encoding proteins that enable the host to distinguish between self and non-self. The term *histocompatibility* is derived from the Greek word for tissue (*histo*) and the ability to get along (*compatibility*). Human MHC is located on chromosome 6 and is called the ***h**uman **l**eukocyte **a**ntigen* **(HLA) complex.** HLA proteins can be divided into three classes: Class I are found on all human cells with a nucleus. Class II appear only on cells that can process antigens and present them to T lymphocytes. Class III molecules include various secreted proteins that have immune functions. Class III molecules are not required for the discrimination between self and non-self and are not discussed further.

Class I MHC molecules identify all healthy nucleated cells in the body as self. Each class I MHC molecule consists of a complex of two protein chains, a larger alpha chain (45,000 Da) and a smaller chain called β_2-microglobulin (12,000 Da). As shown in **figure 32.4*a***, only the alpha chain spans the plasma membrane. Both chains interact to form a pocket that projects from the cell surface. This pocket can bind either self antigen, thereby marking the cell as a host cell, or a peptide extracted from an intracellular pathogen, signaling the immune system that it is an infected host cell. Therefore the identity of the peptide bound to a cell's MHC I molecules proclaims a host cell either as healthy (self) or infected with an intracellular pathogen (non-self).

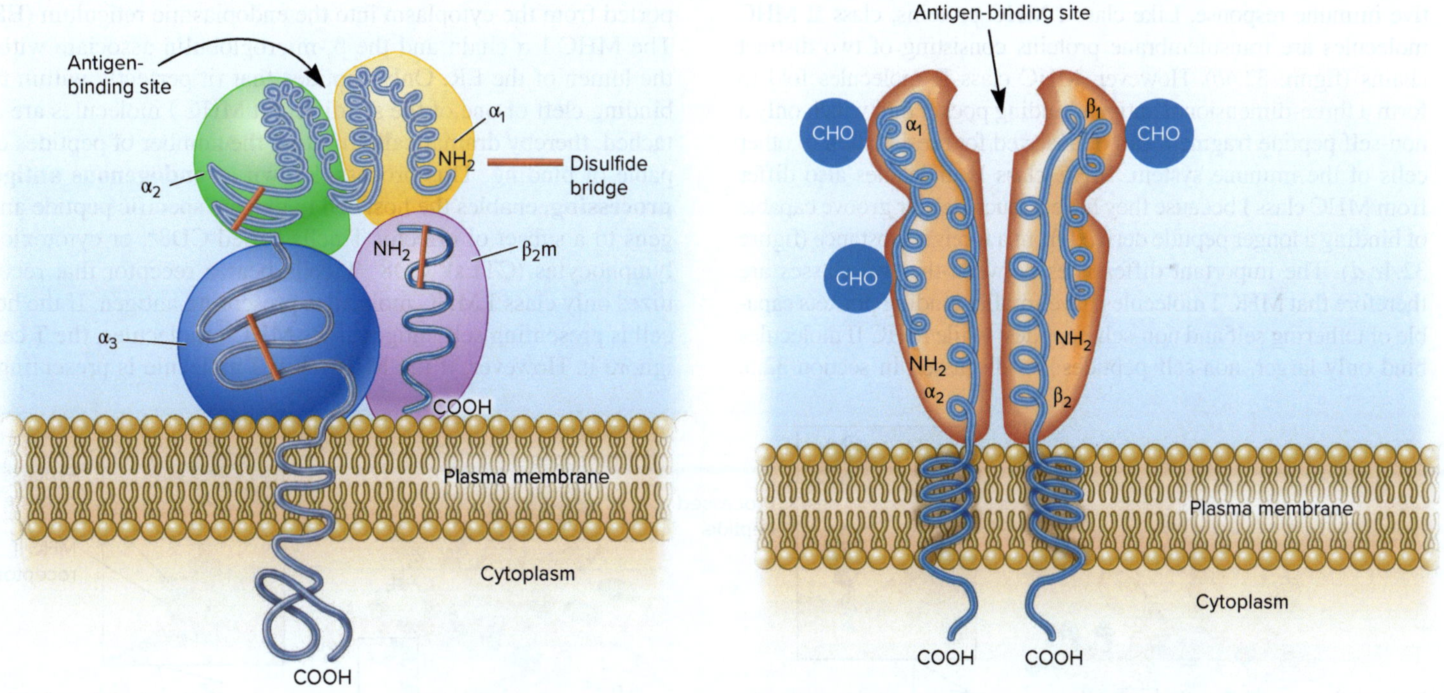

(a) Class I MHC

(b) Class II MHC

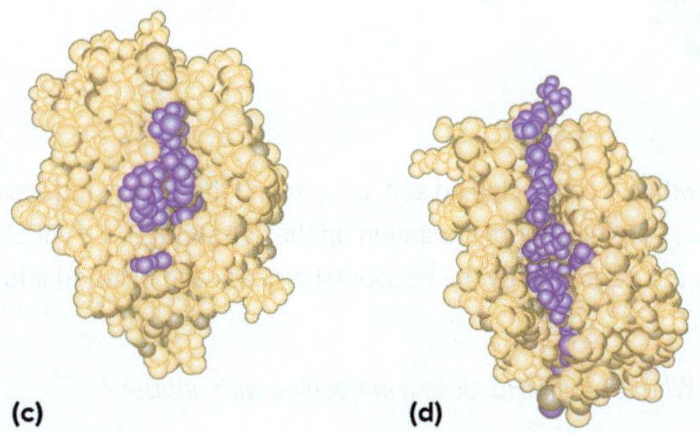

(c)

(d)

Figure 32.4 The Membrane-Bound Class I and Class II Major Histocompatibility Complex Molecules. (a) The class I molecule is a heterodimer composed of the alpha protein, which is divided into three domains—α_1, α_2, and α_3—and the protein β_2 microglobulin (β_2m). (b) The class II molecule is a heterodimer composed of two distinct proteins called alpha and beta. Each is divided into two domains: α_1, α_2 and β_1, β_2, respectively. Both α and β are glycosylated, as indicated by CHO (blue circles). (c) This space-filling model of a class I MHC protein illustrates that its antigen-binding pocket holds shorter peptide antigens (lavender) in a smaller binding pocket than does a class II MHC. (d) This space-filling model of a class II MHC protein shows that the antigen-binding pocket is open on both ends and can accommodate a larger peptide antigen.

MICRO INQUIRY *On what types of cells are MHC class I molecules found? MHC class II?*

There are three types of class I MHC molecules: A, B, and C. These are termed *HLA types* to refer to the genetic locus. HLA proteins differ among individuals; the closer two people are related, the more similar are their HLA molecules. This is because many versions (alleles) of each HLA gene exist. Throughout evolution, multiple alleles of each gene have arisen by mutation, recombination, and other mechanisms. Variation among individuals is further amplified because HLA genes are codominant; thus each person expresses alleles from both birth parents at each A, B, and C locus. Thus a person produces six different class I MHC proteins. Because MHC class I proteins are found on all nucleated cells (i.e., only red blood cells lack them), they stimulate an immune response when cells from one host are introduced into another host with different class I molecules. This is the basis for tissue typing when a patient is being prepared for an organ or bone marrow transplant.

Unlike the broad distribution of class I MHC molecules, **class II MHC molecules** are produced only by certain immune cells, such as dendritic cells (DCs), activated macrophages, mature B cells, and some innate lymphocyte cells (ILCs). As we will learn, class II molecules on DCs and B cells are key to initiating an adaptive immune response. Like class I MHC proteins, class II MHC molecules are transmembrane proteins consisting of two distinct chains (figure 32.4*b*). However, MHC class II molecules fold to form a three-dimensional antigen-binding pocket into which only a non-self peptide fragment can be inserted for presentation to other cells of the immune system. MHC class II molecules also differ from MHC class I because they have a much deeper groove capable of binding a longer peptide derived from a foreign substance (figure 32.4*c,d*). The important differences between the two classes are therefore that MHC I molecules have smaller binding pockets capable of tethering self and non-self peptides, while MHC II molecules bind only larger, non-self peptides. As discussed in section 32.5, non-self peptide antigen fragments in the groove of MHC I and MHC II are needed to activate T cells.

MHCs Present Antigens to T Cells

The mechanism by which peptides become bound to MHC molecules is known as **antigen processing** (**figure 32.5**). Class I molecules bind peptides by sampling proteins in the cytoplasm of all nucleated cells. In healthy cells, self peptide bound to an MHC I molecule communicates to the immune system, "I am a healthy self cell—please leave me alone!" However, if a host cell is infected with an intracellular pathogen, such as a virus, or if the cell has become cancerous, a foreign peptide is placed in the binding groove of the MHC I molecule. This communicates to the immune system, "I am an unhealthy cell—I should be eliminated!" This system thus ensures that all host cells are constantly surveilled for intracellular pathogens and cancerous transformation.

How are self and non-self peptides deposited in the MHC I binding pocket? Recall that intracellular proteins are digested by proteasomes (*see figure 5.8*) as part of the natural process by which a cell continually recycles its protein contents. During this process, specific self and antigenic non-self peptides are transported from the cytoplasm into the endoplasmic reticulum (ER). The MHC I α chain and the β_2-microglobulin associate within the lumen of the ER. Only peptides that fit perfectly within the binding cleft of one of the six different MHC I molecules are attached, thereby dramatically limiting the number of peptides capable of binding. This process, known as **endogenous antigen processing,** enables the host cell to present specific peptide antigens to a subset of effector T cells called $CD8^+$, or cytotoxic T lymphocytes (CTLs). $CD8^+$ T cells bear a receptor that recognizes only class I MHC molecules presenting antigen. If the host cell is presenting self antigen in its MHC I molecules, the T cells ignore it. However, if the MHC class I molecule is presenting a

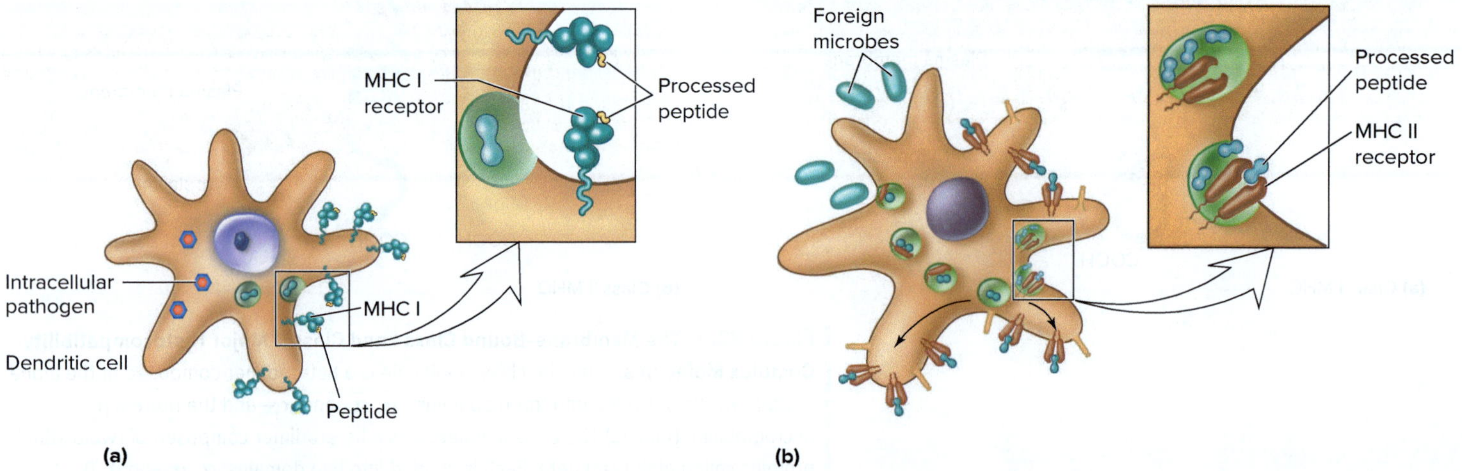

Figure 32.5 Antigen Presentation. (a) Endogenous antigen processing: Antigens arising from within a cell (including self antigen and those from intracellular pathogens) are degraded by proteasomes and inserted into the antigen-binding pocket of the MHC class I proteins for presentation on the cell surface to a variety of immune cells. (b) Exogenous antigen processing: Antigens captured outside the cell by phagocytosis are degraded by the lysosomal process and inserted into the antigen-binding pocket of MHC class II proteins for presentation on the cell surface to T lymphocytes.

MICRO INQUIRY *What cells recognize and bind to MHC class I molecules that carry antigens? What about MHC class II molecules with antigen?*

non-self antigen (e.g., from an intracellular pathogen or a new peptide found only in cancerous cells), these T cells bind and kill the host cell (section 32.5).

Class II MHC molecules differ from class I MHC molecules because they bind fragments that come from antigens outside the cell; thus they undergo **exogenous antigen processing.** Only foreign particles (e.g., bacteria, viruses, toxins) taken up by endo- or phagocytosis can be bound to MHC class II molecules (*see figures 31.19 and 31.20*). Immune cells that place non-self antigen on MHC class II molecules are called **antigen-presenting cells (APCs),** and these include macrophages, DCs, and B cells. Exogenous antigen processing begins with phagocytosis (by macrophages and DCs) or receptor-mediated endocytosis (by B cells). In all cases, degradation in the phagolysosome or autolysosome results in the release of antigenic peptides from the invader. These peptides combine with preformed class II MHC molecules and are delivered to the cell surface. Just as with class I MHCs, only peptides that fit snugly in the class II MHC binding pocket are bound. This peptide can now be recognized by T cells called $CD4^+$ T-helper cells. DCs are particularly adept at presenting foreign peptides to T cells and stimulating them to become activated T-helper cells. Unlike $CD8^+$ T cells, $CD4^+$ T cells do not directly kill target cells. Instead, they secrete cytokines to orchestrate an integrated innate and adaptive immune response. *Several organelles function in the secretory and endocytic pathways (section 5.4); Cytokines are chemical messages between cells (section 31.3); Phagocytosis destroys invaders (section 31.6)*

Comprehension Check

1. Construct a table that lists the differences in structure and function between MHC class I and class II molecules.
2. Why is it important that MHC class I molecules bind and present self antigen?

32.5 T Cells Are Critical for Immune Function

After reading this section, you should be able to:

a. Differentiate between T-cell development and T-cell activation
b. Describe T-cell receptor structure and function
c. Diagram antigen presentation within MHC receptors interacting with the appropriate T-cell type
d. Compare and contrast the activities of T_H1, T_H2, T_H17, Treg cells, and CTLs
e. Follow the path of a common lymphoid progenitor cell from thymus to activation

Adaptive immunity involves the coordinated activity of both B and T cells. However, whereas B cells have a single function, to produce and secrete antibodies, there are multiple subsets of T cells that work together to initiate, orchestrate, and carry out an adaptive immune response. We therefore begin our discussion by learning about T cells.

T-Cell Development

We define *T-cell development* as the maturation of progenitor cells to one of two types of T cells: T helper cells and cytotoxic T lymphocytes. The sole purpose of the thymus is to serve as a site for T-cell development (*see figure 31.14*). This starts with the migration of common lymphoid progenitor (CLP; *see figure 31.9*) cells from the bone marrow to the thymus, a primary lymphoid organ located above the heart (**figure 32.6**). During a process known as **thymic selection,** two major events occur: Positive and negative selection. But first, T cells are sorted into two types based on the structure of the **T-cell receptor (TCR).** The TCR consists of two interacting proteins. Most TCRs are

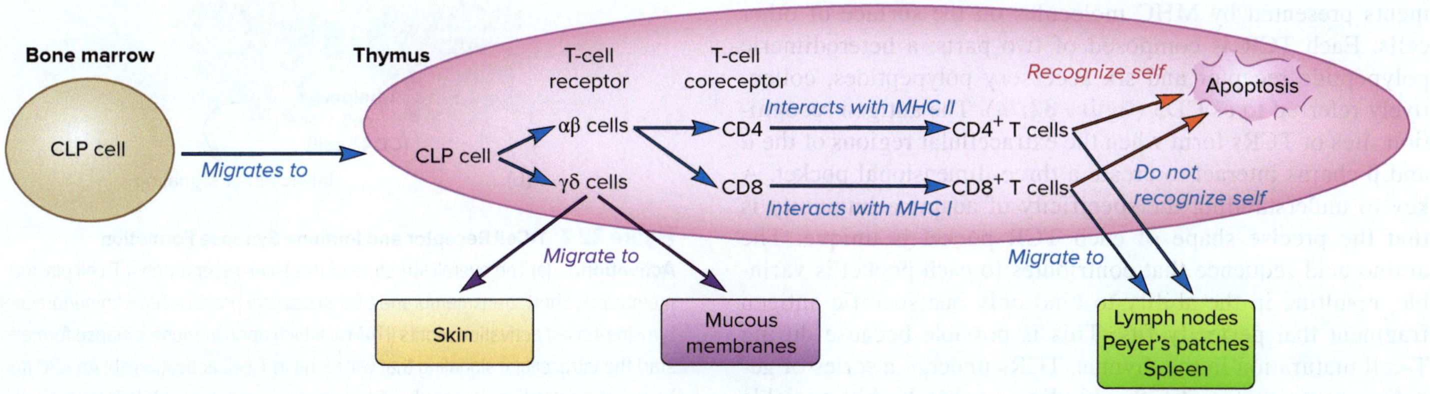

Figure 32.6 Development of T Lymphocytes. Common lymphoid progenitor (CLP) cells in the bone marrow migrate to the thymus, where the structure of the T-cell receptor (TCR) determines their fate. The majority are αβ cells, which undergo positive selection to express one of two coreceptors: CD4 or CD8. T cells with CD4 coreceptors interact with MHC II molecules; those with CD8 coreceptors interact with MHC I molecules. During negative selection, cells that bind to self molecules are eliminated by apoptosis. The remainder leave the thymus and travel to lymph nodes, Peyer's patches, and the spleen to await further instruction during activation. Prior to activation, these cells are said to be mature but naïve.

made up of an α and a β protein and are called **αβ T cells.** A small number of T cells bear TCRs made of γ and δ proteins. These are called **γδ T cells** and they leave the thymus shortly after entering it and migrate to the lymphoid tissue in the skin and mucous membranes. There they principally contribute to innate immunity, despite being T lymphocytes.

The vast majority of cells are αβ T cells, which remain in the thymus, where positive selection commences (figure 32.6). **Positive selection** determines which of two types of TCR coreceptors a mature T cell will have. T-cell coreceptors are members of the family of **cluster of differentiation (CD) molecules.** Cells of the immune system bear many different CDs on their cell surface; here we are concerned with CD4 and CD8. Immature T cells initially bear both CD4 and CD8 but are then selected to have one or the other, but never both. The type of coreceptor (CD4 or CD8) selected is critical to the future development and activity of the T cell.

T cells in the thymus now move on to **negative selection.** In negative selection, all T cells are screened so that any T cell that strongly recognizes self antigens undergoes cell death (apoptosis; figure 32.6). The entire process of T-cell development removes about 98% of CLP cells that enter the thymus. T cells that leave the thymus and migrate to lymph nodes, Peyer's patches in the gut, and the spleen are called mature but naïve. As we will learn, this reflects the fact that they are fully developed (i.e., have matured) but cannot respond to antigen (i.e., are naïve) until instructed by antigen presented on an MHC molecule. By analogy, consider a 19-year-old person who has fully developed (matured) but still requires instruction (is naïve) before launching a career.

T-Cell Activation

T-cell activation occurs in the lymph nodes, Peyer's patches, and spleen where mature, naïve T cells migrate after development in the thymus. T-cell receptors (TCRs) play an essential role in T-cell activation. TCRs evolved to detect antigen fragments presented by MHC molecules on the surface of other cells. Each TCR is composed of two parts: a heterodimeric polypeptide receptor and six accessory polypeptides, collectively referred to as CD3 (**figure 32.7*a***). The antigen recognition sites of TCRs form when the extracellular regions of the α and β chains interact to create a three-dimensional pocket. A key to understanding the specificity of adaptive immunity is that the precise shape of each TCR pocket is unique. The amino acid sequence that contributes to each pocket is variable, resulting in the ability to bind only one specific antigen fragment that perfectly fits. This is possible because during T-cell maturation in the thymus, TCRs undergo a series of genetic events that make the binding pocket highly variable among different T cells (section 32.9). Ultimately, this allows each T-cell receptor to recognize a different, specific antigenic peptide. Importantly, T cells cannot bind free peptides; they must be presented by an APC on its MHC molecule. When a mature T cell binds to a peptide in an MHC molecule, it is no

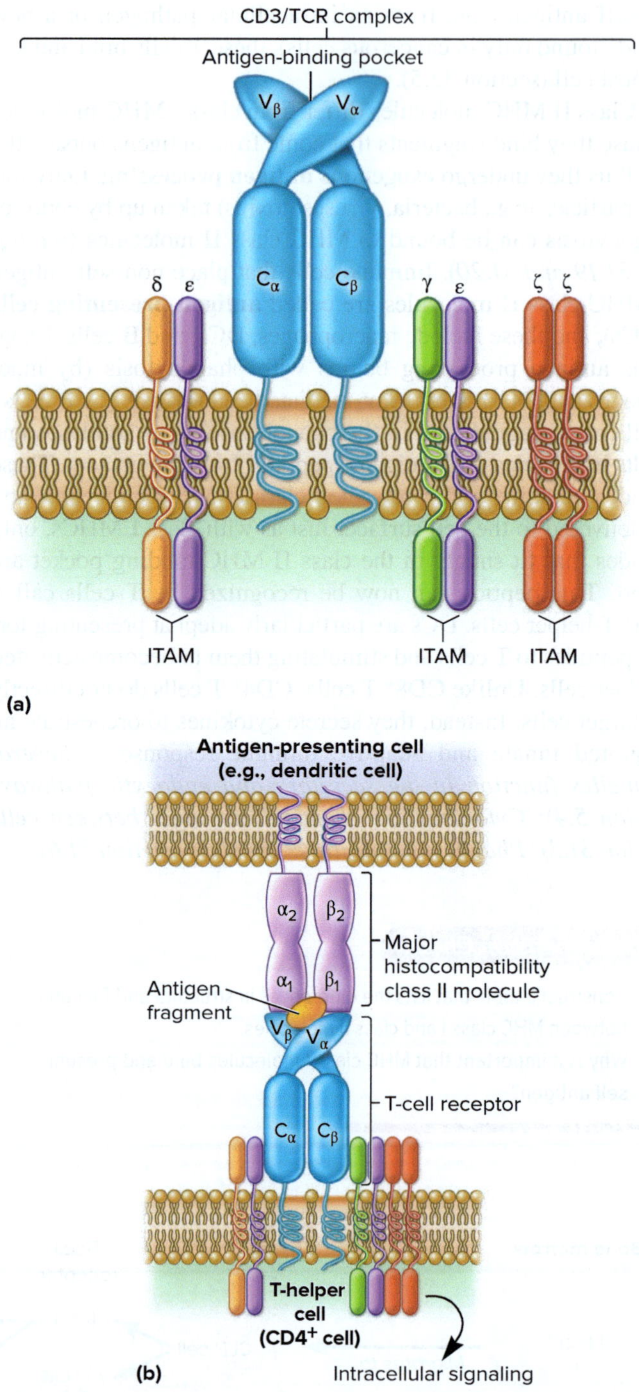

Figure 32.7 T-Cell Receptor and Immune Synapse Formation Activation. (a) The overall structure of the T-cell receptor on a T-cell plasma membrane. Short, transmembrane CD3 accessory proteins have immunoreceptor tyrosine-based activation motifs (ITAMs), which upon immune synapse formation start the intracellular signaling that will result in T-cell activation. (b) An APC (top) begins the activation process by displaying an antigenic peptide (orange) in the binding cleft of its MHC (pink). A T cell (bottom) is activated after the variable region of its receptor (blue, designated V_α and V_β) recognizes and binds the same antigen fragment (orange), creating an immune synapse. Shown here is a T-helper cell, as noted by its interaction with peptide bound to a class II MHC molecule. For simplicity, the CD4 coreceptor is not shown.

longer naïve because it now has found the specific peptide that fits its particular binding pocket and its career path has been determined: It will target this antigen. This is the first step in T-cell activation.

Keep in mind that antigen does not just randomly bump into the naïve T cell. It must be delivered by an antigen-presenting cell (APC). Activation begins with antigen presentation in lymphoid tissue, which is crowded with APCs and naïve T cells. In this process, the antigen held by the MHC of the APC is presented to the T cell. If it binds snugly to the TCR binding pocket, that specific antigen is recognized by the T cell. If the antigen in the APC's MHC pocket does not bind to a T cell's TCR (i.e., it does not fit), the APC will continue to "test" TCRs of other T cells for antigen fit. When the perfect antigen-TCR fit is found, the TCR-peptide-MHC interaction forms a bridge between the APC and T cell called an **immune synapse** (figure 32.7*b*). Immune synapse formation causes a three-dimensional change in the TCR that is communicated to the T-cell nucleus so the appropriate genes can be expressed. However, as inspection of figure 32.7*b* shows, very little of the TCR extends into the cytoplasm. Therefore to transmit the message that antigen has been detected, the cytoplasmic tails of the receptors associate with CD3 accessory polypeptides, which transduce the intracellular signal. The signal transduced by immune synapse formation is called **signal 1.**

In the case of T-helper (T_H) cells, CD4 coreceptors direct the interaction with MHC class II molecules (**figure 32.8*a***). For $CD8^+$ cells to be activated, CD8 coreceptors interact only with MHC class I molecules on the APC. So while both $CD8^+$ and $CD4^+$ cells use their TCRs to form immune synapses, their CD8 or CD4 coreceptors limit interaction with MHC class I or class II molecules, respectively.

T-cell activation must be very carefully regulated. Therefore in addition to signal 1 (immune synapse), two more signals are required. **Signal 2** occurs when the **B7 (CD80) protein** on the surface of an APC binds to the CD28 receptor on the T cell (figure 32.8). Dendritic cells are the APCs that stimulate naïve T cells because they express high levels of B7. Other APCs do not produce high levels of B7 unless stimulated to do so. These APCs (macrophages and B cells) therefore regulate previously activated T cells. T cells that receive signal 1 without signal 2 become **anergic**—that is, they can recognize antigen but are unable to respond to it. *Monocytes, macrophages, and dendritic cells (section 31.4)*

Signal 3 is not as well defined as signals 1 and 2. Signal 3 involves the release of key cytokines by immune cells including the APC and the T cell making the immune synapse. The cytokine IL-2 is particularly critical in stimulating target T cells to proliferate and differentiate into effector cells or memory cells. Thus while signals 1 and 2 involve proteins on an APC interacting with receptors on T cells, signal 3 involves soluble molecules binding to cognate receptors on the target T cell. All three signals trigger a change in T-cell gene expression enabling a specific, adaptive immune response. *Cytokines are chemical messages between cells (section 31.3)*

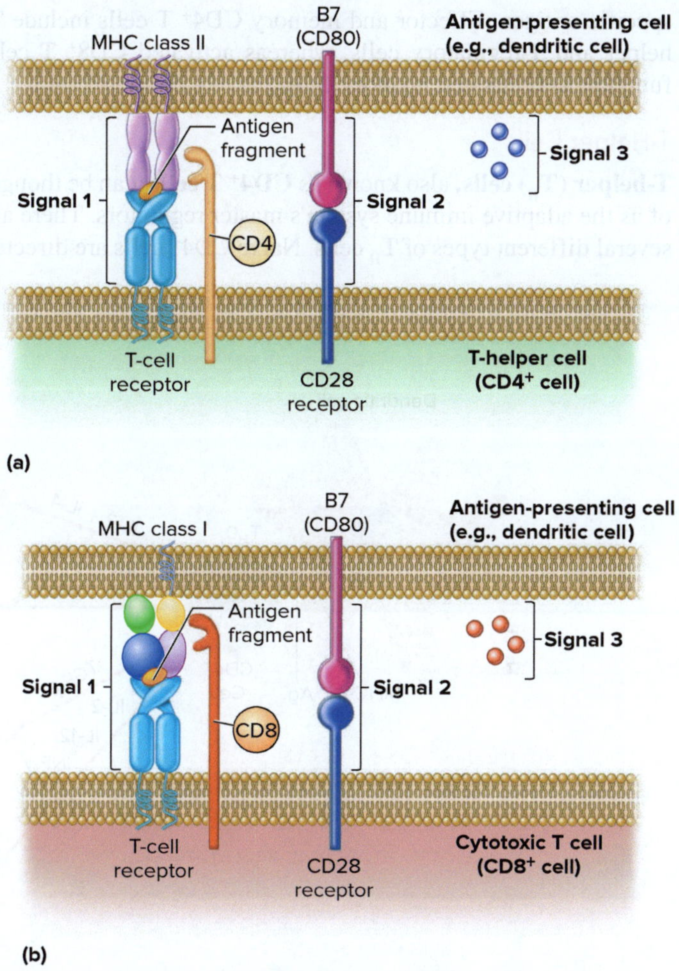

Figure 32.8 Three Signals Are Essential for T-Cell Activation. The first signal is the presentation of the antigen fragment by an APC. Shown here is a dendritic cell, the APC that activates naïve T cells. (a) The CD4 coreceptor on a T-helper cell mediates recognition of an MHC class II molecule on an APC. (b) The MHC class I molecule is recognized by the CD8 coreceptor on cells that will be activated to become CTLs. The second signal occurs when the dendritic cell presents the B7 (CD80) protein to the T helper or CTL with its CD28 protein receptor. The third signal is the release of cytokines by the APC and the T cells themselves, which drives the T cells to differentiate into memory and effector cells, which proliferate and become fully activated. Note that CD3 complexes are not shown for simplicity.

MICRO INQUIRY *How can T-helper cells facilitate $CD8^+$ cell activation?*

Types of T Cells

Recall that αβ T cells migrate to secondary lymphoid organs and tissues (figure 32.6). Mature, naïve T cells in these tissues are awaiting activation by dendritic cell antigen presentation. Upon antigen stimulation, αβ T cells assume their active roles as effector cells and memory cells. Effector T cells carry out specific functions to protect the host against foreign antigen. Memory cells are mature, educated cells (i.e., they "know" their antigen) that are quickly reactivated when the host is reintroduced to their specific antigen. Effector and memory CD4+ T cells include T-helper and T-regulatory cells, whereas activated CD8+ T cells function as CTLs and memory cells.

T-Helper Cells

T-helper (T_H) cells, also known as **CD4+ T cells,** can be thought of as the adaptive immune system's master regulators. There are several different types of T_H cells. Naïve CD4+ cells are directed to become a specific type of T_H cell by the cytokines released by DCs (signal 3) during immune synapse. Different cytokines trigger the activation of specific transcription factors that drive naïve T cells to differentiate into one of several types of CD4+ T-helper cells. The most important types include T_H type 1 (T_H1) cells, T_H type 2 (T_H2) cells, T_H type 17 (T_H17) cells, and regulatory T cells (Treg) (**figure 32.9*a***). **T_H0** is often used to designate mature, naïve cells that have not yet been activated (educated).

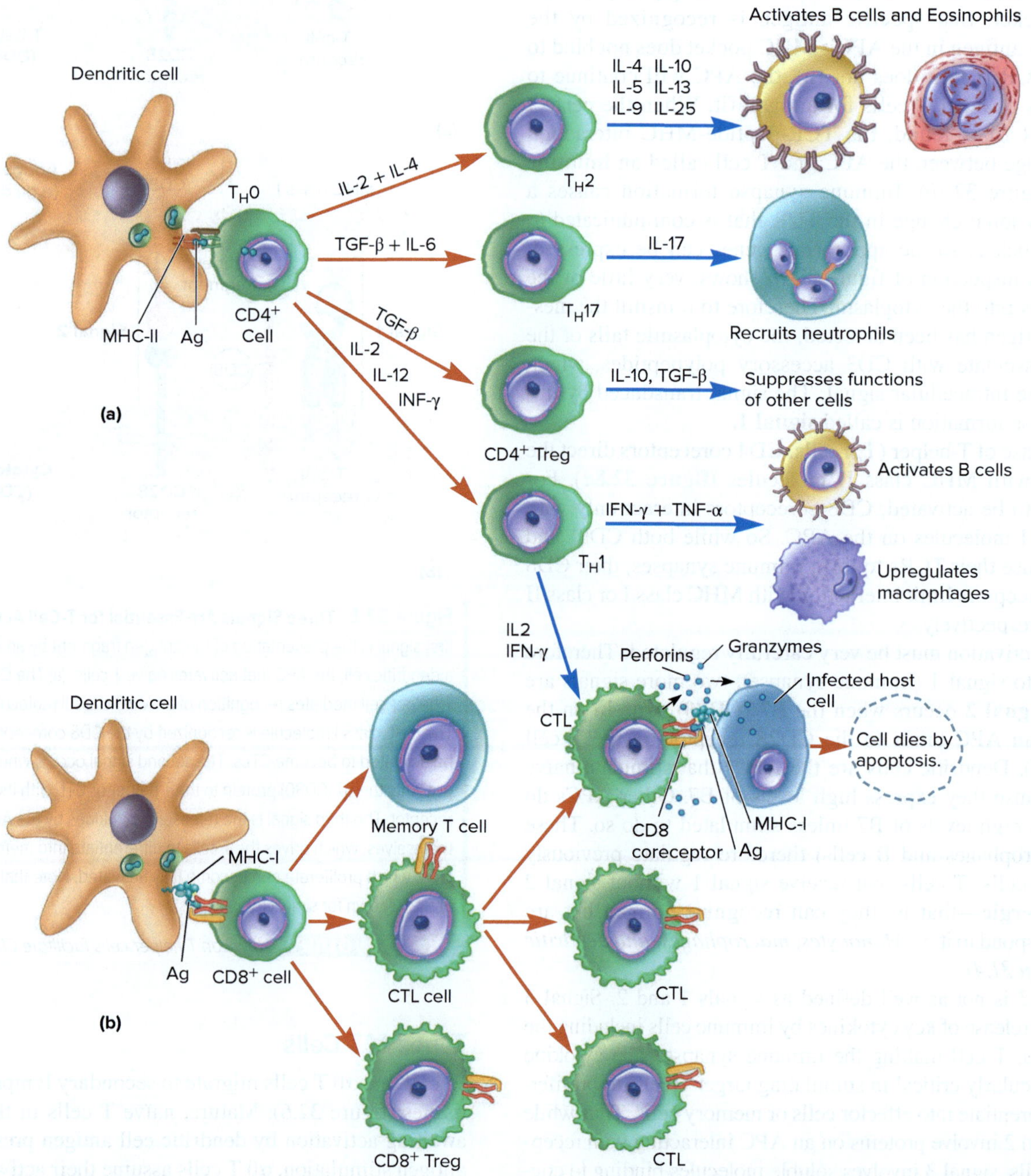

Figure 32.9 T-Cell Activation. (a) During activation, a CD4+ T_H0 cell may differentiate into several kinds of T cells, including T_H1, T_H2, T_H17, or Treg cells. Cytokines near red arrows are produced by DCs during immune synapse and trigger differentiation of T_H0 cells, whereas cytokines near blue arrows are produced by activated T cells to communicate with other types of cells. Memory T cells for each T-cell class are also produced (not shown). (b) CD8+ T cells are activated by DCs presenting non-self antigen on MHC class I molecules. This drives CD8+ T cells to differentiate into activated CTLs. T_H1 cells can also stimulate CTLs by releasing specific cytokines. CTLs induce apoptosis in target cells. A small population of CD8+ cells differentiate into Tregs, but the role of these cells is not well understood.

The release of the cytokines IL-2, interferon-γ (IFN-γ), and interleukin (IL)-12 by DCs induces the activity of the transcription factors STAT4 and T-bet in T_H0 cells. This results in their differentiation into **T_H1 cells.** T_H1 cells mediate inflammation by producing interferon (IFN)-γ and tumor necrosis factor-α (TNF-α). The secretion of these cytokines by T_H1 cells enables their interaction with at least four types of cells: macrophages, DCs, B cells (i.e., APCs), and $CD8^+$ T cells. Interaction with APCs requires that a protein on the surface of T_H1 cells, called CD40L, binds to its receptor (CD40) on the APC surface. CD40L-CD40 engagement in the presence of TNF-α and IFN-γ results in the upregulation of phagocytic activity in macrophages, and DCs are stimulated to interact with $CD8^+$ cells to promote *cytotoxic T lymphocyte (CTL)* activity. B cells are stimulated to produce a particular kind of antibody (IgG) critical for efficient opsonization. Alternatively, if T_H1 cells secrete IL-2 and IFN-γ, they can induce CTL activity independent of DCs. Because T_H1 cells stimulate the activity of so many different types of immune cells, they are critical to the function of a healthy immune response. ⏮ *Innate resistance relies on chemical mediators (section 31.3)*

If, during immune synapse, DCs release IL-4, the T-cell transcription factors STAT6 and GATA-3 are induced, driving T_H0 cells to become **T_H2 cells.** T_H2 cells defend against helminth (worm) parasites by producing cytokines IL-4, IL-5, IL-9, IL-10, IL-13, and IL-25. These cytokines stimulate B cells to produce a specific kind of antibody (IgE) and eosinophils to degranulate. ⏮ *Granulocytes (section 31.4)*

The secretion of two cytokines, transforming growth factor (TGF)-β and IL-6, results in the activation of the T-cell transcription factor RORγt. This triggers T_H0 differentiation into **T_H17 cells.** These cells respond to bacterial invasion by secreting IL-17 and defensins, recruiting neutrophils, and inducing a strong inflammatory response to extracellular pathogens.

Regulatory T cells (Tregs) are derived from approximately 10% of $CD4^+$ T cells and 2% of $CD8^+$ T cells. When DCs release TGF-β during immune synapse, it induces the T-cell transcription factor FOXP3, which is responsible for the development of Tregs. Tregs are important because they dial back an immune response by producing the anti-inflammatory cytokines IL-10 and TGF-β. In addition to damping down acute inflammation and adaptive immune responses, Tregs are critical for maintaining mucosal immunity homeostasis. ⏮ *Inflammation unites all components of immunity (section 31.7)* ⏭ *A functional core microbiome is required for host homeostasis (section 33.3)*

Cytotoxic T Lymphocytes

Recall that the role of **$CD8^+$ T cells** is to destroy unhealthy host cells. Signals 1, 2, and 3 tightly regulate their activation so only unhealthy cells are removed. Signal 1 begins when DCs present non-self antigens in their MHC class I molecules (figure 32.8*b*). If the non-self antigen is recognized by the TCR of a $CD8^+$ T cell, signals 2 and 3 follow and the $CD8^+$ T cell matures into a CTL that functions to destroy host cells infected with intracellular pathogens or presenting cancer neoantigens. This informs the activated CTL to target any host cell that presents the same non-self antigen in its class I MHC receptors as that presented by the DC. Thus during the initial immune synapse, the DC educates the T cell as to what host cells it should seek and destroy.

Activated CTLs kill target cells in at least two ways: More commonly, perforins, granzymes, and granulysins are released, similar to NK cell–mediated cell death (*see figure 31.12*). Perforins generate pores in target cells and granzymes and granulysins stimulate programmed cell death by activating target cell pro-apoptotic proteins. Another apoptotic pathway can be triggered when the apoptotic protein Fas ligand (FasL) engages its receptor, Fas (also called CD95), expressed on the surface of target cells.

T Cell Summary

T-cell development and activation involve many details, so it may be helpful to think about the process from a more holistic point of view. Consider a newborn child: While many of their T cells are undergoing development in the thymus, some have "graduated" because they were selected to possess either a CD4 or CD8 coreceptor and passed the "test" of discriminating self from non-self. These mature, naïve T cells travel to lymphoid tissue to await further education. An infant is exposed to many new microbes, so their DCs are busy processing non-self antigens into their MHC class I (if an intracellular microbe) and MHC class II (if extracellular) molecules. The DCs travel to nearby lymphoid tissue and present the antigen within their MHCs to T cells (MHC I to $CD8^+$ cells; MHC II to $CD4^+$ cells). The DC is persistent in that it continues to introduce each non-self antigen to mature, naïve T cells until it finds the T cell destined for this antigen, as determined by the perfect fit of the antigen in the TCR (it's hard to avoid the Cinderella slipper analogy here). The perfect fit between MHC-antigen-TCR constitutes signal 1. It would be disastrous for the child if T cells were inappropriately activated, so to be fully activated, the T cell must also experience signals 2 and 3. What happens after all three signals have triggered activation depends on what kind of T cell ($CD4^+$ or $CD8^+$) was activated and what path of differentiation it takes (i.e., type of helper cell; memory versus effector cell). It's important to remember that this reflects the kind of pathogen (intracellular, extracellular, worm) that triggered activation. Regardless, further differentiation and proliferation leave the child with a new set of specific T cells that will function now (effector cells) and in the future (memory cells).

Comprehension Check

1. Explain the role of positive and negative selection processes that result in the release of mature, naïve T cells from the thymus.
2. Describe the formation and structure of an immune synapse.
3. Describe the function of the three signals needed for T-cell activation. Why are three signals needed?
4. Contrast and compare the activation of $CD4^+$ and $CD8^+$ T cells.
5. How do T_H1, T_H2, T_H17, and Treg cells differ in function?
6. Briefly describe the two ways in which CTLs destroy target cells.

32.6 B Cells Make Antibodies

After reading this section, you should be able to:

- **a.** Describe B-cell receptor structure and function
- **b.** Diagram the steps required for a B cell to become an antigen-presenting cell (APC)
- **c.** Compare T-cell-dependent and T-cell-independent B-cell activation in terms of mechanism and outcome
- **d.** Follow the path of B cells from bone marrow to plasma cell

Unlike T cells, which have a variety of functions, the sole purpose of B cells is to differentiate into antibody-producing cells as quickly as possible, or, in the case of memory B cells, to become cells producing the same antibodies in the event of reinfection. Like T cells, the development and activation of B cells must be tightly regulated. B-cell development begins in the bone marrow. Like T-cell negative selection in the thymus, negative selection within the bone marrow eliminates developing B cells that recognize self antigens. Upon leaving the bone marrow, mature, naïve B cells (like T cells) migrate between lymph nodes, Peyer's patches, and spleen, where they eventually meet the specific antigen recognized by their B-cell receptor. This results in their activation, proliferation, and differentiation into mature, antibody-secreting cells called **plasma cells** or memory cells. Notice that, like T-cell activation, B-cell activation requires a surface receptor that recognizes and binds specific antigen. However, unlike T cells, B cells can bind antigen directly; antigen is not presented by an APC. In fact, as we learn here, B cells serve as APCs by presenting antigen to T_H cells.

Although we think of antibodies as glycoproteins in serum, **B-cell receptors (BCRs)** are made of membrane-bound antibodies. Mature, naïve B cells attach antibody (IgM or IgD) to their cell membrane so that the part of the antibody that binds to antigen is facing outward, away from the cell (**figure 32.10**). Like TCRs, each BCR binds only a specific molecule that fits in a lock-and-key fashion. These cell-surface, transmembrane antibodies therefore act as receptors for the specific antigen that will activate that particular B cell. Also like TCRs, very little of the membrane-bound antibody extends into the cytoplasm, so for signal transduction to occur, it must associate with other integral membrane proteins, in this case Ig-α/Ig-β heterodimer proteins. Together the transmembrane antibody and the heterodimer protein complex constitutes the BCR. When an antigen is captured by the antibody part of the BCR, the integral membrane components of the receptor communicate this capture to the nucleus through a signal transduction pathway.

When the BCR captures antigen, the B cell must now become activated to proliferate and differentiate into plasma or memory cells. Importantly, each B cell is genetically programmed to make antibody to only one specific antigen. Thus the antigen specificity of the BCR has the same specificity as

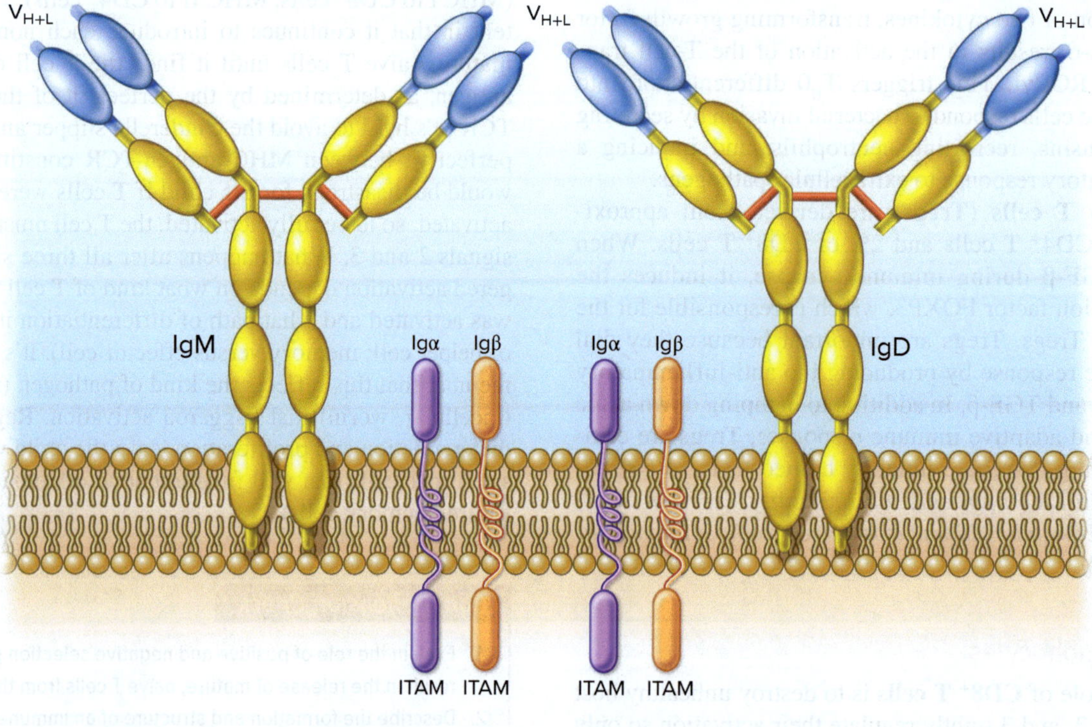

Figure 32.10 The Membrane-Bound B-Cell Receptor (BCR) Complex. The BCR is composed of monomeric IgM and IgD antibody, and the coreceptors Igα and Igβ. B-cell activation begins when the variable region of one receptor (blue; designated V_{H+L}) binds an antigen. The antigen-antibody complex is internalized by endocytosis and processed. A peptide portion of the antigen is presented by the B-cell's class II MHC molecule to a T_H cell that has been activated by the same antigen after immune synapse formation with a DC (T-cell-dependent activation). Activation can also occur if the variable regions of multiple receptors are bridged by an antigen with a repetitive structure, causing the BCRs to cross-link on the B-cell surface (T-cell-independent activation).

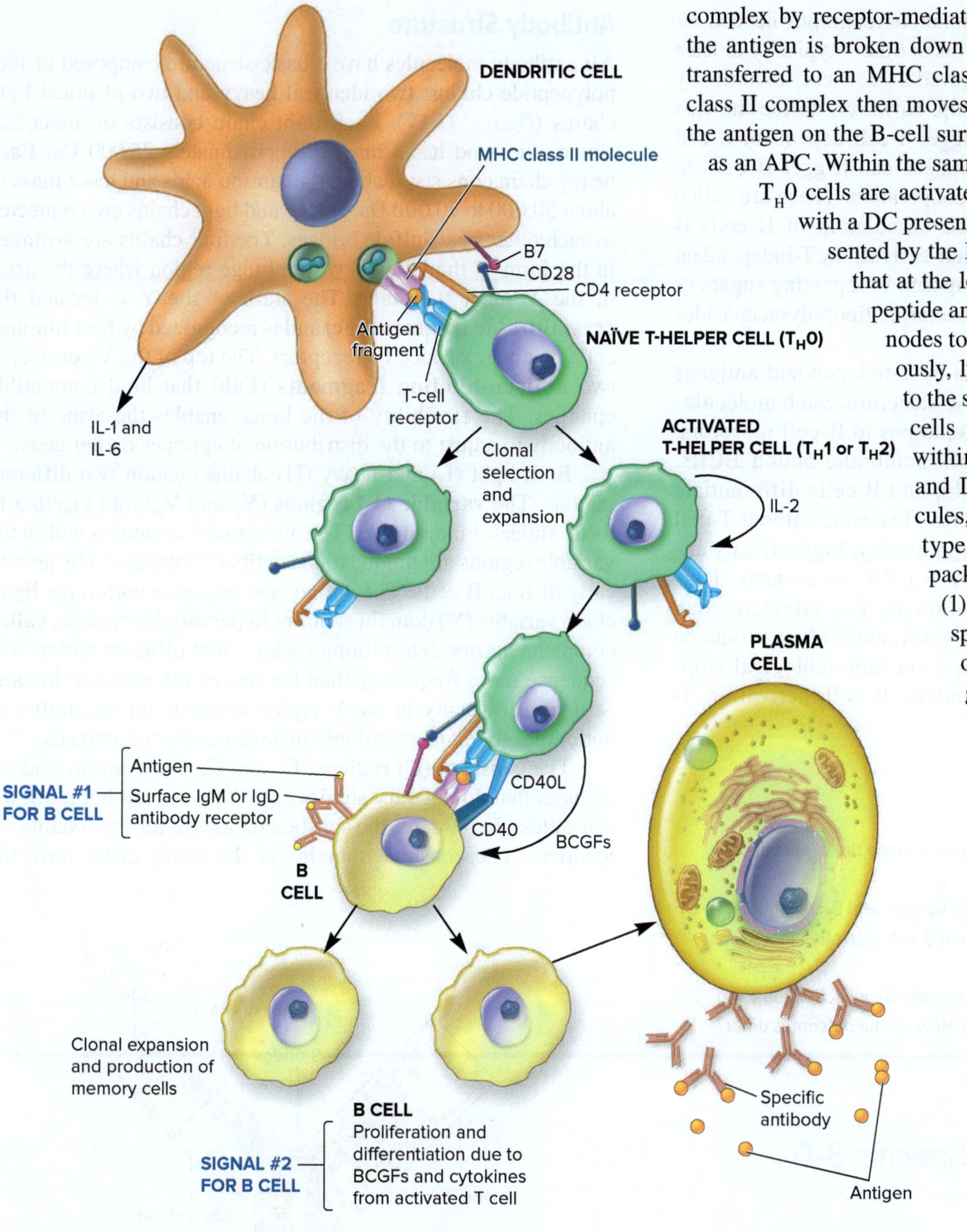

Figure 32.11 T-Dependent Antigen Activation of a B Cell. Many cytokines act as B-cell growth factors (BCGFs) to stimulate B-cell proliferation, whereas others stimulate B-cell differentiation into plasma cells. BCGFs, B-cell growth factors.

MICRO INQUIRY *Which cells function as APCs in this process?*

the antibody that will be secreted once a B cell has differentiated into a plasma cell.

In most cases, B-cell activation involves interaction with T cells, and is thus called **T-cell-dependent B-cell activation.** This process involves the simultaneous activation of both T and B cells with the same antigen (**figure 32.11**). Upon binding antigen snugly in their BCR, B cells internalize the antigen-BCR complex by receptor-mediated endocytosis. Within the B cell, the antigen is broken down and a peptide from the antigen is transferred to an MHC class II molecule. This peptide-MHC class II complex then moves to the B-cell membrane to display the antigen on the B-cell surface. The B cell is now functioning as an APC. Within the same lymph node and at the same time, T_H0 cells are activated by forming an immune synapse with a DC presenting the same antigenic peptide presented by the B cell. How is this possible? Recall that at the local site of infection, DCs package peptide antigens and migrate to nearby lymph nodes to present them to T cells. Simultaneously, lymphatic fluid carries the antigens to the same group of lymph nodes where B cells package and present the antigens within their MHC II molecules. B cells and DCs express the same MHC molecules, as determined by their HLA genotype. The same antigenic peptide is packaged by both types of cells because (1) they share the same lymph node (or spleen) as the peptide antigen and (2) only specific peptides will fit into a given MHC molecule.

The immune synapse between the DC and the T cell results in T-cell activation and differentiation into a T_H1 or T_H2 cell (other T_H cell types may also develop, but these are the types that interact with B cells). Because the B cell is now an APC, displaying the same antigen in its MHC class II molecule as the DC, when the B cell presents the antigen to a T cell, that T cell has already been activated by the DC and is ready to provide help to the B cell (figure 32.11). This help occurs upon formation of an immune synapse between the B cell and T_H cell, when specific cytokines are released by the T_H cell. These cytokines stimulate the expression of genes in the B cell that results in B-cell proliferation followed by differentiation into many antibody-secreting plasma cells and relatively fewer memory B cells.

Antigens that trigger T-cell-dependent B-cell activation are called **T-dependent antigens.** This complex process of B-cell activation requires two signals: antigen-BCR interaction (signal 1) and T-cell cytokines (signal 2). This ensures that a B cell does not differentiate into a potent plasma cell that produces antibody directed against a self antigen. Indeed, one plasma cell can synthesize more than 10 million antibody molecules per hour!

As you might imagine, the cellular choreography needed for T-cell-dependent B-cell activation takes time—typically at least a week. Humans have therefore evolved a "shortcut" to antibody formation that eliminates T-cell help. However, unlike the vast diversity of molecules that can trigger T-cell-dependent B-cell activation, only a few types of antigens can trigger B cells to produce antibodies without T-cell cooperation. These are called **T-independent antigens,** and their stimulation of B cells is known as **T-cell-independent B-cell activation.** T-independent antigens tend to be polymeric, composed of repeating sugars or amino acids. Examples include bacterial lipopolysaccharides and certain polysaccharides.

The mechanism for activation by T-independent antigens relies on their repetitive, polymeric structure. Such molecules present a wide array of identical epitopes to B-cell receptors. The repeating epitopes cross-link membrane-bound BCRs, overriding the need for T cell help, and B cells differentiate into antibody-secreting plasma cells. However, without T-cell help, the B cell cannot produce high-avidity, high-affinity antibodies, and no (or few) memory B cells are formed. Thus T-cell-independent B-cell activation is less effective than T-dependent B-cell activation. However, antibody is produced much more quickly and can be a good immunological stopgap measure until T-cell-dependent B-cell activation is completed.

Comprehension Check

1. What is the structure of B-cell receptors? Trace the events upon BCR-antigen binding.
2. Outline the steps of T-cell activation by a DC and B-cell antigen presentation during T-cell-dependent B-cell activation in a lymph node.
3. How does the process of T-cell-dependent B-cell activation differ from T-cell-independent activation? How do the outcomes differ?

32.7 Antibodies Bind Specific 3-D Antigens

After reading this section, you should be able to:

a. Compare and contrast the structure and function of the five classes of antibody
b. Explain how primary and secondary antibody responses differ

So far we have discussed how B cells are activated to become antibody-secreting plasma cells. We now turn our attention to antibody structure. An **antibody** or **immunoglobulin (Ig)** is a glycoprotein. There are five classes of human antibodies; all share a basic structure that varies to accomplish specific biological functions.

Antibody Structure

All antibody molecules have a basic structure composed of four polypeptide chains: two identical heavy and two identical light chains (**figure 32.12**). Each light chain consists of about 220 amino acids and has a mass of approximately 25,000 Da. Each heavy chain consists of about 440 amino acids and has a mass of about 50,000 to 70,000 Da. Heavy and light chains are connected to each other by disulfide bridges. The four chains are arranged in the form of the letter Y with a hinge region where the arms of the Y meet the stalk. The stalk of the Y is termed the **crystallizable fragment (Fc)** and is recognized by host immune cells that bear surface Fc receptors. The top of the Y consists of two **antigen-binding fragments (Fab)** that bind compatible epitopes. The flexibility of the hinge enables the arms of the antibody to adjust to the distribution of epitopes on antigens.

Both light (L) and heavy (H) chains contain two different regions. The **variable (V) regions** (V_L and V_H) fold together to form antigen-binding sites. The amino acid sequences within the variable regions are unique to the antibodies produced by genetically distinct B cells. The amino acid sequence within the light-chain variable (V) domain contains hypervariable regions, called complementarity-determining regions, that differ in amino acid sequence more frequently than the rest of the variable domain. Structural diversity in the V region accounts for the ability of antibodies to bind a seemingly infinite number of antigens.

The **constant (C) regions** (C_L and C_H) have amino acid sequences that do not vary significantly between antibodies of the same class. In humans, the constant regions of all light chains are identical. The constant domains of the heavy chain form the

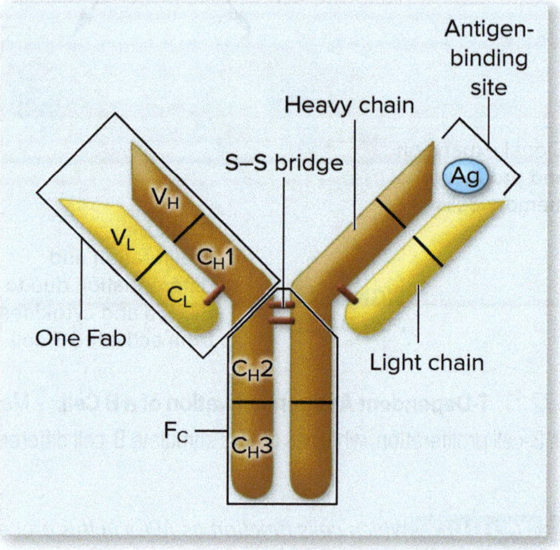

Figure 32.12 Immunoglobulin (Antibody) Structure. Heavy and light chains are connected to each other by disulfide bridges (red, S–S). Variable regions of heavy and light chains (V_H, V_L) form an antigen-binding fragment (Fab). The disulfide-linked heavy chains form the crystallizable fragment (Fc). All light chains contain a single variable domain (V_L) and a single constant domain (C_L). Heavy chains contain a variable domain (V_H) and either three or four constant domains (C_H1, C_H2, and C_H3). Only one bound antigen (Ag) is shown for simplicity.

Table 32.1 Physicochemical Properties of Human Immunoglobulin Classes

Property	IgG[1]	IgM	IgA[2]	IgD	IgE
Heavy chain	γ	μ	α	δ	ε
Percent of total serum antibody	80–85	5–10	5–15	<1	<1
Valency	2	5(10)	2(4)	2	2
Mass of entire molecule (kDa)	146	970	160	184	188
Placental transfer	+	–	–	–	–
Half-life in serum (days)[3]	23	5	6	3	2
Complement activation					
Classical pathway	++	+++	–	–	–
Alternative pathway	–	–	+	–	–
Major characteristics	Most abundant Ig in body fluids; neutralizes toxins; opsonizes bacteria	First to appear after antigen stimulation; agglutinator; monomer functions as part of BCR complex	Secretory antibody; protects mucous membranes and as such is the most abundant antibody type	Serves as part of BCR complex	Anaphylactic-mediating antibody; resistance to helminths

[1] Properties of IgG subclass 1.
[2] Properties of IgA subclass 1.
[3] Time required for half of the antibodies to disappear.

constant (C_H) region. The amino acid sequence of this region determines the classes of heavy chains. In humans, there are five classes of heavy chains designated by lowercase Greek letters: gamma (γ), alpha (α), mu (μ), delta (δ), and epsilon (ε), and usually written as G, A, M, D, or E. The properties of these heavy chains determine, respectively, the five immunoglobulin (Ig) classes: IgG, IgA, IgM, IgD, and IgE (**table 32.1**). Each immunoglobulin class differs in its specific structure, half-life, distribution in the body, and interaction with other components of the host's defensive systems.

Antibody Function

As shown in figure 32.12, antigen binds at the antigen-binding site within the Fab region of the antibody and each antibody has two identical binding sites. Within this region, a pocket is formed by the folding of the V_H and V_L regions. At each site, specific amino acids contact the antigen's epitope and form multiple noncovalent bonds between amino acids of the binding site and the epitope. Because binding is due to noncovalent interactions such as hydrogen bonds and electrostatic attractions, the antigen's shape must exactly match that of the antigen-binding site. If the shape of the epitope and binding site are not truly complementary, the antibody will not effectively bind the antigen.

Whereas the Fab region is concerned with binding antigen, the Fc region binds to receptors (called *Fc receptors*) found on various immune cells such as phagocytes and NK cells. As we learn in chapter 31, antibody is important in opsonizing microbes and initiating the classical complement cascade. The binding of an antibody to an antigen usually opsonizes, or marks, the non-self agent as a target for immunological attack (section 32.8). Immune cells, such as neutrophils, macrophages, and DCs use their Fc receptors to grab the Fc region of the antibody. In this way the antibody serves as a bridge between the opsonized pathogen and the immune cells that will destroy (neutrophils) or destroy and present (macrophages and DCs) the antigen (*see figure 31.5*). NK cells use their Fc receptors in a similar way during antibody-dependent cell-mediated cytotoxicity (*see figure 31.13*). ◀ *Complement is a cascading enzyme system (section 31.3)*

Immunoglobulin (Ig) Classes

Immunoglobulin γ, or **IgG,** is the major immunoglobulin in human serum, where it accounts for 80% of the immunoglobulin pool. Its structure is like that shown in figure 32.12. IgG acts against bacteria and viruses by opsonizing them. It also neutralizes toxins and viruses (section 32.8). It is one of the two immunoglobulin classes that activate complement by the classical pathway (the other is IgM). IgG is the only immunoglobulin molecule able to cross the placenta, providing natural immunity in utero and during the first 6 months of life.

There are four human IgG isotypes (IgG1, IgG2, IgG3, and IgG4). All are monomeric, but IgG isotypes vary chemically in their heavy-chain composition and the number and arrangement of interchain disulfide bonds. Differences in biological function have been noted in these isotypes. For example, IgG2 and IgG4 antibodies bind and neutralize toxins. IgG1 and IgG3 are important for opsonization.

Immunoglobulin μ, or **IgM,** accounts for about 5 to 10% of the serum immunoglobulin pool. In the serum, it is a polymer of five monomeric units (pentamer) (**figure 32.13*a***). The monomers are arranged in a pinwheel array with the Fc ends in the center, held together by disulfide bonds and a protein called the J (joining) chain. Soluble, pentameric IgM is the immunoglobulin made upon initial antigen stimulation. It is the only antibody class produced following T-cell-independent B-cell activation. Pentameric IgM agglutinates (or clumps) bacteria, activates complement by the classical pathway, and enhances the ingestion of pathogens by phagocytic cells. In contrast to pentameric IgM, individual IgM monomers serve as the antibody component of the BCR (figure 32.10).

Immunoglobulin α, or **IgA,** accounts for about 12% of serum immunoglobulin, but is actually the most abundant class of antibody in the human body. This is because most IgA is found in mucous secretions. Whereas serum IgA is monomeric, when secreted into mucous membranes it forms a dimer held together by a J chain (figure 32.13*b*). Transport of IgA from mucosal-associated lymphoid tissue (MALT) to mucosal surfaces requires the attachment of a second protein, the secretory component (*see figure 31.16*). **Secretory IgA (SIgA),** as the modified molecule is called, is the primary immunoglobulin of MALT. Secretory IgA is therefore found in any secretion associated with mucous membranes such as saliva, tears, and intestinal mucus, where it plays a major role in protecting surface tissues against microorganisms by the formation of an immune barrier. For example, when SIgA binds to viruses, bacteria, and protozoa, these microorganisms are blocked from attaching to mucosal surfaces and invading host tissues, a phenomenon known as immune exclusion. In addition, SIgA is found in breast milk, thereby providing a measure of mucosal immunity to breastfed infants. ◀ *Secondary lymphoid organs and tissues (section 31.5)*

Immunoglobulin δ, or **IgD,** has a monomeric structure (figure 32.13*c*). IgD antibodies are abundant in combination with IgM on the surface of B cells and thus are part of the B-cell receptor. Therefore their function is to help signal B cells to start antibody production upon initial antigen binding.

Immunoglobulin ε, or **IgE** (figure 32.13*d*), makes up only a small percent of the total immunoglobulin pool. A number of different cell types, especially mast cells, eosinophils, and basophils, bind IgE molecules with their IgE-specific Fc receptors. When cell-bound IgE molecules become cross-linked by binding to the same antigen, these cells degranulate. Degranulation releases preformed mediators of inflammation (e.g., histamine). It also stimulates production of a large number of eosinophils in the blood (eosinophilia) and increases the rate of movement of the intestinal contents (gut hypermotility). This aids in the elimination of worm parasites. However, the

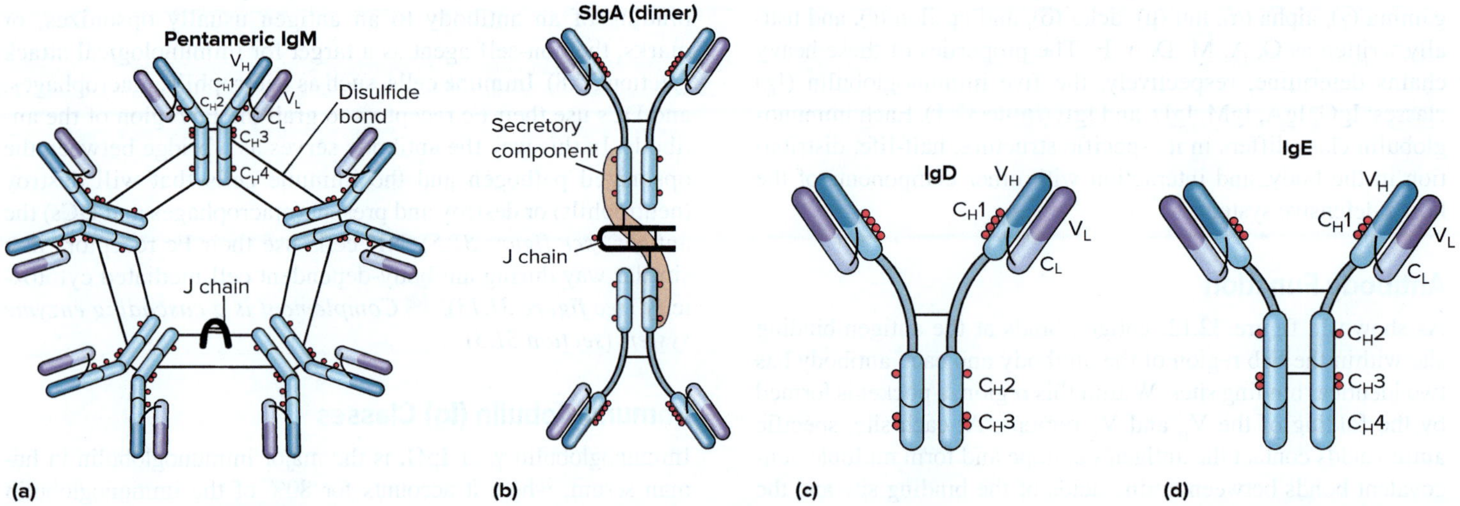

Figure 32.13 Immunoglobulins M, A, D, and E. Antibodies are glycoproteins. Here the carbohydrate side chains are shown in red. Disulfide bridges linking peptide chains are shown in black. (a) The pentameric structure of human IgM generates 10 antigen-binding sites. (b) The dimeric structure of human secretory IgA. Notice the secretory component (tan) wound around the IgA dimer and attached to the constant domain of each IgA monomer. (c) The structure of human IgD. (d) The structure of human IgE.

MICRO INQUIRY *Where is monomeric IgM commonly found? What about secretory IgA?*

release of these mediators, particularly histamine, is also responsible for allergic reactions and anaphylaxis (section 32.10).

Antibody Kinetics

The rate of synthesis and secretion of antibody changes over time. The relatively slow antibody response when an antigen is encountered for the first time is called the primary antibody response. The secondary antibody response occurs upon subsequent exposure to the same antigen. This response is so rapid and efficient, it prevents illness. This demonstrates the ability of the adaptive immune system to remember a pathogen and is the basis for vaccination.

The Primary Antibody Response

When an individual is first exposed to an antigen through either infection or vaccine, there is an initial lag phase, or latent period (up to several weeks), before high levels of antibodies are produced. During this latent period, no antigen-specific antibody can be detected in the blood (**figure 32.14**). During this time, B cells in lymphoid tissues are activated, which triggers their differentiation and proliferation into antibody-secreting plasma cells. IgM appears first, and although it is secreted relatively quickly, IgM affinity for the antigen is relatively low. If T-cell-dependent B-cell activation occurs, B cells will undergo

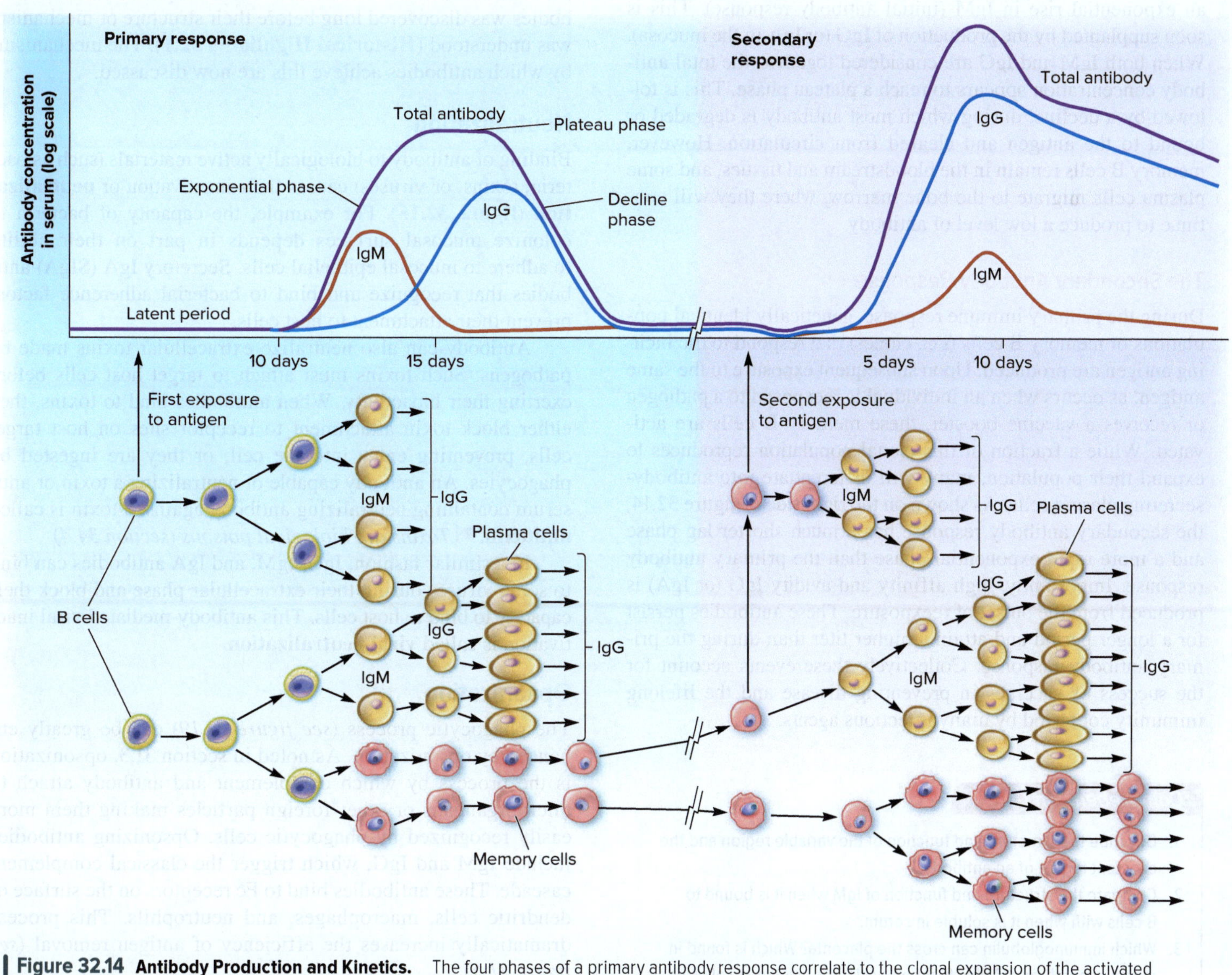

Figure 32.14 Antibody Production and Kinetics. The four phases of a primary antibody response correlate to the clonal expansion of the activated B cell, differentiation into plasma cells, and secretion of antibodies. The secondary response is much more rapid, and total antibody production is nearly 1,000 times greater than that of the primary response.

MICRO INQUIRY *How does the lag in IgM production compare during a primary and secondary response? Contrast this with the lag seen in IgG production during a primary and secondary response. What does this difference imply about the nature of the secondary response?*

antibody class switching. This means they will differentiate into plasma cells that secrete antibody belonging to a different class, usually IgG or IgA. These antibodies have a higher avidity and affinity for the antigen than IgM. The antibody class switching process is irreversible. Therefore once class switching has occurred, that group of activated B cells loses the capacity to produce any other class of antibody. Importantly, B cells will also differentiate into memory cells capable of producing antibody of the specified class against the same antigen in the future.

The **antibody titer,** which is a measurement of serum antibody concentration, reflects these dynamics. An initial period of latency (initial exposure but no antibody response) is followed by an exponential rise in IgM (initial antibody response). This is soon supplanted by the production of IgG (or IgA in the mucosa). When both IgM and IgG are considered together, the total antibody concentration appears to reach a plateau phase. This is followed by a decline, during which most antibody is degraded or bound to the antigen and cleared from circulation. However, memory B cells remain in the bloodstream and tissues, and some plasma cells migrate to the bone marrow, where they will continue to produce a low level of antibody.

The Secondary Antibody Response

During the primary immune response, genetically identical populations of memory B cells (i.e., clones) that respond to the inciting antigen are produced. Upon subsequent exposure to the same antigen, as occurs when an individual is reexposed to a pathogen or receives a vaccine booster, these memory B cells are activated. While a fraction of this clonal population reproduces to expand their population, many will differentiate into antibody-secreting plasma cells. As shown on the right side of figure 32.14, the secondary antibody response has a much shorter lag phase and a more rapid exponential phase than the primary antibody response. Importantly, high affinity and avidity IgG (or IgA) is produced from the outset of reexposure. These antibodies persist for a longer period and attain a higher titer than during the primary antibody response. Collectively these events account for the success of vaccines in preventing disease and the lifelong immunity conferred by many infectious agents.

Comprehension Check

1. Describe the structure and function of the variable region and the constant region of an antibody.
2. Compare the structure and function of IgM when it is bound to B cells with when it is soluble in serum.
3. Which immunoglobulin can cross the placenta? Which is found in mucous membranes? Which antibody class is involved in allergic responses?
4. Relate the graph shown in figure 32.14 with the severity of COVID-19 when the pandemic began in 2020 and with seasonal outbreaks.

32.8 Antibodies Doom Antigens

After reading this section, you should be able to:

a. Explain the consequences of antibody binding of antigen
b. Assess the effectiveness of antigen removal by antibody
c. Predict which antigens will be most susceptible to antibody action

Antigen-antibody interactions exhibit exquisite specificity. The interactions that occur in animals are essential in protecting against the continuous onslaught of microorganisms and their products, as well as cancer cells. The therapeutic value of antibodies was discovered long before their structure or mechanism was understood (**Historical Highlights 32.1**). The mechanisms by which antibodies achieve this are now discussed.

Neutralization

Binding of antibody to biologically active materials (such as bacteria, toxins, or viruses) causes their inactivation or **neutralization** (**figure 32.15**). For example, the capacity of bacteria to colonize mucosal surfaces depends in part on their ability to adhere to mucosal epithelial cells. Secretory IgA (SIgA) antibodies that recognize and bind to bacterial adherence factors prevent their attachment to host cells.

Antibody can also neutralize extracellular toxins made by pathogens. Such toxins must attach to target host cells before exerting their bioactivity. When antibodies bind to toxins, they either block toxin attachment to receptor sites on host target cells, preventing entry into the cell, or they are ingested by phagocytes. An antibody capable of neutralizing a toxin or antiserum containing neutralizing antibody against a toxin is called **antitoxin.** ▶| *Toxins are biological poisons (section 34.4)*

In a similar fashion, IgG, IgM, and IgA antibodies can bind to some viruses during their extracellular phase and block their capacity to bind to host cells. This antibody-mediated viral inactivation is called **viral neutralization.**

Opsonization

The phagocytic process (*see figure 31.19*) can be greatly enhanced by opsonization. As noted in section 31.3, opsonization is the process by which complement and antibody attach to microorganisms or other foreign particles making them more easily recognized by phagocytic cells. Opsonizing antibodies include IgM and IgG, which trigger the classical complement cascade. These antibodies bind to Fc receptors on the surface of dendritic cells, macrophages, and neutrophils. This process dramatically increases the efficiency of antigen removal (*see figure 31.5*).

Immune Complex Formation

Because antibodies have at least two antigen-binding sites (the variable region at the end of each arm) and most antigens have at least two epitopes, cross-linking occurs when a single

HISTORICAL HIGHLIGHTS

32.1 Convalescent Plasma: An Old Treatment for a New Disease

The two greatest microbiologists of the nineteenth century (and maybe ever), Robert Koch and Louis Pasteur were rivals. The source of their deep disdain for each other started with Koch's belief that microbes had unchangeable phenotypes, which contradicted Pasteur's ability to create vaccines from attenuated strains. When two assistants in Koch's lab, Emil Behring and Shibasaburo Kitasato, started using attenuated strains of the bacteria that cause diphtheria (*Corynebacterium diphtheriae*) to develop a diphtheria treatment and vaccine, we can presume Koch was not thrilled.

During the nineteenth century about 50,000 children died of diphtheria per year in Germany alone, so how could Koch object? Behring and Kitasato immunized animals with attenuated *C. diphtheriae* and collected their plasma (the liquid portion of blood, also called serum). They hypothesized that plasma contained what they called antitoxin, which we now know to be antibodies directed against diphtheria toxin. Using Koch's postulates (*see figure 1.17*) to verify that lab animals developed disease when exposed to *C. diphtheriae,* Behring and Kitasato showed that the disease was cured when animals were treated with plasma containing antitoxin. In 1891 Behring and Kitasato demonstrated the effectiveness of antitoxin plasma in children, and the humoral theory of immunity gained traction and the term *antibody* was coined. Behring left Koch's lab and went on to use a similar approach to develop a vaccine against diphtheria. In 1901 Behring (but not Kitasato) received the first Nobel Prize for Physiology or Medicine and was later endowed with the honorific "von," becoming Emil von Behring. Koch was not awarded a Nobel Prize until 1905.

Years later, medical teams trying to combat the 1918 influenza pandemic looked to von Behring's and Kitasato's plasma treatment for diphtheria. However, the causative microbe for influenza was unknown, so animals could not be infected to raise the plasma containing neutralizing antibodies. The only source of such plasma was patients who had recovered from the flu. Thus without knowledge of the structure and mechanism of antibodies (not reported until 1959 by Gerald Edelman and Rodney Porter), so-called *convalescent plasma* (serum) was collected from flu survivors and used as a treatment.

Convalescent plasma was used frequently through the 1930s to treat a variety of infectious diseases such as scarlet fever (caused by *Streptococcus pyogenes*) and pneumococcal pneumonia (caused by *S. pneumoniae*). The development of antibiotics replaced convalescent plasma as a therapy for all but a few viral illnesses. Most notably, convalescent plasma was used to treat Ebola until the development of a vaccine. In 2020 convalescent plasma was used again to treat COVID-19 patients, although its efficacy was confined largely to patients who were immunocompromised (**box figure**). As the explosive growth of the pandemic vastly outpaced the speed by which new drugs could be developed, the world looked once again to an old treatment. This time there was a much better understanding of the molecules and mechanisms involved. In fact, the molecular structures of neutralizing antibodies informed the development of monoclonal antibody treatments. During the COVID-19 pandemic, convalescent plasma was used as a bridge until more effective medicines could be introduced.

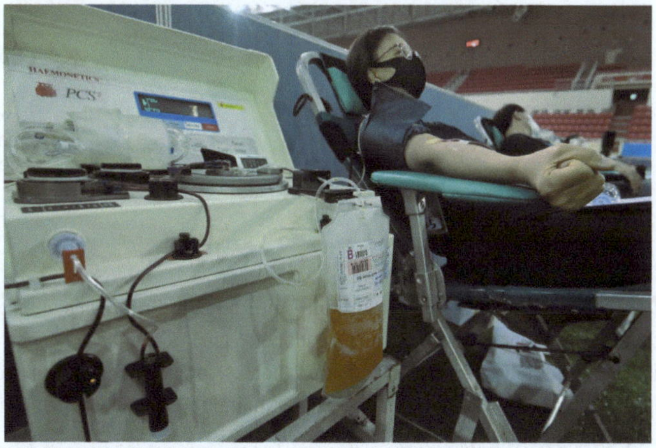

JUNG YEON-JE/AFP/Getty Images

antibody binds two different antigen molecules—one with each arm. When a large number of antigens are bound in this way by multiple antibodies, aggregates termed **immune complexes** are formed (figure 32.15). If the antigens are soluble molecules and the complex becomes large enough to settle out of solution, a **precipitation** (Latin *praecipitare,* to cast down) or a **precipitin reaction** occurs. When the immune complex involves the cross-linking of cells or particles, an **agglutination reaction** occurs. These immune complexes are more rapidly phagocytosed than free antigens. However, immune complex formation can harm the host, as we discuss in section 32.10. Importantly, formation of immune complexes is the basis of some diagnostic tests. *Immune responses can be exploited to detect infections (section 36.3)*

Comprehension Check

1. What is the difference between neutralization and opsonization?
2. How does toxin neutralization occur? Viral neutralization?
3. Draw an immune complex (antigens are circles, antibodies are Y-shaped).

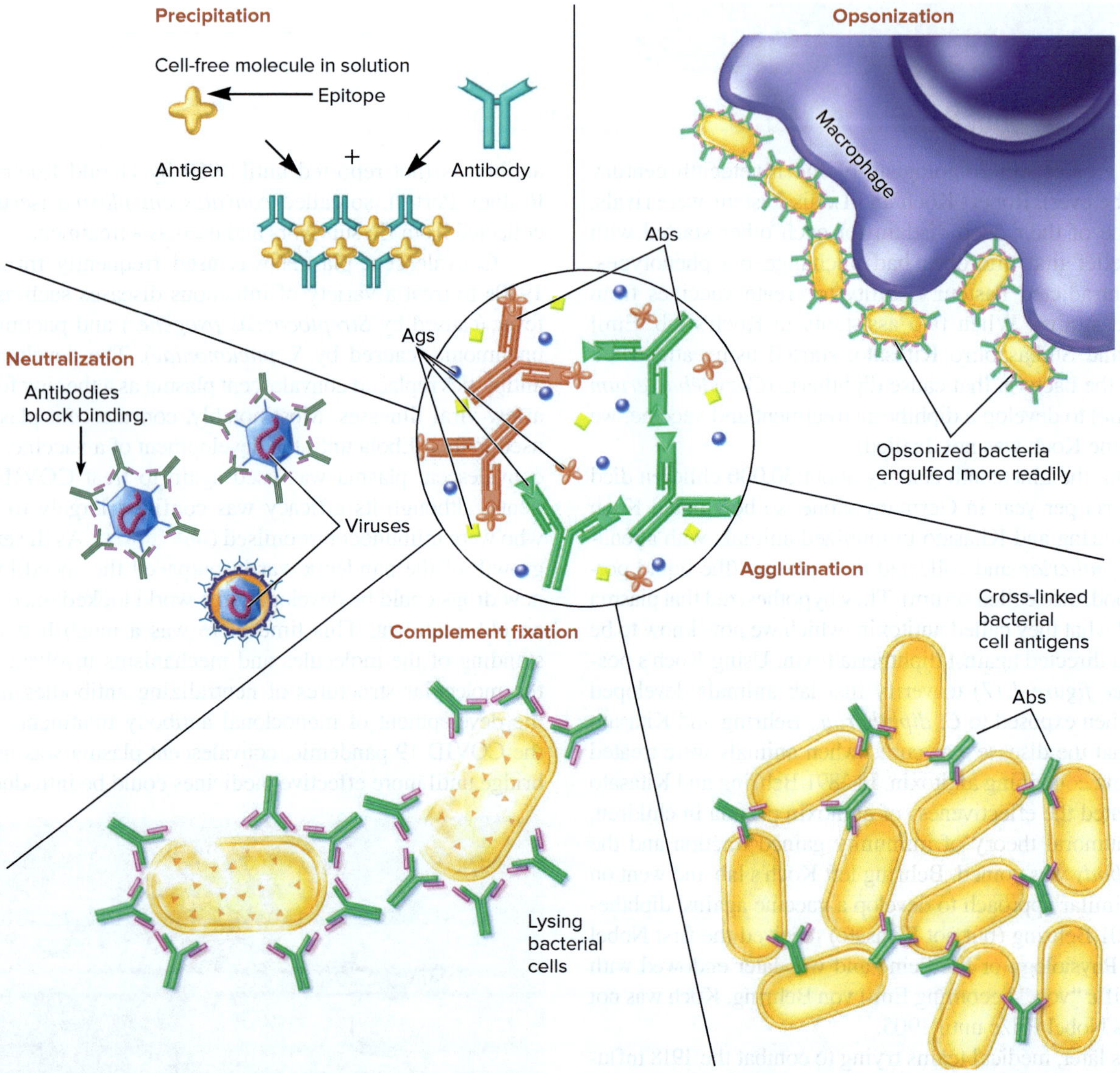

Figure 32.15 Consequences of Antigen-Antibody Binding. Immune complexes form when soluble antigens (Ag) bind soluble antibodies (Ab), resulting in precipitation. Opsonization occurs when antibodies bind to antigens on larger molecules or cells, enhancing their recognition by phagocytic cells. Agglutination results when antigens are cross-linked by antibodies, forming large insoluble complexes. The classical complement cascade can be activated by immune complexes; this is called complement fixation. Neutralization results when antibody binds to toxins or viruses, preventing these antigens from binding to host cells.

MICRO INQUIRY *What is the difference between a precipitation and an agglutination reaction?*

32.9 Generating Populations of Diverse Lymphocytes

After reading this section, you should be able to:

- **a.** Discuss the genetic diversity of BCRs and soluble antibodies that results from recombination, alternative splicing, and somatic hypermutation
- **b.** Compare the development of BCR diversity with the mechanisms used to generate TCR diversity
- **c.** Explain how clonal selection amplifies the populations of genetically distinct lymphocytes

Throughout our discussion of T-cell and B-cell development and activation, we have noted that TCRs and BCRs are both highly specific and variable. For example, consider that each individual human carries BCRs specific for as many as 10^{13} different antigens, which means that every individual produces 10^{13} different B cells, each with a unique BCR structure. Similarly, each human produces about 10^{15} different TCRs. This astounding level of diversity enables the host to respond to antigens for an entire lifetime.

Receptor Diversity

Given that the human genome encodes only about 22,000 genes, how is this astonishing level of lymphocyte diversity possible?

To simplify our discussion, we focus first on the generation of BCR diversity, which involves four mechanisms. The first three occur in the bone marrow: (1) During B-cell development, the genes that encode antibodies rearrange by a cutting and pasting mechanism called *combinatorial joining;* (2) when two ends of an antibody gene join during combinatorial joining, additional nucleotides are inserted, generating additional sequence variability. Also at this time, (3) different codons are created in a process known as *splice-site variability*. The next activity takes place in lymphoid tissue (lymph nodes, Peyer's patches, or spleen) when (4) following T-helper cell stimulation, mutations arise with high frequency in the genes encoding antibodies. This process, known as *somatic hypermutation,* generates B cells that will become plasma cells that produce antibody with increased avidity and affinity for antigen, as compared to the IgM initially produced. Somatic hypermutation occurs only upon T-cell-dependent B-cell activation. We now take a closer look at these mechanisms.

Combinatorial joining is made possible by the organization of immunoglobulin genes. The genes that encode antibody proteins contain a small number of exons close together on the same chromosome that determine the constant (C) region of the light chains. Separated from them but still on the same chromosome is a larger cluster of segments that determines the variable (V) region of the light chains. Between the V and C regions is a cluster of J (joining) regions (**figure 32.16**). During the development of a B cell in the bone marrow, enzymes called RAG-1 and RAG-2 join one V gene segment with one J segment; all other V and J regions are cut out of the DNA and are lost to the cell. Combinatorial joining is thus an irreversible DNA cut-and-paste mechanism that creates many different combinations of the V and J regions. Additional variation is introduced by an enzyme called *t*erminal *d*eoxynucleotidyl *t*ransferase (TdT) that catalyzes the second mechanism that provides antibody diversity. TdT inserts nucleotides at the V-J junction. When the light-chain gene is transcribed, transcription continues through the DNA region that encodes the constant portion of the gene. RNA splicing subsequently joins the V, J, and C regions, creating a unique mRNA.

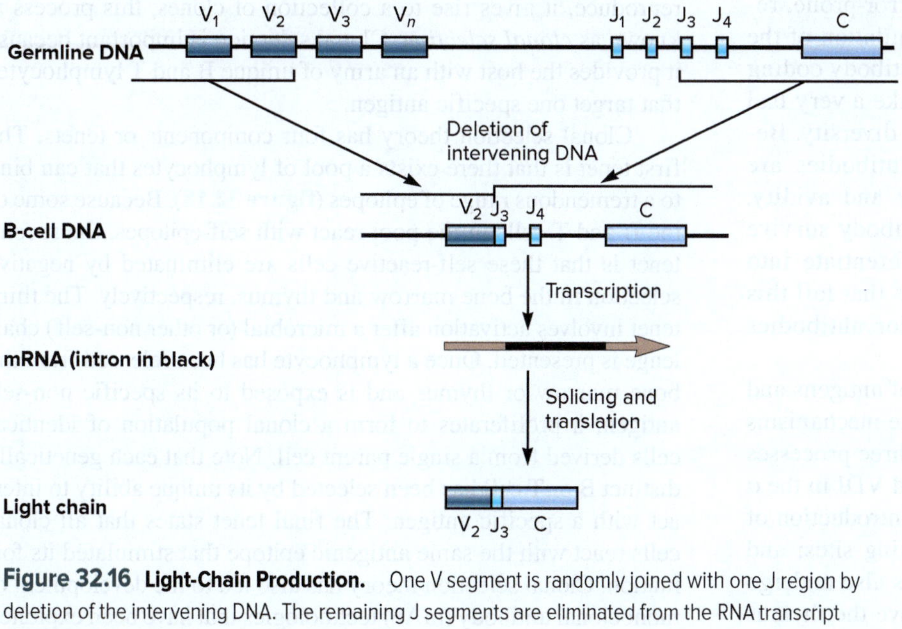

Figure 32.16 Light-Chain Production. One V segment is randomly joined with one J region by deletion of the intervening DNA. The remaining J segments are eliminated from the RNA transcript during RNA processing.

Combinatorial joining of the heavy-chain gene is similar except three different parts of the antibody locus—V, J, and D—must be joined. Therefore heavy-chain development involves cutting and joining the heavy-chain counterparts of V and J DNA, as well as the D (diversity) sequences (**figure 32.17**).

During the process of joining either VJ or VDJ sequences in light- and heavy-chain rearrangement, respectively, the third opportunity to introduce genetic diversity occurs. Because combinatorial joining occurs between different nucleotides, different codons can result in a process known as **splice site variability.** For example, one VJ splicing event can join the V sequence CCTCCC with the J sequence TGGTGG as CCTCCC + TGGTGG = CCGTGG, which codes for proline and tryptophan. Alternatively, the VJ splicing event can eliminate nucleotides, giving rise to the sequence CCTCGG, which codes for proline and arginine. Thus the same VJ joining can produce polypeptides differing in a single amino acid. Importantly, because combinatorial joining, TdT-dependent nucleotide addition, and splice site variability involve changing DNA sequences (as opposed to RNA splicing), each B cell produced has a new and different genotype and its replication will give rise to genetically distinct clones.

Initially, all heavy chains have the μ type of constant region (figure 32.17*b*). This explains why IgM is produced upon initial antigen stimulation (figure 32.14). Upon T-cell-dependent B-cell activation in lymphoid tissue, antibody class switching occurs when the VDJ region is joined with a new constant region that encodes a different class of antibody, usually IgG or IgA (figure 32.17*c*).

A unique editing process is used to switch from IgM to another antibody class. Activated B cells in lymphoid tissue synthesize the enzyme *a*ctivation-*i*nduced cytidine *d*eaminase (AID). AID replaces cytosine residues with uracil. This stimulates DNA repair enzymes to remove uracil residues from the DNA. The resulting abasic nucleotides are then excised, nicking one DNA strand. When nicking occurs adjacent to μ and γ switch regions, recombination occurs; the μ constant region is completely removed, and the γ constant region is joined to the VDJ region. Thus once class switching has occurred, the B cell will produce IgG and never be able to produce IgM again.

In addition to class switching, T cells help trigger B cells to increase antibody diversity through **somatic hypermutation.** This mutational process occurs during activation in lymphoid tissues (lymph nodes, Peyer's patches, and spleen) when B cells are replicating

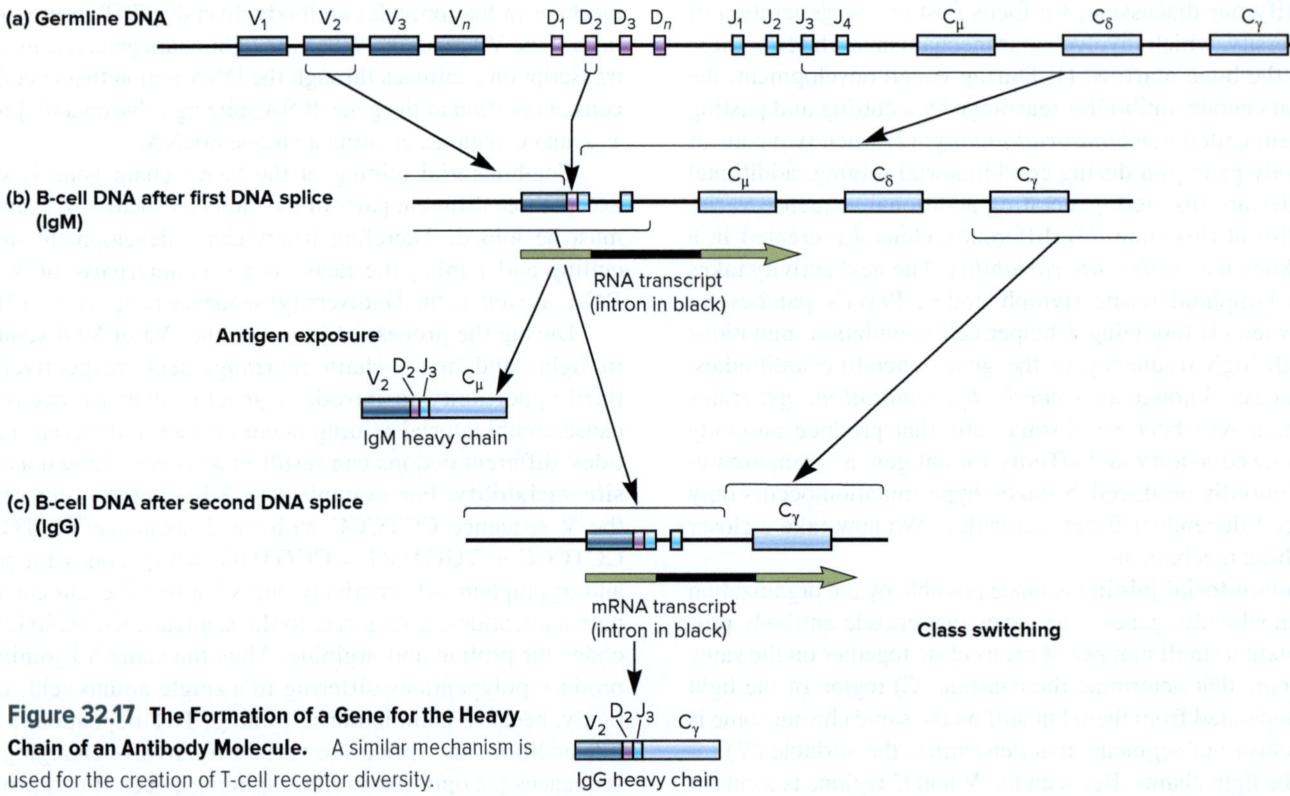

Figure 32.17 The Formation of a Gene for the Heavy Chain of an Antibody Molecule. A similar mechanism is used for the creation of T-cell receptor diversity.

rapidly. When point mutations arise at so called mutational hotspots within the variable regions of the light and heavy chain loci, they are recognized by DNA repair enzymes that attempt to correct the altered bases. AID, the same enzyme that mediates antibody class switching, is one of the enzymes involved. Because AID deaminates DNA at cytosine, base pair mismatches and transition-type mutations occur. Furthermore, repair of the base pair mismatch is error-prone, resulting in even more sequence diversity. Accumulation of the altered bases in the variable regions of B-cell antibody coding genes following T-cell activation would seem like a very bad thing. Instead, it generates significant antibody diversity. Because mutations are random, these variant antibodies are tested within the lymphoid tissue for affinity and avidity. Only B cells producing new and improved antibody survive this selection process and are allowed to differentiate into mature, antibody-secreting plasma cells. B cells that fail this test because the mutations give rise to inferior antibodies undergo apoptosis.

T cell receptors also recognize a vast array of antigens and must therefore also be tremendously variable. The mechanisms by which TCR diversity arises include the first three processes seen in BCRs: (1) combinatorial joining of VJ and VDJ in the α and β chains, respectively, of the TCR; (2) TdT introduction of additional sequence diversity at the V-D-J joining sites; and (3) splice site variability. Note that γ and δ chains also undergo these processes but, as shown in figure 32.6, leave the thymus shortly after TCR formation. The fourth mechanism, somatic hypermutation, only occurs in B cells as it serves to increase the avidity and affinity of antibodies.

Clonal Selection

Each of the processes that lead to antibody diversity changes a cell's DNA and is therefore irreversible. Thus when each genetically distinct B or T cell is stimulated by a specific antigen to reproduce, it gives rise to a collection of clones; this process is known as *clonal selection.* Clonal selection is important because it provides the host with an army of unique B and T lymphocytes that target one specific antigen.

Clonal selection theory has four components or tenets. The first tenet is that there exists a pool of lymphocytes that can bind to a tremendous range of epitopes (**figure 32.18**). Because some of the B and T cells in the pool react with self-epitopes, the second tenet is that these self-reactive cells are eliminated by negative selection in the bone marrow and thymus, respectively. The third tenet involves activation after a microbial (or other non-self) challenge is presented. Once a lymphocyte has been released from the bone marrow or thymus and is exposed to its specific non-self antigen, it proliferates to form a clonal population of identical cells derived from a single parent cell. Note that each genetically distinct B or T cell has been selected by its unique ability to interact with a specific antigen. The final tenet states that all clonal cells react with the same antigenic epitope that stimulated its formation. Clonal selection theory has also led to the development of monoclonal antibody (mAb) technologies that have been exploited to treat disease (**Techniques & Applications 32.2**).

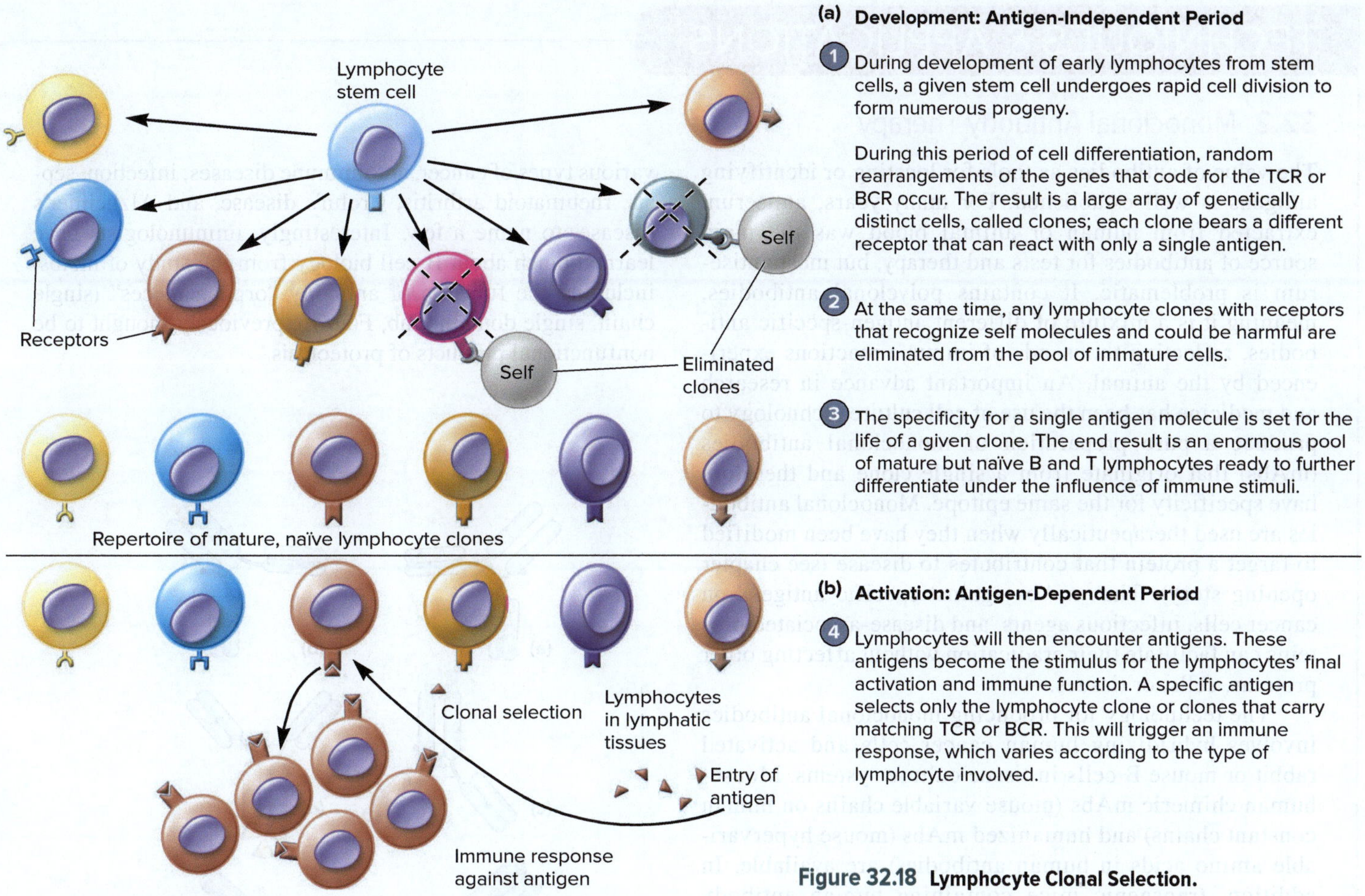

Figure 32.18 Lymphocyte Clonal Selection.

Comprehension Check

1. Compare and contrast combinatorial joining of heavy- and light-chain antibody gene segments.
2. What events lead to B-cell diversity during development in the bone marrow? Compare these with the events that occur only after T-cell activation of B cells in lymphoid tissue.
3. Which mechanisms are used to generate T-cell receptor diversity?
4. How does clonal selection account for the ability of each human to respond to a seemingly infinite number of non-self antigens?

32.10 The Immune System Can Malfunction

After reading this section, you should be able to:

a. Explain the differences between hypersensitivity, autoimmunity, tissue rejection, and immunodeficiency
b. Describe the immunologic basis for donated tissue rejection
c. Predict the impact immunosuppression or immunodeficiency has on the host

As in any body system, disorders also occur in the immune system. Immune disorders can be categorized as hypersensitivities, autoimmune diseases, transplantation (tissue) rejection, and immunodeficiencies. Each of these is now discussed.

Hypersensitivities

Hypersensitivity is an exaggerated adaptive immune response that occurs in an individual on a second or subsequent contact with an antigen. Although we typically think of a hypersensitivity as an allergy like hay fever, hypersensitivity reactions differ by the speed at which they occur and the types of antibodies or immune cells responsible. Realizing this fact in 1963, Peter Gell (1914–2001) and Robert Coombs (1921–2006) developed a classification system for reactions responsible for hypersensitivities. Their system correlates clinical symptoms with the underlying immunologic events that occur during hypersensitivity reactions. The **Gell-Coombs classification** system divides hypersensitivities into four types: I, II, III, and IV.

Type I Hypersensitivity

An **allergy** (Greek *allos,* other; *ergon,* work) is one of the most common kinds of **type I hypersensitivity** reaction.

TECHNIQUES & APPLICATIONS

32.2 Monoclonal Antibody Therapy

The value of antibodies as tools for locating or identifying antigens is well established. For many years, antiserum extracted from human or animal blood was the main source of antibodies for tests and therapy, but most antiserum is problematic. It contains polyclonal antibodies, meaning it is a mixture of different antigen-specific antibodies, reflecting thousands of immune reactions experienced by the animal. An important advance in research and medicine has been the use of cell culture technology to produce a pure preparation of monoclonal antibodies (mAbs) that originate from a single clone and therefore have specificity for the same epitope. Monoclonal antibodies are used therapeutically when they have been modified to target a protein that contributes to disease (see chapter opening story). Thus mAbs against specific antigens on cancer cells, infectious agents, and disease-associated proteins can facilitate their eradication without affecting other proteins, cells, or tissues.

The technology for producing monoclonal antibodies involves hybridizing human cancer cells and activated rabbit or mouse B cells in tissue culture systems. Mouse-human chimeric mAbs (mouse variable chains on human constant chains) and humanized mAbs (mouse hypervariable amino acids in human antibodies) are available. In addition, transgenic mice containing human antibody genes can be immunized with specific antigens to produce completely human mAbs. Additionally, human mAbs can be made in *Escherichia coli* from bacteriophage containing human B-cell DNA.

It is estimated that there are over 500 mAbs presently licensed or in development for the specific diagnosis or treatment of human disease (**box figure**). The mAbs target various types of cancer, autoimmune diseases, infection, sepsis, rheumatoid arthritis, Crohn's disease, and Alzheimer's disease, to name a few. Interestingly, immunologists have learned much about B-cell biology from the study of mAbs, including the function of antibody "orphan pieces" (single chain, single domain, Fab, $F(ab')_2$), previously thought to be nonfunctional products of proteolysis.

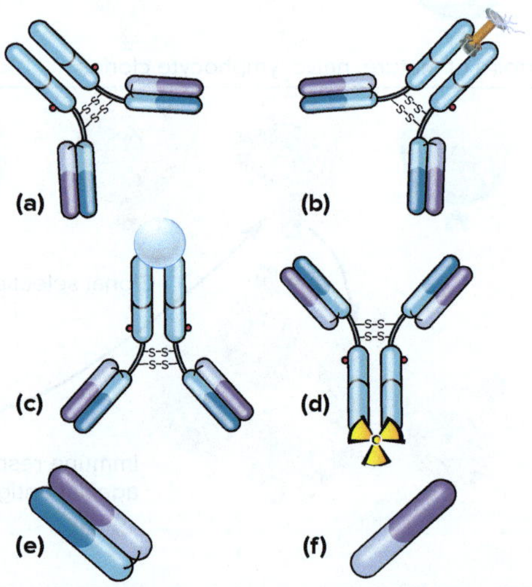

Monoclonal Antibody (mAb) Therapeutics. mAbs that retain their antigen-binding specificity can be used (a) as the naked glycoprotein to stimulate antigen recognition and removal, or they can be altered to deliver (b) cytokines, (c) drugs, or (d) radionucleotides to kill their targets. Additionally, antibody pieces such as (e) the single domain and (f) the single-chain fragments can also be used.

Although IgE is important in the defense against helminths, it also mediates allergic responses. Allergic reactions occur when an individual who has produced IgE antibody in response to the initial exposure to an antigen (in this case called an **allergen**) subsequently encounters the same allergen. Upon initial exposure to a soluble allergen, naïve B cells bind and process the allergen and are stimulated with the help of T_H2 cells to differentiate into plasma cells that produce IgE antibodies against the allergen (**figure 32.19**). The initial exposure to an allergen is called sensitization, and there is a hereditary predisposition for an individual to be sensitized by a given allergen. Once synthesized, IgE binds irreversibly to the Fc receptors of mast cells, and to a lesser extent, basophils and eosinophils. When a subsequent exposure to the allergen occurs, the allergen attaches to the surface-bound IgE on the sensitized cells. This triggers receptor cross-linking that stimulates cell degranulation.

Degranulation of mast cells releases physiological mediators such as histamine, leukotrienes, heparin, prostaglandins, PAF (*p*latelet-*a*ctivating *f*actor), ECF-A (*e*osinophil *c*hemotactic *f*actor of *a*naphylaxis), and proteolytic enzymes. These mediators trigger smooth muscle contractions, vasodilation, increased vascular permeability, and mucus secretion (figure 32.19). The inclusive term for these responses is **anaphylaxis** (Greek *ana,* up, back again; *phylaxis,* protection). Anaphylaxis can be divided into systemic and localized reactions.

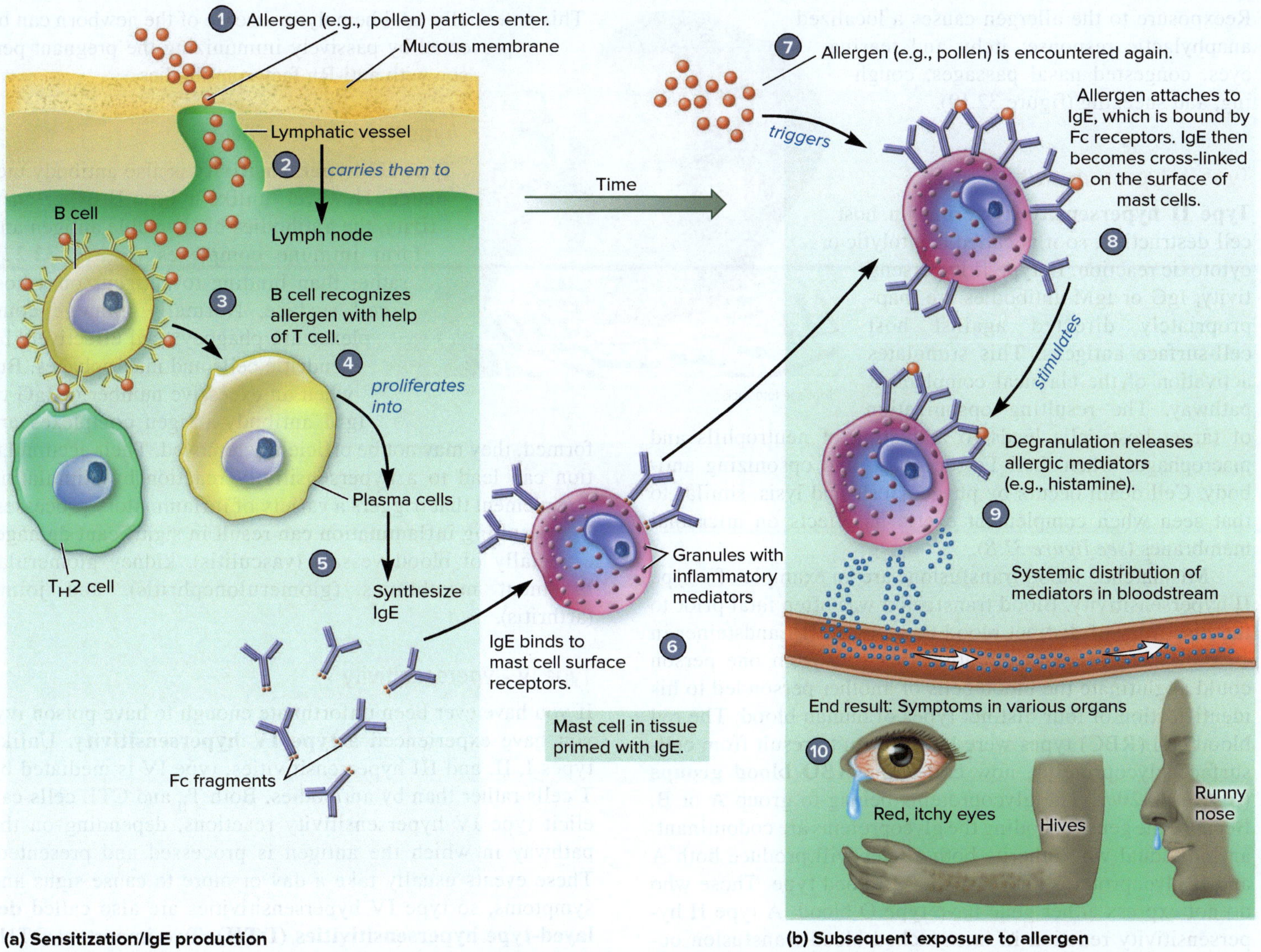

Figure 32.19 Type I Hypersensitivity (Allergic Response). (a) The initial contact (sensitization) of lymphocytes by small-protein allergens at mucous membranes results in T_H2 cell–assisted antibody class switching; plasma cells secrete IgE antibody. The IgE binds to its receptor on tissue mast cells (1–6). (b) Subsequent exposure to the same allergens results in their capture by the cell-bound IgE (7), triggering mast cell degranulation (8, 9). Characteristic signs and symptoms (hives, swelling, itching, etc.) of allergy ensue (10).

MICRO INQUIRY *What are some of the allergic mediators released during mast cell degranulation?*

Systemic anaphylaxis is a medical emergency caused by a sudden burst of mast cell mediators. It is characterized by respiratory distress caused by smooth muscle constriction in the bronchioles. The arterioles dilate, which greatly reduces blood pressure and increases capillary permeability with rapid loss of fluid into the tissue spaces. Systemic anaphylaxis can be rapidly fatal due to asphyxiation, reduced blood flow, low blood pressure, and circulatory shock. Common examples of allergens that can produce systemic anaphylaxis include drugs (penicillin), peanuts, and insect venom from wasps, hornets, or bees.

The more familiar allergic response is **localized anaphylaxis** called an **atopic** (Greek *a,* without; *topos,* place or "out of place") **reaction.** The symptoms depend primarily on the route by which the allergen enters the body. Allergens that enter through the skin stimulate a localized reaction that results in a raised (wheal) and reddened (flare) lesion. Hay fever (allergic rhinitis) is a good example of atopy involving the upper respiratory tract. Initial exposure involves airborne allergens—such as plant pollen, fungal spores, animal dander, and house dust mites—that sensitize mast cells located within the mucous membranes of the respiratory tract.

Reexposure to the allergen causes a localized anaphylactic response: itchy and tearing eyes, congested nasal passages, coughing, and sneezing (figure 32.19).

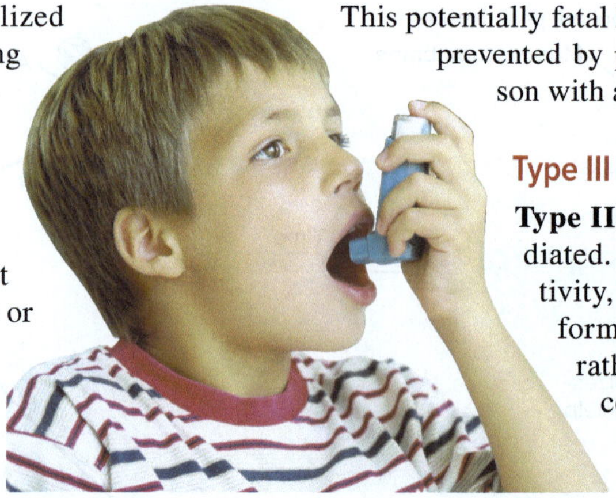
Pixtal/age fotostock

Type II Hypersensitivity

Type II hypersensitivity results in host cell destruction, so it is called a cytolytic or cytotoxic reaction. In type II hypersensitivity, IgG or IgM antibodies are inappropriately directed against host cell-surface antigens. This stimulates activation of the classical complement pathway. The resulting opsonization of target host cells leads to activation of neutrophils and macrophages when their Fc receptors bind opsonizing antibody. Cell death occurs by phagocytosis and lysis, similar to that seen when complement exerts its effects on microbial membranes (*see figure 31.8*).

Mismatched blood transfusions are an example of a type II hypersensitivity. Blood transfusion was often fatal prior to the discovery of distinct blood types by Karl Landsteiner in 1904. Landsteiner's observation that sera from one person could agglutinate the blood cells of another person led to his identification of four distinct types of human blood. The red blood cell (RBC) types were later shown to result from cell-surface glycoproteins, now called the **ABO blood groups** (**figure 32.20*a***). The glycoproteins belong to group A or B. Because the genes encoding the glycoproteins are codominant, an individual who inherits both alleles will produce both A and B glycoproteins and have an AB blood type. Those who do not express either gene have type O blood. A type II hypersensitivity reaction in mismatched blood transfusion occurs as complement is activated by cross-linking antibodies to either the A or B glycoprotein on the surface of RBCs (figure 32.20*b*).

Blood typing can be accomplished by a slightly more sophisticated in vitro method of Landsteiner's process whereby blood from one host is mixed with antibodies specific for type A or type B blood. Agglutination of red cells by antibody (figure 32.20*c*) is used as a diagnostic tool to determine blood type.

After the ABO blood types were discovered, another RBC antigen was discovered as a result of experiments with rhesus monkeys. The so-called rhesus (Rh) factor (or D antigen) is determined by the expression of two alleles, one dominant (coding for the factor) and one recessive (not coding for the factor). Thus an individual who is homozygous or heterozygous for the dominant Rh allele has the antigen and is indicated as Rh^+. Expression of two recessive alleles results in the designation of Rh^- (no Rh factor). Incompatibility between an Rh^- pregnant person and their Rh^+ fetus can result in a type II hypersensitivity in which maternal anti-Rh antibodies destroy fetal blood cells (**figure 32.21**). This is called **erythroblastosis fetalis.** This potentially fatal hemolytic disease of the newborn can be prevented by passively immunizing the pregnant person with anti-Rh factor antibodies.

Type III Hypersensitivity

Type III hypersensitivity is also antibody mediated. However, unlike a Type II hypersensitivity, the antibodies bind soluble antigen and form immune complexes (**figure 32.22**), rather than binding to antigens on a host cell surface. Normally immune complexes are phagocytosed effectively by dendritic cells and macrophages. But when an excessive number of IgG or IgM antibody-antigen complexes are formed, they may not be efficiently removed. Their accumulation can lead to a hypersensitivity reaction by stimulating complement that triggers a variety of inflammatory processes. The resulting inflammation can result in significant damage, especially of blood vessels (vasculitis), kidney glomerular basement membranes (glomerulonephritis), and joints (arthritis).

Type IV Hypersensitivity

If you have ever been unfortunate enough to have poison ivy, you have experienced a **type IV hypersensitivity.** Unlike types I, II, and III hypersensitivities, type IV is mediated by T cells rather than by antibodies. Both T_H and CTL cells can elicit type IV hypersensitivity reactions, depending on the pathway in which the antigen is processed and presented. These events usually take a day or more to cause signs and symptoms, so type IV hypersensitivities are also called **delayed-type hypersensitivities (DTH).** Two important DTHs are tuberculin hypersensitivity and allergic contact dermatitis. Other examples of type IV hypersensitivity reactions include cell and tissue destruction caused by leprosy, tuberculosis, and leishmaniasis. These infectious diseases are discussed in chapters 38 and 39.

Type IV hypersensitivity reactions occur when antigens, especially those binding to serum proteins or tissue cells, are processed and presented to T cells. If the antigen is phagocytosed, it will be presented to T_H cells by class II MHC molecules on the antigen-presenting cell. This activates T_H1 cells, causing proliferation and pro-inflammatory cytokine secretion. If the antigen is lipid soluble, it can cross the cell membrane to be processed within the cytosol and be presented to CTL cells by class I MHC molecules. CTLs are thus stimulated to kill cells presenting the antigen. T_H cell and CTL signals also stimulate the expression of adhesion molecules on local blood vessel endothelium to increase vascular permeability. This enables fluid and cells to enter the tissue space. The cytokines attract and recruit lymphocytes, macrophages, and basophils, exacerbating the inflammatory response. Extensive tissue damage may result.

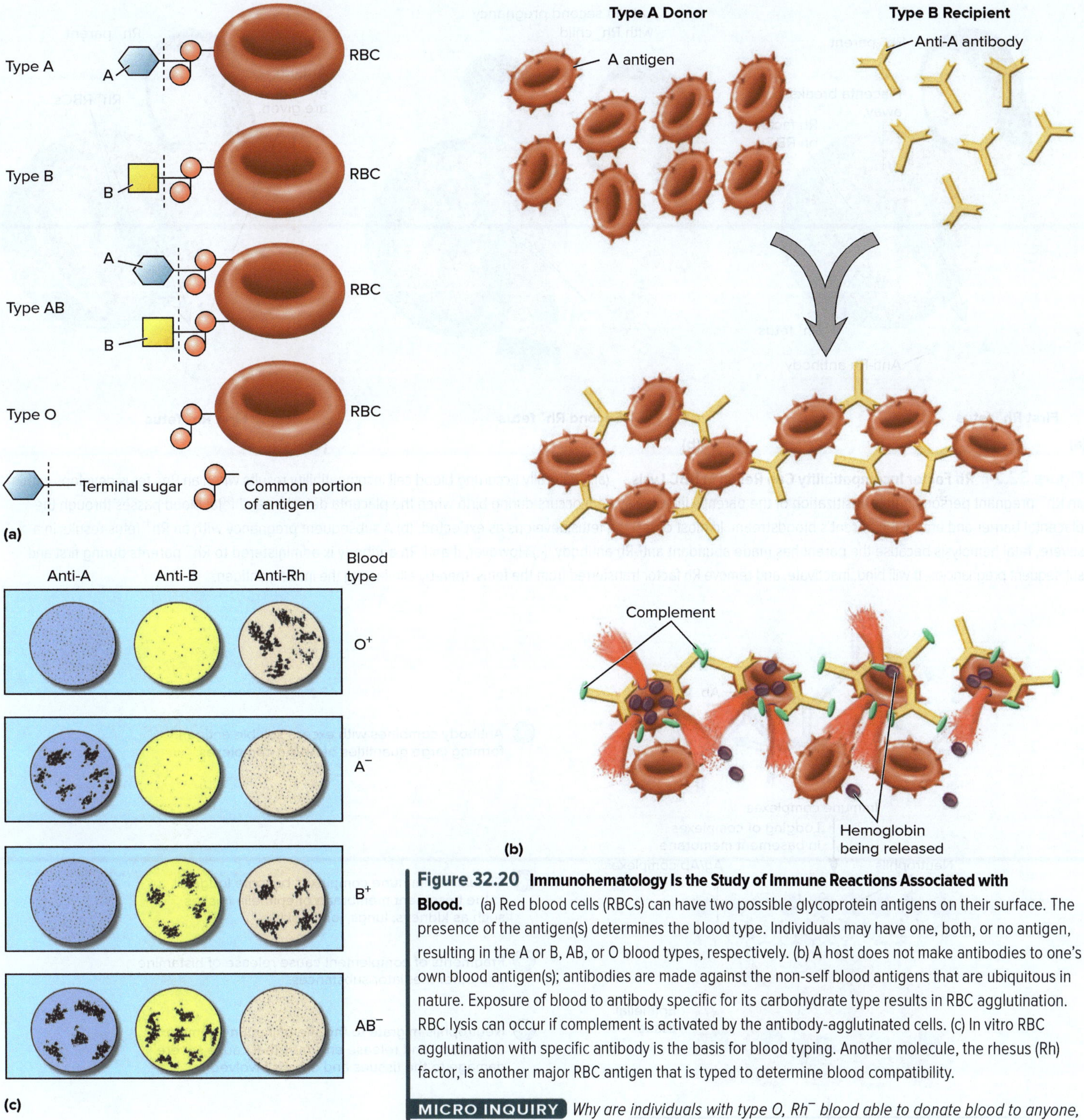

Figure 32.20 Immunohematology Is the Study of Immune Reactions Associated with Blood. (a) Red blood cells (RBCs) can have two possible glycoprotein antigens on their surface. The presence of the antigen(s) determines the blood type. Individuals may have one, both, or no antigen, resulting in the A or B, AB, or O blood types, respectively. (b) A host does not make antibodies to one's own blood antigen(s); antibodies are made against the non-self blood antigens that are ubiquitous in nature. Exposure of blood to antibody specific for its carbohydrate type results in RBC agglutination. RBC lysis can occur if complement is activated by the antibody-agglutinated cells. (c) In vitro RBC agglutination with specific antibody is the basis for blood typing. Another molecule, the rhesus (Rh) factor, is another major RBC antigen that is typed to determine blood compatibility.

MICRO INQUIRY *Why are individuals with type O, Rh^- blood able to donate blood to anyone, regardless of the recipient's blood type?*

In tuberculin hypersensitivity, partially purified proteins collectively called tuberculin are obtained from the bacterium that causes tuberculosis (TB). The proteins are injected into the skin of the forearm. If by 48 to 72 hours the area becomes reddened, firm, and hard (indurated), the test is considered positive. T_H1 cells previously activated by exposure to the tuberculosis bacterium migrate to the injection site and are responsible for the induration. If a person has never been exposed to the bacteria

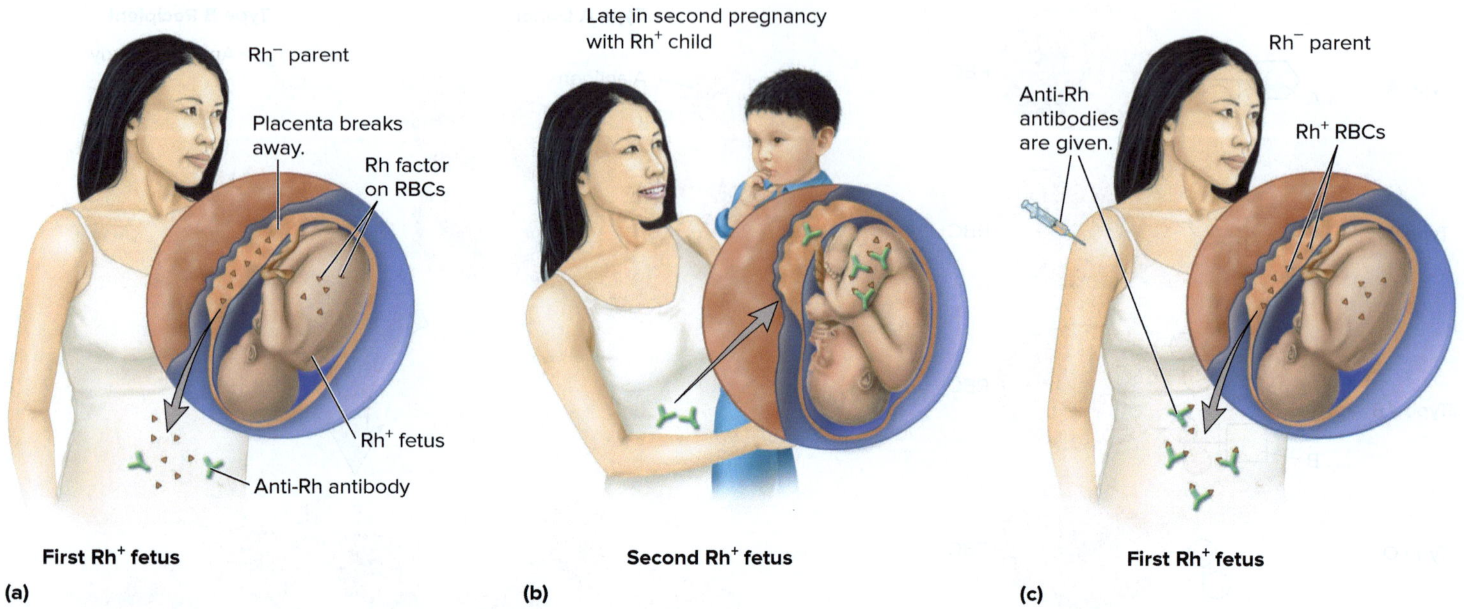

Figure 32.21 Rh Factor Incompatibility Can Result in RBC Lysis. (a) A naturally occurring blood cell incompatibility results when an Rh^+ fetus develops within an Rh^- pregnant person. Initial sensitization of the parent's immune system occurs during birth when the placenta detaches and fetal blood passes through the placental barrier and enters the parent's bloodstream. In most cases, the fetus develops as expected. (b) A subsequent pregnancy with an Rh^+ fetus results in a severe, fetal hemolysis because the parent has made abundant anti-Rh antibody. (c) However, if anti-Rh antibody is administered to Rh^- parents during first and subsequent pregnancies, it will bind, inactivate, and remove Rh factor transferred from the fetus, thereby eliminating the inciting antigen.

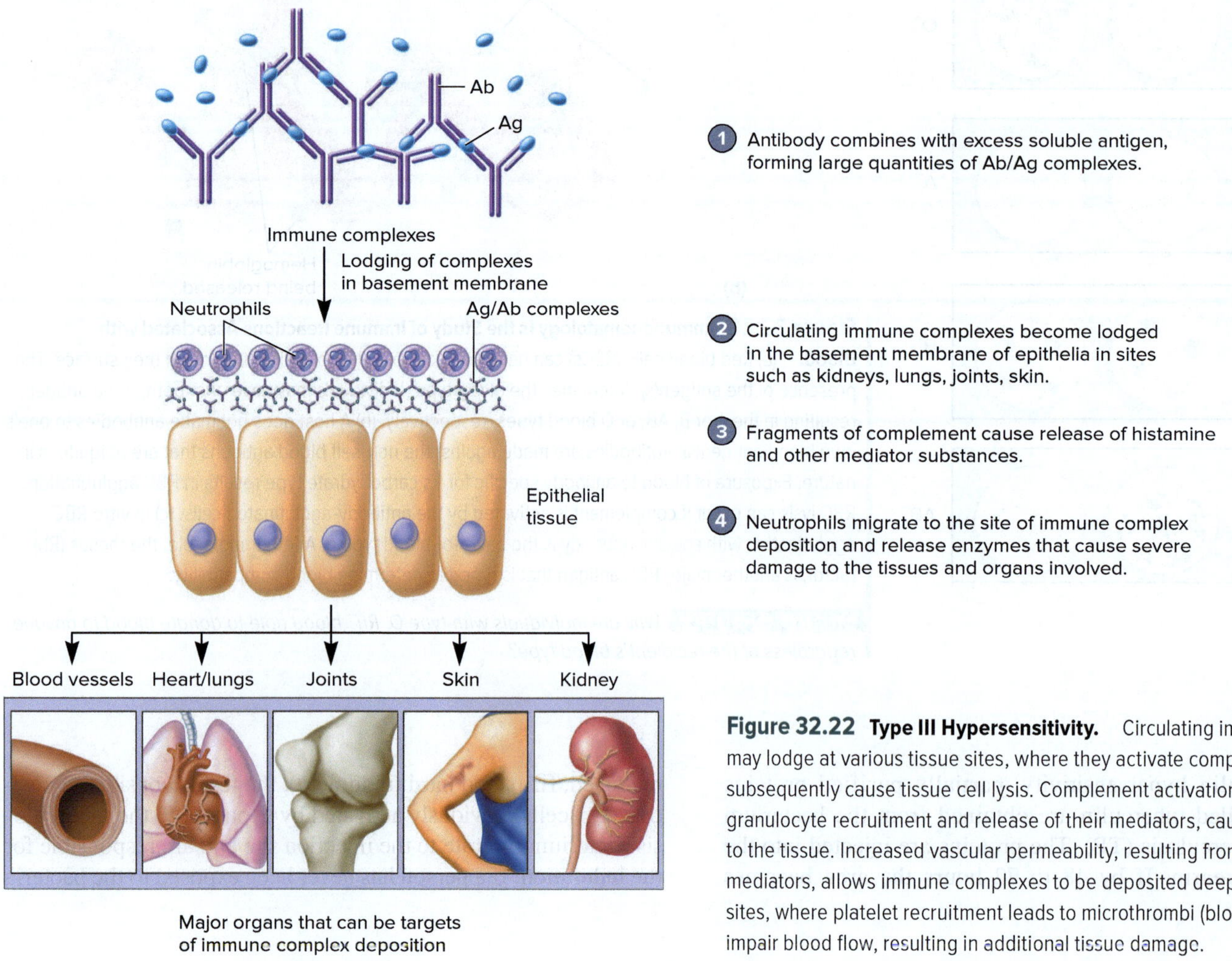

Figure 32.22 Type III Hypersensitivity. Circulating immune complexes may lodge at various tissue sites, where they activate complement and subsequently cause tissue cell lysis. Complement activation also results in granulocyte recruitment and release of their mediators, causing further injury to the tissue. Increased vascular permeability, resulting from granulocyte mediators, allows immune complexes to be deposited deeper into tissue sites, where platelet recruitment leads to microthrombi (blood clots), which impair blood flow, resulting in additional tissue damage.

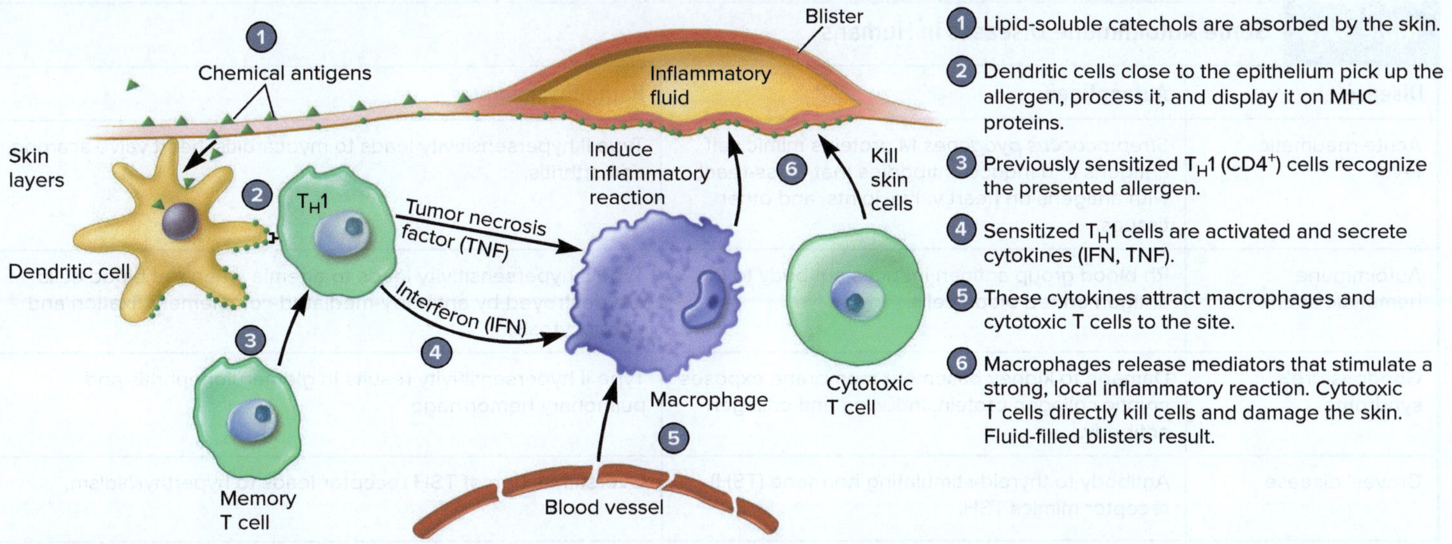

Figure 32.23 Contact Dermatitis. In contact dermatitis to poison ivy, a person initially becomes exposed to the allergen, predominantly 3-n-pentadecyl-catechol found in the resinous sap produced by the leaves, fruit, stems, and bark of the poison ivy plant. The catechol molecules, acting as haptens, combine with high-molecular-weight skin proteins. After 7 to 10 days, sensitized T cells are produced and give rise to memory T cells. Upon second contact, the catechols bind to the same skin proteins, and the memory T cells become activated in only 1 to 2 days, leading to an inflammatory reaction (contact dermatitis).

that cause TB, the injection site may initially become inflamed but does not enlarge and become indurated.

Allergic contact dermatitis is a type IV reaction caused by haptens that combine with carrier proteins in the skin to form the allergen that elicits the immune response (**figure 32.23**). Examples of these haptens include cosmetics, plant materials (catechol molecules from poison ivy and poison oak), topical chemotherapeutic agents, metals, and jewelry (especially jewelry containing nickel).

Comprehension Check

1. Discuss the mechanism of type I hypersensitivity reactions and how these can lead to systemic and localized anaphylaxis.
2. Why are type II hypersensitivity reactions called cytolytic or cytotoxic?
3. What characterizes a type III hypersensitivity reaction? Give an example.
4. Both the TB skin test and a poison ivy rash (figure 32.23) are examples of type IV hypersensitivities. However, only poison ivy causes significant host cell death. Why is this the case?

Autoimmunity and Autoimmune Diseases

The importance of the immune system's ability to distinguish between self and non-self antigens is dramatically illustrated by autoimmune diseases. In these conditions, immune cells fail to recognize certain host antigens as self. This so-called loss of tolerance can be driven by B cells, T cells, or both arms of adaptive immunity. **Autoimmune disease** thus results from the activation of self-reactive T and B cells. Typically, this follows stimulation by genetic or environmental triggers and causes chronic tissue damage. Examples of some autoimmune diseases are listed in **table 32.2.**

There is no single etiology for autoimmune disease. Rather, each disease has its own set of unique and complex immune pathologies (table 32.2). Some autoimmune diseases are largely T-cell driven, others involve autoantibodies, and some involve both mechanisms. In many cases, an infectious trigger, such as a viral or bacterial infection, may be involved. Evidence suggests that a change in the normal host-microbiota relationship (dysbiosis) contributes to immune dysfunction (*see figures 33.6 and 33.7*).

Transplantation (Tissue) Rejection

Transplants between genetically different individuals within a species are termed **allografts** (Greek *allos,* other). Some transplanted tissues do not stimulate an immune response. For example, a transplanted cornea is rarely rejected because lymphocytes do not circulate into the anterior chamber of the eye. This site is considered an *immunologically privileged* site. Another example of a privileged tissue is the heart valve, which can be transplanted from a pig to a human without stimulating an immune response. Transplanted tissue between different species is termed a **xenograft** (Greek *xenos,* strayed).

Most transplants are allographs and involve tissues that are not immunologically privileged. Unless all MHC proteins are identical between organ donor and recipient, as in identical twins, the recipient's cells will recognize the donor's tissues as foreign. This triggers the recipient's innate and adaptive immune responses, which may destroy the donor tissue. Such a response is called **host-versus-graft disease,** or organ rejection. Organ rejection can occur by two different mechanisms. First, foreign MHC molecules on transplanted tissue, or the graft, are recognized as non-self, which results in activation of cytotoxic T cells (**figure 32.24**). Cytotoxic T cells recognize the graft because it bears foreign MHC I molecules. This

Table 32.2 Some Autoimmune Diseases in Humans

Disease	Autoantigen	Pathophysiology
Acute rheumatic fever	*Streptococcus pyogenes* M proteins mimic self antigens and induce antibodies that cross-react with antigens on heart valve, joints, and other tissues.	Type II hypersensitivity leads to myocarditis, heart valve scarring, and arthritis.
Autoimmune hemolytic anemia	Rh blood group antigen induces antibody to Rh antigen on red blood cells.	Type II hypersensitivity leads to anemia when red blood cells are destroyed by antibody-mediated complement fixation and phagocytosis.
Goodpasture's syndrome	Damage to kidney basement membrane exposes cryptic collagen protein, inducing anti-collagen antibody.	Type II hypersensitivity results in glomerulonephritis and pulmonary hemorrhage.
Graves' disease	Antibody to thyroid-stimulating hormone (TSH) receptor mimics TSH.	Overstimulation of TSH receptor leads to hyperthyroidism.
Multiple sclerosis	Antibody and activated T cells to several nervous system antigens	Types II and IV hypersensitivities alter nerve communications, leading to numbness, weakness, spasm, and loss of motor and cognitive function.
Myasthenia gravis	Antibody to acetylcholine receptor in skeletal muscle	Antibody blockade of neurotransmitter receptors results in progressive muscular weakness.
Rheumatoid arthritis	Autoantibodies (IgM, IgG) form against joint tissue; pro-inflammatory cytokines released by T_H1 cells also drive joint destruction.	Type III hypersensitivity from immune complex formation results in joint inflammation and destruction.
Systemic lupus erythematosus	Antibodies to various cellular (DNA, nucleoprotein, cardiolipin) and blood-clotting components	Type III hypersensitivity results in immune complex–induced arthritis, glomerulonephritis, vasculitis, and rash.
Type I diabetes mellitus	Antibody and activated T cells to pancreatic beta cell antigens	Types II and IV hypersensitivities destroy beta cells, resulting in insulin deficiency.

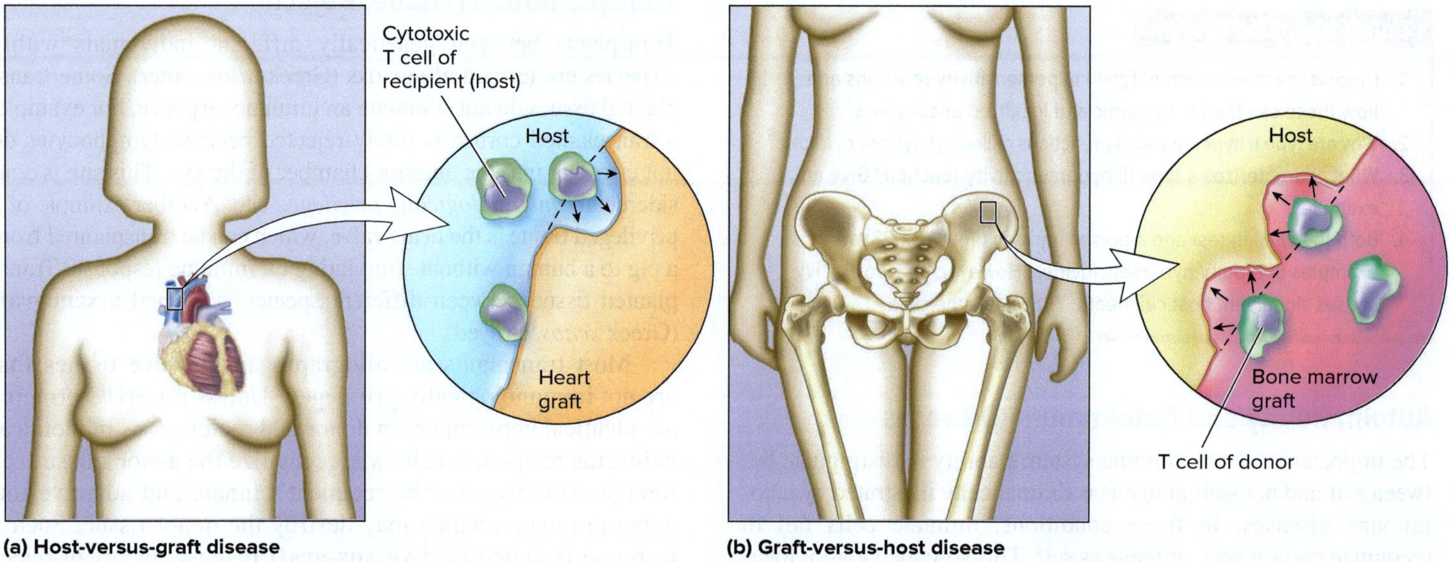

Figure 32.24 Potential Transplantation Reactions. (a) Donated tissues not from an identical twin contain cellular MHC proteins that are recognized as foreign by the recipient host (host-versus-graft disease). The tissue is then attacked by host CTLs, resulting in its damage and rejection. (b) Donated stem cells may contain immune cells that react against host antigens. Recognition of a foreign host by donor CTLs results in graft-versus-host disease.

Table 32.3 Some Congenital Immune Deficiencies in Humans

Condition	Symptoms	Cause
Chronic granulomatous disease	Defective monocytes and neutrophils leading to recurrent bacterial and fungal infections	Failure to produce reactive oxygen intermediates due to defective NADPH oxidase
X-linked agammaglobulinemia	B-cell deficiency and inability to produce adequate antibodies	Defective B-cell differentiation due to loss of tyrosine kinase
DiGeorge syndrome	T-cell deficiency and very poor cell-mediated immunity	Lack of thymus or a poorly developed thymus
Severe combined immunodeficiency disease (SCID)	Both antibody production and cell-mediated immunity impaired due to a great reduction of B- and T-cell levels	Various mechanisms (e.g., defective B- and T-cell maturation because of X-linked gene mutation; absence of RAG enzymes in lymphocytes)

response is much like the activation of CTLs by virus-infected host cells and results in destruction of the donated tissue. A second mechanism involves T-helper cells reacting to the graft and releasing cytokines. The cytokines stimulate macrophages to enter, accumulate within the graft, and destroy it. MHC molecules therefore play a dominant role in tissue rejection reactions because of their unique role in enabling T cells to discriminate self from non-self.

Because class I MHC molecules are central to the development of host-versus-graft disease, it follows that the greater the antigenic difference between class I molecules of the recipient and donor tissues, the more rapid and severe the rejection reaction is likely to be. Class II MHC molecule mismatch can also result in rejection. Thus the goal in matching donor to recipient is to have as many MHC alleles in common as possible. Most recipients are not 100% matched to their donors, so immunosuppressing drugs are used to prevent host-mediated rejection of the graft.

Bone marrow (i.e., stem cell) transplant recipients may develop **graft-versus-host disease,** which is distinct from host-versus-graft seen in organ rejection. Graft-versus-host disease occurs when the transplanted bone marrow contains immune cells that recognize host MHC molecules as foreign and attack the host's cells. Because bone marrow transplant involves the introduction of new immune cells, it requires more rigorous matching between donor and recipient. Some level of mismatch is typical, however, and is controlled by immunosuppressive drugs.

Immunodeficiencies

Defects in one or more components of the immune system can result in its failing to recognize and respond properly to antigens. Such **immunodeficiencies** can make a person more prone to infection than people capable of a complete and active immune response. Many genetic errors associated with immunodeficiency disorders are located on the X chromosome and produce primary or congenital immunodeficiencies (**table 32.3**). Other immunodeficiencies can be acquired through infections by immunosuppressive microorganisms, such as HIV.

Comprehension Check

1. What is an autoimmune disease and how might it develop?
2. What is an immunologically privileged site and how is it related to transplantation success?
3. Why are most patients given immunosuppressive drugs prior to receiving an organ transplant, as well as afterward?
4. Which of the immunodeficiencies listed table 32.3. involve innate immunity? Which are adaptive immunity deficiencies?

Key Concepts

32.1 Adaptive Immunity Relies on Recognition and Memory

- The adaptive immune response system consists of lymphocytes that recognize and respond to specific foreign molecules (antigens) and remember them. Two branches or arms of adaptive immunity are recognized: antibody-mediated immunity and cell-mediated immunity (**figure 32.1**).

32.2 Antigens Elicit Immunity

- An antigen is a substance that stimulates an immune response. Each antigen can have several antigenic determinant sites called epitopes that stimulate production of and combine with specific antibodies (**figure 32.2**).
- Haptens are small molecules that are not antigenic by themselves but can become antigenic if bound to a larger carrier molecule.

32.3 Adaptive Immunity Can Be Earned or Borrowed

- Immunity can be acquired by natural means—actively through infection or passively through receipt of preformed antibodies, as through breast milk (**figure 32.3**).
- Immunity can be acquired by artificial means—actively through immunization or passively through receipt of preformed antibodies, as with antisera.

32.4 Recognition of Foreignness Is Critical for a Strong Defense

- MHC molecules are cell-surface proteins encoded by genes termed the major histocompatibility complex. The human MHC gene products are called the human leukocyte antigens (HLAs) (**figure 32.4**).
- Class I MHC proteins are found on all nucleated cells of mammals. Class II MHC proteins are expressed on antigen-presenting cells (DCs, macrophages, and B cells).
- Class I MHC proteins collect intracellular self and non-self peptides processed by the proteasome and present them to cytotoxic T cells ($CD8^+$), which respond only to non-self peptides (**figure 32.5**).
- Class II MHC proteins collect extracellular foreign peptides processed by the phagosome (macrophages and dendritic cells) or endosome (B cells) and present them to T helper cells ($CD4^+$).

32.5 T Cells Are Critical for Immune Function

- T-cell development in the thymus involves positive selection for the CD4 or CD8 coreceptor and negative selection against T cells that recognize self antigens (**figure 32.6**).
- Antigen-presenting cells take in foreign antigens or pathogens, process them, and present antigenic peptides complexed with MHC class II molecules to T-helper cells ($CD4^+$ cells). Antigens presented in MHC class I are presented to $CD8^+$ T cells.
- When a DC presents antigen to a naïve T cell, an immune synapse is formed. At this time, three signals are required to activate the T cell (**figures 32.7** and **32.8**).
- Subsets of T-helper cells include T_H1, T_H2, T_H17, and Treg, which arise from the undifferentiated T_H0.
- T-helper cells control the development of other cells, including macrophages, neutrophils, B cells, and cytotoxic T cells ($CD8^+$) (**figure 32.9**).
- Regulatory T (Treg) cells arise from both $CD4^+$ and $CD8^+$ T cells and downregulate specific immune reactions.
- Cytotoxic T lymphocytes recognize target host cells such as virus-infected cells with foreign antigens bound to the class I MHC molecules on their surface. The CTLs then attack and destroy the target cells using the perforin pathway, the Fas-FasL apoptotic pathway, or both.

32.6 B Cells Make Antibodies

- B cells have receptor immunoglobulins on their plasma membrane surface specific for three-dimensional epitopes on an antigen (**figure 32.10**).
- B cells activated in lymphoid tissue differentiate into plasma cells that secrete antibodies providing antibody-mediated immunity.
- B-cell differentiation is usually T-cell dependent, which yields plasma cells and memory cells. T-cell-independent B-cell activation results in the formation of only plasma cells that produce low-affinity IgM antibodies (**figure 32.11**).

32.7 Antibodies Bind Specific 3-D Antigens

- Antibodies (immunoglobulins) are a group of glycoproteins present in the blood, tissue fluids, and mucous membranes of vertebrates. All immunoglobulins have a basic structure composed of four polypeptide chains (two light and two heavy) connected to each other by disulfide bonds (**figure 32.12**). In humans, five immunoglobulin classes exist: IgG, IgA, IgM, IgD, and IgE (**figure 32.13** and **table 32.1**). Antibodies bind to antigen epitopes through noncovalent interactions.
- The primary antibody response occurs following initial exposure to the antigen. This response has lag, exponential, plateau, and decline phases. Upon secondary antigen challenge, the B cells mount a heightened and accelerated response (**figure 32.14**).

32.8 Antibodies Doom Antigens

- The complement system can be activated by IgG or IgM, leading to cell lysis, phagocytosis, chemotaxis, or stimulation of the inflammatory response.
- Other defensive antigen-antibody interactions include toxin neutralization, viral neutralization, adherence inhibition, opsonization, and immune complex formation (**figure 32.15**).

32.9 Generating Populations of Diverse Lymphocytes

- Antibody diversity results during B-cell development by the rearrangement and splicing of V, D, and J gene segments on the antibody-coding chromosomes, the generation of different codons during splicing, and the independent assortment of light- and heavy-chain genes. Antibody diversity further increases upon T-cell interaction, when activated B cells undergo somatic hypermutation (**figures 32.16** and **32.17**).
- T-cell receptor diversity arises in the thymus by V-D-J recombination, splice site variability, and independent assortment of the T-cell receptor α and β subunits.
- Clonal selection results in large populations of genetically distinct B and T cells, each of which recognizes a unique antigen (**figure 32.18**).

32.10 The Immune System Can Malfunction

- When the immune response occurs in an exaggerated form and results in tissue damage to the individual, the term *hypersensitivity* is applied. There are four types of hypersensitivity reactions, designated as types I through IV (**figures 32.19** through **32.23**).
- Autoimmune diseases result when self-reactive T and B cells attack the body and cause tissue damage. A variety of factors can influence the development of autoimmune disease (**table 32.2**).
- The immune system rejects tissue transplants because the transplanted tissue expresses different MHC molecules. Organ transplants can result in host-versus-graft disease, whereas bone marrow (stem cell) transplants can cause graft-versus-host disease (**figure 32.24**).
- Immunodeficiency diseases are a diverse group of conditions in which an individual's susceptibility to various infections is increased; several severe diseases can arise because of one or more defects in the adaptive or innate immune responses (**table 32.3**).

Active Learning

1. A rare immunodeficiency disease called bare lymphocyte syndrome-2 is caused by a genetic mutation that prevents the production of MHC class II molecules. Predict the changes in development and function seen in T_H, CTLs, and B cells in a patient with this condition.
2. The diagnosis of mononucleosis, which is caused by the Epstein-Barr virus (EBV), is sometimes made by looking for anti-EBV antibodies, a process called EBV serology. You receive EBV serology reports for two patients: One has a high anti-EBV IgM titer; the other has a high anti-EBV IgG titer. Are you confident that both patients currently have mononucleosis? Explain your answer. (Hint: Refer to figure 32.14.)
3. Very young children have a large pool of mature, naïve T and B cells but very few memory cells; older adults have just the opposite: very few mature, naïve T and B cells but many memory cells. Discuss how this might affect the ability of young children and older adults to respond to the same vaccine or new infectious agent.
4. Discuss reasons that might explain why three signals are required for T-cell activation but only one signal is required for activation of an APC.

CHAPTER

33

The Microbe-Human Ecosystem

Radius Images/Alamy Stock Photo

Embrace Your Gut Flora

Samantha D. was desperate. Over the course of 8 months, she had been battling diarrhea so severe she had lost 60 pounds, was confined to a wheelchair, and was wearing adult diapers. For a period of 6 months, Vanessa R. ran high fevers and had diarrhea about once an hour on her worst days. Her better days were spent sleeping. Both these women suffered from *Clostridioides* (formerly *Clostridium*) *difficile* infection (CDI). Both were eventually cured by a treatment once considered outrageous: fecal microbiota transplant (FMT).

First, the microbe: *C. difficile* (C. diff for short) is a monoderm, endospore-former (*see sections 22.2 and 38.6*) that resides in low numbers in the human gut. The activities of nonpathogenic gut bacteria normally suppress its growth. Many CDIs arise in people undergoing antibiotic therapy. This leads to the inhibition of other gut microbes, enabling the overgrowth of C. diff and its release of toxins, thereby triggering disabling diarrhea and sloughing of the gut epithelium.

Next the cure: Although antibiotic treatment triggers CDI, other antibiotics are the first line of treatment. But such therapy sometimes fails because C. diff has acquired resistance to multiple antimicrobials and antibiotic treatment continues to suppress the growth of gut microorganisms needed to outcompete this virulent pathogen. Instead, recurrent CDI can be treated with FMT. In this procedure, feces collected from healthy donors without a recent history of antibiotic use are screened for pathogens. Prior to 2023, a slurry of this mixture was infused into the patient much like an enema. Thankfully, capsules of fecal microbes are now available. About 90% of patients report a complete cure within 48 hours to a week.

What do experiences with FMT teach us? For a very long time, gut microorganisms have been termed commensals—symbiotic organisms that neither injure nor benefit their host. Clearly, the severity of CDI and its cure by FMT argue that this is a vast oversimplification. In this chapter, we explore the relationship microbes have with their human hosts. As we will see, it is more complex than anyone previously imagined.

Readiness Check:

Based on what you have learned previously, you should be able to:

- ✓ Explain how small subunit ribosomal RNA is used to identify microbes (section 1.2)
- ✓ Compare and contrast the growth and physiology of obligate aerobes, facultative anaerobes, and strict anaerobes (section 7.5)
- ✓ Relate the advances in metagenomic analysis with the development of next-generation sequencing (sections 18.1 to 18.3)
- ✓ Identify and describe the function of professional phagocytes (sections 31.4 and 31.6)
- ✓ Describe the structure and function of mucosal-associated lymphoid tissue (section 31.5)
- ✓ Discuss the process of inflammation (section 31.7)
- ✓ Compare and contrast the major functional differences between T cells and B cells (sections 32.5 through 32.7)

33.1 Humans Are Holobionts

After reading this section, you should be able to:

a. Define the word *microbiome*
b. Explain why humans are considered holobionts

How is it possible that a new human organ could be discovered in the twenty-first century? Dictionaries define the term *organ* as "a group of tissues in a living organism adapted to perform a specific function." By changing the term *tissues* to *genes,* it has been argued that the microbiome is a newly discovered organ. **Microbiome** refers to all the genes found in one's **microbiota**—all the microorganisms that live in and on an organism. Human health depends on their microbial partners—we are **holobionts:** hosts and microbes relying on each other and, importantly, evolving together. After all, the average adult carries about the same number of microbial cells as human somatic cells (an average of about 10^{14} cells). With a combined microbial contribution of over 8 million protein-encoding genes (as compared to our roughly 22,000 genes), we have come to realize that many of these gene products serve essential functions our genome cannot provide.

The acquisition of our microbiota is a selective process. In humans, each microbial niche is related to several factors such as body location, age, sex, diet, and environment. These microorganisms reside in various body sites and were first identified by 16S ribosomal RNA sequencing. Because many of these microbes (e.g., members of Clostridiales and Bacteroidales) survive only within the anoxic and nutritionally unique environments established in or on the human host, their culture requires the development of innovative culture techniques. Indeed, without next-generation nucleotide sequencing, metagenomics, and bioinformatics, we would still know very little about our microbiomes. Through the application of these techniques, the Human Microbiome Project (HMP) sponsored by the National Institutes of Health and citizen science projects like the American Gut Project and Asian Gut Project have produced thousands of metagenomes derived from multiple body sites from thousands of human volunteers. The HMP showed that each person harbors a unique collection of microorganisms. It would therefore be of little use to list all the microbial species associated with health. Instead, the healthy microbiome is defined by microbial functions determined by expression of key genes distributed across many taxa. In this chapter, we begin our exploration of the **human microbiome** with the story of human colonization by our microbial partners, and then we turn our attention to their impact on health and disease. ◀ *DNA sequencing by synthesis (section 18.1); Next-generation genomic sequencing (section 18.2); Metagenomics provides access to uncultured microbes (section 18.3)*

Comprehension Check

1. Differentiate between the terms *microbiome* and *microbiota.*
2. How have advances in molecular biology techniques contributed to our definition of the human microbiome?

33.2 The Microbiome Develops from Birth to Adulthood

After reading this section, you should be able to:

a. Discuss the role of bifidobacteria in the infant microbiome
b. Predict environmental host conditions that favor the association of acidophilic, anaerobic, lactic acid fermenting, and halophilic bacteria
c. Discuss factors that help mediate microbial diversity in and on the human body

Development of a Stable Microbiome

The community of microbiota we host is not static. It begins developing at birth and changes as we age. It is perhaps not surprising that during infancy our microbiota are most sensitive to the environment. This early time in our lives also appears to be a window of opportunity for the development of the kinds of microbes that will be most beneficial for the maturation of our immune systems. By the time we are 3 years old, we have adopted a more stable community of microbes (**figure 33.1**). As we further explore, it is important to develop a microbiome rich in diversity.

Newborns are initially colonized by microorganisms from the immediate environment. For example, babies born vaginally acquire most of their microorganisms from their mother's birth canal, whereas babies born by cesarean delivery acquire microorganisms from the skin of their initial caretakers (nurses, doctors, midwives, parents, etc.). The importance of newborn colonization is now understood to be so critical that investigations seeking to determine the potential benefit of swabbing the mouths and skin of newborns with maternal vaginal secretions after cesarean delivery are underway.

Early in life, human milk acts as a selective medium for nonpathogenic bacteria. Human milk oligosaccharides include complex carbohydrates that select for the growth of intestinal bifidobacteria. Bifidobacteria transport some of the polymeric sugars found in human breast milk directly across their plasma membranes; the transport of such large molecules is a remarkable feat for a bacterium. This helps explain bifidobacterial dominance in a breastfed (and to a lesser extent, bottle-fed) infant's gut. Fermentation of these sugars produces acetate and lactate, which provide calories for the growing baby and lower the gut pH, helping to limit the growth of certain pathogens. Switching to cow's milk or solid food (mostly polysaccharide) appears to result in the loss of bifidobacteria predominance and enterobacteria, enterococci, lactobacilli, clostridia, and *Bacteroides* spp. increase and outcompete bifidobacteria.

An infant's microbial community changes with age until a stable resident microbiota becomes established (**table 33.1**). Each individual cultivates their own microbial communities that are heavily influenced by genetics, developmental changes (e.g., puberty, menopause), diet, personal hygiene, anatomical site, and life events (antimicrobial use, travel, occupation, sexual partners, etc.).

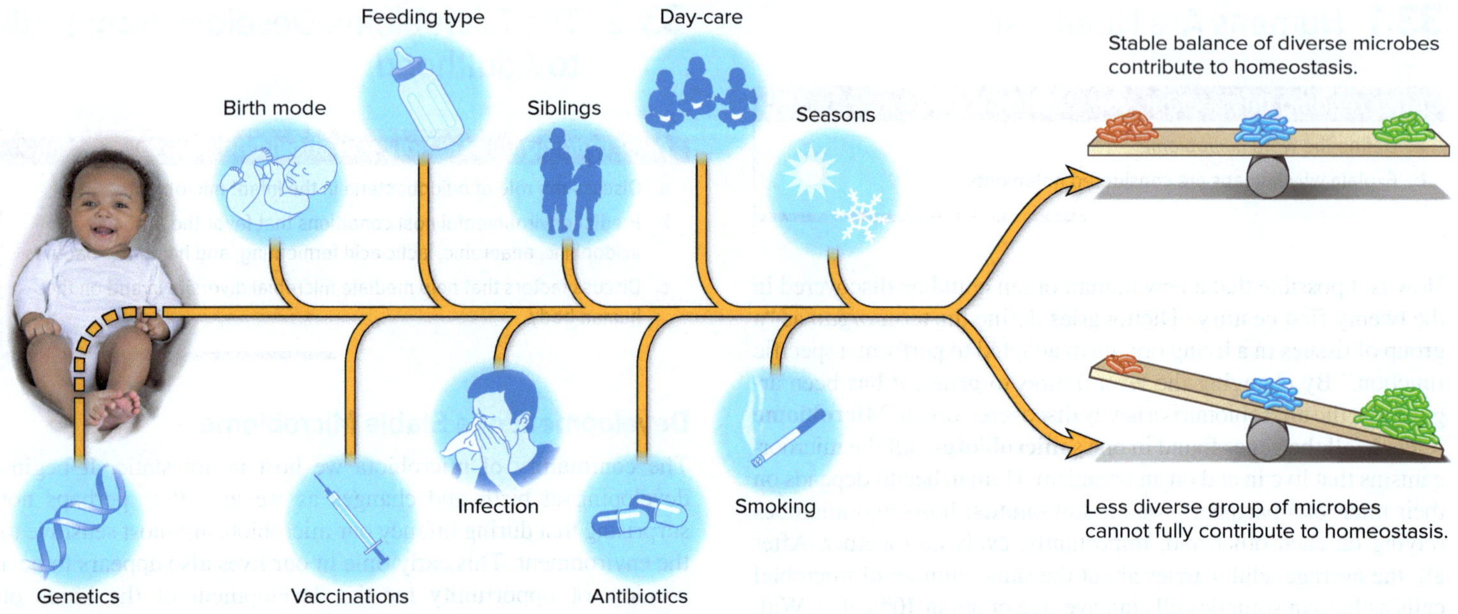

Figure 33.1 Many Factors Contribute to the Development of a Stable Microbiome. A diverse microbial community is linked to better health. Some factors have a clear negative influence (antibiotics) or positive impact (breast milk), but it is the complex interaction of many factors that determines microbial diversity.
(photo) Michel74100/iStock/Getty Images (Text) Source: Man, Wing Ho, Piters, Wouter A. A. de Steenhuijsen and Bogaert, Debby. "The microbiota of the respiratory tract: gatekeeper to respiratory health," *Nature Reviews*, March 20, 2017.

Interestingly, once established, the adult microbiota usually undergoes only transient changes in diversity. These changes are the result of substantial host physical or lifestyle changes (e.g., puberty, adopting a long-term vegetarian diet). For example, acute infection and brief antimicrobial chemotherapy result in short-term microbial population shifts but typically do not alter the long-term composition of the microbiota. However, long-term shifts in microbiota can occur with chronic infection and sustained use of antimicrobial agents.

Table 33.1 Innate and Environmental Factors that Impact Microbial Diversity and Human Immunity

Relative Age	Innate Factors[1]	Environmental Factors	Microbial Diversity[2]
Birth	Genetics	Multiple caregivers, breast milk/formula, immunizations	Mother's vaginal and intestinal flora, transient species from caregivers
Infant	Developing immunity	Solid foods, crawling, mouth contact, first infections, first antibiotics	40–100 species, 7–8 phyla, transient fungi and virus
Child	Body growth	Changing diet, socialization/contact with peers, play, adventure	40–100 species, 7–8 phyla, rare fungi, rare virus
Adolescent	Puberty, sexual activity	Travel, sports, fitness	Species change, 7–8 phyla, increasing fungi, increasing virus
Adult	Weight gain, pregnancy, disease history, stress	Stable cohabitation, job changes, shift work, relocations, children, medications	Stable population, ~500–1,000 species, 7–8 phyla, 1–10% fungi, 10–40% virus
Senior	Muscle atrophy, declining immunity, system failures	Diet change, menopause, declining fitness, loss of mobility, incontinence, cardiopulmonary disorders, age-related infections, hospitalization, residential care	Less stable population, 7–8 phyla, variable fungi, variable virus

[1] Each factor included in subsequent age groups.
[2] Type and relative frequency of taxa.

Sources: Spor, A.; Koren, O.; and Ley, R. 2011. "Unravelling the effects of the environment and host genotype on the gut microbiome." *Nature Rev. Microbiol.* 9: 279–290; and Belkaid, Y., and Segre, J. A. 2014. "Dialog between skin microbiota and immunity." *Science* 346: 954–959.

Ziggy_Mars/Shutterstock

While an adult human's microbiota is relatively stable over time, it is highly variable from person to person and at different sites within the same person. In other words, there are relatively few species of microorganisms common to all humans; each human has a relatively unique microbiota. Based on 16S rRNA and whole genome metagenomic data, bacteria common to human skin, the intestinal tract, and the other mucosal surfaces include six major phyla: Actinobacteriota, Bacteroidota, Firmicutes, Fusobacteriota, Proteobacteria, and Verrucomicrobiota (**figure 33.2**). Additionally, a number of archaea, fungi, and viruses are also present. Despite limited diversity at the phylum level, an enormous number of species are present. In fact, the average adult gut is host to 500 to 1,000 different microbial species.

Microbiota Vary by Body Site

In a healthy human, regardless of age, internal organs and tissues (e.g., brain, blood, muscles) are normally free of microorganisms. Conversely, surface tissues (e.g., skin and mucous membranes) are constantly in contact with the environment and are colonized by various microbes. Because bacterial species make up most of the microbiota, we emphasize them over archaea, fungi, protists, and viruses (including phages) that are also part of the microbial assemblage.

Skin

The skin surface or epidermis has a slightly acidic pH and a high concentration of sodium chloride. Some areas lack moisture, whereas others are bathed in sebum, a moisturizing substance produced by sebaceous glands, and antimicrobial peptides (AMPs). It has been estimated that the average adult hosts about 10^{10} bacteria in these different skin microenvironments (figure 33.2).

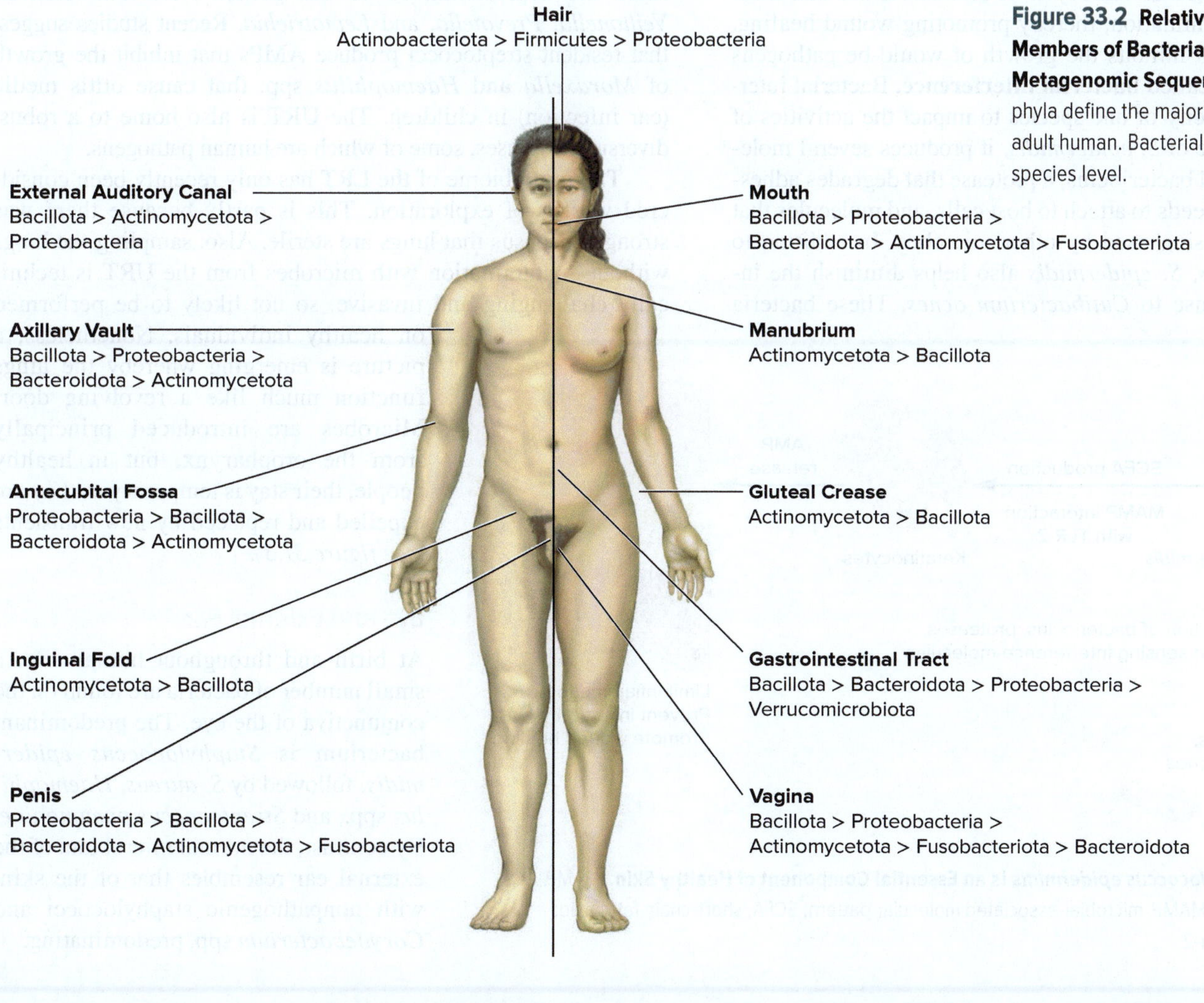

Figure 33.2 Relative Proportion of Members of Bacterial Phyla Identified by Metagenomic Sequencing. Bacteria in six phyla define the majority of microbes in the adult human. Bacterial diversity occurs at the species level.

Some microorganisms are temporarily present and are typically unable to multiply on the skin.

The skin surface can be divided into three environmental niches: dry, moist, and sebaceous (containing sebum). In general, bacterial diversity is greatest at dry sites (e.g., forearms, buttocks, hands), harboring a mixture of organisms from the phyla Actinomycetota, Bacteroidota, Bacillota, and Proteobacteria. Moist areas (e.g., umbilicus, underarms, inguinal and gluteal creases) exhibit less diversity, supporting mostly members of Bacillota and Actinomycetota (e.g., *Staphylococcus* and *Corynebacterium* spp., respectively), in addition to low numbers of other bacteria. The skin surfaces with the lowest bacterial diversity are oily sebaceous sites (e.g., forehead, behind the ear, and the back), where *Cutibacterium* spp. and other Actinomycetota dominate (figure 33.2).

Staphylococcus epidermidis had long been thought to be just hitching a ride by colonizing the skin (thus the species name) where it is universally found. It is generally nonpathogenic, unlike *S. aureus,* which can cause skin infections. We now know that *S. epidermidis* is a key component of healthy skin, where it closely associates with keratinocytes (*see figure 31.15*). *S. epidermidis* modulates keratinocyte gene expression in at least two ways: It secretes products of fermentation called short-chain fatty acids (SCFAs) and its antigens bind to the pattern recognition receptor TLR-2 (**figure 33.3**; *also see figure 31.17*). This interaction stimulates keratinocytes to release AMPs that limit infection and inflammation, thereby promoting wound healing. *S. epidermidis* also inhibits the growth of would-be pathogens through a process called **bacterial interference.** Bacterial interference is the capacity of one species to impact the activities of another. In the case of *S. epidermidis,* it produces several molecules: AMPs called bacteriocins, a protease that degrades adhesins that *S. aureus* needs to attach to host cells, and molecules that derail quorum sensing used by other microbes. In addition to targeting *S. aureus, S. epidermidis* also helps diminish the inflammatory response to *Cutibacterium acnes.* These bacteria usually are harmless; however, they are associated with the skin condition acne vulgaris. ◀ *Bacteriocins (section 31.3); Toll-like receptors (section 31.6);* ▶ *Staphylococcal diseases (section 38.3)*

Respiratory Tract

The respiratory tract is divided into the upper respiratory tract (URT)—the nostrils (also called nares), sinuses, pharynx, and oropharynx—and the lower respiratory tract (LRT)—the larynx below the vocal cords, trachea, bronchi, and lungs. While it has long been understood that the URT is colonized by a diverse group of microbes, we are only beginning to explore the LRT as a site that is not consistently sterile in healthy individuals, as previously thought.

The regions of the URT closest to the environment are the anterior nares, which, like skin, include host cells that produce sebum. Thus like skin, the anterior nares are colonized by the monoderm, lipophilic genera *Staphylococcus, Corynebacterium,* and *Cutibacterium.* Here we see the same interaction between *S. epidermidis* and *S. aureus,* which reside in the nares (figure 33.3). Deeper into the nasal cavity we find additional microbial genera, including members of the monoderm genera *Streptococcus* and *Dolosigranulum* and diderm *Moraxella* and *Haemophilus* spp. Finally, the oropharynx is most diverse, where, in addition to bacteria belonging to genera found in the nares and nasopharynx, there are representatives of the genera *Neisseria, Rothia, Veillonella, Prevotella,* and *Leptotrichia.* Recent studies suggest that resident streptococci produce AMPs that inhibit the growth of *Moraxella* and *Haemophilus* spp. that cause otitis media (ear infection) in children. The URT is also home to a robust diversity of viruses, some of which are human pathogens.

The microbiome of the LRT has only recently been considered worthy of exploration. This is partly because there was strong consensus that lungs are sterile. Also, sampling the lungs without contamination with microbes from the URT is technically challenging and invasive, so not likely to be performed on healthy individuals. Nonetheless, a picture is emerging whereby the lungs function much like a revolving door: Microbes are introduced principally from the oropharynx, but in healthy people, their stay is temporary as they are expelled and replaced by new transients (*see figure 31.3*).

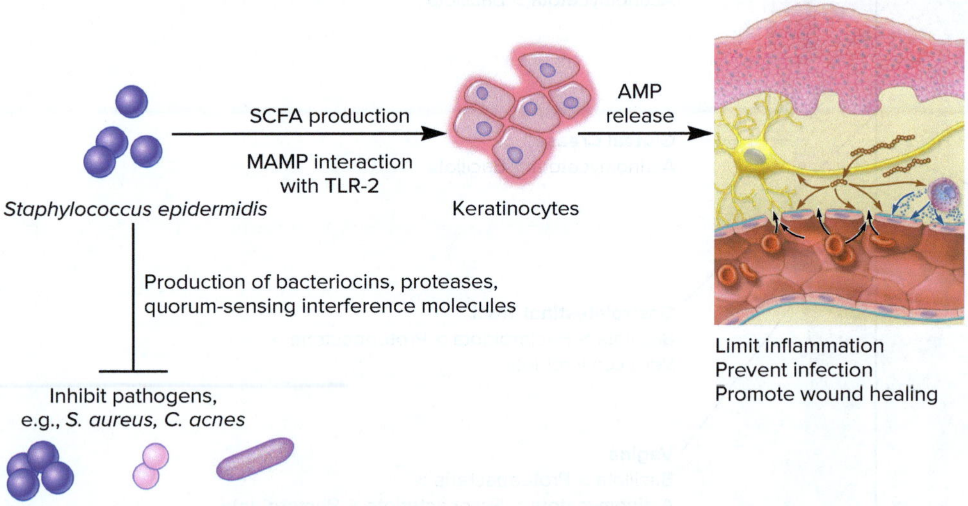

Figure 33.3 ***Staphylococcus epidermidis* Is an Essential Component of Healthy Skin.** AMP, antimicrobial peptide; MAMP, microbial-associated molecular pattern; SCFA, short-chain fatty acid; TLR-2, toll-like receptor-2.

Eye and External Ear

At birth and throughout human life, a small number of bacteria are found on the conjunctiva of the eye. The predominant bacterium is *Staphylococcus epidermidis,* followed by *S. aureus, Haemophilus* spp., and *Streptococcus pneumoniae.* By contrast, the normal microbiota of the external ear resembles that of the skin, with nonpathogenic staphylococci and *Corynebacterium* spp. predominating.

Gastrointestinal Tract

The gastrointestinal (GI) tract starts at the mouth and ends at the anus. Along the way, there are several distinct environments. Soon after an infant is born, the mouth is colonized by microorganisms from the surrounding environment. Most microorganisms that initially invade the oral cavity are aerobes and obligate anaerobes. When the first teeth erupt, anaerobes (*Porphyromonas, Prevotella,* and *Fusobacterium* spp.) become dominant due to the anoxic nature of the space between the teeth and gums. As teeth grow, *Streptococcus parasanguis* and *S. mutans* attach to enamel surfaces; *S. salivarius* attaches to the buccal (e.g., inside the cheeks) and gum epithelial surfaces and colonizes the saliva. These streptococci produce a glycocalyx and various other adherence factors that enable biofilm formation on oral surfaces. The presence of these bacteria contributes to the eventual formation of dental plaque, caries, gingivitis, and periodontal disease. ▶| *Dental diseases (section 38.6)*

From the mouth, food and microbes travel through the esophagus to the stomach. With a pH of 2 to 3, the stomach kills most microorganisms. As a result, the stomach usually contains less than 10 viable bacteria per milliliter of gastric fluid. These are mainly *Streptococcus, Staphylococcus, Lactobacillus, Peptostreptococcus* spp., and yeasts such as *Candida* spp. Microorganisms may survive if they pass rapidly through the stomach or if the organisms ingested with food are particularly resistant to gastric pH.

The small intestine is divided into three regions: duodenum, jejunum, and ileum. The duodenum (the first 25 cm of the small intestine) contains few microorganisms because of the combined influence of the stomach's acidic juices and the inhibitory action of bile and pancreatic secretions that enter here. Monoderm bacteria comprise most of the microbiota. *Enterococcus faecalis,* lactobacilli, diphtheroids, and the yeast *Candida albicans* are found in the jejunum. The distal portion of the small intestine (ileum) becomes more alkaline and the microbiota begins to take on the characteristics of the colon microbiota. Recall that the ileum is the site of Peyer's patches, an important part of the gut-associated lymphoid tissue (GALT) for maintaining tolerance to, as well as defending against, microbial antigens. |◀ *Secondary lymphoid organs and tissues (section 31.5)*

The human large intestine (colon) is one of the most densely packed microbial ecosystems on Earth, with 10^{12} cells/gram of stool. Over a thousand different species have been detected; many of them are not yet cultured. At the phylum level, Bacillota and Bacteroidota dominate healthy gut microbiota (figure 33.2). At the genus level, there is tremendous individual variability. This has led to the development of the notion of a core microbiome. A *core microbiome* predicts that rather than possessing a particular group of microbes, the genes that constitute the microbiome collectively accomplish specific, essential metabolic and regulatory functions. Our core microbiomes are influenced by diet. For example, metagenomic sequencing studies reveal that people who consume the low-fiber, high-fat diet typical of most industrialized nations have a different core microbiome from that of individuals who consume a plant-based (high-fiber, low-fat) diet. In essence, these two groups of people are feeding their microbiomes differently. Interestingly, while the microbiome rapidly responds to large changes in diet, changes are highly variable and endure only as long as the new dietary habits are maintained.

Analyses of large metagenomic data sets have attempted to define the core gut microbiome. Among people living in industrialized nations who consume a low-fiber, high-protein diet, genera that constitute the core microbiome include (but certainly are not limited to) *Bacteroides, Faecalibacterium, Clostridium, Prevotella, Coprococcus,* and *Ruminococcus*. Phylogenetic analysis of gut bacteria that have only recently been cultured has led to the reassignment of some former members of the genus *Bacteroides* to the genera *Alistipes, Prevotella, Paraprevotella, Parabacteroides,* and *Odoribacter.*

What are all these microbes doing? Collectively our gut microbiota functions as an organ that has metabolic, immunological, and endocrine roles. As we explore in more detail, our gut microbes perform an enormous number of tasks, including (1) degrading and fermenting foods that we cannot digest because we lack the genes that encode the necessary enzymes, (2) synthesizing micronutrients including vitamins, (3) metabolizing dietary toxins and carcinogens, (4) ensuring immune system maturation and function both locally and at distant body sites, (5) impacting growth and differentiation of host cells that line the colon (enterocytes), (6) regulating the development of our intestinal blood supply, (7) protecting against enteric pathogens, and (8) converting steroids such as cholesterol and bile acids to other compounds. For example, about one-third to one-half of the cholesterol you consume is transformed by gut microbes into a compound called coprostanol, most of which is then excreted. Likewise, bile acids produced by the liver, called *primary bile acids,* are chemically modified (e.g., via dehydroxylation, oxidation, and epimerization) by our gut microbes to about 20 structurally different compounds called *secondary bile acids.* Secondary bile acids impact the growth of other microbes in the gut (including inhibiting the germination of *Clostridioides difficile* spores—see chapter opening story) and travel to the liver, kidney, and heart, where they are thought to modulate lipid, glucose, and energy metabolism. Different microbes are responsible for different bile acid modifications and not all secondary bile acids have the same effects. Thus some microbes, and the secondary bile acids they produce, promote health better than others. This is just one example by which the composition of the microbiome has far-reaching impacts on human health and homeostasis.

Genitourinary Tract

The upper urinary tract (kidneys, ureters, and urinary bladder) is usually free of microorganisms. The bladder had long been thought to contain microorganisms only during infection, but recent evidence suggests microbial communities may maintain normal bladder homeostasis. In both males and females, a few common bacteria (*S. epidermidis, Enterococcus faecalis,* and *Corynebacterium* spp.) can be cultured from the distal portion of the urethra.

The adult female genital tract, because of its large surface area and mucous secretions, has a complex microbiota that changes with the menstrual cycle. The culturable microorganisms are acid-tolerant lactobacilli, including *Lactobacillus acidophilus,* also called Döderlein's bacillus. They ferment glycogen produced by the vaginal epithelium, forming lactic acid. As a result, the pH of the vagina and cervix is maintained between 4.4 and 4.6. A variety of anaerobic bacteria, as well as *Candida* yeast, also reside in the vagina and cervix. By maintaining an acidic pH, *L. acidophilus* limits their growth. If this bacterial interference is perturbed by antibiotic treatment or hormonal fluctuation, bacterial vaginosis may develop, marked by an overgrowth of diderm anaerobes, or a vaginal yeast infection caused by unchecked *Candida albicans* growth. ▶| *Bacterial vaginosis (section 38.6); Candidiasis (section 39.6)*

Comprehension Check

1. Discuss three factors that have affected the development of your gut microbiota.
2. What is the impact of pH in determining the microbiota of the stomach?
3. What is the role of colon microbes in cholesterol metabolism? What about bile acids?
4. How can antibiotic treatment lead to vaginal infections?

33.3 A Functional Core Microbiome Is Required for Host Homeostasis

After reading this section, you should be able to:

a. Explain the contribution the gut microbiome makes to human metabolism
b. Analyze the importance of germfree mice to our understanding of the human microbiome
c. Differentiate between direct and indirect mechanisms responsible for colonization resistance
d. List three specific reasons the immune system depends on the microbiome to maintain homeostasis
e. Compare and contrast three routes by which gut microbes communicate with the central nervous system

The person-to-person variability in the species composition of our microbiota makes it hard to establish broad generalizations about which microbes contribute to good health. But one theme that repeatedly emerges is that taxonomically diverse microbiota are essential for well-being. As discussed with regard to the GI tract, rather than trying to define a specific group of microbial genera that constitute healthy microbiota, a more realistic accounting is the concept of functional core microbiome. A **functional core microbiome** includes those microbial genes that provide the host with a suite of activities required for health and homeostasis. Some of these activities, like the provision of vitamin K by *Escherichia coli,* have been known for many years. Other activities, like the emerging role of gut microbiota in human behavior, are surprising and just beginning to be understood. The notion of a functional core microbiome means that we need to turn our attention to characterizing microbial genes and metabolites rather than individual microbial species. Here we discuss some of the functions of a healthy microbiome with the caveat that this is a rapidly advancing field, and we present just some of the highlights.

Host Metabolism

Body weight is most closely related not to the number of calories we eat, but to the number of calories we absorb. Our cells are very good at taking up simple sugars and a few polysaccharides, but for the most part, we rely on our gut microbiota to convert our food into calories we can use. This was shown in the early 2000s in a series of experiments that used germfree mice. **Germfree (GF)** mice are born by cesarean delivery, reared in sterile enclosures, and fed sterile food and water. Although GF animals are nowhere near representative of normal life, comparing GF mice with conventional mice allows scientists to test many of the complex relationships between microorganisms, hosts, and environmental factors in a controlled fashion.

These early studies showed that GF mice could eat more but gain less weight than conventional mice. For example, in one 8-week study, GF mice gained 50% less than conventional mice after both groups consumed a diet high in fat. While this may sound like an ad for a weight-loss plan ("Eat more! Weigh less!"), this study and others like it showed that gut microbiota convert much of the food eaten into smaller, digestible products that contribute to total caloric intake of the host. The causal nature of microbes and weight gain has been demonstrated by fecal microbiota transplant (FMT) experiments (see chapter opening story). When FMT is performed using conventional, obese mice as donors and GF mice as recipients, the formerly lean GF recipient mice become obese as well, despite no change to their diet or exercise regime. Conversely, when lean mice are the FMT donors and obese mice are the recipients, the latter lose weight. These results have been extended and confirmed using human microbiota-associated mice, in which human feces is transplanted to mice.

It is frequently (but not always) observed that obese mice, like people who are obese, have higher relative concentrations of gut bacteria belonging to the phylum Bacillota compared to bacteria belonging to the phylum Bacteroidota, along with several other changes in gut microbiota population that make it less diverse. Why is this important? Simply put, we cannot break down fiber very well, but our bacteria can. Although the term *fiber* is not very specific, all fiber consists of a variety of linear and branch chained sugars. Because we can digest only a few complex polysaccharides, we feed most of them to our microbiota. Unlike our own cells, the typical gut microbiome encodes thousands of enzymes that evolved to break down complex polysaccharides into their constituent monomers. Importantly many

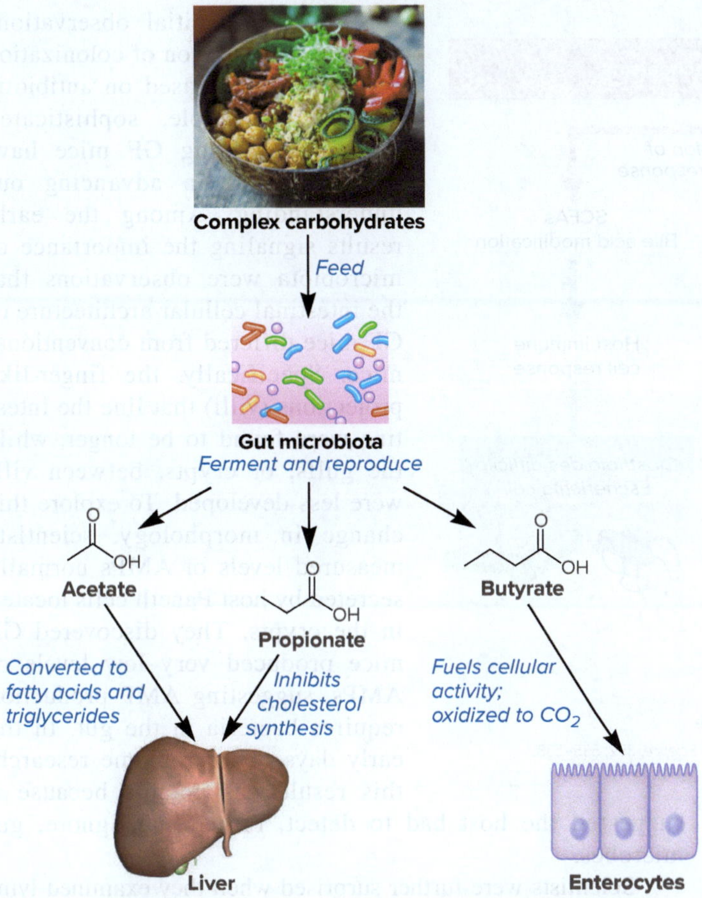

Figure 33.4 Concept Map of the Impact Fermentation End Products Have on Host Metabolism. The relative abundance of butyrate-, propionate-, and acetate-producing bacteria helps determine how many calories are absorbed. These short-chain fatty acids are also distributed systemically (not shown).
(top photo) Magdanatka/Shutterstock; (bottom photo) MedicalRF.com

bacteria ferment these monomers into short-chain fatty acids (SCFAs). Three key SCFAs are butyrate, propionate, and acetate (**figure 33.4**). Butyrate is the principal source of calories for enterocytes lining the gut. In fact, butyrate increases enterocyte mitochondrial activity so that many carbon substrates are oxidized to CO_2—thus calories consumed are literally vaporized. Propionate travels to the liver, where it inhibits cholesterol synthesis and, along with butyrate, helps regulate weight by increasing the release of intestinal hormones that suppress hunger. On the other hand, acetate is readily absorbed by a variety of host cells and acts as a precursor for lipid synthesis in liver and fat cells. For this reason, acetate in this context is sometimes called *obesogenic*. The shift in the population of gut microbiota observed in individuals who are obese results in less diversity. This in turn leads to a higher percentage of bacteria that produce acetate, which promotes weight gain and fewer bacteria that produce butyrate and propionate, the SCFAs that protect against obesity.

Another group of organisms that influence our ability to absorb calories is methanogenic archaea. The initial observation that both mice and people with obesity have a higher concentration of methanogens than their lean counterparts was followed by studies suggesting two potential mechanisms. First, methanogens consume H_2, which may increase fermentation by other bacteria that produce obesogenic SCFAs. Others have found that the presence of methane slows the transit of food through the GI tract, thereby enabling more absorption. Methanogen prevalence is also correlated with irritable bowel syndrome—an example of how our gut microbiota have a role in disease as well as health. ◄ *Syntrophy (section 27.2)*

Perhaps one of the most remarkable aspects of the microbiome's impact on metabolism is that it is heritable, at least in mice. This has been demonstrated by comparing the growth of pups (baby mice) born to GF mice with that of pups born to conventional mice. The SCFAs (particularly propionate) produced by conventional pregnant mice cross the placenta and impact neural cell development, which in turn influences nerve cell growth in the heart and the ability of the pancreas to mature normally. As pups, there is no discernable size difference between those delivered from conventional or GF mice. But as they age, the offspring from the GF mice gain significantly more weight despite the same diet and exercise regimen. These mice also have elevated markers of metabolic disease (e.g., blood glucose, triglyceride, and cholesterol levels). Amazingly, supplementing maternal diets with propionate reverses these effects on their offspring. Most recently, the gut microbiome of male mice has been shown to alter sperm gene expression through epigenetic changes (e.g., methylation), which impacts the birth weight and development of pups they father. This too is reversible by restoring microbiome diversity to the paternal mice.

The role of SCFAs in weight regulation and maintaining host homeostasis has refocused attention from the individual species that constitute the gut microbiota to the products these microorganisms secrete, called the **metabolome.** As we discuss, SCFAs and the microbes that produce them are key players in immune modulation as well.

Immunity

The first clue that the microbiome was critical to host immunity was gleaned from an observation made over 50 years ago. At that time it was noted that following antibiotic treatment, people had a much higher risk of GI tract infection by members of the Enterobacteriaceae, such as *E. coli* or *Salmonella enterica*. It was rightly assumed that antibiotics disrupted the gut microbial community, and this phenomenon was explained by *colonization resistance*. Fifty years ago, it was assumed that the bacteria in the healthy gut simply outcompeted pathogens for nutrients and space. We now understand that bacterial interference is much more complex and many of our microbiota are *mutualists*, as both the host and microbes derive benefit. Although our mutualistic microbial partners do take up space and nutrients, they also protect us through direct pathogen interference and indirect interactions with the host. ◄ *Many types of microbial interactions exist (section 27.1)*

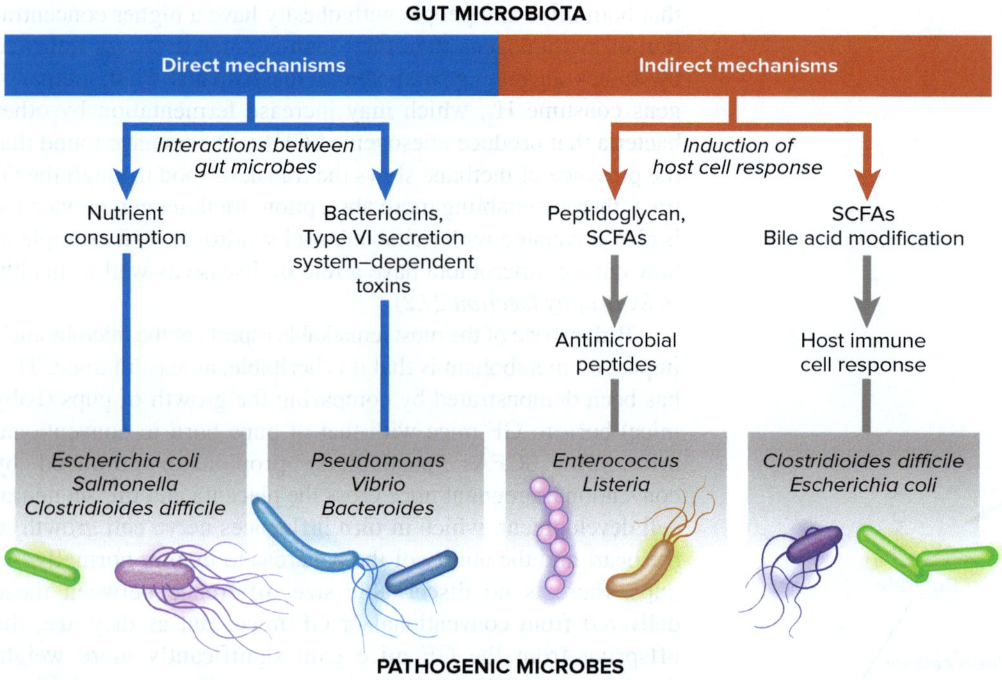

Figure 33.5 Gut Microbiota Directly and Indirectly Impact Pathogen Growth.

Source: Pamer, Eric G. 2016. "Resurrecting the intestinal microbiota to combat antibiotic-resistant pathogens." *Science*, 352: 535–538.

Among well-characterized direct microbiota-pathogen interactions is the release of toxic peptides by one group of bacteria that specifically targets others (**figure 33.5**). These peptides include bacteriocins, microcins, and colicins. Bacteriocins called lantibiotics are produced by monoderm bacteria and target closely related strains or species. Microcins are made by a wide range of Enterobacteriaceae; colicins are *E. coli* products. Like bacteriocins, microcins and colicins target bacteria taxonomically related to the producing microbes. Some diderm bacteria deliver bacteriocidal toxins directly through the cell envelope of competing bacteria through the needle of their type VI secretion system (T6SS). ◀ *Protein translocation and secretion (section 13.9); Competition: microbes vying for common goods (section 27.4); Bacteriocins (section 31.3)*

Examples of indirect, microbe-host-pathogen interaction include the ability of peptidoglycan and SCFAs released by microbiota to induce host cell production of antimicrobial peptides (AMPs) and a C-type lectin (figure 33.5). C-type lectins are a class of carbohydrate-binding proteins that require calcium (thus the C designation). C-type lectins selectively kill Gram-positive microbes, like pathogenic strains of *C. difficile* and *Enterococcus faecalis*. Another example is microbial production of secondary bile acids, which limit inflammation and C. diff germination. Finally, the fermentation product butyrate stimulates host enterocytes to consume oxygen through β-oxidation of fatty acids. This in turn limits oxygen diffusion into the lumen. By keeping the oxygen tension low, the growth of anaerobes is favored over other, less desirable facultative anaerobes such as *E. coli* and *Salmonella enterica*. ◀ *Acute phase proteins (section 31.1)*

While the initial observations that led to the notion of colonization resistance were based on antibiotic therapy in people, sophisticated experiments using GF mice have been essential in advancing our understanding. Among the early results signaling the importance of microbiota were observations that the intestinal cellular architecture of GF mice differed from conventional mice. Specifically, the finger-like projections (villi) that line the intestine were found to be longer, while the gulfs, or crypts, between villi were less developed. To explore this change in morphology, scientists measured levels of AMPs normally secreted by host Paneth cells located in the crypts. They discovered GF mice produced very low levels of AMPs, suggesting AMP production requires bacteria in the gut. In the early days of microbiome research, this result was notable because it suggested the host had to detect, rather than ignore, gut microbes.

Scientists were further surprised when they examined lymphoid tissue outside (e.g., spleen, thymus, lymph nodes) in addition to the intestines of GF mice. These tissues were quite underdeveloped, indicating that microorganisms in the lumen of the gut were influencing distant immune cell and tissue development. Additional studies showed that GF mice have limited development of white blood cells of the innate immune system (**figure 33.6**). These cells include neutrophils, monocytes, macrophages, and dendritic cells—all cells that mature in the bone marrow (*see figure 31.9*). This prompted investigators to look for soluble factors produced by gut microbiota in the bloodstream that could drive immune cell maturation. In what has now become a recurrent theme, it was discovered that soluble SCFAs as well as cytokines released by intestinal immune cells are needed for immune cell maturation throughout the body. Interestingly, cytokine release is triggered by stimulation of toll-like receptors (TLR) when gut immune cells detect microbial-associated molecular patterns (MAMPs) found on our normal microbiota (*see figure 31.17*). ◀ *Recognition of foreignness is key for survival (section 31.6)*

Immune cell maturation is not the end of the story. After immune cells leave the bone marrow and enter the bloodstream, the presence of gut microbiota and their soluble products are thought to prime neutrophils, keeping them ready to defend against pathogens. Macrophages residing in tissues such as the central nervous system, skin, and gut also require a functional microbiome to maintain normal morphology. Although we are still learning about the mechanisms responsible for these effects, it is clear that the

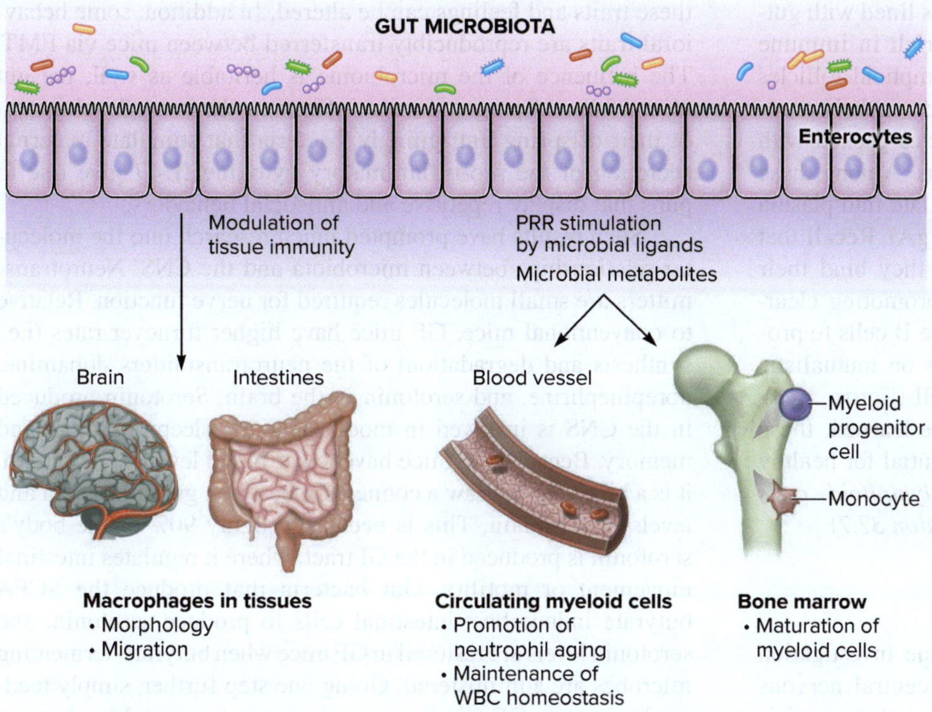

Figure 33.6 Microbiota Signals Influence Immune Cell Function at Sites Distant from the Gut. Studies in GF mice demonstrate that the maturation and function of innate immune cells depend on signals from gut microbiota. PRR, pattern recognition receptor; WBC, white blood cell.

(left) GraphicsRF/Shutterstock. Source: Thaiss, C. A.; Zmora, N.; Levy, M.; and Elinav, E. 2016. "The microbiome and innate immunity." *Nature*, 535(7610): 65–74.

MICRO INQUIRY *How might frequent antibiotic treatment in infants and toddlers affect their immune systems?*

microbiome has a profound effect on the development of the immune system. *Each type of innate immune cell has a specific function (section 31.4)*

These observations argue strongly against the long-held notion that the single layer of host enterocytes that line the intestinal wall forms an impenetrable barrier that only pathogens can cross. This convenient belief helped explain the concept of **mucosal tolerance,** or how we could host so many microbes and foreign food antigens in our guts without responding with an overwhelming inflammatory response. However, the discovery of host cell pattern recognition receptors (PRRs), such as TLRs on enterocytes, combined with GF mice experiments and a plethora of metagenomic microbiome data led to a true paradigm shift: Enterocytes and the immune cells beneath and between them recognize and respond to our mutualistic microbiota in addition to pathogens. While our understanding of mucosal tolerance is incomplete, it has been well documented that SCFAs, especially butyrate, inhibit the release of pro-inflammatory cytokines by enterocytes and macrophages, in part by inhibiting the host enzyme histone deacetylase, thereby regulating host cell gene transcription.

Mucosal tolerance requires a balance between the activity of pro-inflammatory T_H17 cells and Treg cells, which produce the anti-inflammatory cytokine IL-10 (**figure 33.7**). But how is this balance determined and maintained? One of the early, exciting discoveries was that the bacterium *Bacteroides fragilis* produces a molecule called polysaccharide A (PSA) that reverses inflammation by suppressing the release of the pro-inflammatory cytokine IL-17 by T_H17 cells and promoting the production of IL-10 by binding to a PRR (toll-like receptor 2) on Treg cells. Another early discovery was the capacity of SCFAs to increase both the number and activity of intestinal Treg cells. Remarkably, butyrate directly signals Treg cells by binding to their G-protein coupled receptors. In contrast to SCFAs, long-chain fatty acids, 16–18 carbons in length, appear to promote T_H1 and T_H17 differentiation and proliferation. Based on microbiome studies in model mouse systems and patients with inflammatory bowel disease, it would appear that gut microbial communities with a disproportionate number of bacteria belonging to the phylum Proteobacteria promote inflammation at least in part through the production of long-chain fatty acids. Interestingly, many inflammatory bowel conditions such as necrotizing enterocolitis in infants and Crohn's disease in adults are marked by a disproportionate increase in Proteobacteria. *Types of T cells (section 32.5)*

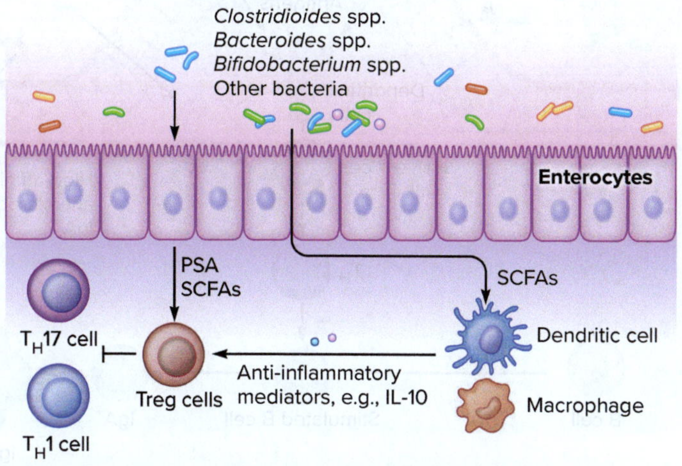

Figure 33.7 Gut Microbiota Metabolites Regulate Inflammation. Inflammation reflects the balance of pro- and anti-inflammatory cytokines released by T_H17 cells and Treg cells, respectively. Bacterial metabolites such as PSA from *Bacteroides fragilis* and SCFAs from clostridia and other fermentative organisms directly promote Treg activity. SCFAs also limit inflammation by interacting with macrophages and dendritic cells, which in turn stimulate Treg cells. PSA, polysaccharide A; SCFAs, short-chain fatty acids.

Source: Thaiss, C. A., Zmora, N., Levy, M., and Elinav, E. 2016. "The microbiome and innate immunity." *Nature*, 535(7610): 65–74.

As discussed in chapter 31, the gut mucosa is lined with gut-associated lymphoid tissue (GALT)—regions rich in immune cells, including Peyer's patches and isolated lymphoid follicles (*see figure 31.16*). When stimulated by microbes or their products, these regions develop into lymphoid follicles packed with cells of the innate and adaptive immune systems. Among these cells are B cells that when activated to differentiate into plasma cells produce antigen-specific secretory IgA (SIgA). Recall that SIgA molecules coat mucosal surfaces where they bind their microbial targets, preventing attachment and promoting clearance. Innate lymphoid cells (ILC3s) help activate B cells to produce SIgA. Because ILC development depends on mutualistic microbiota, the production of SIgA does as well (**figure 33.8**). Although this is just a sample of what we have learned, these examples illustrate how gut microbiota are essential for healthy innate and adaptive immunity. ◀ *Innate lymphoid cells (section 31.4); Immunoglobulin (Ig) classes (section 32.7)*

Gut-Brain Axis

One of the most intriguing areas of microbiome investigation concerns how our gut microbiota affects our central nervous system (CNS). A variety of studies have shown that specific behavioral traits (e.g., inquisitiveness, sociability) and feelings (e.g., anxiety, depression) differ when comparing GF mice and conventional mice. Amazingly, when GF mice are fed bacteria, these traits and feelings can be altered. In addition, some behavioral traits are reproducibly transferred between mice via FMT. The influence of the microbiome is heritable as well; the gut microbiome of pregnant mice influences the neurodevelopment of their offspring. Intriguingly, bacteria that stimulate maternal secretion of the pro-inflammatory cytokine IL-17 give rise to pups that display repetitive and antisocial behaviors.

Such results have prompted much research into the molecular relationships between microbiota and the CNS. Neurotransmitters are small molecules required for nerve function. Relative to conventional mice, GF mice have higher turnover rates (i.e., synthesis and degradation) of the neurotransmitters dopamine, norepinephrine, and serotonin in the brain. Serotonin produced in the CNS is involved in mood, appetite, sleep, learning, and memory. Because GF mice have lower blood levels of serotonin, it is a bit easier to draw a connection between gut microbiota and levels of serotonin. This is because roughly 90% of the body's serotonin is produced in the GI tract, where it regulates intestinal movement or motility. Gut bacteria that produce the SCFA butyrate induce host intestinal cells to produce serotonin, and serotonin levels are restored in GF mice when butyrate-fermenting microbes are administered. Going one step further, simply feeding butyrate to GF mice restores serotonin to normal levels.

However, these observations do not provide a complete explanation of how gut bacteria interact with the brain. In this regard, we can predict at least three ways the microbiome can

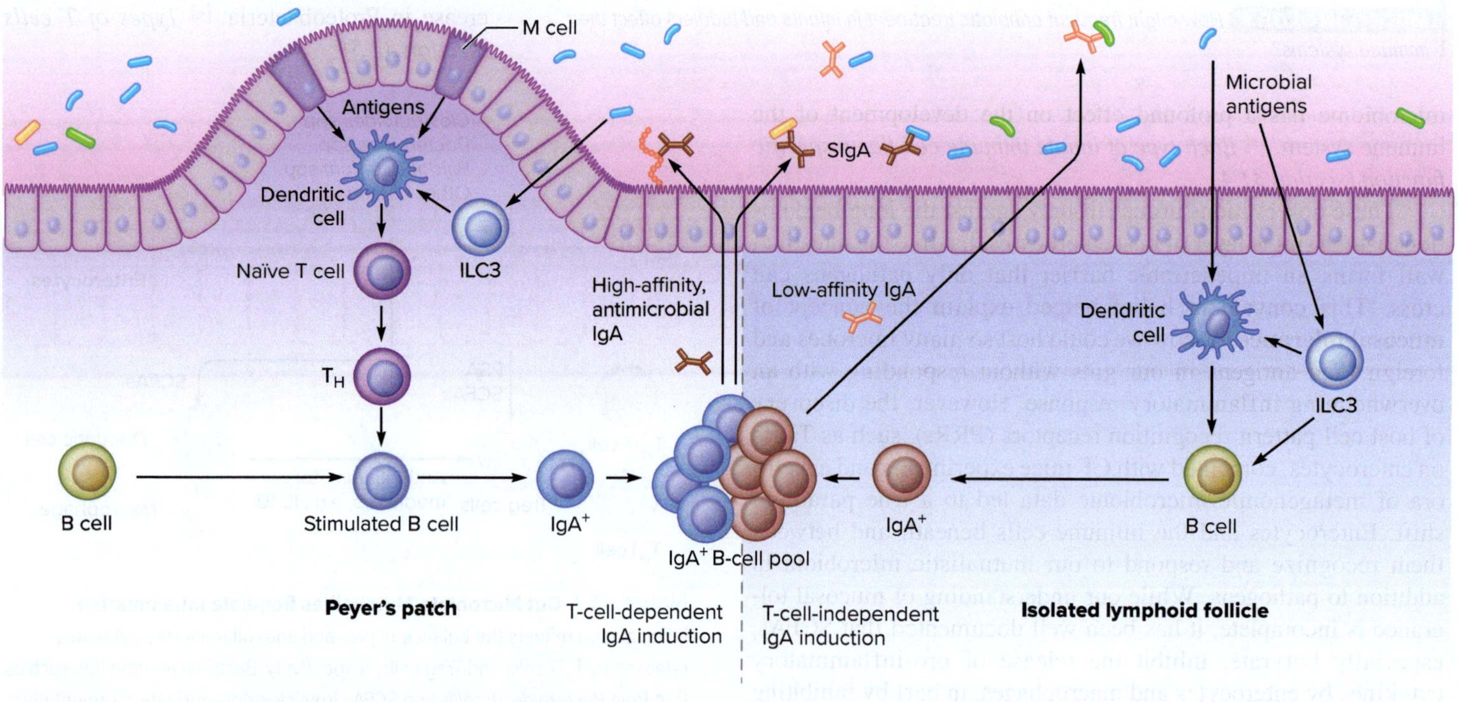

Figure 33.8 Production of Secretory (SIgA) Requires ILC3 and Microbial Stimulation. B-cell stimulation can be T-cell dependent (left, Peyer's patch) or T-cell independent (right, isolated lymphoid follicle). Innate lymphoid cells-3 (ILC3s) interact with gut microbes or their products and optimize dendritic cell (DC) activity. ILC3s can interact directly with B cells or influence B cell activity through DCs. ILC3s' interaction with macrophages also promotes Treg development (not shown).

Source: Honda, K., and Littman, D. R. "The microbiota in adaptive immune homeostasis and disease." 2016. *Nature*, 535(7610): 75–84.

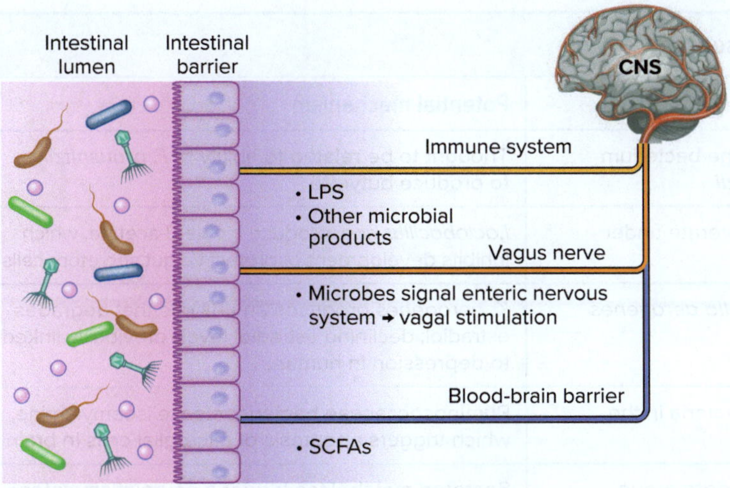

Figure 33.9 The Gut-Brain Axis. Gut microbes influence the central nervous system using at least three mechanisms. LPS, lipopolysaccharide; SCFAs, short-chain fatty acids.

(right top) GraphicsRF/Shutterstock. Source: Sampson, T. R., and Mazmanian, S. K. 2015. "Control of brain development, function, and behavior by the microbiome." *Cell Host Microbe*, 17(5): 565–576.

influence the CNS (**figure 33.9**). First, as we have seen, the microbiome has a profound impact on the immune system, which in turn modulates the CNS. One documented way in which this occurs is through microbial regulation of the balance between pro- and anti-inflammatory cytokines (figure 33.7). When pro-inflammatory cytokines reach the brain, they alter the function of neurons and CNS macrophages, called microglia. For example, when lipopolysaccharide (LPS) crosses the gut, it triggers the production of pro-inflammatory cytokines, which in turn are responsible for the behaviors we associate with being sick: loss of appetite, decreased motor activity, loss of sociability, and altered cognition ("brain fog").

Another mechanism that helps mediate the gut-brain axis is a direct pathway from gut to brain. This is possible because the GI tract is lined with a network of nerves, called the enteric or intrinsic nervous system, that is connected to the CNS via the vagus nerve. Gut bacteria directly stimulate neurons within the enteric nervous system, which then communicate with the vagus nerve. The vagus nerve in turn transmits signals to the brain. The vagus nerve can differentiate between stimuli from normal microbiota and pathogens even in the absence of inflammation, although how this fascinating feat is possible remains under investigation.

Finally, as suggested by experiments in which butyrate and FMT had similar effects on the behavior of GF mice, soluble microbial products can impact the CNS. To understand this, we must first consider the CNS as a site that is set apart from the remainder of the body by the blood-brain barrier (BBB). The BBB is lined with cells that protect the CNS from normal blood components as well as toxins and infectious agents. This is largely due to impermeable connections, called tight junctions, between the cells that make up the BBB. Once again, a flurry of research was triggered by an observation made in GF mice: Their BBBs have the undesirable quality of being more permeable than those of conventional mice. And once again, it is SCFAs, particularly butyrate, that help maintain the BBB as an impenetrable barricade. Because SCFAs are most abundant when mice or humans eat a high-fiber diet, the broader implication is that one's diet can govern the entry of metabolites and other blood-borne compounds to the CNS.

The unexpected influence gut microbiota has on behavior, documented in animals and some human trials, has unleashed a torrent of studies into how the microbiome may mediate neurodegenerative diseases like Parkinson's and Alzheimer's diseases. Autism spectrum disorder (ASD) has also garnered a great deal of attention. This reflects evidence that the establishment of a functional microbiome has long-lasting and important effects on behavior and perhaps on neural development within the first 3 years of life, when most children are diagnosed with ASD. How our microbiomes impact neurodevelopment on the one hand and degeneration on the other are intriguing areas of investigation that will no doubt continue to reveal unexpected findings.

Comprehension Check

1. Describe an experiment performed using GF mice that broadened our understanding of the role of the gut microbiome in maintaining physiologic homeostasis.
2. Compare and contrast a direct mechanism and an indirect mechanism that impact pathogen growth in the gut.
3. What foods have you eaten in the last 24 hours that promoted the microbial production of short-chain fatty acids and why is this desirable?
4. What is mucosal tolerance and how does the microbiome contribute to maintaining it?
5. Describe one way the microbiome impacts innate immunity and one way it impacts adaptive immunity.
6. Define the blood-brain barrier and how it may be involved in the gut-brain axis.

33.4 Many Diseases Have a Connection with Dysbiosis

After reading this section, you should be able to:

a. Describe metabolic syndrome and metabolic endotoxemia
b. Discuss the microbial link between a diet rich in red meat and atherosclerosis
c. Discuss in broad terms the relationship between the microbiome and cancer

A casual review of recent literature might lead one to think that all human disease is linked to the microbiome. A quick internet search reveals hundreds of reports describing the role of our microbiomes in the development of a variety of illnesses such as autoimmune disease (e.g., multiple sclerosis, lupus erythematosus, type 1 diabetes), asthma and allergies, inflammatory bowel

Table 33.2 Some Potential Impacts of Gut Microbiota on Disease

Condition	Microbiome features	Potential mechanism
Exacerbation of Crohn's disease, psoriasis, chronic fatigue syndrome	Decreased abundance of the bacterium *Faecalibacterium prausnitzii*	Thought to be related to ability of *F. prausnitzii* to produce butyrate
Flares of stress-induced irritable bowel syndrome	Gut *Lactobacillus* spp. proliferate under stress	*Lactobacillus* spp. produce indole-3-acetate, which inhibits development of protective gut secretory cells
Depression in premenopausal women	High abundance of *Klebsiella aerogenes*	*K. aerogenes* produces an enzyme that degrades estradiol; declining estradiol levels previously linked to depression in humans
Cognitive decline in aging	Increased abundance of bacteria in the Ruminococcaceae family	Ruminococcaceae bacteria release isoamylamine, which triggers apoptosis of microglial cells in brain
Risk of developing Alzheimer's disease	*Adlercreutzia, Hornefia nodata* group, *Eisenbergiella, Faecalicatena* spp., *Gordonibacter*, and *Prevotella* are thought to be protective; *Collinsella, Bacteroides, Lachnospira*, and *Veillonella* are considered risk factors	Secreted metabolites trigger a neuroinflammatory response
Development of multiple sclerosis	Many changes, including decreased abundance of *Prevotella, Butyricimonas*, and *Bacteroides*	Modulation of host immunity through regulation of multiple metabolic pathways (e.g., SCFAs, bile acid modification) and increased mucin degradation

disease, and neurologic conditions ranging from Parkinson's disease to ASD. **Table 33.2** lists some examples that have been observed in humans and mechanistically studied in mice. Some of these conditions are inflammatory disorders, and it has become clear within the last decade or so that conditions involving chronic inflammation are influenced and in some cases caused by an altered microbiome. When the balance of microbial phyla is disrupted to the point that host homeostasis is altered, it is called **dysbiosis (figure 33.10)**. Here we focus on three conditions that together account for the majority of illness and death in industrialized nations: metabolic syndrome, cardiovascular disease, and cancer.

Metabolic Syndrome

Metabolic syndrome is a condition characterized by at least three of the following: large waist circumference, high blood triglyceride level, high blood pressure, elevated low-density lipoprotein, and high fasting blood glucose levels. The pathophysiology of metabolic syndrome involves insulin resistance, which is the inability of the body's cells to take up insulin from the bloodstream as they should. Because insulin is needed to move glucose from the bloodstream into cells, insulin resistance leads to high blood sugar levels—the hallmark of diabetes mellitus. Insulin resistance precedes the development of type 2 diabetes and both conditions are associated with obesity (by contrast, type 1 diabetes is an autoimmune disease whereby host immunity attacks the pancreas, the site of insulin production). Metabolic syndrome also increases the risk of stroke and heart disease. Importantly the World Health Organization reports that global obesity has more than doubled since 1980, and metabolic syndrome is now a global epidemic.

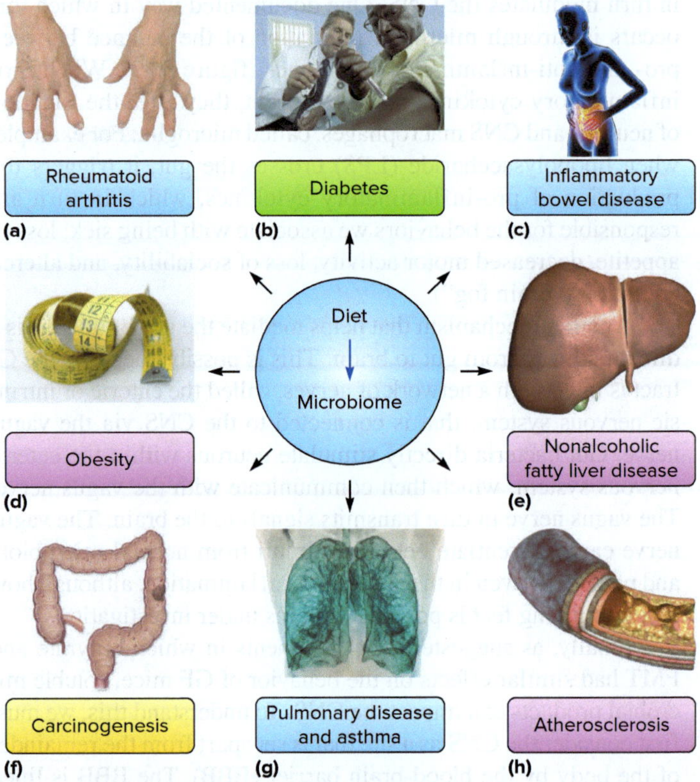

Figure 33.10 Loss of Microbial Diversity Leads to Dysbiosis and a Variety of Diseases That Involve Inflammation.

(a) Suze777/iStock/Getty Images; (b) Adam Gault/Science Photo Library RF/Science Source; (c) Roger Harris/Science Photo Library RF/Science Source; (d) gitanna/123RF; (e) MedicalRF.com; (f) Electronic Publishing Services, Inc., NY/McGraw Hill; (g) Zephyr/Science Photo Library/Alamy Stock Photo; (h) MedicalRF.com. Source: Thaiss, C. A.; Zmora, N.; Levy, M.; and Elinav, E. 2016. "The microbiome and innate immunity." *Nature*, 535(7610): 65–74.

Metabolic syndrome and obesity are associated with chronic, low-level inflammation, and this appears to be linked to microbiota as explained by a phenomenon known as **metabolic endotoxemia** (**figure 33.11**). Metabolic endotoxemia is caused by a series of events that begins with a diet high in fat that induces a change in gut microbiota leading to limited microbial diversity and dysbiosis. Dysbiosis in turn results in an increase in the permeability of tight junctions between enterocytes, making the mucosal barrier leaky. This enables LPS, which is sloughed from diderm bacteria as they grow, to breach the mucosal barrier. In fact, blood levels of LPS in people who are obese can be two to three times higher than that measured in individuals who are lean. LPS is a very potent immune-modulating molecule that, when detected by immune and other cells including adipocytes (fat cells), readily stimulates the release of pro-inflammatory cytokines. Over time, this is thought to result in the low-level, chronic inflammation that characterizes obesity and metabolic syndrome. In addition, LPS in the liver promotes insulin resistance, accumulation of triglycerides, and the development of fatty liver disease. LPS in the blood also induces the expression of pro-inflammatory and adhesion molecules by cells that line blood vessels, promoting the development of cardiovascular disease through atherosclerosis. Atherosclerosis is marked by the accumulation of fat, cholesterol, calcium, and other compounds collectively called plaques on the insides of blood vessels.

While metabolic endotoxemia helps explain the chronic inflammation associated with obesity and metabolic syndrome, it does not address the composition of the microbiome itself. Compelling evidence for the role of the microbiome in metabolic syndrome is obtained by the reproducible observation that insulin resistance can be transferred by FMT. Although this is not the whole story, microbes that contribute to dysbiosis result in the increased production of acetate over butyrate. Acetate is

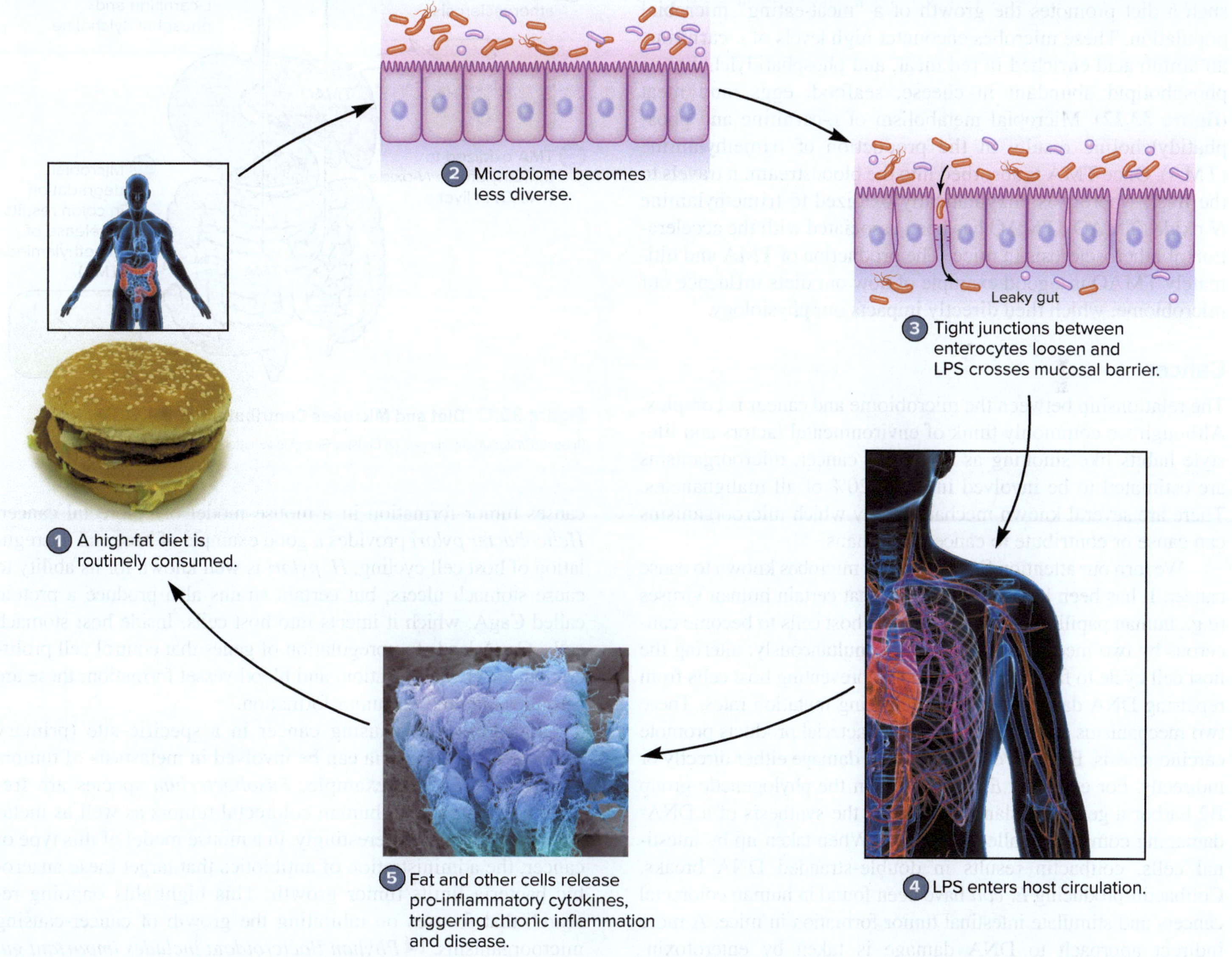

Figure 33.11 The Mechanism of Metabolic Endotoxemia. LPS, lipopolysaccharide.

(1 top) Sebastian Kaulitzki/Eraxion/iStockphoto/Getty Images; (1 bottom) Mark Dierker/McGraw Hill; (4) Science Photo Library -SCIEPRO/Brand X Pictures/Getty Images; (5) Steve Gschmeissner/Science Photo Library/Alamy Stock Photo

converted to fatty acids and triglycerides (figure 33.4). High levels of acetate diminish the positive effects butyrate has on the immune system and promote the production of the appetite-stimulating hormone ghrelin. Numerous studies have shown that a high-fat diet selects for a less diverse microbiome that produces acetate and other products such as branched-chain amino acids, thereby increasing caloric uptake by the host.

Cardiovascular Disease

Although cardiovascular disease and metabolic syndrome are linked by the common finding of atherosclerosis, let's now consider the stereotypical diet consumed by a person with cardiovascular disease: lots of red meat and high-fat foods. From a microbiome point of view, such individuals are feeding their gut microbes little of the fiber that will foster microbial diversification and the production of anti-inflammatory SCFAs. Instead, such a diet promotes the growth of a "meat-eating" microbial population. These microbes encounter high levels of L-carnitine, an amino acid enriched in red meat, and phosphatidylcholine, a phospholipid abundant in cheese, seafood, eggs, and meat (**figure 33.12**). Microbial metabolism of L-carnitine and phosphatidylcholine results in the production of trimethylamine (TMA). Once TMA is absorbed into the bloodstream, it travels to the liver, where it is enzymatically oxidized to trimethylamine *N*-oxide (TMAO). TMAO has been associated with the acceleration of atherosclerosis in mice. The production of TMA and ultimately TMAO is a good example of how our diets influence our microbiome, which then directly impacts our physiology.

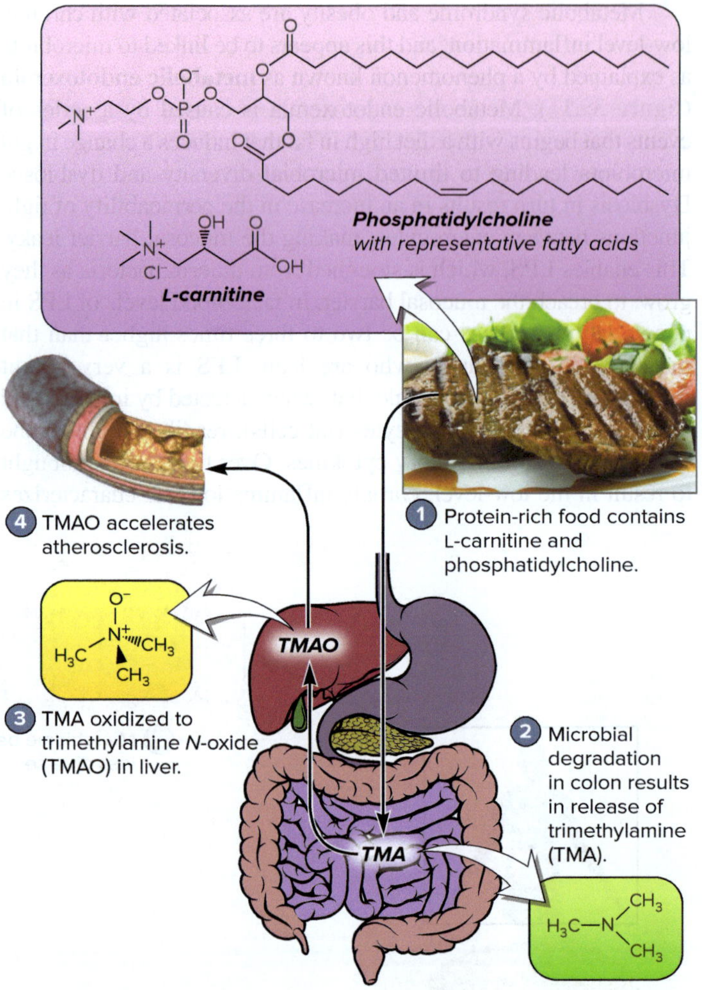

Figure 33.12 Diet and Microbes Contribute to Atherosclerosis.
(1) Gbh007/iStock/Getty Images; (2) Christos Georghiou/Shutterstock; (4) MedicalRF.com

Cancer

The relationship between the microbiome and cancer is complex. Although we commonly think of environmental factors and lifestyle habits like smoking as drivers of cancer, microorganisms are estimated to be involved in about 20% of all malignancies. There are several known mechanisms by which microorganisms can cause or contribute to cancer in humans.

We turn our attention first to specific microbes known to cause cancer. It has been known for decades that certain human viruses (e.g., human papillomavirus) cause their host cells to become cancerous by two mechanisms operating simultaneously: altering the host cell cycle to favor proliferation and preventing host cells from repairing DNA damage, thereby increasing mutation rates. These two mechanisms also account for how bacterial products promote carcinogenesis. Bacteria can cause DNA damage either directly or indirectly. For example, *E. coli* strains in the phylogenetic group B2 harbor a genomic island that directs the synthesis of a DNA-damaging compound called colibactin. When taken up by intestinal cells, colibactin results in double-stranded DNA breaks. Colibactin-producing *E. coli* have been found in human colorectal cancers and stimulate intestinal tumor formation in mice. A more indirect approach to DNA damage is taken by enterotoxin-producing strains of *Bacteroides fragilis* that secrete a metalloprotease called Bft. Bft stimulates the production of high levels of mutagenic radical oxygen species by host cells. Its role in carcinogenesis has been confirmed by experiments in which purified Bft causes tumor formation in a mouse model of colorectal cancer. *Helicobacter pylori* provides a good example of bacterial dysregulation of host cell cycling. *H. pylori* is well known for its ability to cause stomach ulcers, but certain strains also produce a protein called CagA, which it injects into host cells. Inside host stomach cells, CagA leads to upregulation of genes that control cell proliferation, survival, migration, and blood vessel formation; these are activities required for tumor formation.

In addition to causing cancer in a specific site (primary tumorigenesis), bacteria can be involved in metastasis of tumors to distant sites. For example, *Fusobacterium* species are frequently isolated from human colorectal tumors as well as metastatic liver tumors. Interestingly, in a mouse model of this type of cancer, the administration of antibiotics that target these anaerobic bacteria limits tumor growth. This highlights ongoing research that focuses on inhibiting the growth of cancer-causing microorganisms. ◀ *Phylum Bacteroidota includes important gut microbiota (section 20.6); Phylum Fusobacteriota: commensal anaerobes (section 20.7)*

In addition to single microbial species causing cancer, many cancers linked to microorganisms are driven by the inflammatory state associated with dysbiosis (figure 33.10). Low-level chronic inflammation provoked by obesity and metabolic endotoxemia is associated with an increased risk to a variety of cancers. Epidemiology studies reveal that people with obesity are at a higher risk of developing esophageal, liver, kidney, pancreatic, colorectal, endometrial, breast, ovarian, and thyroid cancers. In fact, about one in five of all cancer deaths in the United States is linked to obesity.

The microbiome even impacts the host response to chemotherapy. As discussed, a functional core microbiome is needed for optimal immunity. Because response to both traditional chemotherapy and newer immunotherapies requires an active, functional immune system within the tumor environment, the microbiome can have a significant effect on treatment. In addition, there is evidence suggesting that consumption of *Bifidobacterium* spp. promotes immunity to tumors. This microbe is abundant in yogurt and other probiotic foods and products, as discussed next.

Comprehension Check

1. What is metabolic endotoxemia and how is it related to dysbiosis?
2. Compare the microbial products released after eating a fiber-rich salad to that after a hamburger and fries.
3. Describe the relationship between microbes, obesity, and cancer.

33.5 Microbiome Manipulation Can Be Therapeutic

After reading this section, you should be able to:

a. Describe how a product is classified as a probiotic
b. Predict uses of probiotics in the prevention of illness

Many years ago, a television commercial for yogurt featured a town full of centenarians (at least they were advertised as 100-year-old people) living in Eastern Europe who attributed their longevity to eating yogurt of a specific brand. It turns out that in 1908, microbiologist and Nobel laureate Élie Metchnikov also knew about these villagers and postulated that lactic acid–producing bacteria promoted their health. In the microbiome era, the probiotic industry has blossomed into a $60 billion enterprise that continues to expand.

Despite this burgeoning market, many consumers are not really certain what the term *probiotic* means. In an effort to standardize the use of the term **probiotic,** the Food and Agricultural Organization of the United Nations–World Health Organization (FAO-WHO) has defined it as "live microorganisms, which, when administered in adequate amounts, confer a health benefit to the host." FAO-WHO guidelines provide specific requirements that should be met for foods and supplements to meet this standard.

Despite these guidelines, the U.S. Food and Drug Administration (FDA) does not regulate probiotic foods and supplements, so the health benefits the manufacturers of such products claim have not been rigorously tested. By contrast, if microbes are to be used as a medicine, the FDA must approve their use. Examples include the use of weakened *Mycobacterium bovis* bacteria in BCG to treat superficial bladder cancer and attenuated viruses in vaccines (e.g., measles and rubella vaccines). These products have a long track record of safe, therapeutic success and are regulated as drugs. More recently, the FDA has approved defined mixtures of fecal bacteria to treat recurrent *Clostridoides difficile* infection (see chapter opening story).

Another confounding factor is that until recently it has been difficult to evaluate whether probiotic bacteria actually colonize the gut and are retained long enough to have an effect. The discovery that many so-called probiotic microorganisms do not effectively colonize the intestines has led to the development of synbiotics. **Synbiotics** are foods or supplements that include both a prebiotic and a probiotic. A **prebiotic** is a compound(s) added to enhance the colonization and positive health benefits of probiotic microbes.

Although it is much harder to control the uniformity and effectiveness of a product that contains microbes than a drug with a unique chemical structure, the recognition that the microbiome is, in essence, a newly discovered organ needed for human homeostasis is propelling the unstoppable development of probiotics and synbiotics for preventative medicine. It is likely that hospitalized patients will routinely be given an oral dose of bacteria designed to protect against the development of infection with antibiotic-resistant pathogens. Similarly, outpatients requiring antibiotic therapy may also be prescribed a synbiotic cocktail for microbiome reconstitution following treatment. Tourists may soon be packing probiotics to prevent travelers' diarrhea along with their sunscreen.

While we await further development of probiotic therapies, it is instructive to consider the successful use of probiotics in animals. For example, *Lactobacillus acidophilus* is used in beef cattle feed. When cattle consume feed that has been sprayed with these bacteria, they have markedly lower carriage of the pathogenic *E. coli* strain O157:H7. This can make it easier to produce beef that meets current standards for microbiological quality at the time of slaughter. Probiotics are also used successfully with poultry. For instance, a probiotic *Bacillus subtilis* strain is fed to chickens to increase body weight. There is also a reduction in coliforms and *Campylobacter* spp. in the processed carcasses. This probiotic may decrease the need for antibiotics in poultry production and pathogen levels on farms. *Salmonella enterica* can be controlled by spraying a patented blend of 29 bacteria, isolated from the chicken cecum, on day-old chickens. As they preen themselves, the chicks ingest the bacterial mixture, establishing a functional microbiome in the cecum that limits *Salmonella* colonization of the gut. In an age of increasing antimicrobial resistance, the use of probiotics may prove to be valuable for both humans and the animals they consume.

Comprehension Check

1. Describe the specific use of a currently used probiotic product and one under development.
2. How are probiotics used in agriculture?

Key Concepts

33.1 Humans Are Holobionts

- The average human adult carries about the same number of microbial cells as human cells. Microorganisms that take up residence on a human host without causing disease are called the normal microbiota.
- The Human Microbiome Project and similar projects worldwide are evaluating the role microbes, their genes, and their metabolites have in maintaining homeostasis and causing disease in the host.
- Metagenomic and bioinformatic analysis of the human microbiome has revealed an enormous diversity of microorganisms living on and in the human body. These microbes are essential in establishing and maintaining the health of the host.

33.2 The Microbiome Develops from Birth to Adulthood

- Various microbes have adapted to specific niches found on the human host. These niches are uniquely able to support microbial growth (**figure 33.2**).
- Microorganisms living on or in the skin can be characterized as either transients or residents. *Staphylococcus epidermidis,* a common skin inhabitant, uses several mechanisms to keep skin healthy (**figure 33.3**).
- The oral cavity provides a nutrient-rich habitat but subjects microbes to physical processes that can dislodge them. The normal microbiota of the oral cavity form biofilms that resist this mechanical removal.
- The stomach contains very few microorganisms due to its acidic pH.
- The distal portion of the small intestine and the entire large intestine have the largest microbial community in the body. A person may host between 500 and 1,000 species, of which the vast majority are anaerobic; many have not been cultured.
- The upper urinary tract is usually free of microorganisms. The tip of the urethra (a mucous membrane) harbors numerous skin bacteria. The adult female genital tract has a more complex microbiota that prevents other microorganisms from colonizing it.

33.3 A Functional Core Microbiome Is Required for Host Homeostasis

- A functional core microbiome provides the host with many enzymes capable of digesting complex polysaccharides that would otherwise be indigestible, as well as a variety of metabolites that help maintain homeostasis (**figure 33.4**).
- Gut microbiota exclude pathogens using both direct and indirect mechanisms (**figure 33.5**).
- Germfree mice do not develop a mature, functional immune system, demonstrating the need for microbial colonization for optimal host immunity (**figure 33.6**).
- Fiber-rich foods promote the development of a microbiome that includes microorganisms that produce the short-chain fatty acids acetate, butyrate, and propionate. Butyrate and propionate have positive effects on innate and adaptive immune cells.
- Mucosal tolerance is marked by the absence of an inflammatory response to the many foreign antigens that interact with our mucous membranes, in particular in our GI tracts. Microbial products are needed to maintain the appropriate balance of pro-inflammatory T_H17 cells and anti-inflammatory Treg cells (**figure 33.7**).
- The gut microbiome influences the central nervous system in a system known as the gut-brain axis. Three mechanisms have been proposed to help explain this interesting phenomenon (**figure 33.9**).

33.4 Many Diseases Have a Connection with Dysbiosis

- Lack of microbial diversity and altered ratios of specific bacterial phyla are the hallmarks of dysbiosis, which is linked to a variety of diseases (**figure 33.10**).
- Obesity and metabolic syndrome are associated with dysbiosis and involve chronic inflammation, thought to be caused by metabolic endotoxemia (**figure 33.11**).
- Foods rich in L-carnitine and phosphatidylcholine promote the development of microbiota that produce trimethylamine, which, when metabolized in the liver, accelerates the development of atherosclerosis (**figure 33.12**).
- One in five cancers is linked to obesity and the inflammatory state it causes through dysbiosis.

33.5 Microbiome Manipulation Can Be Therapeutic

- Foods containing microorganisms thought to confer a health benefit have been consumed for hundreds of years and are termed probiotics. The recent appreciation of the human gut microbiome has helped engender more interest in probiotic foods, supplements, and medicines.
- Probiotic products are also used in agriculture, where it is hoped they will diminish reliance on antibiotics.

Active Learning

1. A group of obese mice and a group of lean mice are fed an identical amount of the same chow each day. The mice are carefully monitored so there is no difference in the activity level in the two groups. Discuss why neither group of mice gains or loses weight.
2. Butyrate is known to increase mitochondrial activity in enterocytes, priopionate is known to contribute to gluconeogenesis in the liver, while acetate contributes to lipid biosynthesis in the liver and other tissues. Refer to chapters 11 and 12 and trace means by which these fatty acids are produced by bacteria and then used by the human host (these pathways are conserved across domains). Based on your review, why does acetate serve as a lipid substrate, thereby making it "obesogenic" as compared to butyrate or propionate?
3. Yo-yo dieting is all too common; after losing weight, about 80% of people regain every kilogram within a year or less. Then comes another diet, and another round of weight loss and gain. Although weight loss is associated with a rapid decline in metabolic markers of obesity (e.g., blood triglyceride levels), changes in the community of gut microbiota lags far behind in returning to that of people who are lean. How might the development of "microbiome-targeted approaches" help limit yo-yo dieting? Discuss the information you need to develop such a new approach to weight loss.

CHAPTER

34

Infection and Pathogenicity

Channarong Pherngjanda/Shutterstock; Lunx/Shutterstock; SUPREEYA.CH/Shutterstock

The Unlikely Tale of Miasmas, Bras, and Masks

This may seem like the start of a bad joke: "What's the difference between an N95 mask and a sports bra?" The surprising answer is: Not as much as you would think! The history of face masks is a remarkable tale of misconception, persistence, and ingenuity.

Long before bacteria and viruses were discovered, people intuitively covered their nose and mouth for protection, believing that diseases were derived from "pestilential miasma": putrid gases emanating from the ground, a misconception that persisted until the 1800s. The engraving from the 1600s shown above illustrates the characteristic herb-filled mask used by physicians at that time. Its long, birdlike beak is unsettling and has come to represent the terror of the medieval Black Death (bubonic plague).

In the fall of 1910, a plague epidemic arose in Asia. It had an almost 100% mortality rate and triggered intense competition to find the cause of the disease and stop its spread. A young Chinese doctor named Lien-teh Wu determined that the plague is not always the result of a flea bite, as was the accepted theory, but could also be transmitted through the air. Using inspiration from surgical masks, Wu designed a hardier mask made with gauze and several layers of cotton to filter inhaled air. His idea was initially discredited, but the evidence was conclusive: Those without a mask died; those wearing Wu's mask survived. The mask was cheap and easy to produce, and its efficacy was undeniable, resulting in the first successful use of widespread mask-wearing to prevent disease transmission.

The need for protection from inhaled microscopic particles was well understood by the twentieth century. During World Wars I and II, gas masks were cumbersome with their fiberglass filters, as were those developed to prevent black lung in miners. The need for a lightweight yet effective mask was obvious, and the solution came from an unexpected source. Former *House Beautiful* editor Sara Little Turnbull collaborated with the 3M company to use their melted-polymer fabric in the fashion industry. In 1958, after delivering a presentation suggesting over 100 uses for the fabric, she was tasked with designing a molded bra. While working on her design, Turnbull, who was also caring for three sick family members, noticed health-care professionals struggling with flat, tie-on masks. Turnbull adapted the design for her molded bra cup and the N95 mask was born. As the proverb states, "necessity is the mother of invention." Who would have anticipated that the history of face masks and bras would be intertwined?

Readiness Check:

Based on what you have learned previously, you should be able to:

- ✓ Describe the structure and function of the main cellular components of bacteria, viruses, protists, and fungi, as well as their replication processes and energy requirements (chapters 3, 5, 6, 7, 10, 12, 13, and 15)
- ✓ Explain symbiosis and differentiate the true definition of parasitism from several common uses of the word (section 27.4)
- ✓ Relate the principles of symbiosis to human-microbe interactions (chapter 33)

34.1 The Process of Infection

After reading this section, you should be able to:

a. Differentiate between pathogenicity and virulence
b. Compare opportunistic, extracellular, and intracellular pathogens and discuss their requirements for survival
c. Relate the infectious disease process to time, identifying events associated with each stage of the process
d. Create a concept map to summarize the infectious process

When a pathogenic microorganism is growing and multiplying on or within a larger organism, the latter, known as the **host,** is said to have an **infection.** The nature of an infection can vary widely with respect to severity, location, and type of organisms involved. An infection may or may not result in overt disease. Any organism that causes disease is known as a **pathogen** (Greek *patho,* disease; *gennan,* to produce), and its ability to cause disease is called **pathogenicity.** An organism's **virulence** is the *degree* of harm (pathogenicity) inflicted on its host. An **infectious disease** is any change from a state of health in which part or all of the host is incapable of carrying on its normal functions due to the presence of a pathogen or its products. Even resident microbiota, such as those associated with the gut or skin, can become **opportunistic pathogens** if they infect a host away from their typical niche, especially in a host with a weakened immune system (a compromised host). Host and tissue specificity offer microorganisms shelter and access to resources; however, relationships between organisms can be complex.

To cause disease, microorganisms must not only contact the host but must also be able to survive in it. Bacteria, viruses, protists, and fungi need a suitable environment in which they can grow and reproduce. The range of factors required—temperature, pH, moisture, oxygen concentration, etc.—are provided by the human host. These conditions can be met outside or inside host cells, depending on the pathogen. Infectious microorganisms are considered **extracellular pathogens** if they remain in tissues and fluids but never enter host cells during the course of disease. For instance, some bacteria and fungi can actively grow and multiply in the blood or tissue spaces. Microbes that grow and multiply within host cells are called **intracellular pathogens.** These can be further subdivided into two groups. **Facultative intracellular pathogens** are those organisms that reside within the cells of the host or in the environment but can also be grown in pure culture without host-cell support. *Brucella abortus* is a bacterium that grows and replicates within macrophages, neutrophils, and trophoblast cells (cells that surround the developing embryo) and *Histoplasma capsulatum* is a fungus that grows within macrophages, yet both can be grown in vitro. In contrast, **obligate intracellular pathogens** are incapable of growth and multiplication outside a host cell. By definition, all viruses are obligate intracellular pathogens in that they require a host cell for replication, often to the detriment of that cell. Some bacteria, such as *Chlamydia* spp. and rickettsias (which cause diseases such as Rocky Mountain spotted fever and typhus), are also obligate intracellular pathogens. These microbes cannot be grown in the laboratory outside of their host cells.

Infectious diseases are typically not detected until distinct physical and/or physiological events alert the host to a pathogen's presence. Infectious diseases often have characteristic signs and symptoms. **Signs** are objective changes that can be directly observed and measured, such as fever or rash. **Symptoms** are subjective changes, such as pain and loss of appetite. The term *symptom* is often used in a broader scope to include the clinical signs. A **disease syndrome** is a set of signs and symptoms that are characteristic of the disease.

Clinically, the course of an infectious disease usually has a characteristic pattern and can be divided into several phases (**figure 34.1**). The **incubation period** is the time between pathogen entry and development of signs and symptoms. The pathogen is reproducing but has not reached a sufficient level to cause clinical manifestations. The length of this period varies, depending on the pathogen involved and the number of microbes initially transmitted to the host. Next, the **prodromal stage** occurs with mild, nonspecific signs and symptoms that are not yet defined enough to make a clear diagnosis. However, the infected person is often contagious. This is followed by the **illness period,** when the disease is most severe and characteristic signs and symptoms are apparent. The host adaptive immune response is typically triggered by this stage. Finally, during the period of decline, the signs and symptoms begin to disappear. The recovery stage is often referred to as **convalescence.**

The fundamental process of infection is essentially a competition for resources. The host provides protection, nutrients, and energy for the pathogen. Infectious agents must therefore

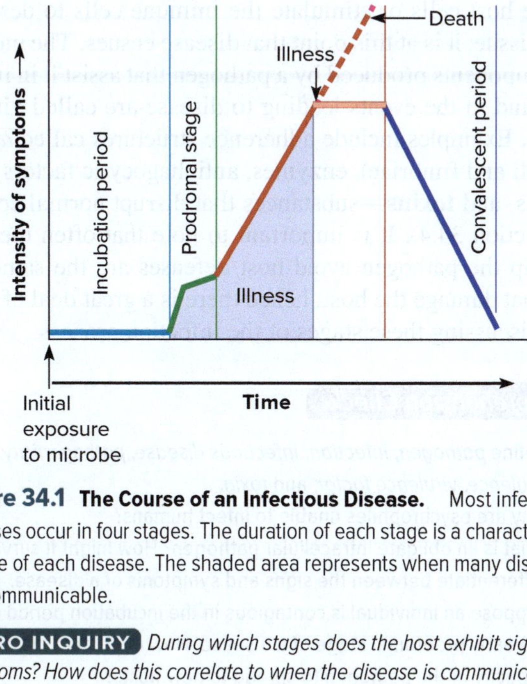

Figure 34.1 The Course of an Infectious Disease. Most infectious diseases occur in four stages. The duration of each stage is a characteristic feature of each disease. The shaded area represents when many diseases are communicable.

MICRO INQUIRY *During which stages does the host exhibit signs and symptoms? How does this correlate to when the disease is communicable?*

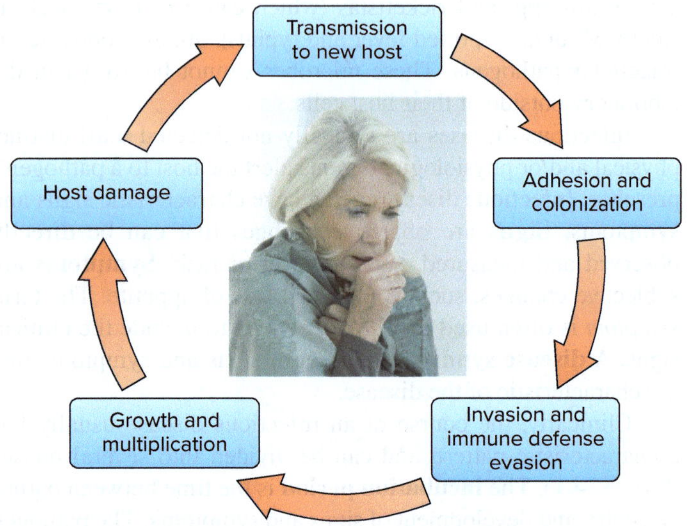

Figure 34.2 Cycle of Bacterial Pathogen Infection. Several steps are required for a pathogen to infect a host and cause disease.

JPC-PROD/Shutterstock

develop mechanisms to access and exploit their hosts while simultaneously avoiding the host's immune defenses. For a pathogen to infect a host and subsequently cause disease, specific events must take place (**figure 34.2**). The organism must be transmitted from the previous host and enter the new host, where it adheres and colonizes. The success of these processes is affected by the virulence of the organism, the number of invading organisms, and the immunocompetence of the host. Next, the organism must be able to outcompete the resident microbiota for resources and survive myriad host defense mechanisms by producing a variety of molecules as a means of counterattack. Finally, the organism must either produce molecules that directly damage host cells or stimulate the immune cells to destroy infected tissue; it is at this point that disease ensues. The molecules and components produced by a pathogen that assist it in invading a host and in the events leading to disease are called **virulence factors.** Examples include adherence structures called *adhesins* (e.g., pili and fimbriae), enzymes, antiphagocytic factors such as capsules, and **toxins**—substances that disrupt normal cell function (section 34.4). It is important to note that often the factors that help the pathogen avoid host defenses are the same molecules that damage the host; hence there is a great deal of overlap when discussing these stages of the infection process.

Comprehension Check

1. Define *pathogen, infection, infectious disease, pathogenicity, virulence, virulence factor,* and *toxin.*
2. Why are psychrophiles unable to infect humans?
3. What is an obligate intracellular pathogen? How might it survive?
4. Differentiate between the signs and symptoms of a disease.
5. Suppose an individual is contagious in the incubation period or in the early prodromal stage of a disease. How might that affect the capacity of the pathogen to spread to new hosts?

34.2 Transmission and Entry into the Host

After reading this section, you should be able to:

a. List and describe the means by which microorganisms access human cells and tissues to cause disease
b. Correlate initial microbial numbers and replication rates to infection and lethality
c. Compare the molecular mechanisms by which microorganisms adhere to and invade human cells and tissues

Before a microbe can cause disease in a human, the two organisms must come into contact. Pathogens are transmitted from a *source,* which is the location from which the pathogen is transferred to the host, either directly or indirectly through an intermediate agent. The source can be either animate (e.g., humans or animals) or inanimate (e.g., water or food). The period of infectivity is the time during which the source is disseminating the pathogen.

A **reservoir** is the natural host or environment in which the pathogen normally resides. Pathogens in a reservoir can be transmitted directly from the reservoir to a human host, or to a source and then to the human; thus a reservoir sometimes functions as a source. An example of direct transmission from a reservoir that is also a source is drinking water that, when contaminated with *Vibrio cholerae,* infects humans. By contrast, West Nile virus is spread from birds (reservoir) to mosquitoes, and then on to humans.

When infectious agents are transmitted from animals to humans, the diseases that ensue are termed **zoonoses** (Greek *zoon,* animal; *nosos,* disease). For example, a dog with rabies can transmit the causative virus while biting a human. Similarly, biting by arthropod **vectors** (organisms such as mosquitoes, ticks and fleas, that transmit pathogens from one host to another) can also spread pathogens, leading to infection (e.g., Zika virus disease, malaria, Lyme disease, and plague).

Transmission and Virulence Are Interrelated

Transmission is a critical component of the infection process. It occurs either directly or indirectly by four main routes: airborne, contact, vehicle, or vector-borne. Females who are pregnant can also pass on a pathogen to their unborn child via vertical transmission. A pathogen's virulence may be strongly influenced by its mode of transmission and ability to live outside its host (**Historical Highlights 34.1**).

When a pathogen can survive for long periods outside its host, it can afford to leave the host and simply wait for a new one to come along. Examples of such pathogens include *Mycobacterium tuberculosis, Corynebacterium diphtheriae,* and hepatitis B virus, the causative agents of tuberculosis, diphtheria and hepatitis B, respectively. These microbes survive for a relatively long time—from hours to days and weeks—outside

HISTORICAL HIGHLIGHTS

34.1 The First Indications of Person-to-Person Spread of an Infectious Disease

In 1773 Charles White, an English surgeon and obstetrician, published his "Treatise on the Management of Pregnant and Lying-in Women." In it, he appealed for surgical cleanliness to combat childbed or puerperal fever, an acute febrile condition that can follow childbirth and is caused by streptococcal infection of the uterus or adjacent regions. In 1795 Alexander Gordon, a Scottish obstetrician, published his "Treatise on the Epidemic Puerperal Fever of Aberdeen," which demonstrated for the first time the contagiousness of the disease. In 1843, Oliver Wendell Holmes, a noted physician and anatomist in the United States, published a paper entitled "On the Contagiousness of Puerperal Fever" and also appealed for surgical cleanliness to combat this disease.

However, the first person to realize that a pathogen could be transmitted from one person to another was the Hungarian physician Ignaz Philipp Semmelweis. Between 1847 and 1849, Semmelweis observed that females who had their babies at the hospital with the help of medical students and physicians were four times as likely to contract puerperal fever as those who gave birth with the help of midwives. He concluded that the physicians and students were infecting the females with material remaining on their hands after autopsies and other activities (amazingly, these individuals did not wash their hands after such procedures!). Semmelweis thus began washing his hands with a calcium hypochlorite solution before examining patients or delivering babies. This simple procedure led to a dramatic decrease in the number of cases of puerperal fever and saved the lives of many females. As a result, Semmelweis is credited with being the pioneer of antisepsis in obstetrics. Unfortunately, in his own time, the medical establishment refused to acknowledge his contribution and adopt his procedures. After years of rejection, Semmelweis had a nervous breakdown in 1865. He died a short time later of a wound infection. It is very probable that it was a streptococcal infection, arising from the same pathogen he had struggled against his whole professional life.

human hosts. Many bacterial virulence factors that facilitate survival are encoded in unique sequences of DNA that are readily transferred between bacteria through conjugation, transduction, or transformation (horizontal gene transfer). ◀ *Horizontal gene transfer: creating genetic variability the asexual way (section 16.3)*

Exposure alone is not sufficient for infection to occur. Rather, the pathogen must also make contact with the appropriate host tissue. For instance, rhinoviruses, which cause the common cold, are spread by airborne transmission from one host to another. Once the virus is inhaled, it makes contact with the upper respiratory tract epithelium, where it invades, causing disease only in that tissue; rhinoviruses do not usually infect any other cell type in the host. This specificity is called a **tropism** (Greek *trope,* turning). Many pathogens exhibit cell, tissue, and organ specificities. A tropism by a specific microbe usually reflects the presence of specific cell surface receptors for that microbe on the host cell. ◀ *Attachment (adsorption) (section 6.3)*

Airborne Transmission

Because air is not a suitable medium for the growth of pathogens, any pathogen that is airborne must have originated from a source such as humans, other animals, plants, soil, food, or water. In **airborne transmission,** the pathogen is suspended in the air in droplets, droplet nuclei, or dust.

Droplets form from saliva, mucus, and other body fluids. They can measure up to 2 mm in diameter and are produced when liquids are placed under force. Because of their size, droplets require close proximity for transfer between hosts (less than 1 meter) and are therefore considered to be transmitted directly. **Droplet nuclei** are smaller particles, 1 to 5 μm in diameter, that result from the evaporation of the larger droplets. Because of their small size, droplet nuclei can remain airborne for hours or days and travel long distances. Particles smaller than 1 μm can become **aerosolized** and disperse even farther. Some of the most contagious diseases, including measles and tuberculosis, are transmitted this way.

Dust is an important factor in airborne transmission. When dust is resuspended by some disturbance, microorganisms adhering to the particles become airborne. If the dust carries endospores or pathogens that can survive for relatively long periods outside a host cell, this creates an epidemiological problem, particularly in clinical environments where hospital-acquired infections can arise. Because of the distance droplet nuclei and dust particles can travel, the term "airborne transmission" is often applied here, referring more specifically to particles traveling farther than 1 meter. This is considered transmission by indirect means. ◀ *Healthcare facilities harbor infection agents (section 35.5)*

Contact Transmission

Contact transmission occurs when the host touches the source or reservoir of the pathogen, either directly or indirectly. Direct contact transmission requires an actual physical interaction with the infectious source (**figure 34.3**). This route, frequently called person-to-person contact, occurs primarily by touching, kissing, or sexual contact (e.g., gonorrhea, mononucleosis); by contact with oral secretions or body lesions (e.g., herpes and boils); by nursing mothers (e.g., HIV, staphylococcal infections); and through the placenta (e.g., syphilis; see Vertical Transmission) or

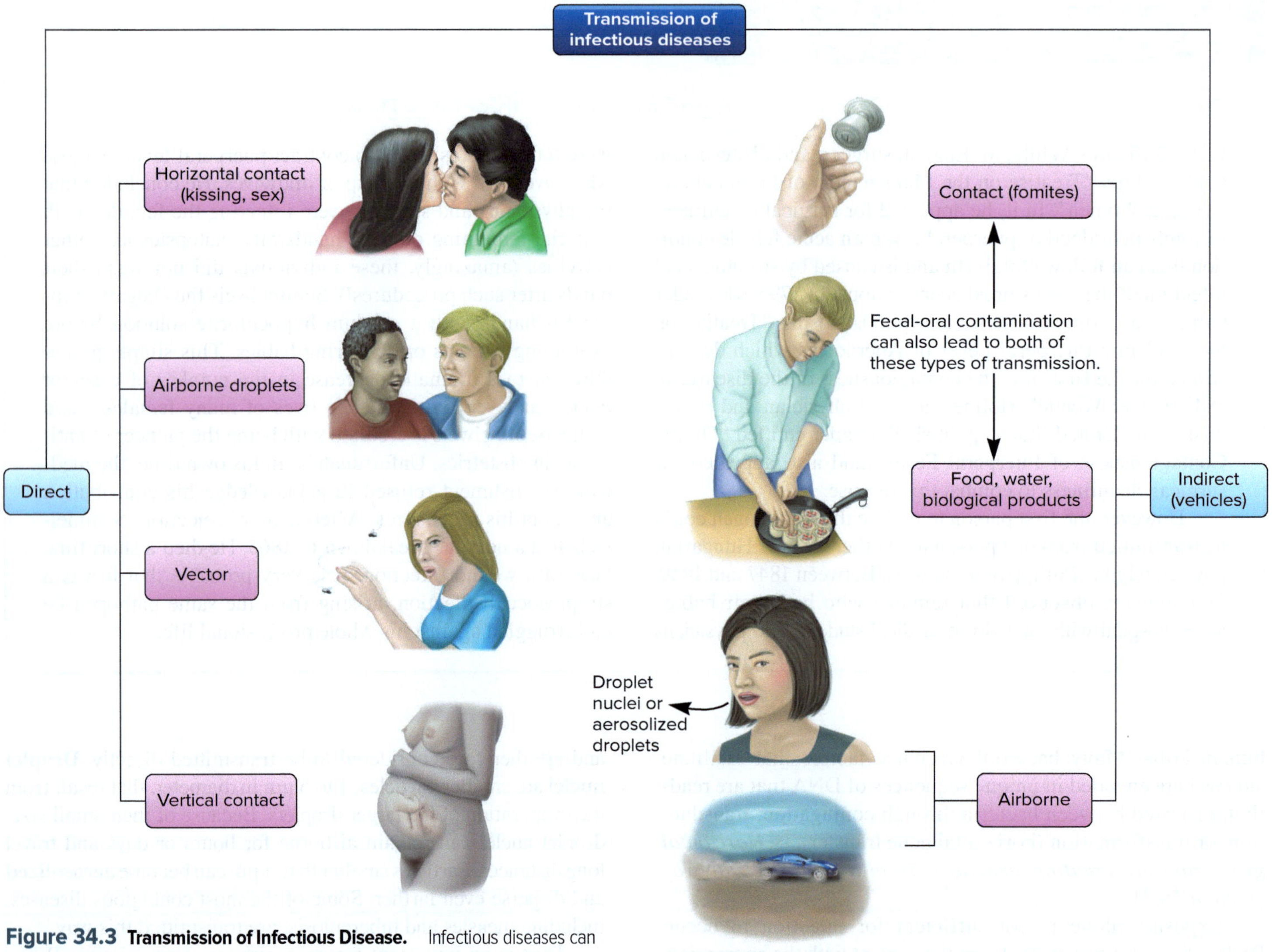

Figure 34.3 Transmission of Infectious Disease. Infectious diseases can be transmitted by various direct and indirect methods.

during birth (e.g., group B streptococcus). Some pathogens also can be transmitted by direct contact with animals or animal products (e.g., members of the genera *Salmonella* and *Campylobacter*). Indirect contact transmission occurs when an inanimate object called a **fomite** transfers the infectious agent between hosts, discussed next.

Vehicle Transmission

Materials that indirectly transmit pathogens are called **vehicles.** In contact vehicle transmission, the vehicle, often a fomite, spreads the pathogen to multiple hosts. Examples of fomites include surgical instruments, drinking vessels, stethoscopes, bedding, eating utensils, and clothing. A single source containing pathogens (e.g., blood, drugs, intravenous fluids) can contaminate a common vehicle that causes multiple infections. Food, water, and biological materials (fluids and tissues) are important common vehicles for many human diseases; moreover they often support pathogen reproduction.

Vector-Borne Transmission

Living organisms that can transmit a pathogen are called vectors. Most vectors are arthropods (e.g., insects, ticks, mites, fleas) or vertebrates (e.g., dogs, cats, skunks, bats). **Vector-borne transmission** involving insects is often categorized as mechanical or biological. In mechanical transmission, the vector carries a pathogen from one host to another, without being infected itself (e.g., a fly carrying bacteria on its legs). With biological transmission, the vector spreads a pathogen that multiplies or carries out part of its life cycle within it (e.g., a mosquito carrying the malaria protozoan). Vector-borne transmission often benefits the

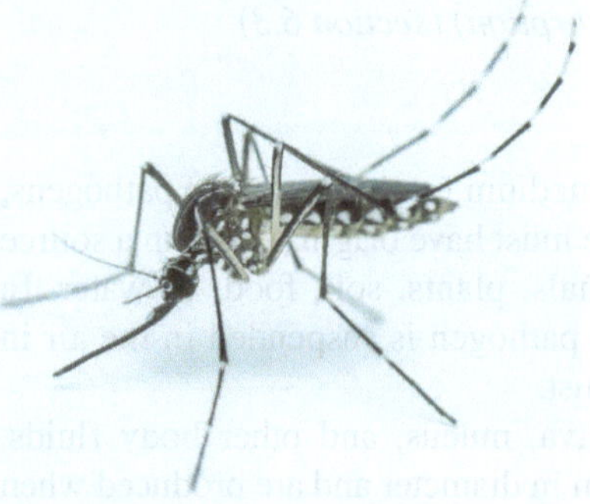

Akil Rolle-Rowan/Shutterstock

pathogen by permitting extensive reproduction and spread across hosts. If pathogen levels are very high in the host, a vector such as a biting insect has a better chance of picking up the pathogen and transferring it to a new host. Indeed, pathogens transmitted by biting arthropods often are highly virulent, causing diseases such as malaria, typhus, and sleeping sickness. It is important that such pathogens do not harm their vectors and the vector generally remains healthy, at least long enough for pathogen transmission.

Vertical Transmission

Direct transmission of pathogens is not only mediated by horizontal, person-to-person passage but by vertical passage as well. **Vertical transmission** of pathogens occurs when a fetus acquires a pathogen from its infected mother. While vertical transmission does not occur as frequently as horizontal transmission, several microorganisms exploit this route, extending their host range. Babies born with an infectious disease are said to have a *congenital* (Latin *congenitus;* born together) infection. Examples of congenital infections include gonorrhea, syphilis, herpes, German measles, and toxoplasmosis. These diseases are discussed in chapters 37 through 39.

The Number of Invading Organisms Differs Among Infections

The rate at which an infection proceeds (the timing of the incubation period and prodromal stage, for example) and its severity are functions of the initial inoculum of microorganisms and their virulence (in addition to the host's resistance). Some microorganisms can establish an infection even when entering the host in low numbers. Conversely, other microorganisms are required in high numbers to establish an infection. The inocula can be measured experimentally to determine the **infectious dose 50 (ID_{50})**—that is, the number of microorganisms required to cause disease in 50% of the inoculated hosts (**figure 34.4**). The ID_{50} varies among pathogens; for example, only 10 *Mycobacterium tuberculosis* cells need be transmitted to cause tuberculosis in the new host, whereas, depending on the species, development of salmonellosis requires transmission of at least 1,000 bacteria. The lower the infectious dose (i.e., the fewer organisms needed for infection), the higher the risk of infection.

Another measure of infectivity, and thus virulence, is the experimental determination of the number of microorganisms required to be lethal to a host species. This is known as the **lethal dose 50 (LD_{50})** and refers to the dose or number of pathogens that kills 50% of a group of experimental hosts within a specified period. Highly virulent pathogens kill at a lower dose when compared to less virulent organisms.

It is important to note that the ID_{50} and LD_{50} for some organisms can also be influenced by the susceptibility of the host. If the host has a robust immune system, the likelihood of developing a disease may be diminished. Factors such as nutrition, cleanliness, emotional health, and genetic predisposition play a role; thus those with a weakened immune system due to medications, stress, or lack of nutrients, for example, may be more likely to contract an infectious disease.

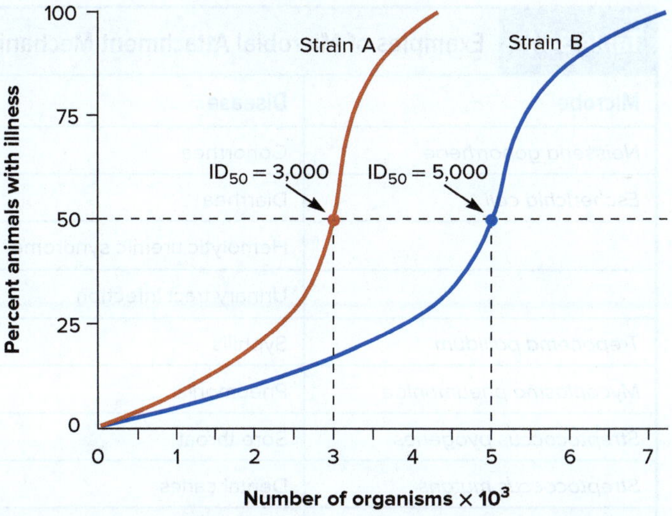

Figure 34.4 Determination of the ID_{50} of a Pathogenic Microorganism. Various doses of a specific pathogen are introduced into experimental host animals. Infections are recorded and a graph constructed. In this example, the graph represents the susceptibility of host animals to two different strains of a pathogen—strain A and strain B. For strain A, the ID_{50} is 3,000, and for strain B, it is 5,000. Hence strain A is more virulent than strain B.

Adhesion and Invasion Factors Initiate Disease

The need to adhere to a host cell, invade the underlying tissues, and colonize is a major obstacle that must be overcome by a pathogen. To accomplish these feats, microbes use molecules—typically proteins or glycoproteins—that are anchored to their cell surface. These surface molecules serve a range of functions to facilitate access to the host. In addition to mediating adhesion and invasion, they allow the pathogen to evade the host immune responses and aid in the formation of biofilms.

Entry and Adhesion Establish Colonization

Once a pathogen has been transmitted from the reservoir or source to a susceptible host in sufficient numbers, access is gained through the appropriate **portal of entry:** the skin, respiratory system, gastrointestinal system, urogenital system, or the conjunctiva of the eye.

After entering the host, the pathogen must be able to adhere to and colonize host cells or tissues. In this context, colonization means the establishment of a site of microbial replication on or within a host. It does not necessarily result in tissue invasion or damage. Adherence can be thought of as a two-step process. Step 1 is nonspecific and often reversible. It results from hydrophobic, electrostatic, and vibrational forces that encourage docking of the microbe on the target cell. Step 2 is specific and permanent, as complementary molecules (cognate receptors) fulfill a lock-and-key style bonding. This latter process is an example of coevolution, permitting pathogen exploitation of the characteristics of host-cell surfaces. Thus pathogens adhere with a high degree of specificity to

Table 34.1 Examples of Microbial Attachment Mechanisms

Microbe	Disease	Adhesion Mechanism	Host Receptor
Neisseria gonorrheae	Gonorrhea	Type IV fimbriae	Mannose on urethral epithelium
Escherichia coli	Diarrhea	Type I fimbriae	Sugar residue on intestinal epithelium
	Hemolytic uremic syndrome	P pili	Sugar residue on kidney cell
	Urinary tract infection	P pili	Sugar residue on urethral epithelium
Treponema pallidum	Syphilis	Outer membrane protein	Fibronectin on mucosal cell
Mycoplasma pneumoniae	Pneumonia	Membrane protein	Sialic acid on lung cell
Streptococcus pyogenes	Sore throat	Protein F	Fibronectin on upper respiratory tract cell
Streptococcus mutans	Dental caries	Sugar residue	Salivary glycoprotein on tooth
Influenza virus	Influenza	Hemagglutinin spike protein	Protein residue on upper respiratory tract cell
Human immunodeficiency virus -1	AIDS	gp120 protein	CD4 receptor on T cells
Human enterovirus C	Poliomyelitis	Capsid protein VP1	CD155 protein on intestinal and nerve cells
SARS-CoV-2	COVID-19	Spike protein	Angiotensin-converting enzyme 2

target tissues. Adhesins such as pili and fimbriae (**table 34.1**), membrane and capsular materials, and specialized adhesion molecules on the surface of the invading microbe bind to host sites (**figure 34.5**). These structures both facilitate infection and contribute to the virulence of the organism, and are thus classified as virulence factors. Examples of pathogen adherence are also discussed in the context of specific diseases in chapters 37 through 39.

Invasion Disseminates Pathogens

Pathogens can be described in terms of their infectivity and invasiveness. **Infectivity** is the ability of the organism to establish a discrete focal point of infection. Attachment to specific host-cell surface molecules permits microbe-mediated infection but may also induce cell-mediated microbe internalization. Viruses are adept at these methods of infectivity. **Invasiveness** is the ability of the organism to spread to adjacent or other tissues. After entry into a host, some pathogens remain localized to a specific site, whereas others have the capacity to invade other tissues.

The first step of invasion is penetration (either active or passive) through the host's surfaces. These include mucous membranes in the respiratory, gastrointestinal, and genitourinary tracts, and other surfaces made up of epithelial cells (epithelium). Active penetration may be accomplished through production of lytic substances that alter the host tissue by (1) attacking the extracellular matrix and basement membranes of body surfaces (integuments) and intestinal linings, (2) degrading carbohydrate-protein complexes between cells or on the cell surface, or (3) disrupting the host-cell surface. Passive mechanisms of penetration are not related to the pathogen. Examples include (1) small breaks, lesions, or ulcers in a mucous membrane that permit initial entry; (2) wounds, abrasions, or burns on the skin's surface; (3) arthropod vectors that create small wounds while feeding; (4) separation of

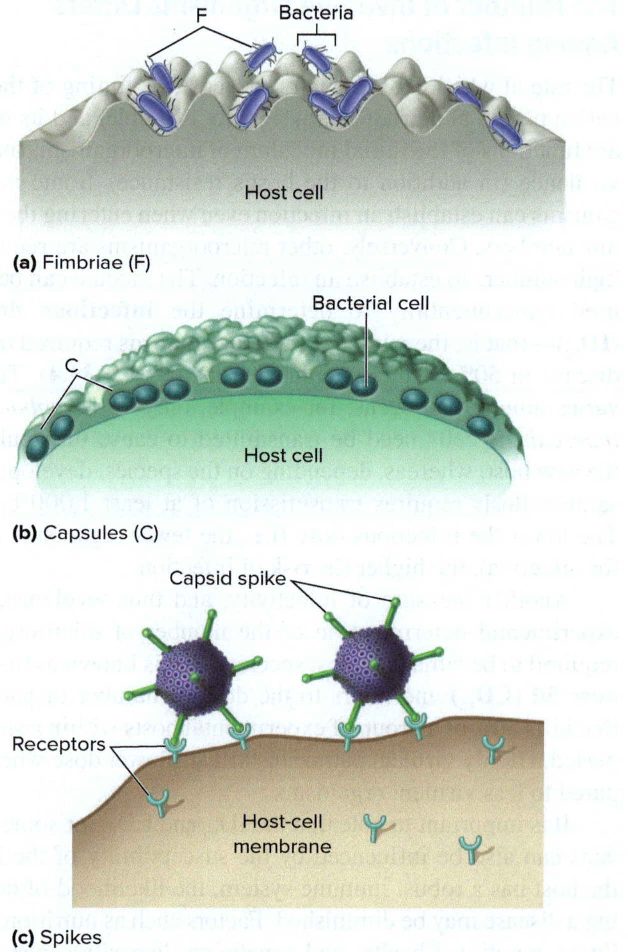

Figure 34.5 Microbial Adherence Mechanisms. (a) Bacterial fimbriae. (b) Bacterial capsules. (c) Viral protein spikes.

tight junctions between cells during inflammation; and (5) tissue damage caused by other organisms (e.g., a dog bite).

Once under the mucous membrane or other barrier, a pathogen may penetrate to deeper tissues and continue disseminating throughout the body of the host. One way the pathogen accomplishes this is by producing specific structures or enzymes that promote spreading (**table 34.2**). These products are also considered virulence factors because they contribute to pathogen success within the host. Pathogens may also enter the small terminal lymphatic capillaries that surround epithelial cells. These capillaries merge into large lymphatic vessels that eventually drain into the circulatory system. Once the circulatory system is reached, the microbes have access to all organs and systems of the host. The presence of viable bacteria in the bloodstream is called **bacteremia.** The infectious disease process caused by bacterial or fungal toxins in the blood is termed **septicemia** (Greek *septikos,* produced by putrefaction; *haima,* blood).

Invasiveness varies greatly among pathogens. For example, *Clostridium tetani* (the cause of tetanus) is considered noninvasive because it does not spread from one tissue to another; it causes disease by producing a toxin that spreads systemically via the blood. Other bacteria, such as *Bacillus anthracis* (the cause of anthrax), also produce toxins, but unlike *C. tetani,* they are also highly invasive. Members of the genus *Streptococcus* span the spectrum of virulence factors and invasiveness with some causing localized strep throat and others causing severe skin and soft-tissue ("flesh-eating") infections. Another method of invasion is seen with *Listeria monocytogenes, Shigella* spp., and *Rickettsia* spp. These intracellular pathogens use cell-surface proteins to bind and polymerize host actin at one pole of the bacterium. The resulting actin tails propel the bacteria within mammalian host cells and allow them to spread from one host cell to the next (**figure 34.6**). To move to an adjacent cell, the bacteria push out against the plasma membrane and form

Table 34.2 Virulence Factors Involved in Bacterial Pathogen Invasion and Dissemination

Product	Organism Involved	Mechanism of Action
Coagulase	*Staphylococcus aureus*	Coagulates (clots) the fibrinogen in plasma. The clot protects the pathogen from phagocytosis and other host defenses.
Collagenase	*Clostridium* spp.	Breaks down collagen that forms the framework of connective tissues; allows the pathogen to spread
Deoxyribonuclease	Group A streptococci, staphylococci, *Clostridium perfringens*	Degrades DNA and thereby lowers viscosity of exudates, giving the pathogen more mobility
Elastase and alkaline protease	*Pseudomonas aeruginosa*	Cleaves laminin associated with basement membranes
Hemolysins	Staphylococci, streptococci, *Escherichia coli, Clostridium perfringens*	Lyse erythrocytes; make iron available for microbial growth
Hyaluronidase	Groups A, B, C, and G streptococci, staphylococci, clostridia	Hydrolyzes hyaluronic acid, a constituent of the extracellular matrix that cements cells together. Its hydrolysis renders the intercellular spaces amenable to passage by the pathogen.
Hydrogen peroxide (H_2O_2) and ammonia (NH_3)	*Mycoplasma* spp., *Ureaplasma* spp.	These metabolic wastes are toxic and damage epithelia in respiratory and urogenital systems.
Immunoglobulin A (IgA) protease	*Streptococcus pneumoniae*	Cleaves immunoglobulin A into Fab and Fc fragments
Lecithinase or phospholipase	*Clostridium* spp.	Destroys the lecithin (phosphatidylcholine) component of plasma membranes, allowing pathogen to spread
Leukocidins	Staphylococci, pneumococci, and other streptococci	Pore-forming exotoxins that kill leukocytes; cause degranulation of lysosomes within leukocytes, which decreases host resistance
Protein A Protein G	*Staphylococcus aureus* *Streptococcus pyogenes*	Located on cell wall. Immunoglobulin G (IgG) binds to either protein A or protein G by its Fc end, thereby preventing complement from interacting with bound IgG.
Pyrogenic exotoxin B (cysteine protease)	Group A streptococci (*Streptococcus pyogenes*)	Degrades host serum proteins
Streptokinase (fibrinolysin) Staphylokinase	Groups A, C, and G streptococci *S. aureus*	A protein that binds to plasminogen and activates the production of plasmin, thus digesting fibrin clots; this allows the pathogen to move from the clotted area.

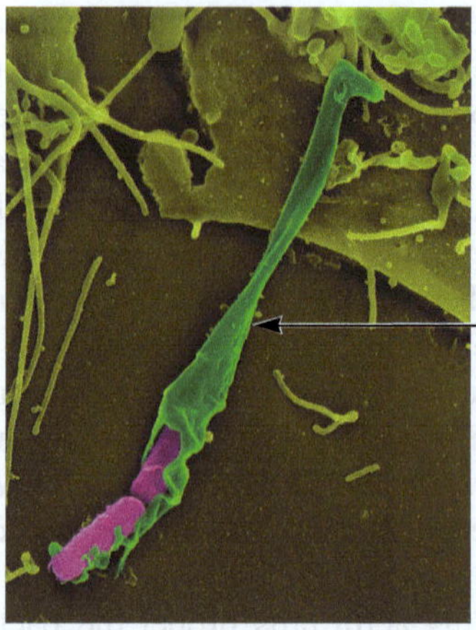

Figure 34.6 Formation of Actin Tails by Intracellular Bacterial Pathogens. Colorized scanning electron micrograph of *Listeria monocytogenes* in a host macrophage. The bacterium (pink) has polymerized host actin into a long tail (green) at only one pole that it uses for intracellular propulsion and to move from one host cell to another.

SPL/Science Source

protrusions. The protrusions are engulfed by the adjacent cell, and the bacteria enter the unsuspecting neighboring host cell, thereby spreading without exposure to the extracellular environment, where they would be easily detected by immune defenses. *Microsporidia* spp. use a novel polar tubule to bore into host cells (*see figure 39.23*), whereas many other fungi use hydrolytic enzymes to invade cells and tissues. These virulence factors are discussed in more detail in chapters 38 and 39.

Comprehension Check

1. Define *source* and *reservoir.* How are they related?
2. Describe the four main types of infectious disease transmission methods and give examples of each.
3. Define *droplet nuclei, vehicle, fomite,* and *vector.*
4. Why are pathogens that are transmitted person-to-person often less virulent than those transmitted by arthropod vectors?
5. How would you determine the ID_{50} of a suspected pathogen?
6. Describe several specific adhesins by which bacterial pathogens attach to host cells.

34.3 Surviving the Host Defenses

After reading this section, you should be able to:

a. Describe features that allow microorganisms to overcome host resistance and immunity
b. Discuss strategies microorganisms have evolved to exploit human cells and tissues as resources for their survival
c. Differentiate between the characteristics of microbes in a biofilm and their planktonic counterparts

The vast majority of microorganisms do not cause disease, in part because they are antagonized by the host's normal microbiota or eliminated by its immune system before infection begins. Competition between invading microbes and normal microbiota along with innate and specific immune defenses are paramount in preventing further proliferation of infectious agents. However, pathogens have evolved multiple mechanisms that enable them to overcome competition and elude host responses.

Before a pathogen encounters specific host immune defense, it must contend with the commensal and mutualistic microbiota that inhabit the host. Certain bacteria produce a Type VI secretion system (T6SS) to directly deliver lethal effector molecules (often a toxin) from its cytoplasm to that of its competitor (section 27.4; *see figure 27.11*). These secretion systems are especially effective in Gram-negative (diderm) enteric pathogens, for which competition for resources in the gut is especially fierce. Other pathogens use microbiota-derived metabolites to boost their growth and increase their virulence. *One-step secretion systems (section 13.9); A functional core microbiome is required for human homeostasis (section 33.3)*

Survival of the microorganism in the host depends on its ability to find shelter and avoid recognition by host defenses; pathogens have evolved myriad ways to accomplish this feat. For example, when hepatitis B virus replicates in host cells, large amounts of protein not associated with the complete virus particle are produced. These enter the bloodstream and surrounding tissues, where they act as decoys, binding to antimicrobial proteins—including antibodies against the virus—so that the virus goes unnoticed. Other pathogens can survive and replicate within the very immune cells that are supposed to destroy them (macrophages and neutrophils); an example is *Mycobacterium tuberculosis,* which withstands destruction by alveolar macrophages. Some microbes use a variety of genetic mechanisms to change or reduce the number of their surface antigens, thus avoiding recognition by immune cells, a process called **antigenic variation.** For example, *Neisseria gonorrhoeae* switches among different genes that encode pilin protein so that different types of pili are expressed. This change in pilin renders specific host responses useless against the new pili so that adherence of the pathogen to host tissue is not obstructed. Another mechanism of changing gene expression, and thus surface antigens, is **phase variation,** in which gene expression is switched on or off. For example, some pathogenic strains of *E. coli* only express genes for the production of fimbriae when they are required for attachment to host tissues. Other strategies for avoiding immune detection include squeezing between host cells (e.g., *Listeria monocytogenes*), burrowing under mucus (e.g., *Helicobacter pylori*), or secreting exopolysaccharides to form communal shelters within biofilms (e.g., *Pseudomonas aeruginosa*). *Gram-negative cell envelopes include a second membrane (section 3.3); Biofilms are common in nature (section 7.6)*

Many pathogenic bacteria (e.g., *Streptococcus pneumoniae* and *Neisseria meningitidis*) produce slippery mucoid capsules that allow them to avoid host defenses. Some capsules resemble host components and thus the pathogen remains undetected, an example of molecular mimicry. The capsules of some other bacteria (e.g., *Haemophilus influenzae*) prevent the deposition of host antimicrobial peptides and complement proteins, thereby protecting the cell from phagocytosis and lysis. Many microorganisms also produce enzymes that inactivate innate resistance mechanisms. In some cases, they produce proteases that degrade host proteins such as immunoglobulin A (IgA). In other cases, bacteria produce surface proteins (e.g., staphylococcal protein A) that bind the Fc region of antibodies, thereby coating themselves with antibodies in the upside-down orientation and thus remaining undetected by phagocytes that recognize that region. Certain pathogens, including respiratory syncytial virus (RSV) and cytomegalovirus (CMV), cause host cell fusion, allowing them to spread from an infected cell to an uninfected cell with minimal exposure to host defenses. ◀ *Capsules and slime layers (section 3.5); Complement is a cascading enzyme system (section 31.3); Antimicrobial peptides and proteins (section 31.3)*

Some pathogens do not avoid immune defenses; instead they have evolved a variety of ways to suppress that response. Human immunodeficiency virus (HIV) infects and kills cells of the adaptive immune system (macrophages and T cells), while ensuring its own survival. Immunosuppression results when the number of lymphocytes in the host falls below a critical level. Measles virus also infects immune cells, as well as epithelial cells and endothelial cells. The virus depletes memory cells, temporarily erasing memory of previous infections, a situation called *immune amnesia*. ◀ *T cells are critical for immune function (section 32.5)*

Biofilms Provide Microbial Protection

Many bacteria live within stable, sessile communities surrounded by a hydrated polymeric matrix of their own production. These **biofilm** bacteria are protected from nutrient deprivation, predators, environmental shifts, antimicrobial agents, and host immune cells. In contrast to free-swimming (planktonic) bacteria of the same species, biofilm bacteria are phenotypically different, expressing genes that favor survival when attached and in poor nutrient conditions. However, the role of the biofilm is much more complex, especially for biofilms that cause infections in humans. Using high throughput sequencing, fluorescent in situ hybridization, and global gene analysis, studies have revealed that some pathogenic bacteria within biofilms exchange plasmids, nutrients, and quorum-sensing molecules, altering their behavior compared to planktonic forms. Biofilm bacteria coordinate gene expression to upregulate mechanisms that make the biofilm community less sensitive to antibiotics and host defense mechanisms.

For example, studies evaluating *Pseudomonas aeruginosa* mixed with antibiotics, antibodies, and phagocytic white blood cells (macrophages and neutrophils) demonstrate that while planktonic *P. aeruginosa* are readily killed by these agents, *P. aeruginosa* in biofilms are not. The bacteria survive by activating global response regulators, increasing the activity of multidrug efflux pumps, activating starvation or stress responses, and stimulating exopolysaccharide secretion. Not only do these activities appear to prevent penetration of antibiotics and antibodies, but they also cause phagocytes to downregulate killing mechanisms. The efficacy of host hydrolytic enzymes is decreased, and enzymes that create super- and nitric oxides are inhibited. Simply put, organisms within biofilms "frustrate" phagocytes by reducing the magnitude of some antibacterial responses and preventing others altogether (**figure 34.7**). ◀ *Microbial growth in natural environments (section 7.6); Phagocytosis destroys invaders (section 31.6)*

Comprehension Check

1. How does a capsule enhance pathogenicity?
2. How does a type VI secretion system aid pathogen survival?
3. Give examples of organisms that can evade and suppress host immune responses.
4. What is a biofilm? How is the ability to form a biofilm a virulence factor?

34.4 Damage to the Host

After reading this section, you should be able to:

a. Predict the microbial virulence factors and host-cell responses that result in host damage
b. Discuss the origin and nature of pathogenicity islands
c. Illustrate the mechanisms by which microbial toxins impact human cells

Recall that pathogenicity is a general term that refers to an organism's potential to cause disease, while virulence refers more specifically to the magnitude of harm (pathogenicity) caused. Some microorganisms are pathogenic due to their rapid replication and explosive departure from host cells. Others are pathogens because they alter normal cell function; many microorganisms, particularly bacteria, use toxins to poison host cells. In many cases of infection, tissue damage is caused indirectly when the pathogen stimulates an excessive or inappropriate immune response by the host. Pathogenicity is thus not exclusively due to a trait inherent in microorganisms, but may be a consequence of the interplay between microbe and host.

Whatever the mechanism, recall that virulence factors are molecules produced by the pathogen that assist in the adherence, colonization, or invasion of the host, including those that help the microbe avoid host immunity. These factors can be encoded in the microbial chromosome or as extrachromosomal elements that are often acquired by horizontal gene transfer (HGT). Synthesis of virulence factors is upregulated by specific host triggers, enabling the microorganism to better survive in the host and to avoid the energetic cost of factor production when in the environment. ◀ *Horizontal gene transfer: creating genetic variability the asexual way (section 16.3)*

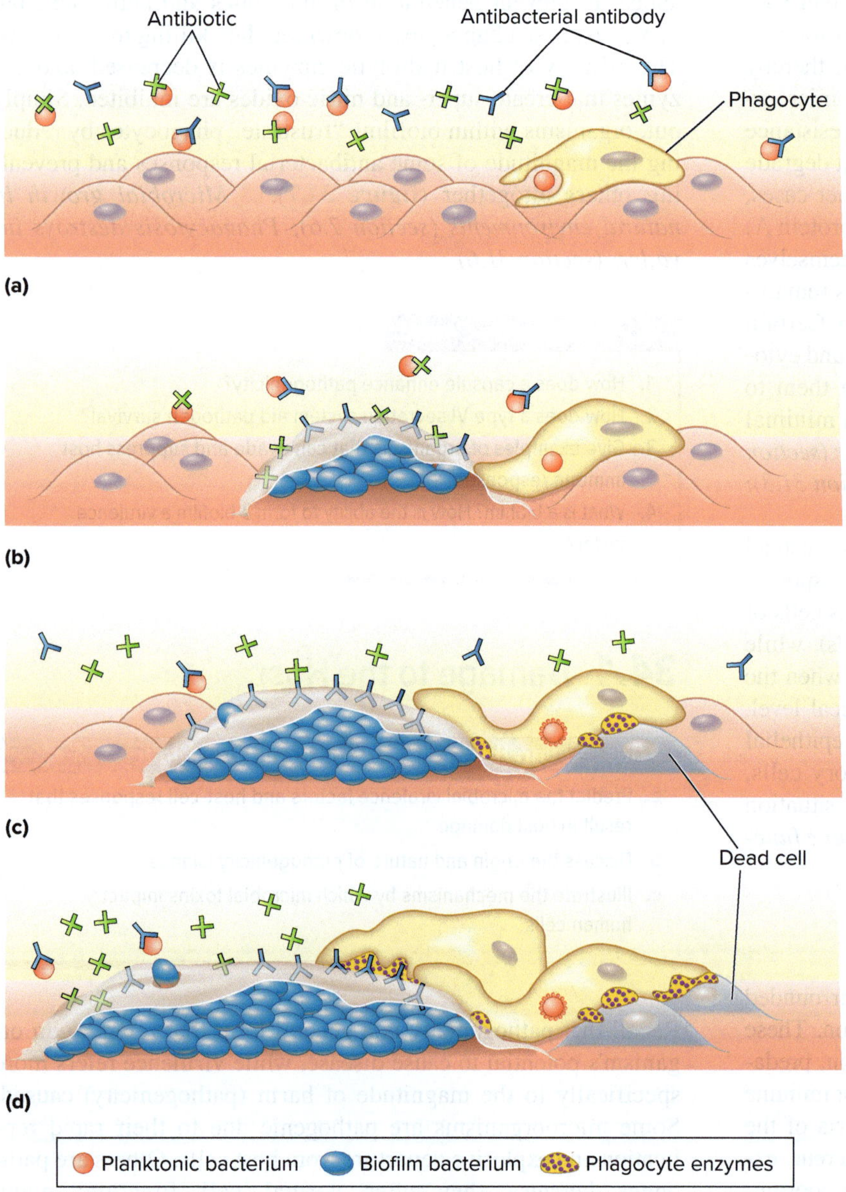

Figure 34.7 Frustrated Phagocytosis. (a) Planktonic cells are sensitive to antibiotics and are detected by host phagocytes and antibodies. (b) Planktonic cells settle to form a biofilm and become biofilm cells. (c) Biofilm cells resist antibiotics and antibody detection. Host phagocytes detect and attempt to destroy biofilm cells. (d) Unable to capture biofilm cells, phagocytes release antimicrobial products that kill host cells but not biofilm cells.

Pathogenicity Islands Encode Virulence Factors

Large segments (10 to 200 kilobases) of bacterial chromosomal and plasmid DNA encoding virulence factors, called **pathogenicity islands,** appear to have been inserted into the existing DNA during the speciation of pathogens from their nonpathogenic ancestors, likely via HGT. Many bacteria (e.g., *Yersinia* spp., *P. aeruginosa, Shigella flexneri, Salmonella* spp., enteropathogenic *Escherichia coli*) carry at least one pathogenicity island. Pathogenicity islands increase bacterial virulence and are absent in nonpathogenic members of the same genus or species (**table 34.3**). For example, the *Staphylococcus aureus* pathogenicity island encodes several superantigen genes, including the gene for the toxin that causes toxic shock syndrome. The movement of genes from one bacterial cell to another is discussed in chapter 16, while the detection of pathogenicity islands within a microbe's genome is explained in chapter 18. ◀ *Comparative genomics (section 18.7)*

Toxins Are Biological Poisons

Recall that a toxin (Latin *toxicum,* poison) is a substance that disrupts the normal metabolism of host cells with damaging effects on the host. **Toxigenicity** is the pathogen's ability to produce toxins, and **intoxications** are diseases that result from a specific toxin produced by the pathogen. Intoxication diseases do not require the presence of the actively growing pathogen—just its toxin, as in the case of tetanus. Bacteria produce two structurally different types of toxins: exotoxins (proteins) and endotoxin (lipopolysaccharide). Some fungi produce potent mycotoxins.

Exotoxins

Exotoxins are soluble, heat-labile proteins (inactivated at 60° to 80°C) usually released into host tissues as the bacterial pathogen grows or when the bacteria die and lyse. Exotoxins often travel from the site of infection to other body tissues or target cells, where they exert their effects. They are frequently encoded by genes carried on plasmids or prophages. Exotoxins are associated with specific diseases and often are named for the disease they produce (e.g., the diphtheria toxin). Some are among the most lethal substances known—toxic in nanogram-per-kilogram of body weight concentrations (e.g., the botulinum toxin).

Exotoxins exert their biological activity by specific mechanisms and are grouped by either mechanism of action (e.g., a cytotoxin kills cells) or their protein structure (**figure 34.8**). A common structural type is the **AB toxin,** which gets its name from the fact that it has two distinct toxin subunit types, an "A" (or active) component and a "B" (or binding) component. The B portion of the toxin binds to a host-cell receptor and triggers endocytosis. Thus the B component determines the cell type infected. Once internalized, the A and B components dissociate from one another. The A component, which functions as an enzyme, is now free to catalyze a reaction that will cause host cell death (figure 34.8*a*). AB toxins act on cells by different mechanisms. Many A subunits have ADP ribosylation activity, which catalyzes the transfer of adenosine diphosphate and ribose moieties of host NAD^+ to target host molecules (*see figure 10.7*).

Table 34.3	**Examples of Pathogenicity Islands and the Products They Encode**		
Organism	**Pathogenicity Island**[1]	**Gene Product**	**Function**
Escherichia coli	PAI-III	Type 1 pili Secreted protein	Attachment Hemolysin, cytotoxic necrosing factor, uropathogenic protein
Helicobacter pylori	cagPI	Type IV secretion proteins	Cytotoxin
Legionella pneumophila	icm/dot	Type IV secretion proteins	Intracellular survival
Salmonella enterica	SPI-1, SPI-2	Type III secretion proteins	Cytotoxin
Shigella flexneri	SHI-1, SHI-2, SHI-3, SHI-O	Type III secretion proteins	Cytotoxin
Staphylococcus aureus	SaPI	Secreted proteins	Superantigens
Vibrio cholerae	VPI-1	20+ proteins	Colonization, enterotoxin production
Yersinia pestis	HPI-1	Siderophore synthesis	Iron uptake and storage

[1] PI and PAI are both common abbreviations for pathogenicity island. HPI = high-pathogenicity island.

Certain exotoxins that are grouped by mechanism of action have the ability to disrupt membranes. Examples of this type of functional exotoxin are the pore-forming toxins (figure 34.8*b*). They destabilize membrane integrity so that the host cell lyses.

A third group of exotoxins is able to stimulate as many as 30% of host T cells to overexpress genes that encode cytokines, resulting in the release of massive amounts of pro-inflammatory molecules. They accomplish this by binding antigen-presenting cells nonspecifically to T cells, that is, in the absence of their

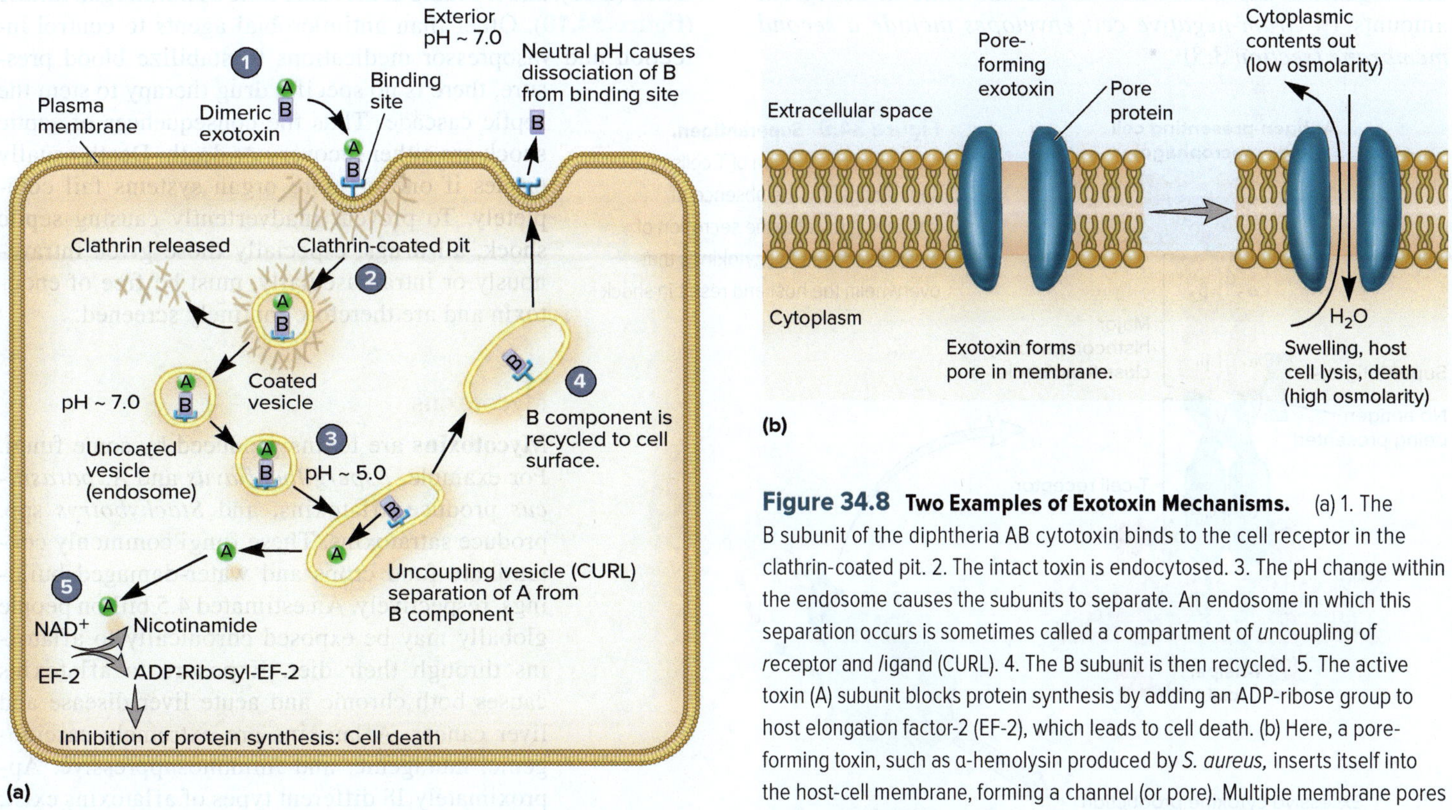

Figure 34.8 Two Examples of Exotoxin Mechanisms. (a) 1. The B subunit of the diphtheria AB cytotoxin binds to the cell receptor in the clathrin-coated pit. 2. The intact toxin is endocytosed. 3. The pH change within the endosome causes the subunits to separate. An endosome in which this separation occurs is sometimes called a compartment of *u*ncoupling of *r*eceptor and *l*igand (CURL). 4. The B subunit is then recycled. 5. The active toxin (A) subunit blocks protein synthesis by adding an ADP-ribose group to host elongation factor-2 (EF-2), which leads to cell death. (b) Here, a pore-forming toxin, such as α-hemolysin produced by *S. aureus,* inserts itself into the host-cell membrane, forming a channel (or pore). Multiple membrane pores result in an osmolarity shift, as water enters the cell and cytoplasmic contents move out. The resulting effect of this toxin is cell lysis.

target antigen. This extreme response earned them the name **superantigens** (figure 34.9). The excessive concentration of cytokines causes multiple host organs to fail, giving the pathogen time to disseminate. By triggering what is called a *cytokine storm,* superantigens cause life-threatening disease; fever, fluid loss, and low blood pressure result in shock and death. The host response is very similar to that described later in this section for endotoxin-mediated septic shock.

The general properties of several exotoxins are presented in **table 34.4.** As proteins, toxins are recognized by the host immune system, which produces antitoxin—antibodies that recognize and bind to their target toxins—to inactivate and remove them. Examples of each specific toxin type are discussed as part of the disease process in chapters 37 through 39. *Cytokines are chemical messages between cells (section 31.3)*

Endotoxin

The lipopolysaccharide (LPS) in the outer membrane of Gram-negative bacteria is toxic to mammals. Thus LPS is called an **endotoxin** because it is bound to the bacterium and is released when the microorganism lyses, and to a lesser degree during cell division. The toxic component of LPS is the lipid portion, called **lipid A** (*see figure 3.22*). Lipid A is not a single macromolecule; rather, it is a complex array of lipid residues that vary among bacterial species in number, length, and attachment position, which affects the pathogenicity of that organism. Lipid A is heat stable and toxic in nanogram amounts. *Gram-negative cell envelopes include a second membrane (section 3.3)*

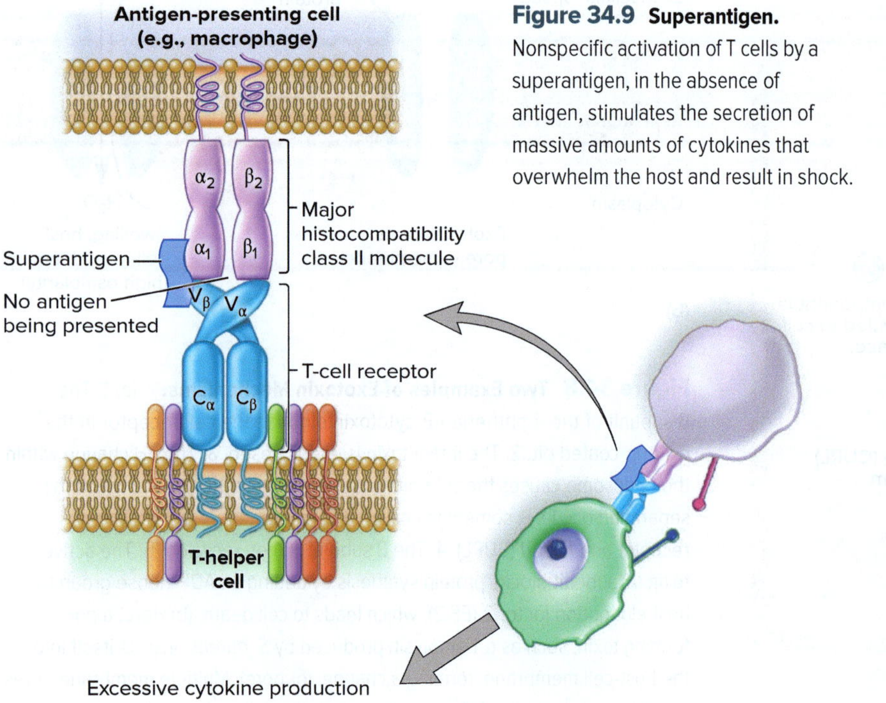

Figure 34.9 Superantigen. Nonspecific activation of T cells by a superantigen, in the absence of antigen, stimulates the secretion of massive amounts of cytokines that overwhelm the host and result in shock.

Unlike the structural and functional diversity of exotoxins, the lipid A of various Gram-negative bacteria produces similar systemic effects regardless of the microbe from which it is derived. These include fever (i.e., endotoxin is pyrogenic), shock (drop in blood pressure, increase in heart rate), damage to the lining of blood vessels, weakness, diarrhea, inflammation, intestinal hemorrhage, and fibrinolysis (enzymatic breakdown of fibrin, the major protein component of blood clots).

Lipid A is the primary initiator of **septic shock,** a culmination of systemic responses to the toxin. The biological effect of lipid A is an indirect one, mediated by host molecules and systems rather than by lipid A itself. Endotoxin indirectly induces a fever in the host by causing macrophages to release the pyrogenic cytokines IL-1 (interleukin-1) and TNF-α (tumor necrosis factor-α) that reset the hypothalamic thermostat. It also stimulates the release of pro-inflammatory cytokines including TNF-α, IL-1, IL-6, and IL-18 from damaged endothelial cells, monocytes, macrophages, neutrophils, and plasma cell precursors. The ensuing cytokine storm causes rapid dilation of the blood vessels accompanied by a drastic drop in blood pressure that has profound physiological effects on the heart, vasculature, and other body organs. Lipid A also damages endothelial cells that line blood vessels, resulting in the exposure of the mediator tissue factor (TF), that in this context ultimately triggers both unregulated clotting and internal bleeding. This condition is called disseminated intravascular coagulation (DIC), and it occurs at the same time as multiorgan failure (**figure 34.10**). Other than antimicrobial agents to control infection and vasopressor medications to stabilize blood pressure, there is no specific drug therapy to stem the septic cascade. Thus the consequences of septic shock are either recovery or death. Death usually ensues if one or more organ systems fail completely. To prevent inadvertently causing septic shock, all drugs, especially those given intravenously or intramuscularly, must be free of endotoxin and are therefore routinely screened.

Mycotoxins

Mycotoxins are toxins produced by some fungi. For example, *Aspergillus flavus* and *A. parasiticus* produce aflatoxins, and *Stachybotrys* spp. produce satratoxins. These fungi commonly contaminate food crops and water-damaged buildings, respectively. An estimated 4.5 billion people globally may be exposed chronically to aflatoxins through their diet. Exposure to aflatoxins causes both chronic and acute liver disease and liver cancer. Aflatoxins are extremely carcinogenic, mutagenic, and immunosuppressive. Approximately 18 different types of aflatoxins exist;

Table 34.4 Exotoxins Produced by Human Bacterial Pathogens

Toxin	Organism	Gene Location	Toxin Type	Mechanism of Action
Edema factor (EF) Lethal factor (LF) Protective antigen (PA)	*Bacillus anthracis*	Plasmid	Tripartite AB	EF causes edema. LF is a cytotoxin. PA is a B component.
Pertussis toxin	*Bordetella pertussis*	Chromosome	AB	↓ATP, ↑cAMP alter cell function, leading to death.
Botulinum toxin	*Clostridium botulinum*	Prophage	AB	Blocks acetylcholine release, leading to flaccid paralysis
CPE enterotoxin	*Clostridium perfringens*	Chromosome	Cytotoxin	Hemolysis
Tetanospasmin	*Clostridium tetani*	Plasmid	AB	Blocks release of inhibitory neurotransmitter, leading to spastic paralysis
Diphtheria toxin	*Corynebacterium diphtheriae*	Phage	AB	Alters translation, leading to protein synthesis inhibition
Enterotoxin Shiga-like toxin	*Escherichia coli* *E. coli* O157:H7	Plasmid Phage gene integrated into chromosome	AB AB	↑cAMP, leading to water secretion from cell Inhibits protein synthesis, leading to death
Cytolysin	*Salmonella* spp.	Chromosome	Cytotoxin	↑cAMP, leading to water secretion from cell
Shiga toxin	*Shigella dysenteriae*	Chromosome	AB	Inhibits protein synthesis, leading to death
Exfoliative toxin Toxic shock syndrome toxin-1 Panton-Valentine leukocidin	*Staphylococcus aureus*	Chromosome Chromosome Phage	Protease Superantigen Cytotoxin	Skin peeling Cytokine-induced shock Lyses white blood cells
Streptolysin O Erythrogenic toxin	*Streptococcus pyogenes*	Chromosome Phage	Cytolysin Superantigen	Hemolysis Cytokine-induced shock
Cholera toxin	*Vibrio cholerae*	Phage	AB	↑cAMP, leading to water secretion from cell

these are classified into two broad groups according to their chemical structure (*see figure 40.3*). *Stachybotrys trichothecene* mycotoxins are potent inhibitors of DNA, RNA, and protein synthesis. They induce inflammation, disrupt surfactant phospholipids in the lungs, and may lead to pathological changes in tissues.

The fungus *Claviceps purpurea* also produces toxic substances generically referred to as ergots, reflecting the name of the tuberlike structure of the fungi (*see figure 24.12*). Ergots contain high concentrations of alkaloids that have varying physiological effects on humans. One such alkaloid is lysergic acid, the psychotropic hallucinogen LSD. Ergotism describes the pathological condition resulting from ingested ergot-contaminated plant matter, and is the earliest recorded example of mycotoxosis, dating as far back as 600 BCE.

Comprehension Check

1. What are pathogenicity islands and why are they important?
2. Describe some general characteristics of exotoxins.
3. Discuss the role of lipid A in the development of septic shock.

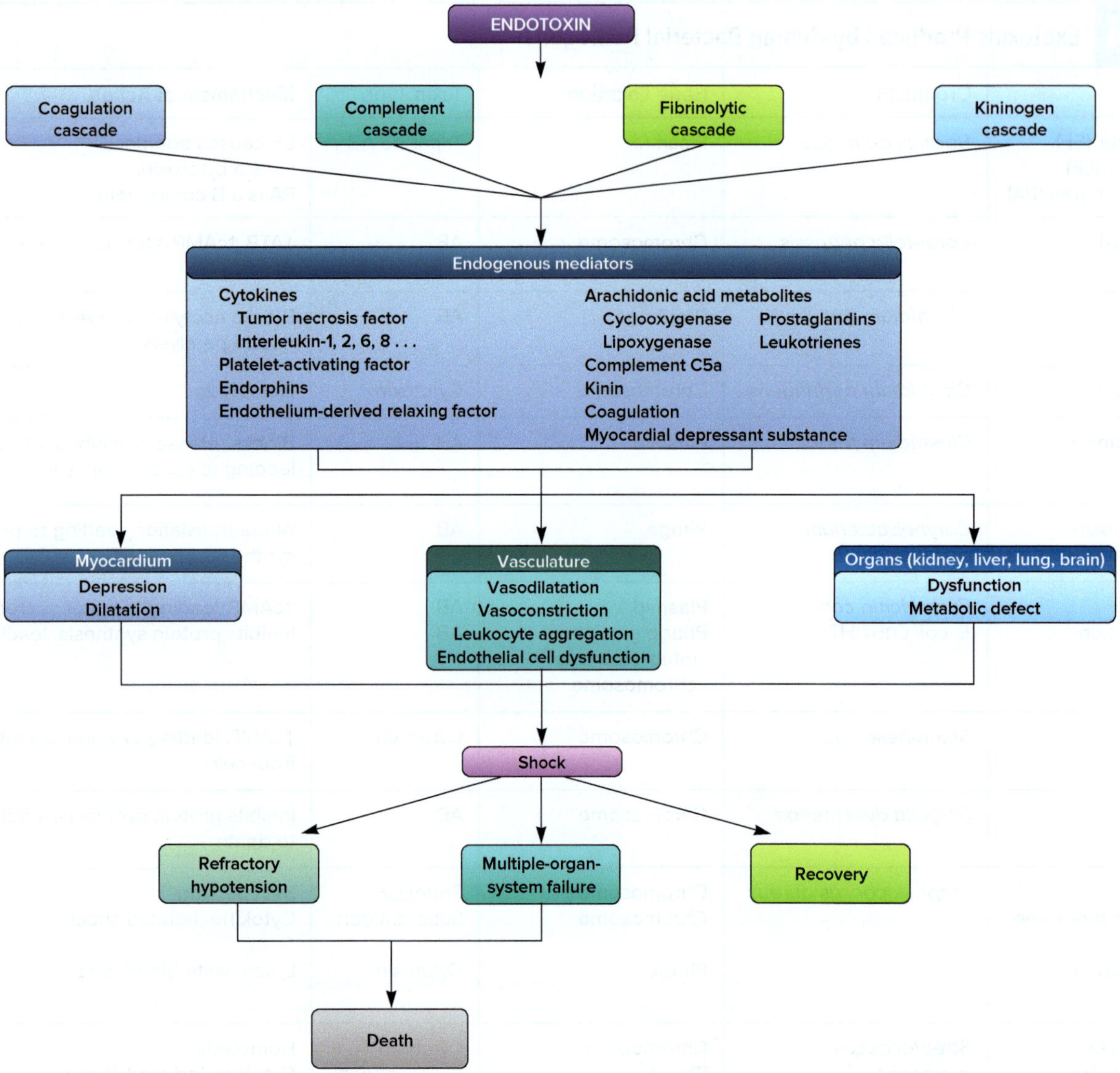

Figure 34.10 The Septic Shock Cascade. Gram-negative bacterial endotoxin triggers biochemical events that lead to such serious complications as shock, acute respiratory distress syndrome, and disseminated intravascular coagulation.

Key Concepts

34.1 The Process of Infection

- Infections occur when microorganisms enter a larger organism, the host, and compete for resources.
- Virulence refers to the magnitude of harm that a pathogen can exert on its host.
- The course of an infection follows a characteristic pattern: the incubation period, the prodromal stage, the illness period, and convalescence (**figure 34.1**).
- Signs of an infectious disease are objective changes in the body. Symptoms are subjective.

34.2 Transmission and Entry into the Host

- Pathogens reside in various environments, termed reservoirs. They infect the host from the reservoir and can return to it.
- Pathogens are transmitted to and between hosts by several common routes, including air, direct contact, arthropod vectors, and vehicles (**figure 34.3**).
- Infectious diseases passed from animals to humans are called zoonoses.

- The number of invading organisms required to cause disease can be quantified using the ID_{50} or LD_{50} (**figure 34.4**).
- Pathogens require virulence factors to assist them in accessing a host. These include the ability to adhere to host cells, chemicals that permit cell and tissue degradation, and invasion of other tissues (**tables 34.1** and **34.2**).

34.3 Surviving the Host Defenses

- Microorganisms have evolved many strategies for avoiding host defenses.
- Proteins synthesized to aid microbial survival in the host are collectively known as virulence factors.
- Biofilms provide a unique environment for bacteria to exchange genetic information and collectively resist antimicrobial agents and host defense mechanisms (**figure 34.7**).

34.4 Damage to the Host

- Specific gene sequences called pathogenicity islands encode virulence factors (**table 34.3**).
- Exotoxins are secreted proteins that have biological activity (**figure 34.8** and **table 34.4**).
- Endotoxins are the lipid A components of the lipopolysaccharide found in most Gram-negative bacterial envelopes that can cause septic shock (**figure 34.10**).
- Mycotoxins are metabolites of fungi. Their effects include hallucinations, cancer, and immunosuppression.

Active Learning

1. In general, infectious diseases that are commonly fatal are newly evolved relationships between the pathogenic organism and the host. Why is this so?
2. Explain the observation that different pathogens infect different host tissues.
3. Intracellular bacterial infections present a particular difficulty for the host. Why is it harder to defend against these infections than against viral infections and extracellular bacterial infections?
4. The Ebola virus disease (EVD) outbreak of 2013 was unprecedented in its geographic distribution and number of infections. Yet the incubation period, duration of illness, disease symptomology (i.e., clinical course of infection), and virus transmissibility were similar to those in earlier EVD outbreaks stemming back to the 1970s. Direct contact with viable virus has been understood to be the route of transmission, especially when contact is made with sick or dead relatives. All individuals who contracted EVD had direct contact with highly infectious body fluids from other patients with the disease. Given its spread, speculation regarding the airborne transmission of Ebola virus has been voiced. Experimentally, Ebola virus particles can be aerosolized. Of note, however, is that family members residing with the first U.S. patient with EVD in Dallas in 2013 did not acquire the disease. Given that the genomes of several strains of Ebola virus have been sequenced, what experiments could you do to test whether airborne Ebola virus is infectious?
5. During late Fall 2020, a mutant strain of SARS-CoV-2 that was rapidly spreading throughout the United Kingdom was analyzed. The Alpha variant was determined to have an asparagine (N) to tyrosine (Y) amino acid substitution in the 501 position of the spike protein (N501Y), a residue within the receptor binding domain. By December 2020, Alpha was the dominant strain in the country, reported to have greater ease of transmission, yet having no greater virulence compared to the original strain. Moreover, it is successfully targeted by host antibodies generated post-vaccination. We know that the SARS-CoV-2 spike protein binds to the ACE2 receptor triggering endocytosis and is the target of vaccines. Predict why transmission is enhanced without appreciable impact on virulence in this variant and, using your knowledge of adaptive immunity, why the efficacy of the vaccine persists.

CHAPTER

35

Epidemiology and Public Health Microbiology

Katherine Welles/Shutterstock

Protecting the Herd

The COVID-19 pandemic was a reminder to the world that new and potentially fatal diseases can arise seemingly without warning. The pandemic took its toll on physical and mental health, the economy, and education. While countries struggled to keep infection rates down and toilet paper supplies up, one field of science had its moment in the spotlight: epidemiology. As the pandemic progressed, epidemiologists kept us informed about disease statistics and recommended measures such as social distancing in an effort to curb the spread of the causative virus, SARS-CoV-2. As the country focused on the pandemic, universities began to report a surge in applications for public health programs. The Association of Schools and Programs of Public Health reported an average 20% increase in applications for master's programs for 2021 compared to previous years, with some schools seeing increases as high as 75%. This increased interest in public health programs has continued even after the COVID-19 pandemic waned.

The emergence and reemergence of infectious diseases are perpetual threats, as we saw in August 2024, when the World Health Organization (WHO) declared the escalating incidence of mpox in several African countries a public health emergency of international concern (PHEIC). By mid-August 2024, the mpox outbreak in Africa involved over 18,000 suspected cases and more than 600 deaths. In declaring the PHEIC, the WHO indicated that there was grave potential for international spread of that disease, requiring a coordinated international response. The PHEIC was ended in May 2023, after a significant decline in global cases. However, the disease remains a public health concern.

Although cases of mpox are still occurring, it is unlikely that the disease will result in a global lockdown comparable to that of COVID-19. This is because mpox, while reemerging, is not new and two preventative vaccines are available. Nonetheless, the disease will continue to be closely tracked so that those at risk can be protected. In this chapter, we describe the practical goal of epidemiology: to establish effective disease recognition, control, prevention, and eradication measures for a given population.

Readiness Check:

Based on what you have learned previously, you should be able to:

- ✓ Describe innate human host resistance mechanisms (chapter 31)
- ✓ Discuss the human adaptive immune response to pathogens (chapter 32)
- ✓ Explain pathogenicity and the infectious disease process (chapter 34)

35.1 Epidemiology Is an Evidence-Based Science

After reading this section, you should be able to:

a. Describe the agencies responsible for disease prevention and control
b. Define the basic vocabulary used in the science of epidemiology

By definition, **epidemiology** (Greek *epi,* upon; *demos,* people or population; *logy,* study) is the science that evaluates the occurrence, determinants, distribution, and control of health and disease in a defined human population (**figure 35.1**). An individual who practices epidemiology is an **epidemiologist.** Their major concerns are the discovery of the factors essential to disease occurrence and the development of methods for disease prevention. In the United States, the **Centers for Disease Control and Prevention (CDC),** headquartered in Atlanta, Georgia, serves as the national agency for developing and carrying out disease prevention and control, environmental health, and health promotion and education activities. Its international counterpart is the **World Health Organization (WHO),** located in Geneva, Switzerland.

The science of epidemiology originated and developed in response to great epidemic diseases such as cholera, typhoid fever, smallpox, influenza, and yellow fever (**Historical Highlights 35.1**). Today, epidemiology encompasses all public health issues: infectious diseases, genetic abnormalities, metabolic dysfunction, malnutrition, neoplasms (abnormal tissue masses), psychiatric disorders, obesity, and aging. This chapter discusses only infectious disease epidemiology.

To explore epidemiology, we first need to define some key terms. The **attack rate** is the proportional number of cases that develop in a population exposed to an infectious agent. Morbidity and mortality refer to illness and death respectively. **Morbidity rate** tracks the number of cases of an illness in a particular population within a specified period of time. Mortality rate indicates the proportion of deaths, either in general or assigned to a particular cause, within a defined population in a given time. The **case fatality rate (CFR)** measures the severity of a disease in a population by defining the proportion of individuals diagnosed with a specific disease who die from that disease; it does not consider the size of the entire population and may change as case definitions are altered and treatments become available. When a disease occurs occasionally or at irregular intervals in a human population, it is a **sporadic disease** (e.g., bacterial meningitis). When it maintains a steady, low-level frequency at a moderately regular interval, it is an **endemic** (Greek *endemos,* dwelling in the same people) **disease** (e.g., the common cold). Hyperendemic diseases gradually increase in frequency beyond the endemic level but not to the epidemic level (e.g., the common cold during winter months). An **outbreak** is the sudden, unexpected occurrence of a disease, usually in a limited segment of a population (e.g., food-borne illnesses resulting from consumption of contaminated produce). An **epidemic** (Greek *epidemios,* upon the people), on the other hand, is an outbreak affecting many people at once. Influenza is an example of a disease that often achieves epidemic status in a community. The influenza virus resides in a **reservoir host,** an organism that remains healthy while the virus thrives, until it is transmitted to a human host. The first person identified with a specific epidemic disease is called the **index case.** A **pandemic** (Greek *pan,* all) is an increase in disease occurrence within a large population over at least two countries around the world. Usually pandemic diseases spread among continents as we saw with COVID-19.

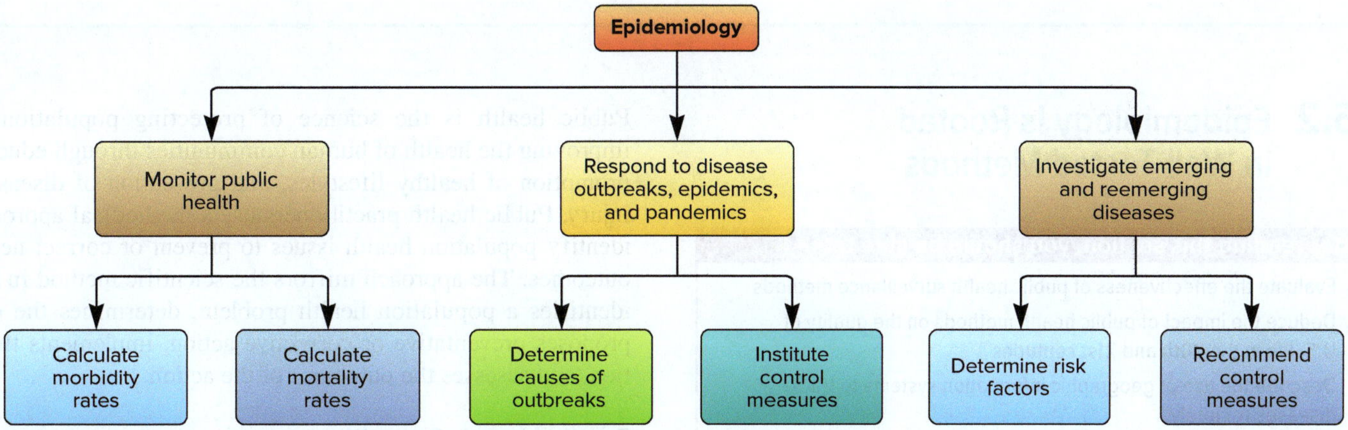

Figure 35.1 Epidemiology. Epidemiology is a multifaceted science that investigates diseases to discover their origin, evaluates diseases to assess their risk, and controls diseases to prevent future outbreaks.

HISTORICAL HIGHLIGHTS

35.1 John Snow, the First Epidemiologist

Much of what we know today about the epidemiology of cholera is based on the classic studies conducted between 1849 and 1854 by the British physician John Snow. During this period, a series of cholera outbreaks occurred in London, England, and Snow set out to find the source of the disease.

When the cholera outbreak of 1854 struck London, Snow suspected that it was caused by contaminated water. With the assistance of Reverend Henry Whitehead, Snow spoke to local residents and identified the source of the outbreak as a pump on Broad Street. He constructed a dot map that confirmed clusters of cholera deaths around the pump and deduced that the disease was being spread because the water from the pump was contaminated with raw sewage containing the disease agent. Snow persuaded officials to disable the pump by removing its handle, and the number of cholera cases in that area dropped dramatically.

Snow went on to conduct his "grand experiment" during this time. Part of London's drinking water came from two different suppliers: the Southwark & Vauxhall Company and the Lambeth Company. Snow compiled data on cholera deaths in the city and found that most of them occurred in households that purchased their drinking water from the Southwark & Vauxhall Company which obtained water from the Thames River below locations at which Londoners discharged sewage. In contrast, the cholera death rate in homes using water from the Lambeth Company, which took water from the Thames upstream of the city, was over eightfold lower. Snow deduced that the cause of the disease was spread in sewage-contaminated water. However, it wasn't until 1883 that Robert Koch identified the causative bacterium as *Vibrio cholerae*.

The John Snow Pub (**box figure**) now stands at the site of the old Broad Street pump, commemorating Snow's achievements.

The John Snow Public House. Soho, London, the United Kingdom.
Dorothy Wood/McGraw Hill

35.2 Epidemiology Is Rooted in Well-Tested Methods

After reading this section, you should be able to:

- **a.** Evaluate the effectiveness of public health surveillance methods
- **b.** Deduce the impact of public health methods on the quality of U.S. life in the 20th and 21st centuries
- **c.** Describe the use of geographic information systems to track diseases remotely
- **d.** Use standardized surveillance data to measure infectious disease frequency
- **e.** Calculate rates of incidence, prevalence, morbidity, and mortality

Public health is the science of protecting populations and improving the health of human communities through education, promotion of healthy lifestyles, and prevention of disease and injury. Public health practitioners use a methodical approach to identify population health issues to prevent or correct negative outcomes. The approach mirrors the scientific method in that it identifies a population health problem, determines the cause, proposes preventative or corrective action, implements that action, and assesses the outcomes of the action.

Public Health Surveillance

The identification of a population health issue is rooted in public health surveillance. This surveillance involves the proactive evaluation of human and microbial genetics, environmental conditions, human behaviors and lifestyle choices, emerging infectious agents, and microbial responses to chemotherapeutic agents to monitor the health of a population. In other words, public health practitioners look for cause-and-effect relationships to determine risk. The impact of public health surveillance is obvious when one considers the change in the leading causes of death in the United States from 1900 to 2022 (**figure 35.2**). The primary cause of death in 1900 was infectious disease. Surveillance identified those diseases and their associated risks. These were remedied by water treatment, by strict sanitation guidelines, and later by the use of antimicrobial agents and vaccines. The result was that by the 1960s very few deaths were due to infectious disease. Ironically, the modernization that helped prevent the leading causes of mortality in the 1900s and led to a longer life expectancy also led to a more sedentary lifestyle in the twenty-first century. Consequently, metabolic syndrome—a group of conditions that increase the risk of developing type 2 diabetes, heart disease, and stroke—became the primary cause of death rather than infectious disease. *Metabolic syndrome (section 33.4)*

Public health surveillance has its roots in the fourteenth century as attempts were made to control the spread of bubonic plague. However, it was not until after the germ theory of disease was widely accepted (late 1800s) that a scientifically based monitoring process was used as a means to track communicable disease. This was not the confidential monitoring that we expect today. The first disease tracking practices were intrusive; people were inspected for signs of disease and those found to be infected were quarantined. The current emphasis on monitoring disease progression in a population and disseminating information in advance of the disease has radically improved the impact of public health. Importantly, disease data come from a number of sources, including mandated clinical reports, laboratory specimens, records and databases, sentinel surveillance, disease registries, and field surveys (**figure 35.3**). These data are used most often for the generation of reports informing public health officials of (1) population morbidity and mortality, (2) disease effect on school attendance, (3) impact of epidemics on employee absenteeism, and (4) animal and vector control policies. If a pandemic is declared, effective management involves implementation of four coordinated control measures: (1) testing the population with speed and accuracy, (2) isolation of individuals who are infected, (3) contact tracing—identifying those who have been exposed to the infectious agent, and (4) quarantine—restricting the movement of those who were exposed. Swift development and distribution of an effective test for the pathogen is vital (*see chapter 36 opening story*), as well as appointing a workforce to execute contact tracing. A breakdown in any of these four actions will spur pathogen spread and challenge public health infrastructure.

Within the CDC, the National Notifiable Disease Surveillance System (NNDSS) maintains a list of notifiable diseases that is regularly updated to reflect those that have newly appeared (*emerging infectious diseases*, or EIDs) or have increased in incidence (*reemerging infectious diseases*).

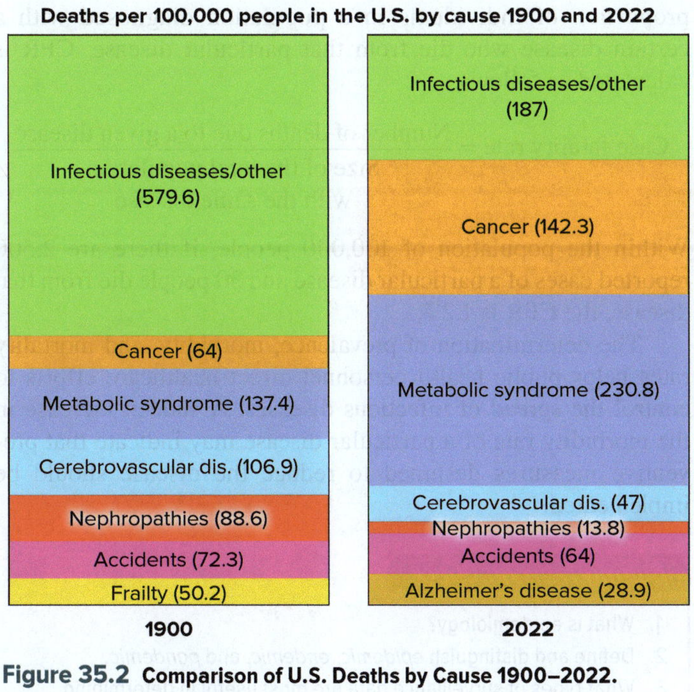

Figure 35.2 Comparison of U.S. Deaths by Cause 1900–2022. Infectious diseases were the major burden in 1900, which has shifted to metabolic syndrome in 2022.

Jones, D. S., Podolsky, S. H., and Greene, J. A. 2012. "The burden of disease and the changing task of medicine." *N. Engl. J. Med.* 366: 2333–2338; Vos, T., et al. 2020. "Global burden of 369 diseases and injuries in 204 countries and territories, 1990–2019: a systematic analysis for the Global Burden of Disease Study 2019." *The Lancet.* 396(10258): 1204–1222.

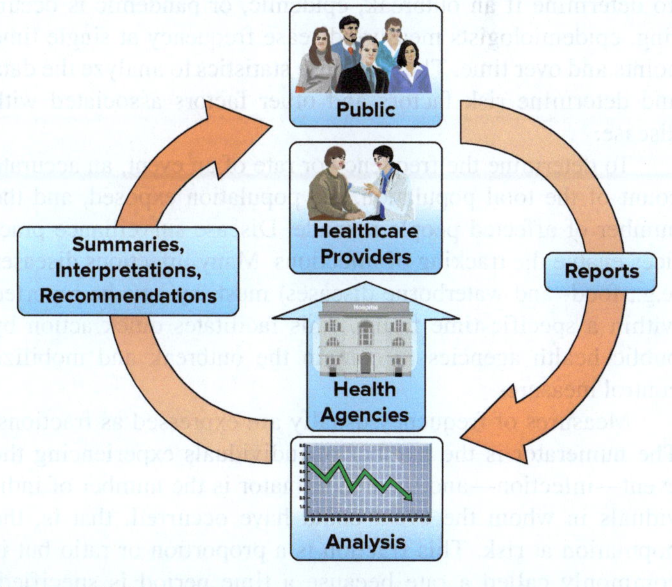

Figure 35.3 Information Loop of Public Health Surveillance. The Centers for Disease Control and Prevention use a number of different data sources to maintain surveillance of the infectious diseases that may erupt into localized epidemics or more widespread pandemics.

MICRO INQUIRY *How might biased data collection impact public health recommendations? Explain.*

Notifiable diseases must be reported to officials when diagnosed because they are of public concern due to their contagiousness, frequency, or severity. The NNDSS currently monitors 79 infectious diseases; the most recent addition to the list is Rift Valley fever virus, which was included in 2025.

Remote Sensing and Geographic Information Systems: Charting Infectious Disease Data

Remote sensing and geographic information systems are map-based tools that can be used to study the distribution, dynamics, and environmental correlates of microbial diseases. **Remote sensing (RS)** gathers digital images of the Earth's surface from satellites and output from biological sensors and transforms the data into maps. A **geographic information system (GIS)** is a data management system that organizes and displays digital map data from RS and facilitates the analysis of relationships between mapped features. Statistical relationships often exist between mapped features and diseases in natural host or human populations. Examples include the location of the habitats of the malaria parasite and mosquito vectors in Mexico and Asia, ticks that transmit Lyme disease in the United States, and African trypanosomes in both humans and livestock in Africa. RS and GIS are most useful if disease dynamics and distributions are clearly related to mapped environmental variables. For example, if a microbial disease is associated with certain vegetation types or physical characteristics (e.g., elevation, precipitation), RS and GIS can identify regions in which risk is relatively high.

Measuring Infectious Disease Frequency

To determine if an outbreak, epidemic, or pandemic is occurring, epidemiologists measure disease frequency at single time points and over time. They then use statistics to analyze the data and determine risk factors and other factors associated with disease.

To determine the frequency or rate of an event, an accurate count of the total population, the population exposed, and the number of affected people is made. Disease surveillance practices enable the tracking of infections. Many infectious diseases (e.g., food- and waterborne diseases) must, by law, be reported within a specific time frame. This facilitates quick action by public health agencies to contain the outbreak and mobilize control measures.

Measures of frequency usually are expressed as fractions. The numerator is the number of individuals experiencing the event—infection—and the denominator is the number of individuals in whom the event could have occurred, that is, the population at risk. This fraction is a proportion or ratio but is commonly called a rate because a time period is specified. (A rate can also be expressed as a percentage.) In population statistics, rates usually are stated per 1,000 individuals, although other powers of 10 may be used for particular diseases (e.g., per 100 for very common diseases and per 10,000 or 100,000 for uncommon diseases). Disease **incidence** measures the occurrence of new cases of a disease during a defined time period, as compared to the total (healthy) population. The incidence of a disease denotes both the rate of occurrence and the relative risk. While the incidence of a disease reflects a rate of new infections over time, the **prevalence** refers to the total number of individuals infected in a population at a specific point in time, no matter when the disease began. The prevalence rate depends on both the incidence rate and the duration of the illness. Prevalence is calculated as follows:

$$\text{Prevalence} = \frac{\text{Total number of cases in population}}{\text{Total population}} \times 100$$

Recall that morbidity rate reflects a change in health status over time. The rate is commonly determined when the number of new cases of illness in the general population is known from clinical reports. It is calculated as follows:

$$\text{Morbidity rate} = \frac{\text{Number of new cases of a disease during a specified period}}{\text{Number of individuals in the population}}$$

For example, if the population of a city is 100,000 people and 5,000 individuals in it become ill, the morbidity rate is 5%.

The mortality rate, or death rate, is a simple statement of the proportion of all deaths that are assigned to a single cause. If, in the same population of 100,000 people, 750 individuals die in a year, the mortality rate is 0.75% for that time period. Compare this with the case fatality rate (CFR), which is the proportion of individuals in a population diagnosed with a certain disease who die from that particular disease. CFR is calculated as follows:

$$\text{Case fatality rate} = \frac{\text{Number of deaths due to a given disease}}{\text{Size of the total population with the same disease}}$$

Within the population of 100,000 people, if there are 2,500 reported cases of a particular disease and 30 people die from that disease, the CFR is 1.2%.

The determination of prevalence, morbidity, and mortality rates helps public health personnel direct healthcare efforts to control the spread of infectious diseases. A sudden increase in the morbidity rate of a particular disease may indicate that preventive measures designed to reduce the disease should be implemented.

Comprehension Check

1. What is epidemiology?
2. Define and distinguish *epidemic, endemic,* and *pandemic*.
3. What types of surveillance data are most useful in determining infectious disease penetration into a population?
4. Define *morbidity rate, incidence, prevalence, mortality rate,* and *case fatality rate.*

35.3 Infectious Disease Is Revealed Through Patterns Within a Population

After reading this section, you should be able to:

a. Compare and contrast communicable and noncommunicable diseases
b. Interpret population infection data to define epidemic and pandemic events
c. Apply the concept of herd immunity to public health in your community
d. Discuss the relationship between R_0 and herd immunity

An **infectious disease** is an illness caused by a microbial agent such as a virus, bacterium, fungus, protozoan, or helminth (worm). A **communicable disease** is an infectious disease that can be transmitted from person to person. Not all infectious diseases are communicable; for example, Lyme disease is acquired through a tick bite but cannot be transmitted between humans. Similarly some toxin-mediated diseases, such as tetanus and botulism, are *noncommunicable*. An epidemiologist studying an infectious disease is concerned with the causative agent, the source or reservoir of the disease agent, how that agent is transmitted, what host and environmental factors can aid development of the disease within a defined population, and how best to control or eliminate the disease. These factors describe the natural history or cycle of an infectious disease.

Epidemiologists recognize an infectious disease in a population by using various surveillance methods. Surveillance is a dynamic activity that includes gathering information on the development and occurrence of a disease, collating and analyzing the data, summarizing the findings, and using the information to select control methods (**figure 35.4**). Surveillance may not always require the direct examination of cases. However, to accurately interpret surveillance data and study the course of a disease in individuals, epidemiologists and other medical professionals must be aware of the pattern of infectious diseases that reflects how the disease spreads through a population. Epidemiologists study disease patterns so they can rapidly identify new outbreaks before substantial morbidity and mortality occur **Historical Highlights 35.2**. ◀ *The process of infection (section 34.1)*

Two major types of epidemics are recognized: common source (noncommunicable) and propagated (communicable). A **common-source epidemic** is characterized as having reached a peak level within a short period of time (1 to 2 weeks), followed by a moderately rapid decline in the number of patients who are infected (**figure 35.5*a***). This type of epidemic usually results from a single, common contaminated source such as food (food poisoning) or water (cholera). By contrast, a **propagated epidemic** is characterized by a relatively slow and prolonged rise and then a gradual decline in the number of individuals infected (figure 35.5*b*). This type of epidemic usually results from the introduction of a single individual with the infection into a susceptible population. The initial infection is then propagated (spread) to others in a gradual fashion until many individuals within the population are infected. An example is the increase in influenza cases that coincides with the return of college students after their winter break. Droplet and contact transmission of flu virus spreads rapidly among susceptible, and unvaccinated, students. One infected student is sufficient to initiate the epidemic, after which person-to-person transmission of the influenza virus extends the epidemic.

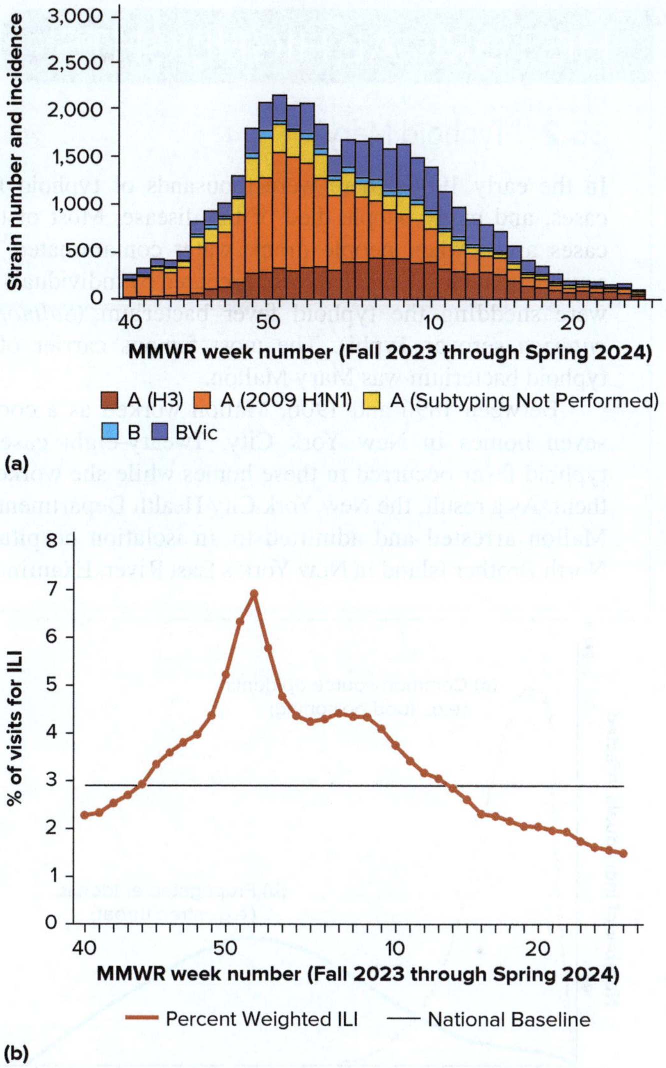

Figure 35.4 Graphic Representation of Epidemiological Data. The Centers for Disease Control and Prevention collect and evaluate various parameters related to human health and disease. These data reflect the cumulative number of confirmed human cases of influenza for the 2023–2024 flu season by (a) strain number and incidence per week and (b) number of hospitalizations. ILI, influenza-like illness.

CDC. FluView Interactive. https://gis.cdc.gov/grasp/fluview/fluportaldashboard.html

HISTORICAL HIGHLIGHTS

35.2 "Typhoid Mary"

In the early 1900s there were thousands of typhoid fever cases, and many people died of the disease. Most of these cases arose when people drank water contaminated with sewage or ate food handled by or prepared by individuals who were shedding the typhoid fever bacterium (*Salmonella enterica* serovar Typhi). The most famous carrier of the typhoid bacterium was Mary Mallon.

Between 1896 and 1906, Mallon worked as a cook in seven homes in New York City. Twenty-eight cases of typhoid fever occurred in these homes while she worked in them. As a result, the New York City Health Department had Mallon arrested and admitted to an isolation hospital on North Brother Island in New York's East River. Examination of her stools showed that she was shedding large numbers of typhoid bacteria, though she exhibited no symptoms of the disease. An article published in 1908 in the *Journal of the American Medical Association* referred to Mallon as "Typhoid Mary," an epithet by which she is still known today. She was released when she pledged not to cook for others or serve food to them, but she changed her name and began to work as a cook again. For 5 years she managed to avoid capture while continuing to spread typhoid fever. Eventually the authorities tracked her down. She was held in custody for 23 years, until she died in 1938. As a lifetime carrier, Mary Mallon was positively linked with 10 outbreaks of typhoid fever, 53 cases, and 3 deaths.

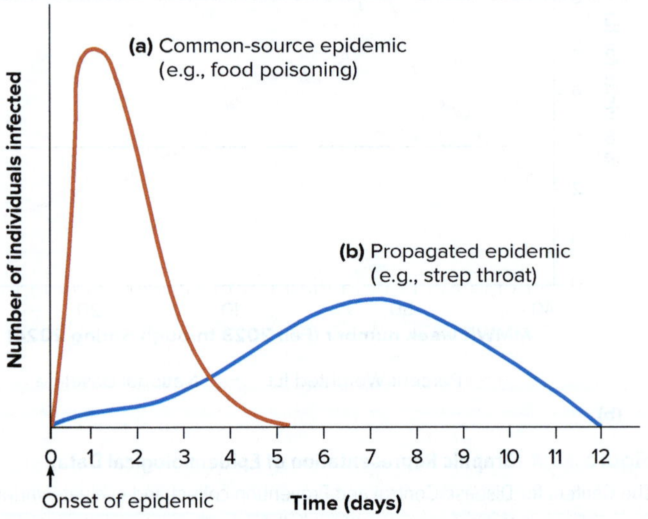

Figure 35.5 Epidemic Curves. (a) In a common-source epidemic, there is a rapid increase up to a peak in the number of individuals infected and then a rapid but more gradual decline. Cases usually are reported for a period that equals approximately one incubation period of the disease. (b) In a propagated epidemic, the curve has a gradual rise and then a gradual decline. Cases usually are reported over a time interval equivalent to several incubation periods of the disease.

MICRO INQUIRY *Is an influenza epidemic an example of a common-source or a propagated epidemic? What about a Salmonella food-poisoning outbreak? Explain your answers.*

To understand how epidemics are propagated, consider **figure 35.6**. At time 0, all individuals in this population are susceptible to a hypothetical pathogen. The introduction of an individual with the infection initiates the epidemic outbreak (lower curve), which spreads to reach a peak by day 15. As individuals recover from the disease, they become immune and no longer transmit the pathogen (upper curve). The number of susceptible individuals therefore decreases. The decline in the number of susceptible individuals to the threshold density (the minimum number of individuals necessary to continue propagating the disease) coincides with the peak of the epidemic wave, and the incidence of new cases declines because the pathogen cannot propagate itself. If the percentage of susceptible people is above the threshold density, the level of protection will decrease and the morbidity rate will increase. For example, the morbidity rates of influenza among schoolchildren may reach epidemic levels if the number of susceptible individuals rises above 30% for the whole population.

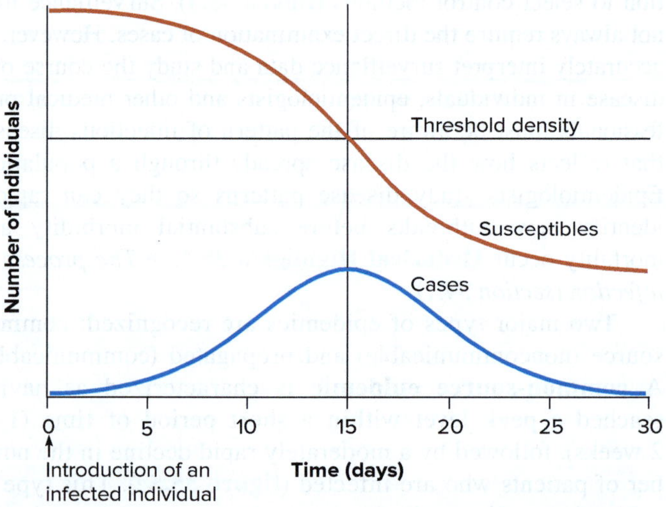

Figure 35.6 The Spread of a Hypothetical Propagated Epidemic. The peak of the epidemic wave coincides with the threshold density of people who are infected.

MICRO INQUIRY *Why does the number of susceptible individuals decline after the epidemic has reached its peak?*

As **figure 35.7** illustrates, the goal from a public health standpoint is to achieve a level of **herd immunity:** a threshold percentage of the population having immunity so that even when isolated cases of a disease reemerge, there is no escalation of the disease throughout that population. Epidemiologists strive to determine the capacity of an infectious agent to spread. This parameter can be quantified; it is called the **basic reproduction number** or $\mathbf{R_0}$ (R naught), which represents the number of susceptible individuals that can be infected from one person with that infection. The R_0 for any given infectious agent is not a constant number; it varies depending on biological, behavioral, and environmental factors. For example, population density can lead to a different R_0 value in different areas. Think of the capacity of a contagious pathogen to spread in a crowded city versus a small town, coupled with the behavior of the population (e.g., riding a crowded subway versus traveling in a private car). Not only does the R_0 assess the transmissibility and thus the severity of the outbreak, it also implies the strength of the medical and behavioral interventions required for its control.

The math to calculate the R_0 is complex, and as is the case when dealing with living organisms, the data do not always fit neatly into the prescribed theory. An important consideration when calculating the R_0 value is that it assumes *all* members of a population are susceptible to the infectious agent. A highly contagious disease has a high R_0 value. For example, measles, which is transmitted via aerosolized particles, has an $R_0 = 12–18$, meaning one person who is infected can potentially infect up to 18 others. Compare this to a disease that does not spread easily from person to person. Human immunodeficiency virus (HIV), for example, must be transmitted directly through blood, blood products, or body fluids, and has an $R_0 = 2–5$. Generally, the higher the R_0 value, the harder it is to control an epidemic, while an $R_0 < 1$ suggests a disease will ultimately die out in a population.

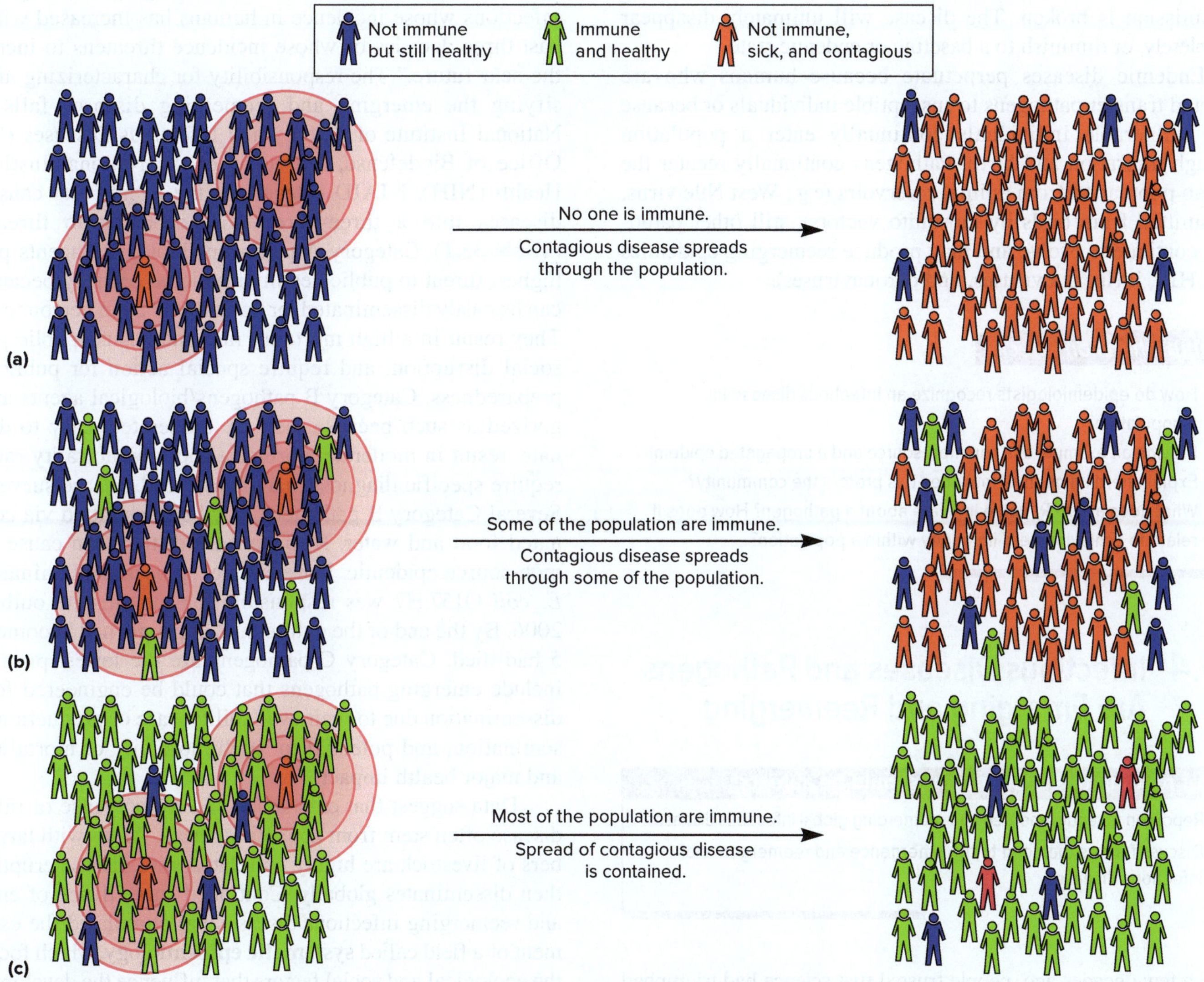

Figure 35.7 Herd Immunity. Immunity can arise from previous exposure or vaccination. Susceptible members of an immune population enjoy a protection that is not of their own making. When (a) no one in the original population is immune, the disease spreads. If a small number are immune (b), the disease still spreads; however, if most of the population are immune (c), the susceptible few are protected by herd immunity.

Knowing the R_0 is crucial to determine what percentage of the population must be immune to achieve herd immunity and stop disease spread. With an R_0 of 5, for example, epidemiologists calculate that approximately 80% of the population must be immune to prevent an epidemic. For a pathogen with a high case fatality rate, herd immunity is best achieved by protection rather than exposure; otherwise hospitals become overwhelmed and there are many unnecessary deaths. Vaccination is the most viable solution. Once a vaccine has been developed and implemented, the goal is to lower the **effective reproduction number** or $\mathbf{R_e}$. Unlike the R_0, which is calculated in an unprotected population, the R_e is calculated as the effective reproduction rate in a population in which some individuals are immune. Ideally, a value of less than 1 is desired, which implies an individual is unlikely to pass the disease to any susceptible individuals. Under these conditions, the rate of infection slows, and although isolated cases may reemerge from time to time, as long as the threshold percentage of vaccinated individuals is maintained, the chain of infectious disease transmission is broken. The disease will ultimately disappear completely, or diminish to a baseline or endemic state.

Endemic diseases perpetuate because humans who are infected transfer pathogens to susceptible individuals or because new, vulnerable individuals continually enter a population through migration and birth. Pathogens continually reenter the human population from animal reservoirs (e.g., West Nile virus, transmitted from birds by mosquito vectors); still other pathogens continue to evolve and may produce reemerging epidemics (e.g., HIV, influenza viruses, and coronaviruses).

Comprehension Check

1. How do epidemiologists recognize an infectious disease in a population?
2. Provide an example of a common-source and a propagated epidemic.
3. Explain herd immunity. How does this protect the community?
4. What does a high R_0 value indicate about a pathogen? How does it relate to achieving herd immunity within a population?

35.4 Infectious Diseases and Pathogens Are Emerging and Reemerging

After reading this section, you should be able to:

a. Report on recently emerged and reemerging global infectious diseases
b. Discuss factors that lead to the emergence and reemergence of infectious diseases

Only a few decades ago, people trusted that science had triumphed over infectious diseases. Antibiotics, vaccines, and public health campaigns had yielded a string of victories over diseases such as whooping cough (pertussis), pneumonia, polio, and smallpox. In many countries, people were lulled into believing that microbial threats were a thing of the past. Trends in the number of deaths caused by infectious diseases in the United States from 1900 through 2022 supported this conclusion (figure 35.2). However, as COVID-19 has made clear, despite this overall downward trend, the incidence of infectious diseases resulting from emerging and reemerging microbial pathogens continues to impact or claim the lives of millions of people. In addition to SARS-CoV-2, of note are HIV/AIDS, Zika virus disease, Ebola virus disease, cholera, mpox, hepatitis C and E, H5N1 avian influenza, *Candida auris* infections, Lyme disease, tuberculosis, and hemolytic uremic syndrome caused by Shiga toxin-producing strains of *E. coli*. Compounding the problem is the rise of the multidrug-resistant bacteria.

It is clear that, for the time being, humans will continue to be faced with both new infectious diseases and the reemergence of older diseases once thought to be conquered. The CDC has defined these diseases as "new, reemerging, or drug-resistant infections whose incidence in humans has increased within the past three decades or whose incidence threatens to increase in the near future." The responsibility for characterizing and classifying the emerging and reemerging diseases falls to the National Institute of Allergy and Infectious Diseases (NIAID) Office of Biodefense, a division of the National Institutes of Health (NIH). NIAID categorizes the agents that cause these diseases into a three-tiered list according to threat level (**Table 35.1**). Category A pathogens/biological agents pose the highest threat to public health or national security because they can be easily disseminated, or transmitted from person to person. They result in a high mortality rate, may cause public panic or social disruption, and require special action for public health preparedness. Category B pathogens/biological agents are categorized as such because they are moderately easy to disseminate, result in moderate morbidity and low mortality rates, and require specific diagnostics and enhanced disease surveillance. Several Category B pathogens may be transmitted via contaminated food and water. If disseminated, they can cause a common-source epidemic. For example, spinach contaminated with *E. coli* O157:H7 was responsible for a multistate outbreak in 2006. By the end of the outbreak, 225 people had become ill and 5 had died. Category C pathogens are the lowest priority and include emerging pathogens that could be engineered for mass dissemination due to their availability, ease of production or dissemination, and potential for high morbidity or mortality rates and major health impact.

Data suggest that emergence and reemergence of infectious disease often stem from zoonotic sources. Areas with large numbers of livestock are highly correlated with disease eruption that then disseminates globally. Continued surveillance of emerging and reemerging infectious diseases has stimulated the establishment of a field called **systematic epidemiology,** which focuses on the ecological and social factors that influence the development of these diseases. Current technology and global cooperation have resulted in an improved ability to track emerging and reemerging infectious diseases in real time. ▶| *Chapter 39 opener*

Table 35.1 National Institute of Allergy and Infectious Diseases Three-Tiered Pathogen Priority List (Partial)

Category A
Bacillus anthracis (anthrax)
Clostridium botulinum (botulism)
Yersinia pestis (plague)
Variola major (smallpox) and other pox viruses
Francisella tularensis (tularemia)
Hemorrhagic fever viruses (including dengue, Ebola, Marburg)
Category B
Burkholderia spp.
Coxiella burnetii (Q fever)
Brucella spp. (brucellosis)
Chlamydia psittaci (psittacosis)
E. coli, Salmonella, Shigella, Listeria, Campylobacter (diarrheal disease agents)
Hepatitis A virus (hepatitis)
Mosquito-borne viruses (e.g., West Nile, Eastern equine encephalitis, Zika, chikungunya)
Protozoans—*Cryptosporidium, Giardia, Toxoplasma, Naegleria fowleri*
Fungi—Microsporidia
Toxins—ricin, *Clostridium perfringens* epsilon, staphylococcal enterotoxin B
Category C
Hantavirus (pulmonary syndrome)
Mycobacterium tuberculosis (TB/extensively drug resistant (XDR)-TB)
Influenza virus (flu)
Rabies virus (rabies)
Prions (neurologic diseases; e.g., Creutzfeldt-Jacob)
Antibiotic-resistant bacteria

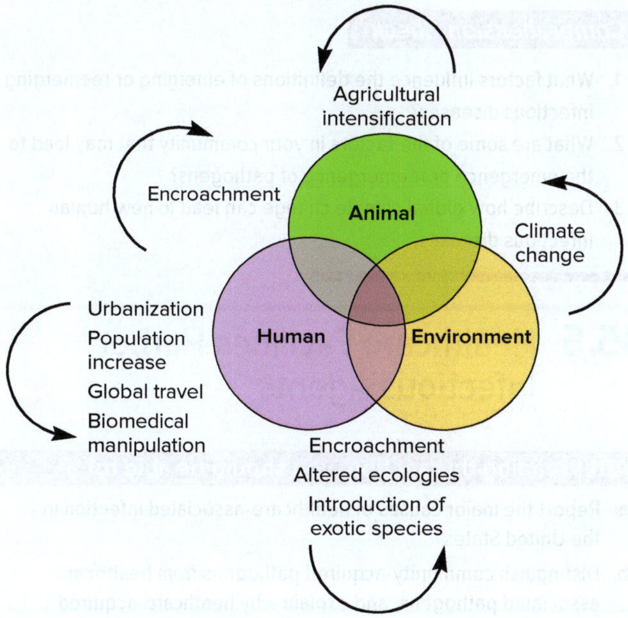

Figure 35.8 Disease Emergence and Reemergence Involve Multiple Factors and Mechanisms. The emergence and reemergence of disease-causing pathogens involve a dynamic, multifactorial interplay between humans, animals, and the environment.

Source: Daszak, P., Cunningham, A. A., and Hyatt, A. D. 2000. "Emerging infectious diseases of wildlife—threats to biodiversity and human health." *Science.* 287(5452): 443–449.

Why are pathogens posing such a problem, despite dramatic advances in medical research, drug discovery, technology development, and sanitation? Many factors characteristic of the modern world undoubtedly favor the development and spread of these microorganisms and their diseases (**figure 35.8**).

As population density increases in cities, the dynamics of microbial exposure and evolution increase in humans. Urbanization often crowds humans and increases exposure to microorganisms. Crowding leads to unsanitary conditions and hinders the effective implementation of adequate medical care, enabling more widespread transmission and propagation of pathogens. In modern societies, crowded workplaces, communal living facilities, day-care centers, large hospitals, and public transportation all facilitate microbial transmission. In addition, the speed and volume of international travel are major factors contributing to the global emergence of infectious diseases. Because air travel essentially eliminates time between exposure and disease outbreak, a person can spread an infectious disease in a matter of hours.

When changes in climate or ecology occur, it is not surprising to also find changes in both beneficial and detrimental microorganisms. Climate change also affects microorganism survival and the distribution of insect vectors. Mass migrations of refugees, workers, and displaced persons have led to a steady growth of urban centers at the expense of rural areas. Furthermore, land development and the exploration and destruction of natural habitats have increased the likelihood of human exposure to new pathogens and may select for pathogens better able to adapt to new hosts and changing environments. The introduction of pathogens to a new environment or host can alter transmission and exposure patterns, leading to sudden proliferation of disease. For example, the spread of Ebola virus disease in 2013–2016 across West Africa resulted when a suspected bat pathogen infected the encroaching human population. Migration of persons who are infected along established roadways facilitated viral transmission. ◀ *Global climate change (section 28.3)*

Emerging and reemerging pathogens and the diseases they cause are therefore the outcome of many different factors. Because the world is so interconnected, we cannot isolate ourselves from other countries and continents. Changes in the disease status of one part of the world or the misuse of antibiotics in another part may well affect health around the planet. As Nobel laureate Joshua Lederberg so eloquently stated, "The microbe that felled one child in a distant continent yesterday can reach yours today and seed a global pandemic tomorrow."

Comprehension Check

1. What factors influence the definitions of emerging or reemerging infectious diseases?
2. What are some of the factors in your community that may lead to the emergence or reemergence of pathogens?
3. Describe how global climate change can lead to new human infectious diseases.

35.5 Healthcare Facilities Harbor Infectious Agents

After reading this section, you should be able to:

a. Report the major causes of healthcare-associated infection in the United States
b. Distinguish community-acquired pathogens from healthcare-associated pathogens, and explain why heathcare-acquired pathogens are of concern to hospital personnel
c. Recommend measures for preventing and controlling healthcare-associated infections

Healthcare-associated infections (HAIs), also called **healthcare-acquired** or **nosocomial infections** (Greek *nosos,* disease; *komeion,* to take care of), result from infections acquired by patients while in a hospital or other clinical care facility. Many potential sources of infection exist in a hospital. Animate sources are the hospital staff, other patients, and visitors. Some examples of inanimate sources are food, plants and flowers, computer keyboards, intravenous and respiratory therapy equipment, and water systems (e.g., dialysis units and hydrotherapy equipment). Besides harming patients, HAIs can affect nurses, physicians, aides, visitors, custodians, and anyone else who has contact with the hospital. Most HAIs become clinically apparent while patients are still hospitalized; however, disease onset can occur after patients have been discharged. Infections that are incubating when patients are admitted to a hospital are not healthcare associated; they are community acquired.

Healthcare-associated diseases are usually caused by bacteria, most of which are noninvasive and part of the human microbiota. Today the most common HAIs are (1) catheter-associated urinary tract infections, (2) surgical site infections, (3) central line (catheters)–associated bloodstream infections, and (4) ventilator-associated pneumonias (**figure 35.9**). In addition to *Clostridioides difficile,* recently recognized healthcare-associated, Gram-positive species include *Paeniclostridium sordellii, Corynebacterium jeikeium,* and *Rhodococcus equi.* The incidence of infections by the Gram-negative pathogens *Pseudomonas aeruginosa, Acinetobacter* spp., *Burkholderia cepacia* complex, and *Stenotrophomonas maltophilia* has also increased. Many Gram-negative bacilli are resistant to β-lactam antibiotics. These bacteria include *E. coli,* *Klebsiella pneumoniae,* other *Klebsiella* spp., *Proteus* spp., *Morganella* spp., *Citrobacter* spp., *Salmonella* spp., and *Serratia marcescens,* which are resistant to penicillins, older cephalosporins, aztreonam, and, most alarmingly, the carbapenems. The β-lactamase-resistant members of Enterobacteriaceae are collectively referred to as extended-spectrum β-lactamase-producing Enterobacteriaceae. ◀ *Antimicrobial drug resistance is a public health threat (section 9.8)*

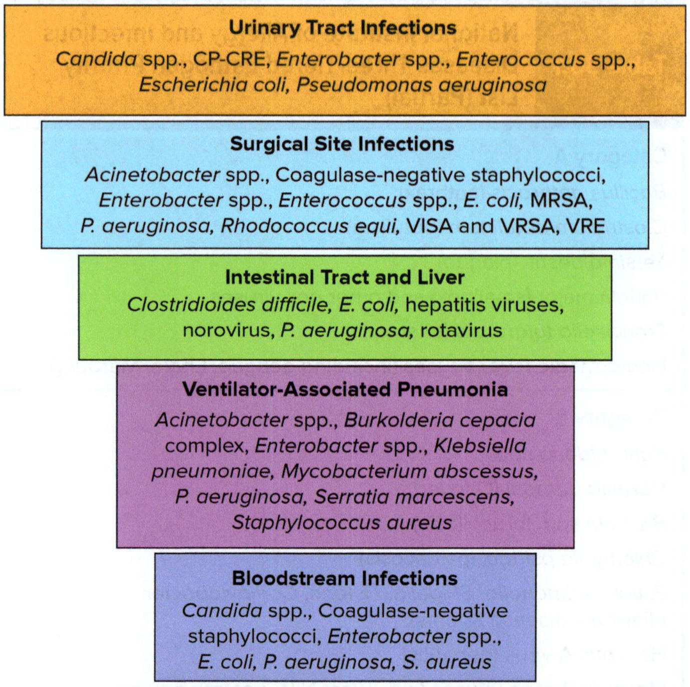

Figure 35.9 Leading Types of Healthcare-Associated Infections (HAIs) in the United States. The most common HAIs are caused by members of the patient's microbiota or by bacteria, fungi, and viruses that are transferred to a patient from a healthcare worker, contaminated instruments, plants and flowers, or fresh foods. CP-CRE: Carbapenemase-producing carbapenem-resistant Enterobacteriaceae; MRSA: methicillin-resistant *Staphylococcus aureus;* VISA and VRSA: vancomycin-intermediate and -resistant *S. aureus* respectively; VRE: vancomycin-resistant Enterococci

The CDC estimates that in the United States alone, 1 in every 31 patients (nearly 1.7 million people) acquires some type of HAI every year. These infections prolong hospital stays by 4 to 14 days, result in an additional $28 to $33 billion per year to overall direct healthcare costs, and lead to approximately 99,000 deaths annually. The enormity of this problem has led hospitals to allocate substantial resources to developing methods and programs for the surveillance, prevention, and control of HAIs.

Comprehension Check

1. Describe a healthcare-associated infection and explain why such infections are significant.
2. What two general sources are responsible for healthcare-associated infections? Give some specific examples of each.

35.6 Coordinated Efforts Are Required to Prevent and Control Epidemics

After reading this section, you should be able to:

a. Discuss the control measures used by epidemiologists to break the chain of infection
b. Describe the chain of infection and explain how the weakest link in it can be exploited to control or prevent disease
c. Describe DNA, RNA, and whole-pathogen vaccines, and give advantages and disadvantages for each

The control of an infectious disease relies on a well-defined network of clinical microbiologists, nurses, physicians, epidemiologists, and infection control personnel who supply epidemiological information to a network of local, state, national, and international organizations. These individuals and organizations comprise the public health system. Each state has a public health laboratory responsible for overseeing infection surveillance and control. The communicable disease section of a state laboratory includes specialized services for the examination of specimens or cultures submitted by physicians, local health departments, hospital laboratories, sanitarians, epidemiologists, and others. These groups share their findings with other health agencies in the state, the CDC, and the WHO.

Control Starts with Breaking the Chain of Infection

Just as the development of an infectious disease is a complex process involving many factors, so too is the design of specific epidemiological control measures. Epidemiologists must consider available resources and time constraints, adverse effects of potential control measures, and human activities that might influence the spread of the infection. To proceed strategically, the components of the infectious disease chain that are primarily responsible for a particular epidemic must be identified: the infectious agent, its reservoir or source, the portal of exit from that reservoir or source, the mode of transmission of the pathogen, the portal of entry into a host, and the susceptibility of that host. Control measures are directed toward that part of the chain that is most susceptible to control—the weakest link in the chain.

One type of control measure is directed toward reducing or eliminating the source or reservoir of infection: social distancing and isolation of carriers, destruction of animal or arthropod reservoirs of infection, treatment of water and sewage to reduce contamination (*see chapter 42*), and therapy that reduces or eliminates infectivity of the individual.

An alternative control measure reduces the number of susceptible individuals and raises the general level of herd immunity. This is achieved by immunization (*see figure 32.3*) as well as prophylactic treatment to prevent infection (e.g., use of chloroquine when traveling to malaria-endemic countries). Antimicrobial chemotherapy is discussed in chapter 9. We discuss the use of vaccines and immunization next.

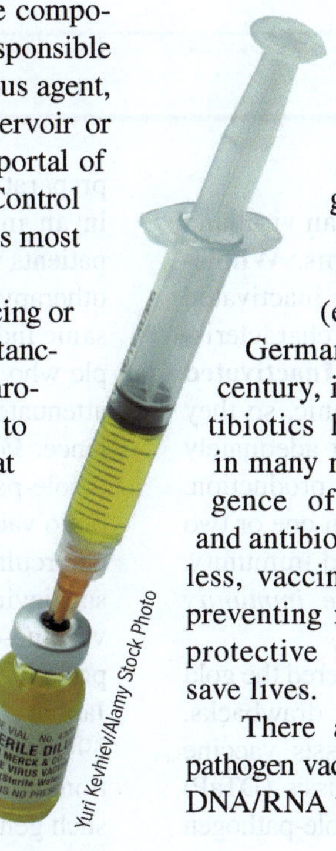
Yuri Kewhiev/Alamy Stock Photo

Vaccines Immunize Susceptible Populations

A **vaccine** (Latin *vacca,* cow) is a preparation of one or more microbial antigens that induce protective immunity in the host. **Immunization** occurs when the host's immune system has successfully responded to the vaccine. The specific goal of vaccination is to induce antibodies and activated T cells to protect a host from future infection. Many epidemics have been prevented by mass prophylactic immunization. Vaccine recommendations and schedules are available on the CDC website; these are dynamic as new formulations become available and the ecology of specific diseases changes, and are updated annually. To date, vaccines have eradicated smallpox, pushed polio to the brink of extinction, and protected millions of individuals from hepatitis A and B, influenza, measles, rotavirus disease, tetanus, typhus, and other dangerous diseases. **Vaccinomics,** the integration of genomics, bioinformatics, and personalized medicine, strives to identify genetic and other mechanisms and pathways that determine immune responses, and thereby provide innovative approaches to vaccine development. ▶| *Vaccines (section 41.1)*

To promote a more efficient immune response, antigens in vaccines can be mixed with an **adjuvant** (Latin *adjuvans,* aiding). Adjuvants are any nontoxic material that assists in the processing of antigens by antigen-presenting cells, or otherwise nonspecifically stimulates the immune response to the antigen; examples include oil in water emulsions (Freund's incomplete adjuvant), aluminum hydroxide salts (alum), beeswax, and various combinations of bacteria (live or killed). |◀ *Recognition of foreignness is critical for a strong defense (section 32.4)*

The modern era of vaccines and immunization began in 1798 with Edward Jenner's use of cowpox as a vaccine against smallpox (**Historical Highlights 35.3**) and in 1881 with Louis Pasteur's rabies vaccine. Vaccines for other diseases did not emerge until the twentieth century when methods for inactivating and attenuating microorganisms were improved. Vaccines were eventually developed against most of the epidemic diseases that plagued Western Europe and North America (e.g., diphtheria, measles, mumps, pertussis, German measles, and polio). Indeed, by the late twentieth century, it seemed that the combination of vaccines and antibiotics had solved the problem of microbial infections in many nations. Such optimism was cut short by the emergence of new and previously unrecognized pathogens and antibiotic resistance among existing pathogens. Nevertheless, vaccination is still the most cost-effective weapon for preventing microbial disease. The role of immunization as a protective measure cannot be overstated: Immunizations save lives.

There are three general categories of vaccines: whole-pathogen vaccines, acellular or subunit vaccines, and recombinant/DNA/RNA vaccines. We discuss examples of each type next.

HISTORICAL HIGHLIGHTS

35.3 The First Immunizations

Since the time of the ancient Greeks, it has been recognized that people who have recovered from plague, smallpox, yellow fever, and various other infectious diseases rarely contract the diseases again. The earliest records of inoculation are against smallpox; accounts from the 1500s describe the practice of grinding up smallpox scabs and scratching them onto the skin or blowing them up the nostrils. This process is known as variolation, and it was introduced to Europe in 1721 by Lady Mary Wortley Montagu after she witnessed it while living in Turkey. Unfortunately, variolation was not risk-free and patients sometimes contracted and died from the disease.

The first scientific attempts at smallpox immunizations were made by Edward Jenner (1749–1823), a country doctor from England. Jenner investigated the basis for the widespread belief of the time that anyone who had vaccinia (cowpox) never contracted smallpox. Smallpox was often fatal and those who recovered had disfiguring scars. Dairy workers who were previously infected with cowpox rarely contracted smallpox.

In 1796, in an experiment that by today's standards was completely unethical, Jenner extracted the contents of a pustule from the arm of a dairy worker infected with cowpox and injected it into the arm of 8-year-old James Phipps. As Jenner expected, immunization with the cowpox virus caused only mild symptoms in the boy. When he subsequently inoculated the boy with smallpox virus, the boy showed no symptoms of the disease. Jenner then inoculated large numbers of his patients with cowpox pus, as did other physicians in England and Europe (**box figure**). Jenner coined this much safer approach "vaccination," from the Latin word *vacca* for cow.

Further work on immunization was carried out by Louis Pasteur (1822–1895). His work led to the first laboratory-developed cholera vaccine. Other scientists developed vaccines for rabies, typhoid, and diphtheria within the next two decades.

Edward Jenner Vaccinating Child, 1796
Photo Researchers/Science History Images/Alamy Stock Photo

Whole-Pathogen Vaccines

Many of the current vaccines effective against human viral and bacterial diseases consist of whole microorganisms. **Whole-pathogen vaccines** contain agents that are either inactivated (killed) or attenuated (live but avirulent). The major characteristics of these vaccines are compared in **table 35.2**. **Inactivated vaccines** are effective, but they are less immunogenic, so they often require several boosters and normally do not adequately stimulate cell-mediated immunity or secretory IgA production. In contrast, **attenuated vaccines** usually are given in one or two doses and stimulate both humoral and cell-mediated immunity. ◀ *Innate host resistance (chapter 31); Adaptive immunity (chapter 32)*

Even though whole-pathogen vaccines are considered the gold standard of vaccines, they are not without their drawbacks. In the 1990s the combined whole-pathogen pertussis vaccine (DTwP) was replaced with combined acellular pertussis (DTaP) due to the risk of serious allergic reactions from the whole-pathogen preparation. Attenuated vaccines can also cause full-blown illness in an individual whose immune system is compromised (e.g., patients with AIDS, those with cancer who are undergoing chemotherapy, females who are pregnant, and older individuals). These same individuals may also contract the disease from healthy people who have recently received an attenuated vaccine. Moreover, attenuated viruses can at times mutate in ways that restore virulence. For this reason, new vaccine technologies rarely include whole-pathogen preparations. For example, in some countries, the polio vaccine is in the form of a live, attenuated virus. If the virus cocirculates with nonpolio enteric (intestinal) viruses such as coxsackievirus, the two viruses can recombine to form pathogenic vaccine-derived poliovirus. (VDPV). Several large outbreaks of polio have been caused by circulating VDPV (cVDPV); between January 2023 and June 2024, nearly 700 cases were reported in 39 countries. Children in the United States and several other nations receive the inactivated polio vaccine, which is incapable of such genetic events.

Table 35.2 A Comparison of Vaccine Types[1]

Major Characteristic	Whole-Pathogen Vaccines		Acellular or Subunit Vaccines	Recombinant or DNA/RNA Vaccines	
	Live, attenuated	Inactivated		Recombinant	DNA or RNA
Typical Immunogen	Viruses	Bacteria or viruses	1–20 antigens from microorganism	Virus or bacterial vector	Antigen(s) encoded by DNA or RNA
T-cell Response	Strong	Weak	Very strong	Very strong	Very strong
B-cell Response	Strong	Strong	Very strong	Very strong	Very strong
Stability	Decreased without refrigeration	Longer shelf life, freeze-dried	Longer shelf life, freeze-dried	Longer shelf life, freeze-dried	
Potential Problems	Reversion to pathogen possible, not given to immunocompromised	Weaker response, requires booster doses	Identification of antigens is costly and research intensive	Only effective against protein immunogens; DNA vaccines have efficacy issues and risk disrupting genomic DNA	

[1] Source: U.S. Department of Health & Human Services. "Vaccine Types."

txking/Shutterstock

Acellular or Subunit Vaccines

Some of the common risks associated with whole-cell vaccines can be avoided by using only specific, purified macromolecules derived from pathogenic microorganisms. These are **subunit vaccines,** of which there are three general forms: (1) capsular polysaccharides, (2) recombinant surface antigens, and (3) inactivated exotoxins called **toxoids.** The purified microbial subunits or their secreted products can be prepared as nontoxic antigens to be used in the formulation of vaccines (**table 35.3**).

Recombinant-Vector, DNA, and RNA Vaccines

Genes isolated from a pathogen that encode major antigens can be inserted into nonvirulent viruses or bacteria. Such recombinant microorganisms serve as vectors, either replicating within the host and expressing the gene product of the pathogen-encoded antigenic proteins or depositing the DNA into the host nucleus, where it is transcribed. The antigens elicit antibody-mediated immunity when they are released, and they also elicit cellular immunity when they are broken down and properly displayed on the cell surface (just as occurs when host cells harbor an active pathogen). Several pathogens, such as hepatitis B virus and attenuated *Salmonella enterica,* have been used in the production of these **recombinant-vector vaccines.**

DNA vaccines introduce fragments of pathogen DNA directly into the host cell. When injected into muscle cells, the DNA is taken into the nucleus and the pathogen's DNA is transiently expressed, generating foreign proteins to which the host's immune system responds. DNA vaccines are very stable between production and use; refrigeration is often unnecessary. The Johnson & Johnson and AstraZeneca/Oxford COVID-19 vaccines deliver SARS-CoV-2

Table 35.3 Example Subunit Vaccines

Microorganism or Toxin	Vaccine Subunit
Capsular polysaccharide	
Haemophilus influenzae type b	Polysaccharide-protein conjugate (HbCV) or bacterial polysaccharide (HbPV)
Neisseria meningitidis	Polysaccharides of serotypes A/C/Y/W-135
Streptococcus pneumoniae	23 distinct capsular polysaccharides
Surface antigen	
Hepatitis B virus (HBV)	Recombinant surface antigen (HBsAg)
Human papillomavirus (HPV)	Recombinant protein subunits
Toxoids	
Corynebacterium diphtheriae toxin	Inactivated exotoxin
Clostridium tetani toxin	Inactivated exotoxin

spike protein DNA to host cells using a modified adenovirus. At present, human trials are underway for several other DNA vaccines against malaria, HIV, influenza, hepatitis B and C, rabies, HPV, and Ebola and Zika virus diseases. *Constructing Vaccines*

The COVID-19 pandemic brought **mRNA vaccines** to the forefront of public awareness. The rapid spread and high incidence of COVID-19 revealed an obvious need for a vaccine. Two mRNA vaccines were quickly produced, and initially used under an Emergency Use Authorization. Both vaccines were fully approved by the FDA in 2021 and 2022 respectively. While it may have felt like the critical steps of research, development, clinical trials, and FDA approval were rushed, RNA vaccine development has been in progress and in veterinary use since the 1990s. Work was pioneered by Katalin Karikó and Drew Weissman, who won the Nobel Prize for Physiology or Medicine in 2023 for their contributions to COVID-19 mRNA vaccine development. Major hurdles to overcome in mRNA vaccine research included (1) minimizing mRNA degradation, (2) limiting detrimental innate immunogenicity, and (3) designing a delivery system that would facilitate entry into the cell. Creating an in vitro–transcribed mRNA that mimics a fully processed mature mRNA resolved the first issue; using nucleoside-modified, purified mRNA solved the second; and inserting the mRNA into liposomes enhanced delivery across the cell membrane and took care of the third issue. The vaccine was constructed by inserting the mRNA encoding the spike protein of SARS-CoV-2 into the lipid vesicle *(see Chapter 25)*. **Figure 35.10** illustrates the mechanism of action of mRNA vaccines. The rapid development of the COVID-19 vaccines was facilitated by government and private-sector funds, an increased workforce, and prior research. Scientists and public health officials agree: Safety and efficacy are never compromised in favor of speed. mRNA vaccine

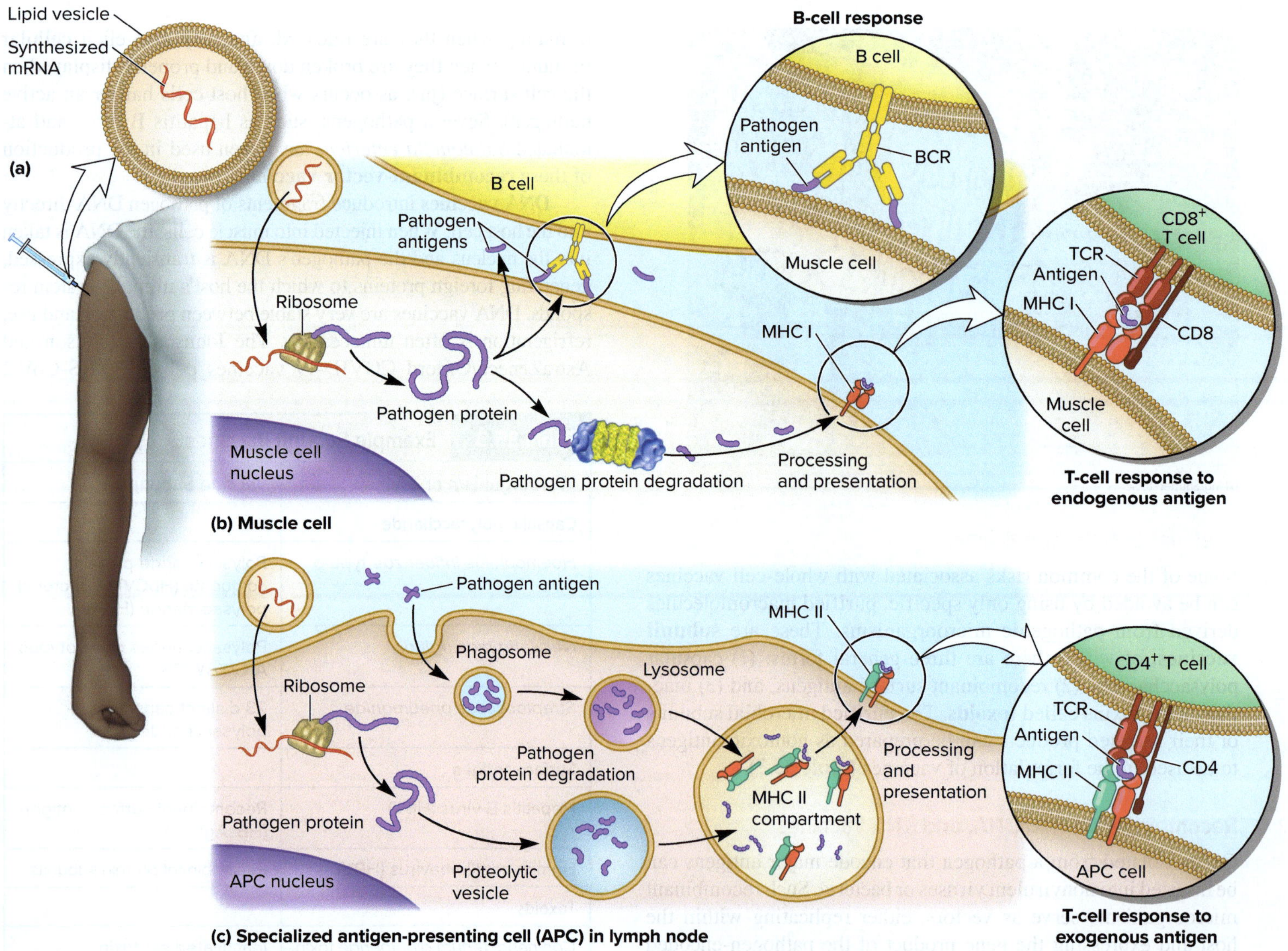

Figure 35.10 mRNA Vaccine. Once injected (a), mRNA is translated into pathogen proteins that behave as both endogenous and exogenous antigens, stimulating cell-mediated and humoral immune responses within muscle cells (b) and antigen-presenting cells (c). BCR: B-cell receptor; TCR: T-cell receptor; MHC: major histocompatibility complex.

technology promises to be robust, cheap, safe, and efficient, and easily adapted for many diseases, which is advantageous when faced with the threat of rapidly evolving pathogens.

Comprehension Check

1. What advantages do DNA and RNA vaccines offer with reference to new diseases and booster shots?
2. What risks are involved when using live, attenuated viruses in vaccine preparations?

35.7 Bioterrorism Readiness Is an Integral Component of Public Health Microbiology

After reading this section, you should be able to:

a. Discuss the efforts necessary to be prepared for a bioterrorism attack
b. Prioritize microorganisms based on public health threat
c. Construct tables of prevention and control information relative to an infectious agent
d. Identify appropriate authorities that respond to suspected acts of bioterrorism

Bioterrorism is defined as "the intentional or threatened use of viruses, bacteria, fungi, or toxins from living organisms to produce death or disease in humans, animals, and plants." The use of biological agents to effect personal or political outcomes is not new, and the modern use of biological agents is a reality (**Historical Highlights 35.4**). Cases of bioterrorism in the United States have been rare. The most serious to date involved the use of *Bacillus anthracis* endospores delivered through the U.S. postal system in 2001. The spores infected 22 people and caused five deaths. NIAID's three-tiered pathogen priority list, shown in table 35.1, was created by the Office of Biodefense, indicating that several pathogens are of national concern and are carefully monitored for their potential as agents of bioterrorism.

Among weapons of mass destruction, biological weapons can be more destructive than chemical weapons. Biological agents are likely to be chosen as a means of localized attack (biocrime) or mass casualty (bioterrorism) for several reasons. They are mostly invisible, odorless, tasteless, and difficult to detect. Furthermore, perpetrators may escape because it may take hours to days before people exhibit signs and symptoms of disease caused by the infectious agent. Additionally, the general public is not likely to be protected immunologically against agents that might be used in bioterrorism.

In 1998 the U.S. government launched the first national effort to create a biological weapons defense. The initiatives included (1) the first-ever procurement of specialized vaccines and medicines for a national civilian protection stockpile; (2) invigoration of research and development in the science of biodefense; (3) investment of more time and money in genome sequencing, new vaccine research, and new therapeutic research; (4) development of improved detection and diagnostic systems; and (5) preparation of clinical microbiologists and clinical microbiology laboratories as members of first-responder teams in the event of a bioterror attack. In 2002 Congress enacted the Public Health Security and Bioterrorism Preparedness and Response Act, which identified select agents whose use is

HISTORICAL HIGHLIGHTS

35.4 1346—Early Biological Warfare Attack

The Black Death, which swept through Europe, Asia, and North Africa in the mid-fourteenth century, was one of the greatest public health disasters in recorded history. Europe alone lost an estimated quarter to a third of its population during this time. This is not only of great historical interest but also relevant to current efforts to evaluate the threat of military or terrorist use of biological weapons. ▶ *Plague (section 38.2)*

Some believe that evidence for the origin of the Black Death in Europe is found in a memoir by Gabriele de' Mussi of Genoa. According to this fourteenth-century memoir, the Black Death reached Europe from the Crimea (a region of Ukraine) in 1346 as a result of a biological warfare attack. The Mongol army hurled plague-infected cadavers into the besieged Crimean city of Caffa (now Feodosija, Ukraine), thereby transmitting the disease to the inhabitants; fleeing survivors then spread the plague from Caffa to the Mediterranean Basin. Such transmission was especially likely at Caffa, where cadavers would have been badly mangled by being hurled and the defenders probably suffered cut or abraded hands from coping with the bombardment. Because many cadavers were involved, the opportunity for disease transmission was greatly increased. Disposal of victims' bodies in a major disease outbreak is always a problem, and the Mongol army used their hurling machines as a solution to limited mortuary facilities. It is possible that thousands of cadavers were disposed of this way; de' Mussi's description of "mountains of dead" might have been quite literally true. Indeed, Caffa could be the site of the most spectacular incident of biological warfare ever, with the Black Death as its disastrous consequence. It is a powerful reminder of the horrific consequences that can result when disease is successfully used as a weapon.

now tightly regulated. The CDC and USDA Federal Select Agent Program's Tier 1 subset overlaps with the NIAID's list to classify those agents that (1) could produce a mass casualty event or economic devastation, (2) were highly communicable, (3) had a low infectious dose, and (4) might be readily weaponized.

In 2003 Congress established the Department of Homeland Security (DHS) to coordinate the defense of the United States against terrorist attacks. As one of many duties, the secretary of the DHS is responsible for maintaining the National Incident Management System to monitor large-scale hazardous events. Bioterrorism and other public health incidents are managed within this system. The Department of Health and Human Services (HHS) responds and deploys assets as needed within the areas of its statutory responsibility (e.g., the Public Health Service Act and the Federal Food, Drug, and Cosmetic Act), while apprising the secretary of DHS of the nature of the response. The HHS secretary directs the CDC to effect the necessary integration of public health activities.

In the United States, the CDC partners with academic institutions across the country to educate, train, and drill public health employees, first responders, and numerous environmental and healthcare providers. Centers for Public Health Preparedness were established to bolster the overall response capability to bioterrorism. Another CDC-managed program, the Laboratory Response Network (LRN) serves to ensure an effective laboratory response to bioterrorism by helping to improve the nation's public health laboratory infrastructure through its partnership with the FBI and the Association of Public Health Laboratories (APHL). The LRN maintains an integrated network that links state and local public health, federal, military, and international laboratories so that a rapid and coordinated response to bioterrorism or other public health emergencies (including veterinary, agriculture, military, and water- and food-related) can occur.

In the absence of overt terrorist threats and without the ability to rapidly detect bioterrorism agents, it is likely that a bioterrorism act would be defined by a sudden spike in an unusual (nonendemic) disease reported to the public health system. Likewise, suddenly increased numbers of zoonoses, diseased animals, or vehicle-borne illnesses may indicate bioterrorism. Important guidelines and standardized protocols prepared for all sentinel laboratories to assist in the management of clinical specimens containing select agents have been established by the American Society for Microbiology in coordination with the CDC and the APHL. It is clear that surveillance and response to a potential act of bioterrorism requires communication and coordination across a number of government agencies. The diseases and microbiology associated with specific select agents are discussed in chapters 37 to 39.

Comprehension Check

1. In what three general ways can epidemics be controlled? Give one or two specific examples of each type of control measure.
2. Name some of the microorganisms that can be used to commit biocrimes. From this list, which pose the greatest risk for causing large numbers of casualties?
3. Why are biological weapons more destructive than chemical weapons?
4. What is the Public Health Security and Bioterrorism Preparedness and Response Act designed to do?

Key Concepts

35.1 Epidemiology Is an Evidence-Based Science

- Epidemiology is the science that evaluates the determinants, occurrence, distribution, and control of health and disease in a defined population.
- Specific epidemiological terminology is used to communicate disease incidence in a given population. Frequently used terms include sporadic disease, endemic disease, hyperendemic disease, epidemic, index case, outbreak, and pandemic.

35.2 Epidemiology Is Rooted in Well-Tested Methods

- Public health surveillance is necessary for recognizing a specific infectious disease within a given population. This consists of gathering data on the occurrence of the disease, collating and analyzing the data, summarizing the findings, and applying the information to control measures.
- Population- and case-based surveillance data are used to track infections within a population.
- Remote sensing and geographic information systems are used to gather data from natural environments.
- Statistical analysis is an important tool used in the study of modern epidemiology.
- Epidemiological data can be obtained from such factors as morbidity, prevalence, mortality, and case fatality rates.

35.3 Infectious Disease Is Revealed Through Patterns Within a Population

- A common-source epidemic is characterized by a sharp rise to a peak and then a rapid but not-as-pronounced decline in the number of individuals infected (**figure 35.5**). A propagated epidemic is characterized by a relatively slow and prolonged rise and then a gradual decline in the number of individuals infected.
- The basic reproduction number, R_0, is an indication of transmissibility. It is defined as the number of infections that can arise from an individual with the infection within a susceptible population.
- Herd immunity is the resistance of a population to infection and pathogen spread because of the immunity of a large percentage of the individuals within the population.

35.4 Infectious Diseases and Pathogens Are Emerging and Reemerging

- Humans will continually be faced with both new infectious diseases and the reemergence of older diseases once thought to be conquered (**table 35.1**).
- The CDC has defined these diseases as "new, reemerging, or drug-resistant infections whose incidence in humans has increased within the past two decades or whose incidence threatens to increase in the near future."
- Many factors characteristic of the modern world undoubtedly favor the development and spread of these microorganisms and their diseases (such as transportation, commerce, and housing development).

35.5 Healthcare Facilities Harbor Infectious Agents

- Healthcare-associated (nosocomial) infections are infections acquired within a hospital or other clinical care facility and are produced by a pathogen acquired during a patient's stay. These infections come from either endogenous or exogenous sources (**figure 35.9**).
- Healthcare-associated diseases are usually caused by bacteria, most of which are noninvasive and part of the human microbiota.

35.6 Coordinated Efforts Are Required to Prevent and Control Epidemics

- The public health system consists of individuals and organizations that function in the control of infectious diseases and epidemics.
- Vaccination is one of the most cost-effective weapons for preventing microbial disease, and vaccines constitute one of the greatest achievements of modern medicine.
- Many of the current vaccines in use for humans consist of whole organisms that are either inactivated (killed) or attenuated (capable of replicating but avirulent).
- Some of the risks associated with whole-cell vaccines can be avoided by using only specific purified macromolecules derived from pathogenic microorganisms. The three general forms of subunit or acellular vaccines are capsular polysaccharides, recombinant surface antigens, and inactivated exotoxins (toxoids) (**table 35.3**).
- A number of microorganisms have been used for recombinant-vector vaccines. The attenuated microorganism serves as a vector, replicating within the host and expressing the gene product of the pathogen-encoded antigenic proteins. The proteins can elicit humoral immunity when the proteins escape from the cells and cellular immunity when they are broken down and properly displayed on the cell surface.
- DNA and RNA vaccines elicit protective immunity against a pathogen by activating both branches of the immune system: antibody-mediated and cellular (**figure 35.10**).
- Epidemiological control measures can be directed toward reducing or eliminating infection sources, breaking the connection between sources and susceptible individuals, or isolating susceptible individuals and raising the general level of herd immunity by immunization.

35.7 Bioterrorism Readiness Is an Integral Component of Public Health Microbiology

- Among weapons of mass destruction, biological weapons can be more destructive than chemical weapons. The list of biological agents that could pose the greatest public health risk in the event of a bioterrorism attack is short and includes viruses, bacteria, parasites, and toxins.

Active Learning

1. Why is international cooperation a necessity in the field of epidemiology?
2. What common sources of infectious disease are found in your community? How can the etiologic agents of such infectious diseases spread from their source or reservoir to members of your community?
3. How could you prove that an epidemic of a given infectious disease was occurring?
4. College dormitories are notorious for outbreaks of flu and other infectious diseases. These are particularly prevalent during final exam weeks. Using your knowledge of the immune response and epidemiology, suggest practices that could be adopted to minimize the risks at such a critical time.
5. Why is an RNA vaccine delivered intramuscularly and not by intravenous or oral routes?
6. Early in the COVID-19 pandemic, epidemiologists raced to calculate the R_0 so that a measure of transmissibility could be gleaned, as well as an insight into what level of herd immunity must be achieved before we could establish protection. Estimates of the R_0 made by various institutions ranged from 1.4 to 6.68, with no definitive agreement on a specific number. What makes the R_0 such an elusive figure? What components must be factored into quantifying transmissibility and why does it vary by locale? Can the threshold level of herd immunity be achieved without a precise calculation for the R_0? What additional factors must be considered in order to attain herd immunity in a population?

CHAPTER

36

Clinical Microbiology and Immunology

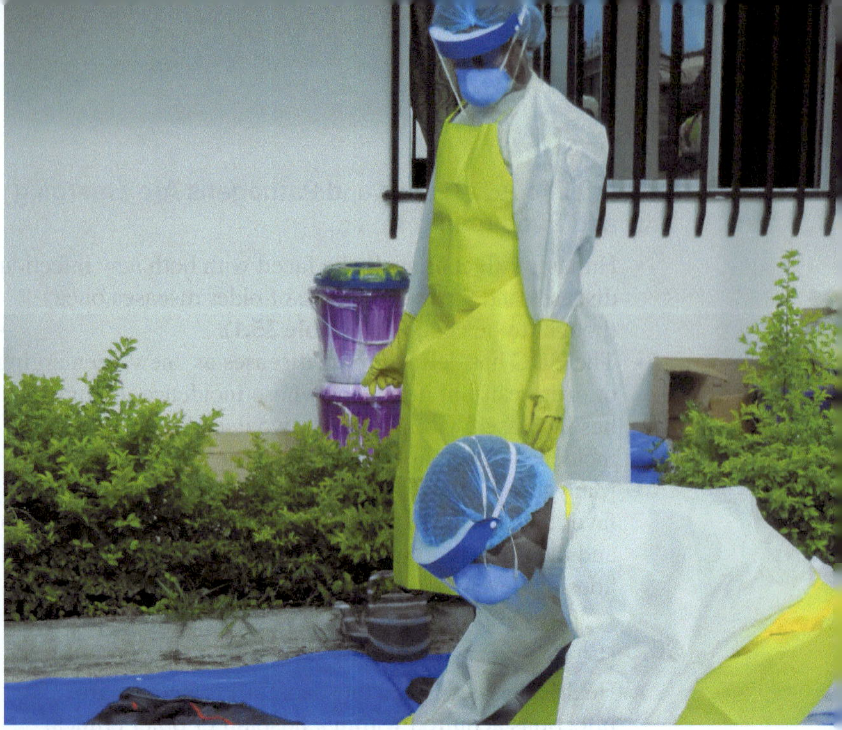

Dr. Heidi Soeters/CDC

Ebola and Global Health Security

The 2014–2016 Ebola virus disease (EVD) epidemic in West Africa was the longest and largest outbreak of its kind. Unlike previous EVD epidemics, which were largely rural, this outbreak rapidly spread via close contacts in urban environments, leading to massive death tolls in Guinea, Liberia, and Sierra Leone. As the epidemic began to wind down, it was clear that there was much to be learned from this tragic event.

The EVD epidemic put a spotlight on the need for rapid, reliable laboratory testing of samples from patients suspected of disease and those already infected. Without timely, accurate diagnostic testing, it is impossible to identify patients in the early stages of disease. Moreover, persons who may have been exposed cannot be traced. Early response to an outbreak, when it might be contained, is impossible without diagnostic tests that can be rapidly deployed on a large scale, skilled workers, mechanisms for sample collection and transport, understanding of biosafety, and supply chain for personal protective equipment.

In response to the 2014 EVD outbreak, the Centers for Disease Control and Prevention (CDC), the World Health Organization (WHO), and other governmental and nongovernmental organizations coordinated efforts with African governments to establish field laboratories and personnel training. New assays were developed and used to diagnose patients, determine when they were safe to return to their families, and trace contacts. These initial tests were complex, but by early 2015, a lateral flow immunoassay was approved. This rapid diagnostic test, which uses the same technology as that of home pregnancy tests, eased the burden on personnel and accelerated response time. Training protocols were codified ensuring that laboratory personnel could accurately perform them using appropriate biosafety precautions.

In 2020, the world again found itself in need of cheap, rapid, and reliable testing, this time for SARS-CoV-2, the virus that causes COVID-19. By the end of that year, the first tests for detecting the virus were produced and widely distributed. Both the EVD epidemic and the COVID-19 pandemic illustrate how important diagnostic testing is to effective healthcare. In this chapter we review some basic microbiological and immunological techniques used to diagnose infectious disease. But first, we discuss the common denominator in all diagnostic testing: biosafety.

Readiness Check:

Based on what you have learned previously, you should be able to:

- ✓ Discuss the tools and techniques used to culture and characterize microorganisms (chapters 2, 7, 17, 18, and 26)
- ✓ Outline important aspects of microbial structure, function, survival requirements, and replication strategies (chapters 3, 4, 5, 6, 7, 13, and 15)
- ✓ Describe the role of complement and antibodies in host defense (sections 31.3, 32.7, and 32.8)
- ✓ Describe the infectious disease process (chapter 34)

36.1 The Clinical Microbiology Laboratory Detects Infectious Agents and Protects Its Workers

After reading this section, you should be able to:

a. List the various scientific disciplines involved in clinical microbiology
b. Describe the risk assessment process leading to a safe laboratory
c. Paraphrase the standard microbiological practices that guide lab safety
d. Differentiate between biosafety levels and risk groups and provide examples of microorganisms assigned to risk groups

The clinical microbiology laboratory is the primary site for the diagnosis of human infectious disease. It also supports public health infrastructure, acting as a sentinel in the rapid detection of naturally occurring outbreaks of infection and bioterrorism. The major goals of the **clinical microbiologist** are (1) rapid and accurate identification of pathogens from clinical specimens and (2) antimicrobial susceptibility testing of these organisms. Clinical microbiology is interdisciplinary, and the clinical microbiologist must have working knowledge of microbial morphology, metabolism, growth, reproduction, biochemistry, and physiology. Additionally, the clinical microbiologist needs to understand the principles of microscopy, immunology, molecular biology, genomics, aseptic technique, sterilization, disinfection, and the dynamics of host-parasite relationships (**figure 36.1**).

Testing a clinical sample must follow a prescribed process to avoid contamination and waste. Thus, a workflow protocol is established in the laboratory to move the specimen rapidly through testing procedures while maintaining its integrity. Throughout this process, the safety of clinical laboratory workers is paramount, as we now discuss.

The Need for Safety Guidance

Although appalling today, not so long ago mouth pipetting, smoking, and eating in the microbiology laboratory were not uncommon. Over the years, several key events changed laboratory culture to promote safety. Laboratories now refer to the *Biosafety in Microbiological and Biomedical Laboratories,* or BMBL, by the Centers for Disease Control and Prevention (CDC) for guidelines on best practices. The current edition of the BMBL (2024) is prefaced with the caveat that "the intent was and is to establish a voluntary code of practice, one that all members of a laboratory community will together embrace to safeguard themselves and their colleagues, and to protect the public health and environment."

To help prevent laboratory-acquired infections (LAIs), microbiologists today follow strict procedures, often called *standard microbiological practices,* codified in the BMBL (**table 36.1**). In addition, some events have led to legislation affecting how microbiologists practice laboratory safety and security. For example, the recognition that human immunodeficiency virus (HIV) and other viruses are transmitted by blood led to the 1991 Bloodborne Pathogens Act. In 2001 the use of anthrax as a bioterrorism agent

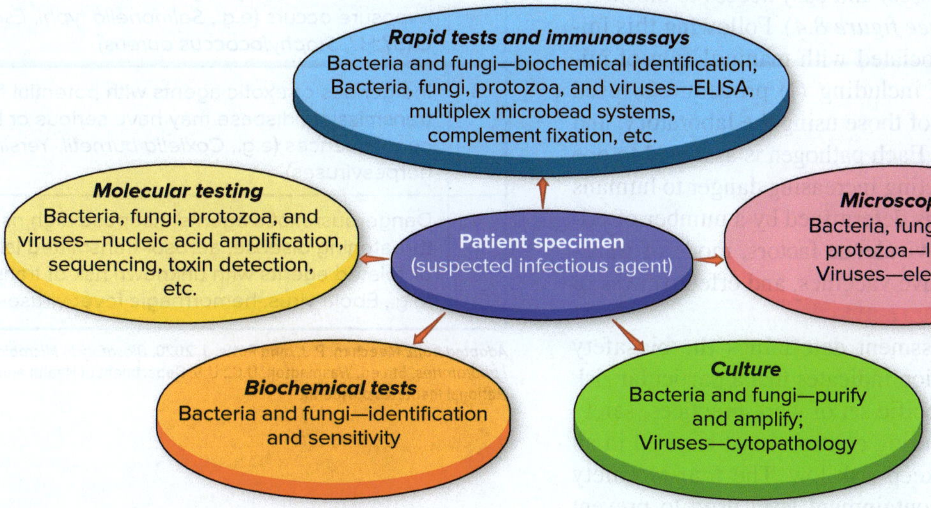

Figure 36.1 The Patient Specimen Is Evaluated by Various Techniques. The specimen source and patient history guide the decision for use of tests and techniques. Not all tests are performed on all specimens, but all laboratories need to be able to perform these tests or refer the specimen to other laboratories.

Table 36.1	Standard Microbiological Practices
1	Persons must wash their hands after working with potentially hazardous materials and before leaving the laboratory.
2	Eating, drinking, smoking, handling contact lenses, applying cosmetics, and storing food for human consumption is not permitted in the laboratory.
3	Mouth pipetting is prohibited; mechanical pipetting devices must be used.
4	Policies for the safe handling of sharps (such as needles, scalpels, pipettes, and broken glassware) must be developed and implemented.
5	Procedures must be developed and implemented to minimize the creation of splashes and/or aerosols.
6	Work surfaces must be decontaminated after completion of work and after any spill or splash of potentially infectious material.
7	All cultures, stocks, and other potentially infectious materials must be sterilized using an effective method before disposal.
8	A sign incorporating the universal biohazard symbol must be posted at the entrance to the laboratory when infectious agents are present. The sign may include the biosafety level, name of the agent(s) in use, and the name and phone number of the laboratory supervisor or other responsible personnel. Information should be posted according to institutional policy.
9	Persons must report all injuries incurred in the laboratory to the laboratory supervisor as soon as possible.

Adapted from: Meechan, P. J., and Potts, J. 2020. *Biosafety in Microbiological and Biomedical Laboratories*, 6th ed. Washington, D.C.: U.S. Department of Health and Human Services, CDC, National Institutes of Health; and Emmert, E. 2013. "Biosafety guidelines for handling microorganisms in the teaching laboratory: Development and rationale." *J. Microbiol. & Biol. Educ.* 14(1): 78–83. doi:10.1128/jmbe.v14i1.531

led to the Public Health Security and Bioterrorism Preparedness and Response Act of 2002. ◀ *Bioterrorism readiness is an integral component of public health microbiology (section 35.7)*

Practicing Biosafety

Safety planning starts with laboratory design (e.g., airflow, sink access, decontamination processes) and easy access to biosafety cabinets to minimize hazard (*see figure 8.4*). Following this initial survey, potential risks associated with manipulation of microorganisms are considered, including (1) procedure(s) used with each agent, (2) skill level of those using the laboratory, and (3) infectious agent risk group. Each pathogen is assigned to one of four risk groups (RGs) reflecting increasing danger to humans (**table 36.2**). A pathogen's RG is determined by a number of criteria, including pathogenicity, virulence factors, mode of transmission, availability of protective vaccines, and effectiveness of chemotherapeutic agents.

The rest of the risk assessment determines the biosafety level (BSL). The BSL designation indicates that a particular risk is acknowledged and that a specific set of work practices, safety equipment, and facility engineering controls must be used in an assigned laboratory space to keep risk low. The four biosafety levels reflect an appropriate containment level used to prevent LAI and environmental contamination (**table 36.3**). Each subsequent level builds on the previous. The common component to all levels is the assumption that standard microbiology practices will be followed.

Table 36.2	Microbiology Risk Group (RG) Characterization with Example Organisms
1	Not known to consistently cause disease in healthy adults (e.g., *Lacticaseibacillus casei, Aliivibrio fischeri*)
2	Associated with human disease; potential hazard if percutaneous injury, ingestion, or mucous membrane exposure occurs (e.g., *Salmonella typhi, Escherichia coli* O157:H7, *Staphylococcus aureus*)
3	Indigenous or exotic agents with potential for aerosol transmission; disease may have serious or lethal consequences (e.g., *Coxiella burnetii, Yersinia pestis,* herpesviruses)
4	Dangerous/exotic agents that pose high risk of life-threatening disease, aerosol-transmitted lab infections, or related agents with unknown risk of transmission (e.g., Ebola virus, hemorrhagic fever viruses)

Adapted from: Meechan, P. J., and Potts, J. 2020. *Biosafety in Microbiological and Biomedical Laboratories*, 6th ed. Washington, D.C.: U.S. Department of Health and Human Services, CDC, National Institutes of Health.

In addition to accepted laboratory behaviors, specific safety equipment is often recommended for each BSL. Personal protective equipment (PPE) includes (1) protective coverings (coats, gowns, or uniforms), (2) protective eyewear, and (3) gloves. PPE

Table 36.3 Biosafety Recommendations for BSL 1–4 Microbiology Laboratories

Biosafety Level (BSL)	Work Practices	Safety Equipment	Facilities
BSL-1	Standard microbiology practices	No barriers required. Personal protective equipment (PPE): laboratory coats and gloves; eye, face protection, as needed	Laboratory bench and sink required
BSL-2	BSL-1 practice plus: Limited lab access Biohazard warning signs "Sharps" precautions Biosafety manual defining any needed waste decontamination or medical surveillance policies	Biosafety cabinets (BSCs) or other physical containment devices used for all manipulations of agents that cause splashes or aerosols of infectious materials PPE: Laboratory coats, gloves, face and eye protection, as needed	BSL-1 facilities plus autoclave available
BSL-3	BSL-2 practice plus: Controlled lab access All waste decontaminated Lab clothing decontaminated before laundering Baseline serum testing	BSCs or other physical containment devices used for all open manipulations of agents PPE: Protective lab clothing, gloves, respiratory protection as needed	BSL-2 facilities plus self-closing double door, negative air pressure, room exhaust not recirculated, away from access corridors
BSL-4	BSL-3 practice plus: Different clothing to enter workspace Shower to exit Decontaminate all material on exit	Class III (glove box) cabinet-based work or Class II (laminar flow) BSC with full-body, positive pressure suit	BSL-3 facilities plus separate building or zone; dedicated heat, ventilation, and air conditioning system, vacuum, and decontamination systems

Adapted from: Meechan, P. J., and Potts, J. 2020. *Biosafety in Microbiological and Biomedical Laboratories*, 6th ed. Washington, D.C.: U.S. Department of Health and Human Services, CDC, National Institutes of Health.

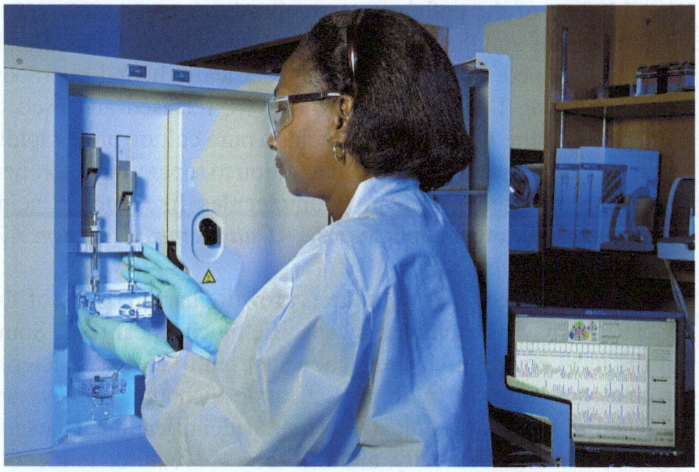

Figure 36.2 Personal Protective Equipment. This scientist is wearing a lab coat, safety glasses, and gloves to prevent contamination of the laboratory and herself from infectious agents.
Hsi Liu, Ph.D., MBA; James Gathany/CDC

prevents transmission of infectious agents by blocking their access to entry portals (**figure 36.2**). Additional PPE (e.g., face shield, respiratory protection) may be necessary as determined by the risk assessment.

36.2 Identification of Microorganisms from Specimens

After reading this section, you should be able to:

- **a.** Provide examples of appropriately collected specimens
- **b.** Explain common methods for culturing bacteria and fungi isolated from patient specimens
- **c.** Describe the use of a preliminary Gram-stain and morphology report
- **d.** Discuss the use of biochemical assays in identifying common bacterial pathogens
- **e.** Compare the speed and accuracy of culture-based and nucleic acid–based methods for microbe identification.

The clinical specimen is an unknown, and microbiology laboratory technicians are detectives who identify the microorganism(s) it contains. In doing so, they determine the probable cause of a patient's infection. Broadly speaking, there are two types of clinical specimens: those from normally sterile body sites, such as blood and cerebrospinal fluid, and those from nonsterile sites, such as skin or the intestines. An important consideration for the clinical microbiologist is that any microorganism found in a sterile-site specimen is a potential problem for the patient and

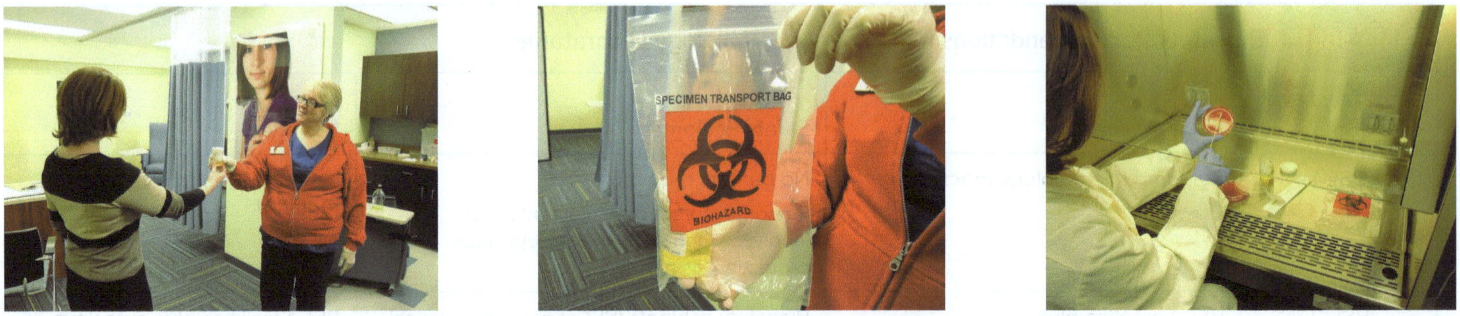

(a) (b) (c)

Figure 36.3 Collection of Patient Specimen for Microorganism Testing. (a) The identification of the microorganism begins with proper instruction for specimen collection. (b) The specimen is protected by transport media and transferred in a biohazard containment bag to prevent contamination and spillage. (c) The specimen is processed in a biosafety cabinet. In this case, the specimen is being cultured to identify the infectious agent.

James Redfearn/McGraw Hill

should be rapidly identified. Assessment of nonsterile-site specimens is no less important but requires a working knowledge of the microbiota normally associated with that site (*see figure 33.2*). The challenge for the microbiologist is to identify the source of the infection amid the background of these microbes.

The Clinical Specimen

A clinical specimen represents a portion or quantity of human material to be examined to determine the presence or absence of particular microorganisms (**figure 36.3**). In general, the clinical specimen should:

1. Adequately represent the diseased area and may include additional sites (e.g., urine and blood specimens) to isolate and identify potential agents of the particular disease process.
2. Be of sufficient quality and quantity to allow a variety of diagnostic tests, should they be necessary.
3. Be free of preventable contamination from the normal microbiota and environment.
4. Be collected in appropriate containers that may include a specific kind of transfer medium, kept at an appropriate temperature, and forwarded promptly to the clinical laboratory.
5. Be obtained before antimicrobial agents have been administered to the patient, if possible.
6. Be accompanied by a putative diagnosis.

Choosing the appropriate test or procedure is determined by the type of specimen and the suspected pathogen(s) based on the patient's signs and symptoms. Most laboratories have prescribed algorithms that dictate how each specimen is processed when it enters the laboratory. In some cases, the specimen is evaluated for microorganisms by direct techniques, including methods that first require growth of the microbe prior to testing. Indirect techniques, such as detection of antibody against a specific pathogen or amplification of pathogen genes, are used to identify a pathogen that is difficult to culture. Molecular tests can provide sensitive and specific results in a short period of time. We now highlight some of the microbiological methods commonly used.

Culture-Based Methods

For many bacteria and fungi, the process of obtaining pure cultures involves their growth on selective or differential media that also provide the first clues to the identity of the microbe. The choice of media used depends in part on the source of the specimen and knowledge of likely causes of the infection. Some examples of media used are described in **table 36.4.** In many cases, classical culture has been replaced by immunomagnetic separation (IMS) (**figure 36.4**). IMS uses immunomagnetic beads (IMBs), which are small magnetic particles coated with antibody targeting a particular microbe. When a clinical sample is mixed with IMBs specific for a suspected pathogen, the antibody mediates binding of the pathogen to the beads. A magnet is used to pull the beads and their attached microbes out of solution. Beads are then washed of debris and nonspecifically bound microbes. The remaining IMB-bound microbes are released and used to inoculate culture medium. This yields a pure culture more rapidly than with enrichment cultures. In addition to separating microbes, IMS can be used to isolate biomolecules like nucleic acids and proteins for other applications. *Culture media (section 7.7)* *Mannitol Salt Agar*

Once isolated in pure culture, growth characteristics of the bacterium may be examined by specific biochemical tests. Some of the most common biochemical tests used to identify bacterial isolates are listed in **table 36.5.** Classic dichotomous keys are coupled with the biochemical tests for the identification of bacteria from specimens. Generally, multiple tests are required to identify clinical bacterial isolates to the species level (**figure 36.5**). *Catalase; Citrate Agar Test; Coagulase; MRVP; Nitrate Test; Oxidase; Phenol Red Carbohydrate; SIM; TSI; Urea*

Commercial kits and robotic automation of biochemical tests have made identification of pathogens in specimens much more efficient. An example is the API 20E test, used for the identification of members of the family Enterobacteriaceae and other Gram-negative bacteria. The API 20E consists of a plastic strip with 20 microtubes containing substrates, each of which detects a different biochemical activity (**figure 36.6**). The microtubes are inoculated with bacteria previously isolated in pure culture.

Table 36.4 Media Commonly Used for Isolation of Pure Bacterial Cultures from Clinical Specimens
Mannitol salt agar (MS) is used for the isolation of staphylococci. Selectivity is obtained by the high (7.5%) salt concentration that inhibits growth of many groups of bacteria. This medium is also differential: Mannitol helps differentiate pathogenic from nonpathogenic staphylococci, as the former ferment mannitol to form acid, which reduces the pH of the medium, causing the agar to change color from red to yellow.
Bismuth sulfite agar (BS) is used for the isolation of *Salmonella enterica* serovar Typhi, especially from stool and food specimens. It is both selective—bismuth suppresses the growth of Gram positives—and differential. *S. enterica* serovar Typhi reduces the sulfite to sulfide, resulting in black colonies with a metallic sheen.
Hektoen enteric agar is used to increase the yield of *Salmonella* and *Shigella* spp. relative to other microbiota. Selectivity is obtained by a high bile salt concentration that inhibits the growth of Gram-positive bacteria and slows the growth of many coliform strains. The medium is differential because *Salmonella* forms black colonies on this medium due to production of sulfide while *Shigella* spp. do not.
Eosin methylene blue agar (EMB) contains lactose, salts, and two dyes—eosin and methylene blue. The medium is selective and differential; the dyes inhibit the growth of Gram-positive bacteria and also change color in the presence of acid, allowing differentiation between lactose fermenters (dark colonies) and nonfermenters (colorless colonies). Fecal coliforms such as *E. coli* may produce colonies with a metallic sheen.
MacConkey agar is used for the selection and recovery of members of Enterobacteriaceae and related Gram-negative rods. The bile salts and crystal violet in this medium inhibit the growth of Gram-positive bacteria and some fastidious Gram-negative bacteria. Because lactose is the sole carbohydrate, lactose-fermenting bacteria produce colonies that are various shades of red, whereas nonlactose fermenters produce colorless colonies.
Blood agar: addition of citrated blood to tryptic soy agar makes possible hemolysis visible, enabling the differentiation of some species of bacteria. The following three hemolytic patterns can be observed on blood agar (*see figure 22.19*). 1. α-hemolysis—greenish to brownish halo around the colony (e.g., *Streptococcus gordonii* and *S. pneumoniae*) 2. β-hemolysis—complete lysis of blood cells resulting in clearing around the colony (e.g., *Staphylococcus aureus* and *Streptococcus pyogenes*) 3. Nonhemolytic—no change in medium (e.g., *Staphylococcus epidermidis* and *S. saprophyticus*)

After 5 to 12 hours of incubation, the 20 test results are converted to a seven- or nine-digit profile. This profile number is matched to a database called the *API Profile Index* to identify the bacterium.

In addition to traditional techniques and kits, larger laboratories also use semiautomated robotic systems to identify microorganisms. Some of these are culture-independent systems, reporting infectious agents by directly identifying microbial protein (e.g., toxins) or DNA within the patient specimen. Several of these technologies couple pure culture isolates with electrospray ionization-mass spectroscopy (ESI-MS), peptide nucleic acid fluorescence in situ hybridization (PNA-FISH), or matrix-assisted laser desorption/ionization–time of flight (MALDI-TOF)—mass spectrometry to provide high-throughput, high-reliability identification capabilities. ◀ *Polymerase chain reaction amplifies targeted DNA (section 17.2); Genomic and metagenomic libraries: cloning genomes in pieces (section 17.3); Molecular characteristics (section 26.2)*

Bacterial and fungal antimicrobial susceptibility is determined by evaluating organism growth when exposed to various antimicrobial agents and reported by similar means. The gold standard for antimicrobial susceptibility testing is still the agar dilution method: Bacteria are inoculated onto media containing increasing concentrations of the appropriate antimicrobial agent to determine the minimum concentration needed for microbial inhibition. However, the most common practice is automated

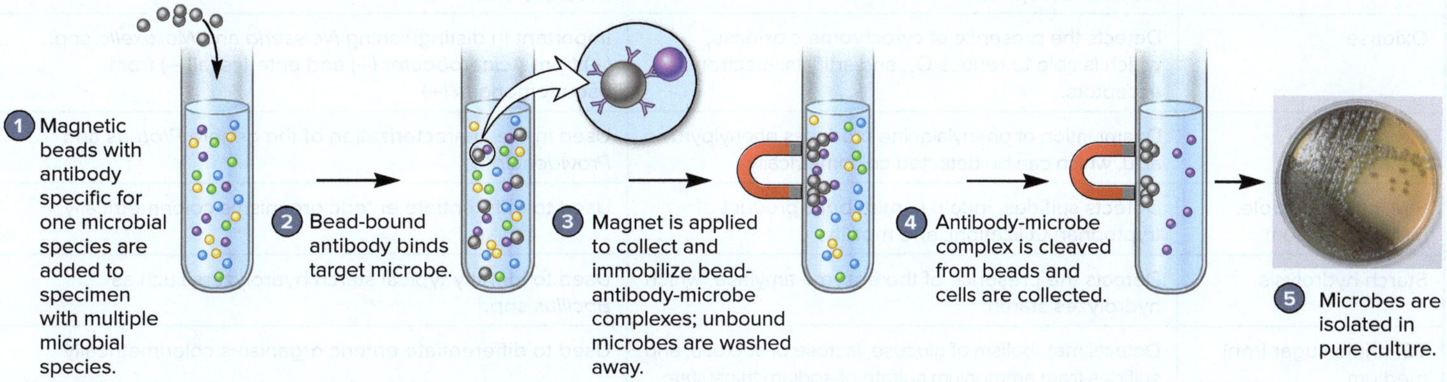

Figure 36.4 Immunomagnetic Beads Can Be Used to Isolate Specific Microbes. Magnetic beads are coated with antibodies against common pathogens. A specific pathogen in a sample can be isolated by the procedure illustrated here.
Lisa Burgess/McGraw Hill

Table 36.5 Some Common Biochemical Tests Used in the Identification and Differentiation of Bacteria

Biochemical Test	Description	Laboratory Application
Carbohydrate fermentation	Acid and/or gas are produced during fermentative growth with sugars or sugar alcohols.	Fermentation of specific sugars used to differentiate enteric bacteria as well as other genera or species
Casein hydrolysis	Detects caseinase, an enzyme that hydrolyzes the milk protein casein. Bacteria that produce caseinase appear as colonies surrounded by a clear zone.	Used to cultivate and differentiate aerobic actinomycetes based on casein utilization. For example, *Streptomyces* uses casein and *Nocardia* does not.
Catalase	Detects the presence of catalase, which converts hydrogen peroxide to water and O_2.	Used to differentiate *Streptococcus* (–) from *Staphylococcus* (+) and *Bacillus* (+) from *Clostridium* (–)
Citrate utilization	When citrate is used as the sole carbon source, the medium is alkalinized, causing a color change from green to blue.	Used in the identification of enteric bacteria. *Klebsiella* (+), *Enterobacter* (+), *Salmonella* (often +); *Escherichia* (–), *Edwardsiella* (–)
Coagulase	Detects the presence of coagulase. Coagulase causes plasma to clot.	This is an important test to differentiate *Staphylococcus aureus* (+) from other staphylococcal species, such as *S. epidermidis* and *S. saprophyticus* (–).
Decarboxylases (arginine, lysine, ornithine)	The decarboxylation of amino acids releases CO_2 and amine.	Used in the identification of enteric bacteria
Esculin hydrolysis	Tests for the cleavage of a glycoside that causes development of a brown-black color.	Used to differentiate *Staphylococcus aureus, Streptococcus mitis,* and others (–) from *Staphylococcus saprophyticus, Streptococcus mutans,* and enterococci (+)
β-galactosidase (ONPG) test	Detects the presence of an enzyme that cleaves lactose to glucose and galactose.	Used to differentiate enterics (*Citrobacter* +, *Salmonella* –) and to identify pseudomonads
Gelatin liquefaction	Detects the production of proteases that hydrolyze gelatin and liquefy solid gelatin medium.	Used in the identification of *Clostridium, Serratia, Pseudomonas,* and *Flavobacterium*
Hydrogen sulfide (H_2S)	Detects the formation of H_2S from cysteine due to cysteine desulfurase, causing the medium to blacken.	Important in the identification of *Edwardsiella, Proteus,* and *Salmonella*
IMViC (indole; methyl red; Voges-Proskauer; citrate)	The indole test detects the production of indole from tryptophan. Methyl red is a pH indicator that shows if the bacterium carries out mixed acid fermentation. Voges-Proskauer (VP) detects acetoin production. The citrate test determines if the bacterium can use citrate as a sole carbon source.	Used to separate *Escherichia* (MR+, VP–, indole+) from *Enterobacter* (MR–, VP+, indole–) and *Klebsiella pneumoniae* (MR–, VP+, indole–); also used to characterize members of the genus *Bacillus*
Lipid hydrolysis	Detects the presence of lipase, which breaks down lipids into simple fatty acids and glycerol.	Used in the differentiation of clostridia
Nitrate reduction	Detects whether a bacterium can use nitrate as an electron acceptor.	Used in the identification of enteric bacteria, which are usually +
Oxidase	Detects the presence of cytochrome *c* oxidase, which is able to reduce O_2, and artificial electron acceptors.	Important in distinguishing *Neisseria* and *Moraxella* spp. (+) from *Acinetobacter* (–) and enterics (all –) from pseudomonads (+)
Phenylalanine deaminase	Deamination of phenylalanine produces phenylpyruvic acid, which can be detected colorimetrically.	Used in the characterization of the genera *Proteus* and *Providencia*
SIM (sulfide, indole, motility) medium	Detects sulfides, indole (a metabolic product of tryptophan utilization), and motility.	Used to differentiate enteric organisms colorimetrically
Starch hydrolysis	Detects the presence of the enzyme amylase, which hydrolyzes starch.	Used to identify typical starch hydrolyzers such as *Bacillus* spp.
TSI (triple sugar iron) medium	Detects metabolism of glucose, lactose or sucrose, and sulfides from ammonium sulfate or sodium thiosulfate.	Used to differentiate enteric organisms colorimetrically
Urease	Detects the enzyme that splits urea to NH_3 and CO_2.	Used to distinguish *Proteus, Providencia rettgeri,* and *Klebsiella pneumoniae* (+) from *Salmonella, Shigella,* and *Escherichia* (–)

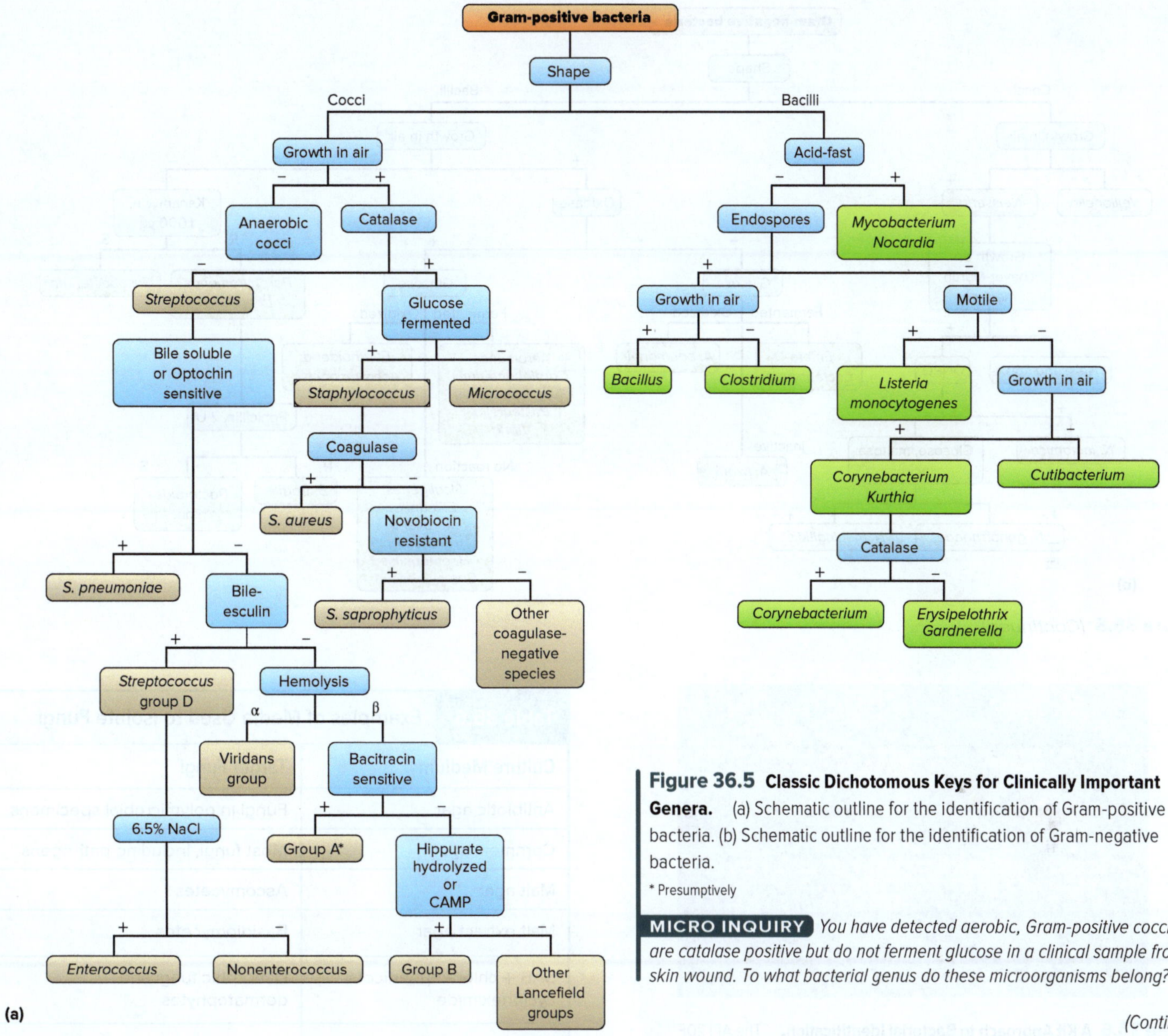

Figure 36.5 Classic Dichotomous Keys for Clinically Important Genera. (a) Schematic outline for the identification of Gram-positive bacteria. (b) Schematic outline for the identification of Gram-negative bacteria.

* Presumptively

MICRO INQUIRY *You have detected aerobic, Gram-positive cocci that are catalase positive but do not ferment glucose in a clinical sample from a skin wound. To what bacterial genus do these microorganisms belong?*

(Continued)

microbroth dilution antibiotic susceptibility tests and the manual Etest® and Kirby-Bauer tests (*see figures 9.3 and 9.4*). Together, the organism's identity and its sensitivity to a panel of antibiotics are determined, often within hours. ◀ *Antimicrobial activity can be measured by specific tests (section 9.3)*

Cultures remain the standard for the recovery of fungi from patient specimens; however, the time needed to culture fungi varies anywhere from a few days to several weeks, depending on the organism. Fungal cultures are evaluated for growth rate and appearance on at least one selective and one nonselective agar medium, with careful examination of colony morphology, color, and dimorphism (i.e., yeast form at 37°C; mold form at room temperature). Typically, the isolation of fungi is accomplished by concurrent culture of the specimen on media that are respectively supplemented and unsupplemented with antibiotics and cycloheximide. Antibiotics inhibit bacteria that may be in the specimen and cycloheximide inhibits saprophytic molds. However, a number of media formulations are routinely used to culture specific fungi (**table 36.6**). If a fungal infection is suspected, serological tests for antibodies to specific fungal pathogens may also be performed. Culture of viruses and protozoa from clinical specimens is not routine. Their detection and identification are by microscopy, molecular techniques, or both.

Microscopy

Sometimes microscopic visualization of microorganisms in a specimen (e.g., urine or sputum sample) or in culture can aid the

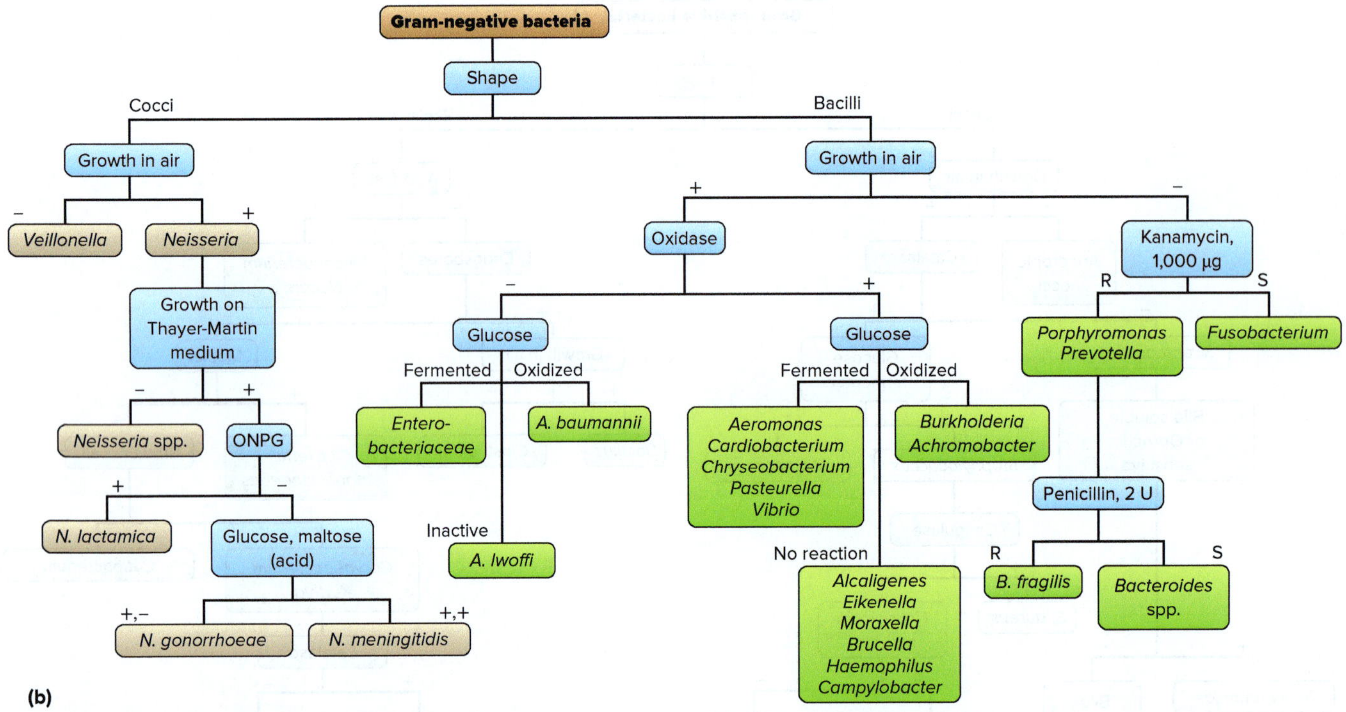

Figure 36.5 *(Continued)*

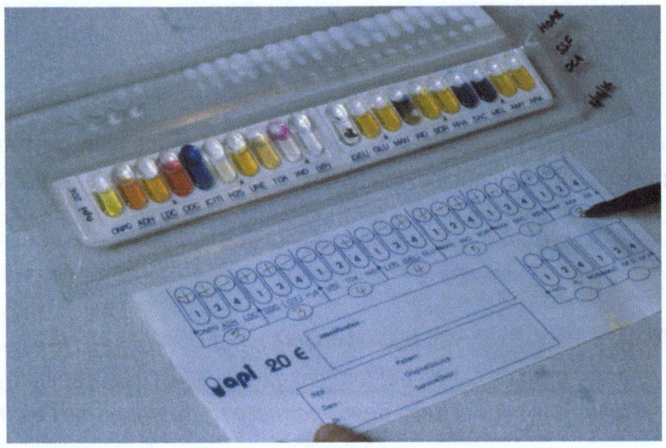

Figure 36.6 A Kit Approach to Bacterial Identification. The API 20E manual biochemical system for microbial identification. Following inoculation and incubation, positive reactions are discerned by a color change. The pattern of color changes is converted to a numerical value and used to identify the bacterial isolate.

Science History Images/Alamy Stock Photo

identification process. Wet-mount, heat-fixed, or chemically fixed specimens are examined with bright-field or phase-contrast microscopy. Among the many stains used to examine specimens, the Gram stain is the most common (*see figure 2.17*). Although it does not identify bacteria, preliminary reports indicate the Gram morphology, such as "Gram-positive cocci in pairs and chains." Such information is often helpful in ensuring appropriate antibiotic therapy. Likewise, the Gram reaction and shape of a bacterium are critical pieces of information used to decide which biochemical tests are appropriate to perform (figure 36.5).

Table 36.6 Examples of Media Used to Isolate Fungi

Culture Medium	Target Fungi
Antibiotic agar	Fungi in polymicrobial specimens
Cornmeal agar	Most fungi, including pathogens
Malt agar	Ascomycetes
Malt extract agar	Basidiomycetes
SAB + chloramphenicol + cycloheximide	Dimorphic fungi and dermatophytes
Sabouraud dextrose agar (SAB)	Most fungi

Definitive identification of most fungi is based on the morphology of hyphae or conidia (spores). Calcofluor white is a fluorescent stain often used to detect fungi because it binds to the chitin in their cell walls. It is often used in combination with potassium hydroxide, which dissolves host tissue, thereby enhancing visualization of fungi.

Protozoa are most often identified based on the morphology of their ova, trophozoites, and cysts. This is typically accomplished by direct microscopic evaluation of the clinical specimen. Helminths (worms) are often visualized in concentrated wet mounts of stool or urine specimens examined microscopically for the presence of eggs, larvae, or adult worms. Viruses are too small to visualize with the light microscope, but some can be diagnosed using fluorescence microscopy and immunofluorescence assays, discussed in section 36.3.

Molecular Methods

The identification of microorganisms through the analysis of specific proteins and nucleic acids is rapid, accurate, and sensitive. Details of techniques that use microbial DNA, RNA, or protein are presented in chapters 17, 18, and 26.

Nucleic acid–based diagnostic methods for the detection and identification of microorganisms have become routine in clinical microbiology laboratories. Many of them rely on increasing the amount of a particular region of a nucleic acid; these tests are collectively referred to as **nucleic acid amplification tests** or **NAATs,** and they include DNA sequencing (*see figures 18.2–18.4*), DNA fingerprinting, and various types of polymerase chain reactions (PCR). Real-time PCR (also called quantitative or qPCR) provides rapid, accurate, and quantitative results by analyzing fluorescent signals of amplification products without the need for gel electrophoresis (**figure 36.7;** also *see figure 17.7*). In digital droplet PCR (ddPCR), a sample is divided into many tiny drops of liquid, each containing one or a few target sequences. A separate PCR reaction occurs in every drop, allowing for the detection of rare targets. By comparison, multiplex PCR (mPCR) uses multiple primers targeting different microbial genomes, enabling a single sample to be screened for several suspected pathogens simultaneously. For example, mPCR for respiratory viruses assays a patient sample for all common respiratory viral pathogens (e.g., adenovirus, influenza, enteroviruses, parainfluenza virus, etc.). *Polymerase chain reaction amplifies targeted DNA (section 17.2)*

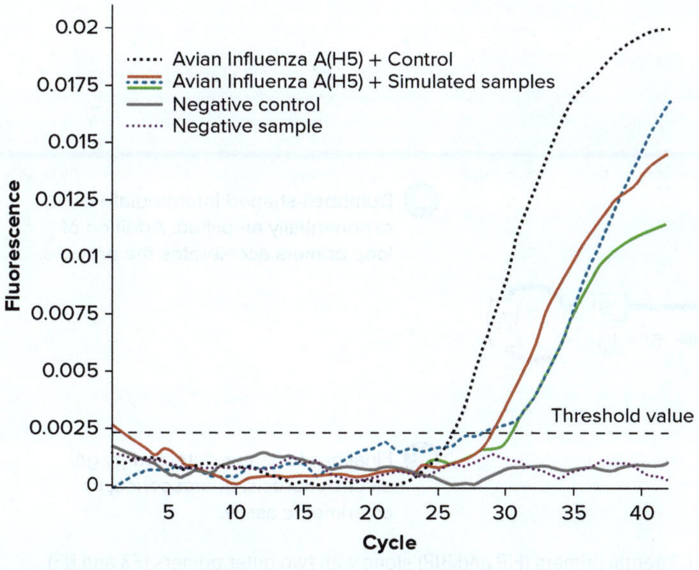

Figure 36.7 PCR Kit Results. Real-time detection of avian influenza is possible using a PCR kit. Note that sufficient DNA amplification occurs after 25 to 30 cycles. Only results above the threshold value are considered positive. The number of cycles required to exceed the threshold is a function of template (sample) concentration such that higher levels of template require fewer cycles.

The gene most commonly used to identify bacteria in blood and other types of cultures is the 16S ribosomal RNA (rRNA) (*see figure 1.8*). Typically the rRNA gene or gene fragment is amplified by PCR. After nucleotide sequencing, the sequence is compared with those in international databases. This method of bacterial identification, called **ribotyping,** is based on the high level of 16S rRNA conservation among bacteria. Multiple housekeeping genes, instead of 16S rRNA genes, may be sequenced and analyzed in a technique called **multilocus sequence typing (MLST).** *Microbial identification is largely based on molecular classification (section 26.2)*

In a method called **isothermal amplification (IA),** nucleic acid amplification is carried out at a constant temperature, unlike PCR, which requires repeated thermal cycles and thus specialized equipment. The enzyme commonly used is a thermostable strand-displacing DNA polymerase from *Geobacillus stearothermophilus* (referred to as *Bst*). There are multiple IA techniques, the most widely used of which is *loop-mediated isothermal amplification,* or *LAMP.* This approach employs two to three pairs of primers specific to distinct regions of the target nucleic acid. The *Bst* polymerase is used to start the reaction and two internal primers (FIP and BIP) along with two outer primers (F3 and B3) then create a dumbbell-shaped DNA structure that acts as a template for further rounds of amplification (**figure 36.8**). Addition of loop primers (loop F and loop B) increases the number of initiation points for the amplification and accelerates the process. Products of IA can be detected in a variety of ways, including color changes, fluorescence detection, and gel electrophoresis. LAMP can be combined with a reverse transcription step (RT-LAMP), facilitating amplification of RNA targets. It is thus effective for diagnosing a variety of bacterial, viral, and protozoan diseases, e.g., tuberculosis, hepatitis B, H5N1 influenza, and malaria.

Comprehension Check

1. Examine table 36.4. What is the difference between selective and differential media? Describe two media that are both selective and differential.
2. Describe in general how biochemical tests are used in the API 20E system to identify bacteria.
3. Why are most viruses identified by molecular methods, such as PCR, rather than microscopy methods?
4. Using the dichotomous key shown in figure 36.5, how would you differentiate between *Staphylococcus saprophyticus* and *S. aureus?*
5. Identify two advantages of isothermal amplification compared with conventional PCR.

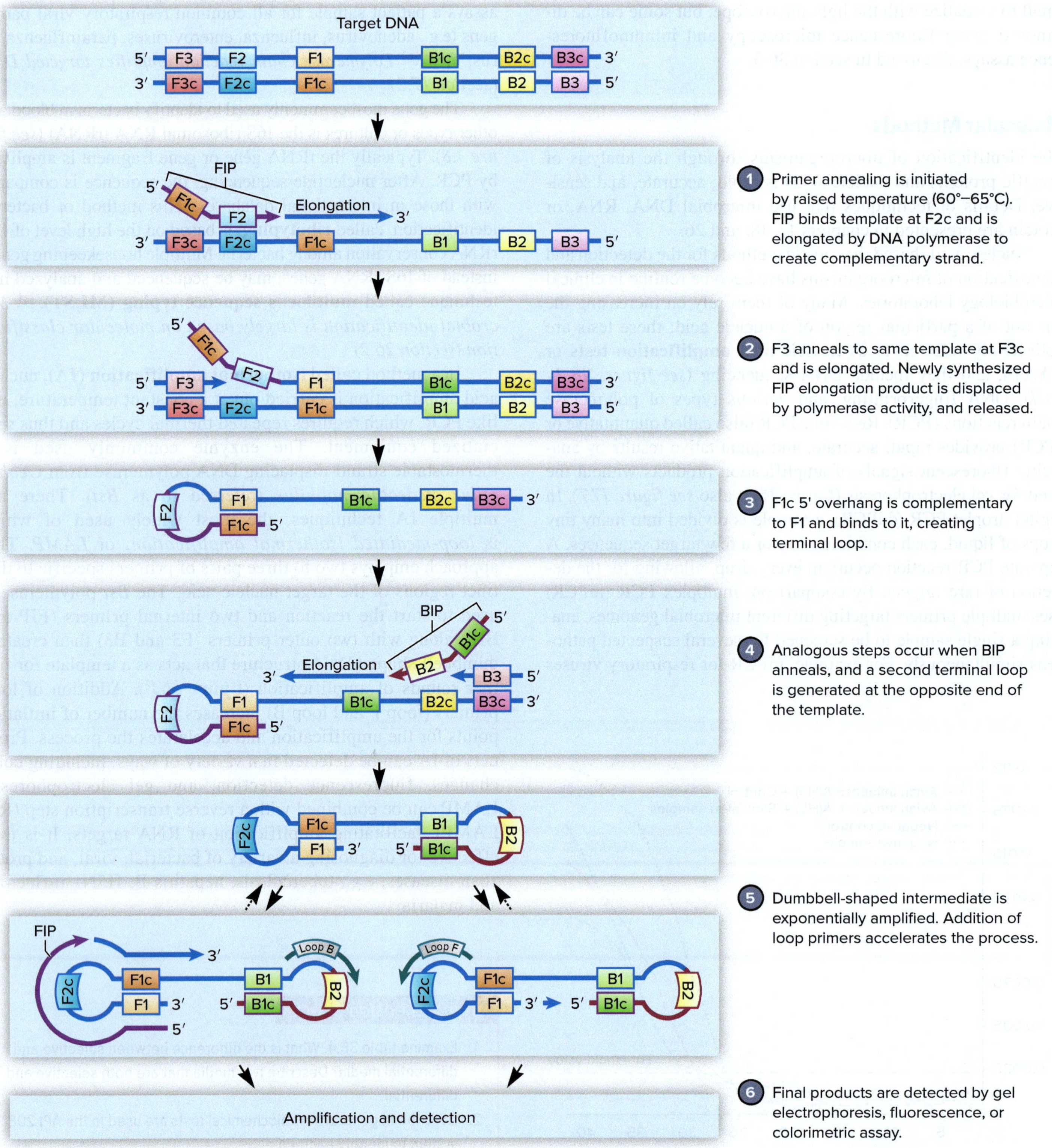

Figure 36.8 Loop-Mediated Isothermal Amplification. Forward and backward internal primers (FIP and BIP) along with two outer primers (F3 and B3) create a dumbbell-shaped DNA structure with end loops that act as a template for amplification. Optional loop primers (loop B and loop F) accelerate continued amplification. The process is generally completed within an hour, faster than conventional PCR.

36.3 Immune Responses Can Be Exploited to Detect Infections

After reading this section, you should be able to:

a. Diagram the methods by which antibodies are used to identify microorganisms
b. Differentiate between direct and indirect immunofluorescence
c. Interpret immunologic data used to identify microorganisms from patient specimens
d. Design an antibody-based test for the identification of a specific microorganism from a patient specimen

It may not be possible to culture certain microorganisms from clinical specimens because the organism is difficult to grow in pure culture or is unculturable (e.g., *Mycobacterium leprae;* hepatitis A, B, and C viruses; and Epstein-Barr virus), the microbe is unsafe (e.g., rickettsias and HIV), or it is impractical for all but a few microbiology laboratories (e.g., strict anaerobes). If cultures were taken after the patient started antimicrobial therapy, growth of the infecting bacterium may be inhibited. Under these circumstances, detection of antibodies or antigens can be quite valuable.

Serology is the identification or measurement of antibodies in the patient's serum directed against a specific microbe. Such assays often distinguish between IgM and IgG. The presence of IgM indicates a recent, active infection, whereas IgG is evidence of a past or chronic infection or of vaccination. The detection of both IgM and IgG against a specific microbe in a patient's serum suggests the patient was recently infected and is in the process of class switching antibody (*see figure 32.15*). In this case, the ratio of IgM to IgG may indicate how long ago the patient was infected.

Advances in clinical immunology have given rise to a marked increase in the number, sensitivity, and specificity of serological tests. This reflects the availability of highly specific monoclonal antibodies with very high affinity and avidity, and the development of sensitive antibody-binding reporter systems (*see Techniques & Applications 32.1*). Although most of these tests are relatively simple to use, the validity of test results depends on proper test selection, use, and timing of specimen collection. Unfortunately, lack of a measurable antibody titer does not always mean the absence of infection. Rather it may reflect a pathogen's lack of immunogenicity or insufficient time for an antibody response to develop following the onset of the infectious disease. Some patients are immunosuppressed due to other disease processes or treatment procedures (e.g., patients with AIDS or cancer) and therefore do not respond. In this section, some of the more common antibody-based techniques used in the diagnosis of microbial and immunological diseases are discussed. ◀ *Antibody structure (section 32.7)*

Serotyping

Serotyping is used to differentiate strains (serovars or serotypes) of microorganisms that differ in key surface molecules that elicit an antibody response. This has traditionally been performed by testing the capacity of different antigen-specific antibodies to bind to the surface of the microbe. For example, there are over 2,500 serotypes of *Salmonella enterica,* of which about 100 cause most human disease. Serotype identification can help trace the source of food-borne disease outbreaks, although it is being replaced by whole-genome sequencing. ◀ *Whole genome sequencing (section 18.2)*

Agglutination

An **agglutination reaction** occurs when an immune complex is formed by cross-linking cells or particles with specific antibodies (*see figure 32.19*). Agglutination reactions form visible aggregates or clumps, called **agglutinates,** that can be seen with the unaided eye. In direct agglutination tests, an antibody-containing suspension (such as patient serum) is mixed with an antigen of interest. An example is the Rapid Plasma Reagin test used to screen for syphilis. Antigenic cardiolipin released from spirochete-damaged host cells stimulates the production of antibodies. These antibodies will agglutinate syphilis bacterial antigens in vitro, yielding a positive test. ◀ *Immune complex formation (section 32.8)* ▶ *Syphilis (section 38.3)*

Indirect or passive agglutination tests use inert particles such as latex beads to which soluble antigens are attached. These coated microspheres bind antibodies in a patient's serum specimen, creating large, easily visible aggregates. Latex agglutination tests are used to detect antibodies that develop during certain bacterial, viral, mycotic (fungal), and helminthic (worm) infections, as well as in drug testing. Microspheres can also be coated with monoclonal antibodies to capture specific diagnostic antigens from patient specimens.

Hemagglutination assays use antibodies to cross-link red blood cells (RBCs) by attaching to surface antigens. Hemagglutination is routinely used in blood typing (*see figure 32.21*). Certain viruses can also mediate RBC clumping. This process, termed **viral hemagglutination,** can be used to detect certain viral infections. For example, if measles virus particles are mixed with RBCs, hemagglutination occurs (**figure 36.9*a***). However, if a person is infected with measles virus, they will have antibodies to that pathogen in their serum. Mixing that serum with measles virions results in neutralization of the virions when the person's antibodies bind to them. When RBCs are then added to the test, the neutralized virions do not attach to the RBCs and hemagglutination is inhibited, indicating a positive result (figure 36.9*b*).

While viral agglutination tests have been largely replaced by molecular methods (e.g., PCR), agglutination tests are used to measure antibody titer (relative concentration). In tube or well agglutination tests, a specific amount of antigen is added to a series of tubes or wells in a microtiter plate (**figure 36.10**). Serial dilutions of serum (1/20, 1/40, 1/80, 1/160, etc.) containing the antibody are then added to each tube or well. The greatest dilution of serum showing an agglutination reaction is determined, and the reciprocal of this dilution is the serum antibody titer.

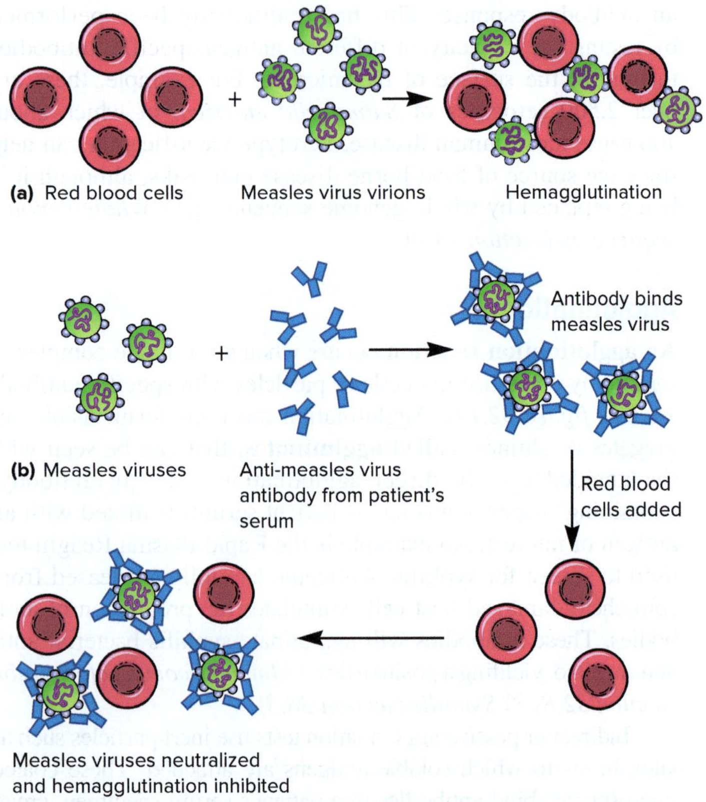

Figure 36.9 Viral Hemagglutination. (a) The virions of certain viruses bind to red blood cells, causing hemagglutination. (b) If patient serum containing specific antibodies to the virus is mixed with the red blood cells, the antibodies will neutralize the virus and inhibit hemagglutination (a positive test).

MICRO INQUIRY *Is the presence of cellular clumping a positive or negative result in this case?*

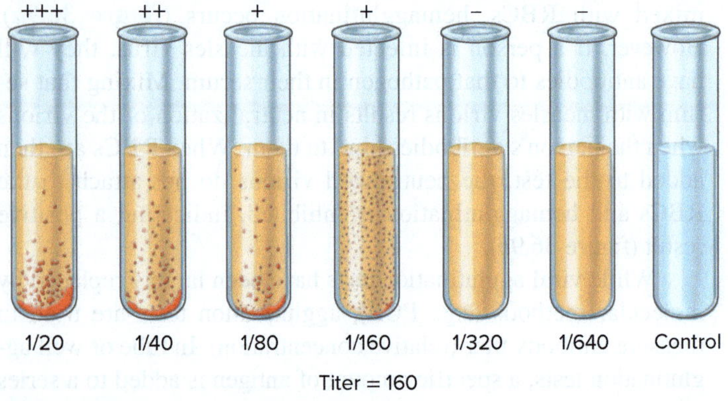

Figure 36.10 Tube Agglutination Test. Agglutination tests for determining antibody titers can be performed in tubes or in microtiter plates. The principle by which the test works is the same in both. The titer in this example is 160 because there is no agglutination in the next tube in the dilution series (1/320). The buff color in the dilution tubes indicates the presence of the patient's serum.

Immunofluorescence Assays

Immunofluorescence assays use fluorescence microscopy to detect either antigens (such as viruses) or pathogen-specific antibodies; direct assays detect antigen, and indirect assays detect patient antibody to the pathogen. **Direct immunofluorescence** starts with fixing the specimen containing the pathogen to be detected onto a slide (**figure 36.11*a***). Fluorochrome-labeled antibodies are added and the slide is incubated. After washing to remove any unbound antibody, the slide is examined with the fluorescence microscope (*see figures 2.12 and 2.13*). If fluorescence is detected, the presence of the pathogen is confirmed. Example uses of direct immunofluorescence are the identification of the bacteria *Treponema pallidum* and *Rickettsia rickettsii,* the fungi *Candida albicans* and *Pneumocystis jirovecii,* as well as some viruses including adenovirus and herpes simplex viruses.

Indirect immunofluorescence is used to detect the presence of antibodies in serum or other body fluid following an individual's exposure to microorganisms (figure 36.11*b*). In this technique, a known antigen is fixed onto a slide. The patient serum is added, and if antibody specific to the fixed antigen is present, it reacts with it to form a complex. When a second, commercially prepared fluorochrome-labeled antibody directed against the constant region of human antibody is added, it binds the patient's antibody in the complex. After incubation and washing to remove unbound antibodies, the slide is examined. Fluorescence detection confirms the presence of the pathogen (figure 36.11c). ◀ *Fluorescence microscopes use emitted light to create images (section 2.2); Antibodies bind to specific 3 D antigens (section 32.7)*

Enzyme-Linked Immunosorbent Assay

The ***e**nzyme-**l**inked **i**mmuno**s**orbent **a**ssay* **(ELISA)** is one of the most widely used serological tests. The science behind the technique is similar to that described for fluorescence microscopy and involves the detection of proteins using antibodies. There are two types of ELISAs: direct and indirect. Direct ELISAs detect protein antigens (e.g., virus capsid proteins) in patient samples; indirect ELISAs detect specific antibodies in patient samples. Thus indirect ELISAs identify pathogens by the presence of antibodies directed against a specific pathogen, not by the presence of the pathogen itself.

The indirect ELISA starts with the addition of patient serum (with its soluble antibodies) to the wells of a microtiter plate that have been previously coated with the antigen of interest. If the patient's serum contains antibody specific for this antigen, the antibody will bind to it (**figure 36.12*a***). Excess serum, containing antibody to other antigens, is removed by washing. A second antibody directed against the constant region of human antibodies is added to each well. The secondary antibody has an enzyme or small molecule (e.g., biotin) attached (conjugated) to it. Only if patient antibody has bound to the known antigen will the labeled secondary antibody bind. The presence of the secondary antibody (and thus patient antibody) is detected by adding the

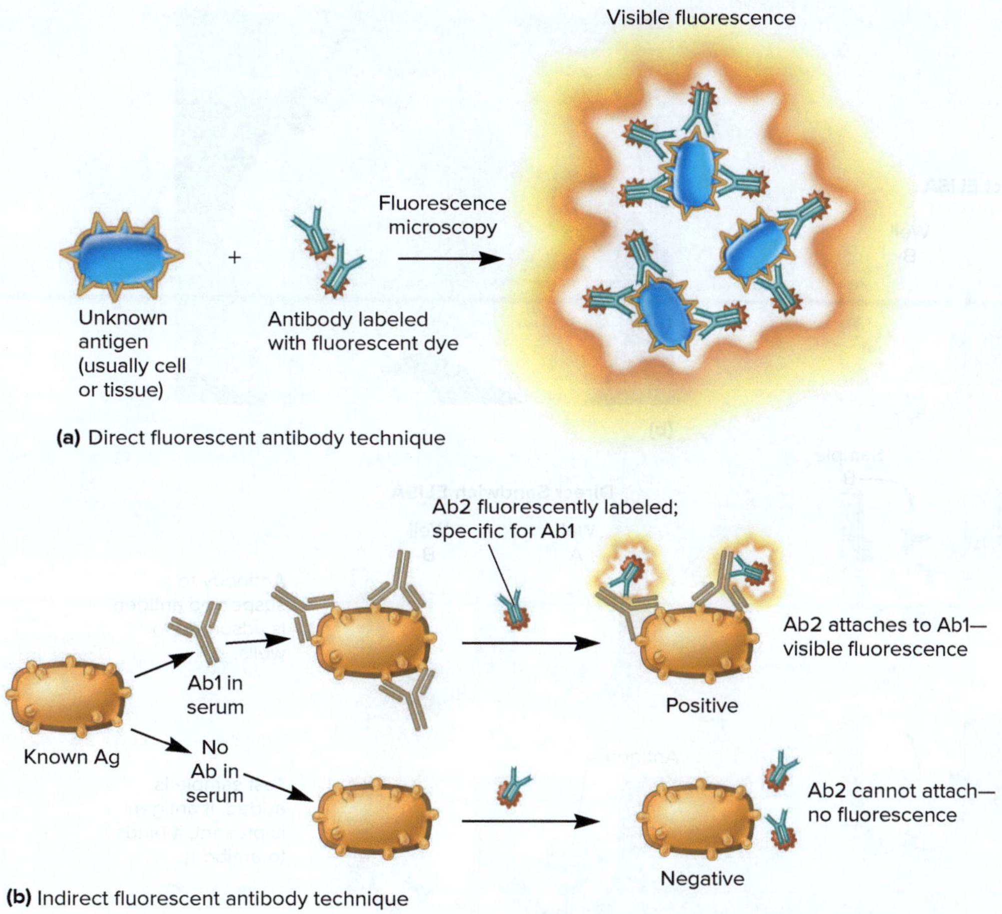

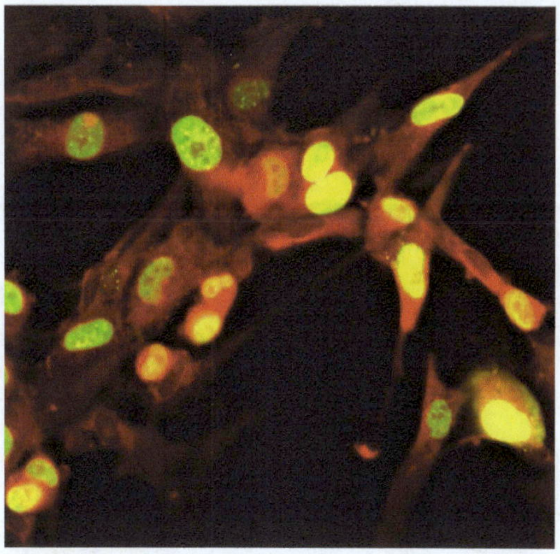

(c)

Figure 36.11 Direct and Indirect Immunofluorescence. (a) The direct fluorescent antibody technique detects the presence of antigen (Ag). Fluorescently labeled antibodies that recognize the antigen are mixed with the sample and examined with a fluorescence microscope. (b) The indirect fluorescent antibody technique detects an antibody (Ab) that recognizes a known antigen (e.g., virus or bacterium) affixed to a slide. The antigen-antibody complex is located with a fluorescently labeled secondary Ab that recognizes the bound antibody's constant region. (c) A photomicrograph of cells infected with cytomegalovirus (green fluorescence) and adenovirus (yellow fluorescence) using an indirect immunofluorescent staining procedure.

(c) Millipore Corporation

MICRO INQUIRY *Which type of immunofluorescence would most likely be used to test a clinical sample for the presence of an intracellular pathogen such as hepatitis C virus?*

substrate that reacts with the labeling molecule to generate either a flash of light (called *chemiluminescence*) or color (figure 36.12*b*). Because secondary antibody, typically made by injecting rabbits or mice with human antibody, can be formed against either human IgM or IgG, it is possible to assay for each antibody class.

In the direct ELISA, samples are added to the wells of a microtiter plate and incubated to allow adherence of antigens to the plate surface. After washing away unbound material, antibody specific to the antigen of interest is added to the wells. As with the indirect ELISA, the antibody is labeled with an enzyme or small molecule and color or light is measured and recorded after addition of the appropriate substrate for the labeling molecule. A more sensitive and commonly used version of this assay is the sandwich method, in which the antigen to be detected is bound by two different antibodies (sandwiched between them); these are referred to as *matched antibody pairs* because they bind to different epitopes on the same antigen. Patient samples are added to microtiter plates with wells coated with commercially prepared monoclonal antibody (Mab) that captures the antigen of interest in that sample (figure 36.12c). Any unbound antigens are removed by a washing step, after which a labeled secondary antibody is added to the wells, where it binds to a different epitope on the captured antigen. After addition of the substrate for the labeling molecule, results are measured and recorded as previously described. The sandwich method can also be used in the indirect ELISA assay. In both direct and indirect ELISAs, the presence and magnitude of signal are determined by placing the microtiter plate in a plate reader, an instrument designed to measure and record color or light in microtiter plate wells in units relative to standard reactions. *ELISA: Enzyme-Linked Immunosorbent Assay*

The principle of the ELISA methodology has been applied to a multitude of quick tests collectively called **lateral flow immunoassays (LFIAs).** These assays, like ELISAs, can be direct (antigen is detected) or indirect (antibody is detected). Patient sample is placed at one end of a filter paper, typically mounted in a plastic or cardboard case. The liquid sample is drawn up the filter paper (i.e., flowing laterally) where

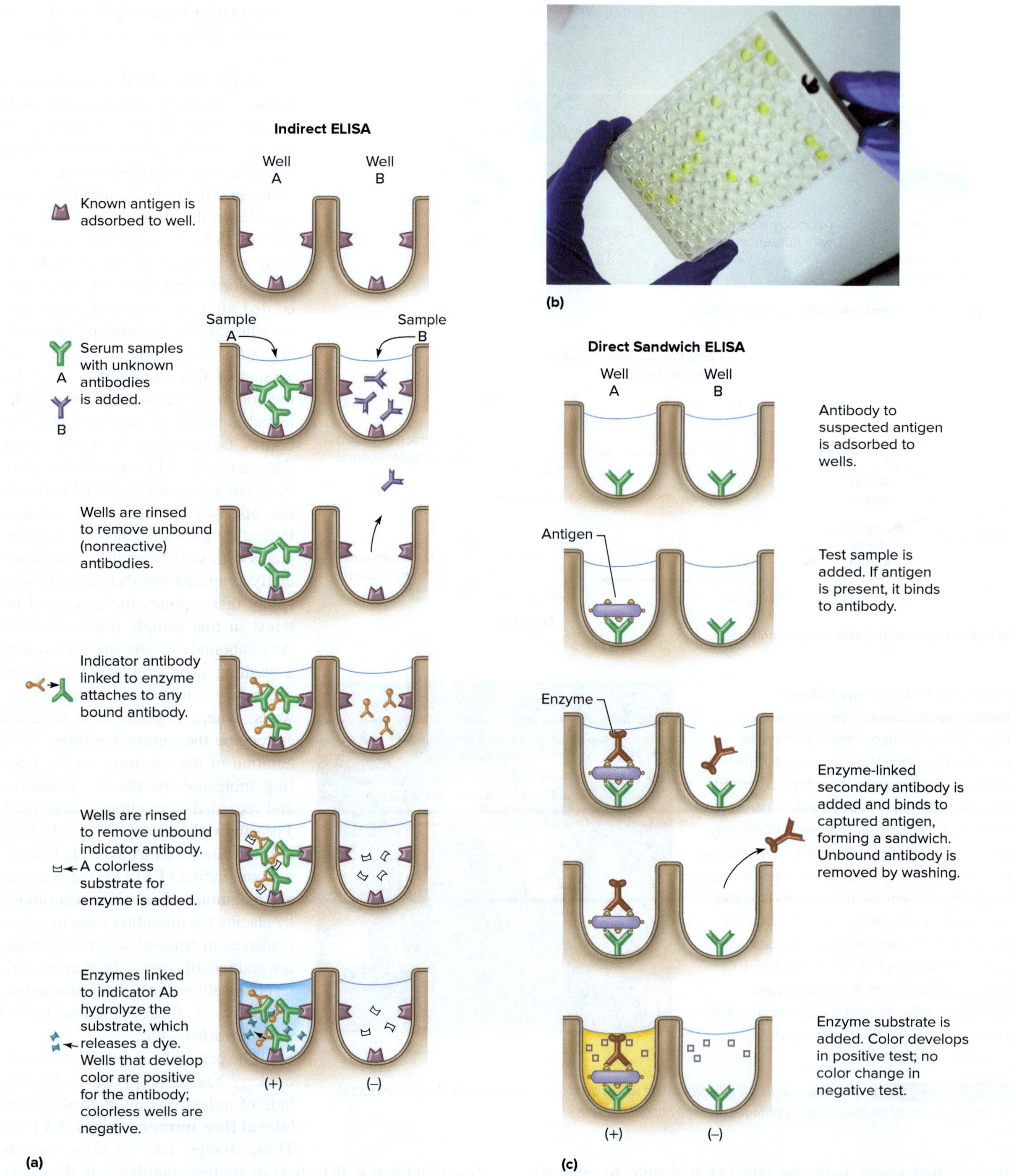

Figure 36.12 The ELISA Test. (a) The indirect ELISA for detecting antibodies. (b) ELISA test results. (c) The direct sandwich method for detecting antigens.

(b) James Redfearn/McGraw Hill

antibody-antigen interaction takes place. As outlined in **figure 36.13,** the detection lines include chromogenic substrates. Examples of pathogens and microbial products detected by direct LFIAs include *Streptococcus pyogenes* (strep throat), SARS-CoV-2 (COVID-19), rotavirus, *Giardia duodenalis,* and toxins produced by *Clostridioides difficile.* Indirect LFIA are used to detect antibodies to viruses such as dengue and chikungunya and *Anaplasma* bacteria in serum.

Immunoblotting (Western Blotting)

Another immunologic technique used in the microbiology laboratory is immunoblotting, also known as Western blotting. **Western blotting** involves polyacrylamide gel electrophoresis to separate proteins in a patient specimen. The proteins are then transferred to sheets of filter paper, thereby immobilizing them on a sturdy substrate. Specific proteins of interest are visualized by treating the filter papers with solutions of enzyme-tagged antibodies and adding appropriate substrate for that enzyme. This procedure was considered more specific than ELISA and lateral flow assays for many applications. However, the use of monoclonal antibodies in these other assays for antigen capture or antibody detection, along with other technological advances, has reduced the need for Western blotting, a more complex and time-consuming process. Western blotting is still used to confirm certain infections, such as Lyme disease.

Immunoprecipitation

The **immunoprecipitation** technique detects antibodies that when bound to antigens are called **precipitins.** The precipitin reaction occurs when antibodies and antigens are mixed in the proper proportions. The antibodies link the antigen to form an antibody-antigen network or lattice that settles out of solution when it becomes sufficiently large (**figure 36.14*a***). Immunoprecipitation reactions occur at what is known as the zone of equivalence. This describes conditions within the assay tube or plate when there is an optimal ratio of antigen to antibody so that almost

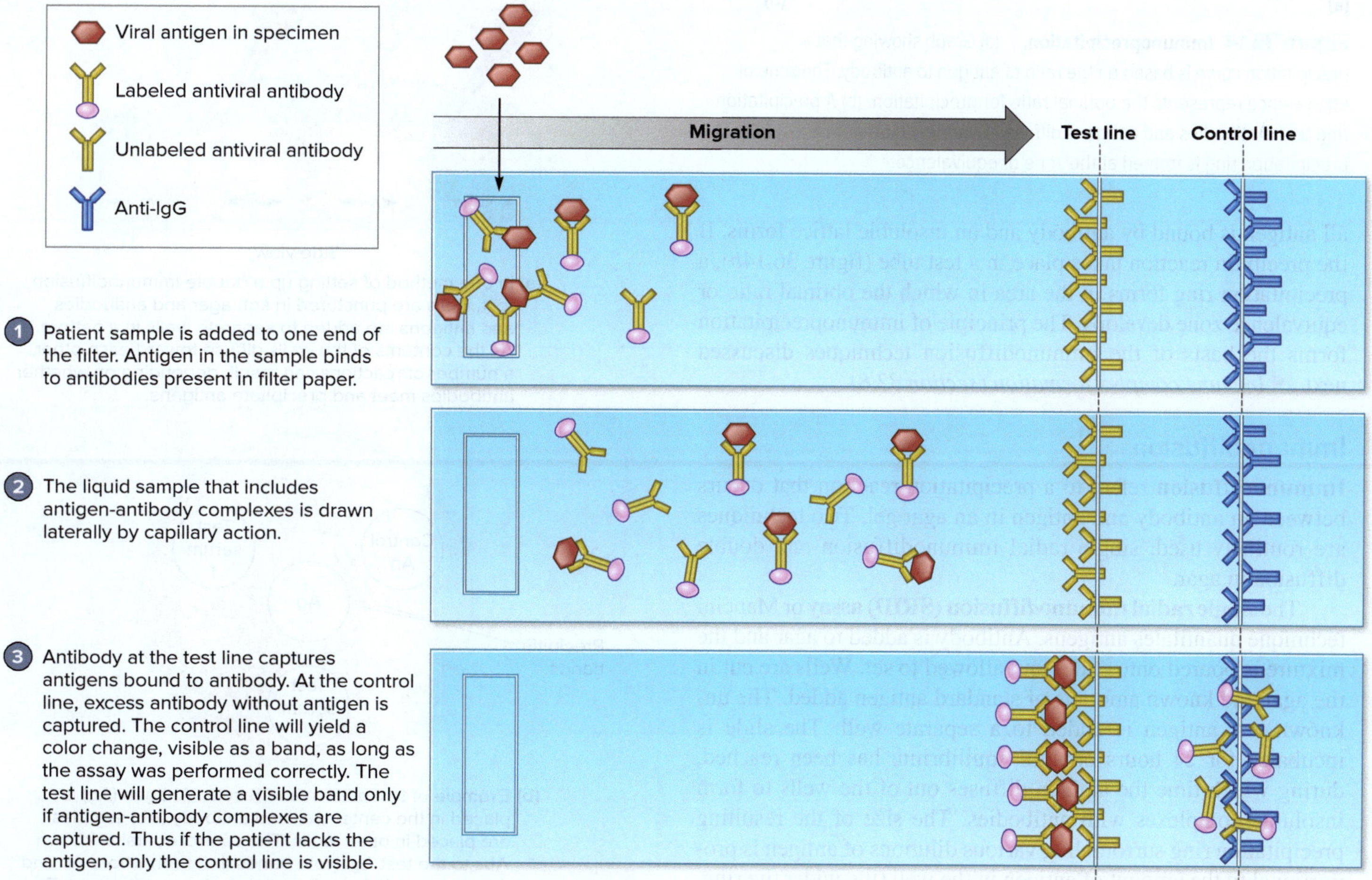

Figure 36.13 Lateral Flow Immunoassay. In this example, patient antigen is detected. Other lateral flow immunoassays detect patient antibody by providing antigen at the site where sample is added.

MICRO INQUIRY *Why must an excess of antibody be present on the filter?*

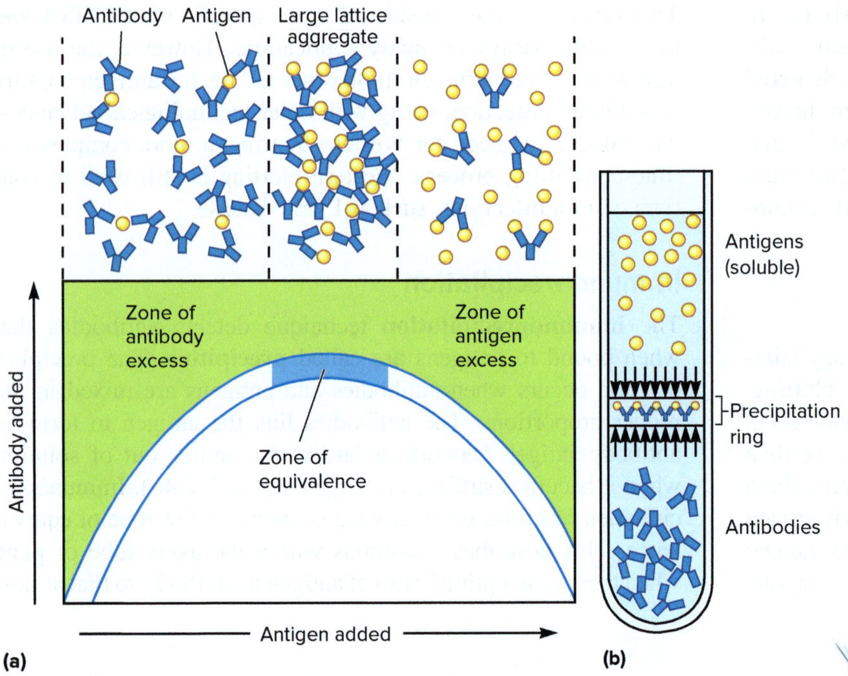

Figure 36.14 Immunoprecipitation. (a) Graph showing that a precipitation curve is based on the ratio of antigen to antibody. The zone of equivalence represents the optimal ratio for precipitation. (b) A precipitation ring test. Antibodies and antigens diffuse toward each other in a test tube. A precipitation ring is formed at the zone of equivalence.

all antigen is bound by antibody and an insoluble lattice forms. If the precipitin reaction takes place in a test tube (figure 36.14*b*), a precipitation ring forms in the area in which the optimal ratio or equivalence zone develops. The principle of immunoprecipitation forms the basis of the immunodiffusion techniques discussed next. *Immune complex formation (section 32.8)*

Immunodiffusion

Immunodiffusion refers to a precipitation reaction that occurs between an antibody and antigen in an agar gel. Two techniques are routinely used: single radial immunodiffusion and double diffusion in agar.

The ***s*ingle *r*adial *i*mmuno*d*iffusion (SRID) assay** or Mancini technique quantitates antigens. Antibody is added to agar and the mixture is poured onto slides and allowed to set. Wells are cut in the agar and known amounts of standard antigen added. The unknown test antigen is added to a separate well. The slide is incubated for 24 hours or until equilibrium has been reached, during which time the antigen diffuses out of the wells to form insoluble complexes with antibodies. The size of the resulting precipitation ring surrounding various dilutions of antigen is proportional to the amount of antigen in the well (the wider the ring, the greater the antigen concentration). This is because antigen concentration drops as it diffuses farther out into the agar. The antigen forms a precipitin ring in the agar when its level has decreased sufficiently to reach equivalence and combine with the antibody to produce a large, insoluble network. SRIDs are most commonly used today to measure potency of influenza vaccines.

The **double immunodiffusion assay (Öuchterlony technique)** is based on the principle that diffusion of both antibody and antigen (hence double diffusion) through agar forms stable and easily observable immune complexes. Test solutions of antigen and antibody are added to separate wells punched in agar. The solutions diffuse outward, and when antigen and the appropriate antibody meet, they combine and precipitate at the equivalence zone, producing an indicator line (or lines) (**figure 36.15**). The visible line of precipitation allows a comparison of antigens for identity (same antigenic determinants), partial identity, or nonidentity against a given selected antibody. Although less commonly applied today, this assay is still used to diagnose certain fungal infections such as

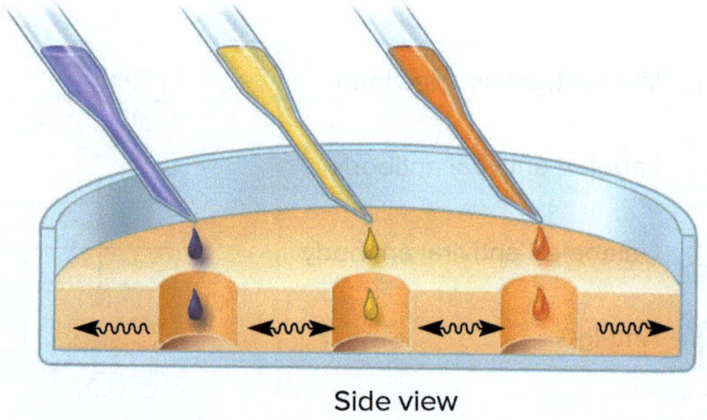

(a) In one method of setting up a double-immunodiffusion test, wells are punctured in soft agar and antibodies and antigens are added to separate wells in a pattern. As the contents of the wells diffuse toward each other, a number of reactions can result, depending on whether antibodies meet and precipitate antigens.

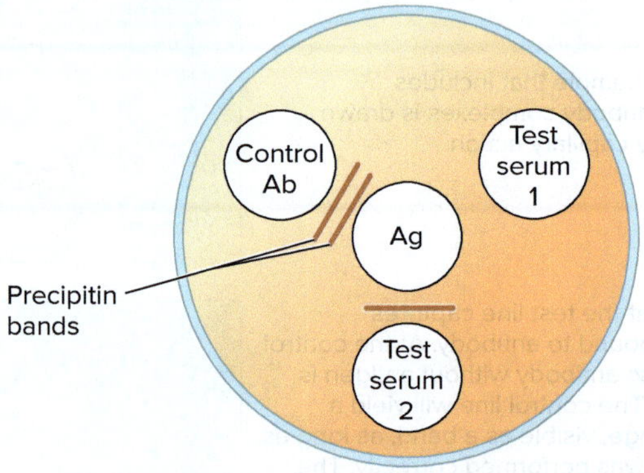

(b) Example of test pattern and results. Antigen (Ag) is placed in the center well and antibody (Ab) samples are placed in outer wells. The control contains known Abs to the test Ag. Note bands that form where Ab and Ag meet. Wells 1 and 2 contain unknown test sera. Test serum 2 reacts with the antigen, while test serum 1 does not. Double bands indicate more than one antigen and antibody that can react.

Figure 36.15 Double Immunodiffusion Assay.

coccidioidomycosis and histoplasmosis. ▶| *Fungi can be transmitted by airborne routes (section 39.2)*

Flow Cytometry

Flow cytometry allows detection of specific lymphocyte subsets ($CD4^+$ T cells, for example, in HIV-infected blood) in an easy, reliable, fast way. Flow cytometry is described in chapter 26 (*see figure 26.4*). Here we emphasize that it has become a routine technique for counting $CD4^+$ and $CD8^+$ lymphocytes (needed for monitoring patients infected with HIV), as well as lymphocytes in varying stages of development as needed in the diagnosis of leukemia and lymphoma. This technique has also enabled the development of quantitative methods to assess antimicrobial susceptibility and drug cytotoxicity in a rapid, accurate, and highly reproducible way.

Comprehension Check

1. Why is serology an indirect means of detecting infectious disease?
2. What is the difference between viral and antibody-mediated hemagglutination?
3. Compare and contrast direct and indirect immunofluorescence assays. What is the difference in terms of what is detected in the patient's sample?
4. What is the zone of equivalence and why is it important in immunoprecipitation reactions?
5. Name two types of immunodiffusion tests and describe how they are performed.

Key Concepts

36.1 The Clinical Microbiology Laboratory Detects Infectious Agents and Protects Its Workers

- Clinical microbiologists provide rapid and accurate identification of disease-causing microorganisms from clinical specimens and the accurate antimicrobial susceptibility testing of isolated organisms.
- A variety of laboratory tests and techniques are used (**figure 36.1**). Many traditional bench tests have been miniaturized, automated, or replaced by faster and more sensitive tests.
- Over 50 years of pioneering work led to protocols and equipment to protect laboratory workers.
- To prevent laboratory-associated infections, scientists follow standard microbiology practices and wear personal protective equipment (**figure 36.2** and **tables 36.1–36.3**).

36.2 Identification of Microorganisms from Specimens

- Detection of disease-causing organisms begins with the collection of patient specimens. The correct selection of body fluid or tissue and subsequent collection, transport, and pre-analytical processing of the specimen are paramount for accurately identifying infecting agents.
- The clinical microbiology laboratory can provide preliminary or definitive identification of microorganisms from sterile site and nonsterile site specimens. A thorough knowledge of the normal human microbiota is necessary to identify pathogens from host organisms.
- The type of suspected microorganisms dictates which tests are performed. Bacteria and fungi are often grown in culture to purify and amplify the pathogen for identification. Viruses and protozoa are typically identified by immunologic or molecular tests.
- Identification of pathogens is accomplished by (1) direct recovery and (2) indirect or circumstantial evidence of a pathogen. Direct recovery of infectious agents is by culture and biochemical analysis, as well as molecular techniques. The culturing of many types of bacteria and fungi (filamentous and yeast forms) is routine; however, a few require special consideration.
- Identification of bacteria, fungi, and protozoa often can be made by concentrating and staining the organisms. Gram stain for bacteria and calcofluor white for fungi permit detection using light microscopy.
- The initial identity of a bacterial organism may be suggested by (1) the source of the culture specimen; (2) its microscopic appearance; (3) its pattern of growth on selective, differential, enrichment, or characteristic media; and (4) its hemolytic, metabolic, and fermentative properties. A simple dichotomous key is used to facilitate the identification (**figure 36.5**).
- Various molecular methods are commonly used to identify microorganisms directly from specimens. Complementary methods like PCR, ribotyping, and isothermal amplification are used on culture isolates to speed identification (**figures 36.7** and **36.8**).

36.3 Immune Responses Can Be Exploited to Detect Infections

- Serotyping refers to procedures used to differentiate strains (serovars or serotypes) of microorganisms that have differences in surface antigenic composition.
- In vitro agglutination reactions usually form aggregates or clumps (agglutinates) visible with the naked eye. Tests have been developed to detect antigen as well as to determine antibody titer. These include the Rapid Plasma Reagin test for syphilis, latex microsphere agglutination

reaction, hemagglutination, and viral hemagglutination (**figures 36.9** and **36.10**).

- Immunofluorescence is a process in which fluorochromes are excited with light to make them fluoresce. These dyes can be coupled to an antibody. There are two main kinds of fluorescent antibody assays: direct and indirect (**figure 36.11**).
- The enzyme-linked immunosorbent assay (ELISA) involves linking indicator enzymes to secondary antibodies. Two basic methods are involved: the double antibody sandwich method and the indirect immunosorbent assay (**figure 36.12**). The first method detects antigens and the latter, antibodies.
- Lateral flow assays are inexpensive easy tests that use antigen-antibody reactions to rapidly detect either antibody or antigen in patient samples (**figure 36.13**).
- Immunoblotting involves polyacrylamide gel electrophoresis of a protein specimen followed by transfer of the separated proteins to filter paper and identification of specific bands by labeled antibodies.
- Immunoprecipitation reactions occur only when there is an optimal ratio of antigen and antibody to produce a lattice at the zone of equivalence, which is evidenced by a visible precipitate (**figure 36.14**).
- Immunodiffusion refers to a precipitation reaction that occurs between antibody and antigen in an agar gel medium. Two techniques are routinely used: single radial immunodiffusion and double immunodiffusion (**figure 36.15**).
- Flow cytometry enables detection of specific cell types, such as $CD4^+$ T cells in a patient infected with HIV, based on their cytometric parameters or by means of fluorochromes.

Active Learning

1. ELISA tests usually use primary and secondary antibodies. Why? What are the necessary controls one would need to perform to ensure that the antibody specificities are valid (i.e., no false-positive or false-negative reactions)?
2. *Legionella pneumophila* is a bacterium that is often found in water systems (e.g., shower heads, air-cooling towers). The dispersal of these bacteria can lead to the development of Legionnaires' disease, a particularly virulent pneumonia in people who are older and those who are immunocompromised *(see Chapter 21 opening story).* Water systems are routinely monitored for *L. pneumophila* contamination by attempting to culture the organism on GVPC medium, which contains glycine and the antibiotics vancomycin, polymyxin B, and cycloheximide. However, this method is known to underestimate the number of *L. pneumophila* cells in the environment. Why do you think these antimicrobial agents are added to the medium? Suggest another method to monitor *L. pneumophila* in the environment. Discuss how you would collect your samples and compare your results to those obtained by plating on GVPC medium. Remember to discuss your positive and negative controls.
3. *Coccidioides immitis* is a common cause of community-acquired fungal pneumonia in the southwestern United States. Design an experiment that would compare the accuracy of a lateral flow assay (figure 36.13) that measures anti-*Coccidioides* antibodies in patient serum to the double immunodiffusion assay (figure 36.15). Keep in mind your selection of positive and negative controls.
4. As a staff scientist in a state health department lab, you are called upon to be a part of a team working to address a sudden increase in the number of dead migratory birds. What laboratory safety precautions will you implement to handle the bird carcasses and materials used to identify the causative agent?

CHAPTER

37

Human Diseases Caused by Viruses and Prions

Carol M. Highsmith/Library of Congress

Remembering HIV/AIDS

The quilt shown here is the AIDS Memorial Quilt, conceived in 1985 by human rights activist Cleve Jones. Each of the nearly 50,000 panels in the quilt was made in memory of persons who died of AIDS-related illnesses. The goals of the quilt were to heighten awareness about HIV/AIDS, to raise funds for education and prevention, and to offer emotional support for those affected by the disease. The quilt toured the United States until 2019. Today it is in San Fransisco, cared for by the AIDS National Memorial, but it can still be viewed virtually.

If one compares the early days of the HIV pandemic, which started in 1981 with the first reports of otherwise healthy homosexual and bisexual men presenting with unusual opportunistic infections and cancers, to today, it is obvious that a lot has changed. In 1996, highly effective drug cocktails were introduced to treat HIV infection, and HIV is now considered a chronic disease instead of a terminal one.

HIV has also brought changes that are not so obvious. For example, HIV research led to much of our current understanding of the immune system, which in turn is yielding new and promising cancer treatments. HIV disrupted the pharmaceutical industry as developing nations began manufacturing their own life-saving antiretroviral drugs that were otherwise too expensive to provide to their citizens. The global commitment to address HIV grew from $250 million in 1996 to over $22 billion by 2024, and there has been a 39% decline in new HIV cases since its peak in 2005. By the end of 2023, 30.7 million people of the roughly 39.9 million infected were receiving treatment. So there is still work to be done.

The HIV pandemic tells us a lot about viruses that seemingly emerge from nowhere. This was a lesson repeated in 2020 with SARS-CoV-2, the virus that causes COVID-19. Like all viruses, the destruction they cause seems incongruent with their size and relative simplicity.

Chapters 6 and 25 review the general biology of viruses and introduce basic virology. Here we continue this coverage by discussing some of the most important viruses that are human pathogens. We group viral diseases according to their mode of transmission; viral diseases that occur in the United States are emphasized.

Readiness Check:

Based on what you have learned previously, you should be able to:

- ✓ Explain the basic principles of virology (sections 6.1–6.6) and prion biochemistry (section 6.7)
- ✓ List the major features of each group of viruses in the Baltimore system of viral classification (chapter 25)
- ✓ Explain the infection process (chapter 34)

37.1 Viruses Can Be Transmitted by Airborne Routes

After reading this section, you should be able to:

a. Discuss the viruses that cause common diseases spread by airborne transmission
b. Identify typical signs and symptoms of viral diseases spread by airborne transmission
c. Correlate airborne viral infection and disease severity with viral life cycles

Because air does not support virus multiplication, any virus that is airborne originated from a living source. When humans are the source of the airborne virus, it usually is propelled from the respiratory tract by coughing, sneezing, or vocalizing.

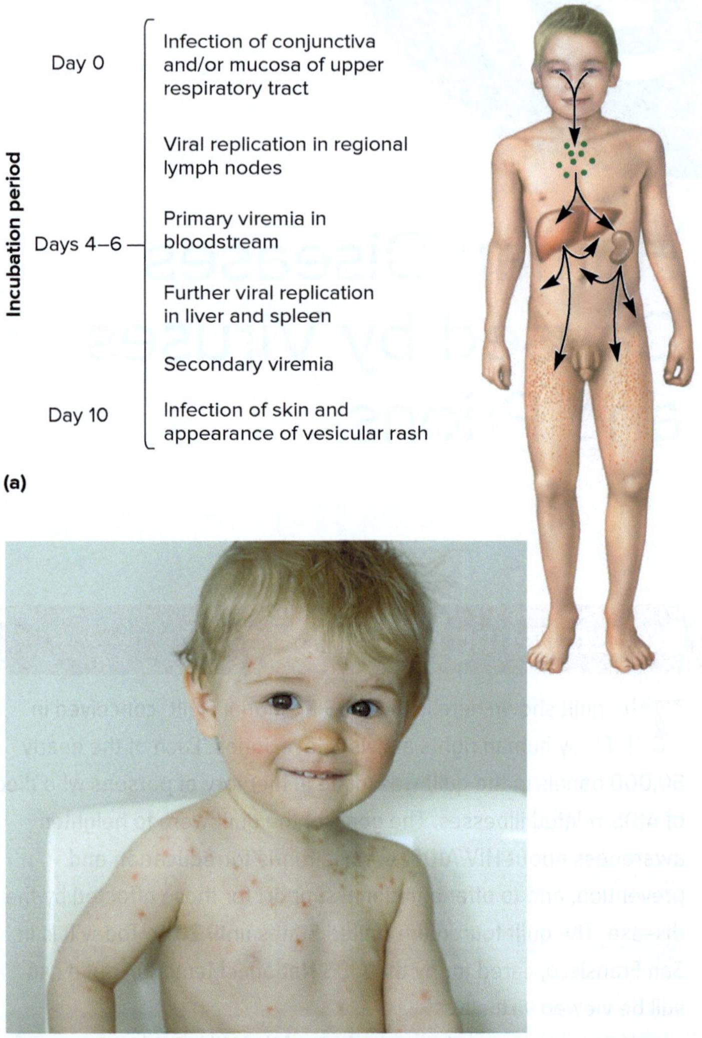

Figure 37.1 Chickenpox (Varicella). (a) Course of infection. Primary viremia occurs when virions first enter the bloodstream from the initial site of infection (in this case, the respiratory tract). Secondary viremia refers to the spread of virions to other body sites. (b) Typical vesicular skin rash. This rash occurs all over the body but is heaviest on the trunk.

(b) Asisyaj/Shutterstock

Chickenpox (Varicella) and Shingles (Herpes Zoster)

Chickenpox (varicella) is a highly contagious disease primarily of children 2 to 7 years of age. Humans are the reservoir and source for this virus, which is acquired by droplet inhalation (**figure 37.1*a***). The virus is highly infectious, with transmission rates in susceptible individuals of 65 to 86%. Prior to 1995, when a vaccine became available, about 4 million cases of chickenpox occurred annually in the United States, resulting in 100 to 150 deaths. Today there are fewer than 150,000 cases and less than 30 deaths each year.

The causative agent is the varicella-zoster virus (VZV), a member of the Herpesviridae family classified as *Human alphaherpesvirus 3* (*see table 25.3*). Like all members of this family, it is enveloped and has a double-stranded (ds) DNA genome. VZV produces at least six surface glycoproteins that play a role in viral attachment to specific receptors on respiratory epithelial cells. Following an incubation period of 10 to 23 days, small vesicles erupt on the face or upper trunk, fill with pus, rupture, and become covered by scabs (figure 37.1*b*). Healing of the vesicles occurs in about 10 days. During this time, intense itching often occurs.

Individuals who recover from chickenpox are subsequently immune to this disease; however, they are not free of the virus, as viral DNA resides as a circular episome in a dormant (latent) state within the nuclei of cranial nerves and sensory neurons in the dorsal root ganglia. Although viral DNA is maintained in infected cells, virions cannot be detected (**figure 37.2*a***). If the person who is infected becomes immunocompromised by such factors as age, cancer, organ transplants, HIV, or psychological or physiological stress, the viral genome may become activated and begin replication (figure 37.2*b*). Virions are produced that migrate down sensory nerves, replicate further, and generate painful vesicles and sensory nerve damage (figure 37.2*c*). This condition, known as **shingles** or **herpes zoster,** is characterized by vesicles that usually occur on only one side of the body, commonly the chest or abdomen, but sometimes on a limb or the face (which may affect the eyes). This pattern reflects the infected nerve and is described as a dermatomal distribution. Shingles is most common in people over age 50. In some cases, pain persists even after the vesicles have healed, a disabling complication called **postherpetic neuralgia.**

The antiviral drug acyclovir or its derivative valacyclovir can be used to shorten the duration of a shingles outbreak Analgesics are given to manage pain. Antiviral medication is used to treat chickenpox only in severe cases or in patients who are immunocompromised. In addition to the chickenpox vaccine for children, a vaccine for shingles is recommended for people 50 years and older.

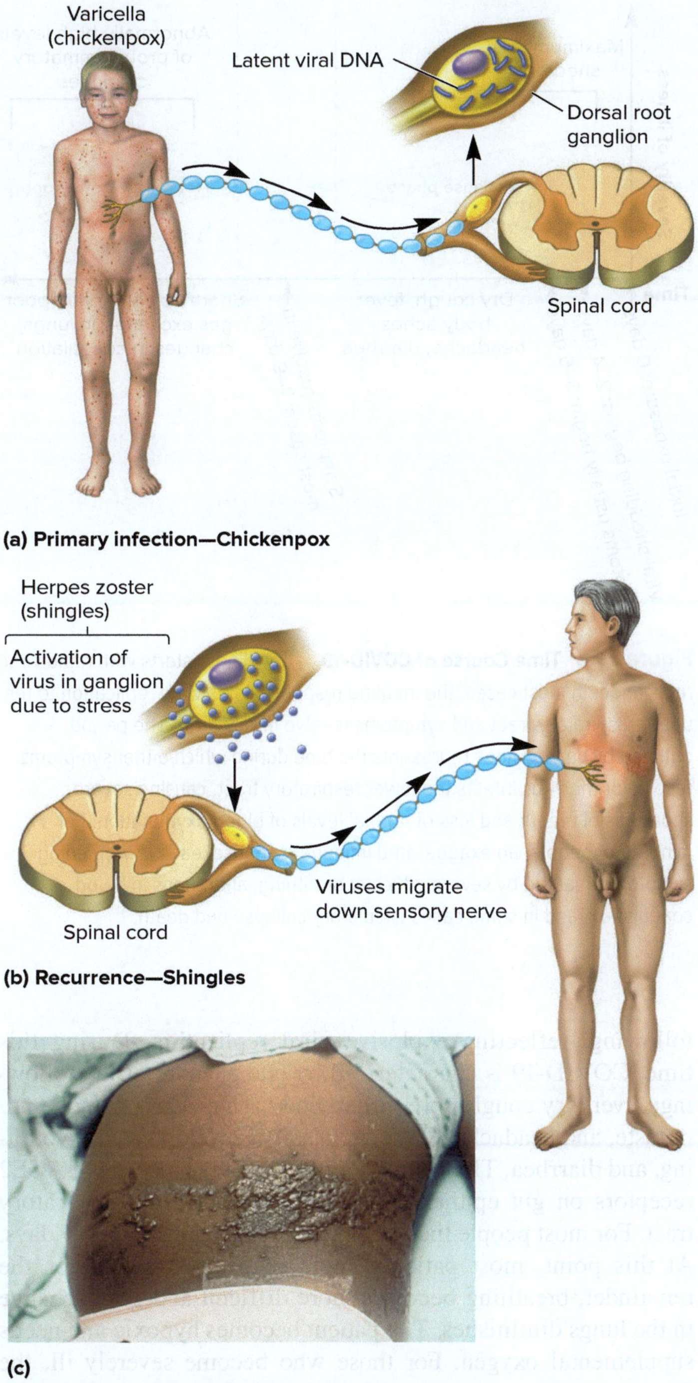

Figure 37.2 Pathogenesis of the Varicella-Zoster Virus. (a) After initial infection with varicella (chickenpox), the viruses migrate up sensory peripheral nerves to the dorsal root ganglia, establishing a latent infection. (b) When a person becomes immunocompromised or is under psychological or physiological stress, the viral genome may be activated. (c) Viruses migrate down sensory nerve axons, replicate further, and produce painful vesicles. Because these vesicles usually appear around the trunk of the body, the name *zoster* (Greek *zoster*, girdle) is used.

(c) Centers for Disease Control and Prevention

MICRO INQUIRY *Where does viral DNA reside during latent infection? Can virions be detected at this time?*

Coronavirus Infections

All coronaviruses (CoVs) belong to the family Coronaviridae. They are relatively large (120–150 nm), enveloped viruses with positive-strand RNA genomes within helical nucleocapsids (*see figures 6.7 and 25.18*). When viewed with an electron microscope, large spikes protruding from the envelope give the illusion of a crown, or corona; these are the source of the name *coronavirus* (**figure 37.3**). ◀ *Coronavirus (section 25.5)*

There are seven coronavirus species that infect people; four cause the common cold (types 229E, NL63, OC43, and HKU1). The other three can cause severe, sometimes lethal, respiratory syndromes. SARS-CoV is the virus that triggered the severe acute respiratory syndrome (SARS) epidemic in 2003 with 8,098 confirmed cases and 774 deaths (mortality rate about 10%). The last case of SARS was reported in 2004. In 2012, Middle East respiratory syndrome (MERS) emerged as a new coronavirus disease. It is caused by MERS-CoV, which is transmitted from dromedary (one hump) camels to humans, although it can also rarely spread person-to-person. Since 2012, there have been 2,613 laboratory-confirmed cases of MERS with 943 deaths. Most cases have occurred in Saudi Arabia. The seventh known coronavirus to infect humans is SARS-CoV-2. In 2025, the World Health Organization (WHO) reported that there had been over 778 million cases of COVID-19 with more than 7 million deaths worldwide since the pandemic began in early 2020.

Like all coronavirus genomes, the SARS-CoV-2 genome is large for an RNA virus—about 30,000 nucleotides (*see figure 25.17*).

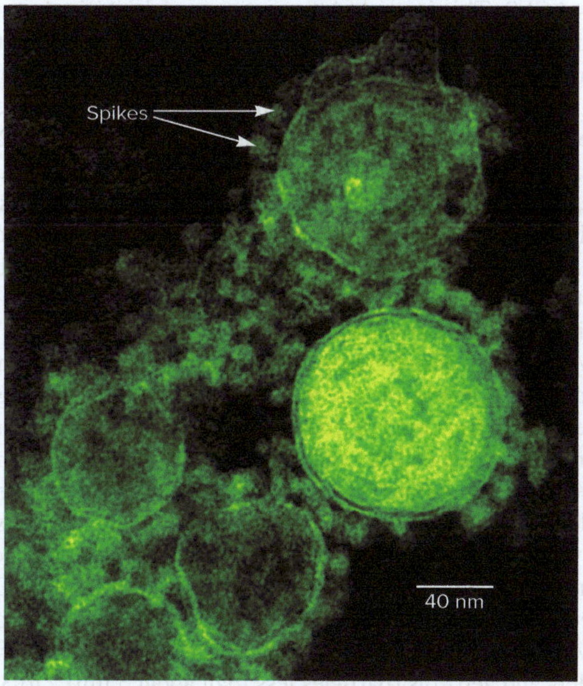

Figure 37.3 Coronavirus. All coronaviruses are large (120–150 nm) and bear a crown of spikes. Shown here is a false colorized transmission electron micrograph of the Middle East respiratory syndrome virus (MERS-CoV).

National Institute of Allergy and Infectious Diseases (NIAID)/CDC

Key proteins include the spike protein, the nucleocapsid protein, and two proteases, which are essential for maturation of newly translated polypeptides into functional proteins. The nucleocapsid protein and spike protein are immunogenic. The spike protein, which binds to the host *a*ngiotensin-*c*onverting *e*nzyme 2 (ACE2) receptor, is targeted by antibodies and is the basis of mRNA vaccines (*see figure 35.10*). Because the proteases are essential to the production of mature virions, they represent important drug targets. Another essential protein is the RNA-dependent RNA polymerase (RdRp), the target of the drug remdesivir—the first drug to be approved specifically for COVID-19.

The mutation rate of SARS-CoV-2 is relatively low because the RdRps of coronaviruses have proofreading activity, setting them apart from the RdRps of other RNA viruses. But the astonishing number of COVID-19 cases provided ample opportunity for mutants to arise. Mutations are catalogued and used to construct phylogenetic trees. Initially, two major clades evolved: one circulating in Asia and Australia, and one in Europe and the Americas. In the spring of 2020, a single mutation within the gene coding for the spike protein made the virus more transmissible but did not change its virulence. The mutation, D614G, replaced an aspartic acid residue (D) with a glycine (G) in the region of the spike involved in membrane fusion. In late 2020 and early 2021, additional highly transmissible variants were discovered in the United Kingdom, South Africa, Brazil, and elsewhere. These isolates had multiple mutations, including in the spike protein gene, which made them more contagious. One key mutation in some of these variants, N501Y, changed an asparagine (N) to a tyrosine (Y) in the region of the spike that binds ACE2. This increased the virus transmissibility because it could bind host receptors more tightly. The delta variant that became dominant in the summer of 2021 included a mutation that increased the efficiency by which a host protease cleaves the spike protein, enabling viral entry into the cell. In early 2022 the omicron strain and several subvariants arose, causing a new wave of COVID-19 infections. Each of these variants had multiple mutations in the spike protein gene. There is no doubt that as the virus keeps replicating, new phenotypes will be selected that confer an advantage in transmission as well as in evading host immune responses.

About 40% of individuals with COVID-19 remain asymptomatic. Symptomatic disease can be thought of as having two phases with a transition period in between. In the first phase, the host responds to the virus, and in the second, the host responds to their own inflammatory response (**figure 37.4**). Not everyone who develops symptomatic COVID-19 advances to the second stage. The viral replication stage begins 3 to 14 days after exposure (with an average of about 4–5 days). At this time, SARS-CoV-2 infects host cells in the nasopharynx and undergoes replication. Individuals who are infected start to shed virus in respiratory droplets and aerosols (particles less than about 5 μm) 2 to 3 days before developing symptoms. These people are therefore *presymptomatic,* which is distinct from those who are *asymptomatic* who are similarly infected but never become ill. Maximum viral shedding occurs as symptoms begin and for several days

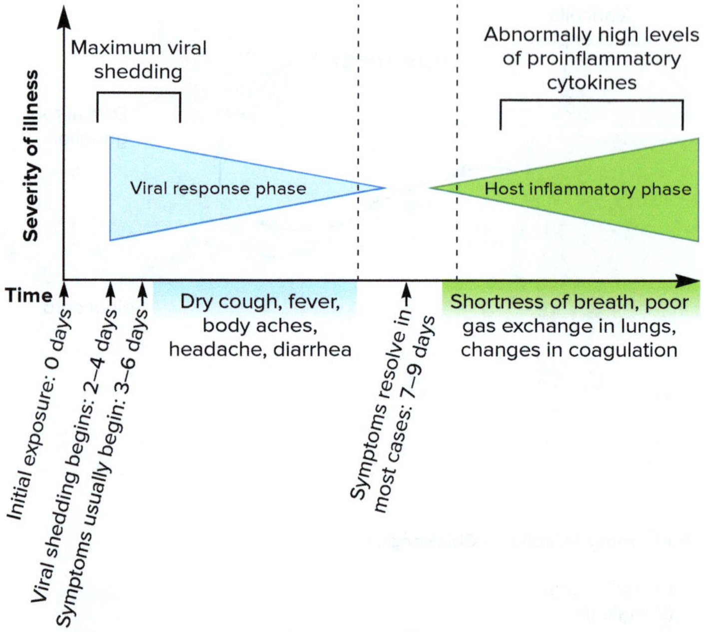

Figure 37.4 Time Course of COVID-19. The illness starts with rapid viral replication. In most cases, the immune response limits virus replication to the upper respiratory tract and symptoms resolve in 7–9 days. The period outlined by dashed lines represents the time during which either symptoms resolve or the virus infects the lower respiratory tract, causing severe shortness of breath and loss of normal levels of blood oxygenation. For a minority of people, an exaggerated immune response leads to worsening symptoms, marked by severe difficulty breathing, alterations in blood coagulation, and in some cases circulatory collapse and death.

following, reflecting explosive viral replication. During this time, COVID-19 is characterized by one or more of the following: fever, dry cough, sore throat, body aches, loss of smell and/or taste, and headache. Some people also develop nausea, vomiting, and diarrhea. These symptoms reflect the presence of ACE2 receptors on gut epithelial cells, as well as in the respiratory tract. For most people these signs and symptoms last 7 to 9 days. At this point, most patients begin to get better, but for the remainder, breathing becomes more difficult and gas exchange in the lungs diminishes. The patient becomes hypoxic and needs supplemental oxygen. For those who become severely ill, the disease is no longer driven by viral replication, but rather their own exaggerated immune response marked by an overproduction of proinflammatory cytokines such as IL-1 and IL-6 (*see table 31.2*). Neutrophils may infiltrate the lungs, as does fluid, limiting gas exchange. This may be accompanied by changes in blood coagulation so that microthrombi (tiny blood clots) clog the small vessels of lungs, kidneys, and other organs. In the most severe cases, blood pressure drops and circulatory collapse leads to death.

It is difficult to calculate a mortality rate for COVID-19, more specifically the *infection fatality rate* (IFR). Recall that mortality rate is the number of people who die of a disease divided by the total number who have the disease. IFR takes into account deaths in all infected individuals, including those who are asymptomatic or undiagnosed; it is impossible to determine the total number of people who are infected (i.e., the denominator). In addition, the mortality rate is drastically different among different populations. The case fatality rate (CFR), which measures the proportion of deaths among confirmed cases, is on average 1 to 1.5% but may be higher in certain populations such as older adults. ⏮ *Measuring infectious disease frequency (section 35.2)*

The most vexing question is why do some people become so ill while others don't even know they are infected? One answer focuses on people who are most at risk of developing severe COVID-19 illness. This includes those who have cardiovascular disease, have type 2 diabetes, or are obese. As discussed in chapter 33, people with these conditions are in a chronic state of inflammation (*see figure 33.11*). It is therefore thought that this higher baseline level of pro-inflammatory cytokines more readily propels people into an exaggerated host inflammatory response during infection. Older people are most at risk because it is believed they are not able to mount a robust early immune response to viral replication. This enables rapid viral growth and tissue injury. Just the opposite may explain why children are largely spared from severe disease. It has been suggested that children are protected by a more robust innate immune response. ⏮ *Innate host resistance (chapter 31)*

COVID-19 is most reliably diagnosed by reverse transcriptase quantitative polymerase chain reaction (RT-PCR). This is a sensitive and specific test because it can detect the presence of a single viral particle and rarely yields a false positive result. Lateral flow assays that detect viral antigen (e.g., protein) are also available. Although these tests are cheaper and easier to perform, they are not quite as sensitive or specific as the RT-PCR-based assays.

Several medications are now approved by the Food and Drug Administration (FDA) for treating those at risk of developing severe disease. Remdesivir targets the SARS-CoV-2 RdRp, thereby interfering with viral replication. Paxlovid combines two medications: Nirmatrelvir inhibits a viral protease and ritonavir prevents the destruction of nirmatrelvir in the liver. Other medications—for example, molnupiravir, which also targets RdRp—are under emergency use authorization (EUA) and are used in certain circumstances.

Currently there is no cure for COVID-19, so vaccination is key. mRNA-based vaccines introduce a short piece of engineered mRNA into the cytoplasm of recipient cells (*see figure 35.10*). The mRNA is translated by those cells into SARS-CoV-2 spike proteins, to which the vaccinated individual makes antibodies. These are the first nucleic acid vaccines to be used for human disease. Booster mRNA vaccines are bivalent and encode two versions of the spike protein, including one of the most recent circulating variants. In 2024, EUA was awarded for a subunit vaccine that contains SARS-CoV-2 spike protein synthesized in moth cells. By mid-2024, approximately 71% of the global population (nearly 5.2 billion people) had received at least one dose of a COVID vaccine. ⏮ *Vaccines immunize susceptible populations (section 35.6)*

Influenza

Influenza A virus, Influenza B virus, and Influenza C virus are members of Orthomyxoviridae; only types A and B cause significant human **influenza** (Italian *un influenza di freddo,* to be influenced by the cold) disease and are the focus here. Influenza viruses are acquired by inhalation or ingestion of virus-contaminated respiratory secretions. During an incubation period of 1 to 2 days, viral particles adhere to host respiratory epithelium and initiate the replication cycle (*see figure 25.23*). Two envelope spikes, hemagglutinin (HA) and neuraminidase (NA), play critical roles in gaining access to the host cell (*see figure 25.22*). HA is the spike protein that binds host cells and thereby triggers receptor-mediated endocytosis. This is facilitated by NA, which is thought to hydrolyze the mucus that covers the epithelium. Once enclosed in an endosome, HA undergoes a dramatic conformational change when the endosomal pH decreases. The hydrophobic ends of the hemagglutinin spring outward and extend toward the endosomal membrane. After they contact the membrane, fusion occurs and the RNA nucleocapsids are released into the cytosol. Each influenza virus nucleocapsid consists of a single negative-strand RNA molecule associated with nucleoprotein (NP) and three subunits of the RdRp (PA, PB1, and PB2). Replication is unusual because influenza viruses are RNA viruses that replicate in the host nucleus rather than in the cytosol. Therefore, each RNA segment is imported into the nucleus to be transcribed (*see figure 25.23*). At the end of the replication cycle, NA has an additional essential role. It aids in the release of newly assembled influenza virus virions from infected cells by cleaving sialic acid residues between virus and host membranes. ⏮ *Negative-strand RNA viruses (section 25.6)*

Influenza viruses are classified into subtypes based on the structure of their HA and NA spikes. There are 18 HA (H1–H18) and 11 NA (N1–N11) forms of the proteins; they are found in various HA/NA combinations. Host specificity is explained in part by the difference in receptor binding. The HA proteins of human influenza viruses preferentially attach to receptors on human tracheal epithelial cells. In contrast, the HA proteins of avian influenza viruses preferentially bind receptors found on intestinal epithelial cells of waterfowl, the main viral replication site of influenza viruses in birds.

The genomes of influenza viruses are linear RNA segments that change frequently due to point mutations and reassortment (**figure 37.5**). They have a high mutation rate because their RdRps lack proofreading capabilities. These smaller genetic changes often result in subtle protein (antigen) changes, known as **antigenic drift.** Antigenic drift results from the accumulation of mutations in a single strain of flu virus and explains why new flu vaccines must be developed each year. **Antigenic shift,** however, is a larger protein change made possible by the linear, segmented nature of the RNA genome. It occurs when two different strains of flu viruses (e.g., one from birds and one from humans) infect the same host cell (e.g., that of a pig). A single pig cell that hosts both bird and human flu (or swine and human flu) strains packages influenza RNA segments into a single new capsid. If a

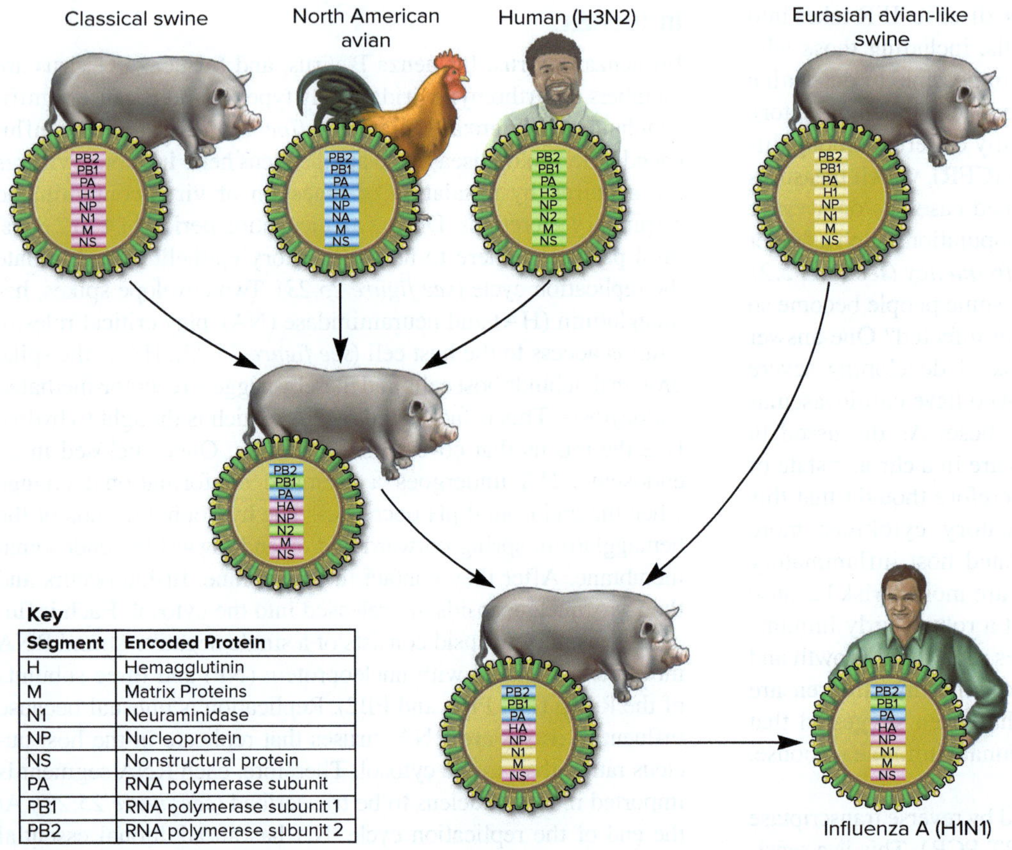

Segment	Encoded Protein
H	Hemagglutinin
M	Matrix Proteins
N1	Neuraminidase
NP	Nucleoprotein
NS	Nonstructural protein
PA	RNA polymerase subunit
PB1	RNA polymerase subunit 1
PB2	RNA polymerase subunit 2

Figure 37.5 Influenza Infection. Influenza A viruses infect birds, pigs, and humans. Reassortment of RNA from more than one viral subtype results in novel subtypes, such as the H1N1 cause of the 2009–2010 influenza pandemic.

capsid encloses RNA segments from human and avian viruses or human and swine viruses, the new virus is said to be *recombinant.* Because each segment encodes a different gene, recombinant viruses have new genes rather than mutated genes, as seen in antigenic drift. For example, antigenic shift may substitute an H3 hemagglutinin gene (and therefore protein) for an H1. Thus, a larger genetic change occurs with antigenic shift than with antigenic drift. Indeed, antigenic shifts are responsible for major epidemics and pandemics. ◀ *Infectious disease is revealed through patterns within a population (section 35.3)*

The first well-documented global epidemic of influenza-like disease occurred in 1580. Since then, 31 possible influenza pandemics have been documented, with four occurring in the twentieth century. The worst pandemic on record occurred in 1918 and killed 50 to 100 million people around the world. This disaster, traced to the Spanish influenza virus, was followed by pandemics of Asian flu (1957), Hong Kong flu (1968), Russian flu (1977), and swine flu (2009). These geographic names reflect popular impressions of where the episodes began.

To track the various influenza subtypes, a standard nomenclature system has been developed. It includes the following information: group (A, B, or C), host of origin, geographic location, strain number, and year of original isolation. Antigenic descriptions of the HA and NA surface proteins are given in parentheses for type A viruses only. The host of origin is indicated for viruses obtained from nonhuman hosts; for example, A/swine/Iowa/15/30 (H1N1). However, the host origin is not given for human isolates; for example, A/Hong Kong/03/68 (H3N2). Influenza A viruses having H1, H2, and H3 HA antigens, along with N1 and N2 NA antigens, predominate in nature, infecting humans since the early 1900s. H1N1 viruses appeared in 1918 and were replaced in 1957 by H2N2 viruses as the predominant subtype. The H2N2 viruses were subsequently replaced by H3N2 as the principal subtype in 1968. The H1N1 subtype reappeared in 1977 and cocirculates today with H2N1, H3N2, H5N1, H5N2, H7N2, H7N3, H7N7, H7N9, H9N2, and H10N7 viruses. Importantly, for avian influenza viruses to infect humans, their surface HA and NA proteins must gain the capacity to recognize human receptors. Indeed, genome sequences of the earliest isolates of the 1918, 1957, and 1968 pandemics demonstrated that otherwise avian influenza A viruses possess HAs usually seen in human influenza strains.

Highly pathogenic H7N9 and H5N1 avian flu viruses that also infect humans are of concern for both the Centers for Disease Control and Prevention (CDC) and the World Health Organization (WHO). H7N9 emerged in China 2013, where the majority of cases occurred; the virus has not spread globally. A novel H5N1 variant emerged in 2021, quickly spreading in wild birds worldwide. In the United States, it has caused outbreaks in poultry and dairy cows. As of early 2025, 67 human cases and one death had been reported by the CDC. There has been no person-to-person transmission thus far. Millions of domestic birds have been culled (destroyed) to prevent transmission of the virus, and the CDC is carefully monitoring the situation.

As with many other viral diseases, only the symptoms of influenza can be treated. The CDC recommends the neuraminidase inhibitors oseltamivir, zanamivir, and peramivir; these can reduce the duration and symptoms of type A influenza if administered during the first 2 days of illness. Neuraminidase inhibitors block the catalytic site of the enzyme. Thus virions do not detach from the host cell membrane and therefore cannot travel to infect another cell. Baloxavir marboxil is a more recently introduced anti-influenza drug. It prevents so-called *cap snatching*—the process by which an influenza encoded exonuclease excises the 5′ cap from host transcripts and places it on its own mRNA to ensure translation (*see figure 25.23*). ◀ *Antiviral drugs (section 9.5)*

Prevention of influenza is by vaccination. Although the flu vaccine is recommended for everyone, it is particularly important for those at risk of severe disease including the chronically ill, individuals under five and over age sixty-five, and residents of nursing homes. Because of influenza's high genetic variability, efforts are made each year to incorporate new virus antigens into the vaccine. The vaccine protection typically lasts only 1 to 2 years because of continuous antigenic drift in flu viruses. Flu vaccines used in the United States are tri- or quadrivalent, meaning that they are designed to protect against three or four viral strains. A variety of vaccines are available, including inactivated (IIV), recombinant (RIV), and live attenuated influenza vaccines (LAIV). Newer vaccines that would stimulate the production of host antibodies that recognize a conserved region of the HA protein are in clinical trials. The goal of these vaccines is to target many different influenza strains and therefore eliminate the need for seasonal immunization.

Measles (Rubeola)

Measles (**rubeola:** Latin *rubeus,* red) is the most contagious disease endemic worldwide. Despite being preventable, in 2023 there were an estimated 10.3 million cases and over 107,500 deaths, primarily in children under the age of five. Although measles is no longer endemic in the United States, sporadic cases have occurred since 2011. Alarmingly, the number of measles cases in the country is rising; between January and May 2025, 1,127 confirmed cases and three deaths were reported. The majority of these cases were outbreak-related and occurred in unvaccinated individuals. There is much misinformation about the perceived dangers of vaccines, but it should be remembered that vaccines save lives; the information that certain vaccines cause autism has been thoroughly and convincingly debunked.

Measles virus (MeV or measles morbillivirus) is a negative-strand, enveloped RNA virus in the genus *Morbillivirus* and the family Paramyxoviridae (**figure 37.6*a***). The MeV genome encodes eight proteins, six of which are structural. When MeV enters the body through the respiratory tract or the conjunctiva of the eyes, the viral surface hemagglutination (H) glycoprotein attaches to receptors on activated B and T cells, as well as antigen-presenting cells (figure 37.6*c*). This prompts fusion of the viral envelope and the host cell's plasma membrane and the nucleocapsid is released into the host cytosol. Like many other viruses, sequential transcription of viral genes occurs and viral mRNAs are capped and polyadenylated in the cytoplasm. Following genome replication and packaging, new virus particles interact with the viral matrix (M) protein on the inner side of the host plasma membrane, from which MeV buds.

The incubation period for measles is usually 10 to 14 days, and visible symptoms begin about the tenth day (figure 37.6*d*). The measles prodrome is characterized by nasal discharge, cough, fever, headache, and conjunctivitis, which intensify several days prior to the onset of rash. Within 3 to 5 days, a rash appears as erythematous (red) maculopapular (small bumps) lesions that are at first discrete but gradually become confluent (figure 37.6*e*). The rash normally lasts about 5 to 7 days. **Koplik's spots,** red lesions with a bluish-white speck in the center of each, are sometimes seen on the mucosal lining of the mouth in the early stages of disease. Recovery from measles begins soon after the rash is reported. Rare complications of measles include pneumonia, otitis media, peri- or endocarditis, and encephalitis. Following recovery, an individual may also experience immune amnesia for several years: MeV kills memory lymphocytes resulting in loss of recognition of past infections, and making the affected person more vulnerable to other pathogens. A rare, devastating complication of measles is a progressive degeneration of the central nervous system called **subacute sclerosing panencephalitis (SSPE)** that is nearly always fatal within three years.

No specific treatment is available for measles, but the disease is preventable through use of attenuated vaccine. The measles vaccine is typically administered in combination with mumps and rubella (MMR vaccine) or as a combined measles, mumps, rubella, and varicella vaccine (MMRV), which is recommended for all children. Measles infection or immunization with measles vaccine typically provides lifelong immunity.

Mumps

Mumps is an acute, highly contagious disease. The annual global incidence of mumps is around 500,000, although many cases likely go unreported. U.S. cases range from a total of several hundred to several thousand every year, the low rates largely due to vaccination. Periodic mumps outbreaks in developed nations are not uncommon. They are typically initiated by individuals who were not vaccinated, but may also occur in previously vaccinated people because unlike the measles and rubella components of the MMR vaccine, immunity to mumps following vaccination often wanes.

Mumps virus (MuV) is a member of the genus *Orthorubulavirus* in the family Paramyxoviridae. MuV virions are pleomorphic and enveloped, and contain a helical nucleocapsid enclosing a negative-strand RNA. Like the measles virus, MuV is transmitted in saliva and respiratory droplets, but it is less contagious than measles or chickenpox. The most prominent manifestations of mumps are swelling and tenderness of the salivary (parotid) glands 16 to 18 days after infection of the host by the virus. The swelling usually lasts for 1 to 2 weeks and is accompanied by a low-grade fever. Severe complications of mumps are rare; however, neurological complications include deafness, meningitis, and encephalitis. In postpubescent males, inflammation of the epididymis and testes (orchitis) that sometimes results in infertility or reduced fertility can occur. Rarely, in postpubescent females, oophoritis (swelling of ovaries) involvement characterized by pain, nausea, and vomiting, occurs. Therapy for mumps is limited to symptomatic and supportive measures.

Respiratory Syndromes and Viral Pneumonia

Acute viral infections of the respiratory system are among the most common causes of human disease. The infectious agents

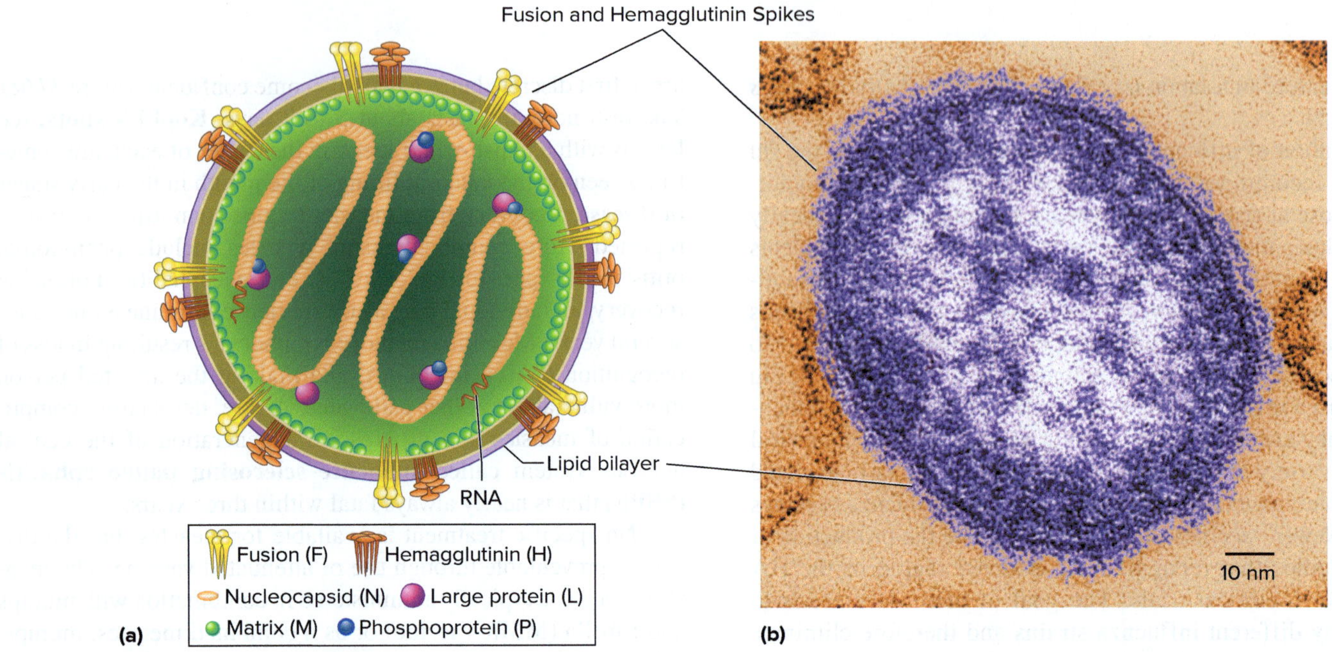

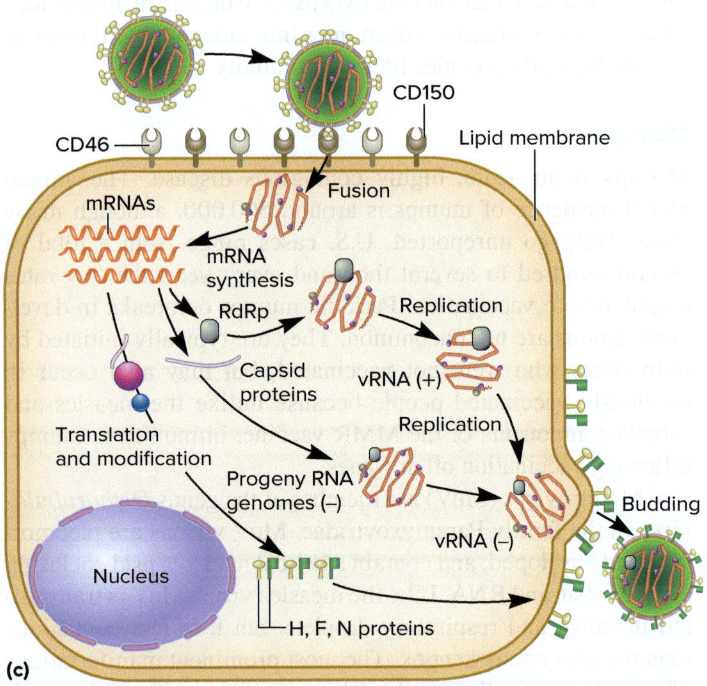

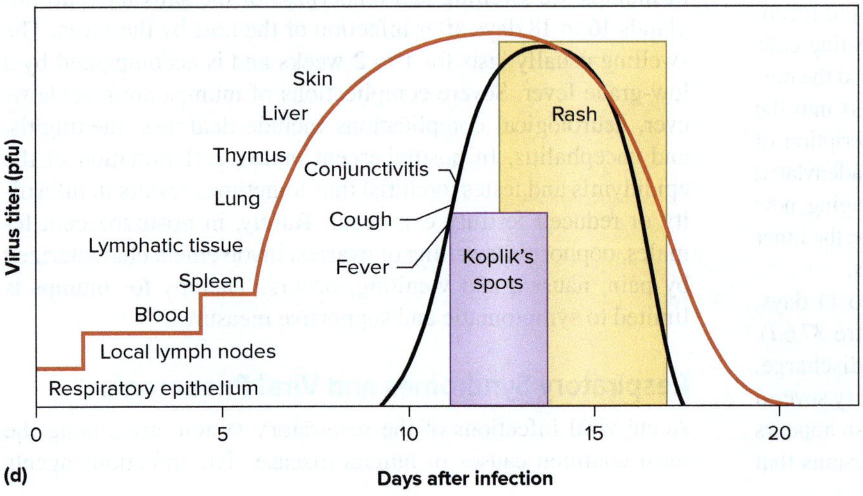

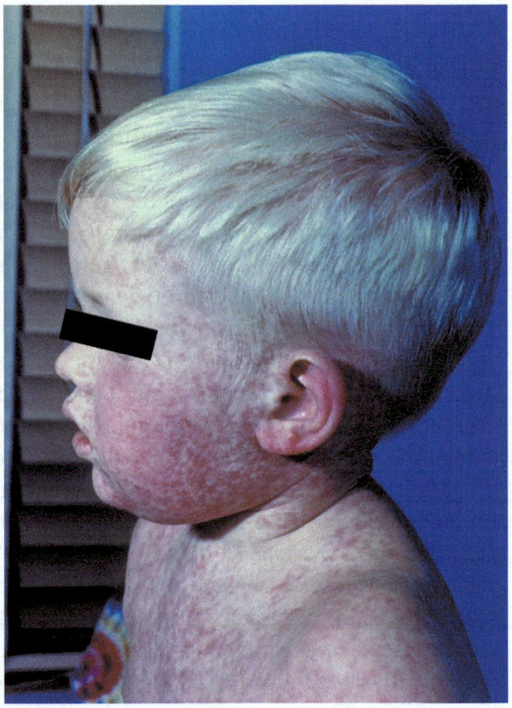

Figure 37.6 Measles (Rubeola). (a) Schematic of the spherical, RNA morbillivirus that causes measles. (b) Colorized transmission electron micrograph of a measles virus (MeV) virion. (c) Life cycle of MeV. MeV virions exit the host by budding. (d) Severity of measles symptoms (black line) corresponds to the viral load (red line) in the tissue sites. (e) The rash of small, raised spots is typical of measles. The rash usually begins on the face and moves downward to the trunk.

(b) Cynthia Goldsmith/CDC; (e) Centers for Disease Control and Prevention are called acute respiratory viruses, and they collectively produce a variety of clinical manifestations, including rhinitis (inflammation of the mucous membranes of the nose), tonsillitis, laryngitis, and bronchitis. Adenoviruses, four species of coronaviruses, picornaviruses (rhinoviruses, coxsackieviruses A and B, and echovirus), influenza viruses, parainfluenza viruses, respiratory syncytial virus, and human metapneumovirus can all cause respiratory illness. About 50% of cases are caused by rhinoviruses (Greek *rhinos,* nose), which are nonenveloped, positive-strand RNA viruses in the family Picornaviridae. There are over 115 distinct serotypes, and each of these antigenic types has a varying capacity to infect the nasal mucosa and cause a cold. In addition, immunity to many of these viruses is not lifelong. Thus colds are common because of the diversity of rhinoviruses, the involvement of other respiratory viruses, and the lack of a durable immunity. The best treatment is rest with symptom management. ◀ *Positive-strand RNA viruses: genomes that are translated upon entry (section 25.5)*

Some viral infections of the respiratory system culminate in pneumonia. The clinical picture of viral pneumonia is nonspecific. Symptoms may be mild, or there may be severe illness and death. Importantly, secondary (subsequent) bacterial infections may also contribute to respiratory illness initiated by viruses.

Respiratory syncytial virus (RSV) often is described as the most dangerous cause of lower respiratory infections in infants (less than one year of age) and, prior to COVID-19, older people (age sixty-five and older). The virus is found worldwide and causes seasonal outbreaks. In the United States, over 2.1 million children (under five years old) require outpatient visits, and about 58,000 children are hospitalized each year for RSV. Additionally, there are 177,000 hospitalizations and 14,000 deaths among those sixty-five years or older.

RSV is a member of the negative-strand RNA virus family Pneumoviridae. The virion is variable in shape and size (average diameter 120–300 nm). It is enveloped with two virus-specific glycoproteins as part of the structure. One of the two glycoproteins, G, is responsible for binding the virus to the host cell; the other, the fusion or F protein, facilitates fusion of the viral envelope with the host-cell plasma membrane, leading to entry of the virus. The F protein also induces fusion of infected cells to form multinuclear masses of cells called *syncytia,* thus giving the virus its name. The syncytia are responsible for inflammation, alveolar thickening, and fluid accumulation in alveolar spaces. RSV is transmitted by direct contact with respiratory secretions of humans. Virions are unstable in the environment (surviving only a few hours on environmental surfaces) and are readily inactivated with soap and water, as well as disinfectants.

Clinical manifestations of RSV infection include acute onset of fever, cough, rhinitis, and nasal congestion. In infants and young children, this can progress to severe respiratory distress. Diagnosis is by either RSV rapid antigen test kit or RSV detection by reverse transcriptase PCR. Treatment is supportive and includes fluid replacement and fever reducers. Prevention is by vaccination: An RSV vaccine given to pregnant people protects infants for approximately six months after birth. That vaccine, or either of two others, is also recommended for older adults. A monoclonal antibody that targets the F protein (nirsevimab) is given as a preventative measure to all infants during their first RSV season.

Rubella (German Measles)

Rubella (Latin *rubellus,* little red) was first described in Germany in the 1800s, giving rise to its moniker German measles. It is a contagious disease that occurs primarily in children five to nine years of age. It is caused by rubella virus (RuV) (*Rubivirus rubellae*), an enveloped, positive-strand RNA virus that is a member of the family Matonaviridae. RuV is spread in droplets shed from the respiratory secretions of individuals who are infected. Once the virus is inside the body, the incubation period ranges from 12 to 23 days. A rash of small red spots (**figure 37.7**), usually lasting no more than 3 days, and a low-grade fever are the normal symptoms. The rash appears as immunity develops and the virus disappears from the blood, suggesting that the rash is immunologically mediated.

Although rubella is typically a mild disease, it can be disastrous if contracted by a female in the first trimester of pregnancy. It can lead to fetal death, premature delivery, or a wide array of congenital anomalies called **congenital rubella syndrome.** These anomalies include damage to the heart, eyes, and ears, as well as brain injuries that can lead to cognitive disabilities. Because there is no effective treatment for rubella, all children and females of childbearing age who have not been previously exposed to rubella should be vaccinated. The United States was declared rubella free in 2009. The live attenuated rubella vaccine (part of MMR or MMRV) continues to be recommended.

Figure 37.7 German Measles (Rubella). This disease is characterized by a rash of red spots. Spots are not raised above the surrounding skin as in measles (rubeola; see figure 37.6*e*).

Centers for Disease Control and Prevention

MICRO INQUIRY *Why is the characteristic rash not caused by viral infection of skin cells?*

Smallpox (Variola)

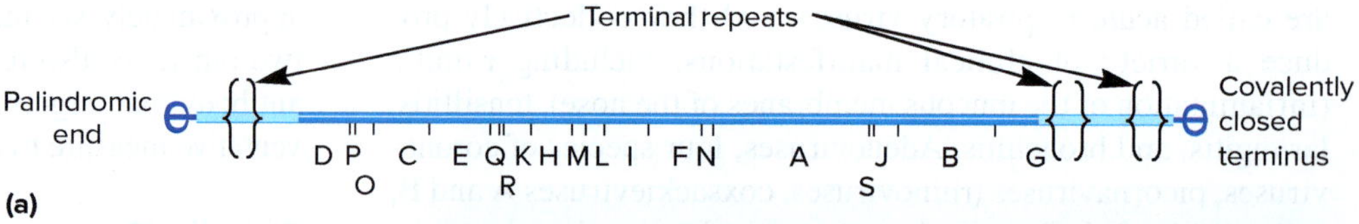

(a)

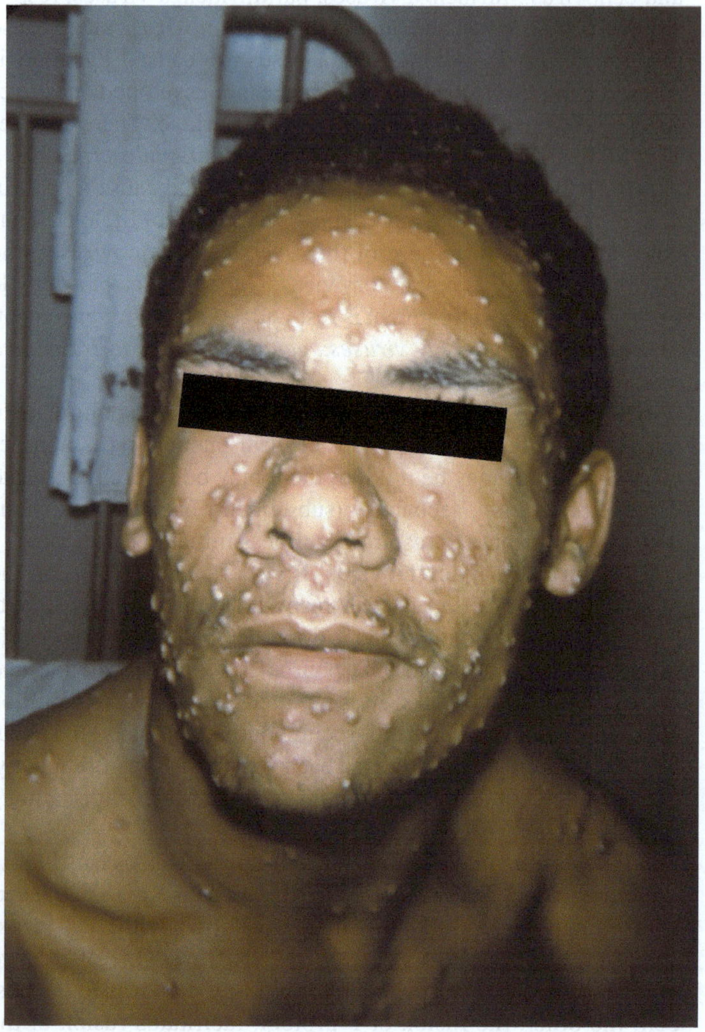

(b)

Figure 37.8 Variola Virus (VARV) Causes Smallpox. (a) Genome architecture of VARV showing its linear, 186-kb covalently closed DNA. Each letter along the genome represents a protein-coding gene. (b) Smallpox rash demonstrating the virus-filled pustules. Pustules erupt at the same time all over the body.

(b) Dr. Noble/CDC

MICRO INQUIRY *What specific clinical and biological features of smallpox made its eradication possible?*

Smallpox (variola) is a highly contagious disease that is to date the only human disease eradicated by vaccination. Smallpox was once one of the most prevalent of all diseases. It was a universally dreaded scourge for more than 3 millennia, with case fatality rates of 20 to 50%; it caused disabilities and changes to physical appearance in those who survived.

Smallpox is caused by orthopoxviruses belonging to the species *Variola virus.* Variola viruses belong to the family Poxviridae, which also includes the species *Mpox virus, Molluscum contagiosum virus,* and *Vaccinia virus. Vaccinia virus* is the virus that serves as the smallpox vaccine. Poxviruses are large and brick-shaped and contain a dumbbell-shaped core (*see figure 6.5*). The genome is large for a virus (about 186,000 bp), consisting of a single, linear molecule of double-stranded DNA (**figure 37.8*a***). It is an unusual DNA virus in that it replicates in the host cytosol rather than the nucleus.

Variola virus enters the respiratory tract, invading the mucous membranes before passing rapidly into regional lymph nodes. The average incubation period is 12 to 14 days. During this time, the virus multiplies in the host, but the infected individual is not contagious. Another brief period of viremia precedes the prodromal phase. The prodromal phase lasts for 2 to 4 days and is characterized by the abrupt onset of malaise, severe head and body aches, occasional vomiting, and high fever. The mucous membranes in the mouth and pharynx become infected, resulting in a rash of small, red spots. These spots progress into open sores that spread large amounts of the virus into the mouth and throat, which allows the virus to be efficiently transmitted from person to person. During this secondary viremia, the virus invades skin cells, leading to the development of lesions that undergo a regular progression of eruptions (rash), papules (bumps), vesicles (fluid-filled blisters), pustules (pus-filled blisters), crusts, and finally desquamation (figure 37.8*b*). The pustules often have a depression in the center that looks like a belly button. This is a major distinguishing characteristic of smallpox, as compared to other diseases that exhibit rashes with lesions of varying stages of development. The fever usually declines as the rash appears but rises and is sustained from the time the vesicles form until they crust over. Oropharyngeal and skin lesions contain abundant virus particles. Death from smallpox is due to toxemia associated with immune-mediated blood clots and elevated blood pressure.

Through the concerted efforts by the WHO, smallpox has been eradicated throughout the world—the greatest public health achievement ever. The last case from a natural infection occurred in Somalia in 1977. Eradication was possible because smallpox has obvious clinical features, very few asymptomatic carriers, only human hosts as reservoirs, and a short period of infectivity (3 to 4 weeks). Furthermore, variola virus is primarily transmitted by direct and fairly prolonged face-to-face contact. It also can be spread through direct contact with infected bodily fluids or contaminated objects such as bedding or clothing. Smallpox has occasionally been reported to spread through the air in enclosed settings such as buildings, buses, and trains.

Routine immunization for smallpox was discontinued in the United States in 1972. An accidental or deliberate release of the virus would be catastrophic in an unimmunized population and could cause a major pandemic. Thus if an outbreak occurred, prompt recognition and institution of control measures would be paramount as there is no FDA-approved treatment for smallpox. *Bioterrorism readiness is an integral component of public health microbiology (section 35.7)*

Comprehension Check

1. Compare and contrast the clinical presentation and epidemiology of chickenpox and shingles.
2. In addition to COVID-19, what other illnesses are caused by coronaviruses?
3. How does the mechanism of disease differ in early COVID-19 and later severe COVID-19?
4. What is the difference between antigenic shift and antigenic drift?
5. What are some common symptoms of measles? When are individuals most contagious?
6. What is one clinical manifestation mumps can cause in a young postpubescent male?
7. When is rubella most dangerous and why?
8. Both chickenpox and smallpox are called pox diseases. Compare the biology of the causative viruses and their disease manifestations.

37.2 Arthropods Can Transmit Viral Diseases

After reading this section, you should be able to:

a. Discuss the viruses that cause common arthropod-borne diseases
b. Identify typical signs and symptoms of two arthropod-borne viral diseases
c. Correlate arthropod-borne virus infection with geography and time of year

*Ar*thropod-*bo*rne viruses (arboviruses) are transmitted by blood-sucking arthropods (e.g., ticks and mosquitoes). They multiply in the tissues of their arthropod hosts without producing disease. Approximately 150 arboviruses cause human illnesses. Arboviral infections are tracked by ArboNET, managed by the CDC and state health departments. Diseases produced by arboviruses can be divided into three clinical syndromes: (1) fevers with or without a rash; (2) encephalitis (inflammation of the brain), often with a high fatality rate; and (3) hemorrhagic fevers, also frequently severe and fatal. No antiviral therapy is effective in treating arbovirus diseases, although supportive treatment is beneficial. Detection of these viruses is by nucleic acid amplification and serology.

Chikungunya

Chikungunya virus (CHIKV), a member of Togaviridae, is a positive-strand, enveloped RNA virus. In 2007 it spread from Africa to Europe, and in 2013 it was found on Caribbean Islands, from where it spread to the U.S., Puerto Rico, and the U.S. Virgin Islands. The virus is transmitted by *Aedes aegypti* and *A. albopictus* mosquitoes. There have been no locally acquired cases in the U.S. since 2019.

Acute CHIKV infection in humans often includes fever and joint pain. Other symptoms may include headache, muscle pain, joint swelling, or rash. These symptoms often resolve in 7 to 10 days. However, many patients report prolonged joint pain persisting for months to years. Rare complications include eye inflammation, myocarditis, hepatitis, nephritis, hemorrhage, encephalitis, and Guillain-Barré syndrome (GBS). GBS is thought to arise when the immune system mistakenly attacks peripheral nerves instead of the infecting pathogen (it is also occasionally seen after influenza infection and vaccination, Zika virus infection, and infection with the bacterium *Campylobacter jejuni*).

There is no therapy for CHIKV infection, but in 2023 the FDA approved an attenuated vaccine. Travelers to endemic areas should follow guidelines to prevent mosquito-borne infectious disease; these include using insect repellent, wearing light-colored clothes with long sleeves and pants, and staying in places with mosquito netting, window and door screens, and fans.

Dengue and Severe Dengue

Dengue virus (DENV) is a member of the Flaviviridae family, and is an enveloped, positive-strand RNA virus. Its distribution is worldwide and dengue is endemic in over 100 countries where its mosquito vector, *Aedes aegypti,* is found. It can also be transmitted by *A. albopictus,* although not as efficiently. The WHO estimates that there are about 400 million cases of dengue annually worldwide, although many cases are asymptomatic and thus unreported. Each year, severe dengue is fatal in approximately 22,000 people, mostly children.

There are four antigenically distinct serotypes of dengue virus: DENV-1 through DENV-4. All have a genome of about 110,000 bases encoding a single polypeptide that is cleaved into seven nonstructural proteins and three structural proteins—the capsid protein, the envelope glycoprotein (E), and the membrane protein. The E protein is responsible for attachment of the virion to host cells, which is followed by endocytosis and entry into the host. After replication and packaging, the virus buds from the host cell's endoplasmic reticulum prior to exiting the cell.

Signs and symptoms of dengue appear about 4 to 7 days after being bitten by an infected mosquito. The individual develops fever, rash, joint and bone pain (earning it the name breakbone fever), muscle aches, and sometimes mild bleeding of the gums and easy bruising. This lasts between 3 and 10 days. The patient recovers thanks to the formation of neutralizing antibodies that bind to DENV E protein, thereby preventing infection of newly made virions into additional host cells. Severe dengue may develop if this same individual is later bitten by a mosquito carrying any of the other DENV serotypes, caused by the phenomenon of **antibody-dependent enhancement (ADE).** This is because each DENV serotype has an antigenically distinct

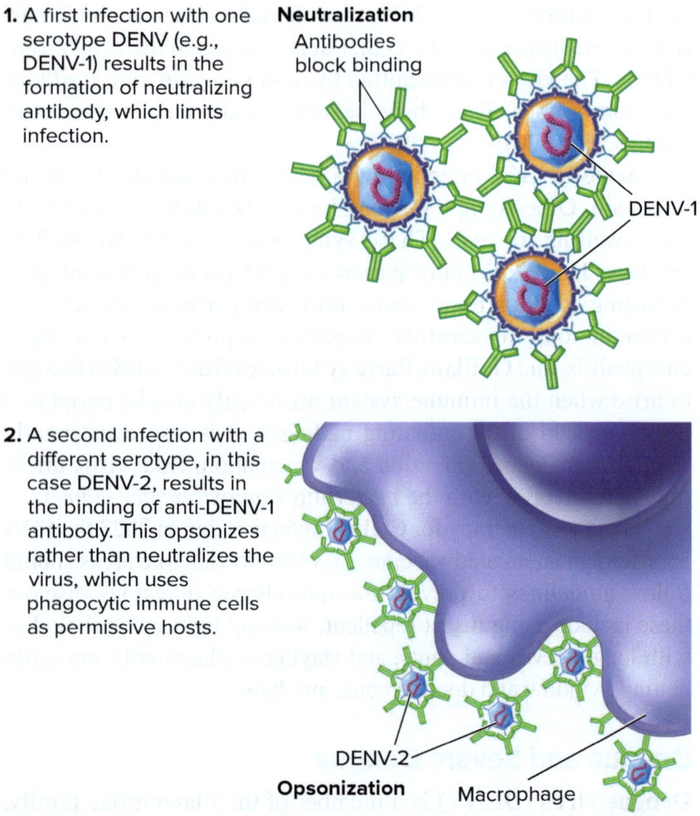

Figure 37.9 Antibody-Dependent Enhancement Leads to Severe Dengue.

E protein and so antibodies against one serotype will recognize the E protein on the second serotype even though it is structurally distinct (**figure 37.9**). Unfortunately in this case, rather than neutralizing the virus, the bound antibodies serve to opsonize (tag) the virus for phagocytosis by white blood cells. These cells, including dendritic cells, monocytes, and macrophages, serve as permissive hosts for DENV, meaning that they allow the second dengue virus to enter and replicate in them. Because these cells are now so efficient at taking up DENV, very high levels of virus are made and severe dengue may ensue. Symptoms of severe dengue begin like dengue, but after the initial weeklong febrile illness, the patient experiences severe abdominal pain and persistent vomiting. They may have difficulty breathing as fluid leaks from capillaries into the lung (pleural) cavity. This crisis can lead to dengue shock syndrome, marked by circulatory system failure and death unless the patient is given supportive care. In 2015 a dengue vaccine was introduced and is recommended for people between 9 and 45 years of age in dengue endemic countries. The vaccine is administered only to those with a history of dengue illness because it can trigger ADE in those whose first natural infection occurs after being vaccinated. A new attenuated vaccine approved by the European Medicine Agency in 2022 can be administered to individuals not previously exposed to Dengue virus. It is not approved for use by the United States. ◀ *Monocytes, macrophages, and dendritic cells (section 31.4); Antibodies doom antigens (section 32.8)*

Equine Encephalitis

Equine encephalitis is caused by several positive-strand, enveloped RNA viruses. As the name implies, this is a disease of horses, where it can be fatal. However, in humans, the disease can present as a spectrum from fever and headache to meningitis and encephalitis. Human disease can progress to include seizures, paralysis, coma, and death. The viruses are transmitted to humans by *Aedes* and *Culex* spp. mosquitoes. Various geographic descriptors are used to define the disease caused by genetically distinct species of these arboviruses. These include eastern equine encephalitis, which occurs along the eastern Atlantic coast from Canada to South America; western equine encephalitis, found from Canada to South America along the western coast; Venezuelan equine encephalitis, in central and southern parts of the United States into South America; and Japanese equine encephalitis (JEE), found in Asia. Reservoir hosts are important in the replication, maintenance, and dissemination of these arboviruses. Treatment consists of supportive care. Currently a vaccine is available only to prevent JEE.

West Nile Fever (Encephalitis)

West Nile fever (encephalitis) is caused by a positive-strand RNA flavivirus (West Nile virus, WNV) that once occurred primarily in the Middle East, Africa, and Southwest Asia. In 1999 the virus appeared in the United States, where it caused 62 confirmed encephalitis cases and seven deaths. Currently, 47 U.S. states and the District of Columbia report West Nile virus (WNV) infections in humans. In 2024 there were nearly 1,500 confirmed cases and 18 deaths in the United States.

WNV is transmitted to humans predominantly by *Culex* spp. mosquitoes that feed on infected wild birds. Humans and other mammals (e.g., horses) infected with WNV are considered *dead-end hosts* because the level of virus in the bloodstream is not sufficient to transmit the virus to uninfected mosquitoes. Mosquitoes harbor the greatest concentration of virus in the late summer and early fall, and corresponding to this, there is a peak of disease in late August to early September. The risk of disease then decreases as the mosquitoes die when the weather becomes colder. Although many people are bitten by WNV-infected mosquitoes, most do not know they have been exposed because they remain asymptomatic or exhibit only mild, flulike symptoms. In rare cases, the virus becomes neuroinvasive, causing seizures, paralysis, altered mental status, and coma. The fatality rate in these cases is high.

There is no treatment other than hospitalization and intravenous fluids. There is no human vaccine to prevent WNV.

Zika Virus Disease

Zika virus (ZIKV) took the world by surprise in the spring of 2015 when people became ill with a virus about which little was known. In the roughly 75 years since its discovery, it had caused only mild febrile illness in fewer than 3,000 people in Asia and Africa. Like dengue, ZIKV is a member of the Flaviviridae, so it has an enveloped capsid and a positive-strand RNA genome. It is transmitted by many species of *Aedes* mosquitoes including *A. africanus, A. aegypti,* and *A. albopictus*. It can also be transmitted from person to person in body fluids, specifically breast milk and semen; males are advised to use condoms for at least 6 months after infection. There are two genetically distinct virus lineages. In both, the genome is about 10,800 bases and encodes a single polypeptide of about 3,400 amino acids that is subsequently cleaved into three structural and seven nonstructural proteins. Permissive hosts in culture include human neurons, skin cells, and immature dendritic cells; more in vivo studies are needed to reveal how exactly the virus causes the severe congenital anomalies, including microcephaly, that have come to define ZIKV infection. However, evidence suggests that a mutation in the gene encoding the envelope protein increases mother-to-fetal transmission, the capacity of the virus to infect neurons, and the amount of virus in the bloodstream (viremia).

ZIKV infection has an incubation period of about 10 days and is symptomatic in only about 18% of people. When it does cause illness, it manifests as a mild fever, rash, joint pain, and conjunctivitis (pink eye), which resolves in 3 to 7 days. It has been associated with the development of GBS in a small minority of patients. ZIKV infection during pregnancy may result in a variety of neurological and cognitive disabilities in the baby. Microcephaly is characterized by diminished brain tissue with calcifications; the skull of some infants is also collapsed. These children also often have damage to the eyes, atypically formed brain structures, and hydrocephalus (atypically high levels of fluid within the skull). Unfortunately, there is no treatment, only prevention. Several vaccines are currently in clinical trials.

The rise and fall of the prevalence of Zika infection illustrates how quickly a virus can expand in an immunologically naïve population—that is, people who have never been exposed to the infectious agent. ZIKV infections peaked in the spring of 2016 with over 35,000 verified infections per week in South and Central America and the Caribbean. Since that time, the prevalence of infections has declined precipitously. Although no one knows for sure why this is the case, herd immunity certainly comes into play. Because roughly 80% of people who are exposed are unaware of infection, yet they nonetheless develop immunity, the transmission cycle from mosquito to human and back to mosquito becomes interrupted. ◀ *Epidemiology is an evidence-based science (section 35.1)*

Comprehension Check

1. Describe the recent biogeographic changes in chikungunya virus. What other arboviruses have undergone significant shifts in their global distribution?
2. Describe how antibody-dependent enhancement is responsible for the development of severe dengue.
3. What is a dead-end host?
4. Why do you think the incidence of Zika and West Nile virus infections has been declining in recent years?

37.3 Direct Contact Diseases Can Be Caused by Viruses

After reading this section, you should be able to:

a. Discuss the viruses that cause common diseases spread by direct contact
b. Identify typical signs and symptoms of common viral diseases spread by direct contact
c. Correlate direct contact virus infection and disease severity with viral life cycles

Transmission of disease from one person to another often requires close personal contact. For example, transfer of microorganisms between people occurs by direct contact with the person who is infected through touching, kissing, sexual contact, contact with body fluids and secretions, or contact with open wounds. We now consider several viral diseases that are transmitted by direct contact.

Acquired Immune Deficiency Syndrome (AIDS)

The devastating global toll **AIDS (acquired immune deficiency syndrome)** has taken across the world cannot be understated (see chapter opening story). More than 42 million people have died from it since the pandemic began in 1981, and in 2024, an estimated 39.9 million people were infected and living with the causative agent, **human immunodeficiency virus (HIV)** (**figure 37.10**). About a million people die each year from HIV/AIDS, but an estimated 1.2 million lives are saved each year thanks to antiretroviral therapy (ART). In 2023, 89% of individuals infected with HIV were accessing treatment and the number of global AIDS-related deaths that year was approximately 680,000. The fatality rate of HIV/AIDS has declined by 51% since 2010. In the United States, AIDS is caused primarily by HIV-1; the HIV-2 species predominates in certain West African countries. Most AIDS cases result from infection with HIV-1, the focus here.

HIV is transmitted from one person to another when infected blood, semen, vaginal secretions, or breast milk comes in contact with an uninfected person's mucous membranes. In sub-Saharan Africa and parts of Asia, AIDS affects as many females as males. In the United States, the groups most at risk for acquiring HIV are (in descending order) males who have unprotected sex with males; people who use intravenous drugs; males and females who engage in unprotected heterosexual sex with partners who are infected; and children born of mothers who are infected. Transmission through semen increases HIV virulence as much as 100,000-fold relative to transmission by other routes. If untreated, HIV disease progresses to AIDS and is almost 100% fatal 9 to 11 years post infection. The use of combined antiviral medications has transformed HIV infection into a chronic disease.

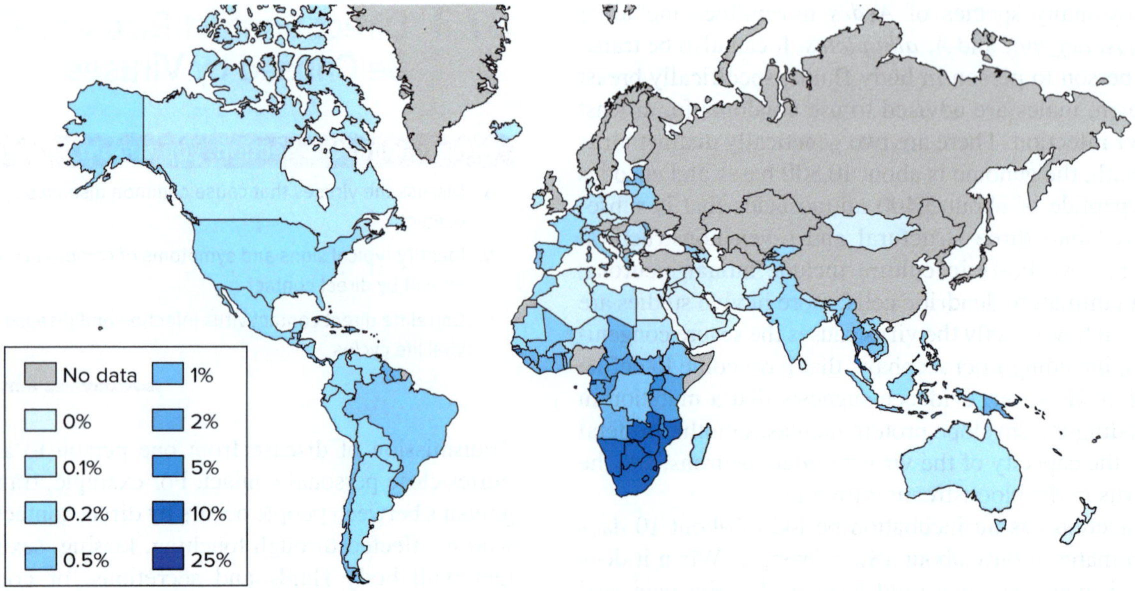

Figure 37.10 The Global HIV/AIDS View. This figure shows percentage of the adult population aged 15–49 living with HIV in 2023.

Data source: Joint United Nations Programme on HIV/AIDS (2023). OurWorldinData.org/hiv-aids

HIV is a positive-strand, enveloped RNA virus within the family Retroviridae. HIV-1 virions have a cylindrical core capsid containing two identical copies of the RNA genome and several enzymes, including three virally encoded enzymes that are targets for anti-HIV therapy: reverse transcriptase, integrase, and protease (**figure 37.11*a***). Among the 10 virus-specific proteins, the gp120 envelope protein facilitates attachment to a susceptible host cell.

To infect a host cell, the virus's gp120 envelope protein (figure 37.11*b*) binds the CD4 glycoprotein in the plasma membranes of $CD4^+$ T cells, macrophages, dendritic cells, and monocytes. The virus requires a coreceptor in addition to CD4. Strains that infect macrophages (macrophage-tropic strains), which predominate early in the disease, require CCR5 (a chemokine coreceptor) as well as CD4. A second chemokine coreceptor, called CXCR-4, is used by T-cell–tropic strains active at later stages of infection. These strains induce the formation of syncytia (multinucleated masses of fused cells). Rare individuals with two defective copies of the CCR5 gene do not develop AIDS because the virus cannot infect their macrophages and T cells. People with one good copy of the CCR5 gene are infected but have an altered disease presentation. ◀ *Retroviruses: positive-strand viruses that use reverse transcriptase in their life cycles (section 25.7); T cells are critical for immune function (section 32.5)*

HIV enters host cells by endocytosis or membrane fusion. When gaining entrance by endocytosis, the viral envelope fuses with the endosome membrane, releasing the nucleocapsid into the cytosol (**figure 37.12*a*; *also see figure 25.25***). When HIV uses membrane fusion to enter host cells, a conformational change in gp120 permits another protein, gp41, to extend, coil, and insert into the host cellular membrane. This process pulls the viral and cellular membranes together, allowing them to fuse.

Once inside the host cell, HIV does not immediately translate its positive-strand RNA genome like other viruses. Instead, it uses its reverse transcriptase (RT) to generate a dsDNA copy of its genome, which is integrated into the host chromosome prior to gene expression. In the first step of this process, RT uses the viral RNA genome as template to generate an RNA:DNA heteroduplex. The RNA template is degraded by another RT component, ribonuclease H. The newly synthesized DNA strand now serves as template for the RT to form a double-stranded DNA copy of the original RNA genome. The nucleocapsid, now containing a double-stranded DNA with the integrase enzyme attached to it, moves into the nucleus, where evidence indicates final uncoating occurs. Integrase catalyzes the integration of viral DNA into the cell's DNA; once integrated the viral genome is known as a *provirus*. The provirus can remain latent, giving no clinical sign of its presence. Alternatively the provirus can force the cell to synthesize viral mRNA (figure 37.12*b*). Some of the RNA is translated by host-cell ribosomes to produce a long polypeptide that is cleaved into individual proteins by a viral protease. These proteins complete the assembly of new virions that bud from the infected host cell (figure 37.12*c*). Eventually the host cell dies.

Once a person becomes infected with HIV, the course of disease can vary greatly. The acute infection stage occurs 2 to 8 weeks after HIV infection. About 70% of individuals in this stage experience a brief illness, sometimes called acute *retroviral syndrome* (ARS), with symptoms that may include fever, malaise, headache, macular (small, red, spotty) rash, weight loss, and

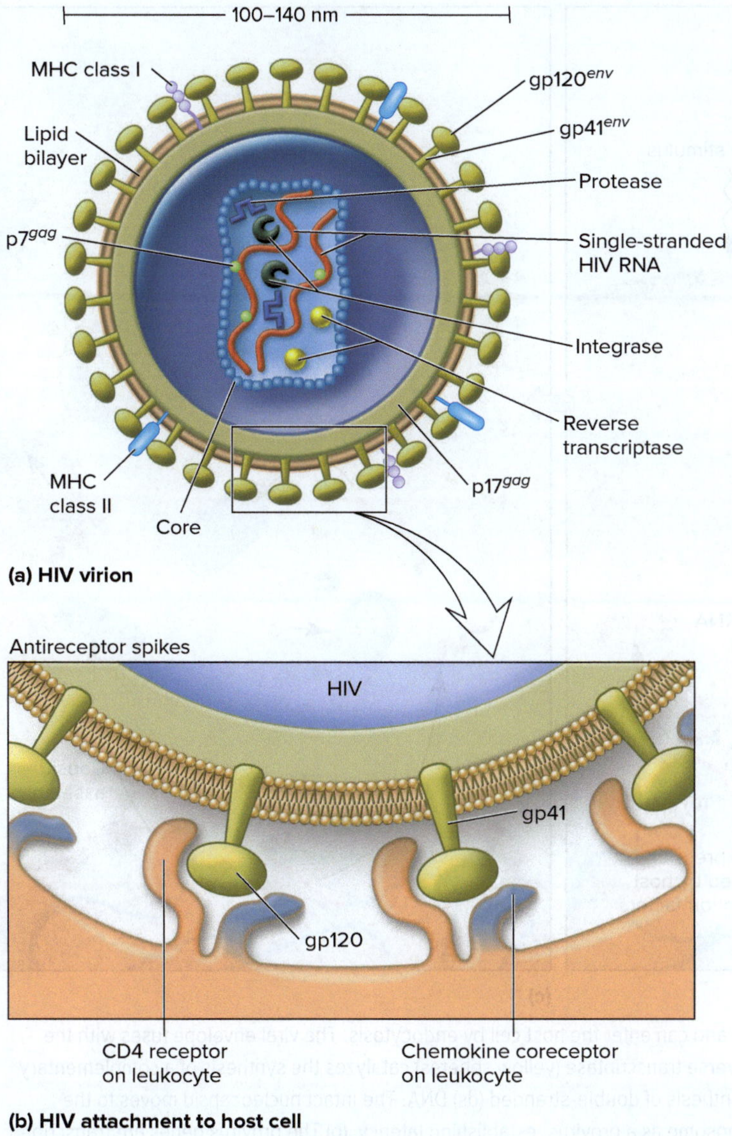

Figure 37.11 Schematic Diagram of the HIV Virion. (a) The HIV virion is enveloped and contains spikes formed by the two major viral envelope proteins, gp120 and gp41 (gp stands for glycoprotein and the number refers to the mass of the protein, in kilodaltons). The HIV lipid bilayer is also studded with various host proteins, including class I and class II major histocompatibility complex molecules, acquired during virion budding. The HIV capsid contains four nucleocapsid proteins (p24, p17, p9, p7), each of which is proteolytically cleaved from a 53 kDa Gag precursor by the HIV protease. The phosphorylated p24 polypeptide forms the main component of the inner shell of the nucleocapsid, whereas the p17 protein is associated with the inner surface of the lipid bilayer and stabilizes the exterior and interior components of the virion. The p7 protein binds directly to genomic RNA and together with p9 forms the core. The retroviral core contains two copies of the single-stranded HIV genomic RNA to which preformed viral reverse transcriptase, integrase, and protease enzymes are attached. (b) The snug attachment of HIV glycoprotein molecules (gp41 and gp120) to their specific receptors on a human cell membrane. These receptors are CD4 and chemokine coreceptors CCR5 and CXCR-4 (sometimes called fusin) that permit docking with the host cell and fusion with the cell membrane.

lymph node enlargement (lymphadenopathy). During this stage, the virus multiplies rapidly and disseminates to lymphoid tissues throughout the body, until an acquired immune response (antibodies and cytotoxic T cells) is generated to reduce virus multiplication (**figure 37.13**). Blood levels of HIV increase rapidly to 10^5 to 10^6 copies of viral RNA per milliliter of plasma. It is believed that the extent to which the immune response is able to control this initial burst of virus multiplication may determine the amount of time required for progression to the next clinical stage.

In untreated individuals, the asymptomatic stage of HIV infection may last from 6 months to 10 years (or longer in some individuals). During this time, the levels of detectable HIV in the blood decrease because the virus continues to replicate, particularly in lymphoid tissues. Even before any decline in $CD4^+$ T cells (T-helper cells) can be detected, the virus may compromise certain immune functions (e.g., memory cell responses to common antigens such as tetanus toxoid or *Candida albicans*).

During the chronic symptomatic stage, which can last for months to years, virus multiplication continues and the number of $CD4^+$ T cells begins to significantly decrease. Because T-helper cells are critically important in the generation of acquired immunity, individuals at this stage develop a variety of symptoms including fever, weight loss, malaise, fatigue, anorexia, abdominal pain, diarrhea, headaches, and lymphadenopathy. Individuals are diagnosed with AIDS when their $CD4^+$ T-cell count drops below 200 cells/mm^3. At this stage, patients develop opportunistic infections (figure 37.13*b*).

In addition to its devastating effects on the immune system, HIV infection can also lead to disease of the central nervous system when virus-infected macrophages cross the blood-brain barrier. The classical symptoms of central nervous system disease include headaches, fevers, subtle cognitive changes, atypical reflexes, and ataxia (irregularity of muscular action). Dementia and severe sensory and motor changes characterize more advanced stages of the disease. Another potential complication of HIV infection is cancer. Individuals infected with HIV-1 have an increased risk of three types of tumors: (1) Kaposi's sarcoma, caused by activation of latent *Human gammaherpesvirus 8* (**figure 37.14**); (2) carcinomas of the mouth and rectum; and (3) B-cell lymphomas or lymphoproliferative disorders. Severe immune suppression enables other tumor-causing agents to initiate these cancers.

Laboratory diagnosis of HIV infection is most commonly accomplished through the detection of specific anti-HIV antibodies and antigens in the blood. Once a patient is confirmed to be HIV positive, the viral load (number of viral copies per milliliter of blood) is determined by real-time polymerase chain reaction (RT-PCR). Viral load and the concentration of $CD4^+$ cells, as determined by flow cytometry (*see figure 26.4*), are used to monitor the patient's response to therapy. ◀ *Polymerase chain reaction amplifies targeted DNA (section 17.2)*

Prevention and control of HIV are achieved primarily through education and screening individuals for HIV infection. Understanding risk factors and practicing strategies to reduce risk are essential in the fight against HIV. There is no cure for

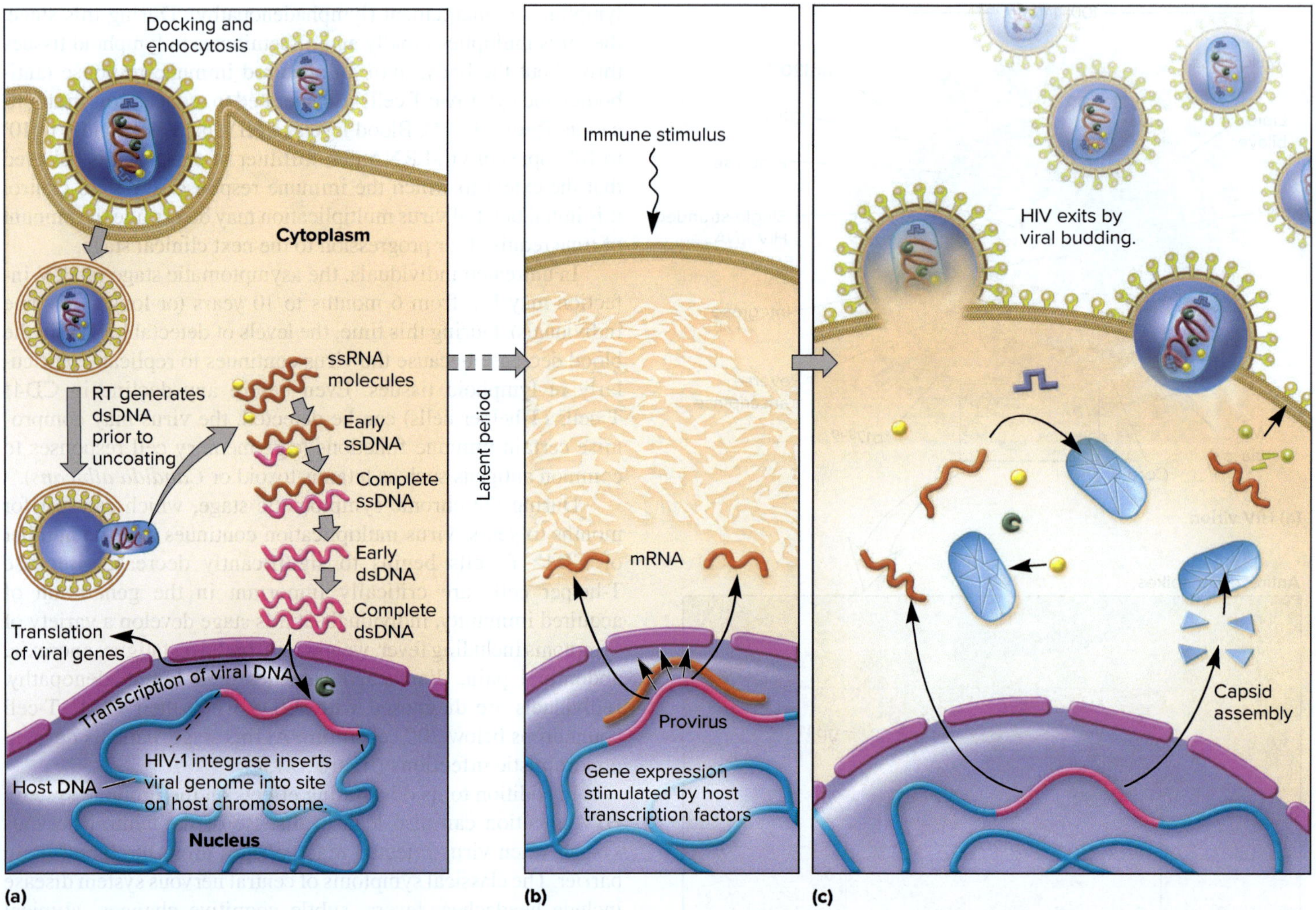

Figure 37.12 HIV Life Cycle. (a) The HIV virion binds to the CD4+ host cell and can enter the host cell by endocytosis. The viral envelope fuses with the endocytic compartment releasing the HIV nucleocapsid into the cytoplasm. Reverse transcriptase (yellow spheres) catalyzes the synthesis of a complementary strand of DNA from the RNA viral genome. This DNA serves as a template for synthesis of double-stranded (ds) DNA. The intact nucleocapsid moves to the nucleus, where uncoating occurs and the dsDNA is inserted into the host chromosome as a provirus, establishing latency. (b) The provirus genes are transcribed. (c) Viral mRNA is translated into virion components (capsid, reverse transcriptase, spike proteins), and the virion is assembled. Maturation is completed as the virus particles bud from the host cell, taking host membrane as their envelope. Note that steps show activity of one strand of viral DNA.

MICRO INQUIRY *What is the function of integrase and how is its activity related to latency?*

HIV infection. The virus cannot be eradicated from long-lived resting memory CD4+ T cells. The antiviral drugs currently approved for use in HIV disease are described in detail in chapter 9. The most successful antiretroviral therapy (ART) is to use a combination of three antiviral agents simultaneously, which prevents the development of drug resistance. Importantly, if a patient stops taking antiviral medications, the reservoir of provirus within quiescent CD4+ T cells will become activated and the viral load will rebound to high levels. The goal of treatment is to reduce viral load to undetectable levels, which eliminates disease burden on the patient and prevents transmission. Providing ART to all individuals infected with HIV could eliminate viral spread, as patients with undetectable levels of virus cannot transmit it. This key strategy is referred to as *Undetectable = Untransmittable* (U=U) or *Treatment as Prevention* (TasP). With appropriate antiviral treatment, HIV is undetectable in an individual within 1 to 6 months. If the virus is still undetectable six months after the first negative HIV test, and remains so, the individual cannot transmit it, as long as they continue with their ART regime.

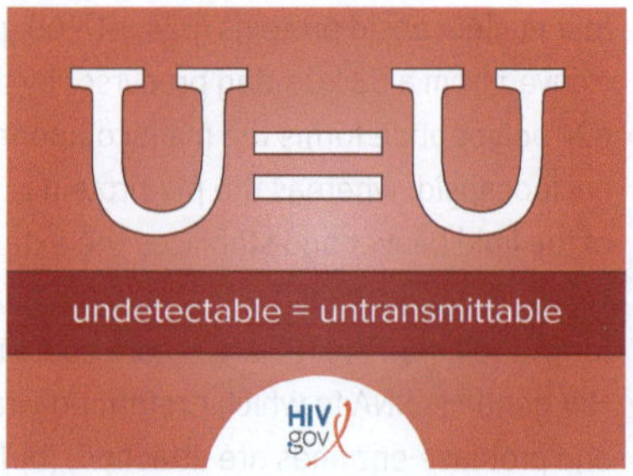

Source: hiv.gov

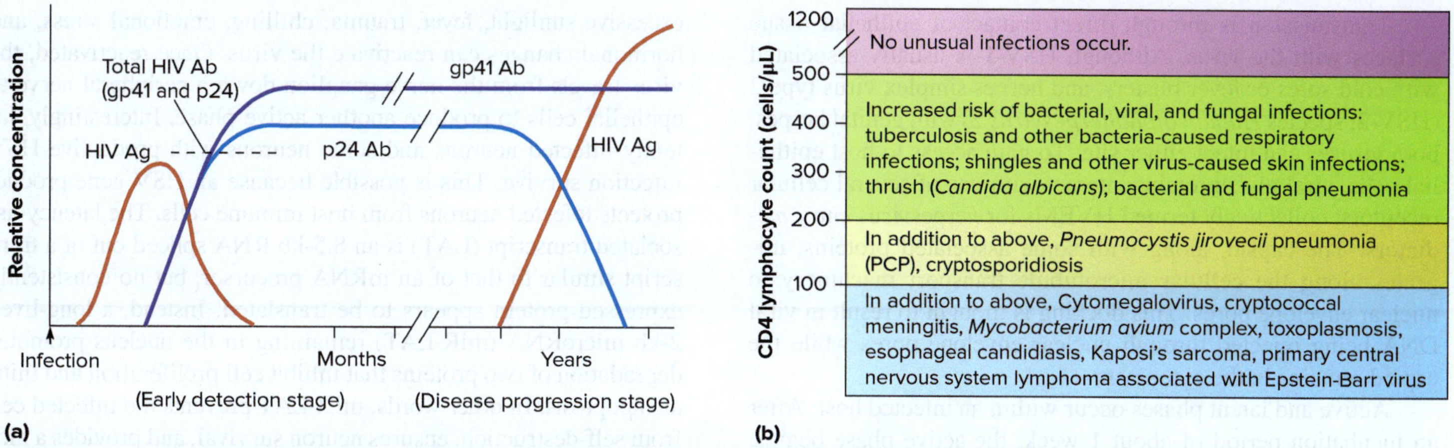

Figure 37.13 The Typical Immunological Pattern in an HIV-1 Infection. (a) HIV-1 antigen (HIV Ag) is detectable as early as 2 weeks after infection and typically occurs before IgG antibodies against gp41 and p24 develop. Seroconversion occurs when HIV-1 antibodies have risen to detectable levels. This usually takes place several weeks to months after infection. The period between HIV-1 infection and seroconversion often is associated with an acute flulike illness. HIV-1 antigen usually disappears following seroconversion but reappears in later stages of the disease. The reappearance of antigen usually indicates impending clinical deterioration. HIV-2 progresses more slowly and may not progress to AIDS. (b) As CD4 lymphocytes are destroyed by HIV, their declining numbers result in increased host susceptibility to specific infections.

Other mitigation strategies are also effective. Blood and blood products are screened for HIV prior to use. Those using intravenous drugs or medications should be educated about the risks of sharing needles and syringes; needle exchange programs reduce some of these risks. Transmission of the virus by pregnant individuals who are HIV-positive to their infant can be prevented by prenatal ART and the use of infant formula rather than breast-feeding (if that is a viable option). Sexual transmission of the virus can be greatly limited by condom use as barrier protection from blood and body fluids. Male circumcision can lower virus acquisition by 60 to 70%; voluntary medical male circumcision (VMMC) is endorsed by WHO and UNAIDS for inclusion as an HIV-transmission prevention strategy for males in areas with high prevalence of heterosexual transmission. Individuals at high risk of HIV exposure can take preexposure prophylaxis (PrEP), a combination of two antiviral medications that dramatically reduce the probability of transmission. Recent PrEP trials using biannual injections of the drug lencapavir, which interferes with HIV uncoating and capsid formation, show exciting promise in the prevention of HIV transmission; trial data in 2024 indicated an almost 100% prevention of transmission.

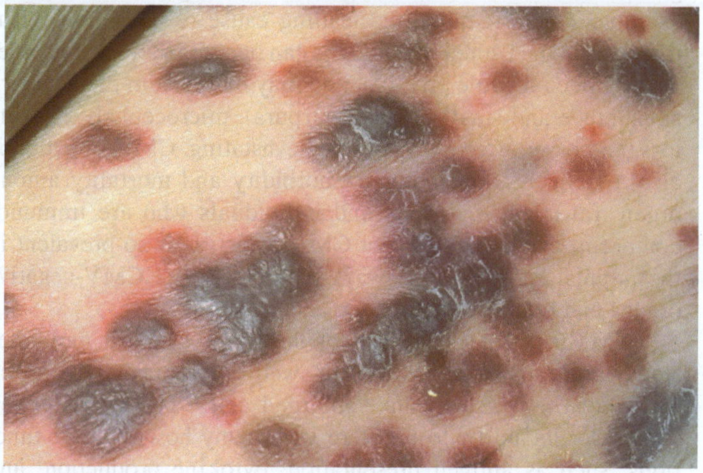

Figure 37.14 Kaposi's Sarcoma on the Arm of a Patient with AIDS. The flat purple tumors can occur in almost any tissue and are a result of activation of latent Human *gammaherpesvirus* 8.

National Institutes of Health/Stocktrek Images/Getty Images

The development of a vaccine for AIDS has been a long-sought goal. Among the many problems encountered in developing an HIV vaccine is the fact that the viral envelope proteins continually change their antigenic properties; HIV has a high mutation rate because reverse transcriptase lacks proofreading capabilities. Despite significant progress in treating individuals infected with HIV, creating a successful vaccine remains a challenge.

Cold Sores (Oral Herpes)

A cold sore, also known as a fever blister, is a fluid-filled lesion typically found on the lips or inside the nostrils. Cold sores cause pain, itching, or a burning sensation because they occur on or near nerve endings. They are caused by infection with the double-stranded DNA virus species *Human alphaherpesvirus 1,* more commonly referred to as herpes simplex virus type 1 (HSV-1). Like all members of the Herpesviridae, HSV-1 has an icosahedral capsid and is enveloped (*see table 25.3*). The genome encodes approximately 100 proteins.

Transmission is through direct contact of epithelial tissue surfaces with the virus. Although HSV-1 is usually associated with cold sores or fever blisters, and herpes simplex virus type 2 (HSV-2; species *Human alphaherpesvirus 2*) with genital herpes, both viruses can infect either site. To gain access to host epithelial cells, viral glycoproteins interact with one of several cellular receptors, collectively termed HVEMs for *h*erpes*v*irus *e*ntry *m*ediators. The capsid, along with some associated proteins, migrates along the cellular microtubule transport machinery to nuclear envelope pores. This docking is thought to result in viral DNA being injected through nuclear envelope pores while the capsid remains in the cytoplasm.

Active and latent phases occur within an infected host. After an incubation period of about 1 week, the active phase begins. During this phase, the virus multiplies explosively; between 50,000 and 200,000 new virions are produced from each infected cell. During this replication cycle, herpesvirus inhibits its host cell's metabolism and degrades host DNA, inducing apoptosis. As a result, the cell dies, releasing viral progeny to infect other cells. Such an active infection may be symptom-free, or painful blisters may arise as a result of a local inflammatory response to cellular lysis. Blisters contain fluid and infectious virions. Importantly, apparently healthy people can transmit HSV to other hosts. Blisters involving the epidermis and surface mucous membranes of the lips, mouth, and gums are referred to as herpes labialis (**figure 37.15**). Primary and recurring HSV infections also may occur in the eyes, causing **herpetic keratitis** (inflammation of the cornea)—the most common cause of infectious blindness in developed nations.

Blisters usually heal within 1 week, and then a very interesting virulence factor initiates the latent survival of HSV-1 in infected neurons. As the active infection is curtailed by a vigorous cellular immune response, viruses released from infected oral epithelia travel to the trigeminal nerve ganglion (cluster of nerve cells in the face and jaw). There the virus will remain in a latent state for the lifetime of the infected host. However, stressful stimuli such as

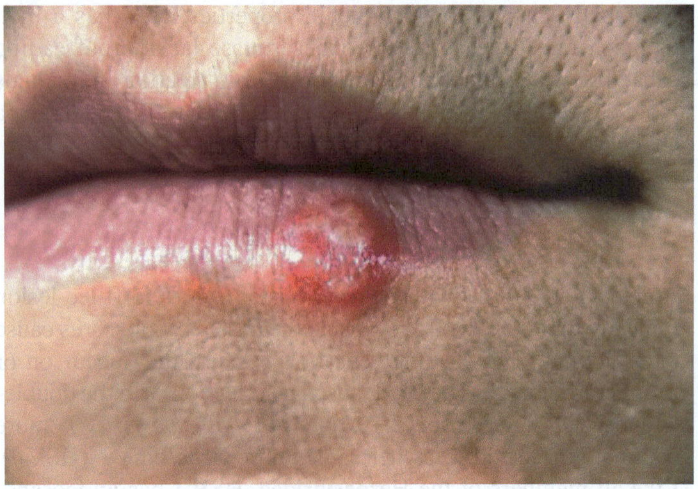

Figure 37.15 Cold Sores. Herpes simplex fever blisters on the lip, caused by herpes simplex virus-1.

Centers for Disease Control and Prevention

excessive sunlight, fever, trauma, chilling, emotional stress, and hormonal changes can reactivate the virus. Once reactivated, the virus travels from the nerve ganglion down a peripheral nerve to epithelial cells to produce another active phase. Interestingly, latently infected neurons and some neurons with productive HSV infection survive. This is possible because an HSV gene product protects infected neurons from host immune cells. The latency-associated transcript (LAT) is an 8.5-kb RNA spliced out of a transcript similar to that of an mRNA precursor, but no consistently expressed protein appears to be translated. Instead, a long-lived 2-kb microRNA (miR-LAT) remaining in the nucleus promotes degradation of two proteins that inhibit cell proliferation and initiate apoptosis. In other words, miR-LAT prevents the infected cell from self-destruction, ensures neuron survival, and provides a hiding place for the virus.

The drugs acyclovir, valacyclovir, and famciclovir are effective against cold sores. Idoxuridine and trifluridine are used to treat herpes infections of the eye. By adulthood, 70 to 90% of people in the United States have been infected and have anti-HSV-1 antibodies. Diagnosis of HSV-1 infection is typically by clinical evaluation of the lesions. However, in rare cases, HSV-1 can cause encephalitis. When this is suspected, the virus is typically detected by nucleic acid amplification from cerebrospinal fluid.

Cytomegalovirus Inclusion Disease

Cytomegalovirus inclusion disease is caused by cytomegalovirus (CMV). As a member of the family Herpesviridae (species *Human betaherpesvirus 5; see table 25.3*), CMV is an enveloped, double-stranded DNA virus with an icosahedral capsid. Most people become infected with this virus at some time during their life; in the United States, about 80% of individuals older than 35 years have been exposed and carry a lifelong infection. Although most CMV infections are asymptomatic, certain groups are at risk for serious illness and long-term effects. For example, CMV remains the leading cause of congenital virus infection in the United States. This can occur if a female becomes newly infected while pregnant and may manifest in the newborn as jaundice and small bruises (purpura), microcephaly, hearing loss, and other effects that can be devastating. CMV infection is also a frequent contributor to morbidity and mortality among organ transplant recipients and individuals who are immunocompromised. Because latent CMV infection is so prevalent in the population, it is not always possible to use CMV-negative organs and blood products.

CMV can infect any cell of the body, where it multiplies slowly and causes the host cell to swell in size; hence the prefix *cytomegalo-*, which means "enlarged cell." The virus is well known for its ability to interfere with many host immune functions, such as antigen presentation, cytokine production, and natural killer cell activity. Infected cells contain unique **intranuclear inclusion bodies** and cytoplasmic inclusions (**figure 37.16**). In fatal cases, cell damage is seen in the gastrointestinal tract, lungs, liver, spleen, and kidneys. In less severe cases, acute cytomegalovirus disease is similar to infectious mononucleosis.

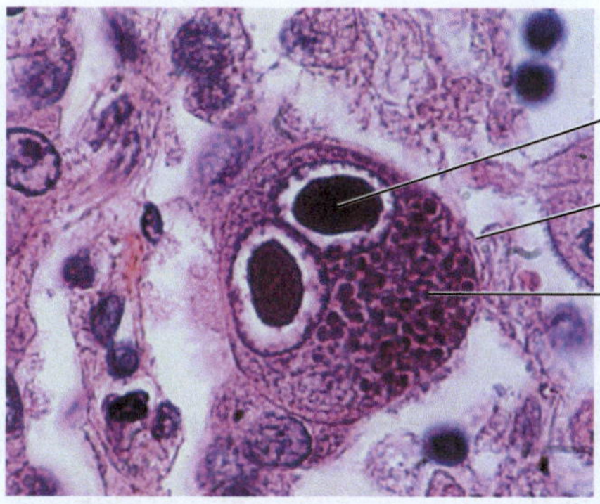

Figure 37.16 Cytomegalovirus Inclusion Bodies. Light micrograph of a giant cell in a lung section infected with cytomegalovirus (×480). Cells containing intranuclear inclusion bodies have a typical "owl-eyed" appearance because of the clear halo that surrounds the inclusion. Sometimes inclusions also are visible in the cytoplasm.

Dan Wiedbrauk, Ph.D/Warde Medical Laboratory Ann Arbor, Michigan

Laboratory diagnosis is by the detection of anti-CMV IgM and IgG in patient serum or amplification of CMV nucleic acid by PCR. The virus may also be cultured.

CMV has a worldwide distribution. The antivirals ganciclovir, valganciclovir, and foscarnet are used to treat high-risk patients. The nucleotide analogue cidofovir is used to treat CMV retinitis, which can occur in individuals who are immunosuppressed.

Genital Herpes

Genital herpes is a lifelong infection caused by herpes simplex virus, predominantly type 2 (HSV-2, species *Human alphaherpesvirus 2*) (**figure 37.17*a***). HSV-2 is usually sexually transmitted, although it can be transmitted nonsexually as well. Infection begins when the virus is introduced into a break in the skin or mucous membranes. The virus infects the epithelial cells of the external genitalia, the urethra, and the cervix (figure 37.17*b*). Rectal and pharyngeal (oral) herpes are also transmitted by sexual contact.

The details of HSV-2 infection are very similar to those of HSV-1 discussed in the section on cold sores. In the case of genital HSV-2, viruses retreat to nerve cells in the sacral plexus of the spinal cord, where they remain latent. Because viral genes are not expressed, the person who is infected is symptom-free. It should be noted that both primary infection and reactivation can occur without any symptoms, and apparently healthy people can transmit HSV-2 to their sexual partners. The virus can also be transmitted to newborns during vaginal delivery, leading to **congenital (neonatal) herpes.** Congenital herpes is a life-threatening infection in newborns, affecting between 5 and 30 of every 100,000 babies per year in the United States. It can involve skin, eyes, and mouth; internal organs; or the central nervous system. Neurological involvement may cause developmental delays, as well as blindness. As a result, any female who is pregnant who has active or new genital herpes should have a cesarean section instead of delivering vaginally. HSV-2 is also associated with a higher-than-normal rate of cervical cancer and miscarriages.

Diagnosis of HSV-2 infection is by screening of blood or serum for anti-HSV-2 antibodies, direct fluorescent antibody testing of tissue, and/or PCR. Although there is no cure for genital herpes, oral use of the antiviral drugs acyclovir, valacyclovir, or famciclovir has proven to be effective in ameliorating recurrent blister outbreaks. Topical acyclovir is more effective than famciclovir in reducing virus shedding.

In the United States, the incidence of genital herpes has increased during the past several decades. Currently about 16% of individuals aged 14 to 49 are infected with HSV-2, with roughly 576,000 new cases developing annually.

Human Betaherpesvirus 6 Infection

Human betaherpesvirus 6 (HHV-6) is the etiologic agent of **exanthem subitum** (Greek *exanthema,* rash), also called *roseola infantum,* which occurs in infants and toddlers. As a member of the family Herpesviridae, it is an enveloped virus with an icosahedral capsid and a core containing double-stranded DNA that establishes latency in host cells (*see table 25.3*). Exanthem subitum is a short-lived disease characterized by a high fever of 3 to 4 days' duration. When the temperature suddenly drops to normal, a macular rash appears on the trunk and spreads to other areas of the body. HHV-6 infects over 95% of the U.S. infant population, and most children are seropositive for HHV-6 by

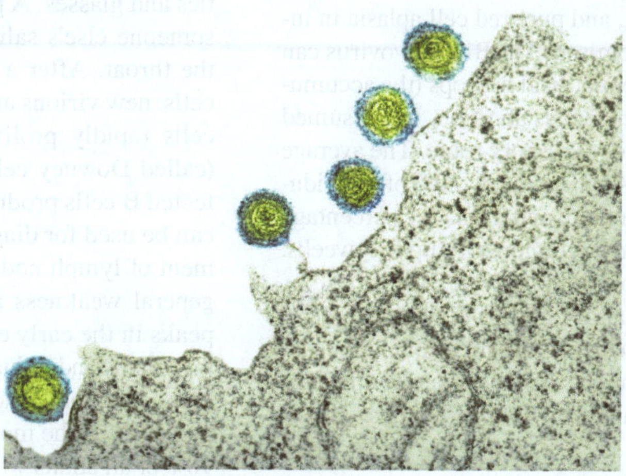
(a)

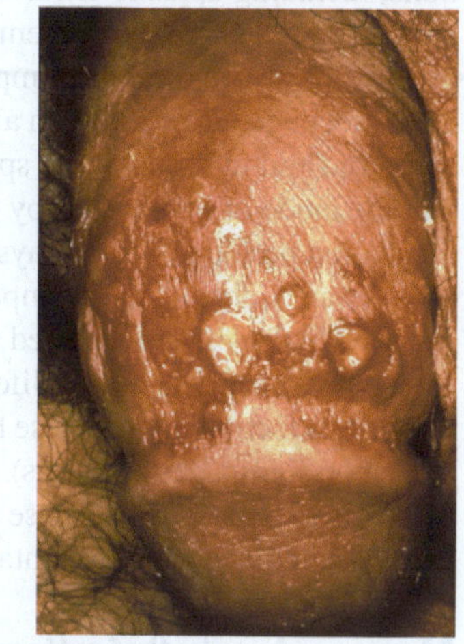
(b)

Figure 37.17 Genital Herpes. (a) Herpes simplex virus virions budding from an infected cell (colorized transmission electron micrograph). (b) Herpes vesicles on the penis. The vesicles contain fluid that is infectious.

(a) National Cancer Institute/Science Source; (b) Dr. N.J. Flumara and Dr. Gavin Hart/Centers for Disease Control and Prevention

MICRO INQUIRY *Where does HSV-2 reside during latency?*

three years of age. $CD4^+$ T cells are the main site of viral replication, whereas monocytes are in an infected, latent state. In adults, HHV-6 is commonly found in peripheral-blood mononuclear cells and saliva. Since the salivary glands are the major site of latent infection, transmission is probably by way of saliva.

Because HHV-6 produces latent infections, it occasionally reactivates in hosts who are immunocompromised, leading to pneumonitis. Diagnosis is by immunofluorescence, enzyme immunoassay, or PCR. There is no antiviral therapy or prevention.

Human Parvovirus B19 Infection

Since its serendipitous discovery in 1974, **human parvovirus B19** (family Parvoviridae, genus *Erythroparvovirus*) has emerged as a significant human pathogen. B19 virions are uniform, icosahedral, naked particles approximately 23 nm in diameter. Parvoviruses have a genome composed of one ssDNA molecule of only about 5,000 bases. Parvoviruses are among the simplest of the DNA viruses. The genome is so small in part because it uses overlapping genes to encode just a few proteins. The genome does not code for any enzymes and the virus uses host-cell enzymes for all biosynthetic processes. Thus viral DNA can only be replicated in the nucleus during the S phase of the cell cycle, when the host cell replicates its own DNA. Because the viral genome is single stranded and linear, host DNA polymerase must be tricked into copying it. By using a self-complementary sequence at the ends of the viral DNA, the parvovirus genome folds back on itself to form a primer for replication (*see figure 25.13*). This is recognized by the host DNA polymerase and DNA replication ensues. ◀ *Parvoviruses (section 25.3)*

Parvovirus B19 infection causes **erythema infectiosum** (**figure 37.18**) (also called **fifth disease**) in children. Because the virus can infect bone marrow and cause a temporary cessation of red blood cell development, it can cause more serious complications, including aplastic crisis in persons with sickle cell disease or autoimmune hemolytic anemia, and pure red cell aplasia in individuals who are immunocompromised. The B19 parvovirus can also infect a fetus, resulting in anemia, fetal hydrops (the accumulation of fluid in tissues), and spontaneous abortion. It is assumed that the mode of infection is by the respiratory route. The average incubation period is 4 to 14 days. Approximately 20% of individuals who are infected are asymptomatic, and a smaller percentage of individuals who are infected have symptoms for up to 3 weeks. Infection typically results in lifelong immunity to B19.

B19 infection in otherwise healthy children is made clinically (based on signs and symptoms). In high-risk patients, it is detected by serology and/or PCR. These patients may be treated with commercial immunoglobulins containing anti-B19 antibodies.

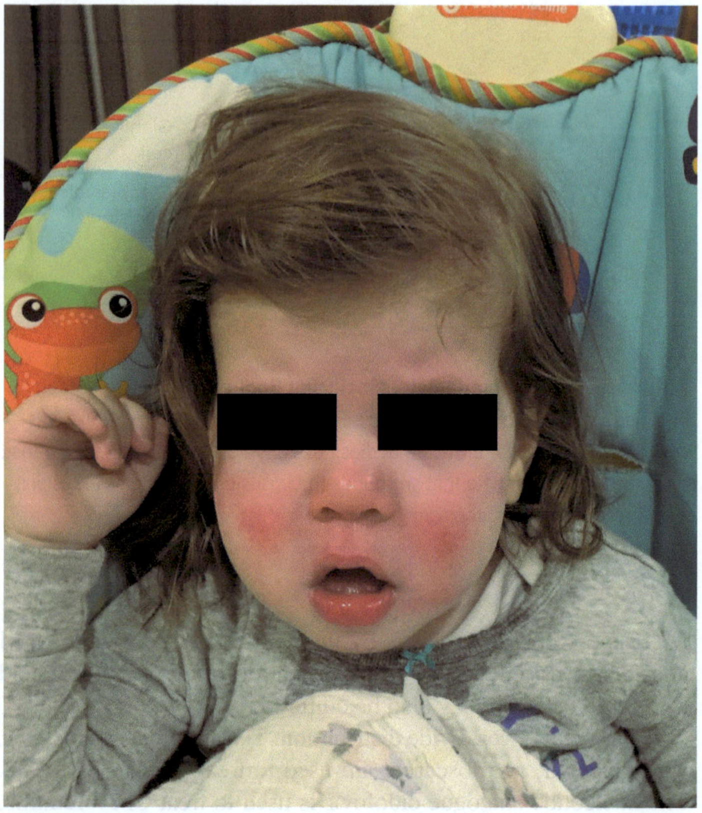

Figure 37.18 Erythema Infectiosum. The facial rash of fifth disease, sometimes called "slapped face syndrome."

Devora Lichtman

Mononucleosis (Infectious)

Infectious mononucleosis is caused by Epstein-Barr virus (EBV; species *Human gammaherpesvirus 4*), a member of the family Herpesviridae (*see table 25.3*). EBV exhibits the characteristic herpesvirus morphology with an enveloped, icosahedral capsid. Like all members of Herpesviridae, its dsDNA genome exists as a linear form in the mature virion and a circular episomal form in latently infected host cells. Because EBV occurs in oropharyngeal secretions, it can be spread by mouth-to-mouth contact (hence the term kissing disease) or shared drinking bottles and glasses. A person becomes infected when the virus from someone else's saliva makes its way into epithelial cells lining the throat. After a brief bout of multiplication in the epithelial cells, new virions are shed and infect memory B cells. Infected B cells rapidly proliferate and take on an atypical appearance (called Downey cells) (**figure 37.19**). In over 90% of cases, infected B cells produce heterophile (multispecific) antibodies that can be used for diagnosis. The disease is manifested by enlargement of lymph nodes and spleen, sore throat, headache, nausea, general weakness and tiredness, and a mild fever that usually peaks in the early evening.

Many individuals acquire immunity to the disease following early childhood infection, which is typically asymptomatic. EBV may well be the most common virus in humans, as it infects 80 to 90% of all adults worldwide. The peak incidence of mononucleosis diagnoses occurs in people fifteen to twenty-five years of age. Treatment is largely supportive and includes plenty of rest. Diagnosis is usually confirmed by demonstration of an increase in circulating mononuclear cells, along with a serological test for heterophile antibodies, specific antiviral antibodies, or identification of viral nucleic acid. Several rapid tests are available.

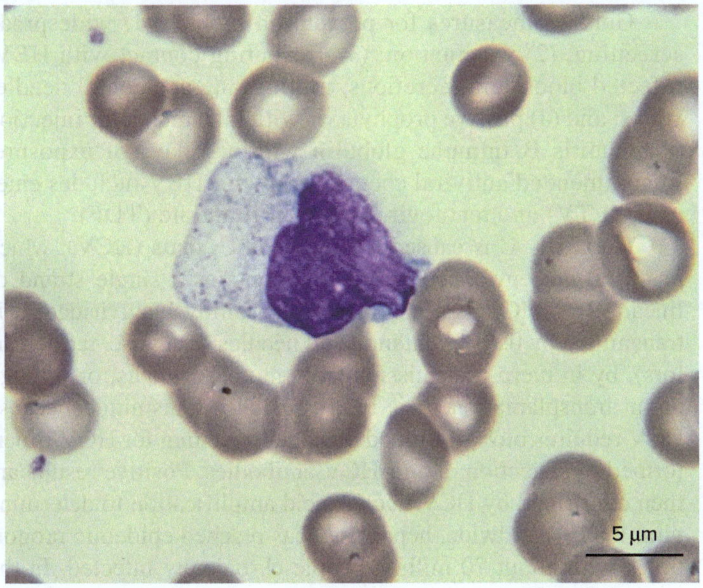

Figure 37.19 Evidence of Epstein-Barr Virus Infection in the Blood Smear of a Patient with Infectious Mononucleosis. Note the abnormally shaped lymphocyte (called a Downey cell) and indented nucleus.

Nature's Faces/Science Source

EBV infections are associated with the cancers Burkitt lymphoma in tropical Africa and nasopharyngeal carcinoma in Southeast Asia, in East and North Africa, and in Inuit populations. Although these associations are well documented, the pathophysiological mechanisms remain unclear.

Viral Hepatitides

Hepatitis (pl., hepatitides; Greek *hepaticus,* liver) is a general term for infection and inflammation of liver cells. Currently 11 viruses are recognized as causing hepatitis. Two are herpesviruses (CMV and EBV, which infect a variety of other cells) and nine are hepatotropic viruses that specifically target liver cells, or hepatocytes. EBV and CMV cause mononucleosis syndromes that can sometimes lead to mild, self-resolving forms of hepatitis with no permanent liver damage. Only five hepatotropic viruses are well characterized (**table 37.1**). Hepatitis A and hepatitis E viruses are transmitted by fecal-oral contamination and are discussed in the section on food-borne and waterborne diseases (section 37.4). The other major types include hepatitis B, hepatitis C, and hepatitis D viruses. Hepatitis G virus, now designated HPgV-1, does not cause hepatitis.

Hepatitis B is caused by **hepatitis B virus** (HBV), an enveloped virus with a partially double-stranded circular DNA genome of complex structure. HBV is classified as an *Orthohepadnavirus* within the family Hepadnaviridae. HBV is a reverse-transcribing DNA virus, so named because it uses a reverse transcriptase enzyme to replicate its genome. *Reverse transcribing DNA viruses (section 25.8)*

Serum from individuals infected with HBV contains three distinct antigenic particles, only one of which is the mature virion, sometimes also called a **Dane particle.** The two other antigenic particles, a smaller spherical and larger filamentous one, serve as decoys for host antibodies (**figure 37.20**). The viral genome is 3.2 kb in length, consisting of four partially overlapping, open-reading frames that encode viral proteins. Viral multiplication takes place predominantly in hepatocytes. The HBV envelope is coated with hepatitis B surface antigen (HBsAg). The inner nucleocapsid core antigen (HBcAg) encloses a single molecule of double-stranded HBV DNA and polymerase. HBsAg in body fluids is (1) an indicator of active hepatitis B infection, (2) used in the large-scale screening of blood for HBV, and (3) the basis for the HBV vaccine, which was the first vaccine for human use developed by recombinant DNA technology. Diagnosis of HBV is made by serology, which can be complex. Most straightforward is the detection of HBsAg in unimmunized individuals; the presence of IgM or IgG anti-HBcAg antibody helps determine if the infection is acute or chronic. If a patient is

Table 37.1 Characteristics of Hepatitides Caused by Hepatotropic Viruses

Disease	Genome	Classification	Transmission	Characteristic	Prevention
Hepatitis A	RNA	Picornaviridae, Hepatovirus	Fecal-oral	Subclinical, acute infection	Killed HAV vaccine
Hepatitis B	DNA	Hepadnaviridae, Orthohepadnavirus	Blood, needles, body secretions, placenta, sexually	Subclinical, acute, or chronic infection; cirrhosis; primary hepatocarcinoma	Recombinant HBV vaccine
Hepatitis C	RNA	Flaviviridae, Hepacivirus	Blood, sexually	Subclinical, acute, or chronic infection; primary hepatocarcinoma	Routine screening of blood
Hepatitis D	RNA	Satellite	Blood, sexually	Coinfection with HBV required for satellite replication	HBV vaccine
Hepatitis E	RNA	Hepeviridae, Orthohepevirus	Fecal-oral	Subclinical, acute infection (but high mortality in females who are pregnant)	Improve sanitary conditions

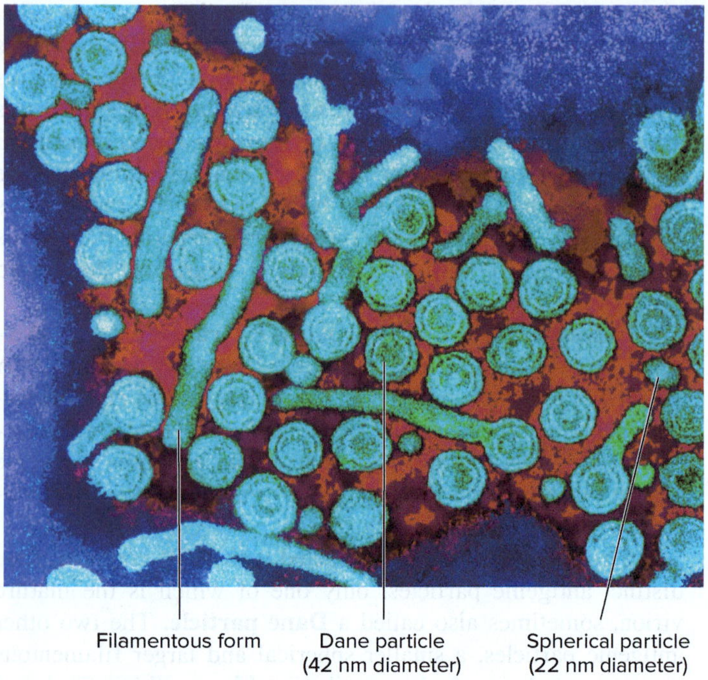

Figure 37.20 Hepatitis B Virus in Serum. Colorized electron micrograph showing the three distinct types of hepatitis B antigenic particles. The spherical particles and filamentous forms lack an internal structure, and only two of the three characteristic viral envelope proteins appear on their surface. Dane particles are the complete, infectious virion.

Dr. Linda M. Stannard, University of Cape Town/Science Source

serologically positive, viral load is determined using PCR to measure HBV nucleic acid.

Hepatitis B virus is transmitted through blood, body-fluid–contaminated equipment (including shared intravenous needles), and sexual contact. The virus can be transmitted vertically, usually during delivery when the baby contacts the mother's vaginal secretions. It is estimated that 296 million persons worldwide are chronically infected with HBV; however, only about 10% know their status. The number of children becoming infected has dramatically declined thanks to vaccination campaigns. In the United States, there are currently between 880,000 and 1.89 million confirmed individuals who are chronically infected. Worldwide, about 1.1 million persons die yearly from HBV-related complications, including liver cancer.

The clinical signs of hepatitis B vary widely. About 50 to 70% of cases are asymptomatic. For others, fever, loss of appetite, abdominal discomfort, nausea, fatigue, and other symptoms gradually appear following an incubation period of 1 to 3 months. The virus causes liver tissue degeneration and the release of liver-associated enzymes (transaminases) into the bloodstream. This is followed by jaundice, caused by the accumulation of bilirubin (a breakdown product of hemoglobin) in the skin and other tissues with a resulting yellow appearance. Chronic hepatitis B infection may also cause the development of hepatocellular carcinoma (liver cancer).

General measures for prevention include (1) widespread screening; (2) vaccination; (3) preventing contact with HBV-infected blood and secretions, minimizing accidental needlesticks; and (4) passive prophylaxis with intramuscular injection of hepatitis B immune globulin within 7 days of exposure. Recommended antiviral chemotherapy for HBV includes entecavir (ETV) and tenofovir disoproxil fumarate (TDF).

Hepatitis C is caused by **hepatitis C virus** (HCV), which has a 50-nm enveloped virion that contains a single strand of linear RNA. HCV is a member of the family Flaviviridae. It is transmitted by use of contaminated needles (drug use, tattoo parlors), by in utero transmission from mother to fetus, or through organ transplantation. It is rarely sexually transmitted because HCV requires blood-to-blood contact. Screening for HCV is performed by detection of anti-HCV antibodies. Positive results are then confirmed by HCV nucleic acid amplification to determine viral load. Worldwide, hepatitis C has reached epidemic proportions, with about 70 million people chronically infected. In the United States, HCV is the most common chronic bloodborne infection; it is estimated that between 2 and 2.9 million persons are chronically infected and approximately 17,000 new acute cases occur annually. Like HBV, HCV infection can cause hepatocellular carcinoma.

There are eight HCV genotypes, with genotype 1 the most common, followed by genotype 3. Treatment is with direct-acting antiviral (DAA) medications, which include the drug combination of sofosbuvir (a nucleotide analogue polymerase inhibitor) and velpatasvir (an HCV replication inhibitor). A 12-week treatment with this drug combination results in complete cure of the virus from the host.

Hepatitis delta virus (HDV) was discovered in 1977 and the disease hepatitis D was designated. There are eight known genotypes of HDV: Genotype 1 has a worldwide distribution; genotypes 2 and 4 exist in East Asia; genotype 3 is found in South America; and genotypes 5 through 8 are in Africa. All genotypes can cause acute as well as chronic liver disease. HDV is a satellite virus that is dependent on hepatitis B virus to provide the surface protein (HBsAg) for its own replication. HDV's dependence on HBV means that HDV only replicates in liver cells coinfected with actively replicating HBV. HDV is spread only to persons who are already infected with HBV (superinfection) or when HBV and the satellite are transmitted together (coinfection). The negative-strand RNA of HDV is smaller than the RNA of the smallest picornaviruses, and its circular conformation differs from the linear structure typical of animal negative-strand RNA viruses. ◀ *Section 6.6. Viroids and satellites: nucleic-acid-based subviral agents*

The primary laboratory tools for diagnosis of an HDV infection are serological tests for anti-HDV antibodies. Treatment of patients mirrors that of those infected with HBV. About 5% of all those with hepatitis B are coinfected with HDV. Prevention and control involve the widespread use of the hepatitis B vaccine. ◀ *Viroids and satellites: nucleic acid–based subviral agents (section 6.6)*

Warts

Warts, or verrucae (Latin *verruca,* wart), are horny projections on the skin caused by human papillomaviruses. Human papillomavirus (HPV) is the name given to a group of DNA viruses that includes more than 100 different strains, some of which are oncogenic (cancer-causing) (**figure 37.21*a***). Strains differ in the types of epithelium they infect; some infect cutaneous sites, whereas others infect mucous membranes. More than 30 of these viruses are sexually transmitted and cause genital warts—soft, pink cauliflower-like growths that occur on genitalia and around the rectum (figure 37.21*b,c*). Genital warts are the most common sexually transmitted infection.

Papillomaviruses are placed in the family Papillomaviridae. These viruses have nonenveloped icosahedral capsids with a double-stranded, supercoiled, circular DNA genome. Warts occur principally in children and young adults and are limited to the skin and mucous membranes. The viruses are spread between people by direct contact; autoinoculation occurs through scratching. Four major kinds of warts are **plantar warts, verrucae vulgaris, flat** or **plane warts,** and **anogenital condylomata (genital warts).** Treatment includes physical destruction of the wart by electrosurgery, cryosurgery with liquid nitrogen or solid CO_2, laser fulguration (drying), direct application of the drug podophyllum to the wart, or injection of IFN-α.

Once HPV enters the body, the incubation period is 1 to 6 months. Infection with specific types of genital HPV plays a major role in the pathogenesis of various cancers. HPV strains are designated by numbers and can be divided into high-risk (oncogenic) and low-risk (nononcogenic) strains. At least eight distinct genotypes produce benign epithelial tumors that vary in respect to their location, clinical appearance, and histopathologic features. Four HPV strains are responsible for 70% of cervical cancers (HPV 16 and 18) and 90% of genital warts (HPV 6 and 11). HPV 16 and 18 have been found in association with invasive cancers of the cervix, vulva, penis, or anus (and other sites, including the mouth, throat, and lung). Other high-risk strains cause benign or low-grade cervical cell changes and genital warts but are rarely, if ever, associated with invasive cancer.

At least 42 million Americans are currently infected with HPV and roughly 14 million become newly infected annually. HPV is typically transmitted through direct contact, especially

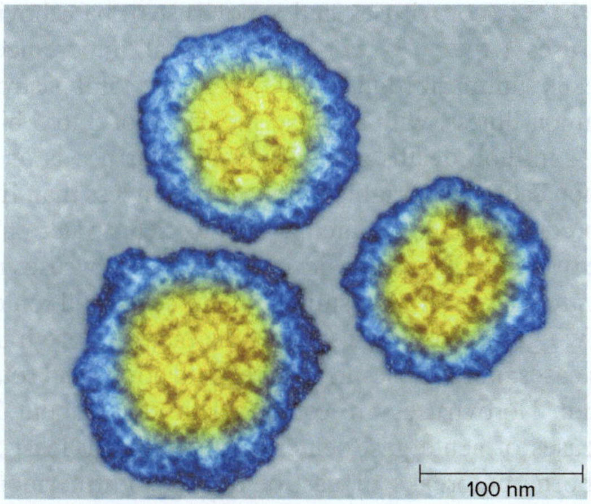

(a)

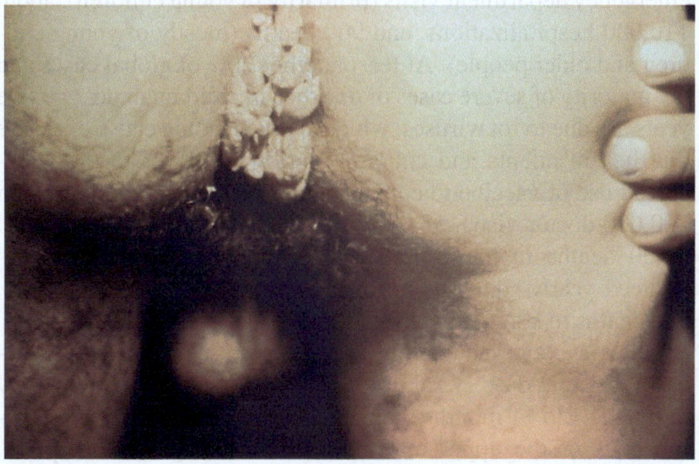

(c)

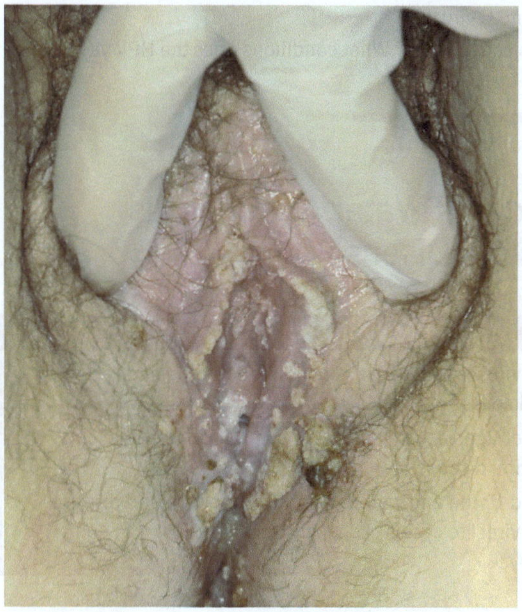

(b)

Figure 37.21 Human Papillomavirus (HPV). (a) Colorized transmission electron micrograph of HPV virions. (b) Genital warts of the vaginal labia. (c) Genital warts in the anal region.

(a) James Cavallini/Science Source; (b) Joe Miller/Centers for Disease Control and Prevention; (c) Dr. Wiesner/Centers for Disease Control and Prevention during vaginal and anal sex. Other types of genital contact in the absence of penetration (oral-genital, manual-genital, and genital-genital contact) can also lead to infection with HPV. Most HPV infections are asymptomatic and transient and clear on their own without treatment. However, in some individuals, HPV infections result in genital warts weeks or months after infection.

Gardasil-9 (9vHPV), a noninfectious recombinant vaccine that protects against nine different HPV strains (6, 11, 16, 18, 31, 33, 45, 52, and 58), is recommended for all preteen males and females in the United States; the vaccine can be taken up to age 26. This multivalent vaccine prevents against both HPV-driven cancer and genital warts.

Comprehension Check

1. Describe HIV structure and its specific interaction with cells of the immune system. Why is HIV a latent infection?
2. Why do people periodically get cold sores? Describe the causative agent. If you get cold sores, what precipitates their emergence?
3. Describe two major ways in which herpes simplex virus type 2 is spread.
4. Describe the causative agent and some symptoms of mononucleosis and exanthem subitum.
5. Compare and contrast the virology and diseases caused by the hepatitis B and C viruses.
6. What viruses cause warts? What conditions does the HPV vaccine prevent?

37.4 Food and Water Are Vehicles for Viral Diseases

After reading this section, you should be able to:

a. Describe the viruses that cause common food-borne and waterborne diseases
b. Identify typical signs and symptoms of common food-borne and waterborne viral diseases
c. Correlate food-borne and waterborne virus infection and disease severity with viral life cycles

Food and water have been recognized as potential carriers (vehicles) of disease since the beginning of recorded history. Collectively more infectious diseases are transmitted by these two routes than any other. A few of the many human viral diseases that are food- and waterborne are now discussed. ▶| *Purification and sanitary analysis ensure safe drinking water (section 42.1)*

Gastroenteritis (Viral)

Acute viral gastroenteritis, inflammation of the stomach or intestines, is caused by six major categories of viruses: adenoviruses, astroviruses, bocaparvoviruses (bocaviruses and parvoviruses), calciviruses (noroviruses and sapoviruses), parechoviruses, and rotaviruses (**figure 37.22**). The medical importance of these viruses is summarized in **table 37.2**. Acute viral gastroenteritis is a common illness, with a global incidence of 3 to 5 billion cases and 1.4 million deaths annually. Viral diarrhea is an especially common cause of mortality among children less than five years of age in countries with limited access to clean water or sanitation, where over 500,000 deaths occur annually. All viruses responsible for gastroenteritis are transmitted by the fecal-oral route.

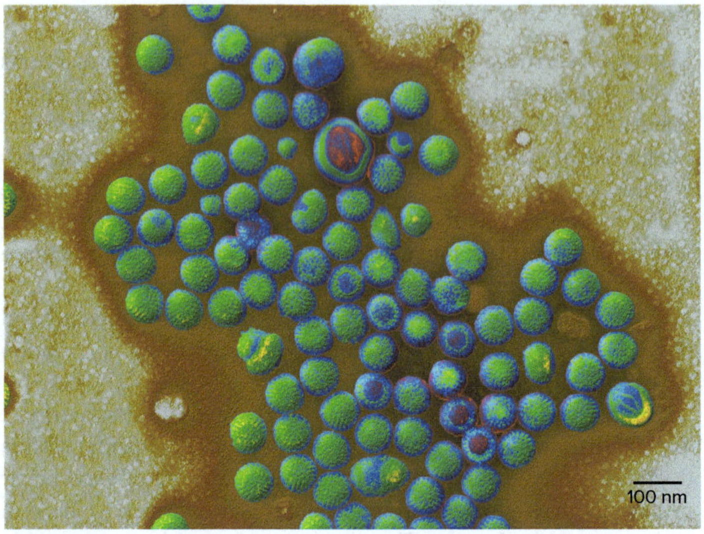

Figure 37.22 Rotavirus Virions. Colorized electron micrograph of rotavirus virions (family Reoviridae) in a human gastroenteritis stool filtrate.
Eye of Science/Science Source

Two groups of viruses cause the majority of viral gastroenteritis cases: noroviruses in children and adults and rotaviruses in children. Noroviruses are estimated to cause about 20 million U.S. cases of acute gastroenteritis annually (at least 50% of all food-borne outbreaks of gastroenteritis). Furthermore, noroviruses are responsible for 2.3 million outpatient visits, 465,000 emergency department visits (primarily in young children), about 110,000 hospitalizations, and 900 deaths (mostly of young children and older people). At least another 20% of global cases and the majority of severe cases of diarrhea in children under age five years are due to rotaviruses, which are the leading cause of severe diarrhea in infants and children worldwide. Prior to the widespread use of vaccine to counter rotavirus, there were more than 400,000 doctor visits, 55,000 to 70,000 hospitalizations, and 20 to 60 deaths in the United States annually. Vaccination has reduced yearly rotavirus hospitalizations to fewer than 20,000, and deaths to single digits.

The average incubation period for most of these viral diseases is 1 to 2 days. Viral gastroenteritis is seen most frequently in infants 1 to 11 months of age, where viruses attack the epithelial cells of the upper intestinal villi, causing malabsorption, impairment of sodium transport, and diarrhea. The clinical manifestations typically range from asymptomatic to a relatively mild

Table 37.2 Medically Important Gastroenteritis Viruses

Family	Virus	Characteristics
Adenoviridae	Adenoviruses (subgroup F, serotypes 40 and 41)	Primarily pediatric cases, outbreaks on military bases, more severe in the immunocompromised
Astroviridae	Human astroviruses	Primarily pediatric cases and outbreaks in nursing homes, lasts 1–3 days
Caliciviridae	Noroviruses (genotypes I and II)	Pediatric epidemics, outbreaks common in older children and adults (families, communities, social events); often associated with shellfish, food handlers with poor hand hygiene, and cruise ships
	Sapoviruses (genogroups 1, 2, 4, and 5)	Primarily pediatric cases with mild diarrhea, rarely adult cases; associated with shellfish and other foods
Parvoviridae	Human bocaparvovirus 2	Primarily pediatric cases; mild to severe diarrhea, rarely with acute flaccid paralysis
Picornaviridae	Human parechoviruses	Primarily pediatric cases; mild diarrhea, rarely with myocarditis and encephalitis complications
Reoviridae	Rotaviruses (Groups A–C)	Primarily pediatric cases; dehydrating diarrhea (5–7 days) with cramps, fever, nausea, and vomiting; 3–5 days diarrhea in adult cases

diarrhea with headache and fever, to a severe, watery, nonbloody diarrhea with abdominal cramps. Fatal dehydration is most common in young children. Vomiting is often present, especially in children. Viral gastroenteritis is usually self-limiting. Treatment is designed to provide relief through the use of oral fluid replacement with isotonic liquids, analgesics, and antiperistaltic agents. Symptoms usually last for 1 to 5 days, and recovery often results in protective immunity to subsequent infection.

Hepatitis A

Hepatitis A (infectious hepatitis) usually is transmitted by fecal contamination of food or drink, or shellfish that live in contaminated water. The disease is caused by hepatitis A virus (HAV) of the genus *Hepatovirus* in the family Picornaviridae. Recall that while all hepatitis viruses cause liver disease, they are not taxonomically related (table 37.1). HAV is an icosahedral, linear, positive-strand RNA virus that lacks an envelope. Once in the digestive system, the virus multiplies within the intestinal epithelium, typically causing only mild intestinal symptoms. Occasionally viruses are found in the blood (viremia) and may spread to the liver. The virus reproduces in the liver, enters the bile, and is released into the small intestine, making the feces highly infectious. After about a 4-week incubation period, symptoms develop that include anorexia, general malaise, nausea, diarrhea, fever, chills, and jaundice. Most cases resolve in 4 to 6 weeks and yield a strong immunity, although some patients relapse, exhibiting symptoms for 6 months or more. The mortality rate is low (less than 1%), and about 40 to 80% of the U.S. population has antibodies, though few are aware of having had the disease. Control of infection is by simple hygienic measures, the sanitary disposal of excreta, and the HAV vaccine. The number of new cases has been dramatically reduced since the introduction of the hepatitis A vaccine in the 1990s.

Hepatitis E

Hepatitis E caused by hepatitis E virus (HEV) genotypes 1 and 2 is implicated in many epidemics in countries with limited clean water and sanitation, while genotypes 3 and 4 are found in Europe and, to a lesser extent, the United States. The single, positive-strand RNA viral genome (7,900 nucleotides) is linear. The virion is spherical, nonenveloped, and 32 to 34 nm in diameter (table 37.1).

Infection with HEV genotypes 1 and 2 usually is associated with feces-contaminated drinking water, whereas genotypes 3 and 4 are transmitted by contaminated food (particularly undercooked pork). HEV enters the blood from the gastrointestinal tract, replicates in the liver, is released from hepatocytes into the bile, and is subsequently excreted in the feces. Like hepatitis A, an HEV infection usually runs a benign course and is self-limiting. The incubation period varies from 15 to 60 days, with an average of 40 days. The disease is most often seen in patients who are 15 to 40 years of age. Children are typically asymptomatic or present mild signs and symptoms, including abdominal pain, anorexia, dark urine, fever, enlarged liver, jaundice, malaise, nausea, and vomiting. Case fatality rates are low (1 to 3%), except for females during pregnancy (15 to 25%), who risk death from fulminant hepatic failure. Diagnosis of HEV infection is by detection of anti-HEV antibodies or reverse transcriptase PCR. There are no specific measures for preventing HEV infections, other than those aimed at improving the level of sanitation in affected areas.

Poliomyelitis

Poliomyelitis (Greek *polios,* gray; *myelos,* marrow or spinal cord), **polio,** or **infantile paralysis** is caused by poliovirus (species *Human enterovirus C*) and was first described in England in 1789 as a leg-wasting disease of children. However, a pictograph from ancient Egypt clearly depicts a man with what appears to be polio (**Historical Highlights 37.1**). The virus is an enterovirus—a transient inhabitant of the gastrointestinal tract—and as a member of the family Picornaviridae, is a nonenveloped, positive-strand RNA virus. Three different poliovirus subtypes have been identified. Strain 2 was declared eradicated in 2015 and strain 3 was declared eradicated in 2019, having not been detected since 2012. Poliovirus virions are stable, especially at acidic pH, and can remain infectious for relatively long periods in food and water, its main routes of transmission.

The incubation period for polio ranges from 6 to 20 days. Once ingested, the virus multiplies in the mucosa of the throat or small intestine. From these sites, the virus invades the tonsils and lymph nodes of the neck and terminal portion of the small intestine. Generally there are either no symptoms or a brief illness characterized by fever, headache, sore throat, vomiting, and loss of appetite. The virus sometimes enters the bloodstream, causing viremia. In more than 99% of cases, the viremia is transient and clinical disease does not result. However, in a minority of cases, the viremia persists and the virus enters the central nervous system, resulting in paralytic polio. The virus enters motor neurons, where it multiplies, destroying the cells and resulting in paralysis of muscles innervated by those neurons.

Two types of polio vaccine are used to prevent the disease; since their development, the incidence of polio has dropped

HISTORICAL HIGHLIGHTS

37.1 A Brief History of Polio

Like many other infectious diseases, polio is probably of ancient origin. Various Egyptian hieroglyphics dated approximately 2000 BCE depict individuals with wasting, withered legs and arms (**box figure**). In 1840 the German orthopedist Jacob von Heine described the clinical features of poliomyelitis and identified the spinal cord as the problem area. Little further progress was made until 1890, when Oskar Medin, a Swedish pediatrician, portrayed the natural history of the disease as epidemic in form. He also recognized that a systemic phase, characterized by minor symptoms and fever, occurred early and was complicated by paralysis only occasionally. Major progress occurred in 1908, when Karl Landsteiner and William Popper successfully transmitted the disease to monkeys. The permanent paralysis of U.S. President Franklin D. Roosevelt from polio increased public awareness in the 1930s. This led to the founding of the March of Dimes campaign in 1938; its sole purpose was to collect money for polio research. In 1949 John Enders, Thomas Weller, and Frederick Robbins discovered that poliovirus could be propagated in vitro in cultures of human embryonic tissues of nonneural origin. This was the breakthrough that later led to the development of vaccines.

In 1952 David Bodian recognized that there were three distinct serotypes of poliovirus. Jonas Salk successfully immunized humans with formalin-inactivated poliovirus in 1952, and this vaccine (IPV) was licensed in 1955. The live attenuated poliovirus vaccine (oral polio vaccine, OPV) developed by Albert Sabin and others was employed in Europe in 1960 and was licensed for U.S. use in 1962. Although Salk and Sabin were professional rivals who didn't much like each other, their vaccines led to a dramatic decline of paralytic poliomyelitis. Polio is likely to be the second human disease eradicated through global vaccination efforts.

Ancient Egyptian with Polio. Note the withered leg.

Bettmann/Getty Images

dramatically. The Salk inactivated vaccine (IVP) is a trivalent vaccine that contains all three poliovirus strains; it is the only vaccine used in the United States. Other countries also use the Sabin attenuated oral polio vaccines (OPV), which are either monovalent or bivalent. As with all attenuated vaccines, OPV cannot be given to individuals who are immunocompromised or pregnant; rarely, the attenuated virus can revert to virulence, either spontaneously or by combining with nonpolio enteric (intestinal) viruses such as coxsackievirus. The pathogenic strain circulates as vaccine-derived poliovirus (VDPV), the most common of which is type 2 (cVDPV2). cVDPV2 has recently caused several significant polio outbreaks (*see chapter 36*). In 2021, a novel OPV (nOPV2) was introduced for use in outbreak response; it is more stable than previous OPVs and 80% less likely to cause type 2 variant poliovirus outbreaks. An ongoing polio eradication effort (the Global Polio Eradication Initiative) aims at eliminating polio by 2030. Currently, wild type polio only exists in Afghanistan and Pakistan; in 2024, there were 62 cases in these countries.

In addition to polio, there are over 100 other enteroviruses that all have the potential to invade the human central nervous system. But like poliovirus, this is very rare. In 2014 there was an outbreak of one such virus, enterovirus D68, that targeted young children with respiratory problems, such as asthma, causing life-threatening respiratory distress and, in a few cases, paralysis. Occasional, sporadic cases of enterovirus D68 continue to be diagnosed.

Comprehension Check

1. What virus groups are associated with acute viral gastroenteritis? How do they cause the disease's symptoms?
2. Describe some symptoms of hepatitis A.
3. Compare and contrast measures that prevent transmission of hepatitis A and E.
4. At what specific sites within the body can the poliomyelitis virus multiply? What is the usual outcome of an infection?

37.5 Zoonotic Diseases Arise from Human-Animal Interactions

After reading this section, you should be able to:

a. Discuss the viruses that cause common diseases spread by contact with animals
b. Identify typical signs and symptoms of zoonotic viral diseases
c. Correlate zoonotic virus infection and disease severity with viral attachment factors

The diseases discussed here are caused by viruses that are normally zoonotic (animal-borne). Members of the RNA virus families Arenaviridae, Bunyaviridae, Flaviviridae, Filoviridae, and Picornaviridae are notable examples of viruses found in animal reservoirs before transmission to and between humans. Although many different animals harbor viruses that also infect people, bats appear to be the most common zoonotic reservoir.

Ebola and Marburg Virus Diseases

Ebola virus disease (EVD) is caused by Ebola viruses, first recognized near the Ebola River in the Democratic Republic of the Congo in Africa. They are members of the genus *Ebolavirus* in the family Filoviridae, a group of filamentous, negative-strand RNA viruses (**figure 37.23**). Six Ebola virus species are known, of which four cause disease in humans: *Sudan ebolavirus, Taï Forest ebolavirus, Bundibugyo ebolavirus, and Zaire ebolavirus*. All Ebola viruses that infect humans are transmitted by contact with body fluids. Convalescing patients continue to harbor virus in body fluids and in rare instances there has been sexual transmission through infected semen up to 17 months after recovery. Bats are thought to be the natural reservoir for Ebola viruses, although the exact species is not known.

Viral hemorrhagic fever (VHF) is the term used to describe the severe, multisystem syndrome seen in Ebola epidemics. Because the host vascular system is damaged, vascular leakage of fluids into body tissues accompanied by the blood's diminished ability to clot (coagulopathy) occurs. The incubation period for EVD ranges from 2 to 21 days and is characterized by abrupt fever, headache, joint and muscle aches, sore throat, and weakness, followed by diarrhea, vomiting, and stomach pain. Signs of infection include fever, rash, red eyes, bleeding, and hiccups—symptoms alerting of internal hemorrhage. There is no standard treatment for Ebola infection. Patients receive supportive therapy consisting of balancing patients' fluids and electrolytes, maintaining their oxygen status and blood pressure, and treating any complicating infections. Two monoclonal antibody treatments are approved by the FDA for treating EVD caused by Zaire ebolavirus.

Marburg virus disease (MVD) is caused by a genetically unique RNA virus also in the Filoviridae family. *Marburgvirus* (MARV) similarly causes a rare, severe type of hemorrhagic

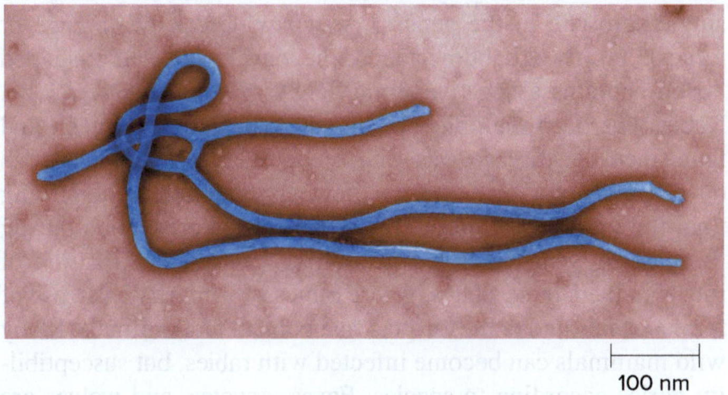

Figure 37.23 **Ebola Virus Virions.**
Cynthia Goldsmith/Centers for Disease Control

fever that affects both humans and nonhuman primates. MARV was first recognized in 1967, when outbreaks of hemorrhagic fever occurred simultaneously in laboratories in Marburg and Frankfurt, Germany, and in Belgrade, Serbia. The first people infected had been exposed to African green monkeys or their tissues. MARV is indigenous to Africa, and the reservoir host of MARV is the African fruit bat, *Rousettus aegyptiacus*. The average incubation period for Marburg hemorrhagic fever is 5 to 10 days. The disease symptoms are abrupt, marked by fever, chills, headache, myalgia, and a maculopapular rash (i.e., discolored with bumps). Nausea, vomiting, chest pain, sore throat, abdominal pain, and diarrhea may also occur in patients with MVD. Symptoms become increasingly severe and may include jaundice, delirium, liver failure, pancreatitis, severe weight loss, shock, and multiorgan dysfunction. Treatment for this disease is limited to supportive hospital therapy.

Hantavirus Pulmonary Syndrome

Hantavirus pulmonary syndrome (HPS) is a disease caused by a negative-strand RNA virus of the family Hantaviridae. HPS is typically transmitted to humans by inhalation of viral particles shed in urine, feces, or saliva of infected rodents. Because it does not cause disease in its rodent hosts, they are considered the reservoir. Although rare in the United States, HPS is potentially deadly, with a mortality rate of 30 to 50%. Rodent control remains the primary strategy for preventing hantavirus infection. HPS in the United States is not transmitted from person to person, nor is it transmitted by rodents purchased from pet stores.

In the environment, virus particles can remain infective for 2 to 3 days. Hantaviruses have lipid envelopes susceptible to most disinfectants, so prevention involves cleaning areas that may be contaminated with rodent droppings. There is no specific treatment for HPS. Supportive therapy is used to treat symptoms, including balancing the patient's fluids and electrolytes, maintaining oxygen status and blood pressure, replacing lost blood and clotting factors, and treating other complicating infections.

Rabies

Rabies (Latin *rabere*, rage or madness) is caused by a number of different strains of highly neurotropic viruses. Most belong to a single serotype in the genus *Lyssavirus* (Greek *lyssa*, rage or rabies), family Rhabdoviridae. The enveloped, bullet-shaped virion contains a negative-strand RNA genome (**figure 37.24**). Rabies has been the object of human fascination, torment, and fear since the disease was first recognized. Prior to Pasteur's development of a rabies vaccine, few words were more terrifying than the cry of "mad dog!" Improvements in prevention during the past 50 years have led to almost complete elimination of rabies in the United States, where rabies is primarily a disease of feral animals and domestic pets that contact feral animals. Many wild mammals can become infected with rabies, but susceptibility varies according to species. Foxes, coyotes, and wolves are the most susceptible; intermediate are skunks, raccoons, insectivorous bats, and bobcats. The virus is primarily transmitted to humans or other animals by biting. Extremely rarely it is transmitted by aerosols in caves where bats dwell or by contamination of scratches, abrasions, open wounds, and mucous membranes with saliva from an infected animal. Worldwide, almost all cases of human rabies are attributed to dog bites, while in the United States most cases result from bat and raccoon bites. Rabid animals may exhibit signs of agitation and aggression (known as furious rabies), although lethargy or paralysis (dumb rabies) is the more common sign. Rabies is rare in the United States, with fewer than 10 deaths reported annually.

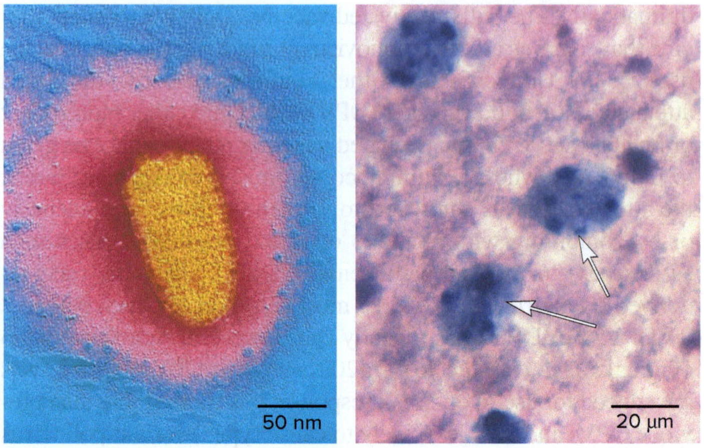

(a) (b)

Figure 37.24 Rabies. (a) Colorized electron micrograph of a rabies virus virion (yellow). Note the bullet shape. (b) Negri bodies (arrows) in brain tissue of dog infected with rabies virus.

(a): CNRI/Science Source; (b): Image courtesy of the Centers for Disease Control and Prevention/Mr G Heid/Centers for Disease Control and Prevention (CDC)

MICRO INQUIRY *What type of host cell does the rabies virus initially infect? To where does the virus then travel?*

Rabies virus multiplies in the salivary glands of an infected host. It travels to the brain, where the virus may change the animal's behavior. The virus also damages the pharyngeal nerve, making swallowing difficult. This causes the animal to foam at the mouth with saliva laden with virions. After inoculation, the virion's envelope spike attaches to the plasma membrane of nearby skeletal muscle cells, which the virus enters. Multiplication of the virus then occurs. When the concentration of the virus in the muscle is sufficient, the virus enters the nervous system through unmyelinated sensory and motor terminals; the binding site is the nicotinic acetylcholine receptor.

The virus spreads relatively slowly to the spinal cord, when the first specific symptoms of the disease—pain or paresthesia (tingling) at the wound site—may occur. A rapidly progressive encephalitis develops as the virus disseminates through the central nervous system. The virus then spreads throughout the body along the peripheral nerves, including those in the salivary glands, where it is shed in the saliva.

Within brain neurons, the virus produces characteristic Negri bodies (figure 37.24*b*), which are masses of virus proteins and RNA that are visible in the light microscope. In the past, diagnosis of rabies consisted solely of examining tissue for the presence of these bodies. Today diagnosis is based on *d*irect immuno*f*luorescent *a*ntibody (DFA), serology for anti-rabies antibody, and PCR testing. *Identification of microorganisms from specimens (section 36.3)*

Symptoms of rabies in humans usually begin 2 to 16 weeks after exposure and include anxiety, irritability, depression, fatigue, loss of appetite, fever, and a sensitivity to light and sound. The disease quickly progresses to paralysis. In about 50% of all cases, intense and painful spasms of the throat and chest muscles occur when the victim swallows liquids. The mere sight, thought, or smell of water can set off spasms. Consequently, rabies has been called hydrophobia (fear of water). Once symptoms of rabies develop in a human, the disease has a 100% mortality rate. Death results from destruction of the regions of the brain that regulate breathing. Safe and effective vaccines (*h*uman *d*iploid-*c*ell rabies *v*accine [HDCV] or *r*abies *v*accine *a*dsorbed [RVA]) against rabies are available. Prevention and control involve preexposure vaccination of dogs and cats and preexposure vaccination of humans at special risk, including veterinarians and persons spending a month or more in countries where rabies is common in dogs.

Individuals suspected of rabies exposure are given rabies immune globulin for passive immunity and rabies vaccine for active immunity, exploiting the relatively long incubation period of the disease. Rabies immune globulin is typically administered directly into the wound after cleaning. The vaccine is administered as four doses given on days 0, 3, 7, and 14. This is usually recommended for anyone bitten by one of the common reservoir species (raccoons, skunks, foxes, and bats), unless it is proven that the animal was uninfected.

A global initiative to make rabies vaccines more available and less expensive is underway. The goal is to vaccinate at least 70% of dogs in all nations by 2030. Because rabies has a variety of animal hosts, it cannot be eradicated.

Comprehension Check

1. What factors make the Ebola and Marburg hemorrhagic fever diseases so contagious and deadly?
2. What precautions can be taken to prevent hantavirus and rabies virus transmission to humans?
3. Most diseases cannot be treated with vaccination. Why is this the case with rabies?

37.6 Prion Proteins Transmit Disease

After reading this section, you should be able to:

a. Describe diseases caused by prions
b. Differentiate prion disease from viral infection

Prion diseases, also called **transmissible spongiform encephalopathies (TSEs),** are fatal neurodegenerative disorders that have attracted enormous attention not only for their unique biological features but also for their impact on public health. Prions consist of misfolded proteins (PrP^{Sc}) that induce normal forms of the protein (PrP^{C}) to also fold abnormally (*see figure 6.20*). This group of diseases includes scrapie, kuru, chronic wasting disease (CWD) in deer, Creutzfeldt-Jakob disease (CJD), variant Creutzfeldt-Jakob disease (vCJD), Gerstmann-Sträussler-Scheinker disease (GSD), and fatal familial insomnia (FFI). Scrapie occurs primarily in sheep. The first of these diseases to be studied in humans was kuru, discovered in the Fore tribe of New Guinea. Carlton Gajdusek and others showed that the disease was transmitted by ritual cannibalism (especially where brains and spinal cords were eaten). The primary symptom of the human disorders is dementia, usually accompanied by manifestations of motor dysfunction such as cerebral ataxia (inability to coordinate muscle activity) and myoclonus (shocklike contractions of muscle groups). Additionally, FFI is characterized by abnormal functioning of the autonomic nervous system and sleep disturbances. Symptoms of TSEs appear insidiously in middle to late adult life and last from months to years prior to death. *Prions are composed only of protein (section 6.7)* *Prion Diseases*

Neuropathologically these disorders produce a characteristic spongiform degeneration of the brain, as well as deposition of amyloid plaques (**figure 37.25**). Prion diseases thus share important clinical, neuropathological, and cell biological features with another, more common cerebral amyloidosis, Alzheimer's disease. Prion diseases are organized into three subcategories: familial (inherited) types, sporadic types, and acquired types. Familial CJD is the result of an inherited mutation within the prion-encoding gene. Certain residues when mutated are known to give rise to CJD, but it is not clear why these amino acid substitutions confer inducible misfolding to the (PrP^{Sc}) conformation.

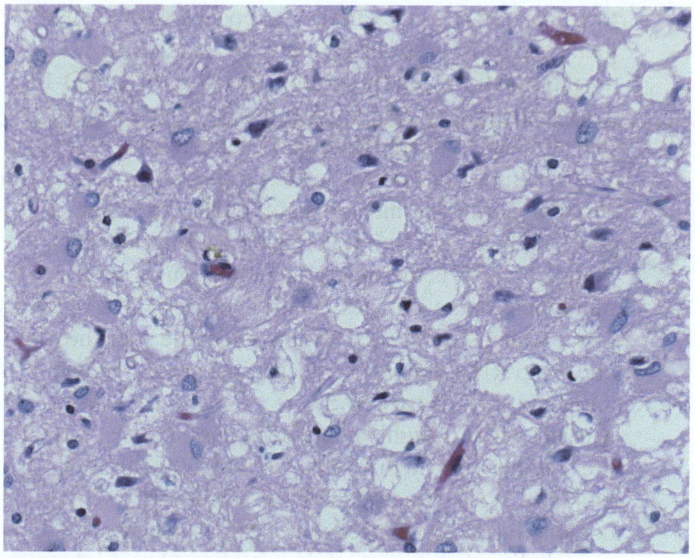

Figure 37.25 Variant Creutzfeldt-Jakob Disease (vCJD). vCJD is a prion-dependent disease that results in prominent spongiotic (holey) changes in the cortex and loss of neurons. Note the spongiotic appearance of this brain tissue photomicrograph (magnified 100× and stained with hematoxylin and eosin).

Teresa Hammett/Centers for Disease Control and Prevention

Familial CJD is very rare, accounting for 5 to 15% of all CJD cases. It progresses rapidly and is always fatal, usually within 1 year of onset of illness.

Sporadic CJD (sCJD) is caused by a spontaneous mutation in the PrP^C protein-coding gene. It has a global incidence of one case per million people per year and accounts for 75% of global CJD cases. There are no known sex, environmental, or lifestyle associations that would induce sCJD. This form of the disease is also rapidly progressive; symptoms move from depression and loss of coordination and walking, to loss of vision and speech, to incontinence and immobility, over an average of 4 to 5 months. Seventy percent of patients die within 6 months of symptom onset.

Variant CJD (vCJD) is acquired when a human consumes meat products from a cow with the prion disease bovine spongiform encephalopathy. Such transmission occurred during the 1980s and early 1990s in the United Kingdom, resulting in 175 cases of vCJD reported between 1996 and 2011. These data suggest a 5-year incubation period. Iatrogenic CJD is acquired from a physician or surgeon, a medical treatment, or diagnostic procedures. The most common source of iatrogenic CJD has been prion-contaminated human growth hormone, corneal grafts, and grafts of dura mater (tissue surrounding the brain). Donor screening and more thorough testing of grafts have decreased the frequency of prion transmission. Iatrogenic CJD represents less than 5% of global CJD cases.

Comprehension Check

1. How are prions different from viruses? How are they similar?
2. In what ways are spongiform encephalopathies commonly acquired?

Key Concepts

37.1 Viruses Can Be Transmitted by Airborne Routes

- Chickenpox and shingles are caused by the same herpesvirus, varicella zoster virus (VZV). Like all members of the Herpesviridae, VZV establishes a latent infection, as it can give rise to shingles many years later. Both chickenpox and shingles are vaccine preventable (**figures 37.1** and **37.2**).
- COVID-19 is caused by SARS-CoV-2, one of seven coronaviruses known to infect humans. Coronaviruses are large, enveloped viruses with positive-strand RNA genomes. Most cases of SARS-CoV-2 are either asymptomatic or relatively mild, but severe COVID-19 causes respiratory distress and altered blood coagulation that can lead to death (**figures 37.3** and **37.4**).
- Viruses in two genera cause the flu, *Influenzavirus A* and *Influenzavirus B,* which are distinguished by their spike proteins, hemagglutinin (HA) and neuraminidase (NA). Antigenic drift is the steady accumulation of point mutations. Large changes in host specificity and virulence occur when two different strains infect the same host cell and swap RNA genome segments in a process called antigenic shift (**figure 37.5**).
- Measles virus is highly contagious; measles is a vaccine-preventable disease that kills over 200,000 children annually. It is characterized by a rash, fever, conjunctivitis, and congestion; it can also cause encephalitis that may result in permanent brain injury or death (**figure 37.6**).
- A number of viruses cause respiratory syndromes that range from the common cold (rhinovirus, adenovirus, coronavirus) to pneumonia and other respiratory syndromes. Respiratory syncytial virus (RSV) causes a respiratory syndrome in infants and toddlers that can be severe.
- Rubella is a vaccine-preventable disease that usually causes mild illness except when contracted during the first trimester of pregnancy, when it can cause congenital rubella syndrome (**figure 37.7**).
- Smallpox is caused by variola virus and was eradicated in 1977 (**figure 37.8**). Until that time, smallpox extracted a large global burden with a 20 to 50% mortality rate; those who survived were affected by lifelong scars from the pustular rash.

37.2 Arthropods Can Transmit Viral Diseases

- The arthropod-borne viral diseases are transmitted by arthropod vectors from human to human or animal to human. In many cases, the majority of infections are mild or go unnoticed. However, for each disease discussed, a small population of patients develops severe, sometimes deadly, disease.
- Chikungunya, dengue, and Zika virus disease are mosquito-borne (*Aedes* spp.) diseases that cause rash, headache, and muscle and joint aches. An individual with a history of dengue who is infected with a different serotype may develop dengue hemorrhagic fever as a result of antibody-dependent enhancement of virus uptake (**figure 37.9**). Zika was almost unknown until 2015 when an outbreak occurred in Brazil and other American countries. If contracted during pregnancy, it can cause severe congenital anomalies, most notably microcephaly.
- Since its introduction to New York in 1999, West Nile virus has spread across the continental United States.

West Nile fever is mosquito-borne (*Culex* spp.); birds are its natural host. Humans are considered dead-end hosts. There is no vaccine or antiviral treatment.

37.3 Direct Contact Diseases Can Be Caused by Viruses

- HIV is a sexually transmitted retrovirus. If infection is left untreated, AIDS will develop after a variable period of latency. HIV targets immune cells by using the host's CD4 protein as the viral receptor. AIDS is characterized by the loss of immune function and the patient is overwhelmed with opportunistic infections and certain cancers. HIV infection cannot be cured and there is no vaccine. A combination of antiretroviral drugs has made living with HIV a chronic condition (**figures 37.10–37.14**).
- Both cold sores (**figure 37.15**) and genital herpes (**figure 37.17**) are caused by herpesviruses: Herpes simplex virus (HSV)-1 and HSV-2, respectively. They cause painful vesicular rashes when active and reside in neurons when latent. Physical and psychological stress can trigger activation of latent viruses.
- Viruses that cause hepatitis are not taxonomically related (**table 37.1**). Those transmitted by direct contact include hepatitis B virus (HBV, a DNA virus), hepatitis C virus (HCV, an RNA virus), and hepatitis delta virus (HDV, a satellite virus dependent on HBV). Hepatitis caused by HBV (with or without HDV) or HCV can be acute or chronic and can lead to hepatocellular carcinoma. Hepatitis caused by HBV (and therefore HDV) is vaccine preventable; recently developed antiviral drugs can cure most hepatitis C infections.
- Cytomegalovirus (CMV), human beta herpesvirus 6 (HHV-6) infections, and human parvovirus B19 are typically infections of childhood. Epstein-Barr virus (EBV) infects children asymptomatically, but it causes infectious mononucleosis in young adults. CMV, HHV-6, and EBV are members of the Herpesviridae, and thus are enveloped DNA viruses that develop latency in all those who have been infected.

37.4 Food and Water Are Vehicles for Viral Diseases

- Viruses transmitted in food and water usually grow in the intestinal system and leave the body in feces (**table 37.2**).
- Viral gastroenteritis is the most common form of food-borne illness. Norovirus is the most common causative virus. Rotavirus is common in young children globally; it is vaccine preventable (**figure 37.22**).
- Hepatitis acquired by fecal contamination of food or water is caused by hepatitis A, which is vaccine preventable, and hepatitis E.
- Poliovirus is an enterovirus and, like other enteroviruses, usually causes mild gastrointestinal upset. In a small minority of cases, poliovirus and other enteroviruses infect the central nervous system. A global polio eradication program has made great strides in eliminating polio.

37.5 Zoonotic Diseases Arise from Human-Animal Interactions

- Diseases transmitted from animals are zoonotic.
- Examples of viral zoonoses include Ebola (**figure 37.23**) and Marburg virus diseases, hantavirus pulmonary syndrome, and rabies (**figure 37.24**).
- These viruses cause severe disease with high mortality rates. Rabies can be treated with a combination of active and passive immunization.

37.6 Prion Proteins Transmit Disease

- A prion disease is a pathological process caused by a transmissible protein (a prion) that remains clinically silent for a prolonged period, after which the clinical disease becomes apparent.
- Examples of prion diseases include Creutzfeldt-Jakob disease, variant Creutzfeldt-Jakob disease, kuru, chronic wasting disease, Gerstmann-Sträussler-Scheinker disease, and fatal familial insomnia. These diseases are chronic infections of the central nervous system that result in progressive degenerative changes and eventual death.

Active Learning

1. You are a county health commissioner who has been informed of a new mosquito-borne disease. Mosquitoes in your county just tested positive for the virus. Most people are asymptomatic if infected, but about 2% of individuals may develop encephalitis (brain infection). Discuss the epidemiological data you would track and efforts you will put into place to protect and educate the citizens of your county.
2. Flaviviruses like West Nile and dengue viruses are often referred to as arboviruses. Hepatitis C virus is also in the genus *Flavivirus* but is not considered an arbovirus. Why is this the case?
3. Infection with some viruses, like measles virus and hepatitis A virus, confers strong lifelong immunity. However, this is not the case with many other viruses such as dengue virus, rhinovirus, coronavirus, and influenza virus. Discuss the factors that contribute to this difference.
4. Discuss the biological, sociological, and political factors that made it possible to eradicate smallpox. Consider these factors in the context of polio eradication; how do they

compare? Do you predict eradication efforts for any other viruses?

5. In 2005 a reconstructed influenza virus containing eight genes from the 1918 Spanish influenza virus was introduced into human lung epithelial cells and mice. This virus was able to rapidly reproduce and cause sudden illness and death in the animals. In 2008 these genes were replaced on an individual basis with homologues from a contemporary H1N1 influenza virus and tested for virulence. Substitution of the 1918 gene with the H1N1 hemagglutinin (HA), neuraminidase (NA), or polymerase PB1 subunit severely diminished the virulence of the 1918 virus. Explain why each of these genes is so critical for infection and virulence. Do you predict an additive effect if more than one gene were replaced? Explain your answer.

6. Avian influenza viruses (AIVs) pose a public health threat if they gain the capacity to infect humans. The H5N6 AIV emerged in Southeast Asia in 2014, where it has caused several human deaths. The nucleotide sequences of the genomes of 175 H5N6 AIVs reveal six different genotypes that arose through genetic reassortment. Refer to figure 37.5 and discuss how the scientists were able to arrive at this conclusion. Will all changes make the virus more likely to infect humans? What changes do you think would be most alarming?

CHAPTER 38

Human Diseases Caused by Bacteria

VCG Wilson/Corbis/Getty Images

The Plague Family Tree

Almost everyone has heard of the Black Death—the outbreak of plague in Europe from 1347 to 1351 when as much as 40% of the population died. Caused by the Gram-negative bacterium *Yersinia pestis,* it was called Black Death because victims bled under the skin and experienced necrosis (cell death) of the extremities, which turned skin purple-black.

For almost as long as the Black Death has been studied, it has been believed that the outbreak in 1347 and other waves of plague that followed over the next several hundred years were independent introductions of *Y. pestis* from Asia. But new evidence suggests that this may not be the case. Europe may have reciprocated and provided Asia with a more virulent *Y. pestis* strain.

How do you study a disease outbreak that happened over 500 years ago? You pair a team of archaeologists, who can find, document, and exhume graves, with a team of forensic anthropologists and bacteriologists, who can extract, sequence, and analyze *Y. pestis* DNA from teeth and bones. Studies of plague victims in Spain, England, Germany, and Russia now give a more complete, albeit complex, story of *Y. pestis* evolution. It is broadly agreed that the story starts about 1320 with a plague outbreak in Mongolia; from there it spread to China in the 1330s. It came to Europe in 1347 on ships that docked in Sicily, where accounts from the time describe dead sailors covered in black boils. From Sicily, the disease quickly migrated north, reaching Russia by 1351.

After the plague epidemic receded in the mid-1350s, it reemerged every few generations for the next three centuries, as depicted in the adjacent painting by Josse Lieferinxe dated 1497. If Europeans were the victims of independent waves of *Y. pestis* from Asia, bacterial DNA from victims of each outbreak should demonstrate genetically different strains. However, *Y. pestis* DNA from victims spanning the fourteenth to seventeenth centuries is highly similar, suggesting it was the same strain that persisted right up until the last outbreak in France in 1722. Some scientists believe that the Black Death *Y. pestis* strain became more virulent and made its way back to China, where it caused outbreaks in the nineteenth and twentieth centuries.

Why does anyone even care about a disease that happened so long ago? COVID-19 has certainly shown us that diseases emerge and move. We need to understand the past to respond in the future. In this chapter we learn about bacteria that cause human disease. Although it is good to remember that only a tiny fraction of bacteria are pathogens, it is wise to become acquainted with those that adversely impact humans.

Readiness Check:

Based on what you have learned previously, you should be able to:

- ✓ Describe basic bacterial cell biology (sections 3.2–3.10)
- ✓ Describe the mechanisms of action of the major classes of antibiotics (section 9.4)
- ✓ Compare and contrast the general principles of innate and adaptive immunity (chapter 31; sections 32.1–32.8)
- ✓ Explain how key pathogens cause infection (chapter 34)
- ✓ Differentiate between different types of vaccines (section 35.6)
- ✓ Explain (in general) methods by which pathogenic bacteria are identified (chapter 36)

38.1 Bacteria Can Be Transmitted by Airborne Routes

After reading this section, you should be able to:

a. Discuss bacteria that cause common airborne diseases
b. Identify typical signs and symptoms of airborne bacterial diseases
c. Correlate airborne bacterial infection and disease severity with bacterial virulence factors

Most airborne bacterial pathogens infect the respiratory tract. Inhalation of infectious particles or self-inoculation can cause disease of the upper tract (sinuses and throat) or, more seriously, the lower tract (bronchi and lungs). From the lungs, some bacteria can disseminate into the bloodstream, particularly in older adults and individuals who are immunocompromised. Some of the most common of these diseases are now discussed.

Chlamydial Pneumonia

Chlamydial pneumonia is caused by *Chlamydophila pneumoniae.* The chlamydiae are obligate intracellular parasites; therefore they must grow and reproduce within host cells. These bacteria exist in two distinct forms: an infectious elementary body (EB) and a noninfectious reticulate body (RB) that grows within host cells (*see figure 20.10*). Infection begins when EBs are inhaled. Host cells endocytose the EBs, which remain in inclusion bodies, where they reorganize to form reticulate bodies (RBs). About 8 to 10 hours after infection, RBs begin binary fission, and then differentiate into infectious EBs. The host cell lyses and releases EBs 48 to 72 hours after infection. ◀ *Phylum Verrucomicrobiota includes human symbionts and obligate intracellular parasites (section 20.4)*

Clinically, chlamydial pneumonia infections are generally mild and may be asymptomatic. When symptoms do occur, they include pharyngitis, bronchitis, sinusitis, fever, sore throat, hoarseness, and pain on swallowing. Mild pneumonia develops after 1 to 4 weeks, characterized by a persistent nonproductive cough (respiratory secretions are not brought up during coughing) and shortness of breath. Approximately 5 to 10% of all cases of adult pneumonia and bronchitis are caused by *C. pneumoniae,* and about 50% of adults have antibodies against this pathogen, indicating prior exposure. *C. pneumoniae* appears to be primarily a human pathogen directly transmitted from human to human by droplet (respiratory) secretions. Diagnosis of chlamydial pneumonia is based on symptoms and serology to detect IgM antibody or nucleic acid amplification. Macrolides are the first line of treatment; tetracyclines are also effective.

Diphtheria

Diphtheria (Greek *diphthera,* membrane; *-ia,* condition) is a vaccine-preventable disease caused by the Gram-positive bacterium *Corynebacterium diphtheriae (see figure 22.8). C. diphtheriae* is well adapted to airborne transmission by way of nasopharyngeal secretions because it is very resistant to drying. Diphtheria is a *toxin-mediated disease,* meaning the illness can be reproduced in animals (or people) by the toxin alone. The *tox* gene that encodes diphtheria toxin is found on a prophage, so only *C. diphtheriae* lysogens that carry this prophage cause disease. This is an example of *lysogenic conversion.*

Upon colonization of the upper respiratory tract (or rarely skin wounds), *C. diphtheriae* releases diphtheria toxin, which enters the circulatory system and is distributed throughout the body. It may cause destruction of cardiac, kidney, and nervous tissues by inhibiting protein synthesis (**figure 38.1**). The toxin is an AB toxin encoded by a single gene that gives rise to a protein with two subunits, A and B (*see figure 34.8*). The B subunit binds to a receptor on the surface of various eukaryotic cells. Once bound, the toxin enters the cytoplasm by endocytosis. The B subunit contains a transmembrane region that embeds itself in the endosome membrane. Acidification of the endosome results in the reduction of a disulfide bond holding the A and B subunits together and releases the A subunit into the cytoplasm. There it catalyzes the attachment of ADP-ribose (from NAD^+) to translation elongation factor-2 (EF-2). A single toxin enzyme (i.e., catalytic domain) can exhaust the entire supply of cellular EF-2 within hours, resulting in protein synthesis inhibition and cell death. ◀ *Translation in bacteria (section 13.7); Order Mycobacteriales includes important human pathogens (section 22.1); Exotoxins (section 34.4)*

Typical signs and symptoms of diphtheria include a thick nasal discharge containing both mucus and pus, pharyngitis, fever, cough, and paralysis of the throat and respiratory muscles. The cardinal sign of diphtheria is a thick, grayish pseudomembrane on the pharynx and respiratory mucosa consisting of dead host cells and *C. diphtheriae* bacteria. Most current diphtheria cases involve children or older individuals who are unvaccinated or undervaccinated, and live in areas where the disease is endemic. Diphtheria is treated with a combination of antitoxin—antibodies that neutralize circulating diphtheria toxin—and macrolides. The preventative vaccine (DTaP) consists of inactivated toxin, making it a toxoid vaccine. Fewer than 10 diphtheria cases are reported annually in the United States, and most occur in nonimmunized individuals. Globally, there are between 10,000 and 20,000 cases of diphtheria annually, with outbreaks in several countries emphasizing the need for vaccination. ◀ *Vaccines immunize susceptible populations (section 35.6)*

Legionnaires' Disease

In 1976 the term **Legionnaires' disease,** or **legionellosis,** was coined to describe an outbreak of pneumonia that occurred at an American Legion convention in Philadelphia. At the time the cause of the outbreak was mysterious and led to multiple conspiracy theories (**figure 38.2*a***). The following year it was discovered that the bacterium responsible was *Legionella pneumophila,* a nutritionally fastidious, aerobic, Gram-negative rod (figure 38.2*b*). The

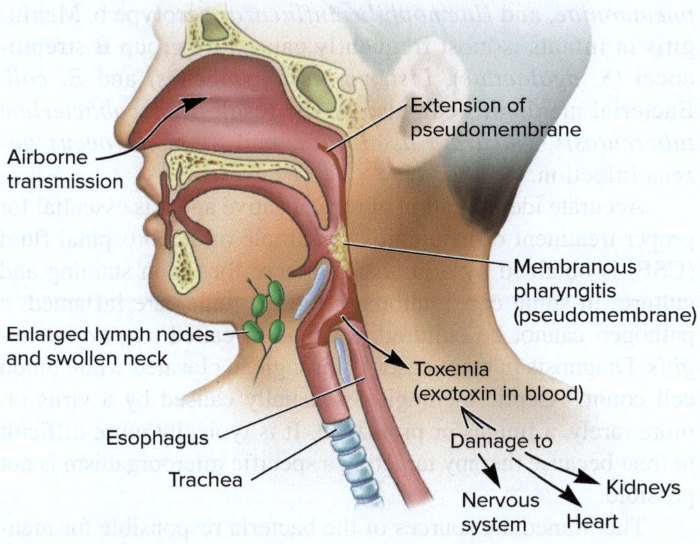

(a)

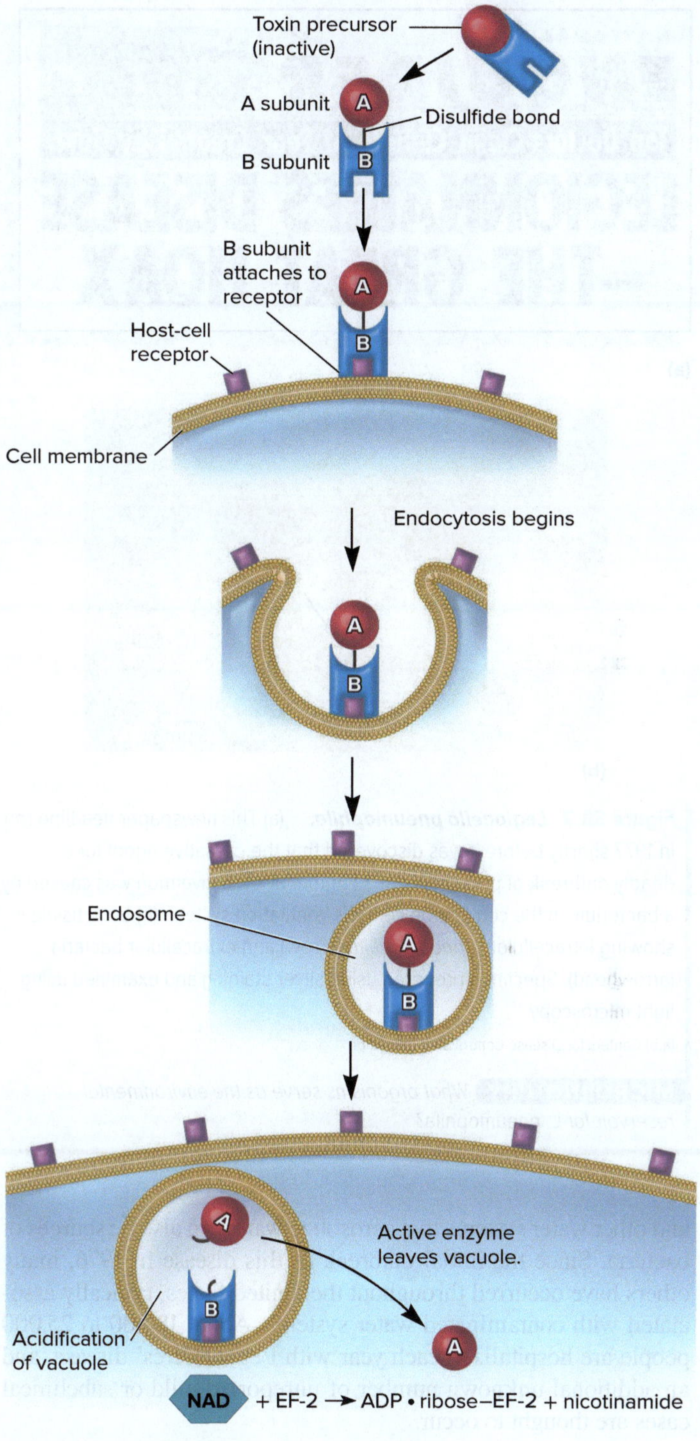

(b) Modified EF-2 prevents protein synthesis.

Figure 38.1 Diphtheria Pathogenesis. (a) Diphtheria is a toxin-mediated infectious disease caused by *Corynebacterium diphtheriae.* The disease is an acute, contagious, febrile illness characterized by local oropharyngeal inflammation and pseudomembrane formation. If the toxin enters the blood, it is disseminated and can damage the peripheral nerves, heart, and kidneys. (b) The B subunit of the diphtheria toxin binds to the host-cell membrane, facilitating entry of the A subunit, which ADP-ribosylates EF-2 to inhibit protein synthesis.

organism is a facultative intracellular parasite found in a variety of free-living fresh-water amoebae and ciliated protozoa that serve as its natural reservoir. These protists also inhabit cooling water found in air-conditioning systems, where they often form biofilms. In fact, most outbreaks are traced to contaminated cooling water systems. *Orders Legionellales and Coxiellales include intracellular pathogens (section 21.2)*

L. pneumophila enters the host by inhalation of contaminated water drops; there is no human-to-human spread of the organism. Most healthy people infected with *L. pneumophila* develop a mild respiratory illness known as *Pontiac fever.* When Legionnaires' disease develops, it is most often seen in individuals who are immunocompromised, including older adults. After infecting the host's lungs, the bacteria reside within the phagosomes of alveolar macrophages, where they multiply and cause localized tissue destruction through export of a cytotoxic exoprotease. Symptoms start 2 to 14 days after exposure and include a high fever, nonproductive cough, headache, neurological manifestations, severe bronchopneumonia, nausea, vomiting, and diarrhea. Although a rapid test can detect *L. pneumophila* antigens in the urine, diagnosis must be confirmed by either isolation of the bacterium or documentation of a rise in antibody titer over time. Treatment begins with supportive measures and the administration of azithromycin or a fluoroquinolone. Death occurs in 5 to 30% of symptomatic cases.

Legionnaires' disease can be prevented by eliminating the environmental source of *L. pneumophila* through chlorination, ozonation, and heating water, and cleaning building water supplies. A variety of regulatory control measures are now in place to reduce the incidence of Legionnaires' disease in hospitals, hotels, apartment buildings, prisons, etc. However, private hot tubs, fountains,

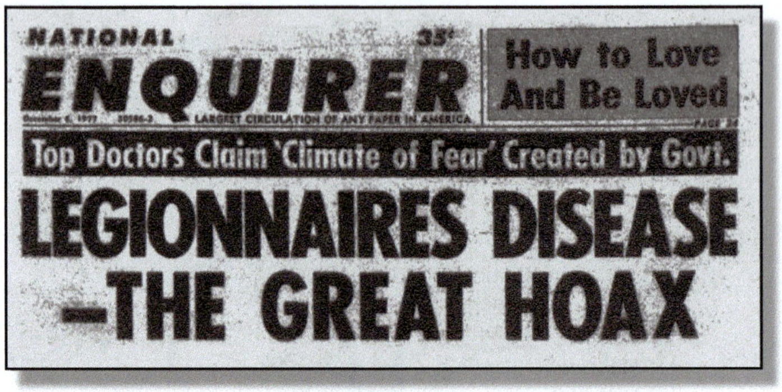
NATIONAL ENQUIRER 35¢

How to Love And Be Loved

LARGEST CIRCULATION OF ANY PAPER IN AMERICA

Top Doctors Claim 'Climate of Fear' Created by Govt.

LEGIONNAIRES DISEASE —THE GREAT HOAX

(a)

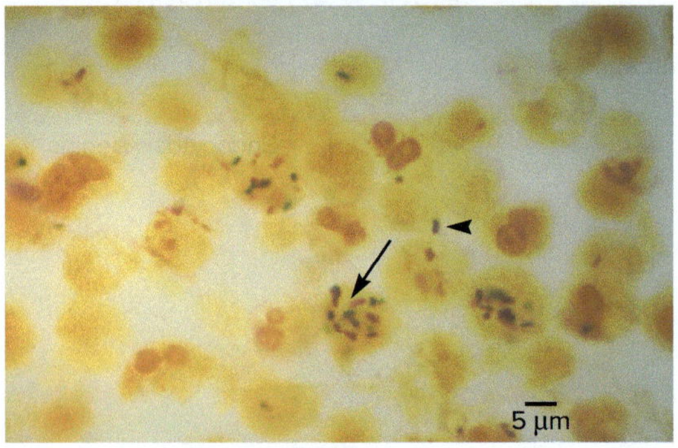

(b)

Figure 38.2 ***Legionella pneumophila.*** (a) This newspaper headline ran in 1977 shortly before it was discovered that the causative agent for a deadly outbreak of pneumonia at a Legionnaires' convention was caused by a bacterium in the convention center's ventilation system. (b) Lung tissue showing intracellular *L. pneumophila* (arrow) and extracellular bacteria (arrowhead). Specimen prepared using silver staining and examined using light microscopy.

(a,b) Centers for Disease Control and Prevention

MICRO INQUIRY *What organisms serve as the environmental reservoir for* L. pneumophila?

and other water systems that aerosolize water can also be sources of bacteria. Since the initial outbreak of this disease in 1976, many others have occurred throughout the United States, typically associated with contaminated water systems. About 18,000 to 25,000 people are hospitalized each year with Legionnaires' disease, and an additional unknown number of unreported mild or subclinical cases are thought to occur.

Meningitis

Meningitis (Greek *meninx,* membrane; *-itis,* inflammation) is an inflammation of the brain or spinal cord meninges (membranes). In addition to many viruses, fungi, and protozoa, a variety of bacteria can cause meningitis. In adults and children over the age of about two years, the most common bacterial agents are *Neisseria meningitidis, Streptococcus pneumoniae,* and *Haemophilus influenzae* serotype b. Meningitis in infants is most frequently caused by group B streptococci (*S. agalactiae*), *Listeria monocytogenes,* and *E. coli.* Bacterial meningitis can also be the result of *Mycobacterium tuberculosis, Nocardia asteroides,* and *Staphylococcus aureus* infection.

Accurate identification of the causative agent is essential for proper treatment of meningitis. A sample of cerebrospinal fluid (CSF) is obtained by a lumbar puncture for Gram staining and culture. In some cases, although the meninges are inflamed, a pathogen cannot be cultured, a situation called *aseptic meningitis*. Diagnosis in these cases is through an elevated white blood cell count. Aseptic meningitis is usually caused by a virus or, more rarely, a fungus or protozoan. It is typically more difficult to treat because therapy targeting a specific microorganism is not possible.

The immediate sources of the bacteria responsible for meningitis are respiratory secretions from carriers. *N. meningitidis,* often referred to as *meningococcus,* is a normal inhabitant of the human nasopharynx (5 to 10% of humans carry nonpathogenic serotypes). Most disease-causing *N. meningitidis* strains belong to serotypes A, B, C, Y, and W-135. In general, serotype A, and to a lesser extent W-135, strains are the cause of epidemic disease in Africa, whereas serotype B, C, Y, and W-135 strains are responsible for meningitis outbreaks in other regions. The disease process is initiated by pili-mediated colonization of the nasopharynx by pathogenic bacteria. The bacteria cross the nasopharyngeal epithelium and the blood-brain barrier (typically through endocytosis) to enter the central nervous system. Often, *N. meningitidis* also invades the bloodstream (meningococcemia). Because meningococcal meningitis and meningococcemia commonly occur simultaneously, the term *meningococcal disease* is often used to describe them.

Symptoms caused by *N. meningitidis* are variable, depending on whether meningococcemia is established. The usual symptoms of meningitis include the meningeal syndromes: fever, vomiting, headache, lethargy, confusion, photophobia (sensitivity to light), and stiff neck. If bacteria have spread to the bloodstream, a characteristic, rapidly progressing, purple rash develops along with septic shock (*see figure 34.10*) and disseminated intravascular coagulation (abnormal blood clotting followed by excessive bleeding throughout the body). Once bacterial meningitis is suspected, ceftriaxone, a third-generation cephalosporin, is administered immediately. Penicillin may be used after it is established that the causative strain has not acquired penicillin resistance. Antibiotics (ciprofloxacin or rifampin) should also be administered prophylactically to patient contacts. Meningococcal disease is rapidly fatal if left untreated; even with treatment, up to 15% of cases are fatal, and up to 20% of those who recover are left with permanent cognitive and physiological impairments.

Prevention of *N. meningitidis* infection is with vaccination (**figure 38.3**). Three conjugate vaccines are effective against serotypes A, C, Y, and W-135. Two vaccines that use recombinant

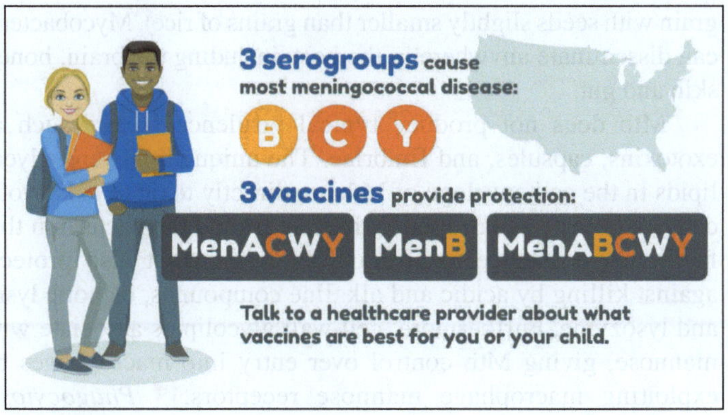

Figure 38.3 Meningitis Is Vaccine Preventable. The most common causes of bacterial meningitis in young people and adults are *Streptococcus pneumoniae*, *Haemophilus influenzae*, and *Neisseria meningitidis*. Vaccines for all three agents are available; there are two vaccines for *Neisseria meningitidis* that cover different serotypes.

CDC. Meningococcal Disease. https://www.cdc.gov/meningococcal/vaccine-info.html

proteins (antigens) are available for protection against serogroup B. A pentavalent vaccine (MenABCWY) protects against serotypes A, B, C, W-135, and Y. *Vaccines (section 41.1)*

Another agent of meningitis is *H. influenzae* serotype b (Hib), a Gram-negative bacterium. Transmission is by inhalation of droplet nuclei shed by infectious individuals or carriers. Hib infects mucous membranes, resulting in sinusitis, pneumonia, bronchitis, and epiglottitis in young, unvaccinated children. In addition to gaining access to the CSF, it can disseminate to the bloodstream and cause bacteremia. *H. influenzae* disease has declined to fewer than 50 cases per year in the United States. Globally, it is estimated that Hib causes at least 3 million cases of serious disease and about 400,000 vaccine-preventable deaths each year.

Three to 6% of all *H. influenzae* infections are fatal. Furthermore, as with meningococcal disease, up to 20% of surviving patients have permanent hearing loss or other long-term sequelae. Therefore it is important that all children begin vaccination with the Hib conjugate vaccine at the age of two months.

Mycobacterium Infections

Mycobacteria are Gram-positive bacteria that normally inhabit soil, water, and house dust. They cause a number of human infections that are difficult to treat. Recall that mycobacteria have cell walls containing mycolic acids. These complex fatty acids have 60 to 90 carbon atoms with a shorter β-hydroxy chain and a longer α-alkyl side chain (*see figure 22.7*). The presence of mycolic acids and other lipids outside the peptidoglycan layer makes mycobacteria acid-fast (*see figure 2.19*). *Differential staining (section 2.3)*

Mycobacterium avium Complex

Mycobacterium avium and *M. intracellulare* are so closely related that they are referred to as the *M. avium* complex (MAC). MAC has become the most common cause of nontuberculous mycobacterial (NTM) infections in the United States. *Order Mycobacteriales includes important human pathogens (section 22.1)*

MAC is found worldwide and infects a variety of insects, birds, and animals. In humans, both the respiratory and the gastrointestinal tracts are entry portals for MAC; however, person-to-person transmission is not very efficient. MAC causes a pulmonary infection in humans similar to that caused by *M. tuberculosis*. Pulmonary MAC is more common in older persons with preexisting lung disease and individuals with suppressed immune systems. The gastrointestinal tract is the most common site of colonization and dissemination in individuals who are HIV-positive. It produces disabling symptoms, including fever, malaise, weight loss, night sweats, and diarrhea.

MAC can be isolated from sputum, blood, and aspirates of bone marrow. Cultures and acid-fast stains are of value in making a diagnosis. Like all mycobacterial infections, MAC is treated with a combination of drugs. The U.S. Food and Drug Administration (FDA) has not approved a specific drug combination for treating MAC infection because each bacterial isolate needs to be tested for antibiotic sensitivity. Infected individuals are usually given a daily regimen of a macrolide and an aminoglycoside along with ethambutol paired with rifamycin for an extended period of time.

Mycobacterium tuberculosis

More than a century ago, Robert Koch identified *Mycobacterium tuberculosis* (Mtb) as the causative agent of **tuberculosis (TB).** We now know TB can also be caused by *M. bovis* (which can be transmitted by drinking unpasteurized milk) and *M. africanum*. The latter two species have recently been reclassified as *M. tuberculosis* and these three organisms make up *M. tuberculosis* complex (MTBC). TB is an enormous global health problem. It is estimated that one-quarter of the world's human population is infected, with about 2 billion people having latent infections. Globally, there were 10 million people newly infected with TB in 2021, and 1.6 million deaths. In 2024 there were nearly 10,350 new TB cases in the United States, where TB occurs most commonly among populations who are unhoused, are malnourished, have alcohol use disorder, are incarcerated, or are from countries with high TB incidence. Additionally, TB cases in American Indians and Alaska Native People are over 8-fold higher than in the general population.

Infection results when the bacteria are inhaled in droplets and aerosols. The infectious dose is quite low, around 10 bacterial cells. The incubation period is about 4 to 12 weeks, and the disease develops slowly. TB is frequently asymptomatic as the immune system controls the disease, keeping the bacteria in a relatively dormant state. The individual has **latent TB infection** (LTBI), but the pathogen is not eliminated. If the host's immune system later becomes stressed, **reactivation tuberculosis** (also called *tuberculosis disease*) may result. In this case, bacteria at the initial site of infection (i.e., the lungs) begin dividing and the infected person develops signs and symptoms, including fever, fatigue, night sweats, weight loss, and a cough that often produces bloody sputum.

Tuberculosis infection begins when inhaled Mtb cells infect resident macrophages in the lungs. Although innate immune defenses are raised, mycobacteria survive macrophage phagocytosis by blocking lysosome fusion with the phagosome. Infected macrophages fuse and a focus of infection begins. Upon activation of the adaptive arm of the immune system, T_H1 and cytotoxic T lymphocytes (CTLs) are recruited to the site of infection. These cells typically do little to clear the infection, but their additional cytokines establish a stalemate between pathogen and host in the form of **granulomas,** whereby infected macrophages are surrounded by other immune cells. As granulomas persist, two things happen. First, infected macrophages in the center die, imparting a cheesy appearance, and the granuloma is said to be caseating (**figure 38.4**). Second, calcium is deposited in and around the granuloma. These appear on X rays and are called **Ghon complexes.** As the granulomas enlarge, the degree of calcification also increases, and these regions are called **tubercles.** Tubercles may be seen in latent infections. Sometimes the tubercle lesions liquefy and form air-filled **tuberculous cavities.** From these cavities, the bacteria can be disseminated to new foci throughout the body (**figure 38.5**). This spreading, seen most commonly in children and people who are immunocompromised, is called **miliary tuberculosis** due to the development of many tubercles the size of millet seeds (millet is a grain with seeds slightly smaller than grains of rice). Mycobacteria can disseminate anywhere in the host, including the brain, bones, skin and gut.

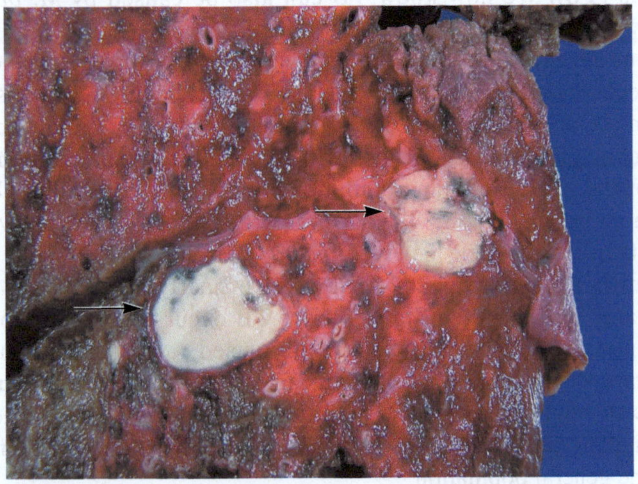

Figure 38.4 ***Mycobacterium tuberculosis*** **Resides in Granulomas.** Tuberculosis is characterized by the development of granulomas. Cells within the granuloma die, causing debris with a cheesy appearance (arrows) called caseous necrosis.

McGraw Hill

Mtb does not produce typical virulence factors such as exotoxins, capsules, and fimbriae. The unique lipids and glycolipids in the cell envelope of Mtb are directly toxic to eukaryotic cells and create a hydrophobic barrier around the bacterium that facilitates resistance to antimicrobial agents. It also protects against killing by acidic and alkaline compounds, osmotic lysis, and lysozyme. Furthermore, cell wall glycolipids associate with mannose, giving Mtb control over entry into macrophages by exploiting macrophage mannose receptors. ◀ *Phagocytosis destroys invaders (section 31.6)*

TB is treated with combination antimicrobial therapy. Four drugs are initially administered simultaneously: isoniazid (INH), rifampin (RIF), ethambutol (EMB), and pyrazinamide (PZA). Rifapentine (RPT), a newer derivative of rifampin, is sometimes used. After 6 to 8 weeks, only INH and RIF are continued for a total of 6 months as a way of decreasing the possibility that the bacterium develops drug resistance. Despite combination therapy, ***m*****ulti*****d*****rug-*****r*****esistant *****tu*****berculosis (MDR-TB)** strains have developed. A multidrug-resistant strain is an Mtb strain resistant to INH and RIF, with or without resistance to other drugs. As alarming as MDR-TB strains are, **e*****x*****tensively *****d*****rug-*****r*****esistant *****tu*****berculosis (XDR-TB)** strains are worse. These are resistant to INH, RIF, and second-line antibiotics. XDR-TB is a unique challenge to public health as first-line and second-line drug resistance severely limits treatment options. For this reason, MDR-TB and XDR-TB have a higher mortality rate than drug-sensitive TB.

Drug-resistant TB arises because Mtb cells within granulomas have spontaneous, predictable rates of chromosomal mutations that confer resistance to drugs. These mutations are unlinked; hence resistance to one drug is not associated with resistance to an unrelated drug. Drug resistance usually arises because the patient does not comply with treatment and patients who have previously undergone therapy are presumed to harbor MDR-TB until proven otherwise. If treatment guidelines are not followed, a susceptible strain of Mtb may become resistant to multiple drugs within a matter of months. To manage noncompliance, many public health organizations worldwide practice *d*irectly *o*bserved *t*herapy (DOT), in which the person with TB takes each medication dose in the presence of a healthcare worker. DOT conducted remotely via live or prerecorded video is called eDOT.

Persons infected with Mtb develop cell-mediated immunity because the bacteria stimulate a strong T_H1-cell response. These sensitized T cells are the basis for the tuberculin skin test, also called the Mantoux skin test. In this test, a *p*urified *p*rotein *d*erivative (PPD) of Mtb is injected intradermally. If the person has had TB or has been exposed to Mtb, sensitized T cells react with these proteins and a delayed hypersensitivity reaction occurs within 48 hours. This positive skin reaction appears as an induration (hardening) and reddening of the area around the injection site (**figure 38.6**). A positive skin test cannot

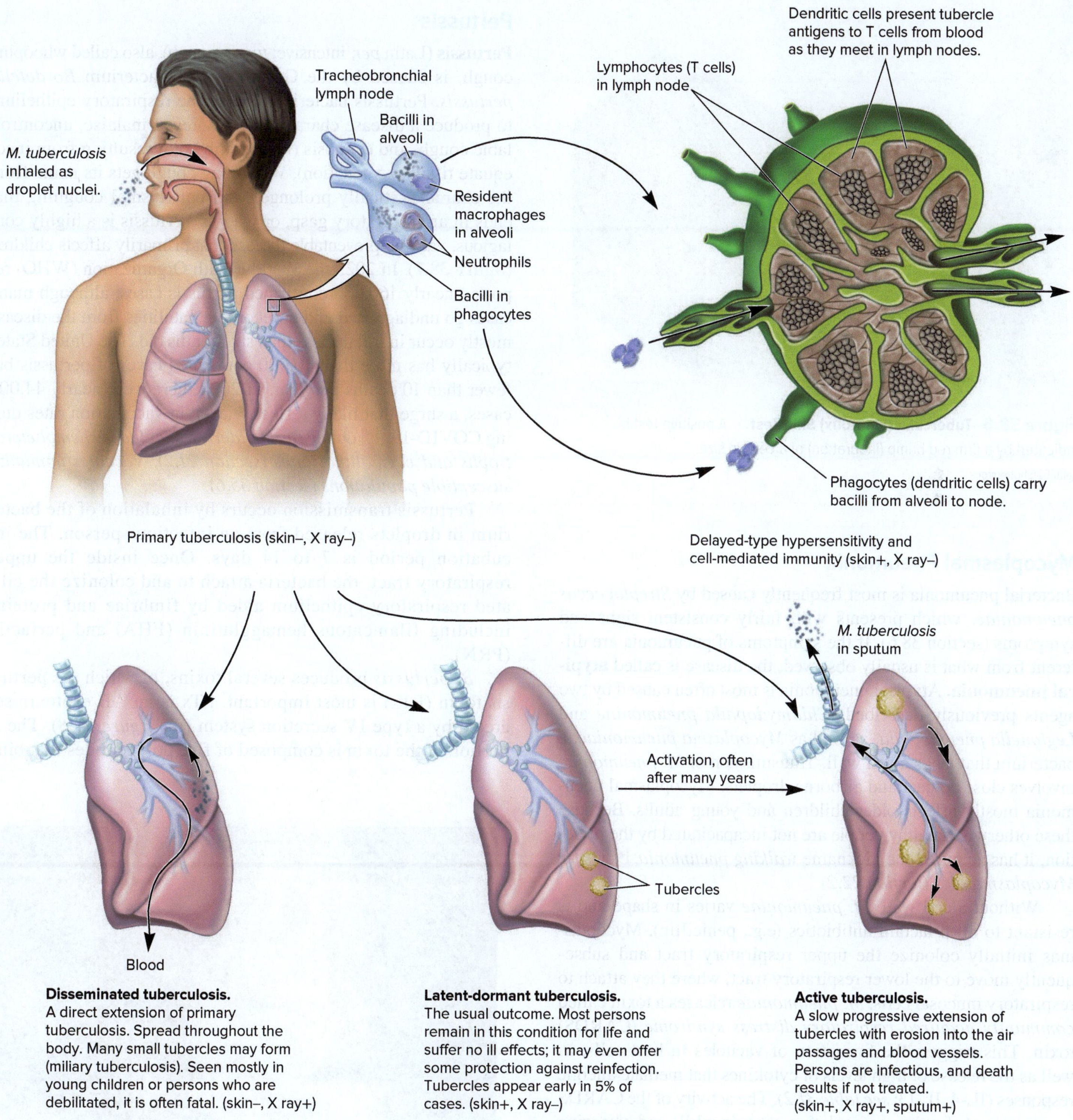

Figure 38.5 Tuberculosis. Tuberculin skin tests, X rays, and sputum results vary during disease progression.

differentiate between active or latent tuberculosis, or prior immunization with the BCG vaccine (not used in the United States). Therefore, a positive PPD test needs to be confirmed with a chest X ray. Laboratory tests include microscopy of the acid-fast organism, culture, and interferon-gamma release assays (IGRAs) that measure cytokine release by a patient's immune cells in response to Mtb antigens. Commercially available DNA assays that detect Mtb and genes for resistance to rifampicin are also used. ◀ *Type IV hypersensitivity (section 32.9)*

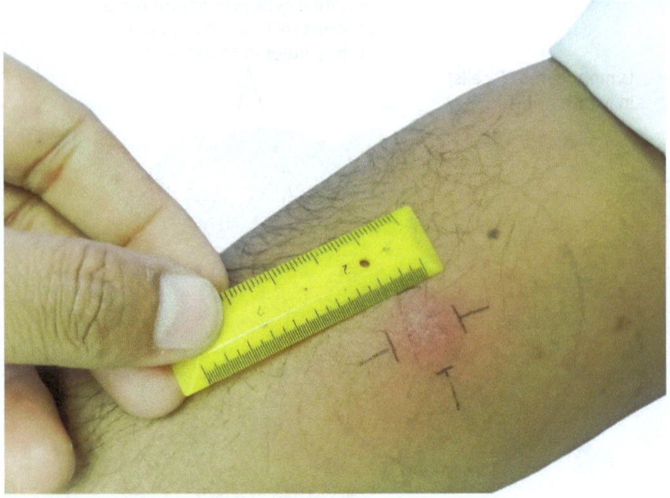

Figure 38.6 Tuberculin (Mantoux) Skin Test. A positive test is indicated by a firm red bump (induration) of a certain size.
Saiful52/Shutterstock

Mycoplasmal Pneumonia

Bacterial pneumonia is most frequently caused by *Streptococcus pneumoniae,* which presents with fairly consistent signs and symptoms (section 38.6). If the symptoms of pneumonia are different from what is usually observed, the disease is called **atypical pneumonia.** Atypical pneumonia is most often caused by two agents previously described, *Chlamydophila pneumoniae* and *Legionella pneumophila,* as well as *Mycoplasma pneumoniae,* a bacterium that lacks a cell wall. Transmission of *M. pneumoniae* involves close contact and airborne droplets. Mycoplasmal pneumonia mostly affects older children and young adults. Because these otherwise healthy people are not incapacitated by the infection, it has acquired the nickname *walking pneumonia.* ⏮ *Order Mycoplasmatales (section 22.2)*

Without a cell wall, *M. pneumoniae* varies in shape and is resistant to all β-lactam antibiotics (e.g., penicillin). Mycoplasmas initially colonize the upper respiratory tract and subsequently move to the lower respiratory tract, where they attach to respiratory mucosal cells. *M. pneumoniae* releases a toxin called *community acquired respiratory distress syndrome* (CARDS) toxin. This triggers the formation of vacuoles in host cells, as well as the release of high levels of cytokines that mediate allergic responses (IL-4, IL-13; *see table 31.2*). The activity of the CARDS toxin accounts for the accumulation of eosinophils and wheezing seen in walking pneumonia. Symptoms also include headache, weakness, a low-grade fever, and a characteristic dry cough. The disease and its symptoms usually persist for weeks. The mortality rate is less than 1%.

Several rapid tests that detect *M. pneumoniae* antibodies are available for diagnosis of mycoplasmal pneumonia; it can also be detected by nucleic acid amplification. Tetracycline or macrolides are effective in treatment.

Pertussis

Pertussis (Latin *per,* intensive; *tussis,* cough), also called whooping cough, is caused by the Gram-negative bacterium *Bordetella pertussis.* Pertussis bacteria colonize the respiratory epithelium to produce a disease characterized by fever, malaise, uncontrollable cough, and cyanosis (bluish skin color resulting from inadequate tissue oxygenation). Whooping cough gets its name from the characteristically prolonged and paroxysmal coughing that ends in an inspiratory gasp, or whoop. Pertussis is a highly contagious, vaccine-preventable disease that primarily affects children (**figure 38.7**). In 2023 the World Health Organization (WHO) reported nearly 160,000 confirmed pertussis cases, although many cases go undiagnosed and unrecorded. Fatalities from the disease mostly occur in infants less than six months old. The United States typically has more than 10,000 confirmed cases of pertussis but fewer than 10 deaths per year. In 2024 there were nearly 44,000 cases, a surge that likely reflects a drop in vaccination rates during COVID-19. ⏮ *Order Burkholderiales includes chemoheterotrophs and chemolithotrophs (section 21.2); Vaccines immunize susceptible populations (section 35.6)*

Pertussis transmission occurs by inhalation of the bacterium in droplets released from an infectious person. The incubation period is 7 to 14 days. Once inside the upper respiratory tract, the bacteria attach to and colonize the ciliated respiratory epithelium aided by fimbriae and proteins including filamentous hemagglutinin (FHA) and pertactin (PRN).

B. pertussis produces several toxins, of which the pertussis toxin (PTx) is most important. PTx is an AB exotoxin secreted by a type IV secretion system (*see figure 34.8*). The B portion of the toxin is composed of five polypeptides that bind

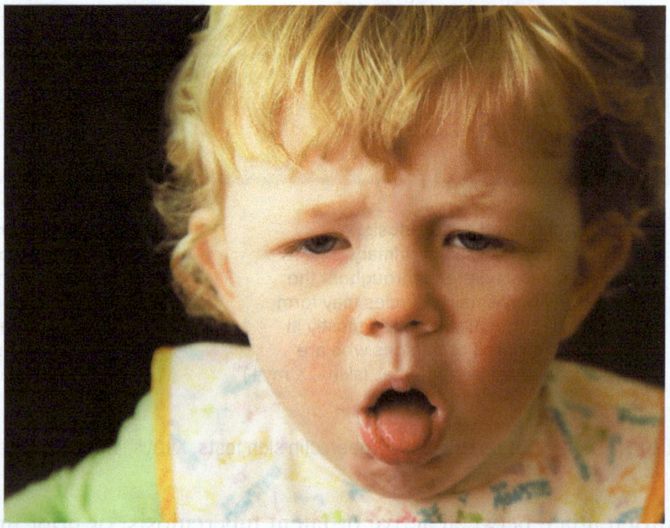

Figure 38.7 Pertussis Causes Paroxysms of Coughing. Pertussis, or whooping cough, results in fits of coughing that end in a sharp inspiratory gasp. Children infected with pertussis grow exhausted simply from coughing.
DavidWhalen/Getty Images

to specific carbohydrates on host-cell surfaces and mediate transport of the A subunit from the plasma membrane to the cytoplasm. The A subunit is an ADP-ribosyl transferase, similar to the diphtheria toxin. However, PTx adds ADP-ribose moieties to the host-cell's G protein–coupled receptor, Gi. This disables Gi, which normally inhibits host adenyl cyclase. Without this molecular brake, adenyl cyclase is locked in the active state and ATP is continuously converted to cyclic AMP (cAMP). The resulting increase in intracellular levels of cAMP accelerates mucin secretion and alters water transport, affecting electrolyte balance. *B. pertussis* also produces adenylate cyclase toxin (CyaA)—which enters the cell and further increases intracellular cAMP levels—and tracheal cytotoxin (TCT), which destroys epithelial tissue. Working together, these toxins provoke secretory cells in the respiratory tract to produce nitric oxide, which kills nearby ciliated cells, inhibiting removal of bacteria and mucus. The secretion of thick mucus also impedes ciliary action. Breathing becomes so difficult that coughing is followed by the whooping sound as patients try to catch their breath.

Pertussis is divided into three stages: (1) The catarrhal (Latin *catarrhus,* flow down) stage is so named because the mucous membrane inflammation resembles the common cold. (2) The paroxysmal stage is characterized by prolonged coughing sieges. During this stage, the person who is infected tries to cough up mucous secretions by making 5 to 15 rapidly consecutive coughs followed by the characteristic whoop. Coughing can be so violent it is not uncommon for patients to fracture a rib. The catarrhal and paroxysmal stages last about 6 weeks. (3) The convalescent stage, when final recovery occurs, may take several months.

Laboratory diagnosis of pertussis is by culture of the bacterium and nucleic acid amplification. The development of a strong, lasting immunity takes place after infection. Treatment is with a macrolide; this ameliorates clinical illness when begun during the catarrhal phase and may also reduce the severity of the disease when begun within 2 weeks of the onset of the paroxysmal cough. After this time, antibiotic therapy has only minimal impact. Pertussis can be prevented by vaccination with the acellular vaccine, which requires periodic boosters. It is important for both young children and their caregivers (including grandparents) to be up to date with pertussis vaccination.

Streptococcal Diseases (Group A)

Pathogenic streptococci, commonly called strep, are a heterogeneous group of Gram-positive bacteria. *Streptococcus pyogenes* are in Lancefield group A. These β-hemolytic streptococci are of major importance (*see figure 22.19*). The different serotypes of **group A streptococci (GAS)** produce a number of virulence factors, including (1) extracellular enzymes that break down host molecules; (2) streptokinases, enzymes which activate a host blood factor that dissolves blood clots; (3) the cytolysins streptolysin O and streptolysin S, which kill host leukocytes; and (4) capsules and M protein, which limit phagocytosis (**figure 38.8**). M protein, an alpha-helical coiled-coil anchored in the GAS plasma membrane, facilitates attachment to host cells and prevents opsonization by complement. The amino acid sequence of M proteins varies among serotypes of *S. pyogenes,* and different M proteins mediate interaction with different host receptors.

Order Lactobacillales (section 22.2); Complement is a cascading enzyme system (section 31.3)

S. pyogenes has a capsule made of hyaluronic acid, a molecule also found in humans. There is no vaccine for *S. pyogenes.* The best antigen for a vaccine against other streptococci is the capsule. However, because hyaluronic acid is also found in the host, it is not immunogenic.

Streptococcal Pharyngitis

Streptococcal pharyngitis is one of the most frequent bacterial infections of humans and is commonly called *strep throat*

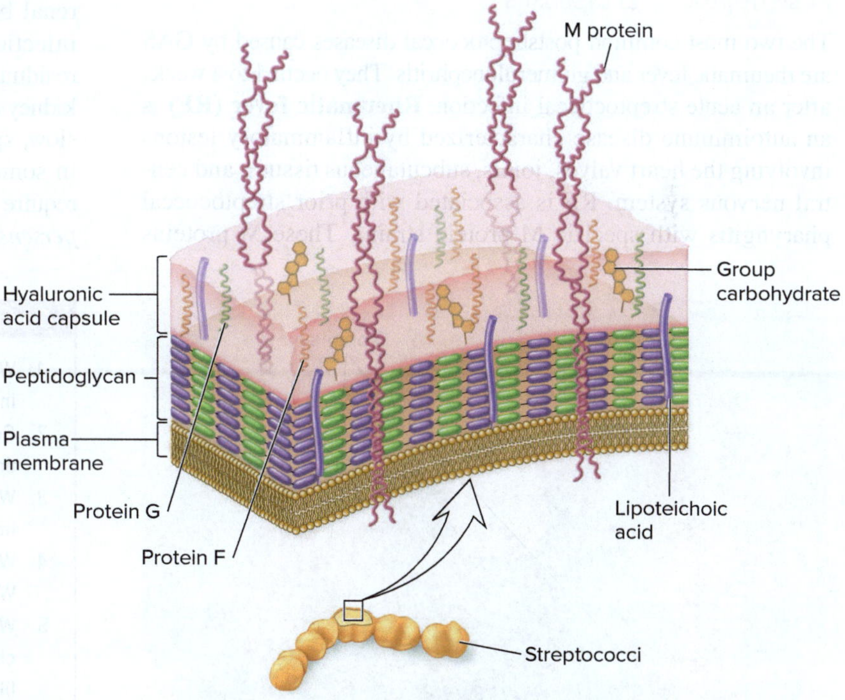

Figure 38.8 ***Streptococcus pyogenes* Cell Envelope.** The M protein is a major virulence factor for group A streptococci. It facilitates bacterial attachment to host cells and has antiphagocytic activity. Protein G prevents attack by antibodies because it binds to the Fc portion of antibody molecules, preventing the antigen-binding site from interacting with the bacterium. Protein F is an epithelial cell attachment factor. The Lancefield group is based on the antigenicity of the group carbohydrate.

MICRO INQUIRY *How is the M protein thought to be involved in poststreptococcal diseases?*

(**figure 38.9**). GAS are spread by droplets of saliva or nasal secretions. The incubation period in humans is 2 to 4 days.

The action of the strep bacteria in the throat (**pharyngitis**) or on the tonsils (**tonsillitis**) stimulates an inflammatory response and lysis of white and red blood cells. An exudate consisting of immune cells and fluid is released from blood vessels and deposited in the surrounding tissue in about 50% of patients with streptococcal pharyngitis. This is accompanied by a general feeling of discomfort or malaise, fever, and headache. Prominent physical manifestations include redness, edema, and lymph node enlargement in the throat. Signs and symptoms alone are not diagnostic because they are manifested by several other infections as well. Rapid test kits are commonly used to diagnose strep throat (*see figure 36.13*). In the absence of complications, the disease can be self-limiting and may disappear within 1 week. However, antibiotic treatment (penicillin or a macrolide antibiotic) can shorten the infection and clinical syndromes, and is especially important in children for the prevention of poststreptococcal diseases, as discussed next. Infections in older children and adults tend to be milder and less frequent due in part to the immunity developed against the many serotypes encountered in childhood.

Poststreptococcal Diseases

The two most common poststreptococcal diseases caused by GAS are rheumatic fever and glomerulonephritis. They occur 1 to 4 weeks after an acute streptococcal infection. **Rheumatic fever (RF)** is an autoimmune disease characterized by inflammatory lesions involving the heart valves, joints, subcutaneous tissues, and central nervous system. RF is associated with prior streptococcal pharyngitis with specific M protein strains. These M proteins trigger the production of autoantibodies that target cardiac proteins. The disease occurs most frequently among children six to fifteen years of age and manifests as a constellation of signs and symptoms including joint pain, heart inflammation (myocarditis), and a ring like rash. In the United States, rheumatic fever has become very rare (less than 0.05% of streptococcal infections), but it occurs more frequently in countries where health care is under-resourced. Therapy is directed at decreasing inflammation and fever, and controlling cardiac failure. Salicylates and corticosteroids are the mainstays of treatment. Although rheumatic fever is rare, it is still the most common cause of permanent heart valve damage in children.

Poststreptococcal glomerulonephritis (PSGN) follows GAS skin infections. It is an inflammatory disease (type III hypersensitivity reaction) of the renal glomeruli—membranous structures within the kidney where blood is filtered. Damage results from the deposition of antigen-antibody complexes in the glomeruli. Evidence suggests that the antigens involved are *S*treptococcal *p*yrogenic *e*xotoxin *B* (SpeB) and nephritis-associated plasmin receptor (NAPlr). The complexes cause destruction of the glomerular membrane, allowing proteins and blood to leak into the urine. Clinically, the person who is affected exhibits edema, fever, hypertension, and hematuria (blood in the urine). Diagnosis is based on clinical history, blood and urine analyses, renal biopsy, and confirmatory evidence of prior streptococcal infection. Penicillin G or azithromycin can be given for any residual streptococci. However, there is no specific therapy once kidney damage has occurred. About 80 to 90% of all cases undergo slow, spontaneous healing of the damaged glomeruli; however, in some cases, a chronic form of the disease develops. This may require a kidney transplant or lifelong renal dialysis. *Type III hypersensitivity (section 32.9)*

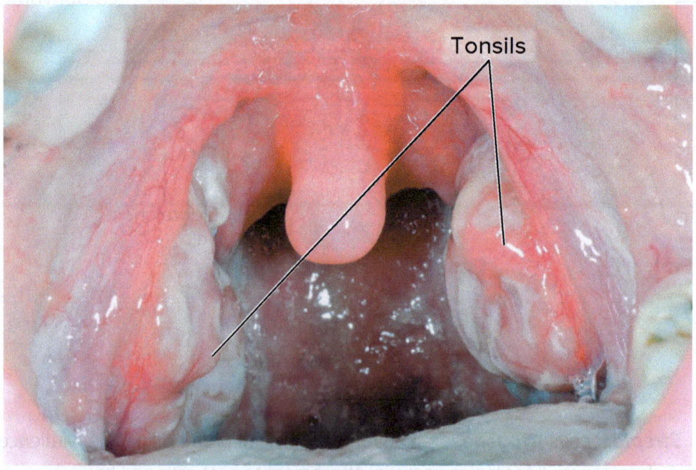

Figure 38.9 Streptococcal Pharyngitis (Strep Throat). Purulent exudate (pus containing white blood cells and bacteria) is most pronounced on the tonsils.

DR P. MARAZZI/Science Source

MICRO INQUIRY *What causes the exudate seen in strep throat? Is it seen in all cases?*

Comprehension Check

1. Why do chlamydiae differentiate into specialized cell types for infection and reproduction?
2. Compare and contrast how people contract chlamydial pneumonia and Legionnaires' disease.
3. What causes the typical symptoms of diphtheria? How are individuals protected against this disease?
4. What are the three major causes of bacterial meningitis in adults? Why is it important to determine which type a person has?
5. What immune cells would you find in a granuloma? Why are children and individuals who are immunocompromised more likely to develop miliary TB?
6. What are multidrug-resistant strains of tuberculosis and how do they develop?
7. Why is pneumonia caused by *M. pneumoniae* also called walking pneumonia?
8. What is the mechanism by which PTx kills host cells?
9. Describe the three stages of pertussis.
10. Which GAS disease is most prevalent? Describe the two major complications of a strep infection.

38.2 Arthropods Can Transmit Bacterial Diseases

After reading this section, you should be able to:

a. Discuss bacteria that cause common arthropod-borne diseases
b. Identify typical signs and symptoms of arthropod-borne bacterial diseases
c. Correlate arthropod-borne bacterial infection and disease severity with bacterial virulence factors

Bacteria cause fewer arthropod-borne diseases than do viruses and protozoa. However, they are of interest either historically (plague) or because they have emerged into human populations (ehrlichiosis, Lyme disease).

Lyme Disease

Lyme disease (Lyme borreliosis) was first documented in the United States in 1975 among residents of Old Lyme, Connecticut. It has become the most common tick-borne zoonosis in the United States. Estimates of confirmed cases of Lyme disease vary depending on surveillance methods, but an average of 30,000 are reported to the Centers for Disease Control and Prevention (CDC), with 95% of these located in only 14 states. The disease is also present in Europe and forested areas of Asia.

At least three spirochete species are responsible for Lyme disease. *Borreliella burgdorferi* is the primary cause in the United States, whereas *B. garinii* and *B. afzelii* cause the disease in Europe. Rodents and other wild animals are the natural hosts for the bacteria, which are transmitted by the bite of infected ticks. In the northeastern United States, the reservoir of *B. burgdorferi* is the white-footed mouse (**figure 38.10*a***). Spirochetes are transmitted to humans by the bite of infected black-legged (or deer) ticks (*Ixodes scapularis;* figure 38.10*b*). Although adult deer ticks feed on deer, these animals do not become infected. On the Pacific Coast, especially in California, the reservoir is a dusky-footed woodrat and the tick, *I. pacificus.* ◀ *Phylum Spirochaetota: bacteria with a corkscrew morphology (section 20.5)*

Clinically, Lyme disease is a complex illness, which if untreated has three major stages. The initial, localized stage occurs 7 to 10 days after an infectious tick bite. The illness sometimes begins with erythema migrans, an expanding, ring-shaped skin lesion with a red outer border and partial central clearing (figure 38.10*c*). However, often the tick bite is unnoticed, or the skin lesion is missed due to skin tone or its obscure location, such as on the scalp. Other signs and symptoms are flulike: malaise and fatigue, headache, fever, and chills.

The second, disseminated stage appears weeks or months after the initial infection. It consists of several symptoms such as neurological abnormalities, heart inflammation, and bouts of

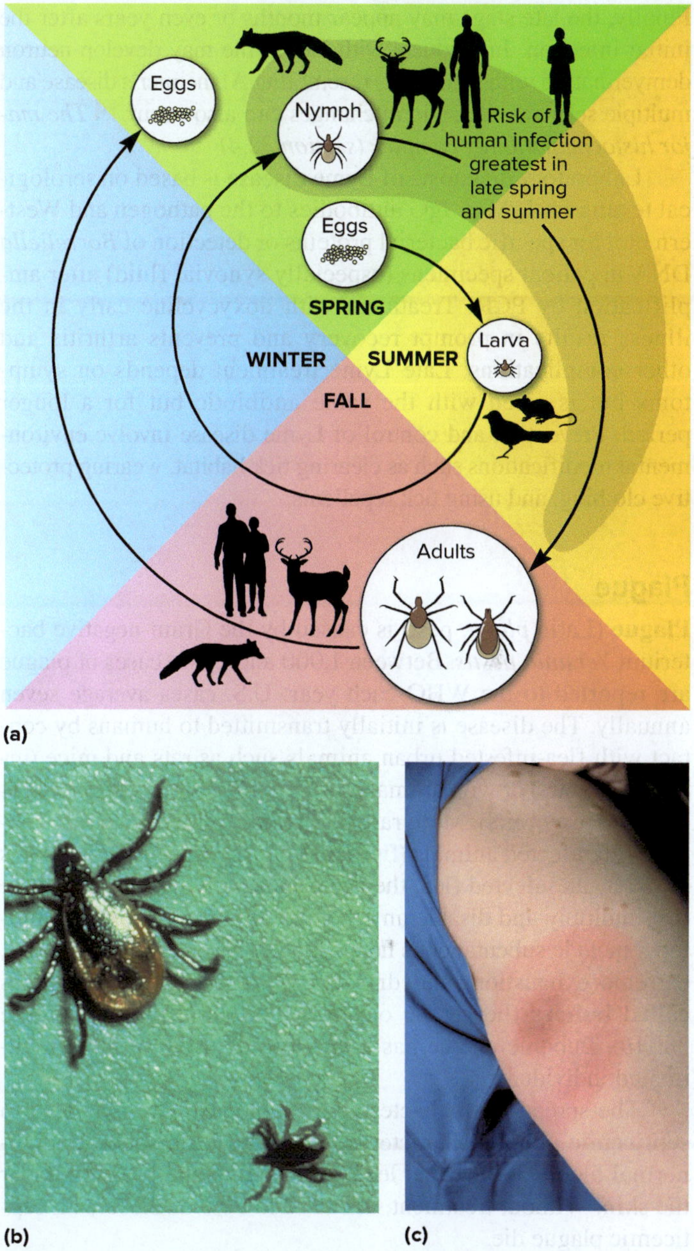

Figure 38.10 Lyme Disease. (a) *Borreliella burgdorferi* is transmitted by the tick *Ixodes scapularis,* which has a two-year life cycle. The life cycle begins with eggs laid in the spring (center). Most people are infected when nymphs take a blood meal. (b) An unengorged adult tick (*Ixodes scapularis;* bottom) is about the size of a pinhead, and an engorged adult (top) can reach the size of a jelly bean. (c) The typical rash (erythema migrans, sometimes called a bull's-eye rash) showing concentric rings around the initial site of the tick bite.

(a) CDC. Lyme Disease. https://www.cdc.gov/lyme/transmission/blacklegged.html; (b) Michael L. Levin, Ph. D./CDC; (c) James Gathany/Centers for Disease Control and Prevention

MICRO INQUIRY *What are the three stages of Lyme disease?*

arthritis, usually in major joints such as elbows or knees. Research indicates that Lyme arthritis might be an autoimmune response to major histocompatibility (MHC) molecules on cells in the synovium (joint) that are similar to bacterial antigens.

Finally, the late stage may appear months or even years after the initial infection. Individuals with late Lyme may develop neuron demyelination with symptoms resembling Alzheimer's disease and multiple sclerosis. Behavioral changes can also occur. *The major histocompatibility complex (section 32.4)*

Laboratory diagnosis of Lyme disease is based on serological testing for IgM or IgG antibodies to the pathogen and Western blot for specific bacterial proteins or detection of *Borreliella* DNA in patient specimens (especially synovial fluid) after amplification by PCR. Treatment with doxycycline early in the illness results in prompt recovery and prevents arthritis and other complications. Late Lyme treatment depends on symptoms but is often with the same antibiotic but for a longer period. Prevention and control of Lyme disease involve environmental modifications such as clearing tick habitat, wearing protective clothing, and using tick repellents.

Plague

Plague (Latin *plaga,* pest) is caused by the Gram-negative bacterium *Yersinia pestis.* Between 1,000 and 2,000 cases of plague are reported to the WHO each year; U.S. cases average seven annually. The disease is initially transmitted to humans by contact with flea-infested urban animals such as rats and mice (urban reservoirs) or wild animals such as coyotes and other rodents (sylvatic reservoirs). More rarely, it is transmitted by direct contact with infected animals (**figure 38.11*a***). When an individual is bitten by an infected flea, the bacteria enter lymph nodes, where they multiply and disable an effective immune response. Symptoms include subcutaneous hemorrhages, fever, chills, headache, extreme exhaustion, and dramatically enlarged lymph nodes called **buboes** (hence the old name, **bubonic plague**) (figure 38.11*b*). Bubonic plague has a fatality rate of 50 to 80% in untreated individuals.

The spread of the bacteria in the bloodstream may lead to **septicemic plague,** characterized by high fever, shock, and abnormal blood clotting that leads to bleeding and bruising under the skin. Without treatment, about 70 to 90% of those with septicemic plague die.

Bubonic plague can give rise to the more dangerous **pneumonic plague** when the bacteria travel through the bloodstream to the lungs (figure 38.11*a*). Pneumonic plague contributed to the epidemic spread in Europe and elsewhere because it is transmitted from person to person through respiratory droplets (see chapter opening story). The mortality rate for pneumonic plague is almost 100% if untreated.

Like many other Gram-negative pathogens, *Y. pestis* uses a type III secretion system to deliver effector proteins directly into targeted host cells. *Y. pestis* secretes plasmid-encoded *y*ersinial *o*uter *p*roteins (YOPS) into phagocytic cells to counteract natural defense mechanisms and help the bacteria multiply and disseminate in the host (**figure 38.12**). Additionally, when *Y. pestis* is phagocytosed by immune cells, it is able to survive and proliferate. *Protein maturation and secretion (section 13.9); Order Enterobacterales is a large, diverse taxon (section 21.2); Pathogenicity islands encode virulence factors (section 34.4)*

Because *Y. pestis* is listed as a potential bioterrorism agent, sentinel laboratory identification is restricted to initial staining and culture tests. Confirmation of the identity of *Y. pestis* by PCR, advanced serological studies, and other approved practices is restricted to national reference laboratories. Treatment is with streptomycin and recovery from the disease provides immunity. Prevention involves flea and rodent control, isolation of human patients, prophylaxis therapy of exposed persons, and vaccination of persons at high risk (e.g., park rangers, veterinarians).

Rocky Mountain Spotted Fever

Rocky Mountain spotted fever (RMSF) is caused by the rickettsia *Rickettsia rickettsii.* Although this disease was originally detected in the Rocky Mountain area, most cases occur east of the Mississippi River. The disease is principally transmitted by two tick vectors: *Dermacentor andersoni,* the wood tick, which is active during the spring and early summer in the Rocky Mountain states, and *D. variabilis,* the dog tick, which has assumed greater importance and is almost exclusively confined to the eastern half of the United States. Unlike other rickettsias, *R. rickettsii* can pass from generation to generation of ticks through their eggs in a process known as transovarian passage. Therefore no animals are needed as reservoirs for the continued propagation of this rickettsia.

The rickettsia are transmitted to humans from the saliva of an infected tick during feeding. The bacteria induce endocytosis by endothelial cells of small blood vessels, in which they multiply. The ensuing endothelial damage leads to leakage of blood into the extravascular space, resulting in the characteristic rash of RMSF. The rash is non blanching because it does not fade when pressed upon. As rickettsia reproduce within the endothelial cells, they polymerize host actin to make a tail at one pole that propels them from one endothelial cell to the next (*see figure 34.6*). This further damages blood vessels and also limits rickettsial exposure to immune cells and host antibodies. *Order Ricketsiales includes obligate intracellular bacteria (section 21.1)*

RMSF is characterized by severe headache, high fever, chills, and a skin rash (**figure 38.13**). The rash initially appears on the hands and feet (including palms and soles, respectively) and then spreads to the trunk of the body. If RMSF is not treated, the rickettsias destroy blood vessels in the heart, lungs, or kidneys and can cause death about 8 days after infection. Severe pathological changes are avoided by antibiotic therapy (doxycycline) and supportive therapy. Diagnosis is made through observation of symptoms and signs such as the characteristic rash and by serological and nucleic acid amplification tests. Prevention is by avoiding tick-infested habitats and animals. The CDC reports an increase in RMSF cases during the last several decades, with approximately 2,000 in the United

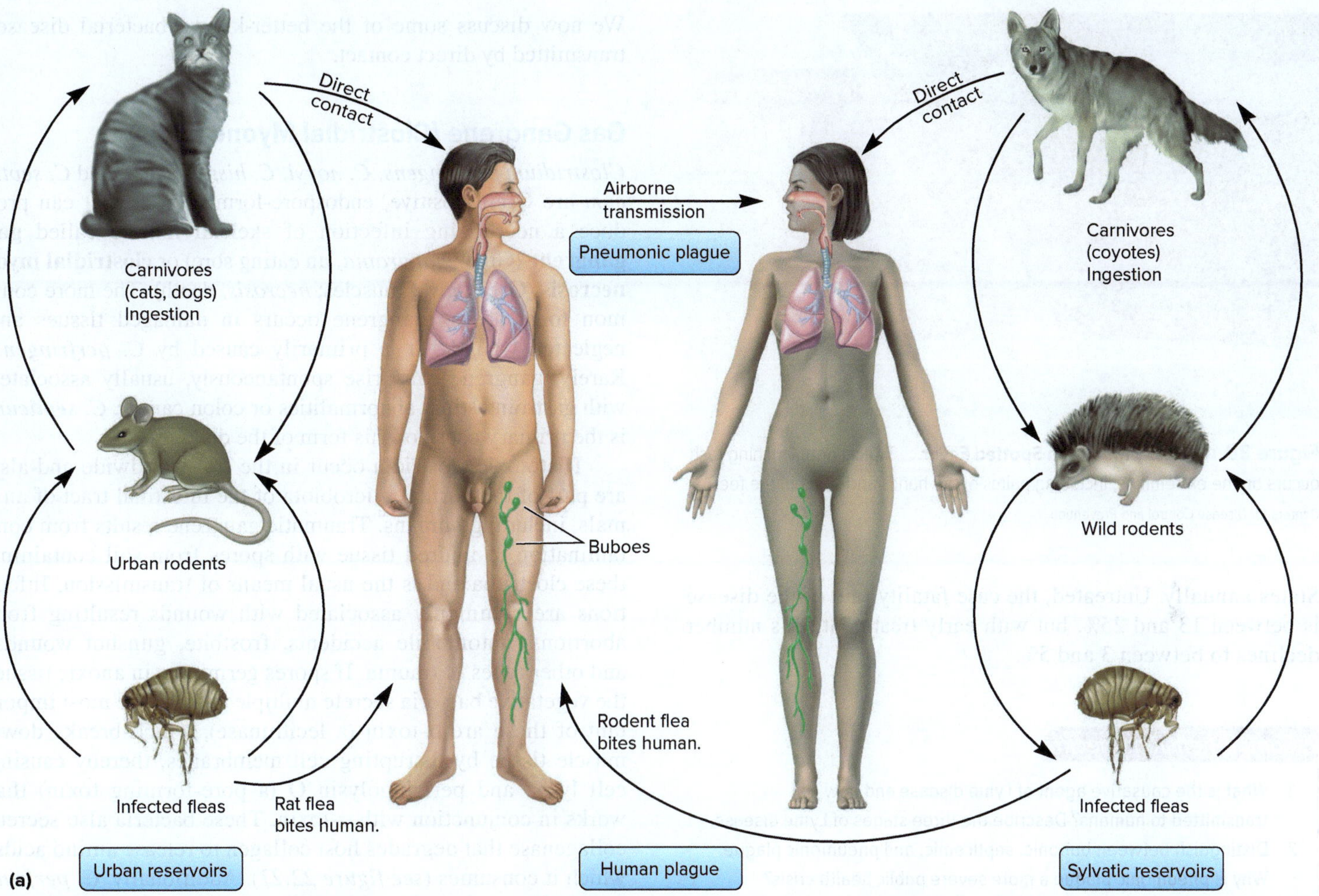

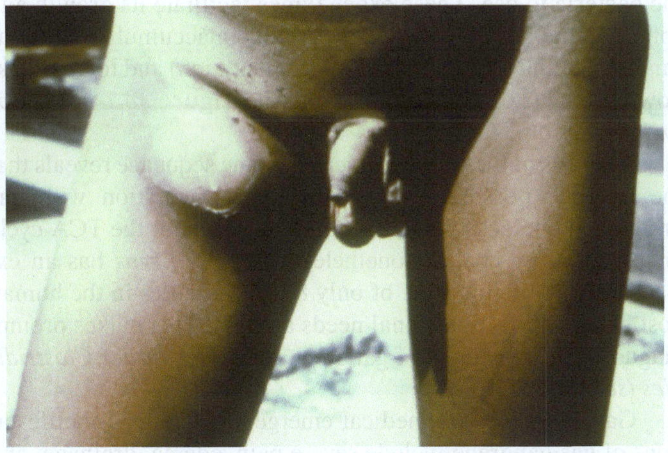

Figure 38.11 Plague. (a) Plague is spread to humans through (1) bites by infected city rodent fleas (urban cycle), (2) bites by infected wild rodent fleas (sylvatic cycle), or (3) airborne transmission from a person who is infected, leading to pneumonic plague. Dogs, cats, and coyotes can also acquire the bacterium by ingestion of infected animals. (b) Enlarged lymph nodes called buboes are characteristic of *Yersinia* infection.

(b) Dr. H.G. Scott/CDC

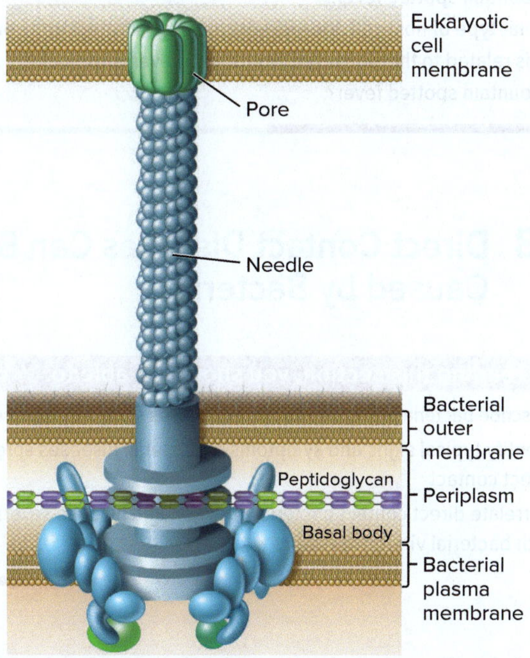

Figure 38.12 Type III Secretion System. X-ray fiber diffraction studies resolve the injectisome (needle) as a helical structure.

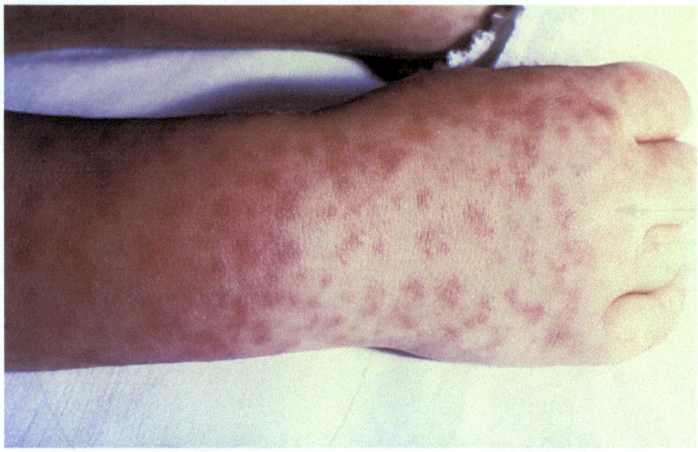

Figure 38.13 Rocky Mountain Spotted Fever. Typical nonblanching rash occurs on the extremities, including palms of the hands and soles of the feet.
Centers for Disease Control and Prevention

States annually. Untreated, the case fatality rate of the disease is between 13 and 25%, but with early treatment, this number declines to between 3 and 5%.

Comprehension Check

1. What is the causative agent of Lyme disease and how is it transmitted to humans? Describe the three stages of Lyme disease.
2. Distinguish between bubonic, septicemic, and pneumonic plague. Why is pneumonic plague a more severe public health crisis?
3. Compare the rash seen in Lyme disease with that seen in Rocky Mountain spotted fever.
4. What type of host cell does *Rickettsia rickettsii* infect and how is this related to the signs, symptoms, and severity of Rocky Mountain spotted fever?

38.3 Direct Contact Diseases Can Be Caused by Bacteria

After reading this section, you should be able to:

a. Describe bacteria that cause common diseases spread by direct contact
b. Identify typical signs and symptoms of bacterial diseases spread by direct contact
c. Correlate direct contact bacterial infection and disease severity with bacterial virulence factors

Most of the direct contact bacterial diseases involve the skin, underlying tissues, or mucous membranes. Occasionally, bacteria penetrate beyond these regions to cause disseminated disease. We now discuss some of the better-known bacterial diseases transmitted by direct contact.

Gas Gangrene (Clostridial Myonecrosis)

Clostridium perfringens, C. novyi, C. histolyticum, and *C. septicum* are Gram-positive, endospore-forming rods that can produce a necrotizing infection of skeletal muscle called **gas gangrene** (Greek *gangraina,* an eating sore) or **clostridial myonecrosis** (Greek *myo,* muscles; *necrosis,* death). The more common form of gas gangrene occurs in damaged tissues and neglected wounds. It is primarily caused by *C. perfringens.* Rarely, gangrene can arise spontaneously, usually associated with gastrointestinal abnormalities or colon cancer. *C. septicum* is the primary cause of this form of the disease.

Histotoxic clostridia occur in the soil worldwide and also are part of the normal microbiota of the intestinal tract of animals, including humans. Traumatic gangrene results from contamination of injured tissue with spores from soil containing these clostridia and is the usual means of transmission. Infections are commonly associated with wounds resulting from abortions, automobile accidents, frostbite, gunshot wounds, and other types of trauma. If spores germinate in anoxic tissue, the vegetative bacteria secrete multiple toxins. The most important of these are α-toxin (a lecithinase), which breaks down muscle tissue by disrupting cell membranes, thereby causing cell lysis, and perfringolysin O (a pore-forming toxin) that works in conjunction with α-toxin. These bacteria also secrete collagenase that degrades host collagen to release amino acids, which it consumes (*see figure 22.21*). Additionally, *C. perfringens* secretes the enzyme sialidase that breaks down sialic acid on the surface of host cells, releasing carbon and nitrogen for the bacteria to use. These exoenzymes facilitate its growth and spread in the host. Growth often results in accumulation of gas (mainly H_2 and CO_2 as a result of fermentation) and toxic breakdown products of skeletal muscle tissue (**figure 38.14**). ⏮ *Exotoxins (section 34.4)*

Analysis of the *C. perfringens* genome sequence reveals that the microbe possesses the genes for fermentation with gas production but lacks genes encoding enzymes for the TCA cycle or a respiratory chain. Nonetheless, *C. perfringens* has an extraordinary doubling time of only 8 to 10 minutes in the human host, where all its nutritional needs are met. This makes prompt diagnosis and treatment of gangrene critical. ⏮ *Order Clostridiales (section 22.2)*

Gas gangrene is a medical emergency! Clinical manifestations of gas gangrene include severe pain, edema, drainage, and muscle necrosis. Diagnosis is through culture of the causative species of clostridia and evaluation of patient signs and symptoms. Treatment requires extensive surgical removal of all dead tissue (debridement) and antimicrobial therapy with a β-lactam antibiotic and a macrolide. Interestingly, laboratory-raised, sterile maggots can also be used to remove dead and infected tissue. Amputation of limbs may be necessary to prevent further spread of the disease.

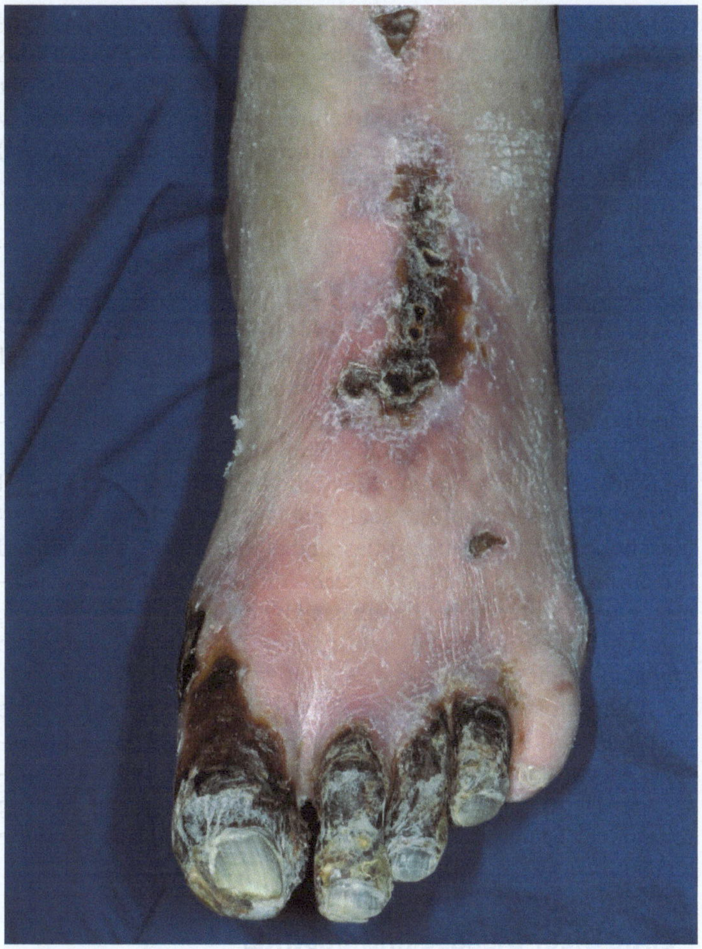

Figure 38.14 Gas Gangrene (Clostridial Myonecrosis). Necrosis of muscle and other tissues results from the numerous toxins produced by *Clostridium perfringens.*

Scott Camazine/Alamy Stock Photo

Group B Streptococcal Disease

Streptococcus agalactiae, or **group B streptococcus (GBS),** is a Gram-positive bacterium that causes illness in older people and adults who are immunocompromised, as well as newborn babies. In susceptible adults, GBS causes sepsis, pneumonia, and skin and bone infections. In newborns, it causes meningitis, bacteremia, and pneumonia. If infected newborns present symptoms in the first week of life, it is called early-onset disease. GBS can also develop in infants who are one week to three months old (late-onset disease). Although early-onset disease is known to be transmitted maternally (mothers can be asymptomatic carriers), only about half of the infants with late-onset GBS disease are associated with a mother who is a GBS carrier. The source of infection for many newborns with late-onset GBS disease is unknown. Infant mortality due to GBS disease is about 5%. Babies who survive GBS disease, particularly those with meningitis, may have permanent disabilities, such as hearing or vision loss or developmental disabilities. The CDC reports approximately 30,000 cases of invasive GBS annually in all age groups. The number of infants with GBS has dropped sharply since routine screening of all females who are pregnant began in the mid-1990s.

Many people are asymptomatic GBS carriers—they are colonized by GBS but do not become ill. GBS carriage is temporary, meaning that carriers do not harbor the bacteria for life. Adults can carry GBS in the bowel, vagina, bladder, or throat. Twenty-five percent of females who are pregnant carry GBS in the rectum or vagina. In these cases, a fetus may be exposed to GBS during birth. Carriage is detected during pregnancy by the presence of the bacterium in culture specimens of either the vagina or the rectum. The CDC recommends that females who are pregnant have vaginal and rectal specimens cultured for GBS in late pregnancy. A positive culture means that the mother carries GBS—not that the mother or the baby will definitely become ill. Administration of antibiotics during labor is effective in preventing the spread of GBS from mother to baby.

GBS disease is diagnosed by culturing the Gram-positive, β-hemolytic streptococcus (*see figure 22.19*). An FDA-approved DNA test for GBS is also available if rapid diagnosis of disease is necessary or culture results are equivocal. GBS infections in both newborns and adults are usually treated with intravenous penicillin, unless resistance or hypersensitivity is indicated. A preventative vaccine is under development. In 2022, the FDA awarded Breakthrough Therapy Designation (BTD) for development of the GBS6 vaccine that protects against the six most prominent GBS strains. BTD status ensures quicker development and review of the vaccine. Nonetheless, GBS vaccine development is problematic due to multiple serotypes and a capsule containing sialic acid, a carbohydrate also found on many human cells. *Vaccines (section 41.1)*

Mycobacterial Skin Infections

Many pathogenic species of nontuberculous mycobacteria cause skin or soft-tissue infections. These diseases often are hard to diagnose as they may cause diverse and misleading signs and symptoms. Two skin-associated diseases caused by nontuberculous mycobacteria are presented here.

Leprosy

Leprosy (Greek *lepros,* scaly, scabby, rough), or Hansen's disease, is a skin disease caused by *Mycobacterium leprae* and *M. lepromatosis,* together referred to as *Mycobacterium leprae complex.* They are acid-fast organisms with a generation time of about 12 to 14 days. Leprosy can cause extreme physical changes in affected individuals. The only reservoirs of proven significance are humans and armadillos (*see chapter 7 opening story*). About 200,000 cases are reported annually worldwide, primarily in countries with warm climates. The 2021–2030 WHO Global Leprosy Strategy has the long-term goal of "zero infection and disease, zero disability, and zero stigma and discrimination" in countries with endemic disease. *Order Mycobacteriales includes important human pathogens (section 22.1)*

Transmission of leprosy is most likely to occur after prolonged exposure to infected individuals who shed large numbers of the causative bacteria, probably in nasal secretions. The incubation period for leprosy is about 3 to 5 years but may be even longer, and the disease progresses slowly. The bacterium invades peripheral nerve and skin cells and reproduces as an obligate intracellular parasite. It is most frequently found in Schwann cells that surround peripheral nerve axons and in mononuclear phagocytes. The earliest symptom is usually a slightly pigmented skin eruption several centimeters in diameter. Approximately 75% of all individuals with this early, solitary lesion heal spontaneously because they mount an effective cell-mediated immune response. However, in some individuals, this immune response fails to clear the mycobacteria and one of two distinct forms of the disease occurs: tuberculoid or lepromatous leprosy.

Tuberculoid (paucibacillary) leprosy is a mild, nonprogressive form of leprosy associated with a delayed-type hypersensitivity reaction to antigens on the surface of *M. leprae* complex. It is characterized by damaged nerves and regions of the skin that have lost sensation and are surrounded by a border of nodules (**figure 38.15**). Individuals who are affected and who do not develop hypersensitivity have a relentlessly progressive form of the disease called **lepromatous (multibacillary) leprosy.** In this form of the disease, large numbers of *M. leprae* complex develop in skin cells, particularly in extremities where the temperature is slightly cooler. The bacteria kill skin tissue, leading to a progressive loss of facial features, fingers, toes, and other structures. Moreover, nodules form all over the body. Nerves are also infected but usually are less damaged than in tuberculoid leprosy. *Type IV hypersensitivity (section 32.9)*

Because *M. leprae* and *M. lepramatosis* cannot easily be cultured, laboratory diagnosis is supported by the demonstration of the bacterium in biopsy specimens using direct fluorescent antibody staining. Serodiagnostic methods, such as the fluorescent leprosy antibody absorption test and ELISA, have been developed.

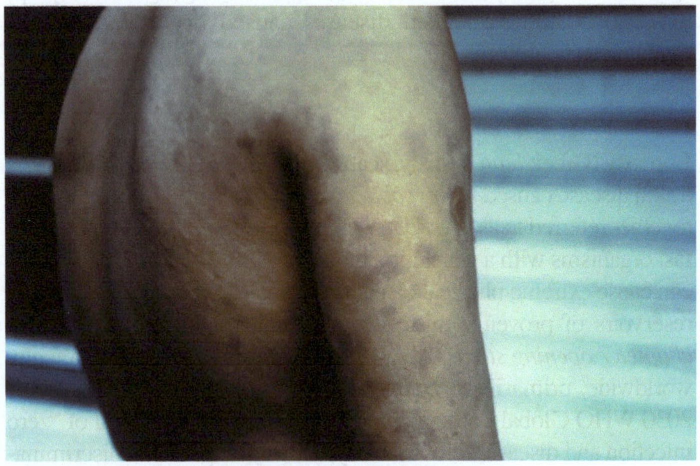

Figure 38.15 Leprosy. In tuberculoid leprosy, the skin manifests shallow, painless lesions.

Arthur E. Kaye/CDC

MICRO INQUIRY *What are the main differences between tuberculoid and lepromatous leprosy?*

Treatment of tuberculoid leprosy is with rifampicin combined with dapsone for 12 months. Clofazimine is added to this regimen for lepramatous leprosy, and treatment is continued for two years. Identification and treatment of patients with leprosy are key to its control. Close contacts (e.g., family members) should be given prophylactic drugs.

Tattoo-Associated Mycobacterial Infections

Sporadic cases and outbreaks of tattoo-associated mycobacterial infections remind us that emerging diseases do not have to be exotic. Rapidly growing mycobacteria (RGM) are a group of nontuberculous mycobacteria commonly found in water, soil, and dust. RGM have been identified as the cause of tattoo-associated skin infections. Tattoo-associated mycobacterial infections feature a range of skin lesions, including pink, red, and purple papules (pimples); scaly papules; granulomatous papules; lichenoid papules; pustules; and plaques. Tattoo-associated infections are caused by *M. haemophilum,* as well as *M. chelonae,* which contaminate the ink used by the artist. In most cases, it takes 6 to 9 months of multicourse antibiotic regimens to eradicate the infection.

Laboratory diagnosis for RGM is by culture and routine bacterial staining. Antibiotic sensitivity testing is also important. As with other mycobacterial infections, multiple antibiotics are used concurrently to prevent the development of drug resistance. Topical steroids may also be indicated to control the lesions.

Peptic Ulcer Disease and Gastritis

Peptic ulcer disease is caused by the multiflagellated, Gram-negative, spiral bacterium *Helicobacter pylori.* This organism is responsible for most cases of chronic **gastritis** not associated with another known primary cause. Importantly, untreated *H. pylori* infection can lead to gastric cancer and it has been classified as a Class I carcinogen by the WHO. Barry Marshall and Robin Warren won the 2005 Nobel Prize in Physiology or Medicine for discovering the link between *H. pylori* and gastritis. *Phylum Campylobacterota: human and animal commensals (section 20.10)*

H. pylori colonizes mucus-secreting cells, beneath the gastric mucous layers using fimbriae as adhesins (**figure 38.16**). It binds to Lewis b antigens, which are on the surface of red blood cells. *H. pylori* also binds to sialic acid, a common sugar component of glycoproteins on the surface of gastric epithelial cells. *H. pylori* produces urease, which creates a localized alkaline environment when it hydrolyzes urea to produce ammonium. The increased pH protects the bacterium from gastric acid as it grows under the layer of mucus in the stomach. The potential virulence factors responsible for epithelial cell damage and inflammation include proteases, phospholipases, and cytotoxins, including VacA and CagA.

Laboratory identification of *H. pylori* is by culture of gastric biopsy specimens, examination of stained biopsies for the presence of bacteria, detection of antigen in stool, and a test that

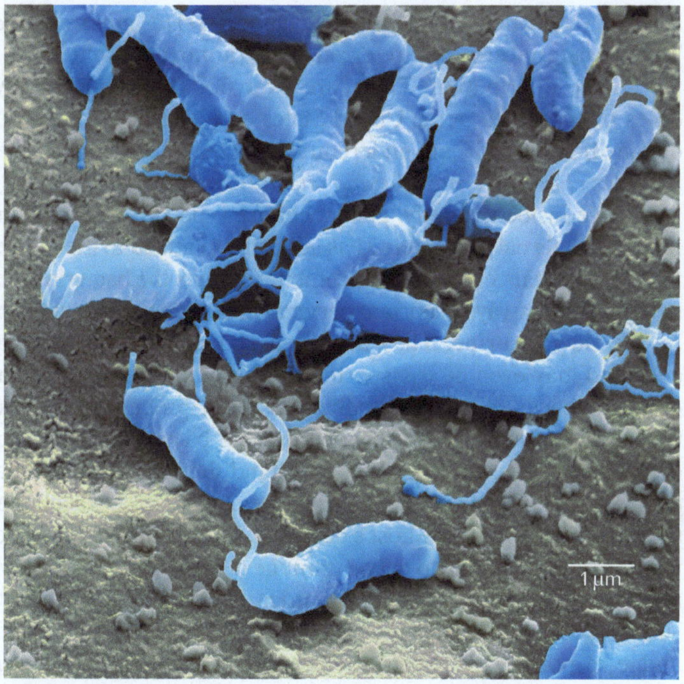

Figure 38.16 *Helicobacter pylori*, the Causative Agent of Peptic Ulcer Disease. Colorized scanning electron micrograph of *H. pylori* adhering to gastric cells.

Juergen Berger/Science Source

MICRO INQUIRY *How does* H. pylori *increase the local pH in its gastric microenvironment?*

detects CO_2 in the patient's breath after they ingest urea. The goal of *H. pylori* treatment is complete elimination of the organism and is two-pronged: use of drugs to decrease stomach acid (proton pump inhibitors) and antibiotics (metronidazole and tetracycline, amoxicillin, or clarithromycin) to kill the bacteria.

Approximately 45% of the world's population is estimated to be infected with *H. pylori*. It is most likely transmitted from person to person (usually acquired in childhood), although infection from a common exogenous source cannot be ruled out, and some hypothesize that it is also spread by food or water. Support for person-to-person transmission comes from evidence of clustering within families and from reports of higher-than-expected prevalences in residents of custodial institutions such as nursing homes.

Sexually Transmitted Infections

Sexually transmitted infections (STIs) represent a worldwide public health problem. There are over 30 different bacteria, viruses, and protozoa transmitted between humans through sexual contact. Viruses that cause STIs are presented in chapter 37, and the protozoa in chapter 39. The most common bacterial conditions include chlamydial infections, gonorrhea, and syphilis. These and other sexually transmitted bacteria, and the diseases they cause, are presented in **table 38.1**.

STIs were previously referred to as sexually transmitted diseases (STDs). Although they occur most frequently in the most sexually active age group, anyone who has sexual contact with an

Table 38.1 Summary of the Major Sexually Transmitted Infections (STIs) Caused by Bacteria

Microorganism	Disease	Comments	Treatment
Chlamydia trachomatis	Nongonococcal urethritis (NGU); cervicitis, pelvic inflammatory disease (PID), lymphogranuloma venereum	Serovars D–K cause most of the STIs in the U.S.; lymphogranuloma venereum rare in the U.S., caused by serovars L1, L2, L3.	Doxycycline and azithromycin
Haemophilus ducreyi	Chancroid soft chancre	Open sores on the genitals can lead to scarring without treatment; on the rise in the U.S.	Azithromycin and ceftriaxone
Helicobacter cinaedi, H. fennelliae	Diarrhea and rectal inflammation in males who have sex with males	Common in individuals who are immunocompromised	Metronidazole, macrolides
Metamycoplasma hominis	Implicated in some cases of PID	Widespread, often asymptomatic but can cause PID in females.	Azithromycin
Mycoplasmoides genitalium	Implicated in some cases of NGU	Only recently described as an STI	Azithromycin
Neisseria gonorrhoeae	Gonorrhea, PID	Usually symptomatic in males and asymptomatic in females; antibiotic-resistant strains are increasing in prevalence.	Ceftriaxone with either azithromycin or doxycycline
Treponema pallidum	Syphilis, congenital syphilis	Manifests many clinical syndromes, depending on stage of infection (primary, secondary, tertiary)	Benzathine penicillin G or doxycycline if allergic
Ureaplasma urealyticum	Urethritis	Widespread, often asymptomatic but can cause PID in females and NGU in males; associated with premature birth	Azithromycin or doxycycline

individual who is infected is at increased risk of acquiring an STI. In general, the more sexual partners a person has, the more likely the person will acquire an STI.

The spread of most STIs is currently at an all-time high. The WHO estimates that over 1 million new STI cases occur each day, with most infections occurring in fifteen- to thirty-year-old individuals. In the United States, the CDC reported more than 2.4 million new cases of syphilis, chlamydia, and gonorrhea in 2024. However, this is an underestimate because many cases go undiagnosed; others are not reported at all. This is particularly problematic as females with untreated chlamydia or gonorrhea infections risk infertility, and both males and females with papillomavirus infection have a greater risk of various cancers.

Some microorganisms that cause STIs can also be transmitted by nonsexual means. Examples include transmission by contaminated hypodermic needles and syringes shared among people who use intravenous drugs and transmission from infected mothers to their infants. Preventive measures are based mainly on education, barrier protection, and treatment of individuals who are infected and their partners with chemotherapeutic agents.

Chlamydial Diseases

Chlamydia is the most frequently reported sexually transmitted bacterial disease in the United States. Over 1.6 million cases of chlamydia were reported to the CDC in 2023. However, this number is thought to be grossly underestimated as most people with chlamydial infections have no symptoms and thus go undiagnosed. The most frequently isolated causative agent is *C. trachomatis;* eight different serotypes of the pathogen cause genital infections. It is transmitted through anal, oral, and vaginal sex and can be transmitted from mother to child during vaginal childbirth. Symptoms of chlamydial disease vary widely. Males and females can be asymptomatic. If symptoms occur in males, they include a urethral discharge, burning during urination, and itching. Symptoms in females include a severe infection of the cervix and urethra, with vaginal discharge and burning on urination. The infection may spread from the cervix to the fallopian tubes, leading to **pelvic inflammatory disease (PID),** with symptoms of abdominal pain, back pain, fever, nausea, and unusual vaginal bleeding. Chlamydial infections of the throat can occur in those who have oral sex with partners who are infected. In females who are pregnant, chlamydial infections are especially serious because they are directly related to miscarriage, stillbirth, newborn conjunctivitis, and infant pneumonia. Chlamydia is easily treated with a macrolide or doxycycline; because it is frequently cotransmitted with *N. gonorrheae,* patients treated for gonorrhea are also treated for chlamydia.

Gonorrhea

Gonorrhea (Greek *gono,* seed; *rhein,* to flow) is an acute sexually transmitted disease of the mucous membranes of the genitourinary tract, eye, rectum, and throat (table 38.1). It is caused by the Gram-negative, diplococcus *Neisseria gonorrhoeae.* These bacteria are also referred to as **gonococci** and have a worldwide distribution. There are over 600,000 reported new cases of gonorrhea in the United States each year, which, as with chlamydia cases, is an undercount. ◀ *Order Burkholderiales includes chemoheterotrophs and chemolithotrophs (section 21.2)*

Once inside the body, gonococci use type IV pili and a surface molecule called protein II as adhesins to attach to the microvilli of mucosal cells. This attachment prevents the bacteria from being washed away by normal cervical and vaginal discharges or by the flow of urine. They are often endocytosed by mucosal cells and may even be transported through the cells to the intercellular spaces and subepithelial tissue, a process known as transcytosis. Neutrophils phagocytose the bacterium but gonococci survive because they can withstand the harsh phagosomal environment (**figure 38.17*a***). When gonococci are intracellular, they are hidden from the host's defenses. Extracellular gonococci use antigenic variation to evade immune detection. *Antigenic variation* is a mechanism by which a microbe randomly changes its surface proteins, forcing immune cells to repeatedly re-identify and re-respond to the bacteria. Through the use of antigenic variation, *N. gonorrhoeae* can display over 1,000 phenotypes to the

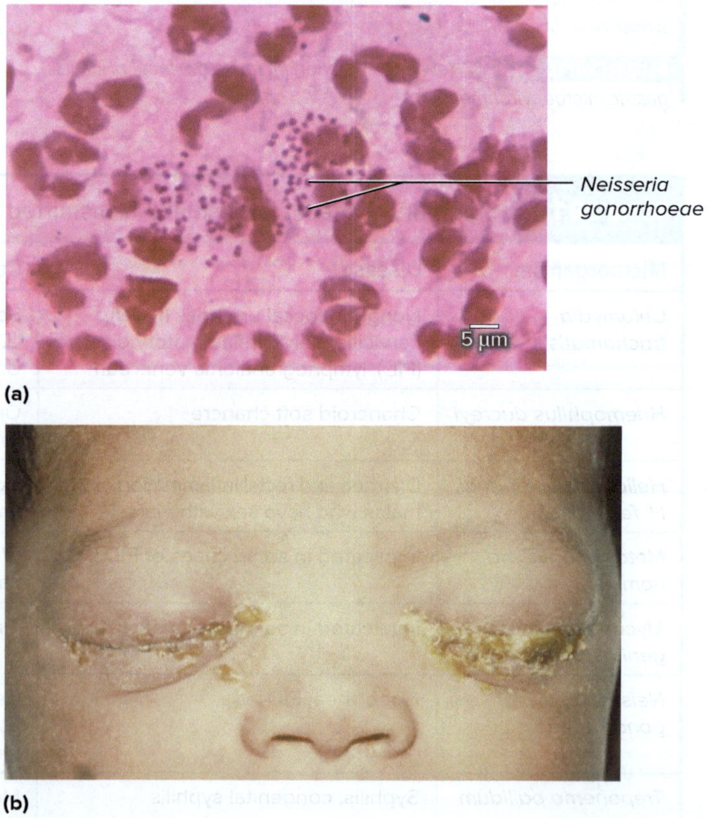

Figure 38.17 Gonorrhea. (a) Gram stain showing *Neisseria gonorrhoeae* (diplococci) inside neutrophils; light micrograph. (b) Gonococcal ophthalmia neonatorum in a one-week-old infant.

(a) Bill Schwartz/Centers for Disease Control and Prevention; (b) J. Pledger/Centers for Disease Control and Prevention host's immune system. Therefore, gonorrhea infection does not lead to immunity to reinfection. ◀ *Granulocytes (section 31.4); Phagocytosis destroys invaders (section 31.6); Surviving the host defenses (section 34.3)*

Gonorrhea is frequently asymptomatic in both males and females, but if symptoms do occur, they appear within 14 days or less. In males, they include urethral discharge of yellow, creamy pus and frequent, and painful urination with a burning sensation. In females, the cervix is the principal site infected, with few individuals showing symptoms. In those that do, symptoms include vaginal discharge and itchiness and abnormal vaginal bleeding. Gonococci also infect the fallopian tubes and surrounding tissues, leading to PID in 10 to 20% of cases. Gonococcal PID is a major cause of sterility and ectopic pregnancies because scar formation in the fallopian tubes prevents passage of the egg to the uterus. Gonococci disseminate most often during menstruation, when there is an increased concentration of free iron available to the bacteria. In both sexes, disseminated gonococcal infection with bacteremia may occur. This can lead to involvement of the joints (gonorrheal arthritis), heart (gonorrheal endocarditis), or pharynx (gonorrheal pharyngitis).

Gonorrheal eye infections can occur in newborns as they pass through an infected birth canal. The resulting disease is called **ophthalmia neonatorum,** or **conjunctivitis of the newborn** (figure 38.17*b*). This was once a leading cause of blindness. To prevent this disease, as well as chlamydial conjunctivitis, erythromycin is placed in the eyes of all newborns. This treatment is required by law in many nations.

Laboratory diagnosis of gonorrhea is by detecting *N. gonorrhoeae* using nucleic acid amplification tests such as PCR. If results are positive, they are followed by culture to determine oxidase reaction, Gram-stain reaction, and colony and cell morphology. Gonorrhea is currently treated with a single injection of ceftriaxone. Because *N. gonorrheae* and *C. trachomatis* are often cotransmitted, current treatment for gonorrhea also includes a macrolide or doxycycline to treat chlamydia as well. Later stages of gonorrhea are more difficult to treat and may require larger drug doses over longer periods of time. Alarmingly, strains resistant to all classes of antibiotics have been isolated, and there is risk that gonorrhea may become untreatable. It is listed by the CDC as an urgent public health threat.

Syphilis

Syphilis (Greek *syn,* together; *philein,* to love) is an ulcerative disease (table 38.1) caused by the spirochete *Treponema pallidum.* Infection occurs through intimate contact with the infected lesion of a sexual partner. Syphilis has many nicknames, including the "great imitator" as its many signs and symptoms mimic those of other diseases. The signs and symptoms of syphilis vary depending on the stage of infection (primary, secondary, latent, and tertiary). **Congenital syphilis** occurs when the disease is transmitted in utero from mother to child. ◀ *Phylum Spirochaetota: bacteria with a corkscrew morphology (section 20.5)*

As a spirochete, *T. pallidum* is actively motile, allowing it to easily enter the body through mucous membranes and minor breaks or abrasions of the skin. It replicates locally, inducing ulceration, before it migrates to the regional lymph nodes, where it rapidly spreads throughout the body. Although *T. pallidum* is a successful pathogen, it is not equipped with many virulence factors. Rather it has few surface molecules that are antigenic, and like *N. gonorrhoeae,* it undergoes antigenic variation.

The signs and symptoms of syphilis vary, depending on the stage of infection. In the *primary stage,* after an incubation period of about 10 days to 3 weeks, the initial symptom is a small, painless, reddened ulcer, or **chancre** (French from Latin, *cancer,* creeping ulcer) with a hard ridge that appears at the infection site and contains spirochetes (**figure 38.18*a***). Contact with the chancre during sexual contact may result in disease transmission. In about one-third of the cases, the disease does not progress further and the chancre disappears. Serological tests are positive in about 80% of the infected individuals during this stage (**figure 38.19**).

Within 2 to 10 weeks after the primary lesion appears, the individual may develop *secondary syphilis,* characterized by a highly variable skin rash (figure 38.18*b*). By this time, 100% of individuals who are infected are serologically positive. Other symptoms during this stage include the loss of hair in patches, malaise, and fever. Like chancres, rash lesions are infectious.

After several weeks, the disease becomes latent. During the latent period, the disease is rarely infectious, except for transmission from mother to fetus. After many years, *tertiary syphilis* develops in about 40% of individuals who are untreated. This stage is characterized by a great reduction in the number of spirochetes in the body. At this time, masses of dead, fibrous tissue called **gummas** (figure 38.18*c*) may form in the skin, bone, and nervous system as the result of hypersensitivity reactions. Involvement of the central nervous system (neurosyphilis) can occur at any stage, and may result in tissue loss that can lead to cognitive deficits, blindness, a shuffle walk (tabes), or mental illness. Newborns with congenital syphilis may appear asymptomatic but develop widespread signs and symptoms early on, including liver and spleen enlargement, a blistering skin rash, watery nasal discharge, irritability, and fever. Without treatment, additional disease manifestations occur as the infant gets older; these include impaired vision and hearing, gastrointestinal disorders, joint pain, and malformations of bone, cartilage, and teeth.

Sadly much of what we know about the progression and prognosis of syphilis is based on shameful studies performed in the United States during the twentieth century (**Disease 38.1**).

Diagnosis of syphilis is through clinical history, physical examination, and immunoassay. Two tests, the venereal disease research laboratory (VDRL) and rapid plasmin reagin (RPR), measure the production of antibody made in response to host cells damaged by *T. pallidum.* These are called nontreponemal tests because they do not detect spirochete-specific antibody. Diagnosis is made by pairing a nontreponemal assay (i.e., VDRL or RPR) with a treponemal specific test, such as the *T. pallidum*

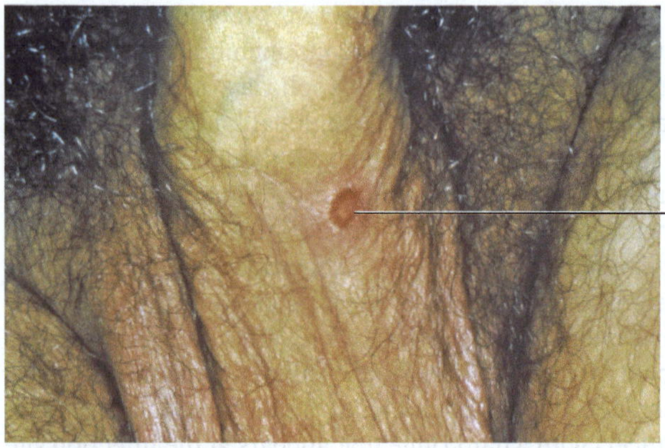

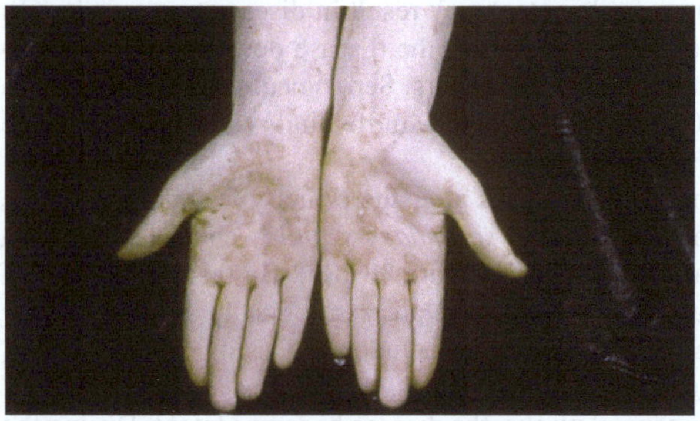

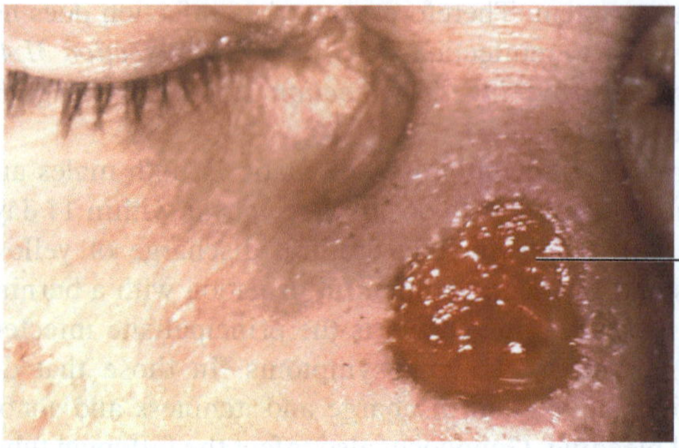

Figure 38.18 Syphilis. (a) Primary syphilitic chancre of the penis. (b) Syphilis can lead to the development of a rash anywhere on the body, including the palms of the hands and soles of the feet. (c) Late-stage tertiary syphilis marked by masses of swollen and dead, fiberlike tissue called gummas.

(a) Dr. N.J. Fiumara, Dr. Gavin Hart/CDC; (b) Robert Sumpter/CDC; (c) J. Pledger/CDC

MICRO INQUIRY *Which of these clinical manifestations are associated with syphilis transmission? Explain.*

particle agglutination (TPPA) test or the fluorescent treponemal antibody absorption test (FTA-ABS), which detect antibodies made against the pathogen. ◀ *Immune complex formation (section 32.8)*

Treatment in the early stages of the disease is easily accomplished with a single dose of long-acting benzathine penicillin G or a 2-week course of doxycycline, for individuals who are allergic to penicillin. Later stages of syphilis require larger doses over a longer period. Immunity to syphilis is not complete, and subsequent re-infections can occur.

Prevention and control of syphilis depend on (1) public education, (2) prompt and adequate treatment of all new cases, (3) follow-up on sources of infection and contacts so they can be treated, and (4) prophylaxis (barrier protection) to prevent exposure. Alarmingly, the CDC has reported a tenfold increase in the number of congenital syphilis cases between 2012 and 2022.

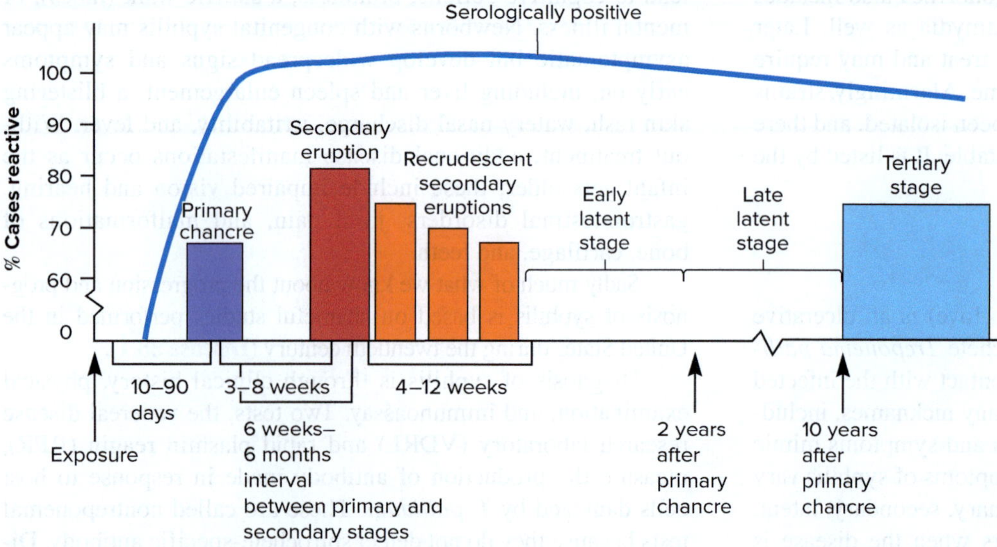

Figure 38.19 The Course of Untreated Syphilis. Recrudescent secondary eruptions are skin lesions that return weeks or months after they appear to have resolved.

Staphylococcal Diseases

Staphylococci are among the most important bacteria that cause disease in humans. They are normal inhabitants of the upper respiratory tract, skin, intestine, and vagina. Staphylococci are members of a group of invasive Gram-positive bacteria known as the **pyogenic** (pus-producing) cocci. ◀ *Order Staphylococcales (section 22.2)*

Staphylococci can be divided into pathogenic and relatively nonpathogenic strains based on synthesis of the enzyme coagulase (*Table 38.3*). The coagulase-positive species *Staphylococcus aureus* is the most important human pathogen in this genus, while coagulase-negative staphylococci

DISEASE

38.1 Syphilis and the Tuskegee Study

A research investigation named "Tuskegee Study of Untreated Syphilis in the Negro Male" would be unthinkable today. But it was the reality in 1932 Macon County, Alabama, when the federal Public Health Service began the study on 600 Black men (399 with syphilis, 201 without the disease). The tale of this study and its participants is a stain on the history of U.S. public health, for which President Bill Clinton formally apologized in 1997.

The "Tuskegee Study," as it was known, started with a racist objective—to develop syphilis treatment for Black people. The enrollees were provided free medical checkups, meals, and burial assistance but were not told the study had anything to do with syphilis. Rather, they were informed that they were being treated for "bad blood" (**box figure**). Except they weren't being treated. Even after penicillin was shown to be a highly effective cure for syphilis in 1947, treatment was withheld. The project was supposed to last 6 months; it went on for 40 years.

Finally in July 1972, a newspaper story broke the news that men were unknowingly enrolled in this highly unethical study, and a government panel confirmed that study participants had been misled and appropriate medical treatment had been withheld. At this time, it was also revealed that the men were never given the opportunity to quit the study. Three months later, the panel shut down the study.

The following year, a class-action lawsuit was filed on behalf of the study participants. In 1974, a $10 million settlement was reached. The U.S. government also promised to provide lifetime medical benefits and burial services for all enrollees. In 1975 the wives and children of the men participants were added. The last study participant died in 2004 and there are currently 10 offspring receiving benefits.

The unethical Tuskekee Experiment lasted 40 years.

CDC

(e.g., *S. epidermidis*) are less virulent. However, they have increasingly been associated with serious healthcare-acquired infections.

Staphylococci are further classified into slime producers and slime nonproducers. Slime is a viscous, extracellular glycoconjugate that allows these bacteria to adhere to smooth surfaces such as prosthetic medical devices and catheters. The ability to produce slime has been proposed as a marker for pathogenic strains of both coagulase-positive and negative staphylococci (**figure 38.20*a***). Scanning electron microscopy clearly demonstrates that biofilms (figure 38.20*b*) consisting of staphylococci encased in a slimy matrix are formed in association with biomaterial-related infections (**Disease 38.2**). Slime also inhibits neutrophil chemotaxis, phagocytosis, and antimicrobial agents (*see figure 34.7*). *Biofilms are common in nature (section 7.6); Biofilms provide microbial protection (section 34.3)*

Staphylococci are common in nature. About one-third of people harbor *S. aureus* on their nasal mucosa as asymptomatic carriers. Staphylococci can be spread by the hands, expelled from the respiratory tract, or transported in or on animate and inanimate objects. *S. aureus* can produce disease in almost every organ and tissue of the body.

S. aureus produces disease through the ability to multiply and spread widely in tissues and produce many virulence factors (**table 38.2**). Some of these factors are exotoxins, and others are enzymes thought to be involved in invasiveness *(see figure 22.15b)*. Many toxin genes are carried on plasmids; in some cases, their genes reside on both a plasmid and the bacterial chromosome, often in pathogenicity islands. The pathogenic capacity of a particular *S. aureus* strain is due to the combined effect of extracellular factors and toxins. At one end of the disease spectrum is staphylococcal food poisoning, a brief, self-limiting condition caused solely by the ingestion of preformed enterotoxin (table 38.2). At the other end of the spectrum are staphylococcal bacteremia and disseminated abscesses in most organs of the body.

The classic example of a staphylococcal lesion is the localized abscess of a domed pustule (**figure 38.21**). When *S. aureus* becomes established in a hair follicle, tissue necrosis results. Coagulase is produced and forms a fibrin wall around the lesion that limits bacterial spread. Within the center of the lesion,

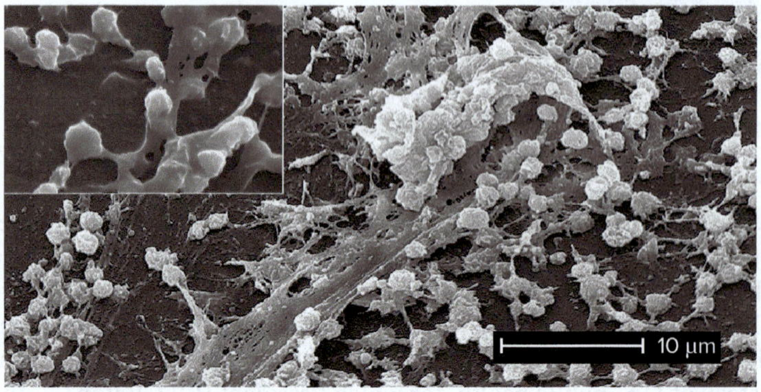

Figure 38.20 Slime and Biofilm. (a) Cells of *S. aureus* producing slime, seen as the black halo surrounding each cell (colorized transmission electron microscopy). (b) A biofilm on the inner surface of an intravenous catheter. Extracellular polymeric substances, mostly polysaccharides, surround and encase the staphylococci (scanning electron micrograph). Higher magnification inset at upper left shows extracellular slime encasing *S. aureus* cells.

(a) Dr. Kari Lounatmaa/Science Source; (b) (1) --Rodney M. Donlan, Ph.D.; Janice Haney Carr/CDC; (b) (2) Janice Carr/Centers for Disease Control and Prevention

DISEASE

38.2 Biofilms

Biofilms consist of microorganisms embedded in an organic polymer matrix of microbial origin. They develop on virtually all surfaces immersed in natural aqueous environments, including both biological (aquatic plants and animals) and abiological (concrete, metal, plastics, stones). Biofilms are particularly detrimental to individuals when they develop on implanted medical devices such as pacemakers and artificial joints.

It is estimated that at least 80% of all human bacterial infections involve biofilms. The microorganisms primarily involved are *Staphylococcus aureus, S. epidermidis,* and other coagulase-negative staphylococci. These normal skin inhabitants possess the ability to adhere tenaciously to the surfaces of inanimate prosthetic devices. Gram-negative bacteria such as *Pseudomonas aeruginosa* and *Klebsiella pneumoniae* also form robust biofilms but are not typically associated with medical devices. Biofilms may be polymicrobial, with many different species, as shown in the colorized scanning electron micrograph (**box figure**). Within biofilms, bacteria are protected from the body's normal defense mechanisms and also from antibiotics. The biofilm also provides a source of infection for other parts of the body as bacteria detach during biofilm sloughing. ◀ *Biofilms are common in nature (section 7.6)*

Some examples of biofilms of medical importance include:

- Infections of artificial knee and hip prostheses that can require replacement of the prosthetic device
- Patients with cystic fibrosis harboring great numbers of *Pseudomonas aeruginosa* in their lungs, which produce large amounts of alginate polymers
- Teeth, where biofilm forms plaque that leads to tooth decay
- Contact lenses, where bacteria may produce severe eye irritation, inflammation, and infection
- Air-conditioning and other water retention systems where potentially pathogenic bacteria, such as *Legionella* species, may be protected by biofilms from the effects of chlorination

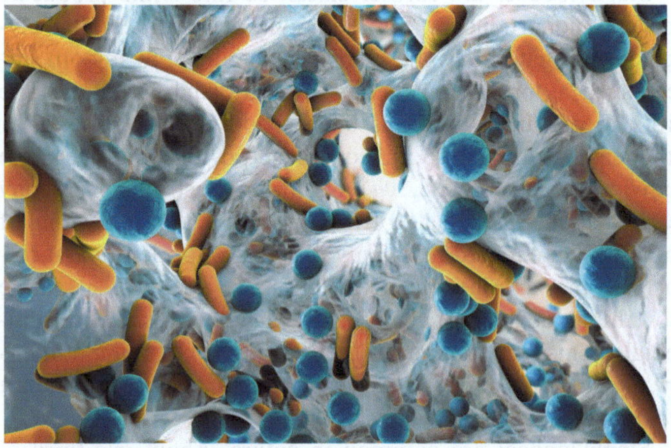

Many biofilms are polymicrobial

Kateryna Kon/Shutterstock

Table 38.2 Some of the Enzymes and Toxins Produced by Staphylococci

Product	Physiological Action
β-lactamase	Breaks down β-lactam antibiotics
Catalase	Converts hydrogen peroxide into water and oxygen to reduce killing by phagocytosis
Coagulase	Reacts with prothrombin to form a complex that can cleave fibrinogen and cause the formation of a fibrin clot; fibrin may also be deposited on the surface of staphylococci, which may protect them from destruction by phagocytic cells; coagulase production is synonymous with invasive pathogenic potential.
DNase	Hydrolyzes DNA
Enterotoxins	Proteins that are divided into heat-stable toxins of six known types (A, B, C1, C2, D, E); responsible for the gastrointestinal upset typical of food poisoning.
Exfoliative toxins A and B (superantigens)	Cause loss of the surface layers of the skin in scalded-skin syndrome
Hemolysins	Alpha hemolysin lyses erythrocytes and causes skin destruction. Beta hemolysin lyses erythrocytes and damages sphingomyelin around nerves.
Hyaluronidase	Also known as spreading factor; breaks down hyaluronic acid located between cells, allowing for penetration and spread of bacteria
Lipases	Break down lipids
Panton-Valentine leukocidin	Inhibits phagocytosis by neutrophils and can destroy these cells by forming pores in their phagosomal membranes
Phenol soluble modulins	Peptides that contribute to biofilm formation and survival within phagosome
Proteases	Break down proteins
Protein A	Surface protein that binds the Fc portion of antibodies so they are upside down on bacterial surface and no longer able to serve as opsonins.
Toxic shock syndrome toxin-1	A superantigen associated with the fever, shock, and multisystem involvement of toxic shock syndrome.

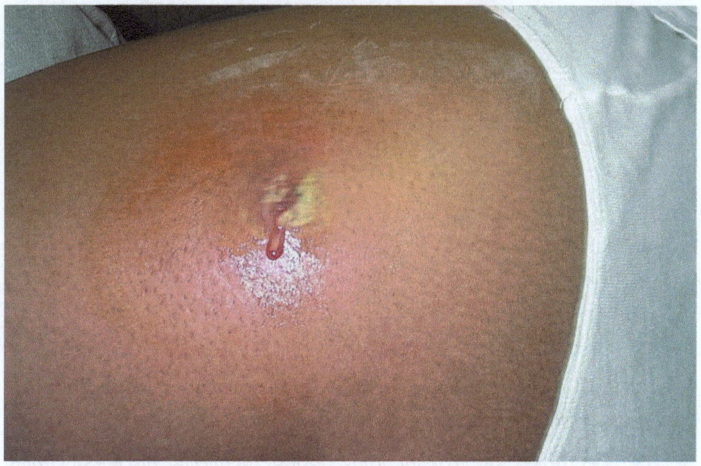

Figure 38.21 Staphylococcal Skin Infections. Draining abscess on the hip.
Centers for Disease Control and Prevention

liquefaction of necrotic tissue occurs. The abscess may be either a furuncle (boil) or a carbuncle (cluster of subdermal boils). The central necrotic tissue drains, and healing eventually occurs. However, the bacteria may spread through lymph or blood to other parts of the body.

Infants sometimes develop **staphylococcal scalded skin syndrome (SSSS)** caused by strains of *S. aureus* that produce **exfoliative toxin (exfoliatin),** a protein that is usually plasmid-encoded. Exfoliative toxin clips proteins that keep epidermal cells attached to the underlying skin layer, so the epidermis peels off to reveal inflamed dermis underneath—thus the name of the disease.

Toxic shock syndrome (TSS) is a staphylococcal disease with potentially serious consequences. TSS was initially described in the 1970s and was linked to the use of superabsorbent tampons. However, non menstrual cases also occur. TSS is caused by toxin-producing strains of *S. aureus*. The disease is

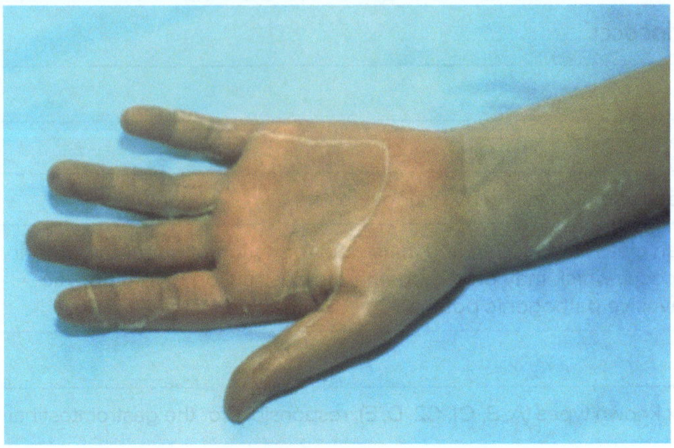

Figure 38.22 Skin Manifestations of Toxic Shock Toxin. Patients with toxic shock syndrome often exhibit significant peeling of skin (desquamation).
Centers for Disease Control and Prevention

characterized by low blood pressure, fever, diarrhea, an extensive skin rash, and shedding of the skin (**figure 38.22**). Signs and symptoms are caused by TSST-1, a **superantigen** that triggers massive overstimulation of T cells, resulting in the release of extraordinary levels of inflammatory cytokines, particularly TNF-α, IL-1, and IL-6 (*see figure 34.10*). This leads to circulatory collapse, shock, and multiorgan failure. ◀ *Toxins are biological poisons (section 34.4)*

The definitive diagnosis of staphylococcal disease can be made only by isolation and identification of the staphylococcus involved. This requires culture, catalase, coagulase, and other biochemical tests. Commercial rapid test kits are also available. There is no specific prevention for staphylococcal disease. The mainstay of treatment is the administration of specific antibiotics: Antistaphylococcal β-lactams (e.g., nafcillin) and first- and second-generation cephalosporins are first-line treatments. Because of the prevalence of drug-resistant strains, all staphylococcal isolates must be tested for antimicrobial susceptibility. Cleanliness, good hygiene, and aseptic management of lesions are the best means of control.

Methicillin-Resistant *Staphylococcus aureus* (MRSA)

Isolates of *Staphylococcus aureus* resistant to all β-lactam antibiotics are known as ***methicillin-resistant Staphylococcus aureus* (MRSA).** These include penicillins (e.g., penicillin, amoxicillin), "antistaphylococcal" penicillins (e.g., methicillin, oxacillin), cephalosporins (except the fifth-generation drug ceftaroline), and carbapenems. Antibiotic resistance is driven by the *mecA* (and in some cases, the *mecC*) gene encoding an altered penicillin-binding protein (transpeptidase) called PBP2a. Otherwise healthy people can acquire MRSA infections, usually on or in the skin, such as abscesses, boils, and other pus-filled lesions. About 100,000 infections and 20,000 deaths due to MRSA occur annually in the United States. When the infection is acquired by people who have not recently been hospitalized or had a medical procedure (e.g., dialysis, surgery, catheters), the infection is referred to as a *c*ommunity-*a*ssociated (CA)-MRSA infection. However, MRSA is often a nosocomial infection; such infections are called hospital-acquired MRSA (HA-MRSA). When they first emerged, CA-MRSA strains were distinct from HA-MRSA strains. But CA-MRSA strains are now commonly reported inside health-care settings and the designation CA- and HA-MRSA is helpful only in distinguishing their general phenotypes. In general, HA-MRSA strains are also resistant to several other classes of antibiotics. By contrast, CA-MRSA strains tend to remain sensitive to other classes of antibiotics but are often more virulent as they produce additional toxins, such as the Panton-Valentine leukocidin (table 38.2). ◀ *Peptidoglycan structure (section 3.3); Antibacterial drugs (section 9.4)*

Similar to other infections caused by staphylococci, MRSA skin infections begin as a bump or reddened area that looks like a large pimple. But unlike carbuncles or boils caused by methicillin-sensitive *S. aureus* (MSSA), MRSA infections may result in the production of more pus (**figure 38.23**). MRSA also causes the same spectrum of diseases as MSSA.

Treatment includes antibiotic therapy. Determining the antibiotic sensitivity of the cultured pathogen is essential. Doxycycline, trimethoprim/sulfamethoxazole, and clindamycin are options for uncomplicated cases. More severe cases are

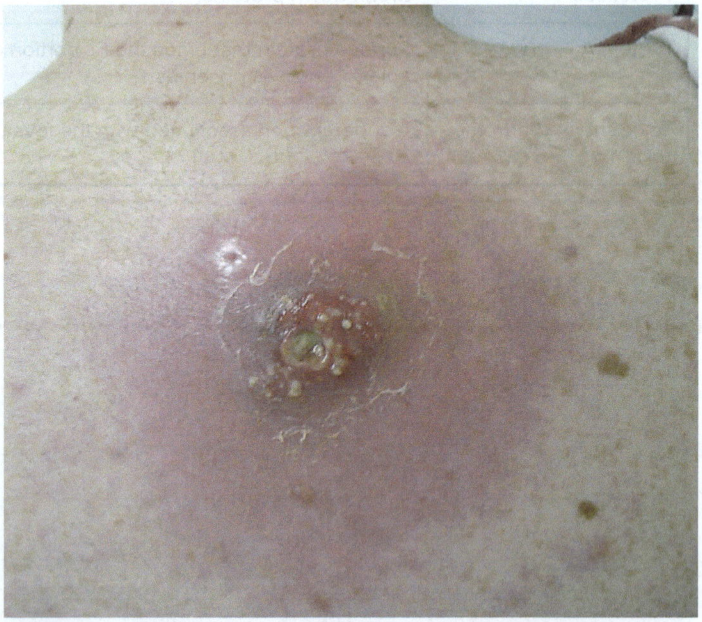

Figure 38.23 Methicillin-Resistant *Staphylococcus aureus* Wound.
Gregory Moran, M.D./Centers for Disease Control and Prevention

MICRO INQUIRY *To what drugs, besides methicillin, are MRSA strains resistant?*

treated with vancomycin, although other medications are used for vancomycin-insensitive or -resistant strains (VISA and VRSA).

Streptococcal Diseases

Skin and soft tissue infections can also be caused by group A streptococci (*S. pyogenes*). In addition, streptococcal disease can invasively infect the skin, spreading deeper into the tissue, often reaching the underlying muscle. This tissue is a very rich growth medium for streptococci, providing the nutrients for rapid replication. We discuss both the superficial and invasive streptococcal diseases next. Airborne transmission of *S. pyogenes* is discussed in section 38.1.

Cellulitis, Impetigo, and Erysipelas

Cellulitis is a diffuse, spreading infection of subcutaneous skin tissue, caused by either *S. pyogenes* or *S. aureus*. The resulting inflammation is characterized by a defined area of redness (erythema) and the accumulation of fluid (edema).

A frequently diagnosed skin infection caused by *S. pyogenes* and *S. aureus* is **impetigo.** Impetigo is a superficial infection, most commonly seen in children, usually located on the face, and characterized by crusty lesions and vesicles surrounded by a red border. It is contagious and can spread rapidly through a nursery, day-care, or school. Treatment of impetigo is with topical mupirocin. In severe cases, systemic antibiotic therapy is also prescribed. Penicillin or a first-generation cephalosporin is the drug of choice; azithromycin is prescribed for individuals allergic to β-lactams.

Unlike cellulitis and impetigo, which can be caused by either *S. pyogenes* or *S. aureus*, **erysipelas** (Greek *erythros*, red; *pella*, skin) is almost always caused by *S. pyogenes*. It is an acute infection of the dermal layer of the skin. It occurs primarily in infants and people over thirty years of age; it sometimes follows a streptococcal sore throat. The skin develops painful, reddish patches that enlarge and thicken with a sharply defined edge. Recovery usually takes a week or longer if no treatment is given. The drugs of choice for the treatment of erysipelas are penicillin or a first-generation cephalosporin. Erysipelas may recur periodically at the same body site for years.

Invasive Streptococcal Infections

The development of invasive disease initiated by *S. pyogenes* is linked to specific serotypes, as determined by M protein structure (figure 38.8), and predisposing host factors (surgical or nonsurgical wounds, diabetes, and other underlying medical problems). A life-threatening infection begins when GAS strains penetrate a mucous membrane or colonize a wound such as a bruise. This infection can quickly lead either to **necrotizing fasciitis** (Greek *nekrosis*, death; Latin *fascis*, band or bandage; *itis*, inflammation), which destroys the sheath covering skeletal muscles, or to **myositis** (Greek *myos*, muscle; *-itis*), the inflammation and destruction of skeletal muscle and fat tissue (**figure 38.24**). The speed at which the infection spreads has earned these strains the name *flesh-eating bacteria*.

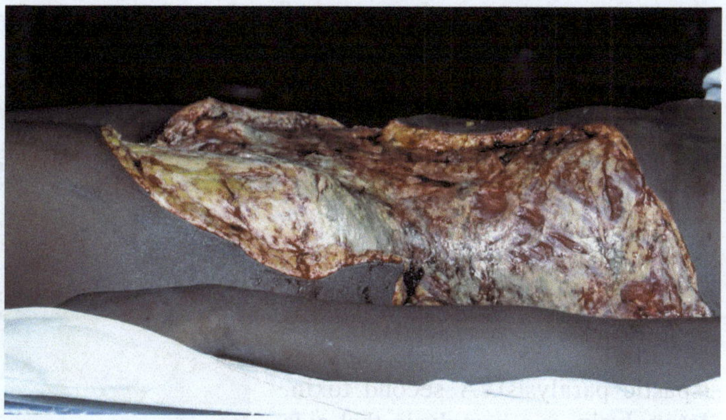

Figure 38.24 Necrotizing Fasciitis. Rapidly advancing streptococcal disease can lead to large necrotic sites that require surgical removal, called debridement, as shown here. Necrotizing fasciitis and myositis are sometimes called flesh-eating disease.

Dr. M.A. Ansary/Science Source

MICRO INQUIRY *What virulence factors enable the rapid invasion of certain strep A strains?*

It is estimated that approximately 11,000 to 24,000 cases of invasive GAS infections occur annually in the United States (1,200 to 1,900 deaths), and 6 to 7% of these are associated with necrotizing conditions. Rapid treatment is necessary to reduce the risk of death, and penicillin G remains the treatment of choice. In addition, surgical removal of dead and dying tissue is needed in more advanced cases.

One reason invasive GAS strains are so deadly is that about 85% of them carry the genes for the production of streptococcal pyrogenic exotoxins A and B (Spe exotoxins). Exotoxin B is a protease that hydrolyzes proteins found in the extracellular matrix of tissues, thereby facilitating bacterial invasion. Exotoxin A is a superantigen, so GAS skin and soft-tissue infections can lead to severe sepsis and septic shock, a condition called streptococcal toxic shock syndrome (STSS). Like staphylococcal toxic shock syndrome, STSS is very difficult to treat and has a mortality rate ranging from 20 to 40%. ◀ *Toxins are biological poisons (section 34.4)*

Tetanus

Tetanus (Greek *tetanos*, to stretch) is caused by *Clostridium tetani*, an anaerobic, Gram-positive, endospore-forming rod (*see figure 22.20*). The endospores of *C. tetani* are commonly found in soil and dust and in the feces of many farm animals and humans. Transmission to humans occurs through a break in the skin, which allows *C. tetani* endospores entry. If the oxygen tension is low enough, the endospores germinate and release the neurotoxin tetanospasmin. **Tetanospasmin** is an endopeptidase that selectively cleaves the synaptic vesicle membrane protein synaptobrevin. This prevents release of inhibitory neurotransmitters (γ-aminobutyric acid and glycine) at synapses within spinal cord motor nerves. The result is uncontrolled stimulation of skeletal muscles (spastic paralysis). A second toxin, **tetanolysin,** is a hemolysin that aids in tissue destruction. *Order Clostridiales (section 22.2)*

(a) (b)

Figure 38.25 The Toll of Tetanus. (a) A patient with tetanus with risus sardonicus. (b) Sir Charles Bell's portrait (c. 1821) of a soldier wounded in the Peninsular War in Spain shows opisthotonus resulting from tetanus.

(a) C Farmer/AFIP/Centers for Disease Control and Prevention; (b) Everett Collection Historical/Alamy Stock Photo-

Early in the course of the disease, tetanospasmin causes tension or cramping and twisting in skeletal muscles surrounding the wound and tightness of the jaw muscles. With more advanced disease, there is trismus ("lockjaw"), an inability to open the mouth because of the spasm of the masseter muscles. Facial muscles may go into spasms, producing the characteristic expression known as *risus sardonicus* (**figure 38.25*a***). Spasms or contractions of the trunk and extremity muscles may be so severe that there is boardlike rigidity, painful convulsions, and backward bowing of the back (opisthotonus; figure 38.25*b*). Death results from respiratory failure caused by spasms of the diaphragm and intercostal respiratory muscles. Tetanus in newborns, called *neonatal tetanus,* can result from the use of nonsterile instruments in cutting the umbilical cord. Symptoms include muscle rigidity and difficulty in breathing and feeding.

Tetanus prevention depends on (1) active immunization with toxoid, with boosters every 10 years; (2) proper care of wounds contaminated with soil; (3) prophylactic use of antitoxin; and (4) administration of penicillin. However, neonatal tetanus is almost 100% fatal, causing as many as 15,000 infant deaths annually. Neonatal tetanus can be prevented by immunization of pregnant females because maternally transmitted antibodies protect the newborn. In resource-rich countries, tetanus is rare.

Trachoma

Trachoma (Greek *trachoma,* roughness) is a highly contagious disease caused by *Chlamydia trachomatis* serovars A, B, Ba, and C. The disease is the greatest single cause of blindness throughout the world: In 2023 about 32.9 million people were treated for trachoma, but this is only about 30% of those who needed treatment. This explains why globally trachoma is responsible for blindness or visual impairment in 1.9 million individuals today. Although trachoma is uncommon in the United States, except among American Indians, it is widespread in Asia, Africa, and South America. *C. trachomatis* can also cause "pink eye," a milder form of eye infection.

Trachoma is transmitted by contact with inanimate objects such as soap and towels, by hand-to-eye contact, or by flies. The disease begins abruptly with an inflamed conjunctiva. This leads to an inflammatory cell exudate and necrotic eyelash follicles (**figure 38.26**). The disease usually heals spontaneously. However, reinfection or secondary infections with other microbes cause fibrosis and scarring. Scar contraction may cause inward turning of the eyelids so that eyelashes scratch the cornea, resulting in ulceration and eventual blindness.

Diagnosis of trachoma is by serological methods. The infection is treated with doxycycline or macrolides. Inverted eyelids and eyelashes may be surgically repaired. Prevention and control of trachoma depend on health education, personal hygiene, and sanitation, including access to clean water.

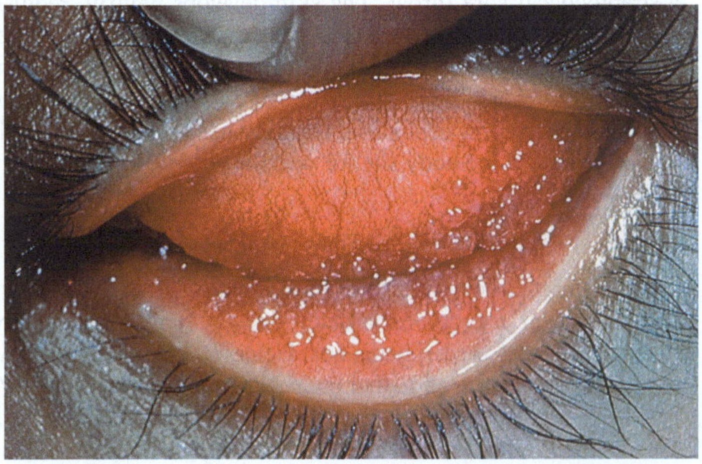

Figure 38.26 Trachoma. This patient's eyelid has been retracted to reveal red inflammatory nodules covering the conjunctival membranes of the eye.

Susan Lindsley/CDC

Comprehension Check

1. How can humans acquire gas gangrene? Group B strep? Leprosy? Describe the major symptoms of each.
2. Why do you think the slow growth rate of *M. leprae* makes leprosy treatment more difficult?
3. Apart from gastric ulcers, what other conditions can *H. pylori* cause?
4. What is ophthalmia neonatorum and how is it transmitted?
5. Describe the three stages of syphilis. How does *T. pallidum* interact with the immune system?
6. Infection with *N. gonorrhoeae* or *T. pallidum* does not confer lifelong protection from reinfection. Why is this the case?
7. What is the difference between CA-MRSA and HA-MRSA?
8. What conditions are caused by superantigens? Why are these toxins so dangerous?
9. Explain why tetanus is potentially life-threatening. When was the last time you had a tetanus booster?
10. How is trachoma transmitted, and how does it cause blindness?

38.4 Food and Water Are Vehicles for Bacterial Diseases

After reading this section, you should be able to:

a. Describe bacteria that cause the common food-borne and waterborne diseases
b. Identify typical signs and symptoms of food-borne and waterborne bacterial diseases
c. Correlate food-borne and waterborne bacterial infection and disease severity with bacterial virulence factors

Some microorganisms that contaminate food and water can cause acute gastroenteritis—inflammation of the stomach and intestinal lining. When food is the source of the pathogen, the condition is often called **food poisoning.** Gastroenteritis can arise in two ways. Some bacteria cause an infection following consumption of tainted food. Alternatively, the pathogen may secrete an exotoxin that contaminates the food and is then ingested by the host. This is referred to as a **food intoxication** because the toxin is ingested but the presence of living microorganisms is not required. These toxins disrupt the functioning of the intestinal mucosa and are called **enterotoxins.**

Worldwide, there are nearly 1.7 billion cases of childhood diarrheal disease annually. In fact, diarrheal diseases are second only to pneumonia as the leading cause of childhood death. For example, the WHO estimates that approximately 500,000 children die of diarrheal disease every year; the majority of deaths are in children under the age of five. In the United States, the CDC estimates that there are about 48 million cases of food-borne illness, leading to 128,000 hospitalizations and 3,000 deaths each year.

This section describes several of the more common bacterial gastrointestinal infections, food intoxications, and waterborne diseases. **Table 38.3** summarizes many of the bacterial pathogens responsible for food poisoning, and **table 38.4** lists many important waterborne bacterial pathogens. The protozoa responsible for food-borne and waterborne diseases are covered in chapter 39. ▶| *Food-borne illnesses (section 40.3)*

Botulism

Food-borne **botulism** (Latin *botulus,* sausage) is a food intoxication because disease results after ingesting preformed botulinum toxin. A common source of disease is home-canned food contaminated with *Clostridium botulinum* endospores. The endospores germinate, and a toxin is produced during anaerobic vegetative growth. If the food is later eaten without adequate heating to denature all toxin molecules, the toxin causes disease.

Botulinum toxin (BTX or BoNT) is an AB neurotoxin that binds to the same target as tetanus toxin, the synaptobrevin of motor neurons, but with opposite effects (**figure 38.27**). The A

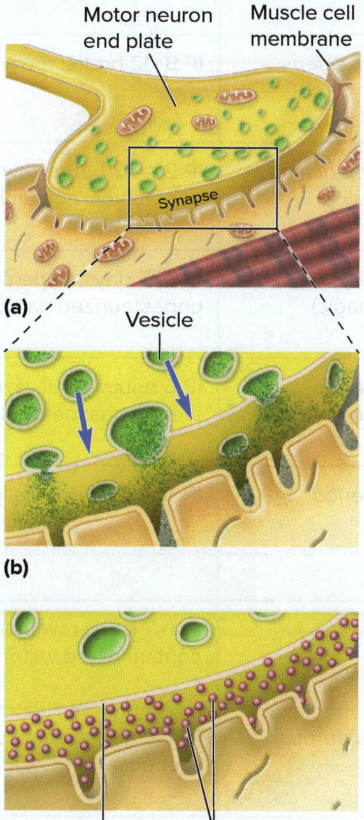

Figure 38.27 The Physiological Effects of Botulinum Toxin. (a) The relationship between the motor neuron and the muscle at the neuromuscular junction. (b) In the normal state, acetylcholine (green) released at the synapse crosses to the muscle cell membrane, binds a receptor, and creates an impulse that stimulates muscle contraction. (c) In botulism, the toxin enters the motor neuron end plate and attaches to the presynaptic membrane, where it blocks release of the acetylcholine. This prevents impulse transmission and prevents the muscle from contracting.

MICRO INQUIRY *How do the neurological effects of botulinum toxin differ from those of tetanus toxin?*

Table 38.3 Bacteria That Cause Acute Diarrhea and Food Poisoning

Organism	Incubation Period (IP) and Common Source	Clinical Features and Pathogenesis
Bacillus cereus	IP 8–16 hours (infection); reheated fried rice, pasta, meat, dairy products	Vomiting and diarrhea. Caused by enterotoxins formed in food or gut.
Campylobacter jejuni	IP 2–10 days; undercooked poultry, meat, seafood; eggs, unpasteurized milk	Fever and diarrhea that may become bloody; PMNs and blood in stool. Invasion of mucus membrane; cytolethal distending toxin produced. Azithromycin given in severe cases.
Clostridioides difficile	IP days to weeks after antibiotic therapy; fecally contaminated food	Watery diarrhea, fever, abdominal cramping. Toxins cause epithelial necrosis in colon. Treated with vancomycin or fidaxomicin, or fecal microbiota transplant (FMT) if recurrent.
Clostridium botulinum	IP 12–72 hours; anoxic food (e.g., canned goods); raw honey in infants	Sometimes vomiting; diarrhea rare; difficulty breathing and speaking, muscle weakness, nausea and vomiting. Exotoxin prevents muscle contraction. Treated with antitoxin and supportive care.
Clostridium perfringens	IP 8–12 hours; rewarmed meat dishes	Abrupt onset of profuse watery diarrhea, occasionally vomiting. Enterotoxins made during sporulation in gut cause hypersecretion.
Enterotoxic Escherichia coli (ETEC)	IP 24–72 hours; raw fruits and vegetables; contaminated water	Abrupt onset of watery diarrhea; vomiting rare. Heat-labile (LT) and heat-stable (ST) enterotoxins cause hypersecretion in small intestine.
Escherichia coli (enterohemorrhagic)	IP 3–5 days; undercooked ground beef; unpasteurized dairy, juices	Mild diarrhea to severe bloody diarrhea; Shiga toxin release can cause hemolytic uremic syndrome (HUS) that may lead to kidney failure.
Salmonella enterica Nontyphoidal serovars	IP 6 hours– 6 days; undercooked poultry, meat, eggs; pet reptiles	Gradual or abrupt onset of watery or bloody diarrhea, low-grade fever; nausea, headache, and muscle aches common. Superficial infection of gut, little invasion. Usually self-limiting.
Salmonella enterica serovar Typhi	IP 10–14 days; fecally contaminated food	Fever (may persist for 2 or more weeks), headache, diarrhea and/or constipation, abdominal pain; rash after 2–4 weeks. Symptoms caused by exotoxin. Requires antibiotic therapy based on individual stain sensitivities due to widespread drug resistance.
Shigella spp.	IP 1–7 days; raw fruits and vegetables; contaminated water	Abrupt onset of diarrhea, often with blood and pus; cramps, lethargy. Organisms invade epithelial cells. Severe cases treated with azithromycin or third-generation cephalosporins.
Staphylococcus aureus	IP 1–6 hours; cold cuts, dairy and bakery products	Abrupt onset; severe cramps, diarrhea, explosive vomiting for up to 24 hours; recovery after 24–48 hours. Symptoms caused by enterotoxins (SE).
Vibrio cholerae	IP 3–72 hours; fecally contaminated water, raw or undercooked shellfish	Abrupt onset; large volumes of liquid diarrhea; exotoxin causes hypersecretion. Treatment is prompt fluid and electrolyte replacement; antibiotics (e.g., doxycycline) reduce pathogen excretion.
Vibrio parahaemolyticus	IP 6–24 hours; seafood, particularly shellfish and crustaceans	Abrupt onset of diarrhea, sometimes bloody; recovery within 1–3 days. Toxin causes hypersecretion.
Yersinia enterocolitica	IP 4–7 days; most commonly undercooked pork; unpasteurized milk	Severe abdominal pain that may mimic appendicitis, diarrhea, fever; PMNs and blood in stool; polyarthritis; gastroenteritis. Treated with antibiotics in severe cases.

Source: Brooks, G. F., et al. 1998. *Medical microbiology,* 21st ed. Norwalk, CT: Appleton & Lange.

Table 38.4 Waterborne Bacterial Pathogens

Organism	Reservoir	Comments
Aeromonas hydrophila	Free-living	Sometimes associated with gastroenteritis, cellulitis, and other diseases
Helicobacter pylori	Free-living	Can cause gastritis, peptic ulcers, gastric adenocarcinomas
Legionella pneumophila	Free-living and associated with protozoa	Found in cooling towers, evaporators, condensers, showers, and other water sources
Leptospira spp.	Infected animals	Hemorrhagic effects, jaundice
Mycobacterium spp.	Infected animals and free-living	Complex recovery procedure required
Pseudomonas aeruginosa	Free-living	Swimmer's ear and skin infections
Salmonella enteritidis	Animal intestinal tracts	Common water contaminant
Vibrio cholerae	Free-living	Found in many waters, including estuaries
Vibrio parahaemolyticus	Free-living in coastal waters	Causes diarrhea in shellfish consumers
Yersinia enterocolitica	Frequent in animals and in the environment	Waterborne gastroenteritis

subunit is a protease that selectively cleaves the synaptic vesicle membrane protein synaptobrevin, thereby preventing release of the neurotransmitter acetylcholine. As a consequence, muscles do not contract in response to motor neuron activity, and flaccid paralysis results (**Techniques & Applications 38.3**). Symptoms of botulism occur within 12 to 72 hours of toxin ingestion and include blurred vision, difficulty swallowing and speaking, bilateral muscle weakness, nausea, and vomiting. Treatment relies on the administration of antitoxin and supportive care. If untreated, up to 5% of patients die of either respiratory or cardiac failure within a few days.

Infant botulism is the most common form of botulism in the United States with about 75 cases reported each year. Ingested endospores, which may be naturally present in raw honey or house dust, germinate in the infant's intestine. *C. botulinum* then multiplies and produces the toxin. The infant becomes constipated, listless and generally weak, and eats poorly. Death may result from respiratory failure. It is recommended that infants are not fed raw honey.

Campylobacter jejuni Gastroenteritis

Campylobacter jejuni is the most commonly reported bacterial cause of food-borne illness in the United States. The CDC estimates there are about 1.5 million cases of **Campylobacter gastroenteritis (campylobacteriosis)** and subsequent diarrhea in the United States each year. Studies with chickens, turkeys, and cattle have shown that 50 to 100% of these animals excrete *C. jejuni*. It is transmitted to humans by contaminated food and water, contact with infected animals, and anal-oral sexual activity. ◀ *Phylum Campylobacterota: human and animal commensals (section 20.10)*

The incubation period for campylobacteriosis is 2 to 10 days. *C. jejuni* invades the epithelium of the small intestine, causing inflammation. It also secretes a cytolethal distending toxin that cleaves host DNA. Symptoms include diarrhea, high fever, severe inflammation, and ulceration of the intestine, which can cause bloody stools. The structural resemblance between *C. jejuni* surface polysaccharides and a sugar moiety found on nerve cell glycoproteins has been linked to the development of Guillain-Barré syndrome in rare cases following *C. jejuni* infection through a mechanism termed **molecular mimicry** (**figure 38.28**). Guillain-Barré syndrome is a disorder in which the body's immune system attacks peripheral nerves, resulting in life-threatening paralysis.

Laboratory diagnosis is accomplished through stool cultures. The disease is usually self-limited, and treatment is supportive, including fluid and electrolyte replacement. Azithromycin may be given to those with additional health problems. Recovery usually takes 5 to 8 days. Prevention and control involve food-handling precautions, pasteurization of milk, and thorough cooking of poultry.

Cholera

Cholera is an acute diarrheal disease caused by infection of the intestine with the Gram-negative, comma-shaped bacterium *Vibrio cholerae*. Although there are many *V. cholerae* serogroups, only the toxigenic O1 and O139 (encapsulated) organisms have caused epidemics. *V. cholerae* O1 is divided into two biotypes, Classical and El Tor. Each biotype is divided into three serotypes—Inaba, Ogawa, and Hikojima—based on the structure of the O-antigen. Throughout recorded history, cholera (Greek *chole*, bile) has caused seven pandemics in various areas of the

TECHNIQUES & APPLICATIONS

38.3 Clostridial Toxins as Therapeutic Agents: Benefits of Nature's Most Toxic Proteins

Some toxins are currently used for the treatment of human disease. The most famous of such toxins, botulinum toxin, is also the most poisonous biological substance known. It is used for the treatment of specific neuromuscular disorders characterized by involuntary muscle contractions. Since approval of type-A botulinum toxin (Botox) by the FDA in 1989 for three disorders (strabismus [crossing of the eyes], blepharospasm [spasmotic contractions of the eye muscles], and hemifacial spasm [contractions of one side of the face]), the number of neuromuscular problems treated has increased to include other tremors, migraine and tension headaches, and other maladies. In 2000 dermatologists and plastic surgeons began using Botox to temporarily eradicate wrinkles caused by repeated muscle contractions as we laugh, smile, or frown (**box figure**). The remarkable therapeutic utility of botulinum toxin lies in its ability to specifically and potently inhibit muscle activity for an extended duration. Overall, clostridia produce more protein toxins than any other known bacterial genus, and are a rich reservoir of toxins for research and medicinal uses. For example, research is underway to use clostridial toxins or toxin domains for drug delivery, prevention of food poisoning, and treatment of cancer and other diseases.

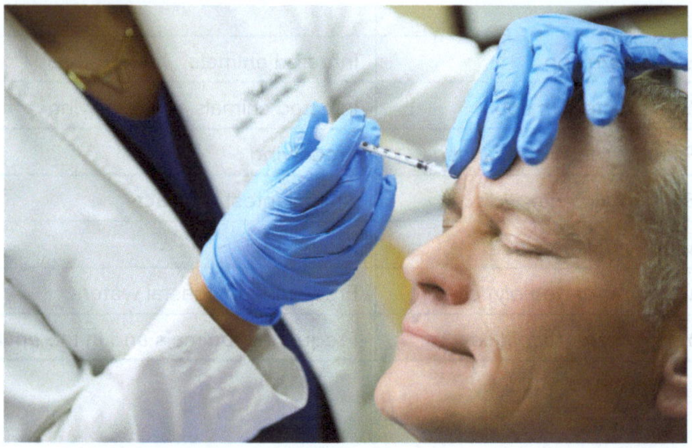

Botox has medical as well as cosmetic uses.
MBI/Alamy Stock Photo

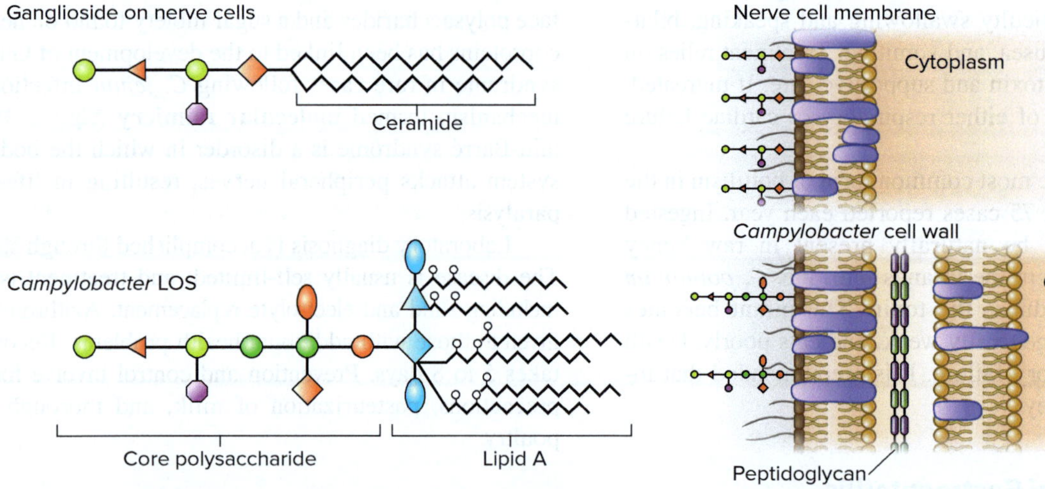

Figure 38.28 Guillian-Barré Syndrome Following *Campylobacter jejuni* Infection Arises Through Molecular Mimicry. The sugars (ganglioside) on the surface of nerve cells are structurally similar to the core polysaccharide of *C. jejuni* lipoligosaccharide (LOS). LOS is similar to lipopolysaccharide but lacks an O-antigen. Antibodies directed against *C. jejuni* LOS bind nerve cell gangliosides and complement deposition leads to targeted destruction of the nerve cells by the host's immune response.

world. In 2016, the worst cholera epidemic ever recorded started in Yemen; by late 2022, it had resulted in an estimated 2.5 million cases and nearly 4,000 deaths. Cholera epidemics are a testament to how quickly bacteria can invade and infect.

Cholera is transmitted by ingesting food or water contaminated by fecal material from individuals who are infected. The incubation period is 3 to 72 hours and infection with serotype O1 or O139 results in mild to severe diarrhea. Approximately 20% of those with severe diarrhea produce large amounts of "rice-water-stool," so named because it is milky white with flecks of mucus floating in it. The bacteria adhere to the intestinal mucosa of the small intestine, where they are not invasive but secrete cholera toxin, an AB toxin (*see figure 34.8*). The A subunit enters the intestinal epithelial cells and adds an ADP-ribosyl group to the G protein–coupled receptor, $G_{s\alpha}$, which becomes locked in the activated state. This results in constant stimulation of host adenyl cyclase and overproduction of cAMP. This leads to hypersecretion of water and chloride ions while inhibiting absorption of sodium ions. The person who is infected loses massive quantities of fluid and electrolytes, causing abdominal muscle cramps, vomiting, fever, and diarrhea so profuse that a person can lose up to 20 liters of fluid per day (**figure 38.29**). Death results from dehydration and electrolyte imbalance.

Interestingly, the cholera toxin gene (*ctxAB*) is carried by the temperate CTX filamentous bacteriophage, which exists as a prophage within the genome of toxin-producing strains of *V. cholerae*. The receptor for the phage is the toxin coregulated pilus—the same structure used to colonize the host's gut. *V. cholerae* is an excellent example of how horizontal transfer of genes can confer pathogenicity.

Treatment of cholera is by oral rehydration therapy with NaCl plus glucose to stimulate water uptake by the intestine; If vomiting occurs, intravenous fluids are given. In severe cases, antibiotics (e.g., doxycycline) may also be given. The most reliable control methods are based on proper sanitation, especially of water supplies. The mortality rate without treatment is often over 50%, but with treatment and supportive care, it is less than 1%.

Escherichia coli Gastroenteritis

A major causative agent of food-borne diarrhea is *Escherichia coli*. The vast majority of *E. coli* strains are harmless and provide assistance in the metabolism of the food we eat. However, certain *E. coli* strains have become pathogenic by acquiring genes through horizontal gene transfer. These strains of *E. coli* are readily tracked by serotyping. Two of the most commonly used serotyping antigens are the H (flagellar) and the O (O-antigen portion of LPS) antigens.

Five categories (pathotypes) of diarrheagenic *E. coli* are recognized (**figure 38.30**): enterotoxigenic *E. coli* (ETEC), enteroinvasive *E. coli* (EIEC), enteropathogenic *E. coli* (EPEC), enteroaggregative *E. coli* (EAEC), and enterohemorrhagic *E. coli* (EHEC). All except EHEC are found in human feces; transmission is from poor hand hygiene or sanitation. Each pathotype uses a different mechanism to cause disease, as we now discuss. *Order Enterobacterales is a large, diverse taxon (section 21.2)*

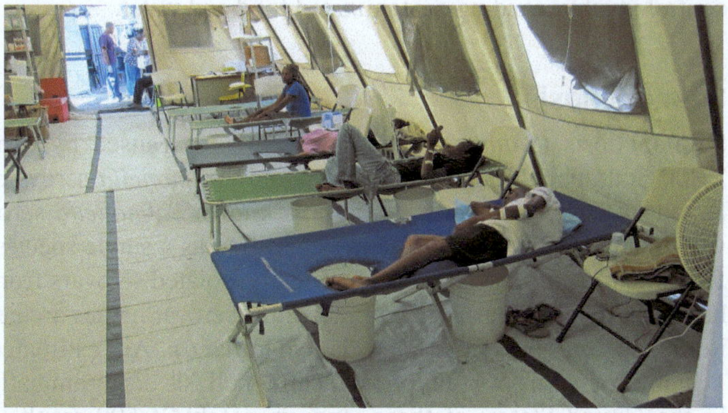

Figure 38.29 Cholera Cots. Cholera, caused by *V. cholerae*, causes such profuse diarrhea that special cots are used during outbreaks to keep those affected clean.

Centers for Disease Control and Prevention

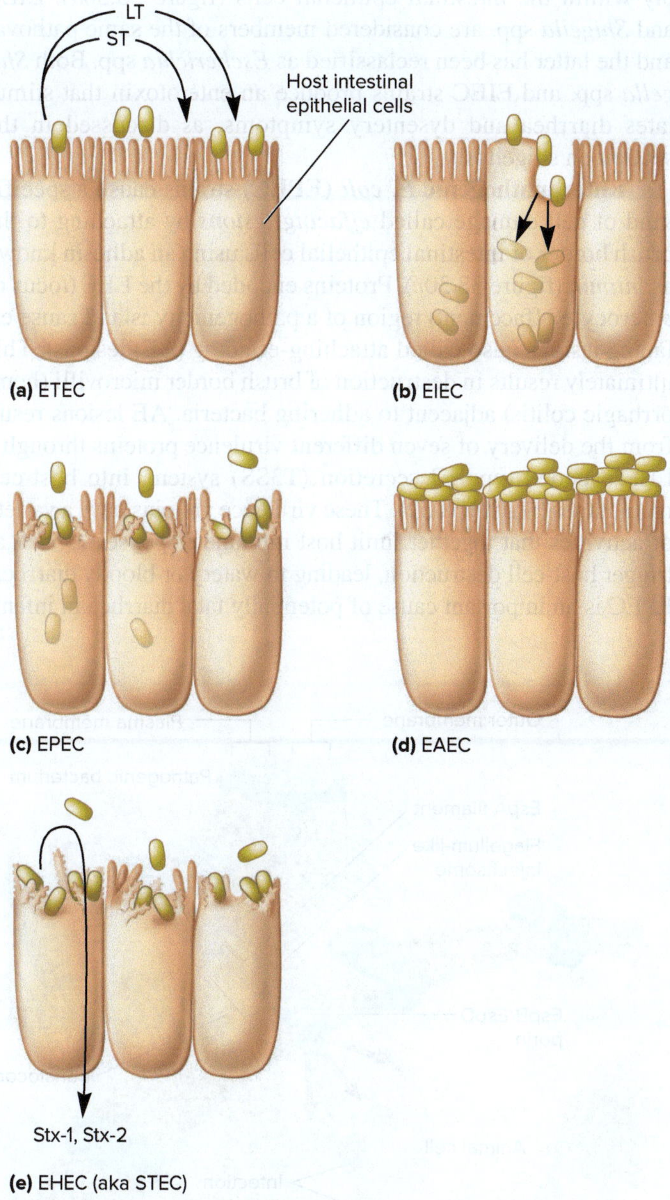

Figure 38.30 Five Strains of Diarrheagenic *E. coli*. Each strain of diarrhea-causing *E. coli* can be classified by the nature of its interaction with host intestinal epithelial cells.

MICRO INQUIRY *To which* E. coli *class does the strain O157:H7 belong?*

Enterotoxic *E. coli* (ETEC) strains produce one or both of two distinct enterotoxins responsible for diarrhea. The toxins are distinguished by their heat stability: heat-stable enterotoxin (ST) and heat-labile enterotoxin (LT) (figure 38.30*a*). The genes for ST and LT production and for colonization factors are usually plasmid-borne and therefore acquired by horizontal gene transfer. ST is a small protein that increases host cGMP concentrations. LT has a structure and activity very similar to cholera toxin. Both toxins induce hypersecretion of electrolytes and water into the intestinal lumen, triggering watery diarrhea. ETEC is the most common cause of travelers' diarrhea.

Enteroinvasive *E. coli* (EIEC) strains penetrate and multiply within the intestinal epithelial cells (figure 38.30*b*). EIEC and *Shigella* spp. are considered members of the same pathovar and the latter has been reclassified as *Escherichia* spp. Both *Shigella* spp. and EIEC strains produce an enterotoxin that stimulates diarrhea and dysentery symptoms, as discussed in the section on shigellosis.

Enteropathogenic *E. coli* (EPEC) strains cause a specific kind of cell damage called *effacing lesions* by attaching to the brush border of intestinal epithelial cells using an adhesin known as *intimin* (figure 38.30*c*). Proteins encoded by the LEE (*l*ocus of *e*nterocyte *e*ffacement) region of a pathogenicity island cause effacing lesions (also called attaching-effacing (AE) lesions). This ultimately results in destruction of brush border microvilli (hemorrhagic colitis) adjacent to adhering bacteria. AE lesions result from the delivery of seven different virulence proteins through a LEE-encoded type III secretion (T3SS) system, into host-cell membranes (**figure 38.31**). These virulence proteins have a variety of activities that together limit host immune responses as well as trigger host-cell destruction, leading to watery or bloody diarrhea. EPEC is an important cause of potentially fatal diarrhea in infants and children living in countries without adequate sanitation infrastructure.

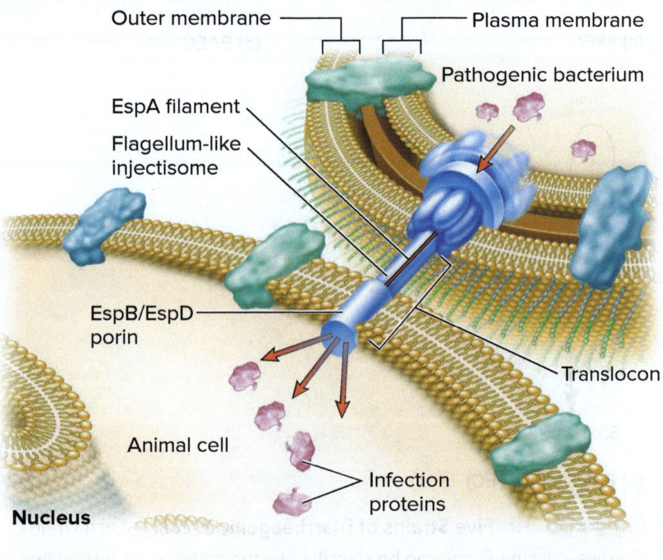

Figure 38.31 Type III Secretion System (T3SS) of Enteropathogenic *E. coli* (EPEC). The T3SS proteins function like a syringe to inject virulence proteins into a host cell.

Enteroaggregative *E. coli* (EAEC) strains adhere to epithelial cells in localized regions, forming clumps of bacteria with a stacked brick appearance (figure 38.30*d*). EAEC can also produce several plasmid-encoded toxins that cause diarrhea, particularly in children in regions lacking good sanitation. It is the second most common cause of travelers' diarrhea.

Cattle serve as the reservoir for **enterohemorrhagic *E. coli* (EHEC)** or **Shiga toxin–producing *E. coli* (STEC),** which is transferred to humans by water or food; outbreaks traced to vegetables reveal that the crops were irrigated with water contaminated with STEC (primarily from cattle or other ruminants such as wild boar). There are over 400 serotypes that carry the bacteriophage-encoded genetic determinants for Shiga toxin (also called Shiga-like toxin Stx-1 and/or Stx-2 proteins; figure 38.30*e*). Some EHEC strains also carry the LEE gene sequence to produce AE lesions, causing hemorrhagic colitis (HC) with severe abdominal pain and cramps followed by bloody diarrhea. Stx-1 and Stx-2 can cause the renal disease hemolytic uremic syndrome (HUS), a severe hemolytic anemia that can lead to kidney failure. The A subunit of these toxins enters renal cells and depurinates 28S rRNA molecules, thereby disabling the ribosome, eliminating protein synthesis, and causing cell death.

▶| *Food-borne illnesses (section 40.3)*

A major serotype of STEC (EHEC) in North America is *E. coli* O157:H7 (STEC O157), which has caused many outbreaks of HC and HUS in the United States since it was first recognized in 1982. A variety of other STEC serotypes cause similar disease. Currently there are an estimated 265,000 STEC (36% of which are STEC O157) cases in the United States each year, resulting in an average of 30 deaths.

Diagnosis of diarrhea caused by *E. coli* is based on travel history and symptoms. Laboratory diagnosis is by culture of the specific type of *E. coli* from feces and immunoassays to detect Shiga toxins. Treatment is primarily by fluid and electrolyte replacement. Antibiotics should not be given if the strain produces Shiga toxin because this can cause an increase in toxin release. Prevention and control involve avoiding contaminated food and water.

Salmonellosis

Salmonellosis (*Salmonella* gastroenteritis) is caused by over 2,000 nontyphoidal (not causing typhoid) *Salmonella* serovars. Based on DNA homology studies, all known *Salmonella* serovars that cause human disease are classified as a single species, *Salmonella enterica*. The most frequently isolated serovars from humans are Typhimurium and Enteritidis (serovar names are not italicized and the first letter is capitalized). An estimated 1.35 million cases of nontyphoidal salmonellosis occur annually in the United States; of these, approximately 26,500 result in hospitalization and 420 result in death.

The initial source of the nontyphoidal *Salmonella* serovars is the intestinal tracts of birds and other animals. Humans acquire the bacteria from contaminated water or foods such as beef products, poultry, eggs, or egg products. Once the bacteria are in the body, the incubation time is between 12 and 72 hours. Disease results when the bacteria multiply and invade the intestinal mucosa. Abdominal pain, cramps, diarrhea, nausea, vomiting, and fever are the most prominent symptoms; these usually persist for 4 to 7 days but can last for several weeks. During the acute phase of the disease, as many as 1 billion *Salmonella* cells can be found per gram of feces. Most adult patients recover, but the loss of fluids can quickly dehydrate children and older people. Those with severe diarrhea may therefore require rehydration therapy with intravenous fluids. Antibiotics are not usually given unless the infection spreads from the intestines.

A variety of serovars have been reported as contaminants of cereal products, chicken, tahini sesame paste, and peanut butter. The ongoing concern over food safety has prompted greater diligence for the detection and reporting of food contamination. Of note is the increasing concern regarding infection resulting from antibiotic-resistant *Salmonella* serovars.

Enteric fever, comprised of typhoid and paratyphoid fevers, is caused by *S. enterica* serovars Typhi and Paratyphi and is acquired by ingestion of food or water contaminated by feces of infected humans or by person-to-person contact. Humans are the sole reservoir for *S.* Typhi and the majority of enteric fever cases are caused by this pathogen. Unlike salmonellosis, enteric fever can result in multiorgan failure and death. *S.* Typhi and *S.* Paratyphi produce an unusual AB toxin with heterodimeric A (catalytic) subunits. One A subunit possesses nuclease activity. Although the second A subunit is known to have ADP-ribosylating activity, its cellular target has not yet been discovered.

The incubation period for enteric fever is about 5 to 21 days. Initially, fever is present with a paradoxically slow heart rate. This is followed by a characteristic rash that features small pink lesions called "rose spots." The bacteria colonize the small intestine, penetrate the epithelium, and spread to lymphoid tissue, blood, liver, and gallbladder. In some patients, infection is dire and can cause septic shock. In others, symptoms are less severe. Following acute illness, it typically takes about 3 months for most individuals to stop shedding bacteria in their feces. However, about 4% of individuals continue to shed the pathogen for extended periods but show no symptoms. In these carriers, the bacteria continue to grow in the gallbladder and reach the intestine through the bile duct. This was almost certainly the case for "Typhoid Mary" (*see Historical Highlights 35.2*). Treatment is with azithromycin, a fluoroquinolone, or a cephalosporin. There are around 400 confirmed cases of enteric fever in the United States every year, mostly associated with travel; many more cases are mild and go unreported. Those going to enteric fever–endemic regions are advised to get the typhoid vaccine that protects against *S.* Typhi.

Shigellosis

Shigellosis, or bacillary dysentery, is a diarrheal illness resulting from an acute inflammatory reaction of the intestinal tract caused by the four species of the genus *Escherichia,* formerly *Shigella*. This nomenclature transition is still in flux, and in this section we retain the use of the term *Shigella*. Globally, these bacteria are estimated to cause up to 165 million cases of disease, with about 600,000 deaths, per year, most of which are in children under the age of five. About 450,000 cases a year are reported in the United States. However, real numbers may be greater because mild cases of disease typically are not reported. The predominant disease-causing species in the United States is *S. sonnei*. *S. flexneri* predominates in many tropical countries. *S. boydii* and *S. dysenteriae* infections are uncommon.

The organism is transmitted by the fecal-oral route and is most prevalent among young children. The infectious dose is only around 10 to 100 bacteria in large part because the bacteria are very tolerant of stomach acidity. Infection is marked by fever, abdominal cramps, and a bloody, mucoidy diarrhea. In severe cases, the colon can become ulcerated.

Shigellae are facultative intracellular parasites that penetrate the gut mucosal barrier. The bacteria are phagocytosed by macrophages within Peyer's patches. After being ingested by macrophages, the bacteria disrupt the phagosome membrane and enter the cytosol, reproduce, and kill the macrophage. They then invade mucosal epithelial cells, where they polymerize actin and use it to propel their intracellular motility, moving from one cell to the next. Both LPS (endotoxin) and exotoxins (some secreted by a T3SS) participate in disease progression, but bacteria do not usually spread beyond the colonic epithelium. *S. dysenteriae* serotype 1 produces Shiga toxin (Stx). This is very similar to Shiga toxins previously described for EHEC. These toxins target the glomerular endothelium leading to hemolytic uremic syndrome and kidney failure. ◀ *Phagocytosis destroys invaders (section 31.6); Exotoxins (section 34.4)*

The incubation period for shigellosis usually ranges from 1 to 7 days, and the organisms are shed over a period of 1 to 2 weeks. Unless the causative agent is *S. dysenteriae,* disease normally is self-limiting in adults and lasts an average of 4 to 7 days; in infants and young children, shigellosis may be fatal regardless of causative species. Fluid and electrolyte replacement are usually sufficient to treat disease. Antibiotics are not generally required in mild cases, but in severe cases azithromycin or a third-generation cephalosporin is used to reduce the duration of symptoms and transmission of the bacteria. Unfortunately, many strains of *Shigella* are becoming resistant to antibiotics.

Staphylococcal Food Poisoning

Staphylococcal food poisoning (SFP) is the major type of food intoxication in the United States. It is caused by ingestion of improperly stored or cooked food (particularly foods such as ham, processed meats, chicken salad, pastries, creamy sauces, and even ice cream) in which *S. aureus* has been allowed to reproduce and release enterotoxins. Because these enterotoxins are heat stable, even if food has been properly cooked and all bacteria killed, active toxin remains. Staphylococcal enterotoxins (SE) are primarily produced by *S. aureus,* although other staphylococci produce them. SE are structurally related, single-chain proteins but are antigenically distinct. Over 20 SE have been identified; of these, SEA, SEB, SEC, SED, and SEE are most commonly associated with SFP outbreaks. Typical symptoms include severe abdominal pain, cramps, diarrhea, explosive vomiting, and nausea. The onset of symptoms is rapid (usually 1–6 hours after eating contaminated food) and usually lasts less than 24 hours. The mortality rate of SFP is negligible among healthy individuals.

Treatment of SFP primarily involves maintaining hydration while letting the toxin flush from the body. The toxemia commonly resolves without medication. Prevention is through thorough hand washing, careful food preparation, and safe dining practices.

Comprehension Check

1. How can *Clostridium botulinum* cause disease without the presence of viable bacteria?
2. Compare the gastroenteritis caused by *Vibrio cholerae* with that of *Campylobacter jejuni.*
3. What are the mechanisms by which *E. coli* can cause disease?
4. Compare and contrast ETEC and EHEC in terms of toxin activity and severity of disease.
5. How do the mechanisms of pathogenicity differ between nontyphoidal and typhoidal *Salmonella* serovars?
6. Compare the intestinal diseases caused by *Salmonella* and *Shigella* spp.
7. Why is staphylococcal food poisoning considered an intoxication? How is it treated?

38.5 Zoonotic Diseases Arise from Human-Animal Interactions

After reading this section, you should be able to:

a. Discuss the bacteria that cause common diseases spread by contact with animals
b. Identify typical signs and symptoms of zoonotic bacterial diseases
c. Correlate zoonotic bacterial infection and disease severity with bacterial virulence factors

Diseases transmitted from animals to humans are called zoonotic diseases, or zoonoses. Many are considered **select agents,** indicating they represent a bioterrorism threat. ◀ *Bioterrorism readiness is an integral component of public health microbiology (section 35.7)*

Anthrax

Anthrax (Greek *anthrax,* coal) is a highly infectious animal disease caused by the Gram-positive, endospore-forming *Bacillus anthracis*. *B. anthracis* (also now known as *B. cereus* Biovar *anthracis*) is found worldwide and can be transmitted to humans by direct contact with infected animals (cattle, goats, sheep) or their products, especially hides. *B. anthracis* endospores remain viable in soil and animal products for decades. There are three forms of anthrax disease. When the bacterial spores enter through a cut or abrasion in the skin, **cutaneous anthrax** results. This produces a lesion called an *eschar,* characterized by black, necrotic tissue with minimal inflammation of the surrounding area. Most cases of anthrax are this form. Inhaling endospores may result in **pulmonary (inhalation) anthrax.** The endospores are engulfed by alveolar macrophages in the lungs; from there they are carried to lymph nodes, where they germinate. Initial symptoms appear rapidly and are flulike; they include fever, chest pain, muscle aches, shortness of breath, and sometimes cyanosis. As the disease progresses, organ failure and shock develop. If endospores are consumed (typically with the ingestion of undercooked meat containing endospores), **gastrointestinal anthrax** may result. Symptoms include fever, nausea, abdominal pain, and vomiting and diarrhea (either or both of which may be bloody). *B. anthracis* bacteremia can develop from any form of anthrax, with pulmonary anthrax almost always resulting in bacteremia.

B. anthracis carries two plasmids encoding virulence factors. One plasmid has genes involved in the synthesis of a polyglutamyl capsule that inhibits phagocytosis, and the other has genes for the synthesis of anthrax exotoxin. This exotoxin is composed of three proteins: protective antigen (PA), edema factor (EF), and lethal factor (LF). Macrophages have two different receptors to which PA attaches. Successful attachment requires the assembly of seven PA-receptor subunits into a doughnut-shaped ring (**figure 38.32**). Toxin proteins EF and LF bind to the PA complex, and PA with bound EF or LF enters the macrophage within an endosome (*see figure 31.20*). PA molecules form a pore that pierces the endosomal membrane, and EF and LF enter the cytosol. EF toxin has adenyl cyclase activity, which causes the cell to lose water, resulting in tissue swelling (edema). LF prevents the transcription factor nuclear factor kappa B (NF-κB) from regulating genes needed for an effective immune response. As thousands of macrophages die, they release their lysosomal contents, leading to fever, internal bleeding, septic shock, and rapid death. Without antibiotic treatment and supportive therapy, mortality rates are between 20 and 50% for cutaneous anthrax, and approach 100% for inhalational and gastrointestinal anthrax.

Between 20,000 and 100,000 cases of anthrax are estimated to occur worldwide annually. In the United States, the annual incidence is less than one case per year. However, anthrax is potentially a weapon of bioterrorism (*see chapter 2 opening story*).

Treatment of each form of anthrax is the same: Ciprofloxacin, penicillin, or doxycycline is successful if begun before

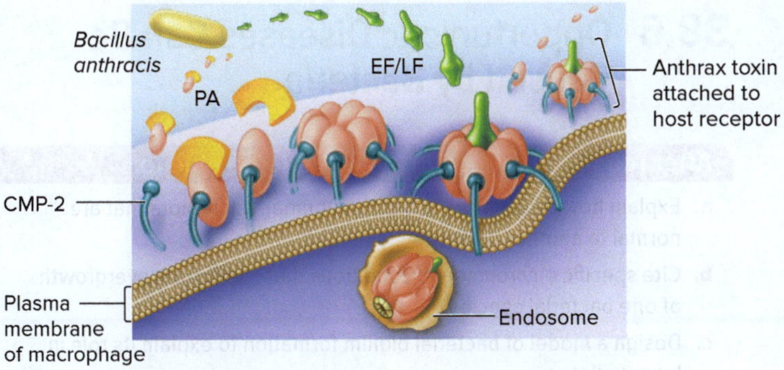

Figure 38.32 Anthrax. Protective antigen (PA) assembles in a barrel-shaped structure on target macrophages. Two proteins, edema factor (EF) and lethal factor (LF), are transported to an endosome. PA then delivers EF and LF from the endosome into the cytoplasm of the macrophage, where they exert their toxic effects.

MICRO INQUIRY *By what mechanisms do EF and LF kill macrophages?*

a critical concentration of toxin accumulates. Although antibiotics may kill the bacterium or suppress its growth, the exotoxin can still eventually kill the patient. Thus supportive care is essential in preventing death. A monoconal antibody that binds PA and prevents entry of EF and LF into host cells can also be administered. Vaccination of animals, primarily cattle, is an important control measure. However, people with a high occupational risk, such as those who handle infected animals or their products, including hides and wool, should be immunized.

Brucellosis (Undulant Fever)

The genus *Brucella* contains important human and animal pathogens. **Brucellosis,** also called *undulant fever,* is caused by faintly staining Gram-negative coccobacilli *B. melitensis* (which now includes microbes formerly classified as *B. abortus, B. suis,* and *B. canis*). These bacteria are commonly transmitted through consumption of contaminated animal products or through abrasions of the skin while handling infected mammals (cattle, sheep, goats, pigs, and, rarely, dogs). The most common route is ingestion of contaminated or unpasteurized milk products. Direct person-to-person spread of brucellosis is extremely rare. However, infants may be infected through their mother's breast milk; sexual transmission of brucellosis has also been reported. A controversy in the northwestern United States over the transmission of brucellosis, endemic in wild bison and elk populations, to otherwise *Brucella*-free cattle has continued for many years.

Brucellosis in humans is rare in the United States, with about 100 cases per year. In contrast to the United States, human brucellosis is a common zoonosis worldwide. About 500,000 cases are reported annually, but it is suspected that the real incidence may be up to 25 times greater.

Brucellosis, in the acute (8 weeks from onset) form, presents as nonspecific, flulike symptoms, including fever, sweats, malaise, anorexia, headache, myalgia (muscle pain), and back pain. In the undulant (rising and falling) form, symptoms of brucellosis include cyclic fevers, arthritis, and testicular inflammation. Neurologic symptoms occur in up to 5% of the cases. In the chronic form (1 year from onset), brucellosis symptoms include chronic fatigue, depression, and arthritis. Mortality is low, less than 2%. However, it is known to trigger spontaneous abortion, premature delivery, and fetal death. Treatment is usually with doxycycline plus streptomycin or rifampin for 6 weeks to prevent recurring infection. The insidious disease, potential for food supply disruption, and significant morbidity associated with *Brucella* infection have prompted its restriction as a select agent; the public health impact, should it be used as a bioweapon, would be substantial.

Psittacosis (Ornithosis)

Psittacosis (ornithosis) is a worldwide infectious disease of birds that is transmissible to humans. Fewer than 50 confirmed cases are reported in the United States each year. Psittacosis (pronounced sit-a-co′-sis) was first described in association with parrots and parakeets, both of which are psittacine birds. The disease is now recognized in many other birds—among them, pigeons, chickens, ducks, and turkeys—and the general term *ornithosis* (Latin *ornis,* bird) is also used.

Psittacosis is caused by *Chlamydophila psittaci*. Humans contract this disease either by handling infected birds or by inhaling dried bird excreta that contains viable *C. psittaci*. Incubation is typically 5 to 19 days. After entering the respiratory tract, the bacteria are transported to the cells of the liver and spleen. The bacteria multiply within these cells and then invade the lungs, where they cause inflammation, hemorrhage, and pneumonia. Endocarditis (inflammation of the heart muscle), hepatitis, and neurologic complications may occasionally occur. With doxycycline or tetracycline therapy, the mortality rate is about 2%.

Q Fever

The Q in **Q fever** stands for "query" because the cause of the fever was long unknown. The disease is now known to be caused by *Coxiella burnetii*. Although it is an intracellular pathogen, *C. burnetii* survives outside host cells by forming resistant, endospore-like bodies. This bacterium infects both wild animals and livestock. Cattle, sheep, and goats are the primary reservoirs. It does not need an arthropod vector for transmission, but in animals, many species of ticks can transmit *C. burnetii*. By contrast, human transmission is primarily by inhalation of dust contaminated with bacteria from dried animal feces or urine, or consumption of unpasteurized milk. Importantly, organisms are shed in high numbers within amniotic fluids and the placenta during birthing. The organisms are resistant to heat, drying, and many common disinfectants. These features enable the bacteria to survive for long periods in the environment. The disease can occur in epidemic form among slaughterhouse workers and sporadically among farmers and veterinarians. Each year, about 100 cases of Q fever are reported in the United States. However, approximately 3% of U.S. adults and an additional 10 to 20% of high-risk populations (farmers, veterinarians, etc.) have antibodies suggesting prior exposure to *C. burnetii*. *Orders Legionellales and Coxiellales include intracellular pathogens (section 21.2)*

In humans, Q fever is an acute illness characterized by the sudden onset of severe headache, malaise, confusion, sore throat, chills, sweats, nausea, chest pain, myalgia, and high fever. It is rarely fatal, but endocarditis occurs in about 10% of the cases. Five to 10 years may elapse between the initial infection and the appearance of endocarditis. During this interval, the bacteria reside in the liver and may cause hepatitis. The fatality rate is approximately 4%.

Diagnosis of Q fever is made by national reference laboratories because it is a select agent. PCR or culture of the bacterium and fluorescent antibody and agglutination tests are used. Treatment is with doxycycline and is most effective when initiated within the first 3 days of illness.

Tularemia

Another select agent is the Gram-negative bacterium *Francisella tularensis,* which causes the disease **tularemia.** It can be transmitted to humans by biting arthropods (ticks and deer flies), direct contact with infected tissue (rabbits), inhalation of aerosolized bacteria, or ingestion of contaminated food or water. However, tularemia is most often transmitted through contact with infected animals; it is called rabbit fever in the central United States because it can occur in hunters who skin their catch. After an incubation period of 2 to 10 days, it causes a disease that resembles plague but is not as severe. The patient develops a primary ulcerative lesion at the infection site, then lymph nodes enlarge, and a high fever develops. Without treatment, tularemia has a 30 to 60% mortality rate.

Diagnosis is made by national reference laboratories using PCR or culture of the bacterium and fluorescent antibody and agglutination tests. Treatment is with streptomycin, tetracycline, or aminoglycoside antibiotics for 10 to 21 days. An attenuated vaccine is available from the U.S. Army for high-risk laboratory workers. Typically 100 to 200 cases of tularemia are reported annually in the United States.

Comprehension Check

1. How do humans acquire anthrax? Brucellosis?
2. Compare and contrast the symptoms of anthrax, brucellosis, and tularemia.
3. How is psittacosis transmitted?
4. Describe the disease of tularemia. Why is *F. tularensis* considered a potential agent of bioterrorism?

38.6 Opportunistic Diseases Can Be Caused by Bacteria

After reading this section, you should be able to:

a. Explain how disease can result from human microbiota that are normal to another body site
b. Cite specific environmental conditions that result from overgrowth of one bacterial species
c. Design a model of bacterial biofilm formation to explain its role in human disease

Opportunistic infections can be divided into two categories. The first involves microorganisms that are part of the normal microbiota but cause disease if they overgrow other members of the microbial community or gain access to otherwise sterile sites in the host. The second are microbes found in the environment that cause disease only in hosts who are immunologically compromised and cannot mount the robust immune defenses that would otherwise eliminate these microbes. We discuss a few examples of opportunistic disease next.

Antibiotic-Associated Colitis

Antibiotic-associated colitis is a spectrum of disease caused by the overgrowth of *Clostridioides* (formerly *Clostridium*) *difficile* that affects about 500,000 patients annually in the United States. *C. difficile* is sometimes part of the human intestinal microbiota but is normally kept at low numbers by other bacteria. The disease spectrum (*C. difficile* infection, or CDI) includes uncomplicated diarrhea, pseudomembranous colitis (a viscous collection of inflammatory cells, dead cells, necrotic tissue, and fibrin that obstructs the intestine), and toxic megacolon (inflammation that results in intestinal tissue necrosis), which is often lethal (**figure 38.33**). The most common symptom of CDI is watery diarrhea, consisting of three or more bowel movements per day for 2 or more days accompanied by fever, loss of appetite, nausea, and abdominal cramping. In severe cases, patients have as many as 20 diarrhea episodes per day, which leads to extreme weight loss, dehydration, and debilitation.

C. difficile is an anaerobic, endospore-forming rod; endospores are often found in health-care facilities. Although therapy with any antibiotic can trigger CDI, the most common antibiotic associated with CDI is clindamycin, a protein synthesis inhibitor that targets anaerobes that normally chemically modify host bile acids to compounds that inhibit *C. difficile* growth. All pathogenic strains of *C. difficile* produce two exotoxins: enterotoxin (TcdA) and a cytotoxin (TcdB), although TcdB appears to be responsible for the inflammation and diarrhea. These toxins disable regulation of the cytoskeleton within host gut epithelial cells, causing loss of barrier integrity. Some of the more aggressive strains also secrete an additional toxin called binary toxin, which targets host-cell actin. Strains that do not produce TcdB do not cause disease—that is, they are avirulent.

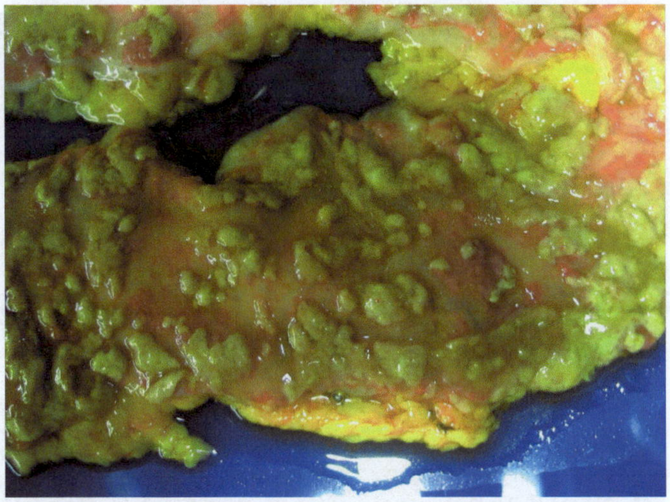

Figure 38.33 ***Clostridioides difficile*** **Infection.** This region of surgically removed bowel shows greenish-yellow sloughing membrane and bacterial overgrowth.

McGraw Hill

MICRO INQUIRY *What cells and tissues make up the pseudomembrane?*

C. difficile is the most commonly recognized cause of diarrhea in hospitalized patients, as endospores from one patient can be transmitted to others. Antibiotic use is the main modifiable risk factor for CDI. In addition to suppressing other bacteria and thereby promoting *C. difficile* overgrowth, antibiotics increase the risk of selecting for antimicrobial-resistant strains. Ironically, treatment of CDI is by oral vancomycin. If disease recurs, a different antibiotic is administered, sometimes with a monoclonal antibody against TcdB. Fecal microbiome transplant (FMT), in which prescreened feces from a healthy donor are transplanted to the CDI patient's colon, may be used to treat recurrent cases (*see chapter 33 opening story*). This reestablishes a healthy microbiome and has a cure rate of about 92%. ◀ *The microbe-human ecosystem (chapter 33)*

Bacterial Vaginosis

Bacterial vaginosis (BV) is a polymicrobial infection that results from an imbalance in the vaginal normal microbiota. It is characterized by a decrease in the *Lactobacillus* population and an overgrowth of *Gardnerella vaginalis, Mobiluncus* spp., *Prevotella* spp., *Peptostreptococcus* spp., and various uncultured BV-associated bacteria (BVAB). These microorganisms inhabit the vagina and rectum of 20 to 40% of healthy females, but their growth is inhibited by vaginal lactobacilli (**figure 38.34**).

BV is typically a mild disease, although it is a risk factor for obstetric infections, adverse outcomes of pregnancy, pelvic inflammatory disease (PID), and HIV infection. The disease is frequently asymptomatic; when symptomatic, it is characterized by a copious, frothy, fishy-smelling discharge with varying degrees of pain or itching. Diagnosis is based on the microscopic observation of clue cells in vaginal discharge. Clue cells are sloughed-off vaginal epithelial cells covered with bacteria, mostly *G. vaginalis* (figure 38.34*b*). Treatment for BV is with metronidazole, a drug that kills anaerobic bacteria and protozoa.

Dental Diseases

Only a few species of bacteria in the oral cavity can be considered true opportunistic dental pathogens, or odontopathogens. These odontopathogens are responsible for the most common bacterial diseases in humans: tooth decay and periodontal disease.

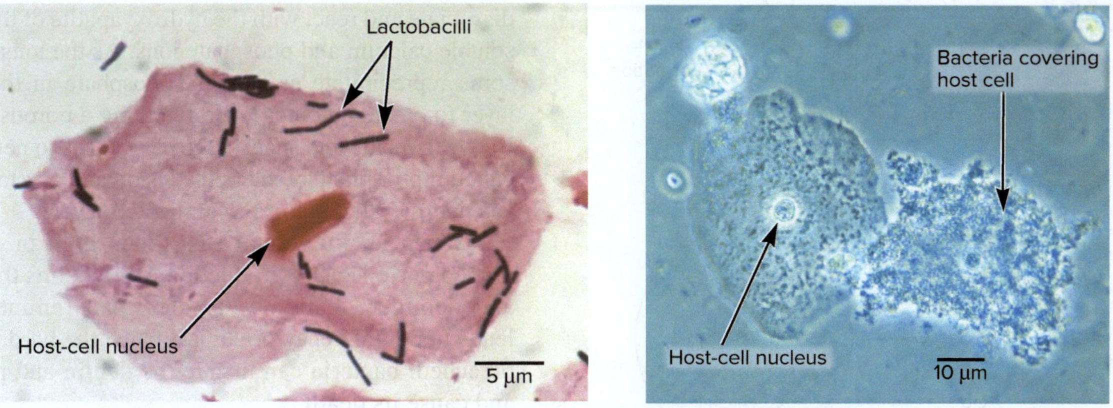

(a) Vaginal lactobacilli

(b) Clue cell

Figure 38.34 Bacterial Vaginosis Is Diagnosed by the Presence of Clue Cells. (a) In the healthy host, Gram-positive lactobacilli keep the vaginal mucosa at an acidic pH, thereby preventing the growth of many other microorganisms (Gram stain). (b) When vaginal lactobacilli growth is inhibited by antibiotic therapy, hormones, or other factors, Gram-negative and Gram-variable anaerobes predominate. Host epithelial cells to which these bacteria attach are called clue cells. (Unstained, phase contrast light microscopy)

(a) Dr. Mike Miller/CDC; (b) M. Rein/CDC

Dental Plaque

The human tooth has a natural defense mechanism against bacterial colonization that complements the protective role of saliva. The hard enamel surface selectively absorbs acidic glycoproteins (mucins) from saliva, forming a membranous layer called the *acquired enamel pellicle.* This pellicle, or organic covering, contains many sulfate (SO_4^{2-}) and carboxylate (COO^-) groups that confer a net negative charge to the tooth surface. Because most bacteria also have a net negative charge, there is a natural repulsion between the tooth surface and bacteria in the oral cavity. Unfortunately, this natural defense mechanism breaks down when dental plaque forms.

Dental plaque is one of the densest collections of bacteria in the body—perhaps one source of the microorganisms seen under a microscope by Antony van Leeuwenhoek in the seventeenth century. Dental plaque formation begins with colonization of the pellicle by *Streptococcus gordonii, S. oralis,* and *S. mitis* (**figure 38.35**). These bacteria selectively adhere to the pellicle by specific ionic, hydrophobic, and lectin like interactions. Once the tooth surface is colonized, subsequent attachment of other bacteria results from a variety of specific coaggregation reactions.

S. mutans and *S. sobrinus* become established on the tooth surface by attaching to these initial colonizers (figure 38.35). *S. mutans* and *S. sobrinus* produce extracellular enzymes that polymerize the glucose moiety of sucrose into a heterogeneous group of extracellular, water-soluble, and water-insoluble polysaccharides. These act like a cement that binds bacterial cells together, forming a plaque ecosystem within a biofilm. The fructose by-product can be used in fermentation. Once plaque becomes established, the surface of the tooth becomes anoxic. This leads to the growth of strictly anaerobic bacteria (*Prevotella melaninogenica, P. oralis,* and *Veillonella alkalescens*), especially between teeth and in the dental-gingival crevices (**figure 38.36**). In addition, bacteria produce lactic and possibly acetic and formic acids from sucrose and other sugars. Because plaque is not permeable to saliva, the acids are not diluted or neutralized, and they demineralize the enamel to produce a lesion on the tooth. It is this chemical lesion that initiates dental decay. ◀ *Biofilms are common in nature (section 7.6)*

Figure 38.36 The Microscopic Appearance of Plaque. Colorized scanning electron micrograph of plaque shows a polymicrobial biofilm.
Steve Gschmeissner/Science Photo Library/Brand X Pictures/Getty Images

MICRO INQUIRY *How is tooth enamel demineralized by bacterial growth?*

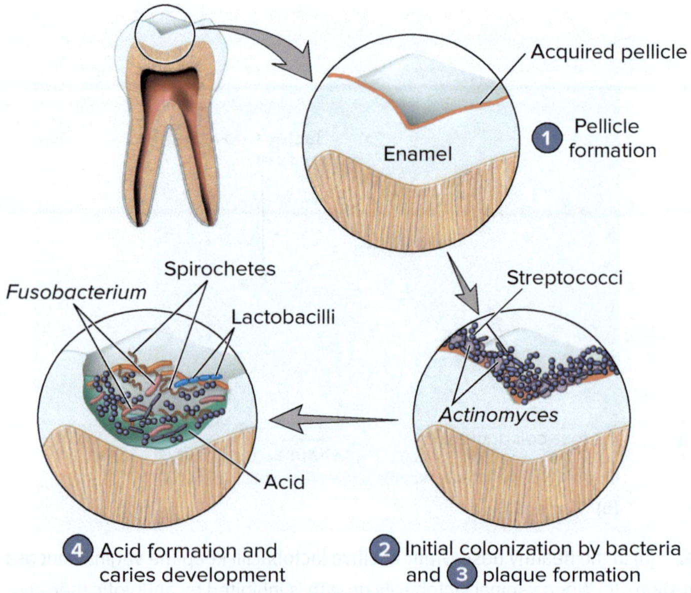

Figure 38.35 Stages in Plaque Development and Cariogenesis. A drawing of a microscopic view of pellicle and plaque formation, acidification, and destruction of tooth enamel.

Dental Decay (Caries)

As fermentation acids move below the enamel surface, they dissociate and react with the hydroxyapatite of the enamel to form soluble calcium and phosphate ions. As the ions diffuse outward, some reprecipitate as calcium phosphate in the tooth's surface layer to create an outer layer overlying a porous subsurface area. Between meals and snacks, the pH returns to neutrality and some calcium phosphate reenters the lesion and crystallizes. The result is a demineralization-remineralization cycle.

When an individual eats foods high in sucrose for prolonged periods, acid production overwhelms the repair process, and demineralization is greater than remineralization. This leads to dental decay or **caries.** Once the hard enamel has been breached, bacteria can invade the dentin and pulp of the tooth and cause its death.

Periodontal Disease

Periodontal disease is the most common chronic infection in adults. It refers to a diverse group of inflammatory diseases that affect the periodontium, which is the supporting structure of a tooth and includes the cementum (a substance that attaches teeth to the underlying bone), the periodontal membrane, the bones of the jaw, and the gingivae (gums). The gingiva is dense, fibrous tissue and the overlying mucous membrane that surrounds the necks of the teeth. The gingiva helps to hold the teeth in place.

Disease is initiated by the formation of plaque that forms at the dentogingival (tooth-gum) margin and extends down into the gingival tissue. A number of bacterial species contribute to tissue damage. The result is an initial inflammatory reaction known as **periodontitis,** which is caused by the host's immune response to both plaque bacteria and tissue destruction. This leads to swelling of the tissue and formation of periodontal pockets. Bacteria colonize these pockets and cause more inflammation, which leads to formation of a periodontal abscess, bone destruction (periodontosis), inflammation of the gingiva (gingivitis), and general tissue necrosis. If the condition is not treated, teeth may be lost.

The main strategies for preventing dental caries and periodontal disease include reduced consumption of sucrose; daily brushing, flossing, and rinsing with mouthwashes; and professional cleaning at least twice a year to remove plaque. The use of fluorides in toothpaste, drinking water, and mouthwashes or fluoride and sealants applied professionally to the teeth protects against lactic and acetic acids and reduces tooth decay.

Streptococcal Pneumonia

Streptococcal pneumonia, also called *pneumococcal pneumonia,* is considered an opportunistic infection because it is contracted from the host's normal microbiota. It is caused by Gram-positive *Streptococcus pneumoniae,* normally found in the upper respiratory tract. Only those individuals with predisposing factors such as viral infections of the respiratory tract, physical injury to the tract, diminished immune function (as found in older adults), alcoholism, diabetes, or smoking are susceptible to this disease. In addition to pneumonia, *S. pneumoniae* causes acute middle ear infection, bacteremia, and meningitis. About 60 to 80% of all respiratory diseases known as pneumonia are caused by *S. pneumoniae*. About 150,000 individuals are hospitalized with pneumococcal pneumonia in the United States annually; the mortality rate is 5–7%.

The primary virulence factor of *S. pneumoniae* is its capsule. Each of the over 90 serotypes of *S. pneumoniae* constructs its capsule from different polysaccharides. The capsule plays an important role in protecting the organism from ingestion and killing by phagocytes. Pathogenesis is due to the rapid multiplication of the bacteria in alveolar spaces. The bacteria also produce a variety of virulence factors including the toxin pneumolysin, which destroys host cells. As host cells are killed, alveoli fill with blood, bacteria, neutrophils, and fluid and become inflamed. Blood in sputum imparts a rust color. The onset of clinical symptoms is usually abrupt, with chills, labored breathing, and chest pain. Diagnosis is by chest X ray, Gram stain, culture, and tests for metabolic products. Treatment needs to be based on antibiotic susceptibility because some strains are resistant to β-lactam and macrolide antibiotics.

Two types of pneumococcal vaccines are available for people who are at risk for exposure. The pneumococcal conjugate vaccines (PCV) cover 13, 15, or 20 pneumococcal serotypes, while the pneumococcal polysaccharide vaccine (PPSV, also called PPSV23) covers 23 serotypes of pneumococcal bacteria. These vaccines are effective because they generate antibodies to the strains that cause about 90% of pneumococcal disease. When these antibodies are deposited on the surface of the bacteria, the microbes become opsonized and phagocytosis is enhanced (*see figure 31.5*).

Urinary Tract Infection and Pyelonephritis

Urinary tract infection (UTI), also called **cystitis,** is one of the most common infections, with an estimated global prevalence of over 400 million cases annually. In the United States, over 12 million people seek help for UTIs in doctor offices and emergency departments every year. The causative bacteria reside in the gut and, when introduced into the urethra, ascend to reach the bladder. If left untreated, the bacteria may move up the ureters to the kidneys; kidney infections are called **pyelonephritis.** Females have a much shorter urethra, which accounts for the higher incidence of UTIs in that population, but UTIs are also not uncommon among infant and older males. A variety of bacteria cause UTIs including *Klebsiella pneumoniae, Staphylococcus saprophyticus, Enterococcus faecalis, Proteus mirabilis,* group B *Streptococcus,* and *Pseudomonas aeruginosa,* but roughly 90% of all UTIs are caused by uropathogenic *E. coli* (UPEC), which is our focus.

UPEC is unique among *E. coli* strains in its ability to infect the bladder and avoid shedding during urination. UPEC localizes the protein FimH at the tip of type 1 pili, where it serves as an adhesin that attaches to receptors on bladder cells. Such binding triggers host-cell actin rearrangement so that UPEC cells are internalized by these normally nonphagocytic cells. Inside bladder cells, UPEC replicates and establishes an intracellular biofilm community, called a *pod.* UPEC also produces the cytolytic toxin HlyA, which contributes to host-cell invasion. As UPEC cells flux in and out of pods, neutrophils invade the bladder and cause acute inflammation, which accounts for the characteristic symptoms of frequency of urination, burning when urinating, abdominal cramping, and blood in the urine.

As long as the infection remains in the bladder, the patient will not develop a fever. However, if the bacteria ascend a ureter and reach a kidney (this is often not bilateral), pyelonephritis develops. This is a much more serious condition as it can cause permanent renal damage. UPEC regulates type 1 pili and flagella in a reciprocal fashion so that when motility is needed to move up the ureter to the kidney, type 1 pili and FimH are not produced. Once UPEC has entered the kidney, it produces fimbriae to mediate attachment. In addition to fever and chills, pyelonephritis is characterized by pain over the affected kidney.

Uncomplicated UTIs are typically diagnosed clinically with the collection of urine. A simple dipstick test will reveal the presence of leukocyte esterase, indicating neutrophil invasion of the bladder. If the causative agent is UPEC, the dipstick will also indicate the presence of nitrite in the urine. Nitrite is produced by UPEC cells that have switched to anaerobic respiration with the use of nitrate as the terminal electron acceptor. The causative agent of pyelonephritis must be determined by urine culture. Obtaining antibiotic sensitivity profile of the causative bacterium is essential. Uncomplicated UTIs are treated with antibiotics that reach a high level in the bladder including sulfamethoxazole/trimethoprim, nitrofurantoin, or fosfomycin. Individuals with pyelonephritis in the outpatient setting can be treated with a fluoroquinolone.

Comprehension Check

1. In both *C. difficile*–associated disease and bacterial vaginosis, the disease begins when the normal microbiota of the tissue site is altered. How do these pathogens become dominant?
2. Name some common odontopathogens responsible for dental caries, dental plaque, and periodontal disease.
3. Discuss reasons cigarette smoking increases the risk of streptococcal pneumonia to the extent that people who smoke are advised to take pneumococcal vaccines.
4. Discuss two adaptations used by UPEC that enable bladder and kidney infection.

Key Concepts

38.1 Bacteria Can Be Transmitted by Airborne Routes

- The vaccine-preventable diseases diphtheria and pertussis are toxin-mediated. Diphtheria toxin blocks host-cell protein synthesis by ADP-ribosylating elongation factor (EF)-2 (**figure 38.1**). *Bordetella pertussis* produces several exotoxins, including pertussis toxin, which stimulates host adenyl cyclase activity, and another toxin that has adenyl cyclase activity. Thus overproduction of cAMP is a key component of *B. pertussis* virulence.
- Atypical pneumonia can be caused by *Chlamydia pneumoniae, Legionella pneumophila,* or *Mycoplasma pneumoniae*. Infections are considered atypical because they are not usually associated with a productive cough. *L. pneumophila* infection causes Legionnaires' disease, which is the most serious of these infections.
- About one-quarter of the global population has tuberculosis (TB) in its latent form. Caused by *Mycobacterium tuberculosis,* TB usually infects the lungs but can also infect other body systems. Strong cell-mediated immunity can keep TB in its latent form. Children, individuals who are HIV-positive, and others who are immunosuppressed often develop severe, life-threatening infections. Treatment is complex, involving multiple antibiotics administered for at least 6 months (**figures 38.4** and **38.5**).
- Vaccine-preventable meningitis is caused by *Neisseria meningitidis, Haemophilus influenzae* serotype b, and *Streptococcus pneumoniae*. However, meningitis in neonates is typically caused by group B streptococcus (*S. agalactiae*), *E. coli,* and *Listeria monocytogenes*. When *N. meningitidis* (meningococcus) causes bacteremia, it is termed meningococcemia. Aggressive antibiotic therapy and supportive care are needed for this medical emergency.
- Pertussis is characterized by uncontrollable coughing and a characteristic "whooping" sound with inhalation. Primarily affecting infants and children, the disease can be prevented by vaccination (DTaP).
- Streptococcal pharyngitis (strep throat) is caused by *S. pyogenes,* or group A streptococcus (GAS). The M protein on the surface of GAS is strain-specific and is one of many virulence factors (**figure 38.8**).

38.2 Arthropods Can Transmit Bacterial Diseases

- Bacteria can be transmitted to humans by arthropod vectors, such as ticks, lice, and fleas.
- Lyme disease is a tick-borne disease caused by spirochetes in the genus *Borreliella*. In the northeast United States, the tick vector *Ixodes scapularis* transmits *B. burgdorferi*. Treatment with doxycycline prevents the progression of disease through three stages: characteristic rash and flulike illness, arthritis and cardiac inflammation, and neurological deficits (**figure 38.10**).
- Plague, caused by *Yersinia pestis,* remains endemic in some regions. The flea vector feeds on rodents; humans are infected by flea bites or direct contact with infected animals. Humans can transmit plague by droplet transmission between individuals (pneumonic plague). Toxic proteins inserted into macrophages and other immune cells enable the survival and growth of *Y. pestis*. Lymph tissue forms bubos, while blood-borne *Y. pestis* causes mottling of skin and necrosis of extremities, thus the terms Bubonic Plague and Black Death, respectively. *Y. pestis* mediates disease by insertion of toxic proteins directly into host cells via a type III secretion system. (**figures 38.11** and **38.12**).
- The majority of Rocky Mountain spotted fever (RMSF) cases occur east of the Mississippi River. *Rickettsia rickettsii* is transmitted by the ticks *Dermacentor andersoni* and *D. variabilis*. *R. rickettsii* invades endothelial cells, causing vasculitis marked by a nonblanching rash. Without treatment (doxycycline), RMSF can be fatal (**figure 38.13**).

38.3 Direct Contact Diseases Can Be Caused by Bacteria

- *Staphylococcus aureus* and group A streptococcus (*S. pyogenes*) cause the superficial skin infections cellulitis and impetigo. *S. pyogenes* and *Clostridium perfringens* cause the life-threatening skin infections necrotizing fasciitis (**figure 38.24**) and gas gangrene (**figure 38.14**), respectively. These are medical emergencies that require surgical excision of infected tissue and aggressive antibiotic therapy.
- In infants, group B streptococcus (*S. agalactiae*) can cause meningitis, bacteremia, and pneumonia; in older people and individuals who are immunocompromised, it can cause bacteremia, pneumonia, and skin and bone infections.
- *Mycobacterium leprae* causes leprosy, which in its tuberculoid form is mild and nonprogressive. Lepromatous leprosy results if cell-mediated immune responses are inadequate. It is marked by necrosis of skin and nerve damage (**figure 38.15**).
- *Helicobacter pylori* infection can result in peptic ulcer disease and gastric cancer. This bacterium survives stomach acidity by enclosing itself in the mucosa and producing ammonium via urease activity.
- *Chlamydia trachomatis* is sexually transmitted and can cause pelvic inflammatory disease (PID) in females and the eye infection trachoma in neonates. Trachoma, which is not sexually transmitted, is the leading cause of preventable blindness globally (**figure 38.26**).
- Gonorrhea, caused by *Neisseria gonorrhoeae,* is another cause of PID, partly because it is easier to diagnose in males than females, who typically have fewer symptoms. *N. gonorrhoeae* can cause eye infection (ophthalmia neonatorum) in newborns during birth (**figure 38.17**). Because of frequent coinfection, gonorrhea and chlamydia are treated simultaneously with ceftriaxone as well as either azithromycin or doxycycline.
- The spirochete *Treponema pallidum* causes syphilis, which progresses through three stages: painless chancre; skin rash, malaise, and fever; and neurosyphilis and the development of gummas (**figure 38.18**). Benzathine penicillin is an effective cure, but like gonorrhea, infection with syphilis does not lead to lifelong immunity.
- Methicillin-resistant *Staphylococcus aureus* (MRSA) can be community or hospital acquired (CA and HA, respectively). MRSA can infect any organ system and is resistant to all β-lactam antibiotics. CA-MRSA is typically sensitive to a wider range of other antibiotics than HA-MRSA but produces more toxins.
- Tetanus, caused by *Clostridium tetani,* is vaccine preventable. Tetanus toxin prevents the release of inhibitory neurotransmitters, thereby causing spastic paralysis. Newborns delivered without sterile precautions are at risk for tetanus (**figure 38.25**).

38.4 Food and Water Are Vehicles for Bacterial Diseases

- Food-borne illnesses are classified as infections or intoxications, which are toxin mediated so presence of the bacterium is not needed (**table 38.3**).
- *Clostridium botulinum* produces botulinum toxin, which blocks the release of the neurotransmitter acetylcholine, thereby causing flaccid paralysis (**figure 38.27**).
- *Campylobacter jejuni* is the most frequently reported cause of food-borne bacterial diarrhea in the United States. It produces a toxin with DNase activity, cytolethal distending toxin, but is not considered an intoxication. In rare cases, *C. jejuni* infection is associated with Guillain-Barré syndrome through a mechanism called molecular mimicry (**figure 38.28**).
- Cholera is caused by the aquatic bacterium *Vibrio cholerae. V. cholerae* produces an AB toxin that ADP-ribosylates the stimulator of adenyl cyclase, which becomes locked in the active state. This leads to high levels of host cell cAMP. This results in the loss of ions and water; patients can lose 20 liters of fluid in diarrhea daily. Rehydration with a salt solution prevents death.
- There are five major strains of pathogenic, food-borne *E. coli,* each classified by the nature of its interaction with host intestinal epithelium (**figure 38.30**). Enterotoxigenic and enterohemorrhagic *E. coli* (ETEC and EHEC, respectively) produce toxins. EHEC also causes hemolytic uremic syndrome.
- Serovars of *Salmonella enterica* are classified as nontyphoidal or as causing enteric fever. *S.* Typhi and *S.* Paratyphi, which cause enteric fever, produce an exotoxin. Nontyphoidal strains cause salmonellosis, a self-limiting gastroenteritis, while enteric fever can be life threatening and is vaccine preventable.
- Shigellosis is caused by one of four species of *Escherichia* (formerly *Shigella*). Shiga toxin (Stx) blocks protein synthesis by depurinating host 28S rRNA. Stx has been horizontally transferred to EHEC strains.
- Staphylococcal food poisoning, usually caused by *S. aureus,* is an intoxication with staphylococcal enterotoxins. Vomiting begins within hours of ingesting the tainted food and resolves within 24 hours.

38.5 Zoonotic Diseases Arise from Human-Animal Interactions

- Anthrax is a toxin-mediated disease that can manifest in one of three ways depending on mode of transmission: cutaneous, pulmonary, and gastrointestinal. The toxin consists of two effector proteins (lethal factor and edema factor) and protective antigen, which binds to host cells (**figure 38.32**).
- Human brucellosis is caused by *Brucella melitensis.* Infection is most frequently transmitted by cattle and can be acute or insidious. Brucellosis is a nonspecific febrile illness that is particularly dangerous for females who are pregnant.

- Psittacosis is a rare pulmonary infection acquired from birds. The causative agent is the obligate intracellular pathogen *Chlamydophila psittaci*. Q fever is also rare. It is caused by *Coxiella burnetii,* also an intracellular pathogen transmitted by cows, sheep, and goats.
- Tularemia, caused by *Francisella tularensis,* can be transmitted by infected animals (e.g., rabbits), arthropods (ticks and deer flies), or direct inhalation of aerosolized bacteria. Signs and symptoms vary depending on how the pathogen enters the host.

38.6 Opportunistic Diseases Can Be Caused by Bacteria

- Normal microbiota can cause disease, especially when they invade normally sterile tissues. Opportunistic pathogens also include those that only infect individuals with compromised immune function.
- Opportunistic diseases that correspond to a loss of normal microbiota include antibiotic-associated colitis (*Clostridioides difficile;* **figure 38.33**) and bacterial vaginosis (*Gardnerella vaginalis;* **figure 38.34**).
- Dental plaque formation begins with the initial colonization of the acquired enamel pellicle by *Streptococcus gordonii, S. oralis,* and *S. mitis*. Other bacteria then become attached and form a plaque ecosystem (**figures 38.35** and **38.36**). The bacteria produce acids that cause a chemical lesion on the tooth and initiate dental decay or caries.
- Periodontal disease is a group of diverse clinical entities that affect the periodontium. Disease is initiated by the formation of subgingival plaque, which leads to tissue inflammation known as periodontitis and to periodontal pockets. Bacteria that colonize these pockets can cause an abscess, periodontosis, gingivitis, and general tissue necrosis.
- *Streptococcus pneumoniae,* a normal inhabitant of the nasopharynx, is a common cause of pneumonia, particularly in older people. There are over 90 serotypes, each with a unique polysaccharide capsule. There are two types of vaccines protecting against the serotypes that commonly cause disease.
- Urinary tract infections are one of the most common infections for which people seek treatment. The most frequent cause is uropathogenic *E. coli* (UPEC). Prompt antibiotic treatment prevents UPEC from ascending the ureters and causing kidney infection (pyelonephritis), a more severe condition.

Active Learning

1. Construct a table of pathogens discussed in this chapter that infect immune cells. Compare the diseases they cause with several other diseases caused by pathogens that do not infect immune cells. Can you see any trends or make any inferences about these two types of pathogens and the diseases they cause?
2. Worldwide, pneumonia and diarrhea claim the lives of more children than any other reason. Review the UNICEF Pneumonia and Diarrhea Report (available online) and provide the three most important steps that can be taken to prevent these avoidable deaths.
3. The incidence of syphilis continues to climb worldwide, including in the United States. Review the three stages of syphilis and discuss why the disease may go unnoticed and untreated.
4. In 2011 Germany saw a large outbreak of EHEC that was eventually traced to contaminated bean sprouts. This outbreak was unusual in several ways. First, the rate of hemolytic uremic syndrome (HUS) was more than double that seen in previous EHEC outbreaks (20% rather than about 8%), and HUS was most commonly seen in adults rather than children. The causative strain, O104:H4, was found to be genetically complex, with elements seen in both EHEC and EAEC strains. Discuss features of both EHEC and EAEC that might have combined to make this strain so virulent. Refer to chapter 16 to discuss the mechanisms of horizontal gene transfer that resulted in the evolution of this unusual strain.
5. When the Lyme disease bacterium *B. burgdorferi* is in the gut of its tick vector, its outer surface protein A (OspA) is expressed at high levels. Once it has been transmitted to a human, OspA production diminishes. However, to complete its infection cycle and return to the tick, *B. burgdorferi* cells must increase OspA synthesis. To accomplish this, *B. burgdorferi* binds the host neuroendocrine stress hormones epinephrine and norepinephrine, which results in upregulation of OspA. Discuss the role of coevolution in the development of *B. burgdorferi*'s ability to sense and respond to these host compounds. How would you show that among all the mediators of inflammation at the site of a tick bite, the spirochete responds specifically to epinephrine and norepinephrine? Your answer should include some carefully considered control experiments.

CHAPTER

39

Human Diseases Caused by Fungi and Protists

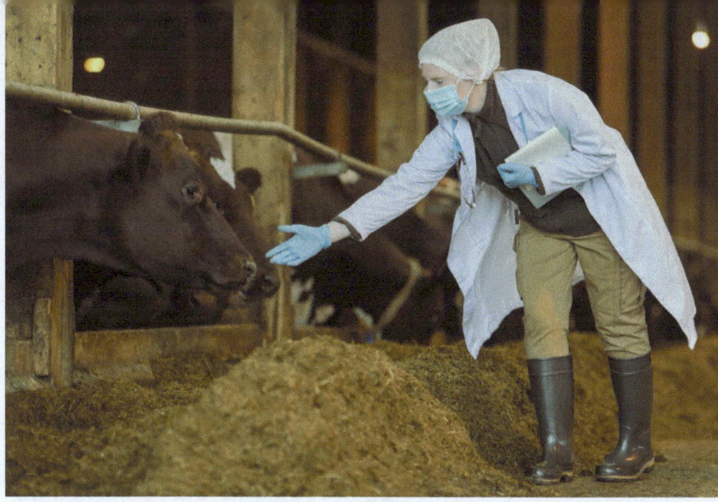

SeventyFour/Shutterstock

One Health for Tropical Diseases That Should Not Be Neglected

As you go about your daily activities, it's often easy to forget that humans are vastly outnumbered by animals and insects. While these nonhuman organisms have many laudable qualities, they also represent a pathogen reservoir so vast it's almost unimaginable. Among the hundreds of zoonotic and arthropod-borne diseases, 23 are considered by the World Health Organization (WHO) to be *Neglected Tropical Diseases* (NTDs).[1] These diseases are neglected not because they have a low global impact—between a fifth and a quarter of the world's population, mostly in tropical regions, live in communities where these diseases are endemic—but because research and development of vaccines and treatments have been historically under-resourced. The majority of NTDs are caused by eukaryotic pathogens, including protozoa and fungi.

Early in 2021, the WHO launched a new, ambitious multidisciplinary roadmap to address NTDs over the next 9 years. NTDs, such as Chagas disease and leishmaniasis, are transmitted through zoonotic pathways or arthropod vectors, making animal and environmental health critical components of disease control. Key to addressing these NTDs is the *One Health Initiative,* a comprehensive approach recognizing the interconnectedness of humans, animals, and the environment.

By integrating efforts across multiple disciplines, One Health aims to create sustainable, holistic solutions for the pervasive health challenges NTDs present. By promoting collaboration among veterinarians, physicians, ecologists, and other professionals, diseases across species and ecosystems can be monitored and managed. Multidisciplinary cooperation enhances early detection and prevention strategies, reducing the incidence and spread of NTDs.

Moreover, One Health emphasizes the importance of improving sanitation, access to clean water, and education in affected communities, addressing the environmental and socioeconomic factors that contribute to the persistence of NTDs. By fostering community engagement and building local capacity, One Health ensures that interventions are culturally appropriate and sustainable.

The One Health Initiative demonstrates that tackling global infectious diseases requires a team approach. So, if your interest is in epidemiology or education or whether you want to become a healthcare provider to humans or animals, an engineer, a scientist, or a sociologist, there is room for you in the One Health Initiative. Only through collaboration will One Health achieve significant, long-lasting reductions in NTD prevalence, improving health outcomes for vulnerable populations worldwide.

In this chapter, we describe some of the fungi and protists that are human pathogens and discuss the clinical manifestations, diagnosis, epidemiology, pathogenesis, and treatment of the diseases caused by them. The biology and diversity of these organisms are covered in chapters 5, 23, and 24. This chapter follows the format of the previous two chapters in categorizing their diseases by route of transmission (**figure 39.1**).

Readiness Check:

Based on what you have learned previously, you should be able to:

- ✓ Describe basic eukaryote cell biology (chapter 5)
- ✓ Identify the major factors supporting growth and reproduction of protists and fungi (sections 23.1 and 24.1, respectively)
- ✓ Distinguish between yeast and mold (section 24.1)
- ✓ Explain pathogenicity and the infection process (chapter 34)
- ✓ Discuss methods by which pathogenic fungi and protozoa are identified (chapter 36)

[1] The complete list of NTDs includes Buruli ulcer, Chagas disease, chromoblastomycosis, dengue and severe dengue, dracunculiasis, echinococcosis, food-borne trematodiases, human African trypanosomiasis, leishmaniasis, leprosy, lymphatic filariasis, mycetoma, noma, onchocerciasis, podoconiosis, rabies; scabies, schistosomiasis, snakebite envenoming, soil-transmitted helminthiases, sporotrichosis, taeniasis/cysticercosis, trachoma, tungiasis, and yaws.

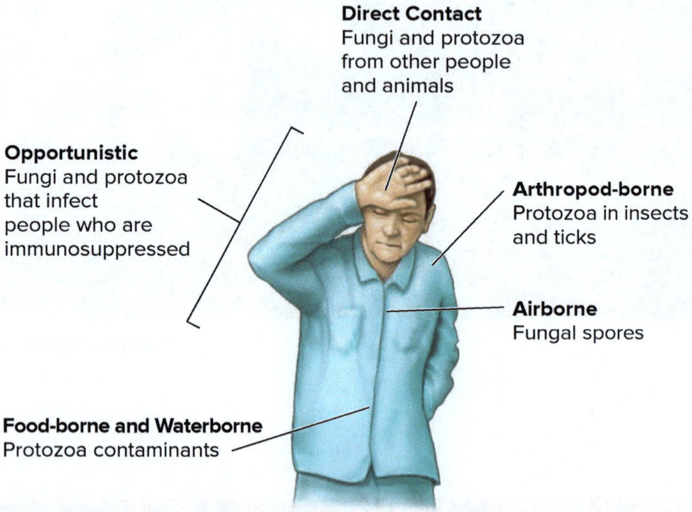

Figure 39.1 Common Routes of Fungal and Protozoan Disease Transmission. Examples of disease-causing fungi and protists are presented in this chapter according to route of transmission.

39.1 Relatively Few Fungi and Protists Are Human Pathogens

After reading this section, you should be able to:

a. Describe the system used to categorize fungal diseases
b. Appreciate the burden of protozoan disease worldwide

While there are millions of species of fungi on the Earth, only about 300 are known to produce disease in humans. Medical mycology is the discipline concerned with fungi that cause human disease. Fungal diseases, known as **mycoses** (s., mycosis; Greek *mykes,* fungus), can be divided into five groups: superficial, cutaneous, subcutaneous, systemic, and opportunistic mycoses (**table 39.1**). Superficial, cutaneous, and subcutaneous mycoses are direct contact infections of the skin, hair, and nails. Systemic mycoses are fungal infections that can disseminate to visceral tissues. The fungi that cause systemic mycoses are

Table 39.1 Examples of Some Medically Important Fungi

Disease Category	Pathogen	Location	Disease
Superficial mycoses	*Piedraia hortae*	Scalp	Black piedra
	Trichosporon beigelii	Beard, mustache	White piedra
	Malassezia furfur	Trunk, neck, face, arms	Tinea versicolor
Cutaneous mycoses	*Trichophyton mentagrophytes, T. verrucosum, T. rubrum*	Beard hair	Tinea barbae
	Trichophyton, Microsporum canis	Scalp hair	Tinea capitis
	T. rubrum	Smooth or bare parts of the skin	Tinea corporis (ringworm)
	Epidermophyton floccosum, T. mentagrophytes, T. rubrum	Groin, buttocks	Tinea cruris (jock itch)
	T. rubrum, T. mentagrophytes, E. floccosum	Feet	Tinea pedis (athlete's foot)
	T. rubrum, T. mentagrophytes, E. floccosum	Nails	Tinea unguium (onychomycosis)
Subcutaneous mycoses	*Phialophora verrucosa, Fonsecaea pedrosoi*	Legs, feet	Chromoblastomycosis
	Madurella mycetomatis	Feet, other areas of body	Maduromycosis
	Sporothrix schenckii	Puncture wounds	Sporotrichosis
Systemic mycoses	*Blastomyces dermatitidis*	Lungs, skin	Blastomycosis
	Coccidioides immitis	Lungs, other parts of body	Coccidioidomycosis
	Histoplasma capsulatum	Within phagocytes	Histoplasmosis
Opportunistic mycoses	*Aspergillus fumigatus, A. flavus*	Respiratory system	Aspergillosis
	Candida albicans	Skin, mucous membranes, bloodstream	Candidiasis
	C. auris	Bloodstream	Candidiasis/sepsis
	Cryptococcus neoformans	Lungs, skin, bones, viscera, central nervous system (CNS)	Cryptococcosis
	Pneumocystis jirovecii	Lungs, sometimes brain	*Pneumocystis* pneumonia
	Microsporidium, Enterocytozoon	Lungs, gastrointestinal (GI) tract, sometimes brain	Microsporidiosis

dimorphic; that is, they exhibit a yeast (single-cell) phase (Y) and a saprophytic mold or mycelial (filamentous) phase (M). The switch from growth as a mycelial form in the environment at cooler temperatures to growth as a yeast at human body temperatures is called the **YM shift** and contributes to virulence. Systemic mycoses are acquired by inhalation of spores from soil. If a susceptible person inhales enough spores, infection begins as a lung infection that may reach the bloodstream and spread to other organs. Opportunistic mycoses are infections seen in individuals who are immunocompromised. Finally, some fungi produce toxins (mycotoxins) harmful to humans. We discuss them with other toxins in chapters 34 and 40 (*see figure 40.3*).

Protozoa have adapted to practically every type of habitat on Earth, including the human body. Pathogenic protozoa can be transmitted to humans by arthropod (i.e., insects and ticks) vectors, from contaminated food or water, or by direct contact. Although members of fewer than 20 genera of protists commonly cause disease in humans (**table 39.2**), their impact is formidable. For example, in 2023, there were about 249 million new cases of malaria and about 597,000 deaths, with children under the age of five bearing most of the burden. In the Americas, Chagas disease (caused by a trypanosome) affects about 7 million people, with about 12,000 deaths annually. Older adults and individuals who are immunosuppressed are at increased risk for some fungal and protozoan infections. We now discuss diseases caused by fungi and protists based on how they are acquired by the human host.

39.2 Fungi Can Be Transmitted by Airborne Routes

After reading this section, you should be able to:

a. Describe fungi that cause common diseases spread by airborne transmission
b. Identify typical signs and symptoms of common airborne fungal diseases

Airborne transmission of fungi occurs by inhaling spores from the environment. Few human protozoan diseases are spread by airborne routes. We therefore present several examples of pathogenic fungi acquired through airborne transmission and the diseases they cause.

Blastomycosis

Blastomycosis is a systemic mycosis caused by *Blastomyces dermatitidis,* a dimorphic fungus that grows as a budding yeast in humans but as a mold on culture media and in the environment. It is found predominately in moist soil enriched with decomposing organic debris. *B. dermatitidis* is endemic in the Mississippi and Ohio River basins and in some regions in Central and South America and Africa. The disease occurs in three clinical stages: pulmonary, disseminated, and cutaneous. The

Table 39.2 Examples of Medically Important Protozoan Infections

Initial Infection Site	Pathogen	Disease
Blood	*Plasmodium* spp. *Trypanosoma brucei* *T. cruzi*	Malaria African sleeping sickness Chagas or American trypanosomiasis
Intestinal Tract	*Cryptosporidium* spp. *Cyclospora* spp. *Entamoeba histolytica* *Giardia intestinalis* *Toxoplasma gondii*	Cryptosporidiosis Cyclosporidiosis Amebiasis Giardiasis Toxoplasmosis
Lungs	*Balamuthia mandrillaris*	*Balamuthia* amebic encephalitis
Mucosal Membranes Ears, Eyes, Nose Genitals	*Acanthamoeba* spp. *Naegleria fowleri* *Trichomonas vaginalis*	Keratitis, Granulomatous amebic encephalitis Primary amebic meningoencephalitis Trichomoniasis
Skin	*Leishmania braziliensis* *L. donovani* *L. tropica*	Mucocutaneous leishmaniasis Visceral leishmaniasis Cutaneous leishmaniasis

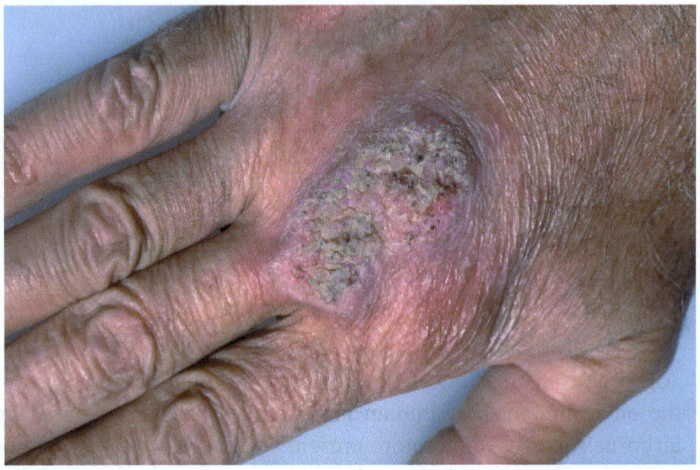

Figure 39.2 Systemic Mycosis: Blastomycosis. Blastomycosis of the skin caused by *Blastomyces dermatitidis*.

Scott Camazine/Science Source

MICRO INQUIRY *What is the morphology of* B. dermatitidis *in the environment? In the human host?*

initial infection begins when blastospores are inhaled into the lungs. If it spreads systemically from the lungs, it becomes disseminated. It often spreads from blood to the skin, causing cutaneous ulcers (**figure 39.2**). *B. dermatitidis* can be isolated from sputum, wound exudates, and biopsy sections. Diagnosis requires the demonstration of thick-walled, yeast-like cells 8 to 15 µm in diameter. Fungal histology, serology, and antigen tests are also useful. Amphotericin B is used for systemic infection, and itraconazole is used for pulmonary infections. Mortality is about 5%. There are no preventive or control measures. *Antifungal drugs (section 9.6)*

Coccidioidomycosis

Coccidioidomycosis is also known as valley fever, San Joaquin fever, or desert rheumatism because of the geographical distribution of the fungus. It is caused by *Coccidioides immitis,* which thrives in the semiarid, highly alkaline soils of the southwestern United States and parts of Mexico and South America. In the United States the number of cases per year has recently been increasing, with about 20,000 people infected annually, resulting in approximately 200 deaths. Endemic areas have been defined by massive skin testing with the antigen coccidioidin, where up to 15% of the test population is positive; endemic areas include Arizona, California, Nevada, New Mexico, and Utah. In these areas coccidioidomycosis causes up to 30% of community-acquired pneumonia. In the soil and on culture media, this fungus grows as a mold that forms arthroconidia at the tips of hyphae (*see figure 24.3*). Because arthroconidia are so abundant, individuals who are immunocompromised can acquire the disease by inhalation as they simply move through these endemic areas. Wind turbulence and even construction of outdoor

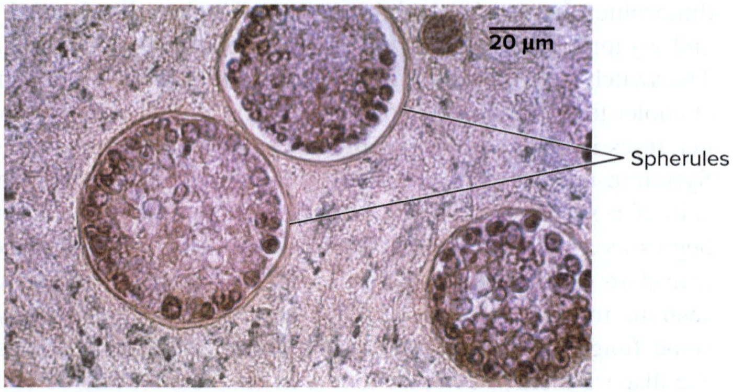

Figure 39.3 Systemic Mycosis: Coccidioidomycosis. *Coccidioides immitis* mature spherules filled with spores within a tissue section; light micrograph.

Lucille K. Georg/CDC

structures have been associated with infection. In humans, the fungus forms thick-walled structures called *spherules* filled with spores about 2 to 5 µm in diameter (**figure 39.3**). Most cases of coccidioidomycosis are asymptomatic or indistinguishable from ordinary upper respiratory tract infections and resolve in a few weeks, with lasting immunity. In rare cases, infection results in progressive chronic pulmonary disease or disseminated infections of other tissues. The fungus can spread throughout the body, involving almost any organ or site.

Diagnosis is accomplished by identification of the large spherules (~60 µm in diameter) in pus, sputum, and aspirates. Culturing clinical samples in the presence of penicillin and streptomycin on the fungal selective medium Sabouraud agar also is diagnostic. Rapid confirmation methods include serology for IgM and IgG titers and skin testing. Fluconazole, itraconazole, and amphotericin B (for systemic infections) are the drugs of choice for treatment. Prevention involves reducing exposure to dust (soil) in endemic areas.

Histoplasmosis

Histoplasmosis is primarily a disease of the lungs caused by *Histoplasma capsulatum* var. *capsulatum*. Like *B. dermatitidis* and *C. immitis, H. capsulatum* can spread from the lungs to cause systemic disease. It differs from other common fungal pathogens in that it grows intracellularly in host macrophages. Symptoms include flulike responses such as fever, cough, headaches, and muscle aches. However, infection is often asymptomatic. *H. capsulatum* is a small, budding yeast in humans and on culture media at 37°C. At 25°C it grows as a mold, with septate hyphae, microconidia, and macroconidia formed on conidiophores (**figure 39.4**; *also see figure 24.3*). Microconidia, which are smaller and less complex conidia that macroconidia, can become airborne when contaminated soil is disturbed. Infection ensues when a large dose of microconidia is inhaled. Microconidia are most prevalent where bird and bat droppings have accumulated. It is noteworthy that birds themselves are not infected; their droppings simply

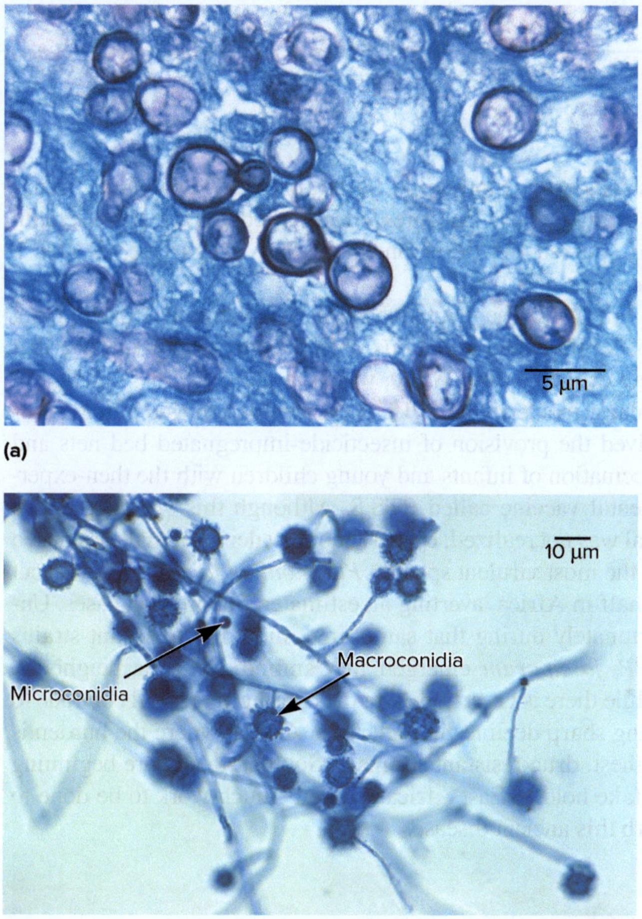

Figure 39.4 Morphology of *Histoplasma capsulatum* var. *capsulatum*. Light micrographs. (a) In the host, *H. capsulatum* grows as a yeast (dark blue spheres). (b) Hyphae, microconidia, and macroconidia as found in soil.
Dr. Libero Ajello/CDC

provide the nutrients for this fungus. Only bats and humans demonstrate the disease and harbor the fungus.

Histoplasmosis is found worldwide, with Africa, Australia, India, and Malaysia being endemic regions. Within the United States, histoplasmosis is endemic within the Mississippi, Ohio, and Rio Grande River basins; however, its range has recently been expanding throughout the eastern United States. In endemic areas of the United States, there are about a 1,000 reported cases of symptomic histoplasmosis with around 100 deaths per year. Histoplasmosis is more common among poultry farmers, spelunkers (people who explore caves), and bat guano miners (bat guano is used as fertilizer).

Because *H. capsulatum* grows within macrophages, the fungus can be disseminated to many organs and is often found in bone marrow. Lesions may appear in the lungs and show calcification, thus the disease may resemble tuberculosis. Laboratory diagnosis is accomplished by isolation of the fungus from tissue specimens, microscopy using Calcofluor White or other fungal stains, and by serology. Severe infections are treated with amphotericin B, whereas the less toxic drugs itraconazole and fluconazole are used to treat mild to moderate disease. Prevention and control involve wearing protective clothing and masks before entering or working in infested habitats.

Comprehension Check

1. Apart from mode of transmission, how can diseases caused by fungi be classified?
2. Compare and contrast the severity of disease caused by the three fungi discussed in this section.
3. Given your geographic location, are you at risk for any of these mycoses?

39.3 Arthropods Can Transmit Protozoal Disease

After reading this section, you should be able to:

a. Discuss the protozoa that cause common diseases spread by arthropod transmission
b. Correlate the signs and symptoms of arthropod-borne protozoal diseases to the life cycle of the causative agent

Protozoa transmitted by insects interact with the human host in a manner that is different from arthropod-borne bacteria and viruses. Bacteria and viruses replicate rapidly in the host, triggering a robust immune response. By contrast, protozoa grow more slowly and have evolved an array of mechanisms to prevent clearance by the host's immune system. Thus when considering arthropod-borne protozoan infections, it is important to keep in mind that the immune system may prevent clinical disease but the protozoan may never be fully cleared from the host. This is seen in malaria; people who live in endemic areas may be referred to as "immune" when, in fact, most carry a low parasite load.

Malaria

The most important human pathogenic protozoa belong to the genus *Plasmodium* and cause **malaria** (**Disease 39.1**). The World Health Organization (WHO) estimates 3.3 billion people (about 40% of the world's population) across 85 countries are at risk for malaria (**figure 39.5**). Although mortality rates have fallen by more than 50% since 2000, disruption to healthcare delivery during and since the COVID-19 pandemic has resulted in an increase in both malaria incidence and mortality. The vast majority (94%) of malaria deaths occur in Africa, with children under the age of five accounting for 80% of deaths.

DISEASE

39.1 A Brief History of Malaria

No other single infectious disease has had more impact on humans than malaria. The first references to its periodic fever and chills can be found in early pre-Babylonian, Chinese, and Hindu writings. In the late fifth century BCE, Hippocrates described malaria symptoms. In the fourth century BCE, the Greeks noted an association between individuals exposed to swamp environments and the subsequent development of periodic fever and enlargement of the spleen. In the seventeenth century, the Italians named the disease *mal' aria* (bad air) because of its association with ill-smelling vapors from swamps near Rome. At about the same time, the bark of the quinaquina (cinchona) tree of South America was used to treat the intermittent fevers, although it was not until the mid-nineteenth century that quinine was identified as the active alkaloid. The major epidemiological breakthrough came in 1880, when French army surgeon Charles Louis Alphonse Laveran observed protozoan gametocytes in his patients' blood. Five years later, the Italian histologist Camillo Golgi observed the multiplication of the asexual blood forms. In the late 1890s, Patrick Manson postulated that malaria was transmitted by mosquitoes. Sir Ronald Ross, a British army surgeon in the Indian Medical Service, subsequently observed developing plasmodia in the intestine of mosquitoes, supporting Manson's theory. Using birds as experimental models, Ross definitively established the major features of the life cycle of plasmodia and received the Nobel Prize for Physiology or Medicine in 1902.

Human malaria is known to have contributed to the fall of the ancient Greek and Roman empires. Troops in both the U.S. Civil War and the Spanish-American War also were severely incapacitated by the disease. During World War II, malaria epidemics severely threatened both the Japanese and Allied forces in the Pacific. The same can be said for the military conflicts in Korea and Vietnam.

In 1955 the WHO began a worldwide malarial eradication program that collapsed by 1976 (**box figure**). Among the major reasons for failure were the development of resistance to the insecticide DDT by mosquitoes and the development of resistance to chloroquine by strains of *Plasmodium.* Drug-resistant malaria parasites encountered during the Vietnam war spurred the development of new drugs (see *chapter 9 opening story*).

In 2000 the United Nations Millennium Development Goals articulated ambitious objectives to reduce malaria that were echoed by the WHO, which targeted a 75% reduction in malaria incidence by 2015. A cornerstone to this initiative involved the provision of insecticide-impregnated bed nets and vaccination of infants and young children with the then-experimental vaccine called RTS,S. Although this very aggressive goal was not realized, by 2015 the prevalence of malaria caused by the most virulent species, *Plasmodium falciparum,* was cut in half in Africa, averting an estimated 663 million cases. Unfortunately during that same time, multidrug-resistant strains of *P. falciparum* emerged in Cambodia and its neighbors. While there is great hope that the coming years will continue to bring sharp declines in malaria in Africa, where the burden is highest, drug-resistant strains in Southeast Asia are beginning to take hold in East Africa. There is much work to be done to curb this ancient disease.

Attempts to control malaria through insecticide use failed.
The Marine Corps History Division

Human malaria is caused by five species of *Plasmodium.* In descending order of incidence, they are *P. falciparum, P. vivax, P. malariae, P. ovale,* and *P. knowlesi.* The life cycle of plasmodial protists is shown in **figure 39.6.** It is divided into three stages: the exoerythrocytic cycle (within the liver), the erythrocytic (red blood cell) cycle, and the mosquito or sporogonic cycle.

The *exoerythrocytic cycle* begins when the protozoan enters the bloodstream through the bite of an infected female *Anopheles* mosquito. As she feeds, the mosquito injects a small amount of saliva containing an anticoagulant and other proteins along with hundreds to thousands of haploid *sporozoites* **(table 39.3).** Within the first hour, the motile sporozoites migrate from the bloodstream to the liver, where they enter hepatocytes (liver cells). There the sporozoites undergo multiple rounds of asexual fission in a process called *schizogony* to form cells called *schizonts.* Each schizont has 6 to 24 nuclei and each nucleus gives rise to a *merozoite.* Over a period of days or weeks, depending on the species, thousands to millions of merozoites are released from

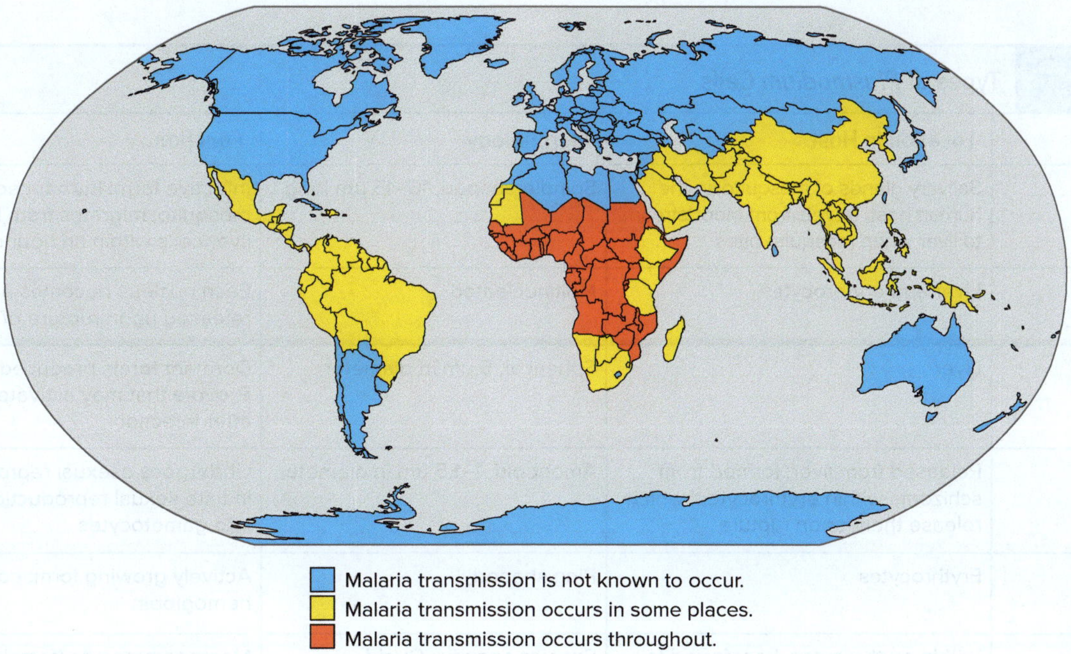

Figure 39.5 Geographic Distribution of Malaria. Temperature is a critical factor. Below 20°C, *Plasmodium falciparum* cannot complete its life cycle in the *Anopheles* mosquito.

Source: CDC, "Where Malaria, Occurs."

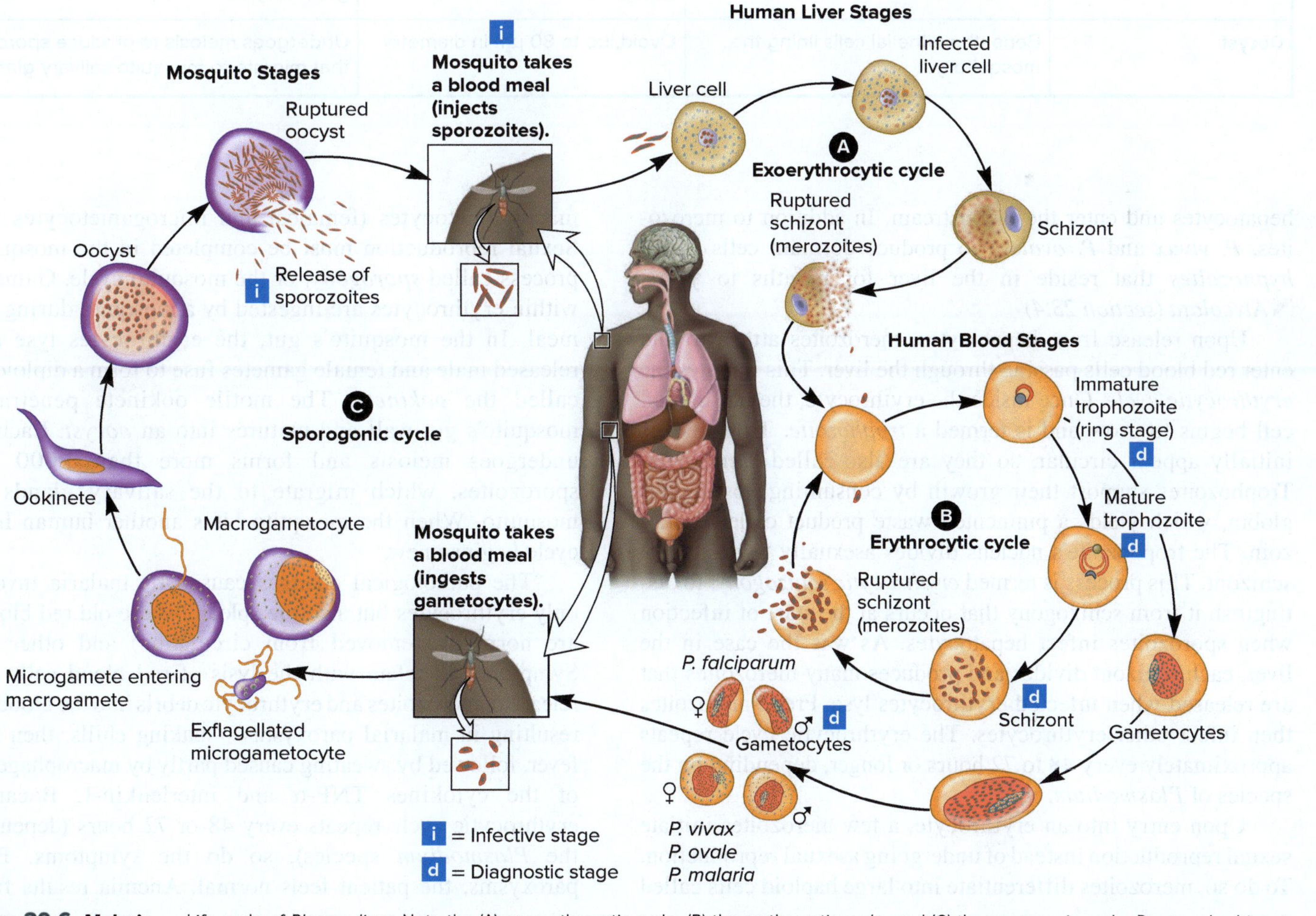

Figure 39.6 Malaria. Life cycle of *Plasmodium.* Note the (A) exoerythrocytic cycle, (B) the erythrocytic cycle, and (C) the sporogonic cycle. Ruptured schizonts release merozoites.

MICRO INQUIRY *How is a schizont formed? What is the cell type that is released from erythrocytes, and where does the protist go next?*

Table 39.3 Types of *Plasmodium* Cells

Cell Type	Location in Host	Morphology	Function
Sporozoite	Salivary glands of mosquito; in the human host, travels from bloodstream to liver when mosquito bites	Spindle-shaped, 10–15 μm long	Infective form: Introduced into host by mosquito; migrates from bloodstream to liver cells within an hour of insect biting
Schizont	Liver and erythrocytes	Multinucleated	Each nucleus becomes a merozoite that is released upon rupture of the schizont.
Hypnozoite	Liver	Spherical, 5 μm in diameter	Dormant forms produced by *P. vivax* and *P. ovale* that may activate months to years after infection
Merozoite	Released from liver; formed from schizonts within erythrocytes, which release them upon rupture	Amoeboid, 1–1.5 μm in diameter	Undergoes asexual reproduction and can initiate sexual reproduction by developing into gametocytes
Trophozoite	Erythrocytes	Ring shaped	Actively growing form; consumes host hemoglobin
Gametocytes	Within erythrocytes, transferred to mosquito	Species specific: Ovoid (*P. ovale*) to banana-shaped (*P. falciparum*)	Macrogametocyte (female) and microgametocyte (male) fuse in mosquito to form zygote.
Ookinete	Mosquito gut	Motile, changes from ovoid to elongated	Zygote that forms upon fusion of gametocytes
Oocyst	Beneath epithelial cells lining the mosquito gut	Ovoid, up to 80 μm in diameter	Undergoes meiosis to produce sporozoites that migrate to mosquito salivary glands

hepatocytes and enter the bloodstream. In addition to merozoites, *P. vivax* and *P. ovale* also produce dormant cells called *hypnozoites* that reside in the liver for months to years. ◀ *Alveolata (section 23.4)*

Upon release from hepatocytes, merozoites attach to and enter red blood cells passing through the liver. This initiates the *erythrocytic cycle*. Once inside the erythrocyte, the plasmodial cell begins to grow and is termed a *trophozoite*. Trophozoites initially appear circular, so they are also called *ring forms*. Trophozoites support their growth by consuming host hemoglobin, which yields a pigmented waste product called hemozoin. The trophozoite's nucleus divides asexually to produce a schizont. This process is termed *erythrocytic schizogony,* to distinguish it from schizogony that occurs at the start of infection when sporozoites infect hepatocytes. As was the case in the liver, each schizont divides and produces many merozoites that are released when infected erythrocytes lyse. Freed merozoites then infect other erythrocytes. The erythrocytic cycle repeats approximately every 48 to 72 hours or longer, depending on the species of *Plasmodium*.

Upon entry into an erythrocyte, a few merozoites initiate sexual reproduction instead of undergoing asexual reproduction. To do so, merozoites differentiate into large haploid cells called macrogametocytes (females) and microgametocytes (males). Sexual reproduction must be completed in the mosquito in a process called *sporogony,* or the mosquito cycle. Gametocytes within erythrocytes are ingested by a mosquito during a blood meal. In the mosquito's gut, the erythrocytes lyse and the released male and female gametes fuse to form a diploid zygote called the *ookinete*. The motile ookinete penetrates the mosquito's gut wall and matures into an *oocyst*. Each oocyst undergoes meiosis and forms more than 1,000 haploid sporozoites, which migrate to the salivary glands of the mosquito. When the mosquito bites another human host, the cycle begins anew.

The pathological changes caused by malaria involve not only erythrocytes but also the spleen (where old red blood cells are normally removed from circulation) and other organs. Symptoms correlate with the lysis of red blood cells and the release of merozoites and erythrocyte debris into the bloodstream, resulting in malarial paroxysms—shaking chills, then burning fever, followed by sweating caused partly by macrophage release of the cytokines TNF-α and interleukin-1. Because the erythrocytic cycle repeats every 48 or 72 hours (depending on the *Plasmodium* species), so do the symptoms. Between paroxysms, the patient feels normal. Anemia results from the loss of erythrocytes, and the spleen and liver often become enlarged. Cerebral malaria is the most dangerous manifestation of malaria, with a mortality rate of about 20%. It is most frequently caused by *P. falciparum* and occurs when protozoa enter the CNS. Children, individuals who are not immune, and individuals who are immunocompromised, including females who are pregnant, are most susceptible. Blackwater fever is another serious outcome, marked by severe kidney injury and cola-colored urine. This occurs when red blood cells lyse within the kidney tubules and release hemoglobin and hemozoin, which is spilled into the urine.

Diagnosis of malaria is made by identifying parasites within Wright- or Giemsa-stained erythrocytes. When blood smears are negative, serological testing can establish a diagnosis of malaria. Outside the United States, rapid diagnostic tests using species-specific antibodies are commonly used to diagnose malaria. PCR-based malarial identification and drug-resistance testing may also be performed.

Specific recommendations for treatment of malaria depend on the geographic region and include administration of chloroquine, amodiaquine, or mefloquine. In the regions where *P. falciparum* is endemic, artemisinin-based combination therapy (ACT)—combining artemisinin with another antimalarial drug—is recommended (*see chapter 9 opening story*). These drugs suppress protozoan reproduction and are effective in eradicating erythrocytic asexual stages. Those infected with *P. vivax* or *P. ovale* must be treated with primaquine to eradicate hypnozoites. Otherwise, individuals infected with these species will relapse after months or years when the hypnozoites activate in the liver. It is worth noting that individuals who are traveling to areas where malaria is endemic (figure 39.5) should receive chemoprophylactic treatment as recommended by the Centers for Disease Control and Prevention (CDC).

Control of malaria is an urgent problem. In 2021, the WHO approved the use of the recombinant vaccine RTS,S in children living in sub-Saharan Africa. This vaccine stimulates the host immune system to neutralize sporozoites as they transit from the skin to the liver. In 2023, the R21/Matrix-M vaccine, which is similar to RTS,S but includes a potent adjuvant, was approved. These vaccines reduce rates of malaria infection in children by about 65 to 75%. This is part of a multipronged global effort to extend the success in malaria control accomplished between 2000 and 2015 through the use of insecticide-treated bednets, ACT, and indoor insecticide spraying.

Leishmaniasis

Leishmanias are trypanosomes that cause a group of human diseases collectively called **leishmaniasis.** There are three forms of leishmaniasis: cutaneous (skin-associated ulcers), mucocutaneous (ulcers of the skin, mouth, nose, and throat), and visceral (affecting liver, spleen, and blood). Worldwide, leishmaniasis is a disease of poverty associated with malnutrition, displacement, poor housing, and discrimination. The WHO estimates there are over 1 million new cases of the cutaneous disease and 50,000 to 90,000 cases of the visceral disease diagnosed around the globe each year. One billion people are at risk of infection across 98 countries in Asia, Africa, and Central and South America. About half of all cases occur in children, who may develop permanent, stigmatizing changes to their appearance (**figure 39.7**).

The primary reservoirs of these parasites are canines and rodents. All species of *Leishmania* use female sand flies such as those of the genera *Lutzomyia* and *Phlebotomus* as intermediate hosts. Sand flies are about one-third the size of a mosquito, so they are hard to see and hear. The leishmanias are transmitted from animals to humans or between humans by sand flies. When an infected sand fly takes a human blood meal, it introduces flagellated promastigotes into the skin of the host (**figure 39.8**). The promastigotes are then engulfed by resident (stationary) macrophages, in which they multiply by binary fission and form small, nonmotile cells called amastigotes. These destroy the host cell; released amastigotes are engulfed by other macrophages, in which they continue to multiply. ◀ *Discoba (section 23.2)*

Over 20 species of *Leishmania* are human pathogens. Most cause cutaneous or mucocutaneous leishmaniasis. Cutaneous leishmaniasis begins with a relatively small, red papule at the site of each insect bite—the inoculation site. These

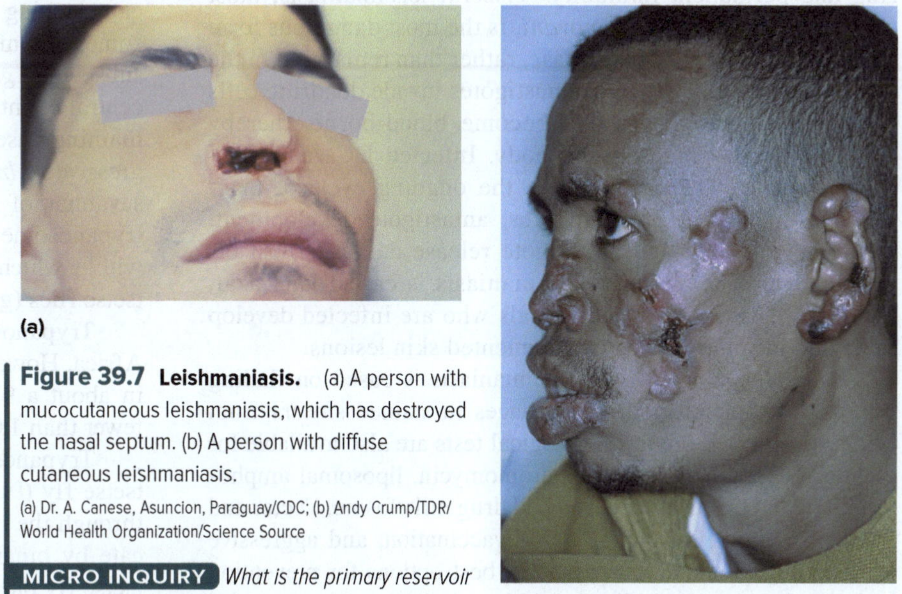

(a) (b)

Figure 39.7 Leishmaniasis. (a) A person with mucocutaneous leishmaniasis, which has destroyed the nasal septum. (b) A person with diffuse cutaneous leishmaniasis.

(a) Dr. A. Canese, Asuncion, Paraguay/CDC; (b) Andy Crump/TDR/World Health Organization/Science Source

MICRO INQUIRY *What is the primary reservoir of leishmanias? What is the intermediate host?*

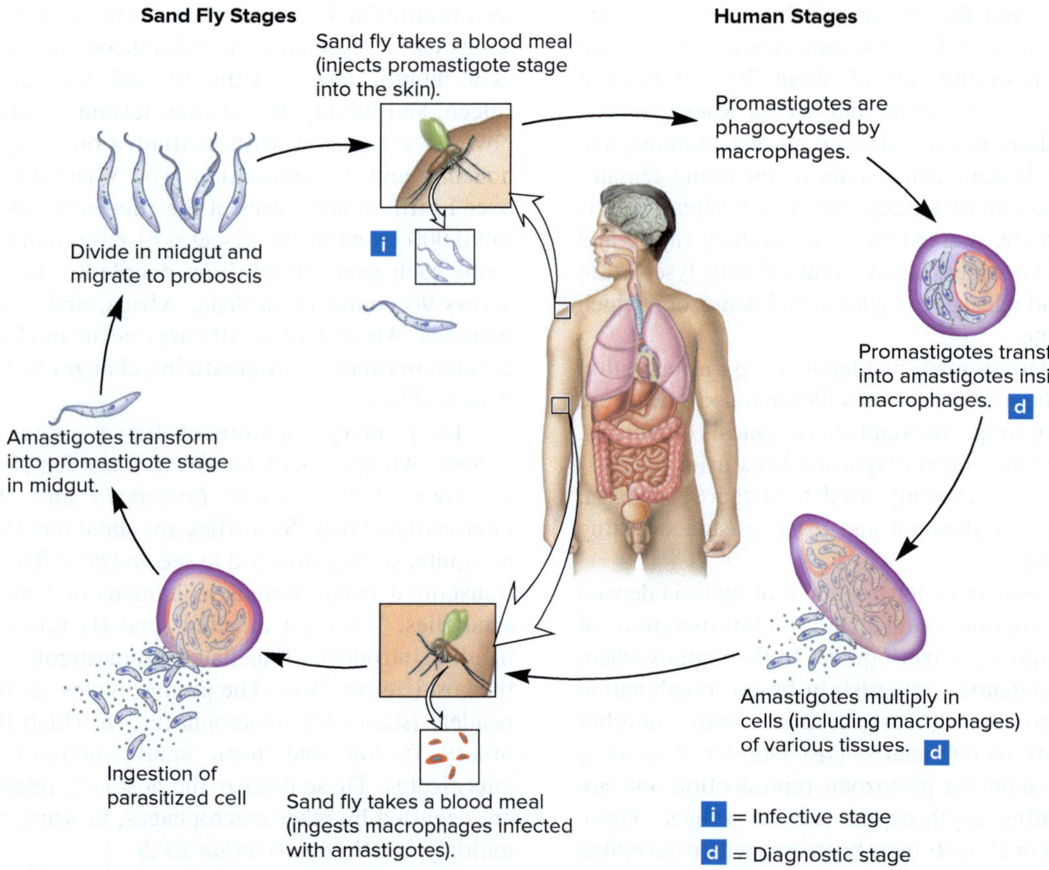

Figure 39.8 ***Leishmania*** **spp. Life Cycle.** Note the promastigote-to-amastigote cycle.

papules are frequently found on the face and ears. They eventually develop into crustated ulcers. Healing occurs with scarring and permanent immunity. Visceral leishmaniasis, most commonly caused by *L. donovani,* is the most dangerous form of the disease. In visceral disease, rather than remaining in the location of the sand fly bite, amastigotes invade dendritic cells (DCs) and macrophages and become blood-borne, thereby traveling to new areas in the body. Infected DCs and macrophages lodge in organs where the ongoing cycle of macrophage capture of promastigotes, amastigote development, macrophage death, and amastigote release damages host organs and tissues. Visceral leishmaniasis is called kala azar, meaning black fever, as individuals who are infected develop fever and in some cases hyperpigmented skin lesions.

Laboratory diagnosis of leishmaniasis is based on finding protozoa within infected macrophages in stained smears from lesions or infected organs. Serological tests are also available for diagnosis. Treatment includes paromomycin, liposomal amphotericin B, and the recently developed drug miltefosine. Vector and reservoir control, including canine vaccination, and aggressive epidemiological surveillance are the best options for prevention and containment of this disease. Although there are no approved vaccines for humans against leishmaniasis, several recombinant protein and nucleic acid vaccines are in clinical trials.

Trypanosomiasis

Human African **Trypanosomiasis** (HAT), also known as African sleeping sickness, is caused by two subspecies of trypanosomes. Gambiense HAT is caused by *Trypanosoma brucei* subspecies *gambiense,* which is distributed over western and central countries and causes over 98% of reported cases. The remaining cases (rhodesiense HAT) are caused by the more aggressive *T. brucei* subspecies *rhodesiense,* found in the upland savannas of eastern and southern Africa. Reservoirs for both trypanosome subspecies are domestic cattle and wild animals, within which they cause severe malnutrition. Both species use tsetse flies (genus *Glossina*) as intermediate hosts.

Trypanosomiasis is found throughout much of sub-Saharan Africa. However, measures to control the tsetse fly have resulted in about a 97% decrease in HAT incidence since 2000, with fewer than 1,000 cases across Africa reported in 2023.

Trypanosomes are transmitted to humans by the bite of the tsetse fly (**figure 39.9;** *also see figure 23.5*). The protists pass through the lymphatic system, enter the bloodstream, and replicate by binary fission as they pass to other body fluids. The tsetse fly bite is painful and can develop into a chancre. Patients may exhibit symptoms of fever, severe headaches, extreme fatigue, muscle and joint aches, irritability, and swollen lymph

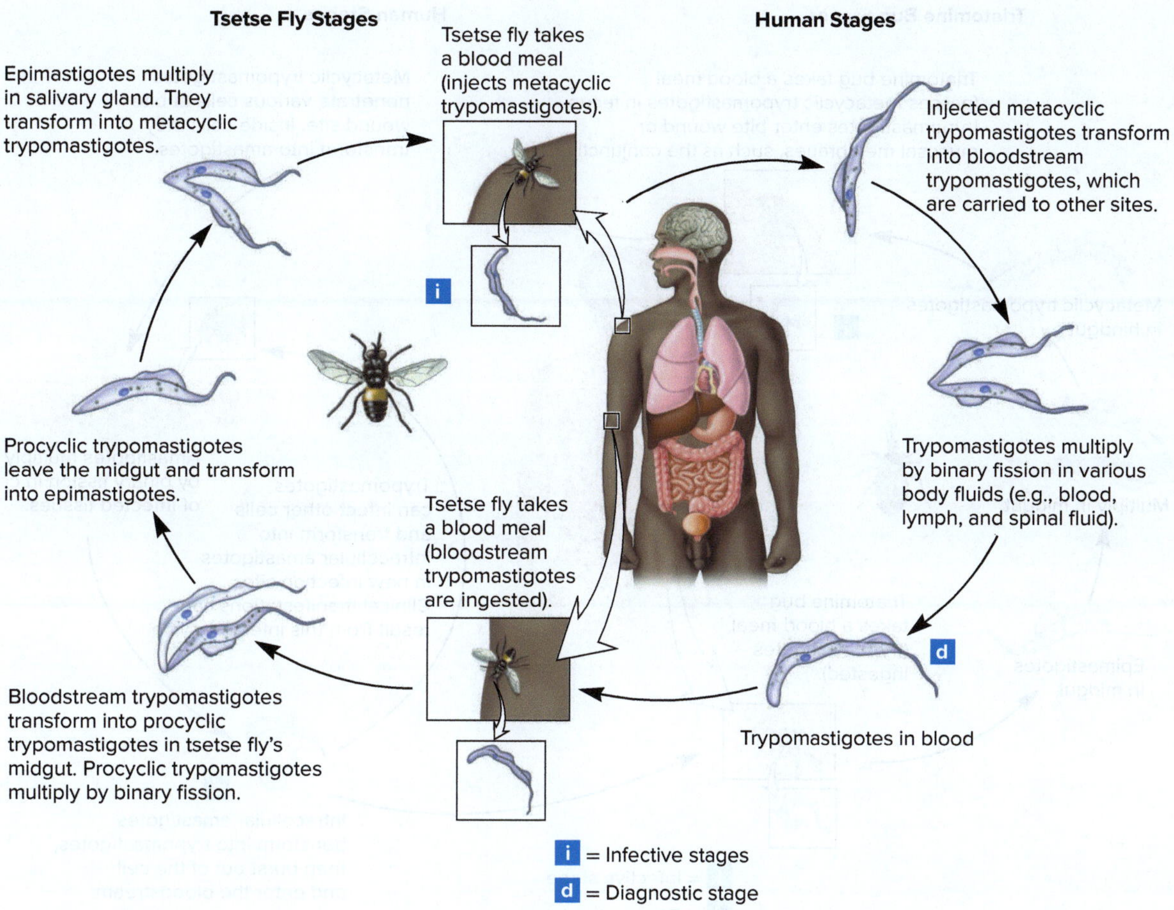

Figure 39.9 Life Cycle of *Trypanosoma brucei*. Trypomastigotes are morphologically similar but are distinguished by their location: metacyclic trypomastigotes are deposited in the skin; procyclic are in the fly's midgut.

nodes. Some patients develop a skin rash. The disease gets its name from the fact that patients exhibit lethargy—characteristically lying prostrate, drooling, and insensitive to pain; they also exhibit progressive confusion, slurred speech, seizures, and personality changes due to infection of the CNS. The protists cause inflammation and necrosis within lymph nodes and the small blood vessels of the brain and heart. Currently treatment depends on the stage of infection. In 2020, the drug fexinidazole was introduced to treat early stages of HAT. Treatment of late-stage disease is more difficult, but clinical trials are promising for a new agent that appears to be effective. If left untreated, HAT is fatal within 1 to 3 years.

T. cruzi causes **American trypanosomiasis (Chagas disease),** which occurs in the tropics and subtropics of the continental Americas (southern United States, Mexico, Central America, and northern South America). *T. cruzi* uses the triatomine (kissing) bug as a vector (**figure 39.10**). As the triatomine bug takes a blood meal, trypanosomes are discharged in the insect's feces, which contain an irritant that stimulates the victim to scratch. This enables trypanosomes to enter the bloodstream and invade the liver, spleen, lymph nodes, and CNS; disease occurs immediately after infection, lasting weeks or months (acute phase). The bloodstream *trypomastigotes* do not replicate until they enter a host cell, where they lose their flagella and become *amastigotes,* or are ingested by the arthropod vector (becoming *epimastigotes*). A chronic indeterminate (asymptomatic) phase of the disease follows the acute phase. During this phase, heart and gastrointestinal damage occurs due to amastigotes within cardiac and intestinal muscle cells, respectively. Years after initial infection, patients can develop severe heart and/or intestinal disease. An estimated 6 to 7 million people are infected, mostly in Latin America, with an annual incidence of about 30,000 new cases and 12,000 deaths each year; however, it is estimated that fewer than 10% of cases are reported.

Trypanosomiasis is diagnosed by finding motile protists in blood, spinal fluid, or muscle biopsy, which reveals amastigotes, and by serological testing. Acute Chagas is treated with benznidazole or nifurtimox; chronic Chagas is more difficult to treat as there is often significant cardiac and/or gastrointestinal pathology. Vaccine development for both HAT and Chagas has been stymied by **antigenic variation,** which is the ability of these protozoa to change surface proteins thereby evading the host's immunologic response. ◀ *Discoba (section 23.2)*

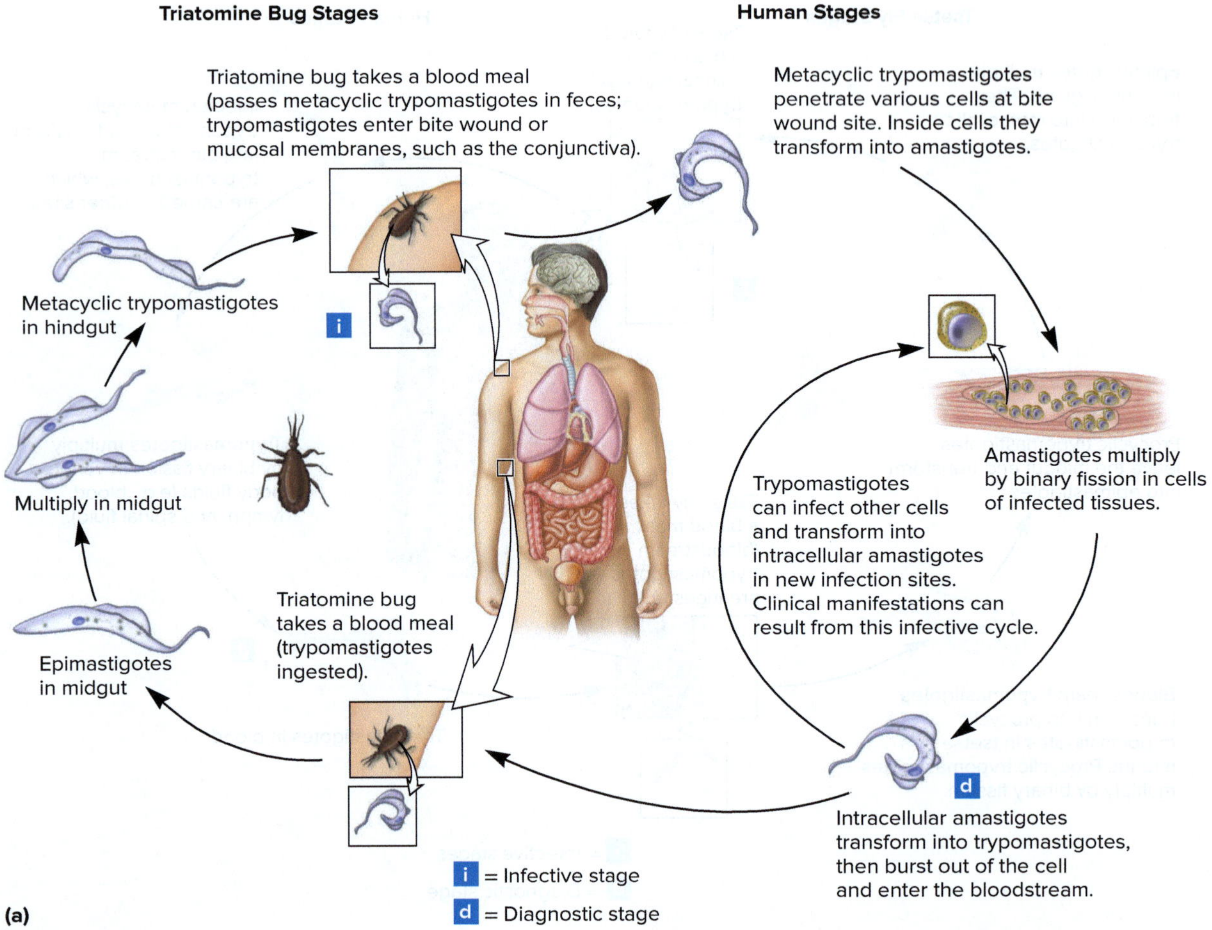

Comprehension Check

1. Describe the *Plasmodium* spp. life cycle. What stages of the life cycle occur in humans? What is the role of merozoites?
2. What are the three types of leishmaniasis? Compare the morphology of promastigotes and amastigotes. What is the location of each in the host?
3. Compare and contrast the life cycles of the trypanosomes that cause African sleeping sickness and Chagas disease. Why has vaccine development been difficult?

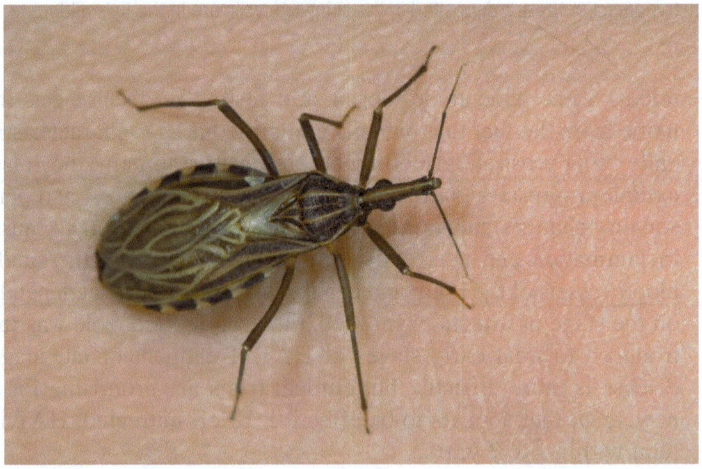
(b)

Figure 39.10 ***Trypanosoma cruzi*** **Infection.** (a) *T. cruzi* life cycle. (b) The triatomine bug.

Ray Wilson/Alamy Stock Photo

39.4 Direct Contact Diseases Can Be Caused by Fungi and Protists

After reading this section, you should be able to:

a. Discuss the microorganisms that cause common fungal and protozoal diseases spread by direct contact
b. Identify typical signs and symptoms of fungal and protozoal diseases spread by direct contact

Transmission of eukaryotic pathogens by direct contact results in a number of infectious diseases of the skin and reproductive organs. The direct contact diseases caused by fungi include cutaneous and subcutaneous mycoses. Diseases caused by protists can occur through hand-to-mouth/nose exchange, as well as their transfer through body fluids. We discuss fungal pathogens first and then a common pathogenic protozoan.

Superficial and Cutaneous Fungal Infections

Fungi responsible for infections limited to the outer surface of hair and skin cause superficial mycoses. These fungi remain above the basement membrane of skin, consuming keratin as their primary carbon source (*see figure 31.2*). Infections of the hair shaft are collectively called **piedras** (Spanish for stone because fungal hyphae form hard nodules on the hair shaft). For example, black piedra caused by *Piedraia hortae* forms hard, black nodules on the hairs of the scalp. White piedra is caused by the yeast *Trichosporon beigelii* and forms light-colored nodules on the beard and mustache. Some superficial and cutaneous mycoses are called **tineas** and involve the outer layers of the skin, nails, and hair. The specific location of tinea is designated by a modifying term. Tinea versicolor is a common mycosis caused by the yeast *Malassezia furfur,* which forms hypopigmented patches on the skin of the trunk, neck, face, and arms (**figure 39.11**). These mycoses are treated with topical antifungal agents.

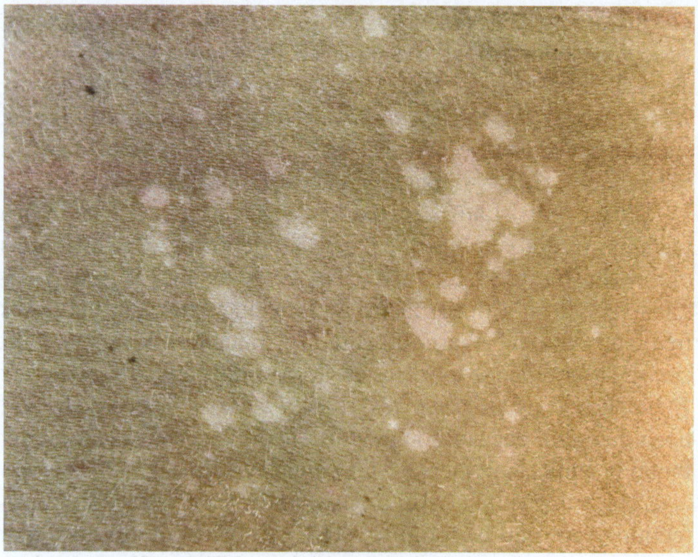

Figure 39.11 Superficial Mycosis: Tinea Versicolor. Infection with *Malassezia furfur* results in hypopigmented patches most commonly on the chest and back.

4FR/iStock/Getty Images

Tinea barbae (Latin *barba,* the beard) is an infection of beard hair caused by *Trichophyton mentagrophytes* or *T. verrucosum.* It is most commonly seen in males who live in rural areas and acquire the fungus from infected animals. Tinea capitis (Latin *capita,* the head) is an infection of the scalp hair (**figure 39.12*a***). It is characterized by hair loss, inflammation, and scaling. Tinea capitis is primarily a childhood disease caused by *Trichophyton* or *Microsporum* species. Person-to-person transmission of the fungus occurs frequently in overcrowded conditions. The fungus also occurs in domestic animals, which can transmit it to humans. An ultraviolet (UV) light called a Wood's lamp can help with the diagnosis of tinea capitis because some species of fungi fluoresce when illuminated by UV radiation (figure 39.12*b*).

Members of three genera of fungi, or *dermatophytes,* are involved in cutaneous mycoses: *Epidermophyton, Microsporum,* and *Trichophyton.* Tinea corporis (Latin *corpus,* the body) is an infection that can occur on any part of the skin (**figure 39.13*a***).

(a) **(b)**

Figure 39.12 Superficial Mycosis: Tinea Capitis. (a) Fungal infection of the head caused by *Microsporum canis.* (b) Close-up using a Wood's lamp (a UV light).

(a) CDC; (b) Mediscan/Alamy Stock Photo

MICRO INQUIRY *What is the difference between a piedra and a tinea? Which is shown here?*

The disease is characterized by circular, red, well-demarcated, scaly, vesiculopustular lesions accompanied by itching. Tinea corporis, also called *ringworm,* is caused by the fungi *Trichophyton rubrum, T. mentagrophytes, T. concentricum,* and *Microsporum canis.* Fungi are transmitted by direct contact with infected animals and humans or by indirect contact through fomites (inanimate objects).

Tinea pedis (Latin *pes,* the foot), or athlete's foot, and tinea manuum (Latin *mannus,* the hand) are dermatophytic infections of the feet (figure 39.13*b*) and hands, respectively. Clinical symptoms vary from a fine scale to a pimplelike eruption. Itching is frequently present. Warmth, humidity, trauma, and occlusion increase susceptibility to infection. Most infections are caused by *T. rubrum, T. mentagrophytes,* or *E. floccosum.* Tinea pedis and tinea manuum occur throughout the world, are most commonly found in adults, and increase in frequency with age.

Tinea unguium (Latin *unguis,* nail) is a dermatophytic infection of the nail bed (figure 39.13*c*). In this condition, the nail becomes discolored and then thickens. The nail plate rises and separates from the nail bed. *Trichophyton rubrum* or *T. mentagrophytes* is the causative fungus.

Diagnosis is usually made through assessment of the clinical presentation since all causative fungi have the same treatment. However, if the causative fungus is to be determined, microscopic examination of biopsied areas of the skin cleared with 10% potassium hydroxide and culture on Sabouraud dextrose agar are performed. Treatment is with topical ointments such as miconazole, tolnaftate, or clotrimazole for 2 to 4 weeks. Oral terbinafine is used to treat nail infections. Dermatomycoses may be insidious; the desquamated (shed) skin cells can contain viable fungi for months or years. Thus any treatment plan should also include thorough cleaning of bedding, carpets, and any other substrates that could hold desquamated skin cells.

Subcutaneous Fungal Infections

Dermatophytes that infect underneath the skin cause subcutaneous mycoses and include sporotrichosis, chromoblastomycosis, and maduromycosis. These fungi are normally saprophytic inhabitants of soil and decaying vegetation. Because they are unable to penetrate the skin, they must be introduced into the subcutaneous tissue by a puncture wound. Once in the subcutaneous tissue, disease develops slowly—often over a period of years. During this time, the fungi produce painless nodules that eventually ulcerates. At times, the nodules drain to the skin surface. Diagnosis is accomplished by culture of the infected tissue.

Sporotrichosis is the subcutaneous mycosis caused by the dimorphic fungus *Sporothrix schenckii.* The disease occurs throughout the world and is the most common subcutaneous mycosis in the United States. The fungus can be found in the soil; on living plants, such as roses; or in plant debris, such as sphagnum moss, baled hay, and pine-bark mulch. Infection occurs by a puncture wound from a thorn or splinter contaminated with the fungus. It is not spread from person to person. After an incubation period of 1 to 12 weeks, a small, red papule arises and begins to ulcerate. If left untreated, in about 60% of cases the infection spreads along

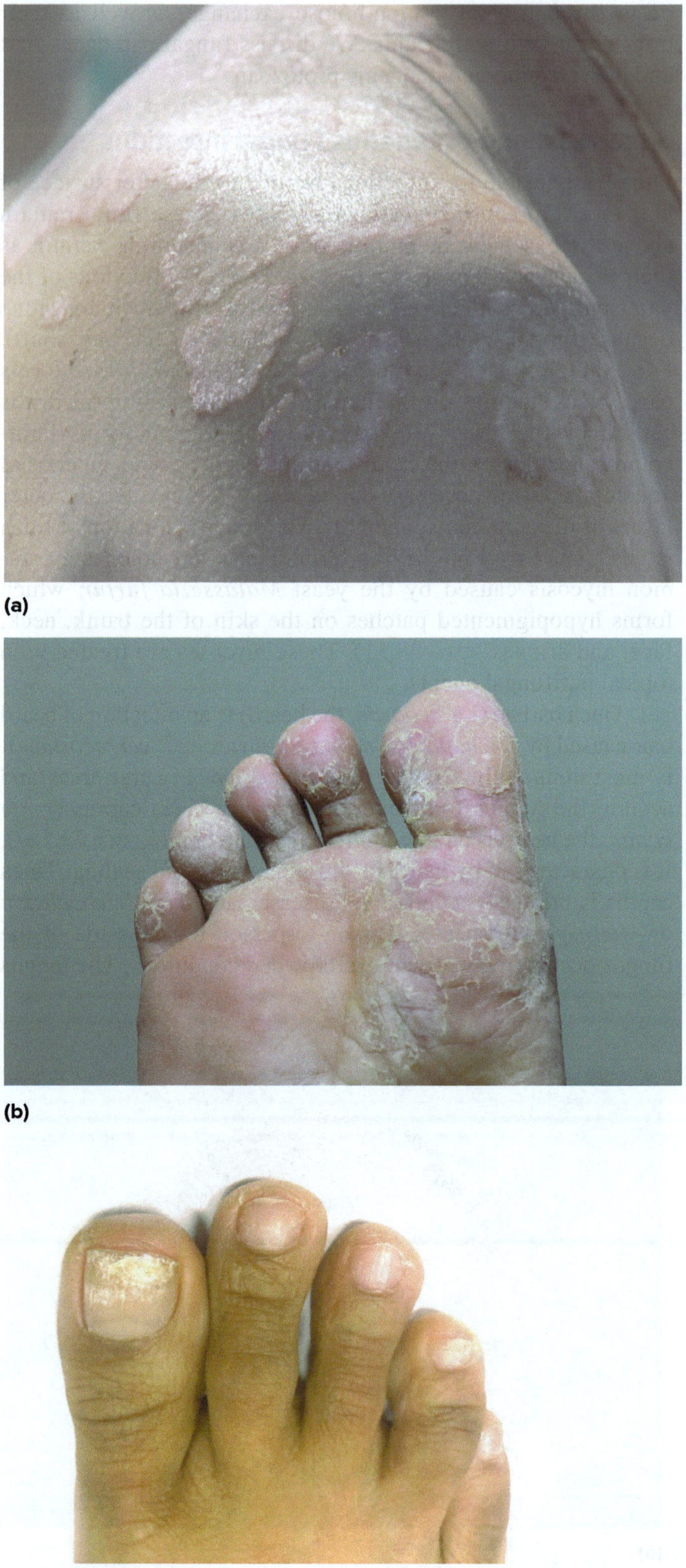

(a) (b) (c)

Figure 39.13 Cutaneous Mycosis. (a) Tinea corporis. Ringworm of the body. (b) Tinea pedis. (c) Tinea unguium of the nails.

(a) schankz/Shutterstock; (b) chaipanya/Shutterstock; (c) plo/Shutterstock

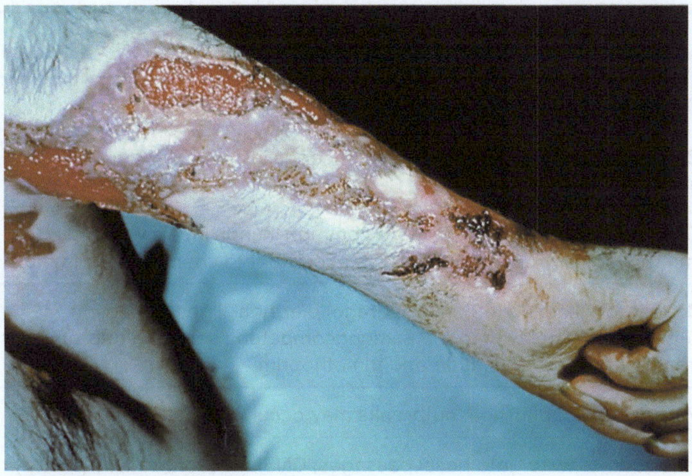

Figure 39.14 Subcutaneous Mycosis: Sporotrichosis. Sporotrichosis of the arm caused by *Sporothrix schenckii.*

Centers for Disease Control and Prevention

lymphatic vessels, causing ulceration (**figure 39.14**). In rare cases, fungi spread throughout the body, causing systemic infection.

Sporotrichosis is typically treated by oral itraconazole for 3 to 6 months. Systemic disease is treated with a lipid formulation of amphotericin B followed by long-term itraconazole.

Chromoblastomycosis, a disease in which nodules are formed, is caused by the black molds *Phialophora verrucosa* and *Fonsecaea pedrosoi*. These fungi exist worldwide, especially in tropical and subtropical regions. Infections develop slowly and often cause disabling changes to the feet and lower legs, the most common site of infection (**figure 39.15**). **Maduromycosis** is caused by *Madurella mycetomatis,* which is distributed worldwide and is especially prevalent in the tropics. Because the fungus destroys subcutaneous tissue and produces serious deformities, the resulting infection is often called a **eumycotic mycetoma,** or fungal tumor. One form of mycetoma, known as Madura foot, occurs through skin abrasions

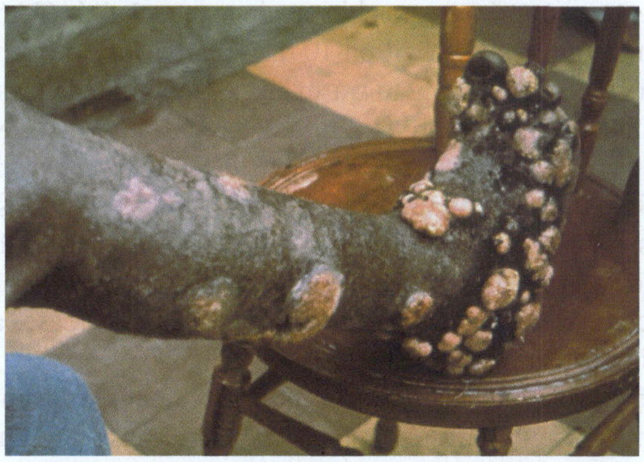

Figure 39.15 Subcutaneous Mycosis: Chromoblastomycosis.

CDC/Dr. Brinkman

MICRO INQUIRY *How do fungi causing subcutaneous mycoses gain access to tissue below the skin surface?*

acquired while walking barefoot on contaminated soil. Griseofulvin, terbinafine, itraconazole, and fluconazole are common therapeutic options; however, these infections are very difficult to cure.

Trichomoniasis

Very few human pathogenic protists are transmitted through direct contact. Thus we only discuss one such disease, **trichomoniasis,** which is a sexually transmitted disease caused by the flagellated protozoan *Trichomonas vaginalis* (*see figure 23.6*). Humans are the only known host of *T. vaginalis*. Infection with this protist increases the risk of acquiring other sexually transmitted infections (STIs). It is one of the most common STIs, with an estimated 3.7 million cases annually in the United States and 156 million cases annually worldwide. In response to the protist, the body accumulates leukocytes at the site of infection. In females, this usually results in a profuse, purulent vaginal discharge that is yellowish to light cream in color and characterized by a disagreeable odor. The discharge is accompanied by itching. The cervical mucosa is covered with punctate (pinpoint) hemorrhages, papules, and vesicles. Males are generally asymptomatic; however, a burning sensation may occur during urination. Prostatic secretions are naturally toxic to this protist, which accounts for the disparity in disease prevalence: 5.3% in females, 0.6% in males according to the WHO. Diagnosis is made by microscopic identification of the protozoan. Treatment is by administration of metronidazole to all sexual contacts. ◀ *Metamonada (section 23.2)*

Comprehension Check

1. Describe two piedras that infect humans.
2. Briefly describe the major tineas that occur in humans.
3. Describe the three types of subcutaneous mycoses that affect humans.
4. Why do the signs and symptoms of trichomoniasis differ in males and females?

39.5 Food and Water Are Vehicles of Protozoal Diseases

After reading this section, you should be able to:

a. Discuss the microorganisms that cause common food-borne and waterborne fungal and protozoal diseases
b. Identify typical signs and symptoms of common food-borne and waterborne fungal and protozoal diseases
c. Explain in specific terms why access to clean water is considered a basic human right

An estimated 4% of deaths per year worldwide are caused by waterborne infectious diseases, and many of them are caused by protozoa (**table 39.4**). Fungal and protozoal contamination of food, particularly as it is shipped great distances, is a problem

Table 39.4 Water-Based Protozoan Reservoirs

Organism	Reservoir	Comments
Acanthamoeba spp.	Sewage sludge disposal areas	Can cause granulomatous amebic encephalitis, keratitis, corneal ulcers
Cryptosporidium spp.	Many species of domestic and wild animals	Causes acute enterocolitis; more common in individuals who are immunologically compromised; cysts resistant to chemical disinfection
Cyclospora cayetanensis	Water—does not withstand drying; possibly other reservoirs	Causes long-lasting (43 days average) diarrheal illness; infection self-limiting in hosts who are immunocompetent; sensitive to prompt treatment with sulfonamide and trimethoprim
Giardia intestinalis	Livestock, dogs, cats, humans	Major cause of diarrhea; outbreaks may occur in day-care centers
Naegleria fowleri	Warm ponds and lakes	Inhalation of water into nasal passages; CNS infection; causes primary amebic meningoencephalitis

that affects all nations. We now discuss a few of the more common protozoan diseases transmitted by ingesting tainted food or water.

Amebiasis

Amebiasis (amebic dysentery) is caused by the amoeba *Entamoeba histolytica*. Worldwide, there are about 50 million cases of invasive (severe) amebiasis with approximately 100,000 deaths per year. ◀ *Entamoebida (section 23.3)*

Infection occurs by ingestion of mature cysts from fecally contaminated water or food. After excystation in the lower region of the small intestine, a form called the metacyst divides rapidly to produce eight small trophozoites (**figure 39.16**). These trophozoites move to the large intestine, where they can invade host tissue, live as commensals in the lumen of the intestine, or undergo encystment. In some cases, the trophozoites remain in the intestinal lumen, resulting in an asymptomatic carrier state with cysts shed in feces.

If the trophozoites invade intestinal tissues, they multiply rapidly and spread laterally within the tissue supporting the intestines, while feeding on erythrocytes, bacteria, and yeasts. The invading trophozoites destroy the epithelial lining of the large intestine by producing a protease that degrades the extracellular matrix and circumvents the host immune response by cleaving secretory immunoglobulin A (SIgA), IgG, and complement factors. Lesions (ulcers) are characterized by minute points of entry into the mucosa and extensive enlargement of the lesion after penetration into the submucosa. *E. histolytica* also may invade and produce lesions in other tissues, especially the liver, to cause hepatic amebiasis. The symptoms of amebiasis are highly variable, ranging from asymptomatic carriage to fulminating dysentery (exhaustive diarrhea accompanied by blood and mucus), appendicitis, and abscesses in the liver, lungs, or brain. ◀ *Antibodies bind specific 3-D molecules (section 32.7)*

Laboratory diagnosis of amebiasis is based on finding cysts in stools. Serological testing for *E. histolytica* is also available. Therapy for amebiasis depends on the location of the infection within the host and the host's condition. Asymptomatic carriers who are passing cysts are treated with paromomycin because they represent the most important reservoir of the protozoan. In symptomatic intestinal amebiasis, paromomycin and metronidazole are the drugs of choice. Prevention and control of amebiasis are achieved by practicing good personal hygiene and avoiding water or food that might be contaminated with human feces. Viable cysts in water can be destroyed by hyperchlorination or iodination. ◀ *Antiprotozoan drugs (section 9.7)*

Amoebic Meningoencephalitis and Keratitis

Free-living amoebae of the genera *Naegleria* and *Acanthamoeba* directly invade the CNS. Therefore unlike other waterborne protozoal pathogens, they are not transmitted by ingestion. *Naegleria fowleri* causes primary amebic meningoencephalitis (PAM), which is almost always fatal within 10 days. *N. fowleri* infection usually occurs when swimming in warm lakes. Contaminated water enters the sinuses and the protist crosses directly into the CNS. There have also been occasional deaths attributed to *Naegleria* infections acquired through the use of neti pots to rinse the sinuses. In these cases, tap water was reported to be contaminated with the protist. While still rare, the incidence of *N. fowleri* infection has expanded in recent years as warming of lakes and ponds increases as a result of climate change.

Several species of *Acanthamoeba* cause granulomatous amebic encephalitis and infect the eye, causing a chronically progressive, ulcerative *Acanthamoeba* keratitis (inflammation of the cornea) that may result in blindness. Wearers of soft contact lenses may be predisposed to this infection and should take care to prevent contamination of lens-cleaning and soaking solutions. Diagnosis of these infections is by detection of amoebae in clinical specimens. Most freshwater amoebae are resistant to

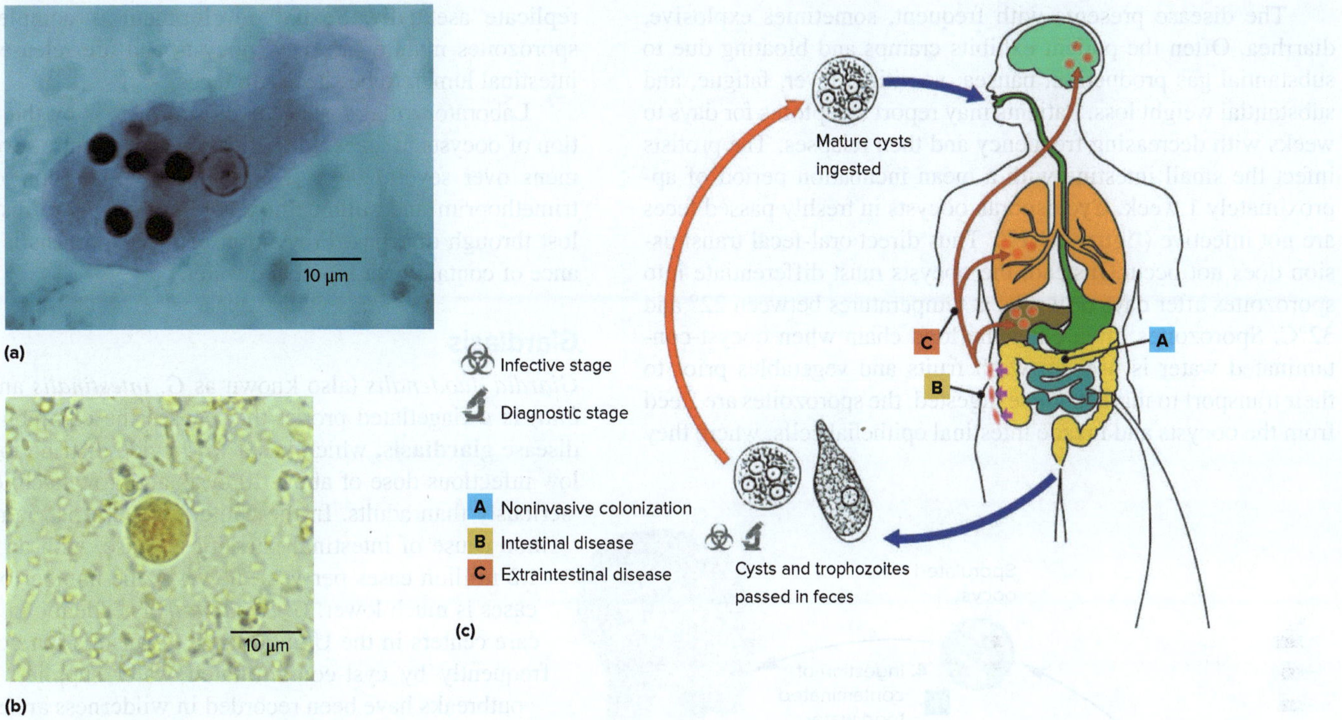

Figure 39.16 Amebiasis Caused by *Entamoeba histolytica*. Light micrographs of (a) a trophozoite and (b) a cyst. (c) Life cycle. Infection occurs by ingestion of a mature cyst. Excystment occurs in the lower region of the small intestine, and the metacyst rapidly divides to give rise to eight small trophozoites (not shown). These enter the large intestine, undergo binary fission, and may (A) live in the lumen of the large intestine without invasion, encyst and pass out of the host in feces, or (B) invade the intestine and ulcerate the underlying tissue, or (C) migrate from the intestine to other sites such as the liver or lungs, causing extraintestinal disease.

(a) Dr. Mae Melvin; Dr. Greene/CDC; (b) Centers for Disease Control and Prevention; (c) CDC parasites A to Z: https://www.cdc.gov/dpdx/amebiasis/index.html

commonly used antimicrobial agents. These amoebae are reported in fewer than 100 human disease cases annually in the United States, although the incidence (especially of *Acanthamoeba* keratitis) is likely higher.

Cryptosporidiosis

A number of cryptosporidial species infect humans and other animals. *Cryptosporidium* (Greek *kruptos,* hidden) spp. are found in about 90% of sewage samples, in 75% of river waters, and in 28% of drinking waters. *C. parvum* is a common protist found in the intestine of many birds and mammals. When these animals defecate, oocysts are shed into the environment. If a human ingests food or water contaminated with oocysts, excystment occurs within the small intestine, and sporozoites enter epithelial cells and develop into merozoites. Some of the merozoites subsequently undergo sexual reproduction to produce zygotes, which differentiate into thick-walled oocysts. Oocysts released into the environment begin the life cycle again. ◀ *Alveolata (section 23.4)*

Major problems for public health arise from the fact that (1) oocysts are small, only 4 to 6 μm in diameter—too small to be easily removed by sand filters used to purify drinking water; (2) they have a low infectious dose, around 10 to 100 oocysts; (3) *Cryptosporidium* spp. are extremely resistant to disinfectants such as chlorine; and (4) oocysts may remain viable for 2 to 6 months in moist environments.

The incubation period for cryptosporidiosis ranges from 5 to 28 days. Diarrhea, which characteristically may be cholera-like (i.e., very watery, without blood), is the most common symptom. Other symptoms include abdominal pain, nausea, fever, and fatigue. The pathogen is routinely diagnosed by fecal concentration and acid-fast stained smears. Treatment of people with healthy immune systems is with nitazoxanide along with rehydration. Although the disease usually is self-limiting in healthy individuals, patients who are immunocompromised may develop prolonged, severe, and life-threatening diarrhea.

Cyclosporiasis

Cyclosporiasis is caused by the protist *Cyclospora cayetanensis*. The disease is most common in tropical and subtropical environments, although it has been reported in most countries. Each year, the CDC reports between 300 and 650 cases of cyclosporiasis in the United States. Many of these are the result of outbreaks linked to contaminated produce.

The disease presents with frequent, sometimes explosive, diarrhea. Often the patient exhibits cramps and bloating due to substantial gas production, nausea, vomiting, fever, fatigue, and substantial weight loss. Patients may report symptoms for days to weeks with decreasing frequency and then relapses. The protists infect the small intestine with a mean incubation period of approximately 1 week. Cyclosporan oocysts in freshly passed feces are not infective (**figure 39.17**). Thus direct oral-fecal transmission does not occur. Instead, the oocysts must differentiate into sporozoites after days or weeks at temperatures between 22° and 32°C. Sporozoites often enter the food chain when oocyst-contaminated water is used to wash fruits and vegetables prior to their transport to market. Once ingested, the sporozoites are freed from the oocysts and invade intestinal epithelial cells, where they replicate asexually. Sexual development is completed when sporozoites mature into new oocysts and are released into the intestinal lumen to be shed with feces.

Laboratory diagnosis of cyclosporiasis is by the identification of oocysts in feces. Identification may require several specimens over several days. Treatment is with a combination of trimethoprim and sulfamethoxazole and fluids to restore water lost through diarrhea. Prevention of cyclosporiasis is by avoidance of contaminated food and water.

Giardiasis

Giardia duodenalis (also known as *G. intestinalis* and *G. lamblia*) is a flagellated protist that causes the common intestinal disease **giardiasis,** which has a worldwide distribution. It has a low infectious dose of about 10 cysts and affects children more seriously than adults. In the United States, it is the most common cause of intestinal parasitic disease, with an estimated 1.2 million cases per year, however the number of reported cases is much lower. *G. duodenalis* is endemic in child daycare centers in the United States. Transmission occurs most frequently by cyst-contaminated water supplies. Epidemic outbreaks have been recorded in wilderness areas, suggesting that humans may be infected from stream water with *G. duodenalis* harbored by livestock or household pets. Although this implies that infection also can be a zoonosis, it is estimated that about 7% of the U.S. population are asymptomatic carriers who shed cysts in their feces. As many as 200 million humans may be infected worldwide.

Following ingestion, *Giardia* cysts undergo excystment in the duodenum, forming trophozoites. The trophozoites inhabit the upper portions of the small intestine, where they attach to the intestinal mucosa by means of their suction disks (**figure 39.18;** *also see figure 23.6*). The trophozoites feed on mucous secretions and reproduce to become such a large population that they interfere with nutrient absorption by the intestinal epithelium. Symptoms of giardiasis typically begin 1 week after infection

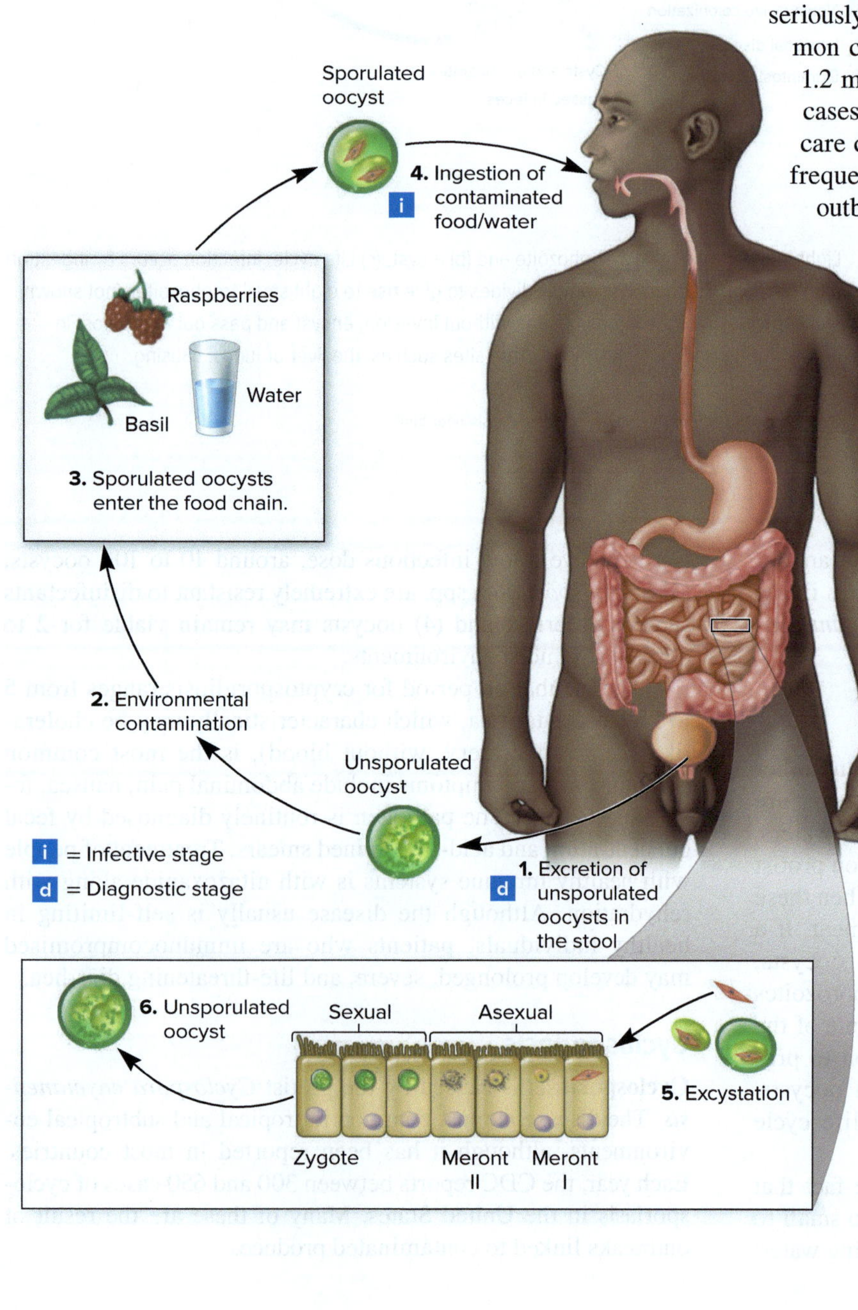

Figure 39.17 Life Cycle of *Cyclospora cayetanensis*. Unsporulated oocysts, shed in the feces of infected animals, sporulate and contaminate food and water. Once ingested, the sporulated oocysts excyst, penetrate host cells, and reproduce by binary fission. Sexual stages unite to form zygotes, resulting in new oocysts (unsporulated) that are shed in feces.
Source: CDC, "Biology."

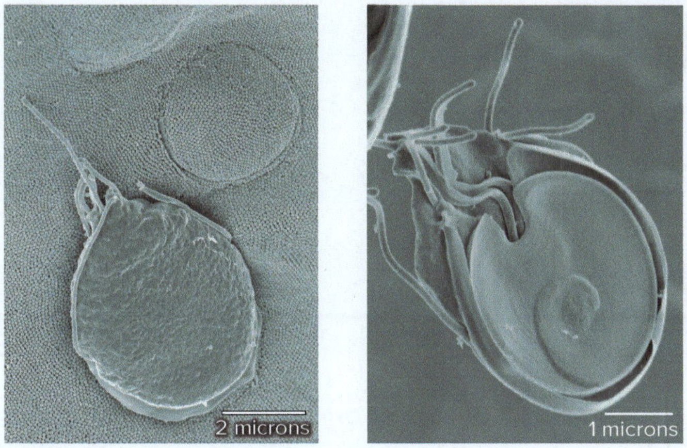

(a) (b)

Figure 39.18 Giardiasis. (a) Scanning electron micrograph (SEM) depicting a *Giardia muris* protist adhering to the microvillus border of a mouse intestinal epithelial cell. The circular lesion above the protozoan is an impression made by the ventral adhesive disk of another *G. muris* organism. (b) This SEM reveals the ultrastructural morphology of the ventral adhesive disk. This disk acts like a suction cup, facilitating the organism's attachment to the intestinal surface.
Dr. Stan Erlandsen/CDC

and can last 6 to 8 weeks or longer if untreated. The disease can be acute or chronic. Acute giardiasis is characterized by severe diarrhea, epigastric pain, cramps, voluminous flatulence, and anorexia. Chronic giardiasis causes intermittent diarrhea, with periodic appearance and remission of symptoms.

Laboratory diagnosis is based on the identification of trophozoites or cysts in stools. A commercial immunoassay is also available for the detection of *G. duodenalis* antigen in stool specimens. Tinidazole, nitazoxanide, and metronidazole are the drugs of choice. Prevention and control involve proper treatment of community water supplies, especially the use of slow sand filtration because cysts are resistant to chlorine treatment. ▶| *Purification and sanitary analysis ensure safe drinking water (section 42.1); Wastewater treatment processes (section 42.2)*

Toxoplasmosis

Toxoplasmosis is a worldwide disease caused by the protist *Toxoplasma gondii.* In some regions of the world, as much as 95% of the population has been infected, whereas in the United States, about 10% of people over the age of 6 have been infected. The natural reservoir of *T. gondii* is wild rodents, birds, and other small animals. Cats are the definitive host, meaning they are required for completion of the sexual cycle (**figure 39.19**). Cats (especially kittens) shed oocysts in their feces for several weeks after infection. Human infection can result through one of three modes of transmission: (1) zoonotic, through the accidental ingestion of oocytes while cleaning a litter box or handling a cat; (2) food-borne via the consumption of raw or undercooked infected meat; and (3) maternal-fetal (congenital) transmission. Congenital transmission can cause serious fetal anomalies or death and females who are pregnant are advised not to empty cat litter boxes.

Most human hosts are unaware of *T. gondii* infection because a strong cell-mediated immune response is stimulated. Absent this immune response, acute toxoplasmosis occurs and may be accompanied by lymph node swelling (lymphadenopathy). Pulmonary necrosis, myocarditis, and hepatitis caused by tissue necrosis may occur. Retinitis (inflammation of the retina of the eye) is associated with necrosis due to proliferation of the protist within retinal cells. *T. gondii* in people who are immunocompromised can produce a unique encephalitis with necrotizing lesions in the brain accompanied by inflammatory infiltrates. *T. gondii* causes at least 400 congenital infections per year in the United States. Toxoplasmosis in newborns can result in a wide array of problems involving many different organ systems, but most commonly affects the CNS. |◀ *Alveolata (section 23.4)*

Laboratory diagnosis of toxoplasmosis is by serological tests. Treatment is with a combination of pyrimethamine and sulfadiazine. Prevention and control require minimizing exposure by the following: avoiding eating raw or undercooked meat and eggs, washing hands after working in soil and emptying cat litter boxes, cleaning litter boxes daily, keeping household cats indoors if possible, and feeding cats commercial food.

Hemera Technologies/Getty Images

39.6 Opportunistic Diseases Can Be Caused by Fungi and Protists

After reading this section, you should be able to:

a. Predict the impact on the host when normal fungal microbiota are introduced to a new host environment
b. Describe growth and reproduction challenges experienced by normal fungal microbiota and environmental fungi when attempting to infect a healthy human
c. Cite specific host conditions that promote fungal infection

Some protists we have discussed can be considered opportunistic because they most frequently infect individuals who are immunocompromised. For example, fewer cysts of *Cryptosporidium parvum* are needed to cause disease in the absence of a robust immune response. The dormant cysts of *Toxoplasma gondii* (bradyzoites, figure 39.19) may become activated without constant immune vigilance. Certain fungi also pose a specific risk for individuals who are immunosuppressed. The mycoses they

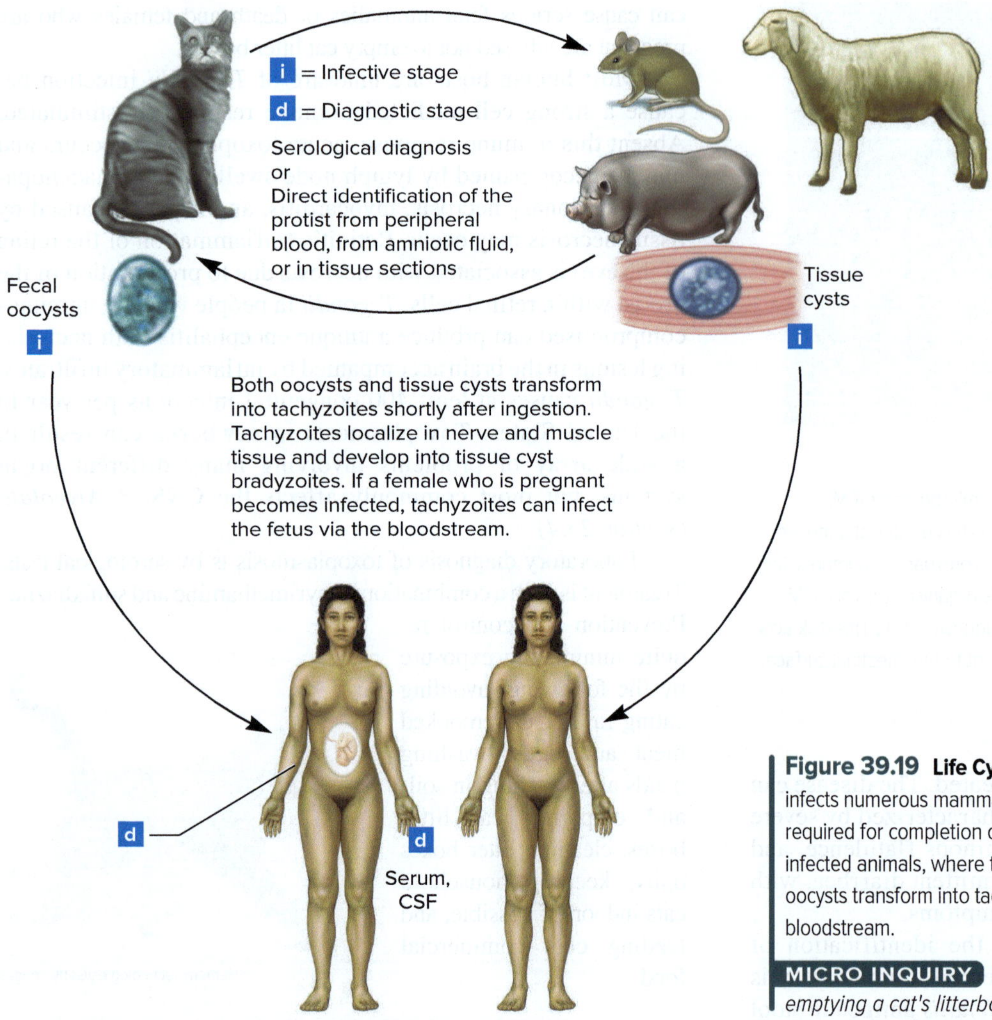

Figure 39.19 Life Cycle of *Toxoplasma gondii*. *Toxoplasma gondii* infects numerous mammals and birds. However, a cat, the definitive host, is required for completion of the sexual cycle. Oocysts are shed in the feces of infected animals, where they may be ingested by another host. Ingested oocysts transform into tachyzoites, which migrate to various tissue sites via the bloodstream.

MICRO INQUIRY *Why are people who are pregnant advised against emptying a cat's litterbox?*

cause are rarely seen in healthy people, but for those whose immune systems have become compromised, they can cause life-threatening infections. Immunosuppression has myriad causes, including malnutrition, alcohol use disorder, cancer, diabetes, another infectious disease (e.g., HIV/AIDS), trauma from surgery or injury, an altered microbiota from prolonged use of antibiotics, genetic deficiencies, cancer chemotherapy, and old age. Here we discuss a few common opportunistic fungal pathogens.

Aspergillosis

Of all the fungi that cause disease in human hosts, none is as widely distributed in nature as members of the genus *Aspergillus*. These filamentous fungi are omnipresent: They are found wherever organic debris occurs, especially in soil, decomposing plant matter, household dust, building materials, some foods, and water. In fact, it is almost impossible to avoid daily inhalation of *Aspergillus* spores (**figure 39.20**). However, infection results only when healthy immune functions deteriorate.

The major portal of entry for *Aspergillus* spp. is the respiratory tract. Inhalation of conidiospores can lead to several types of aspergillosis. *Aspergillus fumigatus* is the species commonly involved in pulmonary **aspergillosis** marked by fever, chest pain, and cough. It also causes both superficial infection and invasive disease in patients who are immunosuppressed. Invasive disease typically results when pulmonary infection disseminates to the brain, kidney, liver, bone, or skin. Invasive aspergillosis has a mortality rate of 50 to 100%. ◀ *Ascomycota includes yeasts and molds (section 24.4)*

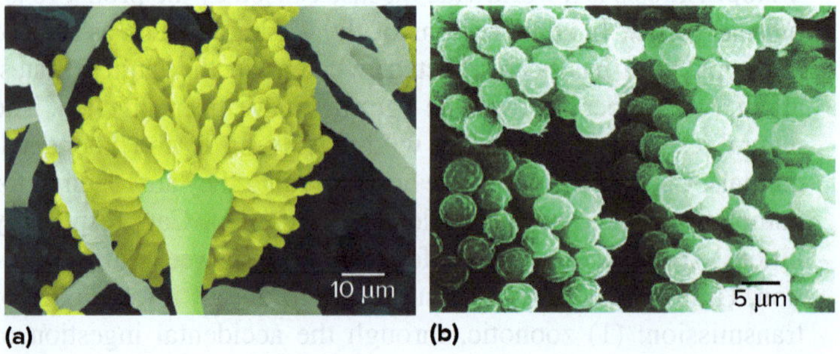

Figure 39.20 *Aspergillus* spp. Cause Opportunistic Mycoses. Colorized scanning electron micrographs of (a) conidiophore of *Aspergillus flavus* and (b) *Aspergillus* conidiospores.

(a) Thierry Berrod, Mona Lisa Production/SPL/Science Source; (b) Robert Simmons/CDC

These fungi also cause allergic aspergillosis whereby individuals may develop an immediate allergic response and experience asthma attacks when exposed to fungal antigens. Another response is bronchopulmonary aspergillosis, in which patients with underlying pulmonary disease, such as tuberculosis, asthma, or cystic fibrosis, develop bronchitis resulting from both type I and type III hypersensitivities. Finally, pulmonary infection may result in the development of an aspergilloma. This occurs when fungal colonies form within the lungs and develop into so-called fungus balls consisting of a tangled mass of hyphae growing in a circumscribed area. From the lung, the fungus may spread, producing disseminated aspergillosis in a variety of tissues and organs. In patients whose resistance is severely compromised, invasive aspergillosis may occur and fill the lung with fungal hyphae. ◀ *Hypersensitivities (section 32.10)*

Laboratory diagnosis of aspergillosis depends on identification either by direct examination of pathological specimens or by isolation and characterization of the fungus. An enzyme immunoassay that detects galactomannan (an antigen of *Aspergillus* spp.) can be used to screen suspected cases of aspergillosis. Treatment for invasive disease is with voriconazole, while allergic-type aspergillosis is treated with itraconazole and corticosteroids. However, successful therapy depends on treatment of the underlying disease so that host resistance increases. ◀ *Antifungal drugs (section 9.6)*

Candidiasis

Candidiasis is a mycosis caused by fungi in the genus *Candida*. *Candida* spp. disseminate within a human host depending on their need to invade through body fluids as yeast or tissues as pseudohyphae. As the term suggests, *pseudohyphae* are not true hyphae; rather they are structures that arise from incomplete budding (**figure 39.21*a***). Cells elongate but remain attached after division. There are over 20 species of *Candida* that can cause infection in humans, most notably *C. albicans, C. glabrata,* and *C. auris*. In contrast to other opportunistic fungi, *Candida* spp. are members of the normal microbiota within the gastrointestinal tract, respiratory tract, vaginal area, and mouth. In healthy individuals, they do not produce disease because growth is suppressed by other microbiota and host resistance mechanisms. However, in the absence of competing microbiota, populations of *Candida* spp. grow. As *C. albicans* gains access to host epithelial cells, it secretes a peptide toxin called candidalysin that permeabilizes host-cell membranes. This provides additional nutrients, which further increases their numbers.

No other mycotic pathogens produce as diverse a spectrum of disease in humans as do *Candida* spp. **Oropharyngeal candidiasis,** or **thrush** (figure 39.22*b*), is a fairly common disease in newborns, persons with dentures, and people who use inhaled steroids. Thrush appears as many small, white flecks that cover the tongue and mouth. More severe colonization of the esophagus occurs in individuals who are immunosuppressed. **Intertriginous candidiasis** involves those areas of the body, usually opposed skin surfaces, that are warm and moist: axillae, groin, skin folds (figure 39.22*c*). **Candidal vaginitis** (vaginal yeast infection) can result as a complication of diabetes, antibiotic therapy, oral contraceptives, pregnancy, or any other factor that disrupts vaginal microbiota. Normally the omnipresent vaginal lactobacilli create a low pH, which limits fungal growth. However, if the numbers of lactobacilli are decreased by any of these factors, yeast may proliferate, causing a curdlike, yellow-white vaginal discharge. *C. albicans* and *C. glabrata* can be transmitted to males during intercourse and lead to **balanitis.** Balanitis is an infection of the male glans penis and occurs primarily in uncircumcised males. In both sexes, candidiasis causes severe itching and burning. *Candida* spp. can also cause urinary tract infections and blood-borne infections. ◀ *Development of a stable microbiome (section 33.2).*

In recent years, *Candida auris* has emerged as an important healthcare-acquired pathogen. This fungus is of particular importance for several reasons. *C. auris* is usually multidrug resistant with some strains resistant to all antifungal agents. This accounts for its high (25%) mortality rate when it causes bloodstream infections. Unlike *C. albicans,* this fungus thrives in the

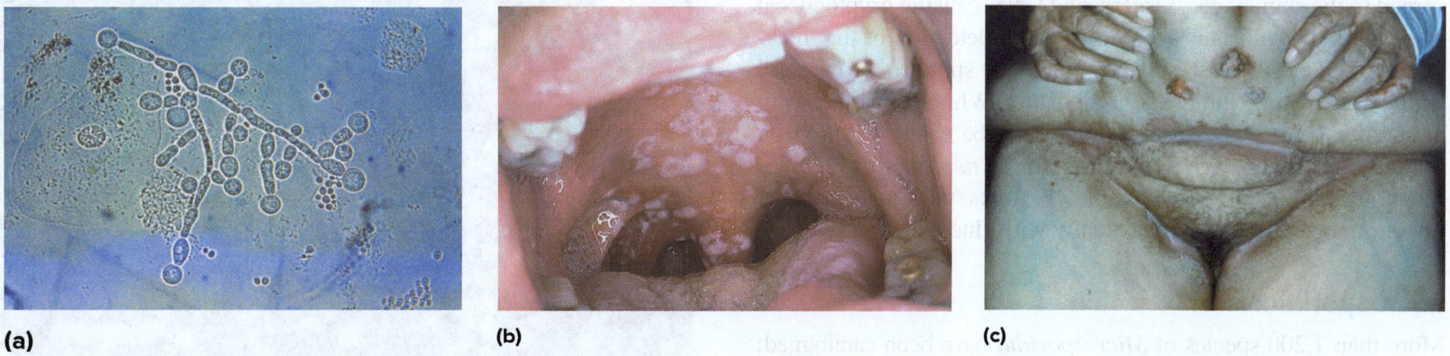

Figure 39.21 Opportunistic Mycoses Caused by *Candida albicans*. (a) Light micrograph showing pseudohyphae. (b) Thrush, or oral candidiasis, is characterized by the formation of white patches on the mucous membranes of the mouth and tongue. These patches form a pseudomembrane composed of spherical yeast cells, leukocytes, and cellular debris. (c) Intertriginous candidiasis involves those areas of the body, usually opposed skin surfaces, that cause irritation, inflammation, and areas of cutaneous ulceration.

(a) CDC/Dr. Brinkman; (b) Sol Silverman, Jr, DDS./Centers for Disease Control and Prevention; (c) Dr. Hardin/Centers for Disease Control and Prevention environment, where it forms biofilms; once it is found in a healthcare facility, it can be very difficult to eliminate. Recall that antibiotic therapy is a significant risk factor for infection with *Candida* yeast, including *C. auris*. This is just one of many reasons that antibiotics must be prescribed judiciously.

Diagnosis of candidiasis is made by a combination of clinical observation and, in more invasive cases, by direct microscopic observation. There is no single satisfactory treatment for candidiasis. Cutaneous lesions can be treated with topical agents such as nystatin, miconazole, and trichomycin. Clotrimazole lozenges and nystatin swishing can be used to treat oropharyngeal candidiasis. Similar topical drugs are used to treat genital/vulvovaginal and rectal candidiasis. Fluconazole can also be used for local infections, while amphotericin B is used to treat systemic candidiasis. *C. auris* is treated with capsofungin.

Cryptococcosis

Cryptococcosis is a mycosis caused by *Cryptococcus gattii* and *C. neoformans*. These fungi grow as large, budding, encapsulated yeasts in the human host but as smaller yeast cells that undergo monokaryotic or dikaryotic reproduction in the environment, where they are saprophytes with a worldwide distribution. *C. gattii* is found in tropical and subtropical regions, as well as the U.S. Pacific Northwest and British Columbia. *C. neoformans* has a broader geographic distribution and therefore is the more common pathogen. Cryptococcosis manifests as pneumonia and/or meningitis with a mortality rate of about 33%. Typically, cryptococci cause disease in those who are immunocompromised, particularly in people who have HIV/AIDS or cancer or have experienced a recent organ transplant.

C. neoformans is found in the droppings of pigeons and several other birds that serve as reservoirs. In people who are healthy, fungi enter the body by the respiratory tract and are immediately cleared by innate immunity; or a minor, transitory pulmonary infection may occur. In contrast, if an individual who is immunosuppressed inhales the fungi, a pneumonia-like disease with coughing and fever occurs. Some pulmonary infections spread to the skin, bones, viscera, and CNS, causing cryptococcal meningitis. Diagnosis is accomplished by detection of the thick-walled, spherical yeast cells from blood, sputum, or cerebral spinal fluid using India ink, Calcofluor White, or fluorescent antibodies (**figure 39.22**). The fungi can be easily cultured on Sabouraud dextrose agar. Treatment for *C. neoformans* includes amphotericin B plus flucytosine. Patients with a history of infection are treated prophylactically with fluconazole.

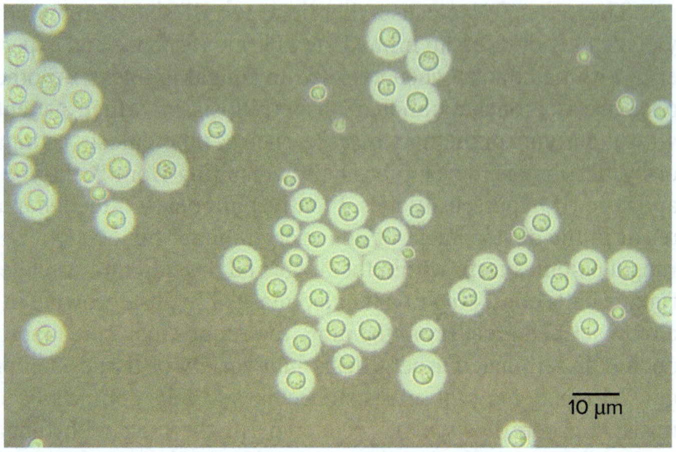

Figure 39.22 Systemic Mycosis: Cryptococcosis. India ink preparation showing *Cryptococcus neoformans*. Note the distinctly outlined capsules surrounding all cells and refractile inclusions in the cytoplasm; light micrograph.

Dr. Leanor Haley/CDC

MICRO INQUIRY *Once this microbe is acquired by inhalation, to what other body systems can it spread?*

Microsporidiosis

More than 1,200 species of *Microsporidia* have been catalogued; at least 14 species are known to be human pathogens causing **microsporidiosis.** Several domestic and feral animals appear to be reservoirs for species that infect humans. Increasingly recognized as opportunistic infectious fungi, microsporidia infect a wide range of vertebrate and invertebrate hosts. One unifying characteristic of microsporidia is their production of a highly resistant spore, capable of surviving long periods of time in the environment. Spore morphology varies with species. Microsporidial spores recovered from human infections are oval to rodlike, measuring 1 to 4 µm. Microsporidia possess a unique organelle known as the polar tubule, which is coiled within the spore (**figure 39.23;** *also see figure 24.4*). *Microsporidia are intracellular parasites (section 24.2)*

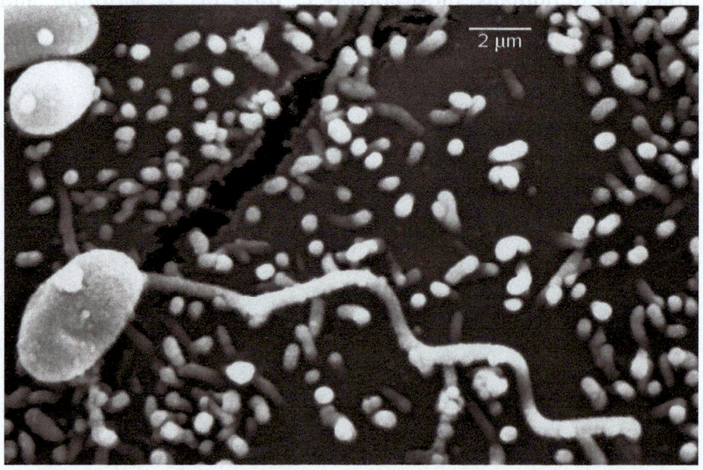

Figure 39.23 Microsporidia. Scanning electron micrograph of a microsporidian spore with an extruded polar tubule inserted into a eukaryotic cell. The spore injects the infective sporoplasms through its polar tubule.

CDC

Infection of a host cell results when a microsporidial spore extrudes its polar tubule, which bores through the target cell's plasma membrane. A sudden increase in spore calcium results in the injection of the sporoplasm (cytoplasm-like contents) through the polar tubule into the host cell. Inside the cell, the sporoplasm condenses and undergoes asexual multiplication. The location of multiplication is species-specific, occurring within the host cytoplasm or within a vacuole, and is usually completed through binary (merogony) or multiple (schizogony) fission. Once the sporoplasm has multiplied, it directs the development of new spores (sporogony) by first encapsulating meronts or schizonts (products of asexual multiplication) within a thick spore coating. New spores increase until the host-cell membrane ruptures, releasing the progeny.

Microsporidia fungal infections can result in a wide variety of conditions. These include hepatitis, pneumonia, skin lesions, diarrhea, weight loss, and wasting syndrome. Diagnosis is based on clinical manifestations and identification of microsporidia in fecal smears. Where possible, identification can be made based on the characteristic polar tubule coiled within a spore visualized by electron microscopy. Microsporidiosis is difficult to treat; albendazole (which inhibits tubulin and ATP synthesis in helminths) is often used for gastrointestinal or disseminated infections while fumagillin (an amebicide) is used for eye infections.

Pneumocystis Pneumonia

Pneumocystis jirovecii is a fungus found in the lungs of a wide variety of mammals. The life cycle of *P. jirovecii* is presented in **figure 39.24.** The disease that this fungus causes was originally termed *Pneumocystis carinii* pneumonia (PCP). The name of the fungus was changed to *Pneumocystis jirovecii* in honor of the Czech microbiologist, Otto Jirovec, who first described this pathogen in humans. The acronym PCP for the disease has been retained; PCP now stands for *p*neumo*c*ystis *p*neumonia.

Serological data indicate that most humans are exposed to *P. jirovecii* by age three or four. However, PCP occurs almost exclusively in hosts who are immunocompromised. Extensive use of immunosuppressive drugs and irradiation for the treatment of cancers and following organ transplants accounts for the formidable prevalence rates for PCP. This pneumonia also occurs in premature, malnourished infants and in more than 80% of patients who have HIV that has progressed to AIDS.

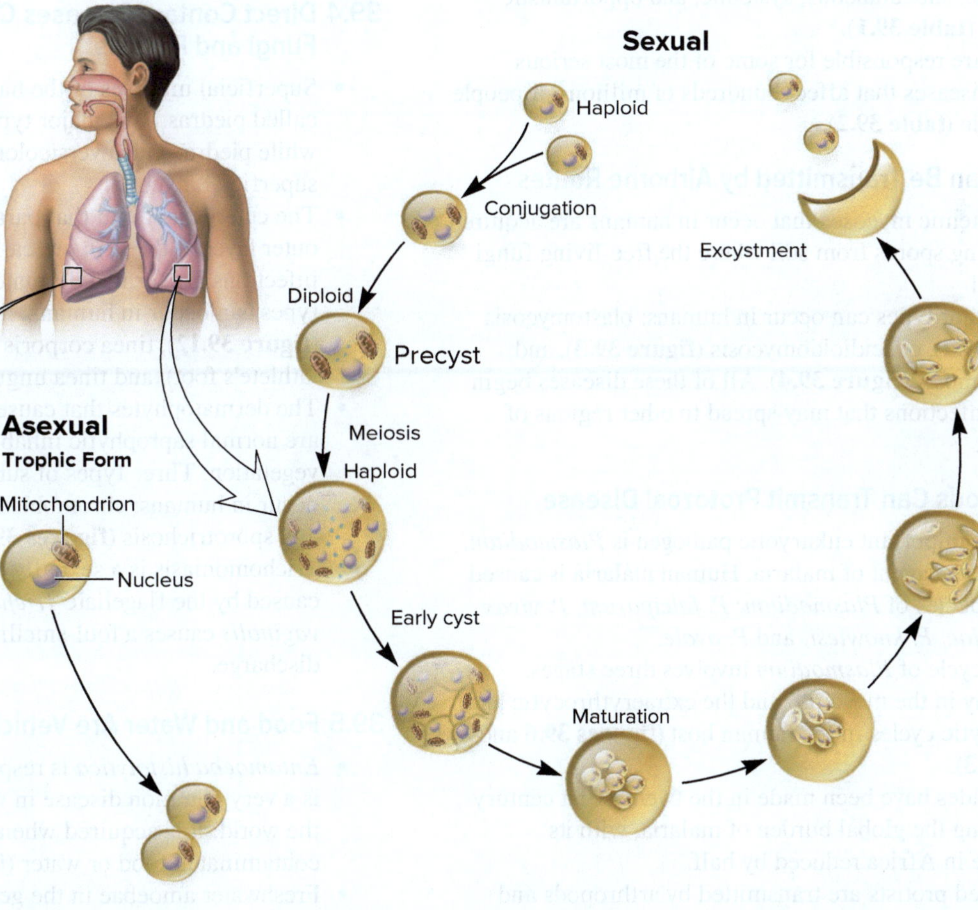

Figure 39.24 Life Cycle of *Pneumocystis jirovecii*. This fungus exhibits both sexual and asexual reproductive stages. Sexual reproduction occurs when haploid cells undergo conjugation followed by meiosis. Asexual reproduction occurs by binary fission. Both forms of reproduction can occur within the human host.

Both the organism and the disease remain localized in the lungs—even in fatal cases. Within the lungs, *P. jirovecii* causes the alveoli to fill with a frothy exudate that diminishes gas exchange.

Laboratory diagnosis of *Pneumocystis* pneumonia can be made by demonstrating the presence of the microorganisms in infected lung material or by PCR analysis. Treatment is by means of oxygen therapy and a combination of trimethoprim and sulfamethoxazole (TMP-SMX). Prevention and control are through prophylaxis with TMP-SMX in persons who are susceptible.

Comprehension Check

1. Compare and contrast infections caused by *Entamoeba histolytica* and *Giardia intestinalis*.
2. How is *Toxoplasma gondii* transmitted?
3. Why are some mycotic diseases of humans called opportunistic mycoses?
4. What parts of the human body can be affected by *Candida* infections?
5. Note that antifungal agents are not used to treat infections caused by microsporidia and *P. jirovecii*. These microorganisms have been difficult to classify taxonomically. How do you think this may relate to their unusual antimicrobial sensitivities?

Key Concepts

39.1 Relatively Few Fungi and Protists Are Human Pathogens

- Human fungal diseases, or mycoses, can be divided into five groups according to the level and mode of entry into the host (**figure 39.1**). These are the superficial, cutaneous, subcutaneous, systemic, and opportunistic mycoses (**table 39.1**).
- Protists are responsible for some of the most serious human diseases that affect hundreds of millions of people worldwide (**table 39.2**).

39.2 Fungi Can Be Transmitted by Airborne Routes

- Most systemic mycoses that occur in humans are acquired by inhaling spores from soil where the free-living fungi are found.
- Three major types can occur in humans: blastomycosis (**figure 39.2**), coccidioidomycosis (**figure 39.3**), and histoplasmosis (**figure 39.4**). All of these diseases begin as lung infections that may spread to other regions of the body.

39.3 Arthropods Can Transmit Protozoal Disease

- The most important eukaryotic pathogen is *Plasmodium,* the causative agent of malaria. Human malaria is caused by five species of *Plasmodium: P. falciparum, P. vivax, P. malariae, P. knowlesi,* and *P. ovale.*
- The life cycle of *Plasmodium* involves three stages: sporogony in the mosquito and the extraerythrocytic and erythrocytic cycles in the human host (**figures 39.6** and **table 39.3**).
- Great strides have been made in the twenty-first century in reducing the global burden of malaria, with its incidence in Africa reduced by half.
- Flagellated protists are transmitted by arthropods and infect the blood and tissues of humans. Two major groups occur: the leishmanias, which cause the diseases collectively termed leishmaniasis (**figures 39.7** and **39.8**), and the trypanosomes (**figures 39.9** and **39.10**), which cause African trypanosomiasis and Chagas disease (American trypanosomiasis).

39.4 Direct Contact Diseases Can Be Caused by Fungi and Protists

- Superficial mycoses of the hair shaft are collectively called piedras. Two major types are black piedra and white piedra. Tinea versicolor is a third common superficial mycosis.
- The cutaneous fungi that parasitize the hair, nails, and outer layer of the skin are called dermatophytes, and their infections are termed dermatophytoses or tineas. Many types can occur in humans: tinea barbae, tinea capitis (**figure 39.12**), tinea corporis (ringworm), tinea pedis (athlete's foot), and tinea unguium (**figure 39.13**).
- The dermatophytes that cause the subcutaneous mycoses are normal saprophytic inhabitants of soil and decaying vegetation. Three types of subcutaneous mycoses can occur in humans: chromoblastomycosis, maduromycosis, and sporotrichosis (**figures 39.14** and **39.15**).
- Trichomoniasis is a sexually transmitted infection caused by the flagellate *Trichomonas vaginalis. T. vaginalis* causes a foul-smelling, mucopurulent vaginal discharge.

39.5 Food and Water Are Vehicles of Protozoal Diseases

- *Entamoeba histolytica* is responsible for amebiasis. This is a very common disease in warm climates throughout the world. It is acquired when cysts are ingested with contaminated food or water (**figure 39.16**).
- Freshwater amoebae in the genera *Naegleria* and *Acanthamoeba* can cause primary amebic

meningoencephalitis and granulomatous amebic encephalitis and keratitis, respectively.

- *Cryptosporidium parvum* is a common apicomplexan protist that causes severe diarrheal disease. It is acquired from contaminated food or water.
- *Cyclospora cayetanensis* is shed in the feces of a current host and infects the gastrointestinal tract of a new host after it has developed at 22° to 32°C for several days or weeks. It can be acquired from contaminated water used to rinse fruit or vegetables (**figure 39.17**).
- *Giardia duodenalis* is a flagellated protozoan that causes the common intestinal disease giardiasis (**figure 39.18**). This disease is distributed throughout the world, and in the United States, it is the most common cause of waterborne diarrheal disease.
- Toxoplasmosis is a disease caused by the apicomplexan protist *Toxoplasma gondii*. It can be acquired by handling cats or their litter boxes, by consuming undercooked or raw meat, or by maternal-fetal transmission (**figure 39.19**).

39.6 Opportunistic Diseases Can Be Caused by Fungi and Protists

- An opportunistic organism is one that is generally harmless in its normal environment but can become pathogenic in a compromised host.
- The most important opportunistic mycoses affecting humans include systemic aspergillosis (**figure 39.20**), candidiasis (**figure 39.21**), cryptococcosis (**figure 39.22**), and pneumocystis pneumonia caused by *Pneumocystis jirovecii* (**figure 39.24**).
- An emerging opportunistic fungal disease, especially of those who are immunocompromised, is caused by fungi known as microsporidia. They live as a spore form and have a unique organelle used for infecting new host cells (**figure 39.23**).

Active Learning

1. *Toxoplasma gondii* manipulates host cell gene expression by modulating the level of key host transcription factors; however, the mechanisms by which this is accomplished are unclear. Proteins delivered from its rhoptry and dense granules are secreted into the host environment (*see figure 23.16*). Discuss how you might determine the function of one such protein.
2. When entering the host's skin, most *Leishmania* spp. are killed by professional phagocytes (neutrophils, macrophages, and dendritic cells). Promastigotes that lead to disease form a parasitophorous vacuole (PV) in macrophages that enables their growth. It is within PVs that promastigotes transform into amastigotes (figure 39.8). Discuss the molecular cues that might trigger the formation of PVs and differentiation of promastigotes into amastigotes. How would you test your predictions?
3. Trypanosomes are notorious for their ability to change their surface antigens frequently. Given the kinetics of a primary immune response (primary antibody production; *see figure 32.14*), how often would surface antigens need to change to stay "ahead" of host antibody specificity?
4. Prior to the COVID-19 pandemic, there was an unprecedented reduction in the rate of malaria infection across Africa. To help resume this progress, epidemiologists must assess the impact of changes in healthcare delivery, preventative measures, medical interventions, as well as population, geographic, and environmental factors. If you were part of such a public health team, what specific data would you want to consider? Explain why your choice is important in the effort to combat malaria.
5. In the effort to reduce the spread of *Candida auris* in healthcare settings, the capacity of silver nanoparticles to inhibit *C. auris* biofilm formation has been explored. Medical equipment and bandages coated with silver nanoparticles have been developed and tested. Propose an assay that could be deployed to determine the level of *C. auris* growth and biofilm formation on these items with and without silver nanoparticle treatment. Be sure to give careful consideration of your positive and negative controls.

CHAPTER

40

Microbiology of Food

View Apart/Shutterstock

The Art, Science, and Genetics of Brewing Beer

Beer is one of the most popular beverages and perhaps part of its popularity has to do with its long history. Prehistoric nomads probably learned to make beer from grains and water before they figured out how to bake bread. Six-thousand-year-old Babylonian tablets describe beer recipes. Ancient Chinese, Egyptians, and Romans brewed beer. It is even rumored that the Pilgrims made landfall in Plymouth in 1620 because their beer supply was running low.

Throughout the ages, several essential steps have defined beer production. First, barley is soaked in warm water to trigger germination. Once germinated, the grain produces two enzymes: gluconase, which breaks down plant cell walls, and amylase, which degrades plant starch to a sugar fermentable by yeast. Next the sprouted grain is heated and dried, denaturing the enzymes. This mixture is called malt, and it is flushed with warm water so yeast can begin fermenting the starch to ethanol and CO_2.

Predictably, many modifications have been made to the art and science of beer brewing. Since the mid-1990s, this has included the development of genetically modified (GM) ingredients. Most breweries find it more cost-effective to add gluconases and amylases collected from bacteria that have been genetically engineered to produce these enzymes. Rice and corn genetically modified to be herbicide-resistant have been used for many years. And there are several GM barley varieties, including those designed to resist pathogenic fungi.

Until recently, however, the use of GM yeast seemed to be entirely off limits. Yeast geneticists developing new strains for brewers do it the old-fashioned way, by mating strains and selecting for new traits. However, some labs are using CRISPR to introduce variation. In fact, CRISPR technology has been used to develop a "hoppy yeast" that imparts the hops flavor to beer, eliminating the need to use actual hops. If this proves to be cost-effective, it may overcome the taboo on GM yeast.

As the history of beer illustrates, humankind long ago figured out how to harness microorganisms to transform raw foods into beverages and foods. It was not until the late nineteenth century, however, that the role microorganisms play in food spoilage was appreciated. This illustrates the two opposing roles of microorganisms in food production and preservation. This chapter begins with an overview of microbial spoilage of food; the second half discusses the importance of other microbes in food and beverage production.

Readiness Check:

Based on what you have learned previously, you should be able to:

- ✓ Describe the capacity of a bacterial endospore to survive in nature, and list example bacteria that form endospores (sections 3.8 and 22.2)
- ✓ Compare and contrast the growth and physiology of obligate aerobes, facultative anaerobes, and strict anaerobes (section 7.5)
- ✓ Review physical and chemical agents commonly used to control microbial growth (sections 8.2 and 8.3)
- ✓ Summarize the possible fates of pyruvate in fermentation (section 11.6)
- ✓ Explain the principles of a toxin-mediated disease (section 34.4)

40.1 Microbial Growth Can Cause Food Spoilage

After reading this section, you should be able to:

a. Define intrinsic factors that influence food spoilage
b. Explain how the composition of a food item helps determine the kind of microbial growth that might occur

The food supply is not sterile; there is an ongoing competition between humans and microbes to see who can consume it first. When the microbes win, we call it food spoilage. Microbial growth is partly controlled by factors inherent in the food itself, called *intrinsic factors*. These include the biochemical composition of the food (**table 40.1**), its pH, redox potential, and the availability of water. *Solutes affect osmosis and water activity (section 7.5)*

Carbohydrate-rich foods with high water activity (fruits and vegetables) first show spoilage by fungi. Molds also produce enzymes that weaken and penetrate the protective outer skin of fruits and vegetables. Once molds degrade carbohydrates within the skin, bacteria can colonize the food. Colonizing bacteria include those that cause soft rots, such as *Pectobacterium carotovorum,* which produces hydrolytic enzymes such as pectinase. Molds also grow on denser carbohydrates such as grains and corn, particularly when these products are stored in moist conditions (**figure 40.1**). Contamination of grains by the ascomycete *Claviceps purpurea* causes ergotism, a toxic condition (*see figure 24.12*). Hallucinogenic alkaloids produced by this fungus can lead to altered behavior, abortion, and death, if infected grains are eaten.

In contrast to carbohydrates, meat and dairy products are rich in proteins and fats that support bacterial growth. The production of short-chained fatty acids from fats renders butter rancid. The anaerobic breakdown of proteins is called **putrefaction.** It yields foul-smelling amine compounds such as cadaverine (imagine the origin of that name) and putrescine.

Raw (unpasteurized) milk is a good example of a complex food containing fats, protein, and carbohydrates. This complex ecosystem features a succession of microbes during spoilage: Acid production by *Lactococcus lactis* is followed by additional acid production associated with the growth of acid-tolerant organisms such as *Lactobacillus.* Next, yeasts and molds dominate and degrade the accumulated lactic acid, gradually decreasing the acidity. Eventually protein-digesting bacteria become active,

(a)

(b)

Figure 40.1 Food Spoilage. Typical examples of mold spoilage include (a) bread, most likely supporting the growth of *Penicillium* sp. (green hyphae), and (b) ear rot in corn, often caused by *Aspergillus* spp. that may produce aflatoxins.

(a) martinfredy/123RF; (b) Daleen Loest/iStock/Getty Images

Table 40.1 Spoilage Processes in Relation to Food Characteristics

Substrate	Food Example	Chemical Reactions or Processes[1]	Typical Products (and Effects)
Pectin	Fruits	Pectinolysis	Methanol, uronic acids (loss of fruit structure, soft rots)
Proteins	Meat	Hydrolysis, deamination	Amino acids, peptides, amines, H_2S, ammonia, indole (bitterness, souring, bad odor, sliminess)
Carbohydrates	Starchy foods	Hydrolysis, fermentation	Organic acids, CO_2, mixed alcohols (souring, acidification)
Fats	Butter	Hydrolysis, fatty acid degradation	Glycerol and mixed fatty acids (rancidity, bitterness)

[1] Other reactions also occur during the spoilage of these substrates.

resulting in a putrid odor and bitter flavor. The milk, originally opaque, eventually becomes clear because the proteins and fats have coagulated.

The pH and oxidation-reduction (redox) potential of a food are also critical intrinsic factors. A low pH favors the growth of yeasts and molds. For neutral or alkaline pH foods, such as meats, bacteria are more dominant in spoilage and putrefaction. Furthermore, when meat products are cooked, they present a reducing environment for microbial growth. These products, with their readily available amino acids, peptides, and growth factors, are ideal media for the growth of anaerobes, including clostridia.

The physical structure of a food can also affect the course and extent of spoilage. Grinding and mixing of foods such as sausage and hamburger increase the food surface area and distribute contaminating microorganisms throughout the food. This can result in rapid spoilage if such foods are stored improperly.

Many foods contain natural antimicrobial substances, including complex chemical inhibitors and enzymes. Coumarins found in fruits and vegetables have antimicrobial activity. Eggs are rich in lysozyme, which lyses cell walls of contaminating monoderm bacteria (*see figure 31.4*). Herbs and spices, once the mainstay of food preservation, often possess significant antimicrobial substances. For example, antimicrobial aldehyde and phenolic compounds are found in cinnamon, mustard, and oregano. Fungi are more sensitive than bacteria to these compounds.

Comprehension Check

1. What are some intrinsic factors that influence food spoilage and how do they exert their effects?
2. How would the spoilage of bread differ from that of a high-fat food like butter?
3. Why do sausage and other ground meat products provide a better environment for the growth of food spoilage organisms than solid cuts of meats?

40.2 Environmental Factors Control Food Spoilage

After reading this section, you should be able to:

a. Review the limitations of refrigeration
b. Compare and contrast two pasteurization methods
c. Describe how the principle of water availability can be used to preserve foods
d. Explain how commonly used chemicals preserve food
e. Defend an informed opinion on the safety and efficacy of food irradiation
f. Examine the packaging of food you frequently consume and assess its capacity to prevent spoilage

In addition to the intrinsic factors that impact food spoilage, there are a variety of environmental, or *extrinsic,* factors that govern microbial growth on and in foods. Unlike intrinsic factors, these can be manipulated to slow or prevent spoilage. They therefore form the basis of methods designed to eliminate or reduce the populations of spoilage microorganisms while maintaining food quality.

Temperature

Refrigeration at 4°C (the temperature of a household refrigerator) inhibits microbial growth, although with extended storage, microorganisms eventually grow and produce spoilage. Of particular concern is the growth of *Listeria monocytogenes,* which grows at refrigeration temperatures. Refrigeration slows the metabolic activity of most microbes, but it does not lead to significant decreases in existing microbial populations.

Controlling microbial populations in foods by means of high temperatures kills microbes and significantly limits disease transmission and spoilage. Heating processes, first used by Nicholas Appert in 1809, provide a safe means of preserving foods, particularly when carried out in commercial canning operations (**figure 40.2**). In the process of *appertization,* canned food is heated in special containers called retorts at about 115°C for intervals ranging from 25 to over 100 minutes. The precise time and temperature depend on the nature of the food. Sometimes canning does not kill all microorganisms but only those that will spoil the food, because growth of remaining bacteria is inhibited by an intrinsic property of the food, such as pH. After heat treatment, cans are cooled as rapidly as possible, usually with cold water. ◀ *Microbes can be controlled by physical means (section 8.2)*

Figure 40.2 Steam Retorts. A production line for canned foods illustrates appertization.

SURAKIT SAWANGCHIT/Shutterstock

MICRO INQUIRY *Canned products such as vegetable soup often contain salt. Other than serving as a flavoring, what is the purpose of salt in the canned product?*

Table 40.2 Methods of Pasteurization

Type of Pasteurization	Temperature and Duration of Heating	Products Pasteurized
Low-temperature holding (LTH)	68.2°C, 30 minutes	Beer, fruit juice, milk for making yogurt
High-temperature short-time (HTST)	72°C, 15 seconds	Industry standard for milk and other dairy products in the United States
Higher heat shorter time (HHST)	89°C, 1 second	Fruit juices, milk
Ultra-high temperature (UHT)	138°C, 2 seconds	Milk that does not require refrigeration until opened

Pasteurization involves heating food to a temperature that kills disease-causing microorganisms and substantially reduces the levels of spoilage organisms. There are several types of pasteurization, which differ by temperature and length of heating time (**table 40.2**). The duration of pasteurization is based on the statistical probability that the number of remaining viable microorganisms will be below a certain level after a particular heating time at a specific temperature. *Food Pathogens and Temperature*

Water Availability

The presence and availability of water also affect the ability of microorganisms to colonize foods. Simply drying a food can slow or prevent spoilage; this is commonly seen in dried fruit, which does not require refrigeration. Water can be made less available by adding solutes such as sugar and salt. Water availability is measured in terms of **water activity (a_w).** This represents the ratio of relative humidity of the air over a test food compared with that of distilled water, which has an a_w of 1. When large quantities of salt or sugar are added to food, most microorganisms are dehydrated by the hypertonic conditions and cannot grow. Dehydration or freeze-drying is a common means of preventing microbial growth. It increases the osmolarity of the product while reducing the a_w so microbial growth is restricted. However, even under these conditions, osmophilic and xerophilic microorganisms may spoil food. **Osmophilic** (Greek *osmus,* impulse; *philein,* to love) microbes grow best in media with a high

Jacques Cornell/McGraw Hill

osmotic concentration (e.g., jams and jellies), whereas **xerophilic** (Greek *xerosis,* dry) microorganisms thrive in low a_w environments (e.g., cereals) and may not grow under high a_w conditions. In contrast to most spoilage bacteria that thrive at an a_w of about 0.9, these microbes grow best at an a_w of 0.6 to 0.7. *Solutes affect osmosis and water activity (section 7.5)*

Chemical-Based Preservation

Various chemical agents are used to preserve foods; these substances are closely regulated by the U.S. Food and Drug Administration and are listed as *g*enerally *r*ecognized *a*s *s*afe or **GRAS.** They include simple organic acids, sulfite, ethylene oxide as a gas sterilant, sodium nitrite, and ethyl formate. These chemical agents damage the microbial plasma membrane or denature proteins. Other compounds interfere with the functioning of nucleic acids, thereby inhibiting microbial reproduction.

Sodium nitrite is an important chemical used to preserve ham, sausage, bacon, and other cured meats by inhibiting the growth of *Clostridium botulinum* and the germination of its endospores (*see figure 38.27*). This protects against botulism and reduces the rate of spoilage. In addition to increasing meat safety, nitrite decomposes to nitric acid, which reacts with heme pigments to keep the meat red in color. However, there is concern about nitrite because it reacts with amines to form carcinogenic nitrosamines.

Low pH can also be used to hinder spoilage. For example, acetic and lactic acids inhibit growth of *Listeria monocytogenes.* In addition, low pH can increase the activity of other chemical preservatives. Sodium propionate is most effective at lower pH values, where it is primarily undissociated. Breads, with their low pH values, often contain sodium propionate as a preservative. The use of citric acid is also common.

High Hydrostatic Pressure

The use of high hydrostatic pressure (HHP), also called *pascalization,* is an alternative way to preserve food. The food product is packaged and sealed, then immersed in water. Pressures from 100 to 1200 megapascals (MPa) are applied without significant changes in temperature. HHP is most detrimental to cell membranes; thus eukaryotic microbes are more susceptible. Predictably, monoderm bacteria, particularly endospores, are less susceptible to HHP than diderm bacteria.

Radiation

Although the public is wary of food radiation, it is an effective way to prevent food spoilage. Unfortunately, there is a misperception that food becomes radioactive upon irradiation. This would be similar to thinking that you become radioactive following a chest or dental X ray. ◀ *Radiation (section 8.2)*

Three sources of radiation are used for food preservation: X rays and electron beams send a stream of high-energy electrons reflected from a heavy metal (X ray) or propelled from an accelerator (electron beam). The third is gamma irradiation, which uses cobalt-60 or cesium-137. Gamma radiation has excellent penetrating power but must be used with moist foods because radiation is effective only if it generates reactive oxygen species from water, resulting in oxidation of nucleic acids, lipids, and proteins. This process of *radappertization,* named after Nicholas Appert, can extend the shelf life of meats, seafoods, fruits, and vegetables. Unlike canning, radappertization does not heat the food product.

Filtration

Microorganisms can be removed from water, wine, beer, juices, soft drinks, and other liquids by filtration. This keeps bacterial populations low or eliminates them entirely. Several major brands of beer are filtered rather than pasteurized to better preserve flavor and aroma.

Atmosphere

Gases present in the atmosphere within a food package can have a significant effect on microbial growth. This has led to the development of **modified atmosphere packaging (MAP).** Modern shrink-wrap materials and vacuum technology make it possible to package foods with controlled atmospheres. These materials are largely impermeable to oxygen. This prolongs shelf life by a factor of two to five times compared to the same product packaged in air. With a CO_2 content of 60% or greater in the atmosphere surrounding a food, spoilage fungi will not grow, even if low levels of oxygen are present. High-oxygen MAP is also effective because it triggers the formation of superoxide anions within microbial cells. Products packaged using MAP technology include delicatessen meats and cheeses, pizza, bakery items, and dried products such as coffee.

Comprehension Check

1. Construct a table that lists the chemical, physical, and biological approaches to food preservation.
2. Keep a list of all the chemicals used to preserve the foods and beverages you consume in a single day.
3. What cellular component does HHP target? What is radappertization?
4. What are the major gases involved in MAP? How are their concentrations varied to inhibit microbial growth?

40.3 Food-Borne Illnesses

After reading this section, you should be able to:

a. Summarize a major food-borne infection that had substantial health and economic consequences
b. Describe the origin and consequences of a major food intoxication event in the United States
c. Describe the types of organisms that release toxins into foods and how those toxins act

Most food-borne illnesses are neither recognized nor reported. The source and the identity of a food pathogen are only investigated if it causes an illness dire enough that the victim visits a physician or hospital. That said, the Centers for Disease Control and Prevention (CDC) estimates there are about 48 million cases of food-related illness annually in the United States, 128,000 hospitalizations, and about 3,000 deaths. *Salmonella enterica* serovars, noroviruses, *Campylobacter jejuni,* enterohemorrhagic *Escherichia coli* (e.g., strain O157:H7), hepatitis A virus, and *Listeria monocytogenes* are major causes of food-borne diseases.

Many food-borne illnesses are transmitted by the fecal-oral route. The 5-Fs—feces, fingers, food, fomites, and flies—are critical for fecal-oral transmission. In this context, *fomites* often include faucets, drinking cups, and cutting boards. Individual food-borne diseases are covered in chapters 37 through 39 and common microbes involved are summarized in table 38.3.

Noam Armonn/Alamy Stock Photo

Food-Borne Infection

A **food-borne infection** involves ingestion of the pathogen, followed by its growth in the host, including tissue invasion or the release of toxins.

Listeriosis, caused by *Listeria monocytogenes,* was responsible for one of the largest meat recalls in U.S. history. In 2002 a seven-state listeriosis outbreak was traced to deli meats and hot dogs produced at a meat-processing plant. Females who are pregnant, babies, older people, and individuals who are immunocompromised are especially vulnerable to *L. monocytogenes* infections. This outbreak resulted in 7 deaths, 3 stillbirths, and 46 illnesses. Microbiologists matched the strain of *L. monocytogenes* in the contaminated food with samples from floor drains in a packaging plant. This prompted the recall of 27.4 million pounds of meats that had been distributed over a 5-month period to stores, restaurants, and school lunch programs. This episode prompted the U.S. Department of Agriculture to step up its environmental testing program for *L. monocytogenes*. Physicians now routinely advise vulnerable people to avoid eating soft cheeses (e.g., feta, Brie, Camembert), refrigerated smoked meats such as lox, and deli meats and undercooked hot dogs.

Outbreaks involving the enterohemorrhagic *E. coli* (EHEC) strain O157:H7 also illustrate the impact of food-borne infections. *E. coli* O157:H7 produces a toxin, Shiga toxin, that is responsible for the life-threatening complication of hemolytic uremic syndrome, most commonly seen in children and older people. Although EHEC produces a toxin, the disease it causes is not an intoxication because the bacterium must be present to produce the toxin in the host. The first U.S. outbreaks of *E. coli* O157:H7 disease (in the 1980s) occurred in individuals who had consumed undercooked hamburgers. This microbe inhabits the intestines of a variety of mammals, and ground beef contaminated with cattle feces during slaughter and processing continues to be a common cause of these outbreaks. ◀ *Escherichia coli gastroenteritis (section 38.4)*

A more recent example of *E. coli* O157:H7 outbreaks involve minimally processed, ready-to-eat fresh salad greens, which are a popular convenience. The produce is repeatedly washed, then dried and packed. Contamination occurs while the crop is in the field. Surveillance studies determine that likely sources of *E. coli* are nearby wild and domestic animals that contaminate irrigation water with their feces. This finding raises more questions than it answers, including: How can thousands of agricultural fields be protected? Why doesn't the postharvest washing procedure always remove *E. coli?* What conditions allow the initial bacterial inoculum on the plants to grow by at least a factor of 10? Regular outbreaks continue to prompt both government and industry to ensure that minimally processed produce remains a convenience, not a sporadic source of disease.

Food Intoxication

Microbial growth in food products can result in **food intoxication** (*see table 38.3*). Intoxication produces symptoms shortly after the food is consumed because the toxin is already present in the food. The classic example is picnickers vomiting a few hours after consuming potato salad contaminated with *Staphylococcus aureus* enterotoxins. The consumption of *S. aureus* cells is not necessary, as the illness results from the toxins.

Two endospore-forming rods are important sources of food intoxications and infections: *Clostridium botulinum* and *Bacillus cereus*. There are typically a few dozen cases of botulism reported each year in the United States, and most are associated with home canning. *B. cereus* causes two types of intoxication, each the result of a different toxin. Ingestion of *B. cereus*–contaminated meats, fish, or vegetables can cause a diarrheal illness. A different *B. cereus* intoxication, characterized by acute (less than 24 hours) vomiting, is associated with consumption of contaminated starchy foods such as rice or pastries. ◀ *Botulism (section 38.4)*

Some fungus-derived toxins have more insidious health effects. The most potent are the fungal carcinogens aflatoxins and fumonisins. **Aflatoxins** are produced by *Aspergillus flavus* and *A. parasiticus*. These fungi grow and release toxin in moist grains and nuts. Aflatoxin consumption is most common in tropical countries where these products are stored in suboptimal conditions and consumed with little or no postharvest processing. Aflatoxins are planar, ringed compounds that wedge between DNA base pairs (intercalate), thereby causing frameshift mutations. About 20 different aflatoxins are known; the most important are shown in **figure 40.3.** Aflatoxin B_1 is the most common and the most potent carcinogen. After ingestion by dairy cows, aflatoxin B_1 is modified in the liver to yield

Katarzyna Bialasiewicz/123RF

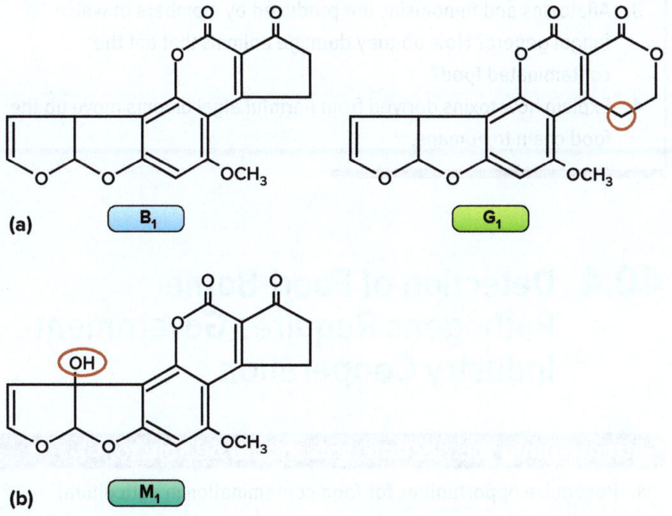

Figure 40.3 Aflatoxins. (a) The letter designations refer to the color of the compounds under ultraviolet (UV) light after extraction from the grain. Toxin B fluoresces blue and toxin G fluoresces green. (b) Type M aflatoxin is found in the milk of lactating animals that have ingested type B aflatoxins. Red circles indicate the chemical differences from toxin B.

MICRO INQUIRY *What foods are most prone to aflatoxin contamination?*

aflatoxin M_1. If cattle consume aflatoxin-contaminated feeds, the toxin contaminates milk and dairy products derived from that milk. Aflatoxins are also found in beer, cocoa, raisins, peanut butter, and soybean meal. ◀ *Exogenous mutations are caused by environmental factors (section 16.1)*

The **fumonisins** are produced by *Fusarium verticillioides* growing on corn and corn products not stored under sufficiently dry conditions. These fungal toxins cause leukoencephalomalacia in horses (also called blind staggers, fatal in 2–3 days), pulmonary edema in pigs, and esophageal cancer in humans. Fumonisins inhibit enzymes that affect synthesis and metabolism of sphingolipids, important compounds that influence a wide variety of cell functions.

Seafood may be contaminated with toxins produced by freshwater and marine microbes that dominate ecosystems during a harmful algal bloom. Cyanobacteria in freshwater ecosystems produce microcystins and several neurotoxins (*see Microbial Diversity and Ecology 29.1*). Marine toxin producers include dinoflagellates and diatoms. These toxins contaminate fish and thereby affect the health of marine animals higher in the food chain. They also contaminate shellfish and finfish, which are later consumed by humans.

Comprehension Check

1. What is the difference between a food-borne infection and a food intoxication? Which is harder to prevent? Explain your answer.
2. What practical recommendations regarding food preparation could you make to someone concerned about *E. coli* O157:H7 contamination?
3. Aflatoxins and fumonisins are produced by members of which fungal genera? How do they damage animals that eat the contaminated food?
4. Explain how toxins derived from harmful algal blooms move up the food chain to humans.

40.4 Detection of Food-Borne Pathogens Requires Government-Industry Cooperation

After reading this section, you should be able to:

a. Recognize opportunities for food contamination in agricultural practices
b. Describe methods used to counter contamination during food production
c. List the principles that guide the development of food testing strategies
d. Describe how a food-borne pathogen consumed by an individual might be traced back to the food source

In 1905 Upton Sinclair's novel *The Jungle* exposed the horrific state of the meat packaging industry. The following year, the first in a series of federal and state laws was enacted in the United States aimed at improving food safety. However, between 1938 and 2011, when the Food and Drug Administration (FDA) Food Safety Modernization Act was passed, no major legislation was enacted. Importantly, the goal of the 2011 act was to change the focus of federal regulators from responding to food contamination to preventing it. But while we may look to the government to keep food safe, food safety is a shared responsibility, as precautions must be defined and in place during production, transport, sale, preparation, and consumption of any food item. The length and complexity of the chain from producer to consumer have increased dramatically with the rise in food imports and exports, making food safety efforts even more complicated. Here we provide a broad overview of safety measures and food pathogen detection.

Farm-to-Market Safety Measures

The microbes responsible for most food-borne diseases originate in two places: from the microbiota normally associated with livestock or produce or from the food processing environment. Produce is monitored beginning with field conditions through harvest and packing. Soil disruption from activities such as driving vehicles around fields or water splashed by a sprinkler risks contamination with soil organisms, many of which are spore formers. Agricultural fields attract wildlife (deer, rabbits, raccoons) that may graze on crops and leave behind feces. Finally, the portion of the plant that is eaten (root, leaves, fruit) has a distinct microbiota, some of which will cause spoilage during packing and transport.

We use chicken to illustrate the challenges and controls in place to provide one of the most popular meats (**figure 40.4**). The pathogens most associated with poultry are *Campylobacter, Salmonella,* and *Clostridium perfringens;* these are part of the normal poultry microbiota. A fourth poultry-associated pathogen, *Listeria monocytogenes,* persists in cold, damp processing facilities. Routine surveillance for these four bacteria is key to detecting and eliminating outbreaks.

Normal body temperature for a chicken is 41°C, the optimal growth temperature for *Campylobacter* and *C. perfringens.* Control strategies for these pathogens rely on decreasing fecal contamination and rapidly cooling carcasses after slaughter. Feed is withheld from the animals for 8 to 12 hours prior to transport and harvesting to clear the gastrointestinal (GI) tract. Following slaughter, the carcasses are scalded in 60°C water for a few minutes. This facilitates feather removal and eliminates some non-spore-forming bacteria, but the water is a major source of cross contamination because of the number of birds immersed at one time. Removal of the animal's GI tract has the greatest potential for contamination by fecal microbes (*Salmonella*), and this step is often automated for consistency. Next, the carcass is cooled by immersion in water to 4.4°C. The cooling process time (less than 4–8 hours, depending on the size of the bird) is specified by federal law. Of note is the fact that feather follicles on the skin may

- Feed and supplements
- Health of flock
- Distress during transport

- Speed of processing
- Chlorine, peracetic acid treatments
- Pathogen surveillance

- Preservative addition
- Packaging
- Transport time

Figure 40.4 Steps in the Chicken Supply Chain That Influence Microbial Composition.

(left) Vedaant Sethia/Shutterstock; (middle) Glow Images; (right) Kathleen Sandman/McGraw Hill

be colonized by *Campylobacter* during the chilling step because of close contact between carcasses. To reduce contamination, both chlorine and peracetic acid are used to treat water used for processing. Depending on the ultimate destination of the meat, the chicken may be cooked for ready-to-eat products or treated with additional preservatives for transport to the market.

Food Pathogen Detection

Several principles guide the development of the technologies, protocols, and policies used to keep foods safe. These include (1) *specificity* and *sensitivity* of the method for the given pathogen—any given test should rarely yield false positives (specificity) and the pathogen in question should be detected even in low concentrations (sensitivity); (2) speed, which is particularly important for fresh produce awaiting distribution; and (3) simplicity—that is, foods should be tested with limited sample preparation. Furthermore, the goal of each detection strategy should be clearly defined. Ideally "testing to prevent"—that is, to confirm food is safe before it leaves the farm or processing plant—is the goal. When this is unreasonable due to logistics and cost, the next best approach is "testing to protect," which involves analysis before the food is accessible for consumption. These strategies are designed to avoid "testing to recover," when an outbreak has occurred and the origin of the contaminated food must be identified.

Growth of contaminating food-borne pathogens on selective media has long been the bedrock of the food safety lab. With the advent of molecular techniques, these approaches have been replaced by methods with excellent sensitivity, specificity, and turnaround time. While some molecular techniques still rely on culturing the pathogen first, many others are culture independent.

In general, three types of molecular methods are used in testing food: nucleic acid–based methods, immunological methods, and biosensors. Nucleic acid–based methods include real-time PCR, multiplex PCR, whole-genome sequencing (WGS), and programmable nuclease-based (CRISPR) detection. For example, multiplex PCR is used to test a food for several pathogens at once using primers specific for each microorganism of interest. Immunoassays to detect pathogen products include lateral flow assays that are simple and inexpensive and, importantly, can be done in the field (*see figure 36.13*). The use of immunomagnetic beads is a more rapid approach than enrichment culture to separate a suspected pathogen from a food item (*see figure 36.4*). Biosensors couple a biological material (e.g., cell receptor, enzyme, synthetic peptide) that detects a pathogen to a transducer that provides an optical, electrical, magnetic, or mechanical readout. Biosensors are valuable in food safety testing because they provide real-time pathogen detection and are typically portable so they can be used in the field. ◀ *Polymerase chain reaction amplifies targeted DNA (section 17.2); Next-generation genomic sequencing (section 18.2)* ▶ *Biosensors (section 41.2)*

Whole-genome sequencing (WGS) is the method of choice for identifying food-borne pathogens (**figure 40.5*a***). WGS comes in two forms, *short read platforms* and *long read platforms,* that give genome fragments of less than and greater than 1,000 base pairs (bp), respectively. While the long reads are essential for the assembly of a complete genome, the shorter reads are sufficient for subtyping. **Subtyping** is the identification of an organism to a level below species or subspecies. Computer alignment of a

(a) (b)

Figure 40.5 Food Pathogen Detection. (a) A public health researcher prepares a sample for whole genome sequencing. (b) A bioinformatician compares whole genome data from a food-borne disease outbreak.

(a) Lauren Bishop/ Courtney Wheeler/CDC; (b) James Gathany/Daniel Drapeau/CDC

MICRO INQUIRY *Why is it important that PulseNet coordinate the identities of food-borne pathogens on a national level?*

short read with an existing genome, known as a *reference sequence* (*see figure 18.6*), highlights differences like single nucleotide polymorphisms and indels. The sum of the differences identified by many short reads describes the unique genotype of the organism in question. These short read sequences may also be compared to other isolates. For example, public health officials want to know how the sequence from a *Campylobacter* isolate that caused a patient's illness compares with isolates from a regional processing facility or isolates from other cases of illness. Complete genome sequencing is necessary to identify and evaluate regions of DNA that are not found in the reference genome; such sequences often represent recent horizontal gene transfers. ⏮ *Whole genome comparison (section 26.2)*

The CDC oversees a program called PulseNet for the early detection of food-borne pathogens, particularly during an outbreak. PulseNet provides molecular tools and resources to document strains of food-borne pathogens, allowing them to be tracked over space and time (figure 40.5*b*). Since 2019 PulseNet has adopted WGS as the gold standard for identifying food-borne pathogens.

The importance of WGS for identifying the source of food-borne outbreaks was demonstrated in 2010 in a nationwide outbreak of *Salmonella enteritidis* in eggs. Laboratories using subtyping identified a pattern in patient samples and samples from egg producers, but that pattern was known to be common among many strains. WGS provided a greater level of resolution, allowing investigators to match patients with samples from a specific egg supplier. ⏮ *Salmonellosis (section 38.4)*

Comprehension Check

1. Consider the food supply chain for an imported vegetable. When and where do you think the vegetable should be screened for pathogens? Explain.
2. What are the molecular approaches used in food safety testing? Which can be used in the field?
3. You are on a team investigating an outbreak of *Salmonella enterica* food poisoning. You have PulseNet and WGS information from several patients and several contaminated foods. What questions can be answered with these data?

40.5 Microbiology of Fermented Foods: Beer, Cheese, and Much More

After reading this section, you should be able to:

a. Compare and contrast processes to ferment milk
b. Outline the process by which cheese is made
c. Provide an overview of the production of wine and beer
d. Describe a fermented food item not commonly eaten in the United States

Fermentation has been used in food production for thousands of years; the earliest records date as far back as 6000 BCE. Microbial growth is used to generate chemical or textural changes to form a product that can be stored for extended periods. The fermentation process is used to create pleasing food flavors and odors—such as cheese, yogurt, and chocolate (**Techniques & Applications 40.1**). Two types of fermentation exist—those performed by the addition of starter cultures and those performed by the microbes inherent to the food being fermented. Here we discuss the microbiology of the most common fermented foods. ⏮ *Fermentation does not involve an electron transport chain (section 11.6)*

Fermented Milks

Although we are most familiar with yogurt, at least 400 different fermented milks are produced around the world. The majority of these rely on **lactic acid bacteria (LAB),** which include species belonging to the genera *Lactobacillus, Lactococcus,* and *Leuconostoc* (**figure 40.6**). These bacteria tolerate acidic conditions, are nonsporing, and are aerotolerant with a strictly fermentative metabolism. ⏮ *Order Lactobacillales (section 22.2)*

Regardless of the type of milk product, **starter cultures** of LAB are added to begin fermentation. *Mesophilic milk fermentation* results in the production of buttermilk and sour cream. This type of fermentation relies on the acidic products of fermentation to denature milk protein. Milk is typically inoculated with the desired starter culture and incubated at 20° to 30°C. Microbial growth is stopped by cooling. *Lactobacillus* spp. and *Lactococcus lactis* starter cultures are used for aroma and acid production.

Despite the name, *thermophilic fermentations* are conducted around 45°C. An important example is yogurt production. In commercial production, milk is pasteurized, cooled to 43°C or lower, and inoculated with a 1:1 ratio of *Streptococcus salivarius* subspecies *thermophilus* (*S. thermophilus*) and *Lactobacillus delbrueckii* subspecies *bulgaricus* (*L. bulgaricus*). *S. thermophilus* grows more rapidly at first and renders the milk anoxic and weakly acidic. *L. bulgaricus* then acidifies the milk even more. Acting together, the two species ferment almost all of the lactose to lactic acid and flavor the yogurt with diacetyl (*S. thermophilus*) and acetaldehyde (*L. bulgaricus*). Fruits or fruit flavors are pasteurized separately and combined with the yogurt. Freshly prepared yogurt contains about 10^9 bacteria per gram.

Yeast-lactic fermentations include kefir, a product that can contain up to 2% ethanol. This fermented milk originated in the Caucasus Mountains, located where Asia and Europe intersect. Kefir products tend to be foamy due to microbial carbon dioxide production. This process is based on the use of kefir grains as a starter culture. Not actual grains, these are coagulated lumps of casein (milk protein) that contain yeasts, lactic acid bacteria, and acetic acid bacteria, which are recovered at the end of the fermentation. Alternately, kefir can be produced by **backslopping,** a common technique in food fermentations in which a sample of each batch is withheld to inoculate the next batch. Originally

TECHNIQUES & APPLICATIONS

40.1 Chocolate: The Sweet Side of Fermentation

Chocolate has been called the "world's favorite food," but few people realize that fermentation is an essential part of chocolate production. The Aztecs were the first to develop chocolate fermentation, serving a chocolate drink made from the seeds of the chocolate tree, *Theobroma cacao* (Greek *theos,* god; *broma,* food, or "food of the gods"). Chocolate trees now grow in Africa as well as South America.

The process of chocolate fermentation has changed very little over the past 500 years. Ripe cocoa pods are harvested from trees and slashed open to release the pulp and seeds (**box figure**). The sooner the fermentation begins, the better the product, so fermentation is set up on the farm. The seeds and pulp are placed in sweat boxes or in heaps in the ground and covered, usually with banana leaves.

Like most fermentations, this process involves a succession of microbes. First, a community of yeasts hydrolyzes the pectin that covers the seeds and releases simple carbohydrates. As the temperature increases, the yeasts are inhibited and lactic acid bacteria increase in number. Lactic acid bacteria ferment the sugars to release lactate, ethanol, and CO_2. Acid production drives the pH down, promoting the growth of bacteria that produce acetic acid as a fermentation end product. Acetic acid is critical to the production of fine chocolate because it kills the sprout inside the seed, which releases enzymes that further degrade proteins and carbohydrates, contributing to the overall taste of the chocolate. Fermentation takes 5 to 7 days. An experienced cocoa grower knows when fermentation is complete: If it is stopped too soon, the chocolate will be bitter and astringent; if it lasts too long, microbes grow on the seeds instead of in the pulp, resulting in off tastes.

After fermentation, the seeds, now called beans, are spread out to dry. The dried beans are brown and lack pulp. They are bagged and sold to chocolate manufacturers, who first roast the beans to further reduce the bitter taste and kill most of the microbes. The beans are then ground and the nibs—the inner part of each bean—are removed. The nibs are crushed into a thick paste called a chocolate liquor, which contains cocoa solids and cocoa butter. Cocoa solids are brown and have a rich flavor, and cocoa butter has a high fat content and is off-white in color. The two components are separated and reunited in controlled ratios, and sugar, vanilla, and other flavors are added. The better the fermentation, the less sugar needs to be added (and the more expensive the chocolate).

The final product, delicious chocolate, is a combination of over 300 different chemical compounds. This mixture is so complex that no one has yet been able to make synthetic chocolate that can compete with the natural fermented plant. Microbiologists and food scientists are studying the fermentation process to determine the role of each microbe. But they have had little luck in replicating the complex, imprecise fermentation that occurs on cocoa farms. In fact, the finest, most expensive chocolate starts as cocoa on farms where the details of fermentation have been handed down through generations. Chocolate production is truly an art as well as a science, while eating it is simply divine.

(a)

(b)

Cocoa Fermentation. (a) Each cocoa pod is 13 to 15 cm in length and contains 30 to 40 seeds in a sticky white pulp. (b) Seeds and pulp are fermented in large heaps for 5 to 7 days and then dried in the sun. Chocolate cannot be produced without fermentation.

(a) Valentyn Volkov/Shutterstock; (b) Rodrigo Buendia/AFP/Getty Images

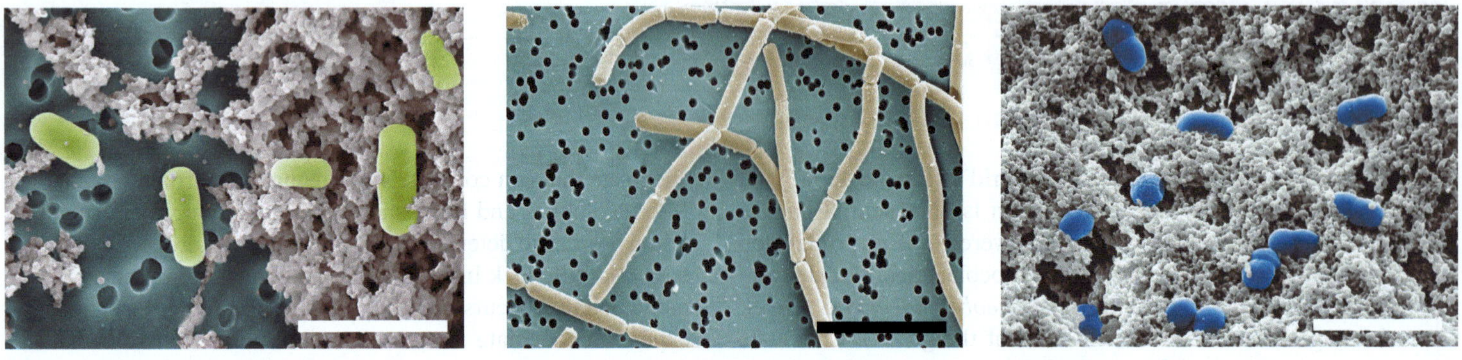

Figure 40.6 **Lactic Acid Bacteria (LAB).** Colorized scanning electron micrographs of LAB used as starter cultures. (a) *Lactobacillus helveticus.* (b) *L. delbrueckii.* (c) *Lactococcus lactis.* The bacteria are supported by filters, seen as holes in the background. Scale bar = 5 μm.

Jeffery Broadbent/Utah State University

MICRO INQUIRY *Can you name at least two features that make these bacteria well suited for milk fermentations?*

kefir was produced in leather sacks hung by the front door during the day, and passersby were expected to push and knead the sack to mix and stimulate the fermentation. Kefir has gained popularity and is now mass-produced and sold in supermarkets worldwide.

Cheese

Cheese is one of the oldest foods, probably developed roughly 8,000 years ago. About 2,000 distinct varieties of cheese are produced throughout the world, representing approximately 20 general types.

Cheese production begins when the milk is pasteurized to remove spoilage microorganisms. It is then cooled to 32°C so that starter cultures can be added. The mixture is held at this temperature for 30 minutes to allow fermentation to begin, and the pH declines as acidic fermentation products accumulate. Next, rennet is added; rennet is a mixture of enzymes produced by mammals to aid in the digestion of milk. Rennin is the key enzyme in rennet; it catalyzes a complex reaction that results in hydrolysis of milk protein (casein), converting the milk to solid curds. The curds set for about 30 minutes to allow the formation of a firm coagulum. Further fermentation of curds continues until the pH drops to 6.4, at which point the cheese is cut into small pieces and heated to 38°C, which helps separate remaining liquid (whey) from the curd. After the whey is completely drained from the curd, the process of cheddaring begins (**figure 40.7**). Here curd is stacked and occasionally flipped; this process expels residual whey while fermentation continues. When the curd mats, now knitted together, reach a pH of 5.1 to 5.5, they are cut into smaller pieces. Depending on what kind of cheese is being made, either salt will be added (e.g., cheddar cheese) or the cheese will be rolled in loaves and placed in a brine (e.g., mozzarella). The next step is to form the salted cheese into blocks so it can age for weeks to years. This will, in part, determine the final hardness and texture of the cheese.

Lactococcus lactis is used as a starter culture for a number of cheeses. Starter culture density is often over 10^9 colony-forming units (CFU) per gram of cheese curd before ripening. However, the high salt, low pH, and temperatures that characterize the cheese microenvironment quickly reduce these numbers. This enables other bacteria, sometimes called nonstarter lactic acid bacteria, to grow; their numbers can reach 10^7 to 10^9 CFU/g after several months of aging. Thus both starter and nonstarter LAB contribute to the final taste, texture, aroma, and appearance of the cheese.

Figure 40.7 **Cheddar Cheese Production.** Cheddar, a village in England, has given its name to a cheese made in many parts of the world. Cheddaring is the process of turning and piling the curd to expel whey and develop the desired texture. It is used in the production of all cheeses.

Robert Estall photo agency/Alamy Stock Photo

In some cases, molds are used to further enhance the cheese. For Roquefort and blue cheeses, *Penicillium roqueforti* spores are added to the curds just before final processing. Sometimes the surface of an already formed cheese is inoculated at the start of ripening; for example, Camembert cheese is inoculated with spores of *Penicillium camemberti.*

Wines

Enology (Greek *oinos,* wine; *ology,* the science of) is the focus of wine production, which starts when the liquid extracted from crushed grapes, called must, is fermented (**figure 40.8**). Wines can be produced by using the natural grape skin microorganisms, but this mixture of bacteria and yeasts gives unpredictable results. To avoid this, fresh must is treated with a sulfur dioxide fumigant to kill naturally occurring microbes and a desired strain of the yeast *Saccharomyces cerevisiae* or *S. ellipsoideus* is added. The juice is fermented for 3 to 5 days at temperatures between 20° and 28°C. Depending on the alcohol tolerance of the yeast strain (the alcohol eventually kills the yeast that produces it), the final product may contain 10 to 14% alcohol. Clearing and development of flavor occur during the aging process.

Grape juice contains high levels of organic acids, including malic and tartaric acids. If the levels of these acids do not decrease during the fermentation process, the wine will be too acidic and have poor stability and "mouth feel." Malic acid levels decrease when fermented by *Oenococcus, Lactobacillus,* and *Pediococcus* species, which transform malic acid (a four-carbon dicarboxylic acid) to lactic acid (a three-carbon monocarboxylic acid) and carbon dioxide. This malolactic fermentation results in deacidification, improvement of flavor stability, and, in some cases, the accumulation of bacteriocins in the wines. *Ascomycota includes yeasts and molds (section 24.4)*

(a) (b)

Figure 40.8 Wine Making. (a) Once grapes are pressed, the sugars in the juice (the must) are immediately fermented to produce wine. (b) Fermentation and aging occur in barrels.

(a) Cavan Images/Alamy Stock Photo; (b) Corbis/VCG/Corbis/Getty Images

A critical part of wine making involves the choice of whether to produce a dry (no remaining free sugar) or a sweeter (varying amounts of free sugar) wine. This can be controlled by regulating the initial must sugar concentration. With higher levels of sugar, alcohol will accumulate and inhibit the fermentation before the sugar is completely consumed, producing a sweeter wine.

Beers and Ales

Beer and ale production uses cereal grains such as barley, wheat, and rice. The first step in beer production, mashing, involves germination of the barley grains and activation of their enzymes to produce a malt (**figure 40.9**). The malt is mixed with water and the desired grains, and the mixture is transferred to the mash tun or cask where the active enzymes hydrolyze starch to usable carbohydrates (*see chapter opening story*). Once this process is completed, the mash is heated with hops (dried flowers of the female vine *Humulus lupulus*). Hops were originally added to inhibit spoilage microorganisms, but they also provide flavor and assist in clarification of the liquid portion called the wort. In this heating step, the hydrolytic enzymes are inactivated and the wort is inoculated with the desired yeast.

Most beers are fermented with bottom yeasts, related to *Saccharomyces pastorianus,* which settle at the bottom of the fermentation vat. Bottom yeasts require 7 to 12 days of fermentation to produce beer with a pH of 4.1 to 4.2. With a top yeast, such as *Saccharomyces cerevisiae,* the pH is lowered to 3.8 to produce ales. Freshly fermented (green) beers are aged, and when they are bottled, CO_2 is usually added. Beer can be pasteurized at 40°C or higher or sterilized by passage through membrane filters to minimize flavor changes.

Bread

Bread is one of the most ancient of human foods. The use of yeasts to leaven bread is depicted in paintings from ancient Egypt, and a bakery at the Giza Pyramid, from the year 2575 BCE, has been excavated. In bread making, yeast growth is carried out under oxic conditions. This results in increased CO_2 production and minimum alcohol accumulation. Bread production involves several steps: moistened flour releases α- and β-amylases and these enzymes produce maltose and glucose from the starch in the flour. The baker's strain of the yeast *Saccharomyces cerevisiae* is added and synthesizes maltase, invertase, and zymase. These consume the mono- and disaccharides. Two flour proteins combine to form gluten, which is not a substrate for yeast, but traps the CO_2 produced in fermentation, causing the bread to rise (**figure 40.10**). Traces of fermentation products contribute to the final flavor. Usually bakers add sufficient yeast to allow the bread to rise within 2 hours. Longer rising times allow contaminating bacteria and fungi to grow.

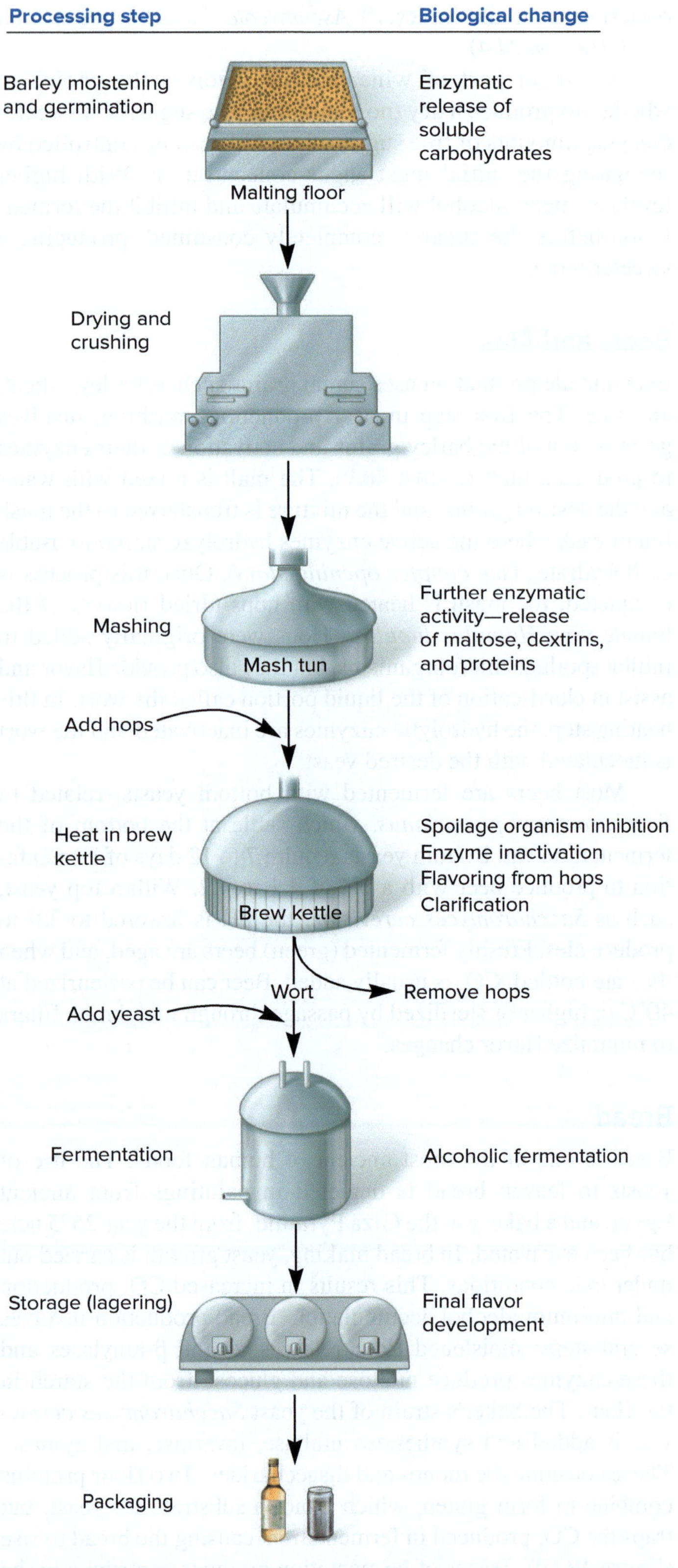

Figure 40.9 Producing Beer. To make beer, the complex carbohydrates in the grain must be converted into a fermentable substrate. Beer production requires mashing, the use of hops, boiling for clarification, flavor development, and inactivation of malting enzymes to produce the wort.

Figure 40.10 Bread Rising. Homemade sourdough increases in volume, and CO_2 production is evident from the bubbles in the dough.
Zagorulko Inka/Shutterstock

Other Fermented Foods

Many other plant products can be fermented, as summarized in **table 40.3.** These include mold-fermented foods such as tempeh and sufu, which is produced by the fermentation of tofu (a chemically coagulated soybean milk product).

Bacteria-fermented foods include sauerkraut and pickles. Sauerkraut or sour cabbage is produced from shredded cabbage. A concentration of 2.2 to 2.8% sodium chloride restricts the growth of many bacteria while favoring the development of lactic acid bacteria. The primary microorganisms contributing to this product are *Leuconostoc mesenteroides* and *Lactiplantibacillus plantarum*. A predictable microbial succession occurs in the development of sauerkraut. The activities of the lactic acid–producing cocci usually cease when the acid content reaches 0.7 to 1%. At this point, *L. plantarum* and *Levilactobacillus brevis* continue to function. The final acidity is generally about pH 1.7, with lactic acid comprising 1 to 1.3% of the total acid in a satisfactory product.

Pickles are produced by placing cucumbers and flavorings such as dill seeds in casks filled with a brine. The sodium chloride concentration begins at 5% and more is added to reach about 16% in 6 to 9 weeks. The salt not only inhibits the growth of undesirable bacteria but also extracts water and water-soluble constituents from the cucumbers. These soluble carbohydrates are fermented to lactic acid. The fermentation, which can require 10 to 12 days, involves the growth of the monoderm bacteria *L. mesenteroides, Enterococcus faecalis, Pediococcus acidilactici, L. brevis,* and *L. plantarum.*

Iconotec/Glow Images

Table 40.3 Fermented Foods Produced from Fruits, Vegetables, Beans, and Related Substrates

Foods	Raw Ingredients	Fermenting Microorganisms	Location
Bacteria-Fermented Foods			
Coffee	Coffee beans	*Enterobacter cloacae, Saccharomyces* spp.	Brazil, Congo, Hawaii, India
Gari	Cassava	*Corynebacterium manihot, Geotrichum* spp.	West Africa
Kimchi	Cabbage and other vegetables	Lactic acid bacteria	Korea
Ogi	Corn	Lactic acid bacteria, *Zygosaccharomyces rouxii*	Nigeria
Olives	Green olives	*Leuconostoc mesenteroides, Lactiplantibacillus plantarum*	Worldwide
Pickles	Cucumbers	Lactic acid bacteria	Worldwide
Poi	Taro roots	Lactic acid bacteria	Hawaii
Sauerkraut	Cabbage	*L. mesenteroides, L. plantarum, Levilactobacillus brevis*	Worldwide
Mold-Fermented Foods			
Kenkey	Corn	*Aspergillus* spp., *Penicillium* spp., lactobacilli, yeasts	Ghana, Nigeria
Miso	Soybeans	*Aspergillus oryzae, Zygosaccharomyces rouxii*	Japan
Soy sauce	Soybeans	*A. oryzae* or *A. soyae, Z. rouxii, Lactobacillus delbrueckii*	Japan
Sufu	Soybeans	*Actinimucor elegans, Mucor* spp.	China
Tempeh	Soybeans	*Rhizopus oligosporus, R. oryzae*	Indonesia, New Guinea, Suriname

Source: Jay, James M.; Loessner, Martin J.; and Golden, David A. 2000. Modern Food Microbiology, *6th ed. New York: Springer.*

L. plantarum plays the dominant role in this fermentation process. Sometimes, to achieve more uniform pickle quality, natural microorganisms are first destroyed and the cucumbers are fermented using pure cultures of *P. acidilactici* and *L. plantarum.*

Finally, fermentation is important in agriculture. When grass, chopped corn, and other fresh animal feeds are stored under moist anoxic conditions, they undergo a mixed acid fermentation that produces pleasant-smelling **silage.** Silage is used to feed livestock during the winter. Trenches or more traditional vertical steel or concrete silos are used to store silage.

Comprehension Check

1. Describe the major types of milk fermentations.
2. What steps are used to produce cheese? How is the cheese curd formed in this process?
3. Which fungal genus is often used in cheese making?
4. Describe and contrast the processes of wine and beer production.
5. Describe the fermentation of one food described here that is unfamiliar to you.

Key Concepts

40.1 Microbial Growth Can Cause Food Spoilage

- Most foods, especially when raw, provide an excellent environment for microbial growth. This growth can lead to spoilage or preservation, depending on the microorganisms present and environmental conditions.
- The course of microbial development in a food is influenced by the intrinsic characteristics of the food itself—its composition (protein, lipid, carbohydrate), pH, salt content, substrates present, and water availability (**table 40.1**).

40.2 Environmental Factors Control Food Spoilage

- Foods can be preserved in a variety of physical and chemical ways, including alteration of temperature (cooling, pasteurization, sterilization), drying, addition of chemical preservatives, radiation, filtration, and fermentation (**table 40.2**).
- Modified atmosphere packaging (MAP) is used to control microbial growth in foods and to extend product shelf life. This process involves decreased oxygen and increased carbon dioxide levels in the space between the food surface and the wrapping material.

40.3 Food-Borne Illnesses

- Foods can be contaminated by pathogens at any point in their production, storage, or preparation. Pathogens that are ingested and grow in the host are said to cause food-borne infections.
- If the pathogen grows in the food before consumption and forms toxins that affect consumers without further microbial growth, the disease is a food intoxication. Fungi that grow in foods, especially cereals and grains, can produce carcinogens aflatoxins (**figure 40.3**) and fumonisins. Seafood is susceptible to contamination from cyanobacterial and dinoflagellate toxins.

40.4 Detection of Food-Borne Pathogens Requires Government-Industry Cooperation

- Processing livestock or produce from farm to market includes many steps where microbial contamination may corrupt a food item. Food processing facilities follow strict protocols to minimize contamination.
- Molecular techniques have largely replaced culture-based approaches to detect food-borne pathogens. Nucleic acid–based techniques, immunological methods, and biosensors are all employed to detect food pathogens. PulseNet is used to respond to food-borne illness outbreaks.

40.5 Microbiology of Fermented Foods: Beer, Cheese, and Much More

- Dairy products are fermented to yield cultured milk products. These include mesophilic, thermophilic, and yeast-lactic products.
- Growth of lactic-acid-forming bacteria, often with the additional use of rennin, coagulates milk solids. These solids are processed to yield a wide variety of cheeses. Both bacteria and fungi are used in cheese production processes.
- Wines are produced from pressed grapes and are either dry or sweet, depending on the level of free sugar that remains at the end of the alcoholic fermentation (**figure 40.8**).
- Beer and ale are produced from cereals and grains. The starches in these substrates are hydrolyzed, in the processes of malting and mashing, to produce a fermentable wort. *Saccharomyces cerevisiae* and *S. pastorianus* are yeasts commonly used in the production of beer and ale (**figure 40.9**).
- Many plant products can be fermented with bacteria, yeasts, and molds. Important products are breads, soy sauce, sufu, and tempeh (**table 40.3**). Sauerkraut and pickles are produced in a fermentation process in which natural populations of lactobacilli play a major role.

Active Learning

1. Compare the "sell-by" date of a package of hot dogs with that of a fresh cut of meat. What are the intrinsic factors that contribute to the difference? What role does packaging play?
2. Keep a record of what you eat for a day or two. Determine if any of the food and beverages you ate were produced with the aid of microorganisms. Indicate at what levels microorganisms were deliberately used. Be sure to consider ingredients such as citric acid, which is produced industrially by several species of fungi. (*See Organic acids, section 41.1*)
3. During cheese production, lactic acid bacteria (LAB) convert lactose to lactate and casein (milk protein) to amino acids. Lactate and amino acids then become the substrates for further microbial growth, which results in aroma production and deacidification of the cheese. The yeast *Yarrowia lipolytica* grows on the surface of many cheeses; it is capable of both lactate and amino acid catabolism. When grown on a lactate plus amino acid medium, *Y. lipolytica* preferentially consumes amino acids. Amino acid degradation results in the release of ammonia, which increases the pH. Draw a flow chart that shows the LAB fermentation of milk, followed by the growth of *Y. lipolytica*. Indicate which substrates are consumed first and what happens to the pH. Based on this simplified scenario, why do you think most cheeses involve the activity of more than one yeast species?
4. Review the discussion of in-use testing in section 8.4. Imagine that you have just been hired as the staff microbiologist at a processing plant that cans tomato puree. How could you verify that each batch is free of *Salmonella?* How could you verify that your facility is free of *Listeria?*
5. Consumer demand for ready-to-eat foods and natural foods has brought preassembled salads to the grocery store. How would you assess the potential for spoilage for each ingredient in the product illustrated in this figure?

artem evdokimov/Shutterstock

6. "Hurdle technology" for food preservation uses combinations of extrinsic factors to force potential contaminants to mount successive hurdles. How would a combination of high-pressure treatment, exposure to UV light, and modified atmospheric packaging affect the microbes that may be present in seafood?

CHAPTER

41

Biotechnology and Industrial Microbiology

Rocketclips/123RF

Where Are the New Antibiotics?

Keisha's dad takes a statin drug to lower his cholesterol, and her grandmother takes a beta-blocker for her heart. Her mom has a steroid inhaler to keep her asthma under control. When Keisha comes down with strep throat, her physician prescribes a week-long course of amoxicillin and she gets better.

Keisha's relatives are benefiting from drugs pharmaceutical companies love to develop. These companies aren't so wild about helping Keisha, though. It's not that they don't want to; they believe they can't afford it. Consider this: It costs about 1.3 billion U.S. dollars and at least a decade to bring a drug from discovery to market. Then the patent lasts for about another 10 years. Once the patent expires, the cost of the drug (and the profit therein) drops precipitously. Whereas Keisha's relatives will very likely take their medicines for the rest of their lives, Keisha takes her antibiotic for a week. When she's done: no more strep throat, no more profit for the drug manufacturer.

But in the meantime, amoxicillin (like all antibiotics) is losing its effectiveness as more and more bacteria develop resistance. So when new antibiotics are developed, they cannot be widely prescribed, lest resistance quickly develop. An antibiotic will never be a blockbuster drug.

Despite these disincentives, the need for new antibiotics grows more urgent with each passing year. Recall that two-thirds of antibiotics are natural microbial products. However, because less than 5% of all bacteria and archaea—and even fewer fungi—have been cultured, the environment holds great untapped resources. But history tells us we need new strategies to access this potential antibiotic reservoir. Between 1935 and 1962, seven classes of antibiotics[1] still in use today were developed. But from 1962 to 2000, no new drug classes were discovered. Instead, while astronauts went to the moon and the internet was born, existing antibiotics were chemically modified, generating multiple generations of existing drugs. It was not until the twenty-first century that three new classes were introduced to clinical practice: oxazolidinones, lipopeptides, and mutilins. Currently, antibiotic discovery is becoming synonymous with microbial discovery: looking for novel genes in new habitats, such as deep-sea trenches and the symbionts of other animals, or inventing new ways to coax microbes from common environments (like soil) to grow in the lab.

Industrial microbiology harnesses the capacity of microbes to synthesize compounds that have important applications in medicine, agriculture, food preparation, and other industrial processes. These compounds are commonly called **natural products.** In this chapter, we discuss these products and how microorganisms are optimized for industrial manufacturing.

Readiness Check:

Based on what you have learned previously, you should be able to:

- ✓ Describe endospore formation (section 3.8)
- ✓ Describe the difference between batch and continuous culture techniques (sections 7.7 and 7.9)
- ✓ Compare and contrast at least five different classes of antibiotics (section 9.4)
- ✓ Explain transcriptional regulation by riboswitches (section 14.3)
- ✓ Explain the mechanisms of horizontal gene transfer (sections 16.3–16.5)
- ✓ Discuss the principles by which vaccines confer immunity (section 35.6)

[1] In chronological order of discovery, these seven drug classes are sulfa drugs, β-lactams, chloramphenicol/tetracyclines, aminoglycosides, macrolides, glycopeptides, and quinolones/streptogramins.

41.1 Microbes Are the Source of Many Products of Industrial Importance

After reading this section, you should be able to:

a. Describe at least five compounds of industrial importance made by microbes
b. Define biocatalysis and explain its advantages over chemical synthesis
c. Describe molecular design strategies for vaccines

Most people are not aware that the majority of antibiotics and other medicines, food additives, enzymes, and biofuels are examples of natural products (**table 41.1**). Here we introduce a few of the more common microbial products that have become an integral part of modern life. Microbial products are typically produced and extracted from organisms that have been engineered to optimize their synthesis in industrial settings.

Antibiotics

More than 65% of all antibiotics are produced by members of the genus *Streptomyces* and by filamentous fungi (*see table 22.1*). Antibiotics are secreted by these microbes and either are used clinically in their native state or are chemically modified to produce semisynthetic derivatives. Semisynthetic derivatives were developed to treat a broader spectrum of bacteria or bacteria that had become resistant to the parent drug.

The synthesis of penicillin and its derivatives illustrates how an antibiotic is commercially produced. As with any industrially important natural product, the manufacturer determines how to maximize product yield. In this case, the growth of the fungus *Penicillium chrysogenum* requires careful adjustment of the medium so that the slowly hydrolyzed disaccharide lactose is used in combination with a limited supply of nitrogen to stimulate a greater accumulation of penicillin after growth has stopped (**figure 41.1**). If a particular penicillin is needed, a specific precursor is added to the medium. For example, phenylacetic acid is added to maximize production of penicillin G, which has a benzyl side chain (*see figure 9.5*). After about a week, the broth is separated from the fungal hyphae and processed by absorption, precipitation, and crystallization to yield the final product. Manufacture of semisynthetic β-lactam antibiotics (e.g., ampicillin and amoxicillin) begins with natural penicillin. The β-lactam

Table 41.1 Major Microbial Products and Processes of Industrial Importance

Substances	Microorganisms
Industrial Products	
Ethanol	*Saccharomyces cerevisiae*
Acetone and butanol	*Clostridium acetobutylicum*
Food Additives	
Amino acids (e.g., lysine)	*Corynebacterium glutamicum*
Organic acids (citric acid)	*Aspergillus niger*
Vitamins B_1, B_2, B_3, B_5, B_6, B_7, B_9, B_{12}, and C	Engineered *Bacillus subtilis, Escherichia coli, Aspergillus* spp.
Polysaccharides	*Xanthomonas*
Medical Products	
Antibiotics	*Penicillium, Streptomyces, Bacillus*
Steroid transformations	Actinomycetota spp.
Insulin, human growth hormone, somatostatin, interferons, vaccines	Engineered *E. coli, S. cerevisiae,* and others
Biofuels	
Methane	Hydrogenotrophic methanogens
Ethanol	*Zymomonas, Thermoanaerobacter*

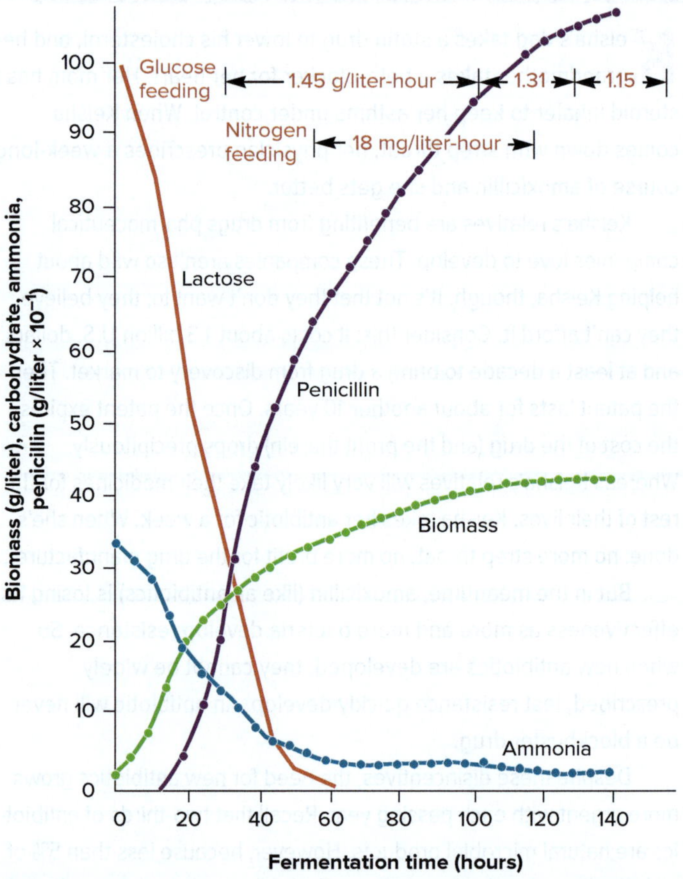

Figure 41.1 Penicillin Fermentation Involves Precise Control of Nutrients. Penicillin synthesis begins when nitrogen from ammonia becomes limiting. After most of the lactose has been degraded, glucose is added along with a low level of nitrogen. This stimulates maximum transformation of the carbon sources to penicillin.

MICRO INQUIRY *Why are the cells first fed lactose and then glucose?*

ring is preserved, but the side chain is modified by chemical procedures. ◀ *Penicillins (section 9.4)*

Amino Acids

Amino acids such as lysine and glutamic acid are used in the food industry as nutritional supplements and as flavor-enhancing compounds such as monosodium glutamate (MSG). Production of glutamate and several other amino acids uses mutants of *Corynebacterium glutamicum* that lack or have only a limited ability to convert the TCA cycle intermediate α-ketoglutarate to succinyl-CoA (*see figure 11.8*). The impaired bacteria use the glyoxylate cycle (*see figure 12.21*) to meet their needs for essential biochemical intermediates, especially during exponential phase. By the time nutrients are depleted and growth stops, an almost complete molar conversion of isocitrate to glutamate occurs. By keeping the biotin level low and adding fatty acid derivatives, membrane permeability increases, resulting in high concentrations of glutamate released to the medium. ◀ *Synthesis of amino acids consumes many precursor metabolites (section 12.5)*

Organic Acids

Simply reading the ingredient list on most processed foods illustrates the widespread use of organic acids such as citric, acetic, and lactic acids. They are principally used as preservatives.

Organic acid production exemplifies how manipulating metabolism influences product yield. One example is citric acid (citrate), commonly added to food as a preservative or to impart a tart flavor. Its production involves limiting trace metals such as manganese and iron to stop the fungus *Aspergillus niger* growth at a specific point. Generally, high sugar concentrations (15–18%) are used. This reflects the regulation of glycolysis and the TCA cycle (recall that citric acid is an intermediate of the TCA cycle). If the sugar levels are high after exponential phase, citrate synthase activity increases and the activities of aconitase and isocitrate dehydrogenase decrease. This results in citric acid accumulation and excretion by the stressed microorganism.

Enzymes

Industrial enzymes are used in a variety of processes in the pharmaceutical, agricultural, food, and consumer products industries, with a global market of about $13 billion annually. The vast majority of the enzymes are bacterial and fungal in origin. In many industrial biocatalysis processes, enzyme-catalyzed reactions substitute for chemical treatments, offering benefits like reduced cost, reduced toxic by-products, and improved quality and safety. Often, these enzymes have been genetically modified to operate in different pH and temperature conditions. ◀ *Directed evolution (section 17.6)*

The food industry uses biocatalysis treatments of raw materials to provide consistent quality, and most processed foods have been treated with at least one enzyme. For example, pectinases are superior to mechanical methods to enhance juice recovery from fruits. Cheese production has traditionally used rennin, extracted from animal stomachs, to coagulate milk. Chymosin, the active enzyme in rennin, is commercially sold as a recombinant protein produced in bacteria. Microbial enzymes also remove ingredients that trigger allergic responses. For instance, lactase removes lactose from dairy products. ◀ *Cheese (section 40.5)*

Two consumer products have been reformulated in recent years by including enzymes like proteases and lipases. Both laundry and dishwasher detergents rely on enzymes to remove stains and soil from clothing and dishes. Blood and grease stains on cloth can be removed without the use of harsh chemicals such as bleach. In addition, phosphates that contribute to eutrophication have been eliminated from these products. ◀ *Microorganisms in lakes can cause seasonal blooms (section 29.3)*

Lipids

Microalgae, single-celled autotrophs, can produce 40 to 70% of their biomass as lipids. Carotenoids (*see figure 11.29*), polyunsaturated fatty acids, and triacylglycerols (*see figure 12.26*) are the primary products obtained from microalgae; they are used in the food and pharmaceutical industries. Microalgae grow well in large-scale fermentation vessels, and strategies to induce lipid overproduction include increasing light intensity, CO_2, and temperature.

Archaeal lipids are distinct from bacterial and eukaryotic lipids and have been used to synthesize novel liposomes termed *archaeosomes*. Archaeosomes exhibit increased stability to temperature, pH, and enzymatic degradation, a reflection of the environmental conditions that their archaeal source organisms inhabit. ◀ *Archaea plasma membranes are composed of unique lipids (section 4.2)*

Vaccines

Vaccines are the single most important public health weapon. New technologies are improving vaccines by using genomic and molecular approaches to identify the best antigens to elicit an immune response. **Reverse vaccinology** mines the genomic sequence of the pathogen to assemble a list of antigens that may be good vaccine targets. Recall that an antigen is a foreign molecule targeted by the host immune system. Rather than whole organisms or viruses, purified antigens or the DNA or RNA that encodes them can be used in vaccines. Such antigens must (1) be expressed by the pathogen during infection; (2) be secreted or found on the surface of the pathogen; (3) be found in all strains of the pathogen; (4) elicit a host immune response; and (5) be essential for the survival of the pathogen, at least while it is in the host.

Reverse vaccinology considers the thousands of genes revealed in each annotated genome to find those whose products meet all of these criteria. For example, *Neisseria meningitidis* serovar B (MenB), has a capsule whose polysaccharides are similar to those on the surface of many human cells, rendering it ineffective for vaccine development. In a reverse vaccinology approach, 25 surface proteins were analyzed, which led to the discovery of several antigens that in combination induce a robust immune response. In 2014 the first MenB vaccine was approved for use in the United States.

As nucleic acids that encode products that trigger an immune response have been developed, researchers are exploiting viral mechanisms to improve the efficacy of these mRNA vaccines. An mRNA vaccine encodes an antigen and upon entry to the cytoplasm is translated to synthesize the antigen of interest (AOI), which provokes an immune response (**figure 41.2*a*;** *see figure 35.10*). Replicon mRNA delivery aims to enhance the AOI immunogenicity. A **replicon** is a 9 to 12 kbp mRNA molecule based on positive-strand RNA viruses that encodes both the AOI and viral nonstructural proteins (NSPs). The NSPs include the replicase protein that synthesizes a negative-strand RNA that in turn is used as the template to synthesize large quantities of the positive-strand RNA (figure 41.2*b*). This amplification of the original replicon dramatically increases the number of mRNAs for translation. Because viral structural proteins are absent from the replicon, a functional viral particle does not form, and thus it cannot spread to other cells or cause disease. ◀ *Positive-strand RNA viruses: genomes that are translated upon entry (section 25.5)*

Comprehension Check

1. What critical limiting factors are used in the production of penicillin? Why do you think the microbe must be stressed to produce high levels of drug?
2. Why are regulatory mutants used to increase the production of glutamic acid by *Corynebacterium glutamicum*?
3. What kinds of molecular changes result from biocatalysis?
4. How do reverse vaccinology and replicon vaccine design differ from conventional vaccine development?

41.2 Microbial Cells Are Customized for Specific Tasks

After reading this section, you should be able to:

a. Distinguish a cell factory from a natural isolate
b. Define a microbial chassis and describe the features desired in such a construct
c. Explain the utility of biosensors

The ease with which microbial cells and genomes can be manipulated in the laboratory has opened many applications. Here we highlight microbial cells as factories and as biosensors.

Cell Factories

Evolution selects microbial cells adapted to their biological niche, but their use in efficient bioproduction typically requires adjustments. **Cell factories** are strains of bacteria developed in the laboratory that synthesize specific products (metabolite or protein) of industrial importance. In an analogy to automobile manufacturing, the cells are built on a **chassis,** the basic support structure onto which the components customized for the final product are added. A microbial chassis is the hardware: the cell envelope, ribosomes, and enzymes of central metabolism. The software that operates the cell factory includes the genetic circuits that regulate and synthesize the desired products. ◀ *Biotechnology develops custom microbes for industrial use (section 17.6)*

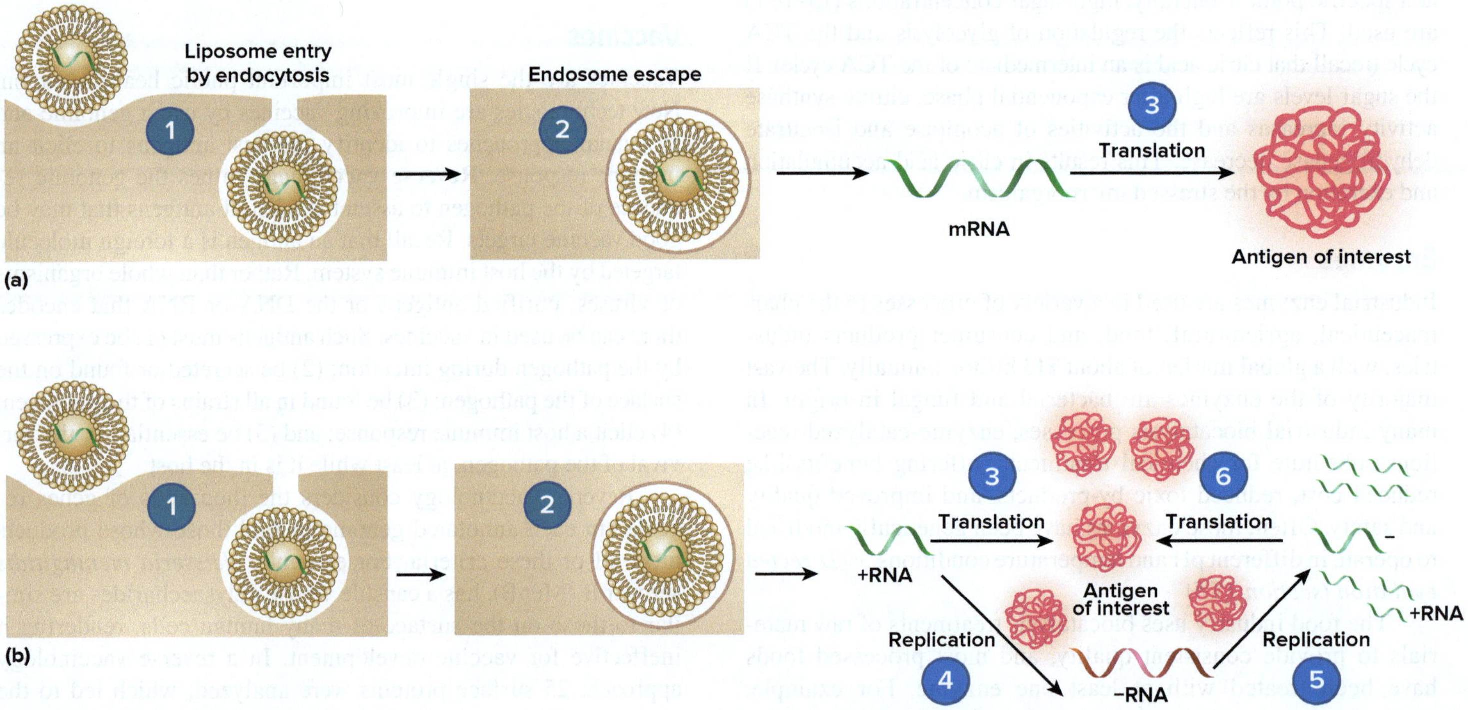

Figure 41.2 Replicon Vaccine. (a) An mRNA vaccine enters a cell and undergoes translation to generate an antigen. (b) A replicon RNA enters a cell and, in addition to translation, undergoes replication to generate more template for translation, resulting in increased antigen production.

Table 41.2 Selected Chassis Organisms for Industrial Production of Microbial Compounds

Microorganism	Advantages for Industrial Production
Escherichia coli	Best-studied bacterium with extensive genetic and genomic tools and a history of producing recombinant proteins
Bacillus subtilis	Free of endotoxin (LPS), generally regarded as safe (GRAS), efficient at metabolite and enzyme secretion
Pseudomonas putida	Tolerant of many industrial solvents, grows on a wide range of carbon sources, GRAS
Saccharomyces cerevisiae	Capable of performing many eukaryotic posttranslational protein modifications

The microorganisms listed in **table 41.2** are common chassis organisms. Each has features that make it an optimal starting point for particular applications, and these organisms have been in industrial use for decades. For example, endotoxin, the lipopolysaccharide component of the diderm outer membrane, can provoke immune responses if even small amounts are present in products intended for human or veterinary use. Monoderm bacteria like *Bacillus subtilis* are a more appropriate chassis than *E. coli* a diderm, for synthesizing many such reagents. *Endotoxin (section 34.4)*

Microorganisms with novel capabilities are in development as chassis organisms; these must be efficient in bioproduction but tolerant of the conditions imposed by industrial processes. A chassis organism should grow quickly and efficiently in simple media, be moderately resistant to physical and chemical stresses, have tools available for genetic and genome manipulation, and have efficient secretion systems for purification of the desired product. Notably, the yeast *Saccharomyces cerevisiae* has been completely rebuilt from scratch to enhance its value in bioproduction.

Biosensors

A recurring theme in this book is the ability of microbes to detect changes in their environment and respond through changes in gene expression. What if we could harness the microbial power of detection? They could alert us to changes in our environment (the presence of toxins in drinking water) and changes in our bodies (disease processes). By combining microbial detection with an output that humans can interpret, microbes have been engineered as **biosensors** to alert us to conditions that we cannot see directly. The examples of gene regulation discussed in chapter 14 illustrate how microbes convert sensory information to responses. By exploring the molecular details of how these detectors work, scientists have repurposed and engineered microbes as sensitive biosensors (*see chapter 14 opening story*).

Riboswitches are regulatory elements found at the 5′ end of RNA transcripts. The RNA folding pattern depends on the presence or absence of its effector molecule. In one configuration, transcription is halted or the ribosome binding site is hidden, thereby preventing translation. In the alternate configuration, gene expression results. Because the riboswitch folds as it is synthesized by RNA polymerase, it can function as a detector regardless of the downstream gene. Riboswitches have been documented as detectors of a wide range of molecules, including amino acids, ions, vitamins, and nucleotides. **Figure 41.3** illustrates how a fluoride-detecting riboswitch has been fused to green fluorescent protein to monitor fluoride levels in water supplies. *Riboswitches: effector-mRNA interaction regulates transcription (section 14.3)*

A simple repressor/inducer model system has been adapted to act as a biosensor for heme, as an indicator of blood. An operator sequence and its repressor, originally from *Lactococcus lactis,* control gene expression by the presence or absence of heme (**figure 41.4**). Heme binding to the repressor releases

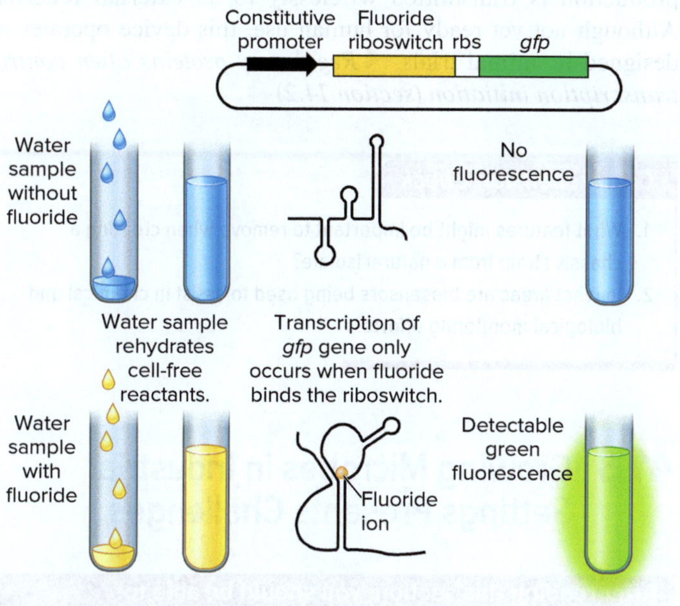

Figure 41.3 Biosensor to Detect Fluoride in Water. A cell-free biosensor is hydrated by the addition of the water sample to be tested. Transcription of the constitutive promoter produces mRNA with a riboswitch. The presence of fluoride permits transcription of the *gfp* gene, whose product, green fluorescent protein, is easily detectable. Rbs, ribosome binding site.

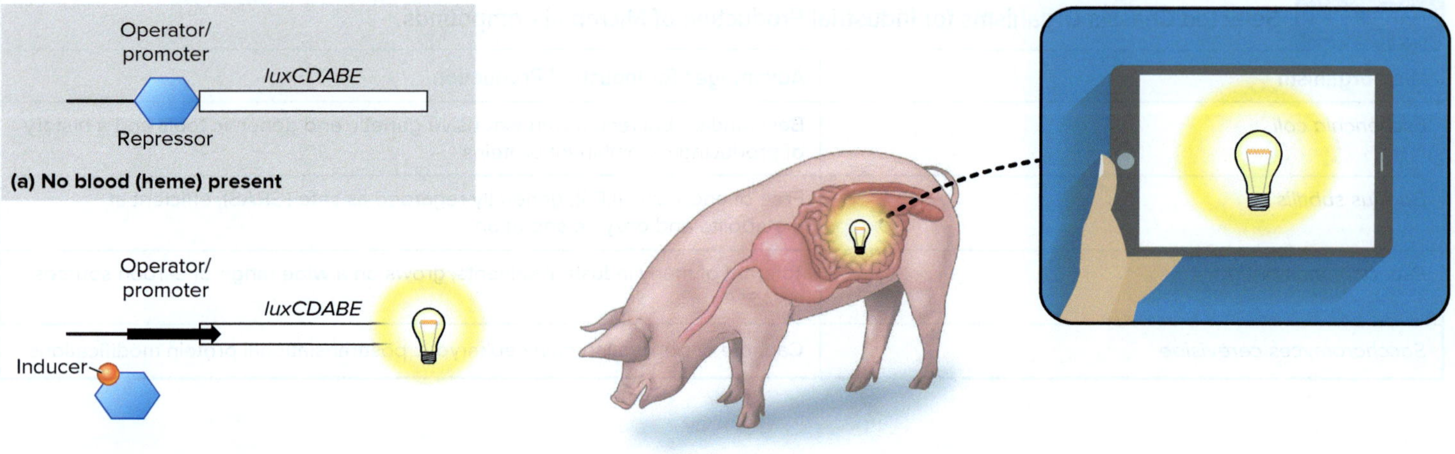

Figure 41.4 Biosensor to Detect Blood in Situ. (a) In the absence of heme, the repressor binds the operator, and the *lux* operon is not expressed. (b) Heme induces transcription by binding the repressor. The operon encodes the enzyme luciferase, which produces light. (c) In animal trials, the biosensor is ingested and can detect intestinal blood. Its accompanying microelectronics transmit a signal to a nearby detector.

it from the operator/promoter region of a structural gene. This genetic control switch has been fused to the *lux* operon that encodes luciferase, an enzyme that produces light, and the recombinant operon is propagated in *E. coli*. The resulting biosensor lights up in the presence of blood (heme). By combining this biosensor with miniaturized electronics, it can be swallowed and deployed to the gastrointestinal tract. Upon detecting blood, as might be present if there is cancer or an ulcer, light production is transmitted wirelessly to an external receiver. Although not yet ready for human use, this device operates as designed in animal trials. ◀ *Regulatory proteins often control transcription initiation (section 14.2)*

Comprehension Check

1. What features might be important to remove when creating a chassis strain from a natural isolate?
2. In what areas are biosensors being used to assist in chemical and biological monitoring efforts?

41.3 Growing Microbes in Industrial Settings Presents Challenges

After reading this section, you should be able to:

a. Explain the use of the term *fermenter* as it is used in an industrial setting
b. Describe two types of fermenters

The development of appropriate culture media and conditions is critical to the manufacture of microbial products. Culturing microbes in industrial settings is much different from their growth in either nature or the laboratory. The term **fermentation** is applied to the mass culture of microorganisms, plant, or animal cells. Industrial fermentations require the development of appropriate culture media and the process of **scale-up.** This is when a procedure developed in a small flask is modified for use in a large fermenter. The microenvironment of the small culture must be maintained despite increases in culture volume. Although the process of transitioning a culture originally developed in a 250-milliliter Erlenmeyer flask to a 100,000-liter reactor seems like it should be relatively straightforward, it is often difficult to obtain the same yield of the desired product.

Microorganisms are often grown in stirred **fermenters** that can range in size from 3 liters to over 100,000 liters, depending on production requirements. A typical industrial stirred fermentation unit is illustrated in **figure 41.5;** such units are best for culturing planktonic cells. Upon inoculation of microbes into sterile growth medium, aeration, pH and temperature adjustments, and sampling are carried out under rigorously controlled conditions. When required, agents to control foaming are added, especially with high-protein media. Probes monitor biomass, levels of critical metabolic products, pH, input and exhaust gas composition, and other parameters. Environmental conditions can be changed or held constant over time, depending on the requirements of the particular process.

Frequently a critical component in the medium is added continuously—a process called **continuous feed**—so that the microorganisms will not have excess substrate available at any given time. This is particularly important with glucose and

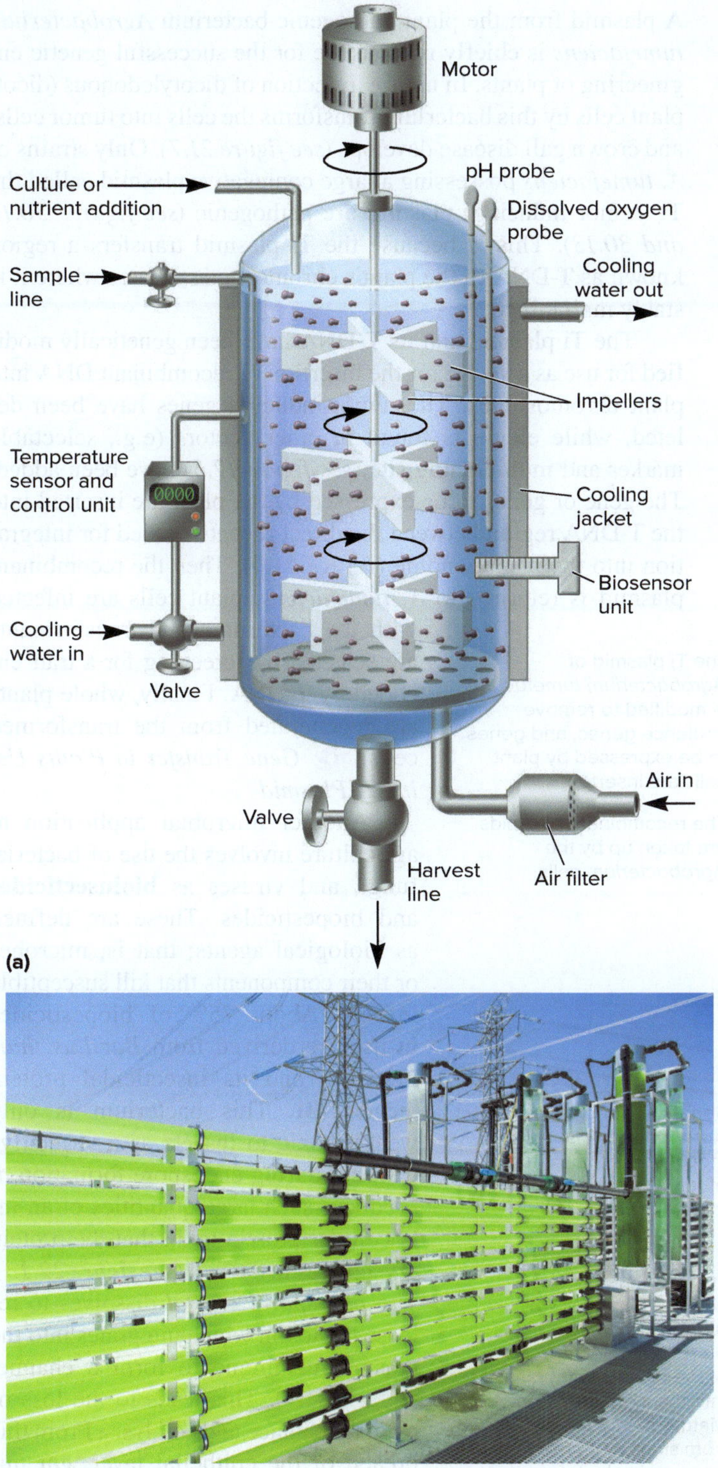

Figure 41.5 Industrial Fermentation. (a) This stirred reactor can be run under oxic or anoxic conditions, with nutrient additions, sampling, and fermentation monitoring carried out under aseptic conditions. Biosensors and infrared monitoring provide real-time information on the course of the fermentation. (b) Tubular bioreactors increase the surface available for light penetration when culturing phototrophs as biofilms.

(b) Santiago Urquijo/Getty Images

other carbohydrates. If excess glucose is present at the beginning of a fermentation, it can be catabolized and lost as volatile end products, reducing the final yield of the desired product.

Microbial products can be classified as primary and secondary metabolites. **Primary metabolites** consist of compounds related to the synthesis of microbial cells during exponential growth. The use of chemostats can sometimes markedly improve rates of substrate use and product yield because microorganisms can be maintained in exponential phase. Example primary metabolites include amino acids, nucleotides, and products such as ethanol and organic acids. In addition, industrially useful enzymes, associated with either the microbial cells or exoenzymes, often are synthesized by microorganisms during exponential growth. **Secondary metabolites** are usually synthesized when nutrients become limited or waste products accumulate following the active growth phase, and thus are not produced in chemostats. These compounds are sometimes considered part of a stress response as nutrients become scarce. Most antibiotics fall into this category. ◀ *Growth curves consist of five phases (section 7.4); Chemostats and turbidostats are used for continuous culture of microorganisms (section 7.9)*

Some microbial processes occur when the producing microorganisms are in the **biofilm** state. Adapting such processes for industrial-scale cultivation requires increasing the surface area to volume ratio in the cultivation vessel. Recall that biofilms develop on a **substratum** for anchoring (*see figure 7.23*). Liquid nutrient medium is passed through biofilm-coated tubes (figure 41.5*b*) or trickled over microbe-covered substrata (*see figures 42.5 and 42.6*). In fact, most wastewater treatment processes are performed by biofilms. ◀ *Wastewater treatment maintains human and environmental health (section 42.2)*

One advantage to biofilm cultivation is that once a mature biofilm develops, the producing organisms (either in axenic culture or as a community) may persist for extended periods while continuously metabolizing. Soluble products can be purified from the effluent. The flow rate of the nutrient medium through a stationary biofilm reactor is optimized for maintenance of the biofilm and production of the desired metabolite.

Comprehension Check

1. What is the objective of the scale-up process and why is it so critical?
2. What parameters can be monitored in a large-scale industrial fermentation?
3. Which of the following are secondary metabolites: pyruvate, a secreted amylase, an antibiotic, leucine?
4. What kind of system (fermenter vs. chemostat) would most likely be used for the production of the streptomycete antibiotic streptomycin? Explain your answer.

41.4 Agricultural Biotechnology Repurposes Plant and Insect Pathogens

After reading this section, you should be able to:

a. Explain how the natural infection process used by *Agrobacterium tumefaciens* has been adapted for the genetic modification of plants
b. Describe the Bt toxin mechanism of action
c. Defend the use of Bt as a pesticide

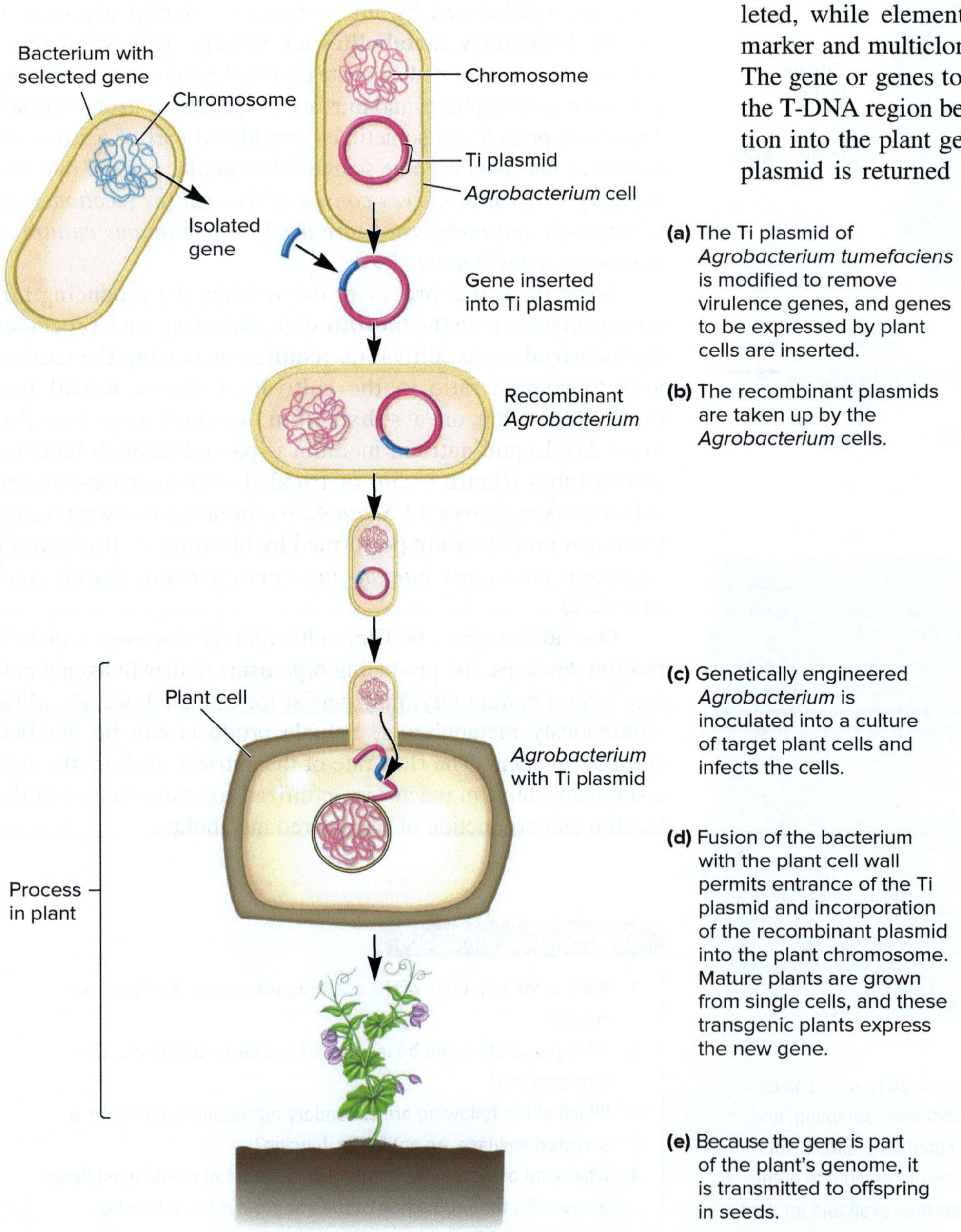

Figure 41.6 Bioengineering of Plants. Most techniques employ a genetically modified strain of the natural tumor-producing bacterium *Agrobacterium tumefaciens.*

A plasmid from the plant pathogenic bacterium *Agrobacterium tumefaciens* is chiefly responsible for the successful genetic engineering of plants. In nature, infection of dicotyledonous (dicot) plant cells by this bacterium transforms the cells into tumor cells, and crown gall disease develops (*see figure 21.7*). Only strains of *A. tumefaciens* possessing a large conjugative plasmid called the Ti (*t*umor *i*nducing) plasmid are pathogenic (*see figures 30.12 and 30.13*). This is because the Ti plasmid transfers a region known as T-DNA to the plant's chromosomes at sites where it is stably maintained.

The Ti plasmid and its T-DNA have been genetically modified for use as a vector for the insertion of recombinant DNA into plant chromosomes. The tumor-inducing genes have been deleted, while elements found in other vectors (e.g., selectable marker and multicloning site; *see figure 17.5*) have been added. The gene or genes to be expressed by the plant are inserted into the T-DNA region between the direct repeats needed for integration into the plant genome (**figure 41.6**). Then the recombinant plasmid is returned to *A. tumefaciens,* plant cells are infected with the bacterium, and transformants are selected by screening for a trait encoded by T-DNA. Finally, whole plants are regenerated from the transformed cells. *Gene Transfer in Plants Using Ti Plasmid*

Another microbial application to agriculture involves the use of bacteria, fungi, and viruses as **bioinsecticides** and biopesticides. These are defined as biological agents; that is, microbes or their components that kill susceptible insects. About 75% of biopesticides in use are derived from *Bacillus thuringiensis* and its insecticidal protein termed **Bt.** This bacterium is only weakly toxic to insects as a vegetative cell, but during endospore formation, it produces such large quantities of an intracellular protein toxin that the protein crystallizes in the cell (*see figure 22.13*). The *parasporal crystal* dissolves to release the toxin, which integrates into the plasma membrane to form a channel (**figure 41.7**). This leads to the loss of osmotic balance and cell lysis. From this breach of the epithelial layer, gut microbes proliferate in the insect's tissues, leading to death. *Order Bacillales (section 22.2)*

The manufacture of Bt begins by growing *B. thuringiensis* in fermenters through vegetative growth and sporulation. The endospores are released into the medium when the cells lyse to release the endospore and the crystals

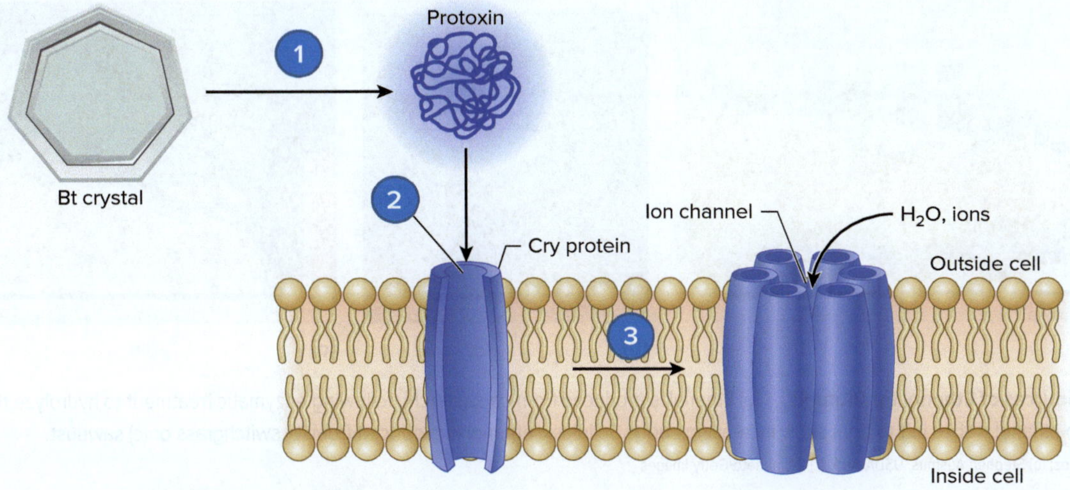

Figure 41.7 The Mode of Action of the *Bacillus thuringiensis* Cry Toxin in the Target Insect. (1) The parasporal crystal dissolves in the insect gut to release the protein as a protoxin. (2) Protoxin monomers are proteolyzed, bind to cell receptors on midgut epithelial cells, and enter the membrane. (3) Further proteolysis triggers toxin oligomerization to form a membrane channel.

(*see figure 3.39*), then dried to a powder for application to plants. Bt has been used on a worldwide basis for over 50 years. Unlike chemical insecticides, Bt does not accumulate in the soil or in nontarget animals. Rather, it is readily degraded in the environment.

The gene encoding the Bt toxin (*cry* genes, for *cry*stal) can be introduced into plants (figure 41.6). Unlike some genetically modified organisms, transgenic plants expressing the *cry* gene have generally been well accepted, reflecting the history of safe application of Bt as an insecticide without adverse environmental or health impacts. Long-term studies have shown that Bt is nontoxic to mammals and is not an allergen in humans.

Comprehension Check

1. What is the Ti plasmid and how is it modified for the genetic modification of plants?
2. How is Bt toxin produced and why is it so widely accepted?
3. Trace the steps needed to construct a transgenic plant that produces the Cry protein.

41.5 Biofuel Production Is a Dynamic Field

After reading this section, you should be able to:

a. Compare the value of biofuel use with the challenges presented in its manufacture and transport
b. Discuss at least two types of biofuels

As population and economic growth increases energy demands to new levels, most nations have pledged to take steps to limit atmospheric CO_2 (*see figure 28.10*). A key step in this effort is to develop economically competitive alternatives to fossil fuel. **Microbial energy conversion** includes the microbial transformation of organic materials into **biofuels,** such as bioethanol and hydrogen, that can fuel car, truck, and bus engines (table 41.1). In this section, we introduce the microbial production of these biofuels. ◀ *Global climate change: biogeochemical cycling out of balance (section 28.3)*

Ethanol is currently used as an additive to gasoline because it is easy to store and can be burned in cars designed to use pure gasoline. Most common is E10, a mixture of 10% ethanol/90% gasoline. Corn is the substrate most commonly fermented to ethanol in the United States (**figure 41.8*a***), while sugar cane is used in other countries. Ethanol production involves the degradation of plant starch using microbial amylases and amyloglucosidases. The resulting sugars are then fermented to ethanol by microbes. However, the use of corn and other foods to make bioethanol has unintended consequences, including increased pesticide and fertilizer use and inflating worldwide food prices. As an alternative to plant starch from corn, cellulose from crop residue is being developed for biofuel production. Crop residues are plant materials left in the field after harvest, as well as sawdust and plants like switchgrass (figure 41.8*b,c*) that grow on land not fertile enough for crops. This material consists of cellulose and hemicellulose—polymers of five different hexoses and pentoses: glucose, xylose, mannose, galactose, and arabinose. Degradation of cellulose and hemicellulose to release these monomers is commonly done by heating the plant material and treating it with acid, which is energy-intensive and corrosive. Ruminants, however, perform cellulolytic degradation efficiently (*see figure 27.5*). This has sparked interest in replacing chemical pretreatment with microbial degradation of these

(a) (b) (c)

Figure 41.8 Plant Sources of Ethanol for Biofuels. (a) Ethanol is produced from corn (starch), following enzymatic treatment to hydrolyze the starch to sugars, which are fermented to ethanol. Tubes left to right show increasing purity. Cellulosic ethanol is produced from (b) switchgrass or (c) sawdust.

(a) ThamKC/iStock/Getty Images; (b) Stephen Ausmus, USDA-ARS; (c) Reimphoto/Getty Images

polymers. Newly isolated microbes and genetically engineered organisms that can degrade cellulose at temperatures in excess of 100°C have been developed.

In many ways, hydrogen (H_2) as a biofuel compares favorably to other fuels. With about three times more potential energy per unit weight than gasoline, H_2 is the highest energy-content fuel available. A diverse group of microbes releases H_2, making its commercial production quite flexible. Like ethanol, H_2 is a product of fermentation. It is also produced by oxygenic photosynthetic microbes (cyanobacteria and algae) and anoxygenic photoheterotrophs. Two different enzymes (thus processes) generate H_2: hydrogenase and nitrogenase. Both are keenly sensitive to oxygen, so H_2 production is an anaerobic process.

Oxygenic photosynthetic microbes produce H_2 at night, when they oxidize cellular storage products such as poly-β-hydroxybutyrate. The use of aquatic phototrophs to generate H_2 holds promise because there is plenty of substrate (light and CO_2). However, hydrogenase is strongly inhibited by its product, so H_2 must be removed from the system as soon as it is created. By contrast, H_2 production by anoxygenic photoheterotrophic bacteria can occur via cellular hydrogenases or nitrogenases. This is of interest to industrial microbiologists because in an atmosphere artificially purged of all N_2, nitrogenase will form only H_2. This process requires a tremendous amount of ATP and reductant, but unlike hydrogenase, nitrogenase is not inhibited by the accumulation of H_2, and electron conversion to H_2 reaches efficiencies near 100%.

Although the future of specific biofuels remains to be determined, several things are clear. The growing desire for energy will eventually deplete fossil fuels. At the same time, increased food production will make it untenable to grow feedstocks for biofuel production on arable land. It may be that instead of drilling for oil, we will be harvesting it from bioreactors.

Comprehension Check

1. Why is there interest in developing alternatives to ethanol as the biofuel of choice in the United States?
2. What types of microbes are good candidates for the industrial production of H_2?

Key Concepts

41.1 Microbes Are the Source of Many Products of Industrial Importance

- A wide variety of compounds are produced in industrial microbiology that impact our lives in many ways. These include antibiotics, amino acids, organic acids, lipids, and enzymes (**table 41.1**).
- Microbial enzymes are used as biocatalysts to carry out specific chemical reactions.
- Vaccine development continues to challenge biotechnologists; reverse vaccinology identifies potential targets and uses their structure and function to inform vaccine production. Replicon vaccines use viral replication to amplify the number of templates for translation of mRNA vaccines (**figure 41.2**).

41.2 Microbial Cells Are Customized for Specific Tasks

- Cell factories are microbial cells developed in the laboratory to be efficient at bioproduction
- Chassis strains use cell hardware to support genetic circuits that regulate and synthesize microbial products
- Biosensors exploit the detection capabilities of microbes to convert environmental conditions to an output that can be easily measured (**figures 41.3** and **41.4**).

41.3 Growing Microbes in Industrial Settings Presents Challenges

- Microorganisms can be grown in controlled environments of various types using fermenters and other culture systems (**figure 41.5**).
- If defined constituents are used, growth parameters can be chosen and varied in the course of growing a microorganism. This approach is used particularly for the production of amino acids, organic acids, and antibiotics.

41.4 Agricultural Biotechnology Repurposes Plant and Insect Pathogens

- Many plants are genetically modified through the use of the Ti plasmid of *Agrobacterium tumefaciens*. The plasmid can be engineered to introduce genes whose products confer new traits to plants (**figure 41.6**).
- *Bacillus thuringiensis* is an important biopesticide, and the Bt *cry* gene has been incorporated into several important crop plants (**figure 41.7**).

41.5 Biofuel Production Is a Dynamic Field

- The field of microbial energy conversion includes the production of biofuels and the development of microbial sources of H_2.
- Although corn is currently the most common substrate used for the microbial production of ethanol, methods to extract carbohydrates from cellulose are in development.

Active Learning

1. People who are affected by kidney stones are advised to consume less oxalate, a common ingredient in kidney stones that is found in many foods like spinach. *Bacillus subtilis* has an enzyme, oxalate decarboxylase, that metabolizes oxalate. Devise a strategy to use this enzyme in a pharmaceutical or probiotic to help decrease the frequency of kidney stone formation.
2. Review the challenges to food preservation in chapter 40. Suggest a biosensor design that could be incorporated into food packaging to assure consumers that the product is safe.
3. Some cell factories are used *in situ,* for example, directly in an agricultural setting. What types of biocontainment features might be effective in preventing such cells from spreading through the environment?

CHAPTER

42

Applied Environmental Microbiology

Chris Sattlberger/Blend Images LLC

Viruses Down the Drain

City dwellers share infrastructure: We move along shared roadways, we inhale city air, and we all send sewage to the same water treatment plant when we flush. In urban areas, sewage is collected and treated at municipal facilities before being discharged to the environment. Can the molecules and microbes in sewage reveal anything about the population that produces it?

One legacy of the COVID-19 pandemic is renewed interest in wastewater surveillance and its application to community public health. Early in the pandemic it was established that people infected with the SARS-CoV-2 coronavirus shed virus in their stool and that the viral RNA could be detected in sewage. When cities began to track SARS-CoV-2 in sewage, it became clear that there was a good correlation between wastewater virus levels and the number of disease cases in the community. Assaying sewage for that virus gave public health officials a sense of how widespread viral infection was in the community, especially when many cases were undiagnosed and widespread testing was unavailable. While the exact number of cases can't be predicted from virus levels, comparisons from the same location over time illustrate trends in the community health status. Rising or falling viral numbers bear out infection levels.

The Centers for Disease Control and Prevention (CDC) maintains a National Wastewater Surveillance System that has monitored public systems around the country for several years. With the success of the program in tracking SARS-CoV-2, the CDC has expanded the surveillance to include other viruses like influenza A and mpox. Because we all contribute to sewage when we flush, everyone participates, making wastewater epidemiology a valid early warning system for public health preparedness.

Readiness Check:

Based on what you have learned previously, you should be able to:

- ✓ List environmental factors that influence microbial growth (section 7.5)
- ✓ Distinguish between mineralization and immobilization of an element (section 28.1)
- ✓ Compare and contrast carbon substrates that are readily degraded with those that are not (section 28.2)
- ✓ Sketch generalized carbon and nitrogen cycles (section 28.2)
- ✓ Describe several waterborne diseases (sections 38.4 and 39.5)

42.1 Purification and Sanitary Analysis Ensure Safe Drinking Water

After reading this section, you should be able to:

a. Trace water purification steps from source to tap
b. Describe how the U.S. Environmental Protection Agency regulates water quality
c. Discuss strategies for water quality monitoring

Access to clean, drinkable water is a fundamental human right, yet 2–4 billion people worldwide do not enjoy a reliable source of potable water (Latin, *potabilis*, fit to drink). Not only must water be purified before drinking, it must be monitored for purity. The U.S. Environmental Protection Agency (EPA) has designated certain bacterial, viral, and protozoan pathogens that can survive (or thrive) in water as "contaminant candidates" because they represent particularly severe health risks (**table 42.1**).

Water purification is therefore a critical link in controlling disease transmission. The choice of purification method depends on the volume, source, and initial quality of the water. Many municipalities draw their water from surface sources, such as reservoirs. Surface water is usually purified by a process that consists of several steps (**figure 42.1**). First, the water may be held for a specified duration to allow larger particles to settle out. When the water is drawn into the purification system, chemical coagulants such as alum (aluminum sulfate) and lime are added in a procedure called **coagulation** or flocculation. The coagulated particles are called *flocs*. The water is then moved to a **sedimentation basin** where the flocs settle by gravity. The water is further purified by passing it through **rapid sand filters.** Here sand grains roughly 1 mm in diameter physically trap fine particles and flocs to remove up to 99% of bacteria. After filtration, the water is disinfected. This step usually involves chlorination or ozonation.

These purification processes remove or inactivate most bacteria. Unfortunately, these steps often do not remove *Giardia* cysts, *Cryptosporidium* oocysts, *Cyclospora* spp., or viruses. *Giardia duodenalis* (*see figure 23.6*) and *Cryptosporidium* spp. are significant causes of waterborne diarrheal diseases in the United States. Both organisms are resistant to chlorine and most other disinfectants. One source of *G. duodenalis* and *Cryptosporidium* spp. contamination is the Canada goose, which has an ever-expanding population. Water supplies that receive terrestrial runoff are therefore at particular risk of contamination by protists. ◀ *Food and water are vehicles of fungal and protozoal diseases (section 39.5)*

In the United States, the EPA has established policies for all public water systems that draw from surface waters or groundwater fed by surface waters. It sets a maximum containment level goal (MCLG), defined as "the maximum level at which no known or anticipated adverse health effects would

Table 42.1 Microbial Water Contaminant Candidate List (USEPA CCL5, 2022)

Organism	Concern
Adenoviruses	Viruses that cause respiratory illness and gastrointestinal disease
Caliciviruses	Viruses that cause self-limiting nausea, vomiting, diarrhea; includes norovirus
Campylobacter jejuni	Bacterium that causes nausea, vomiting, and diarrhea; usually self-limiting
Enteroviruses	Viruses that infect the gastrointestinal tract; includes poliovirus, coxsackie viruses, and echoviruses
Escherichia coli O157	Enterohemorrhagic *E. coli* that causes severe gastrointestinal (GI) illness; can also cause kidney failure
Helicobacter pylori	Bacterium that causes gastric ulcers and cancer
Legionella pneumophila	Bacterium found in cooling towers; causes bacterial pneumonia when inhaled.
Mycobacterium abscessus	Bacterium found in plumbing biofilms that causes respiratory symptoms
Mycobacterium avium complex (MAC)	Bacterium found in plumbing biofilms; can cause pulmonary infection; causes GI illness in people who are immunocompromised
Naegleria fowleri	Protozoan found in warm surface and groundwater; causes primary amoebic meningoencephalitis
Pseudomonas aeruginosa	Bacterium common in plumbing biofilms; causes pneumonia in patients with cystic fibrosis and septicemia in people who are immunocompromised
Shigella sonnei	Bacterium that causes GI illness, including bloody diarrhea

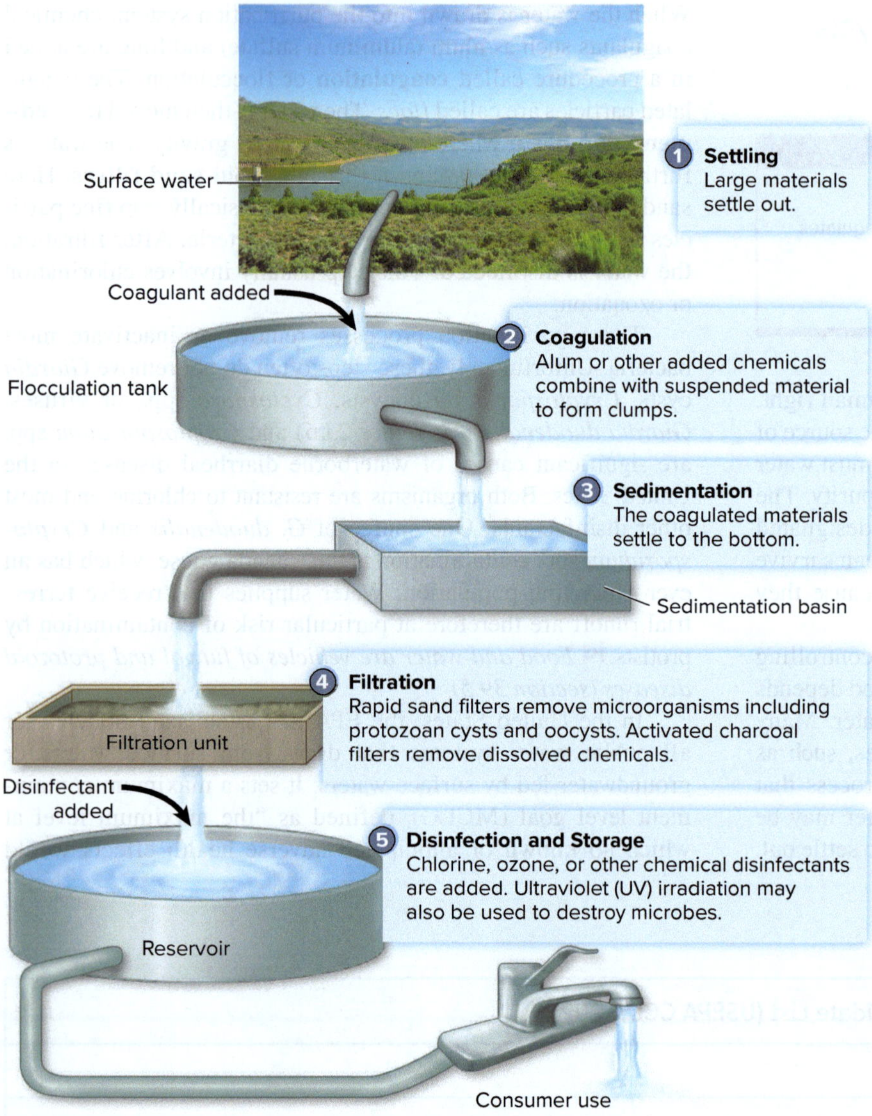

Figure 42.1 Water Purification. Preparing water for drinking involves several steps to physically remove particles (including microbes) and disinfection.

Mark Dierker/Bear Dancer Studios

occur, allowing an adequate margin of safety." MCLGs are set at zero for *Giardia, Cryptosporidium, Legionella* spp., and enteric viruses. In addition, the Revised Total Coliform Rule, which applies to all public water systems regardless of supply source, sets an MCLG of zero for total and fecal coliform bacteria. Water systems found to be contaminated must develop additional water treatment strategies to provide 99.9% inactivation of the microbes.

Once water has been purified, it must be deemed potable. The standard approach to monitoring drinking water safety is the detection of **indicator organisms.** Indicator organisms permit a quick estimate of water quality. Researchers are still searching for the ideal indicator organisms, and among the suggested criteria for such a microbe are:

1. The indicator organism should be suitable for analysis of all types of water: tap, river, ground, reservoir, recreational, estuary, sea, and waste.
2. It should be present whenever enteric pathogens are present.
3. It should survive longer than the hardiest enteric pathogen.
4. It should not reproduce in the contaminated water, as this would produce an inflated value.
5. It should be harmless to humans.
6. Its level in contaminated water should have some direct relationship to the degree of fecal pollution.
7. The assay procedure for the indicator should be highly specific; in other words, the presence of other, nonpathogenic bacteria should not generate positive results. In addition, the procedure should have high sensitivity and detect low levels of the indicator microbe.
8. The testing method should be easy to perform.

Coliforms, including *Escherichia coli,* are members of the order Enterobacterales. These bacteria are commonly found in the gut of humans and other animals and are used widely as indicator organisms. They lose viability in freshwater at slower rates than most of the major intestinal bacterial pathogens. When such enteric indicator bacteria are not detectable in a specific volume of water, the water is considered potable. *Order Enterobacterales is a large, diverse taxon (section 21.2)*

However, coliforms include a wide range of bacteria that do not primarily come from the intestinal tract. To address this, tests for the presence of **fecal coliforms** are used. These are coliforms from the guts of warm-blooded animals, which grow at the more restrictive temperature of 44.5°C. Other indicator microorganisms include **fecal enterococci.** These bacteria are used as an indicator of fecal contamination in brackish and marine water. In saltwater, these bacteria die more slowly than fecal coliforms, providing a more reliable indicator of pollution. *Order Lactobacillales (section 22.2)*

The **membrane filter technique** is a common method to culture and count the number of indicator organisms in water. The sample is passed through a membrane filter, which is then transferred to the surface of a solid medium or to an absorptive pad containing the desired liquid medium (*see figure 7.33*). Specific media enable the rapid detection of total coliforms, fecal coliforms, or fecal enterococci by the growth of their characteristic colonies.

To test for both coliforms and *E. coli,* the indicator organism for fecal coliforms, the **Colilert defined substrate test** can be used. A water sample is added to a specialized medium containing *o*-nitrophenyl-β-D-galactopyranoside (ONPG) and 4-methylumbelliferyl-β-D-glucuronide (MUG) as the only nutrients. ONPG detects the presence of β-galactosidase, an enzyme common in coliforms, while MUG detects β-glucosidase, found specifically in *E. coli.* If coliforms are present, the medium will turn yellow within 24 hours at 35°C due to hydrolysis of ONPG, which releases *o*-nitrophenol (**figure 42.2**). To check for *E. coli,* the medium is observed for fluorescence under ultraviolet (UV) light. When *E. coli* is present, MUG is hydrolyzed to yield a fluorescent product. If the test is negative for the presence of coliforms, the water is considered acceptable for human consumption. If coliforms are present, tests for fecal coliforms or *E. coli* must be performed.

The use of indicator organisms as a means of detecting all pathogenic microbes is limited. Some pathogens simply do not co-occur with indicator organisms. Another problem is that indicator organisms can be more sensitive to disinfectants; thus when drinking water is tested after such treatment, indicator-based assays may be misleading. In addition, culture-based methods generally require at least 18 hours to complete.

It is clear that effective, reliable, rapid, and quantitative methods for determining the presence or absence of specific pathogens are needed. Molecular approaches can be all these things—with the ability to quickly detect multiple, specific

Figure 42.2 The Defined Substrate Test. This test detects coliforms and fecal coliforms and has also been miniaturized for rapid testing of multiple samples. (a) Sterile control. (b) Yellow color due to the presence of coliforms. (c) Fluorescence resulting from the presence of fecal coliforms.

(a-b) Donald Klein/McGraw Hill; (c) Ethan Shelkey and Joanne M. Willey, Ph.D.

pathogens in a single water sample. Some of the methodologies that have been applied to water testing include flow cytometry (*section 26.3*), fluorescent in situ hybridization (FISH; *section 26.4*), quantitative PCR (*section 17.2*), microarrays (*section 18.5*), and metagenomics (*sections 18.3 and 26.4*). Targeted amplification of conserved regions within a metagenomic sample is a rapid means of determining the presence or absence of certain microbes in a water sample.

Merely identifying the presence of microbes in a body of water doesn't address their source. For example, fecal coliforms in a lake could indicate failure of a residential septic system, runoff from a livestock pasture, or a flock of migrating birds. **Microbial source tracking (MST)** methods identify the origin of contaminants. Enumeration of fecal coliforms (FC) and fecal streptococci (FS) is performed by culturing. The ratio of FC/FS in a sample indicates the source: FC/FS of >4 indicates human origin, while FC/FS <0.7 indicates animal origin. In addition, molecular techniques can identify bacteria at a level below species. *Library-dependent methods* compare an isolate to a collection of strains previously derived from a locale; *library-independent methods* use PCR to amplify target genes correlated with specific animal sources.

Comprehension Check

1. How can *Giardia* and *Cryptosporidium* cysts be removed from a water supply?
2. Why is it important that an indicator microorganism survive but not grow in the water to be tested?
3. Why do you think the age of pipes used in municipal water systems is a factor in introducing waterborne pathogens into the water supply?

42.2 Wastewater Treatment Maintains Human and Environmental Health

After reading this section, you should be able to:

a. Compare and contrast the processes that occur during primary, secondary, and tertiary wastewater treatment
b. Explain the general operation of septic systems
c. Identify at least two challenges for the future of wastewater treatment

The flip side of ensuring safe drinking water is wastewater management. Wastewater includes sewage, defined as feces and dirty water that are released into the public sewer system. Dirty water, sometimes called *gray water,* includes water from showers, laundry, and dishwashers. In addition, municipal sewer systems also receive industrial and agricultural effluent and street runoff collected in storm sewers. These waters often contain high levels of organic matter, heavy metals, nutrients, and particulates.

We begin our discussion of wastewater treatment with an overview of large-scale wastewater treatment processes, followed by the means by which water quality is monitored. We then review home treatment systems.

Wastewater Treatment Processes

Wastewater treatment involves at least two spatially segregated steps: primary and secondary, with an optional tertiary treatment (**figure 42.3**). At the end of the process, water is usually chlorinated or treated with ozone before it is released into the environment.

Primary treatment prepares wastewater for further treatment by physically removing much of the solid material. This can be accomplished in several ways, including passing the wastewater through screens to remove large items, precipitation of small particulates, and settling in basins or tanks. The water (effluent) is then further purified by biological treatment. The resulting solid material, called **sludge,** is treated before disposal.

Secondary treatment relies on microbial growth to convert dissolved organic matter (DOM) into microbial biomass and carbon dioxide. It involves aerobic or anaerobic processes or both. This step reduces degradable DOM by about 90 to 95%. When microbial growth is complete, microorganisms should aggregate and form stable flocs, also called **granular sludge,** that settle. As shown in **figure 42.4,** the addition of commercial coagulants to secondary treatments increases the rate of flocculation and the amount of material that settles. Alternatively, bacteria that have been genetically modified to create a stable, granular sludge may be added. Efficient flocculation is important because a sewage treatment facility that generates stable granular sludge requires less space and reduces cost.

Figure 42.3 Aerial View of a Conventional Sewage Treatment Plant. Sewage treatment plants allow natural processes of self-purification that occur in rivers and lakes to be carried out under more intense, managed conditions in large concrete vessels.

Mariusz Szczygiel/Shutterstock

MICRO INQUIRY *This plant includes anaerobic digesters. What are the advantages of this additional step?*

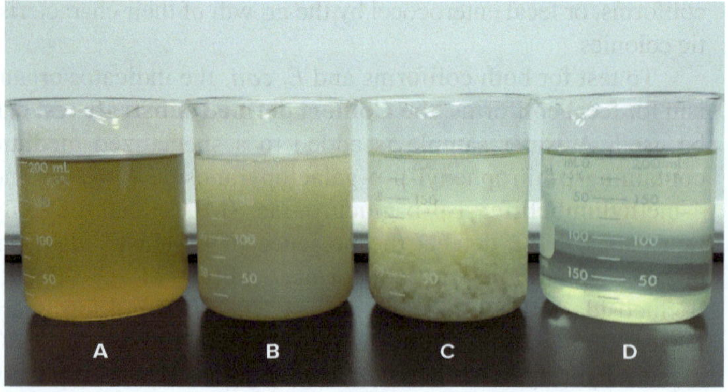

Figure 42.4 Proper Floc Formation in Activated Sludge. Microorganisms play a critical role in the functioning of activated sludge systems. Wastewater in beaker A contains bacteria and other organic materials that resist settling. A commercial coagulant containing ferric chloride has been added to beaker B. This accelerates the formation of stable flocs, which settle at the bottom of beaker C. Beaker D shows the effluent after the settled material has been removed.

Ecologix Environmental Systems, LLC./Eli Gruber

MICRO INQUIRY *Why is proper floc formation critical to the function of a treatment facility?*

One type of aerobic secondary treatment, an **activated sludge** system (**figure 42.5*a***), involves the horizontal flow of materials with recycling of sludge. Activated sludge systems can be designed with variations in mixing and with variations in the ratio of organic matter added to the active microbial biomass. A low-rate system (low nutrient input per unit of microbial biomass), with slower-growing microorganisms, produces an effluent with low residual levels of DOM. A high-rate system (high nutrient input per unit of microbial biomass), with faster-growing microorganisms, removes more DOM per unit time but produces a poorer-quality effluent. Another type of aerobic secondary treatment is carried out with a **trickling filter** (figure 42.5*b*). The effluent from primary treatment is passed over rocks or other solid materials coated with microbial biofilms, and the microbial community degrades the organic waste.

All aerobic processes produce excess microbial biomass, or sewage sludge, which contains many recalcitrant organic chemicals. Often the sludge from aerobic sewage treatment, together with the materials that settled in primary treatment, are further treated by anaerobic digestion. **Anaerobic digestion** occurs in large tanks designed to operate with continuous input of untreated sludge and removal of the final, stabilized sludge product (**figure 42.6**). A second product of anaerobic digestion, methane, the major component of natural gas, is vented and often burned for heat and electricity production. Within the anaerobic digester, at least three metabolic processes occur: (1) fermentation of sludge components to form organic acids, including acetate; (2) production of the other methanogenic substrates CO_2 and hydrogen; and (3) methanogenesis. These methanogenic processes involve critical balances between electron acceptors and donors (**table 42.2**). To function most efficiently, the hydrogen concentration must be maintained

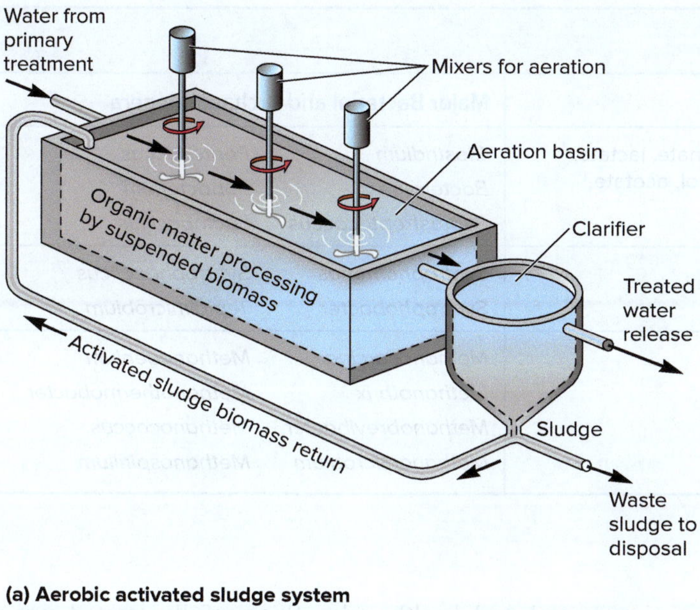

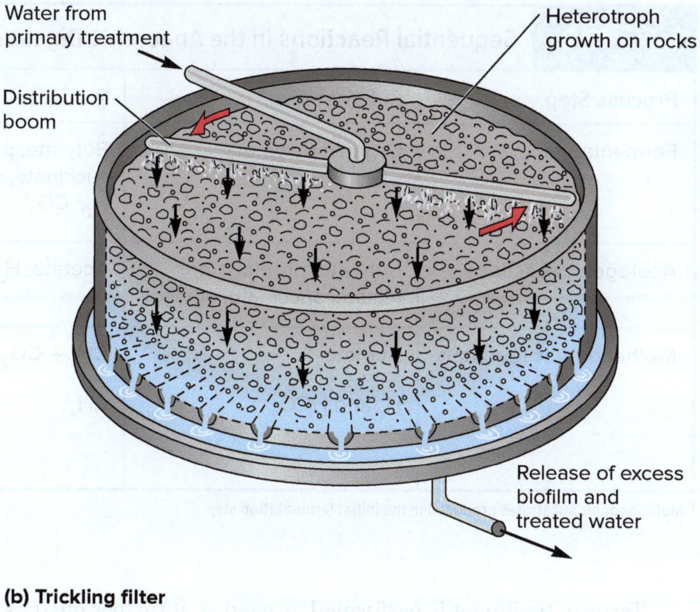

Figure 42.5 Aerobic Secondary Sewage Treatment. (a) Activated sludge with microbial biomass recycling. The biomass is maintained in a suspended state to maximize oxygen, nutrient, and waste transfer processes. (b) Trickling filter, where wastewater flows over biofilms attached to rocks or other solid supports, resulting in transformation of dissolved organic matter to new biofilm biomass and carbon dioxide. Excess biomass and treated water flow to a final clarifier.

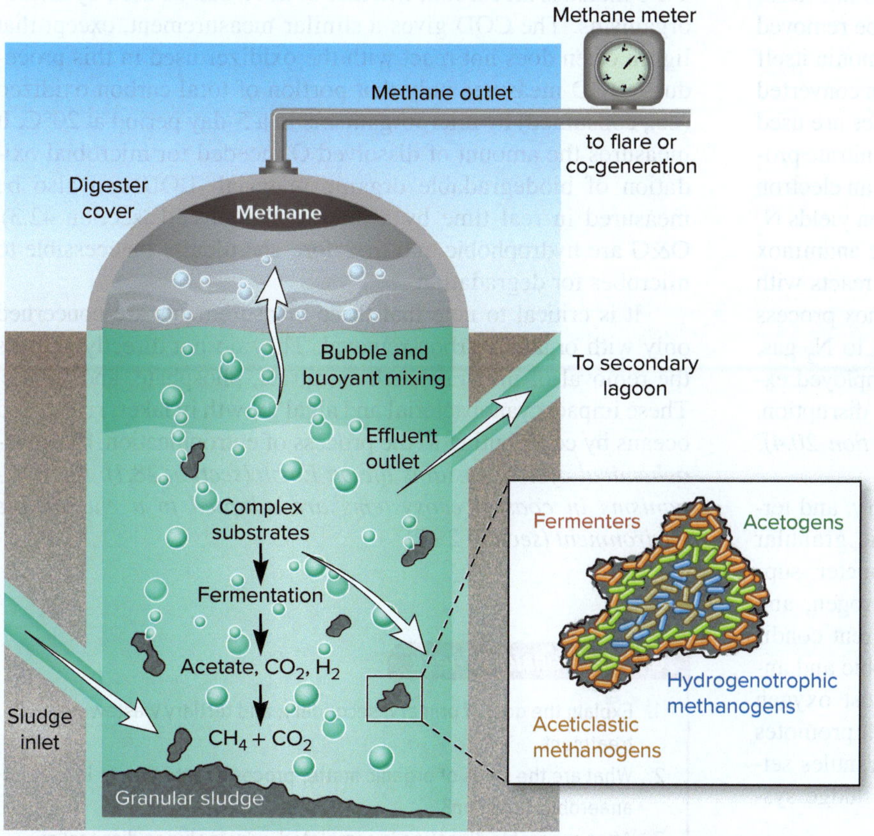

Figure 42.6 Anaerobic Secondary Sewage Treatment Generates Methane. The basic design of an anaerobic digester includes an inlet for the delivery of waste. Microbial consortia ferment complex substrates to yield acetate, CO_2, and H_2, which methanogens convert to CH_4. The effluent outlet optimizes the retention time to maximize CH_4 capture.

at a low level, and this is ensured by the presence of methanogens that consume H_2 and CO_2. The microbial consortia that carry out anaerobic digestion aggregate into methanogenic granules, a type of biofilm that promotes syntrophic interactions (figure 42.6). ◀ *Methanogens (section 19.4); Syntrophy (section 27.2)*

It has been discovered that the efficiency of anaerobic digestion can be increased by **direct interspecies electron transfer (DIET).** In syntrophic systems, electron transport occurs via hydrogen transfer. In DIET, *Geobacter* spp. and other syntrophic bacteria form conductive pili (*see figure 20.16*) that connect to methanogens in anaerobic digesters. Scientists have also found that conductive materials like magnetite or activated carbon participate in DIET by providing a path for electrons. Addition of such conductive materials to anaerobic digesters results in increased methane yield and a faster rate of methanogenesis.

Anaerobic digestion has many advantages, including the production of methane. Most of the microbial biomass produced in aerobic growth is degraded in the anaerobic digester. Also, because methanogenesis is energetically inefficient, methanogenic archaea consume about twice as many nutrients to produce an equivalent biomass as that of aerobic systems. Consequently less sludge is produced and it can be easily dried and used as organic fertilizer.

Table 42.2 Sequential Reactions in the Anaerobic Digester

Process Step	Substrates	Products	Major Bacterial and Archaeal Genera	
Fermentation	Organic polymers	Butyrate, propionate, lactate, succinate, ethanol, acetate,[1] H_2,[1] CO_2[1]	*Clostridium* *Bacteroides* *Peptostreptococcus*	*Peptococcus* *Eubacterium* *Lactobacillus*
Acetogenic reactions	Butyrate, propionate, lactate, succinate, ethanol	Acetate, H_2, CO_2	*Syntrophomonas* *Syntrophobacter*	*Syntrophaceticus* *Tepidimicrobium*
Methanogenic reactions	Acetate H_2 and HCO_3^-	$CH_4 + CO_2$ CH_4	*Methanosarcina* *Methanothrix* *Methanobrevibacter* *Methanomicrobium*	*Methanogenium* *Methanothermobacter* *Methanococcus* *Methanospirillum*

[1] Methanogenic substrates produced in the initial fermentation step.

Tertiary treatment is performed as needed; it further purifies wastewater by removing nitrogen and phosphorus compounds that promote eutrophication. Phosphate usually is precipitated as calcium or iron phosphate (e.g., by the addition of lime). To remove phosphorus, oxic and anoxic conditions are used alternately in a series of treatments, and phosphorus accumulates in microbial biomass as polyphosphate. Excess nitrogen may be removed by "stripping," volatilization of NH_3 at high pH. Ammonia itself can be chlorinated to form dichloramine, which is then converted to nitrogen gas (N_2). In some cases, microbial processes are used to remove nitrogen and phosphorus. In denitrification, nitrate produced by microbes under aerobic conditions serves as an electron acceptor during anaerobic respiration. Nitrate reduction yields N_2 and nitrous oxide (N_2O) as the major products. In the anammox process, ammonium ion (used as the electron donor) reacts with nitrite (the electron acceptor) to yield N_2. The anammox process can convert up to 80% of the initial ammonium ion to N_2 gas. Tertiary treatment is expensive and is usually not employed except where necessary to prevent obvious ecological disruption. ◀ *Phylum Planctomycetota: anammox bacteria (section 20.4); Nitrogen cycle (section 28.2)*

A recent innovation combines primary, secondary, and tertiary treatments into a single process in an aerobic granular sludge system. Granules, several millimeters in diameter, support bacterial communities that remove carbon, nitrogen, and phosphate. Because these communities require different conditions for their metabolism, the reactor alternates aerobic and anaerobic growth. Aerobic heterotrophs rapidly exhaust oxygen and organic matter, and the shift to anaerobic growth promotes nitrifiers and phosphate-accumulating organisms. Granules settle much faster than traditional flocs, and granular sludge systems can treat a greater volume of wastewater.

Measuring Treated Wastewater Quality

The process of wastewater treatment must be monitored to ensure that waters released into the environment do not pose environmental and health risks. Successfully treated water should have little organic carbon, and four measurements are typically made after wastewater treatment: **total organic carbon (TOC), chemical oxygen demand (COD), biochemical oxygen demand (BOD),** and **oil and grease (O&G).** The TOC includes all carbon, whether or not it can be used by microorganisms. The COD gives a similar measurement, except that lignin often does not react with the oxidizer used in this procedure. BOD measures only that portion of total carbon oxidized (i.e., consumed) by microorganisms in a 5-day period at 20°C. It measures the amount of dissolved O_2 needed for microbial oxidation of biodegradable organic material. BOD can also be measured in real time by a microbial fuel cell (section 42.3). O&G are hydrophobic and therefore chemically inaccessible to microbes for degradation.

It is critical to note that these measurements are concerned only with organic carbon removal. They do not directly address the removal of minerals such as nitrate, phosphate, and sulfate. These impact cyanobacterial and algal growth in lakes, rivers, and oceans by contributing to the process of eutrophication. ◀ *Biogeochemical cycling sustains life on Earth (section 28.1); Microorganisms in coastal ecosystems are adapted to a changeable environment (section 29.2)*

Comprehension Check

1. Explain the goal of primary, secondary, and tertiary wastewater treatment.
2. What are the steps of organic matter processing that occur in anaerobic digestion?
3. After anaerobic digestion is completed, why is sludge disposal of concern?
4. Draw a Venn diagram to illustrate the relationships between TOC, COD, BOD, and O&G.

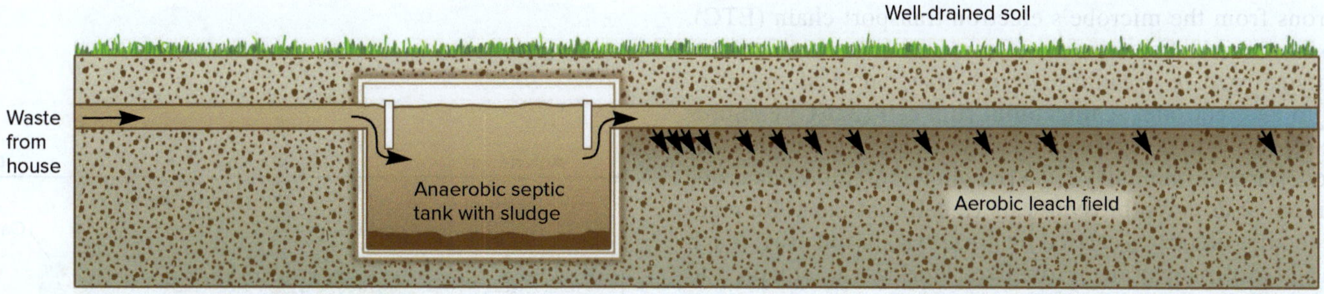

Figure 42.7 The Conventional Septic Tank Home Treatment System. This system combines an anaerobic waste liquefaction unit (the septic tank) with an aerobic leach field. Biological oxidation of the liquefied waste takes place in the leach field, unless the soil becomes flooded.
Ecologix Environmental Systems, LLC.

MICRO INQUIRY *Why would flooding halt biological oxidation of waste in the leach field?*

Home Treatment Systems

In communities without municipal sewage systems, **septic tanks** are most commonly used. About 20% of American households rely on this system to dispose of wastewater. A septic tank functions as a simple anaerobic digester, with an anaerobic liquefaction and digestion step. This is followed by organic matter adsorption and entrapment of microorganisms in an aerobic leach field, where biological oxidation occurs. Pathogenic microorganisms and dissolved organic matter are removed from the wastewater during subsurface passage through adsorption and trapping by fine sandy materials, clays, and organic matter (**figure 42.7**). Microorganisms associated with these materials—including predators such as protozoa—use the trapped pathogens as food. This results in purified water with a lower microbial population.

It is imperative that septic systems be located far above groundwater, often an important source of potable water. Thus not all sites are suitable for septic systems. Prior to installation, a percolation test (perc test) must be performed to determine if the surrounding soil can absorb the liquid produced by the septic tank. Variable groundwater levels, often related to climate change, are rendering many home septic systems ineffective.

A septic tank may not operate correctly for several reasons. If the waste retention time in the septic tank is too short, undigested solids move into the leach field, plugging the system. If the leach field floods and becomes anoxic, biological oxidation does not occur, and effective treatment ceases. Other problems can occur, especially when a suitable soil is not present and the septic tank outflow drains too rapidly to the deeper subsurface. Fractured rocks and coarse gravel materials do not effectively adsorb or filter wastewater. In addition, nitrogen and phosphorus from waste can pollute groundwater. This leads to nutrient enrichment of ponds, lakes, rivers, and estuaries as the subsurface water enters these environmentally sensitive ecosystems.

Future Challenges

Our planet now has over 8 billion people, and the needs for clean drinking water and wastewater treatment have become ever more pressing. In an effort to address both at once, water reuse or recycling has become a reality for many. Water reuse is the purification of wastewater so that it can serve as drinking water. Not surprisingly, water reuse is complex. Not only must the obvious contaminants be removed (i.e., pathogens, heavy metals, nutrients that cause eutrophication), but twenty-first-century sewage also carries alarming levels of pharmaceuticals, introduced by both industry and the households. This has stimulated the development of methods to detect and remove the vast array of drugs common in sewage. In addition to recycling water, recent technologies enable the reclamation of biopolymers, plastics, and cellulose fibers in an economically viable fashion. The current challenge is the widespread application of these technologies.

Comprehension Check

1. How, in principle, does a conventional septic tank and leach field system work? What factors can reduce the effectiveness of this system?
2. What type of wastewater treatment system does your hometown use?
3. List two areas of active research in the wastewater treatment field. Which seem more pressing in your community? Explain your answer.

42.3 Microbial Fuel Cells: Batteries Powered by Microbes

After reading this section, you should be able to:

a. Diagram the basic design of a microbial fuel cell
b. Explain potential applications of microbial fuel cells

An exciting and emerging field in microbiology is the use of bacteria to generate electricity, not through the production of methane (as in anaerobic digesters, figure 42.6) but by capturing electrons from the microbe's electron transport chain (ETC). This is possible because when some heterotrophic microorganisms oxidize organic material, they can pass the electrons to an electrode. A **microbial fuel cell (MFC)** captures these electrons to generate electricity. To accomplish this, the microbes are continuously fed a rich diet of organic substrates to minimize de novo biosynthesis (recall that anabolic reactions require electrons carried by NADPH). In this way, much of the organic substrate is oxidized during catabolism and electrons are donated to the ETC. The basic design of an MFC is relatively simple (**figure 42.8**). It consists of two chambers, one oxic and the other anoxic, divided by a membrane through which protons can pass. Nutrients are delivered to the bacteria in the anoxic side. The oxidation of this organic matter yields protons and electrons. The protons diffuse across the membrane to the oxic chamber and the electrons are deposited on an anode. As electrons flow from the anode toward the cathode, they generate electricity. A catalyst in the oxic chamber combines the protons, electrons, and oxygen to generate water.

Many different types of heterotrophic microbes can produce power in MFCs. The only requirement is that they oxidize organic matter anaerobically. In fact, communities of microbes like those found in wastewater can be employed. The largest constraint on maximum efficiency appears to be delivery of electrons to the anode. There are several ways this can be accomplished. Often a chemical mediator is used to shuttle electrons from inside the cell, across the cellular membrane, to the anode (figure 42.8). Some microbes, such as *Geobacter* and *Shewanella* spp., produce nanowires to transmit electrons to the anode (in nature, these nanowires shuttle electrons to external electron acceptors such as Fe^{3+}; *see figure 20.16*).

MFC development is ongoing, and great strides have been made to maximize power output. Applications of MFCs include drinking water purification and the treatment of landfill leachate: In both cases, unwanted organic material is used to feed the bacteria in the MFC. MFCs can also be used when energy requirements are small but batteries cannot be used. Environmental monitoring instruments that need to be left in the field for years at a time are good examples. In addition, because MFCs are self-contained, they are attractive for regions that lack well-developed power grids.

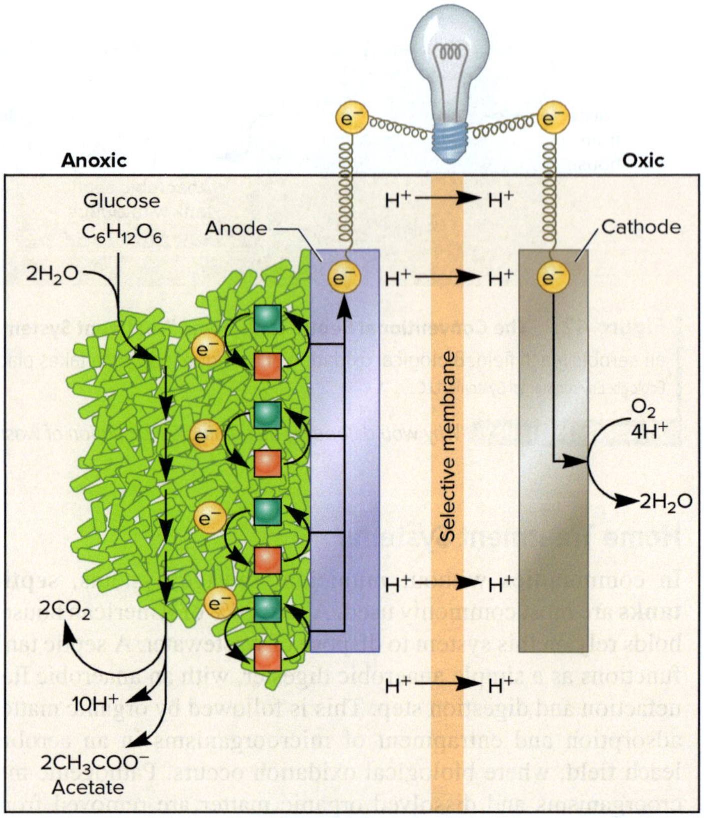

(a)

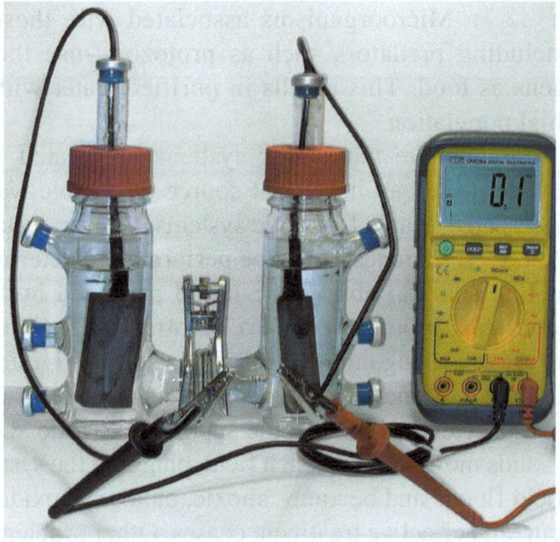

(b)

Figure 42.8 Microbial Fuel Cells. (a) Nutrients are catabolized anaerobically by microbes in the anoxic chamber. The resulting protons diffuse to the oxic chamber, and the electrons are shuttled from the cell to an anode, where they travel to the oxic chamber. Green boxes represent oxidized electron shuttles and red boxes represent reduced shuttles. (b) A two-chambered microbial fuel cell.

(b) Kelly P. Nevin

MICRO INQUIRY *Why must the membrane between the two chambers be selectively permeable?*

Comprehension Check

1. How does a microbial fuel cell work?
2. What would be the outcome if microbes in a fuel cell were carbon limited?
3. How might microbial fuel cells be involved in wastewater management?

42.4 Biodegradation and Bioremediation Harness Microbes to Clean the Environment

After reading this section, you should be able to:

a. List the steps of biodegradation
b. Describe the importance of reductive dehalogenation in bioremediation
c. Discuss factors that influence the rate of biodegradation and bioremediation of a compound
d. Discuss some of the challenges in biodegradation of chemical contaminants

The metabolic activities of microbes can be exploited in complex natural environments for beneficial outcomes. Examples are the use of resident microbial communities to carry out biodegradation, bioremediation, and environmental maintenance processes.

Biodegradation is the process of breaking down an organic chemical through the action of microbes. In the simplest cases, enzymes like proteases hydrolyze peptide bonds when plants or animals decompose. Complex substrates aren't typically degraded by a single enzyme, but instead, several enzymes act synergistically. Importantly, biodegradation is a community process. In an analogy to the human gut, the total sum of the genes in the community constitutes a **core microbiome.** The enzymes to degrade a complex substrate are distributed among the individuals in the community, also called a *consortium*.

Figure 42.9 highlights the three general steps in biodegradation. **Biofragmentation** relies on exoenzymes, secreted enzymes that degrade large organic compounds to smaller chemical forms. The degradation products are metabolized in **bioassimilation,** and the end point is **mineralization,** the conversion of organic matter to inorganic forms. Here we discuss the biodegradation of major organic compounds that pollute soils and water.

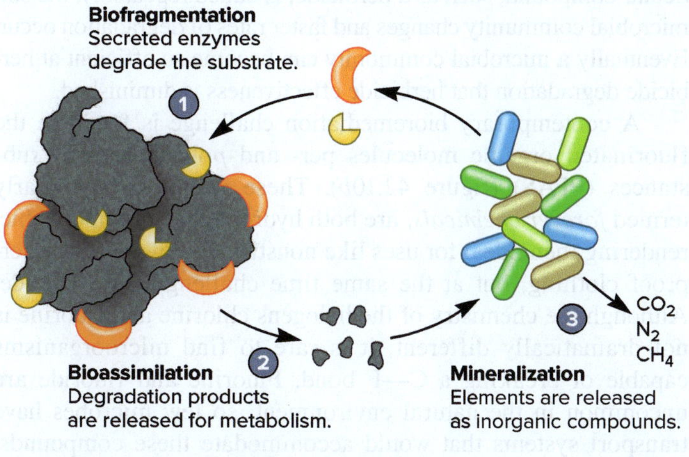

Figure 42.9 Three Steps of Biodegradation.

Petroleum Hydrocarbons

Oil spills have become recurring events, as our dependence on oil has increased exploration and refinement. Crude oil is a complex mix of hydrocarbons that can serve as a source of carbon and energy. Oil-degrading microbes are part of the **rare biosphere**—that is, they are ubiquitous in soil, marine, and freshwater ecosystems, but they are not abundant. When petroleum is available, for example, around underwater petroleum seeps, these microbes exploit the resource and bloom. This requires enzymes to convert the hydrocarbons to organic intermediates that enter the cell's central carbon pathways.

Alkanes comprise about 50% of crude oil. Microbes that degrade hydrocarbons encode enzymes that use alkanes as substrates. Alkane monooxygenase, for example, catalyzes the oxidation of alkanes to alcohols (**figure 42.10**). Following this initial activation step, other enzymes further oxidize the carbon or cleave the carbon chain. Metagenomic profiles of the marine microbial community following the Deepwater Horizon accident in 2010 off the coast of Louisiana documented an increased abundance of genes for hydrocarbon metabolism.

Phosphorus and nitrogen are limiting elements for microbial growth on hydrocarbons, and the addition of these elements stimulates the microbial community. This was demonstrated following an earlier oil spill in Alaska in 1989 (**figure 42.11**). Simple fertilization with phosphorus and nitrogen promoted the growth of oil-degrading microbes and significantly diminished the oil in coastal sediments. Following the Deepwater Horizon incident, metagenomics documented parallel increases in genes for nitrogen fixation with genes for petroleum degradation. Oil-degrading microbes with the ability to reduce atmospheric nitrogen have a clear competitive advantage under these conditions.

Halogenated Organic Molecules

The removal of toxic industrial products in soils and aquatic environments is a daunting and necessary task. Compounds such as

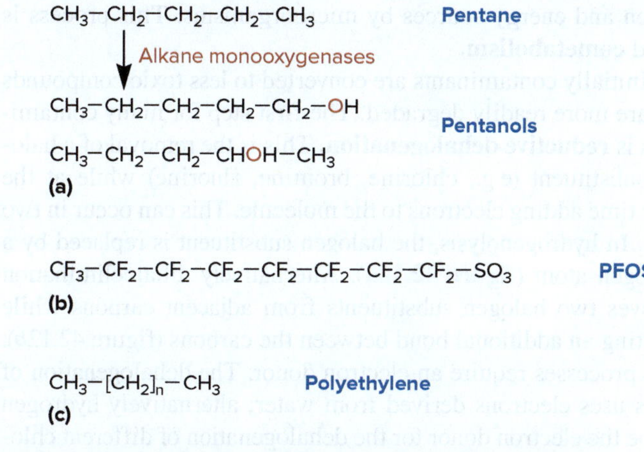

Figure 42.10 Chemical Structures of Linear Carbon Molecules. (a) Pentanes are typical of the alkanes found in petroleum. Alkane monooxygenases add an oxygen atom to make primary or secondary alcohols. (b) Perfluorooctane sulfonate (PFOS). (c) Polyethylene.

Figure 42.11 Bioremediation of an Oil Spill. Workers sprayed the coastline with nitrogen and phosphorus–containing fertilizer to provide limiting nutrients to hydrocarbon-degrading microorganisms. This increased the rate at which oil was degraded.
Accent Alaska.com/Alamy Stock Photo

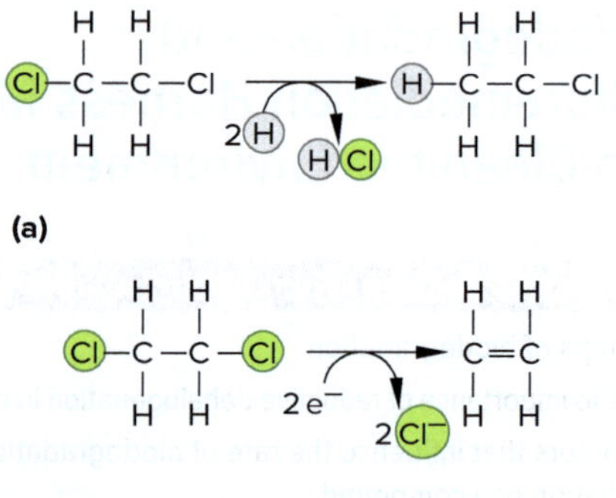

Figure 42.12 Reductive Dehalogenation. In these two examples, 1,2-dichloroethane is dehalogenated by (a) hydrogenolysis and (b) dihaloelimination. Note that although the substrate is the same, these strategies result in different products.

perchloroethylene (PCE), trichloroethylene (TCE), and polychlorinated biphenyls (PCBs) are common contaminants. The use of microbes to transform these contaminants to nontoxic degradation products is called **bioremediation.**

Bioremediation usually involves stimulating the degradative activities of microorganisms already present in contaminated waters or soils. These existing microbial communities typically cannot carry out biodegradation processes at the desired rate due to limiting physical or nutritional factors. For example, biodegradation may be limited by low levels of oxygen, nitrogen, phosphorus, or other nutrients. In these cases, it is necessary to determine the limiting factors and supply the needed materials or modify the environment. Often the addition of easily metabolized organic matter such as glucose increases biodegradation of recalcitrant compounds that otherwise would not be used as carbon and energy sources by microorganisms. This process is called **cometabolism.**

Initially contaminants are converted to less toxic compounds that are more readily degraded. The first step for many contaminants is **reductive dehalogenation.** This is the removal of a halogen substituent (e.g., chlorine, bromine, fluorine) while at the same time adding electrons to the molecule. This can occur in two ways. In hydrogenolysis, the halogen substituent is replaced by a hydrogen atom (**figure 42.12*a***). Alternatively dihaloelimination removes two halogen substituents from adjacent carbons while inserting an additional bond between the carbons (figure 42.12*b*). Both processes require an electron donor. The dehalogenation of PCBs uses electrons derived from water; alternatively hydrogen can be the electron donor for the dehalogenation of different chlorinated compounds. Major genera with species that carry out this process include *Desulfitobacterium, Dehalococcoides,* and *Desulfomonile.* Some of these organisms are considered obligate dehalogenators—that is, their only known terminal electron acceptors are organic halides.

Reductive dehalogenation usually occurs under anoxic conditions. Once the anaerobic dehalogenation steps are complete, degradation of the main structure of many pesticides and other xenobiotics (foreign compounds) often proceeds more rapidly in the presence of O_2.

Structure and stereochemistry are critical in predicting the fate of a specific chemical in nature. When a constituent is in the *meta* as opposed to the *ortho* position, the compound will be degraded at a much slower rate. This stereochemical difference is the reason the herbicide 2,4-dichlorophenoxyacetic acid (2,4-D), with chlorine in the *ortho* position, will be largely degraded in a single summer. In contrast, 2,4,5-trichlorophenoxyacetic acid, with a constituent in the *meta* position, will persist in the soils for several years and thus is used for long-term brush control.

Another important factor that influences biodegradation is a compound's chirality or handedness (*see figure AI.8*). Microorganisms often degrade one isomer of a substance but not the other. About 30% of herbicides are chiral. Thus it is critical to add the herbicide isomer that is effective and also degradable. However, populations that make up a microbial community can change in response to the addition of inorganic or organic substrates. If a particular compound, such as a herbicide, is added repeatedly, the soil microbial community changes and faster rates of degradation occur. Eventually a microbial community can become so efficient at herbicide degradation that herbicide effectiveness is diminished.

A contemporary bioremediation challenge is found in the fluorinated organic molecules per- and *p*oly*f*luoro*a*lkyl *s*ubstances (PFAS) (figure 42.10*b*). These chemicals, popularly termed *forever chemicals,* are both hydrophobic and lipophobic, rendering them ideal for uses like nonstick cookware and waterproof clothing, but at the same time challenging to degrade. Although the chemistry of the halogens chlorine and fluorine is not dramatically different, it is rare to find microorganisms capable of breaking a C—F bond. Fluorine and fluoride are uncommon in the natural environment, so few microbes have transport systems that would accommodate these compounds. In addition, removal of the fluorine atom from PFAS would result in a fluoride ion. F^- is highly toxic and inhibits many cellular enzymes including ATPases. This contrasts with Cl^-, which is common in the bacterial cytoplasm. The multifaceted bioremediation challenge described in **Microbial Diversity and Ecology 42.1** involves both Cl^- and F^- decontamination.

Plastics

Plastics are perhaps the most abundant organic polymer contaminant on Earth. It should not be surprising that materials designed to withstand degradation have accumulated in the environment and now present a planetary scale disposal problem (**figure 42.13**). Chemically, plastics are carbon polymers; the density and degree of cross-linking determines the properties of the final product.

As plastics have accumulated, a new term, the **plastisphere,** was coined to describe the microbial environment on plastic debris. Plastic exposed to heat or ultraviolet radiation has a weakened polymer structure that can provide an entry for microbial degradation. For example, floating marine plastic is colonized by photoautotrophs and photoheterotrophs that form surface biofilms and initiate degradation. The rate of decomposition, however, is extremely slow, far slower than the rate at which plastics are added to the oceans for disposal.

What makes plastics so recalcitrant to biodegradation? First, plastics are generally hydrophobic, and microbes require their nutrients in solution. Second, long repeating hydrocarbon chains are difficult for enzymes to access. Finally and most importantly, there are other more easily digestible carbon sources available in most ecosystems.

Polyethylene (PE) exemplifies the challenges to degradation. The structural similarities between PE and hydrocarbons (figure 42.10*c*) hint at potential mechanisms. Organisms like *Alcanivorax* spp. that degrade hydrocarbons in oil spills use mono- and dioxygenases to add oxygen atoms to carbon polymers, thus forming alcohol or peroxyl groups. Further oxidation of the terminal carbon generates carboxylated hydrocarbons (fatty acids) that are metabolized through the β-oxidation pathway (*see figure 11.21*). PE is a mix of crystalline and amorphous regions; so although the chemical composition is the same, the polymer organization differs (**figure 42.14**). The amorphous

MICROBIAL DIVERSITY & ECOLOGY

42.1 Microbial First Responders

In early 2023, a train assembled from 150 individual cars partially derailed in the town of East Palestine, Ohio. Of the 51 cars piled up, 11 warranted concern as they carried a mix of hazardous chemicals. Over the following week, those chemicals escaped the rail cars, from both damage upon the initial impact and a controlled release and burn. Contaminants, which included vinyl chloride and butyl acrylate, permeated the soil, entered local streams, and blew through the atmosphere.

Although many of the local residents were initially evacuated, as chemical odors persisted in the air and from residential water supplies, health symptoms continue in the residents and animals exposed to the contaminated air. Members of soil and aquatic animal communities could not be evacuated, and the consequences of chemical contamination are apparent. Tens of thousands of aquatic animals died.

Meanwhile, scientists have sampled the microbial communities in the soil and water. Preliminary reports track the changes in these communities as they adapt following the incident. In laboratory studies, vinyl chloride was degraded efficiently in soil and sediment samples where community analysis indicates the presence of *Dehalococcoides* spp. The most likely mechanism for vinyl chloride (C_2H_3Cl) biodegradation is reductive dehalogenation to ethylene (C_2H_4) (figure 42.12). Similarly, butyl acrylate was degraded, with the resulting butanol metabolized as described for hydrocarbons.

Adding another layer of tragedy to this chemical spill, the fire resulting from the controlled release and burn of 100,000 gallons of vinyl chloride was extinguished with firefighting foam composed of polyfluoroalkyl substances (PFAS; figure 42.10*b*). This so-called forever chemical resists microbial degradation and now further contaminates the soil and potentially the groundwater at the accident site.

(a)

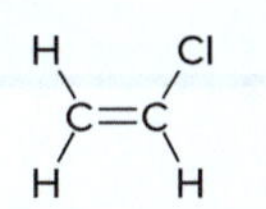

Vinyl chloride

(b)

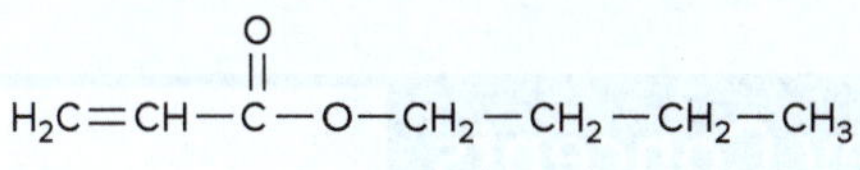

Butyl acrylate

(c)

Chemical Contamination at the East Palestine, Ohio, Train Derailment. (a) Hazardous chemicals carried in train cars leaked chemicals. Structures of vinyl chloride (b) and butyl acrylate (c), two chemicals that contaminated the soil and stream sediments.

Gene J. Puskar/AP Photos

Figure 42.13 Plastic Debris Littering a Shoreline.
LCDR Eric Johnson, NOAA Corps, NOAA/Dept. of Commerce

regions are more biodegradable such that, at a small scale, a degraded sample of PE resembles Swiss cheese.

The bacterium *Ideonella sakaensis* demonstrates that one type of plastic, polyethylene terephthalate (PET), can be completely depolymerized to monomers ethylene glycol and terephthalic acid. *I. sakaensis* was isolated from soil around a plastic recycling plant, and two secreted enzymes catalyze the reactions. Although it is tempting to declare, "Problem solved!" there are many steps between isolating an organism that can use PET as its carbon and energy source and tackling mountains of plastic.

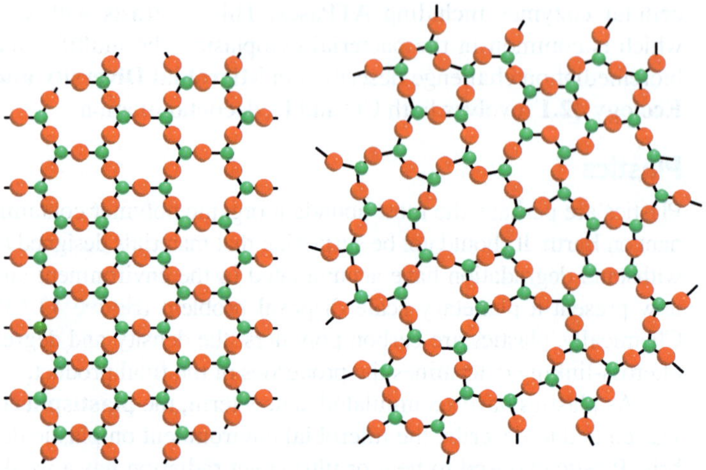

Figure 42.14 Crystalline and Amorphous Structures. Plastics have varying degrees of crystalline and amorphous structure. Microbes are best able to degrade amorphous regions.

An important early step is protein engineering to stabilize the enzymes for high temperature, where PET shifts from crystalline to amorphous and is therefore more amenable to hydrolysis.

There are currently thousands of synthetic polymers used to make different types of plastics, and their varied properties will undoubtedly require varied degradation methods. In 1952 microbial ecologist Ernest Gale proposed the microbial infallibility hypothesis, which states that there is a strong selective pressure for microbes to exploit chemicals in their environment as growth substrates. In the twenty-first century, we are undoubtedly testing the limits of this hypothesis.

Comprehension Check

1. List the steps in biodegradation.
2. What components are commonly added to microbes used to treat petrochemical contamination? Why are these added?
3. Explain the two mechanisms of reductive dehalogenation.
4. Discuss chirality and isomerization and their importance for understanding biodegradation.
5. Why is it challenging for microorganisms to degrade plastic?

Key Concepts

42.1 Purification and Sanitary Analysis Ensure Safe Drinking Water

- Water purification typically involves sedimentation, coagulation, chlorination, and sand filtration (**figure 42.1**).
- *Cryptosporidium, Cyclospora, Giardia* spp., and viruses are of concern, as conventional water purification and chlorination do not always ensure their removal or inactivation to acceptable limits.
- Indicator organisms are used to assess the presence of pathogenic microorganisms. Membrane filtration procedures are employed to estimate the number of indicator organisms present. Defined substrate tests are used for coliforms and *E. coli* (**figure 42.2**).

42.2 Wastewater Treatment Maintains Human and Environmental Health

- Conventional sewage treatment is a controlled intensification of natural self-purification processes and can involve primary, secondary, and tertiary treatment steps (**figures 42.3** through **42.6**).
- Water quality is assessed by measuring total organic carbon, chemical oxygen demand, biological oxygen demand, and oil and grease.
- Home treatment systems operate on general self-purification principles. The conventional septic tank provides anaerobic liquefaction and digestion, whereas the aerobic leach field allows oxidation of the soluble effluent. These systems are now designed to provide nitrogen and phosphorus removal to lessen impacts of on-site sewage treatment systems on vulnerable marine and freshwaters (**figure 42.7**).

42.3 Microbial Fuel Cells: Batteries Powered by Microbes

- Microbial fuel cells capture electrons donated to an anode during heterotrophic bacterial respiration. MFCs have a variety of applications, including the purification of waters while generating power and providing energy for monitoring environmental processes (**figure 42.8**).

42.4 Biodegradation and Bioremediation Harness Microbes to Clean the Environment

- Biodegradation is a critical part of natural systems mediated largely by microorganisms. This involves biofragmentation, bioassimilation, and mineralization (**figure 42.9**).
- Biodegradation is typically carried out by a consortia of microbes that secrete enzymes to break down complex substrates.
- Biodegradation can be influenced by many factors, including oxygen levels and the presence of readily usable organic matter. Reductive dehalogenation occurs best under anoxic conditions (**figure 42.12**), and the presence of organic matter can facilitate modification of recalcitrant compounds in the process of cometabolism.
- The plastisphere is the microbial community that colonizes plastic waste in the environment. Although inroads have been made in biodegradation, many polymers remain incalcitrant.

Active Learning

1. You wish to build a house in a rural community where you will need to install a septic system and drill a well for drinking water. Where will you site the well in relation to the septic system? Following the perc test, you discover that local regulations require you to install a sand filter in addition to your septic tank. Why might this be the case?
2. You are responsible for the bioremediation of soil contaminated by jet fuel. Your coworker suggests you fertilize the soil with nitrogen and phosphorus to promote the growth of naturally occurring hydrocarbon-degrading microbes. However, you just read about a new mixture of microbes sold in a slurry with nutrients and a surfactant. What data do you need to collect to decide which approach would be best? How will you obtain these data?
3. Biofilms form on surfaces in drinking water distribution networks and in some cases have been shown to be reservoirs of protists in the genera *Giardia* and *Cryptosporidium,* as well as a strain of *Poliovirus.* Discuss how you would assess the attachment to, growth within, and detachment from model biofilms in tap water in a model water distribution network. What specific variables would you consider and what microbial, chemical, and physical parameters would you measure?
4. How might studies to characterize the human gut microbiome impact the selection of indicator bacteria for tracking pollution?
5. Some lifestyle influencers are recommending the consumption of "raw water," water obtained directly from environmental sources like springs, lakes, and rain. They claim that this is healthier than water treated with chemicals before distribution through tap water systems. What health consequences might arise from consuming raw water?

Appendix I
A Review of the Chemistry of Biological Molecules

Appendix I provides a brief summary of the chemistry of organic molecules with particular emphasis on the molecules present in microbial cells. Only basic concepts and terminology are presented; introductory textbooks in biology and chemistry should be consulted for a more extensive treatment of these topics.

Atoms and Molecules

Matter is made of elements that are composed of atoms. An element contains only one kind of atom and cannot be broken down to simpler components by chemical reactions. An atom is the smallest unit characteristic of an element and can exist alone or in combination with other atoms. When atoms combine, they form molecules. Molecules are the smallest particles of a substance. They have all the properties of the substance and are composed of two or more atoms.

Although atoms contain many subatomic particles, three directly influence their chemical behavior: protons, neutrons, and electrons. The atom's nucleus is located at its center and contains varying numbers of protons and neutrons (**figure AI.1**). Protons have a positive charge, and neutrons are uncharged. The mass of these particles and the atoms that they compose are given in terms of the atomic mass unit (AMU), which is $\frac{1}{2}$ the mass of the most abundant carbon isotope. Often the term *dalton* (Da) is used to express the mass of molecules. It also is $\frac{1}{2}$ the mass of an atom of ^{12}C or 1.661×10^{-24} gram. Both protons and neutrons have a mass of about 1 Da. The atomic weight is the actual measured weight of an element and is almost identical to the mass number for the element, the total number of protons and neutrons in its nucleus. The mass number is indicated by a superscripted number preceding the element's symbol (e.g., ^{12}C, ^{16}O, and ^{14}N).

Negatively charged particles called electrons circle the atomic nucleus (figure AI.1). The number of electrons in a neutral atom equals the number of its protons and is given by the atomic number, the number of protons in an atomic nucleus. The atomic number is characteristic of a particular type of atom. For example, carbon has an atomic number of six, hydrogen's number is one, and oxygen's is eight (**table AI.1**).

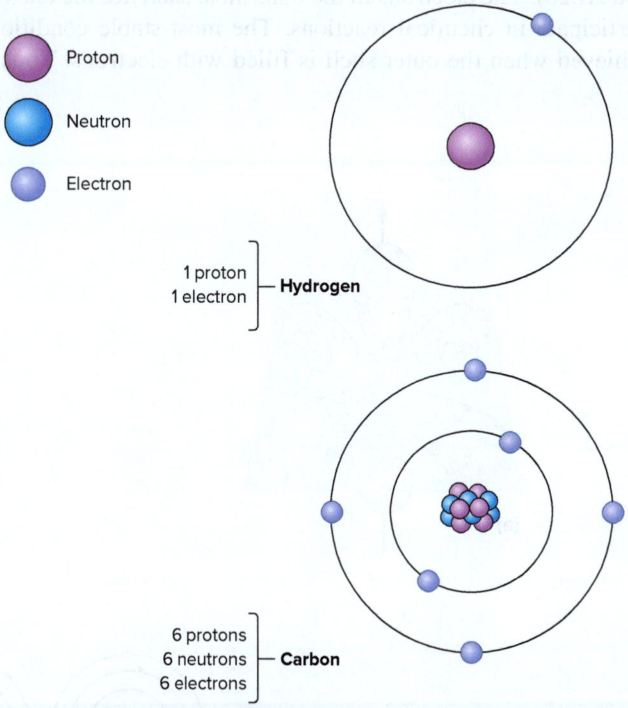

Figure AI.1 Diagrams of Hydrogen and Carbon Atoms. The electron orbitals are represented as concentric circles.

Table AI.1 Atoms Commonly Present in Organic Molecules

Atom	Symbol	Atomic Number	Atomic Weight	Number of Chemical Bonds
Hydrogen	H	1	1.01	1
Carbon	C	6	12.01	4
Nitrogen	N	7	14.01	3
Oxygen	O	8	16.00	2
Phosphorus	P	15	30.97	5
Sulfur	S	16	32.06	2

The electrons move constantly within a volume of space surrounding the nucleus, even though their precise location in this volume cannot be determined accurately. This volume of space in which an electron is located is called its orbital. Each orbital can contain two electrons. Orbitals are grouped into shells of different energy that surround the nucleus. The first shell is closest to the nucleus and has the lowest energy; it contains only one orbital. The second shell contains four orbitals, one circular and three shaped like dumbbells (**figure AI.2*a***). It can contain up to eight electrons. The third shell has even higher energy and holds more than eight electrons. Shells are filled beginning with the innermost and moving outward. For example, carbon has six electrons, two in its first shell and four in the second (figures AI.1 and AI.2*b*). The electrons in the outermost shell are the ones that participate in chemical reactions. The most stable condition is achieved when the outer shell is filled with electrons. Thus the number of bonds an element can form depends on the number of electrons required to fill the outer shell. Because carbon has four electrons in its outer shell and the shell is filled when it contains eight electrons, it can form four covalent bonds (table AI.1).

Chemical Bonds

Molecules are formed when two or more atoms associate through chemical bonding. Chemical bonds are attractive forces that hold together atoms, ions, or groups of atoms in a molecule or other substance. Many types of chemical bonds are present in organic molecules; three of the most important are covalent bonds, ionic bonds, and hydrogen bonds.

In covalent bonds, atoms are joined together by sharing pairs of electrons (**figure AI.3**). If the electrons are equally shared between identical atoms (e.g., in a carbon–carbon bond), the covalent bond is strong and nonpolar. When two different

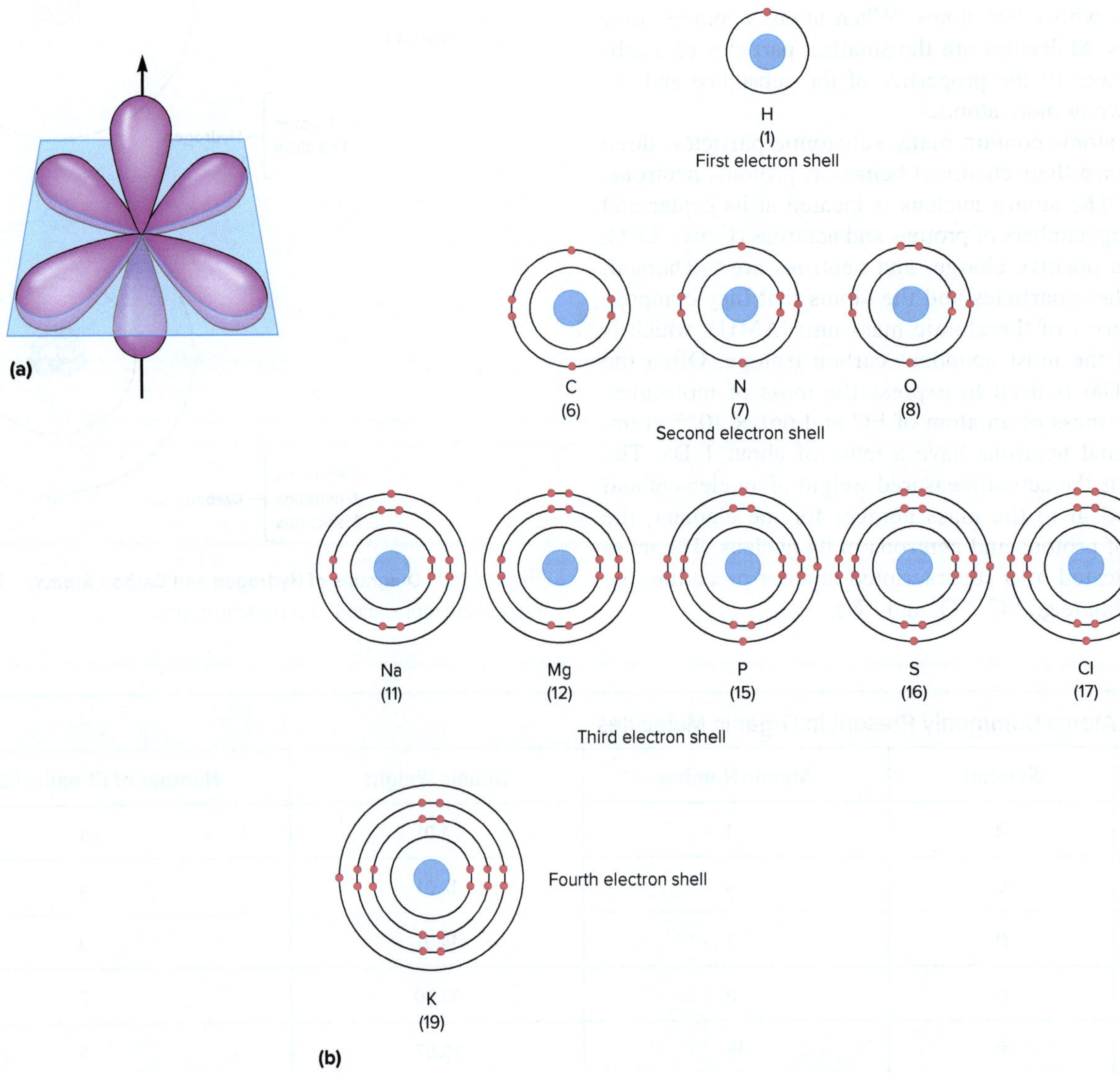

Figure AI.2 Electron Orbitals. (a) The three dumbbell-shaped orbitals of the second shell. The orbitals lie at right angles to each other. (b) The distribution of electrons in some common elements. Atomic numbers are given in parentheses.

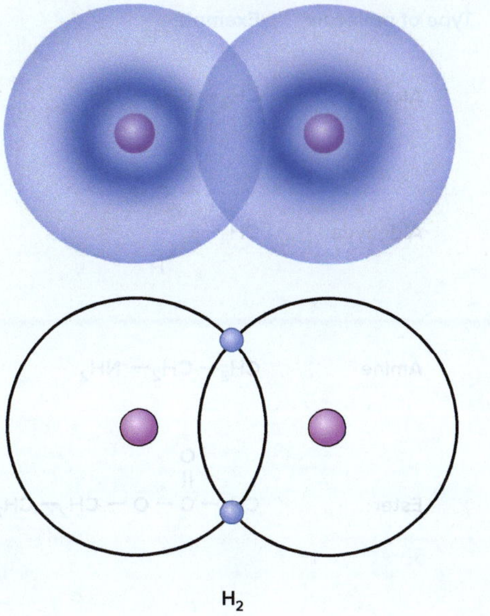

Figure AI.3 The Covalent Bond. A hydrogen molecule is formed when two hydrogen atoms share electrons.

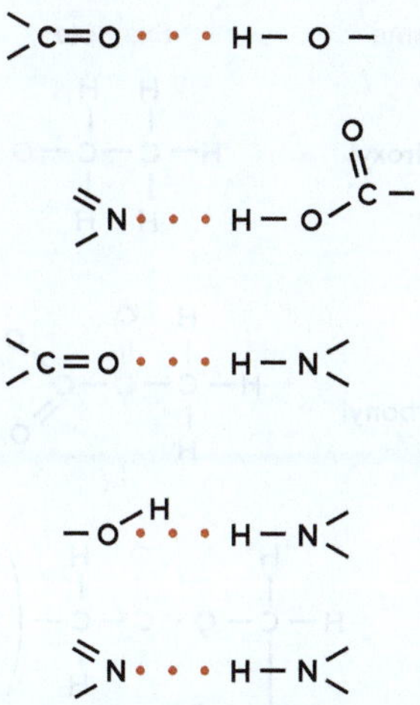

Figure AI.4 Hydrogen Bonds. Representative examples of hydrogen bonds present in biological molecules.

atoms such as carbon and oxygen share electrons, the covalent bond formed is polar because the electrons are pulled toward the more electronegative atom, the atom that more strongly attracts electrons (the oxygen atom). A single pair of electrons is shared in a single bond; a double bond is formed when two pairs of electrons are shared.

Atoms often contain either more or fewer electrons than the number of protons in their nuclei. When this is the case, they carry a net negative or positive charge and are called ions. Cations carry positive charges and anions have a net negative charge. When a cation and an anion approach each other, they are attracted by their opposite charges. This ionic attraction that holds two groups together is called an ionic bond. While it is difficult to make a direct comparison between the strength of ionic and covalent bonds, in an aqueous environment ionic bonds are easily disrupted, making them an important component of biological molecules.

When a hydrogen atom is covalently bonded to a more electronegative atom such as oxygen or nitrogen, the electrons are unequally shared and the hydrogen atom carries a partial positive charge. It will be attracted to an electronegative atom such as oxygen or nitrogen, which carries an unshared pair of electrons; this attraction is called a hydrogen bond (**figure AI.4**). Although an individual hydrogen bond is weak, there are so many hydrogen bonds in proteins and nucleic acids that they play a major role in determining protein and nucleic acid structure.

Organic Molecules

Most molecules in cells are organic molecules that contain carbon. Because carbon has four electrons in its outer shell, it forms four covalent bonds to fill its outer shell with eight electrons. This property makes it possible to form chains and rings of carbon atoms that also can bond with hydrogen and other atoms (**figure AI.5**). Although adjacent carbons usually are connected by single bonds, they may be joined by double or triple bonds. Rings that have alternating single and double bonds, such as the benzene ring, are called aromatic rings. The hydrocarbon chain or ring provides a chemically inactive skeleton to which more reactive groups of atoms may be attached. These reactive groups with specific properties are known as functional groups. They usually contain atoms of oxygen, nitrogen, phosphorus, or sulfur (**figure AI.6**) and are largely responsible for most characteristic chemical properties of organic molecules.

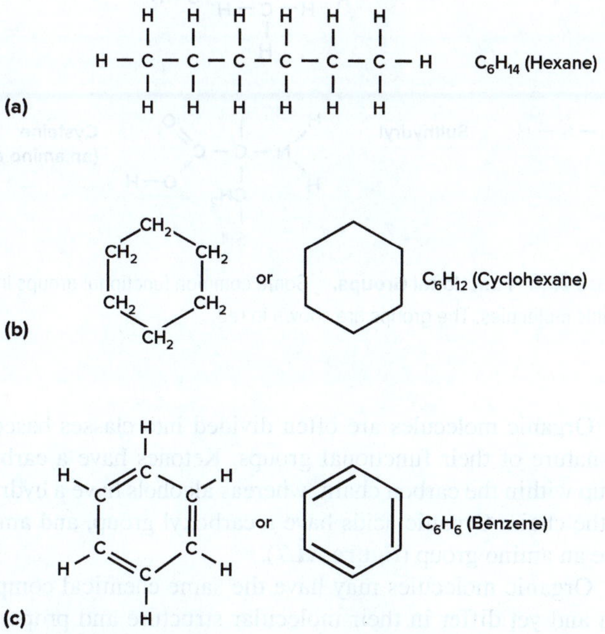

Figure AI.5 Hydrocarbons. Examples of hydrocarbons that are (a) linear, (b) cyclic, and (c) aromatic.

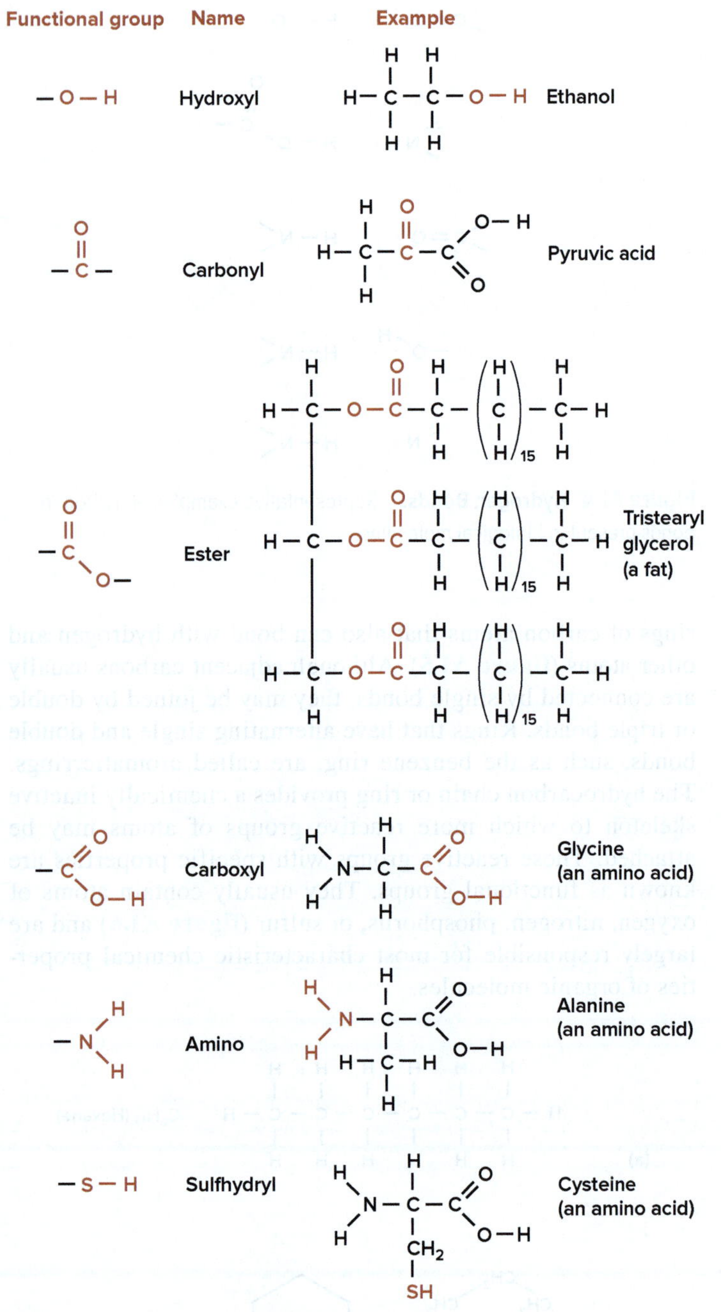

Figure AI.6 Functional Groups. Some common functional groups in organic molecules. The groups are shown in red.

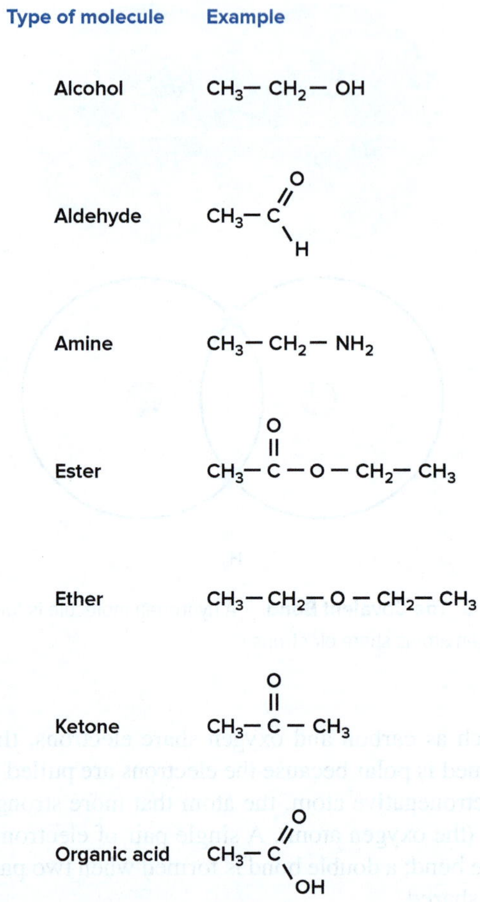

Figure AI.7 Types of Organic Molecules. These are classified on the basis of their functional groups.

Organic molecules are often divided into classes based on the nature of their functional groups. Ketones have a carbonyl group within the carbon chain, whereas alcohols have a hydroxyl on the chain. Organic acids have a carboxyl group, and amines have an amino group (**figure AI.7**).

Organic molecules may have the same chemical composition and yet differ in their molecular structure and properties. Such molecules are called isomers. One important class of isomers is the stereoisomers. Stereoisomers have the same atoms arranged in the same nucleus-to-nucleus sequence but differ in the spatial arrangement of their atoms. For example, an amino acid such as alanine can form stereoisomers (**figure AI.8**). L-Alanine and other L-amino acids are the stereoisomer forms normally present in proteins.

Carbohydrates

Carbohydrates are aldehyde or ketone derivatives of polyhydroxy alcohols. The smallest and least complex carbohydrates are the simple sugars or monosaccharides. The most common sugars have five or six carbons. A sugar in its ring form has two isomeric structures, the α and β forms, that differ in the orientation of the hydroxyl on the aldehyde or ketone carbon, which is called the anomeric or glycosidic carbon (**figure AI.9**). Microorganisms have many sugar derivatives in which a hydroxyl is replaced by an amino group or some other functional group (e.g., glucosamine).

Two monosaccharides can be joined by a bond between the anomeric carbon of one sugar and a hydroxyl or the anomeric

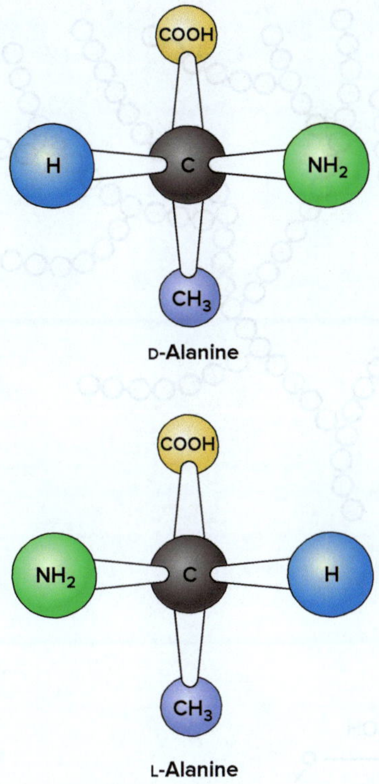

Figure AI.8 The Stereoisomers of Alanine. The α-carbon is in gray. L-Alanine is the form usually present in proteins.

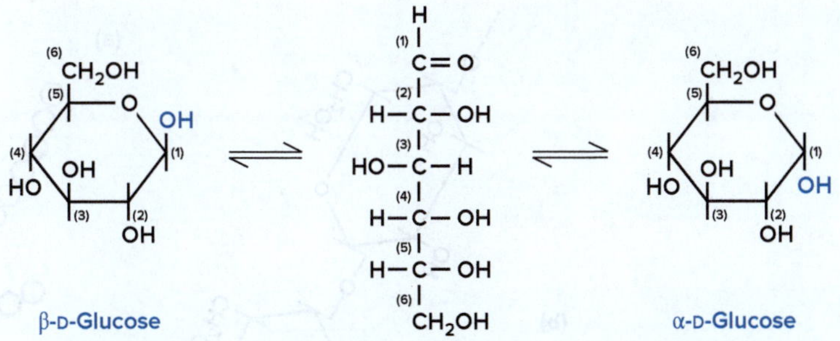

Figure AI.9 The Interconversion of Monosaccharide Structures. The open chain form of glucose and other sugars is in equilibrium with closed ring structures (depicted here with Haworth projections). Aldehyde sugars form cyclic hemiacetals, and keto sugars produce cyclic hemiketals. When the hydroxyl on carbon 1 of cyclic hemiacetals projects above the ring, the form is known as a β form. The α form has a hydroxyl that lies below the plane of the ring. The same convention is used in showing the α and β forms of hemiketals such as those formed by fructose.

carbon of the second (**figure AI.10**). The bond joining sugars is a glycosidic bond and may be either α or β depending on the orientation of the anomeric carbon. Two sugars linked in this way constitute a disaccharide. Some common disaccharides are maltose (two glucose molecules), lactose (glucose and galactose),

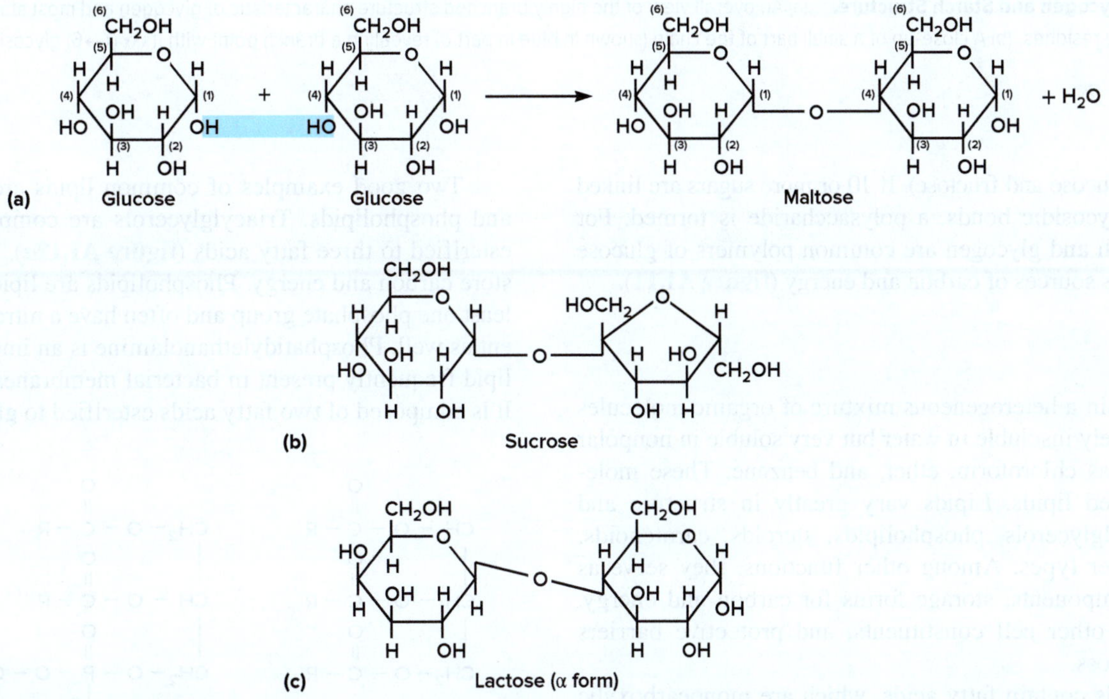

Figure AI.10 Common Disaccharides. (a) The formation of maltose from two molecules of α-glucose. The bond connecting the glucose extends between carbons 1 and 4 and involves the α form of the anomeric carbon. Therefore it is called an α (1→4) glycosidic bond. (b) Sucrose is composed of a glucose and a fructose joined to each other through their anomeric carbons, an αβ (1→2) bond. (c) The milk sugar lactose contains galactose and glucose joined by a β (1→4) glycosidic bond.

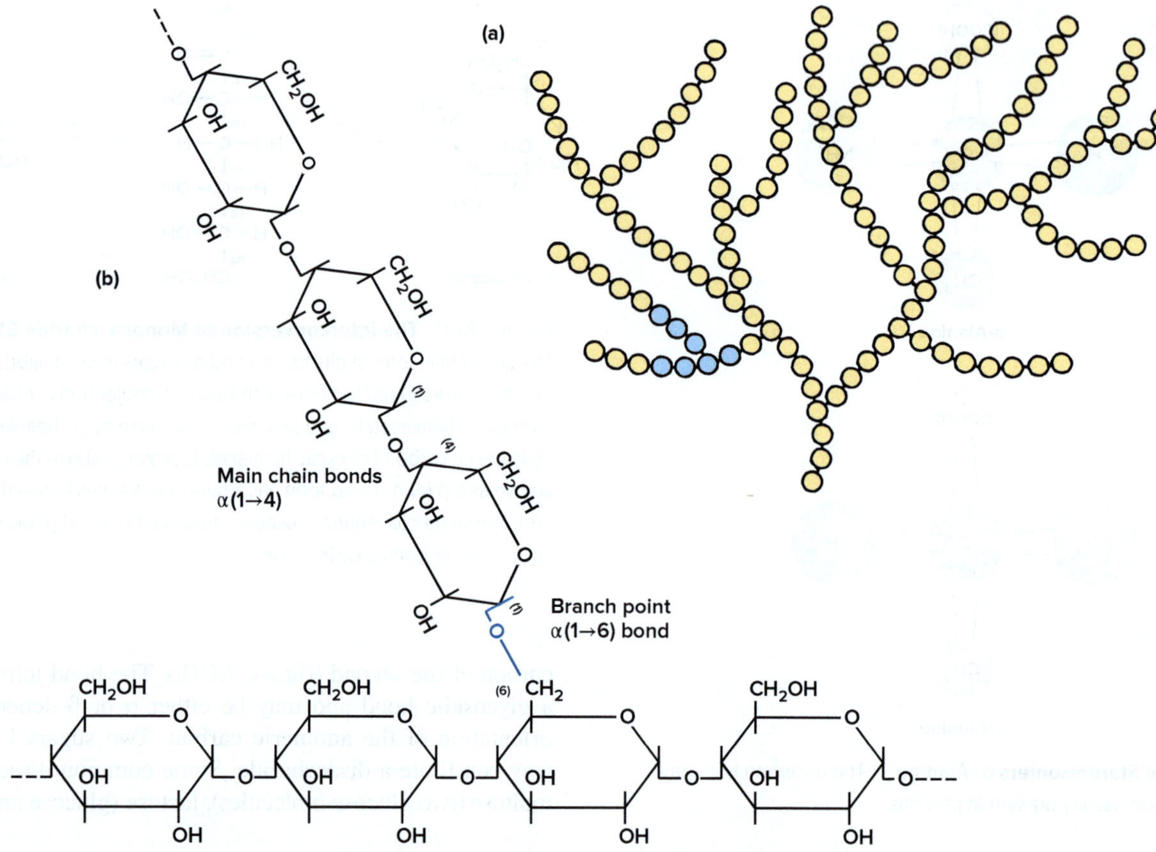

Figure AI.11 Glycogen and Starch Structure. (a) An overall view of the highly branched structure characteristic of glycogen and most starch. The circles represent glucose residues. (b) A close-up of a small part of the chain (shown in blue in part *a*) revealing a branch point with its α (1→6) glycosidic bond, which is colored blue.

and sucrose (glucose and fructose). If 10 or more sugars are linked together by glycosidic bonds, a polysaccharide is formed. For example, starch and glycogen are common polymers of glucose that are used as sources of carbon and energy (**figure AI.11**).

Lipids

All cells contain a heterogeneous mixture of organic molecules that are relatively insoluble in water but very soluble in nonpolar solvents such as chloroform, ether, and benzene. These molecules are called lipids. Lipids vary greatly in structure and include triacylglycerols, phospholipids, steroids, carotenoids, and many other types. Among other functions, they serve as membrane components, storage forms for carbon and energy, precursors of other cell constituents, and protective barriers against water loss.

Most lipids contain fatty acids, which are monocarboxylic acids that often are straight chained but may be branched. Saturated fatty acids lack double bonds in their carbon chains, whereas unsaturated fatty acids have double bonds. The most common fatty acids are 16 or 18 carbons long.

Two good examples of common lipids are triacylglycerols and phospholipids. Triacylglycerols are composed of glycerol esterified to three fatty acids (**figure AI.12*a***). They are used to store carbon and energy. Phospholipids are lipids that contain at least one phosphate group and often have a nitrogenous constituent as well. Phosphatidylethanolamine is an important phospholipid frequently present in bacterial membranes (figure AI.12*b*). It is composed of two fatty acids esterified to glycerol. The third

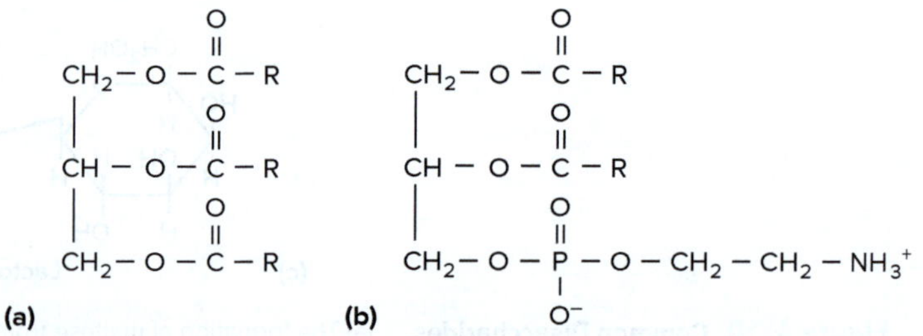

Figure AI.12 Examples of Common Lipids. (a) A triacylglycerol or neutral fat. (b) The phospholipid phosphatidylethanolamine. The R groups represent fatty acid side chains.

glycerol hydroxyl is joined with a phosphate group, and ethanolamine is attached to the phosphate. The resulting lipid is very asymmetric with a hydrophobic nonpolar end contributed by the fatty acids and a polar, hydrophilic end. In cell membranes, the hydrophobic end is buried in the interior of the membrane, while the polar-charged end is at the membrane surface and exposed to water.

Proteins

The basic building blocks of proteins are amino acids. An amino acid contains a carboxyl group and an amino group on its alpha carbon (**figure AI.13**). About 20 amino acids are normally found in proteins; they differ from each other with respect to their side chains. In proteins, amino acids are linked together by peptide bonds between their carboxyls and α-amino groups to form linear polymers called peptides. A small protein is called a peptide; a protein that contains more than 50 amino acids is called a polypeptide. Each protein is composed of one or more polypeptide chains and has a molecular weight greater than about 6,000 to 7,000.

Proteins have three or four levels of structural organization and complexity. The primary structure of a protein is the sequence of the amino acids in its polypeptide chain or chains. The structure of the polypeptide chain backbone is also considered part of the primary structure. Each different polypeptide has its own amino acid sequence dictated by the nucleotide sequence in the gene that codes for its synthesis. The polypeptide chain can coil along one axis in space into various shapes such as the α-helix (**figure AI.14**). This arrangement of the polypeptide in space around a single axis is called the secondary structure. Secondary structure is formed and stabilized by the interactions of amino acids that are fairly close to one another on the polypeptide chain. The polypeptide with its primary and secondary structures can be coiled or organized in space along three axes to form a more complex, three-dimensional shape (**figure AI.15**). This level of organization is the tertiary structure (**figure AI.16**). Amino acids more distant from one another on the polypeptide chain contribute to tertiary structure. Secondary and tertiary structures are examples of conformation, molecular shape that can be changed by bond rotation and without breaking covalent bonds. When a protein contains more than one polypeptide chain, each chain with its own primary, secondary, and tertiary structures associates with the other chains to form the final molecule. The way in which polypeptides associate with each other in space to form the final protein is called the protein's quaternary structure (**figure AI.17**). The final conformation of a protein is ultimately determined by the amino acid sequence of its polypeptide chains.

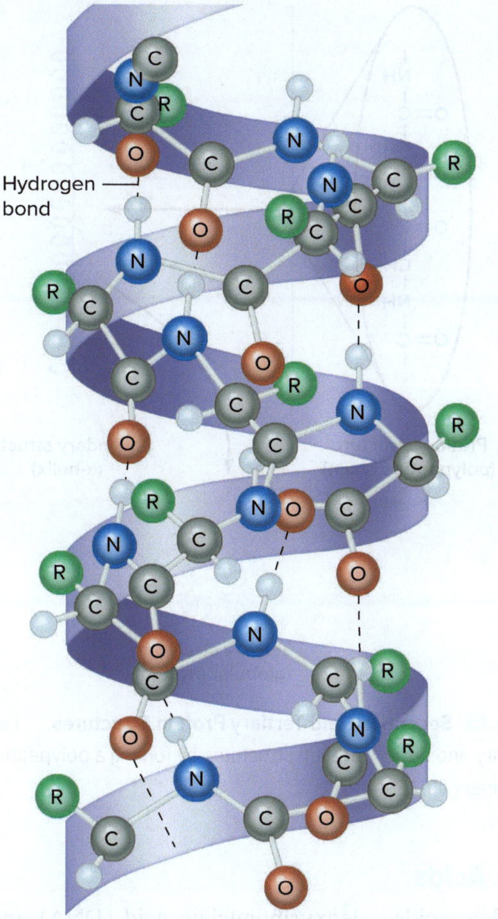

Figure AI.14 The α-Helix. A polypeptide twisted into one type of secondary structure, the α-helix. The helix is stabilized by hydrogen bonds joining peptide bonds that are separated by three amino acids.

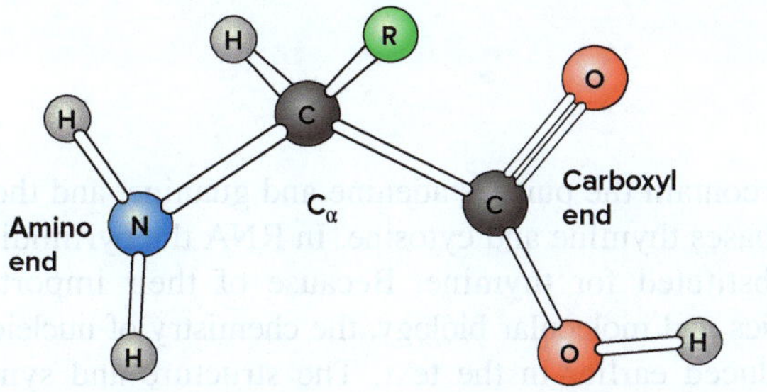

Figure AI.13 L-Amino Acid Structure. The uncharged form is shown.

Protein secondary, tertiary, and quaternary structures are largely determined and stabilized by many weak noncovalent forces such as hydrogen bonds and ionic bonds. Because of this, protein shape often is very flexible and easily changed. This flexibility is very important in protein function and in the regulation of enzyme activity. Because of their flexibility, however, proteins readily lose their proper shape and activity when exposed to harsh conditions. The only covalent bond commonly involved in the secondary and tertiary structures of proteins is the disulfide bond. The disulfide bond is formed when two cysteines are linked through their sulfhydryl groups. Disulfide bonds generally strengthen or stabilize protein structure.

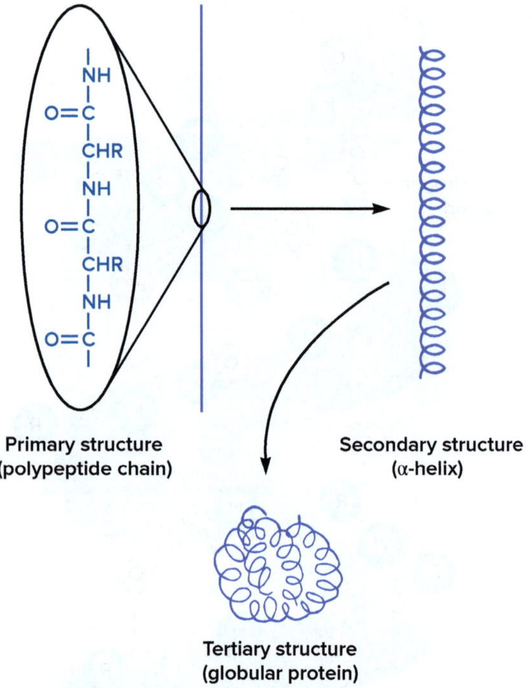

Figure AI.15 Secondary and Tertiary Protein Structures. The formation of secondary and tertiary protein structures by folding a polypeptide chain with its primary structure.

Nucleic Acids

The nucleic acids—deoxyribonucleic acid (DNA) and ribonucleic acid (RNA)—are polymers of deoxyribonucleosides and ribonucleosides joined by phosphate groups. The nucleosides in DNA contain the purines adenine and guanine, and the pyrimidine bases thymine and cytosine. In RNA the pyrimidine uracil is substituted for thymine. Because of their importance for genetics and molecular biology, the chemistry of nucleic acids is introduced earlier in the text. The structure and synthesis of purines and pyrimidines are discussed in chapter 12 (*section 12.6*). The structures of DNA and RNA are described in chapter 13 (*section 13.2*).

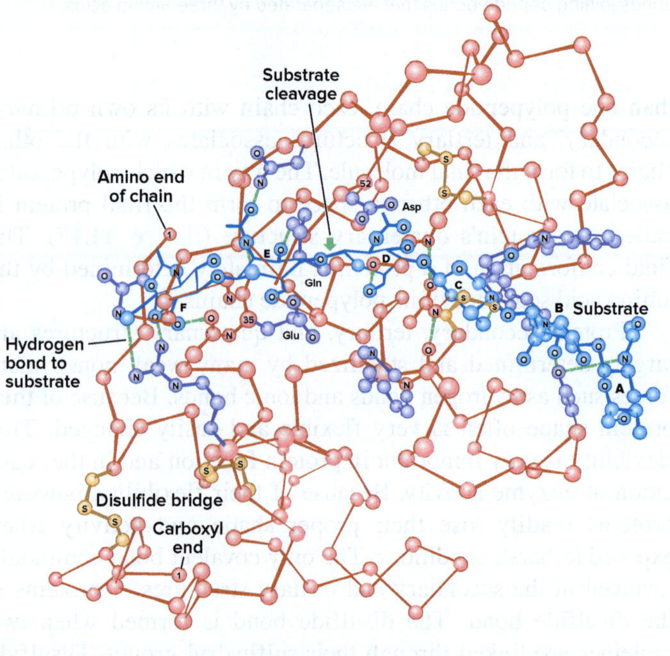

Figure AI.16 Lysozyme. The tertiary structure of the enzyme lysozyme. A diagram of the protein's polypeptide backbone with the substrate hexasaccharide shown in blue. The point of substrate cleavage is indicated.

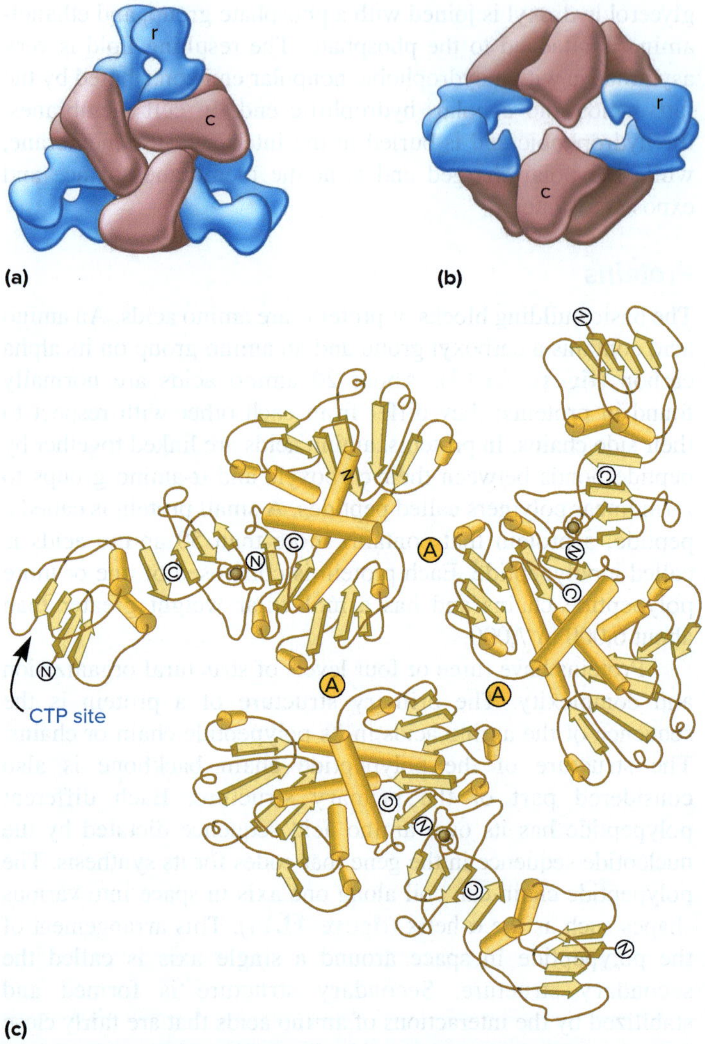

Figure AI.17 An Example of Quaternary Structure. The enzyme aspartate carbamoyltransferase from *Escherichia coli* has two types of subunits: catalytic and regulatory. The association between the two types of subunits is shown: (a) a top view and (b) a side view of the enzyme. The catalytic (c) and regulator (r) subunits are shown in different colors. (c) The peptide chains shown when viewed from the top as in (a). The active sites of the enzyme are located at the positions indicated by A.

Source: Kantrowitz, et al., in Trends in Biochemical Science, V. 5, 1980, as appeared in *Biochemistry*, 3d edition, by Lubert Stryer.

Appendix II

Common Metabolic Pathways

Appendix II contains a few of the more important pathways discussed in the text, particularly those involved in carbohydrate catabolism. Enzyme names and final end products are given in color. Consult the text for a description of each pathway and its physiologic role.

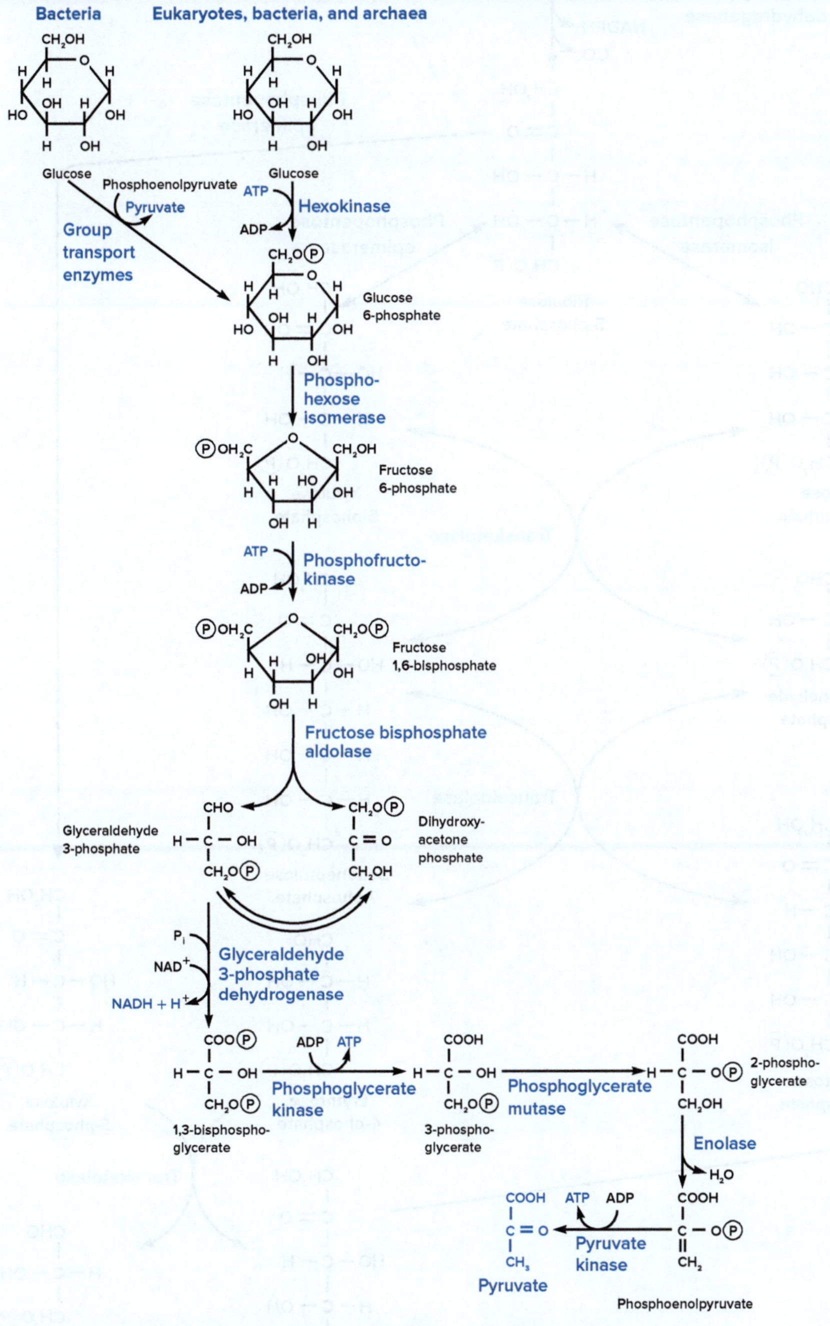

Figure AII.1 The Embden-Meyerhof Pathway. This pathway converts glucose and other sugars to pyruvate and generates NADH and ATP. In some bacteria, glucose is phosphorylated to glucose 6-phosphate during group translocation transport across the plasma membrane.

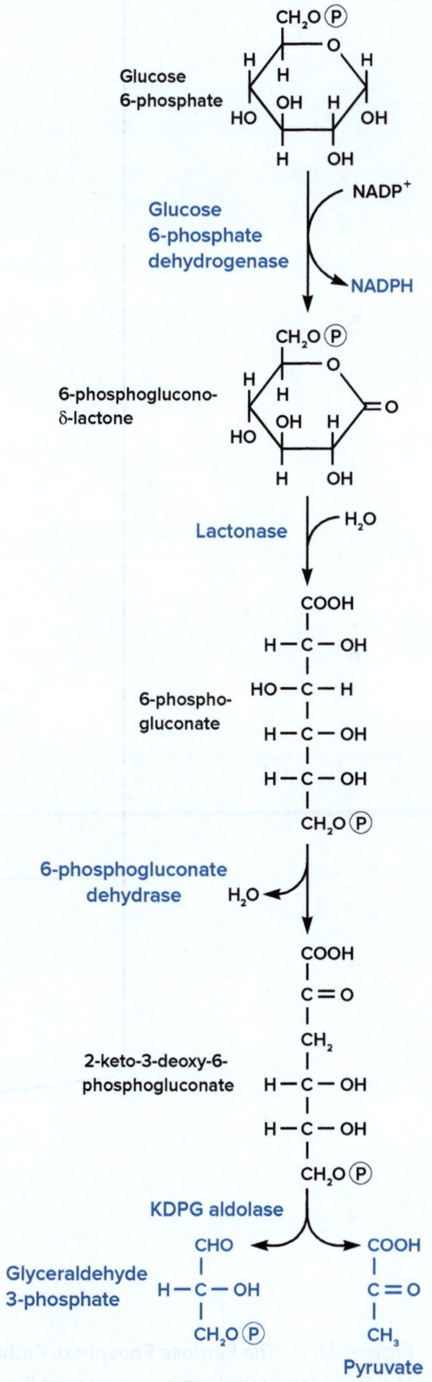

Figure AII.2 The Entner-Doudoroff Pathway.

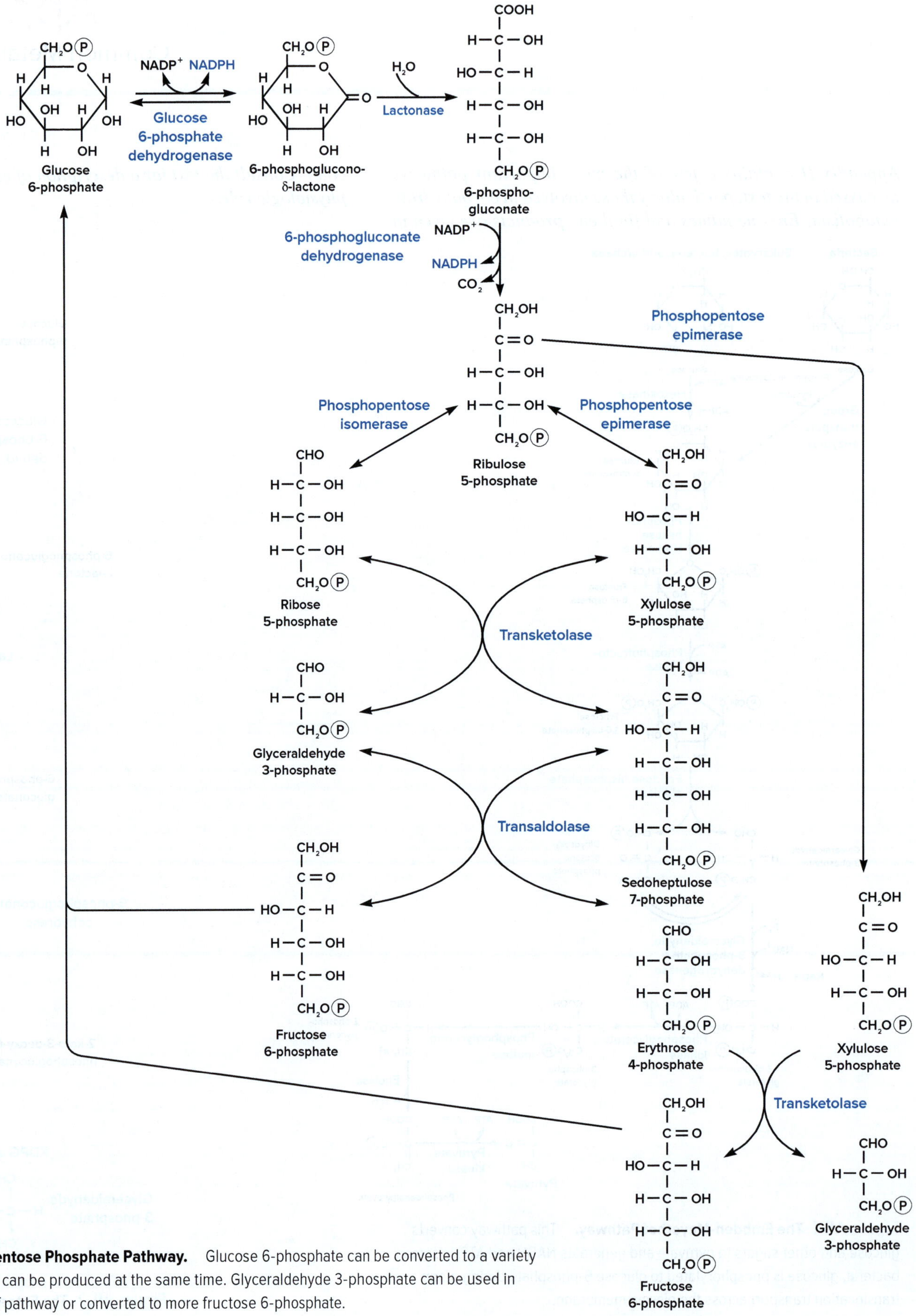

Figure AII.3 The Pentose Phosphate Pathway. Glucose 6-phosphate can be converted to a variety of sugars, and NADPH can be produced at the same time. Glyceraldehyde 3-phosphate can be used in the Embden-Meyerhof pathway or converted to more fructose 6-phosphate.

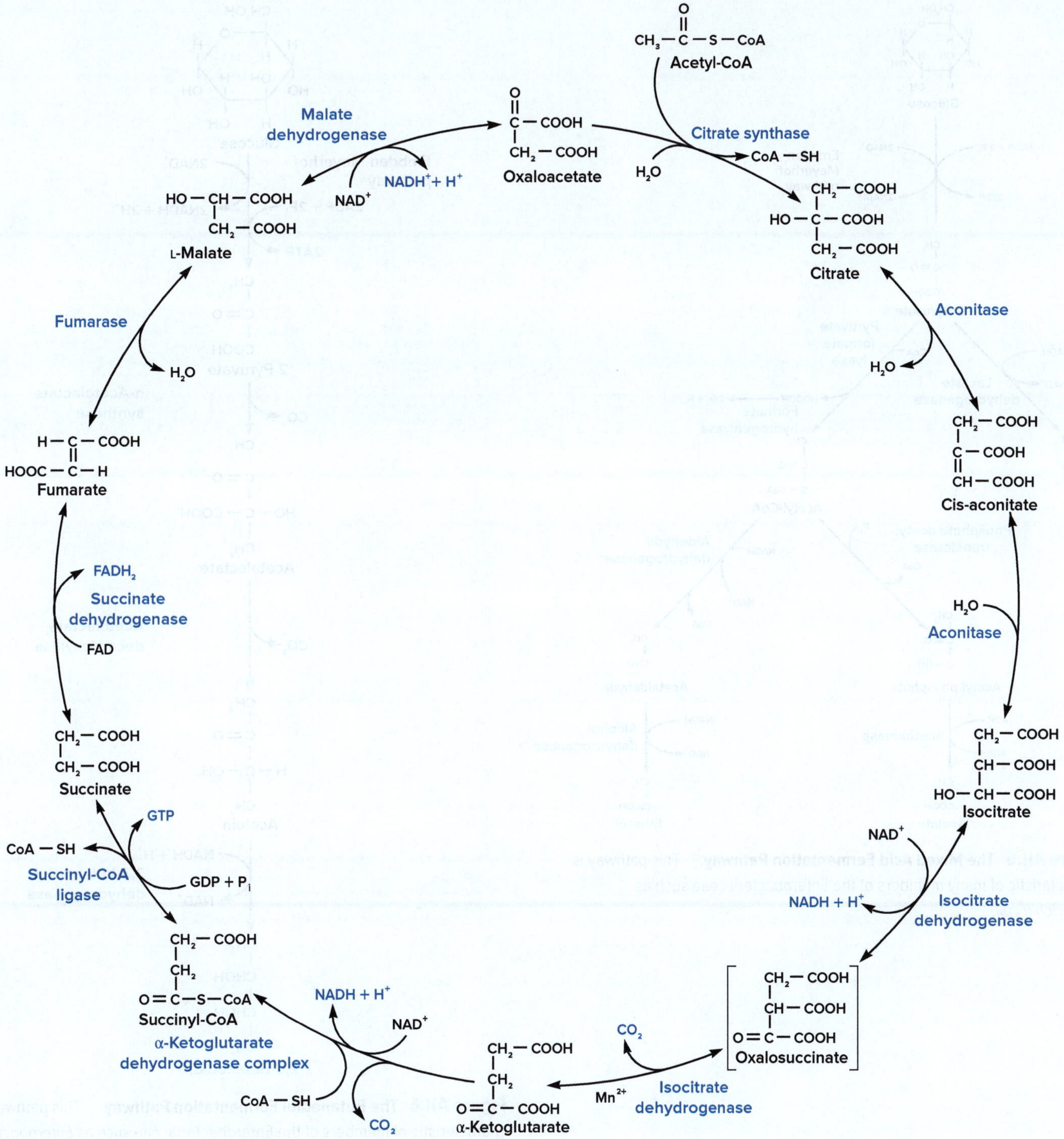

Figure AII.4 The Tricarboxylic Acid Cycle. Cis-aconitate and oxalosuccinate remain bound to aconitase and isocitrate dehydrogenase, respectively. Oxalosuccinate has been placed in brackets because it is so unstable.

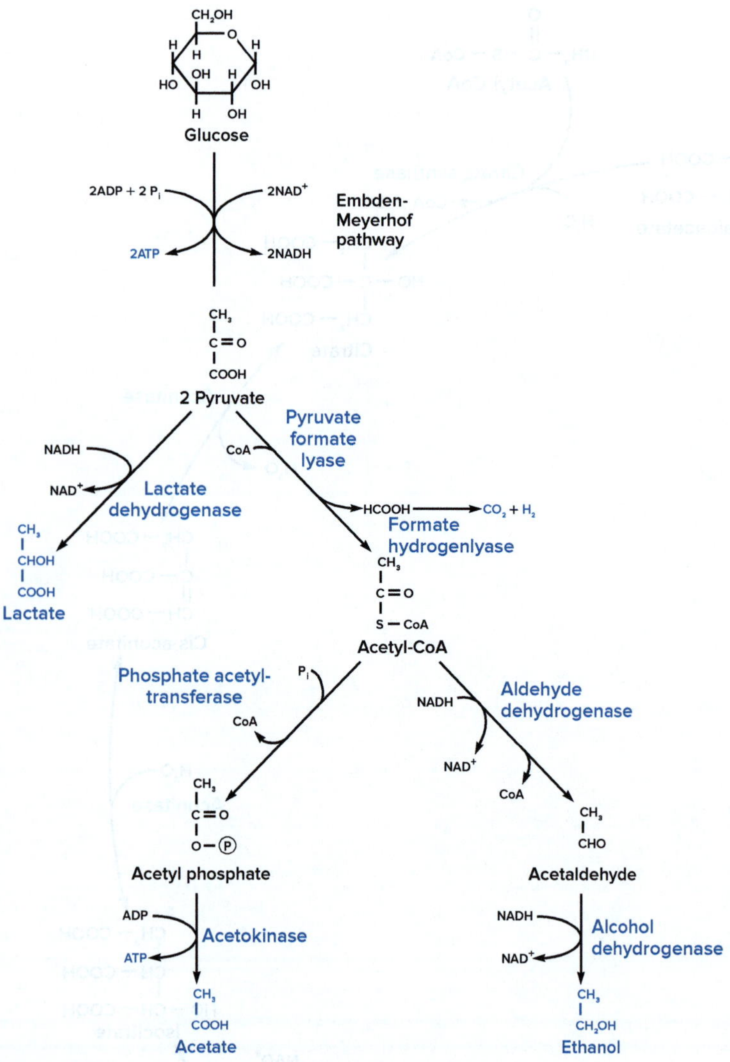

Figure AII.5 The Mixed Acid Fermentation Pathway. This pathway is characteristic of many members of the Enterobacteriaceae such as *Escherichia coli.*

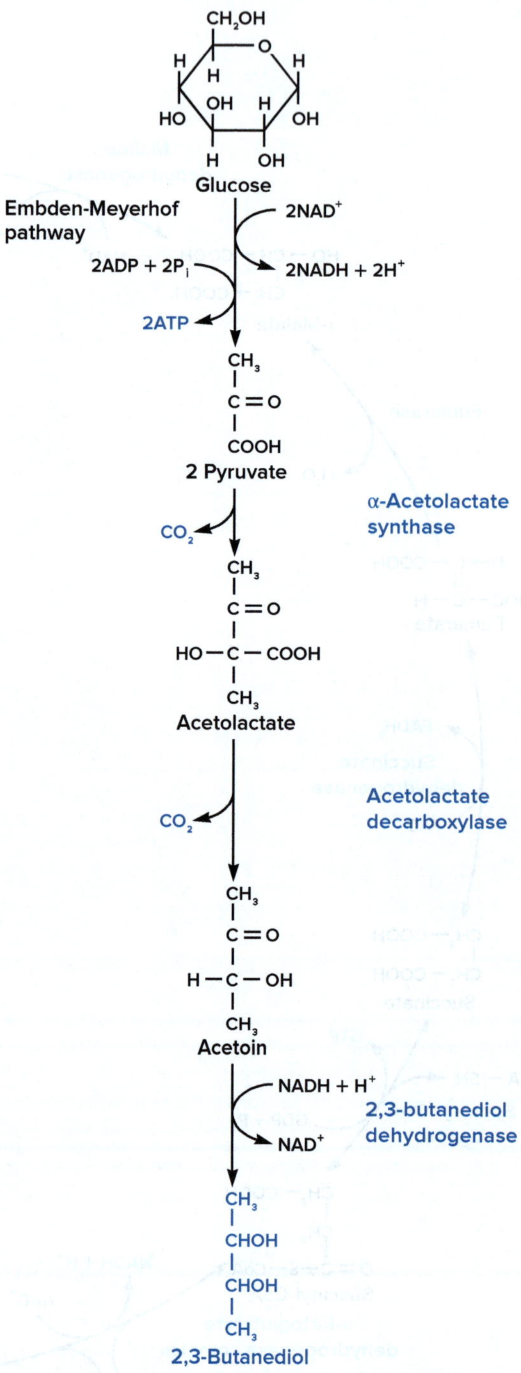

Figure AII.6 The Butanediol Fermentation Pathway. This pathway is characteristic of members of the Enterobacteriaceae such as *Enterobacter.* Other products may also be formed during butanediol fermentation.

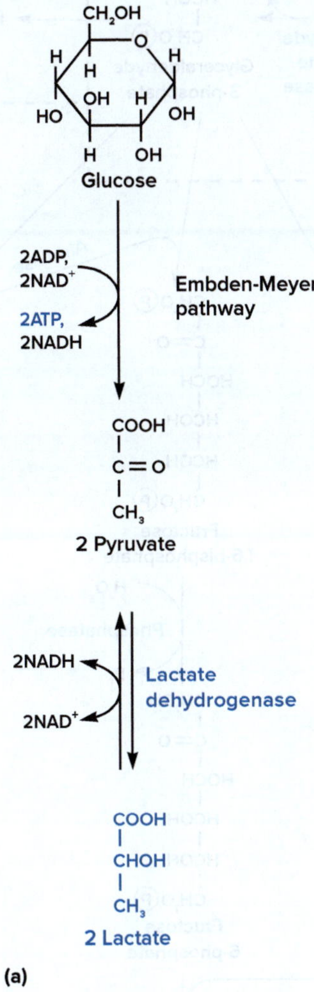

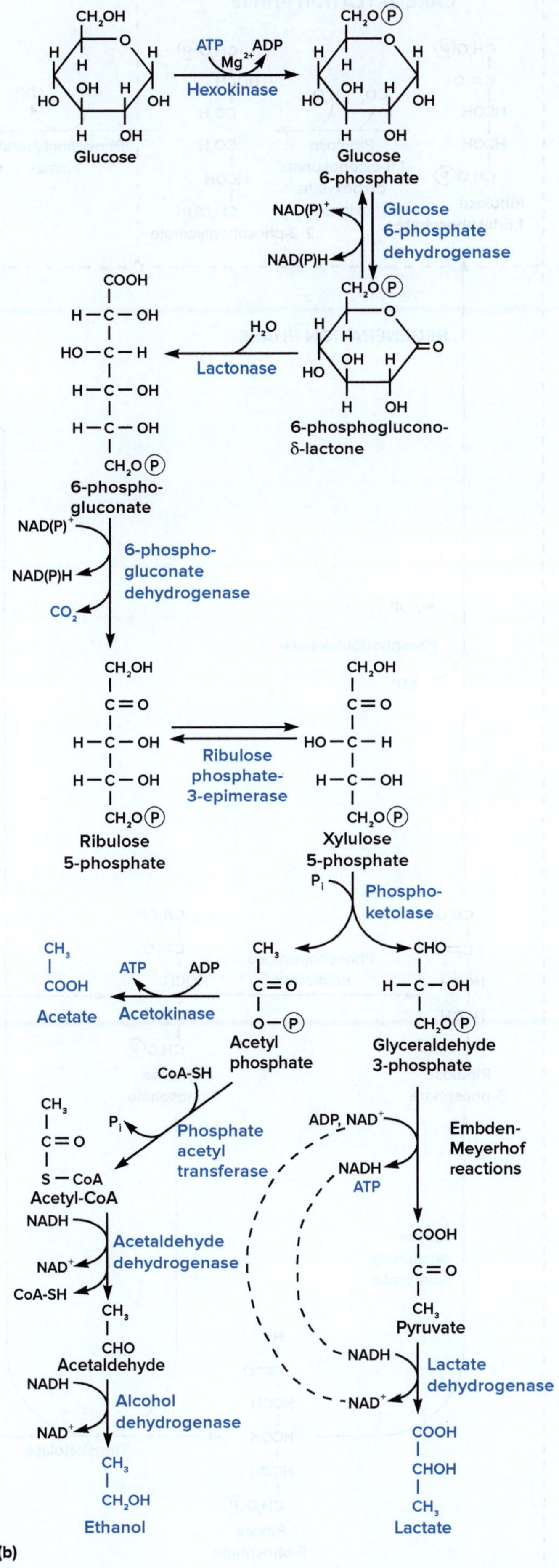

Figure AII.7 Lactic Acid Fermentations. (a) Homolactic fermentation pathway. (b) Heterolactic fermentation pathways.

Figure AII.8 The Calvin-Benson Cycle.

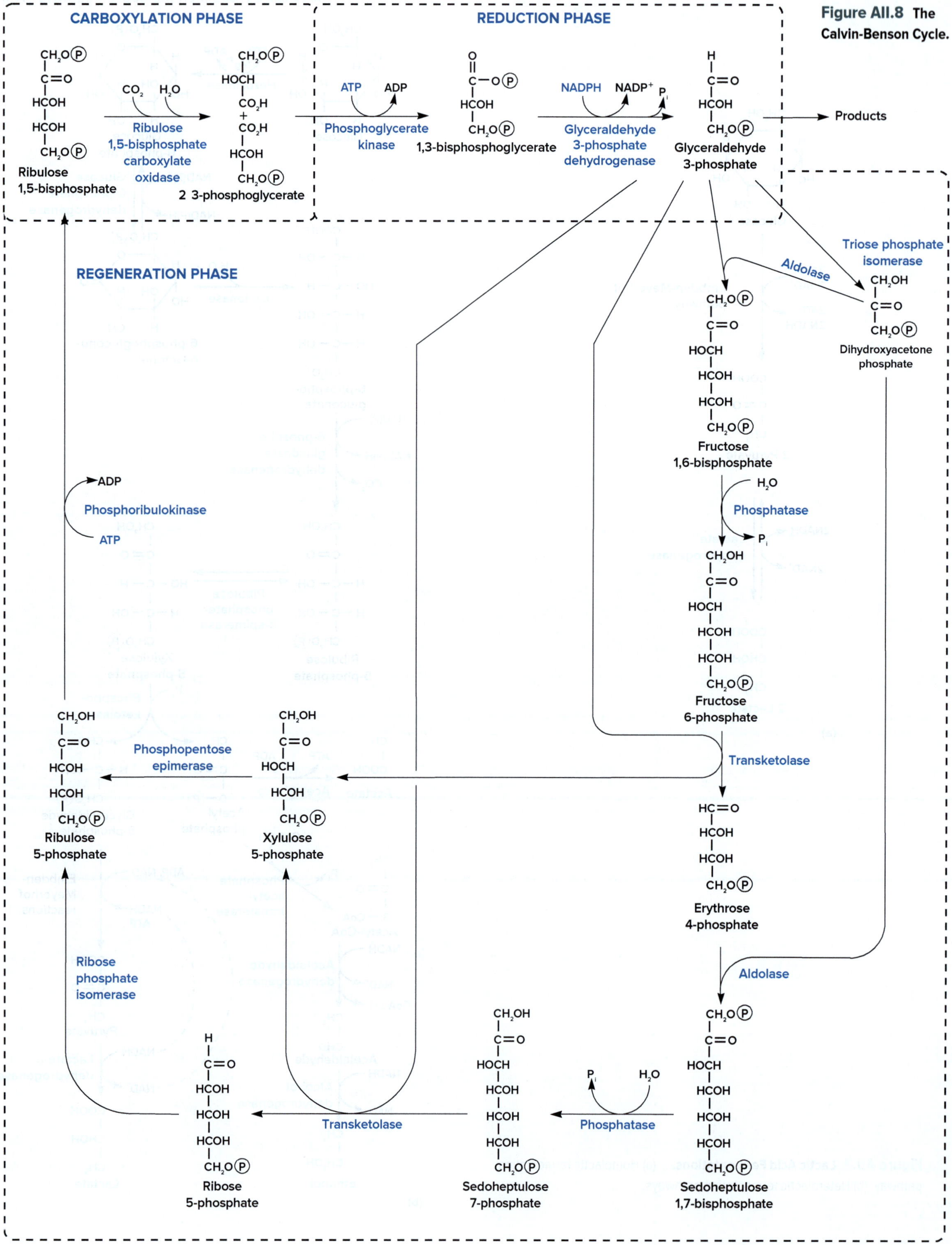

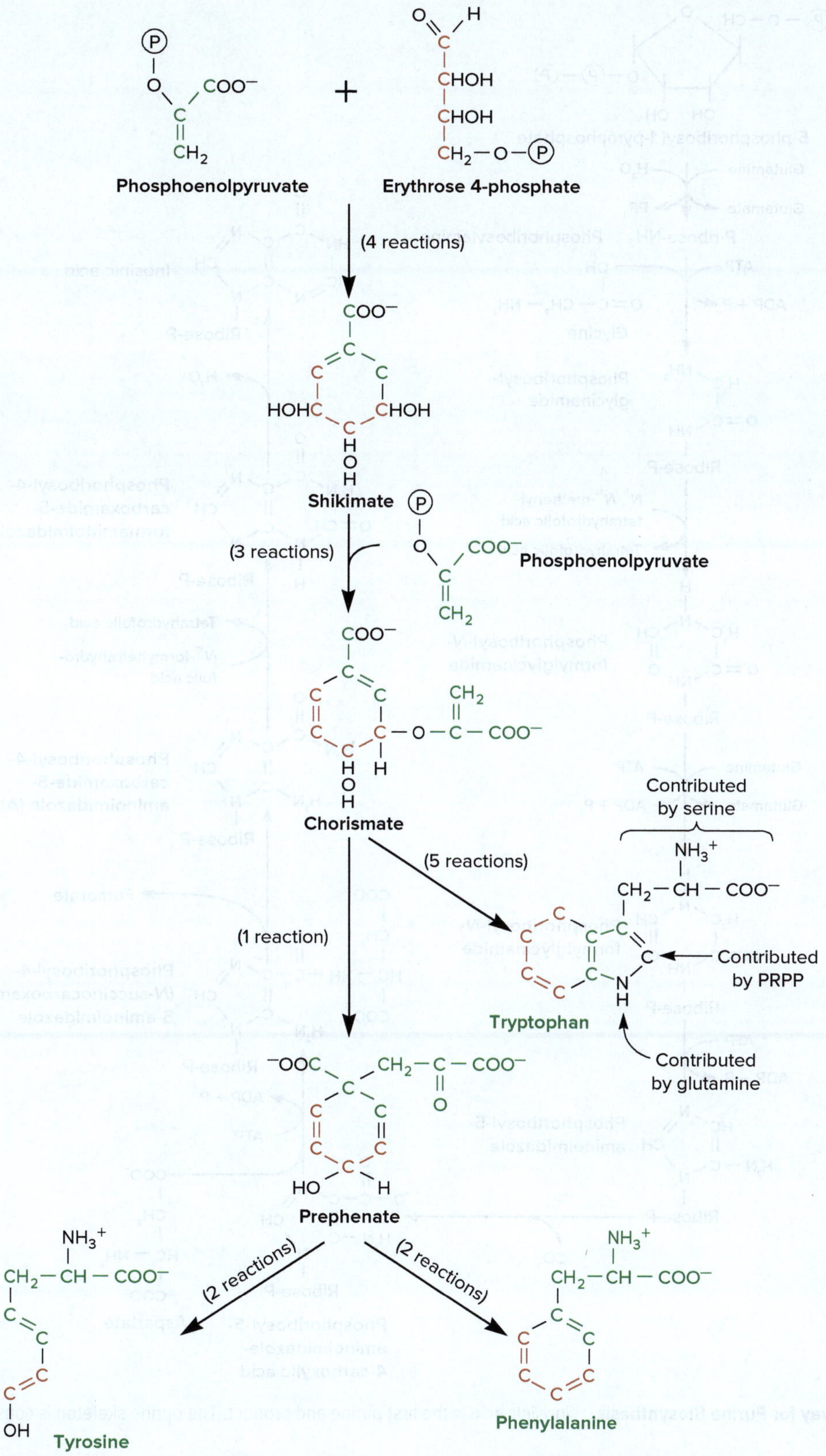

Figure AII.9 Aromatic Amino Acid Synthesis. The carbons arising from phosphoenolpyruvate (green) and erythrose 4-phosphate (red) are shown. The remaining carbons present in tryptophan are provided by 5-phosphoribosyl-1-pyrophosphate (PRPP) and the amino acid serine. PRPP is also important in purine biosynthesis (figure AII.10).

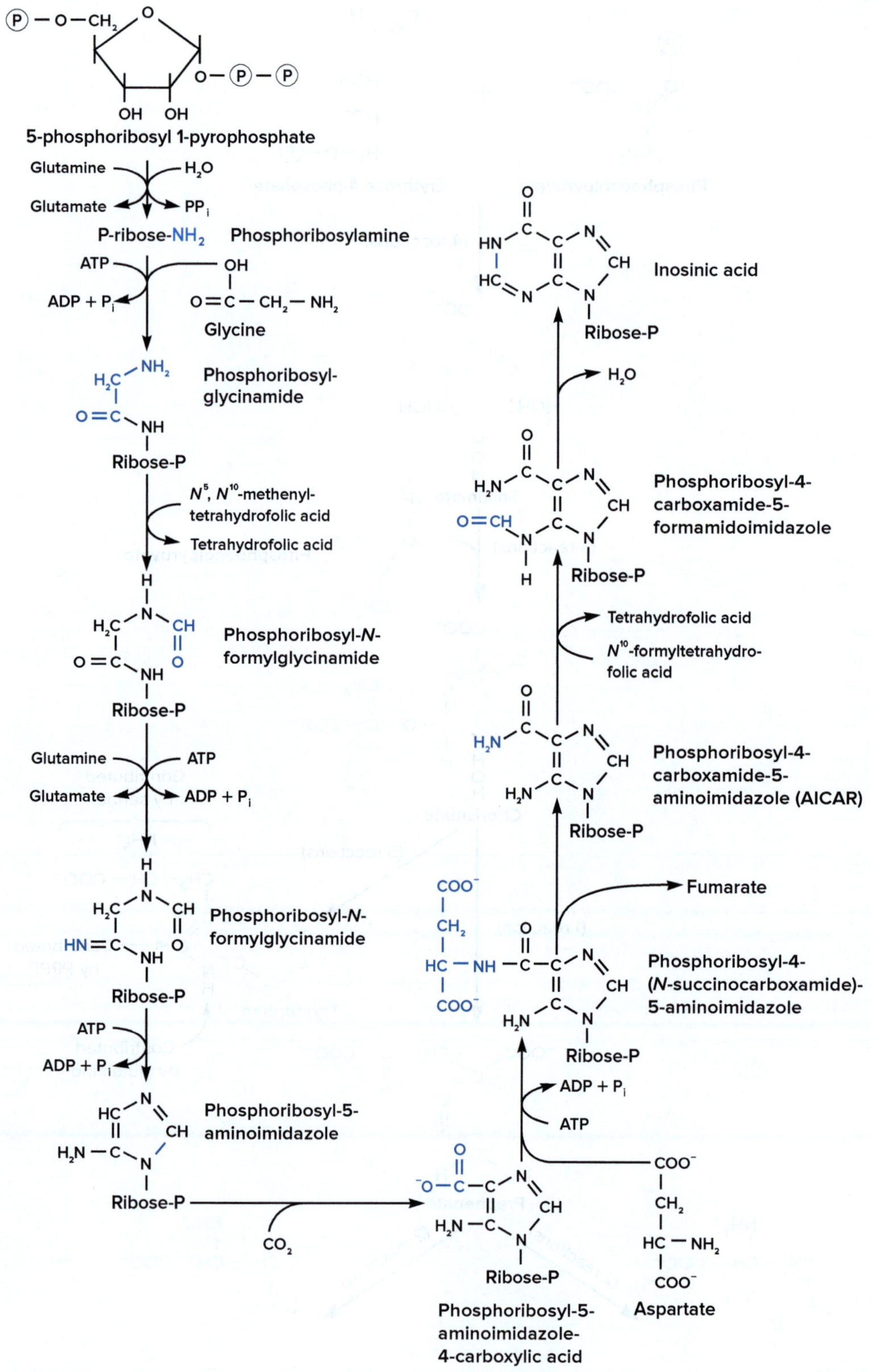

Figure AII.10 The Pathway for Purine Biosynthesis. Inosinic acid is the first purine end product. The purine skeleton is constructed while attached to a ribose phosphate.

Appendix III

Microorganism Pronunciation Guide

The pronunciation of each name is given in parentheses. The phonetic spelling system is explained at the beginning of the glossary.

Bacteria and Archaea

Acetobacter (ah-se″to-bak′ter)
Acidithiobacillus (ah-sid″ah-thi″o-bah-sil′us)
Acinetobacter (as″ĭ-net″o-bak′ter)
Actinomyces (ak″tĭ-no-mi′sēz)
Agrobacterium (ag″ro-bak-te′re-um)
Akkermansia (ak″er-manz′e-ah)
Alcaligenes (al″kah-lij′ĕ-nēz)
Altiarchaeum (al″te-ark′e-um)
Anabaena (ah-nab′ē-nah)
Arthrobacter (ar″thro-bak′ter)
Bacillus (bah-sil′lus)
Bacteroides (bak″tĕ-roi′dēz)
Bdellovibrio (del″o-vib′re-o)
Beggiatoa (bej″je-ah-to′ah)
Bifidobacterium (bi″fid-o-bak-te′re-um)
Bordetella (bor″dĕ-tel′lah)
Borreliella (bo-rel″e-el′-ah)
Burkholderia (berk-hōl-der′e-ah)
Campylobacter (kam″pī-lo-bak′ter)
Caulobacter (kaw″lo-bak′ter)
Chlamydia (klah-mid′e-ah)
Chlorobium (klo-ro′be-um)
Chromatium (kro-ma′te-um)
Chroococcus (kro″o-kok′us)
Citrobacter (sit″ro-bak′ter)
Clostridioides (klo-strid′e-oi″dēz)
Clostridium (klo-strid′e-um)
Corynebacterium (ko-ri″ne-bak-te′re-um)
Coxiella (kok″se-el′lah)
Cytophaga (si-tof′ah-gah)
Deinococcus (di′no-kok″us)
Desulfovibrio (de-sul″fo-vib′re-o)
Endoriftia (en″do-rif′te-ah)
Enterobacter (en″ter-o-bak′ter)
Erwinia (er-win′e-ah)
Escherichia (esh″er-i′ke-ah)
Francisella (fran-sĭ-sel′ah)
Frankia (frank′e-ah)
Haemophilus (he-mof′ĭ-lus)
Klebsiella (kleb″se-el′lah)
Lactobacillus (lak″to-bah-sil′lus)
Legionella (le″jun-el′ah)
Leptospira (lep″to-spi′rah)
Leptothrix (lep′to-thriks)
Leuconostoc (loo″ko-nos′tok)
Listeria (lis-te′re-ah)
Lokiarchaeum (lo″ke-ark′e-um)
Methanothermobacter (meth″ah-no-ther″mo-bak′ter)
Methylococcus (meth″il-o-kok′-us)
Micrococcus (mi″kro-kok′us)
Mycobacterium (mi″ko-bak-te′re-um)
Mycoplasma (mi″ko-plaz′mah)
Neisseria (nīs-se′re-ah)
Nitrobacter (ni″tro-bak′ter)
Nitrosomonas (ni-tro″so-mo′nas)
Nocardia (no-kar′de-ah)
Oenococcus (e″no-kok′us)
Pelagibacter (pel-a″ji-bak′ter)
Propionibacterium (pro″pe-on″e-bak-te′re-um)
Proteus (pro′te-us)
Pseudomonas (soo″do-mo′nas)
Rhizobium (ri-zo′be-um)
Rhodopseudomonas (ro″do-soo″do-mo′nas)
Rickettsia (rĭ-ket′se-ah)
Salmonella (sal″mo-nel′ah)
Serratia (sĕ-ra′she-ah)
Shewanella (shu″en-el′ah)
Shigella (shĭ-gel′ah)
Sphaerotilus (sfe-ro′tĭ-lus)
Spirochaeta (spi″ro-ke′tah)
Spiroplasma (spi″ro-plaz′mah)
Staphylococcus (staf″ĭ-lo-kok′us)
Streptococcus (strep″to-kok′us)
Streptomyces (strep″to-mi′sēz)
Sulfolobus (sul″fo-lo′bus)
Synechococcus (sin′-eh-ko-kok′-us)
Thermoactinomyces (ther″mo-ak″tĭ-no-mi′sēz)
Thiothrix (thi′o-thriks)
Treponema (trep″o-ne′mah)
Veillonella (va″yon-el′ah)
Vibrio (vib′re-o)
Wolbachia (wol-bak′e-ah)
Xanthomonas (zan″tho-mo′nas)
Xenorhabdus (ze″no-rab′dus)
Yersinia (yer-sin′e-ah)

Viruses

Vernacular, nonscientific virus names are not written in italics, and thus the following names are not italicized.

adenovirus (ad″ĕ-no-vi′rus)
arbovirus (ar″bo-vi′rus)
baculovirus (bak″u-lo-vi′rus)
coronavirus (kor-o″nah-vi′rus)
cytomegalovirus (si″to-meg″ah-lo-vi′rus)
Epstein-Barr virus (ep′stīn-bar′)
hepadnavirus (hep-ad″nə-vi′rus)
hepatitis virus (hep″ah-ti′tis)
herpesvirus (her″pēz-vi′rus)
influenza virus (in″flu-en′zah)
measles virus (me′zelz)
mimivirus (me′me-vi′rus)
mumps virus (mumps)
orthomyxovirus (or″tho-mik″so-vi′rus)
papillomavirus (pap″ĭ-lo″mah-vi′rus)
paramyxovirus (par″ah-mik″so-vi′rus)
parvovirus (par″vo-vi′rus)
picornavirus (pi-kor″nah-vi′rus)
poxvirus (poks-vi′rus)
rabies virus (ra′bēz)
reovirus (re″o-vi′rus)
retrovirus (re″tro-vi′rus)
rhabdovirus (rab″do-vi′rus)
rhinovirus (ri″no-vi′rus)
rotavirus (ro′tah-vi″rus)
rubella virus (roo-bel′ah)
togavirus (to″gah-vi′rus)
varicella-zoster virus (var″ĭ-sel′ah zos′ter)
variola virus (vah-re-o′-lah)

Fungi

Amanita (am″ah-ni′tah)
Arthrobotrys (ar″thro-bo′tris)
Aspergillus (as″per-jil′us)
Blastomyces (blas″to-mi′sēz)
Candida (kan′dĭ-dah)
Cephalosporium (sef″ah-lo-spo′re-um)
Claviceps (klav′ĭ-seps)
Coccidioides (kok-sid″e-oi′dēz)
Cryptococcus (krip″to-kok′us)
Dictyostelium (dik″te-o-stel′e-um)
Epidermophyton (ep″ĭ-der-mof′ĭ-ton)
Fusarium (fu-sa′re-um)
Histoplasma (his″to-plaz′mah)
Microsporum (mi-kros′po-rum)
Mucor (mu′kor)
Neurospora (nu-ros′po-rah)
Penicillium (pen″ĭ-sil′e-um)
Phytophthora (fi-tof′tho-rah)
Pichia (pik′e-uh)
Pneumocystis (noo″mo-sis′tis)
Rhizopus (ri-zo′pus)
Saccharomyces (sak″ah-ro-mi′sēz)
Sporothrix (spo′ro-thriks)
Trichoderma (trik-o-der′mah)
Trichophyton (tri-kof′ĭ-ton)

Protists

Acanthamoeba (ah-kan″thah-me′bah)
Amoeba (ah-me′bah)
Balantidium (bal″an-tid′e-um)
Chlamydomonas (klah-mid″do-mo′nas)
Chlorella (klo-rel′ah)
Cryptosporidium (krip″to-spo-rid′e-um)
Entamoeba (en″tah-me′bah)
Euglena (u-gle′nah)
Giardia (je-ar′de-ah)
Gonyaulax (gon″e-aw′laks)
Leishmania (lēsh-ma′ne-ah)
Naegleria (na-gle′re-ah)
Paramecium (par″ah-me′se-um)
Plasmodium (plaz-mo′de-um)
Prototheca (pro″to-the′kah)
Spirogyra (spi″ro-ji′rah)
Tetrahymena (tet″rah-hi′mē-nah)
Toxoplasma (toks″o-plaz′mah)
Trichomonas (trik″o-mo′nas)
Trichonympha (trik″o-nim′fuh)
Trypanosoma (tri″pan-o-so′mah)
Volvox (vol′voks)

Glossary

Pronunciation Guide

Some of the boldface terms in this glossary are followed by a phonetic spelling in parentheses. These pronunciation aids usually come from *Dorland's Illustrated Medical Dictionary.* The following rules are taken from this dictionary and will help in using its phonetic spelling system.

1. An unmarked vowel ending a syllable (an open syllable) is long; thus, *ma* represents the pronunciation of *may; ne,* that of *knee; ri,* of *wry; so,* of *sew; too,* of *two;* and *vu,* of *view.*
2. An unmarked vowel in a syllable ending with a consonant (a closed syllable) is short; thus, *kat* represents *cat; bed, bed; hit, hit; not, knot; foot, foot;* and *kusp, cusp.*
3. A long vowel in a closed syllable is indicated by a macron; thus, *māt* stands for *mate; sēd,* for *seed; bīl,* for *bile; mōl,* for *mole; fūm,* for *fume;* and *fo͞ol,* for *fool.*
4. A short vowel that ends or itself constitutes a syllable is indicated by a breve; thus, *ĕ-fekt′* for *effect; ĭ-mūn′* for *immune;* and *ŭ-klo͞od′* for *occlude.*

Primary (′) and secondary (″) accents are shown in polysyllabic words. Unstressed syllables are followed by hyphens.

Some common vowels are pronounced as indicated here.

ə	sofa	ĕ	met	ŏ	got
ā	mate	ī	bite	ū	fuel
ă	bat	ĭ	bit	ŭ	but
ē	beam	ō	home		

Source: Dorland, W. A. Newman. 1998. *Dorland's Illustrated Medical Dictionary.* Philadelphia: Saunders.

A

+1 nucleosome The nucleosome that covers a transcriptional template and is positioned closest to the promoter.

AB toxins Exotoxins composed of two parts (A and B). The B portion is responsible for toxin binding to a cell but does not directly harm it; the A portion enters the cell and disrupts its function.

ABC protein secretion pathway *See* ATP-binding cassette transporters.

ABO blood group The classification of human red blood cells based on inherited surface antigens.

accessory pigments Photosynthetic pigments such as carotenoids and phycobiliproteins that aid chlorophyll in trapping light energy.

acellular slime mold *See* plasmodial slime mold.

acetyl-coenzyme A (acetyl-CoA) A molecule made of acetic acid and coenzyme A that is energy rich; it is produced by many catabolic pathways and is the substrate for the tricarboxylic acid cycle, fatty acid biosynthesis, and other pathways.

acid-fast Refers to bacteria (e.g., mycobacteria) that cannot be easily decolorized with acid alcohol after being stained with dyes such as basic fuchsin.

acid-fast staining A staining procedure that differentiates between bacteria based on their ability to retain a dye when washed with an acid alcohol solution.

acidophile (a-sid′o-fīl) A microorganism that has its growth optimum between about pH 0 and 5.5.

acquired immunity *See* adaptive immunity.

actin filaments One of the proteinaceous components of the eukaryotic cytoskeleton; also called microfilaments.

actinomycete (ak″tĭ-no-mi′sēt) An aerobic, monoderm bacterium that forms branching filaments (hyphae) and asexual spores.

actinorhizae Associations between actinomycetes and plant roots.

activated sludge Solid matter or sediment composed of actively growing microorganisms that participate in the aerobic portion of a biological sewage treatment process.

activation energy The energy required to bring reacting molecules together to reach the transition state in a chemical reaction.

activator protein A transcriptional regulatory protein that binds to a specific site on DNA (activator-binding site) and enhances transcription initiation.

activator-binding site *See* activator protein.

active (catalytic) site The part of an enzyme that binds the substrate to form an enzyme-substrate complex and catalyze the reaction. Also called the catalytic site.

active transport The transport of solute molecules across a membrane against a gradient; it requires a carrier protein and the input of energy. Three major types are primary active transport, which uses hydrolysis of ATP to power transport; secondary active transport, which uses ion gradients across a membrane to power active transport; and group translocation. *See also* group translocation.

acute-phase proteins Liver proteins that assist in the prevention of blood loss and ready the host for microbial invasion.

adaptive immunity Immunity that develops in response to a specific antigen; involves T and B lymphocyte differentiation and antibody production. Immunologic memory is formed to each antigen.

adenine (ad′e-nēn) A purine derivative, 6-aminopurine, found in nucleosides, nucleotides, coenzymes, and nucleic acids.

adenosine 5′-triphosphate (ATP) A high-energy molecule that serves as the cell's major form of energy currency.

adhesin (ad-he′zin) A molecule on the surface of a microorganism involved in adhesion to a substratum (e.g., host tissue).

adjuvant Material added to an antigen in a vaccine preparation that increases its immunogenicity.

aerial mycelium The mat of hyphae formed by actinomycetes or fungi that grows above the substrate, imparting a fuzzy appearance to colonies.

aerobic anoxygenic phototrophy (AAnP) Phototrophic process by which electron donors such as organic matter or sulfide are used under aerobic conditions.

aerobic respiration A metabolic process in which molecules, often organic, are oxidized with oxygen as the final electron acceptor.

aerosolized microbes Infectious bacteria and viruses that are shed in tiny liquid droplets emanating from the nasopharyngeal tract.

aerotolerant anaerobes Microbes that grow equally well whether or not oxygen is present.

aflatoxin (af″lah-tok′sin) A polyketide secondary fungal metabolite that can cause cancer.

agar (ahg′ar) A complex sulfated polysaccharide, usually from red algae, that is used to solidify culture media.

agglutinates The visible aggregates or clumps formed by an agglutination reaction.

agglutination reaction The formation of an insoluble immune complex by the cross-linking of cells or particles.

airborne transmission The spread of a pathogen as it travels suspended in air over 1 meter or more from the source to the host.

akinetes (ā-kin′ēts) Specialized, nonmotile, dormant, thick-walled resting cells formed by some cyanobacteria.

alarmone A signaling molecule synthesized in response to cellular stressors that regulates gene activity so the cell survives the stress.

alcoholic fermentation A fermentation process that produces ethanol and CO_2 from sugars.

alga (al′gah) A common term for several unrelated groups of photosynthetic eukaryotic microorganisms. Most are now considered protists.

alignment The comparison of gene, RNA, and protein sequences in a base-by-base (or amino acid by amino acid) fashion; used to calculate the degree of similarity between sequences.

alkaliphile A microorganism that grows best at pHs from about 8.5 to 11.5.

allele An alternative form of a gene.

allergen An antigen that induces an allergic response.

allergy *See* type I hypersensitivity.

allochthonous (ăl′ək-thə-nəs) Substances not native to a given environment (e.g., nutrient influx into freshwater ecosystems).

allograft A transplant between genetically different individuals of the same species.

allosteric effector *See* allosteric enzyme.

allosteric enzyme An enzyme whose activity is altered by the noncovalent binding of a small molecule (allosteric effector) at a regulatory site separate from the catalytic site; effector binding prompts a conformational change in the enzyme's catalytic site, causing enzyme activation or inhibition.

αβ T cells CD4 or CD8 T cells that possess alpha/beta T-cell receptors.

alpha diversity A statistical quantity that describes the diversity within a community.

α-hemolysis *See* hemolysis.

alternate sigma factors Sigma is the subunit of bacterial RNA polymerase required for transcription initiation; alternate sigma factors direct the core RNA polymerase to transcribe distinct sets of inducible genes or operons, and is thereby a mechanism for global regulation.

alternative complement pathway An antibody-independent pathway of complement activation.

alternative splicing The use of different exons during RNA splicing to generate different polypeptides from the same gene.

alveolar macrophage A vigorously phagocytic macrophage located on the epithelial surface of the lung alveoli, where it ingests inhaled particulate matter and microorganisms.

amino acid activation The preparatory step of protein synthesis in which amino acids are attached to transfer RNA molecules. The reaction is catalyzed by aminoacyl-tRNA synthetases.

aminoacyl or **acceptor site (A site)** The ribosomal site that contains an aminoacyl-tRNA at the beginning of the elongation cycle during protein synthesis.

aminoacyl-tRNA synthetases *See* amino acid activation.

aminoglycoside antibiotics A group of antibiotics that contain a cyclohexane ring and amino sugars; all aminoglycoside antibiotics inhibit protein synthesis by binding to the small ribosomal subunit.

amphibolic pathways Metabolic pathways that function both catabolically and anabolically.

amphipathic Term describing a molecule that has both hydrophilic and hydrophobic regions (e.g., phospholipids).

amphitrichous (am-fit′rĕ-kus) A cell with a single flagellum at each end.

amplification The increase in the number of DNA fragments that occurs during the polymerase chain reaction.

amplified sequence variants A set of DNA sequences derived by amplification using a community nucleic acid preparation as the template.

anabolism The synthesis of complex molecules from simpler molecules with the input of energy and reducing power.

anaerobic digestion (an″a-er-o′bik) The microbiological treatment of sewage wastes under anoxic conditions to produce methane.

anaerobic respiration An energy-conserving process in which the terminal electron transport chain acceptor is a molecule other than oxygen.

anammox reaction The coupled use of nitrite as an electron acceptor and ammonium ion as an electron donor under anoxic conditions to yield nitrogen gas.

anammoxosome The compartment within a planctomycete cell in which the anammox reaction occurs.

anaphylaxis (an″ah-fĭ-lak′sis) A severe, immediate type I hypersensitivity reaction following exposure of a sensitized individual to the appropriate antigen.

anaplerotic reactions (an′ah-plĕ-rot′ik) Reactions that replenish depleted tricarboxylic acid cycle intermediates.

anergy (an′ər-je) A state of immune cell unresponsiveness to antigens.

anoxic Without oxygen present.

anoxygenic photosynthesis Photosynthesis that does not oxidize water to produce oxygen.

antagonism A type of symbiotic relationship in which one partner negatively impacts the other.

antibiotic A microbial product or its derivative that kills susceptible microorganisms or inhibits their growth.

antibody (immunoglobulin) A glycoprotein made by plasma cells in response to the introduction of an antigen.

antibody affinity The strength of binding between an antigen and an antibody.

antibody class switching The process by which plasma cells change their expression from one isotype of immunoglobulin to a second isotype.

antibody titer An approximation of the antibody concentration.

antibody-dependent cell-mediated cytotoxicity (ADCC) The killing of antibody-coated target cells by immune cells with Fc receptors that recognize the Fc region of the bound antibody.

antibody-dependent enhancement (ADE) A phenomenon whereby the production of neutralizing antibodies to one serotype of dengue virus serves to opsonize rather than neutralize other serotypes.

antibody-mediated (humoral) immunity The type of immunity that results from the presence of soluble antibodies in blood and lymph.

anticodon The three bases on a tRNA that are complementary to the codon on mRNA.

antigen A substance (such as a protein, nucleoprotein, polysaccharide, or sometimes a glycolipid) to which lymphocytes respond.

antigen processing The hydrolytic digestion of antigens to produce peptide fragments, which are collected and presented by class I or class II MHC molecules on the surface of an antigen-presenting cell.

antigen-binding fragment (Fab) "Fragment antigen binding." A monovalent antigen-binding fragment of an immunoglobulin molecule that consists of one light chain and part of one heavy chain, linked by interchain disulfide bonds.

antigenic determinant site *See* epitope.

antigenic drift A small change due to mutation in the antigenic character of an organism.

antigenic shift A major change in the antigenic character of an organism that makes it unrecognized by host immune mechanisms. This term is most often used to describe an antigenically distinct virus that results when two different strains of virus recombine genetic material.

antigenic variation The capacity of some microbes to change surface proteins when in a host organism, thereby evading the host immune system.

antigen-presenting cells (APCs) Cells that take in protein antigens, process them, and present peptide fragments bound to MHC molecules to T cells; includes macrophages, dendritic cells, and B cells.

antimetabolite A compound that blocks metabolic pathway function by competitively inhibiting a key enzyme because it closely resembles the normal enzyme substrate.

antimicrobial agent A drug that inhibits the growth of or is lethal to any kind of microbe (e.g., virus, bacterium, fungus).

antiport Coupled transport of two molecules in which one molecule enters the cell as the other leaves the cell.

antisense RNA A single-stranded RNA with a base sequence complementary to a segment of a target RNA molecule; when bound to the target RNA, the target's activity is altered.

antisepsis The prevention of infection or sepsis.

antiseptic Chemical agent applied to tissue to prevent infection by killing or inhibiting pathogens.

antitoxin An antibody to a microbial toxin, usually a bacterial exotoxin, that binds and neutralizes the toxin.

apical complex A set of organelles characteristic of members of the protist subdivision Apicomplexa: includes polar rings, subpellicular microtubules, conoids, rhoptries, and micronemes.

apicomplexan (a′pĭ-kom-plek′san) A protist that has an apical complex and a spore-forming stage. It is either an intra- or extracellular parasite of animals; a member of the taxon Apicomplexa.

apicoplast (a′pĭ-ko-plast) Organelles in apicomplexan protists derived from cyanobacterial endosymbionts. Apicoplasts, however, are not involved in phototrophy; rather, they are the site of fatty acid, isoprenoid, and heme biosythesis.

apoenzyme (ap″o-en′zīm) The protein part of an enzyme that also has a nonprotein component.

appressorium (pl., **appressoria**) A flattened region of hypha found in some plant-infecting fungi that aids in penetrating the host plant cell wall. Occurs in both pathogenic fungi and nonpathogenic mycorrhizal fungi.

aptamer (ap′tăm-er) A single-stranded DNA or RNA molecule engineered to bind to a target protein to modify the activity of the protein.

arbuscular mycorrhizae (AM) The mycorrhizal fungi in a fungus-root association that penetrate the outer layer of the root, grow intracellularly, and form characteristic much-branched hyphal structures called arbuscules.

arbuscules *See* arbuscular mycorrhizal fungi.

archaea (ar′ke-ah) Members of the domain Archaea.

archaella (ar′ke-el″uh) Appendages on archaeal cells that function in motility; functional equivalent of bacteria flagella.

archaeol A lipid molecule that is a component of archaeal plasma membranes.

archaerhodopsin A transmembrane protein to which retinal is bound; it functions as a light-driven proton pump resulting in photophosphorylation without chlorophyll or bacteriochlorophyll. Found in the purple membrane of halophilic archaea.

artemisinin An antimalarial drug used in combination with at least one other antimalarial drug for the treatment for infections caused by *Plasmodium falciparum*.

arthroconidia (arthrospores) Fungal spores formed by fragmentation.

artificially acquired active immunity The type of immunity that results from immunizing an animal with a vaccine. The immunized animal produces its own antibodies and activated lymphocytes.

artificially acquired passive immunity The temporary immunity that results from introducing into an animal antibodies that have been produced either in another animal or by in vitro methods.

ascocarp A multicellular structure in ascomycete fungi lined with specialized cells called asci.

ascogenous hypha A specialized hypha that gives rise to one or more asci.

ascomycetes (as″ko-mi-se′tēz) A group of fungi that form ascospores.

ascospore (as′ko-spor) A spore contained or produced in an ascus.

ascus (pl., **asci**) A specialized cell, characteristic of the ascomycetes, in which two haploid nuclei fuse to produce a zygote, which often immediately divides by meiosis; at maturity, an ascus contains ascospores.

aseptate (coenocytic) hypha A multinucleate hypha (filament) that lacks cross walls.

assimilatory reduction (e.g., **assimilation, assimilatory nitrate reduction,** and **assimilatory sulfate reduction**) The reduction of an inorganic molecule to incorporate it into organic material. No energy is conserved during this process.

atomic force microscope *See* scanning probe microscope.

atopic reaction *See* localized anaphylaxis.

ATP *See* adenosine 5′-triphosphate.

ATP synthase An enzyme that catalyzes synthesis of ATP from ADP and P_i, using energy derived from the proton motive force.

ATP-binding cassette transporters (ABC transporters) Transport systems that use ATP hydrolysis to drive translocation across the plasma membrane; can be used for nutrient uptake (ABC importer) or export of substances (ABC exporter), including protein secretion (ABC protein secretion pathway).

attack rate The proportional number of cases that develop in a population exposed to an infectious agent.

attenuated vaccine Live, nonpathogenic organisms used to activate adaptive immunity.

attenuation (1) A mechanism for the regulation of transcription termination of some bacterial operons by aminoacyl-tRNAs. (2) A procedure that reduces or abolishes the virulence of a pathogen without altering its immunogenicity.

attenuator A factor-independent transcription termination site in the leader sequence that is involved in attenuation.

autochthonous (ô-tŏk′-thə-nəs) Substances (nutrients) that originate in a given environment.

autoclave An apparatus for sterilizing objects by the use of steam under pressure.

autoimmune disease A disease produced by the immune system attacking self antigens.

autoinducer (AI) With respect to quorum sensing, a small molecule that induces synthesis of the enzyme responsible for the autoinducer's synthesis. *See also* quorum sensing.

autoinducing peptide (AIP) Short (5–17 amino acids) peptides used by some monoderm bacteria for cellular communication.

autoinduction In general terms, any molecule that increases its own level of production.

autolysins (aw-tol′ĭ-sins or aw″to-lye′sins) Enzymes that partially digest peptidoglycan in growing bacteria so that the cell wall can be enlarged.

autophagocytosis Ancient means of homeostasis whereby intracellular components are recycled.

autophagosome *See* macroautophagy.

autophagy *See* autophagocytosis and macroautophagy.

autotroph (aw′to-trōf) An organism that uses CO_2 as its sole or principal source of carbon.

auxotroph (awk′so-trōf) An organism with a mutation that causes it to lose the ability to synthesize an essential nutrient; because of the mutation, the organism must obtain the nutrient or a precursor from its surroundings.

average nucleotide identity (ANI) A pairwise comparison between the whole or partial genome sequences of two microorganisms; expressed as percent identity.

avidity The combined strength of binding between an antigen and all the antibody-binding sites.

axenic (a-zen′ik) Not contaminated by any foreign organisms; the term is used in reference to pure microbial cultures. *See also* pure culture.

axopodium A thin, needlelike type of pseudopodium with a central core of microtubules.

B

B cell (B lymphocyte) A type of lymphocyte that, following interaction with antigen, becomes a plasma cell, which synthesizes and secretes antibody molecules involved in humoral immunity.

B lymphocyte *See* B cell.

B7 (CD80, CD86) proteins Glycoproteins on the surface of antigen-presenting cells (B cells, macrophages, dendritic cells, as well as T cells); it binds CD28 on T cells, resulting in second signal activation.

bacillus (pl., **bacilli**) (bah-sil′lus) A rod-shaped bacterium or archaeon.

backslopping A technique in food fermentation in which a sample of each batch is withheld to inoculate the next batch.

bacteremia The presence of viable bacteria in the blood.

bacteria Members of the domain Bacteria.

bacterial artificial chromosome (BAC) A cloning vector constructed from the *Escherichia coli* F-factor plasmid.

bacterial interference The capacity of one species to impact the activities of another.

bactericide An agent that kills bacteria.

bacteriochlorophyll (bak-te″re-o-klo′ro-fil) A modified chlorophyll that serves as the primary light-trapping pigment in purple and green photosynthetic bacteria and heliobacteria.

bacteriocin (bak-te′re-o-sin) A secreted bacterial protein that kills other closely related bacteria.

bacteriocyte Specialized eukaryotic cell that contains symbiotic bacteria.

bacteriophage (bak-te′re-o-fāj″) A virus that uses bacteria as its host; often called a phage.

bacteriostatic Inhibiting the growth and reproduction of bacteria.

bacteroid A modified, often pleomorphic, bacterial cell within the root nodule cells of legumes; it carries out nitrogen fixation.

bactoprenol A lipid molecule used to carry peptidoglycan precursors from the cytoplasm to the periplasm.

baeocytes Small, spherical, reproductive cells produced by some cyanobacteria through multiple fission.

barcodes Short conserved regions of a gene that are specific to phylogenetically defined groups of organisms.

barophilic (bar″o-fil′ik) or **barophile** *See* piezophilic.

barotolerant Organisms that can grow and reproduce at high pressures but do not require them.

base excision repair *See* excision repair.

basic reproduction number (R_0) Defines the number of susceptible individuals that can be infected from one infected person assuming all members of the population are vulnerable.

basidiocarp (bah-sid′e-o-karp″) The fruiting body of a basidiomycete that contains the basidia.

basidiomycetes (bah-sid″e-o-mi-se′tēz) A group of fungi in which the spores are borne on club-shaped organs called basidia.

basidiospore (bah-sid′e-ō-spor) A spore borne on the outside of a basidium.

basidium (bah-sid′e-um; pl., **basidia**) A structure formed by basidiomycetes; it bears on its surface basidiospores (typically four) that are formed following karyogamy and meiosis.

basophil A weakly phagocytic white blood cell in the granulocyte lineage. It synthesizes and stores vasoactive molecules (e.g., histamine) that are released in response to external triggers.

batch culture Growth of micro-organisms in a closed culture vessel without adding fresh or removing old (spent) medium.

B-cell receptor (BCR) A trans-membrane immunoglobulin complex on the surface of a B cell that binds an antigen and stimulates the B cell. It is composed of a membrane-bound immunoglobulin, usually IgD or a modified IgM, complexed with the Ig-α/Ig-β heterodimer.

benthic Pertaining to the bottom of the sea or other body of water.

beta diversity A statistical measure of the diversity between two communities or two conditions.

β-hemolysis *See* hemolysis.

β-lactam Referring to antibiotics containing a β-lactam ring (e.g., penicillins and cephalosporins).

β-lactam ring The cyclic chemical structure composed of one nitrogen and three carbon atoms. It has antibacterial activity, interfering with bacterial cell wall synthesis.

β-lactamase An enzyme that hydrolyzes the β-lactam ring, rendering the antibiotic inactive.

β-oxidation pathway The major pathway of fatty acid oxidation to produce NADH, $FADH_2$, and acetyl coenzyme A.

binal symmetry The symmetry of some virus capsids (e.g., those of complex phages) that is a combination of icosahedral and helical symmetry.

binary fission Asexual reproduction in which a cell separates into two identical daughter cells.

binomial system The nomenclature system in which an organism is given two names: The first is the capitalized generic name and the second is the uncapitalized specific epithet.

bioassimilation A component of biodegradation in which the degradation products are metabolized.

biochemical oxygen demand (BOD) The amount of oxygen used by organisms in water under standard conditions; it provides an index of the amount of microbially oxidizable organic matter present.

biochemical pathways Sets of chemical reactions performed by organisms that convert a starting substrate into one or more products.

biocide Chemical or physical agent, usually with a broad spectrum of biological activity, that inactivates microorganisms.

biodegradation The breakdown of a complex chemical through biological processes that results in minor loss of functional groups, fragmentation into smaller constituents, or complete breakdown to CO_2 and minerals.

biofilms Organized microbial communities encased in extracellular polymeric substances and associated with surfaces, often with complex structural and functional characteristics.

biofragmentation The degradative process in which large organic polymers are hydrolyzed by exoenzymes.

biofuels Energy-dense compounds (i.e., fuels) produced by microbes, usually by the degradation of plant substrates.

biogeochemical cycling The oxidation and reduction of substances carried out by living organisms and abiotic processes that results in the cycling of elements within and between different parts of the ecosystem and the atmosphere.

bioinformatics The interdisciplinary field that manages and analyzes large biological data sets, including genome and protein sequences.

bioinsecticide A microbe or microbial product used to kill or disable unwanted insect pests.

biomagnification The increase in concentration of a substance in higher-level consumer organisms.

biomolecular condensate A protein complex separated from the cytoplasm by its biophysical properties, not by a membrane.

bioremediation The use of biologically mediated processes to remove or degrade pollutants from specific environments.

biosensor A device for the detection of a particular substance (an analyte) that combines a biological receptor with a physicochemical detector.

biosynthetic gene clusters (BGCs) Large operons whose products are the enzymes that comprise the biosynthetic pathway to generate secondary metabolites like antibiotics.

bioterrorism The intentional or threatened use of microorganisms or organic toxins to produce death or disease in humans, animals, and plants.

biotrophic fungi Plant pathogenic fungi that do not kill their plant hosts

biovar Variant strains of microbes characterized by biological or physiological differences.

BLAST (basic local alignment search tool) Software that identifies regions of similarity between nucleotide or amino acid sequences. It compares input sequences with databases and determines the statistical significance of matches. Homologies can be used to infer gene or protein function and evolutionary relationships.

blastospores Fungal spores produced from a vegetative mother cell by budding.

bloom The growth of a single microbial species or a limited number of species in an aquatic environment, usually in response to a sudden pulse of nutrients. *See also* harmful algal bloom.

BONCAT (bioorthogonal noncanonical amino acid tagging) A method to identify translationally active cells within a population by incorporating analogs of amino acids.

bright-field microscope A micro-scope that illuminates the specimen directly with bright light and forms a dark image on a brighter background.

broad-spectrum drugs Chemo-therapeutic agents that are effective against many different kinds of pathogens.

bronchial-associated lymphoid tissue (BALT) *See* mucosal-associated lymphoid tissue.

Bt An insecticide that consists of a toxin produced by *Bacillus thuringiensis*.

bubo (bu′bo) A tender, inflamed, enlarged lymph node that results from a variety of infections.

budding A vegetative outgrowth of yeast and some bacteria as a means of asexual reproduction; the daughter cell is smaller than the parent.

butanediol fermentation A type of fermentation most often found in members of the family Enterobac-teriaceae in which 2,3-butanediol is a major product; acetoin is an inter-mediate in the pathway and may be detected by the Voges-Proskauer test.

C

5′ cap *See* pre-mRNA.

cable bacteria Bacteria that grow in long chains and transport electrons over millimeter distances across hundreds of cells.

calorie The amount of heat needed to raise one gram of water from 14.5° to 15.5°C.

Calvin-Benson cycle The main pathway for the fixation (reduction and incorporation) of CO_2 into organic material by photoautotrophs and chemolithoautotrophs.

cannulae Tubelike structures extending from the surface of some archaeal cells. They are thought to function in attachment to surfaces.

cap snatching A viral process in which the 5′ cap of cellular mRNA is removed and used to prime viral RNA synthesis.

capsid The protein coat or shell that surrounds a virion's nucleic acid.

capsomere The ring-shaped morphological unit of icosahedral capsids.

capsule A layer of well-organized material, not easily washed off, lying outside the cell wall.

capsule staining A technique that reveals the presence of capsules

carbapenems A group of broad-spectrum β-lactam antibiotics typically resistant to β-lactamases.

carbon to nitrogen ratio (C:N ratio) The ratio of carbon to nitrogen in a soil; it can be used to predict rates of organic decomposition and thus the cycling of nutrients in the system.

carbonate equilibrium system The interchange among CO_2, HCO_3^-, and CO_3^{2-} that keeps oceans buffered between pH 7.6 and 8.2.

carbonosome A bacterial compartment for storage of carbon in the form of polyhydroxyalkanoate.

carboxysomes Polyhedral inclusions (microcompartments) where CO_2 fixation occurs; contain the CO_2-fixation enzyme ribulose 1,5-bisphosphate carboxylase/oxygenase.

cardinal temperatures The minimal, maximal, and optimum temperatures for growth.

carriers Membrane-bound proteins that transport specific substances across the membrane.

Cas9 genome editing (CRISPR or CRISPR/CAS9) A procedure that directs the nuclease activity of Cas9 to specific DNA sequences in a genome for the purpose of genome modification.

case fatality rate A measure of the severity of a disease in a population by defining the proportion of individuals diagnosed with a specific disease who die from that disease.

catabolism That part of metabolism in which larger, more complex molecules are broken down into smaller, simpler molecules with the release of energy.

catabolite activator protein (CAP) A protein that regulates the expression of genes encoding catabolite repressible enzymes; also called cyclic AMP receptor protein (CRP).

catabolite repression Inhibition of the synthesis of several catabolic enzymes by a preferred carbon and energy source (e.g., glucose).

catalase An enzyme that catalyzes the destruction of hydrogen peroxide.

catalyst A substance that accelerates a reaction without being permanently changed itself.

catalytic site *See* active site.

catenanes (kăt′ə-nāns′) Circular, covalently closed nucleic acid molecules that are interlocked like the links of a chain.

cathelicidins Antimicrobial cationic peptides that are produced by a variety of cells (e.g., neutrophils, respiratory epithelial cells, and alveolar macrophages).

caveolin-dependent endocytosis *See* endocytosis.

CD4⁺ cell *See* T-helper cell.

CD8⁺ cell *See* cytotoxic T lymphocyte.

cell cycle The sequence of events in a cell's growth-division cycle between the end of one division and the end of the next.

cell envelope The plasma membrane plus all other external layers.

cell factory A strain of bacteria that synthesizes a specific product of industrial importance.

cell wall The strong structure that lies outside the plasma membrane; it supports and protects the membrane and helps maintain cell shape.

cellular (cell-mediated) immunity The type of immunity mediated by T cells.

cellular slime molds Protists with a vegetative phase consisting of amoeboid cells that aggregate to form a multicellular pseudoplasmodium. They belong to the taxon *Dictyostelium;* formerly considered fungi.

Centers for Disease Control and Prevention (CDC) Public health agency of the United States, headquartered in Atlanta, Georgia; serves as the national agency for developing and carrying out disease prevention and control, environmental health, and health promotion and education.

central metabolic pathways Those pathways central to the metabolism of an organism because they function catabolically and anabolically (e.g., glycolytic pathways and tricarboxylic acid cycle).

central tolerance The process by which self-reactive B cells are eliminated in the bone marrow and self-reactive T cells are eliminated in the thymus.

chain-termination DNA sequencing method A DNA-sequencing method that uses dideoxynucleotides to cause termination of DNA replication at random sites.

chancre (shang′ker) The primary lesion of syphilis occurring at the site of entry of the pathogen.

channels Membrane-bound proteins that provide a portal through which molecules can pass.

chaperone Proteins that assist in the folding and stabilization of other proteins. Some are also involved in directing newly synthesized proteins to protein secretion systems or to other locations in the cell.

chassis The components of a cell factory onto which are added the genetic elements for synthesizing an industrial product.

cheaters Members of a bacterial population who benefit from a community action but do not contribute to it—for example, antibiotic production.

chemical fixation *See* fixation.

chemical oxygen demand (COD) The amount of chemical oxidation required to convert organic matter in water and wastewater to CO_2.

chemiosmotic hypothesis (kem″e-os-mot′ik) The hypothesis that proton and electrochemical gradients are generated by electron transport and then used to drive ATP synthesis by oxidative phosphorylation or photophosphorylation.

chemoheterotroph (ke″mo-het′er-o-trōf″) An organism that uses organic compounds as sources of energy, electrons, and carbon for biosynthesis.

chemokine A type of cytokine that stimulates chemotaxis and chemokinesis of immune cells.

chemolithoautotroph A microorganism that oxidizes reduced inorganic compounds to derive both energy and electrons; CO_2 is the carbon source.

chemolithoheterotroph A microorganism that uses reduced inorganic compounds to derive both energy and electrons; organic molecules are used as the carbon source.

chemolithotroph A microorganism that uses reduced inorganic compounds as a source of energy and electrons.

chemoreceptors Proteins in the plasma membrane or periplasmic space that bind chemicals and trigger the appropriate chemotactic response.

chemostat A continuous culture apparatus that feeds medium into the culture vessel at the same rate as medium containing microorganisms is removed; the medium in a chemostat contains one essential nutrient in a limiting quantity.

chemotaxins Agents that cause white blood cells to migrate up its concentration gradient.

chemotaxis The pattern of cellular behavior in which the cell moves toward chemical attractants and away from repellents.

chemotherapy The use of chemical agents to treat disease.

chemotrophs Organisms that obtain energy from the oxidation of chemical compounds.

chlamydiae (klə-mid′e-e) Members of the genera *Chlamydia* and *Chlamydophila:* diderm, obligate intracellular pathogens.

chlorophyll The green photosynthetic pigment that consists of a large tetrapyrrole ring with a magnesium atom in the center.

chloroplast A eukaryotic plastid that contains chlorophyll and is the site of photosynthesis.

chlorosomes Elongated, intramembranous vesicles found in the green sulfur and nonsulfur bacteria; they contain light-harvesting pigments.

chromatic acclimation The ability of cyanobacteria to change the relative concentrations of light-harvesting pigments in response to changes in wavelengths of incident light.

chromatin The complex of DNA and proteins, including histones, from which eukaryotic and some archaeal chromosomes are made.

chromatin immunoprecipitation (ChIP) A technique that enables the identification of protein-DNA interactions.

chromogen (kro′mo-jen) A colorless substrate that is acted on by an enzyme to produce a colored end product.

chromophore group A chemical group with double bonds that absorbs visible light and gives a dye its color.

chromosome interaction domains Portions of a bacterial or archaeal chromosome that are organized within the nucleoid.

chromosomes The bodies that have most or all of the cell's DNA and contain most of its genetic information (mitochondria and chloroplasts also contain DNA).

chronic inflammation Inflammation that is characterized by a dense tissue infiltration of lymphocytes and macrophages into the affected site and the formation of new connective tissue, which usually causes permanent tissue damage.

chytrids A term used to describe members of Chytridiomycota, fungi that produce motile zoospores with single, posterior, whiplash flagella.

cidal Causing death.

cilia Threadlike appendages extending from the surface of some protists that beat rhythmically to propel them; cilia are membrane-bound cylinders with a complex internal array of microtubules, usually in a 9 + 2 pattern.

ciliates Protists that move by rapidly beating cilia.

cistron A segment of DNA that encodes a single polypeptide.

citric acid cycle *See* tricarboxylic acid cycle.

clade A group of organisms derived from a common ancestor.

class I MHC molecule *See* major histocompatibility complex.

class II MHC molecule *See* major histocompatibility complex.

classical complement pathway The antibody-dependent pathway of complement activation; it stimulates lysis of pathogens, phagocytosis, and other host defenses.

clathrin-dependent endocytosis *See* endocytosis.

clonal selection The process by which an antigen binds to the best-fitting B-cell receptor, activating that B cell, resulting in the synthesis of that specific antibody and clonal expansion.

cloning vector A DNA molecule that can replicate independently of the host chromosome and maintain a piece of inserted foreign DNA, such as a gene, in a recipient cell. It may be a plasmid, phage, cosmid, or artificial chromosome.

club fungi Common name for members of Basidiomycota.

cluster of differentiation (CD) molecules Functional cell surface proteins that can be used to identify leukocyte subpopulations (e.g., interleukin-2 receptor, CD4, and CD8).

coagulation The process of adding chemicals to precipitate impurities from water during purification; also called flocculation.

coccolithophores Photosynthetic protists belonging to the group Stramenopila; characterized by coccoliths—intricate cell walls made of calcite.

coccus (kok′us, pl., **cocci**, kok′si) A roughly spherical bacterial or archaeal cell.

code degeneracy The presence of more than one codon for each amino acid.

coding sequences (CDS) In genomic analysis, open-reading frames presumed to encode proteins but not tRNA or rRNA.

codon (ko′don) A sequence of three nucleotides in mRNA that directs the incorporation of an amino acid during protein synthesis or signals the stop of translation.

coenocytic (se″no-sit′ik) *See* aseptate hypha.

coenzyme A loosely bound cofactor that often dissociates from the enzyme active site after product has been formed.

coenzyme Q (CoQ, ubiquinone) *See* electron transport chain.

coevolution The evolution of two interacting organisms such that they optimize their survival with one another; that is, they develop reciprocal adaptations.

cofactor The nonprotein component of an enzyme required for catalytic activity.

cold plasma A method of microbial control that uses unstable radicals to disrupt lipids, proteins, and nucleic acids in microbes. Also known as nonthermal plasma or nonequilibrium plasma.

colicin (kol′ĭ-sin) A plasmid-encoded protein produced by enteric bacteria; binds to receptors on the cell envelope of sensitive target bacteria, where it may cause lysis or attack specific intracellular sites such as ribosomes.

coliform A diderm, nonsporing, facultative rod that ferments lactose with gas formation within 48 hours at 35°C.

Colilert defined substrate test An assay designed to detect the presence of coliforms in water.

collectins A family of innate immunity proteins that are α-helical with a collagen-like motif. They bind foreign sugars and lipids, serving to activate complement, aggregate, or opsonize antigens.

colony An assemblage of microorganisms growing on a solid surface; the assemblage often is directly visible.

colony forming units (CFU) The number of microorganisms that form colonies when cultured using spread plates or pour plates, an indication of the number of viable microorganisms in a sample.

colony-stimulating factor (CSF) A protein that stimulates the growth and development of specific cell populations (e.g., granulocyte-CSF stimulates granulocytes to be made from precursor stem cells).

colorless sulfur bacteria A diverse group of nonphotosynthetic bacteria that can oxidize reduced sulfur compounds such as hydrogen sulfide. Many are chemolithotrophs and derive energy from sulfur oxidation.

comammox The capacity of some *Nitrospira* bacteria to perform both steps in the nitrification process, thereby performing *com*plete *amm*onia *ox*idation.

combinatorial joining The genomic recombination of V, D, and J regions during lymphocyte development that gives rise to diverse T-cell receptors, B-cell receptors, and antibodies.

cometabolism The modification of a compound not used for growth by a microorganism, which occurs in the presence of another organic material that serves as a carbon and energy source.

common-source epidemic An epidemic characterized by a sharp rise to a peak and then a rapid but not as pronounced decline in the number of individuals infected; it usually involves a single contaminated source that infects individuals.

communicable disease A disease associated with a pathogen that can be transmitted from one host to another.

community A group of organisms that includes multiple populations that occupy the same ecosystem.

comparative genomics Comparison of genomes from different organisms to discern differences and similarities.

compartmentation The differential distribution of enzymes and metabolites to separate cell structures or organelles.

compatible solute A low molecular weight molecule used to protect cells against changes in solute concentrations (osmolarity) in their habitat; it can exist at high concentrations within the cell without hindering metabolism or growth.

competent cell A cell that can take up free DNA fragments and incorporate them into its genome during transformation.

competition An interaction between two organisms attempting to use the same resource (nutrients, space, etc.).

competitive inhibitor A molecule that inhibits enzyme activity by binding the enzyme's active site.

complement system A group of plasma proteins that plays a major role in innate immunity.

complementary Refers to the matching of two strands of DNA or RNA based on the base-pairing rules.

complementary DNA (cDNA) A DNA copy of an RNA molecule (e.g., a DNA copy of an mRNA).

complex medium Culture medium that contains some ingredients of unknown chemical composition.

concatemer A long DNA molecule consisting of several genomes linked together in a row.

condenser lens Lens that focuses a cone of light on a slide.

conditional mutations Mutations with phenotypes that are expressed only under certain environmental parameters (e.g., temperature).

confocal microscope A light microscope in which monochromatic laser-derived light scans across the specimen at a specific level. Stray light from above and below the plane of focus is blocked out to give an image with excellent contrast and resolution.

conidiospore (ko-nid′e-o-spōr) An asexual, thin-walled spore borne on hyphae and not contained within a sporangium; it may be produced singly or in chains.

conidium (ko-nid′e-um; pl., **conidia**) *See* conidiospore.

conjugate redox pair The electron acceptor and electron donor of a redox half reaction.

conjugation (1) The form of gene transfer and recombination in bacteria and archaea that requires direct cell-to-cell contact. (2) A complex form of sexual reproduction commonly employed by protists.

conjugative plasmid A plasmid that carries the genes that enable its transfer to other bacteria during conjugation (e.g., F plasmid).

consensus sequence A commonly occurring sequence of nucleotides within a genetic element.

conserved hypothetical proteins In genomic analysis, proteins of unknown function for which at least one match exists in a database.

constant (C) region (C_L and C_H) The part of an antibody molecule that does not vary greatly in amino acid sequence among molecules of the same class, subclass, or type.

constitutive gene A gene that is expressed at nearly the same level at all times.

contact transmission Spread of a pathogen by contact of the source or reservoir of the pathogen.

contact-dependent growth inhibition (CDI) A form of competition in which one organism transfers toxins to a competitor to inhibit its growth.

contact-independent growth inhibition A form of competition in which one organism secretes diffusible chemicals to inhibit the growth of a competitor.

contigs In genomic sequencing, fragments of DNA with identical sequences at the ends that have been assembled into a longer, continuous sequence of DNA.

continuous culture system A culture system with constant environmental conditions maintained through continual provision of nutrients and removal of wastes. *See also* chemostat and turbidostat.

continuous feed The constant addition of a carbon source during industrial fermentation.

contractile vacuole (vak′u-ōl) In protists and some animals, a clear fluid-filled vacuole that takes up water from within the cell and then contracts, releasing it to the outside through a pore in a cyclical manner. It functions primarily in osmoregulation and excretion.

convalescence Recovery stage following an illness.

cooperation A positive but not obligatory interaction between two different organisms.

copy number The number of copies of a plasmid molecule present in a cell.

coral bleaching The loss of photosynthetic pigments by either physiological inhibition or expulsion of the coral photosynthetic endosymbiont, a dinoflagellate.

core genome The common set of genes found in all genomes in a species or other taxon.

core microbiome The suite of genes expressed by normal microbiota that are essential to maintain physiologic homeostasis of the human host.

core polysaccharide *See* lipopolysaccharide.

corepressor A small molecule that binds a transcription repressor protein, thereby activating the repressor and inhibiting the synthesis of a repressible enzyme.

cosmid A plasmid vector with lambda phage *cos* sites that can be packaged in a phage capsid; it is useful for cloning large DNA fragments.

cotranslationally Occurring at the same time as translation.

coverage The number of times a nucleotide is sequenced in any given sequencing project. Synonymous with depth of coverage, so depth and coverage are used interchangeably. Breadth of coverage refers to the percentage of genome sequenced.

CRISPR/Cas system An array of DNA sequences and a set of enzymes that recognize invading DNA of plasmid or bacteriophage origin and target them for destruction; found in many bacteria and most archaea.

cristae (kris′te) Infoldings of the inner mitochondrial membrane.

crosslinks A type of DNA damage in which covalent bonds form to link the two molecules.

cryo-electron microscopy (cryo-EM) A specialized electron microscopy procedure that involves rapid freezing of intact specimens, examining the specimen in a frozen state, and imaging from different angles. The information from each angle is used to create a three-dimensional reconstruction.

cryophile *See* psychrophile.

cryptidins Antimicrobial peptides produced by Paneth cells in the intestines.

crystallizable fragment (Fc) The stem of the Y portion of an antibody molecule. Cells such as macrophages bind to the Fc region; it also is involved in complement activation.

culture medium The source of nutrients used to propagate microbes in the laboratory.

culturomics An approach to isolating and identifying microbes that involves cultivation under many different conditions followed by identification by mass spectrometry and SSU rRNA gene sequencing.

cyanobacteria A large group of diderm, morphologically diverse bacteria that carry out oxygenic photosynthesis.

cyanophycin Inclusions that store nitrogen; found in cyanobacteria.

3′, 5′-cyclic adenosine monophosphate (cAMP) A small nucleotide that is involved in regulating a number of cellular processes in bacteria and other organisms. In some bacteria, it functions in catabolite repression.

cyclic dinucleotides Nucleotides linked together by their phosphate that function as second messengers in many bacteria and some eukaryotes.

cyclic photophosphorylation (fo″to-fos″for-ĭ-la′shun) The formation of ATP when light energy is used to move electrons cyclically through an electron transport chain during photosynthesis.

cyst A general term used for a specialized microbial cell enclosed in a wall. Cysts are formed by protists and a few bacteria. They may be dormant, resistant structures formed in response to adverse conditions or reproductive cysts that are a normal stage in the life cycle.

cystitis *See* urinary tract infection.

cytochromes (si′to-krōms) *See* electron transport chain.

cytokine (si′to-kīn) A general term for proteins released by immune cells in response to inducing stimuli; they mediate the activity of other cells.

cytokinesis Processes that apportion the cytoplasm and organelles, synthesize a septum, and divide a cell into two daughter cells during cell division.

cytopathic effect (CPE) The observable change that occurs in cells as a result of viral replication.

cytoplasm All material in the cell enclosed by the plasma membrane, with the exception of the nucleus in eukaryotic cells.

cytoproct A specific site in certain protists (e.g., ciliates) where digested material is expelled.

cytosine (si′to-sēn) A pyrimidine 2-oxy-4-aminopyrimidine found in nucleosides, nucleotides, and nucleic acids.

cytoskeleton A network of structures made from filamentous proteins (e.g., actin and tubulin) and other components in the cytoplasm of cells.

cytosol Liquid component of the cytoplasm.

cytostome A permanent site in a ciliate protist in which food is ingested.

cytotoxic T lymphocyte (CTL) Effector T cell that can lyse target cells bearing class I MHC molecules complexed with antigenic peptides. CTLs arise from antigen-activated Tc cells.

D

D value *See* decimal reduction time.

Dane particle A 42-nm spherical particle that is one of three seen in hepatitis B virus infections. The Dane particle is the complete virion.

dark reaction *See* photosynthesis.

dark-field microscopy Microscopy in which the specimen is brightly illuminated while the background is dark.

deamination The removal of amino groups from amino acids or nucleotides.

death phase During growth in a closed system, the point at which the number of viable cells decreases as cell death occurs at a constant rate.

decimal reduction time (D or D value) The time required to kill 90% of the microorganisms or endospores in a sample at a specified temperature.

defensins A large family of antimicrobial peptides produced by neutrophils, macrophages, lymphocytes, and a variety of other host cells.

defined (synthetic) medium Culture medium made with components that can be described with chemical formulae.

delayed-type hypersensitivity (DTH) *See* type IV hypersensitivity.

denaturation A loss of the three-dimensional protein shape or nucleic acid structure.

dendritic cell An antigen-presenting cell that has long membrane extensions resembling the dendrites of neurons.

denitrification The reduction of nitrate to gaseous products, primarily nitrogen gas, during anaerobic respiration.

deoxyribonucleic acid (DNA) (de-ok″se-ri″bo-nu-kle′ik) A polynucleotide that constitutes the genetic material of all cellular organisms. It is composed of deoxyribonucleotides connected by phosphodiester bonds.

depth filters Filters composed of fibrous or granular materials that are used to decrease microbial load and sometimes to sterilize solutions.

desensitization The process by which a sensitized or hypersensitive individual is made insensitive or nonreactive to a sensitizing agent (e.g., an allergen).

detergent An organic molecule, other than a soap, that serves as a wetting agent and emulsifier; it is normally used as cleanser, but some may be used as antimicrobial agents.

diatoms Photosynthetic protists with siliceous cell walls called frustules belonging to the group Stramenopila.

diauxic growth (di-awk′sik) Biphasic growth response in which a microorganism, when exposed to two nutrients, initially uses one of them for growth and then alters its metabolism to make use of the second.

diazotroph An organism capable of fixing nitrogen.

diderm A bacterial cell with two membranes: the plasma membrane and an outer membrane.

differential interference contrast (DIC) microscope A light microscope that employs two beams of plane-polarized light. The beams are combined after passing through the specimen and their interference is used to create the image.

differential media Culture media that distinguish between groups of microorganisms based on differences in their growth and metabolic products.

differential staining Staining procedures that divide bacteria into groups based on staining properties.

digital droplet PCR (ddPCR) A quantitative polymerase chain reaction that partitions templates into discrete droplets at a very low dilution, providing higher sensitivity than real-time quantitative PCR.

diglycerol tetraether lipids Archaeal lipids formed when isoprenoid hydrocarbons are linked to two glycerols by ether bonds.

dinoflagellate (di″no-flaj′e-lāt) A photosynthetic protist characterized by two flagella used in swimming in a spinning pattern; belong to the group Alveolata.

diplococcus (dip″lo-kok′us) A pair of cocci.

direct counts Methods for determining population size by examining a sample of the population and counting the cells within the sample.

direct immunofluorescence Technique for imaging of microscopic agents using fluorescently labeled antibodies to bind to the agent.

direct interspecies electron transfer (DIET) Transfer of electrons from one organism to another via a nanowire or other conductive material.

directed evolution A variety of approaches used to mutate a specific gene or genes with the aim of increasing product yield or generating new molecules.

disease syndrome A set of signs and symptoms that are characteristic of a disease.

disinfectant An agent, usually chemical, that disinfects; normally, it is employed only with inanimate objects.

disinfection The killing, inhibition, or removal of microorganisms that may cause disease. It usually refers to the treatment of inanimate objects with chemicals.

dissimilatory reduction (e.g., **dissimilatory nitrate reduction** and **dissimilatory sulfate reduction**) The use of a substance as an electron acceptor for an electron transport chain. The acceptor (e.g., sulfate or nitrate) is reduced but not incorporated into organic matter.

dissolved organic matter (DOM) Nutrients that are available in the soluble, or dissolved, state.

divisome A collection of proteins that aggregate at the region in a dividing microbial cell where a septum will form.

DNA cloning The process of propagating a piece of DNA in a plasmid or phage vector.

DNA ligase An enzyme that joins two DNA fragments through the formation of a new phosphodiester bond.

DNA microarrays Solid supports that have DNA attached in organized arrays and are used to evaluate gene expression.

DNA polymerase (pol-im′er-ās) An enzyme that synthesizes new DNA using a parental nucleic acid strand (usually DNA) as a template.

DNA shuffling A procedure for generating hybrid genes by recombining the genes from homologous organisms.

DNA vaccine Introduces fragments of pathogen DNA directly into the host cell.

DNA-dependent DNA polymerase An enzyme that catalyzes the synthesis of DNA using a DNA molecule as a template.

DNA-dependent RNA polymerase An enzyme that catalyzes the synthesis of RNA using a DNA molecule as a template.

domain A portion of a protein that has a specific structure and function. Some proteins have multiple domains.

double immunodiffusion agar assay (Öuchterlony technique) An immunodiffusion reaction in which both antibody and antigen diffuse through agar to form stable immune complexes, which can be observed visually.

double membrane vesicles (DMVs) Vesicles formed by the rearrangement of the endoplasmic reticulum in animal cells infected by certain viruses. These are the sites of viral replication and gene expression.

doubling time *See* generation time.

droplet nuclei Small particles (0 to 4 μm in diameter) that represent what is left from the evaporation of larger particles (10 μm or more in diameter) called droplets.

dysbiosis An alteration in the microbiome that has a negative impact on the health of the host.

E

ecosystem An environment that includes all living and nonliving components.

ectomycorrhizae A mutualistic association between fungi and plant roots in which the fungus surrounds the root with a sheath.

ectoplasm In some protists, the cytoplasm directly under the cell membrane (plasmalemma) is divided into an outer gelatinous region, the ectoplasm, and an inner fluid region, the endoplasm.

effective reproduction number (R_e) The number of people in a population who can be infected by an individual at any given time.

effector proteins Proteins secreted by pathogens that are targeted to host cells via a secretion system.

effector response The mechanisms by which the immune system eliminates or neutralizes invading pathogens.

efflux pumps Membrane-associated protein assemblies that transport amphiphilic molecules from cytoplasm to the extracellular environment.

electron acceptor A compound that accepts electrons in an oxidation-reduction reaction. Often called an oxidizing agent or oxidant.

electron donor A molecule that becomes oxidized by giving up an electron during an oxidation-reduction reaction. Often called a reducing agent or reductant.

electron shuttle A substance or structure along which electrons are passed from the electron transport chain to an external terminal electron acceptor.

electron transport chain (ETC) A series of electron carriers that operate together to transfer electrons from donors to acceptors such as oxygen. Molecules involved in electron transport include nicotinamide adenine dinucleotide (NAD^+), NAD phosphate ($NADP^+$), cytochromes, heme proteins, nonheme proteins (e.g., iron-sulfur proteins and ferredoxin), coenzyme Q, flavin adenine dinucleotide (FAD), and flavin mononucleotide (FMN). Also called an electron transport system.

electroporation The application of an electric field to create temporary pores in the plasma membrane to render a cell temporarily transformation competent.

elementary body (EB) A small, dormant body that serves as the agent of transmission between host cells in the chlamydial life cycle.

elongasome The complex of enzymes that synthesize peptidoglycan during cell growth.

elongation factors (EF) Proteins that function in the elongation cycle of protein synthesis.

Embden-Meyerhof pathway (EMP) A pathway that degrades glucose to pyruvate (i.e., a glycolytic pathway).

emergent properties Properties of a system that cannot be inferred from the components of the system. For example, cells in biofilms are protected from UV light, unlike individual cells.

encystment The formation of a cyst.

end point PCR A type of polymerase chain reaction (PCR) in which the reaction goes to completion.

end product inhibition *See* feedback inhibition.

endemic disease A disease that exists at a steady, low-level frequency at a moderate, regular interval.

endergonic reaction (end″er-gon′ik) A reaction that does not spontaneously go to completion as written; the standard free energy change is positive, and the equilibrium constant is less than one.

endocytic pathways *See* endocytosis.

endocytosis The process in which a cell takes up solutes or particles by enclosing them in vesicles pinched off from its plasma membrane. It often occurs at regions of the plasma membrane coated by proteins such as clathrin and caveolin. Endocytosis involving these proteins is called clathrin-dependent endocytosis and caveolin-dependent endocytosis, respectively. When the substance endocytosed first binds to a receptor, the process is called receptor-mediated endocytosis.

endogenous antigen processing Antigen degradation by cytoplasmic proteasomes and subsequent complexing with class I MHC molecules for presentation on the processing cell's surface.

endomycorrhizae Referring to a mutualistic association of fungi and plant roots in which the fungus penetrates into the root cells.

endophyte A microorganism living within a plant but not necessarily parasitic on it.

endoplasm *See* ectoplasm.

endoplasmic reticulum (ER) A system of membranous tubules and flattened sacs (cisternae) in eukaryotic cells. Rough endoplasmic reticulum (RER) bears ribosomes on its surface; smooth endoplasmic reticulum (SER) lacks them.

endosome A membranous vesicle formed by endocytosis. It undergoes a maturational process that starts with early endosomes and proceeds to late endosomes and finally lysosomes.

endospore An extremely heat- and chemical-resistant, dormant, thick-walled spore that develops within some monoderm bacteria. It has a complex structure that includes (from outermost to innermost) exosporium, spore coat, cortex, spore cell wall, and spore core.

endospore staining A differential staining technique that allows the detection of endospores.

endosymbiont An organism that lives entirely within another organism.

endosymbiotic hypothesis The hypothesis that mitochondria, and related organelles (e.g., hydrogenosomes), and chloroplasts arose from bacterial endosymbionts of ancestral eukaryotic cells.

endotoxin The lipid A component of lipopolysaccharide (LPS) from Gram-negative bacterial cell walls. Nanogram quantities can induce fever, activate complement and coagulation cascades, act as a mitogen to B cells, and stimulate cytokine release from a variety of cells. Systemic effects of endotoxin are referred to as septic shock.

enhancer A site in DNA to which a eukaryotic activator protein binds.

enology The science of wine making.

enriched media Media that contain nutrients to encourage growth of fastidious microbes.

enrichment culture The growth of specific microbes from natural samples by including media components or manipulating environmental conditions to promote growth of the desired microbes while suppressing the growth of other microbes.

enteric bacteria (enterobacteria) Members of the family Enterobacteriaceae; also can refer to bacteria that live in the intestinal tract.

enteric fever Gastrointestinal bacterial infection comprised of typhoid and paratyphoid fevers.

enterotoxin A toxin specifically affecting the cells of the intestinal mucosa, causing vomiting and diarrhea.

enthalpy Heat content of a system.

Entner-Doudoroff pathway (EDP) A pathway that converts glucose to pyruvate and glyceraldehyde 3-phosphate by producing 6-phosphogluconate and then dehydrating it (i.e., a glycolytic pathway).

entropy A measure of the randomness or disorder of a system; a measure of that part of the total energy in a system that is unavailable for useful work.

envelope In virology, an outer membranous layer that surrounds the nucleocapsid of some viruses.

enveloped virus A virus having virions with a nucleocapsid enclosed within a lipid bilayer.

enzyme A protein catalyst with specificity for the reaction catalyzed and its substrates.

enzyme-linked immunosorbent assay (ELISA) A serological assay in which bound antigen or antibody is detected by another antibody that is conjugated to an enzyme. The enzyme converts a colorless substrate to a colored product reporting the antibody capture of the antigen.

eosinophil (e″o-sin′o-fil) A leukocyte that has a two-lobed nucleus and cytoplasmic granules that release chemical mediators in response to parasite infections.

epidemic A disease that suddenly increases in occurrence above the normal level in a given population.

epidemiologist A person who specializes in epidemiology.

epidemiology The study of the factors determining and influencing the frequency and distribution of disease, injury, and other health-related events and their causes in defined human populations.

epifluorescence microscopy The illumination of a microscope sample with a high-intensity light from above. The light is passed through an exciter filter so that when light of a specific wavelength is focused on the specimen, it emits light, thereby making the sample visible as a specific color on a black background.

epilimnion The upper, warmer layer of water in a stratified lake.

epiphyte An organism that grows on the surface of plants.

episome A plasmid that can either exist independently of the host cell's chromosome or be integrated into it.

epitope (ep′ĭ-tōp) An area of an antigen that stimulates the production of, and combines with, specific antibodies; also known as the antigenic determinant site.

equilibrium The state of a system in which no net change is occurring and free energy is at a minimum; in a chemical reaction at equilibrium, the rates in the forward and reverse directions exactly balance each other out.

equilibrium constant (K_{eq}) A value that relates the concentration of reactants and products to each other when a reaction is at equilibrium.

ergot The dried sclerotium of *Claviceps purpurea.* Also, an ascomycete that parasitizes rye and other plants; production of an alkaloid causes an animal disease called ergotism.

erythroblastosis fetalis Hemolytic disease of the newborn whereby red blood cells are destroyed by maternal antibody.

eukaryotic cells Cells that have a membrane-delimited nucleus; all plants, fungi, protists, and animals are eukaryotic.

eukaryotic signature proteins Proteins present almost exclusively in eukaryotic cells but not in most bacterial or archaeal cells.

eutrophic Nutrient enriched.

eutrophication The enrichment of an aquatic environment with nutrients.

evenness A measure of microbial communities that evaluates the relative abundance of different populations.

excision repair A type of DNA repair mechanism in which a section of a strand of damaged DNA is excised and replaced, using the complementary strand as a template. Two types are recognized: base excision repair and nucleotide excision repair.

excystment The escape of one or more cells or organisms from a cyst.

exergonic reaction (ek″ser-gon′ik) A reaction that spontaneously goes to completion as written; the standard free energy change is negative, and the equilibrium constant is greater than one.

exfoliative toxin (exfoliatin) An exotoxin produced by *Staphylococcus aureus* that causes the separation of epidermal layers and the loss of skin surface layers; produces scalded skin syndrome.

exit site (E site) The location on a ribosome to which an empty (uncharged) tRNA moves from the P site before it finally leaves during protein synthesis.

exocytosis The release of vacuolated cellular contents by fusion of the vacuole with the cell membrane.

exogenous antigen processing Antigen degradation by phagolysosome enzymes and subsequent complexing with class II MHC molecules for presentation on the processing cell's surface.

exon The region in a gene that is retained in mRNA after introns are removed by splicing.

exospore A bacterial spore that does not develop within a mother cell, but rather differentiates and constricts from a hypha.

exotoxin A heat-labile, toxic protein produced by a bacterium and usually released into the bacterium's surroundings.

exponential phase The phase of the growth curve during which the microbial population is growing at a constant and maximum rate, dividing and doubling at regular intervals.

expression vector A plasmid into which a recombinant gene is cloned; when in host cells, the gene is transcribed and its protein synthesized.

extant organism An organism that exists on Earth today.

extensively drug-resistant TB (XDR-TB) Strains of Mtb resistant to isoniazid and rifampin and second-line antibiotics.

extinction culture technique A method for obtaining an axenic (pure) culture in which natural samples are serial diluted so that the final dilution has between 1 and 10 cells (i.e., diluted to extinction).

extracellular pathogen A microbe that causes disease without entering the host's cells.

extracellular polymeric substances (EPS) Carbohydrate and nucleic acid polymers secreted by biofilm bacteria that comprise the biofilm matrix.

extracellular vesicles (EVs) Small vesicles with a lipid bilayer that form by pinching off a portion of a cell membrane. These contain a sample of the cytoplasm or periplasm.

extreme environment An environment in which physical factors such as temperature, pH, salinity, and pressure are outside of the normal range for growth of most microorganisms; these conditions allow unique organisms to survive and function.

extreme halophiles *See* haloarchaea.

eyepiece Lens of a compound microscope that is closest to the eye; also called the ocular lens.

F

F (fertility) factor The fertility factor, a plasmid that carries genes for bacterial conjugation and makes its *E. coli* host the gene donor during conjugation.

F′ plasmid An F plasmid that carries some bacterial genes and transmits them to recipient cells when the F′ cell carries out conjugation.

facilitated diffusion Diffusion across the plasma membrane that is aided by a channel protein or a carrier protein.

factor Protein that alters chromatin structure by inserting or removing variant histone proteins from a nucleosome.

facultative A term describing an interaction or condition that is not obligatory.

facultative anaerobes Microorganisms that do not require oxygen for growth, but grow better in its presence.

facultative intracellular pathogen A microbe that when infecting the host, resides within host cells but can reproduce outside host cells as well.

fatty acid synthase The multienzyme complex that makes fatty acids.

fecal coliform Diderm, lactose-fermenting bacteria whose normal habitat is the intestinal tract and that can grow at 44.5°C. They are used as indicators of fecal pollution of water.

fecal enterococci (en″ter-o-kok′si) Enterococci found in the intestine of humans and other warm-blooded animals.

feedback inhibition A negative feedback mechanism in which an end product inhibits the activity of an enzyme in the pathway leading to its formation.

fermentation (1) An energy-yielding process in which an organic molecule is oxidized without an exogenous electron acceptor. Usually pyruvate or a pyruvate derivative serves as the electron acceptor. (2) The growth of microbes in very large volumes for the production of industrially important products.

fermenter In industrial microbiology, the large vessel used to culture microorganisms.

ferredoxin *See* electron transport chain.

filopodia Long, narrow pseudopodia found in certain amoeboid protists.

fimbria (fim′bre-ah; pl., **fimbriae**) *See* pilus.

first law of thermodynamics Energy can be neither created nor destroyed (although it can be changed in form or redistributed).

fixation (1) The process in which the internal and external structures of cells and organisms are preserved and fixed in position. (2) The conversion of inorganic, gaseous elements such as carbon (CO_2) and nitrogen (N_2) to organic forms.

flagella staining A staining technique that provides information on the presence and distribution of flagella on bacterial and archaeal cells.

flagellin (flaj′ĕ-lin) A family of related proteins in motile bacteria used to construct the filament of a flagellum.

flagellum (flah-jel′um; pl., **flagella**) A threadlike appendage that is responsible for bacterial motility. Flagella are composed of a basal body, which attaches it to the cell, and a hook, which connects the basal body to the filament. The filament is the part of the flagellum that rotates and moves the bacterium.

flavin adenine dinucleotide (FAD) (fla′vin ad′ĕ-nēn) *See* electron transport chain.

flavin mononucleotide (FMN) *See* electron transport chain.

flavin-based electron bifurcation (FBEB) The conservation of energy by linking an unfavorable, endothermic reaction with a favorable exothermic reaction.

flotillin Integral membrane protein that assembles large protein complexes in the plasma membrane.

flow cytometry A method for defining and enumerating cells by sending them in single file past a laser to detect cell size and morphology.

fluorescence activated cell sorting (FACS) A form of flow cytometry that results in separating cells according to size or other quantifiable trait (e.g., DNA content).

fluorescence microscope A microscope that exposes a specimen to light of a specific wavelength and then forms an image from the fluorescent light produced.

fluorescent in situ hybridization (FISH) A technique for identifying certain genes or organisms in which single-stranded DNA fragments are labeled with fluorescent dye and hybridized to genes of interest.

fluorescent light The light emitted by a substance when it is irradiated with light of a shorter wavelength.

fluorochrome A fluorescent dye.

fluoroquinolones A group of broad-spectrum bacteriocidal agents that block DNA gyrase and topoisomerase IV.

fomite Common inanimate materials that transmit pathogens to humans.

food intoxication Food poisoning caused by microbial toxins produced in a food prior to consumption. The presence of living bacteria is not required.

food poisoning A general term usually referring to a gastrointestinal disease caused by the ingestion of food contaminated by pathogens or their toxins.

food-borne infection Gastrointestinal illness caused by ingestion of microorganisms, followed by their growth within the host.

formate dehydrogenase (FDH) An enzyme system that cleaves formic acid into H_2 and CO_2. It is used by enteric bacteria that perform mixed acid fermentation.

forward mutation A mutation from the wild type to a mutant form.

frameshift mutations Mutations arising from the loss or gain of a base or DNA segment, leading to a change in the codon reading frame and thus a change in the amino acids incorporated into protein.

free energy change The total energy change in a system that is available to do useful work as the system goes from its initial state to its final state at constant temperature and pressure.

fruiting body A specialized structure that holds sexually or asexually produced spores; found in fungi, some protists, and some bacteria (e.g., the myxobacteria).

fueling reactions Chemical reactions that supply the ATP, precursor metabolites, and reducing power needed for biosynthesis.

fumonisins A family of toxins produced by molds belonging to the genus *Fusarium*. It primarily affects corn and is known to be hepato- and nephrotoxic in animals.

functional core microbiome A group of microbes that provides activities and metabolites required for health and homeostasis of the host.

functional genomics Genomic analysis concerned with determining the way a genome functions.

functional membrane microdomains A portion of the plasma membrane comprised of distinct lipids that localizes protein complexes like secretion systems.

fungal highway A network of fungal hyphae that provides a physical platform for bacterial motility.

fungicide An agent that kills fungi.

fungistatic Inhibiting the growth and reproduction of fungi.

fungus (pl., **fungi**) Achlorophyllous, heterotrophic, spore-bearing eukaryotes with absorptive nutrition and a walled thallus.

G

G + C content The percent of guanines and cytosines present in a genome; this percent is used for taxonomic purposes.

gametangium (gam-ĕ-tan′je-um; pl., **gametangia**) A structure that contains gametes or in which gametes are formed.

γδ T cells CD4 or CD8 T cells that possess gamma/delta T-cell receptors.

gamonts Gametic cells formed by protists when they undergo sexual reproduction.

gas vacuole A gas-filled vacuole found in cyanobacteria and some other aquatic bacteria and archaea that provides flotation. It is composed of proteinaceous gas vesicles.

gas vesicle *See* gas vacuole.

Gell-Coombs classification System used to codify hypersensitivity reactions.

gene A DNA segment or sequence that codes for a polypeptide, rRNA, or tRNA.

gene ontology (GO) A structured vocabulary adopted for annotating functional gene product data.

gene transfer agents Phage-like particles that package and transfer DNA.

generalized transduction The transfer of any part of a bacterial or archaeal genome when the DNA fragment is packaged within a virus's capsid by mistake.

generation (doubling) time (*g*) The time required for a microbial population to double in number.

genetic complementation The capacity of a gene, when introduced into a cell lacking a functional copy of that gene, to restore the wild-type phenotype to the cell. *See also* phenotypic rescue.

genetic drift Changes in the genotype of an organism that collect over time. These changes are generally neutral mutations that occur and are maintained without selection.

genome The full set of genes present in a cell or virus; all the genetic material in an organism.

genome annotation The process of determining the location and potential function of specific genes and genetic elements in a genome sequence.

genomic island A region in a genome introduced in an ancestral microbe by horizontal gene transfer. Genomic islands often bring new phenotypic traits to the microbe; *see also* pathogenicity island.

genomic library A collection of clones that contains fragments that represent the complete genome of an organism.

genomic reduction The decrease in genomic information that occurs over evolutionary time as an organism or organelle becomes increasingly dependent on another cell or a host organism.

genomics The study of the molecular organization of genomes, their information content, and the gene products they encode.

genus A well-defined group of one or more species that is clearly separate from other organisms.

geographic information system (GIS) Use of hardware, software, and data for the capture, management, and display of geographically referenced information.

germfree (GF) State of having no microbial flora; describing animals born by cesarean delivery and reared in sterile environments.

germination The stage following spore activation in which the spore breaks its dormant state. Germination is followed by outgrowth.

Ghon complex The initial focus of infection in primary pulmonary tuberculosis.

gliding motility A type of motility in which a microbial cell glides smoothly along a solid surface without the aid of flagella.

global regulatory protein A protein that affects the transcription of numerous operons or genes simultaneously.

global regulatory systems Regulatory systems that simultaneously affect many genes.

gluconeogenesis (gloo″ko-ne″o-jen′e-sis) The synthesis of glucose 6-phosphate from noncarbohydrate precursors such as lactate and amino acids.

glutamine synthetase–glutamate synthase (GS-GOGAT) system A mechanism used by many microbes to incorporate ammonia into carbon skeletons.

glycan A type of polysaccharide.

glycerol diether lipids Archaeal lipids formed when isoprenoid hydrocarbons are linked to glycerol by ether bonds.

glycocalyx (gli″ko-kal′iks) A network of polysaccharides extending from the surface of bacteria and other cells.

glycogen (gli′ko-jen) A highly branched polysaccharide containing glucose, which is used to store carbon and energy.

glycolysis (gli-kol′ĭ-sis) The conversion of glucose to pyruvic acid by use of the Embden-Meyerhof pathway, pentose phosphate pathway, or Entner-Doudoroff pathway.

glycolytic pathway A pathway that converts glucose to pyruvic acid.

glycomics Analysis of all the carbohydrates produced by a cell under specific circumstances.

glyoxylate cycle A modified tricarboxylic acid cycle in which the decarboxylation reactions are bypassed by the enzymes isocitrate lyase and malate synthase; it is used to convert acetyl-CoA to succinate and other metabolites.

Golgi apparatus (gol′je) A membranous eukaryotic organelle composed of stacks (dictyosomes) of flattened sacs (cisternae) involved in packaging and modifying materials for secretion and many other processes.

gonococci (gon′o-kok′si) Bacteria of the species *Neisseria gonorrhoeae,* the organism causing gonorrhea.

Gram stain A differential staining procedure that divides bacteria into Gram-positive and Gram-negative groups based on their ability to retain crystal violet when decolorized with an organic solvent such as ethanol.

granular sludge In wastewater management, the stable aggregation of bacteria that readily settle, thereby facilitating recovery of the effluent.

granulocyte A type of white blood cell that stores preformed enzymes and antimicrobial proteins in vacuoles near the cell membrane.

granuloma Term applied to nodular inflammatory lesions containing phagocytic cells.

granzyme Enzyme found in the granules of cytotoxic T cells and natural killer cells that causes targets of these cells to undergo apoptosis.

GRAS An acronym for *g*enerally *r*egarded *a*s *s*afe: a category of food additives recognized in the United States and by the United Nations.

grazing The feeding habit of protozoans consuming bacteria.

great plate count anomaly The discrepancy between the number of viable microbial cells and the number of colonies that can be cultivated from the same natural sample.

green fluorescent protein (GFP) A protein from the jellyfish *Aequorea victoria* that fluoresces when illuminated at a specific wavelength. The gene for GFP can be fused to gene promoters to determine when a gene is expressed (transcriptional fusion). When fused to the coding sequence of a gene so that a chimeric protein is produced, GFP fluorescence reveals the localization of the protein (translational fusion).

green nonsulfur bacteria Anoxygenic photosynthetic bacteria that contain bacteriochlorophyll *a* and *c;* usually photoheterotrophic and display gliding motility. Include members of the phylum Chloroflexota.

green sulfur bacteria Anoxygenic photosynthetic bacteria that contain bacteriochlorophyll *a,* plus *c, d,* or *e;* photolithoautotrophic; use H_2, H_2S, or S as electron donor.

greenhouse gases Gases (e.g., CO_2, CH_4) released from Earth's surface through chemical and biological processes that interact with the chemicals in the stratosphere to decrease radiational cooling of Earth.

group A streptococcus (GAS) A monoderm, coccus-shaped bacterium often found in the throat and on the skin of humans, having the Lancefield group A surface carbohydrate.

group B streptococcus (GBS) A monoderm, coccus-shaped bacterium found occasionally on mucous membranes of humans, having the Lancefield group B surface carbohydrate.

group translocation An active transport process in which a molecule is moved across a membrane by carrier proteins while being chemically altered at the same time (e.g., phosphoenolpyruvate: sugar phosphotransferase system).

growth arrest A phenomenon in which bacterial cells are neither actively dividing nor dead. They are stressed by lack of nutrients or by deviation from optimal conditions.

growth factors Organic compounds that must be supplied in the diet for growth because they are essential cell components or precursors of such components and cannot be synthesized by the organism.

growth rate constant (*k*) The rate of microbial population growth expressed in terms of the number of generations per unit time.

guanine (gwan′in) A purine derivative, 2-amino-6-oxypurine found in nucleosides, nucleotides, and nucleic acids.

guanosine tetraphosphate (ppGpp) A small nucleotide that functions as a second messenger for the stringent response. *See also* stringent response.

guide RNA (gRNA) A short oligoribonucleotide that directs Cas9 nuclease to a specific region of the genome by base-pairing with the target.

gumma (gum′ah) A soft, gummy, benign tumor occurring in tertiary syphilis.

gut-associated lymphoid tissue (GALT) *See* mucosal-associated lymphoid tissue.

H

haloarchaea (extreme halophiles) A group of archaea that depend on high NaCl concentrations for growth and do not survive at a concentration below about 1.5 M NaCl.

halophile A microorganism that requires high levels of sodium chloride for growth.

halotolerant The ability to withstand large changes in salt concentration.

hami Structures extending from the surface of some archaeal cells that terminate in a grappling hooklike apparatus. They are thought to function in attachment to surfaces.

hapten A molecule not immunogenic by itself that, when coupled to a macromolecular carrier, can elicit antibodies directed against itself.

harmful algal bloom (HAB) In aquatic ecosystems, the growth of a single population of phototroph, either a protist (e.g., diatom, dinoflagellate) or a cyanobacterium, that produces a toxin poisonous to other organisms, sometimes including humans. Alternatively, in the absence of toxin production, the concentration of bloom microbe reaches levels that are inherently harmful to other organisms, such as filter-feeding bivalves.

Hartig net The area of nutrient exchange between ectomycorrhizal fungal hyphae and plant host cells.

healthcare-associated (acquired) infection (HAI) An infection that is acquired during a patient's stay in a hospital or other type of clinical care facility.

heat fixation *See* fixation.

helical capsid A viral capsid in the form of a helix.

helicases Enzymes that use ATP energy to unwind DNA ahead of the replication fork.

hemagglutination assay A testing procedure based on a hemagglutination reaction.

hemagglutinin (HA) (1) Proteins that cause red blood cells to clump together (hemagglutination). (2) One of the envelope spikes of influenza and some other viruses. They are the basis for identifying different influenza virus strains.

hematopoiesis The process by which blood cells develop into specific lineages from stem cells. Red and white blood cells and platelets develop from this process.

hemolysis The disruption of red blood cells and release of their hemoglobin. In α-hemolysis, a greenish zone of incomplete hemolysis forms around a colony grown on blood agar. A clear zone of complete hemolysis without any obvious color change is formed during β-hemolysis.

herd immunity The resistance of a population to infection and spread of an infectious agent due to the immunity of a high percentage of the population.

heterocysts Specialized cells of cyanobacteria that are the sites of nitrogen fixation.

heterokont flagella A pattern of flagellation found in the members of the protist taxon Stramenopila, featuring two flagella, one extending anteriorly and the other posteriorly.

heterolactic fermenters Microorganisms that ferment sugars to form lactate and other products such as ethanol and CO_2.

heterologous gene expression The cloning, transcription, and translation of a gene that has been introduced (cloned) into an organism that normally does not possess the gene.

heterotroph An organism that uses reduced, preformed organic molecules as its principal carbon source.

hexose monophosphate pathway *See* pentose phosphate pathway.

Hfr conjugation Conjugation involving an Hfr strain and an F^- strain.

Hfr strain A strain of *E. coli* that donates its chromosomal genes with high frequency to a recipient cell during conjugation because the F factor is integrated into the donor's chromosome.

hierarchical cluster analysis The organization of gene expression data such that induced and repressed genes, or genes of similar function, are grouped separately.

high phosphate transfer potential A characteristic of a phosphorylated compound such that it readily transfers a phosphoryl group to another molecule concomitant with a large energy release.

high-efficiency particulate (HEPA) filter Thick, fibrous filter constructed to remove 99.97% of particles that are 0.3 μm or larger.

high-throughput screening (HTS) A system that combines liquid handling devices, robotics, computers, data processing, and a sensitive detection system to screen thousands of compounds for a single capability.

histatin An antimicrobial peptide composed of 24 to 38 amino acids, heavily enriched with histidine, that targets fungal mitochondria.

histone A small basic protein with large amounts of lysine and arginine that is associated with eukaryotic DNA in chromatin. Related proteins are observed in many archaeal species.

holdfast A structure produced by some bacteria (e.g., *Caulobacter*) that attaches them to a solid object.

holobiont A host organism and its microbes that together form a single functional unit.

holoenzyme A complete enzyme consisting of the apoenzyme plus a cofactor. Also refers to a complete enzyme consisting of all protein subunits (e.g., RNA polymerase holoenzyme).

holozoic nutrition Acquisition of nutrients (e.g., bacteria) by endocytosis and the subsequent formation of a food vacuole or phagosome.

homolactic fermenters Organisms that ferment sugars almost completely to lactic acid.

homologous recombination Recombination involving two DNA molecules that are very similar in nucleotide sequence.

hopanoids Lipids found in bacterial membranes that are similar in structure and function to the sterols found in eukaryotic membranes.

horizontal gene transfer (HGT) The process by which genes are transferred from one mature, independent organism to another. In bacteria and archaea, transformation, conjugation, transduction, and vesiduction are the major mechanisms by which HGT can occur.

hormogonia Small motile fragments produced by fragmentation of filamentous cyanobacteria; used for asexual reproduction and dispersal.

hospital-acquired infections *See* healthcare-associated infections.

host An organism that harbors another organism.

housekeeping genes Genes that encode proteins that function throughout most of the life cycle of an organism (e.g., enzymes of the Embden-Meyerhof pathway).

human leukocyte antigen (HLA) complex A collection of genes on chromosome 6 that encodes proteins involved in host defenses; also called major histocompatibility complex.

hydrogen hypothesis A hypothesis that considers the origin of the eukaryotes through the development of the hydrogenosome. It suggests the organelle arose as the result of an endosymbiotic anaerobic bacterium that produced CO_2 and H_2 as the products of fermentation.

hydrogenosome One member of the mitochondrial family of organelles, it is found in some anaerobic protists that produce ATP by fermentation.

hydrophilic A polar substance that has a strong affinity for water (or is readily soluble in water).

hydrophobic A nonpolar substance lacking affinity for water (or which is not readily soluble in water).

hyperthermophile (hi″per-ther′mo-fīl) A microbe that has its growth optimum between 85° and about 120°C. Hyperthermophiles usually do not grow well below 55°C.

hypha (hi′fah; pl., **hyphae**) The unit of structure of most fungi and some bacteria; a tubular filament.

hypolimnion The colder, bottom layer of water in a stratified lake.

I

icosahedral capsid A viral capsid that has the shape of a regular polyhedron having 20 equilateral triangular faces and 12 corners.

IgA Immunoglobulin A; the class of dimeric immunoglobulins that is present in mucous membranes and in many body secretions.

IgD Immunoglobulin D; the class of immunoglobulins found on the surface of many B lymphocytes, where it serves as an antigen receptor in the stimulation of antibody synthesis.

IgE Immunoglobulin E; the immunoglobulin class that binds to mast cells and basophils and is responsible for type I or anaphylactic hypersensitivity reactions such as hay fever and asthma. IgE is also involved in resistance to helminth parasites.

IgG Immunoglobulin G; the predominant immunoglobulin class in serum. It serves to neutralize toxins, opsonize bacteria, activate complement, and cross the placenta to protect the fetus and neonate.

IgM Immunoglobulin M; the class of serum antibody produced first upon infection. It is a large, pentameric molecule that agglutinates pathogens and activates complement. The monomeric form is present on the surface of some B lymphocytes.

Illness period When a disease is most severe and characteristic signs and symptoms are apparent.

immobilization The incorporation of a simple, soluble substance into the body of an organism, making it unavailable for use by other organisms.

immune complex The product of antibody binding to antigen; may also contain components of the complement system.

immune synapse The binding of a T cell, via its T-cell receptor and coreceptor (CD4 or CD8), to a major histocompatibility complex (MHC)–peptide on the surface of an antigen-presenting cell; it is the first step in T-cell activation.

immune system The defensive system in an animal consisting of the innate and adaptive immune responses. It is composed of widely distributed cells, tissues, and organs that recognize foreign substances and microorganisms and acts to neutralize or destroy them.

immunity Refers to the overall general ability of a host to resist a particular disease; the condition of being immune.

immunization The deliberate introduction of foreign materials into a host to stimulate an adaptive immune response. *See also* vaccine.

immunoblotting The electrophoretic transfer of proteins from polyacrylamide gels to filters to demonstrate the presence of specific proteins through reaction with labeled antibodies. Also called Western blotting.

immunodeficiency The inability to produce a normal innate and/or adaptive immune response.

immunodiffusion A technique involving the diffusion of antigen or antibody within a semisolid gel to produce a precipitin reaction.

immunofluorescence assay A technique used to identify particular antigens microscopically in cells or tissues by binding a fluorescent antibody conjugate.

immunoglobulin (Ig) *See* antibody.

immunology The study of host defenses against invading foreign materials, including pathogenic microorganisms, transformed or cancerous cells, and tissue transplants from other sources.

immunoprecipitation A reaction involving soluble antigens reacting with antibodies to form an aggregate that precipitates out of solution.

in silico analysis The study of biology through the examination of nucleic acid and amino acid sequence. *See also* bioinformatics.

inactivated vaccines Nonviable infectious agents prepared to stimulate immunity but not cause infectious disease; also known as killed vaccines.

incidence The number of new cases of a disease in a population at risk.

inclusions (1) Granules, crystals, or globules of organic or inorganic material in the cytosol of cells. (2) Clusters of viral proteins or virions within the nucleus or cytoplasm of virus-infected cells.

incubation period The period after pathogen entry into a host and before signs and symptoms appear.

indel A blending of the words *insertion* and *deletion;* these genetic mutations can be taxonomically useful.

index case The first disease case in an outbreak or epidemic.

indicator organism An organism whose presence indicates the condition of a substance or environment—for example, the potential presence of pathogens. Coliforms are used as indicators of fecal pollution.

indirect immunofluorescence (IIF) Technique for imaging microscopic agents using antibodies to target the agent and a second, fluorochrome-labeled reagent to bind to the antibodies.

induced mutations Mutations caused by exposure to a mutagen.

inducer A small molecule that stimulates the synthesis of an inducible enzyme.

inducible gene A gene whose expression level can be increased by a regulatory molecule.

induction (1) In virology, the events that trigger a virus to switch from a lysogenic mode to a lytic pathway. (2) In genetics, an increase in gene expression.

infection The invasion of a host by a microorganism with subsequent establishment and multiplication of the agent. An infection may or may not lead to overt disease.

infection thread A tubular structure formed during the infection of a root by nitrogen-fixing bacteria. The bacteria enter the root by way of the infection thread and stimulate the formation of the root nodule.

infectious disease Any change from a state of health in which part or all of the host's body cannot carry on its normal functions because of the presence of an infectious agent or its products.

infectious dose 50 (ID_{50}) Refers to the number of organisms that infect 50% of an experimental group of hosts within a specified time period.

infectivity Infectiousness; the state or quality of being infectious or communicable.

inflammasomes A complex formed within the cytoplasm of certain immune cells that promotes inflammation; composed of oligomerized NOD-like receptors and formed in response to the capture of microbe-associated molecular patterns and danger-associated molecular patterns.

inflammation A localized protective response to tissue injury or destruction. Acute inflammation is characterized by pain, heat, swelling, and redness in the injured area.

initiation factors Proteins that function during initiation of protein synthesis.

initiator tRNA The tRNA that binds to the start codon during initiation of translation.

innate (natural) immunity Refers to nonspecific defenses that are mounted within minutes to hours after challenge by a microbial or other foreign antigen. Sometimes also called nonspecific immunity.

innate lymphoid cell (ILC) A lymphocyte that functions as a bridge between the innate and adaptive arms of immunity, amplifying and relaying information about infection and tissue damage.

innate resistance mechanisms *See* innate (natural) immunity.

insertion sequence A mobile genetic element that contains genes only for the enzymes required for transposition.

integral membrane protein *See* plasma membrane.

integrase An enzyme observed in some viruses; it catalyzes the integration of provirus DNA into the host chromosome.

integrative conjugative elements (ICEs) *See* conjugative transposon.

integrins (in′ta-grinz) Cellular adhesion receptors that mediate cell-cell and cell-substratum interactions, usually by recognizing linear amino acid sequences on protein ligands.

integron A genetic element with an attachment site for site-specific recombination and an integrase gene. It can capture genes and gene cassettes.

interferon (IFN) A cytokine that stimulates cells to produce antiviral proteins (type 1 IFNs) and regulates growth, differentiation, or function of immune system cells (type 2 IFNs).

interleukin (in″tər-loo′kin) A cytokine that regulates growth and differentiation, particularly of lymphocytes.

intermediate filaments One of the proteinaceous components of the eukaryotic cytoskeleton.

interspecies hydrogen transfer The linkage of H_2 production by fermentative microorganisms to the use of H_2 by archaea during methanogenesis.

intoxication A disease that results from the entrance of a specific toxin into the body of a host. The toxin can induce the disease in the absence of the toxin-producing organism.

intracellular pathogen A disease-causing microbe that infects host cells.

intraepidermal lymphocytes T cells found in the epidermis of the skin.

intranuclear inclusion body A structure found within cells infected with cytomegalovirus.

intrinsically disordered region A segment of a polypeptide that does not have a defined three-dimensional structure.

intron A noncoding intervening sequence in a gene that codes for pre-mRNA and is removed (spliced) from the final RNA product.

in-use testing Method for evaluating disinfectant effectiveness under real-world conditions.

invasiveness The ability of a microorganism to enter a host, grow and reproduce within the host, and spread throughout its body.

iodophor An antimicrobial agent consisting of an organic compound complexed with iodine.

ionizing radiation Radiation of very short wavelength and high energy that causes atoms to lose electrons (i.e., ionize).

iron-sulfur (Fe-S) protein *See* electron transport chain.

isoelectric focusing An electrophoretic technique in which proteins are separated based on their isoelectric point.

isoelectric point The pH value at which a protein or other molecule no longer carries a net charge.

isoenzyme (isozyme) An enzyme that carries out the same catalytic function but differs in terms of its amino acid sequence, regulatory properties, or other characteristics.

isolated lymphoid follicles (ILFs) Highly organized lymphoid tissue distributed throughout the small intestine.

isothermal amplification Nucleic acid amplification carried out at a constant temperature.

isotope fractionation The preferential use of one stable isotope over another by microbes carrying out metabolic processes.

J

joule The SI (International System of Units) unit of measure for energy or work. One calorie is equivalent to 4.1840 joules.

K

kallikrein An enzyme that acts on kininogen, releasing the active bradykinin protein.

keratinocytes (kĕ-rat′ĭ-no-sīt) Cells of the skin that express keratin.

Kirby-Bauer method A disk diffusion test to determine the susceptibility of a microorganism to chemotherapeutic agents.

Koch's postulates (koks) A set of rules for proving that a microorganism causes a particular disease.

Koplik's spots White lesions that appear on mucous membranes in the mouth (buccal mucosa) characteristic of measles infection.

Krebs cycle *See* tricarboxylic acid (TCA) cycle.

L

labyrinthulids (lab-ah-rin-thu′lids) A subgroup of stramenopiles characterized by having heterokont flagellated zoospores.

lactic acid bacteria (LAB) Strictly fermentative, monoderm bacteria that produce lactic acid as the primary fermentation end product; used in the fermentation of dairy products.

lactoferrin An iron-sequestering protein released from macrophages and neutrophils into plasma.

lag phase A period following the introduction of microorganisms into fresh culture medium when there is no increase in cell numbers or mass during batch culture.

lagging strand The strand of DNA synthesized discontinuously during DNA replication.

Lancefield system (group) One of the serologically distinguishable groups (e.g., group A and group B) into which streptococci can be divided.

Langerhans cell Cell found in the skin that internalizes antigen and moves in the lymph to lymph nodes, where it differentiates into a dendritic cell.

last universal common ancestor (LUCA) The most recent organism to which all extant organisms are related.

lateral flow immunoassay (LFIA) A simple, inexpensive diagnostic test in which patient sample is applied to a filter paper. The sample is wicked along the filter paper (lateral flow) where an antigen-antibody interaction is detected.

leader A sequence in a gene that lies between the promoter and the start codon. It is transcribed, becoming a nontranslated sequence at the 5′ end of mRNA. It often aids in initiation and regulation of transcription and translation.

leading strand The strand of DNA that is synthesized continuously during DNA replication.

lectin complement pathway An antibody-independent pathway of complement activation that is initiated by microbial lectins (proteins that bind carbohydrates).

leghemoglobin A heme-containing protein produced in leguminous plants. It protects the nitrogenase enzyme from oxygen in nodule-forming, nitrogen-fixing bacteria.

leishmanias (lēsh″ma′ne-ăs) Trypanosomal protists of the genus *Leishmania*.

lentic An aquatic system that features slow-moving or still waters.

lethal dose 50 (LD_{50}) Refers to the number of organisms that kills 50% of an experimental group of hosts within a specified time period.

leukocyte (loo′ko-sīt) Any white blood cell.

ligand *See* receptor.

light reactions *See* photosynthesis.

lignin An irregularly branched molecule made of phenylpropene units; an important structural component of woody plants.

lignocellulose A biopolymer found in plants and soils consisting of cellulose, hemicellulose, and lignin.

limnology The study of the biological, chemical, and physical aspects of freshwater systems.

lipid A The lipid component of a lipopolysaccharide; also called endotoxin.

lipidomics Study of all the lipids produced by an organism under specific circumstances.

lipopolysaccharide (LPS) (lip″o-pol″e-sak′ah-rīd) A molecule containing both lipid and polysaccharide, found in the outer membrane of the typical diderm cell wall. In many bacteria, it consists of three components: lipid A, core polysaccharide, and O antigen.

liquid-liquid phase separation (LLPS) A phenomenon in which two soluble protein structures form distinct phases based on the aggregation of disordered protein regions.

lithoheterotrophy The metabolic strategy characterized by the use of inorganic chemicals as an electron and energy source and organic carbon as a source of carbon. *See also* chemolithoheterotroph.

lithotroph An organism that uses reduced inorganic compounds as its electron source.

littoral zone The coastal region of a lake, stream, or ocean.

lobopodia Rounded pseudopodia found in some amoeboid protists.

localized anaphylaxis (atopic reaction) Type I hypersensitivity that remains contained in one tissue space (e.g., hay fever).

long-term stationary phase The phase following the death phase of a growth curve in which the population size remains at a more or less constant low level for an extended period.

lophotrichous (lo-fot′rĭ-kus) A cell with a cluster of flagella at one or both ends.

lotic An aquatic system that features fast-moving, free-running waters.

lymph node A small secondary lymphoid organ that contains lymphocytes, macrophages, and dendritic cells. It serves as a site for (1) filtration and removal of foreign antigens and (2) activation and proliferation of lymphocytes.

lymphocyte A nonphagocytic, mononuclear leukocyte that is an immunologically competent cell, or its precursor. Lymphocytes are present in the blood, lymph, and lymphoid tissues. *See also* B cell and T cell.

lymphoid tissue Tissues that house large numbers of immune cells such as lymph nodes and Peyer's patches.

lymphoid tissue inducer (LTi) cells Innate lymphoid cells that stimulate the production of secondary lymphoid tissue including lymph nodes and Peyer's patches.

lysis (li′sis) The rupture or physical disintegration of a cell.

lysogenic (li-so-jen′ik) *See* lysogens.

lysogenic conversion A change in the phenotype of a bacterium due to the presence of a prophage.

lysogenic cycle The phase of a temperate virus's life cycle in which it establishes and maintains lysogeny.

lysogens (li′so-jens) Bacterial and archaeal cells that carry a provirus and can produce viruses under the proper conditions.

lysogeny (li-soj′e-ne) The state in which a viral genome remains within a bacterial or archaeal cell after infection and reproduces along with it, rather than taking control of the host cell and destroying it.

lysosome A spherical membranous eukaryotic organelle that contains hydrolytic enzymes and is responsible for the intracellular digestion of substances.

lysozyme An enzyme that degrades peptidoglycan by hydrolyzing the $\beta(1 \rightarrow 4)$ bond that joins *N*-acetylmuramic acid and *N*-acetylglucosamine.

lytic cycle (lit′ik) A viral life cycle that results in the lysis of the host cell.

M

M cell Specialized cell of the intestinal mucosa and other sites (e.g., urogenital tract) that delivers antigen from its apical face to lymphocytes clustered within the pocket in its basolateral face.

macroautophagy Digestion of cytoplasmic components that involves enclosing the material (e.g., an organelle) in a double-membrane structure called an autophagosome. The autophagosome delivers the material to a lysosome for digestion.

macrolide antibiotic An antibiotic containing a macrolide ring—a large lactone ring with multiple keto and hydroxyl groups—linked to one or more sugars.

macromolecular crowding A phenomenon that occurs in the cytoplasm from the high concentration of solutes, resulting in reduced solvent.

macromolecule A large molecule that is a polymer of smaller units joined together.

macronucleus The larger of the two nuclei in ciliate protists. It is normally polyploid and directs the routine activities of the cell.

macronutrient A nutrient that is required in relatively large amounts (e.g., carbon and nitrogen).

macrophage The name for a large, mononuclear phagocytic antigen-presenting cell, present in blood, lymph, and other tissues.

magnetosomes Magnetic greigite or magnetite particles in magnetotactic bacteria that allow the bacteria to orient themselves in magnetic fields.

maintenance energy The energy a cell requires simply to maintain itself or remain alive and functioning properly. It does not include the energy needed for either growth or reproduction.

major histocompatibility complex (MHC) A chromosome locus encoding the histocompatibility antigens and other components of the immune system. Class I MHC molecules are cell-surface glycoproteins present on all nucleated cells; class II MHC glycoproteins are on antigen-presenting cells.

mannose-binding protein (MBP) Also known as mannose-binding lectin, a serum protein that binds to mannose residues in microbial cell walls, thereby initiating complement activation.

marine snow Organic matter that sinks out of the photic zone of the ocean. These flocculent particles are composed of fecal pellets, diatom frustules, and other materials that are not rapidly degraded.

marker gene analysis A method for analyzing community structure by focusing on a single gene present in most members.

mass spectrometry (MS) A type of spectrometry that determines the mass-to-charge ratio of ions formed from the molecule being analyzed. The ratio can be used to identify structures and determine sequences of proteins.

massively parallel nucleotide sequencing *See* next-generation DNA sequencing.

mast cell A white blood cell that produces vasoactive molecules (e.g., histamine) and stores them in vacuoles near the cell membrane where they are released upon cell stimulation by external triggers.

meiosis (mi-o′sis) The type of cell division by which a diploid cell divides and forms four haploid cells.

membrane attack complex (MAC) The complex of complement proteins (C5b–C9) that creates a pore in the plasma membrane of a target cell and leads to cell lysis.

membrane filter Porous material that retains microorganisms as the suspending liquid passes through the pores.

membrane filter technique The use of a thin porous filter to collect microorganisms from water, air, and food.

memory cell An inactive lymphocyte derived from a sensitized B or T cell capable of a rapid response to subsequent antigen exposure.

mesophile A microorganism with a growth optimum around 20° to 45°C, a minimum of 15° to 20°C, and a maximum about 45°C or lower.

messenger RNA (mRNA) Single-stranded RNA synthesized from a nucleic acid template (DNA in cellular organisms, RNA in some viruses) during transcription; mRNA binds to ribosomes and directs the synthesis of protein.

metabolic channeling The localization of metabolites and enzymes in different parts of a cell.

metabolic endotoxemia Increased permeability of the gut wall to lipopolysaccharide (LPS) observed in obese individuals that leads to a state of chronic inflammation, fatty liver disease, and atherosclerosis.

metabolic engineering The use of molecular techniques to develop novel biochemical pathways that either promote product yield or generate new products.

metabolism The total of all chemical reactions in the cell; almost all are enzyme catalyzed.

metabolite flux The rate of turnover of a metabolite.

metabolites Chemical compounds produced by the metabolic activities of organisms.

metabolome All the small molecules produced by a cell, including cytoplasmic and secreted molecules.

metabolomics Analysis of all the small metabolites present in a cell under specific circumstances.

metagenome-assembled genomes (MAGs) A genome that has been assembled from the DNA sequence derived from an environmental sample, not from a pure culture of microbes.

metagenomic library A genomic library constructed of DNA (or cDNA derived from mRNA) extracted directly from the environment.

metagenomics The study of genomes recovered from environmental samples (including the human body) without first isolating members of the microbial community and growing them in cultures.

metalimnion In lakes, the region of the water column between the hypolimnion and the epilimnion.

metaproteomics The emerging field of examining all the proteins produced in a given microbial habitat at a specific time.

metatranscriptomics Large-scale sequencing of mRNAs collected from natural environments as a means to assess microbial activities.

methane hydrate Pools of trapped methane that accumulates in latticelike cages of crystalline water 500 m or more below the sediment surface in many regions of the world's oceans.

methanogenesis The production of methane by certain members of the archaeal phylum Methanobacteriota.

methanogens (meth′ə-no-jens″) Strictly anaerobic archaea that conserve energy by converting CO_2, H_2, formate, acetate, and other compounds to either methane or methane and CO_2.

methanotroph A microbe that has the ability to grow on methane as its sole carbon and energy source.

methylotroph A bacterium that uses reduced one-carbon compounds such as methane and methanol as its sole source of carbon and energy.

metronidazole Antimicrobial chemotherapeutic agent of the nitromidazole family having biological activity on anaerobic bacteria and protozoa.

Michaelis constant (K_m) A value that represents the substrate concentration required for an enzyme to operate at half maximal velocity.

microaerophile (mi″kro-ār′o-fīl) A microorganism that requires low levels of oxygen for growth, around 2 to 10%, but is damaged by normal atmospheric oxygen levels.

microautoradiography (MAR) A technique whereby the uptake of a radioactive substrate by single cells is determined.

microbe-associated molecular pattern (MAMP) Conserved molecular structures that occur in patterns on microbial surfaces. The structures and their patterns are unique to particular types of microorganisms and invariant among members of a given microbial group.

microbial dark matter Microorganisms that have never been cultivated in the laboratory. Their properties are inferred from their genome sequence.

microbial ecology The study of microbial populations and communities in natural habitats, such as soils, water, food, etc.

microbial energy conversion The processes by which microbes are used to generate useful forms of energy, such as biofuels (ethanol, H_2, methanol), or electricity through microbial fuel cells.

microbial flora (microbiota) *See* microbiota.

microbial fuel cell (MFC) An apparatus designed to capture electrons produced during microbial respiration and convert these electrons to electricity.

microbial loop The cycling of organic matter synthesized by photosynthetic microorganisms among other microbes, such as bacteria and protozoa. This process "loops" organic nutrients, minerals, and carbon dioxide back for reuse by the primary producers and makes the organic matter unavailable to higher consumers.

microbial mat A firm structure of layered microorganisms with complementary physiological activities that can develop on surfaces in aquatic environments.

microbial source tracking (MST) Methods for bacterial identification to determine likely sources of contamination in water systems.

microbiology The study of organisms that are usually too small to be seen with the naked eye.

microbiome The totality of microorganisms and microbial genomes that constitute a host's normal microbiota.

microbiota The microbes found in a habitat.

microcosms Incubation chambers with conditions that mimic those in a natural setting.

microdomain Regions of the plasma membrane demarcated by phospholipids, farnesol, hopanoids, and carotenoids; these serve as platforms for the assembly of large protein complexes.

microelectrode An electrode with a tip that is sufficiently small to perform nondestructive measurements at intervals of less than 1 mm.

micronucleus The smaller of the two nuclei in ciliate protists. Micronuclei are diploid and involved only in genetic recombination and the regeneration of macronuclei.

micronutrients Nutrients required in very small quantities for growth and reproduction. Also called trace elements.

microorganism An organism that is usually too small to be seen clearly with the naked eye and is often unicellular, or if multicellular, does not exhibit a high degree of differentiation.

microtubules (mi″kro-tu′būls) One of the proteinaceous components of the eukaryotic cytoskeleton and cilia; composed of tubulin proteins.

mineral soil Soil that contains less than 20% organic carbon.

mineralization The conversion of organic nutrients into inorganic material during microbial growth and metabolism.

minimal inhibitory concentration (MIC) The lowest concentration of a drug that will prevent the growth of a particular microorganism.

mismatch repair A type of DNA repair in which a portion of a newly synthesized strand of DNA containing mismatched base pairs is removed and replaced, using the parental strand as a template.

missense mutation A single base substitution in DNA that changes a codon for one amino acid into a codon for another.

mitochondrion (mi″to-kon′dre-on) The eukaryotic organelle that is the site of electron transport, oxidative phosphorylation, and pathways such as the TCA cycle; it provides most of a nonphotosynthetic cell's energy under aerobic conditions.

mitosis A process that takes place in the nucleus of a eukaryotic cell and results in the formation of two new nuclei, each with the same number of chromosomes as the parent.

mitosome A member of the mitochondria family of organelles found in certain protists that lack mitochondria or hydrogenosomes.

mixed acid fermentation A type of fermentation carried out by members of the family Enterobacteriaceae in which ethanol and a complex mixture of organic acids are produced.

mixotroph An organism that uses a variety of different sources of energy, electrons, and carbon rather than a single trophic mode of growth.

mixotrophy A nutritional strategy that includes different types of metabolism (e.g., phototrophy and chemoorganotrophy).

mobile genetic element (MGE) A small DNA molecule that carries the genes for transposition and thus can move around the genome.

mobilome The mobile genetic elements within a microbial population.

modified atmosphere packaging (MAP) Addition of gases such as nitrogen and carbon dioxide to packaged foods to inhibit the growth of spoilage organisms.

mold Any of a large group of fungi that exist as multicellular filamentous colonies; typically lack macroscopic fruiting bodies.

molecular chaperones *See* chaperone.

molecular mimicry When the molecular structure between a host and non-self antigen is so similar that antibodies made against the non-self antigen also target the host (self) antigen.

monocistronic mRNA An mRNA containing a single coding region.

monocyte A mononuclear phagocytic leukocyte; precursor of macrophages and dendritic cells.

monoderm A bacterial cell with a single membrane, the plasma membrane.

monotrichous (mon-ot′rĭ-kus) Having a single flagellum.

morbidity rate Measures the number of individuals who become ill as a result of a particular disease within a susceptible population during a specific time period.

mortality rate The ratio of the number of deaths from a given disease to the total number of cases of the disease.

mRNA vaccine Vaccine in which mRNA is introduced directly into the host cell.

mucociliary escalator The layer of cilia and mucus that lines certain portions of the respiratory system; it traps microorganisms and then transports them by ciliary action away from the lungs.

mucosal tolerance The ability of the immune system to not respond to, or tolerate, antigens (including microbes) present in and on mucosal tissues.

mucosal-associated lymphoid tissue (MALT) Organized and diffuse immune tissues found as part of the mucosal epithelium. It can be specialized to the gut (GALT) or the bronchial system (BALT).

multicloning site (MCS) A region of DNA on a cloning vector that has a number of restriction enzyme recognition sequences to facilitate the cloning of a gene.

multidrug-resistant tuberculosis (MDR-TB) Strains of Mtb resistant to isoniazid and rifampin.

multilocus sequence analysis (MLSA) A method for genotypic classification of bacteria or archaea within a single genus using nucleotide differences among five to seven housekeeping genes.

multiple displacement amplification (MDA) A method of in vitro DNA synthesis using random primers for rapid amplification of genomic templates, typically using phi29 DNA polymerase.

murein *See* peptidoglycan.

mutagen (mu′tah-jen) A chemical or physical agent that causes mutations.

mutation A heritable change in the genetic material.

mutualism A type of symbiosis in which both partners gain from the association and are dependent on each other.

Myc factor (mik) A lipochitooligosaccharide produced by some mycorrhizal fungi that prompts their uptake by the roots of host plants.

mycelium (mi-se′le-um) A mass of branching hyphae found in fungi and some bacteria.

mycobiont The fungal partner in a lichen.

mycolic acids Complex 60 to 90 carbon fatty acids with a hydroxyl on the β-carbon and an aliphatic chain on the α-carbon; found in the cell walls of mycobacteria.

mycologist A person specializing in mycology; a student of mycology.

mycology The science and study of fungi.

mycomembrane The distinctive mycolic acid–containing outer membrane present in *Mycobacterium*.

mycoplasmas Common name for bacteria that are members of the order Mycoplasmatales. They lack cell walls and cannot synthesize peptidoglycan precursors.

mycorrhizal fungi Fungi that form stable, mutualistic relationships on (ectomycorrhizal) or in (endomycorrhizal) the roots of vascular plants.

mycosis (mi-ko′sis; pl., **mycoses**) Any disease caused by a fungus.

mycotoxicology The branch of mycology that studies the toxins produced by fungi.

mycotoxin A toxin produced by fungi.

myositis (mi″o-si′tis) Inflammation of a striated or voluntary muscle.

myxobacteria A group of diderm, aerobic soil bacteria characterized by gliding motility, a complex life cycle with the production of fruiting bodies, and the formation of myxospores.

myxospores (mik′so-spōrs) Special dormant spores formed by myxobacteria.

N

***N*-acyl homoserine lactone (AHL)** *See* quorum sensing.

naked amoeba An amoeboid protist that lacks a cell wall or other supporting structures.

naked viruses *See* nonenveloped virus.

nanowires Threadlike extensions of plasma membrane that transfer electrons from the terminal point in the electron transport chain to an external metal surface.

narrow-spectrum drugs Chemotherapeutic agents effective only against a limited range of microorganisms.

natural immunity *See* innate immunity.

natural killer (NK) cell A type of white blood cell that destroys host cells infected with intracellular pathogens and tumor cells.

natural products Industrially important compounds, such as antibiotics, produced by microbes.

naturally acquired active immunity The type of active immunity that develops when an individual's immune system encounters an appropriate antigen during the course of normal activities; it often arises as the result of recovering from an infection and lasts a long time.

naturally acquired passive immunity The type of temporary immunity that involves the transfer of antibodies from one individual to another.

necrotrophic fungi Plant pathogenic fungi that kill their host.

negative control Regulation of transcription by a repressor protein. When bound to the repressor-binding site, transcription is inhibited.

negative selection In immunology, the process by which lymphocytes that recognize host (self) antigens undergo apoptosis or become anergic (inactive).

negative staining A staining procedure in which a dye is used to make the background dark while the specimen is unstained.

negative strand A viral single-stranded nucleic acid that is complementary to the viral mRNA.

netosis Neutrophil cell death and simultaneous neutralization of microbes by the release of web like structures called neutrophil extracellular traps (NETs) composed of DNA, histones, and antimicrobial proteins.

neuraminidase An enzyme that cleaves the chemical bond linking neuraminic acids to the sugars present on the surface of animal cells; in virology, one type of envelope spike on influenza viruses has neuraminidase activity and is used to identify different strains.

neutral mutation A missense mutation that does not affect the function of the protein (e.g., a polar amino acid is substituted for a polar amino acid).

neutralization Binding of specific immunoglobulins to toxins or viruses, inhibiting their biological activity.

neutrophil A mature white blood cell in the granulocyte lineage. It has a nucleus with three to five lobes and is very phagocytic.

neutrophile A microorganism that grows best at a neutral pH range between pH 5.5 and 8.0.

next-generation sequencing (NGS) Methodologies that allow rapid genome sequencing without the prior construction of a genomic library. Instead, fragments of genomic DNA are tethered to a solid substrate, and each fragment is sequenced simultaneously or in a massively parallel fashion.

nicotinamide adenine dinucleotide (NAD^+) (nik″o-tin′ah-mīd) An electron-carrying coenzyme; it is particularly important in catabolic processes and usually transfers electrons from an electron source to an electron transport chain.

nicotinamide adenine dinucleotide phosphate ($NADP^+$) An electron-carrying coenzyme that most often participates as an electron carrier in biosynthetic metabolism.

nitrification The oxidation of ammonia to nitrate.

nitrifying bacteria Chemolithotrophic, diderm bacteria that either oxidize ammonia to nitrite or nitrite to nitrate.

nitrogen fixation The metabolic process in which atmospheric molecular nitrogen (N_2) is reduced to ammonia; carried out by nitrogen-fixing bacteria and archaea.

nitrogenase (ni′tro-jen-ās) The enzyme that catalyzes biological nitrogen fixation.

nitroplast An organelle found in marine algae that contains the enzymes required for nitrogen fixation.

Nod factors Signaling compounds that alter gene expression in the plant host of a rhizobium.

NOD-like receptors (NLRs) Receptors similar to nucleotide-binding and oligomerization domain (NOD) receptors that sense endogenous metabolites and regulate the production of inflammatory cytokines in response to microbe-associated molecular patterns and danger-associated molecular patterns.

noncompetitive inhibitor A chemical that inhibits enzyme activity by a mechanism that does not involve binding the active site of the enzyme.

noncyclic photophosphorylation The process in which light energy is used to make ATP when electrons are moved from water to $NADP^+$ during oxygenic photosynthesis; both photosystem I and photosystem II are involved.

nonenveloped (naked) virus A virus having virions lacking an envelope.

nonheme iron protein *See* electron transport chain.

nonhomologous end joining (NHEJ) A genetic repair phenomenon in which two linear chromosomal fragments are rejoined despite no sequence homology.

nonsense (stop) codon A codon that does not code for an amino acid but is a signal to terminate protein synthesis.

nonsense mutation A mutation that converts a sense codon to a nonsense (stop) codon.

nosocomial infection (nos′o-ko″me-al) *See* healthcare-associated infection.

nuclear envelope The complex double-membrane structure forming the outer boundary of the eukaryotic nucleus. It is covered by nuclear pores through which substances enter and leave the nucleus.

nuclear pore complex The nuclear pore plus about 30 proteins that form the pore and are involved in moving materials across the nuclear envelope.

nucleic acid amplification tests (NAATs) Molecular tests that rely on increasing the amount of a particular region of a nucleic acid.

nucleocapsid (nu″kle-o-kap′sid) The viral nucleic acid and its surrounding capsid; the basic unit of virion structure.

nucleoid An irregularly shaped region in a bacterial or archaeal cell containing genetic material.

nucleoid occlusion A mechanism used by many bacteria to prevent premature formation of a septum during binary fission.

nucleoid-associated protein (NAP) Proteins that serve an architectural role in the nucleoid and contribute to chromosome structure.

nucleolus (nu-kle′o-lus) The organelle, located within the nucleus and not bounded by a membrane, that is the location of ribosomal RNA synthesis and the assembly of ribosomal subunits.

nucleoside (nu′kle-o-sīd″) A combination of ribose or deoxyribose with a purine or pyrimidine base but lacking phosphate.

nucleosome (nu′kle-o-sōm″) A complex of histones and DNA found in eukaryotic chromatin and some archaea; the DNA is wrapped around the surface of the histone complex.

nucleotide (nu′kle-o-tīd) A combination of ribose or deoxyribose with phosphate and a purine or pyrimidine base; a nucleoside plus one or more phosphates.

nucleotide excision repair *See* excision repair.

nucleotide-binding and oligermization domain (NOD)–like receptors Intracellular sensors of microbe and damage associated molecular pattern molecules.

nucleus The eukaryotic organelle enclosed by a double-membrane envelope that contains the cell's chromosomes.

numerical aperture The property of a microscope lens that determines how much light can enter and how great a resolution the lens can provide.

O

O antigen The terminal polysaccharide of the lipopolysaccharide found on some diderm bacterial cell walls.

objective lens Lens closest to the specimen that forms a magnified image.

obligate aerobes Organisms that grow only in the presence of oxygen.

obligate anaerobes Microorganisms that cannot tolerate oxygen and die when exposed to it.

obligate intracellular pathogen A pathogen that requires a host cell for replication.

obligatory Term describing an interaction or condition that is required for bacterial growth.

ocean acidification The decline in pH in ocean waters as a result of changes to the carbonate equilibrium system.

oceanography The study of the biological, chemical, and physical aspects of the oceans.

ocular lens Lens of a compound microscope that is closest to the eye; also called the eyepiece.

oil and grease (O&G) One component of total organic carbon that is quantified in wastewater systems.

Okazaki fragments Short stretches of polynucleotides produced during discontinuous DNA replication.

oligonucleotide primer A short fragment of DNA or RNA, usually artificially synthesized, used in a number of molecular genetic techniques such as DNA sequencing and polymerase chain reaction.

oligonucleotide signature sequence *See* barcode.

oligotrophic environment (ol″ĭ-go-trof′ik) An environment containing low levels of nutrients that support microbial growth.

oocyst (o′o-sist) Cyst formed around a zygote of certain protozoa.

öomycetes (o″o-mi-se′tēz) A collective name for protists also known as water molds. Formerly thought to be fungi.

open reading frame (ORF) A sequence of DNA thought to encode a protein because it is not interrupted by a stop codon and has an apparent promoter and ribosome-binding site at the 5′ end and a terminator at the 3′ end.

operator The segment of DNA in an operon to which the repressor protein binds; it controls the expression of the genes adjacent to it.

operon In bacteria, the sequence of bases in DNA that contains a promoter and one or more structural genes and often an operator or activator-binding site that controls their expression.

opportunistic microorganism or pathogen A microorganism that is usually free living or a part of the host's normal microbiota but may become pathogenic under certain circumstances, such as when the immune system is compromised.

opsonin Proteins, including antibodies and complement proteins, that bind to cell surfaces and mark them for destruction by phagocytosis.

opsonization (op″so-ni-za′shun) The coating of foreign substances by antibody, complement proteins, or fibronectin to make the substances more readily recognized by phagocytic cells.

opsonize The process by which opsonins bind to target cell surfaces (*see* opsonin).

organelle A structure within or on a cell that performs specific functions and is related to the cell in a way similar to that of an organ to the body of a multicellular organism.

organic soil A soil that contains at least 20% organic carbon.

organotrophs Organisms that use reduced organic compounds as their electron source.

origin of replication A site on a chromosome or plasmid where DNA replication is initiated.

origin recognition complex (ORC) A protein complex that marks the origins of replication in eukaryotic chromosomes.

orthologue A gene found in the genomes of two or more different organisms that share a common ancestry. The products of orthologous genes are presumed to have similar functions.

osmophiles Microorganisms that grow best in or on media of high solute concentration.

osmotolerant Organisms that grow over a fairly wide range of water activity or solute concentration.

osmotrophy A form of nutrition in which soluble nutrients are absorbed through the cytoplasmic membrane; found in bacteria, archaea, fungi, and some protists.

Öuchterlony technique *See* double immunodiffusion agar assay.

outbreak The sudden, unexpected occurrence of a disease in a given population.

outer membrane A membrane located outside the peptidoglycan layer in the cell envelope of typical diderm bacteria.

outer membrane protein (OMP) A protein found spanning the outer membrane of diderm bacteria that adopts a barrel-like structure composed of beta-sheets.

outer membrane vesicle (OMV) A spherical portion of the diderm outer membrane that pinches off to leave the cell. OMVs may carry proteins, toxins, and nucleic acids.

oxazolidinones Synthetic drugs, characterized by a heterocyclic five-membered ring, that inhibit protein synthesis.

oxidation-reduction (redox) reactions Reactions involving electron transfers; the electron donor (reductant) gives electrons to an electron acceptor (oxidant).

oxidative burst *See* respiratory burst.

oxidative phosphorylation (fos″for-ĭ-la′-shun) The synthesis of ATP from ADP using energy made available during electron transport initiated by the oxidation of a chemical energy source.

oxygenic photosynthesis Photosynthesis that oxidizes water to form oxygen; the form of photosynthesis characteristic of plants, protists, and cyanobacteria.

P

palisade arrangement Angular arrangements of bacteria whereby cells remain partially attached after snapping division.

pandemic An increase in the occurrence of a disease within a large and geographically widespread population (often refers to a worldwide epidemic).

Paneth cell (pah′net) A specialized epithelial cell of the intestine that secretes hydrolytic enzymes and antimicrobial proteins and peptides.

pan-genome The collection of genes found in all strains that belong to a single species or other taxonomic group.

paralogue Two or more genes in the genome of a single organism that arose through duplication of a common ancestral gene. The products of paralogous genes may have slightly different functions.

parasite An organism that lives on or within another organism (the host) and benefits from the association while harming its host. Often the parasite obtains nutrients from the host.

parasitism A type of symbiosis in which one organism benefits from the other and the host is usually harmed.

parasporal body An intracellular, solid protein crystal made by the bacterium *Bacillus thuringiensis*. It is the basis of the bacterial insecticide Bt.

parfocal A microscope that retains proper focus when the objectives are changed.

particulate organic matter (POM) Nutrients that are not dissolved or soluble, generally referring to aquatic ecosystems. This includes microorganisms and their cellular debris after senescence or viral lysis.

passive diffusion The movement of molecules from a region of higher concentration to one of lower concentration as a result of random thermal agitation.

pasteurization The process of heating liquids to destroy microorganisms that can cause spoilage or disease.

pathobionts A component of the normal microbiota that may be pathogenic under some circumstances.

pathogen (path′o-jən) Any organism that causes disease.

pathogenicity (path″o-je-nis′ĭ-te) The condition or quality of being pathogenic, or the ability to cause disease.

pathogenicity island A segment of DNA in some pathogens that contains genes responsible for virulence; often codes for a type III secretion system that allows the pathogen to secrete virulence proteins and damage host cells.

pattern recognition receptor (PRR) A receptor found on macrophages and other phagocytic cells that binds to microbe-associated molecular patterns on microbial surfaces.

pelagic zone The central region of a lake or ocean not influenced by events or nutrients found in the coastal regions.

pellicle (pel′ĭ-kuhl) (1) A relatively rigid layer of proteinaceous elements just beneath the plasma membrane in many protists. The plasma membrane is sometimes considered part of the pellicle. (2) A biofilm that forms at an air-liquid interface.

penicillin A β-lactam antibiotic. The first true antibiotic to be discovered and used clinically.

penicillinase A β-lactamase enzyme that digests the active moiety of penicillin.

pentose phosphate (hexose monophosphate) pathway A glycolytic pathway that oxidizes glucose 6-phosphate to ribulose 5-phosphate and then converts it to a variety of three to seven carbon sugars.

peplomer (spike) (pep′lo-mer) A protein or protein complex that extends from the viral envelope and often is important in virion attachment to the host cell surface.

peptide interbridge A short peptide chain that connects the tetrapeptide chains in the peptidoglycan of some bacteria.

peptidoglycan (pep″tĭ-do-gli′kan) A large polymer composed of long chains of alternating *N*-acetylglucosamine and *N*-acetylmuramic acid residues. The polysaccharide chains are linked to each other through connections between tetrapeptide chains attached to the *N*-acetylmuramic acids. It provides much of the strength and rigidity possessed by bacterial cell walls; also called murein.

peptidyl or donor site (P site) The site on the ribosome that contains the peptidyl-tRNA at the beginning of the elongation cycle during protein synthesis.

peptidyl transferase The ribozyme that catalyzes the transpeptidation reaction in protein synthesis; in this reaction, an amino acid is added to the growing peptide chain.

perforin Pore-forming proteins secreted along with granzymes by cytotoxic T lymphocytes and NK cells.

peripheral membrane protein *See* plasma membrane.

periplasm (per′ĭ-plaz-əm) The substance that fills the periplasmic space.

periplasmic flagella The flagella that lie under the outer sheath and extend from both ends of a spirochete cell to overlap in the middle and form the axial filament. Also called endoflagella.

periplasmic space The space between the plasma membrane and the outer membrane in typical diderm bacteria, and between the plasma membrane and the cell wall in typical monoderm bacteria. A similar space is sometimes observed between the plasma membrane and the cell wall of some archaea.

peristalsis The muscular contractions of the gut that propel digested foods and waste through the intestinal tract.

peritrichous (pĕ-rit′rĭ-kus) A cell with flagella distributed over its surface.

peronosporomycetes Name for öomycetes, a subgroup of Stramenopila, distinguished by the formation of large egg cells when they reproduce.

peroxidase Enzyme that detoxifies hydrogen peroxide.

persisters Variants of microbial cells in a population that are resistant to antibiotics.

Peyer's patches Highly organized lymphoid tissue in the small intestine.

phage (fāj) *See* bacteriophage.

phage-plasmids Hybrid genetic elements that transfer vertically as plasmids and horizontally as phages.

phagocytic vacuole A membrane-delimited vacuole produced by cells carrying out phagocytosis. It is formed by the invagination of the plasma membrane and contains solid material.

phagocytosis The endocytotic process in which a cell encloses large particles in a phagocytic vacuole (phagosome) and engulfs them.

phagolysosome The vacuole that results from the fusion of a phagosome with a lysosome.

phagosome A membrane-enclosed vacuole formed by the invagination of the cell membrane during endocytosis.

phase variation A reversible bacterial process that facilitates rapid genetic changes in response to a changing environment.

phase-contrast microscope A microscope that converts slight differences in refractive index and cell density into easily observed differences in light intensity.

phenol coefficient test A measurement of the effectiveness of disinfectants that compares their activity against test bacteria with that of phenol.

phenotype Observable characteristics of an organism.

phenotypic rescue The identification of a cloned gene based on its ability to complement a genetic deficiency in the host cell.

phosphatase (fos′fah-tās″) An enzyme that catalyzes the hydrolytic removal of phosphate from molecules.

phosphoenolpyruvate:sugar phosphotransferase system (PTS) A group translocation system used by many bacteria. As a sugar is transported into the cell, the hydrolysis of a high-energy phosphate bond fuels its import, and the sugar is modified by the covalent attachment of the phosphoryl group.

phosphorelay system A set of proteins involved in the transfer of phosphate from one protein in the set to another. It is often used to regulate protein activity or transcription.

photic zone The illuminated area of an aquatic habitat from the surface to the depth to which the rate of photosynthesis equals that of respiration.

photoautotroph *See* photolithoautotroph.

photobiont (fo″to-bi′ont) The photosynthetic protist or cyanobacterial partner in a lichen.

photolithoautotroph An organism that uses light energy, an inorganic electron source (e.g., H_2O, H_2, H_2S), and CO_2 as its carbon source. Also called photolithotrophic autotroph or a photoautotroph.

photoorganoheterotroph A microorganism that uses light energy, organic electron sources, and organic molecules as a carbon source. Also called photoorganotrophic heterotroph.

photophosphorylation The synthesis of ATP from ADP using energy made available by the absorption of light.

photosynthate Nutrient material released from phototrophic organisms that contributes to the pool of dissolved organic matter.

photosynthesis The trapping of light energy and its conversion to chemical energy (light reactions), which is then used to reduce CO_2 and incorporate it into organic molecules (dark reactions).

photosystem I The photosystem in eukaryotic cells and cyanobacteria that absorbs longer wavelength light, usually greater than about 680 nm, and transfers the energy to chlorophyll P700 during photosynthesis; it is involved in both cyclic and noncyclic photophosphorylation.

photosystem II The photosystem in eukaryotic cells and cyanobacteria that absorbs shorter wavelength light, usually less than 680 nm, and transfers the energy to chlorophyll P680 during photosynthesis; it participates in noncyclic photophosphorylation.

phototaxis The ability of certain phototrophic organisms to move in response to a light source.

phototrophs Organisms that use light as their energy source.

phycobiliproteins Photosynthetic pigments found in cyanobacteria that are composed of proteins with attached tetrapyrroles.

phycobilisomes Particles on the membranes of cyanobacteria that contain photosynthetic pigments.

phycocyanin (fi″ko-si′an-in) A blue phycobiliprotein pigment used to trap light energy during photosynthesis.

phycoerythrin (fi″ko-er′i-thrin) A red photosynthetic phycobiliprotein pigment used to trap light energy.

phycology (fi-kol′o-je) The study of algae.

phyletic classification system *See* phylogenetic classification system.

phyllosphere The surface of plant leaves.

phylogenetic classification system A classification system based on evolutionary relationships, rather than the general similarity of characteristics.

phylogenetic tree A graph made of nodes and branches, much like a tree in shape, that shows phylogenetic relationships between groups of organisms and sometimes also indicates the evolutionary development of groups.

phylogenomics The use of comparative genomics to determine the evolutionary history of organisms.

phylogeny The evolutionary development of a species.

physical map In genomics, this refers to a diagram of a chromosome, gene, or other genetic element that shows restriction endonuclease recognition sites. Specific genes and other genetic elements (e.g., origin of replication) may also be noted.

phytoplankton A community of floating photosynthetic organisms, composed of photosynthetic protists and cyanobacteria.

picoplankton Planktonic microbes between 0.2 and 2.0 μm in size, including the cyanobacterial genera *Prochlorococcus* and *Synechococcus*.

piezophilic (pi-e-zo-fil′ic) Describing organisms that prefer or require high pressures for growth and reproduction.

pili (s., **pilus**) A fine, hairlike protein appendage on many bacteria, archaea, and some fungi. They attach cells to surfaces, and some are involved in twitching motility.

pilin The protein monomer that comprises a pilus.

planktonic Describes microbes that are free-living and not attached to a substrate.

plant microbiome All microorganisms that inhabit above- and below-ground parts of plants.

plaque A clear area in a lawn of host cells that results from their lysis by viruses.

plaque assay A method used to determine the number of infectious virions.

plaque-forming unit (PFU) The unit of measure of a plaque assay. It usually represents a single infectious virion.

plasma cell A mature, differentiated B lymphocyte that synthesizes and secretes antibody.

plasma membrane The selectively permeable membrane surrounding the cell's cytoplasm; also called the cell membrane, plasmalemma, or cytoplasmic membrane. For most cells it is a lipid bilayer (some archaea have a lipid monolayer) with proteins embedded in it (integral membrane proteins) and associated with the surface (peripheral membrane proteins).

plasmalemma The plasma membrane in protists.

plasmid A double-stranded DNA molecule that can exist and replicate independently of the chromosome. A plasmid is stably inherited but is not required for the host cell's growth and reproduction.

plasmolysis (plaz-mol′ĭ-sis) The process in which water osmotically leaves a cell, which causes the cytoplasm to shrivel up and pull the plasma membrane away from the cell wall.

plastid A cytoplasmic organelle of algae and higher plants that contains pigments such as chlorophyll, stores food reserves, and often carries out processes such as photosynthesis.

plastisphere The microbial niche found in plastic waste.

pleomorphic (ple″o-mor′fik) Refers to cells or viruses that are variable in shape and lack a single, characteristic form.

point mutation A mutation that affects only a single base pair.

polar flagellum A flagellum located at one end of an elongated cell.

3′ poly-A tail *See* pre-mRNA.

polycistronic mRNA An mRNA that has more than one coding region; formed when an operon is transcribed.

polyhistidine tagging (His-tagging) A method by which six consecutive histidines are incorporated in a protein, thereby facilitating the protein's purification through affinity chromatography.

polymerase chain reaction (PCR) An in vitro technique used to synthesize large quantities of specific nucleotide sequences from small amounts of DNA. It employs oligonucleotide primers complementary to specific sequences in the target gene and special heat-stable DNA polymerases (e.g., *Taq* polymerase).

polymorphonuclear cell (PMN) (pol″e-mor″fo-noo′kle-ər) A term used to describe basophils, eosinophils, and neutrophils.

polyprotein A large polypeptide that contains several proteins, which are released when the polyprotein is cut by proteases. Polyproteins are often produced by viruses having positive-strand RNA genomes.

polyribosome (pol″e-ri′bo-sōm) A complex of several ribosomes translating a single messenger RNA.

population A group of organisms of the same species within a community.

porin proteins Proteins that form channels across the outer membrane of typical diderm bacterial cell walls through which small molecules enter the periplasm.

portal of entry Body surface that is the initial site of pathogen access to the host.

positive control Control of transcription by an activator protein. When the activator is bound to the activator-binding site, the level of transcription increases.

positive selection The process by which only those immature T cells in the thymus that appropriately recognize MHC class I or II molecules bearing self-peptides are signaled to continue development into either CD4 or CD8 T cells.

positive-strand (plus-strand) DNA A viral nucleic acid strand that is equivalent in base sequence to the viral mRNA.

postherpetic neuralgia The severe pain after a herpes infection.

posttranslational regulation Alteration of the activity of a protein after it has been synthesized by a structural modification, such as phosphorylation or a conformational change. *See also* allosteric enzyme and reversible covalent modification.

posttranslationally Occurring after translation has terminated.

pour plate A Petri dish of solid culture medium with isolated microbial colonies growing both on its surface and within the medium that has been prepared by mixing microorganisms with cooled, still-liquid medium and then allowing the medium to harden.

prebiotic Compound(s) added to foods and supplements to enhance the colonization of probiotic microbes.

precipitation (precipitin) reaction The reaction of an antibody with a soluble antigen to form an insoluble precipitate.

precipitin The antibody responsible for a precipitation reaction.

precursor metabolites Intermediates of glycolytic pathways, TCA cycle, and other pathways that serve as starting molecules for biosynthetic pathways that generate monomers and other building blocks needed for synthesis of macromolecules.

pre-mRNA In eukaryotes, the RNA transcript of DNA made by RNA polymerase II prior to processing to form mRNA by the addition of the 5′ cap and 3′ poly-A tail, and the removal of introns.

prevalence Refers to the total number of individuals infected at any one time in a given population regardless of when the disease began.

primary active transport *See* active transport.

primary metabolites Microbial metabolites produced during active growth of an organism.

primary mRNA *See* pre-mRNA.

primary producer Photoautotrophic and chemoautotrophic organisms that incorporate carbon dioxide into organic carbon and thus provide new biomass for the ecosystem.

primary production The incorporation of carbon dioxide into organic matter by photosynthetic organisms and chemoautotrophic organisms.

primers *See* polymerase chain reaction.

prion (pre′on) An infectious agent consisting only of protein; prions cause a variety of transmissible spongiform encephalopathies such as scrapie in sheep.

probe A labeled single-stranded oligonucleotide designed to hybridize with a microorganism of interest.

probiotic A living microorganism that may provide health benefits beyond its nutritional value when ingested.

prochlorophytes A group of cyanobacteria having both chlorophyll *a* and *b* but lacking phycobilins.

prodromal stage (pro-dro′məl) The period during the course of a disease when signs and symptoms are present, but they are not yet distinctive enough for an accurate diagnosis.

production strains Microbial strains used in industrial processes that have been engineered to generate a high yield of the desired product.

professional phagocyte Any one of the three types of immune cells that phagocytose microbes. These include neutrophils, macrophages, and dendritic cells.

promoter The region on DNA at the start of a gene that RNA polymerase binds to before beginning transcription.

proofreading The ability of enzymes to check their products to ensure that the correct product is made. For instance, DNA polymerase checks newly synthesized DNA and replaces an incorrect nucleotide with the correct one prior to further synthesis.

propagated epidemic An epidemic that is characterized by a relatively slow and prolonged rise and then a gradual decline in the number of individuals infected. It usually results from the introduction of an infected individual into a susceptible population and transmission of the pathogen from person to person.

prophage (pro′fāj) **(provirus)** (1) The latent form of a temperate bacterial or archaeal virus that remains within the lysogen, usually integrated into the host chromosome. (2) The form of a eukaryotic virus that remains within the host cell during a latent infection. Also refers to a retroviral genome after it has been integrated into the host's chromosome.

prostheca (pros-the′kah) An extension of a bacterial cell, including the plasma membrane and cell wall, that is narrower than the mature cell.

prosthetic group A tightly bound cofactor that remains at the active site of an enzyme during its catalytic activity.

protease (pro′te-ās) An enzyme that hydrolyzes proteins. Also called a proteinase.

proteasome A large, cylindrical protein complex observed in eukaryotic cells that degrades ubiquitin-labeled proteins to peptides in an ATP-dependent process. Also called 26S proteasome. Similar protein-degrading machinery has been observed in some archaea and bacteria.

protein disulfide isomerase (PDI) A periplasmic enzyme that catalyzes the formation of disulfide bonds as a polypeptide folds into its functional form.

protein modeling The process by which the amino acid sequence of a protein is analyzed using software designed to predict the protein's three-dimensional structure.

proteome The complete collection of proteins that an organism produces.

proteomics The study of the structure and function of cellular proteins.

proteorhodopsin A rhodopsin molecule that enables rhodopsin-based phototrophy. *See also* archaerhodopsin.

protist Unicellular and sometimes colonial eukaryotic organisms that lack cellular differentiation into tissues. Many chemoorganotrophic protists are referred to as protozoa; many phototrophic protists are referred to as algae.

protistology The study of protists.

protomer An individual protein of a viral capsid; a capsomer is made of protomers.

proton motive force (PMF) The potential energy arising from a gradient of protons and a membrane potential that powers ATP synthesis and other processes.

protoplast (1) The plasma membrane and everything within it. (2) A bacterial, archaeal, or fungal cell with its cell wall completely removed. It is spherical in shape and osmotically sensitive.

protospacer adjacent motif (PAM) A series of nucleotides recognized by the Cas9 component in genome editing.

prototroph A microorganism that requires the same nutrients as the majority of naturally occurring members of its species.

protozoa (pro″to-zo′a) A common term for a group of unrelated unicellular, chemoorganotrophic protists.

protozoology The study of protozoa.

provirus *See* prophage.

pseudogene A degraded, nonfunctional gene.

pseudomurein A complex polysaccharide observed in the cell walls of some archaea; it resembles peptidoglycan (murein) in structure and chemical makeup.

pseudopodium/pseudopod (soo″do-po′de-um) A nonpermanent cytoplasmic extension of the cell by which amoeboid protists move and feed.

psychrophile (si′kro-fīl) A microorganism that has an optimum growth temperature of 15°C or lower and a temperature maximum around 20°C. Synonymous with cryophile.

psychrotolerant (psychrotroph) A microorganism that has a growth optimum between 20° and 30°C, and a maximum of about 35°C.

pure (axenic) culture A population of cells that are identical because they arise from a single cell.

purine (pu′rēn) A heterocyclic, nitrogen-containing molecule with two joined rings that occurs in nucleic acids and other cell constituents (e.g., adenine and guanine).

purple membrane An area of the plasma membrane of archaeal halophiles that contains archaerhodopsin and is active in trapping light energy.

purple nonsulfur bacteria Anoxygenic photosynthetic Alphaproteobacteria.

purple sulfur bacteria Anoxygenic photosynthetic Gammaproteobacteria.

putrefaction The microbial decomposition of organic matter, especially the anaerobic breakdown of proteins, with the production of foul-smelling compounds such as hydrogen sulfide and amines.

pyelonephritis An infection of one or both kidneys, often as a result of an untreated urinary tract infection.

pyogenic Pus-producing.

pyrenoid (pi′rĕ-noid) The differentiated region of the chloroplast that is a center of starch formation in some photosynthetic protists.

pyrimidine (pi-rim′i-dēn) A nitrogen-containing molecule with one ring that occurs in nucleic acids and other cell constituents; the most important pyrimidines are cytosine, thymine, and uracil.

pyruvate dehydrogenase complex (PDH) A multienzyme complex that catalyzes pyruvate oxidation and decarboxylation to acetyl-CoA, which then enters the tricarboxylic acid cycle.

pyruvate formate-lyase (PFL) An enzyme system that cleaves pyruvate into formic acid and acetyl-CoA. It is used by enteric bacteria that perform mixed-acid fermentation.

Q

quinine drugs Antimicrobial chemotherapeutic agents derived from the bark of the cinchona tree and used to treat malaria.

quorum sensing The process in which bacteria monitor their own population density or the presence of other species of bacteria by sensing the levels of signal molecules (e.g., *N*-acyl homoserine lactone) released by the microorganisms. When these signal molecules reach a threshold concentration, quorum-dependent genes are expressed.

R

rapid sand filters A step in drinking water purification that removes particulates by passage over sand.

rare biosphere Organisms that are ubiquitous but not abundant.

reaction center A protein-pigment complex in the photosynthesis apparatus that converts light energy into chemical energy by initiating electron transfer.

reaction-center chlorophyll pair The two chlorophyll molecules in a photosystem that are energized by the absorption of light and release electrons to an associated electron transport chain, thus initiating energy conservation by photophosphorylation.

reactive nitrogen intermediate (RNI) Charged nitrogen radicals.

reactive oxygen species (ROS) Superoxide radical (O_2^-), hydrogen peroxide (H_2O_2), singlet oxygen ($^1O_2$), and hydroxyl radical (OH^-) derivatives of oxygen.

reading frame The way in which nucleotides in DNA and mRNA are grouped into codons for reading the message contained in the nucleotide sequence.

readthrough A phenomenon in which a ribosome ignores a stop codon and continues protein synthesis. This generates a different protein than that produced if the ribosome had stopped at the stop codon.

realm The highest classification level for viruses, equivalent to domain for cellular organisms.

real-time PCR (qPCR) A type of polymerase chain reaction (PCR) that quantitatively measures the amount of template in a sample as the amount of fluorescently labeled amplified product.

RecA A protein that functions in DNA repair and recombination processes. Homologues of RecA have been found in all three domains of life.

receptor-mediated endocytosis *See* endocytosis.

receptors Proteins that bind signaling molecules (ligands), thereby initiating cellular responses.

recombinants Organisms produced following a recombination event.

recombinant-vector vaccine A vaccine produced by the introduction of one or more of a pathogen's genes into attenuated viruses or bacteria, which serves as a vector, replicating within the vertebrate host and expressing the gene(s) of the pathogen. The pathogen's antigens induce an immune response.

recombinases Enzymes that catalyze integration and excision of

DNA segments during site-specific recombination.

recombination The process in which a new recombinant chromosome is formed by combining genetic material from two different nucleic acids. In some cases, the nucleic acids are from two different organisms.

recombinational repair A DNA repair process that repairs damaged DNA when there is no remaining template; a piece of DNA from a sister molecule is used.

red tides Population blooms of dinoflagellates that release red pigments and toxins, which can lead to paralytic shellfish poisoning. *See also* harmful algal blooms.

redox potential The tendency of a chemical compound or system to accept electrons; more formally called oxidation-reduction potential.

reducing power Molecules such as NADH and NADPH that temporarily store electrons. The stored electrons are used in anabolic reactions such as CO_2 fixation and the synthesis of monomers (e.g., amino acids).

reductive amination pathway A mechanism used by many microbes to incorporate ammonia.

reductive dehalogenation The cleavage of carbon-halogen bonds by anaerobic bacteria that creates a strong electron-donating environment.

reductive TCA cycle A pathway used by some chemolithotrophs and some anoxygenic phototrophs to fix CO_2. It involves the TCA cycle running in the reverse direction; also called reverse citric acid cycle.

refractive index A measure of how much a substance deflects a light ray from a straight path as it passes from one medium (e.g., glass) to another (e.g., air). It is calculated as the ratio of the velocity of light passing through the first medium to that of light passing through the second medium.

regulatory site *See* allosteric enzyme.

regulatory T cells (Tregs) T cells that maintain immune homeostasis by suppressing the activity of other immune cells.

regulon A collection of genes or operons that is controlled by a common regulatory protein.

remote sensing (RS) The gathering of data and transformation of data into maps.

replicase An RNA-dependent RNA polymerase used to replicate the genome of an RNA virus.

replication The process in which an exact copy of parental DNA (or viral RNA) is made with the parental molecule serving as a template.

replication fork The Y-shaped structure where DNA is replicated.

replicative form (RF) A double-stranded nucleic acid that is formed from a single-stranded viral genome and used to synthesize mRNA and new copies of the genome.

replicon A unit of the genome that contains an origin of replication and in which DNA is replicated.

replisome A large protein complex that copies the DNA double helix to form two daughter chromosomes.

repressible gene A gene that encodes a protein whose level drops in the presence of a small molecule. If the gene product is a biosynthetic enzyme, the small molecule is often an end product of the metabolic pathway in which it functions.

repressor protein A protein that can bind to a repressor-binding site and inhibit transcription.

reservoir A site, alternate host, or carrier that normally harbors pathogenic organisms and serves as a source from which other individuals can be infected.

reservoir host An organism other than a human infected with a pathogen that can also infect humans.

residual body A lysosome after digestion of its contents has occurred. It contains undigested material.

resolution The ability of a microscope to distinguish between small objects that are close together.

respiration An energy-yielding process in which the energy substrate is oxidized using an exogenous or externally derived electron acceptor.

respiratory burst An increase in oxygen consumption by an activated phagocytic cell to generate reactive oxygen species within the phagolysosome.

response regulator *See* two-component signal transduction system.

restriction A process used by bacteria to defend the cell against viral infection; accomplished by enzymes that cleave the viral DNA; also called host restriction.

restriction endonucleases (restriction enzymes) (REs) Enzymes produced by bacterial cells that cleave DNA at specific nucleotide sequences. They evolved to protect bacteria from viral infection and are used in genetic engineering.

reticulate body (RB) The cellular form in the chlamydial life cycle that grows and reproduces within the host cell.

reticulopodia Netlike pseudopodia found in certain amoeboid protists.

retroviruses A group of viruses with RNA genomes that carry the enzyme reverse transcriptase and form a DNA copy of their genome during their life cycle.

reverse electron flow An energy-consuming process used by some chemolithotrophs and phototrophs to generate reducing power.

reverse transcriptase (RT) A multifunctional enzyme used by retroviruses and reverse transcribing DNA viruses during their life cycles. It has RNA-dependent DNA polymerase, DNA-dependent DNA polymerase, and RNAase activity. It synthesizes double-stranded DNA from single-stranded RNA.

reverse vaccinology The use of genomics to identify candidate target molecules on a pathogen for vaccine development.

reversible chain termination sequencing A next-generation DNA sequencing technique that incorporates fluorescently labeled nucleotides that bear a small molecule on the 3′ OH to block addition of the next nucleotide triphosphate. Once the incorporation of the modified base is recorded, the blocking molecule is removed, leaving an unmodified 3′ OH so the next labeled, modified base can be incorporated and recorded; the process continues in an iterative fashion.

reversion mutation A mutation in a mutant organism that changes its phenotype back to the wild type; the second mutation is at the same site as the initial mutation.

rhizobia Any one of a number of α-proteobacteria that form symbiotic nitrogen-fixing nodules on the roots of leguminous plants.

rhizomorph A macroscopic, densely packed thread consisting of individual cells formed by some fungi. A rhizomorph can remain dormant or serve as a means of fungal dissemination.

rhizoplane The surface of a plant root.

rhizosphere A region around the plant root where materials released from the root increase the microbial population and its activities.

ribonucleic acid (RNA) A polynucleotide composed of ribonucleotides joined by phosphodiester bonds.

ribosomal frameshifting A phenomenon in which a ribosome shifts the reading frame of the message, thus generating a different protein than it would have if it had stopped at the stop codon.

ribosomal RNA (rRNA) The RNA present in ribosomes; contributes to ribosome structure and also directly involved in the mechanism of protein synthesis.

ribosome The organelle where protein synthesis occurs; the message encoded in mRNA is translated here.

ribosome-binding site (RBS) Nucleotide sequences on mRNA recognized by a ribosome so that it orients properly on the mRNA and begins translation at the appropriate codon.

riboswitch A site in the leader of an mRNA molecule that interacts with a metabolite or other small molecule, causing the leader to change its folding pattern. In some riboswitches, this change can alter transcription; in others, it affects translation.

ribozyme An RNA molecule with catalytic activity.

ribulose 1,5-bisphosphate carboxylase/oxygenase (RubisCO) The enzyme that catalyzes incorporation of CO_2 in the Calvin-Benson cycle.

richness A term that describes a microbial community in terms of the number of different taxa present.

rifamycins Semisynthetic antibacterial drugs that block transcription by binding to bacterial RNA polymerase.

RNA-dependent DNA polymerase An enzyme that synthesizes DNA using an RNA template; also called reverse transcriptase.

RNA-dependent RNA polymerase An enzyme that synthesizes RNA using an RNA template.

RNA interference A eukaryotic gene regulatory mechanism in which short RNA molecules destabilize or degrade mRNAs.

RNA polymerase An enzyme that catalyzes the synthesis of RNA.

RNA polymerase core enzyme The portion of bacterial RNA polymerase that synthesizes RNA. It consists of all subunits except sigma factor.

RNA polymerase holoenzyme Bacterial RNA polymerase core enzyme plus sigma factor.

RNA thermometers Regions of RNA that adopt different configurations based on temperature.

RNA world The theory that posits that the first self-replicating molecule was RNA and this led to the evolution of the first primitive cell.

RNA-seq Direct sequencing of total cellular mRNA.

rolling-circle replication A mode of DNA replication in which the replication fork moves around a circular DNA molecule, displacing a strand to give a 5′ tail that is also copied to produce new double-stranded DNA.

root nodule Gall-like structures on roots that contain endosymbiotic nitrogen-fixing bacteria.

rough endoplasmic reticulum (RER) *See* endoplasmic reticulum.

S

sac fungi Common name for the Ascomycota.

sanitization Reduction of the microbial population on an inanimate object to levels judged safe by public health standards.

saprophyte (sap′ro-fīt) An organism that takes up nonliving organic nutrients in dissolved form and usually grows on decomposing organic matter.

SAR11 The most abundant group of microbes on Earth, this lineage of α-proteobacteria has been found in almost all marine ecosystems studied.

satellites Subviral infectious agents composed of DNA or RNA encapsidated with the aid of an unrelated helper virus.

scale-up The process of increasing the volume of a culture from a small, experimental system to a large-volume system for the production of an industrially important compound.

scanning electron microscope (SEM) An electron microscope that passes a beam of electrons over the surface of a specimen; an image is formed from the electrons that are emitted by the specimen.

scanning probe microscope (SPM) A microscope used to study surface features by moving a sharp probe over the object's surface (e.g., atomic force and scanning tunneling microscopes).

scanning tunneling microscope (STM) *See* scanning probe microscope.

sclerotia (sklĕ-ro′she-ə) Compact masses of hyphae produced by some filamentous fungi that permit overwintering. In the spring, they germinate to produce either additional hyphae or conidia.

Sec system A system found in all domains of life that transports a protein through or inserts it into a membrane.

second law of thermodynamics Physical and chemical processes proceed in such a way that the entropy of the universe (the system and its surroundings) increases to the maximum possible.

second messengers Small molecules that are made in response to an extracellular signal (the first messenger). They pass the message to effector molecules that bring about the appropriate response to the signal.

secondary active transport *See* active transport.

secondary metabolites Products of metabolism that are synthesized after growth has been completed. Antibiotics are considered secondary metabolites.

secretion The process by which proteins are transported through the plasma membrane of a bacterial cell.

secretory IgA (sIgA) The class of immunoglobulin associated with mucous membranes. *See also* IgA.

secretory pathway The process used by eukaryotic cells to synthesize proteins and lipids, followed by secretion or delivery to organelles or the plasma membrane.

sedimentation basin A containment vessel used during surface-water purification. Following the addition of material to facilitate the formation of coagulated impurities, water is left until such material has precipitated, forming sediment.

segmented genome A viral genome that is divided into several parts or fragments, each usually coding for a single polypeptide.

select agent Biological entities (bacteria, viruses, toxins, etc.) designated by the United States as having the potential to pose great harm to the public; as such, laboratory use is highly regulated.

selectable marker A gene conferring a wild-type or mutant phenotype that can be determined by growth on specific media.

selectins A family of cell adhesion molecules displayed on activated endothelial cells; mediate leukocyte binding to the vascular endothelium.

selective media Culture media that favor the growth of specific microorganisms; this may be accomplished by inhibiting the growth of undesired microorganisms.

selective toxicity The ability of a chemotherapeutic agent to kill or inhibit a microbial pathogen while damaging the host as little as possible.

sense codon A codon that specifies an amino acid.

sense strand The strand of DNA that is not used as the transcription template.

sensor kinase *See* two-component signal transduction system.

sensory rhodopsin Microbial rhodopsin that senses the spectral quality of light. *See also* archaerhodopsin.

septate hyphae Fungal hyphae (filaments) having cross walls.

septation The process of forming a cross wall between two daughter cells during cell division.

septic shock Sepsis associated with severe hypotension despite adequate fluid resuscitation, along with the presence of perfusion abnormalities that may include but are not limited to lactic acidosis, oliguria, or an acute alteration in mental status.

septic tank A tank used to process domestic sewage. Solid material settles out and is partially degraded by anaerobic bacteria as sewage slowly flows through the tank. The outflow is treated or dispersed in oxic soil.

septicemia (sep″tĭ-se′me-ah) A disease associated with the presence in the blood of pathogens or bacterial toxins.

septum (pl., **septa**) A partition that is used to separate cellular material. Septa are formed during cell division to divide the two daughter cells. They also separate cells in filamentous organisms and a developing endospore from its mother cell.

sequencing by synthesis Any DNA sequencing technology that determines the nucleotide sequence as labeled nucleotides are incorporated.

serology The branch of immunology that is concerned with in vitro reactions involving one or more serum constituents (e.g., antibodies and complement).

serotyping A technique or serological procedure used to differentiate between strains (serovars or serotypes) of microorganisms that have differences in the antigenic composition of a structure or product.

sessile Describes a pattern of growth in which microbial cells are attached to a solid substrate.

sex pilus (pĭ′lus) A thin protein appendage required for bacterial conjugation. The cell with sex pili donates DNA to recipient cells.

sheath A hollow, tubelike structure surrounding a cell; present in some genera of bacteria.

Shine-Dalgarno sequence A segment in the leader of bacterial mRNA and some archaeal mRNA that binds to a sequence on the 16S rRNA of the small ribosomal subunit. This helps orient the mRNA on the ribosome.

shuttle vector A DNA vector (e.g., plasmid, cosmid) that has two origins of replication, each recognized by a different microorganism. Thus the vector can replicate in both microbes.

siderophore (sid′er-o-for″) A small molecule that complexes with ferric iron and supplies it to a cell by aiding in its transport across the plasma membrane.

sigma factor A protein that helps bacterial RNA polymerase core enzyme recognize the promoter at the start of a gene. Thus it is a transcription factor.

sign An objective change in a diseased body that can be directly observed (e.g., a fever or rash).

signal 1 Primary lymphocyte activation that occurs when antigen is presented to the lymphocyte by its appropriate MHC molecule on the presenting cell.

signal 2 Confirmatory event causing lymphocytes stimulated by signal 1 to complete activation and thus become effector cells.

signal 3 The release of cytokines by an antigen-presenting cell during immune synapse required to complete T-cell activation.

signal peptidase The membrane-bound enzyme that cleaves the signal peptide from a polypeptide as it is transported through the plasma membrane.

signal peptide The amino terminal sequence on a protein destined for transport through or into a membrane. It delays protein folding and is recognized by the Sec system.

silage Fermented plant material with increased palatability and nutritional value for animals, which can be stored for extended periods.

silent mutation A mutation that does not change an organism's proteins or phenotype even though the DNA base sequence has been changed.

simple staining Technique in which single dye is used to stain bacteria.

single nucleotide polymorphism (SNP) Difference of one base in genes or genetic elements that are otherwise highly conserved among members of a given taxon.

single radial immunodiffusion (SRID) assay An immunodiffusion technique that quantitates antigens by following their diffusion through a gel containing antibodies directed against the test antigens.

single-cell genomic sequencing A method whereby the genome of a single microbial cell is copied many times by the multiple displacement amplification (MDA) technique. The genome is then sequenced by a next-generation technique.

site-directed mutagenesis The in vitro process by which a specific nucleotide change is made to a particular gene.

site-specific recombination Recombination between two DNA molecules that does not involve extensive areas of homology between the two molecules. It is catalyzed by enzymes called recombinases distinct from RecA, and it is the mechanism by which transposition occurs.

skin-associated lymphoid tissue (SALT) The lymphoid tissue in the skin that forms a first-line defense as a part of innate immunity.

S-layer A regularly structured layer composed of protein or glycoprotein that lies on the surface of many bacteria and archaea.

slime layer A layer of diffuse, unorganized, easily removed material lying outside an archaeal or bacterial cell wall.

sludge The solid material remaining after the liquid phase of sewage has been treated and removed.

small RNAs (sRNAs) Small regulatory RNA molecules that do not function as messenger, ribosomal, or transfer RNAs.

small subunit rRNA (SSU rRNA) The rRNA associated with the small ribosomal subunit: 16S rRNA in bacterial and archaeal cells and 18S rRNA in eukaryotes.

smooth endoplasmic reticulum (SER) *See* endoplasmic reticulum.

snapping division A distinctive type of binary fission resulting in an angular or a palisade arrangement of cells; characteristic of *Corynebacterium*.

soil microbiome All microorganisms that inhabit the soil, including bacteria, archaea, fungi, protists, and viruses.

soil organic matter (SOM) The organic material within a soil; organic material is important because it helps to retain nutrients, maintain soil structure, and hold water for plant use.

somatic hypermutation A high rate of mutation within antibody-encoding genes that occurs in B cells following T-cell activation. This mechanism generates antibodies that are then screened for high avidity and affinity.

sortase An enzyme that catalyzes the attachment of a secreted protein to the surface peptidoglycan of a monoderm bacterium.

sorting signal A sequence at the C-terminus of some monoderm secreted proteins that directs the protein to sortase enzymes for anchoring to the peptidoglycan.

SOS response A complex, inducible process that allows bacterial cells with extensive DNA damage to survive, although often in a mutated form; it involves cessation of cell division, upregulation of several DNA repair systems, and induction of translesion DNA synthesis.

specialized transduction A transduction process in which only a specific set of bacterial or archaeal genes is carried to a recipient cell by a temperate virus; the cell's genes are acquired because of a mistake in the excision of a provirus during the lysogenic life cycle.

species Species of higher organisms are groups of interbreeding or potentially interbreeding natural populations that are reproductively isolated. Archaeal and bacterial species are often defined as collections of strains that have many stable properties in common and differ significantly from other strains. However, there is currently considerable debate about the best definition of archaeal and bacterial species.

spheroplast (sfēr′o-plast) A spherical cell formed by the weakening or partial removal of the rigid cell wall component.

spike *See* peplomer.

spirillum (spi-ril′um) A rigid, spiral-shaped bacterium.

spirochete (spi′ro-kēt) A flexible, spiral-shaped bacterium with periplasmic flagella.

spleen A secondary lymphoid organ where old erythrocytes are destroyed and blood-borne antigens are trapped and presented to lymphocytes.

splice site variability Variations in DNA sequence within the boundaries between exons and introns. During transcription of eukaryotic DNA, it can lead to different mRNA products; in the context of immunology, it can result in nucleotide sequence variation following VDJ recombination.

spliceosome A complex of proteins and small RNAs that carries out RNA splicing in eukaryotes.

spontaneous generation An early belief that living organisms could develop from nonliving matter.

spontaneous mutations Mutations that occur due to errors in replication or the occurrence of lesions in the DNA (e.g., apurinic sites).

sporangiospore (spo-ran′je-o-spōr) A spore borne within a sporangium.

sporangium (spo-ran′je-um; pl., **sporangia**) A saclike structure or cell, the contents of which are converted into one or more spores.

spore A differentiated, specialized cell that can be used for dissemination, for survival in adverse conditions because of its heat and desiccation resistance, and/or for reproduction. Spores are usually unicellular and may develop into vegetative organisms or gametes. They may be produced asexually or sexually.

sporozoite The motile, infective stage of apicomplexan protists.

sporulation The process of spore formation.

spread plate A Petri dish of solid culture medium with isolated microbial colonies growing on its surface that has been prepared by spreading a dilute microbial suspension evenly over the agar surface.

stable isotope analysis The use of stable (nonradioactive) isotopes to determine the fate of a given nutrient or compound in the environment.

stable isotope probing A technique that uses stable isotopes to examine nutrient cycling and identify the microbes involved.

stalk A bacterial appendage produced by the cell and extending from it.

standard free energy change The free energy change of a reaction at 1 atmosphere pressure when all reactants and products are present in their standard states; usually the temperature is 25°C.

standard plate counts *See* viable counting methods.

standard reduction potential A measure of the tendency of an electron donor to lose electrons in an oxidation-reduction (redox) reaction. The more negative the reduction potential of a compound, the better electron donor it is.

start codon The first codon in an mRNA that is translated; it signifies the translation start site.

starter culture An inoculum, consisting of a mixture of microorganisms, used to start a commercial fermentation.

static Inhibiting or retarding growth.

stationary phase The phase of microbial growth in a batch culture when population growth ceases and the growth curve levels off.

sterilization The process by which all living cells, spores, viruses, and viroids are either destroyed or removed from an object or habitat.

Stickland reaction A type of fermentation in which an amino acid serves as the energy source and another amino acid serves as the electron acceptor.

sticky ends The complementary single-stranded ends of double-stranded DNA that result from cleavage with certain restriction endonuclease enzymes.

stigma A photosensitive region on the surface of certain protists that is used in phototaxis.

stop codon *See* nonsense codon.

strain A population of organisms that descends from a single organism or pure culture isolate.

streak plate A Petri dish of solid culture medium with isolated microbial colonies growing on its surface that has been prepared by spreading a microbial mixture over the agar surface, using an inoculating loop.

streptomycete Any high G + C monoderm bacterium of the genera *Kitasatospora* and *Streptomyces.*

strict anaerobes *See* obligate anaerobes.

stringent response A decrease in tRNA and rRNA, and an increase in specific amino acid biosynthetic enzymes in response to amino acid starvation.

structural gene A gene that codes for the synthesis of a polypeptide or polynucleotide (i.e., rRNA, tRNA) with a nonregulatory function.

structural proteomics Analysis of the three-dimensional structure of many proteins to predict the structure of other proteins and protein complexes.

substrate (1) A reacting molecule in a chemical reaction that is bound by an enzyme. (2) The first molecule of a biochemical pathway. (3) A surface upon which microbes grow.

substrate analog probing A method to study microbial communities by incorporating synthetic analogs of common biochemicals.

substrate mycelium In actinobacteria and fungi, a mat of hyphae that penetrates the surface into which the microbes are growing.

substrate-level phosphorylation The synthesis of ATP from ADP by phosphorylation coupled with the exergonic breakdown of a high-energy organic substrate molecule.

substratum The basal layer of a biofilm that is attached to an inanimate platform.

subunit vaccines Immunogenic materials derived from whole infectious agents; e.g., bacterial capsular polysaccharide.

sulfate- or sulfur-reducing bacteria (SRB) Anaerobic bacteria that use sulfate or sulfur or both as a terminal electron acceptor during respiration, thereby performing dissimilatory sulfate reduction.

sulfate reduction The use of sulfate as an electron acceptor, which results in the accumulation of reduced forms of sulfur such as sulfide or incorporation of sulfur into organic molecules, usually as sulfhydryl groups.

superantigens Toxic bacterial proteins that overstimulate the immune system.

superoxide dismutase (SOD) (dis-mu′tās) An enzyme found in all aerobes, facultative anaerobes, and microaerophiles that catalyzes the conversion of the radical oxygen species superoxide to hydrogen peroxide and molecular oxygen (O_2).

superphylum A classification of organisms below the level of domain but above phylum; includes several phyla that are united by one or more distinct characteristics.

super-resolution microscopy Microscopy that achieves much higher resolution than that of conventional optical microscopes.

supportive media Culture media that are able to sustain the growth of many different kinds of microorganisms.

suppressor mutation A mutation at a different site that overcomes the effect of a forward mutation and produces the normal phenotype. The different site can be within the same gene (intragenic suppressor) or different gene (extragenic suppressor) as the forward mutation.

Svedberg unit (sfed′berg) The unit used in expressing the sedimentation coefficient; the greater a particle's Svedberg value, the faster it travels in a centrifuge.

symbiont Any organism that has a specific relationship with another that can be characterized as mutualism, cooperation, commensalism, predation, parasitism, amensalism, or commensalism.

symbiosis The living together or close association of two dissimilar organisms, each of these organisms being known as a symbiont.

symbiosome The organelle-like compartment that houses bacteroids that fix nitrogen during rhizobium nodulation.

symport Linked transport of two substances across a plasma membrane in the same direction.

symptom A change during a disease that a person subjectively experiences (e.g., pain, fatigue, or loss of appetite). Sometimes the term *symptom* is used more broadly to include any observed signs.

synbiotic A food or supplement that includes a compound(s) added to enhance the colonization and positive health benefits of probiotic microbes also found in the product.

syncytial growth A type of growth common in actinomycetes in which cells elongate without septation.

syndrome *See* disease syndrome.

syngamy The fusion of haploid gametes.

synteny Partial or complete conservation of gene order observed when the genomes of two closely related organisms are compared.

synthetic biology The construction of biological systems (e.g., bacterial strains) with unique, new capabilities such as the production of a specific molecule of industrial importance.

synthetic medium *See* defined medium.

syntrophism (sin′trōf-izəm) The association in which the growth of one organism either depends on or is improved by the provision of one or more growth factors or nutrients by a neighboring organism. Sometimes both organisms benefit.

systematic epidemiology The field of epidemiology that focuses on the ecological and social factors that influence the development of emerging and reemerging infectious diseases.

systemic anaphylaxis Type 1 (immediate) hypersensitivity that involves multiple tissue sites and results in shock. It is often fatal if not treated.

systems biology The application of mathematical and computational approaches to understand biological systems holistically by exploring regulatory and physiological networks.

systems metabolic engineering The use of *-omic* technologies (genomics, metabolomics, etc.) to optimize cellular production of a desired product.

T

T cell (T lymphocyte) A type of lymphocyte that matures into an immunologically competent cell within the thymus.

T lymphocyte *See* T cell.

tandem mass spectrometry The analysis of molecules using two mass spectrometers sequentially.

***Taq* polymerase** A thermostable DNA polymerase used in the polymerase chain reaction; derived from *Thermus aquaticus*.

Tat system A system used to transport folded proteins across the plasma membrane in bacteria and some archaea; some eukaryotes have homologues of the Tat system in their chloroplasts and mitochondria.

taxon A group into which related organisms are classified.

taxonomy The science of biological classification; it consists of three parts: classification, nomenclature, and identification.

T-cell receptor (TCR) The receptor on the T-cell surface consisting of two antigen-binding peptide chains and associated with several other glycoproteins. Binding of antigen to the TCR in association with MHC activates the T cell.

T-cell-independent B-cell activation The stimulation of B cells by antigen without T-cell help, thus class switching and somatic hypermutation are absent and plasma cells produce only IgM antibody.

T-dependent antigen An antigen that activates B-cell response only with the aid of T-helper cells.

T-DNA A portion of the *Agrobacterium tumefaciens* Ti plasmid integrated into the host plant's chromosome.

teichoic acids (ti-ko′ik) Polymers of glycerol or ribitol joined by phosphates; found in the cell walls of most monoderm bacteria.

teliospores Diploid spores formed by fungi of *Ustilaginomycota*.

telomerase An enzyme in eukaryotes that replicates the ends of chromosomes.

telomeres Complexes of DNA and proteins at the ends of eukaryotic chromosomes.

temperate bacteriophages (viruses) Bacterial and archaeal viruses that can establish a lysogenic relationship, rather than immediately lysing their hosts.

template strand A DNA or RNA strand that specifies the base sequence of a new complementary strand of DNA or RNA.

terminator A sequence that marks the end of a gene and stops transcription.

terminus The site on the chromosome where DNA replication halts.

testate amoeba An amoeba with a covering (test) made either by the protist or from materials collected from the environment.

tetanolysin (tet″ah-nol′ĭ-sin) A hemolysin that aids in tissue destruction; produced by *Clostridium tetani*.

tetanospasmin (tet″ah-no-spaz′min) The neurotoxic component of the tetanus toxin, which causes the muscle spasms of tetanus.

tetracyclines (tet″rah-si′klēns) A family of antibiotics with a common four-ring structure that is isolated from the genus *Streptomyces* or produced semisynthetically.

T_H0 cell Precursor to T_H1, T_H17, and T_H2 cells.

T_H1 cell A CD4$^+$ T-helper cell that secretes interferon-gamma, interleukin-2, and tumor necrosis factor.

T_H2 cell A $CD4^+$ T-helper cell that secretes interleukin (IL)-4, IL-5, IL-6, IL-9, IL-10, and IL-13, influencing growth and differentiation of B cells.

T_H17 cells T_H cells found predominantly in the skin and intestinal epithelia, where their release of IL-17 and IL-22 stimulates neutrophils in response to bacterial invasion.

thallus (thal′us) A type of body that is devoid of root, stem, or leaf; characteristic of fungi.

thaumarchaeol Cyclopentane ring-containing lipids unique to certain mesophilic archaea.

T-helper (T_H) cell A T lymphocyte that bears the CD4 coreceptor. Subtypes amplify innate immune responses and promote cell-mediated and antibody-mediated immune responses.

therapeutic index The ratio between the toxic dose and the therapeutic dose of a drug, used as a measure of the drug's relative safety.

thermoacidophiles Microbes that grow best at acidic pHs and temperatures 55°C or higher.

thermocline A water column characterized by a steep temperature gradient.

thermodynamics A science that analyzes energy changes in a system.

thermophile (ther′mo-fīl) A microorganism that can grow at temperatures of 55°C or higher; the minimum is usually around 45°C.

thylakoid (thi′lah-koid) A flattened sac in the chloroplast stroma that contains photosynthetic pigments and the photosynthetic electron transport chain. Similar structures are observed in the cytoplasm of cyanobacteria.

thymic selection A process occurring in the thymus in which T-cell precursors are evaluated for functionality.

thymine (thi′min) The pyrimidine 5-methyluracil found in nucleosides, nucleotides, and DNA.

thymus (thi′məs) A primary lymphoid organ in the chest that is necessary in early life for the development of immunological functions, including T-cell maturation.

Ti (tumor-inducing) plasmid Plasmid found in plant pathogenic species of *Agrobacterium.*

T-independent antigen An antigen that triggers a B cell to produce immunoglobulin without T-cell cooperation.

toll-like receptor (TLR) A type of pattern recognition molecule (PRM) on phagocytes and other cells that when bound to a microbial ligand, triggers the production of transcription factor NF-κB, which stimulates formation of cytokines, chemokines, and other defense molecules.

topoisomerases Enzymes that change the topology of DNA molecules by transiently breaking one or both strands. They play an important role in DNA replication and transcription.

total organic carbon (TOC) The amount of particulate and dissolved carbon in an environmental sample.

toxemia (tok-se′me-ah) The condition caused by toxins in the blood of the host.

toxigenicity (tok″sĭ-jĕ-nis′i-tē) The capacity of an organism to produce a toxin.

toxin Substance that disrupts normal cell function.

toxin neutralization *See* neutralization.

toxoid A bacterial exotoxin that has been inactivated but still stimulates antitoxin formation when injected into a person or animal.

trace elements *See* micronutrients.

trailer That portion of a gene downstream of the coding region. It is transcribed but not translated.

transaminase An enzyme that catalyzes a transamination reaction.

transamination The transfer of an amino group from an amino acid to an α-keto acid acceptor. Transamination is used both for degradation of amino acids and in their synthesis.

transcriptase (trans-krip′tās) An enzyme that catalyzes transcription.

transcription The process in which RNA with a base sequence complementary to the template strand of DNA or RNA is synthesized.

transcriptome All the messenger RNA transcribed from the genome of an organism under a given set of circumstances.

transcriptomics The study of all the mRNA expressed by a cell at the time of sampling.

transduction The transfer of genes between bacterial or archaeal cells by viruses.

transfer RNA (tRNA) A small RNA that binds an amino acid and delivers it to the ribosome for incorporation into a polypeptide chain during protein synthesis.

transfer RNA fragments (tRFs) Small pieces of tRNA, often of microbial origin, that regulate gene expression in plants by binding to complementary mRNA, thereby preventing translation.

transferrin A protein found in blood serum that binds iron, thereby limiting its availability to microbial pathogens.

transformation (1) A mode of gene transfer in bacteria and archaea in which a piece of free DNA is taken up by a cell and integrated into its genome. (2) In recombinant DNA technology, the uptake of free DNA by a cell. (3) The conversion of a cell into a malignant, cancerous cell.

transition mutations Mutations that involve the substitution of a different purine base for the purine present at the site of the mutation or the substitution of a different pyrimidine for the normal pyrimidine.

translation Protein synthesis; the process by which the genetic message carried by mRNA directs the synthesis of polypeptides with the aid of ribosomes and other cell constituents.

translesion DNA synthesis A type of DNA synthesis that occurs during the SOS response. DNA synthesis occurs without an intact template and generates mutations as a result.

translocation (1) During the elongation cycle of protein synthesis, the movement of the ribosome relative to the mRNA that repositions the A site of the ribosome over the next codon to be translated. (2) The movement of a protein across or into a membrane.

transmissible spongiform encephalopathies (TSEs) Degenerative central nervous system diseases in which the brain has a spongy appearance; due to prions.

transmission electron microscope (TEM) A microscope in which an image is formed by passing an electron beam through a specimen and focusing the scattered electrons with magnetic lenses.

transpeptidation (1) The reaction that forms the peptide cross-links during peptidoglycan synthesis. (2) The reaction that forms a peptide bond during the elongation cycle of protein synthesis.

transposable elements (TE) Another term for transposons. *See* mobile genetic element.

transposase An enzyme that catalyzes the movement of a mobile genetic element.

transposition The movement of a piece of DNA within or between chromosomes.

transposon (tranz-po′zon) *See* mobile genetic element.

tricarboxylic acid (TCA) cycle The cycle that oxidizes acetyl coenzyme A to CO_2 and generates NADH and $FADH_2$ for oxidation in an electron transport chain; also supplies precursor metabolites for biosynthesis. Also called citric acid cycle and Krebs cycle.

trichocyst (trĭk′o-sist) A small organelle beneath the pellicle of some protists. It can be extruded to function as either an anchoring device or a form of defense.

trichome (tri′kōm) A filament of microbial cells that are in close contact with one another over a large area.

trickling filter A bed of rocks covered with a microbial biofilm that aerobically degrades organic waste during secondary sewage treatment.

trigger factor (TF) A protein associated with the ribosome exit tunnel that facilitates proper folding of a nascent polypeptide.

trimethoprim A synthetic antibiotic that inhibits production of folic acid by binding to dihydrofolate reductase. Trimethoprim has a wide spectrum of activity and is bacteriostatic.

tRNA fragment (tRF) Small, noncoding RNA resulting from enzymatic cleavage of tRNA that contribute to gene regulation.

trophozoite (trof″o-zo′īt) The active, motile feeding stage of a protozoan organism.

tropism (1) The movement of living organisms toward or away from a focus of heat, light, or other stimulus. (2) The selective infection of certain organisms or host tissues by a virus due to the presence of the specific receptor to which the virus binds.

trypanosome (tri-pan′o-sōm) A flagellated protozoan of the genus *Trypanosoma* or *Leishmania;* often lives in the blood of humans and other vertebrates and is transmitted by insect bites.

tubercle (too′ber-kuhl) A small, rounded nodular lesion produced by *Mycobacterium tuberculosis*.

tuberculous cavity An air-filled cavity that results from a tubercle lesion caused by *M. tuberculosis*.

tumble Random turning or tumbling movements made by flagellated bacteria when they stop swimming in a straight line (a run).

tumor necrosis factor (TNF) A pro-inflammatory cytokine, so named because of initial observations.

turbidostat A continuous culture system equipped with a photocell that adjusts the flow of medium through the culture vessel to maintain a constant cell density or turbidity.

twitching motility A type of bacterial motility characterized by short, intermittent jerky motions due to extension and retraction of type IV pili.

two-component signal transduction system A regulatory system that uses the transfer of phosphoryl groups to control gene transcription and protein activity. It has two major components: a sensor kinase and a response regulator protein.

two-dimensional gel electrophoresis An electrophoretic procedure that separates proteins using two procedures: isoelectric focusing and SDS-PAGE.

tyndallization The process of repeated heating and incubation to destroy bacterial endospores.

type I hypersensitivity A form of immediate hypersensitivity arising from the binding of antigen to IgE attached to mast cells, which then release anaphylaxis mediators such as histamine. Examples: hay fever, asthma, and food allergies.

type II hypersensitivity A form of immediate hypersensitivity involving the binding of antibodies to antigens on cell surfaces followed by destruction of the target cells (e.g., through complement attack, phagocytosis, or agglutination).

type III hypersensitivity A form of immediate hypersensitivity resulting from the exposure to excessive amounts of antigens to which antibodies bind. These antibody-antigen complexes activate complement and trigger an acute inflammatory response with subsequent tissue damage.

type IV hypersensitivity A delayed hypersensitivity response that results from the binding of antigen to activated T lymphocytes, which then release cytokines to trigger inflammation that damages tissue. Type IV hypersensitivity is seen in contact dermatitis from poison ivy, leprosy, and tertiary syphilis.

type I secretion system (T1SS) A structure to transport proteins from the periplasm through the outer membrane.

type II secretion system (T2SS) A two-component system in proteobacteria for the export of proteins. A pseudopilus provides the thrust for the movement of proteins through the outer membrane.

type III secretion system (T3SS) A system in diderm bacteria that secretes virulence factors and injects them into host cells.

type IV secretion system (T4SS) A system in diderm bacteria that secretes proteins; some monoderm and diderm bacteria use this secretion system to transfer DNA during bacterial conjugation.

type V secretion system (T5SS) A pathway in diderm bacteria that translocates proteins across the outer membrane after they have been translocated across the plasma membrane by the Sec system.

type VI secretion system (T6SS) A protein secretion system in diderm bacteria that is similar in structure to the machinery used by some bacteriophages to insert their DNA into host cells.

type VII secretion system (T7SS) A protein secretion system in monoderm bacteria that translocates proteins across the plasma membrane.

type IX secretion system (T9SS) A two-step protein secretion system in the Bacteroidota that guides proteins to the outer surface of the cell.

type X secretion systems (T10SS) A two-step secretion machine composed of proteins similar to those used by bacteriophages to escape from their bacterial hosts during a lytic infection.

U

ubiquinone *See* electron transport chain.

ultraviolet (UV) radiation High-energy radiation of short wavelength, about 10 to 400 nm.

uniporters Carrier proteins that move a single solute across a membrane.

universal phylogenetic tree A phylogenetic tree that considers the evolutionary relationships among organisms from all three domains of life: Bacteria, Archaea, and Eukarya.

uracil (u′rah-sil) The pyrimidine 2,4-dioxypyrimidine, which is found in nucleosides, nucleotides, and RNA.

urinary tract infection (UTI) Infection of the bladder most commonly caused by uropathogenic *E. coli*.

V

vaccine A preparation of antigens and adjuvant administered to induce development of an adaptive immune response that will protect the individual against a specific pathogen or a toxin.

vaccinomics The application of genomics to vaccine development.

valence The number of antigenic determinant sites on the surface of an antigen or the number of antigen-binding sites possessed by an antibody molecule.

vancomycin A glycopeptide antibiotic that is effective only against monoderm bacteria and inhibits peptidoglycan synthesis.

variable (V) region (V_L and V_H) The region at the N-terminal end of immunoglobulin heavy and light chains whose amino acid sequence varies between antibodies of different specificity. Variable regions form the antigen-binding site.

vasoactive mediator Compound released by cells that causes blood vessels to contract or dilate, influencing blood flow and permeability of blood vessels.

vector (1) In genetic engineering, a DNA molecule that can replicate and transport a piece of inserted foreign DNA, such as a gene, into a recipient cell. It may be a plasmid, phage, cosmid, or artificial chromosome. (2) In epidemiology, it is a living organism, usually an arthropod or other animal, that transfers an infective agent between hosts.

vector-borne transmission The transmission of an infectious pathogen between hosts by means of a vector.

vehicle An inanimate substance or medium that transmits a pathogen.

vertical gene transfer The transfer of genes from a parent organism to its offspring.

vertical transmission Pathogen transfer from mother to unborn child (in utero).

vesicular transport The movement of materials in small membrane-bound vesicles between the endoplasmic reticulum, Golgi apparatus, lysosomes, plasma membrane, and other organelles of the endocytic and secretory pathways in eukaryotic cells.

vesiduction A mechanism of horizontal gene transfer mediated by the inclusion of nucleic acids in extracellular vesicles.

viable but nonculturable (VBNC) Microbes known to be growing in a specific environment but that cannot be cultured under standard laboratory conditions.

viable counting methods Methods for determining population size that depend on a microbe's ability to reproduce when cultured. Also called standard plate counts.

vibrio (vib′re-o) A rod-shaped bacterial cell that is curved to form a commalike shape.

viral hemagglutination The clumping of red blood cells caused by the attachment of some viruses.

viral neutralization An antibody-mediated process in which IgG, IgM, and IgA antibodies bind to some viruses during their extracellular phase and inactivate or neutralize them.

viricide (vir′i-sīd) An agent that inactivates viruses so that they cannot reproduce within host cells.

virion A complete virus particle; at the simplest, it consists of a protein capsid surrounding a single nucleic acid molecule.

virioplankton Viruses that occur in waters; high levels are found in marine and freshwater environments.

viroid An infectious agent that is a single-stranded RNA not associated with any protein; the RNA does not code for any proteins and is not translated.

virology The branch of microbiology that is concerned with viruses and viral diseases.

virulence The degree or intensity of pathogenicity of an organism as indicated by case fatality rates and/or ability to invade host tissues and cause disease.

virulence factor A product, usually a protein or carbohydrate, that contributes to virulence or pathogenicity.

virulent viruses (bacteriophages) Viruses that lyse their host cells at the end of the viral life cycle.

virus An infectious agent having a simple acellular organization with a protein coat and a nucleic acid genome, lacking independent metabolism, and multiplying only within living host cells.

viruslike particles (VLP) Presumptive viruses detected by electron microscopy or fluorescence microscopy; because these approaches do not demonstrate that these entities can infect host cells, they cannot be considered viruses.

W

water activity (a_w) A quantitative measure of water availability in the habitat; the water activity of a solution is one-hundredth its relative humidity.

water molds *See* öomycetes.

Western blotting *See* immunoblotting.

white blood cell (WBC) Blood cell having innate or acquired immune function. These cells are named for the white or buffy layer in which they are found when blood is centrifuged.

whole-genome shotgun sequencing Genome sequencing in which random fragments of a complete genome are individually sequenced. The nucelotide sequences of the fragments are placed in the proper order based on overlapping identical sequences.

whole-genome shuffling A procedure to generate novel genomes by recombination between the genomes of homologous organisms.

whole-pathogen vaccine A vaccine made from complete pathogens, which can be either killed microorganisms or live, attenuated microbes.

wild type Prevalent form of a gene or phenotype.

Winogradsky column A glass column with an anoxic lower zone and an oxic upper zone, which allows growth of microorganisms under conditions similar to those found in a nutrient-rich lake.

wobble The loose base pairing between an anticodon and a codon at the third position of the codon.

Wolfe cycle Biochemical cycle for the reduction of carbon dioxide to methane. *See* methanogenesis.

World Health Organization (WHO) An international health agency located in Geneva, Switzerland.

X

xenograft (zen″o-graft) A tissue graft between animals of different species.

xerophilic Describes microorganisms that thrive in environments with lower water activity.

xerotolerant Microorganisms that grow best under low water activity (a_w) conditions, and may not be able to grow at high a_w values.

Y

yeast A unicellular, uninuclear fungus that reproduces either asexually by budding or fission or sexually through spore formation.

yeast artificial chromosome (YAC) Engineered DNA that contains all the elements required to propagate a chromosome in yeast and is used to clone foreign DNA fragments in yeast cells.

YM shift The change from yeast (Y) morphology in the host to mold (M) morphology in the environment.

Z

zoonosis (zo″o-no′sis; pl., **zoonoses**) A disease that can be transmitted from animals to humans.

zooxanthella (zo″o-zan-thel′ah) A dinoflagellate found living symbiotically within cnidarians (corals) and other invertebrates.

zygospore A thick-walled, sexual, resting spore characteristic of the zygomycetous fungi.

Index

Page numbers followed by t and f designate tables and figures, respectively; page numbers in *italics* refer to definitions; page numbers in **boldface** refer to major discussions.

A

AAnPs (aerobic anoxygenic phototrophs), *438,* 577
Abbé equation, 26
Abbé, Ernst, 26
ABC (ATP-binding cassette) transporters, *51,* 51f
ABO blood groups, *658*
AB toxin, *694*
Abundance, 3
Acanthamoeba keratitis (cornea inflammation), 826
Acarbose (type 2 diabetes drug), 459
Acceptor (A) tRNA binding site, *286*
Accessory pigments in photosynthesis, 235, 237f
Acellular slime molds, *476,* 476f
Acellular vaccines, 713
Acetyl-coenzyme A, *217*
Acid-fast cell walls, *458*
Acid-fast staining, 34, 35f
Acidithiobacillus ferrooxidans, 446, 564
Acidithiobacillus species, 232
Acid mine drainage, 445, 446
Acidobacteria, 239
Acidophile microorganisms, *133*
Acne vulgaris, 460
Acquired immunity. *See* Adaptive immunity
Acquired resistance, antimicrobial drugs, 185
ACT (artemisinin-based combination therapy), 819
Actin filaments, 88, 89f
Actinobacteria, **455–462**
  Actinomycetales, 456
  Bifidobacteriales, 457
  cell wall types, 455t
  characteristics of, 455t, 457
  Corynebacteriales, 457–458
  exospores, *456,* 456f
  Frankiales, 460
  Micrococcales, 456
  Micromonosporales, 459–460
  as monoderms, *53*
  overview, 459–460
  Propionibacteriales, 460
  Streptomycetales, 460–461
Actinomyces, 456, 457f
Actinomycetales, 456
Actinomycetoma, 461
Actinoplanes, 459–460
Actinoplanetes
  Micromonosporales, 459
Actinorhizae, *599,* 599f
Actin tails, 691, 691f
Activation energy for reactions, *200*
Activator proteins, 301, 338–339
Active sites for reactions, *201*
Active transport, *50*
  in archaea, 79
  in eukaryotic cells, 88
  primary and secondary, **50–51**
Acute inflammatory response, *629,* 630f
Acute-phase proteins, *614*
Acute viral gastroenteritis, 760–761, 761t
Acyclovir antiviral drug, 181, 738, 754
Adaptive immunity, **633–663,** 634f. *See also* Host resistance, innate
  activation, *634*
  antigen-antibody interactions, **650–652,** 652f
  antigens to elicit, **634–635,** *635*
  autoimmunity and autoimmune diseases, *661,* 662t
  B cells in, **644–646**
  cancer treatment, 633
  3-D antigens, antibodies, **646–650**
  foreignness, recognition of, **636–639**
  hypersensitivities, 655–661
  immune system, **655–663**
  immunoglobulin (*See* Antibodies)
  lymphocytes, **652–655**
  recognition and memory in, 634
  T cells in, **639–643**
  transplantation (tissue) rejection, 661–663
ADCC (antibody-dependent cell-mediated cytotoxicity), *620,* 620f, 647
ADE (antibody-dependent enhancement), *747*
Adenine, *257*
Adenoids, 623
Adenosine diphosphate (ADP), *193*
Adenosine monophosphate, 257
Adenosine 5′-phosphosulfate (APS), 232, 429, 430f, 446f
Adhesion factors, 689–692, 691t
Adhesion molecules, 440
Adhesion of pathogens, 689–690
Adjuvant, *711*
ADP (adenosine diphosphate), *193*
Aequorea, 31
Aerial mycelium, *456*
Aerobes, 136, 145
Aerobic anoxygenic phototrophy (AAnPs), 577
Aerobic respiration, **212–218**
  ATP yield, 221–224
  electron transport chains, 219–220, 219f
  electron transport to oxidative phosphorylation, 220–221
  glucose to pyruvate, 213–215
  overview, *213*
  oxidative phosphorylation process, 220–221
  pyruvate to carbon dioxide, 215
Aerosolized, *687*
Aerotolerant anaerobes, *136*
A-factor in streptomycin production, 454
Aflatoxins, 500, 696, *841,* 841f
African sleeping sickness, 474, 813t
Agar
  for cultures, *144*
  dilution method for antimicrobial susceptibility, 723
  Mueller-Hinton, 173
Agaricus campestris, 501
Agglutinates, *729*
Agglutination reaction, *651, 729,* 730
Agriculture
  biotechnology in, **858–859,** 858f
  microbiology in, 14
  runoff from, 582
Agrobacterium, 441, 441f, 599–600, 600f
AHL (*N*-acyl homoserine lactone), *142*–143, *319,* 554

Airborne transmission of disease
of bacteria, **770–778**
blastomycosis, *813*–814
chickenpox (varicella), 738, 738f
chlamydial pneumonia, *770*
coccidioidomycosis, *814*
description of, *687*
diphtheria, *770,* 771f
of fungi, **813–815**
histoplasmosis, *814*–815, 815f
Legionnaires' disease, *770*–772, 772f
measles (rubeola), *743*
meningitis, *772*–773
mumps, *743*
mycobacterial infections, 773–776
mycoplasmal pneumonia, 776
pertussis (whooping cough), *776*–*777*
respiratory syndromes, 743, 745
rubella (German measles), *745,* 745f
shingles (herpes zoster), 738, 739f
smallpox (variola), *746*–747, 746f
streptococcal diseases, 777–778
viral pneumonia, 743, 745
of viruses, **738–747**
Air conditioning systems, bacterial biofilms in, 436
Akinetes, *421*
Akkermansia muciniphila, 424
Alarmones, *139, 314*
Alcoholic fermentation, *228*
Alcohols for microbe control, 163, 163f
Aldehydes for microbe control, 164, 165f
Ales and beers, 847
Alexandrium, 572
Algae, 208
green, 486–488
overview, *3*
peptidoglycan in chloroplasts of, 9
as photosynthetic protists, *470*
Algicides, 163
Alignment in genome sequencing, *389*
Alkaliphile microorganisms, *133*
Alleles, *352*
Allergen, *656*
Allergies. *See also* Hypersensitivities
allergic-type aspergillosis, 831
eosinophils and, *616*
overview, *655*
Allochthonous nutrients, 581–*582*
Allografts, *661*
Allosteric regulation of enzymes, *204,* 204f
Alpha-amanitin toxin, 501
αβ T cells, *640*
Alpha diversity, *536*
α-hemolysis, *466,* 467
Alphaproteobacteria, **437–443**
anoxygenic photosynthesis, *418*
Caulobacteraceae, 439–440
characteristics of, 437t
nitrifying, 441–442, 441f
origin of mitochondria, 443
Pelagibacterales, 442, 442f
purple nonsulfur, 437–438
Rhizobiales for N_2 fixation, 440–441, 441f
Rhodospirillales, 437–439
Rickettsia, 442–443, 443f
SAR11 clade abundance, *576*
Alternate sigma (σ) factors, *311*
Alternative complement pathway, *611*
Alternative splicing, *332*
Altman, Sidney, 203
Alveolar macrophages, *609*
Alveolata protists, 479–483
Alzheimer's disease, 765, 780
Amanita phalloides, 501
Amebiasis, 813t, *826*
Amebic dysentery, 185, 813t, *826*
Amebic meningoencephalitis and keratitis, 826–827
American Society for Microbiology, 716
American trypanosomiasis, 474, *821*
Amino acid biosynthetic pathways, 254–256
ammonia incorporation, 252
anaplerotic reactions, 255–256, 256f
assimilatory nitrate reduction, 252–253, 253f
biosynthetic pathways for, 254–256
inorganic nitrogen assimilation, 252–253
nitrogen fixation, 253
precursor metabolites, 251–256
sulfur assimilation, 253–254
Amino acids
activation in translation, *285*–286
catabolism of, 230
in industrial microbiology, 853
proteins as polymers of, 269–271, 271f
Aminoacyl (A) tRNA binding site, *286*
Aminoacyl-tRNA synthases, *285*
Aminoglycoside antibiotics, *178*
Ammonia
in amino acid synthesis, 254–256
anammox reaction (*See* Anammox reaction)
Ammonia monooxygenase (AMO), 442
Amoebozoa clade, 475–477
Amoebozoa supergroup, 475–477
Amphibolic pathways, *215*
Amphipathic membrane-associated lipids, *48*
Amphitrichous bacteria, *66*
Amplified sequence variants (ASVs), *534*
Amycolatopsis orientalis, 177
Anabolism, 244–262
amino acid synthesis, **251–256** (*See also* Amino acids)
biosynthesis principles, **245**
carbohydrate synthesis, **248–251**
CO_2 fixation, **247–248**
lipid synthesis, **258–262**
nucleotide synthesis, **257–258**
pentose phosphate pathway for, 213, 214f
phosphorus assimilation, **257**
precursor metabolites, ***246***
purine biosynthesis, **257–258**
pyrimidine synthesis, **258**
Anaerobes, 136, 145, 466
Anaerobic bacteria, 259
Anaerobic digestion, *866,* 867f
Anaerobic methanotrophic (ANME) archaea, 411, 411f
Anaerobic respiration, 212–218
Anaerobic sulfate/sulfur reducers, **429–431**
Anaerobic workstation and incubator, 145f
Anammox bacteria, *232,* 424, 424f
Anammoxosome, *424*
Anammox reaction, *424,* 424f, *561,* 576
Anaphylaxis, *656*
Anaplerotic reactions, 255–256, 256f
Andromeda Galaxy, 1
Angiotensin-converting enzyme 2 (ACE2) receptor, 740
ANI (average nucleotide identity), *531*
Animal cells, viral infections, 115f, 116
Animal-free foods, 377
Animalia kingdom, 2
Animal reservoirs of infection, 708
Anogenital condylomata, *759*
Anopheles mosquitoes, 816. *See also* Malaria
Anoxic, 402
Anoxygenic photosynthesis, *239*–240, 418, *418,* 447
Anoxygenic phototrophy, *239*
Antagonism, *545*
Antarctic lakes, 581, 581f
Anthrax
bioterrorism attack using (2001), 23
Koch's work on, 17–18

Public Health Security and Bioterrorism Preparedness and Response Act of 2002 and, 715
transmission of, *802*–803, 803f
vaccine, 19, 803
Antibacterial drugs, **175–180**. *See also* specific antibacterial drugs
Antibiotic-associated colitis, *804*–805
Antibiotic resistance, 186f
to biocides, 157
β-lactamases in, *175*
genetic variation for, 343, 363–364
horizontal gene transfer (HGT) to spread, 352–353
intrinsic, 185
mechanisms, 185–187, 186t
methicillin-resistant *Staphylococcus aureus* (MRSA), 185, *792*–793, 792f
multidrug-resistant strains of tuberculosis (MDR-TB), *774*
to penicillin, 363
penicillinases in, *175*
in persisters, *140*
Antibiotics, *156*. *See also* specific antibiotics
antibacterial drugs, **175–180**
biotechnology and industrial microbiology for, 852–853
common targets of, 172, 172f
definition of, *170*
development disincentives, 851
early developments, 244
persisters resistant to, *140*
Antibiotic stewardship programs, 187
Antibodies, *646*
binding to targets, **646–650**
clonal selection, 654, 655f
diversity of, 652–654
function of, 647
immune complex formation, 650–*651*
immunoglobulin classes of, 647–649, 647t
kinetics of, 649–650, 649f
neutralization, *650*
opsonization, 650
reagin (IgE antibodies), *648*, 648f
structure of, 646–647, 646f
in type III hypersensitivity, *658*, 660f
Antibody affinity, *635*
Antibody class switching, *650*
Antibody-dependent cell-mediated cytotoxicity (ADCC), *620*, 620f
Antibody-dependent enhancement (ADE), *747*
Antibody-mediated immunity, *634*
Antibody titer, *650*
Anticodon, *285*
Antifungal drugs, **182–183**
Antigenic determinant sites, *635*
Antigenic drift, *741*
Antigenic shift, *741*
Antigenic variation, *692, 821*
Antigen-presenting cells (APCs), *629, 639*
Antigen processing, *638*
Antigens
adaptive immunity elicited by, ***634–635**, 635*, *664*–665, 665f
description of, *607*
Antimalaria drugs, 396
Antimetabolitic drugs, *179*
Antimicrobial activity, 173–174
Antimicrobial agents, *170*, 186t
naturally occurring in food, 847
resistance to (*See* Antibiotic resistance)
Antimicrobial drug resistance, public health threat, **185–187**, 188f
Antimicrobial drugs
selective toxicity of, ***172–173***
Antimicrobial peptides (AMP)
gut bacteria and, 674
Antiport, *51*
Antiprotozoan drugs, **183–185**
Antiretroviral drugs, 737
Antisense RNAs, *309*
Antisepsis, 156f, *157*, **171**
Antiseptics, *157*, 163f
Antitoxins, 19, *650*
Antiviral drugs, **181–182**, 182f
Acyclovir, 181
in chemotherapy, **181–182**, 182f
cidofovir, 181
Foscarnet, 181
oseltamivir, 742
peramivir, 181, 742
valacyclovir, 738
zanamivir, 181, 742
Apical complex, *482*
Apicomplexan cell, *482*, 482f
Apicoplasts, *482*
API 20E kit, 722, 726f
API Profile Index, 722
Aplastic crisis, 756
Apoenzymes, *200, 374*
Apoptosis, 621
Appert, Nicholas, 838, 840
Appressoria hyphae, *496*
APS (Adenosine 5′-phosphosulfate), 232, 429, 430f, 446f
Aptamers (engineered RNAs), *377*
Aquaporin protein, 41, 41f
Aquatic ecosystems, **569–585**
benthic marine sediments, 578–580
coastal, 571–573
freshwater, 580–585
as microbial habitat, 570–571
open ocean, 573–578
Arabinose operon, 305, 305f
Arber, Werner, 363
Arbuscular mycorrhizae (AM), 593
Arbuscules, 593
Archaea, **401–412**
bacteria and eukaryotic properties *vs.*, 402t
bacteria phylogenetic relationship to, 403f
bacteria *vs.*, 2, 8–9
chemoorganotrophic pathways, 404
CO_2 fixation, 402–404
DNA replication, 272–275
eukaryotic cell cycle similarities to, 127
horizontal gene transfer (HGT) and, 9
lytic and lysogenic infections for, 113–114, 114f
metabolism of, 402–405
mosaic nature of genome of, 9
overview, **3**, 402
regulation of gene expression, 203–204
role in iron cycle, 563
taxonomy of, 402
thermophilic, 563
Archaeal cell structure, **74–83**
attachment and motility structures, 81–83
bacterial cell structure *vs.*, 75t, 76f, 81
cell envelopes of, **77–80**
common features, 75–76
cytoplasm of, 80–81
eukaryotic and bacterial cell structure *vs.*, 99–100
Archaeal protein translocation systems, 337–338
Archaea, transcription, 333–334
Archaella, 82–83, 82f
Archaeol, *259*
Archaeplastida clade, green/red algae, 486–488
Archaeplastida supergroup, 486–488
Archaerhodopsin protein, *240, 412*, 412f
Arsenic treatment of syphilis, 171
Arsphenamine, 171
Artemisinin, 170, 185

Artemisinin-based combination therapy (ACT), 170, 819
Arthrobacter, 456, 457f
Arthroconidia fungal spores, *494*
Arthropod transmission of disease
by bacteria, **779–782**
Chikungunya, *747*
dengue, *747*–748
equine encephalitis, *748*
leishmaniasis, *819*–820
Lyme disease, *779*–780, 779f
malaria, *815*–819
plague, *780*
pneumonic plague, *780*
by protists, **815–822**
Rocky Mountain spotted fever, *780,* 782, 782f
trypanosomiasis, *820*–822
by viruses, **747–749**
West Nile fever, *748*
Zika, *748*–749
Artificial chromosomes, *367,* 368f
Artificially acquired active immunity, 636, 636f
Artificially acquired passive immunity, 636, 636f
Ascocarp, *498*
Ascogenous hyphae, *498*
Ascomycetes, *497*
Ascomycota, 497–500, 497f
Ascospores, *498, 499*
Ascus, *497*
ASD (autism spectrum disorder), 677
Aseptate hyphae, *492*
Asgard archaea, 533
Asgardarchaeota, 406
Asian influenza virus, 742
Aspergillosis, *830,* 830f
Aspergillus flavus, 696
Aspergillus parasiticus, 696
Aspergillus species, 498
Assembly, 111–112, 111f
Assimilatory nitrate reduction, *252, 561*
Assimilatory sulfate reduction, *562*
Assimilatory sulfur reduction, 253–254
Associative nitrogen fixation, 592
Atelocyanobacterium thalassa, 423
Atherosclerosis, 680
Athlete's foot, 824
Atmosphere as extrinsic factor in food spoilage, 840
Atomic force microscopy, *41,* 41f
Atopic reactions, *657*
ATP (adenosine triphosphate)
in acidic tolerance response, 134
aerobic respiration yield of, 194, 197f
archaella rotation powered by, 82
as bacterial energy source, 49
as coupling agent, 193f
electron transport and oxidative phosphorylation yields of, 220–224, 222f
Embden-Meyerhof pathway yields of, 213–215
Entner-Doudoroff pathway yields of, 215
in metabolism, 193–194, 193f
nitrogen fixation use of, 253
proton gradients in synthesis of, 223f
proton motive force for synthesis of, 221
26S proteasome broken down by, 91
ATP-binding cassette (ABC) transporters, *51,* 51f
ATP synthase, *221,* 223f
AT-rich segments of DNA, 272
A (aminoacyl or acceptor) tRNA binding site, 286
Attachment
archaeal, 82–83
bacterial structures for, **66–68**
in life cycle of viruses, 108–109
Attack rate of diseases, *701*
Attenuated cultures, 19
Attenuated vaccines, *712*
Attenuation, *305*–308, 306f
Attenuator sequences, *306*
Atypical pneumonia, *776*
Autism spectrum disorder (ASD), 677
Autochthonous carbon, *581*
Autoclave, *160,* 161f
Autoimmune diseases
and autoimmunity, *661,* 662t
monoclonal antibody therapy for, 656, 656f
rheumatic fever, *778*
Autoimmune hemolytic anemia, 756
Autoimmunity, 661
Autoinducer, *142*
Autoinducing peptides, *143*
Autoinduction, *319*
Automated blood culture system, 722
Automated microbroth dilution susceptibility test, 723, 725
Automated Sanger method of DNA sequencing, 381, 382f
Autophagocytosis, *624*
Autophagosomes, *93*
Autophagy, *93,* 628, 628f
Autotrophs, *209,* 232
Auxotroph mutants, *347*
Average nucleotide identity (ANI), *531*
Avery, Oswald, 267–268, 267f
Avian influenza virus, 742f
Avidity of antibodies, *635*
Avsunviroidae viroids, 117, 117f
Axenic culture, 14, *145, 531. See also* Pure culture
Axopodia, 89
Azidothymidine (AZT), 181
Azole antifungal drugs, 183
Azospirillum, 438

B

Bacillary dysentery, 801
Bacilli, *45,* 45f, **462–468**
Bacillales, 462–463
Bacillus thuringiensis, 858, 859f
biofilm formation by, 464f
characteristics of, 463t
Lactobacillales, 464–467
Listeriaceae, 467
rod shape of, 45
Staphylococcaceae, 464
Bacillota
Gram-positive bacteria in phylum, 55
Bacillus alcalophilus, 134
Bacillus anthracis, 691
anthrax connection, discovery of, 17
as bioterrorism agent, 802
capsule of, 59, 59f
spores of, 69
Bacillus cereus, 841
Bacillus infernus, 131
Bacillus spp., 462–468, 588
Bacillus subtilis
bioform formation by, 464f
cell cycle of, *123*
cell walls of, 69
characteristics of, 462–463
endospores of, 68f, 69
oligopeptide communication by, 143
riboswitch for, 305–308, 307f
sporulation, 69
sporulation in, 320–321, 321f
transformation in, 358–360
Bacillus thuringiensis, 858, 859f
Bacitracin, 251, 466
Backslopping, *844*
Bacteremia, *691*
Bacteria.
airborne transmission of diseases, **770–778**

in Antarctic, 581
antibacterial drugs, **175–180**
archaea and eukarya *vs.,* 3, 8–9, 402t
arthropod transmission of disease, **779–782**
conjugation of, 355–358
diarrhea from, 796t
direct contact transmission of, **782–795**
food and water as vehicles for disease, **795–802**
in food spoilage, 838
horizontal gene transfer (HGT) and, 9
human diseases caused by, **769–808** (*See also* specific diseases or pathogens)
lactic acid, 844–846, 846f
lysogenic, *113*–114, 114f
lytic and lysogenic infections for, 113–114, 114f
magnetotactic, 563
mosaic nature of genome of, 9
nitrifying, *231*
nitrogen-fixing, 19
opportunistic diseases, **804–808**
overview, 3
oxygen and growth of, 137f
rapid transport, 44
role in iron cycle, 563
structural features
  cell organization, 47–48
  rapid transport, 44
  shape, arrangement and size, 45–47
sulfate-reducing, 19, 411–412, 411f
transformation from free DNA uptake, 360–363
viroids and viruses *vs.,* 117
vomiting from, 796t
zoonotic diseases, **802–804**
Bacterial artificial chromosomes (BACs), *367*
Bacterial cell cycle, **123–127**
  cell growth and shape determination, 126–127
  chromosome replication and partitioning, 124
  cytokinesis, *124*–126
  overview, 123–127
Bacterial cell structure, **44–71,** 47f
  for attachment and motility, **66–68**
  bacterial plasma membranes, **48–54**
    nutrients, 49–52
    plasma membrane structure, 48–49, 48f
  cell envelope, *47,* 53f, **59–60**
  cell walls of, **52–58**
    gram-negative cell, 55–58
    overview of, 53
    peptidoglycan structure, 53–55
  common features of, **45–47**
  cytoplasm, **60–65**
  endospores as survival strategy, **68–71,** 68f
  eukaryotic and archaeal cell structure *vs.,* 99–100
  fluid mosaic model of membrane structure, 49–50, 48f
  functions of, **52–58**
  Gram-negative, 55–58
  Gram-positive, 55
  Gram staining mechanism, 34
  lacking peptidoglycan or cell walls, 58
  lipids in membranes of, 49, 56
  nutrient transport mechanisms, 49–52, 49t
  osmotic protection, 58
  peptidoglycan structure, **55**
  size, 45–47
Bacterial cellular processes, regulation of, **299–323**
  attenuation, 306–307
  catabolite repression in *Escherichia coli, 311,* 313, 314
  chemotaxis in *Escherichia coli,* 316–317
  global regulation mechanisms, 309–316
  overview, 300, 300f
  quorum sensing in *Vibrio* species, 318–319
  riboswitches, 308
  small RNA, translation control by, 308–309
  sporulation in *Bacillus subtilis,* 319–320, 320f
  stringent response in *Escherichia coli,* 313–315
  transcription initiation, **301–302**
    arabinose operon for control of, 305
    induction and repression of enzyme synthesis, 300
    lactose operon, 302–303
    regulatory proteins for, 301–302
    tryptophan operon, 303–304
  translation control, 308–309
Bacterial genome replication and expression, **265–295**. *See also* Eukaryotic and archaeal genome replication and expression
  accuracy in translation, 288
  amino acid activation, 285
  coding regions and other sequences in, 277–279
  daughter chromosome separation for termination, 275
  DNA as genetic material, 266–268
  DNA replication, **271–276**
  genetic code, 282–285
  nucleic acid and protein structure, 268–271
  origin of replication, 271–272
  polypeptide chain elongation, 286–288
  protein maturation and secretion, 292–295
  pyrrolysine insertion, 288
  replication fork, 271–274
  replisomes for, 272–273
  ribosomal tRNA binding sites, 286
  selenocysteine insertion, 288
  70S initiation complex for protein synthesis, 288
  stop codon to end, 288
  transcription in, **279–282**
  translation in, **285–290**
Bacterial identification, *726,* 727
Bacterial interference, *670*
Bacterial pathogen infection, cycle of, 686f
Bacterial taxonomy
  and cell envelopes, 415
Bacterial vaginosis (BV), *805,* 805f
Bactericidal permeability-increasing (BPI) protein, 627
Bactericide, *157*
Bacteriochlorophylls, 240, 577
Bacteriocins, 610, 847
Bacteriocytes, *545*
λ Bacteriophage
  genome, 512t
  overview, **508–511,** 509f
  transduction mechanism, 361, 361f
Bacteriophages
  complex capsids of, 105–106
  description of, 103, *104*
  φ6, 516
  φX174, 514, 514f
  lambda, 508–511, 509f–511f
  Pf1, 514
  T4, 506–508
  temperate, 113, 114, 114f, *361*
  as therapies, 168
  to treat bacterial disease, 187
  vectors, 365
  virulent, *113, 361*

Bacterioplankton, 563
Bacteriostatic substances, *157*
Bacteroid, 440, *598*
Bacteroides fragilis, 675
Bacteroidetes
type IX secretion systems (T9SS), 295
Bactoprenol, *126*
Bactoprenol phosphate, 249
Baeocytes, *122, 421*
Baker's yeast, 497
Balantidium coli, 482
BALT (bronchial-associated lymphoid tissue), *623*
Baltimore, David, 365, 505
Baltimore system of virus classification, 505, 506t, 515
Bam complex, 296, 296f
Banded iron formation, 563
Barcodes, *530*
Barophilic microbes, *137*
Barotolerant microbes, *137*
Basal body, 67
Base analogues, *345*
Base excision repair, *350*
Basic local alignment search tool (BLAST), *389*
Basic reproduction number (R_0), *707*
Basidiocarps, *501*
Basidiomycetes, *501*–502, 587, 593t
Basidiomycota, 501–502, 501f
Basidiospores, *501*
Basidium, *501*–502
Basophils, *616*
Bassi, Agostino, 17
Batch culture, *128*
Batrachochytrium dendrobatidis, 495
Bats, 500
Batteries, microbe-powered, **869–870**
Bauer, A.W., 173
Bayesian inference, 11
BBB (blood-brain barrier), 677, 751
B-cell receptor (BCR) complex, *644,* 644f
B cells
humoral immunity, *617*
lymphomas of, 751
receptor complex, 644, 644f
BCR (B-cell receptor) complex, *644,* 644f
Bdellovibrio, 431–433
life cycle of, 432f
Bdellovibrio killing, 168
Beers and ales, 847
Beggiatoa alba, 447, 447f
Beggiatoa spp., 447–448
Beijerinck, Martinus, 18, 146
Benthic marine sediments, 578–580
Bergey's Manual of Systematic Bacteriology, 415
Berg, Paul, 365
Berkeley, M. J., 17
Betadine iodophor for microbe control, 163
Beta diversity, *536*
β-Galactosidase, 300, 301f, 303
β-Hemolysis, 466, 467f
β-Hydroxyaspartate cycle, 442, 442f
β-Lactam antibiotic, 175–176, 176f
antibiotics block transpeptidation, peptidoglycan chains, 176f
β-Lactamases, 365
β-lactamases inhibitors, 176–177
β-lactam ring, *175,* 186f
β-Oxidation pathway, *229*
Betaproteobacteria. See also Proteobacteria
β-tubulin, 89
Burkholderiales, 444–446
Bifidobacteria, 456, 457f, 667
Bifidobacteriales, 457
Bigo, M., 17
Bilirubin, 758
Binal symmetry of T4 phage of *Escherichia coli, 106,* 106f
Binary fission, *122,* 122f, 472, 472f
Binnig, Gerd, 41
Binomial system of classification (Linnaeus), 13
Biochemical assessment of microbial community activity, 530
Biochemical characteristics, 530
Biochemical oxygen demand (BOD), *868*
Biochemical pathways, 198, 199f
Biocides, *156,* 162, 164t
Biodegradation and bioremediation
chemoorganotrophic metabolism and, 211
definitions, *871*
microbiology applications, 14, 871–874
oil spill cleanup, 871, 872f
Bioencryption, 265
Bioengineering, 858f
Biofilms
antimicrobial agent resistance of, 142
Bacillus subtilis formation of, 462, 464f
c-di-GMP signaling, 312f, 315
common nature of, *140*–142
definition of, *140*
dental plaque (*See* Dental plaque)
examples of, 141f
formation of, *140*–142, 141f
in freshwater ecosystems, 582t
Fusobacterium, 429
heterogeneity of, 141f
of microorganisms, 790
pathogenicity and, *693*
Pseudomonas aeruginosa, 315, 315f, 450
of staphylococci, 790, 790f
Biofuels
biotechnology and, **858–859**
from food waste, 414
methanogenesis, 409
synthetic biology effort in, 377
Biogeochemical cycling
carbon cycle, 558–559
carbon fixation, 565, 570, 570f
climate change and, 565–566
elements important in, 557t
harmful algal blooms and, 572
interaction between cycles, 565
iron cycle, 563–564
manganese cycle, 564–565
mercury cycle, 564–565
nitrogen cycle, 560–561
out of balance, 565–567
phosphorus cycle, 565–567
sulfur cycle, 561–563
Bioinformatics, 20, **389–390**
Bioinsecticides, *858*
Biological control of microorganisms, 168
Biological entities, microbiological studies of, 1–4, 2f
Bioluminescence, 143, 318
Biomagnification, *565*
Biomass of marine microbes, 577f, 578
Biomolecular condensates, *63,* 327
Biopesticides, *858*
Biopolymers, 869
Biosafety in Microbiological and Biomedical Laboratories (BMBL), 719
Biosafety level (BSL), 720, 721t
Biosafety practices, 720–721, 721t
Biosensors
electronic applications, 299
food safety applications, 843
industrial applications, 851
Biostorage, 265
Biosynthesis
amino acid synthesis pathways in, 254–256

in *Escherichia coli,* 245t
principles of, **245**
Biotechnology
custom microbes, 375–376
and industrial microbiology, 20, **851–860**
agricultural, **858–859,** 858f
amino acids, 853
antibiotics, 852–853
biofuels, ***859–860,*** 860f
challenges of, **856–857**
microbes as products, **852–854,** 852t
microbial cells, **854–856**
organic acids, 853
vaccines, 853–854
Bioterrorism
anthrax attack, 23
anthrax in, 802
Brucellosis potential in, 803
epidemiology and, ***715–716***
Biotrophic fungi, *600*
Biovars, 13
Black Death
description of, 769
Yersinia as cause of, 13
Black piedra, 823
BLAST (basic local alignment search tool), *389*
Blastomycosis, *813*–814
Blastospores, *494*
Bleach for microbe control, 163
Blindness, 754, 787
Blood agar, 144
Bloodborne Pathogens Act of 1991, 719
Blood-brain barrier (BBB), 677
Blooms
coccolithophore *Emiliania huxleyi,* 486, 486f
description of, 618
B lymphocytes, *634*
BMBL *(Biosafety in Microbiological and Biomedical Laboratories),* 719
Bodian, David, 762
BOD (biochemical oxygen demand) test, *868*
Body's sterile sites, 155
Bone marrow
hematopoiesis process in, *616*
in lymphoid system, 622f, 624
in lymph system, 621
mast cells derived from, *616*
Bootstrapping of phylogenetic trees, 11
Bordetella pertussis, 445, 776
Borlaug, Norman, 587
Borreliella burgdorferi, 427
chromosomes in, 65
"end replication problem" solved by, 329
Lyme disease from, 427
shape of, 127
Bosch, Karl, 566
Botox (type-A botulinum toxin), 798
Botrytis cinerea, 600
Botulism, 69, 469, *795,* 795f, 797
Bovine spongiform encephalopathy (BSE), 118, 766
BPI (bactericidal permeability-increasing protein), 627
B7 (CD80) protein, *641*
Bradford coccus, 513
Bradykinin, 629, 630f
Brain-gut axis, 676–677, 677f
Braun's lipoprotein, 55
Bright-field microscopy, ***24–25,*** 25f
Broad-spectrum antimicrobial drugs, *172*
Bronchial-associated lymphoid tissue (BALT), *623*
Bronchopulmonary aspergillosis, 831
Brown, Robert, 94
Brown rot fungi, 588
Brucellosis *(Brucella abortus), 803*
BSE (bovine spongiform encephalopathy), 118, 766
BSL (biosafety level), 720, 721t
Buboes (enlarged lymph nodes), *780*
Buchnera aphidicola, 545, 552
Budding
in bacteria, 122f
reproduction by, 472, 472f
for virion release, 112, 112f
Bulk soil, 592
Burkholderia cepacia, 143
Burkholderiales, 444–446
Burkitt lymphoma, 757
Butanediol fermentation, *228*

C

Cable bacteria, *431,* 431f
Cadaverine (amine compounds), 837
*Ca*lcium and *dipicolinic acid* (Ca-DPA), 69, 70
Calcium spiking, 598
Calories, *191*
Calvin-Benson cycle, 447
action of, *247*–248, 247f
cAMP (3′, 5′-cyclic adenosine monophosphate), *313*
Campylobacteraceae, 433
Campylobacter jejuni, 162, 433, *797*
Cancer
aflatoxins and, 841
cervical, 759
dysbiosis, 680–681
Epstein-Barr virus (EBV) associated with, 756, 757f
Helicobacter pylori and, 433, 784
as HIV/AIDS complication, 751
human papillomaviruses associated with, 759
immunological treatment of, 633
liver, 500
microbes as cause of, 680
monoclonal antibody therapy for, 656, 656f
stomach, 126
viruses and, 103, 115f
Candida albicans, 143, 671
Candida auris, 183
Candidal vaginitis, *831*
Candidatus Carsonella ruddii, 397
Candidiasis, *831*–832
Cannulae, 82, 82f
CAP (catabolite activator protein), *303, 313,* 313f, 314f
Capsid, ***104–107,*** 104f
Cap snatching, 181
Capsomers, *105*
Capsules, *59,* 59f, 690f, 693
Capsule staining, *35,* 35f
Carbapenem antibiotics, *176*
Carbapenem-resistant *Enterobacteriaceae* (CRE), 185
Carbenicillin, 176
Carbohydrates
anabolic synthesis of, 248–251
catabolism of, 228–229, 229f
food spoilage, 837
Carbon
to nitrogen ratio (C/N ratio), *590*
PHA to store, 62
sources of, 210, 210t
Carbonate equilibrium system, *570,* 570f
Carbon cycle
description of, 558–559, 558f
diatom importance in, 485
prochlorophyte cyanobacteria effect on, 422
Carbon dioxide, 74, 209
Carbon dioxide fixation
in anabolism, 248
in archaea, 402–404

Carbon fixation
by phytoplankton, 573, 574f
primary production, 570
Carboxysomes, *62,* 63f, 247, *420*
Carcinogen testing systems, 346
CARD-FISH, 537
Cardinal temperatures, *135*
Cardiovascular disease
dysbiosis, 680
Caries, 671, *806*
Carotenoids, 138
Carriers, *50,* 51f
Case fatality rate (CFR), *701*
Cas9 genome editing, *373*–375, 374f
Catabolism, **208–240**. *See also* Anabolism; Metabolism
aerobic respiration
ATP yield, 221–224
electron transport chains, 219–220
electron transport to oxidative phosphorylation, **219–224**
glucose to pyruvate, 213–215
overview, 219–224
oxidative phosphorylation process, 220–221
pyruvate to carbon dioxide, 219
anabolism *vs., 211,* 245
anaerobic respiration, 224–225
chemolithotrophy, 230–232
chemoorganotrophic fueling processes, 211–212
fermentation, *211*
metabolic diversity and nutritional types, 209–210
phototrophy
anoxygenic photosynthesis reactions, *239*–240
oxygenic photosynthesis reactions, *234*–239
phototrophy, ***245–252***
rhodopsin-based, 240
Catabolite activator protein (CAP), *303, 313,* 313f , 314f
Catabolite repression, *311–313*
Catalase, *137,* 137f
Catalysts, *200*
Catalytic proteins, 5. *See also* Enzymes
Catalytic sites for reactions, *201*
Catalyzed reported deposition-FISH (CARD-FISH), 537
Catenanes, *275,* 276f
Caterpillar fungus, 490
Cathelicidin, *610*
Cationic peptides, 610
Caulobacterales, 439–440, 440f
Caulobacter crescentus, 60, 63
Caulobacter vibrioides, 60, 63
cell cycle, *123*–124
cell shape, 126, 127f
holdfast, *439*
life cycle, 440, 440f
Caveolin-dependent endocytosis, *92*
Cavities
dental caries (*See* Caries)
tuberculous, *774*
CD (cluster of differentiation) molecules
cDNA (complementary DNA), *365,* 367f, 375
CDS (coding sequences), *410*
$CD4^+$ T cells, 642, 643
$CD8^+$ T cells, *643*
Cecal coccidiosis, 483
Cech, Thomas, 5, 203
Cell-cell communication, 142–143, 142f
Cell envelopes, *47*
archaeal, **75–76,** 76f
bacterial, *47,* 53f, **59–60**
and bacterial taxonomy, 415
eukaryotic, 87–88
of Gram-negative bacteria, 53f
of Gram-positive bacteria, 53f
Cell-mediated immunity, *634*
Cells
energy cycle of, 194f
of immune system, 616–620
measuring mass of, 149–150
membranes (*See* Membranes)
work of, *191*
Cell shapes
Halophiles, 412
Cell suspensions, 267
Cellular microbes, **9**
Cellular slime molds, *476*–477
Cellulitis, *793*
Cell walls
in *Actinobacteria,* 456
archaeal, 79–80
bacterial, *53,* 53f
biosynthesis of, 127f
of diatoms, 484
drugs inhibiting synthesis of, **175–177**
in fungal hyphae, 492
lysozyme action on, 610
mycobacterial, 457, 458f
naked amoebae lacking, *475*
and osmotic protection, **58**
Centers for Disease Control and Prevention (CDC), *701,* 719
antiviral drugs recommended by, 742
food-borne illnesses, 840–842
on global health security, 718
on HIV/AIDS therapy, 181
PulseNet program of, 843f, 844
role in epidemiology, 718
on toxoplasmosis treatment, 185
on water quality, 862
Central metabolic pathways, 246
Central nervous system, complications from HIV/AIDS, 751
Cephalosporin antibiotics, 176, 177f
Cerebral amyloidosis, 765
Cerebrospinal fluid (CSF), 23, 773
Cervical cancer, 755
CFU (colony forming units), *149*
Chagas' disease, 473, *821*
Chain, Ernst, 171, 244
Chain termination, 181
Chain-termination DNA sequencing, *381*
Chalfie, Martin, 31
Chamberland, Charles, 18, 160
Chancre, *787*
Channeling, metabolic, 203
Channel-mediated facilitated diffusion, *50*
Chaperone, *136*
Charpentier, Emmanuelle, 374
Chase, Martha, 267, 268f
Cheaters, *553*
Cheese production, 846–847, 846f
Chemical-based food preservation, 839
Chemical control of microorganisms, 162–165
Chemical fixation, *33*
Chemical oxygen demand (COD) test, *868*
Chemical work of cells, *191*
Chemiosmotic hypothesis, *220*
Chemoheterotrophs, 444–445
Chemokines, *614*
Chemolithoautotrophy, *209,* 210t
Chemolithotrophs, *230,* 230f, 444–445
Chemoorganoheterotrophs, *210,* 210t
Chemoorganotrophic fueling processes, 211–212, 212f
Chemoorganotrophic pathways, 404–405
Chemostats for continuous microbial culture, *151*–152, 151f
Chemotaxins, *629*
Chemotaxis
Escherichia coli, 316–317, 317f
Chemotherapy, **170–188**
antibacterial drugs, **175–180**
antifungal drugs, **182–183**
antiprotozoan drugs, **183–185**
antisepsis efforts leading to, *157*

antiviral drugs, **181–182,** 182f
compounds from *Streptomyces,* 460–461, 461t
measurement of activity, 173–174
overview, *157*
selective toxicity of, **172–173**
Chemotrophs, *209*
Chestnut blight, 497, 587
Chickenpox (varicella), 738, 738f
Chikungunya virus, *747*
Childbed fever, 687
Chinese medicine, 170
ChIP (chromatin immunoprecipitation), *395*
Chlamydiae, *424*
Chlamydial life cycle, 425f
Chlamydial pneumonia, *770*
Chlamydial sexually transmitted diseases, 786
Chlamydia pneumoniae, 425, 425f
Chlamydia psittaci, 425
Chlamydia spp., 425
Chlamydia trachomatis, 425
Chlamydomonas, 487–488, 487f
Chlamydomonas nivalis, 135
Chlorine dioxide sterilizing gas, 165
Chlorine for microbe control, 163–165, 872
Chlorobi, light reactions in photosynthesis, 239
Chloroflexus spp., 419
Chlorophyll
for anoxygenic photosynthesis, *239*–240
in bacteria, 416
in diatoms, 484
for oxygenic photosynthesis, *234*–239
Chloroplastida, 486–488
Chloroplasts
action of, *97,* 97f
from endosymbiosis, 8–9
evolution of, 8–9
pyrenoid, in protist, *472*
Chloroquine, 184, 816
in COVID-19, 184
Chlorosomes, *418*
Chlortetracycline antibiotics, 178
Chocolate, fermentation in, 845
Cholera
description of, *797*–799
treatment of, 799f
Vibrio cholerae as causative agent of, 67
Chromatiales, 447
Chromatic acclimation, *420*
Chromatic adaptations in cyanobacteria, *421*
Chromatin, *94*
Chromatin immunoprecipitation (ChIP), *395,* 395f
Chromatin remodeling and modification, 338–339
Chromatin remodeling factors, *339*
Chromoblastomycosis, *825,* 825f
Chromophore groups, *33*
Chromophores, 412, 412f
Chromosome interaction domains, 64f, 65
Chromosomes
archaeal, 81
artificial, 367, 368f
bidirectional replication of, 272f
catenanes and dimerized, 275, 276f
daughters separate for termination, 275
end protection of, 274
"end replication problem" of, 276f
eukaryotic, *94,* 327, 328f
Hfr conjugation transfer of DNA in, 356–357
replication and partitioning, 124
telomerase replicating telomeric DNA of, 330f
Chronic inflammation, 630–631
Chytridiomycota, 495, 495f
Chytrids, 495
Cidal chemotherapeutic agents, *172*
Cidofovir, 181
Cilia, *99,* 99f, 100
Ciliates, 479–480, 480f
Ciliophora, 479–480, 480f
Citric acid cycle, 217
Citric acid fermentation, 853
Citrus greening, 603
Classical approaches to taxonomy, 529
Classical complement pathway, 613
Classification
genotypic, 344
Linnaeus system of, 13
of organisms, 2
phyletic, 9
Class II MHC molecules, *638*
Class I MHC molecules, *637*
Clathrin-dependent endocytosis, *92*
Claviceps purpurea, 499
Climate change
biogeochemical cycling and, 565–566
harvest of caterpillar fungus and, 490
methanogenesis and, 409
microbial ecology and, 19
Clinical laboratory scientists (CLS), 20
Clinical microbiology, **718–735**
agglutination, *729,* 730
biosafety practices, 720–721, 721t
complement fixation, 719f
definition of, *719*
Ebola epidemic and, 718
enzyme-linked immunosorbent assay, *730*–733, 732f
flow cytometry, *735*
immunoblotting (Western blotting), *733*
immunoprecipitation, *733*–734
laboratory detection of infectious diseases, 729
measurement or exploitation of immune responses, **729–735**
serotyping, *729*
specimens for microorganism identification, **721–728,** 722f (*See also* Specimens for microorganism identification)
Clonal selection, lymphocyte, 654, 655f
Cloning
artificial chromosomes as vectors, 367
bacteriophages as vectors, 365
cDNA synthesis and, 365, 366f
cosmids as vectors, 367
genes, 362, 363f
genomic libraries, 371–372
Gibson assembly, 370, 370f
host cells, 368
plasmids as vectors, 365
process, 368f
seamless, 370–371, 370f
vectors, 365
vectors for recombinant DNA, 365, 366f, 367f
viruses as vectors, 365
ClO_2 sterilizing gas, 164
Clostridia, 469–470, 469f, 470f
Clostridial myonecrosis, *782,* 783f
Clostridial toxins, 798
Clostridioides (Clostridium) difficile, 69, 666, 671
antibiotic-associated colitis, *804*–805
vancomycin effective against, 177
Clostridium botulinum, 69, 469, 795
Clostridium perfringens, 69, 469, 782
Clostridium tetani
endospores, 69
overview, 469, 469f, 470f
CLR-MAMP binding, 624
CLRs (C-type lectin receptors), 624
Club fungi, *501*–502

Cluster analysis, 392
Clustered, regularly interspaced short palindromic repeats (CRISPR), 321, 836, 843
Cluster of differentiation (CD) molecules, *640*
C1-metabolizing bacteria, 448–449
Coagulation in water purification, *863*
Coal mines, 446
Coastal ecosystems, 571–573
Cocci, *45,* 45f, *75,* 76f
Coccidioidomycosis, *814,* 814f
Coccolithophores, *486,* 571, 571f
Code degeneracy, *283*
Coding regions, *277,* 277–279
Coding sequences (CDS), *389,* 390f
Codons, *283*
COD (chemical oxygen demand) test, *868*
Coenocytic hyphae, 493f, 495–496
Coenzymes, 195f, *196,* 409, 410f
Coevolution, *545*
Cofactors, *200,* 245
CO_2-fixation pathway, **247–248,** 402–404
Cohn, Ferdinand, 16
Cold plasma, *161*
Cold sores, 753–754
Cold sterilization using radiation, 162
Colicins, *610*
Coliforms, water quality, *864*
Colilert defined substrate test, *865,* 865f
Colitis, *804*
Colonies
  on membrane filter, 150f
  of microorganisms, *146,* 147f
Colonization, adhesion, 689–690
Colonization resistance, 673, 674
Colon microbiota. *See* Gut microbiome
Colony forming units (CFU), *149*
Colony-stimulating factors (CSFs), *614*
Colorless sulfur bacteria, *445*
Comammox, *232, 442, 561*
Combinatorial joining, 653–654
Cometabolism, *872*
Commensal anaerobes, 429
Common cold, 739
Common-source epidemic, *705*
Communicable diseases, *705*
Communication, cell-to-cell, 142–143, 142f
Communities, 543–554
Community activity assessment, 538–541
Comparative genomics, *396*–399, 398f
Compartmentalization as metabolic channeling, *203*
Compartmentation of anabolism and catabolism, 245
Compatible solutes, *132,* 571
Competent cells, *359*
Competition, *552*–553, 553f
Competitive inhibitor of enzymes, *202*
Complementary DNA (cDNA), 277, *365,* 367f, 391
Complement fixation, 719f
Complement system of cascading enzymes, *611*–613, 611f, 612t
Complex culture media, *144,* 144t
Concatemers, *507*–509
Condenser lens, *25*
Conditional mutations, *347*
Confocal microscopy, three-dimensional images, ***31–32,*** 32f
Confocal scanning laser microscope (CSLM), 31
Congenital herpes, *755*
Congenital immunodeficiencies, 663, 663t
Congenital rubella syndrome, *745*
Congenital syphilis, *787*
Conidia, *498*
Conidiospores, *494*
Coniophora, 588
Conjugate redox pair, *194*
Conjugation, *355*–358, 355f, *472,* 481f
Conjugative plasmids, *355*
Conjugative transposons, 364
Conjunctivitis of newborns, *787*
Consensus sequences, *280*
Conserved hypothetical proteins, *389*
Constitutive genes, *300*
Contact-dependent growth inhibition (CDI), *553*
Contact dermatitis, 661, 661f
Contact-independent growth inhibition, *552*
Contact lenses, microbe growth on, 790, 826
Contact time of antimicrobial exposure, 167
Contact transmission of disease, *687*–688. *See also* Direct contact transmission of disease
Containment to prevent contamination, 720
Contigs, *385*
Continuous culture systems, *151*
Continuous feed, *856*
Contractile vacuoles, *471*
Control of microorganisms, **155–168**
  biological methods of, 167–168
  chemical agents for, 162–165
  clean rooms, 159
  death and growth patterns, 157–158
  evaluation of effectiveness, 166–167
  mechanical removal, 158–162
  targets for, 156–158
Convalescent period of disease, *685*
Convalescent plasma, 651
Cooperation, *544*
Copper sulfate, 163
Copy number, *365*
CoQ, 195f, *196*
Coral bleaching, 546
Coral holobiont, 546–547, 547f
Cordyceps, 490
Core enzyme, *280*
Core genome, *397,* 399f, *536*
Core microbiome, 671
Core polysaccharide, *56*
Corepressor proteins, *300*
Corneal inflammation (keratitis), *754,* 826
Coronaviruses, *518*–520, 518f. *See also* COVID-19 pandemic
Corrosion, 430
Corynebacteriales, 457–458, 459f
Corynebacterium glutamicum, 853
Cosmids, 367
Cotranslational reactions, *292*
Cotransporters, 51f
Coughing reflex, 609
Coulter counter, 149
Coumarins, 838
Counting chamber, 148, 148f
Coverage in genomic sequencing, *386*
COVID-19 pandemic, 20
  chloroquine, 184
  deaths, 776
  electron microscopy (EM), 38
  remdesivir, 740
  reverse transcriptase PCR, 370
  reverse transcriptase quantitative polymerase chain reaction (RT-PCR), 741
  SARS-CoV-2
    genome, 739
    RNA polymerase, 349
    time course, 740f
  virus spreading, 397
Coxiella burnetii, 804
  within a host cell vacuole, 450f
Coxiellales, 449–450

CPEs (cytopathic effects), *114*
CRE (carbapenem-resistant Enterobacteriaceae), 185
C-reactive protein (CRP), 616
Creutzfeldt-Jakob disease (CJD), 118, 765
CRISPR-Cas9 genome editing, *373*–375, 374f
CRISPR-Cas system, 321–322, 377
CRISPR RNA (crRNA), 321
Cristae, *96*
Crithidia fasciculata, 135
Crocosphaera watsonii, 573, 576
Cross-linking, 54, 55f
Crown gall disease, 441
CRP (C-reactive protein), 616
crRNA (CRISPR), 321
Cryophiles, *135*
Cryptococcal pneumonia, 753f
Cryptococcosis meningitis, 832, 832f
Cryptosporidiosis, 483, *827*
Cryptosporidium contamination of food and water, 863
CSF (cerebral spinal fluid), 23, 773
CSFs (colony-stimulating factors), *614*
C-type lectin receptors (CLRs), 624
Culture-based methods, **14–19**
  developments in, 15f
  disease and, 17
  immunology, 19
  Koch's postulates, 17–18
  microbial ecology, 19
  in microbial ecology studies, 529
  for microorganism identification, 723, 723t, 727
  pure, 14
  spontaneous generation theories, 14–17
Culture media, 19, ***144–145,*** 144t
Culturomics, *147, 532*
Custom microbes, biotechnology, 375–376
Cut-and-paste transpositions, *354*
Cutaneous anthrax, *802*
Cutaneous mycosis, 824f
Cyanobacteria
  description of, *420*–422, 421f
  dissimilatory sulfate reduction by, 562
  evolution of, 8
  in fossilized stromatolites, 8, 8f
  in lake ecosystems, 583
  multiple fission of, 122, 122f
  Prochloron, 9
Cyanobacterial thylakoids, 420f
Cyanophycin granules, *421*
Cyclic adenosine monophosphate (cAMP), 311–314, 312f
Cyclic AMP receptor protein (CRP), *313*
Cyclic dimeric GMP (c-di-GMP), *311,* 312f, 315–316
Cyclic photophosphorylation, 240f
Cyclosporiasis, *827*–828
Cyclosporine, 492
Cyro-electron microscopy (cryo-EM), ***39–40,*** 40f
Cysteine, 254
Cyst formation, *438,* 439f, 474
Cystic fibrosis, 444, 790
Cystitis, *807*
Cysts, *156*
Cytidine 5′-triphosphate (CTP), 194
Cytochromes, *196*
Cytocidal infection, 114
Cytokines
  as cellular chemical messages, *614*
  functions of, 615t
  pro-inflammatory, 614
Cytokinesis, *124*–126
Cytomegalovirus (CMV)
  fusion of host cells from, 693
  in HIV/AIDS complications, 754
  inclusion disease, 754–755
  size of, 397
Cytopathic effects (CPEs), *114*
Cytophaga spp., 588
Cytoplasm, *60*
  archaeal, 80–81
  bacterial, **60–65**
  eukaryotic, 88–89, 89f
  peptidoglycan synthesis in, 249–251
Cytoproct, *471*
Cytosine, *257*
Cytoskeleton
  archaeal, *80,* 81t
  bacterial, **60–61,** 61f
  eukaryotic, 88–89, 89f
Cytosol, *60,* 88
Cytostome, *471*
Cytotoxic T lymphocytes (CTLs), *643*

D

Damage-associated molecular patterns (DAMPs), 626, 626f
Dane particle, *757*
DAPI (4′, 6-diamido-2-phenylindole) stain, 149, 529f
  nucleic acids, 532f
Dapsone, 185
Dark-field microscope, 14, ***27,*** 27f, 28f
Dark reactions in photosynthesis, *234*
Davis, Bernard, 356
DC/HB (dicarboxylate/ 4-hydroxybutyrate) cycle, 402, 402f
DDT, 816
Dead zones in oceans, 569, 582
Deamination, *230*
Death and growth patterns, 157–158, 158f
Death phase in microbial growth curve, *129*
Death trajectory, 139f
de Bary, Heinrich, 17
Decimal reduction time (D value), *157*
Deep hot biosphere, 604
Deep-ocean sediments, 578–579, 580f
Deep sequencing, 386
Deepwater Horizon oil spill, 871
Defensin peptides, *610*
Defined culture media, *144*–145, 144t
Deforestation, 565
Degranulation, 616
Dehydration, fatal, 761
Deinococcota, 415–416, 416f
Deinococcus, 416, 416f
Deinococcus radiodurans, 138, 392, 392f
Delayed-type hypersensitivities (DTH), *658*
Delbrück, Max, 108
Delta agent (hepatitis D virus), *758*
Denaturation of enzymes, *202*
Dendritic cells (DCs), 615f, *618*
Dengue fever, 555, *747*–748
Denitrification, 225, *445, 561,* 576
Dental diseases, 671, 805–807
Dental plaque, 671, *806,* 806f
Deoxyribonucleoside triphosphates (dNTPs), 368
Depth filters as barrier to microorganisms, *158*
Dermatomycoses, 824
Desert rheumatism, 814
Destructive interference, 28
Desulfonema species, 562
Desulfotomaculum, 563
Desulfovibrio species, 560f, 561, 562
Detergents to control microbes, *163*
DFA (direct immunofluorescent antibody) diagnosis, 765
DGCs (diguanylate cyclases), 315
d'Herelle, Felix, 168, 187

DHFR (dihydrofolate reductatase), 180, 180f
Diabetes
acarbose treatment, 459
type 1, 606
type 2, 678
Diarrhea
Dientamoeba fragilis as cause of, 475
Enterocytozoon bieneusi as cause of, 494
rotaviruses as cause of, 516–517, 760, 760f
Diatoms
in algal blooms, 572
description of, *484*
life cycle of, 485f
Diauxic growth of *Escherichia coli*, *313,* 313f
Diazotrophs, *253,* 423
Dicarboxylate/4-hydroxybutyrate (DC/HB) cycle, 402, 402f
DIC (differential interference contrast) microscopy, ***29,*** 30f
Dictyostelia, *476*
Dictyostelium discoideum, 476, 477, 477f
Diderm bacteria, 294
Diderm cell, *53,* 53f, 415
Dientamoeba fragilis, 475
Differential interference contrast (DIC) microscopy, ***29,*** 30f
Differential media, *144,* 145f
Differential staining, ***34–35***
Digital droplet PCR (ddPCR), 536
Diglycerol tetraether lipids, *77*
Diguanylate cyclases (DGCs), 315, 316
Dihydrofolate reductase (DHFR), 180, 180f
Dikarya, 497–502
Dikaryotic stage, *497*
Dilution susceptibility test, **173**
Dimerized chromosome, 275, 276f
Dimethylallyl pyrophosphate (DMAPP), 259
Dimethylsulfoniopropionate (DMSP), 563
Dinoflagellates, ***479–480,*** 480f
Alexandrium, 572
photosynthetic, 98
Symbiodinium, 546–547
Diphtheria, 19, 459, *770,* 771f
Diplococci, *45*
Direct contact transmission of disease
by bacteria, **782–795**
cellulitis, *793*
chlamydial diseases, 786
chromoblastomycosis, *825,* 825f
cold sores, 753–754
common cold, 739
cytomegalovirus inclusion disease, 754–755
erysipelas, *793*
eumycotic mycetoma, **825**
by fungi and protists, **822–825**
gas gangrene, *782,* 783f
genital herpes, *755*
gonorrhea, *786*–787, 786f
group B streptococcal disease, *783*
HIV/AIDs, 749–753
human betaherpesvirus 5 infection, 754
human parvovirus B19 infection, *756*
impetigo, *793*
invasive streptococcal infections, 793
leprosy, *783*–784, 784f
maduromycosis, *825*
methicillin-resistant *Staphylococcus aureus, 792*
mononucleosis, 757, 757f
nectrotizing fasciitis, *793,* 793f
peptic ulcer disease and gastritis, 784, *784*–785
piedras, *823*
sexually transmitted diseases, 785–788, 785t, 825
sporotrichosis, *824*
staphylococcal diseases, 788–793
streptococcal diseases, 793
syphilis, *787*–788, 788f
tattoo-associated myobacterial infections, 784
tetanus, *793*–794, 794f
tineas, *823*
toxic shock syndrome, 793
trachoma, *794,* 794f
trichomoniasis, *825*
viral hepatitides, 757–758
by viruses, **749–760**
warts, *759*–760
Direct counts, *148*
Directed evolution, 376, 378t
nucleic acid, 378f
Direct immunofluorescence, *730*
Direct immunofluorescent antibody (DFA) diagnosis, 765
Direct interspecies electron transport (DIET), *867*
Directly observed therapy (DOT), 774
Direct repair, *350*
Discoba-Metamonada clade, 473–475
Discriminatory ability, *634*
Disease, *701*
Disease ecology, 1
Disease syndrome, *685*
Disinfectants, *156*
Disinfection, *156,* 156f, 165, 167f
Disk diffusion tests for rapidly growing microbes, **173**
Dissimilatory iron reduction, 563
Dissimilatory metal reduction, 429–431
Dissimilatory nitrate reduction, *225, 252*
Dissimilatory sulfate reduction, 429–430, 430f
Dissolved organic carbon (DOC), *573,* 574f, 582
Dissolved organic matter (DOM), *573*
Dissolved oxygen sag curve, 582, 583f
Distilled water, 839
Diversity
in adaptive immunity, *634*
antibiotic development and, 852–853
of Archaea, 402
human microbiome, 667
photosynthetic bacteria, 438
plant pathogens, 600–602
soil environment, 588
Divisome, *125,* 125f
DMSP (dimethylsulfoniopropionate), 563
DNA (deoxyribonucleic acid). *See also* Bacterial genome replication and expression; Microbial genomics; Recombinant DNA technology
archaeal, *81*
as bacterial genetic material, 266–268
in bacterial nucleoid, *64,* 64f
daughter chromosome separation for termination, 275
as deoxyribonucleotide polymer, *268*–269, 269f
endospore resistance, **70**
Hfr conjugation transfer of chromosomal, 356–357
mitochondrial, 96–97, 96f
origin of replication, 123f, 271–272
protein interactions with, 395
repair of, 356–357
replication fork, 272
replication of, **272–276, 327–330**
replisomes, 272–275
replisome synthesizing machinery for, *124*
DNA amplification, 370
DNA as genetic material, 266–268
DNA-binding proteins, 300, 301

DNA-dependent DNA polymerase, 506
DNA-dependent RNA polymerase, 509
DNA fingerprinting, 370, 394, 727
DNA gyrase, 136
DNA ligase, 274, 275f
DNA microarrays, 391
DNA-modifying agents, 345–346
DNA polymerases, 329
  families, 329, 329t
  holoenzyme, 280f
  reaction, 273f
  in replisomes, 272–273
DNA sequencing
  next-generation methods of, 381–385
  overview, 381
  Sanger method of, 381
DNA shuffling, *376*
DNA vaccines, *713*–715
DNA viruses
  bacterial transformation from free, 358–359
  bacteriophage lambda, 508–511
  bacteriophage T4, 506–509
  double-stranded, **506–513**
  herpesviruses, 512
  multiplication strategy of, 506f
  negative-strand, 516
  nucleocytoplasmic large DNA, 513
  plus-strand, 505
  reverse transcribing, 524–525
  single-strand, 514–515
  transduction for virus-mediated transfer of, 360–363
dNTPs (deoxyribonucleoside triphosphates), 368
DOC (dissolved organic carbon), *573,* 574f, 582
DOM (dissolved organic matter), *573*
Domagk, Gerhard, 171
Domains of life, 8–9, 292–293
Donor tRNA binding site, 286
DOT (directly observed therapy), 774
Double diffusion agar assay, *734,* 734f
Double-stranded RNA viruses, 515–517, 516f
Doudna, Jennifer, 374
Drinking water quality, **863–865**
Droplet nuclei, *687*
Drug resistance, **185–187,** 186t. *See also* Antibiotic resistance
  bacteria with, 363–364
  gene expression systems for detecting, 187
  overcoming, 187
  public health threat, **185–187,** 188f
Drugs
  fungi used for, 492
  probiotics, 681
  use for microbe control (*See* Chemotherapy)
Dry heat incineration for microbe control, 159–161, 161f
Dubochet, James, 40
Duchesne, Ernest, 171
Dunaliella viridis, 133
Dutch elm disease, 497, 587
Dyes
  fluorescent, 31f
  ionizable, 33, 34t
  for staining, **33–34**
Dysbiosis, **677–681**
  microbial diversity, loss of, 678f
Dysentery
  amebic, 813t, *826*
  bacillary, 801
  bacteriophage therapy, 168

E

EAEC (enteroaggregative *Escherichia coli*), 799f, 800
Early life, RNA basis of, 5–8
Ear microbiota, 670
East Coast fever, 483
Eastern equine encephalitis (EEE), 748
EB (elementary body), *425*
Ebola virus disease (EVD)
  epidemic, 718
  global climate change and, 555
  overview, 763
  virion, 763f
ECF-A (eosinophil chemotactic factor of anaphylaxis), *656*
ECF (extracytoplasmic function) sigmas, *311*
Ecological characteristics in microbial taxonomy, 529
Ecology, microbial
  field of microbiology, 19
  rhodopsins, 240
  study of, 30–31
Economics of caterpillar fungus, 490
Ectomycorrhizae (ECM), 593, 594f
Ectoplasm, *471*
Ectosymbionts, *544*
Ectosymbiosis, 544
Editing in genome sequencing, 385–387, 386f
EEE (eastern equine encephalitis), 748
Eelgrass, 486
EF (elongation factors), *288*
Effective reproduction number (R_e), *708*
Effector proteins, *294*
Effector response, *634*
Efflux pumps, *186*
EHEC (enterohemorrhagic *Escherichia coli*), 799f, 800
Ehrlich, Paul, 171
EIEC (enteroinvasive *Escherichia coli*), 799f, 800
Electromagnetic spectrum, 138f
Electron acceptors, 444–445
Electron cryotomography, *39,* 39f
Electron donor to electron acceptor reactions (redox), 194–196
Electronema, 431
Electron gun, 36
Electron microscopy, **35–40,** 36f, 37f, 38t
  electron cryotomography, *39*–40, 40f
  light microscopy *vs.*, 36f
  scanning, *38*–39, 39f
  transmission, 36–38
Electron shuttles, *430*
Electron sources, 209, 210t
Electron transport chains (ETC)
  membrane-associated, 136, 196, 197f
  in metabolism, 194–196
Electrophoresis. *See* Gel electrophoresis
Electrophoretic mobility shift assays (EMSA), 395
Electroporation, *368*
Electrospray ionization-mass spectroscopy, 723
Electrothrix, 431
Elementary body (EB), *425*
Ellis, Emory, 108
Elongasome, *126*
Elongation cycle, *288,* 289f
Elongation factors (EF), *288*
Embden-Meyerhof (EM) pathway, 427
  as amphibolic pathway, 214f
  Borreliella burgdorferi use of, 427
  Calvin-Benson cycle connection to, *247*–248, 247f
  carbohydrates catabolized by, 228
  in chemoorganotrophic archaea, 404, 405f
  gluconeogenic pathway connection to, *248*
  glucose to pyruvate catabolized in, *213*–215, 214f
  homolactic fermentation in, *227,* 465
  lipids catabolized by, 229–230
  in *Plasmodium,* 482
  reversible enzymes in, 215

Emergent properties, *142*
Emiliania huxleyi, 486, 486f, 571f, 578, 578f
EMSA (electrophoretic mobility shift assays), 395
Encephalitis
  amoebic meningoencephalitis, 826–827
  as Chikungunya virus complication, 747
  Encephalitozoon cuniculi as cause of, 494
  equine, *748*
  from rabies virus, 764
  West Nile fever, *748*
Encephalitozoon cuniculi, 397, 494, 552
Encrypting data in biostorage, 265
Encystment, *472*
Endemic diseases, *701*
Endergonic reactions, *192*
Enders, John, 762
Endobiotic predators, *551*
Endocytic
  and secretory pathways, **90–93,** 92f, 109
Endogenous antigen processing, *638*
Endomycorrhizae, 592
Endophytes, *591*
Endoplasm, *471*
Endoplasmic reticulum (ER), *90,* 95
Endoriftia persephone, 547
Endosomes, *93,* 109
Endospores, *68, 156*
  from Bacilli, 462–463
  as bacterial survival strategy, **68–71,** 68f
  discovery of, 14
  formation of, 70f
  location and size, 69f
  resistance of, **70**
  schematic of, 68f
  staining of, *35,* 35f
  structure of, 70, 70f
  vegetative cell, 70–71
Endosymbionts, *545*
Endosymbiosis, 545
Endosymbiotic hypothesis, **8–9,** 9f, 533
Endosymbiotic theory, 9
Endotoxins, *696*
End-point PCR, *369*
End product inhibition to regulate enzymes, *204*
End replication problem, 329
Energy. *See also* Catabolism; Metabolism
  anabolism consumption of, 245
  conservation in eukaryotic cells, **95–97**
  cycle of, in cells, 194f
  definition of, *191*
  sources of, 209t, 210
Enhancers as transcription activator bindings sites, *339*
Enology, *847*
Enriched media, *144,* 145f
Enrichment culture, *146, 532*
Entamoebida, 476
Enteroaggregative *Escherichia coli* (EAEC), 799f, 800
Enterobacterales, 451–452
Enterobacteria/enteric bacteria, 451
Enterococcaceae, 463t, 466
Enterococcus, 177, 464
Enterococcus faecalis, 126, 364, 671, 674
Enterocytozoon bieneusi, 494
Enterohemorrhagic *Escherichia coli* (EHEC), 799f, 800
Enteroinvasive *Escherichia coli* (EIEC), 799f, 800
Enteropathogenic *Escherichia coli* (EPEC), 799f, 800
Enterotoxic *Escherichia coli* (ETEC), 799f, 800
Enterotoxins, *795*
Enthalpy, *192*
Entner-Doudoroff (ED) pathway
  in chemoorganotrophic archaea, 404, 405f
  glucose to pyruvate catabolized in, 213–214, 214f
  modified, in *P. ubique,* 442f
Entomopathogenic fungi, 490
Entropy, *191*
Entry of viruses into host, 109, 110f
Enveloped viruses, 104f, *105,* 107f, 109
Environmental microbiology, 17, *555,* **862–874**
  biodegradation and bioremediation, ***871–874***
  drinking water safety, **863–865**
  microbially powered fuel cells, **869–870,** 870f
  wastewater treatment, **865–869** (*See also* Wastewater treatment)
Enzyme-linked immunosorbent assay (ELISA), *730*–733, 732f
Enzymes
  activation energy lowered by, 201f
  amphibolic, 215
  β-lactamases (*See* β-lactamases)
  catalase, *137,* 137f
  catalytic proteins as, 5
  classification of, *200,* 200t
  in complement system, 611–614
  core, of RNA polymerases, *280*
  DNA polymerases, 272–273, 273f
  exoenzymes, 93
  induced fit model of function of, 201, 201f
  induction and repression of synthesis of, 300–301
  lysozyme, *58*
  in metabolism, 199–203
  reverse transcriptase (RT), *365*
  substrate concentration and, 201, 202f
  superoxide dismutase (SOD), 137f
  transaminase, *252*
Enzymology, 363–367
Eosin methylene blue agar, 145
Eosinophil chemotactic factor of anaphylaxis (ECF-A), *656*
Eosinophils, *616*
EPA (U.S. Environmental Protection Agency), 166, 863, 863t
EPEC (enteropathogenic *Escherichia coli*), 799f, 800
Epibiotic predators, *551,* 551f
Epidemics, *701*
  control and prevention of, **711–715**
Epidemiologist, *701*
Epidemiology, **700–716,** 701f
  bioterrorism readiness, ***715–716***
  evidence basis of, **701,** 701f
  methods in, **702–704**
  population patterns of infectious disease, **705–708**
Epifluorescence microscopy, 30, 30f, *536*
Epilimnion, *584*
Epiphytes, *591*
Episomes, *65, 356*
Epitopes, *635*
EPS (extracellular polymeric substances), *141*
Epstein-Barr virus (EBV), 512, 756, 757f
Epulopiscium fishelsoni, 46, 46f
Equilibrium, *192,* 192f
Equilibrium constant (K_{eq}), *192*
Equine encephalitis, *748*
ER (endoplasmic reticulum), *90,* 95

Ergot, *499,* 499f
Erysipelas, *793*
Erythema infectiosum, *756,* 756f
Erythema migrans, 779
Erythroblastosis fetalis, *658*
Erythrocytic cycle, 817f, 818
Erythromycin, 178, 179f
Escherichia coli, 121
acidic tolerance response of, 134
average size of, 46
bacteriocin produced by, 610
catabolite repression in, 311–313
cell cycle of, *123,* 123f, 124–125, 125f
chemotaxis in, 316–317
diauxic growth of, 313f
divisome of, *125,* 125f
fecal contamination of water and, 451
food-borne diseases caused by, 840
food intoxication by, 841–842
gastroenteritis from, 799–800
gastroenterits caused by, 841
glutamine synthetase regulatory enzyme of, 204, 205f
irradiation to control, 162
metabolism of pyruvate and formate by, 451f
methyl-directed mismatch repair in, 350, 350f
nucleoid and chromosomes of, 64f, 65, 65f
nutrient transport systems of, 49–52, 49t
specialized transduction in phage lambda of, 365
stringent response in, 314–315
T4 phage of, *106,* 106f
as water purification indicator organism, *864*
ESCRT proteins, 152
Estuaries, 571–572
ETEC (enterotoxic *Escherichia coli*), 799f, 800
Etest®, **173–174,** 174f
Etest for antibiotic susceptibility testing, 725
Ethylene oxide (EtO) to control microbes, 164, 165f
E (exit) tRNA binding site, *286*
Euglenozoa, 473–474, 473f
Eukarya domain, **3–4**
Eukaryotes, *2,* 533. *See also* Microbial growth
archaea cell cycle similarities to, 127
bacteria and archaea *vs.,* 402t
cell structure, **85–100**
ESCRT proteins of, 128
evolution of, 533
genome replication and expression, **326–340**
infection of, 114, 115f
replisomes in, 328t
signature proteins, *406*
Eukaryotic and archaeal genome replication and expression, **326–340**
in archaea, 340
bacteria protein folding similar to, 335
bacteria translation similar to, 335
cellular process regulation, **338–340**
commonalities, 327
DNA replications, **327–330**
localization and secretion, 335, 337
overview, 326
transcriptions, **331–334**
translation and protein maturation and localization, **335–338**
Eukaryotic cell structure, **85–100**
bacterial and archaeal cells *vs.,* 95
bacterial membranes *vs.,* 49
cell envelopes, 87–88
common features, 86–87
cytoplasm of, 88–89
energy conservation in, 95–97
microorganism examples, 86, 86f
motility structures, 99–100
nucleus, 94–95
overview, 2
ribosomes, 94–95
secretory and endocytic pathways, 90–93
ultrastructure of, 87f
Eukaryotic signature proteins, 406
Eumycetozoa, 476–477
Eumycotic mycetoma, *825*
Eutrophication, *566*–567, 572
Eutrophic environments, *139*
E-value, 389
Evolution of microorganisms, **1–21**
biological entities, 2–4, 2f
culture-based methods, 14–19
developments in, 14–19
disease and, 17
immunology, 19
Koch's postulates, 17
microbial ecology, 19
pure, 18
microbiology fields, 19–21
microscopy for study of, 14
molecular and genomic methods, 19
origin of life theories, 4–5
RNA basis of early life, 5–8
spontaneous generation theories, 14–17
three domains of life, 8–9
Exanthem subitum, *755*
Excision repair, *350,* 351f
Excystment, *472,* 472f
Exergonic reactions, *192*
Exfoliative toxin, *791*
Exit (E) tRNA binding site, *286*
Exocytosis, *628*–629
Exoenzymes, 228, 461, 559, 573, 782
Exogenous antigen processing, *639*
Exon sequences, *331*
Exospores, 432, 456, 456f
Exotoxins, *694*–696, 695f
human bacterial pathogens, 697t
Exponential phase in microbial growth curve, *128*–130, 130f, 130t
Exposporium, 69
Expression cassette, 427
Expression vectors, *371*
Extant organisms in origins of life studies, *5*
Extinction culture technique, *532*
Extracellular membrane vesicles, 59
Extracellular pathogens, *685*
Extracellular polymeric substances (EPS), *141*
Extracellular vesicles (EVs), ***58–59,*** 59f, 93
Extracytoplasmic function (ECF) sigma factors, *311*
Extreme environments, *131*
Extreme habitats, 75
Extreme halophiles, *412*
Extreme thermophilic S^0-reducers, 413
Extremophilic microorganisms, *131,* 402
Extrinsic factors in food spoilage, 838
Eyepiece, *25*
Eyes
corneal inflammation, *754,* 826
microbiota adaptations in, 670

F

5Fs (feces, fingers, food, fomites, and flies), 840
Facilitated diffusion, *50,* 50f, 88
FACS (fluorescence activated cell sorting), *534*

Factor-dependent transcription
termination, 282
Facultative anaerobes, *136*
Facultative intracellular pathogens, *685*
Facultative predators, 551
FAD (flavin adenine dinucleotide), *196*
Famciclovir, 755
Fatal familial insomnia (FFI), 118, 765
Fatty acids
synthesis of, 258–259
Fatty acid synthase complex, *258*
F′ conjugation, 357–358, 358f
Fecal coliform assays, *864*
Fecal enterococci, *864*
Fecal microbiome transplant, 805
Fecal microbiota transplant (FMT), 672
Fecal-oral route of food-borne illness,
840
Feedback inhibition to regulate enzymes,
204–205, 205f
Feline leukemia virus (FeLV), 504
Fermentation, 844
citric acid, 853
by *Clostridium tetani,* 470f
electron transport chain, 225–228
endogenous electron acceptor
for, *214*
heterolactic, *465*
homolactic, *465*
in industrial microbiology, 851
mixed acid, *228*
Pasteur's work on, 16
penicillin, 852, 852f
yeasts for, 492
Fermented foods, 849t
beer and ales, 847
bread, 847, 848f
cheese, 846–847, 846f
chocolate, 845
milk, 844–846
plant-based foods, 848
wine, 847
Fermenters, *856,* 857f
Ferredoxin Fe-S protein, *196*
Fe-S proteins, *196*
Feulgen procedure for DNA
staining, 33
Fever blisters, 753, 754
F factor conjugative plasmids, *355,*
355f
FFI (fatal familial insomnia), 118,
765
$F^+ \times F^-$ mating: plasmid only transfer,
355–356, 356f
Fifth disease, *756,* 756f
Filamentous bacteria, 447–448
Filaments
actin, *89*
in actinobacteria, 456
in bacterial flagella, 67, 68f
flimmer filaments on flagella, 99
hormogonia, *421*
intermediate, *89*
tricome, *421*
Filopodia, *475*
Filtration to control food spoilage, 840
Fimbriae, ***66***
First law of thermodynamics, *191*
FIs (fusion inhibitors), 181
Fish and meat, fermented, 841
FISH (fluorescent in situ hybridization)
technique, *536*
Fixation
5′ cap nucleotide, *332*
associative nitrogen, 592
ATP (adenosine triphosphate) in
nitrogen, 253
in carbon cycle, 558
chemical, *33*
CO_2 (*See* Carbon dioxide fixation)
complement, 719f
heat, *33*
nitrogen (*See* Nitrogen fixation)
primary production, *570*
Flagella, *47, 66*
in bacteria, **66–68,** 68f
distribution, 66f
eukaryotic, 99–100, 99f
heterokont, *483*
lophotrichous, *66,* 66f
in Micromonosporales, 459
microtubules, 89
periplasmic, *427*
peritrichous, 66f, *67*
staining of, *35,* 35f
whiplash and tinsel, *99,* 99f
Flagellin protein, *67,* 67f
Flagellum, *66*
Flat warts, *759*
Flavin adenine dinucleotide (FAD),
196
Flavin mononucleotide (FMN), *196,*
307
Flavobacterium, 295
Flavobacterium johnsoniae, 428
Fleming, Alexander, 20, 171, 244
Flimmer filaments, 99
Flocculation in water purification, 863
Florey, Howard, 171
Flotillins, *49*
Flow cytometry, 149, *534,* 534f, *735*
Fluorescence activated cell sorting
(FACS), *534*
Fluorescence labeling, 371, 373f
Fluorescence microscopy, *30,* **30–31**
Fluorescent dyes and tags, 31f
Fluorescent in situ hybridization (FISH)
technique, 536, 537, 537f, 540
Fluorescent light, *30*
Fluorescent stains, 726
Fluorochromes, *30,* 30t
Fluoroquinolone drugs, *180,* 180f
FMN (flavin mononucleotide), *196,*
307
Focal length, 24
Folic acid, 179, 180, 183
Fomite, *688,* 840
Food and Agricultural Organization
of the United Nations–World
Health Organization (FAO–WHO),
681
Food and water as disease vehicles.
See also Food microbiology
acute bacterial diarrhea, 796t
for bacteria, **795–802**
bacteriophage spray to disinfect,
168
botulism, *795,* 795f, 797
Campylobacter jejuni gastroenteritis,
797
cholera, *797*–799
Escherichia coli gastroenteritis,
799–800
food poisoning (*See* Food poisoning)
gastroenteritis, 760–761, 761t
hepatitis A, *761*
hepatitis E, *761*
irradiation to disinfect, 162
poliomyelitis, 762–763
Salmonella gastroenteritis, 800
Shigellosis, *801*
Staphylococcal food poisoning,
801–802
typhoid, 801
for viruses, **760–763**
waterborne pathogens, 797t
Food intoxication, *795,* 841–842
Food microbiology, **836–849**
beer and ales, 847
bread, 847, 848f
cheese, 846–847, 846f
chocolate, 845
disease outbreaks from, 842, 843f
distilled water, 839
fermented foods, **844–849**

field of, 19
government-industry cooperation for detection, **842–844**
irradiation, 162
meat and fish, 841
plant-based foods, fermentation of, 848
probiotics, 681
spoilage, control of, **838–840**
spoilage from microbial growth, **837–838**
wine, 847
Food poisoning
Clostridium perfringens as cause of, 69
overview, *795,* 796t
staphylococcal, 801–802
Food Safety Modernization Act of 2001, 842
Foraminifera, 478–479, 479f
Foreignness, recognition of, 624–626, **636–639**
Forespore, 69
Formaldehyde to control microbes, 164
Formate dehydrogenase (FDH), 451
Forward mutations, *346,* 346t
Foscarnet antiviral drug, 181
Fossil fuel combustion, 558
F′ plasmid, *357*
Fracastoro, Girolamo, 14
Fragment alignment in genome sequencing, *385*
Frameshift mutations, *347*
Frankia, 599
Free energy change, *192*
Freeze-etching procedure, 38, 38f
Freshwater ecosystems, 570–571
Fruiting bodies, *432*
in bacteria, 45f, 46
in mycobacteria, *457*–460, 458f
Frustrated phagocytosis, 694f
Frustule, *484*
FtsZ, 124–126
Fucoxanthin, 484
Fuel cells, microbially powered, **869–870,** 870f
Fueling reactions, *209,* 210, 211f
Fuels, food waste, 414
Fumonisins, 500, *842*
Functional genomics, ***390–395***
DNA-protein interactions and, 395
in mutation screening, 348
overview, 390
proteomics, *393*–395
transcriptome analysis, *391*–393
Functional membrane microdomains, *49*
Fungal distribution and importance, 491–492
Fungal highway, *549,* 550f
Fungal reproduction, 493–494
Fungi, 2, **490–502**
airborne transmission of diseases, 813–815
antifungal drugs, **182–183**
Ascomycota, 497–499
Basidiomycota, 501–502
branches of, 491f
Candida albicans, 143
Chytridiomycota, 495
classification of, 492t
culture-based methods to identify, 725, 726t
diversity of, 491–494
as eukaryotic group, 85
example of, 85
fluorescence microscopy to identify, 730
in food spoilage, 837
Glomeromycota, 496
growth and reproduction, 493–494
human diseases from, 811–834 (*See also* specific diseases and pathogens)
medically important fungi, 812t
opportunistic diseases, 829–834
overview, 812–813
routes of transmission, 812f
media to isolate, 723t
Microsporidia, 494–495
mycotoxins from, *696*–697
overview, *3*
plant diseases caused by, 602t
silkworm disease caused by, 17
in soils, 588
structure, 492–493
Zygomycota, 495
Fungicides, *157*
Fungistatic substances, *157*
Fusion inhibitors (FIs), 181
Fusobacterium, 429
spindle shape, 429f
Fusobacterium nucleatum genome, 429

G

γδ T cells, *640*
Galen, 17
Galileo, 14
GALT (gut-associated lymphoid tissue), 609, *623,* 671, 676
Gambiense human African trypanosomiasis, 820
Gametangia, *494*
Gamma-butyrolactone (A-factor), 454
Gamma irradiation, 138, 162
Gammaproteobacteria, **443–452**
Burkholderiales, 444–446
characteristics of, 444t
Enterobacteriales, 451–452
in iron cycle, 561
Legionellales, 449–450
Methylococcales, 448–449
Pseudomonadales, 450–451
Thiotrichales, 447–448
Vibrionaceae, 451
Gamonts, *472*
Ganciclovir, 181
Gap closure in genome sequencing, *385*
GAS (group A streptococci), *777*
Gas gangrene, *782,* 783f
Gastric cancer, 433
Gastric ulcers, 126
Gastritis, 433, *784*
Gastroenteritis
campylobacters in, 433, 797
Escherichia coli, 610, 799–800
overview, 799
Salmonella, 800
Vibrio parahaemolyticus, 797t
viral, 760–761, 761t
Gastrointestinal anthrax, *802*
Gastrointestinal tract, 671, 773. *See also* Gut microbiome
Gas vacuoles, 62f, *63*
Gas vesicles, *63*
G + C content
in DNA pairing, *531*
Gel electrophoresis
in recombinant DNA technology, 364, *364*
two-dimensional, *393*
Gell-Coombs classification, *655*
Gene expression, 187, 203–204, 299
in archaea, 340
colliding polymerases, 290–291
coupled transcription/translation, 291–292
Gene ontology, *389*
Generalized transduction, *361,* 362f
Generally recognized as safe (GRAS) chemical food preservatives, *839*
Generation time, *129,* 131f, 131t

Genes
community function and metabolic diversity assessed by, 539
constitutive, *300*
housekeeping, *300,* 536
inducible, *300,* 302f
monocistronic, *331*
as polynucleotide sequence, *277*
protein-coding, 277–278
repressible, *300,* 300f
Genetically modified (GM) materials, 836
Genetic code, 282–285, 284t
Genetic complementation, *370–371*
Genetic variation, **343–359**
antibiotic resistance in bacteria, 343, 359
conjugation, 355–358
DNA repair, 347–350
environmental DNA, 350–352
microbes, 350–352
mobile genetic elements (MGEs), *353*–358
mutation as heritable change, 344–347
transduction for virus-mediated DNA transfer, 354–355
transformation from free DNA uptake, 352, 352f
transposable elements to move genes, 358
Gene transfer agents (GTAs), 358
Genital herpes, *755*
Genitourinary tract, 609, 671–672
Genome annotation, *389*
Genome editing
with Cas9 Nuclease, 374
Genomes
segmented, *107*
Streptomyces coelicolor, 454
viral, 107
Genome sequencing, 415
automated Sanger method of, 381, 382f
bridge amplification in, 383
chain-termination DNA, *381*
coverage in, *386*
deep sequencing, 386
editing in, 374–376, 375f
food-borne pathogen detection, 842–844, 843f
fragment alignment and gap closure in, *385*
horizontal gene transfer (HGT) shown by, 9
massively parallel sequencing techniques, *381*
in microbial genomics, 385–388
next-generation, *381*–384, 539
open reading frame (ORF) in, *389*
overview, 380
proofreading in, 386
Sanger method of, 381
single-cell, *386*
whole-genome shotgun sequencing, *385*
Genomic analysis, 13
Genomic Encyclopedia of Bacteria and Archaea Project, 388
Genomic islands, *397*
Genomic libraries, *370,* 370f
Genomic magnetosome island (GMI), 439
Genomic reduction, *551*
Gentamicin, 460
Geobacter metallireducens, 430, 561
Geobacter spp., 563, 867
Geographic information systems (GIS), *704*
German measles, *745,* 745f
Germfree animals, *672*
Germinant receptors, 70
Germination of spores, *70*
Germ theory of disease
Lister's experiments on, 17
Spallanzani's work on, 14
Tyndall's experiments on, 16–17
Gerstmann-Strässler-Scheinker syndrome (GSS), 118
GFP (green fluorescent protein), 299, *371*
GHG (greenhouse gas), 74, *565*–*567,* 566f
Ghon complexes, *774*
Giardia intestinalis, 474–475, 474f
Giardiasis, 828–829
Gibson assembly, 370, 370f
Gibson, Daniel, 370, 370f
Giemsa-stained erythrocytes, 819
Gilbert, Walter, 5, 381
Gingivitis, 671
Glaciers, 581
Glide (Gld) proteins, 428
Gliding motility
of filamentous cyanobacteria, *428*
of *Flavobacterium johnsoniae,* 428
of *Mycoplasma mobile,* 468, 468f
of *Myxococcus xanthus,* 432
Global climate change, 565
Global regulatory systems for cell processes, *310*
Glomeromycetes, *496*
Glomeromycota, 496, 594
Glomerulonephritis, 466, *778*
Gluconeogenesis, *248*
Glucose to pyruvate pathway
aerobic respiration, 212–218
in chemoorganotrophic archaea, 404
Glutamine synthetase-glutamate synthase (GS-GOGAT) system, *252,* 595
Glutaraldehyde to control microbes, 164, 165f
Glycans, 428
Glycerol diether lipids, *77*
Glycocalyx, *60,* 88
Glycogen, *420*
Glycolytic pathways, 212, 213, 246
Glycomics, 394
Glyoxylate cycle, *256,* 256f, 404, 404f
GM (genetically modified) materials, 836
GOGAT (glutamine synthetase-glutamate synthase) system, 595
Gold, Thomas, 604
Golgi apparatus, *90,* 90f
Golgi, Camillo, 816
Gonorrhea, *786*–*787,* 786f
Graft-*vs.*-host disease, 662f, *663*
Gram, Christian, 34, 53
Gramicidin, 463
Gram-negative bacteria (diderm bacteria)
ampicillin effective against, 176
for biological control of microbes, 168
cell envelope of, 53f
cell structure of, 53, 55–58
Chlamydiae, *424*
differential staining for, 34–35
DNA uptake systems of, 359, 361f
facultatively anaerobic rods, 451
lipopolysaccharides in membranes of, 260–262, 262f
meropenem effective against, 176
monoderms and diderms, 57f
outer membrane of, *55*
pertussis, 776
phosphorelay signal transduction system in, 319
photosynthesis by, 417t
plague, *780*
protein secretion in, 292–295, 293f, 294f
quorum sensing, 318–319
riboswitches in translation in, 309
second membrane, **55–58**
selective media for, *144*
septic shock induced by, *696,* 698f

Gram-positive bacteria, 34–35. *See also* Actinobacteria; (monoderm bacteria)
- Actinobacteria, 455–461 (*See also* Actinobacteria)
- bacterial conjugation in, 364
- cell envelope of, 53f, 56f
- cell structure of, 55
- differential staining for, 34–35
- diphtheria caused by, 770
- DNA uptake systems of, 359, 361f
- food intoxication by, 841–842
- monoderms and diderms, *53*
- penicillin G effective against, 176
- phosphorelay signal transduction system in, 319
- in phylum Actinomycetota and Bacillota, 55
- protein secretion in, 292–295, 293f, 294f
- riboswitches in transcription in, 309
- selective media for, *144*
- streptococcal pneumonia, *807*
- streptococci, 777

Gram stain, *34*, 34f
- of cerebral spinal fluid, 23
- as differential stain, 34f
- process of, 34–35, 34f

Gram staining
- antibiotic therapy determined by, 726
- of bacteria, 53

Granular sludge, *866*
Granulocytes, 616
Granulomas, 631, 631f
GRAS (generally recognized as safe) chemical food preservatives, *839*
Grazing, 551
Great plate count anomaly, *532*
Green algae, 486–487, 487f
Green fluorescent protein (GFP), 31, 31f, 299, *371*
Greenhouse gas (GHG), 74, *565*–567, 566f
Green nonsulfur bacteria, *419*
Grey mold fungus, 601
Griffith, Fred, 266, 266f, 359
gRNA (guide RNA), *373*
Group A streptococcal (GAS) disease, 44, *777*
Group B streptococcal (GBS) disease, *783*
Group translocation, ***51–52***
Growth arrest, *139*
Growth curves, **128–131**
Growth factors, *50*
Growth rate constant *(k)*, *130*, 130f
GS-GOGAT (glutamine synthetase-glutamate synthase) system, *252*, 595
GSS (Gerstmann-Strässler-Scheinker syndrome), 118
Guanidine pentaphosphate (pppGpp), 311, 312f, 314–315
Guanine, *257*
Guanosine monophosphate, 257, 257f
Guanosine tetraphosphate (ppGpp), 311, 312f, 314–315
Guanosine 5′-triphosphate (GTP), 194
Guide RNA (gRNA), *373*
Guillain-Barré syndrome (GBS), 433, 747, 797
Gummas, *787*
Gut-associated lymphoid tissue (GALT), 609, *623*
Gut-brain axis, 676–677, 677f
Gut microbiome, 678t
- adaptations of, 667
- *Akkermansia muciniphila* in, 424
- bacteria in, 3
- Bacteroidota, 427–429
- *Bifidobacteria* in, 456–457
- *Clostridioides difficile* in, 666, 671
- colonization resistance, 673, 674
- immunity and, 673–676
- neurodevelopment and, 676, 677
- obesity related to, 681
- Verrucomicrobiota, 423–425

H

HAB (harmful algal bloom), *572*, 572f
Haber-Bosch reaction, 566
Haber, Fritz, 566
Haemophilus influenzae
- capsules, 693
- immune system impact of, 693
- in pharynx, 670
- rifampin effective against, 180

Haemophilus influenzae serotype b (Hib), 772–773
Hair shaft infections, 823
Hallucinogenic alkaloids from fungus, 501, 837
Haloarchaea, *411*, 412f
Halobacterium NRC-1, 412, 412f
Halobacterium salinarum, 240, 412
Halogens for microbe control, 163
Halophiles, 76, *132*, 412–413
Haloquadratum walsbyi, 75
Halorhodopsin, 412
Halothiobacillus neapolitanus, 63f
Halotolerant microbes, *571*
Hami, 82, 82f
Hantavirus pulmonary syndrome (HPS), *764*
Haptens, *635*
Haptista clade, 486
Harmful algal bloom (HAB), *572*, 572f, 584
Hartig net (hyphae meshwork), *593*
HAT (Human African Trypanosomiasis), 820–822
Hata, Sahachiro, 171
Haustorium, 600
Hayes, William, 356
HDCV (human diploid-cell rabies vaccine), 765
Health-care-acquired infections, *710*
Healthcare-associated infections (HAI), *710*
Heat fixation, *33*
Heat killing of microbes, 157t
Heatley, Norman, 171
Heat-shock proteins (HSPs), 311, 335
Heat-stable protein (HPr), 52
Heat to control microorganisms, 159–161, 160t
Heine, Jacob von, 762
Helical capsids, *105*
Helicases, *272*
Helicobacteraceae, 433
Helicobacter pylori, 433
- as cause of disease, 784–785, 785f
- cell shape, 126
- as Class I carcinogen, 784
- pathogenicity, 433

Heliophilum, 470
Helix-turn-helix DNA-binding proteins, 301
Hemagglutination, *729*
Hematopoiesis process, *614*
Hemolytic uremic syndrome (HUS), 800
Hemorrhage, 747, 763
Henderson, Richard, 39
Henle, Jacob, 17
Hepadnaviridae, 757, 757t
HEPA (high-efficiency particulate air) filters, *159*
Hepatitis
- as Chikungunya virus complication, 747
- hepatitis A (infectious), *761*
- hepatitis B, 525, *757*, 757t, 758
- hepatitis C, 757t, *758*
- hepatitis D, 757t, 758
- hepatitis E, *761*
- viral, 757–758

Hepatitis C virus (HCV) infection, 182
Hepatitis delta virus (HDV), 757t, 758, *758*
Herbs and spices for food preservation, 838
Herd immunity, *707,* 707f, 708, 711, 749
Hermatypic (reef-building) corals, 546–548
Herpes viruses
  acyclovir to treat, 181, 754
  autophagy elusion, 628
  cold sores, 753–754
  congenital (neonatal) herpes, *755*
  corneal inflammation from, *754*
  Epstein-Barr virus (EBV), 756, 757f
  eye infections from, 754
  genital herpes, *755*
  herpes labialis, 754
  herpes simplex, 754
  Human gammaherpesvirus 8-induced Kaposi's sarcoma, 751
  idoxuridine, for herpes eye infections, 754
  keratitis, *754*
  miscarriage and, 755
  shingles (herpes zoster), 738, 739f
  trifluridine, for herpes eye infections, 754
Herpetic keratitis, *754*
Hershey, Alfred, 267, 268f
Heterocysts, *421,* 422f
Heterokont flagella, *483*
Heterolactic fermentation, *228, 465, 466*f
Heterologous gene expression, *371*
Heterotrophs, *209*
Hexamida salmonis, 475
Hfr conjugation, 356–*357,* 359f
HHP (high hydrostatic pressure) for food preservation, 839
HHV-5 (Human betaherpesvirus 5) infection, 754. *See also* Herpes viruses
Hierarchical cluster analysis, *392*–393, 392f
High-efficiency particulate air (HEPA) filters, *159*
High hydrostatic pressure (HHP) for food preservation, 839
High multiplicity of infection (MOI), 114
High phosphate transfer potential, 193, 193f
Hippocrates, 816
His-tagging, *371*
Histamine, 616, 629
Histatin peptides, *610*
Histones, *81,* 94
Histoplasmosis, *814*–815, 815f
Historical microbiology
  person-to-person spread of disease, 687
  polio, 763
HIV/AIDS
  anti-HIV agents to block replication, 181, 182f
  Bloodborne Pathogens Act of 1991 and, 719
  cryptococcosis and, 832
  global spread of, 750f
  host defenses suppressed by, 692–693
  infection stages in, 753f
  Kaposi's sarcoma in, 512t
  life cycle, 752f
  opportunistic infections in, 751
  pandemic, 737
  public health concern, 20
  as retrovirus example, 523–524, 523f
  transmission of, *749*–753
  virion of, 751f, 752f
  virus attachment to host cells, 108–109
H1N1 influenza pandemic (2009), 742
H7N9 influenza virus, 742
Holdfast, *439*
Holobionts, 546
  humans as, ***667***
Holoenzymes, *200*
Holozoic nutrition, *471*
Homeostasis
  metabolism regulated for, **203–205**
Homolactic fermentation, *227, 465*
Homologous recombination of genes, *353, 374*
Hong Kong influenza virus, 742
Hook, 67
Hooke, Robert, 14
Hopane molecules, 5
Hopanoids, *49,* 49f
Hops, 847
Horizontal gene transfer (HGT)
  for asexual-based variability, 9, ***352–353,*** 353f, 397
  bacterial and archaeal diversity increased by, 9
  drug resistance transmitted by, 363–364
  heritable genetic changes from, 344–348, 344f
  levels of, *397*
  in marine ecosystems, 577
  natural transformation of, 359
Hormogonia filaments, *421*
Horseradish peroxidase (HRP), 537
Host cells, 371–373
Host damage, **693–698**
Host defenses, surviving, **692–693**
Host metabolism, microbiome, 672–673
  impact fermentation end products, concept map of, 673f
Host resistance, innate, **606–631**
  adaptive immunity (*See* Adaptive immunity)
  barriers as, 607–609
  cell-specific functionality, 616–620
  chemical mediators for, 610–616
  immune system cells, 616–620
  immunity from, 607
  inflammation, 629–631
  organs and tissues as sites of defense, 621–623
  phagocytosis, 624–629 (*See also* Phagocytosis)
Host susceptibility to disease, *685*
Host-*vs.*-graft disease, *661,* 662f
Housekeeping genes, *300,* 536
HP/HB (3-hydroxypropionate/4-hydroxybutyrate) cycle, 402–403, 403f
HPS (hantavirus pulmonary syndrome), *764*
HRP (horseradish peroxidase), 537
HSP (heat-shock protein), 311, 335
Huanglongbing (HLB), 603
Human African Trypanosomiasis (HAT), 820–822
Human betaherpesvirus 5 (HHV-5) infection. *See* Herpes viruses
Human *Brucella* infections, 803
Human cytomegalovirus, 754–755
Human diploid-cell rabies vaccine (HDCV), 765
Human enterovirus C, 762
Human immunoglobulin classes, physiochemical properties of, 647, 647t
Human leukocyte antigen (HLA) complex, *637*
Human Microbiome Project (HMP), 667
Human papillomavirus, 759–760, 759f
Human parvovirus B19 infection, *756*
Humans
  gut microbiome (*See* Gut microbiome)
  microbe ecosystem in, 666–681 (*See also* Microbiome)
Human symbionts, 424–425

Humidity as extrinsic factor in food spoilage, 839
HUS (hemolytic uremic syndrome), 800
Hydrocarbon-degrading microorganisms, 871, 872f
Hydrocarbons, bioremediation, 871
Hydrogen hypothesis, *9*
Hydrogen, interspecies transfer, *545*
Hydrogenosomes, *97,* 97f, 474
Hydrophilic, *48*
Hydrophobia, 765
Hydrophobic, *48*
Hydrothermal vents, 7
  microbial communities, 578
  Pyrolobus fumarii at, 407
Hydroxylamine oxidoreductase (HAO), 442
3-Hydroxypropionate bi-cycle, 419, 419f
3-Hydroxyproprionate/4-hydroxybutyrate (HP/HB) cycle, 402–403, 403f
Hypersensitivities, 655–661
  bronchopulmonary aspergillosis, 831
  type I, 655–658, 657f, 831
  type II, *658*
  type III, *658,* 660f, 831
  type IV, *658,* 659, 661
Hyperthermophilic archaea and bacteria, 136
Hyphae, *45,* 45f
  in Actinobacteria, 455t, *456*
  appressoria, *496*
  ascogenous, *498*
  fungal, *492,* 493f
Hyphomicrobium
  budding, 122f
Hypolimnion, *584*
Hypothetical propagated epidemic, 706, 706f
Hypovirus, 602

I

Iatrogenic Creutzfeldt-Jakob disease (CJD), 765
Ice-dwelling psychrophilic microbes, 581
ICEs (integrative conjugative elements), 363
ICMs (intracellular membranes), 437, 443
Icosahedral capsids, *105*
ICTV (International Committee on Taxonomy of Viruses), 505
ID_{50} (infectious dose 50%), *689*
Idoxuridine, 754
IgA immunoglobulin, *648,* 648f
IgD immunoglobulin, *648,* 648f
IgE immunoglobulin, *648,* 648f
IgG immunoglobulin, *647*
IgM immunoglobulin, *648,* 648f
Ignicoccus hospitalis, 406f, 407
ILC (Innate lymphoid cells), *618*–620
Illness period, *685*
IMB (immunomagnetic bead) technology, 722, 723f
Immobilization of nutrients, *558*
Immune complex formation, 650–*651*
Immune synapse, *641*
Immune system
  adaptive immunity, **655–663**
Immunity. *See also* Adaptive immunity; Host resistance, innate
  cells of immune system, 19, **616–620**
  gut microbiome and, 673–676
  herd, *707,* 707f, 708, 711, 749
  identification of microorganism by immune response, **729–735**
  innate and environmental factors affecting, 668t
  organs and tissues of, 621–623
  overview, *607*
Immunization through vaccines, *742–745, 711,* 712. *See also* Vaccines
Immunoblotting (Western blotting), *733*
Immunodeficiencies, *663. See also* HIV/AIDS
Immunodiffusion, *734*
Immunofluorescence, 731–732, 731f, 733
Immunogen, 635
Immunoglobulin (Ig), *646. See also* Antibodies
Immunoglobulin γ (IgG), *647*
Immunohematology, 659f
Immunology, 19. *See also* Clinical microbiology
  culture-based methods and, 14
  as major field in microbiology, 19
  overview, 607, *607,* 608f
Immunomagnetic bead (IMB) technology, 722, 723f
Immunoprecipitation, *733*–734, 734f
Immunosuppressive drugs, 492
Impetigo, *793*
Inactivated vaccines, *712*
Incidence of disease, *704*
Inclusions in bacterial cells, *47,* **62–64,** ***62–64,*** 62f
Incubation period, *685*
Indels (insertion/deletion), *531*
Index case in epidemics, *701*
Indian Medical Service, 816
Indicator organisms in water purification, 864
Indirect evidence of origins of life, 4–5
Indirect immunofluorescence, *730*
Induced fit model of enzyme activity, 201, 201f
Induced mutations, *344,* 345–346
Inducers, enzyme levels affected by, *300*
Inducible genes, *300,* 302f
Induction and repression of enzyme synthesis, 300
Industrial microbiology and biotechnology, 20, **851–860**
  agricultural, **858–859,** 858f
  amino acids, 853
  antibiotics, 852–853
  biofuels, ***859–860,*** 860f
  challenges of, **856–857**
  microbes as products, **852–854,** 852t
  microbial cells, **854–856**
  organic acids, 853
  vaccines, 853–854
Infantile paralysis (polio), *762*
Infant microbiome development, 667
Infections, *685. See also* Epidemiology; Pathogenicity; specific disease or pathogen
  diagnosis of (*See* Clinical microbiology)
  of eukaryotic cells, 114, 115f
  food-borne, *840*–841
  hospital-acquired (*See* Health-care-acquired infections)
  invading organisms, 689
  laboratory-acquired, 720
  monoclonal antibody therapy for, 656, 656f
  mycobacterial, 773–776
  process overview, 685–686, 685f
  sexually transmitted (STIs) (*See* Sexually transmitted diseases (STDs))
  treatment of (*See* Chemotherapy)
  urinary tract (*See* Urinary tract infections (UTIs))
  viral, 113–115, 115f
Infection thread, *598*
Infectious diseases, *705. See also* Epidemiology
  chain, 711
  description of, *685*
  emerging and reemerging, **708–709,** 709f
  in health-care facilities, **710**

Infectious diseases (*continued*)
measuring frequency of, 704
person-to-person spread of, 687
population patterns in, **705–708**
remote sensing for, 704
transmission of, 688f
Infectious dose 50% (ID_{50}), *689*
pathogenic microorganism, 689f
Infectious hepatitis (hepatitis A), *761*
Infectious mononucleosis, 757, 757f
Infectivity, *690*
Inflammasome, *626*
Inflammation
chronic, *629*–631
as defense reaction to tissue injury, *629*
eye, 747
in innate host resistance, **610–616**
innate lymphoid cells to regulate, *618*–620
Inflammatory response, 629
Influenza
budding, for virion release, 112, 112f
enveloped virus, with spikes, 104f, 107, 110f
virions of, 113f
viruses of, 521–523, 522f
Inhalation anthrax, 23
Inhibition zone diameter of chemotherapeutic drugs, 173, 174t
Initiation factors for protein synthesis, *286*
Initiator tRNA, *287*
Innate lymphoid cells (ILC), *618*–619
Inner membrane, 69
Inorganic nitrogen assimilation, 252–253
Insecticides, 466
Insect-microorganism mutualism, 545–546
Insertion sequences (IS elements), *354*
In silico analysis, *389*
Insulin production, 363
Integral membrane proteins, *49*
Integrase, *509*
Integrase inhibitors, 181
Integrative conjugative elements (ICEs), 363
Integrins (selectin receptors), *629*
Intercalating agents, *346*
Interferons, *614*
Interleukins (ILs), *614,* 818
Intermediate filaments (IFs), *89*
International Committee on Taxonomy of Viruses (ICTV), 505
Interspecies hydrogen transfer, *411,* 545
Intertriginous candidiasis, 831, *831*
Intoxications, *694*
Intracellular bacteria, 442–443
Intracellular digestion in phagocytosis, 626–628
Intracellular membranes (ICMs), 437, 443
Intracellular pathogens, *685*
Intracytoplasmic membranes, **61–62,** 61f, 196
Intraepidermal lymphocyte, *623*
Intranuclear inclusion bodies, *754,* 755f
Intrinsic factors in food spoilage, 837–838, 837f
Introns, *331,* 332, 334f
In-use testing, *167*
Invasion factors, 689–692, 691t
Invasiveness, *690*
Iodophors for microbe control, *163*
Ionizable dyes, 33, 34t
Ionizing radiation, *138, 162,* 162f
Irish Potato Famine of 1846–1847, 601
Iron cycle, 563–564, 563f
Iron reduction, 563
Iron-sulfur proteins, *196,* 198f
Iron uptake in bacteria, **52**
Irradiated food, 162
IS elements (insertion sequences), *354*
Isoelectric focusing, *393*
Isoelectric point, *393*
Isoenzymes (isozymes), *205*
Isoleucine, 255, 255f
Isopentenyl pyrophosphate (IPP), 259
Isoprene, 77, 77f
Isoprenoid lipid synthesis, 259–260
Isotope fractionation, *539*
Ivanowski, Dmitri, 18

J

Jackson, David, 365
Jacob, François, 302
Jaundice, 758
Jenner, Edward, 19
Jirovec, Otto, 833
Joules, *191*
Jungle, The (Sinclair), 842

K

Kallikrein, *629,* 630f
Kanamycin antibiotics, 178
Kaposi's sarcoma, 751
Karenia brevis, 572
Kefir products, 844
Kennel cough, 445
K_{eq} (equilibrium constant), *192*
Keratinocytes, *608*
Keratitis, *754,* 826–827
Ketolides, 178
Kirby-Bauer method disk diffusion test, *173,* 174f, 725
Kirby, William, 173
Kissing bugs, 474, 821
Kitasato, Shibasaburo, 19
Koch, Robert, 14–15, 15f, 17, 146, 773
Koch's postulates, 17, 18f, 146
Koplik's spots, *743*
Krebs cycle, *217. See also* Tricarboxcylic acid (TCA) cycle
Kuru, 118, 765

L

Laboratory-acquired infections (LAIs), 720
Laboratory culture of microbes, **143–148**
aerobes and anaerobes cultivation, 145
culture media, 144–145
new approaches, 147–148
pure cultures, enrichment and isolation of, *145*–147
on solid media, 144
Laboratory detection of infectious diseases, 719–721. *See also* Clinical microbiology
Labyrinthulids, 486
Lactic acid bacteria (LAB), 465, *844,* 845
Lactobacillales, 464–467
Lactobacillus species, 463t, 464
Lactococcus, 45
Lactococcus lactis, 846, 846f
Lactoferrin, 610
Lactose operon, 302–303, 302f–303f
Ladderane lipids, 424, 424f
Lagging strand of DNA, 274, 274f, 329
Lag phase in microbial growth curve, *128*
LAIs (laboratory-acquired infections), 720
Lakes
blooms in, 572, 584f
microbial community in, 582t
permanently frozen, 581
Laminar flow biological safety cabinets or hoods, 159, 160f
Lancefield grouping system, 466
Landsteiner, Karl, 762
Langerhans cells, 621
Last universal common ancestor (LUCA), 8
Lateral flow assays, 733, 733f
Latex agglutination tests, 729
Laveran, Charles Louis Alphonse, 816
Leader portion of gene, 279, 286
Leading strand of DNA, 274, 329

Lectin complement pathway, 613
Lederberg, Joshua, 356
Leeuwenhoek, Antony van, 14, 16f, 473, 806
Leghemoglobin (LegH), *377, 598*
Legionella-containing vacuole (LCV), 449
Legionellales, 449–450, *770*
Legionella pneumophila, 45, 121, 449
Legionnaires' disease, *770–772,* 772f
Leguminous plants, rhizobia and, 597–599
Leibig, Justus van, 559
Leibig's law of the minimum, 559
Leishmania, 474
Leishmaniasis
  arthropod transmission of, *819*–820
  effects of, 819f
  life cycle of, 820f
Lenses, **24,** 26t
  function of, 24f
Lentic systems, *581*
Lentivirus, 523
Lepromatous (progressive) leprosy, *784*
Leprosy
  comparative genomics and, 396
  description of, *783*–784, 784f
  Mycobacterium leprae as causative agent of, 18, 458
Leptothrix sheath, 445, 445f
Lethal dose 50% (LD_{50}), *689*
Leuconostoc, 466, 466f
Leucothrix mucor, 448f
  life cycle and morphology of, 448f
Leukocytes, 614, 616
Lichens, 549, 550f
Life cycle of viruses, **108–113**. *See also* specific pathogens
  assembly, 111–112, 111f
  attachment, 108–109
  entry into host, 109, 110f
  influenza virus, 521, 522f
  rotaviruses, 517f
  synthesis, 109, 111
  tobacco mosaic virus, 520
  virion release, 112, 112f
Ligand molecules in endocytosis, *92*
Light-emitting diode (LED), 30
Light microscope, **24–33**
  bright-field, 24–25
  confocal, 31
  confocal microscopy *vs.,* 32f
  dark-field, 27
  differential interference contrast (DIC), 29–30
  electron microscopy *vs.,* 36f
  fluorescence, 30–31
  operation of, 26–27
  phase-contrast, 27–29
  resolution of, 27–29
  transmission electron microscopy *vs.,* 38t
Light reactions
  in anoxygenic photosynthesis, *239*–240
  of green sulfur bacteria, 234t, 239
  in oxygenic photosynthesis, *234*–239
Lignin, 559, *588*
Lignocellulose, *589*
Limnology, *569*
Lincosamide antibiotics, 179
Linker proteins, 420
Linnaeus, Carolus, 13
Lipid A, *56,* 260–262, *696*
Lipidomics, *394*
Lipid rafts, 88
Lipids
  anabolic synthesis of, 260–262
  in archaeal membranes, **77–79**
  in bacterial membranes, 49, 56
  bacterial *vs.* archaeal synthesis, 261t
  catabolism of, *229*–230
  in eukaryotic nuclear envelope, 94
  ladderane, 424, 424f
Lipopolysaccharides (LPSs), *56,* 57f, 260–262, 262f
Liposomes, 5
Liquid-liquid phase separation (LLPS), *63*
Listeriaceae, 467
Listeria monocytogenes
  actin tail of, 691, 691f
  disease outbreak due to contamination, 842
  food spoilage and, 838–840
  invasiveness, 691
  as pathogen, 466
Listeriosis, 840
Lister, Joseph, 14, 162
Lithoheterotrophy, *577*
Lithotrophs, *209*
Lithotrophy, 546
Littoral zone, *583*
Lobar pneumonia, 466
Lobopodia, *475*
Localization, 335–338
Localized anaphylaxis, *657*
Lockjaw, 794
Long-term stationary phase in microbial growth curve, *129*
Lophotrichous flagella, *66,* 66f
Lotic systems, *581*
Lower respiratory tract (LRT), 670
LPS biosynthetic pathway, 260–262, 262f
LSD (lysergic acid diethyamide), 499, 697
LUCA (last universal common ancestor), 8
Lucretius, 14
Luminescent bacteria, 142–143, 142f
Lumpy jaw disease, 456
Luria, Salvador, 320
Lyme disease
  Borrelia *burgdorferi* as cause of, 45, 45f, 127, 427
  climate change effect on, 555
  description of, *779*–780, 779f
Lymph nodes, *621*
Lymphocytes
  adaptive immunity, **652–655**
  B lymphocytes (*See* B cells)
  innate host resistance, *610*–616, 610f
  T lymphocytes (*See* T cells)
Lymphoid system, 621–623, 622f
Lymphoid tissues, *621*
Lymphoproliferative disorders, 751
Lysergic acid diethylamide (LSD), 499, 697
Lysine, 255, 255f
Lysing host cells, 116f
Lysis of protoplasts, *58*
Lysobacter, 591
Lysogenic conversion, *113*
Lysogenic cycle, *113*–114, 114f
Lysogens, *113, 354*
Lysogeny, *113, 354,* 509–510, 510f
Lysol disinfectant, 162
Lysosomes, ***90–91,*** 90f, *626*
Lysozyme, *58, 610,* 610f
Lytic cycle, *113, 354*
Lytic infections, 113–114, 114f

M

mAb (monoclonal antibody) therapies, 656, 656f, 745
MacConkey agar, 145, 145f
MacLeod, C. M., 266, 267f
Macroautophagy, *93*
Macrolide antibiotics, *178,* 179f
Macromolecular crowding, 60
Macromolecules, *245*
Macronucleus, *481*
Macronutrients, *49,* 49t
Macrophages, 609, *618, 623*
Mad cow disease, 4, 118

Madura foot, 825
Maduromycosis, *825*
Magnetic lenses, 36
Magnetosomes, *63,* 63f, 439, 439f
Magnetotactic bacteria, 563
Maintenance energy, *152*
Malaria
climate change and, 555
description of, *815*–819
drugs to treat, 183–184
geographical distribution of, 817f
history of, 816
impact of, 813
Plasmodium as cause of, 482–483, 483f
protists causing, *815*–819
MALDI-ToF (matrix-assisted laser desorption/ionization-time of flight) mass spectrometry, 530, 723
Malolactic fermentation in wine making, 847
MALT (mucosal-associated lymphoid tissue), *623*
Malt in beer brewing, 847
MamK, 64
MAMPs (microbe-associated molecular patterns), *624,* 625f
Mancini technique, 734
Manganese cycle, 564–565, 564f
Mannose-binding protein (MBP), *613*
Manson, Patrick, 816
MAP (modified atmosphere packaging) for food preservation, *840*
Marburg disease (MARV), *763*–764
March of Dimes, 762
MAR-FISH, 540
Mariana Trench system surface sediments, 579, 579t
Marine ecosystems
carbonate equilibrium system, 570, 570f
coastal, 571–573
as microbial habitat, 570–571
ocean acidification, 570–571, 571f
overview, 571
Marine snow, *573,* 574f
Marker gene analysis, *534*
MARV (Marburg disease), *763*–764
Mashing in beer brewing, 847
Massively parallel sequencing techniques, *381*
Mass spectrometry, *394,* 540
Mast cells, *616*
Mastigonemes, 99
Matrix-assisted laser desorption/ionization-time of flight (MALDI-ToF) mass spectrometry, 530, 723
Maturation of proteins, 292
Maxam, Alan, 381
Maximum contaminant level goal (MCLG) in water purification, 863–864
Maximum likelihood in phylogeny, 11
Maximum parsimony in phylogeny, 11
MBP (mannose-binding protein), *613*
McCarty, M. J., 267, 267f
McClintock, Barbara, 354
M cells, *623,* 624f
MCLG (maximum contaminant level goal) in water purification, 863–864
MCM (minichromosome maintenance), 327
MCS (multicloning sites), 365, 365f
MDA (multiple displacement amplification), *386*–387
MDR-TB (multidrug-resistant strains of tuberculosis), *774*
Measles (rubeola)
overview, *743*
schematic, life cycle, and symptoms, 744f
Measles, mumps, and rubella (MMR) vaccine, 743
Mechanical work of cells, *191*
Mechanosensitive channels, 132
Mediator, 610–616
Medical insurance, 140
Medical microbiology, 20
Medin, Oskar, 762
Mefloquine, 170
Megaviruses, 513
Meiosis, 353
Meister, Joseph, 19
Membrane-associated lipids, 48
Membrane attack complex (MAC), *612,* 614f
Membrane filters
procedure, 150f, *159*
results, 150f
sterilization, *159f*
water analysis, *864*
Membrane hopanoids, 49, 49f
Membranes
electron transport chain associated with, 136
Gram-negative bacteria, 53
plasma (*See* Plasma membranes)
Membrane steroids, 49, 49f
Membrane vesicles, 59
Memory cells, *634*
Memory in adaptive immunity, *634*
Meningitis
bacterial, *772*–773
meningococcal, *772*–773
Meningoencephalitis, 826–827
Mercury cycle, 564–565, 564f
Mercury poisoning, 565
MERS-CoV (Middle East respiratory syndrome coronavirus), *518,* 739
Mesophiles, *135*
Mesophilic ammonia oxidizers, 408
Mesophilic fermentation, 844
Metabolic antagonist drugs, **179–180**
Metabolic channeling, 203
Metabolic endotoxemia, *679,* 679f
Metabolic engineering, *371,* 396
Metabolic flux (metabolite turnover), *198*
Metabolic syndrome, 678–680
Metabolism, **190–205**. *See also* Anabolism; Catabolism
of archaea, 402–404
ATP, 193–194
biochemical pathways, 198, 199f
electron transport chains, 194–196
Embden-Meyerhof (EM) pathway (*See* Embden-Meyerhof (EM) pathway)
Entner-Doudoroff (ED) pathway (*See* Entner-Doudoroff (ED) pathway)
enzymes, 199–203
homeostasis and waste prevention, **203–205**
as microbial taxonomy characteristic, 529
nutritional types of, 209–210
overview, *191*
pentose phosphate pathway (*See* Pentose phosphate pathway)
principles of, 191
redox reactions, 194–196
ribozyme, 199–203
Metabolites
anaplerotic reactions to replace, 255–256, 256f
in industrial microbiology, 857
precursor, for anabolism, *210*
as reaction products, *369*
Metabolome, *673*
Metabolomics, *394*
Metagenomics, 405–406
human microbiome, 20
library, *370,* 370f

microbial community activity assessment, **534–538**
microbial interactions found by, 544
screening of libraries, 370f
uncultured microbe access through, *387*–388, 388f
Metalimnion, *584*
Metal instruments, 155
Metals to control microbes, 163
Metaorganism, 544
Metatranscriptomics, *393, 539*
Metchnikoff, Elie, 19
Methane
as greenhouse gas, 565
ruminant animal production of, 74
wastewater treatment production of, 867, 867f
Methane hydrates, *580*
Methanochondroitin, 79
Methanogenesis
biogeochemical cycling, 556, 557f
overview, ***409–411***
synthesis, 411f
wastewater treatment, 867
Methanogens, 401, 408–411
and methanotrophs
archaea as, 74, 401
coenzymes, 410f
Haloarchaea, *412*
Methanosarcina species, 79
Methanospirillum species, 79
Methanotrophs, 411, *448,* 449f
Methicillin-resistant *Staphylococcus aureus* (MRSA), 179, 185
antibiotic resistance, 185
description, *792*–793, 792f
Methionine, 255, 255f
Methyl-accepting chemotaxis proteins (MCPs), *316,* 317f
Methylaspartate pathway, 404, 404f
Methylococcales, 448–449
Methylotrophs, *448,* 449f
Verrucomicrobia, 423–425
Metronidazole, 185
MHC (major histocompatibility complex), *637*–638, 637f
antigen presentation, 638–639, 638f
Michaelis constant (K_m), *201*
Michaelis-Menten kinetics of ribozymes, 202f, 203
Microaerophiles, *136*
Microbe-associated molecular patterns (MAMPs), *624,* 625f
Microbes
filtration, 158–159
heat, 159–161
as products, **852–854,** 852t, 864
radiation, 161
Microbial adherence mechanisms, 690f
Microbial attachment mechanisms, 690t
Microbial DNA technologies
biotechnology, custom microbes, 375–376
Cas9 Nuclease, genome editing, 373–374
cloning technologies, DNA structure and enzymology, 363–367
expressing foreign genes in host cells, 371–373
genomic and metagenomic libraries, 370–371
polymerase chain reaction amplifies targeted DNA, 367–369
Microbial ecology, 20, ***528–541***
acid mine drainage, 445, 446
biogeochemical study of, 529
chloroplast evolution, *97*
clean rooms, 159
community activity assessment, **534–541**
cultures for, 14, **531–532,** 534
genetic methods, 536–537
Microbial energy conversion, *859*
Microbial fuel cells (MFC), *906,* 906f, ***869–870,*** 870f
Microbial genomics, 19, **380–398**
bioinformatics, 389–390
comparative, 396–399
DNA-protein interactions, 395
DNA sequencing methods, 381–385
genome sequencing, 385–387
Halobacterium NRC-1, 412, 412f
metagenomics, *387*–388
next-generation DNA sequencing methods, *381*–385
proteomics, 393–395
Sanger DNA sequencing methods, 381
Sulfolobus solfataricus, 407, 407f
systems biology, 396
transcriptome analysis, 391–393
Microbial growth, **138–143, 156–158**
aerobes and anaerobes cultivation, 145
archaeal and eukaryotic cell cycle similarities, 127
bacterial cell cycle, **123–127**
binary fission, *122*
cell growth and shape determination, 126–127
chemostats, *151*–152
chromosome replication and partitioning, 124
culture media, 144–145
cytokinesis, *124*–126
environmental factors affecting, **131–138**
growth curves, 128–130
laboratory culture, **143–148**
in natural environment, 138–143
new approaches, 147–148
oxygen concentration and, 136–137
pH and, 133–134
pressure and, 133
pure cultures, enrichment and isolation of, 145–147
radiation and, 137–138
rod-coccus cycle, 457, 457f
solutes, 132–133
temperature, 131t, 135–136, 135f
turbidostats, *152*
Microbial identification, **529–531**
Microbial interactions, **543–553**
antagonistic interactions, **550–553**
cooperation, **549–550**
human microbiome, **666–681** (*See also* Microbiome)
mutualism, **545–549** (*See also* Mutualism)
overview, 544, 544f
parasitism, 551
predation, 551
types, 544–545
Microbial loop, *573,* 574f, 578f
Microbial mats, 421, 580
Microbial populations, 529, 536
antimicrobial agent effectiveness and size of, 166
measuring size of, 149–150
Microbial product-based inhibition, 838
Microbiology. *See also* Clinical microbiology; Environmental microbiology; Food microbiology
development of, 14
fields in, 20–21
tools for, *14*
Microbiome, **666–681**
birth to adulthood development, **667–672**
cancer and, 680
cardiovascular disease and, 680
development of, 667–669, 668f
dysbiosis, **677–681**
in ears, 670
in eyes, 670

Microbiome (*continued*)
factors influencing, 678t
functional core microbiome, ***672–677***
in gastrointestinal tract, 671
(*See also* Gut microbiome)
in genitourinary tract, 671–672
gut-brain axis, 676–677, 677f
homeostasis and, 671, 672
host metabolism and, 672–673
humans as holobionts, ***667***
immunity and, 673–676
metabolic syndrome and, 678–680
microbiota adaptations in, *667*
normal microbiota, 670, 674
overview, *667*
in pharynx, 670
in respiratory tract, 670
on skin, 669–670
therapeutic manipulation, **681**
variation by body site, 669–672
Micrococcales, 456–457, 457f
Micrococcus, 45, 456, 457f
Microcompartments in bacteria, 62
Microcosms (incubation chambers), *539*
Microdomains, 48f, 49
of eukaryotic membranes, 88
Micrographia (Hooke), 14
Micromonospora echinospora, 178
Micromonosporales, 459–460, 459f
Picomonas judraskeda, 488
of Spirochaetes, 426–427
Micronucleus, *481*
Micronutrients, *50*
Microorganism-insect mutualisms, 545–560
Microorganisms, ***1–4***
building, 379
Microscopic resolution, limits of, 36f
Microscopy, **23–41**
bright-field, 24–25, 27, 28f
confocal, 31–32
dark-field, 27f
early, 11, 13f
electron, **35–40**
electron cryotomography, *39,* 39f
epifluorescence, 30f
fluorescence, 29–31, 730
lenses bending light for, 24, 24f
of living, unstained microbes, 27–29
for microorganism identification, 730
numerical aperture, 26f
phase-contrast, 28f, 29f
resolution of, 25–27, 35f
scanning electron, *37*–39
scanning probe, 39–40
staining specimens for, 33–34
transmission electron, 35–37
Microsporidia, 491f, *494*–495
Microsporidiosis, 832–833, 832f
Microtubules, *89,* 89f
Microvesicles, 93
Middle East respiratory syndrome coronavirus (MERS-CoV), *518,* 739
Miliary tuberculosis, *774*
Milk
fermented, 844–846
spoilage of, 846
Milky Way Galaxy, 1
Mimivirus, 513
Mineralization, *556*
Mineral soil, *588*
Minichromosome maintenance (MCM), 327
Minimal inhibitory concentration (MIC), *173,* 365
Minus-strand RNA viruses, *505,* 505f
Miscarriage, 755
Mismatch repair, *350,* 350f
Missense mutations, *346*
Mitchell, Peter, 220
Mitochondria
from endosymbiosis, **8–9**
in energy conservation, **95–97,** 96f
hydrogenosomes similar to, *97*
protein translocation from cytosol to, 335
Mitochondrial contact site and cristae organizing system (MICOS), 443
Mitosomes, *474*
Mixed acid fermentation, *228*
Mixotrophs, *581*
Mixotrophy, *471*
Microbial dark matter, 402
MMR vaccine (measles, mumps, and rubella), 743
Mobile genetic elements (MGEs), *353*–358
Mobilome, 358
Modified atmosphere packaging (MAP) for food preservation, *840*
MOI (multiplicity of infection), 114, 116
Molds. *See also* Antibiotics; Food microbiology; Fungi
in cheeses, 846
food spoilage, 837, 837f
thallus, *492*
Molecular chaperones, 95, 292
Molecular characterization in microbial taxonomy, 530–531
Molecular methods, 18
Molecular mimicry, 433, *797*
Monera kingdom, 2
Monobactam antibiotics, *176*
Monocistronic genes, *279*
Monocistronic mRNA (messenger), *279,* 280f
Monoclonal antibody (mAb) therapies, 656, 656f, 745
Monocytes, *618*
Monoderms, *53,* 53f, 59f, 415
Monod, Jacques, 302
Mononucleosis, 512, 757, 757f
Monosaccharide synthesis, 249
Monosodium glutamate (MSG), 853
Monotrichous bacteria, *66,* 66f
Morbidity rate, *701*
Mordant substance, *35*
Morphological characteristics, in microbial taxonomy, 529
Mosquitoes
Chikungunya virus transmitted by, 747
Dengue fever transmitted by, 747
West Nile fever (encephalitis) transmitted by, *748*
Mother cell, 69
Motility
actin filaments for, 89
actinobacterial, 456
archaeal, 81–83
bacterial cell structures for, **66–68**
eukaryotic, 99–100
gliding (*See* Gliding motility)
Mountain pine beetle, 85
mRNA (messenger)
monocistronic, *279,* 279f
pre-, 332
in protein construction, 7
riboswitches and, 308
subgenomic, *524*
total cellular (RNA-Seq), *392*
transcription for, 281f, 282f
transcription generating, *282*
mRNA vaccines, *714,* 714f
MSG (monosodium glutamate), 853
Mucociliary blanket, 609
Mucor fungus, 14
Mucoromycota, 495–496
Mucosal-associated lymphoid tissue (MALT), *623*
Mucosal tolerance, *675*
Mucous membranes as immune barrier, 608
Mueller-Hinton agar, 173
Mullis, Kary, 368

Multicloning sites (MCS), 365, 365f
Multidrug-resistant strains of tuberculosis (MDR-TB), *774*
Multiple displacement amplification (MDA), *386*–387
Multiplicity of infection (MOI), 114, 116
Multivesicular endosome (MVE), 93
Mumps, *743*
Must (crushed grape liquid), 847, 847f
Mutagens, *344,* 346t
Mutations. *See also* specific types of mutation
  Creutzfeldt-Jakob disease from, 765
  detection and isolation, 348–349
  in evolution, 9
  as heritable changes, 344–348
  industrial fermentation yields and, **856–857,** 857f
  from ionizing radiation, *138*
  tautomerization and transition, 344f
Mutualism, ***545–549,*** 547f
  microorganism-insect, 545–546
  mountain pine beetle and blue stain fungus, 85
  mycorrhizal fungi and plants, *496*
  with primary producers, 546–548
  in rumen ecosystem, 548–549
  sulfide-based, 547
  zooxanthellae, 479
Mutualists, *545*
Mycelium, *45, 456, 492*
Myc factors, *594,* 596f
Mycobacterial infections, 773–776
  related to tattooing, 784
Mycobacterium avium complex, 773
Mycobacterium bovis, 458
Mycobacterium leprae, 15, 121, 398, 783
Mycobacterium tuberculosis (Mtb), 121
  comparative genomics and, 396
  description of, 773–776, 776f
  Koch's work on, 17, 17f
  as pathogen, 458–459
  secretion, 302
  SEM, 39f
Mycobionts, *550*
Mycolic acids, *458*
Mycology, *490. See also* Fungi
Mycoplasma genitalium, 379, 552
Mycoplasmal pneumonia, 776
Mycoplasma mobile, 468
Mycoplasma mycoides, 379
Mycoplasma pneumoniae, 468, 776
Mycorrhizae, 592–596
  associations, *496,* 593t
  nitrogen and carbon exchange, 596f
  plant growth facilitated by, *592*
  root colonization by, 3, 493, 594f
Mycorrhizal fungi, 496, 592
Mycoses, *490. See also* Fungi
Mycotoxicology, *490. See also* Fungi
Mycotoxins, *696*–697
Myc protein, 595
Myocarditis, 747
Myositis, *793*
Myxobacteria, 45, 45f, *432*
Myxococcota, bacterial predators, 431–433
Myxococcus, 432, 551, 551f
Myxococcus xanthus
  life cycle of, 432f
Myxogastria, 476
Myxospores, *432*

N

NAC (nascent-chain-associated chaperone), 335
N-acetylglucosamine (NAG), 53, 54, 260, 261, 427, 610, 610f
N-acetylmuramic acid (NAM), 53, 54
  lysozyme action on, 610, 610f
  peptidoglycan synthesis, 249–251
N-acyl homoserine lactone (AHL), *142*–143, 318
NADH
  in catabolism, 245
  Embden-Meyerhof pathway yields of, 213–214
  Entner-Doudoroff pathway yields of, 215
NAD^+ (nicotinamide adenine dinucleotide)/NADH conjugate redox pair, 194
NADPH (nicotinamide adenine dinucleotide phosphate)
  in anabolism, *211,* 245
  Entner-Doudoroff pathway yields of, 215
  overview, *196*
  in photosynthesis, 239
NA (neuraminadase) envelope spikes, 741
NAG (*N*-acetylglucosamine), 249, 610, 610f
NAG-NAM-peptidoglycan building block, 126, 126f
Naked amoebae, 475
Naked viruses, *105*
Naming system of Linnaeus, 12
Nanoarchaeum equitans, 406, 406f
Nanosurfaces
  to repel microbes, 166
Nanowire, 430, 431f
NAPs (nucleoid-associated proteins), 65, 81
Narrow-spectrum antimicrobial drugs, *172*
Nascent-chain-associated chaperone (NAC), 335
Nasopharyngeal carcinoma, 757
National Aeronautics and Space Administration (NASA), 5
Natural gas, 401, 580
Natural immunity, 636, 636f. *See also* Host resistance, innate
Natural killer (NK) cells, 619, 619f, 620f
Naturally acquired active immunity, 636, 636f
Naturally acquired passive immunity, 636, 636f
Natural products, 851
NCR (nodule-specific cysteine-rich) peptide signals, 597
Necrotic lesions on plants, 116
Necrotizing fasciitis, *793,* 793f
Necrotrophic fungi, *600*
Needham, John, 14
Negative selection, *640*
Negative staining, *34*
Negative-strand DNA viruses, 516
Negative-strand RNA viruses, 515, 521–523, 521f
Negative transcriptional control, 302
Negativicutes, 470–471, 470f
Neglected Tropical Diseases (NTDs), 811
Neighbor joining in phylogeny, 11
Neisseria gonorrhoeae, 135, 445, 445f, 786
Neisseria meningitidis
  in pharynx, 670
  rifampin treatment, 180
Neisseria meningitidis serovar B (MenB), 853
Neomycin, 171, 178
Neonatal herpes, *755*
Nephritis, 494
Neuraminidase (NA), 741
  enzyme in viruses, 181
  inhibitors, 742
Neuromuscular disorders, 798
Neuron demyelination, 780
Neutralization of biologically active materials, *650*
Neutral mutations, *346*
Neutrophils, *133, 616,* 618, 627, 629

Next-generation DNA sequencing (NGS) methods, 381, *385–386*
N-formylmethionyl-tRNA, 286, 287f
NHEJ (nonhomologous end joining), *374*
Nitrate reductase, 252–253
Nitrate reduction
assimilatory, 252–253
dissimilatory, *225, 561*
Nitrification
Alphaproteobacteria for, 441, 441f
nitrifying bacteria, *231,* 441–442, 441f
in nitrogen cycle, *560*
Nitrogen, 252–253
Nitrogenase, *253*
Nitrogen cycle, 560–561, 560f, 566f
Nitrogen fixation, 576
agriculture, 596–599
associative, 592
by cyanobacteria, 421
discovery of bacteria for, 18
by Frankiales, 460
in nitrogen cycle, *560*
process of, 253
in terrestrial ecosystems, 592
Trichodesmium for, in open ocean, 576f
Nitroplast, 423
NLRs (NOD-like receptors), *626*
NNRTIs (nonnucleoside reverse transcriptase inhibitors), 181
Nobel Prize, 17
to Binnig and Rohrer, 41
to Borlaug, Norman, 587
to Domagk, 171
to Ehrlich, 171
to Fleming, Florey, and Chain, 171
to Koch, Robert, 17
to McClintock, Barbara, 354
to Metchnikov, 681
to Ross, Ronald, 816
to Shimomura, Chalfie, and Tsien, 31
Nocard, Edmond, 459
Nocardia, 459
Nocardioform bacteria, 459
Node, 11
Nod factors, *594*
NOD-like receptors (NLRs), 626
Nodule-specific cysteine-rich (NCR) peptide signals, 597
Nomenclature, 12
Noncompetitive inhibitor of enzymes, *203*
Noncyclic photophosphorylation, *236*
Nonenveloped viruses, 104f, *105,* 109, 110f
Nonheme iron proteins, *196*
Nonhomologous end joining (NHEJ), 375
Nonnucleoside reverse transcriptase inhibitors (NNRTIs), 181
Nonobligatory positive interaction, 549–550
Nonphosphorylative Entner-Doudoroff (ED) pathway, 408
Nonpoint source pollution, 582
Nonproteobacterial Gram-negative bacteria, 414–433
anammox bacteria, 424, 424f
commensal anaerobes, 429
Deinococcus-Thermus phylum, 416
Fusobacterium, 429
green nonsulfur, *419*
green sulfur, 418
in human gut microbiota, 424
human symbionts, 423–425
obligate intracelluar parasites *(Chlamydia), 424*
oxygenic, *420*–422
photosynthetic bacteria, 416–422
Planctomycetota, 423–425
Spirochaetes, 426–427
Verrucomicrobia, 423–425
Nonsense codons, 283
Nonsense mutations, 346
Noroviruses, 760
Nosocomial infections, *710*
NRTIs (nucleoside reverse transcriptase inhibitors), 181
NTPs (nucleoside triphosphates), 194
Nuclear envelope, *94*
Nuclear pore complex, *94,* 94f
Nucleic acids
bacterial, 268–271
composition of, 269f
in diagnostic methods, 727
drugs for synthesis inhibition of, **180**
in microbial taxonomy, 529
subviral agents based on, 117
in viruses, 107
Nucleocapsid, *104*
Nucleocytoplasmic large DNA viruses (NCLD), 513
Nucleoids, *47*
archaeal, *81*
bacterial, **64–65,** 64f
nucleoid-associated proteins (NAPs), 65
occlusion of, in septation, *125*
Nucleolus, *95*
Nucleoprotein (NP), 741
Nucleoside diphosphate, *257*
Nucleoside diphosphate carriers, 249
Nucleoside reverse transcriptase inhibitors (NRTIs), 181
Nucleoside triphosphates (NTPs), 194
Nucleosomes, *94,* 94f
Nucleotide-binding and oligomerization domain (NOD)-like receptors (NLRs), *626*
Nucleotide excision repair, *350,* 351f
Nucleotide synthesis, *257*–258
Nucleus, eukaryotic, 94–95, 94f
Numerical aperture, *25,* 26f
Nutrient concentration, 129f
Nutrient immobilization, *556*
Nutrient transport mechanisms, 49–52, 49t
Nutritional types, 209–210, 210t

O

O antigen, *56,* 260
Obesity
cancer and, 680
gut microbiome and, 609, 693
metabolic syndrome, 678–680
Objective lenses, *24,* 25f, 26t
Obligate aerobes, *136*
Obligate anaerobes, *136*
Obligate intracellular parasites, 424–425
Obligate intracellular pathogens, *685*
Ocean-based climate change solutions, 575, 576t
Oceans
acidification, 546–547, 571, 571f
ecosystems, 571–580
study of, 569
Ocular lense, *25*
OEC (oxygen-evolving complex), 238
Oil immersion objective lens, 26, 26f
Oil spill bioremediation, 871, 872f
Okazaki fragments of DNA, *274*
Okazaki, Reiji, 274
Oligonucleotides, 368
primers, 368–370
signature sequences, 10, *530*
Oligotrophic environments, *139*
Oligotrophs, **437–443**
One-step growth experiment on viruses, 108, 108f
Öomycetes, *485*
Open ocean ecosystem, 573–578
Open reading frame (ORF) in genome sequencing, *389,* 427
Operating room, 155
Operons
arabinose, 305, 305f
definition, *302*

description of, *302*
lactose, 302–303, 303f–305f
tryptophan, 303–304, 304f
Ophiocordyceps sinensis, 490
Ophthalmia neonatorum, *787*
Opines, 600
Opportunistic diseases
antibiotic-associated colitis, *804*–805
aspergillosis, *830,* 830f
bacterial, **804–808**
candidiasis, *831*–832
cryptococcosis, 832, 832f
dental caries, 806
dental diseases, 805–807
dental plaque, *806,* 806f
from fungi, 829–834
in HIV/AIDS patients, 751
microsporidiosis, 832–833, 832f
odontopathogens, 805
pathogens, *685*
periodontal disease, *806*–807
pneumocystis pneumonia, 833–834
pyelonephritis, *807*–808
streptococcal pneumonia, *807*
urinary tract infection, *807*–808
vaginosis, *805,* 805f
Opsonization, 611, 611f, 650
Orchids, 595
ORF (open reading frame) in genome sequencing, *389*
Organelles
eukaryotic, *87*
mitochondrial family of, **95–97,** 96f
in secretory and endocytic pathways, **90–93**
Organic acids, 492, 853
Organic soil, *588*
Organotrophs, *209*
Origin of life theories, 4–5, 6f, 7f
Origin of replication complex (ORC), *327*
definition, *124*
eukaryote replication, 327–330, 329f
initiation of replication, *272*
plasmids, *365*
Ornithosis (psittacosis), *803*
Oropharyngeal candidiasis, *831*
Orthologues (genes with similar ORFs), *389,* 400
Oseltamivir, 181, 742
Osmophiles, *132*
osmotolerant microorganisms, *133*
Osmosis, 69
osmotic stress in halophiles, 412
osmotrophy, *471*
osmotrophy in fungi, *491*
solutes and, 132–134
Osmotolerant microorganisms, *133*
Otitis media (middle ear infection), 466
OTU (operational taxonomic unit), *397*
Outbreak, *701*
Outer membrane, *55*
Gram-negative bacteria, *53,* 69
Negativicutes Gram-positive bacteria, 470–471
Outer membrane proteins (OMPs), *296*
Outer membrane vesicles (OMVs), *59*
Oxazolidinones, *179*
Oxidation-reduction potential, 556, 557f
Oxidation-reduction (redox) reactions in metabolism, 194–196
Oxidative burst, *627*
Oxygen concentration, microbial growth affected by, 136–137, 137f
Oxygen-evolving complex (OEC), 238
Oxygenic photosynthesis, *234*–239
Oxygenic photosynthetic bacteria, *420*–422
Oxytetracycline antibiotics, 603
Ozonation in water purification, 863

P

PABPs (poly-A binding proteins), 335
Pacemaker enzymes, 204
Packaging for food preservation, 842, 843f
Packasome, 112, 507–508
PAF (platelet-activating factor), *656*
Palisade arrangement of cells, *459,* 459f
p-aminobenzoic acid (PABA), 179
Pandemics, *701,* 737
Paneth cells, 608
Pan-genome, *397,* 399f, *536*
Panspermia theory, 7
Papillomaviruses, 759, 759f
PAPS (phosphoadenosine 5′-phosphosulfate), 254, 254f
Paralogues (genes with similar sequences), *389*
Paralytic polio, 762
Paramecium, 46
Paramecium caudatum, 86f, 100, 481, 481f
Paramecium tetraurelia, 397
Paramyxovirus, 521
Parasitic Metamonada, 474
Parasitism. *See also* Pathogenicity
by ciliates, 479
as microbial interaction, *551*–552
Parasporal body, *463,* 464f
Parfocal microscopes, *25*
Parkinson's disease, 677
Particulate methane monooxygenase (pMMO), 448, 448f
Particulate organic carbon (POC), 582
Particulate organic matter (POM), *573*
Parvoviruses, 514, 515f, *756*
Passive diffusion, *50,* 50f, 88
Pasteur Institute (France), 19, 168
Pasteurization, 17, *161, 839,* 839t
Pasteur, Louis, 15f
fermentation research of, 17
immunological studies of, 19
rabies vaccine of, 764
spontaneous generation theory refuted by, 14
swan-neck flask experiments of, 16f
wine spoilage prevention process of, 161
Pathogenicity, **684–698**. *See also* Infections
adhesion of pathogen, 689–690
airborne transmission, *687*
biofilm development, *693*
contact transmission, *687*–688
damage to host, **693–698**
endotoxins, *696*
entry of pathogen, 689–690
exotoxins, *694*–696, 695f
factors initiating disease, 689–692, 691t
infectious disease driven by, *685*
infectivity and virulence, 686–687
invasion factors, 689–692, 691t
mycotoxins, *696*–697
pathogen defined, *685*
person-to-person transmission, 687
surviving host defenses, **692–693**
transmission of pathogens, **686–692**
vector-borne transmission, *688*–689
vehicle transmission, *688*
vertical transmission, *689*
virulence, *685, 686*
Pathogenicity islands, 353, 364, *694,* 695t
Pathogens
innate host resistance, *606*
microbial plant, 600, 602t
Pattern recognition receptors (PRRs), *624,* 625f
Pause loop to slow RNA polymerase, 306
PBPs (penicillin-binding proteins), 126
PCP *(Pneumocystis carinii pneumonia),* 833
PCV (pneumoncoccal conjugate vaccine), 807

PDI (protein disulfide isomerase), *293*
Pelagibacter ubique, 442, 442f
Pelagic zone, *583*
Pellicle, *471*
Pelvic inflammatory disease (PID), *786*
Penicillinases in penicillin-resistant bacteria, *175*
Penicillin-binding proteins (PBPs), 126, 175
Penicillins, 175–176
  bactericidal action of, 171f
  as cell wall synthesis inhibitors, *171*
  development of, 244
  discovery of, 171
  as first antibiotic, 20
  horizontal gene transfer of resistance to, 352
  structures of, 175f
  transpeptidation reaction blocked by, 251
Penicillium chrysogenum spore, 171
Pentose phosphate pathway
  for anabolic reactions, *215*–217, 216f
  Helicobacter pylori, 433
Peplomers of enveloped viruses, *107*
PEP (phosphoenolpyruvate): sugar PTS (phosphotransferase system)
  in archaea, 79
  in bacteria, 313, 314f
  group translocation, *51,* 52, 52f
Peptic ulcer disease, 459, 784, *784*–785
Peptide interbridge, *54,* 55f
Peptidoglycan (murein), *53*
  in Actinobacteria, 456, 457f
  archaeal cell walls lacking, 79
  in bacterial cell structure, 53–58
  cell shape determined by, 126, 126f
  in chloroplasts, 8
  composition of, 54f
  cross-links in, 54, 55f
  diamino acids in, 55f
  in Gram-positive bacteria, 53f
  helical strand of, 54f
  mycoplasmas lacking, 58
  synthesis of, 249–251, 250f
Peptidyl transferase, *288*
Peptidyl tRNA binding site, *286*
Peramivir antiviral drug, 181, 742
Periodontal disease, 671, *806*–807
Peripheral membrane proteins, *49*
Periplasm, *53*
Periplasmic flagella, *427*
Periplasmic space, *53,* 249–251
Peristalsis, *609*
Peritrichous flagella, 66f, *67*
Peronosporomycetes, *485*–486
Personal protective equipment (PPE), 720–721, 721f
Pertussis (whooping cough), *776*–*777*
Petri dish, 171
Petroff-Hausser counting chamber, 148, 148f
Peyer's patches, *623*
PFUs (plaque-forming units), *116*
PGA (3-phosphoglycerate), 247
PGE_2, 629
PGF_2, 629
pH
  enzyme activity and, 201–202
  food spoilage and, 838
  microbial growth affected by, 133–134, 133f
PHA (polyhydroxyalkonates), 62, 62f
Phaeospirillum cells, 439
Phage-plasmids, 358
Phagocytic vacuoles, *471*
Phagocytosis, 624, **624–629,** 625f
  biofilms and, *693*
  exocytosis in, 628–629
  foreignness recognition in, 624–626, 625f
  holozoic nutrition and, 471
  intracellular digestion in, 626–628
  opsonization by, 611f
  phagosomes in, *92*
  white blood cell action as, 19
Phagophores, 628
Phalloidin toxin, 501
Pharyngitis, *778*
Pharynx microbiota, 670
PHAs (polyhydroxyalkonates), 404
Phase-contrast microscopes, **27–29,** 28f, 29f
Phase variation, *427, 692*
  in *Borreliella,* 427f
PHB (Polyhydroxybutyrate), 62, 62f
Phenetic system, 529
Phenol coefficient test of disinfectant potency, *167*
Phenolics for microbe control, 162–163, 163f
Phenotype, **390–396**
  DNA-protein interactions and, 395–396
  in mutation screening, 346
  overview, 390
  proteomics, 393–395
  transcriptome analysis, 391–393
Phenotypic rescue, *371*
Phenylpropene units, 588, 590f
φ6 bacteriophage, 516
φX174 bacteriophage, 514, 514f
Phosphatases, *257*
Phosphate transfer potential, *193*
Phosphoadenosine 5′-phosphosulfate (PAPS), 254, 254f
Phosphoenolpyruvate (PEP): sugar phosphotransferase system (PTS)
  in archaea, 79
  in bacteria, 313, 314f
  group translocation, *51,* 52, 52f
Phospholipids, 48, 48f
  in bacterial membranes, 87, 88f
  synthesis of, 258–259
Phosphorelay systems, *52, 311,* 319
Phosphorus assimilation, *257*
Phosphorus cycle, 561, 562f
Phosphorus in amino acid synthesis, 257
Phosphorus to oxygen (P/O) ratio, 221–222
Phosphorylative Entner-Doudoroff (ED) pathway, 404, 405f
Photic zone, *573*
Photobiont, *550*
Photolithoautotrophs, *209,* 418
Photoorganoheterotrophs, 210
Photoorganotrophs, 573
Photophosphorylation, *234,* 235f
Photoreactivation, *350*
Photoreceptors, 412
Photosensitizer pigments, 138
Photosynthate, *573*
Photosynthesis
  anoxygenic, *239*–240
  chloroplast reactions, 97, 97f
  evolution of, 7–8
  Gram-negative bacteria, 416–422, 417t
  green nonsulfur bacteria, 419
  green sulfur bacteria, *418*–419
  oxygenic, *234*–239, *420*–422
  pigments for, 418f
  purple nonsulfur bacteria, 437–438
Photosynthetic purple sulfur bacteria, 447, 447f
Photosystems I and II in photosynthesis, *236*
Phototaxis, 82, *420*
Phototrophy, 234–241, 546
  anoxygenic photosynthesis reactions, *239*–240
  definition of, *234*
  diversity of microorganisms, 234t
  fueling reactions in, *209*
  oxygenic photosynthesis reactions, *234*–239
  rhodopsin-based, 240

Phycobiliproteins, *235, 420*
Phycobilisome protein complexes, *420,* 420f
Phycobionts, 550
Phycocyanin pigment, *420*
Phycoerythrin pigment, *420*
Phycology, 470. *See also* Protists
Phyletic classification, *9*
Phyllosphere microorganisms, *592*
Phylogenetic classification systems, 9
Phylogenetic diversity, in microbial communities, 535–536
Phylogenetic trees
  distance method of construction, 12f
  evolutionary relationships shown by, *8*
  topologies, 12
  universal, 4, 4f
Phylogeny, *9*
Phylum Bacteroidota, 427–429
Phylum Bdellovibriota, 431–433
Phylum Campylobacterota, 433
Phylum Chloroflexota, 419
Phylum Cyanobacteriota, 420–422
Phylum Desulfobacterota, **429–431**
Phylum Fusobacteriota, 429
Phylum Planctomycetota, 424
Phylum Spirochaetota, 426–427
Phylum Verrucomicrobiota, 424–425
Physical map of genome, *390*
Physiological characteristics in microbial taxonomy, 2
Phytohormones, 592
Phytophthora infestans, 485, 601
Phytoplankton, *573,* 574f, 575, 580, 581
Pichia pastoris, 377
Pickles, 848, 849t
Picoplankton, *573*
Picozoa, 488
PID (pelvic inflammatory disease), *786*
Piedras, *823*
Piezophilic microbes, *137, 578*
Pili, *47,* **66,** *81*
PIs (protease inhibitors), 181
Planctomyces, 424
Planctomycetota, 423–425
Plane warts, *759*
Planktonic, *140*
Plantae kingdom, 2
Plantar warts, *759*
Plant microbiome, ***591–601***
Plant pathogens, 599–600
Plant roots, microbial communities, 592
Plaque assays of cultured virus, *116*
Plaque-forming units (PFUs), *116*
Plasma cells, *644*
Plasmalemma, *471*
Plasma membranes
  archaeal, **77–79**
  bacterial, *47*
  eukaryotic, 87
  methyl-accepting chemotaxis proteins in, 316
  peptidoglycan synthesis in, 249–251
  purple membrane region of, 412
Plasmids
  in bacteria, ***65***
  as cloning vectors, 365–367, 367f
  conjugative, *355,* 364
  F′, *358*
  recombinant, 366f
  Ti (tumor-inducing), 600
Plasmodium
  cell types of, 818t
  description of, 482–483
  life cycle of, 817f
  as malaria causative agent, 815–819
  TCA cycle in, 482, 483f
Plasmodium falciparum, 170
Plasmolysis, *58*
Plastids, *97*
Platelet-activating factor (PAF), *656*
Pleomorphic bacteria, *45*
Plus-strand DNA viruses, *505*
Plus-strand RNA genomes, 505
Plus-strand RNA viruses, *505*
PMNs (polymorphonuclear neutrophils), *618*
PNA-FISH (peptide nucleic acid fluorescence *in situ* hybridization), 723
Pneumococcal conjugate vaccine (PCV), 807
Pneumocystis carinii pneumonia (PCP), 833
Pneumocystis jirovecii, 183, 730, 753, 833–834, 833f
Pneumocystis pneumonia, 833–834
Pneumonia
  atypical, *776*
  chlamydial, *770*
  Enterocytozoon bieneusi as cause of, 494
  mycoplasmal, 776
  Pneumocystis, 833–834
  streptococcal, *807*
Pneumonic plague, *780*
POC (particulate organic carbon), 582
Point mutations, *344,* 346t
Polar flagellum, *66,* 66f
Poliomyelitis, 762–763
Polio vaccine, 762
Poliovirus, 763
Poly-A binding proteins (PABPs), 335
Poly-β-hydroxybutyrate (PHB), 860
Polycistronic mRNA (messenger), *279,* 280f
Polyene antifungal drugs, 183
Polyhistidine tagging, *371*
Polyhydroxyalkonates (PHAs), 62
Polymerase chain reaction (PCR)
  drug-resistance gene detection by, 187
  food-borne pathogen detection, 842–844, 843f
  for HIV/AIDS assay, 751
  in microorganism identification, 727
  in next-generation DNA sequencing (NGS) methods, 385
  process of, 368–370
  in recombinant DNA technology, 363, 365–367
  rRNA gene regions amplified by, 488
  seamless cloning, 369
Polymorphonuclear cells (PMNs), 618
Polymyxin, 463
Polypeptides
  elongation of, 292–295
  modifications to N-terminus, 292
Polyphosphate granules, 62
Polyporus squamosus, 501
Polyprotein, *519*
Polyribosome, *291*
Polysaccharide synthesis, 249
Polysaccharide utilization loci (PULs), 428
Polysomes, eukaryotic, 335
POM (particulate organic matter), *573*
Pontiac fever, 771
Popper, William, 762
Populations
  human infectious disease patterns, **705–708**
  microbial (*See* Microbial populations)
Porin proteins, 309
Porins, *56,* 58f
Portal of entry, *689*
Positive selection, *640*
Positive-strand RNA viruses, *505,* 505f
Positive transcriptional control, *305*
Pospiviroidae viroids, 117, 117f
Postherpetic neuralgia, *738*
Poststreptococcal diseases, 778
Poststreptococcal glomerulonephritis (PSGN), *778*

Posttranslational reactions, 300
Posttranslational regulation, *204*
Pouchet, Felix, 16
Pour plate method for isolating pure cultures, *147*
Powdery mildew, 497
Poxviruses, 105–106, 746
PPE (personal protective equipment), 720–721, 721f
ppGpp (guanosine tetraphosphate), 139, *311,* 312f, 314–315
Prebiotic, *681*
Precipitation, *651*
Precipitin reaction, *651*
Precipitins, *733*
Precursor metabolites, *210,* 211f
Predation, 551–552
Pre-mRNA, 332
Pre-replicative complex (pre-RC), 327
Presence-absence (P-A) test for coliforms, 865
Preservatives, organic acids as, 853
Pressure, microbial growth affected by, 137
Prevalence rate, *704*
Primaquine, 184, 819
Primary active transport
  description of, *51*
  in eukaryotic cells, 88
Primary antibody response, 649–650, 649f
Primary metabolites in industrial microbiology, *857*
Primary producers, *570*
Primary treatment of wastewater, 866
Primases, *329,* 330f
Primosome complex, *273*
Prions, 4, 118, 162, 765
Prism, bending of light, 24, 24f
Probes, 391
Probionts, 5
Probiotics
  Bifidobacterium, 457
  development of, 681
  overview, *681*
  in yogurt, 465
Prochlorococcus, 234
Prochloron cyanobacteria, 9
Prochlorophyte cyanobacteria, 422
Prodromal stage of disease, *685*
Production strains, 376
Professional phagocytes, 618
Programmed cell death, 620
Progressive leprosy, 784
Prokaryotes, *2,* 15
Prokaryotic cells, 2
Promoters
  in biosensors, 299
  eukaryotic and archaeal, 332
  gene sites for, 278
  sigma factors to recognize, 280, 281f
Proofreading, 350
  by DNA polymerase, *350*
  in genome sequencing, 385
Propagated epidemics, *705*
Prophage, *113, 361,* 508
Propionibacteriales, 460
Prostaglandins, 629
Prosthetic devices, bacteria adhering to, 790
Prosthetic groups, cofactors as, *200*
Protease inhibitors (PIs), 181
Proteases, *230*
Proteasome degradation of proteins, 91, 91f
Protein disulfide isomerase (PDI), *293*
Protein modeling, *394*
Proteins
  actin, 89
  activator, *301*
  acute-phase, *614–616*
  as amino acid polymers, 269–271
  archaerhodopsin, *412*
  bacterial, 268–271
  in bacterial cytoskeleton, **60–61,** 61f
  bacterial transport, 44
  bactericidal permeability-increasing, 627
  carrier, *50*
  catabolism of, *230*
  of complement system, 611–614
  conserved hypothetical, *389*
  C-reactive, 616
  DNA interactions with, 395
  drugs inhibiting synthesis of, 170
  ESCRT, 128
  eukaryotic and archaea folding of, 335
  flagellin, *67*
  flavoproteins, 196
  fluorescence labeling, 371, 373f
  FtsZ, 124–126
  genes coding for, 278–279
  global regulatory, *310*
  heat-shock, 311, 335, 368
  hemagglutinin, 113f
  initiation of synthesis of, 288
  integral membrane, 48f, 49
  iron-sulfur proteins, *196,* 198f, 219f
  linker, 420
  mannose-binding, *613*
  maturation, 297
  maturation and localization in eukaryotic and archaeal cells, 335–338
  methyl-accepting chemotaxis, 317f
  nonheme iron, *196*
  nucleoid-associated, 65, 81
  penicillin-binding, 126
  peripheral membrane, *49*
  phycobilisome protein complexes, *420,* 420f
  poly-A binding, 335
  prion, 765
  proteasome degradation of, 91, 91f
  purification, 371
  RecA, *510*
  recombinant, 371–373
  replisome, 274f
  repressor, *300–302,* 339
  RpoS, 139
  scaffolding, in virus assembly, 112
  secretion, 295
  silencing, 346
  70S initiation complex for synthesis of, 288
  starvation, 140
  stop codon to end synthesis of, 288
  surfactants A and D, 616
  TATA-binding, 331, 333, 334f
  thioredoxin, 258
  transcriptional regulatory, 301
  translation for synthesis of, 109, 111
  translocation, 293–294
  trichocyst, in dinoflagellates, *479*
  tumor suppressor, 115f
  of unknown function, *389*
Protein synthesis inhibitors, **177–179**
Proteobacteria, **436–452**
  Alphaproteobacteria, 441, 441f
  dissimilatory iron reduction by, 563
  Enterobacteriales, 451–452
  Gammaproteobacteria, **443–452**
  Legionellales, 449–450
  Methylococcales, 448–449
  nitrifying, 441–442, 441f
  Pseudomonadales, 450–451
  purple nonsulfur, 437–438
  purple sulfur, 447
  Rhizobiales for N_2 fixation, 440–441, 441f
  Rickettsia, 442–443, 443f
  SAR11 clade abundance, 576
  Thiotrichales, 447–448
  type II secretion systems (T2SS), 514
  Vibrionales, 451

Proteolytic enzymes, 394
Proteomics, *393–395*
Proteorhodopsin, 576
Protists, 3, 471–473
  Alveolata, 479–483
  Amoebozoa supergroup, 475–477
  in Antarctic lakes, 581, 581f
  antiprotozoan drugs, **183–185**
  Archaeplastida supergroup, 486–488
  arthropod transmission of diseases, **815–822**
  direct contact transmission of diseases, **822–825**
  diversity of, 471–473
  as eukaryotic group, 85
  example of, 85
  food and water as vehicles for diseases, **825–829**
  human diseases from, 811–834 (*See also* specific diseases and pathogens)
  medically important protists, 813t
  morphology, 471
  overview, *3,* 812–813
  with pseudopodia, 475–477
  Rhizaria, 478–479
  routes of transmission, 812f
  Stramenopila, 483–486
  structure of, 471f
  water-based reservoirs, 826t
Protomers, *105*
Proton gradients in ATP synthesis, 220
Proton motive force (PMF)
  from anammox reaction, 424
  description of, *221,* 222f
Proto-oncogenes, 115f
Protoplasts, *58, 60*
Protospacer adjacent motif (PAM), *373*
Prototheca moriformis, 488
Protothecosis, 488
Prototroph, mutants from, *347*
Protozoology, *470*
PRRs (pattern recognition receptors), 624
Pseudogenes, *397, 442*
Pseudogymnoascus destructans, 500
Pseudomonadales, 450–451
  opportunistic pathogens, 450–451
Pseudomonas aeruginosa
  c-di-GMP signaling, 310–311, 312f, 315
  as Gram-negative bacteria, 143
  host immune responses suppressed by, 692–693
Pseudomonas denitrificans, 220, 221, 225, 226f
Pseudomonas fluorescens, 38
Pseudomonas phage φ6, 516
Pseudomonas phage Pf1, 514
Pseudomurein, *79,* 80f
Pseudonitzschia, 572
Pseudonocardia, 543
Pseudopodia, 89, *475*–477
Psittacosis (ornithosis), *803*
PSP (paralytic shellfish poisoning), 572
Psychrophiles, *135*
Psychrotolerants, *135*
Psychrotrophs, *135*
P (peptidyl or donor) tRNA binding site, *286*
Public health. *See also* Epidemiology
  antimicrobial drug resistance as threat to, **185–187**, 188f
  microbiology in, 13
Public Health Security and Bioterrorism Preparedness and Response Act of 2002, 715, 720
Public health surveillance, 702–704, 703f
Puccinia graminis subsp. tritici, 587
Puerperal fever, 687
Pulmonary anthrax, *802*
Pulmonary MAC, 773
Pure culture, 14
Pure red cell aplasia, 756
Purine synthesis, *257*
Purple membrane, *412*
Purple nonsulfur *Alphaproteobacteria,* 437–438
Purple sulfur bacteria, 62, 447
Putrefaction, *837*
Putrescine, 837
PVC superphylum, 423–425
Pyelonephritis, *807*–808
Pyrenoid, *472*
Pyrimethamine, 185, 829
Pyrimidine synthesis, *257*
Pyrodictium occultum, 338
Pyrolobus fumarii, 135, 407
Pyrrolysine, 283, 285, 288
Pyruvate
  to carbon dioxide, in aerobic respiration, 213
  from glucose, in aerobic respiration, 212–218
Pyruvate formate-lyase (PFL), 451

Q

Q fever, *803*–804
Quality-assurance mechanism in secretory pathway, 91
Quaternary ammonium compounds, 163–164
Quinine, 183, 816
Quinolone drugs, 180
Quorum sensing
  bacterial bioluminescence as model for, 451
  as cell-cell communication, *142*–143
  for cellular regulation, 316–317
  in *Vibrio* species, 318–319

R

Rabies
  description of, *764*–765
  Pasteur's vaccine for, 19, 19f
  virion, 764
Rabies vaccine adsorbed (RVA), 765
RAC (ribosome-associated complex), 335
Radappertization, 840
Radial immunodiffusion (RID) assay, 734
Radiation
  to control microorganisms, 161–162
  for food preservation, *840*
  microbial growth affected by, 137–138
Radiolaria, 478, 478f
Random sequencing, *385*
Rapidly growing mycobacteria (RGM), 784
Rapid Plasma Reagin test, 729
Rapid sand filters, *863*
Rare biosphere, *536*
RB (reticulate body), *425*
RBS (ribosome-binding site), *306*
RDR (recombination-driven DNA replication), 330
Reaction center, *418*
Reaction-center chlorophyll pair, *236*
Reactivation tuberculosis, *774*
Reactive nitrogen species (RNS), *627*
Reactive oxygen species (ROS), *136,* 610
Reading frame for codons, *283,* 284f
Readthrough, *520*
Reagin (IgE antibodies), 729
Real-time PCR, *370*
RecA protein, *350,* 351, 353, *510*
Receptor diversity, lymphocyte, 652–654
Receptor-mediated endocytosis, *92*
Receptors on viral host cells, *108*
Recombinant DNA technology, 363, 365, 366f
  cloning vectors, *365*
  foreign gene expression in host cells, 372–374
  gel electrophoresis, 363
  genetic cloning and cDNA synthesis, 365

Recombinant DNA technology (*continued*)
genomic libraries, 371
host-cell introduction, 367–368
polymerase chain reaction, 368–370
restriction enzymes, 363–364
Recombinants, *352*
Recombinant-vector vaccines, *713*–715
Recombinases, *353*
Recombinational repair, *352*f
Recombination-driven DNA replication (RDR), 330
Recombinations of genes, *352*
Redi, Francesco, 14
Redox (oxidation-reduction) reactions
biogeochemical cycles and, 556, 604
food spoilage and, 838
metabolism and, 194–196
Red tides, *572,* 572f
Reducing power, *210*
Reductive acetyl-CoA pathway, 402, 402t
Reductive amination pathway, 252
Reductive dehalogenation, *872,* 872f
Reductive TCA cycle, 248, 248f
Reef-building corals, 546
Refractive index, *24*
Regulatory sites for allosteric regulation of enzymes, *204,* 204f
Regulatory T cells (Tregs), *643*
Regulon, *310*
Relaxosome, 356, 357f
Release of virions, 112, 112f
Remdesivir, in COVID-19, 740
Remote sensing for charting infectious diseases, *704*
Rennet in cheese production, 846–847, 846f
Replica plating, *348,* 349f
Replicase, *516*
Replication fork
bacterial, *271*–276
Replicative forms (RFs), *514*
Replicative transposition, *354*
Replicons, *276, 327*
Replisomes, *272*–276, 290–291
Repressible genes, *300,* 303
Repression of enzyme synthesis, 300–301
Repressor proteins, *301,* 347
Reproduction
in *Chlamydomonas,* 487
in diatoms, 484f
in *Dictyostelium discoideum,* 476
in filamentous ascomycetes, 498
in fungi, 492–493
mutualism necessary for, 544
in protists, 470
in *Saccharomyces cerevisiae,* 497f
RER (rough endoplasmic reticulum), *90*
REs (restriction endonucleases), *322*
Reservoir for pathogens, *686*
Reservoir host, *701*
Residual body, *93*
Resolution of microscopes, ***25–27,*** 26f
Respiration
aerobic, 212–218 (*See also* Aerobic respiration)
anaerobic, 224–225
electron acceptors in, 224
overview, *211*
Respiratory syncytial virus (RSV), *745*
Respiratory system
as immune barrier, 608, 609f
viral syndromes, 743, 745
Respiratory tract
microbiota, 670
Response regulator, DNA-binding proteins as, *310*
Restriction, *507*
Restriction endonucleases (REs), *320*
action of, *364*f
cloning technology, *363*–364
structure, 364f
viral infection and, *507*
Restriction-modification system to combat viral infection, *321*
Reticulate body (RB), *425*
Reticulopodia, *475*
Retinoic-acid-inducible gene I (RIG-I)-like receptors (RLRs), 626
Retroviruses, 365, 516
Reverse DNA gyrase in hyperthermophiles, *136*
Reverse electron flow, *232*
Reverse transcribing DNA viruses, 524–525
Reverse transcriptase (RT), *364, 523*
Reverse transcriptase PCR, in COVID-19 test, 370
Reverse vaccinology, *853*
Reversible chain termination sequencing, *381,* 383f
Reversible covalent modification of enzymes, 204
Reversion mutations, 347, 349
RFs (replicative forms), *514*
RGM (rapidly growing mycobacteria), 784
RGs (risk groups), 720, 720t
Rhesus (Rh) factor, 660f
Rheumatic fever, 466, *778*
Rhizaria protists, 478–479
Rhizobia
leguminous plants and, *596*–598
stem-nodulating, 598–599, 599f
Rhizobium, 440, 441f
Rhizomorphs, 591
Rhizoplane, *592*
Rhizopus stolonifer, 495, 496f
Rhizosphere, *592*
Rhodesiense human African trypanosomiasis, 820
Rhodococcus, 457, 459f
Rhodopseudomonas palustris, 559
Rhodopsin-based phototrophy, 234, 240
Rhodospirillum
metabolic flexibility of, 438, 439f
Rhodospirillum centenum
cyst formation in, 439
Rho termination factor, 282–283
Ribosomal frameshifting, *524*
Ribosomal RNA (rRNA), 5
nucleic acid sequences, 3
Ribosome-associated complex (RAC), 335
Ribosome-binding site (RBS), *286*
Ribosomes, *47,* ***64,*** 64f
archaeal, 80–81, 95
bacterial, 75t, 81
cellular pool of RNA in, 5
eukaryotic, 95, 95f
nucleolus in synthesis of, 95
tRNA binding sites, 288
Riboswitches (sensory RNAs), 7, 269, 278, 305, 307–308, 855, 855f
Ribotyping, *727*
Ribozymes, *5,* 199–203, 203t
Ribulose 1,5-bisphosphate carboxylase/oxygenase (RuBisCO), 62, 63, *247*
Rickettsia
parasitic or mutualistic, 443
Rocky Mountain spotted fever caused by, 443, *780,* 782, 782f
RID (radial immunodiffusion) assay, 734
Rifampin, 180
Rifamycin, *180*
Riftia worms, 547
RIG-I (retinoic-acid-inducible gene I)-like receptors (RLRs), 626
Ringworm, 824
Risk groups (RGs), 720, 720t
Risus sardonicus, 794, 794f
River ecosystems, 581–582
RNA (ribonucleic acid)
aptamers, *377*
as basis of early life, 5–8
catalytic, 5, 203
DNA, protein and, 8f

engineered (aptamers), 377
guide RNA, *375*
messenger (*See* mRNA (messenger))
as ribonucleotide polymer, *269*
small RNAs to control translation, *308*–309
subgenomic, 524
transfer (*See* tRNA (transfer))
in viroids, 4
RNA-dependent DNA polymerase, *524*
RNA-dependent RNA polymerase, *516*, 521–523
RNA interference (RNAi), *117*
RNAPII promoter, 331–333
RNA polymerase holoenzyme, 280, 303
RNA-Seq, *390*–393
RNA vaccines, 713–715
RNA viruses
double-stranded, 515–517
minus-strand, *505*, 521–523
plus-strand, *505*, 517–521
RNA world hypothesis, 4–7, 8f
RNS (reactive nitrogen species), *627*
Robbins, Frederick, 762
Robotic automation of biochemical tests, 722
Rocky Mountain spotted fever
description, *780*, 782, 782f
rickettsias as cause of, 443
Rod-shaped cells, 126
Rohrer, Heinrich, 41
Rolling-circle replication, *356*, 358f
Rolling-hairpin replication by parvoviruses, 515, 515f
Roosevelt, Franklin D., 762
Root nodule, *598*
ROS (reactive oxygen species), *136*–137, 610, 616, 626
ROS detoxification, 416
Roseobacter, *366*, 577
Roseola infantum, 755
Ross, Ronald, 816
Rotaviruses, 516–517, 760, 760f
Rough endoplasmic reticulum (RER), *90*
Roux, Pierre, 19
Royal Society of London, 14
R plasmids, *363*, 363f
RpoS protein, 139, 396
RSV (respiratory syncytial virus), *745*
RT (reverse transcriptase), *365*, 367f, *523*
Rubella, *745*, 745f
RuBisCO (ribulose 1,5-bisphosphate carboxylase/oxygenase), 62, 63, *247*
Ruminant animals
methane production of, 74
mutualism in, 544
stomachs of, 549f
Runoff, agricultural, 582
Ruska, Ernst, 41
Russian influenza virus, 742
Rusts in wheat, 501, 587
Rut (rho-utilization site), 283
RVA (rabies vaccine adsorbed), 765

S

Sabin, Albert, 762
Sabin vaccine for polio, 762
Saccharomyces cerevisiae, 90, 93, 852t, 855, 855t
antimalaria drug from, 396
clathrin-dependent endocytosis in, *92*
GAL gene regulation, 339, 339f
life cycle of, 497f
as model organism, 498
Saccharomyces pastorianus, 847
Sac fungi, *497*
Salem witchcraft trials, 499
Salivation, 610
Salk, Jonas, 762
Salk vaccine for polio, 763
Salmonella enterica serovar Typhimurium
pathogenicity islands, *694*, 695t
Salmonella gastroenteritis, 800
Salmonella spp.
lysogenic conversion in, *113*
Salmonellosis, *800*–801
SALT (skin-associated lymphoid tissue), *621*, 623f
Salt marshes, 572
Salvarsan, 171
Sanger, Frederick, 381
Sanger method of DNA sequencing, 381
Sanitization, *157*
San Joaquin fever, 814
Saprophytes, *471*, *491*
Sarcina, 45
SAR11 (Sargasso Sea) clade, *576*
SARS-CoV (severe acute respiratory syndrome coronavirus), 739–741
spike protein of, 690f, 690t
strains of, 397
Satellite viruses, *4*, *117*
Satratoxins, 696
Sauerkraut, 848, 849t
Scaffold in DNA sequence, 385
Scaffolding proteins in virus assembly, 112
Scale-up, *856*
Scanning electron microscopy (SEM), ***39***, 39f, 45f
Scanning probe microscopy (SPM), ***40–41***
Scanning tunneling microscopy, *40*, 41f
Schiff's reagent, 33
Scientific method, 4
Sclerotia, 498
Sclerotinia sclerotiorum, 601
Scrapie, 4, 118, 765
SDS polyacrylamide gel electrophoresis (SDS-PAGE), 394
Seamless cloning, 370–371
Secondary active transport, *51*, 79, 88
Secondary antibody response, 649f, 650
Secondary metabolites, *456*, *857*
Secondary treatment of wastewater, 866, 867f
Second law of thermodynamics, *191*
Second messengers in global control systems, 311–316
Secretion
eukaryotic, 335, 337
of proteins, *292*–295
Secretory IgA (sIgA) immunoglobulin, *648*
Secretory pathways, eukaryotic, *91*–92
Sec system for protein secretion, *293*, 295f, *337*
Sedimentation basins in water purification, *863*
Seedling blight in rice, 496
Segmented genomes, *107*, *516*
Selectable markers, *365*
Selectins (cell adhesion molecules), *629*
Selective media, *144*, 149
Selective toxicity of chemotherapy, *172*
Selenocysteine, 283, 285, 288
SELEX (systematic evolution of ligands by exponential enrichment), 377, 378
Self-assembly, 67, 245
SEM (scanning electron microscopy), *38*–39, 39f
Sense codons, *283*
Sense strand of DNA, 284
Sensor kinase, *310*
Sensory rhodopsins, *412*
Sentinel cells, 477
Septa, *492*
Septate hyphae, *492*
Septation, *124*
Septicemia, 456, *691*
Septic shock, *696*, 698f, 772
Septic tanks, *869*, 869f

Sequencing by synthesis, *381*–385
SER (smooth endoplasmic reticulum), *90*
Serology, 729
Serotyping, *729*
Serovars, 729
Sessile, *140*
Settling basins in water purification, 863
Severe acute respiratory syndrome (SARS), 737, 739
70S initiation complex for protein synthesis, 288
Sex pili, *66*
Sex pilus, *356*
Sexually transmitted diseases (STDs), 785–788, 785t
 chlamydial, 786
 gonorrhea (*See* Gonorrhea)
 overview, 785
 summary of, 785t
 syphilis (*See* Syphilis)
 trichomoniasis, *825*
Sexual reproduction, 352
Sheaths, 444, 445f
Shewanella cells, 430, 431
 in manganese cycle, 564–565
Shiga toxin-producing *Escherichia coli* (STEC), *800*
Shigella spp., *691*
Shigellosis, *801*
Shimomura, Osamu, 31
Short-chain fatty acids (SCFAs), 673
 immunity and, 673–676
Shuttle vectors, plasmids as, *365*
Sick building syndrome, 500, 500f
Sickle cell disease, 756
Siderophores, *52,* 52f, 563
Sigma factor (σ) of RNA polymerases, *280*
Signal 1 in T-cell activation, *641*
Signal 2 in T-cell activation, *641*
Signal 3 in T-cell activation, *641*
Signal peptidase, *293*–294
Signal peptide, *293, 337, 347*
Signal recognition particle (SRP), *293*
Signature rRNA, 11
Signature sequences, 530–531
Signs of infectious disease, *685*
Silage, *849*
Silent mutations, *346*
Silk fibers, 362
Silver sulfadiazine, 163
Simple staining, ***33–34,*** 34f
Simple transpositions, *354,* 355f
Sinclair, Upton, 842
Single-cell sequencing, *386*–387
Single radial immunodiffusion (RID) assay, *734*
Single-stranded DNA, 514–515
Single-stranded DNA-binding proteins (SSB), *273,* 274f
Sinorhizobium meliloti, 398
Site-directed mutagenesis, *375,* 375f
Site-specific recombination of genes, *353*
Size of microbial genomes, 397
Skin
 erysipelas, *793*
 as immune barrier, 607, 608f
 microbiota on, 669–670
 mycobacterial infections of, 783–784
 staphylococcal infections of, 789, 791f
 streptococcal diseases of, 793
Skin-associated lymphoid tissue (SALT), *621,* 623f
S-layer, *60,* 60f
Slime
 as archaea cell envelope, 77
 in bacteria, *60*
 staphylococci as producer of, 789, 791t
Slime molds (Myxogastria), *2,* 476–477, 476f–477f
Slow sand filters in water purification, 863
Sludge in wastewater treatment, *866*
Small cell variant (SCV), 449
Smallpox
 cowpox-based vaccine against, 15
 description of, *746*–747, 746f
 rash in, 746f
Small RNAs (sRNAs), *308,* 315f
Small subunit (SSU) ribosomal RNA (rRNA), 9
 evolutionary relationships assessed by, 505
 mitochondria and chloroplasts analysis based on, 9
 in phylogenetic trees, 488
Small subunit ribosomal RNAs (SSU rRNAs), 530
Smith, Hamilton, 363, 385
Smooth endoplasmic reticulum (SER), *90*
Smuts, 501
Snapping division, *458,* 459f
Sneezing reflexes, 609
Snow, John, 702
SOD (superoxide dismutase), *137,* 137f
Sodium chloride impact on microbial growth, 132f
Sodium motive force in *Clostridium tetani,* 469
Sofosbuvir, 758
Soil microbiome, **590–591**
Soil microbiota, 19, 141, 143, 461, 566, ***588–590,*** 589f, 872
Soil organic matter (SOM), 588–590, 589f
Solid media for laboratory culture of microbes, 144
Soluble substances in disease prevention, 19
Solutes, microbial growth affected by, 132–133
SOM (soil organic matter), 588–590, 589f
Somatic hypermutation, *653*
Sortase, *294*
Sorting signal, *294*
SOS response as DNA global control, *350*–352
Source of pathogens, 494
Soybeans, 496
Spallanzani, Lazzaro, 14
Spanish influenza virus, 742
Specialized transduction, *355*
Species
 definition of, *13*
 microbial, 9
Specificity in adaptive immunity, *634*
Specimens for microorganism identification, **721–728,** 722f
 clinical specimens, 722, 722f
 culture-based methods, 722–723, 723t, 725
 microscopy, 725–727
 molecular methods, 727
Spectrophotometry, 150f
Spheroplasts, *58*
Sphingolipids, 87, 88f
Spices and herbs for food preservation, 838
Spider silk fibers, 362
Spidroin protein, 362
Spikes of enveloped viruses, 104f, 107, 110f
Spirilla, *45*
Spirochetes, *45,* 45f, 426–427
 characteristics of, 426t
 morphology, 426
Spiroplasma, 127, 468
Spleen, *621*
Spliceosome, *333,* 334f
Splice-site variability, *653*
SPM (scanning probe microscopy), *40*
Spoilage of food
 control of, **838–840**
 disease outbreaks from, 842, 843f
 government-industry cooperation for detection, **842–844**
 from microbial growth, **837–838**

Spongiform encephalopathies, 118, 765
Spontaneous generation theories, *14*–17
Spontaneous mutations, *344*–345
Sporadic disease, *701*
Sporangiospores, *456, 494*
Sporangium, 69, *493*
Spores
  arthroconidia, *493*
  fungal, 494
  from *Micromonospora,* 459
Sporosarcina, 463
Sporotrichosis, *824*
Sporozoite, *482*
Sporulation, ***69,*** 320–321, 321f
Spread plate method for isolating pure cultures, *147*
Squid/*Vibrio fisheri* symbiosis, 319
sRNAs (small RNAs), 308–310
SRP (signal recognition particle), *293*
SSB (single-stranded DNA-binding proteins), *273,* 274f
SSSS (staphylococcal scalded skin syndrome), *791*
Stable isotope analysis, *539*
Stable microbiome development, 667–669, 668f
Stachybotrys chartarum, 500
Staining procedures
  differential staining, ***34–35,*** 35f
  dyes and simple staining, **33–34**
  fixation, **33**
Staining techniques, 32–35, 529, 532f
Standard free energy change, *192*
Standard plate counts, *149*
Standard reduction potential (E0), *194*
St. Anthony's fire (ergotism), 499
Staphylococcaceae, 464, 467f
Staphylococcal scalded skin syndrome (SSSS), *791*
Staphylococcus aureus, 121, 465f, 670
  growth inhibition of, 126
  halotolerance of, 133
  irradiation to control, 162
  methicillin-resistant, 186t, *792*–793, 792f
  vancomycin resistance of, 177
  virulence factors acquired by, 464, 465f
Staphylococcus epidermidis, 670, 670f
Staphylococcus species
  diseases from, 788–793
  enzymes and toxins produced by, 791t
  food poisoning from, 801–802
  shape of, 45
Start codons, *283*
Starter cultures, *844,* 846
Starvation proteins, 140
Static chemotherapeutic agents, *172*
Stationary phase in microbial growth curve, *129*
STEC (Shiga toxin-producing *Escherichia coli*), *800*
Stelluti, Francesco, 14
Stem-loop structure in mRNA, 306
Stem-nodulating rhizobia, 598–599, 599f
Sterility, 159, 609
Sterilization
  definition, *156*
  dry heat incineration, 160
  gases, 164–165
  heat, 159–161
  microbial control, 156f
Sterilizing, with ultraviolet radiation, 162f
Sterol synthesis, 259–260
Stickland reaction, 230, *469,* 469f
Sticky ends of endonucleases, *363*
Stigma (eyespot), *473*
Stirred fermenters, 856
Stomach cancer, 126
Stop codons, 278f, 279, 283, *285f,* 288, 307
Storage inclusions in bacterial cells, 62
Strain, definition of, *13,* 183
Stramenopila, 483–486
Stramenopiles (photosynthetic protists), 98, 477
Streak plate method for isolating pure cultures, *146,* 146f
Stream ecosystems, 581–582
Strep throat, 778
Streptococcaceae, 466
Streptococcal diseases
  group A (airborne transmission), 777–778
  group B (direct contact), *783*
  pharyngitis, *778*
  pneumonia, *807*
  poststreptococcal diseases, 778
  sulfa drugs to treat, **179–180,** 179f
Streptococcus parasanguis, 671
Streptococcus pneumoniae
  capsule of, 59
  meningitis caused by, *772*–773
  in pharynx, 670
  transformation in, 359, 361f
  virulence transfer in, 266
Streptococcus pyogenes, 376
Streptococcus salivarius, 844
Streptomyces
  antibiotics isolated from, *178*
  bioactive compounds from, 461
Streptomyces coelicolor
  hyphae, 461f
  life cycle, 460
  spore formation by, 122, 122f
Streptomyces griseus, 143, 171
Streptomyces scabies, 461
Streptomyces somaliensis, 461
Streptomyces sudanensis, 461
Streptomycetales, 460–462
  antibiotics from, 454, 460–462
  life cycle of, 460f
  natural products from, 461t
Streptomycetes, 461
Streptomycin, 171, 174t
  aminoglycoside antibiotic, 178f
  description of, 178, 454
  from *Streptomyces griseus,* 461f
  for tuberculosis treatment, 454
Stress response, 139, 693
Stringent response for *E. coli,* 314–315
Stroma, 97
Stromatolites, 8, 8f
Structural proteins, 5
Structural proteomics, 394
Subacute sclerosing panencephalitis, *743*
Subgenomic RNAs, 518f, 519, 524
Substrate analog probing, *539*
Substrate-level phosphorylation, *194,* 211, 213, 232
Substrate mycelium, 456
Substrates, enzymes and, *200*
Subsurface biosphere, 601–604
Subtherapeutic antibiotic use, 343
Subunit vaccines, *713,* 713t
Sudan III (Sudan Black) stain, 33
Sufu, 496
Sulfadiazine, 185
Sulfa drugs, **179–180,** 179f
Sulfate-or sulfur-reducing bacteria (SRB), 411, 411f, 412, 429
Sulfolobus acidocaldarius, 134
Sulfolobus solfataricus, genomic reconstruction of, 407, 407f
Sulfolobus species, 75, 79, 81, 127, 564
Sulfonamide-resistant bacteria, 187
Sulfonamides/sulfa drugs, **179–180,** 179f
Sulfur
  in amino acid synthesis, 253–254
  sulfur cycle, 561–563
  sulfur-oxidizing microbes, 232
Sulfur-dependent thermophiles, 407–408
Sulfur globules, *62,* 62f

Sunlight as radiation source, 137
Superantigens, *696,* 696f, *792*
Supercoiling of DNA, 81, 273, 275, 281
Superorganism, 544
Superoxide dismutase (SOD), *137*
Superphylum, 13
Super-resolution microscopy (SRM), **32–33**
Supportive media, *144*
Suppressor mutations, 347, 347t
Surface area-to-volume (S/V) ratio, 46, 76
Surfactant proteins A and D, 616
Sustainability of caterpillar fungus, 490
Sutcliffe, Iain, 53
Svedberg unit, *64*
S/V (surface area-to-volume) ratio, 46, 76
Sweet wormwood plant, 185
Swimming motion by bacteria, 316
Swine flu, 742
Symbiodinium, 479, 546
Symbionts, 46, 406, 424–425
Symbiosis
  description of, *544*
  leguminous plants and rhizobia, 597–599
  squid/*Vibrio fisheri,* 318, 451, 544
Symbiosome, *598*
Symons, Robert, 365
Symport, *51*
Symptoms of infectious disease, 171, *685*
Synbiotics, *681*
Syncytial growth, 461
Syngamy, *472*
Synovium, 779
Synteny in comparative genomics, *398,* 398f
Synthesis
  in life cycle of viruses, 109, 111
  sequencing by, *381*–385
Synthetic biology, *376,* 396. *See also* Control of microorganisms
Synthetic culture media, *144*
Syntrophy, *545*
Syphilis
  arsenic-based chemicals to treat, 171
  description of, *787*–788, 788f
  Rapid Plasma Reagin test, 729
  treatment of, 171
  Treponema pallidum as causative agent of, 390, 391f
Systematic epidemiology, *708*
Systematic evolution of ligands by exponential enrichment (SELEX), 377, 378f
Systematics, 394
Systemic anaphylaxis, *657*
Systemic fungal infections, 183
Systems biology, 396
Systems metabolic engineering, *371*
Szostak, Jack, 5

T

Tandem mass spectrometry, 394, 394f
Taq polymerase, 369
TATA-binding protein (TBP), 331, 333, 334f
Tat (twin-arginine translocation) system, 294, 337
Tattoo-associated myobacterial infections, 784
Tatum, Edward, 356
Taxon, 12
Taxonomists, 2
Taxonomy, 12
  and cell envelopes, 415
  of archaea, 402–405
  hierarchical arrangement, 13f
  protist, 473
TBP (TATA-binding protein), 331, 333, 334f
T-cell-dependent B-cell activation, *645,* 645f
T-cell-independent B-cell activation, *646*
T-cell receptor (TCR), *639*
T cells, 643
  αβ T cells, *640*
  activation, 640–641, 641f, 642f
  development, 639–640, 639f
  γδ T cells, *640*
  types of, 641–643
T-dependent antigens, *645,* 645f
T DNA region of Ti (tumor-inducing) plasmid, *599*
Teichoic acids in bacteria cell walls, *55,* 56f
Teliospores, 502
Telomeres and telomerases, 329–330, 330f
Temin, Howard, 365
Tempeh, 496
Temperate bacteriophages, *113,* 114, 114f, 354, 508–512
Temperature
  to control food spoilage, 838–839
  to control microorganisms, 158–162
  enzyme activity and, 202
  as extrinsic factor in food spoilage, 838
  for heat killing of microbes, 157t
  microbial control by, 167
  microbial growth affected by, 135–136, 135f
  syphilis treatment, 171
Template strand of DNA, 278
Terminator loop, 306
Terminator region of gene, 278, 279
Terminators in transcription, 280–282
Terminus, *124*
Termites, 475, 545
Terrestrial ecosystems, **587–604**
  Agrobacterium, 599–600, 600f
  microbial plant pathogens, 600, 602t
  nitrogen-fixing, 596–599
  phyllosphere microorganisms, *592*
  in rhizoplane and rhizosphere, 592
  in soils, 588–590, 589f
  subsurface biosphere, 603–604
Tertiary treatment of wastewater, 866, 868
Testate amoebae, 475
Tetanolysin, *794*
Tetanospasmin, *794*
Tetanus, 69, 469, *793*–794, 794f
Tetracycline antibiotics, *178,* 178f
Tetrahymena, 5
TFB (transcription factor B), 334, 334f
TFIIB (transcription factor B for RNA polymerase II), 332
T4 bacteriophage
  binal symmetry, *106,* 106f
  virulence, 506–508
Thaumarchaeol, 408, 408f
T_H0 cells, *642*
T_H1 cells, *643*
T_H2 cells, *643*
T_H17 cells, *643*
T-helper (T_H) cells, 642, 643
Therapeutic index, *172*
Therapeutic manipulation of microbiome, **681**
Thermoacidophiles, *407*
Thermoactinomyces, 463–464
Thermoclines, *570*
Thermococcus kodakarensis, 330, 413
Thermodynamics, *191*
Thermophile microorganisms, *136,* **406–408**
Thermophiles, 78, 80
Thermophilic fermentation, 844
Thermophilic S^0-reducers, extreme, 413
Thermoplasma species, 80
Thermoproteota, 407–408
Thermoproteus tenax, 75, 76f
Thermosome chaperones, 338

Thermotoga bacteria
dissimilatory iron reduction by, 563
fuel from food waste, 414
horizontal gene transfer in, 390
Thermotoga maritima, 414–416, 415f
Thermus aquaticus, 369
Thermus spp., 416
Thiobacillus denitrificans, 445
Thiobacillus spp., 232
sulfur oxidation in, 445, 446f
Thiomargarita namibiensis, 46
Thiomargarita spp., 447, 448f
Thioredoxin, 258
Thiothrix, 447
Thiotrichales, 447–448
Threonine, 255, 255f
3-phosphoglycerate (PGA), 247
3′ poly-A tail nucleotide, *332*
Thrush *(Candida albicans),* 753f
Thrush (oral candidiasis), *831*
Thylakoids, *97,* 97f
Thymic selection, *639*
Thymine, 345, 346f, 350
Thymus, 616, 621
Ticks, 779
Time of contact to antimicrobial agents, 166–168
Timing, in adaptive immunity, *634*
T-independent antigens, *646*
Tineas mycoses, *823*
Ti (tumor-inducing) plasmid, *599,* 858
TIR (toll/interleukin-1 receptor) domain, 625
Tissues of immune system, 621–624
TLRs (toll-like receptors), 606, 625, 627f, 628f
T lymphocytes, 621, 629, 631, *634*
TMA (trimethylamine), 680
TMAO (trimethylamine *N*-oxide), 680
TNF (tumor necrosis factor), 614
TNT-GFP biosensor, 299
Tobacco mosaic virus (TMV)
assembly, 520f
capsid of, 105, 105f
early work on, 18
life cycle of, 520
as plant pathogen, *600*
virions of, 105, 105f
Tobramycin antibiotics, 178
TOC (total organic carbon) as water quality measure, *868*
Togaviridae, 747
Toll/interleukin-1 receptor (TIR) domain, 625
Toll-like receptors (TLRs), 625
Tonsillitis, *778*
Topoisomerases, 273
Total organic carbon (TOC) as water quality measure, *868*
Toxic shock syndrome (TSS), *791,* 792f
Toxic waste storage, 603
Toxigenicity, *694*
Toxin-mediated killing of microbes, 168
Toxins, 469, *686,* 694–697
Toxoids, *713*
Toxoplasmosis, 483, 689, *829*
Trace elements, *50*
Trachoma, *794,* 794f
Traditional chinese medicine, 170
Trailer region of gene, 278
Transaminase enzymes, *252*
Transamination, *230,* 230f
Transcriptase, 516, 523–524
Transcription
activator and repressor proteins for, 339
arabinose operon for, 305
in Archaea, 333–334
in bacterial genome replication and expression, **279–282**
basal factors of, 333
in eukaryotes, 331–333
initiation control, **301–302**
induction and repression of enzyme synthesis, 300–301
lactose operon, 302–303
overview, 300–301
regulatory proteins for, 301–302
tryptophan operon, 303–305
initiation of, 331–332
RNA molecules from, 281–282
stages of, 286–288
translation coupled with, 291–292
Transcriptional fusion, 371
Transcriptional regulatory proteins, 301, 301f
Transcription factor B (TFB), 334, 334f
Transcription factor B for RNA polymerase II (TFIIB), 332
Transcriptome analysis, 391–393
Transduction, 360–363
Transferrin, 610
Transformation
bacterial, 363–365, 364f
Griffith's experiments on, 266, 266f
Transition mutations, 344, 344f
Translational fusion, 371
Translation in genome replication and expression
accuracy in, 288–290
amino acid activation, 285–286
archaeal, 335–338
bacterial, **285–290**
eukaryotic, 335
overview of control of, 300–305
polypeptide chain elongation, 286–288
pyrrolysine insertion, 288
ribosomal structure for, ***64***
ribosomal tRNA binding sites, 288
riboswitches to control, 305–306
selenocysteine insertion, 288
70S initiation complex for protein synthesis, 287f, 288
small RNAs to control, 308
stop codon to end, 288
Translesion DNA synthesis, *351*
Translocation, *288, 292,* 293f
Transmissible spongiform encephalopathies (TSEs), *765*
Transmission electron microscopy (TEM), 36–39, 37f, 48
light microscopy *vs.,* 37f
overview, 36
staining for, 38f
Transmission of infectious disease, **686–692,** 688f. *See also* Airborne transmission of disease; Arthropod transmission of disease; Direct contact transmission of disease; Zoonoses
Transpeptidation, 250, 251f
Transplantation (tissue) rejection, 661–663
Transport, endocytic, **90–93,** 92f
Transport proteins, 44, 49, 50, 55
Transport work of cells, *191*
Transposable elements to move genes, 353–354
Transposase, *353*
Transposons, *353,* 363–364
Transversion mutations, *344*
Travelers' diarrhea, 681, 800
Treatise on the Epidemic Puerperal Fever of Aberdeen (Gordon), 687
Treatise on the Management of Pregnant and Lying-in Women (White), 687
Trefouel, Jacques, 171
Trefouel, Therese, 171
Treponema pallidum, 27, 28f, 427
as cause of syphilis, 787
Triatomine bugs, 822f
Tricarboxcylic acid (TCA) cycle
acetyl-coenzyme A in, *217,* 218f
anaplerotic reactions and, 255–256, 256f

Tricarboxcylic acid (TCA) cycle (*continued*)
in catabolism, 213
precursor metabolites in, 263
reductive, 263
Trichocyst proteins, *479*
Trichoderma reesei, 377
Trichodesmium, 422, 560, 575, 576f
Trichome filaments in cyanobacteria, *421*
Trichomonadida, 475
Trichomoniasis, *825*
Trichonymphida, 475
Trickling filter in wastewater treatment, *866*
Triclosan resistance, 157
Trifluridine, 754
Trigger factor chaperone, *292*
Trimethoprim synthetic antibiotic, *180,* 180f
Trimethylamine (TMA), 680
Trimethylamine *N*-oxide (TMAO), 680
tRNA (transfer). *See also* mRNA (messenger)
amino acid attachment to, 283
binding sites for, 286
genes for, 282, 284t
initiator, *287*
mutations in, 348
in protein construction, 7
ribosomal binding sites for, 288–289
structure, 290f
transcription for, 277f
transcription generating, 277
wobble impact on, 285f
tRNA fragments (tRFs), *598*
Tropical lakes, 584
Tropism, *108, 687*
Truffles, 497
Trypanosama cruzi, 473
Trypanosomes, *473*
Trypanosomiasis
arthropod transmission, *820–822*
impact of, 820–821
life cycle of, 820–821
Tryptic soy broth as complex culture media, *144*
Tryptophan operon, 303–304, 304f
TSAR supergroup, 477–486
Alveolata, 479–483
Rhizaria, 478–479
Stramenopila, 483–486
TSEs (transmissible spongiform encephalopathies), *765*
Tsetse flies, 820
Tsien, Roger, 31
TSS (toxic shock syndrome), *791,* 792f
Tubercles, *774*
Tuberculoid leprosy, *784*
Tuberculosis
drug-resistant strains of, *774*
in HIV/AIDS, 753f
Koch's work on, 17, 17f
Mycobacteria as causative agent of, 34, 180, 631, 773–776, 776f
streptomycin to treat, 171
Tube worm holobiont, 547–548, 548f
Tubulinea, 475
Tularemia, *804*
Tumor necrosis factor (TNF), 614, 792
Tunneling current, 40
Turbidity to measure cell mass, 150, 150f
Turbidostats for continuous microbial culture, *152*
Tuskegee Study of Untreated Syphilis in the Negro Male, 789
Twin-arginine translocation (TAT) system, 337
26S proteasome, *91*
Two-component signal transduction system, *310*–311
Two-dimensional gel electrophoresis, *393*–394, 394f
Tyndallization (intermittent sterilization), *161*
2D-polyacrylamide (2D-PAGE), 540
2D-polyacrylamide (2D-PAGE), 540
Tyndall, John, 16, 161
Type 1 hypersensitivity, 617
Type III secretion systems (T3SS)
in enteropathogenic *E. coli,* 800, 800f
in plague, *780,* 781f
Type II secretion systems (T2SS), *295*
Type I secretion systems (T1SS), *294,* 294f
Type IV pili, *66,* 82
Type IV secretion systems (T4SS), *295,* 449
Type IX secretion systems (T9SS), *295*
Type VII secretion systems (T7SS), *296*
Type VI secretion systems (T6SS), *295–296,* 674
Type V secretion systems (T5SS), *295*
Type X secretion systems (T10SS), *295*
Typhoid Mary, 706, 801
Typhus, 685, 689

U

Ubiquinone, *196*
UCYN-A, 423
UDPG (uridine diphosphate glucose), 249, 249f
Ultraviolet (UV) radiation, *138, 161,* 346, 346f
Uniporters, *51*
Universal phylogenetic tree, 4f, *8*
Unsaturated fatty acid synthesis, 135, 259
Unstained microbes, 27
Upper respiratory tract (URT), 670
Uracil, *257*
Ureaplasma urealyticum, 468, 468f
Urediniomycetes, 501
Uridine diphosphate glucose (UDPG), 249, 249f
Uridine 5′-triphosphate (UTP), 194
Urinary tract infections (UTIs), 177, 180, 466, 468, *807*–808
U.S. Department of Agriculture (USDA), 840
U.S. Department of Energy, 603
U.S. Environmental Protection Agency (EPA), 166, 863, 863t
U.S. Food and Drug Administration (FDA)
bacteriophage sprays approved by, 168
Botox (type-A botulinum toxin) approved by, 798
Food Safety Modernization Act of 2001, 842
"generally recognized as safe" chemical food preservation, *839*
hepatitis C drugs approved by, 757t, *758*
irradiated food, approval of, 162
microbe control agents regulated by, 156
regulation of probiotics, 681
U.S. Public Health Service, 789
Ustilaginomycetes, 492t, 501
Ustilago maydis, 501, 502, 502f
Utilisomes, 428, 428f
UTP (Uridine 5′-triphosphate), 194

V

Vaccines, *711*
anthrax, *802*–803, 803f
attenuated cultures as, 19
biotechnology and, 853–854
hepatitis A (infectious), *761*
human papillomavirus, 759–760, 759f
influenza, 741
malaria, strategies in, 819
MMR (measles, mumps, and rubella), 743
Neisseria meningitidis serogroup B (MenB), 853

overiew, *636*
pneumoncoccal conjugate, 807
poliomyelitis, 762–763
protective immunity from, 711–715
rabies, *764*–765
rubella (German measles), *745,* 745f
streptococcal pneumonia, *807*
types of, 713t
Vaccinia virus (VACV), 112, *746*
Vaccinomics, *711*
Vaginitis
bacterial, 465
candidal, *831*
Vaginosis, bacterial, *805,* 805f
Valacyclovir antiviral drug, 738, 754
Valence, *635*
Valley fever, 814
Vampirococcus, 551, 551f
Vancomycin, *177,* 177f
Vancomycin-resistant enterococci (VRE), 185
Vaporized hydrogen peroxide (VHP) sterilizing gas, 164
Variant Creutzfeldt-Jakob disease (vCJD), 765, 765f
Varicella (chicken pox), 738, 738f
Vasoactive mediators, *616*
VBNC (viable but nonculturable cells), *140, 532*
vCJD (variant Creutzfeldt-Jakob disease), 765, 765f
Vector-borne transmission of disease, *688*–689
Vectors, arthropod, *686*
Vectors, cloning, *365*–367
Vegetative cell, **70–71**
Vehicle transmission of disease, *688*
Venereal warts, *759*
Venter, J. Craig, 379, 385
Verrucae vulgaris, *759*
Verrucomicrobiota, 423–425
Vertical gene transfer, *352,* 353f
Vertical transmission of disease, *689*
Vesicular transport, *335*
Vesiduction, *352*
VHF (viral hemorrhagic fever), *763*
VHP (vaporized hydrogen peroxide) sterilizing gas, 164t, 165
Viable but nonculturable (VBNC) cells, *140, 158, 532*
Viable counting methods, *149*
Vibrio cholerae, 67, 140, 295, 451, 686, 697t
Vibrio fischeri, 142, 143f, 319
Vibrionaceae, 451
Vibrionales, 451
Vibrios, *45*
Vibrio species, quorum sensing in, *142,* 319–320
Vibroid cell shape, 127f
Vinegar for microbe control, 163
Viral hemagglutination, *729*
Viral hemorrhagic fever (VHF), *763*
Viral hepatitides, 757–758
Viral-mediated lysis of microbes, 168
Viral neutralization, *650*
Viral pneumonia, 743, 745
Viral replication complex, 111, 111f
Viricides, *157*
Virions, *104,* 104f, 105, 105f, 106, 106f
Virioplankton, *577*
Viroids, 2f, *4, 117,* 156–157
Virulence, *685, 686*
factors, bacterial pathogen invasion and dissemination, 691t
and transmission, 686–687
Virulent bacteriophages, 114, *354,* 506–508
Viruses, **103–118, 504–525**. *See also* Bacterial genome replication and expression; Microbial genomics
acellular organization of, **104**
airborne transmission of human diseases, **738–747**
antiviral drugs, **181–182,** 182f
archaeal, 104, 107, 108, 109f, 113–114
arthropod transmission of, **747–749**
assembly, 111–112, 111f
attachment step, 108–109
bacteria and viroids *vs.,* 117
bacteriophage lambda, 505–512
bacteriophage T4, 506–508
cellular responses to infection, 321–323
Chamberland's work on, 18
cultivation and enumeration of, **115–116**
direct contact diseases, **749–760**
disruption of life cycle, 519
double-stranded DNA, **506–514**
bacteriophage, 506–509
herpesviruses, 512–513
nucleocytoplasmic large DNA, 513–514
double-stranded RNA, 515–517
entry into host, 109, 110f
food and water vehicles for, **760–763**
herpesviruses, 512–513
human diseases caused by, **737–766** (*See also* specific diseases and viruses)
infections from, **113–115**
life cycle of, **108–113**
marine environment, 571–580
minus-strand RNA, 505, 521–523
nucleic acid–based subviral agents, **117**
nucleocytoplasmic large DNA, 513–514
overview, *4*
phylogeny of, 505–506
plus-strand RNA, 519–521
prion proteins for disease transmission, 765–766
prions *vs.,* **118**
release of virions, 112, 112f
retroviruses, *523*–524
reverse transcribing DNA, 525
single-stranded DNA, 514–515
structure of, **104–107**
synthesis, 109, 111
transduction for DNA transfer in, 360–363
zoonotic diseases, **763–765**
Viruslike particles (VLPs), *577*
Vitreous ice, 40
vlsE gene, 427
Volta, Alessandro, 409
Volvox, 28f
Vomiting, bacteria causing, 841
von Behring, Emil, 19
VRE (Vancomycin-resistant enterococci), 185

W

Waksman, Selman, 171, 244, 461
Walking pneumonia, 468
Warts, *759*–760
Wastewater treatment, 190, **865–869**
challenges of, 869
common contaminants, 863t
home treatment systems, 869, 869f
methanogenic archaea, 411
methods in, 863t
plant facilities, 866, 866f
processes, 866–868
quality measurement, 868
water reclamation, 869
Water. *See also* Wastewater treatment
bacterial biofilms in retention systems for, 790
drinking water, safety of, **863–865**
microorganism growth affected by, 131–138
pathogens carried in, 797t
reclamation, 869
viruses transmitted by, **760–763**

Water activity (a_w), *133,* 837
Water molds, 227
Water-splitting complex, 238
WEE (western equine encephalitis), 748
Weller, Thomas, 762
Wescodyne iodophor for microbe control, 163
Western blotting method, *733*
Western equine encephalitis (WEE), 748
West Nile fever (encephalitis), 686, *748*
WGS (whole-genome sequencing), 843
White blood cells (leukocytes), 19, 611, *617*
White, Charles, 687
White cliffs of Dover, 479, 479f
White mold fungus, 601
White-nose syndrome in bats, 500
White piedra, 823
White pine blister rust, 587
White rot fungi, 588
Whole-genome sequencing (WGS), 531–532, 534, 843
Whole-genome shotgun sequencing, *385*
Whole-genome shuffling, *376*
Whole-pathogen vaccines, *712*–713, 713t
Whooping cough (pertussis), 445, 776–777
Wild-type organisms
- inhibiting growth of, 348–349
- phenotype, 346
- prototrophs, 347

Wine, 847
Winogradsky column, *572*–573, 572f
Winogradsky, Sergei, 19, 146, 572
Witchcraft trials in Salem, 499
Wobble in base pairing, *283,* 285f
Woese, Carl, 3, 415, 505
Wolbachia, 552, 552f
Wood-Ljungdahl pathway, 402–404, 409, 424
Wood's lamp UV light, 823
World Health Organization (WHO), 811
- on climate change and disease, 555
- Ebola epidemic response, 718
- *Helicobacter pylori* as Class I carcinogen, 784
- on HIV/AIDS, 753
- influenza prevention effort coordinated by, 742
- irradiated food, approval of, 162
- malarial eradication program of, 816
- on malarial risk, 815
- on malaria treatment, 183–184
- on pertussis, 776
- on probiotics, 681
- role in epidemiology, *701*
- smallpox eradication by, *746–747,* 746f
- viruses causing Ebola and COVID-19, 555

Wright-stained erythrocytes, 819

X

Xenografts, *661*
Xenorhabdus nematophila–Steinernema carpocapsae mutualism, 549, 549f
Xerophillic, *839*
Xerotolerant, *133*
X-ray crystallography, 39
X rays as ionizing radiation, *138*

Y

YAC (yeast artificial chromosomes), *367*
Yanofsky, Charles, 305
Yeast artificial chromosomes (YACs), *367*
Yeast-lactic fermentation, 844
Yeasts
- as Ascomycota fungi, 497–500
- in beer brewing, 847
- description of, 492
- life cycle of, 498f
- *Saccharomyces cerevisiae,* 90
- top and bottom, in beer brewing, 847

Yellow fever, 555
Yellowstone National Park, 131
Yersinial outer membrane proteins (YOPS), 780
YM shift, *813*
Yogurt, 839t, 844
- production, 465

YOPS (yersinial outer membrane proteins), 780

Z

Zanamivir, 181, 742
Zernike, Frits, 29
Zhang, Feng, 375
Ziehl-Neelsen method, 34
Zika virus, 20, *748*–749
Zoonoses
- animals as reservoirs of, *686*
- anthrax, *802*–803, 803f
- bacterial, **802–804**
- Brucellosis (*See* Brucellosis)
- Ebola virus disease, *763,* 763f
- hantavirus pulmonary syndrome, *764*
- Marburg disease, *763*–764
- ornithosis, *803*
- psittacosis, *803*
- Q fever, *803*–804
- rabies, *764–765*
- tularemia, *804*
- viral, **763–765**
- viral hemorrhagic fever, *763*

Zoosporic fungi form motile spores
- Chytridiomycota, 495
- microsporidia, intracellular parasites, 494–495

Zooxanthellae, *479, 546*
Z-ring assembly in septation, 125
Z-ring formation, 125
Zygomycetes, *495,* **495–496**
Zygomycota, 495
Zygosaccharomyces rouxii, 133
Zygospores, *495*